Lutz D. Schmadel

Dictionary of Minor Planet Names

Fourth Revised and Enlarged Edition

With a CD-ROM of the Catalogue of Minor Planet Names and Discovery Circumstances

Springer

Dr. Lutz D. Schmadel
Astronomisches Rechen-Institut
Mönchhofstrasse 12–14
D-69120 Heidelberg
Germany
e-mail: s21@ix.urz.uni-heidelberg.de

ISBN 3-540-66292-8 4th Edition Springer-Verlag Berlin Heidelberg New York

ISBN 3-540-61747-7 3rd Edition Springer-Verlag Berlin Heidelberg New York

Library of Congress Cataloging-in-Publication Data applied for.

Die Deutsche Bibliothek – CIP-Einheitsaufnahme

Dictionary of minor planet names [Medienkombination] / Lutz D. Schmadel. –
Berlin; Heidelberg; New York; Barcelona; Hong Kong; London; Milan; Paris; Singapore; Tokyo: Springer
ISBN 3-540-66292-8 Buch. – 4., rev. and enl. ed. – 1999 Gb. CD-ROM zur 4. ed. 1999

Typesetting: Camera-ready copy by the author
Cover design: Erich Kirchner, Heidelberg
Computer-to-film, printing and binding: Konrad Triltsch, Graphischer Betrieb Würzburg

SPIN 10717081 55/3144/tr – 5 4 3 2 1 0 – Printed on acid-free paper

Dedicated to

my colleagues, dear friends and everlasting supporters of this project

Ingrid van Houten-Groeneveld

Senior Staff Astronomer, Leiden Observatory, The Netherlands

and

Brian G. Marsden

Director, IAU Minor Planet Center, Cambridge, MA, USA

Foreword
to the Fourth Edition

People love to name things. Parents name their children. Children name their pets. Why? Otherwise rational human beings put an inordinate effort into this naming activity. Some names are selected to remind the namer of some other person, place or event. In other instances, the choice of a name means something that "sounds good", or is easily spelled. "What's the baby's name?" is much more likely to be asked than some question about its state of health, its weight or the color of its eyes. People are often named according to religious tradition, exemplified in the custom, in some countries, of speaking of a "Christian" name. In other countries, it is a "given" name, often the name of some favored relative, particularly a father, as in the system of patronymics. In some parts of the world a name may be more practical, making it clear that this person is "number one son", for example. But in such a case, why not simply give numbers to the children in order of their birth? One might presume that other animals number their children: take one away, and the mother will realize that the count is wrong, possibly even to the point of knowing precisely which infant has been removed.

The most basic part of speech is a "noun", a word that itself means "name", particularly in its dominant use as the subject, i.e., in the "nominative case". Communication is essential to an intelligent species, and the human race simply would not have advanced if it didn't have, in its numerous languages, generic words for "tree", "bird", "food", "rock", "star" and "computer". One clearly also needs to be specific, introducing further nouns in each of these categories, such as "apple", "kookaburra", "apple", "punk", "quasar" and "apple". But where does that specificity end? Should every individual member of a species, type or class end up with an obscure number or alphanumeric designation? Or should it have a name or descriptor like "The One that Fell Down in the Hurricane Last Week", "The Very Noisy One that Somehow Found its Way into Our Barbecue", "The 10-Meter Cake", "Hope Diamond", "Arcturus" and, well, "Arcturus".

Astronomers named stars – and groupings of stars – long ago. There were truly imaginative names for Arcturus that translate into English as phrases like "Leg of the Lance Bearer", "Patriarch Mentor of the Train" (or, in the original Arabic, "Al Harris al Simak"). Nowadays, stars are named, for a fee, and with no obvious benefit to astronomy, by "The International Star Registry (ISR)". After their children have left the roost, and their children's pets have been buried 'neath the apple tree, ordinary people with money to burn can still name stars. The International Astronomical Union (IAU) takes a dim view of the ISR, and the ISR takes advantage of human irrationality. Astronomers, trained to cultivate at least an air of rationality, nowadays give to the objects of their quests more useful labels like "Alpha Bootis", PPM 130442, BD +19°2777, PSR 1257-12 B and GRO J1744-28.

But in the solar system, names still – so to speak – have a place. Galileo knew what he was about when he referred to the "Medician Stars", as did Herschel with "Georgium Sidus". A broader, more traditional view won out, however, and Piazzi's "Ferdinandea" appendix was soon dropped from his choice of a name for the small planet he discovered between Mars and Jupiter. But when it was established that Ceres was not alone, the practice of acknowledging each new discovery with the name of a goddess somehow seemed appropriate and international – at least for discoveries made in different parts of Europe and with the goddess names from the "classical" tradition. But nationalist squabbles quickly arose, and soon the supply of names of classical goddesses was exhausted. Other traditions were tapped, more general female names were used, and the discoverers started concocting feminizations of other names and words. At the same time, the objects were simply given sequential numbers. The advent of photography also brought systems of provisional designations, the use of which greatly facilitated the bookkeeping of linking together independent discoveries of the same object, which could then receive a sequential number. And still the minor planets also received names. In a few cases, the names were inspired mnemonically by the provisional designations; and in a few other cases, particular classes of names were applied to members of particular dynamical classes of minor planets. One could say that such names were helpful. Even when there was no obvious connection, to have both a name and a number could provide some useful redundancy when typographical errors occurred. From time to time, a few astronomers would deprecate the naming practice, but it prevailed, with the various monikers proposed by the discoverers being adopted essentially automatically. Perhaps unlike their colleagues in the star and galaxy businesses, planetary astronomers do evidently cultivate that air of irrationality common to the general population.

Which are my own particular favorites among the names of minor planets? Certainly, I like the ones where there is some particular "connection" between planet and name. Some discoverers have their lists of numberings and the lists of the names they want to use. As each new numbering comes along, they simply apply the next name on the list, whatever interesting characteristics the object may have. Most proposers of names could usefully learn from Paul Wild, veteran Swiss discoverer of minor planets: he gave to the minor planet with provisional designation 1968 HB the name "Swissair", HB being the international designator for that airline; and the WK in another provisional designation reminded him of the Swiss-German word for annual military exercises, a grueling maneuver through the Alps and – "Hannibal". He gave the name "Cucula" to a rare discovery in May, the month when the cuckoos call incessantly in the woods near his observatory; and the name "Tripaxeptalis" went to minor planet (2037), because 2037 = 3 x 679 = 7 x 291 – and given (679) Pax and (291) Alice. If one wants to name a minor planet for the twelfth-century mathematician Fibonacci, the least he can do is ensure that its number is a number in the Fibonacci sequence: so this name was therefore given recently to minor planet (6765).

When the Minor Planet Center was established in 1947, the numbering of minor planets extended to (1564). At a meeting of IAU Commission 20 in 1952 it was remarked that to reach (3000) would represent a "reasonable upper limit". When (3000) was finally reached, in 1984, that remark had clearly been forgotten, and further doubling could be expected in about a decade. When the first edition of the Dictionary of Minor Planet Names was prepared, in late 1991, (5000) had just been numbered. It was then a foregone conclusion that (10000) would be reached around the end of the century. That milestone was in fact reached as early as March 1999, a little more than 198 years after Piazzi first sighted Ceres. There are already predictions that (20000) will be attained as early as 2004!

Anticipating the surge of activity that began in the late 1970s, with many more professional and amateur astronomers entering the field, some members of Commission 20 felt that oversight of the names being proposed was desirable. So a committee of three was formed... It grew to a committee of seven, and in 1994 it became the nine-member Small Bodies Names Committee, also taking on the responsibility for naming comets. In 1997 the membership increased to 11, and the Committee was directly attached to IAU Division III, in which Commission 20 was one of six IAU commissions. Those discoverers whose pet names have been rejected might not agree, but the SBNC has good international representation and has generally acted in a fair and effective manner. The main difficulties arise with names considered "too nearly similar" to others, although a precise definition of what is meant by this has so far been elusive.

Quite as impressive as the numbering rate has been the progression of provisional designations. The modern system of these designations was introduced in 1925 with the idea that a simple sequence of letters could be used each halfmonth, it being considered unlikely that there would be more than 25 new discoveries in any such interval. The first complement of 25 was already achieved in the second half of April 1926 with the assignment of 1926 GZ (the letters I and J being considered the same). Then, in the second half of February 1928 (well, it was a leap year), a 26th discovery was accommodated in the form 1928 DA_1 and a 30th as 1928 DE_1. (The utility of the system is that further designations can be added at any time, and these two halfmonths currently terminate with 1926 GF_1 and 1928 DK_1.) The 58 designations of the first half of October 1931, extending to 1931 TH_2, remained a special accomplishment for a long time, particularly when the examination of plates obtained in the continuation of the Pluto search at the Lowell Observatory a few years later took this sequence to 1931 TJ_4; at present, it runs to 1931 TQ_4, a count not superseded until 1950 TU_4. The next record was 1969 TG_8, and double digits were reached for the first time three years later with what now stands at 1972 TE_{11}, i.e., a total of 280 discoveries. Of course, these advances are quite artificial, because 1972 also saw the publication of the Palomar-Leiden survey; the latter did not in fact utilize the system of provisional designations, but had it done so, there would have been designations extending to something like 1960 SR_{83} – a total of 2092 objects not surpassed until the second half of March 1993, when the designations reached 1993 FR_{84}. That record remained only until 1998, when the subscripts first reached three digits, new peaks of 1998 FT_{144}, 1998 HH_{151} and 1998 SL_{165} (this last corresponding to 4136 objects!) following in quick succession. There is no shortage of multiple-opposition identifications that will indeed assure the numbering of the 20 000th minor planet in only a few more years.

But, in the meantime, there is certainly cause to celebrate the myriad (or decachiliad) of numbered objects. Indeed, (10000) has just been given the name "Myriostos", Greek for ten-thousandth, this winning entry again being supplied by Paul Wild. The citation dedicates this minor planet to all the astronomers, from Piazzi on, who participated in this achievement. Of course, Piazzi himself was directly honored with (1000), the subsequent "millennia" honoring William Herschel, Leonardo da Vinci, Hipparchus, the IAU, the United Nations, and Marie and Pierre Curie. That was the status with the publication of the third edition of the Dictionary, in 1996. At (8000), Isaac Newton now also joins this list, while (9000) has just become – Hal (what else?). There are now names for every one of the first 3336 numberings; but after that, numerous minor planets remain unnamed. Only 63 percent of the numbered objects are currently named, a fraction that is well below the 75 percent of the third edition and the 80 percent of the first and second editions.

Given the rapid rates of both designation and numbering these days, some tailing off in the naming rate is probably inevitable. Indeed, there was another recent suggestion that the practice should be terminated. Others think that, as long as a sizable fraction of the discoverers want to name their finds, it is still reasonable and appropriate to recognize their industry in this manner. The danger is, of course, that future names will become more and more trivial. Yet if the naming is not done by the IAU, in a general cooperation with the discoverers, some "International Asteroid Registry" will surely appear somewhere to carry out the task – undoubtedly for a hefty fee. Not all the minor planets need to be named, they certainly don't have to be named immediately, and the tailing off is not necessarily such a bad thing. To my mind, the main problem is in fact the preparation of the citations, and in ensuring that they are concise and well written and – yes – interesting to read. Relatively few astronomers seem to have developed the art of writing a good citation. It should not be a "potted biography" or a "travelogue". It should be something that arrests the attention of the reader, perhaps by pointing out some unusual fact about the person or place being honored, or why this particular minor planet is being selected to have that name. One thing that can be guaranteed is that, in the future, citations are going to have to be a lot shorter than they are now!

As with the earlier editions, the principal responsibility for this volume rests with Lutz Schmadel, who has done a superb job putting together a truly authoritative piece of work. It is not just a catalogue or dictionary in the usual sense. One can in fact read the book like a novel, from beginning to end. In so doing, one could certainly draw some conclusions about the psychology of astronomers. One can also learn a great deal about the world and its customs. The book is also a marvelous source of information about mythology, classical and otherwise.

Cambridge, MA, May 1999

Brian G. Marsden
Director
IAU Minor Planet Center

Foreword
to the First Edition

According to a long-standing astronomical tradition, the naming of minor planets in the solar system is the privilege of the discoverers. Contrary to most other kinds of celestial objects which receive complex alpha-numerical designations, the names of minor planets often say more about the discoverers than about the object in question. There is a rich and colourful variety of ingenious names, from those of heavenly goddesses in the nineteenth century, to the more prosaic and sometimes very specific names of observatories, towns and mountains, computers and persons, given by present-day discoverers.

Commission 20 of the International Astronomical Union, under whose auspices the naming of minor planets falls, has long been concerned with the need to establish a complete catalogue of these names, as well as of the interpretation of their meanings. For this purpose, a Study Group on the Origin of Minor Planet Names was set up at the time of the IAU General Assembly in Baltimore in August 1988. The Working Group immediately started to collect information about these matters from all available sources, including some earlier, incomplete compilations made in the U.S. and in Europe, and also by personal interaction with living discoverers of minor planets.

The present book contains the names and their meanings of all named minor planets. It informs about the discoverers as well as the circumstances of discovery and also gives an interesting insight into the minds of these astronomers, who over two centuries vested their affinities in these names. Moreover, this unique volume illustrates the history of a particular branch of modern astronomy which deals with our immediate environment in space. Although the minor planets in the solar system may seem remote to us at the moment, there is little doubt that technological advances will bring them closer to us in the future. They are also the largest representatives of a vast population of celestial objects, many of which, as we now know, must in the past have collided with Earth and thereby had a significant influence on life conditions on our planet.

On behalf of Commission 20, it is a pleasure to thank most cordially the members of the Study Group and its chairman, Lutz D. Schmadel, and to congratulate them on the excellent result of their hard labours. I am sure that this book will be studied with great interest, not only by astronomer colleagues, but also by historians of our science as well as all others interested in the more philosophical aspects of space research.

Garching, June 1991

Richard M. West
President
IAU Commission 20

Preface
to the First Edition

In the history of natural sciences it is a common procedure to give any newly detected object an individual proper name. But the increase of similar objects in the course of time makes it more and more difficult to choose adequate proper names. So any new object is given an individual number. This system, however, will soon cause confusion and even provoke mistakes. To prevent this, it is best to combine both the name and the number of an object, thus getting a sufficient redundancy no matter how many objects there may be. This system has been realized in an exemplary manner in the assignment of names to the minor planets.

It was the first night of the 19th century when Giuseppe Piazzi established a new era in the exploration of the Solar System. His discovery proved the existence of a body between the orbits of Mars and Jupiter which had been expected for a long time. The naming of this body, however, caused the first controversies. During the second half of the last century the number of discoveries grew dramatically. The naming or the assigning of a special sign as had been done in the beginning could no longer be continued. The practice was started to add an ordinal number to the planets related to the dates of their discovery. However, this procedure was to fail as well, because more and more discoveries were made. Numbers had to be corrected subsequently, and there were fierce arguments about some of the names proposed by the discoverers. The honor of science seemed to be endangered. The violent discussion about the assignment of names did decrease with the exorbitant increase in discoveries, but on the other hand the amount of numbered yet unnamed objects grew rapidly. Great confusion was to come.

The necessity of assigning an adequate name became as apparent as the request for a short explanation of its origin. Not very often could one learn from literature who should be honored and for what reason. After the end of World War II, Antonio Paluzíe-Borrell, a librarian from Barcelona and General Secretary of the "Sociedad Astronómico de España y America", started a first investigation on the origin of some of the names of the first 1650 numbered planets. It is entirely owing to Paul Herget, however, that after the foundation of the Minor Planet Center at the Cincinnati Observatory he demanded of the discoverer that the assignment of each name should be accompanied by an explanation of its meaning. He also suggested that a compilation should be prepared which would indicate the meanings of the names which had already been assigned in the past. A first short list was published in 1955, a more explicit one in 1968. The first newly numbered minor planet after the interruption of World War II was (1565). Since then all newly assigned names have been more or less appropriately described in the Minor Planet Circulars along with their announcement. Herget's successor Brian G. Marsden has strived in an unprecedented manner to meet this demand. His work, based in part on some resolutions of IAU Commission 20, became more and more complex with the constantly growing flood of proposals for naming.

During the XXth IAU General Assembly in 1988 in Baltimore, U.S.A., Edward Bowell (chairman 1988/89) and some other colleagues suggested the establishment of a "Study Group on the Origin of Minor Planet Names", with the aim of thus not only obtaining a database containing the names of all numbered planets, but also, and in particular, of learning about the meaning of many names from the early beginnings of our science. Members of the Study Group (1988-1991) were

V. K. Abalakin, Pulkovo (U.S.S.R.)
E. L. G. Bowell, Flagstaff, AZ (U.S.A.)
F. K. Edmondson, Bloomington, IN (U.S.A.)
H. F. Haupt, Graz (Austria)
L. K. Kristensen, Aarhus (Denmark)
B. G. Marsden, Cambridge, MA (U.S.A.)
P. M. Millman (†), Ottawa, ON (Canada)
J. D. Mulholland, Gainesville, FL (U.S.A.)
E. Roemer, Tucson, AZ (U.S.A.)
L. D. Schmadel, Heidelberg (Germany), chairman
K. Tomita, Tokyo (Japan)
I. van Houten-Groeneveld, Leiden (The Netherlands)
J.-x. Zhang, Nanjing (China)

I want to express my deepest appreciation to Ingrid van Houten-Groeneveld, who has worked for decades on the nomenclature of minor planets, for the many months she spent going over many entries in this compilation. Some other colleagues and amateur astronomers contributed to this work or served as Consultants to the Study Group:

C. M. Bardwell, Cambridge, MA (U.S.A.)
R. L. Branham Jr., Mendoza (Argentina)
R. Bremer, St. Charles, MO (U.S.A.)
N. S. Chernykh, Nauchnyj (U.S.S.R.)
M.-A. Combes, Paris (France)
C. J. Cunningham, Kitchener, ON (Canada)
J. B. Gibson, Pasadena, CA (U.S.A.)
E. Goffin, Hoboken (Belgium)
J. U. Gunther, Durham, NC (U.S.A.)
A. W. Harris, Pasadena, CA (U.S.A.)
Z. Knežević, Belgrade (Yugoslavia)
J. Meeus, Erps-Kwerps (Belgium)
A. Schnell, Vienna (Austria)
V. A. Shor, St. Petersburg (U.S.S.R.)
N. Solovaya, Moscow (U.S.S.R.)

I am grateful to all colleagues who supported this project by manifold advice and investigations. I would also like to thank my colleague Gernot Burkhardt who helped to realize this work by complex and time-consuming programming in TEX.

It is my special pleasure to thank Richard M. West, 1988-1991 President of Commission 20, for his encouragement throughout the project. I am deeply indebted to the past General Secretary Derek McNally for the financial support of this book by the IAU.

Heidelberg, December 1991 **Lutz D. Schmadel**

Preface

to the Fourth Edition

The history and development of minor planet discoveries is a fascinating story with a rather breathtaking evolution. The figure of numbered planets exceeded the magic cornerstone of 10000 objects in January 1999. A comparison with the corresponding numbers in the first three editions of this book – 5012 (Dec. 1991), 5655 (Sept. 1993) and 7041 (June 1996) – yields the astonishing fact of a complete doubling of the material in only seven years. This must be compared with the huge time span of 190 years which was necessary to detect and to refine the orbits of the first half of all numbered minor planets! This fourth edition comprises over 50% more material than its predecessor and serious estimates foresee a further doubling to minor planet (20000) in 2004.

This adventurous enhancement of data in even shorter time intervals requires a new publication policy. Size and the prize of a book obviously are strongly correlated items. With this fourth edition the technical possibilities of the bookbinder are already endangered; at least a two-volume copy will be needed for possible future printed editions. A second disadvantage is given by the fact that new information requires permanent updating, which overtaxes any print medium. Springer-Verlag and the author therefore decided to create a parallel CD-ROM version of this "Dictionary". It is the aim of this new service to inform the user of new results as quickly as possible. The CD-ROM accompanying this edition contains all the information given in the printed part of the catalogue. A special search engine offers manifold possibilities, including a full-text survey. Many links to somehow related entries can be used in the stand-alone mode of the CD-ROM. In addition, the licensee is allowed to extract up-to-date information via the Internet by means of a dedicated server at Springer-Verlag during the continuance of this agreement.

The fundamental structure of the "Dictionary" has been left unchanged. It is, of course, a thoroughly revised and considerably enlarged data collection and every effort has been made to check and correct each single piece of information again.

As author I owe a great debt of gratitude to Ingrid van Houten-Groeneveld for her enthusiastic encouragement and to Brian G. Marsden for his everlasting support of this common project of all interested astronomers. My gratitude is also due to Wolf Beiglböck and Petra Treiber from the Springer-Verlag for their excellent cooperation. Finally, I would like to thank again my colleague Gernot Burkhardt for his support with the TeX processing and his indefatigable endeavour in preparing the CD-ROM.

Heidelberg, June 1999 **Lutz D. Schmadel**

Contents

Introduction

Introductory Remarks

In a period when discoveries flourish, more and more minor planets are being discovered; it is inevitable that this dictionary can be no more than an inventory of the situation at this moment. A compilation of this nature is by definition bound to be out of date before it appears in print. There are, however, two reasons that justify this effort. After having numbered and named the 10000th minor planet, a historic goal was achieved. It seems reasonable to avail ourselves of this opportunity to summarize an interesting aspect of this narrow branch of astronomy during the past two centuries. The second reason is purely pragmatic: this seems to be the last chance to handle the enormous amount of minor-planet data in one volume.

At first glance, this book is simply a compilation of data. For the historians of our science, however, the naming of an object is a complex and revealing matter. Piazzi's discovery of Ceres was the beginning of an absolutely new branch of science that gave an extraordinary impulse to the development of mathematical tools for astronomy. The brilliant work in the period between Gauss and Poincaré made celestial mechanics prosper to the extent that it is now applied in many very different branches of science. The entire second half of the nineteenth century was characterized by intensive observational and numerical work on minor planets. As the importance of this field decreased, the process of name giving changed in a very obvious way. Mystification changed to profanation. The previously dramatic and rare discovery of new members of our Solar System changed to the finding of ever fainter and smaller objects in large numbers. These changes were reflected in the nomenclature of minor planets.

Interest in questions of nomenclature originated not only from the perspective of cultural history but also from a very pragmatic consideration. The large number of new discoveries made it more and more difficult to give a newly numbered planet an adequate name. Only recently the same name was given to two different minor planets; this oversight had to be corrected as soon as possible. The 'Minor Planet Names Committee' of IAU Commission 20 was founded, and 1994 enlarged to the 'Small Bodies Names Committee', in order to study the proposals for names (of minor planets and comets) and to investigate whether the suggested names do not resemble too strongly any other names previously given - a work that is not always easy to manage! Discoverers of minor planets might also find this book useful. On the other hand, they will still have to seek advice in the literature of classical antiquity if, for instance, a newly discovered Trojan is to be named. Some people have already had the awkward vision of a 'first Trojan soldier', 'second Trojan soldier', ...; we can but hope that this will not become customary!

Everyone has their own specific ideas as to what to include in a book and what not to include. It is not easy to reach a general consensus. We have mainly used original quotations in order that the character of this compilation be evident. From number (1565) Lemaître, which was the first object to be numbered after World War II, this policy has been applied to every minor planet. From that time onward, the 'Minor Planet Center' - then in Cincinnati, Ohio, now in Cambridge, Massachusetts - took the task of nomenclature over from the Astronomisches Rechen-Institut (*ARI*) - then in Berlin, now in Heidelberg. The 'Minor Planet Circulars' (*M.P.C.*) report not only on new name assignments but also on many names that were assigned subsequently for numbers earlier than (1565). The names of these older planets were in most cases taken from the lists of Herget (1955, 1968). The explanations of these older names could, in most cases, be found in the contemporary literature, including the 'Astronomische Nachrichten' (*AN*), the 'Zirkulare des Rechen-Instituts' (*RI*), etc. These explanations often have the character of an official announcement. These older explanations differ widely regarding accuracy, completeness, and the aesthetic use of language; thus they reveal much about the proposer of the name, mostly the discoverer, and his cultural background. In this compilation we deviate from the English language in only those few cases in which important original contributions exist.

The minor-planet sky, like the surface of the Moon or of Mars, has became a sort of 'astronomer's cemetery'. In those cases where a minor planet has been named for a deceased astronomer, we have supplemented the relevant name quotations with biographical data. The references given to published obituaries may be helpful for detailed study. Many names are correlated in various ways, so the reader will find cross references throughout the catalogue. In this way, entire genealogical tables could be set up; there is ample evidence for specific preferences on the part of particular proposers of the names. Comparisons can show a change in the 'Zeitgeist', and there are also more or less subtle allusions to events of worldwide importance. From the very beginning of this project, it was clear that not only the names and their explanations should be mentioned but also the most important circumstances under which the discoveries were made. There is only a slight correlation between the discovery date of a minor planet and its definitive number. In many cases, an object could be numbered only some decades after its discovery; yet more time might have elapsed before a name was assigned to the numbered minor planet. It must furthermore be taken into account that in many cases the discoverer is not the patron who names the object. All these facts must be considered when one wants to use this book not only as a mere dictionary but as a resource on cultural and astronomical history as well.

The author and colleagues who were involved in compiling this dictionary took pains to handle the vast amount of data carefully. Nevertheless there may remain inaccuracies or omissions for which the author accepts the blame. In the future the remaining gaps will be filled and many new names added. We are convinced that there will be more studies on names with a meaning so far unknown or doubtful as well as further research on the existing nomenclature. We would welcome any corrections or comments from the users of this compilation.

Minor Planet Designations

The choice of an appropriate name for a celestial body obviously presents difficulties when the total number of objects increases considerably. Eventually one is forced to introduce a numbering system. Very large numbers of objects, however, require the simultaneous use of numbers and names in order to avoid mistakes. In particular, the rules of minor-planet nomenclature require such a procedure.

Until the middle of the nineteenth century, nomenclature problems did not exist. Ceres, Pallas, Juno, and Vesta were mentioned in scientific literature without associated numbers. They were handled in the same way as the major planets known at the time. Problems only arose with the dramatic increase of minor-planet discoveries which began about 1850. It became customary to assign a special symbol to the name and number of a minor planet, following the custom of the traditional symbols associated with the major planets. This procedure, however, soon failed. On the one hand, it was difficult to print these symbols; on the other hand, it soon became impossible to remember all the different symbols assigned. It seems that Luther (1855) was the last astronomer who assigned a special symbol to a minor planet, namely to (37) Fides.

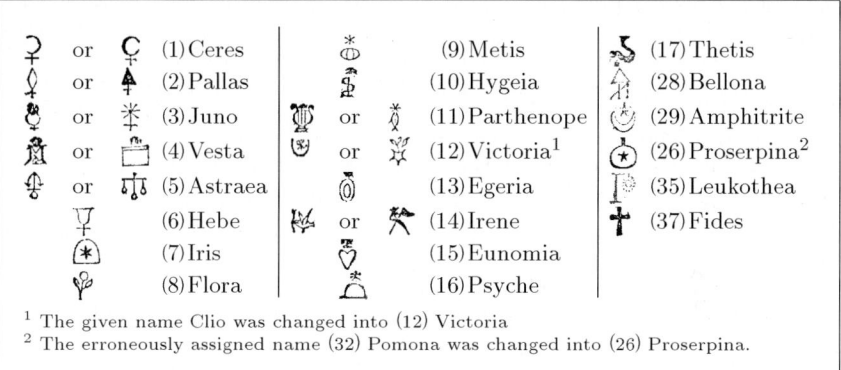

or (1) Ceres	(9) Metis	(17) Thetis
or (2) Pallas	(10) Hygeia	(28) Bellona
or (3) Juno	or (11) Parthenope	(29) Amphitrite
or (4) Vesta	or (12) Victoria[1]	(26) Proserpina[2]
or (5) Astraea	(13) Egeria	(35) Leukothea
(6) Hebe	or (14) Irene	(37) Fides
(7) Iris	(15) Eunomia	
(8) Flora	(16) Psyche	

[1] The given name Clio was changed into (12) Victoria

[2] The erroneously assigned name (32) Pomona was changed into (26) Proserpina.

The minor planet symbols are adopted from Webster's *A Dictionary of the English Language* , G. & C. Merriam & Co., Springfield, MA, USA, p. 1780 (1884). (Courtesy of R. W. Sinnott, Cambridge, MA, USA)

In place of symbols, the system of ordinal numbers was introduced. Ferguson (1852) initiated this development with (16) Psyche. He used an encircled number rather than setting the number in parentheses as is practiced today. Obviously, Ferguson adopted this style of designation from Encke (1851) who declared in the *BAJ* for 1854: "Endlich füge ich noch hinzu, daß bei der Verwickelung und Schwierigkeit der neueren Planetenzeichen ich mir erlaubt habe, statt der Zeichen Zahlen in einen Kreis eingeschlossen einzuführen." [Finally, I want to add that – in view of the complications and difficulties with the recently used planetary symbols – I took the liberty to introduce encircled numbers instead of symbols.] Wolf (1892) stated that he and Gould introduced the encircled numbers as early as 1851.

A number was given by the editor of the *AN* upon publication of the announcement of the discovery of a minor planet. This custom soon resulted in awkward consequences. By the end of the year 1857, some fifty planets had been discovered - on October 9th Ferguson had detected (50) Virginia. The numbering, until then a chronological one, became confused because Goldschmidt had already found a new planet on September 9th, which was later named (56) Melete. The method of numbering had been criticized before. Laugier (1859) proposed that the numbering should be absolutely chronological. Indeed, new discoveries were given names very soon, but as for the number, the chronological order was to be followed strictly. Foerster (1861) gave a rather sarcastic outline of this policy of the 'Berliner Astronomisches Jahrbuch' (*BAJ*): "Was die Benennungen der

Planeten betrifft, so werden dieselben von Berlin aus vom Planeten (60) an nur mit Nummern bezeichnet werden ... Die Namensgebung könnte fortan nur eine Quelle von Lächerlichkeiten werden." [As for naming planets, Berlin defines them by numbers only from planet (60) onwards ... The naming process could become a source of ridicule.]

The problems of nomenclature and numbering were further complicated by the question of who was the true discoverer and who had the right to propose a name and act as the 'patron'. Wolf (1859) wrote a noteworthy comment: "So wenig man Flamsteed die Entdeckung des Uranus, oder Lalande die Entdeckung Neptun's zugeschrieben hat, so wenig darf man Goldschmidt die Entdeckung von (56) zuschreiben, - nicht wer zuerst gesehen oder beobachtet, sondern wer zuerst erkannt hat, ist der Entdecker." [The discovery of Uranus cannot be ascribed to Flamsteed, the one of Neptune not to Lalande, accordingly you cannot ascribe the discovery of (56) to Goldschmidt, - the discoverer of a planet is not the one who first saw or observed it, but the one who first recognized it as a new object.] This was the first time that the patronage of a new discovery could be transferred to the person who computed the orbit if the first observer had not realized the nature of the object. Nowadays, this problem remains of some importance.

The sharp increase in the rate of discovery mandated that the *BAJ* or the *AN* assign numbers promptly. Although the basic idea of chronological numbering remained, other deficiencies in the nomenclature system soon emerged. In quite a few cases subsequent confirming observations of a "discovery" could not be made; hence the number of spurious minor planets accumulated. What should then be done with the numbers which had been assigned to them? Tietjen, editor of the *BAJ*, proposed a way out of this dilemma. Krueger (1892) had suggested that all subsequent new discoveries should be given a provisional designation: "...der Herausgeber der Astronomischen Nachrichten wird die neuen Planeten von jetzt an zunächst nur mit einer provisorischen Bezeichnung, 18.. A, B, C, ..., nach dem Datum der Anmeldung bei der Centralstelle für Astronomische Telegramme versehen. Die definitive Numerierung wird der Herausgeber des Berliner Astronomischen Jahrbuchs erst später zu geeigneter Zeit vornehmen und hierbei alle diejenigen Planeten, bei denen ein genügendes Material zur Berechnung der Bahnelemente nicht vorhanden sein sollte, von der Numerierung ausschließen." [From now on, the editor of the *AN* will first give the new planet a provisional designation, 18.. A, B, C, ..., according to the date of registration at the Central Bureau for Astronomical Telegrams. The definitive number will be given only later by the editor of the *BAJ*. This procedure will make it possible to exclude all planets from numbering whose orbital elements could not be calculated due to a lack of material.] Only one year later, in 1893, it was realized that one single capital letter in the provisional designation was not enough - the letter 'A', in fact the entire alphabet, was taken once more. So 1893 Z was followed by 1893 AA, 1893 AB, and so on. Krueger (1893) noted that the assignment of the letter should be done without respect to the interruption by the beginning of a new year. During World War I some additional systems for provisional designations were introduced, raising new problems. For example, the astronomers in Simeïs did not have a regular transmission line during this time, and found it necessary to introduce their own provisional numbering.

The solution of the designation problem goes back to a suggestion by Bower (1924) and this system is still in use: "To produce homogeneity of nomenclature, the following suggestion is made. Assign as a provisional designation, the year of discovery and two capital letters. The first letter will indicate the time in the year of discovery; A, the interval Jan. 1-15, B Jan. 16-31, C Febr. 1-15, D Febr. 16-29, etc. The second letter will indicate the order in the discovery interval that notice came to the Recheninstitut; A = first, B = second, etc. This designation is of the same brevity as the one most used, accurately indicates time of discovery, approximately indicates the right ascension at discovery, and has a sufficient expansibility to make unnecessary more than one system of designation." The Berlin Recheninstitut adopted this suggestion and, since 1925, has acted accordingly. Kopff (1924) suggested the following refinement to the system of Bower: "Diese Vorschläge sind noch dahin zu ergänzen, daß bei Entdeckung von mehr als 25 Planeten in einem halben Monat, ein Fall, der allerdings kaum vorkommt, das Alphabet an zweiter Stelle von neuem wieder zu laufen beginnt unter Hinzufügung des Index 1, 2 ... an den zweiten Buchstaben; so wäre z.B. der 26. Planet in der ersten Januarhälfte 1925 mit 1925 AA_1, der 27. mit 1925 AB_1, der 51. mit 1925 AA_2 zu bezeichnen." [These suggestions are to be complemented as follows: In the improbable case (*sic!*) of more than 25 discoveries within half a month, the alphabet for the second letter starts running once more followed by an index number; so the 26th planet in the first half of January 1925 should be designated 1925 AA_1, the 27th planet 1925 AB_1, the 51th planet 1925 AA_2.] In this way the demand for a chronological ordering was finally met. This dictionary lists the principal designations following the new-style system with any entry. A comprehensive index of these preliminary principal designations is published elsewhere in this book.

In the course of time, a permanent number was given only if a certain number of observations, appropriately distributed over time, was assured. The *ARI* assigned a definitive number in practically all cases if a first elliptical orbit from three observations yielded sufficiently small residuals for all the observations which were not taken into account. Herget (1952) sharpened these conditions considerably: "An unnumbered minor planet will be assigned a permanent number when elements have been derived which: a) depend upon observations in at least two oppositions, b) include perturbations, and c) produce satisfactory residuals for all known observations. Condition a) will not be required if the perihelion distance is less than 1.67 a.u." During the decades following Herget's refinements, the conditions became even stricter. Except for very unusual, earth-approaching planets, observations from three oppositions were demanded, including one longer-observed arc in one opposition. After having numbered a new record of 384 minor planets in 1990, Marsden (1991) even announced: "However, the general increase in the accuracy of the orbits of the numbered minor planets would seem to dictate that the standards for new numberings should be more stringent than hitherto. Although earth-approaching objects may be numbered after two oppositions and uniformly well-observed objects after three, it seems not unreasonable generally to delay numbering until there are four or even more oppositions..." Precise quantitative guidelines are discussed by Marsden (1996).

Evolution of Naming Rules

Let us now turn to the historical evolution of the process of giving names to minor planets. Ceres, Pallas, Juno, and Vesta are terms taken from classical mythology. With the increase of discoveries beginning around 1850 it seemed only natural to follow these examples. The name given to the early minor planet (12) Victoria, however, was the first of many to initiate a long controversy. One of the most pugnacious astronomers in this field was the founder of the Bilk Observatory near Düsseldorf, Luther (1861), who vehemently adhered to classical names. He dictated: "So lange als für andere Himmelskörper, z.B. für Sterne, Cometen, die Trabanten des Saturn und Uranus und sogar für die Mondgebirge besondere Namen für zweckmäßig erachtet werden, dürfte es auch zweckmäßig sein, die auf der classischen Mythologie beruhenden Planeten-Namen beizubehalten, indem die alleinige Bezeichnung durch Nummern leicht eine ergiebige Quelle von Fehlern und Verwechslungen werden könnte. Unclassische Namen sind freilich ... auf die Dauer nicht haltbar, so daß es rathsam sein wird, statt solcher ausschließlich die Nummern zu gebrauchen." [As long as people believe it appropriate to give special names to celestial bodies like stars, comets, the moons of Saturn and Uranus and even for the mountains of the Moon, it seems also appropriate to adhere to names from classical mythology, since a mere number could easily lead to mistakes and misunderstandings. Unclassical names, however, ... are not tenable in long run; instead of such an alternative, one should rather turn to the numbers alone.] His demand culminated in the rule: "Classische Namen werden gebraucht, unclassische verworfen; für die unclassischen Namen haben die Berechner das Recht, classische Namen zu substituiren." [Classical names are necessary, unclassical names are rejected; (orbit) computers have the right to substitute classical names for unclassical ones.]

This dogmatic attitude was immediately and strongly criticized. Steinheil (1861) noted smugly: "Welchen Vortheil aber soll es bieten ... nur classische Namen zu wählen? Sollen uns die neuen Planeten daran erinnern, daß wir einst im Gymnasium gewesen sind? Hat die Astronomie der Philologie so viel zu verdanken, daß sie keiner anderen Erinnerung, als dieser, Raum geben soll? Ich glaube nicht, daß dieses nachgewiesen werden kann." [What advantage should there be ... in choosing only classical names? Are the new planets to remind us of having gone to a classical secondary school? Does astronomy owe so much to philology as to remember nothing but this? I do not think that this can be proved.] And he goes on: "Herr Dr. Luther scheint aber auch die practische Seite nicht genau erwogen zu haben, denn wenn zwei oder mehr Rechner einen Planeten bearbeiten, welcher soll das Recht der Umtaufe haben, oder soll es allen bleiben?" [It seems that Dr. Luther has not taken into account the practical side of his demand. Who should have the right to substitute unclassical names if two or even more orbit computers are working on the same planet; each of them?] Attempts to give up names entirely in favor of only numbering failed as well. Goldschmidt (1861) agreed to the defenders of a classical line such as the one taken by Luther, Hind, and the editor of the *AN*, Peters: "Ich stimme ... gegen die Abschaffung der Namen, da die größten Verwirrungen daraus entstehen würden. Haben sich einige missfällige Namen eingestellt, so ist es nicht der Mühe werth, den Pathen deshalb Verdruß zu machen. An einem Planeten ist alles Ziffer, das einzige, was poetisch bleibt, ist der Name, und der soll auch wieder Ziffer werden, ohne daß eigentlich eine besondere Ursache dazu da wäre." [I vote ... against the repeal of names, which would cause great confusion. Some displeasing names are not worth the trouble of annoying their patrons. A planet is all number - the name is the only poetic part of it, and this would now become again a number without there being a real cause to do so.] Peters (1861) tried to put an end to this controversy by standing on his authority as editor of the *AN*: "... glaube ich den Wunsch äußern zu dürfen, dass die Controverse über den, in wissenschaftlicher Beziehung unerheblichen Gegenstand, in diesen Blättern nicht weiter geführt werde." [... I do not like this controversy, which being of no scientific importance will not be discussed further in this journal.]

The determination of the classicists remain unbroken, however. A decade and a half later Luther (1878) attacked again: "In Bezug auf die neuerdings wieder allzubunt werdenden Benennungen ... wäre es zu Gunsten der Würde der Wissenschaft gewiß sehr zu wünschen, daß die Entdecker dem älteren Brauche gemäß vorzugsweise klassisch mythologische Namen wählen und neuere Anspielungen jeglicher Art vermeiden möchten... Durch strengeres Festhalten an den Namen des klassischen Alterthums ... wird es hoffentlich gelingen, die wachsende Schaar der kleinen Planeten vor zunehmender Gleichgültigkeit zu bewahren." [The names having now become a more than colorful mixture ... it seems very advisable to return to the old usage of preferring classical, mythological names. Allusions of any kind should be avoided - for the sake of the honor of science... A stricter adherence to classical names will hopefully help to shield the growing number of minor planets from increasing indifference.] Bruhns (1878) concurred with Luther: "... da einmal klassische Namen üblich und allgemein adoptirt sind, ist es das Vorteilhafteste, von dem Gebrauche nicht abzuweichen und alle Namen, welche sich auf lebende Personen oder auf vorübergehende Ereignisse beziehen, zu vermeiden... Nur durch die Wahl klassischer Namen wird es auch ferner möglich sein, den gewählten Namen allgemeine Anerkennung zu verschaffen." [... classical names being in use and commonly adopted, it seems to be best not to deviate from this usage and to avoid all names referring to living people or current events... Only the choice of classical names will further win general recognition.] At the end of the 19th century Holden (1896) still denounced some unclassical, female names by reproaching: "Many of them, at least, read like the Christian names in a girl's school."

After the number of minor planets had reached some 400, however, the classical line could no longer be maintained. The rule was restricted to the choice of female names. Bauschinger (1899) gave the opinion of the *ARI* by threatening: "Es ist Anlaß gegeben die Herren Entdecker zu ersuchen, bei der Namensgebung der kleinen Planeten von dem herkömmlichen Gebrauch, weibliche Namen zu wählen, nicht abzugehen; es ist dieser Gebrauch bisher nur einmal aus guten Gründen bei (433) Eros durchbrochen worden. Männliche Personennamen werden im Berliner Jahrbuch keine Aufname finden." [There is reason to ask the discoverers not to deviate from the rule of choosing female names; so far this rule has only once been offended - and for a good reason - with (433) Eros. Male names will not be accepted in the *BAJ*.] Kreutz (1899) agreed and concluded that male names would not be admitted by the *AN*. This policy could, of course, not be long maintained. First, names of cities were given the feminine suffix 'a' or 'ia', until finally more and more male names were transformed in this way to follow the rule. This was practiced for a long time. But without a clear cut, World War II seems to have made an end to this procedure.

At the beginning of this century the nature of the nomenclature problem changed notably. First there was a gradual abandoning of classical names - the large numbers forced a pragmatic approach. The importance of naming, however, has never been denied. Bauschinger (1901) argued for the policy of the *ARI* as expressed in the *BAJ*, edited by the Recheninstitut, as well as in the *AN*: "... darf nicht übersehen werden, daß auch die Namen ihre volle Berechtigung haben. Abgesehen davon, daß ein durch 100 Jahre üblicher Gebrauch nicht ohne Nothwendigkeit aufgegeben werden sollte, bieten die Namen ein werthvolles mnemotechnisches Huelfsmittel ... Nummern und Buchstaben werden leicht verwechselt, der Namen dagegen prägt sich mit der ganzen Geschichte des Planeten leicht dem Gedächtnis ein." [...it may not be ignored that names have their justification, too. Other than the fact that a usage that has been practiced for a century should not be given up without necessity, names offer a good mnemotechnical tool ... numbers and letters are mixed up easily; a name, however, stamps upon the memory all the history of a planet.]

There was an appeal made to the discoverers to make use of their naming rights within a reasonable span of time. On behalf of the *ARI*, Bauschinger asserted that the right of giving a name should be withdrawn from the discoverer if a name had not been assigned to a planet after observations at a second opposition. This system has worked until the present time. In order to make this dictionary as complete as possible, the Minor Planet Center gave the order in *M.P.C.* 17249 (1990 Dec. 2) that the resolution of IAU Commission 20 from 1979 would be applied: "... if the discoverer does not exercise his established right within ten years after the numbering of a minor planet, that right will be lost." As a consequence of the adoption of this resolution, the discoverers' rights of naming all planets with a number lower than (2378) was terminated effective 1991 May 1. A similar notice was announced in *M.P.C.* 22089 (1993 June 4) in order to support the second edition of this dictionary: "... This applies to minor planets up to (2892), for which, under the "ten-year rule", discoverers will lose their naming privileges." For the benefit of the third edition this procedure was announced once more in *M.P.C.* 26205 (1996 Jan. 5) and applied now to all minor planets up to (3414). The Editorial Notice in *M.P.C.* 33151 (1998 Dec. 8) announced a similar procedure for the fourth edition which "... particularly applies to minor planets up to (4044)".

After the World War II, the rule of assigning names with female endings was finally given up. The note in *M.P.C.* 837 leaves no doubt on this matter: "The custom of attaching feminine endings to masculine names has had numerous exceptions in the past. Names which are submitted will not be rejected or modified if they are masculine." This rule still exists, although quite recently some exceptions have again followed the traditional rules.

Compulsory regulations concerning the assignment of names were first given by Herget (1952) in *M.P.C.* 837. The Minor Planet Center retained the decisive authority: "The discoverer may propose the name for each numbered planet, and this name shall be recognized only after it has been announced in the *M.P.C.* The announcement shall also contain an explanation of the significance of the name and the reason for assigning the honor." Without this rule, the meaning of many of the names and their patrons would tend to vanish in the dark of the history of astronomy. At the same time as announcing the regulations, Herget justified this procedure as well as his liberal attitude towards the assignment of names: "The principal justification for exercising control over the assignment of names is to avoid names which are too similar to others as to cause confusion, and to prevent names which are deliberately offensive or in bad taste. In the past, names have been rejected on the grounds of political connotations. This policy will not be continued in the future... The discoverer is usually motivated, with complete sincerity, to assign what he considers an honor, and his privilege should not be restricted because someone else holds a different opinion."

Herget's views did not remain unchallenged. Occasionally there were discussions about the admissibility of suggested names. These discussions were mainly about questions of taste and about eventual political implications. Regulations were discussed and resolutions were passed in various IAU General Assemblies. The Minor Planet Names Committee is the decisive authority. Before 1991, the Names Committee was composed of the President and Vice President of Commission 20 and the Director of the Minor Planet Center. In case of controversy the proposer of a name should have the right to appeal to the entire Commission 20. This right has been exercised in several cases, with differing results. The regulation in use at the present time was passed at the 1985 IAU General Assembly held in New Delhi. The resolution, published e.g. in *M.P.C.* 10194 (1985 Dec. 27), reads as follows: "Names proposed for minor planets will not be accepted if, in the opinion of the Minor Planet Names Committee, they are too nearly similar to those of other minor or major planets or natural satellites, or are in questionable taste. Names should be pronounceable, preferably expressible as a single word, and no more than sixteen characters long. Names glorifying individuals or events principally known for their political or military activities or implications are considered unsuitable unless at least one hundred years have elapsed since the individuals died or the events took place. Objects involved with the Jovian triangular libration points should be named in accordance with the tradition of honoring heroes of the Trojan War. In a disputed case, the proposer may appeal the committee's decision at a general meeting of Commission 20, provided that due written notice is given to the President of the Commission." In view of the increasing problems in the field of nomenclature the suggestion to enlarge the 'Minor Planet Names Committee' to seven members was accepted at the 1991 IAU General Assembly held in Bueños Aires. Subsequent actions by Commission 20 increased the size, renamed it the 'Small Bodies Names Committee' at the 1994 The Hague General Assembly and required the group of nine (see Appendix 9) to judge on the names of both minor planets and comets. In the past, it has happened that names were chosen for natural satellites of the major planets that had already been assigned to minor planets; an agreement with the 'Working Group on Planetary System Nomenclature' was found in order to avoid an unnecessary duplication of names. After consultations, policies were also set down for naming transjovian bodies.

The rapid evolution in naming minor planets also called for detailed decisions on the form of the citations. Marsden (1995, unpublished) formulated the following guidelines: "Citations should be concise, to the point and devoid of remarks that, if tied directly to the name proposal they support, might cause some to think that the name should be rejected for some ... reasons ... Citations supporting names of persons should not be "potted biographies" (but they should include birth and death dates), nor those supporting names of places "travelogues". A name may have no obvious connection with astronomy in general or minor planets in particular; a more obscure connection would therefore be of interest to the readers and should be mentioned. Propaganda, whether political or for the aggrandisement of the proposer, should not be included in a citation. A name is often of very personal significance to the proposer; if the proposer chooses to acknowledge this in the citation, this should be done very specifically, rather than in terms of generalities. Final editing of the citations is done by the Minor Planet Center staff. The actual text of a citation will not exceed a maximum of ten printed lines in the *M.P.C.s.*" Principles, guidelines and rules, necessarily, became important constituents in this particular field of astronomy.

This dictionary cannot discuss in detail the evolution of specific nomenclature. Above all it cannot enter into particulars concerning the delicate question of who is the discoverer of a planet and who, after a discoverer's death, should have the moral right to assign a name to a planet. Recent discussions on such cases leave no doubt that many of our colleagues attach as much importance to the problem of nomenclature as was done more than a century ago.

Information Content of the Catalogue

This entire catalogue results from the analysis of a computerized data base set up by the author. Data fields were defined as narrowly as possible, so that the data could satisfy a wide range of uses. Some data fields are obviously obligatory, such as the definitive number or the date of discovery, others, such as the explanation of the name, are not. The data fields depend logically on each other in ways which can be used for plausibility checks or completeness control. For example, wherever a naming citation is given, the data field must also contain the relevant references to the citation. The provisional designation of a planet enables the control of its date of discovery, and vice versa. The data base contains additional information such as classification marks. These are not given along with every entry, but some of the appendices were set up with their help.

The definitive number of the planet and its official name form the most important, obligatory information. But about one fourth of all of the planets which have so far been numbered are yet to be named. In these cases the dictionary gives the provisional designation assigned by the Minor Planet Center. This systematic representation gives the discovery date of a planet with an accuracy of half a month. Prior to 1925 - see e.g. Krüger (1892, 1893) - the provisional designations did not follow a strict rule. For this reason the old designations are not given in this dictionary. In some exceptional cases after 1925 the provisional designations did not follow Bower's (1924) system. The catalogue also lists designations assigned according to the modern method to planets discovered before 1925. These designations are marked with the letter 'A' instead of the number '1' in the date of the year. A third group of provisional designations comes from the generally known Palomar-Leiden Survey made by van Houten et al. (1970). It consists of an ordinal number followed by the suffix 'P-L'. Further investigations describing searches for Trojan planets are reported in the papers by van Houten et al. (1984, 1991) and van Houten-Groeneveld et al. (1989). The suffixes 'T-1', 'T-2', and 'T-3' refer to the three surveys for Trojans. The total number of planets with modern-style provisional designations amounts to 9613 entries (90%). A complete index of all these preliminary (principal) designations in relation to the permanent planet number is given just ahead of the main names index.

The obligatory entries concerning the discovery circumstances report three pieces of information: the date of discovery, the name of the discoverer, and the place of discovery. The date refers to UTC, so it does not depend on the meridian of the place of discovery. The name of the discoverer is given in about 91% of all cases. If more than one name is given this indicates that there was a team working at the same observatory. In 922 cases an individual discoverer was not named, so the name of the observatory is given as the corporate discoverer. The place of discovery is always a geographical mark like a city or nearby locality, or a mountain. This procedure was preferred rather than quoting particular observatories; these often have branches elsewhere, or have moved to other places or closed down. Many planets were discovered at small, private observatories. A list of concordances in Appendix 4 shows in which way place names and observatories are related.

During the first third of this century, in particular, the discovery rate was so rapid that in several cases more than one independent report on the same object was given to the *AN* or to the *ARI*, before it could be published. These independent discoveries are also listed, together with the relevant dates, discoverers, and places.

The explanations of names constitute the main part of the catalogue. We attempt to give complete original quotations. This was possible with all entries originating from the *M.P.C.* All insertions by the author, such as life dates or cross references to other planets are indicated by braces. Names from mythology are often interpreted differently by different poets from the classics; in such cases, several interpretations are given.

We have attempted to classify the explanations of names with respect to their reliability by a flag in the data base. Approximately 2% of all background information on particular names is indicated as uncertain or questionable by using adverbs like 'probably' or 'possibly'. A total of 6475 of all planet names, corresponding to more than 96% of all names, are interpreted as absolutely correct. This is not very remarkable, since the great majority of explanations could be quoted from literature or from the *M.P.C.*'s. The meanings of 134 names are still in question; these names are appropriately indicated in the dictionary. Most of such cases involve female names which could not be assigned to a specific woman. A small number of names – only 2% of the total – remains that are still completely inexplicable; the smallness of this percentage indicates the great success of efforts of the 'Study Group on the Origin of Minor Planet Names'.

Ever since the discovery of the first planets the right of naming has often been transferred to other astronomers; this practice is still being followed. If there was any information on the proposer it is given following the explanation of the name. Sometimes we were able to give quite detailed information on the problems of naming, on the discovery circumstances, or on the discoverers themselves. Astronomer colleagues are honored in many cases. The naming citations by themselves give but few biographical facts; more explicit descriptions of the lives of deceased astronomers are found in the literature. Therefore, we took many bibliographical remarks from obituaries published in the volumes of 'Astronomy and Astrophysics Abstracts' during the past decades and placed them at the end of the respective entries.

The dictionary contains additional useful information in the appendices. Alphabetical lists of all discoverers and of all observatories involved with the discoveries are given. The discoverers are listed with the total number of discoveries, and, in parentheses, the number of co-discoveries involved. The time interval between the first and the last discovery is also given. Appendix 11 lists the planet names that could not be interpreted so far, together with the year of the discovery and the discoverer. There is also a table that gives the names of minor planets of special types that cannot be inferred from the data base (Appendix 10). The objects with special characteristics were compiled from a data base of the orbital elements of all numbered planets. The Hilda-type minor planets are characterized by a period of revolution around the sun almost exactly 2/3 that of Jupiter. Whereas the Trojans are defined by semi-major axes of mean Jovian distance, the Aten, Amor, and Apollo planets are Earth approaching, thus deserving the special interest of astronomers. In many assignments one can see how this special type of orbit influenced the names.

Statistics and Classification of the Names

This dictionary contains information on all 10666 planets which had been numbered until May 1999. The object (1) Ceres was discovered on January 1, 1801 and is thus the 'eldest' numbered planet; (10645) 1999 ES_4 was discovered only on March 14, 1999. The chronological distribution of the discoveries during this period of more than 198 years can be taken from the cumulative diagram Fig. 1. The immense increase of the rate of discovery during the last decades is obvious. At the turn of the century, only some 4% of the planets that are now numbered had been discovered. The number 1000 was exceeded in 1921; the number 2000, shortly after World War II.

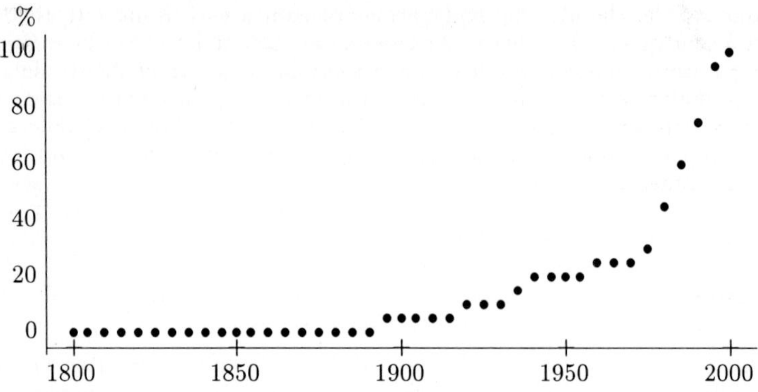

Fig. 1: Cumulative numbers of minor planet discoveries

The number of planets reached one half of the entries listed in this dictionary in the year 1992. Marsden (1979), Director of the Minor Planet Center and a leading expert in this field, enormously underrated this evolution when he said: "At the present and anticipated future rate of growth, it is not unreasonable to suppose that there will be 4000 numbered minor planets by the end of the century." The rate of growth becomes evident by noting that during the past 7 years as many minor planets were numbered as during the span of 191 years before that. This rate of growth is all the more impressive when one notes that the standards applied to the quality of the orbital elements are now much stricter than before. Further, the numbers assigned during the most recent years are affected by a selection effect, since many of the planets discovered now will only be numbered in forthcoming years.

In the 19th century it was a common use to number a minor planet immediately after his discovery. This procedure, however, has proved to be a dangerous one – many planets were lost just after the publication of the discovery. The discovery rate of numbered planets is discussed in detail by Bowell et al. (1989) and more recently by Marsden (1996). Actually we are faced with a doubling rate of minor planet numberings of approximately 14 years. The first $4096 = 2^{12}$ objects were numbered until 1989 and one can predict that there will be 2^{14} numbered planets by the end of 2003. It is very interesting to compare this exponential data growth given by the numbering procedure with the real discovery dates of the now numbered planets. There is practically no difference between planet 2^0 and 2^9, corresponding to the time span 1801–1903. A remarkable gap developed in the middle of this century. In 1978, 2^{11} planets were numbered but a look into the discovery statistics (of now numbered planets) shows that this figure was already reached in 1948! This fact is readily explained by disruption effects due to World War II and the more and more stringent requirements for a definitive numbering, e.g. the numbering was heavily delayed during the decades 1940–1980. The situation has changed mainly by the use of modern computing facilities. Therefore, the gap between the discovery and the numbering times has shortened considerably: 2^{13} (now numbered) planets which were discovered until 1992 already reached their numbering status in 1998.

The data base also reveals the discovery frequency according to seasons in the year (see Fig. 2). Because of the weather conditions, there are discovery maxima in spring and in autumn. The differing heights of the maxima reflect the fact that by far most of the planets were found from the Northern Hemisphere, where relatively good conditions for observations away from the band of the Milky Way are found in autumn. The ranking list of discovery places given in Appendix 5 confirms the leading role of the Northern observatories – only two stations in the Southern Hemisphere rank among the first ten places. Due to the establishment of special minor planet survey telescopes this situation significantly will evolve towards an even larger asymmetry in the near future.

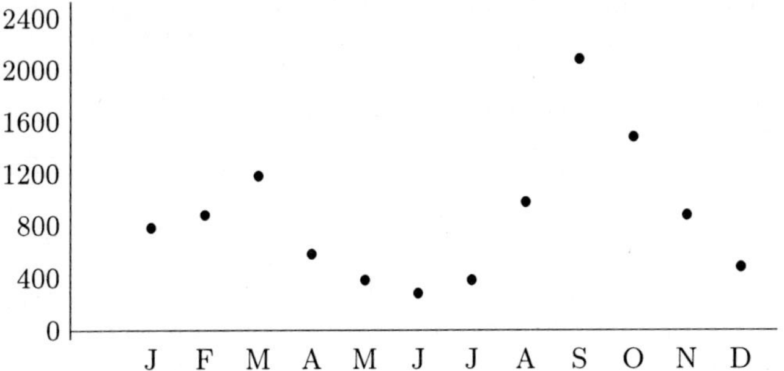

Fig. 2: Minor Planet discovery frequency by months

At May 1999, out of a total of 10666 planets which have been given numbers, only 6730 or 63.1% have been named. In general, the earlier a planet was discovered, the sooner it was named. For example, each of the first 3300 planets were assigned a name. Unfortunately, the naming ratio seriously dropped off during the last few years. Fig. 3 demonstrates that the completeness of names declines considerably with higher numbers.

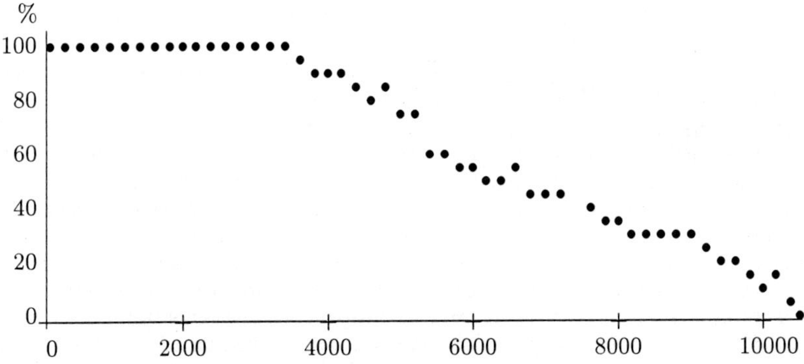

Fig. 3: Fraction of named minor planets

Up to planet (6000) more than half of every 100 subsequent objects have been named, but this fraction declines rapidly for planets with higher numbers. The Minor Planet Center does not like this delay in naming, since the increase of the number of unnamed objects increases the potential for confusion. The combination of a definitive number and name gives the security of redundancy.

There has always been a preference for names consisting of a single word. This rule could not be maintained, however, since the names of some of the people to be honored consist of two or even three words. Thus 120 out of 6730 names combined of more than one word had to be tolerated. A total of 54 designations contain a hyphen, and 36 an apostrophe. For some time a diacritical mark, an umlaut, etc. could not be adequately represented by data processing machines. The use of diacritical marks causes no such problem with machine-readable data now, and consequently such marks are found in 295 names. The length of a chosen name is also interesting. Names consisting of a single character have never been accepted. For reasons of data processing, however, a maximum length of 16 characters was adopted. Unfortunately, this rule was violated in a special case. The following table shows the present distribution of the length of names:

Number of characters	Minor Planet names	Number of characters	Minor Planet names
2	6	10	495
3	79	11	319
4	412	12	186
5	806	13	81
6	1203	14	46
7	1216	15	21
8	1081	16	10
9	768	17	1

More than half of the names consist of six to eight characters. There is a certain overabundance of names consisting of four characters. This is due to an old rule to assign planets a four-letter name if their orbital elements have exceptional characteristics. This rule is now rarely applied. With an increasing number of names difficulties arise with names considered too nearly similar. We therefore observe a slight trend towards longer names.

The names may be classified, of course, according to many different criteria. We decided to arrange groups of countries and groups of free classification. Names dedicated to countries, cities, rivers, buildings, etc., can be classified very easily. There may be difficulties, for example, in assigning names of astronomers, artists, or other people to only one country if they happen to have worked in several different ones. Names from mythology cannot be assigned to a particular country. Names given in honor of a discoverer's relatives were, as a rule, not classified by countries. Nevertheless, as can be seen in Appendix 7, the statistics on some 3900 classifications is quite informative. There is a clear predominance of the 'great powers' in the field of minor planets: names that can be definitely assigned to a country come mainly from the former USSR or the USA, followed by Japan, Germany, France and Finland. The USA and especially Japan became outstanding only during the last decades. The USSR, Germany, and Finland owe their ranking mainly to the observatories at Simeïs/CrAO, Heidelberg, and Turku, respectively. The preponderance of countries from the Northern Hemisphere is striking.

After the publication of the first edition of this "Dictionary" some very useful papers dealing with name classifications have been appeared. Combes (1993) published a very interesting list with free classification criteria. Schmadel (1992) compiled a special catalogue of astronomers belonging to the European Southern Observatory (ESO). Special investigations concerning names with close relations to certain countries have been undertaken by Meeus (1988), Denoyelle (1995/96) and by Schnell and Haupt (1996).

Classification according to other criteria is not so straightforward. After several attempts we found 19 categories which came up rather automatically. In this way more than 95% of all names could be classified, compared to only 58% according to a classification based on country. A first analysis shows that the names of minor planets do not reflect an 'astronomers' cemetery', a 'female sky', or even an 'Iliad sphere', as has often been suggested. There are, however, periods during which the naming could be so described.

Mythological names predominated only in the first three quarters of the 19th century. Later, classical names were almost exclusively invoked for the Trojans. These are divided (with some unfortunate mistakes) into groups of Greek besiegers and their Trojan opponents. The Trojan planets are distinguished by their 1:1 resonance with Jupiter. They were given male names in contrast to the normal, 'female' planets. This anti-feminine character shows as well in the clear preponderance of male names to female names, in the ratio of 4 to 1: as far as naming minor planets is concerned, emancipation has not yet been very successful.

The suspicion that a cemetery for astronomers has been founded cannot be substantiated either, since only a bare quarter of all names refers to this profession. Appendix 6 shows that many other scientists have been honored, and the list of amateur astronomers is also remarkable. A long list of names refers to relatives of the discoverer; that list remains longer than the one of important names from occidental culture. As an analogy to characters from classical mythology, genealogical tables can be reproduced. An example with the names of relatives suggested by the discoverer couple Carolyn and Eugene Shoemaker is given in Fig. 4.

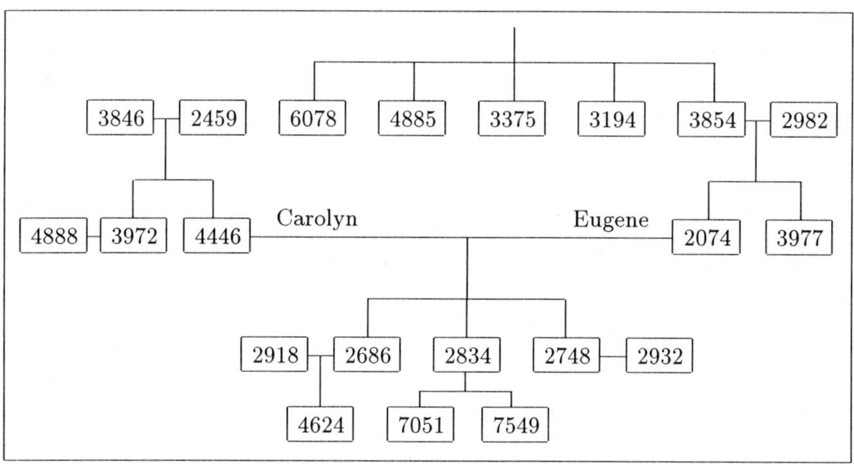

Fig. 4: The Shoemaker dynasty

Without going into details we see that spouses of the patrons prevail over their children as well as over their parents. It is much easier for a grandchild to become astronomically immortal than it is for a grandparent. And the uncles and aunts of the discoverers are much less represented than their friends.

Distinguished musicians, painters, and writers constitute a substantial group of names. Dramatists and novelists are still prevailing in the group of intellectuals, but musicians and composers are coming up rapidly. One can also try to find out if a particular discoverer tends towards names in particular categories. One can see personal preferences very clearly, but trends common to many discoverers only rarely. Today plants and animals are chosen seldom, whereas acronyms and abbreviations are in fashion. Somewhat to our regret, there are not many entries in the category of curiosities. This is a wide field for imaginative discoverers.

Planets named in honor of famous contemporaries can be further analyzed. These names can be arranged into subcategories. This is shown for two groups. Appendices 8 and 9 give lists of Nobel laureates and of IAU officers who gave reason for being honored with a planet's name. Remarkably, not only prize-holders for physics were honored. Considering the great number of all Nobel laureates, this list is rather short. This is very different from the list of astronomers who once were officers in the IAU, the discoverers' professional representation. Here we find nearly no gaps and one can assume that existing ones will soon be filled, too.

Sources of Information

During the second half of the 19th century, the *AN* were the leading resource in this regard, because they appeared promptly and were distributed widely. The *AN* were both a scientific journal and a data base. Contributions taken from this source are referred to here by '*AN*', followed by the number of the volume and the column, as well as by the year. All other journals are referred to by the system of abbreviations used in 'Astronomy and Astrophysics Abstracts'. Concurrently with the *AN*, the *BAJ* gained a central importance. Reports on current affairs were often published in the Circulars of the *BAJ*; for these, we use the abbreviation 'BAJ Circ.', followed by the serial number and the year.

Two series of informative papers have taken over the role of these journals as distributors of data about minor planets since 1926, namely the *RI* and its successor, the *M.P.C.* From 1926 until the end of World War II, the *ARI*, Berlin, edited special circulars publishing observations, orbital elements, namings, etc. Every astronomer working in this field knows these circulars as '*RI*', an abbreviation for 'Rechen-Institut'. When the Minor Planet Center was founded after World War II, the function of the *RI* was taken over by the Minor Planet Circulars. This dictionary refers to the *M.P.C.* by the letter '*M*', followed by the appropriate number.

An important source of information has been the compilation 'The Names of the Minor Planets', edited by Paul Herget and published by the Cincinnati Observatory (1955, 1968). It reports not only on discovery circumstances but also gives explanations to the names of numerous minor planets, with reference to the astronomers who contributed to the explanations of these names. We took over, partly revised and completed, a great part of the notes published by Herget. This is shown in the dictionary by the letter '*H*', followed by the page number. On the matter of names from classical mythology, we sometimes quote Zimmermann (1964), referred to by the letter 'Z' followed by the appropriate page number.

The bibliographical data contained in the references include other important publications which were very helpful for the description and explanation of names. In many cases, colleagues gave helpful comments. Such private communications to the author (LDS) are not mentioned explicitly; instead, the names of these colleagues are given in parentheses, following the relevant citations. Generally, all informations given which do not exist in the original sources have been included into braces.

Catalogue of Minor Planet Names and Discovery Circumstances

DELLA SCOPERTA

DEL NUOVO PIANETA

CERERE FERDINANDEA

OTTAVO TRA I PRIMARJ DEL NOSTRO SISTEMA SOLARE.

PALERMO

1802

NELLA STAMPERIA REALE.

Title page of Giuseppe Piazzi's book "On the discovery of the new planet CERES FERDINANDEA, the eighth of those known in our solar system". The vignette, against the background of Monte Pelegrini and the city of Palermo, shows an angel observing the goddess Ceres sitting in a carriage drawn by two snakes. The inscription on the telescope "CERES ADDITA COELI" (Ceres was added to the heavens) celebrates this epoch-making discovery of the first of the minor planets. (Courtesy of A. Baldi, Bologne)

(1) Ceres

Discovered 1801 January 1 by G. Piazzi at Palermo.

Named after the Roman goddess of corn and harvests, daughter of Rhea {see planet (577)} and Saturn, sister of Juno, Vesta {see planets (3) and (4)}, Jupiter, Neptune and Pluto, mother of Proserpina {see planet (26)} by Jupiter. (H 1)
The symbol used for Ceres in ancient times was a reaper's hook.

The planet was originally named *Ceres Ferdinandea*, honoring King Ferdinand IV (1751-1825) of Sicily, who has also been King of Naples, and who was taking refuge in Palermo at the time of the discovery. In Germany the planet was called Hera {this is now planet (103)} for a short time. F. X. von Zach {see planet (999)} first discussed the original right of the discoverer to propose a name for a newly discovered object. In the naming controversy about this planet he stated: "Da Herr Prof. Piazzi nunmehr sein eigenes Kind getauft, und *Ceres Ferdinandea* benannt hat, wozu er als erster Entdecker das offenbar Recht hat, auch alle seine Correspondenten zu dieser Benennung von ihm aufgefordert sind: so unterschreiben wir auch unserer Seits diese recht schickliche Benennung mit wahrem und desto grösserem Vergnügen, weil dem Könige von Neapel unstreitig als eifrigem Beschützer und Beförderer der Sternkunde, und als grossmüthigem Stifter einer neuen stattlichen Sternwarte, unsere dankbarste Erkenntlichkeit um so mehr gebührt, da er eine Sternwarte zu bauen nicht nur angefangen, sondern auch vollendet hat; nicht bloss die prächtigsten und kostbarsten Englischen Werkzeuge angekauft hat, und in Kisten und Verschlägen auf Rumpelkammern aufbewahrt, sondern, wohin sie gehören, setzen lässt: diese vortrefflichen Instrumente nicht ungeschickten und unfleissigen Händen, sondern einem Gelehrten von anerkannten Verdiensten und Geschicklichkeit anvertraut, und diesen ein für allemahl in den Stand gesetzt, seine Arbeiten und Beobachtungen auf königl. Kosten zum Druck zu befördern. Daher denn auch in so kurzer Zeit die nützlichsten und glänzendsten Früchte aus der Palermer Sternwarte hervorgegangen, die gelehrte Welt mit mehreren Bänden der schätzbarsten Beobachtungen beschenkt, und dieser Tempel der Sicilianischen Urania durch die merkwürdigste Entdeckung, mit dem pünctlich eintretenden neuen Jahrhundert, auf Jahrtausende mit ihrem Stifter und Priester ist verewigt worden. Mit Recht sagt daher Piazzi in seiner Abhandlung, dass Ferdinand IV mit grösserem Rechte eine Stelle am Himmel, als manchem andern Protector der Sternkunde gebühre." (Mon. Corr., Band IV, p. 577-578 (1804)). The naming controversy is discussed in some detail by A. Manara (Mem. Soc. Astron. Ital., Vol. 68, No. 3, p. 679-686 (1997)).

(2) Pallas

Discovered 1802 March 28 by H. W. Olbers at Bremen.

War-like appellation of Athene {see planet (881)} or Minerva {see planet (93)}, goddess of wisdom, war and the liberal arts, who sprang from Zeus' {see planet (5731)} head. (H 1)
The symbol used for Pallas in ancient times was a head of a spear.

The planet was discovered accidentally while looking for (1) Ceres. This is also the name of a lunar crater.

(3) Juno

Discovered 1804 September 1 by K. Harding at Lilienthal.

Named after the queen of all the gods, daughter of Saturn and Rhea {see planet (577)}, wife of Jupiter, mother of Mars, Hebe {see planet (6)}, Lucina {see planet (146)} and Vulcan, sister of Ceres {see planet (1)}. (H 1)

The symbol used for Juno in ancient times was a scepter surmounted with a star.

(4) Vesta

Discovered 1807 March 29 by H. W. Olbers at Bremen.

Named for the patroness of the vestal virgins, goddess of fire and sister of Ceres {see planet (1)}. (H 1)

Named by C. F. Gauss. The symbol used for Vesta in ancient times was an altar with fire upon it.

(5) Astraea

Discovered 1845 December 8 by K. L. Hencke at Driesen.

This is the Latin name for the Greek Astraia. This was the goddess of justice, daughter of Zeus and Themis {see planets (5731) and (24)}. Astraea lived on earth during the Golden Age, but the wickedness and impiety of mankind during the Brazen and Iron ages drove her to heaven, where she was placed among the constellations of the zodiac under the name Virgo. (H 2; Z 34)

Named by J. F. Encke (Am. J. Sci., Vol. 2, p. 435 (1846)). The symbol used for Astraea in ancient times was a pair of balances.
The planet was accidentally discovered while looking for (4) Vesta.

(6) Hebe

Discovered 1847 July 1 by K. L. Hencke at Driesen.

Named after the Greek goddess of youth and cup-bearer to all the gods, daughter of Zeus and Hera {see planets (5731) and (103)}, wife of Hercules. (H 2)

Named by C. F. Gauss (AN 26, 155 (1847)). The symbol used for Hebe in ancient times was a beaker.

(7) Iris

Discovered 1847 August 13 by J. R. Hind at London.

Named after the messenger of the gods, especially Juno {see planet (3)}, daughter of Thaumas and Electra {see planet (130)}, sister of the Harpies. Synonym with the term rainbow. W. Bishop (Astronomical Observations, 1839-1851) stated: "The name of Iris, an attendant upon Juno, appeared peculiarly appropriate under the circumstances of the discovery. Juno was then in the 18th hour of right ascension and Iris closely followed in the 19th." The planet was discovered at Mr. Bishop's private observatory. The rainbow symbol is shown in AN 26, 207 (1847). (H 2)

(8) Flora

Discovered 1847 October 18 by J. R. Hind at London.
Named for the goddess of flowers and gardens, wife of Zephyrus. (H 2)

Named by J. Herschel (AN 26, 254 (1847)): "Pallas, Juno, Ceres and Vesta, as sober and majestic Duennas will abundantly provide for the respectability of the group between Mars and Jupiter, while Astraea, Iris, Hebe, and Flora will attract all eyes and fill all imaginations with sweet and graceful images." The symbol used for Flora in ancient times was the rose of England.

(9) Metis

Discovered 1848 April 25 by A. Graham at Markree.

Named after an Oceanid, daughter of Tethys and Oceanus, and first wife of Jupiter. He devoured her lest she bear a child more powerful than he. She personalizes prudence. (H 2)

The planet was discovered at the private observatory of Mr. Cooper who delegated to the discoverer the choice of a name from either Metis, put forward by Robinson, Armagh or Thetis suggested by J. Herschel. The name Thetis was later given to planet (17).

The name Metis has also been given to the satellite Jupiter XVI, discovered 1979 by S. P. Synnott. The symbol used for Metis in ancient times was an eye and a star.

(10) Hygiea

Discovered 1849 April 12 by A. de Gasparis at Naples.

Named for the goddess of health, daughter of Aesculapius {see planet (1027)}. (H 2)

Named by M. Capocci, director of the Naples observatory. The symbol used for Hygiea in ancient times was a star and a serpent.

(11) Parthenope

Discovered 1850 May 11 by A. de Gasparis at Naples.

Named for a Siren, sometimes described as having a fish's body, who, cast upon the shore, founded the city of Parthenope, now Naples. (H 2)

The symbol used for Parthenope in ancient times was a star and a fis h. Alternatively, a harp was used as a proper sign.

The discoverer "used his utmost endeavors to realize a Parthenope in the heavens, such being the name suggested by Sir John Herschel on the occasion of the discovery of (10) Hygiea." (Mon. Not. R. Astron. Soc., Vol. 10, p. 145 (1850)).

(12) Victoria

Discovered 1850 September 13 by J. R. Hind at London.

Named after the Roman goddess of victory, daughter of the giant Pallas {see planet (2)}, or Titan and Styx; also the reigning queen of England. The corresponding Greek goddess Nike {see planet (307)} was added some 40 years later to the asteroid sky. The adopted sign for this planet is described in AN 31, 191 (1850): "Es ist ein Stern mit einem Lorbeerzweige." (H 3)

A celebrated controversy arose over the naming, as some astronomers objected to the use of the name of a reigning sovereign. B. A. Gould, editor of the Astronomical Journal adopted Clio, an alternative name proposed by the discoverer. W. C. Bond {see planet (767)}, Harvard College Observatory, held that the mythological condition was fulfilled, and the name justified, an opinion concurred in by the great majority of astronomers.

(13) Egeria

Discovered 1850 November 2 by A. de Gasparis at Naples.

Named for the nymph of Aricia in Italy, wife of Numa Pompilius, second king of Rome. (H 3)

Named by U. J. J. Leverrier. The symbol used for Egeria in ancient times was a star and a plate.

(14) Irene

Discovered 1851 May 19 by J. R. Hind at London.

Named for one of the Hours, personification of peace, daughter of Jupiter and Themis {see planet (24)}. The others are Eunomia and Dike {see planets (15) and (99)}. They were represented as opening the gates of heaven and of Olympus. The discoverer states in AN 32, 277 (1851): "You will readily discover that this name, properly Eirene (peace), has some relation to this event (the Great Industrial Exhibition) which is now filling our Metropolis with the talent of art and science, in which all mankind must feel an interest." (H 3)

Named by J. Herschel. The symbol used for Irene in ancient times was a dove carrying an olive branch and having a star on its head. Sometimes an alternative sign, representing an olive branch, a flag of truce and a star, was in use.

(15) Eunomia

Discovered 1851 July 29 by A. de Gasparis at Naples.

Named for one of the Hours, personification of order and the law. The others are Irene and Dike {see planets (14) and (99)}. (H 3)

The symbol used for Eunomia in ancient times was a heart, surmounted with a star.

(16) Psyche

Discovered 1852 March 17 by A. de Gasparis at Naples.

Named after the nymph whom Cupid married. Venus put her to death because she had robbed the world of her son; but Jupiter at the request of Cupid granted immortality to Psyche. The word signified "the soul" which she personifies. (H 3)

The symbol used for Psyche in ancient times was a butterfly's wing, a personification of the soul.

(17) Thetis

Discovered 1852 April 17 by R. Luther at Düsseldorf.

Named after the wife of Peleus, king of Thessaly, daughter of Nereus and Doris {see planet (48)}, mother of Achilles {see planet (588)}. (H 3)

Named by F. W. A. Argelander. The symbol used for Thetis in ancient times was a dolphin and a star.

(18) Melpomene

Discovered 1852 June 24 by J. R. Hind at London.

Named for the Muse of tragedy. All the Muses were offspring of Jupiter and Mnemosyne {see planet (57)}. The others are Klio, Euterpe, Thalia, Terpsichore, Erato, Polyhymnia, Kalliope and Urania {see, respectively, planets (84), (27), (23), (81), (62), (33), (22), (30)}. (H 4; AN 34, 391 (1852))

Named by the Astronomer Royal G. B. Airy who, in a letter to the Cape astronomer D. Gill, stated: "I look upon her as my planet for the following reason which you will not find in books. On 1839 June 24 I lost my noble boy Arthur. On 1852 June 24 (just 13 years later) I lost my dear daughter Elizabeth. And, while feeling that day of sorrow, I learnt that on that day a planet was discovered which I was requested to name. So I fixed on the name of the Muse of sadness." (cf. G. Forbes: David Gill, man and astronomer, p. 91 (London, 1916))
For the first time, no sign was adopted for this planet.

(19) Fortuna

Discovered 1852 August 22 by J. R. Hind at London.

Named after a powerful deity among the ancients, goddess of fortune, dispenser of riches and poverty, pleasures and misfortunes. (H 4)

Named by the discoverer "at the deputation of Mr. Bishop". The planet was discovered at W. Bishop's private observatory.

(20) Massalia

Discovered 1852 September 19 by A. de Gasparis at Naples. Independently discovered 1852 September 20 by J. Chacornac at Marseilles.

This is the Greek name of the city of Marseilles. The Latin spelling Massilia was also used sometimes. (H 4)

Named by B. Valz without knowledge of the Naples discovery. The planet was detected independently by J. Chacornac at Marseilles 1852 September 20. De Gasparis acquiesced in the matter. Valz was one of the first to suggest the encircled number of the planet as its symbol which is now modified to parentheses (AN 35, 194 (1852)). A. de Gasparis had chosen the name Themis for this planet {see also planet (24)} as proposed by Herschel (AN 35, 279 (1852)).

This is the first minor planet name which does not originate from ancient mythologies.

(21) Lutetia

Discovered 1852 November 15 by H. Goldschmidt at Paris.

This planet is named for the city of Paris by its Latin name. The discoverer was a German painter living in Paris who discovered 14 minor planets, some from his window above the Cafe Procope. (H 4)
Named (AN 35, 343 (1852)) by F. J. D. Arago, director of the Paris Observatory. This is the first minor planet discovered by an amateur astronomer.

(22) Kalliope

Discovered 1852 November 16 by J. R. Hind at London.
Named for the Muse of heroic poetry {see also planet (18)}. (H 4)
Named by J. C. Adams, president of the Royal Astronomical Society.

(23) Thalia

Discovered 1852 December 15 by J. R. Hind at London.

Named for the Muse of comedy {see also planet (18)}. Thalia is also one of the three Graces, daughters of Venus by Jupiter or Bacchus {see planet (2063)}. The other Graces are Euphrosyne and Aglaja {see planets (31) and (47)}. (H 4)

Named by W. Bishop at whose private observatory the planet was discovered (AN 35, 380 (1852)).

(24) Themis

Discovered 1853 April 5 by A. de Gasparis at Naples.

Named for one of the Titans, goddess of justice, daughter of Uranus and Gaea {see planet (1184)}, mother of the Hours {see planets (14) Irene, (15) Eunomia, and (99) Dike} and Fates {see planets (97) Klotho, (120) Lachesis, and (273) Atropos} by Zeus {see planet (5731)}. Themis is known as the goddess of law. (H 4)

Named by A. Secchi. This is the same name that A. de Gasparis had planned to call planet (20).

(25) Phocaea

Discovered 1853 April 6 by J. Chacornac at Marseilles.

Named after a maritime town of Ionia in Asia Minor with colonies in Italy, Spain, and France. Beset by the tyrant Haspages a group of the inhabitants built Massalia (Marseilles, see planet (20)), sometimes designated Phocaica, about 600 B.C. (H 5; AN 36, 279 (1853))
Named by B. Valz.

(26) Proserpina

Discovered 1853 May 5 by R. Luther at Düsseldorf.

Named after the daughter of Ceres {see planet (1)} and Jupiter, carried away by Pluto to become queen of the infernal regions. Her mother after a lengthy search was unable even with the aid of Jupiter to secure her daughter's return to the upper regions since she had partaken of food in the Elysian fields - a pomegranate. A compromise was brought about whereby she spent half the year in the upper world. Proserpina presided over the death of mankind. See also the citation for planet (399) Persephone which is the Greek name for Proserpina. (H 5)

Named by "Sr. Excellenz dem wirklichen Geheimen Rath" Alexander von Humboldt (AN 36, 26 (1853)). The sign of the planet is "ein Granatapfel mit einem Stern im Innern."

(27) Euterpe

Discovered 1853 November 8 by J. R. Hind at London.

Named for the Muse of music and lyric poetry {see also planet (18)}. Her symbol was the flute. Some legends say she invented the flute and all the wind instruments. She loved wild melodies and associated more with Bacchus than with Apollo {see planets (2063) and (1862)}. (Z 105)

(28) Bellona

Discovered 1854 March 1 by R. Luther at Düsseldorf.
Named for the war goddess, wife, sister, nurse or mother of Mars. (H 5)
Named by J. F. Encke.

The planet was named at the outset of the Crimean War (1854-1856). See also the remarks to planets (40) and (494). The sign of the planet is given in AN 38, 143 (1854): "die Geissel und der Spiess der kriegerischen Schwester des Mars."

(29) Amphitrite

Discovered 1854 March 1 by A. Marth at London. Independently discovered 1854 March 2 by N. R. Pogson at Oxford and March 3 by J. Chacornac at Paris.

Named after an Oceanid, wife of Poseidon {see planet (4341)} and mother of Triton. (H 5)

Named by W. Bishop at whose private observatory the planet was discovered. The symbol used for Amphitrite in ancient times was a shell and a star.

(30) Urania

Discovered 1854 July 22 by J. R. Hind at London.
Named for the Muse of astronomy {see also planet (18)}. (H 5; AN 39, 92 (1854))
Named by A. de Morgan, secretary of the Royal Astronomical Society.

(31) Euphrosyne

Discovered 1854 September 1 by J. Ferguson at Washington.

Named for one of the three Graces. The others are Thalia and Aglaja {see planets (23) and (47)}. (H 5)
This is the first numbered minor planet discovered in North America.

(32) Pomona

Discovered 1854 October 26 by H. Goldschmidt at Paris.

Named for a nymph at Rome, goddess of the fruit trees and gardens and wife of Vertumnus. (H 5)

Named by U. J. J. Leverrier. Wrongly, the symbol of an apple and a star was used sometimes. This sign, however, belongs to (26) Proserpina.

(33) Polyhymnia

Discovered 1854 October 28 by J. Chacornac at Paris.
Named for the Muse of singing and rhetoric {see also planet (18)}. (H 6)
Named by U. J. J. Leverrier.

(34) Circe

Discovered 1855 April 6 by J. Chacornac at Paris.

Named for the enchantress, daughter of the Sun, celebrated for her knowledge of magic and venomous herbs. Circe changed the companions of Odysseus {see planet (1143)} into pigs. She had no influence on Odysseus because Hermes protected him. Odysseus lived a year with Circe, his friends were retransformed into men. (H 6)
The planet was named by the members of Paris observatory.

(35) Leukothea

Discovered 1855 April 19 by R. Luther at Düsseldorf.

Named for the daughter of Cadmus and Hermione {see planet (121)} who was also called Ino {see planet (173)}. She married Athamas, king of Thebes, and was later changed to a sea deity. (H 6)

Named by G. Rümker and C. A. F. Peters (AN 40, 373 (1855)) for the "Beschützerin der Seefahrer". The sign adopted for the planet consisted of an ancient lighthouse.

(36) Atalante

Discovered 1855 October 5 by H. Goldschmidt at Paris.

Named for a girl of Arcadia {see planet (1020)} whose hand was won by Hippomenes who defeated her in a race by dropping three golden apples given him by Aphrodite {see planet (1388)}. (H 6)

Named by U. J. J. Leverrier.

(37) Fides

Discovered 1855 October 5 by R. Luther at Düsseldorf.

Named after the Roman goddess of faith, oaths and honesty. (H 6)

Named by the Düsseldorf municipal council (AN 42, 107 (1855)): "... dass der Düsseldorfer Gemeinderath in seiner gestrigen Sitzung, aus einer vom Curatorium der Sternwarte zu diesem Zweck entworfenen Liste, für meinen neuesten Planeten den Namen ... gewählt hat." A cross was adopted as the sign for this planet.

(38) Leda

Discovered 1856 January 12 by J. Chacornac at Paris.

Named for the wife of the king of Sparta. By Jupiter, who had taken the form of a swan, she produced two eggs from one of which came Pollux and Helena {see planet (101)} and from the other Castor and Klytemnaestra {see planet (179)}. (H 6)

Named by U. J. J. Leverrier.

This planet could not be observed in its second apparition. It was rediscovered in its third opposition by W. Förster. The name Leda has also been given to the satellite Jupiter XIII, discovered 1974 by C. Kowal.

(39) Laetitia

Discovered 1856 February 8 by J. Chacornac at Paris.

Named for the secondary Roman goddess of gaiety. (H 6)

Named by U. J. J. Leverrier.

(40) Harmonia

Discovered 1856 March 31 by H. Goldschmidt at Paris.

Named after the daughter of Ares and Aphrodite {see planet (1388)}, wife of Cadmus {see planet (7092)}, the founder of Thebes. She introduced art and music into Greece. (H 6)

Named by U. J. J. Leverrier.

The name was selected in order to "place a lasting monument to the happy re-establishment of peace". The Crimean War (1854-1856) came to a close with the signing of peace in Paris, February 1856. See also the citation for planet (28).

(41) Daphne

Discovered 1856 May 22 by H. Goldschmidt at Paris.

Named after the daughter of Terra. She entreated the gods to save her from the pursuit of Apollo {see planet (1862)} and she was changed into a laurel tree. (H 6)

Named by U. J. J. Leverrier.

(42) Isis

Discovered 1856 May 23 by N. R. Pogson at Oxford.

Named after the Egyptian goddess. Isis (the Moon) and her brother-husband Osiris {the Sun, see planet (1923)} comprehended all nature and were the most beneficial of the Egyptian gods. (H 7)

Named by Prof. Manuel J. Johnson, director of the Radcliffe Observatory, Oxford.

The planet is probably also named for the discoverer's daughter, Elizabeth Isis Pogson, who was a member of the Royal Astronomical Society.

(43) Ariadne

Discovered 1857 April 15 by N. R. Pogson at Oxford.

Named for the daughter of Minos {see planet (6239)}, second king of Crete, and Pasiphae. Ariadne fell in love with Theseus who was shut up in the labyrinth to be devoured by the Minotaur. She aided his escape with a thread and was later married and forsaken by him. (H 7)

(44) Nysa

Discovered 1857 May 27 by H. Goldschmidt at Paris.

It was to the nymphs of Nysa that the education of the young Bacchus {see planet (2063)} was entrusted. No less than ten places had the name of Nysa including those in Ethiopia, Arabia, India {see planets (1432) and (1157)}, Thrace and Euboea {see planet (1119)}. (H 7)

Named by A. von Humboldt.

(45) Eugenia

Discovered 1857 June 27 by H. Goldschmidt at Paris.

Named by the discoverer in honor of the French empress and Spanish noblewoman Eugenia de Montijo de Guzmán (1826-1920), wife of Napoleon III. (H 7)

This is the first example of assigning the name of a human being to a minor planet. William J. Merline et al. reported the discovery of a satellite of (45) on 1998 Nov. 1. The moon with the preliminary designation S/1998 (45) 1 orbits the minor planet in a near-circular orbit every 4.7 days (IAU Circ. No. 7129).

(46) Hestia

Discovered 1857 August 16 by N. R. Pogson at Oxford.

This is the Greek name for Vesta {see planet (4)}, here considered as one of the Hesperides, daughters of Hesperis and Atlas, guardians of the golden apples given to Juno {see planet (3)} by Jupiter on their wedding day. The other Hesperides are Aegle and Arethusa {see planets (96) and (95)}. (H 7)

Named by Admiral W. H. Smyth, "who oblingly complied with my request to stand God-father to this first offspring of his former telescope" (AN 47, 213 (1857)).

(47) Aglaja

Discovered 1857 September 15 by R. Luther at Düsseldorf.

Named for one of the three Graces. The others are Thalia and Euphrosyne {see planets (23) and (31)}. (H 7)

This planet was named by the members of the Philosophical Faculty of the University of Bonn (AN 47, 47 (1858)).

(48) Doris

Discovered 1857 September 19 by H. Goldschmidt at Paris.

Named for one of the sea nymphs, called Oceanides, daughter of Oceanus and Tethys. Doris married her brother Nereus {see planet (4660)} by whom she bore the sea nymphs called Nereids, about fifty in number. (H 7)

Named by E. de Beaumont.

The name was chosen by the French geologist from a list of suggestions by Babinet, Secretary of the Paris Academy of Sciences. Beaumont further suggested the appellation *deux Jumelles* (the Twins) for this planet and (49) Pales, both discovered by Goldschmidt on the same night - a unique occurrence at that time.

(49) Pales

Discovered 1857 September 19 by H. Goldschmidt at Paris.
Named for the Roman goddess of sheepfolds and pastures. (H 8)
See also the remarks on planet (48).

(50) Virginia

Discovered 1857 October 4 by J. Ferguson at Washington. Independently discovered 1857 October 19 by R. Luther at Düsseldorf.

Two possible interpretations for this name are published. Virginia is perhaps the famous Roman girl, stabbed by her father, the centurion Virginius, to save her from the decemvir Appius Claudius in 448 B.C. The name could be also an allusion to the contiguous state of Virginia. (H 8)

(51) Nemausa

Discovered 1858 January 22 by A. Laurent at Nimes.

The planet was named "in memory and honor of the city and the fountain of the god Nemausus". Nemausus was the ancient name of Nimes. (H 8)
Named (AN 47, 349 (1858)) by B. Valz with permission of the discoverer.

(52) Europa

Discovered 1858 February 4 by H. Goldschmidt at Paris.

Named for the daughter of Agenor {see planet (1873)}, king of Phoenicia, and sister of Cadmus {see planet (7092)}. She was carried off across the water to Crete on the back of a bull, which form had been assumed by Jupiter who was enamored of her. Europa became the mother of Minos, Sarpedon {see planets (6239) and (2223), respectively} and Rhadamanthus. (H 8)
Named by Marshal Vaillant.

The name Europa has also been given to the second satellite of Jupiter, discovered 1610 by Galilei.

(53) Kalypso

Discovered 1858 April 4 by R. Luther at Düsseldorf.

Named for one of the Oceanides, goddess of silence and daughter of Oceanus and Tethis. Kalypso received Odysseus {see planet (1143)} with hospitality when he was shipwrecked on her shores on the island of Ogygia and became enamored of him. (H 8)
Named by E. Schönfeld.

The name Calypso has also been given to the satellite Saturn XIV, discovered 1980 by B. Smith.

(54) Alexandra

Discovered 1858 September 10 by H. Goldschmidt at Paris.

Named in honor of Baron Alexander von Humboldt (1769-1859), renowned naturalist and explorer of South America, Mexico and Siberia. His name is also associated with the Humboldt stream in the Pacific Ocean. (H 8)
Named by Moigno.

The departure in the naming usage from the broad mythological category was probably allowed because of Alexandra, daughter of Priamus {see planet (884)}, although the discoverer is clear in his intent (Comptes Rendus 47, p. 446).

(55) Pandora

Discovered 1858 September 10 by G. Searle at Albany.

Named after the first mortal female. Pandora was made from clay by Vulcan at the request of Jupiter as a curse on Prometheus {see planet (1809)}. She was given all the necessary gifts by the gods - beauty by Venus, eloquence by Mercury, etc., hence the name. Then she was given a beautiful box to present to the man who married her. Prometheus saw through the deceit but his brother Epimetheus {see planet (1810)} married Pandora. He opened the box and there issued a multitude of evils which have not ceased to this day to afflict the human race. Only hope remained at the bottom of the box. In another version she was not to open the box but curiosity overruled. (H 8)

A famous controversy of the last century was that which raged between B. A. Gould and the trustees of the Dudley Observatory at Albany, from which place he issued the Astronomical Journal. The published charges and countercharges can be found today in astronomical libraries. He was eventually ejected by "hired ruffians". Mrs. Dudley has proposed the name Pandora, "the apt significance of which will be obvious to all" (Gould). The name Pandora has also been given to the satellite Saturn XVII, discovered 1980 by the Voyager 1 spacecraft.

(56) Melete

Discovered 1857 September 9 by H. Goldschmidt at Paris.

Named for one of the three ancient Muses worshipped on Mount Helicon, daughter of Uranus. (H 9)

Named by E. Schubert, Ann Arbor, the orbit computer, at the request of the discoverer (AN 56, 349 (1861)).

The planet was also known as Pseudo-Daphne due to a confusion with planet (41).

(57) Mnemosyne

Discovered 1859 September 22 by R. Luther at Düsseldorf.

Named after the daughter of Coleus and Terra, mother of the nine Muses {see also planet (18)} by Jupiter. The word signifies "memory" without which endowment progress in the spheres of the Muses would not take place. (H 9)
Named by M. Hoek, director of the Utrecht Observatory (AN 51, 255 (1859)).

(58) Concordia

Discovered 1860 March 24 by R. Luther at Düsseldorf.

Named for the Roman goddess of peace and concord, daughter of Jupiter and Themis {see planet (24)}. (H 9)

Named (AN 53, 58 (1860)) by C. Bruhns, Leipzig at the request of the discoverer.

(59) Elpis

Discovered 1860 September 12 by J. Chacornac at Paris.

Named after the Greek word for hope, also the personification of hope. (AN 56, 229 (1861))

Named by K. von Littrow on the request of E. Weiss. Weiss stated: "Da ich die Bearbeitung dieses Planeten übernommen und der Herr Entdecker keine Anstalten getroffen hat, ihn zu benennen, obwohl er schon vor mehr als einem Jahre aufgefunden wurde, habe ich Herrn Director v. Littrow ersucht, einen Namen für denselben auszusuchen. Er wählte 'Elpis' zur Erinnerung an die politische Stimmung der Epoche, in welche die Entdeckung fiel, oder, wenn man will, als Anspielung darauf, dass der Planet eben seinen Namen so lange zu erwarten hatte..."

Leverrier, director of the Paris Observatory did not want to have this planet named, attempting to introduce a new nomenclature - attaching the name of the discoverer to the number. Hind, Goldschmidt, Luther, Airy, Herschel, Argelander and others pronounced against the innovation. Dr. Weiss of the Vienna Observatory, who had taken particular charge of the observations of the planet, requested Prof. K. von Littrow, director of that observatory to name it. He chose Elpis in allusion to the political condition in Europe at the time of discovery and the name was adopted in Germany. In January 1862 Leverrier finally allowed the discoverer to select a name. The latter delegated the office to J. R. Hind who selected Olympia {now the name of planet (582)}, and the dual appellation was in use for a short time.

(60) Echo

Discovered 1860 September 14 by J. Ferguson at Washington.

Named for an Oread, daughter of the Air and Tellus, condemned by Juno {see planet (3)} for her chattering to the loss of voice except for purposes of reply. She subsequently fell in love with Narcissus who was driven away by her manner of speech. Gradually she faded away till nothing was left but her voice. (H 9)

The planet was called Titania at first without knowledge of the previous appropriation of the name for a satellite of Uranus by J. Herschel.

(61) Danaë

Discovered 1860 September 9 by H. Goldschmidt at Paris.

Named after the daughter of Acrisius, king of Argos, and Eurydike {see planet (75)}. Danaë was confined by her father to a brazen tower which was entered by Zeus {see planet (5731)} in the form of rain. By him she had Perseus. (H 9)
Named (AN 54, 93 (1860)) by R. Luther at the request of the discoverer.
The discoverer conceded the naming because of illness.

(62) Erato

Discovered 1860 September 14 by O. Lesser and W. Foerster at Berlin.
Named after the muse of lyric poetry {see also planet (18)}. (H 9)
Named by J. F. Encke, director of the Berlin Observatory.

(63) Ausonia

Discovered 1861 February 10 by A. de Gasparis at Naples.

Named for a part or all of Italy after its king Auson, a son of Odysseus {see planet (1143)} and Kallisto {see planet (204)}. (H 10)
Named by M. Capocci (AN 55, 79 (1861)).

The planet was first called Italia, a name which nowadays is assigned to planet (477).

(64) Angelina

Discovered 1861 March 4 by E. W. Tempel at Marseilles.

Named in remembrance of the astronomical station of Baron F. X. von Zach (1754-1832) {see planet (999)} at Notre-Dame-des-Anges on the mountain of Mimet near Marseilles, hermitage and convent of the Pères de l'Oratoire. The aim of von Zach's observations was to determine the plumb deviation caused by mountains. He published his results in his monumental work *L'attraction des montagnes* (Avignon, 1814). Zach travelled extensively in his early life, promoted international cooperation, was active as editor of an astronomical publication, and helped organize the search for the missing planet between Mars and Jupiter. (H 10)
Named by B. Valz.

A storm of protest raised by Herschel, Airy and Argelander arose over the naming of this planet. The discoverer stated in AN 71, 96 (1868): "Der Planet, welcher von Herrn Valz ... Angelina getauft wurde und Anfechtungen erhielt, wäre durch Weglassung des letzten 'n' classisch mythologisch geworden und hätte gewiss beide Theile befriedigt."

(65) Cybele

Discovered 1861 March 8 by E. W. Tempel at Marseilles.

Named for the Phrygian goddess often identified with Rhea {see planet (577)}. (H 10)

The planet was originally named Maximiliana by the Bavarian Ministerialrat K. A. von Steinheil in allusion to Maximilian II, the king of Bavaria. The name was regarded as non-classical and was changed after protest by Herschel, Airy, Argelander and others by the *Berliner Astronomisches Jahrbuch*. The discoverer transferred the naming rights to Steinheil "aus Anerkennung für die ausgezeichneten Leistungen seines aus dessen Werkstätte bezogenen Fernrohrs." (AN 55, 93 (1861))

(66) Maja

Discovered 1861 April 9 by H. P. Tuttle at Cambridge, Massachusetts.

Named for one of the Pleiades, daughters of Atlas and Pleione. Maja was the mother of Mercury by Jupiter. (H 10)

Named by J. Quincy, former president of Harvard University, friend of the Observatory and member of its Visiting Committee.

Further three minor planets are named for one of the Pleiades: Elektra, Asterope, and Merope {see planets (130), (233), and (1051), respectively}. G. P. Bond, director of the Harvard College Observatory, in AN 55, 299 (1861) pointed out: "It may perhaps be objected that the same name has been applied to one of the Plejades, but there can scarcely be any risk of confusions from the double application."

(67) Asia

Discovered 1861 April 17 by N. R. Pogson at Madras.

Asia was the wife of Iapetus and the mother of Atlas and Prometheus {see planet (1809)}. The discoverer stated in AN 55, 285 (1861)): "The name "Asia" has been selected in consequence of this being the first discovery made in this quarter of the globe. As one of the Oceanides, Asia has just as much right to a celestial celebrity as her sisters Europa, Doris {see planets (52) and (48)} and many others of the same family long since adopted." (H 10)
This is the first numbered minor planet discovered in Asia.

(68) Leto

Discovered 1861 April 29 by R. Luther at Düsseldorf.

Named for the daughter of Coeus and Phoebe and mother of Artemis {see planet (105)} and Apollo {see planet (1862)} by Zeus {see planet (5731)}. (H 10)
Named by F. W. A. Argelander, C. N. A. Krueger and Tiele at Bonn.
The Latin name for Leto is Latona - see planet (639).

(69) Hesperia

Discovered 1861 April 26 by G. Schiaparelli at Milan.

Hesperia is the Greek name for Italy derived from Hespera or Vesper, the setting Sun, the evening or the evening star in which direction Italy lay. (H 10)

(70) Panopaea

Discovered 1861 May 5 by H. Goldschmidt at Paris.

Named for a sea nymph, one of the Nereids, whom sailors invoked during storms. (H 11)

Named (AN 55, 235 (1861)) by R. Main, Oxford Observatory, president of the Royal Astronomical Society, who had the occasion to present the Gold Medal of the RAS to the discoverer for his planetary discoveries.

(71) Niobe

Discovered 1861 August 13 by R. Luther at Düsseldorf.

Named for the daughter of Tantalus {see planet (2102)} and sister of Pelops, the king of Phrygia. She presumed to be greater than Latona {see planet (639)}. Apollo {see planet (1862)} punished her by killing her seven sons and seven daughters and changing her into a rock. (H 11)

The name was chosen by a majority of astronomers who met in Dresden on August 20/21, 1861.

(72) Feronia

Discovered 1861 May 29 by C. H. F. Peters at Clinton.

Named for the Roman goddess of groves and freedmen. (H 11)

Named by T. H. Safford, assistant at Harvard, who discovered that this was a new planet and not (66) as originally identified.

(73) Klytia

Discovered 1862 April 7 by H. P. Tuttle at Cambridge, Massachusetts.

Named for the nymph loved by and later despised by Apollo {see planet (1862)}. She was changed into a sunflower. (H 11)

Named by J. I. Bowditch of Boston, member of the Harvard Observatory Visiting Committee.

(74) Galatea

Discovered 1862 August 29 by E. W. Tempel at Marseilles.

Two possible interpretations of the name are published. Galatea probably refers to the name given by the Cypriot sculptor Pygmalion to his marble statue of which he had become enamored. The other possibility is that the planet was named after the Nereid loved by Acis who was destroyed by the jealous Cyclops Polyphemus. (H 11)

Named by K. von Littrow (AN 58, 255 (1862)) at the request of the discoverer (AN 58, 207 (1862)).

The name Galatea has also been given to the satellite Neptune VI.

(75) Eurydike

Discovered 1862 September 22 by C. H. F. Peters at Clinton.

Named for the wife of Orpheus {see planet (3361)} who descended into Hades to secure her return which he obtained on the condition that he would not look back at her until reaching the upper world. She was returned to Hades when he looked back at the last moment. (H 11)

(76) Freia

Discovered 1862 October 21 by H. d'Arrest at Copenhagen.

Named after the Norse goddess of love and beauty. (AN 59, 16 (1862))

(77) Frigga

Discovered 1862 November 12 by C. H. F. Peters at Clinton.

Named after the Norse goddess and queen of all the gods. Frigga is the wife of Odin {see planet (3989)}. (H 11)

The discoverer stated in AN 60, 107 (1863): "... erlaube ich mir den Namen Frigga vorzuschlagen, da Frigga und Freia {see planet (76)} in der nordischen Mythologie so häufig in Gesellschaft sich finden..."

(78) Diana

Discovered 1863 March 15 by R. Luther at Düsseldorf.

Named for the goddess of the hunt and daughter of Jupiter and Latona {see planet (639)}. The Greek name of Diana is Artemis {see planet (105)}. (H 11)

Named by Privy Councilor of Justice W. Luther of Naumburg, uncle of the discoverer.

(79) Eurynome

Discovered 1863 September 14 by J. C. Watson at Ann Arbor.

Named after a nymph, daughter of the Ocean and Thetis {see planet (17)} and mother of Leukothea {see planet (35)}. (H 12)

(80) Sappho

Discovered 1864 May 2 by N. R. Pogson at Madras.

Named in honor of the renowned Greek lyric poetess (610 B.C.) who threw herself into the ocean because of her unrequited love for the young Phaon. (H 12)

(81) Terpsichore

Discovered 1864 September 30 by E. W. Tempel at Marseilles.
Named for the Muse of choral dance and song {see also planet (18)}. (H 12)
Named (AN 63, 127 (1874)) by C. H. F. Peters.

(82) Alkmene

Discovered 1864 November 27 by R. Luther at Düsseldorf.

Named after the mother of Hercules by Jupiter who had assumed the shape of her husband Amphitryon. (H 12)

Named by C. von Littrow, E. Weiss and Th. von Oppolzer in Vienna at the request of the discoverer (AN 63, 269 (1864)).

(83) Beatrix

Discovered 1865 April 26 by A. de Gasparis at Naples.

Named for a Florentine lady of the noble family of Portinari, immortalized by Dante {see planet (2999)} in his "Divine Comedy". Beatrix is also the name of the discoverer's daughter. (H 12)

The remark in Mon. Not. R. Astron. Soc., Vol. 25, p. 221 (1875) states "in honor to Dante". The discoverer in AN 64, 256 (1875) stated: "Sauf précédence dans la découverte, en hommage au Dante je propose pour cette planète le nom de Béatrix".

(84) Klio

Discovered 1865 August 25 by R. Luther at Düsseldorf.
Named for the Muse of history {see also planet (18)}. (H 12)
Named by Leipzig astronomers at the meeting on August 31, 1865.

(85) Io

Discovered 1865 September 19 by C. H. F. Peters at Clinton.

Named after the daughter of the river god Inachus, son of Oceanus. She was changed into a heifer by Jupiter at the approach of Juno {see planet (3)} who saw through the deception and had her guarded by Argus who had a hundred eyes and slept with them closed only two at a time. The latter was killed by Mercury at Jupiter's order and Io escaped through the Ionian Sea, named for her, to Egypt where some say she became Isis, the wife of Osiris {see planets (42) and (1923)}. (H 12)

The name Io has also been given to the first satellite of Jupiter, discovered 1610 by Galilei.

(86) Semele

Discovered 1866 January 4 by F. Tietjen at Berlin.

Named for the daughter of Cadmus and mother of Bacchus {see planet (2063)} by Jupiter. (H 12)

(87) Sylvia

Discovered 1866 May 16 by N. R. Pogson at Madras.

Named possibly in honor of Sylvie Petiaux-Hugo Flammarion, first wife of Camille Flammarion {see planet (1021)}, French popularizer of astronomy. This interpretation seems very questionable. Paluzíe-Borrell quoted the book of Camille Flammarion, *Astronomie Populaire*, Lib. IV, Chap. V: "La 87e, la 107e, la 141e, la 154e, la 169e, ont été nommées en l'honneur d'un jeune astronome qui consacrait ses meilleures années au culte de l'astronomie et à l'apostolat de cette belle science." (H 12)

The discoverer stated: "The name Sylvia as in the case of Sappho {see planet (80)} was chosen out of a list supplied to me some years back by Sir John Herschel." Herschel undoubtedly was an advocate for using classical names. A very similar way of thinking must be assumed for Pogson who exclusively used classical names for former discoveries {see planets (29), (42), (43), (46), (67), and (80), respectively}.

(88) Thisbe

Discovered 1866 June 15 by C. H. F. Peters at Clinton.

Pyramus and Thisbe were two Babylonian lovers prominent in Shakespeare's *A Midsummer Night's Dream.* (H 13)

(89) Julia

Discovered 1866 August 6 by E. Stephan at Marseilles.

Named probably after the patron saint Julia of Corsica who died as a martyr in the fifth century. At the time of the discovery, the island in the vicinity of Marseilles already was a part of France. (LDS)

(90) Antiope

Discovered 1866 October 1 by R. Luther at Düsseldorf.

The name appears twice in the mythology. Antiope was the daughter of Nycteus, king of Thebes, and Polyxo {see planet (308)} and mother of two children by Zeus {see planet (5731)}. Antiope is also mentioned as one of the queens of the Amazones and daughter of Ares. She was taken prisoner by Hercules and given in marriage to Theseus. (H 13)
Named by von Kühlwetter, district president of Düsseldorf.

(91) Aegina

Discovered 1866 November 4 by A. Borrelly at Marseilles.

Named after the daughter of Asopus, a son of Poseidon {see planet (4341)}. She was changed by Zeus {see planet (5731)}, father of one of her sons, into the island bearing her name. (H 13)
Named by C. Bruhns (AN 71, 37 (1868)).

(92) Undina

Discovered 1867 July 7 by C. H. F. Peters at Clinton.

Named for the heroine of the early 19th century novel *Undine* by the German romantic writer F. H. K. de la Motte Fouqué (1777-1843) who wrote many chivalric romances, tales and plays based on Norse mythology. *Undine* is one of a group of female water spirits. When an *Undine* married a mortal and born a child she received a soul. If, however, her lover proved to be unfaithful, she would be forced to return to the sea. (LDS)

The discoverer explained the naming in AN 69, 349 (1867): "Der Planet ... hat den Namen Undina erhalten, welcher sich vielleicht nicht in den alten Classikern findet, aber jedenfalls klassisch richtig gebildet, und ausserdem durch die liebliche Dichtung von de la Motte Fouqué verherrlicht worden ist." Two romantic operas by A. Lortzing (1801-1851) and by E. T. A. Hoffmann (1776-1822) are based on the tale.

(93) Minerva

Discovered 1867 August 24 by J. C. Watson at Ann Arbor.

Minerva is the Latin name of the goddess Athene or Pallas {see planets (881) and (2)}. (H 13)

Named by the Astronomische Gesellschaft.

(94) Aurora

Discovered 1867 September 6 by J. C. Watson at Ann Arbor.

Aurora is the Latin name for Eos {see planet (221)}, the goddess of the dawn. (H 13)

Named by the Astronomische Gesellschaft.

(95) Arethusa

Discovered 1867 November 23 by R. Luther at Düsseldorf.

Named after one of the Hesperides {see also planet (46)}. (H 13)

Named by J. Galle, the discoverer of Neptune, and Günther in Breslau (AN 70, 237 (1867)).

(96) Aegle

Discovered 1868 February 17 by J. Coggia at Marseilles.

Named after one of the Hesperides {see also planet (46)}. (H 13)

(97) Klotho

Discovered 1868 February 17 by E. W. Tempel at Marseilles.

Named after one of the three Fates. The others are Lachesis and Atropos {see planets (120) and (273)}. Klotho presides over birth and spins destinies. (H 13)

Named Clotho by the Société Impériale des Sciences naturelles de Cherbourg.

Tempel proposed also names for the next two planets (AN 71, 97 (1868)): "Il me semble qu'il serait un peu poétique, de donner aux trois dernières planètes (97), (98) et (99) les noms de 3 Parques: Clotho, Lachésis et Atropos, non pas avec l'idée que le ciseau d'Atropos dût couper le fil de cette recherche, mais avec la pensée de diviser la première centaine des petites planètes, qui représentent assurément un grand et noble travail astronomique." His naming proposals, however, were not realized. The two other Fates received the numbers (120) and (273), respectively.

(98) Ianthe

Discovered 1868 April 18 by C. H. F. Peters at Clinton.

Named for the young girl who became betrothed to Iphis, a Cretan girl who was changed by Isis {see planet (42)} into a man. (H 14)

(99) Dike

Discovered 1868 May 28 by A. Borrelly at Marseilles.

Dike represents the human justice who with the divine justice Themis {see planet (24)} sat beside Zeus {see planet (5731)} on Olympus. (H 14)

Named by E. Stephan.

(100) Hekate

Discovered 1868 July 11 by J. C. Watson at Ann Arbor. Independently discovered 1868 July 18 by C. Wolf at Paris.

Named after the goddess of the lower world and of darkness in the upper world. (H 14)

The naming was delegated to the National Academy of Sciences, Washington, D.C. The name was submitted by a committee of four including B. Peirce and B. A. Gould.

J. Meeus suggested the interesting interpretation that the choice of this name could have been influenced by the fact that the Greek word 'hekaton' means 100.

(101) Helena

Discovered 1868 August 15 by J. C. Watson at Ann Arbor.

Named after the daughter of Zeus and Leda {see planets (5731) and (38)}, and cause for the Trojan War. (H 14)

The naming was delegated to the U.S. National Academy of Sciences, of which the discoverer was a member. The naming was redelegated to B. A. Gould who selected the name (AN 72, 191 (1868)).

(102) Miriam

Discovered 1868 August 22 by C. H. F. Peters at Clinton.

Named after the sister of Moses, the Hebrew prophet who led the Israelites out of Egyptian slavery and at Mt. Sinai delivered to them the Law establishing God's covenant with them. (H 14)

The naming of this planet produced a controversy. Edward S. Holden {see planet (2974)} stated: "The name of the asteroid Miriam was chosen in defiance of a rule, and of malice aforethought; so that the discoverer could tell a theological professor, whom he thought to be too pious, that Miriam also was a 'mythological personage'."

(103) Hera

Discovered 1868 September 7 by J. C. Watson at Ann Arbor.

Named after the Greek goddess, daughter of Cronus and Rhea {see planet (577)}, sister and at the same time wife of Zeus {see planet (5731)}, and mother of Ares, Hephaistos {see planet (2212)}, Hebe {see planet (6)}, and Eileithyia. Hera is the Greek name for the Latin Juno {see planet (3)}. (H 14)

(104) Klymene

Discovered 1868 September 13 by J. C. Watson at Ann Arbor.

There are about ten mythological characters by this name. In the Iliad Clymene is a female servant of Helen {see planet (101)}, who accompanied her mistress when Paris {see planet (3317)} abducted her. Clymene is also known as the daughter of Oceanus and Tethys and mother of Epimetheus and Prometheus {see planets (1810) and (1809)}. (Z 65)

(105) Artemis

Discovered 1868 September 16 by J. C. Watson at Ann Arbor.

Artemis (Roman name Diana, see planet (78)) was the daughter of Zeus by Leto {see planets (5731) and (68)} and the twin sister of Apollo {see planet (1862)}. She was the Moon goddess and goddess of hunting. In Roman mythology she was also the goddess of childbirth as Lucina {see planet (146)} and the patroness of unmarried girls and of chastity. (Z 32)

(106) Dione

Discovered 1868 October 10 by J. C. Watson at Ann Arbor.

Named for the mother of Aphrodite by Zeus {see planets (1388) and (5731)}. When Aphrodite was wounded by Diomedes {see planet (1437)}, Dione comforted her daughter. Dione foretold the death of Diomedes for daring to wage war against the immortals. Aphrodite is sometimes called Dione. (Z 87)

The name Dione has also been given to the fourth satellite of Saturn, discovered 1684 by J. D. Cassini.

(107) Camilla

Discovered 1868 November 17 by N. R. Pogson at Madras.

Named possibly for a daughter of Metabus. When she was young, her father dedicated her to the service of Artemis {see planet (105)}. She became a virgin warrior and queen of the Volscians. She assisted Turnus against Aeneas {see planet (1172)}. Camilla fought with one breast bared to give freedom to her bow arm. She was a swift runner, comparable to Atalanta {see planet (36)}. She was killed by the spear of Aruns. (LDS, Z 49)

Other sources - see the remarks to planet (87) - give the alternative interpretation that the name refers to the French astronomer Camille Flammarion (see planet (1021)). This seems unlikely, however, because Pogson frequently used mythological names for his discoveries.

(108) Hecuba

Discovered 1869 April 2 by R. Luther at Düsseldorf.

Named for the wife of Priam {see planet (884)} at the time of the Trojan War. She was a most unfortunate mother of fifty sons and twelve daughters. Many of her sons were slain, Polyxena {see planet (595)} was sacrified, Polydorus {see planet (4708)} drowned, Cassandra {see planet (114)} was murdered, Hecuba herself changed into a bitch. (Z 118)

Named by Dr. Franz Heinen, director of the Realschule in Düsseldorf and trustee of the Bilk Observatory (AN 74, 31 (1869)).

(109) Felicitas

Discovered 1869 October 9 by C. H. F. Peters at Clinton.

Named for the divinity personifying happiness. She is portrayed seated in a throne with a caduceus and a horn of plenty in her hands. (H 15)

(110) Lydia

Discovered 1870 April 19 by A. Borrelly at Marseilles.

Named after the country in Asia Minor whose early inhabitants were called Phrygians. About the time of the Trojan War, the Heraclidae reigned there. (Z 156)

(111) Ate

Discovered 1870 August 14 by C. H. F. Peters at Clinton.

Named for the goddess of all evil, infatuation, and mischief. Ate (or Eris) was a daughter of Zeus {see planet (5731)}. Not invited to the marriage of Peleus and Thetis {see planet (17)}, she rolled the Golden Apple that led to the Judgement of Paris {see planet (3317)}, a main cause for the Trojan War. Hesiod says she was a daughter of Eris. (Z 35)

(112) Iphigenia

Discovered 1870 September 19 by C. H. F. Peters at Clinton.

Named for the daughter of Agamemnon {see planet (911)} and Clytemnestra {see planet (179)} and sister of Orestes and Electra {see planet (130)}. The Greeks, detained at Aulis by contrary winds and unable to sail for Troy, were informed by a soothsayer that Iphigenia must be sacrificed to appease the gods, for Agamemnon had provoked Artemis {see planet (105)} by killing her favorite stag. As Agamemnon was about to strike the fatal blow, a large and beautiful stag appeared in Iphigenia's place. The winds became favorable and the Greeks sailed for Troy. In Iliad, Iphigenia is called Chrysothemis {see planet (637)}. (Z 138)

(113) Amalthea

Discovered 1871 March 12 by R. Luther at Düsseldorf.

Named for the daughter of King Melissus of Crete, who nourished Zeus {see planet (5731)} with the milk of a goat. (H 15)

Named (AN 77, 217 (1871)) by the Berlin astronomers by a vote of 5 to 3.

The name Amalthea has also been given to the fifth satellite of Jupiter, discovered 1892 by E. Barnard.

(114) Kassandra

Discovered 1871 July 23 by C. H. F. Peters at Clinton.

Named after one of the twelve daughters of Priam {see planet (884)} and Hecuba {see planet (108)} of Troy. The twin of Helenus {see planet (1872)}, she had fifty brothers. She was the most beautiful of Priam's daughters and was courted by many princes. Apollo {see planet (1862)} loved her and promised to give her what she desired if she would gratify his passion. She asked for the power of prophecy, but as soon as she obtained the gift, refused to fulfill the bargain. Apollo could not withdraw the gift of prophecy he had granted her, but he wet her lips with his tongue, thus effecting a curse that no reliance would ever be placed upon her predictions. She would indeed be able to foretell the future, but no credence would ever be given to her prophecy. Her name has become synonymous with prophets of doom whose warnings are heeded too late. (Z 51)

(115) Thyra

Discovered 1871 August 6 by J. C. Watson at Ann Arbor.

Named after an early Danish queen. Queen Thyri was the wife of King Gorm the Old of Denmark (?-940) and the mother of the Dane Harald Bluetooth (935-986), who died as a Christian. In 935 King Harald wrote on a runic stone, that he had unified all Denmark and Norway under himself and made the Danes Christians. This unique monument Harald erected to himself and his parents at Jellings, Jutland. These two vast runestones are the largest and most impressing in the world. The smaller of the two is dedicated by Gorm to honor his wife Thyri, restorer of Denmark. (H 15; R. Bremer)

At the time that this planet was discovered, the discoverer had just returned from a visit to Europe, and probably this particular name was chosen as a memento of his stay in Denmark.

(116) Sirona

Discovered 1871 September 8 by C. H. F. Peters at Clinton.

Named after the Celtic mother goddess of Belgian and Roman Gaul, whose worship was in connection with the thermal fountains. (H 15)

(117) Lomia

Discovered 1871 September 12 by A. Borrelly at Marseilles.

Very probably, the name 'Lamia' originally was misspelled to 'Lomia'. Lamia, queen of Lybia, was beloved by Zeus {see planet (5731)} and made insane by a jealous Hera {see planet (103)}. (I. van Houten-Groeneveld)
Lamia, or Greek Lameia, exists as planet (248).

(118) Peitho

Discovered 1872 March 15 by R. Luther at Düsseldorf.

Named for a daughter of Hermes and Aphrodite {see planet (1388)}, goddess of persuasion among the Greeks and Romans. Peitho (also Pitho, Suada, Suadela) was an attendant of Eros {see planet (433)}. (Z 195)

Named by Becker, Leiden and G. Rümker, director of the Hamburg Observatory.

(119) Althaea

Discovered 1872 April 3 by J. C. Watson at Ann Arbor.

Named after the mother of Meleager, who, having killed his mother's two brothers, was himself killed when she threw on the fire the log on whose existence his life depended. (H 16)

(120) Lachesis

Discovered 1872 April 10 by A. Borrelly at Marseilles. Independently discovered 1872 April 11 by C. H. F. Peters at Clinton.

Named for one the three Fates. The other two are Clotho {see planet (97)} and

Atropos {see planet (273)}. They are daughters of Erebus and Nyx {see planet (3908)}. Lachesis carries the globe or scroll and determines the length of the thread of life. (H 16)

(121) Hermione

Discovered 1872 May 12 by J. C. Watson at Ann Arbor.

Named after the daughter of Menelaus and Helen {see planets (1647) and (101)}. She was promised in marriage to Orestes, but Menelaus, ignorant of this, gave her to Pyrrhus {see planet (5283)} for the services he rendered in the Trojan War. (Z 125)

(122) Gerda

Discovered 1872 July 31 by C. H. F. Peters at Clinton.

Named for the wife of Freyr in Norse mythology. (H 16)

(123) Brunhild

Discovered 1872 July 31 by C. H. F. Peters at Clinton.

Named for a character of the *Nibelungenlied*, the Volsung Cycle, and the Thidrek Saga. Brunhild was a queen in German legend, won by Siegfried for Gunther. Brunhild was the daughter of Odin {or Wodan, see planets (3989) and (2155), respectively} and Erda {see planet (894)}, tha eldest of the Valkyries. (H 16; I. van Houten-Groeneveld)

(124) Alkeste

Discovered 1872 August 23 by C. H. F. Peters at Clinton.

Named after the daughter of Pelias and Anaxibia and wife of Admetus. She volunteered to die for her husband on the promise of Apollo {see planet (1862)} that Admetus should never die if someone were found to die in his stead. Some legends say Heracles brought Alcestis back from Hades. (Z 15)

Named by Adelinde Weiss, wife of the Austrian astronomer E. Weiss, at the request of the discoverer.

(125) Liberatrix

Discovered 1872 September 11 by P. M. Henry at Paris.

This name very probably is an allusion to the liberation of France at the time of the discovery. (Littrow, 7. ed., p. 445 (1886))

Another interpretation is (H 16) that this planet is possibly named for the first president of the French Republic, Adolphus Thiers, who arranged a loan of 5,000,000,000 golden francs, by means of which the Prussian troops were evacuated from the French territory after the France-Prussian War, 1870/71.

(126) Velleda

Discovered 1872 November 5 by P. P. Henry at Paris.

Named for a priestess of the Germans in the first century A.D. who was worshiped as a deity. She provoked the rebellion of the Bataves and other tribes against the Romans. (H 17)

(127) Johanna

Discovered 1872 November 5 by P. M. Henry at Paris.

Named probably for Jeanne d'Arc (1412-1431), saint and French national heroine. (Littrow, 7. ed., p. 445 (1886))

(128) Nemesis

Discovered 1872 November 25 by J. C. Watson at Ann Arbor. Independently discovered 1872 December 5 by A. Borrelly at Marseilles.

Named for the goddess of vengeance. Nemesis was the daughter of Erebus and Nyx (the Night) {see planet (3908)}. (H 16)

(129) Antigone

Discovered 1873 February 5 by C. H. F. Peters at Clinton.

Named after the daughter of Oedipus, king of Thebes. She guided her father when he was blind and exiled by her uncle Creon. (H 17)

(130) Elektra

Discovered 1873 February 17 by C. H. F. Peters at Clinton.

There are several meanings known in classical mythology. After Aeschylus, Sophocles and Euripides, Electra was the daughter of Agamemnon and Clytemnestra {see, respectively, planets (911) and (179)} and sister of Orestes, Iphigenia {see planet (112)}, and Chrysothemis {see planet (637)}. Electra incited her brother Orestes to avenge their father's murder by killing their mother and her lover Aegisthus. (Z 92)

(131) Vala

Discovered 1873 May 24 by C. H. F. Peters at Clinton.
Named for a prophetess in Scandinavian mythology. (AN 82, 130 (1873))

(132) Aethra

Discovered 1873 June 13 by J. C. Watson at Ann Arbor.

Named (in the Iliad) after the mother of Theseus, the Greek hero. After another legend Aethra was one of the seven Oceanids. They died of grief over the death of their brother Hyas and were changed into stars called the Hyades. (Z 10)

(133) Cyrene

Discovered 1873 August 16 by J. C. Watson at Ann Arbor.

Named after a daughter of Hypseus, king of the Lapithae. She loved to hunt and wrestle with lions. Apollo {see planet (1862)} saw her wrestle with a lion and fell in love with her. He was the father of her two sons Aristaeus {see planet (2135)} and Idmon, the latter a famous seer. (Z 78)

(134) Sophrosyne

Discovered 1873 September 27 by R. Luther at Düsseldorf.

Named for one of the virtues in Plato's (427?-347 B.C.) {see planet (5451)} system: equanimity, healthy mind and impartiality. (H 17)
Named (AN 82, 287 (1873)) by F. W. A. Argelander and Theodor Wolff, Bonn.

(135) Hertha

Discovered 1874 February 18 by C. H. F. Peters at Clinton.

Named for the Teutonic and Scandinavian goddess of fertility (also Nerthus, see planet (601)). The discoverer was born in southern Schleswig, at that time a region of Denmark. A few years prior to the discovery he visited Scandinavia. (H 17)

E. Weiss stated in Littrow, Wunder des Himmels, 7. ed., p. 446 (1886): "Möge ihr Erscheinen Freude und Frieden wie in der deutschen Mythe, nur für längere Zeit als dort, bedeuten."

(136) Austria

Discovered 1874 March 18 by J. Palisa at Pola.

Named by the discoverer in honor of his native country. This was the first (of a total of 121) numbered minor planet discovered by Palisa and also the first one discovered in Austria. (H 17)

(137) Meliboea

Discovered 1874 April 21 by J. Palisa at Pola.

Named for a daughter of Oceanus. Meliboea married Pelasgus by whom she bore Lycaon {see planet (4792)}. (A. Schnell)
Named (AN 85, 108 (1875)) by K. von Littrow.

There are two further figures in Greek mythology with this name. The explanation given above, however, was published by K. von Littrow in an appendix to the Wiener 'Kalender für alle Stände' for 1876, p. 21.

(138) Tolosa

Discovered 1874 May 19 by J. Perrotin at Toulouse.

Named for the city where this planet was discovered. Tolosa is the Latin name for Toulouse. It became a Roman colony under Augustus and was afterwards celebrated for the cultivation of the sciences. (H 17)

(139) Juewa

Discovered 1874 October 10 by J. C. Watson at Peking.

The name stands as Juewa, or more fully, Jue-wa-sing, which means literally the "Star of China's Fortune". (H 18)

This planet was discovered while preparing for the transit of Venus of December 9, 1874. The discoverer stated (Observation of the Transit of Venus, Part II, p. 108 (1881)): "On the night of the 10th October, while observing in the constellation Pisces, with the 5-inch equatorial, I came across a star of the 11th magnitude in a region of the heavens with which I was very familiar, and where I had not hitherto seen any such star. Subsequent observations the same night by means of a micrometer, extemporized for the purpose, showed that the star was slowly retrograding, and that it was a new member of the group of planets between Mars and Jupiter. The discovery was duly announced to astronomers in other lands, and it became also speedily known in Peking. Some mandarins of high rank came to our station to see the stranger with their own eyes, and upon observing the change of configuration with neighboring stars on two successive nights, they gave free expression to their astonishment and delight. This being the first planet discovered in China, I requested Prince Kung, regent of the Empire, to give it a suitable name. In due time, a mandarin of high rank brought to me the document containing the name by which the planet should be known, coupled with a request - communicated verbally - that I would not publish the name in China until the astronomical board had communicated to the Emperor an account of the discovery and the name which had been given to the planet. This request was of course promptly acceded to; and I afterwards learned upon inquiry that if the knowledge had come to the Emperor otherwise than through the astronomical board, organized specially for his guidance in celestial matters, some of the ministers would have been disgraced."

(140) Siwa

Discovered 1874 October 13 by J. Palisa at Pola.

Named for the Slavic goddess of fertility. This naming refers to the numerous Illyric population of Istria. Pola is situated at the southern coast of this peninsula. (A. Schnell)

Named by Marine-Commandant Vice-Admiral Freiherr F. von Pöck. The planet is erroneously published as (139) in AN 85, 108 (1875). The meaning of the name is published in the Wiener 'Kalender für alle Stände' for 1876, p. 20.

(141) Lumen

Discovered 1875 January 13 by P. P. Henry at Paris.

Named after a literary work of the French astronomer Camille Flammarion {see planet (1021)}, *Lumen: Récits de l'infini.* (H 18)

(142) Polana

Discovered 1875 January 28 by J. Palisa at Pola.

Named for the city of discovery, located on the northern Adriatic sea. Since Palisa's days Pola has become first part of Italy and now Yugoslavia. Pola was the main naval port of Austria since 1850. The Pola Naval Observatory was a department of the Austrian Hydrographic Service. Palisa served as head of this department. (H 18; A. Schnell)

(143) Adria

Discovered 1875 February 23 by J. Palisa at Pola.

Named for the Adriatic sea. The city of Pola {see planet (142)}, at the time of discovery the chief naval station of Austria-Hungary, is an Adriatic seaport. (H 19)

(144) Vibilia

Discovered 1875 June 3 by C. H. F. Peters at Clinton.

Named for the Roman goddess and patroness of journeyings. The planet was discovered immediately after the return of Peters from the transit of Venus expedition of 1874. (H 19)

(145) Adeona

Discovered 1875 June 3 by C. H. F. Peters at Clinton.

Named for the Roman divinity and patroness of homecomings. See also the remarks to planet (144) which was discovered at the same night. (H 19)

(146) Lucina

Discovered 1875 June 8 by A. Borrelly at Marseilles.

Named for the Roman goddess of the travails of women and of childbirth. For the Greek counterpart see the citation for planet (105) Artemis. See also the remarks to planet (494). (Z 153)

(147) Protogeneia

Discovered 1875 July 10 by L. Schulhof at Vienna.

The name, translated from Greek, means first-born. (A. Schnell)

Named by K. von Littrow (AN 86, 122 (1875)) on request of the discoverer in allusion to the first discovery of a minor planet by an astronomer which in other fields of astronomy was already well known. Schulhof at the time of the discovery was assistant at the Vienna Observatory.

K. von Littrow (Wunder des Himmels, 6. ed., Berlin 1878) stated: "Erhielt ihren Namen auf Einladung des Entdeckers..., um die Erstlingsschaft des Fundes für diesen Entdecker anzudeuten."

(148) Gallia

Discovered 1875 August 7 by P. M. Henry at Paris.

Named in honor of the discoverer's home country. Gallia is the Latin name for France. (H 19)

(149) Medusa

Discovered 1875 September 21 by J. Perrotin at Toulouse.

Named for one of the three Gorgons. Medusa was mortal whereas the other two, Euryale and Stheno, were immortal. The eyes of the three Gorgons had the power of killing or turning onlookers into stone. Perseus killed Medusa, cut off her head, and placed it on the shield of Athena {see planet (881)}, where it had the same petrifying power as when Medusa was alive. From the blood of Medusa sprang the winged horse Pegasus. (Z 161)

(150) Nuwa

Discovered 1875 October 18 by J. C. Watson at Ann Arbor.

Named for a character in Chinese mythology. This planet was possibly named in remembrance of the discoverer's trip to China. For more details see the remarks to planet (139). (H 19)

(151) Abundantia

Discovered 1875 November 1 by J. Palisa at Pola.

This name is the personification of fulness, its main attribute is the horn of plenty. The naming refers to the large number of newly discovered bodies between Mars and Jupiter. Von Littrow (Wunder des Himmels, 6. ed., p. 434, Berlin 1878) complained: "Benannt ... nach einer Göttin, die mit den armen Astronomen auf diesem Felde seit Jahren erbarmungsloses Spiel treibt und namentlich in dieser Epoche ihr Füllhorn über uns ausschüttete." (A. Schnell)
Named by Prof. E. Weiss, director of the Vienna Observatory.

(152) Atala

Discovered 1875 November 2 by P. P. Henry at Paris.

Named for the character in a novel of the same name by the French author François René de Chateaubriand (1768-1848). (H 19)

(153) Hilda

Discovered 1875 November 2 by J. Palisa at Pola.

Named in honor of (probably the eldest) daughter of the Austrian astronomer Theodor von Oppolzer (1841-1886) {see planet (1492)}. She died some years prior to her father. (Vierteljahrsschr. Astron. Ges., 22. Jahrg., p. 191 (1887))
Named by Th. von Oppolzer.

(154) Bertha

Discovered 1875 November 4 by P. M. Henry at Paris.

Named probably in honor of Berthe Martin-Flammarion (1844-1936), sister of Camille Flammarion {see planet (1021)}. (H 20)

(155) Scylla

Discovered 1875 November 8 by J. Palisa at Pola.

Named for the daughter of Phorcys and Ceto and sister of Gorgons, Sirens and Graeae. The sea nymph Scylla and Charybdis {see planet (388)} guarded the strait between Sicily and Italy, expanding her long necks and six heads to remove one sailor from each passing ship. (H 20)

The planet was lost for many decades - see the detective recovering story in Sky Telesc., Vol. 40, No. 6, p. 361-362 (1970).

(156) Xanthippe

Discovered 1875 November 22 by J. Palisa at Pola.

Named for the wife of the Greek philosopher Socrates (470?-399 B.C.) {see planet (5450)} whose peevish disposition and ill humor have become proverbial. (H 20)
Named by Amalia Palisa, (first) wife of the discoverer.

(157) Dejanira

Discovered 1875 December 1 by A. Borrelly at Marseilles.

Named for the second wife Dejanira (Greek: Deianeira) of Heracles; Megara was the first. She unwittingly killed Heracles by sending him a garment steeped in the poisoned blood of the centaur Nessus {see planet (7066)}. This garment, Nessus has said, had power to reclaim a husband from unlawful loves. (Z 82)

(158) Koronis

Discovered 1876 January 4 by V. Knorre at Berlin.

There exist several interpretations of this name among mythologists. After Pausanias, Coronis was the mother of Aesculapius by Apollo {see, respectively, planets (1027) and (1862)}. Coronis is also the name of a Hyades member, originally a nymph, daughter of Atlas and the Oceanid Aethra {see planet (132)}. (H 20; LDS)
Named by Dr. Maywald.

(159) Aemilia

Discovered 1876 January 26 by P. P. Henry at Paris.

Named probably after the famous Roman road 'Via Aemilia' which went from Piacenza to Rimini. (I. van Houten-Groeneveld)

K. von Littrow (Wunder des Himmels, 6. ed., Berlin 1878) ironically stated: "Mit dem Namen begegnet man auf mythologischem Gebiete nur einer Dienerin der Vesta, die vom Paten wohl kaum gemeint war."

(160) Una

Discovered 1876 February 20 by C. H. F. Peters at Clinton.

Named after the heroine of the poetic work *Faerie Queene* by the English writer Edmund Spenser (1552-1599). (H 20)

(161) Athor

Discovered 1876 April 19 by J. C. Watson at Ann Arbor.

Named for the Egyptian goddess, often identified with Isis {see planet (42)}. She had the head of a cow and wore the Sun's disk and possessed the characteristics of Aphrodite {see planet (1388)}. (H 20)
Athor is the same goddess as Hathor {see planet (2340)}.

(162) Laurentia

Discovered 1876 April 21 by P. M. Henry at Paris.

Named in honor of the amateur astronomer A. Laurent, Nimes, the discoverer of (51) Nemausa. (H 20)

(163) Erigone

Discovered 1876 April 26 by J. Perrotin at Toulouse.

Apollodorus mentioned Erigone as daughter of Icarus {see planet (1566)}, who hanged herself when she learned that her father had been killed by drunken shepherds. She was changed into the constellation Virgo. According to Pausanias, Erigone was a daughter of Aegisthus and Clytemnestra {see planet (179)} and has a son by Orestes. She brought Orestes to trial for the murder of her mother and hanged herself when Orestes was acquitted before the Areopagus. (Z 98)

(164) Eva

Discovered 1876 July 12 by P. P. Henry at Paris.
Any reference of this name to a person or occurence is unknown.

K. von Littrow (Wunder des Himmels, 6. ed., Berlin 1878) characterized the naming practice of the discoverer ironically: "Mit dem Namen könnten wir wie bei Miriam {see planet (102)} wieder den biblischen Boden zu betreten glauben, wenn wir bei diesem Entdecker nicht an Taufen weltlichen Ursprungs gewöhnt wären."

(165) Loreley

Discovered 1876 August 9 by C. H. F. Peters at Clinton.

Named after a siren of Germanic legend who by her songs lures Rhine river boatmen to destruction on a reef. Sirenes appear in the *Nibelungenlied* and other tales. (H 21)

(166) Rhodope

Discovered 1876 August 15 by C. H. F. Peters at Clinton.

Named after an attendant of Artemis {see planet (105)}. She thought herself more beautiful than Hera {see planet (103)}, and the goddess changed her into the high mountain in Thrace that bears her name. (Z 229)

(167) Urda

Discovered 1876 August 28 by C. H. F. Peters at Clinton.

Named for one of the three Norns in Norse mythology. Urda stands for the Past. (H 21)

The other Norns are Werdandi (the Present) and Skuld (the Future) {see planets (621) and (1130)}.

(168) Sibylla

Discovered 1876 September 28 by J. C. Watson at Ann Arbor.

Named probably for the Sibyls collectively. The Sibyls were women, young and old, inspired by heaven and endowed with prophetic powers, who interceded with the gods on behalf of men. The most famous of the (presumably ten) Sibyls was that of Cumae in Italy, a woman of wisdom and vision. (Z 239)

(169) Zelia

Discovered 1876 September 28 by P. M. Henry at Paris.

Named in honor of a niece of Camille Flammarion {see planet (1021)} and daughter of his sister Berthe {see planet (154)}. (H 21)

(170) Maria

Discovered 1877 January 10 by J. Perrotin at Toulouse. Independently discovered 1877 January 22 by C. H. F. Peters at Clinton.

Named in honor of the sister of Antonio Abetti who computed the orbit, and aunt of Giorgio Abetti {see planet (2646)}. She died in her youth. (H 21)

(171) Ophelia

Discovered 1877 January 13 by A. Borrelly at Marseilles.

Named for the daughter of Polonius in Shakespeare's {see planet (2985)} play *Hamlet*. (H 21)

The name Ophelia has also been given to the satellite Uranus VI, discovered 1986 by the Voyager 2 spacecraft.

(172) Baucis

Discovered 1877 February 5 by A. Borrelly at Marseilles.

Baucis and her husband Philemon entertained the gods Zeus {see planet (5731)} and Hermes most hospitably in their humble cottage in Phrygia. As a reward for this kindness their cottage was changed into a beautiful golden-roofed marble temple. Baucis and Philemon desired to die at the same time, and their bodies were changed into trees, an oak and a linden. (Z 40)

(173) Ino

Discovered 1877 August 1 by A. Borrelly at Marseilles.

Named after the daughter of Cadmus and Harmonia {see planet (40)}, sister of Agave, Autonoe, Polydorus {see planet (4708)}, and Semele {see planet (86)} and second wife of Athamas, king of Thebes. She was the wicked stepmother of Helle and Phrixus, the boy who was saved from death by the ram of the Golden Fleece. Ino later became a sea goddess who saved Odysseus {see planet (1143)} from drowning when his raft was shattered. Her name was even later changed to Leukothea {see planet (35)}. (H 21)

(174) Phaedra

Discovered 1877 September 2 by J. C. Watson at Ann Arbor.

nd Pasiphae, sister of Ariadne {see planet (43)} and wife of Theseus. Phaedra became infatuated with her stepson Hippolytus and made overtures to him, but was rejected. In anger and humiliation she hanged herself, but left a message accusing Hippolytus of having attacked her. The outraged Theseus, without hearing Hippolytus' side of the story, appealed to Poseidon {see planet (4341)}, god of sea, for appropriate revenge. Poseidon sent a sea monster which terrified Hippolytus' horses so that they bolted and dragged their master to his death under the wheels of his own chariot. (Z 202)

(175) Andromache

Discovered 1877 October 1 by J. C. Watson at Ann Arbor.

Named for the devoted and faithful wife of Hector {see planet (624)}. Andromache warned Hector against engaging in battle during the Trojan War. After the war she became a concubine of Neoptolemus {see planet (2260)} and later married Helenus {see planet (1872)}. Aeneas {see planet (1172)} found her at Epirus on his way to Italy. (Z 22)

(176) Iduna

Discovered 1877 October 14 by C. H. F. Peters at Clinton.

Named by the discoverer after a club in Stockholm, Sweden on the occasion of a meeting of the Astronomische Gesellschaft, 1877. In Norse mythology, Iduna was keeper of the apples which preserve the youth of the gods. (H 22)

The planet originally received the name Idunna and the number 175 (AN 91, 63 (1877): "Ich lege diesem Planeten den Namen Idunna bei, womit wenigstens diejenigen Mitglieder der Astronomischen Gesellschaft, die in Stockholm an der Gastfreundschaft der "Ydun" Theil nahmen, einverstanden sein werden."

(177) Irma

Discovered 1877 November 5 by P. P. Henry at Paris.

Any reference of this name to a person or occurence is unknown.

(178) Belisana

Discovered 1877 November 6 by J. Palisa at Pola.

Named for Belisana, the name of Athene or Minerva {see planets (881) and (93)} among the Gauls, signifying the "queen of heaven". Belisana is also the most warlike goddess among British Celts. (H 22; A. Schnell)

Named by M. Loewy at the request of the discoverer. The first interpretation seems more likely because the naming was performed after Loewy's naturalization in France.

(179) Klytaemnestra

Discovered 1877 November 11 by J. C. Watson at Ann Arbor.

Named after the daughter of Tyndareus, king of Sparta, by Leda {see planet (38)}. She became the wife of Agamemnon {see planet (911)} and the mother of Orestes, Electra, Iphigenia and Chrysothemis {see, respectively, planets (130), (112), (637)}. When Agamemnon went to Trojan War, Clytemnestra became the paramour of Aegisthus. When Agamemnon returned from Trojan War, Aegisthus and Clytemnestra murdered him. Orestes later killed the murderers of his father. (Z 65)

(180) Garumna

Discovered 1878 January 29 by J. Perrotin at Toulouse.

Named for the Garonne river on which the city of discovery is situated. Garumna is the ancient name. (H 22)

(181) Eucharis

Discovered 1878 February 2 by P. Cottenot at Marseilles.

Named for a nymph of the goddess Calypso {see planet (53)}. (H 22)

(182) Elsa

Discovered 1878 February 7 by J. Palisa at Pola.

Named after the Austrian form 'Elsbeth' of the most common feminine Christian name 'Elisabeth'. It was also the name of the Austrian empress. This name was later changed in agreement with the discoverer into the more melodious 'Elsa'. (A. Schnell)

Named by Admiral Bourgignon at the request of the discoverer. Bourgignon was the military superior of Palisa in Pola.

(183) Istria

Discovered 1878 February 8 by J. Palisa at Pola.

Named for the (now Croatian) peninsula at the northern end of the Adriatic sea, containing Trieste and the city of discovery. (H 183; A. Schnell)

Named by Vice-Admiral B. Freiherr von Wüllerstorf who was the commander of the first Austrian circumnavigatory adventure with the frigate 'Novara'.

(184) Dejopeja

Discovered 1878 February 28 by J. Palisa at Pola.

Named for the fairest of the 14 nymphs who attended Juno {see planet (3)} who promised her in marriage of Aeolus, god of the winds, if he would destroy the fleet of Aeneas {see planet (1172)} which was sailing for Italy. Some sources call Dejopeja an attendant nymph of Cyrene {see planet (133)}. (H 23)
Named by R. Müller, director of the Hydrographic Service in Pola.

(185) Eunike

Discovered 1878 March 1 by C. H. F. Peters at Clinton.

Named for one of the Nereids meaning "happy victory", commemorating the treaty of San Stefano, signed by Russia and Turkey, two days after the discovery of the asteroid, at the Turkish village of Yesilkoy (San Stefano), concluding the Russo-Turkish War (1877-1878). The treaty gave enormous gains to the Russians. (AN 92, 207 (1878); H 23)

(186) Celuta

Discovered 1878 April 6 by P. M. Henry at Paris.
Any reference of this name to a person or occurence is unknown.

(187) Lamberta

Discovered 1878 April 11 by J. Coggia at Marseilles.

Named in honor of Johann Heinrich Lambert (1728-1777), who described the Milky Way as "the ecliptic of the fixed stars" and made other remarkably good cosmological guesses. His law of exponential absorption of light and his theorem regarding inferences from the curvature of the apparent paths of comets are still in use. He first proved the irrationality of π (1768) and he introduced the present hyperbolic function symbols and elaborated the new subject. (H 23)

Lambert is also honored by craters on the Moon and on Mars. The cgs unit of brightness equal to the brightness of a perfectly diffusing surface that radiates or reflects one lumen per square centimeter bears his name.

(188) Menippe

Discovered 1878 June 18 by C. H. F. Peters at Clinton.

Named after a daughter of the celebrated hunter Orion who offered to die with her sister Metioche when a pestilence was raging in the Greek province Boeotia and the oracle had demanded the sacrifice of two virgins. Their voluntary sacrifice was rewarded by a place in the heavens as comets. Menippe is also the name of one of the Nereids. (H 23)

(189) Phthia

Discovered 1878 September 9 by C. H. F. Peters at Clinton.

Many explanations are given in Greek mythology: (1) Phthia was a daughter of Niobe {see planet (71)} and was killed by Artemis {see planet (105)}. (2) Phthia was a mistress of Zeus, Apollo {see planets (5731), (1862)}, or Phoenix. It is also the name of a town in Thessaly where Achilles {see planet (588)} was born. (Z 209)

(190) Ismene

Discovered 1878 September 22 by C. H. F. Peters at Clinton.

Named for the daughter of Oedipus and Jocaste {see planet (899)}, sister of Antigone {see planet (129)}, Eteocles, and Polynices. When her sister Antigone was condemned by Creon to be entombed alive for giving burial to their brother Polynices, Ismene declared herself as guilty as her sister and demanded to be punished equally, but Antigone did not want Ismene to be involved in her calamities. (Z 141)

(191) Kolga

Discovered 1878 September 30 by C. H. F. Peters at Clinton.

Named for the daughter of Aeger in Norse mythology. All names of his daughters are personifications of waves. (H 24)

(192) Nausikaa

Discovered 1879 February 17 by J. Palisa at Pola.

Named after the daughter of Alcinous, king of the Phaeacians. She befriended the shipwrecked Odysseus {see planet (1143)} and persuaded her father to be kind to the visitor. Some accounts say Nausicaa later married Telemachus, Odysseus' son. (Z 172)

Named by the Berlin astronomers on the occasion of the September 1879 assembly of the Astronomische Gesellschaft. These astronomers also proposed well-known names from the Odyssey for planets (195), (197), and (201) (Astronomischer Kalender für 1882, p. 106).

(193) Ambrosia

Discovered 1879 February 28 by J. Coggia at Marseilles.

Named for the food of the gods. Their drink was nectar. Ambrosia gives the gods immortality. This is also the name of one of the Hyades which have been originally nymphs, daughters of Atlas and Aethra {see planet (132)}. The second interpretation seems more likely because Coggia named another planet after one of these nymphs {see planet (217)}. (H 24; LDS)

(194) Prokne

Discovered 1879 March 21 by C. H. F. Peters at Clinton.

Named for the daughter of Pandion, king of Athens, and wife of Tereus, king of Thrace. She sent her husband to Athens to bring her sister Philomela {see planet (196)}, to whom she was attached, to Thrace. Tereus became enamored of Philomela and she was carried of to a castle and her tongue removed; reporting back, Tereus claimed she had died. When the infamy became known Prokne served her son Itys to her husband at a feast. Upon this disclosure he drew his sword but was changed into a hoopoe; Philomela, also present, into a nightingale, and Prokne into a swallow. The discoverer stated: "Prokne, found on the day of the vernal equinox, was suggested by the swallow coming with spring; in May followed Philomela, the nightingale." (H 24, AN 96, 336 (1880))

(195) Eurykleia

Discovered 1879 April 19 by J. Palisa at Pola.

Named for Odysseus' {see planet (1143)} aged nurse. When, after twenty years away, Odysseus returned home disguised as a beggar, she recognized him by the scar of a wound given him by a wild boar. (Z 103)

Named by the Berlin astronomers - see the remarks to planet (192).

(196) Philomela

Discovered 1879 May 14 by C. H. F. Peters at Clinton.

Named for the sister of Prokne {see planet (194)}. Philomela was changed into a nightingale. See the remarks to (194). (H 24)

(197) Arete

Discovered 1879 May 21 by J. Palisa at Pola.

Named for the wife of Alcinous, king of the Phaeacians and mother of Nausicaa {see planet (192)}. Arete is also the name for the goddess of virtue. (H 24) Named by the Berlin astronomers - see the remarks to planet (192).

(198) Ampella

Discovered 1879 June 13 by A. Borrelly at Marseilles.

Named probably after Ampelos, a friend of Dionysus {see planet (3671)}, god of wine and revelry. (R. R. de Freitas Mourão)

(199) Byblis

Discovered 1879 July 9 by C. H. F. Peters at Clinton.

Named after the daughter of Miletus and Idothea, who was in love with her brother Caunus who she pursued through various lands, till at length, worn out with sorrow, she was changed into a fountain. (H 24)

(200) Dynamene

Discovered 1879 July 27 by C. H. F. Peters at Clinton.

The meaning is twofold: Dynamene is one of the Nereids and the Greek prefix "di" stands for number two. The discoverer wrotes (AN 96, 336 (1880)): "The name of Dynamene may aid the memory in writing its number, 200." (H 25)

(201) Penelope

Discovered 1879 August 7 by J. Palisa at Pola.

Named after the celebrated princess of Greece, daughter of Icarius and Periboea. She was the wife of Odysseus {see planet (1143)} and mother of Telemachus. During the absence of her husband, she was beset with 108 suitors whom she treated with coldness and disdain. The three most persistent were Amphinomus, Antinous {see planet (1863)}, and Eurymachus, but she was relieved of their unwelcome attentions when Odysseus returned home after an absence of twenty years. Odysseus killed his rivals. Penelope is described as a model of feminine virtue and chastity by Homer. (Z 198)
Named by the Berlin astronomers - see the remarks to planet (192).

(202) Chryseïs

Discovered 1879 September 11 by C. H. F. Peters at Clinton.

Named for the daughter of Chryses, a priest of Apollo {see planet (1862)}. She was captured by Achilles {see planet (588)} but awarded to Agamemnon {see planet (911)}. Apollo sent a plague to avenge her capture, and the Greeks were compelled to return Chryseïs to her father. Then Agamemnon took Briseïs {see planet (655)}, another captive of Achilles, and the famous quarrel resulted. Achilles, angry because he had lost both Briseïs and Chryseïs, withdrew from the Trojan War. (Z 61)

(203) Pompeja

Discovered 1879 September 25 by C. H. F. Peters at Clinton.

The discoverer wrote: "For the planet, whose discovery very coincided with the day, when at Pompeji was commemorated the destruction of that city by Mt. Vesuvius 18 centuries ago, I believed the name Pompeja eminently proper." (AN 96, 336 (1880))

(204) Kallisto

Discovered 1879 October 8 by J. Palisa at Pola.

Named for the mother of Arcas, king of Arcadia, by Zeus {see planets (1020) and (5731)}. She was changed by Zeus into the constellation Ursa, and Arcas was translated into a bear keeper, located in the heavens behind his mother as her guardian. (Z 28)

Named by Th. von Oppolzer at the request of the discoverer.

This name was also given by Simon Marius to the fourth satellite of Jupiter, discovered 1610 by Galilei.

(205) Martha

Discovered 1879 October 13 by J. Palisa at Pola.

Named for the biblical Martha (Book of Luke of the New Testament), sister of Mary and Lazarus of Bethany. Martha's hospitality to Jesus is documented. Palisa selected this name in remembrance to the hospitality of his Berlin colleagues on the occasion of the September 1879 assembly of the Astronomische Gesellschaft. (Astronomischer Kalender für 1883, p. 108)

(206) Hersilia

Discovered 1879 October 13 by C. H. F. Peters at Clinton.

Named after one of the Sabines carried away by the Romans. She was given and married to Romulus. After death she was deified by Juno {see planet (3)} and became Hora, a goddess of beauty. (H 25)

(207) Hedda

Discovered 1879 October 17 by J. Palisa at Pola.

Named in honor of Hedwig Winnecke, wife of Friedrich A. T. Winnecke, director of the Strassburg Observatory. Hedwig Winnecke, née Dell, was a niece of Otto Struve {see planet (2227)}. (Vierteljahrsschr. Astron. Ges., 16. Jahrg., p. 281 (1881))

Named during the meeting of the Astronomische Gesellschaft at Strassburg, September 1881 (BAJ Circ., No. 174 (1882)). The nordic form 'Hedda' was proposed by J. Gyldén.

(208) Lacrimosa

Discovered 1879 October 21 by J. Palisa at Pola.

The name means whining, deplorable. Very probably, this name should reflect the sorrow about the difficulties associated with a recovery of this planet. Lacrimosa could have been observed only four times by Palisa. (Astronomischer Kalender für 1883, p. 108)

Named by L. Schulhof, astronomer in Paris (BAJ Circ., No. 174 (1882)), at the request of the discoverer.

(209) Dido

Discovered 1879 October 22 by C. H. F. Peters at Clinton.

Dido (also named Elissa) was the daughter of Belus, king of Tyre. Her husband Sichaeus was secretly murdered for his money by her brother Pygmalion. She went to Africa, founded Carthage, and became its queen. When Aeneas {see planet (1172)} came to Carthage on his way to Italy after the Trojan War, Dido fell in love with him. When he left for Italy she uttered a curse against the Trojans and then stabbed herself with her sword. (Z 86)

(210) Isabella

Discovered 1879 November 12 by J. Palisa at Pola.

Any reference of this name to a person or occurence is unknown.

(211) Isolda

Discovered 1879 December 10 by J. Palisa at Pola.

Named possibly for the character in the early mediaeval Arthurian legend perpetuated in the opera *Tristan und Isolde* (1865) by Richard Wagner {see planet (3992)}. (H 26; A. Schnell)

Named by the wife of K. Friesach, professor of mathematics in Graz.

(212) Medea

Discovered 1880 February 6 by J. Palisa at Pola.

Named for the daughter of Aeetes, king of Colchis {see planet (1135)} and niece of Circe {see planet (34)}. Medea was a powerful enchantress, she aided Jason {see planet (6063)} in obtaining the Golden Fleece, she restored Aeson, Jason's aged father, to the vigor of youth, and she caused the death of Glauce {see planet (288)}. When Jason abandoned her, she killed their two children and fled to Athens. There she married Aegeus, king of Athens, became jealous of his son Theseus, and finally left him and returned to Colchis. (Z 161)

Named (BAC Circ., No. 176 (1882)) by the Vienna astronomers.

The naming history is handed down in the (Wiener) Astronomischer Kalender für 1883, p. 108: "Nun waren von den in Pola entdeckten Planeten noch die folgenden vier 212, 216, 218 und 219 zu benennen, deren Taufe der Entdecker den Wiener Astronomen überliess. Zu diesem Zwecke wurde eine kleine Conferenz veranstaltet, an welcher auch Dr. H. Kreutz {editor of the Astronomische Nachrichten}, der sich damals in Wien befand, und Frau Direktor Weiss theilnahmen. Jeder der Anwesenden nannte mehrere ihm passend erscheinende Namen, so dass im Ganzen etwa 30 vorgeschlagen wurden, aus denen nach mehreren Wahlgängen Medea, Kleopatra, Bianca und Thusnelda als gewählt hervorgingen. Den Schluss der Conferenz bildete ein fröhliches Mahl, bei dem auch auf die Wiederauffindung der neu benannten Himmelsbürger ein Glas geleert wurde."

(213) Lilaea

Discovered 1880 February 16 by C. H. F. Peters at Clinton.

Named for one of the Naiads. The Naiads are water nymphs, female beings granted very long life but still mortal. The Naiads incarnate the divinity of the spring or stream which they inhabit. (E. Weiss in Littrow, Wunder des Himmels, 7. ed., p. 458 (1886); Z 301)

(214) Aschera

Discovered 1880 February 29 by J. Palisa at Pola.

Named for a goddess of the Sidonians, identified by the Phoenicians with Astarte {see planet (672)} or Ashtoreth. (H 26, A. Schnell)

Named by R. Müller, director of the Hydrographic Service in Pola. The Astronomischer Kalender für 1883, p. 108 stated: "Bei der Taufe von Dejopeja {planet (184)} hatte sich Dir. Müller in Pola scherzweise auch die Taufe des 25. Fundes von Palisa vorbehalten."

(215) Oenone

Discovered 1880 April 7 by V. Knorre at Berlin.

Named for a nymph of Mount Ida in Phrygia. Oenone was the wife of Paris {see planet (3317)}. She foretold that her husband would desert her, abduct Helen {see planet (101)}, and bring ruin to Troy. (Z 181)

(216) Kleopatra

Discovered 1880 April 10 by J. Palisa at Pola.

Named for the Egyptian queen (69-30 B.C.), famed as mistress of Julius Caesar and of Mark Antony. (H 26)

Named (BAJ Circ., No. 176 (1882)) by the Vienna astronomers. See also the remarks to planets (212) and (494).

(217) Eudora

Discovered 1880 August 30 by J. Coggia at Marseilles.

Eudora is the name for (1) one of the Nereides, (2) one of the Atlantides, and (3) one of the Hyades in Greek mythology. The last interpretation seems most likely because Coggia also named another planet {see planet (193)} after a Hyades member. (H 26; LDS)

(218) Bianca

Discovered 1880 September 4 by J. Palisa at Pola.

Named by the Vienna astronomers in honor of the famous coloratura soprano Bianca Bianchi (stage name; Bertha Schwarz (1858-1947), born in Mannheim, Germany), prima donna at the Vienna opera. (A. Schnell; BAJ Circ., No. 176 (1882))

The naming circumstances were published in great detail by several Vienna newspapers in spring 1882. The Fremdenblatt, Abendblatt, 36. Jahrg., Nr. 74 (15 March 1882) stated: "Bianca Bianchi unter den Sternen. Vor Kurzem fand auf der Sternwarte eine recht interessante Sitzung der Mitglieder dieses Institutes statt. Auf der Tagesordnung derselben stand die Taufe jener Asteroiden, welche der fixige Herr Adjunkt Palisa in der letzten Zeit wieder entdeckt, ja sozusagen aus dem Aermel des Weltenraumes geschüttelt hat. Die Adjunkten und Assistenten der Sternwarte zerbrachen sich ihre gelehrten Köpfe... Da erhob sich Dr. Hepperger, ein junger Astronom, und stellte in Anbetracht der Verdienste, welche sich die Primadonna Bianca Bianchi um die schöne Kunst des Gesanges erworben, den Antrag, einen der neuentdeckten Asteroiden Bianca zu taufen. Der Antrag wurde unter stürmischer Akklamation zum Beschlusse erhoben, und so schwebt denn der erste Stern unserer Oper als kleiner Stern in unendlichen Sphären." See also the remarks to planet (212).

The name Bianca has also been given to the satellite Uranus VIII, discovered 1986 by the Voyager 2 spacecraft.

(219) Thusnelda

Discovered 1880 September 30 by J. Palisa at Pola.

Named probably after the daughter of the German chief Segestes. She was taken to wife by force by the German hero Arminius (17 B.C.?-21) who maintained a bloody war against the Romans. He was finally defeated by general Germanicus (15 B.C.-19). Thusnelda was brought to Rome to be seen in the triumph procession of Germanicus. (H 26)

Named (BAJ Circ., No. 176 (1882)) by the Vienna astronomers - see the remarks to planet (212).

(220) Stephania

Discovered 1881 May 19 by J. Palisa at Vienna.

Named in honor of Princess Stephanie of Belgium who in the year of discovery married Archduke Rudolf (1858-1889), Crown Prince of Austria and only son of Emperor Franz Joseph I (1830-1916). Rudolf's later romance with a court beauty, Baroness Maria Vetsera, led to one of the most discussed tragedies of the day when they were found dead in his hunting lodge at Mayerling, near Vienna. (H 27)
Named (BAJ Circ., No. 220 (1884)) by His Majesty the Emperor of Austria.
This was Palisa's first discovery after transferring to Vienna from Pola.

(221) Eos

Discovered 1882 January 18 by J. Palisa at Vienna.

Named for the Greek goddess of the dawn (also Aurora, see planet (94)). (H 27; A. Schnell)

Named (BAJ Circ., No. 220 (1884)) by Staatsrath A. von Braun in Vienna. The Vienna Astronomischer Kalender für 1885, p. 88 stated the reason for this choice: "... eine sehr sinnige Anspielung, um der Hoffnung Ausdruck zu geben, dass mit der Vollendung der neuen Sternwarte die Morgenröte einer besseren Zeit für die Astronomie in Österreich angebrochen sei."

(222) Lucia

Discovered 1882 February 9 by J. Palisa at Vienna.
Named probably after the youngest daughter of Graf Wilczek. (A. Schnell)

Named (BAJ Circ., No. 220 (1884)) by Graf Johann N. Wilczek (1837-1922) who sponsored and participated in the preliminary voyages of the Austro-Hungarian Arctic expedition of 1872-1874 and the Austrian polar expedition to Jan Mayen as part of the First International Polar Year 1882-1883.

(223) Rosa

Discovered 1882 March 9 by J. Palisa at Vienna.
Any reference of this name to a person or occurence is unknown.
Named (BAJ Circ., No. 218 (1883)) by Prof. E. Vincent of Vienna.

(224) Oceana

Discovered 1882 March 30 by J. Palisa at Vienna.
Named for the Pacific Ocean. (H 27)

The discoverer communicated from Honolulu on return from the solar eclipse expedition of May 6, 1883 that Governor von Dessarts of Tahiti has named this planet (BAJ Circ., No. 210 (1883)).

(225) Henrietta

Discovered 1882 April 19 by J. Palisa at Vienna.

Named in honor of the wife of the French astronomer Pierre J. C. Janssen (1824-1907), pioneer of solar spectroscopy and director of the Meudon Observatory. (H 27)
Named (BAJ Circ., No. 213 (1883)) by P. J. C. Janssen.

(226) Weringia

Discovered 1882 July 19 by J. Palisa at Vienna.

Named for Währing, a part of Vienna in which the University Observatory is situated. (H 27)

Named (BAJ Circ., No. 212 (1883)) by Ferdinand Oberwimmer, architect of the Vienna Observatory.

(227) Philosophia

Discovered 1882 August 12 by P. P. Henry at Paris.
The name personifies philosophy. (H 27)

(228) Agathe

Discovered 1882 August 19 by J. Palisa at Vienna.

Named in honor of the youngest daughter of Theodor von Oppolzer (1841-1886), professor of astronomy in Vienna. (Vierteljahrsschr. Astron. Ges., 22. Jahrg., p. 191 (1887))
Named (BAJ Circ., No. 218 (1883)) by Th. von Oppolzer.

(229) Adelinda

Discovered 1882 August 22 by J. Palisa at Vienna.

Named in honor of the wife of E. Weiss, director of the Vienna Observatory. (Astronomischer Kalender für 1884, p. 112)

Named (BAJ Circ., No. 213 (1883)) at the September 1883 Vienna meeting of the Astronomische Gesellschaft by the participants.

(230) Athamantis

Discovered 1882 September 3 by L. de Ball at Bothkamp.

Athamantis or Helle was the daughter of Athamas, king of Thebes, and Nephele {see planet (431)} and the sister of Phrixus. (A. Paluzie-Borrell)

This planet was accidentally discovered during the observation of the nearby planet (12) Victoria. The planet was very bright. The naming obviously was influenced by this fact because the German word 'Helle' means brightness.

(231) Vindobona

Discovered 1882 September 10 by J. Palisa at Vienna.

Named for Vienna, the place of discovery. Vindobona is the Latin name of Vienna. (H 28)

Named (BAJ Circ., No. 213 (1883)) by the participants of the September 1883 Vienna meeting of the Astronomische Gesellschaft.

(232) Russia

Discovered 1883 January 31 by J. Palisa at Vienna.

Named after the former empire in eastern Europe and northern Asia which is nearly coextensive with the present U.S.S.R. (H 28)

Named (BAJ Circ., No. 200 (1883)) by Baron Basil von Engelhardt of Dresden. The Engelhardt Observatory near Kasan, U.S.S.R. resulted from a gift of the instruments of Engelhardt's private observatory in Dresden. Engelhardt was born in Russia.

Times are changing – Russia -> U.S.S.R. -> "new" Russia!

(233) Asterope

Discovered 1883 May 11 by A. Borrelly at Marseilles.

Named for one the seven Pleiades (also Sterope), daughters of Atlas by Pleione, greatly beloved by the gods. Asterope is also commemorated by a Pleiades star. (H 28)

(234) Barbara

Discovered 1883 August 12 by C. H. F. Peters at Clinton.

Named probably for the Saint Barbara and also for a relative of Prof. Ibbotson of Vienna. (R. R. de Freitas Mourão)

Named (BAJ Circ., No. 213 (1883)) at the Vienna meeting of the Astronomische Gesellschaft.

(235) Carolina

Discovered 1883 November 28 by J. Palisa at Vienna.

Named for an atoll of the Line Islands, 450 miles northwest of Papeete, Tahiti, where the discoverer observed the solar eclipse of May 6, 1883. Palisa observed the solar neighborhood in order to find an intra-Mercurian planet. (H 28)

Named by the discoverer (BAJ Circ., No. 218 (1883)) in remembrance of his voyage to this island.

(236) Honoria

Discovered 1884 April 26 by J. Palisa at Vienna.

Named possibly for the goddess personifying honor, worshiped at Rome. (H 28)

Named (BAJ Circ., No. 227 (1884)) by Edward Singleton Holden (1846-1914) {see planet (2974)}, director of the Washburn Observatory of the University of Wisconsin in Madison and later director of the Lick Observatory. Holden was head of the American expedition to Caroline Island {see also the citation for planet (235)} where he met the discoverer. Holden searched visually for an intra-Mercurian planet.

(237) Coelestina

Discovered 1884 June 27 by J. Palisa at Vienna.

Named in honor of Coelestine, née Mauthner von Markhof, wife of Theodor von Oppolzer, professor of astronomy in Vienna. Coelestine was the daughter of the well-known Viennese industrialist family Mauthner von Markhof. (Vierteljahrsschr. Astron. Ges., 22. Jahrg., p. 191 (1887))

Named by Th. von Oppolzer.

(238) Hypatia

Discovered 1884 July 1 by V. Knorre at Berlin.

Named probably after the daughter of the famous mathematician and philosopher Theon of Alexandria, whom she is said to have surpassed in erudition. She was celebrated for her beauty. Among Hypatia's (370?-415 B.C.) writings were commentaries on ancient astronomical works. She was murdered by a mob of Christian fanatics. (H 29)

Hypatia is also honored by a lunar crater.

(239) Adrastea

Discovered 1884 August 18 by J. Palisa at Vienna.

Named for a nymph of Crete to whose care Rhea {see planet (577)} entrusted the infant Zeus {see planet (5731)}. (H 29)

Named (BAJ Circ., No. 242 (1884)) by Mrs. Bertha Schneider. In that source, planet (243) is named for another nurse of Zeus.

The name Adrastea has also been given to the satellite Jupiter XV.

(240) Vanadis

Discovered 1884 August 27 by A. Borrelly at Marseilles.

Named for Freya {see planet (76)}, the Norse goddess of love and beauty. Vanadis is one of the names of Freya. She drives forth with a chariot drawn by cats. (H 29)

(241) Germania

Discovered 1884 September 12 by R. Luther at Düsseldorf.

This is the Latin name for Germany, the native country of the discoverer. (H 29)

(242) Kriemhild

Discovered 1884 September 22 by J. Palisa at Vienna.

Named probably after the sister of Gunther in the medieval German *Nibelungenlied* who married the hero Siegfried. (H 29)

Named (BAJ Circ., No. 241 (1884)) by Moritz von Kuffner, owner of an important private observatory in Ottakring.

(243) Ida

Discovered 1884 September 29 by J. Palisa at Vienna.

Named for a nymph of Crete who nursed the young Zeus {see planet (5731)}. See also the remarks to planet (239). (H 29)

Named (BAJ Circ., No. 242 (1884)) by Moritz von Kuffner.

The planet was imaged by the Galileo spacecraft in 1994. It measures 56 x 24 x 21 km and is heavily cratered. Ida has a satellite which was named Dactyl. The moon is remarkably spherical with a mean diameter of 1.4 km. The Dactyls of Mount Ida were Daemons, Cretan or Phrygian in origin. Their name means 'the fingers'. The Dactyls were magicians, credited with the spread, and sometimes the invention, of the Mysteries. To amuse the infant Zeus, they organized the first Olympic Games. (Z 124; LDS)

(244) Sita

Discovered 1884 October 14 by J. Palisa at Vienna.

Named possibly for the wife of Rama in the Sanskrit epic *The Ramayana*. It is a symbol of the ideal spouse and of everlasting faith. (H 29)
Named by Baron von Pfungen (BAJ Circ., No. 247 (1885)).

(245) Vera

Discovered 1885 February 6 by N. R. Pogson at Madras.
Any reference of this name to a person or occurence is unknown.
Name suggested by the wife of the discoverer (AN 111, 318 (1885)).

(246) Asporina

Discovered 1885 March 6 by A. Borrelly at Marseilles.

Named for the mother of the gods worshiped at the Asporenus mountain in Asia Minor. (H 30)

(247) Eukrate

Discovered 1885 March 14 by R. Luther at Düsseldorf.
Named for one of the Nereids. (H 30)
Named by Geh. Rath. Schönfeld, Bonn (AN 111, 335 (1885)).

(248) Lameia

Discovered 1885 June 5 by J. Palisa at Vienna.

Named for the daughter of Belus loved by Zeus {see planet (5731)}. Hera {see planet (103)}, out of jealousy, deformed Lameia and killed all her children but one. Lameia had the face and breasts of a woman and the body of a serpent. Because she could not avenge herself on Hera, Lameia lured strangers so that she might devour them. (Z 146)

(249) Ilse

Discovered 1885 August 16 by C. H. F. Peters at Clinton.

Named probably after a legendary princess of the Harz mountains in Germany. It is also the name of a small river flowing from the Harz. (H 30)

(250) Bettina

Discovered 1885 September 3 by J. Palisa at Vienna.

Named for Baroness Bettina von Rothschild of the Austrian plutocratic family. In Observatory, Vol. 8, p. 63 (1885) the following information is published: "Herr Palisa, being desirous to raise funds for his intended expedition to observe the total solar eclipse of Aug. 29, 1886 will sell the right of naming the minor planet No. 244 for 50 pounds." The bright idea seems to have struck Palisa, who had already discovered many planets and begun to find difficulties in assigning suitable names, that he might turn his difficulty into a source of profit in a good cause. The offer was not responded to immediately, nor until Palisa had discovered two more planets (Nos. (248) and (250)). He found names for two, leaving, however, the last discovered always open for a patron. A note in Observatory, Vol. 9, p. 142 (1886) informs: "Minor planet No. 250 has been named "Bettina" by Baron Albert von Rothschild." (H 30)

(251) Sophia

Discovered 1885 October 4 by J. Palisa at Vienna.

Named in honor of Sophie von Seeliger, née Stoeltzel, wife of the German as-

tronomer Hugo von Seeliger {see planet (892)}. Palisa was informed on the marriage which took place in 1885, the year of the meeting of the Astronomische Gesellschaft in Geneva. (A. Schnell)
Name proposed by H. von Seeliger.

(252) Clementina

Discovered 1885 October 11 by J. Perrotin at Nice.
Any reference of this name to a person or occurence is unknown.
Named (BAJ Circ., No. 265 (1885)) by the discoverer.

(253) Mathilde

Discovered 1885 November 12 by J. Palisa at Vienna.

Named probably in honor of Mathilde, née Worms, wife of the astronomer Moritz Loewy (1833-1907). (A. Schnell)

Named by V. A. Lebeuf who computed the orbit. Lebeuf (1859-1929) was a staff member of the Paris Observatory. Loewy served as a vice-director of the observatory.

(254) Augusta

Discovered 1886 March 31 by J. Palisa at Vienna.

Named in honor of the widow of Carl Ludwig von Littrow (1811-1877) who succeeded his father Johann Joseph von Littrow (1781-1840) as director of the Vienna Observatory in 1842. Auguste von Littrow (1819-1890) worked as an author with the pen name Otto August. Her home was one of the intellectual centers in Vienna. (H 30; A. Schnell)

(255) Oppavia

Discovered 1886 March 31 by J. Palisa at Vienna.

Named for the city of Troppau in northern Silesia {see planet 257)}, Austria, where the discoverer was born. The city now belongs to Czechoslovakia. Oppavia is the Czech name of Troppau. (H 31)

(256) Walpurga

Discovered 1886 April 3 by J. Palisa at Vienna.

Named after St. Walpurga (?-777), princess of Wessex, England who with her brothers went to Germany to convert the heathen. Her feast day is May 1 (hence Walpurgis Night) when the witches ride on broomsticks to the ancient places of sacrifice to revel with Satan. The best known of this places is the Brocken {see planet (4724)}, highest peak in the Harz mountains, immortalized by the scene of the witches' sabbath in Goethe's {see planet (3047)} *Faust*. (H 31)
Named by Baroness Bettina von Rothschild (BAJ Circ., No. 291 (1887)).

(257) Silesia

Discovered 1886 April 5 by J. Palisa at Vienna.

Named for the native province of the discoverer. There was an Austrian and a German part of Silesia, bordering the Sudeten mountains and the valley of the upper Oder river. The region nowadays chiefly belongs to northern Czechoslovakia and southwestern Poland. (H 31)

(258) Tyche

Discovered 1886 May 4 by R. Luther at Düsseldorf.

Named for the Greek goddess of fortune. Tyche is also the name of one of the Oceanids. (H 31)
Name proposed (AN 114, 383 (1886)) by G. Rümker.

(259) Aletheia

Discovered 1886 June 28 by C. H. F. Peters at Clinton.

Named for a daughter of Zeus {see planet (5731)}, one of the nurses of Apollo {see planet (1862)}. Aletheia is the Greek goddess of truth. (Z 18)

(260) Huberta

Discovered 1886 October 3 by J. Palisa at Vienna.

Named for St. Hubertus (656?-727) of Liège, patron of the hunters. (A. Schnell)

The 'Beilage zum (Wiener) Astronomischer Kalender für 1888', p. 87 ironically stated: "... Palisa ... Huberta nannte, offenbar nach dem heiligen Hubertus, der sich aber der Aufnahme unter die Gestirne durch eine Veränderung der Endsylbe us in a, und damit durch die Verwandlung aus einem Patrone in eine Patronin der Jäger erkaufen musste."

(261) Prymno

Discovered 1886 October 31 by C. H. F. Peters at Clinton.

Named after an Oceanid nymph. (H 31)

(262) Valda

Discovered 1886 November 3 by J. Palisa at Vienna.

Any reference of this name to a person or occurence is unknown.

Name proposed by the Baroness Bettina von Rothschild {see the remarks to planet (250)}.

(263) Dresda

Discovered 1886 November 3 by J. Palisa at Vienna.

The planet is named to honor the German city of Dresden {see also planet (3053)}. (H 31; A. Schnell)

Named by Baron B. von Engelhardt, owner of a private observatory in the city of Dresden.

(264) Libussa

Discovered 1886 December 22 by C. H. F. Peters at Clinton.

Named after the princess who founded the city of Prague {see planet (2367)} in the 8th century. (H 31)

(265) Anna

Discovered 1887 February 25 by J. Palisa at Vienna.

Named probably in honor of Anny Weiss, née Kretschmar, daughter-in-law of Prof. E. Weiss, director of the Vienna Observatory. (A. Schnell)

In the 'Beilage zum astronomischen Kalender 1888', p. 88, the contributor E. Weiss laconically noted: "... erhielten die Namen Anna und Aline, die eines weiteren Commentares nicht bedürfen."

(266) Aline

Discovered 1887 May 17 by J. Palisa at Vienna.

Named probably in honor of Linda von Schuster, daughter of Prof. E. Weiss, director of the Vienna Observatory. (A. Schnell)

See also the remark to planet (265).

(267) Tirza

Discovered 1887 May 27 by A. Charlois at Nice.

Named after the biblical Tirzah (Solomon's Song 6, 4). (H 32)

(268) Adorea

Discovered 1887 June 8 by A. Borrelly at Marseilles.

Named after the flat cake produced from meal and salt which was offered by the Romans as a sacrifice to their gods. (Beilage zum astronomischen Kalender 1889, p. 86)

The naming was violently criticized in the 'Astronomischer Kalender': "Der Asteroid hat mittlerweile den Namen Adorea erhalten, womit die Römer die leichten flachen Kuchen aus Mehl und Salz bezeichneten, welche sie den Göttern opferten. Wir haben also jetzt neben einem Büchertitel {planet (141) Lumen)}, einem Romanhelden {planet (152) Atala} und der Götternahrung {planet (193) Ambrosia} auch noch eine Opferspeise durch die Asteroiden am Himmel verewigt erhalten: fürwahr wohl die beste denkbare Satire auf die Grundsätze, die in den letzten Jahren beim Benennen dieser Himmelskörper massgebend geworden sind."

(269) Justitia

Discovered 1887 September 21 by J. Palisa at Vienna.

Named for the goddess of justice, daughter of Jupiter and Astraea {see planet (5)}. She and her mother went to heaven when the Golden Age was ended. She carries a balance and a sword. (H 32)

Named by Hofrath August Biela, "einem eifrigen Freunde der Astronomie und Besitzer einer sehr hübschen kleinen Privatsternwarte" (Beilage zum astronomischen Kalender für 1889, p. 83)

(270) Anahita

Discovered 1887 October 8 by C. H. F. Peters at Clinton.

Named for the Persian Great Mother, goddess of fertility. (H 32)

The discoverer proposed the name and, simultaneously, provided a quotation in AN 118, 44 (1887): "Ich schlage für diesen Planeten den Namen Anahita vor, von der altpersischen Göttin, bei den Occidentalen zuweilen auch Anaitis übersetzt. (Man mag darüber z.B. nachlesen: Ebers, Aegyptische Königstochter, Bd. 2, die Anmerkung 38)." Prof. E. Weiss, director of the Vienna Observatory, stated that this planet is also named in honor of Anita, daughter of the famous American astronomer S. Newcomb {see planet (855)}. (Beilage zum astronomischen Kalender 1889, p. 83)

(271) Penthesilea

Discovered 1887 October 13 by V. Knorre at Berlin.

Named for the queen of the Amazones {see planet (1042)} and ally of Troy in the Trojan War. She fought against Achilles {see planet (588)} by whom she was slain. (H 32)

(272) Antonia

Discovered 1888 February 4 by A. Charlois at Nice.

Any reference of this name to a person or occurence is unknown.

(273) Atropos

Discovered 1888 March 8 by J. Palisa at Vienna.

Named after one of the three Fates. The others are Clotho and Lachesis {see planets (97) and (120), respectively}. Atropos carries the shears and cuts the thread of life. (Z 37)

Named by the Berlin astronomers (BAJ Circ., No. 317 (1888)).

The Beilage zum astronomischen Kalender 1889, p. 86 gives a connection between the choice of this particular name with the death of the German emperor Wilhelm I.: "Der Planet wurde am 8. März, wenige Stunden vor dem Hinscheiden Kaiser Wilhelm I., aufgefunden und in Folge dessen von den Berliner Astronomen für ihn sehr sinnig der Name der Parze Atropos gewählt."

(274) Philagoria

Discovered 1888 April 3 by J. Palisa at Vienna.
Named by the discoverer for a recreation club in Olmütz. (A. Schnell)

(275) Sapientia

Discovered 1888 April 15 by J. Palisa at Vienna.
This is the Latin name for wisdom. (H 32)

(276) Adelheid

Discovered 1888 April 17 by J. Palisa at Vienna.
Any reference of this name to a person or occurence is unknown.

(277) Elvira

Discovered 1888 May 3 by A. Charlois at Nice.

Named possibly for the principal character in "*Méditations poétiques*" (1820) and "*Harmonies poétiques et religieuses*" (1830) of the French statesman and poet Alphonse de Lamartine (1790-1869). (H 32)

(278) Paulina

Discovered 1888 May 16 by J. Palisa at Vienna.
Any reference of this name to a person or occurence is unknown.

(279) Thule

Discovered 1888 October 25 by J. Palisa at Vienna.

Named for an island in the northernmost part of the North Sea to which the ancients gave the name Ultima Thule. Its location has never been accurately ascertained. Some writers have thought it to be Iceland, Greenland, or the Shetland Islands. To the ancients, it was the northern limit of the habitable world. (Z 271)

The name is exceptionally appropriate because the planet's orbit was the farthest from the Sun then known. Except for the special dynamical case of the Trojan asteroids and some other spectacular objects, which probably are cometary nuclei, Thule still occupies this position. It is very near the 3:4 commensurability of period with Jupiter.

(280) Philia

Discovered 1888 October 29 by J. Palisa at Vienna.

This name belongs to the nymph who, with Koronis {see planet (158)} and Clyda educated the young Dionysos {see planet (3671)} on Naxos. See also the citation for planet (44). (H 33)

Named by A. Berberich.

The planet was accidentally discovered while searching for (255) Oppavia. The ephemeris position, computed by Berberich, however was in error by 1 degree in declination. A spelling mistake thus led to the discovery of a new planet. Palisa therefore transferred the naming right to the Berlin astronomer.

(281) Lucretia

Discovered 1888 October 31 by J. Palisa at Vienna.

Named in honor of Lucretia Caroline Herschel (1750-1848), sister of the discoverer (1781) of Uranus Sir William Herschel (1738-1822) {see planet (2000)}, whom she assisted, beginning in 1772. She independently discovered seven or eight comets. After her brother's death she returned from England to Hannover, Germany and constructed a catalogue of the nebulae and clusters discovered by him. She received the Gold Medal of the Royal Astronomical Society in 1828. (AN 120, 335 (1889))

Named by Mrs. Šafařik, wife of the astronomer Adalbert Šafařik, professor of astronomy at the Prague University.

(282) Clorinde

Discovered 1889 January 28 by A. Charlois at Nice.

Named probably after the heroine of the epic poem *Jerusalem Delivered* by the Italian writer Torquato Tasso (1544-1595). (H 33)

(283) Emma

Discovered 1889 February 8 by A. Charlois at Nice.
Any reference of this name to a person or occurence is unknown.
See also the comment to planet (295).

(284) Amalia

Discovered 1889 May 29 by A. Charlois at Nice.
Any reference of this name to a person or occurence is unknown.
See also the comment to planet (295).

(285) Regina

Discovered 1889 August 3 by A. Charlois at Nice.
Any reference of this name to a person or occurence is unknown.
See also the comment to planet (295).

(286) Iclea

Discovered 1889 August 3 by J. Palisa at Vienna.

Named "in honor of the heroine of the astronomical romance *Uranie*, published last year by Camille Flammarion" (l'Astronomie, Vol. 9, p. 392 (1890)). (H 33)

(287) Nephthys

Discovered 1889 August 25 by C. H. F. Peters at Clinton.

Named for the Egyptian goddess of the dead, sister of Isis {see planet (42)} and wife of Seth. (H 34)

The discoverer wrote in AN 122, 393 (1889): "Ich schlage den Namen Nephthys vor - bei den Ägyptern repräsentirt Neb-ta vielleicht den Planeten Venus ..."

(288) Glauke

Discovered 1890 February 20 by R. Luther at Düsseldorf.

Named for the daughter of Creon, king of Corinth, whom Jason {see planet (6063)} planned to marry. Glauke is also the name of one of the Danaides and of one of the Nereides. (Z 111)

Named by the son of the discoverer. Luther stated in AN 124, 288 (1890): "Nachdem ich mich von der Neuheit des Planeten hinlänglich überzeugt zu haben glaubte, schickte ich an meinen Sohn in Hamburg, welcher den Planeten seit dem 24. Februar {1890} daselbst eifrigst beobachtet hat, eine Namen-Liste zur Auswahl eines Namens, worunter sich auch drei seiner Zeit von Argelander gewünschte Namen befanden. Nach einem Briefe meines Sohnes vom 9. März hat einer dieser drei Namen "Glauke" in Hamburg besonders gefallen, so dass ich denselben als gewählt betrachte."

(289) Nenetta

Discovered 1890 March 10 by A. Charlois at Nice.

The name originates from a French argot which denotes a more or less frivolous woman. (M.-A. Combes)

Une "nénette" est un mot français argotique qui signifie tout simplement une fille, pas forcément frivole. C'est un mot plutôt sympathique. See also the comments to planets (295) and (494), respectively.

(290) Bruna

Discovered 1890 March 20 by J. Palisa at Vienna.

Named by Hofrath August Biela for his home town Brünn, now Brno, Czechoslovakia {see also planet (2889)}. (A. Schnell; Beilage zum Astronomischen Kalender 1892, p. 92)

(291) Alice

Discovered 1890 April 25 by J. Palisa at Vienna.
Any reference of this name to a person or occurence is unknown.

Named by the Société Astronomique de France "at the gracious invitation of the discoverer" (l'Astronomie, Vol. 10, p. 112 (1891)). The planets (291) and (292) were independently discovered by A. Charlois at Nice one night later.

(292) Ludovica

Discovered 1890 April 25 by J. Palisa at Vienna.
Any reference of this name to a person or occurence is unknown.
Named by the Société Astronomique de France. See the note to planet (291).

(293) Brasilia

Discovered 1890 May 20 by A. Charlois at Nice.

Named for the South American country. Dom Pedro II (1825-1891), former emperor of Brazil who had withdrawn to France after the revolution of 1889 may have occasioned the naming. He was known as an enthusiastic patron of astronomy. Dom Pedro was an admirer and friend of Camille Flammarion who invited him to Juvisy (see planets (1021), (605)). (H 34; M.-A. Combes)

(294) Felicia

Discovered 1890 July 15 by A. Charlois at Nice.
Any reference of this name to a person or occurence is unknown.
See also the comment to planet (295).

(295) Theresia

Discovered 1890 August 17 by J. Palisa at Vienna.
Any reference of this name to a person or occurence is unknown.

The name was erroneously attributed to Maria Theresia (1717-1780), queen of Hungary and Bohemia, and wife of emperor Franz I. As was mentioned in the Beilage zum Astronomischen Kalender 1892, p. 91, this interpretation obviuosly is wrong: "Die Namen der neun Planeten 283, 284, 285, 289, 294, 295, 297, 302 und 303, welche der Reihe nach lauten: Emma, Amelia, Regina, Nenetta, Felicia, Theresia, Cäcilia, Clarissa und Josefina und von den Entdeckern oder Freunden derselben gewählt wurden, gehören, wie man leicht erkennt, nicht antiken, sondern lebenden Göttinnen an: es gebietet daher schon die Courtoisie, sich weiterer Commentare über dieselben zu enthalten." (Communicated by A. Schnell)

(296) Phaëtusa

Discovered 1890 August 19 by A. Charlois at Nice.

Named for one of the daughters of Apollo and Klymene {see planets (1862) and (104)}, changed by Zeus {see planet (5731)} into poplars after the death of their brother Phaethon {see planet (3200)}. (H 34)

(297) Caecilia

Discovered 1890 September 9 by A. Charlois at Nice.
Any reference of this name to a person or occurence is unknown.
See also the comment to planet (295).

(298) Baptistina

Discovered 1890 September 9 by A. Charlois at Nice.
Any reference of this name to a person or occurence is unknown.

(299) Thora

Discovered 1890 October 6 by J. Palisa at Vienna.
Named for the Norse god of thunder, weather and crops. (A. Schnell)

Named by Geheimrat Prof. Scheibler in Berlin. In Norse mythology this name repeatedly exists as spouse of Helge, spouse of Ragnar Lodbrok, and as a girlfriend of Gudrun (Beilage zum Astronomischen Kalender 1892, p. 92).

(300) Geraldina

Discovered 1890 October 3 by A. Charlois at Nice.
Any reference of this name to a person or occurence is unknown.

(301) Bavaria

Discovered 1890 November 16 by J. Palisa at Vienna.

Named to commemorate the August 1891 meeting of the Astronomische Gesellschaft in Munich, the capital city of Bavaria. (AN 128, 197 (1891))
Name proposed by B. A. Gould.

(302) Clarissa

Discovered 1890 November 14 by A. Charlois at Nice.
Any reference of this name to a person or occurence is unknown.
See also the comment on planet (295).

(303) Josephina

Discovered 1891 February 12 by E. Millosevich at Rome.

The discoverer states in the Memoirs of the Collegio Romano College that the planet was named "... in homage to a person dear to me ...". (H 35)
See also the comment on planet (295).

(304) Olga

Discovered 1891 February 14 by J. Palisa at Vienna.
Named by F. W. Argelander in honor of his niece. (H 35)

(305) Gordonia

Discovered 1891 February 16 by A. Charlois at Nice.

Named probably in honor of James Gordon Bennett Jr. (1841-1918), editor of the *New York Herald*, founded by his father. He sent Stanley to Africa to find Livingstone (1869-1871). Bennett gave monetary aid to the Juvisy {see planet (605)} Observatory of Flammarion {see planet (1021)} by whom he is praised in l'Astronomie, p. 307 (1890). (H 35)

(306) Unitas

Discovered 1891 March 1 by E. Millosevich at Rome.

Named in honor of the Italian astronomer Angelo Pietro Secchi (1818-1878) {see planet (4705)} and also for the unity of Italy. The citation reads: "Al pianeta, scoperto ... dal E. Millosevich, e da lui pregato di denominarlo, do il nome di Unitas, associando in questo nome due idee, la prima il ricordo d'un libro classico del mio illustre predecessore ed amico A. Secchi, la seconda l'unità della patria." (AN 127, 167 (1891))

Named by P. Tacchini (1838-1905), director of the Modena Observatory in 1859. He went to Palermo in 1863 and succeeded Secchi in 1879 as director of the Osservatorio del Collegio Romano. He was a pioneer of solar spectroscopy, paying particular attention to solar prominences which he showed to obey the 11-year period (H 35).

(307) Nike

Discovered 1891 March 5 by A. Charlois at Nice.

Named for the Greek goddess of victory corresponding to the Roman Victoria {see planet (12)}. The name also denotes the Greek name of the city of discovery. (H 35)

(308) Polyxo

Discovered 1891 March 31 by A. Borrelly at Marseilles.

Named for a priestress of Apollo's {see planet (1862)} temple in Lemnos and the nurse of Queen Hypsipyle {see planet (587)}. She advised the Lemnian women to murder their husbands, and suggested to Hypsipyle welcoming the Argonauts so that they would father sons by the Lemnian women, in order that the race would not become extinct. Polyxo is also the name of one of the Hyades which were said to have been originally nymphs, daughters of Atlas and the Oceanid Aethra {see planet (132)}. This second interpretation seems more likely because Borrelly named another planet after a Hyades member - see planet (322). (Z 218; LDS)

(309) Fraternitas

Discovered 1891 April 6 by J. Palisa at Vienna.
This is the Latin word for fraternity. (H 35; A. Schnell)

The Beilage zum Astronomischen Kalender 1892, p. 92 reported the occasion of the naming: "... wurde bei einer geselligen Zusammenkunft der Studiengenossen Palisa's zur Feier des 25. Jahrestages der Ablegung der Maturitätsprüfung Fraternitas getauft."

(310) Margarita

Discovered 1891 May 16 by A. Charlois at Nice.
Any reference of this name to a person or occurence is unknown.

(311) Claudia

Discovered 1891 June 11 by A. Charlois at Nice.
Any reference of this name to a person or occurence is unknown.

(312) Pierretta

Discovered 1891 August 28 by A. Charlois at Nice.
Any reference of this name to a person or occurence is unknown.

(313) Chaldaea

Discovered 1891 August 30 by J. Palisa at Vienna.

Named for the country in Asia between the Tigris and Euphrates whose capital was Babylon. The Chaldaeans were famous for originating astrology as it is known today. (H 36)

Name assigned by Miss Catherine W. Bruce of New York at the invitation of Dr. Palisa, in token of the gratitude of astronomers (Astron. J., Vol. 11, No. 8, p. 64 (1891)).

(314) Rosalia

Discovered 1891 September 1 by A. Charlois at Nice.
Any reference of this name to a person or occurence is unknown.

(315) Constantia

Discovered 1891 September 4 by J. Palisa at Vienna.

This name describes a special attribute: "Mot latin qui signifie constance et persévérance, qualité jugée essentielle pour faire un bon astronome par Camille Flammarion qui proposa le nom." (J. Meeus)
This interpretation originates from M.-A. Combes.

(316) Goberta

Discovered 1891 September 8 by A. Charlois at Nice.
Any reference of this name to a person or occurence is unknown.

(317) Roxane

Discovered 1891 September 11 by A. Charlois at Nice.

Named for the daughter of the Persian king Oxyartes and wife of the Macedonian king Alexander the Great (356-323 B.C.). (A. Schnell; I. van Houten-Groeneveld)

Named by F. Bidschof, assistant at the Vienna Observatory, at the request of the discoverer. (Kalender für alle Stände, Vienna 1894). The planet originally (AN 132, 176 (1893)) was named 'Roxana'.

(318) Magdalena

Discovered 1891 September 24 by A. Charlois at Nice.
Any reference of this name to a person or occurence is unknown.

(319) Leona

Discovered 1891 October 8 by A. Charlois at Nice.
Any reference of this name to a person or occurence is unknown.

(320) Katharina

Discovered 1891 October 11 by J. Palisa at Vienna.
Named in honor of the mother of the discoverer, Katharina. (H 36)

(321) Florentina

Discovered 1891 October 15 by J. Palisa at Vienna.
Named in honor of the daughter of the discoverer, Florentine. (H 36)

(322) Phaeo

Discovered 1891 November 27 by A. Borrelly at Marseilles.

Named for one of the Hyades which were said to have been originally nymphs, daughters of Atlas and an oceanid named Aethra {see planet (132)}. Their number varies from two to seven, and their names were no less variable. The most usual seem to have been Ambrosia {see planet (193)}, Eudora {(217)}, Aesyle, Coronis {(158)}, Dione {(106)}, Polyxo {(308)} and Phaeo. (M.-A. Combes)

(323) Brucia

Discovered 1891 December 22 by M. Wolf at Heidelberg.

Named in honor of Miss Catherine Wolfe Bruce (1816-1900), noted American patroness of astronomy "who presented Dr. Wolf with the means to secure the largest and best photographic telescope that could be made for his especial research in the domain of planet and nebula photography." (J. A. Brashear, Pop. Astron., Vol. 11, p. 548 (1903)). (H 36)

This is the first numbered minor planet discovered photographically. With the Bruce telescope of the Heidelberg Observatory more now numbered minor planets were discovered than with any other telescope in the world. Miss Bruce is also honored by a lunar crater.

(324) Bamberga

Discovered 1892 February 25 by J. Palisa at Vienna.

"Der Planet ... hat bei Gelegenheit der Astronomenversammlung in Bamberg den Namen Bamberga erhalten." Bamberg is a leading city of Bavaria, Germany. (AN 141, 405 (1896))

Named by Dr. von Brandt, mayor of the city of Bamberg, at the occasion of the September 1896 AG meeting.

(325) Heidelberga

Discovered 1892 March 4 by M. Wolf at Heidelberg.

Named for the famous German city on the Neckar {see planet (1223)} river. This planet was discovered at Wolf's private observatory in the Märzgasse in the old part of the city of Heidelberg. (H 36)

(326) Tamara

Discovered 1892 March 19 by J. Palisa at Vienna.

Named for a queen of Georgia (1184-1212) who became a legendary figure in late medieval romance. She had her lovers killed and thrown into a river from her palace window when she was tired of them. (H 37)
Name proposed by the Russian Grand Duke Georg Alexandrovich.

(327) Columbia

Discovered 1892 March 22 by A. Charlois at Nice.

Named in honor of Christopher Columbus (1446?-1506) to commemorate the 400th anniversary of his first voyage to the New World. See also the citation for planet (334). (H 37)
Columbus is also honored by craters on the Moon and on Mars.

(328) Gudrun

Discovered 1892 March 18 by M. Wolf at Heidelberg.

Named for the wife of Sigurd and later of Atli in the Gudrun Saga in Norse mythology. (I. van Houten-Groeneveld)

(329) Svea

Discovered 1892 March 21 by M. Wolf at Heidelberg.

This name is an allusion to Sweden. The discoverer has made studies in Sweden in 1889. (H 37)

(330) Adalberta

A910 CB. Discovered 1910 February 2 by M. Wolf at Heidelberg.

Named in honor of Adalbert Merx {see planet (808)}, father-in-law of the discoverer, or, possibly, for Adalbert Krüger, editor of the *Astronomische Nachrichten*. (H 37)

Another planet also discovered by Max Wolf had been called (330) Ilmatar (AN 130, 159 (1892)). This planet was identified with (298) Baptistina and the name was transferred (Veröff. Rechen-Inst. 16, p. 27 (1901)) to (385). The "original" Adalberta was discovered by Wolf 1892 March 18. There exist no other observations than those from the discovery apparition. A reexamination of the original plates by R. M. West, C. Madsen and L. D. Schmadel (Astron. Astrophys., Vol. 110, p. 198-202 (1982)) indicates that Wolf measured two stellar images and that therefore this planet is non-existent. The number and name were therefore assigned to another minor planet (M 6939).

(331) Etheridgea

Discovered 1892 April 1 by A. Charlois at Nice.
Any reference of this name to a person or occurence is unknown.

(332) Siri

Discovered 1892 March 19 by M. Wolf at Heidelberg.
Any reference of this name to a person or occurence is unknown.

(333) Badenia

Discovered 1892 August 22 by M. Wolf at Heidelberg.

Named for the region in southwestern Germany, former German grand-duchy, in which Heidelberg {see planet (325)} is situated. It is now part of the German state of Baden-Württemberg. (H 37)

(334) Chicago

Discovered 1892 August 23 by M. Wolf at Heidelberg.

Named by the discoverer at an astronomical congress in Chicago to honor this great city. (AN 134, 167 (1893))

This congress was in connection with the World's Columbian Exposition (May to October 1893) to commemorate the 400th anniversary of Columbus' discovery of America. The name Columbia was to be chosen but it had been appropriated shortly before by A. Charlois for (327). The newly constructed mounting and tube of the Yerkes 40-inch refractor in operating position were exhibited at the Exposition by the builders Warner and Swasey (see also planets (990) and (992)) (H 37).

(335) Roberta

Discovered 1892 September 1 by A. Staus at Heidelberg.

Named by the discoverer in honor of Robert von der Osten-Sacken who lived in Heidelberg as a private scientist. He was an entomologist and his research trips led him many times to the U.S.A. From one of these trips he brought back a 3-inch refractor which he gave to Staus as a present. The planet was named in recognition and gratitude for his friendship. (H 38)

Staus discovered this planet with the 6-inch telescope of Max Wolf's private observatory in the Märzgasse in Heidelberg.

(336) Lacadiera

Discovered 1892 September 19 by A. Charlois at Nice.

Named for the village La Cadière-d'Azur in the departement Var in southern France where the discoverer held some property. (H 38)

(337) Devosa

Discovered 1892 September 22 by A. Charlois at Nice.

Any reference of this name to a person or occurence is unknown.

(338) Budrosa

Discovered 1892 September 25 by A. Charlois at Nice.

Any reference of this name to a person or occurence is unknown.

(339) Dorothea

Discovered 1892 September 25 by M. Wolf at Heidelberg.

Named in honor of Dorothea Klumpke Roberts (1861-1943) {see also planet (1040)}. She was the first lady to gain the degree Doctor of Mathematical Sciences at the Sorbonne. She assisted her husband Isaac Roberts (1829-1904), a building contractor who took up astronomy as a hobby when about 50, built two private observatories at Liverpool and Crowborough, and became a pioneer in the photography of clusters and nebulae. (H 38)

(340) Eduarda

Discovered 1892 September 25 by M. Wolf at Heidelberg.

Named in honor of Eduard von Lade, amateur astronomer at Geisenheim on the Rhine. (H 38)

(341) California

Discovered 1892 September 25 by M. Wolf at Heidelberg.

Named by the discoverer for the state of California in order to commemorate his visit in 1893. (H 38)

(342) Endymion

Discovered 1892 October 17 by M. Wolf at Heidelberg.

Named for the youthful shepherd who so enchanted Selene (the Moon) by his unsurpassed beauty as he lay sleeping on Mt. Latmos that she came down every night to visit him. (H 38)

Named by F. Bidschof (AN 141, 71 (1896)), who was an orbit computer and served as assistant at the Vienna Observatory. For the first time, the name of a male god was used for a minor planet. (A. Schnell)

Endymion is also commemorated by a lunar crater.

(343) Ostara

Discovered 1892 November 15 by M. Wolf at Heidelberg.

Named after the early Norse goddess of Spring. The Easter rabbit was the escort of Ostara who thus contributed the name Easter. The German word is "Ostern". (H 39)

(344) Desiderata

Discovered 1892 November 15 by A. Charlois at Nice.

Named probably in honor of Désirée Clary of Marseilles who became the wife of General Jean-Baptiste Jules Bernadotte (1763-1844). He later became King Charles XIV John of Sweden (1818-1844) and was the founder of the present Swedish dynasty. Désirée was a former fiancée of Napoleon. (H 39)

(345) Tercidina

Discovered 1892 November 23 by A. Charlois at Nice.

Any reference of this name to a person or occurence is unknown.

(346) Hermentaria

Discovered 1892 November 25 by A. Charlois at Nice.

Named probably for the village of Herment in the département du Puy-de-Dôme (Auvergne region). (H 39; M.-A. Combes)

(347) Pariana

Discovered 1892 November 28 by A. Charlois at Nice.

Any reference of this name to a person or occurence is unknown.

(348) May

Discovered 1892 November 28 by A. Charlois at Nice.

Named possibly in honor of the contemporal German writer Karl May (1842-1912), well-known for his many novels dealing with adventures among the Indians of the wild West (*Winnetou*). May also published a lot of exciting travelogues from the Ottoman Empire. (G. Zech)

The books of Karl May were widely distributed in many languages - partly in large editions - in central Europe. French translations appeared in journals (starting 1881) as well as in books (by Mame et Fils, Tours, since 1885).

(349) Dembowska

Discovered 1892 December 9 by A. Charlois at Nice.

Named in honor of Ercole Dembowski (1812-1881), native of Milan, Italy who established a private observatory at Naples which was later transferred to Milan. His careful measurements of double stars marked an important stage in this branch of astronomy. (H 39)

Dembowski is also honored by a lunar crater.

(350) Ornamenta

Discovered 1892 December 14 by A. Charlois at Nice.

Named "in remembrance of the mariner Hornemann, of Holland, whose son is a very zealous member of the Société Astronomique de France". (l'Astronomie, Vol. 11, p. 332 (1897))

(351) Yrsa

Discovered 1892 December 16 by M. Wolf at Heidelberg.
Any reference of this name to a person or occurence is unknown.

(352) Gisela

Discovered 1893 January 12 by M. Wolf at Heidelberg.

Named in honor of Gisela Wolf, neé Merx {see planet (808)}, wife of the discoverer. (H 39)

(353) Ruperto-Carola

Discovered 1893 January 16 by M. Wolf at Heidelberg.

The discoverer stated in AN 154, 15 (1900): "Der Planet hat durch die Versammlung der A.G. in Heidelberg den Namen Ruperto-Carola erhalten." This is a reference to Heidelberg {see planet (325)} University, one of the oldest and most famous of German universities, which was founded by Elector Ruprecht I (1309-1390) in 1386. Grand Duke Karl Friedrich von Baden reestablished the university in 1803. (H 39)

(354) Eleonora

Discovered 1893 January 17 by A. Charlois at Nice.
Any reference of this name to a person or occurence is unknown.

(355) Gabriella

Discovered 1893 January 20 by A. Charlois at Nice.

Named in honor of Gabrielle Renaudot (1877-1962), later Mme. Camille Flammarion {see planet (1021)}, in recognition of her zeal as a student of astronomy and an observer at the Observatory of Juvisy {see planet (605)}. Later she was editor-in-chief of *l'Astronomie*, Laureate of the Académie des Sciences de France, and officier de la Légion d'Honneur. (H 39)

(356) Liguria

Discovered 1893 January 21 by A. Charlois at Nice.

Named after the region in northwest Italy and southern France between Genoa and Nice, bordering on the Ligurian sea north of the island of Corsica. (H 40)

In AN 156, 239 (1901), 34 names of planets between numbers (356) and (451) are published by J. Bauschinger. The text only states: "Nach Zustimmung des Herrn Charlois haben folgende Planeten ... Namen erhalten." It therefore seems obvious that the names were selected by the Astronomisches Rechen-Institut.

(357) Ninina

Discovered 1893 February 11 by A. Charlois at Nice.
Any reference of this name to a person or occurence is unknown.

(358) Apollonia

Discovered 1893 March 8 by A. Charlois at Nice.

The planet was probably named for a "ville ancienne de l'Illyrie, à l'embouchure de l'Aoüs, réputée à l'époque gréco-romaine comme centre culturel et commercial." This interpretation was given by M.-A. Combes. (J. Meeus)
See also the remarks to planet (356).

(359) Georgia

Discovered 1893 March 10 by A. Charlois at Nice.

This planet was named at the 1902 meeting of the Astronomische Gesellschaft in Göttingen. It honors Georg August, later King George II of England (1683-1760), who founded the University of Göttingen in 1737. (H 40; AN 159, 279 (1902)) Named by Miss Klein, daughter of the famous mathematician.

(360) Carlova

Discovered 1893 March 11 by A. Charlois at Nice.
Any reference of this name to a person or occurence is unknown.

(361) Bononia

Discovered 1893 March 11 by A. Charlois at Nice.

This is the Latin name for the city of Bologna {see planet (2601)} in north central Italy and of Boulogne, the French seaport on the English Channel. (H 40)
See also the remarks to planet (356).

(362) Havnia

Discovered 1893 March 12 by A. Charlois at Nice.

This planet is probably named for the city of Copenhagen, capital of Denmark. The Latin name is Hafnia. (H 40)
See also the remarks to planet (356).

(363) Padua

Discovered 1893 March 17 by A. Charlois at Nice.
Named after the city of Padova near Venice, Italy. (H 40)
See also the remarks to planet (356).

(364) Isara

Discovered 1893 March 19 by A. Charlois at Nice.

Isara is an early name of the Isère river which joins the Rhone river in southern France. (H 40)
See also the remarks to planet (356).

(365) Corduba

Discovered 1893 March 21 by A. Charlois at Nice.

This planet is possibly named for the city of Córdoba, Spain on the Guadalquivir river. Corduba is the ancient name. (H 40)
See also the remarks to planet (356).

(366) Vincentina

Discovered 1893 March 21 by A. Charlois at Nice.

Named by the orbit computer Prof. G. Boccardi to honor the Italian astronomer Vincenzo Cerulli (1859-1927). Boccardi worked for many years in the private observatory of Cerulli in Teramo. See also the citation for planet (704). (H 40)

This planet was studied by E. Strömgren on its approach of only 4.5 million miles to (386) Siegena in January, 1921.

(367) Amicitia

Discovered 1893 May 19 by A. Charlois at Nice.
This planet is named with the Latin word for friendship. (H 40)
See also the remarks to planet (356).

(368) Haidea

Discovered 1893 May 19 by A. Charlois at Nice.
Any reference of this name to a person or occurence is unknown.

(369) Aëria

Discovered 1893 July 4 by A. Borrelly at Marseilles.

Named presumably after one of the four elements of the ancients. E. Weiss stated in the Astronomischer Kalender für 1896, p. 155: "... während wir wohl kaum fehlgehen dürften, wenn wir annehmen, dass mit Aëria eines der vier Elemente (Feuer, Luft, Wasser und Erde) der Alten am Himmel verewigt werden sollte..." (A. Schnell)

The naming may be influenced by the two letters of the provisional designation 1893 AE. See also the remarks to planet (579).

(370) Modestia

Discovered 1893 July 14 by A. Charlois at Nice.

Named for the quality of modesty, the freedom from conceit or vanity. (H 41) See also the remarks to planet (356).

(371) Bohemia

Discovered 1893 July 16 by A. Charlois at Nice.

Named for the region which today lies in western Czechoslovakia. With Moravia and Silesia {see planets (1901) and (257)}, Bohemia now forms one of the two constituent states of Czechoslovakia. At the time of the discovery, Bohemia was a part of Austria. (H 41)

Named by A. Abetti.

The planet's orbit was computed by H. Mader (AN 153, 447 (1900)) whose native country is Bohemia.

(372) Palma

Discovered 1893 August 19 by A. Charlois at Nice.

Named probably for the capital of Mallorca, the largest of the Balearic islands in the Mediterranean south of France. (H 41)

See also the remarks to planet (356).

(373) Melusina

Discovered 1893 September 15 by A. Charlois at Nice.

Named probably for the legendary ancestress of the noble family of Lusignan of the 10th century, a heroine of French romance. She married a knight on the condition that he was never to see her on Saturday. But his curiosity led to the discovery that she was half snake. Thereupon she took the form of a dragon and flew away to reappear as an omen before the death of her descendants. The town of Lusignan is supposed to be named for her. (H 41)

See also the remarks to planet (356).

(374) Burgundia

Discovered 1893 September 18 by A. Charlois at Nice.

Named for the former kingdom, duchy and province in eastern France. (H 41) See also the remarks to planet (356).

(375) Ursula

Discovered 1893 September 18 by A. Charlois at Nice.

Any reference of this name to a person or occurence is unknown.

See also the remarks to planet (356).

(376) Geometria

Discovered 1893 September 18 by A. Charlois at Nice.

The Latin word originally means "to measure the Earth". It concerns a branch of mathematics that deals with the measurement, properties and relationships of points, lines, angles, surfaces, and solids. (H 41)

See also the remarks to planet (356).

(377) Campania
Discovered 1893 September 20 by A. Charlois at Nice.

Named after a region in southern Italy bordering on the Tyrrhenian sea and containing the city of Naples. (H 41)
See also the remarks to planet (356).

(378) Holmia
Discovered 1893 December 6 by A. Charlois at Nice.
This is the ancient name of Stockholm, the capital of Sweden. (H 41)
See also the remarks to planet (356).

(379) Huenna
Discovered 1894 January 8 by A. Charlois at Nice.

This is the Latin form for Hveen. Tycho Brahe {see planet (1677)} observed for more than 20 years on this small island in the Sound between Denmark and Sweden. See also the citations for planets (499) and (1678). This interpretation was given by P. Colombier who found "Huenne" under the entry Tycho-Brahé in the *Grand Vocabulaire Français* (1773). (J. Meeus)
See also the remarks to planet (356).

(380) Fiducia
Discovered 1894 January 8 by A. Charlois at Nice.
This is the Latin name for confidence. (H 42)
See also the remarks to planet (356).

(381) Myrrha
Discovered 1894 January 10 by A. Charlois at Nice.

Named for a woman in Greek mythology who was changed into a myrrh tree. Her son Adonis {see planet (2101)} was born out of this tree. (H 42)
See also the remarks to planet (356).

(382) Dodona
Discovered 1894 January 29 by A. Charlois at Nice.

Named for a famous ancient town with a celebrated oracle of Zeus {see planet (5731)} on a nearby hill. The town and temple were built by Deucalion after the universal flood. Dodona was destroyed in 219 B.C. (H 42)
See also the remarks to planet (356).

(383) Janina
Discovered 1894 January 29 by A. Charlois at Nice.
Any reference of this name to a person or occurence is unknown.

(384) Burdigala
Discovered 1894 February 11 by F. Courty at Bordeaux.

Named in honor of the city of the discovery of this planet. Burdigala is the Latin name of Bordeaux. (H 42)

(385) Ilmatar
Discovered 1894 March 1 by M. Wolf at Heidelberg.

Named for a figure in Finnish mythology which symbolizes the daughter of the Air, who brought forth the Earth, Sky, Sun, Moon and Stars. (AN 130, 159 (1892))
Named by A. Krüger, editor of the Astronomische Nachrichten.

Krüger stated: "In Bezug auf (330) wünscht der Entdecker, dass ich den Namen vorschlagen solle." This planet, however, was identified as (298) Baptistina. The planet, also discovered by M. Wolf, which eventually became (330) was named

"Adalberta". The name "Ilmatar" was transferred to (385). Krüger (1832-1896) spent several years in Finland, having gone to Helsingfors in 1862, hence the choice of the name. Earlier he had been assistant to Argelander {see planet (1551)} in Bonn. In 1876 he went to Gotha. He determined a number of stellar parallaxes and in 1893 published a catalogue of over 2,000 red stars. (H 42). See also the remarks to (330).

(386) Siegena

Discovered 1894 March 1 by M. Wolf at Heidelberg.

Named for the city of Siegen in Westphalia, Germany, home town of Prof. Kreutz, by whom it was named. (H 42; AN 154, 15 (1900))

(387) Aquitania

Discovered 1894 March 5 by F. Courty at Bordeaux.

Named for a Roman division of southwestern Gaul to which Bordeaux {Burdigala, see planet (384)} belonged. Under Caesar the region consisted of the country between the Pyrenees mountains and Garonne river, and under Augustus expanded to the Loire and Allier rivers. (H 42)

(388) Charybdis

Discovered 1894 March 7 by A. Charlois at Nice.

Named for a daughter of Poseidon and Gaea who was thrown into the sea off Sicily by Zeus {see planet (5731)} where by swallowing and spewing water she created a whirlpool. Odysseus {see planet (1143)}, concerned with avoiding Charybdis, lost six of his men to the monster Scylla {see planet (155)}. Charybdis is an island, close to Scylla, where a whirlpool was. A giant woman was spying water in and out three times per day. (H 42)
See also the remarks to planet (356).

(389) Industria

Discovered 1894 March 8 by A. Charlois at Nice.
This is the Latin name for diligence. (J. Reichert)
See also the remarks to planet (356).

(390) Alma

Discovered 1894 March 24 by G. Bigourdan at Paris.

This planet is probably named for the river in southwest Russia and Crimea which gives its name to a famous victory gained over the Russians (1854) by the French and English in the Crimean War. (H 43)
See also the remarks to planet (356).

(391) Ingeborg

Discovered 1894 November 1 by M. Wolf at Heidelberg.
Any reference of this name to a person or occurence is unknown.

(392) Wilhelmina

Discovered 1894 November 4 by M. Wolf at Heidelberg.

Concerning this name there exist different explanations. E. Weiss (Astronomischer Kalender für 1896, p. 155) interpreted the name as "eine Ehrenbezeugung für Kaiser Wilhelm", the German emperor. T. Gehrels gives the more likely explanation that this planet was named in honor of Queen Wilhelmina of the Netherlands (1880-1962): "At the time of discovery, it must have drawn quite a bit of attention in western Europe that Princess Wilhelmina was so young and, because of the death of her father became nominally Queen already at the age of 10! There must have been sympathetic publicity because the Dutch Royal Family was and is highly respected." Queen Wilhelmina abducated 1948 in favor of her daughter Juliana {see also the citation for planet (816)}. (LDS)

(393) Lampetia

Discovered 1894 November 4 by M. Wolf at Heidelberg.

There are two characters in Greek mythology with this name: (1) Lampetia was a daughter of Apollo {see planet (1862)} and Neaera and the sister of Phaetusa {see planet (296)} with whom she guarded her father's sacred cattle on the island of Thrinacia, where Odysseus {see planet (1143)} and his men arrived. Although warned by Odysseus to keep their hands off the sacred cattle of Apollo, the men carried away and killed some of the sacred beasts. Lampetia informed her father of the sacrilege. Apollo requested that Zeus {see planet (5731)} punish the offenders. As the men sailed away from Thrinacia, a terrible storm rose and all of the Greeks perished - except Odysseus, who held on to a piece of broken mast and saved himself from the shipwreck. (2) Lampetia is the name of one of the Heliades, who, with her sisters was changed into a poplar tree at the death of her brother Phaethon {see planet (3200)}. (Z 146)

Named by A. Berberich (AN 156, 127 (1901)).

(394) Arduina

Discovered 1894 November 19 by A. Borrelly at Marseilles.

Named probably for Arduenna, goddess of hunting among the Gauls. Silva Arduenna was the vast Ardennes Forest in northeastern Gaul. (H 43)

The planet was erroneously called Darwina in Bull. Astron., Vol. 21, p. 267 (AN 166, 319 (1904)). Darwin is honored by planet (1991).

(395) Delia

Discovered 1894 November 30 by A. Charlois at Nice.

Delia is another name for Artemis, Cynthia, Diana, Hecate, Luna, Phoebe, or Selene - the goddess of the Moon. Artemis {see planet (105)} was called Delia because she was born on the island of Delos. Delia was also the name of a festival, instituted by Theseus in honor of Apollo {see planet (1862)}, celebrated on the island of Delos. (Z 83)

Named by the discoverer (AN 162, 54 (1903)).

(396) Aeolia

Discovered 1894 December 1 by A. Charlois at Nice.

Named for the ancient country of northwestern Asia Minor near the Aegean Sea. Aeolus was the god of storms and winds and supposedly a great astronomer. Another possible interpretation belongs to the Lipari (or Aeolian) islands in the Tyrrhenian sea northeast of Sicily. (H 43)

Named by the discoverer (AN 162, 54 (1903)).

(397) Vienna

Discovered 1894 December 19 by A. Charlois at Nice.

The most probable explanation was given by M.-A. Combes: "Nommée probablement en l'honneur de la ville de Vienna en Autriche où observa Johann Palisa de 1881 à sa mort et pour lequel Auguste Charlois, le découvreur, avait une grande admiration." (M.-A. Combes; A. Schnell)

Named by J. Bauschinger.

See also the remarks to planet (356).

(398) Admete

Discovered 1894 December 28 by A. Charlois at Nice.

Named for a daughter of Eurystheus who appointed the Twelve Labors of Heracles {see planet (5143)}. Obtaining the golden girdle of Hippolyta, queen of the Amazons, was one of the labors because Admete coveted it. (Z 43)

(399) Persephone

Discovered 1895 February 23 by M. Wolf at Heidelberg.

The Greek word means maiden. Persephone was the daughter of Zeus and Demeter {see planets (5731) and (1108)}, the wife of Pluto, and the queen of Hades. Pluto abducted her while she was gathering flowers on the plain of Enna in Sicily. Demeter sought her everywhere and threatened destruction for all mankind by withdrawing fertility from the Earth if she could not find her. Zeus promised to restore Persephone to her mother provided Persephone had eaten nothing in Hades; but she had eaten some pomegranate seeds and was compelled to spend six months each year with Pluto but allowed six months with her mother. The Latin name for Persephone is Proserpina {see planet (26)}. (Z 200)
Named by A. Berberich (AN 156, 127 (1901)).

(400) Ducrosa

Discovered 1895 March 15 by A. Charlois at Nice.

Named by the discoverer (AN 170, 375 (1906)) in honor of J. Ducros, mechanician at the Nice Observatory. (H 44)

(401) Ottilia

Discovered 1895 March 16 by M. Wolf at Heidelberg.

"Für den Planeten wurde der in der Schwarzwald- und Vogesensage eine Rolle spielende Name Ottilia gewählt." The Black Forest runs north and south along the eastern side of the Rhine river in Germany, the Vosges along the western side of the river in France. (AN 138, 31 (1895))
Named by A. Berberich.

Berberich was interested in the planet because of the commensurability with Jupiter.

(402) Chloë

Discovered 1895 March 21 by A. Charlois at Nice.

Named for the pretty, sportive shepherdess in an old Greek pastoral romance of Daphnis and Chloë attributed to Longus. (H 44)
See also the remarks to planet (356).

(403) Cyane

Discovered 1895 May 18 by A. Charlois at Nice.

Named for a nymph of Sicily who tried to prevent Hades from taking Persephone (or Proserpina, see planet (26)) to the underworld. Hades changed her into a fountain. (H 44)
See also the remarks to planet (356).

(404) Arsinoë

Discovered 1895 June 20 by A. Charlois at Nice.

Named for the mother of Asclepius (or Aesculapius, see planet (1027)) by Apollo {see planet (1862)}, according to some mythologists, although Coronis {see planet (158)} is usually accepted as the mother of Asclepius. The name is also in use for a nurse of Orestes. Arsinoë put her own son in the bed of Orestes, where he was killed by Aegisthus, and Orestes was saved. (H 44)
See also the remarks to planet (356).

(405) Thia

Discovered 1895 July 23 by A. Charlois at Nice.

Named for one of the Titans, daughter of Uranus and Gaea {see planet (1184)}.

Her attribute was light and she was the mother of Helios (the Sun), Selene (the Moon, see planet (580)), and Eos (the dawn, see planet (221)). She married her brother Hyperion. After another source, Thia (or Thea, Theia) was a daughter of Chiron {see planet (2060)} the centaur. She was ravished by Aeolus and feared that her father would know that she was pregnant. She was changed by Poseidon into a mare named Euippe and gave birth to a foal named Melanippe. Euippe became a constellation named the Horse. The foal was changed into a girl. (Z 262)
See also the remarks to planet (356).

(406) Erna

Discovered 1895 August 22 by A. Charlois at Nice.

Named probably in honor of the grand-daughter of Palisa, Erna Bidschof. Erna is the daughter of the astronomer Friedrich Bidschof (1864-1915) and his wife Helene, née Palisa. (A. Schnell)
Named (AN 169, 363 (1905)) by J. Palisa.

(407) Arachne

Discovered 1895 October 13 by M. Wolf at Heidelberg.

Named for a daughter of Idmon, a dyer. She was so skillful in weaving that she challenged Athena {see planet (881)} to a weaving contest. When Athena ripped the cloth, Arachne hanged herself and was changed into a spider. (Z 27)
Named by A. Berberich (AN 156, 127 (1901)).

(408) Fama

Discovered 1895 October 13 by M. Wolf at Heidelberg.

Named for a daughter of Titan and the Earth. Fama was a powerful goddess worshipped by the ancients, usually represented winged and blowing a trumpet. Fama, in Greek Pheme, is the goddess of fame. (H 44)
Named by A. Berberich (AN 156, 127 (1901)).

(409) Aspasia

Discovered 1895 December 9 by A. Charlois at Nice.

Named for the Greek adventuress (470?-410 B.C.) and consort of Pericles. Aspasia of Miletus was a famous courtesan. Her house was a literary and philosophical center. (H 45)
See also the remarks to planet (356).

(410) Chloris

Discovered 1896 January 7 by A. Charlois at Nice.

Named for the goddess of flowers who married Zephyrus. Chloris was daughter of Amphion and Niobe {see planet (71)}. Some legends say she married Neleus, king of Pylos, was the mother of Nestor {see planet (659)}, eleven other sons, and one daughter. All her sons except Nestor were killed by Heracles {see planet (5143)}. (Z 60)

(411) Xanthe

Discovered 1896 January 7 by A. Charlois at Nice.

Xanthe is the name of an Oceanid, one of more than 40 daughters of Oceanus and Tethys. (H 45; R. Bremer)

(412) Elisabetha

Discovered 1896 January 7 by M. Wolf at Heidelberg.

Named probably after Elise Wolf (1840-1924), née Helwerth {see planet (801)}, the mother of the discoverer. (I. van Houten-Groeneveld)

(413) Edburga
Discovered 1896 January 7 by M. Wolf at Heidelberg.
Any reference of this name to a person or occurence is unknown.

(414) Liriope
Discovered 1896 January 16 by A. Charlois at Nice.
Named for an Oceanid, wife of Cephissus and mother of Narcissus. (Z 152)

(415) Palatia
Discovered 1896 February 7 by M. Wolf at Heidelberg.

Named for the Pfalz, a region in southwestern Germany in which Heidelberg {see planet (325)} is situated. (H 45)
Named by A. Berberich (AN 156, 127 (1901)).

(416) Vaticana
Discovered 1896 May 4 by A. Charlois at Nice.

Named for a hill near the city of Rome. It is now part of the Vatican City State. (H 45)

(417) Suevia
Discovered 1896 May 6 by M. Wolf at Heidelberg.
Named for a Heidelberg student fraternity. (H 45)
Named by A. Berberich (AN 156, 127 (1901)).

(418) Alemannia
Discovered 1896 September 7 by M. Wolf at Heidelberg.
Named for a Heidelberg student fraternity. (H 45)
Named by A. Berberich (AN 156, 127 (1901)).

(419) Aurelia
Discovered 1896 September 7 by M. Wolf at Heidelberg.
Any reference of this name to a person or occurence is unknown.
Named by A. Berberich (AN 156, 127 (1901)).

(420) Bertholda
Discovered 1896 September 7 by M. Wolf at Heidelberg.

Named in memory of Markgraf Berthold von Baden, ancestor of the Grand Dukes of Baden {see also planet (333)}. He built the Zähringen castle in the 11th century (see also planet (421)). (H 46)

(421) Zähringia
Discovered 1896 September 7 by M. Wolf at Heidelberg.
Named for a family of Grand Dukes of Baden (see also planet (420)). (H 46)

(422) Berolina
Discovered 1896 October 8 by G. Witt at Berlin.

This is the Latin name of the city of Berlin, Germany, where this planet was discovered. (H 46)

(423) Diotima
Discovered 1896 December 7 by A. Charlois at Nice.

Named for a priestess and teacher of the Greek philosopher Socrates (470?-399 B.C.) {see planet (5450)}. (H 46)
See also the remarks to planet (356).

(424) Gratia
Discovered 1896 December 31 by A. Charlois at Nice.

This is the Latin name of grace. The three Graces Aglaia, Euphrosyne, and Thalia {see planets (47), (31), and (23), respectively} were called Gratiae or Charites. (H 46)
See also the remarks to planet (356).

(425) Cornelia
Discovered 1896 December 28 by A. Charlois at Nice.
Named probably after Cornelia (ca. 180 B.C.), daughter of Scipio Africanus, mother of Tiberius Sempronius Gracchus and Gaius Sempronius Gracchus. Cornelia is also the name of the first wife of the Roman statesman Gaius Julius Caesar (100-44 B.C.). (H 46)
See also the remarks to planet (356).

(426) Hippo
Discovered 1897 August 25 by A. Charlois at Nice.
Named for the ancient city in northern Africa and chief town of Numidia (also Hippo Regius). (H 46)

(427) Galene
Discovered 1897 August 27 by A. Charlois at Nice.
Named for a Nereid, one of the many daughters of Nereus and Doris (see planet (48)), and grand-daughter of Oceanus. (H46)

(428) Monachia
Discovered 1897 November 18 by W. Villiger at Munich.
This is the Latin name of the discovery city of Munich, the capital of the German country of Bavaria {see planet (301)}. (H 46)

(429) Lotis
Discovered 1897 November 23 by A. Charlois at Nice.
Named for a beautiful nymph and daughter of Poseidon {see planet (4341)}. Pursued by Priapus, she fled from his lust and prayed to the gods to help her. She was changed into a tree called the lotus. (Z 153)

(430) Hybris
Discovered 1897 December 18 by A. Charlois at Nice.
Named for the personification of insolent justice in Greek mythology. (H 46)

(431) Nephele
Discovered 1897 December 18 by A. Charlois at Nice.
Named after the first wife of Athamas, king of Thebes. She was the mother of Phrixus and Helle. See also the notes to planet (173). (H 47)

(432) Pythia
Discovered 1897 December 18 by A. Charlois at Nice.
Named after the famous priestess of Apollo {see planet (1862)} at Delphi (which is another name for Pythia). She was fifty years old before she assumed the office. She delivered the answers of Apollo to those who came to consult the famous oracle. The oracle could be consulted only one month in the year, and rich presents were required for Apollo, hence the opulence, splendor, and magnificence of the celebrated temple at Delphi. (Z 226)
See also the remarks to planet (356).

(433) Eros
Discovered 1898 August 13 by G. Witt at Berlin. Independently discovered 1898 August 13 by A. Charlois at Nice.

Named after the Greek god of love (in Latin called Cupido or Amor), son of Mercury and Venus. He protected the beautiful Psyche {see planet (16)} from the vengeance of his mother and when he later married her she became immortal. See also the citation for planet (763). (H 47)

(434) Hungaria

Discovered 1898 September 11 by M. Wolf at Heidelberg.

This is the Latin name for Hungary. The planet was named in remembrance of an astronomical meeting at Budapest in 1898. (H 47; AN 147, 335 (1898))

(435) Ella

Discovered 1898 September 11 by M. Wolf and A. Schwassmann at Heidelberg.
Any reference of this name to a person or occurence is unknown.
Named by A. Schwassmann (AN 156, 239 (1901)).

(436) Patricia

Discovered 1898 September 13 by M. Wolf and A. Schwassmann at Heidelberg.
Any reference of this name to a person or occurence is unknown.
Named by A. Schwassmann (AN 156, 239 (1901)).

(437) Rhodia

Discovered 1898 July 16 by A. Charlois at Nice.
Named for one of the Oceanid nymphs, daughter of Oceanus and Tethys. (H 48)

(438) Zeuxo

Discovered 1898 November 8 by A. Charlois at Nice.
Named for one of the Oceanid nymphs, daughter of Oceanus and Tethys. (H 48)

(439) Ohio

Discovered 1898 October 13 by E. F. Coddington at Mount Hamilton.

Named for the state in the eastern part of central U.S.A. and for the river flowing from the junction of Allegheny and Monongahela rivers into the Mississippi. (H 48)

(440) Theodora

Discovered 1898 October 13 by E. F. Coddington at Mount Hamilton.

Named in honor of the daughter of Mr. Julius F. Stone who was a benefactor and long time trustee of The Ohio State University. She became the wife of Charles Sutton, professor of landscape architecture at that university. (H 48; AN 152, 175 (1900))
Named by J. F. Stone.

(441) Bathilde

Discovered 1898 December 8 by A. Charlois at Nice.
Any reference of this name to a person or occurence is unknown.

(442) Eichsfeldia

Discovered 1899 February 15 by M. Wolf and A. Schwassmann at Heidelberg.

Named for the region of Eichsfeld in central Germany. (AN 154, 443 (1901); LDS)

Named by Rev. Anton K. Thraen {1843-1902} {see planet (4098)} who improved the elements of this planet. Thraen was born in Holungen und died in Dingelstädt, both villages in the Eichsfeld region.

(443) Photographica

Discovered 1899 February 17 by M. Wolf and A. Schwassmann at Heidelberg.
The name refers to the method found by M. Wolf to discover asteroids. (H 48)
Named (AN 156, 239 (1901)) by A. Schwassmann.

(444) Gyptis

Discovered 1899 March 31 by J. Coggia at Marseilles.

Named for the wife of Protis, chief of the expedition from Phocaea {see planet (25)} which founded Marseilles, the most ancient city in France, about 600 B.C. She was the daughter of the king of the Segobrigs who assigned the ground where Marseilles (or Massalia, see planet (20)) was built. (H 48)

(445) Edna

Discovered 1899 October 2 by E. F. Coddington at Mount Hamilton.

Named in honor of the wife of Julius F. Stone who has made outstanding gifts to promote scientific work at The Ohio State University (see also the remarks to planet (440)). (H 48; AN 152, 175 (1900))
Named by J. F. Stone.

(446) Aeternitas

Discovered 1899 October 27 by M. Wolf and A. Schwassmann at Heidelberg.

This name is a personification of eternity among the Romans, that is represented by a snake biting its tail and by the phoenix reborn from its ashes. (H 48)
Named by A. Schwassmann (AN 156, 239 (1901)).

(447) Valentine

Discovered 1899 October 27 by M. Wolf and A. Schwassmann at Heidelberg.

Named probably for Valentine Noëmi von Rothschild (1886-1969), the only daughter of Baron Albert von Rothschild, a benefactor of the Vienna Observatory {see also planets (250), and (719), respectively}. (A. Schnell)
Named by J. Palisa (AN 156, 143 (1901)) at the request of M. Wolf.

The name is given as 'Valentina' in 'Wiener Astronomischer Kalender für 1903'. This can also be interpreted as the feminine form of the German Christian name 'Valentin'.

(448) Natalie

Discovered 1899 October 27 by M. Wolf and A. Schwassmann at Heidelberg.
Any reference of this name to a person or occurence is unknown.
Named by J. Palisa (AN 156, 143 (1901)) at the request of M. Wolf.

(449) Hamburga

Discovered 1899 October 31 by M. Wolf and A. Schwassmann at Heidelberg.

Named by A. Schwassmann, Seewarte Hamburg, to honor the city of Hamburg, Germany, during the annual festival of the Mathematical Society of Hamburg in 1901. (AN 156, 239 (1901))
The city of Hamburg is also honored by planet (723).

(450) Brigitta

Discovered 1899 October 10 by M. Wolf and A. Schwassmann at Heidelberg.
Any reference of this name to a person or occurence is unknown.
Named by A. Berberich (AN 156, 127 (1901)).

(451) Patientia

Discovered 1899 December 4 by A. Charlois at Nice.
This is the Latin name for patience. (H 49)
See also the remarks to planet (356).

(452) Hamiltonia

Discovered 1899 December 6 by J. E. Keeler at Mount Hamilton.
Named for Mt. Hamilton in California, the site of the Lick Observatory. (H 49)

(453) Tea

Discovered 1900 February 22 by A. Charlois at Nice.
Any reference of this name to a person or occurence is unknown.

(454) Mathesis

Discovered 1900 March 28 by A. Schwassmann at Heidelberg.

This Greek word for science was given to this planet at the anniversary of the founding of the Mathematische Gesellschaft in Hamburg in 1901. (AN 156, 239 (1901))

(455) Bruchsalia

Discovered 1900 May 22 by M. Wolf and A. Schwassmann at Heidelberg.

Named for the city of Bruchsal in southwestern Germany, the native city of Secretary Nokk, who promoted the erection of Heidelberg Observatory. The discoverer stated in AN 156, 157 (1901): "Zur Erinnerung an unseren in den Ruhestand getretenen Staatsminister Dr. Nokk, der unserer Wissenschaft in Baden durch die Errichtung der Observatorien auf dem Königstuhl eine ausgezeichnete Arbeitsstätte geschaffen hat, {habe ich} dem Planeten den Namen "Bruchsalia" gegeben, so dass durch die Erinnerung an den Geburtsort eine dauernde Erinnerung an unseren Wohlthäter gegeben sein möge." (H 49)

(456) Abnoba

Discovered 1900 June 4 by M. Wolf and A. Schwassmann at Heidelberg.

This name was used by the Roman troops in southern Germany for the mountainous region of the Black Forest. 'Diana Abnoba' is the Celtic-Roman deity of the Black Forest as was found in old inscriptions. (LDS)

(457) Alleghenia

Discovered 1900 September 15 by M. Wolf and A. Schwassmann at Heidelberg.

Named by the first discoverer in honor of the famous optician John A. Brashear {see planet (5502)}: "Dem ersten mit meinem neuen Fernrohr entdeckten kleinen Planeten habe ich in meiner Freude über die ausgezeichneten Linsen, die mir mein Freund Brashear in Allegheny dazu geschliffen hat, den Namen Alleghenia gegeben. (AN 154, 443 (1901))

See also the citation of planet (484). Brashear is also honored by craters on Mars and the Moon.

(458) Hercynia

Discovered 1900 September 21 by M. Wolf and A. Schwassmann at Heidelberg.

This is the Latin name for the mountainous forests in Germany from the upper part of the Rhine to the Carpathian Mountains in southeastern Europe, which, according to Caesar, required nine days journey to cross it. The name (Hercynia silva) was later used only for parts of the entire region. (H 50; AN 162, 195 (1903))

Named by J. Riem of Astronomisches Rechen-Institut, Berlin, who computed the orbit.

(459) Signe

Discovered 1900 October 22 by M. Wolf at Heidelberg.

Named probably after the character of the Volsunga Saga, daughter of Volsung and sister of Sigmund. (H 50)

Named at the meeting of the Astronomische Gesellschaft in Lund, Sweden (AN 166, 207 (1904)).

(460) Scania

Discovered 1900 October 22 by M. Wolf at Heidelberg.

Named at the meeting of the Astronomische Gesellschaft in Lund, Sweden, in

1904. This is the Latin name of the province in southern Sweden in which Lund is situated. (H 50; AN 166, 207 (1904))

(461) Saskia

Discovered 1900 October 22 by M. Wolf at Heidelberg.

Named probably in honor of Saskia van Uylenburgh, wife (in 1634) of the famous Dutch painter Rembrandt van Rijn (1606-1669) {see planet (4511)}. (H 50)

(462) Eriphyla

Discovered 1900 October 22 by M. Wolf at Heidelberg.

Named for the sister of Adrastus, king of Argos, and wife of Amphiaraus who foretold the disastrous end of the expedition of the Seven against Thebes. Eriphyle caused her husband to accompany the expedition and was killed by her son Alcmaeon on the news of his father's death. (H 50)

(463) Lola

Discovered 1900 October 31 by M. Wolf at Heidelberg.

Named conceivably for a character in the opera *Cavalleria Rusticana* (1890) by the Italian composer Pietro Mascagni (1863-1945). (H 50)

(464) Megaira

Discovered 1901 January 9 by M. Wolf at Heidelberg.

Named for one of the three Erinyes or Furiae, daughters of Nox and Acheron, avenging spirits who bring retribution on those guilty of a wide variety of sins including homicide, perjury, and lack of hospitality. The three are Alecto (Unresting), Megaera (Jealous), and Tisiphone (Avenger). (H 50)
This is the first numbered minor planet detected in the 20th century.

(465) Alekto

Discovered 1901 January 13 by M. Wolf at Heidelberg.

Named for one of the three Furies. Alecto is represented with flaming torches, her head covered with serpents, and breathing vengeance, war and pestilence. See also planet (464). (H 50)

(466) Tisiphone

Discovered 1901 January 17 by M. Wolf and L. Carnera at Heidelberg.

Named for one of the three Furies. Tisiphone was represented with a whip in her hand and with snakes for bracelets. See also planet (464). (H 50)

(467) Laura

Discovered 1901 January 9 by M. Wolf at Heidelberg.

Named probably after a character in the opera *La Gioconda* by the Italian composer Amilcare Ponchielli (1834-1886). The opera is based on a play by the French poet Victor Hugo (1802-1885) {see planet (2106)}. This planet could also have been named after the character in *Sonnets to Laura* by the Italian poet Francesco Petrarca (1304-1374). (H 51)

(468) Lina

Discovered 1901 January 18 by M. Wolf at Heidelberg.
Named for a housemaid of the discoverer's family. (H 51)

(469) Argentina

Discovered 1901 February 20 by L. Carnera at Heidelberg.

Named by the discoverer, who was at that time in Oncativo, Argentina, for the country in South America. (AN 173, 63 (1906))

(470) Kilia

Discovered 1901 April 21 by L. Carnera at Heidelberg.
Named for the city of Kiel in northern Germany. Kilia is the Latin form. (H 51)

Named by Dr. J. Möller (AN 156, 127 (1901)) who was honorary director of the "Seefahrtsschule Elsfleth" in Kiel.

(471) Papagena

Discovered 1901 June 7 by M. Wolf at Heidelberg.

Named for the girl friend of Papageno in the opera *Die Zauberflöte* (1791) by Wolfgang Amadeus Mozart {see planet (1034)}. (H 51)

(472) Roma

Discovered 1901 July 11 by L. Carnera at Heidelberg.

Named for the capital city of Italy, the native country of the discoverer. (AN 158, 255 (1902))
Named by A. Abetti.

(473) Nolli

Discovered 1901 February 13 by M. Wolf at Heidelberg.
This is the pet name of a small child in the Wolf family. (H 51)

(474) Prudentia

Discovered 1901 February 13 by M. Wolf at Heidelberg.

Named for the allegorical divinity personifying prudence, represented by a mirror surrounded by a snake. (H 51)

(475) Ocllo

Discovered 1901 August 14 by D. Stewart at Arequipa.

Named by the discoverer for the first Inca queen, by tradition daughter of the Sun. (AN 159, 129 (1902))

This is the first minor planet discovered in South America and also the first discovered in the southern hemisphere.

(476) Hedwig

Discovered 1901 August 17 by L. Carnera at Heidelberg.

Named in honor of the wife of the Danish astronomer Elis Strömgren (1870-1947) who computed the orbit of this planet {see also planet (1422)}. (H 51)
Named by E. Strömgren (AN 158, 48 (1902)).

(477) Italia

Discovered 1901 August 23 by L. Carnera at Heidelberg.
Named by the discoverer in honor of his native country. (P. Sicoli)

See also the remarks to planets (63) and (69), respectively. For the naming source see the citation for planet (478).

(478) Tergeste

Discovered 1901 September 21 by L. Carnera at Heidelberg.

This is the ancient name of Trieste, the city at the northeast of the Adriatic sea. Luigi Carnera (1875-1962) was born in Trieste. (P. Sicoli)
Named by the discoverer (Mem. Soc. Spettrosc. Ital., Vol. 32, p. 164 (1903).

(479) Caprera

Discovered 1901 November 12 by L. Carnera at Heidelberg.

Named by the discoverer (AN 173, 63 (1906)) for the island off the northeast coast of Sardinia. Caprera was the place where Garibaldi {see planet (4317)} retired and died, after his brilliant military exploits that helped bring about the unification of Italy. (P. G. Comba)

(480) Hansa

Discovered 1901 May 21 by M. Wolf and L. Carnera at Heidelberg.

Named in memory of the medieval Hanse, a merchant trading association. The Hanseatic League was constituted of merchants of various free German cities and reached the height of its power in the 14th and 15th centuries. (H 52)

Named by H. Kreutz (AN 172, 127 (1906)), editor of the Astronomische Nachrichten at Kiel, an old Hanse city.

(481) Emita

Discovered 1902 February 12 by L. Carnera at Heidelberg.
Any reference of this name to a person or occurence is unknown.

(482) Petrina

Discovered 1902 March 3 by M. Wolf at Heidelberg.
Named after the dog Peter of the discoverer. (H 52)

(483) Seppina

Discovered 1902 March 4 by M. Wolf at Heidelberg.
Named after the dog Sepp of the discoverer. (H 52)

(484) Pittsburghia

Discovered 1902 April 29 by M. Wolf at Heidelberg.

Named for the city of Pittsburgh, Pennsylvania, home of John A. Brashear, who figured the lenses of the discoverer's 16-inch photographic doublet. (H 52)

Named by J. A. Brashear {see planet (5502)}, who stated in Pop. Astron., Vol. 11, No. 10, p. 549 (1903) that the discoverer presents him the honor of naming the planet. "Desiring to share this honor with the good people of this great city, he proposed the name, which has been accepted by Dr. Wolf, and henceforth planet No. 484 will be recorded in astronomical annals with the name of our city as it is engraved upon its great seal, except that the Greek suffix has been attached to the last syllable to make the name more euphonious, as has been the custom in all names permitting this ending. Some time in the far-distant future our great city may lie silent and dead, as Mycenae or Pompeii, but her name, now recorded in the skies, will be remembered as long as the beautiful science of astronomy has its votaries on Old Mother Earth."

The optics of this famous Bruce {see planet (323)} telescope were later removed and replaced by two four-lens Zeiss objectives. See also the citation for planet (457).

(485) Genua

Discovered 1902 May 7 by L. Carnera at Heidelberg.

Named for the city of Genova, Italy. Genua is the ancient name. (H 52, AN 168, 307 (1905))

Name suggested by Prof. F. Porro.

(486) Cremona

Discovered 1902 May 11 by L. Carnera at Heidelberg.

Named after the city in northern Italy in Lombardy on the Po. (H 52, AN 168, 307 (1905))

Name suggested by Prof. F. Porro.

(487) Venetia

Discovered 1902 July 9 by L. Carnera at Heidelberg.

Named for the area in northeast Italy and northwest Yugoslavia including the territory between lower Po river and the Alps. The city of Venice lies in this region. (H 52; P. Sicoli)

Name suggested by Prof. E. Millosevich (Mem. Oss. Astron. Coll. Romano, Ser. III, Vol. IV, part 1, p. 27 (1904)).

(488) Kreusa

Discovered 1902 June 26 by M. Wolf and L. Carnera at Heidelberg.

The name Kreusa exists in some variations in Greek mythology: (1) Kreusa (also Glauke, see planet (288)) was the daughter of Creon, king of Corinth. (2) Kreusa is one of the Nereides or, respectively, one of the Danaides. (3) She was the daughter of Priam by Hecuba and the wife of Aeneas {see planets (884), (108), and (1172), respectively}. (4) Kreusa was the daughter of Erechtheus, king of Athens, and mother of Janus by Apollo {see planet (1862)}. Some accounts say she was the mother of Ion by Apollo. She deserted her son Ion but when she later married Xuthus she was reunited with Ion. Apollodorus and Pausanias say Ion was the son of Xuthus by Kreusa. (Z 74)

(489) Comacina

Discovered 1902 September 2 by L. Carnera at Heidelberg.

Named for a little island in Lake Como in northern Italy. The name is conceivably related to the Comacine masters, the Lombard master builders of the Middle Ages who influenced the architecture of that period. (H 53)
Named by the discoverer (AN 169, 95 (1905)) at the suggestion of L. Volta.

(490) Veritas

Discovered 1902 September 3 by M. Wolf at Heidelberg.

The word means "truth". It was personified by the ancients as a deity and called the daughter of Cronos and the mother of Virtue. Democritus used to say that Veritas hides herself at the bottom of a well, she is so difficult to find. (H 53; Z 286)

(491) Carina

Discovered 1902 September 3 by M. Wolf at Heidelberg.
Any reference of this name to a person or occurence is unknown.

Named by Premierlieutenant Th. Lassen, Odense, Denmark, who computed an orbit for this planet (AN 162, 63 (1903)).

(492) Gismonda

Discovered 1902 September 3 by M. Wolf at Heidelberg.

Named after the daughter of Tancred, prince of Salerno, from the *Decamerone* of the Italian author Giovanni Boccacio (1313-1375). (R. Bremer)

(493) Griseldis

Discovered 1902 September 7 by M. Wolf at Heidelberg.

Named for the lady proverbial for her virtue and patience. She appears in the *Decamerone* by the Italian author Giovanni Boccaccio (1313-1375) and in an opera by Paer (1771-1839). (H 53)

(494) Virtus

Discovered 1902 September 7 by M. Wolf at Heidelberg.

Named for the personification of virtue. The ancient Romans made deities of all the major virtues and built temples to virtue and to honor. The statues of the most important virtues - temperance, honesty, modesty, and liberty - were characterized by their dress. Among the Romans, virtues came to include many of the manly virtues - manliness, courage, integrity, strength, and fortitude. (H 53; AN 168, 307 (1905))

Named at the request of the discoverer by C. Flammarion (l'Astronomie, Vol. 19, p. 309 (1905)) who states that by an oversight the astronomers had neglected to place Virtue in the skies and that if it disappeared from the Earth, it would be nice to find it in the heavens. He expressed his regret that the name Virtus was chosen long time after names such as (28) Bellona, (146) Lucina, (216) Kleopatra, and (289) Nenetta.

(495) Eulalia

Discovered 1902 October 25 by M. Wolf at Heidelberg.
Named in honor of the grandmother of Mrs. Wolf. (H 53)

(496) Gryphia

Discovered 1902 October 25 by M. Wolf at Heidelberg.

Named in honor of the German baroque poet Andreas Gryphius (1616-1664). (H 53, AN 167, 223 (1905))

Named with permission of the discoverer by W. Ebert, Paris, who computed an early orbit for this planet.

(497) Iva

Discovered 1902 November 4 by R. S. Dugan at Heidelberg.

Named after Iva Shores, a little daughter in the household where the discoverer lived while in Amherst College. (H 53)

The naming information was provided by Mrs. Edith Eveleth, sister of the discoverer.

(498) Tokio

Discovered 1902 December 2 by A. Charlois at Nice.

Named for the capital of Japan. An earlier planet discovered 1900 March 6 by S. Hirayama in Tokio for which circular elements had been computed was shown to be identical with the newly discovered planet. (H 54)

The first photograph was taken by Hirayama, assisted by K. Toda. The circular elements were computed by K. Saotome. The planet, temporarily designated 1990 FF, was identified by P. V. Neugebauer with the Nice object. Tokio is the German transliteration form of Tokyo. Tokyo means eastern metropolis. (K. Tomita).

(499) Venusia

Discovered 1902 December 24 by M. Wolf at Heidelberg.

Named for the insula Venusia, or Hveen, island in the Sound between Denmark and Sweden, provided by King Frederick II of Denmark in 1576 to Tycho Brahe {see planet (1677)} for his observatories Uraniborg and Stjerneborg. (H 54; AN 166, 319 (1904))

Named on the occasion of a visit of the participants of the 1904 meeting of the Astronomische Gesellschaft in Lund, Sweden, by Miss Bruhns, daughter of C. Bruhns, director of the Leipzig Observatory.
The island is also commemorated by planets (379) and (1678).

(500) Selinur

Discovered 1903 January 16 by M. Wolf at Heidelberg.

Named after a character in the novel *Auch Einer* (1879) by the German writer Friedrich Theodor Vischer (1807-1887). Selinur represents a Celtic Moon goddess whose symbol is the half moon. She is opposed by the evil spirit "Grippo" who is responsible for having inflicted the "Grippe" (influenza) on the villager's male population. In order to ban evil Grippo and his helpers the villagers stage all kinds of coughing contests, coughing worships, coughing procedures, etc. (H 54; R. Bremer)

(501) Urhixidur

Discovered 1903 January 18 by M. Wolf at Heidelberg.

Named after a character in the novel *Auch Einer* (1879) by the German writer Friedrich Theodor Vischer (1807-1887). Urhixidur is housekeeper, nurse and female companion of the priest Angus. He, a progressive, opposes her conservative views. She eventually dominates him since he cannot live without her. (H 54; R. Bremer)

(502) Sigune

Discovered 1903 January 19 by M. Wolf at Heidelberg.

Named after a character in the novel *Auch Einer* (1879) by the German writer Friedrich Theodor Vischer (1807-1887). (I. van Houten-Groeneveld)

(503) Evelyn

Discovered 1903 January 19 by R. S. Dugan at Heidelberg.
Named by the discoverer in honor of his mother, Evelyn Smith Dugan. (H 54)
See the remark for planet (497).

(504) Cora

Discovered 1902 June 30 by S. I. Bailey at Arequipa.
Named for a figure from old Peruvian mythology. (AN 169, 95 (1905))

(505) Cava

Discovered 1902 August 21 by R. H. Frost at Arequipa.
Named for a figure from old Peruvian mythology. (AN 169, 95 (1905))

(506) Marion

Discovered 1903 February 17 by R. S. Dugan at Heidelberg.
Named by the discoverer in honor of his cousin Marion Orcutt. (H 54)
See the remark for planet (497).

(507) Laodica

Discovered 1903 February 19 by R. S. Dugan at Heidelberg.

Named for a daughter of Priam and Hecuba {see planets (884) and (108)}. She fell in love with Acamas {see planet (2594)}, son of Theseus and Phaedra {see planet (174)}, when he and Diomedes {see planet (1437)} came from the Greeks to demand the restoration of Hellen. (H 54)

The planet is not named by the discoverer. It had previously been observed by A. Charlois in Nice, but lost. The name appeared when an ephemeris was published.

(508) Princetonia

Discovered 1903 April 20 by R. S. Dugan at Heidelberg.

Named by the discoverer in honor of the Princeton University in Princeton, New Jersey. (H 55)

See also the remarks for planets (497), and (534). Princeton University is also honored by planets (534) and (6151).

(509) Iolanda

Discovered 1903 April 28 by M. Wolf at Heidelberg.
Any reference of this name to a person or occurence is unknown.

(510) Mabella

Discovered 1903 May 20 by R. S. Dugan at Heidelberg.

Named by the discoverer in honor of Mabel Loomis Todd, daughter of the mathematician and astronomer Elias Loomis. She was the wife of David P. Todd

{see planet (511)}, the discoverer's professor of astronomy at Amherst College. (H 55)
See the remark for planet (497).

(511) Davida

Discovered 1903 May 30 by R. S. Dugan at Heidelberg.

Named by the discoverer in honor of David P. Todd (1855-1939), professor of astronomy and director of the Amherst College Observatory (1881-1920). (H 55)
See the remark for planet (497).

(512) Taurinensis

Discovered 1903 June 23 by M. Wolf at Heidelberg.
This is the ancient name of the city of Turin, Italy. (H 55)

The planet was named (AN 167, 239 (1905)) with permission of the discoverer by the Turin astronomers.

(513) Centesima

Discovered 1903 August 24 by M. Wolf at Heidelberg.
The naming commemorates the 100th planet discovery by Max Wolf. (H 55)

(514) Armida

Discovered 1903 August 24 by M. Wolf at Heidelberg.

Named for the beautiful legendary sorceress in Torquato Tasso's (1544-1595) *Jerusalem Delivered.* She is the leading character in the opera *Armida* (composed 1777) by Christoph Willibald Gluck (1714-1787). (H 55)
See also the remarks to planet (579).

(515) Athalia

Discovered 1903 September 20 by M. Wolf at Heidelberg.

Named for the impious and murderous queen of Judah, daughter of Ahab and Jezebel (II Kings 8, 26; 11, 1-20). Athalia was the only woman who ever occupied the throne of either of the Hebrew kingdoms. (H 55)

(516) Amherstia

Discovered 1903 September 20 by R. S. Dugan at Heidelberg.

Named in honor of the discoverer's Alma Mater, the Amherst College in Amherst, Massachusetts. (H 55)
See the remark for planet (497).

(517) Edith

Discovered 1903 September 22 by R. S. Dugan at Heidelberg.

Named by the discoverer in honor of his sister Mrs. Edith Dugan Eveleth. (H 55)
See the remark for planet (497).

(518) Halawe

Discovered 1903 October 20 by R. S. Dugan at Heidelberg.

Named after an Arabic sweetmeat that was a favorite of the discoverer. The origin of this taste may have been during Dugan's sojourn as instructor in the American University, Beirut, Lebanon (1899-1902). (H 55)
See the remark for planet (497).

(519) Sylvania

Discovered 1903 October 20 by R. S. Dugan at Heidelberg.

This planet is named for the large forests that the discoverer enjoyed tramping through even as a small boy. Mrs. Edith Eveleth, sister of the discoverer, stated:

"In Heidelberg he climbed up and down from the city to the Königstuhl through deep woods; for holidays he tramped through the Black Forest, the Odenwald and all the forest within reach." (H 56)
See the remark for planet (497).

(520) Franziska

Discovered 1903 October 27 by M. Wolf and P. Götz at Heidelberg.
Any reference of this name to a person or occurence is unknown.
Named by Götz (AN 169, 363 (1905)).

(521) Brixia

Discovered 1904 January 10 by R. S. Dugan at Heidelberg.

This is the Latin name of Brescia, Italy, the native town of Prof. E. Bianchi, who computed the orbit. It was at his request his birthplace was honored. (H 56)
See the remark for planet (497).

(522) Helga

Discovered 1904 January 10 by M. Wolf at Heidelberg.
Any reference of this name to a person or occurence is unknown.
Named (AN 169, 363 (1905)) by Lieutenant Th. Lassen who computed the orbit.

(523) Ada

Discovered 1904 January 27 by R. S. Dugan at Heidelberg.

Named by the discoverer in honor of Ada Helme, a school friend and neighbor of the discoverer in Montague, Massachusetts {see planet (535)}. (H 56)
See the remark for planet (497).

(524) Fidelio

Discovered 1904 March 14 by M. Wolf at Heidelberg.

This is the name of Leonora when disguised as a man in the opera *Fidelio* (composed 1805) by the German composer Ludwig van Beethoven {see planet (1815)}. (H 56)

(525) Adelaide

Discovered 1908 October 21 by J. H. Metcalf at Taunton.

Named by Max Wolf probably for the German-born Queen Adelaide, consort of the British king William IV (1765-1837). The capital of Southern Australia (founded 1836) was named for her. (LDS; B. G. Marsden)

The number and name were originally given to a minor planet discovered 1904 by Max Wolf at Heidelberg. A. Patry (MPC 1831) established the identity of (525) with (1171) Rusthawelia on the basis of well known elements of (1171). There are only a few references to (525) in the literature. The Minor Planet Center retained the number and name of (1171) and vacated the number (525). In MPC 1935 the vacated number and name were assigned to the 1908 Metcalf planet.

(526) Jena

Discovered 1904 March 14 by M. Wolf at Heidelberg.

Named for the city of Jena in eastern Germany at the occasion of the September 1906 meeting of the Astronomische Gesellschaft in Jena. (AN 172, 287 (1906))
Name proposed by J. Palisa.

(527) Euryanthe

Discovered 1904 March 20 by M. Wolf at Heidelberg.

Named after the character *Euryanthe* of Savoy in the opera *Euryanthe* by the German composer Carl Maria von Weber (1786-1826) {see planet (4152)}. (H 56)

(528) Rezia

Discovered 1904 March 20 by M. Wolf at Heidelberg.

Named for a character in the opera *Oberon* by the German composer Carl Maria von Weber (1786-1826) {see planet (4152)}. (H 56)

(529) Preziosa

Discovered 1904 March 20 by M. Wolf at Heidelberg.

This is the title of a play (1810) by Pius Alexander Wolff, set to music (1820) by Carl Maria von Weber {see planet (4152)}. It is also a character in the young gipsy novel *La Gitanilla* by the Spanish novelist, dramatist and poet Miguel de Cervantes Saavedra (1547-1616). (R. R. de Freitas Mourão; LDS; R. Bremer)

(530) Turandot

Discovered 1904 April 11 by M. Wolf at Heidelberg.

Named for the daughter of the Emperor of China in the opera *Turandot* by the Italian composer Giacomo Puccini (1858-1924) {see planet (4579)}. (H 57)

(531) Zerlina

Discovered 1904 April 12 by M. Wolf at Heidelberg.

Named for a character in the opera *Don Giovanni* (1787) by Wolfgang Amadeus Mozart {see planet (1034)}. (H 57)

(532) Herculina

Discovered 1904 April 20 by M. Wolf at Heidelberg.

Any reference of this name to a person or occurence is unknown.

Named by E. Millosevich, Osservatorio al Collegio Romano (AN 167, 45 (1905)).

(533) Sara

Discovered 1904 April 19 by R. S. Dugan at Heidelberg.

Named in honor of a friend of the discoverer. (H 57)

(534) Nassovia

Discovered 1904 April 19 by R. S. Dugan at Heidelberg.

Named in honor of Princeton University. Lyman Spitzer {see planet (2160)}, Director of Princeton University Observatory stated: "Nassovia is presumably a Latinized version of Nassau. When the first building in the town of Princeton was built for what was then the College of New Jersey, this building was named Nassau Hall "in memory of the glorious King William III, who was a branch of the illustrious House of Nassau". The quotation marks enclose a recommendation of the Governor of New Jersey who was a member of the Board of Trustees. Nassau Hall has been the center of the university ever since it was built, and the name Nassau has become associated with Princeton University." (H 57)

Princeton University is also honored by minor planets (508) and (6151), respectively.

(535) Montague

Discovered 1904 May 7 by R. S. Dugan at Heidelberg.

Named by the discoverer in honor of his native town in Massachusetts. (H 57)

See the remark for planet (497).

(536) Merapi

Discovered 1904 May 11 by G. H. Peters at Washington.

Named after a mountain in west central Sumatra, site of the U.S. Naval Observatory and other expeditions to the total solar eclipse of May 17, 1901. It gives off smoke more or less continuously and the name means "with fire". It should not to be confused with the nearby active volcano of the same name in central Java. The discoverer was a member of the eclipse expedition. (H 57)

(537) Pauly

Discovered 1904 July 7 by A. Charlois at Nice.

Named in memory of Max Pauly (1849-1917) of the Jena Carl Zeiss optical works. The genial Pauly was a good friend of Max Wolf. Pauly ground the 10-inch lens of the Bruce telescope, "das Juwel des Königstuhls". Under his supervision, many large telescopes were produced for European as well as American Observatories. (H 57)

Named by Max Wolf.

Obituary published in Astron. Nachr., Band 207, Nr. 4963, p. 219-224 (1918).

(538) Friederike

Discovered 1904 July 18 by P. Götz at Heidelberg.

Named in honor of a friend of the discoverer in Heidelberg. (H 57)

(539) Pamina

Discovered 1904 August 2 by M. Wolf at Heidelberg.

Named for the daughter of the Queen of the Night in the opera *Die Zauberflöte* by Wolfgang Amadeus Mozart {see planet (1034)}. (H 57)

(540) Rosamunde

Discovered 1904 August 3 by M. Wolf at Heidelberg.

Named for the character in the opera of the same name by the Austrian composer Franz Peter Schubert (1797-1828) (H 58)

(541) Deborah

Discovered 1904 August 4 by M. Wolf at Heidelberg.

Named for the Biblic prophetess in Israel who helped to free the Israelites from the Canaanites, and celebrated the victory by famous verses (Judges 4, 4). (H 58)

(542) Susanna

Discovered 1904 August 15 by P. Götz and A. Kopff at Heidelberg.

Named by the first discoverer in honor of a friend in Heidelberg. (H 58)

(543) Charlotte

Discovered 1904 September 11 by P. Götz at Heidelberg.

Named in honor of a friend of the discoverer in Heidelberg. (H 58)

(544) Jetta

Discovered 1904 September 11 by P. Götz at Heidelberg.

Named by the discoverer after a legendary figure of Heidelberg. The name of the site on which the Heidelberg castle was built in the 15th century is called Jettenbühl. (H 58; R. Bremer)

(545) Messalina

Discovered 1904 October 3 by P. Götz at Heidelberg.

Named after Valeria Messalina (c. 25-48), the third wife of the Roman emperor Claudius (10 B.C.-54). Messalina was the great-granddaughter of Marcus Antonius. (H 58)

(546) Herodias

Discovered 1904 October 10 by P. Götz at Heidelberg.

Named for the consort of Herod Antipas. She and her daughter Salome {see planet (562)} caused the death of John the Baptist. (H 58)

(547) Praxedis

Discovered 1904 October 14 by P. Götz at Heidelberg.

Named after a character in the narration *Ekkehard* by the German author Victor von Scheffel (1826-1886). (H 58)

(548) Kressida

Discovered 1904 October 14 by P. Götz at Heidelberg.

Named for a beautiful girl, proverbial for her infidelity in Shakespeare's {see planet (2985)} *Troilus and Cressida* (1601). The figure belongs to a medieval, not ancient, myth. The name is derived from Chryseis {see also planet (202)}. (H 58) The name Cressida has also been given to the satellite Uranus IX.

(549) Jessonda

Discovered 1904 November 15 by M. Wolf at Heidelberg.

Named presumably after the character in the opera of the same name by the German composer, conductor and violinist Ludwig Spohr (1784-1859), one of the leading composers in the early romantic period. (H 58)

(550) Senta

Discovered 1904 November 16 by M. Wolf at Heidelberg.

Named probably for the daughter of Daland, Norse sea captain in the opera *Der fliegende Holländer* (composed 1841) by Richard Wagner {see planet (3992)}, which is based on Heinrich Heine's (1797-1856) version of the legend. (H 58)

(551) Ortrud

Discovered 1904 November 16 by M. Wolf at Heidelberg.

Named for the wife of Frederick of Telramund in the opera *Lohengrin* (1878) by Richard Wagner {see planet (3992)}. (H 58)

(552) Sigelinde

Discovered 1904 December 14 by M. Wolf at Heidelberg.

Named after a character in the opera *Die Walküre* {see planet (877)} by Richard Wagner {see planet (3992)}. (H 59)

(553) Kundry

Discovered 1904 December 27 by M. Wolf at Heidelberg.

Named for a girl, part sorceress, part mortal woman, in the opera *Parsifal* {see planet (2095)} by Richard Wagner {see planet (3992)}. The opera is based on the epic by Wolfram von Eschenbach (who is a leading character in Wagner's opera *Tannhäuser*). (H 59)

(554) Peraga

Discovered 1905 January 8 by P. Götz at Heidelberg.
Named for a village between Padua and Venice. (H 59)

Named (AN 168, 141 (1905)) with permission of the discoverer by G. Abetti, Padua, who computed the orbit, because his relatives had a country place there.

(555) Norma

Discovered 1905 January 14 by M. Wolf at Heidelberg.

Named for the high priestess of the druids in the opera *Norma* by the Italian composer Vincenzo Bellini (1801-1835). (H 59)

(556) Phyllis

Discovered 1905 January 8 by P. Götz at Heidelberg.

Named for the daughter of Lycurgus, king of Thrace. She fell in love with Demophon, son of Theseus, who stopped in Thrace on his return from the Trojan War. He married her and became king. Shortly thereafter, Demophon went to Athens, and when he failed to return after one month, Phyllis threw herself into the sea. (Z 210)

(557) Violetta

Discovered 1905 January 26 by M. Wolf at Heidelberg.

Named for the frivolous woman and leading character in the opera *La Traviata* by Giuseppe Verdi {see planet (3975)}, which is based on *La Dame aux Camelias* by Alexandre Dumas the younger (1824-1895). (H 59)

(558) Carmen

Discovered 1905 February 9 by M. Wolf at Heidelberg.

Named for the leading character in the opera *Carmen* by the French composer Alexandre César Léopold (named Georges) Bizet (1838-1875), which is based on the novel by the French writer Prosper Mérimée (1803-1870). (H 59)

(559) Nanon

Discovered 1905 March 8 by M. Wolf at Heidelberg.

Named after the operetta *Nanon, die Wirtin vom Goldenen Lamm* (1877) by the German-Austrian composer Richard Genée (1823-1895) which is based on a French comedy, set in Paris of Louis XIV. (R. Bremer; LDS)

(560) Delila

Discovered 1905 March 13 by M. Wolf at Heidelberg.

Named for the mistress and betrayer of Samson in the book of Judges, set to music by the French composer Camille Saint-Saëns (1835-1877) {see planet (5210)} in his opera *Samson et Delila* (1877). (H 59)

(561) Ingwelde

Discovered 1905 March 26 by M. Wolf at Heidelberg.

Named probably after the opera by the German composer Max von Schillings (1868-1933) who became well-known by his opera *Mona Lisa* (1915). (R. Bremer; LDS)

(562) Salome

Discovered 1905 April 3 by M. Wolf at Heidelberg.

Named for the daughter of Herod Antipas who was given the head of John the Baptist as a reward for her dancing. She is the leading figure in the opera *Salome* (1905) by the German composer Richard Strauss (1864-1949) {see planet (5039)}, which is based on the romance by the Irish-born writer Oscar Wilde (1854-1900). (H 59)

(563) Suleika

Discovered 1905 April 6 by P. Götz at Heidelberg.

Named for a character in *Also sprach Zarathustra* (1885) by the German philospher Friedrich Wilhelm Nietzsche (1844-1900). Suleika and Dudu {see planet (564)} are mentioned in Part IV in the chapter *Unter Töchtern der Wüste*. These are the only feminine names in Zarathustra. (I. van Houten-Groeneveld)

(564) Dudu

Discovered 1905 May 9 by P. Götz at Heidelberg.

Named for a character in *Also sprach Zarathustra* (1885) by the German philosopher Friedrich Wilhelm Nietzsche (1844-1900). (I. van Houten-Groeneveld)
See also the remarks to planet (563).

(565) Marbachia

Discovered 1905 May 9 by M. Wolf at Heidelberg.

Named for the town Marbach in southwest Germany, the native city of the German poet Friedrich Schiller {see planet (3079)}. (H 60)

(566) Stereoskopia

Discovered 1905 May 28 by P. Götz at Heidelberg.

This name is an allusion to C. Pulfrich's stereo-comparator (see also his paper in

AN 159, 83 (1902)), by the aid of which he discovered a planet which had previously been overlooked on older photographic plates. It was the first planet discovered by the method and was later identified with Stereoskopia. (H 60)
Named (AN 169, 363 (1905)) by C. Pulfrich.

(567) Eleutheria

Discovered 1905 May 28 by P. Götz at Heidelberg.
Named after the Greek goddess of liberty. (H 60)

(568) Cheruskia

Discovered 1905 July 26 by P. Götz at Heidelberg.
Named for a Heidelberg student fraternity. (H 60)

(569) Misa

Discovered 1905 July 27 by J. Palisa at Vienna.

Named after Misa, or Mise, a mytic divinity that appears in the Orphic mysteries. Misa is known as the mother of Bacchus {see planet (2063)}. (H 60; A. Schnell)

(570) Kythera

Discovered 1905 July 30 by M. Wolf at Heidelberg.

Named for the famous island in the Ionian Sea which contends with Cyprus for the honor of being the island to which Aphrodite {see planet (1388)} floated when she rose from the foam. Kythera is also a name for Aphrodite. (Z 78)

(571) Dulcinea

Discovered 1905 September 4 by P. Götz at Heidelberg.

Named for the beautiful girl, object of the attention of Don Quixote in the famous novel *El Ingenioso Hidalgo Don Quijote de La Mancha* by the Spanish writer Miguel de Cervantes Saavedra (1547-1616). (H 60)

(572) Rebekka

Discovered 1905 September 19 by P. Götz at Heidelberg.

Named by the discoverer for a "Heidelberger Bürgerstochter" as he stated to I. van Houten-Groeneveld. (I. van Houten-Groeneveld)

The naming might be influenced by the two letters of the provisional designation 1905 RB. See also the remarks to planet (579).

(573) Recha

Discovered 1905 September 19 by M. Wolf at Heidelberg.

Named after the feminine character Recha in the drama *Nathan der Weise* by the German dramatist and critic Gotthold Ephraim Lessing (1729-1781), a strong advocate of religious tolerance. (H 60; R. Bremer)

The naming might be influenced by the two letters of the provisional designation 1905 RC. See also the remarks to planet (579).

(574) Reginhild

Discovered 1905 September 19 by M. Wolf at Heidelberg.
Any reference of this name to a person or occurence is unknown.

The naming might be influenced by the two letters of the provisional designation 1905 RD. See also the remarks to planet (579).

(575) Renate

Discovered 1905 September 19 by M. Wolf at Heidelberg.
Any reference of this name to a person or occurence is unknown.

The naming might be influenced by the two letters of the provisional designation 1905 RE. See also the remarks to planet (579).

(576) Emanuela

Discovered 1905 September 22 by P. Götz at Heidelberg.
Named for a friend of the discoverer. (H 61, AN 180, 183 (1909))

(577) Rhea

Discovered 1905 October 20 by M. Wolf at Heidelberg.

Named for the daughter of Uranus and Gaea {see planet (1184)}, sister and wife of Cronus, and mother of Zeus, Poseidon {see planets (5731), (4341)}, Hades, Demeter {see planet (1108)}, Hera {see planet (103)}, and Hestia {see planet (46)}. Rhea was called "the mother of the gods" or the "great mother goddess". Her Roman name was Tellus (or Terra) and her Phrygian name was Cybele {see planet (65)}. (Z 228)

The naming might be influenced by the two letters of the provisional designation 1905 RH. See also the remarks to planet (579). Rhea is also the name of the fifth satellite of Saturn, discovered 1672 by J. D. Cassini.

(578) Happelia

Discovered 1905 November 1 by M. Wolf at Heidelberg.

Named in honor of Carl Happel (1820-1914), painter and great benefactor to the Heidelberg Observatory. From his financial support the Happel-Labor on the Königstuhl was erected. (H 61; LDS)

(579) Sidonia

Discovered 1905 November 3 by A. Kopff at Heidelberg.

Named for a character in the opera *Armida* (composed 1777) by the German composer Christoph Willibald Gluck (1714-1787) which is based on Torquato Tasso's (1544-1595) poem *Jerusalem Delivered*. See also the citation for planet (514). (H 61)

In 1913, the Astronomisches Rechen-Institut insisted to name a number of planets in order to avoid possible errors. F. Cohn (AN 196, 137 (1913)) stated: "Seit längerer Zeit hat eine regelmässige Benennung der Kleinen Planeten nicht mehr stattgefunden. Die Unbequemlichkeit und Möglichkeit der Verwechselungen ... gab den Anlass zu einer an die Entdecker gerichteten Aufforderung, wenigstens die Planeten bis zur Nr. 700 benennen zu wollen. Eine ihnen zugesandte Vorschlagsliste enthielt in üblicher Art weibliche Vornamen, insbesondere aus Mythologie und Geschichte. Nach verschiedenen Abänderungsvorschlägen seitens der Herren Götz, Helffrich, Kopff und Lohnert sind die Namen vereinbart worden ...". A list of 108 names between (570) and (727) follows. An inspection reveals that, obviously, a lot of names were chosen according to the two letters of the provisional designation assigned to the planets. There is a strong correlation in this way for about 30 planets which demonstrates a statistical significance. The provisional designation for (579) Sidonia was 1905 SD.

(580) Selene

Discovered 1905 December 17 by M. Wolf at Heidelberg.

This name personified the Moon among the Greeks. See also the remarks to planet (105). (H 61)

The naming might be influenced by the two letters of the provisional designation 1905 SE. See also the remarks to planet (579).

(581) Tauntonia

Discovered 1905 December 24 by J. H. Metcalf at Taunton.

Named after the city in southeast Massachusetts and place of the discovery. (H 61)

Named (AN 172, 391 (1906)) by the orbit computer Herbert R. Morgan. Morgan was sometime director of the observatory in Glasgow, Missouri and for many years director of the nine-inch transit circle division at the U.S. Naval Observatory.

(582) Olympia

Discovered 1906 January 23 by A. Kopff at Heidelberg.

Named for the town in the northwestern part of the Peloponnese where Zeus {see planet (5731)} had a famous temple with a celebrated statue by Phidias {see planet (4753)}. The statue was reckoned one of the Seven Wonders of the Ancient World. At Olympia the Olympic games originated in 776 B.C. (Z 182)

(583) Klotilde

Discovered 1905 December 31 by J. Palisa at Vienna.

Named in honor of the daughter of the Austrian astronomer Hofrat Edmund Weiss (1837-1917) who founded with Karl von Littrow the Vienna Imperial Observatory in 1875. Later, he was director of the observatory where this planet was discovered. (H 61)

Named (AN 173, 31 (1907)) by Adelinde Weiss, wife of E. Weiss.

(584) Semiramis

Discovered 1906 January 15 by A. Kopff at Heidelberg.

Named for the mythical Assyrian queen, wife of Ninus, founder of Niniveh, whom she succeeded as ruler. She was famed for her beauty, wisdom, and voluptuousness. She is said to have built Babylon with its hanging gardens and many other cities. Semiramis conquered Egypt and much of Asia and Ethiopia and unsuccessfully attacked India. She was queen of Babylon during the days of Pyramus and Thisbe {see planet (88)}. (Z 237)

(585) Bilkis

Discovered 1906 February 16 by A. Kopff at Heidelberg.

Bilkis, or Balkis, is the name in the Koran of the queen of Sheba {see planet (1196)} who visited Solomon. Ancient Abyssinian tradition says that all the rulers of that country are descendants of Menelik, son of Solomon and the queen of Sheba and legendary first king of Abyssinia. (H 62)

(586) Thekla

Discovered 1906 February 21 by M. Wolf at Heidelberg.

Named probably after St. Thekla, who, according to Apocryphical sources, accompanied Paul on some of his missionary journeys. The name was selected from the list mentioned in the remarks to planet (579). (H 62)

(587) Hypsipyle

Discovered 1906 February 22 by M. Wolf at Heidelberg.

Named after the queen of Lemnos. The women of Lemnos put to death all other males, but Hypsipyle saved the life of her father Thoas {see planet (4834)}, king of Lemnos. When the Argonauts landed at Lemnos, all the women of Lemnos became pregnant. By Jason {see planet (6063)}, Hypsipyle became the mother of twins. Jason forgot his vows of fidelity to Hypsipyle, abandoned her, and the women of Lemnos banished the queen of Nemea. (Z 132)

(588) Achilles

Discovered 1906 February 22 by M. Wolf at Heidelberg.

Named for the bravest of the Greeks in the Trojan War. As an infant he was plunged in the River Styx by his mother Thetis {see planet (17)}, thus rendering his body invulnerable excepting the heel by which he was held. He slew Hector {see planet (624)}, the greatest Trojan warrior. He was eventually killed by an

arrow in the heel by Paris {see planet (3317)}. Achilles is the central figure in Homer's *Iliad.* (H 62)
Name suggested (AN 175, 191 (1907)) by J. Palisa.

This planet oscillates around the leading equilateral libration point formed with the Sun and Jupiter. It is first known example of the stable solution of the three-body problem worked out by Lagrange {see planet (1006)} in 1772. Palisa also named planets (617) and (624) which have similar orbital characteristics to (588) after heroes from the Trojan War. Henceforth such planets have been known as Trojan asteroids. It might seem natural to place the Greeks in one equilateral point and the Trojans in the other. Crommelin {see planet (1899)} remarked in Observatory, Vol. 30, p. 328 (1907) that "it seems strange that Achilles and Patroclus should be placed on opposite sides of Jupiter".

(589) Croatia

Discovered 1906 March 3 by A. Kopff at Heidelberg.

Named for the region in southeast Europe which was a part of Yugoslavia and, since 1991, is an independent state. (H 62)

(590) Tomyris

Discovered 1906 March 4 by M. Wolf at Heidelberg.

Named after the queen of the Massagets in Scythia who sold and killed the Persian king Cyrus the Great {see planet (7209)} 529 B.C. (H 62)

The naming might be influenced by the two letters of the provisional designation 1906 TO. See also the remarks to planet (579).

(591) Irmgard

Discovered 1906 March 14 by A. Kopff at Heidelberg.
Any reference of this name to a person or occurence is unknown.

(592) Bathseba

Discovered 1906 March 18 by M. Wolf at Heidelberg.

Named for the Biblic character, wife of Urias and later king David, who was the mother of Solomon. (H 62)

(593) Titania

Discovered 1906 March 20 by A. Kopff at Heidelberg.

Named after the wife of Oberon, queen of fairies and gnomes in Shakespeare's {see planet (2985)} *A Midsummer Night's Dream.* (H 63)

The naming might be influenced by the two letters of the provisional designation 1906 TT. See also the remarks to planet (579). Titania is also the name of the third satellite of Uranus, discovered 1787 by W. Herschel.

(594) Mireille

Discovered 1906 March 27 by M. Wolf at Heidelberg.

Named after the poem of the Provencal poet Frédéric Mistral (1830-1914). (H 63)
Name proposed by C. Flammarion (l'Astronomie, p. 118 (1913)).

(595) Polyxena

Discovered 1906 March 27 by A. Kopff at Heidelberg.

Named for the daughter of Priam and Hecuba {see planets (884) and (108)}. Beautiful and accomplished, she was courted by Achilles, but the marriage was opposed by Hector {see planets (588) and (624)}. She accompanied her father when he went to Achilles to claim the body of Hector. When Achilles was killed by Paris {see planet (3317)}, Polyxena sacrificed herself at his tomb. Some accounts say the sacrifice was not voluntary but that the spirit of Achilles appeared to the Greeks and demanded her sacrifice. (Z 218)

(596) Scheila

Discovered 1906 February 21 by A. Kopff at Heidelberg.

Named in honor of an acquaintance of the discoverer, an English woman student in Heidelberg. (H 63)

(597) Bandusia

Discovered 1906 April 16 by M. Wolf at Heidelberg.

Named after a fountain near Polezzo, Apulia, Italy. The Roman poet and satirist Horace (65-8 B.C.) {see planet (4294)} had a cottage near this fountain. (H 63)

(598) Octavia

Discovered 1906 April 13 by M. Wolf at Heidelberg.

Named for Octavia (70?-11 B.C.), sister of the first Roman emperor Augustus (63 B.C.-14), and later the second wife of Marcus Antonius (83?-30 B.C.). (H 63)

(599) Luisa

Discovered 1906 April 25 by J. H. Metcalf at Taunton.

Any reference of this name to a person or occurence is unknown.

(600) Musa

Discovered 1906 June 14 by J. H. Metcalf at Taunton.

Named for the nine goddesses in Greek mythology presiding over song and poetry and the arts and sciences. The muses are daughters of Zeus by Mnemosyne {see planets (5731) and (57)}. They were Erato (amatory verses and hymnes), Euterpe (music), Kalliope (heroic verses), Klio (history), Melpomene (tragedy), Polyhymnia (solemn songs), Thalia (comedy), Terpsichore (dance), and Urania (astronomy). All muses are existent at the minor planet sky {see planets (62), (27), (22), (84), (18), (33), (23), (81), and (30), respectively}. (H 63)

(601) Nerthus

Discovered 1906 June 21 by M. Wolf at Heidelberg.

Named for the Nordic goddess, wife of the Heaven's god. (H 63)

(602) Marianna

Discovered 1906 February 16 by J. H. Metcalf at Taunton. Independently discovered 1906 February 22 by A. Kopff at Heidelberg.

Any reference of this name to a person or occurence is unknown.

(603) Timandra

Discovered 1906 February 16 by J. H. Metcalf at Taunton.

Named for the daughter of Tindareus and Leda in Greek mythology. Timandra was the wife of Equemus and the mother of Evandrus. (H 63)

(604) Tekmessa

Discovered 1906 February 16 by J. H. Metcalf at Taunton.

Named for the daughter of the Phrygian prince Teubrantes, captive of Ajax {see planet (1404)}, by whom she had a son Eurysaces {see planet (8317)}. (H 64)

The naming might be influenced by the two letters of the provisional designation 1906 TK. See also the remarks to planet (579).

(605) Juvisia

Discovered 1906 August 27 by M. Wolf at Heidelberg.

Named for the city of Juvisy-sur-Orge in the department of Essonne, near Paris,

France, where the French astronomer Camille Flammarion {see planet (1021)} had his observatory. He lived there between 1882 and 1925. (H 64)

(606) Brangäne

Discovered 1906 September 18 by A. Kopff at Heidelberg.

Named for a maid servant of Isolde in the opera *Tristan und Isolde* (composed 1859) {see planets (1966) and (211)} by the German composer Richard Wagner {see planet (3992)}. (H 64)

(607) Jenny

Discovered 1906 September 18 by A. Kopff at Heidelberg.

Named in honor of the wife of a friend of the discoverer on the occasion of her engagement. Her full name was Jenny Adolfine Kessler (see also planet (608)). (H 64)

(608) Adolfine

Discovered 1906 September 18 by A. Kopff at Heidelberg.

See the name citation of planet (607). (H 64)

(609) Fulvia

Discovered 1906 September 24 by M. Wolf at Heidelberg.

Named for Fulvia, first wife of the orator, triumvir and general Marcus Aurelius Antonius (83?-30 B.C.). (H 64)

(610) Valeska

Discovered 1906 September 26 by M. Wolf at Heidelberg.

Any reference of this name to a person or occurence is unknown.

The naming might be influenced by the two letters of the provisional designation 1906 VK. See also the remarks to planet (579).

(611) Valeria

Discovered 1906 September 24 by J. H. Metcalf at Taunton.

Any reference of this name to a person or occurence is unknown.

The naming might be influenced by the two letters of the provisional designation 1906 VL. See also the remarks to planet (579).

(612) Veronika

Discovered 1906 October 8 by A. Kopff at Heidelberg.

Any reference of this name to a person or occurence is unknown.

The naming might be influenced by the two letters of the provisional designation 1906 VN. See also the remarks to planet (579).

(613) Ginevra

Discovered 1906 October 11 by A. Kopff at Heidelberg.

Named very probably after Guinevere, wife of King Arthur {see planet (2597)}. In the novel *Le morte d'Arthur* {orig.: Le Morte Darthur} (1469) by Thomas Malory her adulterous love, which Arthur is reluctant to recognize, causes Lancelot's {see planet (2041)} estrangement from Arthur and leads indirectly to Arthur's death and the destruction of the fellowship of the Round Table. (G. Cibis; LDS)

(614) Pia

Discovered 1906 October 11 by A. Kopff at Heidelberg.

Named probably after the Pia Observatory at Trieste, Italy, which was named by the astronomer Krieger after his wife. (R. Bremer)

(615) Roswitha
Discovered 1906 October 11 by A. Kopff at Heidelberg.
Named in honor of the German poetess Roswitha von Gandersheim. (H 64)

(616) Elly
Discovered 1906 October 17 by A. Kopff at Heidelberg.

Named in honor of Mrs. Elly Boehm, wife of Prof. K. Boehm, who was professor of mathematics at the Technical University at Karlsruhe, Germany. (H 64)

(617) Patroclus
Discovered 1906 October 17 by A. Kopff at Heidelberg.

Named for the Greek hero and friend of Achilles slain by Hector (see also planets (588) and (624)) at the Trojan War. (H 65)
Name suggested (AN 175, 191 (1907)) by J. Palisa.

(618) Elfriede
Discovered 1906 October 17 by K. Lohnert at Heidelberg.
Any reference of this name to a person or occurence is unknown.

(619) Triberga
Discovered 1906 October 22 by A. Kopff at Heidelberg.

Named for the small town of Triberg, near the city of Villingen, in the Black Forest in Germany because of the discoverer's love for that region. (H 65)

(620) Drakonia
Discovered 1906 October 26 by J. H. Metcalf at Taunton.

Named probably in honor of the Drake University, Des Moines, Iowa. (LDS; AN 175, 305 (1907))

Named by D. W. Morehouse and E. B. Stouffer, who calculated the orbit, both of the Drake University.

(621) Werdandi
Discovered 1906 November 11 by A. Kopff at Heidelberg.

Named for one of the three Norns in Norse mythology. Werdandi stands for the Present. (H 21)

The other Norns are Urda (the Past) and Skuld (the Future) {see planets (167) and (1130)}.

(622) Esther
Discovered 1906 November 13 by J. H. Metcalf at Taunton.

Named probably for the Jewish heroine of the Old Testament book of Esther. (H 65)

(623) Chimaera
Discovered 1907 January 22 by K. Lohnert at Heidelberg.

Named for the mount in Lycia whose summit emits flames and produces lions. Half way up there are pastures where goates graze. The foot is infested by snakes. Bellerophon {see planet (1808)} made this mount habitable. This is the basis of the fable in which Chimaera is a monster that emits flames, has the lion's head, the goat's body and the dragon's tail and that was killed by Bellerophon. Others say that the Chimaera had three heads, one of the lion, one of the goat and one of the snake. (H 65)

(624) Hektor
Discovered 1907 February 10 by A. Kopff at Heidelberg.

Named for a son of Priam, husband of Andromache, and Trojan champion slain by Achilles {see also planets (884), (175), and (588), respectively}. (H 65)

Name suggested (AN 175, 191 (1907)) by J. Palisa. The original spelling was "Hector".

(625) Xenia

Discovered 1907 February 11 by A. Kopff at Heidelberg.

Any reference of this name to a person or occurence is unknown. (H 65)

The naming might be influenced by the two letters of the provisional designation 1907 XN. See also the remarks to planet (579).

(626) Notburga

Discovered 1907 February 11 by A. Kopff at Heidelberg.

Named after the Catholic Saint Notburga von Hochhausen (ca. 1517), a small town in the Neckar Valley. (H 65)

(627) Charis

Discovered 1907 March 4 by A. Kopff at Heidelberg.

Named for the Greek goddess and wife of Hephaistos {see planet (2212)}. The Charites were the three Greek goddesses of charm and grace. The three graces (47) Aglaja, (31) Euphrosyne, and (23) Thalia are all members of the asteroid sky. (H 65)

(628) Christine

Discovered 1907 March 7 by A. Kopff at Heidelberg.

Any reference of this name to a person or occurence is unknown.

(629) Bernardina

Discovered 1907 March 7 by A. Kopff at Heidelberg.

Any reference of this name to a person or occurence is unknown.

(630) Euphemia

Discovered 1907 March 7 by A. Kopff at Heidelberg.

Named after a Christian holy figure whose festival is on the 16th of September. The name also means a good omen. (H 66)

(631) Philippina

Discovered 1907 March 21 by A. Kopff at Heidelberg.

Named in honor of Philipp Kessler, a friend of the discoverer, on the occasion of his engagement. (H 66)

(632) Pyrrha

Discovered 1907 April 5 by A. Kopff at Heidelberg.

Named after the daughter of Epimetheus and Pandora {see planets (1810) and (55), respectively}. She married Deucalion, a son of Prometheus {see planet (1809)}. Deucalion and Pyrrha were the only survivors of the deluge which Zeus {see planet (5731)} sent to destroy mankind. The couple replaced the loss of mankind by throwing stones behind their backs - those Deucalion threw became men, those Pyrrha threw became women. (Z 226)

(633) Zelima

Discovered 1907 May 12 by A. Kopff at Heidelberg.

Any reference of this name to a person or occurence is unknown. (H 66)

The naming might be influenced by the two letters of the provisional designation 1907 ZM. See also the remarks to planet (579).

(634) Ute

Discovered 1907 May 12 by A. Kopff at Heidelberg.

Named by the discoverer in honor of a friend on the occasion of her engagement. (H 66)

(635) Vundtia

Discovered 1907 June 9 by K. Lohnert at Heidelberg.

Named in honor of the German physicist Vundt. (H 66)

(636) Erika

Discovered 1907 February 8 by J. H. Metcalf at Taunton.

Any reference of this name to a person or occurence is unknown.

(637) Chrysothemis

Discovered 1907 March 11 by J. H. Metcalf at Taunton.

Named for the daughter of Agamemnon and Klytaemnestra in Greek mythology {see planets (911) and (179)}. (LDS)

(638) Moira

Discovered 1907 May 5 by J. H. Metcalf at Taunton.

Named for the Greek goddess of fate. Hesiod mentioned three goddesses of fate, daughters of Zeus, Klotho, Lachesis and Atropos {see planets (5731), (97), (120) and (273), respectively}. (H 66)

(639) Latona

Discovered 1907 July 19 by K. Lohnert at Heidelberg.

This is the Latin name for Leto {see planet (68)}, the mother of Apollo {see planet (1862)} and Artemis {see planet (105)}. She was the daughter of the Titan Coeus and Phoebe. (H 66)

(640) Brambilla

Discovered 1907 August 29 by A. Kopff at Heidelberg.

Named after the novel *Prinzessin Brambilla* (written 1821) by the German writer and composer E. T. A. Hoffmann (1776-1822), set to music by Walter Braunfels (1882-1954). (H 66)

(641) Agnes

Discovered 1907 September 8 by M. Wolf at Heidelberg.

Any reference of this name to a person or occurence is unknown.

(642) Clara

Discovered 1907 September 8 by M. Wolf at Heidelberg.

Named by the discoverer for a housekeeper of the Wolf family. (H 67)

(643) Scheherezade

Discovered 1907 September 8 by A. Kopff at Heidelberg.

Named for the fictional wife of an oriental king and the narrator of the tales in *Arabian Nights.* (H 67)

(644) Cosima

Discovered 1907 September 7 by A. Kopff at Heidelberg.

Named in honor of Cosima Wagner (1837-1930). She was the daughter of the Hungarian composer Franz Liszt {see planet (3910)} and the Countess d'Agoult and the second wife of the German composer Richard Wagner {see planet (3992)}. (H 67)

(645) Agrippina

Discovered 1907 September 13 by J. H. Metcalf at Taunton.

Named for two characters of the Roman history. Agrippina the elder (13? B.C.-33) was the daughter of the Roman statesman Marcus Vipsanius Agrippa, the wife of Germanicus and the mother of Caligula. Her daughter Agrippina the younger (15?-59) was the mother of Emperor Nero. (LDS)

The naming might be influenced by the two letters of the provisional designation 1907 AG. See also the remarks to planet (579).

(646) Kastalia

Discovered 1907 September 11 by A. Kopff at Heidelberg.

Named for the fountain in the footing of the Parnassus mountain, near Delphi, consecrated to Apollo {see planet (1862)} and the Muses. In Greek mythology Kastalia is the symbol for poetry. (H 67)
Another spelling of Kastalia is used with planet (4769) Castalia.

(647) Adelgunde

Discovered 1907 September 11 by A. Kopff at Heidelberg.
Any reference of this name to a person or occurence is unknown.

The naming might be influenced by the two letters of the provisional designation 1907 AD. See also the remarks to planet (579).

(648) Pippa

Discovered 1907 September 11 by A. Kopff at Heidelberg.

Named for the main character in *Und Pippa tanzt* by the German writer Gerhart Hauptmann (1862-1946). (H 67)

(649) Josefa

Discovered 1907 September 11 by A. Kopff at Heidelberg.
Any reference of this name to a person or occurence is unknown.

(650) Amalasuntha

Discovered 1907 October 4 by A. Kopff at Heidelberg.

Named after the daughter of the Ostrogoth king Theodoric the Great (474-526) and regent for his grandson Athalarich during his whole reign (526-534). Amalasuntha (also Amalaswintha) was killed 535 by her cousin Theodahat who succeeded as Ostrogoth king. The assassination gave rise to the intervention in Italy by the Byzantine emperor Justinian I. (483-565). (LDS)

The naming might be influenced by the two letters of the provisional designation 1907 AM. See also the remarks to planet (579).

(651) Antikleia

Discovered 1907 October 4 by A. Kopff at Heidelberg.

Named for the wife of Laertes, king of Ithaca, who died of an affliction caused by the prolonged absence of her son, Odysseus {see planet (1143)}. (H 67)

The naming might be influenced by the two letters of the provisional designation 1907 AN. See also the remarks to planet (579).

(652) Jubilatrix

Discovered 1907 November 4 by J. Palisa at Vienna.

This planet was named at the occasion of the 1908 meeting of the Astronomische Gesellschaft in Vienna (AN 179, 47 (1908)). It honors the 60th year reign jubilee of the Austrian Emperor Franz Joseph (1830-1916). (H 67)

Named by E. Weiss. A report on the meeting is published in Observatory, Vol. 31, p. 448 (1908): "The evening before the opening of the Congress a great number of foreign astronomers were present at the University Observatory at the formal unveiling of a bust of the Emperor Francis Joseph I. on the occasion of the sixtieth anniversary of his succession to the throne... Lastly, Prof. Palisa, of the Vienna Observatory, presented to the Assembly the new photographic celestial charts, publication of which had been undertaken by himself and Prof. Wolf... It seems that on these occasions it is customary to christen a small planet. This year, 1905 AU, discovered by Prof. Palisa, was named Jubilatrix, in honour of the Jubilee of the Emperor..."

(653) Berenike

Discovered 1907 November 27 by J. H. Metcalf at Taunton.

Named after the daughter of king Magas of Cyrene (273?-221 B.C.) and wife (247 B.C.) of Ptolemaeus III. Berenike sacrified her magnificent hair at the temple of Venus after the safely return of Ptolemaeus from a war against Syria. The hair was transferred to the skies and formed a constellation (Coma Berenices). (LDS)

The naming might be influenced by the two letters of the provisional designation 1907 BK. See also the remarks to planet (579).

(654) Zelinda

Discovered 1908 January 4 by A. Kopff at Heidelberg.

Named by Elia Millosevich with the permission of the discoverer in honor of Zelinda, sister of the famous Italian mathematician Ulisse Dini (1845-1918). Dini was a good friend of Millosevich. (Mem. Soc. Spettrosc. Italiani, Vol. XXXVII, 1908 (E. Colombini))

(655) Briseïs

Discovered 1907 November 4 by J. H. Metcalf at Taunton.

Named after the Trojan girl captured by the Greeks and given to Achilles {see planet (588)}. When Agamemnon {see planet (911)} demanded her for himself, Achilles refused to take further part in the Trojan War. (Z 43)

(656) Beagle

Discovered 1908 January 22 by A. Kopff at Heidelberg.

Named for the ship with which Charles Darwin {see planet (1991)} sailed around the world (1831-1836). (H 68)

(657) Gunlöd

Discovered 1908 January 23 by A. Kopff at Heidelberg.

Named after the daughter of the giant Sutung who took care of the Scandinavian hydromel. Odin {see planet (3989)} seduced her and stole three sips of the liquor. Gunlöd, changed into an eagle, pursued Odin who let fall some drops. (A. Paluzie-Borrell)

(658) Asteria

Discovered 1908 January 23 by A. Kopff at Heidelberg.

The name appears with different meanings in Greek mythology: (1) Asteria was the daughter of the Titans Coeus and Phoebe, the sister of Leto {see planet (68)}, and by Perseus the mother of Hecate {see planet (100)}. She was changed into a quail. (2) Asteria was one of the fifty daughters of Danaus. (3) Apollodorus says that Asteria was a daughter of Atlas. (4) Asteria is also a name of a town in Greece. (Z 34)

(659) Nestor

Discovered 1908 March 23 by M. Wolf at Heidelberg.

Named for the king of Pylos who served in his old age as a counselor to the Greeks at Troy. (LDS; AN 186, 223 (1910))
Named at the meeting of the Astronomische Gesellschaft in Breslau in 1910.

(660) Crescentia

Discovered 1908 January 8 by J. H. Metcalf at Taunton.
Any reference of this name to a person or occurence is unknown.

The naming might be influenced by the two letters of the provisional designation 1908 CC. See also the remarks to planet (579).

(661) Cloelia

Discovered 1908 February 22 by J. H. Metcalf at Taunton.

Named probably for a Roman noble girl. With nine other girls she was an hostage given to Porsenna who had besieged Rome. But during the night she escaped and swimming crossed the Tiber river. (A. Paluzie-Borrell)

The naming might be influenced by the two letters of the provisional designation 1908 CL. See also the remarks to planet (579).

(662) Newtonia

Discovered 1908 March 30 by J. H. Metcalf at Taunton.

Named for the city of Newton, Massachusetts. (B. G. Marsden; AN 182, 332 (1909))
Named by Z. Daniel, Princeton University Observatory.

Previously, the name was interpreted as an honor for the great English physicist Isaac Newton. The correct interpretation was now given by Brian G. Marsden who met Zaccheus Daniel at the April 1960 meeting of the American Astronomical Society in Pittsburgh, PA. Marsden states: "As I interpreted it, the point was that Z. Daniel did not know whether to honor Isaac or H. A. Newton, so he got around the matter by choosing Newton, Massachusetts, instead." The Small Bodies Names Committee in 1998 corrected the situation by naming planet (8000) to the great physicist.

(663) Gerlinde

Discovered 1908 June 24 by A. Kopff at Heidelberg.
Any reference of this name to a person or occurence is unknown.

(664) Judith

Discovered 1908 June 24 by A. Kopff at Heidelberg.

Named after the figure from the play (1839) of the German dramatist Friedrich Hebbel (1813-1863). Judith is a book of Scripture included in the Roman Catholic canon of the Old Testament and in the Protestant Apocrypha. (H 68)

(665) Sabine

Discovered 1908 July 22 by W. Lorenz at Heidelberg.
Any reference of this name to a person or occurence is unknown.

(666) Desdemona

Discovered 1908 July 23 by A. Kopff at Heidelberg.

Named for the wife of Othello in William Shakespeare's {see planet (2985)} tragedy (written 1604). (H 68)

The naming might be influenced by the two letters of the provisional designation 1908 DM. See also the remarks to planet (579). The name Desdemona has also been given to the satellite Uranus X, discovered 1986 by the Voyager 2 spacecraft.

(667) Denise

Discovered 1908 July 23 by A. Kopff at Heidelberg.
Any reference of this name to a person or occurence is unknown.

The naming might be influenced by the two letters of the provisional designation 1908 DN. See also the remarks to planet (579).

(668) Dora

Discovered 1908 July 27 by A. Kopff at Heidelberg.
Named after a girl friend of the discoverer's wife. (H 68)

The naming might be influenced by the two letters of the provisional designation 1908 DO. See also the remarks to planet (579).

(669) Kypria

Discovered 1908 August 20 by A. Kopff at Heidelberg.

Named after a poem, in early days ascribed to Homer {see planet (5700)} but denied to him by Herodotus {see planet (3092)}. It detailed the causes of the Trojan War and served as a sort of introduction to the *Iliad*. Kypria is also a surname of Aphrodite {see planet (1388)}, who was born from the foam of the sea and came to land near the island of Cyprus. (H 68; I. van Houten-Groeneveld)

(670) Ottegebe

Discovered 1908 August 20 by A. Kopff at Heidelberg.

Named for a woman character in the drama *Der arme Heinrich* by the German writer Gerhart Hauptmann (1862-1946). (H 69)

(671) Carnegia

Discovered 1908 September 21 by J. Palisa at Vienna.

Named by the discoverer (AN 184, 159 (1910)) in honor of the American philanthropist Andrew Carnegie (1835-1919), founder of the Carnegie Institution in Washington. D.C. (H 69)

(672) Astarte

Discovered 1908 September 21 by A. Kopff at Heidelberg.
Named for the Phoenician goddess of love and fertility. (H 69)

(673) Edda

Discovered 1908 September 20 by J. H. Metcalf at Taunton. Independently discovered 1908 September 21 by A. Kopff at Heidelberg.

Named for the Norse Edda which is a 13th century collection of mythological, heroic, and aphoristic poems in alliterative verse. (H 69)

The naming might be influenced by the two letters of the provisional designation 1908 EA. See also the remarks to planet (579).

(674) Rachele

Discovered 1908 October 28 by W. Lorenz at Heidelberg.

Named in honor of the wife of the orbit computer E. Bianchi. (H 69; AN 181, 292 (1909))
Named by E. Bianchi.

(675) Ludmilla

Discovered 1908 August 30 by J. H. Metcalf at Taunton.

Named probably after the title figure of the opera *Ruslan and Ludmilla* by the Russian composer Mikhail Ivanovich Glinka (1804-1857) {see planet (2205)}. (I. van Houten-Groeneveld)

(676) Melitta

Discovered 1909 January 16 by P. J. Melotte at Greenwich.
Melitta is the Attic form of the Greek Melissa, daughter of the Cretan king Melissus and sister of Amalthea {see planet (113)}. She was changed into a bee. At the same time the name is a play upon words for its resemblance to the discoverer's name. (H 69)

(677) Aaltje

Discovered 1909 January 18 by A. Kopff at Heidelberg.
Named in honor of the Dutch singer Aaltje Noordewier Reddingius. (H 69)

(678) Fredegundis

Discovered 1909 January 22 by W. Lorenz at Heidelberg.

Named after an unfinished opera by Ernest Guirand which was completed by the French composer Camille Saint-Saëns (1835-1921) {see planet (5210)}. (H 69)

The naming might be influenced by the two letters of the provisional designation 1909 FS. See also the remarks to planet (579).

(679) Pax

Discovered 1909 January 28 by A. Kopff at Heidelberg.
Named for the Roman goddess of peace. (H 69)
Named (AN 186, 223 (1910)) by G. Zappa, who computed the orbit.

(680) Genoveva

Discovered 1909 April 22 by A. Kopff at Heidelberg.

Named probably after the main figure in the drama *Genoveva* by the German dramatist Friedrich Hebbel (1813-1863). See also planet (664). (H 69)

(681) Gorgo

Discovered 1909 May 13 by A. Kopff at Heidelberg.

Gorgo is the German word for Gorgon. In the Greek mythology by Homer {see planet (5700)} only one Gorgon exists as a monster. Later there are three Gorgons, the three daughters of Ceto and Phorcys. Two of them, Euryale and Stheno, were immortal, the third, Medusa (see planet (149)), was mortal. (I. van Houten-Groeneveld)

(682) Hagar

Discovered 1909 June 17 by A. Kopff at Heidelberg.

Named for the concubine of Abraham driven into the desert with her son Ishmael because of Sarah's jealousy according to the account in the Old Testament (Genesis XXI, 14). (H 70)

The naming might be influenced by the two letters of the provisional designation 1909 HA. See also the remarks to planet (579).

(683) Lanzia

Discovered 1909 July 23 by M. Wolf at Heidelberg.

Named in honor of Karl Lanz, founder of the Heidelberg Academy of Sciences. (H 70)

(684) Hildburg

Discovered 1909 August 8 by A. Kopff at Heidelberg.
Any reference of this name to a person or occurence is unknown.

The naming might be influenced by the two letters of the provisional designation 1909 HD. See also the remarks to planet (579).

(685) Hermia

Discovered 1909 August 12 by W. Lorenz at Heidelberg.

Any reference of this name to a person or occurence is unknown.

The naming might be influenced by the two letters of the provisional designation 1909 HE. See also the remarks to planet (579).

(686) Gersuind

Discovered 1909 August 15 by A. Kopff at Heidelberg.

Named for a woman character out of a drama of the German writer Gerhart Hauptmann (1862-1946). (H 70)

(687) Tinette

Discovered 1909 August 16 by J. Palisa at Vienna.

Any reference of this name to a person or occurence is unknown.

The name was first published (AN 184, 320 (1910)) by R. Coniel, Paris.

(688) Melanie

Discovered 1909 August 25 by J. Palisa at Vienna.

Any reference of this name to a person or occurence is unknown.

Named (AN 186, 15 (1910)) by Dr. O. Prelinger, co-worker at the Wolf-Palisa photographic star charts.

(689) Zita

Discovered 1909 September 12 by J. Palisa at Vienna.

Named in honor of Empress Zita von Bourbon-Parma (1892-1989), wife of Emperor Karl I of Austria (and King Karl IV of Hungary) (1887-1922). (H 70)

The name was assigned on the occasion of the marriage of Zita 1911. After planet (220) this is the second planet which was given at a wedding present.

(690) Wratislavia

Discovered 1909 October 16 by J. H. Metcalf at Taunton.

Named for the (then German) city of Breslau in the province of Silesia {see planet (257)}. Wratislavia is the Latin form of the name. (H 70)

Named (AN 186, 223 (1910)) on the occasion of the meeting of the Astronomische Gesellschaft in Breslau in 1910.

(691) Lehigh

Discovered 1909 December 11 by J. H. Metcalf at Taunton.

Named for the Lehigh University, Bethlehem, Pennsylvania. (H 70)

Named in accordance with the discoverer by the orbit computer J. B. Reynolds, Lehigh University.

(692) Hippodamia

Discovered 1901 November 5 by M. Wolf at Heidelberg.

Named for a daughter of Oenomaus, king of Pisa, wife of Pelops and mother of Atreus and Thyestes. She bribed Myrtilus, her father's charioteer, to remove a spoke from the royal chariot wheels so that Pelops could win her. Oenomaus had already defeated and killed 13 other suitors whom he had challenged to chariot races. After killing Oenomaus, Pelops murdered Myrtilus. These murders were primal sins, all paid for later by the many troubles of the house of Atreus. (Z 128)

The naming might be influenced by the two letters of the provisional designation 1901 HD. See also the remarks to planet (579). The planet was rediscovered by A. Kopff at the 1910 apparition.

(693) Zerbinetta

Discovered 1909 September 21 by A. Kopff at Heidelberg.

Named after a character from the opera *Ariadne auf Naxos* by the German composer Richard Strauss (1864-1949) {see planet (5039)}. (H 70)

(694) Ekard

Discovered 1909 November 7 by J. H. Metcalf at Taunton. Independently discovered 1909 November 9 by J. Helffrich at Heidelberg.

Named by the American astronomer Seth Barnes Nicholson {see planet (1831)} and his wife while they were at the Drake University, Des Moines, Iowa. They were the first to compute its orbit and the name is "Drake" spelled backwards. (H 70)

(695) Bella

Discovered 1909 November 7 by J. H. Metcalf at Taunton.
Any reference of this name to a person or occurence is unknown.

(696) Leonora

Discovered 1910 January 10 by J. H. Metcalf at Taunton.

Named by Arthur Snow of the Nautical Almanac Office, Washington, D.C., who computed the orbit for this planet, in honor of his wife, Mary Leonora Snow. (Astron. J., Vol. 27, p. 16 (1911))

(697) Galilea

Discovered 1910 February 14 by J. Helffrich at Heidelberg.

Named for the discovery of the Jupiter satellites by Galileo Galilei: "Im besonderen soll der Name Galilea daran erinnern, dass der Planet am Tage der 300jährigen Wiederkehr der Entdeckung der Jupitersatelliten durch Galilei aufgefunden wurde." (AN 196, 139 (1913))
Galilei is also honored by a lunar crater.

(698) Ernestina

Discovered 1910 March 5 by J. Helffrich at Heidelberg.
Named in honor of Ernst Wolf, son of Max Wolf. (H 71)

(699) Hela

Discovered 1910 June 5 by J. Helffrich at Heidelberg.

Named possibly after Hel {see also planet (949)}, the Norse goddess of the dead. Yggdrasil was a tree, one of whose roots grew in Asgard, the second in Nifflheim and the third in the realm of Hela or Death. The lower world was ruled by the goddess Hel and to it were consigned those who had not died in battle. (H 71; R. Bremer)

(700) Auravictrix

Discovered 1910 June 5 by J. Helffrich at Heidelberg.

This planet is named in memory of the trips of the Schutte-Lanz Zeppelin No. 1. It is a Latin word meaning "victory against the wind". The first flights were performed in the years 1911 and 1912. (H 71)

(701) Oriola

Discovered 1910 July 12 by J. Helffrich at Heidelberg.

Named after a family of usually bright colored Old World passerine birds related to the crows. The English name for the most known bird of this family is 'golden oriole'. (H 71)

(702) Alauda

Discovered 1910 July 16 by J. Helffrich at Heidelberg.

Named for the numerous singing birds of the Alaudidae family (the lark) mostly of Europe, Asia, and northern Africa. (H 71)

(703) Noëmi

Discovered 1910 October 3 by J. Palisa at Vienna.

Probably named for Valentine Noëmi von Rothschild at the occasion of her marriage with Baron Sigismund von Springer in 1911. At that time Baron Rothschild donated a Zeiss Stereocomparator to the Vienna Observatory. {See also the citations for planets (250), (447), and, respectively (719)}. (A. Schnell)

(704) Interamnia

Discovered 1910 October 2 by V. Cerulli at Teramo.

Ancient name of the city of Teramo where the discoverer was born and where this planet was discovered. (J. Meeus)

Vincenzo Cerulli (1859-1927) in 1890 built a private observatory with a 40-cm Cooke telescope which he left to the Italian nation. He extensively studied minor planet (433) Eros at its 1900 opposition. The name Interamnia was sometimes erroneously interpreted as the ancient name of Terni in Umbria, Italy. The ancient name of Terni {see planet (5654)}, however, is Interamna. This clarification was given by S. De Meis.

(705) Erminia

Discovered 1910 October 6 by E. Ernst at Heidelberg.

Named after an opera (1885) by Jacobowsky termed the "most successful operetta of modern times" and based on an old melodrama *Robert Macaire.* (H 71)

(706) Hirundo

Discovered 1910 October 9 by J. Helffrich at Heidelberg.

Named for the numerous small long-winged passerine birds of the Hirundinidae family (the swallow) that are noted for their graceful flight and regular migrations, have a short bill with a wide gape, small weak feet, and often a deeply forked tail, occur in all parts of the world except New Zealand and the polar regions, and feed on insects caught on the wing. (H 71)

(707) Steina

Discovered 1910 December 22 by M. Wolf at Heidelberg.

Named after Mr. Stein, benefactor of the Breslau Observatory. (H 71; AN 214, 69 (1921))

Stein was honored twice: "Am 30. Juli 1921 fand ... die Einweihung des neuen Institutes (of the Breslau Observatory) statt; dem hervorragendsten Freund der Sternwarte, Herrn Fabrikdirektor Stein, wurde bei dieser Gelegenheit die Ehrenbürgerschaft der Universität verliehen. Dank der grossen Liebenswürdigkeit von Herrn Geheimrat Wolf in Heidelberg wurde Herr Stein in dankbarer Anerkennung seines Verdienstes um die Förderung der Astronomie an den Himmel versetzt, indem der Planet 707 auf den Namen Steina getauft wurde." (Vierteljahrsschr. Astron. Ges., 57. Jahrg., p. 78 (1922))

(708) Raphaela

Discovered 1911 February 3 by J. Helffrich at Heidelberg.

Named in honor of Raphael von Bischoffsheim, founder of the Nice Observatory. The name was chosen because A. Charlois, Nice, discovered the planet 1892 S which was identical to (708). (AN 196 139 (1913))

(709) Fringilla

Discovered 1911 February 3 by J. Helffrich at Heidelberg.

Named after the numerous singbirds of the Fringillidae family (the finch) having a short stout conical bill adapted for crushing seeds. (H 72)

(710) Gertrud

Discovered 1911 February 28 by J. Palisa at Vienna.

Named probably in honor of Gertrud Rheden, daughter of the astronomer J. Rheden and granddaughter of J. Palisa {see planet (914)}. (A. Schnell)
Named (AN 190, 223 (1912)) by E. Anding in Gotha.

(711) Marmulla

Discovered 1911 March 1 by J. Palisa at Vienna.

Named possibly after the German word "Marmel" or "Murmel" ("Marmul" in the medieval form) which means a marble - surely a nice allusion to a minor planet. (P. Wild)

The name was spelled 'Marmula' in the naming announcement (AN 191, 216 (1912)).

(712) Boliviana

Discovered 1911 March 19 by M. Wolf at Heidelberg.

Named in honor of Simon Bolivar (1783-1830), the liberator of South America. (H 72)

Name proposed by C. Flammarion.

(713) Luscinia

Discovered 1911 April 18 by J. Helffrich at Heidelberg.

Named after several Old World thrushes of genus Luscinia (the nightingale) noted for the sweet and usually nocturnal song of the male. (H 72)

The naming might be influenced by the two letters of the provisional designation 1911 LS. See also the remarks to planet (579).

(714) Ulula

Discovered 1911 May 18 by J. Helffrich at Heidelberg.

Named for an order of birds (the owl) of prey with large head and eyes, short hooked bill, strong talons, and more or less nocturnal habits. Ulula comes from the Latin word ululare which is an imitation of the call of the owl. (H 72)

(715) Transvaalia

Discovered 1911 April 22 by H. E. Wood at Johannesburg.

Named after the province in the northeastern part of the Republic of South Africa between the Vaal and the Limpopo {see planet (1490)} rivers. (H 72)

This is the first numbered minor planet discovered in Africa.

(716) Berkeley

Discovered 1911 July 30 by J. Palisa at Vienna.

This planet was named after the city of Berkeley, California, where Armin O. Leuschner (see planet (1361)) studied the orbits of minor planets. Leuschner especially worked on an orbit determination of the lost Palisa planet (719) Albert. (A. Schnell)

(717) Wisibada

Discovered 1911 August 26 by F. Kaiser at Heidelberg.

This is the Latin name of the city of Wiesbaden, the home of the discoverer. (H 72)

The city of Wiesbaden, Germany is also honored by the discoverer by planet (765) Mattiaca.

(718) Erida

Discovered 1911 September 29 by J. Palisa at Vienna.

Named in honor of the daughter of A. O. Leuschner {see planet (1361)}, Erida. (H 72; A. Schnell)

The friendship between Palisa and his colleague Leuschner is also documented by the naming of planet (716).

(719) Albert

Discovered 1911 October 3 by J. Palisa at Vienna.

Named by the Vienna astronomers in honor of Albert Freiherr von Rothschild, a benefactor of the Vienna Observatory. (AN 193, 334 (1913))

The naming was commented on in Observatory, Vol. 36, p. 153 (1913): "Masculine names are used only in the case of remarkable orbits like Eros, Albert, and the Trojan group." This is the only numbered minor planet that is supposed to be lost.

(720) Bohlinia

Discovered 1911 October 18 by F. Kaiser at Heidelberg.

Named in honor of the Swedish astronomer Karl Petrus Theodor Bohlin (1860-1939), author of a study of Jupiter's perturbations, on the occasion of his 65th birthday. B. Asplind stated in AN 225, 354 (1925): "Dieser zur Koronis-Gruppe gehörende Planet ... hat mit Zustimmung des Entdeckers als Ehrung für Prof. Karl Bohlin in Stockholm zu seinem 65. Geburtstage (30. Okt. 1925) für seine großen Verdienste auf dem Gebiet der Kleinen Planeten, insbesondere für seine Arbeiten zur gruppenweisen Berechnung der allgemeinen Störungen, den Namen Bohlinia erhalten." (H 72)

Named by M. Wolf at the meeting of the Astronomische Gesellschaft without the knowledge of the discoverer.

(721) Tabora

Discovered 1911 October 18 by F. Kaiser at Heidelberg.

This planet was named by M. Wolf after the ocean liner Tabora which was inspected by the participants in a meeting of the Astronomische Gesellschaft in Hamburg in 1913. The discoverer adds that Tabora also was the name of a coastal town in former German East Africa. (H 73; AN 196, 139 (1913))

(722) Frieda

Discovered 1911 October 18 by J. Palisa at Vienna.

Named probably in honor of Frieda Hillebrand, grand-daughter of Prof. Weiss, director of the Vienna Observatory. Frieda is the daughter of Prof. Karl Hillebrand (1861-1939), astronomer at the Graz University, and his wife Irene, née Weiss {see planet (794)}. (A. Schnell)

(723) Hammonia

Discovered 1911 October 21 by J. Palisa at Vienna.

The planet was named (AN 196, 139 (1913)) on the occasion of the 24th meeting of the Astronomische Gesellschaft in Hamburg in 1913. It is an allusion to the city. (H 73)

The city of Hamburg is also honored by planet (449).

(724) Hapag

Discovered 1911 October 21 by J. Palisa at Vienna.

This planet honors by the initials the "Hamburg-Amerika Paketfahrt Aktien-Gesellschaft". The planet was named on the occasion of the meeting of the Astronomische Gesellschaft in Hamburg, 1913. (H 73; AN 196, 139 (1913))

At the Hamburg AG meeting the planets (721), (723), (724) and (725) were named. Turner (Observatory, Vol. 36, No. 466, p. 415-416 (1913)) published verses, written by Palisa, which elucidate the naming history.

(725) Amanda

Discovered 1911 October 21 by J. Palisa at Vienna.

Named in honor of the wife of the Hamburg astronomer Richard Schorr {see planet (1235)}. (H 73)

The planet was named at the occasion of the 24th meeting of the Astronomische Gesellschaft in Hamburg in 1913 (AN 196, 139 (1913)). Turner noted in Observatory, Vol. 36, No. 466, p. 414 (1913): "Before the end of the meeting we had another christening of a planet: Prof. Palisa announced the name "Amanda". It is difficult for us who have not yet seen the planet to judge of the suitability of the celestial application: but Prof. Palisa was kind enough to give us a terrestrial parallel, the suitability of which was recognized with hearty applause."

(726) Joëlla

Discovered 1911 November 22 by J. H. Metcalf at Winchester.
Named in honor of the discoverer, Joel H. Metcalf (1866-1925). (H 73)

The discoverer is also honored by planet (792). See also the remarks to planet (1052).

(727) Nipponia

Discovered 1912 February 11 by A. Massinger at Heidelberg.

Named after the country of Japan where this planet was discovered accidently (by Hirayama, Tokyo) twice in 1900 and 1908, respectively. (AN 196, 140 (1913))

The planet was first photographed on March 3, 1900 by S. Hirayama under the assistance of K. Toda. K. Saotome computed first circular elements for 1900 FE. This is the first minor planet discovered in Japan. Nippon is the official transliteration of Japan. (K. Tomita).

(728) Leonisis

Discovered 1912 February 16 by J. Palisa at Vienna.

Named in honor of Geheimrat Leo Gans (1843-1935), president of the Physical Society at Frankfurt a.M., Germany on the occasion of his 70th birthday. The emblem of the Society is an Isis, thus the combined name for this planet. (AN 197, 416 (1914))
Named by M. Brendel.

(729) Watsonia

Discovered 1912 February 9 by J. H. Metcalf at Winchester.

Named in honor of the American astronomer James Craig Watson (1838-1880), director of the Ann Arbor Observatory, Michigan, who detected some comets and 22 minor planets. He left an endowment with the National Academy for the care of the "Watson" asteroids. (H 73)
Watson is also honored by a lunar crater. See also the remarks to planet (1052).

(730) Athanasia

Discovered 1912 April 10 by J. Palisa at Vienna.

Athanasia is the Greek word for immortality. Any reference to a person or occurrence is unknown. (H 73)
Named by friends of the discoverer (Astronomischer Kalender für 1917).

(731) Sorga

Discovered 1912 April 15 by A. Massinger at Heidelberg.

The name 'surga' means 'the heavens' in the Indonesian language. The paradise on earth was named 'surga dunia'. Both, 'surga' and 'sorga' are possible transliterations from Indonesian. (B. Hidayat)

(732) Tjilaki

Discovered 1912 April 15 by A. Massinger at Heidelberg.

Named for a river in Indonesia. "Tji" means river. The river rises in the mountains where the city of Malabar {see planet (754)} is situated. (H 74; B. Hidayat)

(733) Mocia

Discovered 1912 September 16 by M. Wolf at Heidelberg.

Named in honor of Max Wolf's youngest son, Werner Wolf, who in his childhood had the nickname "Mok". (I. van Houten-Groeneveld)

(734) Benda

Discovered 1912 October 11 by J. Palisa at Vienna.

Named in honor of Anna Benda, second wife of the discoverer, which he married in 1902. (M. G. Firneis, Astron. Ges., Abstr. Ser., No. 13, p. 143 (1997))

This name erroneously was attributed to the Czech composer Karl Bendl (1838-1897) in previous editions of this Dictionary.
Named by friends of the discoverer (Astronomischer Kalender für 1917).

(735) Marghanna

Discovered 1912 December 9 by H. Vogt at Heidelberg.

Named in honor of the discoverer's mother Margarete Vogt and after Hanna, who was a relative of H. Vogt {see planet (1439)}. (H 74)

(736) Harvard

Discovered 1912 November 16 by J. H. Metcalf at Winchester.
Named for the famous University at Cambridge, Massachusetts. (H 74)
See also the remarks to planet (1052).

(737) Arequipa

Discovered 1912 December 7 by J. H. Metcalf at Winchester.

Named for the city in Peru where the Harvard Observatory had a southern observing station. (H 74)
See also the remarks to planet (1052).

(738) Alagasta

Discovered 1913 January 7 by F. Kaiser at Heidelberg.

Named after the first German name of the city of Gaualgesheim on the Rhine river from which the family of the discoverer originated. The city originally was named Alagastesheim after Alagast, modified from Alberich, the trustee of the "Hort der Nibelungen". (H 74)

(739) Mandeville

Discovered 1913 February 7 by J. H. Metcalf at Winchester.

Named for the city of Mandeville, Jamaica, where Edward C. Pickering {see planet (784)} observed. (H 74)
See also the remarks to planet (1052).

(740) Cantabia

Discovered 1913 February 10 by J. H. Metcalf at Winchester.

This name is probably a contraction of Cantabridgia, the Latin form of Cambridge, Massachusetts, site of the Harvard Observatory. (H 74)
See also the remarks to planet (1052).

(741) Botolphia

Discovered 1913 February 10 by J. H. Metcalf at Winchester.

Named for the city of Boston, Massachusetts. The city derives its name from Saint Botolph who founded a monastery in the year 654. (H 74)
See also the remarks to planet (1052).

(742) Edisona

Discovered 1913 February 23 by F. Kaiser at Heidelberg.

Named in honor of the great American scientist and inventor Thomas Alva Edison (1847-1931) immediately after his death. The "Astronomical Society Urania, Wiesbaden", which Franz Kaiser founded, sent an artistically arranged scroll of dedication to the heirs of Edison by the help of the American Consulate at Frankfurt a.M. (H 75)
Edison is also honored by a lunar crater.

(743) Eugenisis

Discovered 1913 February 25 by F. Kaiser at Heidelberg.

This name is composed from the two Greek words "eu" (good, well) and "genesis" (creation). Obviously, there is a close connection with the birth of the daughter of the discoverer (see also the citation for planet (746)). (M. Kretlow; I. van Houten-Groeneveld)

The naming information was given by Dr. M.-L. Kaiser, daughter of the discoverer.

(744) Aguntina

Discovered 1913 February 26 by J. Rheden at Vienna.

Named for Aguntinum, a Roman town in the province of Noricum close to Lienz, the birthplace of the discoverer. (H 75)

The naming information was given by Mrs. Rheden, widow of the discoverer and daughter of J. Palisa.

(745) Mauritia

Discovered 1913 March 1 by F. Kaiser at Heidelberg.

Named by the discoverer after the patron St. Mauritius of a church of Wiesbaden {see planet (717)}, Germany. (H 75)

(746) Marlu

Discovered 1913 March 1 by F. Kaiser at Heidelberg.

Named by the discoverer in honor of his daughter, the physician Dr. Marie-Louise Kaiser. See also the citation for planet (743). (H 75)

(747) Winchester

Discovered 1913 March 7 by J. H. Metcalf at Winchester.

Named by the discoverer for the city of Winchester, Massachusetts, where he made this discovery. (H 75)

(748) Simeïsa

Discovered 1913 March 14 by G. N. Neujmin at Simeïs.

Named for the Simeïs Observatory and the town of Simeïs, Crimea, where the planet was first discovered. (H 75)

This is the first minor planet discovered in Russia.

(749) Malzovia

Discovered 1913 April 5 by S. I. Belyavskij at Simeïs.

Named in honor of the Russian amateur astronomer S. I. Malzov. In 1911, he deeded his private observatory at Simeïs as a gift to the Pulkovo Observatory. Malzov was a close friend of the discoverer and of J. O. Backlund {see planet (856)}. After 1918, Malzov resided at Menton, French Riviera. (H 75)

(750) Oskar

Discovered 1913 April 28 by J. Palisa at Vienna.

Named very probably in remembrance of Oskar Ruben von Rothschild (1888-1909), the youngest son of Baron Albert von Rothschild {see planet (719)}. The reconstruction and improvement of the telescope drive of the so-called Rothschild Coudé Telescope of the Vienna Observatory was payed by the Oskar Freiherr von Rothschild-Stiftung in 1913. (A. Schnell)

Named by friends of the discoverer (Astronomischer Kalender für 1917).

(751) Faïna

Discovered 1913 April 28 by G. N. Neujmin at Simeïs.

Named in honor of Mrs. F. M. Neujmin, the first wife of the discoverer. (N. S. Samojlova-Yakhontova)

(752) Sulamitis

Discovered 1913 April 30 by G. N. Neujmin at Simeïs.

Named for Sulamith, a woman in the book "Solomon's Song" 7, 1 of the Old Testament. The writer describes her great beauty. Sulamith probably was the queen of Sheba. (I. van Houten-Groeneveld)

(753) Tiflis

Discovered 1913 April 30 by G. N. Neujmin at Simeïs.

Named for the native city of the discoverer. Tiflis (or Tbilisi) is the capital city of Georgia on the Kura river. (H 75)

(754) Malabar

Discovered 1906 August 22 by A. Kopff at Heidelberg.

Named in remembrance of the Dutch-German solar eclipse expedition to Christmas Island in 1922. Malabar is a city and mountain on Java. (I. van Houten-Groeneveld; B. Hidayat)

The naming is described in AN 218, 253 (1923): "Aus Anlass der holländisch-deutschen Sonnenfinsternis-Expedition nach Christmas Island wurde der Planet der Niederländisch-Indischen Sternkundigen-Vereinigung zur Benennung überlassen als Zeichen des Dankes für die der Expedition zuteil gewordene Förderung. Herr K. A. R. Bosscha auf Malabar (Java), der Vorsitzende der Vereinigung, erteilte ihm den Namen." Bosscha and his friend Kerkhoven determined that after their death a great part of their capital should be put into a fund to serve astronomy in the Netherlands and Indonesia. In 1954 the Kerkhoven-Bosscha-Fund was established in Leiden. In the course of the years this fund has grown and helps today the Dutch and Indonesian astronomy in many ways. Bosscha dedicated the name to the Malabar mountain, 40 km south of the city of Bandung, location of his beautifully cultivated tea plantations.

(755) Quintilla

Discovered 1908 April 6 by J. H. Metcalf at Taunton.

This is the name of an Italian girl. The name was suggested by Miss Arville D. Walker, secretary to H. Shapley {see planet (1123)}, because at the time of the naming (in 1926) no other planet name began with the letter "Q". (H 76)
See also the remarks to planet (1052).

(756) Lilliana

Discovered 1908 April 26 by J. H. Metcalf at Taunton.
Named in honor of the sister of H. Shapley {see planet (1123)}. (H 76)
Name suggested by H. Shapley in 1926.
See also the remarks to planet (1052).

(757) Portlandia

Discovered 1908 September 30 by J. H. Metcalf at Taunton.

Named for the city of Portland, Maine, where the discoverer was minister of a church at the time of his death. (H 76)

(758) Mancunia

Discovered 1912 May 18 by H. E. Wood at Johannesburg.

Named after the city of Manchester, England, the native city of the discoverer. Mancunia is the Latin name of Manchester. (H 76)

(759) Vinifera

Discovered 1913 August 26 by F. Kaiser at Heidelberg.

Named by the discoverer after the vine from which his ancestors made a living. (H 76)

(760) Massinga

Discovered 1913 August 28 by F. Kaiser at Heidelberg. Independently discovered 1913 August 28 by G. N. Neujmin at Simeïs.

Named in honor of A. Massinger, assistant astronomer at the Heidelberg Königstuhl Observatory who died in the World War I as a soldier. (H 76)

(761) Brendelia

Discovered 1913 September 8 by F. Kaiser at Heidelberg.

Named by the discoverer in honor of the late theoreticist of the minor planets, Otto Rudolph Martin Brendel (1862-1939), who at that time was director of the "Planeteninstitut" at the University of Frankfurt am Main. (H 76; AN 215, 471 (1922))

Brendel himself chose for the planet to be named the object with the smallest inclination from the planets discovered by Kaiser and not previously named. Obituary published in Astron. Nachr., Band 270, p. 248 (1940).

(762) Pulcova

Discovered 1913 September 3 by G. N. Neujmin at Simeïs.

Named for Pulkovo, the locality near St. Petersburg (the former Leningrad), where the famous observatory was erected in 1839. (H 76)

(763) Cupido

Discovered 1913 September 25 by F. Kaiser at Heidelberg.

Named after the Roman god of erotic love because of its relatively small distance from the Sun. See also the citation for planet (433). (H 76)
Named probably by B. Asplind.

(764) Gedania

Discovered 1913 September 26 by F. Kaiser at Heidelberg.

Named by the discoverer after the Latin name of the free city of Danzig (now Gdansk, Poland) in which he was first assistant at the Observatory 1921-1925. (H 77)
The city of Danzig is also honored by planet (1419).

(765) Mattiaca

Discovered 1913 September 26 by F. Kaiser at Heidelberg.

Named by the discoverer for his home town. Mattiacum was the Latin name for the city of Wiesbaden, Germany. (H 77)

The city of Wiesbaden was also honored by the discoverer with planet (717) Wisibada.

(766) Moguntia

Discovered 1913 September 29 by F. Kaiser at Heidelberg.

Named by the discoverer in honor of the city of Mainz (Latin name Moguntia), Germany. He was active for more than three years at the University as an instructor in astronomy and as the first who gave lectures within the natural sciences faculty. (H 77)

(767) Bondia

Discovered 1913 September 23 by J. H. Metcalf at Winchester.

Named in honor of William Cranch Bond (1789-1859) and his son George Phillips Bond (1825-1865) who both were directors of the Harvard College Observatory in Cambridge, Massachusetts. W. C. Bond was codiscoverer of comet 1850 II and was one of the first astronomers who used the photographic technique in astronomy. G. P. Bond discovered in 1848 the Saturn satellite Hyperion and (1850) Saturn's faint crepe ring. (H 77)

Both astronomers are also honored by lunar craters, G. P. Bond also by a crater on Mars. See also the remarks to planet (1052).

(768) Struveana

Discovered 1913 October 4 by G. N. Neujmin at Simeïs.

Named in honor of Friedrich Georg Wilhelm Struve (1793-1864), Otto Wilhelm Struve (1819-1905), directors of the Pulkovo Observatory and famous for their double star studies, and Hermann Struve (1854-1920), director of the Königsberg and Berlin Observatories. (H 77)

The Struves are also honored by a lunar crater.

(769) Tatjana

Discovered 1913 October 6 by G. N. Neujmin at Simeïs.

Named possibly in honor of a former scientific collaborator at the Pulkovo Observatory. Another interpretation says that this planet is named for the heroine of the celebrated poem *Evgeni Onigin* by the Russian poet Aleksander Sergeevich Pushkin (1799-1837, see planet (2208)). (H 77)

The naming might be influenced by the two letters of the provisional designation 1913 TA.

(770) Bali

Discovered 1913 October 31 by A. Massinger at Heidelberg.

This planet was named after the Indonesian island east of Java. The discoverer repeatedly named planets after geographic items of Indonesia (cf. planets (732) and (772)). (LDS; B. Hidayat)

(771) Libera

Discovered 1913 November 21 by J. Rheden at Vienna.

Named by Mrs. Hedwig Rheden in honor of a friend of the discoverer. (H 77)

(772) Tanete

Discovered 1913 December 19 by A. Massinger at Heidelberg.

Named very probably after the city of Tanete on the southwest coast of Celebes, Indonesia. (R. Bremer)

See also the citation for planet (770).

(773) Irmintraud

Discovered 1913 December 22 by F. Kaiser at Heidelberg.

Named by the discoverer for an old German woman's name which frequently appears in old songs and sagas. (H 77)

(774) Armor

Discovered 1913 December 19 by C. Le Morvan at Paris.

Armor or Armorica is the Celtic name of the Normandy and Brittany coastal regions in France. Brittany is the native region of the discoverer. (H 77)

(775) Lumière

Discovered 1914 January 6 by J. Lagrula at Nice.

Named in honor of Auguste Lumière (1862-1954) and his brother Louis Lumière (1864-1948), noted French physicists and inventors of photography and cinematography. (H 78)

(776) Berbericia

Discovered 1914 January 24 by A. Massinger at Heidelberg.

Named in honor of the German astronomer Adolf Berberich (1861-1920). (H 78)

Named by Max Wolf, who (AN 211, 44 (1920)) wrote: "Dem Planeten 776 möchte ich den Namen Berbericia, dem Planeten 914 den Namen Palisana beilegen, um so für die beiden Astronomen Adolf Berberich und Johann Palisa, denen die Erforschung der Planetoiden in den letzten Dezennien wohl am meisten verdankte, eine Erinnerung zu stiften."

(777) Gutemberga

Discovered 1914 January 24 by F. Kaiser at Heidelberg.

Named by the discoverer in memory of Johann Gensfleisch (Gutenberg) (1400?-1468) who discovered the art of printing from movable types. He lived and died in Mainz and the neighboring Eltville. (H 78)
Gutenberg is also honored by a lunar crater.

(778) Theobalda

Discovered 1914 January 25 by F. Kaiser at Heidelberg.
Named by the discoverer in honor of his father Theobald Kaiser. (H 78)

(779) Nina

Discovered 1914 January 25 by G. N. Neujmin at Simeïs.

Named in honor of the sister of the discoverer, the mathematician Nina Nikolaevna Neujmina (1877-1956). (N. S. Samojlova-Yakhontova; N. Solovaya; N. S. Chernykh)

(780) Armenia

Discovered 1914 January 25 by G. N. Neujmin at Simeïs.

Named for the region in western Asia southeast of the Black sea and southwest of the Caspian sea. The area is now divided between the USSR, Turkey and Iran. (H 78)

(781) Kartvelia

Discovered 1914 January 25 by G. N. Neujmin at Simeïs.

This name honors the people of the Russian region of Georgia. Kartveli is the ancient name of the inhabitants. (H 78)

(782) Montefiore

Discovered 1914 March 18 by J. Palisa at Vienna.

Probably named in honor of the wife of Alphonse Mayer Rothschild (1878-1942), the second son of Baron Albert von Rothschild {see planet (719)}, Clarice Sebag-Montefiore. The marriage took place in 1912. (A. Schnell)
Named by friends of the discoverer (Astronomischer Kalender für 1917).

(783) Nora

Discovered 1914 March 18 by J. Palisa at Vienna.

Named probably for the drama of the Norwegian poet Henrik Ibsen (1828-1906) {see planet (5696)}. (H 78)
Named by friends of the discoverer (Astronomischer Kalender für 1917).

(784) Pickeringia

Discovered 1914 March 20 by J. H. Metcalf at Winchester.

Named in honor of the brothers Edward Charles Pickering (1846-1919), director of Harvard Observatory, and William Henry Pickering (1858-1930), director of Arequipa {see planet (737)} Observatory and discoverer (1898) of the ninth satellite of Saturn. (H 78)

The Pickerings are also honored by craters on the Moon and on Mars. See also the remarks to planet (1052).

(785) Zwetana

Discovered 1914 March 30 by A. Massinger at Heidelberg.
Named in honor of the daughter of Prof. K. Pophoff, Sofia. (H 78)
Named by A. Kopff.

(786) Bredichina

Discovered 1914 April 20 by F. Kaiser at Heidelberg.

Named in honor of the Russian astronomer Feodor Alexandrovich Bredichin (1831-1904) who was director of the Pulkovo Observatory. He made important contributions to the study of comets. (H 78)
Bredichin is also honored by a lunar crater.

(787) Moskva

Discovered 1914 April 20 by G. N. Neujmin at Simeïs.
This planet was named for the city of Moscow, USSR. (H 79)

(788) Hohensteina

Discovered 1914 April 28 by F. Kaiser at Heidelberg.

Named by the discoverer after the castle Hohenstein in the Taunus mountains near Bad Schwalbach, the original residence of the knights "Breder von Hohenstein" which is synonymous with the family name (Broeder or Breder) of his wife, who's ancestors came from that region. (H 79)

(789) Lena

Discovered 1914 June 24 by G. N. Neujmin at Simeïs.

Named in honor of Lena (Elena) Petrovna Neujmina (1860-1942), mother of the discoverer. (N. S. Chernykh)

(790) Pretoria

Discovered 1912 January 16 by H. E. Wood at Johannesburg.

Named after the capital city of Transvaal {see planet (715)} in South Africa. (H 79)

(791) Ani

Discovered 1914 June 29 by G. N. Neujmin at Simeïs.

Named for the ruined city in the Erivan district which was destroyed by an earthquake in 1319. Ani is situated 30 miles southeast of Kars, northeast Turkey, near the border to Armenia. Ani had so many churches that Sir John Mandeville (?-1372) called it 'the city of 1001 churches'. (H 79; R. Bremer)

(792) Metcalfia

Discovered 1907 March 20 by J. H. Metcalf at Taunton.

Named in honor of Rev. Joel Hastings Metcalf (1866-1925). A member of the Committee of Harvard Observatory, he discovered six comets and 41 minor planets. He designed two objectives of 10 and 16 inch aperture. (H 79)
Metcalf is also honored by planet (726). See also the remarks to planet (1052).

(793) Arizona

Discovered 1907 April 9 by P. Lowell at Flagstaff. Independently discovered 1907 April 17 by J. H. Metcalf at Taunton.

Named for the state in which the Lowell Observatory is located at Flagstaff {see planet (2118)}. (H 79)

(794) Irenaea

Discovered 1914 August 27 by J. Palisa at Vienna.

Named probably in honor of Irene Hillebrand, née Weiss, daughter of the director of the Vienna Observatory E. Weiss. The name had to be given a 'classical' form because of the existence of planet (14). (A. Schnell)

(795) Fini

Discovered 1914 September 26 by J. Palisa at Vienna.
Any reference of this name to a person or occurence is unknown.
In Austria, this nick name is a short form of Josefine.

(796) Sarita

Discovered 1914 October 15 by K. Reinmuth at Heidelberg.
Any reference of this name to a person or occurence is unknown.

(797) Montana

Discovered 1914 November 17 by H. Thiele at Bergedorf.

This is the first asteroid discovered at the Bergedorf Observatory and is named by this Latin word in honor of Bergedorf ("mountain village"). (H 79)

(798) Ruth

Discovered 1914 November 21 by M. Wolf at Heidelberg.

This planet is possibly named for the biblical heroine in the Old Testament (Book of Ruth). (H 79)

(799) Gudula

Discovered 1915 March 9 by K. Reinmuth at Heidelberg.
See the remarks to planet (913).

(800) Kressmannia

Discovered 1915 March 20 by M. Wolf at Heidelberg.

Named in honor of Major A. Kressmann who donated the Kressmann refractor for the Heidelberg Königstuhl Observatory. (H 80; AN 214, 69 (1921))

(801) Helwerthia

Discovered 1915 March 20 by M. Wolf at Heidelberg.

Named in honor of Mrs. Elise Helwerth-Wolf (1840-1924), the mother of the discoverer. See also planet (412). (H 80)

(802) Epyaxa

Discovered 1915 March 20 by M. Wolf at Heidelberg.
Any reference of this name to a person or occurence is unknown.

(803) Picka

Discovered 1915 March 21 by J. Palisa at Vienna.

Named in honor of the Czech physician Friedrich Pick (1867-1921). He was the first physician in Prague who introduced endoscopic methods in medicine. (H 80; A. Schnell)

Named by friends of the discoverer (Astronomischer Kalender für 1917).

(804) Hispania

Discovered 1915 March 20 by J. Comas Solá at Barcelona.

This is the Latin name of Spain. Hispania is the first planet discovered in this country. (H 80)

(805) Hormuthia

Discovered 1915 April 17 by M. Wolf at Heidelberg.

Named in honor of Mrs. Hormuth Kopff, the wife of A. Kopff {see planet (1631)}. (H 80)

(806) Gyldénia

Discovered 1915 April 18 by M. Wolf at Heidelberg.

Named in honor of the Swedish astronomer Johan August Hugo Gyldén (1841-1896), director of the Stockholm Observatory and author of a new method for

computing perturbations of planets and comets. (H 80)
Gyldén is also honored by a lunar crater.

(807) Ceraskia

Discovered 1915 April 18 by M. Wolf at Heidelberg.

Named in honor of the Russian astronomer Vitold Karlovich Ceraski (1849-1925). Born in Slutsk (now in Byelorussia), he graduated from Moscow University and worked for 45 years at the Moscow Observatory. He was director of that observatory for 25 years. Ceraski worked extensively on stellar and solar photometry. (N. S. Chernykh)

Ceraski is also honored by a lunar crater. The citation (H 80) given in the first three editions of this 'Dictionary' is erroneous.

(808) Merxia

Discovered 1901 October 11 by L. Carnera at Heidelberg.
Named in honor of Adalbert Merx, the father-in-law of Max Wolf. (H 80)

(809) Lundia

Discovered 1915 August 11 by M. Wolf at Heidelberg.

Named for the city of Lund in southwestern Sweden and for the Lund Observatory. (H 80)

(810) Atossa

Discovered 1915 September 8 by M. Wolf at Heidelberg.

Named after the Persian queen, daughter of king Cyrus and wife of Darius I {see, respectively, planets (7209) and (7210)}. (H 80)

(811) Nauheima

Discovered 1915 September 8 by M. Wolf at Heidelberg.
Named for the city of Bad Nauheim in Hessen, Germany. (H 80)

(812) Adele

Discovered 1915 September 8 by S. I. Belyavskij at Simeïs. Independently discovered 1915 September 11 by M. Wolf at Heidelberg.

Named probably after a figure from the opera *Die Fledermaus* by Johann Strauss (1825-1899) {see planet (4559)}. (R. Bremer; I. van Houten-Groeneveld)
Named by the independent discoverer M. Wolf.

(813) Baumeia

Discovered 1915 November 28 by M. Wolf at Heidelberg.

Named by the discoverer in honor of H. Baum, student of astronomy at Heidelberg, who died in World War I. (H 81; AN 214, 69 (1921))

(814) Tauris

Discovered 1916 January 2 by G. N. Neujmin at Simeïs.

Named for the Tauric Mount on the southwest coast of Crimea. It is also the ancient name of Crimea. Gustav L. W. von Struve (1858-1920) taught astronomy at the Tauris University in Simferopol, Crimea. (H 81; R. Bremer)

(815) Coppelia

Discovered 1916 February 2 by M. Wolf at Heidelberg.

Named for the short ballet in two acts by Léo Delibes (1836-1891) which is based on a tale by E. T. A. Hoffmann (1776-1822). (H 81)

(816) Juliana

Discovered 1916 February 8 by M. Wolf at Heidelberg.

Named presumably in honor of Queen Juliana (1909-) of the Netherlands. T. Gehrels, in connection with his interpretation of the meaning of the name

Wilhelmina (see planet (392)) pointed out that it seems very consistent that Max Wolf named an asteroid after Queen Wilhelmina's daughter, too. (LDS)

(817) Annika

Discovered 1916 February 6 by M. Wolf at Heidelberg.
Any reference of this name to a person or occurence is unknown.

(818) Kapteynia

Discovered 1916 February 21 by M. Wolf at Heidelberg.

Named in honor of the late Prof. Jacobus Cornelius Kapteyn {1851-1922}, who was director of the Groningen Astronomical Laboratory from 1878 till 1921. He made pioneering researches in the field of galactic structure, initiated the plan of Selected Areas and made a major contribution to the Cape Photographic Durchmusterung. (M 2822)

See also the remark to (1694) Kaiser. Kapteyn is also honored by a lunar crater.

(819) Barnardiana

Discovered 1916 March 3 by M. Wolf at Heidelberg.

Named in honor of the American astronomer Edward Emerson Barnard (1857-1923), great observer and star photographer who discovered many double stars, 17 comets, the Jupiter satellite Amalthea (1892) and the high proper motion star which is named after him. (H 81; AN 215, 471 (1922))

Barnard is also honored by a lunar crater, a crater on the Mars surface and a region on the Jupiter satellite Ganymede.

(820) Adriana

Discovered 1916 March 30 by M. Wolf at Heidelberg.
Any reference of this name to a person or occurence is unknown.

(821) Fanny

Discovered 1916 March 31 by M. Wolf at Heidelberg.
Any reference of this name to a person or occurence is unknown.

(822) Lalage

Discovered 1916 March 31 by M. Wolf at Heidelberg.
Any reference of this name to a person or occurence is unknown.

(823) Sisigambis

Discovered 1916 March 31 by M. Wolf at Heidelberg.

Named for Sisigambis, the mother of Darius III, the last Persian king. During the Issus Battle (333 B.C.) she was taken prisoner by Alexander the Great along with the wife Roxane {see planet (317)} and daughter of Darius. (H 81)

(824) Anastasia

Discovered 1916 March 25 by G. N. Neujmin at Simeïs. Independently discovered 1916 April 1 by M. Wolf at Heidelberg.
Named in memory of Mrs. Anastasia Semenoff. (AN 223, 103 (1925))

(825) Tanina

Discovered 1916 March 27 by G. N. Neujmin at Simeïs. Independently discovered 1916 April 3 by M. Wolf at Heidelberg.
Any reference of this name to a person or occurence is unknown.

(826) Henrika

Discovered 1916 April 28 by M. Wolf at Heidelberg.
Any reference of this name to a person or occurence is unknown.

(827) Wolfiana

Discovered 1916 August 29 by J. Palisa at Vienna.

Named in honor of Max Wolf (1863-1932), professor of astronomy at the University of Heidelberg and founder and director of the Königstuhl Observatory. Wolf discovered several novae, comets and hundreds of minor planets. The original citation for this planet and (828) is published in AN 211, 441 (1920): "Den Planeten 827 widme ich meinem Freunde Wolf, dem Begründer der photographischen Entdekkungsmethode Kleiner Planeten; den Planeten 828 widme ich Herrn Lindemann, dem uneigennützigen und großherzigen Förderer astronomischer Forschung." (H 82)

Wolf is also honored by a lunar crater and by planet (1217).

(828) Lindemannia

Discovered 1916 August 29 by J. Palisa at Vienna.

Named in honor of Prof. Adolf F. Lindemann (1846-1931) of Sidmouth, England, who invented, together with his son F. A. Lindemann, the Lindemann Electrometer. Wolf visited Lindemann's private observatory during a stay in England in 1896. See also the citation for planet (827). (H 82; A. Schnell)

(829) Academia

Discovered 1916 August 25 by G. N. Neujmin at Simeïs. Independently discovered 1916 August 31 by M. Wolf at Heidelberg.

This planet honors the 200th anniversary of the founding of the Academy of Sciences in St. Petersburg (see also planet (830)). (H 82)

(830) Petropolitana

Discovered 1916 August 25 by G. N. Neujmin at Simeïs. Independently discovered 1916 September 3 by M. Wolf at Heidelberg.

This planet is named on the same occasion as planet (829) which was discovered at the same night. It honors the city of St. Petersburg (the former Leningrad). (H 82)

(831) Stateira

Discovered 1916 September 20 by M. Wolf at Heidelberg.

Named after the wife of king Artaxerxes II of Persia who died following poisoning by her mother-in-law Parysatis {see planet (888)}. (H 82)

(832) Karin

Discovered 1916 September 20 by M. Wolf at Heidelberg.

Named after Karin Mansdotter from Swedish history. She was the mistress of Erik XIV of Sweden in the 16th century. Karin married Erik 1567, but he was pushed from his throne because of this marriage. (H 82)

(833) Monica

Discovered 1916 September 20 by M. Wolf at Heidelberg.

Any reference of this name to a person or occurence is unknown.

(834) Burnhamia

Discovered 1916 September 20 by M. Wolf at Heidelberg.

Named in honor of the American astronomer Sherburne Wesley Burnham (1838-1921). From 1869 he observed as an amateur astronomer and discovered many visual binaries. Later he observed in Chicago (1877), Lick (1888) and Yerkes (1897) Observatories. Burnham is the author of the famous *General Catalogue of Double Stars* of 13,665 binaries. (H 82; AN 214, 69 (1921))

Burnham is also honored by a lunar crater.

(835) Olivia

Discovered 1916 September 23 by M. Wolf at Heidelberg.
Any reference of this name to a person or occurence is unknown.

(836) Jole

Discovered 1916 September 23 by M. Wolf at Heidelberg.

Named for Jole or Iole, the daughter of Eurytos and by force wife of Hercules. (LDS)

(837) Schwarzschilda

Discovered 1916 September 23 by M. Wolf at Heidelberg.

Named in honor of the German astronomer Karl Schwarzschild (1873-1916), director of the Göttingen (1901) and Potsdam (1909) Observatories. He worked in photometry, geometrical optics, stellar statistics and theoretical astrophysics. (H 82)

Schwarzschild is also honored by a lunar crater. The Schwarzschild black hole, the Schwarzschild radius and the Schwarzschild telescope bear his name.

(838) Seraphina

Discovered 1916 September 24 by M. Wolf at Heidelberg.
Any reference of this name to a person or occurence is unknown.

(839) Valborg

Discovered 1916 September 24 by M. Wolf at Heidelberg.

Named for the heroine in the drama *Axel e Valborg* of the Danish writer Adam Gottlob Oehlenschläger (1779-1850). (H 82)

(840) Zenobia

Discovered 1916 September 25 by M. Wolf at Heidelberg.

The planet is probably named for the Slavic goddess of hunting. It is also the name of a queen of Palmyra. Zenobia enlarged her country with Egypt and parts of Arabia and Asia Minor. She was defeated by Marcus Aurelius in 272 near Antiochia. (H 83; I. van Houten-Groeneveld)

(841) Arabella

Discovered 1916 October 1 by M. Wolf at Heidelberg.

Named after the opera of the same name by the German composer Richard Strauss (1864-1949) {see planet (5039)}. (H 83)

(842) Kerstin

Discovered 1916 October 1 by M. Wolf at Heidelberg.
Any reference of this name to a person or occurence is unknown.

(843) Nicolaia

Discovered 1916 September 30 by H. Thiele at Bergedorf.

Named in honor of Torvald Nicolai Thiele (1838-1910), the father of the discoverer. (H 83)

(844) Leontina

Discovered 1916 October 1 by J. Rheden at Vienna.

Named by the discoverer for his home town Lienz in the province of Tirol, Austria. (AN 204, 63 (1917))

(845) Naëma

Discovered 1916 November 16 by M. Wolf at Heidelberg.
Any reference of this name to a person or occurence is unknown.

(846) Lipperta
Discovered 1916 November 26 by K. Gyllenberg at Bergedorf.

Named in honor of Eduard Lippert, donor of the Lippert Astrograph of the Bergedorf Observatory. (H 83)

(847) Agnia
Discovered 1915 September 2 by G. N. Neujmin at Simeïs.

Named in honor of a Simeïs physician, Agnia Ivanovna Bad'ina (1877-1956). (N. S. Samojlova-Yakhontova; N. Solovaya; N. S. Chernykh)

(848) Inna
Discovered 1915 September 5 by G. N. Neujmin at Simeïs. Independently discovered 1915 September 9 by M. Wolf at Heidelberg.

Named in honor of Inna Nikolaevna Leman-Balanovskaya (1881-1945), astronomer at the Pulkovo Observatory. (N. S. Chernykh)

(849) Ara
Discovered 1912 February 9 by S. I. Belyavskij at Simeïs.

Named in appreciation of the help rendered by ARA (American Relief Administration) under Herbert Hoover during the famine of 1922-1923. (H 83)

(850) Altona
Discovered 1916 March 27 by S. I. Belyavskij at Simeïs.

Named for the German city on the Elbe river near Hamburg. Altona is the home-town of W. Struve, founder of the Pulkovo Observatory. (H 83)

(851) Zeissia
Discovered 1916 April 2 by S. I. Belyavskij at Simeïs.

Named in honor of the famous German optician and mechanician Carl Zeiss (1816-1888) who founded 1846 the Zeiss optical works in Jena. (H 83)

(852) Wladilena
Discovered 1916 April 2 by S. I. Belyavskij at Simeïs.

Named in honor of the Russian Communist leader Vladimir Ilyich Ulyanov (Lenin) (1870-1924). The name comprises the first syllables of Lenin's name. (H 84)

(853) Nansenia
Discovered 1916 April 2 by S. I. Belyavskij at Simeïs.

Named in honor of the Norwegian Arctic explorer Fridtjov Nansen (1861-1930). With his ship "Fram" he explored the vicinity of the North Pole (1893-1896). Nansen is also known as statesman who, in 1922, received the Nobel Peace Prize. (H 84)

Nansen is also honored by craters on the Moon and on Mars.

(854) Frostia
Discovered 1916 April 3 by S. I. Belyavskij at Simeïs.

Named in honor of the American astronomer Edwin Brant Frost (1866-1935) in gratitude for his generous help during the difficult years after World War I (see also planet (849)). Frost was professor of astrophysics at the University of Chicago and director of the Yerkes Observatory. He also edited the *Astrophysical Journal.* (H 84)

Frost is also honored by a lunar crater.

(855) Newcombia
Discovered 1916 April 3 by S. I. Belyavskij at Simeïs. Independently discovered 1916 April 28 by M. Wolf at Heidelberg.

Named in honor of the American astronomer Simon Newcomb (1835-1909), professor of astronomy and director of the U.S. Nautical Almanac Office. Newcomb

worked on cometary and planetary orbits and on the theory of the orbit of the Earth. He measured the velocity of light and determined the astronomical unit anew. (H 84)
Newcomb is also honored by craters on the Moon and on Mars, respectively.

(856) Backlunda

Discovered 1916 April 3 by S. I. Belyavskij at Simeïs.

Named in honor of the Russian astronomer Johann Oskar Backlund (1846-1916). Backlund, who was born in Sweden, worked on the orbits of comets and especially on comet Encke. (H 84)
Backlund is also honored by a lunar crater.

(857) Glasenappia

Discovered 1916 April 6 by S. I. Belyavskij at Simeïs.

Named in honor of Sergej Pavlov Glasenapp (1848-1937), director of the Observatories in Pulkovo (1870-1877) and St. Petersburg (1893). Glasenapp worked on visual binaries and on the satellites of Jupiter. He observed the transits of Venus and Mercury and several solar eclipses. Glasenapp was a founder of the Russian Astronomical Society. (H 84)
Glasenapp is also honored by a lunar crater.

(858) El Djezaïr

Discovered 1916 May 26 by F. Sy at Algiers.

This is the Arabian name for the city of Algiers which means "the islands". (H 84)
This is the first minor planet whose name is composed of two words.

(859) Bouzaréah

Discovered 1916 October 2 by F. Sy at Algiers.

This planet was named after the locality where the Algiers Observatory was erected. (H 84)

(860) Ursina

Discovered 1917 January 22 by M. Wolf at Heidelberg.
Any reference of this name to a person or occurence is unknown.

(861) Aïda

Discovered 1917 January 22 by M. Wolf at Heidelberg.

Named for the opera of the same name by the Italian composer Giuseppe Verdi (1813-1901) {see planet (3975)}. (H 84)

(862) Franzia

Discovered 1917 January 28 by M. Wolf at Heidelberg.
Named in honor of Franz Wolf, son of the discoverer. (H 84)

(863) Benkoela

Discovered 1917 February 9 by M. Wolf at Heidelberg.
Named possibly for the city of Benkoelen on the island of Sumatra. (H 84)

(864) Aase

A921 SB. Discovered 1921 September 30 by K. Reinmuth at Heidelberg.

Named for a character in the play *Peer Gynt* of the Norwegian poet and dramatist Henrik Ibsen (1828-1906) {see planet (5696)}. (LDS)

The French astronomer A. Patry, Nice, detected in 1958 the identity between the numbered planets (864) Aase and (1078) Mentha (MPC 1763). (864) has never been re-observed after the discovery opposition of 1917 until the identification with (1078). Therefore, the Minor Planet Center retained the number and name for (1078) Mentha and vacated the number (864). This number was later assigned to the Reinmuth discovery.

(865) Zubaida

Discovered 1917 February 15 by M. Wolf at Heidelberg.

Named after a character from *Abu Hassan* by the German composer Carl Maria Friedrich Ernst von Weber (1786-1826) {see planet (4152)}. (H 85)

(866) Fatme

Discovered 1917 February 25 by M. Wolf at Heidelberg.

Named after a character from *Abu Hassan* by the German composer Carl Maria Friedrich Ernst von Weber (1786-1826) {see planet (4152)}. (H 85)

(867) Kovacia

Discovered 1917 February 25 by J. Palisa at Vienna.

Named in honor of Prof. Friedrich Kovacs (1861-1931), a Vienna physician who restored Mrs. Palisa's health. (H 85)

(868) Lova

Discovered 1917 April 26 by M. Wolf at Heidelberg.

Any reference of this name to a person or occurence is unknown.

(869) Mellena

Discovered 1917 May 9 by R. Schorr at Bergedorf. Independently discovered 1917 May 10 by M. Wolf at Heidelberg.

Named in honor of the Bürgermeister (mayor) of Hamburg, von Melle, founder of the Hamburg Observatory. (H 85)

(870) Manto

Discovered 1917 May 12 by M. Wolf at Heidelberg.

Named for the famous soothsayer and daughter of Teresias Thebanus. She erected Apollo's {see planet (1862)} oracle in Delphi. (H 85)

(871) Amneris

Discovered 1917 May 14 by M. Wolf at Heidelberg.

Named for a character from the opera *Aida* {see planet (861)} by the Italian composer Giuseppe Verdi (1813-1901) {see planet (3975)}. (H 85)

(872) Holda

Discovered 1917 May 21 by M. Wolf at Heidelberg.

Named presumably in honor of Edward Singleton Holden (1846-1914), American astronomer at Lick Observatory, which Wolf had visited before World War I. (I. van Houten-Groeneveld)

Holden, definitively, is honored by planet (2974).

(873) Mechthild

Discovered 1917 May 21 by M. Wolf at Heidelberg.

Any reference of this name to a person or occurence is unknown.

(874) Rotraut

Discovered 1917 May 25 by M. Wolf at Heidelberg.

Named probably after the poem *Schön Rotraut* by the German poet Eduard Möricke (1804-1875). (I. van Houten-Groeneveld)

(875) Nymphe

Discovered 1917 May 19 by M. Wolf at Heidelberg.

Named for the minor divinities of nature in ancient mythology represented as beautiful maidens dwelling in the mountains, forests, trees and waters. (H 85)

(876) Scott

Discovered 1917 June 20 by J. Palisa at Vienna.

This name had been attributed erroneously to the British polar explorer Robert Fulcon Scott in earlier publications. The real motive is published in the (Vienna) Astronomischer Kalender für 1923: "Herr Palisa hat dem von ihm entdeckten Planeten den Namen Scott zur dankbaren Erinnerung an Miss E. Scott gegeben, welche lange Zeit hindurch in selbstlosester Weise die Hilfeleistung der Society of Friends in Wien geleitet und sich ganz besonders um die Verbesserung der Lage von Mitgliedern der Hochschulen Österreichs bemüht hat." The name therefore honors Miss E. Scott in recognition of her help and assistance to the members of the Austrian universities after World War I. (A. Schnell)

(877) Walküre

Discovered 1915 September 13 by G. N. Neujmin at Simeïs.

he heroes to be slain in battle and conduct them to Valhalla {see planet (1260)}. It is also the title of an opera by the German composer Richard Wagner (1813-1883) {see planet (3992)} in *Ring der Nibelungen*. (LDS)

Named (AN 219, 401 (1923)) by S. I. Belyavskij.

(878) Mildred

Discovered 1916 September 6 by S. B. Nicholson at Mount Wilson.

Named in honor of Mildred Shapley Matthews {1915- }, the daughter of Harlow Shapley {see planet (1123)}. (H 85)

(879) Ricarda

Discovered 1917 July 22 by M. Wolf at Heidelberg.

Named in honor of the German poetess Ricarda Huch (1864-1947). (H 85)

(880) Herba

Discovered 1917 July 22 by M. Wolf at Heidelberg.

Named for the Greek god of misery and poverty. (H 86)

(881) Athene

Discovered 1917 July 22 by M. Wolf at Heidelberg.

Named for the Greek goddess of wisdom, comparable to the Roman Minerva {see planet (93)}. Athene is often called Pallas Athene. (H 86)

(882) Swetlana

Discovered 1917 August 15 by G. N. Neujmin at Simeïs. Independently discovered 1917 August 18 by M. Wolf at Heidelberg.

Any reference of this name to a person or occurence is unknown.

(883) Matterania

Discovered 1917 September 14 by M. Wolf at Heidelberg. Independently discovered 1917 September 14 by R. Schorr at Bergedorf.

Named for the producer of photographic plates who donated many photographic plates for the Heidelberg Observatory. Max Wolf in June 1921 wrote: "In Anerkennung der Verdienste, die sich Herr August Matter in Mannheim um unsere wissenschaftliche Arbeit erworben hat, dadurch dass er jahrelang seine Kraft und Geschicklichkeit an die Herstellung unserer Aufnahmeplatten setzte, sodass dadurch zahlreiche Entdeckungen und Grundlagen für die künftige Forschung ermöglicht worden sind, geben wir dem Planeten (883) 1917 CP, entdeckt am 14. Sept. 1917 mit Matterplatten, den Namen Matterania zur dauernden Erinnerung an unseren Wohltäter, dem wir dadurch unseren Dank ausdrücken wollen." Matter's factory was destroyed in World War II. (LDS; AN 214, 69 (1921))

(884) Priamus

Discovered 1917 September 22 by M. Wolf at Heidelberg.

Named for the king of Troy during the Trojan War, father of Hector and Paris {see planets (624) and (3317), respectively}. (H 86)

(885) Ulrike

Discovered 1917 September 23 by S. I. Belyavskij at Simeïs. Independently discovered 1917 September 26 by M. Wolf at Heidelberg.

Named presumably in honor of Ulrike von Levetzow (1804-1899). The eminent German poet Johann Wolfgang von Goethe {see planet (3047)} met Ulrike during summertime 1821-1823 in Marienbad, Bohemia. Goethe proposed to her but she denied his offer of marriage. His grief found an expression in the famous *Marienbader Elegie* (1823). (A. N. Deutsch; LDS)

This interpretation seems very likely in view of the fact that the discoverer studied in Göttingen, Germany, and was an expert in German literature.

(886) Washingtonia

Discovered 1917 November 16 by G. H. Peters at Washington. Independently discovered 1917 November 12 by M. Harwood at Nantucket.

Named in honor of George Washington (1732-1799), the first president (1789-1797) of the United States of America. (H 86)

For a more detailed description of the discovery circumstances see the the citations for planets (7040) and (7041), respectively.

(887) Alinda

Discovered 1918 January 3 by M. Wolf at Heidelberg.

According to Australian aboriginal mythology, Alinda is the man in the moon. He drowned his two sons after they returned from fishing and lied that they had caught nothing. His two wives then burned him to death. Alinda proclaimed that although death would henceforth be the fate of everyone else, except for three days each month he would live forever. (From John Gunther's *Inside Australia*, Harper & Row, New York, 1971, p. 81). R. Bremer in an alternative explanation suggested that Alinda is a name of an ancient city and recent archaeological site in Asia Minor in the province of ancient Caria. (B. G. Marsden)
Name proposed by H. Kobold (AN 250, 145 (1933)).

(888) Parysatis

Discovered 1918 February 2 by M. Wolf at Heidelberg.

Named after the wife of the Persian king Darius II, mother of Artaxerxes II {see planet (831)} and Cyrus the Younger. (H 86)

(889) Erynia

Discovered 1918 March 5 by M. Wolf at Heidelberg.

Named after one of the avenging deities who according to Greek mythology tormented criminals and inflicted plagues. In Roman mythology they are called furies. (H 86)

(890) Waltraut

Discovered 1918 March 11 by M. Wolf at Heidelberg.

Named after a character in the opera *Götterdämmerung* (the fourth and last part of *Der Ring der Nibelungen*) of the German composer Richard Wagner (1813-1883) {see planet (3992)}. (H 86)

(891) Gunhild

Discovered 1918 May 17 by M. Wolf at Heidelberg.
Any reference of this name to a person or occurence is unknown.

(892) Seeligeria

Discovered 1918 May 31 by M. Wolf at Heidelberg.

Named in honor of the German astronomer Hugo Hans von Seeliger (1849-1924), director of the Munich Observatory. Von Seeliger investigated the movement of the perihelion of Mercury and was one of the founders of stellar astronomy. He published books on the motion of binaries and on the theory of the heliometer. (H 86)

Seeliger is also honored by a lunar crater.

(893) Leopoldina

Discovered 1918 May 31 by M. Wolf at Heidelberg.

Named in honor of the "Carolinisch-Leopoldinische Akademie der Naturforscher in Halle", Germany. (H 86)

(894) Erda

Discovered 1918 June 4 by M. Wolf at Heidelberg.

Named for the Norse goddess, incarnation of the nature, a seer who knows the origin and the destination of all things. (H 87)

(895) Helio

Discovered 1918 July 11 by M. Wolf at Heidelberg.

Named by the German physicist Friedrich Paschen (1865-1940) in remembrance of common investigations about the helium spectrum together with Runge. (H 87)

(896) Sphinx

Discovered 1918 August 1 by M. Wolf at Heidelberg.

Named for a female monster having, according to Greek mythology, a winged lion's body and a human head and having the habit of asking a riddle and killing anyone who failed to answer it. (H 87)

(897) Lysistrata

Discovered 1918 August 3 by M. Wolf at Heidelberg.

Named after the comedy of the Athenian dramatist Aristophanes (448?-380? B.C.) {see planet (2934)}. Lysistrata organized a general strike of women in order to prevent a war. (H 87)

(898) Hildegard

Discovered 1918 August 3 by M. Wolf at Heidelberg.

Named probably after the St. Hildegard von Bingen (1098-1179). (H 87)

(899) Jokaste

Discovered 1918 August 3 by M. Wolf at Heidelberg.

Named for the mother of Oedipus. She married her son without recognizing him, and by him had sons Polynice and Eteocles. (H 87)

(900) Rosalinde

Discovered 1918 August 10 by M. Wolf at Heidelberg.

Named probably for a figure in the opera *Die Fledermaus* by Johann Strauss (1825-1899) {see planet (4559)}. (I. van Houten-Groeneveld)

See also the citation for planet (812).

(901) Brunsia

Discovered 1918 August 30 by M. Wolf at Heidelberg.

Named in honor of Ernst Heinrich Bruns (1848-1919), professor of astronomy and director of the Leipzig Observatory. (H 87; AN 214, 69 (1921))

(902) Probitas

Discovered 1918 September 3 by J. Palisa at Vienna.

This name stands for the adherence to the highest principles and ideals. This has been assigned in honor of the discoverer after his death because uprightness was considered to be one of the essential traits of his character. (H 87)

Named by the Austrian astronomer J. Rheden in agreement with Anna Palisa, the second wife of the discoverer (AN 240, 303 (1930)).

The planets (975) and (996) have been named in a similar way after the death of Palisa.

(903) Nealley

Discovered 1918 September 13 by J. Palisa at Vienna.

Named in honor of the New York amateur astronomer Nealley who supported the edition of the Wolf-Palisa photographic star charts. (Beobachtungszirk. AN 3, 61 (1921))

(904) Rockefellia

Discovered 1918 October 29 by M. Wolf at Heidelberg.

Named in honor of John David Rockefeller (1839-1937), founder of the Rockefeller Foundation. (H 87)

(905) Universitas

Discovered 1918 October 30 by A. Schwassmann at Bergedorf.

Named by the discoverer in memory of the founding of the University of Hamburg, Germany. (H 88)

(906) Repsolda

Discovered 1918 October 30 by A. Schwassmann at Bergedorf.

Named in honor of Johann Georg Repsold (1771-1830), famous manufacturer of optics and precision mechanics for astrometric instruments. (H 88)
Repsold is also honored by a lunar crater.

(907) Rhoda

Discovered 1918 November 12 by M. Wolf at Heidelberg.

Named in honor of the wife of the American astronomer Edward Emerson Barnard {see planet (819)}. (H 88; AN 215, 471 (1922))

(908) Buda

Discovered 1918 November 30 by M. Wolf at Heidelberg.
Named for a part of the Hungarian city of Budapest on the Danube. (H 88)

(909) Ulla

Discovered 1919 February 7 by K. Reinmuth at Heidelberg.

Named in honor of Ulla Ahrens, a daughter of a friend of the discoverer. See also the citation for planet (950). (H 88)

(910) Anneliese

Discovered 1919 March 1 by K. Reinmuth at Heidelberg.

Named for an acqaintançe of the German astronomer Julius Dick from the Babelsberg Observatory. (H 88)
Named by J. Dick.

(911) Agamemnon

Discovered 1919 March 19 by K. Reinmuth at Heidelberg.

Named for the king of Mycenae who was the leader of the Greeks in the Trojan War. (H 88)

(912) Maritima

Discovered 1919 April 27 by A. Schwassmann at Bergedorf.

Probably named during the first marine travel of Hamburg University. Before World War II, once a year the University closed during the summer semester and went out on the North Sea for three to four days on one of the larger ocean liners which were put at the disposal of the University by the big Hamburg shipping companies. (H 88)

Named by R. Schorr.

See also the citation for planet (947).

(913) Otila

Discovered 1919 May 19 by K. Reinmuth at Heidelberg.

Karl Reinmuth discovered so many minor planets that he often had difficulties to find proper names for the numbered ones. He therefore frequently used girl's names from the calendar *Der Lahrer hinkende Bote*. These names need not to have any connections to his contemporaries. (I. van Houten-Groeneveld)

(914) Palisana

Discovered 1919 July 4 by M. Wolf at Heidelberg.

Named in honor of the famous Austrian astronomer Johann Palisa (1848-1925), observer at the Marine Observatory in Pola and vice-director of the Vienna Observatory. Palisa discovered more than 120 minor planets and the comet 1879 V. In collaboration with M. Wolf he produced the famous Wolf-Palisa star charts and he initiated the Carte du Ciel endeavor. (H 88)

Palisa is also honored by a lunar crater and by the planets (902), (975), and (996). See also the remarks to planet (776).

(915) Cosette

Discovered 1918 December 14 by F. Gonnessiat at Algiers.

Named in honor of the youngest daughter of the discoverer. (H 88)

(916) America

Discovered 1915 August 7 by G. N. Neujmin at Simeïs.

Named for the American continent possibly in appreciation of the help rendered by the American Relief Administration under President Hoover during the famine in Crimea. (H 89)

Named (AN 219, 59 (1923)) by the Pulkovo astronomers.

(917) Lyka

Discovered 1915 September 5 by G. N. Neujmin at Simeïs.

Named after a girl-friend of the sister of the discoverer. (N. S. Chernykh)

(918) Itha

Discovered 1919 August 22 by K. Reinmuth at Heidelberg.

See the remarks to planet (913).

(919) Ilsebill

Discovered 1918 October 30 by M. Wolf at Heidelberg.

Named for the famous character appearing in the German fairy tale *Von den Fischer un sijne Fru* (The Fisherman and His Wife) by J. and W. Grimm. (H 89)

Name proposed by Mrs. G. Wolf (RI 1013).

(920) Rogeria

Discovered 1919 September 1 by K. Reinmuth at Heidelberg.

See the remarks to planet (913).

(921) Jovita

Discovered 1919 September 4 by K. Reinmuth at Heidelberg.
See the remarks to planet (913).

(922) Schlutia

Discovered 1919 September 18 by K. Reinmuth at Heidelberg.

Named in honor of the important businessmen Edgar Schlubach (Hamburg) and Mr. Tiarks (London) who together supported the Dutch-German solar eclipse expedition to Christmas Island in 1922. (H 89)
Named (AN 218, 253 (1923)) by Schlubach and Tiarks.

(923) Herluga

Discovered 1919 September 30 by K. Reinmuth at Heidelberg.
See the remarks to planet (913).

(924) Toni

Discovered 1919 October 20 by K. Reinmuth at Heidelberg.
See the remarks to planet (913).

(925) Alphonsina

Discovered 1920 January 13 by J. Comas Solá at Barcelona.

Named in honor of two kings of Spain: Alphonsus X (1223-1284), King of Castille, and Alphonsus XIII (1886-1941), King of Spain. The original citation reads: "La petite planète ... a reçu le nom de Alphonsina, comme double hommage au roi Alphonse X, dit le Savant, roi de Castille et de Léon (siècle XIII), et à qui on doit de si importantes inspirations sur l'astronomie pendant le moyen âge, et au roi Alphonse XIII, grand enthousiaste pour le développement des Sciences dans son pays. Le nom de Alphonsina a été accepté avec plaisir par S.M. le roi d'Espagne." (AN 211, 223 (1920))

(926) Imhilde

Discovered 1920 February 15 by K. Reinmuth at Heidelberg.
See the remarks to planet (913).

(927) Ratisbona

Discovered 1920 February 16 by M. Wolf at Heidelberg.

Named for the German city of Regensburg (Ratisbona is the Latin name). The famous astronomer Johannes Kepler {see planet (1134)} died in Regensburg in 1630. (AN 240, 135 (1930))

(928) Hildrun

Discovered 1920 February 23 by K. Reinmuth at Heidelberg.
See the remarks to planet (913).

(929) Algunde

Discovered 1920 March 10 by K. Reinmuth at Heidelberg.
See the remarks to planet (913).

(930) Westphalia

Discovered 1920 March 10 by W. Baade at Bergedorf.

Named by the discoverer for the region Westfalen in western Germany bordering the Netherlands and the Rhine river. Baade was born in Schröttinghausen in Westfalen. (H 90)

(931) Whittemora

Discovered 1920 March 19 by F. Gonnessiat at Algiers. Independently discovered 1920 March 21 by K. Reinmuth at Heidelberg.

Named in honor of Thomas Whittemore who was professor at both Harvard and Columbia Universities. (H 90)

(932) Hooveria

Discovered 1920 March 23 by J. Palisa at Vienna.

Named in honor of Herbert Clark Hoover (1874-1964) in recognition of his help and assistance to Austria after World War I. (AN 216, 192 (1922))
Named by the Academic Senate of the Vienna University.
Hoover, later president of the United States, was also honored by planet (1363).

(933) Susi

1927 CH. Discovered 1927 February 10 by K. Reinmuth at Heidelberg.

Named in honor of Mrs. S. Graff, wife of the director of the Vienna Observatory. (H 90)

(934) Thüringia

Discovered 1920 August 15 by W. Baade at Bergedorf.

Thüringia was one of the ships of the Hamburg-America Line. The discoverer made two trips from Hamburg to New York on this ship during the 1920's. The captain of the ship was an amateur astronomer and was invited by the discoverer to name one of his minor planets. (H 90)
The planet and the ship are named for the German province of Thüringen.

(935) Clivia

Discovered 1920 September 7 by K. Reinmuth at Heidelberg.
Named after a flower of the Amaryllis family. (H 90)

(936) Kunigunde

Discovered 1920 September 6 by K. Reinmuth at Heidelberg.
See the remarks to planet (913).

(937) Bethgea

Discovered 1920 September 12 by K. Reinmuth at Heidelberg.
Named in honor of the German lyrical poet Hans Bethge (1876-1946). (H 90)

(938) Chlosinde

Discovered 1920 September 9 by K. Reinmuth at Heidelberg.
See the remarks to planet (913).

(939) Isberga

Discovered 1920 October 4 by K. Reinmuth at Heidelberg.
See the remarks to planet (913).

(940) Kordula

Discovered 1920 October 10 by K. Reinmuth at Heidelberg.
See the remarks to planet (913).

(941) Murray

Discovered 1920 October 10 by J. Palisa at Vienna.

Named in honor of the family of the American professor Gilbert Murray (1866-1957) who helped Austria after World War I in 1920. (H 91)

(942) Romilda

Discovered 1920 October 11 by K. Reinmuth at Heidelberg. Independently discovered 1920 October 11 by A. Schwassmann at Bergedorf.
See the remarks to planet (913).

(943) Begonia
Discovered 1920 October 20 by K. Reinmuth at Heidelberg.

Named after a genus of tropical herbs having asymmetrical leaves and being widely cultivated as ornamentals. (H 91)

(944) Hidalgo
Discovered 1920 October 31 by W. Baade at Bergedorf.

German astronomers observed the total solar eclipse 1923 September 10 in Mexico. After the eclipse they had an audience with the president of Mexico and asked permission to call this planet after Miguel Hidalgo y Costilla (1753-1811) who proclaimed the Mexican independence in 1810. R. Schorr wrote: "Zur Erinnerung an die Deutsche Sonnenfinster-Expedition nach Mexiko im Sommer 1923 und die ihr durch die mexikanische Regierung erwiesene gastliche Aufnahme ist unter der Zustimmung des Staatspräsidenten Don Alvaro Obregón der ... am Bergedorfer Spiegelteleskop entdeckte, durch seine Bahn besonders interessante Planet ... nach dem mexikanischen Nationalhelden Hidalgo benannt worden." (AN 221, 159 (1924))

(945) Barcelona
Discovered 1921 February 3 by J. Comas Solá at Barcelona.

Named for the birthplace of the discoverer and the place were this minor planet was discovered. (H 91)

(946) Poësia
Discovered 1921 February 11 by M. Wolf at Heidelberg.
Named after the goddess of poetry. (H 91)
Name proposed by N. Komendantov.

(947) Monterosa
Discovered 1921 February 8 by A. Schwassmann at Bergedorf.

Probably named for the ship "Monterosa" on the occasion of a marine travel by the Hamburg University to the North Sea. (RI 929; H 91)
Named by R. Schorr.
See also the citation for planet (912).

(948) Jucunda
Discovered 1921 March 3 by K. Reinmuth at Heidelberg.
See the remarks to planet (913).

(949) Hel
Discovered 1921 March 11 by M. Wolf at Heidelberg.

Named for the Norse goddess of the dead and the queen of the underworld. (LDS)

Name proposed by Mrs. G. Wolf, the widow of the discoverer, two years after his death (RI 1013).

(950) Ahrensa
Discovered 1921 April 1 by K. Reinmuth at Heidelberg.

Named in honor of the Ahrens family, friends of the discoverer who helped financially at the Königstuhl Observatory. See also the citation for planet (909). (H 91)

(951) Gaspra
Discovered 1916 July 30 by G. N. Neujmin at Simeïs.

Named after the resort on the southern shore of Crimea in which the famous Russian writer Lev Nikolaevich Tolstoj (1828-1910) spent many years of his life. (H 91)

This planet was the first space mission asteroid target with the Galileo spacecraft. Galileo submitted the first detailed photograph of a minor planet from the flyby of 1991 October 29.

(952) Caia

Discovered 1916 October 27 by G. N. Neujmin at Simeïs.

Named after a heroine in the novel *Quo Vadis* of the Polish writer Henryk Sienkiewicz (1846-1916). (N. S. Samojlova-Yakhontova)

(953) Painleva

Discovered 1921 April 29 by B. Jekhovsky at Algiers.

Named in honor of Paul Painlevé (1863-1933), French mathematician and several times French minister (1917-1933). His scientific works refer to analysis and aerodynamics. (H 92; M.-A. Combes)

(954) Li

Discovered 1921 August 4 by K. Reinmuth at Heidelberg.

Named in honor of Mrs. Lina Alstede Reinmuth, wife of the discoverer. (H 92)
Mrs. Reinmuth is also honored by planet (955).

(955) Alstede

Discovered 1921 August 5 by K. Reinmuth at Heidelberg.

Named in honor of Mrs. Lina Alstede Reinmuth, wife of the discoverer. (H 92)
Mrs. Reinmuth is also honored by planet (954).

(956) Elisa

Discovered 1921 August 8 by K. Reinmuth at Heidelberg.

Named in honor of Elisa Reinmuth, mother of the discoverer. (H 92)

(957) Camelia

Discovered 1921 September 7 by K. Reinmuth at Heidelberg.

Named for a genus of shrubs or trees of the tea family, especially a ornamental greenhouse shrub with glossy evergreen leaves and showy roselike flowers. (H 92)

(958) Asplinda

Discovered 1921 September 28 by K. Reinmuth at Heidelberg.

Named in honor of Bror Ansgar Asplind {1890-1954}, the Swedish computer of planetary orbits. (H 92)

(959) Arne

Discovered 1921 September 30 by K. Reinmuth at Heidelberg.

Named in honor of a son of B. Asplind {see planet (958)}. (H 92)

(960) Birgit

Discovered 1921 October 1 by K. Reinmuth at Heidelberg.

Named in honor of a daughter of B. Asplind {see planet (958)}. (H 92)

(961) Gunnie

Discovered 1921 October 10 by K. Reinmuth at Heidelberg.

Named in honor of a daughter of B. Asplind {see planet (958)}. (H 92)

(962) Aslög

Discovered 1921 October 25 by K. Reinmuth at Heidelberg.

This is a Christian name from Norse mythology. Aslög followed her foster-father hidden in his harp. Her fiancé Ragnar wanted to test Aslög's intelligence and asked her to come to him: naked but dressed. She used a fishnet as dress and so she was accepted as his wife. (H 92)
Name proposed by B. Asplind.

(963) Iduberga

Discovered 1921 October 26 by K. Reinmuth at Heidelberg.
See the remarks to planet (913).

(964) Subamara

Discovered 1921 October 27 by J. Palisa at Vienna.

Amarus is the Latin word for bitter and the addition of the prefix sub yield the meaning 'very bitter'. This name very probably reflects the bad observing situation from the Vienna Observatory during the last years of Palisa's life. This interpretation is supported by his paper in AN 222, 172 (1924). (A. Schnell)

(965) Angelica

Discovered 1921 November 4 by J. Hartmann at La Plata.
Named in honor of Angelica Hartmann, wife of the discoverer. (H 92)

(966) Muschi

Discovered 1921 November 9 by W. Baade at Bergedorf.
Named in honor of the wife of the discoverer. Muschi is a nickname. (H 93)

(967) Helionape

Discovered 1921 November 9 by W. Baade at Bergedorf.

Named in honor of the theatrical artist Sonnenthal. Helionape is the Greek translation (Sonne = helio, Tal = nape). Adolf Ritter von Sonnenthal (1834-1909), famous actor at the Vienna Burgtheater and later director of this institution, was known world-wide by many guest performances in Europe and America. (AN 223, 247 (1925))
Named by C. Wirtz.

(968) Petunia

Discovered 1921 November 24 by K. Reinmuth at Heidelberg.

Named for a genus of tropical American herbs of the nightshade family with funnel-shaped corollas. (H 93)

(969) Leocadia

Discovered 1921 November 5 by S. I. Belyavskij at Simeïs. Independently discovered 1921 November 25 by K. Reinmuth at Heidelberg.
Any reference of this name to a person or occurence is unknown.

(970) Primula

Discovered 1921 November 29 by K. Reinmuth at Heidelberg.

Named for a genus of perennial herbs with large tufted basal leaves and showy variously colored flowers. The English name is primrose. (H 93)

(971) Alsatia

Discovered 1921 November 23 by A. Schaumasse at Nice.

Named after the province Alsace in northeast France between the Rhine river and the Vosges mountains. (H 93)

The discoverer proposed the name Alsace (J. Obs., Tome V, p. 6, 100 (1922)). The Astronomisches Rechen-Institut changed the name in Alsatia.

(972) Cohnia

Discovered 1922 January 18 by M. Wolf at Heidelberg.

Named in honor of the German astronomer F. Cohn (1866-1921) who was director of the Astronomisches Rechen-Institut in Berlin. (H 93)

Named by M. Wolf and G. Stracke (AN 219, 401 (1923)) who wrote: "Der Planet 972 hat zum Andenken an den verstorbenen Direktor des Astronomischen Rechen-Instituts zu Berlin-Dahlem, Professor F. Cohn, den Namen erhalten. Es ist dies nur ein schwacher Ausdruck für die großen Verdienste, die sich Cohn gerade auf dem Gebiete der Kleinen Planeten erworben hat."

(973) Aralia

Discovered 1922 March 18 by K. Reinmuth at Heidelberg.

Named after a widely cultivated ornamental climbing or prostrate or sometimes shrubby Eurasian vine of the ginseng family with evergreen leaves, small yellowish flowers, and black berries. (H 93)

(974) Lioba

Discovered 1922 March 18 by K. Reinmuth at Heidelberg.

Named after the Saint Lioba, abbess in Tauberbischofsheim, Germany. She was buried at the side of Bonifatius in 782 in Fulda, Germany. (I. van Houten-Groeneveld)

(975) Perseverantia

Discovered 1922 March 27 by J. Palisa at Vienna.

Named in honor of the discoverer after his death (1925). Perseverance was an essential trait of Palisa's character. (H 93)

Named by the Austrian astronomer J. Rheden in agreement with Anna Palisa, the second wife of the discoverer (AN 240, 303 (1930)).

The planets (902) and (996) have been named in a similar way after the death of Palisa.

(976) Benjamina

Discovered 1922 March 27 by B. Jekhovsky at Algiers.
Named in honor of the discoverer's son Benjamin. (H 93)

(977) Philippa

Discovered 1922 April 6 by B. Jekhovsky at Algiers.
Named in honor of the financier Philipp von Rothschild. (H 94)

(978) Aidamina

Discovered 1922 May 18 by S. I. Belyavskij at Simeïs. Independently discovered 1922 May 30 by M. Wolf at Heidelberg.

Named in honor of Aida Minaevna, friend of the discoverer's family. (N. S. Samojlova-Yakhontova; N. Solovaya; N. S. Chernykh)

(979) Ilsewa

Discovered 1922 June 29 by K. Reinmuth at Heidelberg.
Named in honor of Ilse Walldorf, an acquaintance of the discoverer. (H 94)

(980) Anacostia

Discovered 1921 November 21 by G. H. Peters at Washington.

Named for an urban district in the city of Washington, D.C. and also for a river in the region. (H 94)

(981) Martina

Discovered 1917 September 23 by S. I. Belyavskij at Simeïs.
Named in honor of the French revolutionary Henri Martin. (H 94)

(982) Franklina

Discovered 1922 May 21 by H. E. Wood at Johannesburg.

Named by the discoverer in honor of the late John Franklin Adams (1843-1912), British amateur astronomer and author of stellar maps. (H 94)
Franklin Adams is also honored by planet (1925).

(983) Gunila

Discovered 1922 July 30 by K. Reinmuth at Heidelberg.
See the remarks to planet (913).

(984) Gretia

Discovered 1922 August 27 by K. Reinmuth at Heidelberg.
Named for a sister-in-law of A. Kahrstedt {see planet (1587)}. (H 94)
See also the remarks to planet (1026).

(985) Rosina

Discovered 1922 October 14 by K. Reinmuth at Heidelberg.
See the remarks to planet (913).

(986) Amelia

Discovered 1922 October 19 by J. Comas Solá at Barcelona.
Named in honor of the discoverer's wife. (H 94)

(987) Wallia

Discovered 1922 October 23 by K. Reinmuth at Heidelberg.
See the remarks to planet (913).

(988) Appella

Discovered 1922 November 10 by B. Jekhovsky at Algiers.

Named in honor of Paul Émile Appell (1855-1930), president of the Academy of Sciences and of the Société Astronomique de France. Appell is the author of the magisterial *Traité de Mécanique Rationnelle.* (H 94)
Obituary published in l'Astronomie, Vol. 45, p. 42-43 (1931).

(989) Schwassmannia

Discovered 1922 November 18 by A. Schwassmann at Bergedorf.

Named in honor of the German astronomer Arnold Schwassmann (1870-1964) who discovered four comets and 22 minor planets. Schwassmann worked at the Potsdam and Hamburg-Bergedorf Observatories. (H 94)

(990) Yerkes

Discovered 1922 November 23 by G. Van Biesbroeck at Williams Bay. Independently discovered 1922 December 14 by M. Wolf at Heidelberg.

Named by the discoverer for the observatory where this planet was discovered. It was the first planet, Van Biesbroeck discovered at the Yerkes Observatory. (H 95)
The name Yerkes also appear as a lunar crater.

(991) McDonalda

Discovered 1922 October 24 by O. Struve at Williams Bay.

Named for the observatory erected on Mount Locke, Fort Davis, Texas, whose 82-inch reflector was made up with the legacy of Mr. W. J. McDonald who is also honored by this planet. (H 95)

(992) Swasey

Discovered 1922 November 24 by O. Struve at Williams Bay.

Named in honor of the optician Dr. Swasey of the Warner and Swasey Works who built the 82-inch reflector of the McDonald Observatory. (H 95)

(993) Moultona

Discovered 1923 January 12 by G. Van Biesbroeck at Williams Bay.

Named by the discoverer in honor of the American astronomer and mathematician Forest Ray Moulton (1872-1952) who is wellknown for his work on celestial

mechanics. (H 95)
Moulton is also honored by a lunar crater.

(994) Otthild

Discovered 1923 March 18 by K. Reinmuth at Heidelberg.
See the remarks to planet (913).

(995) Sternberga

Discovered 1923 June 8 by S. I. Belyavskij at Simeïs.

Named in honor of the Russian astronomer Pavel K. Sternberg (1865-1920) from the Moscow University. (H 95)
Sternberg is also honored by a lunar crater.

(996) Hilaritas

Discovered 1923 March 21 by J. Palisa at Vienna.

The name stands for a happy or contented mind. This was an important quality of Palisa's character who is honored by this planet. (H 95)

Named by the Austrian astronomer J. Rheden in agreement with Anna Palisa, the second wife of the discoverer (AN 240, 303 (1930)).

The planets (902) and (975) have been named in a similar way after the death of Palisa.

(997) Priska

Discovered 1923 July 12 by K. Reinmuth at Heidelberg.
See the remarks to planet (913).

(998) Bodea

Discovered 1923 August 6 by K. Reinmuth at Heidelberg.

Named in honor of the German astronomer and mathematician Johann Elert Bode (1747-1826). He was the founder of the *Berliner Astronomisches Jahrbuch* (1774) and director of the Berlin Observatory in 1780. He is also known for the empiric law of planetary distances (1766) which was originally found by J. D. Titius {see planet (1998)}. (H 95)
Name proposed by B. Asplind.
Bode is also honored by a lunar crater.

(999) Zachia

Discovered 1923 August 9 by K. Reinmuth at Heidelberg.

Named in honor of the astronomer and mathematician Franz Xaver Freiherr von Zach (1754-1832). He was the director of the Seeberg Observatory near Gotha, Germany. Von Zach recovered planet (1) Ceres profiting the Gauss computings because the planet was lost in consequence of Piazzi's serious illness. He founded the journal *Monatliche Correspondenz zur Beförderung der Erd- und Himmelskunde* which was the main medium for exchange and improvements in observation and data treatment of the first asteroids. (H 95)
Zach is also honored by a lunar crater.

(1000) Piazzia

Discovered 1923 August 12 by K. Reinmuth at Heidelberg.

Named in honor of Giuseppe Piazzi (1746-1826), Theatin monk and director of the observatories of Palermo and Naples. He discovered (1) Ceres and elaborated a catalog of 7,646 stars. (H 96)
Piazzi is also honored by a lunar crater.

(1001) Gaussia

Discovered 1923 August 8 by S. I. Belyavskij at Simeïs.

Named in honor of Carl Friedrich Gauss (1777-1855), director of the Göttingen

Observatory. With his new computing method F. X. von Zach {see planet (999)} rediscovered (1) Ceres. (H 96)
Named by B. Asplind.
Gauss is also honored by a lunar crater.

(1002) Olbersia

Discovered 1923 August 15 by V. A. Albitskij at Simeïs.
Named in honor of Heinrich Wilhelm Matthias Olbers (1758-1840) who was a physician in Bremen and a keen amateur astronomer. Olbers discovered (2) Pallas, (4) Vesta and six comets and was the author of the first method for computing cometary orbits. (H 96)
Named by B. Asplind.
Olbers is also honored by a lunar crater.

(1003) Lilofee

Discovered 1923 September 13 by K. Reinmuth at Heidelberg.
Named by the discoverer (RI 402) for a legendary character which is the title figure in an old German folk-song *Die schöne junge Lilofee* by August Schnezler. Lilofee was a mermaid of the Mummelsee in the Black Forest. (H 96; R. Bremer)
Name proposed by J. Riem, astronomer at the Astronomisches Rechen-Institut in Berlin.

(1004) Belopolskya

Discovered 1923 September 5 by S. I. Belyavskij at Simeïs. Independently discovered 1923 September 13 by K. Reinmuth at Heidelberg.
Named in honor of A. A. Belopolsky (1854-1934), astrophysicist at the Pulkovo Observatory. (H 96)
Belopolsky is also honored by a lunar crater.

(1005) Arago

Discovered 1923 September 5 by S. I. Belyavskij at Simeïs.
Named in honor of François Arago (1786-1853), since 1843 director of the Paris Observatory, life-secretary of the Academy of Sciences, politician and author of the four volume *Astronomie Populaire* (1854-1857). (H 96)
Arago is also honored by craters on Mars and the Moon.

(1006) Lagrangea

Discovered 1923 September 12 by S. I. Belyavskij at Simeïs.
Named in honor of the astronomer and mathematician Joseph Louis Lagrange (1736-1813). He was professor of mathematics in Turin (1753), Berlin (1766) and Paris (1787) and author of the famous *Mécanique analytique* (1788) and *Théories des fonctions analytiques* (1797). (H 96)
Lagrange is also honored by a lunar crater.

(1007) Pawlowia

Discovered 1923 October 5 by V. A. Albitskij at Simeïs.
Named in honor of the Russian biologist Ivan Petrovich Pavlov (1849-1936), the 1904 Nobel Prize winner for physiology and medicine. (H 96)
Pavlov is also honored by a lunar crater.

(1008) La Paz

Discovered 1923 October 31 by M. Wolf at Heidelberg.
Named for the capital of Bolivia, South America. (H 96)

(1009) Sirene

Discovered 1923 October 31 by K. Reinmuth at Heidelberg.
Named for the sirens in Greek mythology. (H 96)

(1010) Marlene

Discovered 1923 November 12 by K. Reinmuth at Heidelberg.

The name honors the Berlin born motion picture star Maria Magdalena von Losch alias Marlene Dietrich {1901-1992}. (H 97)

Name suggested by G. Stracke.

(1011) Laodamia

Discovered 1924 January 5 by K. Reinmuth at Heidelberg.

This name has two meanings in the Greek mythology: (1) Laodamia, the daughter of Akastos, was the wife of Protesilaos {see planet (3540)}.who was killed in the Trojan War. Her fate is described in a tragedy of Euripides {see planet (2930)}. (2) Laodamia was the daughter of Bellerophon {see planet (1808)} and the wife of Sarpedon {see planet (2223)}. She was killed by the arrows of Artemis {see planet (105)}. (LDS)

Name suggested by N. Komendantov (RI 740).

(1012) Sarema

Discovered 1924 January 12 by K. Reinmuth at Heidelberg.

Named after a character in one of the poems of Aleksandr Pushkin {see planet (2208)}. (H 97)

Name suggested by N. Komendantov (RI 740).

(1013) Tombecka

Discovered 1924 January 17 by B. Jekhovsky at Algiers.

Named in honor of D. Tombeck, secretary of the Faculty of Sciences of Paris. (H 97)

(1014) Semphyra

Discovered 1924 January 29 by K. Reinmuth at Heidelberg.

Named after a character in one of the poems of Aleksandr Pushkin {see planet (2208)}. (H 97)

Name proposed by N. Komendantov (RI 740).

(1015) Christa

Discovered 1924 January 31 by K. Reinmuth at Heidelberg.

Any reference of this name to a person or occurence is unknown.

Name possibly proposed by G. Stracke.

(1016) Anitra

Discovered 1924 January 31 by K. Reinmuth at Heidelberg.

Named probably after the Arabian dancer Anitra in Henrik Ibsen's (1828-1906) {see planet (5696)} drama *Peer Gynt*. The Norwegian Edvard Grieg (1843-1907) {see planet (4872)} composed the *Peer Gynt Suite No. 1* of which one of the pieces is called *Anitra's Dance*. (R. Bremer; I. van Houten-Groeneveld)

(1017) Jacqueline

Discovered 1924 February 4 by B. Jekhovsky at Algiers.

Named in honor of Jacqueline Zadoc-Kahn, an old disciple of the discoverer. (H 97)

(1018) Arnolda

Discovered 1924 March 3 by K. Reinmuth at Heidelberg.

Der Planet hat aus Anlass des 70. Geburtstags des Herausgebers der *Naturwissenschaften*, Dr.Ing. e.h. Dr.phil. Arnold Berliner den Namen "Arnolda" erhalten. (AN 247, 295 (1933))

(1019) Strackea

Discovered 1924 March 3 by K. Reinmuth at Heidelberg.

Named in honor of the German astronomer Gustav Stracke (1887-1943). Stracke for decades until his death was in charge of the minor planet department at the Astronomisches Rechen-Institut in Berlin-Dahlem. (H 97)

The original citation (RI 2501) reads: "Der Planet ... ist vor Jahren von dem Entdecker ... "Strackea" getauft worden. Auf Wunsch von Prof. Stracke ist bisher die Benennung unterblieben. Zum Andenken an den langjährigen Leiter der Arbeiten an den Kleinen Planeten soll nun der Planet 1019 den Namen "Strackea" tragen." Stracke was also honored by a series of planets - see the citation of planet (1234).

(1020) Arcadia

Discovered 1924 March 7 by K. Reinmuth at Heidelberg.

The planet was named for the Greek province on the Peloponnese peninsula. (H 97)

(1021) Flammario

Discovered 1924 March 11 by M. Wolf at Heidelberg.

Named in honor of the French astronomer Camille Flammarion (1842-1925), author of many astronomical books. Flammarion founded the journal *l'Astronomie* (1882) and the "Société Astronomique de France" (1887). (H 98)
Flammarion is also honored by craters on Mars and the Moon.

(1022) Olympiada

Discovered 1924 June 23 by V. A. Albitskij at Simeïs.

Named in honor of Olimpiada Albitskaya, the mother of the discoverer. (N. S. Chernykh)

(1023) Thomana

Discovered 1924 June 25 by K. Reinmuth at Heidelberg.

Named in honor of the famous boys' choir (Thomaner) in the Thomas-Kirche in Leipzig, Germany. Reinmuth stated in a letter: "Auf der Leipziger Astronomentagung erhielt er zu Ehren des Thomanerchores dann den Namen..." (H 98, LDS)

(1024) Hale

Discovered 1923 December 12 by G. Van Biesbroeck at Williams Bay.

Named in honor of George Ellery Hale (1868-1938), founder and first director of Yerkes and Mt. Wilson Observatories and inspirer of the observatory at Mt. Palomar. With J. E. Keeler {see planet (2261)} he founded the *Astrophysical Journal*, and he invented the spectroheliograph. (H 98)
Hale is also honored by craters on Mars and the Moon.

(1025) Riema

Discovered 1923 August 12 by K. Reinmuth at Heidelberg.

Named in honor of Johannes Riem {1868-1945}, a German astronomer at the Astronomisches Rechen-Institut in Berlin-Dahlem. (H 98)
Name proposed by the Astronomisches Rechen-Institut (RI 762).

(1026) Ingrid

Discovered 1923 August 13 by K. Reinmuth at Heidelberg.

Named after a niece of A. Kahrstedt {see planet (1587)} as a present on the day of baptism. (H 98; LDS)

The naming of this planet and that of (984) Gretia is well documented by a letter from Kahrstedt to the discoverer from Feb. 24, 1926: "...Ich möchte Sie bitten, mir zu gestatten, dass ich zwei der von Ihnen entdeckten Planetoiden "Gretia" und "Ingrid" taufe, Namen, die entschieden vernünftiger sind, als mancher Unsinn, der schon auf unserer Namensliste steht. Im Laufe des Sommers soll nämlich das Töchterchen meines Bruders "Ingrid" getauft werden, ich soll Pate stehen und würde mich sehr freuen, wenn ich Mutter und Kind je einen kleinen Planeten als Patengeschenk mitbringen könnte..."

(1027) Aesculapia

Discovered 1923 November 11 by G. Van Biesbroeck at Williams Bay.

Named for Aesculapius, the Roman god of medicine, son of Apollo {see planet (1862)} and Koronis {see planet (158)}. (H 98)

(1028) Lydina

Discovered 1923 November 6 by V. A. Albitskij at Simeïs.

Named in honor of Lydia Il'inichna Albitskaya, the wife of the discoverer. (N. S. Chernykh)

(1029) La Plata

Discovered 1924 April 28 by J. Hartmann at La Plata.

Named after the city of La Plata, Argentina. (H 98)

(1030) Vitja

Discovered 1924 May 25 by V. A. Albitskij at Simeïs.

Named in honor of Viktor (Vitja) Viktorovich Zaslavskij (1925-1944), a nephew of Spiridon Zaslavskij {see planet (1330)}, the brother-in-law of the discoverer. (LDS)

(1031) Arctica

Discovered 1924 June 6 by S. I. Belyavskij at Simeïs.

Named for the Arctic sea. (H 98)

(1032) Pafuri

Discovered 1924 May 30 by H. E. Wood at Johannesburg.

Named for a river in northern Transvaal, South Africa. (H 98)

(1033) Simona

Discovered 1924 September 4 by G. Van Biesbroeck at Williams Bay. Independently discovered 1924 September 5 by S. I. Belyavskij at Simeïs.

Named in honor of the discoverer's daughter who is now Mrs. John Titus. (H 98)

(1034) Mozartia

Discovered 1924 September 7 by V. A. Albitskij at Simeïs.

Named in honor of the composer Wolfgang Amadeus Mozart {1756-1791}. (M 837)

(1035) Amata

Discovered 1924 September 29 by K. Reinmuth at Heidelberg.

Named possibly after the wife of king Latinus and mother of Lavinia, the wife of Aeneas {see planet (1172)} in Latium. (Z 19; R. Bremer; LDS)

(1036) Ganymed

Discovered 1924 October 23 by W. Baade at Bergedorf.

Named after the beautiful youth in classical mythology carried off to Olympus to be the cupbearer of the gods. (LDS)

The name Ganymede has also been given to the third satellite of Jupiter, discovered 1610 by Galilei.

(1037) Davidweilla

Discovered 1924 October 29 by B. Jekhovsky at Algiers.

Named in honor of David Weill, a member of the Academy of Sciences and benefactor of the Sorbonne University, Paris. (H 99)

(1038) Tuckia

Discovered 1924 November 24 by M. Wolf at Heidelberg.

Named in honor of Mr. and Mrs. Edward Tuck, philantropists. He is the son of the founder of the Republican Party in the United States. (H 99)

Name proposed by Mme. G. Camille Flammarion (l'Astronomie, Vol. 41, p. 229 (1927)).

(1039) Sonneberga

Discovered 1924 November 24 by M. Wolf at Heidelberg.

Named for the city of Sonneberg in Thüringen, Germany, location of the Sonneberg Observatory. (H 99)

(1040) Klumpkea

1925 BD. Discovered 1925 January 20 by B. Jekhovsky at Algiers.

Named in honor of the amateur astronomer Dorothea Klumpke {1861-1943} {see also planet (339)}, wife of the English astronomer Isaac Roberts. She was the first woman to receive the degree of Doctor of Mathematical Sciences at the Sorbonne University. (H 99)

(1041) Asta

1925 FA. Discovered 1925 March 22 by K. Reinmuth at Heidelberg.

This planet is possibly named in honor of the Danish actress Asta Nielsen (1881-1972). (LDS)

Name proposed by G. Stracke.

(1042) Amazone

1925 HA. Discovered 1925 April 22 by K. Reinmuth at Heidelberg.

Named for the women of Scythia near the rivers Tanais and Thermodon. They had no men and joined with their neighbours, killed the sons and educated the daughters of whom the teats were burnt so that they could discharge the arrows more rapidly. The name means "without teat". Heracles {see planet (5143)} and Theseus fought against the Amazones. The most famous queens are Penthesileia {see planet (271)} and Hippolyte. (I. van Houten-Groeneveld)

Name proposed by G. Stracke.

(1043) Beate

1925 HB. Discovered 1925 April 22 by K. Reinmuth at Heidelberg.

Any reference of this name to a person or occurence is unknown.

Name proposed by G. Stracke.

(1044) Teutonia

Discovered 1924 May 10 by K. Reinmuth at Heidelberg.

Named after the region, now in Denmark and Germany, that was once inhabited by the tribe of Teutons. (I. van Houten-Groeneveld)

Name proposed by G. Stracke.

(1045) Michela

Discovered 1924 November 19 by G. Van Biesbroeck at Williams Bay.

Named in honor of the discoverer's daughter Micheline. (H 99)

(1046) Edwin

Discovered 1924 December 1 by G. Van Biesbroeck at Williams Bay.

Named in honor of the discoverer's son. (H 99)

(1047) Geisha

Discovered 1924 November 17 by K. Reinmuth at Heidelberg.

Named for the opera of the same name by Jones. The dictionary defines: a Japanese girl who is trained to provide entertaining and lighthearted company especially for a man or a group of men. (H 100)

(1048) Feodosia

Discovered 1924 November 29 by K. Reinmuth at Heidelberg.
Named for the city Feodosija in the Crimea. (A. Paluzíe-Borrell)
Named in accordance with a proposal by the computer I. Putilin (RI 628).

(1049) Gotho

1925 RB. Discovered 1925 September 14 by K. Reinmuth at Heidelberg.
Any reference of this name to a person or occurence is unknown.

(1050) Meta

1925 RC. Discovered 1925 September 14 by K. Reinmuth at Heidelberg.
Any reference of this name to a person or occurence is unknown.

(1051) Merope

1925 SA. Discovered 1925 September 16 by K. Reinmuth at Heidelberg.
Named for one of the Pleiades, daughters of Atlas and Pleione.

(1052) Belgica

1925 VD. Discovered 1925 November 15 by E. Delporte at Uccle.

Named in honor of the state of Belgium. This is the first minor planet to be discovered in Uccle {see planet (1276)}, Belgium. (H 100)

Brian G. Marsden (private communication) discovered a remarkable piece of history involving minor planet names and World War I. He detected a correspondence between Prof. W. W. Campbell, director of the Lick Observatory, and Rev. J. Metcalf which proves that Metcalf submitted the name "Belgica", evidently during 1915-1917 (after the Belgian relief effort but before the United States entered the war), and had it turned down by Prof. F. Cohn, director of the Astronomisches Rechen-Institut in Berlin.

In a letter from October 1, 1919 Prof. Campbell wrote to Metcalf: "Professor Russell of Princeton told a group of astronomers in Washington last June about a most remarkable incident of which he said you were the hero. The circumstances were reported to be as follows: Before the United States entered the great war you forwarded to the Rechen-Institut in Berlin a list of four asteroids discovered by you, with proposals for the names which they should bear. One of these names was Belgica. The head of the Recheninstitut responded refusing to utilize this name and expressing the hope that you were not casting political aspersions, or something of that kind. Would you be so good as to let me have the precise facts in the case? I think the incident is most remarkable..." The {undated} answer by Metcalf says the following: "...The facts of the would be Belgica are briefly these. At the Allied Bazaar Prof. {Henry Norris} Russell asked me for an unnamed asteroid that the naming of it might be sold to the highest bidder. I sent him four. The people paid the price (about $50) and sent me through Prof. Russell the names. One of them was Belgica. I saw the possibility of trouble but I sent it along as I had agreed to. Prof. Cohn of the Rechen-Institut sent backward that it was "impossible" *at that moment anyway.* I think myself it was poor taste. Of course he blamed me. I never took the trouble to tell him how it happened. The other names were very long queer ones that they objected also. In fact, the only successful part of it was that we obtained some money for the Allied Bazaar. I am afraid Belgium was a pretty sore spot with them about the time I wrote the

letter..." Unfortunately, our search did not show any surviving ARI documents that shed more light on this incident.
Marsden suspects that all minor planets discovered by Metcalf from (726) Joëlla onward, with the possible exception of (747) Winchester {home town of Rev. Metcalf}, were named by Harlow Shapley soon after Metcalf's death in 1925. Metcalf moved from Winchester, Mass. to Portland, Maine around 1920, two years before Shapley came from Mt. Wilson to direct the Harvard College Observatory.

(1053) Vigdis

1925 WA. Discovered 1925 November 16 by M. Wolf at Heidelberg.
Any reference of this name to a person or occurence is unknown.

(1054) Forsytia

1925 WD. Discovered 1925 November 20 by K. Reinmuth at Heidelberg.

Named after a genus of ornamental shrubs of the olive family with opposite leaves and yellow bell-shaped flowers appearing before the leaves in early spring. (H 100)

Reinmuth published (RI 803) a large list of new names for 66 planets between (1009) and (1200). In this compilation there is a number of 28 consecutive names which belong to flowers and other plants.

(1055) Tynka

1925 WG. Discovered 1925 November 17 by E. Buchar at Algiers.

Named in honor of the mother of the discoverer who worked at the Institute of Astronomy and Geophysics at the Technical University, Prague. (H 100)

(1056) Azalea

Discovered 1924 January 31 by K. Reinmuth at Heidelberg.

Named after a genus of rhododendrons with funnel-shaped corollas and usually deciduous leaves including many species and hybrid forms cultivated as ornamentals. (H 100)
See also the remark to planet (1054).

(1057) Wanda

1925 QB. Discovered 1925 August 16 by G. Shajn at Simeïs. Independently discovered 1925 August 19 by K. Reinmuth at Heidelberg.

Named in honor of the well-known Polish and Soviet writer Wanda Wasilewska (1905-1964). In the USSR her name is Wasilewskaya. Another interpretation says that Wanda was a mythic person in ancient Polish annals. Daughter of Krakus or Krak, she was the legendary founder of Craccow. (N. S. Chernykh; A. Palužie-Borrell)

(1058) Grubba

1925 MA. Discovered 1925 June 22 by G. Shajn at Simeïs.

Named in honor of Sir Howard Grubb of Grubb Parson and Co., Newcastle on Tyne, England, manufacturer of the 40-inch reflector of the Simeïs Observatory {see planet (748)}. (H 100)

(1059) Mussorgskia

1925 OA. Discovered 1925 July 19 by V. A. Albitskij at Simeïs.

Named in honor of the Russian composer Modest Mussorgsky {1839-1881}. (M 837)

(1060) Magnolia

1925 PA. Discovered 1925 August 13 by K. Reinmuth at Heidelberg.

Named after a genus of North American and Asian shrubs and trees with entire

evergreen or deciduous leaves and usually showy white, yellow, rose, or purple flowers appearing in early spring. (H 101)
See also the remark to planet (1054).

(1061) Paeonia

1925 TB. Discovered 1925 October 10 by K. Reinmuth at Heidelberg.
Named after a flower. (H 101)
See also the remark to planet (1054).

(1062) Ljuba

1925 TD. Discovered 1925 October 11 by S. I. Belyavskij at Simeïs.

Named in honor of the perished lady-parachutist Ljuba Berlin (1915-1936). See also planets (1084) and (1086). (N. S. Samojlova-Yakhontova; N. Solovaya)

(1063) Aquilegia

1925 XA. Discovered 1925 December 6 by K. Reinmuth at Heidelberg.

Named after a genus of plants of the buttercup family with irregular showy spurred flowers. The English name of the flower is columbine. (H 101)
See also the remark to planet (1054).

(1064) Aethusa

1926 PA. Discovered 1926 August 2 by K. Reinmuth at Heidelberg.

Named after a plant of the carrot family. The fool's parsley belongs to the Aethusa family. (H 101)
See also the remark to planet (1054).

(1065) Amundsenia

1926 PD. Discovered 1926 August 4 by S. I. Belyavskij at Simeïs.

Named in honor of the Norwegian polar explorer Roald E. Amundsen (1872-1928). He explored the Arctic regions in 1901-1906, 1918-1920, and the Antarctic in 1897 and 1910, the north pole by aeroplane in 1925 and by dirigible in 1926. Amundsen perished in searching for the wreckage of the dirigible *Italia*. (H 101)
Amundsen is also honored by a lunar crater.

(1066) Lobelia

1926 RA. Discovered 1926 September 1 by K. Reinmuth at Heidelberg.

Named for a genus of widely distributed herbaceous plants cultivated for their terminal clusters of showy lipped flowers. (H 101)
See also the remark to planet (1054).

(1067) Lunaria

1926 RG. Discovered 1926 September 9 by K. Reinmuth at Heidelberg.

Named after any of a genus of European plants of the mustard family with cordate leaves and broad siliques (e.g. honesty). (H 101)
See also the remark to planet (1054).

(1068) Nofretete

1926 RK. Discovered 1926 September 13 by E. Delporte at Uccle.

Named for the Egyptian queen and wife of Amenhotep IV {see also planet (4847)}. (H 101)
Named by G. Stracke.

Nofretete is honored twice because the name Nefertiti {see planet (3199)} is only a different spelling.

(1069) Planckia

1927 BC. Discovered 1927 January 28 by M. Wolf at Heidelberg.

Named in honor of the famous German physicist Max Karl Ernst Ludwig Planck (1858-1947), Nobel prize winner in 1918, on the occasion of his 80th birthday anniversary. He was a professor of physics at Berlin University and the discoverer of the quantum nature of radiation. (H 101)
Planck is also honored by a lunar crater.

(1070) Tunica

1926 RB. Discovered 1926 September 1 by K. Reinmuth at Heidelberg.

Named after one of numerous cultivated usually double-flowered pinks derived from the common gillyflower. (LDS)
See also the remark to planet (1054).

(1071) Brita

Discovered 1924 March 3 by V. A. Albitskij at Simeïs.

Named by the discoverer in honor of Great Britain as a mark of gratitude to the country where the 1-m telescope for the Simeïs Observatory {see planet (748)} was made. (N. Solovaya; N. S. Chernykh)

(1072) Malva

1926 TA. Discovered 1926 October 4 by K. Reinmuth at Heidelberg.

Named after any of a genus of herbs with palmately lobed or dissected leaves, usually showy flowers, and a disk-shaped fruit. The English name for the genus is mallow. (H 101)
See also the remark to planet (1054).

(1073) Gellivara

Discovered 1923 September 14 by J. Palisa at Vienna.

Named for the small town Gällivare in Swedish Lapland where in the year 1927 astronomers from several countries observed the total solar eclipse of 1927 June 29. (H 101)
Named by the Austrian astronomer J. Rheden and endorsed by Anna Palisa.

(1074) Beljawskya

1925 BE. Discovered 1925 January 26 by S. I. Belyavskij at Simeïs.

This planet is named in honor of the discoverer S. I. Belyavskij (1883-1953), astronomer at the Simeïs Observatory. He discovered 36 numbered minor planets between 1912 and 1927 and the brilliant comet 1911 IV. (H 102)
Name proposed by the Simeïs Observatory staff.

(1075) Helina

1926 SC. Discovered 1926 September 29 by G. N. Neujmin at Simeïs.

Named in honor of Helij Grigor'evich Neujmin (1910-1982), a son of the discoverer. (N. S. Chernykh; I. I. Neyachenko)

(1076) Viola

1926 TE. Discovered 1926 October 5 by K. Reinmuth at Heidelberg.

Named after the family of garden hybrids with solitary white, yellow, or purple often variegated flowers resembling but smaller than typical pansies. (H 102)
See also the remark to planet (1054).

(1077) Campanula

1926 TK. Discovered 1926 October 6 by K. Reinmuth at Heidelberg.
Named after the bellflower. (H 102)
See also the remark to planet (1054).

(1078) Mentha

1926 XB. Discovered 1926 December 7 by K. Reinmuth at Heidelberg.
Named after a planet of the mint family. (H 102)

See also the remark to planet (1054). A. Patry, Nice, in 1958 found an identity between this planet and (864) Aase (see the remarks for that planet, too).

(1079) Mimosa

1927 AD. Discovered 1927 January 14 by G. Van Biesbroeck at Williams Bay.

Named after a genus of leguminous trees, shrubs, and herbs of tropical and warm regions with usually bipinnate often prickly leaves and globular heads of small white or pink flowers. (H 102)

(1080) Orchis

1927 QB. Discovered 1927 August 30 by K. Reinmuth at Heidelberg.
Named after a flower of a genus with fleshy roots and a spurred lip. (H 102)
See also the remark to planet (1054).

(1081) Reseda

1927 QF. Discovered 1927 August 31 by K. Reinmuth at Heidelberg.

Named after a genus of Old World herbs of the mignonette family having racemose flowers with cleft petals and numerous stamens. (H 102)
See also the remark to planet (1054).

(1082) Pirola

1927 UC. Discovered 1927 October 28 by K. Reinmuth at Heidelberg.

Named after a member of the plant family of flowering herbs, mostly evergreen. (H 102)
See also the remark to planet (1054).

(1083) Salvia

1928 BC. Discovered 1928 January 26 by K. Reinmuth at Heidelberg.

Named after a largely and widely distributed genus of herbs or shrubs of the mint family having a two-lipped open calyx and two anthers. The English name of the flower is sage. (H 102)
See also the remark to planet (1054).

(1084) Tamariwa

1926 CC. Discovered 1926 February 12 by S. I. Belyavskij at Simeïs.

This is a combination of the feminine name Tamara Ivanova (1912-1936), a perished lady-parachutist. See also planets (1062) and (1086). (N. S. Samojlova-Yakhontova)

(1085) Amaryllis

1927 QH. Discovered 1927 August 31 by K. Reinmuth at Heidelberg.

Named after a genus of bulbous African herbs with showy umbellate flowers. (H 102)
See also the remark to planet (1054).

(1086) Nata

1927 QL. Discovered 1927 August 25 by S. I. Belyavskij and N. Ivanov at Simeïs.

Named in honor of the perished lady-parachutist Nata (Nadezhda) Babushkina (1915-1936). See also planets (1062) and (1084). (N. S. Samojlova-Yakhontova; N. Solovaya; N. S. Chernykh)

(1087) Arabis

1927 RD. Discovered 1927 September 2 by K. Reinmuth at Heidelberg. Independently discovered 1927 September 2 by S. I. Belyavskij and N. Ivanov at Simeïs.
Named after a large genus of herbs from the mustard family. (H 102)
See also the remark to planet (1054).

(1088) Mitaka

1927 WA. Discovered 1927 November 17 by O. Oikawa at Tokyo.

Named for a little village near Tokyo, the site of the Tokyo Astronomical Observatory. (H 103)

(1089) Tama

1927 WB. Discovered 1927 November 17 by O. Oikawa at Tokyo.
Named for a river near the Tokyo Astronomical Observatory. (H 103)

(1090) Sumida

1928 DG. Discovered 1928 February 20 by O. Oikawa at Tokyo. Independently discovered 1928 February 24 by K. Reinmuth at Heidelberg.
Named for a river near Tokyo, Japan. (H 103)

(1091) Spiraea

1928 DT. Discovered 1928 February 26 by K. Reinmuth at Heidelberg.

Named after a genus of shrubs of the rose family with small perfect white or pink flowers in dense racemes, corymbs, cymes, or panicles. (H 103)
See also the remark to planet (1054).

(1092) Lilium

Discovered 1924 January 12 by K. Reinmuth at Heidelberg.

Named after a genus of erect perennial leafy-stemmed bulbous herbs that are native to the northern hemisphere and are widely culticated for their showy flowers. The English name of the flower is lily. (H 103)
See also the remark to planet (1054).

(1093) Freda

1925 LA. Discovered 1925 June 15 by B. Jekhovsky at Algiers.

Named in memory of Fred Prévost, a civil engineer of mines and benefactor of the Faculty of Sciences of Bordeaux, France. (H 103)

(1094) Siberia

1926 CB. Discovered 1926 February 12 by S. I. Belyavskij at Simeïs.
Named for a region of Russia located in northern Asia. (H 103)

(1095) Tulipa

1926 GS. Discovered 1926 April 14 by K. Reinmuth at Heidelberg.

Named after a genus of Eurasian bulbous herbs of the lily family that have linear or broadely lanceolate leaves and are widely grown for their showy flowers. The English name of the flower is tulip. (H 103)

See also the remark to planet (1054). The name was first assigned to planet 1928 DC, discovered 1928 February 24 by K. Reinmuth at Heidelberg. This object, however, was proofed to be identical with planet (1449) Virtanen. Therefore, the name Tulipa has been reassigned to the above object.

(1096) Reunerta

1928 OB. Discovered 1928 July 21 by H. E. Wood at Johannesburg.

Named by the discoverer (RI 636) in honor of Dr. Reunert, Johannesburg, an engineer and friend. (H 103)

(1097) Vicia

1928 PC. Discovered 1928 August 11 by K. Reinmuth at Heidelberg. Independently discovered 1928 August 15 by P. F. Shajn at Simeïs and August 22 by H. E. Wood at Johannesburg.

Named after a flower of the Papilionaceae family. The English name of the flower is vetch. (H 103)

See also the remark to planet (1054).

(1098) Hakone

1928 RJ. Discovered 1928 September 5 by O. Oikawa at Tokyo. Independently discovered 1928 September 9 by M. Wolf at Heidelberg and September 11 by G. N. Neujmin at Simeïs.

Named for a well-known composite volcanic mountain about 80 km from the observatory where this planet was discovered. It is a typical mountain resort of Japan with a lake and numerous hot springs. It provides a wonderful view of Mt. Fuji (see planet (1584)) and attracts a lot of tourists all over the year. (K. Tomita)

(1099) Figneria

1928 RQ. Discovered 1928 September 13 by G. N. Neujmin at Simeïs. Independently discovered 1928 September 13 by M. Wolf at Heidelberg.

Named by the discoverer in honor of the well-known Russian writer and revolutionary Vera N. Figner (1852-1942). (RI 789)

(1100) Arnica

1928 SD. Discovered 1928 September 22 by K. Reinmuth at Heidelberg. Independently discovered 1928 October 14 by A. Schwassmann and A. A. Wachmann at Bergedorf.

Named after a genus of composite herbs including some with bright yellow ray flowers. (H 103)

See also the remark to planet (1054).

(1101) Clematis

1928 SJ. Discovered 1928 September 22 by K. Reinmuth at Heidelberg.

Named after a vine or herb of the buttercup family having three leaflets on each leaf and usually white or purple flowers. (H 103)

See also the remark to planet (1054).

(1102) Pepita

1928 VA. Discovered 1928 November 5 by J. Comas Solá at Barcelona.

Named in honor of the discoverer José Comas Solá (1868-1937), astronomer at Barcelona and founder of the "Sociedad Astronomica de España y America". Pepita is the feminine form of Pepito, the familiar name of the discoverer. The discoverer is also honored by planet (1655). (H 104)

Comas Solá is also honored by a crater on Mars.

(1103) Sequoia

1928 VB. Discovered 1928 November 9 by W. Baade at Bergedorf.

Named by the discoverer for the Sequoia National Park where he spent vacations. The park is named after the huge coniferous California trees of the pine family that reach a height of over 100 meters. (H 104)

(1104) Syringa

1928 XA. Discovered 1928 December 9 by K. Reinmuth at Heidelberg.

Named after a genus of ornamental shrubs of the saxifrage family of which several are widely known in temperate regions for their showy white flowers. The

genus Syringa is also known as philadelphus. The English name of the plant is lilac. (H 104)
See also the remark to planet (1054).

(1105) Fragaria

1929 AB. Discovered 1929 January 1 by K. Reinmuth at Heidelberg.

Named after a genus of the rose family that is technically an enlarged pulpy receptacle bearing numerous achenes (strawberry). (H 104)
See also the remark to planet (1054).

(1106) Cydonia

1929 CW. Discovered 1929 February 5 by K. Reinmuth at Heidelberg.
Named after a tree (English name quince) belonging to the apple family. (H 104)
See also the remark to planet (1054).

(1107) Lictoria

1929 FB. Discovered 1929 March 30 by L. Volta at Pino Torinese. Independently discovered 1929 March 17 by K. Reinmuth at Heidelberg.

The name is strictly connected with the symbol of the fascist party, which in Italian was called "Fasci Littori", derived from the Latin "Fasces Lictores". During the fascist regime it became common-place to apply that name to events or objects ranging from festivals ("littoriali") to fast trains ("littorine"). When the Pontine Marshes were being reclaimed, the first two towns established there were Littoria and Sabaudia {see planet (1115)}. It is noteworthy that after the demise of the fascist regime the town of Littoria was renamed "Latina". (P. G. Comba)

(1108) Demeter

1929 KA. Discovered 1929 May 31 by K. Reinmuth at Heidelberg.

Named for the Greek goddess of the fruitful soil and of agriculture. The Romans identified Demeter with Ceres {see planet (1)}. (H 104)

(1109) Tata

1929 CU. Discovered 1929 February 5 by K. Reinmuth at Heidelberg.
Any reference of this name to a person or occurence is unknown.

(1110) Jaroslawa

1928 PD. Discovered 1928 August 10 by G. N. Neujmin at Simeïs. Independently discovered 1928 August 18 by E. Delporte at Uccle.

Named in honor of Jaroslav Grigor'evich Neujmin (1928-   ), a son of the discoverer. (N. S. Chernykh; I. I. Neyachenko)

(1111) Reinmuthia

1927 CO. Discovered 1927 February 11 by K. Reinmuth at Heidelberg.

Named in honor of Karl Reinmuth {1892-1979}, a tireless observer on the staff of the Heidelberg Königstuhl Observatory. Reinmuth discovered more than 380 now numbered minor planets - a unique record for many years. (H 104)

Obituaries published in Sterne Weltraum, Jahrg. 18, p. 202 (1979); Minor Planet Bull., Vol. 7, p. 10 (1979); Mitt. Astron. Ges., Nr. 50, p. 7-8 (1980).

(1112) Polonia

1928 PE. Discovered 1928 August 15 by P. F. Shajn at Simeïs. Independently discovered 1928 August 16 by G. N. Neujmin at Simeïs.
Named for the country of Poland. Polonia is the Latin name. (H 104)

Named in accordance with the discoverer by L. Matkiewicz (RI 304), who computed the orbit of this planet.
This is the first minor planet that was discovered by a woman.

(1113) Katja

1928 QC. Discovered 1928 August 15 by P. F. Shajn at Simeïs. Independently discovered 1928 August 24 by M. Wolf at Heidelberg.

Named in honor of the laboratory assistant and calculator Katja (Ekaterina) Ios'ko of the discovery observatory. She was the daughter of the Simeïs mechanician Iosif Gavrilovich Ios'ko. (N. S. Samojlova-Yakhontova; N. Solovaya; N. S. Chernykh; V. A. Shor)

(1114) Lorraine

1928 WA. Discovered 1928 November 17 by A. Schaumasse at Nice. Independently discovered 1928 November 18 by L. Volta at Pino Torinese.

Named after the region and former duchy in northeastern France, remnant of the medieval kingdom of Lotharingia. (R. Bremer)

(1115) Sabauda

1928 XC. Discovered 1928 December 13 by L. Volta at Pino Torinese. Independently discovered 1928 December 18 by J. Comas Solá at Barcelona.

Sabauda, or Sapauda, is the Latin name of the House of Savoy, the former rulers of Italy. The planet probably could also have been named after the new established town Sabauda in the Pontine Marshes - see the remarks to planet (1107). (H 104)

(1116) Catriona

1929 GD. Discovered 1929 April 5 by C. Jackson at Johannesburg.

Named probably after the novel (1893) by the Scottish poet and novelist Robert Louis Balfour Stevenson (1850-1894). (R. Bremer)

(1117) Reginita

1927 KA. Discovered 1927 May 24 by J. Comas Solá at Barcelona.
Named in honor of the niece of the discoverer. (H 105)

(1118) Hanskya

1927 QD. Discovered 1927 August 29 by S. I. Belyavskij and N. Ivanov at Simeïs. Independently discovered 1927 August 30 by K. Reinmuth at Heidelberg and September 17 by E. Delporte at Uccle.

The planet was named "zum Andenken an den vor 25 Jahren verstorbenen ersten Astronomen der Sternwarte Simeïs, Alexis Hansky {1872-1908}, dessen Initiative diese Sternwarte ihre Entstehung verdankt." (RI 858)
Hansky is also honored by a lunar crater.

(1119) Euboea

1927 UB. Discovered 1927 October 27 by K. Reinmuth at Heidelberg. Independently discovered 1927 October 29 by S. I. Belyavskij at Simeïs.
Euboea (or Negropont) is the largest island of Greece in the Aegean. (H 105)

(1120) Cannonia

1928 RV. Discovered 1928 September 11 by P. F. Shajn at Simeïs. Independently discovered 1928 September 13 by G. N. Neujmin at Simeïs and September 23 by E. Delporte at Uccle.

Named by the discoverer (RI 584) in honor of Miss Annie Jump Cannon (1863-1941). At Harvard University she classified the spectra of about 225,000 stars for the *Henry Draper Catalogue* in a system of spectral types which she had developed. (H 105)
Cannon is also honored by a lunar crater.

(1121) Natascha

1928 RZ. Discovered 1928 September 11 by P. F. Shajn at Simeïs.

Named - as a birthday present - in honor of the hydrogeologist Natasha (Natalia) Tichomirova, daughter of the Simeïs astronomer G. N. Neujmin. (N. S. Samojlova-Yakhontova; N. Solovaya; N. S. Chernykh)

(1122) Neith

1928 SB. Discovered 1928 September 17 by E. Delporte at Uccle.

Named for the Egyptian goddess of Libyan origin. She passed for the mother of the Sun, Râ. (H 105)

(1123) Shapleya

1928 ST. Discovered 1928 September 21 by G. N. Neujmin at Simeïs.

Named by the discoverer (RI 509) in honor of the American astronomer and director of Harvard Observatory Harlow Shapley (1885-1972). (H 105)

Obituaries published in Nature, Vol. 240, p. 429-430 (1972); Orion, 30. Jahrg., p. 186 (1972); Sky Telesc., Vol. 44, p. 354-357 (1972); Sterne Weltraum, Vol. 11, p. 296 (1972); Astrophys. Space Sci., Vol. 18, p. 258-266 (1972); Coelum, Vol. 41, p. 13-15 (1973); J.R. Astron. Soc. Can., Vol. 67, p. 31-33 (1973); Irish Astron. J., Vol. 10, p. 302-303 (1972); C.R. Acad. Sci., Vie Acad., Vol. 277, p. 119-120 (1973); Vesmír, Vol. 52, p. 254 (1973); Alm. Österr. Akad Wiss., 123. Jahrg., p. 315-321 (1973).

(1124) Stroobantia

1928 TB. Discovered 1928 October 6 by E. Delporte at Uccle.

Named in honor of Paul Stroobant (1868-1936), director of the Royal Observatory at Uccle, Belgium. His work concerned the number, mass and distribution of the minor planets. (H 105)

(1125) China

1957 UN_1. Discovered 1957 October 30 by Y. C. Chang at Nanking.
Named for the country in which this planet was discovered. (M 13179)
See also the citation for minor planet (3789).

(1126) Otero

1929 AC. Discovered 1929 January 11 by K. Reinmuth at Heidelberg.

Named probably after Caroline Otéro (1868-1965), known as "la belle Otéro". The whole world admired her for being charming and beautiful; for a long period she embodied "la femme de grand luxe" and "la Belle Epoque". At the turn of the century she was the most demanded of the Parisian courtesans and led an excessive life. Thanks to her numerous rich and famous lovers, who fought each other and ruined themselves in favor of her, she was able to amass immense riches. Unfortunately, she was bound to lose them, because she was passionately fond of gambling. She died in poverty at the age of 97 after a life of fulfillment. (M.-A. Combes)
Named by the discoverer (RI 803).

(1127) Mimi

1929 AJ. Discovered 1929 January 13 by S. Arend at Uccle.
Named in honor of the wife of E. Delporte. (H 105)

The planets (1127) and (1145) have been discovered at the Uccle Observatory by S. Arend and E. Delporte, respectively. Arend proposed the name Robelmonte for his planet, while Delporte chose the name Mimi for his own. Through an error, planet (1127) received the name Mimi and the name Robelmonte was assigned to (1145).

(1128) Astrid

1929 EB. Discovered 1929 March 10 by E. Delporte at Uccle.
Named in honor of H.M. Astrid (1905-1935), Queen of the Belgians. (H 106)

(1129) Neujmina

1929 PH. Discovered 1929 August 8 by P. Parchomenko at Simeïs.

Named by the discoverer (RI 369) in honor of Grigory N. Neujmin (1885-1946), astronomer at the Pulkovo and Simeïs Observatories. Neujmin discovered seven comets and more than 70 minor planets. (H 106)
Neujmin is also honored by a lunar crater.

(1130) Skuld

1929 RC. Discovered 1929 September 2 by K. Reinmuth at Heidelberg. Independently discovered 1929 September 12 by A. Schwassmann and A. A. Wachmann at Bergedorf.

Named for one of the three Norns in Norse mythology. Skuld stands for the Future. (H 106)

The other Norns are Urda (the Past) and Werdandi (the Present) {see planets (167) and (621)}.

(1131) Porzia

1929 RO. Discovered 1929 September 10 by K. Reinmuth at Heidelberg.
Named for a character in Shakespeare's play *Julius Caesar* (1599). (H 106)

(1132) Hollandia

1929 RB_1. Discovered 1929 September 13 by H. van Gent at Johannesburg.

This is the Latin name for Holland, a synonymous name for the Netherlands. (H 106)
Named by the discoverer and the orbit computer G. Pels.

(1133) Lugduna

1929 RC_1. Discovered 1929 September 13 by H. van Gent at Johannesburg.

Lugdunum Batavorum is the Latin name for the city of Leiden, The Netherlands. (I. van Houten-Groeneveld)
Named by the discoverer and the orbit computer G. Pels.

(1134) Kepler

1929 SA. Discovered 1929 September 25 by M. Wolf at Heidelberg.

Named in honor of the famous German astronomer Johannes Kepler (1571-1630) on the occasion of the 300th anniversary of his death. After incredible labor, Kepler discovered the elliptical orbits of the planets, the law of equal areas, and the relation between the period and the size of an orbit. (AN 240, 135 (1930))

Kepler is also honored by craters on the Moon and Mars and by a mountain on Phobos.

(1135) Colchis

1929 TA. Discovered 1929 October 3 by G. N. Neujmin at Simeïs.

Named for the ancient country bordering on Black Sea south of the Caucasus mountains. The area now constitutes the western part of the Georgian Republic. (H 106)

(1136) Mercedes

1929 UA. Discovered 1929 October 30 by J. Comas Solá at Barcelona.
Named in honor of the sister-in-law of the discoverer. (H 106)

(1137) Raïssa

1929 WB. Discovered 1929 October 27 by G. N. Neujmin at Simeïs. Independently discovered 1929 November 21 by K. Reinmuth at Heidelberg.

Named in honor of Raïssa Izrailevna Maseeva (1900-1930), a former scientific collaborator at the Pulkovo Observatory. (H 106)

(1138) Attica

1929 WF. Discovered 1929 November 22 by K. Reinmuth at Heidelberg.
Named after the region in eastern Greece with the capital Athens. (H 106)

(1139) Atami

1929 XE. Discovered 1929 December 1 by O. Oikawa and K. Kubokawa at Tokyo.
Named after a harbor near Tokyo, Japan. (H 106)

(1140) Crimea

1929 YC. Discovered 1929 December 30 by G. N. Neujmin at Simeïs.

Named for the southern peninsula of Russia extending into the Black Sea where Simeïs is located. (H 106)

(1141) Bohmia

1930 AA. Discovered 1930 January 4 by M. Wolf at Heidelberg.

Named in honor of Mrs. Katharina Bohm-Waltz (? -1901) who donated the 72-cm Waltz reflector to the Heidelberg Observatory. (H 107; LDS)

Mrs. Bohm donated a sum of 25,000 M. The invoice of Zeiss Jena for this superb instrument from July 1906 amounts to 20,000 M.

(1142) Aetolia

1930 BC. Discovered 1930 January 24 by K. Reinmuth at Heidelberg.
Named for the Greek province north of the Gulf of Patras. (H 107)

(1143) Odysseus

1930 BH. Discovered 1930 January 28 by K. Reinmuth at Heidelberg.

Named after the king of Ithaca and Greek hero in the Trojan War who after the war wandered 10 years before reaching home. (H 107)

(1144) Oda

1930 BJ. Discovered 1930 January 28 by K. Reinmuth at Heidelberg.
See the remarks to planet (913).

(1145) Robelmonte

1929 CC. Discovered 1929 February 3 by E. Delporte at Uccle.

Named for the city of Robelmont, Belgium, the birthplace of S. Arend {see planet (1502)}. Robelmonte is the feminine form. (H 107)
See also the remarks to planet (1127).

(1146) Biarmia

1929 JF. Discovered 1929 May 7 by G. N. Neujmin at Simeïs.

Named by the discoverer after a legendary country in northern Russia near to the Finnish border. (H 107)

(1147) Stavropolis

1929 LF. Discovered 1929 June 11 by G. N. Neujmin at Simeïs.

Named by the discoverer for the city of Stavropol' in northern Caucasus region. Between 1936 and 1946 the city was named Woroschilowsk. (H 107)

(1148) Rarahu

1929 NA. Discovered 1929 July 5 by A. N. Deutsch at Simeïs. Independently discovered 1929 July 28 by C. Jackson and H. E. Wood at Johannesburg.

This is a Tahitian name for a girl. The name is taken from the novel by Louis Marie Julien Viaud (pseudonym Pierre Loti, 1850-1923) *Mariage de Loti.* (H 107)

For many decades the wrong spelling "Raraju" was used. J. Meeus showed that an erroneous transliteration from French to Russian to German caused the name of this minor planet to be spelled incorrectly in the literature. The correct name Rarahu has officially been adopted in 1985 (M 10194).

(1149) Volga

1929 PF. Discovered 1929 August 1 by E. Skvortsov at Simeïs.

This planet is named for the large river in the European part of the USSR. (M 2740)

Name proposed by the Institute of Theoretical Astronomy, Leningrad.

(1150) Achaia

1929 RB. Discovered 1929 September 2 by K. Reinmuth at Heidelberg. Independently discovered 1929 September 12 by A. Schwassmann and A. A. Wachmann at Bergedorf.

Named after the Greek region in northern Peloponnese bordering on the gulfs of Corinth and Patras. The spelling Achaea for the province is also in use. (H 107)

(1151) Ithaka

1929 RK. Discovered 1929 September 8 by K. Reinmuth at Heidelberg. Independently discovered 1929 September 13 by G. N. Neujmin at Simeïs.

Named after the Greek island, the island of Odysseus, in the Ionian Sea. (H 107)

(1152) Pawona

1930 AD. Discovered 1930 January 8 by K. Reinmuth at Heidelberg. Independently discovered 1930 January 19 by L. Volta at Pino Torinese and January 21 by G. N. Neujmin at Simeïs.

This name stands for a combination of the names of J. Palisa and M. Wolf in honor of the cooperation between these two important astronomers. (H 107)

Name proposed by B. Asplind who found the identity of (1152) with 1925 SF which was the last minor planet discovered by Palisa (AN 240, 406 (1930)).

(1153) Wallenbergia

Discovered 1924 September 5 by S. I. Belyavskij at Simeïs.

Named in honor of the German mathematician Georg James Wallenberg (1864-1924). (H 107)

(1154) Astronomia

1927 CB. Discovered 1927 February 8 by K. Reinmuth at Heidelberg.

Named for the science which explores the celestial bodies. (H 108)

(1155) Aënna

1928 BD. Discovered 1928 January 26 by K. Reinmuth at Heidelberg.

This artificial name contains in the German pronunciation the initials "A" and "N" followed by the feminine ending. It honors the *Astronomische Nachrichten.* The naming was published together with that for planet (1164). (H 108)

Name proposed by the Astronomisches Rechen-Institut (RI 709).

(1156) Kira

1928 DA. Discovered 1928 February 22 by K. Reinmuth at Heidelberg.

Any reference of this name to a person or occurence is unknown.

The naming of this planet was granted to the Heidelberg astronomer M. Mündler.

(1157) Arabia

1929 QC. Discovered 1929 August 31 by K. Reinmuth at Heidelberg.
Named for the peninsula in southwest Asia. (H 108)

(1158) Luda

1929 QF. Discovered 1929 August 31 by G. N. Neujmin at Simeïs.

Named in honor of the sister of the discoverer. Luda is a diminutive of Ludmilla. (N. S. Samojlova-Yakhontova)

(1159) Granada

1929 RD. Discovered 1929 September 2 by K. Reinmuth at Heidelberg.

Named after the Spanish city and province in Andalusia in southern Spain. (H 108)

(1160) Illyria

1929 RL. Discovered 1929 September 9 by K. Reinmuth at Heidelberg.

Named for the ancient region in the Balkan peninsula bordering on the Adriatic Sea. (H 108)

(1161) Thessalia

1929 SF. Discovered 1929 September 29 by K. Reinmuth at Heidelberg.

Named for the region in eastern Greece between Pindus mountains and the Aegean Sea. (H 108)

(1162) Larissa

1930 AC. Discovered 1930 January 5 by K. Reinmuth at Heidelberg.
Named for the city in eastern Thessaly {see planet (1161)}, Greece. (H 108)
The name Larissa has also been given to the satellite Neptune VII.

(1163) Saga

1930 BA. Discovered 1930 January 20 by K. Reinmuth at Heidelberg.

The planet is named after a detailed recital of events, historical or legendary or both, of Icelandic heroes, most often set down in prose. (H 108)

(1164) Kobolda

1930 FB. Discovered 1930 March 19 by K. Reinmuth at Heidelberg. Independently discovered March 20 by W. Baade at Bergedorf.

Named in honor of Hermann Albert Kobold {1858-1942}, astronomer at Kiel and editor of the *Astronomische Nachrichten* {see also planet (1155)} from 1907 to 1938. (H 108)

(1165) Imprinetta

1930 HM. Discovered 1930 April 24 by H. van Gent at Johannesburg.
Named in honor of the wife of the discoverer. (H 108)
Named by the discoverer and the orbit computer G. Pels.

(1166) Sakuntala

1930 MA. Discovered 1930 June 27 by P. Parchomenko at Simeïs. Independently discovered 1930 June 29 by K. Reinmuth at Heidelberg.

Named by the discoverer (RI 895) for the heroine of a famous sanskrit drama by the Indian poet Kalidasa. The Hindu poet and dramatist lived sometime between the fourth and sixth century, was a native of Ujjain and lived at the court of one King Vikramaditya. Sakuntala is a dramatization of part of the *Mahabharata*. (H 108; LDS)

(1167) Dubiago

1930 PB. Discovered 1930 August 3 by E. Skvortsov at Simeïs.

This planet is named in honor of the late Prof. Alexander D. Dubiago (1903-1959), an eminent astronomer of the USSR. (M 2740)
Name proposed by the Institute of Theoretical Astronomy, Leningrad.
Dubiago is also honored by a lunar crater.

(1168) Brandia

1930 QA. Discovered 1930 August 25 by E. Delporte at Uccle. Independently discovered 1930 August 31 by G. N. Neujmin at Simeïs.

Named in honor of Eugène Brand, Professor of mathematics at the University of Brussels, Belgium. (H 109)

(1169) Alwine

1930 QH. Discovered 1930 August 30 by M. Wolf and M. Ferrero at Heidelberg.
Any reference of this name to a person or occurence is unknown.

(1170) Siva

1930 SQ. Discovered 1930 September 29 by E. Delporte at Uccle.

Named for the god of destruction and regeneration in the Hindu sacred triad (with Brahma and Vishnu). A legendary, colorful personality and delight to the intellectuals of the Hindu priesthood, Siva was the divine dancer who could interpret the mathematical law of the universe in 108 different movements, and he was the lord of knowledge, the center about whom the universe revolved. (H 109)

(1171) Rusthawelia

1930 TA. Discovered 1930 October 3 by S. Arend at Uccle. Independently discovered 1930 September 29 by G. N. Neujmin at Simeïs.

Named in honor of the Georgian poet Schota Rusthaweli (c. 1300). See also the remarks to planet (525). (H 109)
Name proposed by Simeïs astronomers.

(1172) Äneas

1930 UA. Discovered 1930 October 17 by K. Reinmuth at Heidelberg.

Named for the defender of Troy and hero of Vergil's Aeneid, son of Anchises {see planet (1173)} and Aphrodite {see planet (1388)}. (H 109)

(1173) Anchises

1930 UB. Discovered 1930 October 17 by K. Reinmuth at Heidelberg.

Named for the father of Aeneas {see planet (1172)}, rescued by his son from the burning city of Troy. (H 109)

(1174) Marmara

1930 UC. Discovered 1930 October 17 by K. Reinmuth at Heidelberg.

Named by the discoverer (RI 803) for the Marmara Sea which is connected with the Black Sea by the Bosporus and with the Aegean Sea by the Dardanelles. (H 109)

(1175) Margo

1930 UD. Discovered 1930 October 17 by K. Reinmuth at Heidelberg.
Any reference of this name to a person or occurence is unknown.
Named by the discoverer (RI 803).

(1176) Lucidor

1930 VE. Discovered 1930 November 15 by E. Delporte at Uccle. Independently discovered 1930 November 15 by M. Wolf at Heidelberg and November 30 by G. N. Neujmin at Simeïs.

Named in honor of an amateur astronomer and feminine friend of the discoverer. (H 109)

(1177) Gonnessia

1930 WA. Discovered 1930 November 24 by L. Boyer at Algiers.

Named in honor of the late François Gonnessiat (1856-1934), director of the Algiers Bouzaréah and Quito Observatories. (H 109)

(1178) Irmela

1931 EC. Discovered 1931 March 13 by M. Wolf at Heidelberg.

Named in honor of Mrs. Irmela Ruska, the wife of Ernest Ruska who invented the electron microscope and received the Nobel Prize in 1986. The prize had to be shared with B. Binnig and H. Rohrer. (H 109)

The naming was granted by the discoverer to A. Kopff.

(1179) Mally

1931 FD. Discovered 1931 March 19 by M. Wolf at Heidelberg.

Named by the discoverer in honor of his daughter-in-law, the wife of Franz Wolf. (H 110)

(1180) Rita

1931 GE. Discovered 1931 April 9 by K. Reinmuth at Heidelberg.

Any reference of this name to a person or occurence is unknown.

(1181) Lilith

1927 CQ. Discovered 1927 February 11 by B. Jekhovsky at Algiers.

Named by the discoverer in honor of Mme. L. Boulanger. (H 110)

(1182) Ilona

1927 EA. Discovered 1927 March 3 by K. Reinmuth at Heidelberg.

Any reference of this name to a person or occurence is unknown.

Name proposed by G. Stracke.

(1183) Jutta

1930 DC. Discovered 1930 February 22 by K. Reinmuth at Heidelberg.

Any reference of this name to a person or occurence is unknown.

Name proposed by G. Stracke.

(1184) Gaea

1926 RE. Discovered 1926 September 5 by K. Reinmuth at Heidelberg.

Named after the Greek Earth goddess. After Chaos, Gaea appeared and bore Uranus, the personification of Heaven that covered the Earth. Gaea united with her son Uranus to produce the first race, the twelve Titans, six males and six females. Uranus and Gaea were also the parents of the Cyclops, the Centimani, and the Giants. (LDS; Z 108)

Name proposed by G. Stracke.

(1185) Nikko

1927 WC. Discovered 1927 November 17 by O. Oikawa at Tokyo.

Named for a popular city and tourist resort in Japan noted for its shrines. (H 110)

(1186) Turnera

1929 PL. Discovered 1929 August 1 by C. Jackson at Johannesburg.

Named in honor of the British astronomer Herbert Hall Turner (1861-1930), director of the Oxford University Observatory. (H 110)

(1187) Afra

1929 XC. Discovered 1929 December 6 by K. Reinmuth at Heidelberg.

Any reference of this name to a person or occurence is unknown.

Name proposed by G. Stracke.

(1188) Gothlandia

1930 SB. Discovered 1930 September 30 by J. Comas Solá at Barcelona. Independently discovered 1930 October 17 by G. N. Neujmin at Simeïs and October 18 by K. Nakamura at Kyoto.

This is the ancient name of Catalonia which has the city of Barcelona {see planet (945)} as the capital. (H 110)

(1189) Terentia

1930 SG. Discovered 1930 September 17 by G. N. Neujmin at Simeïs.

Named in remembrance of the deceased collaborator at the Simeïs Observatory, Mrs. Lidiya Ivanovna Terent'eva (1879-1933), an orbit computer. (RI 762)

Name proposed by the scientific collaborators of the Simeïs Observatory.

(1190) Pelagia

1930 SL. Discovered 1930 September 20 by G. N. Neujmin at Simeïs.

Named in honor of the Simeïs astronomer Pelageya Fedorovna Shajn (1894-1956), the first woman who discovered a minor planet (planet (1112)). (RI 895) See also the citation for planet (1648).

(1191) Alfaterna

1931 CA. Discovered 1931 February 11 by L. Volta at Pino Torinese.

The ancient city founded by the Oschi about 1000 B.C. between Pompeii and Salerno. It is entombed beneath the present Nocera Superiore, birthplace of A. Fresa, who proposed the name (in 1957). (M 2882)

(1192) Prisma

1931 FE. Discovered 1931 March 17 by A. Schwassmann at Bergedorf.

This planet was named in honor of the *Bergedorf Spectral Catalogue*. (H 111)

(1193) Africa

1931 HB. Discovered 1931 April 24 by C. Jackson at Johannesburg.

Named for the large continent in which Johannesburg is located. (H 111)

(1194) Aletta

1931 JG. Discovered 1931 May 13 by C. Jackson at Johannesburg.

Named in honor of the wife of the discoverer. (H 111)

(1195) Orangia

1931 KD. Discovered 1931 May 24 by C. Jackson at Johannesburg.

Named in honor of the South African province of Orange Free State. (H 111)

(1196) Sheba

1931 KE. Discovered 1931 May 21 by C. Jackson at Johannesburg.

This planet is named for the biblical queen of Sheba, who visited Solomon. (H 111)

(1197) Rhodesia

1931 LD. Discovered 1931 June 9 by C. Jackson at Johannesburg.

Named after the country in southern Africa which now has the name Zimbabwe. (H 111)

(1198) Atlantis

1931 RA. Discovered 1931 September 7 by K. Reinmuth at Heidelberg.
This planet is named for the legendary island which sank.
Name proposed by G. Stracke.

(1199) Geldonia

1931 RF. Discovered 1931 September 14 by E. Delporte at Uccle.

Geldonia is the Latin name of Geldenaken (Jodoigne), Belgium, the native town of the discoverer. (H 111)

(1200) Imperatrix

1931 RH. Discovered 1931 September 14 by K. Reinmuth at Heidelberg.

Imperatrix is the Latin name for empress. Any reference to a person is unknown. (I. van Houten-Groeneveld)
The name was proposed by G. Stracke.

(1201) Strenua

1931 RK. Discovered 1931 September 14 by K. Reinmuth at Heidelberg.

This name was assigned in honor of Prof. Gustav Stracke (1887-1943), famous German astronomer at the Berlin Rechen-Institut, without his knowledge of the meaning. The name is from the Latin word "strenuus" which means strong, careful, and exemplifies virtues of G. Stracke. The German substantive is "Stärke" which sounds quite similar as "Stracke". (H 111)

G. Stracke is also honored by minor planet (1019) and the eight planets (1227) to (1234).

(1202) Marina

1931 RL. Discovered 1931 September 13 by G. N. Neujmin at Simeïs. Independently discovered 1931 September 15 by K. Reinmuth at Heidelberg.

Named in honor of Mrs. Marina Davidovna Lavrova-Berg (1898-1943), scientific collaborator at the Pulkovo Observatory. (N. S. Samojlova-Yakhontova; N. Solovaya; N. S. Chernykh)

(1203) Nanna

1931 TA. Discovered 1931 October 5 by M. Wolf at Heidelberg.

This planet is named honoring the many paintings named "Nanna" by the German painter Anselm Feuerbach (1829-1880). One such painting was in the possession of the Wolf family. (H 111)

(1204) Renzia

1931 TE. Discovered 1931 October 6 by K. Reinmuth at Heidelberg.

Named in honor of Franz Robert Renz (1860-1942), astronomer at Dorpat and Pulkowo. (H 112)

(1205) Ebella

1931 TB_1. Discovered 1931 October 6 by K. Reinmuth at Heidelberg.

Named in honor of Carl Wilhelm Ludwig Martin Ebell (1871-1944) who was an astronomer in Kiel, Germany on the staff of the *Astronomische Nachrichten*. (H 112)

(1206) Numerowia

1931 UH. Discovered 1931 October 18 by K. Reinmuth at Heidelberg.

Named in honor of the Russian astronomer Boris Vasil'evich Numerov (1891-1941), the founder and director of the Institute of Theoretical Astronomy in Leningrad. (H 112)
Numerov is also honored by a lunar crater.

(1207) Ostenia

1931 VT. Discovered 1931 November 15 by K. Reinmuth at Heidelberg.

Named in honor of the amateur astronomer, orbital computer, and business man, Hans Osten. (H 112)

(1208) Troilus

1931 YA. Discovered 1931 December 31 by K. Reinmuth at Heidelberg.

Named for a son of Priam {see planet (884)} who in a medieval legend loved Cressida {see planet (548)} and lost her to Diomedes {see planet (1437)}. Troilus was killed by Achilles {see planet (588)}. (H 112)

(1209) Pumma

1927 HA. Discovered 1927 April 22 by K. Reinmuth at Heidelberg.
This is the nickname of the niece of A. Kahrstedt {see planet (1587)}. (H 112)
Name proposed by A. Kahrstedt.

(1210) Morosovia

1931 LB. Discovered 1931 June 6 by G. N. Neujmin at Simeïs.

Named in honor of the Russian revolutionist and scientist Nikolaj A. Morozov (1854-1946). (H 112)
Morozov is also honored by a lunar crater.

(1211) Bressole

1931 XA. Discovered 1931 December 2 by L. Boyer at Algiers.
Named in honor of a nephew of the discoverer. (H 112)

(1212) Francette

1931 XC. Discovered 1931 December 3 by L. Boyer at Algiers.
Named in honor of the wife of the discoverer. (H 112)

(1213) Algeria

1931 XD. Discovered 1931 December 5 by G. Reiss at Algiers. Independently discovered 1931 December 8 by G. Van Biesbroeck at Williams Bay.

Named for the country in North Africa and its capital city Algiers in which this planet was discovered. (H 112)

(1214) Richilde

1932 AA. Discovered 1932 January 1 by M. Wolf at Heidelberg. Independently discovered 1932 January 6 by K. Nakamura at Kyoto.
Any reference of this name to a person or occurence is unknown.

(1215) Boyer

1932 BA. Discovered 1932 January 19 by A. Schmitt at Algiers. Independently discovered 1932 January 27 by K. Reinmuth at Heidelberg.

Named in honor of Louis Boyer, astronomer at the Algiers Observatory and later at the Nice Observatory who has worked extensively on minor planets and comets. At Algiers he discovered 40 minor planets and at Nice has worked on identifications. (M 4418)

(1216) Askania

1932 BL. Discovered 1932 January 29 by K. Reinmuth at Heidelberg.

Named after the Askania works in Berlin, manufacturers of optical and astronomical instruments. (H 112)

(1217) Maximiliana

1932 EC. Discovered 1932 March 13 by E. Delporte at Uccle. Independently discovered 1932 March 12 by M. Wolf at Heidelberg.

Named in honor of Max Wolf (1863-1932), founder and director of the Heidelberg Königstuhl Observatory. Wolf is also honored by planet (827). (H 112)

The citation (RI 695) reads: "E. Delporte hat den Wunsch geäussert, den von ihm entdeckten Planeten, der auch von Max Wolf entdeckt ist, zu Ehren des Verstorbenen zu benennen. Auf Vorschlag von Frau G. Wolf hat der Planet den Namen erhalten."

(1218) Aster

1932 BJ. Discovered 1932 January 29 by K. Reinmuth at Heidelberg. Independently discovered 1932 January 31 by M. Ferrero at Pino Torinese.

Named after various chiefly fall-blooming leafy-stemmed composite herbs with often showy heads containing tubular flowers or both tubular and ray flowers. (H 113)

(1219) Britta

1932 CJ. Discovered 1932 February 6 by M. Wolf at Heidelberg.
Any reference of this name to a person or occurence is unknown.

(1220) Crocus

1932 CU. Discovered 1932 February 11 by K. Reinmuth at Heidelberg.

Named for a genus of herbs of the iris family having solitary long-tubed flowers and slender linear leaves. (H 113)

(1221) Amor

1932 EA_1. Discovered 1932 March 12 by E. Delporte at Uccle.

Amor is the Latin name for the Greek Eros, the god of love. Like (433) Eros this planet makes close approaches to the Earth. (H 113)

(1222) Tina

1932 LA. Discovered 1932 June 11 by E. Delporte at Uccle.

Tina is the Christian name of a woman, amateur astronomer, and friend of the discoverer. (H 113)

(1223) Neckar

1931 TG. Discovered 1931 October 6 by K. Reinmuth at Heidelberg. Independently discovered 1931 October 11 by F. Rigaux at Uccle.

Named for the Neckar river which has its origin in the Black Forest, crosses the city of Heidelberg, and flows into the Rhine. (H 113)

(1224) Fantasia

1927 SD. Discovered 1927 August 29 by S. I. Belyavskij and N. Ivanov at Simeïs. Independently discovered 1927 September 17 by E. Delporte at Uccle.

Possibly named for a work as a poem or play in which the author's fancy roves unrestricted. Fantasia is the Latin word for fantasy. (LDS)

(1225) Ariane

1930 HK. Discovered 1930 April 23 by H. van Gent at Johannesburg.

Named for the principal role, Ariane Leprieur in the play *Le Chemin de Crête* by the French dramatist Gabriel Marcel (1889-1973). (H 113)

(1226) Golia

1930 HL. Discovered 1930 April 22 by H. van Gent at Johannesburg.

Named in honor of Golius, the first professor of astronomy and founder of the Leiden Observatory in 1633. He also bought the quadrant, used by Snellius for his trigonometry of the Netherlands, which is now in the Museum of History and Sciences in Leiden. (I. van Houten-Groeneveld)

(1227) Geranium

1931 TD. Discovered 1931 October 5 by K. Reinmuth at Heidelberg.

Named after a genus of plants having regular flowers without spurs and with glands that alternate with the petals. (H 113)
See also the citation for planet (1234).

(1228) Scabiosa

1931 TU. Discovered 1931 October 5 by K. Reinmuth at Heidelberg.

Named after a genus of herbs of the teasel family with terminal flower heads subtended by a leafy involucre. (H 113)
See also the citation for planet (1234).

(1229) Tilia

1931 TP_1. Discovered 1931 October 9 by K. Reinmuth at Heidelberg.
Named after a plant (limetree, basswood) from the Tiliaceae family. (H 113)
See also the citation for planet (1234).

(1230) Riceia

1931 TX_1. Discovered 1931 October 9 by K. Reinmuth at Heidelberg.

Named in honor of Hugh Rice, amateur astronomer of New York and director of the Museum of Natural Sciences. See also the citation for planet (1234). (H 113; LDS)

Name proposed by I. L. Meyer and endorsed by G. Stracke. See also the remarks to planet (1249).

(1231) Auricula

1931 TE_2. Discovered 1931 October 10 by K. Reinmuth at Heidelberg. Independently discovered 1931 October 11 by F. Rigaux at Uccle.
Named after the yellow flowered Alpine primrose. (H 113)
See also the citation for planet (1234).

(1232) Cortusa

1931 TF_2. Discovered 1931 October 10 by K. Reinmuth at Heidelberg.
Named after a plant of the primrose family. (H 114)
See also the citation for planet (1234).

(1233) Kobresia

1931 TG_2. Discovered 1931 October 10 by K. Reinmuth at Heidelberg.
Kobresia is the name of a grass. (I. van Houten-Groeneveld)
See also the citation for planet (1234).

(1234) Elyna

1931 UF. Discovered 1931 October 18 by K. Reinmuth at Heidelberg.

This is a name of a plant. The first letters of planets (1227) through (1234) form the name G. Stracke. Due to the desire of Stracke that no planet be named after him it was in this manner that the discoverer was able to circumvent and to honor the noted German astronomer and diligent orbit computer. Stracke later was also honored by planet (1019). (H 114)

The same method to honor an astronomer by using consecutive initial letters was used again in 1995 - see the citation of planet (5699).

(1235) Schorria

1931 UJ. Discovered 1931 October 18 by K. Reinmuth at Heidelberg.

Named in honor of the German astronomer Richard Schorr (1867-1951), staff member of the Hamburg Bergedorf Observatory. He discovered the periodic comet 1918 III and the planets (869) and (1240). (H 114)

Named by the Astronomisches Rechen-Institut in accordance with the discoverer (RI 862).

Schorr is also honored by a lunar crater.

(1236) Thaïs

1931 VX. Discovered 1931 November 6 by G. N. Neujmin at Simeïs.

Named for the famous Athenian hetaera who accompanied Alexander the Great (356-323 B.C.) in Asia. She later joined king Ptolemy I, from whom she had three children. Thaïs also is the title figure of a novel (1890) of the French novelist and satirist and 1921 Nobel laureate Anatole France (1844-1924, pseudonym of Jacques Anatole François Thibault). (N. Solovaya; A. Paluzíe-Borrell)

(1237) Geneviève

1931 XB. Discovered 1931 December 2 by G. Reiss at Algiers. Independently discovered 1931 December 3 by G. N. Neujmin at Simeïs.

Named in honor of the eldest daughter of the discoverer. (H 114)

(1238) Predappia

1932 CA. Discovered 1932 February 4 by L. Volta at Pino Torinese.

Named after the small village of Predappio near Forli, Italy. Predappio was the birthplace of "il Duce" Mussolini and, obviously, this is another clear instance of homage to him. (P. G. Comba)

(1239) Queteleta

1932 CB. Discovered 1932 February 4 by E. Delporte at Uccle. Independently discovered 1932 February 4 by L. Boyer at Algiers and February 13 by G. Van Biesbroeck at Williams Bay.

Named in honor of Lambert Adolphe Jacques Quetelet (1796-1874), Belgian astronomer, mathematician, meteorologist, statistician and in 1828 first director of the Royal Observatory of Belgium which he helped to found. (H 114)

Quetelet is also honored by a lunar crater.

(1240) Centenaria

1932 CD. Discovered 1932 February 5 by R. Schorr at Bergedorf.

Named on the occasion of the 100th anniversary of the Hamburg Observatory (1933 Oct. 31). (H 114)

(1241) Dysona

1932 EB_1. Discovered 1932 March 4 by H. E. Wood at Johannesburg.

Named in honor of Sir Frank Watson Dyson (1868-1939), Astronomer Royal of England, director of the Greenwich Observatory and president of the International Astronomical Union 1928-1932. (RI 814, H 114)

Dyson is also honored by a lunar crater.

(1242) Zambesia

1932 HL. Discovered 1932 April 28 by C. Jackson at Johannesburg.

The name applies to the former British territories in the Zambezi Basin, southern Africa. (H 114)

(1243) Pamela

1932 JE. Discovered 1932 May 7 by C. Jackson at Johannesburg.

Named in honor of the daughter of the discoverer. (H 114)

(1244) Deira

1932 KE. Discovered 1932 May 25 by C. Jackson at Johannesburg.
Named after the ancient name of the birthplace of the discoverer. (H 115) Jackson (see planet (2193)) was born in the town of Ossett, Yorkshire.

(1245) Calvinia

1932 KF. Discovered 1932 May 26 by C. Jackson at Johannesburg.
Named for a village in the Cape Province, South Africa. (H 115)

(1246) Chaka

1932 OA. Discovered 1932 July 23 by C. Jackson at Johannesburg.

Named for Chaka (or Tchaka), king of the Zulu tribe and founder of the Zulu empire in 1812. (H 115)

(1247) Memoria

1932 QA. Discovered 1932 August 30 by M. Laugier at Uccle. Independently discovered 1932 August 30 by G. N. Neujmin at Simeïs and September 6 by K. Reinmuth at Heidelberg.

This is the Latin word for remembrance. The discoverer was often reminded of her pleasant relationship while in Uccle in 1932. (H 115)

(1248) Jugurtha

1932 RO. Discovered 1932 September 1 by C. Jackson at Johannesburg. Independently discovered 1932 September 29 by G. N. Neujmin at Simeïs.

Named after the Numidian king (160-104 B.C.) and enemy of Rome. Jugurtha was throttled to death in Rome. (H 115)

(1249) Rutherfordia

1932 VB. Discovered 1932 November 4 by K. Reinmuth at Heidelberg. Independently discovered 1932 November 29 by E. Delporte at Uccle.
Named after the city of Rutherford, New York. (LDS)

Name proposed by Irving L. Meyer, Rutherford, N.Y. and endorsed by G. Strakke (RI 1534).

The name is erroneously interpreted in many sources as belonging to the 1908 Nobel Prize winner Lord Ernest Rutherford (1871-1937). The correct interpretation, however, has been found on a postcard, dated Feb. 2, 1937, from G. Stracke to the discoverer: "Ein amerikanischer Kollege, der selber keinen Planeten entdeckt hat, möchte gern zwei Planeten mit dem Namen "Riceia" {see planet (1230)} (Leiter Rice vom Amerik. Museum der Naturwissenschaften in New York) und "Rutherfordia" (Rutherford ist der Ort einer Privatsternwarte) taufen..."

(1250) Galanthus

1933 BD. Discovered 1933 January 25 by K. Reinmuth at Heidelberg.
Named after a flower, the snowdrop. (H 115)

(1251) Hedera

1933 BE. Discovered 1933 January 25 by K. Reinmuth at Heidelberg.
Named after a plant of the ivy family. (H 115)

(1252) Celestia

1933 DG. Discovered 1933 February 19 by F. L. Whipple at Cambridge, Massachusetts.

Named by the discoverer in honor of his mother. For the naming of this planet the spelling needed no change. (H 115)

(1253) Frisia

1931 TV_1. Discovered 1931 October 9 by K. Reinmuth at Heidelberg. Independently discovered 1931 November 6 by P. F. Shajn at Simeïs.

Named for the people that inhabit principally the Netherlands province of Friesland and the Friesian islands in the North sea. (H 115)

(1254) Erfordia

1932 JA. Discovered 1932 May 10 by J. Hartmann at La Plata.

Named (RI 835) by J. Hartmann for the city of Erfurt, Germany, which was the birthplace of the discoverer. (H 115)

(1255) Schilowa

1932 NC. Discovered 1932 July 8 by G. N. Neujmin at Simeïs.

Named in honor of M. W. Schilowa (1870-1934), astronomer and orbit computer at the Pulkovo Observatory. (H 115)
The spelling of the name originally was "Shilowa" (RI 1160).

(1256) Normannia

1932 PD. Discovered 1932 August 8 by K. Reinmuth at Heidelberg.
This planet is probably named for the inhabitants of Normandy. (H 115)

(1257) Móra

1932 PE. Discovered 1932 August 8 by K. Reinmuth at Heidelberg.

Named in honor of the Hungarian astronomer Károly Móra (1899-1938). (H 116; I. van Houten-Groeneveld)

The number and name of this planet were later used for another object in a rather unusual manner - see the citation for planet (2517).

(1258) Sicilia

1932 PG. Discovered 1932 August 8 by K. Reinmuth at Heidelberg.
Named for the island of Sicily in the Mediterranean sea. (H 116)

(1259) Ógyalla

1933 BT. Discovered 1933 January 29 by K. Reinmuth at Heidelberg.

Named for the Hungarian seismological, meteorological and astronomical observatory. (H 116)

(1260) Walhalla

1933 BW. Discovered 1933 January 29 by K. Reinmuth at Heidelberg.

Named for the memorial hall near Regensburg, Germany. In Norse mythology it was the hall of Odin {see planet (3989)} where warriors who have died in battle are received. (H 116)

(1261) Legia

1933 FB. Discovered 1933 March 23 by E. Delporte at Uccle.
Legia is the Latin name for the city of Liège (Luik), Belgium. (H 116)

(1262) Sniadeckia

1933 FE. Discovered 1933 March 23 by S. Arend at Uccle.

Named in honor of Jan Sniadecki (1756-1830), a Polish scholar, professor of mathematics and astronomy, and founder of the Cracow Observatory. (H 116)
Named by T. Banachiewicz (RI 843).
Sniadecki is also honored by a lunar crater.

(1263) Varsavia

1933 FF. Discovered 1933 March 23 by S. Arend at Uccle.

Varsavia is the Latin name for Warsaw, Poland. The citation reads: "zum Dank der Stadt Warschau für ihre der Sternwarte Lubomir (Beskiden) geleistete Hilfe." (RI 843)
Named by T. Banachiewicz.

(1264) Letaba

1933 HG. Discovered 1933 April 21 by C. Jackson at Johannesburg.
Named for a river in Transvaal, South Africa. (H 116)

(1265) Schweikarda

Discovered 1911 October 18 by F. Kaiser at Heidelberg.
Named by the discoverer for his mother's family name, Schweikard. (H 116)

(1266) Tone

1927 BD. Discovered 1927 January 23 by O. Oikawa at Tokyo. Independently discovered 1927 January 24 by G. N. Neujmin at Simeïs.
Named for the largest river in Japan. (H 116)

(1267) Geertruida

1930 HD. Discovered 1930 April 23 by H. van Gent at Johannesburg. Independently discovered 1930 April 28 by K. Reinmuth at Heidelberg.

Named in honor of the daughter of the Mrs. Hamerslag, a sister of G. Pels. (I. van Houten-Groeneveld)
Named by G. Pels {see planet (1667)} who computed the orbit.

(1268) Libya

1930 HJ. Discovered 1930 April 29 by C. Jackson at Johannesburg.

Named for the country in northern Africa bordering the Mediterranean sea. (H 116)

(1269) Rollandia

1930 SH. Discovered 1930 September 20 by G. N. Neujmin at Simeïs.

Named in honor of the French writer and 1915 Nobel laureate Romain Rolland (1866-1944). (H 116)

(1270) Datura

1930 YE. Discovered 1930 December 17 by G. Van Biesbroeck at Williams Bay. Independently discovered 1930 December 20 by M. Wolf at Heidelberg.

Named for a genus of widely distributed strong-scented herbs, shrubs, or trees of the nightshade family (e.g. hawthorn). (H 116)

(1271) Isergina

1931 TN. Discovered 1931 October 10 by G. N. Neujmin at Simeïs.

Named in honor of the friend and physician of the discoverer, Pyotr Vasil'evich Isergin (1870-1936) who treated him. (H 117; G. R. Kastel'; I. I. Neyachenko)

(1272) Gefion

1931 TZ_1. Discovered 1931 October 10 by K. Reinmuth at Heidelberg.

Named for a figure in Norse mythology and for the Gefion spring in Copenhagen. (H 117)

(1273) Helma

1932 PF. Discovered 1932 August 8 by K. Reinmuth at Heidelberg.

Named in honor of an acquaintance of the German astronomer W. Schaub. (H 117)

(1274) Delportia

1932 WC. Discovered 1932 November 28 by E. Delporte at Uccle.

Named in honor of the discoverer, Eugène J. Delporte (1882-1955). He observed and discovered more than 60 minor planets at the Uccle Observatory where he was director from 1936 to 1947. (H 117)
Name proposed by G. Stracke.
Delporte is also honored by a lunar crater.

(1275) Cimbria

1932 WG. Discovered 1932 November 30 by K. Reinmuth at Heidelberg.

Named for the people encountered by the Romans in Noricum about 113 B.C. At first victorious, they were destroyed by Marius in 101 B.C. (H 117)

(1276) Ucclia

1933 BA. Discovered 1933 January 24 by E. Delporte at Uccle. Independently discovered 1933 January 26 by R. Schorr at Bergedorf.

Named in honor of Uccle, both the city and the observatory, where this planet was discovered. (H 117)

(1277) Dolores

1933 HA. Discovered 1933 April 18 by G. N. Neujmin at Simeïs.

Named in honor of Dolores Gómez Ibarruri ("La Pasionária") (1895-1989), a leading figure of the Spanish communists. In 1920, she was co-founder of the communist party in Spain. She served as general secretary (1942-1960) and chairwoman (1960-1967) of the exile party. She returned to her home country in 1977 and was selected as member of the parliament. (H 117)

(1278) Kenya

1933 LA. Discovered 1933 June 15 by C. Jackson at Johannesburg.
Named for the country in eastern Africa. (H 117)

(1279) Uganda

1933 LB. Discovered 1933 June 15 by C. Jackson at Johannesburg.
Named for the country in central Africa, north of Lake Victoria. (H 117)

(1280) Baillauda

1933 QB. Discovered 1933 August 18 by E. Delporte at Uccle. Independently discovered 1933 August 19 by G. N. Neujmin at Simeïs.

Named in honor of Jules Baillaud (1876-1960), astronomer at the Paris Observatory and director of the Pic-du-Midi Observatory from 1937 to 1947. (H 117)
Baillaud is also honored by a lunar crater.

(1281) Jeanne

1933 QJ. Discovered 1933 August 25 by S. Arend at Uccle.
Named in honor of the daughter of the discoverer. (H 117)

(1282) Utopia

1933 QM_1. Discovered 1933 August 17 by C. Jackson at Johannesburg.

Named for that imaginary country, a place of ideal perfection especially in laws, government, and social conditions. It was described 1516 by the English statesman and author Sir Thomas More (1478-1535). (H 117)

(1283) Komsomolia

1925 SC. Discovered 1925 September 25 by V. A. Albitskij at Simeïs. Independently discovered 1925 October 10 by K. Reinmuth at Heidelberg.

The name of the communistic youth organization of {the former} USSR. (M 838)

(1284) Latvia

1933 OP. Discovered 1933 July 27 by K. Reinmuth at Heidelberg. Independently discovered 1933 August 19 by G. N. Neujmin at Simeïs.

Named for the country of Latvia. Since 1991 Latvia again is an independent state. (H 118)
Name proposed by J. Stein.

(1285) Julietta

1933 QF. Discovered 1933 August 21 by E. Delporte at Uccle.
Named in honor of the discoverer's mother. (J. Meeus)

(1286) Banachiewicza

1933 QH. Discovered 1933 August 25 by S. Arend at Uccle.

Named in honor of the Polish astronomer Tadeusz Banachiewicz (1882-1954), director of the Cracow Observatory. (H 118)

Obituary published in l'Astronomie, Vol. 69, p. 290 (1955). Banachiewicz is also honored by a lunar crater.

(1287) Lorcia

1933 QL. Discovered 1933 August 25 by S. Arend at Uccle.
Named in honor of the wife of T. Banachiewicz. (H 118)
Named by T. Banachiewicz.

(1288) Santa

1933 QM. Discovered 1933 August 26 by E. Delporte at Uccle.
Any reference of this name to a person or occurence is unknown.
Named by De Caro who computed the orbit.

(1289) Kutaïssi

1933 QR. Discovered 1933 August 19 by G. N. Neujmin at Simeïs. Independently discovered 1933 August 25 by E. Delporte at Uccle and September 11 by C. Jackson at Johannesburg.
Named for a city in the Georgian Republic. (H 118)

(1290) Albertine

1933 QL_1. Discovered 1933 August 21 by E. Delporte at Uccle.
Named in honor of Albert I (1875-1934), King of the Belgians. (H 118)

(1291) Phryne

1933 RA. Discovered 1933 September 15 by E. Delporte at Uccle.

Named for a hetaera of the 4th century B.C. celebrated for her beauty. She was the model for several statues including the Cnidian Aphrodite of Praxiteles {see planet (5983)} and also, it is said, for the Aphrodite Anadyomene of Apelles. (H 118)

(1292) Luce

1933 SH. Discovered 1933 September 17 by F. Rigaux at Uccle.
Named in honor of the wife of the discoverer. (H 118)

(1293) Sonja

1933 SO. Discovered 1933 September 26 by E. Delporte at Uccle. Independently discovered 1933 September 28 by G. N. Neujmin at Simeïs.
Any reference of this name to a person or occurence is unknown.
Name proposed by the Astronomisches Rechen-Institut (RI 1039).

The naming might be influenced by the two letters of the provisional designation 1933 SO. See also the remarks to planet (579).

(1294) Antwerpia

1933 UB_1. Discovered 1933 October 24 by E. Delporte at Uccle.
Named for the city of Antwerp, Belgium. (H 118)

(1295) Deflotte

1933 WD. Discovered 1933 November 25 by L. Boyer at Algiers.
Named by the discoverer in honor of his nephew. (H 118)

(1296) Andrée

1933 WE. Discovered 1933 November 25 by L. Boyer at Algiers.
Named by the discoverer in honor of his niece. (H 118)

(1297) Quadea

1934 AD. Discovered 1934 January 7 by K. Reinmuth at Heidelberg.

Named by the discoverer in honor of the parents-in-law of his brother, Prof. E. Reinmuth. (H 119)

(1298) Nocturna

1934 AE. Discovered 1934 January 7 by K. Reinmuth at Heidelberg.
Nocturna is the feminine adjective of nocturnus and means nightly. (H 119)
Name proposed by G. Stracke.

(1299) Mertona

1934 BA. Discovered 1934 January 18 by G. Reiss at Algiers.
Named for the English astronomer Gerald Merton {1893-1983}. (H 119)
Obituary published in J. Br. Astron. Assoc., Vol. 94, No. 1, p. 28-30 (1983).

(1300) Marcelle

1934 CL. Discovered 1934 February 10 by G. Reiss at Algiers. Independently discovered 1934 February 14 by S. Arend at Uccle.
Named in honor of the second daughter of the discoverer. (H 119)

(1301) Yvonne

1934 EA. Discovered 1934 March 7 by L. Boyer at Algiers.
Named in honor of the sister of the discoverer. (H 119)

(1302) Werra

Discovered 1924 September 28 by K. Reinmuth at Heidelberg.

Named for the German river which joins the Fulda at Münden to form the Weser. (H 119)

(1303) Luthera

1928 FP. Discovered 1928 March 16 by A. Schwassmann at Bergedorf.

Named in honor of the German astronomer Karl Theodor Robert Luther (1822-1900) who, at the Bilk Observatory near Düsseldorf, discovered 24 minor planets between 1852 and 1890. (H 119)
Luther is also honored by a lunar crater.

(1304) Arosa

1928 KC. Discovered 1928 May 21 by K. Reinmuth at Heidelberg.
Named for the small city in Switzerland. (H 119)

(1305) Pongola

1928 OC. Discovered 1928 July 19 by H. E. Wood at Johannesburg.
Named by the discoverer for a river in South Africa. (H 119)

(1306) Scythia

1930 OB. Discovered 1930 July 22 by G. N. Neujmin at Simeïs.

Named for the country of the ancient Scythians comprising parts of Europe and Asia now in the U.S.S.R. in regions north of the Black sea and east of the Aral sea. (H 119)

(1307) Cimmeria

1930 UF. Discovered 1930 October 17 by G. N. Neujmin at Simeïs.

Named after the ancient inhabitants of the Crimea peninsula. The Cimmerians were expelled by the Scythians {see planet (1306)} in the 7th century B.C. (A. N. Deutsch)

(1308) Halleria

1931 EB. Discovered 1931 March 12 by K. Reinmuth at Heidelberg.

Named in honor of Albrecht von Haller (1708-1777), physician, botanist, and poet at the 1935 meeting of the Astronomische Gesellschaft in Berne, Switzerland. (I. van Houten-Groeneveld)

(1309) Hyperborea

1931 TO. Discovered 1931 October 11 by G. N. Neujmin at Simeïs. Independently discovered 1931 October 13 by E. Delporte at Uccle.

Named for the region in Greek mythology which is the home of a group of people placed by Herodotus {see planet (3092)} in the extreme north who were especially associated with the cult of Apollo {see planet (1862)}. The land was so far north that it was regarded as the region of perpetual sunshine as it was beyond the north wind. (N. S. Samojlova-Yakhontova)

(1310) Villigera

1932 DB. Discovered 1932 February 28 by A. Schwassmann at Bergedorf.

Named in honor of the Swiss astronomer Walter Augustin Villiger (1872-1938) who was head of the department for astronomical instruments of Carl Zeiss, Jena. (H 120)

(1311) Knopfia

1933 FF_1. Discovered 1933 March 24 by K. Reinmuth at Heidelberg.

Named in honor of the German astronomer O. Knopf {1856-1945} who worked at Jena. (H 120)

Name proposed by H. Vogt (RI 1445).

(1312) Vassar

1933 OT. Discovered 1933 July 27 by G. Van Biesbroeck at Williams Bay.

Named by Mrs. Maud W. Makemson who computed the orbit while teaching in the women's college of Vassar, Poughkeepsie, N.Y. (H 120)

(1313) Berna

1933 QG. Discovered 1933 August 24 by S. Arend at Uccle.

Named for the capital of Switzerland. (H 120)

Name proposed by S. Mauderli of the University of Berne who computed the definitive orbit of this planet.

The planet was first announced (RI 1156) as "Bernia" but Mauderli (RI 1161) requested the actual name.

(1314) Paula

1933 SC. Discovered 1933 September 16 by S. Arend at Uccle.

Named by the discoverer in honor of his wife. (H 120)

(1315) Bronislawa

1933 SF_1. Discovered 1933 September 16 by S. Arend at Uccle.

Named for the Polish saint Bronislava (?-1259), a cousin of St. Hyazinth of Poland. She was a professed Premonstratensian nun, but died a recluse. (H 120; R. Bremer)

Name proposed (RI 1254) by Miss L. Stankiewicz.

(1316) Kasan

1933 WC. Discovered 1933 November 17 by G. N. Neujmin at Simeïs.

Named for the city on the Volga river east of Moscow and for the Kasan Observatory. (H 120)

(1317) Silvretta

1935 RC. Discovered 1935 September 1 by K. Reinmuth at Heidelberg. Independently discovered 1935 September 19 by C. Jackson at Johannesburg and P. F. Shajn at Simeïs.
Named for a peak (3282 m) and a lake in Graubünden, Swiss Alps. (H 120)

(1318) Nerina

1934 FG. Discovered 1934 March 24 by C. Jackson at Johannesburg.

Named after a genus of south African bulbous herbs from the amaryllidaceae family. (H 120)

(1319) Disa

1934 FO. Discovered 1934 March 19 by C. Jackson at Johannesburg.

Named for a large genus of tropical African terrestrial orchids, perhaps the most showy of all orchids. (H 120)

(1320) Impala

1934 JG. Discovered 1934 May 13 by C. Jackson at Johannesburg.

Named after the large brownish African antelope. The male has slender lyrate horns. (H 120)

(1321) Majuba

1934 JH. Discovered 1934 May 7 by C. Jackson at Johannesburg.
Named for a mountain in northwest Natal, South Africa. (H 120)

(1322) Coppernicus

1934 LA. Discovered 1934 June 15 by K. Reinmuth at Heidelberg. Independently discovered 1934 June 15 by E. Delporte at Uccle.

Named for the great Polish astronomer Nicolaus Copernicus (1473-1543), the founder of the heliocentric planetary system and of modern astronomy. (H 120)
Copernicus is also honored by a lunar and a Martian crater.

(1323) Tugela

1934 LD. Discovered 1934 May 19 by C. Jackson at Johannesburg.
Named after a river in Natal, South Africa. (H 121)

(1324) Knysna

1934 LL. Discovered 1934 June 15 by C. Jackson at Johannesburg.

Named by the discoverer for the village in the Cape Province, South Africa. (H 121)

(1325) Inanda

1934 NR. Discovered 1934 July 14 by C. Jackson at Johannesburg.
This is a name of a village community inhabited by the Zulus. (H 121)

(1326) Losaka

1934 NS. Discovered 1934 July 14 by C. Jackson at Johannesburg.

Named for the capital of northern Rhodesia. Lusaka today is the capital of Zambia. (H 121)

(1327) Namaqua

1934 RT. Discovered 1934 September 7 by C. Jackson at Johannesburg.
Named after a coastal region in Namibia, southwest Africa. (H 121)

(1328) Devota

1925 UA. Discovered 1925 October 21 by B. Jekhovsky at Algiers.

Named by the discoverer in honor of his friend Fortunato Devoto, director of the La Plata Observatory and president of the National Council of Observatories of Argentina. (H 121)

(1329) Eliane

1933 FL. Discovered 1933 March 23 by E. Delporte at Uccle.

Named in honor of the daughter of Prof. Paul Bourgeois {see planet (1543)} of the Royal Observatory at Uccle, Belgium. (H 121)

(1330) Spiridonia

1925 DB. Discovered 1925 February 17 by V. A. Albitskij at Simeïs.

Named in honor of Spiridon Il'ich Zaslavskij (1883-1942), the brother of Mrs. Albitskij. Zaslavskij was the uncle of Vitja {see planet (1030)}. (M 2882; LDS)

(1331) Solvejg

1933 QS. Discovered 1933 August 25 by G. N. Neujmin at Simeïs. Independently discovered 1933 September 11 by C. Jackson at Johannesburg.

Named for the heroine of Henrik Ibsen's (1828-1906) {see planet (5696)} drama *Peer Gynt*. (N. S. Samojlova-Yakhontova)

(1332) Marconia

1934 AA. Discovered 1934 January 9 by L. Volta at Pino Torinese.

Named in honor of Marchese Guglielmo Marconi (1874-1937), Italian electrical engineer and inventor and winner of the physics Nobel prize in 1909 (together with K. F. Braun). (H 121)

Marconi is also honored by a lunar crater.

(1333) Cevenola

1934 DA. Discovered 1934 February 20 by O. Bancilhon at Algiers.

Named for the Cévennes mountain range in southern France at the eastern edge of the Massif Central. (H 121)

(1334) Lundmarka

1934 OB. Discovered 1934 July 16 by K. Reinmuth at Heidelberg.

Named in honor of the Swedish astronomer Knut Emil Lundmark (1889-1958). Lundmark studied extensively globular clusters and galaxies and determined absolute stellar magnitudes as well as galactic distances. He also contributed to our knowledge in the field of history of astronomy. (H 121)

Lundmark is also honored by a lunar crater.

(1335) Demoulina

1934 RE. Discovered 1934 September 7 by K. Reinmuth at Heidelberg. Independently discovered 1934 September 13 by E. Delporte at Uccle.

Named in honor of Prof. Demoulin of the University of Ghent, Belgium. (H 121)

(1336) Zeelandia

1934 RW. Discovered 1934 September 9 by H. van Gent at Johannesburg.

Named for a province in southwestern Netherlands. (H 121)

(1337) Gerarda

1934 RA_1. Discovered 1934 September 9 by H. van Gent at Johannesburg.

Named in honor of Mrs. G. Prins, wife of Mr. Prins who was a computer at the Leiden Observatory. (I. van Houten-Groeneveld)

Named by G. Pels who computed the orbit.

(1338) Duponta

1934 XA. Discovered 1934 December 4 by L. Boyer at Algiers.
Named by the discoverer in honor of his nephew Marc Dupont. (H 122)

(1339) Désagneauxa

1934 XB. Discovered 1934 December 4 by L. Boyer at Algiers. Independently discovered 1934 December 8 by G. N. Neujmin at Simeïs and December 9 by E. Delporte at Uccle.
Named in honor of the brother-in-law of the discoverer. (H 122)

(1340) Yvette

1934 YA. Discovered 1934 December 27 by L. Boyer at Algiers.
Named by the discoverer in honor of his niece. (H 122)

(1341) Edmée

1935 BA. Discovered 1935 January 27 by E. Delporte at Uccle.
Named in honor of Mrs. Edmée Chandon who was an astronomer at the Paris Observatory. (H 122)

(1342) Brabantia

1935 CV. Discovered 1935 February 13 by H. van Gent at Johannesburg.
Named for the northern part of Brabant which is a province of the Netherlands. (I. van Houten-Groeneveld)

(1343) Nicole

1935 FC. Discovered 1935 March 29 by L. Boyer at Algiers. Independently discovered 1935 April 3 by E. Delporte at Uccle.
Named by the discoverer in honor of his niece. (H 122)

(1344) Caubeta

1935 GA. Discovered 1935 April 1 by L. Boyer at Algiers.
Named in honor of Paul Caubet (? -1942), astronomer at the Toulouse Observatory. (H 122)

(1345) Potomac

Discovered 1908 February 4 by J. H. Metcalf at Taunton.
Named for the river on which Washington, D.C. is located. The river flows from West Virginia into the Chesapeake Bay and forms the southern boundary of Maryland. (H 122)
See also the remarks to planet (1052).

(1346) Gotha

1929 CY. Discovered 1929 February 5 by K. Reinmuth at Heidelberg.
Named for the city of Gotha in the district of Erfurt in Thüringen {see, respectively, planets (1254) and (934)}, Germany. Gotha is wellknown in astronomy by his famous observatory and the work of F. X. von Zach {see planet (999)}. (H 122)

(1347) Patria

1931 VW. Discovered 1931 November 6 by G. N. Neujmin at Simeïs.
This is the Latin word for native country or fatherland. (H 122)

(1348) Michel

1933 FD. Discovered 1933 March 23 by S. Arend at Uccle.
Named in honor of the elder son of the discoverer. (H 122)

(1349) Bechuana

1934 LJ. Discovered 1934 June 13 by C. Jackson at Johannesburg.
Named for the region south of the Molopo river which became a province of

the Union of South Africa in 1895. It is now the independent state of Botswana. (H 122)

(1350) Rosselia

1934 TA. Discovered 1934 October 3 by E. Delporte at Uccle. Independently discovered 1934 November 3 by R. Schorr at Bergedorf.

Named in honor of Miss Rossel, editor of the Belgian newspaper *Le Soir*. (H 122)

(1351) Uzbekistania

1934 TF. Discovered 1934 October 5 by G. N. Neujmin at Simeïs.

Named in honor of the {former} Uzbek Soviet Socialist Republic, where the discoverer resided during World War II. (M 2882; M 2903; N. S. Chernykh)

This name was found in the discoverer's handwriting in his personal volume of "Kleine Planeten für 1941" by Dr. S. G. Makover.

(1352) Wawel

1935 CE. Discovered 1935 February 3 by S. Arend at Uccle.
Named for a castle of the ancient kings of Poland at Cracow. (H 123)
Name proposed by T. Banachiewicz.

(1353) Maartje

1935 CU. Discovered 1935 February 13 by H. van Gent at Johannesburg.

Named in honor of the daughter of B. G. Mekking who was a computer at the Leiden Observatory. (H 123)

(1354) Botha

1935 GK. Discovered 1935 April 3 by C. Jackson at Johannesburg.

Named after a distinguished South African patriot who was also one of the prime movers in the formation of the Union of South Africa. {Louis Botha (1862-1919) was the first prime minister of Transvaal (1907) and of the Union of South Africa (1910-1919)}. (M 908)

(1355) Magoeba

1935 HE. Discovered 1935 April 30 by C. Jackson at Johannesburg.

The name of a native chief of the North Transvaal, South Africa, who has given his name also to one of the beauty spots of Transvaal, viz. MagoebasKloof. (M 908)

(1356) Nyanza

1935 JH. Discovered 1935 May 3 by C. Jackson at Johannesburg.
Named for a region in southwestern Kenya with the capital Kisumu. (H 123)

(1357) Khama

1935 ND. Discovered 1935 July 2 by C. Jackson at Johannesburg.

Name of the late Paramount Chief of the Bechuana tribe and a distinguished native leader. (M 908)

(1358) Gaika

1935 OB. Discovered 1935 July 21 by C. Jackson at Johannesburg.

Name of a native chief of the Transkei, Cape Province, who figures largely in South African history. (M 909)

(1359) Prieska

1935 OC. Discovered 1935 July 22 by C. Jackson at Johannesburg.
Named for a village in the Cape Province, South Africa. (H 123)

(1360) Tarka

1935 OD. Discovered 1935 July 22 by C. Jackson at Johannesburg.

Same as above {cf. citation for planet (1358)}. His name is given to the South African town of Tarkastad. (M 909)

(1361) Leuschneria

1935 QA. Discovered 1935 August 30 by E. Delporte at Uccle.

Named in honor of the American astronomer Armin Otto Leuschner (1868-1953), director of Students' Observatory (now Leuschner Observatory), Berkeley, University of California. Leuschner worked extensively on celestial mechanics and is wellknown for his books *Celestial Mechanics* and *The Minor Planets of the Hecuba Group.* (H 123)

Named at the suggestion of S. Arend during his visit to Berkeley.

Leuschner is also honored by a lunar crater.

(1362) Griqua

1935 QG_1. Discovered 1935 July 31 by C. Jackson at Johannesburg.

Named for the tribe of mixed people of Bushman and Hottentot descent in Griqualand which is the district around the city of Kimberley, north of the Orange river. (H 124)

(1363) Herberta

1935 RA. Discovered 1935 August 30 by E. Delporte at Uccle.

Named in honor of Herbert Clark Hoover (1874-1964), 31st president of the United States from 1929 to 1933. He was also president of the Commission for Relief in Belgium. This planet was named for the president after his visit to Belgium in 1938. (H 124)

Hoover is also honored by planet (932).

(1364) Safara

1935 VB. Discovered 1935 November 18 by L. Boyer at Algiers.

Named in honor of André Safar, Algiers. (H 124)

(1365) Henyey

1928 RK. Discovered 1928 September 9 by M. Wolf at Heidelberg.

Named in honor of Dr. Louis G. Henyey {1910-1970}, late professor of astronomy at the University of California, Berkeley, a pioneer in the calculation of stellar interiors and stellar evolution, and a member of the U.S. National Academy of Sciences. (M 3143)

Name proposed by William Westbrooke.

Obituaries published in Publ. Astron. Soc. Pac., Vol. 82, p. 367-369 (1970); Sky Telesc., Vol. 39, p. 290 (1970). Henyey is also honored by a lunar crater.

(1366) Piccolo

1932 WA. Discovered 1932 November 29 by E. Delporte at Uccle.

Named in honor of M. d'Arsac, editor-in-chief of the Brussels newspaper *Le Soir* who used this pseudonym. (H 124)

(1367) Nongoma

1934 NA. Discovered 1934 July 3 by C. Jackson at Johannesburg. Independently discovered 1934 July 6 by K. Reinmuth at Heidelberg.

Named for the capital city of the Kwa-Zulu homeland in South Africa. (M 5181)

(1368) Numidia

1935 HD. Discovered 1935 April 30 by C. Jackson at Johannesburg.

Named for the ancient country in northern Africa, east of Mauretania, in modern Algeria. (H 124)

(1369) Ostanina

1935 QB. Discovered 1935 August 27 by P. F. Shajn at Simeïs. Independently discovered 1935 August 31 by K. Reinmuth at Heidelberg.

Named for a small town in the Perm district, USSR, the birthplace of the astronomer P. F. Shajn {see planet (1190)}. (M 838; LDS)

(1370) Hella

1935 QG. Discovered 1935 August 31 by K. Reinmuth at Heidelberg.

Named in honor of Helene Nowacki {1904-1972}, an astronomer at the Astronomisches Rechen-Institut, Heidelberg. (H 124)
Name proposed by G. Stracke (RI 1644).
Obituary published in Astron. Nachr., Band 294, p. 191 (1973).

(1371) Resi

1935 QJ. Discovered 1935 August 31 by K. Reinmuth at Heidelberg.
Named for a cousin of Mrs. Schaub, an acquaintance of the discoverer. (H 124)
Name proposed by W. Schaub (RI 1644).

(1372) Haremari

1935 QK. Discovered 1935 August 31 by K. Reinmuth at Heidelberg.

The name jointly honors all the women on the staff of the Astronomisches Rechen-Institut. It is combined from the word "harem" and the abbreviation "ARI". This often published version for the meaning of the name is not fully correct. Reinmuth has been asked by some collaborators of the ARI to 'donate' planets for their girl-friends, popular actresses etc. He compiled all these suggestions to this peculiar term. However, Reinmuth did not want to publish the original meaning and he, therefore, devised the interpretation of the first sentence in 1948. (I. van Houten-Groeneveld)

(1373) Cincinnati

1935 QN. Discovered 1935 August 30 by E. Hubble at Mount Wilson.

Since most of the orbit computations for this planet have been provided by the staff of the Cincinnati Observatory, this name has been recommended by the Minor Planet Center. (M 2116)

(1374) Isora

1935 UA. Discovered 1935 October 21 by E. Delporte at Uccle.

This name is a combination of the word "Isor" (which is the backward form of the feminine name "Rosi") and the traditional "a". (H 125)
Name proposed by G. Stracke.

(1375) Alfreda

1935 UB. Discovered 1935 October 22 by E. Delporte at Uccle.
Named in honor of a friend of the discoverer. (H 125)

(1376) Michelle

1935 UH. Discovered 1935 October 29 by G. Reiss at Algiers.
Named in honor of the third daughter of the discoverer. (H 125)

(1377) Roberbauxa

1936 CD. Discovered 1936 February 14 by L. Boyer at Algiers. Independently discovered 1936 February 20 by E. Delporte at Uccle.

Nommée en l'honneur de l'ingénieur français Robert Baux (1900-1987), ami d'enfance du découvreur Louis Boyer, qui proposa un nom soudé pour cet astéroïde. Robert Baux était spécialiste dans les études des productivité des grandes entreprises et devient directeur et administrateur des sociétés. C'était aussi un alpiniste au titre de capitaine de réserve des Chasseurs alpins et il participa à de nombreuses ascensions. (B. G. Marsden; J. Baux; M.-A. Combes)

The original naming citation bei Herget (H 125) read 'Named in honor of the engineer Mr. Roberbaux'. The correct name was first mentioned by Marsden. M.-A. Combes elucidates the full story through a contact with the widow of the honoree, Mrs. Jacqueline Baux. Citation by M.-A. Combes.

(1378) Leonce

1936 DB. Discovered 1936 February 21 by F. Rigaux at Uccle.
Named in honor of the father of the discoverer. (H 125)

(1379) Lomonosowa

1936 FC. Discovered 1936 March 19 by G. N. Neujmin at Simeïs. Independently discovered 1936 March 19 by P. Djurkovic at Uccle.

Named in honor of the first famous Russian physicist and astronomer, Mikhail Vasilevich Lomonosov, 1711-1765. (M 1252)

Name submitted by N. Yakhontova, Institute of Theoretical Astronomy, Leningrad.
Lomonosov is also honored by craters on Mars and the Moon.

(1380) Volodia

1936 FM. Discovered 1936 March 16 by L. Boyer at Algiers. Independently discovered 1936 March 21 by E. Delporte at Uccle.

Named in honor of Vladimir Vesselovskij (diminuted Volodia) who was born the same night the minor planet was discovered. (H 125)

(1381) Danubia

1930 QJ. Discovered 1930 August 20 by E. Skvortsov at Simeïs.

Named for the great European river Danube (or Donau) flowing from southern Germany into the Black sea. (H 125)
Name proposed by the orbit computer C. Popovici (RI 1534).

(1382) Gerti

1925 BB. Discovered 1925 January 21 by K. Reinmuth at Heidelberg.

Named in honor of Miss Gertrud Höhne who was a secretary at the Astronomisches Rechen-Institut in Berlin. (H 125)

Name proposed by L. Ringelmann (RI 1642). In the RI copy of Reinmuth, this name is framed by two hearts with red pencil.

(1383) Limburgia

1934 RV. Discovered 1934 September 9 by H. van Gent at Johannesburg.
Named for the province of Limburg in southern Netherlands. (H 125)

(1384) Kniertje

1934 RX. Discovered 1934 September 9 by H. van Gent at Johannesburg.

Named after the principal character in the Dutch play *Op Hoop van Zegen* by Herman Heyermans (1864-1924). (H 125)

(1385) Gelria

1935 MJ. Discovered 1935 May 24 by H. van Gent at Johannesburg.

Named for the province Gelderland in the eastern Netherlands bordering on the Ijsselmeer. Gelre is an old name of Gelderland. (H 125)

(1386) Storeria

1935 PA. Discovered 1935 July 28 by G. N. Neujmin at Simeïs. Independently discovered 1935 August 2 by E. Delporte at Uccle.

The orbit was computed from Van Biesbroeck's observations by a student (Miss Brenton) at the University of Kansas, and she named it after her professor, Dr. N. Wyman Storer. (RI 1602; H 125)

(1387) Kama

1935 QD. Discovered 1935 August 27 by P. F. Shajn at Simeïs. Independently discovered 1935 August 31 by K. Reinmuth at Heidelberg.

The name of a large river in USSR east of Kasan, and a tributary to the Volga. (M 838)

(1388) Aphrodite

1935 SS. Discovered 1935 September 24 by E. Delporte at Uccle.

Named for the Greek goddess of love and beauty, daughter of Zeus and Dione {see planets (5731) and (106)} in the Iliad but in later poems it is said that she sprung from the foam of the sea near Cyprus. According to later poems she was the mother of Eros {see planet (433)}. (H 126)

Name proposed by the Astronomisches Rechen-Institut (RI 1702).

(1389) Onnie

1935 SS_1. Discovered 1935 September 28 by H. van Gent at Johannesburg.

Named in honor of Mrs. A. Kruyt, a sister-in-law of G. Pels. (I. van Houten-Groeneveld)

Named by G. Pels {see planet (1667)}who computed the orbit.

(1390) Abastumani

1935 TA. Discovered 1935 October 3 by P. F. Shajn at Simeïs. Independently discovered 1935 October 3 by C. Jackson at Johannesburg.

The name of a town in the Georgian Caucasus; more recently the site of a new astronomical observatory. (M 838)

(1391) Carelia

1936 DA. Discovered 1936 February 16 by Y. Väisälä at Turku.

Named for the region in northeastern Europe between the Gulf of Finland and the White sea in the U.S.S.R. (H 126)

(1392) Pierre

1936 FO. Discovered 1936 March 16 by L. Boyer at Algiers.

Named in honor of a nephew of the discoverer. (H 126)

(1393) Sofala

1936 KD. Discovered 1936 May 25 by C. Jackson at Johannesburg.

Name of the largest province of the former Portuguese African Territory of Mozambique. (M 909)

(1394) Algoa

1936 LK. Discovered 1936 June 12 by C. Jackson at Johannesburg.

Name of a South African Bay which has historical associations. (M 909)

(1395) Aribeda

1936 OB. Discovered 1936 July 16 by K. Reinmuth at Heidelberg.

This name is an abbreviation formed by the first letters of "Astronomisches Rechen-Institut, Berlin-Dahlem". (H 126)

(1396) Outeniqua

1936 PF. Discovered 1936 August 9 by C. Jackson at Johannesburg.

Name of a range of mountains in the south western Cape Province and also home of a now extinct race of Hottentots. (M 909)

(1397) Umtata

1936 PG. Discovered 1936 August 9 by C. Jackson at Johannesburg.

Name of the capital town of the native province of the Transkei and seat of the native administration. (M 909)

(1398) Donnera

1936 QL. Discovered 1936 August 26 by Y. Väisälä at Turku.

Named in honor of the Finnish astronomer Anders Severin Donner (1854-1939), director of the Helsinki Observatory. (H 127)

Donner is also honored by a lunar crater.

(1399) Teneriffa

1936 QY. Discovered 1936 August 23 by K. Reinmuth at Heidelberg.

Named for the largest of the Spanish Canary islands. (H 127)

(1400) Tirela

1936 WA. Discovered 1936 November 17 by L. Boyer at Algiers.

Named in honor of Charles Tirel, a friend of the discoverer. (H 127)

(1401) Lavonne

1935 UD. Discovered 1935 October 22 by E. Delporte at Uccle.

Named in honor of the granddaughter of Mrs. Maud W. Makemson who computed the orbit of this planet and received permission by the discoverer to name it. (H 127)

(1402) Eri

1936 OC. Discovered 1936 July 16 by K. Reinmuth at Heidelberg.

Named in honor of Erika Kollnig-Schattschneider {1913-1978}, astronomer at the Heidelberg Königstuhl Observatory. (H 127)

(1403) Idelsonia

1936 QA. Discovered 1936 August 13 by G. N. Neujmin at Simeïs. Independently discovered 1936 August 17 by K. Reinmuth at Heidelberg.

Named in honor of the late, well-known Soviet astronomer, Naum I. Idelson {1885-1951}. He was a member of the staff of the Pulkovo Observatory. (M 1252)

Name submitted by N. Yakhontova, Institute of Theoretical Astronomy, Leningrad.

Idelson is also honored by a lunar crater.

(1404) Ajax

1936 QW. Discovered 1936 August 17 by K. Reinmuth at Heidelberg.

Named for the Greek hero in the Trojan War and son of Telamon {see planet (1749)} who kills himself because the armor of Achilles {see planet (588)} was awarded to Odysseus {see planet (1143)}. (H 127)

(1405) Sibelius

1936 RE. Discovered 1936 September 12 by Y. Väisälä at Turku.

Named in honor of the great Finnish composer Jean Sibelius (1865-1957). (M 3928)

(1406) Komppa

1936 RF. Discovered 1936 September 13 by Y. Väisälä at Turku.

Named in honor of G. Komppa, a chancellor of Turku University and one of the founders of the Turku Observatory. (H 127)

(1407) Lindelöf

1936 WC. Discovered 1936 November 21 by Y. Väisälä at Turku.

Named in honor of the late Ernest L. Lindelöf, professor of mathematics at the Helsinki University. (H 127)

(1408) Trusanda

1936 WF. Discovered 1936 November 23 by K. Reinmuth at Heidelberg.

Named in honor of Trude Hochgesand, an acquaintance of the Heidelberg astronomer H. Vogt {see planet (1439)}. (H 127)

(1409) Isko

1937 AK. Discovered 1937 January 8 by K. Reinmuth at Heidelberg.
Named in honor of Ise Koch, wife of the astronomer Kubach. (H 127)

(1410) Margret

1937 AL. Discovered 1937 January 8 by K. Reinmuth at Heidelberg.

Named in honor of Margret Braun (?-1991), wife of the Heidelberg astronomer H. Vogt {see planet (1439)}. (H 127)
Mrs. Vogt is also honored by planet (1411).

(1411) Brauna

1937 AM. Discovered 1937 January 8 by K. Reinmuth at Heidelberg.

Named in honor of Margret Braun, wife of H. Vogt (see also the citation for planet (1410)). (H 128)

(1412) Lagrula

1937 BA. Discovered 1937 January 19 by L. Boyer at Algiers.

Named in honor of J. Ph. Lagrula {1906-1988}, director of the Algiers Observatory. (H 128)

Obituary published in C.R. Acad. Sci., Sér. Gén., Vie Sci., Tome 6, No. 6, p. 583-589 (1989).

(1413) Roucarie

1937 CD. Discovered 1937 February 12 by L. Boyer at Algiers.
Named in honor of the mother of the discoverer. (H 128)

(1414) Jérôme

1937 CE. Discovered 1937 February 12 by L. Boyer at Algiers.
Named in honor of the father of the discoverer. (H 128)

(1415) Malautra

1937 EA. Discovered 1937 March 4 by L. Boyer at Algiers.
Named in honor of the wife of the discoverer. (H 128)

(1416) Renauxa

1937 EC. Discovered 1937 March 4 by L. Boyer at Algiers.

Named in honor of P. Renaux, assistant astronomer at the Algiers Observatory. (H 128)

(1417) Walinskia

1937 GH. Discovered 1937 April 1 by K. Reinmuth at Heidelberg. Independently discovered 1937 April 9 by G. N. Neujmin at Simeïs.

Named in honor of an acqaintance of an astronomer at the Astronomisches Rechen-Institut, Berlin.
Name proposed by W. Schaub.

(1418) Fayeta

Discovered 1903 September 22 by P. Götz at Heidelberg.

Named in honor of Gaston-Jules Fayet (1874-1967), director of the Nice Observatory who computed extensive tables of minor planets and searched for mutual close encounters. (H 128)
Name proposed by the orbit computer Ch. Cailliatte.

Obituaries published in C.R. Acad. Sci., Vol. 266, p. 16, Vol. 267, p. 103-104 (1968).

(1419) Danzig

1929 RF. Discovered 1929 September 5 by K. Reinmuth at Heidelberg.
Named for the city (now Gdansk) and port on the Baltic sea. (H 128)
The city of Danzig is also honored by planet (764).

(1420) Radcliffe

1931 RJ. Discovered 1931 September 14 by K. Reinmuth at Heidelberg. Independently discovered 1931 October 10 by F. Rigaux at Uccle.

Named by the orbit computer, Mrs. Maud W. Makemson, in honor of the class of 1912 of Radcliffe College, Cambridge, Mass. (H 128)

(1421) Esperanto

1936 FQ. Discovered 1936 March 18 by Y. Väisälä at Turku.

Named to commemorate the international language which was proposed by L. L. Zamenhof {see planet (1462)} in 1887 for common scientific usage. The inventor used the pseudonym "doctoro Esperanto". (M 1350)

(1422) Strömgrenia

1936 QF. Discovered 1936 August 23 by K. Reinmuth at Heidelberg.

Named in honor of the Danish astronomer Prof. Elis Strömgren (1870-1947) who was director of the Copenhagen University Observatory, and director of the Bureau Central des Télégrammes Astronomiques. (H 128)
Strömgren is also honored by a lunar crater.

(1423) Jose

1936 QM. Discovered 1936 August 28 by J. Hunaerts at Uccle.

Named in honor of Giuseppina Bianchi, deceased young daughter of the Italian astronomer E. Bianchi. (H 128)
Name proposed by the orbit computer Cesare Lombardi.

(1424) Sundmania

1937 AJ. Discovered 1937 January 9 by Y. Väisälä at Turku.

Named in honor of K. F. Sundman (1873-1949), director of the Helsingfors Observatory, who intensively worked on the n-body problem. (H 129)
Sundman is also honored by a lunar crater.

(1425) Tuorla

1937 GB. Discovered 1937 April 3 by K. Inkeri at Turku.

Named in honor of the Research Institute for Astronomy and Optics near Turku, Finland. (M 2277)

(1426) Riviera

1937 GF. Discovered 1937 April 1 by M. Laugier at Nice. Independently discovered 1937 April 3 by C. Jackson at Johannesburg.

Named for the southeast Mediterranean coast of France, noted for its mild weather, which favors the observatory at which this planet was discovered. (M 6831)
Name proposed by F. Pilcher.

(1427) Ruvuma

1937 KB. Discovered 1937 May 16 by C. Jackson at Johannesburg.

Name of the most important river in southern Tanganyika, now Tanzania. (M 909)

(1428) Mombasa

1937 ND. Discovered 1937 July 5 by C. Jackson at Johannesburg.

Name of the chief port in Kenya, East Africa, which has also ancient historical associations. (M 909)

(1429) Pemba

1937 NH. Discovered 1937 July 2 by C. Jackson at Johannesburg.

A large island off the East Coast of Africa which was under the rule of the Sultan of Zanzibar. (M 909)

(1430) Somalia

1937 NK. Discovered 1937 July 5 by C. Jackson at Johannesburg.
Named for the state in the northeast of Africa. (M 5181)

(1431) Luanda

1937 OB. Discovered 1937 July 29 by C. Jackson at Johannesburg.
Named for the capital city of Angola. (M 5182)

(1432) Ethiopia

1937 PG. Discovered 1937 August 1 by C. Jackson at Johannesburg.
Ancient name of Abyssinia; still called the empire of Ethiopia. (M 909)

(1433) Geramtina

1937 UC. Discovered 1937 October 30 by E. Delporte at Uccle.

This is a compound name which has been formed to recall and to honor Miss Asplind, sister of B. Asplind {see planet (958)}. (H 129)

Named by B. Asplind who computed several orbits of planets found at Uccle on the occasion of the 1938 Stockholm IAU meeting.

(1434) Margot

1936 FD_1. Discovered 1936 March 19 by G. N. Neujmin at Simeïs.

Named in honor of Miss Gertrud Margot Görsdorf (1915-1990), later Mrs. Zottmann. She was for some years in Berlin friend and schoolfellow of the German astronomer W. Gliese {see planet (1823)} who proposed this name. (LDS)

(1435) Garlena

1936 WE. Discovered 1936 November 23 by K. Reinmuth at Heidelberg. Independently discovered 1936 December 13 by G. N. Neujmin at Simeïs.

Named in honor of an acquaintance of the German astronomer W. Schaub. (H 129)

(1436) Salonta

1936 YA. Discovered 1936 December 11 by G. Kulin at Budapest. Independently discovered 1936 December 14 by A. Patry at Nice.
Named for the birthplace of the discoverer. (M 5182)

(1437) Diomedes

1937 PB. Discovered 1937 August 3 by K. Reinmuth at Heidelberg.

Named for Diomedes, king of Argos, one of the Greek heroes of the Trojan War. (H 129)

(1438) Wendeline

1937 TC. Discovered 1937 October 11 by K. Reinmuth at Heidelberg.
Any reference of this name to a person or occurence is unknown.
Name proposed by W. Schaub.

(1439) Vogtia

1937 TE. Discovered 1937 October 11 by K. Reinmuth at Heidelberg.

Named in honor of the Heidelberg astronomer Heinrich Vogt (1890-1968). In 1929, Vogt became professor and director of the Jena University Observatory. From 1933 he was director of the Heidelberg Königstuhl Observatory. Vogt is especially well known for his fundamental work on the theory of stellar interiors.

He detected the Vogt-Russell theorem which uniquely describes the structure of a star by using only the mass and the chemical composition. (G. Klare)

Obituaries published in Astron. Nachr., Vol. 292, p. 45-46 (1970); Jahrb. Heidelb. Akad. Wiss., p. 47-49 (1970).

(1440) Rostia

1937 TF. Discovered 1937 October 11 by K. Reinmuth at Heidelberg.

Named probably in honor of the German amateur astronomer, poet, and popularizer Johann Leonhard Rost (1688-1727) whose *Astronomisches Handbuch* (1718) has found a wide distribution. (LDS)

Rost is also honored by a lunar crater.

(1441) Bolyai

1937 WA. Discovered 1937 November 26 by G. Kulin at Budapest.

Name proposed by the discoverer in honor of Farkas Wolfgang Bolyai (1775-1856), a Hungarian astronomer and computer. (H 130)

(1442) Corvina

1937 YF. Discovered 1937 December 29 by G. Kulin at Budapest. Independently discovered 1937 December 29 by K. Reinmuth at Heidelberg.

Named by the discoverer probably in memory of the Hungarian king Matthias I. Corvinus (1440-1490) who conquered Vienna in 1485 and who was the founder of the Pressburg University. Corvinus was the surname of the Hungarian kings. His library at Buda {see planet (908)}, the Corvina, was one of the finest in Europe. (LDS)

(1443) Ruppina

1937 YG. Discovered 1937 December 29 by K. Reinmuth at Heidelberg.

Named for the German city of Ruppin, the birthplace of the astronomer M. Ebell {see planet (1205)}, who proposed this name (RI 2247). (H 130)

(1444) Pannonia

1938 AE. Discovered 1938 January 6 by G. Kulin at Budapest.

Named for the province of the Roman empire that includes much of modern Hungary. (M 5182)

(1445) Konkolya

1938 AF. Discovered 1938 January 6 by G. Kulin at Budapest.

Named in honor of the founder of the Ógyalla {see planet (1259)} Observatory, Miklós (Nicolaus) von Konkoly Thege {1842-1916}. He was chiefly interested in the new methods of celestial photography and astrophysics - especially spectroscopy. In 1898, Konkoly Thege presented his observatory to the Hungarian government, together with the funds necessary to ensure its continuation. The instruments became the basis for the new Budapest observatory. From 1890 Konkoly Thege directed the Hungarian Meteorological Service until his retirement in 1911. (LDS)

(1446) Sillanpää

1938 BA. Discovered 1938 January 26 by Y. Väisälä at Turku.

The great Finnish writer, Frans Eemil Sillanpää (1888-1964), winner of the Nobel prize for literature in 1939. (M 3928)

(1447) Utra

1938 BB. Discovered 1938 January 26 by Y. Väisälä at Turku.

Named in honor of the birthplace of the discoverer in northeastern Finland. (M 2277)

(1448) Lindbladia

1938 DF. Discovered 1938 February 16 by Y. Väisälä at Turku.

Named in honor of the Swedish astronomer Bertil Lindblad (1895-1965), director

of the Stockholm Observatory at Salsjöbaden, and president of the I.A.U. during the critical, post-war years 1948-52. (M 1350)
Lindblad is also honored by a lunar crater.

(1449) Virtanen

1938 DO. Discovered 1938 February 20 by Y. Väisälä at Turku.

Named by the discoverer in honor of Artturi Ilmari Virtanen, famous Finnish biochemist, on the occasion of his 75th birthday. He was awarded the Nobel prize in chemistry in 1945, and was for many years president of the Academy of Finland. (M 3023)
See also the remark to planet (1095).

(1450) Raimonda

1938 DP. Discovered 1938 February 20 by Y. Väisälä at Turku.

Named in honor of Dr. Jean Jacques Raimond (1903-1961), president of the Dutch Astronomical Society, Nederlandse Vereniging voor Weer- en Sterrenkunde and director of the Zeiss planetarium at The Hague. (M 2347)
Name suggested by Jean Meeus.
Raimond is also honored by a lunar crater.

(1451) Granö

1938 DT. Discovered 1938 February 22 by Y. Väisälä at Turku.

Named in honor of J. G. Granö (1882-1956), first professor of geography, later chancellor of the Turku University. (M 2277)

(1452) Hunnia

1938 DZ_1. Discovered 1938 February 26 by G. Kulin at Budapest.
Named in honor of the Hungarian nation. (M 5182)

(1453) Fennia

1938 ED_1. Discovered 1938 March 8 by Y. Väisälä at Turku. Independently discovered 1938 March 23 by G. N. Neujmin at Simeïs.
This is the Latin word for the country of Finland. (H 130)

(1454) Kalevala

1936 DO. Discovered 1936 February 16 by Y. Väisälä at Turku.

Named for the Finnish national epic, a source of inspiration for the music of Sibelius {see planet (1405)}. (M 3928)

(1455) Mitchella

1937 LF. Discovered 1937 June 5 by A. Bohrmann at Heidelberg.

Named in honor of the American woman astronomer Maria Mitchell (1818-1889) who was professor of astronomy and director of Vassar College Observatory. (H 131)

Mitchell is also honored by a lunar crater. The observatory in Nantucket, Massachusetts bears her name.

(1456) Saldanha

1937 NG. Discovered 1937 July 2 by C. Jackson at Johannesburg.

Named for the newly-discovered harbor on the southwest tip of South Africa. (M 5182)

(1457) Ankara

1937 PA. Discovered 1937 August 3 by K. Reinmuth at Heidelberg.
Named for the Turkish capital. (H 131)
Name proposed by W. Gleissberg.

(1458) Mineura

1937 RC. Discovered 1937 September 1 by F. Rigaux at Uccle.

Named in honor of Adolphe Mineur who was professor of mathematics at the University of Brussels, Belgium. (H 131)

(1459) Magnya

1937 VA. Discovered 1937 November 4 by G. N. Neujmin at Simeïs. Independently discovered 1937 November 6 by A. Patry at Nice.

The word, translated from Latin to Russian, means "clear, bright, wonderful". (N. Solovaya; N. S. Chernykh)
Name suggested by A. Patry who computed a first orbit.

(1460) Haltia

1937 WC. Discovered 1937 November 24 by Y. Väisälä at Turku.
Named for the highest mountain in Finland. (M 3928)

(1461) Jean-Jacques

1937 YL. Discovered 1937 December 30 by M. Laugier at Nice.
This planet is named by the discoverer in honor of her son. (M 2116)

(1462) Zamenhof

1938 CA. Discovered 1938 February 6 by Y. Väisälä at Turku.

Named in honor of L. L. Zamenhof (1859-1917), a Polish physician and inventor of the international language, Esperanto {see minor planet (1421)}. (M 1350)

(1463) Nordenmarkia

1938 CB. Discovered 1938 February 6 by Y. Väisälä at Turku. Independently discovered 1938 February 19 by E. Delporte at Uccle.

Named in honor of the Swedish astronomer N. V. E. Nordenmark (1867-1962) whose writings created increased interest in astronomy. (M 1350)

(1464) Armisticia

1939 VO. Discovered 1939 November 11 by G. Van Biesbroeck at Williams Bay.

Since this was the 21st anniversary of the signing of the armistice of World War I, the name was given in the hope for a continuation of world peace, in the presence of the then ominous conditions. (M 2196)

(1465) Autonoma

1938 FA. Discovered 1938 March 20 by A. A. Wachmann at Bergedorf.

According to the wishes of the discoverer and of O. Heckmann {see planet (1650)}, director of the Hamburg-Bergedorf Observatory, the planet is named "Autonoma". This name has been chosen in honor of the "Universidad Autonoma de El Salvador", in recognition of the hospitality granted by this university to the Hamburg Observatory. (M 698)

(1466) Mündleria

1938 KA. Discovered 1938 May 31 by K. Reinmuth at Heidelberg.

Named in honor of the German astronomer Max Mündler (1876-1969) who worked at the Heidelberg Königstuhl Observatory. (H 131)
Name proposed by H. Vogt (RI 2284).

Obituaries published in Astron. Nachr., Vol. 291, p. 224 (1969); Mitt. Astron. Ges., Nr. 27, p. 236 (1969); Astron. Nachr., Vol. 292, p. 143-144 (1970).

(1467) Mashona

1938 OE. Discovered 1938 July 30 by C. Jackson at Johannesburg.
Name of a large race of native people who inhabit Rhodesia. (M 909)

(1468) Zomba

1938 PA. Discovered 1938 July 23 by C. Jackson at Johannesburg. Independently discovered 1938 August 2 by L. Boyer at Algiers.
Name of an important town in Nyassaland. (M 909)

(1469) Linzia

1938 QD. Discovered 1938 August 19 by K. Reinmuth at Heidelberg. Independently discovered 1938 August 31 by G. N. Neujmin at Simeïs.
Named after the city of Linz, Austria, on the Danube river. (H 132)
Name proposed by A. Wersig (RI 2319).

(1470) Carla

1938 SD. Discovered 1938 September 17 by A. Bohrmann at Heidelberg.

This planet is named in honor of Mrs. Carla Ziegler, Heidelberg, a friend of the Bohrmann family. (M 1129)

(1471) Tornio

1938 SL_1. Discovered 1938 September 16 by Y. Väisälä at Turku.

Named for a Finnish river and town at the northern end of the Gulf of Bothnia. (M 3928)

(1472) Muonio

1938 UQ. Discovered 1938 October 18 by Y. Väisälä at Turku.
One of the principal rivers of Finland. (M 3928)

(1473) Ounas

1938 UT. Discovered 1938 October 22 by Y. Väisälä at Turku.
One of the principal rivers of Finland. (M 3928)

(1474) Beira

1935 QY. Discovered 1935 August 20 by C. Jackson at Johannesburg.
Name of an important port in Mozambique. (M 909)

(1475) Yalta

1935 SM. Discovered 1935 September 21 by P. F. Shajn at Simeïs.

A resort city in Crimea; site of an important political conference during World War II. (M 838)

(1476) Cox

1936 RA. Discovered 1936 September 10 by E. Delporte at Uccle.

Named in honor of Jacques Cox (1898-1972), professor of astronomy at the University of Brussels. He has been rector of the university, permanent secretary of the Royal Academy of Belgium, president of the Belgian National Committee of Astronomy, and president of the Belgian Society of Astronomy, Meteorology and Physics of the Globe. He is well known for his researches concerning minor planets and the origin of comets. (M 3824)

(1477) Bonsdorffia

1938 CC. Discovered 1938 February 6 by Y. Väisälä at Turku.

Named in honor of Ilmari Bonsdorff {1879-1950}, Finnish astronomer and founder and director of the Geodetic Institute of Finland. (H 132)

(1478) Vihuri

1938 CF. Discovered 1938 February 6 by Y. Väisälä at Turku.

Named in honor of A. Vihuri, a Finnish ship owner, and generous patron of science and the arts. (M 2882)

(1479) Inkeri

1938 DE. Discovered 1938 February 16 by Y. Väisälä at Turku.

A popular feminine Christian name in Finland; name of the discoverer's granddaughter and niece. Also a province inhabited by Finns. (M 1350)

(1480) Aunus

1938 DK. Discovered 1938 February 18 by Y. Väisälä at Turku.

A popular masculine Christian name in Finland; name of the discoverer's grandson. Also, a province inhabited by Finns. (M 1350)

(1481) Tübingia

1938 DR. Discovered 1938 February 7 by K. Reinmuth at Heidelberg. Independently discovered 1938 February 22 by Y. Väisälä at Turku.

Named for the German city of Tübingen on the Neckar river south of Stuttgart, where Kepler {see planet (1134)} has worked for many years, and renowned for its university. (H 133)

Name proposed by A. Wersig (RI 2319).

(1482) Sebastiana

1938 DA_1. Discovered 1938 February 20 by K. Reinmuth at Heidelberg. Independently discovered 1938 February 24 by Y. Väisälä at Turku.

Named in honor of the German scientist Sebastian Finsterwalder, Munich. (RI 2417)

Name proposed by O. Volk.

(1483) Hakoila

1938 DJ_1. Discovered 1938 February 19 by Y. Väisälä at Turku. Independently discovered 1938 March 6 by M. B. Protitch at Belgrade.

Named by the discoverer in honor of his former assistant in physics, K. J. Hakoila, Ph. D., D. Eng. (1898-). His technical assistance has been of great value in the construction of the new Turku Observatory. (M 2278)

(1484) Postrema

1938 HC. Discovered 1938 April 29 by G. N. Neujmin at Simeïs.

The name signifies the last link of the last of a group. It is assigned to the last minor planet discovered by Neujmin which has been numbered. (M 1252)

Name submitted by N. Yakhontova, Institute of Theoretical Astronomy, Leningrad.

The citation statement holds for the time of publication of MPC 1252 (June 24, 1955). Until now, (4420) 1936 PB is the Neujmin discovery with the highest number. The latest numbered discovery of Neujmin is planet (2536), discovered 1939 August 15.

(1485) Isa

1938 OB. Discovered 1938 July 28 by K. Reinmuth at Heidelberg.

Isa is a diminutive of the Italian name Marisa. (H 133)

The naming was granted by the discoverer to Massimo Cimino, astronomer at the Monte Mario Observatory, Rome who calculated the first orbit.

(1486) Marilyn

1938 QA. Discovered 1938 August 23 by E. Delporte at Uccle.

Named in honor of the daughter of Paul Herget {see planet (1751)} who computed the orbit. (H 133)

(1487) Boda

1938 WC. Discovered 1938 November 17 by K. Reinmuth at Heidelberg.

Named in honor of the German astronomer Karl Boda {1889-1942} who at the Frankfurt Planeten-Institut worked on minor planet dynamics. (H 133)

(1488) Aura

1938 XE. Discovered 1938 December 15 by Y. Väisälä at Turku.
Named for the river on which Turku stands. (M 3928)

(1489) Attila

1939 GC. Discovered 1939 April 12 by G. Kulin at Budapest.
Named for the fifth-century king of the Huns. (M 5182)

(1490) Limpopo

1936 LB. Discovered 1936 June 14 by C. Jackson at Johannesburg.

Name of the large river which flows round the northern edge of the Transvaal and through Portuguese East Africa, now Mozambique. (M 909)

(1491) Balduinus

1938 EJ. Discovered 1938 February 23 by E. Delporte at Uccle.

This planet is named in honor of Baudouin {1930-1993}, King of the Belgians. It was the intention of the discoverer, long time observer at the Royal Observatory of Belgium, to honor his king and his patron in this way. (M 1860)

(1492) Oppolzer

1938 FL. Discovered 1938 March 23 by Y. Väisälä at Turku.

Named in honor of Hofrath Professor Theodor Ritter von Oppolzer (1841-1886), professor of astronomy in Vienna and author of the monumental *Canon der Finsternisse.* (M 2278)
Name suggested by Jean Meeus.
Oppolzer is also honored by a lunar crater.

(1493) Sigrid

1938 QB. Discovered 1938 August 26 by E. Delporte at Uccle.

Named in honor of Sigrid Strömgren, wife of the Danish astronomer Bengt Strömgren {see planet (1846)}. (H 134; RI 2297)

(1494) Savo

1938 SJ. Discovered 1938 September 16 by Y. Väisälä at Turku. Independently discovered 1938 September 18 by A. A. Wachmann at Bergedorf.
Named after a province in Finland. (M 1350)

(1495) Helsinki

1938 SW. Discovered 1938 September 21 by Y. Väisälä at Turku.
Named for the capital city of Finland. (M 3928)

(1496) Turku

1938 SA_1. Discovered 1938 September 22 by Y. Väisälä at Turku.

Named in honor of the city which in ancient times was the capital of Finland, is now the home of the discoverer, and was the home of Argelander {see planet (1551)} when he began his monumental Durchmusterung. (M 1350)

(1497) Tampere

1938 SB_1. Discovered 1938 September 22 by Y. Väisälä at Turku.
Named for the second largest city in Finland. (M 3928)

(1498) Lahti

1938 SK_1. Discovered 1938 September 16 by Y. Väisälä at Turku.
Named for a town in Finland. (M 3928)

(1499) Pori

1938 UF. Discovered 1938 October 16 by Y. Väisälä at Turku.

Named for an important Finnish town and port on the Gulf of Bothnia. (M 3928)

(1500) Jyväskylä

1938 UH. Discovered 1938 October 16 by Y. Väisälä at Turku.

Named for a town in central Finland. (M 3928)

(1501) Baade

1938 UJ. Discovered 1938 October 20 by A. A. Wachmann at Bergedorf.

Named in memory of Wilhelm Heinrich Walter Baade (1893-1960), German-born astronomer who joined the staff of Mount Wilson Observatory in 1931 and whose observations of M31 and its companion led to the classification of stars in two discrete populations. While at Bergedorf in 1920 he discovered (944) Hidalgo, and at Palomar in 1949 he discovered (1566) Icarus. These are the minor planets of largest aphelion and smallest perihelion distance, respectively. (M 4236)

Obituaries published in Astron. Nachr., Band 285, p. 286 (1960); Nordisk Astron. Tidsskr., p. 115-116 (1960); Nature, Vol. 187, p. 1075 (1960); Observatory, Vol. 80, p. 166 (1960); Phys. Bl., Vol. 16, p. 428 (1960); Publ. Astron. Soc. Pac., Vol. 72, p. 434 (1960); Science, Vol. 132, p. 137 (1960); Sky Telesc., Vol. 20, p. 76 (1960); Sterne, Band 36, p. 204-207 (1960); J. R. Astron. Soc. Canada, Vol. 55, p. 113-116 (1961); Mitt. Astron. Ges., p. 5-9 (1960); Orion, Vol. 6, p. 135-138 (1961); Q.J.R. Astron. Soc., Vol. 2, p. 118-121 (1961). Baade is also honored by a lunar crater. Presently, the record holding planets are (5145) Pholus and (3200) Phaethon, respectively.

(1502) Arenda

1938 WB. Discovered 1938 November 17 by K. Reinmuth at Heidelberg.

Named in honor of Sylvain Arend {1902-1992}, astronomer at the Uccle Observatory near Brussels, Belgium, who observed and discovered more than 50 numbered minor planets and a bright comet. (H 134)

Name proposed by T. Banachiewicz (RI 2333).

Obituaries published in Ciel Terre, Vol. 108, No. 4, p. 101-102 (1992); Astronomie, Vol. 106, p. 25 (1993); Inf. Circ. IAU Comm. 26, No. 122, p. 4 (1994).

(1503) Kuopio

1938 XD. Discovered 1938 December 15 by Y. Väisälä at Turku.

Named for a town in central Finland. (M 3928)

(1504) Lappeenranta

1939 FM. Discovered 1939 March 23 by L. Oterma at Turku.

Named for a town in south-eastern Finland. (M 3928)

(1505) Koranna

1939 HH. Discovered 1939 April 21 by C. Jackson at Johannesburg.

Name of a tribe of wandering bushmen who inhabit the southern part of the Kalahari Desert in southern Africa. (M 909)

(1506) Xosa

1939 JC. Discovered 1939 May 15 by C. Jackson at Johannesburg.

Name of a tribe of native peoples of the Cape Province who early came into contact with the white people. Nowadays the name Xosa is spelled Xhosa. (M 909; I. van Houten-Groeneveld)

(1507) Vaasa

1939 RD. Discovered 1939 September 12 by L. Oterma at Turku.

Named for an important Finnish town and port on the Gulf of Bothnia. (M 3928)

(1508) Kemi

1938 UP. Discovered 1938 October 21 by H. Alikoski at Turku. Independently discovered 1938 October 30 by G. Kulin at Budapest.

Named for a Finnish river and town at the northern end of the Gulf of Bothnia. The four-letter name is in keeping with the names of several other minor planets of large orbital eccentricity and inclination. (M 3928)

(1509) Esclangona

1938 YG. Discovered 1938 December 21 by A. Patry at Nice.

Named in honor of Ernest Esclangon (1876-1954), director of the Paris Observatory. He served as president of the International Astronomical Union from 1935 to 1938. (H 134)

(1510) Charlois

1939 DC. Discovered 1939 February 22 by A. Patry at Nice.

Named in memory of Auguste Charlois (1864-1910), assistant at the Nice Observatory from 1881 and subsequently first observer until his early and tragic death. An indefatigable observer of comets and minor planets, he discovered 99 numbered minor planets. Although the first 27 of his discoveries were visual ones, he started making photographic searches in 1892, a few months after the pioneering work in that field at Heidelberg. (M 4190)

(1511) Daléra

1939 FB. Discovered 1939 March 22 by L. Boyer at Algiers.

Named in honor of Paul Daléra, a friend of the discoverer. (H 135)

(1512) Oulu

1939 FE. Discovered 1939 March 18 by H. Alikoski at Turku.

Named in honor of the birthplace of the discoverer, a town in northern Finland. (M 2278)

(1513) Mátra

1940 EB. Discovered 1940 March 10 by G. Kulin at Budapest.

Named for the mountain in northern Hungary where an outstation of the Konkoly {see planet (1445)} Observatory is located. (M 5182)

(1514) Ricouxa

Discovered 1906 August 22 by M. Wolf at Heidelberg.

Any reference of this name to a person or occurence is unknown. (H 135) Named by the French astronomer A. Patry.

(1515) Perrotin

1936 VG. Discovered 1936 November 15 by A. Patry at Nice.

Named in memory of Henri Joseph Anastase Perrotin (1845-1904), the first director of the Nice Observatory and discoverer of (252) Clementina, the first minor planet found there. While working at Toulouse he had previously discovered five other minor planets and developed a precise theory of the motion of (4) Vesta. (M 4358)

(1516) Henry

1938 BG. Discovered 1938 January 28 by A. Patry at Nice.

Named in memory of the brothers Paul Pierre Henry (1848-1905) and Prosper Matthieu Henry (1849-1903), who constructed several telescopes, including the 76-cm refractor at Nice and most of the Carte du Ciel telescopes. In the course of their program of mapping the ecliptic they each visually discovered seven minor planets. (M 4358)

Both brothers are also honored by lunar craters. Paul is honored by a crater on Mars.

(1517) Beograd

1938 FD. Discovered 1938 March 20 by M. B. Protitch at Belgrade.

Named by the discoverer in honor of his native city and the capital of his country. (M 2277)

(1518) Rovaniemi

1938 UA. Discovered 1938 October 15 by Y. Väisälä at Turku. Independently discovered 1938 October 21 by E. Delporte at Uccle.

Named for a town in Finland, located nearly on the arctic circle. (M 3929)

(1519) Kajaani

1938 UB. Discovered 1938 October 15 by Y. Väisälä at Turku.

Named for a town in central Finland. (M 3929)

(1520) Imatra

1938 UY. Discovered 1938 October 22 by Y. Väisälä at Turku.

Named for a town in south-eastern Finland. (M 3929)

(1521) Seinäjoki

1938 UB_1. Discovered 1938 October 22 by Y. Väisälä at Turku.

Named for a town in western Finland. (M 3929)

(1522) Kokkola

1938 WO. Discovered 1938 November 18 by L. Oterma at Turku.

Named for a Finnish town and port on the Gulf of Bothnia. (M 3929)

(1523) Pieksämäki

1939 BC. Discovered 1939 January 18 by Y. Väisälä at Turku.

Named for a town in central Finland. (M 3929)

(1524) Joensuu

1939 SB. Discovered 1939 September 18 by Y. Väisälä at Turku.

Named for the town in eastern Finland where the discoverer received his early schooling. (M 3929)

(1525) Savonlinna

1939 SC. Discovered 1939 September 18 by Y. Väisälä at Turku.

Named for a town in eastern Finland. (M 3929)

(1526) Mikkeli

1939 TF. Discovered 1939 October 7 by L. Oterma at Turku.

Named for a town in eastern Finland. (M 3929)

(1527) Malmquista

1939 UG. Discovered 1939 October 18 by Y. Väisälä at Turku.

Named in honor of the Swedish astronomer G. Malmquist (1893-1982), director of the Uppsala Observatory. (M 1350)

Obituary published in Astron. Tidsskr., Arg. 15, p. 176-177 (1982).

(1528) Conrada

1940 CA. Discovered 1940 February 10 by K. Reinmuth at Heidelberg.

Named in honor of Fritz Conrad (1883-1944), rear-admiral of the German navy and well-known nautical expert and oceanographer. (LDS; RI 2485)

Name presumably proposed by A. Kohlschütter.

A lot of detailed informations concerning the life and work of Admiral Conrad were provided by Mrs. Ilse Palm, née Conrad. During World War I, Conrad commanded two cruisers. After the war he studied astronomy, geodesy, mathematics and physics at the Berlin University and got his Ph.D. in 1933 with *Astronomische Ortsbestimmung und Kimmtiefenmessung auf See.* Conrad was founder and head of the Naval Meteorology Service and of the Nautic and Hydrographic Department of the Deutsche Seewarte in Hamburg as well as of the Naval Observatory Wilhelmshaven. He is also honored by the "Conrad Mountains" in Neu-Schwabenland which were discovered by the German Antarctic Expedition 1938/39.

(1529) Oterma

1938 BC. Discovered 1938 January 26 by Y. Väisälä at Turku.

This member of the Hilda Group is named in honor of Liisi Oterma (1915-), astronomer at, and since 1971 director of, the Astronomical-Optical Institute at Turku. She is well known for her discoveries and observations of comets and minor planets and also for her work on their orbits. (M 3929)
Name proposed by B. G. Marsden and E. Roemer.

(1530) Rantaseppä

1938 SG. Discovered 1938 September 16 by Y. Väisälä at Turku. Independently discovered 1938 September 18 by E. Delporte at Uccle.

Named in memory of Hilkka Rantaseppä-Helenius (1925-1975), astronomer at the Turku Astronomical-Optical Institute, observer of comets and minor planets. (M 3929)

(1531) Hartmut

1938 SH. Discovered 1938 September 17 by A. Bohrmann at Heidelberg.

Named by the discoverer in honor of his grandson, Hartmut Neckel, whose father is an astronomer at the Hamburg-Bergedorf Observatory. (M 3929)

(1532) Inari

1938 SM. Discovered 1938 September 16 by Y. Väisälä at Turku.
One of the principal lakes of Finland. (M 3929)

(1533) Saimaa

1939 BD. Discovered 1939 January 19 by Y. Väisälä at Turku.
One of the principal lakes of Finland. (M 3929)

(1534) Näsi

1939 BK. Discovered 1939 January 20 by Y. Väisälä at Turku.
One of the principal lakes of Finland. (M 3929)

(1535) Päijänne

1939 RC. Discovered 1939 September 9 by Y. Väisälä at Turku.
One of the principal lakes of Finland. (M 3929)

(1536) Pielinen

1939 SE. Discovered 1939 September 18 by Y. Väisälä at Turku.
One of the principal lakes of Finland. (M 3930)

(1537) Transylvania

1940 QA. Discovered 1940 August 27 by G. Strommer at Budapest.

Named for the section of the old Hungarian empire in which the discoverer was born. (M 5182)

(1538) Detre

1940 RF. Discovered 1940 September 8 by G. Kulin at Budapest.

Named in memory of Laszlo Detre (1906-1974), Hungarian astronomer well known for his work on variable stars, director of the Konkoly {see planet (1445)} Observatory for many years. (M 5182)

Obituaries published in BAV Rundbrief, 24. Jahrg., p. 40 (1975); Coelum, Vol. 43, p. 113 (1975); Sterne Weltraum, Vol. 14, p. 83 (1975); Mitt. Astron. Ges., Nr. 38, p. 7-9 (1976).

(1539) Borrelly

1940 UB. Discovered 1940 October 29 by A. Patry at Nice.

Named in honor of Alphonse Louis Nicolas Borrelly (1842-1926), astronomer at the Marseilles Observatory who discovered 19 minor planets between 1866 and 1894 and at least 16 comets between 1871 and 1919. (M 4358)

(1540) Kevola

1938 WK. Discovered 1938 November 16 by L. Oterma at Turku.

Named for one of the observing stations of the Turku Astronomical-Optical Institute. (M 3930)

(1541) Estonia

1939 CK. Discovered 1939 February 12 by Y. Väisälä at Turku.

Named in honor of the country south of Finland, formerly populated by the Finno-Ugric tribe of Esthonians on the east shores of the Baltic. Since 1991 Estonia is again an independent state. (M 1350)

(1542) Schalén

1941 QE. Discovered 1941 August 26 by Y. Väisälä at Turku.

Named in honor of Carl Adam Wilhelm Schalén {1902-1993} of the Lund Institute of Astronomy. Sometime director of the Lund Observatory, he is noted for his work on interstellar reddening. Some of his research was made using plates taken with the Schmidt-Väisälä telescope at Uppsala. (M 3930)
Obituary published in Astron. Tidsskr., Arg. 27, Nr. 2, p. 84-85 (1994).

(1543) Bourgeois

1941 SJ. Discovered 1941 September 21 by E. Delporte at Uccle.

Named in honor of Paul E.-E. Bourgeois {1898-1974}, late director of the Royal Observatory of Belgium, professor at the Free University of Brussels, author of many publications in astrometry, meridian astronomy, astrophysics and stellar statistics. (M 3930)
Obituary published in Ciel Terre, Vol. 91, p. 1-4 (1975).

(1544) Vinterhansenia

1941 UK. Discovered 1941 October 15 by L. Oterma at Turku.

Named in honor of Miss Julie M. Vinterhansen (1890-1960), astronomer at the Copenhagen Observatory. She was for a long time editor of the IAU circulars and telegrams. (M 1350)

(1545) Thernöe

1941 UW. Discovered 1941 October 15 by L. Oterma at Turku.

Named in honor of Karl August Thernöe, formerly an astronomer at the Copenhagen Observatory who worked in celestial mechanics. He has for some time been well known as a popularizer of astronomy in Denmark and from 1950 to 1964 was director of the IAU Central Telegram Bureau. (M 3930)

(1546) Izsák

1941 SG_1. Discovered 1941 September 28 by G. Kulin at Budapest.

Named in memory of Imre Izsák (1929-1965), Hungarian-born astronomer and celestial mechanician, who made important contributions to the study of artificial-

satellite motion. In his later years he worked at the Cincinnati Observatory and the Smithsonian Astrophysical Observatory. (M 5182)
Izsák is also honored by a lunar crater.

(1547) Nele

1929 CZ. Discovered 1929 February 12 by P. Bourgeois at Uccle.

Named for the wife of Till Eulenspiegel, the roguish fourteenth-century German folk-hero. The earliest extant edition of the stories about him was published in Antwerp in 1515 by Charles Théodore Henri de Coster in *La Légende d'Ulenspiegel.* (M 6831; I. van Houten-Groeneveld)
Name proposed by the wife of J. Meeus.
Nele is a Frisian name, an abbreviation of Cornelia.

(1548) Palomaa

1935 FK. Discovered 1935 March 26 by Y. Väisälä at Turku.

Named in honor of Prof. M. H. Palomaa (1871-1947), the first professor of chemistry in the University of Turku, and for many years the dean of the faculty of sciences. (M 1351)

(1549) Mikko

1937 GA. Discovered 1937 April 2 by Y. Väisälä at Turku.

Named in honor of Mikko Arthur Levander, a Finnish pastor, amateur astronomer and the father-in-law of the discoverer. (H 136)

(1550) Tito

1937 WD. Discovered 1937 November 29 by M. B. Protitch at Belgrade.

Named in honor of Josip Tito (1892-1980), president of Yugoslavia, the leader of Yugoslavian resistance against aggression during World War II, and an early enthusiast of the United Nations. (M 2277)

(1551) Argelander

1938 DC_1. Discovered 1938 February 24 by Y. Väisälä at Turku.

Named in honor of F. W. A. Argelander (1799-1875), director of the ancient observatory of Turku (Abo), and later director of the Bonn Observatory, and author of the famous *Bonner Durchmusterung.* (M 2278)
Argelander is also honored by a lunar crater.

(1552) Bessel

1938 DE_1. Discovered 1938 February 24 by Y. Väisälä at Turku.

Named in honor of the eminent German astronomer F. W. Bessel (1789-1846), who measured the first stellar parallax (61 Cyg). (M 2278)
Bessel is also honored by a lunar crater.

(1553) Bauersfelda

1940 AD. Discovered 1940 January 13 by K. Reinmuth at Heidelberg.

This planet is named in honor of Prof. Dr.Ing. W. Bauersfeld {1879-1959} on the occasion of his 75th birthday anniversary, 1954 January 23. Prof. Bauersfeld, formerly with Zeiss-Opton at Oberkochen/Württemberg (Germany), is best known as the designer of the famous Zeiss planetaria {see also planet (851)}. (M 994)

(1554) Yugoslavia

1940 RE. Discovered 1940 September 6 by M. B. Protitch at Belgrade.

Named by the discoverer in honor of his country, which has been defended by martyrs during two World Wars. (M 2277)

(1555) Dejan

1941 SA. Discovered 1941 September 15 by F. Rigaux at Uccle.

Named in honor of the son of the Yugoslav astronomer P. M. Djurkovic (1908-1981) from the Belgrade Observatory. (H 137)

(1556) Wingolfia

1942 AA. Discovered 1942 January 14 by K. Reinmuth at Heidelberg.

Named in honor of one of the Heidelberg student fraternities on the occasion of its 104th anniversary celebrated 1955 June 17. Reinmuth writes, "Dem Kleinen Planeten (1556) 1942 AA gebe ich den Namen "Wingolfia" zu Ehren der alten, christlichen, in der Hitlerzeit verbotenen und nach dem 2. Weltkriege wieder erstandenen Heidelberger Studentenverbindung "Wingolf", aus Anlass ihres 104. Stiftungsfestes am 17. Juni 1955." (M 1221)

(1557) Roehla

1942 AD. Discovered 1942 January 14 by K. Reinmuth at Heidelberg.

This planet is named in honor of Dr. Lars Roehl of Lund, Sweden, presently the chief physician of the surgical clinic in Heidelberg, to whom the discoverer is grateful for the restoration of his health. (M 2277)

(1558) Järnefelt

1942 BD. Discovered 1942 January 20 by L. Oterma at Turku.

Named in honor of Gustaf J. Järnefelt, director of the Helsinki Observatory from 1945 to 1969, who has worked on relativity theory and has published extensive catalogues of observations of artificial satellites. (M 3930)

(1559) Kustaanheimo

1942 BF. Discovered 1942 January 20 by L. Oterma at Turku.

Named in honor of Paul H. Kustaanheimo {1924- }, an astronomer at the Helsinki Observatory who has made important contributions to celestial mechanics and relativity theory. (M 3930)

(1560) Strattonia

1942 XB. Discovered 1942 December 3 by E. Delporte at Uccle.

Named in honor of Professor Frederick J. M. Stratton {1881-1960}, Director of the Solar Physics Observatory, Cambridge, England, and long time member of the International Astronomical Union {General Secretary 1925-1938}, upon the occasion of the presentation of a commemorative volume at the R.A.S. meeting of 1955 August 23. (M 1255)

The dedication by the discoverer appears at the beginning of the volume "Vistas in Astronomy" (A. Beer, ed., Pergamon Press, 1955).
Stratton is also honored by a lunar crater.

(1561) Fricke

1941 CG. Discovered 1941 February 15 by K. Reinmuth at Heidelberg.

Named in honor of Walter Ernst Fricke {1915-1988}, Director of the Astronomisches Rechen-Institut in Heidelberg since 1955. The principal author of the FK4, he has also worked extensively on the system of astronomical constants. He served as president of IAU Commissions 4 (1958-1964) and 8 (1970-1973) and vice president of the IAU (1964-1967). (M 3930)

Obituaries published in Mitt. Astron. Ges., Nr. 72, p. 4-14 (1989); Q.J.R. Astron. Soc., Vol. 31, No. 3, p. 515-517 (1990); Astron. Nachr., Vol. 309, No. 3, p. 226 (1988); Sterne Weltraum, 27. Jahrg., Nr. 5, p. 279 (1988); Alm. Österr. Akad. Wiss., 138. Jahrg., p. 377-383 (1988); Observatory, Vol. 108, No. 1087, p. 251 (1988); Heidelb. Akad. Wiss. 1989, p. 89-91 (1990).

(1562) Gondolatsch

1943 EE. Discovered 1943 March 9 by K. Reinmuth at Heidelberg.

Named in honor of Friedrich Gondolatsch {1904- }, astronomer at the Heidelberg Rechen-Institut since 1928, who has worked on the orbits and ephemerides of minor planets, notably of Hermes. He was also in charge of the preparation of the Astronomisch-Geodätisches Jahrbuch. (M 3930)

(1563) Noël

1943 EG. Discovered 1943 March 7 by S. Arend at Uccle.
Named in honor of Emanuel Arend, son of the discoverer. (H 138)

(1564) Srbija

1936 TB. Discovered 1936 October 15 by M. B. Protitch at Belgrade. Independently discovered 1936 October 15 by M. Laugier at Nice.

The name has been given at the request of Prof. V. V. Michkovitch {see planet (2348)} in honor of the land of Serbia in Yugoslavia. This is the first minor planet to be discovered in Belgrade. The spelling is Serbian. (M 844)

(1565) Lemaître

1948 WA. Discovered 1948 November 25 by S. Arend at Uccle.

Named in honor of Canon Georges Lemaître (1894-1966), professor of astronomy and celestial mechanics at the University of Louvain. He was a recognized authority on relativity, the three-body problem and cosmology. He was president of the Pontifical Academy, and he received the first Franqui Prize in Belgium. (M 3824)

This is the first newly numbered planet after World War II. Since then, all newly assigned names have been appropriately described in the MPC's when they are announced. Lemaître is also honored by a lunar crater.

(1566) Icarus

1949 MA. Discovered 1949 June 27 by W. Baade at Palomar.

The discoverer has assigned to this unusual planet the name of the mythological boy who escaped imprisonment on wings of wax, but he flew too near the Sun and they melted, so that he fell into the sea. (M 347)
Name suggested by R. C. Cameron and Dr. Folkman.

This is the first minor planet which was observed (in 1968) by radar. Icarus, like his father Daedalus {see planet (1864)}, is also honored by a lunar crater.

(1567) Alikoski

1941 HN. Discovered 1941 April 22 by Y. Väisälä at Turku.

Named in honor of Heikki A. Alikoski, assistant to the discoverer from 1937 to 1956 and observer and discoverer of minor planets. He helped greatly in establishing the Turku Astronomical-Optical Institute. (M 3930)

(1568) Aisleen

1946 QB. Discovered 1946 August 21 by E. L. Johnson at Johannesburg.
Named by the discoverer in honor of his wife. (M 941)

(1569) Evita

1948 PA. Discovered 1948 August 3 by M. Itzigsohn at La Plata.

The discoverer has assigned the name in honor of the wife of President Peron {1895-1974} of Argentina. (M 519)

Eva Peron (1919-1952) is also honored by planets (1581), (1582), (1588), and (1589), respectively.

(1570) Brunonia

1948 TX. Discovered 1948 October 9 by S. Arend at Uccle.

This planet is named in honor of Brown University, Providence, Rhode Island. Chartered in 1764, it is the seventh oldest university in the United States. Its astronomical history dates back to the transit of Venus in 1769, observed by Prof. Benjamin West. Two local streets are named Planet and Transit. (M 1040)

(1571) Cesco

1950 FJ. Discovered 1950 March 20 by M. Itzigsohn at La Plata.

Named in memory of Ronaldo P. Cesco, a former director of the La Plata Observatory, an outstanding mathematician and celestial mechanician. The planet also honors his brother, Carlos U. Cesco, in charge of the program of astrometric observations of comets and minor planets at El Leoncito. (M 6954)

The Felix Aguilar (see planet (1800)) Observatory at El Leoncito (see planet (2311)) was renamed into Carlos U. Cesco Observatory.

(1572) Posnania

1949 SC. Discovered 1949 September 22 by J. Dobrzycki and A. Kwiek at Poznan.

This planet is named in honor of the city and the observatory at which it was discovered. (M 877)

(1573) Väisälä

1949 UA. Discovered 1949 October 27 by S. Arend at Uccle.

This planet is named in honor of Prof. Yrjö Väisälä {1891-1971}, Director of the Turku Observatory. (M 2116)
Väisälä is also honored by planet (2804).

Obituaries published in Bull. Geod., Nouv. Ser., No. 102, p. 348 (1971); Sky Telesc., Vol. 42, p. 273 (1971); Bull. Geod., Nouv. Ser., No. 103, p. 4 (1972); Astron. Tidsskr., Arg. 5, p. 100 (1972); Sci. Rev. Beograd, Vol. 22, p. 133-140 (1971); Irish Astron. J., Vol. 11, p. 159 (1973).

(1574) Meyer

1949 FD. Discovered 1949 March 22 by L. Boyer at Algiers.

Named in honor of M. Georges Meyer {1894- }, director of the Algiers Observatory. (M 837)

(1575) Winifred

1950 HH. Discovered 1950 April 20 by R. C. Cameron at Brooklyn, Indiana.

This planet is named in honor of Miss Winifred Sawtelle, a staff member of the U.S. Naval Observatory, at the request of Mr. R. C. Cameron, whose discovery and observations of this planet in 1950 lead to its being numbered. (M 844)

(1576) Fabiola

1948 SA. Discovered 1948 September 30 by S. Arend at Uccle.

This planet is named in honor of H. M. Fabiola {1928- }, Queen of the Belgians. (M 2116)

(1577) Reiss

1949 BA. Discovered 1949 January 19 by L. Boyer at Algiers.

Named in honor of Guy Reiss (1904-1964), astronomer at the Algiers Observatory and later at Nice, discoverer of five minor planets. (M 4418)

(1578) Kirkwood

1951 AT. Discovered 1951 January 10 at the Goethe Link Observatory at Brooklyn, Indiana.

This planet has been named in honor of Daniel Kirkwood {1814-1895}, professor

of mathematics at Indiana University from 1856 to 1886. In 1866 he discovered the gaps in the distribution of the mean distances of the minor planets which bear his names: Kirkwood's Gaps. This is the first numbered planet to result from the minor planet observing program at Indiana University. (M 738)
Kirkwood is also honored by a lunar crater.

(1579) Herrick

1948 SB. Discovered 1948 September 30 by S. Arend at Uccle.

Named in honor of Samuel Herrick (1911-1974), well-known for his many contributions to celestial mechanics and astrodynamics, particularly for his perturbation method in vector elements and for his concept of universal variables. Dr. Herrick also worked extensively on the orbits of the Earth-approaching minor planets (1580) Betulia, (1685) Toro (each of which was named for Mrs. Herrick), (1566) Icarus and (1620) Geographos. (M 3824)

Obituaries published in Sky Telesc., Vol. 48, No. 1, p. 3 (1974); Celest. Mech., Vol. 10, p. 2-3 (1974); J. Navig., Vol. 27, p. 402-403 (1974); Science, Vol. 186, p. 461 (1974); Spaceflight, Vol. 16, p. 472 (1974); Q.J.R. Astron. Soc., Vol. 16, p. 321-322 (1975).

(1580) Betulia

1950 KA. Discovered 1950 May 22 by E. L. Johnson at Johannesburg.

This planet has been named at the request of Dr. S. J. Herrick in honor of his wife. (M 768)

(1581) Abanderada

1950 LA_1. Discovered 1950 June 15 by M. Itzigsohn at La Plata.

This name signifies a leader who carries a banner, and it is bestowed to honor the late Eva Peron {1919-1952} {see planet (1569)} for her labors in behalf of social development. (M 877)

(1582) Martir

1950 LY. Discovered 1950 June 15 by M. Itzigsohn at La Plata.

The Spanish spelling of martyr, this name is also assigned as an homage to Eva Peron {see planet (1569)}. (M 877)

(1583) Antilochus

1950 SA. Discovered 1950 September 19 by S. Arend at Uccle.

The discoverer has assigned to this Trojan the name Antilochus, who was the youngest son of Nestor, intimate friend of Achilles {see planets (659) and (588)}, and commander of the Pylians in the Trojan war. (M 770)

(1584) Fuji

1927 CR. Discovered 1927 February 7 by O. Oikawa at Tokyo.

Named in honor of Mt. Fuji, the highest mountain in Japan and a place of sacred worship. (M 2882)
The proposal of the discoverer was transmitted by S. Kanda.

(1585) Union

1947 RG. Discovered 1947 September 7 by E. L. Johnson at Johannesburg.
Named in honor of the Union Observatory at Johannesburg. (M 941)

(1586) Thiele

1939 CJ. Discovered 1939 February 13 by A. A. Wachmann at Bergedorf.

Named in memory of Thorvald Nicolai Thiele (1838-1910), sometime director of the Copenhagen Observatory, who worked on theoretical celestial mechanics. His son Holger Thiele, the discoverer of comet 1906 VII, later observed minor planets at Bergedorf and subsequently moved to Leuschner Observatory in California. (M 4236)

(1587) Kahrstedt

1933 FS_1. Discovered 1933 March 25 by K. Reinmuth at Heidelberg.

Named in memory of Albrecht Kahrstedt (1897-1971), on the staff of the Astronomisches Rechen-Institut since 1922 and later in charge of the section at Potsdam. When that section combined with the Babelsberg Observatory, he became director. He was an indefatigable computer of minor planet orbits, notably of (1221) Amor, and he also prepared the right-ascension system of the FK3. (M 3930)

Obituaries published in Mitt. Astron. Ges., Nr. 30, p. 166 (1971); Astron. Nachr., Vol. 292, p. 282 (1971), Vol. 294, p. 147-148 (1973).

(1588) Descamisada

1951 MH. Discovered 1951 June 27 by M. Itzigsohn at La Plata.

Literally a naked woman. As a symbol of the Argentine trade unions, it means a shirtless worker. Eva Peron {see planet (1569)} adopted this symbol of the very poor people and she was called The First Descamisada. (M 1069)

(1589) Fanatica

1950 RK. Discovered 1950 September 13 by M. Itzigsohn at La Plata.

Literally a fanatical woman or a feminine zealot. Named as a tribute to Eva Peron {see planet (1569)} whose devotion and enthusiasm for the welfare of the Argentine people led her to champion the cause of the workers. (M 1069)

(1590) Tsiolkovskaja

1933 NA. Discovered 1933 July 1 by G. N. Neujmin at Simeïs.

This planet is named in honor of the well-known Russian physicist and teacher Konstantin Tsiolkovsky (1857-1935), designer of liquid fuel rockets and space vehicles. (M 2116)

Tsiolkovsky is also honored by a lunar crater.

(1591) Baize

1951 KA. Discovered 1951 May 31 by S. Arend at Uccle.

This planet is named in honor of Dr. Paul Baize {1901-1995} of Paris who {was} a practicing physician and at the same time an assiduous amateur double star observer and computer at the Paris Observatory. This planet was recovered by Van Biesbroeck in 1952. Both he and the discoverer are ardent double star observers and they join in signifying their high esteem for their colleague. (M 943)

Obituaries published in Int. Astron. Union Comm. 26, Inf. Circ., No. 127, p. 2 (1995); Ciel Terre, Vol. 112, No. 3, p. 119-120 (1996); Astronomie, Vol. 110, p. 32 (1996).

(1592) Mathieu

1951 LA. Discovered 1951 June 1 by S. Arend at Uccle.

Named after one of the grand-children of the discoverer. (M 3825)

(1593) Fagnes

1951 LB. Discovered 1951 June 1 by S. Arend at Uccle.

Named after the "Plateau des Hautes Fagnes", largest national park of Belgium. (M 3825)

Name proposed by Jean Meeus.

(1594) Danjon

1949 WA. Discovered 1949 November 23 by L. Boyer at Algiers.

Named in honor of Prof. André Danjon {1890-1967}, Director of the Paris Observatory, and former president of the International Astronomical Union {1955-1958}. (M 2196)
Danjon is also honored by a lunar crater.

(1595) Tanga

1930 ME. Discovered 1930 June 19 by C. Jackson and H. E. Wood at Johannesburg.
Named for a port on the eastern seabord of Tanzania. (M 5182)

(1596) Itzigsohn

1951 EV. Discovered 1951 March 8 by M. Itzigsohn at La Plata.

Named by the La Plata Observatory in memory of Miguel Itzigsohn, some time professor of spherical and practical astronomy and head of the department of extrameridian astronomy at the La Plata Observatory. He was in charge of the work on minor planets at La Plata and contributed extensively to the development of astrometry in Argentina. (M 5449)

(1597) Laugier

1949 EB. Discovered 1949 March 7 by L. Boyer at Algiers.

Named in memory of Margueritte Laugier (1896-1976), astronomer at the Nice Observatory and discoverer of {19} minor planets. (M 4418).

(1598) Paloque

1950 CA. Discovered 1950 February 11 by L. Boyer at Algiers.

Named in honor of E. Paloque, Director of the Observatory of Toulouse. (M 1221)

Name assigned by the discoverer, following a suggestion made by P. Pretre, Toulouse, who computed the orbit of this planet.

(1599) Giomus

1950 WA. Discovered 1950 November 17 by L. Boyer at Algiers.

Named in honor of the birthplace of P. Pretre. This is the sixth-century name of the present town of Gien (Loiret), which has a rich history and suffered terribly during the occupation of 1940. (M 1221)

Name assigned by the discoverer, following a suggestion made by P. Pretre, Toulouse, who computed the orbit of this planet.

(1600) Vyssotsky

1947 UC. Discovered 1947 October 22 by C. A. Wirtanen at Mount Hamilton.

Named in honor of Prof. Alexander Vyssotsky (1888-1973) who joined the faculty of the University of Virginia in 1923 and was active in astrometry, photometry and spectral classification at the Leander McCormick Observatory. (M 3931)
Obituary published in Sky Telesc., Vol. 47, p. 166 (1974).

(1601) Patry

1942 KA. Discovered 1942 May 18 by L. Boyer at Algiers.

Named in honor of André Patry {1902-1960}, astronomer at the Nice Observatory, especially distinguished for his research in establishing identities of minor planets from observations in widely separated years, of which this planet is one example. (M 2196)

(1602) Indiana

1950 GF. Discovered 1950 March 14 at the Goethe Link Observatory at Brooklyn, Indiana.

Named in honor of the state of Indiana and Indiana University, the parent institution of the astronomy department and the observatory in which the planet was discovered. (M 1171)

The discovery was credited (MPC 1171) to Beryl H. Potter {see planet (1729)}. Frank K. Edmondson {see planet (1761)}, however, stated that 'during the first years there were 10 or 12 persons involved in blinking the plates'. The discovery is therefore credited to Indiana University.

(1603) Neva

1926 VH. Discovered 1926 November 4 by G. N. Neujmin at Simeïs.

This planet is named after the large river in Russia upon whose banks the city of Leningrad is situated. Leningrad was renamed in St. Petersburg in 1991. (M 2117)

(1604) Tombaugh

1931 FH. Discovered 1931 March 24 by C. W. Tombaugh at Flagstaff.

Named by the Lowell Observatory for Clyde W. Tombaugh {1906-1997}, the discoverer of Pluto, on the occasion of a symposium on Pluto, held on the fiftieth anniversary of its discovery, 1980 Feb. 18. Tombaugh marked, during the course of his blink examination, over 4000 minor planets on plates obtained with the 0.33-m photographic telescope during the trans-Saturnian search program at the Lowell Observatory. (M 5280)

Obituaries published in Zenit, Jaarg. 24, Nr. 7/8, p. 310-311 (1997); Nature, Vol. 385, No. 6619, p. 778 (1997); Icarus, Vol. 127, No. 1, p. V (1997); Astronomy, Vol. 25, No. 4, p. 28, 30 (1997); Sidereal Times, Vol. 47, No. 4, p. 12 (1997); Mon. Notes Astron. S. Afr., Vol. 56, No. 3-4, p. 20 (1997); Planet. Rep., Vol. 17, No. 4, p. 14 (1997); Phys. Today, Vol. 50, No. 7, p. 77 (1997); Orion, Jahrg. 55, Nr. 282, p. 7 (1997); G. Astron., Vol. 23, N. 1, p. 55-56 (1997); Astron. Geophys., Vol. 38, Issue 2, p. 38 (1997); J. R. Astron. Soc. Canada, Vol. 91, No. 2, p. 91-92 (1997); Sky Telesc., Vol. 93, No. 4, p. 102-103 (1997).

(1605) Milankovitch

1936 GA. Discovered 1936 April 13 by P. Djurkovic at Uccle. Independently discovered 1936 April 15 by J. Piegza at Cracow and by T. Banachiewicz at Warsaw.

Named in memory of M. Milankovitch (1879-1958), famous Yugoslav astronomer, celestial mechanician and mathematician, known for his astronomical theory of long-term changes in the Earth's climate and for numerous papers on the history of astronomy. (M 5449)

Milankovitch is also honored by craters on Mars and the Moon.

(1606) Jekhovsky

1950 RH. Discovered 1950 September 14 by L. Boyer at Algiers.

Named in memory of Benjamin Jekhovsky, astronomer at the Algiers and later the Bordeaux and Toulouse observatories, discoverer of {12} minor planets between 1921 and 1927. (M 4418)

(1607) Mavis

1950 RA. Discovered 1950 September 3 by E. L. Johnson at Johannesburg.

Named in honor of the wife of J. A. Bruwer {see planet (1811)}, astronomer at the observatory in Johannesburg. (M 3931)

(1608) Muñoz

1951 RZ. Discovered 1951 September 1 by M. Itzigsohn at La Plata.

Named in memory of F. A. Muñoz, an assistant in the department of extra-meridian astronomy at the La Plata Observatory and involved for many years in computational and observational work on minor planets at La Plata; he also took an active part in site testing for the 2.15-m Argentine telescope. (M 5449)

(1609) Brenda

1951 NL. Discovered 1951 July 10 by E. L. Johnson at Johannesburg.
Named by the discoverer in honor of his granddaughter. (M 3931)

(1610) Mirnaya

1928 RT. Discovered 1928 September 11 by P. F. Shajn at Simeïs.
The name means peaceful. (M 2740)
Name proposed by the Institute of Theoretical Astronomy, Leningrad.

(1611) Beyer

1950 DJ. Discovered 1950 February 17 by K. Reinmuth at Heidelberg.

This planet is named by the discoverer in honor of Dr. h.c. Max Beyer {1894-1982}, astronomer at the Hamburg-Bergedorf Observatory. (M 1948)

Obituaries published in BAV Rundbrief, 32. Jahrg., Nr. 2, p. 51 (1983); Mitt. Astron. Ges., Nr. 59, p. 5-6 (1983).

(1612) Hirose

1950 BJ. Discovered 1950 January 23 by K. Reinmuth at Heidelberg.

Named in honor of the Japanese astronomer, Prof. H. Hirose, under whose leadership the observations of minor planets and the computation of their orbits at the Tokyo-Mitaka astronomical observatory have been so successfully conducted. (M 1342)

(1613) Smiley

1950 SD. Discovered 1950 September 16 by S. Arend at Uccle.

Named in honor of Charles Hugh Smiley {1903-1977}, interested mainly in orbit computations by Leuschner's method, director of the Ladd Observatory and professor at Brown University, Providence, R.I. (M 3931)

Smiley not only worked on the field of minor planet dynamics. He led 14 solar eclipse expeditions, in Peru, Canada, Brazil, Thailand, Pakistan, and the U.S.A., and did much work on the study of the Mayan calendar. Obituaries are published in J.R. Astron. Soc. Canada, Vol. 72, p. 46-47 (1978); Q.J.R. Astron. Soc., Vol. 19, p. 510-511 (1978).

(1614) Goldschmidt

1952 HA. Discovered 1952 April 18 by A. Schmitt at Uccle.

Named in memory of Hermann Goldschmidt (1802-1866), German-born astronomer who observed in Paris and discovered there 14 of the first 70 minor planets. (M 4418)
Goldschmidt is also honored by a lunar crater.

(1615) Bardwell

1950 BW. Discovered 1950 January 28 at the Goethe Link Observatory at Brooklyn, Indiana.

Named in honor of Conrad M. Bardwell {1926- }, research associate in the Minor Planet Center at the Cincinnati Observatory, in recognition of his close cooperation with the observing program of the Goethe Link Observatory. He has provided observers with numerous reliable ephemerides and has had repeated successes in establishing identifications of observations in widely separated oppositions. (M 3643)
Name proposed by Dr. F. K. Edmondson and Mrs. Deloris J. Owings.

(1616) Filipoff

1950 EA. Discovered 1950 March 15 by L. Boyer at Algiers.

Named in memory of Lionel Filipoff (1893-1940), astronomer at the Algiers Observatory and later at Paris, assiduous observer of minor planets and comets. (M 4418)

(1617) Alschmitt

1952 FB. Discovered 1952 March 20 by L. Boyer at Algiers.

Named in memory of Alfred Schmitt (1907-1973), astronomer at the Algiers and Strasbourg observatories and some time director of the Quito Observatory. He made numerous observations of minor planets at Algiers and Uccle and at Strasbourg was concerned with astrometric problems involving artificial Earth satellites. (M 4418)

(1618) Dawn

1948 NF. Discovered 1948 July 5 by E. L. Johnson at Johannesburg.

Named by the discoverer in honor of his granddaughter. (M 3931)

(1619) Ueta

1953 TA. Discovered 1953 October 11 by T. Mitani at Kwasan.

Named by the discoverer after the former Director of Kwasan Observatory who encouraged him to keep on with his observations of asteroids and comets. (M 2347)

(1620) Geographos

1951 RA. Discovered 1951 September 14 by A. G. Wilson and R. Minkowski at Palomar.

The name, meaning Geographer, has been assigned by the discoverers to honor the National Geographic Society for its contribution to astronomy by the support of the National Geographic Society - Palomar Mountain Sky Survey. (M 1468)

(1621) Druzhba

1926 TM. Discovered 1926 October 1 by S. I. Belyavskij at Simeïs.

Druzhba means friendship. (M 2740)

Name proposed by the Institute of Theoretical Astronomy, Leningrad.

(1622) Chacornac

1952 EA. Discovered 1952 March 15 by A. Schmitt at Uccle.

Named in memory of Jean Chacornac (1823-1873), astronomer at Marseilles and later Paris, co-discoverer of (20) Massalia and discoverer of (25) Phocaea and five other minor planets. (M 4418)

Chacornac is also honored by a lunar crater.

(1623) Vivian

1948 PL. Discovered 1948 August 9 by E. L. Johnson at Johannesburg.

Named in honor of the daughter of W. P. Hirst {see planet (3172)}, who calculated the preliminary orbit for this and several other planets discovered by Johnson. (M 3569)

This name had previously been assigned to (1605) 1952 OH, until that planet was found to be identical with (735) Marghanna {see MPC 1116, 1119}.

(1624) Rabe

1931 TT_1. Discovered 1931 October 9 by K. Reinmuth at Heidelberg.

Named in memory of Eugene Rabe (1911-1974), on the staff of the Astronomisches Rechen-Institut from 1937 to 1948 and then at the Cincinnati Observatory. He worked extensively on the theory of the motions of the Trojan minor planets and especially the orbit of (433) Eros. (M 3931)

Obituaries published in Science, Vol. 186, p. 719 (1974); Sky Telesc., Vol. 48, p. 157 (1974).

(1625) The NORC

1953 RB. Discovered 1953 September 1 by S. Arend at Uccle.

This planet is named in honor of the Naval Ordnance Research Calculator at Dahlgren, Virginia. This was the most powerful electronic calculator ever built

up to the time the planet was named. It was conceived by Mr. Byron Havens and developed at the Watson Scientific Computation Laboratory under the direction of Dr. W. J. Eckert {see planet (1750)}. The NORC has already served to provide a vast amount of computations of minor planet orbits, and it is to be expected that more will follow. This has been made possible through the assistance of the Naval Proving Grounds, the Office of Naval Research and the National Science Foundation. (M 1591)
The assignment of the name was granted by the discoverer to Paul Herget.

(1626) Sadeya

1927 AA. Discovered 1927 January 10 by J. Comas Solá at Barcelona.
Named in honor of the Sociedad Astronomica de España y America. (M 2277)
Name communicated by J. F. Carbo, president of the Society.

(1627) Ivar

1929 SH. Discovered 1929 September 25 by E. Hertzsprung at Johannesburg.

This planet is named by the discoverer in honor of his late brother Ivar. (M 1860)

(1628) Strobel

Discovered 1923 September 11 by K. Reinmuth at Heidelberg.

Named in honor of Willi Strobel {1909-1988}, at the Astronomisches Rechen-Institut since 1938 and author of the 1963 edition of the *Identifizierungsnachweis der Kleinen Planeten.* (M 3931)

(1629) Pecker

1952 DB. Discovered 1952 February 28 by L. Boyer at Algiers.

Named by the discoverer in honor of Prof. Jean-Claude Pecker {1923- }, director of the Nice Observatory. (M 2740)

Pecker served as general secretary of the International Astronomical Union from 1964 to 1967.

(1630) Milet

1952 DA. Discovered 1952 February 28 by L. Boyer at Algiers.

Named in honor of Bernard Milet, astronomer at the Nice Observatory who conducts a large program of observations of minor planets and comets. (M 4418)

(1631) Kopff

1936 UC. Discovered 1936 October 11 by Y. Väisälä at Turku.

Named in memory of August Kopff (1882-1960), who as Wolf's assistant in Heidelberg discovered and observed many minor planets. In 1924 he became Director of the Astronomisches Rechen-Institut in Berlin, and after the western section moved to Heidelberg he also became director of Heidelberg-Königstuhl Observatory. He was responsible for constructing the FK3 and initiated work on the FK4. (M 3931)

Karl Reinmuth discovered the planet as 1926 TH on 1926 October 5. The planet could only have been observed on two nights which was not sufficient to derive a reliable orbit. The discovery is therefore credited to Y. Väisälä. Elliptical elements were derived by H. Walter (MPC 1897) from five positions of 1936 UC. The identity between both apparitions was found by A. Patry (MPC 1451). Kopff is also honored by a lunar crater.

(1632) Sieböhme

1941 DF. Discovered 1941 February 26 by K. Reinmuth at Heidelberg.

Named in honor of Siegfried Böhme {1909-1996}, an astronomer at the Heidelberg Rechen-Institut since 1949, who has improved the orbits of many minor planets, notably (919) Ilsebill. (M 3931)

(1633) Chimay

1929 EC. Discovered 1929 March 3 by S. Arend at Uccle. Independently discovered 1929 March 8 by M. Wolf at Heidelberg.

The name of the native town of Georges Roland, co-discoverer of the comet Arend-Roland. (M 3931)

(1634) Ndola

1935 QP. Discovered 1935 August 19 by C. Jackson at Johannesburg.

Named for the most important mining town in Zambia. (M 5182)

(1635) Bohrmann

Discovered 1924 March 7 by K. Reinmuth at Heidelberg.

Named in honor of Alfred Bohrmann (1904-), at the Königstuhl Observatory from 1924 to 1969. Well known as an observer of minor planets, he was responsible for the publication of 700 accurate observations of these bodies. (M 3931)

(1636) Porter

1950 BH. Discovered 1950 January 23 by K. Reinmuth at Heidelberg.

Named for the U.S. astronomer Jermain Gildersleeve Porter (1852-1933), Director of the Cincinnati Observatory from 1884 to 1930 and well known for his work on star positions. Also named for John Guy Porter (1900-1981), for many years principal scientific officer at H.M. Nautical Almanac Office, Director of the Computing Section of the British Astronomical Association, and well-known popularizer of astronomy. Both J. G. Porters have contributed extensively to the astronomy of comets and minor planets, the first one as an observer, the second as a computer and a compiler of a catalogue of cometary orbits. (M 3932)

Obituaries published in Int. Comet. Q., Vol. 4, p. 55 (1982); Q.J.R. Astron. Soc., Vol. 24, No. 3, p. 364-367 (1983); Q.J.R. Astron. Soc., Vol. 24, No. 3, p. 364-367 (1983).

(1637) Swings

1936 QO. Discovered 1936 August 28 by J. Hunaerts at Uccle. Independently discovered 1936 September 15 by C. Jackson at Johannesburg.

Named in honor of Pol Swings {1906-1983}, president of the I.A.U. during 1964-1967. He has contributed immensely to our knowledge of the physics of comets, especially the analysis and understanding of their spectra. (M 3932)

Obituaries published in Ciel, Vol. 45, p. 252-253 (1983); C.R. Séances Acad. Sci., Suppl., Tome 297, No. 13, p. 91 (1983); Sky Telesc., Vol. 67, No. 4, p. 325 (1984); Astrophys. Space Sci., Vol. 102, No. 1, p. 1-2 (1984); Ciel Terre, Vol. 100, No. 4, p. 135-138 (1984); C.R. Acad Sci., Sér. Gén., Vie Sci., Tome 2, No. 6, p. 657-662 (1986); Q.J.R. Astron. Soc., Vol. 27, No. 2, p. 305-308 (1986); Postepy Astron., Tom 32, Zesz. 1, p. 87-98 (1984).

(1638) Ruanda

1935 JF. Discovered 1935 May 3 by C. Jackson at Johannesburg.

Named for the state of Ruanda-Urundi. (M 5182)

(1639) Bower

1951 RB. Discovered 1951 September 12 by S. Arend at Uccle.

Named in honor of Ernest Clare Bower who worked intensively on the Laplacian method of orbit computation and the preliminary orbit of Pluto at the Students Observatory, Berkeley. He proposed the system of provisional designations which has been in use since 1925. (M 3932)

(1640) Nemo

1951 QA. Discovered 1951 August 31 by S. Arend at Uccle.

Named for the fictional builder of advanced technology in the novel by Jules Verne {see planet (5231)}, *Twenty Thousand Leagues under the Sea* {1870}. (M 6831) Name proposed by J. Meeus.

(1641) Tana

1935 OJ. Discovered 1935 July 25 by C. Jackson at Johannesburg.
Named for a river in Kenya. (M 5183)

(1642) Hill

1951 RU. Discovered 1951 September 4 by K. Reinmuth at Heidelberg.

Named in memory of George William Hill (1838-1914), American celestial mechanician whose works include the development of precepts for the Hill-Brown lunar theory and the derivation of second-order theories for the motions of Jupiter and Saturn. (M 3932)

(1643) Brown

1951 RQ. Discovered 1951 September 4 by K. Reinmuth at Heidelberg.

Named in memory of Ernest William Brown (1866-1938), the Cambridge-Yale mathematical astronomer who spent much of the last fifty years of his life working on the theory of the motion of the Moon, developing the method of G. W. Hill {see planet (1642)} with spectacular success. He also concerned himself with the problem of the Earth's variable rotation, studied the motions of minor planets on resonant orbits (particularly the Trojans) and wrote treatises on lunar theory and planetary theory. (M 3932)

(1644) Rafita

1935 YA. Discovered 1935 December 16 by R. Carrasco at Madrid.
Named by the discoverer in honor of his late son. (M 2277)

(1645) Waterfield

1933 OJ. Discovered 1933 July 24 by K. Reinmuth at Heidelberg.

Named in honor of Reginald Lawson Waterfield {1900-1986}, for many years a hemotologist at Guy's Hospital in London and one of England's best known amateur astronomers. He has been photographing comets since 1927 and still maintains an extensive program of astrometric observations. The minor planet also honors his cousin, William Francis Herschel Waterfield (1886-1933), a grandson of Sir John Herschel, a prolific observer of variable stars, a staff member of the Harvard Observatory during 1926-1928 and subsequently chief assistant at the Boyden Station. (M 3932)

Obituaries published in Q.J.R. Astron. Soc., Vol. 28, No. 4, p. 544-546, p. 546-551 (1987); J. Br. Astron. Assoc., Vol. 97, No. 4, p. 211-214 (1987).

(1646) Rosseland

1939 BG. Discovered 1939 January 19 by Y. Väisälä at Turku.

Named in honor of Svein Rosseland {1894-1985}, the renowned Norwegian Astrophysicist who founded the Institute of Theoretical Astrophysics at Oslo and served as its first director. His work on the theory of stellar interiors included studies of stellar rotation and stability and the derivation of the "Rosseland Mean". (M 3932)

Obituaries published in Astron. Tidsskr., Årg. 18, Nr. 2, p. 71-72 (1985); Inst. Teor. Astrofys., Blindern-Oslo, Småtrykk, Nr. 106 (1985); Q.J.R. Astron. Soc., Vol. 27, No. 3, p. 512-514 (1986).

(1647) Menelaus

1957 MK. Discovered 1957 June 23 by S. B. Nicholson at Mount Wilson.

The name was selected by the discoverer to enable another of the famous Greek

heroes to join the majority of his compatriots in the camp to the east of Jupiter. (M 2019)
Menelaus is also commemorated by a lunar crater.

(1648) Shajna

1935 RF. Discovered 1935 September 5 by P. F. Shajn at Simeïs. Independently discovered 1935 September 19 by C. Jackson at Johannesburg.

The planet is named in honor of the two late Russian astronomers Mr. G. A. Shajn {1892-1956} and Mrs. P. F. Shajn {1894-1956} {see planet (1190)}. (M 2117)
G. A. Shajn is also honored by a lunar crater.

(1649) Fabre

1951 DE. Discovered 1951 February 27 by L. Boyer at Algiers.

Named in honor of Hervé Fabre {1905-1995}, astronomer at the Nice Observatory and author of numerous works in celestial mechanics. (M 4418)
Obituary published in J. Astron. Fr., No. 50, p. 4-6 (1996).

(1650) Heckmann

1937 TG. Discovered 1937 October 11 by K. Reinmuth at Heidelberg.

Named in honor of Otto Heckmann {1901-1983}, Director of the Hamburg-Bergedorf Observatory from 1941 to 1962 and subsequently the first Director of the European Southern Observatory, the foundation of which had been initiated by him. His research activities cover cosmology and several aspects of fundamental astronomy. He was president of the IAU from 1967 to 1970. (M 3932)

Obituaries published in Messenger, No. 33, p. 1 (1983); Mitt. Astron. Ges., Nr. 60, p. 9-12 (1983); C.R. Séances Acad. Sci., Suppl., Tome 297, No. 13, p. 91-92 (1983); Q.J.R. Astron. Soc., Vol. 25, No. 3, p. 374-376 (1984); C.R. Acad. Sci., Sér. Gén., Vie Sci., Tome 1, No. 6, p. 591-593 (1984).

(1651) Behrens

1936 HD. Discovered 1936 April 23 by M. Laugier at Nice. Independently discovered 1936 May 18 by K. Reinmuth at Heidelberg.

Named in memory of Johann Gerhard Behrens (1889-1978), pastor at Detern, well known for his orbit computations on comets and minor planets. (M 5523)

Name proposed by O. Kippes, who found some of the identifications involving this planet.

(1652) Hergé

1953 PA. Discovered 1953 August 9 by S. Arend at Uccle.

Named in honor of Georges Remi, better known under his pseudonym Hergé, on the occasion of his seventy-fifth birthday. Considered by many as the father of the comic strip, he created his hero Tintin in 1929. (M 6831)
Name proposed by J. Meeus.

(1653) Yakhontovia

1937 RA. Discovered 1937 August 30 by G. N. Neujmin at Simeïs. Independently discovered 1937 August 31 by H. Alikoski at Turku.

Named in honor of Prof. N. S. Yakhontova. Mrs. Yakhontova was in charge of the Minor Planet Department of the Institute of Theoretical Astronomy for about 30 years. (M 2740)
Name proposed by the Institute of Theoretical Astronomy, Leningrad.

(1654) Bojeva

1931 TL. Discovered 1931 October 8 by P. F. Shajn at Simeïs.

This planet is named in honor of Dr. Nina Fedorovna Bojeva (1890-1956 who was

a valuable member of the staff of the Institute of Theoretical Astronomy for many years. Miss Bojeva served as computer of minor planet orbits and perturbations. (M 2740)
Name proposed by the Institute of Theoretical Astronomy, Leningrad.

(1655) Comas Solá

1929 WG. Discovered 1929 November 28 by J. Comas Solá at Barcelona.

Named by the Fabra Observatory in memory of José Comas Solá (1868-1937), first director of the Fabra Observatory, Barcelona, and a well-known observer of minor planets and comets for many years. Among his discoveries were the short-period comet Comas Solá and 11 numbered minor planets. The discoverer is also honored by planet (1102). (M 5357)
Comas Solá is also honored by a crater on Mars.

(1656) Suomi

1942 EC. Discovered 1942 March 11 by Y. Väisälä at Turku.

The name of this Hungaria-type minor planet, like that of (1453) Fennia, honors the country in which it was discovered. (M 3932)

(1657) Roemera

1961 EA. Discovered 1961 March 6 by P. Wild at Zimmerwald.

Named by the discoverer in honor of Dr. Elizabeth Roemer {1929- }, U.S. Naval Observatory, in appreciation of her untiring and successful efforts to advance the knowledge of the motions and physical properties of comets and minor planets. (M 2347)

(1658) Innes

1953 NA. Discovered 1953 July 13 by J. A. Bruwer at Johannesburg.

Named in honor of R. T. A. Innes (1861-1933), first Director of the Republic Observatory (originally Transvaal Observatory, later Union Observatory), from 1903 to 1927. He was not only a first-class observer, well-known for his observations of double stars and his deliberate search for Proxima Centauri, but he also made important theoretical and computational contributions to celestial mechanics, including the first convincing demonstration of the irregular rotation of the Earth. (M 3297)
Innes is also honored by a lunar crater.

(1659) Punkaharju

1940 YL. Discovered 1940 December 28 by Y. Väisälä at Turku.

Named for a well-known and beautiful isthmus in southeastern Finland. (M 3933)

(1660) Wood

1953 GA. Discovered 1953 April 7 by J. A. Bruwer at Johannesburg.

Named in honor of H. E. Wood (1881-1946), second Director of the Republic (then Union) Observatory (1928-1941). He had the prime responsibility for the Franklin-Adams camera since its acquisition in 1909 and inaugurated the Observatory's program for observing minor planets. (M 3297)

(1661) Granule

A916 FA. Discovered 1916 March 31 by M. Wolf at Heidelberg.

Named in honor of Edward A. Gall, M.D., internationally renowned pathologist, to commemorate his discovery of Gall's granule in lymphocytes, and his career as an inspiring teacher, on the occasion of his retirement as Director of the Medical Center of the University of Cincinnati. (M 3757)
The name was suggested by Mrs. Anne L. Herget of the Cincinnati Observatory.

(1662) Hoffmann

A923 RB. Discovered 1923 September 11 by K. Reinmuth at Heidelberg.

Named for Irmtraud Hoffmann who is the daughter-in-law of the discoverer. (M 2901)

(1663) van den Bos

1926 PE. Discovered 1926 August 4 by H. E. Wood at Johannesburg.

Named in honor of W. H. van den Bos {1896-1974}, Director of the Republic (then Union) Observatory from 1941 to 1956. He has made an outstanding series of visual micrometric observations of double stars and is responsible for the pre-eminence of the Republic Observatory in this work. (M 3297)

Obituaries published in Circ. Inf., No. 63 (1974); Mon. Not. Astron. Soc. S. Afr., Vol. 33, p. 60-61 (1974); l'Astronomie, 88. année, p. 305, 309-311 (1974).

(1664) Felix

1929 CD. Discovered 1929 February 4 by E. Delporte at Uccle.

Named in memory of Felix Timmermans (1886-1947), popular Belgian writer. (M 6831)

Name proposed by J. Meeus.

(1665) Gaby

1930 DQ. Discovered 1930 February 27 by K. Reinmuth at Heidelberg.

Named by the discoverer for his daughter-in-law, Gaby Reinmuth. (M 2901)

(1666) van Gent

1930 OG. Discovered 1930 July 22 by H. van Gent at Johannesburg.

This planet is named in honor of the late Dr. H. van Gent {1900-1947}, who was a Leiden observer in South Africa from 1928 to 1946, during which time he discovered many new minor planets. (M 2740)

Van Gent is also honored by a lunar crater.

(1667) Pels

1930 SY. Discovered 1930 September 16 by H. van Gent at Johannesburg.

This planet is named in honor of the late Mr. G. Pels {1893-1966}, a life long member of the scientific staff of the Leiden Observatory. He observed minor planets at the Leiden Observatory and in addition he computed orbital elements for many of the minor planets found by van Gent. (M 2740)

(1668) Hanna

1933 OK. Discovered 1933 July 24 by K. Reinmuth at Heidelberg.

Named by the discoverer for his daughter-in-law, Hanna Reinmuth. (M 2901)

(1669) Dagmar

1934 RS. Discovered 1934 September 7 by K. Reinmuth at Heidelberg.

Named by the discoverer, a German feminine name. No special meaning is assigned to this name. (M 2901)

(1670) Minnaert

1934 RZ. Discovered 1934 September 9 by H. van Gent at Johannesburg.

Named in honor of the late Prof. Marcel G. J. Minnaert {1893-1970}, who was Director of the Utrecht Observatory from 1937 until 1963. He made major contributions to solar research and prepared (with Mulders and Houtgast) the *Photometric Atlas of the Solar Spectrum.* He was an extraordinarily effective lecturer and writer in the popularization of astronomy. (M 3185)

Minnaert is also honored by a lunar crater. Obituaries published in Astron. Nachr., Vol. 292, p. 192 (1970); Orion, 28. Jahrg., p. 195 (1970); Hemel en Dampkring, Vol. 68, p. 289-292 (1970); l'Astronomie, Vol. 84, p. 525 (1970); Observatory, Vol. 90, p. 272 (1970); Sky Telesc., Vol. 40, p. 344 (1970); Nature, Vol. 229, p. 214 (1971); Astrophys. Space Sci., Vol. 10, p. 183-185 (1971); Solar Phys., Vol. 17, p. 3-5 (1971); Astron. Tidsskr., Vol. 3, p. 199-200 (1970); Icarus, Vol. 15, p. 147-148 (1971); Q.J.R. Astron. Soc., Vol. 12, p. 338-341 (1971); Irish Astron. J., Vol. 11, p. 161 (1973).

(1671) Chaika

1934 TD. Discovered 1934 October 3 by G. N. Neujmin at Simeïs.

The name means sea gull. This planet honors V. V. Nikolajeva-Tereshkova who was given the name the "sea gull" as she was the first woman to fly in space. (M 2740)

Name proposed by the Institute of Theoretical Astronomy, Leningrad.

(1672) Gezelle

1935 BD. Discovered 1935 January 29 by E. Delporte at Uccle.

Named in memory of Guido Gezelle (1830-1899), famous Flemish poet, who wrote extensively on religion and nature. (M 6832)

Name proposed by J. Meeus.

(1673) van Houten

1937 TH. Discovered 1937 October 11 by K. Reinmuth at Heidelberg.

Named by the discoverer in honor of C. J. van Houten, who is an astronomer at the Leiden Observatory. (M 2901)

(1674) Groeneveld

1938 DS. Discovered 1938 February 7 by K. Reinmuth at Heidelberg. Independently discovered 1938 February 22 by Y. Väisälä at Turku and G. Kulin at Budapest.

Named by the discoverer in honor of I. van Houten-Groeneveld, who is an astronomer at the Leiden Observatory. (M 2901)

(1675) Simonida

1938 FB. Discovered 1938 March 20 by M. B. Protitch at Belgrade. Independently discovered 1938 March 20 by F. Rigaux at Uccle.

In Yugoslavia this name is a symbol of beauty and is related to the personality of a gracious Serbian princess from the middle ages. (M 3359)

(1676) Kariba

1939 LC. Discovered 1939 June 15 by C. Jackson at Johannesburg.

Named for a large manmade lake between Zambia and Zimbabwe-Rhodesia. (M 5183)

(1677) Tycho Brahe

1940 RO. Discovered 1940 September 6 by Y. Väisälä at Turku.

Named for the great Danish-born astronomer Tycho Brahe (1546-1601). (M 4236)

Brahe is also honored by craters on Mars and the Moon.

(1678) Hveen

1940 YH. Discovered 1940 December 28 by Y. Väisälä at Turku.

Named for the island of Hveen, where Tycho Brahe's {see planet (1677)} most productive observations were made between 1576 and 1597. (M 4236)

The island is also commemorated by planets (379) and (499).

(1679) Nevanlinna

1941 FR. Discovered 1941 March 18 by L. Oterma at Turku.

Named in honor of Rolf H. Nevanlinna {1895- }, great Finnish mathematician, on the occasion of his 80th birthday, 1975 Oct. 22. He was a member of the Academy of Finland from 1948 to 1965 and chancellor of Turku University from 1965 to 1970. (M 3933)

(1680) Per Brahe

1942 CH. Discovered 1942 February 12 by L. Oterma at Turku.

Named in memory of the Swedish count Per Brahe (1602-1680), who was governor general of Finland. The "count's epoch" was a happy era, with the establishment of Academia Aboensis, the first university in Finland, the construction of various new towns and many schools, and the publication of the first Finnish Bible. (M 5280)

(1681) Steinmetz

1948 WE. Discovered 1948 November 23 by M. Laugier at Nice.

Named in memory of Julius Steinmetz (1893-1965), pastor at Gerolfingen; orbit computer. (M 5523)

Name proposed by O. Kippes, who found one of the identifications involving this planet.

(1682) Karel

1949 PH. Discovered 1949 August 2 by K. Reinmuth at Heidelberg.

Named by the discoverer for the son of Dr. and Mrs. van Houten of the Leiden Observatory. (M 2901)

See also the citations for (1673) and (1674).

(1683) Castafiore

1950 SL. Discovered 1950 September 19 by S. Arend at Uccle.

Named for Bianca Castafiore, a comic-strip character created by Hergé {see planet (1652)}. (M 6832)

Name proposed by J. Meeus.

(1684) Iguassú

1951 QE. Discovered 1951 August 23 by M. Itzigsohn at La Plata.

Named for the large waterfall 60 meters high and more than 1 km wide on a river of the same name. The Iguassú river forms part of the boundary between Argentina and Brazil. (M 6832)

Name proposed by F. Pilcher.

(1685) Toro

1948 OA. Discovered 1948 July 17 by C. A. Wirtanen at Mount Hamilton.

Named, as companion to (1580) Betulia, to honor the maiden name of Mrs. Herrick, and because of its phonetic, masculine, and legendary appropriateness to the group of close minor planets. (M 2504)

Naming delegated to Samuel Herrick and his associates, especially Kenneth C. Ford, whose predictions led to its recovery in 1956 and 1964.

(1686) De Sitter

1935 SR_1. Discovered 1935 September 28 by H. van Gent at Johannesburg.

Named in honor of the late Prof. Willem de Sitter {1872-1934}, who was director of the Leiden Observatory from 1918 till 1934. He made major contributions in the theory of the Jupiter satellites, astronomical constants and cosmology. (M 2822)

Name proposed by the Leiden Observatory.

De Sitter is also honored by a lunar crater.

(1687) Glarona

1965 SC. Discovered 1965 September 19 by P. Wild at Zimmerwald.

Named by the discoverer in honor of his home valley, the Swiss canton and town of Glarus. (M 2971)

(1688) Wilkens

1951 EQ_1. Discovered 1951 March 3 by M. Itzigsohn at La Plata.

Named in memory of the German astronomer Alexander Wilkens {1881-1968}, researcher in many branches of astronomy, most notably celestial mechanics. He worked for many years in Germany, then at the La Plata Observatory, where he produced two generations of celestial mechanicians before returning to his native country. (M 5449)

Obituary published in Astron. Nachr., Vol. 291, p. 87-88 (1969).

(1689) Floris-Jan

1930 SO. Discovered 1930 September 16 by H. van Gent at Johannesburg. Independently discovered 1930 September 21 by E. F. Skvortsov at Simeïs.

This planet is named after Floris-Jan van der Meulen, the 5000th visitor to a 14-day astronomical exhibition at the Leiden Observatory. (M 3470)

(1690) Mayrhofer

1948 VB. Discovered 1948 November 8 by M. Laugier at Nice.

Named in honor of Karl Mayrhofer, Austrian astronomer, currently living at Ried im Innkreis, well known for his orbit computations on minor planets. (M 5523)

Name proposed by O. Kippes, who found some of the identifications involving this planet. Mayrhofer died 1982 at the age of 79.

(1691) Oort

1956 RB. Discovered 1956 September 9 by K. Reinmuth and I. Groeneveld at Heidelberg.

Named in honor of Jan Hendrik Oort {1900-1992}, director of the Leiden Observatory 1945-1970, former president of the International Astronomical Union {1958-1961}, and a well-known authority on stellar statistics and galactic structure. (M 3023)

Obituaries published in Mercury, Vol. 21, No. 6, p. 196 (1992); Messenger, No. 70, p. 1-2 (1992); Mon. Notes Astron. Soc. S. Afr., Vol. 51, Nos. 11-12, p. 107-108 (1992); Phys. Today, Vol. 46, No. 11, p. 104-105 (1993); C.R. Acad. Sci., Sér. Gén, Vie Sci., Tome 10, No. 5, p. 535-540 (1993); Bull. Am. Astron. Soc., Vol. 25, No. 4, p. 1499 (1993); Q.J.R. Astron. Soc., Vol. 35, No. 2, p. 237-242 (1994); J. R. Astron. Soc. Can., Vol. 87, No. 2, p. 73-76 (1993); IAU Inf. Bull., No. 69, p. VII (1993); Astronomie, Vol. 107, p. 60-64 (1993); Zenit, Jaarg. 20, Nr. 5, p. 196-210 (1993); Zemlya Vselennaya, No. 2, p. 53-56 (1993);Sky Telesc., Vol. 85, No. 4, p. 44-45 (1993); Int. Astrom. Union Symp., No. 169, p. XV-XVI (1996).

(1692) Subbotina

1936 QD. Discovered 1936 August 16 by G. N. Neujmin at Simeïs. Independently discovered 1936 August 17 by K. Reinmuth at Heidelberg.

This planet is named in honor of the late Prof. Mikhail F. Subbotin (1893-1966) who was Director of the Institute of Theoretical Astronomy for about 20 years. (M 2740)

Name proposed by the Institute of Theoretical Astronomy, Leningrad.
Subbotin is also honored by a lunar crater.

(1693) Hertzsprung

1935 LA. Discovered 1935 May 5 by H. van Gent at Johannesburg.

Named in honor of the late Prof. E. Hertzsprung {1873-1967}, who was Director of the Leiden Observatory from 1934 till 1945. A well-known authority in the field of astronomical photometry, he initiated the Leiden Variable Star Survey of the southern Milky Way, during which survey many asteroids and some comets were found. (M 2822)
Name proposed by the Leiden Observatory.

Hertzsprung is also honored by a lunar crater.Obituary published in Astron. Nachr., Vol. 291, p. 85-87 (1969).

(1694) Kaiser

1934 SB. Discovered 1934 September 29 by H. van Gent at Johannesburg.

Named in honor of the late Prof. Frederick Kaiser {1808-1872}, who was Director of the Leiden Observatory from 1837-1872. He is the founder of the new Leiden Observatory and stimulated Dutch astronomical research. (M 2883)

Due to an oversight, the name of Kapteyn was assigned to this planet in MPC 2822, when it had already been assigned to (818). Kaiser is also honored by craters on Mars and the Moon.

(1695) Walbeck

1941 UO. Discovered 1941 October 15 by L. Oterma at Turku.

Named in memory of H. J. Walbeck (1793-1822), astronomer at the old Academia Aboensis who used the method of least squares to derive a good value for the Earth's flattening. (M 5281)

(1696) Nurmela

1939 FF. Discovered 1939 March 18 by Y. Väisälä at Turku.

Named in honor of Finnish academician Tauno Kalervo Nurmela, some time professor of Romanic philology and later chancellor of Turku University. (M 5281)

(1697) Koskenniemi

1940 RM. Discovered 1940 September 8 by H. Alikoski at Turku.

Named in memory of Veikko Antero Koskenniemi (1885-1962), the famous Finnish poet, a member of the Academy of Finland, professor of literature at Turku University. In various poems he wrote about the stars, and he was a founder member of the society Turun Ursa for amateur astronomers. (M 5281)

(1698) Christophe

1934 CS. Discovered 1934 February 10 by E. Delporte at Uccle.

Named in honor of a grand-nephew of G. Roland, administrator of the Royal Observatory, Uccle, co-discoverer of comet Arend-Roland (1957 III). (M 6832)

(1699) Honkasalo

1941 QD. Discovered 1941 August 26 by Y. Väisälä at Turku.

Named in memory of Tauno Bruno Honkasalo (1912-1975), a disciple of Y. Väisälä who measured geodesic standard base lines in various countries with the Väisälä interference comparator. (M 5281, M 5392)

(1700) Zvezdara

1940 QC. Discovered 1940 August 26 by P. Djurkovic at Belgrade.

The Serbian word for observatory, this is the name of the section of the city of Belgrade in which the Observatory, founded in 1934, is located. (M 5449)

(1701) Okavango

1953 NJ. Discovered 1953 July 6 by J. Churms at Johannesburg.

Named for a large river in southern Africa. It flows into a swamp of the same name in Botswana. The swamp is an important wildlife preserve and has no outlet to the ocean. (M 7156)
Name proposed by F. Pilcher.

(1702) Kalahari

A924 NC. Discovered 1924 July 7 by E. Hertzsprung at Johannesburg.

Named for the great desert of Namibia {see planet (1718)} and adjoining region of the Republic of South Africa. It is continuous with the steppe country that extends to the vicinity of the discovery site. (M 7156)
Name proposed by F. Pilcher.

(1703) Barry

1930 RB. Discovered 1930 September 2 by M. Wolf at Heidelberg. Independently discovered 1930 September 16 by H. van Gent at Johannesburg and September 21 by E. F. Skvortsov at Simeïs.

Named for Roger Barry (1752-1813), who became an astronomer at the Mannheim Observatory in 1788. The observatory at Heidelberg-Königstuhl is a direct successor to the old Mannheim Observatory. (M 3933)

(1704) Wachmann

A924 EE. Discovered 1924 March 7 by K. Reinmuth at Heidelberg.

Named in honor of Arno Arthur Wachmann {1902-1990}, an astronomer at Hamburg-Bergedorf since 1927. He discovered and observed several minor planets, and together with Schwassmann, found four comets. He is well-known for his work on stellar spectroscopy and especially on variable and binary stars. (M 3933)
Obituary published in Mitt. Astron. Ges., Nr. 74, p. 5-6 (1991).

(1705) Tapio

1941 SL_1. Discovered 1941 September 26 by L. Oterma at Turku.

Named for the guardian spirit of the forest in Kalevala {see planet (1454)}, the Finnish national epic. Tapio is also a common boy's name in Finland. (M 5281)

(1706) Dieckvoss

1931 TS. Discovered 1931 October 5 by K. Reinmuth at Heidelberg.

Named in honor of Wilhelm Dieckvoss {1908-1982}, Hamburg-Bergedorf astronomer since 1935 who has worked extensively on stellar positions and proper motions, especially on the catalogues AGK2 and AGK3. (M 3933)
Obituary published in Mitt. Astron. Ges., Nr. 59, p. 7-8 (1983).

(1707) Chantal

1932 RL. Discovered 1932 September 8 by E. Delporte at Uccle.
Named in honor of a niece of Uccle astronomer G. Roland. (M 6832)

(1708) Pólit

1929 XA. Discovered 1929 December 1 by J. Comas Solá at Barcelona.

Named in memory of Isidre Pólit (1880-1958), second director of the astronomical section of the Fabra Observatory and an assiduous observer of minor planets and comets. (M 5357)

(1709) Ukraina

1925 QA. Discovered 1925 August 16 by G. Shajn at Simeïs.

The planet is named for the {former} Ukrainian Soviet Socialist Republic. (M 2740)
Name proposed by the Institute of Theoretical Astronomy, Leningrad.

(1710) Gothard

1941 UF. Discovered 1941 October 20 by G. Kulin at Budapest.

Named in memory of Jeno Gothard (1857-1909), a Hungarian astronomer who discovered the central star in M57. (M 5183)

(1711) Sandrine

1935 BB. Discovered 1935 January 29 by E. Delporte at Uccle.

Named in honor of a grand-niece of Uccle astronomer G. Roland. (M 6832)

(1712) Angola

1935 KC. Discovered 1935 May 28 by C. Jackson at Johannesburg.

Named for the state on the southwestern coast of Africa. (M 5183)

(1713) Bancilhon

1951 SC. Discovered 1951 September 27 by L. Boyer at Algiers.

Named in honor of Madame Alfred Schmitt, née Odette Bancilhon, astronomer at the Algiers and later the Strasbourg observatories, discoverer of (1333) Cevenola. (M 4419)

(1714) Sy

1951 OA. Discovered 1951 July 25 by L. Boyer at Algiers.

Named in memory of Frederic Sy, a computer at the Paris Observatory during 1879-1887 and an assistant astronomer at the Algiers Observatory during 1887-1918. An avid observer of minor planets and comets, he made the first discovery of a numbered minor planet at Algiers, (858) El Djezair, in 1916. (M 4419)

(1715) Salli

1938 GK. Discovered 1938 April 9 by H. Alikoski at Turku.

Named in honor of the wife of the discoverer. (M 5281)

(1716) Peter

1934 GF. Discovered 1934 April 4 by K. Reinmuth at Heidelberg.

Named by the discoverer in honor of his grandson. (M 3933)

(1717) Arlon

1954 AC. Discovered 1954 January 8 by S. Arend at Uccle.

Named for the principal town of the province of Belgian Luxembourg, situated on a hill above the headwaters of the Semoise. The Orolaunum of the Romans was a station on the Antoninian way connecting Reims and Trier. (M 8150)

(1718) Namibia

1942 RX. Discovered 1942 September 14 by M. Väisälä at Turku.

Named for the {now independent} African country, where the discoverer worked for many years, teaching the children of Finnish missionaries. (M 5281)

(1719) Jens

1950 DP. Discovered 1950 February 17 by K. Reinmuth at Heidelberg.

Named by the discoverer in honor of his grandson. (M 3933)

(1720) Niels

1935 CQ. Discovered 1935 February 7 by K. Reinmuth at Heidelberg.

Named by the discoverer in honor of his grandson. (M 3933)

(1721) Wells

1953 TD_3. Discovered 1953 October 3 at the Goethe Link Observatory at Brooklyn, Indiana.

Named for Herman B. Wells, president of Indiana University for 25 years from

1937 to 1962. Following his retirement as president, he was given the new title of university chancellor, and has continued to be active in university affairs. Almost single handed, he changed Indiana University from a good university to a great university. Over the years he also served the cause of higher education on the national and international level with great distinction and statesmanship. (M 3508)

(1722) Goffin

1938 EG. Discovered 1938 February 23 by E. Delporte at Uccle.

Named in honor of the Belgian amateur astronomer Edwin Goffin, who has made extensive computations involving minor-planet orbits, and whose initials are indicated by this object's provisional designation. (M 6832)
Name proposed by J. Meeus.

(1723) Klemola

1936 FX. Discovered 1936 March 18 by Y. Väisälä at Turku.

Named in honor of Irja Klemola, formerly rector of a school in Turku, one of the founders and for many years secretary of the society Turun Ursa for amateur astronomers; she also frequently participated in the minor planet program at the Turku Observatory. This planet also honors Arnold R. Klemola, currently in charge of the proper-motion program at the Lick Observatory, also well known for his astrometric observations of comets and minor planets. (M 5281)

(1724) Vladimir

1932 DC. Discovered 1932 February 28 by E. Delporte at Uccle.

Named by M. B. Protitch, whose rediscovery of this object in 1952 made the permanent numbering possible, in honor of his grandson. (M 5281)

(1725) CrAO

1930 SK. Discovered 1930 September 20 by G. N. Neujmin at Simeïs.

The name is an abbreviation for the Crimean Astrophysical Observatory, which succeeded the Simeïs Observatory. (M 2740)
Name proposed by the Institute of Theoretical Astronomy, Leningrad.

(1726) Hoffmeister

1933 OE. Discovered 1933 July 24 by K. Reinmuth at Heidelberg.

Named in memory of Cuno Hoffmeister (1892-1968), founder and later director of the Sonneberg {see planet (1039)} Observatory. Best known as a prolific discoverer and observer of variable stars, he also did extensive work on meteors and discovered a comet and several minor planets. (M 3933)

Hoffmeister is also honored by minor planet (4183) and by a lunar crater. Obituaries published in Irish Astron J., Vol. 9, p. 169 (1969); IAU Colloq. No. 15, p. 315-316 (1972).

(1727) Mette

1965 BA. Discovered 1965 January 25 by A. D. Andrews at Bloemfontein.

Named by the discoverer in honor of his wife "for her tolerance of my nocturnal working hours and for the many months spent away from home". (M 5183)

(1728) Goethe Link

1964 TO. Discovered 1964 October 12 at the Goethe Link Observatory at Brooklyn, Indiana.

Named in honor of Dr. Goethe Link, eminent surgeon of Indianapolis, Indiana, enthusiastic amateur astronomer, generous patron of astronomy at Indiana University, and donor of the Goethe Link Observatory. (M 2882)

(1729) Beryl

1963 SL. Discovered 1963 September 19 at the Goethe Link Observatory at Brooklyn, Indiana.

Named in honor of Mrs. Beryl H. Potter {1901-1985}, whose devoted work from 1949 to 1966 contributed immensely to the program of minor planet observations at Indiana University. (M 2883)

An editorial notice concerning the retirement of Mrs. Potter is published in MPC 2561. Obituary published in Phys. Today, Vol. 39, No. 2, p. 92 (1986).

(1730) Marceline

1936 UA. Discovered 1936 October 17 by M. Laugier at Nice. Independently discovered 1936 October 24 by G. N. Neujmin at Simeïs.

Named for the heroine of André Gide's novel *L'Immoraliste.* As a beautiful and devoted young wife Marceline nursed her husband from all the brink of death to robust health. When soon afterward Marceline became ill her husband benignly neglected her. Marceline suffered much physical and mental anguish and finally died needlessly. (M 6832)
Name proposed by F. Pilcher.

(1731) Smuts

1948 PH. Discovered 1948 August 9 by E. L. Johnson at Johannesburg.

Field-marshal Jan Christiaan Smuts {1870-1950}, a distinguished South African under whom the discoverer fought in both World Wars. He was prime minister of South Africa from 1919 to 1924 and again from 1938 to 1948. (M 3933)

(1732) Heike

1943 EY. Discovered 1943 March 9 by K. Reinmuth at Heidelberg.
Named in honor of Heike Neckel, a granddaughter of A. Bohrmann. (M 3933)

(1733) Silke

1938 DL_1. Discovered 1938 February 19 by A. Bohrmann at Heidelberg.
Named by the discoverer for his granddaughter, Silke Neckel. (M 3933)

(1734) Zhongolovich

1928 TJ. Discovered 1928 October 11 by G. N. Neujmin at Simeïs.

Named in honor of Prof. Ivan Danilovich Zhongolovich, a distinguished astronomer and geodesist, on the staff of the ITA {Institute of Theoretical Astronomy, Leningrad} since 1920, the head of the Special Ephemeris Department, world renowned for his fundamental ideas in satellite geodesy. (M 3933)

(1735) ITA

1948 RJ_1. Discovered 1948 September 10 by P. F. Shajn at Simeïs.

Named in 1979 on the occasion of the 60th anniversary of the founding of the Institute for Theoretical Astronomy, U.S.S.R. Academy of Sciences. (M 5357)

(1736) Floirac

1967 RA. Discovered 1967 September 6 by G. Soulie at Bordeaux.

Named by the discoverer for the suburb in Bordeaux in which the observatory is located. (M 2883)

(1737) Severny

1966 TJ. Discovered 1966 October 13 by L. I. Chernykh at Nauchnyj.

Named by the discoverer in honor of Prof. A. B. Severny {1913-1987}, who was the Director of the Crimean Astrophysical Observatory. (M 2971)

Obituaries published in Astron. Zh., Tom 64, Vyp. 4, p. 891-892 (1987); Pis'ma Astron. Zh., Tom 13, No. 7, p. 638-640 (1987); Zemlya Vselennaya, No. 4, p. 33-34

(1987); Sov. Astron., Vol. 31, No. 4, p. 469-470 (1987); Sov. Astron. Lett., Vol. 13, No. 4, p. 268-269 (1987); Vestn. Akad. Nauk SSSR, No. 7, p. 131-132 (1987); Q.J.R. Astron. Soc., Vol. 29, No. 4, p. 581-584 (1988); Sol. Phys., Vol. 115, No. 1, p. 1-3 (1988); Izv. Krym. Astrofiz. Obs., Tom 78, p. 160-227 (1988).

(1738) Oosterhoff

1930 SP. Discovered 1930 September 16 by H. van Gent at Johannesburg.

Named in memory of P. Th. Oosterhoff (1904-1978), professor of astronomy at Leiden University from 1948 until his retirement in 1972, Assistant General Secretary of the IAU during 1951-1952 and General Secretary during 1952-1958. He was well known for his contributions to photographic photometry and for his investigations of Cepheids with double periods. (M 4785)

(1739) Meyermann

1939 PF. Discovered 1939 August 15 by K. Reinmuth at Heidelberg.

Named in memory of Bruno Meyermann (1876-1963), who spent most of his life as a classical astronomer and academic teacher at Göttingen Observatory. His fields of interest included polar motion and relativistic effects. (M 4155)

(1740) Paavo Nurmi

1939 UA. Discovered 1939 October 18 by Y. Väisälä at Turku.

Named in memory of Turku-born Paavo Nurmi, the famous Finnish long-distance runner who won seven gold and three silver medals in the Olympic Games and broke 15 world records. (M 5281)

(1741) Giclas

1960 BC. Discovered 1960 January 29 at the Goethe Link Observatory at Brooklyn, Indiana.

Named for Henry L. Giclas {1910- }, longtime staff member of the Lowell Observatory, who has had the major responsibility for the programs of minor planet positions and stellar proper motions carried out with the 13-inch Lawrence Lowell Telescope. (M 3934)

Name proposed by Frank K. Edmondson and Paul Herget.

(1742) Schaifers

1934 RO. Discovered 1934 September 7 by K. Reinmuth at Heidelberg.

Named in honor of Karl Schaifers {1921- }, astronomer on the Heidelberg-Königstuhl staff, editor of *Sterne und Weltraum* and well-known popularizer of astronomy in Germany. (M 4358)

(1743) Schmidt

4109 P-L. Discovered 1960 September 24 by C. J. van Houten and I. van Houten-Groeneveld at Palomar.

Named in honor of Bernhard Schmidt (1879-1935), the inventor of the Schmidt telescope. (M 3086)

This name has been proposed by Dr. Paul Herget.

A description of his life and work appeared in Sky Telesc., November 1955. Schmidt is also honored by a lunar crater.

(1744) Harriet

6557 P-L. Discovered 1960 September 24 by C. J. van Houten and I. van Houten-Groeneveld at Palomar.

Named in honor of the wife of Paul Herget {see planet (1751)}, Director of the Cincinnati Observatory. (M 3023)

(1745) Ferguson

1941 SY_1. Discovered 1941 September 17 by J. E. Willis at Washington.

Named in honor of James Ferguson (1797-1867), a civil engineer, member of the Northwest Boundary Survey, assistant in the U.S. Coast Survey, and from 1848-1867 an astronomer at the U.S. Naval Observatory on the 9.6 inch refractor. He made the first discovery of a minor planet in America in Sept. 1854, namely (31) Euphrosyne, and later discovered (50) Virginia and (60) Echo. He was a skilled computer, widely read, and an admirable conversationalist. (M 4155)

(1746) Brouwer

1963 RF. Discovered 1963 September 14 at the Goethe Link Observatory at Brooklyn, Indiana.

Named in honor of the late Prof. Dirk Brouwer (1902-1966), Director of the Yale University Observatory since 1941, dean of celestial mechanicians in the U.S.A., president of Commission 20, I.A.U. (1948-1955), and an enthusiastic, inspiring leader in astrometry and dynamical astronomy. (M 2883)

The name was suggested by B. G. Marsden (who independently established the identity 1963 RF = 1947 QA) to memorialize Prof. Brouwer's studies of the minor planet families. Following (1578) Kirkwood, this is the twenty-first permanently numbered member of the Hilda group.

Brouwer is also honored by a lunar crater.

(1747) Wright

1947 NH. Discovered 1947 July 14 by C. A. Wirtanen at Mount Hamilton.

Named in memory of William H. Wright (1871-1959), a pioneer in astrophysics who was on the Lick Observatory staff from 1897 to 1944 and the Director from 1935 to 1942. His last project was the design and construction of the 20-inch Carnegie double astrograph with which the recent discoveries of asteroids and comets are by-products of the proper motion program which he had conceived. (M 3934)

Wright is also honored by craters on Mars and the Moon.

(1748) Mauderli

1966 RA. Discovered 1966 September 7 by P. Wild at Zimmerwald.

Named by the discoverer in honor of Prof. Sigmund Mauderli (1876-1962) who was Director of the Astronomical Institute of the University of Berne from 1921-1946. He devoted much of his time to orbit determination and perturbation computing of minor planets for the Astronomisches Rechen-Institut. (M 2971)

(1749) Telamon

1949 SB. Discovered 1949 September 23 by K. Reinmuth at Heidelberg.

The name for this Trojan planet was selected by the discoverer. (M 3023)

Telamon was a son of Aecus and Endeis and the father of Ajax and Teucer {see planets (1404) and (2797)}. Ajax committed suicide, Teucer buried him in the sands of Troy, and when he returned home with this news and without the bones of Ajax, Telamon banished him as he had been banished by his own father. Telamon participated in the Calydonian boar hunt and also sailed the Argo to Colchis {see planet (1135)} in search of the Golden Fleece. (Z 256).

(1750) Eckert

1950 NA_1. Discovered 1950 July 15 by K. Reinmuth at Heidelberg.

Named in memory of Wallace J. Eckert (1902-1971), Director of the U.S. Nautical Almanac Office from 1940 to 1945 and a pioneer in the use of automatic computing machines. On two occasions he dedicated the most powerful computing machine ever built (SSEC in 1948 and the NORC {see planet (1625)} in 1954).

With Brouwer and Clemence {see planets (1746) and (1919)} he produced the integration of the orbits of the five outer planets. By use of sophisticated computing techniques he was able to check and extend Brown's {see planet (1643)} lunar theory. He was president of the IAU Commission 7 from 1967 to 1970. (M 3934)

Obituaries published in Phys. Today, Vol. 24, No. 11, p. 73 (1971); Science, Vol. 173, p. 1115 (1971); Sky Telesc., Vol. 42, p. 207 (1971); Celest. Mech., Vol. 6, p. 3 (1972).

(1751) Herget

1955 OC. Discovered 1955 July 27 at the Goethe Link Observatory at Brooklyn, Indiana.

Named in honor of Paul Herget {1908-1981}, Director of the Cincinnati Observatory and distinguished service professor in the University of Cincinnati. He was founder of the Minor Planet Center in 1947, and a pioneer in the application of high speed computers to astronomical problems and one of the most skilled practitioners of this art at the present time. He was a member of the U.S. National Academy of Sciences, and past president of Commission 20 of the International Astronomical Union. (M 3143)

Proposed by F. K. Edmondson and approved by acclamation at the meeting of Comm. 20, University of Sussex, England, 1970 August 26.

Obituaries published in MPC 6221 (1981); Sky Telesc., Vol. 62, p. 531 (1981); Phys. Today, Vol. 35, No. 1, p. 86-87 (1982).

(1752) van Herk

1930 OK. Discovered 1930 July 22 by H. van Gent at Johannesburg.

Named in honor of G. van Herk {1907- }, former staff member of the Leiden Observatory and a well-known authority on astrometry. (M 5357)

Name proposed by the Leiden Observatory.

(1753) Mieke

1934 JM. Discovered 1934 May 10 by H. van Gent at Johannesburg.

Named in honor of the wife {Mieke Oort-Graadt van Roggen, 1906-1993} of former director of the Leiden Observatory, Jan Hendrik Oort {see planet (1691)}. (M 5357)

Name proposed by the Leiden Observatory.

(1754) Cunningham

1935 FE. Discovered 1935 March 29 by E. Delporte at Uccle.

Named in honor of Leland E. Cunningham {1904-1989}, a prolific computer of comet orbits with F. L. Whipple {see planet (1940)} at Harvard during the 1930s, observer of faint comets, including the discovery of 1941 I and the recovery of comet Gale in 1938, and since 1946 on the staff of the Leuschner Observatory, University of California. (M 3934)

(1755) Lorbach

1936 VD. Discovered 1936 November 8 by M. Laugier at Nice.

Named in honor of Anne Lorbach Herget, assistant at the Cincinnati Observatory since 1962 and wife of Paul Herget {see planet (1751)}. She has been responsible for the assignment of provisional designations of minor planets and for key-punching almost all the material in the MPC's, work that was accomplished to a high standard of accuracy. (M 4419)

Name suggested by C. M. Bardwell and B. G. Marsden.

(1756) Giacobini

1937 YA. Discovered 1937 December 24 by A. Patry at Nice.

Named in memory of Michel Giacobini (1873-1938), astronomer at the Nice Observatory who discovered 13 comets. (M 4358)

(1757) Porvoo

1939 FC. Discovered 1939 March 17 by Y. Väisälä at Turku.

Named for the second oldest (after Turku) town in Finland. At the Porvoo Diet in 1809 the Russian czar confirmed that Finland was annexed to the Russian empire as an autonomous nation. (M 5449)

(1758) Naantali

1942 DK. Discovered 1942 February 18 by L. Oterma at Turku.

Named for an idyllic small town near Turku, founded in the early fifteenth century around a convent and monastery. The summer residence of the Finnish president is located there. (M 5449)

(1759) Kienle

1942 RF. Discovered 1942 September 11 by K. Reinmuth at Heidelberg.

Named in memory of Hans Kienle (1895-1975), an astrophysicist, who served as Director of several German observatories. He was director of the Heidelberg-Königstuhl Observatory from 1950-1962. His work on spectrophotometry is well-known. He was president of IAU Commission 36 from 1955 to 1958. (M 4155)

Obituaries published in Phys. Bl., 31. Jahrg., p. 222-223 (1975); Sky Telesc., Vol. 49, p. 368 (1975); Sterne Weltraum, Vol. 14, p. 83, p. 184-186 (1975); Orion, 33. Jahrg., p. 111 (1975); Ruperto Carola, Univ. Heidelb., 27. Jahrg., p. 114-115 (1975); Astron. Nachr., Vol. 297, p. 99-105 (1976); Mitt. Astron. Ges., Nr. 38, p. 9-11 (1976).

(1760) Sandra

1950 GB. Discovered 1950 April 10 by E. L. Johnson at Johannesburg.

Named by the discoverer in honor of his granddaughter. (M 3934; M 3988; M 5391)

(1761) Edmondson

1952 FN. Discovered 1952 March 30 at the Goethe Link Observatory at Brooklyn, Indiana.

Named in honor of Prof. Frank K. Edmondson {1912- }, chairman of the Astronomy Department of Indiana University since 1944, treasurer of the American Astronomical Society (AAS) since 1954, program Director for astronomy of the National Science Foundation (NSF), 1956-57, and chairman of Section D (Astronomy) of the American Assoc. for the Advancement of Science (AAAS), 1962. He was chairman, U.S. National Committee of the International Astronomical Union (IAU), 1962-64. He was vice president, 1957-61, president, 1962-65 of the Association of Universities for Research in Astronomy, Inc. (AURA), and president of Comm. 20 of the International Astronomical Union (IAU), 1970. In 1964 he was decorated with the "Order of Merit" by the government of Chile. (M 3143)

The proposal of this name was presented by the entire departmental staff.

(1762) Russell

1953 TZ. Discovered 1953 October 8 at the Goethe Link Observatory at Brooklyn, Indiana.

Named in honor of Henry Norris Russell {1877-1957}, one of the world's most distinguished astronomers, noted for the H-R diagram and his brilliant research on a variety of topics in fundamental astronomy, astrophysics, and the analysis of atomic spectra (RS- or Russell-Saunders coupling). His advice and counsel were

sought by many astronomers who valued his scientific judgement and respected his wide range of interests and encyclopedic knowledge of many subjects. His Ph.D. thesis at Princeton was on the general perturbations of the major axis of the orbit of {433} Eros by the action of Mars. (M 3143)
Proposed by F. K. Edmondson.
Russell is also honored by craters on Mars and the Moon.

(1763) Williams

1953 TN_2. Discovered 1953 October 13 at the Goethe Link Observatory at Brooklyn, Indiana.

Named in honor of K. P. Williams, professor of mathematics at Indiana University (1909-58). He was known for his textbook *The calculation of the orbits of asteroids and comets* and his detailed analysis of the transits of Mercury from 1723 to 1927. He also wrote *Lincoln finds a general*, a definitive military history of the U.S. Civil War. (M 3143)
Proposed by F. K. Edmondson.

(1764) Cogshall

1953 VM_1. Discovered 1953 November 7 at the Goethe Link Observatory at Brooklyn, Indiana.

Named in honor of W. A. Cogshall, professor of astronomy at Indiana University (1900-44). He was known for his work on visual binary stars, photography of solar eclipses, and as a teacher of many who followed professional careers in astronomy. (M 3143)
Proposed by F. K. Edmondson.

(1765) Wrubel

1957 XB. Discovered 1957 December 15 at the Goethe Link Observatory at Brooklyn, Indiana.

Named in honor of Marshal H. Wrubel, faculty member at Indiana University (1950-68) and university professor of astronomy (1966-68). He was co-founder of the Indiana University Research Computing Center and a pioneer in the use of high speed computers for astrophysical computations, and a brilliant teacher whose profound influence continues to show in the work of his 14 Ph.D. students. (M 3143)
Proposed by F. K. Edmondson.

(1766) Slipher

1962 RF. Discovered 1962 September 7 at the Goethe Link Observatory at Brooklyn, Indiana.

Named in honor of Vesto Melvin Slipher {1876-1969} and E. C. Slipher {1883-1964}, graduates of Indiana University in 1901 and 1906. V. M. Slipher was a pioneer investigator of the spectra of the planets, and was the first to measure the redshifts of galaxies. E. C. Slipher developed the art of direct photography of the planets to the highest level, and his photographs are the only continuous and systematic record of the appearance of the planets for a period of more than half a century. (M 3144)
Proposed by F. K. Edmondson.

E. C. Slipher is also honored by a crater on Mars. Both brothers are honored by a lunar crater. Obituaries (V.M.S.) published in Publ. Astron. Soc. Pac., Vol. 81, p. 922-923 (1969); Science, Vol. 166, p. 1608 (1969); Phys. Today, Vol. 23, No. 2, p. 101 (1970); Year Book Am. Phil. Soc. 1970, p. 161-166.

(1767) Lampland

1962 RJ. Discovered 1962 September 7 at the Goethe Link Observatory at Brooklyn, Indiana.

Named in honor of C. O. Lampland {1873-1951}, a graduate of Indiana University in 1902. His best known work was the radiometric measurement of planetary temperatures in collaboration with W. W. Coblentz, and the long photographic history of the changes in NGC 2261 (Hubble's variable nebula). (M 3144)
Proposed by F. K. Edmondson.
Lampland is also honored by craters on Mars and the Moon.

(1768) Appenzella

1965 SA. Discovered 1965 September 23 by P. Wild at Zimmerwald.

Appenzella, named by the discoverer in honor of the Swiss Canton of Appenzell, on the occasion of the 150th anniversary of the Kantonsschule Trogen (Appenzell-Ausserrhoden). (M 3297)

(1769) Carlostorres

1966 QP. Discovered 1966 August 25 by Z. M. Pereyra at Córdoba.

Named in memory of Carlos Guillermo Torres (1910-1965), astronomer at the Córdoba Observatory, an active observer with the meridian circle and of comets and minor planets. The planet also honors Carlos Torres R., who is in charge of the program for observing comets and minor planets at Santiago. (M 6954)

(1770) Schlesinger

1967 JR. Discovered 1967 May 10 by C. U. Cesco and A. R. Klemola at El Leoncito.

Named in memory of Frank Schlesinger (1871-1943), Director of the Allegheny Observatory from 1905 to 1920 and then Director of the Yale University Observatory until his retirement in 1941. His pioneering work in photographic astrometry revolutionized the determination of stellar parallaxes and the preparation of zone catalogues. His invention of the method of dependences for parallax reductions has also found extensive application in obtaining minor planet positions. He was prominent in the establishment of the IAU and served as its president from 1932 to 1935. (M 3934)
Schlesinger is also honored by a lunar crater.

(1771) Makover

1968 BD. Discovered 1968 January 24 by L. I. Chernykh at Nauchnyj.

Named in honor of the late Samuel Gdalevich Makover {1908-1970}, scientist in the I.T.A. since 1946. He studied extensively the orbit of comet Encke-Backlund, and he was a pioneer in the use of electronic calculators for computing planetary perturbations and orbit improvements. In 1963 he became head of the Department of Minor Planets and Comets, and editor of the annual volume of Minor Planet Ephemerides. He was vice president of Comm. 20, I.A.U. (1964-67). (M 3185)

Obituaries published in Astron. Tsirk., No. 579, p. 7-8 (1970); Byull. Inst. Teor. Astron. Leningrad, Vol. 12, p. 437-438 (1970).

(1772) Gagarin

1968 CB. Discovered 1968 February 6 by L. I. Chernykh at Nauchnyj.

Named in honor of the late Yuri Gagarin {1934-1968}, the illustrious Soviet cosmonaut, and the first man to circumnavigate the Earth in outer space. (M 3185)
Gagarin is also honored by a lunar crater.

(1773) Rumpelstilz

1968 HE. Discovered 1968 April 17 by P. Wild at Zimmerwald.
The leading character in one of Grimm's fairy tales. (M 4155)

(1774) Kulikov

1968 UG_1. Discovered 1968 October 22 by T. M. Smirnova at Nauchnyj.

Named in honor of the late Dmitri Kuzmich Kulikov {1912-1964}, specialist in geodesy and fundamental astronomy. During 1934-1941 he led five geodetic expeditions. He was decorated for his service in World War II with two Orders of the Red Star. During 1949-1956 he served as scientific secretary of I.T.A. and since then as head of the Department of the Astronomical Yearbook. He worked on the orbits of comets and Jupiter VIII. In 1952 he was awarded a prize for his treatment of Zinger star pairs. (M 3185)

(1775) Zimmerwald

1969 JA. Discovered 1969 May 13 by P. Wild at Zimmerwald.

The charming village that harbours the new observing station, seven miles south of Berne {see planet (1313)}. (M 4155)

(1776) Kuiper

2520 P-L. Discovered 1960 September 24 by C. J. van Houten and I. van Houten-Groeneveld at Palomar.

Named in honor of G. P. Kuiper {1905-1973}, former Director of the Lunar and Planetary Laboratory at Tucson, and former Director of the Yerkes Observatory. Dr. Kuiper was a well known authority on the solar system and initiated both the McDonald Survey and the Palomar-Leiden Survey of minor planets. (M 3185)

Kuiper is also honored by a crater on Mars. Obituaries published in Icarus, Vol. 22, p. 117-118 (1974); Nature, Vol. 248, p. 539-540 (1974); Observatory, Vol. 94, p. 94 (1974); Phys. Today, Vol. 27, No. 3, p. 85, 87 (1974); Science, Vol. 183, p. 1323 (1974); Sky Telesc., Vol. 47, p. 83 (1974); Zenit, Vol. 1, No. 1, p. 13 (1974); Irish Astron. J., Vol. 11, p. 158 (1973); Zemlya Vselennaya, No. 5, p. 56-58 (1974); l'Astronomie, 89. année, p. 84-85 (1975).

(1777) Gehrels

4007 P-L. Discovered 1960 September 24 by C. J. van Houten and I. van Houten-Groeneveld at Palomar.

Named in honor of Tom Gehrels {1925- }, staff member of the Lunar and Planetary Laboratory at Tucson. Dr. Gehrels is well known for his photometric and polarimetric observations of minor planets and the Moon. (M 3185)

(1778) Alfvén

4506 P-L. Discovered 1960 September 26 by C. J. van Houten and I. van Houten-Groeneveld at Palomar.

Named in honor of Nobel prize winner Hannes Alfvén {1908-1995} {1970, together with L. E. N. Néel}, who has stimulated physical studies of asteroids to be made with telescopes on the Earth and on future spacecraft. (M 3643)
This name has been proposed by Dr. T. Gehrels.

Obituaries published in Ciel Terre, Vol. 111, No. 3, p. 67-74 (1995); Phys. Today, Vol. 48, No. 9, p. 118-119 (1995); COSPAR Inf. Bull., No. 133, p. 57-58 (1995); Astrophys. Space Sci., Vol. 234, No. 2, p. 173-175 (1995); IEEE Trans. Plasma Sci., Vol. 25, No. 3, p. 409-414 (1997); Nature, Vol. 375, No. 6528, p. 187 (1995); Postepy Astron., Tom 44, No. 2, p. 91-92 (1996); Q.J.R. Astron. Soc., Vol. 37, No. 2, p. 259-260 (1996); Vasiona, Vol. 44, No. 1-2, p. 24-25 (1996).

(1779) Paraná

1950 LZ. Discovered 1950 June 15 by M. Itzigsohn at La Plata.

Named for a large river in northern Argentina, a major tributary to the La Plata {see planet (1029)} river, near which this minor planet was discovered. (M 6832)
Name proposed by F. Pilcher.

(1780) Kippes

A906 RA. Discovered 1906 September 12 by A. Kopff at Heidelberg.

This name is bestowed by the Minor Planet Center in honor of Pastor Otto Kippes {1905- }, an instructor in religion in the Parochial School of Reckendorf, Germany. He is a diligent and meticulous worker, and an amateur astronomer in the finest sense, who established hundreds of identifications of minor planets in widely separated oppositions. (M 3508)
See also the Scylla story in Sky Telesc., Vol. 40, No. 6, p. 361-362 (1970).

(1781) Van Biesbroeck

A906 UB. Discovered 1906 October 17 by A. Kopff at Heidelberg.

Named for Georges Van Biesbroeck {1880-1974} in recognition of, and appreciation for, many years of devoted service to astronomy through observations and discoveries of minor planets, comets, satellites, and double stars. (M 3569)

Obituaries published in Circ. Inf., No. 62 (1974); Phys. Today, Vol. 27, No. 7, p. 59 (1974); Sky Telesc., Vol. 47, p. 215 (1974); Ciel Terre, Vol. 90, p. 321-325 (1974); Icarus, Vol. 23, p. 134-135 (1974); J.R. Astron. Soc. Can., Vol. 68, p. 202-204, p. L23-L24 (1974); l'Astronomie, 88. année, p. 305-308 (1974); Q.J.R. Astron. Soc., Vol. 16, p. 104-105 (1975).

(1782) Schneller

1931 TL_1. Discovered 1931 October 6 by K. Reinmuth at Heidelberg.

Named in memory of Heribert Schneller (1901-1967), a prolific observer of variable stars, who worked at Berlin-Babelsberg Observatory, and later at the Astrophysical Observatory in Potsdam. He published many catalogues and collections of references on variable stars. (M 4156)
Schneller is also honored by a lunar crater.

(1783) Albitskij

1935 FJ. Discovered 1935 March 24 by G. N. Neujmin at Simeïs.

Named in memory of Vladimir Aleksandrovich Albitskij (1891-1952), head of the Simeïs department of the Pulkovo Observatory from 1934 onward. He discovered ten numbered minor planets and is well known for his research on radial velocities and variable stars. (M 5357)

(1784) Benguella

1935 MG. Discovered 1935 June 30 by C. Jackson at Johannesburg.
Named for the chief port of Angola. (M 5183)

(1785) Wurm

1941 CD. Discovered 1941 February 15 by K. Reinmuth at Heidelberg.

Named in memory of Karl Wurm (1899-1975), who is well-known for his work on interstellar matter and cometary physics. He worked at Potsdam, the Yerkes Observatory, and later at the Hamburg-Bergedorf Observatory. He was president of IAU Commission 15 from 1958 to 1964. (M 4156)

Obituaries published in Coelum, Vol. 43, p. 114-115 (1975); Orion, 33. Jahrg., p. 82 (1975); Sterne Weltraum, Vol. 14, p. 113 (1975); Astrophys. Space Sci., Vol. 35, p. 221-222 (1975); Mitt. Astron. Ges., Nr. 38, p. 14-16 (1976).

(1786) Raahe

1948 TL. Discovered 1948 October 9 by H. Alikoski at Turku.

Named for a Finnish town, founded by Per Brahe {see planet (1680)} in 1649 on an ancient market place near Oulu {see planet (1512)}. (M 5449)

(1787) Chiny

1950 SK. Discovered 1950 September 19 by S. Arend at Uccle.

Named for the principal town in the ancient county of the same name, situated on the beautiful river Semoise. (M 8402)

(1788) Kiess

1952 OZ. Discovered 1952 July 25 at the Goethe Link Observatory at Brooklyn, Indiana.

Named for Carl C. Kiess, a graduate of Indiana University (AB 1910, Honorary D.Sc. 1963). Dr. Kiess made distinguished contributions both in astronomy and spectroscopy. His spectroscopic research at the U.S. National Bureau of Standards began in 1917 and spanned a period of 40 years. He was a member of several eclipse expeditions, and conducted two expeditions to Hawaii to observe planetary spectra with high dispersion under good seeing conditions. He was serving on the Astronomy Faculty of Georgetown University at the time of his death in 1967. (M 3508)

(1789) Dobrovolsky

1966 QC. Discovered 1966 August 19 by L. I. Chernykh at Nauchnyj.

Named in honor of Lt. Col. Georgi Timofeyevich Dobrovolsky, the commander. (M 3296)

This name and the following two names are a memorial to the cosmonauts of the spacecraft Soyuz 11, who perished on 30 June 1971 during the return of the vehicle to the Earth after completing the flight program of the first manned orbital station, Salyut.

(1790) Volkov

1967 ER. Discovered 1967 March 9 by L. I. Chernykh at Nauchnyj.
Named in honor of Vladislav Nikolayevich Volkov, the flight engineer. (M 3296)
See the remark to (1789).

(1791) Patsayev

1967 RE. Discovered 1967 September 4 by T. M. Smirnova at Nauchnyj.
Named in honor of Victor Ivanovich Patsayev, the test engineer. (M 3296)
See the remark to (1789).

(1792) Reni

1968 BG. Discovered 1968 January 24 by L. I. Chernykh at Nauchnyj.

Named in honor of the town Reni (Ukrainian Soviet Socialist Republic), the birthplace of A. N. Deutsch, the well-known investigator of minor planets and authority in the field of astrophotometry and stellar astronomy. (M 3297)

(1793) Zoya

1968 DW. Discovered 1968 February 28 by T. M. Smirnova at Nauchnyj.

Named in honor of Zoya Kosmodemyanskaya {see planet (2072)}, the young heroine of the Soviet people, who perished during the Great Patriotic War. (M 3297)
See also the citations for minor planets (1977) and (2072).

(1794) Finsen

1970 GA. Discovered 1970 April 7 by J. A. Bruwer at Hartbeespoort.

Named in honor of W. S. Finsen {1905-1979}, Director of the Republic Observatory from 1957 to 1965. He has contributed immensely to the discovery and observation of double stars both micrometrically and by means of the eyepiece interferometer he invented. (M 3297)

Obituaries published in Circ. Inf., No. 79 (1979); Sky Telesc., Vol. 58, p. 137 (1979).

(1795) Woltjer

4010 P-L. Discovered 1960 September 24 by C. J. van Houten and I. van Houten-Groeneveld at Palomar.

Named in honor of Dr. Jan J. Woltjer Jr. {1891-1946}, who was assistant professor at the Leiden Observatory from 1941-1946. He was an authority in the field of celestial mechanics and Cepheid pulsations. (M 3569)
This name has been proposed by Dr. G. P. Kuiper.
Woltjer is also honored by a lunar crater.

(1796) Riga

1966 KB. Discovered 1966 May 16 by N. S. Chernykh at Nauchnyj.

Named in honor of the capital of Latvia, site of the Astronomical Observatory of the Latvian State University. (M 3185)
Named at the request of Prof. M. Dirikis.

(1797) Schaumasse

1936 VH. Discovered 1936 November 15 by A. Patry at Nice.

Named in memory of Alexandre Schaumasse (1882-1958), astronomer at the Nice Observatory who discovered (971) Alsatia and three comets. (M 4358)
Schaumasse also discovered minor planet (1114) Lorraine.

(1798) Watts

1949 GC. Discovered 1949 April 4 at the Goethe Link Observatory at Brooklyn, Indiana.

Named for Chester B. Watts {1889-1971}, a graduate of Indiana University (AB 1915, Honorary D.Sc. 1953). Dr. Watts made distinguished contributions in the field of positional astronomy during his 44 years at the U.S. Naval Observatory. He pioneered in the field of automation of transit circle observations, leading to results of the highest systematic accuracy. Culminating his career was the monumental task of mapping every feature on the marginal zone of the Moon, which he commenced in the late 1940's and completed in 1963, four years after his normal retirement. (M 3508)

Obituaries published in Sky Telesc., Vol. 42, p. 131 (1971); Q.J.R. Astron. Soc., Vol. 13, p. 110-112 (1972); Moon, Vol. 6, p. 233-234 (1973).

(1799) Koussevitzky

1950 OE. Discovered 1950 July 25 at the Goethe Link Observatory at Brooklyn, Indiana.

Named in honor of Serge Koussevitzky, distinguished conductor of the Boston Symphony Orchestra during its golden years. His 25 year tenure was noteworthy for his efforts to encourage young American composers as well as for his masterly interpretations of the classic repertoire. (M 3569)

Proposed by Frank K. Edmondson to celebrate the centenary of the birth of Dr. Koussevitzky on July 26, 1974.

(1800) Aguilar

1950 RJ. Discovered 1950 September 12 by M. Itzigsohn at La Plata.

Named in memory of Eng. Felix Aguilar { -1943}, former director of the La Plata Observatory, founder of the University School of Astronomy and Geophysics, the most outstanding scientific worker for the development of Argentine astronomy in the first half of this century. (M 3934)

(1801) Titicaca

1952 SP_1. Discovered 1952 September 23 by M. Itzigsohn at La Plata.

Named for the large lake, at 3800 meters the highest in the world, that forms

part of the border between Peru and Bolivia. (M 6832)
Name proposed by F. Pilcher.

(1802) Zhang Heng

1964 TW_1. Discovered 1964 October 9 at the Purple Mountain Observatory at Nanking.

Named for Zhang Heng (78-139), a prominent scientist of the eastern Han dynasty who devised and constructed armillary spheres and seismographs. He also invented a celestial sphere driven by water power that is in many respects the forerunner of the modern planetarium. (M 4419)
Zhang Heng (or Chang Heng) is also honored by a lunar crater.

(1803) Zwicky

1967 CA. Discovered 1967 February 6 by P. Wild at Zimmerwald.

Named in memory of Fritz Zwicky (1898-1974), a Swiss citizen, who as a professor at Caltech, Pasadena, did pioneer work in many fields, notably in the study of supernovae and of clusters of galaxies, in high-energy astrophysics, and also in developing jet propulsion for airplanes and spacecraft. (M 4156)

Obituaries published in l'Astronomie, 88. année, p. 221-224 (1974); Observatory, Vol. 94, p. 94 (1974); Orion, 32. Jahrg., p. 63, p. 113-114 (1974); Phys. Today, Vol. 27, No. 6, p. 70-71 (1974); Sky Telesc., Vol. 47, p. 215 (1974); Sterne Weltraum, Vol. 13, p. 77-78 (1974); Acta Astronaut., Vol. 1, No. 7-8, p. IX (1974); Science, Vol. 186, p. 284 (1974); Q.J.R. Astron. Soc., Vol. 16, p. 106-108 (1975).

(1804) Chebotarev

1967 GG. Discovered 1967 April 6 by T. M. Smirnova at Nauchnyj.

Named in honor of Prof. G. A. Chebotarev {1913-1975}, Director of the Institute of Theoretical Astronomy since 1964 and past president of Commission 20 of the International Astronomical Union. Prof. Chebotarev is well known for his intensive work on different problems of celestial mechanics concerned with the asteroids, comets and satellites. (M 3569)

Obituaries published in Zvaigžnotā Debess, p. 52-53 (1976); Byull. Inst. Teor. Astron., Vol. 14, p. 133-135 (1975); Celest. Mech., Vol. 12, p. 395-396 (1975); Zemlya Vselennaya, No. 6, p. 40-41 (1975).

(1805) Dirikis

1970 GD. Discovered 1970 April 1 by L. I. Chernykh at Nauchnyj.

Named in honor of Dr. M{atiss} A. Dirikis {1923-1993}, staff member of the Astronomical Observatory of the Latvian State University, chairman of the Latvian branch of the Astronomical-Geodetical Society of the U.S.S.R. His work on the motion of small bodies of the solar system contributes much to this field of theoretical astronomy. (M 3569)
Obituary published in Zvaigžnotā Debess, No. 143, p. 60-61 (1994).

(1806) Derice

1971 LC. Discovered 1971 June 13 at the Perth Observatory at Bickley.

Named in honor of the wife of Dennis Harwood, member of the astrometric team at the Perth Observatory. (M 6530)
This is the first numbered minor planet discovered in Oceania.

(1807) Slovakia

1971 QA. Discovered 1971 August 20 by M. Antal at Skalnaté Pleso.

Named in honor of one of the two socialistic republics of Czechoslovakia {see planet (2315)}, the country in which the discoverer's observatory is situated. (M 3508)
The Slovak Republic now is an independent state.

(1808) Bellerophon

2517 P-L. Discovered 1960 September 24 by C. J. van Houten and I. van Houten-Groeneveld at Palomar.

Named after the hero of a Greek saga, who vanquished the Chimaera {see planet (623)}, a monster which was partly lion, partly goat, partly serpent. (M 3934)

(1809) Prometheus

2522 P-L. Discovered 1960 September 24 by C. J. van Houten and I. van Houten-Groeneveld at Palomar.

Named after the hero of a Greek saga, who stole the fire from the gods. (M 3934)

The name Prometheus has also been given to the satellite Saturn XVI, discovered 1980 by the Voyager 1 spacecraft.

(1810) Epimetheus

4196 P-L. Discovered 1960 September 24 by C. J. van Houten and I. van Houten-Groeneveld at Palomar.

Epimetheus, the brother of Prometheus {see planet (1809)}, opened the Pandora {see planet (55)} box, which contained all the illnesses and ailments of mankind. (M 3935)

The name Epimetheus has also been given to the satellite Saturn XI, discovered 1980 by the Voyager 1 spacecraft.

(1811) Bruwer

4576 P-L. Discovered 1960 September 24 by C. J. van Houten and I. van Houten-Groeneveld at Palomar.

Named in honor of Jacobus Albertus Bruwer {1915- }, former staff member of the South African Astronomical Observatory in Johannesburg. He is well-known for his observational work on positions of minor planets with the Franklin-Adams camera at Hartbeespoort. (M 4547)

(1812) Gilgamesh

4645 P-L. Discovered 1960 September 24 by C. J. van Houten and I. van Houten-Groeneveld at Palomar.

Named after the hero of an old-Babylonian saga. (M 4547)

(1813) Imhotep

7589 P-L. Discovered 1960 October 17 by C. J. van Houten and I. van Houten-Groeneveld at Palomar.

Named for a famous physician and architect in the third dynasty of the ancient Egyptian Kingdom, builder of the step-pyramid near Sakkara. (M 4547)

(1814) Bach

1931 TW_1. Discovered 1931 October 9 by K. Reinmuth at Heidelberg.

The German composer, Johann Sebastian Bach (1685-1750). (M 3935)

(1815) Beethoven

1932 CE_1. Discovered 1932 January 27 by K. Reinmuth at Heidelberg.

The German composer, Ludwig van Beethoven (1770-1827). (M 3935)

(1816) Liberia

1936 BD. Discovered 1936 January 29 by C. Jackson at Johannesburg.

Named for the state on the western coast of Africa. (M 5183)

(1817) Katanga

1939 MB. Discovered 1939 June 20 by C. Jackson at Johannesburg.

Named for the chief mining area of the Congo, today named Zaïre. (M 5183)

(1818) Brahms

1939 PE. Discovered 1939 August 15 by K. Reinmuth at Heidelberg.
The German composer, Johannes Brahms (1833-1897). (M 3935)

(1819) Laputa

1948 PC. Discovered 1948 August 9 by E. L. Johnson at Johannesburg.

Named for the floating island in *Gulliver's Travels*, a novel by the English satirist, poet, political writer and clergyman Jonathan Swift (1667-1745). (M 3935)

(1820) Lohmann

1949 PO. Discovered 1949 August 2 by K. Reinmuth at Heidelberg.

Named in honor of Werner Lohmann {1911-1983}, astronomer at Heidelberg, who worked at the Königstuhl Observatory, and later, since 1949, at the Rechen-Institut. He was editor or co-editor of the *Astronomischer Jahresbericht*, indexing the astronomical literature of the period 1943-1968. (M 4156)

(1821) Aconcagua

1950 MB. Discovered 1950 June 24 by M. Itzigsohn at La Plata.

Named for one of the highest mountains in South America. It has an elevation of 6960 meters and is situated on the Argentina-Chile border. The western slope of the mountain is the headwaters of a river of the same name whose outlet to the Pacific Ocean is at Valparaiso. (M 7156)
Name proposed by F. Pilcher.

(1822) Waterman

1950 OO. Discovered 1950 July 25 at the Goethe Link Observatory at Brooklyn, Indiana.

Named to honor the memory of Alan T. Waterman {1892-1967}, first director of the U.S. National Science Foundation, in the year which marks the 25th anniversary of the foundation. After 25 years as an academic physicist, he went to Washington to serve with OSRD (1941-45), ONR (1946-51), and NSF (1951-63). Among the many honors he received was the Karl Taylor Compton Gold Medal for distinguished statesmanship in science, awarded by the American Institute of Physics two months before his death. (M 3825)
The name is proposed by Frank K. Edmondson.
Waterman is also honored by a lunar crater.

(1823) Gliese

1951 RD. Discovered 1951 September 4 by K. Reinmuth at Heidelberg.

Named in honor of Wilhelm Gliese {1915-1993}, an astronomer at the Rechen-Institut since 1943, who is well-known for his work on the nearby stars and the FK4. (M 4156)
Obituary published in Mitt. Astron. Ges., Nr. 77, p. 5-7 (1994).

(1824) Haworth

1952 FM. Discovered 1952 March 30 at the Goethe Link Observatory at Brooklyn, Indiana.

Named in honor of Leland J. Haworth, a graduate of Indiana University and second director of the National Science Foundation. His long and varied career included high school teaching, academic appointments at Wisconsin, MIT, and Illinois, director of the Brookhaven National Laboratory, vice-president and president of Associated Universities, Inc. (AUI), member of the Atomic Energy Commission, and director of the National Science Foundation from 1963 to 1969. He conducted the negotiations which led to the joint funding by NSF and the Ford Foundation of a 4 meter telescope at the Cerro Tololo Interamerican Observatory. (M 4156)
Proposed by Frank K. Edmondson.

(1825) Klare

1954 QH. Discovered 1954 August 31 by K. Reinmuth at Heidelberg.

Named in honor of Gerhard Klare {1932- }, an observing astronomer at Heidelberg-Königstuhl Observatory since 1960, whose fields of interest include minor planets. (M 4156)

(1826) Miller

1955 RC_1. Discovered 1955 September 14 at the Goethe Link Observatory at Brooklyn, Indiana.

Named in honor of John A. Miller, founder of the Astronomy Department at Indiana University and first director of the Kirkwood {see planet (1578)} Observatory which he built and named for his former teacher. Later he built the Sproul Observatory at Swarthmore College to continue and expand the double star program he had started in Indiana. (M 4236)
Proposed by Frank K. Edmondson.

(1827) Atkinson

1962 RK. Discovered 1962 September 7 at the Goethe Link Observatory at Brooklyn, Indiana.

Named in honor of Robert d'Escourt Atkinson {1898-1982}, noted for his contributions to fundamental astronomy. He also pioneered in studying nuclear energy-generation in the Sun and stars, in collaboration with Houtermans in the late 1920's. During his time as chief assistant at the Royal Observatory, Greenwich, he handled the many details of the move to Herstmonceux, Sussex. He was awarded the Eddington Medal by the Royal Astronomical Society in 1960. Following his retirement from the Royal Greenwich Observatory he served on the Faculty of Indiana University for several years. (M 4236)
Proposed by Frank K. Edmondson

Obituaries published in J. Br. Astron. Assoc., Vol. 93, No. 4, p. 172-173 (1983); Phys. Today, Vol. 36, No. 5, p. 81-82 (1983); Sky Telesc., Vol. 65, No. 5, p. 417 (1983); Astronomie, Vol. 97, p. 442 (1983); Q.J.R. Astron. Soc., Vol. 25, No. 1, p. 100-104 (1984).

(1828) Kashirina

1966 PH. Discovered 1966 August 14 by L. I. Chernykh at Nauchnyj.

Named in honor of Valentin Semenovich Kashirin, physician from Simferopol (Crimea). (M 3825)
Name proposed by L. Chernykh.

(1829) Dawson

1967 JJ. Discovered 1967 May 6 by C. U. Cesco and A. R. Klemola at El Leoncito.

Named in memory of Bernhard H. Dawson (1890-1960), a U.S.-born astronomer who worked at the La Plata Observatory in Argentina from 1913 onward. From 1948 to 1955 he was a professor at the Faculdad de Ingeniera de San Juan. He was well known for his visual discoveries and observations of double stars and was also an active observer of minor planets and comets and computer of their orbits. (M 3935)
Dawson is also honored by a lunar crater.

(1830) Pogson

1968 HA. Discovered 1968 April 17 by P. Wild at Zimmerwald.

Named in memory of Norman Robert Pogson (1829-1891), astronomer at Oxford and later at Madras, discoverer of eight minor planets, including (42) Isis and (67) Asia, and inventor of the astronomical magnitude scale. (M 4236)
Pogson is also honored by a lunar crater.

(1831) Nicholson

1968 HC. Discovered 1968 April 17 by P. Wild at Zimmerwald.

Named in memory of Seth Barnes Nicholson (1891-1963), discoverer of four of Jupiter's satellites and pioneer in several branches of observational research on members of the solar system. (M 4236)

Nicholson is also honored by craters on the Moon and Mars and by a region on the Jupiter satellite Ganymed.

(1832) Mrkos

1969 PC. Discovered 1969 August 11 by L. I. Chernykh at Nauchnyj.

Named in honor of Antonin Mrkos {1918-1996}, Director of the Kleť Observatory in Czechoslovakia, well known for his contributions to cometary astronomy. {He was the founder and for many years head of Kleť Observatory who initiated the Kleť minor planet survey, the first one in Czechoslovakia.} (M 3825; N. S. Chernykh)

Name proposed by L. Chernykh.

Obituaries published in Heavens, Vol. 77, p. 272-273 (1996); Int. Comet Q., Vol. 18, No. 4, p. 182-183 (1996).

(1833) Shmakova

1969 PN. Discovered 1969 August 11 by L. I. Chernykh at Nauchnyj.

Named in honor of Mrs. M. Ya. Shmakova (1910-1971), staff member of the I.T.A. {Institute for Theoretical Astronomy, Leningrad} who contributed much to planetary and cometary orbit computations. (M 3825)

(1834) Palach

1969 QP. Discovered 1969 August 22 by L. Kohoutek at Bergedorf.

Named in memory of the Czech student of philosophy Jan Palach, who burned himself to death on 1969 Jan. 16 in Wenceslas Square, Prague, as a protest against the occupation of Czechoslovakia the previous August, resulting in the violent end of the reform movement known as Prague Spring. (M 18643)

(1835) Gajdariya

1970 OE. Discovered 1970 July 30 by T. M. Smirnova at Nauchnyj.

Named in honor of Arkadij Gajdar (1904-1941), Russian writer. (M 3825)

(1836) Komarov

1971 OT. Discovered 1971 July 26 by N. S. Chernykh at Nauchnyj.

Named in honor of the Soviet cosmonaut Vladimir Mikhajlovich Komarov (1927-1967). In 1964 he headed the manned flight on the Voskhod spacecraft. He perished on 24 April 1967 when completing the Soyuz-1 spacecraft test-flight. (M 3825)

Komarov is also honored by a lunar crater.

(1837) Osita

1971 QZ_1. Discovered 1971 August 16 by J. Gibson at El Leoncito.

Named by the discoverer in honor of his wife Ursula, of which Osita is the Spanish equivalent. While the discoverer was located at the Yale-Columbia Southern Station she measured and/or reduced more than 150 positions of comets and minor planets. Her volunteer services as assistant, visitor's guide and secretary were performed in addition to the daily work of making a pleasant home at an isolated location. (M 3935)

(1838) Ursa

1971 UC. Discovered 1971 October 20 by P. Wild at Zimmerwald.

Named for the discoverer's wife, Ursula, their son, Urs, and for the bears of Berne. (M 4156)

(1839) Ragazza

1971 UF. Discovered 1971 October 20 by P. Wild at Zimmerwald.

The Italian word for girl. Also the name alludes to Bad Ragaz, a resort town with hot springs, in Switzerland. (M 4156)

(1840) Hus

1971 UY. Discovered 1971 October 26 by L. Kohoutek at Bergedorf.

Named in honor of Jan Hus (1372-1415), Czech religious and linguistic reformer, rector of Charles University in Prague. He was condemned to death by the Council of Constance and burned at the stake for his reformation ideas. (M 3757)

(1841) Masaryk

1971 UO_1. Discovered 1971 October 26 by L. Kohoutek at Bergedorf.

Named in honor of Thomas G. Masaryk (1850-1937), Czechoslovak statesman and philosopher, known for his humanistic ideas. The first president of the independent Czechoslovak Republic. (M 3757)

(1842) Hynek

1972 AA. Discovered 1972 January 14 by L. Kohoutek at Bergedorf.

Named by the discoverer in honor of his father, Hynek Kohoutek, on the occasion of his 70th birthday. (M 3757)

(1843) Jarmila

1972 AB. Discovered 1972 January 14 by L. Kohoutek at Bergedorf.

Named by the discoverer in honor of his mother, Jarmila Kohoutkova, on the occasion of her 70th birthday. (M 3757)

(1844) Susilva

1972 UB. Discovered 1972 October 30 by P. Wild at Zimmerwald.

See {the citation} for (1845) which follows. (M 4156)

(1845) Helewalda

1972 UC. Discovered 1972 October 30 by P. Wild at Zimmerwald.

Named for two charming former schoolmates of the discoverer, Susi and Helen, both from the town of Wald (Canton Zurich) Switzerland. (M 4156)

(1846) Bengt

6553 P-L. Discovered 1960 September 24 by C. J. van Houten and I. van Houten-Groeneveld at Palomar.

Named in honor of Bengt Strömgren {1908-1987} on the occasion of his 70th birthday. This renowned Danish astronomer was an authority in the field of stellar structure and stellar evolution. He served as director of the Yerkes Observatory during 1951-1957 and as president of the IAU during 1970-1973. (M 4547)

Obituaries published in Astron. Tidsskr., Årg. 20, Nr. 4, p. 175-177 (1987); Messenger, No. 49, p. 1, 43-44 (1987); Sterne Weltraum, 26. Jahrg., Nr. 12, p. 677 (1987); Phys. Today, Vol. 41, No. 3, p. 112-114 (1988); Q.J.R. Astron. Soc., Vol. 29, No. 2, p. 282-284 (1988); Mitt. Astron. Ges., Nr. 73, p. 5-7 (1990); C.R. Acad. Sci., Sér. Gén., Vie Sci., Tome 6, p. 609-610 (1989); IAU Today (20th General Assembly, Baltimore, 1988), No. 3, p. 2, 8 (1988).

(1847) Stobbe

A916 CA. Discovered 1916 February 1 by H. Thiele at Bergedorf.

Named in honor of Joachim Otto Stobbe (1900-1943), astronomer at the Hamburg-Bergedorf Observatory during 1925-1927 who observed comets and minor planets. As chief assistant at the Berlin-Babelsberg Observatory he later did an extensive study of the light curve of (433) Eros. From 1941 until his death he was director of the Poznan Observatory. (M 4236)

(1848) Delvaux

1933 QD. Discovered 1933 August 18 by E. Delporte at Uccle.
Named in honor of the sister-in-law of Uccle astronomer G. Roland. (M 6832)

(1849) Kresák

1942 AB. Discovered 1942 January 14 by K. Reinmuth at Heidelberg.

Named in honor of Luboš Kresák {1927-1994}, astronomer at the Slovak Academy of Sciences in Bratislava. Well-known for his theoretical work on meteors and the question of their relationship with comets and minor planets, he has also been an observer at the Skalnaté Pleso {see planet (2619)} Observatory, where in 1951 he rediscovered the short-period comet now known as Tuttle-Giacobini-Kresák {see planets (5036) and (1756) for the codiscoverers}. He was president of IAU Commission 20 for the period 1973-1976. (M 3935)

Obituaries published in Icarus, Vol. 109, No. 1, p. 1 (1994); WGN, Vol. 22, No. 1, p. 1-2 (1994); IAU Symp. No. 160, p. 75-76 (1994); Planet. Space Sci., Vol. 42, No. 2, 9. 99 (1994); Q.J.R. Astron. Soc., Vol. 35, No. 4, p. 579 (1994).

(1850) Kohoutek

1942 EN. Discovered 1942 March 23 by K. Reinmuth at Heidelberg.

Named in honor of the Czech astronomer, Lubos Kohoutek {1935- }, on the staff of the observatory at Hamburg-Bergedorf since 1969. He has contributed much to our knowledge of planetary nebulae and emission-line stars and is also widely known as a discoverer and observer of comets and minor planets. (M 3935)

(1851) Lacroute

1950 VA. Discovered 1950 November 9 by L. Boyer at Algiers.

Named in honor of Pierre Lacroute {1906-1993}, well-known French astrometrist who made an independent reduction of the AGK3 using a plate-overlapping technique. As director of the Strasbourg Observatory he was instrumental in the establishment of the Stellar Data Center there, and he served as president of IAU Commission 24 during 1973-1976. (M 4419)

Obituaries published in Bull. Am. Astron. Soc., Vol. 25, No. 4, p. 1498 (1993); Astronomie, Vol. 107, p. 128-129 (1993).

(1852) Carpenter

1955 GA. Discovered 1955 April 1 at the Goethe Link Observatory at Brooklyn, Indiana.

Named in memory of Edwin F. Carpenter (1898-1963), second director of the Steward Observatory and a director-at-large on the AURA board. His primary research interests were spectroscopic binaries, photometry and interacting galaxies. Despite extremely limited financial resources he succeeded in keeping Steward Observatory an active research center during the nearly three decades of his directorship, and he played a major role in persuading the Papago Tribal Council to lease the top of Kitt Peak to the AURA Board, thereby enabling the construction of the National Observatory on Kitt Peak. (M 5282)

Name proposed by F. K. Edmondson and E. Roemer. Citation written by W. S. Fitch.

(1853) McElroy

1957 XE. Discovered 1957 December 15 at the Goethe Link Observatory at Brooklyn, Indiana.

Named in honor of William David McElroy, distinguished biologist and biochemist, chairman of the biology department at Johns Hopkins University during 1956-1969, later director of the National Science Foundation (1969-1972) and chancellor of the University of California at San Diego (1972-1980). During his tenure as director of N.S.F. the U.S. government decided to fund the Very Large Array, which will be the pre-eminent instrument for radio astronomy for many years to come. (M 5450)

(1854) Skvortsov

1968 UE_1. Discovered 1968 October 22 by T. M. Smirnova at Nauchnyj.

Named in honor of Prof. E. F. Skvortsov (1882-1952), an instructor of astronomy in the Simferopol Pedagogical Institute, an active observer of minor planets at Simeïs, and the discoverer of several new planets. (M 3825)

(1855) Korolev

1969 TU_1. Discovered 1969 October 8 by L. I. Chernykh at Nauchnyj.

Named in honor of academician Sergej Pavlovich Korolev (1907-1966), an outstanding designer of Soviet space rockets and spacecraft. (M 3825)
Korolev is also honored by craters on Mars and the Moon.

(1856) Ružena

1969 TW_1. Discovered 1969 October 8 by L. I. Chernykh at Nauchnyj.

Named in honor of Miss Ružena Petrovicova, a staff member of the Kleť Observatory, and an observer of comets and minor planets. (M 3825)
Name proposed by L. Chernykh.

(1857) Parchomenko

1971 QS_1. Discovered 1971 August 30 by T. M. Smirnova at Nauchnyj.

Named in honor of Praskoviya Georgievna Parchomenko (1886-1970), astronomer. In 1930-1940 she observed minor planets at Simeïs and discovered several new ones. (M 3826)

(1858) Lobachevskij

1972 QL. Discovered 1972 August 18 by L. V. Zhuravleva at Nauchnyj.

Named in honor of Nikolaj Ivanovich Lobachevskij (1792-1856), Russian mathematician, the creator of the first comprehensive system of non-Euclidian geometry. (M 3826)
Lobachevskij is also honored by a lunar crater.

(1859) Kovalevskaya

1972 RS_2. Discovered 1972 September 4 by L. V. Zhuravleva at Nauchnyj.

Named in honor of Sophie Kovalevskaya (1850-1891), the first Russian woman mathematician, associate member of the St. Petersburg Academy of Sciences, widely known for her researches on differential equations, Abelian integrals and the rotational motion of solid bodies. (M 3826)
Kovalevskaya is also honored by a lunar crater.

(1860) Barbarossa

1973 SK. Discovered 1973 September 28 by P. Wild at Zimmerwald.

The name of the famous German emperor was the nickname (Barba) of the discoverer's favorite mathematics teacher, Dr. Jakob Stauber, (1880-1952) in Trogen, Switzerland. (M 4157)

(1861) Komenský

1970 WB. Discovered 1970 November 24 by L. Kohoutek at Bergedorf.

Named in honor of Jan Amos Komenský (Comenius, 1592-1670), Czech educational reformer and theologian, sometimes called the grandfather of modern education. He had to spend most of his life in exile. (M 3757)

(1862) Apollo

1932 HA. Discovered 1932 April 24 by K. Reinmuth at Heidelberg.

This object is named for the god of the Sun, child of Zeus and Leto {see planets (5731) and (68)}. (M 3758)
See also the naming citation of planet (4544).

(1863) Antinous

1948 EA. Discovered 1948 March 7 by C. A. Wirtanen at Mount Hamilton.

Antinous was one of the many unwelcome suitors for Penelope's {see planet (201)} hand while her husband, Odysseus {see planet (1143)}, was away on his travels. Antinous, being the most insolent of all, was the first to be killed by Odysseus on his return. (M 3935)

(1864) Daedalus

1971 FA. Discovered 1971 March 24 by T. Gehrels at Palomar.

Named for the builder of King Minos' {see planet (6239)} labyrinth, who was subsequently imprisoned there with his son Icarus. Both escaped, on wings of feathers and wax, but whereas Icarus was drowned when the wax in his wings melted {see planet (1566)}, Daedalus went on to Sicily and built there a temple to Apollo {see planet (1862)}. (M 3758)
This mythological figure also exists as a lunar crater.

(1865) Cerberus

1971 UA. Discovered 1971 October 26 by L. Kohoutek at Bergedorf.

Named for the three-headed dog that guarded the entrance to Hades. His capture (and subsequent return) marked the last of the twelve labors of Hercules. Cerberus is also the name of an extinct constellation now contained in the eastern part of Hercules. (M 3758)

(1866) Sisyphus

1972 XA. Discovered 1972 December 5 by P. Wild at Zimmerwald.

Named for the cruel king of Corinth, punished by being given the task of rolling a large stone up to a hill in Hades, only to have it roll down again each time he neared the top. (M 3758)

(1867) Deiphobus

1971 EA. Discovered 1971 March 3 by C. U. Cesco at El Leoncito.

Named after the Trojan warrior, son of Priamus {see planet (884)} and Hecuba {see planet (108)}. (M 3935)

(1868) Thersites

2008 P-L. Discovered 1960 September 24 by C. J. van Houten and I. van Houten-Groeneveld at Palomar.

Named after a Greek warrior who wanted to stop the siege of Troy and go home. This is a fitting name for this asteroid which was found farthest from the Trojan libration point. (M 3826)

(1869) Philoctetes

4596 P-L. Discovered 1960 September 24 by C. J. van Houten and I. van Houten-Groeneveld at Palomar.

Named after a Greek archer and armor-bearer of Hercules, wounded and left to die on the way to the Trojan War. Since it had been prophesied that Troy could not be taken without the arrows of Hercules, Philoctetes was later taken to Troy, where he killed Paris {see planet (3317)}. (M 3826)

(1870) Glaukos

1971 FE. Discovered 1971 March 24 by C. J. van Houten and I. van Houten-Groeneveld at Palomar.
Glaukos, king of the Lycians, was an ally of Troy during the Trojan War. (M 3826)

(1871) Astyanax

1971 FF. Discovered 1971 March 24 by C. J. van Houten and I. van Houten-Groeneveld at Palomar.

Named after a son of Hector and Andromache {see planets (624) and (175)}, killed after the capture of Troy. (M 3826)

(1872) Helenos

1971 FG. Discovered 1971 March 24 by C. J. van Houten and I. van Houten-Groeneveld at Palomar.

Named after a son of Priamus and Hecuba {see planets (884) and (108)}, who was taken prisoner by the Greeks during the Trojan War. (M 3826)

(1873) Agenor

1971 FH. Discovered 1971 March 25 by C. J. van Houten and I. van Houten-Groeneveld at Palomar.

Named after a Trojan warrior, who fought with and wounded Achilles {see planet (588)}. Apollo {see planet (1862)} assumed his form in order to lead Achilles away from the retreating Trojans. (M 3826)
Name proposed by Brian G. Marsden.

(1874) Kacivelia

A924 RC. Discovered 1924 September 5 by S. I. Belyavskij at Simeïs.

Named in honor of the Black Sea Hydrographical Station (now the Marine Hydrographical Institute). Kaciveli is a village near Simeïs {see planet (748)}, the site of the institute. (M 3936)
The name had been proposed in the 1930's by the late Prof. E. Skvortsov.

(1875) Neruda

1969 QQ. Discovered 1969 August 22 by L. Kohoutek at Bergedorf.

Named in memory of Jan Neruda (1834-1891), Czech lyric poet, novelist and journalist. He published several volumes of poems, notably *Cosmic Songs*, as well as collections of essays, plays and stories, of which the best known is *Stories from Malá Strana*. (M 18446)

(1876) Napolitania

1970 BA. Discovered 1970 January 31 by C. T. Kowal at Palomar.
Named in honor of the city of Naples, the birthplace of Mrs. Kowal. (M 3826)

(1877) Marsden

1971 FC. Discovered 1971 March 24 by C. J. van Houten and I. van Houten-Groeneveld at Palomar.

Named in honor of Brian G. Marsden {1937- }, Smithsonian Astrophysical Observatory, in recognition of his numerous contributions in the field of orbit calculations for comets and minor planets, his improved versions of the Catalogue of Cometary Orbits, and his activities in the Central Bureau and in Commission 20 of the I.A.U. (M 3826)

Proposed by his colleagues, Elizabeth Roemer, Frank Edmondson, Tom Gehrels and Paul Herget.

Marsden was also honored by a series of minor planets - see the citation for planet (5699).

(1878) Hughes

1933 QC. Discovered 1933 August 18 by E. Delporte at Uccle.

Named in honor of the son of Mireille (née Delporte) Demiddelaer {see planet (1926)}, granddaughter of the discoverer. (M 8402)

(1879) Broederstroom

1935 UN. Discovered 1935 October 16 by H. van Gent at Johannesburg.

Named for the village in South Africa near which the Leiden Southern Observatory was located from 1957 to 1982. (M 6833)

(1880) McCrosky

1940 AN. Discovered 1940 January 13 by K. Reinmuth at Heidelberg.

Named in honor of Richard E. McCrosky {1924- }, in appreciation of his role in the recovery of (1862) Apollo. On the staff of the Harvard-Smithsonian Center for Astrophysics, he has been in charge of the Prairie Network for the observations of fireballs, which led to the recovery of the Lost City meteorite in 1970. From 1970 to 1973 he was president of IAU Commission 22. (M 3936)

(1881) Shao

1940 PC. Discovered 1940 August 3 by K. Reinmuth at Heidelberg.

Named in honor of Cheng-yuan Shao {1927- }, an assistant to McCrosky {see planet (1880)} in several observing projects at the Harvard-Smithsonian Center for Astrophysics, including the recovery of (1862) Apollo. (M 3936)

(1882) Rauma

1941 UJ. Discovered 1941 October 15 by L. Oterma at Turku.

Named for a Finnish town, famous for its seafaring and lace-making, that developed around a monastery in the first part of the fifteenth century. The area was already populated in the bronze age. (M 5450)

(1883) Rimito

1942 XA. Discovered 1942 December 4 by Y. Väisälä at Turku.

Named for Rymattyla, a Finnish country commune near Turku, already inhabited in the bronze age. Comprising about 400 islands with numerous summer cottages, it is also the center for winter seine-fishing in Finland. (M 5450)

(1884) Skip

1943 EB_1. Discovered 1943 March 2 by M. Laugier at Nice.

Named in honor of Gunther (Skip) Schwartz, manager of and observer at Harvard Observatory Agassiz {see planet (2267)} Station and previously field manager of various other Harvard and Smithsonian optical and radio observing stations. He is best known for his role in the recovery of the Lost City meteorite in 1970. (M 4645)

Name proposed by R. E. McCrosky and B. G. Marsden.

(1885) Herero

1948 PJ. Discovered 1948 August 9 by E. L. Johnson at Johannesburg.

Named for a tribe in the northern highlands of South-West Africa who fought a disastrous war with the Nama Hottentots in 1930. They came under German rule in 1884 but were ruthless suppressed after a revolt in 1904. (M 3936)

(1886) Lowell

1949 MP. Discovered 1949 June 21 by H. L. Giclas and R. D. Schaldach at Flagstaff.

Named in honor of Percival Lowell {1855-1916}, the founder, director and endower of Lowell Observatory at Flagstaff, Arizona. Lowell, during his lifetime and beginning in 1905 directed three searches of the ecliptic for a trans-Neptunian planet. In all these surveys positions for all the minor planets found were measured and he was the discoverer of (793) Arizona. (M 4157)

The planet was found with the 13-inch astrograph with which Pluto, the PlanetX predicted by Lowell, was discovered. Lowell is also honored by craters on Mars and the Moon.

(1887) Virton

1950 TD. Discovered 1950 October 5 by S. Arend at Uccle.

Named for the capital of the Gaume (or Belgian Lorraine), very close to Robelmont {see planet (1145)}, the discoverer's birthplace. (M 8151; M 8183)

(1888) Zu Chong-Zhi

1964 VO_1. Discovered 1964 November 9 at the Purple Mountain Observatory at Nanking.

Named for Zu Chong-Zhi (492-500), a distinguished astronomer-mathematician during the period of the Liu-Sung dynasty. He improved the calendrical system of his time. He constructed a bronze compass that can rotate freely while pointing in the same direction. He also ascertained that the ratio of the circumference to the diameter of a circle is between 3.1415926 and 3.1415927. (M 4419)

(1889) Pakhmutova

1968 BE. Discovered 1968 January 24 by L. I. Chernykh at Nauchnyj.

Named in honor of Aleksandra Nikolaevna Pakhmutova, Soviet composer. (M 3936)

(1890) Konoshenkova

1968 CD. Discovered 1968 February 6 by L. I. Chernykh at Nauchnyj.

Named in honor of Olga Petrovna Konoshenkova (1919-1975), the schoolmistress at the Crimean Observatory school. (M 3936)

(1891) Gondola

1969 RA. Discovered 1969 September 11 by P. Wild at Zimmerwald.

Selected as a beautiful sounding word, well suited to an object moving smoothly and silently across the sky. (M 4157)

(1892) Lucienne

1971 SD. Discovered 1971 September 16 by P. Wild at Zimmerwald.

Named in honor of Mlle. Lucienne Divan of the Institut d'Astrophysique in Paris, who is a spectroscopist working mainly on the spectral and luminosity classification of stars and on interstellar absorption. (M 4157)

(1893) Jakoba

1971 UD. Discovered 1971 October 20 by P. Wild at Zimmerwald.

Named in honor of the discoverer's grandfather, Dr. Jakob Oberholzer (1862-1939) in Glarus, Switzerland, who was one of the outstanding alpine geologists of his time. (M 4157)

(1894) Haffner

1971 UH. Discovered 1971 October 26 by L. Kohoutek at Bergedorf.

Named in honor of the German astronomer Hans Haffner (1912-1977), who worked in Göttingen from 1941 to 1953 prior to the Hamburg Observatory in Bergedorf. Since 1967 he has been Director at the Astronomical Institute of the University in Würzburg. His activity in astronomical photography and photometry is mainly directed towards the study of open clusters. Best known is his photographic photometry of Praesepe (together with Heckmann, see planet (1650)). In addition he discovered 26 open clusters. (M 3827)

Obituaries published in Mitt. Astron. Ges., Nr. 42, p. 5-8 (1977); Sterne Weltraum, 16. Jahrg., p. 121 (1977).

(1895) Larink

1971 UZ. Discovered 1971 October 26 by L. Kohoutek at Bergedorf.

Named in honor of the German astronomer Johannes Larink (1893-1988), who worked at the Hamburg Observatory in Bergedorf as senior observer from 1922 until 1958. He determined the periods of Cepheids in M3, but his main field was classical astronomy. He carried out astrometric programs using the Bergedorf meridian circle, for instance making measurements of reference stars for the AGK2 and AGK3 and a catalogue of faint stars. (M 3827)

(1896) Beer

1971 UC_1. Discovered 1971 October 26 by L. Kohoutek at Bergedorf.

Named in honor of the astronomer Arthur Beer {1900-1980}, born in Bohemia in 1900, who worked at the Naval Observatory in Hamburg during 1929-1934, prior to moving to England. He was senior observer at the Cambridge Observatories from 1947 until his retirement in 1967. Although perhaps best known for his on-going series *Vistas in Astronomy*, he has worked extensively on spectroscopic binaries, on the nova DQ Herculis and on the spectrophotometry of O and B stars and galactic structure. (M 3827)

Name suggested by B. G. Marsden.

Beer is also honored by craters on Mars and the Moon. Obituaries published in Vistas Astron., Vol. 24, p. I-II (1980); Observatory, Vol. 101, p. 24 (1981).

(1897) Hind

1971 UE_1. Discovered 1971 October 26 by L. Kohoutek at Bergedorf.

Named in honor of John Russell Hind (1823-1895), discoverer of (7) Iris, (8) Flora and eight other minor planets. Hind worked for many years at George Bishop's Observatory near London and served also as superintendent of the British Nautical Almanac Office from 1853 to 1891. (M 3827)

Name suggested by B. G. Marsden.

Hind is also honored by a lunar crater.

(1898) Cowell

1971 UF_1. Discovered 1971 October 26 by L. Kohoutek at Bergedorf.

Named in honor of Philip Herbert Cowell (1870-1949), superintendent of the British Nautical Almanac Office from 1910-1930 and inventor of the well-known method used for numerical integration of the orbits of comets and minor planets. (M 3827)

Name suggested by B. G. Marsden.

(1899) Crommelin

1971 UR_1. Discovered 1971 October 26 by L. Kohoutek at Bergedorf.

Named in honor of Andrew Claude de la Cherois Crommelin (1865-1939), who worked and wrote extensively on the orbits of comets and minor planets. Together with P. H. Cowell {see minor planet (1898)}, Crommelin computed the orbit of Jupiter VIII and identified appearances of Halley's comet back to the year -239. (M 3827)

Name suggested by B. G. Marsden.

Crommelin is also honored by craters on Mars and the Moon.

(1900) Katyusha

1971 YB. Discovered 1971 December 16 by T. M. Smirnova at Nauchnyj.

Named in honor of Ekaterina Ivanovna Zelenko, airwoman who valorously perished at the front in Sept. 1941. Katyusha is a petname for Ekaterina. (M 3936)

(1901) Moravia

1972 AD. Discovered 1972 January 14 by L. Kohoutek at Bergedorf.

Central region of Czechoslovakia, formerly a province. Other regions are Bohemia (371), Silesia (257) and Slovakia (1807), each with a distinct history and own folk culture. Native province of the discoverer. (M 3827)

(1902) Shaposhnikov

1972 HU. Discovered 1972 April 18 by T. M. Smirnova at Nauchnyj.

Named in honor of Vladimir Grigorevich Shaposhnikov (1905-1942), an expert in astrometry. Before the Great Patriotic War he worked at the Simeïs Observatory. He perished at the front. (M 3936)

(1903) Adzhimushkaj

1972 JL. Discovered 1972 May 9 by T. M. Smirnova at Nauchnyj.

Named in honor of the heroic warriors of the Soviet Army and the Crimean guerillas who fought in the Adzhimushkaj quarries during the Great Patriotic War. The discoverer's mother was among the courageous defenders of the underground fortress. (M 3936)

(1904) Massevitch

1972 JM. Discovered 1972 May 9 by T. M. Smirnova at Nauchnyj.

Honoring Dr. Alla Genrichovna Massevitch {1918- }, a well-known woman astronomer-astrophysicist, vice-president of the Astronomical Council of the USSR Academy of Sciences, the organizer of optical tracking of artificial Earth satellites in the USSR. (M 3936)

(1905) Ambartsumian

1972 JZ. Discovered 1972 May 14 by T. M. Smirnova at Nauchnyj.

Named in honor of Viktor Amazaspovich Ambartsumian {1908-1996}, a world renowned scientist, the founder of the Soviet School for Astrophysics, president of the Academy of Sciences of the Armenian SSR, served as president of the IAU in 1961-1964, Director of the Byurakan Astrophysical Observatory. (M 3937)

Obituaries published in Izv. Nats. Akad. Nauk Arm., Fiz., Tom 32, No. 1, p. 54-59 (1997); J. Astrophys. Astron., Vol. 18, No. 1, p. 1-2 (1997); Ir. Astron. J., Vol. 24, No. 1, p. 4-6 (1997); Astron. Geophys., Vol. 38, Issue 2, p. 37 (1997); South. Stars, Vol. 37, No. 3, p. 94-96 (1997); Postepy Astron., Tom 45, No. 2, p. 34-35 (1997); Zvaigznota Debess, No. 155, p. 18-27 (1997); IAU Inf. Bull., No. 79, p. V-VI (1997); Astron. Rep., Vol. 41, No. 2, p. 278-279 (1997); Astrophysics, Vol. 39, No. 4, p. 307-309 (1996); Sonderdr. Almanach Österreich. Akad. Wiss., 147. Jahrg., p. 497-504 (1996/97).

(1906) Naef

1972 RC. Discovered 1972 September 5 by P. Wild at Zimmerwald.

Named in memory of Robert A. Naef (1907-1975) of Zurich, an ardent amateur astronomer, who since 1940 spent practically all his spare time (from the banking business) to produce the yearly observers almanac, *Der Sternenhimmel.* (M 4157)

Obituaries published in G. Ass. Astrofili Bologn., No. 38, p. 8 (1975); Orion, 33. Jahrg., p. 84-85 (1975); Sterne Weltraum, Vol. 14, p. 113, p. 209 (1975).

(1907) Rudneva

1972 RC_2. Discovered 1972 September 11 by N. S. Chernykh at Nauchnyj.

Named in honor of Evgeniya Maksimovna Rudneva, hero of the Soviet Union, member of Moscow branch of the Astronomical-Geodetical Society of the USSR, head of the solar department. During the Great Patriotic War she voluntarily joined the army. Being assigned to the Woman's Night Bombers Regiment, she fought at the front and perished valiantly in April 1944 while flying a mission. (M 3937)

(1908) Pobeda

1972 RL_2. Discovered 1972 September 11 by N. S. Chernykh at Nauchnyj.

The name pobeda (victory) is given on the occasion of the thirtieth anniversary of the victory of the Soviet people in the Great Patriotic War (1941-1945). (M 3937)

(1909) Alekhin

1972 RW_2. Discovered 1972 September 4 by L. V. Zhuravleva at Nauchnyj.

Named in honor of Aleksandr Aleksandrovich Alekhin (1892-1946), an outstanding Russian chess-player, the world chess champion (1927-1935, 1937-1946). (M 3937)

(1910) Mikhailov

1972 TZ_1. Discovered 1972 October 8 by L. V. Zhuravleva at Nauchnyj.

Named in honor of Aleksandr Aleksandrovich Mikhailov {1888-1983}, a prominent Soviet astronomer and gravimetrist, academician, president of the Astronomical Council of the USSR Academy of Sciences (1939-1962), vice-president of the International Astronomical Union (1946-1948), director of the Pulkovo Observatory (1947-1964). (M 3937)

Obituaries published in Astron. Zh., Tom 61, Vyp. 2, p. 412-413 (1984); Geod. i kartogr., No. 12, p. 55-56 (1983); Vesmír, Vol. 63, No. 4, p. 126 (1984); Zemlya Vselennaya, No. 2, p. 59-62 (1984); Sov. Astron., Vol. 28, No. 2, p. 243-244 (1984); Tsirk. Vses. astron.-geod. o-vo Akad. Nauk SSSR, No. 37, p. 78-88 (1984); Vestn. AN SSSR, No. 2, p. 99 (1984); Q.J.R. Astron. Soc., Vol. 26, No. 3, p. 209 (1985).

(1911) Schubart

1973 UD. Discovered 1973 October 25 by P. Wild at Zimmerwald.

Named in honor of Joachim Schubart {1928- }, Astronomisches Rechen-Institut, who has developed averaging techniques for studying the long-term motions of minor planets and used them to investigate in detail the theory of the Hilda {see planet (153)} group of minor planets, of which this is a member. With P. Stumpff he has produced a widely-used computer program for the numerical integration of n bodies and has utilized it in determinations of the masses of (1) Ceres and (2) Pallas. (M 3937)

(1912) Anubis

6534 P-L. Discovered 1960 September 24 by C. J. van Houten and I. van Houten-Groeneveld at Palomar.

Named for the Egyptian god who had the shape of a jackal and was the protector of the dead. (M 5013)

(1913) Sekanina

1928 SF. Discovered 1928 September 22 by K. Reinmuth at Heidelberg.

Named in honor of Zdenek Sekanina {1936- }, an astronomer at the Harvard-Smithsonian Center for Astrophysics, who has made many contributions to cometary astronomy, notably the calculation of orbits, the development of models for icy nuclei, and a detailed investigation of dust trails. (M 3937)

(1914) Hartbeespoortdam

1930 SB_1. Discovered 1930 September 28 by H. van Gent at Johannesburg.

Named for the lake in South Africa near which the Leiden Southern Observatory was located from 1957 to 1982. (M 6954)

(1915) Quetzálcoatl

1953 EA. Discovered 1953 March 9 by A. G. Wilson at Palomar.

Named for the god of wisdom and culture who brought learning to the Toltec people and has been familiarized in D. H. Lawrence's novel *The plumed serpent.* (M 3827)

(1916) Boreas

1953 RA. Discovered 1953 September 1 by S. Arend at Uccle.

Named for the god of the north wind, this Amor-type {see planet (1221)} object

was discovered as it moved rapidly northward after passing the ascending node of its orbit. (M 6833)
Name proposed by J. Meeus.

(1917) Cuyo

1968 AA. Discovered 1968 January 1 by C. U. Cesco and A. G. Samuel at El Leoncito.

Named in honor of the Universidad Nacional de Cuyo, which operated the observatory at El Leoncito {see planet (2311)} in association with Yale and Columbia Universities. The name is also that of the region formed by three provinces in western Argentina, namely, Mendoza, San Juan and San Luis. (M 3828)

(1918) Aiguillon

1968 UA. Discovered 1968 October 19 by G. Soulie at Bordeaux.

Named for the discoverer's birthplace, a small town on the Garonne river. (M 5038)

(1919) Clemence

1971 SA. Discovered 1971 September 16 by J. Gibson and C. U. Cesco at El Leoncito.

Named in memory of Gerald M. Clemence (1908-1974), Director of the U.S. Nautical Almanac Office from 1945 to 1958 and then the first scientific director of the U.S. Naval Observatory. Later he became professor of astronomy at the Yale Observatory. Although best known for his work on the theory of the motion of Mars, he made numerous other contributions in celestial mechanics, notably on the motion of Mercury and on the system of astronomical constants. He served as president of the American Astronomical Society and of IAU Commissions 4 and 7. Among his many honors were the awards of the Gold Medal of the R.A.S. and honorary doctorates from Case Institute of Technology and from Cuyo University in Western Argentina, where the Yale-Columbia Station was located. (M 3937)

Obituaries published in Phys. Today, Vol. 28, No. 3, p. 59, 61 (1975); Q.J.R. Astron. Soc., Vol. 16, p. 210-214 (1975); Sky Telesc., Vol. 49, p. 93, p. 215-216 (1975); J.R. Astron. Soc. Can., Vol. 69, p. 136-138 (1975).

(1920) Sarmiento

1971 VO. Discovered 1971 November 11 by J. Gibson and C. U. Cesco at El Leoncito.

Named for Domingo F. Sarmiento, president of Argentina between 1868 and 1874 who promoted education, particularly in scientific fields, in his country. Born in San Juan, he grew to maturity in the province and knew the region where the El Leoncito Observatory now stands. With the aid of B. A. Gould he was responsible for founding the Córdoba Observatory, dedicating it with the words 'it is said that an observatory is premature or superfluous for a rising country whose public treasury is exhausted or overburdened. But I say that we must renounce the rank of nation, or the title of civilized people, if we do not take our part in the progress and movement of the natural sciences'. (M 3938)

(1921) Pala

1973 SE. Discovered 1973 September 20 by T. Gehrels at Palomar.

Named for the Indian reservation at the base of Palomar Mountain. The name is believed to apply to an Indian tribe whose members have lived in the area for many centuries. (M 3938)

(1922) Zulu

1949 HC. Discovered 1949 April 25 by E. L. Johnson at Johannesburg.

Named for the well-known South African tribe, to recognize some of the tribesmen who were faithful workmen at the Johannesburg Observatory. Coupled with two other planets having tribal names, (1362) Griqua and (1921) Pala, these planets presently librate in the 2/1 ratio of Jupiter's mean motion. (M 3938)

(1923) Osiris

4011 P-L. Discovered 1960 September 24 by C. J. van Houten and I. van Houten-Groeneveld at Palomar.

Named for the Egyptian god of vegetation, of the waxing and waning Moon and of the annual flooding of the Nile. (M 5013)

(1924) Horus

4023 P-L. Discovered 1960 September 24 by C. J. van Houten and I. van Houten-Groeneveld at Palomar.

Named for the son of Osiris {see minor planet (1923)}, an Egyptian god in the shape of a falcon, king of the sky and the stars. (M 5013)

(1925) Franklin-Adams

1934 RY. Discovered 1934 September 9 by H. van Gent at Johannesburg.

Named in memory of John Franklin-Adams (1843-1912), the British amateur astronomer who acquired and originally owned the 0.25-m photographic refractor he later donated to the Union Observatory in Johannesburg. Initially utilized for the pioneering photographic survey of the Milky Way, the Franklin-Adams Star Camera was employed for observations of minor planets in Johannesburg and Hartbeespoort for almost 70 years. Among the many discoveries with this celebrated instrument was Proxima Centauri. (M 8402)
Franklin-Adams is also honored by planet (982).

(1926) Demiddelaer

1935 JA. Discovered 1935 May 2 by E. Delporte at Uccle.

Named in honor of the family of Mireille Demiddelaer, granddaughter of the discoverer. (M 8402)

(1927) Suvanto

1936 FP. Discovered 1936 March 18 by R. Suvanto at Turku.

Named in memory of Rafael Suvanto, an assistant of Y. Väisälä {see planets (1573) and (2804)}. Later, as rector of a school in Naantali {see planet (1758)}, he continued to participate in orbit calculations. He fell near Summa {see planet (1928)} in the last days of the Finnish winter war of 1939-40. (M 5450)

(1928) Summa

1938 SO. Discovered 1938 September 21 by Y. Väisälä at Turku.

Named for a village on the Karelian isthmus, scene of violent battles during the Finnish winter war {1939/40}. (M 5450)

(1929) Kollaa

1939 BS. Discovered 1939 January 20 by Y. Väisälä at Turku.

Named for a river in Karelia, scene of violent battles during the Finnish winter war {1939/40}. (M 5450)

(1930) Lucifer

1964 UA. Discovered 1964 October 29 by E. Roemer at Flagstaff.

Named for the proud, rebellious archangel, identified with Satan, who was expelled from heaven. (M 4419)

(1931) Čapek

1969 QB. Discovered 1969 August 22 by L. Kohoutek at Bergedorf.

Named in memory of Karel Čapek (1890-1938), Czech dramatist and novelist, best known for his allegorical plays "R.U.R." (*Rossum's Universal Robots*) and *The Insect Comedy* and for his novel *Krakatit*. He anticipated both the destructive possibilities of nuclear physics and the moral problems these possibilities would raise. (M 18447)

(1932) Jansky

1971 UB_1. Discovered 1971 October 26 by L. Kohoutek at Bergedorf.

Named for Karl Guthe Jansky (1905-1950), the father of radio astronomy, the research engineer at Bell Telephone Laboratories who in 1932 made the first detection of extraterrestrial radio emission. (M 4785)

Jansky is also honored by a lunar crater.

(1933) Tinchen

1972 AC. Discovered 1972 January 14 by L. Kohoutek at Bergedorf.

Named in honor of Christine Kohoutek, wife of the discoverer. (M 3938)

(1934) Jeffers

1972 XB. Discovered 1972 December 2 by A. R. Klemola at Mount Hamilton.

Named in honor of Hamilton M. Jeffers {1893-1976}, an astronomer on the staff of the Lick Observatory from 1924 until his retirement in 1961. He has conducted an extensive program of astrometric observations of comets and faint minor planets and has made important contributions to double-star astronomy, notably his collaboration on the Index Catalogue of Double-star Observations. (M 3938)

Obituaries published in Q.J.R. Astron. Soc., Vol. 21, p. 69-70 (1980); Phys. Today, Vol. 29, No. 10, p. 69 (1976), Sky Telesc., Vol. 52, p. 91 (1976).

(1935) Lucerna

1973 RB. Discovered 1973 September 2 by P. Wild at Zimmerwald.

This and the following three minor planets are named for a Swiss lake and a city on its shores. (M 4358)

(1936) Lugano

1973 WD. Discovered 1973 November 24 by P. Wild at Zimmerwald.

See the citation for minor planet (1935). (M 4358)

(1937) Locarno

1973 YA. Discovered 1973 December 19 by P. Wild at Zimmerwald.

See the citation for minor planet (1935). (M 4358)

(1938) Lausanna

1974 HC. Discovered 1974 April 19 by P. Wild at Zimmerwald.

See the citation for minor planet (1935). (M 4358)

(1939) Loretta

1974 UC. Discovered 1974 October 17 by C. T. Kowal at Palomar.

Named by the discoverer in honor of his daughter. (M 3828)

(1940) Whipple

1975 CA. Discovered 1975 February 2 at the Harvard College Observatory at Harvard.

Named in honor of Fred L.Whipple {1906- }, Harvard astronomer since 1931, professor since 1950 and director of the Smithsonian Astrophysical Observatory from 1955 to 1973. His countless contributions to our knowledge of the smaller bodies of the solar system include his icy-conglomerate model for cometary nuclei, and the development of modern techniques for the photographic observations of meteors. He has served as president of IAU Commissions 6, 15, and 22, and is now active on the NASA panel of space missions to comets and minor planets. (M 3828)

(1941) Wild

1931 TN_1. Discovered 1931 October 6 by K. Reinmuth at Heidelberg.

Named in honor of Paul Wild, of the Astronomical Institute, Berne University, whose principal work involves the discovery and observation of supernovae in other galaxies. His program has also produced discoveries of comets and minor planets, several of which he subsequently recovered as a result of his own unpublished orbit computations. (M 3938)

(1942) Jablunka

1972 SA. Discovered 1972 September 30 by L. Kohoutek at Bergedorf.

Named after a village in eastern Moravia {see planet (1901)}, Czechoslovakia, which is alive in the discoverer's memory as the site of summer camps where he made his first astronomical observations. Dedicated to all friends of Jablunka. (M 4157)

(1943) Anteros

1973 EC. Discovered 1973 March 13 by J. Gibson at El Leoncito.

Anteros was in attendance upon Eros and was sometimes said to be the avenger of slighted love, sometimes the one who opposes love, and was said by other authorities to be the twin brother of Eros. Although briefly used for the minor planet {2101} Adonis shortly after its discovery, the name is particularly suitable for (1943) because of the similarity of its orbit to that of (433) Eros. (M 4237)
Name proposed by M.-A. Combes.

(1944) Günter

1925 RA. Discovered 1925 September 14 by K. Reinmuth at Heidelberg.

Named by the discoverer in honor of his son, Günter Reinmuth. (M 4157; M 26599)

(1945) Wesselink

1930 OL. Discovered 1930 July 22 by H. van Gent at Johannesburg.

Named in honor of A. J. Wesselink {1909-1995}, astronomer at the Leiden, Radcliffe and Yale Observatories. From 1946 to 1950 he was Leiden observer at the Union Observatory, Johannesburg. (M 10311)
Name proposed by the Leiden Observatory.

Obituaries published in Zenith, No. 11, p. 483 (1995); Astron. Soc. Pac. Conf. Ser., Vol. 83, p. XXII (1995); Q. J. R. Astron. Soc., Vol. 37, p. 95-97 (1996); Bull. Am. Astron. Soc., Vol. 27, No. 4, p. 1486-1488, (1995); Mon. Notes Astron. Soc. S. Afr., Vol. 54, No. 9/10, p. 84-85 (1995).

(1946) Walraven

1931 PH. Discovered 1931 August 8 by H. van Gent at Johannesburg.

Named in honor of Th. Walraven {1916- }, retired professor of astronomy at the University of Leiden and for many years resident astronomer at the former Leiden Southern Station near Hartbeespoortdam, South Africa. He constructed special photometers for the telescopes at the station, including the 5-color photometer for which he developed the Walraven photometric system. (M 12968)

Name proposed by the Leiden Observatory, following a suggestion by C. J. van Houten.

(1947) Iso-Heikkilä

1935 EA. Discovered 1935 March 4 by Y. Väisälä at Turku.

Named for the farm, owned by Turku University, that became the site of the Turku Observatory. This minor planet was the first one to be discovered there. (M 5450)

(1948) Kampala

1935 GL. Discovered 1935 April 3 by C. Jackson at Johannesburg.
Named for the capital of Uganda. (M 5183)

(1949) Messina

1936 NE. Discovered 1936 July 8 by C. Jackson at Johannesburg.

Named for a copper-mining town on the border between South Africa and Zimbabwe-Rhodesia. (M 5183)

(1950) Wempe

1942 EO. Discovered 1942 March 23 by K. Reinmuth at Heidelberg.

Named in honor of Johann Wempe {1906-1980}, astronomer at the Astrophysical Observatory in Potsdam since 1944, and editor of *Astronomische Nachrichten* from 1951 to 1973. In 1936 and 1937 he observed and discovered minor planets at Heidelberg-Königstuhl. (M 4157)

Obituaries published in Astron. Nachr., Band 302, p. 59-60 (1981); Sterne, 57. Band, p. 109-112 (1981); Mitt. Astron. Ges., Nr. 53, p. I-II (1981).

(1951) Lick

1949 OA. Discovered 1949 July 26 by C. A. Wirtanen at Mount Hamilton.

Named in honor of James Lick (1796-1876) the founder of the Lick Observatory of the University of California. (M 3938)
Lick is also honored by a lunar crater.

(1952) Hesburgh

1951 JC. Discovered 1951 May 3 at the Goethe Link Observatory at Brooklyn, Indiana.

Named in honor of the Very Reverend Theodore M. Hesburgh, C.S.C., president of the University of Notre Dame since 1952. Father Hesburgh has an extraordinary record of public service in areas ranging from the humanitarian to the technical and scientific. During his 12 years as a member of the National Science Board he played a very important role in the founding of both the Kitt Peak National Observatory and the Cerro Tololo Interamerican Observatory. His support for CTIO during the 1964 period of inflation in Chile was decisive. (M 5688)
Name proposed by F. K. Edmondson.

(1953) Rupertwildt

1951 UK. Discovered 1951 October 29 at the Goethe Link Observatory at Brooklyn, Indiana.

Named in memory of Rupert Wildt (1905-1976), who was awarded the Eddington Medal by the Royal Astronomical Society in 1966 for his discovery in 1939 that the negative hydrogen ion is an important contributor to the opacity of the solar atmosphere. He identified the absorption bands in the red part of the spectra of the outer planets as due to methane and ammonia, and he made pioneer calculations of models for the interiors of the giant planets. A professor in Yale University's Department of Astronomy for many years, he was Yale's first scientific representative on the Board of Directors of the Association of Universities for Research in Astronomy and served two terms as AURA president and chairman of the board (1965-1968, 1971-1974). (M 6954)

Obituaries published in Q.J.R. Astron. Soc., Vol. 17, p. 522 (1976); Strolling

Astron., Vol. 26, p. 46 (1976); Phys. Today, Vol. 29, No. 4, p. 89 (1976); Sky Telesc., Vol. 51, p. 156 (1976); Icarus, Vol. 30, p. 441-445 (1977).

(1954) Kukarkin

1952 PH. Discovered 1952 August 15 by P. F. Shajn at Simeïs.

Named in memory of Boris Vasil'evich Kukarkin (1909-1977), professor of stellar astronomy and astrometry at Moscow State University for many years and a prominent specialist in the fields of variable stars and the structure of stellar systems. He was the initiator and one of the compilers of the *General Catalogue of Variable Stars*, served as Vice President of the Astronomical Council of the U.S.S.R. Academy of Sciences from 1947 to 1960, as Vice President of the IAU from 1955 to 1961 and as President of IAU Commission 27 from 1951 to 1958. (M 5358)

Obituaries published in Perem. Zvezdy, Tom 20, p. 493-498, 499-500 (1978); Priroda, No. 10, p. 155 (1977), Astron. Tsirk., No. 969, p. 1-3 (1977), Astron. Zh., Vol. 55, p. 202-204 (1978), Coelum, Vol. 46, p. 72-75 (1978), Q.J.R. Astron. Soc., Vol. 19, p. 247-248 (1978), Zemlya i Vselennaya, No. 1, p. 59-60 (1978).

(1955) McMath

1963 SR. Discovered 1963 September 22 at the Goethe Link Observatory at Brooklyn, Indiana.

Named in memory of Robert R. McMath (1891-1962), astronomer, engineer and businessman. Co-donor of the McMath-Hulbert Observatory to the University of Michigan in 1931, he served as its director during 1931-1961. He was an adviser to the National Science Foundation in its early years and chaired the panel that advised NSF on the need for a national observatory and conducted the site survey leading to the selection of Kitt Peak. He played a leading role in the incorporation of AURA, served as its first president (1957-1958) and then as chairman of the AURA board. (M 5848)

Name proposed by F. K. Edmondson.

McMath is also honored by a lunar crater.

(1956) Artek

1969 TX_1. Discovered 1969 October 8 by L. I. Chernykh at Nauchnyj.

Named for an all-Union young pioneer's camp in Crimea. (M 4190)

(1957) Angara

1970 GF. Discovered 1970 April 1 by L. I. Chernykh at Nauchnyj.

Named for a river in Siberia. (M 4190)

(1958) Chandra

1970 SB. Discovered 1970 September 24 by C. U. Cesco at El Leoncito.

Named for Subrahmanyan Chandrasekhar {1910-1995}, the Indian-born distinguished service professor at the University of Chicago, generally regarded as the world's leading theoretical astrophysicist. (M 5013)

Chandrasekhar, with A. Fowler, received the 1983 Nobel prize for physics.

Obituaries published in Khagol, No. 24, p. 2 (1995); Orion, Jahrg. 53, Nr. 271, p. 323 (1995); Phys. Today, Vol. 48, No. 11, p. 106-108 (1995); Astrophys. J., Vol. 454, No. 2, p. 551 (1995); Bull. Am. Astron. Soc., Vol. 28, No. 4, p. 1448 (1996); Nature, Vol. 377, No. 6549, p. 484 (1995); Phys. Bl., Jahrg. 51, Heft 12, p. 1193 (1995); Zenit, Jaarg. 22, Nr. 12, p. 518-519 (1995); Zemlya Vselennaya, No. 1, p. 57-60 (1996); J. Astrophys. Astron., Vol. 17, No. 3-4, p. I-VII (1996); Zvaigznota Debess, No. 157, p. 30-34 (1997); Observatory, Vol. 116, No. 1131, p. 121-124 (1996); Q.J.R. Astron. Soc., Vol. 37, No. 2, p. 261-263 (1996); Ciel Terre, Vol. 112, No. 3, p. 117-119 (1996).

(1959) Karbyshev

1972 NB. Discovered 1972 July 14 by L. V. Zhuravleva at Nauchnyj.

Named in honor of Dmitrij Michajlovich Karbyshev (1880-1945), Soviet military leader, hero of the Soviet Union, who had been taken prisoner at the beginning of the Great Patriotic War and was atrociously tortured to death by fascists at the extermination camp at Mauthausen. (M 4190)

(1960) Guisan

1973 UA. Discovered 1973 October 25 by P. Wild at Zimmerwald.

Named in memory of Henri Guisan (1874-1960), the general of the Swiss army from 1939 to 1945. His integrity and courage were essential to the strength of Switzerland during the Second World War. (M 4157)

(1961) Dufour

1973 WA. Discovered 1973 November 19 by P. Wild at Zimmerwald.

Named in memory of Henri Dufour (1787-1875), citizen of Geneva, who as a general in 1847 swiftly and with almost no bloodshed, subdued the catholic Sonderbund, and who later as an engineer, initiated and directed the first complete geodetic survey of Switzerland. (M 4157)

(1962) Dunant

1973 WE. Discovered 1973 November 24 by P. Wild at Zimmerwald.

Named in memory of Henri Dunant (1828-1910), citizen of Geneva, and founder of the Red Cross. (M 4158)

Dunant, together with F. Passey, received the first Peace Nobel Prize in 1901.

(1963) Bezovec

1975 CB. Discovered 1975 February 9 by L. Kohoutek at Bergedorf.

Named after a mountain in western Slovakia, Czechoslovakia, in remembrance of the numerous meteor expeditions that have been organized in that area since 1958. Dedicated to all participants, mostly amateur astronomers. (M 4158)

(1964) Luyten

2007 P-L. Discovered 1960 September 24 by C. J. van Houten and I. van Houten-Groeneveld at Palomar.

Named in honor of Willem Jacob Luyten {1899-1994}, emeritus professor of astronomy at the University of Minnesota, well known for his investigations on stars with large proper motion. (M 5038)

Obituaries published in I.A.P.P.P. Commun., No. 58, p. 60-61 (1994); Bull. Am. Astron. Soc., Vol. 27, No. 4, p. 1480-1481 (1995); Q.J.R. Astron. Soc., Vol. 37, No. 3, p. 453-456 (1996); Phys. Today, Vol. 68, No. 4, p. 107 (1995).

(1965) van de Kamp

2521 P-L. Discovered 1960 September 24 by C. J. van Houten and I. van Houten-Groeneveld at Palomar.

Named in honor of Peter van de Kamp {1901-1995}, director of Sproul Observatory from 1937 until 1972, well known for his investigations on astrometric binary stars. (M 5038)

Obituaries published in Phys. Today, Vol. 48, No. 8, p. 70-71 (1995); Bull. Am. Astron. Soc., Vol. 27, No. 4, p. 1483-1484 (1995); Publ. Astron. Soc. Pac., Vol. 108, No. 725, p. 556-559 (1996).

(1966) Tristan

2552 P-L. Discovered 1960 September 24 by C. J. van Houten and I. van Houten-Groeneveld at Palomar.

Named for one of the knights of the round table. (M 6421)

(1967) Menzel

A905 VC. Discovered 1905 November 1 by M. Wolf at Heidelberg.

Named in honor of Donald Howard Menzel (1901-1976), Harvard astronomer since 1932 and director of the Harvard College Observatory from 1954 to 1966. Both a theoretical and observational pioneering solar and stellar astrophysicist, he calculated atomic transition probabilities, chemically analysed stars and nebulae, and made fundamental contributions to our understanding of physical processes in gaseous nebulae, the solar chromosphere, and interpretation of stellar spectra. He observed 15 solar eclipses, determined spectroscopically the rotation rates of Uranus and Neptune and helped educate a number of prominent astronomers. On the lighter side he is well known for his doodling and for debunking of UFO's. (M 4158)

Obituaries published in Astronomie, Vol. 91, p. 50 (1977); Nature, Vol. 267, p. 189 (1977); Phys. Today, Vol. 30, No. 5, p. 96, 98 (1977); Sky Telesc., Vol. 53, p. 244-251 (1977).

(1968) Mehltretter

1932 BK. Discovered 1932 January 29 by K. Reinmuth at Heidelberg.

Named in honor of Johannes Peter Mehltretter {1934-1982}, who became a solar astronomer at the Fraunhofer-Institut in Freiburg in 1962, after he started with work at Königstuhl Observatory. (M 4158)

Obituary published in Mitt. Astron. Ges., Nr. 59, p. 9-10 (1983).

(1969) Alain

1935 CG. Discovered 1935 February 3 by S. Arend at Uccle.

Named in honor of Alain Vanheste, husband of Isabelle Dubois, granddaughter of the discoverer. (M 8151)

(1970) Sumeria

1954 ER. Discovered 1954 March 12 by M. Itzigsohn at La Plata.

Named in honor of one of the oldest known civilizations, located in the southernmost part of Mesopotamia between the Tigris and Euphrates rivers. It was first settled between 4500 to 4000 B.C., probably shortly after JD 0, by Ubaidans. The Sumerians themselves arrived around 3300 B.C. and lost their separate identity around 1900 B.C. when conquered by the Amorites. The Sumerians contributed many technological and cultural innovations, among them the first wheeled vehicles and potter's wheels, the first system of writing, the first written codes of law, and the first city-state. (M 21128)

Named by the Minor Planet Names Committee following a suggestion by F. Pilcher.

(1971) Hagihara

1955 RD_1. Discovered 1955 September 14 at the Goethe Link Observatory at Brooklyn, Indiana.

Named in honor of Yusuke Hagihara {1897-1979}, professor of astronomy at the University of Tokyo from 1935 to 1957, Director of the Tokyo Observatory from 1946 to 1957, and long an inspiring leader and teacher, principally in celestial mechanics and also in several other branches of astronomy and astrophysics. Best known for his theory of libratory motions and general discussion of stability problems in celestial mechanics, he has also made important contributions to the study of the velocity distribution of free electrons in planetary nebulae. In recent years he has produced a monumental five-volume treatise on celestial mechanics. He served as a vice-president of the IAU during 1961-1967 and as president of Commission 7 during 1964-1967. (M 4419)

This minor planet was named at the IAU Symposium No. 81, The Dynamics of the Solar System, held in Tokyo in May 1978 on the occasion of his 81st birthday.

Obituaries published in Phys. Today, Vol. 32, No. 6, p. 71 (1979); Q.J.R. Astron. Soc., Vol. 20, p. 325-328 (1979).

(1972) Yi Xing

1964 VQ_1. Discovered 1964 November 9 at the Purple Mountain Observatory at Nanking.

Named for Yi Xing or I Hsing (683-727), also known as Zhang Sui. This famous Chinese astronomer of the Tang dynasty constructed a bronze armillary sphere for measuring stellar altitudes, and he conducted a large-scale latitude-measurement project in many places across China. He also reformed the Chinese calendrical system. (M 4420)

(1973) Colocolo

1968 OA. Discovered 1968 July 18 by C. Torres at Cerro El Roble.

Named for the old and wise Araucanian chief who succeeded in unifying the Araucanian Indian tribes against the Spanish conquerors. (M 5358)

(1974) Caupolican

1968 OE. Discovered 1968 July 18 by C. Torres at Cerro El Roble.

Named for the Great Chief of the unified Araucanian tribes. He was selected after winning a competition in which a trunk had to be carried on the shoulders for as long as possible; he carried it for three days and nights. (M 5358)

(1975) Pikelner

1969 PH. Discovered 1969 August 11 by L. I. Chernykh at Nauchnyj.

Named in honor of Solomon Borisovich Pikelner (1921-1975), renowned astronomer, professor at the University of Moscow, founder of the Soviet school of cosmic electrodynamics. He was among the first to understand the significance of plasma physics and magnetohydrodynamics for astrophysics. His research activities cover physics of the Sun, interstellar matter, gaseous nebulae, remnants of supernova explosions, galactic structure, formation of stars, cosmology. He was chairman of the Commission 34 (Interstellar matter) of IAU (1964-1967). (M 4190)

Obituaries published in Q.J.R. Astron. Soc., Vol. 17, p. 352-355 (1976); Astron. Tsirk., No. 896, p. 7-9 (1975); Astron. Zh., Vol. 53, p. 233-235 (1976); Astrophys. Space Sci., Vol. 40, p. 241-242 (1976); Phys. Today, Vol. 29, No. 2, p. 65 (1976); Sky Telesc., Vol. 51, p. 98 (1976); Zemlya Vselennaya, p. 71-72 (1976).

(1976) Kaverin

1970 GC. Discovered 1970 April 1 by L. I. Chernykh at Nauchnyj.

Named in memory of Aleksej Aleksandrovich Kaverin (1904-1976), an instructor in astronomy at Irkutsk Pedagogical Institute, expert in the field of the theory of eclipses. (M 4190)

(1977) Shura

1970 QY. Discovered 1970 August 30 by T. M. Smirnova at Nauchnyj.

Named in honor of Aleksandr Anatolevich Kosmodemyanskij (1925-1945), hero of the Soviet Union, died valiantly at the front during the Great Patriotic War. Shura is a pet name for Aleksandr. (M 4190)

See also the citations for minor planets (1793) and (2072).

(1978) Patrice

1971 LD. Discovered 1971 June 13 at the Perth Observatory at Bickley.

Named in honor of the daughter of Dennis Harwood, member of the astrometric team at the Perth Observatory. (M 6530)

(1979) Sakharov

2006 P-L. Discovered 1960 September 24 by C. J. van Houten and I. van Houten-Groeneveld at Palomar.

Named in honor of Academician Andrej Dmitrievich Sakharov, distinguished physicist and mathematician. (M 6207)
Sakharov received the 1975 Nobel peace prize.

(1980) Tezcatlipoca

1950 LA. Discovered 1950 June 19 by A. G. Wilson and A. A. E. Wallenquist at Palomar.

Named for the Toltec god of matter, "the smoking mirror", the counterpart to (1915) Quetzálcoatl, "the plumed serpent". The two gods are the twin serpents that entwine to produce time, the opposites that display constancy and change. (M 4237)

(1981) Midas

1973 EA. Discovered 1973 March 6 by C. T. Kowal at Palomar.

Named for the Phrygian king supposed to have the gift of transforming all that he touched to gold. Although he had chosen this power as a reward for kind treatment of his captive, Silenus, he soon realized his folly and was relieved of his power by bathing in the Pactolus River. (M 4237)

(1982) Cline

1975 VA. Discovered 1975 November 4 by E. F. Helin at Palomar.

Named to honor the memory of a friend, Edwin Lee Cline. Cline was a distinguished inventor in the automotive field who looked to space as the new frontier. (M 4158)

(1983) Bok

1975 LB. Discovered 1975 June 9 by E. Roemer at Kitt Peak.

Named for Bart J. {1906-1983} and Priscilla F. Bok {1896-1975}. This is the first minor planet discovered with the 229-cm telescope of the Steward Observatory to be numbered. It is named for the Boks in recognition of their role in bringing the large reflector of the Steward Observatory into operation and for their support of the astrometric work on comets, minor planets, and satellites. This minor planet was at far southern declination at discovery (and observations at Wellington were important in obtaining a reliable orbit). This is also in fitting association with the Boks, who have devoted much effort to study of the structure of the southern Milky Way and to encouragement of astronomy in the southern hemisphere. (M 4158)
Name proposed by E. Roemer and A. C. Gilmore.

Priscilla F. Bok is also honored by minor planet (2137). Obituaries (B.J.B) published in Sky Telesc., Vol. 66, No. 4, p. 303-306 (1983); Phys. Today, Vol. 36, No. 12, p. 73 (1983); Mercury, Vol. 12, No. 5, p. 155 (1983); ESA Spec. Publ., ESA SP-201, p. 14-15 (1983); J.R. Astron. Soc. Can., Vol. 78, No. 1, p. 8-9 (1984); South. Stars, Vol. 30, No. 6, p. 424-426 (1984); Zenit, 11. Jaarg., No. 2, p. 38-42 (1984); Proc. Astron. Soc. Aust., Vol. 5, No. 4, p. 608-610 (1984); Q.J.R. Astron. Soc., Vol. 28, No. 4, p. 539-542 (1987). Obituary (P.F.B.) published in Sky Telesc., Vol. 51, p. 25 (1976).

(1984) Fedynskij

1926 TN. Discovered 1926 October 10 by S. I. Belyavskij at Simeïs.

Named in memory of Vsevolod Vladimirovich Fedynskij (1908-1978), an outstanding expert in reconnaissance geophysics, physics of the Earth and meteor astronomy, a capable organizer of scientific research, Vice President of IAU Commission 22 (1958-1964), and an honorary member and Vice President of the All-Union Astronomy and Geodesy Association. (M 5358)
Obituary published in Meteoritika, Vyp. 38, p. 157-158 (1979).

(1985) Hopmann

1929 AE. Discovered 1929 January 13 by K. Reinmuth at Heidelberg.

Named in memory of Josef Hopmann (1890-1975), the former director of the Universitäts-Sternwarte in Vienna, who was active as an observer of binary and variable stars. During 1930/31 he took part in an international program to observe {433} Eros. (M 4237)

Obituaries published in Mitt. Astron. Ges., Nr. 40, p. 7-9 (1976), Sternenbote, 18. Jahrg., p. 189 (1975), 19. Jahrg., p. 22-30 (1976), Sterne Weltraum, 15. Jahrg., p. 21 (1976).

(1986) Plaut

1935 SV_1. Discovered 1935 September 28 by H. van Gent at Johannesburg.

Named in honor of Lukas Plaut (1910-1984), Dutch astronomer, who worked at the Leiden Observatory from 1933 to 1940 and at the Kapteyn Laboratory in Groningen from 1940 to 1975. He is well known for his investigations of RR Lyrae variables near the Galactic Center. Born in Germany, he emigrated to Holland in 1933. During part of World War II he was interned in a working camp near the German town of Fürstenau under an assumed name, because of his Jewish descent; he escaped from there and returned to Holland. (M 17026)

Name proposed by the Leiden Observatory.

Obituary published in Zenit, 12. Jaarg., No. 4, p. 152-153 (1985).

(1987) Kaplan

1952 RH. Discovered 1952 September 11 by P. F. Shajn at Simeïs.

Named in memory of Samuil Aronovich Kaplan (1921-1978), head of the astrophysics department of the Lvov Observatory from 1948 to 1961 and subsequently a staff member of the Scientific Research Radiophysics Institute in Gorkij. He contributed extensively to a wide range of astrophysical topics, including white dwarfs, interstellar matter, radiative transfer, solar radiation, pulsars and galactic nuclei. (M 5358)

Obituaries published in Astrofizika, Vol. 14, p. 695-696 (1978); Zemlya Vselennaya, No. 1, p. 46-47 (1979).

(1988) Delores

1952 SV. Discovered 1952 September 28 at the Goethe Link Observatory at Brooklyn, Indiana.

Named in honor of Mrs. Delores Owings, who has worked diligently in the minor planet program of Indiana University, first as a co-worker with Dr. T. Gehrels {see planet (1777)} in the absolute magnitude determinations and then as the supervisor of all plate measurements for accurate positions. (M 4190)

Name proposed by Paul Herget, Director of the Minor Planet Center.

(1989) Tatry

1955 FG. Discovered 1955 March 20 by A. Paroubek and R. Podstanicka at Skalnaté Pleso.

Named for Vysoke Tatry, the High Tatras, the highest mountain range in Czechoslovakia. The Skalnaté Pleso Observatory {see planet (2619)} is located in this range. (M 5183)

(1990) Pilcher

1956 EE. Discovered 1956 March 9 by K. Reinmuth at Heidelberg.

Named in honor of Frederick Pilcher, associate professor of physics at Illinois College, Jacksonville, who has done much to promote interest in minor planets among amateur astronomers. He has himself made visual observations of almost 1100 different minor planets, an all-time record. (M 6833)

Name proposed by J. U. Gunter and J. Meeus, endorsed by C. M. Bardwell and B. G. Marsden.

(1991) Darwin

1967 JL. Discovered 1967 May 6 by C. U. Cesco and A. R. Klemola at El Leoncito.

Named in memory of Charles Robert Darwin (1809-1882), the English naturalist who first established the theory of organic evolution; much of his research was done in Argentina, and he crossed the Andes at a pass located some 100 km south of El Leoncito. This planet also honors his second son, George Howard Darwin (1845-1912), the astronomer noted for his pioneering application of detailed dynamical analyses to problems of cosmogony and geology. (M 5282)

The Darwins are also honored by craters on the Moon and on Mars (G.H.D.)

(1992) Galvarino

1968 OD. Discovered 1968 July 18 by C. Torres at Cerro El Roble.

Named for the heroic Araucanian chief condemned by the Spanish soldiers to have his hands cut off and then to be freed as a living lesson to other Indians. Since his request for death was not granted, he promised revenge. He continued to fight the conquerors until recaptured and condemned to the gallows. (M 5358)

(1993) Guacolda

1968 OH_1. Discovered 1968 July 25 by H. Wroblewski at Cerro El Roble.

Named for the beautiful and heroic wife of the Araucanian chief Lautaro {see planet (2976)}, formerly a servant in a Spanish home. She accompanied her husband in battle, fighting side by side with him. (M 5358)

(1994) Shane

1961 TE. Discovered 1961 October 4 at the Goethe Link Observatory at Brooklyn, Indiana.

Named in honor of Charles D. Shane {1895-1983}, the second president of AURA (1958-1962). He played a major role in the planning and construction of the first telescopes and buildings on Kitt Peak and also the Tucson headquarters building. Even more important was his initiative in arranging for AURA to take on the responsibility for the construction of a major observatory in the southern hemisphere; this led to the establishment of the Cerro Tololo Interamerican Observatory. His long and distinguished career at the University of California spanned a period of more than 50 years; he was director of the Lick Observatory during 1945-1958 and was responsible for modernizing and expanding its facilities. (M 5848)

Name proposed by F. K. Edmondson.

Obituaries published in Phys. Today, Vol. 37, No. 2, p. 80 (1984); South. Stars, Vol. 30, No. 6, p. 427-428 (1984); Q.J.R. Astron. Soc., Vol. 25, No. 4, p. 532-533 (1984).

(1995) Hajek

1971 UP_1. Discovered 1971 October 26 by L. Kohoutek at Bergedorf.

Named in honor of Tadeas Hajek (Hagecius, 1525-1600), Czech humanist astronomer regarded today as one of the founders of modern stellar and cometary astronomy. Known especially for his studies on the great comet of 1577 and on the supernova of 1572. He also worked in other scientific fields, notably medicine and botany. (M 4358)

(1996) Adams

1961 UA. Discovered 1961 October 16 at the Goethe Link Observatory at Brooklyn, Indiana.

Named in memory of John Couch Adams (1819-1892), the British mathematical astronomer who predicted the existence of the planet Neptune. (M 4237)
Adams is also honored by a lunar crater.

(1997) Leverrier

1963 RC. Discovered 1963 September 14 at the Goethe Link Observatory at Brooklyn, Indiana.

Named in memory of Urbain Joseph Le Verrier (1811-1877), the French mathematical astronomer who predicted the existence of the planet Neptune. (M 4237)
Leverrier is also honored by craters on Mars and the Moon.

(1998) Titius

1938 DX_1. Discovered 1938 February 24 by A. Bohrmann at Heidelberg.

Named in memory of Johann Daniel Titius (1729-1796), who made the initial formulation of the Titius-Bode Law of planetary distances. (M 4237)
Titius is also honored by a lunar crater.

(1999) Hirayama

1973 DR. Discovered 1973 February 27 by L. Kohoutek at Bergedorf.

Named in memory of Kiyotsugu Hirayama (1874-1943), discoverer of the so-called Hirayama families of minor planets. (M 4237)
Hirayama is also honored by a lunar crater.

(2000) Herschel

1960 OA. Discovered 1960 July 29 by J. Schubart at Sonneberg.

Just as (1000) honors Giuseppe Piazzi, the discoverer of the first minor planet, so (2000) honors William Herschel, (1738-1822), the discoverer of the first telescopic major planet. (M 4237)
Herschel is also honored by craters on Mars and the Moon.

(2001) Einstein

1973 EB. Discovered 1973 March 5 by P. Wild at Zimmerwald.

Named in memory of Albert Einstein (1879-1955), the greatest scientist of the twentieth century. Although Einstein was born in Germany and died in the U.S., it is particularly fitting that his name be given to a Bernese minor planet, for he laid the foundations of his revolutionary scientific thoughts while working as an examiner in the Swiss Patent Office in Berne. (M 4237)

Einstein received the Nobel prize for physics in 1921. He is also honored by a lunar crater.

(2002) Euler

1973 QQ_1. Discovered 1973 August 29 by T. M. Smirnova at Nauchnyj.

Named for the great Swiss-born mathematician and astronomer Leonhard Euler (1707-1783), who spent much of his time in St. Petersburg and was associated with the Russian Academy of Sciences from the age of 20 until his death. Among the many contributions to astronomy were two theories for the motion of the Moon. (M 4238)
Euler is also honored by a lunar crater.

(2003) Harding

6559 P-L. Discovered 1960 September 24 by C. J. van Houten and I. van Houten-Groeneveld at Palomar.

Named in memory of Karl Ludwig Harding (1765-1834), the discoverer of (3) Juno. (M 4238)
Harding is also honored by a lunar crater.

(2004) Lexell

1973 SV_2. Discovered 1973 September 22 by N. S. Chernykh at Nauchnyj.

St. Petersburg and for a short period Euler's {see planet (2002)} successor as professor of mathematics. He was the first to realize that the object discovered by Herschel {see planet (2000)} was a new planet. His name is also associated with comet 1770 I, for which he established the revolution period as less than six years. (M 4238)
Lexell is also honored by a lunar crater.

(2005) Hencke

1973 RA. Discovered 1973 September 2 by P. Wild at Zimmerwald.

Named in honor of Karl Ludwig Hencke (1793-1866), postmaster at Driesen, the discoverer of (5) Astraea and (6) Hebe. (M 4238)

(2006) Polonskaya

1973 SB_3. Discovered 1973 September 22 by N. S. Chernykh at Nauchnyj.

Named in honor of Elena Ivanovna Kazimirchak-Polonskaya {1902-1992}, a prominent researcher on cometary motion and the evolution of cometary orbits. She has demonstrated with numerous examples the possibility of the formation of typical short-period cometary orbits from the capture of comets by major planets. In 1968 she was awarded the F. A. Bredikhin {see planet (786)} prize by the U.S.S.R. Academy of Sciences for her considerable contributions to cometary astronomy. She is a member of the organizing committee of IAU Commission 20. (M 4481)

(2007) McCuskey

1963 SQ. Discovered 1963 September 22 at the Goethe Link Observatory at Brooklyn, Indiana.

Named in memory of Sidney W. McCuskey (1907-1979), director of the Warner and Swasey Observatory during 1959-1970. His papers on the stellar luminosity function and its variations form a major contribution to the understanding of these problems of galactic structure. He served as president of IAU Commission 33 (Structure and Dynamics of the Galactic System). (M 6954)
Citation prepared by F. D. Miller.

Obituaries published in Sky Telesc., Vol. 58, p. 117 (1979); Q.J.R. Astron. Soc., Vol. 21, p. 214-215 (1980).

(2008) Konstitutsiya

1973 SV_4. Discovered 1973 September 27 by L. I. Chernykh at Nauchnyj.

Named on the occasion of the adoption of the new constitution of the U.S.S.R. in 1977. (M 4481)

(2009) Voloshina

1968 UL. Discovered 1968 October 22 by T. M. Smirnova at Nauchnyj.

Named in honor of Vera Danilovna Voloshina (1919-1941), a partisan who perished valiantly during the Great Patriotic War (1941-1945). (M 4481)

(2010) Chebyshev

1969 TL_4. Discovered 1969 October 13 by B. A. Burnasheva at Nauchnyj.

Named in honor of Academician Pafnutij Lvovich Chebyshev (1821-1894), the great Russian mathematician and mechanician. (M 4481)
Chebyshev is also honored by a lunar crater.

(2011) Veteraniya

1970 QB_1. Discovered 1970 August 30 by T. M. Smirnova at Nauchnyj.
Named in honor of the Soviet veterans of the Great Patriotic War. (M 4481)

(2012) Guo Shou-Jing

1964 TE_2. Discovered 1964 October 9 at the Purple Mountain Observatory at Nanking.

Named for Guo Shou-Jing (1231-1316), a prominent Chinese astronomer of the Yuan dynasty. Believing that observations are fundamental to calendrical work and that high-quality instruments are indispensable for making satisfactory observations, he devised and constructed numerous ingenious astronomical instruments. He also conducted an expedition to measure latitudes and derived an accurate value for the obliquity of the ecliptic. (M 4420)

(2013) Tucapel

1971 UH_4. Discovered 1971 October 22 by C. Torres at Cerro El Roble.

Named for one of the brave Araucanian chiefs who, with his wife Gualeva, victoriously entered the city of Imperial. He died in 1560, fighting against the Spanish soldiers. (M 5359)

(2014) Vasilevskis

1973 JA. Discovered 1973 May 2 by A. R. Klemola at Mount Hamilton.

Named in honor of Stanislavs Vasilevskis { -1988}, an astronomer on the staff of the Lick Observatory from 1949 until his retirement in 1974. He has conducted extensive programs of astrometric observations for stellar proper motions and parallaxes and has made important contributions to astrometric instrumentation, notably his success in establishing a system of automatic plate measurement coupled with the use of electronic computers. (M 4190)

(2015) Kachuevskaya

1972 RA_3. Discovered 1972 September 4 by L. V. Zhuravleva at Nauchnyj.

Named in honor of Natasha Kachuevskaya, a heroine of the Great Patriotic War, killed in the battle of Stalingrad. (M 4481)

(2016) Heinemann

1938 SE. Discovered 1938 September 18 by A. Bohrmann at Heidelberg.

Named in memory of Karl Heinemann (1898-1970), astronomer at the Astronomisches Rechen-Institut from 1927 to 1963. His activities included fundamental astronomy and the editing of the *Astronomischer Jahresbericht* for the period 1934-1958. (M 4238)
Obituary published in Astron. Nachr., Vol. 292, p. 190 (1970).

(2017) Wesson

A903 SC. Discovered 1903 September 20 by M. Wolf at Heidelberg.

Named in honor of Mary Joan Wesson Bardwell, wife of Conrad M. Bardwell {see planet (1615)}. The latter established the identifications for this minor planet. (M 4358)

(2018) Schuster

1931 UC. Discovered 1931 October 17 by K. Reinmuth at Heidelberg.

Named in honor of Hans-Emil Schuster {1934- }, astronomer at the European Southern Observatory, who is active as observer and discoverer of minor planets and comets. (M 4358)

(2019) van Albada

1935 SX_1. Discovered 1935 September 28 by H. van Gent at Johannesburg.

Named in memory of G. B. van Albada (1911-1972), Dutch astronomer, director of the Bosscha Observatory in Lembang from 1948 to 1958 and of the Astronomical Institute of the Municipal University in Amsterdam from 1959 until his death. (M 18135)

(2020) Ukko

1936 FR. Discovered 1936 March 18 by Y. Väisälä at Turku.

Named for the supreme god in Finnish folklore. Ukko means "old man"; ukkonen means "thunder". (M 5450)

(2021) Poincaré

1936 MA. Discovered 1936 June 26 by L. Boyer at Algiers.

Named in honor of Henri Poincaré (1854-1912), distinguished French mathematician and celestial mechanician, contributor to the theory of functions, modern algebra, algebraic topology, number theory, and *Les méthodes nouvelles de la mécanique céleste.* (M 4420)

Poincaré is also honored by a lunar crater.

(2022) West

1938 CK. Discovered 1938 February 7 by K. Reinmuth at Heidelberg.

Named in honor of Richard M. West, astronomer at the European Southern Observatory, well known as the discoverer of a bright comet, and active as an observer and discoverer of minor planets. (M 4359)

West served as general secretary of the International Astronomical Union from 1982 to 1985, and as president of IAU Commission 20 from 1988 to 1991.

(2023) Asaph

1952 SA. Discovered 1952 September 16 at the Goethe Link Observatory at Brooklyn, Indiana.

Named in memory of Asaph Hall (1829-1907), discoverer of Phobos and Deimos, the satellites of Mars. (M 4238)

This minor planet was named at a special meeting held at the U.S. Naval Observatory in Washington on 1977 Aug. 11, the one-hundredth anniversary of the discovery of Deimos.

(2024) McLaughlin

1952 UR. Discovered 1952 October 23 at the Goethe Link Observatory at Brooklyn, Indiana.

Named in memory of Dean Benjamin McLaughlin (1901-1965), astronomical spectroscopist and geologist at Swarthmore College and the University of Michigan. He made the first rigorous spectroscopic measurement of stellar rotation - the rotation of Algol. He also applied his skill as a structural geologist to an interpretation of the telescopically observable markings on Mars, and this was later confirmed by direct observation from space vehicles. (M 6955)

Citation prepared by O. C. Mohler.

McLaughlin is also honored by craters on Mars and the Moon.

(2025) Nortia

1953 LG. Discovered 1953 June 6 by J. Churms at Johannesburg.

Nortia was the Etruscan goddess of fortune who was worshipped at Volsinii. Each year a nail was driven into the wall of her temple, for the purpose of counting the years. (M 21128)

Named by the Minor Planet Names Committee following a suggestion by F. Pilcher.

(2026) Cottrell

1955 FF. Discovered 1955 March 30 at the Goethe Link Observatory at Brooklyn, Indiana.

Named in memory of Frederick Gardner Cottrell in the centennial year of his birth. Cottrell organized the Research Corporation in 1912 and provided an initial endowment consisting of his valuable electrostatic precipitation rights for the United States east of the Rocky Mountains. Grants from the Research Corporation have been of great assistance in many fields of science, and they have made possible the purchase of the measuring engine used in the minor planet program at the Goethe Link Observatory. (M 4547)

(2027) Shen Guo

1964 VR_1. Discovered 1964 November 9 at the Purple Mountain Observatory at Nanking.

Named for the versatile Chinese scholar Shen Guo (1031-1095), who lived during the Sung dynasty and had many achievements in astronomy. He determined the distance of Polaris from the north celestial pole to be 3° and devised an experiment to demonstrate that the Moon is a dark, spherical body illuminated by sunlight. He also proposed that the traditional lunar calendar be abandoned in favor of one based on solar reckoning. (M 4785)

(2028) Janequeo

1968 OB_1. Discovered 1968 July 18 by C. Torres at Cerro El Roble.

Named for the wife of the Araucanian chief Guepotan. After her husband's death in battle, she took command and won. Later, she brought together Indians from various tribes and commanded them successfully in several further battles. (M 5359)

(2029) Binomi

1969 RB. Discovered 1969 September 11 by P. Wild at Zimmerwald.

This minor planet, five ahead of (2034) Bernoulli, is named for the mythical inventor of mathematical formulae. It seems that a student, on being asked when Binomi lived, credited him with having been a contemporary of Newton {see planet (8000)}, a response that became notorious around the university. A few years later another student, on being asked when Bernoulli lived, immediately answered: "I'm not going to fall into the trap; it is well known that the man never existed!" (M 6208)

(2030) Belyaev

1969 TA_2. Discovered 1969 October 8 by L. I. Chernykh at Nauchnyj.

Named in honor of Colonel Pavel Ivanovich Belyaev (1925-1970), hero of the Soviet Union, Soviet cosmonaut, commander of the spaceship Voskhod 2. (M 4482) Belyaev is also honored by a lunar crater.

(2031) BAM

1969 TG_2. Discovered 1969 October 8 by L. I. Chernykh at Nauchnyj.

Named in honor of the heroic builders of the Baikal-Amur-Main railroad line. (M 4482)

(2032) Ethel

1970 OH. Discovered 1970 July 30 by T. M. Smirnova at Nauchnyj.

Named in honor of the English writer Ethel Lilian Voynich (1864-1960). (M 4482)

(2033) Basilea

1973 CA. Discovered 1973 February 6 by P. Wild at Zimmerwald.

Named for the old city of Basel and the 50th anniversary of the Astronomical Institute at Basel University. (M 5359)

(2034) Bernoulli

1973 EE. Discovered 1973 March 5 by P. Wild at Zimmerwald.

Named for the great dynasty of mathematicians of Basel, notably Jakob (1654-1705), founder of the calculus of variations; Johann (1667-1748), contributor to integral calculus and the teacher of Euler {see planet (2002)}; and Daniel (1700-1782), cofounder of hydrodynamics. (M 5359)
The name Bernoulli is also commemorated by a lunar crater.

(2035) Stearns

1973 SC. Discovered 1973 September 21 by J. Gibson at El Leoncito.

Named in memory of Carl Leo Stearns (1892-1972), Fisk professor of astronomy at Wesleyan University and second director of the Van Vleck Observatory. His primary research consisted of the determination of more than 200 trigonometric parallaxes. He also participated in the world-wide {433} Eros campaign in 1931 but is perhaps best known as the discoverer of comet 1927 IV, which holds the record for the maximum heliocentric distance at which a comet has ever been observed. (M 4548)
Obituary published in Science, Vol. 179, p. 1212 (1973).

(2036) Sheragul

1973 SY_2. Discovered 1973 September 22 by N. S. Chernykh at Nauchnyj.

Named in honor of the Siberian village where the discoverer spent his school years. The name also honors his school-mates and relatives. (M 4482)

(2037) Tripaxeptalis

1973 UB. Discovered 1973 October 25 by P. Wild at Zimmerwald.

A whimsical name, suggested by the fact that (2037) = 3 x (679) Pax = 7 x (291) Alice. (M 5359)

(2038) Bistro

1973 WF. Discovered 1973 November 24 by P. Wild at Zimmerwald.

The French name for a small, cozy restaurant. The name is also suggested by the fact that (2038) = 2 x (1019) Strackea. (M 5359)

(2039) Payne-Gaposchkin

1974 CA. Discovered 1974 February 14 at the Harvard College Observatory at Harvard.

Named in honor of Cecilia Payne-Gaposchkin {1900-1979}, Harvard astronomer since 1923, professor of astronomy since 1956, and a staff member of the Smithsonian Astrophysical Observatory since 1962. One of the pioneers of modern astrophysics, she was engaged in research on the physical states and chemical composition of stars and nebulae in the 1920s. She is a leading authority on variable stars, supergiants, novae and supernovae, and the author of seven important and influential books whose topics include stars of high luminosity, stars in the making, novae, and variable stars. (M 4238)

Obituaries published in Phys. Today, Vol. 33, No. 6, p. 64-65 (1980); Irish Astron. J., Vol. 14, p. 69 (1979); Q.J.R. Astron. Soc., Vol. 23, p. 450-451 (1982).

(2040) Chalonge

1974 HA. Discovered 1974 April 19 by P. Wild at Zimmerwald.

Named in honor of Daniel Chalonge (1895-1977), astronome titulaire at the Paris Observatory, a co-founder of the Institut d'Astrophysique, well known for his fundamental spectrophotometric work, especially on the Sun and on early-type stars. He was the author of the first three-dimensional stellar classification. A native of Grenoble, he was a passionate mountaineer and had some first ascents to his credit. (M 4785)
Obituary published in Q.J.R. Astron. Soc., Vol. 21, p. 481-483 (1980).

(2041) Lancelot

2523 P-L. Discovered 1960 September 24 by C. J. van Houten and I. van Houten-Groeneveld at Palomar.
Named for one of the knights of the round table. (M 6421)

(2042) Sitarski

4633 P-L. Discovered 1960 September 24 by C. J. van Houten and I. van Houten-Groeneveld at Palomar.

Named in honor of Grzegorz Sitarski {1932- }, astronomer at the Polish Academy of Sciences, well known for his studies of the motions of comets, including consideration of nongravitational effects. He has recently succeeded in dynamically linking the observations made of (1862) Apollo and (2101) Adonis in the 1930s and the 1970s. Currently involved in an extensive program of new orbit determinations of one-apparition comets, he was (1979-1982) President of IAU Commission 20. (M 5523)

(2043) Ortutay

1936 TH. Discovered 1936 November 12 by G. Kulin at Budapest.

Named in memory of Gyula Ortutay, a professor of ethnography and a Hungarian cultural leader who patronized the popularization of astronomy in his country. (M 5183)

(2044) Wirt

1950 VE. Discovered 1950 November 8 by C. A. Wirtanen at Mount Hamilton.

Named in honor of Carl A. Wirtanen {1910-1990}, known for his discovery of several comets and minor planets and for his contribution to the Shane-Wirtanen survey of galaxies. He contributed to astrometric and other programs of the Lick Observatory for 33 years, including the proper-motion program with respect to galaxies. (M 5688)
Name proposed by A. R. Klemola.

Obituary published in Bull. Am. Astron. Soc., Vol. 23, No. 4, p. 1495-1496 (1991).

(2045) Peking

1964 TB_1. Discovered 1964 October 8 at the Purple Mountain Observatory at Nanking.

Named for the largest city in northern China and the capital of the People's Republic of China today. Peking has served as the Chinese capital for most of the time since the early 13th century. (M 4786)

(2046) Leningrad

1968 UD_1. Discovered 1968 October 22 by T. M. Smirnova at Nauchnyj.

Named for one of the largest industrial, cultural and scientific centers of the U.S.S.R. (M 5282)
The city was renamed into St. Petersburg in 1991.

(2047) Smetana

1971 UA_1. Discovered 1971 October 26 by L. Kohoutek at Bergedorf.

Named for the great Czech national composer Bedřich Smetana (1824-1884) who wrote songs, piano pieces, chamber music, operas and instrumental works. Best known are the opera *The Bartered Bride*, the cycle of six symphonic poems *My country* and the string quartet *From my life*. (M 4786)

(2048) Dwornik

1973 QA. Discovered 1973 August 27 by E. F. Helin at Palomar.

Named by the discoverer for Stephen E. Dwornik, a geologist, noted for his role in the exploration of the Moon and planets by unmanned spacecraft. As chief of the planetology programs of NASA, he was instrumental in the application of systematic geologic mapping to Mars and to Mercury. (M 4359)

(2049) Grietje

1973 SH. Discovered 1973 September 29 by T. Gehrels at Palomar.

Named in honor of Mrs. G. A. M. Haring-Gehrels, an exceptional person. (M 5282)

(2050) Francis

1974 KA. Discovered 1974 May 28 by E. F. Helin at Palomar.

Named by the discoverer in honor of her parents, Fred and Kay Francis. (M 4359)

(2051) Chang

1976 UC. Discovered 1976 October 23 at the Harvard College Observatory at Harvard.

Named in honor of Yu-che Chang {1902-1986}, one of the leading astronomers in the Peoples Republic of China, Director of the Purple Mountain Observatory, Nanking, since 1950, and long an active observer of minor planets and comets and computer of their orbits. He has also determined rotation periods of minor planets and has worked on spectroscopic binaries. From 1927 to 1929 he was on the staff of the Yerkes Observatory. (M 4420)

(2052) Tamriko

1976 UN. Discovered 1976 October 24 by R. M. West at La Silla.
Named in honor of Tamara West, wife of the discoverer. (M 4786)

(2053) Nuki

1976 UO. Discovered 1976 October 24 by R. M. West at La Silla.
Named in honor of Nodari West, son of the discoverer. (M 4786)

(2054) Gawain

4097 P-L. Discovered 1960 September 24 by C. J. van Houten and I. van Houten-Groeneveld at Palomar.
Named for one of the knights of the round table. (M 6421)

(2055) Dvořák

1974 DB. Discovered 1974 February 19 by L. Kohoutek at Bergedorf.

Named for Antonin Dvořák (1841-1904), the great Czech composer who wrote in virtually all the musical genres and received worldwide recognition. Symphony No. 8 *From the New World*, *Slavonic Dances* and the opera *Rusalka* are his best known works. (M 4786)

(2056) Nancy

A909 TB. Discovered 1909 October 15 by J. Helffrich at Heidelberg.

Named in honor of Nancy Lou Zissell Marsden, wife of Brian G. Marsden {see planet (1877)}. The latter established the identifications for this minor planet. (M 4359)

(2057) Rosemary

1934 RQ. Discovered 1934 September 7 by K. Reinmuth at Heidelberg.

Named in honor of Rosemary Birky Hoffmann Scholl, {first} wife of Hans Scholl {see planet (2959)}. The latter was an astronomer at the Astronomisches Rechen-Institut in Heidelberg {who now works at the Observatoire de Nice}. (M 4359)

(2058) Róka

1938 BH. Discovered 1938 January 22 by G. Kulin at Budapest.

Named in memory of Gedeon Róka, a well-known popularizer of astronomy in Hungary for three decades. (M 5183)

(2059) Baboquivari

1963 UA. Discovered 1963 October 16 at the Goethe Link Observatory at Brooklyn, Indiana.

This Amor-type object, recovered with the Steward Observatory's 2.3-m reflector on Kitt Peak in 1976, is named in honor of the Papago Indian Tribe, which made available the land on which the AURA observatories are based. Baboquivari, located a few kilometers south of Kitt Peak, is the highest and most sacred peak in Papago country. It figures prominently in Papago mythology as the home of the culture hero I'itoi. (M 5038)
Name proposed by F. K. Edmondson.

(2060) Chiron

1977 UB. Discovered 1977 October 18 by C. T. Kowal at Palomar.

The name is that of the wisest and most just of the Centaurs, son of Kronos and the ocean nymph Philyra. Beloved by Apollo and Diana {see planets (1862) and (78), respectively}, he learned from them botany, music, astronomy, divination and medicine and served in turn as instructor of the Greek heroes. (M 4359)

(2061) Anza

1960 UA. Discovered 1960 October 22 by H. L. Giclas at Flagstaff.

Named for Captain Juan Bautista de Anza, born at the Presidio of Fronteras, 250 km southeast of Tucson, Arizona, in 1736. He became the commander at the Presidio of Tubac and in 1774-1776 explored the first overland route from southern Arizona to Monterey, California. (M 4420)

(2062) Aten

1976 AA. Discovered 1976 January 7 by E. F. Helin at Palomar.

Named for the Egyptian sun god. This object is distinguished among the Apollo asteroids as the first discovered to have a semimajor axis less than 1 AU and a period less than one year. (M 4420)

(2063) Bacchus

1977 HB. Discovered 1977 April 24 by C. T. Kowal at Palomar.

This Apollo-type object is named for the god of wine and merriment, son of Jupiter and Semele {see planet (86)}. Bacchus was honored in various festivals, such as the Bacchanalia, characterized by drink and debauchery. (M 4421)

(2064) Thomsen

1942 RQ. Discovered 1942 September 8 by L. Oterma at Turku.

Named in memory of Ivan Leslie Thomsen (1910-1969), director of the Carter Observatory, Wellington, from 1945 until his appointment as astronomer-in-charge of the Mount John Observatory only two months before his death. He was an enthusiastic coordinator of astronomy in New Zealand, and his efforts eventually led to the minor-planet observing program with the Carter Observatory 41-cm reflector. It was the 1977 rediscovery at the Carter Observatory that allowed this minor planet to be numbered. (M 4421)

(2065) Spicer

1959 RN. Discovered 1959 September 9 at the Goethe Link Observatory at Brooklyn, Indiana.

Named in memory of Edward H. Spicer (1906-1983), professor of anthropology at the University of Arizona, and a former president of the American Anthropological Association. His assistance was a major factor in the success of the negotiations with the Schuk Toak District Council and the Papago Tribal Council in 1955 to obtain permission for the site testing of Kitt Peak as a possible location for an astronomical observatory. (M 7944)
Name proposed by F. K. Edmondson.

(2066) Palala

1934 LB. Discovered 1934 June 4 by C. Jackson at Johannesburg.

Named for a river, a tributary of the Limpopo {see planet (1490)}, in the northwestern Transvaal, South Africa. (M 5184)

(2067) Aksnes

1936 DD. Discovered 1936 February 23 by Y. Väisälä at Turku.

Named in honor of Kaare Aksnes {1938- }, celestial mechanician at the Smithsonian Astrophysical Observatory from 1971 to 1978, well-known for his work on both artificial and natural satellites, especially for his detailed analyses of observations of the mutual phenomena of the Galilean satellites. He is currently on the staff of the Norwegian Defense Research Establishment. (M 4482)

(2068) Dangreen

1948 AD. Discovered 1948 January 8 by M. Laugier at Nice.

Named in honor of Daniel W. E.Green {1958- }, student aide at the Smithsonian Astrophysical Observatory during 1978 June-August, in appreciation of his invaluable assistance during the transition of the Minor Planet Center from Cincinnati to Cambridge. (M 4482)
Name proposed by B. G. Marsden and C. M. Bardwell.

(2069) Hubble

1955 FT. Discovered 1955 March 29 at the Goethe Link Observatory at Brooklyn, Indiana.

Named in memory of Edwin P. Hubble (1889-1953), who provided the first comprehensive exploration of the universe beyond our own galaxy. He established a self-consistent distance scale as far as the 2.5-m Mount Wilson reflector could reach, and his classification scheme for galaxies is still the standard. He discovered the unique minor planet (1373) Cincinnati. His greatest achievement, however, known as Hubble's law of redshifts, can be interpreted as observational basis for the expanding universe. (M 8403)
Name proposed by F. K. Edmondson. Citation written by N. U. Mayall.
Hubble is also honored by a lunar crater.

(2070) Humason

1964 TQ. Discovered 1964 October 14 at the Goethe Link Observatory at Brooklyn, Indiana.

Named in memory of Milton L. Humason (1891-1972), who provided all the larger redshifts in the Hubble-Humason epoch-making program. Unrivaled in obtaining spectrograms of very faint galaxies, he correctly identified the few available features. Originally a mule skinner who brought up material for the construction of the Mount Wilson Observatory, he rose through the ranks as a janitor, night assistant and telescope user for colleagues. Appointed a regular staff member with only grammar-school education, he was eventually awarded an honorary degree by the University of Lund. (M 8403)
Name proposed by F. K. Edmondson. Citation written by N. U. Mayall.

Obituaries published in Mercury, Vol. 1, No. 5, p. 12 (1972); Phys. Today, Vol. 25, No. 10, p. 59 (1972); Sky Telesc., Vol. 44, p. 87 (1972); Orion, 31. Jahrg., p. 15 (1973); Q.J.R. Astron. Soc., Vol. 14, p. 235-236 (1973).

(2071) Nadezhda

1971 QS. Discovered 1971 August 18 by T. M. Smirnova at Nauchnyj.

Named in memory of Nadezhda Konstantinova Krupskaya (1869-1939), one of the creators of the Soviet public education system, wife of V. I. Lenin {see planet (852)}. (M 5282)

(2072) Kosmodemyanskaya

1973 QE_2. Discovered 1973 August 31 by T. M. Smirnova at Nauchnyj.

Named in memory of Lubov' Timofeevna Kosmodemyanskaya (1900-1978), social worker, mother of Soviet heroes Zoya and Aleksandr Kosmodemyanskij. (M 5282)

See also the citations for minor planets (1793) and (1977).

(2073) Janáček

1974 DK. Discovered 1974 February 19 by L. Kohoutek at Bergedorf.

Named for the Moravian musician Leos Janáček (1854-1928), the leading Czech composer of the early twentieth century. His chief works are *Sinfonietta* and the opera *Jenufa*. (M 4786)

(2074) Shoemaker

1974 UA. Discovered 1974 October 17 by E. F. Helin at Palomar.

Named by the discoverer to honor her colleague and friend Eugene M. Shoemaker {1928-1997}. An outstanding scientist, he has played a principal role in the Lunar Ranger, Surveyor and Apollo missions. His definitive work on meteor craters is a classic. His more recent areas of interest and contributions range from paleomagnetism to a planet-crossing asteroid survey. (M 4548)

This name has also been proposed by B. G. Marsden.

Obituaries published in Publ. Astron. Soc. Pac., Vol. 110, p. 490-492 (1998); Meteorit. Planet. Sci., Vol. 32, No. 6, p. 985-986 (1997); Planet. Space Sci., Vol. 45, No. 8, p. 1049-1050 (1997); Ir. Astron. J., Vol. 25, No. 1, p. 3 (1998); J. R. Astron. Soc. Can., Vol. 91, No. 5, p. 238-239 (1997); Heavens, Vol. 78, No. 9, p. 26-28 (1997); Nature, Vol. 389, No. 6647, p. 132 (1997); Int. Comet Q., Vol. 19, No. 3, p. 153-154 (1997); Astron. Raumfahrt Unterr., Jahrg. 34, Heft 6, p. 41 (1997); WGN, Vol. 25, No. 4, p. 145-146 (1997); Sterne Weltraum, Jahrg. 36, Nr. 10, p. 898 (1997).

(2075) Martinez

1974 VA. Discovered 1974 November 9 at the Felix Aguilar Observatory at El Leoncito.

Named in memory of Hugo Arturo Martinez (1890-1976), astronomer at the La Plata Observatory for many years. (M 5013)

(2076) Levin

1974 WA. Discovered 1974 November 16 at the Harvard College Observatory at Harvard.

Named in honor of Boris Julevich Levin {1912-1989}, Moscow astronomer and geophysicist who has contributed much to our knowledge of the evolution of the solar system. He has made many important specific studies in the areas of meteors and meteorites, and his desorption model played a crucial role in understanding the physical processes in comets. (M 4421)

Obituaries published in Pis'ma Astron., Tom 15, No. 8, p. 767 (1989); Astron. Vestn., Tom 23, No. 4, p. 345-346 (1989); Zemlya Vselennaya, No. 5, p. 45-46 (1989); Komety Meteory, No. 39, p. 58 (1989); Meteoritics, Vol. 24, No. 4, p. 353 (1989); Icarus, Vol. 84, No. 2, p. 277-279 (1990).

(2077) Kiangsu

1974 YA. Discovered 1974 December 18 at the Purple Mountain Observatory at Nanking.

Named for a maritime province in eastern China. In Kiangsu are located two major cities, Shanghai, the largest in China and a major seaport, and Nanking, the provincial capital. (M 4786)

(2078) Nanking

1975 AD. Discovered 1975 January 12 at the Purple Mountain Observatory at Nanking.

Named for the city on whose outskirts the Purple Mountain {see planet (3444)} Observatory is located. On the southern bank of the Yangtze River in central eastern China, Nanking served as China's capital during several imperial periods prior to the 15th century. It is the capital of Kiangsu {see planet (2077)} province of the People's Republic today. (M 4786)

(2079) Jacchia

1976 DB. Discovered 1976 February 23 at the Harvard College Observatory at Harvard.

Named in honor of Luigi Giuseppe Jacchia {1911-1996}, research associate at the Harvard College Observatory since 1939 and a physicist at the Smithsonian Astrophysical Observatory. In his work on meteors and subsequently on the motions of artificial satellites he has made pioneering investigations of the Earth's upper atmosphere. His interest in variable stars dates from his student days in Bologna, and he is well known as an extraordinary linguist. (M 4421)

Obituaries published in Bull. Am. Astron. Soc., Vol. 28, No. 4, p. 1453-1454 (1996); G. Astron., Vol. 23, N. 1, p. 54 (1997).

(2080) Jihlava

1976 DG. Discovered 1976 February 27 by P. Wild at Zimmerwald.

Named for a Moravian town, by the river of the same name. Founded in the 11th century, the community prospered from rich silver deposits. The municipal and mining laws of Jihlava were to become a model for analogous regulations all over the world. (M 4786)

This minor planet was named by I. Bauersima, a geodesist at Berne University, in honor of his native town.

(2081) Sázava

1976 DH. Discovered 1976 February 27 by P. Wild at Zimmerwald.

Named for a river in Bohemia, running from the Zdar Mountains into the Vltava {see also (2123)} south of Prague. (M 4787)

Name proposed by I. Bauersima.

(2082) Galahad

7588 P-L. Discovered 1960 October 17 by C. J. van Houten and I. van Houten-Groeneveld at Palomar.

Named for one of the knights of the round table. (M 6421)

(2083) Smither

1973 WB. Discovered 1973 November 29 by E. F. Helin at Palomar.

Named by the discoverer and E. M. Shoemaker for John C. Smith in appreciation and recognition of his dedicated assistance in the Palomar planet-crossing minor planet program. As an undergraduate student at Caltech he was known as Smither. (M 4548)

(2084) Okayama

1935 CK. Discovered 1935 February 7 by S. Arend at Uccle.

Named for the station of the Tokyo Astronomical Observatory at which the Japanese 1.9-m reflector is located. (M 8151)

Name proposed by the discoverer, following a suggestion by K. Tomita, who made an accidental discovery of this object as 1965 HA.

(2085) Henan
1965 YA. Discovered 1965 December 20 at the Purple Mountain Observatory at Nanking.

Named for the province situated in the lower stretch of the Yellow River, the cradle of civilization in ancient China. (M 5184)

(2086) Newell
1966 BC. Discovered 1966 January 20 at the Goethe Link Observatory at Brooklyn, Indiana.

Named in memory of Homer E. Newell (1915-1983), one of the most important organizers of the U.S. space program. He was a member of the advisory committee that helped to establish the National Aeronautics and Space Administration, and he became one of the original staff members when NASA came into being in 1958. Newell served in a number of leading positions and was Associate Administrator of NASA at the time of his retirement in 1973. (M 8151)
Name proposed by F. K. Edmondson.

(2087) Kochera
1975 YC. Discovered 1975 December 28 by P. Wild at Zimmerwald.

Named in memory of Theodor Kocher (1841-1917), surgeon of world fame, Nobel laureate in 1909, benefactor of Berne University. (M 4787)

(2088) Sahlia
1976 DJ. Discovered 1976 February 27 by P. Wild at Zimmerwald.

Named in memory of Hermann Sahli (1856-1933), professor of internal medicine at Berne University, philosophically minded teacher, inventor of many new methods and improved instruments of diagnosis. (M 4787)

(2089) Cetacea
1977 VF. Discovered 1977 November 9 by N. G. Thomas at Flagstaff.

Named for the order of mammals commonly called whales, many species of which are currently endangered. (M 4787)

(2090) Mizuho
1978 EA. Discovered 1978 March 12 by T. Urata at Yakiimo.
Named by the discoverer in honor of his daughter. (M 4482)

(2091) Sampo
1941 HO. Discovered 1941 April 26 by Y. Väisälä at Turku.

Named for the wonder-object in Kalevala {see planet (1454)}, the Finnish national epic. Sampo was to produce every kind of fortune. When Kalevala and Pohjola {the North, see planet (3606)}) were fighting for its possession, out at sea, it broke into pieces, and only tiny fragments could be found. (M 5450)

(2092) Sumiana
1969 UP. Discovered 1969 October 16 by L. I. Chernykh at Nauchnyj.
Named for the town of Sumy, in the Ukrainian S.S.R. (M 5282)

(2093) Genichesk
1971 HX. Discovered 1971 April 28 by T. M. Smirnova at Nauchnyj.
Named for the discoverer's birthplace, a town in the Ukrainian S.S.R. (M 5282)

(2094) Magnitka
1971 TC_2. Discovered 1971 October 12 at the Crimean Astrophysical Observatory at Nauchnyj.

Named for the town of Magnitogorsk, one of the largest centers of metallurgy in the U.S.S.R. (M 5282)

(2095) Parsifal

6036 P-L. Discovered 1960 September 24 by C. J. van Houten and I. van Houten-Groeneveld at Palomar.
Named for one of the knights of the round table. (M 6421)

(2096) Väinö

1939 UC. Discovered 1939 October 18 by Y. Väisälä at Turku.

Named for Väinämöinen, an old and wise magician, a central figure in Finnish folklore and Kalevala {see planet (1454)}. This minor planet also honors Väinö Väisälä, brother of the discoverer. (M 5451)

(2097) Galle

1953 PV. Discovered 1953 August 11 by K. Reinmuth at Heidelberg.

Named in memory of Johann Gottfried Galle (1812-1910), famous German astronomer who in 1846 discovered the planet Neptune, the existence of which had been theoretically demonstrated by J. C. Adams and U. J. Leverrier {see planets (1996) and (1997), respectively}. Galle studied mathematics and natural sciences in Berlin and was appointed staff astronomer at the Berlin Observatory by its director J. F. Encke. He then went to the University of Breslau and in 1851 became professor of astronomy and director of the Breslau Observatory. The discoverer of three comets within only three months in 1839-40, Galle was also known for his catalogues of cometary orbits that culminated with the definitive edition of 1894. In 1872, more than three decades before its realization, Galle proposed the use of minor planet observations for the determination of the solar parallax. He was also engaged in meteorological problems. (M 18304)

Name suggested and citation prepared by L. D. Schmadel, endorsed by the Heidelberg-Königstuhl Observatory.
Galle is also honored by craters on Mars and the Moon.

(2098) Zyskin

1972 QE. Discovered 1972 August 18 by L. V. Zhuravleva at Nauchnyj.

Named in honor of Lev Yur'evich Zyskin, professor at the Crimean Medical Institute, head of the Pulmonary Surgery Center. (M 5283)

(2099) Öpik

1977 VB. Discovered 1977 November 8 by E. F. Helin and E. M. Shoemaker at Palomar.

Named in honor of Ernst J. Öpik {1893-1985} on the occasion of his 85th birthday. Over a career spanning more than 60 years he has influenced almost every branch of astronomy. In particular he has developed the discipline of statistical celestial mechanics and of methods for estimating the lifetimes of small bodies on planet-crossing orbits. In 1951 he computed the probability of collision with Mars of the Mars-crossing minor planets recognized at that time, and 14 years before their discovery by the Mariner 4 spacecraft he inferred that it would be worthwhile to search for impact craters on Mars. It is therefore fitting that (2099) Öpik should be a Mars-crosser. (M 4548)

Obituaries published in Icarus, Vol. 66, No. 2, 9. 193-194 (1986); Sky Telesc., Vol. 71, No. 2, p. 149 (1986); Astrophys. Space Sci., Vol. 124, No. 1, p. 1-3 (1986); Q.J.R. Astron. Soc., Vol. 27, No. 3, p. 508-512 (1986).

(2100) Ra-Shalom

1978 RA. Discovered 1978 September 10 by E. F. Helin at Palomar.

Named by the discoverer for the Egyptian sun-god Ra, who symbolizes enlightenment and life, and for Shalom, the traditional Hebrew greeting meaning peace. This name is chosen to commemorate the Camp David mid-east peace conference, at which time this unusual body was found. May it stand as a symbol for the universal hope for peace. (M 4548)

(2101) Adonis

1936 CA. Discovered 1936 February 12 by E. Delporte at Uccle.

This Apollo-type minor planet was named previously for the lover of Aphrodite {see planet (1388)}. Adonis was ordered by Zeus {see planet (5731)} to spend part of each year with Aphrodite and part with Persephone {see planet (399)}. He is regarded as a god of vegetation and of swine and was killed by a boar sent by Artemis {see planet (105)}. (M 4548)

(2102) Tantalus

1975 YA. Discovered 1975 December 27 by C. T. Kowal at Palomar.

Named for the king of Sipylus, who was punished by being immersed to his neck in water and prevented from drinking the water or gathering fruit hung above his head. (M 4645)

(2103) Laverna

1960 FL. Discovered 1960 March 20 at the La Plata Observatory at La Plata.

Laverna was the protecting divinity of thieves and imposters. Her name was probably connected with the Latin word levator, meaning thief. (M 21128)
Named by the Minor Planet Names Committee following a suggestion by F. Pilcher.

(2104) Toronto

1963 PD. Discovered 1963 August 15 by K. W. Kamper at Tautenburg.

This minor planet, the first to be found at a Canadian Observatory, is named in honor of the University of Toronto, during whose sesquicentennial celebration it was discovered, to acknowledge the central role played by the university in the development of astronomy in Canada. (M 4645)

This planet was detected at the David Dunlap Observatory on plates taken by S. van den Bergh at the Tautenburg Observatory, East Germany.

(2105) Gudy

1976 DA. Discovered 1976 February 29 by H.-E. Schuster at La Silla.

Named by the discoverer for Mrs. Gudrun Werner of Hamburg in sentimental reminiscence of college days. This Phocaea-type {see planet (25)} minor planet was found on the same blue survey plates as the large-perihelion comet 1976c = 1975 II. (M 4645)

(2106) Hugo

1936 UF. Discovered 1936 October 21 by M. Laugier at Nice.
Named for the celebrated French writer Victor Hugo (1802-1885). (M 6833)
Name proposed by J. Meeus.

(2107) Ilmari

1941 VA. Discovered 1941 November 12 by L. Oterma at Turku.

Named for the master-smith Ilmarinen, who forged Sampo {see planet (2091)}, the wonder-object in Kalevala {see planet (1454)}. Ilmari is a common boy's name in Finland. (M 5451)

(2108) Otto Schmidt

1948 TR_1. Discovered 1948 October 4 by P. F. Shajn at Simeïs.

Named in memory of Otto Yul'evich Schmidt (1891-1956), a prominent Soviet scientist famous for his investigations in mathematics, astronomy and geophysics. He was also an outstanding Arctic researcher, academician and statesman. His cosmogonic theories contributed extensively to the evolution of concepts on the formation of the Earth and other planets. (M 5359)

(2109) Dhotel

1950 TH_2. Discovered 1950 October 13 by S. Arend at Uccle.

Named in honor of André d'Hotel, recipient of the 1974 "Grand Prix de Littérature de l'Académie Française" and a friend of the Arend family. (M 8323)
The originally adopted name d'Hotel (MPC 8151) was later changed.

(2110) Moore-Sitterly

1962 RD. Discovered 1962 September 7 at the Goethe Link Observatory at Brooklyn, Indiana.

Named to honor Charlotte E. Moore (Mrs. Bancroft W. Sitterly) {1898-1990}, whose long-time collaboration with Henry Norris Russell {see planet (1762)} included analyses of many atomic spectra and the important work on dynamical parallaxes that led to their landmark treatise, *The masses of the stars*, in 1940. She joined the staff of the U.S. National Bureau of Standards in 1945 and continued her work on atomic spectra of astrophysical interest. She served as president of IAU Commission 14 during 1961-1967. Her formal retirement in 1968 did not stop her activities, and she continues to be an international resource in this field. (M 8797)
Name proposed by F. K. Edmondson.

Obituaries published in Q.J.R. Astron. Soc., Vol. 32, No. 2, p. 209-210 (1991); Bull. Am. Astron. Soc., Vol. 23, No. 4, p. 1492-1494 (1991); Phys. Today, Vol. 44, No. 4, p. 128, 130 (1991).

(2111) Tselina

1969 LG. Discovered 1969 June 13 by T. M. Smirnova at Nauchnyj.

Named on the 25th anniversary of the development of tselina (virgin soil) and long fallow lands in the U.S.S.R. (M 5283)

(2112) Ulyanov

1972 NP. Discovered 1972 July 13 by T. M. Smirnova at Nauchnyj.

Named in memory of Aleksandr Ulyanov (1866-1887), eldest brother of V. I. Lenin {see planet (852). (M 5283)

(2113) Ehrdni

1972 RJ_2. Discovered 1972 September 11 by N. S. Chernykh at Nauchnyj.

Named in memory of Ehrdni Tel'dzhievich Delikov (1922-1942), a hero from the Kalmyk S.S.R. who was killed in the Great Patriotic War (1941-1945). (M 5283)

(2114) Wallenquist

1976 HA. Discovered 1976 April 19 by C.-I. Lagerkvist at Mount Stromlo.

Named in honor of Åke A. F. Wallenquist {1904-1994}, retired director of the Kvistaberg {see planet (3331)} Station but still active at the Uppsala Observatory, where he was a codiscoverer of (1980) Tezcatlipoca. (M 4645)
Obituary published in Astron. Tidsskr., Arg. 27, Nr. 3, p. 123-124 (1994).

(2115) Irakli

1976 UD. Discovered 1976 October 24 by R. M. West at La Silla.
Named in honor of Irakli West, son of the discoverer. (M 4787)

(2116) Mtskheta

1976 UM. Discovered 1976 October 24 by R. M. West at La Silla.

Named after the ancient capital of the Kingdom of Georgia, at the confluence of the rivers Mtkvari and Aragvi, rich in architectural treasures and the scene of many decisive events in the history of Georgia. The name is proposed by the discoverer in admiration of a beautiful country and its valiant people. (M 4787)

(2117) Danmark

1978 AC. Discovered 1978 January 9 by R. M. West at La Silla.
Named in honor of the native country of the discoverer. (M 4787)

(2118) Flagstaff

1978 PB. Discovered 1978 August 5 by H. L. Giclas at Anderson Mesa.

Named of the city of Flagstaff, Arizona, where Percival Lowell {see planet (1886)} founded the Lowell Observatory in 1894. In more recent times several other astronomical facilities have been set up in Flagstaff. (M 5184)

(2119) Schwall

1930 QG. Discovered 1930 August 30 by M. Wolf and M. Ferrero at Heidelberg.

Named in memory of August Schwall (1877-1947), mechanician at the Heidelberg-Königstuhl Observatory for almost half a century, night assistant to Wolf during 1914-1932, guiding exposures with the 0.72-m reflector. (M 5524)
Name proposed by A. Bohrmann.

(2120) Tyumenia

1967 RM. Discovered 1967 September 9 by T. M. Smirnova at Nauchnyj.

Named for the Tyumen' district of the R.S.F.S.R., the center of the west Siberian oil-gas basin that has become in recent years the main source of power in the U.S.S.R. (M 5283)

(2121) Sevastopol

1971 ME. Discovered 1971 June 27 by T. M. Smirnova at Nauchnyj.

Named for this Crimean city on the 200th anniversary of its foundation. (M 7616)

(2122) Pyatiletka

1971 XB. Discovered 1971 December 14 by T. M. Smirnova at Nauchnyj.

Named on the 50th anniversary of the adoption of the first five-year plan for the development of the national economy of the U.S.S.R. (M 5283)

(2123) Vltava

1973 SL_2. Discovered 1973 September 22 by N. S. Chernykh at Nauchnyj.
Named for the river on which the city of Prague is situated. (M 5283)

(2124) Nissen

1974 MK. Discovered 1974 June 20 at the Felix Aguilar Observatory at El Leoncito.

Named in memory of Juan Jose Nissen (1901-1978), first director of the Felix Aguilar Observatory, previously director of the Córdoba Observatory and a department head at the La Plata Observatory. (M 5013)

(2125) Karl-Ontjes

2005 P-L. Discovered 1960 September 24 by C. J. van Houten and I. van Houten-Groeneveld at Palomar.

Named in honor of Karl-Ontjes Groeneveld {1935- }, a well-known physicist at the Institut für Kernphysik, Frankfurt-am-Main, and a brother of one of the discoverers. (M 7471)

(2126) Gerasimovich

1970 QZ. Discovered 1970 August 30 by T. M. Smirnova at Nauchnyj.

Named in memory of Boris Petrovich Gerasimovich (1889-1937), professor at Kharkov University (1922-1931), then chief of the astrophysics section and from 1933 director of the Pulkovo Observatory. His scientific papers cover a large range of astrophysical problems, and he was a member of many scientific societies. (M 5359)
Gerasimovich is also honored by a lunar crater.

(2127) Tanya

1971 KB_1. Discovered 1971 May 29 by L. I. Chernykh at Nauchnyj.

Named in memory of Tanya Savicheva, a 12-year-old schoolgirl who perished during the 1941-1944 blockade of Leningrad. In her diary she made a record as her parents and other relatives died, one by one. Her last note was "... all the Savichev's have died, Tanya is left alone". (M 5283)

(2128) Wetherill

1973 SB. Discovered 1973 September 26 by E. F. Helin at Palomar.

Named in honor of George W. Wetherill {1925- }, director of the Department of Terrestrial Magnetism, Carnegie Institution of Washington. Noted for his work on isotopic age determination, meteorites, and the origin of the terrestrial planets, he has been a leading student of the dynamical evolution of orbits of meteorites and planet-crossing asteroids. (2128) is a Mars crosser that lies close to a secular resonance shown by Wetherill to be of importance in the delivery of meteorites and asteroids into Earth-crossing orbits. (M 4787)

(2129) Cosicosi

1973 SJ. Discovered 1973 September 27 by P. Wild at Zimmerwald.

The Italian characterization of indifference. (M 5359)

(2130) Evdokiya

1974 QH_1. Discovered 1974 August 22 by L. V. Zhuravleva at Nauchnyj.

Named by the discoverer in honor of her mother, Evdokiya Efimovna Shchelokova. (M 5283)

(2131) Mayall

1975 RA. Discovered 1975 September 3 by A. R. Klemola at Mount Hamilton.

Named in honor of Nicholas U. Mayall {1906-1993}, director of the Kitt Peak National Observatory during 1960-1971. It was under his direction that the planning and construction of a major portion of the astronomical facilities at the Kitt Peak and Cerro Tololo observatories were accomplished. He was on the Lick Observatory staff from 1945 to 1960. His major research contribution spans the domains of nebular spectroscopy, radial velocities of globular clusters and planetary nebulae, and the redshifts and internal motions of galaxies. (M 5848) Obituary published in Bull. Am. Astron. Soc., Vol. 25, No. 4, p. 1499 (1993).

(2132) Zhukov

1975 TW_3. Discovered 1975 October 3 by L. I. Chernykh at Nauchnyj.

Named in memory of Georgij Konstantinovich Zhukov (1896-1974), Marshal of the Soviet Union and leader during the Great Patriotic War. (M 5283)

(2133) Franceswright

1976 WB. Discovered 1976 November 20 at the Harvard College Observatory at Harvard.

Named in honor of Frances Woodworth Wright {1898-1990}, long time staff member of the Harvard College Observatory, known for her research on photographic meteors, extraterrestrial dust and variable stars, and revered for teaching generations of navigators to fare forth and return safely to port. (M 5039)

(2134) Dennispalm

1976 YB. Discovered 1976 December 24 by C. T. Kowal at Palomar.

Named in memory of C. Dennis Palm (1945-1974), a night assistant at Palomar Observatory, an active amateur astronomer, and a dear friend to those who were privileged to work with him. (M 4788)

(2135) Aristaeus

1977 HA. Discovered 1977 April 17 by E. F. Helin and S. J. Bus at Palomar.

This Apollo object is named for the son of Apollo and the nymph Cyrene {see planets (1862) and (133), respectively}. (M 5014)

(2136) Jugta

1933 OC. Discovered 1933 July 24 by K. Reinmuth at Heidelberg.

Named in honor of J. U. Gunter {1911-1994} and his publication *Tonight's Asteroids*. This publication has been extraordinary successful at interesting the general public in minor planets and has inspired countless amateur and several professional astronomers to study these objects. (M 5284)

Name proposed by E. Fogelin; endorsed by C. M. Bardwell, D. W. E. Green and B. G. Marsden.

Obituary published in Minor Planet Bull., Vol. 22, No. 1, p. 1 (1995).

(2137) Priscilla

1936 QZ. Discovered 1936 August 24 by K. Reinmuth at Heidelberg.

Named in memory of Priscilla Fairfield Bok (1896-1975). With her husband, Bart J. Bok, she participated in four decades of fruitful galactic research, first at the Harvard College Observatory, then at the Mt. Stromlo Observatory, Steward Observatory, with briefer tours elsewhere. She also enjoyed teaching astronomy courses successively at Smith College, Wellesley College and Connecticut College for Women. Together with her husband she prepared several editions of *The Milky Way*, a popular title in the Harvard Books on Astronomy. (M 7616)

Name proposed by F. Pilcher.

Priscilla F. and Bart J. Bok are also honored by minor planet (1983).

(2138) Swissair

1968 HB. Discovered 1968 April 17 by P. Wild at Zimmerwald.

Named for the Swiss national airline, for which HB is the international designation. (M 5360)

(2139) Makharadze

1970 MC. Discovered 1970 June 30 by T. M. Smirnova at Nauchnyj.

Named in memory of the friendship between the Georgian and Ukrainian peoples. Makharadze is the Georgian twin city of Genichesk {see planet (2093)} in the Ukraine. (M 6647)

(2140) Kemerovo

1970 PE. Discovered 1970 August 3 by T. M. Smirnova at Nauchnyj.

Named for the regional center of the Kemerovo district of Russia, an important industrial center of Siberia. (M 6647)

(2141) Simferopol

1970 QC_1. Discovered 1970 August 30 by T. M. Smirnova at Nauchnyj.

Named for the central city of the Crimean district of the Ukrainian SSR on the occasion of its two-hundredth anniversary. (M 8911)

(2142) Landau

1972 GA. Discovered 1972 April 3 by L. I. Chernykh at Nauchnyj.

Named in memory of Lev Davydovich Landau (1908-1968), founder of the study of modern theoretical physics in the U.S.S.R. Among his accomplishments was fundamental research in quantum electrodynamics, superconductivity and neutrino theory. (M 5284)

Landau is also honored by a lunar crater.

(2143) Jimarnold

1973 SA. Discovered 1973 September 26 by E. F. Helin at Palomar.

Named in honor of James R. Arnold, professor of chemistry at the University

of California at San Diego, distinguished for his applications of nuclear chemistry to the interpretation of meteorites, the history of cosmic radiation, the lunar regolith and compositional mapping of the Moon. In order to elucidate the origin of meteorites he developed a Monte Carlo method for studying the effects of close planetary encounters on the orbital evolution of small planet-crossing bodies. The method has become the primary technique for investigating the dynamical history of planet-crossing asteroids, of which (2143) is an example. (M 4788)
Name proposed by E. F. Helin and E. M. Shoemaker.

(2144) Marietta

1975 BC_1. Discovered 1975 January 18 by L. I. Chernykh at Nauchnyj.

Named in honor of Marietta Sergeevna Shaginyan, Soviet writer, doctor of philology, a member of the Armenian Academy of Sciences. (M 5284)

(2145) Blaauw

1976 UF. Discovered 1976 October 24 by R. M. West at La Silla.

Named by the discoverer in honor of Adriaan Blaauw {1914- }, director of ESO (1970-74), president of the IAU (1976-79) and professor at the Leiden Observatory (1975-1981). He has made many important contributions to stellar kinematics, the structure of the Galaxy and the study of stellar associations. He has been closely associated with ESO since its beginning and has been very active in the furthering of collaboration in European astronomy. He is one of the founders of the European journal *Astronomy and Astrophysics* and has been Chairman of the Board of Directors since 1969. (M 4788)

(2146) Stentor

1976 UQ. Discovered 1976 October 24 by R. M. West at La Silla.

This Trojan planet is named for the Greek warrior with the famous voice as loud as fifty men together. (M 4788)

(2147) Kharadze

1976 US. Discovered 1976 October 25 by R. M. West at La Silla.

Named by the discoverer in honor of Academician Evgenij Kirillovich Kharadze {1907- }, founder and director since 1932 of the Abastumani {see planet (1390)} Astrophysical Observatory, Vice-President (1972-1978) and now President of the Academy of Sciences of the Georgian S.S.R., and Vice-President of the IAU (1976-1982). He has made numerous contributions to the study of interstellar matter and the structure of the Galaxy. He is the author of several astronomical textbooks in the Georgian language and has developed a modern Georgian astronomical terminology. (M 4788)

(2148) Epeios

1976 UW. Discovered 1976 October 24 by R. M. West at La Silla.

This Trojan planet is named for the leader of the troops from Phocis, who also constructed the Wooden Horse that became the fate of Troy. (M 4788)

(2149) Schwambraniya

1977 FX. Discovered 1977 March 22 by N. S. Chernykh at Nauchnyj.

Named for the wonderland created by the characters in L. A. Kassil's children's novel *Conduite and Schwambraniya*. (M 5284)

(2150) Nyctimene

1977 TA. Discovered 1977 October 13 by W. Sebok at Palomar.

Nyctimene was a daughter of Epopeus, King of Lesbos. Pursued and raped by her father, she concealed herself in the shade of forests. Athena {see planet (881)} rescued her there by metamorphosing her into an owl. (M 21128)

Named by the Minor Planet Names Committee following a suggestion by F. Pilcher.

(2151) Hadwiger

1977 VX. Discovered 1977 November 3 by P. Wild at Zimmerwald.

Named in memory of Hugo Hadwiger (1908-1981), for more than 40 years professor of mathematics at the University of Berne, very popular for the human and historical touch of his lectures, and admired by students and colleagues alike for his refined art of presentation. Known as a "master of beautiful formulae", his course on special functions deservedly drew crowds of students of all ages. (M 8151)

(2152) Hannibal

1978 WK. Discovered 1978 November 19 by P. Wild at Zimmerwald.

Named for the great Carthaginian general who, on his way to Rome in 217 B.C., crossed the Alps with his army, including some elephants to frighten the enemy. For the Swiss, WK is the abbreviation for "Wiederholungskurs", the yearly repetition of military exercises, and the 1978 WK for a colleague of the discoverer consisted of a large Alpine maneuver called Hannibal. (M 8151)

(2153) Akiyama

1978 XD. Discovered 1978 December 1 at the Harvard College Observatory at Harvard.

Named in memory of Kaoru Akiyama (1901-1970), professor at Hosei University, Tokyo, during 1930-1940 and 1949-1970, well known for his work on minor planets. Initially in collaboration with K. Hirayama {see minor planet (1999)}, he made the first detailed study of the orbit of the 3/2 commensurability object (153) Hilda. (M 5014)

Name proposed by K. Tomita.

(2154) Underhill

2015 P-L. Discovered 1960 September 24 by C. J. van Houten and I. van Houten-Groeneveld at Palomar.

Named in honor of Anne B. Underhill, a well-known astrophysicist and an expert in the field of the structure of early-type stars. (M 7471)

(2155) Wodan

6542 P-L. Discovered 1960 September 24 by C. J. van Houten and I. van Houten-Groeneveld at Palomar.

Named for the chief god of Teutonic mythology, also honored in some languages by the fourth day of the week. (M 7782)

(2156) Kate

A917 SH. Discovered 1917 September 23 by S. I. Belyavskij at Simeïs.

Named by L. K. Kristensen, who found several of the identifications involving this planet, in honor of his wife. (M 5284)

(2157) Ashbrook

A924 EF. Discovered 1924 March 7 by K. Reinmuth at Heidelberg.

Named in honor of Joseph Ashbrook {1918-1980}, Editor of "Sky and Telescope", co-discoverer of periodic comet Ashbrook-Jackson. His extensive professional background in astronomy, together with his encyclopedic knowledge of intriguing and obscure astronomical facts and fancies, have secured for "Sky and Telescope" a unique position as a vital source of information, not only for professional astronomers and historians of science, but also for amateur astronomers of all levels of expertise. (M 5284)

Name proposed by N. Sperling; endorsed by C. M. Bardwell and B. G. Marsden.

Obituaries published in J. Hist. Astron., Vol. 11, p. 201 (1980); Sky Telesc., Vol. 60, p. 281-284 (1980); J. Am. Assoc. Variable Star Obs., Vol. 9, p. 43-44 (1980); Strolling Astron., Vol. 28, p. 203 (1980).

(2158) Tietjen

1933 OS. Discovered 1933 July 24 by K. Reinmuth at Heidelberg.

Named in memory of Friedrich Tietjen (1832-1895), professor of astronomy at Berlin University and director of the Astronomisches Rechen-Institut since 1874. For many years, he served as an editor of the respected *Berliner Astronomisches Jahrbuch* and *Nautisches Jahrbuch*. After a 13-year period of observations of minor planets and comets, Tietjen became the most prolific orbit computer of the ARI. (M 18447)
Name proposed and citation prepared by L. D. Schmadel.

(2159) Kukkamäki

1941 UX. Discovered 1941 October 16 by L. Oterma at Turku.

Named in honor of T. J. Kukkamäki {1909- } on the occasion of his 70th birthday, 1979 Oct. 11. A distinguished disciple of Y. Väisälä {see planet (1573)}, this leading geodesist was for many years director of the Finnish Geodetic Institute. He is an active member of several international scientific organizations and has served as president of the International Association of Geodesy since 1975. (M 5014)

(2160) Spitzer

1956 RL. Discovered 1956 September 7 at the Goethe Link Observatory at Brooklyn, Indiana.

Named in honor of Lyman Spitzer Jr. {1914-1997}, director of the Princeton University Observatory from 1947 to 1979 and one of the most influential pioneers in recognizing and promoting the importance of making astronomical observations from space vehicles. In addition to performing outstanding astronomical research, he was involved in the Princeton project for the controlled release of thermonuclear energy as director of Project Matterhorn (1953-1961) and chairman of the executive committee of the Plasma Physics Laboratory (1961-1966). A member of the AURA board of directors during 1959-1969, he became chairman of the new AURA Space Telescope Institute Council in 1981. (M 8798)
Name proposed by F. K. Edmondson.

Obituaries published in Nature, Vol. 387, No. 6630, p. 244 (1997); Astron. Geophys., Vol. 38, Issue 6, p. 36 (1997); Vasiona, Vol. 46, No. 3, p. 79-80 (1998); Phys. Today, Vol. 50, No. 10, p. 123-124 (1997); Bull. Astron. Soc. India, Vol. 25, No. 2, p. 281-283 (1997); Pub. Astron. Soc. Pac., Vol. 110, No. 745, p. 215-222 (1998); Space Telesc. Sci. Inst., Newsl., Vol. 14, No. 2, p. 2 (1997).

(2161) Grissom

1963 UD. Discovered 1963 October 17 at the Goethe Link Observatory at Brooklyn, Indiana.

Named in memory of Virgil I. "Gus" Grissom (1926-1967), one of the seven Mercury astronauts. On 1961 July 21, as pilot of the Mercury-Redstone 4 suborbital mission, he became the second American in space. In 1965 he commanded a three-orbit mission, the first manned Gemini flight. As command pilot of Apollo 1, he lost his life in the tragic fire in that spacecraft. (M 5848)

Name proposed by D. Owings and F. K. Edmondson. Citation prepared by Muriel M. Thorne, NASA.
Grissom is also honored by a lunar crater.

(2162) Anhui

1966 BE. Discovered 1966 January 30 at the Purple Mountain Observatory at Nanking.

Named for a province in the northeastern part of East China that straddles the Yangtze and Huai Rivers. (M 5184)

(2163) Korczak

1971 SP_1. Discovered 1971 September 16 at the Crimean Astrophysical Observatory at Nauchnyj.

Named in memory of Janusz Korczak (1878-1942), Polish writer, teacher and doctor who perished with his 200 pupils at Treblinka. (M 5284)

(2164) Lyalya

1972 RM_2. Discovered 1972 September 11 by N. S. Chernykh at Nauchnyj.

Named in memory of Elena (Lyalya) Konstantinova Ubijvovk (1918-1942), a student in astronomy at Kharkov University who perished with other members of the Resistance during the Great Patriotic War. (M 5285)

(2165) Young

1956 RJ. Discovered 1956 September 7 at the Goethe Link Observatory at Brooklyn, Indiana.

Named in memory of Charles Augustus Young (1834-1908), known affectionately as "Twinkle" Young by the Princeton students. He accepted the professorship of astronomy at Princeton in 1877, the year that his most famous student, Henry Norris Russell {see planet (1762)}, was born. Earlier he was a professor at Dartmouth, as his father and grandfather had been. He discovered the green line (λ5304) in the solar corona in 1869, and the following year he was the first both to observe the "flash spectrum" and to photograph a prominence. In 1876 he made the first use of the grating spectroscope in astronomy for the determination of the Sun's rotation period. The last of his three successful textbooks, *Manual of astronomy*, was updated by Russell, Dugan {see planet (2772)} and Stewart in 1926. (M 8798)

Name proposed by F. K. Edmondson.

(2166) Handahl

1936 QB. Discovered 1936 August 13 by G. N. Neujmin at Simeïs. Independently discovered 1936 August 17 by K. Reinmuth at Heidelberg.

Named by D. W. E. Green, who found identifications for this planet, in honor of his mother, Violet Handahl Green. (M 5285)

(2167) Erin

1971 LA. Discovered 1971 June 1 at the Perth Observatory at Bickley.

Named in honor of the daughter of George Punko, member of the astrometric team at the Perth Observatory. (M 6530)

(2168) Swope

1955 RF_1. Discovered 1955 September 14 at the Goethe Link Observatory at Brooklyn, Indiana.

Named in memory of Henrietta Hill Swope (1902-1980), best known for her work in establishing distance ratios for nearby galaxies. One of her most significant contributions was the calibration of the distance of the Andromeda galaxy. She worked as an assistant at the Harvard College Observatory during 1928-1942 and was on the staff of the Mount Wilson and Palomar Observatories during 1952-1968. She received the Annie J. Cannon {see planet (1120)} prize of the American Astronomical Society in 1968 and an honorary Ph.D. from the University of Basel in 1975. The 1-m Swope telescope at the Las Campañas Observatory was named in her honor by the Carnegie Institution of Washington. (M 8798)

Name proposed by F. K. Edmondson.
Obituary published in Phys. Today, Vol. 34, No. 3, p. 88 (1981).

(2169) Taiwan
1964 VP_1. Discovered 1964 November 9 at the Purple Mountain Observatory at Nanking.
Named for an insular province of China. (M 5184)

(2170) Byelorussia
1971 SZ. Discovered 1971 September 16 at the Crimean Astrophysical Observatory at Nauchnyj.
Named for one of the Soviet Socialist Republics. (M 5285)
Since 1991 Byelorussia is an independent state.

(2171) Kiev
1973 QD_1. Discovered 1973 August 28 by T. M. Smirnova at Nauchnyj.

Named on the occasion of the 1500th anniversary of this city, the capital of the Ukraine and one of the largest industrial, cultural and scientific centers of the {former} U.S.S.R. (M 6647)

(2172) Plavsk
1973 QA_2. Discovered 1973 August 31 by T. M. Smirnova at Nauchnyj.

Named for the district center of the Tula region of the R.S.F.S.R. Plavsk is the twin town of the Ukrainian town Genichesk {see the citation for minor planet (2093)}, the birthplace of the discoverer. (M 7616)

(2173) Maresjev
1974 QG_1. Discovered 1974 August 22 by L. V. Zhuravleva at Nauchnyj.

Named in honor of Alexej Petrovich Maresjev, a war veteran whose heroic deed is described in B. Polevoj's novel *Story about a True Man.* (M 5285)

(2174) Asmodeus
1975 TA. Discovered 1975 October 8 by S. J. Bus and J. P. Huchra at Palomar.
Named for the Babylonian god of lust. (M 5039)

(2175) Andrea Doria
1977 TY. Discovered 1977 October 12 by P. Wild at Zimmerwald.

Named for the head of the Genoan family prominent in Schiller's {see planet (3079)} tragedy *Die Verschwörung des Fiesco zu Genua.* In combination with (2176) Donar, the name suggests the repeated curse of Giannettino Doria (nephew of Andrea) against the Republican conspirators; "Donner und Doria" is now used as an exclamation of dismay. (M 8151)
Name proposed by the discoverer, following a suggestion by W. Gurtner.

This name became known worldwide through the Italian ship which sunk in the Atlantic ocean on 25 July 1956.

(2176) Donar
2529 P-L. Discovered 1960 September 24 by C. J. van Houten and I. van Houten-Groeneveld at Palomar.

Named for the god of thunderstorms in Teutonic mythology, also honored in some languages by the fifth day of the week. (M 7782)

(2177) Oliver
6551 P-L. Discovered 1960 September 24 by C. J. van Houten and I. van Houten-Groeneveld at Palomar.

Named in honor of Bernard M. Oliver {1916-1995}, retired Vice President for Research and Development of the Hewlett-Packard Company, where he was an

innovator in the field of electronic instrumentation. He has made major contributions toward the search for extraterrestrial intelligence, the Monterey Institute for Research in Astronomy, and sky surveying with charge-coupled devices. (M 8541)
Obituary published in Bull. Am. Astron. Soc., Vol. 28, No. 4, p. 1459-1461 (1996).

(2178) Kazakhstania

1972 RA_2. Discovered 1972 September 11 by N. S. Chernykh at Nauchnyj.
Named for one of the Soviet Socialist Republics. (M 5285)
Since 1991 the Kazakh Republic is an independent state.

(2179) Platzeck

1965 MA. Discovered 1965 June 28 by C. U. Cesco and A. R. Klemola at El Leoncito.
Named in honor of Ricardo Pablo Platzeck, director of the Córdoba Observatory; the Córdoba Institute of Mathematics, Physics and Astronomy; and the Balseiro Institute of Atomic Physics at Cuyo University. An authority on the construction, testing and treatment of telescope mirrors, he played an important role in the completion of the 1.5-m reflector at Bosque Alegre in 1941. (M 8798)

(2180) Marjaleena

1940 RJ. Discovered 1940 September 8 by H. Alikoski at Turku.
Named by the discoverer in honor of his daughter, Marjaleena Johnsson. (M 7944)

(2181) Fogelin

1942 YA. Discovered 1942 December 28 by K. Reinmuth at Heidelberg.
Named in honor of Eric S. Fogelin {1961- }, an assistant at the Minor Planet Center during 1979-1980 who has helped extensively in the production of the Minor Planet Circulars and in the preparation of computerized data relevant to the Minor Planet Center. (M 5451)
Name proposed by C. M. Bardwell and B. G. Marsden.

(2182) Semirot

1953 FH_1. Discovered 1953 March 21 at the Goethe Link Observatory at Brooklyn, Indiana.
Named in memory of Pierre Sémirot (1907-1972), who joined the staff of the Bordeaux Observatory in 1931, spent the war years at the Paris Observatory, and served as director of the Bordeaux Observatory from 1947 to 1970. His work in positional astronomy included collaboration with Paul Herget {see planet (1751)} to improve the plate constants of the Bordeaux zone of the *Carte du Ciel*. He was president of IAU Commission 23 during 1961-1967. (M 8798)
Name proposed by F. K. Edmondson.
Obituary published on l'Astronomie, 87. année, p. 295-297 (1973).

(2183) Neufang

1959 OB. Discovered 1959 July 26 by C. Hoffmeister at Bloemfontein.
Named for a village in Thuringia where the discoverer lived and worked for many decades. The village is in the vicinity of the Sonneberg {see planet (1039)} Observatory, which was established by the discoverer, and it is the current staff members of that Observatory who have proposed the name for this minor planet. (M 18135)

(2184) Fujian

1964 TV_2. Discovered 1964 October 9 at the Purple Mountain Observatory at Nanking.
Named for a coastal province in southeastern China. (M 5184)

(2185) Guangdong

1965 WO. Discovered 1965 November 20 at the Purple Mountain Observatory at Nanking.
Named for a coastal province in southern China. (M 5184)

(2186) Keldysh

1973 SQ_4. Discovered 1973 September 27 by L. I. Chernykh at Nauchnyj.

Named in memory of Academician Mstislav Vsevolodovich Keldysh (1911-1978), famous Soviet scientist and mathematician who made a valuable contribution to space science and engineering. He was president of the U.S.S.R. Academy of Sciences from 1961 to 1975. (M 5849)
Obituary published in Icarus, Vol. 41, p. 320-322 (1980).

(2187) La Silla

1976 UH. Discovered 1976 October 24 by R. M. West at La Silla.

Named for the mountain in the Chilean Atacama desert on the top of which the European Southern Observatory is situated. (M 5039)

(2188) Orlenok

1976 UL_4. Discovered 1976 October 28 by L. V. Zhuravleva at Nauchnyj.

Named on the occasion of the 20th anniversary of the all-Union pioneer camp in the Tuapse region. An orlenok, or an eaglet, is the symbol of the young revolutionary. (M 5849)

(2189) Zaragoza

1975 QK. Discovered 1975 August 30 at the Felix Aguilar Observatory at El Leoncito.

Named in memory of Aldo Zaragoza (1924-1979), a member of the staff of the Felix Aguilar Observatory for over 20 years. As head of the Calculations Office he was responsible for the analysis and reduction of the plates taken at El Leoncito, and he calculated orbits for the comets and minor planets observed. (M 5524)

(2190) Coubertin

1976 GV_3. Discovered 1976 April 2 by N. S. Chernykh at Nauchnyj.

Named in memory of Pierre de Coubertin (1863-1937), a prominent French public figure, teacher, historian and man of letters. He was responsible for the renaissance of the Olympic Games, and the name is suggested by the discoverer in connection with the 22nd Olympic Games in Moscow. (M 5360)

(2191) Uppsala

1977 PA_1. Discovered 1977 August 6 by C.-I. Lagerkvist at Mount Stromlo.
Named for the ancient Swedish city and university. (M 5285)

(2192) Pyatigoriya

1972 HP. Discovered 1972 April 18 by T. M. Smirnova at Nauchnyj.

Named on the occasion of the 200th anniversary of the town Pyatigorsk, in the Caucasus. (M 5849)

(2193) Jackson

1926 KB. Discovered 1926 May 18 by H. E. Wood at Johannesburg.

Named in honor of Cyril V. Jackson {1903-1988}, whose {70} discoveries of numbered minor planets constitute a record for the southern hemisphere. After working at the former Union Observatory in Johannesburg he was subsequently associated with the Yale-Columbia Southern Station at its successive locations. (M 5451)
Name proposed by C. M. Bardwell and B. G. Marsden.

Obituaries published in Mon. Notes Astron. Soc. S. Afr., Vol. 48, No. 7/8, p. 66 (1989); Q.J.R. Astron. Soc., Vol. 32, No. 2, p. 211 (1991).

(2194) Arpola

1940 GE. Discovered 1940 April 3 by Y. Väisälä at Turku.

Named for the discoverer's summer cottage, located in the archipelago off Turku. A keen sailor, the discoverer yachted from Turku to Arpola about a hundred times each year. (M 7944)

(2195) Tengström

1941 SP_1. Discovered 1941 September 27 by L. Oterma at Turku.

Named in honor of Erik Tengström, emeritus professor at the University of Uppsala, on the occasion of his 70th birthday, 1983 Apr. 3. Initiator of the study of modern geodesy at Uppsala, he directed research in a wide number of subjects, including astronomy. One of his ancestors, Jacob Tengström, lived in Turku, was vice chancellor of the old Academia Aboensis (1803-1817) and the first archbishop of Finland (1817-1832). (M 7782)

(2196) Ellicott

1965 BC. Discovered 1965 January 29 at the Goethe Link Observatory at Brooklyn, Indiana.

Named in memory of Andrew Ellicott Douglass (1867-1962), American astronomer and founder in 1901 of the science of dendrochronology. As Percival Lowell's {see planet (1886)} principal assistant from 1894 to 1901, Douglass had investigated observing sites in Arizona and Mexico, and upon joining the faculty of the University of Arizona in Tucson in 1906 he mounted a campaign to bring a major observatory to southern Arizona. Through a gift from Lavinia Steward this effort met success with the completion in 1921 of a 0.9-m telescope, and Douglass served as director of the Steward Observatory until 1937. Having already made a mark in the use of dendrochronology for the dating of archaeological ruins, he then founded and directed until his 91st year the University's Laboratory for Tree Ring Research. (M 8799)
Name proposed by F. K. Edmondson. Citation prepared by E. Roemer.

(2197) Shanghai

1965 YN. Discovered 1965 December 30 at the Purple Mountain Observatory at Nanking.

Named for the largest city in China, one of the best known ports in the world, located at the mouth of the Yangtze River. (M 6059)

(2198) Ceplecha

1975 VF. Discovered 1975 November 7 at the Harvard College Observatory at Harvard.

Named in honor of the Czechoslovak astronomer Zdeněk Ceplecha, well known for his work on meteors. His investigation of the Pribram meteorite represents the first time that an accurate orbit could be derived for such an object. (M 5285)

(2199) Kleť

1978 LA. Discovered 1978 June 6 by A. Mrkos at Kleť.

Named for the observatory at which this minor planet was discovered and for the mountain in southern Bohemia on which the observatory is located. (M 5524)

(2200) Pasadena

6090 P-L. Discovered 1960 September 24 by C. J. van Houten and I. van Houten-Groeneveld at Palomar.

Named for a suburb of Los Angeles, California. Prominent astronomical insti-

tutions in Pasadena are the California Institute of Technology (which operates Palomar Observatory), the Mount Wilson Observatory of the Carnegie Institution of Washington, and the Jet Propulsion Laboratory. (M 5524)
Name proposed by E. Bowell.

(2201) Oljato

1947 XC. Discovered 1947 December 12 by H. L. Giclas at Flagstaff.

Named for the place of Moonlight Water near Monument Valley, Utah, on the Navajo Indian Reservation. During the day the hot Sun evaporates the small amount of spring water that seeps up through the sand. During the night, with less evaporation, the sand becomes moist enough to reflect the light of the Moon. (M 7782; M 8025)

This Apollo object was rediscovered as 1979 XA on 1979 Dec. 13 by Q. Passey and S. J. Bus at Palomar.

(2202) Pele

1972 RA. Discovered 1972 September 7 by A. R. Klemola at Mount Hamilton.

Named for the goddess of fire in Hawaiian legend. She made her home in the volcanoes Kilauea, after being driven out of the western sea by her angry sister, the sea goddess. (M 5360)

(2203) van Rhijn

1935 SQ_1. Discovered 1935 September 28 by H. van Gent at Johannesburg.

Named in memory of Pieter J. van Rhijn (1886-1960), Dutch astronomer, former director of the Kapteyn Astronomical Laboratory at Groningen. His main activity was the investigation of the Galactic structure. (M 18135)
Van Rhijn is also honored by a lunar crater.

(2204) Lyyli

1943 EQ. Discovered 1943 March 3 by Y. Väisälä at Turku.

Named in honor of Lyyli Heinanen (née Hartonen), a disciple of the discoverer. In 1927, as a student, she assisted in the first attempt to measure a length of 192 m by the Väisälä interference method. It was winter, the weather was very cold, and it seemed that the experiment would fail. She urged that they try once more, and the interferences were found. Later she taught mathematics in secondary school and was for many years chairman of the society Mikkelin Ursa for amateur astronomers. (M 7944)

(2205) Glinka

1973 SU_4. Discovered 1973 September 27 by L. I. Chernykh at Nauchnyj.

Named for Mikhail Ivanovich Glinka (1804-1857), celebrated composer and the acknowledged founder of Russian classical music. (M 5849)

(2206) Gabrova

1976 GR_3. Discovered 1976 April 1 by N. S. Chernykh at Nauchnyj.

Discovered on All Fools' Day, this minor planet is named for the Bulgarian town known for its humor and high-spirited merriment. (M 5849)

(2207) Antenor

1977 QH_1. Discovered 1977 August 19 by N. S. Chernykh at Nauchnyj.

This Trojan planet is named for the Trojan hero and sage who called for peace with the Greeks. He advised the citizens of Troy to put an end to the war by returning the kidnapped Helena to her husband Menelaus {see planets (101) and (1647)}. (M 5849)

(2208) Pushkin

1977 QL_3. Discovered 1977 August 22 by N. S. Chernykh at Nauchnyj.

Named for the great Russian poet Aleksandr Sergeevich Pushkin (1799-1837). (M 5849)

(2209) Tianjin

1978 US_1. Discovered 1978 October 28 at the Purple Mountain Observatory at Nanking.

Named for the largest port city in northern China, site of the latitude station of the Peking Observatory. (M 6059)

(2210) Lois

9597 P-L. Discovered 1960 September 24 by C. J. van Houten and I. van Houten-Groeneveld at Palomar.

Named in honor of Lois J. Baldwin, spouse of Ralph B. Baldwin, author of several books, including the pioneering studies *The face of the Moon* and *The measure of the Moon.* (M 11155)

(2211) Hanuman

1951 WO_2. Discovered 1951 November 26 by L. E. Cunningham at Mount Wilson.

Named for the monkey-god in the mythology of India. Son of Vayu, Hanuman was gifted with almost unlimited physical strength, powerful virility and total chastity, a combination that made him the perfect worshipper of Rama. After a great victory, Rama gave a necklace of fine pearls to Hanuman, who proceeded to break them apart one after another with his teeth. After inspection Hanuman threw away the pieces. "They do not contain the name of Rama", he declared, "and have no value". Then he tore open his own chest, and on his heart in letters of flame stood out the name of Rama. (M 21605)

Named by the Minor Planet Names Committee following a suggestion by F. Pilcher.

(2212) Hephaistos

1978 SB. Discovered 1978 September 27 by L. I. Chernykh at Nauchnyj.

This Apollo-type object is named for the god of fire and blacksmiths. (M 5849)

(2213) Meeus

1935 SO_1. Discovered 1935 September 24 by E. Delporte at Uccle.

Named in honor of the Belgian amateur astronomer and professional meteorologist Jean Meeus, among whose numerous contributions to astronomy are the original 1963 edition of *Tables des Petites Planètes* and improved and updated versions of Oppolzer's canon of eclipses. His other writings range from dynamical studies of long-term motions of Apollo objects, through examinations of geometrical configurations of the planets, to the debunking of the idea that there exist observations of faint natural satellites of the Earth. (M 6208)

Name proposed by E. Bowell, who made the key identification for this planet, and also by J. U. Gunter, following a suggestion by E. Fogelin.

(2214) Carol

1953 GF. Discovered 1953 April 7 by K. Reinmuth at Heidelberg.

Named by the Minor Planet Center for Carol D. Valenti, in honor of her service as secretarial assistant to the Central Bureau for Astronomical Telegrams since 1974 and to the Minor Planet Center since 1978. (M 6833)

(2215) Sichuan

1964 VX_2. Discovered 1964 November 12 at the Purple Mountain Observatory at Nanking.

Named for the most populous province of China, noted for its agricultural products. (M 6059)

(2216) Kerch

1971 LF. Discovered 1971 June 12 by T. M. Smirnova at Nauchnyj.
Named for the hero city in the Crimea. (M 5850)

(2217) Eltigen

1971 SK_2. Discovered 1971 September 26 by T. M. Smirnova at Nauchnyj.

Named for the site of the heroic landing of Soviet troops in the Crimea in November 1943. (M 5850)

(2218) Wotho

1975 AK. Discovered 1975 January 10 by P. Wild at Zimmerwald.

Named for an atoll in the Ralik Chain in the Pacific, perhaps an appropriate variation of (1282) Utopia. (M 8152)

Name proposed by the discoverer following a suggestion by R. Fiedler, formerly director of the public relations office at the California Institute of Technology.

(2219) Mannucci

1975 LU. Discovered 1975 June 13 at the Felix Aguilar Observatory at El Leoncito.

Named in memory of Edgardo Mannucci, outstanding precision mechanic, whose work at the Felix Aguilar Observatory benefited astronomy in Argentina for many years. He played an important part in setting up the meridian circle and Danjon {see planet (1594)} astrolabe and the adaption of the K-50 geodesic camera for observations of artificial satellites. Chief of precision work in the department of geodesy, topography and cartography of San Juan, and a valued member of the faculty of engineering at the Cuyo National University, he had previously served as assistant chief for the light and buoy workshop in the Argentine Naval Hydrographic Service. (M 8152)

(2220) Hicks

1975 VB. Discovered 1975 November 4 by E. F. Helin at Palomar.

Named in honor of William B. Hicks, alumnus of the California Institute of Technology, distinguished engineer, businessman and supporter of the sciences. (M 5285)

Name proposed by E. F. Helin and E. M. Shoemaker.

(2221) Chilton

1976 QC. Discovered 1976 August 25 at the Harvard College Observatory at Harvard.

Named in honor of Jean Chilton McCrosky, wife of Harvard-Smithsonian astronomer Richard E. McCrosky {see planet (1880)}. (M 5285)

(2222) Lermontov

1977 ST_1. Discovered 1977 September 19 by N. S. Chernykh at Nauchnyj.

Named for the great Russian poet Mikhail Yur'evich Lermontov (1814-1841). (M 5850)

(2223) Sarpedon

1977 TL_3. Discovered 1977 October 4 at the Purple Mountain Observatory at Nanking.

One of the leaders of the Lycians, Sarpedon was killed by Patroclus {see planet (617)} in the Trojan War. At the command of Zeus {see planet (5731)}, his body was seized by Apollo {see planet (1862)} and returned to Lycia. (M 6208)

(2224) Tucson

2528 P-L. Discovered 1960 September 24 by C. J. van Houten and I. van Houten-Groeneveld at Palomar.

Named for the city in Arizona. Prominent astronomical institutions in Tucson are the Lunar and Planetary Laboratory of the University of Arizona, Kitt Peak

National Observatory, and Steward Observatory. (M 5524)
Name proposed by E. Bowell.

(2225) Serkowski

6546 P-L. Discovered 1960 September 24 by C. J. van Houten and I. van Houten-Groeneveld at Palomar.

Named in memory of Krzysztof M. Serkowski (1930-1981), an astronomer distinguished particularly in the fields of interstellar polarization and the development of astronomical instrumentation. Despite a debilitating illness during the last eight years of his life, he remained actively working and maintained a serene and courageous attitude. (M 8541)
Obituary published in Astron. Q., Vol. 4, p. 76-81 (1982).

(2226) Cunitza

1936 QC_1. Discovered 1936 August 26 by A. Bohrmann at Heidelberg.

Named by the discoverer for the family of his sister-in-law, Lydia Cunitz. (M 5451)

(2227) Otto Struve

1955 RX. Discovered 1955 September 13 at the Goethe Link Observatory at Brooklyn, Indiana.

Named in memory of Otto Struve (1897-1963), last of a remarkable astronomical lineage. His great-grandfather, Wilhelm Struve {see planet (768)}, founded the Pulkovo Observatory in 1839; his grandfather (Otto), uncle (Hermann) and father (Ludwig) were also distinguished astronomers. Following a period of great privation and misery after World War I, he was invited by Edwin B. Frost {see planet (854)} to come to Yerkes Observatory in 1921. He started working in spectroscopy and remained a spectroscopist to the end of his days. He succeeded Frost as Yerkes director in 1932 and was the major force responsible for the establishment of the McDonald Observatory in 1933. Managing editor of the Astrophysical Journal from 1932 to 1947, he raised it to the preeminent position it now occupies. He became head of the astronomy department of the University of California in Berkeley in 1950, and he served as director of the National Radio Astronomy Observatory from 1960 to 1962. He served as president of the IAU during 1952-1955. (768) Struveana is named for three of his ancestors; following in their footsteps, Otto Struve received the Royal Astronomical Society's gold medal in 1944. (M 8912)
Name proposed by F. K. Edmondson.
Struve is also honored by a lunar crater.

(2228) Soyuz-Apollo

1977 OH. Discovered 1977 July 19 by N. S. Chernykh at Nauchnyj.
Named in honor of the joint Soviet-American space flight in 1975. (M 5850)

(2229) Mezzarco

1977 RO. Discovered 1977 September 7 by P. Wild at Zimmerwald.

Italian for "half arch, half vault", the name is that of the house and studio, on Lake Maggiore, of a friend of the discoverer. A painting of hers, *The Separation of the Elements*, is from a cycle *Creation*. The name is therefore suggested for this planet because (2229) = 3 x (743) Eugenisis. (M 8152)

(2230) Yunnan

1978 UT_1. Discovered 1978 October 29 at the Purple Mountain Observatory at Nanking.

Named for a province along the southwestern border of China. The Yunnan Observatory is situated in its capital city of Kunming. (M 6059)

(2231) Durrell

1941 SG. Discovered 1941 September 21 by S. Arend at Uccle.

Named in honor of the British writer Lawrence Durrell, author of the Alexandria Quartet (*Justine, Balthazar, Mountolive* and *Clea*) and a friend of the Arend family. (M 8152)

(2232) Altaj

1969 RD_2. Discovered 1969 September 15 by B. A. Burnasheva at Nauchnyj.

Named by the discoverer for the place of residence of her mother, Elena Andreevna Vasil'eva. (M 5850)

(2233) Kuznetsov

1972 XE_1. Discovered 1972 December 3 by L. V. Zhuravleva at Nauchnyj.

Named in memory of Nikolaj Ivanovich Kuznetsov (1911-1944), hero of the Soviet Union, renowned participant in the partisan movement during the Second World War. (M 5850)

(2234) Schmadel

1977 HD. Discovered 1977 April 27 by H.-E. Schuster at La Silla.

Named in honor of Lutz D. Schmadel {1942- }, Astronomisches Rechen-Institut, computer of orbits and ephemerides of minor planets, and Editor of *Astronomy and Astrophysics Abstracts*. The lost objects (1206) Numerowia and (1370) Hella were recovered as the result of his calculations. (M 5285)

(2235) Vittore

A924 GA. Discovered 1924 April 5 by K. Reinmuth at Heidelberg.

Named in honor of the Osservatorio S. Vittore, Bologna, where this object was rediscovered in 1979. It was a result of the extensive and careful series of observations there by a dedicated group of amateur astronomers that the earlier observations could be identified and the object permanently numbered. (M 5524)

(2236) Austrasia

1933 FX. Discovered 1933 March 23 by K. Reinmuth at Heidelberg.

Named for the eastern kingdom of the Merovingian Franks from the sixth to the eighth centuries. Austrasia embodied an extensive region on both sides of the Rhine, with Metz as its capital. (M 5524)

Name proposed by E. Bowell, who found one of the identifications involving this planet.

(2237) Melnikov

1938 TB. Discovered 1938 October 2 by G. N. Neujmin at Simeïs.

Named in memory of Oleg Aleksandrovich Melnikov (1912-1982), on the staff of the Pulkovo Observatory since 1933 and a professor at Leningrad University since 1947. His scientific research was centered on the study of the Sun, stellar astronomy and interstellar matter by spectroscopic methods. He was also concerned with astronomical instruments and served as president of IAU Commission 9. (M 8912)

Obituaries published in Astron. Zh., Tom 59, p. 1036-1037 (1982); Astrofizika, Tom 18, Vyp. 3, p. 498-500 (1982); Zemlya Vselennaya, No. 1, p. 46-47 (1983).

(2238) Steshenko

1972 RQ_1. Discovered 1972 September 11 by N. S. Chernykh at Nauchnyj.

Named in honor of Nikolaj Vladimirovich Steshenko {1927- }, deputy director of the Crimean Astrophysical Observatory, whose comprehensive support has contributed to the success of the program for the discovery and observation of minor planets. Well-known for his work in solar physics, he is in charge of the program of solar observations from space, and he is the author of the design for the Soviet 25-m-diameter mosaic optical telescope. (M 5850)

(2239) Paracelsus

1978 RC. Discovered 1978 September 13 by P. Wild at Zimmerwald.

Named for the great physician and natural philosopher whose real name was Theophrastus Bombastus von Hohenheim (1493?-1541). Because of his vehement fight against traditions, he was hounded throughout Europe during his lifetime, but with his concept of dosages of medicaments, he clearly set the foundations for modern medicine and pharmacology. (M 8152)

Paracelsus is also honored by a lunar crater.

(2240) Tsai

1978 YA. Discovered 1978 December 30 at the Harvard College Observatory at Harvard.

Named in honor of Tsai Chang-hsien, director of the Taipei Observatory since World War II, an active observer of planets and variable stars, and a long-time popularizer of astronomy. He has enlightened the public and directed amateur activity in astronomy for more than three decades with great patience and dedication. (M 5360)

(2241) Alcathous

1979 WM. Discovered 1979 November 22 by C. T. Kowal at Palomar.

Alcathous proved himself the noblest suitor in all of Troy by marrying Hippodamia {see planet (692)}, eldest and dearest daughter of Anchises {see planet (1173)}, who excelled all other girls of her age in beauty, skill and wit. As the Trojans achieved their greatest advance toward the Greek ships, Alcathous, previously uninjured in the Trojan War, was struck by Poseidon {see planet (4341)} with blindness and paralysis. In this hapless state he was killed easily by a spear thrown by Idomeneus {see planet (2759)}. (M 21128)

Named by the Minor Planet Names Committee following a suggestion by F. Pilcher.

(2242) Balaton

1936 TG. Discovered 1936 October 13 by G. Kulin at Budapest.

Named for the largest lake in Hungary, located in the western part of the country. (M 21605)

Named by the Minor Planet Names Committee following a suggestion of F. Pilcher.

(2243) Lönnrot

1941 SA_1. Discovered 1941 September 25 by Y. Väisälä at Turku.

Named for Elias Lönnrot (1802-1884), a physician in Kajaani and later professor of the Finnish language in Helsinki. While on extensive walking tours he wrote down a great number of runes and other traditional poems. He published the Finnish national epic Kalevala {see planet (1454)} and numerous medical, philosophical and other works. (M 7944)

(2244) Tesla

1952 UW_1. Discovered 1952 October 22 by M. B. Protitch at Belgrade.

Named in memory of Nikola Tesla (1856-1943), the famous Yugoslav-born physicist well known for his numerous scientific researches and discoveries in the field of multi-phase and high frequency currents and radio electro-communications. (M 9477)

Tesla is also honored by a lunar crater.

(2245) Hekatostos

1968 BC. Discovered 1968 January 24 by L. I. Chernykh at Nauchnyj.

The Greek ordinal number acknowledges this as the 100th minor planet to be numbered as a result of the joint observational program of the Institute for Theoretical Astronomy in Leningrad and the Crimean Astrophysical Observatory at Nauchnyj. (M 5850)

(2246) Bowell

1979 XH. Discovered 1979 December 14 by E. Bowell at Anderson Mesa.

Named in honor of Edward L. G. Bowell, Lowell Observatory astronomer who has made and who continues to make impressive contributions in many areas of minor planet astronomy. Following his comprehensive UBV photoelectric photometry of minor planets, he has revived and augmented the Lowell photographic astrometric program and has discovered several new objects. He also does extensive orbital work on minor planets, including the establishment of identifications and the prediction of occultations. (M 5688)
Name proposed by B. G. Marsden.

(2247) Hiroshima

6512 P-L. Discovered 1960 September 24 by C. J. van Houten and I. van Houten-Groeneveld at Palomar.

Named for the Japanese city Hiroshima in the hope that this planet may be a symbol for world peace. (M 14207)

(2248) Kanda

1933 DE. Discovered 1933 February 27 by K. Reinmuth at Heidelberg.

Named in memory of Shigeru Kanda (1894-1974), a staff member of the Tokyo Astronomical Observatory from 1920 to 1943 and a leading source of inspiration and encouragement to amateur astronomers in Japan. He published papers on observations of minor planets, comets and variable stars, identifications of minor planets, orbits of comets, and the history of Japanese and Chinese astronomy. He wrote several popular books on astronomy and in 1926 began publication of the *Kanda Circular* for amateur astronomers. After leaving Tokyo Observatory he founded the Kanda Astronomical Society, which later became the Japan Astronomy Study Group. (M 6647)
Name proposed by T. Urata, who found the identifications for this planet.

(2249) Yamamoto

1942 GA. Discovered 1942 April 6 by K. Reinmuth at Heidelberg.

Named in memory of Issei Yamamoto (1889-1959), professor of astronomy at Kyoto University, director of the Kwasan Observatory, and first president of the IAU commission on the zodiacal light. The most well-known popularizer of astronomy in Japan, he founded the Oriental Astronomical Association in 1920 and was a source of encouragement for many amateur astronomers in Japan. (M 6648)
Name proposed by S. Nakano, who determined the orbit for this planet.
Yamamoto is also honored by a lunar crater.

(2250) Stalingrad

1972 HN. Discovered 1972 April 18 by T. M. Smirnova at Nauchnyj.

Named in commemoration of the fierce battle for the city. The victory by the defenders was an important turning point in World War II. The city is now called Volgograd. (M 6648)

(2251) Tikhov

1977 SU_1. Discovered 1977 September 19 by N. S. Chernykh at Nauchnyj.

Named in memory of Gavriil Adrianovich Tikhov (1875-1960), active on the staff of the Pulkovo Observatory during 1906-1941 and from 1947 head of the astrobotanical department of the Kazakh Academy of Sciences. His principal scientific work was concerned with stellar and planetary photometry and colorimetry and with atmospheric optics. He was also known for his research on the physical nature of Mars. (M 5451)
Tikhov is also honored by craters on Mars and the Moon.

(2252) CERGA

1978 VT. Discovered 1978 November 1 by K. Tomita at Caussols.

Named for the Centre d'Etudes et de Recherches Géodynamiques et Astronomiques, which operates the 0.9-m Schmidt telescope at Caussols-Cipieres with which this minor planet was discovered. (M 5451)

(2253) Espinette

1932 PB. Discovered 1932 July 30 by G. Van Biesbroeck at Williams Bay. Independently discovered 1932 July 31 by C. Jackson at Johannesburg and August 4 by G. N. Neujmin at Simeïs.

Named for the discoverer's residence in Williams Bay, Wisconsin. For many of the 45 years the discoverer worked at the Yerkes Observatory. The Van Biesbroecks fed and housed students and visiting astronomers from all over the world. The name originally comes from a small cafe in one of the parks in Brussels. (M 6059)

Name proposed by the discoverer's children, Edwin Van Biesbroeck, Simone Van Biesbroeck Titus and Micheline Van Biesbroeck Wilson.

(2254) Requiem

1977 QJ_1. Discovered 1977 August 19 by N. S. Chernykh at Nauchnyj.

The name is dedicated to the memory of the discoverer's mother, Melaniya Petrovna Chernykh, who died on 1977 Aug. 19, the day that this minor planet was discovered. (M 5524)

(2255) Qinghai

1977 VK_1. Discovered 1977 November 3 at the Purple Mountain Observatory at Nanking.

Named for a large province in the western part of China. The sources of the two largest rivers in China, the Yellow River and the Yangtze, are in this province. (M 6059)

(2256) Wisniewski

4519 P-L. Discovered 1960 September 24 by C. J. van Houten and I. van Houten-Groeneveld at Palomar.

Named in honor of Wieslaw Z. Wisniewski {1931-1994}, astronomer at the University of Arizona. Wisniewski worked on the lightcurves of comets and minor planets, especially Earth-approaching objects. He was a leading authority on photometric standards and on photometry with CCDs as well as with photomultipliers. (M 16040)

Obituaries published in Minor Planet. Bull., Vol. 21, No. 2, p. 20 (1994); Icarus, Vol. 112, No. 2, p. 300-301 (1994); Q.J.R. Astron. Soc., Vol. 37, No. 4, p. 845 (1996); Bull. Am. Astron. Soc., Vol. 26, No. 4, p. 1611-1612 (1994); Postepy Astron., Tom 43, No. 1, p. 41-43 (1995).

(2257) Kaarina

1939 QB. Discovered 1939 August 18 by H. Alikoski at Turku.
Named by the discoverer in honor of his daughter, Kaarina Soini. (M 7944)

(2258) Viipuri

1939 TA. Discovered 1939 October 7 by Y. Väisälä at Turku.

Named for an old Finnish town on the Gulf of Finland. Runic inscriptions mention Viipuri as a market place in the eleventh century. The Scandinavian name is Viborg. At present the town belongs to Russia. (M 7945)

(2259) Sofievka

1971 OG. Discovered 1971 July 19 by B. A. Burnasheva at Nauchnyj.

Named for a dendrological park in Uman', in the Cherkasskaya district of the {former} Ukrainian SSR. The park is a monument to the landscaping and architecture of the 18th and 19th centuries. (M 8912)

(2260) Neoptolemus

1975 WM_1. Discovered 1975 November 26 at the Purple Mountain Observatory at Nanking.

The son of Achilles and Deidameia, Neoptolemus was brought to Troy by Odysseus in the last year of the Trojan War. He was one of the warriors in the wooden horse and was responsible for the death of Priam. (M 6208)

See also the citations for (588) Achilles, (3912) Troja, (1143) Odysseus, and (884) Priamus, respectively.

(2261) Keeler

1977 HC. Discovered 1977 April 20 by A. R. Klemola at Mount Hamilton.

Named in memory of James Edward Keeler (1857-1900), pioneer American astrophysicist and second director of the Lick Observatory. He measured accurately the wavelengths of the chief nebular lines, confirmed spectroscopically the particle nature of Saturn's rings and recognized the spiral "nebulae" as numerous, important constituents of the universe. In 1899 he also discovered photographically with the Crossley reflector the faint minor planet (452) Hamiltonia. (M 13172)
Citation prepared by D. E. Osterbrock.
Keeler is also honored by craters on Mars and the Moon.

(2262) Mitidika

1978 RB. Discovered 1978 September 10 by P. Wild at Zimmerwald.

Named for the gypsy girl of great charm and courage, one of the central figures in Clemens Brentano's novel *Die mehreren Wehmüller und ungarischen Nationalgesichter*. (M 8152)

(2263) Shaanxi

1978 UW_1. Discovered 1978 October 30 at the Purple Mountain Observatory at Nanking.

Named for a province in central China, near the upper reaches of the Yellow River. Its capital, Xi'an {see planet (2387)}, is one of the best known ancient cities in the country. (M 6059)

(2264) Sabrina

1979 YK. Discovered 1979 December 16 by E. Bowell at Anderson Mesa.

Named for a legendary English princess, daughter of King Locrine. Along with her mother, Sabrina was drowned in the River Severn by Locrine's angry widow. Sabrina became the Roman name for the River Severn. (M 5451)

(2265) Verbaandert

1950 DB. Discovered 1950 February 17 by S. Arend at Uccle.

Named in memory of Jean Verbaandert (1901-1974), a Belgian astronomer who was a colleague of the discoverer for many years. (M 8403)

(2266) Tchaikovsky

1974 VK. Discovered 1974 November 12 by L. I. Chernykh at Nauchnyj.

Named in memory of the celebrated Russian composer Petr Il'ich Chajkovskij (1840-1893). (M 7782)

(2267) Agassiz

1977 RF. Discovered 1977 September 9 at the Harvard College Observatory at Harvard.

Named in memory of Jean Louis Rodolphe Agassiz (1807-1873), Swiss-born naturalist, later a professor at Harvard, where he was the leading U.S. opponent of Darwin; his son Alexander Agassiz (1835-1910), marine zoologist and oceanographer; and his grandson George Russell Agassiz (1862-1951), for many years a friend and benefactor of the Harvard Observatory and for whom the Agassiz Station is named. (M 5451)

Agassiz is also honored by a crater on Mars.

(2268) Szmytowna

1942 VW. Discovered 1942 November 6 by L. Oterma at Turku.

Named in honor of the distinguished Polish scientist Maria Szmytowna on the occasion of her eightieth birthday, 1983 March 26. A professor of inorganic and analytical chemistry at the Akademia Medyczna w Poznaniu, she has done extensive research work for half a century on waters and mineral springs and is an active member of many societies in the areas of balneology and hydrobiology. (M 7945)

(2269) Efremiana

1976 JA_2. Discovered 1976 May 2 by N. S. Chernykh at Nauchnyj.

Named in memory of Ivan Antonovich Efremov (1907-1972), renowned paleontologist and the author of historical, science-fiction and adventure novels. (M 7782)

(2270) Yazhi

1980 ED. Discovered 1980 March 14 by E. Bowell at Anderson Mesa.

The name comes from the Navajo Indian language and means 'little one'. (M 5525)

The name was suggested by Debbie Geoffrion.

(2271) Kiso

1976 UV_5. Discovered 1976 October 22 by H. Kosai and K. Hurukawa at Kiso.

Named for the station of the Tokyo Observatory at which this minor planet was discovered. (M 5525)

(2272) Montezuma

1972 FA. Discovered 1972 March 16 by T. Gehrels at Palomar.

Named in honor of the ninth emperor of the Aztec empire, who reigned 1502-1520. He believed that his land rightly belonged to the serpent god Quetzálcoatl {see planet (1915)}, who one day would return from the East. When the first Spanish invaders landed in Central America, Montezuma interpreted them as the followers of Quetzálcoatl returning. After only feeble negotiation, he allowed Cortez to occupy his capital city peacefully and rule through him. During Cortez' temporary absence hostilities developed between Spaniards and Aztecs in which Montezuma was among the first killed. (M 21605)

Named by the Minor Planet Names Committee following a suggestion by F. Pilcher.

(2273) Yarilo

1975 EV_1. Discovered 1975 March 6 by L. I. Chernykh at Nauchnyj.

Named for the ancient Slavic god of the Sun, spring, fertility and love. (M 7783)

(2274) Ehrsson

1976 EA. Discovered 1976 March 2 by C.-I. Lagerkvist at Kvistaberg.
Named in honor of a friend of the discoverer. (M 5525)

(2275) Cuitlahuac

1979 MH. Discovered 1979 June 16 by H.-E. Schuster at La Silla.

Named in honor of the tenth emperor of the Aztec empire, who reigned briefly in 1520. Brother of Montezuma {see planet (2272)}, he was released from captivity by Cortez to negotiate a dispute. Immediately he rallied the Aztecs to battle the Spaniards. After many days of fierce fighting in which thousands of Aztec warriors were killed, Cuitlahuac drove Cortez from the city with the loss of over half his men and all his treasure. Within a few months Cuitlahuac and uncounted thousands of his fellow Aztecs died from smallpox brought by an occupying Spanish soldier. (M 21605)

Named by the Minor Planet Names Committee following a suggestion by F. Pilcher.

(2276) Warck

1933 QA. Discovered 1933 August 18 by E. Delporte at Uccle. Independently discovered 1933 August 19 by G. N. Neujmin at Simeïs.

Named in honor of the family of Evelyne (née Delporte) Warck, granddaughter of the discoverer. (M 8403)

(2277) Moreau

1950 DS. Discovered 1950 February 18 by S. Arend at Uccle.

Named in memory of Fernand Moreau (1888-1979), a long-time colleague of the discoverer at the Uccle Observatory. (M 8403)
Obituary published in Ciel Terre, Vol. 96, p. 123-124 (1980).

(2278) Götz

1953 GE. Discovered 1953 April 7 by K. Reinmuth at Heidelberg.

Named in memory of Paul Götz, first assistant of Max Wolf {see planets (827) and (1217)} at the Heidelberg-Königstuhl Observatory during 1903-1905. Götz was a skillful and diligent observer with the Bruce telescope and the 0.15-m astrograph. He discovered 20 minor planets. (M 18447)
Name proposed and citation prepared by G. Klare and L. D. Schmadel.

(2279) Barto

1968 DL. Discovered 1968 February 25 by L. I. Chernykh at Nauchnyj.

Named in memory of Agniya Lvovna Barto (1906-1982), a famous Soviet poetess, who dedicated all her works to children. (M 7783)

(2280) Kunikov

1971 SL_2. Discovered 1971 September 26 by T. M. Smirnova at Nauchnyj.

Named in memory of Tzezar' L'vovich Kunikov (1909-1943), commander of the landing party that seized Malaya Zemlya in the Novorossijsk region in February 1943. (M 6059)

(2281) Biela

1971 UQ_1. Discovered 1971 October 26 by L. Kohoutek at Bergedorf.

Named in memory of Wilhelm von Biela (1782-1850), Austrian military officer and astronomer, who discovered comet 1826 I in Josefov, Bohemia. The comet is famous for having split and disappeared and for helping establish the connection between comets and meteors. In 1971 the discoverer of this minor planet tried unsuccessfully to recover the comet. (M 22244)
Biela is also honored by a lunar crater.

(2282) Andrés Bello

1974 FE. Discovered 1974 March 22 by C. Torres at Cerro El Roble.

Named for Andrés Bello, Venezuelan intellectual, first rector of the University of Chile, who helped establish the Observatorio Astronomico Nacional in Santiago in 1852. Bello persuaded the Chilean government to purchase the equipment brought to Chile by a U.S. Naval Observatory team headed by J. M. Gillis. This homage is made in 1981 to commemorate the bicentennial of Bello's birth. (M 6531)

(2283) Bunke

1974 SV_4. Discovered 1974 September 26 by L. V. Zhuravleva at Nauchnyj.
Named in memory of the German patriot Tamara Bunke (1937-1967). (M 7616)

(2284) San Juan

1974 TG_1. Discovered 1974 October 10 at the Felix Aguilar Observatory at El Leoncito.

Named for the university of which the Felix Aguilar {see planet (1800)} Observatory is a part and for the Argentine state in which it is located. (M 8152)

(2285) Ron Helin

1976 QB. Discovered 1976 August 27 by S. J. Bus at Palomar.

Named in honor of Ronald P. Helin, husband of Eleanor F. Helin. His never-ending friendship and encouragement have been invaluable to the success of the Palomar planet-crossing asteroid survey. (M 6531)

(2286) Fesenkov

1977 NH. Discovered 1977 July 14 by N. S. Chernykh at Nauchnyj.

Named in memory of Vasilej Grigor'evich Fesenkov (1889-1972), one of the founders of the study of astrophysics in the U.S.S.R. His scientific activities covered a wide range of topics, including solar and stellar physics, the Moon and planets, atmospheric optics, meteoritics and cosmogony. From 1924 to 1964 he was editor of the *Astronomicheskij Zhurnal* and from 1945 Chairman of the Committee on Meteorites of the U.S.S.R. Academy of Sciences. (M 5525)

Fesenkov is also honored by a crater on Mars. Obituaries published in Astron. Zhurn., Vol. 49, p. 678-682 (1972); Sky Telesc., Vol. 43, p. 284 (1972); Zemlya Vselennaya, No. 3, p. 46-47 (1972); Astron. Tsirk., No. 713, p. 4-7 (1972); Meteoritika, No. 3, p. 162-163 (1972); Trudy Astrofiz. Inst. Alma Ata, Vol. 19, p. 129 (1972); Vestn. AN KazSSR, No. 4, p. 67 (1972); Irish Astron. J., Vol. 11, p. 162 (1973).

(2287) Kalmykia

1977 QK_3. Discovered 1977 August 22 by N. S. Chernykh at Nauchnyj.
Named for one of the autonomous republics of the {former} R.S.F.S.R. (M 5688)

(2288) Karolinum

1979 UZ. Discovered 1979 October 19 by L. Brožek at Kleť.

Named for the original main building, still in use, of the Charles University, founded in Prague in 1348. (M 5525)
Charles University is also honored by planet (4339).

(2289) McMillan

6567 P-L. Discovered 1960 September 24 by C. J. van Houten and I. van Houten-Groeneveld at Palomar.

Named in honor of Robert S. McMillan, astronomer at the University of Arizona. McMillan heads the radial velocity program that searches for planets of other stars. He also works for Spacewatch, being instrumental in that project's development for the discovery of asteroids and comets. (M 16040)

(2290) Helffrich

1932 CD_1. Discovered 1932 February 14 by K. Reinmuth at Heidelberg.

Named in memory of J. Helffrich { -1971}, on the staff of the Heidelberg-Königstuhl Observatory between 1909 and 1911. As an assistant of Max Wolf {see planets (827) and (1217)}, he discovered 13 numbered minor planets. (M 18447) Name proposed and citation prepared by G. Klare and L. D. Schmadel.

(2291) Kevo

1941 FS. Discovered 1941 March 19 by L. Oterma at Turku.

Named for the field station of the University of Turku's Kevo Subarctic Research Institute. Situated in the valley of the river Kevo in far Lappland at latitude 69°45′, the station cooperates with other research centers above the Arctic circle in the study of biology and Earth sciences. (M 7945)

(2292) Seili

1942 RM. Discovered 1942 September 7 by Y. Väisälä at Turku.

Named for an island, some 30 km from Turku, originally a leper colony, and in 1619 the site of the first hospital in Finland. In 1965 a station of the University of Turku's Archipelago Institute was established there. (M 7945)

(2293) Guernica

1977 EH_1. Discovered 1977 March 13 by N. S. Chernykh at Nauchnyj.

Named for the town Guernica y Luno, Spain, historical center of Basque culture. (M 9767)

(2294) Andronikov

1977 PL_1. Discovered 1977 August 14 by N. S. Chernykh at Nauchnyj.

Named in memory of Iraklij Luarsabovich Andronikov (Andronikashvili, 1908-1990), Soviet writer and literary scholar. He was a brilliant teller of his own stories and reminiscences about writers, artists and other men of art. His main research was devoted to the life and creative work of the celebrated Russian poet M. Yu. Lermontov {see planet (2222)}. (M 17465)

(2295) Matusovskij

1977 QD_1. Discovered 1977 August 19 by N. S. Chernykh at Nauchnyj.

Named in memory of Mikhail L'vovich Matusovskij (1915-1990), well known Soviet poet. (M 17465)

(2296) Kugultinov

1975 BA_1. Discovered 1975 January 18 by L. I. Chernykh at Nauchnyj.

Named in honor of David Nikitich Kugul'tinov, renowned Soviet poet and national poet of Kalmykia {see planet (2287)}. (M 17465)

(2297) Daghestan

1978 RE. Discovered 1978 September 1 by N. S. Chernykh at Nauchnyj.
Named for one of the autonomous republics of the {former} R.S.F.S.R. (M 5688)

(2298) Cindijon

A915 TA. Discovered 1915 October 2 by M. Wolf at Heidelberg.

Named by B. G. Marsden {see planet (1877)}, who found the identification for this planet, in honor of his children, Cynthia Louise and Jonathan Brian. (M 6833)

(2299) Hanko

1941 SZ. Discovered 1941 September 25 by Y. Väisälä at Turku.

Named for a Finnish town on the Gulf of Finland, famous for its beaches and regattas. Its harbor was already known in the thirteenth century. (M 7945)

(2300) Stebbins

1953 TG_2. Discovered 1953 October 10 at the Goethe Link Observatory at Brooklyn, Indiana.

Named in memory of Joel Stebbins (1878-1966), who pioneered the development of photoelectric methods for measuring astronomical radiation, first at the University of Illinois, then at the University of Wisconsin. He applied the method in work on eclipsing binaries, interstellar reddening and the integrated magnitudes of galaxies. Others have used the technique to study the shapes, rotation periods and surface properties of asteroids. (M 8403)
Name proposed by F. K. Edmondson. Citation written by A. E. Whitford. Stebbins ia also honored by a lunar crater.

(2301) Whitford

1965 WJ. Discovered 1965 November 20 at the Goethe Link Observatory at Brooklyn, Indiana.

Named in honor of Albert E. Whitford, director of the Washburn Observatory (1945-1958), director of the Lick Observatory (1958-1968), and a past president of the American Astronomical Society. As a pioneer in photoelectric photometry he developed, in collaboration with Stebbins {see planet (2300)}, the first precise multicolor photometric system. He made major contributions in both galactic and extragalactic research. His studies of the wavelength dependence of interstellar extinction are fundamental to the understanding of interstellar particles, and the standard galactic extinction curve is referred to as the Whitford curve. (M 8403)
Name proposed by F. K. Edmondson. Citation written by A. D. Code.

(2302) Florya

1972 TL_2. Discovered 1972 October 2 by N. E. Kurochkin at Nauchnyj.

Named in memory of Nikolaj Fyodorovich Florya (1912-1941), astronomer at the Sternberg State Astronomical Institute, skilled observer and prominent researcher on variable stars. (M 12208)

(2303) Retsina

1979 FK. Discovered 1979 March 24 by P. Wild at Zimmerwald.

Named on the occasion of the Patras IAU General Assembly in honor of the resined wine of Greece. (2303) = (47) Aglaja x (49) Pales, the product of one of the three Graces and the goddess of pastures, and a reminder of the discoverer's tour of the Peloponnese with three charming companions. (M 8152)

(2304) Slavia

1979 KB. Discovered 1979 May 18 by A. Mrkos at Kleť.

This is an ancient Slavonic name, now also that of a famous sports club in Prague. (M 6060)

(2305) King

1980 RJ_1. Discovered 1980 September 12 at the Harvard College Observatory at Harvard.

Named in memory of Martin Luther King (1929-1968), the American civil rights leader who worked to bring about social, political and economic equality for blacks by peaceful means. A Baptist minister who received the 1964 Nobel peace prize, he preached 'non-violent resistance' to achieve full civil rights for all. (M 6208)

(2306) Bauschinger

1939 PM. Discovered 1939 August 15 by K. Reinmuth at Heidelberg.

Named in memory of Julius Bauschinger (1860-1934), eminent German astronomer, professor of astronomy and director of the Astronomisches Rechen-Institut,

Berlin, and the Leipzig University Observatory. For almost 15 years, Bauschinger worked on classical astrometric problems at the Munich Observatory. His extensive meridian observations resulted in two large catalogues, *Münchener Sternverzeichnisse*, including mean places of almost 50,000 stars. He also discussed the orbit of periodic comet Brooks 2 (1889 V) in great detail. Successor of F. Tietjen {see planet (2158)}, Bauschinger in 1897 managed the separation of the ARI from the Berlin Observatory. For many years, he was editor of the famous *Berliner Astronomisches Jahrbuch*. With P. V. Neugebauer, Bauschinger elaborated tables of the history and statistics of minor planets (1901) and his well-known textbook on orbit determination *Die Bahnbestimmung der Himmelskörper* (1906). The decade 1909-1919 was a period of practical work at the Strasbourg Observatory 49-cm refractor, with which he photometrically observed NGC nebulae and measured double stars. As director of the Leipzig Observatory (1920-1930), Bauschinger heavily supported a repetition of the AG zones project. (M 18447)
Name proposed and citation prepared by L. D. Schmadel.

(2307) Garuda

1957 HJ. Discovered 1957 April 18 at the La Plata Observatory at La Plata.

Named for a son of the sage Kasyapa and Vinata {see planet (2347)} in the mythology of India. Vinata laid two eggs, and after a thousand years one hatched into Garuda. He symbolizes aspiration to spiritual truth on the higher planes, and feels hatred toward the serpents who are guardians of spiritual truth on the material and terrestrial planes. He was the mount chosen by Vishnu, protector and conservator of the universe. (M 21605)

Named by the Minor Planet Names Committee following a suggestion by F. Pilcher.

(2308) Schilt

1967 JM. Discovered 1967 May 6 by C. U. Cesco and A. R. Klemola at El Leoncito.

Named in honor of Jan Schilt {1894-1982}, Columbia University astronomer who has made significant contributions to the study of galactic structure and dynamics and the problems of stellar luminosity calibrations. The cooperative venture by Schilt at Columbia and Brouwer {see planet (1746)} at Yale, together with the Argentine astronomers at San Juan, led to the establishment of the Yale-Columbia Southern Station in the early 1960s. (M 6531)

Obituary published in Phys. Today, Vol. 35, No. 6, p. 67-68 (1982); Zenit, 9. Jaarg., p. 375 (1982).

(2309) Mr. Spock

1971 QX_1. Discovered 1971 August 16 by J. Gibson at El Leoncito.

Named for the ginger short-haired tabby cat (1967-) who selected the discoverer and his soon-to-be wife at a cat show in California and accompanied them to Connecticut, South Africa and Argentina. At El Leoncito he provided endless hours of amusement, brought home his trophies, dead or alive, and was a figure of interest to everyone who knew him. He was named after the character in the television program *Star Trek* who was also imperturbable, logical, intelligent and had pointed ears. (M 10042)

(2310) Olshaniya

1974 SU_4. Discovered 1974 September 26 by L. V. Zhuravleva at Nauchnyj.

Named in honor of Konstantin Olshanskij and the other daring fighters who entered the occupied city of Nikolaev in March 1944. (M 6648)

(2311) El Leoncito

1974 TA_1. Discovered 1974 October 10 at the Felix Aguilar Observatory at El Leoncito.

Named for the observing station at which this minor planet was discovered. (M 8153)

(2312) Duboshin

1976 GU_2. Discovered 1976 April 1 by N. S. Chernykh at Nauchnyj.

Named in honor of Georgij Nikolaevich Duboshin {1904-1986}, distinguished authority on celestial mechanics, author of several textbooks, president of IAU Commission 7 during 1970-1973. (M 7471)
Obituary published in Celest. Mech., Vol. 39, No. 2, p. 115 (1986).

(2313) Aruna

1976 TA. Discovered 1976 October 15 by H. L. Giclas at Anderson Mesa.

Named for a son of the sage Kasyapa and Vinata {see planet (2347)} in the mythology of India. Vinata laid two eggs. When after five hundred years neither had hatched, she broke one open to find Aruna with only his upper half developed. He went into the sky to become the red glow of dawn, where ever since he has driven the suns's chariot each morning. (M 21605)

Named by the Minor Planet Names Committee following a suggestion by F. Pilcher.

(2314) Field

1977 VD. Discovered 1977 November 12 at the Harvard College Observatory at Harvard.

Named in honor of George B. Field {1929- }, director of the Harvard College Observatory and Smithsonian Astrophysical Observatory since 1973. A well-known theoretician who has made significant contributions in many areas of astrophysics, most notably concerning processes in the interstellar medium and intergalactic medium, he was also serving as chairman of the U.S. Astronomy Survey committee on the needs of astronomy in the 1980s. (M 6648)

(2315) Czechoslovakia

1980 DZ. Discovered 1980 February 19 by Z. Vávrová at Kleť.

Named in honor of the discoverer's country, renowned for its research activity on minor planets, comets, meteorites and meteor streams. (M 6060)

(2316) Jo-Ann

1980 RH. Discovered 1980 September 2 by E. Bowell at Anderson Mesa.

Named in honor of Jo-Ann Bowell, wife of the discoverer {see planet (2246)}, on the occasion of her nth birthday, 1982 May 13. (M 6833)

(2317) Galya

2524 P-L. Discovered 1960 September 24 by C. J. van Houten and I. van Houten-Groeneveld at Palomar.

Named in honor of Galya Lubarsky, a friend of T. Gehrels. {See also the citation for planet (2318)}. (M 9214)

(2318) Lubarsky

6521 P-L. Discovered 1960 September 24 by C. J. van Houten and I. van Houten-Groeneveld at Palomar.

Named in honor of Cronid Lubarsky, a friend of T. Gehrels. {See also the citation for planet (2317)}. (M 9214)

(2319) Aristides

7631 P-L. Discovered 1960 October 17 by C. J. van Houten and I. van Houten-Groeneveld at Palomar.

Named for the Athenian politician Aristides (fl. c. 500 B.C.), famous for his just determination of the contribution from Athens' allies. (M 16590)

(2320) Blarney

1979 QJ. Discovered 1979 August 29 by P. Wild at Zimmerwald.

Named for the famous Irish castle near Cork. What will happen to the first astronaut to kiss this stone remains to be seen. (M 17465)

(2321) Lužnice

1980 DB_1. Discovered 1980 February 19 by Z. Vávrová at Kleť.

Named for the small river running through the idyllic southern Bohemian countryside very close to the fifteenth meridian. (M 6060; M 30493)

(2322) Kitt Peak

1954 UQ_2. Discovered 1954 October 28 at the Goethe Link Observatory at Brooklyn, Indiana.

Named to commemorate two important events in the history of the Kitt Peak National Observatory. Members of the Papago Tribal Council and the Schuk Toak District Council visited the Steward Observatory on the evening of 1955 Oct. 28, just one year after the discovery of this minor planet. This was the first step in arranging for the establishment of an astronomical observatory on Kitt Peak. The Association of Universities for Research in Astronomy (AURA) was incorporated on 1957 Oct. 28. (M 9079)
Name proposed by F. K. Edmondson.

(2323) Zverev

1976 SF_2. Discovered 1976 September 24 by N. S. Chernykh at Nauchnyj.

Named in honor of Mitrofan Stepanovich Zverev {1903-1991}, specialist in fundamental astrometry, initiator of the international program to compile the Catalogue of Faint Stars (KSZ). This Pulkovo astrometrist organized expeditions in the 1960s for the observation of stars in the southern hemisphere. He is also popular as a fine pianist. (M 7783)
Obituary published in Zemlya Vselennaya, No. 4, p. 44-45 (1992).

(2324) Janice

1978 VS_4. Discovered 1978 November 7 by E. F. Helin and S. J. Bus at Palomar.

Named in honor of Janice Cline, who for many years has encouraged astrometric studies of minor planets at Caltech. Her kindness and support are greatly appreciated. (M 5850)

(2325) Chernykh

1979 SP. Discovered 1979 September 25 by A. Mrkos at Kleť.

Named in honor of Ludmilla I. Chernykh {1935- } and Nikolaj S. Chernykh {1931- }, astronomers at the Crimean Astrophysical Observatory who lead the impressive program of observations and discoveries of minor planets there. (M 6060)

(2326) Tololo

1965 QC. Discovered 1965 August 29 at the Goethe Link Observatory at Brooklyn, Indiana.

Named to commemorate the founding of the Cerro Tololo Inter-American Observatory in northern Chile on 1962 Nov. 23. CTIO is funded by the National Science Foundation and operated by AURA. (M 9079)
Name proposed by F. K. Edmondson.

(2327) Gershberg

1969 TQ_4. Discovered 1969 October 13 by B. A. Burnasheva at Nauchnyj.

Named in honor of the Soviet astronomer Roald Evgenevich Gershberg {1933- }, who has made significant contributions to the study of processes occurring in gaseous nebulae and of the nature of stellar outbursts. (M 8912)

(2328) Robeson

1972 HW. Discovered 1972 April 19 by T. M. Smirnova at Nauchnyj.

Named in memory of the American singer and actor Paul Robeson (1898-1976). (M 6648)

(2329) Orthos

1976 WA. Discovered 1976 November 19 by H.-E. Schuster at La Silla.

Named for Orthos, the two-headed dog from Greek mythology. Together with his master Eurytion, this dog looked after the cattle of Geryones. One of the twelve labors of Heracles {see planet (5143)} was to steal these cattle. (M 5688)

(2330) Ontake

1977 DS_3. Discovered 1977 February 18 by H. Kosai and K. Hurukawa at Kiso.

Named for the conical volcano near the Kiso station. In October 1979, for the first time in its recorded history, the volcano, which is 3063 m high, exhibited small eruptions. (M 5851)

(2331) Parvulesco

1936 EA. Discovered 1936 March 12 by E. Delporte at Uccle.

Named in memory of the Roumanian professor Constantin Parvulesco (1890-1945), who worked with the discoverer at the Uccle Observatory in the early 1930s. The planet also honors his daughter, Carina Parvulesco, who was born at Uccle, and who, as a professor of astronomy at San Mateo College, California, has made contributions in stellar and galactic dynamics; she has more recently served as vice-president of two corporations and as president of the U.N. Association of San Mateo County. The planet also honors her brother, Antares Parvulesco, mathematical physicist and acoustician, formerly at Columbia University, now at the University of Hawaii. (M 8799)

(2332) Kalm

1940 GH. Discovered 1940 April 4 by L. Oterma at Turku.

Named for Pehr Kalm (1716-1779), a Finnish naturalist, and from 1747 the first professor of economics at the old Academia Aboensis. His numerous explorations took him as far as North America. He also founded a botanical garden for the Academia and wrote the pioneering work *Flora Fennica*. (M 7945)

(2333) Porthan

1943 EP. Discovered 1943 March 3 by Y. Väisälä at Turku.

Named for Henrik Gabriel Porthan (1739-1804), historian and professor of oratory at the Academia Aboensis. This 'Father of Finnish History' pioneered the study of history based on facts, rather than on fantasy. (M 7945)

(2334) Cuffey

1962 HD. Discovered 1962 April 27 at the Goethe Link Observatory at Brooklyn, Indiana.

Named in honor of James Cuffey, a member of the Indiana University faculty from 1946 to 1966 and then of the New Mexico State University faculty until 1976. He played a major role in arranging for the transfer of the 10-inch Cooke triplet from the Cincinnati Observatory to the Goethe Link Observatory in 1948 and thus the start of the Indiana minor-planet program. (M 9079)
Name proposed by F. K. Edmondson.

(2335) James
1974 UB. Discovered 1974 October 17 by E. F. Helin at Palomar.
Named in honor of James G. Williams, celestial mechanician who extended the theory of secular perturbations to orbits of high inclination. This minor planet lies very near one of the secular resonances discovered by Williams, and its argument of perihelion librates around the value of 90 degrees. (M 5688)

(2336) Xinjiang
1975 WL_1. Discovered 1975 November 26 at the Purple Mountain Observatory at Nanking.
Named for the Xinjiang Uygur Autonomous Region, the largest Chinese region by area, situated at the northwestern border of the country. (M 6060)

(2337) Boubín
1976 UH_1. Discovered 1976 October 22 by P. Wild at Zimmerwald.
Named for a mountain in Bohemia that offers a wonderful panorama and still has a virgin forest on its slopes. (M 17465; M 30493)
Name suggested by I. Bauersima.

(2338) Bokhan
1977 QA_3. Discovered 1977 August 22 by N. S. Chernykh at Nauchnyj.
Named in honor of Nadezhda Antonovna Bokhan {1916- }, a staff member of the Institute for Theoretical Astronomy in Leningrad during 1944-1957 and 1965-1974, known for her valuable contributions to the study of minor planets and for her investigation on the motion of periodic comet Encke. (M 6060)

(2339) Anacreon
2509 P-L. Discovered 1960 September 24 by C. J. van Houten and I. van Houten-Groeneveld at Palomar.
Named for the Greek poet Anacreon (fl. c. 550 B.C.), whose poems celebrate wine, love and friendship. (M 16590)

(2340) Hathor
1976 UA. Discovered 1976 October 22 by C. T. Kowal at Palomar.
Like the other objects of Aten type, (2340) is named for an Egyptian deity. Known as a sky-goddess and the daughter of Ra, Hathor was also identified with Aphrodite {see planet (1388)}. (M 6060)
Name proposed by E. Helin, who made an independent discovery of the object and who also made crucial recovery observations in 1981.
Hathor is the same goddess as Athor {see planet (161)}.

(2341) Aoluta
1976 YU_1. Discovered 1976 December 16 by L. I. Chernykh at Nauchnyj.
Named on the occasion of the centennial, in 1981, of the founding of the Astronomical Observatory of Leningrad University. The first part of the name is an acronym for that institution. (M 6648)

(2342) Lebedev
1968 UQ. Discovered 1968 October 22 by T. M. Smirnova at Nauchnyj.
Named in memory of Nikolaj Aleksandrovich Lebedev (1914-1942), a valiant tank officer who fell in the battle of Stalingrad. (M 6060)

(2343) Siding Spring
1979 MD_4. Discovered 1979 June 25 by E. F. Helin and S. J. Bus at Siding Spring.
Named for the New South Wales community in which the Australian National Observatory, Anglo-Australian Observatory and U.K. Schmidt telescope are lo-

cated. This is the first minor planet discovered with the U.K. Schmidt telescope to be numbered. (M 5851)

(2344) Xizang

1979 SC_1. Discovered 1979 September 27 at the Purple Mountain Observatory at Nanking.

Named for the autonomous region, also known as Tibet, on the southwestern border of China. (M 6060)

(2345) Fučik

1974 OS. Discovered 1974 July 25 by T. M. Smirnova at Nauchnyj.

Named in memory of Julius Fučik (1903-1943), a Czechoslovak national hero and writer. (M 6648)

(2346) Lilio

1934 CB. Discovered 1934 February 5 by K. Reinmuth at Heidelberg.

Named on the occasion of the 400th anniversary of the establishment of the Gregorian Calendar in memory of its inventor, Luigi Lilio (ca. 1510-1576), an obscure physician from southern Italy. (M 7471)

Name proposed by B. G. Marsden, who found the identifications involving this planet.

(2347) Vinata

1936 TK. Discovered 1936 October 7 by H. L. Giclas at Flagstaff.

Named for a daughter of Prajapati and wife of the sage Kasyapa in the mythology of India. She laid only two eggs, while her sister and Kasyapa's other wife laid a thousand. After five hundred years her sister's eggs hatched into serpents. Jealous and impatient, she broke the shell of one of her two eggs to find Aruna {see planet (2313)} inside with only his upper half developed. The second egg hatched into Garuda {see planet (2307)}, who had such spiritual purity as to exceed that of all his thousand half-brothers. (M 21606)

Named by the Minor Planet Names Committee following a suggestion by F. Pilcher.

(2348) Michkovitch

1939 AA. Discovered 1939 January 10 by M. B. Protitch at Belgrade.

Named by the discoverer as a sign of much appreciation and in memory of his professor, Vojislav V. Michkovitch (1892-1976), the first director of the new Belgrade Astronomical Observatory, which, by his efforts, was built in 1932. He was also founder of the Astronomical Institute of the Serbian Academy of Sciences and was well known for his works on minor planets, especially their identifications. (M 9477)

Obituary published in Bull. Obs. Astron. Belgrade, No. 129, p. III-VI (1978).

(2349) Kurchenko

1970 OG. Discovered 1970 July 30 by T. M. Smirnova at Nauchnyj.

Named in memory of Nadezhda Kurchenko (1950-1970), an airline stewardess killed in a clash with two armed bandits. (M 6648)

(2350) von Lüde

1938 CG. Discovered 1938 February 6 by A. Bohrmann at Heidelberg.

Named in memory of Heinz von Lüde (1914-1974), astronomer at the Astronomisches Rechen-Institut who calculated many preliminary orbits of minor planets. He also studied a fictitious example of 3/1 libration in the restricted three-body problem and was involved with the meridian-circle program at the Heidelberg Observatory. (M 6060)

(2351) O'Higgins

1964 VD. Discovered 1964 November 3 at the Goethe Link Observatory at Brooklyn, Indiana.

Named in memory of Bernardo O'Higgins (1778-1842) and to honor the people of Chile, who regard him as the "Father of the Country". Chile declared independence from Spanish rule on 1810 Sept. 18. This became a reality on 1817 Feb. 22, when the Spaniards were defeated by General O'Higgins and his forces in a decisive battle at Chacabuco. (M 9079)
Name proposed by F. K. Edmondson.

(2352) Kurchatov

1969 RY. Discovered 1969 September 10 by L. I. Chernykh at Nauchnyj.

Named in memory of Igor' Vasil'evich Kurchatov (1902-1960), great atomic physicist and organizer of science, founder and first director of the Institute of Atomic Energy of the U.S.S.R. Academy of Sciences. (M 7617)
Kurchatov is also honored by a lunar crater.

(2353) Alva

1975 UD. Discovered 1975 October 27 by P. Wild at Zimmerwald.
Named for an American former girlfriend of the discoverer. (M 17465)

(2354) Lavrov

1978 PZ_3. Discovered 1978 August 9 by L. I. Chernykh and N. S. Chernykh at Nauchnyj.

Named in honor of Svyatoslav Sergeevich Lavrov {1923- }, director of the Institute for Theoretical Astronomy, Leningrad, a corresponding member of the U.S.S.R. Academy of Sciences, and a well-known authority on computer mathematics and computer languages. (M 7617)

(2355) Nei Monggol

1978 UV_1. Discovered 1978 October 30 at the Purple Mountain Observatory at Nanking.

Named for an autonomous region, also known as Inner Mongolia, on the northern border of China, noted for its extensive prairie lands. (M 6061)

(2356) Hirons

1979 UJ. Discovered 1979 October 17 by E. Bowell at Anderson Mesa.

Named in honor of the discoverer's parents-in-law, Charles and Ann Hirons. (M 6208)

(2357) Phereclos

1981 AC. Discovered 1981 January 1 by E. Bowell at Anderson Mesa.

This Trojan planet is named for the skilled craftsman and builder of the ships that took Paris to Sparta. Phereclos was killed in battle by Meriones {see planet (3596)}. (M 6208)

(2358) Bahner

1929 RE. Discovered 1929 September 2 by K. Reinmuth at Heidelberg.

Named in honor of Klaus Bahner {1921- }, staff member of the Heidelberg-Königstuhl Observatory who has made outstanding contributions to the design of large telescopes. The design of the 1.2-m, 2.2-m and 3.5-m large telescopes of the Max-Planck-Institut für Astronomie is mainly due to his efforts. (M 6649)

Name proposed by L. D. Schmadel; endorsed by O. Kippes, who found the identifications for this planet.

(2359) Debehogne

1931 TV. Discovered 1931 October 5 by K. Reinmuth at Heidelberg.

Named in honor of Henri Debehogne {1928- }, astronomer at the Royal Observatory, Uccle, noted for his astrometric work on comets and minor planets. In recent years he has trained and collaborated with several astronomers in South America, and it was during one of these collaborations that he rediscovered this minor planet in 1979. (M 6209)

Name proposed by E. Bowell, who made the identification.

(2360) Volgo-Don

1975 VD_3. Discovered 1975 November 2 by T. M. Smirnova at Nauchnyj.

Named on the occasion of the 30th anniversary of the construction of the Volgo-Don Navigation Canal, which connects the Volga and Don rivers at their closest approach. (M 7617)

(2361) Gogol

1976 GQ_1. Discovered 1976 April 1 by N. S. Chernykh at Nauchnyj.

Named for the celebrated writer Nikolaj Vasil'evich Gogol' (1809-1852). (M 7783)

(2362) Mark Twain

1976 SH_2. Discovered 1976 September 24 by N. S. Chernykh at Nauchnyj.

Named for Mark Twain, pen name of Samuel Langhorne Clemens (1835-1910), world-famous American writer. (M 9214)

(2363) Cebriones

1977 TJ_3. Discovered 1977 October 4 at the Purple Mountain Observatory at Nanking.

This Trojan planet is named for the charioteer of Hektor {see planet (624)}, wounded in the conflict between Hektor and Patroclus {see planet (617)} that led to the latter's death. (M 6209)

(2364) Seillier

1978 GD. Discovered 1978 April 14 by H. Debehogne at La Silla.

Named by the discoverer in honor of his mother and her family. (M 9079)

(2365) Interkosmos

1980 YQ. Discovered 1980 December 30 by Z. Vávrová at Kleť.

Named for the eastern European organization for space exploration. (M 6649)

(2366) Aaryn

1981 AC_1. Discovered 1981 January 10 by N. G. Thomas at Anderson Mesa.

Named in honor of Aaryn G. Baltutis, grandson of the discoverer. (M 7617)

(2367) Praha

1981 AK_1. Discovered 1981 January 8 by A. Mrkos at Kleť.

Named for the capital of {the former} Czechoslovakia, which has a longstanding astronomical tradition going back to the time of Tycho Brahe and Johannes Kepler. (M 6649)

(2368) Beltrovata

1977 RA. Discovered 1977 September 4 by P. Wild at Zimmerwald.

This Amor object shares the name given by the Zürich-born writer Gottfried Keller to his friend Betty Tendering. Keller's novel *Der grüne Heinrich* contains a comparable character in Dortchen Schönfund. (M 6209)

(2369) Chekhov

1976 GC_8. Discovered 1976 April 4 by N. S. Chernykh at Nauchnyj.

Named for the celebrated writer Anton Pavlovich Chekhov (1860-1904). (M 7783)

(2370) van Altena

1965 LA. Discovered 1965 June 10 by A. R. Klemola at El Leoncito.

Named in honor of William F. van Altena, well-known astrometrist at the Yale University Observatory. His contributions include works on stellar proper motions and parallaxes, including a revised edition of the Yale Catalogue of Stellar Trigonometric Parallaxes. The southern proper motion program with respect to galaxies is conducted under his direction with the 0.5-m astrograph at El Leoncito. (M 9477)

(2371) Dimitrov

1975 VR_3. Discovered 1975 November 2 by T. M. Smirnova at Nauchnyj.

Named in memory of Georgij Mikhailovich Dimitrov (1882-1949), founder and leader of the Bulgarian Patriotic Front and chairman of the council of ministers of Bulgaria. (M 6649)

(2372) Proskurin

1977 RA_8. Discovered 1977 September 13 by N. S. Chernykh at Nauchnyj.

Named in memory of Vitalij Fedorovich Proskurin (1919-1964), celestial mechanician at the Institute for Theoretical Astronomy, known for his work on the motions of Ceres, Jupiter VI and VIII and artificial satellites. (M 7783)

(2373) Immo

1929 PC. Discovered 1929 August 4 by M. Wolf at Heidelberg.

Named in honor of Immo Appenzeller (1940-), professor of astronomy at Heidelberg University and since 1975 director of the observatory at Königstuhl. He has made important contributions to the fields of star formation, stellar evolution, interstellar magnetic fields and active galaxies. Appenzeller is also deeply involved in the development of astronomical instrumention. (M 18447)
Name proposed and citation prepared by G. Klare and L. D. Schmadel.

(2374) Vladvysotskij

1974 QE_1. Discovered 1974 August 22 by L. V. Zhuravleva at Nauchnyj.

Named in memory of Vladimir Semenovich Vysotskij (1938-1980), a notable dramatic actor, composer and poet, particularly popular as a music-hall singer who performed songs of his own composition. (M 7617)

(2375) Radek

1975 AA. Discovered 1975 January 8 by L. Kohoutek at Bergedorf.

Named by the discoverer in honor of his brother, Ctirad Kohoutek (1929-), Czech composer, who has been professor for music at the academies in Brno and Prague as well as director of the Czech Philharmonic Orchestra. (M 22244)

(2376) Martynov

1977 QG_3. Discovered 1977 August 22 by N. S. Chernykh at Nauchnyj.

Named in honor of Dmitrij Yakovlevich Martynov {1904-1989}, outstanding astrophysicist, for many years the director of the Sternberg Astronomical Institute in Moscow, author of widely used texts on astrophysics. (M 7783)

Obituaries published in Astron. Calendar Bulg., p. 124-126 (1991); Astron. Vestn., Tom 24, No. 2, p. 173-175 (1990); Astron. Zh., Tom 67, Vyp. 2, p. 440-442 (1990); Zemlya Vselennaya, No. 1, p. 45-48 (1990); Astron. Tsirk., No. 1541, p. 39-40 (1989).

(2377) Shcheglov

1978 QT_1. Discovered 1978 August 31 by N. S. Chernykh at Nauchnyj.

Named in honor of Vladimir Petrovich Shcheglov {1904-1985}, since 1941 the director of the Astronomical Institute of the {former} Uzbek SSR in Tashkent, investigator in astrometry and the history of astronomy, renowned popularizer of astronomy. (M 7783)

(2378) Pannekoek

1935 CY. Discovered 1935 February 13 by H. van Gent at Johannesburg.

Named in memory of A. Pannekoek (1873-1960), Dutch astronomer, director of the Astronomical Institute of the Municipal University in Amsterdam from 1921 to 1941. His main activity was stellar astrophysics. (M 18135)
Pannekoek is also honored by a lunar crater.

(2379) Heiskanen

1941 ST. Discovered 1941 September 21 by Y. Väisälä at Turku.

Named in memory of V. A. Heiskanen (1895-1971), professor of geodesy at the Helsinki Technological University and director of the Finnish Geodetic Institute. After 1950 he worked mainly at the Ohio State University in Columbus, where he began the program in geodesic sciences. He also wrote several Finnish textbooks on astronomy. (M 7945)

(2380) Heilongjiang

1965 SN. Discovered 1965 September 18 at the Purple Mountain Observatory at Nanking.

Named for a forested province in northeastern China, rich in petroleum, coal and gold. (M 7617)

(2381) Landi

1976 AF. Discovered 1976 January 3 at the Felix Aguilar Observatory at El Leoncito.

Named in honor of Jorge Landi Dessy, formerly director of the Córdoba Observatory and professor at the Córdoba Institute of Mathematics, Physics and Astronomy; also vice-president of the Argentine Association of Astronomy. His principal fields of study involved astronomical optics and photometry, and the structure of the Magellanic Clouds. (M 8799)

(2382) Nonie

1977 GA. Discovered 1977 April 13 at the Perth Observatory at Bickley.

Named in honor of the daughter of Peter Jekabsons {see planet (3188)}, member of the astrometric team at the Perth Observatory. (M 6531)

(2383) Bradley

1981 GN. Discovered 1981 April 5 by E. Bowell at Anderson Mesa.

Named in honor of Martin and Maud Bradley, friends of the discoverer. (M 6209)

(2384) Schulhof

1943 EC_1. Discovered 1943 March 2 by M. Laugier at Nice.

Named in memory of Leopold Schulhof (1847-1921), Hungarian-born astronomer who spent most of his working life in Paris, although he discovered (147) Protogeneia while a student in Vienna. His extensive computations on the orbits of comets and minor planets frequently involved the interplay of perturbations and determinacy, and he contributed much toward the recovery of lost objects. His very detailed study of the comet now known as P/Crommelin is acknowledged as that object comes now again to perihelion. (M 8541)

Name proposed by B. G. Marsden, who found the identifications involving this planet.

(2385) Mustel

1969 VW. Discovered 1969 November 11 by L. I. Chernykh at Nauchnyj.

Named in honor of Evald Rudolfovich Mustel' {1911-1988}, chairman of the Astronomical Council of the U.S.S.R. Academy of Sciences, editor of the *Astronomicheskij Zhurnal*, and a vice-president of the IAU during 1970-1976. His research activities involve several aspects of solar and stellar physics and the correlation of geophysical phenomena and solar activity. (M 7617)

Obituaries published in Astron. Zh., Tom 65, Vyp. 4, p. 891-892 (1988); Pis'ma Astron. Zh., Tom 14, No. 8, p. 764 (1988); Zemlya Vselennaya, No. 5, p. 42-45 (1988); Astrophys. Space Sci., Vol. 155, No. 1, p. 1 (1989); Sov. Astron. Lett., Vol. 14, No. 4, p. 326 (1988); Sov. Astron., Vol. 32, No. 4, p. 466-467 (1988).

(2386) Nikonov

1974 SN_1. Discovered 1974 September 19 by N. S. Chernykh at Nauchnyj.

Named in honor of Vladimir Borisovich Nikonov {1905- }, chief of the stellar department of the Crimean Astrophysical Observatory, a pioneer in stellar photoelectric photometry. (M 7471)

(2387) Xi'an

1975 FX. Discovered 1975 March 17 at the Purple Mountain Observatory at Nanking.

Named for the famous historical city, origin of Chinese splendid culture and home of a rich treasure of human civilization. Known in ancient times as Chang'an, it takes first place among the six ancient capitals of China, serving in this capacity for twelve dynasties since the eleventh century B.C. With Athens, Cairo and Rome it ranks among the four leading ancient capitals of the world. The terra cotta warriors and horses at the Museum of Qin Shihuang are praised as an "eighth wonder of the world". Today, Xi'an is the political and economic center of Shaanxi {see planet (2263)} province and the largest central city in the western part of China. It is an important center for national higher education and scientific research, as well as for the aerospace, aviation, electronic, textile and power industries. (M 17655)

(2388) Gase

1977 EA_2. Discovered 1977 March 13 by N. S. Chernykh at Nauchnyj.

Named in memory of Vera Fedorovna Gaze (1899-1954), who worked at the Pulkovo and Simeïs observatories on stellar spectroscopy and the study of diffuse nebulae. As a young assistant at the Astronomical Institute (predecessor of ITA) in Leningrad she determined the orbits of minor planets. (M 7783)

(2389) Dibaj

1977 QC_1. Discovered 1977 August 19 by N. S. Chernykh at Nauchnyj.

Named in memory of Ernest Apushevich Dibaj (1931-1983), an astrophysicist known for his research on the interstellar medium, variable stars and extragalactic systems. A professor at Moscow University, he was head of the Crimean Station of the Sternberg Astronomical Institute from 1962 to 1977. (M 8542)

Obituary published in Pis'ma Astron. Zh., Tom 9, No. 12, p. 755 (1983); Astron. Zh., Tom 61, Vyp. 1, p. 204-205 (1984); Sov. Astron. Lett., Vol. 9, No. 6, p. 390-391 (1983); Sov. Astron., Vol. 28, No. 1, p. 121-122 (1984).

(2390) Nežárka

1980 PA_1. Discovered 1980 August 14 by Z. Vávrová at Kleť.

Named for a small river running through the town of Veseli {see planet (2599)} in southern Bohemia where the discoverer lives. (M 6649)

(2391) Tomita

1957 AA. Discovered 1957 January 9 by K. Reinmuth at Heidelberg.

Named in honor of Koichiro Tomita, astronomer at the Tokyo Astronomical

Observatory during 1947-1985, prominent observer of comets and minor planets, discoverer of comet 1964 VI and recoverer of almost all the comets that returned in 1967. Also a leading popularizer of astronomy in Japan, Tomita has been a source of inspiration for H. Oishi {see planet (3379)}, who made the key identification for this planet and proposed the name. (M 2391)

(2392) Jonathan Murray

1979 MN_1. Discovered 1979 June 25 by E. F. Helin and S. J. Bus at Siding Spring.

Named for Jonathan Murray, the young son of Bruce and Suzanne Murray, longtime friends of the discoverers. The planet was discovered a few weeks before his birth on 1979 July 19. (M 6209)

(2393) Suzuki

1955 WB. Discovered 1955 November 17 by M. Laugier at Nice.

Named in honor of Keishin Suzuki (1905-), professor of astronomy at Tokyo Gakugei University, an expert on practical astronomy and teaching of astronomy. He played the principal part in first publishing the Japanese Nautical Almanac, when astronomical almanacs could not be imported during the second World War. He has made a great contribution to the teaching and popularization of astronomy in Japan, mainly by writing or translating a great many books and articles over a period of more than fifty years. (M 11441)

Name proposed by T. Urata, who found the identifications involving this planet.

(2394) Nadeev

1973 SZ_2. Discovered 1973 September 22 by N. S. Chernykh at Nauchnyj.

Named in memory of Lev Nikolaevich Nadeev (1902-1974), an astrometrist and geodesist who made extensive geodetic surveys in the northern USSR. He founded the Laboratory of Time and Frequency in Irkutsk {see planet (3224)}, where the discoverer began his own career under Nadeev's guidance. (M 7784)

(2395) Aho

1977 FA. Discovered 1977 March 17 at the Harvard College Observatory at Harvard.

Named in honor of Arne J. Aho by his coworkers in appreciation of his hard work that made their work easier. (M 6833)

(2396) Kochi

1981 CB. Discovered 1981 February 9 by T. Seki at Geisei.

Named by the discoverer in honor of his native city. (M 6421)

See also the citation for planet (3150).

(2397) Lappajärvi

1938 DV. Discovered 1938 February 22 by Y. Väisälä at Turku.

Named for a lake, which proved to be a meteorite crater, in the western part of central Finland. (M 7945)

(2398) Jilin

1965 UD_2. Discovered 1965 October 24 at the Purple Mountain Observatory at Nanking.

Named for the province in northeastern China, famous for its "ginseng". (M 7617)

(2399) Terradas

1971 MA. Discovered 1971 June 17 by C. U. Cesco at El Leoncito.

Named in memory of Esteban Terradas e Illa (1883-1950), outstanding Spanish mathematician, professor at the universities of Zaragoza, Barcelona and Madrid, and later at the Astronomical School of La Plata University. He made important contributions to geodesy and was involved with the fundamental precision marigraph in Puerto Madryn. (M 8799)

(2400) Derevskaya

1972 KJ. Discovered 1972 May 17 by T. M. Smirnova at Nauchnyj.

Named in memory of Alexandra Avramovna Derevskaya (1902-1959), mother heroine, who brought up 48 orphans of different nationalities. (M 7617)

(2401) Aehlita

1975 VM_2. Discovered 1975 November 2 by T. M. Smirnova at Nauchnyj.

Named for the heroine in one of the novels of Aleksej Nikolaevich Tolstoj (1883-1945, see planet (3771)). In the book of the same name Aehlita was a Martian girl who fell in love with an Earth man who went to Mars in a space ship. (M 6649)

(2402) Satpaev

1979 OR_{13}. Discovered 1979 July 31 by N. S. Chernykh at Nauchnyj.

Named in memory of Kanysh Imantaevich Satpaev (1899-1964), outstanding geologist. As president of the Academy of Sciences of the Kazakh SSR he contributed much to the development of astronomy in Kazakhstan. (M 8542)

(2403) Šumava

1979 SQ. Discovered 1979 September 25 by A. Mrkos at Kleť.

Named for the mountains on the southern border of Bohemia in the vicinity of the Kleť Observatory. (M 6649)

(2404) Antarctica

1980 TE. Discovered 1980 October 1 by A. Mrkos at Kleť.

Named for the southernmost continent, in honor of the third Soviet Antarctic Expedition, in which the discoverer participated. (M 6649)

(2405) Welch

1963 UF. Discovered 1963 October 18 at the Goethe Link Observatory at Brooklyn, Indiana.

Named in honor of David F. ("Kelly") Welch, AURA Corporate Staff Executive from 1978 to 1983. He upgraded AURA administrative and operational procedures and brought modern technology into the corporate office. He also played a major role in the preparation of the successful proposal to NASA for AURA to establish and operate the Space Telescope Science Institute. Following (2404) Antarctica, this planet also acknowledges Welch's earlier career in the U.S. Navy, which culminated in his service as commander of the Naval Support Force for the scientific bases in Antarctica during 1969-1971. (M 9079)
Name proposed by F. K. Edmondson.

(2406) Orelskaya

1966 QG. Discovered 1966 August 20 at the Crimean Astrophysical Observatory at Nauchnyj.

Named in honor of Varvara Ivanovna Orel'skaya, a member of the staff of the Institute for Theoretical Astronomy from 1937 to 1982, well-known for her work on determining the equator-equinox correction using observations of minor planets. Her plan, supported by IAU Commission 20, of observations of 20 selected planets during 1974-1990 is continuing. (M 7618)

(2407) Haug

1973 DH. Discovered 1973 February 27 by L. Kohoutek at Bergedorf.

Named in memory of Ulrich Haug (1929-1992), German astronomer, who inves-

tigated interplanetary dust particles in his early years in Tübingen and who is known from later work at the Hamburg Observatory, mainly for his contributions on interstellar extinction and for observations with the Schmidt telescope, moved from Hamburg-Bergedorf to the German-Spanish Astronomical Center at Calar Alto, where he made tests for quasars being the basis for the Hamburg Quasar Survey. (M 22244)
Obituary published in Mitt. Astron. Ges., Nr. 76, p. 5-7 (1993).

(2408) Astapovich

1978 QK_1. Discovered 1978 August 31 by N. S. Chernykh at Nauchnyj.

Named in memory of Igor' Stanislavovich Astapovich (1908-1976), a professor at Kiev University known for his research in meteor astronomy. (M 7784)
Obituary published in Komety Meteory, No. 26, p. 43-48 (1977).

(2409) Chapman

1979 UG. Discovered 1979 October 17 by E. Bowell at Anderson Mesa.

Named in honor of Clark R. Chapman, planetary astronomer at the Planetary Science Institute in Tucson. He has made outstanding contributions to our understanding of asteroid compositions and physical processes, particularly surface mineralogical identification, taxonomy and collisional evolution. (M 6209)

(2410) Morrison

1981 AF. Discovered 1981 January 3 by E. Bowell at Anderson Mesa.

Named in honor of David Morrison, astronomer at the University of Hawaii, Honolulu. His research into the infrared radiometric properties of asteroids has been fundamental in revealing the diversity of asteroid surface albedos and compositions. He has also worked on the definition of a taxonomic system for asteroids and on the distribution of the various taxonomic types in the main belt. (M 6209)

(2411) Zellner

1981 JK. Discovered 1981 May 3 by E. Bowell at Anderson Mesa.

Named in honor of Benjamin H. Zellner, astronomer at the University of Arizona, Tucson. He brought to fruition the polarimetric-photometric method of albedo and diameter determination for asteroids. He has also helped define a taxonomic system for asteroids and has investigated in detail the distribution of the various taxonomic types in the main belt. (M 6209)

(2412) Wil

3537 P-L. Discovered 1960 October 17 by C. J. van Houten and I. van Houten-Groeneveld at Palomar.

Named in honor of Wil van de Hulst, psychotherapist, married to the astronomer Henk C. van de Hulst {see planet (2413)}. (M 9767)

(2413) van de Hulst

6816 P-L. Discovered 1960 September 24 by C. J. van Houten and I. van Houten-Groeneveld at Palomar.

Named in honor of Henk C. van de Hulst {1918- }, husband of Wil van de Hulst (see 2412) and astrophysicist at the Leiden Observatory. A leader in the development of radioastronomy in the Netherlands, van de Hulst predicted the interstellar hydrogen 21-cm line. He has made contributions to the theory of light scattering by small particles, about which topic he wrote two books. He is active in promoting international cooperation in space research. (M 11640)

(2414) Vibeke

1931 UG. Discovered 1931 October 18 by K. Reinmuth at Heidelberg.

Named by L. K. Kristensen {see planet (3455)}, who made the identifications involving this planet, in honor of his daughter. (M 6210)

(2415) Ganesa

1978 UJ. Discovered 1978 October 28 by H. L. Giclas at Anderson Mesa.

In the mythology of India, Ganesa is the son of Siva and Parvati {see planets (1170) and (2847), respectively}, who represents the call to spiritual power. Soon after birth he glanced at the god Sani, who symbolizes obstacles, and the baby's head was reduced to ashes. His head was then replaced with the head of Indra's elephant, Airavata. He procures wealth and encourages the success of all enterprises, and is much revered by the merchant class. (M 21606)

Named by the Minor Planet Names Committee following a suggestion by F. Pilcher.

(2416) Sharonov

1979 OF_{13}. Discovered 1979 July 31 by N. S. Chernykh at Nauchnyj.

Named in memory of Vsevolod Vasil'evich Sharonov (1901-1964), professor at Leningrad University and director of the Leningrad University Observatory, who contributed greatly to physical studies of the Moon and planets. (M 7784) Sharonov is also honored by craters on Mars and the Moon.

(2417) McVittie

1964 CD. Discovered 1964 February 15 at the Goethe Link Observatory at Brooklyn, Indiana.

Named in honor of George C. McVittie {1904-1988} on the occasion of his 80th birthday, 1984 June 5. Born in Turkey, he was educated and spent his early career in the U.K. Moving to the U.S., he served as head of the astronomy department at the University of Illinois from 1952 to 1972, during which time he converted a one-man undergraduate teaching department into a major graduate and research center. His own research was theoretical, involving relativity and cosmology, but he built up observational optical astronomy and added radio astronomy to the department to produce a well-rounded program. From 1961 to 1970 he served as secretary of the American Astronomical Society. Following his formal retirement he returned to the U.K. to a position at the University of Kent. (M 9080) Name proposed by F. K. Edmondson.

Obituaries published in Phys. Today, Vol. 42, No. 3, p. 128-132 (1989); Q.J.R. Astron. Soc., Vol. 30, No. 1, p. 119-122 (1989).

(2418) Voskovec-Werich

1971 UV. Discovered 1971 October 26 by L. Kohoutek at Bergedorf.

Named in memory of Jiří Voskovec (1905-1981) and Jan Werich (1905-1980), Czech actors and dramatists who became famous in the 1930s for several theater plays and reviews, mainly in the avant-garde Liberal Theater and the "V+W" Theater in Prague. After World War II they acted in many theater, film and TV productions, Voskovec in the U.S. and Werich in Czechoslovakia. They were highly respected for their humanistic ideas. (M 27124)

(2419) Moldavia

1974 SJ. Discovered 1974 September 19 by N. S. Chernykh at Nauchnyj.
Named for the {former} Moldavian Soviet Socialist Republic. (M 7471)

(2420) Čiurlionis

1975 TN. Discovered 1975 October 3 by N. S. Chernykh at Nauchnyj.

Named in memory of M. K. Čiurlionis (1875-1911), a well-known Lithuanian painter and composer. (M 8542)

(2421) Nininger

1979 UD. Discovered 1979 October 17 by E. Bowell at Anderson Mesa.

Named on the occasion of the 95th birthday, 1982 Jan. 17, of Harvey Harlow Nininger {1887-1986}, celebrated U.S. meteoriticist. Through field investigations and publicity he was responsible for the recovery of some 200 separate meteorite falls and finds, and his resulting 3000 samples are more than those of any other individual prior to the exploration of the Antarctic blue-ice fields. The author of many popular and technical works, he conducted much of his research at the Arizona meteorite crater, not far from Flagstaff. (M 6531)
Name proposed by N. Sperling.

Obituaries published in Meteoritics, Vol. 21, No. 2, p. 239, No. 4, p. 551-552 (1986).

(2422) Perovskaya

1968 HK_1. Discovered 1968 April 28 by T. M. Smirnova at Nauchnyj.

Named in memory of Sofya Lvovna Perovskaya (1853-1881), executed for organizing the attempt of the life of Tsar Alexander II on 1881 Mar. 1. (M 6649)

(2423) Ibarruri

1972 NC. Discovered 1972 July 14 by L. V. Zhuravleva at Nauchnyj.

Named in memory of Ruben Ibarruri (1920-1942), who died valiantly in the battle of Stalingrad. (M 6649)

(2424) Tautenburg

1973 UT_5. Discovered 1973 October 27 by F. Börngen at Tautenburg.

Named for the village near which the Karl Schwarzschild Observatory is located, some 15 km to the northeast of the city of Jena {see planet (526)}. (M 7784)

(2425) Shenzhen

1975 FW. Discovered 1975 March 17 at the Purple Mountain Observatory at Nanking.

Named for China's first special economic zone, well known at home and abroad. It has often been referred to as an 'overnight' modernized, export-oriented and multifunctional city with industry as its mainstay but combining industry with trade. Located on the east bank of the Pearl river estuary in Guangdong {see planet (2185)} province and neighboring Hong Kong, Shenzhen enjoys geographical advantages and is an ideal tourist attraction. Rich in natural resources and technological expertise, it has a well-developed commodity economy and an efficient transportation and telecommunications system. Shenzhen serves as the nation's major gateway to international trade and exchange. (M 17655)

(2426) Simonov

1976 KV. Discovered 1976 May 26 by N. S. Chernykh at Nauchnyj.

Named in memory of Konstantin Mikhailovich Simonov (1915-1979), Soviet writer and public figure. (M 8064)

(2427) Kobzar

1976 YQ_7. Discovered 1976 December 20 by N. S. Chernykh at Nauchnyj.

Named in memory of the great Ukrainian poet and painter Taras Grigor'evich Shevchenko (1814-1861), who exerted a profound influence on Ukrainian literature. He was popularly known as Kobzar, which in old Ukraine means one who sings to the accompaniment of a kobza, a stringed instrument. (M 7471)

(2428) Kamenyar

1977 RZ_6. Discovered 1977 September 11 by N. S. Chernykh at Nauchnyj.

Named in memory of Ivan Yakovlevich Franko (1856-1916), a Ukrainian writer and scientist whose remarkably rich creative legacy is now being published in 50 volumes. He was sometimes called Kamenyar, or 'stonecrusher', the title of one of his poems. (M 7471)

(2429) Schürer

1977 TZ. Discovered 1977 October 12 by P. Wild at Zimmerwald.

Named in honor of Max Schürer (1910-1997), who from 1947 to 1980 was director of the Astronomical Institute of the University of Berne. The existence of the observatory at Zimmerwald {see planet (1775)} is due to his initiative, endurance and great technical competence. As a pupil of S. Mauderli {see planet (1748)} he did much numerical work on the orbits of minor planets. Then he dealt with stellar dynamics and finally got deeply involved as a pioneer in satellite geodesy. His courses had broad scope and great clarity, and they inspired many young physicists to have a lasting interest in astronomical problems. (M 21129)
Obituary published in Orion, Jahrg. 56, Nr. 287, p. 1-2 (1998).

(2430) Bruce Helin

1977 VC. Discovered 1977 November 8 by E. F. Helin and E. M. Shoemaker at Palomar.

Named in honor of the first discoverer's son, who has for many years tolerated his mother's preoccupation with extraterrestrial objects. (M 6421)

(2431) Skovoroda

1978 PF_3. Discovered 1978 August 8 by N. S. Chernykh at Nauchnyj.

Named in memory of Grigorij Savvich Skovoroda (1722-1794), an outstanding Ukrainian philosopher and poet. He wandered over the Ukraine preaching his learning to the people. (M 7472)

(2432) Soomana

1981 FA. Discovered 1981 March 30 by E. Bowell at Anderson Mesa.

In the Hopi Indian language of northern Arizona, Soomana means "star girl". A Hopi child is traditionally named by its father's mother or sister clan, in this case the Sun clan. (M 6650)
Name proposed by Michael Lomatewama and Ekkehart Malotki.

(2433) Sootiyo

1981 GJ. Discovered 1981 April 5 by E. Bowell at Anderson Mesa.

In the Hopi Indian language, Sootiya means "star boy". The tradition and proposers are as for (2432) Soomana. (M 6650)

(2434) Bateson

1981 KA. Discovered 1981 May 27 by A. C. Gilmore and P. M. Kilmartin at Lake Tekapo.

Named in honor of Frank Maine Bateson {1909- }, director of the Variable Star Section of the Royal Astronomical Society of New Zealand since 1928. He led the site survey of New Zealand that resulted in the establishment of the Mount John University Observatory and was astronomer-in-charge there until 1968. Dr. Bateson and his wife Doris continue to support and encourage astronomy in New Zealand. (M 6210)

(2435) Horemheb

4578 P-L. Discovered 1960 September 24 by C. J. van Houten and I. van Houten-Groeneveld at Palomar.

Named for the last pharaoh of the eighteenth dynasty in ancient Egypt. His tomb was recently excavated by a joint British-Dutch expedition. (M 8153)

(2436) Hatshepsut

6066 P-L. Discovered 1960 September 24 by C. J. van Houten and I. van Houten-Groeneveld at Palomar.
Named for the only female pharaoh to reign over ancient Egypt. (M 8153)

(2437) Amnestia

1942 RZ. Discovered 1942 September 14 by M. Väisälä at Turku.
Named to honor the work of Amnesty International. (M 7946)

(2438) Oleshko

1975 VO_2. Discovered 1975 November 2 by T. M. Smirnova at Nauchnyj.
Named in memory of Valentina Iosifovna Oleshko (1924-1943), who organized an underground resistance group at Lampovo, near Leningrad. (M 6650)

(2439) Ulugbek

1977 QX_2. Discovered 1977 August 21 by N. S. Chernykh at Nauchnyj.
Named for the celebrated Uzbek astronomer and mathematician Ulugbek (Ulugh Beg) Mukhammed Taragaj (1394-1449). (M 7784)
Ulugh Beg is also honored by a lunar crater.

(2440) Educatio

1978 VQ_4. Discovered 1978 November 7 by E. F. Helin and S. J. Bus at Palomar.
The Latin name of one of the most important human endeavors. The education of people on Earth is foremost to the industrialization of space. (M 7618)
Name proposed by M. Bush, of the World Space Foundation, which supports the Asteroid Project at Palomar.

(2441) Hibbs

1979 MN_2. Discovered 1979 June 25 by E. F. Helin and S. J. Bus at Siding Spring.
Named in honor of Al and Marka Hibbs, long time friends of the first discoverer, who wishes to acknowledge the role Al Hibbs played in introducing her to Space Science at Caltech shortly after Sputnik. (M 6422)

(2442) Corbett

1980 TO. Discovered 1980 October 3 by Z. Vávrová at Kleť.
Named for Jim Corbett, hunter, writer and a great friend of animals, born in Nainital, India. (M 6650)

(2443) Tomeileen

A906 BJ. Discovered 1906 January 24 by M. Wolf at Heidelberg.
Named by B. G. Marsden, who found the identifications for this planet, in memory of his parents, Thomas Marsden (1905-1980) and Eileen (née West) Marsden (1905-1981). (M 6833)

(2444) Lederle

1934 CD. Discovered 1934 February 5 by K. Reinmuth at Heidelberg.
Named in honor of Trudpert Lederle {1922- }, astronomer at the Astronomisches Rechen-Institut since 1942. Mainly involved with the program of fundamental star catalogues, he has also worked on the motion of (1036) Ganymed. (M 6833)
Name proposed by B. G. Marsden and J. Schubart.

(2445) Blazhko

1935 TC. Discovered 1935 October 3 by P. F. Shajn at Simeïs.
Named in memory of Sergej Nikolaevich Blazhko (1870-1956), director of the Moscow Observatory during 1920-1931, a professor at Moscow University during 1931-1953 and a corresponding member of the U.S.S.R. Academy of Sciences.

The founder of the Moscow school of variable-star research, he constructed the first general theory of eclipsing variables of the Algol type. He also proposed a new method of photographing minor planets and constructed a number of original instruments, including the blink microscope and slitless stellar spectrograph. (M 6834)
Name proposed by the Institute for Theoretical Astronomy.
Blazhko is also honored by a lunar crater.

(2446) Lunacharsky

1971 TS_2. Discovered 1971 October 14 by L. I. Chernykh at Nauchnyj.
Named in memory of Anatolij Vasil'evich Lunacharskij (1875-1933), Soviet statesman, writer and literary and art critic. (M 7784)

(2447) Kronstadt

1973 QY_1. Discovered 1973 August 31 by T. M. Smirnova at Nauchnyj.
Named for the town, situated on Kotlin Island in the Gulf of Finland, that played an integral part in the defense of Leningrad in World War II. (M 6650)

(2448) Sholokhov

1975 BU. Discovered 1975 January 18 by L. I. Chernykh at Nauchnyj.
Named in memory of Mikhail Aleksandrovich Sholokhov (1905-1984), famous Soviet writer, 1965 Nobel prize-winner. (M 13172)

(2449) Kenos

1978 GC. Discovered 1978 April 8 by W. Liller at Cerro Tololo.
Named for the first man in the mythology of the native Americans of Tierra del Fuego, sent by the Supreme Being to bring order into the world. He created the human race by using peat to make male and female organs, taught them language and instructed them in rules to fashion a harmonious society. (M 21606)
Named by the Minor Planet Names Committee following a suggestion by F. Pilcher.

(2450) Ioannisiani

1978 RP. Discovered 1978 September 1 by N. S. Chernykh at Nauchnyj.
Named in honor of Bagrat Konstantinovich Ioannisiani {1911-1985}, designer of astronomical instruments, including the largest Soviet telescopes. (M 7784)
Obituary published in Astrofizika, Tom 24, Vyp. 1, p. 208 (1986).

(2451) Dollfus

1980 RQ. Discovered 1980 September 2 by E. Bowell at Anderson Mesa.
Named in honor of Audouin Dollfus, astronomer at the Observatoire de Paris, Meudon. Dollfus is renowned for his fundamental work on planetary polarimetry and ground-based imaging, including studies of Saturn's rings, the Martian and lunar surfaces, and the Venus atmosphere. Recently he has worked on the polarimetric properties of minor planets. (M 6531)

(2452) Lyot

1981 FE. Discovered 1981 March 30 by E. Bowell at Anderson Mesa.
Named in memory of the French astronomer Bernard Lyot (1897-1952). One of the outstanding experimental astronomers of the twentieth century, Lyot invented the solar coronagraph and the birefringent filter. He developed the study of the polarization of light from planets to a perfection that has hardly been surpassed. (M 6531)
Lyot is also honored by craters on Mars and the Moon.

(2453) Wabash

A921 SA. Discovered 1921 September 30 by K. Reinmuth at Heidelberg.

Named in honor of Bob (Wabash) Warshow, occasional operator in the central computing facility at the Harvard-Smithsonian Center for Astrophysics. In his spare time he is a spelunker, and among his exploits was an expedition directly underneath the radio telescope at Arecibo. (M 18448)

Name proposed by B. G. Marsden, who found the key identification involving this planet.

(2454) Olaus Magnus

1941 SS. Discovered 1941 September 21 by Y. Väisälä at Turku.

Named for Olaus Magnus (1490-1557), the last Catholic archbishop of Sweden, who spent most of his life in exile in Rome. His 'Carta Marina' was the first map to represent Finland approximately correctly. (M 7946)

(2455) Somville

1950 TO_4. Discovered 1950 October 5 by S. Arend at Uccle.

Named in memory of Oscar Somville (1880-1980), some time head of the seismological section of the Uccle Observatory. (M 8404)

(2456) Palamedes

1966 BA_1. Discovered 1966 January 30 at the Purple Mountain Observatory at Nanking.

Named for the most intelligent of all the Greek commanders who fought in the Trojan War. (M 7618)

(2457) Rublyov

1975 TU_2. Discovered 1975 October 3 by L. I. Chernykh at Nauchnyj.

Named for Andrej Rublyov (ca. 1360-ca. 1430), a world-renowned Russian painter who worked on the paintings of some of the cathedrals in Moscow and elsewhere. (M 13172)

(2458) Veniakaverin

1977 RC_7. Discovered 1977 September 11 by N. S. Chernykh at Nauchnyj.

Named in memory of the famous Soviet writer Veniamin Aleksandrovich Kaverin (1902-1989). (M 18304)

(2459) Spellmann

1980 LB_1. Discovered 1980 June 11 by C. S. Shoemaker at Palomar.

Named in honor of the discoverer's father Leonard Spellmann, who has been a never-failing source of encouragement and support. (M 6532)

(2460) Mitlincoln

1980 TX_4. Discovered 1980 October 1 by L. G. Taff and D. Beatty at Socorro.

The name honors the Lincoln Laboratory and its association with the Massachusetts Institute of Technology. (M 9080)

(2461) Clavel

1981 EC_1. Discovered 1981 March 5 by H. Debehogne and G. DeSanctis at La Silla.

Named in honor of Gustavine Clavel and her family on the occasion of her hundredth birthday, 1991 Apr. 13, in recognition of a life of work and humility. (M 19332)

(2462) Nehalennia

6578 P-L. Discovered 1960 September 24 by C. J. van Houten and I. van Houten-Groeneveld at Palomar.

Named for a Roman goddess of fortune, worshiped mainly in the region that is now the Netherlands. (M 8153)

(2463) Sterpin

1934 FF. Discovered 1934 March 10 by G. Van Biesbroeck at Williams Bay.

Name proposed by the discoverer's daughter, Micheline Van Biesbroeck Wilson, in memory of her mother, Julia Sterpin Van Biesbroeck (1882-1968). (M 14480)

(2464) Nordenskiöld

1939 BF. Discovered 1939 January 19 by Y. Väisälä at Turku.

Named for the Finnish explorer Nils Adolf Erik Nordenskiöld (1832-1901). He conducted many Arctic expeditions and first navigated the northeast passage from the Atlantic to the Pacific Ocean in 1878-1879. (M 7946)

(2465) Wilson

1949 PK. Discovered 1949 August 2 by K. Reinmuth at Heidelberg.

Named in honor of Sir Robert Wilson F.R.S., Perren Professor of Astronomy at University College London, and head of the Department of Physics and Astronomy. Wilson played a leading role in the development of the International Ultraviolet Explorer, one of the most successful astronomical satellites ever launched. He is one of the U.K.'s foremost astronomers and a past vice president of the IAU. (M 18448)

Name proposed by C. M. Bardwell, who made the identifications involving this minor planet, and G. V. Williams, following a suggestion by M. M. Dworetsky, who prepared the citation.

(2466) Golson

1959 RJ. Discovered 1959 September 7 at the Goethe Link Observatory at Brooklyn, Indiana.

Named in memory of John C. Golson (1927-1984), the first employee of the Kitt Peak National Observatory. "J. C." had been a night assistant at the McDonald Observatory before he was hired by Aden B. Meinel {see planet (4065)} to work as an observer on the site survey for a national optical astronomy observatory. The first person hired after Kitt Peak was selected, he received a special service award during the celebration of KPNO's 25th anniversary in February 1983. (M 10043) Name proposed by F. K. Edmondson.

(2467) Kollontai

1966 PJ. Discovered 1966 August 14 by L. I. Chernykh at Nauchnyj.

Named in memory of Aleksandra Mikhailovna Kollontai (1872-1952), who served as Soviet ambassador to Norway, Mexico and Sweden, the first woman of any nation ever to become an accredited minister to a foreign country. (M 8064)

(2468) Repin

1969 TO_1. Discovered 1969 October 8 by L. I. Chernykh at Nauchnyj.

Named in honor of Il'ya Efimovich Repin (1844-1930), famous Russian painter. (M 13172)

(2469) Tadjikistan

1970 HA. Discovered 1970 April 27 by T. M. Smirnova at Nauchnyj.
Named for the {former} Tadjik Soviet Socialist Republic. (M 6650)

(2470) Agematsu

1976 UW_{15}. Discovered 1976 October 22 by H. Kosai and K. Hurukawa at Kiso.

Named for one of three towns to which the Kiso Station administratively belongs. (M 8912)

For the two other cities mentioned, see the citations of planets (2924) Mitake-mura and (2960) Ohtaki.

(2471) Ultrajectum

6545 P-L. Discovered 1960 September 24 by C. J. van Houten and I. van Houten-Groeneveld at Palomar.

Named in honor of the University Observatory at Utrecht, best known for its work on solar physics. Ultrajectum was the Roman name for Utrecht. (M 8799)

(2472) Bradman

1973 DG. Discovered 1973 February 27 by L. Kohoutek at Bergedorf.

Named in honor of the legendary Australian cricketer Donald George Bradman (1908-), whose test-match batting average of 99.94 in 80 innings far exeeds that of his closest meaningful competitor, and whose 42 half-centuries give him the best consistency ratio. Bradman's score of 334 runs in a test match against England in 1930 was then a record, as was his 452 (not out) for New South Wales against Queensland, a record in first-class cricket that was unbroken for 29 years. (M 27329)

Name suggested by B. G. Marsden, D. I. Steel and G. V. Williams.

(2473) Heyerdahl

1977 RX_7. Discovered 1977 September 12 by N. S. Chernykh at Nauchnyj.

Named in honor of Thor Heyerdahl, outstanding Norwegian ethnographer and archaeologist, brave traveller and prominent writer, famous for his expeditions along ancient sea routes. (M 11156)

(2474) Ruby

1979 PB. Discovered 1979 August 14 by Z. Vávrová at Kleť.

Named for the discoverer's dog, which lives at the Kleť Observatory. (M 6650)

(2475) Semenov

1972 TF_2. Discovered 1972 October 8 by L. V. Zhuravleva at Nauchnyj.

Named in memory of Pavel Afanes'evich Semenov (1912-1942), a fighter in the International Brigade in Spain during 1937-1938 and one of the valiant defenders who perished in the battle of Stalingrad. (M 6650)

(2476) Andersen

1976 JF_2. Discovered 1976 May 2 by N. S. Chernykh at Nauchnyj.

Named for Hans Christian Andersen (1805-1875), great Danish writer of fairy tales. (M 9767)

(2477) Biryukov

1977 PY_1. Discovered 1977 August 14 by N. S. Chernykh at Nauchnyj.

Named in memory of Nikolaj Zotovich Biryukov (1912-1966), famous Soviet writer. In a rescue attempt at the age of 18 he rushed into icy water, and as a result was bedridden for the rest of his life, an example of courage and fortitude. (M 7472)

(2478) Tokai

1981 JC. Discovered 1981 May 4 by T. Furuta at Tokai.

Named by the discoverer for the city in which he lives. (M 6650)

(2479) Sodankylä

1942 CB. Discovered 1942 February 6 by Y. Väisälä at Turku.

Named for a large country commune, more than 12,000 square kilometers in area, in southern Lapland. A temporary observatory was erected there during the first Polar Year (1882-1883), and since 1913 the Finnish Academy of Sciences and Letters has maintained its Geophysical Observatory there. (M 7946)

(2480) Papanov

1976 YS_1. Discovered 1976 December 16 by L. I. Chernykh at Nauchnyj.

Named in memory of Anatolij Dmitrievich Papanov (1922-1987), talented Soviet actor. (M 13172)

(2481) Bürgi

1977 UQ. Discovered 1977 October 18 by P. Wild at Zimmerwald.

Named for Jost Bürgi (1552-1632) of Lichtensteig in the Toggenburg valley, who won European fame as a highly skilled maker of precious instruments for astronomy and geometry. As a practical mathematician, he also devised and used refined new methods. Initially in the service of the landgraves Wilhelm and Moritz of Kassel, he later served the emperors Rudolf and Mathias in Prague, where he became a good friend and helper to Kepler {see planet (1134)}. Bürgi on his own invented logarithms, but being slow (or even loath) to publish his system, he lost the priority to John Napier. (M 21129)

(2482) Perkin

1980 CO. Discovered 1980 February 13 at the Harvard College Observatory at Harvard.

Named in honor of Richard S. and Gladys T. Perkin, long-time friends and supporters of astronomical research, particularly at the Harvard College Observatory. Richard Perkin (1906-1969) was a founder of the Perkin-Elmer Corporation, the company that has been responsible for the construction of a number of advanced astronomical instruments, including the Space Telescope. Gladys Perkin served for over a decade as a member of the Committee to Visit the Harvard University Department of Astronomy. (M 6955)

Richard Perkin is also honored by a lunar crater.

(2483) Guinevere

1928 QB. Discovered 1928 August 17 by M. Wolf at Heidelberg.

Named for the heroine of the Arthurian legends, the wife of King Arthur {see planet (2597)}, but the lover of Lancelot {see planet (2041)}. When this affair was made public, civil war developed between King Arthur and Lancelot. Lancelot rescued Guinevere from burning at the stake, but these events initiated the downfall of Arthur's idyllic kingdom. (M 8064)

Following a suggestion by F. Pilcher, the name was proposed by E. Bowell, who found the key identification for this planet.

(2484) Parenago

1928 TK. Discovered 1928 October 7 by G. N. Neujmin at Simeïs.

Named in memory of Pavel Petrovich Parenago (1906-1960), a professor at Moscow University, a corresponding member of the U.S.S.R. Academy of Sciences and the founder of the Moscow school of stellar astronomy. In the early 1940s he developed a method for the determination of the interstellar absorption of light. He established the existence of the subdwarf sequence and its location on the Hertzsprung-Russell diagram. He also determined the galactic orbit of the Sun and the first more-or-less reliable determination of the velocity of the Galaxy relative to its neighbors. (M 6955)

Name proposed by the Institute for Theoretical Astronomy.

Parenago is also honored by a lunar crater.

(2485) Scheffler

1932 BH. Discovered 1932 January 29 by K. Reinmuth at Heidelberg.

Named in honor of Helmut Scheffler (1928-), staff member of the Heidelberg Königstuhl Observatory and professor of astronomy at Heidelberg University (1963-1991), on the occasion of his retirement. He has made important contributions to the fields of radiation transfer in the outer solar atmosphere, atmospheric seeing and the structure of the interstellar medium. In collaboration with H. Elsässer {see planet (4385)}, Scheffler has written the the well-known textbooks *Physik der Sterne und der Sonne* and *Physics of the Galaxy and Interstellar Matter*. (M 18643)

Name proposed and citation prepared by G. Klare and L. D. Schmadel. Endorsed by E. Bowell, who found the key identification involving this planet.

(2486) Metsähovi

1939 FY. Discovered 1939 March 22 by Y. Väisälä at Turku.

Named for a donated farm near Helsinki, where various institutes have established their observing stations: the Finnish Geodetic Institute for space geodesy, the University of Helsinki for astrophysics, and the Helsinki Technological University for radio astronomy. (M 7946)

(2487) Juhani

1940 RL. Discovered 1940 September 8 by H. Alikoski at Turku.
Named by the discoverer in honor of his son. (M 7946)

(2488) Bryan

1952 UT. Discovered 1952 October 23 at the Goethe Link Observatory at Brooklyn, Indiana.

Named in memory of William Lowe Bryan (1860-1955), president of Indiana University from 1902 to 1937, whose pioneering work in experimental psychology was recognized by his election as president of the American Psychological Association in 1903. Appointed to the faculty soon after his graduation from Indiana University in 1884 (two years before the retirement of Daniel Kirkwood {see planet (1578)}), Bryan also served for more than a quarter of a century as vice president and president emeritus, and he was the architect of the present-day structure of the university: the present academic organization, including the department of astronomy, is truly his creation. (M 10043)
Name proposed by F. K. Edmondson.

(2489) Suvorov

1975 NY. Discovered 1975 July 11 by L. I. Chernykh at Nauchnyj.

Named in memory of Alexandr Vasil'evich Suvorov (1729-1800), military theoretician, famous for his progressive ideas, renowned among the common people of Russia. (M 7946)

(2490) Bussolini

1976 AG. Discovered 1976 January 3 at the Felix Aguilar Observatory at El Leoncito.

Named in memory of Juan A. Bussolini, S.J. (1905-1966), solar physicist, director of the Observatorio de Fisica Cosmica de San Miguel and a member of the commission of the International Year of the Quiet Sun. He was also an important benefactor to the Felix Aguilar Observatory. (M 8800)

(2491) Tvashtri

1977 CB. Discovered 1977 February 15 by W. Sebok at Palomar.

Named for the carpenter in the mythology of India, the divine power for construction in the world. He molds all forms and assembles the pieces, and has given to earth and sky the marvelous variety of their forms. (M 21606)

Named by the Minor Planet Names Committee following a suggestion by F. Pilcher.

(2492) Kutuzov

1977 NT. Discovered 1977 July 14 by N. S. Chernykh at Nauchnyj.

Named for Mikhail Illarionovich Kutuzov (1745-1813), military leader, commander-in-chief during the 1812 war against the troops of Napoleon. (M 9767)

(2493) Elmer

1978 XC. Discovered 1978 December 1 at the Harvard College Observatory at Harvard.

Named in memory of Charles Wesley Elmer (1872-1954), whose meeting with Richard S. Perkin {see planet (2482)} at the Harvard Tercentenary celebration in 1936 led to the establishment of the Perkin-Elmer Corporation two years later. Director of the astronomy department at the Brooklyn Academy of Arts and Sciences, he founded the Amateur Astronomers Association in New York City and the Custer Institute in Southold, Long Island. His summer abode in Southold long served as a lending repository of telescopes for amateur astronomers. (M 8404)

(2494) Inge

1981 LF. Discovered 1981 June 4 by E. Bowell at Anderson Mesa.

Named in honor of Jay L. Inge, friend of the discoverer and cartographer at the U.S. Geological Survey's Branch of Astrogeologic Studies, Flagstaff. Inge has been responsible for producing many topographic and geologic maps of planets and satellites. (M 6834)

(2495) Noviomagum

7071 P-L. Discovered 1960 October 17 by C. J. van Houten and I. van Houten-Groeneveld at Palomar.

Named in honor of the Astronomical Institute of the Nijmegen University, where many of the positions for the Palomar-Leiden Trojan survey are being measured. Noviomagus was the Roman name for Nijmegen. (M 8800)

(2496) Fernandus

1953 TC_1. Discovered 1953 October 8 at the Goethe Link Observatory at Brooklyn, Indiana.

Named in memory of Fernandus Payne (1881-1977), dean of the graduate school and head of the department of zoology at Indiana University from 1927 to 1947 and dean of the college of arts and sciences from 1942 to 1947. His support played an important role in expanding the department of astronomy and the establishment of a graduate program leading to the Ph.D. degree. President of the American Society of Zoologists in 1931, Payne was chairman of the Association of American Universities' Committee on Classification for 18 years. Following his formal retirement in 1951, he served for a year as assistant director for biological and medical sciences of the newly-established National Science Foundation and actively continued his research for two more decades. (M 10043)

Name proposed by F. K. Edmondson.

(2497) Kulikovskij

1977 PZ_1. Discovered 1977 August 14 by N. S. Chernykh at Nauchnyj.

Named in honor of Petr Grigor'evich Kulikovskij {1910- }, stellar astronomer and distinguished authority on the history of astronomy. An assistant professor at Moscow University, Kulikovskij is widely known as the author of a reference book for amateur astronomers and also as a musician and composer. (M 7784)

(2498) Tsesevich

1977 QM_3. Discovered 1977 August 23 by N. S. Chernykh at Nauchnyj.

Named in honor of Vladimir Platonovich Tsesevich {1907-1983}, former director of the Odessa University Observatory, renowned for his research on variable stars. He also studied the brightness variations of {433} Eros and is the author of a handbook for amateur astronomers. (M 7784)

Obituaries published in Astron. Tsirk., No. 1319, p. 6-8 (1984); Tsirk Vses astron.-geod. o-vo Akad. Nauk SSSR, No. 37, p. 89-92 (1984); Zemlya Vselennaya, No. 4, p. 58-60 (1984); Rocznik Astron. Obs. Krakowskiego 1986, p. VII (1985); Perem. Zvezdy, Tom 22, No. 6, p. 816-817 (1988).

(2499) Brunk

1978 VJ_7. Discovered 1978 November 7 by E. M. Shoemaker and E. F. Helin and S. J. Bus at Palomar.

Named in honor of William E. Brunk, astronomer with the National Aeronautics and Space Administration. As chief of NASA's Planetary Astronomy Program, he played an important role in the construction of the 2.6-m reflector at the McDonald Observatory and the Infrared Telescope Facility at Mauna Kea. He is also known for his work on the rings of Saturn, the atmosphere of Jupiter, and the feasibility of detecting planets about other stars. (M 6955)

(2500) Alascattalo

1926 GC. Discovered 1926 April 2 by K. Reinmuth at Heidelberg.

Named for Alaska's mythical beast, supposedly a cross between a moose and a walrus and genetically bred by miners during the Alaska Gold Rush around 1900. The alascattalo is also the epitome of Alaska's unique sense of humor, "absurding", which is the Alaskan way of dealing with tourists who ask stupid questions. A parade, lasting just four minutes and extending just one block down an alley, is held each year at three minutes past noon on Alascattalo Day, the first Sunday after the third Saturday in November. (M 19332)

Named by the Minor Planets Names Committee following a suggestion by S. C. Levi.

(2501) Lohja

1942 GD. Discovered 1942 April 14 by L. Oterma at Turku.

Named for a town, ridge and lake in southern Finland. An important market in the Middle Ages, Lohja is famous for its limestone quarries. In the old mine shafts, 150 m under the Earth's surface, the Finnish Geodetic Institute has established a station for studying Earth tides. (M 7946)

(2502) Nummela

1943 EO. Discovered 1943 March 3 by Y. Väisälä at Turku.

Named for a village in southern Finland. On a plateau atop the Lohja {see planet (2501)} Ridge lies the Finnish Geodetic Institute's standard base line, 864 m long. (M 7946)

(2503) Liaoning

1965 UB_1. Discovered 1965 October 16 at the Purple Mountain Observatory at Nanking.

Named for a province in northeastern China, noted for its coal and iron mines. (M 7618)

(2504) Gaviola

1967 JO. Discovered 1967 May 6 by C. U. Cesco and A. R. Klemola at El Leoncito.

Named in honor of Enrique Gaviola {1900-1990}, some time director of the Córdoba Observatory, founder of the Córdoba Institute of Mathematics, Physics and Astronomy. An authority on the construction, testing and treatment of astro-

nomical mirrors, it was he who brought to completion, in 1941, the 1.5-m reflector at Bosque Alegre. (M 7618)
Obituary published in Bol. Asoc. Argent. Astron., No. 35, p. 7-12 (1989).

(2505) Hebei

1975 UJ. Discovered 1975 October 31 at the Purple Mountain Observatory at Nanking.

Named for a province in northern China. The Yellow River and the Great Wall run through the province. (M 7946)

(2506) Pirogov

1976 QG_1. Discovered 1976 August 26 by N. S. Chernykh at Nauchnyj.

Named in memory of Nikolaj Ivanovich Pirogov (1810-1881), distinguished Russian surgeon and anatomist who developed new methods of anatomical research and the basic principles of field surgery. He has been credited with placing surgery on a scientific basis. (M 9214)

(2507) Bobone

1976 WB_1. Discovered 1976 November 18 at the Felix Aguilar Observatory at El Leoncito.

Named in memory of Jorge Bobone (1901-1958), some time director of the Córdoba Observatory, well known for his work on the orbits of comets, minor planets and the satellites of Jupiter. His investigation on the orbit of P/Halley for the 1986 return was unfinished at his death. He also participated in astrometric programs and recovered P/Encke and P/Kopff at their returns in 1931 and 1932, respectively. (M 11156)
Bobone is also honored by a lunar crater.

(2508) Alupka

1977 ET_1. Discovered 1977 March 13 by N. S. Chernykh at Nauchnyj.

Named for a small town on the south coast of the Crimea, famous for its health resorts, also the Vorontsov palace and nearby park. (M 9214)

(2509) Chukotka

1977 NG. Discovered 1977 July 14 by N. S. Chernykh at Nauchnyj.

Named for a National Area of the R.S.F.S.R., situated in the northeastern part of the U.S.S.R. The discoverer participated in an expedition there to observe the 1972 total solar eclipse. (M 7472)

(2510) Shandong

1979 TH. Discovered 1979 October 10 at the Purple Mountain Observatory at Nanking.

Named for a province in eastern China. The Shandong peninsula stretches between the Bohai Sea and the Yellow Sea. (M 7618)

(2511) Patterson

1980 LM. Discovered 1980 June 11 by C. S. Shoemaker at Palomar.

Named in honor of Clair C. Patterson {1922-1995}, geochemist at the California Institute of Technology. Patterson is world-renowned for his determination, through exquisite techniques of isotopic analysis, of the age of the Earth and of meteorites at 4.55×10^9 years; with the aid of additional lines of evidence, this can be shown to be the age of the solar system as a whole. He also played a leading role in revising the estimates of elemental abundances in various solar system materials. (M 6955)
Name proposed by C. S. Shoemaker and E. F. Helin.
Obituary published in Nature, Vol. 379, No. 6565, p. 487 (1996).

(2512) Tavastia

1940 GG. Discovered 1940 April 3 by Y. Väisälä at Turku.

The Latin name for the province of Hame, in the lake district of south-central Finland. Archeological studies show evidence for the presence of man 9,000 years ago. The Finns began to settle there in the fifth century. (M 7947)

(2513) Baetslé

1950 SH. Discovered 1950 September 19 by S. Arend at Uccle.

Named in memory of Paul-Louis Baetslé (1909-1983), professor of astronomy and geodesy at the Brussels Royal Military School and a friend of the discoverer. (M 8404)

Obituary published in Ciel Terre, Vol. 100, No. 1, p. 11-12 (1984).

(2514) Taiyuan

1964 TA_1. Discovered 1964 October 8 at the Purple Mountain Observatory at Nanking.

Named for the capital city of the province of Shaanxi {see planet (2263)}, located in northern China, near the central portion of the Yellow River. (M 7618)

(2515) Gansu

1964 TX_1. Discovered 1964 October 9 at the Purple Mountain Observatory at Nanking.

Named for a province in northwestern China, at the boundary of the Qinghai-Xizang and loess plateaux. (M 7618)

(2516) Roman

1964 VY. Discovered 1964 November 6 at the Goethe Link Observatory at Brooklyn, Indiana.

Named in honor of Nancy Grace Roman, a space-age astronomer who joined the staff of the National Aeronautics and Space Administration during its first year and who served with distinction at NASA headquarters in positions of increasing responsibility for two decades. Her earlier professional experience included positions at the Yerkes Observatory and the U.S. Naval Research Laboratory. Her many honors include the NASA Exceptional Scientific Achievement award in 1969, the NASA Outstanding Leadership award in 1978 and four honorary degrees. (M 10311)

Name proposed by F. K. Edmondson.

(2517) Orma

1968 SB. Discovered 1968 September 28 by P. Wild at Zimmerwald.

The name, the Italian word for a trace or track, seems very appropriate for a minor planet. That it should apply to this particular object follows from the fact that the number and name are an alphanumerical variation of (1257) Móra. (M 17465)

(2518) Rutllant

1974 FG. Discovered 1974 March 22 by C. Torres at Cerro El Roble.

Named in memory of Federico Rutllant Alcina (1904-1971), director of the Observatório Astronómico Nacional of the Universidad de Chile in Santiago from 1950 to 1963, later professor of mathematics at the Universidad Tecnica Federico Santa Maria in Valparaiso. Among his principal contributions to Chilean astronomy were his arrangement for moving the observatory from Lo Espejo to its present location at Cerro Calán, an agreement with the University of Florida that created the Radio Observatório de Maipu, an agreement with the Universities of Chicago and Texas - and later AURA - that produced the Cerro Tololo Interamerican Observatory, and an agreement with the U.S.S.R. Academy of Sciences that produced the Estacion Astronómica de Cerro El Roble - where this minor planet was discovered. (M 10545)

(2519) Annagerman

1975 VD_2. Discovered 1975 November 2 by T. M. Smirnova at Nauchnyj.

Named in memory of Anna German (1936-1982), a popular Polish variety singer. (M 11156)

(2520) Novorossijsk

1976 QF_1. Discovered 1976 August 26 by N. S. Chernykh at Nauchnyj.

Named for a town on the Black Sea coast. (M 7618)

(2521) Heidi

1979 DK. Discovered 1979 February 28 by P. Wild at Zimmerwald.

Named for the heroine of a still highly popular, deeply moving children's tale from the Swiss Alps, by Johanna Spyri (1829-1901). (M 17465)

(2522) Triglav

1980 PP. Discovered 1980 August 6 by Z. Vávrová at Kleť.

Named for a Slavonic deity. (M 21954)

(2523) Ryba

1980 PV. Discovered 1980 August 6 by Z. Vávrová at Kleť.

Named for Jakub Jan Ryba (1765-1815), Czech baroque composer, author of the famous Czech Christmas Mass. (M 21954)

(2524) Budovicium

1981 QB_1. Discovered 1981 August 28 by Z. Vávrová at Kleť.

The ancient Roman name for the town of České Budějovice, the center of South Bohemia near which the Kleť Observatory is located. (M 7618)

(2525) O'Steen

1981 VG. Discovered 1981 November 2 by B. A. Skiff at Anderson Mesa.

Named in honor of the discoverer's mother, Mary Elizabeth O'Steen Skiff. (M 6834)

(2526) Alisary

1979 KX. Discovered 1979 May 19 by R. M. West at La Silla.

Named by the discoverer in honor of his parents, Alice Benedicta (née Loethman) West and Harry Richard West, as an expression of sincere filial gratitude and in thankful recognition of their many sacrifices and unremitting efforts to give their children a good start in life. (M 6834)

(2527) Gregory

1981 RE. Discovered 1981 September 3 by N. G. Thomas at Anderson Mesa.

Named in honor of Bruce Gregory Thomas, youngest son of the discoverer, whose name was taken from that of James Gregory, the Scottish astronomer. (M 6834)

(2528) Mohler

1953 TF_1. Discovered 1953 October 8 at the Goethe Link Observatory at Brooklyn, Indiana.

Named in memory of Orren C. Mohler (1908-1985), solar astronomer, director of the McMath-Hulbert Observatory (1962-1979), chairman of the department of astronomy at the University of Michigan (1962-1970), member of the board of directors of the Association of Universities for Research in Astronomy (1962-1974).

Mohler pioneered the exploration of the infrared solar spectrum with the lead sulphide infrared detector. His development of the vacuum spectrograph at the McMath-Hulbert Observatory led to the discovery of the "wiggly" solar spectral lines and to an understanding of the role of turbulence in the solar structure of the solar photosphere. (M 10546)
Name proposed by F. K. Edmondson. Citation written by W. A. Hiltner.
Obituary published in Phys. Today, Vol. 39, No. 4, p. 74 (1986).

(2529) Rockwell Kent

1977 QL_2. Discovered 1977 August 21 by N. S. Chernykh at Nauchnyj.

Named for Rockwell Kent (1882-1971), American artist and writer of travel books, whose works draw upon the experiences of his varied career as an architectural draftsman, a lobsterman and a carpenter on the coast of Maine. He also illustrated the works of Melville, Shakespeare and Chaucer {see, respectively, planets (2985), (2984)}. (M 9767)

(2530) Shipka

1978 NC_3. Discovered 1978 July 9 by L. I. Chernykh at Nauchnyj.

Named for the mountainous site of the victory of the Russian and Bulgarian troops during the Russo-Turkish war of 1877-1878. (M 7947)

(2531) Cambridge

1980 LD. Discovered 1980 June 11 by E. Bowell at Anderson Mesa.

Named for the centers of learning in England and in Massachusetts, U.S.A. The Minor Planet Center is located at the Smithsonian Astrophysical Observatory in the latter city. (M 6834)

(2532) Sutton

1980 TU_5. Discovered 1980 October 9 by C. S. Shoemaker at Palomar.

Named in honor of Robert L. Sutton, geologist of the U.S. Geological Survey. As a member of the field geology investigation team for the Apollo lunar landing missions, Sutton had the unique responsibility of identifying the exact location, orientation and geologic field relationships of each sample returned from the Moon. This information is the necessary link that ties the detailed analyses of the returned samples to the heavenly body from which the samples were obtained and is crucial to the correct interpretation of many of the laboratory observations. (M 6956)

(2533) Fechtig

A905 VA. Discovered 1905 November 3 by M. Wolf at Heidelberg.

Named in honor of Hugo Fechtig (1929-), director at the Max-Planck Institute for Nuclear Physics in Heidelberg and professor of physics at Heidelberg University. Early in his career Fechtig became interested in cosmophysics. He is a well-known specialist in studies of micrometeoroids and interplanetary dust by rockets and spacecraft and has shared in the organization and development of laboratory and space experiments on meteoritic, lunar, interplanetary and cometary material. This especially refers to the projects on the Helios, Giotto, Vega, Ulysses and Galileo space missions. (M 18643)

Name proposed and citation prepared by L. D. Schmadel and J. Schubart. Endorsed by E. Bowell, who found the key identifications involving this planet.

(2534) Houzeau

1931 VD. Discovered 1931 November 2 by E. Delporte at Uccle.

Named in memory of Jean-Charles Houzeau (1820-1888), second director of the Observatoire Royal de Bruxelles and one of the greatest Belgian astronomers. He

is best known internationally as co-author with A.-B. Lancaster, of the *Bibliographie générale de l'astronomie jusqu'à 1880*; he also wrote the *Vade-mecum de l'astronomie.* (M 8800)
Houzeau is also honored by a lunar crater.

(2535) Hämeenlinna

1939 DH. Discovered 1939 February 17 by Y. Väisälä at Turku.

Named for an old town in the province of Häme. The town grew around a thirteenth-century castle (linna) in an ancient populated area. With its ridges, hills and lakes, Hämeenlinna presents the most fascinating views. It has been an important educational center since 1639, and the renowned Finnish composer Jean Sibelius {see planet (1405)} received his schooling there. (M 7947)

(2536) Kozyrev

1939 PJ. Discovered 1939 August 15 by G. N. Neujmin at Simeïs.

Named in honor of Nikolaj Aleksandrovich Kozyrev (1908-1983), a staff member of the Pulkovo Observatory, distinguished expert on the physics of stars, the Moon and the planets, skillful experimenter and observer. He developed a theory on extended atmospheres and found properties of the radiation emitted from them, and he discovered volcanic activity on the Moon and the presence of hydrogen in the atmosphere of Mercury. (M 10546)
Obituary published in Zemlya Vselennaya, No. 1, p. 50-51 (1984).

(2537) Gilmore

1951 RL. Discovered 1951 September 4 by K. Reinmuth at Heidelberg.

Named in honor of Alan C. and Pamela M. (Kilmartin) Gilmore, whose program of astrometric observations of comets and minor planets has for more than a decade been one of the most productive and rapidly responsive such efforts ever to be undertaken in the southern hemisphere. At the Mount John University Observatory since 1980, they were formerly on the staff of the Carter Observatory. They also serve as co-directors of the Comet and Minor Planet Section of the Royal Astronomical Society of New Zealand. (M 8064)

Name proposed by C. M. Bardwell and B. G. Marsden, identifiers for this planet.
Pamela M. Gilmore is also honored by minor planet (3907) Kilmartin.

(2538) Vanderlinden

1954 UD. Discovered 1954 October 30 by S. Arend at Uccle.

Named in memory of Henri Vanderlinden (1892-1983), a Belgian astronomer and professor at the Ghent State University. (M 8404)

(2539) Ningxia

1964 TS_2. Discovered 1964 October 8 at the Purple Mountain Observatory at Nanking.

Named for a province in northwestern China, along the middle section of the Yellow River. (M 7619)

(2540) Blok

1971 TH_2. Discovered 1971 October 13 by L. I. Chernykh at Nauchnyj.

Named in honor of Aleksandr Aleksandrovich Blok (1880-1921), outstanding Russian poet. (M 13172)

(2541) Edebono

1973 DE. Discovered 1973 February 27 by L. Kohoutek at Bergedorf.

Named in honor of Edward de Bono (1933-   ) who is regarded as the leading international authority in conceptual and creative thinking and in the teaching of

thinking as a skill. He originated the term 'lateral thinking' and is well known for the deliberate creative techniques associated with it and the powerful Six Hats method. He has written over 50 books, published in 28 languages, and has made two TV series. (M 27124)
Name proposed by P. Colemont.

(2542) Calpurnia

1980 CF. Discovered 1980 February 11 by E. Bowell at Anderson Mesa.

Named for the last wife of Julius Caesar. She bade him stay home from the Senate on the Ides of March, but he ignored her advice and was assassinated there. (M 6834)
Name suggested by F. Pilcher.

(2543) Machado

1980 LJ. Discovered 1980 June 1 by H. Debehogne at La Silla.

Named in honor of Luiz Eduardo da Silva Machado { -1992}, director of the Valongo Observatory and professor at the Universidade Federal do Rio de Janeiro, codeveloper with the discoverer of the program of astrometry of minor planets and comets at the European Southern Observatory. He has also promoted the installation of a Zeiss astrograph at Campinas. (M 10844)

(2544) Gubarev

1980 PS. Discovered 1980 August 6 by Z. Vávrová at Kleť.

Named in honor of Soviet cosmonaut Aleksei A. Gubarev, commander of the spacecraft Soyuz 28. (M 7619)

(2545) Verbiest

1933 BB. Discovered 1933 January 26 by E. Delporte at Uccle.

Named in memory of Father Ferdinand Verbiest, Belgian missionary in China and astronomer at the court of Emperor Kang-Hi. (M 8800)

(2546) Libitina

1950 FC. Discovered 1950 March 23 by E. L. Johnson at Johannesburg.

Named for an ancient Roman divinity who presided over the burial of the dead. Roman poets frequently employed her name to mean death itself. (M 21606)

Named by the Minor Planet Names Committee following a suggestion by F. Pilcher.

(2547) Hubei

1964 TC_2. Discovered 1964 October 9 at the Purple Mountain Observatory at Nanking.

Named for a province in the central part of China, located in the middle stretch of the Yangtze River. (M 7947)

(2548) Leloir

1975 DA. Discovered 1975 February 16 at the Felix Aguilar Observatory at El Leoncito.

Named in honor of the outstanding Argentinian biochemist Luis Federico Leloir {1906- }, recipient of the Nobel prize in chemistry in 1970 and many other international honors. Born in France, he worked for several years in the United States. He has served as Extraordinary Research Professor at the Universidad Nacional de Buenos Aires and as director of the Instituto de Investigaciones Bioquimicas. As a member of the board of directors of the Consejo Nacional de Investigaciones Cientificas y Tecnicas during 1958-1964 he had an important influence on scientific research in Argentina. (M 11156)

(2549) Baker

1976 UB. Discovered 1976 October 23 at the Harvard College Observatory at Harvard.

Named in honor of James G. Baker {1914- }, astrophysicist, innovator, advisor. To many he is best known for his legendary designs of very demanding optical cameras and spectrographs. By some he is appreciated for his unselfish contributions to the success of difficult endeavors, both on behalf of his fellow scientists and his country. (M 8800)

(2550) Houssay

1976 UP_{20}. Discovered 1976 October 21 at the Felix Aguilar Observatory at El Leoncito.

Named in memory of the remarkable Argentinian scientist and teacher Bernardo A. Houssay (1887-1971), one of the most influential researchers and teachers in medicine in Latin America, recipient of the Nobel prize for medicine in 1947 and many other international honors. Known in particular for his work in physiology and pharmacology, he wrote extensively on nutrition, internal secreting glands, experimental pathology and toxicology. (M 11156)

The 1947 Nobel prize in medicine was shared between Houssay and C. F. Cori, and G. T. Cori, respectively.

(2551) Decabrina

1976 YX_1. Discovered 1976 December 16 by L. I. Chernykh at Nauchnyj.

Named in memory of the revolutionary group of noblemen that led the uprising against tsarist autocracy and servitude in 1825. (M 7947)

(2552) Remek

1978 SP. Discovered 1978 September 24 by A. Mrkos at Kleť.

Named for the first Czechoslovak astronaut Vladimir Remek, born in South Bohemia, a member of the crew of Soyuz 28. (M 7619)

(2553) Viljev

1979 FS_2. Discovered 1979 March 29 by N. S. Chernykh at Nauchnyj.

Named in honor of Mikhail Anatol'evich Vil'ev (1893-1919), who, because of his exceptional abilities and tireless efforts, and in spite of his short life, made a prodigious number of investigations in the areas of celestial mechanics, theoretical astronomy and the history of astronomy. (M 7785)

Viljev is also honored by a lunar crater.

(2554) Skiff

1980 OB. Discovered 1980 July 17 by E. Bowell at Anderson Mesa.

Named in honor of Brian A. Skiff, who has recently made valuable contributions to the asteroid astrometry program at Lowell Observatory. (M 6834)

(2555) Thomas

1980 OC. Discovered 1980 July 17 by E. Bowell at Anderson Mesa.

Named in honor of Norman G. Thomas, for many years an observer on the Lowell proper motion and asteroid astrometry programs. (M 6835)

(2556) Louise

1981 CS. Discovered 1981 February 8 by N. G. Thomas at Anderson Mesa.

Named in honor of Carol Louise Thomas-Baltutis, youngest daughter of the discoverer. (M 6835)

(2557) Putnam

1981 SL_1. Discovered 1981 September 26 by B. A. Skiff and N. G. Thomas at Anderson Mesa.

Named in appreciation of the long and continuing support of the Lowell Observatory by members of the Putnam family, in particular Roger Lowell Putnam (1893-1972) and Michael C. J. Putnam, father and son, successive trustees of the Lowell Observatory. (M 6835)

(2558) Viv

1981 SP_1. Discovered 1981 September 26 by N. G. Thomas at Anderson Mesa. Named in memory of the discoverer's mother, Vivian Russell Thomas. (M 6835)

(2559) Svoboda

1981 UH. Discovered 1981 October 23 by A. Mrkos at Kleť.

Named in memory of Jindřich Svoboda (1884-1941), professor of astronomy and geodesy at Prague Polytechnic. (M 22828)

(2560) Siegma

1932 CW. Discovered 1932 February 14 by K. Reinmuth at Heidelberg.

Named in honor of Siegfried A. Marx (1934-1995), director of the Karl Schwarzschild Observatory at Tautenburg and professor of astronomy at Jena University. Marx is working on problems of interstellar and intergalactic matter, and he has made important contributions to questions of techniques and proper interpretations of astrophotographic observations. He is well-known for the publication of many textbooks and is one of the great popularizers of astronomy in Germany. His interests include the astrometric work on minor planets and he always supported observational campaigns with the largest Schmidt telescope in the world. (M 18643)

Name proposed and citation prepared by L. D. Schmadel. Endorsed by E. Bowell, who found the key identifications involving this planet.

Obituaries published in Sterne Weltraum, Jahrg. 34, Nr. 11, p. 783 (1995); Astron. Raumfahrt, Jahrg. 32, Heft 29, p. 6 (1995); Mitt. Astron. Ges., Nr. 79, p. 5-7 (1996); Mitt. Astronomiegesch., Nr. 7, p. 5 (1995); Ahnerts Kalender Sternfreunde 1996, p. 6 (1995).

(2561) Margolin

1969 TK_2. Discovered 1969 October 8 by L. I. Chernykh at Nauchnyj.

Named in memory of Mikhail Vladimirovich Margolin (1906-1975), blind designer and inventor, a man of heroic fate. (M 7785)

(2562) Chaliapin

1973 FF_1. Discovered 1973 March 27 by L. V. Zhuravleva at Nauchnyj.

Named in memory of Feodor Ivanovich Chaliapin (1873-1938), outstanding Russian singer and actor. (M 17026)

(2563) Boyarchuk

1977 FZ. Discovered 1977 March 22 by N. S. Chernykh at Nauchnyj.

Named in honor of Aleksandr Alekseevich Boyarchuk {1931- }, deputy director of the Crimean Astrophysical Observatory, well known for his research on stellar physics, president of IAU Commission 29 (Stellar spectra) during 1973-76. (M 7785)

Boyarchuk was elected IAU president for the period 1991-1994.

(2564) Kayala

1977 QX. Discovered 1977 August 19 by N. S. Chernykh at Nauchnyj.

Named for the river in *The Song of Igor's Campaign* on the occasion of the 800th anniversary of the old Russian literary monument. (M 12012)

(2565) Grögler

1977 TB_1. Discovered 1977 October 12 by P. Wild at Zimmerwald.

Named in memory of Norbert Grögler (1928-1983), mineralogist and planetologist who made significant contributions in meteorite research and in the analysis of lunar samples. After graduating from the University of Vienna he followed the invitation of F. G. Houtermans to join an interdisciplinary team of space physicists and planetologists at the University of Berne. He developed sophisticated methods for the quantitative separation of minerals from meteorites and lunar samples and made several significant discoveries on the mineralogy and petrography of meteorites and lunar rocks. He also had strong interests in archaeology and made relevant scientific contributions investigating thermoluminescence of ancient pottery. (M 21129)

(2566) Kirghizia

1979 FR_2. Discovered 1979 March 29 by N. S. Chernykh at Nauchnyj.

Named for the {former} Kirghiz Soviet Socialist Republic. (M 7472)

(2567) Elba

1979 KA. Discovered 1979 May 19 by O. Pizarro and G. Pizarro at La Silla.

Named by the discoverers in memory of their mother, Elba Aguilera de Pizarro (1926-1965). (M 6835)

(2568) Maksutov

1980 GH. Discovered 1980 April 13 by Z. Vávrová at Kleť.

Named in memory of Dmitrij Dmitrievich Maksutov (1896-1964), inventor of catadioptric meniscus optical systems. The observations of minor planets at the Kleť Observatory are made with a telescope of Maksutov design. (M 7619)
Maksutov is also honored by a lunar crater.

(2569) Madeline

1980 MA. Discovered 1980 June 18 by E. Bowell at Anderson Mesa.

Named for the heroine in John Keats' poem 'The Eve of St. Agnes'. According to the legend, virtuous young girls who perform the proper ceremonies on the evening before St. Agnes' Day (January 21) were supposed to dream that night of their future husbands. (M 7156)
Name proposed by F. Pilcher.

(2570) Porphyro

1980 PG. Discovered 1980 August 6 by E. Bowell at Anderson Mesa.

Named for the hero in John Keats' poem *The Eve of St. Agnes*. Porphyro was deeply in love with Madeline (planet 2569) but was thoroughly despised by all of Madeline's family. A friend helped Porphyro hide in Madeline's bedchamber before her bedtime on the Eve of St. Agnes. Porphyro stood over Madeline as she wakened from her dream of him. With Madeline's family in drunken stupor following revelry the evening before, Porphyro was able to carry her away. (M 7157)
Name proposed by F. Pilcher.

(2571) Geisei

1981 UC. Discovered 1981 October 23 by T. Seki at Geisei.

Named for the location of the discoverer's observatory, some 35 km to the east of the city of Kochi {see planet (2396)}. (M 6956)

(2572) Annschnell

1950 DL. Discovered 1950 February 17 by K. Reinmuth at Heidelberg.

Named in honor of Anneliese Schnell (1941-), Austrian astronomer at the Vienna Observatory. Although she mainly studies variable stars, she also works on problems in the history of astronomy, especially on the meaning of the names and on the discovery circumstances of the minor planets discovered by J. Palisa {see planet (914)}. Schnell was the first woman on the board of the Astronomische Gesellschaft - more than one century after its founding. (M 19333)

Name suggested and citation provided by L. D. Schmadel and endorsed by E. Bowell and B. G. Marsden.

(2573) Hannu Olavi

1953 EN. Discovered 1953 March 10 by H. Alikoski at Turku.
Named by the discoverer in honor of his son. (M 7947)

(2574) Ladoga

1968 UP. Discovered 1968 October 22 by T. M. Smirnova at Nauchnyj.
Named for the largest lake in Europe, situated near St. Petersburg. (M 8912)

(2575) Bulgaria

1970 PL. Discovered 1970 August 4 by T. M. Smirnova at Nauchnyj.
Named for the People's Republic of Bulgaria. (M 8912)

(2576) Yesenin

1974 QL. Discovered 1974 August 17 by L. V. Zhuravleva at Nauchnyj.

Named in memory of the lyric poet Sergej Alexandrovich Yesenin (1895-1925). (M 7619)

(2577) Litva

1975 EE_3. Discovered 1975 March 12 by N. S. Chernykh at Nauchnyj.

Named for the Lithuanian Soviet Socialist Republic, since 1991 the independent state of Lithuania. (M 7472)

(2578) Saint-Exupéry

1975 VW_3. Discovered 1975 November 2 by T. M. Smirnova at Nauchnyj.

Named in memory of the French writer Antoine de Saint-Exupéry (1900-1944), some time pilot, author of several novels and the world-famous tale *The Little Prince*. (M 12012)

(2579) Spartacus

1977 PA_2. Discovered 1977 August 14 by N. S. Chernykh at Nauchnyj.

Named for the leader of a large-scale rebellion of the slaves in Rome in 73-71 B.C. (M 9767)

(2580) Smilevskia

1977 QP_4. Discovered 1977 August 18 by N. S. Chernykh at Nauchnyj.

Named in memory of Moisej Vasil'evich Smilevskij (1913-1944), a journalist from the city of Cherson {see planet (2701)} in the Ukraine. (M 12968)

(2581) Radegast

1980 VX. Discovered 1980 November 11 by Z. Vávrová at Kleť.
Named for a Slavonic deity. (M 21954)

(2582) Harimaya-Bashi

1981 SA. Discovered 1981 September 26 by T. Seki at Geisei.

Named for the well-known red bridge in the center of the city of Kochi. (M 6956)

(2583) Fatyanov

1975 XA_3. Discovered 1975 December 3 by T. M. Smirnova at Nauchnyj.

Named in memory of Aleksej Ivanovich Fat'yanov (1919-1959), well-known Soviet poet and writer of popular songs. (M 13172)

(2584) Turkmenia

1979 FG_2. Discovered 1979 March 23 by N. S. Chernykh at Nauchnyj.
Named for the {former} Turkmen Soviet Socialist Republic. (M 7472)

(2585) Irpedina

1979 OJ_{15}. Discovered 1979 July 21 by N. S. Chernykh at Nauchnyj.

Named in honor of the Irkutsk {see planet (3224)} Pedagogical Institute. The name is dedicated to fellow students of the discoverer there. (M 8542)

(2586) Matson

1980 LO. Discovered 1980 June 11 by C. S. Shoemaker at Palomar.

Named in honor of Dennis L. Matson, planetary scientist at the Jet Propulsion Laboratory, who has played a leading role in developing the method for determination of the sizes and albedos of minor planets by means of infrared radiometry. He has also shown that the heat flow from the interior of Jupiter's satellite Io can be measured and monitored from ground-based telescopes. (M 6956)

(2587) Gardner

1980 OH. Discovered 1980 July 17 by E. Bowell at Anderson Mesa.

Named in honor of the American mathematician Martin Gardner, well known for his column on mathematical games in *Scientific American*. His wide interests range from recreational mathematics to philosophy to the debunking of pseudoscience. (M 7157)
Name proposed by J. Meeus.

(2588) Flavia

1981 VQ. Discovered 1981 November 2 by B. A. Skiff at Anderson Mesa.

The name is the feminine form of Flavius, a Roman gens, or patrilineal clan, which included the Emperors Vespasian, Titus and Domitian. The name also appeared in the science-fiction novel *A Torrent of Faces* (J. Blish and N. Knight, 1967), wherein Flavia is an asteroid on a collision course with the Earth. (M 6956)

(2589) Daniel

1979 QU_2. Discovered 1979 August 22 by C.-I. Lagerkvist at La Silla.
Named by the discoverer in honor of his son. (M 7472)

(2590) Mourão

1980 KJ. Discovered 1980 May 22 by H. Debehogne at La Silla.

Named in honor of R. R. de Freitas Mourão {1935- }, astronomer at the National Observatory in Rio de Janeiro, known for his work on double stars, minor planets and comets. He has participated extensively in the discoverer's program of observations of minor planets at the European Southern Observatory, is the author of several astronomical books and is the leader of the Foundation of the Brazilian Museum for Astronomy. (M 9767)

(2591) Dworetsky

1949 PS. Discovered 1949 August 2 by K. Reinmuth at Heidelberg.

Named in honor of Michael M. Dworetsky, senior lecturer at University College, London. His research interests mainly involve the stellar abundances of the mercury group of elements. He has also taken a large part in the development of the undergraduate astronomy degree program, particularly the practical work associated with it. He is currently tutor to astronomy studies and has maintained a wide interest in all matters astronomical. (M 18448)

Name proposed by C. M. Bardwell, who made the identifications involving this minor planet, following a suggestion by G. V. Williams. Citation written by D. McNally.

(2592) Hunan

1966 BW. Discovered 1966 January 30 at the Purple Mountain Observatory at Nanking.

Named for a province in southern China, rich in the production of rice and world-famous for its embroidery. (M 7619)

(2593) Buryatia

1976 GB_8. Discovered 1976 April 2 by N. S. Chernykh at Nauchnyj.

Named for the {former} Buryat Autonomous Soviet Socialist Republic. (M 7472)

(2594) Acamas

1978 TB. Discovered 1978 October 4 by C. T. Kowal at Palomar.

Named for a commander of the Thracians who allied themselves with the Trojans in order to fight the Greeks. (M 21606)

Named by the Minor Planet Names Committee following a suggestion by F. Pilcher.

(2595) Gudiachvili

1979 KL. Discovered 1979 May 19 by R. M. West at La Silla.

Named in memory of Lado Gudiachvili (1896-1980), a great painter and a dear friend of the discoverer. Born in Ducheti, Georgia, he studied in Tbilisi, and after a period in Paris (1919-1927), he returned to his native country, where he remained the rest of his life. He early developed a very personal style, to some extent influenced by early church frescoes. His persistent affection for beauty in all its aspects and his uncompromising pursuit of poetic and sometimes fantastic expressions was only fully recognized at a later time. His drawings and paintings are frequently of a philosophical or allegorical nature, and many are connected with the turbulent history of Georgia. He is one of the founders of twentieth-century Georgian art and became a national legend during his lifetime. (M 6835)

(2596) Vainu Bappu

1979 KN. Discovered 1979 May 19 by R. M. West at La Silla.

Named in memory of Manali Kallat Vainu Bappu (1927-1982), famous Indian astrophysicist and a dear friend of the discoverer. Educated at Harvard and Caltech, he established, under difficult circumstances, the first modern Indian observatory at Naini Tal during the 1950s. Appointed director of the Kodaikanal Observatory in 1960, he subsequently founded and directed the Indian Institute of Astrophysics in Bangalore. He was the initiator and driving force of many projects, among them the 2.3-m Kavalur telescope, entirely designed and built in India. In the multiple roles of brilliant scientist, teacher and administrator, he contributed decisively to the high level of astronomy and astrophysics in India today. He served as vice-president (1967-1973) and president (1979-1982) of the IAU and as chairman of the editorial board of the Indian Journal of Astronomy and Astrophysics. (M 10043)

Obituaries published in Messenger, No. 29, p. 1, No. 30, p. 1-2 (1982); Astrophys. Space Sci., Vol. 89, No. 1, p. 3-4 (1983); J. Astrophys. Astron., Vol. 3, No. 3, p. 217-218 (1983); Patrika, No. 5, p. 13 (1983); Sky Telesc., Vol. 65, No. 2, p. 137 (1983); Trans. IAU, Vol. XVIII B, p. 1-6 (1983); Bull. Astron. Soc. India, Vol. 10, No. 4, p. 275-279 (1982); Q.J.R. Astron. Soc., Vol. 24, No. 3, p. 360-361 (1983).

(2597) Arthur

1980 PN. Discovered 1980 August 8 by E. Bowell at Anderson Mesa.

Named for the central figure of the Arthurian legends of medieval England. In his youth, Arthur pulled the sword Excalibur out of a rock where all others had failed. This feat entitled him to become king. He established a castle at Camelot wherein he placed a Round Table, around which all knights who were worthy sat

as equals with Arthur, and for many years the kingdom flourished in this spirit of democracy. (M 7157)
Name proposed by F. Pilcher.

(2598) Merlin

1980 RY. Discovered 1980 September 7 by E. Bowell at Anderson Mesa.

Named for the sage and sorcerer of the Arthurian legends. Merlin's magic enabled Arthur {see planet (2597)} to pull the sword Excalibur from the rock and thereby become king. Afterward, Arthur continued to rely heavily on Merlin for advice in running his kingdom. When Merlin withdrew from the scene, Arthur could not guide the kingdom successfully on his wisdom alone. (M 7157)
Name proposed by F. Pilcher.

(2599) Veselí

1980 SO. Discovered 1980 September 29 by Z. Vávrová at Kleť.
Named for the town in South Bohemia where the dicoverer resides. (M 7619)

(2600) Lumme

1980 VP. Discovered 1980 November 9 by E. Bowell at Anderson Mesa.

Named in honor of Kari Lumme, of the University of Helsinki. This Finnish astronomer has done much theoretical work on the scattering of light from particles in Saturn's rings. He has recently considered the light-scattering properties of the surfaces of atmosphereless bodies, especially asteroids, and has devised a theory to explain the observed phase curves of these bodies. (M 6956)

(2601) Bologna

1980 XA. Discovered 1980 December 8 at the Osservatorio San Vittore at Bologna.

Named for the famous Italian city, home of one of the oldest universities in the world, in which the Osservatorio San Vittore is located. (M 6835)

(2602) Moore

1982 BR. Discovered 1982 January 24 by E. Bowell at Anderson Mesa.

Named in honor of Patrick Moore {1924- }, astronomer, broadcaster, and writer. For some years director of the Lunar Section of the British Astronomical Association, Moore has been most energetic and successful in popularizing astronomy. He is author of many books and has regularly presented 'The Sky at Night' on BBC television since April 1957. In 1967 he was awarded the Order of the British Empire. (M 7157)

(2603) Taylor

1982 BW_1. Discovered 1982 January 30 by E. Bowell at Anderson Mesa.

Named in honor of Gordon E. Taylor, astronomer at the Nautical Almanac Office, Royal Greenwich Observatory. Taylor has been director of the Computing Section of the British Astronomical Association since 1974. His research has centered on the prediction of occultations of stars by solar system objects, and in recent years he has been instrumental in the successful determination of several asteroid sizes by the occultation method. (M 7157)

(2604) Marshak

1972 LD_1. Discovered 1972 June 13 by T. M. Smirnova at Nauchnyj.

Named in memory of Samuil Yakovlevich Marshak (1887-1964), famous Soviet poet, writer of classic children's literature. His translations of poems by Robert Burns and of sonnets by Shakespeare {see planet (2985)} won him wide recognition. (M 13172)

(2605) Sahade

1974 QA. Discovered 1974 August 16 at the Felix Aguilar Observatory at El Leoncito.

Named in honor of Jorge Sahade, former {1985-1988} president of the International Astronomical Union, well-known Argentinian astrophysicist who has served as director of the Córdoba and La Plata Observatories and was the first director of the Instituto de Astronomia y Fisica del Espacio. During the 1960s he initiated the process that led to the acquisition of the 2.15-m reflector in the Complejo Astronomica El Leoncito. (M 11156)

(2606) Odessa

1976 GX_2. Discovered 1976 April 1 by N. S. Chernykh at Nauchnyj.

Named for the large port and industrial and cultural city on the coast of the Black Sea. (M 7472)

(2607) Yakutia

1977 NR. Discovered 1977 July 14 by N. S. Chernykh at Nauchnyj.

Named for the {former} Yakut Autonomous Soviet Socialist Republic. (M 7473)

(2608) Seneca

1978 DA. Discovered 1978 February 17 by H.-E. Schuster at La Silla.

Named for the great Roman philosopher and statesman Lucius Annaeus Seneca {4 B.C.-65}. (M 6835)

Seneca is also honored by a lunar crater.

(2609) Kiril-Metodi

1978 PB_4. Discovered 1978 August 9 by L. I. Chernykh and N. S. Chernykh at Nauchnyj.

Named for the brothers Kiril (827-869) and Metodi (815-885), also known as Cyril and Methodius, 'the apostles of the Slavs', great Slavic enlighteners, generally credited with the creation of the Slavic alphabet. (M 7785)

(2610) Tuva

1978 RO_1. Discovered 1978 September 5 by N. S. Chernykh at Nauchnyj.

Named for the {former} Tuvinian Autonomous Soviet Socialist Republic. (M 7473)

(2611) Boyce

1978 VQ_5. Discovered 1978 November 7 by E. M. Shoemaker and E. F. Helin and S. J. Bus at Palomar.

Named in honor of Joseph M. Boyce, planetary scientist at the National Aeronautics and Space Administration. A leading investigator of the cratering record of the surfaces of planets and satellites, his detailed studies of the small craters on the lunar plains are the foundation for our present knowledge of the ages and sequence of volcanic eruptions on the Moon. (M 6956)

(2612) Kathryn

1979 DE. Discovered 1979 February 28 by N. G. Thomas at Anderson Mesa.

Named in honor of Kathryn Gail Thomas-Hazelton, daughter of the discoverer. Her birthdate, Dec. 26, is apparent in both the object's number and provisional designation. (M 6957)

(2613) Plzeň

1979 QE. Discovered 1979 August 30 by L. Brožek at Kleť.

Named for the industrial and cultural center of West Bohemia, birthplace of the discoverer. (M 7619)

(2614) Torrence

1980 LP. Discovered 1980 June 11 by C. S. Shoemaker at Palomar.

Named in honor of Torrence V. Johnson, planetary scientist at the Jet Propulsion Laboratory, who is best known for his systematic investigations of the satellites of Jupiter and Saturn, both by means of ground-based telescopes and the data returned from the Voyager missions. He is also well known for his photometric observations of minor planets. (M 6957)

(2615) Saito

1951 RJ. Discovered 1951 September 4 by K. Reinmuth at Heidelberg.

Named in honor of Keiji Saito, astrophysicist at the Tokyo Astronomical Observatory during 1961-1985 who has studied the physics of comets and meteors. While still a college student, he was an independent discoverer of the nova T CrB at its 1946 recurrence and has inspired many amateur astronomers in Japan. (M 11748)

Name proposed by H. Oishi, who made the key identification involving this planet.

(2616) Lesya

1970 QV. Discovered 1970 August 28 by T. M. Smirnova at Nauchnyj.

Named for Lesya Ukrainka, the pen name of Larisa Petrovna Kosach (1871-1913), an outstanding Ukrainian poetess, playwright and literary critic. (M 8912)

(2617) Jiangxi

1975 WO_1. Discovered 1975 November 26 at the Purple Mountain Observatory at Nanking.

Named for a province in southern China, located along the eastern stretch of the Yangtze River. (M 7619)

(2618) Coonabarabran

1979 MX_2. Discovered 1979 June 25 by E. F. Helin and S. J. Bus at Siding Spring.

Named for this typical country town situated at the base of the beautiful Warrumbungle mountain range in New South Wales. The town is home for many of the astronomers and staff who work at the Siding Spring {see planet (2343)} Observatory. (M 7473)

(2619) Skalnaté Pleso

1979 MZ_3. Discovered 1979 June 25 by E. F. Helin and S. J. Bus at Siding Spring.

Named for the observatory in the Tatra mountains of Slovakia {see planet (1807)}. A remarkably successful program of visual comet hunting was conducted there, beginning in 1946, and the observatory is also noted for an excellent series of star charts. (M 7473)

(2620) Santana

1980 TN. Discovered 1980 October 3 by Z. Vávrová at Kleť.

Named in honor of the contemporary composer and singer Carlos Santana, author of magic music. (M 22245)

(2621) Goto

1981 CA. Discovered 1981 February 9 by T. Seki at Geisei.

Named in honor of Seizo Goto, well-known manufacturer of telescopes and planetaria, founder of Goto Optics. In 1981 he greatly aided astronomical education by presenting to his native prefecture of Kochi a 60-cm reflector, the largest made by Goto Optics. (M 6957)

(2622) Bolzano

1981 CM. Discovered 1981 February 9 by L. Brožek at Kleť.

Named on the 200th anniversary of the birth of Bernardo Bolzano (1781-1848), distinguished Czech mathematician and philosopher, professor at the Charles University during 1805-1819. (M 7619)

(2623) Zech

A919 SA. Discovered 1919 September 22 by K. Reinmuth at Heidelberg.

Named in honor of Gert Zech (1941-), astronomer at the Astronomisches Rechen-Institut and editor of *Astronomy and Astrophysics Abstracts* who has also worked on the determination of the mass of the earth and the astronomical unit from observations of (433) Eros. (M 19692)

Name proposed and citation prepared by L. D. Schmadel; endorsed by O. Kippes, who found the key identification involving this planet.

(2624) Samitchell

1962 RE. Discovered 1962 September 7 at the Goethe Link Observatory at Brooklyn, Indiana.

Named in memory of Samuel Alfred Mitchell (1874-1960), a faculty member of Columbia University from 1899 to 1913 and then director of the Leander McCormick Observatory until 1945, known for his work on solar eclipses and stellar parallaxes. His measurements of the flash spectrum at the eclipses in 1900, 1901 and 1905 referred to nearly 3,000 lines. His book *Eclipses of the Sun* went through five editions. His photographic parallax work with the McCormick refractor resulted in the publication of 2,001 parallaxes by 1950. Subsequent work by van de Kamp {see planet (1965)} and Vyssotsky {see planet (1600)}, encouraged and supported by Mitchell, yielded absolute proper motions of 29,000 stars between magnitudes 8 and 12. (M 10844)

Name proposed by F. K. Edmondson.

(2625) Jack London

1976 JQ_2. Discovered 1976 May 2 by N. S. Chernykh at Nauchnyj.

Named for the celebrated American writer Jack London (1876-1916). (M 8542)

(2626) Belnika

1978 PP_2. Discovered 1978 August 8 by N. S. Chernykh at Nauchnyj.

Named in honor of Nikolaj Alekseevich Belyaev, a celestial mechanician at the Institute for Theoretical Astronomy, known for his research on the dynamics of comets and the evolution of cometary orbits. (M 8542)

(2627) Churyumov

1978 PP_3. Discovered 1978 August 8 by N. S. Chernykh at Nauchnyj.

Named in honor of Klim Ivanovich Churyumov {1937- }, an astronomer on the Astronomical Faculty of Kiev University known for his work on comets. He was a codiscoverer of periodic comet Churyumov-Gerasimenko in 1969. (M 8542)

(2628) Kopal

1979 MS_8. Discovered 1979 June 25 by E. F. Helin and S. J. Bus at Siding Spring.

Named in honor of Zdeněk Kopal {1914-1993}, astronomer, Czech-born chairman of the astronomy department of Victoria University, Manchester, during 1951-1981. A world authority on eclipsing variables, the Moon and the terrestrial planets, Kopal played a leading role in the exploration of the Moon in a NASA-sponsored program. (M 7158)

Name proposed by E. Helin, endorsed by E. M. Shoemaker.

Obituaries published in Bull. Am. Astron. Soc., Vol. 25, No. 4, p. 1498 (1993); Q.J.R. Astron. Soc., Vol. 35, No. 2, p. 229-230 (1994); Phys. Today, Vol. 47, No. 3, p. 80 (1994); Astrophys. Space Sci., Vol. 213, No. 2, p. 171-173 (1994).

(2629) Rudra

1980 RB_1. Discovered 1980 September 13 by C. T. Kowal at Palomar.

Named for the destroyer aspect of Siva {see planet (1170)}, who in the mythology of India was both destroyer and regenerator of all things in the universe. His strength was especially unleashed on all sin and evil, and even the gods feared him. (M 27329)
Name suggested by F. Pilcher.

(2630) Hermod

1980 TF_3. Discovered 1980 October 14 at the Haute Provence Observatory at St. Michel.

Named for a son of Odin and Frigga {see planets (3989) and (77)} in Scandinavian mythology. He entered the land of the dead, guarded by Hel {see planet (949)}, in an attempt to rescue his brother Balder {see planet (4059)}. Hel agreed on condition that everyone and everything in the world, even the plants, stones, and metals, should mourn for Balder. All did, except for a single giant witch in a cave, Thokk. (M 22495)

Named by the Minor Planet Names Committee following a suggestion by F. Pilcher.

(2631) Zhejiang

1980 TY_5. Discovered 1980 October 7 at the Purple Mountain Observatory at Nanking.

Named for a province in eastern China. The West Lake in its capital city of Hangzhou attracts visitors from all over the world. (M 7619)

(2632) Guizhou

1980 VJ_1. Discovered 1980 November 6 at the Purple Mountain Observatory at Nanking.

Named for a province in southwestern China. Many minority people live in this province. (M 7620)

(2633) Bishop

1981 WR_1. Discovered 1981 November 24 by E. Bowell at Anderson Mesa.

Named for George Bishop (1785-1861), from whose observatory in Regents Park, London, eleven minor planets were discovered. The proprietor of a wine-making business, Bishop served as president of the Royal Astronomical Society in 1857 and 1858. (M 9768)
Citation prepared by B. Hetherington

(2634) James Bradley

1982 DL. Discovered 1982 February 21 by E. Bowell at Anderson Mesa.

Named for James Bradley (1693-1762), one of the greatest observers of his time and third Astronomer Royal, from 1742 until his death. Discoverer of aberration (1729) and nutation (1748), Bradley pioneered modern methods of determining instrumental effects on position measurements and published positions of some 3,000 stars with an accuracy never before attained. (M 9768)
Name suggested and citation prepared by B. Hetherington.

(2635) Huggins

1982 DS. Discovered 1982 February 21 by E. Bowell at Anderson Mesa.

Named for William Huggins (1824-1910), pioneer in astronomical spectroscopy. Huggins stated that the chemical elements on the Earth also existed in the stars, discovered the gaseous nature of bright nebulae, was the first to study the spectrum of a nova and measured the radial velocity of Sirius. As early as 1875 he had devised methods of photographing spectra, and this resulted in the publication of his Atlas of Representative Stellar Spectra in 1899. (M 9768)

Name suggested and citation prepared by B. Hetherington.
Huggins is also honored by craters on Mars and the Moon.

(2636) Lassell

1982 DZ. Discovered 1982 February 20 by E. Bowell at Anderson Mesa.

Named for William Lassell (1799-1880), discoverer of Neptune's satellite Triton and Uranus' satellites Ariel and Umbriel and an independent discoverer of Saturn's satellite Hyperion. A brewer by trade, he also found some 600 nebulae. (M 9768)
Name suggested and citation prepared by B. Hetherington.
Lassell is also honored by craters on Mars and the Moon.

(2637) Bobrovnikoff

A919 SB. Discovered 1919 September 22 by K. Reinmuth at Heidelberg.

Named in honor of Nicholas T. Bobrovnikoff, director of the Perkins Observatory from 1934 to 1951, who prepared an authoritative history of astronomy. Perhaps best known for his pioneering investigations on the effect of aperture size on the observed total magnitude of comets, he made an exhaustive investigation of the 1909-11 apparition of Halley's Comet and prepared in manuscript form a comprehensive catalogue of physical observations of comets. His spectroscopic study showing differences in the surfaces of minor planets, published in 1929, has been cited as 'so far ahead of its time that it was overlooked'. (M 8064)

Following suggestions by D. D. Meisel and N. Sperling, the name was proposed by B. G. Marsden, who found the identifications involving this planet.

(2638) Gadolin

1939 SG. Discovered 1939 September 19 by Y. Väisälä at Turku.

Named for Jacob Gadolin (1719-1802), the first astronomer at the Academia Aboensis, and Johan Gadolin (1760-1852), the 'Father of Finnish Chemistry'. (M 7947)

(2639) Planman

1940 GN. Discovered 1940 April 9 by Y. Väisälä at Turku.

Named for Anders Planman (1724-1803), an astronomer and from 1763 professor of physics at the Academia Aboensis. A participant in the international program to determine the astronomical unit, he made careful observations in Kajaani {see planet (1519)} of the 1761 and 1769 transits of Venus. (M 7947)

(2640) Hällström

1941 FN. Discovered 1941 March 18 by L. Oterma at Turku.

Named for Gustav Gabriel Hällström (1775-1844), professor of physics and several times rector at the Academia Aboensis. His efforts led to the acquisition by Turku of an astronomical observatory and the regular appointment of an observer. (M 7947)

(2641) Lipschutz

1949 GJ. Discovered 1949 April 4 at the Goethe Link Observatory at Brooklyn, Indiana.

Named in honor of Michael E. Lipschutz, professor of chemistry at Purdue University and author or co-author of more than 100 scientific papers on meteorites, radiochemistry, cosmochemistry, the Antarctic and extraterrestrial materials processing. Recipient of the first annual Nininger {see planet (2421)} Award for meteorite research in 1962, he has also received three NASA achievement awards and in 1986 was given a certificate of appreciation by the National Commission on Space. He has served as treasurer and council member of the Meteoritical Society (1979-1984), was associate editor of the proceedings of the eleventh Lunar and

Planetary Science Conference (1980) and co-convenor of a workshop on Antarctic Glaciology and Meteorites (1982). (M 12012)
Name proposed by F. K. Edmondson.

(2642) Vésale

1961 RA. Discovered 1961 September 14 by S. Arend at Uccle.

Named in memory of André Vésale (1514-1564), physician to Charles V, promoter of the value of the dissection of cadavers to medicine and author of *De humani corporis fabrica*, which brought about a revolution in the study of human anatomy. (M 12209)

(2643) Bernhard

1973 SD. Discovered 1973 September 19 by T. Gehrels at Palomar.

Named in honor of H.R.H. Prince Bernhard of the Netherlands, in recognition of his interest in comets and minor planets. (M 22828)

(2644) Victor Jara

1973 SO_2. Discovered 1973 September 22 by N. S. Chernykh at Nauchnyj.

Named in memory of the internationally-known Chilean singer and composer Victor Jara (1938-1973). (M 8800)

(2645) Daphne Plane

1976 QD. Discovered 1976 August 30 by E. F. Helin at Palomar.

Named in honor of Daphne Plane, librarian in Caltech's geology department for more than a quarter of a century and a long-time friend of the discoverer. (M 7620)

(2646) Abetti

1977 EC_1. Discovered 1977 March 13 by N. S. Chernykh at Nauchnyj.

Named in memory of Antonio Abetti (1846-1928) and of his son Giorgio Abetti (1882-1982), who each served as director of the Asiago Astrophysical Observatory. Collectively they made important contributions to the study of minor planets, solar physics, and the history of astronomy. (M 8542)

Name proposed by the discoverer, following a suggestion from E. Colombini at the Osservatorio San Vittore, where the object was rediscovered as 1982 FB.

Obituaries (G.A.) published in Coelum, Vol. 50, p. 289-291 (1982); Observatory, Vol. 103, No. 1054, p. 184 (1983); Sky Telesc., Vol. 65, No. 1, p. 27 (1983); Vesmír, Vol. 62, p. 94 (1983); Q.J.R. Astron. Soc., Vol. 25, No. 1, p. 98-100 (1984); Zenit, 10. Jaarg., No. 3, p. 128 (1983).

(2647) Sova

1980 SP. Discovered 1980 September 29 by Z. Vávrová at Kleť.

Named in memory of Antonín Sova (1864-1928), Czech poet, who sang of southern Bohemia. (M 22245)

(2648) Owa

1980 VJ. Discovered 1980 November 8 by E. Bowell at Anderson Mesa.

In the Hopi Indian language of northern Arizona owa means 'rock', the almost certain constituent of this minor planet. (M 7473)
Name suggested by E. Malotki.

(2649) Oongaq

1980 WA. Discovered 1980 November 29 by E. Bowell at Anderson Mesa.

In the Hopi Indian language oongaq means 'from up there', being the ablative case form of the locative stem oo- ('up' or 'high'). The word can be used to donate things that come from the sky, although it is to be hoped that this particular

minor planet will remain in orbit. (M 7473)
Name suggested by E. Malotki.

(2650) Elinor

1931 EG. Discovered 1931 March 14 by M. Wolf at Heidelberg.

Named by the Minor Planet Center in honor of Elinor Gates in appreciation of her assistance in preparing and checking observations as a summer student, 1989 May-August. (M 14971)

(2651) Karen

1949 QD. Discovered 1949 August 28 by E. L. Johnson at Johannesburg.

Named by F. N. Bowman, who found the key identification involving this planet, in honor of Karen S. Mayer, his sister-in-law, and also Karen S. Franz, a fellow physics major at the University of Cincinnati. (M 7158)

(2652) Yabuuti

1953 GM. Discovered 1953 April 7 by K. Reinmuth at Heidelberg.

Named in honor of Kiyosi Yabuuti, a member of the Japan Academy, former professor of the Astronomical Institute and former director of the Research Institute for Humanistic Studies, Kyoto University. A pioneer in the study of sinology, he has published various books on Chinese ancient astronomy and technology. The minor planet is named on the occasion of the international conference on the history of science in China (Kyoto, 1987), which was held in his honor. (M 12803)

Name proposed by K. Hurukawa, who was a student of Yabuuti and who found the identifications involving this planet.

(2653) Principia

1964 VP. Discovered 1964 November 4 at the Goethe Link Observatory at Brooklyn, Indiana.

Named in honor of Isaac Newton's {1642-1727, see planet (8000)} *Philosophiae Naturalis Principia Mathematica* on the three-hundredth anniversary of its publication. The culminating work of the seventeenth-century scientific revolution, weaving together many diverse strands into one consistent fabric and establishing the enduring paradigm of modern science, Newton's book was also of major importance for astronomy, including programs for the discovery and observation of minor planets. (M 12012)
Name proposed by F. K. Edmondson. Citation prepared by R. S. Westfall.

(2654) Ristenpart

1968 OG. Discovered 1968 July 18 by C. Torres at Cerro El Roble.

Named in memory of the German astronomer Friedrich Wilhelm Ristenpart (1868-1913), appointed in 1906 as director of the Observatorio Astronomico Nacional in Santiago with the mission to modernize it. He accomplished the move from the downtown site of Quinta Normal to Lo Espejo, about 13 km south of the city. Of the 1260 plates assigned to the Santiago zone of the Astrographic Catalogue, 745 were taken before his untimely death. A series of fifty charts showing southern-hemisphere stars down to tenth magnitude is known as the Carta de Ristenpart. (M 10546)

(2655) Guangxi

1974 XX. Discovered 1974 December 14 at the Purple Mountain Observatory at Nanking.

Guangxi Zhuangzu Zizhiqu is located in southwestern China. Its Guilin district is famous for its beauty. (M 7620)

(2656) Evenkia

1979 HD_5. Discovered 1979 April 25 by N. S. Chernykh at Nauchnyj.
Named for the Evenkian National Area of the R.S.F.S.R. (M 7473)

(2657) Bashkiria

1979 SB_7. Discovered 1979 September 23 by N. S. Chernykh at Nauchnyj.

Named for the {former} Bashkir Autonomous Soviet Socialist Republic. (M 7473)

(2658) Gingerich

1980 CK. Discovered 1980 February 13 at the Harvard College Observatory at Harvard.

Named in honor of Owen Jay Gingerich {1930- }, Harvard professor of astronomy and of the history of science, and an astrophysicist at the Smithsonian Astrophysical Observatory. Although he has made important contributions to the modeling of stellar atmospheres, he is best known for effectively setting the standards of scholarship for modern studies of the history of astronomy and astrophysics. His historical interests are broad, but much of his work centers on Copernicus and Kepler, as is appropriately acknowledged by this planet's provisional designation. He is also celebrated for his basic course on science to non-scientist Harvard students, and he served as director of the IAU Central Bureau for Astronomical Telegrams for the three years following the transfer from Copenhagen at the end of 1964. (M 9477)

(2659) Millis

1981 JX. Discovered 1981 May 5 by E. Bowell at Anderson Mesa.

Named in honor of Robert L. Millis, planetary astronomer at Lowell Observatory, whose work on occultations of stars by solar-system objects has led to the discovery of the rings of Uranus and accurate determination of the sizes of a number of minor planets. He has also used the occultation technique to probe the atmospheres of Mars and Uranus, to refine the ephemerides of the Galilean satellites and to search for possible rings around Neptune. (M 7620)

(2660) Wasserman

1982 FG. Discovered 1982 March 21 by E. Bowell at Anderson Mesa.

Named in honor of Lawrence H. Wasserman, planetary astronomer at Lowell Observatory. Like his colleague Millis {see planet (2659)}, Wasserman has specialized in the study of solar-system objects by the occultation technique. He has analyzed lightcurves of stars occulted by planetary atmospheres and has worked on the prediction, observation and reduction of occultations of stars by minor planets. (M 7620)

(2661) Bydžovský

1982 FC_1. Discovered 1982 March 23 by Z. Vávrová at Kleť.

Named in honor of Academician Bohumil Bydžovský (1880-1969), professor of mathematics and chancellor of the Charles University in Prague. He was born in the small town of Veselínad Lužnicí{see planets (2599) and (2321)} in southern Bohemia and became its most eminent citizen. (M 21607)

(2662) Kandinsky

4021 P-L. Discovered 1960 September 24 by C. J. van Houten and I. van Houten-Groeneveld at Palomar.

Named in memory of the Russian-born painter Vassily Kandinsky (1866-1944), one of the first and best of the abstract painters. (M 9080)

(2663) Miltiades

6561 P-L. Discovered 1960 September 24 by C. J. van Houten and I. van Houten-Groeneveld at Palomar.

Named for the commander of the Athenian troops who conquered the Persian army in the battle near Marathon in 490 B.C. (M 16590)

(2664) Everhart

1934 RR. Discovered 1934 September 7 by K. Reinmuth at Heidelberg.

Named in honor of Edgar Everhart {1920-1990}, since 1969 in the physics-astronomy department at the University of Denver and director of the Chamberlain Observatory. After an impressive career working on atomic cross-sections, he has made equally fundamental contributions to our knowledge of the distribution of comets and the evolution of cometary orbits, including the development of an efficient integration technique for the purpose. Visual discoverer of comets 1964 IX and 1966 IV, he has more recently designed and constructed a measuring engine and used it in a highly successful program of photographic astrometry of comets. (M 8065)

Name proposed by B. G. Marsden, who found the identifications involving this planet.

(2665) Schrutka

1938 DW_1. Discovered 1938 February 24 by A. Bohrmann at Heidelberg.

Named in honor of the Austrian astronomer Guntram Schrutka von Rechtenstamm (1910-1995), who worked as professor of astronomy at Vienna University. In 1936 he became the first astronomer to derive the shape of minor planet (433) Eros, assuming it to be a triaxial ellipsoid. Furthermore, Schrutka is an eminent computer of cometary orbits, and he worked extensively on the difficult cases of P/Tempel 1, P/Tempel-Swift and P/Westphal. His work on positions and heights of lunar formations is also well known. (M 20519, M 20542)
Name proposed and citation prepared by H. Haupt and L. D. Schmadel.
Obituary published in Mitt. Astron. Ges., Nr. 79, p. 9-10 (1996).

(2666) Gramme

1951 TA. Discovered 1951 October 8 by S. Arend at Uccle.

Named in memory of Zénobe Gramme (1826-1901), a joiner by profession, who constructed the first direct-current dynamo and invented the collector that derives direct current from a revolving armature. (M 12209)

(2667) Oikawa

1967 UO. Discovered 1967 October 30 by L. Kohoutek at Bergedorf.

Named in memory of Okuro Oikawa (1896-1980), who joined the staff of the Tokyo Astronomical Observatory while the new observatory at Mitaka {see planet (1088)} was under construction. The Brashear {see planet (5502)} astrograph used at the original Azabu {see planet (3290)} site by S. Hirayama when he made the 1900 prediscovery observations of the objects later named (498) Tokio and (727) Nipponia was also moved to Mitaka, and Oikawa used it to discover eight numbered and seven unnumbered minor planets during 1926–1930. (M 27329)
Name suggested by Y. Kozai.

(2668) Tataria

1976 QV. Discovered 1976 August 26 by N. S. Chernykh at Nauchnyj.
Named for the {former} Tatar Autonomous Soviet Socialist Republic. (M 7473)

(2669) Shostakovich

1976 YQ_2. Discovered 1976 December 16 by L. I. Chernykh at Nauchnyj.

Named in memory of Dmitrij Dmitrievich Shostakovich (1906-1975), outstanding twentieth-century Soviet composer. (M 7785)

(2670) Chuvashia

1977 PW_1. Discovered 1977 August 14 by N. S. Chernykh at Nauchnyj.

Named for the {former} Chuvash Autonomous Soviet Socialist Republic. (M 7473)

(2671) Abkhazia

1977 QR_2. Discovered 1977 August 21 by N. S. Chernykh at Nauchnyj.

Named for the {former} Abkhazian Autonomous Soviet Socialist Republic. (M 7474)

(2672) Písek

1979 KC. Discovered 1979 May 31 by J. Květon at Kleť.

Named for a town in western South Bohemia where many outstanding Czech writers and artists were born and lived. (M 7620)

(2673) Lossignol

1980 KN. Discovered 1980 May 22 by H. Debehogne at La Silla.

The family name of friends of the discoverer. (M 15572)

(2674) Pandarus

1982 BC_3. Discovered 1982 January 27 at the Oak Ridge Observatory at Harvard.

Named for the Lycian warrior whose treacherous wounding of Menelaus {see planet (1647)} broke the truce in the Trojan War. (M 6957)

(2675) Tolkien

1982 GB. Discovered 1982 April 14 by M. Watt at Anderson Mesa.

Named in memory of J. R. R. Tolkien (1892-1973), author and philologist, Merton professor of English language at the University of Oxford. Best known for his imaginative writings, in particular *The Lord of the Rings* and *The Hobbit*, he also had a lifelong interest in astronomy. (M 7474)

(2676) Aarhus

1933 QV. Discovered 1933 August 25 by K. Reinmuth at Heidelberg.

Named by L. K. Kristensen, who found the identifications involving this object, in honor of the Danish city and its university. (M 7158)

(2677) Joan

1935 FF. Discovered 1935 March 25 by M. Laugier at Nice.

Named in honor of Joan Jordan, secretary in the Solar and Stellar Physics division at the Harvard-Smithsonian Center for Astrophysics. (M 22495)

Name proposed by members of the Planetary Sciences division, the identifications for this object having been made by B. G. Marsden.

(2678) Aavasaksa

1938 DF_1. Discovered 1938 February 24 by Y. Väisälä at Turku.

Named for a hill in the Tornio {see planet (1471)} river valley. Just south of the Arctic circle, it offers a fine view of the midnight Sun each June. (M 7947)

(2679) Kittisvaara

1939 TG. Discovered 1939 October 7 by Y. Väisälä at Turku.

Named for the mountain that is the northern terminus of the 'degree measurement' conducted in the Tornio {see planet (1471)} valley in 1736-1737 by Maupertuis {see planet (3281)}. (M 7948)

(2680) Mateo

1975 NF. Discovered 1975 July 1 at the Felix Aguilar Observatory at El Leoncito.

Named in memory of José Mateo (1914-1978), a geophysicist who was for many years in charge of the Gravimetry and Tide Department of the La Plata Observatory. Later, as director of the observatory, he started the final phase of construction of the astronomical complex at El Leoncito. (M 13608)

(2681) Ostrovskij

1975 VF_2. Discovered 1975 November 2 by T. M. Smirnova at Nauchnyj.

Named in memory of Nikolaj Alekseevich Ostrovskij (1904-1936), Soviet writer, known for his book *How the steel was tempered.* (M 10546; I. I. Neyachenko)

(2682) Soromundi

1979 MF_4. Discovered 1979 June 25 by E. F. Helin and S. J. Bus at Siding Spring.

Named by E. Helin in honor of the Los Angeles chapter of the Young Women's Christian Association, 'sisters of the world'. (M 7620)

(2683) Brian

1981 AD_1. Discovered 1981 January 10 by N. G. Thomas at Anderson Mesa.
Named by the discoverer in honor of his eldest son. (M 7474)

(2684) Douglas

1981 AH_1. Discovered 1981 January 3 by N. G. Thomas at Anderson Mesa.

Named by the discoverer in honor of his brother, Douglas B. Thomas, a physicist at the National Bureau of Standards. (M 7474)

(2685) Masursky

1981 JN. Discovered 1981 May 3 by E. Bowell at Anderson Mesa.

Named in honor of Harold Masursky {1923-1990}, planetary geologist at the Branch of Astrogeologic Studies of the U.S. Geological Survey, Flagstaff. He has been active in nearly every U.S. program of lunar and planetary exploration including Ranger, Surveyor, Lunar Orbiter, Apollo, Mariner 9, Viking, Pioneer Venus, and Voyager. He was involved in planning future space missions, including the Galileo mission (Jupiter orbiter probe) and the Venus Radar Mapper. (M 7158)

Obituaries published in Sky Telesc., Vol. 81, No. 2, p. 130-131 (1991); Bull. Am. Astron. Soc., Vol. 25, No. 4, p. 1499 (1993).

(2686) Linda Susan

1981 JW_1. Discovered 1981 May 5 by C. S. Shoemaker at Palomar.

Named in honor of Linda Susan Salazar, youngest daughter of the discoverer. (M 8404)

(2687) Tortali

1982 HG. Discovered 1982 April 18 by M. Watt at Anderson Mesa.

Named for the spirit of daytime and the Sun in the mythology of the Melanesians of the New Hebrides Islands, which recognizes no gods. He was the rival of Ul {see planet (3271)}, spirit of nighttime and the Moon. (M 22496)

Named by the Minor Planet Names Committee following a suggestion by F. Pilcher.

(2688) Halley

1982 HG_1. Discovered 1982 April 25 by E. Bowell at Anderson Mesa.

Named in memory of the great English astronomer Edmond Halley (1656-1742) on the occasion of the recovery, in October 1982, of the comet that bears his name. (M 7474)

Halley is also honored by craters on Mars and the Moon.

(2689) Bruxelles

1935 CF. Discovered 1935 February 3 by S. Arend at Uccle.
Named for the capital of Belgium. (M 12209)

(2690) Ristiina

1938 DG_1. Discovered 1938 February 24 by Y. Väisälä at Turku.

Named for a country commune in the lake district of southern Finland. In 1968 over sixty prehistoric rock-paintings were found there in the largest discovery of this type in Fennoscandia. (M 7948)

(2691) Sersic

1974 KB. Discovered 1974 May 18 at the Felix Aguilar Observatory at El Leoncito.

Named in honor of José Luis Sersic {1933-1993}, well known for his work in extragalactic astronomy and on supernovae. He has served as director of the Córdoba Observatory and is currently head of the section for extragalactic studies. With Jorge Sahade {see planet 2605)} he was responsible for establishing IAFE, the Argentinian Institute for Astronomy and Physics of Space, and he played an important role in the establishment of the El Leoncito {see planet (2311)} Station. (M 14207)

Obituary published in Q.J.R. Astron. Soc., Vol. 36, No. 1, p. 79 (1995).

(2692) Chkalov

1976 YT_3. Discovered 1976 December 16 by L. I. Chernykh at Nauchnyj.

Named in memory of Valerij Pavlovich Chkalov (1904-1938), who participated in the historic flight from Moscow over the North Pole to the U.S.A. in 1937. (M 7785)

(2693) Yan'an

1977 VM_1. Discovered 1977 November 3 at the Purple Mountain Observatory at Nanking.

Named for a town in the northern part of Shaanxi {see planet (2263)} province, along the Yan River. (M 9478)

(2694) Pino Torinese

1979 QL_1. Discovered 1979 August 22 by C.-I. Lagerkvist at La Silla.

Named for the small village in which the Turin Observatory is situated. (M 7474)

(2695) Christabel

1979 UE. Discovered 1979 October 17 by E. Bowell at Anderson Mesa.

Named for a lovely lady in a poem of the same name by the English poet Samuel Taylor Coleridge (1772-1834). While riding at night, Christabel rescued another lovely lady, who turned out to be the daughter of her father's estranged boyhood friend. Christabel sought to reconcile the two fathers. (M 7158)

(2696) Magion

1980 HB. Discovered 1980 April 16 by L. Brožek at Kleť.

Named for the first Czechoslovak artificial satellite, launched together with Interkosmos 18 in 1978. Designed for studying the relationship of the magnetosphere and the ionosphere, Magion examined the special structure of ELF- and E-wave phenomena. (M 7620)

(2697) Albina

1969 TC_3. Discovered 1969 October 9 by B. A. Burnasheva at Nauchnyj.

Named in honor of Albina Alekseevna Serova, Moscow astronomer, friend of the discoverer. (M 11156)

(2698) Azerbajdzhan

1971 TZ. Discovered 1971 October 11 at the Crimean Astrophysical Observatory at Nauchnyj.

Named for the {former} Azerbajdzhan Soviet Socialist Republic. (M 7474)

(2699) Kalinin

1976 YX. Discovered 1976 December 16 by L. I. Chernykh at Nauchnyj.

Named in memory of Mikhail Ivanovich Kalinin (1875-1946), Soviet statesman. (M 8065)

(2700) Baikonur

1976 YP_7. Discovered 1976 December 20 by N. S. Chernykh at Nauchnyj.

Named for the Soviet cosmodrome, where the world's first artificial satellite was launched just 25 years ago {1957}. (M 7474)

(2701) Cherson

1978 RT. Discovered 1978 September 1 by N. S. Chernykh at Nauchnyj.

Named for a town in the Ukraine, center of the Chersonian region, which borders with the Crimea. (M 7474)

(2702) Batrakov

1978 SZ_2. Discovered 1978 September 26 by L. V. Zhuravleva at Nauchnyj.

Named in honor of Yurij Vasil'evich Batrakov {1926- }, deputy director of the Institute for Theoretical Astronomy, Leningrad, head of the Small Bodies of the Solar System Department and editor-in-chief of the annual *Ehfemeridy Malykh Planet* {see planet (5001)}. Soon after joining ITA in 1955 he studied new classes of periodic solutions in celestial mechanics and application of the results to minor planets and the rings of Saturn. Beginning in 1958 he worked extensively on the orbits of artificial satellites, including the development, with V. F. Proskurin {see planet (2372)}, of a widely-used analytical theory of their motions, the use of radio measurements and of synchronous observations, and he was the pioneer in the determination of the earth's gravitational field from resonant satellites. He served as vice president of IAU Commission 20 during 1985-1988. (M 18304)

(2703) Rodari

1979 FT_2. Discovered 1979 March 29 by N. S. Chernykh at Nauchnyj.

Named in memory of Gianni Rodari (1920-1980), Italian writer of children's books. (M 9768)

(2704) Julian Loewe

1979 MR_4. Discovered 1979 June 25 by E. F. Helin and S. J. Bus at Siding Spring.

Named in honor of Julian Loewe, a science and staff writer for the *Pasadena Star News* who has for a number of years written on activities at the California Institute of Technology and Jet Propulsion Laboratory. (M 7620)

(2705) Wu

1980 TD_4. Discovered 1980 October 9 by C. S. Shoemaker at Palomar.

Named in honor of Sherman S. C. Wu, chief photogrammetrist of the Astrogeology Branch of the U.S. Geological Survey. Wu and his group have published highly detailed topographic maps of selected regions of the Moon and of Mars, including Olympus Mons and parts of the Valles Marineris. These maps, obtained by means of analytical photogrammetry, represent the most accurate solutions extant of the morphology of surfaces on extraterrestrial bodies. (M 8153)

(2706) Borovský

1980 VW. Discovered 1980 November 11 by Z. Vávrová at Kleť.

Named in memory of Karel Havlíček Borovský (1821-1856), Czech writer, poet and journalist. (M 16441)

(2707) Ueferji

1981 QS_3. Discovered 1981 August 28 by H. Debehogne at La Silla.

Named in honor of the Universidade Federal do Rio de Janeiro, which, through the Valongo Observatory, supports research in astrometry of minor planets in South America. Staff members have participated in the discoverer's observing program at La Silla. (M 9768)

(2708) Burns

1981 WT. Discovered 1981 November 24 by E. Bowell at Anderson Mesa.

Named in honor of Joseph A. Burns, planetary astronomer at Cornell University. Burns' wide-ranging research on solar-system dynamics includes the study of planetary rings, satellites, orbital evolution and tides, origin of the solar system, dust motions, and planetary and asteroid rotation. He is currently editor of the journal *Icarus*. (M 7158)

(2709) Sagan

1982 FH. Discovered 1982 March 21 by E. Bowell at Anderson Mesa.

Named in honor of Carl Sagan {1934-1996}, planetary scientist at Cornell University. He has been active on a number of planetary space missions, including the Mariner 9 and Viking missions to Mars and the Voyager mission to the outer planets. Sagan's research has included studies of the greenhouse effect on Venus, windblown dust on Mars, the atmosphere and surface of Titan, and the possibility of extraterrestrial intelligent life. He has been editor of the journal *Icarus* and won the 1978 Pulitzer prize for literature. A leading popularizer of astronomy, Sagan is founder of the Planetary Society. (M 7158)

Obituaries published in Astron. Raumfahrt Unterr., Jahrg. 34, Heft 2, p. 7 (1997); Astron. Now, Vol. 11, No. 2, p. 4 (1997); Sidereal Times, Vol. 47, No. 4, p. 12 (1997); Postepy Astron., Tom 45, No. 1, p. 40-41 (1997); Phys. Today, Vol. 50, No. 9, p. 94-95 (1997); Mon. Notes Astron. Soc. S. Afr., Vol. 56, No. 1-2, p. 3-4 (1997); Astronomy, Vol. 25, No. 3, p. 28 (1997); Origins Life Evol. Biosphere, Vol. 27, No. 4, p. 309-310 (1997); Nature, Vol. 385, No. 6615, p. 400 (1997); Vasiona, Vol. 45, No. 1-2, p. 30-31 (1997); Planet. Rep., Vol. 17, No. 2, p. 4-22 (1997); Astronomy, Vol. 25, No. 4, p. 38-39 (1997); Sky Telesc., Vol. 93, No. 3, p. 6-7 (1997).

(2710) Veverka

1982 FQ. Discovered 1982 March 23 by E. Bowell at Anderson Mesa.

Named in honor of Joseph Veverka, planetary astronomer at Cornell University. One of the first to study the polarimetric and photometric properties of asteroids, Veverka has made substantial contributions to our knowledge of other small objects in the solar system, notably in his detailed work on Phobos and Deimos. He has also studied the morphology and motions of wind streaks on the Martian surface and has been a strong advocate of space missions to comets. (M 7159)

(2711) Aleksandrov

1978 QB_2. Discovered 1978 August 31 by N. S. Chernykh at Nauchnyj.

Named in honor of Anatolij Petrovich Aleksandrov, outstanding physicist, president of the USSR Academy of Sciences. (M 8542)

(2712) Keaton

1937 YD. Discovered 1937 December 29 by G. Kulin at Budapest. Independently discovered 1937 December 29 by K. Reinmuth at Heidelberg.

Named for Joseph Francis ("Buster") Keaton (1895-1966), American comedian, film actor and director. He joined his family's acrobatic act in vaudeville when still a child, and he perfected his art both as a comic actor and director producing three classic stony-faced performances in *Sherlock Junior*, *The Navigator* and *The*

General, all masterpieces of the silent screen. Thereafter his career declined, but the film *The Buster Keaton Story*, a 1957 screen biography, restored Keaton to his rightful place as one of the world's greatest comedians. (M 22496)

Name proposed by G. V. and R. Williams, the identifications for this object having been made by C. M. Bardwell.

(2713) Luxembourg

1938 EA. Discovered 1938 February 19 by E. Delporte at Uccle.
Named for the country in western Europe. (M 12209)

(2714) Matti

1938 GC. Discovered 1938 April 5 by H. Alikoski at Turku.
Named by the discoverer in honor of his son. (M 7948)

(2715) Mielikki

1938 US. Discovered 1938 October 22 by Y. Väisälä at Turku.

Named for the mistress of the forest in Kalevala {see planet (1454)}. Her favor was needed for successful hunting. Mielikki is a common girl's name in Finland. (M 7948)

(2716) Tuulikki

1939 TM. Discovered 1939 October 7 by Y. Väisälä at Turku.

Also a common girl's name in Finland, Tuulikki was a fairy of the forest in Kalevala {see planet (1454)}. (M 7948)

(2717) Tellervo

1940 WJ. Discovered 1940 November 29 by L. Oterma at Turku.

Named for the fairy of the forest in Kalevala {see planet (1454)}, whose duty was to protect domestic animals from beasts of prey. Tellervo is another common Finnish girl's name. (M 7948)

(2718) Handley

1951 OM. Discovered 1951 July 30 by E. L. Johnson at Johannesburg.

Named for Tommy Handley (1894-1949), one of Britain's greatest entertainers of the Second World War. He began his career in variety and concert parties, but from about 1926 he found that his very fast, zany style of spoken humor suited the new medium of radio. Tommy Handley's biggest success was the radio show ITMA ("It's That Man Again"), which was first broadcast in 1939. He introduced many colorful characters and catch-phrases, including Mrs. Mopp, the charlady, with her eternal question - "Can I do you now, sir?" (M 22496)

Name proposed by B. G. Marsden and G. V. and R. Williams, the identifications for this object having been made by C. M. Bardwell.

(2719) Suzhou

1965 SU. Discovered 1965 September 22 at the Purple Mountain Observatory at Nanking.

Named for a city in southeastern Jiangsu province, famous at home and abroad for its beautiful scenery and elegant ancient gardens. Long described as "the land of fish and rice", "home of silk" and "Venice of the Orient", Suzhou is a shining pearl inlaid on the Yangtse River delta and is also a center of industry and commerce. (M 18448)

(2720) Pyotr Pervyj

1972 RV_3. Discovered 1972 September 6 by L. V. Zhuravleva at Nauchnyj.

Named in memory of Peter I (1672-1725), the Russian tsar known as Peter the Great for his progressive reforms and military talent. (M 7621)

(2721) Vsekhsvyatskij

1973 SP_2. Discovered 1973 September 22 by N. S. Chernykh at Nauchnyj.

Named in honor of Sergej Konstantinovich Vsekhsvyatskij {1905-1984}, professor at Kiev University, a prominent researcher on comets, the Sun and solar activity. (M 7785)

Obituaries published in Říše hvězd, Vol. 66, No. 5, p. 88 (1985); Zemlya Vselennaya, No. 2, p. 60-63 (1985); Vesmír, Vol. 64, No. 9, p. 538 (1985). - This planet and the following ones are named on the occasion of the fiftieth anniversary of the All-Union Astronomical Geodetical Organization. The initial letters of the four names spell out the Russian abbreviation VAGO (Vsesoyuznoe Astronomo-Geodezicheskoe Obshchestvo) for this body.

(2722) Abalakin

1976 GM_2. Discovered 1976 April 1 by N. S. Chernykh at Nauchnyj.

Named in honor of Viktor Kuz'mich Abalakin {1930- }, leader of the ephemeris group at the Institute for Theoretical Astronomy, editor of the *Astronomicheskij Ezhegodnik*, president of IAU Commission 4 (Ephemerides) during 1976-79, outstanding linguist and friend of astronomers around the world. (M 7785)
See also the citation for minor planet (2721).

(2723) Gorshkov

1978 QL_2. Discovered 1978 August 31 by N. S. Chernykh at Nauchnyj.

Named in memory of Petr Mikhailovich Gorshkov (1883-1975), professor at Leningrad University, prominent geodesist and astronomer, researcher in geodesy, gravimetry, celestial mechanics and the history of astronomy. (M 7786)
See also the citation for minor planet (2721).

(2724) Orlov

1978 RZ_5. Discovered 1978 September 13 by N. S. Chernykh at Nauchnyj.

Named in memory of Sergej Vladimirovich Orlov (1880-1958), professor at Moscow University who contributed much to research on comets. This planet also honors Aleksandr Yakovlevich Orlov (1880-1954), founder and first director of the Golosseevo Astronomical Observatory, known for his research on the Earth's polar motion and on tidal variations in gravity. (M 7786)

See also the citation for minor planet (2721). Aleksandr and Sergej Orlov are also honored by lunar craters.

(2725) David Bender

1978 VG_3. Discovered 1978 November 7 by E. F. Helin and S. J. Bus at Palomar.

Named in honor of David F. Bender, whose analytical techniques and innovations are widely used to evaluate potential space missions to minor planets. A pioneer in the generation of trajectories for both flyby and rendezvous missions, and now a septuagenarian, he is still actively involved in related work at the Jet Propulsion Laboratory, where he has long been a member of the Advance Project Group in the Mission Design Section. (M 8065)

(2726) Kotelnikov

1979 SE_9. Discovered 1979 September 22 by N. S. Chernykh at Nauchnyj.

Named in honor of academician Vladimir Aleksandrovich Kotel'nikov {1908- }, Soviet scientist, radio engineer and vice-president of the U.S.S.R. Academy of Sciences. Radar observations of Mercury, Venus, Mars and Jupiter conducted under his supervision were of help in correcting the value of the astronomical unit, in determining the period and the direction of the rotation of Venus, and in understanding the physics and dynamics of these planets. (M 9214)

(2727) Paton

1979 SO_9. Discovered 1979 September 22 by N. S. Chernykh at Nauchnyj.

Named in memory of Evgenij Oskarovich Paton (1870-1953), Soviet scientist known for bridge building and electric welding. The planet also honors academician Boris Evgen'evich Paton, prominent Soviet metallurgist and president of the Academy of Sciences of the Ukrainian S.S.R. (M 9214)

(2728) Yatskiv

1979 ST_9. Discovered 1979 September 22 by N. S. Chernykh at Nauchnyj.

Named in honor of Yaroslav Stepanovich Yatskiv {1940- }, Soviet astrometrist and geodynamicist, director of the Main Astronomical Observatory of the Ukrainian Academy of Sciences, a vice-president of the IAU, president of IAU Commission 19, and a member of the International Halley Watch Steering Group. (M 9215)

(2729) Urumqi

1979 UA_2. Discovered 1979 October 18 at the Purple Mountain Observatory at Nanking.

Named for an open city in the western part of China, capital of the Xinjiang Uygur {see planet (2336)} autonomous region. The political, economic, cultural, scientific and technological center of Xinjiang, Urumqi is located at the northern foot of the Tianshan Mountains, with the Junggar Basin in the north. It is farther from a sea than any other city in the world. It is a "treasure bowl" rich in natural resources and with a wide cultural diversity. Strategically located on the ancient "Silk Road", Urumqi now occupies an important place on the second Eurasian Land Bridge. (M 18448)

(2730) Barks

1981 QH. Discovered 1981 August 30 by E. Bowell at Anderson Mesa.

Named for Carl Barks, writer and illustrator. His comic-book stories have emphasized exploration and invention, often featuring space adventure, satellites and minor planets. He touched on the idea of 'rubble pile asteroids' more than 20 years ago. (M 7621)

Name proposed by P. Thomas.

(2731) Cucula

1982 KJ. Discovered 1982 May 21 by P. Wild at Zimmerwald.

After (1775) Zimmerwald, this is the second minor planet to be discovered at Zimmerwald in the month of May, when the call of the cuckoo is continually heard from the large woods nearby. (M 8153)

(2732) Witt

1926 FG. Discovered 1926 March 19 by M. Wolf at Heidelberg.

Named in memory of Gustav Witt (1866-1946), astronomer at the Berlin Observatory, discoverer of (433) Eros. (M 8153)

Following a suggestion by M. Gressmann, of Falkensee, the city to which Witt moved in 1943 and in which he died, this planet was named by B. G. Marsden, who found the identifications involving it.

(2733) Hamina

1938 DQ. Discovered 1938 February 22 by Y. Väisälä at Turku.

Named for an old Finnish town on the eastern Gulf of Finland. It was known as market place and harbor as early as the fourteenth century. (M 7948)

(2734) Hašek

1976 GJ_3. Discovered 1976 April 1 by N. S. Chernykh at Nauchnyj.

Named for Jaroslav Hašek (1883-1923), prominent Czech writer. (M 9768)

(2735) Ellen

1977 RB. Discovered 1977 September 13 by S. J. Bus and T. Lauer at Palomar.

Named by the discoverer of comet 1981b (P/Bus) in honor of the discoverer of comet 1981k (P/Howell), then his fiancée, now his wife. (M 7474)

(2736) Ops

1979 OC. Discovered 1979 July 23 by E. Bowell at Anderson Mesa.

Named for the Roman goddess of abundance. (M 7621)

(2737) Kotka

1938 DU. Discovered 1938 February 22 by Y. Väisälä at Turku.

Named for an important Finnish industrial center and harbor, a town constructed mainly on islands in the eastern Gulf of Finland. (M 7948)

(2738) Viracocha

1940 EC. Discovered 1940 March 12 by G. Kulin at Budapest.

Named for the supreme being in Inca mythology. He created the human race and gave them laws for which he required obedience. In his later travels in the Andes Mountains he was accompanied by the Deceiver, Taguacipa {see planet (2739)}, who turns to evil the good men he created. Later, the creation completed, he departed in the direction of the setting Sun. (M 22496)

Named by the Minor Planet Names Committee following a suggestion by F. Pilcher.

(2739) Taguacipa

1952 UZ_1. Discovered 1952 October 17 by J. L. Brady at Mount Wilson.

Named for the great Deceiver in Inca mythology. He is the evil dual of the creator of the human race Viracocha {see planet (2738)}, who turns to evil the good men Viracocha created. The two always traveled in company, and their companionship continues by their being adjacent in the minor planet tables. (M 22496)

Named by the Minor Planet Names Committee following a suggestion by F. Pilcher.

(2740) Tsoj

1974 SY_4. Discovered 1974 September 26 by L. V. Zhuravleva at Nauchnyj.

Named in memory of Victor Robertovich Tsoj (1962-1990), poet, composer and soloist of the rock-group Kino. (M 20834)

(2741) Valdivia

1975 XG. Discovered 1975 December 1 by C. Torres at Cerro El Roble.

Named in memory of the Spanish Captain Pedro de Valdivia (1502-1553), conqueror of Chile, who left Peru accompanied by seven soldiers, one Spanish woman and many Indians, to realize his dreams of being discoverer of new territories and governor of a country, of laying the foundations of new cities, and of mixing the Spanish and native races as the first step for a new identity. (M 10546)

(2742) Gibson

1981 JG_3. Discovered 1981 May 6 by C. S. Shoemaker at Palomar.

Named in honor of James B. Gibson, discoverer of the Amor object (1943) Anteros, one of the most accessible known neighbors of the Earth. His painstaking follow-up observations of periodic comets and newly-discovered minor planets and his search for lost objects have been conducted in California and Argentina during much of the past quarter of a century. (M 7621)

(2743) Chengdu

1965 WR. Discovered 1965 November 21 at the Purple Mountain Observatory at Nanking.

Named for the capital of Sichuan {see planet (2215)} province in southwestern

China. Located in the heart of the Sichuan basin, Chengdu was the capital of the Shu-Han dynasty during the Three Kingdoms period of Chinese history. Today, Chengdu is a major agricultural and industrial center, and it is also the homeland of the Giant Panda. (M 20834)

(2744) Birgitta

1975 RB. Discovered 1975 September 4 by C.-I. Lagerkvist at Kvistaberg.

Named in honor of Anna Birgitta Angelica Lagerkvist, daughter of the discoverer. (M 8800)

(2745) San Martin

1976 SR_{10}. Discovered 1976 September 25 at the Carlos U. Cesco Observatory at El Leoncito.

Named in memory of General José de San Martin (1778-1850), considered the father of Argentina and liberator of Chile and Peru. After fighting for Spain in Bailen against the Napoleonic army he returned to his native Argentina and created the Granaderos a Caballo regiment, which won the battle of San Lorenzo in 1812. As governor intendent of Cuyo {see planet (1917)}, he created the Andes army, crossed those mountains, and by conquering the Spanish army at the battles of Chacabuco and Maipu (1817-1818) assured Chile's independence. His defeat of the Royal army in Peru in 1820 led to the independence of that nation also. He refused to take part in the subsequent civil war in Argentina and moved to Europe. (M 17026)

(2746) Hissao

1979 SJ_9. Discovered 1979 September 22 by N. S. Chernykh at Nauchnyj.

Named for the Hissar (Gissar) Astronomical Observatory, part of the Institute of Astrophysics of the Tadjik Academy of Sciences in Dushanbe, famous for its research activities on meteors, comets and minor planets. (M 7786)

(2747) Český Krumlov

1980 DW. Discovered 1980 February 19 by A. Mrkos at Kleť.

Named for the historical town on the Vltava {see planet (2123)} river. Kleť Observatory is situated 700 m above the town. (M 22828)

(2748) Patrick Gene

1981 JF_2. Discovered 1981 May 5 by C. S. Shoemaker at Palomar.

Named in honor of Patrick Gene Shoemaker, son of the discoverer. (M 8404)

(2749) Walterhorn

1937 TD. Discovered 1937 October 11 by K. Reinmuth at Heidelberg.

Named in memory of Walter Horn (1881-1967), promoter of astronomical knowledge in Nordrhein-Westfalen, Germany's most industrialized area. Horn founded the public observatory in Solingen and, as early as 1924, a society for amateur astronomers. He was considered, not only as an amateur with remarkable didactic skills, but also as a philanthropist who helped many people in the economically unstable mid 1920s. (M 20158)

Name proposed and citation prepared by L. D. Schmadel and W. Schlosser.

(2750) Loviisa

1940 YK. Discovered 1940 December 30 by Y. Väisälä at Turku.

Named for a small Finnish town, an important export harbor, on the eastern Gulf of Finland. About two-thirds of its area is water. (M 7948)

(2751) Campbell

1962 RP. Discovered 1962 September 7 at the Goethe Link Observatory at Brooklyn, Indiana.

Named in memory of W. W. Campbell (1862-1938), observational spectroscopist, director of the Lick Observatory (1901-1930), president of the University of California (1923-1930), president of the International Astronomical Union (1922-1925) and president of the U.S. National Academy of Sciences (1931-1935). He pioneered in conceiving, organizing and carrying out the first large-scale systematic program for the accurate measurement of stellar radial velocities. (M 12457)
Name proposed by F. K. Edmondson. Citation prepared by D. E. Osterbrock.
Campbell is also honored by craters on Mars and the Moon.

(2752) Wu Chien-Shiung

1965 SP. Discovered 1965 September 20 at the Purple Mountain Observatory at Nanking.

Named in honor of Wu Chien-Shiung, born in China and currently Pupin Professor Emerita at Columbia University. She is renowned for her work in nuclear physics, particularly in the experimental study of the beta decay of radioactive atomic nuclei. The precision and elegance of her experiments have earned her the title of greatest living woman physicist. With coworkers she made her most famous contribution in 1956 with a critical experiment on polarized cobalt-60 beta decay. The result of this work, which substantiated the theory of Lee and Yang, shocked the world of physics and overthrew the concept of parity conservation in weak interactions. (M 16040)

(2753) Duncan

1966 DH. Discovered 1966 February 18 at the Goethe Link Observatory at Brooklyn, Indiana.

Named in memory of John Charles Duncan (1882-1967), a graduate of Indiana University, who taught astronomy at Harvard and Radcliffe until 1916 and at Wellesley until his retirement in 1950 - after which he taught for another 14 years at the University of Arizona. The first edition of his widely-used textbook appeared in 1916; the fifth edition appeared in 1955. His Lick Ph.D. thesis on the orbits of two cepheid variables was the first step toward the abandonment of the widely-held binary star theory of cepheid variation. During a 28-year association with the Mount Wilson Observatory he measured the expansion of the Crab Nebula (1921), discovered three variable stars in external galaxies (1922) and took a superb series of pictorial photographs of galaxies, nebulae and Milky Way fields. (M 12457)
Name proposed by F. K. Edmondson.

(2754) Efimov

1966 PD. Discovered 1966 August 13 by T. M. Smirnova at Nauchnyj.

Named in memory of the aviator Mikhail Nikiforovich Efimov (1881-1919), who was the first to realize steep turns and dives. (M 13173)

(2755) Avicenna

1973 SJ_4. Discovered 1973 September 26 by L. I. Chernykh at Nauchnyj.

Named for the medieval scientist, philosopher, physician and poet Ibn Sina, or Abu Ali al Hussein ibn Abdallah (980-1037), known in Europe under the name Avicenna. (M 7786)
Avicenna is also honored by a lunar crater.

(2756) Dzhangar

1974 SG_1. Discovered 1974 September 19 by L. I. Chernykh at Nauchnyj.

Named for the Kalmyk epics, which reflect the hopes, dreams, and struggles of the Kalmyk people. (M 8912)

(2757) Crisser

1977 VN. Discovered 1977 November 11 by S. Barros at Cerro El Roble.

The name is formed from the first halves of the first names of the discoverer's wife, Cristina, and himself, Sergio. (M 11157)

(2758) Cordelia

1978 RF. Discovered 1978 September 1 by N. S. Chernykh at Nauchnyj.
Named for the younger daughter of King Lear in Shakespeare's tragedy. (M 9769)

The name Cordelia has also been given to the satellite Uranus VI, discovered 1986 by the Voyager 2 spacecraft.

(2759) Idomeneus

1980 GC. Discovered 1980 April 14 by E. Bowell at Anderson Mesa.

Named for a king of Crete, renowned for his valor in the Trojan War, during which he slew many Trojans. (M 8404)

(2760) Kacha

1980 TU_6. Discovered 1980 October 8 by L. V. Zhuravleva at Nauchnyj.

Named for the flying school in Kacha, a settlement in the Crimea, where some Soviet cosmonauts receive training. In 1980 the school celebrated its 70th anniversary. (M 8913)

(2761) Eddington

1981 AE. Discovered 1981 January 1 by E. Bowell at Anderson Mesa.

Named in memory of the great English astronomer and physicist Arthur Stanley Eddington (1882-1944) on the occasion of the centennial of his birth. Eddington made fundamental contributions to studies of stellar structure and relativity, and he was also a lucid and undefatigable popularizer of astronomy. (M 7621)

Eddington served as president of the International Astronomical Union from 1938 to 1943. He is also honored by a lunar crater.

(2762) Fowler

1981 AT. Discovered 1981 January 14 by E. Bowell at Anderson Mesa.

Named in memory of Ralph Howard Fowler (1889-1944), well-known English astrophysicist. Among his many discoveries was the recognition of the state of matter in white-dwarf stars. (M 7621)
Fowler is also honored by a lunar crater.

(2763) Jeans

1982 OG. Discovered 1982 July 24 by E. Bowell at Anderson Mesa.

Named in memory of the celebrated English astronomer James Hopwood Jeans (1877-1946), whose wide-ranging research embraced cosmogony, stellar evolution and stellar dynamics. Jeans was also a brilliant writer and popularizer of astronomy. (M 7621)
Jeans is also honored by craters on Mars and the Moon.

(2764) Moeller

1981 CN. Discovered 1981 February 8 by N. G. Thomas at Anderson Mesa.

Named by the discoverer in honor of his mother, Sonia Louise Moeller-Thomas. (M 7621)

(2765) Dinant

1981 EY. Discovered 1981 March 4 by H. Debehogne and G. DeSanctis at La Silla.

Named for a city in southern Belgium, birthplace of Adolphe Sax {see planet (3534)}, inventor of the saxophone, and of the painter Wiertz. (M 19333)

(2766) Leeuwenhoek

1982 FE_1. Discovered 1982 March 23 by Z. Vávrová at Kleť.

Named in memory of Antonius van Leeuwenhoek (1632-1723), inventor of the microscope. (M 16441)
Leeuwenhoek is also honored by a lunar crater.

(2767) Takenouchi

1967 UM. Discovered 1967 October 30 by L. Kohoutek at Bergedorf.

Named in honor of Tadao Takenouchi (1922-), who worked variously in the Department of Astronomy of Tokyo University, the Tokyo Astronomical Observatory, the space agency NASDA and the Institute of Space and Astronautical Sciences. During the 1950s he worked with H. Hirose {see planet (1612)} computing ephemerides of minor planets by hand for the Minor Planet Center. Also by hand, he made a numerical study of the orbit of (279) Thule and examined the behavior of its 4:3 libration with Jupiter over a complete cycle. (M 27329)
Name suggested by Y. Kozai.

(2768) Gorky

1972 RX_3. Discovered 1972 September 6 by L. V. Zhuravleva at Nauchnyj.

Named for Maxim Gorky, pen name of Aleksei Makhsimovich Peshkov (1868-1936), outstanding Russian writer and the founder of Soviet literature. (M 8913)

(2769) Mendeleev

1976 GZ_2. Discovered 1976 April 1 by N. S. Chernykh at Nauchnyj.

Named in memory of Dmitrij Ivanovich Mendeleev (1834-1907), world-famous Russian chemist, discoverer of the periodic law of the elements. (M 9215)
Mendeleev is also honored by a lunar crater.

(2770) Tsvet

1977 SM_1. Discovered 1977 September 19 by N. S. Chernykh at Nauchnyj.

Named in memory of Mikhail Semenovich Tsvet (1872-1919), physiologist and plant biochemist, the founder of chromatography. (M 12012)

(2771) Polzunov

1978 SP_7. Discovered 1978 September 26 by L. V. Zhuravleva at Nauchnyj.

Named for Ivan Ivanovich Polzunov (1728-1766), inventor of the first universal steam-engine in Russia. (M 13173)
Polzunov is also honored by a lunar crater.

(2772) Dugan

1979 XE. Discovered 1979 December 14 by E. Bowell at Anderson Mesa.

Named in memory of the American astronomer Raymond Smith Dugan (1878-1940), perhaps best known as a co-author with Russell and Stewart of an important astronomical textbook. His principal field of research concerned variable stars, but during 1902-1904 he participated in the Heidelberg minor-planet program, discovering 18 planets. (M 7622)
Dugan is also honored by a lunar crater.

(2773) Brooks

1981 JZ_2. Discovered 1981 May 6 by C. S. Shoemaker at Palomar.

Named in honor of William R. Brooks (1844-1921), American astronomer. As the village photographer at Phelps, New York, he independently discovered comet 1881 V with a 5-inch reflector. Although credited to Denning, this find encouraged Brooks to build a 9 1/4-inch telescope with which he found comet Brooks-Swift 1883 I. During the next 28 years, Brooks discovered 20 more comets to which his name has been attached. Three of those were found during a particularly successful 4-week period in 1886. In 1900 he became professor of astronomy at Hobart College in Geneva, New York. (M 17655)
Citation provided by D. H. Levy at the request of the discoverer.

(2774) Tenojoki

1942 TJ. Discovered 1942 October 3 by L. Oterma at Turku.
Named for the border stream between Norway and Finland. (M 7948)

(2775) Odishaw

1953 TX_2. Discovered 1953 October 14 at the Goethe Link Observatory at Brooklyn, Indiana.

Named in memory of Hugh Odishaw (1916-1984), an organizer in international geophysical research during a time of revolution in Earth and space sciences. He served as executive director of the U.S. committee for the International Geophysical Year, director of the IGY World Data Center, executive director of the Space Science Board, executive secretary of the U.S. National Academy of Sciences' division of physical sciences and dean of the University of Arizona's college of Earth and space sciences. His excitement and enthusiasm about the recent flood of discoveries in geophysics led to a television film series, *Planet Earth*, premiered not long after his death. (M 12968)
Name proposed by F. K. Edmondson. Citation prepared by E. Roemer.

(2776) Baikal

1976 SZ_7. Discovered 1976 September 25 by N. S. Chernykh at Nauchnyj.
Named for the large Siberian lake. (M 8543)

(2777) Shukshin

1979 SY_{11}. Discovered 1979 September 24 by N. S. Chernykh at Nauchnyj.

Named in memory of Vasilij Makarovich Shukshin (1929-1974), distinguished Soviet writer, film director and actor. (M 8543)

(2778) Tangshan

1979 XP. Discovered 1979 December 14 at the Purple Mountain Observatory at Nanking.

Named for a city in Hebei {see planet (2505)} province in northern China with a long history and flourishing industry. Tangshan city overlooks Bohai Bay, which is rich in natural resources. The coal, iron and steel and power industries are its economic pillars, and the city is well known as the "granary of eastern Hebei province", the "northern city of coal" and the "northern center of porcelain". (M 20834)

(2779) Mary

1981 CX. Discovered 1981 February 6 by N. G. Thomas at Anderson Mesa.

Named by the discoverer in honor of his wife, Maryanna Ruth Thomas. During the 1960s she measured positions of comets and minor planets in the program then conducted by Elizabeth Roemer {see planet (1657)} at the U.S. Naval Observatory's station in Flagstaff. (M 7622)

(2780) Monnig

1981 DO_2. Discovered 1981 February 28 by S. J. Bus at Siding Spring.

Named in honor of Oscar Monnig, of Fort Worth, Texas, who has devoted a lifetime to the study and popularization of meteoritics. He has made outstanding contributions as the result of countless field investigations of meteorite falls. His tireless efforts have led to the accumulation of one of the world's largest private collections of meteorites. (M 7786)
Name proposed by E. Helin.

(2781) Kleczek

1982 QH. Discovered 1982 August 19 by Z. Vávrová at Kleť.

Named in honor of Josip Kleczek (1923-), distinguished astronomer and solar physicist, known in particular for his outstanding organization of the IAU/-

UNESCO International School for Young Astronomers. He is the author of many books, including *The Six Languages Astronomical Dictionary*. (M 22245)

(2782) Leonidas
2605 P-L. Discovered 1960 September 24 by C. J. van Houten and I. van Houten-Groeneveld at Palomar.
Named for the king of Sparta who defended the Thermophylae pass against the Persians in 480 B.C. (M 16590)

(2783) Chernyshevskij
1974 RA_2. Discovered 1974 September 14 by N. S. Chernykh at Nauchnyj.
Named in memory of Nikolaj Gavrilovich Chernyshevskij (1828-1889), Russian writer and philosopher. (M 9215)

(2784) Domeyko
1975 GA. Discovered 1975 April 15 by C. Torres at Cerro El Roble.
Named in memory of Ignacio Domeyko (1802-1889), a Polish scientist who moved to Chile in 1830 as a teacher of chemistry and mineralogy in La Serena. In 1846 he became professor of mineralogy and later rector at the University of Chile in Santiago. In recognition of his extensive geological investigations throughout the country, the government offered him Chilean nationality. (M 11157)

(2785) Sedov
1978 QN_2. Discovered 1978 August 31 by N. S. Chernykh at Nauchnyj.
Named in memory of Georgij Yakovlevich Sedov (1877-1914), celebrated Russian Arctic explorer who perished on an expedition to the North Pole. (M 8543)

(2786) Grinevia
1978 RR_5. Discovered 1978 September 6 by N. S. Chernykh at Nauchnyj.
Named in memory of the Soviet writer-romanticist Aleksandr Stepanovich Grinevskij (1880-1932). (M 8543)

(2787) Tovarishch
1978 RC_6. Discovered 1978 September 13 by N. S. Chernykh at Nauchnyj.
Named for the Soviet training ship that has participated in many international sailing regattas and has twice been awarded first prize. (M 8800)

(2788) Andenne
1981 EL. Discovered 1981 March 1 by H. Debehogne and G. DeSanctis at La Silla.
Named for an industrial city in southern Belgium, known for its foundries and metallurgic factories. (M 19333)

(2789) Foshan
1956 XA. Discovered 1956 December 6 at the Purple Mountain Observatory at Nanking.
Named for a famous city in the Guangdong {see planet (2185)} province in southern China with a long history and flourishing industry. With its favorable geographical conditions, fertile land and developed commodity economy, and well known for its industry of textiles, electronics, pottery and porcelain, casting, etc., Foshan is a bright pearl of the Pearl River delta. (M 15088)

(2790) Needham
1965 UU_1. Discovered 1965 October 19 at the Purple Mountain Observatory at Nanking.
Named in honor of Joseph Needham, a famous natural scientist and academician of the British Royal Academy of Sciences. For nearly half a century, he was diligent and unremitting in writing *Science and Civilization in China*, a monumental

series that vividly elaborates the historical contributions of China to science and technology. (M 16041)

(2791) Paradise

1977 CA. Discovered 1977 February 13 by S. J. Bus at Palomar.

Named for the town of Paradise, California. Situated on the lower slopes of the Sierras, this old mining town is now a thriving community and the home of Schelte und Alice Bus, parents of the discoverer. (M 7948)

(2792) Ponomarev

1977 EY_1. Discovered 1977 March 13 by N. S. Chernykh at Nauchnyj.

Named in memory of Nikolaj Georgievich Ponomarev (1900-1942), designer of astronomical instruments. (M 11157)

(2793) Valdaj

1977 QV. Discovered 1977 August 19 by N. S. Chernykh at Nauchnyj.

Named for the Valdaj Hills, near Moscow, well known in Russian history. The discoverer als dedicates this name to the memory of his father, Stepan Semenovich Chernykh, who perished there on 1942 Mar. 3 in World War II. (M 9769)

(2794) Kulik

1978 PS_3. Discovered 1978 August 8 by N. S. Chernykh at Nauchnyj.

Named in memory of Leonid Alekseevich Kulik (1883-1942), Soviet mineralogist, researcher of meteorites, and a founder of meteoric research in the U.S.S.R. He is particularly known for his investigation of the place and circumstances of the Tunguska event {see planet (5471)}. (M 9215)

Kulik is also honored by a lunar crater.

(2795) Lepage

1979 YM. Discovered 1979 December 16 by H. Debehogne and E. R. Netto at La Silla.

Named in honor of Theophile Lepage, professor of mathematics at the Université de Liège during 1928-1930 and at the Université de Bruxelles during 1931-1971. Known for his work in analysis, modern algebra, group theory and the theory of geodesic fields, he directed the doctoral studies of the first discoverer. (M 9769)

(2796) Kron

1980 EC. Discovered 1980 March 13 by E. Bowell at Anderson Mesa.

Named in honor of Gerald E. Kron, member of the staff of the Lick Observatory during 1938-1965 and director of the U.S. Naval Observatory's Flagstaff station during 1965-1973. A pioneer in the application of the photomultiplier tube to astronomical photometry, he has published a large quantity of high-precision photometric measurements of stars, clusters and galaxies. In addition, he developed an improved data-retrieval system for the Lallemand electronic camera, which he then applied to a study of globular clusters. (M 8800)

(2797) Teucer

1981 LK. Discovered 1981 June 4 by E. Bowell at Anderson Mesa.

Named for the son of Telamon {see planet (1749)} and half-brother of Ajax {see planet (1404)}. This most renowned of the Greek archers disposed of many Trojan soldiers by springing out from his great shield to loose an arrow, then darting back behind its cover. (M 8543)

(2798) Vergilius

2009 P-L. Discovered 1960 September 24 by C. J. van Houten and I. van Houten-Groeneveld at Palomar.

Named for the Roman poet Virgil (70-18 B.C.), author of the poems *Bucolica, Georgica* and *Aeneis.* (M 16590)

(2799) Justus

3071 P-L. Discovered 1960 September 25 by C. J. van Houten at Palomar.

Named in honor of Justus Cramer, a descendant of H. G. van de Sande Bakhuyzen, a former director of the Leiden Observatory. (M 8404)

(2800) Ovidius

4585 P-L. Discovered 1960 September 24 by C. J. van Houten and I. van Houten-Groeneveld at Palomar.

Named for the Roman poet Ovid (43 B.C.-A.D. 18), whose collections of poems include *Ars Amatoria, Metamorphoses, Tristia* and *Fasti.* (M 16590)

(2801) Huygens

1935 SU_1. Discovered 1935 September 28 by H. van Gent at Johannesburg.

Named in memory of Christian Huygens (1629-1695), celebrated Dutch physicist and astronomer, well known for his wave theory of light and as the discoverer of Saturn's satellite Titan. (M 18135)

Huygens is also honored by a crater on Mars.

(2802) Weisell

1939 BU. Discovered 1939 January 19 by Y. Väisälä at Turku.

Named in memory of the discoverer's father, whose many technical and scientific interests included astronomy. (M 7948)

(2803) Vilho

1940 WG. Discovered 1940 November 29 by L. Oterma at Turku.

Named in memory of Vilho Väisälä (1889-1969), a meteorologist and geophysicist, elder brother of Y. Väisälä {see planet (2804)}. Director of the Finnish Meteorological Institute (1919-1948) and professor of meteorology at the University of Helsinki (1948-1958), he was well known for devising scientific apparatus, including a radiosonde used in many countries. (M 7949)

(2804) Yrjö

1941 HF. Discovered 1941 April 19 by L. Oterma at Turku.

Named in memory of Yrjö Väisälä (1891-1971), originator and leader of the minor-planet program at Turku, and a teacher of the discoverer. He became the first professor of physics at the University of Turku in 1923 and soon afterward he inaugurated instruction in astronomy. He was also well known as a geodesist and an esperantist. An important contributor to astronomical fields as diverse as optics and orbit determination, he planned and constructed the observatory and the telescopes at Iso-Heikkilä {see planet (1947)}. In 1951 he was appointed a member of the Academy of Finland. On his initiative the University of Turku then established the Astronomical-Optical Institute at Tuorla {see planet (1425)}, and he served as its director until his death. (M 7949)

Väisälä is also honored by planet (1573).

(2805) Kalle

1941 UM. Discovered 1941 October 15 by L. Oterma at Turku.

Named in memory of Kalle Väisälä (1893-1968), an algebraist and number theorist, younger brother of Y. Väisälä {see planet (2804)}. Author of various textbooks for secondary school and university, he was professor of mathematics at Tartu University during 1919-1922, at the University of Turku during 1922-1938, and at the Helsinki Technological University during 1939-1960. (M 7949)

(2806) Graz

1953 GG. Discovered 1953 April 7 by K. Reinmuth at Heidelberg.

Named for the capital city of the Austrian province of Styria {see planet (6482)}. With its three universities and a population of 240,000, Graz is a center of culture, music and the arts. The city also hosts and supports the Institute for Space Research of the Austrian Academy of Sciences. Among its distinguished past residents is Johannes Kepler {see planet (1134)}, who lived there from 1594 to 1600. (M 20519)
Name proposed and citation prepared by H. Haupt and L. D. Schmadel.

(2807) Karl Marx

1969 TH_6. Discovered 1969 October 15 by L. I. Chernykh at Nauchnyj.

Named in memory of Heinrich Karl Marx (1818-1883), author of *Das Kapital*, philosopher of history and a student of the theory of socio-economic systems. (M 8065)

(2808) Belgrano

1976 HS. Discovered 1976 April 23 at the Carlos U. Cesco Observatory at El Leoncito.

Named in memory of General Manuel Belgrano (1770-1820), creator of the Argentinian national flag. After studying law in Spain he returned to his native land and served as secretary for the commerce consulate of the viceroyship El Rio de la Plata. A deep believer in freedom and education, he hailed the establishment of the schools of agriculture, commerce and navigation in La Plata and was restless in his fight to improve the social and economic conditions of his people. Pioneer of the May revolution and a member of the first La Plata governmental meeting (1810), he participated in the first military campaign to Paraguay (1811) and defeated the Spanish army in Tucumán and Salta (1812-1813). He traveled to Europe on a diplomatic mission in the hope of gaining independence for these lands. On his return he was put in charge of the army fighting in Peru, and after his death in Buenos Aires he was considered the spiritual martyr of anarchy in his motherland. (M 17027)

(2809) Vernadskij

1978 QW_2. Discovered 1978 August 31 by N. S. Chernykh at Nauchnyj.

Named in memory of Academician Vladimir Ivanovich Vernadskij (1863-1945), distinguished naturalist, mineralogist and crystallographer, founder of the geochemical and radiogeological investigation of the terrestrial biosphere. (M 11157)
Vernadskij is also honored by a lunar crater.

(2810) Lev Tolstoj

1978 RU_5. Discovered 1978 September 13 by N. S. Chernykh at Nauchnyj.

Named for the great Russian writer Lev Nikolaevich Tolstoj (1828-1910). (M 8801)

(2811) Střemchoví

1980 JA. Discovered 1980 May 10 by A. Mrkos at Kleť.

Named for the small village in Moravia {see planet (1901)} where the discoverer was born. (M 22828, 25457)

(2812) Scaltriti

1981 FN. Discovered 1981 March 30 by E. Bowell at Anderson Mesa.

Named in honor of Franco Scaltriti, astronomer at the Pino Torinese Observatory. With his colleague Zappalà {see planet (2813)}, Scaltriti has undertaken an extensive program of minor-planet rotational studies. He has also worked on lightcurves of variable stars. (M 8404)

(2813) Zappalà

1981 WZ. Discovered 1981 November 24 by E. Bowell at Anderson Mesa.

Named in honor of Vincenzo Zappalà, astronomer at the Pino Torinese Observatory. Zappalà's work on minor-planet brightness variation has provided a wealth of data on rotation and shape for the brighter main-belt objects. He has also studied collisional evolution and has been active in a program of minor-planet astrometry. (M 8405)

(2814) Vieira

1982 FA_3. Discovered 1982 March 18 by H. Debehogne at La Silla.

Named in honor of Gilson Vieira, astronomer at the Valongo Observatory and professor at the Universidade Federal do Rio de Janeiro, collaborator in the program that produced the discovery of this object, in particular by adapting computer programs for orbit computations and astrometric reductions. (M 10845)

(2815) Soma

1982 RL. Discovered 1982 September 15 by E. Bowell at Anderson Mesa.

Named for the Soma cube, a three-dimensional mathematical game invented by the Danish writer Piet Hein and popularized in articles by Martin Gardner {see planet (2587)}. (M 9080)

Name proposed by the discoverer, following a suggestion by J. Meeus.

(2816) Pien

1982 SO. Discovered 1982 September 22 by E. Bowell at Anderson Mesa.

Named in honor of Armand Pien, of the Royal Meteorological Institute, Uccle. Well known for his popularization of meteorology and astronomy, he has presented televised weather forecasts in Belgium for more than 30 years. (M 9080)

Name proposed by the discoverer, following a suggestion by J. Meeus.

(2817) Perec

1982 UJ. Discovered 1982 October 17 by E. Bowell at Anderson Mesa.

Named for Georges Perec, who wrote a 300-page novel *La Disparition* (Paris, 1969) without using the letter "e". This "eccentricity" would seem to suit him to studies of minor planets. (M 9080)

Name proposed by the discoverer, following a suggestion by J. Meeus.

(2818) Juvenalis

2580 P-L. Discovered 1960 September 24 by C. J. van Houten and I. van Houten-Groeneveld at Palomar.

Named for the Roman satirical poet Juvenal (fl. c. A.D. 100). (M 16591)

(2819) Ensor

1933 UR. Discovered 1933 October 20 by E. Delporte at Uccle.

Named in memory of Baron James Ensor (1860-1949), renowned painter and sculptor from Ostende, whose works were principally about death or the sea. (M 12209)

(2820) Iisalmi

1942 RU. Discovered 1942 September 8 by Y. Väisälä at Turku.

Named for a town, parts of which date back to the stone age, in the lake district of central Finland. Known for the spa Runni, monuments there memorialize important events in Finnish history. Iisalmi was also the home of famous writers, including Eero Salmelainen, who compiled the celebrated collection of *Finnish Folk-Tales*, published during 1852-1866. (M 7949)

(2821) Slávka

1978 SQ. Discovered 1978 September 24 by Z. Vávrová at Kleť.

Named in memory of Sláva Vávrová (1910-1985), the discoverer's mother. (M 22245)

(2822) Sacajawea

1980 EG. Discovered 1980 March 14 by E. Bowell at Anderson Mesa.

Named for the young Shoshone Indian woman who guided Lewis and Clark on their expedition of discovery across North America to the Pacific Ocean during 1804-1806. She displayed limitless courage and loyalty to both the expedition, in which she faced danger as bravely as any other member, and to her infant son, whom she carried for the entire trip. (M 8065)

(2823) van der Laan

2010 P-L. Discovered 1960 September 24 by C. J. van Houten and I. van Houten-Groeneveld at Palomar.

Named in honor of Harry van der Laan, former professor of astronomy at Leiden Observatory and well-known for his work on radio radiation of gaseous nebulae. He is especially honored for his activity in securing the Dutch participation in the Observatorio de Roque de los Muchachos at La Palma, Canary Islands. Van der Laan is since 1987 director of the European Southern Observatory. (M 10043)

(2824) Franke

1934 CZ. Discovered 1934 February 4 by K. Reinmuth at Heidelberg.

Named in honor of Ernst H. Franke, professor of biophysics at the University of Cincinnati. He gained the respect of his students by being a doer first and a teacher second, facts enhanced by his experience and desire to teach the latest technologies. (M 8065)

Name proposed by F. N. Bowman, who found the key identification involving this planet.

(2825) Crosby

1938 SD_1. Discovered 1938 September 19 by C. Jackson at Johannesburg.

Named for Harry Lillis ("Bing") Crosby (1903-1977), American singer and film actor. Crosby began to sing while studying law at college and, after a spell with the Paul Whiteman Orchestra in a trio called "The Rhythm Boys", appeared in the early talkie *King of Jazz* (1930). Later, turning solo, he began to make films in earnest, notably the 'Road' series of films with Bob Hope {see planet (2829)} and Dorothy Lamour, and won an Academic Award for *Going My Way* (1944). Crosby's 'White Christmas' and 'Silent Night' are two of the biggest selling recordings of all time. (M 22496)

Name proposed by G. V. and R. Williams, the identifications for this object having been made by C. M. Bardwell.

(2826) Ahti

1939 UJ. Discovered 1939 October 18 by Y. Väisälä at Turku.

Named for the male water sprite in Kalevala {see planet (1454)}. Ahti is a common boy's name in Finland. (M 7949)

(2827) Vellamo

1942 CC. Discovered 1942 February 11 by L. Oterma at Turku.

Named for the female water sprite in Kalevala {see planet (1454)}. Vellamo is yet another common girl's name in Finland. (M 7949)

(2828) Iku-Turso

1942 DL. Discovered 1942 February 18 by L. Oterma at Turku.
Named for the deep-sea monster in Kalevala {see planet (1454)}. (M 7949)

(2829) Bobhope

1948 PK. Discovered 1948 August 9 by E. L. Johnson at Johannesburg.

Named in honor of Bob Hope (real name Leslie Townes Hope), American comedian, born Eltham, England, in 1903. Star of numerous radio and television productions, Hope made his mark playing bumbling fools and cowards in such films as *Cat and the Canary* (1939), *Caught in the Draft* (1941) and *Princess and the Pirate* (1944). Hope has been the recipient of four special Academy Awards and more than forty honorary doctorates. (M 22496)

Name proposed by G. V. and R. Williams, the identifications for this object having been made by C. M. Bardwell.

(2830) Greenwich

1980 GA. Discovered 1980 April 14 by E. Bowell at Anderson Mesa.

Named for the Royal Greenwich Observatory on the occasion of the centennial of its adoption as the prime meridian for longitude and time. Founded by King Charles II in 1675 for the determination of longitude at sea, the Royal Observatory was soon established as one of the world's leading astronomical research institutions. The prime meridian was agreed upon, after some dispute, at an international conference in Washington, D.C., on 1884 October 13. (M 8801) Citation material provided by S. R. Malin.

(2831) Stevin

1930 SZ. Discovered 1930 September 17 by H. van Gent at Johannesburg.

Named in memory of Simon Stevin (1548-1620), Dutch mathematician, inventor of the decimal point. (M 18135)

(2832) Lada

1975 EC_1. Discovered 1975 March 6 by N. S. Chernykh at Nauchnyj.

Named for the ancient Slavic goddess of marriage and family happiness. (M 12013)

(2833) Radishchev

1978 PC_4. Discovered 1978 August 9 by L. I. Chernykh and N. S. Chernykh at Nauchnyj.

Named in memory of Aleksandr Nikolaevich Radishchev (1749-1802), Russian writer, philosopher and revolutionary. (M 8913)

(2834) Christy Carol

1980 TB_4. Discovered 1980 October 9 by C. S. Shoemaker at Palomar.

Named in honor of Christine Carol Woodard, eldest daughter of the discoverer. (M 8405)

(2835) Ryoma

1982 WF. Discovered 1982 November 20 by T. Seki at Geisei.

Named in memory of the Japanese revolutionary, Ryoma Sakamoto (1835-1867), born in the discoverer's hometown of Kochi {see planet (2396)}, assassinated for his efforts, eventually successful, to democratize his country. (M 7786)

(2836) Sobolev

1978 YQ. Discovered 1978 December 22 by N. S. Chernykh at Nauchnyj.

Named in honor of Academician Viktor Viktorovich Sobolev {1915- }, professor at Leningrad University, distinguished theoretical astrophysicist, one of the founders of the modern theory of radiative transfer in stellar atmospheres, author of fundamental works on the theory of stellar spectra. (M 8543)

(2837) Griboedov

1971 TJ_2. Discovered 1971 October 13 by L. I. Chernykh at Nauchnyj.

Named in memory of Aleksandr Sergeevich Griboedov (1795-1829), Russian dramatic author and diplomat. His *Gore ot Uma*, or *The mischief of being clever*, is one of the great comedies of European literature. (M 8913)

(2838) Takase

1971 UM_1. Discovered 1971 October 26 by L. Kohoutek at Bergedorf.

Named in honor of Bunshiro Takase (1924-), who worked at the Tokyo Astronomical Observatory and the Department of Astronomy of Tokyo University, where he taught a course on Galactic structure. With the start of the project of building the 105-cm Schmidt telescope at Kiso {see planet (2271)} he returned to the Tokyo Astronomical Observatory and during 1974–84 served as its first director. Early in the 1950s Takase worked with H. Hirose {see planet (1612)} on the computation of ephemerides of minor planets for the Minor Planet Center. (M 27329)
Name suggested by Y. Kozai.

(2839) Annette

1929 TP. Discovered 1929 October 5 by C. W. Tombaugh at Flagstaff.
Named by the discoverer in honor of his daughter. (M 10845)

(2840) Kallavesi

1941 UP. Discovered 1941 October 15 by L. Oterma at Turku.

Named for a large lake, dotted with numerous islands, in south-central Finland. After the ice age the lake was a gulf, and it still contains relics from the Arctic Ocean. (M 7949)

(2841) Puijo

1943 DM. Discovered 1943 February 26 by L. Oterma at Turku.

Named for a hill on Lake Kallavesi {see planet (2840)}. An important tourist and skiing center, it is well known for its beautiful view. (M 7950)

(2842) Unsöld

1950 OD. Discovered 1950 July 25 at the Goethe Link Observatory at Brooklyn, Indiana.

Named in honor of Albrecht Unsöld {1905-1995}, author of *Physik der Sternatmosphären*, long-term professor and director of the Institute for Theoretical Physics at the Christian Albrecht University at Kiel. A student of Sommerfeld's, he pioneered the use of physics in understanding the structure of stellar atmospheres through the analysis of the strengths and profiles of spectral lines, and he sowed the seeds for the new science of laboratory astrophysics as a source of information about the properties of atoms and for the simulation of physical conditions in astronomical objects. Unsöld is respected as much for his vision for the future of astronomy as for his deep sense of appreciation for its history. (M 12969)
Name proposed by F. K. Edmondson. Citation written by G. K. Oertel.

Obituaries published in Publ. Astron. Soc. Pac., Vol. 108, No. 725, p. 553-555 (1996); Phys. Bl., Jahrg. 52, Heft 9, p. 890-891 (1996); Sterne Weltraum, Jahrg. 35, Nr. 3, p. 182-183 (1996); Mitt. Astron. Ges., Nr. 79, p. 11-15 (1996); Astron. Geophys., Vol. 38, Issue 1, p. 37-38 (1997).

(2843) Yeti

1975 XQ. Discovered 1975 December 7 by P. Wild at Zimmerwald.

Named to ensure reality for the mysterious "abominable snowman" that supposedly roams the Himalayan glaciers. (M 17465)

(2844) Hess

1981 JP. Discovered 1981 May 3 by E. Bowell at Anderson Mesa.

Named in honor of Frederick Hess, professor of natural sciences at the State University of New York at Fort Schuyler and long-time lecturer at the Hayden

Planetarium-American Museum in New York City. Hess has directed a number of solar eclipse expeditions and has accumulated more than 30 minutes in the shadow of the Moon. (M 9215)

Name proposed by the discoverer following a suggestion from the Custer Institute. Citation prepared by T. Carey.

(2845) Franklinken

1981 OF. Discovered 1981 July 26 by E. Bowell at Anderson Mesa.

Named in honor of Kenneth Linn Franklin, astronomer at the Hayden Planetarium-American Museum in New York City. At the Carnegie Institution of Washington he was co-discoverer of high-frequency radio emission from Jupiter, and while at the American Museum he directed the Kalbfleisch Research Station on Long Island and designed a watch that displays lunar time. (M 9215)

Name proposed by the discoverer following a suggestion by the Custer Institute. Citation prepared by T. Carey.

(2846) Ylppö

1942 CJ. Discovered 1942 February 12 by L. Oterma at Turku.

Named in honor of the distinguished Finnish physician and scientist Arva Ylppö on the occasion of his hundredth birthday, 1987 October 12. He has been a pioneer in saving premature babies and in the development of pediatrics and mother-and-child welfare in Finland. (M 12803)

(2847) Parvati

1959 CC_1. Discovered 1959 February 1 at the Lowell Observatory at Flagstaff.

Named for the daughter of the Himalayas and wife of Siva {see planet (1170)} in the mythology of India. She was often portrayed as a beautiful young woman at Siva's side and was instructed by him in the arts of asceticism and dancing. (M 22828)

Named by the Minor Planet Names Committee following a suggestion by F. Pilcher.

(2848) ASP

1959 VF. Discovered 1959 November 8 at the Goethe Link Observatory at Brooklyn, Indiana.

Named in honor of the Astronomical Society of the Pacific on the 100th anniversary of its founding, 1889 Feb. 7. The A.S.P., composed of both professional and amateur astronomers from around the world, has worked throughout its existence to promote the field of astronomy and to improve public understanding of it, through the Publications of the A.S.P., Mercury magazine and its newsletter for teachers (The Universe in the Classroom). (M 14029)

Name proposed by F. K. Edmondson. Citation prepared by K. Bracher.

(2849) Shklovskij

1976 GN_3. Discovered 1976 April 1 by N. S. Chernykh at Nauchnyj.

Named in honor of Iosif Samuilovich Shklovskij {1916-1985}, corresponding member of the USSR Academy of Sciences, professor at Moscow University and member of the staff at the Space Research Institute. A brilliant popularizer of astronomy, he has also made substantial contributions to research on the solar corona, galactic radio emission and various cosmic objects. (M 8543)

Obituaries published in Astron. Zh., Tom 62, Vyp. 3, p. 618-619 (1985); Kosm. Issled., Tom 23, Vyp. 3, p. 495 (1985); Pis'ma Astron. Zh., Tom 11, No. 4, p. 319-320 (1985); Říše hvězd, Vol. 66, No. 6, p. 113 (1985); Astron. Vestn., Tom 19, No. 4, p. 359-361 (1985); Sky Telesc., Vol. 70, No. 2, p. 109 (1985); Sov. Astron., Vol. 29, No. 3, p. 364-365 (1985); Sov. Astron. Lett., Vol. 11, No. 2, p. 131-132 (1985); Sterne, 61. Band, Heft 4, p. 232-234 (1985); Sterne Weltraum, 24. Jahrg.,

Nr. 8-9, p. 427 (1985); Zemlya Vselennaya, No. 4, p. 44-46 (1985); Astron. Zh., Tom 63, Vyp. 5, p. 835-838 (1986); Q.J.R. Astron. Soc., Vol. 27, No. 4, p. 700-702 (1986); Sov. Astron., Vol. 30, No. 5, p. 495-497 (1986).

(2850) Mozhaiskij

1978 TM_7. Discovered 1978 October 2 by L. V. Zhuravleva at Nauchnyj.

Named in memory of Aleksandr Fedorovich Mozhaiskij (1825-1890), Russian inventor and pioneer of flight. (M 20834)

(2851) Harbin

1978 UQ_2. Discovered 1978 October 30 at the Purple Mountain Observatory at Nanking.

Named for the city in the north of China, famed as "the jewel of the Eurasia Continent Bridge". Harbin is the capital of Heilongjiang {see planet (2380)} province and is a popular summer resort, on account of its cool and pleasant climate. (M 20834)

(2852) Declercq

1981 QU_2. Discovered 1981 August 23 by H. Debehogne at La Silla.
The family name of the wife of the discoverer. (M 15573)

(2853) Harvill

1963 RG. Discovered 1963 September 14 at the Goethe Link Observatory at Brooklyn, Indiana.

Named in memory of Richard A. Harvill (1905-1988), president of the University of Arizona from 1951 to 1971, the longest term in the history of the university and a period of major growth. During these 20 years the enrollment grew from 5,700 to 26,500, and 45 new buildings were constructed. Harvill provided significant help in many ways to those who were involved in the establishment of the Kitt Peak National Observatory. He was also successful in using the presence of AURA and KPNO to push for increased appropriations from the legislature, following Board of Regents approval, for expansion of the department of astronomy and increased support for the other physical sciences. (M 14207)
Name proposed by F. K. Edmondson.

(2854) Rawson

1964 JE. Discovered 1964 May 6 by D. McLeish at Córdoba.

Named in memory of Guillermo Colesbery Rawson (1825-1890), Argentine physician who worked very hard during the 19th century epidemic in Buenos Aires. He was also a member of the National Parliament on numerous occasions. (M 22496)

(2855) Bastian

1931 TB_2. Discovered 1931 October 10 by K. Reinmuth at Heidelberg.

Named in honor of Ulrich Bastian (1951-), astronomer at the Astronomisches Rechen-Institut, Heidelberg. Together with S. Röser {see planet (2856)}, Bastian produced the *PPM Star Catalogue*, which is a valuable practical tool for minor planet and comet work and constitutes an important aid for the transition from the FK4/B1950 to the FK5/J2000 reference system. He has also worked on other astrometric topics, such as Giotto targeting and the voluminous Hipparcos data-reduction task. (M 20520)

Name proposed and citation prepared by L. D. Schmadel, endorsed by G. Klare and B. G. Marsden.

(2856) Röser

1933 GB. Discovered 1933 April 14 by K. Reinmuth at Heidelberg.

Named in honor of Siegfried Röser (1948-), astronomer at the Astronomisches Rechen-Institut, Heidelberg. Together with U. Bastian {see planet (2855)}, Röser produced the *PPM Star Catalogue*. Röser has also contributed to astrometry by re-reducing all the positional data of Halley's Comet at its 1835/36 and 1909/11 apparitions. (M 20520)

Name proposed and citation prepared by L. D. Schmadel, endorsed by G. Klare and B. G. Marsden.

(2857) NOT

1942 DA. Discovered 1942 February 17 by L. Oterma at Turku.

Named for the 2.56-m Nordic Optical Telescope, erected on La Palma. The optics were made in Y. Väisälä's tunnel laboratory at Tuorla, University of Turku, in a project led by Tapio Korhonen, a student of Väisälä {see planets (2988) and (1573)}. (M 18449)

(2858) Carlosporter

1975 XB. Discovered 1975 December 1 by H. Wroblewski at Cerro El Roble.

Named in memory of Carlos Porter (1867-1942), Chilean zoologist, director of the Valparaiso museum and a professor at Chilean and foreign universities, honored by the Chilean Academy of Natural Sciences and in several other countries. From 1911 to 1928 he was in charge of the invertebrate section of the Museum of Natural History in Santiago. (M 11641)

(2859) Paganini

1978 RW_1. Discovered 1978 September 5 by N. S. Chernykh at Nauchnyj.

Named for the great Italian violinist and composer Niccolo Paganini (1782-1840). (M 11157)

(2860) Pasacentennium

1978 TA. Discovered 1978 October 8 by E. F. Helin at Palomar.

Named as a celestial tribute to the city of Pasadena {see also the citation for planet (2200)}, home of Caltech, and a world center for astronomical research, in honor of the Pasadena Centennial, 1886-1986. (M 10546)

The name was suggested by the city of Pasadena.

(2861) Lambrecht

1981 VL_2. Discovered 1981 November 3 by F. Börngen at Tautenburg.

Named in memory of Hermann Lambrecht (1908-1983), professor at the University of Jena and director of the University Observatory from 1948 to 1968. An authority on the physics and chemistry of interstellar gas and dust particles, he worked on a wide number of astronomical topics and was also well known as a popularizer of astronomy. (M 9080)

Obituaries published in Sterne, 59. Band, Heft 4, p. 195-197 (1983); Mitt. Astron. Ges., Nr. 60, p. 13-14 (1983).

(2862) Vavilov

1977 JP. Discovered 1977 May 15 by N. S. Chernykh at Nauchnyj.

Named in memory of Academician Nikolaj Ivanovich Vavilov (1887-1943), prominent botanist and geneticist, who organized and participated in many botanical expeditions to the various parts of the globe. He established the Soviet collection of plant cultures that includes more than 300,000 samples. The planet also honors his brother, Academician Sergej Ivanovich Vavilov (1891-1951), president of the U.S.S.R. Academy of Sciences from 1945 until his death, renowned for his work in the field of physical optics. (M 11157)

Vavilov is also honored by a lunar crater.

(2863) Ben Mayer

1981 QG_2. Discovered 1981 August 30 by E. Bowell at Anderson Mesa.

Named in honor of the Californian amateur astronomer Ben Mayer, who has

conceived, developed and coordinated the PROBLICOM Sky Survey for novae. This project was inspired following his accidental observations of the rise of Nova Cygni 1975. (M 8065)
Name proposed by the discoverer, following a suggestion by P. L. Dombrowski.

(2864) Soderblom

1983 AZ. Discovered 1983 January 12 by B. A. Skiff at Anderson Mesa.

Named in honor of Lawrence A. Soderblom, planetary geologist with the U.S. Geological Survey. His contributions to planetary science have included research in the geologic histories of both rocky and icy planets and satellites, image processing of spacecraft imagery, and a vigorous program of lunchtime volleyball. A former chief of the U.S.G.S. Branch of Astrogeologic Sciences, he was a member of the Viking and Voyager imaging teams, the Galileo Near Infrared Mapping Spectrometer Team, and the Space Science Advisory Committee. (M 7950)

(2865) Laurel

1935 OK. Discovered 1935 July 31 by C. Jackson at Johannesburg.

Named for Stan Laurel (1890-1965), American comedian (born Arthur Stanley Jefferson in Ulverston, England), the thin half and gag deviser of Hollywood's first great comedy duo. Stan performed in circuses and vaudeville before acting in silent films. He teamed with Oliver Hardy {see planet (2866)} in 1926, and the pair (Stan as the bumbling innocent and pompous Ollie with his long-suffering glances into the camera) made more than 200 timeless slapstick films and shorts. (M 22496)

Name proposed by G. V. and R. Williams, the identifications for this object having been made by C. M. Bardwell.

(2866) Hardy

1961 TA. Discovered 1961 October 7 by S. Arend at Uccle.

Named for Oliver Norville Hardy (1892-1957), American comedian, the fat half of Hollywood's first great comedy duo. Ollie was a former vaudeville actor and singer, who began doing silent-film comedy in 1913. Laurel {see planet (2865)} and Hardy successfully made the transition from silent to sound movies. Among their outstanding films is *Way Out West*, which incorporates Ollie's and Stan's rendition of 'Trail of the Lonesome Pine'. (M 22496)

Name proposed by G. V. and R. Williams, the identifications for this object having been made by C. M. Bardwell.

(2867) Šteins

1969 VC. Discovered 1969 November 4 by N. S. Chernykh at Nauchnyj.

Named in memory of Karlis Augustovich Šteins (1911-1983), director of the Latvian University's Astronomical Observatory from 1959, well known for his work on cometary cosmogony. He also studied the rotation of the Earth and designed astronomical instruments. (M 11157)
Obituary published in Zvaigžnotā Debess, gada ziema, p. 39-40 (1983/84).

(2868) Upupa

1972 UA. Discovered 1972 October 30 by P. Wild at Zimmerwald.

Named for a European nonpasserine bird of the family upupidae, or hoopie, this is a cousin of (2731) Cucula, a rare bird of somewhat peculiar habit and habits. (M 21129)

(2869) Nepryadva

1980 RM_2. Discovered 1980 September 7 by N. S. Chernykh at Nauchnyj.

Named to commemorate the Russian victory over the Tatar-Mongolians in the battle at Kulikovo near the Nepryadva River on 1380 Sept. 8. (M 12013)

(2870) Haupt

1981 LD. Discovered 1981 June 4 by E. Bowell at Anderson Mesa.

Named in honor of Hermann F. Haupt {1926- }, director of the Institute for Astronomy, University of Graz. One of the pioneers of photometry of minor planets, he made photoelectric observations while working at the Lick Observatory in 1951 and 1952. He found the peculiar brightness distribution in the near infrared and the reddening with phase of Vesta. (M 9769)

(2871) Schober

1981 QC_2. Discovered 1981 August 30 by E. Bowell at Anderson Mesa.

Named in honor of Hans Josef Schober {1944- }, astronomer at the Institute for Astronomy, University of Graz {see planet (2806)}. Most of his research has been devoted to photoelectric photometry of minor planets, where he made a major contribution to the study of lightcurves and rotation periods. He is known especially for discoveries of very slow-spinning minor planets and for studies of minor planets with complex lightcurve features. (M 9769)

(2872) Gentelec

1981 RU. Discovered 1981 September 5 at the Oak Ridge Observatory at Harvard.

Named for the GTE Research Laboratories, Waltham, Massachusetts, in recognition of their support of astronomy during the current apparition of Halley's Comet. (M 9478)

(2873) Binzel

1982 FR. Discovered 1982 March 28 by E. Bowell at Anderson Mesa.

Named in honor of Richard P. Binzel {1958- } of the University of Texas at Austin. During his thesis research, Binzel has been one of the most prolific observers of minor-planet lightcurves, especially in determining rotation periods of small, main-belt objects. (M 9215)
Citation prepared by A. W. Harris.

(2874) Jim Young

1982 TH. Discovered 1982 October 13 by E. Bowell at Anderson Mesa.

Named in honor of James W. Young, resident astronomer at the Table Mountain Observatory of the Jet Propulsion Laboratory. In recent years he has obtained photoelectric lightcurves for more than 100 minor planets, thus contributing about a third of the currently known rotation rates. (M 9081)
Name proposed by the discoverer, following a suggestion by A. W. Harris.

(2875) Lagerkvist

1983 CL. Discovered 1983 February 11 by E. Bowell at Anderson Mesa.

Named in honor of Claes-Ingvar Lagerkvist, planetary astronomer at the Uppsala Astronomical Observatory, well known for his observational work on shapes and spin properties of minor planets, particularly small ones. His research has also provided extensive astrometric data and has led to the numbering of six of his discoveries as of July 1984. An inspiring teacher with a great ability to stimulate the interest and research activity of young students, Lagerkvist has made important contributions to the popularization of astronomy in Sweden in recent years. (M 9081)
Citation prepared by H. Rickman.
Lagerkvist is discoverer of 29 numbered planets as of 1996.

(2876) Aeschylus

6558 P-L. Discovered 1960 September 24 by C. J. van Houten and I. van Houten-Groeneveld at Palomar.

Named for the author of tragedies in ancient Athens (525-456 B. C.). Of his more than 90 tragedies only seven survived. (M 10044)

(2877) Likhachev

1969 TR_2. Discovered 1969 October 8 by L. I. Chernykh at Nauchnyj.

Named in honor of Dmitrij Sergeevich Likhachev, prominent Soviet literary scholar and outstanding expert on the history of Slavic culture. (M 8913)

(2878) Panacea

1980 RX. Discovered 1980 September 7 by E. Bowell at Anderson Mesa.

Named for the Roman goddess of health, daughter of Aesculapius {see planet (1027)} and sister of Machaon, Podalirius and Hygiea {see planets (3063), (4086), and (10), respectively}. (M 8405)

Name proposed by the discoverer, following a suggestion by J. Meeus.

(2879) Shimizu

1932 CB_1. Discovered 1932 February 14 by K. Reinmuth at Heidelberg.

Named in honor of Shin-ichi Shimizu (1889-), a pioneer in astrophotography among amateur astronomers in Japan, who in 1937, with only an 80-mm astrocamera, recovered periodic comet Daniel, which had been lost since its discovery apparition of 1909-1910. He also observed many comets and minor planets in response to requests from the Tokyo Observatory. (M 8801)

Name proposed by T. Urata, who found the identifications involving this planet, and who lives in the city of Shimizu, Shizuoka prefecture.

(2880) Nihondaira

1983 CA. Discovered 1983 February 8 by T. Seki at Geisei.

Named for a hill and prominent beauty spot in central Japan. Overlooking the city and harbor of Shimizu, it offers a fine view of Mt. Fuji. The minor planet also honors the Nihondaira Observatory, where T. Urata conducts his observational and orbital work on comets and minor planets. (M 8066)

(2881) Meiden

1983 AA_1. Discovered 1983 January 12 by B. A. Skiff at Anderson Mesa.

Named for the hare-god of ancient Lithuanian mythology, Meiden was also deity of animals and forests. (M 22828)

Named by the Minor Planet Names Committee following a suggestion by F. Pilcher.

(2882) Tedesco

1981 OG. Discovered 1981 July 26 by E. Bowell at Anderson Mesa.

Named in honor of Edward F. Tedesco, planetary scientist at the Jet Propulsion Laboratory, who has made wide-ranging contributions to minor-planet science, including studies of rotational brightness variation, pole and shape determination, and the compositional structure of the belt. He is currently engaged in analyzing observations of minor planets by the Infrared Astronomical Satellite. (M 8543)

(2883) Barabashov

1978 RG_6. Discovered 1978 September 13 by N. S. Chernykh at Nauchnyj.

Named in memory of Nikolaj Pavlovich Barabashov (1894-1971), director of the Kharkov Observatory for 40 years, known for his research on the Moon and planets. (M 9216)

Barabashov is also honored by a crater on Mars.

(2884) Reddish

1981 ES_{22}. Discovered 1981 March 2 by S. J. Bus at Siding Spring.

Named in honor of Vincent C. Reddish, former Astronomer Royal for Scotland and director of the Royal Observatory, Edinburgh. Project scientist for the U.K. Schmidt Telescope, he played a leading role in organizing the survey of the south-

ern sky being undertaken at Siding Spring. (M 8405)
Name proposed by J. A. Dawe, current director of the U.K. Schmidt.

(2885) Palva

1939 TC. Discovered 1939 October 7 by Y. Väisälä at Turku.

Named in honor of Tauno Palva, son-in-law of the discoverer, internationally known ear surgeon and professor of otorhinolaryngology at the University of Helsinki. (M 18449)

(2886) Tinkaping

1965 YG. Discovered 1965 December 20 at the Purple Mountain Observatory at Nanking.

Named in honor of Tinkaping, well-known Hong Kong industrialist, who has made important contributions to health and education. (M 22828)

(2887) Krinov

1977 QD_5. Discovered 1977 August 22 by N. S. Chernykh at Nauchnyj.

Named in memory of Evgenij Leonidovich Krinov (1906-1984), celebrated Soviet meteoriticist, recipient of the Leonard medal of the American Meteoritical Society. (M 11157)

Obituaries published in Meteoritics, Vol. 19, No. 1, p. 67 (1984); Meteoritika, Vyp. 43, p. 186-187 (1984); Zemlya Vselennaya, No. 5, p. 73-75 (1984).

(2888) Hodgson

1982 TO. Discovered 1982 October 13 by E. Bowell at Anderson Mesa.

Named in honor of Richard G. Hodgson, professor at Dordt College, Sioux Center, Iowa. Founder in 1973 of the Minor Planets Section of the Association of Lunar and Planetary Observers, and Section Recorder for more than nine years, Hodgson has inspired many amateur astronomers around the world to make scientifically-valuable observations of minor planets. He continues, as a full-time teacher, to direct students toward minor-planet studies and other astronomical fields. (M 8801)
Citation written by R. P. Binzel.

(2889) Brno

1981 WT_1. Discovered 1981 November 17 by A. Mrkos at Kleť.
Named for the capital of Moravia {see planet (1901)}. (M 22828)
The town of Brno is also honored by minor planet (290).

(2890) Vilyujsk

1978 SY_7. Discovered 1978 September 26 by L. V. Zhuravleva at Nauchnyj.

Named for the central town of the Vilyujsk district of the {former} Yakut Autonomous SSR on the occasion of its 350th anniversary. (M 8913)

(2891) McGetchin

1980 MD. Discovered 1980 June 18 by C. S. Shoemaker at Palomar.

Named in memory of Thomas R. McGetchin (1936-1979), geologist and planetary scientist, from 1977 director of the Lunar and Planetary Institute in Houston. A leading investigator of volcanism and tectonics on the Earth and other solid bodies in the solar system, he was particularly noted for his work on kimberlite pipes and on the dynamics of volcanoes. (M 8153)

Obituaries published in Icarus, Vol. 41, p. 177 (1980); Proc. Tenth Lunar Planet Sci. Conf., p. VII-VIII (1979).

(2892) Filipenko

1983 AX_2. Discovered 1983 January 13 by L. G. Karachkina at Nauchnyj.

Named in honor of Aleksandr Grigorevich Filipenko, chief of surgery at the Crimean regional hospital in Bakhchisaraj. The discoverer owes him for saving the life of a person dear to her. (M 8913)

(2893) Peiroos

1975 QD. Discovered 1975 August 30 at the Felix Aguilar Observatory at El Leoncito.

Named for the Thracian chief who fought courageously to defend Troy. (M 13608)

(2894) Kakhovka

1978 SH_5. Discovered 1978 September 27 by L. I. Chernykh at Nauchnyj.

Named for a town of the {former} Chersonian district of the Ukrainian SSR. (M 8913)

(2895) Memnon

1981 AE_1. Discovered 1981 January 10 by N. G. Thomas at Anderson Mesa.

Named for a king of Ethiopia, who took 10,000 men to Troy to assist his uncle Priam in the Trojan War. Slayer of Antilochos, Nestor's son, Memnon was killed in combat by Achilles. (M 8405)

Name proposed by the discoverer, following a suggestion by E. Bowell.

See also the citations for planets (1432) Ethiopia, (3912) Troja, (884) Priamus, (1583) Antilochus, (659) Nestor, and (588) Achilles, respectively.

(2896) Preiss

1931 RN. Discovered 1931 September 15 by K. Reinmuth at Heidelberg.

Named in honor of Günter Preiss (1929-) on the occasion of his retirement in 1992 from the position of lawyer and administrator of the Max-Planck-Society. He earned great merit through his involvement in the construction and development of the Max-Planck-Institut in Heidelberg and also at its observatory at Calar Alto in southern Spain. The same holds for his involvement in the Institut für Radioastronomie in Bonn. (M 21607)

Name proposed and citation prepared by H. Elsässer, endorsed by G. Klare and L. D. Schmadel.

(2897) Ole Römer

1932 CK. Discovered 1932 February 5 by K. Reinmuth at Heidelberg.

Named for Ole Römer (1644-1710), discoverer of the finite velocity of light and inventor of the transit circle. Born in Aarhus {see planet (2676)}, he did his best astronomical work in Paris. Later professor of mathematics at the University of Copenhagen, he became extensively involved with terrestrial weights and measures, including the establishment of a temperature scale tied to the freezing and boiling points of water, the definition of the pound, and the measurement of spirit strength. He was also an authority on hydraulics and ballistics, and he served in various civil capacities, including mayor of Copenhagen, master of the mint, and head of the Danish state council. (M 8544)

Name proposed by L. K. Kristensen, who found some of the identifications involving this planet.

(2898) Neuvo

1938 DN. Discovered 1938 February 20 by Y. Väisälä at Turku.

Named in honor of Yrjö Neuvo, grandson of the discoverer, professor of digital sound processing at Tampere University of Technology and research professor of the Academy of Finland. (M 18449)

(2899) Runrun Shaw

1964 TR_2. Discovered 1964 October 8 at the Purple Mountain Observatory at Nanking.

Named in honor of Run Run Shaw, a famous entrepreneur of movie and television in Hong Kong. Public-spirited and wishing to perform social and welfare services, he has made important contributions to the development of Chinese education. (M 16041)

(2900) Luboš Perek

1972 AR. Discovered 1972 January 14 by L. Kohoutek at Bergedorf.

Named in honor of Luboš Perek, well known for his work on galactic dynamics and planetary nebulae. He worked at the Purkyne University in Brno and later at the Astronomical Institute of the Czechoslovak Institute of Sciences in Prague, serving as director from 1968 to 1975. Subsequently he has been chief of the space department in the secretariat of the United Nations. He was general secretary of the IAU from 1967 to 1970 and has also served as president of the International Astronautical Federation. (M 9478)

(2901) Bagehot

1973 DP. Discovered 1973 February 27 by L. Kohoutek at Bergedorf.

Named in memory of Walter Bagehot (1826-1877), English economist, political analyst and highly influential editor of *The Economist* from 1860 until his death. He has been described as Victorian England's "most versatile genius". (M 27329; M 27347)

Name suggested by O. Morton.

(2902) Westerlund

1980 FN_3. Discovered 1980 March 16 by C.-I. Lagerkvist at La Silla.

Named in honor of Bengt E. Westerlund, director of the Uppsala Astronomical Observatory, on the occasion of his retirement. Well known for his study of the structure of the Milky Way and for his work on the Magellanic Clouds, he has for many years given strong moral support to the Uppsala program on minor planets and comets, and he has even participated in the observations of minor planets with the Schmidt telescope at the Uppsala Southern Station. (M 12013)

(2903) Zhuhai

1981 UV_9. Discovered 1981 October 23 at the Purple Mountain Observatory at Nanking.

Named for an important open city in southern China, located at the west coast of Zhujiang port in Guangdong {see planet (2185)} province. With its fertile land and scenic beauty, Zhuhai is sometimes called a bright pearl of the South China Sea. It is a new city, developing especially in industry, but also in agriculture, fishing, animal husbandry, commerce and tourism. (M 16041)

(2904) Millman

1981 YB. Discovered 1981 December 20 by E. Bowell at Anderson Mesa.

Named in honor of Peter MacKenzie Millman {1906-1990}, former head of Upper Atmosphere Research and, since his formal retirement, a guest scientist at the Herzberg Institute of Astrophysics of the National Research Council of Canada. Well known for his research on meteors and for his interest in meteorites, comets and planets, he has served as president of the Royal Astronomical Society of Canada and of the Meteoritical Society, as well as president of Commission 22 and chairman of the Working Group on Planetary System Nomenclature of the IAU. (M 9081)

Name proposed by the discoverer, following a suggestion by C. E. Spratt.

Obituaries published in Icarus, Vol. 93, No. 2, p. 181-182 (1990); WGN, Vol. 19, Nr. 1, p. 1 (1991); IAU Colloquium No. 126, p. xxvii (1991); J. R. Astron. Soc. Can., Vol. 85, No. 2, p. 67-78 (1991); Radiant, Jaarg. 13, Nr. 2, p. 27 (1991); Meteoritics, Vol. 26, No. 2, p. 173 (1991).

(2905) Plaskett

1982 BZ_2. Discovered 1982 January 24 by E. Bowell at Anderson Mesa.

Named in memory of John Stanley Plaskett (1865-1941), director of the Dominion Astrophysical Observatory in Victoria from its founding in 1917 to 1935; and of his son, Harry Hemley Plaskett (1893-1980), Savilian professor of astronomy at the University of Oxford and director of the Oxford Observatory from 1932 to 1960. Known for his work on the radial velocities of O- and B-type stars that provided observational confirmation of the theory of galactic rotation, J. S. Plaskett was also particularly associated with the massive Plaskett's star. The younger Plaskett made far-reaching contributions to stellar spectroscopy and spectrophotometry and to solar physics; as president of the Royal Astronomical Society just after World War II he was instrumental in bringing into being the 2.5-m telescope, completed in 1967 and initially installed at the Royal Greenwich Observatory at Herstmonceux. Both father and son were awarded the gold medal of the Royal Astronomical Society. (M 9081)

Name proposed by the discoverer, following a suggestion by C. E. Spratt.

John S. Plaskett is also honored by a lunar crater. Obituaries (H.H.P.) published in Nature, Vol. 285, p. 58 (1980); J.R. Astron. Soc. Canada, Vol. 74, p. 234-236 (1980); Q.J.R. Astron. Soc., Vol. 21, p. 486-488 (1980).

(2906) Caltech

1983 AE_2. Discovered 1983 January 13 by C. S. Shoemaker at Palomar.

Named for the California Institute of Technology, of which the Palomar Observatory is a part. The 0.46-m Schmidt telescope, with which this minor planet was discovered, was the first telescope placed into operation by Caltech on Palomar Mountain. Five Caltech students assisted Caltech staff members C. S. and E. M. Shoemaker in the observations that established the unusual nature of the orbit of this object. (M 8154)

(2907) Nekrasov

1975 TT_2. Discovered 1975 October 3 by L. I. Chernykh at Nauchnyj.

Named in memory of Nikolaj Alekseevich Nekrasov (1821-1878), a poet and revolutionary who expressed in his works the sorrows and sufferings of the common people. (M 8913)

(2908) Shimoyama

1981 WA. Discovered 1981 November 18 by T. Furuta at Tokai.

Named for the village where the Tokai {see planet (2478)} station is located. (M 8801)

(2909) Hoshi-no-ie

1983 JA. Discovered 1983 May 9 by S. Sei at Chirorin.

Named for the discoverer's observatory. The English translation is "a star house". (M 8544)

(2910) Yoshkar-Ola

1980 TK_{13}. Discovered 1980 October 11 by N. S. Chernykh at Nauchnyj.

Named for a city on the Volga basin, capital of the {former} Mari Autonomous Soviet Socialist Republic, on the occasion of the city's 400th anniversary. (M 9216)

(2911) Miahelena

1938 GJ. Discovered 1938 April 8 by H. Alikoski at Turku.
Named by the discoverer in honor of his wife. (M 18449)

(2912) Lapalma

1942 DM. Discovered 1942 February 18 by L. Oterma at Turku.

Named for one of the Canary Islands, site of the Nordic Optical Telescope {see planet (2857)} and telescopes from several other European countries. (M 18449)

(2913) Horta

1931 TK. Discovered 1931 October 12 by E. Delporte at Uccle.

Named in memory of Baron Victor Horta (1861-1947), Belgian architect, who broke with tradition and was one of the first to glimpse the ornamental and calligraphic value of iron. (M 12209)

(2914) Glärnisch

1965 SB. Discovered 1965 September 19 by P. Wild at Zimmerwald.

Named for the most prominent mountain chain near the discoverer's home town of Glarus {see planet (1687)}. Its highest summit is 2914 meters above sea level. In 1926 the famous astronomer Fritz Zwicky {see planet (1803)} and one of his friends (the later Nobel laureate in chemistry, Thadeus Reichstein) became the first to climb the upper half of its north wall. (M 21129)

(2915) Moskvina

1977 QY_2. Discovered 1977 August 22 by N. S. Chernykh at Nauchnyj.

Named in honor of Valentina Nikolaevna Moskvina, a doctor at the Bakchisaraj regional hospital in the Crimea. (M 9216)

(2916) Voronveliya

1978 PW_2. Discovered 1978 August 8 by N. S. Chernykh at Nauchnyj.

Named in honor of Boris Aleksandrovich Vorontsov-Vel'yaminov {1904-1994}, famous Soviet astrophysicist at the Sternberg Astronomical Institute for many decades. His scientific works cover a wide range of topics, including comets, variable stars, galaxies, nebulae and the history of astronomy. He is also well-known as an author of textbooks and popular books on astronomy. (M 9216)
Obituary published in Zemlya Vselennaya, No. 3, p. 36-40 (1994).

(2917) Sawyer Hogg

1980 RR. Discovered 1980 September 2 by E. Bowell at Anderson Mesa.

Named in honor of Helen Sawyer Hogg {1905-1993}, professor emeritus of astronomy at the University of Toronto, known for her research on globular clusters and for her inspiring course in introductory astronomy, which she taught for more than 35 years. She has served as president of the Royal Astronomical Society of Canada, the Canadian Astronomical Society, and the American Association of Variable Star Observers. For nearly 30 years she wrote a weekly astronomical column for the *Toronto Star*, and she conducted a series of television programs for the Ontario Educational Television Authority. She is a Companion of the Order of Canada, a recent recipient of the Klumpke-Roberts award from the Astronomical Society of the Pacific, and the first Canadian and second woman to receive the Rittenhouse silver medal. (M 8801)
Name proposed by C. E. Spratt.

Obituaries published in J.R. Astron. Soc. Canada, Vol. 87, No. 6, p. 351-356 (1993); Bull. Am. Astron. Soc., Vol. 25, No. 4, p. 1497 (1993); J. Am. Assoc. Variable Star Obs., Vol. 22, No. 1, p. 83-86 (1993).

(2918) Salazar

1980 TU_4. Discovered 1980 October 9 by C. S. Shoemaker at Palomar.
Named in honor of Frederick Salazar, son-in-law of the discoverer. (M 9216)

(2919) Dali

1981 EX_{18}. Discovered 1981 March 2 by S. J. Bus at Siding Spring.

Named in honor of the great Spanish painter Salvador Dali (1904-1989), who in a career spanning over six decades has produced countless works dealing in imageries of the subconscious mind. His art has been influenced by surrealist contemporaries, as well as by his love for his wife Gala, but his own hallucinatory style and mastery of multiple illusions have made him a genius of our time. (M 8802)

(2920) Automedon

1981 JR. Discovered 1981 May 3 by E. Bowell at Anderson Mesa.

Named for the son of Diores and the charioteer of Achilles {see planet (588)}. Automedon killed the Trojan Aretus to avenge the death of Patroclus {see planet (617)}. (M 8544)

(2921) Sophocles

6525 P-L. Discovered 1960 September 24 by C. J. van Houten and I. van Houten-Groeneveld at Palomar.

Named for the author of tragedies in acient Athens (496-406 B.C.). Of his 123 tragedies only seven survived. (M 10044)

(2922) Dikan'ka

1976 GY_1. Discovered 1976 April 1 by N. S. Chernykh at Nauchnyj.

Named for the Ukrainian village mentioned in Gogol's {see planet (2361)} novel *Evenings at the farmstead near Dikan'ka*. (M 11158)

(2923) Schuyler

1977 DA. Discovered 1977 February 22 at the Harvard College Observatory at Harvard.

Named in honor of Catherine Schuyler on the occasion of the completion of her studies at Harvard University and in appreciation of her assistance with the administration of the Minor Planet Center and Central Bureau for Astronomical Telegrams during two years. (M 9769)

(2924) Mitake-mura

1977 DJ_2. Discovered 1977 February 18 by H. Kosai and K. Hurukawa at Kiso.

Named for another of the three towns to which the Kiso Station administratively belongs. Although in Roman characters the name is very similar to that of Mitaka, the town in which the Tokyo Observatory's headquarters are located, the names are very different when written in Japanese characters. The endings "take" and "taka" mean mountain and hawk, respectively. (M 8914)

For the other two cities mentioned, see the citations for planets (2470) Agematsu and (2960) Ohtaki.

(2925) Beatty

1978 VC_5. Discovered 1978 November 7 by E. F. Helin and S. J. Bus at Palomar.

Named in honor of J. Kelly Beatty, associate editor of *Sky and Telescope*, on the occasion of his marriage, 1983 Oct. 1. A long-time friend of the first discoverer, and a geologist and planetary scientist in his own right, he has given encouragement to the Palomar planet-crossing asteroid survey for many years. (M 8405)

(2926) Caldeira

1980 KG. Discovered 1980 May 22 by H. Debehogne at La Silla.

Named in honor of Felipe Caldeira, astronomer at the Valongo Observatory and professor at the Universidade Federal do Rio de Janeiro, participant in the minor-planet program at the European Southern Observatory. (M 10845)

(2927) Alamosa

1981 TM. Discovered 1981 October 5 by N. G. Thomas at Anderson Mesa.

Named for the discoverer's birthplace, the central town of the San Luis Valley of southern Colorado, located on the upper Rio Grande River. (M 8405)

(2928) Epstein

1976 GN_8. Discovered 1976 April 5 at the Felix Aguilar Observatory at El Leoncito.

Named in honor of Isadore Epstein, professor emeritus of astronomy at Columbia University and formerly director of Harriman Observatory. He conducted extensive site testing for a southern observatory in Australia, Chile and Argentina during 1957-1962. This resulted in the inauguration of the Yale-Columbia Southern Observatory at El Leoncito on 1965 Mar. 30. (M 12969)

(2929) Harris

1982 BK_1. Discovered 1982 January 24 by E. Bowell at Anderson Mesa.

Named in honor of Alan W. Harris, planetary scientist at the Jet Propulsion Laboratory, whose research has included studies of the origin of the solar system and the dynamics of planetary satellites and ring systems. In recent years he has become the most prolific observer of minor-planet rotational lightcurves. (M 8406)

(2930) Euripides

6554 P-L. Discovered 1960 September 24 by C. J. van Houten and I. van Houten-Groeneveld at Palomar.

Named for the author of tragedies in ancient Athens (480-406 B.C.). Of his 92 plays 19 survived. (M 10044)

(2931) Mayakovsky

1969 UC. Discovered 1969 October 16 by L. I. Chernykh at Nauchnyj.

Named in memory of the Soviet poet Vladimir Vladimirovich Mayakovskij (1893-1930). (M 8914)

(2932) Kempchinsky

1980 TK_4. Discovered 1980 October 9 by C. S. Shoemaker at Palomar.

Named in honor of Paula M. Kempchinsky, daughter-in-law of the discoverer. (M 9216)

(2933) Amber

1983 HN. Discovered 1983 April 18 by N. G. Thomas at Anderson Mesa.

Named in honor of the discoverer's granddaughter, Amber Marie Baltutis. (M 8406)

(2934) Aristophanes

4006 P-L. Discovered 1960 September 25 by C. J. van Houten and I. van Houten-Groeneveld at Palomar.

Named for the author of comedies in ancient Athens (445-385 B.C.). Of his 44 plays 11 survived. (M 10044)

(2935) Naerum

1976 UU. Discovered 1976 October 24 by R. M. West at La Silla.

Named after the small town of Naerum, some 15 km north of Copenhagen, where the discoverer spent his early years in the parental home. Archaeological finds from the stone age and bronze age tumuli bear witness to habitation during several millenia, as does the present name, which is derived from "Njarthar-heim", i.e. the home of Njord, god of the sea and its riches in Norse mythology. (M 12969)

(2936) Nechvíle

1979 SF. Discovered 1979 September 17 by A. Mrkos at Kleť.

Named in memory of V. Nechvíle (1890-1964), professor of astronomy at Charles University. During a stay in France he collaborated with J. W. Ritchey and the Henry brothers. (M 22828)

(2937) Gibbs

1980 LA. Discovered 1980 June 14 by E. Bowell at Anderson Mesa.

Named in memory of Josiah Willard Gibbs (1839-1903), mathematician and physicist, who worked at Yale and Princeton Universities. Although his principal work concerned thermodynamics, theories of optics and vector analysis, he made significant contributions to minor-planet studies through his work on orbits. (M 8544)

Gibbs is also honored by a lunar crater.

(2938) Hopi

1980 LB. Discovered 1980 June 14 by E. Bowell at Anderson Mesa.

Named for the Pueblo Indian people, who have maintained a continuous presence in northeastern Arizona for almost a thousand years. The Hopis' vital cultural heritage is reflected in their elaborate ceremonies, in particular the kachina and snake dances, and in their distinctive arts and crafts. (M 8544)

(2939) Coconino

1982 DP. Discovered 1982 February 21 by E. Bowell at Anderson Mesa.

Named for the county of Arizona of which Flagstaff is the county seat. The word Coconino derives from the Hopi koonina and refers primarily to the Supai and Hualapai Indian peoples living to the west of the Hopi. A secondary meaning applies to Hopi children not yet initiated into the religious practices of the Hopi kachina society. (M 8544)

(2940) Bacon

3042 P-L. Discovered 1960 September 24 by C. J. van Houten and I. van Houten-Groeneveld at Palomar.

Named for the English scholar Francis Bacon (1561-1626), thought by some to have been the author of Shakespeare's plays. (M 10044)

(2941) Alden

1930 YV. Discovered 1930 December 24 by C. W. Tombaugh at Flagstaff.
Named by the discoverer in honor of his son. (M 10845)

(2942) Cordie

1932 BG. Discovered 1932 January 29 by K. Reinmuth at Heidelberg.

Named in honor of Cordula (Cordie) Astrid Robinson, planetary geologist at the Harvard-Smithsonian Center for Astrophysics. Cordula gained her Ph.D. from University College, London, studying the crustal dichotomy of Mars, and she is currently using Magellan data to investigate the surface electrical properties on Venus with respect to geochemical and weathering processes. (M 21607)

Name proposed by C. M. Bardwell, who found the identifications for this planet, following a suggestion by G. V. Williams.

(2943) Heinrich

1933 QU. Discovered 1933 August 25 by K. Reinmuth at Heidelberg.

Named in honor of Inge Heinrich (1941-), astronomer at the Astronomisches Rechen-Institut Heidelberg and editor of *Astronomy and Astrophysics Abstracts*. Her ever-lasting dedication and profound knowledge of the astronomical literature significantly contribute to the worldwide reputation of this bibliography. (M 21607)
Name proposed and citation prepared by L. D. Schmadel.

(2944) Peyo

1935 QF. Discovered 1935 August 31 by K. Reinmuth at Heidelberg.

Named in memory of Pierre Culliford (1928-1992), better known under his pseudonym Peyo. He was known worldwide for his comic strips, where he created the blue dwarfs known as the Smurfs. Together with Georges Remi (Hergé) {see planet (1652)}, the father of Tintin, Peyo raised the drawing of comic strips to a 'Belgian' work of art. (M 21954)

Name proposed and citation prepared by J. Meeus, endorsed by G. Klare and L. D. Schmadel.

(2945) Zanstra

1935 ST_1. Discovered 1935 September 28 by H. van Gent at Johannesburg.

Named in memory of H. Zanstra (1894-1972), Dutch astronomer, director of the Astronomical Institute of the Municipal University in Amsterdam from 1946 to 1959. He is well known for his method for obtaining the surface temperatures of the central stars of planetary nebulae. (M 18135)
Obituary published in 18th Coll. Int. Astrophys. Liège 1972, P. 11-15 (1973).

(2946) Muchachos

1941 UV. Discovered 1941 October 15 by L. Oterma at Turku.

Named for the Roque de los Muchachos, the mountain on La Palma where the Nordic Optical Telescope {see planet (2857)} and other European telescopes are situated. (M 18449)

(2947) Kippenhahn

1955 QP_1. Discovered 1955 August 22 by I. Groeneveld at Heidelberg.

Named in honor of Rudolf Kippenhahn (1926-), German astronomer, director of the Max-Planck-Institut für Physik und Astrophysik at Garching and currently a vice president of the IAU. (M 18136)

(2948) Amosov

1969 TD_2. Discovered 1969 October 8 by L. I. Chernykh at Nauchnyj.

Named in honor of Nikolaj Mikhailovich Amosov, distinguished cardiologist, specialist in medical cybernetics, and member of the Academy of Sciences of the Ukrainian S.S.R. He has performed more than four thousand heart operations. (M 9216)

(2949) Kaverznev

1970 PR. Discovered 1970 August 9 at the Crimean Astrophysical Observatory at Nauchnyj.

Named in memory of Aleksandr Aleksandrovich Kaverznev (1932-1983), Soviet journalist, documentary-film maker and political commentator. (M 8914)

(2950) Rousseau

1974 VQ_2. Discovered 1974 November 9 by P. Wild at Zimmerwald.

Named in honor of Jean-Jacques Rousseau (1712-1778), citizen of Geneva, moralist and writer. The guiding idea of his discourses and novels was that human nature is principally good but gets corrupted by society under the influence of science and art, which he believed to enhance social inequality and discontentment. Hence his emotional exhortations for a return to nature and simple life. Real and also imagined persecution made him seek refuge in many remote places in western Europe. Rousseau's ideas later proved highly influential in inspiring the French Revolution, Storm and Stress, romanticism and socialism. (M 22496)

(2951) Perepadin

1977 RB_8. Discovered 1977 September 13 by N. S. Chernykh at Nauchnyj.

Named in honor of Aleksandr Ivanovich Perepadin, friend of the discoverer, learned agronomist, good manager, reader at the Crimean Agricultural Institute and chairman of the Bakhchisaraj District Council. (M 18136)

(2952) Lilliputia

1979 SF_2. Discovered 1979 September 22 by N. S. Chernykh at Nauchnyj.

Named for the land of tiny people in *Gulliver's Travels*, a novel by the English satirist and poet Jonathan Swift (1667-1745), this minor planet is one of the smallest discovered at this observatory. (M 9216)

(2953) Vysheslavia

1979 SV_{11}. Discovered 1979 September 24 by N. S. Chernykh at Nauchnyj.

Named in honor of Leonid Nikolaevich Vysheslavskij, Soviet writer, author of *Stellar sonnets* and numerous other poetical works. (M 11158)

(2954) Delsemme

1982 BT_1. Discovered 1982 January 30 by E. Bowell at Anderson Mesa.

Named in honor of Armand H. Delsemme, professor of astrophysics at the University of Toledo, Ohio, recognized for his extensive investigations of the chemical, physical, dynamical and evolutionary processes of comets. In 1952 he proposed, with P. Swings {see planet (1637)}, that cometary nuclei contain clathrate hydrates of gases, a theory that has been popular ever since. In 1976 he organized and edited the proceedings of the Lyons IAU colloquium "Comets, Asteroids and Meteorites". (M 11158)

Citation prepared by Z. Sekanina.

(2955) Newburn

1982 BX_1. Discovered 1982 January 30 by E. Bowell at Anderson Mesa.

Named in honor of Ray L. Newburn, astronomer at the Jet Propulsion Laboratory, recognized for his development of physical models of comets based on spectrophotometric observations and for his adroit leadership of the International Halley Watch during the apparition of that comet that began in 1982. (M 11158)

Citation prepared by S. J. Edberg.

(2956) Yeomans

1982 HN_1. Discovered 1982 April 28 by E. Bowell at Anderson Mesa.

Named in honor of Donald K. Yeomans, celestial mechanician at the Jet Propulsion Laboratory and discipline specialist for the astrometric team of the International Halley Watch. Well known for his orbit determinations that take account of the nongravitational forces acting on comets, he has made a detailed study of the motion of Halley's Comet back to the year -1404 and has analyzed the orbits of meteor showers and their relation to the orbits of parent comets. (M 11158)

Citation prepared by S. J. Edberg and Z. Sekanina.

(2957) Tatsuo

1934 CB_1. Discovered 1934 February 5 by K. Reinmuth at Heidelberg.

Named in honor of Tatsuo Yamada (1923-), who is a Japanese observer and researcher of variable stars. Yamada was the director of the variable star section of the Oriental Astronomical Association (O.A.A.), and he is devoted to the diffusion of astronomical knowledge. Yamada has continued to hold monthly meetings of the Nagoya branch of the O.A.A. since 1954. (M 20158)

Name proposed by T. Furuta, who found the identifications involving this planet, and who was inspired by T. Yamada.

(2958) Arpetito

1981 DG. Discovered 1981 February 28 by H. Debehogne and G. DeSanctis at La Silla.

Named in honor of E. Araya, J. Perez, R. Tighe and A. Torrecon, who prepare the plates used in the observing program on minor planets with the Grand Prism Objectif at La Silla. (M 19333)

(2959) Scholl

1983 RE_2. Discovered 1983 September 4 by E. Bowell at Anderson Mesa.

Named in honor of Hans Scholl {1942- }, former astronomer at the Astronomisches Rechen-Institut, Heidelberg, now at Nice Observatory. Well known for his theoretical work on the orbits of minor planets, Scholl has investigated resonant motion in the outer belt and has studied a variety of particularly interesting orbits, including those of Aten and Chiron. His broad range of minor-planet research has also embraced problems from mass determination to asteroid missions and from libration to depletion. (M 8802)

Citation written by J. Schubart.

(2960) Ohtaki

1977 DK_3. Discovered 1977 February 18 by H. Kosai and K. Hurukawa at Kiso.

Named for the third of the towns to which the Kiso Station administratively belongs. In fact, the Observatory is located at the top of the mountain where the boundaries of the three towns converge. (M 8914)

For the other two cities mentioned, see the citations for planets (2470) Agematsu and (2924) Mitake-mura.

(2961) Katsurahama

1982 XA. Discovered 1982 December 7 by T. Seki at Geisei.

Named for the beautiful seashore in the discoverer's home city of KocOhi {see planet (2396)} and one of the most famous tourist resorts in Japan. (M 8802)

(2962) Otto

1940 YF. Discovered 1940 December 28 by Y. Väisälä at Turku.

Named in honor of Otto Oskari Väisälä, great-grandson of the discoverer. (M 18449)

(2963) Chen Jiageng

1964 VM_1. Discovered 1964 November 9 at the Purple Mountain Observatory at Nanking.

Named in memory of Chen Jiageng (1874-1961), a famous Chinese educator who devoted his life and finances to running schools, and who made brilliant contributions to the development of Chinese education. (M 16041)

(2964) Jaschek

1974 OA_1. Discovered 1974 July 16 at the Carlos U. Cesco Observatory at El Leoncito.

Named in honor of Carlos Jaschek, professor of astronomy at the University of Strasbourg and director of the Stellar Data Center. Born in Germany, he was educated in Argentina and served as head of the astrophysics department at La Plata from 1957 to 1973. His earliest work, around 1950, was with the La Plata program for astrometric observations of minor planets. As head of the astrophysics group he inaugurated programs in stellar spectroscopy and arranged for the development of instrumentation, especially for photoelectric photometry. He was also involved with the start of radioastronomy and space activities in Argentina and in 1972 organized the first Latin American conference on astrophysics. After a year in Geneva, he began work in Strasbourg in 1974, strengthening the research there in astrophysics and attempting to make the stellar data center the largest such center in the world. He has also contributed to the creation of data centers in China, Japan, India, the U.S.S.R. and Argentina. A member of the Argentine National Academy of Sciences, he has also served in the CNRS and as president of IAU Commission 45. (M 17221)

(2965) Surikov

1975 BX. Discovered 1975 January 18 by L. I. Chernykh at Nauchnyj.

Named in memory of Vasilij Ivanovich Surikov (1848-1916), famous Russian painter. (M 13173)

(2966) Korsunia

1977 EB_2. Discovered 1977 March 13 by N. S. Chernykh at Nauchnyj.

Named for the ancient Crimean town of Chersonesus, known in medieval Russia as Korsun'. (M 12969)

(2967) Vladisvyat

1977 SS_1. Discovered 1977 September 19 by N. S. Chernykh at Nauchnyj.

Named for Vladimir Svyatoslavich (ca. 950-1015), Kiev Grand Prince who worked for the consolidation of Kiev and introduced Christianity into Russia in 988-989. (M 12969)

(2968) Iliya

1978 QJ. Discovered 1978 August 31 by N. S. Chernykh at Nauchnyj.

Named for the Russian epic hero Il'ya Muromets, defender of the Russian land. (M 12969)

(2969) Mikula

1978 RU_1. Discovered 1978 September 5 by N. S. Chernykh at Nauchnyj.

Named for the Russian epic hero Mikula Selyaninovich, the grain grower. (M 12970)

(2970) Pestalozzi

1978 UC. Discovered 1978 October 27 by P. Wild at Zimmerwald.

Named for Johann Heinrich Pestalozzi (1746-1827), Swiss educator of worldwide influence. Inspired by the ideas of J.-J. Rousseau {see planet (2950)}, he worked unremittingly for better education and schooling of the children of the poor. Some of his highly unselfish actions failed at the time, but in the long term his way of thinking succeeded. Even today people use his name as a hallmark for true altruism. (M 21129)

(2971) Mohr

1980 YL. Discovered 1980 December 30 by A. Mrkos at Kleť.

Named in memory of Josef M. Mohr (1901-1979), professor of astronomy at Charles University, founder of modern stellar astronomy in the former Czechoslovakia. (M 22828)

(2972) Niilo

1939 TB. Discovered 1939 October 7 by Y. Väisälä at Turku.

Named in honor of Niilo Anselmi Väisälä, great-grandson of the discoverer. (M 18449)

(2973) Paola

1951 AJ. Discovered 1951 January 10 by S. Arend at Uccle.

Named in honor of Princess Paola, sister-in-law of King Baudouin of Belgium {see planet (1491)}. (M 12209)

(2974) Holden

1955 QK. Discovered 1955 August 23 at the Goethe Link Observatory at Brooklyn, Indiana.

Named in memory of Edward Singleton Holden (1846-1914), first director of the Lick Observatory and founder of the Astronomical Society of the Pacific {see planet (2848)}. (M 14480)

Name proposed by F. K. Edmondson in the ASP's centennial year and one year after the centennial of the Lick Observatory.

Holden was also director of the Washburn Observatory of the University of Wisconsin in Madison. He is also honored by craters on the Moon and on Mars.

(2975) Spahr

1970 AF_1. Discovered 1970 January 8 by H. Potter and A. Lokalov at Cerro El Roble.

Named in honor of Timothy Bruce Spahr (1970-) of the Bigelow Sky Survey. This photographic survey utilizes the 0.41-m Catalina Schmidt in a search for high-latitude minor planets. (M 27124)

Name proposed by B. G. Marsden, G. V. Williams and S. M. Larson.

(2976) Lautaro

1974 HR. Discovered 1974 April 22 by C. Torres at Cerro El Roble.

Named in honor of the Chilean Indian Levtraru (1534-1557), modified to Lautaro (Swift Hawk) by the Spanish soldiers during the conquest of Chile. Son of an Indian chief, he was made prisoner by Pedro de Valdivia {see planet (2741)}, who named him his horseboy. In this activity, Lautaro learned a great deal about Spanish soldiers, their horses and military science, knowledge he used together with his own strategies when, at the age of 18, he became by acclamation big chief of all tribes to defend their land against the Spanish soldiers. After Lautaro's death his head was brought to Santiago and exhibited for fifteen days at the center of Plaza de Armas. (M 10547)

(2977) Chivilikhin

1974 SP. Discovered 1974 September 19 by L. I. Chernykh at Nauchnyj.

Named in memory of Vladimir Alekseevich Chivilikhin (1928-1984), prominent Soviet writer. (M 9216)

(2978) Roudebush

1978 SR. Discovered 1978 September 26 at the Harvard College Observatory at Harvard.

Named in honor of Susan Horner Roudebush, in appreciation of her outstanding work as administrator for the Planetary Sciences division of the Harvard-Smithsonian Center for Astrophysics, through which both the Minor Planet Center and the Oak Ridge Observatory (formerly the Agassiz Station) are also administered. She has in the past also served with distinction as head of the Smithsonian Astrophysical Observatory's travel office. (M 16244)

(2979) Murmansk

1978 TB_7. Discovered 1978 October 2 by L. V. Zhuravleva at Nauchnyj.

Named for the famous Arctic seaport. (M 8914)

(2980) Cameron

1981 EU_{17}. Discovered 1981 March 2 by S. J. Bus at Siding Spring.

Named in honor of Alastair G. W. Cameron {1925- }, astrophysicist and cosmogonist and currently associate director for theoretical astrophysics at the Harvard-Smithsonian Center for Astrophysics. Known to his colleagues as "Big Al", Cameron has, in his long and distinguished career, been a prolific producer of cosmogonical theories, never hesitating to revise or replace them when confronted with new data. He has consistently emphasized that the origin of planetary systems must be understood in the context of star formation. He was among the first to advocate such concepts as a turbulent accretion disk solar nebula, and the origin of the moon by a giant impact on the proto-earth. Cameron has also contributed greatly to studies of nucleosynthesis in stars and supernovae, and to understanding the significance of the cosmic abundances of nuclides. This work has been valuable for interpreting the meteoritic record of the early history of the solar system. (M 18449)
Citation provided by S. J. Weidenschilling.

(2981) Chagall

1981 EE_{20}. Discovered 1981 March 2 by S. J. Bus at Siding Spring.

Named in honor of the Russian-born painter Marc Chagall (1887-1985), whose dreamlike, often whimsical, representations of people and animals have made him one of the most popular and innovative artists of the twentieth century. Chagall's paintings reflect his strong religious background and an inner, almost childish, joy and love for life and the world. His major works include hundreds of paintings and book illustrations, as well as stage and costume design for theater and ballet. (M 8802)

(2982) Muriel

1981 JA_3. Discovered 1981 May 6 by C. S. Shoemaker at Palomar.

Named in honor of Muriel May Scott Shoemaker, mother-in-law of the discoverer. (M 9217)

(2983) Poltava

1981 RW_2. Discovered 1981 September 2 by N. S. Chernykh at Nauchnyj.
Named for the city in the {former} Ukrainian S.S.R. (M 11158)

(2984) Chaucer

1981 YD. Discovered 1981 December 30 by E. Bowell at Anderson Mesa.
Named for the English poet Geoffrey Chaucer (1340?-1400). (M 10044)
Chaucer is also honored by a lunar crater.

(2985) Shakespeare

1983 TV_1. Discovered 1983 October 12 by E. Bowell at Anderson Mesa.

Named for the English poet and playwright William Shakespeare (1564-1616). (M 10044)

(2986) Mrinalini

2525 P-L. Discovered 1960 September 24 by C. J. van Houten and I. van Houten-Groeneveld at Palomar.

Named in honor of Mrinalini Sarabhai, a distinguished author, choreographer and performer of classical South Indian dances, and deeply involved with the problems of the people of India. {See also the citation for planet (2987)}. (M 9478)

(2987) Sarabhai

4583 P-L. Discovered 1960 September 24 by C. J. van Houten and I. van Houten-Groeneveld at Palomar.

Named in honor of Vikram Ambalal Sarabhai (1919-1971), a cosmic-ray physicist who created several institutions, including the Physical Research Laboratory in Ahmedabad, and directed the Indian programs of space research and atomic energy. He was married to Mrinalini Sarabhai {see planet (2986)}. (M 9478)

(2988) Korhonen

1943 EM. Discovered 1943 March 1 by L. Oterma at Turku.

Named in honor of Tapio Korhonen, well-known telescope maker, especially of telescope optics. Among his successes is the optical system of the Nordic Optical Telescope {see planet (2857)}, for which an image quality of 0".2 is obtainable in principle. (M 18450)

(2989) Imago

1976 UF_1. Discovered 1976 October 22 by P. Wild at Zimmerwald.

The Latin word for image, in various degrees of reality, from full appearance (as e.g. the mature stage of an insect) to a mental picture (e.g. of oneself or another person), to visions and dreams. (M 22496)

(2990) Trimberger

1981 EN_{27}. Discovered 1981 March 2 by S. J. Bus at Siding Spring.

Named in honor of Stephen M. Trimberger, who as a Caltech undergraduate participated in the Palomar Planet-Crossing Asteroid Survey. He works in computer software development and has authored two books on computer-aided design for integrated circuits. (M 13173)

(2991) Bilbo

1982 HV. Discovered 1982 April 21 by M. Watt at Anderson Mesa.

Named for the central character in J. R. R. Tolkien's {see planet (2675)} classic tale of Middle Earth, *The Hobbit.* (M 27124)

The name is proposed by G. V. Williams and is in keeping with the discoverer's first numbered object, (2675) Tolkien.

(2992) Vondel

2540 P-L. Discovered 1960 September 24 by C. J. van Houten and I. van Houten-Groeneveld at Palomar.

Named for the Dutch poet and playwright (1587-1679). (M 10044)

(2993) Wendy

1970 PA. Discovered 1970 August 4 at the Perth Observatory at Bickley.

Named by Peter Birch {see planet (3924)} in honor of his wife. This was the first minor planet discovered at the Bickley site. (M 18644)

(2994) Flynn

1975 PA. Discovered 1975 August 14 at the Perth Observatory at Bickley.

Named by Mike Candy {see planet (3015)} in honor of his wife Vicki Marie Flynn, at one time a staff observer and still an observer on her own time. (M 18644)

(2995) Taratuta

1978 QK. Discovered 1978 August 31 by N. S. Chernykh at Nauchnyj.

Named in honor of Evgeniya Aleksandrovna Taratuta, Soviet writer and literary scholar. (M 11158)

(2996) Bowman

1954 RJ. Discovered 1954 September 5 at the Goethe Link Observatory at Brooklyn, Indiana.

Named in honor of Fred N. Bowman, a volunteer astronomer at the Cincinnati Observatory, born on the day this object was discovered, and who has found several minor-planet identifications. (M 9478)

Name proposed by F. K. Edmondson, following a suggestion by Viola R. Bowman.

(2997) Cabrera

1974 MJ. Discovered 1974 June 17 at the Carlos U. Cesco Observatory at El Leoncito.

Named in honor of Laurentino Ascencion Cabrera (1917-), an outstanding Argentine astronomer, long on the staff of the La Plata Observatory. His early research involved gravimetry, and from 1949 to 1952 he also worked for the International Latitude Service. From 1958 to 1969 he was in charge of the site-survey commission for the 2.15-m reflector of the Argentine National Observatory and since 1986 has been collaborating with that observatory. (M 18305)

(2998) Berendeya

1975 TR_3. Discovered 1975 October 3 by L. I. Chernykh at Nauchnyj.

Named for the wonderland in A. N. Ostrovskij's (1823-1886) fairy tale *The Snow-Maiden*. (M 11159; I. I. Neyachenko)

(2999) Dante

1981 CY. Discovered 1981 February 6 by N. G. Thomas at Anderson Mesa.
Named for Dante Alighieri (1265-1321), greatest of the Italian poets. (M 10044)
Dante is also honored by a lunar crater.

(3000) Leonardo

1981 EG_{19}. Discovered 1981 March 2 by S. J. Bus at Siding Spring.

Named for Leonardo da Vinci (1452-1519), Italian painter, sculptor, architect, musician, enigineer and natural philosopher. (M 10044)
Da Vinci is also honored by a crater on Mars.

(3001) Michelangelo

1982 BC_1. Discovered 1982 January 24 by E. Bowell at Anderson Mesa.
Named for Michelangelo Buonarroti (1475-1564), Italian artist. (M 10045)

(3002) Delasalle

1982 FB_3. Discovered 1982 March 20 by H. Debehogne at La Silla.

Named for St. Jean-Baptiste de la Salle, founder of the Frères des Ecoles Chrétiennes in France during the eighteenth century. The Frères are teachers who prepare pupils for the higher education. The discoverer has both studied and taught in their schools, and he wishes to honor all his fellow teachers and pupils. (M 15573)

(3003) Končeк

1983 YH. Discovered 1983 December 28 by A. Mrkos at Kleť.

Named in memory of academician Mikuláš Konček (1900-1982), founder of the Meteorological Institute in Bratislava. (M 22828)

(3004) Knud

1976 DD. Discovered 1976 February 27 by R. M. West at La Silla.

Named in memory of the Danish-Eskimo explorer and ethnologist Knud (Johan Victor) Rasmussen (1879-1933). He first visited the Polar Eskimos in northwestern Greenland in 1902-04. Later he founded the Thule station, which became the starting point of his legendary "Thule expeditions", during which he eventually visited virtually every Eskimo tribe between Greenland and the Bering Strait. The vast scope and depth of his many-sided scientific studies brought new insights into the ancient Eskimo culture. His many writings opened this fascinating world to the wide public. Knud Rasmussen was a close friend of the discoverer's paternal grandfather and namesake, a captain in the Royal Danish Navy who frequently sailed in Greenland waters in the 1920s. (M 27124)

(3005) Pervictoralex

1979 QK_2. Discovered 1979 August 22 by C.-I. Lagerkvist at La Silla.

Named by the discoverer in honor of his son, Per Victor Alexander Lagerkvist, born on 1987 Apr. 9. (M 12970)

(3006) Livadia

1979 SF_{11}. Discovered 1979 September 24 by N. S. Chernykh at Nauchnyj.

Named for a suburb of the Crimean city of Yalta {see planet (1475)}. (M 9769)

(3007) Reaves

1979 UC. Discovered 1979 October 17 by E. Bowell at Anderson Mesa.

Named in honor of Gibson Reaves, astronomer, historian and educator at the University of Southern California. Himself an expert on dwarf galaxies in clusters, his students have made signal contributions to the study of minor planets. (M 9769)

Citation prepared by D. T. Thompson.

(3008) Nojiri

1938 WA. Discovered 1938 November 17 by K. Reinmuth at Heidelberg.

Named in memory of Hoei Nojiri (1885-1977), the most famous popularizer of astronomy in Japan. Having begun his career as a teacher of English, he wrote many books based on his study of the mythology and ethnology of stars in Japan and abroad. His books have inspired many professional astronomers currently active in Japan. (M 9478)

Name proposed by T. Urata, who found the identifications involving this planet, and endorsed by Y. Kozai.

(3009) Coventry

1973 SM_2. Discovered 1973 September 22 by N. S. Chernykh at Nauchnyj.

Named for the city in England, twin city of Volgograd. (M 9770)

(3010) Ushakov

1978 SB_5. Discovered 1978 September 27 by L. I. Chernykh at Nauchnyj.

Named in memory of the Russian admiral and sea captain Fedor Fedorovich Ushakov (1744-1817). (M 11159)

(3011) Chongqing

1978 WM_{14}. Discovered 1978 November 26 at the Purple Mountain Observatory at Nanking.

Named for a city in southwestern China on the upper reach of the Yangtse River. Founded more than 3,000 years ago, Chongqing was the capital of Ba state in ancient times as well as the provisional capital of China from 1937 to 1946. (M 20834)

(3012) Minsk

1979 QU_9. Discovered 1979 August 27 by N. S. Chernykh at Nauchnyj.

Named for the capital city of the {former} Byelorussian S.S.R. {see the citation for planet (2170)}. (M 9770)

(3013) Dobrovoleva

1979 SD_7. Discovered 1979 September 23 by N. S. Chernykh at Nauchnyj.

Named in honor of Oleg Vail'evich Dobrovol'skij {1914-1989}, head of the Cometary Astronomy Department of the Institute of Astrophysics of the Tadjik S.S.R. Academy of Sciences in Dushanbe, well known for his research on the physics of comets. (M 11159)

Obituaries published in Icarus, Vol. 94, No. 2, p. 259 (1991); Astron. Vestn., Tom 24, No. 3, p. 271-272 (1990); Komet. Tsirk., No. 409, p. 2 (1990).

(3014) Huangsushu

1979 TM. Discovered 1979 October 11 at the Purple Mountain Observatory at Nanking.

Named in memory of Su-Shu Huang (1915-1977), professor of physics and astronomy at Northwestern University and well-known theoretical astrophysicist. He made many pioneering studies and important contributions, especially on close binary systems. Huang died on a visit to his native China. (M 22828)

(3015) Candy

1980 VN. Discovered 1980 November 9 by E. Bowell at Anderson Mesa.

Named in honor of Michael P. Candy {1928-1994}, acting director of Perth Observatory and an active astrometrist and orbit computer for more than thirty years. While on the staff of the Royal Greenwich Observatory he discovered comet 1961 II, and he now fulfills a major role in southern-hemisphere astronomy by obtaining positions, particularly of comets and minor planets, that would be difficult for northern observers. A former director of the comet section of the British Astronomical Association and editor of the B.A.A. Circulars, he served as president of IAU Commission 6 during 1982-1985. (M 10845)

Citation prepared by P. V. Birch and B. G. Marsden.

Obituaries published in Aust. J. Astron., Vol. 5, No. 5, p. 191-192 (1994); Q. J. R. Astron. Soc., Vol. 36, No. 3, p. 285-286 (1995); J. Br. Astron. Assoc., Vol. 105, No. 2, p. 56 (1995).

(3016) Meuse

1981 EK. Discovered 1981 March 1 by H. Debehogne and G. DeSanctis at La Silla.

Named for the river, more than 900 km long, with its sources at Langres that flows through the cities of Verdun, Domremy and Sedan in France, Dinant, Namur {see planets (2765) and (3374)}, Huy and Liège in Belgium, and Maastricht, Nijmegen and Rotterdam in the Netherlands. (M 19333)

(3017) Petrovič

1981 UL. Discovered 1981 October 25 by A. Mrkos at Kleť.

Named in honor of Štefan Petrovič (1906-), organizer of modern climatology in Slovakia. (M 22828)

(3018) Godiva

1982 KM. Discovered 1982 May 21 by E. Bowell at Anderson Mesa.

Named for the wife of Leofric, Earl of Mercia, who, it is said, rode naked, but for her long hair, through the streets of Coventry {see planet (3009)} so that her husband would reduce the oppressive taxes he levied on the people of the city. In a later, embellished version of the legend, the populace was entreated to stay behind shuttered windows; but a tailor named Peeping Tom, who disobeyed, was instantly struck blind. (M 9770)

(3019) Kulin

1940 AC. Discovered 1940 January 7 by G. Kulin at Budapest.

Named in memory of György Kulin (1905-1989), staff astronomer at the Konkoly Observatory from 1935 to 1947. Internationally renowned for his discoveries of minor planets (19 of which have been numbered) with the Konkoly 0.60-m reflector, he also made accurate computations of their orbits. In Hungary he was known for encouraging the development of amateur astronomy: he made almost 7000 telescope mirrors, wrote dozens of books and hundreds of articles to popularize astronomy. The Hungarian "Flammarion" {see planet (1021)}, he directed the Urania Observatory in Budapest from 1947 to 1975 and facilitated in many ways the publication of the Hungarian Astronomical Association's magazine *Meteor*. (M 17465)

Name suggested by Attila Mizser and endorsed by E. Bowell and B. G. Marsden, who found the identifications involving this planet.

(3020) Naudts

1949 PR. Discovered 1949 August 2 by K. Reinmuth at Heidelberg.

Named in memory of Ignace Naudts (1949-1992), active Belgian amateur astronomer, editor of the monthly magazine *Heelal* of the Flemish association 'Vereniging voor Sterrenkunde'. His particular fields of interest were planetary satellites and rings, chaos in the solar system, and theoretical and practical study of sundials. He was a talented mathematician, a dynamic and enthusiastic popularizer of astronomy and science. (M 21955)

Name proposed and citation prepared by C. Steyaert, endorsed by G. Klare and L. D. Schmadel.

(3021) Lucubratio

1967 CB. Discovered 1967 February 6 by P. Wild at Zimmerwald.

The original meaning of this Latin word is "night work" (from lucubrum, meaning candle). Then it seems to have been extended to any laboriously earned study to overlabored work - and even to idle talk. In spite of this pejorative development the discoverer hopes that this naming will be welcomed by astronomical observers and theorists alike. (M 21130)

(3022) Dobermann

1980 SH. Discovered 1980 September 16 by Z. Vávrová at Kleť.

Named in honor of Karl Friedrich Louis Dobermann (1834-1894), German zoologist and amateur astronomer. (M 16441)

(3023) Heard

1981 JS. Discovered 1981 May 5 by E. Bowell at Anderson Mesa.

Named in memory of John Frederick Heard (1907-1976), professor of astronomy at the University of Toronto and fourth director of the David Dunlap Observatory. An outstanding and meticulous spectroscopist, he specialized in spectroscopic binaries and stellar radial-velocity standards. In addition, he was a dedicated teacher who helped train many Canadian astronomers. (M 9770)
Citation prepared by H. Guetter.
Obituary published in J.R. Astron. Soc. Canada, Vol. 71, p. 1-8 (1977).

(3024) Hainan

1981 UW_9. Discovered 1981 October 23 at the Purple Mountain Observatory at Nanking.

Named for a newly established province in southern China. Hainan Island, under its jurisdiction, has rich developing resources of tropical plants and animals, sea aquatic products, petrochemical and metallurgical industry, rare-earth metals, etc. With its bright and colorful scene, Hainan Island, acclaimed as "Eastern Hawaii", is a tourist resort and a bright pearl in the vast South China Sea. (M 15088)

(3025) Higson

1982 QR. Discovered 1982 August 20 by C. S. Shoemaker and E. M. Shoemaker at Palomar.

Named in honor of Roger Higson, night assistant for the 1.2-m Schmidt telescope at Palomar Observatory since 1979. An outstanding and thoroughly dedicated member of the observatory's supporting staff, he has skillfully assisted astronomers in a broad range of research conducted with the 1.2-m Schmidt. His efforts have been especially appreciated by observers of comets and minor planets. (M 9217) Name proposed by the discoverers, endorsed by J. Gibson and C. Kowal.

(3026) Sarastro

1977 TA_1. Discovered 1977 October 12 by P. Wild at Zimmerwald.

Named for the lord or high priest in the Temple of Wisdom, a leading figure in Mozart's {see planet (1034)} opera "*Die Zauberflöte*". (M 21130)

(3027) Shavarsh

1978 PQ_2. Discovered 1978 August 8 by N. S. Chernykh at Nauchnyj.

Named in honor of Shavarsh Vladimirovich Karapetyan, Armenian sportsman, eleven times the world champion in underwater sports. Also known for heroic deeds, he saved twenty persons in a trolley bus submerged in water from a dam in Erevan. (M 11159)

(3028) Zhangguoxi

1978 TA_2. Discovered 1978 October 9 at the Purple Mountain Observatory at Nanking.

Named in honor of Zhang Guoxi, Chinese industrialist, for his contributions to social welfare and public education. (M 21607)

(3029) Sanders

1981 EA_8. Discovered 1981 March 1 by S. J. Bus at Siding Spring.

Named in honor of Jeffrey D. Sanders, who participated in the Palomar Planet-Crossing Asteroid Survey as a Caltech undergraduate student. (M 13173)

(3030) Vehrenberg

1981 EH_{16}. Discovered 1981 March 1 by S. J. Bus at Siding Spring.

Named in honor of Hans Vehrenberg (1910-1991) of Düsseldorf: publisher, amateur astronomer, and creator of atlases of the sky. His *Falkauer Atlas* and *Atlas Stellarum* are widely used by amateur and professional astronomers interested in or conducting research on minor planets and comets. His other publications include the popular *Atlas of Deep-Sky Splendors.* (M 9081)

Name proposed by the discoverer following a suggestion by T. P. Kohman and J. U. Gunter.

Obituaries published in Sterne Weltraum, Jahrg. 30, Nr. 12, p. 766-768 (1991); Mitt. Astron. Ges., Nr. 75, p. 11-12 (1992).

(3031) Houston

1984 CX. Discovered 1984 February 8 by E. Bowell at Anderson Mesa.

Named in honor of Walter Scott Houston, American amateur astronomer well known for his column "Deep Sky Wonders" in the magazine *Sky and Telescope.* Houston has specialized in the visual study of deep-sky objects and has guided countless amateurs to view and marvel at the varied objects within the grasp of small telescopes. (M 10845)
Name proposed by the discoverer following a suggestion by P. L. Dombrowski.

(3032) Evans

1984 CA_1. Discovered 1984 February 8 by E. Bowell at Anderson Mesa.

Named in honor of the Reverend Robert O. Evans, Australian amateur astronomer and discoverer of several extragalactic supernovae. Evans has successfully

used a systematic visual search program to examine selected galaxies for supernova activity. (M 10845)
Name proposed by the discoverer following a suggestion by P. L. Dombrowski.

(3033) Holbaek

1984 EJ. Discovered 1984 March 5 by K. Augustesen and P. Jensen and H. J. Fogh Olsen at Brorfelde.

Named in honor of the town nearest to the Brorfelde {see planet (3309)} Observatory on the occasion of the town's 700th anniversary in 1986. (M 10045) Holbaek is situated on the Danish island of Sjaelland in the Baltic Sea.

(3034) Climenhaga

A917 SE. Discovered 1917 September 24 by M. Wolf at Heidelberg.

Named in honor of John L. Climenhaga, first head of physics at the University of Victoria, on the occasion of his seventieth birthday. Known for his work on the C12/C13 abundance ratio in carbon stars and for studies of line blanketing and microturbulence in late-type stars, he has also long had an interest in cometary spectra. On his retirement in 1982 the University's Observatory was named in his honor, and among the Observatory's activities is the only Canadian program of astrometric observations of comets and minor planets. (M 11441)

Following a suggestion from C. E. Spratt, the name was proposed by B. G. Marsden, who found the identifications involving this minor planet.

(3035) Chambers

A924 EJ. Discovered 1924 March 7 by K. Reinmuth at Heidelberg.

Named in honor of John Eric Chambers (1969-), graduate student at Manchester University, currently a predoctoral fellow at the Harvard-Smithsonian Center for Astrophysics. John's research concentrates on the effect of mean-motion resonances on the orbital evolution of solar system objects, and he has recently completed a study of P/Swift-Tuttle, showing that the motion of this object is dominated by the 1:11 resonance with Jupiter. (M 22497)

Name proposed by members of the Planetary Sciences division, the identifications for this object having been made by B. G. Marsden.

(3036) Krat

1937 TO. Discovered 1937 October 11 by G. N. Neujmin at Simeïs.

Named in memory of Vladimir Alekseevich Krat (1911-1983), corresponding member of the U.S.S.R. Academy of Sciences, a staff member of the Pulkovo Observatory and from 1964 to 1979 its director. His main contributions to astronomy involved solar physics and chromospheric structure, figures of equilibrium of close binaries, classification of eclipsing variables and cosmogony. He initiated and actively participated in the development of the first Soviet stratospheric balloon observatory. (M 10547)

Obituaries published in Zemlya Vselennaya, No. 6, p. 33-34 (1983); Sol. Phys., Vol. 89, No. 1, p. 1-2 (1983); Izv. Glav. Astron. Obs. Pulkovo, Astrometr. Astrofiz., No. 202, p. 3-5 (1984).

(3037) Alku

1944 BA. Discovered 1944 January 17 by Y. Väisälä at Turku.

Named for the boat the discoverer enjoyed in his boyhood. It was built by the father of the discoverer and instilled in him a lifelong love of sailing. The name means "the beginning". (M 18450)

(3038) Bernes

1978 QB_3. Discovered 1978 August 31 by N. S. Chernykh at Nauchnyj.

Named in memory of Mark Naumovich Bernes (1911-1969), popular Soviet film actor and singer. (M 12970)

(3039) Yangel

1978 SP_2. Discovered 1978 September 26 by L. V. Zhuravleva at Nauchnyj.

Named in memory of Mikhail Kuz'mich Yangel' (1911-1971), Soviet designer of space-rocket systems. (M 10547)

(3040) Kozai

1979 BA. Discovered 1979 January 23 by W. Liller at Cerro Tololo.

Named in honor of Yoshihide Kozai, astronomer and celestial mechanician at the Tokyo Observatory, whose interests include natural and artificial satellites, the motions of comets and minor planets, families of minor planets, and the use of satellite and lunar positions for astronomy and geodesy. His investigation of the theory of secular perturbations of minor planets of large orbital inclination or eccentricity led to the recognition of the first object known to exhibit argument-of-perihelion libration. (M 9770)

Name proposed by J. G. Williams, who has found that this object also to be an argument-of-perihelion librator.

Kozai served as IAU president 1988-1991.

(3041) Webb

1980 GD. Discovered 1980 April 15 by E. Bowell at Anderson Mesa.

Named for Thomas William Webb, discoverer of S Ori. Like many other English clergymen of his days, he was a keen amateur astronomer. His observations on many years formed the basis of his *Celestial Objects for Common Telescopes* (1859), a work covering many aspects of astronomical observation and destined to become a classic handbook for the amateur astronomer. (M 9770)

Name suggested and citation prepared by B. Hetherington.

(3042) Zelinsky

1981 EF_{10}. Discovered 1981 March 1 by S. J. Bus at Siding Spring.

Named in honor of David S. Zelinsky, now a mathematician at Brown University, Providence, formerly an active participant in the Palomar Planet-Crossing Asteroid Survey while an undergraduate student at Caltech. (M 13173)

(3043) San Diego

1982 SA. Discovered 1982 September 20 by E. F. Helin at Palomar.

Named as a celestial tribute to the city of San Diego in appreciation of the city's responsiveness and cooperation in the campaign to restore dark skies for astronomers probing the universe. (M 8914)

(3044) Saltykov

1983 RE_3. Discovered 1983 September 2 by N. V. Metlova and N. E. Kurochkin at Nauchnyj.

Named by the first discoverer in memory of her grandfather, Nikita Saltykov (1893-1946), well known as a grower of vegetables in the region of Yaroslavl and the Urals. (M 22245)

(3045) Alois

1984 AW. Discovered 1984 January 8 by J. F. Wagner at Anderson Mesa.

Named by the discoverer in memory of his grandfather, Alois T. Stuczynski. (M 9479)

(3046) Molière

4120 P-L. Discovered 1960 September 24 by C. J. van Houten and I. van Houten-Groeneveld at Palomar.

Named for the French playwright (1622-1673). (M 10045)

(3047) Goethe

6091 P-L. Discovered 1960 September 24 by C. J. van Houten and I. van Houten-Groeneveld at Palomar.

Named for the German poet and playwright {Johann Wolfgang von Goethe} (1749-1832). (M 10045)

(3048) Guangzhou

1964 TH_1. Discovered 1964 October 8 at the Purple Mountain Observatory at Nanking.

Named for the largest open city in southern China and the capital of Guangdong province. The provincial center of culture, finance and technology, and located in the inner part of the Pearl River delta, Guangzhou is celebrated as the city of Flowers, Rams and the Ear of Grain, and it has a sound foundation in industry and commerce and a long history of trade relations with foreign countries. (M 15089)

(3049) Kuzbass

1968 FH. Discovered 1968 March 28 by T. M. Smirnova at Nauchnyj.

Named for the Kuznetskij coal basin, one of the richest coal deposits in the U.S.S.R. and the world. (M 13173)

(3050) Carrera

1972 NW. Discovered 1972 July 13 by C. Torres at Cerro El Roble.

Named in memory of the brothers Carrera, Javiera (1781-1862), Juan José (1782-1818), José Miguel (1785-1821) and Luis (1791-1818), active participants in gaining Chile's independence from Spain, in spite of continuous disagreements with Bernardo O'Higgins {see planet (2351)}, the "Father of the Country", and José de San Martin {see planet (2745)}, an Argentine general and political liberator of Argentina, Chile and Peru. José Miguel Carrera was the first president of Chile. During his government Chile acquired its first political constitution, a law against slavery, a law prohibiting obedience to foreign authorities, a law of civil rights, and the establishment of diplomatic relations with the United States. (M 10547)

(3051) Nantong

1974 YP. Discovered 1974 December 19 at the Purple Mountain Observatory at Nanking.

Named for the industrial port city at the mouth of the Yangtse, on the Yellow Sea. The city has a long history of education and culture, and the first school and the first museum in China were here. (M 20835)

(3052) Herzen

1976 YJ_3. Discovered 1976 December 16 by L. I. Chernykh at Nauchnyj.

Named in memory of Aleksandr Ivanovich Herzen (1812-1870), revolutionary, writer and philosopher, founder of the free Russian press abroad. (M 11159)

(3053) Dresden

1977 QS. Discovered 1977 August 18 by N. S. Chernykh at Nauchnyj.

Named for the city in the {former} German Democratic Republic {see also planet (263)}. (M 9770)
Dresden now is an important town in the Federal Republic of Germany.

(3054) Strugatskia

1977 RE_7. Discovered 1977 September 11 by N. S. Chernykh at Nauchnyj.

Named in honor of the brothers Arkadij Natanovich and Boris Natanovich Strugatskij, well-known Soviet writers of science fiction. (M 9771)

(3055) Annapavlova

1978 TR_3. Discovered 1978 October 4 by T. M. Smirnova at Nauchnyj.

Named in honor of the renowned ballet-dancer Anna Pavlovna Pavlova (1881-1931). (M 13173)

(3056) INAG

1978 VD_1. Discovered 1978 November 1 by K. Tomita at Caussols.

Named in honor of the French Institut National d'Astronomie et de Géophysique, at which the 0.9-m Schmidt telescope used for the discovery of this object was constructed. The 3.6-m Canada-France-Hawaii telescope and the 2-m Pic-du-Midi telescope were also built at INAG. (M 9082)

(3057) Mälaren

1981 EG. Discovered 1981 March 9 by E. Bowell at Anderson Mesa.

Named for the large Swedish lake between Stockholm and Uppsala. A souvenir of a relaxing shipboard evening spent at the conclusion of a most successful conference on minor planets, comets and meteors held in Uppsala in June 1985. (M 10311)

Name suggested by B. G. Marsden following a request by the discoverer.

(3058) Delmary

1981 EO_{17}. Discovered 1981 March 1 by S. J. Bus at Siding Spring.

Named in honor of the American artist Delmary Rose Schanz (1938-), whose seascapes, rendered in the glazed oil technique of the Flemish masters, have inspired art enthusiasts internationally. Highly respected for her sensitive use of light and color, the artist has produced an extensive body of work widely acclaimed by collectors and peers for its spiritual power and technical mastery. (M 9082)

(3059) Pryor

1981 EF_{23}. Discovered 1981 March 3 by S. J. Bus at Siding Spring.

Named in honor of Carlton P. Pryor, who participated in the Palomar Planet-Crossing Asteroid Survey while an undergraduate student at Caltech. Until recently an astronomer at Vanderbilt University, he is now a research fellow at Rutgers University. (M 13173)

(3060) Delcano

1982 RD_1. Discovered 1982 September 12 by P. Wild at Zimmerwald.

Named in memory of Juan Sebastián del Cano, the first to circumnavigate the globe. While Magellan {see planet (4055)} is usually credited as being the first, this is true only by virtue of splicing his first voyage eastward from Europe to the Philippines with his second one westward to those islands where he was killed. The first true continuous circumnavigation was the heroic voyage of Juan del Cano, a young lieutenant of Magellan, who in command of the only remaining ship of the five that had begun the westward voyage navigated it safely for about 30,000 additional kilometers, returning home with his fellow 30 or so survivors more than three years after leaving home. Of astronomical interest is the fact that his log failed to tally by one day with the calendar of those who had not made the voyage - demonstrating the need for an international data line or its equivalent. (M 21130)

Name proposed and citation written by the late Harlan J. Smith.

(3061) Cook

1982 UB_1. Discovered 1982 October 21 by E. Bowell at Anderson Mesa.

Named for James Cook (1728-1779), British circumnavigator and one of the first scientific navigators. He observed the solar eclipse of 1766 Aug. 5 from Newfoundland and in 1769 measured the transit of Venus from Tahiti. In 1761 he assisted the Astronomer Royal, Nevil Maskelyne, in tests of John Harrison's fourth marine chronometer as a means of determining longitude at sea. (M 10846) Name proposed by the discoverer following a suggestion by B. Hetherington. Cook is also honored by a lunar crater.

(3062) Wren

1982 XC. Discovered 1982 December 14 by E. Bowell at Anderson Mesa.

Named for Christopher Wren (1632-1723), British architect and astronomer who was professor of astronomy at Gresham College, London, and later professor of astronomy at Oxford. Wren prepared a scheme for rebuilding London after the Great Fire of 1666, his most noted building being St. Paul's Cathedral. He also designed and built the Royal Observatory at Greenwich in 1675. (M 10846) Name proposed by the discoverer following a suggestion by B. Hetherington.

(3063) Makhaon

1983 PV. Discovered 1983 August 4 by L. G. Karachkina at Nauchnyj. Named for the physician to the Greek troops during the Trojan War. (M 10547)

(3064) Zimmer

1984 BB_1. Discovered 1984 January 28 by E. Bowell at Anderson Mesa.

Named in memory of Louis Zimmer (1888-1970), world-famous Belgian maker of astronomical clocks. Many of his clocks were made for crowned heads and political leaders all over the world. Today he is best remembered for his "wonderclock", exhibited at the 1939 New York World Fair; his "Jubilee clock", donated to his native town of Lier in 1930; and his Astronomical Studio. All three masterpieces were housed in or near the "Zimmer Tower", which has subsequently become one of the main attractions of Lier. (M 11748)

Name proposed by the discoverer, following a suggestion by E. Goffin, who prepared the citation.

(3065) Sarahill

1984 CV. Discovered 1984 February 8 by E. Bowell at Anderson Mesa.

Named in honor of Sarah J. Hill, professor of astronomy at Wellesley College (1952-1974) and chairman of its astronomy department (1952-1971). Her enthusiasm for teaching and for observational astronomy influenced the graduates to earn doctorates and become professional astronomers. Many others among her students earned advanced degrees in astronomy or related fields. (M 12013)

Name proposed by the discoverer following a suggestion by F. Vilas, who provided the citation.

(3066) McFadden

1984 EO. Discovered 1984 March 1 by E. Bowell at Anderson Mesa.

Named in honor of Lucy-Ann A. McFadden, planetary scientist at the University of Maryland, for her studies of the nature of Earth-approaching minor planets and for the comparison of their spectra with those of meteorites. Recently, she has been working on the relationship between cometary nuclei and Earth approachers, and on ultraviolet spectrophotometry of comets. (M 11748)
Name endorsed by M. F. A'Hearn.

(3067) Akhmatova

1982 TE_2. Discovered 1982 October 14 by L. G. Karachkina and L. V. Zhuravleva at Nauchnyj.

Named in honor of Anna Andreevna Akhmatova (1889-1966), outstanding poetess, awarded an honorary doctorate by the University of Oxford. (M 13174)

(3068) Khanina

1982 YJ_1. Discovered 1982 December 23 by L. G. Karachkina at Nauchnyj.

Named in honor of Frida Borisovna Khanina, specialist on orbit computations, a staff member of the Institute for Theoretical Astronomy from 1946 to 1983. She contributed extensively to nearly forty volumes of *Ehfemeridy Malykh Planet* and improved the orbits of many hundreds of minor planets. (M 10547)

(3069) Heyrovský

1982 UG_2. Discovered 1982 October 16 by Z. Vávrová at Kleť.

Named in honor of Jaroslav Heyrovský (1890-1967), Czech physicist and inventor of polarography, recipient of the 1959 Nobel Prize for Chemistry. (M 16441)

(3070) Aitken

1949 GK. Discovered 1949 April 4 at the Goethe Link Observatory at Brooklyn, Indiana.

he Lick Observatory (1930-1935), and as associate director under W. W. Campbell {see planet (2751)} he ran the Observatory when the latter was also president of the University of California (1923-1930). During his 40 years at Lick Aitken became the leading authority on double stars, his work culminating in the publication of his *New General Catalogue of Double Stars within 120 degrees of the North Pole*, published in two large volumes in 1932. His book *The Binary Stars* was published in 1918, and an updated second edition appeared in 1935. (M 14481)

Name proposed by F. K. Edmondson.

Aitken is also honored by a lunar crater.

(3071) Nesterov

1973 FT_1. Discovered 1973 March 28 by T. M. Smirnova at Nauchnyj.

Named in honor of Petr Nikolaevich Nesterov (1887-1914), the Russian airman who was the first to carry out a number of piloting maneuvers, among them the loop. (M 11159)

(3072) Vilnius

1978 RS_1. Discovered 1978 September 5 by N. S. Chernykh at Nauchnyj.

o planet (2577)}. (M9771)

Vilnius is since 1991 the capital of the independent state of Lithuania.

(3073) Kursk

1979 SW_{11}. Discovered 1979 September 24 by N. S. Chernykh at Nauchnyj.

Named for an old Russian city. (M 9771)

(3074) Popov

1979 YE_9. Discovered 1979 December 24 by L. V. Zhuravleva at Nauchnyj.

Named in memory of Aleksandr Stepanovich Popov (1859-1906), the inventor of radio in Russia. (M 13174)

Popov is also honored by a lunar crater.

(3075) Bornmann

1981 EY_{15}. Discovered 1981 March 1 by S. J. Bus at Siding Spring.

Named in honor of Patricia L. Bornmann, who participated in the Palomar Planet-Crossing Asteroid Survey while an undergraduate student at Caltech and now performs solar research at the National Oceanic and Atmospheric Administration in Boulder. (M 13174)

(3076) Garber

1982 RB_1. Discovered 1982 September 13 at the Oak Ridge Observatory at Harvard.

Named in honor of Paul E. Garber, historian emeritus and Ramsey fellow of the

Smithsonian Institution's National Air and Space Museum, on the occasion of his ninetieth birthday and in recognition of his lifelong commitment to aviation and the exploration of both air and space. (M 14971)
Citation prepared by J. C. Cornell.

(3077) Henderson

1982 SK. Discovered 1982 September 22 by E. Bowell at Anderson Mesa.

Named for Thomas Henderson (1798-1844), Scottish astronomer and noted computer. He was appointed Royal Astronomer at the Cape of Good Hope and later Astronomer Royal for Scotland. Henderson computed an improved value for the solar parallax and was the first to measure the distance to a star, Alpha Centauri, in 1839. (M 10846)
Name proposed by the discoverer following a suggestion by B. Hetherington.
Henderson is also honored by a lunar crater.

(3078) Horrocks

1984 FG. Discovered 1984 March 31 by E. Bowell at Anderson Mesa.

Named for Jeremiah Horrocks (1619-1641), the English astronomer who predicted the transit of Venus across the face of the Sun in 1639 Nov. and became the first to see such an event. From his observations he improved the orbital elements and the diameter of Venus. He believed the Moon to have an elliptical orbit with the Earth at one focus - a fact that Newton {see planet (8000)} was later to acknowledge. (M 10846)
Name proposed by the discoverer following a suggestion by B. Hetherington.

(3079) Schiller

2578 P-L. Discovered 1960 September 24 by C. J. van Houten and I. van Houten-Groeneveld at Palomar.
Named for the German poet and playwright (1759-1805). (M 10045)

(3080) Moisseiev

1935 TE. Discovered 1935 October 3 by P. F. Shajn at Simeïs.

Named in memory of Nikolaj Dmitrevich Moisseiev (Moiseev) (1902-1955), professor at Moscow University and founder of the Moscow school of celestial mechanics. He studied the secular and long-period perturbations in the motion of natural celestial bodies, especially minor planets. (M 10548)
Moisseiev is also honored by a lunar crater.

(3081) Martinůboh

1971 UP. Discovered 1971 October 26 by L. Kohoutek at Bergedorf.

Named in memory of the Czech composer Bohuslav Martinů (1890-1959), known in the musical world as the author of remarkable orchestral and scenic works (six symphonies and twelve operas), as well as chamber and vocal music. (M 31609)

(3082) Dzhalil

1972 KE. Discovered 1972 May 17 by T. M. Smirnova at Nauchnyj.

Named in memory of Musa Mustafovich Dzhalil' (1906-1944), outstanding Tatar Soviet poet, author of lyric poetry, poems and opera libretti. (M 10548)

(3083) OAFA

1974 MH. Discovered 1974 June 17 at the Felix Aguilar Observatory at El Leoncito.

Named in honor of the Observatorio Astronomico Felix Aguilar {see also planet (1800)}, established on 1953 Sept. 28 under the auspices of the National San Juan University in San Juan province, Argentina. OAFA has made important contributions to astronomy in the areas of meridian and extrameridian astrometry.

Since 1974 the observing station at El Leoncito {see planet (2311)}, operated in collaboration with Yale University and now known as the Dr. Carlos U. Cesco Observatory, has been a branch of OAFA. (M 19333)

(3084) Kondratyuk

1977 QB_1. Discovered 1977 August 19 by N. S. Chernykh at Nauchnyj.

Named in memory of Yurij Vasil'evich Kondratyuk (1897-1942), one of the Soviet pioneers in rocket technology and cosmonautics, known for his work in electrical engineering. (M 12970)

Kondratyuk is also honored by a lunar crater.

(3085) Donna

1980 DA. Discovered 1980 February 18 at the Harvard College Observatory at Harvard.

Named in honor of Donna Marie Thompson, who as administrative assistant for the Minor Planet Center and the Central Bureau for Astronomical Telegrams is in charge of handling subscriptions. Her official position as secretary for the Planetary Sciences division of the Harvard-Smithsonian Center for Astrophysics scarecely describes her talents, which include serving as timekeeper and deputy administrator, setting up audio-visual aids at meetings and maintaining laser printers. (M 16244)

(3086) Kalbaugh

1980 XE. Discovered 1980 December 4 by E. Bowell at Anderson Mesa.

Named in honor of Carroll Kalbaugh Liller, father of astronomer William Liller {see planet (3222)}, lover of life and nature, friend of all who come in peace. (M 10548)

Named by the discoverer following a suggestion by W. Liller.

(3087) Beatrice Tinsley

1981 QJ_1. Discovered 1981 August 30 by A. C. Gilmore and P. M. Kilmartin at Lake Tekapo.

Named in memory of Beatrice Muriel Tinsley (1941-1981), a graduate of the University of Canterbury, New Zealand, and a professor at Yale University, well known for her work on the evolution of galaxies. (M 9479)

Obituary published in Phys. Today, Vol. 34, No. 9, p. 110, 112 (1981); Q.J.R. Astron. Soc., Vol. 23, p. 162-165 (1982).

(3088) Jinxiuzhonghua

1981 UX_9. Discovered 1981 October 24 at the Purple Mountain Observatory at Nanking.

The name reflects the natural beauty and ancient civilization of China. Jinxiuzhonghua, or "Splendid China", is the largest miniature scenic spot in the world. Situated in Shenzhen, in the southern part of China, and constructed by China Travel Service, it is an attractive place for tourists and well-known for its outstanding scenery, including the Great Wall, the Imperial Palace, the Terra-Cotta Warriors and Horses of Qin Shihuang Mausoleum, and scenic and historical sites such as Guilin, Suzhou {see planet (2719)} and Hangzhou. (M 17221)

(3089) Oujianquan

1981 XK_2. Discovered 1981 December 3 at the Purple Mountain Observatory at Nanking.

Named in honor of Oujianquan, Chinese entrepreneur, for his notable contributions in developing township enterprises. (M 22497)

(3090) Tjossem

1982 AN. Discovered 1982 January 4 by J. Gibson at Palomar.

Named in honor of a pioneer family in central Washington, four generations of whose members have been friends of the discoverer and his family. It honors in particular the memory of Peter Tjossem (1878-1957), millwright, farmer, amateur entomologist, lapidarist, amateur paleobotanist working at a professional level on identification of Miocene fossil woods of central Washington, elder in the Presbyterian Church - and friend to any young person who came to him, in the best traditions of the amateur scientist. (M 10045)

(3091) van den Heuvel

6081 P-L. Discovered 1960 September 24 by C. J. van Houten and I. van Houten-Groeneveld at Palomar.

Named in honor of E. P. J. van den Heuvel, professor of astronomy at the Municipal University of Amsterdam, well known for his studies on the structure and evolution of neutron stars. The name also honors Julia Edith van den Heuvel, niece of the astronomer, for her lively interest in astronomy. (M 11159)

(3092) Herodotus

6550 P-L. Discovered 1960 September 24 by C. J. van Houten and I. van Houten-Groeneveld at Palomar.

Named for the Greek historian who lived in the fifth century B.C. and is known as the "Father of Historiography". (M 11159)

Herodotus is also commemorated by a lunar crater.

(3093) Bergholz

1971 MG. Discovered 1971 June 28 by T. M. Smirnova at Nauchnyj.

Named in memory of Ol'ga Fedorovna Bergholz (1910-1975), talented poetess and writer. (M 13174)

(3094) Chukokkala

1979 FE_2. Discovered 1979 March 23 by N. S. Chernykh at Nauchnyj.

Named in memory of Kornej Ivanovich Chukovskij, pen name of Nikolaj Vasil'evich Kornejchukov (1882-1969), outstanding writer and literary scholar, known especially for his poetry for children. The name comes from Chukovskij's album "Chukokkala". Many prominent writers, artists and other men of culture included their notes, poetry, and paintings in this work. (M 12013)

(3095) Omarkhayyam

1980 RT_2. Discovered 1980 September 8 by L. V. Zhuravleva at Nauchnyj.

Named for the great Tadjik and Persian poet, mathematician and philosopher Omar Khayyam Giyasaddin-abu-L' Fatkh ibn Ibragim (ca. 1048-after 1122). (M 13174)

(3096) Bezruč

1981 QC_1. Discovered 1981 August 28 by Z. Vávrová at Kleť.

Named in honor of Petr Bezruč (1867-1958), popular Silesian poet. (M 21607)

(3097) Tacitus

2011 P-L. Discovered 1960 September 24 by C. J. van Houten and I. van Houten-Groeneveld at Palomar.

Named for the Roman historian who lived in the first century A.D. (M 11159)

Tacitus is also honored by a lunar crater.

(3098) van Sprang

4579 P-L. Discovered 1960 September 24 by C. J. van Houten and I. van Houten-Groeneveld at Palomar.

Named in honor of Bert van Sprang, a Dutch amateur who is very active in popularizing astronomy. An organization of young amateur astronomers, founded by him, has at present 1,400 members. He initiated the International Astronomical Youth Camps, which are held in various European countries each year. He also gives many popular lectures and stimulates the construction of small telescopes by young people. (M 12013)

(3099) Hergenrother

1940 GF. Discovered 1940 April 3 by Y. Väisälä at Turku.

Named in honor of Carl William Hergenrother (1973-) of the Bigelow Sky Survey. This photographic survey has been very successful in discovering new high-inclination minor planets. (M 27124)
Name proposed by B. G. Marsden, G. V. Williams and S. M. Larson.

(3100) Zimmerman

1977 EQ_1. Discovered 1977 March 13 by N. S. Chernykh at Nauchnyj.

Named in memory of Nikolaj Vladimirovich Zimmerman (1890-1942), head of the astrometric department of Pulkovo Observatory from 1938, and professor at Leningrad University from 1937, known for his work on astrometry and star catalogues. (M 12970)

(3101) Goldberger

1978 GB. Discovered 1978 April 11 by E. F. Helin at Palomar.

Named in honor of Marvin L. Goldberger, gifted physicist, teacher and humanitarian, to commemorate his birthday, Oct. 22. Since his inauguration in 1978 as president of the California Institute of Technology, he has dedicated himself to sustaining the Institute's excellence and to opening new opportunities for women in science. His steadfast support for research on minor planets has enhanced the importance and vitality of the subject as a scientific endeavor. (M 9217)

(3102) Krok

1981 QA. Discovered 1981 August 21 by L. Brožek at Kleť.

Krok was a mythical prince of the Slavonic tribes in ancient Bohemia. He had three daughters. The youngest of them was the sibyl Libuše, who married the young Přemysl; together they are considered the mythical founders of the Czech royal dynasty of the Přemyslids. (M 24410)

(3103) Eger

1982 BB. Discovered 1982 January 20 by M. Lovas at Piszkéstetö.

Named for a famous Hungarian city. In 1762 its bishop, Count Karoly Esterhazy, established a university there at his own expense. The university included an observatory, which he furnished on consultation with Maximilian Hell {see planet (3727)} in Vienna, instruments being made in both Vienna and London. The Eger observations began in 1778, and the results were published in Hell's annual astronomical ephemerides. The city is also known for the successful stand of its populace against the besieging Turks in 1552. It is situated in an excellent wine-growing area and is known internationally for its hearty red Egri Bikaver ("bull's blood"). (M 23135)

(3104) Dürer

1982 BB_1. Discovered 1982 January 24 by E. Bowell at Anderson Mesa.

Named for Albrecht Dürer of Nuremberg (1471-1528), master painter, wood-

cutter, engraver, scholar of the Renaissance. Dürer is noted for his meticulous applications of perspective geometry to art, not only for his rendering of the unusual themes of his times, but also for his beautiful and accurate depictions of scientific instruments and of constellations. (M 10846)
Name proposed by the discoverer following a suggestion by G. Reaves.

(3105) Stumpff

A907 PB. Discovered 1907 August 8 by A. Kopff at Heidelberg.

Named in memory of Karl Stumpff (1895-1970), eminent celestial mechanist and professor of astronomy at Berlin, Graz and Göttingen. His method of treatment of periodic processes was a forerunner of the Fast Fourier Analysis. The collected experience of decades of research went into his main scientific heritage, the three-volume *Himmelsmechanik*. (M 22497)

Name proposed by G. Klare and L. D. Schmadel. Citation prepared by P. Brosche.

(3106) Morabito

1981 EE. Discovered 1981 March 9 by E. Bowell at Anderson Mesa.

Named in honor of Linda A. Morabito, member of the Optical Navigation Team for the Voyager 1 spacecraft, who because of her alertness, knowledge of astronomy and skill in image processing discovered an active volcano on the Galilean Satellite Io. (M 10846)
Name proposed by the discoverer following a suggestion by G. Reaves.

(3107) Weaver

1981 JG_2. Discovered 1981 May 5 by C. S. Shoemaker at Palomar.

Named in honor of Kenneth F. Weaver, senior assistant editor for science of the National Geographic magazine. For more than two decades Weaver has followed closely the exploration of the solar system by spacecraft. He has been responsible for the accurate and skillful presentation of new discoveries in space to a large segment of the public. (M 10311)

(3108) Lyubov

1972 QM. Discovered 1972 August 18 by L. V. Zhuravleva at Nauchnyj.

Named in memory of Lyubov Petrovna Orlova (1902-1975), outstanding Soviet actress. (M 13174)

(3109) Machin

1974 DC. Discovered 1974 February 19 by L. Kohoutek at Bergedorf.

Named in memory of renowned sculptor Arnold Machin (1911-1999). His sculpture of the effigy of queen Elizabeth II was used on British coinage from 1968 until 1984. Machin also sculpted the monarch's profile, and this formed the basis for the Machin definitive series of British definitive stamps, first issued in June 1967 and still in use today. (M 34618)

Name suggested by B. G. Marsden, who made identifications for this object, following a prompting by G. V. Williams.

(3110) Wagman

1975 SC. Discovered 1975 September 28 by H. L. Giclas at Anderson Mesa.

Named in memory of Nicholas E. Wagman (1905-1980), American astronomer and leading astrometrist. Associated with the Allegheny Observatory beginning in 1930, he was its director and the chairman of the University of Pittsburgh's Astronomy Department from 1941 until 1970. Under his direction the 0.76-m Thaw refractor was renovated and used to set the standard for parallax determinations, of which were over 1200 made; and a number of astrometric binaries were discovered and characterized. In his honor the Amateur Astronomers Association of Pittsburgh named its Nicholas E. Wagman Observatory in 1976. (M 27733)

Name proposed and citation prepared by T. P. Kohman on behalf of the association.

(3111) Misuzu

1977 DX_8. Discovered 1977 February 19 by H. Kosai and K. Hurukawa at Kiso.

This planet is being given the former name of Nagano prefecture, where the Kiso Station {of the Tokyo Observatory} is located. (M 11441)

(3112) Velimir

1977 QC_5. Discovered 1977 August 22 by N. S. Chernykh at Nauchnyj.

Named for Velimir (Viktor Vladimirovich) Khlebnikov (1885-1922), Russian poet known for his experiments in poesy, who also applied mathematical analysis to history and discovered some cycles among historical events. (M 12970)

(3113) Chizhevskij

1978 RO. Discovered 1978 September 1 by N. S. Chernykh at Nauchnyj.

Named in honor of Aleksandr Leonidovich Chizhevskij (1897-1964), Soviet biologist and one of the founders of heliobiology, who discovered the dependence of many phenomena of the biosphere and atmosphere upon the solar cycle. (M 12970)

(3114) Ercilla

1980 FB_{12}. Discovered 1980 March 19 by C. Torres at Cerro El Roble.

Named in memory of Don Alonso de Ercilla y Zuniga (1533-1594), Spanish poet and soldier who distinguished himself in the campaign in Chile against the Araucanians. He returned to Spain in 1562, taking with him the first fifteen cantos of *La Araucana*, his epic poem about the early part of the 300-year-long war between Spanish soldiers and Chilean Indians. He can be considered the first chronicler of the history of Chile. (M 11160)

(3115) Baily

1981 PL. Discovered 1981 August 3 by E. Bowell at Anderson Mesa.

Named for Francis Baily (1774-1844), English astronomer and one of the founders of the Royal Astronomical Society. During his observation of the total solar eclipse of 1836 he noticed intrusions of sunlight around the Moon's limb, which have become known as Baily's beads. (M 10847)

Name proposed by the discoverer following a suggestion by B. Hetherington.

Baily is also honored by a lunar crater.

(3116) Goodricke

1983 CF. Discovered 1983 February 11 by E. Bowell at Anderson Mesa.

Named for John Goodricke (1764-1786), the deaf and dumb Dutch-English astronomer who studied the light variations of the star Algol and correctly suggested that it is what we now call an eclipsing variable. He also discovered and studied the variable star Delta Cephei. (M 10847)

Name proposed by the discoverer following a suggestion by B. Hetherington.

(3117) Niepce

1983 CM_1. Discovered 1983 February 11 by N. G. Thomas at Anderson Mesa.

Named for the Frenchman Joseph Nicéphore Niepce (1765-1833), who made the first photograph in 1827 using the bitumen heliographic process. (M 10548)

Name suggested by Douglas B. Thomas, brother of the discoverer.

Niepce is also honored by a lunar crater.

(3118) Claytonsmith

1974 OD. Discovered 1974 July 19 at the Felix Aguilar Observatory at El Leoncito.

Named in memory of Clayton Albert Smith (1934-1993), a leading expert on astrometry. Smith was known for his work on stellar catalogues and for improving the celestial coordinate system. He was director of the Yale-Columbia Southern Observatory (now the Dr. Carlos U. Cesco Station) in El Leoncito from 1968 to 1970 and director of the U.S. Naval Observatory's Astrometry Department from 1992. (M 22497)

(3119) Dobronravin

1972 YX. Discovered 1972 December 30 by T. M. Smirnova at Nauchnyj.

Named in honor of Petr Pavlovich Dobronravin {1908- }, well-known Soviet astrophysicist and spectroscopist, who served successively on the staffs of the Leningrad Astronomical Institute, the Leningrad State Optical Institute and the Pulkovo Observatory. As deputy director of the Crimean Astrophysical Observatory during 1952-1969 he made an impressive contribution to the development of that organization, equipping it with powerful optical and radio telescopes. (M 10548)

(3120) Dangrania

1979 RZ. Discovered 1979 September 14 by N. S. Chernykh at Nauchnyj.

Named in honor of Daniil Aleksandrovich Granin, Soviet writer whose work is mainly about scientists and researchers. (M 12970)

(3121) Tamines

1981 EV. Discovered 1981 March 2 by H. Debehogne and G. DeSanctis at La Silla.

Named for an industrial city on the river Sambre {see planet (4016)} in southern Belgium, known for foundries and glass works. (M 19333)

(3122) Florence

1981 ET_3. Discovered 1981 March 2 by S. J. Bus at Siding Spring.

Named in memory of Florence Nightingale (1820-1910), English nurse and hospital reformer, who almost singlehandedly established trained nursing as an honorable profession for women. She transformed the English field hospitals during the Crimean War and was known there for her quiet dignity, her rigorous discipline and her determination. Florence Nigthingale is most remembered as The Lady of the Lamp for her courage, compassion and devotion to the injured troops as she visited the hospital wards after a full day's work. (M 21955)
Citation prepared by C. S. Shoemaker at the request of the discoverer.

(3123) Dunham

1981 QF_2. Discovered 1981 August 30 by E. Bowell at Anderson Mesa.

Named in honor of David W. Dunham, American astronomer and organizer of the International Occultation Timing Association. Dunham has played a cardinal role in collecting and analyzing occultation observations, particularly those involving asteroids and grazing occultations by the Moon. In addition, he has stimulated many observers to make accurate and useful timings of occultation phenomena. (M 10847)

Name proposed by the discoverer following suggestions by E. Goffin and P. L. Dombrowski.

(3124) Kansas

1981 VB. Discovered 1981 November 3 by D. J. Tholen at Kitt Peak.

Named for the discoverer's home state, which derives its name from that of the Kansa Indians who migrated into the northeastern region of the state during the latter portion of the 18th century. The state's motto is "Ad astra per aspera", which means "To the stars through difficulties". The planet is also named for the

University of Kansas, the discoverer's alma mater, to commemorate the centennial of observational astronomy there, which began with the purchase of an Alvan Clark 6-inch refractor in 1885, an instrument still in use today. (M 10045)

(3125) Hay

1982 BJ_1. Discovered 1982 January 24 by E. Bowell at Anderson Mesa.

Named for William Thompson Hay (1888-1949), British music-hall comedian and film star of the 1930s and early 1940s. Will Hay was an accomplished planetary observer who in 1933 discovered a famous white spot on Saturn. (M 10847)
Name proposed by the discoverer following a suggestion by B. Hetherington.

(3126) Davydov

1969 TP_1. Discovered 1969 October 8 by L. I. Chernykh at Nauchnyj.

Named in memory of Denis Vasil'evich Davydov (1784-1839), officer, writer and poet, hero of the war of 1812 in Russia. (M 11160)

(3127) Bagration

1973 ST_4. Discovered 1973 September 27 by L. I. Chernykh at Nauchnyj.

Named in memory of Petr Ivanovich Bagration (1765-1812), Russian general, hero of the war of 1812. (M 11160)

(3128) Obruchev

1979 FJ_2. Discovered 1979 March 23 by N. S. Chernykh at Nauchnyj.

Named in honor of Vladimir Afanasjevich Obruchev (1863-1956), outstanding geologist and geographer who made a valuable contribution to the geologic exploration of Siberia and central Asia. He also wrote popular books on science and science-fiction novels. (M 12013)
Obruchev is also honored by a lunar crater.

(3129) Bonestell

1979 MK_2. Discovered 1979 June 25 by E. F. Helin and S. J. Bus at Siding Spring.

Named in honor of Chesley Bonestell {1888-1986}, whose art has inspired generations of astronomers, space enthusiasts and artists. (M 10548)

Name proposed by the first discoverer, following a suggestion from Ronald Paludan.

Obituaries published in Icarus, Vol. 67, No. 3, p. 343-344 (1986); Spaceflight, Vol. 28, Nos. 9/10, p. 355 (1986).

(3130) Hillary

1981 YO. Discovered 1981 December 20 by A. Mrkos at Kleť.

Named in honor of the New Zealand mountaineer Sir Edmund Percival Hillary (1919-), who, with Sherpa Tenzing Norgay, was the first to conquer Mt. Everest. The discoverer worked with Hillary in Antarctica during the International Geophysical Year, 1957-1958. (M 23135)

(3131) Mason-Dixon

1982 BM_1. Discovered 1982 January 24 by E. Bowell at Anderson Mesa.

Named for Charles Mason (1730-1787) and Jeremiah Dixon (1773-1779), British astronomers who observed the 1761 transit of Venus from the Cape of Good Hope. Between 1763 and 1767 they surveyed the boundary between Pennsylvania and Maryland - the Mason-Dixon Line. (M 10847)
Name proposed by the discoverer following a suggestion by B. Hetherington.

(3132) Landgraf

1940 WL. Discovered 1940 November 29 by L. Oterma at Turku.

Named in honor of Werner Landgraf {1959- }, who found the identifications and computed the orbit for this minor planet, and whose initials appear in the object's principal provisional designation. (M 12457)

(3133) Sendai

A907 TC. Discovered 1907 October 4 by A. Kopff at Heidelberg.

Named for the biggest city in northeastern Japan, home of Tôhuku University and several other institutes of higher learning, sometimes called the "Heidelberg of the East". The Sendai Municipal Astronomical Observatory, established in 1955 at the urging of the Sendai Amateur Astronomical Association, has an active program for astrometric observations of comets. (M 10045)

Name proposed by S. Nakano, who found the identifications involving this planet.

(3134) Kostinsky

A921 VA. Discovered 1921 November 5 by S. I. Belyavskij at Simeïs.

Named in honor of Sergej Konstantinovich Kostinsky (1867-1936), one of the founders of astrophotography and photographical astrometry in Russia, a corresponding member of the U.S.S.R. Academy of Sciences and a staff member of the Pulkovo Observatory. He made numerous determinations of stellar parallaxes and proper motions, studied star clusters and nebulae, as well as planets and their satellites. (M 10548)
Kostinsky is also honored by a lunar crater.

(3135) Lauer

1981 EC_9. Discovered 1981 March 1 by S. J. Bus at Siding Spring.

Named in honor of Tod R. Lauer, who participated in the Palomar Planet-Crossing Asteroid Survey while an undergraduate student at Caltech. Now an astronomer at the Princeton University, he is engaged in extragalactic studies and is a member of the Space Telescope Wide-Field Camera Team. (M 13174)

(3136) Anshan

1981 WD_4. Discovered 1981 November 18 at the Purple Mountain Observatory at Nanking.

Named for the city, known as the "Steel Metropolis", in the Liaodong peninsula. The largest iron and steel producer in China, Anshan is rich in natural resources and is an industrial center. (M 21607)

(3137) Horky

1982 SM_1. Discovered 1982 September 16 by A. Mrkos at Kleť.

Named for the hill where the discoverer installed his first telescope in 1939. (M 22828)

(3138) Ciney

1980 KL. Discovered 1980 May 22 by H. Debehogne at La Silla.

Named for the chief town of the Condroz, in the province of Namur {see planet (3374)}, where the discoverer studied and maintains a residence. Ciney is renowned for its schools, its horse and cattle fairs and its casting houses. (M 15573)

(3139) Shantou

1980 VL_1. Discovered 1980 November 11 at the Purple Mountain Observatory at Nanking.

Named for an open city on the coast of southeast China. A fine seaport, it is also a flourishing industrial and agricultural center. (M 16244)

(3140) Stellafane

1983 AO. Discovered 1983 January 9 by B. A. Skiff at Anderson Mesa.

Named for the regional telescope maker's annual meeting in Vermont that has become recognized worldwide as a preeminent gathering for telescopic design and innovation. The "Shrine to the Stars" has provided a forum to exchange astronomical ideas, to enhance the performance of astronomical instruments and to further good contact between amateurs and professionals. (M 13174)
Name suggested and citation provided by P. L. Dombrowski.

(3141) Buchar

1984 RH. Discovered 1984 September 2 by A. Mrkos at Kleť.

Named in memory of Emil Buchar (1901-1979), professor of astronomy and geodesy at Prague Polytechnic. He was one of the pioneers of satellite geodesy. (M 22828)

(3142) Kilopi

1937 AC. Discovered 1937 January 9 by A. Patry at Nice.

The name acknowledges the fact that the number of this planet is the approximate circumference of a circle of diameter 1,000 units. (M 9771)

Following a suggestion by J. G. Williams, this planet was named by B. G. Marsden, who made the identifications involving it.

(3143) Genecampbell

1980 UA. Discovered 1980 October 31 at the Harvard College Observatory at Harvard.

Named in honor of I. Gene Campbell, systems programmer in the central computing facility at the Harvard-Smithsonian Center for Astrophysics. Ever helpful and always patient, he has quickly and quietly resolved many problems over the years, both for the observing program at Oak Ridge and for the Minor Planet Center and Central Bureau for Astronomical Telegrams. It is a rare magnetic tape that he cannot salvage and decipher, and fixing recalcitrant modems is second nature to him. When it became necessary to dispense with the teletype machines for the MPC/CBAT TWX number, he did the necessary programming for operating EASYLINK from the CFAPS2 MicroVAX and arranged for the automatic receipt there of TWX/telex messages. (M 16244)

(3144) Brosche

1931 TY_1. Discovered 1931 October 10 by K. Reinmuth at Heidelberg.

Named in honor of Peter Brosche (1936-), astronomer at Heidelberg and Bonn. He has developed the determination of systematic differences into orthogonal functions, participated in the foundation of absolute radio interferometry and investigated the history of the Seeberg Observatory and its scientific founder F. X. von Zach {see planet (999)}. (M 22497)
Name proposed by L. D. Schmadel, endorsed by G. Klare.

(3145) Walter Adams

1955 RY. Discovered 1955 September 14 at the Goethe Link Observatory at Brooklyn, Indiana.

Named in memory of Walter S. Adams (1876-1956), whose spectroscopic studies of sunspots and stars led to the discovery, with A. Kohlschütter, of a spectroscopic method for determining stellar distances, the relative intensities of spectral lines being used to determine absolute magnitudes of both giant and main-sequence stars. Adams identified Sirius B as the first white-dwarf star known, and his measurement of its gravitational redshift was taken as confirming evidence for the general theory of relativity. He served as director of the Mount Wilson Observatory from 1923 to 1946. (M 15089)
Name proposed by F. K. Edmondson. Citation prepared by J. Tenn.

(3146) Dato

1972 KG. Discovered 1972 May 17 by T. M. Smirnova at Nauchnyj.

Named in memory of Dato Kratsashvili (1963-1980), gifted Georgian painter. (M 13174)

(3147) Samantha

1976 YU_3. Discovered 1976 December 16 by L. I. Chernykh at Nauchnyj.

Named in memory of Samantha Smith (1971-1985), the American girl who dreamed about the friendship of people all over the world. (M 11160)

(3148) Grechko

1979 SA_{12}. Discovered 1979 September 24 by N. S. Chernykh at Nauchnyj.

Named in honor of Georgij Mikhajlovich Grechko, Soviet cosmonaut and scientist who made three space flights involving astronomical and geophysical observations and research. (M 12971)

(3149) Okudzhava

1981 SH. Discovered 1981 September 22 by Z. Vávrová at Kleť.

Named in honor of the contemporary Soviet writer, poet and singer Bulat Okudzhava. (M 12209, M 12851)

(3150) Tosa

1983 CB. Discovered 1983 February 11 by T. Seki at Geisei.

This planet is being given the old name of Kochi {see planet (2396)} prefecture, where the discoverer resides. (M 10847)

(3151) Talbot

1983 HF. Discovered 1983 April 18 by N. G. Thomas at Anderson Mesa.

Named for the Englishman William Henry Fox Talbot (1800-1877), who made the first silver nitrate photographic negatives in 1834. With Rawlinson and Hincks he was one of the earliest to decipher the cuneiform inscriptions of Nineveh. (M 10548)

(3152) Jones

1983 LF. Discovered 1983 June 7 by A. C. Gilmore and P. M. Kilmartin at Lake Tekapo.

Named in honor of Albert Francis Arthur Lofley Jones, New Zealand amateur astronomer renowned for the quantity and precision of his visual magnitude estimates of variable stars. His record of some 300,000 variable star observations over nearly 50 years is unrivaled worldwide. He is also a noted observer of comets and the discoverer of comet 1946 VI. He was an independent discoverer of supernova 1987A in the Large Magellanic Cloud. (M 13175)

(3153) Lincoln

1984 SH_3. Discovered 1984 September 28 by B. A. Skiff at Anderson Mesa.

Named in memory of Abraham Lincoln (1809-1865), sixteenth president of the United States. Lincoln served tirelessly during the great drama of the American Civil War as commander-in-chief of the Union armies in a struggle to preserve the country. A self-taught prairie lawyer, Lincoln rose to greatness through sheer determination and hard work. His election as president in 1860 sparked the southern secession movement, which had smoldered for decades. Lincoln's brilliance as a politician and army strategist is well documented. His efforts to preserve the Union and emancipate American slaves ended in Union victory just days before he was assassinated at Ford's Theatre in Washington, D.C., on Good Friday, 1865 Apr. 14. (M 16442)

Name proposed and citation prepared by David J. Eicher.

(3154) Grant

1984 SO_3. Discovered 1984 September 28 by B. A. Skiff at Anderson Mesa.

Named in memory of Ulysses Simpson Grant (1822-1885), eighteenth president of the United States and lieutenant general commanding the armies of the Union during the American Civil War. Grant served with distinction as a young lieutenant during the Mexican War and afterward resigned his commission to enter civilian life. Having failed at several dozen occupations, a poor, distraught Grant entered the volunteer service as a colonel at the outbreak of the Civil War. Grant captured Fort Donelson on the Tennessee River and Vicksburg on the Mississippi, and this transformed him from an obscure army commander to the best-known general in Blue. In the spring of 1864 Lincoln {see planet (3153)} promoted him to general-in-chief; his crunching "May Campaign" through Virginia ended in Lee's {see planet (3155)} surrender at Appomattox Court House on 1865 Apr. 9. (M 16442)

Name proposed and citation prepared by David J. Eicher.

(3155) Lee

1984 SP_3. Discovered 1984 September 28 by B. A. Skiff at Anderson Mesa.

Named in memory of Robert Edward Lee (1807-1870), general-in-chief of the Confederate States of America during the American Civil War. The son of Revolutionary War patriot Henry "Lighthorse Harry" Lee and great-grandnephew of George Washington {see planet (886)}, Robert E. Lee attended the U.S. Military Academy of West Point and distinguished himself in the Mexican War. He was a career soldier whose expertise as an engineer served the U.S. Army well until his native state, Virginia, seceded from the Union. Lee's actions throughout the Civil War demonstrated a cunning ability to out-command many of his opponents, despite having a smaller, poorly-fed army. With his grandfatherly bearing and grey-white beard, Lee came to epitomize the southern gentleman, and, after the war was over, he stirred dream-like memories of an Old South that would never return. (M 16442)

Name proposed and citation prepared by David J. Eicher.

(3156) Ellington

1953 EE. Discovered 1953 March 15 by A. Schmitt at Uccle.

Named in honor of Edward Kennedy ("Duke") Ellington (1899-1974), renowned jazz pianist, composer and bandleader. Ellington crafted more than 1,000 orchestrations for big bands, film scores, operas, ballets, musicals and church services. (M 22497)

Name proposed by B. G. Marsden, who made the identifications for this object.

(3157) Novikov

1973 SX_3. Discovered 1973 September 25 by L. V. Zhuravleva at Nauchnyj.

Named in memory of Aleksej Ivanovich Novikov (1916-1986), Soviet aviator and poet. (M 13175)

(3158) Anga

1976 SU_2. Discovered 1976 September 24 by N. S. Chernykh at Nauchnyj.

Named for a Siberian village, birthplace of the Russian ethnographers Ivan Evseevich Venyaminov (1797-1879) and Afanasij Prokopevich Shchapov (1831-1876). The former was the first explorer of the Aleutians and the latter a historian and popularizer of culture. (M 12014)

(3159) Prokof'ev

1976 US_2. Discovered 1976 October 26 by T. M. Smirnova at Nauchnyj.

Named in honor of Vladimir Konstantinovich Prokof'ev {1898- }, well-known expert on atomic spectroscopy and emission spectral analysis, a professor at the Leningrad and Gorkij universities and since 1961 on the staff of the Crimean Astrophysical Observatory. He contributed much to solar ultraviolet spectroscopic research and discovered the presence of oxygen in the atmosphere of Venus. He served as president of IAU Commission 44 during 1970-1973. (M 10847)

(3160) Angerhofer

1980 LE. Discovered 1980 June 14 by E. Bowell at Anderson Mesa.

Named in memory of Philip E. Angerhofer (1950-1986), astronomer at the U.S. Naval Observatory, Washington. He was involved in using the National Radio Astronomy Observatory connected-element interferometer at Green Bank for the determination of Earth-oriented parameters. His other research interests included supernova remnants such as CTB 80 and Cas A. Angerhofer's caring attitude toward his colleagues and family are greatly missed. (M 10848)

Name proposed by the discoverer, following a suggestion by P. K. Seidelmann, who also prepared the citation.

(3161) Beadell

1980 TB_5. Discovered 1980 October 9 by C. S. Shoemaker at Palomar.

Named in honor of Len Beadell, surveyor at Salisbury, South Australia. The last of the geographical explorers of the Australian outback, Beadell laid out more than 6,000 km of graded track, including the well-known "Gun Barrel Highway", through the most remote regions of Australia. (M 11748)

(3162) Nostalgia

1980 YH. Discovered 1980 December 16 by E. Bowell at Anderson Mesa.

Named in remembrance of good things that are no more. (M 11641)

Name proposed by the discoverer, following a suggestion by J. Meeus.

(3163) Randi

1981 QM. Discovered 1981 August 28 by C. T. Kowal at Palomar.

Named in honor of the American magician James Randi (1928-); (real name Randall James Hamilton Zwinge) for his continuing efforts in debunking the claims of the paranormal community and exposing the tricks that charlatans use. His use of scientific techniques in many disciplines has contributed to the refutation of suspicious and fraudulent claims of paranormal results. (M 27124)

Name proposed by A. Dill and J. Meeus.

(3164) Prast

6562 P-L. Discovered 1960 September 24 by C. J. van Houten and I. van Houten-Groeneveld at Palomar.

Named in honor of Martin Prast, friend of one of the discoverers. Severely wounded in Vietnam in 1970, he became paraplegic and confined to a wheelchair. Together with his father he founded "Prast Research Association: Mobility Aids for Handicapped Persons". In 1977 he received the Outstanding Citizen award in the 36th Congressional District of New York State, and in 1982 he parachuted successfully into the Niagara River near Grand Island, N.Y. (M 12457)

(3165) Mikawa

1984 QE. Discovered 1984 August 31 by K. Suzuki and T. Urata at Toyota.

Named for the district in which the Toyota Station is located. Mikawa, the old name of the district, also appears in the title of the Western Mikawa Astronomical Club, of which the first discoverer is a member. (M 10046)

(3166) Klondike

1940 FG. Discovered 1940 March 30 by Y. Väisälä at Turku.

Named in memory of the brothers Karl F. Joutsen and Anton F. Johnson, who during 1901-1905 made a fortune in their mine, Dominion Creek 21, in the Klondike gold rush. Among their benefactions to the University of Turku were the means to construct its library in 1954. (M 18450)

(3167) Babcock

1955 RS. Discovered 1955 September 13 at the Goethe Link Observatory at Brooklyn, Indiana.

Named in memory of Harold D. Babcock (1882-1968) and in honor of his son Horace W. Babcock, astronomers at Mount Wilson Observatory, the latter also serving as director of Palomar Observatory. The elder Babcock's precise laboratory studies of atomic spectra allowed others to identify the first "forbidden" lines in the laboratory and to discover the rare isotopes of oxygen. With C. E. St. John and others, he extended Rowland's tables of the solar spectrum into the ultraviolet and infrared. The Babcocks ruled excellent large gratings, including those used in the coudé spectrographs of the 2.5-m and 5-m telescopes, and they measured the distribution of magnetic fields over the solar surface to unprecedented precision. The younger Babcock invented and built many astronomical instruments, including the solar magnetograph, microphotometers and automatic guiders. By combining his polarization analyzer with the spectrograph he discovered magnetic fields in other stars, and he developed important models of sunspots and their magnetism. (M 15089)
Name proposed by F. K. Edmondson. Citation prepared by J. Tenn.

Obituary published in Q.J.R. Astron. Soc., Vol. 10, p. 68-72 (1969). Harold Babcock is also honored by a lunar crater.

(3168) Lomnický Štít

1980 XM. Discovered 1980 December 1 by A. Mrkos at Kleť.

Named for the meteorological and solar observatory in the High Tatras, where the discoverer worked for some 20 years. (M 23136; M 30819)

(3169) Ostro

1981 LA. Discovered 1981 June 4 by E. Bowell at Anderson Mesa.

Named in honor of Steven J. Ostro, planetary scientist at the Jet Propulsion Laboratory of the California Institute of Technology. Ostro has specialized in radar studies of solar system objects, including the Galilean satellites. He is the leader in radar research on minor planets and has used his observations to deduce their surface roughness and radar reflectivity. He has also combined his radar data with optical rotational lightcurves to study the shapes and axial orientations of minor planets. (M 11749)
Citation prepared by the discoverer and A. W. Harris.

(3170) Dzhanibekov

1979 SS_{11}. Discovered 1979 September 24 by N. S. Chernykh at Nauchnyj.

Named in honor of Vladimir Aleksandrovich Dzhanibekov, famous Soviet cosmonaut who flew into outer space five times and made a valuable contribution to outer space research. (M 12971)

(3171) Wangshouguan

1979 WO. Discovered 1979 November 19 at the Purple Mountain Observatory at Nanking.

Named in honor of Wang Shouguan, member of the Chinese Academy of Sciences and an unfailing friend of the discoverers. Wang has contributed to the development of modern Chinese astronomy and has served as chief editor of Acta Astrophysica Sinica since its inception. He is also an honorary president of the Chinese Astronomical Society and honorary director of Beijing Observatory. (M 21607)

(3172) Hirst

1981 WW. Discovered 1981 November 24 by E. Bowell at Anderson Mesa.

Named in honor of William P. Hirst, computer of the orbits of several of the minor planets discovered in Johannesburg. A Yorkshireman by birth, he worked for many years as Shell Oil's chief chemist in South Africa. He led the Cape Town Moonwatch Team from its inception in 1957, and following his retirement from Shell he served for several years at the Smithsonian Astrophysical Observatory as Moonwatch world coordinator. After returning to South Africa he continued to lecture on celestial mechanics at the University of Cape Town until the age of 83. (M 12014)

Name proposed by the discoverer, following a suggestion by R. Hurly and B. G. Marsden.

(3173) McNaught

1981 WY. Discovered 1981 November 24 by E. Bowell at Anderson Mesa.

Named in honor of Rob McNaught, who has been in charge of the University of Aston's satellite-tracking camera at Herstmonceux and more recently at Siding Spring. In his spare time he successfully conducts patrols for novae, identifies images of prenovae and unusual variable stars on survey plates, measures their positions, makes astrometric observations of comets and minor planets and photometric observations of comets and novae, carries out extensive observational and computational work on meteors, as well as on occultations by minor planets. (M 12457)

Named by the discoverer, following a suggestion by D. A. J. Seargent.

(3174) Alcock

1984 UV. Discovered 1984 October 26 by E. Bowell at Anderson Mesa.

Named in honor of the outstanding British amateur astronomer George E. D. Alcock, visual discoverer of five comets and four novae. (M 12458)

(3175) Netto

1979 YP. Discovered 1979 December 16 by H. Debehogne and E. R. Netto at La Silla.

Named by the first discoverer in honor of the second, Edgar Rangel Netto, astronomer at the Valongo Observatory (Universidade do Rio de Janeiro), measurer of many of the plates obtained in the program that produced this discovery. (M 10848)

(3176) Paolicchi

1980 VR_1. Discovered 1980 November 13 by Z. Knežević at Piszkéstetö.

Named in honor of Paolo Paolicchi, planetary scientist, associate professor of astrophysics at the University of Pisa. His research activity has included studies on the origin of stellar and planetary systems and the dynamical and collisional history of small solar-system bodies. His work on minor planets has focused on the evolution of rotational properties and on the modeling of catastrophic breakup events. (M 12209)

Name proposed by the discoverer following a suggestion by V. Zappalà and P. Farinella, the latter of whom prepared the citation.

(3177) Chillicothe

1934 AK. Discovered 1934 January 8 by H. L. Giclas at Flagstaff.

Named for the southern Ohio city of Chillicothe, located at the confluence of the Scioto River and Paint Creek, on the occasion of the bicentennial of its founding by Nathanial Massie. The name is derived from the Shawnee Indian word meaning "principal town". This area was occupied by the Adena Mound Builders over 2000 years ago and at the time of its founding by the Shawnee Indians. Chillicothe served as the capital of the eastern section of the Northwest Territory and was the first capital of Ohio from 1803 to 1816. (M 27733)

Name proposed by F. N. Bowman, who independently made the principal identification involving this planet and who lives 20 miles southwest of Chillicothe.

(3178) Yoshitsune

1984 WA. Discovered 1984 November 21 by K. Suzuki and T. Urata at Toyota.

Named for one of the most famous and tragic Japanese military commanders, Minamoto Yoshitsune (1159-1189). In 1185, he defeated the Heike in the great sea battle of Dannoura, and afterwards the jealousy of his elder brother Yoritomo {see planet (3902)} forced him to commit suicide by the sword. (M 19692)

(3179) Beruti

1962 FA. Discovered 1962 March 31 at the La Plata Observatory at La Plata.

Named in memory of Arturo Beruti (1862-1938), Argentine composer, author of the operas *Pampa, Kryse, Evangelina* and *Taras Bulba,* among others. (M 22497)

(3180) Morgan

1962 RO. Discovered 1962 September 7 at the Goethe Link Observatory at Brooklyn, Indiana.

Named in honor of William W. Morgan {1906-1994}, whose career at the Yerkes Observatory since 1926 has been devoted to morphology, the classification of objects by their form and structure. He and P. Keenan introduced stellar luminosity classes and the two-dimensional classification of stellar spectra strictly on the basis of the spectra themselves. With D. Osterbrock {see planet (6107)} and S. Sharpless he discovered spiral arms in the Galaxy from spectral studies of O and B stars. Morgan invented the UBV system of magnitudes and colors, and with N. Mayall {see planet (2131)} he developed a spectral classification system for giant galaxies. (M 15089)

Name proposed by F. K. Edmondson. Citation prepared by J. Tenn.

Obituaries published in Phys. Today, Vol. 67, No. 12, p. 82-83 (1994); Bull. Am. Astron. Soc., Vol. 26, No. 4, p. 1606 (1994); Q.J.R. Astron. Soc., Vol. 36, No. 2, p. 175-177 (1995); Publ. Astron. Soc. Pac., Vol. 107,, No. 712, p. 507-512 (1995).

(3181) Ahnert

1964 EC. Discovered 1964 March 8 by F. Börngen at Tautenburg.

Named in honor of Paul Ahnert {1897-1989}, author since 1949 of the annual *Kalender für Sternfreunde* and well known by professional and amateur astronomers alike in many countries. Although his principal aim is the popularization of astronomy, and he has published several books on the art of making observations, he has carried out research since 1938 at the Sonneberg {see planet (1039)} Observatory on variable stars and on the physics of the solar system. (M 9771)

Obituaries published in Ahnerts Kalender Sternfreunde 1990, p. 7 (1989); Sterne Weltraum, Jahrg. 28, Nr. 7/8, p. 480 (1989); Astron. Tidsskr., Årg. 22, Nr. 4, p. 155-156 (1989); Bull. Assoc. Fr. Obs. Etoiles Variables, No. 51, p. 8 (1990).

(3182) Shimanto

1984 WC. Discovered 1984 November 27 by T. Seki at Geisei.

Named for the longest river in the discoverer's home prefecture of Kochi {see planet (2396)}. (M 10848)

(3183) Franzkaiser

1949 PP. Discovered 1949 August 2 by K. Reinmuth at Heidelberg.

Named in memory of Franz Kaiser (1891-1962), astronomer at the Heidelberg-Königstuhl Observatory from 1911 to 1914. In 1914 Kaiser modified and improved the formulae for deriving photographic positions of minor planets by the interpolation method. He discovered 21 numbered minor planets. (M 22497)
Name proposed and citation prepared by G. Klare.

(3184) Raab

1949 QC. Discovered 1949 August 22 by E. L. Johnson at Johannesburg.

Named in honor of Herbert Raab (1969-), author of the widely-used and acclaimed *Astrometrica* software package. *Astrometrica* has enabled many amateur astronomers to participate in their own astrometric programs on comets and minor planets. Raab is a software developer and is also president of the Linzer Astronomische Gemeinschaft, the oldest amateur association in Austria. (M 27124)
Name proposed by B. G. Marsden and G. V. Williams.

(3185) Clintford

1953 VY_1. Discovered 1953 November 11 at the Goethe Link Observatory at Brooklyn, Indiana.

Named in honor of Clinton B. Ford {1913-1992}, secretary of the American Association of Variable Star Observers since 1948 and president in 1961, the year of his retirement from a career in industry. In 1965 he helped found what is now called the Ford Observatory in southern California, an amateur facility with a 0.46-m telescope dedicated to observing variable stars. Since joining the AAVSO at the age of 16 Ford has made more than 60,000 observations, drawn up over 800 new finding charts, and encouraged and assisted many hundreds of amateur observers in monitoring variables. He has also taught classes in astronomy at Brown University and Smith College. In 1987 he received the Amateur Achievement Award of the Astronomical Society of the Pacific for his contributions to science and to the fostering of interest in amateur astronomy. (M 15260)

Name proposed by F. K. Edmondson, following a suggestion by A. G. Fraknoi, who also prepared the citation.

Obituaries published in Bull. Assoc. Fr. Obs. Etoiles Variables, No. 62, p. 18-19 (1992); J. Am. Assoc. Variable Star Obs., Vol. 21, No. 2, p. 144-146 (1992); Sky Telesc., Vol. 85, No. 4, p. 102-103 (1993); Bull. Am. Astron. Soc., Vol. 26, No. 4, p. 1602-1603 (1994).

(3186) Manuilova

1973 SD_3. Discovered 1973 September 22 by N. S. Chernykh at Nauchnyj.

Named in memory of the prominent sculptor Ol'ga Maksimilianovna Manuilova (1893-1984). (M 12014)

(3187) Dalian

1977 TO_3. Discovered 1977 October 10 at the Purple Mountain Observatory at Nanking.

Named for the city in northeastern China, located on the southern tip of the Liao Dong peninsula. A center for industry, technology and education, Dalian is a port city and a tourist center. (M 22497)

(3188) Jekabsons

1978 OM. Discovered 1978 July 28 at the Perth Observatory at Bickley.

Named in memory of Peter Jekabsons (1943-1990), a gifted observer on the Perth staff for sixteen years. His astronomical paintings adorn the walls of the Observatory. His talents as an artist carried over into his scientific work, and he produced some of the Observatory's best plates. (M 18644)

(3189) Penza

1978 RF_6. Discovered 1978 September 13 by N. S. Chernykh at Nauchnyj.

Named for a Russian city, the center of the region where Lermontov, Belinskij, Davydov, Kuprin {see planets (2222), (3747), (3126), and (3618), respectively} and some other Russian writers lived. (M 12971)

(3190) Aposhanskij

1978 SR_6. Discovered 1978 September 26 by L. V. Zhuravleva at Nauchnyj.

Named in memory of Vladimir Mikhailovich Aposhanskij (1910-1943), Soviet poet and journalist. (M 11160)

(3191) Svanetia

1979 SX_9. Discovered 1979 September 22 by N. S. Chernykh at Nauchnyj.

Named for Svanetia, a historical region in Georgia near the Caucasus Mountains. (M 12971)

(3192) A'Hearn

1982 BY_1. Discovered 1982 January 30 by E. Bowell at Anderson Mesa.

Named in honor of Michael F. A'Hearn, professor of astronomy at the University of Maryland. A prominent student of cometary physics, A'Hearn has pursued coordinated spectroscopic and spectrophotometric observations of comets spanning the spectral interval from the vacuum ultraviolet to the radio region. He participated in the 1983 discovery with the IUE spacecraft of diatomic sulphur in the spectrum of Comet IRAS-Araki-Alcock (1983d) and has made many other important contributions to our current understanding of comets. (M 10848)
Citation prepared by R. L. Millis.

(3193) Elliot

1982 DJ. Discovered 1982 February 20 by E. Bowell at Anderson Mesa.

Named in honor of James L. Elliot, professor of physics and astronomy at the Massachusetts Institute of Technology and codiscoverer of the Uranian rings. A pioneer in the study of solar-system bodies through photometric observations of stellar occultations, he has made major contributions to understanding the structure of planetary atmospheres and ring systems. Elliot is currently involved in planning solar-system studies using the Hubble Space Telescope and other Earth-orbiting platforms. (M 10848)
Citation prepared by R. L. Millis.

(3194) Dorsey

1982 KD_1. Discovered 1982 May 27 by C. S. Shoemaker and E. M. Shoemaker at Palomar.

Named in honor of Dorsey Taylor Shoemaker Jr., businessman in Gabbs, Nevada, and uncle of the second discoverer. (M 10311)

(3195) Fedchenko

1978 PT_2. Discovered 1978 August 8 by N. S. Chernykh at Nauchnyj.

Named for the family of scientists, including Aleksej Pavlovich Fedchenko (1844-1873), outstanding Russian naturalist and geographer known for his exploration of central Asia; his wife Ol'ga Aleksandrovna Fedchenko (1845-1921), prominent Russian botanist and plant collector, pioneer in the study of Turkestan flora; and

their son Boris Alekseevich Fedchenko (1872-1947), distinguished botanist and geographer, prolific writer, and one of the initiators and most active contributors to the multi-volume "Flora of the U.S.S.R." (M 12971)
Name proposed by the discoverer, following a suggestion by C. E. Spratt.

(3196) Maklaj

1978 RY. Discovered 1978 September 1 by N. S. Chernykh at Nauchnyj.

Named in honor of Nikolaj Nikolaevich Miklukho-Maklaj (1846-1888), a prominent Russian anthropologist and ethnographer who studied the native populations of southeast Asia, Australia and Oceania and who lived for about three years among the Papuans in New Guinea. (M 12971)

(3197) Weissman

1981 AD. Discovered 1981 January 1 by E. Bowell at Anderson Mesa.

Named in honor of Paul R. Weissman {1947- }, cometary physicist at the Jet Propulsion Laboratory who has made studies of the dynamics of the Oort cloud and of the thermal properties of cometary nuclei. (M 11160)

(3198) Wallonia

1981 YH_1. Discovered 1981 December 30 by F. Dossin at St. Michel.

Named in honor of Wallonia, the French speaking part of Belgium, where the discoverer was born and where the Institut d'Astrophysique is located. (M 10848)

(3199) Nefertiti

1982 RA. Discovered 1982 September 13 by C. S. Shoemaker and E. M. Shoemaker at Palomar.

Named for the beautiful consort of the revolutionary pharaoh Akhenaten of the seventeenth dynasty in Egypt. She is generally believed to have had a major influence on radical changes that occurred in the court and religion of Egypt during the reign of Akhenaten. (M 10311)
The name Nefertiti is only a different spelling of Nofretete {see planet (1068)}.

(3200) Phaethon

1983 TB. Discovered 1983 October 11 with the Infrared Astronomical Satellite.

This object associated with the Geminid meteor stream has the smallest known perihelion distance for a body in a short-period orbit and is named for the son of Helios, who operated the solar chariot for a day, lost control of it and almost set fire to the Earth. (M 9771)

(3201) Sijthoff

6560 P-L. Discovered 1960 September 24 by C. J. van Houten and I. van Houten-Groeneveld at Palomar.

Named in honor of Albert Georg Sijthoff, publisher, owner of the former Zeiss planetarium in The Hague and first president of the board of the Omniversum space theater there. Through his lifelong efforts he greatly contributed to the popularization of astronomy in the Netherlands. He played a key role in founding Omniversum, the main center in the Netherlands for the popularization of science in general and astronomy in particular. (M 12458)

(3202) Graff

A908 AA. Discovered 1908 January 3 by M. Wolf at Heidelberg.

Named in honor of Gareth (Graff) Vaughan Williams {1965- }, of Leighton Buzzard, England, where he has during the past three years identified several low-numbered minor planets among objects that had been given provisional designations. During 1990 January-April he has been at the Minor Planet Center, where he continued this work with ingenuity and success, has made excellent progress on the daunting task of correcting the MPC files of old observations of low-numbered minor planets, and has generally been improving the efficiency of the whole MPC operation. (M 16245)

Name proposed by B. G. Marsden and C. M. Bardwell, the identification of this Hilda-type object having been made by the latter.

(3203) Huth

1938 SL. Discovered 1938 September 18 by C. Hoffmeister at Sonneberg.

Named in memory of Hans Huth (1925-1988), an assiduous observer who obtained more than 100,000 patrol plates at the Sonneberg Observatory. Huth was also known for his bibliographic catalogue of variable stars. (M 15260)

(3204) Lindgren

1978 RH. Discovered 1978 September 1 by N. S. Chernykh at Nauchnyj.

Named in honor of Astrid Anna Emilia Lindgren {1907- }, renowned Swedish writer, author of many fascinating stories for children and recipient of the international H. C. Andersen {see planet (2476)} gold medal. (M 12971)

(3205) Boksenberg

1979 MO_6. Discovered 1979 June 25 by E. F. Helin and S. J. Bus at Siding Spring.

Named in honor of Alexander Boksenberg, director of the Royal Greenwich Observatory, in recognition of his invention of the image-photon counting system and its application to a wide variety of astronomical problems. The first discoverer acknowledges the role he played in introducing her to the U.K. 1.2-m Schmidt facility in New South Wales, where she conducted the program in which (3205) was discovered. (M 13480)

Name endorsed by W. L. W. Sargent.

(3206) Wuhan

1980 VN_1. Discovered 1980 November 13 at the Purple Mountain Observatory at Nanking.

Named for the largest city in central China. Located at the confluence of the Han and Yangtze rivers, Wuhan has served as an inland commercial center since the third century AD. (M 16245)

(3207) Spinrad

1981 EY_{25}. Discovered 1981 March 2 by S. J. Bus at Siding Spring.

Named in honor of Hyron Spinrad {1934- }, a professor of astronomy at the University of California at Berkeley and a specialist in the study of both comets and galaxies. Spinrad has been a pioneer in the use of the metastable oxygen line at 630 nm to determine water production in comets and in the study of faint periodic comets at large heliocentric distances. His image-dissector scanner observations of 25 comets constitute one of the larger, uniform cometary data sets. He is currently adding to cometary knowledge with interpretation of very high dispersion CCD échelle spectroscopy and two-dimensional (interference filter) CCD photometry of brighter comets. In more than a quarter-century of teaching Spinrad has guided many outstanding students in their doctorate research. (M 18450)

Citation provided by Ray L. Newburn at the request of the discoverer.

(3208) Lunn

1981 JM. Discovered 1981 May 3 by E. Bowell at Anderson Mesa.

Named in memory of Borge Lunn (1912-1986), Danish civil engineer and metallurgist who did much to encourage the study of metal and iron meteorites. He invented a particular unmagnetized bronze for deepwater experiments on the marine Galathea expedition in 1950, and for his work on the metallurgy of sleeve bearing metals he was awarded the Hunt Medal of the American Society of Lubrication Engineers. Twice chairman of the Danish Metallurgical Society, he served as head

of the department of metallurgy at the Technical University of Denmark and was permanent censor in metallurgy in that department for 27 years. (M 12014)
Name proposed by the discoverer, following a suggestion of J. Gradie.

(3209) Buchwald

1982 BL_1. Discovered 1982 January 24 by E. Bowell at Anderson Mesa.

Named in honor of Vagn Fabritius Buchwald, associate professor in the department of metallurgy at the Technical University of Denmark. Participant in numerous expeditions, most recently to Antarctica, to search for iron meteorites, he discovered the 20-tonne iron meteorite Agpalilik in Greenland in 1963. Among his international publications is the three-volume Handbook of Iron Meteorites (1975) and a catalogue of meteorites (1965, 1976). (M 12014)
Name proposed by the discoverer, following a suggestion by J. Gradie.

(3210) Lupishko

1983 WH_1. Discovered 1983 November 29 by E. Bowell at Anderson Mesa.

Named in honor of Dmitrij Fedorovich Lupishko {1942- }, deputy director of the Kharkow Astronomical Observatory. An assiduous observer of minor-planet lightcurves, Lupishko has made fundamental contributions to the understanding of how photometric data may be used to derive the shapes and spin rates of these bodies. He has also undertaken photometry of Mars and since its inception in 1984 has been chairman of the Working Group on Asteroids of the U.S.S.R. Academy of Sciences. (M 12014)
Citation prepared by V. A. Shor at the request of the discoverer.

(3211) Louispharailda

1931 CE. Discovered 1931 February 10 by G. Van Biesbroeck at Williams Bay.

Named in memory of the discoverer's parents, Louis Pierre Van Biesbroeck (1839-1919) and Pharailda de Colpaert Van Biesbroeck (1840-1920). (M 14481)
Name proposed by Micheline Van Biesbroeck Wilson.

(3212) Agricola

1938 DH_2. Discovered 1938 February 19 by Y. Väisälä at Turku.

Named for Mikael Agricola (c. 1510-1557), bishop, the reformer of Finland, 'father of Finnish literature'. Around 1538 he issued his *ABC-kiria*, the first book printed in the Finnish language, and in 1548 he translated the New Testament into Finnish. (M 18450)

(3213) Smolensk

1977 NQ. Discovered 1977 July 14 by N. S. Chernykh at Nauchnyj.
Named for the one of the oldest towns in Russia. (M 12971)

(3214) Makarenko

1978 TZ_6. Discovered 1978 October 2 by L. V. Zhuravleva at Nauchnyj.

Named in memory of Anton Semenovich Makarenko (1888-1939), outstanding Soviet teacher and writer. (M 13175)

(3215) Lapko

1980 BQ. Discovered 1980 January 23 by L. G. Karachkina at Nauchnyj.

Named in honor of Konstantin Kuz'mich Lapko, assistant professor at the Crimean medical institute, a surgeon to whom the discoverer owes her recovery. (M 10549)

(3216) Harrington

1980 RB. Discovered 1980 September 4 by E. Bowell at Anderson Mesa.

Named in honor of Robert S. Harrington {1942-1993}, astronomer at the U.S.

Naval Observatory in Washington and director of the program to determine parallaxes and proper motions of faint nearby stars. A leading investigator on the orbital characteristics of Pluto's satellite and the mass of Pluto, he has made observational and theoretical studies of the motions of the planets and satellites and likely evolutions leading to the current distribution of the planets. He has also investigated the possible existence of another principal planet in the solar system. (M 11160)

Obituary published in Bull. Am. Astron. Soc., Vol. 25, No. 4, p. 1496-1497 (1993).

(3217) Seidelmann

1980 RK. Discovered 1980 September 2 by E. Bowell at Anderson Mesa.

Named in honor of P. Kenneth Seidelmann {1937- }, director of the Nautical Almanac Office of the U.S. Naval Observatory, currently vice president of IAU Commission 4, and a past chairman and secretary of the Division on Dynamical Astronomy of the American Astronomical Society. Seidelmann has played an important role in the introduction of new constants, theories, timescales and reference frames into the astronomical ephemerides, particularly a new theory of nutation. He has also explored the use of new techniques for obtaining astrometric data on minor planets and satellites. (M 11161)

(3218) Delphine

6611 P-L. Discovered 1960 September 24 by C. J. van Houten and I. van Houten-Groeneveld at Palomar.

Named in honor of Delphine Jehoulet Delsemme, wife of astronomer Armand Delsemme {see planet (2954)} of the University of Toledo, Ohio. Before her marriage she contributed to astronomical research by investigating the behavior of the population II Cepheid RU Camelopardalis and as a co-recoverer of periodic comet Honda-Mrkos-Pajdušaková at its first predicted return. (M 13175)

(3219) Komaki

1934 CX. Discovered 1934 February 4 by K. Reinmuth at Heidelberg.

Named in memory of Kojiro Komaki (1903-1969), director of the meteor section of the Oriental Astronomical Association from 1928 until 1969 and well known as a meteor observer and a leader of amateur astronomy in Japan. He established the Kii Astronomical Society (1943), the Japan Meteor Committee (1956) and the Japan Meteor Society (1968), and he was the director of these until his death. Together with his wife Sigeyo, he observed more than 30,000 meteors. He was a strong influence on many astronomers now active in Japan, including S. Nakano {see planet (3431)}, who found the identifications involving this planet and proposed the name. (M 10046)

(3220) Murayama

1951 WF. Discovered 1951 November 22 by M. Laugier at Nice.

Named in honor of Sadao Murayama, observer of Mars, authority on meteorites, and director of the Earth-science section of the National Science Museum, at Ueno Park, Tokyo. (M 11441)

Named by one of his former students, S. Nakano, who found the identifications involving this planet.

(3221) Changshi

1981 XF_2. Discovered 1981 December 2 at the Purple Mountain Observatory at Nanking.

Named for a bright pearl city in the lower reaches of the Yangtze river in China, with charming natural scenery and a cultural history dating back for more than

3000 years. It is famed for its abundant products and enjoys another name – the 'land rich in rice and fish'. (M 25229)

(3222) Liller

1983 NJ. Discovered 1983 July 10 by E. Bowell at Anderson Mesa.

Named in honor of William Liller, formerly Robert Wheeler Wilson Professor of Applied Astronomy at Harvard University, on the occasion of his sixtieth birthday. A premier observer, he has made substantial contributions through observations of a broad range of astronomical objects and phenomena: planetary nebulae, minor planets, comets, novae, variable stars, globular clusters, X-ray sources, quasars, solar eclipses and stellar occultations. Now living in Chile, he has in recent years participated in the PROBLICOM survey and has discovered several novae. During the recent passage of Halley's Comet he was a crucial member of the IHW Island Network. He has been a leader in astronomical education and an important supporter of amateur astronomy. His enthusiastic encouragement has been greatly appreciated by his colleagues and students. (M 12015)

(3223) Forsius

1942 RN. Discovered 1942 September 7 by Y. Väisälä at Turku.

Named for Sigfrid Aronus Forsius (c. 1550-1624), pastor, naturalist and astronomer. Among his various life-works he was rector of several parishes, mapped the Tornio river valley (1601-1602), was professor of astronomy at Uppsala University (1608-1610), and prepared almanacs for the horizons of Stockholm and Turku, from 1613 as a monopoly. (M 18450)

(3224) Irkutsk

1977 RL_6. Discovered 1977 September 11 by N. S. Chernykh at Nauchnyj.
Named for the Siberian city on the occasion of its 300th anniversary. (M 11161)

(3225) Hoag

1982 QQ. Discovered 1982 August 20 by C. S. Shoemaker and E. M. Shoemaker at Palomar.

Named in honor of Arthur Allen Hoag, American astronomer and, since 1977, director of the Lowell Observatory at Flagstaff, Arizona. He is specially recognized for his work on both photoelectric and photographic photometry, development of astronomical sites and instruments, and investigations of quasistellar sources. (M 10311)
Name endorsed by E. Bowell and W. A. Baum.

(3226) Plinius

6565 P-L. Discovered 1960 September 24 by C. J. van Houten and I. van Houten-Groeneveld at Palomar.

Named for Pliny (62-114), the Roman author who described the A.D. 79 eruption of Vesuvius. (M 16591)
Plinius is also honored by a lunar crater.

(3227) Hasegawa

1928 DF. Discovered 1928 February 24 by K. Reinmuth at Heidelberg.

Named in honor of Ichiro Hasegawa {1928- }, editor of the Yamamoto Circulars, well known for his research on the origin of comets and for the computation of their orbits. Author of several books, including some on orbit determination, he is also the mentor of S. Nakano {see planet (3431)}, who found the identifications involving this planet. (M 11441)

(3228) Pire

1935 CL. Discovered 1935 February 8 by S. Arend at Uccle.

Named in memory of Dominique Pire (1910-1969), who worked for the underprivileged and won the Nobel peace prize in 1958. (M 12209)

(3229) Solnhofen

A916 PC. Discovered 1916 August 9 by H. Thiele at Bergedorf.

The rocks around the limestone quarry at Solnhofen, Bavaria, consist of fossilized reefs. It seems that during the Upper Jurassic this region was covered by lagoons, in which limy sediments slowly accumulated to form very fine-grained limestones. This slow accumulation permitted the remarkable preservation of fossils – even internal organs and soft-bodied creatures are preserved in exquisite detail. The most famous specimen to come from Solnhofen was one of the few complete examples of Archaeopteryx, the earliest known flying vertebrate. (M 23351)

Name proposed by G. V. Williams, the identifications for this object having been made by C. M. Bardwell.

(3230) Vampilov

1972 LE. Discovered 1972 June 8 by N. S. Chernykh at Nauchnyj.

Named in memory of Aleksandr Valentinovich Vampilov (1937-1972), talented Soviet dramatist. (M 12972)

(3231) Mila

1972 RU_2. Discovered 1972 September 4 by L. V. Zhuravleva at Nauchnyj.

Named in memory of the Soviet ice dancer Ludmila Alekseevna Pakhomova (1946-1986). (M 13175)

(3232) Brest

1974 SL. Discovered 1974 September 19 by L. I. Chernykh at Nauchnyj.
Named for the well-known city in Byelorussia {see planet (2170)}. (M 12015)

(3233) Krišbarons

1977 RA_6. Discovered 1977 September 9 by N. S. Chernykh at Nauchnyj.

Named for Krišjanis Barons (1835-1923), prominent Latvian folklorist, writer and public figure. He collected and published the most complete collection of 'dainas', Latvian folksongs. (M 12015)

(3234) Hergiani

1978 QO_2. Discovered 1978 August 31 by N. S. Chernykh at Nauchnyj.

Named in memory of Mikhail Vissarionovich Hergiani (1932-1969), outstanding Soviet mountaineer. (M 12015)

(3235) Melchior

1981 EL_1. Discovered 1981 March 6 by H. Debehogne and G. DeSanctis at La Silla.

Named in honor of Paul Melchior, well-known geophysicist, general secretary of the IUGG since 1971 and director of the Royal Observatory of Belgium. His demonstration of the relationship of astronomical phenomena (precession-nutation) and geophysical phenomena (earth tides) was the basis of the IAU nutation series adopted in 1979. His development of a world tidal gravitational network permitted the evaluation of the indirect effects of the oceanic tides and revealed a correlation between the amplitudes of the earth tides and tectonic features. (M 16442)

(3236) Strand

1982 BH_1. Discovered 1982 January 24 by E. Bowell at Anderson Mesa.

Named in honor of Kaj Aa. Strand {1907- }, former Scientific Director of the U.S. Naval Observatory. Strand is known for his advances in the photographic astrometry of double stars. He was responsible for the construction and initial programs of the Naval Observatory's 1.55-meter astrometric reflector and its associated semi-automatic measuring machine. These led to the determination of stellar trigonometric parallaxes having an order-of-magnitude greater accuracy than heretofore possible, thus expanding the realm of the nearest stars. (M 11641) Name suggested by and citation prepared by R. S. Harrington.

(3237) Victorplatt

1984 SA_5. Discovered 1984 September 25 by J. Platt at Palomar.
Named in honor of Victor D. Platt, M.D., father of the discoverer. (M 10549)

(3238) Timresovia

1975 VB_9. Discovered 1975 November 8 by N. S. Chernykh at Nauchnyj.

Named in honor of Nikolaj Vladimirovich Timofeev-Resovskij (1900-1981), famous Soviet biologist, one of the founders of radiation genetics and molecular biology. (M 12972)

(3239) Meizhou

1978 UJ_2. Discovered 1978 October 29 at the Purple Mountain Observatory at Nanking.

Named for the historical cultural city in China, situated in northeastern Guangdong province {see planet (2185)}. (M 23792)

(3240) Laocoon

1978 VG_6. Discovered 1978 November 7 by S. J. Bus at Palomar.

Named for the priest of Apollo at Troy who warned the Trojans against the Trojan Horse. He and his two sons were killed by serpents sent by Athena {see planet (881)} or Apollo {see planet (1862)}. (M 12210)

(3241) Yeshuhua

1978 WH_{14}. Discovered 1978 November 28 at the Purple Mountain Observatory at Nanking.

Named in honor of Ye Shuhua, well-known Chinese astronomer, member of the Chinese Academy of Sciences and an unfailing friend of the discoverers. She has contributed to the development of modern Chinese astronomy, is an honorary president of the Chinese Astronomical Society and a vice-president of the IAU (1988-1994). (M 23792)

(3242) Bakhchisaraj

1979 SG_9. Discovered 1979 September 22 by N. S. Chernykh at Nauchnyj.

Named for a town in Crimea, the center of the district in which the Crimean Astrophysical Observatory is located. The name is widely known, thanks to Pushkin's {see planet (2208)} poem *The Bakhchisaraj Fountain*. (M 12015)

(3243) Skytel

1980 DC. Discovered 1980 February 19 at the Harvard College Observatory at Harvard.

Named in honor of the golden anniversary of *Sky and Telescope* magazine, which for 50 years has chronicled the development of astronomy and space exploration, while serving as a vital link between the amateur and professional astronomical communities. (M 17655)

(3244) Petronius

4008 P-L. Discovered 1960 September 24 by C. J. van Houten and I. van Houten-Groeneveld at Palomar.

Named for Petronius (fl. c. A.D. 60), Roman author whose main work is the novel Satyricon. (M 16591)

(3245) Jensch

1973 UL_5. Discovered 1973 October 27 by F. Börngen at Tautenburg.

Named in honor of Alfred Jensch, internationally renowned for his creative genius in the advancement of the design of astronomical instruments at the Jena Zeiss Works for almost 40 years. The 300-mm and 600-mm Jensch Coelostats are particularly innovative, and he was essentially responsible for the design of the 2 m reflectors at Tautenburg, Shemakha, Ondřejov and Rozhen. The new 'support' mounting, perhaps his most important innovation, proved to be of enormous value. Jensch has also represented Zeiss Jena on several national and international panels and holds a number of awards. (M 10046)

(3246) Bidstrup

1976 GQ_3. Discovered 1976 April 1 by N. S. Chernykh at Nauchnyj.

Named in memory of Herluf Bidstrup (1912-1988), well known Danish caricaturist. (M 18136)

(3247) Di Martino

1981 YE. Discovered 1981 December 30 by E. Bowell at Anderson Mesa.

Named in honor of Mario Di Martino, astronomer at the Osservatorio di Torino, who in recent years has been a prolific observer of minor planet lightcurves. Di Martino has specialized in observations for the purpose of shape and pole determinations. (M 11749)

Name proposed by the discoverer, following a suggestion by A. W. Harris, who also prepared the citation.

(3248) Farinella

1982 FK. Discovered 1982 March 21 by E. Bowell at Anderson Mesa.

Named in honor of Paolo Farinella, planetary scientist at the University of Pisa, whose research has included studies of the origin of the solar system and the dynamics of planetary satellites and ring systems. Farinella's work on minor planets has concerned the collisional evolution of the belt and the formation of families, both from a theoretical and an experimental point of view. (M 11749)

Named by the discoverer, following a suggestion by A. W. Harris. Citation prepared by A. Coradini and V. Zappalà.

(3249) Musashino

1977 DT_4. Discovered 1977 February 18 by H. Kosai and K. Hurukawa at Kiso.

Named for the area surrounding Tokyo that includes Mitaka {see planet (1088)}, site of the headquarters of the Tokyo Astronomical Observatory. (M 11442)

(3250) Martebo

1979 EB. Discovered 1979 March 6 by C.-I. Lagerkvist at Mount Stromlo.

Named for a small village on the island of Gotland, where the discoverer spends his summer vacations. (M 12972)

(3251) Eratosthenes

6536 P-L. Discovered 1960 September 24 by C. J. van Houten and I. van Houten-Groeneveld at Palomar.

Named for the Greek scholar Eratosthenes, who lived in Alexandria around 250 B.C. and was the first person to determine the size of the Earth correctly. (M 16591)

Eratosthenes is also honored by a lunar crater.

(3252) Johnny

1981 EM_4. Discovered 1981 March 2 by S. J. Bus at Siding Spring.

Named in honor of Johnny Carson (1925-), entertainer and host of television's

"The Tonight Show" from 1962 to 1992. Carson's witty monologue and unique, relaxed style of interviewing guests helped make this show a popular, late-night tradition in millions of American households. Away from the public spotlight, Johnny Carson is a devoted amateur astronomer. This interest in astronomy was shared with his TV audience through numerous light-hearted chats with guest Carl Sagan {see planet (2709)}. (M 27124)

(3253) Gradie

1982 HQ_1. Discovered 1982 April 28 by E. Bowell at Anderson Mesa.

Named in honor of Jonathan C. Gradie, planetary scientist at the Hawaiian Institute of Geophysics, University of Hawaii. Gradie's research has centered on the physics of minor planets, satellites and comets, particularly on their light-scattering properties. His wide-ranging contributions to minor-planet science include a detailed physical study of the Eos and Koronis {see planets (221) and (158)} families and development of a number of aspects of minor-planet taxonomy. (M 11749)

(3254) Bus

1982 UM. Discovered 1982 October 17 by E. Bowell at Anderson Mesa.

Named in honor of S. J. ("Bobby") Bus {1956- }, currently a research assistant at Lowell Observatory. Bus has been largely responsible for the successful planning and carrying out of the U.K. Schmidt Telescope/California Institute of Technology Asteroid Survey (UCAS). Observations at Siding Spring and subsequent analysis at Caltech, the Jet Propulsion Laboratory and Lowell Observatory have led to the establishment of more than a thousand orbits of faint minor planets, almost all of which are potentially recoverable. (M 9771)

(3255) Tholen

1980 RA. Discovered 1980 September 2 by E. Bowell at Anderson Mesa.

Named in honor of David J. Tholen {1955- }, planetary scientist at the Institute for Astronomy of the University of Hawaii. Tholen's work on the eight-color survey of minor planets led him to devise an improved taxonomy of minor planets. He has considered the physical properties of minor planets, satellites and comets in terms of composition and evolution, and he was among the first to observe events in the series of occultations and transits now occurring between Pluto and its satellite Charon. (M 11749)

(3256) Daguerre

1981 SJ_1. Discovered 1981 September 26 by B. A. Skiff and N. G. Thomas at Anderson Mesa.

Named for the Frenchman Louis Jacques Mande Daguerre (1787-1851), who invented the daguerrotype photographic process in 1835. (M 10549)
Daguerre is also honored by a lunar crater.

(3257) Hanzlík

1982 GG. Discovered 1982 April 15 by A. Mrkos at Kleť.

Named in memory of Stanislav Hanzlík (1878-1956), professor of meteorology and climatology at Charles University, Prague. He founded modern dynamical meteorology in Czechoslovakia. (M 23136)

(3258) Somnium

1983 RJ. Discovered 1983 September 8 by P. Wild at Zimmerwald.

The Latin word for dream or vision. Title of Johannes Kepler's {see planet (1134)} famous posthumously (1643) published science-fiction work expounding astronomy for an observer on the Moon. (M 22497)

(3259) Brownlee

1984 SZ_4. Discovered 1984 September 25 by J. Platt at Palomar.

Named in honor of Donald E. Brownlee, professor of astronomy at the University of Washington. Brownlee is renowned for his successful recovery, identification and chemical and minerological investigation of extraterrestrial particles from the atmosphere - the "Brownlee" particles. He showed that many of these particles very probably are cometary dust. Brownlee also extracted and studied extraterrestrial particles from deep sea sediments and polar ice. His results from the interplanetary dust captured by the earth provide especially important clues about the material from which comets accreted. (M 18450)
Citation provided by E. M. and C. S. Shoemaker.

(3260) Vizbor

1974 SO_2. Discovered 1974 September 20 by L. V. Zhuravleva at Nauchnyj.

Named in memory of Yurij Iosifovich Vizbor (1934-1984), well-known journalist, poet, actor, producer and bard. (M 13175)

(3261) Tvardovskij

1979 SF_9. Discovered 1979 September 22 by N. S. Chernykh at Nauchnyj.

Named in memory of Aleksandr Trifonovich Tvardovskij (1910-1971), famous Soviet poet. (M 12972)

(3262) Miune

1983 WB. Discovered 1983 November 28 by T. Seki at Geisei.

Named for a mountain in the eastern part of Kochi {see planet (2396)} prefecture. (M 10849)

(3263) Bligh

1932 CN. Discovered 1932 February 5 by K. Reinmuth at Heidelberg.

Named for Lt. William Bligh (1754-1817), renowned navigator, who had sailed with Captain Cook {see planet (3061)} on the latter's second voyage to the south Pacific. Despatched to Tahiti in H.M.S. Bounty {see planet (3264)}, Bligh was cast adrift in an open boat by mutineers off the Friendly Islands with 18 of his crew. Without charts, Bligh managed to navigate the tiny vessel over some 4000 miles of ocean to safety in Timor. Commended for his command of the Bounty's launch by the court-martial set up to investigate the loss of his ship, Bligh was cleared of all blame for the mutiny. However, he remains unique in British naval history as the only naval commander to suffer two mutinies - the second being in 1808, while he was Governor of New South Wales. (M 19692)

Name proposed by G. V. Williams, who wrote the citation, and endorsed by E. Bowell, who made the principal identification involving this planet.

(3264) Bounty

1934 AF. Discovered 1934 January 7 by K. Reinmuth at Heidelberg.

Named for His Majesty's Ship Bounty. In 1787 the Admiralty dispatched the ship, under the command of Lt. Bligh {see planet (3263)}, to Tahiti, to collect breadfruit plants for the slave plantations in the West Indies. The Bounty was taken by mutineers to Pitcairn Island, where the ship was destroyed, in a bay now known as Bounty Bay. The mutineers' refuge was accidentally discovered in 1808 by an American whaling vessel, which found only one survivor from the mutineers who went to Pitcairn. (M 19692)

Name proposed by G. V. Williams, who wrote the citation, and endorsed by E. Bowell, who made the principal identification involving this planet.

(3265) Fletcher

1953 VN_2. Discovered 1953 November 9 by K. Reinmuth at Heidelberg.

Named for Fletcher Christian (1764-?), leader of the mutiny aboard H.M.S. Bounty {see planet (3264)}. A long-standing friend of Bligh's {see planet (3263)}, Fletcher was invited by Bligh to accompany him on the voyage to Tahiti as Master's Mate. After the mutiny, the ship returned to Tahiti, where 16 of the mutineers decided to stay. Fletcher took the other mutineers to Pitcairn Island, where they hoped to avoid capture by the Royal Navy. Although the descendants of Fletcher Christian still live on Pitcairn, the fate of Christian himself remains unclear - one account has him being killed on Pitcairn, another has him returning secretly to England. (M 19692)

Name proposed by G. V. Williams, who wrote the citation, and endorsed by E. Bowell, who made the principal identification involving this planet.

(3266) Bernardus

1978 PA. Discovered 1978 August 11 by H.-E. Schuster at La Silla.

Named in honor of Andres Bernardus Muller, astronomer at the Leiden Observatory and leader of the site-testing campaigns for the European Southern Observatory in South Africa and Chile. From 1964 to 1969 he was superintendent at La Silla {see planet (2187)} while ESO was under construction. He was involved with instrumental improvements at ESO, especially for the 1-m Schmidt telescope, with which this minor planet was discovered. (M 27124)

(3267) Glo

1981 AA. Discovered 1981 January 3 by E. Bowell at Anderson Mesa.

Named in honor of Eleanor F. ("Glo") Helin, planetary scientist at the Jet Propulsion Laboratory, in appreciation of her extraordinary contributions to the discovery of near-Earth minor planets. Her finding of 1976 AA (=2062 Aten) heralded the recognition of a new class of planet-crossers, and her initiation of the Palomar planet-crossing asteroid survey has resulted in increased worldwide interest in the observation of minor planets. Helin's education and experience as a geologist and in the analysis of meteorites has provided a unique background for her interest in asteroids and comets. (M 11641)
Citation prepared by R. Helin.

(3268) DeSanctis

1981 DD. Discovered 1981 February 26 by H. Debehogne and G. DeSanctis at La Silla.

Named by the first discoverer in honor of the second, Giovanni DeSanctis, astronomer at the Osservatorio Astronomico di Torino who has also participated in minor-planet programs at the European Southern Observatory, Catania and elsewhere. (M 10849)

(3269) Vibert-Douglas

1981 EX_{16}. Discovered 1981 March 6 by S. J. Bus at Siding Spring.

Named in memory of Alice Vibert Douglas (1894-1988), the pioneer in the teaching of astronomy at McGill University and later dean of women at Queen's University {see planet (5457)}. Her research interests included spectroscopic absolute magnitudes of stars and the Stark effect in stellar atmospheres. An authority on historical astronomy, she was the biographer of Eddington {see planet (2761)}, as well as an early advocate of an increased role for women in science. She was made an Officer of the Order of Canada. (M 13481)

Name suggested and citation prepared by C. J. Cunningham, endorsed by P. M. Millman.

Obituaries published in Phys. Today, Vol. 42, No. 7, p. 88-89 (1989); J. R. Astron. Soc. Can., Vol. 82, No. 6, p. 309-311 (1988).

(3270) Dudley

1982 DA. Discovered 1982 February 18 by C. S. Shoemaker and S. J. Bus at Palomar.

Named in honor of H. Dudley Wright, engineer, inventor, entrepreneur and benefactor of science, education and the arts in California and in Geneva, Switzerland. (M 10312)
Name endorsed by E. M. Shoemaker.

(3271) Ul

1982 RB. Discovered 1982 September 14 by H.-E. Schuster at La Silla.

Named for the spirit of night-time and the moon in the mythology of the Melanesians of the New Hebrides Islands. Ul was the rival of Tortali {see planet (2687)}, spirit of daytime and the sun. The Melanesians recognize no gods. The name is shared by a god character in a best-selling fantasy series by David Eddings. (M 27329)
Name suggested by F. Pilcher.

(3272) Tillandz

1938 DB_1. Discovered 1938 February 24 by Y. Väisälä at Turku.

Named for Elias Tillandz (1640-1693), from 1670 professor of medicine at the old Academia Aboensis. He was a botanist and the only physician in Finland at the time, cultivating his own medicinal plants. (M 18451)

(3273) Drukar

1975 TS_2. Discovered 1975 October 3 by L. I. Chernykh at Nauchnyj.

Named for Ivan Fedorov (ca. 1510-1583), the first printer of books in Russia and the Ukraine. In old Russian (and in modern Ukrainian) the word means 'bookprinter'. (M 12015)

(3274) Maillen

1981 QO_2. Discovered 1981 August 23 by H. Debehogne at La Silla.

Named for the discoverer's birthplace, a village of 1,000 inhabitants in Wallony, 15 km south of Namur. (M 15573)

(3275) Oberndorfer

1982 HE_1. Discovered 1982 April 25 by E. Bowell at Anderson Mesa.

Named in honor of Hans Oberndorfer, director of the Bavarian Public Observatory, which he founded in Munich in 1947. He has contributed greatly to the development of German amateur astronomy and is well known for his publications on telescope construction and for his articles in the periodical *Sterne und Weltraum*. (M 12016)

Name proposed by the discoverer, following a suggestion of L. D. Schmadel, H. Schwaiger and J. Stromeyer, the last of whom prepared the citation.

(3276) Porta Coeli

1982 RZ_1. Discovered 1982 September 15 by A. Mrkos at Kleť.

Named for the twelfth-century convent and church in Tišnov, Moravia. (M 23136)

(3277) Aaronson

1984 AF_1. Discovered 1984 January 8 by E. Bowell at Anderson Mesa.

Named in memory of Marc Aaronson (1950-1987), associate professor of astronomy at the University of Arizona, killed in a tragic accident at the 4-m telescope at Kitt Peak. After completing his dissertation on the infrared properties of spiral galaxies, Aaronson demonstrated the determination of relative distances of galaxies independently of redshifts and detected the decelerative effect of the Virgo cluster on the Hubble flow. Observations of globular clusters in the Magellanic clouds showed carbon stars to be a characteristic of stellar populations of intermediate age, and his measurement of the large velocity dispersion of carbon stars in dwarf spheroidal galaxies suggested that even the smallest galaxies possess halos of cold dark matter. (M 12016)

Obituary published in Nature, Vol. 327, No. 6118, p. 92 (1987); Sterne Weltraum, 26. Jahrg., Nr. 9, p. 465 (1987); IAU Coll. No. 95, p. XVIII-IXX (1987); Phys. Today, Vol. 41, No. 7, p. 91-92 (1988).

(3278) Běhounek

1984 BT. Discovered 1984 January 27 by A. Mrkos at Kleť.

Named in memory of František Běhounek (1898-1973), the first professor of cosmic-ray physics at Charles University, Prague. He was also a member of Nobile's airship expedition to the north pole. (M 23136)

(3279) Solon

9103 P-L. Discovered 1960 October 17 by C. J. van Houten and I. van Houten-Groeneveld at Palomar.

Named for the Athenian legislator who lived around 600 B.C. (M 16591)

(3280) Grétry

1933 SJ. Discovered 1933 September 17 by F. Rigaux at Uccle.

Named in memory of André Grétry (1741-1813), composer from Liège, particularly known for his comic operas, although his compositions included symphonies and a requiem. (M 12210)

(3281) Maupertuis

1938 DZ. Discovered 1938 February 24 by Y. Väisälä at Turku.

Named for Pierre-Louis Moreau de Maupertuis (1698-1759), French mathematician and geodesist, from 1731 a member of the French Academy. During 1736-1737 he conducted the degree measurement in the Tornio {see planet (1471)} river valley in Finland. (M 18451)

Maupertuis is also honored by a lunar crater.

(3282) Spencer Jones

1949 DA. Discovered 1949 February 19 at the Goethe Link Observatory at Brooklyn, Indiana.

Named in memory of Harold Spencer Jones (1890-1960), successively astronomical assistant at the Royal Greenwich Observatory, H.M. astronomer at the Cape of Good Hope, and Astronomer Royal (1933-1955). He also served as president of the IAU (1945-1948). His work was devoted to fundamental positional astronomy, and he conclusively demonstrated that the small residuals in the apparent motions of the planets were due to the irregular rotation of the Earth. He led the worldwide effort to determine the length of the astronomical unit by triangulating the distance to (433) Eros when it passed near the Earth in 1930-31. (M 16041)

Name proposed by F. K. Edmondson. Citation prepared by J. S. Tenn.

Spencer Jones is also honored by a lunar crater.

(3283) Skorina

1979 QA_{10}. Discovered 1979 August 27 by N. S. Chernykh at Nauchnyj.

Named in honor of Francis Skorina, who lived from before 1490 to not later than 1551, and who pioneered printing in Belorus. (M 20835)

(3284) Niebuhr

1953 NB. Discovered 1953 July 13 by J. A. Bruwer at Johannesburg.

Named in memory of Carsten Niebuhr (1733-1815), astronomer-surveyor, the sole survivor of the ill-fated Danish expedition to Arabia Felix (Yemen) in 1760–67. During an adventurous return voyage via India, and at great personal sacrifice, he carefully copied the inscriptions at the ruins of Persepolis. This served as the basis for the subsequent decoding of the cuneiform script and thus our access to source documents of early astronomical observations in that geographical area. (M 27329)

Name suggested by R. M. West.

(3285) Ruth Wolfe

1983 VW_1. Discovered 1983 November 5 by C. S. Shoemaker and E. M. Shoemaker at Palomar.

Named in honor of Ruth Fanton Wolfe {1937- }, mathematician with the U.S. Geological Survey, recognized for her dynamical investigations of the orbital evolution and collision of small bodies in the solar system. (M 10312)

Name endorsed by B. G. Marsden.

(3286) Anatoliya

1980 BV. Discovered 1980 January 23 by L. G. Karachkina at Nauchnyj.

Named in memory of Anatolij Vasilevich Karachkin (1947-1984), brother of the discoverer's husband, a skilled specialist in the construction of industrial establishments. (M 12016)

(3287) Olmstead

1981 DK_1. Discovered 1981 February 28 by S. J. Bus at Siding Spring.

Named in honor of C. Michelle Olmstead for her enthusiastic assistance in the asteroid astrometry programs at Lowell Observatory and the U.S. Geological Survey. As a student at Northern Arizona University, Olmstead is a NASA undergraduate fellow, and she has participated in the observational, discovery, identification and astrometric stages of the asteroid survey work. (M 18451)

Name and citation endorsed by E. Bowell, E. M. and C. S. Shoemaker and H. E. Holt.

(3288) Seleucus

1982 DV. Discovered 1982 February 28 by H.-E. Schuster at La Silla.

Named for one of the generals of Alexander the Great and heir to the largest part of his empire. Seleucus maintained the best of relations with the great Babylonian teachers and priests, and the Chaldean astronomers are said to have predicted his becoming king. (M 10046)

Seleucus is also a name of a lunar crater.

(3289) Mitani

1934 RP. Discovered 1934 September 7 by K. Reinmuth at Heidelberg.

Named in honor of Tesuyasu Mitani, discoverer of (1619) Ueta and independent recoverer of periodic comet Honda-Mrkos-Pajdušaková in 1954. He worked at Ikoma-san Solar Observatory of Kyoto Imperial University during 1944-1945 and under the difficult conditions following World War II he enthusiastically made astrometric observations of Pluto, comets and minor planets at Kyoto University's Kwasan Observatory. From 1957 to 1960 he was engaged in occultation observations for the geodetic survey (U.S. Army Map Service, Far East). (M 15426)

Name proposed by H. Oishi, who found the identifications involving this minor planet, endorsed by I. Hasegawa and K. Hurukawa.

(3290) Azabu

1973 SZ_1. Discovered 1973 September 19 by C. J. van Houten and I. van Houten-Groeneveld at Palomar.
Named for the former site of the Tokyo Astronomical Observatory. (M 11442)
See also the comment for minor planet (3291).

(3291) Dunlap

1982 VX_3. Discovered 1982 November 14 by H. Kosai and K. Hurukawa at Kiso.

Named in honor of Larry Dunlap, research assistant at the Lunar and Planetary Laboratory who has published lightcurves of asteroids and is now teaching high-school and other students the beauty of astronomy at the Flandrau planetarium. (M 11161)

Named in accordance with a proposal by T. Gehrels that naming rights for (3291) and the Palomar discovery (3290) be interchanged, so that this and the following three names can be in sequence.

(3292) Sather

2631 P-L. Discovered 1960 September 24 by C. J. van Houten and I. van Houten-Groeneveld at Palomar.

Named in honor of Lunar and Planetary Laboratory research assistant and high-school teacher Bob Sather, who has published lightcurves of asteroids and discovered a new method of determining the orientation of asteroidal rotational axes in space. (M 11161)

(3293) Rontaylor

4650 P-L. Discovered 1960 September 24 by C. J. van Houten and I. van Houten-Groeneveld at Palomar.

Named in honor of Ron Taylor, internationally known expert in the determination of asteroid shapes, sense of rotation, and orientation of rotational axes in space. A research specialist at the Lunar and Planetary Laboratory, he is also a dedicated high-school teacher. (M 11161)

(3294) Carlvesely

6563 P-L. Discovered 1960 September 24 by C. J. van Houten and I. van Houten-Groeneveld at Palomar.

Named in honor of Carl Vesely, high-school teacher and research assistant at the Lunar and Planetary Laboratory who published lightcurves of asteroids. He also assisted G. Van Biesbroeck {see planet (1781)} in his comet observations and participated extensively in the completion of Van Biesbroeck's work for publication after the latter's death in 1974. (M 11161)

(3295) Murakami

1950 DH. Discovered 1950 February 17 by K. Reinmuth at Heidelberg.

Named in memory of Tadayoshi Murakami (1907-1985), professor of astronomy at the Hiroshima University, president of the Hiroshima Jogakuin College. He studied meteors extensively and encouraged many meteor observers in Japan. He contributed much to the teaching and popularization of astronomy in Japan, not only by lecturing at universities, but also by writing many books and articles over a period of fifty years. His father, Harutaro Murakami, studied lunar theory and is known for his work *Theory of the perturbation of the Moon*. (M 11442)
Name proposed by T. Urata, who found the identifications involving this planet.

(3296) Bosque Alegre

1975 SF. Discovered 1975 September 30 at the Felix Aguilar Observatory at El Leoncito.

Named in honor of the Astrophysical Station of the Córdoba Astronomical Observatory, founded in 1942 some 40 km southwest of Córdoba. (M 22828)

(3297) Hong Kong

1978 WN_{14}. Discovered 1978 November 26 at the Purple Mountain Observatory at Nanking.

Named for the island and city in southeast Asia, a key center of international finance and trade. (M 32787)

(3298) Massandra

1979 OB_{15}. Discovered 1979 July 21 by N. S. Chernykh at Nauchnyj.

Named for a suburb of Yalta {see planet (1475)}, a beautiful site noted for its production of the finest Crimean wines. (M 21608)

(3299) Hall

1980 TX_5. Discovered 1980 October 10 by C. S. Shoemaker at Palomar.

Named in honor of John Scoville Hall {1908-1991}, American astronomer and director of the Lowell Observatory from 1958 to 1977. He was a pioneer in the photoelectric photometry of stars in the infrared region of the spectrum and codiscoverer with W. A. Hiltner of the polarization of starlight. (M 10312)

Obituary published in Bull. Am. Astron. Soc., Vol. 24, No. 4, p. 1323-1325 (1992).

(3300) McGlasson

1928 NA. Discovered 1928 July 10 by H. E. Wood at Johannesburg.

Named in honor of Van McGlasson, head of the Computation Facility at the Harvard-Smithsonian Center for Astrophysics. (M 22497)

Name proposed by B. G. Marsden, who made the identifications for this object.

(3301) Jansje

1978 CT. Discovered 1978 February 6 at the Perth Observatory at Bickley.

Named by Arie Verveer {see planet (3974)} in honor of his mother. (M 18644)

(3302) Schliemann

1977 RS_6. Discovered 1977 September 11 by N. S. Chernykh at Nauchnyj.

Named in memory of Heinrich Schliemann (1822-1890), famous German archaeologist, who undertook, at his own expense, a 12-year excavation at Hissarlik and discovered the ancient site of Troy. (M 18136)

Name proposed following a suggestion by C. E. Spratt.

Schliemann is also honored by a lunar crater.

(3303) Merta

1967 UN. Discovered 1967 October 30 by L. Kohoutek at Bergedorf.

Named by the discoverer in memory of his grandfather, František Merta (1872-1953), elementary-school teacher and pedagogic journalist, a writer and translator who propagated humanistic ideas. (M 29142)

(3304) Pearce

1981 EQ_{21}. Discovered 1981 March 2 by S. J. Bus at Siding Spring.

Named in honor of Joseph A. Pearce {1893-1988}, who joined the staff of the Dominion Astrophysical Observatory in 1924 and served as its director from 1940 to 1951. Together with J. S. Plaskett {see planet (2905)}, he conducted studies of the motions of distant early-type stars that established the reality of galactic rotation. An active promoter of both science and culture, he served as president of the Royal Society of Canada in 1949-50. (M 13481)

Name suggested and citation prepared by J. C. Cunningham, endorsed by P. M. Millman.
Obituary published in J.R. Astron. Soc. Canada, Vol. 83, No. 1, p. 3-7 (1989).

(3305) Ceadams

1985 KB. Discovered 1985 May 21 by A. C. Gilmore and P. M. Kilmartin at Lake Tekapo.

Named in memory of Charles Edward Adams (1870-1945), New Zealand Government Astronomer and Seismologist 1912-36. Among his many innovations was the introduction of the P and Q coefficients for ephemeris computation. He began the transmission of radio time signals in New Zealand, pioneered the use of the cine-camera for astronomical timing and devised one of the first moon cameras for direct measurement of the moon's position against background stars. He was also noted for his contributions to seismology. (M 18305)

(3306) Byron

1979 SM_{11}. Discovered 1979 September 24 by N. S. Chernykh at Nauchnyj.

Named for the great English poet Lord George Noel Gordon Byron (1788-1824). (M 12972)

(3307) Athabasca

1981 DE_1. Discovered 1981 February 28 by S. J. Bus at Siding Spring.

Named for the native North Americans who originally settled in the subarctic region that is now Northwestern Canada and central Alaska. The Athabascan hunter-gatherers comprise a diverse group whose influence spread, about a thousand years ago, as far as the southwestern United States. They are the ancestors of the Navajo {see planet (3688)} and Apache peoples. (M 18644)
Name suggested and citation prepared by J. J. Klavetter.

(3308) Ferreri

1981 EP. Discovered 1981 March 1 by H. Debehogne and G. DeSanctis at La Silla.

Named in honor of Walter Ferreri, who works in the photographic minor-planet program at the Osservatorio Astronomico di Torino and has discovered several minor planets at the European Southern Observatory. (M 11750)

(3309) Brorfelde

1982 BH. Discovered 1982 January 28 by K. S. Jensen and K. Augustesen at Brorfelde.

On the observatory's fortieth anniversary, this first minor planet discovered at the Copenhagen University Observatory in Brorfelde is named for the village in the middle of Zealand where the observatory is situated. (M 12210)

(3310) Patsy

1931 TS_2. Discovered 1931 October 9 by C. W. Tombaugh at Flagstaff.
Named by the discoverer in honor of his wife. (M 10849)

(3311) Podobed

1976 QM_1. Discovered 1976 August 26 by N. S. Chernykh at Nauchnyj.

Named in honor of Vladimir Vladimirovich Podobed {1918-1992}, professor at the Sternberg Astronomical Institute, prominent specialist on meridian and photographic astrometry and author of two textbooks on astrometry. (M 18136)
Obituary published in Zemlya Vselennaya, No. 3, p. 54 (1992).

(3312) Pedersen

1984 SN. Discovered 1984 September 24 by K. Augustesen and P. Jensen and H. J. Fogh Olsen at Brorfelde.

Named in honor of Bodil and Helge Pedersen, who have made a great contribution to the popularization of astronomy in Denmark by donating a planetarium (to be erected in Copenhagen) to the Danish people. (M 10849)

(3313) Mendel

1980 DG. Discovered 1980 February 19 by A. Mrkos at Kleť.

Named in memory of Gregor Johann Mendel (1822-1884), Austrian monk and botanist, discoverer of the law of genetics. (M 23136)
Mendel is also honored by a craters on Mars and on the Moon.

(3314) Beals

1981 FH. Discovered 1981 March 30 by E. Bowell at Anderson Mesa.

Named in memory of Canadian astronomer Carlyle Smith Beals (1899-1979), fourth Dominion Astronomer and the only man who has been both President of the American Astronomical Society and the National President of the Royal Astronomical Society of Canada. Beals made important contributions to the observation and interpretation of emission lines in the spectra of hot stars, to the understanding of the nature of interstellar gas clouds, and to the development of instrumentation for astronomy. He also initiated a program to identify and study meteorite craters in Canada. (M 12210)
Name proposed by the discoverer following a suggestion by P. M. Millman.

Obituaries published in J.R. Astron. Soc. Canada, Vol. 73, p. 325-332 (1979); Q.J.R. Astron. Soc., Vol. 21, p. 212-213 (1980).

(3315) Chant

1984 CZ. Discovered 1984 February 8 by E. Bowell at Anderson Mesa.

Named in memory of Clarence Augustus Chant (1865-1956), generally referred to as the "father of Canadian astronomy". A renowned teacher, Chant organized the Astronomy Department of the University of Toronto and built up the Royal Astronomical Society of Canada. He participated in five solar eclipse expeditions, the most important being the one he led to Australia in 1922 to test Einstein's {see planet (2001)} prediction of the deflection of starlight by a massive body. (M 12210)
Name proposed by the discoverer following a suggestion by P. M. Millman.
Chant is also honored by a lunar crater.

(3316) Herzberg

1984 CN_1. Discovered 1984 February 6 by E. Bowell at Anderson Mesa.

Named in honor of Gerhard Herzberg {1904- }, Canadian Nobel Laureate in Chemistry in 1971 and acknowledged world leader in the study of molecular spectra. Herzberg's specialty has been free radicals, both in the laboratory and in interstellar space. He has identified numerous features in the spectra of comets, planets and interstellar material. In 1975 the National Research Council of Canada's astronomy and spectroscopy units were reorganized as the Herzberg Institute of Astrophysics. (M 12210)
Name proposed by the discoverer following a suggestion by P. M. Millman.

(3317) Paris

1984 KF. Discovered 1984 May 26 by C. S. Shoemaker and E. M. Shoemaker at Palomar.

Named for one of the numerous sons of Priam {see planet (884)}, the king of Troy. Paris, considered by the gods to be the most handsome man on Earth, abducted Helen {see planet (101)}, the most beautiful woman and the wife of Menelaus {see planet (1647)}, thereby precipitating the Trojan War. (M 10312)

(3318) Blixen

1985 HB. Discovered 1985 April 23 by P. Jensen at Brorfelde.

Named in memory of the celebrated Danish writer Karen Blixen (1885-1962) on the hundredth anniversary of her birth. Among her best known writings are *Seven Gothic Tales* (1934) and the memory-novel *Out of Africa* (1937). An American screen version of the latter was produced in 1985. (M 11161)

(3319) Kibi

1977 EJ_5. Discovered 1977 March 12 by H. Kosai and K. Hurukawa at Kiso.

This planet was given the former name of the area in Okayama {see planet (2084)} where co-discoverer Kosai grew up. (M 11442)

(3320) Namba

1982 VZ_4. Discovered 1982 November 14 by H. Kosai and K. Hurukawa at Kiso.

This planet was given the former name of Osaka, where co-discoverer Hurukawa grew up. (M 11442)

For another former name of the city of Osaka see the citation for minor planet (3607).

(3321) Dasha

1975 TZ_2. Discovered 1975 October 3 by L. I. Chernykh at Nauchnyj.

Named in honor of Dar'ya Lavrent'evna Mikhailova, the first Russian army sister of charity during the Crimean War, known as Dasha Sevastopol'skaya. (M 13175)

(3322) Lidiya

1975 XY_1. Discovered 1975 December 1 by T. M. Smirnova at Nauchnyj.

Named in honor of Lidiya Vissarionovna Zvereva (1890-1916), the first Russian female pilot. She began flying in 1911, instructed other pilots and was also involved in constructing airplanes. (M 17027)

(3323) Turgenev

1979 SY_9. Discovered 1979 September 22 by N. S. Chernykh at Nauchnyj.

Named in memory of the famous Russian writer Ivan Sergeevich Turgenev (1818-1883). (M 18136)

(3324) Avsyuk

1983 CW_1. Discovered 1983 February 4 by A. Mrkos at Kleť.

Named in honor of Yurij N. Avsyuk, geodesist and gravimetrist, who worked with the discoverer on the Antarctic plateau during the International Geophysical Year {1957-1958}. (M 23136; M 33817)

(3325) TARDIS

1984 JZ. Discovered 1984 May 3 by B. A. Skiff at Anderson Mesa.

The acronym TARDIS stands for Time And Relative Dimensions In Space and derives from the long-running science-fiction television program 'Dr. Who'. It is the name of the vehicle The Doctor uses for travel through space-time. Whimsically, it has the outward appearance of a British police telephone box but is quite spacious inside. (M 16041)

Name suggested and citation provided by C. J. Cunningham.

(3326) Agafonikov

1985 FL. Discovered 1985 March 20 by A. Mrkos at Kleť.

Named in honor of Askol'd M. Agafonikov, geophysicist and navigator of the third Russian Antarctic expedition, in recognition of his work with the discoverer in the region of the south magnetic pole. (M 23136)

(3327) Campins

1985 PW. Discovered 1985 August 14 by E. Bowell at Anderson Mesa.

Named in honor of Humberto Campins, research scientist at the Planetary Science Institute in Tucson. Well known for his work on the properties of cometary comae, Campins has helped establish pioneering techniques to measure the physical properties of cometary nuclei using simultaneous infrared and visual observations. He has also undertaken infrared searches for intramercurial bodies. (M 12458)

Citation written by R. P. Binzel at the request of the discoverer.

(3328) Interposita

1985 QD_1. Discovered 1985 August 21 by T. Schildknecht at Zimmerwald.

The discovery film was exposed hastily between two satellite laser ranging sessions in the adjacent dome. (M 27125)

(3329) Golay

1985 RT_1. Discovered 1985 September 12 by P. Wild at Zimmerwald.

Named in honor of Marcel Golay (1927-), who from 1956 to 1992 was director of the Observatoire de Genève. He developed a seven-color system of stellar photometry that allows precise and relatively rapid determination of temperature, gravity, rotation, and chemistry. Under his inspiring leadership the observatory staff made these measurements on myriads of stars and cultivated practically all the branches of astrophysics profiting from them. It is due mainly to Golay that Switzerland is adhering firmly to ESO and ESA, giving access to the most modern observing methods and instruments. Golay has been president of IAU commissions 25, 37 and 45. (M 22497)

(3330) Gantrisch

1985 RU_1. Discovered 1985 September 12 by T. Schildknecht at Zimmerwald.

Named for a prominent mountain about 20 km due south of the Zimmerwald {see planet (1775)} Observatory. (M 27125)

(3331) Kvistaberg

1979 QS. Discovered 1979 August 22 by C.-I. Lagerkvist at La Silla.

Named for the location of the Uppsala Observatory's observing station. Several minor planets have been discovered with the Schmidt telescope there. (M 12972)

(3332) Raksha

1978 NT_1. Discovered 1978 July 4 by L. I. Chernykh at Nauchnyj.

Named in memory of Yurij Mikhajlovich Raksha (1937-1980), gifted artist well known for his paintings and for his work in film-making. (M 13175)

(3333) Schaber

1980 TG_5. Discovered 1980 October 9 by C. S. Shoemaker at Palomar.

Named in honor of Gerald G. Schaber, geologist with the U.S. Geological Survey and chief of the branch of astrogeology since 1983. Schaber is specially recognized for his research on the geology of the Moon, Mercury, Venus, Mars and the satellite Io and for the application of radar to the investigation of the terrestrial planets. He was a codiscoverer, with J. F. McCauley and C. S. Breed, of the ancient buried river channels of the western desert of Egypt. (M 10549)

(3334) Somov

1981 YR. Discovered 1981 December 20 by A. Mrkos at Kleť.

Named in memory of Mikhail M. Somov, leader of the first Russian Antarctic expedition during the International Geophysical Year {1957-1958}. (M 23136)

(3335) Quanzhou

1966 AA. Discovered 1966 January 1 at the Purple Mountain Observatory at Nanking.

Named for a well-known historical and cultural city. In ancient times, Quanzhou was a starting point of "the Silk Road at Sea", and its port of Citong was famed as "First Port in the East" from 960 to 1368. As a result of its economic development in recent times Quanzhou has become a star city on the southeastern coast of China. (M 26424)

(3336) Grygar

1971 UX. Discovered 1971 October 26 by L. Kohoutek at Bergedorf.

Named in honor of Jiří Grygar on the occasion of his 60th birthday. A Czech astronomer working in Prague, he has contributed to our knowledge of meteors, comets and variable stars. He is also known in both the Czech and the Slovak republics as a very successful scientific popularizer and as the author of several astronomical books for the public. (M 26424)

(3337) 1971 UG_1

Discovered 1971 October 26 by L. Kohoutek at Bergedorf.

(3338) Richter

1973 UX_5. Discovered 1973 October 28 by F. Börngen at Tautenburg.

Named in memory of Nikolaus B. Richter (1910-1980), first director of the Tautenburg Observatory (1960-1975), renowned for his experiments with meteorites and his comparative study of terrestrial particles and interplanetary matter. Author of the monograph *Statistik und Physik der Kometen* (1954; English edition published in 1963 under the title *The nature of comets*), he served as president of IAU Commission 15 during 1973-1976. He was also interested in the study of blue objects and compact galaxies. (M 10549)

Obituaries published in Astron. Nachr., Band 302, p. 203 (1981); Mitt. Astron. Ges., Nr. 53, p. III-V (1981); Sterne, 57. Band, p. 182-183 (1981).

(3339) Treshnikov

1978 LB. Discovered 1978 June 6 by A. Mrkos at Kleť.

Named in honor of Aleksej F. Treshnikov, leader of the second Russian Antarctic expedition, in recognition of his first meeting with the discoverer on the Antarctic plateau in 1957. (M 23136)

(3340) Yinhai

1979 TK. Discovered 1979 October 12 at the Purple Mountain Observatory at Nanking.

Named for a beautiful city located on Hailing Island in the South China Sea, famous for its scenery. It is a tourist attraction with a pretty beach and known as the 'eastern Hawaii'. (M 25229)

(3341) Hartmann

1980 OD. Discovered 1980 July 17 by E. Bowell at Anderson Mesa.

Named in honor of William K. Hartmann, senior scientist at the Planetary Science Institute in Tucson. Hartmann's contributions to solar system research have ranged from work on planetary cratering rates and the origin of the Moon to studies of comets and Trojan minor planets. He is the author of several textbooks on astronomy and planetary science, as well as popular books on space exploration. Hartmann is also a renowned space artist whose paintings depict scenes predicted by modern research. (M 12458)

Citation written by R. P. Binzel at the request of the discoverer.

(3342) Fivesparks

1982 BD_3. Discovered 1982 January 27 at the Oak Ridge Observatory at Harvard.

Named in honor of Newton and Margaret Mayall, who have enriched the literature for amateur astronomers with their delightful and informative books entitled *Sundials* and *Skyshooting*; who have preserved the papers and memorabilia of Annie Jump Cannon; and who carried the AAVSO through the difficult period of moving its headquarters from the Harvard College Observatory. (M 15573)

The name, which refers to the Mayall's residence in Cambridge, Massachusetts, was suggested by B. L. Welther, who also wrote the citation.

(3343) Nedzel

1982 HS. Discovered 1982 April 28 by L. G. Taff at Socorro.

Named in memory of V. Alexander Nedzel, scholar, gentleman, patron of the sciences and friend. The foremost supporter of the Lincoln Laboratory Earth-Approaching Asteroid Search, he was head of the Laboratory's Aerospace Division, and the Search was carried out under his auspices until his untimely death on 1984 Sept. 6. (M 10849)

(3344) Modena

1982 JA. Discovered 1982 May 15 at the Osservatorio San Vittore at Bologna.

Named for the city near Bologna, famous for its beautiful bell-tower Ghirlandia, the Romanesque cathedral Duomo, and the military academy that was formerly the royal palace of the Estensi, the dynasty that governed the city until 1859. Modena was the home of the astronomers Montanari (1633-1687), Amici (1786-1863) {see planet (3809)} and Bianchi (1791-1866). It is also the birthplace and residence of Ermes Colombini, an amateur astronomer of the group at the Osservatorio San Vittore {see planet (2235)} in Bologna. (M 10549)

(3345) Tarkovskij

1982 YC_1. Discovered 1982 December 23 by L. G. Karachkina at Nauchnyj.

Named in memory of Andrej Arsen'evich Tarkovskij (1932-1986), Soviet theater and film producer. (M 13176)

(3346) Gerla

1951 SD. Discovered 1951 September 27 by S. Arend at Uccle.

Named in honor of Gertrude Lawrence (1898-1952), real name Gertrud Alexandra Dagma Lawrence Klasen, English actress, best remembered for her performances in musicals and Noël Coward plays. (M 22497)

Citation written by G. V. Williams following a suggestion by J. Bardwell, the identifications for this object having been made by C. M. Bardwell.

(3347) Konstantin

1975 VN_1. Discovered 1975 November 2 by T. M. Smirnova at Nauchnyj.

Named in memory of Konstantin Alekseevich Kalinin (1889-1938), a remarkable flier and talented aircraft designer who constructed many aircraft of various types, one of which was awarded a gold medal at the International Aviation Exhibition in Berlin in 1928. (M 17027)

(3348) Pokryshkin

1978 EA_3. Discovered 1978 March 6 by N. S. Chernykh at Nauchnyj.

Named in memory of Aleksandr Ivanovich Pokryshkin (1913-1985), a Soviet pilot known as a writer, the author of four popular books on Soviet aviation. His contribution to the theory and practice of flying and maneuvering small, high-speed aircraft allowed Soviet pilots to win some world aviation records. He was also involved in the selection of the first cosmonauts and the realization of the first Soviet space flights. (M 12972)

(3349) Manas
1979 FH_2. Discovered 1979 March 23 by N. S. Chernykh at Nauchnyj.
Named for an epic poem of the Kirghiz people. (M 18136)

(3350) Scobee
1980 PJ. Discovered 1980 August 8 by E. Bowell at Anderson Mesa.
Named in memory of Francis R. Scobee (1939-1986), commander. (M 10549)

This planet and the following six are named in memory of the astronauts who perished in the flight of the space shuttle Challenger on 1986 Jan. 28.

(3351) Smith
1980 RN_1. Discovered 1980 September 7 by E. Bowell at Anderson Mesa.
Named in memory of Michael J. Smith (1945-1986), pilot. (M 10550)
See the comment to planet (3350).

(3352) McAuliffe
1981 CW. Discovered 1981 February 6 by N. G. Thomas at Anderson Mesa.

Named in memory of S. Christa C. McAuliffe (1948-1986), teacher, observer. (M 10550)
See the comment to planet (3350).

(3353) Jarvis
1981 YC. Discovered 1981 December 20 by E. Bowell at Anderson Mesa.

Named in memory of Gregory B. Jarvis (1944-1986), payload specialist. (M 10550)
See the comment to planet (3350).

(3354) McNair
1984 CW. Discovered 1984 February 8 by E. Bowell at Anderson Mesa.

Named in memory of Ronald E. McNair (1950-1986), mission specialist. (M 10550)
See the comment to planet (3350).

(3355) Onizuka
1984 CC_1. Discovered 1984 February 8 by E. Bowell at Anderson Mesa.

Named in memory of Ellison S. Onizuka (1946-1986), mission specialist. (M 10550)
See the comment to planet (3350).

(3356) Resnik
1984 EU. Discovered 1984 March 6 by E. Bowell at Anderson Mesa.

Named in memory of Judith A. Resnik (1949-1986), mission specialist. (M 10550)
See the comment to planet (3350).

(3357) Tolstikov
1984 FT. Discovered 1984 March 21 by A. Mrkos at Kleť.

Named in memory of Evgenij I. Tolstikov, Russian meteorologist and leader of the third Antarctic expedition during International Geophysical Year {1957-1958}. Tolstikov and the discoverer worked for three years at the Antarctic station Mirny. (M 23136)

(3358) Anikushin
1978 RX. Discovered 1978 September 1 by N. S. Chernykh at Nauchnyj.

Named in honor of Mikhail Konstantinovich Anikushin, member of the U.S.S.R.

Academy of Arts, well known Soviet sculptor, creator of the Pushkin {see planet (2208)} momument in Leningrad and other works. (M 18136)

(3359) Purcari

1978 RA_6. Discovered 1978 September 13 by N. S. Chernykh at Nauchnyj.
Named for the producer of the best Moldavian wines. (M 18136)

(3360) 1981 VA

Discovered 1981 November 4 by E. F. Helin and R. S. Dunbar at Palomar.

(3361) Orpheus

1982 HR. Discovered 1982 April 24 by C. Torres at Cerro El Roble.

Named for the poet and musician of Greek mythology who almost rescued his wife Eurydice from Hades by charming Pluto and Persephone with his lyre. (M 11442)
Name proposed by the discoverer following a suggestion by S. J. Ostro.
See also the citations for minor planets (75) Eurydike and (399) Persephone.

(3362) Khufu

1984 QA. Discovered 1984 August 30 by R. S. Dunbar and M. A. Barucci at Palomar.

Named for the Egyptian god-king Khufu, better known by his Greek name of Cheops, a pharaoh of the 29th century B.C. and builder of the largest of the great pyramids at Giza {see planet (5249)}, one of the seven wonders of the ancient world. (M 10550)

(3363) Bowen

1960 EE. Discovered 1960 March 6 at the Goethe Link Observatory at Brooklyn, Indiana.

Named in memory of Ira S. Bowen (1898-1973), whose investigation of the ultraviolet spectra of highly ionized atoms led to his identification of the mysterious 'nebulium' spectral lines of gaseous nebulae as forbidden lines of ionized oxygen and nitrogen; he soon explained most of the spectral lines of gaseous nebulae. As director of the Mount Wilson and Palomar Observatories from 1948 to 1964 he oversaw the completion of the 5-m reflector and 1.2-m Schmidt and designed many of their instruments, including a novel spectrograph. He also initiated the baking of photographic plates to improve their sensitivity. (M 16245)
Name proposed by F. K. Edmondson. Citation prepared by J. S. Tenn.

Obituaries published in Publ. Astron. Soc. Pac., Vol. 85, p. 174 (1973); Science, Vol. 180, p. 1158 (1973); Sky Telesc., Vol. 45, p. 212-214 (1973); Mercury, Vol. 2, No. 3, p. 3-5 (1973); Q.J.R. Astron. Soc., Vol. 15, p. 193-196 (1974).

(3364) Zdenka

1984 GF. Discovered 1984 April 5 by A. Mrkos at Kleť.

Named in honor of Zdenka Vávrová for her 20 devoted years in the astrometric program at Kleť. (M 23136)

(3365) Recogne

1985 CG_2. Discovered 1985 February 13 by H. Debehogne at La Silla.

Named for a high point in the Ardennes, in the Province de Luxembourg. (M 15573)

(3366) Gödel

1985 SD_1. Discovered 1985 September 22 by T. Schildknecht at Zimmerwald.

Named for Kurt Gödel (1906-1978), American logician of Austrian origin. He is the author of the famous article *Über formal unentscheidbare Sätze der Principia*

Mathematica und verwandter Systeme (1931). His ideas accompany the discoverer through many observing nights. (M 27125)

(3367) Alex

1983 CA_3. Discovered 1983 February 15 by N. G. Thomas at Anderson Mesa. Named by the discoverer in honor of his grandson, Alex R. Baltutis. (M 10550)

(3368) Duncombe

1985 QT. Discovered 1985 August 22 by E. Bowell at Anderson Mesa.

Named in honor of Raynor L. Duncombe {1917- }, astronomer at the University of Texas, on the occasion of his seventieth birthday. A leader in positional and dynamical astronomy, he has pursued wide-ranging research that has included the study of the motions of major planets, minor planets and satellites; he was a pioneer in artificial-satellite orbit determination and in the use of digital computers in astronomy; and he has recently been involved with the astrometric uses of the Hubble Space Telescope as a member of the Astrometry Team. He served as director of the U.S. Nautical Almanac Office from 1963 to 1975, was the first chairman of the American Astronomical Society's Division on Dynamical Astronomy and for a number of years the executive editor of *Celestial Mechanics*. (M 11750)

Name proposed by the discoverer following a suggestion by P. K. Seidelmann, who provided the citation.

(3369) Freuchen

1985 UZ. Discovered 1985 October 18 by K. Augustesen and P. Jensen at Brorfelde.

Named in memory of the well-known Danish polar explorer and writer Peter Freuchen (1886-1957) on the hundredth anniversary of his birth. He participated in several arctic expeditions to Greenland, and in 1910 he was co-founder of the Thule trading station, which he led until 1919. (M 11161)

Name proposed by the second discoverer.

(3370) Kohsai

1934 CU. Discovered 1934 February 4 by K. Reinmuth at Heidelberg.

Named in honor of Hiroki Kosai {1933- }, astronomer at the Tokyo Astronomical Observatory, outstanding observer of comets and minor planets, codiscoverer of comet 1976 XVI and popularizer of astronomy in Japan. (M 11750)

Name proposed by H. Oishi and K. Hurukawa, who found the identifications for this planet.

(3371) Giacconi

1955 RZ. Discovered 1955 September 14 at the Goethe Link Observatory at Brooklyn, Indiana.

Named in honor of Riccardo Giacconi {1931- }, since 1981 the first director of the Space Telescope Science Institute. After serving as a research associate in the cosmic-ray physics group at Indiana University, and then briefly at Princeton, in 1959 he joined American Science and Engineering, where he began work on X-ray astronomy. His team there developed grazing-incidence X-ray telescopes, and in 1962 they discovered Sco X-1, the first known X-ray source outside the solar system. This was followed by the orbiting X-ray observatory UHURU, which made the first surveys of the X-ray sky. Joining the Harvard-Smithsonian Center for Astrophysics in 1973 as associate director for high-energy physics, Giacconi led the construction and successful operation of the powerful X-ray observatory HEAO-2, also known as Einstein, which made detailed images of X-ray sources. (M 16884)

Name proposed by F. K. Edmondson. Citation prepared with the help of J. Tenn.

(3372) Bratijchuk

1976 SP_4. Discovered 1976 September 24 by N. S. Chernykh at Nauchnyj.

Named in honor of Matrena Vasil'evna Bratijchuk, professor of astronomy and founder and head of the Laboratory for Space Research at Uzhgorod University. (M 18137)

(3373) Koktebelia

1978 QQ_2. Discovered 1978 August 31 by N. S. Chernykh at Nauchnyj.

Named for the Crimean settlement of Koktebel', home and workplace of the brilliant Russian poet, painter and water-colorist Maksimilian Aleksandrovich Kirienko-Voloshin (1877-1932), almost all of whose work was devoted to the Crimea. (M 18137)

(3374) Namur

1980 KO. Discovered 1980 May 22 by H. Debehogne at La Silla.

Named for the principal town in the Belgian province of the same name, a place where the discoverer studied. Namur is at the confluence of the Meuse and the Sambre {see planets (3016) and (4016)} rivers, and it was there that Julius Caesar battled Belgian tribes in 59 and 57 B.C. (M 15573)

(3375) Amy

1981 JY_1. Discovered 1981 May 5 by C. S. Shoemaker at Palomar.

Named in honor of Amy Shoemaker Prescott, aunt of E. M. Shoemaker. (M 11750)

(3376) Armandhammer

1982 UJ_8. Discovered 1982 October 21 by L. V. Zhuravleva at Nauchnyj.

Named in honor of the American businessman Armand Hammer, a champion of good trade relations between the U.S.A. and the U.S.S.R. (M 13176)

(3377) Lodewijk

4122 P-L. Discovered 1960 September 24 by C. J. van Houten and I. van Houten-Groeneveld at Palomar.

Named in honor of Lodewijk Woltjer, former editor of the Astronomical Journal and former director of the European Southern Observatory, well known for his studies on the Crab nebula. (M 16591)

Name proposed by J. H. Oort.

(3378) Susanvictoria

A922 WB. Discovered 1922 November 25 by G. Van Biesbroeck at Williams Bay.

Named in honor of the discoverer's granddaughters, Susan Titus and Victoria Van Biesbroeck Streeter. (M 14481)

Name proposed by Micheline Van Biesbroeck Wilson.

(3379) Oishi

1931 TJ_1. Discovered 1931 October 6 by K. Reinmuth at Heidelberg.

Named in honor of Hideo Oishi, orbit computer and identifier of minor planets, editor of the Japan Astronomical Association's Minor Planet Circulars. (M 11442)

Name proposed by S. Nakano, who found the identifications involving this planet.

(3380) Awaji

1940 EF. Discovered 1940 March 15 by G. Kulin at Budapest.

Named for the largest island in the Seto inland sea in Japan. In Japanese mythology, this was the first island created. (M 22497)

Name proposed by H. Oishi and S. Nakano, who found the identifications involving this object.

(3381) Mikkola

1941 UG. Discovered 1941 October 15 by L. Oterma at Turku.

Named in honor of Seppo Mikkola, well-known specialist in celestial mechanics and stellar dynamics, a student of Väisälä {see planet (1573)} who has continued the Turku tradition of computing orbits of minor planets and comets. (M 18451)

(3382) Cassidy

1948 RD. Discovered 1948 September 7 by H. L. Giclas at Flagstaff.

Named in honor of William A. Cassidy (1928-), geologist at the University of Pittsburgh. He was a pioneer investigator of tektite strewn fields and meteorite-impact craters. At the Vampo del Cielo in Argentina he documented one of the first known occurrences of multiple impact craters intimately associated with meteorite masses. For 15 years he led meteorite recovery teams on glacial stranding surfaces in Antarctica and was responsible for the acquisition of several thousand meteorites there, including lunar and martian specimens. He has investigated crystallization in plasmas simulating dust condensation in stellar atmospheres and vapor-deposition processes on surfaces simulating the moon and planets. (M 29142)

Named by the discoverer following a suggestion by T. P. Kohman and B. W. Hapke, who wrote the citation.

(3383) Koyama

1951 AB. Discovered 1951 January 9 by K. Reinmuth at Heidelberg.

Named in honor of Hisako Koyama {1916-1997}, staff member of the National Science Museum in Tokyo for more than 40 years, internationally known for her solar observations. (M 11442)

Name proposed by S. Nakano, who found the identifications involving this planet.

Obituary published in Yamamoto Circ., No. 2279, p. 2 (1997).

(3384) Daliya

1974 SB_1. Discovered 1974 September 19 by L. I. Chernykh at Nauchnyj.

Named for Vladimir Ivanovich Dal' (1801-1872), outstanding lexicologist, ethnographer and writer, compiler of the famous *Explanatory Dictionary of the Living Russian Language*. (M 13176)

(3385) Bronnina

1979 SK_{11}. Discovered 1979 September 24 by N. S. Chernykh at Nauchnyj.

Named in honor of Nina Mikhailovna Bronnikova, astronomer at the Pulkovo Observatory. One of the main observers with the Pulkovo normal astrograph for many years, she is known for her work on star catalogues and for her observations of minor planets and comets. (M 18137)

(3386) Klementinum

1980 FA. Discovered 1980 March 16 by L. Brožek at Kleť.

Named for the Jesuit college of St. Clemens in the Old Town of Prague. At the beginning of the eighteenth century a mathematical museum and astronomical observatory were established there, the observation tower being completed in 1722. Klementinum was long an important center for astronomical, meteorological and geophysical research and measurements. Now it contains the national and university libraries. (M 21130)

(3387) Greenberg

1981 WE. Discovered 1981 November 20 by E. Bowell at Anderson Mesa.

Named in honor of Richard J. Greenberg {1947- }, planetary scientist at the University of Arizona. Greenberg has carried out careful and innovative studies of the evolution of dynamical resonances in outer-planet satellite and ring systems. He has also contributed important new insights into collisional processes between minor planets and into the formation of resulting dust bands. (M 12803)
Citation prepared by W. B. Hubbard following a suggestion by L. A. Lebofsky.

(3388) Tsanghinchi

1981 YR_1. Discovered 1981 December 21 at the Purple Mountain Observatory at Nanking.

Named in honor of Tsanghinchi, well-known Chinese industrialist, honorary principal of the Life Science College of Zhongshan University. He has made important contributions to education and culture. (M 22829)

(3389) Sinzot

1984 DU. Discovered 1984 February 25 by H. Debehogne at La Silla.
The family name of the discoverer's maternal grandmother. (M 15573)

(3390) Demanet

1984 ES_1. Discovered 1984 March 2 by H. Debehogne at La Silla.
The family name of the discoverer's paternal grandmother. (M 15573)

(3391) Sinon

1977 DD_3. Discovered 1977 February 18 by H. Kosai and K. Hurukawa at Kiso.
Named for an ancient Greek warrior, a hero of the Trojan War. (M 11443)

(3392) Setouchi

1979 YB. Discovered 1979 December 17 by H. Kosai and G. Sasaki at Kiso.
Named for a beautiful place around the Seto Inland Sea. (M 11443)

(3393) Štúr

1984 WY_1. Discovered 1984 November 28 by M. Antal at Piszkéstetö.

Named in memory of Ľudovít Velislav Štúr (1815-1856), ideologist and organizer of the Slovak national-liberation movement, codifier of the Slovak literary language, linguist, philosopher, historian, writer, poet, journalist, educationalist and editor. Štúr was also an enthusiastic promoter of Czecho-Slovak contacts, Slavonic solidarity and human coexistence with high respect for democracy. An active authority during the 1848-1849 revolution, he led the revival from the oppression of a millenium that almost caused the Slovaks to lose their identity. (M 13608)

(3394) Banno

1986 DB. Discovered 1986 February 16 by S. Inoda and T. Urata at Karasuyama.

Named in memory of Yoshiaki Banno (1952-1991), a coworker of the discoverers and a computer engineer at the National Laboratory for High-Energy Physics in Tsukuba city. A discoverer of (4200) Shizukagozen, he was one of the pioneers among Japanese amateur astronomers in the use of electronic computers for orbital calculations. He was tragically killed in a traffic accident on his way home from the laboratory. (M 21130)

(3395) Jitka

1985 UN. Discovered 1985 October 20 by A. Mrkos at Kleť.

Named in honor of Jitka Beneš in recognition of her assistance at Kleť during the International Halley Watch. (M 23136)

(3396) Muazzez

A915 TE. Discovered 1915 October 15 by M. Wolf at Heidelberg.

Named in honor of Muazzez K. Lohmiller, weekend operator in the Smithsonian Astrophysical Observatory's Computer Center, who has for several years assisted the work of the Minor Planet Center by loading and unloading the magnetic tapes for archiving the Minor Planet Circulars and attending to massive piles of computer printout. (M 11443)

Name proposed by B. G. Marsden, who found the identifications involving this minor planet.

(3397) 1964 XA

Discovered 1964 December 8 by R. A. Burnham and N. G. Thomas at Flagstaff.

(3398) Stättmayer

1978 PC. Discovered 1978 August 10 by H.-E. Schuster at La Silla.

Named in honor of Peter Stättmayer {1945- }, director of the Munich Public Observatory, on the occasion of his 50th birthday, 1995 Nov. 3. He has been a very active amateur astronomer all his lifetime and is well known for his great ability to communicate with the public. His excellent astrophotographic observations, mostly made from his well-equipped private observatory in Herrsching at the Ammersee, are unsurpassed among amateurs in Germany and beyond. (M 25976)

Proposed by the discoverer and R. M. West, as well as his amateur friends in Munich.

(3399) Kobzon

1979 SZ_9. Discovered 1979 September 22 by N. S. Chernykh at Nauchnyj.

Named in honor of Iosif Davidovich Kobzon, well-known singer in the former Soviet Union. (M 22829)

(3400) Aotearoa

1981 GX. Discovered 1981 April 2 by A. C. Gilmore and P. M. Kilmartin at Lake Tekapo.

Usually taken to be the Maori name for New Zealand, although originally it referred only to the North Island. Commonly accepted to mean 'Land of the Long White Cloud', it is attributed to Hine-te-aparangi, wife of the legendary Maori navigator, Kupe. She is said to have called out 'He, ao' ('A Cloud!') on sighting land when first arriving in the vicinity of the country, and Kupe used this declaration in naming the new land. The name is therefore particularly appropriate for this minor planet, the first to be discovered from New Zealand. (M 18305)

(3401) Vanphilos

1981 PA. Discovered 1981 August 1 at the Harvard College Observatory at Harvard.

Named by G. V. Williams in honor of his friends Vanessa Hall and Philip Osborne, on the occasion of their marriage, 1991 August 3. Both planned a two-year tour of duty with the Voluntary Service Overseas organization. (M 18644)

(3402) Wisdom

1981 PB. Discovered 1981 August 5 by E. Bowell at Anderson Mesa.

Named in honor of Jack Wisdom, a dynamicist at the Massachusetts Institute of Technology. Wisdom is well known for his discovery of a dynamical mechanism for the removal of minor-planet type objects from the 3:1 and other Kirkwood {see planet (1578)} gaps via chaotic behavior, which can also deliver meteorites to the Earth. He and his colleagues have also explored the long-term stability of the solar system, the tidal evolution of planetary satellite systems and the chaotic rotation states of irregularly shaped natural satellites. (M 12803)

Citation prepared by W. C. Tittemore and L. M. French at the request of the discoverer.

(3403) Tammy

1981 SW. Discovered 1981 September 25 by L. G. Taff at Socorro.

Named in honor of the wife of R. L. Irelan, principal night assistant at Lincoln Laboratory's Observatory, as a token of thanks for her patience and support during the years of strange and trying schedules of his effort. (M 12016)

(3404) Hinderer

1934 CY. Discovered 1934 February 4 by K. Reinmuth at Heidelberg.

Named in memory of Fritz Hinderer (1912-1991), German astronomer at the Potsdam and Babelsberg Observatories and professor at the Berlin Free University. For many years, Hinderer was engaged in the study of irregular variables, as well as W UMa stars. By means of more than 850 plates taken at the Babelsberg Observatory, he thoroughly investigated the spectrophotometric behavior of SS Cyg. After World War II, Hinderer served solely as an academic teacher of astronomy in West Berlin. (M 22497)

Name suggested and citation prepared by L. D. Schmadel, who was a student of Hinderer.

Obituary published in Mitt. Astron. Ges., Nr. 75, p. 5-7 (1992).

(3405) Daiwensai

1964 UQ. Discovered 1964 October 30 at the Purple Mountain Observatory at Nanking.

Named in memory of Dai Wensai (1911-1979), one of the founders of modern astronomy in China, and director of the astronomy department at Nanjing University from 1955 to 1979. He worked on stellar spectroscopy, stellar astronomy and the origin of the solar system. (M 21608)

(3406) Omsk

1969 DA. Discovered 1969 February 21 by B. A. Burnasheva at Nauchnyj.

Named for the city of Omsk, the discoverer's birthplace, one of the great industrial, cultural and scientific centers in Siberia {see planet (1094)}. (M 19693)

(3407) Jimmysimms

1973 DT. Discovered 1973 February 28 by L. Kohoutek at Bergedorf.

Named in honor of James A. C. Simms III (1957-), a system administrator in the AXAF Science Center at the Smithsonian Astrophysical Observatory. Beginning in 1980, as an operator in the observatory's central computing facility, he was always particularly helpful in expediting the Minor Planet Center's computing jobs. (M 34618)

Name proposed by B. G. Marsden and C. M. Bardwell, who made the identifications involving this object.

(3408) Shalamov

1977 QG_4. Discovered 1977 August 18 by N. S. Chernykh at Nauchnyj.

Named in memory of the Soviet writer Varlam Tikhonovich Shalamov (1907-1982). (M 18137)

(3409) Abramov

1977 RE_6. Discovered 1977 September 9 by N. S. Chernykh at Nauchnyj.

Named in memory of Fyodor Aleksandrovich Abramov (1920-1983), well-known writer who depicted in his works the life and fates of the ordinary people of northern Russia. (M 22498)

(3410) Vereshchagin

1978 SZ_7. Discovered 1978 September 26 by L. V. Zhuravleva at Nauchnyj.

Named for Vasilij Vasil'evich Vereshchagin (1842-1904), Russian publicist and war artist. (M 20835)

(3411) Debetencourt

1980 LK. Discovered 1980 June 2 by H. Debehogne at La Silla.

The family name of the mother of Georges Roland, codiscoverer of the naked-eye comet Arend-Roland 1957 III. (M 15574)

(3412) Kafka

1983 AU_2. Discovered 1983 January 10 by R. Kirk and D. Rudy at Palomar.

Named in memory of the Bohemian writer Franz Kafka (1883-1924). In his novels and short stories (of which *The Castle*, *The Trial* and *Metamorphosis* are best known), Kafka depicted the fatalistic struggles of ordinary individuals to cope with a world turned surreal and incomprehensibly hostile. (M 11641)

(3413) Andriana

1983 CB_3. Discovered 1983 February 15 by N. G. Thomas at Anderson Mesa.

Named in honor of Andriana Marie Hazelton, granddaughter of the discoverer. (M 12016)

(3414) Champollion

1983 DJ. Discovered 1983 February 19 by E. Bowell at Anderson Mesa.

Named for Jean-François Champollion (1790-1832), French Egyptologist who in 1822 found the key to the decipherment of the Egyptian hieroglyphics on the Rosetta stone. (M 12458)

Named by the discoverer following a suggestion by C. E. Spratt.

Champollion is also honored by a lunar crater.

(3415) Danby

1928 SL. Discovered 1928 September 22 by K. Reinmuth at Heidelberg.

Named in honor of John Michael Anthony Danby {1929- }, celestial mechanician and sometime oboeist in the London Symphony Orchestra. An inspiring teacher, at the Yale University Observatory during the Brouwer {see planet (1746)} era and more recently at the North Carolina State University, he is the author of modern texts on celestial mechanics and differential equations. The name also honors his daughter, Dinah, who while a student at Harvard provided occasional volunteer assistance with address labels for the MPCs and IAU Circulars. (M 11443)

Name proposed by B. G. Marsden, who found the identifications involving this Hilda-type planet.

(3416) Dorrit

1931 VP. Discovered 1931 November 8 by K. Reinmuth at Heidelberg.

Named in honor of Dorrit Hoffleit {1907- }, renowned authority on variable stars, director of the Maria Mitchell Observatory from 1957 to 1978, research astronomer at Harvard and later Yale. The summer research internship program she conducted at the M.M.O. provided training for more than 100 female students, many of whom subsequently became well-known astronomers. Her research at Harvard also included stellar spectra and meteors. Author of the third and fourth editions (1964, 1982) of the Yale Bright Star Catalogue, she was also in charge of the Yale astrometric zone work for several years. (M 11641)

Name proposed by B. G. Marsden, who found the identifications involving this planet. Citation prepared in collaboration with J. A. Mattei.

(3417) Tamblyn

1937 GG. Discovered 1937 April 1 by K. Reinmuth at Heidelberg.

Named in honor of Peter Tamblyn, enthusiastic volunteer at the Minor Planet Center during the summer of 1987. (M 12210)

Name proposed by B. G. Marsden, who found the identifications involving this planet.

(3418) Izvekov

1973 QZ_1. Discovered 1973 August 31 by T. M. Smirnova at Nauchnyj.

Named in honor of Vladimir Andreevich Izvekov, an expert on solar-system dynamics and a staff member of the Institute for Theoretical Astronomy from 1951 to 1988. One of the first in the U.S.S.R. to employ electronic computers for improving orbits of minor planets and computing ephemerides, he contributed extensively to automation and quality control in the preparation of the *Ephemerides of Minor Planets* {see planet (5001)}. (M 17027)

(3419) Guth

1981 JZ. Discovered 1981 May 8 by L. Brožek at Kleť.

Named in memory of Vladimír Guth (1905-1980), outstanding Czech astronomer, founder of contemporary Czechoslovak meteoric astronomy. He started his astronomical work at Ondřejov Observatory and during the 1950s went to Slovakia (to Skalnaté Pleso, then later as first director of the Astronomical Institute of the Slovak Academy of Sciences). After his return to Ondřejov, Guth led the department of interplanetary matter. Guth was president of IAU Commission 22 from 1952 to 1958. (M 22498)
Name proposed by L. Kresák.

(3420) Standish

1984 EB. Discovered 1984 March 1 by E. Bowell at Anderson Mesa.

Named in honor of E. Myles Standish Jr., celestial mechanician at the Jet Propulsion Laboratory, who has continually improved the accuracy of planetary ephemerides by making use of diverse types of data, including optical astrometry, planetary ring occultations, radio interferometry, planetary radar and spacecraft tracking data. His efforts have been important in making the JPL planetary ephemeris an international standard. (M 12016)

Name proposed by the discoverer following a suggestion by D. K. Yeomans, who provided the citation.

(3421) Yangchenning

1975 WK_1. Discovered 1975 November 26 at the Purple Mountain Observatory at Nanking.

Named in honor of the theoretical physicist, Chen Ning Yang (1922-), Albert Einstein Professor of Physics at the State University of New York. The Yang-Mills theory, parity nonconservation in weak interactions and the Yang-Baxter equation are epoch making contributions to physics and mathematics. Yang has received the Nobel Prize (1957), the Benjamin Franklin Medal (1993) and the Bower Award (1994). (M 27125)
Yang received the Nobel prize together with Tsung-Dao Lee, see planet (3443).

(3422) Reid

1978 OJ. Discovered 1978 July 28 by B. Stewart and C. Pratt at Bickley.

Named in honor of Gordon Reid (1923-1989) and his wife Ruth. Gordon Reid was foundation professor of politics at the University of Western Australia and became deputy vice-chancellor of the university in 1978. A Companion of the Order of Australia, he served as governor of Western Australia from 1984 until

his death. Ruth Reid is well known in Western Australia for her warmth and compassion for the disadvantaged in the community. (M 20520)
Citation prepared by M. P. Candy.

(3423) Slouka

1981 CK. Discovered 1981 February 9 by L. Brožek at Kleť.

Named in memory of Hubert Slouka (1903-1973), Czech astronomer, well known as a popularizer of astronomy and as the author of a great number of books, articles and lectures. Slouka was the long-time editor-in-chief of *Říše hvězd*, a Czech journal for amateur astronomers. His wonderful book *Views into the Sky* inspired an interest in astronomy in many Czech and Slovak astronomers. (M 22498)

(3424) Nušl

1982 CD. Discovered 1982 February 14 by L. Brožek at Kleť.

Named in memory of František Nušl (1867-1951), Czech astronomer and mathematician, director of the National Observatory in Prague and for many years the chairman of the Czech Astronomical Society. His main interest consisted in the construction of astronomical instruments for the determination of geographical coordinates (circumzenithal, diazenithal, radiozenithal, etc.) in collaboration with J. J. Frič. (M 22498)
Nušl is also honored by a lunar crater.

(3425) Hurukawa

1929 BD. Discovered 1929 January 29 by K. Reinmuth at Heidelberg.

Named in honor of Kiichiro Hurukawa, astronomer at the Tokyo Astronomical Observatory, known for his identifications and orbit computations and for his participation in the observational program of minor planets with the Kiso {see planet (2271)} Schmidt. (M 11443)

Name proposed by S. Nakano, who found the identifications involving this planet.

(3426) Seki

1932 CQ. Discovered 1932 February 5 by K. Reinmuth at Heidelberg.

Named in honor of Tsutomu Seki, visual discoverer of six comets between 1961 and 1970. He later undertook an extensive program of photographic astrometric observations and has recovered several returning periodic comets. Editor of the Oriental Astronomical Association's Comet Bulletin, he is also an accomplished performer and teacher of classical guitar. (M 11443)

Name proposed by S. Nakano, who found the identifications involving this planet.

(3427) Szentmártoni

1938 AD. Discovered 1938 January 6 by G. Kulin at Budapest.

Named in memory of Belá Szentmártoni (1931-1988), outstanding Hungarian amateur astronomer. He founded the Albireo Amateur Astronomical Society and its magazine 'Albireo' in 1971 and through his articles and translations helped promote amateur astronomy in Hungary. Belá was an all-rounder, although he specialized in deep-sky and double star observing, as well as variable stars. He was also a mirror maker. (M 23792)

(3428) Roberts

1952 JH. Discovered 1952 May 1 at the Goethe Link Observatory at Brooklyn, Indiana.

Named in memory of Walter Orr Roberts (1915-1990), from 1960 to 1970 the founding director of the National Center for Atmospheric Research. In the early

1940s he established the Harvard College Observatory's Fremont Pass Station (which later became the University of Colorado's High Altitude Observatory), where he made solar observations with the first coronagraph in the western hemisphere. His interests soon broadened from solar physics to climatic research in general, and he served as president (and later president emeritus) of the University Corporation for Atmospheric Research. His final major activity was the Greenhouse Glasnost project between scientists in the U.S. and the U.S.S.R. (M 16885) Name proposed by F. K. Edmondson.

Obituaries published in Bill. Am. Astron. Soc., Vol. 24, No. 4, p. 1331-1332 (1992); Q.J.R. Astron. Soc., Vol. 33, No. 1, p. 35-37 (1992).

(3429) Chuvaev

1974 SU_1. Discovered 1974 September 19 by L. I. Chernykh at Nauchnyj.

Named in memory of Konstantin Konstantinovich Chuvaev (1917-1994), astrophysicist and staff member of the Crimean Astrophysical Observatory from 1947. He investigated nocturnal airglow, variable stars and active galaxies. One of the first observers at the 2.6-m Shajn telescope, Chuvaev was a pioneer in the use of electronic imaging with this telescope. (M 27125)

Obituary published in Bull. Crimean Astrophys. Obs., Vol. 93, p. 184-186 (1996).

(3430) Bradfield

1980 TF_4. Discovered 1980 October 9 by C. S. Shoemaker at Palomar.

Named in honor of William A. Bradfield, rocket engineer of Dernancourt, South Australia. Discoverer of twelve comets, Bradfield has been chiefly responsible for the greatly increased rate of discovery of bright comets from the southern hemisphere during the 1970s and 1980s. (M 11750)

(3431) Nakano

1984 QC. Discovered 1984 August 24 by T. Seki at Geisei.

Named in honor of Syuichi Nakano {1947- }, energetic computer of orbits and identifier of minor planets, prolific author of computer software and related astronomical books, catalogues and circulars. (M 10849)

(3432) Kobuchizawa

1986 EE. Discovered 1986 March 7 by M. Inoue and O. Muramatsu and T. Urata at Kobuchizawa.

Named for the observing station, 150 km west of Tokyo, at which this minor planet was the first discovery. (M 10849)

(3433) Fehrenbach

1963 TJ_1. Discovered 1963 October 15 at the Goethe Link Observatory at Brooklyn, Indiana.

Named in honor of Charles Fehrenbach, who pioneered the successful use of the objective prism to measure stellar radial velocities. One of his instruments was used to identify members of the Magellanic Clouds from their radial velocities, starting in 1961 during the ESO site survey in southern Africa and continuing in 1968 on La Silla in Chile. He served as vice president of the IAU from 1973 to 1979, as a member of ESO and president of its Commission on Instruments, and as a member and president of the council of the Canada-France-Hawaii Observatory. His many honors include membership in the French Academy of Sciences, the gold medal of the CNRS and the grand scientific prize of the city of Paris. (M 17027) Name proposed by F. K. Edmondson.

(3434) Hurless

1981 VO. Discovered 1981 November 2 by B. A. Skiff at Anderson Mesa.

Named in memory of Carolyn J. Hurless {1934-1987}, of Lima, Ohio. From 1959 until her death in 1987 she made over 79,000 observations of variable stars for the American Association of Variable Star Observers, which places her among the ten most prolific observers in the AAVSO's history. Besides these observations, Hurless was also active in promoting variable-star observing among hundreds of amateur astronomers by means of a newsletter and other correspondence with observers worldwide. The amateur variable star observing community benefited greatly from her interest and enthusiasm. (M 16041)
Name suggested and citation provided by P. L. Sventek.

Obituary published in J. Am. Assoc. Variable Star Obs., Vol. 16, No. 1, p. 35-36 (1987).

(3435) Boury

1981 XC_2. Discovered 1981 December 2 by F. Dossin at St. Michel.

Named in memory of Arsène Boury (1934-1982), fellow-student, colleague and friend of the discoverer. He was a theoretical astrophysicist known for his work on stellar evolution and stability, mainly concerning massive and population III stars. The discovery of this minor planet was announced around the date of his untimely death. (M 11750)
Obituary published in Proc. 24th Liège Int. Astrophys. Colloq., p. I (1983).

(3436) Ibadinov

1976 SS_3. Discovered 1976 September 24 by N. S. Chernykh at Nauchnyj.

Named in honor of Hursandkul Ibadinov, researcher on comets, founder and head of the laboratory for modeling cometary processes at the Tadjik Academy of Sciences' Institute of Astrophysics in Dushanbe. (M 18137)

(3437) Kapitsa

1982 UZ_5. Discovered 1982 October 20 by L. G. Karachkina at Nauchnyj.

Named in memory of the celebrated physicist Pyotr Leonidovich Kapitsa (1894-1984), recipient of the Nobel prize for physics in 1978 {shared with A. A. Penzias and R. W. Wilson}. (M 12016)

(3438) Inarradas

1974 SD_5. Discovered 1974 September 21 at the Felix Aguilar Observatory at El Leoncito.

Named for the Instituto Argentino de Radioastronomia, founded in 1962, for its outstanding contributions to astronomy. (M 23351)

(3439) Lebofsky

1983 RL_2. Discovered 1983 September 4 by E. Bowell at Anderson Mesa.

Named in honor of Larry A. Lebofsky, planetary scientist at the Lunar and Planetary Laboratory of the University of Arizona, Tucson. Lebofsky was the first to find chemically-bound water and the presence of ice in the regoliths of minor planets and has been a major contributor to the development of minor-planet thermal models. He has also played an important role in the extraction of minor-planet data from IRAS {see planet (3728)} infrared observations. He has undertaken related laboratory spectral studies on icy condensates and the comparison of minor planets with cometary dust, planetary satellites and Pluto. (M 12458)
Citation prepared by J. S. Lewis.

(3440) Stampfer

1950 DD. Discovered 1950 February 17 by K. Reinmuth at Heidelberg.

Named in memory of Simon Stampfer (1790-1864), who despite his humble beginnings became an astronomer and a founding member of the Imperial Academy of Sciences in Vienna. He was the first, in 1851, to compute photometric diameters of minor planets on the basis of the mean albedo derived from observations of the satellites of the giant planets, and in 1862 he was the first to show that the revolution period of P/Swift-Tuttle {see also planets (5035), (5036), respectively} was somewhat in excess of a century. (M 22498)

Named by the Minor Planet Names Committee following a suggestion by H. Haupt.

(3441) Pochaina

1969 TS_1. Discovered 1969 October 8 by L. I. Chernykh at Nauchnyj.

Named for the small river in the old town of Kiev. According to some historians, the inhabitants of Kiev were baptized here in 988 when Christianity was accepted by Russia. (M 13481)

(3442) Yashin

1978 TO_7. Discovered 1978 October 2 by L. V. Zhuravleva at Nauchnyj.

Named in memory of Lev Ivanpvich Yashin (1929-1990), football goalkeeper and trainer. (M 20835)

(3443) Leetsungdao

1979 SB_1. Discovered 1979 September 26 at the Purple Mountain Observatory at Nanking.

Named in honor of the theoretical physicist Tsung-Dao Lee (1926-), professor at Columbia University, winner of the Nobel prize in physics in 1957 {together with C. N. Yang, see planet (3421)}. Among his outstanding works is the discovery of parity and other symmetry violations in weak interactions. From the late 1940s onward, he made many other landmark contributions to weak interactions. In the 1970s and 1980s Lee established the field of nontopological solitons and discovered a large class of new astronomical solutions of general relativity, called soliton stars. (M 27733)

(3444) Stepanian

1980 RJ_2. Discovered 1980 September 7 by N. S. Chernykh at Nauchnyj.

Named in honor of husband-and-wife staff members at the Crimean Astrophysical Observatory for more than 30 years. Natalia Nikolaevna Stepanian, prominent astrophysicist, is head of the observatory's solar physics department. Arnol'd Artashesovich Stepanian, head of the gamma-ray laboratory, initiated the construction of the unique 48-element telescope for the observatory's research on gamma-ray sources. (M 18137)

(3445) 1983 FC

Discovered 1983 March 16 by E. Barr at Anderson Mesa.

(3446) Combes

1942 EB. Discovered 1942 March 12 by K. Reinmuth at Heidelberg.

Named in honor of the French amateur astronomer Michel-Alain Combes (1942-), who for many years has been interested in minor planets, especially the near-earth asteroids and their discovery circumstances. He writes a regular column about minor planets in the French periodical *L'Astronomie*, and he contributed to the *Dictionary of Minor Planet Names* by L. D. Schmadel {see planet (2234)},

who found the identification involving this planet and who proposed the name. (M 21130)
Citation prepared by J. Meeus.

(3447) Burckhalter

1956 SC. Discovered 1956 September 29 at the Goethe Link Observatory at Brooklyn, Indiana.

Named in memory of Charles Burckhalter (1849-1923), well known for his research in solar-eclipse photography, a founder of the Astronomical Society of the Pacific and its first vice president. He became the first full-time director of the two-year-old Chabot Observatory in downtown Oakland, California, in 1885 and built it into an important popular-science institution. Under his direction it was moved to a new building at a darker hill site in 1913 and featured a 50-cm refractor. (M 17028)
Name proposed by N. Sperling with the concurrence of F. K. Edmondson.

(3448) Narbut

1977 QA_5. Discovered 1977 August 22 by N. S. Chernykh at Nauchnyj.

Named in memory of Georgij Ivanovich Narbut (1886-1920), outstanding Ukrainian artist, master of graphic arts, and rector of the Ukrainian Academy of Arts. One of his pictures shows Halley's Comet in 1910. (M 18137)

(3449) Abell

1978 VR_9. Discovered 1978 November 7 by E. F. Helin and S. J. Bus at Palomar.

Named in memory of George O. Abell (1927-1983), astronomer at the University of California at Los Angeles, noted author, lecturer, educator, popularizer of astronomy and long-time director of the Summer Science Program of Thacher School in Ojai, California. As a graduate student, he took a large fraction of the plates for the *Palomar-National Geographic Sky Survey*, and his analysis of them produced the Abell catalogue of rich clusters of galaxies. In addition, he discovered a new class of planetary nebulae and showed that planetary nebulae represent a normal phase in the evolution of solar-mass stars in the disks of galaxies. (M 11162)
Name proposed by the first discoverer following a suggestion from D. A. Pierce.

Obituaries published in Mercury, Vol. 12, No. 6, p. 186 (1983); Phys. Today, Vol. 37, No. 2, p. 76-77 (1984); Sky Telesc., Vol. 67, No. 1, p. 22 (1984); Mercury, Vol. 13, No. 4, p. 108-112 (1984); Q.J.R. Astron. Soc., Vol. 30, No. 2, p. 283-285 (1989).

(3450) Dommanget

1983 QJ. Discovered 1983 August 31 by H. Debehogne at La Silla.

Named in honor of Jean Dommanget, head of the department of astrometry and celestial mechanics at the Royal Observatory of Belgium (1967-89), for his more than 40 years of continuous research activity in double-star astronomy and on related problems concerning astrometry and image quality. He participated in the ESO site survey in South Africa (1955-57) and was acting director of the Boyden Observatory (1964-65). He served as president of IAU Commission 26 during 1970-73 and since 1980 has been coordinator on double stars for the Hipparcos Input Catalogue Consortium. (M 15574)

(3451) Mentor

1984 HA_1. Discovered 1984 April 19 by A. Mrkos at Kleť.

Named for the father of Imbrius and son of Imbrus at Pedaseus, who allied with the Trojans. (M 22829)
Name proposed by A. Mrkos following a suggestion by F. Pilcher.

(3452) Hawke

1980 OA. Discovered 1980 July 17 by E. Bowell at Anderson Mesa.

Named in honor of B. R. Hawke, planetary geologist at the University of Hawaii, Honolulu. Hawke's research has concentrated on the geologic evolution of the Moon as revealed by sample studies, spacecraft photography and infrared spectroscopy. He has also undertaken spectroscopic evaluation of minor planet mineralogy. (M 12804)

Name suggested and citation prepared by J. F. Bell.

(3453) Dostoevsky

1981 SS_5. Discovered 1981 September 27 by L. G. Karachkina at Nauchnyj.

Named for the great Russian writer Fyodor Mikhailovich Dostoevsky (1821-1881). (M 12017)

(3454) Lieske

1981 WB_1. Discovered 1981 November 24 by E. Bowell at Anderson Mesa.

Named in honor of Jay Henry Lieske {1941- }, astronomer at the Jet Propulsion Laboratory, well known for his work on the Galilean satellites, ephemerides and astronomical constants, the precessional formulation in the J2000 system and for an accurate determination of the solar parallax from the motion of (433) Eros. (M 11162)

Name proposed by the discoverer following a suggestion from L. D. Schmadel, who also prepared the citation.

(3455) Kristensen

1985 QC. Discovered 1985 August 20 by E. Bowell at Anderson Mesa.

Named in honor of Leif Kahl Kristensen, teacher in theoretical physics at the University of Aarhus, who has extensively discussed the orbit of (51) Nemausa and has thereby derived equinox and equator corrections to the FK4 system. In addition, he has worked on the recovery of long-lost minor planets, on orbit determinations and identifications. (M 11162)

Name proposed by the discoverer following a suggestion by L. D. Schmadel, who also prepared the citation.

(3456) 1985 RS_2

Discovered 1985 September 5 by H. Debehogne at La Silla.

(3457) 1985 RA_3

Discovered 1985 September 5 by H. Debehogne at La Silla.

(3458) Boduognat

1985 RT_3. Discovered 1985 September 7 by H. Debehogne at La Silla.

Named for Boduognat, chief of the Nerviens, a Belgian tribe in the provinces of Hainaut and Brabant. Boduognat fought against the Roman invaders led by Julius Caesar and was defeated and killed at the battle of Sambre {see planet (4016)} in 59 B.C. (M 23792)

(3459) Bodil

1986 GB. Discovered 1986 April 2 by P. Jensen at Brorfelde.

Named in honor of Bodil Jensen, wife of the discoverer. (M 12211)

(3460) Ashkova

1973 QB_2. Discovered 1973 August 31 by T. M. Smirnova at Nauchnyj.

Named in honor of Nataliya Vladimirovna Ashkova {1932- }, an expert on the dynamics of minor planets on the staff of the Institute for Theoretical Astronomy from 1954 to 1987. She was heavily involved in the automation of computations and to the improvement of orbits of minor planets for the *Ephemerides of Minor Planets* {see planet (5001)}. (M 17028)

(3461) Mandelshtam

1977 SA_1. Discovered 1977 September 18 by N. S. Chernykh at Nauchnyj.

Named in memory of the Soviet poet Osip Emilievich Mandel'shtam (1891-1938). (M 18137)

(3462) Zhouguangzhao

1981 UA_{10}. Discovered 1981 October 25 at the Purple Mountain Observatory at Nanking.

Named in honor of Zhou Guangzhao, president of the Chinese Academy of Sciences and known for his creative work in several major fields of theoretical physics, notably particle physics, quantum-field physics, condensed-matter physics and statistical physics. He is recognized internationally as one of the founding fathers of research on the partially conserved axial vector current, and he introduced the concept of helicity amplitude and its corresponding mathematical description. He has also developed a unified formalism for the study of equilibrium and non-equilibrium states. (M 26424)

(3463) Kaokuen

1981 XJ_2. Discovered 1981 December 3 at the Purple Mountain Observatory at Nanking.

Named in honor of Charles Kuen Kao (1933-), an internationally renowned telecommunications scientist and electronic engineer. In 1966 he was the principal author of the seminal paper on the theory and practice of the use of optical fibers for communication applications, heralding a new era in telecommunications. For his numerous contributions to the field he is referred to as the "Father of Fiber Optics". (M 27125)

(3464) Owensby

1983 BA. Discovered 1983 January 16 by E. Bowell at Anderson Mesa.

Named in honor of Pamela D. Owensby, planetary astronomer at the University of Hawaii, Honolulu. Her hard work and dedication to accurate data analysis have made possible the success of several large observational programs at Mauna Kea Observatory, including the 24-color visual and 52-color infrared spectral surveys of minor planets. (M 12804)
Name suggested and citation prepared by J. F. Bell.

(3465) Trevires

1984 SQ_5. Discovered 1984 September 20 by H. Debehogne at La Silla.

Named for the Trevires, an ancient Belgian tribe, mentioned in Julius Caesar's *Gallic Wars.* (M 23792)

(3466) Ritina

1975 EA_6. Discovered 1975 March 6 by N. S. Chernykh at Nauchnyj.

Named in honor of discoverer's daughter Margarita, an astronomer at the Crimean Astrophysical Observatory. (M 20835)

(3467) Bernheim

1981 SF_2. Discovered 1981 September 26 by N. G. Thomas at Anderson Mesa.
Named in honor of Robert Burnham Jr. {1931-1993}, past staff member of the Lowell Observatory during the Lowell Proper Motion Survey, discoverer or co-discoverer of six comets, and author of the *Celestial Handbook*, a 2,000-page 'wunderwerk' of the universe beyond the solar system. Bernheim is the ancestral Bohemian name of the Burnham family. (M 12017)

Obituaries published in Heavens, Vol. 79, No. 3, p. 12-13 (1998); J. R. Astron. Soc. Can., Vol. 92, No. 1, p. 18-19 (1998); Int. Comet Q., Vol. 19, No. 4, p. 221 (1997).

(3468) Urgenta

1975 AM. Discovered 1975 January 7 by P. Wild at Zimmerwald.

This minor planet is named for a popular, wholesome type of potato, perhaps alluding to the likely form of the object. The name also indicates that the time is near when the discoverer might lose the privilege of naming the object. (M 27125)

(3469) Bulgakov

1982 UL_7. Discovered 1982 October 21 by L. G. Karachkina at Nauchnyj.

Named in memory of Mikhail Afanasevich Bulgakov (1891-1940), Soviet writer and dramatist. (M 12017)

(3470) Yaronika

1975 ES. Discovered 1975 March 6 by N. S. Chernykh at Nauchnyj.

Named in honor of the discoverer's son Yaroslav, who works in electronics at the Crimean Astrophysical Observatory. (M 20835)

(3471) Amelin

1977 QK_2. Discovered 1977 August 21 by N. S. Chernykh at Nauchnyj.

Named in memory of Valentin Mikhajlovich Amelin (1930-1989), an authority on geodesy who lectured on the subject at Leningrad University. A staff member of the Institute for Theoretical Astronomy since 1956, he took part in the preparation of the Naval Astronomical Almanac from 1962 and became supervisor of this work in 1977. (M 17028)

(3472) Upgren

1981 EJ_{10}. Discovered 1981 March 1 by S. J. Bus at Siding Spring.

Named in honor of Arthur R. Upgren Jr., director of Van Vleck Observatory and president of IAU Commission 24. He has contributed to our knowledge of galactic structure by the determination of parallaxes and proper motions of both nearby stars and the halo population. (M 11642)

Name suggested by J. Caruso and K. Gloria.

(3473) Sapporo

A924 EG. Discovered 1924 March 7 by K. Reinmuth at Heidelberg.

Named for the capital of Hokkaido {see planet (3720)}, location of the Winter Olympic games in 1972. One of the many cultural institutions in Sapporo, the Sapporo Science Center, has an active program for astrometric observations of comets and minor planets. (M 12972)

Name suggested by K. Watanabe and proposed by S. Nakano, who found the identifications involving this planet.

(3474) Linsley

1962 HE. Discovered 1962 April 27 at the Goethe Link Observatory at Brooklyn, Indiana.

Named in memory of Earle Garfield Linsley (1882-1969), professor of geography at Mills College, Oakland, California. He was the second director of the Chalbot Observatory, serving from 1923 until his retirement in 1947. He nurtured several young astronomers who went on to distinguished research careers, including John W. Evans, Gibson Reaves {see planet (3007)}, Elizabeth Roemer {see planet (1657)}, Dorothy Locanthi and Elizabeth Scott. (M 18305)

Name proposed by Norman Sperling with the concurrence of F. K. Edmondson.

(3475) Fichte

1972 TD. Discovered 1972 October 4 by L. Kohoutek at Bergedorf.

Named in memory of Hubert Fichte (1935-1986), German writer and an important presence in the post-World War II literature scene. (M 34618)

(3476) Dongguan

1978 UF_2. Discovered 1978 October 28 at the Purple Mountain Observatory at Nanking.

Named for the "Town of Fish and Rice and Fruit", a developing industrial city with modernized construction and natural rural scenery. (M 29142)

(3477) Kazbegi

1979 KH. Discovered 1979 May 19 by R. M. West at La Silla.

Named for one of the highest mountains in Georgia, Kazbegi (also known as Mkinvartsveri), in the Caucasian range. At its foot, on the banks of the wild river Tergi, is situated the village of Kazbegi (formerly Stepantsminda), whence the ancient church of Sameba is seen against the backdrop of eternal snows. The high mountains, the clear skies and the turbulent streams of this beautiful region have been a rich source of inspiration to many poets. It is the home of the Mokheve people, whose ways of life were described by the famous Georgian writer Alexandre Kazbegi (1848-1893), who was born in the village of Stepantsminda into the Chopikachvili family, of which the discoverer's wife is also a member. (M 14632)

(3478) Fanale

1979 XG. Discovered 1979 December 14 by E. Bowell at Anderson Mesa.

Named in honor of Fraser P. Fanale, planetary scientist at the University of Hawaii, Honolulu. Fanale has specialized in the study of volatile compounds in the planets and carried out some of the earliest work in relating C-type minor planets to the carbonaceous chondrites. (M 12804)
Name suggested and citation prepared by J. F. Bell.

(3479) Malaparte

1980 TQ. Discovered 1980 October 3 by Z. Vávrová at Kleť.

Named in memory of the discoverer's favorite writer and poet, Curzio Malaparte {1898-1957}, born Kurt Erich Suckert. (M 12017)

(3480) Abante

1981 GB. Discovered 1981 April 1 by E. Bowell at Anderson Mesa.

Named in honor of Robert Hamilton Brown, planetary astronomer at the Jet Propulsion Laboratory. A pioneer in the study of the compositions of the satellites of Uranus, Brown has also contributed greatly to the development of the radiometric method of diameter determination for minor planets and satellites. He has been involved in several calibrations of the radiometric method using occultation diameters, and he has investigated the geometric assumptions by developing a generalized ellipsoidal radiometric model. "Abante" derives from the name of Brown's Italian grandfather; in English it loosely corresponds to "Bob", a nickname for Robert. (M 12804)

Citation prepared by D. L. Matson and L. A. Lebofsky following a suggestion by the discoverer.

(3481) 1982 DS_6

Discovered 1982 February 19 at the Beijing Observatory at Xinglong.

(3482) Lesnaya

1975 VY_4. Discovered 1975 November 2 by T. M. Smirnova at Nauchnyj.

Named in honor of the village in the Mogilev district (Belorussia) near which the Russian army under the command of Peter the Great defeated the Swedes on 1708 Sept. 28. (M 22498)

(3483) Svetlov

1976 YP_2. Discovered 1976 December 16 by L. I. Chernykh at Nauchnyj.

Named in memory of the Soviet poet and dramatist Mikhail Arkad'evich Svetlov (1903-1964). (M 13481)

(3484) Neugebauer

1978 NE. Discovered 1978 July 10 by E. F. Helin and E. M. Shoemaker at Palomar.

Named in honor of the Neugebauers, prominent family of physicists and mathematicians who have made significant contributions in their chosen fields. Gerry Neugebauer is chairman of the division of physics, mathematics and astronomy, California Institute of Technology, and director of Palomar Observatory; Marcia Neugebauer is project scientist for the Comet Asteroid Flyby mission at the Jet Propulsion Laboratory; and Otto E. Neugebauer, now on the faculty of the Institute for Advanced Study, Princeton, is professor emeritus, Brown University, and celebrated his ninetieth birthday on 1989 May 26. (M 14632)

Obituaries for Otto Neugebauer (1899-1990) are published in Isis, Vol. 82,No.311, p. 87-88 (1991); J. Hist. Astron., Vol. 24, Part 4, p. 289-299 (1993).

(3485) Barucci

1983 NU. Discovered 1983 July 11 by E. Bowell at Anderson Mesa.

Named in honor of M. Antonella Barucci, planetary scientist at the Istituto di Astrofisica Spaziale in Rome. A prolific contributor to the study of the physical properties of minor planets, Barucci has carried out both photometric and astrometric observations at the telescope and has studied minor planet body shapes and surface light-scattering properties in the laboratory. (M 11750)

Citation prepared by the discoverer, with assistance from A. Coradini, A. W. Harris and V. Zappalà.

(3486) Fulchignoni

1984 CR. Discovered 1984 February 5 by E. Bowell at Anderson Mesa.

Named in honor of Marcello Fulchignoni, associate professor of physics at the University of Rome and director of the Istituto di Astrofisica Spaziale. Fulchignoni has played a crucial role in promoting planetary science in Italy, particularly in stimulating Italian participation in space exploration. His research has included studies of the geological and geochemical evolution of the Moon and planetary surfaces. His interest in minor planets has centered on the problem of lightcurve inversion by means of laboratory simulation and numerical modeling. (M 11751) Citation prepared in part by A. Coradini and V. Zappalà.

(3487) Edgeworth

1978 UF. Discovered 1978 October 28 by H. L. Giclas at Anderson Mesa.

Named in memory of Kenneth Essex Edgeworth (1880-1972), Irish engineer, economist, military man and independent theoretical astronomer, who reasoned that the solar system did not end with Neptune. As early as 1943 he pointed out the likely existence of a reservoir of potential comets near the invariable plane. This preceded the discovery of 1992 QB_1 by almost half a century. (M 34618)

(3488) Brahic

1980 PM. Discovered 1980 August 8 by E. Bowell at Anderson Mesa.

Named in honor of André Brahic, astronomer at the Observatoire de Paris, Meudon, and at the Université de Paris. His studies of the dynamics of planetary rings have provided insight into how collisional interactions among particles can control the large-scale behavior of a swarm. He was leader of the team that discovered the apparent "ring arcs" of Neptune and has performed several other ring occultation experiments. Brahic is also a member of the imaging team for the Voyager spacecraft. Through popular articles, public talks and books, he has helped convey the excitement of planetary science to the public in both his native France and abroad. (M 12804)
Citation prepared by R. J. Greenberg at the request of the discoverer.

(3489) Lottie

1983 AT_2. Discovered 1983 January 10 by K. Herkenhoff and G. Ojakangas at Palomar.

Named in honor of Lottie Soll-Herkenhoff, wife of one of the co-discoverers. (M 11751)

(3490) Šolc

1984 SV. Discovered 1984 September 20 by A. Mrkos at Kleť.

Named in honor of Ivan Šolc, well-known Czech inventor of birefringent polarizing filters for research on solar prominences and surface activity. (M 25976)

(3491) Fridolin

1984 SM_4. Discovered 1984 September 30 by P. Wild at Zimmerwald.

Named for the patron saint of the Swiss valley of Glarus, where it is still a popular Christian name, usually abbreviated to Fritz. The number of this object exceeds by one the sum of (1687) Glarona and (1803) Zwicky. (M 17466)

(3492) Petra-Pepi

1985 DQ. Discovered 1985 February 16 by M. Mahrová at Kleť.

Named in honor of the discoverer's daughter on the occasion of her eighteenth birthday. (M 21955)

(3493) Stepanov

1976 GR_6. Discovered 1976 April 3 by N. S. Chernykh at Nauchnyj.

Named in memory of Vladimir Evgen'evich Stepanov (1913-1986), a corresponding member of the former Soviet Academy of Sciences, well-known for his work in solar physics and solar-terrestrial relations. For many years he led the solar researches at the Siberian Institute of Terrestrial Magnetism, Ionosphere and Radio Wave Propagation, and he did much for the development of astronomy in Siberia. (M 20835)

(3494) Purple Mountain

1980 XW. Discovered 1980 December 7 at the Purple Mountain Observatory at Nanking.

Named in honor of the Purple Mountain Observatory (PMO), established in 1934 in the eastern part of Nanjing {see planet (2078)}, near the Yangtse River. Research undertaken at PMO includes celestial mechanics, astrophysics, and radio and space astronomy. Four comets and over one hundred minor planets have been discovered at PMO. The observatory has also played an important role in developing modern Chinese astronomy. (M 22829)

(3495) Colchagua

1981 NU. Discovered 1981 July 2 by L. E. Gonzalez at Cerro El Roble.

Named for the Provincia de Colchagua, situated in the central part of Chile. It is the land of the "huasos", cheerful cattleman and agriculturists. The province includes the city of San Fernando, birthplace of the discoverer. (M 11642)

(3496) Arieso

1977 RC. Discovered 1977 September 5 by H.-E. Schuster at La Silla.

This Pallas-type object was the first to be discovered during the 1977 survey of High-Inclination Minor Planets jointly conducted by the Astronomisches Rechen-Institut and the European Southern Observatory. The name, consisting of the acronyms of the two institutions involved, was suggested by the astronomers involved, L. D. Schmadel, J. Schubart, H.-E. Schuster and R. M. West {see planets (2234), (1911), (2018), and (2022), respectively}. (M 11642)

(3497) Innanen

1941 HJ. Discovered 1941 April 19 by L. Oterma at Turku.

Named in honor of the Finnish-Canadian astronomer Kimmo Innanen, well-known specialist in celestial mechanics and galactic dynamics. With Seppo Mikkola {see planet (3381)} he predicted the existence of "Mars Trojans", the first of which was recently discovered. (M 18451)

(3498) Belton

1981 EG_{14}. Discovered 1981 March 1 by S. J. Bus at Siding Spring.

Named in honor of Michael J. S. Belton {1934- } of the National Optical Astronomy Observatories for his fundamental contributions to our understanding of solar system astrophysics, his contagious enthusiasm for astronomy, and his many outstanding achievements as an internationally highly respected scientist and a motivating teacher who has the ability to convey the excitement of astronomy to specialist and public alike. Belton has tirelessly and selflessly worked to further the causes of space science and space exploration and has repeatedly demonstrated that the U.S. has a future in space that is exciting to the individual and important to the nation and the human race. Among many other things, he has made significant discoveries about the properties of comets and asteroids, as well as the outer solar system. He has worked on a number of planetary and astrophysics missions, most recently as team leader of the Galileo imaging investigation of the Jovian system. (M 18451)

Citation provided by J. H. Rahe at the request of the discoverer.

(3499) Hoppe

1981 VW_1. Discovered 1981 November 3 by F. Börngen at Tautenburg.

Named in honor of Johannes Hoppe (1907-1987), appointed professor of astronomy at the University of Jena in 1959. He studied the physical processes involved with the passage of meteors through the Earth's atmosphere, in particular the differentiation relative to height, the real deceleration and mass loss, vaporization and fragmentation, with conclusions concerning the size and consistency of meteoroidal bodies. He is also the author of the monograph *Planeten-Sterne-Nebel* and of the chapter "Planetensystem" in the *Grundriss der Astrophysik* of Graff-Lambrecht. (M 11642)

Obituaries published in Astron. Raumfahrt, 25. Jahrg., Heft 5, p. 155 (1987); Publ. Astron. Inst. Czech. Acad. Sci., No. 67, p. 139 (1987).

(3500) Kobayashi

A919 SD. Discovered 1919 September 18 by K. Reinmuth at Heidelberg.

Named in honor of Takao Kobayashi, an active computer of cometary orbits and identifier of minor planets. He is also a vice-director of the Computing Section of Oriental Astronomical Association and is currently collecting and verifying almost all the observations of minor planets made by Japanese amateur astronomers. (M 12972)

Name proposed by S. Nakano, who found the identifications involving this planet.

(3501) Olegiya

1971 QU. Discovered 1971 August 18 by T. M. Smirnova at Nauchnyj.

Named in honor of Oleg Nikolaevich Korottsev (1922-), a member of the Astronomical-Geodetical Society since 1946 and a well-known St. Petersburg popularizer of astronomical knowledge, especially on minor planets and comets. He is the author of a number of astronomical books. (M 22498)

(3502) Huangpu

1964 TR_1. Discovered 1964 October 9 at the Purple Mountain Observatory at Nanking.

Named for Huangpu district, located in the center of Shanghai {see planet (2197)}. It is a center for commercial and financial business, as well as foreign trade and culture. (M 21608)

(3503) Brandt

1981 EF_{17}. Discovered 1981 March 1 by S. J. Bus at Siding Spring.

Named in honor of John C. Brandt of the Laboratory for Atmospheric and Space Physics, University of Colorado, for his fundamental contributions to our understanding of solar system astrophysics. Founder of a "Comet Observatory", Brandt has worked on many aspects of cometary science, particularly the interaction of the solar wind with comets. He was Comet Scientist on the International Cometary Explorer mission, which ushered in a new era of in situ exploration by flying through the tail of Comet Giacobini-Zinner in 1985, and thus represented the only tail interception by any of the Giacobini-Zinner or Halley spacecraft. (M 18452)

Citation provided by J. H. Rahe at the request of the discoverer.

(3504) Kholshevnikov

1981 RV_3. Discovered 1981 September 3 by N. S. Chernykh at Nauchnyj.

Named in honor of the celestial mechanician Konstantin Vladislavovich Kholshevnikov {1939- }, professor at Leningrad University. (M 18138)

(3505) Byrd

1983 AM. Discovered 1983 January 9 by B. A. Skiff at Anderson Mesa.

Named in honor of Deborah Byrd, writer and producer of "Star Date", a radio program that since 1978 has broadcast astronomical news and information over several hundred stations in the United States. Byrd began her career providing public information via telephone messages and brochures for the University of Texas McDonald Observatory. This has progressed to include the popular radio program, a series of brief spots for television and more extensive writing for magazines such as *Physics Today*, "Highlights for Children", *Astronomy* and *Sky and Telescope*. In 1979 she founded the "Texas Star Party" {see planet (4932)}, now one of the largest meetings for amateur astronomers in North America, which takes place each year near the McDonald Observatory in west Texas. (M 16443)

Citation provided by D. H. Levy at the request of the discoverer.

(3506) French

1984 CO_1. Discovered 1984 February 6 by E. Bowell at Anderson Mesa.

Named in honor of Linda M. French, planetary scientist at the Department of Earth, Atmospheric and Planetary Sciences of the Massachusetts Institute of

Technology. An active teacher, French has encouraged undergraduate research in planetary astronomy by using a hands-on approach to observing and data analysis and by arranging for students to observe at major facilities. Involved in research on the shapes, spin states and surface compositions of small solar-system bodies, she has emphasized the properties of Trojan asteroids in an effort to understand their origins. (M 12805)
Citation prepared by F. Vilas.

(3507) Vilas

1982 UX. Discovered 1982 October 21 by E. Bowell at Anderson Mesa.

Named in honor of Faith Vilas, planetary scientist at the Johnson Manned Space Center in Houston. Vilas has used high-resolution visual and near-infrared spectral measurements to search for compositional trends among outer-belt minor planets and to investigate the mineralogy of Mercury. She designed and built the coronagraph/spectrograph that was used to image the planetary disk around Beta Pictoris and is currently evaluating the hazard presented by Earth-orbiting debris for future manned missions, including NASA's Space Station. (M 12805)

Citation prepared by M. V. Sykes, with assistance from N. Lebofsky and E. Roemer.

(3508) Pasternak

1980 DO_5. Discovered 1980 February 21 by L. G. Karachkina at Nauchnyj.

Named in memory of Boris Leonidovich Pasternak (1890-1960), famous poet and writer. (M 12017)
Pasternak received the 1958 literature Nobel prize.

(3509) Sanshui

1978 UH_2. Discovered 1978 October 28 at the Purple Mountain Observatory at Nanking.
Named for the Chinese city situated on the delta of the Pearl River. (M 23792)

(3510) Veeder

1982 TP. Discovered 1982 October 13 by E. Bowell at Anderson Mesa.

Named in honor of Glenn J. Veeder, planetary astronomer at the Jet Propulsion Laboratory. Veeder was chiefly responsible for the initiation of minor planet observations through the J, H and K bandpasses. This work resulted in the identification of new minor planet types and led to a new classification scheme. Veeder is also noted for his work on minor planet radiometry for the determination of diameters and albedos, particularly among the Earth-crossers, and he was in charge of the scientific analysis of the characteristics of the IRAS {see planet (3728)} Asteroid and Comet Survey. (M 12805)

Citation prepared by D. L. Matson and L. A. Lebofsky following a suggestion by the discoverer.

(3511) Tsvetaeva

1982 TC_2. Discovered 1982 October 14 by L. G. Karachkina and L. V. Zhuravleva at Nauchnyj.

Named in honor of Marina Ivanovna Tsvetaeva (1892-1941), talented Soviet poetess. (M 13176)

(3512) Eriepa

1984 AC_1. Discovered 1984 January 8 by J. F. Wagner at Anderson Mesa.
Named for the discoverer's home town of Erie, Pennsylvania. (M 14481)

(3513) Quqinyue

1965 UZ. Discovered 1965 October 16 at the Purple Mountain Observatory at Nanking.

Named in honor of Qu Qinyue, professor of astronomy at Nanjing University, known for his research on X-ray and γ-ray sources and neutron stars, as well as for his contributions to the development of high-energy astrophysics in China. During his twelve years of service as president, he raised the academic standard and reputation of Nanjing University considerably. (M 34619)

(3514) 1971 UJ

Discovered 1971 October 26 by L. Kohoutek at Bergedorf.

(3515) Jindra

1982 UH_2. Discovered 1982 October 16 by Z. Vávrová at Kleť.

Named in honor of Lumír Jindra (1936-), an old friend of the discoverer, for his contribution to Czech pharmacology. (M 27125)

(3516) Rusheva

1982 UH_7. Discovered 1982 October 21 by L. G. Karachkina at Nauchnyj.

Named in memory of Nadya Rusheva (1952-1969), a talented painter who died in her youth. (M 12017)

(3517) Tatianicheva

1976 SE_1. Discovered 1976 September 24 by N. S. Chernykh at Nauchnyj.

Named in memory of the lyric poetess Lyudmila Konstantinovna Tatianicheva (1915-1980). (M 20835)

(3518) Florena

1977 QC_4. Discovered 1977 August 18 by N. S. Chernykh at Nauchnyj.

Named in memory of Pavel Aleksandrovich Florenskij (1882-1943), well-known Russian religious philosopher, physicist and mathematician. The name also honors his son, Kirill Pavlovich Florenskij (1915-1982), geochemist and mineralogist. (M 22498)

Obituary (K.P.F.) published in Icarus, Vol. 61, No. 3, p. 351-354 (1985).

(3519) Ambiorix

1984 DO. Discovered 1984 February 23 by H. Debehogne at La Silla.

Named for Ambiorix, chief of the Eburons, an ancient Belgian tribe. Ambiorix revolted against the occupying forces under Julius Caesar and was pursued and killed in the forest. There is a statue of Ambiorix in the town of Tongres. (M 23792)

(3520) Klopsteg

1952 SG. Discovered 1952 September 16 at the Goethe Link Observatory at Brooklyn, Indiana.

Named in memory of Paul E. Klopsteg (1889-1991), who died on 1991 Apr. 28, a month before his 102nd birthday. He left Northwestern University in 1951 to become the National Science Foundation's first assistant director for mathematical, physical and engineering sciences. He was appointed to the new position of associate director in 1952 and retired in 1958. His many honors include election as president of the American Association for the Advancement of Science in 1959, two medals and an award from three different archery organizations and membership in the Archery Hall of Fame. During his time at the National Science Foundation he played an important role in the planning for the National Radio Astronomy Observatory and the Kitt Peak National Observatory. (M 18305)
Name proposed by F. K. Edmondson.

(3521) Comrie

1982 MH. Discovered 1982 June 26 by A. C. Gilmore and P. M. Kilmartin at Lake Tekapo.

Named in memory of Leslie John Comrie (1893-1950), New Zealand-born innovator in computational science and authority on the production of mathematical tables. Superintendent of the British Nautical Almanay Office for six years from 1930, he applied calculating machines to astronomical computing. As president of IAU Commission 4 (Ephemerides) he initiated the annual *Apparent Places of Fundamental Stars* and saw adopted his earlier proposal for the use of standard equinoxes, the first being 1950. Though resident in Britain for most of his working life he actively fostered and assisted astronomy in New Zealand. (M 18306)
Comrie is also honored by a lunar crater.

(3522) Becker

1941 SW. Discovered 1941 September 21 by Y. Väisälä at Turku.

Named for Reinhold von Becker (1788-1858), a member of the old Academia Aboensis from 1813 onward. The weekly *Turun Viikkosanomat*, which he began publishing in 1820, had an essential influence on the evolution of the Finnish literary language. (M 18452)

(3523) Arina

1975 TV_2. Discovered 1975 October 3 by L. I. Chernykh at Nauchnyj.

Named for Arina Rodionova Yakovleva (1758-1828), the nurse of A. S. Pushkin {see planet (2208)}. (M 13481)

(3524) Schulz

1981 EE_{27}. Discovered 1981 March 2 by S. J. Bus at Siding Spring.

Named in honor of Charles M. Schulz (1922-), cartoonist and creator of the syndicated comic strip "Peanuts". Since 1950, Charlie Brown, Snoopy and the rest of the Peanuts gang have reminded us of the innocent, though sometimes frustrating, experiences of childhood. Through tireless devotion to his work, Schulz has managed to capture the hearts of both young and old, making Peanuts one of the most celebrated and enduring comic strips of all time. (M 27125)

(3525) Paul

1983 CX_2. Discovered 1983 February 15 by N. G. Thomas at Anderson Mesa.
Named in honor of Paul J. Baltutis, son-in-law of the discoverer. (M 12017)

(3526) Jeffbell

1984 CN. Discovered 1984 February 5 by E. Bowell at Anderson Mesa.

Named in honor of Jeffrey F. Bell, planetary astronomer at the University of Hawaii, Honolulu. Bell has contributed greatly to our understanding of the composition of minor planets and to their relationship with meteorites. He was responsible for the 0.3 to 2.5 micrometer 52-color minor planet survey, which has led to the spectral characterization of more than 100 minor planets. Recently, he has proposed that the CV and C0 chondrite meteorites may be derived from Eos {see planet (221)} family members. (M 12805)
Citation prepared by L. A. Lebofsky following a suggestion by the discoverer.

(3527) McCord

1985 GE_1. Discovered 1985 April 15 by E. Bowell at Anderson Mesa.

Named in honor of Thomas B. McCord, planetary scientist at the University of Hawaii, Honolulu. McCord developed reflection spectroscopy as a means of probing the mineralogical composition of planetary surfaces and played a major role in early applications of this technique to minor planets and the Moon. (M 12805)
Citation prepared by J. F. Bell at the request of the discoverer.

(3528) Counselman

1981 EW_3. Discovered 1981 March 2 by S. J. Bus at Siding Spring.

Named in honor of Charles C. Counselman III (1943-), professor of planetary science at the Massachusetts Institute of Technology. As a scientist involved with the Pioneer Venus mission, he used interferometric measurements of the spacecraft multiprobes to deduce the vertical distribution of Venusian wind velocities. More recently, he has pioneered the use of Global Positioning System satellites to achieve very high precision geodetic measurements of the surface of the earth. Counselman is also a dedicated teacher and is particularly devoted to student advising and the instruction of planetary astronomy laboratory courses. (M 27125)
Citation prepared by R. P. Binzel.

(3529) Dowling

1981 EQ_{19}. Discovered 1981 March 2 by S. J. Bus at Siding Spring.

Named in honor of Timothy E. Dowling, professor of planetary science at the Massachusetts Institute of Technology. An expert in the dynamics of the atmospheres of the giant planets, Dowling determined a relationship between potential vorticity and the zonal wind on Jupiter and developed explicit planetary isentropic-coordinate models for Jupiter, Saturn, Uranus and Neptune. He is also a dedicated teacher of planetary science at the undergraduate and graduate levels. (M 27125)
Citation prepared by R. P. Binzel.

(3530) Hammel

1981 EC_{20}. Discovered 1981 March 2 by S. J. Bus at Siding Spring.

Named in honor of Heidi Beth Hammel, planetary scientist at the Massachusetts Institute of Technology. An indefatigable observer of the atmospheres of the outer planets, she is best known for her long-term monitoring of Neptune. As an expert in planetary imaging, she was selected as the team leader for the Hubble Space Telescope project to observe the consequences of the July 1994 Jupiter impact by comet D/1993 F2 (Shoemaker-Levy 9). Hammel also devotes substantial effort to public education and is a great communicator of the excitement of planetary science. (M 27125)
Citation prepared by J. l. Elliot and R. P. Binzel.

(3531) Cruikshank

1981 FB. Discovered 1981 March 30 by E. Bowell at Anderson Mesa.

Named in honor of Dale P. Cruikshank, planetary scientist at the University of Hawaii, Honolulu. Cruikshank is well known for his observational work on solar system small bodies, including Trojan asteroids, comets and Pluto. He is especially known for studies of outer-planet satellites, including Triton, Iapetus and Io, through both telescopic and Voyager spacecraft observations. He has been active in developing instrumentation and facilities at Mauna Kea Observatory and has promoted historical studies of planetary science. Through several extended working visits to the Soviet Union and other projects, Cruikshank has also been a leader in furthering international scientific relations. (M 12806)

Citation prepared by W. K. Hartmann and D. J. Tholen at the request of the discoverer.

(3532) Tracie

1983 AS_2. Discovered 1983 January 10 by G. Ojakangas and K. Herkenhoff at Palomar.

Named in honor of Tracie Lynn Ojakangas, wife of one of the co-discoverers. (M 11751)

(3533) Toyota

1986 UE. Discovered 1986 October 30 by K. Suzuki and T. Urata at Toyota.

Named for the city in central Japan, home of the first discoverer and known throughout the world for its car industry. (M 12211)

(3534) Sax

1936 XA. Discovered 1936 December 15 by E. Delporte at Uccle.

Named in memory of Adolphe Sax (1814-1894), inventor of the saxophone. (M 12211)

(3535) Ditte

1979 SN_{11}. Discovered 1979 September 24 by N. S. Chernykh at Nauchnyj.

Named for the principal heroine of *Ditte, a human child*, a novel by the Danish writer Martin Andersen Nex{ø} (1869-1954). (M 18138)

(3536) Schleicher

1981 EV_{20}. Discovered 1981 March 2 by S. J. Bus at Siding Spring.

Named in honor of David G. Schleicher of the Lowell Observatory for his many contributions to the physical study of comets. An enthusiastic observer, Schleicher has pursued spectroscopic and spectrophotometric investigations of cometary comae using a variety of ground-based telescopes and the International Ultraviolet Explorer satellite. Theseies of comets and the identification of a number of peculiar objects with clearly anomalous relative abundances. Schleicher played a key role in the discovery of Comet Halley's periodic variability and in the subsequent interpretation of this behavior in terms of nuclear rotation. In addition to his observational research, Schleicher's detailed calculation of the fluorescence efficiency of OH has been extremely important to the analysis of cometary spectra. (M 18452)

Citation provided by R. L. Millis at the request of the discoverer.

(3537) Jürgen

1982 VT. Discovered 1982 November 15 by E. Bowell at Anderson Mesa.

Named in honor of Jürgen Rahe {1939-1997}, planetary scientist. Rahe was director of the Dr. Remeis Sternwarte and the Astronomisches Institut der Universität Erlangen-Nürnberg, and is discipline scientist in the planetary science branch of NASA's Solar System Exploration Division. He is co-leader of the International Halley Watch and has been for a long time associated with the work of IAU Commission 15. While he has worked in a number of fields of astrophysics, Rahe is especially noted for his work on comets, both from the ground and with the International Ultraviolet Explorer, as well as for his two atlases of cometary morphology. His diplomatic leadership in the IAU and the IHW has been particularly important in projects requiring international cooperation. (M 12973)

Citation prepared by R. L. Newburn and S. J. Edberg at the request of the discoverer.

Obituaries published in Icarus, Vol. 131, No. 1, p. 1-3 (1998); Phys. Today, Vol. 50, No. 12, p. 92 (1997); Int. Comet Q., Vol. 19, No. 3,p. 154-155 (1987).

(3538) Nelsonia

6548 P-L. Discovered 1960 September 24 by C. J. van Houten and I. van Houten-Groeneveld at Palomar.

Named in honor of Elisabeth Nelson, former secretary of the Landessternwarte and the Max-Planck-Institut für Astronomie, both at Heidelberg. She took care of the investigators of the Palomar-Leiden Survey during the blinking of the survey plates in Heidelberg. (M 16591)

(3539) Weimar

1967 GF_1. Discovered 1967 April 11 by F. Börngen at Tautenburg.

Named for the venerable town, 30 km from Tautenburg {see planet (2424)}, renowned for its contributors (Goethe, Schiller {see planets (3047) and (3079)} and others) to 'classical' German literature around the end of the eighteenth century and the home of several famous musicians and painters. The town is also known for the Weimar Republic (1919) and the Weimar Constitution. (M 12017)

(3540) Protesilaos

1973 UF_5. Discovered 1973 October 27 by F. Börngen at Tautenburg.

Named for the first Greek warrior to jump to the shore in the landing at Asia Minor in the Trojan war. He died at the hand of Aeneas {see planet (1172)}. (M 11751)

(3541) Graham

1984 ML. Discovered 1984 June 18 by V. M. Candy and M. P. Candy at Bickley.

Named in honor of Lloyd Wilson Graham (1940-), executive director of the department of state services, for his sympathetic support of the Perth Observatory in 1987, a difficult time in its history. He was also instrumental in the formation of the Perth Astronomy Research Group, which includes staff from the observatory and three local tertiary institutions. (M 20520)

(3542) Tanjiazhen

1964 TN_2. Discovered 1964 October 9 at the Purple Mountain Observatory at Nanking.

Named in honor of Tan Jiazhen (1909-), professor at Fudan University and considered the founder of genetics in China. His creative work leading to the discovery of the mosaic-dominance phenomenon in the color-pattern inheritances of the ladybug and of the role of chromosomal rearrangements and gene changes in Drosophila speciation were recognized as outstanding contributions to the development of classical gene theory and the modern synthetic theory of evolution. (M 34619)

(3543) Ningbo

1964 VA_3. Discovered 1964 November 11 at the Purple Mountain Observatory at Nanking.

Named for the historic Chinese port city, birthplace of the Neolithic Hemudu culture, dating back 7000 years and symbolized by oars and rice. It is also a birthplace of modern Chinese industry. (M 32787)

(3544) Borodino

1977 RD_4. Discovered 1977 September 7 by N. S. Chernykh at Nauchnyj.

Named for a village not far from Moscow where there was a fierce battle between the Russians and Napoleon's troops in 1812. (M 21130)

(3545) Gaffey

1981 WK_2. Discovered 1981 November 20 by E. Bowell at Anderson Mesa.

Named in honor of Michael J. Gaffey, planetary scientist at the Rensselaer Polytechnic Institute. His pioneering and comprehensive study of the spectra of meteorites has become the principal resource in interpreting minor planet spectra. More recently, he has demonstated that careful spectrophotometry can resolve geological units on minor planet surfaces. (M 12806)
Citation prepared by J. F. Bell at the request of the discoverer.

(3546) Atanasoff

1983 SC. Discovered 1983 September 28 by E. F. Helin and V. G. Shkodrov and V. G. Ivanova and A. Georgieva at Rozhen.

Named in honor of the mathematician John Atanasoff (1903-), a pioneer in the development of computers who between 1936 and 1942 created an operating model with regenerative memory that utilized a magnetic drum and some relay logical schemes. At the same time he lectured on mathematics and physics at Iowa

State University. Atanasoff, whose father was a Bulgarian, has been a member of the Bulgarian Academy of Sciences since 1983. (M 12806)

(3547) Serov

1978 TM_6. Discovered 1978 October 2 by L. V. Zhuravleva at Nauchnyj.

Named in memory of Valentin Alexandrovich Serov (1865-1911), a famous Russian painter. (M 24120)

(3548) Eurybates

1973 SO. Discovered 1973 September 19 by C. J. van Houten and I. van Houten-Groeneveld at Palomar.

Eurybates was a herald in the Greek army during the siege of Troy. (M 18138)

(3549) Hapke

1981 YH. Discovered 1981 December 30 by E. Bowell at Anderson Mesa.

Named in honor of Bruce W. Hapke, planetary scientist at the University of Pittsburgh. A pioneer investigator of the physical and chemical nature of the surfaces of minor planets, he was one of the first to recognize the division of minor planets into two principal color classes (later known as C and S types) on the basis of UBV photometry. More recently, he has developed a comprehensive theoretical treatment of the scattering of sunlight in planetary regoliths. (M 12806)
Citation prepared by J. F. Bell at the request of the discoverer.

(3550) Link

1981 YS. Discovered 1981 December 20 by A. Mrkos at Kleť.

Named in memory of František Link (1906-1984), founder and for ten years director of the modern astrophysical observatory at Ondřejov and a specialist in solar astronomy, high-atmosphere physics and the study of interplanetary matter. (M 26424)

Obituaries published in Rise hvezd, Vol. 65, No. 11, p. 238 (1984); Earth, Moon, Planets, Vol. 35, No. 3, p. 201-202 (1986).

(3551) Verenia

1983 RD. Discovered 1983 September 12 by R. S. Dunbar at Palomar.

Named for the first vestal virgin consecrated by the legendary Roman king Numa Pompilius, builder of the temple of Vesta {see planet (4)}, in the eighth to seventh centuries B.C. This Amor object is the first V-Type (Vesta) asteroid known after (4) Vesta itself. It was under consideration for a flyby mission by the Clementine spacecraft in October 1995. (M 23540; M 23697)

(3552) Don Quixote

1983 SA. Discovered 1983 September 26 by P. Wild at Zimmerwald.

Named for the hero of Cervantes' {1547-1616} great romance, the Knight of the Sorrowful Countenance, the noble-minded eccentric who tries, in strange ways and in vain, to bring back the Age of Chivalry. The naming of this object is rather obvious, given the long-term erratic nature of the extended orbit, comparable to that of (944) Hidalgo. (M 17466)

(3553) Mera

1985 JA. Discovered 1985 May 14 by C. S. Shoemaker and E. M. Shoemaker at Palomar.

Named for a daughter of Praetus. She was a follower of Artemis {see planet (105)} and was killed by the goddess for having given herself to Zeus {see planet (5731)}. Before dying, she gave birth to Locri, ancestor of the Locrians. (M 11751)

(3554) Amun

1986 EB. Discovered 1986 March 4 by C. S. Shoemaker and E. M. Shoemaker at Palomar.

Named for the invisible or hidden god of Hermopolis, who later became the local god of Thebes and then the king of the gods of the Egyptian empire. Great temples at Karnak and Luxor were dedicated to Amun. (M 11751)

(3555) Miyasaka

1931 TC_1. Discovered 1931 October 6 by K. Reinmuth at Heidelberg.

Named in honor of Seidai Miyasaka (1955-), an active observer of minor planets and one of the few observers in Japan who devote themselves to follow-up observations. He has observed many minor planets so that they could be numbered. (M 18644)

Name proposed by T. Kobayashi, who found the identifications involving this minor planet, and with whom Miyasaka has been collaborating for many years.

(3556) Lixiaohua

1964 UO. Discovered 1964 October 30 at the Purple Mountain Observatory at Nanking.

Named in honor of Li Xiaohua, a young industrialist in Beijing whose concern for education led him to establish middle and primary schools in remote villages and mountain districts of China. (M 26424)

(3557) Sokolsky

1977 QE_1. Discovered 1977 August 19 by N. S. Chernykh at Nauchnyj.

Named in honor of Andrej Georgievich Sokolskij, director of the Institute of Theoretical Astronomy in St. Petersburg, known for his work on the theory of periodic and quasiperiodic solutions of Hamiltonian systems. As founder and executive director of the International Institute for Problems of the Asteroid Hazard he is much involved in the current international interest in Near-Earth asteroids. (M 20835)

(3558) Shishkin

1978 SQ_2. Discovered 1978 September 26 by L. V. Zhuravleva at Nauchnyj.

Named in memory of Ivan Ivanovich Shishkin (1832-1898), a well-known Russian painter. (M 24120)

(3559) Violaumayer

1980 PH. Discovered 1980 August 8 by E. Bowell at Anderson Mesa.

Named for the village Violau in Bavaria and in honor of Martin Mayer, who there directs the "Bruder-Klaus-Heim", a Catholic educational center. Known for his enthusiastic and didactic style of teaching astronomy, Mayer has erected a well-equipped observatory to show the beauty of the universe to his guests and to the public. International Astronomical Youth Camps and conventions of the German Planetary Observers, a group of amateur astronomers, have been held in Violau. (M 12806)

Name suggested and citation prepared by C. M. Schambeck and G. Marxer.

(3560) Chenqian

1980 RZ_2. Discovered 1980 September 3 at the Purple Mountain Observatory at Nanking.

Named in honor of Chen Qian, director of the History Museum of Chinese Astronomy for some years and a great contributor to the popularization of astronomy in China. (M 26424)

(3561) Devine

1983 HO. Discovered 1983 April 18 by N. G. Thomas at Anderson Mesa.

Named in honor of John Devine Hazelton, son-in-law of the discoverer and grandson of Colonel John Devine, who was with the forces under General Gordon at Khartoum. (M 12017)

(3562) Ignatius

1984 AZ. Discovered 1984 January 8 by J. F. Wagner at Anderson Mesa.

Named in honor of the 500th anniversary of the birth of Ignatius Loyola (1491-1556) {see also planet (3589)}, founder of the Jesuits. (M 17028)

(3563) Canterbury

1985 FE. Discovered 1985 March 23 by A. C. Gilmore and P. M. Kilmartin at Lake Tekapo.

Named for the province of New Zealand on the eastern side of the South Island. The name also honors Canterbury University, based in the city of Christchurch, of which Mount John University Observatory is a field station. The region was named after the Canterbury Association formed in England in 1848 with the purpose of organizing an idealized Anglican settlement in New Zealand. (M 18306)

(3564) Talthybius

1985 TC_1. Discovered 1985 October 15 by E. Bowell at Anderson Mesa.
Named for the chief herald of the Greek forces in the Trojan war. (M 12459)

(3565) Ojima

1986 YD. Discovered 1986 December 22 by T. Niijima and T. Urata at Ojima.

Named for the small town in the central Japan where the observing station is located. Known for the manufacture of Japanes dolls, Ojima is the home of the first discoverer. It is close to the 1828-m volcano Agaki and the river Tone {see planet (1266)}. (M 12211)

(3566) Levitan

1979 YA_9. Discovered 1979 December 24 by L. V. Zhuravleva at Nauchnyj.

Named in memory of Isaac Il'ich Levitan (1860-1900), a famous Russian landscape painter. (M 24120)

(3567) Alvema

1930 VD. Discovered 1930 November 15 by E. Delporte at Uccle.

Named for the three great-granddaughters of the discoverer, Aline De Middelaer, and Véronique and Martine Warck. (M 25652)

Name proposed by M. De Middelaer-Delporte and endorsed by J. Denoyelle, who prepared the citation.

(3568) ASCII

1936 UB. Discovered 1936 October 17 by M. Laugier at Nice.

Named for the computer character code, which is also the name of the principal magazine on microcomputers in Japan. (M 12973)

Name proposed by S. Nakano, who found the identifications involving this planet and whose stay at the Smithsonian Astrophysical Observatory has partly been supported by articles written for this magazine.
ASCII is the acronym for American Standard Code for Information Interchange.

(3569) Kumon

1938 DN_1. Discovered 1938 February 20 by K. Reinmuth at Heidelberg.

Named in honor of Toru Kumon, a pioneer in the education of children in Japan and founder of the Kumon Education Research Center in Osaka. (M 12973)

Name proposed by S. Nakano, who found the identification involving this planet, and whose stay at the Smithsonian Astrophysical Observatory has been partly supported by the Kumon Center.

(3570) Wuyeesun

1979 XO. Discovered 1979 December 14 at the Purple Mountain Observatory at Nanking.

Named in honor of Wu Yeesun, a famous bonsai artist, who has been engaged in research on the art of bonsai for many years and who is regarded as a representative of the Lingnan School. His book *Potted Landscape of Wennong*, published in 1969, is the most highly regarded Chinese-English bilingual volume on the subject, and it has exerted a great influence in the bonsai art circles of the world. (M 32787)

(3571) Milanštefánik

1982 EJ. Discovered 1982 March 15 by A. Mrkos at Kleť.

Named in memory of Milan Rastislav Štefánik (1880-1919), astronomer and meteorologist. He worked at Meudon Observatory during 1905-1907 and proposed to build a new observatory on the island of Tahiti. He received the Janssen award in 1907 and the Wilde award in 1911. (M 25976)

(3572) Leogoldberg

1954 UJ_2. Discovered 1954 October 28 at the Goethe Link Observatory at Brooklyn, Indiana.

Named in memory of Leo Goldberg (1913-1987), one of the three incorporators of the Association of Universities for Research in Astronomy on 1957 Oct. 28, exactly three years after the discovery of this minor planet. Having begun work in solar physics as a student of Donald H. Menzel {see planet (1967)}, he served as director of the University of Michigan's observatory from 1946 to 1960, after which he moved to Harvard University and succeeded Menzel as director of the Harvard College Observatory in 1966. He became the third director of the Kitt Peak National Observatory in 1971 and retired in 1977. He was president of the American Astronomical Society during 1964-1966 and of the International Astronomical Union during 1973-1976. (M 19333)

Name proposed by B. G. Marsden with the concurrence of F. K. Edmondson, who prepared the citation.

Obituaries published in Phys. Today, Vol. 43, No. 2, p. 146-148 (1990); IAU Today (20th General Assembly, Baltimore, 1988), No. 1, p. 4 (1988).

(3573) Holmberg

1982 QO_1. Discovered 1982 August 16 by C.-I. Lagerkvist at La Silla.

Named in honor of the Swedish astronomer Erik Holmberg on the occasion of his eightieth birthday. Internationally renowned for his pioneering work on galaxies, paricularly multiple galaxies, Holmberg served as an assistant and associate professors at Lund during 1937-1951 and as professor at Uppsala and director of the Uppsala Observatory during 1959-1975. Holmberg was an inspiring teacher, and the discoverer is very grateful to him for allowing a young student to start work on minor planets, until then an almost unknown topic for research in Uppsala. (M 12973)

(3574) Rudaux

1982 TQ. Discovered 1982 October 13 by E. Bowell at Anderson Mesa.

Named for Lucien Rudaux (1874-1947), French astronomical painter, writer and amateur astronomer. Rudaux is considered the grandfather of astronomical art. He maintained his own observatory at Donville and used his observations of mountain profiles on the lunar limb to argue that lunar mountains are rounded. His lunar paintings and articles, from about 1910 onward, correctly described lunar topography, whereas the popular misconception was of spiky crags. His paintings also showed various aspects of other planets, satellites, the zodiacal light and comets, using the best available data. His writings include many illustrated

articles about astronomy and space travel and the book *Sur les Autres Mondes* (1937). (M 12807)
Name suggested and citation prepared by W. K. Hartmann.

(3575) Anyuta

1984 DU_2. Discovered 1984 February 26 by N. S. Chernykh at Nauchnyj.

Named in honor of the outstanding sportswoman and pioneer Soviet parachutist Anna Aleksandrovna Shishmareva. See also the citation for planet (3576). (M 18138)

(3576) Galina

1984 DB_3. Discovered 1984 February 26 by N. S. Chernykh at Nauchnyj.

Named in honor of Galina Bogdanova Pyasetskaya, who with Anna Shishmareva {see planet (3575)} accomplished in 1935 a twin parachute jump from an altitude of 7923 meters without oxygen equipment. (M 18138)

(3577) Putilin

1969 TK. Discovered 1969 October 7 by L. I. Chernykh at Nauchnyj.

Named in memory of Ivan Ivanovich Putilin (1893-1954), assistant professor at the Kiev State University, known mainly for his works on minor planets. He observed minor planets and computed their orbits. His book *Malye planety* was for many years a reference source for observers and other researchers on minor planets. (M 18452)

(3578) Carestia

1977 CC. Discovered 1977 February 11 at the Felix Aguilar Observatory at El Leoncito.

Named in memory of Reinaldo Augusto Carestia (1932-1993) outstanding researcher who was, for more than 26 years, in charge of the Repsold Meridian Circle at the Felix Aguilar Observatory in San Juan, Argentina. He published five fundamental catalogues listing about 10,000 stars. He was also a professor of spherical astronomy and geographic determinations at the School of Topography of the National University of San Juan, as well as a member of the National Committee of Scientific and Technological Research of Chile. (M 24120)

(3579) 1977 YA

Discovered 1977 December 18 by M. Lovas at Piszkéstetö.

(3580) Avery

1983 CS_2. Discovered 1983 February 15 by N. G. Thomas at Anderson Mesa.

Named for the discoverer's grandson Avery Jordan Thomas, who was born 1994 March 4. (M 23540)

(3581) Alvarez

1985 HC. Discovered 1985 April 23 by C. S. Shoemaker and E. M. Shoemaker at Palomar.

Named in honor of Luis W. Alvarez {1911-1988}, physicist and Nobel laureate, and his son Walter Alvarez, geologist, both on the faculty of the University of California at Berkeley. The Alvarezes headed a team that discovered a global geochemical anomaly of noble metals at the Cretaceous-Tertiary boundary. They propounded the theory that a mass extinction of living species at the end of the Cretaceous period was triggered by impact of an asteroid or a comet about 10 km in diameter. Their work has stimulated intensive international research on the possible relationships between large body impacts and the evolution of life. (M 12807)
Obituary (L.W.A.) published in Phys. Bl., Jahrg. 45, Heft 1, p. 29 (1989).

(3582) Cyrano

1986 TT_5. Discovered 1986 October 2 by P. Wild at Zimmerwald.

Named for the whimsical French poet and soldier Cyrano de Bergerac (1619-1655), who in some of his comedies made brisk use of fanciful ways of spaceflight. (M 22498)

(3583) Burdett

1929 TQ. Discovered 1929 October 5 by C. W. Tombaugh at Flagstaff.
Named for the discoverer's home town in Kansas {see planet (3124)}. (M 12018)

(3584) Aisha

1981 TW. Discovered 1981 October 5 by N. G. Thomas at Anderson Mesa.

Named for the discoverer's granddaughter Aisha Renee Thomas, who was born 1991 June 13. (M 23540)

(3585) Goshirakawa

1987 BE. Discovered 1987 January 28 by T. Niijima and T. Urata at Ojima.

Named for the 77th emperor of Japan, Goshirakawa (1127-1192), who reigned from 1155 to 1158. He influenced five later emperors and managed to keep the authority of the imperial household during a period of increasing military power of the Genji and the Heike. (M 30798)
Name proposed by the second discoverer.

(3586) Vasnetsov

1978 SW_6. Discovered 1978 September 26 by L. V. Zhuravleva at Nauchnyj.

Named in memory of Viktor Mikhailovich Vasnetsov (1848-1926) and Apollinarij Mikhailovich Vasnetsov (1856-1933), famous Russian painters. (M 22245; M 26439)

(3587) Descartes

1981 RK_5. Discovered 1981 September 8 by L. V. Zhuravleva at Nauchnyj.

Named in honor of René Descartes (1596-1650), outstanding French mathematician and philosopher. (M 22245)
Descartes is also honored by a lunar crater.

(3588) Kirik

1981 TH_4. Discovered 1981 October 8 by L. I. Chernykh at Nauchnyj.

Named for Kirik Novgorodets, a twelfth-century chronicler from the town of Novgorod, author of the first Russian treatise on the luni-solar calendar. "The First Novgorod Chronicle" contains many of his remarks about the various astronomical events he observed. (M 13481)

(3589) Loyola

1984 AB_1. Discovered 1984 January 8 by J. F. Wagner at Anderson Mesa.

Named for the town in Spain, birthplace of Ignatius {see planet (3562)}, founder of the Jesuits. (M 17028)

(3590) Holst

1984 CQ. Discovered 1984 February 5 by E. Bowell at Anderson Mesa.

Named for English composer Gustav Holst (1874-1934). One of Holst's best-known and most popular compositions is the symphonic suite *The Planets*, written between 1914 and 1916. The titles of the seven movements are taken from the name of the seven major planets (excluding the Earth) known at the time the piece was written. Although the ideas for the scores are based on the astrological character of each planet, the work is nevertheless enjoyed by astronomers throughout the world. (M 12807)
Name suggested and citation provided by D. J. Tholen.

(3591) Vladimirskij

1978 QJ_2. Discovered 1978 August 31 by N. S. Chernykh at Nauchnyj.

Named in honor of Boris Mikhajlovich Vladimirskij, astronomer at the Crimean Astrophysical Observatory since 1958, well known for his researches in various fields of astronomy, especially high-energy astrophysics and solar-terrestrial relations. (M 20836)

(3592) Nedbal

1980 CT. Discovered 1980 February 15 by Z. Vávrová at Kleť.

Named in memory of the Czech composer and conductor Oskar Nedbal (1874-1930). (M 16443)

(3593) Osip

1981 EB_{20}. Discovered 1981 March 2 by S. J. Bus at Siding Spring.

Named in honor of David J. Osip. As an undergraduate at the Massachusetts Institute of Technology, Osip was involved in a number of projects concerning solar system occultation research, including both lunar occultations and the prediction of occultations of stars by Saturn's rings. In his current role as a research assistant at Lowell Observatory, he has taken part in asteroid occultation chases and has participated in an extensive program of cometary research. This research has involved the observation of comets using both photoelectric photometry and CCD imaging. (M 18452)

Citation provided by D. G. Schleicher at the request of the discoverer.

(3594) Scotti

1983 CN. Discovered 1983 February 11 by E. Bowell at Anderson Mesa.

Named in honor of James V. Scotti {1960- } of the University of Arizona, Tucson. Scotti works with the SPACEWATCH {see planet (4255)} Telescope, which is the 0.9-m reflector of the Steward Observatory on Kitt Peak. He has developed most of the system's software and has carried out final checks and data reduction for the CCD scanning observations of comets and minor planets. (M 12211)

(3595) Gallagher

1985 TF_1. Discovered 1985 October 15 by E. Bowell at Anderson Mesa.

Named in honor of John S. Gallagher III, director of Lowell Observatory. Gallagher's particular devotion to astronomy has been manifest in his service to the community, energetic and productive research, and teaching. He is a council member of the American Astronomical Society, is an editor of the Astrophysical Journal and has served on numerous committees and boards. His research has included work on luminous stars, novae, a broad variety of extragalactic problems, and cosmology. With a colleague, he discovered the ultraviolet output of novae and developed a method to evaluate star-formation rates at different stages in the evolution of galaxies. (M 12807)

Citation prepared by D. A. Hunter at the request of the discoverer.

(3596) Meriones

1985 VO. Discovered 1985 November 14 by P. Jensen and K. Augustesen at Brorfelde.

Named for the Greek warrior, who with Idomeneus {see planet (2759)} was leader of the Cretans, and who achieved distinction in the Trojan War, especially in the Battle of the Ships. (M 12211)

(3597) Kakkuri

1941 UL. Discovered 1941 October 15 by L. Oterma at Turku.

Named in honor of Juhani Kakkuri, since 1977 director of the Finnish Geodetic

Institute, interested also in astronomy. He developed for practical use the stellar triangulation method first presented by Väisälä {see planet (1573)}. The Institute measured and still measures standard baselines in various continents using the Väisälä interference method. (M 18453)

(3598) Saucier

1977 KK_1. Discovered 1977 May 18 by E. S. Bus at Palomar.

Named in honor of Agnes Elizabeth Saucier, grandmother of the discoverer. (M 12018)

(3599) Basov

1978 PB_3. Discovered 1978 August 8 by N. S. Chernykh at Nauchnyj.

Named in honor of the contemporary physicist Nikolaj Gennadievich Basov {1922- }, Nobel Prize Laureate {in 1964, with A. M. Prochorow and C. H. Townes} and one of the founders of quantum electronics. (M 22498)

(3600) Archimedes

1978 SL_7. Discovered 1978 September 26 by L. V. Zhuravleva at Nauchnyj.

Named for Archimedes (c. 287-212 B.C.), great ancient Greek scientist. (M 22245)

Archimedes is also honored by a lunar crater.

(3601) Velikhov

1979 SP_9. Discovered 1979 September 22 by N. S. Chernykh at Nauchnyj.

Named in honor of Evgenij Pavlovich Velikhov, member of the Russian Academy of Sciences, renowned for his research in plasma physics, magnetohydrodynamics and controllable thermonuclear synthesis. He is the president of the International Center for Scientific Culture-World Laboratory. (M 20836)

(3602) Lazzaro

1981 DQ_2. Discovered 1981 February 28 by S. J. Bus at Siding Spring.

Named in honor of Daniela Lazzaro (1956-), an expert on the dynamics of planetary satellites at the Observatório Nacional, Rio de Janeiro. She has worked on the dynamics of the β Pictoris system, and in 1993 she began working on the physical properties of minor planets and comets. Lazzaro is playing an important role in developing a researh group of planetary scientists in Brazil. (M 27458)

Name suggested and citation provided by M. A. Barucci.

(3603) Gajdušek

1981 RM. Discovered 1981 September 5 by L. Brožek at Kleť.

Named in memory of Vilém Gajdušek (1895-1977), Czech optician and maker of many outstanding telescopes at Czech and Slovak observatories. A 0.57-m f/5.2 reflector with Gajdušek's main mirror is used at Kleť {see planet (2199)} Observatory for CCD observations of minor planets and comets. (M 23792)

(3604) Berkhuijsen

5550 P-L. Discovered 1960 October 17 by C. J. van Houten and I. van Houten-Groeneveld at Palomar.

Named in honor of E. M. Berkhuijsen {1937- } of the Max-Planck-Institut für Radioastronomie, Bonn. Her fields of interest include continuum radio radiation of our Galaxy and other galaxies and the study of supernova remnants. (M 16885)

(3605) Davy

1932 WB. Discovered 1932 November 28 by E. Delporte at Uccle.

Named for Davy DeWinter, son of the current administrator of the Royal Observatory of Belgium, Mrs. Asselberghs. (M 14029)

Name proposed by G. Roland.

(3606) Pohjola

1939 SF. Discovered 1939 September 19 by Y. Väisälä at Turku.

Named for a place in the Finnish national epic Kalevala {see planet (1454)}. Pohjola, ruled by the mistress Louhi {see planet (3897)}, was the home of darkness, coldness and all kinds of badness, an exact opposite of Kalevala. (M 18453)

(3607) Naniwa

1977 DO_4. Discovered 1977 February 18 by H. Kosai and K. Hurukawa at Kiso.

This planet is being given the former name of Osaka, where co-discoverer Hurukawa grew up. (M 12018)

For another former name of the city of Osaka see the citation for minor planet (3320).

(3608) Kataev

1978 SD_1. Discovered 1978 September 27 by L. I. Chernykh at Nauchnyj.

Named in memory of the writer Valentin Petrovich Kataev (1897-1986). (M 13608)

(3609) Liloketai

1980 VM_1. Discovered 1980 November 13 at the Purple Mountain Observatory at Nanking.

Named in honor of Li Loketai, founder of the Zishan College. He has won a high reputation for his efforts to develop education in poor districts of China. (M 26424)

(3610) Decampos

1981 EA_1. Discovered 1981 March 5 by H. Debehogne and G. DeSanctis at La Silla.

Named in honor of Jose Adolfo Snajdauf de Campos, of the Valongo Observatory, Federal University of Rio de Janeiro. A teacher of astronomy and a researcher on the photometry of minor planets, he participates in the first discoverer's group for photometric observations at the European Southern Observatory. (M 19334)

(3611) Dabu

1981 YY_1. Discovered 1981 December 20 at the Purple Mountain Observatory at Nanking.

Named for a county in the northeastern part of Guangdong {see planet (2185)} province. Often called "the land of ceramics and tea", it enjoys a high reputation for its beautiful scenery and unique Hakka culture. It is the birthplace of observer J.-x. Yang. (M 34619)

(3612) Peale

1982 TW. Discovered 1982 October 13 by E. Bowell at Anderson Mesa.

Named in honor of Stanton J. Peale, planetary astronomer at the University of California at Santa Barbara. Peale has made important contributions to the theory of tidal interactions between orbiting bodies and resulting capture into spin-orbit and orbit-orbit resonances, and to chaotic orbital and rotational motion. He is perhaps best known for predicting, almost literally on the eve of the Voyager 1 encounter with Jupiter, that its satellite Io would show extensive evidence of volcanism as a result of tidal heating by Jupiter. (M 12808)
Citation prepared by A. W. Harris at the request of the discoverer.

(3613) Kunlun

1982 VJ_{11}. Discovered 1982 November 10 at the Purple Mountain Observatory at Nanking.
Named for a mountain range in the western part of China. (M 34619)

(3614) Tumilty

1983 AE_1. Discovered 1983 January 12 by N. G. Thomas at Anderson Mesa.

Named in honor of Jodi Anne Tumilty Thomas, daughter-in-law of the discoverer. The name is derived from the Irish 'Tomaltach'. (M 25443)

(3615) Safronov

1983 WZ. Discovered 1983 November 29 by E. Bowell at Anderson Mesa.

Named in honor of Victor S. Safronov {1917- } of the O. Yu. Shmidt Institute of Earth Physics in Moscow. Starting with Shmidt's pioneering work from 1944 to 1960, Safronov developed a comprehensive theory of the origin of planetary bodies from the protoplanetary swarm. This work has well stood the tests of time and scrutiny by other scientists and it is the basis of most modern research on the origin of terrestrial and minor planets. (M 12808)
Citation prepared by A. W. Harris at the request of the discoverer.

(3616) Glazunov

1984 JJ_2. Discovered 1984 May 3 by L. V. Zhuravleva at Nauchnyj.

Named in honor of Il'ya Sergeevich Glazunov (1930-), well-known Russian painter. (M 22245)

(3617) Eicher

1984 LJ. Discovered 1984 June 2 by B. A. Skiff at Anderson Mesa.

Named in honor of David J. Eicher, founder and editor-in-chief of *Deep Sky* magazine. In June 1977, as a high-school student, Eicher began *Deep Sky Monthly*, designed to promote amateur deep-sky observing. After more than fifty issues, the monthly became a quarterly in 1982, and Eicher became an assistant editor of *Astronomy* magazine. Through his magazine and two books on viewing, Eicher has worked tirelessly to encourage amateur astronomers to make high-quality observations of galaxies, clusters and nebulae. (M 16245)
Name proposed and citation provided by D. H. Levy.

(3618) Kuprin

1979 QP_8. Discovered 1979 August 20 by N. S. Chernykh at Nauchnyj.

Named in honor of Aleksandr Ivanovich Kuprin (1870-1938), famous Russian writer. (M 20836)

(3619) Nash

1981 EU_{35}. Discovered 1981 March 2 by S. J. Bus at Siding Spring.

Named in honor of Douglas B. Nash of the San Juan Capistrano Research Institute, California, in recognition of his many contributions to the understanding of the compositions and processes affecting solar system bodies. Nash has achieved these advances by innovation and persistence in designing, executing and applying the results of laboratory investigations. His work on the spectral reflectances of lunar samples, meteorites, rock and mineral samples, and frozen gases has allowed comparison with and interpretation of telescopic data. His investigations of luminescence, ultraviolet irradiation, sputtering by a variety of ion species, and evaporation have led to the identification of new "space weathering" processes and a better understanding of the surfaces of the moon and the Galilean satellites of Jupiter. (M 18453)
Citation provided by Dennis L. Matson at the request of the discoverer.

(3620) Platonov

1981 RU_2. Discovered 1981 September 7 by L. G. Karachkina at Nauchnyj.

Named in memory of the writer Andrej Platonovich Platonov (1899-1951). (M 13176)

(3621) Curtis

1981 SQ_1. Discovered 1981 September 26 by N. G. Thomas at Anderson Mesa.

Named in memory of the discoverer's high-school science teacher Curtis R. Carbutt (1911-1976), who was known for his personal attention to his students. For nearly 40 years, Curtis was a school administrator, teacher and junior-college instructor in Colorado and California. (M 24916)

Assistance with the citation was provided by B. Carbutt.

(3622) Ilinsky

1981 SX_7. Discovered 1981 September 29 by L. V. Zhuravleva at Nauchnyj.

Named in memory of the actor Igor' Vladimirovich Il'insky (1901-1987). (M 13176)

(3623) Chaplin

1981 TG_2. Discovered 1981 October 4 by L. G. Karachkina at Nauchnyj.

Named in memory of the great film actor and producer Charles Spencer Chaplin (1889-1977). (M 13176)

(3624) Mironov

1982 TH_2. Discovered 1982 October 14 by L. G. Karachkina and L. V. Zhuravleva at Nauchnyj.

Named in memory of Soviet actor and producer Andrej Aleksandrovich Mironov (1941-1987). (M 13176)

(3625) Fracastoro

1984 HZ_1. Discovered 1984 April 27 by W. Ferreri at La Silla.

Named in honor of Mario Girolamo Fracastoro, former director of the Catania and the Pino Torinese {see planet (2694)} observatories. His early research (in Florence) included solar physics, astronomical technology and the physics of photometric binaries. In 1963 he began a systematic study of the star RS CVn. More recently, he has been involved in astrometric work on binary systems, serving as president of IAU Commission 26 during 1982-1985 and as a member of the Hipparcos mission definition group. (M 13608)

(3626) Ohsaki

1929 PA. Discovered 1929 August 4 by M. Wolf at Heidelberg.

Named in honor of Shoji Ohsaki (1912-) on the occasion of the publication of his life work of historical research on the Chinese constellations. An amateur astronomer and historian, Ohsaki is well known for his work on the astronomical histories of the far-eastern countries and as a collaborator in S. Kanda's {see planet (2248)} 1935 publication of astronomical phenomena in Japanese historical records before the year 1600. (M 12973)

Name proposed by H. Oishi, who found the identifications involving this minor planet.

(3627) Sayers

1973 DS. Discovered 1973 February 28 by L. Kohoutek at Bergedorf.

Named in memory of Dorothy Leigh Sayers (1893-1957), British author of the Peter Wimsey detective stories, which she wrote during the 1920s and 1930s. In later life she tried hard to play down this notoriety, turning her considerable talents instead to the writing of theological plays and translating the works of Dante {see planet (2999)}. (M 34619)

Name suggested by B. G. Marsden, with whom she consulted extensively during the last year of her life in an attempt to rehabilitate the Roman poet Lucan, whose astronomy and geography had been condemned by other twentieth-century critics.

(3628) Božněmcová

1979 WD. Discovered 1979 November 25 by Z. Vávrová at Kleť.

Named in memory of Božena Němcová (1820-1862), outstanding Czech writer, author of the well-known book *Granny*. (M 22245)

(3629) Lebedinskij

1982 WK. Discovered 1982 November 21 by A. Mrkos at Kleť.

Named in memory of Aleksandr Ignat'evich Lebedinskij (1913-1967), astrophysicist at Moscow State University known for his research on the cosmogony of stars and planets, novae, aurorae and the magnetohydrodynamics of the solar atmosphere, as well as for his design of astronomical and geophysical instruments. He and the discoverer collaborated on spectroscopic research on aurorae in Antarctica during the International Geophysical Year. (M 26424)

(3630) Lubomír

1984 QN. Discovered 1984 August 28 by A. Mrkos at Kleť.

A well-known old Slavic name, widespread in southern Bohemia {see planet (371)}. (M 25976)

(3631) Sigyn

1987 BV_1. Discovered 1987 January 25 by E. W. Elst at La Silla.

Named by the discoverer in honor of his daughter, who has a lively interest in astronomy and who accompanied him on his mission from Belgium to the Bulgarian National Observatory last year. (M 12459)

(3632) Grachevka

1976 SJ_4. Discovered 1976 September 24 by N. S. Chernykh at Nauchnyj.

Named for a village in the Tambov (now Lipetsk) region in Russia, the birthplace of the discoverer's parents, Stepan Semenovich Chernykh (1904-1942) and Melaniya Petrovna Chernykh (1910-1977). (M 20836)

(3633) Mira

1980 EE_2. Discovered 1980 March 13 at the Felix Aguilar Observatory at El Leoncito.

Named in memory of Hugo Mira (1937-1994), outstanding researcher who worked for more than twenty-five years at the discovery site of this minor planet. Mira was an observer at the transit circle of the U.S. Naval Observatory's Yale-Columbia Southern Observatory for four years. He also worked for twenty years on minor planets and proper motions with the double astrograph. (M 24120)

(3634) Iwan

1980 FV. Discovered 1980 March 16 by C.-I. Lagerkvist at La Silla.

Named in honor of Iwan P. Williams, of Queen Mary College, London, in recognition of his well-known work on meteor streams and interest in comets and minor planets. The discoverer appreciates their long and fruitful collaboration. (M 15427)

(3635) 1981 WO_1

Discovered 1981 November 21 by L. Kohoutek at Calar Alto.

(3636) Pajdušáková

1982 UJ_2. Discovered 1982 October 17 by A. Mrkos at Kleť.

Named in memory of Ludmila Pajdušáková (1916-1979), the first Slovak woman astronomer, a specialist in solar astronomy and recognized as the discoverer of five comets. Some time wife of the discoverer, she was the third director of the Skalnaté Pleso Observatory {see planet (2619)} and Astronomical Institute in Tatranská Lomnica. (M 26424)

(3637) O'Meara

1984 UQ. Discovered 1984 October 23 by B. A. Skiff at Anderson Mesa.

Named in honor of Stephen James O'Meara. Well known for his fine drawings of solar system objects, O'Meara joined *Sky and Telescope* {see planet (3243)} in 1979 and in 1989 was promoted to associate editor. His unique visual observations have included pre-Voyager sightings of spokes in Saturn's B ring and a determination of Uranus' rotation period. In January 1985 he made an apparent visual observation of P/Halley at magnitude 19.6 using a 0.6-m telescope on Mauna Kea. O'Meara's writing and remarkable visual skill inspires other observers to improve the quality of their observations. An energetic speaker, he has taught introductory classes in astronomy in Massachusetts and Hawaii and is regularly sought for astronomy talks in the U.S. and abroad. (M 16245)

Name suggested and citation provided by D. H. Levy.

(3638) Davis

1984 WX. Discovered 1984 November 20 by E. Bowell at Anderson Mesa.

Named in honor of Donald R. Davis, senior scientist at the Planetary Science Institute in Tucson. Davis has made fundamental theoretical and experimental contributions to research on the collisional evolution of minor planets. With colleagues, he was the first to propose the "gravitationally bound rubble pile" model for large minor planets. Another of his research interests is infrared searching for intramercurial bodies. (M 12459)

Citation written by R. P. Binzel at the request of the discoverer.

(3639) Weidenschilling

1985 TX. Discovered 1985 October 15 by E. Bowell at Anderson Mesa.

Named in honor of Stuart J. Weidenschilling, research scientist at the Planetary Science Institute in Tucson. Weidenschilling is a noted expert in the study of the origin of the solar system, and his research has also included collisional evolution of minor planets. He and colleagues are conducting a program of "photometric geodesy" to model the shapes of large, rapidly rotating minor planets from extensive lightcurve observations. (M 12459)

Citation written by R. P. Binzel at the request of the discoverer.

(3640) Gostin

1985 TR_3. Discovered 1985 October 11 by C. S. Shoemaker and E. M. Shoemaker at Palomar.

Named in honor of Victor A. Gostin, geologist on the faculty of the University of Adelaide, South Australia. A specialist in sedimentology and stratigraphy, Gostin discovered in the Flinders Ranges of South Australia a deposit of shocked debris ejected from the Lake Acraman impact structure about 300 km to the west. His careful studies of this ancient deposit have provided the first detailed picture of the distant ejecta from a known large terrestrial impact crater. (M 12808)

(3641) Williams Bay

A922 WC. Discovered 1922 November 24 by G. Van Biesbroeck at Williams Bay.

Named for the village where the Yerkes Observatory is situated and where this minor planet was discovered. (M 14632)

Name proposed by Micheline Wilson, daughter of the discoverer.

(3642) Frieden

1953 XL_1. Discovered 1953 December 4 by H. Gessner at Sonneberg.

This planet is given the German form of the goddess Pax {see planet (679)} in the hope for peace around the world. (M 12808)

(3643) 1978 UN_2

Discovered 1978 October 29 at the Purple Mountain Observatory at Nanking.

(3644) Kojitaku

1931 TW. Discovered 1931 October 5 by K. Reinmuth at Heidelberg.

Named for Takuo Kojima, who has made more than 1,000 accurate observations of comets and discovered more than 10 numbered minor planets. He has also written the Comet Observers Guide column in *Gekkan-Tenmon* since 1990. (M 22498)

Name proposed by T. Kobayashi, who found the identifications involving this minor planet. Until 1996, Kojima has discovered 23 numbered minor planets.

(3645) Fabini

1981 QZ. Discovered 1981 August 28 by A. Mrkos at Kleť.

Named in memory of Tatiana Fabini (1943-1989), editor-in-chief of the Slovak popular-astronomy magazine *Kozmos* and a source of inspiration for many amateur astronomers. (M 26424)

Obituary published in Kozmos, Vol. 20, No. 2, p. 39 (1989).

(3646) Aduatiques

1985 RK_4. Discovered 1985 September 11 by H. Debehogne at La Silla.

Named for the Aduatiques, an ancient Belgian tribe from Namur {see planet (3374)}. The tribe is mentioned in Julius Caesar's *Gallic Wars*. (M 23792)

(3647) Dermott

1986 AD_1. Discovered 1986 January 11 by E. Bowell at Anderson Mesa.

Named in honor of Stanley F. Dermott, planetary scientist at Cornell University, Ithaca, New York. Dermott's unusually broad range of contributions includes work on planetary origin, resonance effects on planetary satellites, rings, zodiacal dust bands, tidal interactions, and minor planets, for the last of which he is known for his study of the statistics of spin rates. (M 12808)

Citation prepared by A. W. Harris at the request of the discoverer.

(3648) Raffinetti

1957 HK. Discovered 1957 April 24 at the La Plata Observatory at La Plata.

Named in memory of Virgilio Raffinetti (1869-1946), Argentine astronomer, who was director of La Plata Observatory between 1889 and 1905. Under his personal supervision the Gautier meridian circle was installed at this Observatory in 1903. (M 22498)

(3649) Guillermina

1976 HQ. Discovered 1976 April 26 at the Felix Aguilar Observatory at El Leoncito.

Named in honor of Maria Guillermina Martin de Cesco (1915-), widow of Carlos U. Cesco {see planet (1571)} and mother of M. R. Cesco. She was the first student registered at the Astronomy School of La Plata State University and was a generous collaborator of Ing. Felix Aguilar {see planet (1800)}. (M 24120)

(3650) 1978 UO_2

Discovered 1978 October 30 at the Purple Mountain Observatory at Nanking.

(3651) Friedman

1978 VB_5. Discovered 1978 November 7 by E. F. Helin and S. J. Bus at Palomar.

Named in honor of Louis and Connie Friedman on the occasion of their 25th wedding anniversary. As executive director of the Planetary Society, Louis D. Friedman is a leader in promoting the exploration of the solar system. At the Jet Propulsion Laboratory, he originated the International Halley Watch. Connie Friedman has shared her husband's excitement for planetary exploration. (M 13481)

Name proposed by the first discoverer, endorsed by R. L. Staehle.

(3652) Soros

1981 TC_3. Discovered 1981 October 6 by T. M. Smirnova at Nauchnyj.

Named in honor of George Soros, the founder of the International Science Foundation. This charitable organization provides financial support to scientifically meritorious scientists of the former Soviet Union in the field of fundamental civilian research. (M 23136)
Name proposed by the Institute of Theoretical Astronomy.

(3653) Klimishin

1979 HF_5. Discovered 1979 April 25 by N. S. Chernykh at Nauchnyj.

Named in honor of Ivan Antonovich Klimishin, professor of astronomy at the Ivano-Frankovsk Pedagogical Institute. Prominent in the field of cosmic hydrodynamics, Klimishin is also an expert on the history of astronomy and is the author of some popular astronomical books. (M 22498)

(3654) AAS

1949 QH_1. Discovered 1949 August 21 at the Goethe Link Observatory at Brooklyn, Indiana.

Named in honor of the American Astronomical Society, the professional society of American astronomers and astrophysicists from its founding in 1899 until today. Through its journals, its meetings, and especially its members now numbering over five thousand, it has promoted research and education in every branch of astronomy, including the study of minor planets. (M 19334)

Name proposed by D. E. Osterbrock and F. K. Edmondson. Citation by D. E. Osterbrock.

(3655) Eupraksia

1978 SA_3. Discovered 1978 September 26 by L. V. Zhuravleva at Nauchnyj.

Named for the wife of the thirteenth-century prince Fyodor Ryazansky. She preferred death to being taken prisoner by the Tatar-Mongolians. (M 13176)

(3656) Hemingway

1978 QX. Discovered 1978 August 31 by N. S. Chernykh at Nauchnyj.

Named in honor of Ernest Hemingway (1899-1961), great American writer. (M 22498)
Hemingway received the 1954 Literature Nobel Prize.

(3657) Ermolova

1978 ST_6. Discovered 1978 September 26 by L. V. Zhuravleva at Nauchnyj.

Named in memory of Maria Nikolaevna Ermolova (1853-1928), outstanding Russian actress. (M 22246)

(3658) Feldman

1982 TR. Discovered 1982 October 13 by E. Bowell at Anderson Mesa.

Named in honor of Paul D. Feldman, professor of physics and astronomy at the Johns Hopkins University, Baltimore, for his numerous contributions in ultraviolet spectroscopy, particularly of the Earth's atmosphere, Venus, the outer planets and comets. His design and supervision of a number of spacecraft instruments have led to many advances in our understanding of physical processes in the solar system. The minor planet also honors Paul A. Feldman, a radioastronomer at the Herzberg Institute of Astrophysics, Ottawa, who is currently involved in making submillimeter observations of minor planets. (M 12974)
Name suggested and citation provided by M. F. A'Hearn and B. G. Marsden.

(3659) Bellingshausen

1969 TE_2. Discovered 1969 October 8 by L. I. Chernykh at Nauchnyj.

Named in memory of the Russian admiral Faddej Faddeevich Bellingshausen

(1778-1852), head of the first Russian Antarctic expedition in 1819-1821. (M 13608)
Bellingshausen is also honored by a lunar crater.

(3660) Lazarev

1978 QX_2. Discovered 1978 August 31 by N. S. Chernykh at Nauchnyj.
Named in honor of the Russian admiral Mikhail Petrovich Lazarev (1797-1851), a participant in the first Russian Antarctic expedition and commander of the ship Mirnyj. (M 13608)

(3661) Dolmatovskij

1979 UY_3. Discovered 1979 October 16 by N. S. Chernykh at Nauchnyj.
Named in honor of Evgenij Aronovich Dolmatovskij, poet and publicist. (M 20836)

(3662) Dezhnev

1980 RU_2. Discovered 1980 September 8 by L. V. Zhuravleva at Nauchnyj.
Named in honor of Semen Ivanovich Dezhnev (c. 1605-1673), Russian seafarer. (M 22246)

(3663) Tisserand

1985 GK_1. Discovered 1985 April 15 by E. Bowell at Anderson Mesa.
Named for François Tisserand (1845-1896), contributor to the theory of general perturbations (Tisserand's criterion) and lunar theory, and author of the four-volume compendium *Traité de Mécanique Céleste.* (M 12974)
Name suggested and citation provided by G. Reaves.
Tisserand is also honored by a lunar crater.

(3664) Anneres

4260 P-L. Discovered 1960 September 24 by C. J. van Houten and I. van Houten-Groeneveld at Palomar.
Named in honor of Anna Theresia ("Anneres") Schmadel. She is the wife of Lutz D. Schmadel {see planet (2234)}, a senior staff astronomer at the Astronomisches Rechen-Institut in Heidelberg. (M 16591)

(3665) Fitzgerald

1979 FE. Discovered 1979 March 19 by A. Mrkos at Kleť.
Named in memory of Ella Fitzgerald (1917-1996), renowned jazz singer known internationally as the "First Lady of Song". A phenomenal improvisationalist who used her voice as if it were a musical instrument, she was also known for her albums of definitive performances of the works of many of the leading songwriters of the first half of the twentieth century. (M 27733)
Name proposed by M. Tichý and B. G. Marsden.

(3666) Holman

1979 HP. Discovered 1979 April 19 by J. C. Muzzio at Cerro Tololo.
Named in honor of Matthew J. Holman (1967-), astronomer in the planetary sciences division at the Harvard-Smithsonian Center for Astrophysics. He is particularly known for his long-term investigations on the stability of the outer solar system by means of the sympletic integrator he codeveloped, and he has studied the stability of planets around other stars. Recently, he has also become an active observer of centaurs and transneptunian objects. (M 34619)

(3667) Anne-Marie

1981 EF. Discovered 1981 March 9 by E. Bowell at Anderson Mesa.
Named for Anne-Marie Malotki, a friend of the discoverer. (M 12974)

(3668) Ilfpetrov

1982 UM_7. Discovered 1982 October 21 by L. G. Karachkina at Nauchnyj.

Named for Il'ya Arnol'dovich Feinsilberg (1897-1937) and Eugeny Petrovich Kataev (1903-1942), the authors of the books *Twelve Chairs* and *Golden Calf.* (M 13176)

(3669) Vertinskij

1982 UO_7. Discovered 1982 October 21 by L. G. Karachkina at Nauchnyj.

Named in memory of Aleksandr Nikolaevich Vertinskij (1889-1957), variety actor, composer and poet, famous for his sincere rendition of songs with words by Blok, Akhmatova and Tsvetaeva {see, respectively, planets (2540), (3067) and (3511)}. (M 20836)

(3670) Northcott

1983 BN. Discovered 1983 January 22 by E. Bowell at Anderson Mesa.

Named for the Canadian astronomer Ruth Josephine Northcott (1913-1969), an associate professor at the University of Toronto and a staff member of the David Dunlap Observatory from its inception in 1935. A specialist in stellar radial velocities, Northcott served on IAU Commissions 26 (double stars), 30 (radial velocities) and 41 (history of astronomy). In 1956 she succeeded C. A. Chant {see planet (3315)} as editor of the Observer's Handbook of the Royal Astronomical Society of Canada, a position she held until her untimely death. (M 12974)
Name suggested and citation provided by C. E. Spratt.

Obituaries published in J.R. Astron. Soc. Canada, Vo. 63, p. 225-226 (1969); Sky Telesc., Vol. 38, p. 211 (1969).

(3671) Dionysus

1984 KD. Discovered 1984 May 27 by C. S. Shoemaker and E. M. Shoemaker at Palomar.

Originating in Thrace, where he was revered as a god of wine, the cult of Dionysus spread gradually through Greece. Dionysus became, in turn, the god of vegetation and warm moisture, the god of pleasure, the god of civilization, and ultimately a kind of supreme god and a symbol of rebirth or everlasting life. In some instances he appeared as a Sun deity. (M 19347)

The original name 'Dionysius' (M 12808) was corrected by the Minor Planet Names Committee.

(3672) Stevedberg

1985 QQ. Discovered 1985 August 22 by E. Bowell at Anderson Mesa.

Named in honor of Stephen J. Edberg, planetary scientist at the Jet Propulsion Laboratory, for his long commitment to bringing amateur and professional astronomers together. Although his most noted activity has been his work with the Amateur Observations Net of the International Halley Watch, he has also been a leader in the program for amateur use of the Hubble Space Telescope, in the Mars project of the Planetary Society, and as president of the Western Amateur Astronomers. (M 12974)
Citation prepared by D. H. Levy at the request of the discoverer.

(3673) Levy

1985 QS. Discovered 1985 August 22 by E. Bowell at Anderson Mesa.

Named in honor of David H. Levy {1948- }, comet discoverer and observer, recognized for his perseverance in observing comets using the oldest visual and the newest electronic techniques. Author of several books and articles, he is known for his biographies of astronomers. As an educator Levy has concentrated on bringing observational astronomy to both amateur astronomers and to children, and he has initiated school and camp programs for this purpose. (M 12974)
Citation prepared by S. J. Edberg at the request of the discoverer.

(3674) Erbisbühl

1963 RH. Discovered 1963 September 13 by C. Hoffmeister at Sonneberg.

Named for the mountain on which the Sonneberg {see planet (1039)} Observatory is situated. The discoverer, who founded the observatory, lived and worked there for many decades. (M 12809)

(3675) Kemstach

1982 YP_1. Discovered 1982 December 23 by L. G. Karachkina at Nauchnyj.

Named in memory of Marfa Vladimirovna Kemstach (1888-1971) and Semen Stepanovich Kemstach (1880-1938), grandparents of the discoverer. (M 13177)

(3676) Hahn

1984 GA. Discovered 1984 April 3 by E. Bowell at Anderson Mesa.

Named in honor of Gerhard Hahn, a planetary astronomer at Uppsala Observatory and a member of the research group studying minor planets and comets. Hahn has undertaken extensive photometry and astrometry of minor planets and has been studying the long-term orbital evolution and physical properties of these objects. (M 12975)
Citation prepared by C.-I. Lagerkvist at the request of the discoverer.

(3677) Magnusson

1984 QJ_1. Discovered 1984 August 31 by E. Bowell at Anderson Mesa.

Named in honor of Per Magnusson, a planetary astronomer at Uppsala Observatory and a member of the research group studying minor planets and comets. Well known for his method of determining the spin characteristics of minor planets, Magnusson has estimated pole directions for more than thirty objects. (M 12975)
Citation prepared by C.-I. Lagerkvist at the request of the discoverer.

(3678) Mongmanwai

1966 BO. Discovered 1966 January 20 at the Purple Mountain Observatory at Nanking.

Named in honor of Man Wai Mong, prominent in the Chinese electrical engineering community. Mong takes an active interest in the development of education and sciences at several universities in China. (M 27125)

(3679) Condruses

1984 DT. Discovered 1984 February 24 by H. Debehogne at La Silla.

Named for the Condruses, an ancient Belgian tribe from southeast Namur {see planet (3374)} and southern Liège. The tribe is mentioned in Julius Caesar's *Gallic Wars*. (M 23792)

(3680) Sasha

1987 MY. Discovered 1987 June 28 by E. F. Helin at Palomar.

Named in honor of Alexandra Rachel Druyan Sagan, young daughter of Carl Sagan {see planet (2709)} and Ann Druyan {see planet (4970)}. Sasha's charismatic parents are passionately committed to preserving a peaceful, intact world to their children to inherit. This is well expressed in the dedication of his book *Cosmos*: "for Alexandra, who comes of age with the millenium. May we leave your generation a world better than the one we were given." In the hope that they will continue and expand the quests of their parents, this minor planet is dedicated to Sasha and her contemporaries throughout the world. (M 12459)

(3681) Boyan

1974 QO_2. Discovered 1974 August 27 by L. I. Chernykh at Nauchnyj.

Named for Wise Boyan, the legendary old Russian story-teller and bard. (M 18453)

(3682) Welther

A923 NB. Discovered 1923 July 12 by K. Reinmuth at Heidelberg.

Named in honor of Barbara Welther (1938-), historian of science at the Harvard-Smithsonian Center for Astrophysics. Barbara has recently completed an educational video examining the life and work of Annie Jump Cannon {see planet (1120)}. (M 22499)

Name proposed by members of the Planetary Sciences division, the identifications for this object having been made by B. G. Marsden.

(3683) Baumann

1987 MA. Discovered 1987 June 23 by W. Landgraf at La Silla.

Named in honor of Paul Baumann (1901-1976) and his wife Helene (1899-1986). An amateur astronomer since 1959, Baumann founded the astronomical association (1961) and public observatory (1962) in Mainz, was involved in the establishment of several other associations in this area and was well known throughout Germany and in many other countries. Baumann was also a member of the first parliament of Rheinland-Pfalz (1947-1951) and for a long time a member of the Mainz city council. The Baumann family were good friends of the father of the discoverer for several decades. (M 12459)

(3684) Berry

1983 AK. Discovered 1983 January 9 by B. A. Skiff at Anderson Mesa.

Named in honor of Richard Berry, since 1979 editor-in-chief of *Astronomy.* He has actively promoted amateur astronomy by means of books on beginning astronomy and telescope making as well as editorship of the technical journal *Telescope Making.* He was also a pioneer in increasing awareness of light pollution by mapping its extent in southern Ontario in the 1970s. (M 16246)

Citation prepared by D. H. Levy and S. J. Edberg at the request of the discoverer.

(3685) Derdenye

1981 EH_{14}. Discovered 1981 March 1 by S. J. Bus at Siding Spring.

Named in honor of Derald and Denise Nye, dedicated amateur astronomers and observatory builders in Tucson. For the past thirteen years, Derald, with the assistance of Denise, has served as distributor for the *Minor Planet Bulletin,* the publication of the Minor Planets Section of the Association for Lunar and Planetary Observers. Through this work they have served as a contact point for hundreds of amateur astronomers around the world seeking to contribute to minor planet research. (M 27125)

Name suggested and citation prepared by R. P. Binzel.

(3686) Antoku

1987 EB. Discovered 1987 March 3 by T. Niijima and T. Urata at Ojima.

Named for the 81st emperor of Japan, Antoku (1178-1185), son of Kenreimonin {see planet (5242)}. He reigned from 1180 to 1185, and when the Heike was destroyed at Dannoura, he was drowned in Niinoama's {see planet (4959)} arms. (M 30798)

Name proposed by second discoverer.

(3687) Dzus

A908 TC. Discovered 1908 October 7 by A. Kopff at Heidelberg.

Named in honor of Paul K. Dzus {1969- } in appreciation of his helpful assistance at the Minor Planet Center since October 1987, much of the time as a volunteer. (M 14029)

Named by B. G. Marsden, who found the identifications involving this minor planet, on Dzus' departure for the University of Arizona.

(3688) Navajo

1981 FD. Discovered 1981 March 30 by E. Bowell at Anderson Mesa.

This 2:1 Jupiter librator is named for the indigenous North American people, inhabitants of a 60,000-square kilometer area of Arizona, New Mexico and Utah. The Navajo people probably migrated to the southwest United States in the fifteenth century. Today their flourishing culture is particularly renowned for its artistic achievement, including weaving and silver work. (M 12975)
P. E. Roques contributed to the citation.

(3689) Yeates

1981 JJ_2. Discovered 1981 May 5 by C. S. Shoemaker at Palomar.

Named in honor of Anthony N. Yeates, geologist with the Bureau of Mineral Resources of the Commonwealth of Australia. In the course of regional geologic mapping at the southern edge of the Great Sandy desert of Western Australia, Yeates led a team of geologists that discovered the Veevers meteorite crater. This site, discovered in 1975, is the fifteenth and latest recognized locality where meteorites have been found associated with an impact crater. (M 12809)

(3690) Larson

1981 PM. Discovered 1981 August 3 by E. Bowell at Anderson Mesa.

Named in honor of Stephen M. Larson, planetary scientist at the Lunar and Planetary Laboratory. Well known for his work on comets, Larson has recently digitized many 1910 images of Comet Halley to bring out fresh details of jet structure and has been active in organizing the Near Nucleus Studies Net of the International Halley Watch. His program of monitoring comets by means of spectroscopy and imaging continues to reveal interesting aspects of cometary phenomena. With J. Fountain, he correctly demonstrated the existence of and specified the revolution period of Saturn's eleventh satellite before the extensive 1980 data completely clarified the confusing situation presented at the 1966 ring-plane passage. (M 12975)
Name suggested and citation prepared by D. H. Levy.

(3691) 1982 FT

Discovered 1982 March 29 by L. E. Gonzalez at Cerro El Roble.

(3692) Rickman

1982 HF_1. Discovered 1982 April 25 by E. Bowell at Anderson Mesa.

Named in honor of Hans Rickman, a planetary astronomer at Uppsala Observatory and a member of the research group studying minor planets and comets. Rickman is widely known for his theoretical modelling of cometary nuclei, and he is also interested in the relationships between comets and planet-crossing asteroids. He has for a long time favored the popularization of astronomy, and to this end he has written a large number of articles in popular magazines and has lectured to general audiences. Rickman is a co-organizer of the "Asteroids, Comets, Meteors" meetings that are held regularly in Uppsala. (M 12975)
Citation prepared by C.-I. Lagerkvist at the request of the discoverer.

(3693) Barringer

1982 RU. Discovered 1982 September 15 by E. Bowell at Anderson Mesa.

Named for Daniel Moreau Barringer (1860-1929), pioneer investigator of terrestrial meteorite craters. From intensive field studies, including much subsurface exploration, Barringer demonstrated the impact origin of Meteor Crater, Arizona, now widely known as the Barringer Crater, the first recognized impact crater on the Earth. On the basis of his studies, Barringer strongly supported the hypothesis of impact origin for the craters on the Moon. (M 12976)

Name proposed jointly by E. M. and C. S. Shoemaker.
Barringer is also honored by a lunar crater.

(3694) Sharon

1984 SH_5. Discovered 1984 September 27 by A. Grossman at Palomar.
Named in honor of Sharon Rachel Vinick, a great source of joy and inspiration to the discoverer. (M 15260)

(3695) 1973 UU_4

Discovered 1973 October 21 by H. L. Giclas at Anderson Mesa.

(3696) Herald

1980 OF. Discovered 1980 July 17 by E. Bowell at Anderson Mesa.
Named in honor of the Australian amateur astronomer David Herald for his astrometric observations of comets and his work on occultations by minor planets. The number of his cometary positions is the second highest in the world for an amateur astronomer in modern times, and his prompt and reliable attention to new comets discovered in the southern hemisphere has been invaluable. (M 12809)
Name suggested by B. G. Marsden, who wrote the citation.

(3697) Guyhurst

1984 EV. Discovered 1984 March 6 by E. Bowell at Anderson Mesa.
Named in honor of Guy M. Hurst, editor of the British amateur astronomical journal *The Astronomer* {see planet (6411)} since 1975. An enthusiastic visual observer of comets and variable stars himself, he is in contact with an enormous network of other observers all over western Europe, and his frequent assistance in thereby securing verifications of reports for the IAU Central Telegram Bureau is greatly appreciated. (M 12809)
Name suggested by R. H. McNaught and B. G. Marsden, the citation being prepared by the latter.

(3698) Manning

1984 UA_2. Discovered 1984 October 29 by E. Bowell at Anderson Mesa.
Named in honor of Brian Manning, British amateur cometary astrometrist, whose numerous contributions to the IAU Circulars and Minor Planet Circulars, as well as to the astrometric campaigns of International Halley Watch, have been consistently timely and of the highest accuracy. (M 12809)
Name suggested by D. K. Yeomans and B. G. Marsden, the citation being prepared by the latter.

(3699) Milbourn

1984 UC_2. Discovered 1984 October 29 by E. Bowell at Anderson Mesa.
Named in honor of Stanley William Milbourn {1925-1997}, editor of the circulars of the British Astronomical Association during 1969-1986, director of the comet section during 1968-1977 and currently assistant director of the computing section. Long interested in astronomical computations, particularly those involving returning periodic comets, he has produced a steady stream of accurate predictions over the years, as he has adapted procedures from mechanical calculating machine to pocket calculator to personal computer. (M 12810)
Name suggested by B. G. Marsden, who wrote the citation.
Obituaries published in Int. Comet Q., Vol. 20, No. 1, p. 3 (1998); J. Br. Astron. Assoc., Vol. 108, No. 3, p. 181-182 (1998).

(3700) Geowilliams

1984 UL_2. Discovered 1984 October 23 by C. S. Shoemaker and E. M. Shoemaker at Palomar.

Named in honor of George E. Williams, exploration geologist with Broken Hill Propriety Co., Ltd. at Adelaide, South Australia. Williams discovered the Lake Acraman impact structure of South Australia, the largest such feature found so far on the Australian continent. He also discovered rhythmically layered sedimentary deposits of Precambrian age that exhibit periods closely matching those of the modern solar cycle. (M 12810)

(3701) Purkyně

1985 DW. Discovered 1985 February 20 by A. Mrkos at Kleť.

Named in memory of J. E. Purkyně (1787-1869), professor of physiology in Wroclaw and Prague. He discovered the physiological effect, well known to observers of variable stars, that causes comparable variations in brightness to appear more extreme for red light than for blue light. (M 26424)

(3702) Trubetskaya

1970 NB. Discovered 1970 July 3 by L. I. Chernykh at Nauchnyj.

Named for Ekaterina Ivanovna Trubetskaya (1800-1854), a princess who voluntarily followed her husband, the Decembrist S. P. Trubetskoj, to exile in Siberia. (M 13609)

(3703) Volkonskaya

1978 PU_3. Discovered 1978 August 9 by L. I. Chernykh at Nauchnyj.

Named for Mariya Nikolaevna Volkonskaya (1805-1865), a princess who voluntarily followed her husband, the Decembrist S. G. Volkonskij, to exile in Siberia. (M 13609)

(3704) 1981 YX_1

Discovered 1981 December 20 at the Purple Mountain Observatory at Nanking.

(3705) Hotellasilla

1984 ET_1. Discovered 1984 March 4 by H. Debehogne at La Silla.

Named in honor Erich Schumann and his Chilean team, who, for almost three decades and with great dedication, have secured the accomodation and catering of thousands of staff members and visitors to the European Southern Observatory at the "Hotel La Silla" in Chile. (M 33786)

(3706) Sinnott

1984 SE_3. Discovered 1984 September 28 by B. A. Skiff at Anderson Mesa.

Named in honor of Roger W. Sinnott, associate editor of *Sky and Telescope*. His extensive telescope-making experience began with a first mirror ground when he was aged 13 and has culminated in his conducting the 'Gleanings for ATMs' department of the magazine for many years. He also created (with Alan Hirshfeld) the two-volume *Sky Catalogue 2000.0* and edited NGC 2000.0 using computer databases. This led to the far-ranging magazine column 'Astronomical Computing'. (M 16246)

Citation provided by D. H. Levy, S. J. Edberg and J. K. Beatty at the request of the discoverer.

(3707) Schröter

1934 CC. Discovered 1934 February 5 by K. Reinmuth at Heidelberg.

Named in honor of Egon Horst Schröter (1928-), German solar astronomer and director of the Freiburg Kiepenheuer-Institut, on the occasion of his retirement. In 1976 he succeeded K.-O. Kiepenheuer at the Freiburg Institute for Solar Physics. An important achievement during his directorship consisted in negotiations about the erection of two new tower telescopes at Teide Observatory on Tenerife, Canary Islands. Schröter served as president of the Astronomische Gesellschaft from 1987 to 1990. In 1978 he became a member of the German Committee in COSPAR. (M 22499)

Name proposed and citation prepared by J. Schubart, endorsed by G. Klare and L. D. Schmadel.

(3708) 1974 FV_1

Discovered 1974 March 21 at the University of Chile Observatory at Cerro El Roble.

(3709) Polypoites

1985 TL_3. Discovered 1985 October 14 by C. S. Shoemaker and E. M. Shoemaker at Palomar.

Named after the victor in a contest among the Greeks to see who would win an iron meteorite by throwing it the farthest. Polypoites flung the meteorite "as far as a herdsman throws a cattle staff, rifling it clear across the herd". When it landed it probably made a small impact crater. (M 12976)

Name proposed by Dorothy and Jerome Preston, who suggested that, had the discoverers been present at the Trojan War, they might have studied with some interest the result of Polypoites' throw.

(3710) Bogoslovskij

1978 RD_6. Discovered 1978 September 13 by N. S. Chernykh at Nauchnyj.

Named in honor of Nikita Bogoslovskij, well-known contemporary composer and writer, on the occasion of his eightieth birthday. (M 22499)

(3711) Ellensburg

1983 QD. Discovered 1983 August 31 by J. Gibson at Palomar.

Named by the discoverer to honor the city of his birth, established in the 1880s by the tracks of the Northern Pacific Railroad. Originally a small trading post in what was then Washington Territory, the city is named for Ellen Shoudy, wife of its founder and first mayor. The city is home to Central Washington University, originally Washington State Normal School, which this year celebrates the centennial of its founding. (M 18453)

(3712) Kraft

1984 YC. Discovered 1984 December 22 by E. A. Harlan and A. R. Klemola at Mount Hamilton.

Named in honor of Robert P. Kraft (1927-), professor of astronomy and astrophysics at the University of California at Santa Cruz and Lick Observatory (1967-1993) and director of Lick Observatory (1981-1991). His research spans stellar spectroscopy, galactic structure, and chemical composition and evolution of stars in the galactic field and star clusters and the Magellanic Clouds. Kraft served as president (1974-1976) and vice president (1982-1988) of the American Astronomical Society and is currently president-elect of the International Astronomical Union. (M 26761)

(3713) Pieters

1985 FA_2. Discovered 1985 March 22 by E. Bowell at Anderson Mesa.

Named in honor of Carle M. Pieters, geologist on the faculty of Brown University, Providence. Pieters obtained the first compositionally diagnostic observations of an Earth-approaching minor planet when she observed (433) Eros during its 1974-75 apparition. She has extensively studied the surface composition of geological units on the near side of the Moon using groundbased reflectance spectroscopy combined with laboratory measurements of returned lunar samples. Pieters has also developed and applied remote sensing techniques for the study of the Earth, Mars and Venus using airborne and spacecraft data. (M 13177)
Name suggested and citation provided by L. A. McFadden.

(3714) Kenrussell

1983 TT_1. Discovered 1983 October 12 by E. Bowell at Anderson Mesa.

Named for Ken S. Russell, an astronomer at the U.K. Schmidt Telescope Unit, Siding Spring Observatory. For more than a decade, Russell has been concerned with most aspects of the operation of the 1.2-m U.K. Schmidt, including observation, data analysis and computer programming. He has been involved with the so-called ESO/SERC Southern Sky Survey and was instrumental in securing a very fine series of photographic images of P/Halley during its recent apparition. Russell has discovered five comets - four of them periodic - since 1979, and he worked on the observational phase of the U.K. Schmidt-Caltech Asteroid Survey in 1981. Recently, he began a collaboration with the discoverer on a deep astrometric survey of minor planets, bringing to bear his expertise in image recognition to the problem of semiautomatic identification and measurement of minor-planet images. (M 13177)

Name endorsed by S. J. Bus and B. G. Marsden.

(3715) Štohl

1980 DS. Discovered 1980 February 19 by A. Mrkos at Kleť.

Named in memory of Ján Štohl (1932-1993), well known for his work on the dynamics of meteors, director of the Astronomical Institute at Tatranská Lomnica and president of IAU Commission 22 (Meteors and Interplanetary Dust) during 1991-1993. (M 25976)

Name proposed by the discoverer, following a suggestion by astronomers at Tatranská Lomnica.

Obituaries published in WGN, Vol. 21, No. 3, p. 77-78 (1993); Kozmos, Vol. 24, No. 3, p. 2-3 (1993); International astronomical symposium: Meteoroids and their parent bodies, p. 8 (1993).

(3716) Petzval

1980 TG. Discovered 1980 October 2 by A. Mrkos at Kleť.

Named in memory of József Miska Petzval (1807-1891), mathematician, professor and inventor at the University of Vienna. He proposed the Petzval lens, and many minor planets were discovered by means of this portrait design at the end of the nineteenth century. (M 25976)

Name proposed by the discoverer, following a suggestion by astronomers of the Astronomical Institute at Tatranská Lomnica.

Petzval is also honored by a lunar crater.

(3717) Thorenia

1964 CG. Discovered 1964 February 15 at the Goethe Link Observatory at Brooklyn, Indiana.

Named in memory of Victor Eugene Thoren (1935-1991), one of the foremost contemporary historians of astronomy and a faculty member in Indiana University's department of history and philosophy of science for his entire academic career. Although Thoren's research contributions were wide-ranging, they nonetheless focused on Tycho Brahe {see planet (1677)} and his emphasis on accuracy in a manner never before thought possible. A copy of his monumental study, *The Lord of Uraniborg: A Biography of Tycho Brahe*, reached Thoren a few days before his tragic death, thus affording him an opportunity to enjoy, all too briefly, the fruits of his intellectual labors. (M 19334)

Name proposed by F. K. Edmondson and endorsed by IAU Commission 41. Citation prepared by E. Grant.

Obituaries published in Isis, Vol. 82, No. 314, p. 693-694 (1991); J. Hist. Astron., Vol. 22, Part 3, p. 253-254 (1991); Bull. Am. Astron. Soc., Vol. 24, No. 4, p. 1334 (1992).

(3718) Dunbar

1978 VS_{10}. Discovered 1978 November 7 by E. F. Helin and S. J. Bus at Palomar.

Named in honor of Roy Scott Dunbar, physicist and planetary scientist at the Jet Propulsion Laboratory. The discoverers wish to acknowledge his dedicated commitment and participation in the Palomar planet-crossing asteroid search, which included his own discovery of the Aten-type asteroid (3362) Khufu. His dissertation on the stability of Trojan-type librations in the Earth-Sun system led to an Earth-Trojan search program at Palomar. (M 12976)

(3719) Karamzin

1976 YO_1. Discovered 1976 December 16 by L. I. Chernykh at Nauchnyj.

Named in memory of Nikolaj Mikhailovich Karamzin (1766-1826), author of the celebrated twelve-volume *History of the Russian State*. (M 13609)

(3720) Hokkaido

1987 UR_1. Discovered 1987 October 28 by S. Ueda and H. Kaneda at Kushiro.

Named for the northern island of Japan, which has a population of 5.5 million. This planet was the first to be discovered in Hokkaido. In addition to Kushiro {see planet (4096)}, there are active programs in Sapporo {see planet (3473)} and Kitami {see planet (3785)} concerned with the astrometric observations of minor planets and comets. (M 12976)

(3721) Widorn

1982 TU. Discovered 1982 October 13 by E. Bowell at Anderson Mesa.

Named in memory of Thomas R. Widorn, for several decades an astronomer at the Vienna University Observatory. Widorn is best known for his discovery of an inverse relationship between the geometric albedo and the rate of change of the optical polarization with solar phase angle, which led to improved estimates of the diameters of minor planets. (M 13177)

Name suggested and citation provided by H. F. Haupt.

(3722) Urata

1927 UE. Discovered 1927 October 29 by K. Reinmuth at Heidelberg.

Named in honor of Takeshi Urata (1947-), one of the most active amateur astronomers in Japan and the director of the minor planet section of the Oriental Astronomical Association. Well known for both his astrometric observations and his orbital computations for both comets and minor planets, he pioneered the contributions by Japanese amateur astronomers in both the discovery and the identification of minor planets. His discovery of (2090) Mizuho in 1978 was the first to be made of a minor planet by an amateur astronomer anywhere in more than half a century. He also discovered periodic comet Urata-Niijima in 1986. Details of his work are included in the Nihondaira Observatory Circular, of which some 1700 issues have appeared since he established this publication in 1967. He is also an editor of the Japanese Ephemerides of Minor Planets. (M 16591)

Name proposed and citation prepared by S. Nakano, who found the identifications involving this planet, and who has been his best friend for many years.

(3723) Voznesenskij

1976 GK_2. Discovered 1976 April 1 by N. S. Chernykh at Nauchnyj.

Named in honor of Andrej Andreevich Voznesenskij, distinguished contemporary poet, known also as the author of unique word-and-figure combinations ("vidioms"), and a friend of some of the astronomers of the Crimean Astrophysical Observatory. (M 22499)

(3724) Annenskij

1979 YN_8. Discovered 1979 December 23 by L. V. Zhuravleva at Nauchnyj.

Named for Innokentij Fedorovich Annenskij (1855-1909), well-known Russian poet and writer. (M 22246; M 26439)

(3725) Valsecchi

1981 EA_{11}. Discovered 1981 March 1 by S. J. Bus at Siding Spring.

Named in honor of Giovanni B. Valsecchi, of the Istituto di Astrofisica Spaziale, Rome. Valsecchi's work on the dynamical effects of the outer planets, principally Jupiter, on cometary orbits gave new insight to the existence of temporary captures and the evolution of cometary orbits. This work is summarized in the first *Atlas of Orbital Patterns at Close Encounters*, of which he is a co-author. He was one of the first, together with Andrea Carusi {see planet (4700)}, to recognize the difficulties in reliably identifying asteroid families, and he pointed out many inconsistencies in the existing families claimed by various workers. He also contributed to the theory of the dynamical evolution of a swarm of gravitationally interacting small bodies and the understanding of three-body interactions in terms of a two-body model. His keen insight into the physical nature of problems makes it always fruitful to seek his counsel. (M 18453)
Citation provided by Donald R. Davis at the request of the discoverer.

(3726) Johnadams

1981 LJ. Discovered 1981 June 4 by E. Bowell at Anderson Mesa.

Named in honor of John B. Adams, geologist at the University of Washington, Seattle. Adams experimentally calibrated the relationship between pyroxene chemistry and the wavelength of the pyroxene band present in the absorption spectra of many minor planets, this band being the most important diagnostic feature for interpreting surface composition. He has continued to develop remote sensing techniques that have contributed to our understanding of the composition and evolution of the asteroids and planets, including the Earth. (M 13177)
Name suggested and citation provided by L. A. McFadden.

(3727) Maxhell

1981 PQ. Discovered 1981 August 7 by A. Mrkos at Kleť.

Named in memory of Maximilian Hell (1720-1792), famous for his determination of the solar parallax from his observations of the transit of Venus in 1769. Appointed director of the Imperial Observatory in Vienna in 1755, he prepared and published an important series of astronomical ephemerides. (M 26424)

Name suggested by astronomers at the Astronomical Institute at Tatranská Lomnica.

(3728) IRAS

1983 QF. Discovered 1983 August 23 with the Infrared Astronomical Satellite.

Named for the Infrared Astronomical Satellite (IRAS), a joint U.S.-Netherlands-U.K. spaceborne all-sky infrared survey satellite launched on 1983 Jan. 26. Over its ten-month lifespan, IRAS made observations of more than 250,000 solar-system and extra-solar-system bodies at wavelengths of 12, 25, 60 and 100 μm. IRAS discovered (3200) Phaethon, parent body of the Geminid meteor stream, as well as four long-period and two short-period comets. (M 34619)

(3729) 1983 VP_7

Discovered 1983 November 1 at the Purple Mountain Observatory at Nanking.

(3730) Hurban

1983 XM_1. Discovered 1983 December 4 by M. Antal at Piszkéstetö.

Named in memory of Jozef Miloslav Hurban (1817-1888), poet, writer, journalist, editor, critic, historian and a leading ideologist of the Slovak national revival. A close colleague (and biographer) of Ľudovít Štúr {see planet (3393)}, he promoted the Slovak literary language and organized cultural and educational activities. During the revolution of 1848-1849 he asserted Slovak national and social claims. (M 13609)

(3731) Hancock

1984 DH_1. Discovered 1984 February 20 by V. M. Candy and M. P. Candy at Bickley.

Named in memory of Langley George ("Lang") Hancock (1909-1992), who started life as manager of the family sheep station and became a leading Australian mining prospector, entrepreneur and visionary. He prospected for and developed many minerals, but he will be forever famous for his discovery of the massive iron ore deposits in the Pilbara region of Western Australia. Hancock had a profound influence on the development of Western Australia, establishing several newspapers and writing many books and other publications on mining and free enterprise. The family corporation is now chaired by daughter Georgina Rinehart, who prepared the citation. (M 22246)

(3732) Vávra

1984 SR_1. Discovered 1984 September 27 by Z. Vávrová at Kleť.

Named in honor of Anton Alfred Vávra (1896-), the discoverer's father. (M 22246)

(3733) Yoshitomo

1985 AF. Discovered 1985 January 15 by K. Suzuki and T. Urata at Toyota.

Named for a Japanese military commander, Minamoto Yoshitomo (1123-1160), father of Yoritomo and Yoshitsune {see planets (3902) and (3178)}. Though he won the battle of Hogen (1156) and had a chance to enter the militarist government of Japan, he lost the battle of Heiji (1159) and was put to death. (M 19693)

(3734) Waland

9527 P-L. Discovered 1960 October 17 by C. J. van Houten and I. van Houten-Groeneveld at Palomar.

Named in honor of Scottish optician Robert L. Waland, who developed new techniques for making the optics of Schmidt telescopes. In the 1960s, when he was at the University of Arizona's Lunar and Planetary Laboratory, he made the superb mirrors for the 1.54-meter reflector at the Catalina Station. Waland authored the book *Optics of the Cassegrain Telescope* in 1990. (M 20158; I. van Houten-Groeneveld)

Name proposed by Tom Gehrels.

(3735) Třeboň

1983 XS. Discovered 1983 December 4 by Z. Vávrová at Kleť.

Named in honor of the discoverer's home town in South Bohemia. (M 16443)

(3736) Rokoske

1987 SY_3. Discovered 1987 September 26 by E. Bowell at Anderson Mesa.

Named for Thomas Leo Rokoske, professor of physics and astronomy at Appalachian State University in Boone, North Carolina. A specialist in solid-state physics, he is known for his exceptional teaching skills and has been active in developing and expanding ASU's astronomical instrumentation and teaching staff. He has played an instrumental role in the publication of the *International Comet Quarterly* (and its predecessors) since its inception in 1973, and his unselfish work has helped to create this very useful archive of cometary photometric data. (M 12976)

Name suggested by D. W. E. Green, ICQ editor and one of his former students, who found the identifications involving this minor planet.

(3737) Beckman

1983 PA. Discovered 1983 August 8 by E. F. Helin at Palomar.

Named in honor of Arnold O. Beckman, the inventor of the first practical electrical pH meter. Through his personal efforts and accomplishments, he has created scientific instrumentation of extraordinary consequence to the world in the areas of chemistry and medicine. He is an alumnus, former professor and past chairman of the Board of Trustees of Caltech. Beckman and his wife, Mabel, throughout their 63 years of marriage developed a partnership that has made a lasting impact on scientific research and education. This minor planet is named in part to commemorate their immeasurable philanthropic gestures. (M 14971)

(3738) Ots

1977 QA_1. Discovered 1977 August 19 by N. S. Chernykh at Nauchnyj.

Named in memory of the opera singer Georg Karlovich Ots (1920-1975). (M 22499)

(3739) Rem

1977 RE_2. Discovered 1977 September 8 by N. S. Chernykh at Nauchnyj.

Named in memory of Rem Viktorovich Khokhlov (1926-1977), Soviet physicist and one of the founders of nonlinear optics. An academician, professor and then rector of the Moscow University, Rem was an outstanding alpinist who perished ascending the Pamirs. (M 22499)

(3740) Menge

1981 EM. Discovered 1981 March 1 by H. Debehogne and G. DeSanctis at La Silla.

Named in honor of Sergio Menge de Freitas, vice-director of the Valongo Observatory, Federal University of Rio de Janeiro, during 1983-1990. Known for his research on stellar motions, he has also been a collaborator of the first discoverer. (M 19334)

(3741) Rogerburns

1981 EL_{19}. Discovered 1981 March 2 by S. J. Bus at Siding Spring.

Named in memory of Roger George Burns (1937-1994), professor of mineralogy and geochemistry at the Massachusetts Institute of Technology. Author of the book *Mineralogical Applications of Crystal Field Theory,* first published in 1970, Burns was a pioneer in the field of mineral spectroscopy and in establishing the fundamental basis for interpreting modern remote sensing observations of planetary surfaces. Burns also applied his expertise in geochemistry to understanding chemical weathering processes on Mars. (M 27126)
Name suggested and citation prepared by R. P. Binzel.

(3742) Sunshine

1981 EQ_{27}. Discovered 1981 March 2 by S. J. Bus at Siding Spring.

Named in honor of Jessica M. Sunshine, visiting scientist at the Massachusetts Institute of Technology. She is best known for developing a technique called the "Modified Gaussian Model" that allows reflectance spectroscopy measurements to be deconvolved and interpreted in terms of constituent absorptions from a suite of mineral species. Sunshine has applied this technique to the study of minor planets, satellites, meteorites and the surfaces of Mars and the earth. (M 27126)
Citation prepared by R. P. Binzel.

(3743) 1983 EW

Discovered 1983 March 10 by E. Barr at Anderson Mesa.

(3744) Horn-d'Arturo

1983 VE. Discovered 1983 November 5 at the Osservatorio San Vittore at Bologna.

Named in memory of Guido Horn-d'Arturo (1879-1967), director of the Bologna Observatory for almost half a century and a talented astronomer far ahead of his time. A pioneer in the design and construction of multiple-mirror telescopes, he designed and utilized the world's first such instrument, a 1.80-m zenith telescope at Bologna. In the 1920s he correctly interpreted the shadow-band phenomenon and attributed an important component of stellar scintillation to currents in the Earth's stratosphere. Following a suggestion by Luigi Jacchia {see planet (2079)}, he founded the magazine *Coelum* in 1931. (M 13178)

(3745) Petaev

1949 SF. Discovered 1949 September 23 by K. Reinmuth at Heidelberg.

Named in honor of Michail Ivanovich Petaev, visiting planetary geologist in the Harvard-Smithsonian Center for Astrophysics Planetary Sciences division. Petaev's investigations include the study of the origins of enstatite chondrite and achondrite meteorites, and he has made detailed studies of numerous meteorites. (M 22499)

Name proposed by members of the Planetary Sciences division, the identifications for this object having been made by C. M. Bardwell.

(3746) 1964 TC_1

Discovered 1964 October 8 at the Purple Mountain Observatory at Nanking.

(3747) Belinskij

1975 VY_5. Discovered 1975 November 5 by L. I. Chernykh at Nauchnyj.

Named in honor of Vissarion Grigor'evich Belinskij (1811-1848), Russian literary critic and publicist, revolutionary democrat. (M 13609)

(3748) Tatum

1981 JQ. Discovered 1981 May 3 by E. Bowell at Anderson Mesa.

Named in honor of Jeremy B. Tatum, professor of astronomy at the University of Victoria, British Columbia. An enthusiastic teacher and expert in molecular spectroscopy, especially of comets and interstellar gas clouds, Tatum was the chief founder and is the driving force behind the Climenhaga {see planet (3034)} Observatory's program of astrometry of minor planets and comets, the only one of its kind in Canada. (M 13178)

Name suggested and citation provided by C. Spratt.

(3749) Balam

1982 BG_1. Discovered 1982 January 24 by E. Bowell at Anderson Mesa.

Named in honor of David D. Balam, research assistant at the University of Victoria's Climenhaga Observatory. In addition to being the observatory's principal observer, Balam has developed most of the software for the astrometric program on minor planets and comets. (M 13178)

Name suggested and citation provided by C. Spratt.

(3750) Ilizarov

1982 TD_1. Discovered 1982 October 14 by L. G. Karachkina at Nauchnyj.

Named in honor of Gavriil Abramovich Ilizarov, a remarkable orthopedic surgeon whose new methods have cured more than half a million patients. (M 17028)

(3751) Kiang

1983 NK. Discovered 1983 July 10 by E. Bowell at Anderson Mesa.

Named in honor of Tao Kiang, an astronomer at the Dunsink Observatory, near Dublin. A teacher of the discoverer, Kiang has carried out a wide range of astronomical research, including contributions to the study of minor planets, comets, quasars and cosmology. In particular, he has worked on the structure of the asteroidal belt and the stability of Kirkwood gaps. He has served as editor of *Chinese*

Astronomy and Astrophysics since its inception in 1977 and has published several papers on astronomical research in China, both ancient and modern. (M 12976)

(3752) Camillo

1985 PA. Discovered 1985 August 15 by E. F. Helin and M. A. Barucci at Caussols.

Named for the young son of Turno, king of the earliest Romans. The name also honors the son of the second discoverer. (M 14633)

(3753) Cruithne

1986 TO. Discovered 1986 October 10 by J. D. Waldron at Siding Spring.

Named for the most common term used by the Irish for the Picts – one of the ancient peoples of Scotland. 'Cruithne' was the eponymous ancestor of the Picts, and it was his seven sons who gave their names to the seven divisions of the Pictish Kingdom. (M 31609)

(3754) Kathleen

1931 FM. Discovered 1931 March 16 by C. W. Tombaugh at Flagstaff.

Named in honor of Kathleen Willoughby Clifford, granddaughter of the discoverer. (M 18306)

(3755) Lecointe

1950 SJ. Discovered 1950 September 19 by S. Arend at Uccle.

Named in honor of Georges Lecointe (1869-1929), director of the Royal Observatory of Belgium from 1900 to 1925. Under his leadership, the observatory obtained several instruments and deployed great scientific activity. (M 25976)
Proposed by P. Páquet and N. Arend.

(3756) Ruscannon

1979 MV_6. Discovered 1979 June 25 by E. F. Helin and S. J. Bus at Siding Spring.

Named in honor of Russell Cannon, astrophysicist, director of the Anglo-Australian Observatory, coordinating the activities of the joint observatories. His special interests are the study of stars in globular clusters and in dwarf galaxies of the Local Group. Cannon was officer-in-charge of the U.K. 1.2-m Schmidt from its first operation in 1974 and is responsible for setting the high standards for which it is renowned. The first discoverer wishes to acknowledge his support in helping her initiate special asteroid surveys at the U.K. Schmidt telescope in Siding Spring {see planets (4128) and (2343), respectively}. (M 19334)

(3757) 1982 XB

Discovered 1982 December 14 by E. F. Helin at Palomar.

(3758) Karttunen

1983 WP. Discovered 1983 November 28 by E. Bowell at Anderson Mesa.

Named for Hannu Karttunen, Finnish astronomer and mathematician. Karttunen has developed a rather general method to model the lightcurves of minor planets and, in collaboration with the discoverer and other colleagues, has sought to understand lightcurves and phase curves in terms of body morphology and large-scale albedo features. He has recently served as an editor of a comprehensive college-level text on fundamental astronomy, which has been published in Finnish and English. (M 12977)
Name endorsed by B. A. Skiff, who took the discovery plates.

(3759) Piironen

1984 AP. Discovered 1984 January 8 by E. Bowell at Anderson Mesa.

Named in honor of Jukka Piironen, astronomer at the Finnish Meteorological Institute, Helsinki. Working with the discoverer, Piironen made accurate lightcurve observations of a number of brighter minor planets. Subsequently, he was involved

in the physical interpretation of such lightcurves by studying the behavior of rotating models in the laboratory. In Turku he participated in the grinding and figuring of the 2.5-m mirror for the Nordic Optical Telescope {see planet (2857)}, and he worked on aspects of the Soviet Phobos mission. (M 12977)
Name endorsed by B. A. Skiff, who took the discovery plates.

(3760) Poutanen

1984 AQ. Discovered 1984 January 8 by E. Bowell at Anderson Mesa.

Named in honor of Markku Poutanen, Finnish astronomer. Poutanen has been active in the observation of asteroid lightcurves and, with the discoverer, obtained the most extensive phase curve of a minor planet then observed, especially at small phase angles. Like his colleague Karttunen {see planet (3758)}, Poutanen recently served as coeditor of a textbook on astronomy. In addition, he has enthusiastically sought to bring amateur and professional astronomers together, by means of both teaching and writing. (M 12977)
Name endorsed by B. A. Skiff, who took the discovery plates.

(3761) Romanskaya

1936 OH. Discovered 1936 July 25 by G. N. Neujmin at Simeïs.

Named in memory of Sof'ya Vasil'evna Romanskaya (1886-1969). The first Russian woman concerned with astronomical observations, she made over 20,000 highly precise latitude determinations with a ZTF-135 zenith-telescope. Her major scientific interests were in the fields of the variation of latitude and the earth's polar motion. From 1908 to 1959 she worked at the Pulkovo Observatory. (M 24916)
Name proposed by L. D. Kostina.

(3762) Amaravella

1976 QN_1. Discovered 1976 August 26 by N. S. Chernykh at Nauchnyj.

Named for a group of Russian painters, known for the cosmic themes of their work. (M 20836)

(3763) 1980 TA_6

Discovered 1980 October 14 at the Purple Mountain Observatory at Nanking.

(3764) Holmesacourt

1980 TL_{15}. Discovered 1980 October 10 by A. McNay at Bickley.

Named in honor of Robert Holmes a Court (1937-1990), a leading Australian lawyer, businessman and collector. Holmes a Court was born in South Africa and studied in New Zealand and Australia. He practised law for a number of years before embarking on a business career and developed a world-ranking reputation for original and lateral thinking in every pursuit, from art and bloodstock to corporate management and philantropy. He was described by his peers as a renaissance man and a citizen of the world. (M 22829)

(3765) Texereau

1982 SU_1. Discovered 1982 September 16 by K. Tomita at Caussols.

Named in honor of Jean Texereau, leading figure in the world of astronomical optics. Among the fine telescopes he has constructed is the Caussols 0.9-m Schmidt with which this minor planet was discovered. Wide diffusion of his books and advice has greatly helped others, notably thousands of amateur astronomers, to build their own instruments. (M 14481)

(3766) Junepatterson

1983 BF. Discovered 1983 January 16 by E. Bowell at Anderson Mesa.

Named in memory of June C. Patterson (1923-1988) of Sierra Vista, Arizona. She and her husband David were well-known amateur astronomers. Through their volunteer work with astronomy clubs and the Flandrau Planetarium in Tucson,

the Patterson's inspired many young people to pursue interests in astronomy. (M 13178)
Name proposed by J. V. Scotti and D. H. Levy.

(3767) DiMaggio

1986 LC. Discovered 1986 June 3 by E. F. Helin at Palomar.

Named in memory of the Yankee Clipper, Joe DiMaggio (1914-1999), U.S. baseball legend. DiMaggio's most enduring record was set during the 1941 season, when he had hits in 56 consecutive games. (M 34619)
DiMaggio was married with the movie star Marilyn Monroe {see planet (3768)}.

(3768) Monroe

1937 RB. Discovered 1937 September 5 by C. Jackson at Johannesburg.

Named for Marilyn Monroe (1926-1962), née Norma Jean Mortensen (also known as Baker), renowned glamorous star of the movie screen. (M 34619)
Monroe was married with the baseball player Joe DiMaggio {see planet (3767)}.

(3769) 1967 UV

Discovered 1967 October 30 by L. Kohoutek and A. Kriete at Bergedorf.

(3770) Nizami

1974 QT_1. Discovered 1974 August 24 by L. I. Chernykh at Nauchnyj.

Named in memory of the distinguished Azerbaijani poet and thinker Nizami Ghiandjevi Abu Mukhammed Il'yas ibn Yusub (1141-1209), whose poetry had an major influence on the development of literature in the Middle East. (M 18454)

(3771) Alexejtolstoj

1974 SB_3. Discovered 1974 September 20 by L. V. Zhuravleva at Nauchnyj.

Named in memory of Alexej Nikolaevich Tolstoj (1883-1945), a well-known Russian writer and public figure. (M 24121)

(3772) Piaf

1982 UR_7. Discovered 1982 October 21 by L. G. Karachkina at Nauchnyj.
Named in memory of the great French singer Edith Piaf (1915-1963). (M 17028)

(3773) Smithsonian

1984 YY. Discovered 1984 December 23 at the Oak Ridge Observatory at Harvard.

Named in honor of the Smithsonian Institution on the occasion of the centennial celebration of its Astrophysical Observatory, 1990 Nov. 2-4. (M 17221)

(3774) Megumi

1987 YC. Discovered 1987 December 20 by T. Kojima at Chiyoda.
Named in honor of the discoverer's wife and observing partner. (M 12977)

(3775) Ellenbeth

1931 TC_4. Discovered 1931 October 6 by C. W. Tombaugh at Flagstaff.

Named in honor of Ellen Elizabeth Willoughby, granddaughter of the discoverer. (M 18306)

(3776) Vartiovuori

1938 GG. Discovered 1938 April 5 by H. Alikoski at Turku.

Named for the hill in Turku on which Argelander's {see planet (1551)} old observatory was situated. The building now includes astronomical collections in memory of Väisälä {see planet (1573)}. (M 18454)

(3777) McCauley

1981 JD_2. Discovered 1981 May 5 by C. S. Shoemaker at Palomar.

Named in honor of John Francis McCauley, geologist with the U.S. Geological

Survey and chief of the Branch of Astrogeologic Studies from 1970 to 1974. McCauley is specially recognized for his pioneering research on the geology of the Moon, Mars, and Mercury and for his leading role in using Space Shuttle imaging radar data to describe an ancient trans-African river system that crossed the now hyperarid Sahara Desert. (M 13482)
Citation provided by G. G. Schaber at the request of the discoverer.

(3778) Regge

1984 HK_1. Discovered 1984 April 26 by W. Ferreri at La Silla.

Named in honor of Tullio Regge, professor of the theory of relativity at Turin University. His research has included particle scattering at high energy and the development of the concept of complex angular momentum. Regge is a recipient of the American Physical Society's Heineman award and of the Lewis Strauss Foundation's Einstein medal. (M 13609)

(3779) Kieffer

1985 JV_1. Discovered 1985 May 13 by C. S. Shoemaker and E. M. Shoemaker at Palomar.

Named in honor of Hugh Hartman Kieffer, geophysicist with the U.S. Geological Survey and chief of the Branch of Astrogeology since 1986. Kieffer is specially recognized for his work on the geology of Mars and his participation in spacecraft missions. His laboratory studies of the spectra of water and carbon-dioxide ices, his leadership in the design of the Viking Infrared Thermal Mapper, and his analysis of its data with colleagues, many of whom are his former students, defined the thermal properties of Mars and the seasonal behavior of water and carbon dioxide in the polar caps. (M 13482)

(3780) Maury

1985 RL. Discovered 1985 September 14 by E. Bowell at Anderson Mesa.

Named in honor of Alain Maury, photographic scientist working on the Palomar Sky Survey II at Mount Palomar Observatory. Well known as a discoverer of comets and Earth-approaching minor planets, Maury began his professional career in his native France at the CERGA {see planet (2252)} Schmidt telescope. He joined the Palomar team in 1984. (M 13178)
Citation provided by D. H. Levy at the request of the discoverer.

(3781) Dufek

1986 RG_1. Discovered 1986 September 2 by A. Mrkos at Kleť.

Named in honor of George Dufek, an associate of Richard E. Byrd and for a long time the leader of American expeditions in Antarctica. (M 26424)

(3782) Celle

1986 TE. Discovered 1986 October 3 by K. Augustesen and H. J. Fogh Olsen and P. Jensen at Brorfelde.

Named for the beautiful German city of Celle on the occasion of its seven-hundredth anniversary in 1992. Celle, which is situated in the province of Niedersachsen, is the twin town to Holbaek {see planet (3033)}, the Danish town nearby the Brorfelde {see planet (3309)} Observatory. The Danish queen Caroline Mathilde (1751-1775), banished from the court in 1772, went to Celle, where she died of copper poisoning. She was buried in the old Celle castle. (M 19693)

Name proposed by the discoverers following a suggestion from the town of Holbaek.

(3783) Morris

1986 TW_1. Discovered 1986 October 7 by E. Bowell at Anderson Mesa.

Named for Charles S. Morris, one of the foremost visual observers of comets. In addition to making more than 1,200 photometric observations of more than

90 comets, he made a rare visual recovery of P/Faye in 1983, and he has been most helpful in confirming newly discovered comets at the request of the Central Bureau for Astronomical Telegrams. He has done much analytical work with visual magnitude data and has contributed extensively to the development of the *International Comet Quarterly*, of which he is associate editor, and its photometric archive. (M 13178)
Name suggested and citation prepared by D. W. E. Green and D. H. Levy.

(3784) Chopin

1986 UL_1. Discovered 1986 October 31 by E. W. Elst at St. Michel.

Named in memory of the great composer and pianist Frédéric Chopin (1810-1849). (M 13179)

(3785) Kitami

1986 WM. Discovered 1986 November 30 by T. Seki at Geisei.

Named for a city in Hokkaido {see planet (3720)} that has been since 1983 a "friendship city" of the discoverer's own city of Kochi {see planet (2396)}. Several amateur astronomers there have been conducting an active program of astrometric observations of minor planets and comets. (M 13482)

(3786) Yamada

1988 AE. Discovered 1988 January 10 by T. Kojima at Chiyoda.

Named in memory of Sakao Yamada (1918-1986), an outstanding Japanese engineer who devoted his life to the research and development of aspherical optics. Founder of Japan Special Optics, he made much optical equipment for Japanese amateur astronomers in cooperation with Nobuhisa Kojima {see planet (4351)}, discoverer of two comets and 1973 MA, the first minor planet discovered by a Japanese amateur and one of the identifications for 1988 AE. Yamada, a teacher of both N. Kojima and the discoverer (no relation), pioneered the mass production of Schmidt-type cameras in Japan. With their high performance and capability, ease of handling and precise star images, these Schmidt cameras, which include the 0.25-m Wright Schmidt with which this discovery was made, have produced many splendid astro-photographs and discoveries of new heavenly bodies. (M 17028)

(3787) Aivazovskij

1977 RG_7. Discovered 1977 September 11 by N. S. Chernykh at Nauchnyj.

Named in honor of Ivan Konstantinovich Aivazovskij (1817-1900), great Russian painter of seascapes who lived and worked in Feodosia {see planet (1048)}, in the Crimea. (M 22499)

(3788) 1986 QM_3

Discovered 1986 August 29 by H. Debehogne at La Silla.

(3789) Zhongguo

1928 UF. Discovered 1928 October 25 by Y. C. Chang at Williams Bay.

This minor planet, originally intended to be (1125) China, was accidentally usurped from that number and name by 1957 UN1, discovered by Y. C. Chang {see planet (2051)} at the Purple Mountain Observatory. It seems therefore particularly appropriate that 1928 UF should receive the Chinese form of the name of China. The minor planet is a 2:1 Jupiter librator. (M 13179)

(3790) Raywilson

1937 UE. Discovered 1937 October 26 by K. Reinmuth at Heidelberg.

Named in honor of Raymond N. Wilson (1928-), astronomical optician and pioneer of new-technology telescopes, on the occasion of his retirement. After serving as a department head and chief optician with Carl Zeiss, in 1972 he joined the European Southern Observatory, where he began work on ESO's telescopes and auxiliary instrumentation. His great experience in the optical design of large telescopes, as well as his profound knowledge of test methods led him to develop the concept of deformable, adjustable active optical elements. Wilson was responsible for the optical design and the evaluation of suitable tests for the large MPIA and ESO telescopes. (M 22499)

Name suggested and citation prepared by L. D. Schmadel, endorsed by the Heidelberg Königstuhl Observatory.

(3791) Marci

1981 WV_1. Discovered 1981 November 17 by A. Mrkos at Kleť.

Named in memory of Jan Marcus Marci of Kronland (1595-1667), Czech phycicist, mathematician, astronomer and physician at the Charles University in Prague during the Thirty Years War. Twenty years before Newton, Marci thoroughly described the spectral dispersion and diffraction of light, color effects on thin layers and rainbow colors. He also studied elastic and inelastic collisions of spheres, the motion of a pendulum and tried to solve the problem of squaring the circle. (M 27733)

Name suggested by J. Tichá and M. Šolc.

(3792) Preston

1985 FA. Discovered 1985 March 22 by C. S. Shoemaker and E. M. Shoemaker at Palomar.

Named for Richard Preston, American author and teacher of English literature. Preston's prize-winning book, *First Light*, describes the history of Palomar Observatory, including the personalities and the research of many astronomers who have observed at Palomar. (M 13482)

(3793) Leonteus

1985 TE_3. Discovered 1985 October 11 by C. S. Shoemaker and E. M. Shoemaker at Palomar.

Named for a good fighter who strove in a contest among his fellow Greeks to win a valuable piece of loot - an iron meteorite - by throwing it the farthest, Leonteus threw mightily, but lost. (M 13482)

Name proposed by Dorothy and Jerome Preston.

(3794) Sthenelos

1985 TF_3. Discovered 1985 October 12 by C. S. Shoemaker and E. M. Shoemaker at Palomar.

Named for the fighting companion who accompanied Diomedes {see planet (1437)} during his great rush against the Trojans. Sthenelos pulled an arrow from Diomedes' shoulder and then stole Aeneas' {see planet (1172)} chariot horses and drove them back among the Greeks. (M 13482)

Name proposed by Anna McCann Taggart and Robert Taggart.

(3795) 1986 GV_1

Discovered 1986 April 8 by E. F. Helin at Palomar.

(3796) Lene

1986 XJ. Discovered 1986 December 6 by K. Augustesen and P. Jensen at Brorfelde.

Named in honor of Lene Augustesen, daughter of the first discoverer. (M 15089)

(3797) Ching-Sung Yu

1987 YL. Discovered 1987 December 22 at the Oak Ridge Observatory at Harvard.

Named in memory of Ching-Sung Yu (1897-1978), a Chinese-American astrophysicist, whose spectrophotometric measurements of 91 stars at the Lick Observatory in the early 1920s represented a milestone in research on stellar energy distribution. After returning to China he established the Purple Mountain Observatory and served as its first director. He also developed the site of what is now the Yunnan Observatory, was director of the Academia Sinica's Institute of Astronomy and president of the Chinese Astronomical Society. After World War II Yu worked at the Harvard College Observatory and later became professor of astronomy and director of the Williams Observatory at Hood College in Frederick, Maryland. (M 14481)

(3798) de Jager

2402 T-3. Discovered 1977 October 16 by I. van Houten-Groeneveld at Palomar.

Named in honor of Cornelis de Jager, Dutch astronomer, former director of the observatory at Utrecht and General Secretary of the IAU from 1967 to 1973. His research concentrated on solar physics. He promoted international scientific collaboration, in particular with the Soviet Union. (M 18138)

(3799) Novgorod

1979 SL_9. Discovered 1979 September 22 by N. S. Chernykh at Nauchnyj.
Named for one of the oldest towns in Russia. (M 22499)

(3800) Karayusuf

1984 AB. Discovered 1984 January 4 by E. F. Helin at Palomar.

Named in honor of Alford S. Karayusuf, M.D., in recognition of his enthusiasm and dedication for the extension of space exploration. With a vitality uncommon since the first days of space exploration, he has encouraged the dreams of man's discovery and adventure in the solar system. He has generously supported near-earth asteroid research projects at the Jet Propulsion Laboratory and has spearheaded a World Space Foundation program to investigate the distant regions of the solar system. (M 17221)

(3801) Thrasymedes

1985 VS. Discovered 1985 November 6 by the Spacewatch at Kitt Peak.

Named for the son of Nestor {see planet (659)} and brother of Antilochus. Thrasymedes commanded a fleet of 15 ships to Troy and was involved in the fight against Memnon {see planet (2895)} during which Antilochus was killed. He later was among the warriors hidden inside the wooden horse. It has been noted that the orbits of (1583) Antilochus and (3801) Thrasymedes are quite similar, suggesting a common origin for these two minor planets. (M 13179)

(3802) Dornburg

1986 PJ_4. Discovered 1986 August 7 by F. Börngen at Tautenburg.

Named for a small town, more than 1050 years old, a few kilometers distant from the Tautenburg Observatory. Situated on a limestock rock above the mean Saale {see planet (5409)} river valley, it is known for its ceramics and for the ensemble of three "Dornburg castles" from three architectural epochs. Goethe {see planet (3047)} spent some time in this attractive place. (M 13482)

(3803) Tuchkova

1981 TP_1. Discovered 1981 October 2 by L. V. Zhuravleva at Nauchnyj.

Named in memory of Margarita Mikhailovna Tuchkova (1781-1852), who built the Temple Spas Nerukotvorny in 1820. She founded the Spaso-Borodinskij nunnery in 1839 in honor of the battle at Borodino (1812) {see planet (3544)} where her husband and brother-in-law were killed. Her monastic name was Mother Maria. (M 24121)

(3804) Drunina

1969 TB_2. Discovered 1969 October 8 by L. I. Chernykh at Nauchnyj.
Named in honor of the Soviet poetess Yuliya Vladimirnova Drunina. (M 18454)

(3805) Goldreich

1981 DK_3. Discovered 1981 February 28 by S. J. Bus at Siding Spring.

Named in honor of Peter Goldreich (1939-), of the California Institute of Technology. Goldreich and his students have contributed to a remarkably broad range of topics in planetary science and astrophysics. Goldreich has written influential and much-quoted papers on polar wandering, the origin of the moon, solar oscillations, the cause of radio bursts from Jupiter, tidal evolution of satellite orbits and planetary spins, planetary rings, the formation of the solar system, instabilities in rotating stars, the origin of planetary nebulae, pulsar magnetospheres, accretion disks, astrophysical masers, spiral structure in galaxies and galaxy formation. (M 27458)

Citation provided by S. D. Tremaine.

(3806) Tremaine

1981 EW_{32}. Discovered 1981 March 1 by S. J. Bus at Siding Spring.

Named in honor of Scott D. Tremaine (1950-). Tremaine has made seminal contributions to solar-system and galactic dynamics. He is co-author with James Binney of the leading monograph on galactic dynamics. Since 1985 Tremaine has been director of the Canadian Institute for Theoretic Astrophysics in Toronto, and the success of that organization owes much to his vision and leadership. (M 27458)

Citation provided by P. M. Goldreich following a suggestion by the discoverer and A. W. Harris.

(3807) Pagels

1981 SE_1. Discovered 1981 September 26 by B. A. Skiff and N. G. Thomas at Anderson Mesa.

Named in memory of Heinz R. Pagels (1939-1988), physicist, activist, educator, administrator, editor and author. His books include *The Dreams of Reason: The Computer and the Rise of Sciences of Complexity* (1988), *Perfect Symmetry: The Search for the Beginning of Time* (1985), and *The Cosmic Code: Quantum Physics as the Language of Nature* (1985). Among his numerous offices were the executive directorship of the New York Academy of Sciences and the presidency of the International League for Human Rights. He was an avid mountaineer. (M 26761)

(3808) Tempel

1982 FQ_2. Discovered 1982 March 24 by F. Börngen at Tautenburg.

Named in memory of Wilhelm Ernst Tempel (1821-1889), discoverer of sixteen comets, five minor planets and several diffuse nebulae. Born in the German district of Upper Lusatia, Tempel was an uncommonly successful and enthusiastic observer. His minor planets were discovered while he was in Marseilles. (M 13483)

Name suggested by M. Gressmann on the 100th anniversary of Tempel's death in Florence.

Tempel is also honored by a lunar crater.

(3809) Amici

1984 FA. Discovered 1984 March 26 at the Osservatorio San Vittore at Bologna.

Named in memory of Giovanni Battista Amici (1786-1863), professor of mathematics at the University of Modena {see planet (3344)} and astronomer at the Museum of Physics and Natural History in Florence. Well known for his construction of optical instruments, his famous 0.28-m achromatic objective, one of the largest in the world when it was made, is still in use at the Arcetri Astrophysical Observatory. In Italian the name of this minor planet also means "friends". (M 13483)

Amici is also honored by a lunar crater.

(3810) Aoraki

1985 DX. Discovered 1985 February 20 by A. C. Gilmore and P. M. Kilmartin at Lake Tekapo.

Aoraki, or Aorangi in northern dialects, is the Maori name for New Zealand's highest mountain, 3764 meters high, also known as Mount Cook. The name is also given to the region in which Mount John lies, east of the mountain. (M 18306)

(3811) Karma

1953 TH. Discovered 1953 October 13 by L. Oterma at Turku.

Named in memory of Birger Karma (1899-1943), mathematician, a teacher of the discoverer at school and university and one of the founders of the society Turun Ursa for amateur astronomers. (M 18454)

(3812) Lidaksum

1965 AK_1. Discovered 1965 January 11 at the Purple Mountain Observatory at Nanking.

Named in honor of Li Dak Sum, noted public figure and enthusiastic promoter of Chinese educational and cultural activities. (M 27126)

(3813) Fortov

1970 QA_1. Discovered 1970 August 30 by T. M. Smirnova at Nauchnyj.

Named in honor of Russian academician Vladimir Evgen'evich Fortov (1946-), well-known for his work on the theory of high-velocity processes of burning and impact. He has made valuable contributions to the theory of the interaction of cosmic bodies with the earth's atmosphere and surface. (M 23136)

Name proposed by the International Institute of Problems of the Asteroid Hazard (St. Petersburg).

(3814) Hoshi-no-mura

1981 JA. Discovered 1981 May 4 by T. Furuta at Tokai.

Named for a small vocational training institute not far from the discoverer's home. The main purpose of the institute is to help mentally handicapped people obtain job opportunities and participate positively in various social activities. The name, which means "star village", was suggested by Youichi Hanaki, an amateur astronomer and one of the institute's leaders. (M 16042)

(3815) König

1959 GG. Discovered 1959 April 15 by A. König and G. Jackisch and W. Wenzel at Heidelberg.

Named by the second and third discoverers in memory of the first, Arthur König (1895-1969), who succeeded Reinmuth {see planet (1111)} in 1957 as the leader of the long-standing Heidelberg observing program on minor planets. An astrometrist known also for his work on stellar positions and proper motions, König also held a leading position in the astronomical department of the Zeiss company and was an authority on coordinate measuring engines. (M 13483)

(3816) Chugainov

1975 VG_9. Discovered 1975 November 8 by N. S. Chernykh at Nauchnyj.

Named in memory of Pavel Fedorovich Chugainov (1933-1992), astronomer at the Crimean Astrophysical Observatory for more than 30 years, and a prominent specialist in stellar photometry and physics. Known for his research of red dwarfs, Chugainov was the head of the Working Group on Flare Stars of IAU Commission 27 and he had been a member of the Organizing Committee of IAU Commission 25 for some years. (M 22499)

(3817) Lencarter

1979 MK_1. Discovered 1979 June 25 by E. F. Helin and S. J. Bus at Siding Spring.

Named in honor of Leonard J. Carter, executive secretary of the British Interplanetary Society. For more than 50 years, his efforts have been the basis for the constructive role of the BIS in space advocacy, education and international communications. (M 13483)

Name proposed by the first discoverer following a suggestion by W. I. McLaughlin and endorsed by R. L. Staehle.

(3818) Gorlitsa

1979 QL_8. Discovered 1979 August 20 by N. S. Chernykh at Nauchnyj.

Named in honor of Mariya Avksent'evna Rudenko, village schoolmistress in the Ukrainian region of Mogilov-Podol'skij, collector of Ukrainian folklore, and founder and leader of Gorlitsa. Meaning turtle-dove, this women's amateur chorus is very popular throughout Ukraine. (M 20836)

(3819) Robinson

1983 AR. Discovered 1983 January 12 by B. A. Skiff at Anderson Mesa.

Named in honor of Leif J. Robinson, editor of 'Sky and Telescope'. Robinson's career as an observer began with a series of planetary drawings and observations of the rapidly changing variable stars in the Orion Nebula. He worked at the Griffith Planetarium in Los Angeles before joining the staff of the magazine in 1962 as an editorial assistant, and he succeeded the late Joseph Ashbrook {see planet (2157)} as editor in 1980. Robinson has been an active promoter of professional-amateur cooperation in astronomy, and retains interests in solar-eclipse viewing and bird-watching. (M 16246)

Citation provided by D. H. Levy, S. J. Edberg and J. K. Beatty at the request of the discoverer.

(3820) 1984 DV

Discovered 1984 February 25 by H. Debehogne at La Silla.

(3821) 1985 RC_3

Discovered 1985 September 6 by H. Debehogne at La Silla.

(3822) Segovia

1988 DP_1. Discovered 1988 February 21 by T. Seki at Geisei.

Named in memory of Andreas Segovia (1893-1987), considered by many to be the most celebrated guitarist of all time. The discoverer heard Segovia perform in Japan in 1959, and this greatly inspired his own interest in playing the guitar. (M 14481)

(3823) Yorii

1988 EC_1. Discovered 1988 March 10 by M. Arai and H. Mori at Yorii.

Named for the small town in central Japan where this minor planet was discovered. Known for the attractions of Hachigata Castle and Five Hundred Rakans (small stone statues), Yorii is situated close to the Tokyo Astronomical Observatory's Dodaira Station. (M 13483)

(3824) Brendalee

1929 TK. Discovered 1929 October 5 by C. W. Tombaugh at Flagstaff.

Named in honor of Brenda Willoughby Anderson, granddaughter of the discoverer. (M 18306)

(3825) 1967 UR

Discovered 1967 October 30 by L. Kohoutek at Bergedorf.

(3826) Handel

1973 UV_5. Discovered 1973 October 27 by F. Börngen at Tautenburg.

Named for the illustrious composer Georg Friedrich Handel (1685-1759). Although his greatest works were composed after he moved to England, Handel was born in Halle, only some 60 km from Tautenburg. (M 13483)
Handel is the anglicized form of the original German spelling Händel.

(3827) Zdeněkhorský

1986 VU. Discovered 1986 November 3 by A. Mrkos at Kleť.

Named in memory of Zdeněk Horský (1929-1988), outstanding Czech historian of medieval and renaissance astronomy and authority on Kepler {see planet (1134)}. Horský also studied megalithic localities in Bohemia, explained the cosmological architectural symbolism of the gothic Charles Bridge and the New Town of Prague and dated the origin of the famous astronomical clock on the Old Town Hall in Prague. (M 27733; M 30819)
Name suggested by J. Tichá, M. Tichý and Z. Moravec.
Obituary published in J. Hist. Astron., Vol. 19, Part 3, p. 215 (1988).

(3828) Hoshino

1986 WC. Discovered 1986 November 22 by K. Suzuki and T. Urata at Toyota.

Named in honor of Jiro Hoshino, an amateur astronomer who has ground more than seven hundred telescope mirrors, one of which is in the telescope with which this minor planet was found. Formerly a public officer in Fukuoka prefecture, he is author of *How to make a reflecting telescope* and *Album of astrophotography*. (M 14971)

(3829) Gunma

1988 EM. Discovered 1988 March 10 by T. Kojima at Chiyoda.

Named for the prefecture in central Japan that is home to the discoverer. A public astronomical observatory under construction in Gunma will include a 1.5-m telescope, the largest in Japan for public use. This telescope will also be used for astronomical research and discovery. (M 28089)

(3830) Trelleborg

1986 RL. Discovered 1986 September 11 by P. Jensen and K. Augustesen and H. J. Fogh Olsen at Brorfelde.

Named in honor of an old town in southern Sweden, twinned with Holbaek {see planet (3033)}, the nearest town to the Brorfelde {see planet (3309)} Observatory. (M 22499)
Name proposed by the third discoverer.

(3831) Pettengill

1986 TP_2. Discovered 1986 October 7 by E. Bowell at Anderson Mesa.

Named in honor of Gordon H. Pettengill, planetary physicist at the Massachusetts Institute of Technology and currently director of the MIT Center for Space Research. Pettengill pioneered the development of planetary radar astronomy, beginning in the late 1950s with the first application of coherent Earth-based radar to lunar studies. In 1965 he led delay-Doppler observations that revealed the 3/2 spin-orbit resonance of Mercury. Since then his observations have helped characterize the physical and dynamical properties of the inner planets, the Galilean satellites, Saturn's rings, and several asteroids and comets. Pettengill was principal investigator for the Pioneer Venus radar mapper experiment, which during 1978-1981 provided the first global maps of Venus topography, radar reflectivity and surface slope. He is also PI for the radar experiments on the Magellan Mission to Venus. (M 13610)

Name suggested and citation written by S. J. Ostro, who was a student of Pettengill.

(3832) Shapiro

1981 QJ. Discovered 1981 August 30 by E. Bowell at Anderson Mesa.

Named in honor of Irwin I. Shapiro {1929- }, physicist associated with the Massachusetts Institute of Technology and with Harvard University during the past three decades, and since 1983 the director of the Harvard-Smithsonian Center for Astrophysics. Shapiro has been responsible for major advances in radio and radar science, and he has made fundamental theoretical and observational contributions to celestial mechanics, astrometry, astrophysics, gravitation physics, geophysics and planetary physics. His planetary ephemerides, considered among the best available, have been used for virtually every Arecibo Observatory radar observation of asteroids, comets, planets, and the satellites of Mars and Jupiter. The underlying ephemeris software also constitutes a primary tool for analysis of VLBI measurements, whose diverse applications include high-precision terrestrial geodesy. During the early 1960s Shapiro played a central role in the radar determination of the astronomical unit, and his research since then has led to refined values for planetary orbital elements, masses, dimensions and spin vectors. (M 13610)

Name suggested and citation written by S. J. Ostro, who was a student of Shapiro.

(3833) Calingasta

1971 SC. Discovered 1971 September 27 by J. Gibson and C. U. Cesco at El Leoncito.

Named for the region of Argentina where the discovery site is located. The region has good astronomical seeing, and a large area has been declared as both an astronomical and an ecological reserve in order to preserve the observing conditions. (M 29142)

Name proposed by the first discoverer following a suggestion by J. G. Sanguin.

(3834) Zappafrank

1980 JE. Discovered 1980 May 11 by L. Brožek at Kleť.

Named in memory of Frank Zappa (1940-1993), rock musician and composer of innovative contemporary symphonic, chamber and electronic music. Zappa was an eclectic, self-trained artist and composer with incredible energy and a biting wit, and his music transcends the usual music barriers. Before 1989 he was regarded as a symbol for democracy and freedom by many people in Czechoslovakia. (M 23792)

(3835) Korolenko

1977 SD_3. Discovered 1977 September 23 by N. S. Chernykh at Nauchnyj.

Named in honor of Vladimir Galaktionovich Korolenko (1853-1921), Russian writer and publicist whose works are filled with democratic and humanistic ideas. (M 22499)

(3836) Lem

1979 SR_9. Discovered 1979 September 22 by N. S. Chernykh at Nauchnyj.

Named in honor of Stanislaw Lem (1921-), famous Polish writer and founder of the Polish Astronautical Society. (M 20837)

(3837) Carr

1981 JU_2. Discovered 1981 May 6 by C. S. Shoemaker at Palomar.

Named for Michael Harold Carr, geologist with the U.S. Geological Survey and chief of the Branch of Astrogeologic Studies from 1974 to 1978. As leader of the

Viking Orbiter Imaging Team, Carr is widely recognized for his role in planetary exploration. He is best known for his comprehensive investigations of the geology and climatic history of Mars. (M 13483)

(3838) Epona

1986 WA. Discovered 1986 November 27 by A. Maury at Palomar.

Named for the Gaulish goddess of horses and horseriders, specifically with respect to those on Palomar Mountain, where this object was discovered. The name derives from the ancient Indo-European word "Ekwos", which gave among others the words Epos (in Gaulish), Hippos (in Greek) and Equus (in Latin). Horses have been long used by mankind in their conquest of new worlds, and there is hope that rockets with large "horsepower" will let us soon explore earth-crossing asteroids such as this one. (M 16443)

Citation written by the discoverer with the very knowledgeable help of J. P. Rivet and at the suggestion of many other French amateur astronomers.

(3839) Bogaevskij

1971 OU. Discovered 1971 July 26 by N. S. Chernykh at Nauchnyj.

Named in honor of Konstantin Fedorovich Bogaevskij (1872-1943), Russian painter who lived and worked in the Crimea for many years and who devoted many of his paintings to the eastern part of that peninsula. (M 20837)

(3840) Mimistrobell

1980 TN_4. Discovered 1980 October 9 by C. S. Shoemaker at Palomar.

Named in honor of Mary E. (Mimi) Strobell, geologist with the U.S. Geological Survey. Early in her career she was a member of a team that conducted some of the earliest airborne surveys in the United States. In recent years she has coordinated much of the detailed information used to establish the extensive nomenclature of features on the terrestrial planets and on the satellites of the giant planets. (M 15574)

(3841) Dicicco

1983 VG_7. Discovered 1983 November 4 by B. A. Skiff at Anderson Mesa.

Named in honor of Dennis di Cicco, since 1974 a staff member and since 1983 an associate editor of *Sky and Telescope*. An imaginative and outstanding astrophotographer, he has participated in many expeditions, specifically to observe eclipses and comets, although his best-known work is probably the analemma showing the Sun from the same spot at the same mean time every few days throughout the year. Particularly meticulous and appropriately cautions in all his writings, he regularly conducts the 'Observers Page' column in the magazine. (M 16246) Citation prepared by B. G. Marsden at the request of the discoverer.

(3842) Harlansmith

1985 FC_1. Discovered 1985 March 21 by E. Bowell at Anderson Mesa.

Named in honor of Harlan J. Smith {1924-1991}, the Edward Randall Jr., M.D. Centennial Professor in Astronomy at the University of Texas and director of McDonald Observatory since 1963. Under his leadership, McDonald has become one of the most productive observatories in the world. While serving as chairman of the astronomy department from 1963 to 1978, Smith oversaw its development into the largest and one of the best academic astronomy programs in the United States. (M 13610)

Name suggested and citation prepared by R. P. Binzel in celebration of Harlan Smith's 25th year as observatory director.

Obituaries published in Phys. Today, Vol. 45, No. 8, p. 83 (1992); Bull. Am. Astron. Soc., Vol. 24, No. 4, p. 1332-1334 (1992); Postepy Astron., Tom 42, No. 4,

p. 179-183 (1994); J. Astrophys. Astron., Vol. 13, No. 1, p. 145-150 (1992); Q.J.R. Astron. Soc., Vol. 34, No. 1, p. 109-111 (1993).

(3843) OISCA

1987 DM. Discovered 1987 February 28 by Y. Oshima at Gekko.

Named for the Organization for Industrial, Spiritual and Cultural Advancement, founded by Yonosuke Nakano in 1961 as a sister organization of the International Foundation fur Cultural Harmony, to which Gekko {see planet (4261)} Observatory belongs. OISCA-International is now active in supporting the efforts of people in developing countries to be free from poverty and hunger through promoting education. (M 13610)

(3844) 1966 BZ

Discovered 1966 January 30 at the Purple Mountain Observatory at Nanking.

(3845) Neyachenko

1979 SA_{10}. Discovered 1979 September 22 by N. S. Chernykh at Nauchnyj.

Named in memory of Il'ya Isakovich Neyachenko, journalist and amateur astronomer from Yalta, known for his research on the history of the Simeïs Astronomical Observatory and the names of minor planets discovered there. (M 20837)

(3846) Hazel

1980 TK_5. Discovered 1980 October 9 by C. S. Shoemaker at Palomar.

Named in honor of Hazel Arthur Spellmann (1896-1968), mother of the discoverer. (M 15574)

(3847) Šindel

1982 DY_1. Discovered 1982 February 16 by A. Mrkos at Kleť.

Named in memory of Jan Ondřejův, known as Šindel (c.1375-c.1456), Czech medieval astronomer, mathematician, physician, professor at the universities in Nuremberg, Vienna and Prague, rector of the Charles University in Prague. He designed the famous Prague astronomical clock, which was built by the clockmaster Nicolas of Kadaň in 1410 and is still working today. (M 27733)
Name suggested by J.Tichá and M. Šolc.

(3848) Analucia

1982 FH_3. Discovered 1982 March 21 by H. Debehogne at La Silla.

Named in honor of Ana Lucia Martins, a good friend of the discoverer and who helps him very much when he is working in connection with his observing program at the Universidade Federal do Rio de Janeiro. (M 21955)

Name proposed by the discoverer and by U.F.R.J. professor L. E. Machado, who died on 1992 July 13.

(3849) Incidentia

1984 FC. Discovered 1984 March 31 by E. Bowell at Anderson Mesa.

Named in honor of Roger W. Martin, a graduate of the Electrical and Computer Engineering Department at the University of California, San Diego. He has creatively contributed to asteroid science by his analysis of the spectra of (2201) Oljato and objects near the 3:1 Kirkwood gap, exhibiting both a wide range of technical expertise and an infectiously high-spirited nature. Incidentia is named after the school of music developed by Martin, who is also a talented musician. The name is quite the antithesis of the effects of Martin's efforts in whatever he undertakes. (M 14029)
Name proposed and citation provided by L. A. McFadden.

(3850) Peltier

1986 TK_2. Discovered 1986 October 7 by E. Bowell at Anderson Mesa.

Named in memory of Leslie C. Peltier (1900-1980), one of this century's most highly respected American amateur astronomers. Between 1925 and 1954 he independently discovered 12 comets (of which 10 bear his name), Nova Herculis 1963, and several other novae. Over a span of 62 years he made 132,123 observations of variable stars. In 1965 he published his autobiography, *Starlight Nights*, a moving account of a man who loved the sky. (M 14633)

Name suggested by D. Levy and J. Mattei. Citation provided by Levy.

Obituary published in J. Am. Assoc. Variable Star Obs., Vol. 9, p. 32-34 (1980).

(3851) Alhambra

1986 UZ. Discovered 1986 October 30 by T. Seki at Geisei.

Named for the famous palace of the Moorish kings in Granada. The great guitarist Segovia {see planet (3822)} frequently included "Memory of Alhambra" in his performances. (M 14482)

(3852) 1987 DR_6

Discovered 1987 February 24 by H. Debehogne at La Silla.

(3853) Haas

1981 WG_1. Discovered 1981 November 24 by E. Bowell at Anderson Mesa.

Named in honor of Walter H. Haas of Las Cruces, New Mexico, founder and director, from 1947 until 1985, of the Association of Lunar and Planetary Observers. His leadership resulted in the evolution of A.L.P.O. into a respected organization dedicated to observing solar-system objects. Through the example set by his meticulous observations and his correspondence with members around the world he is responsible for launching the careers of many solar-system students. (M 14633)

Name suggested and citation provided by D. Levy.

(3854) George

1983 EA. Discovered 1983 March 13 by C. S. Shoemaker and E. M. Shoemaker at Palomar.

Named in honor of George Estel Shoemaker (1904-1960), father of the second discoverer. (M 15574)

(3855) Pasasymphonia

1986 NF_1. Discovered 1986 July 4 by E. F. Helin at Palomar.

Named in honor of the Pasadena Symphony, now celebrating its 60th anniversary. The orchestra has brought artistic recognition to honor the community. Under the brilliant musical direction of Maestro Jorge Mester, the symphony has won nationwide acclaim as an ensemble of outstanding musicians dedicated to the performance of a richly diverse repertoire. (M 13484)

Name suggested by Edith Roberts and endorsed by the discoverer.

(3856) Lutskij

1976 QX. Discovered 1976 August 26 by N. S. Chernykh at Nauchnyj.

Named in honor of Valerij Konstantinovich Lutskij, Moscow astronomer and scientific commentator on astronomy and space exploration. An authority on the history of astronomy, he has been a scientific consultant at the Moscow Planetarium for many years. (M 20837)

(3857) Cellino

1984 CD_1. Discovered 1984 February 8 by E. Bowell at Anderson Mesa.

Named in honor of Alberto Cellino, an astronomer at the Osservatorio Astronomico di Torino. Cellino has been involved in several minor planet photometric campaigns. He has also studied in detail the processes connected with catastrophic collisional break-up and how they can affect minor planet evolution. (M 14030)
Name suggested and citation provided by V. Zappalà.

(3858) Dorchester

1986 TG. Discovered 1986 October 3 by K. Augustesen and P. Jensen and H. J. Fogh Olsen at Brorfelde.

Named for the county town of Dorset, England. Founded by the Romans in A.D. 70, Dorchester was made the twin town of Holbaek {see planet (3033)}, Denmark, the nearest town to the observatory at Brorfelde {see planet (3309)}. (M 21608)
Name proposed by the third discoverer.

(3859) Börngen

1987 EW. Discovered 1987 March 4 by E. Bowell at Anderson Mesa.

Named in honor of Freimut Börngen, senior staff astronomer at the Karl Schwarzschild Observatory, Tautenburg. For 27 years Börngen has worked as an observer at the 1.3-m Schmidt telescope. He has been particularly interested in problems of photographic photometry and has published numerous papers on galaxies of a wide variety of types. Eighteen of his many minor planet discoveries have so far been numbered. (M 14207)
Name suggested and citation provided by L. D. Schmadel.

The actual number of minor planets detected by Börngen amounts to 165 in 1999.

(3860) Plovdiv

1986 PM_4. Discovered 1986 August 8 by E. W. Elst and V. G. Ivanova at Rozhen.
Named for an ancient town in Bulgaria. (M 14030)

(3861) Lorenz

A910 FA. Discovered 1910 March 30 by J. Helffrich at Heidelberg.

Named in memory of Konrad Zacharias Lorenz (1903-1989), Austrian ethologist, who established the field of ethology, the basic concept of which is that animal behavior is genetically programmed. He shared the 1973 Nobel Prize for Physiology and Medicine with Nikolaas Tinbergen and Karl von Frisch for their work on animal behavior and its impact on the understanding of human nature. (M 20520)

Name proposed by T. Urata and S. Nakano, who found the identifications involving this planet. Citation prepared by T. Hidaka.

(3862) Agekian

1972 KM. Discovered 1972 May 18 by T. M. Smirnova at Nauchnyj.

Named in honor of Tateos Artemjevich Agekian, professor at the St. Petersburg University. Agekian is a world famous scientist in stellar statistics, kinematics and dynamics. He has found two evolutionary sequences of stellar systems: nearly spherical and strongly flattened. He initiated a numerical study of the dynamics of triple systems, developed a new approach to study the motion in the field of axially symmetric potential, and he suggested an essentially new method to investigate the structure and kinematics of the Milky Way Galaxy using radio observations of neutral hydrogen. His attractive book *Stars, Galaxies, the Metagalaxy*, translated into many languages, is the lodestar for many young astronomers. His manuals on probability theory and the treatment of observations are reference books for astronomers and physicists. (M 24121)

(3863) Gilyarovskij

1978 SJ_3. Discovered 1978 September 26 by L. V. Zhuravleva at Nauchnyj.

Named after the Russian writer Vladimir Alexeevich Gilyarovskij (1853-1935). (M 24121)

(3864) Søren

1986 XF. Discovered 1986 December 6 by K. Augustesen and P. Jensen at Brorfelde.

Named in honor of Søren Augustesen, son of the first discoverer. (M 15090)

(3865) 1988 AY_4

Discovered 1988 January 13 by H. Debehogne at La Silla.

(3866) 1988 BH_4

Discovered 1988 January 20 by H. Debehogne at La Silla.

(3867) Shiretoko

1988 HG. Discovered 1988 April 16 by M. Yanai and K. Watanabe at Kitami.

Shiretoko is the name of the peninsula in northeastern Hokkaido {see planet (3720)}, not far from the city of Kitami {see planet (3785)}. Established in 1964, Shiretoko is a land of mystery, still in a pristine state. (M 14030)

(3868) Mendoza

4575 P-L. Discovered 1960 September 24 by C. J. van Houten and I. van Houten-Groeneveld at Palomar.

Named in honor of Eugenio E. Mendoza V. (1928-) on the occasion of his 65th birthday. After obtaining his Ph.D. at the University of Chicago, he undertook a postdoctoral position at the University of Cambridge. Mendoza taught astronomy at several universities in Mexico, and he is an expert on photometry and spectroscopy. (M 22499)

(3869) Norton

1981 JE. Discovered 1981 May 3 by E. Bowell at Anderson Mesa.

Named in memory of Arthur P. Norton (1876-1955), author of Norton's Star Atlas, the most widely used publication of its type. Born in Cardiff, Wales, Norton was a geography teacher with a lifelong interest in astronomy. The first edition of his atlas, published in 1910, achieved immediate success, largely due to the use of a low-distortion projection of Norton's devising. For the 1933 edition, Norton redrew the maps to incorporate new IAU constellation boundaries, and he extended the magnitude limit to 6.2. Many of today's astronomers, amateur and professional, use the version first published in 1943, which has maps of epoch 1950.0 and limiting magnitude 6.35. The naming of this planet also celebrates the publication of the epoch 2000.0 edition of the atlas, the first edition not drawn by Norton. (M 15090)

Name suggested and citation provided by I. Ridpath, editor of the new atlas.

(3870) Mayré

1988 CG_3. Discovered 1988 February 13 by E. W. Elst at La Silla.

Named in honor of the discoverer's youngest daughter. (M 14030)

(3871) Reiz

1982 DR_2. Discovered 1982 February 18 by R. M. West at La Silla.

Named in honor of Anders Reiz (1915-), professor emeritus at the Copenhagen University Observatory and highly esteemed teacher and friend of the discoverer. He obtained a doctorate in astronomy from the Lund University Observatory in 1941 and worked there until 1958, when he became a full professor in Copenhagen and director of the university observatories in Copenhagen and Brorfelde. With remarkable skills as teacher and administrator, he contributed substantially to the rapid progress in theoretical and observational astrophysics in Denmark that commenced in the early 1960s. In particular, his persistent efforts soon provided the young generation of astronomers with better tools and research opportunities, from the first digital GIER computer at the observatory in 1961 and several medium-size instruments at the Brorfelde observatory, to the large telescopes at La Silla when Denmark became a member of ESO in 1967. He took the initiatives for the construction of the national Danish 1.5-m telescope on La Silla and the 2.5-m Nordic Optical Telescope {see planet (2857)} on La Palma, two first-class instruments which serve astronomers from Scandinavia and beyond. His research interests have ranged from galaxies to stellar evolution, including the first models of red giants in 1947 and of metal-deficient subdwarfs in 1953, as well as the perfection of associated numerical methods. He made more than 30,000 visual observations with the meridian circle in Lund during the war and recently returned to high-precision astrometry with the Danish 1.5-m telescope at ESO. He early conceived the idea of a joint European professional astronomy journal and was instrumental to the mergers that led to the launch of *Astronomy and Astrophysics* in 1968. (M 16592)

(3872) Akirafujii

1983 AV. Discovered 1983 January 12 by B. A. Skiff at Anderson Mesa.

Named for Japan's highly talented and renowned astrophotographer, Akira Fujii of Koriyama. He edited the quarterly magazine *Hoshi No Techou* (star handbook) and is the author of a beautifully illustrated series of popular astronomy books for young people. Fujii has fostered public awareness of astronomy through television broadcasts from his Chiro Observatory in Shirakawa, Fukushima prefecture, and he toured Japan in 1986, offering the public views of Halley's Comet with a trailer-mounted 60-cm reflector. Fujii is best known outside of Japan for his stunning celestial photographs, a hallmark of which is technical perfection. (M 22829)

Name proposed by D. di Cicco and R. W. Sinnott and endorsed by the discoverer.

(3873) Roddy

1984 WB. Discovered 1984 November 21 by C. S. Shoemaker and E. M. Shoemaker at Palomar.

Named in honor of David J. Roddy, geologist with the U.S. Geological Survey. A leading investigator of impact and explosion craters, Roddy is best known for his work on the impact crater of Devonian age at Flynn Creek, Tennessee, the structure of craters produced by large field experiments with high explosives, and numerical modeling of large impacts. (M 15574)

(3874) Stuart

1986 TJ_1. Discovered 1986 October 4 by E. Bowell at Anderson Mesa.

Named in honor of Stuart E. Jones on the occasion of his retirement as Lowell Observatory's photographic specialist. A member of the Lowell staff since 1962, Jones has been responsible for a wide variety of photographic tasks, including, in particular, the processing of high-quality photographic images for the International Planetary Patrol, a decade-long program that led to the production of about 1 million planetary images. He helped develop a method of photographically extracting fine planetary surface features by means of compositing and a method of reproducing images of planets in true color. Jones will be warmly remembered for his multifaceted talents, which range from practical chemistry to the construction of sophisticated electronics, and for his cheerful disposition and dedication. (M 14207)

(3875) Staehle

1988 KE. Discovered 1988 May 17 by E. F. Helin at Palomar.

Named in honor of Robert L. Staehle, astronautical engineer, member of the technical staff of the Jet Propulsion Laboratory, and president and founder of the World Space Foundation. The Foundation, a non-profit corporation, promotes research and the exploration of space and provides funding from private sources. The discoverer wishes to acknowledge Rob Staehle for the role he and the Foundation have played in recognizing the importance of near-earth asteroids and sponsoring some of the research carried out by the Palomar Planet-Crossing Asteroid Survey. The NASA/JPL Asteroid Project has received valuable assistance and encouragement from Staehle and the Foundation since 1981. (M 14482)

(3876) Quaide

1988 KJ. Discovered 1988 May 19 by E. F. Helin at Palomar.

Named in honor of William L. Quaide for his extraordinary record of scientific achievement in furthering the understanding of our solar system; his many years of service to NASA and the scientific community in support of solar system exploration; his dedication to his work, his personal and professional integrity, which have earned him the trust and respect of his colleagues and friends. (M 14482) Name proposed by the discoverer, following a suggestion by J. Rahe.

(3877) Braes

3108 P-L. Discovered 1960 September 24 by C. J. van Houten and I. van Houten-Groeneveld at Palomar.

Named in honor of Luc L. Braes (1936-), born in Belgium and since 1958 living and working in the Netherlands, first in Groningen and then at Leiden Observatory. His main interests are the OB-stars and the determination of stellar proper motions. (M 22500)

(3878) Jyoumon

1982 VR_4. Discovered 1982 November 14 by H. Kosai and K. Hurukawa at Kiso.

Named for the Jyoumon Era, from the eighth to the third centuries B.C., when the early Japanese people used a special type of pottery with rope-print decoration. (M 21608)

(3879) Machar

1983 QA. Discovered 1983 August 16 by Z. Vávrová at Kleť.

Named in honor of the Czech writer and poet, Josef Svatopluk Machar (1864-1942), whose work is close to the discoverer's heart. (M 16443)

(3880) Kaiserman

1984 WK. Discovered 1984 November 21 by C. S. Shoemaker and E. M. Shoemaker at Palomar.

Named in honor of Michael Kaiserman, American aeronautical engineer and enthusiastic supporter of research in astronomy. (M 15574)

(3881) Doumergua

1925 VF. Discovered 1925 November 15 by B. Jekhovsky at Algiers.

Named in memory of Gaston Doumergue (1863-1937), president of France during 1924-1931, and his wife. (M 22500)

This name was proposed for 1925 VF by the discoverer in 1933 in *J. Obs.*, Vol. 17, p. 102, long before the enactment of the rule excluding contemporary politicians from being honored with minor planets.

(3882) Johncox

1962 RN. Discovered 1962 September 7 at the Goethe Link Observatory at Brooklyn, Indiana.

Named in memory of John Paul Cox (1926-1984), whose professional life was devoted to an investigation of the pulsational mechanisms in variable stars. First at Cornell University and later at the University of Colorado, from where he collaborated with workers at the Los Alamos National Laboratories, he made several outstanding contributions in his field. He produced the first exact, linear, non-adiabatic treatment of pulsation. Along with Zhevakin, Cox realized the importance of the second helium ionization zone as the source of the driving power for Cepheids. He also made significant contributions to our understanding of the Beta Cephei and RR Lyrae stars and authored several review papers and two books that are models of clarity in a very complex subject and will be cited for years to come. (M 19335)
Name proposed by H. R. Johnson and F. K. Edmondson.
Obituaries published in Phys. Today, Vol. 38, No. 1, p. 112 (1985); Q.J.R. Astron. Soc., Vol. 27, No. 2, p. 302-305 (1986).

(3883) Verbano

1972 RQ. Discovered 1972 September 7 by N. S. Chernykh at Nauchnyj.
Named after the lake, also known as Lake Maggiore, in Italy and Switzerland between the south spurs of the Alps. IAU Symposium No. 160, 'Asteroids, Comets, Meteors', was held on the western shore of the lake in the resort village of Belgirate {see planet (5110)}, 1993 June 14-18. (M 23136)
Name proposed by the Symposium participants from Russia, Ukraine and Tadjikistan.

(3884) Alferov

1977 EM_1. Discovered 1977 March 13 by N. S. Chernykh at Nauchnyj.
Named in honor of the outstanding physicist and academician Zhores Ivanovich Alferov (1930-), director of the Ioffe Physical and Technical Institute in St. Petersburg. Alferov made fundamental contributions to the physics of solid bodies and semiconductors, semiconductor heterostructures and semiconductor and quantum electronics. (M 29143)
Named by the discoverer following a suggestion by the Institute of Theoretical Astronomy.

(3885) Bogorodskij

1979 HG_5. Discovered 1979 April 25 by N. S. Chernykh at Nauchnyj.
Named in memory of Aleksandr Fyodorovich Bogorodskij (1907-1984), Soviet astrophysicist, director of the Astronomical Observatory of Kiev University, well known for his works on Einsteinian gravitation, solar physics and planetary nebulae. (M 19693)

(3886) Shcherbakovia

1981 RU_3. Discovered 1981 September 3 by N. S. Chernykh at Nauchnyj.
Named in memory of Sergej Vasil'evich Shcherbakov (1856-1931), founder of the Nizhegorodskij circle of amateur physicists and astronomers, author of the first *Manual of Cosmography* in Russia and involved in establishing the *Russian Astronomical Calendar*. The name is also in memory of his granddaughter Marianna Valentinovna Shcherbakova (1910-1991), docent of geography of Kiev University and an active lecturer in the Kiev Planetarium. (M 19693)

(3887) Gerstner

1985 QX. Discovered 1985 August 22 by A. Mrkos at Kleť.
Named in memory of the physicist František Josef Gerstner (1756-1832) and his son František Antonín Gerstner (1795-1840). F. J. Gerstner worked at the observatories in Vienna and Prague but then turned from astronomy to applied mathematics, physics, mechanics and metallurgy. He founded the Prague Poly-

technic School, later Technical University, in 1806. He also designed the horse railroad from České Budějovice to Linz. F. A. Gerstner continued building this horse railroad, which was completed in 1832 and is considered the first railroad in continental Europe. (M 27733)
Name suggested by J. Tichá and M. Šolc.

(3888) Hoyt

1984 FO. Discovered 1984 March 28 by C. S. Shoemaker and E. M. Shoemaker at Palomar.

Named in memory of William Graves Hoyt (1921-1985), American journalist and historian. His books *Planet 'X' and Pluto*, *Lowell and Mars*, and *Coon Mountain Controversies* (about Meteor Crater, Arizona), written while he was resident historian at Lowell Observatory, are widely recognized as major contributions to the history of planetary science. (M 15575)

(3889) Menshikov

1972 RT_3. Discovered 1972 September 6 by L. V. Zhuravleva at Nauchnyj.

Named in honor of Aleksandr Danilovich Menshikov (1673-1729), Russian statesman and military leader during the reign of Peter the Great. (M 22246)

(3890) Bunin

1976 YU_5. Discovered 1976 December 18 by L. I. Chernykh at Nauchnyj.

Named in memory of an outstanding Russian writer and 1933 Nobel laureate Ivan Alekseevich Bunin (1870-1953). (M 18454)

(3891) Werner

1981 EY_{31}. Discovered 1981 March 3 by S. J. Bus at Siding Spring.

Named in honor of Robert A. Werner, graduate student in aerospace engineering at the University of Texas at Austin. Werner's dissertation work on *Polyhedron Gravitation* has direct application to the problem of calculating orbits around small irregularly shaped objects such as minor planets. For the past eleven years Werner has also served as the typesetter and producer of the *Minor Planet Bulletin*, an activity which has fostered scientific interactions on minor planet research between professional and amateur astronomers. (M 27126)
Name suggested and citation prepared by R. P. Binzel.

(3892) Dezsö

1941 HD. Discovered 1941 April 19 by L. Oterma at Turku.

Named in honor of the Hungarian astronomer Dezsö Loránt, an old friend of the discoverer, founder of the Observatory for Solar Physics in Debrecen and its director for many years. (M 18454)

(3893) DeLaeter

1980 FG_{12}. Discovered 1980 March 20 by M. P. Candy at Bickley.

Named in honor of John DeLaeter, retired professor at Curtin University, Western Australia, in recognition of his pioneering application of mass spectrometry to a range of astrophysical, chemical, geological and nuclear problems over many decades. His outstanding scientific achievements and communication skills have been recognized in awards from ANZAAS, the Royal Society and the Australian Geological Society. He is an officer of the Order of Australia. His support of the Perth Observatory during funding crises was instrumental in maintaining the operation of this isolated institution. (M 27733)
Name suggested and citation provided by J. Biggs and P. Birch.

(3894) Williamcooke

1980 PQ_2. Discovered 1980 August 14 by P. Jekabsons and M. P. Candy at Bickley.

Named in memory of William Ernest Cooke (1863-1947), government astronomer for Western Australia during 1896-1912 and first director of the Perth Observatory. His astronomical expertise and willingness to engage in international collaboration paved the way for the establishment of the Observatory and led to its involvement with the 'Carte du Ciel' project and the acquisition of all the plates for the declination zone -31° and -41°. During 1912-1926 Cooke was government astronomer for New South Wales, and he remains the only Sydney Observatory director to hold the post of professor of astronomy at Sydney University at the same time. (M 27733)
Name suggested and citation provided by J. Biggs and P. Birch.

(3895) Earhart

1987 DE. Discovered 1987 February 23 by C. S. Shoemaker and E. M. Shoemaker at Palomar.

Named for Amelia Mary Earhart (1897-1937), a distinguished pioneer in the field of aviation. She was the first woman to fly across the Atlantic, the first person to fly across the Pacific Ocean from Honolulu to California, the first to solo from Mexico City to New York and the first woman to receive the Distinguished Flying Cross. She was an organizer of the National Glider Association, first president of the Ninety-Nines and a charter member of Zonta International, a women's professional organization. Earhart perished in a second attempt to fly around the world, when her plane was lost at sea near Howland Island in the Pacific. (M 24916)

(3896) Pordenone

1987 WM. Discovered 1987 November 18 by J. M. Baur at Chions.

Named in memory of Giovanni Antonio Licinio (1483-1539), also known as Il Pordenone, from the birthplace, not far from the Chaonis {see planet (4630)} Observatory. One of the masters of painting of the sixteenth century, Il Pordenone was declared a prince of the Friuli region, and the Pictor Modernus, for his exploits in drawing, chiaroscuro and relief. He invented an original language that was violently dynamic and expressive. A disciple of Bellini and Giorgione, he studied the works of Raffael and Michelangelo {see planet (3001)}, influenced Tintoretto and was a rival of Titian. (M 14208)

(3897) Louhi

1942 RT. Discovered 1942 September 8 by Y. Väisälä at Turku.

Named for the powerful mistress of the dark and cold Pohjola {see planet (3606)}, a place in the national epic Kalevala {see planet (1454)}. (M 18454)

(3898) Curlewis

1981 SF_9. Discovered 1981 September 26 by M. P. Candy at Bickley.

Named in honor of Harold Burnham Curlewis (1874-1960), government astronomer for Western Australia during 1912-1940 and second director of the Perth Observatory. Because of severe funding cuts, he was sometimes the observatory's only employee, but he managed to keep the observatory operational during the severe years of the depression. His 38 years of continuous service remain the longest of any employee of the Perth Observatory. (M 27734)
Name suggested and citation provided by J. Biggs and P. Birch.

(3899) Wichterle

1982 SN_1. Discovered 1982 September 17 by M. Mahrová at Kleť.

Named in honor of Otto Wichterle (1913-), Czech chemist with a special interest in macromolecular substances. He is well known as the inventor of hydrophilic gels used for the production of contact lenses. (M 22500)

(3900) Knežević

1985 RK. Discovered 1985 September 14 by E. Bowell at Anderson Mesa.

Named in honor of Zoran Knežević, an astronomer at the Astronomical Observatory of Belgrade. Knežević has studied the physical properties of minor planets using both observational and theoretical approaches. He has analyzed the evolution of minor planet families and has investigated high-order perturbation theories with a view to improving the determination of proper elements. (M 14030) Citation prepared by V. Zappalà at the request of the discoverer.

(3901) 1958 GQ

Discovered 1958 April 7 at the Purple Mountain Observatory at Nanking.

(3902) Yoritomo

1986 AL. Discovered 1986 January 14 by S. Inoda and T. Urata at Karasuyama.

Named for Minamoto Yoritomo (1147-1199), the first shogun and founder of the Japanese feudal system. In his boyhood, Yoritomo was confined under the Heike's surveillance for 20 years in a place of exile near the residence of the second discoverer of this asteroid. Yoritomo destroyed the Heike with his brother's {Yoshitsune, see planet (3178)} help and became the first supreme commander of Japan in 1192, marking the beginning of the Kamakura era. (M 19693)

(3903) Kliment Ohridski

1987 SV_2. Discovered 1987 September 20 by E. W. Elst and V. G. Shkodrov and V. G. Ivanova at Rozhen.

Named in memory of Kliment Ohridski (840-916), one of the first Bulgarian philosophers {of Macedonian origin}. A disciple of Konstantin, he took holy orders in Rome in 868. A pupil and collaborator of Kyril and Methodius, he established a school where he taught some 3,500 students the Bulgarian alphabet and contributed to development of the Bulgarian language. This minor planet, named on the occasion of the 100th anniversary of the founding of the University of Sofia, of which he is considered the patron, is testament to Ohridski's influence on Bulgarian science and culture. (M 14208)

(3904) Honda

1988 DQ. Discovered 1988 February 22 by R. H. McNaught at Siding Spring.

Named in honor of Minoru Honda {1917-1990} for his outstanding achievement in the discovery of 12 comets and 12 novae. The comet discoveries were made visually over the period 1940 to 1968, the latter year adding three discoveries. Honda's emphasis then turned to photographic nova searching, with his first two discoveries coming in 1970. The first, FH Ser, displayed a pronounced minimum during its fade, and multifrequency observations made of this object make it a benchmark in nova studies. Although he was chronologically not the first discoverer of the bright nova V1500 Cyg in 1975, it was his observation that was first disseminated to the astronomical community. These discoveries acted as a major influence in the development of the discoverer's astronomical interests. (M 14633) Citation prepared by the discoverer, with assistance from Charles Morgan.

Obituaries published in Yamamoto Circ., No. 2154 (1990); Komet Tsirk., No. 420, p. 16 (1990).

(3905) Doppler

1984 QO. Discovered 1984 August 28 by A. Mrkos at Kleť.

Named in memory of the Austrian physicist Christian Doppler (1803-1853), professor of mathematics and geometry at the Technical University in Prague and later the first director of the Physical Institute in Vienna. The well-known effect relating a shift in observed wavelength to the relative motion of source and ob-

server was formulated by Doppler in Prague in his book *Ueber das farbige Licht der Doppelsterne* (1842). (M 27734)
Name suggested by J. Tichá and M. Šolc.

(3906) Chao

1987 KE_1. Discovered 1987 May 31 by C. S. Shoemaker and E. M. Shoemaker at Palomar.

Named in honor of Edward C. T. Chao, geologist with the U.S. Geological Survey. A discoverer of the high-pressure silica polymorphs coesite and stishovite at Meteor Crater, Arizona, Chao was the pioneer investigator of the shock metamorphism of rocks. He studied numerous terrestrial impact structures, including especially the Ries {see planet (4327)} crater in Germany, and played a prominent role in demonstrating impact effects in the rocks brought back from the moon. Chao also carried out intensive studies of tektites and advanced key evidence that they are also the product of impacts. (M 18454)

(3907) Kilmartin

A904 PC. Discovered 1904 August 14 by M. Wolf at Heidelberg.

Named in honor of Pamela Margaret Kilmartin, co-director with her husband, Alan C. Gilmore {see planet (2537)}, of the Comets and Minor Planets section of the Royal Astronomical Society of New Zealand. Originally employed as the librarian of the Carter Observatory in Wellington, she quickly became an astronomer in her own right and has been solely responsible for the measurement and reduction of the plates taken in the course of the astrometric programs in Wellington and more recently at the Mount John University Observatory. (M 14482)

Name proposed by B. G. Marsden, who found the identifications involving this minor planet, and endorsed by A. C. Gilmore and F. M. Bateson.

(3908) Nyx

1980 PA. Discovered 1980 August 6 by H.-E. Schuster at La Silla.

Named for the goddess and personification of the night. She was a daughter of Chaos and had many children by her brother Erebus, who personifies the subterranean darkness. The realm of Nyx was in the far West, beyond the land of Atlas. (M 34339)

(3909) Gladys

1988 JD_1. Discovered 1988 May 15 by K. W. Zeigler at Anderson Mesa.

Named in memory of the discoverer's mother, Gladys Marie Zeigler (1921-1988), who encouraged him as a youth to pursue his interest in astronomy despite all obstacles. Her continued encouragement was directly responsible for the discovery of this minor planet. (M 15260)

(3910) Liszt

1988 SF. Discovered 1988 September 16 by E. W. Elst at St. Michel.

Named in memory of Franz Liszt (1811-1886), legendary master of the piano and a courageous fighter for progress in the musical art. A grand and many-sided composer, his works ranged from Hungarian rhapsodies to symphonic poems. While a student at the University of Bonn, the discoverer became acqainted with Lady Elisabeth von Loe-Schultz, who was privileged to have known the composer. At her home the discoverer regularly played before a small audience of students several of Liszt's famous piano etudes. (M 15575)

Name endorsed by F. Börngen, Tautenburg, who independently proposed the name for another minor planet, and who notes that from 1848 to 1861 Liszt was the conductor of the court orchestra in Weimar, not far from Tautenburg.

(3911) Otomo

1940 QB. Discovered 1940 August 31 by K. Reinmuth at Heidelberg.

Named in honor of Satoshi Otomo, a dentist living in Kiyosato {see planet (5488)}. Discoverer of numerous minor planets since 1991, he is also an active confirmer of new objects in collaboration with S. Nakano {see planet (3431)}, who proposed the name and made the identifications involving this object. (M 22500)

(3912) Troja

1988 SG. Discovered 1988 September 16 by E. W. Elst at St. Michel.

Named for the ancient legendary city featured in the works of Homer {see planet (5700)}. The historical site of Troy was found in 1872 at Hissarlik, Turkey, by Schliemann {see planet (3302)}. (M 14208)

(3913) Chemin

1986 XO_2. Discovered 1986 December 2 by the CERGA at Caussols.

Named in honor of Henriette and Robert Chemin, who, after many years at the Paris Observatory as librarian and engineer, respectively, observed, discovered and measured many asteroids on plates taken at the Caussols Schmidt Telescope. After Henriette's death in 1991, Robert retired and started popularizing and teaching astronomy. (M 22829)

Name suggested by J.-L. Heudier and C. Pollas.

(3914) Kotogahama

1987 SE. Discovered 1987 September 16 by T. Seki at Geisei.

Named for the beautiful coast situated 2 km south of the Geisei station on the Pacific shore of Shikoku {see planet (4223)} island. There are many pine trees on this coast, and there is a famous place where their leaves play the sound on the Japanese harp Koto in the breeze. (M 15090)

(3915) Fukushima

1988 PA_1. Discovered 1988 August 15 by M. Yanai and K. Watanabe at Kitami.

Named in honor of Hisao Fukushima (1910-1997), professor emeritus at Hokkaido University, famous for his research in hydrodynamics. An active amateur astronomer, he is a representative of the Hokkaido Astronomical Liaison Group. He works particularly on astronomical history and is a great source of inspiration to younger amateur astronomers. (M 14030)

Obituaries published in Heavens, Vol. 79, No. 2, p. 8-18 (1998); Heavens, Vol. 79, No. 3, p. 20-21 (1998).

(3916) Maeva

1981 QA_3. Discovered 1981 August 24 by H. Debehogne at La Silla.

Named in memory of Maeva d'Alloy d'Hocquincourt Vitry, who died tragically on 1992 June 7 at the age of seven. Niece of ESO staff astronomer Patrice Bouchet Vitry {see planet (4313)}, this extraordinary open-hearted little girl was fond of nature, flowers and beauties of the universe, and she showed every promise of becoming an astronomer herself. (M 21131)

(3917) Franz Schubert

1961 CX. Discovered 1961 February 15 by F. Börngen at Tautenburg.

Named for the great composer Franz Schubert (1797-1828). (M 14208)

(3918) Brel

1988 PE_1. Discovered 1988 August 13 by E. W. Elst and G. Sause at St. Michel.

Named in memory of the well-known Belgian artist Jacques Brel (1929-1978), famous also for his songs and poems. The songs *Marieke* and *Le plat pays* are a tribute to Flanders. (M 14030)

(3919) Maryanning

1984 DS. Discovered 1984 February 23 by H. Debehogne at La Silla.

Named in memory of Mary Anning (1799-1847), whose discoveries from the age of 12 of marine fossils were catalytic to the recognition of dinosaurs as the giant extinct monsters of antiquity. From early childhood, she and her brother Joseph followed their father in searches for ammonite fossils in the cliffs around Lyme Regis, England. After the death of her father in 1810, fossil hunting became the means of support for the two surviving children and their mother. She discovered the first complete skeleton of the fish-lizard Ichthyosaurus, which aroused great interest as it passed into the hands of fossil collectors and then to the British Museum a few years later. Her other major discoveries include the Plesiosaurus, Pterodactylus and Squaloraja. (M 34619)

(3920) Aubignan

1948 WF. Discovered 1948 November 28 by S. Arend at Uccle.

Named for the small village in the French Provence, at the foot of the Mont Ventoux, where the discoverer regularly spent his holidays in the home of his youngest son Noel {see planet (1563)}, who proposed this name. (M 25976)
Name endorsed by P. Páquet.

(3921) Klement'ev

1971 OH. Discovered 1971 July 19 by B. A. Burnasheva at Nauchnyj.

Named in memory of Zakhar Ivanovich Klement'ev (1903-1994), professor of mathematics at Tomsk University. For more than 60 years he delivered lectures on mathematical analysis, functional analysis and other topics, imbuing in young minds by the beauty and accessibility of his presentation the most complex problems of mathematics. He authored more than 60 scientific papers and the six-part series *Lectures on Mathematical Analysis*. The fascination of his personality and creative individuality is preserved in the memory of his numerous disciples, scattered now throughout Russia and abroad. (M 24916)

(3922) 1971 SP_3

Discovered 1971 September 26 by C. Torres at Cerro El Roble.

(3923) Radzievskij

1976 SN_3. Discovered 1976 September 24 by N. S. Chernykh at Nauchnyj.

Named in honor of Vladimir Vyacheslavovich Radzievskij, professor at the Nizhegorodskij Pedagogical Institute, known for his work in celestial mechanics, the cosmogony of the solar system and the origin of comets. (M 19695)

(3924) Birch

1977 CU. Discovered 1977 February 11 by E. Bowell at Palomar.

Named in honor of Peter V. Birch. An astronomer at the Perth Observatory since 1970, Birch made many planetary photographic observations as part of Lowell Observatory's International Planetary Patrol Program. In 1977 he was involved in the discovery of the rings of Uranus. He has also carried out a variety of photometric observations, including photoelectric lightcurves of minor planets and comets and CCD work on Comet Halley. (M 14208)
Citation provided by M. P. Candy at the request of the discoverer.

(3925) Tret'yakov

1977 SS_2. Discovered 1977 September 19 by L. V. Zhuravleva at Nauchnyj.

Named in memory of Pavel Mikhailovich Tret'yakov (1832-1898) and his brother Sergej Mikhailovich (1834-1892). Pavel founded the Art Gallery; Sergej presented his Western-European art collection to the city of Moscow. (M 24121)

(3926) Ramirez

1978 VQ_3. Discovered 1978 November 7 by E. F. Helin and S. J. Bus at Palomar.

Named in honor of Abel R. Ramirez, manager and impeccable host of the California Institute of Technology's Athenaeum, on the occasion of his 50th birthday, 1992 May 18. Fondly regarded by faculty, students and guests from all over the world, he has graciously presided over this elegant establishment since 1978, the year that this minor planet was discovered. (M 20520)

(3927) Feliciaplatt

1981 JA_2. Discovered 1981 May 5 by C. S. Shoemaker and E. M. Shoemaker at Palomar.

Named in honor of the mother of John Platt, who discovered (3237) Victorplatt and (3259) Brownlee while a student at Caltech. (M 25443)

(3928) Randa

1981 PG. Discovered 1981 August 4 by P. Wild at Zimmerwald.

Named for a picturesque old village near Zermatt. This minor planet's number is the same as the village's code in the Swiss postal system. (M 17466)

The discoverer informed us that the Italian word 'randa' (which means 'brink') obviously explains the name of the village near the Italian border.

(3929) Carmelmaria

1981 WG_9. Discovered 1981 November 16 by P. Jekabsons at Bickley.

Named in honor of Carmel Maria Borg (nee Materazzo), secretary at the Perth Observatory and administrative assistant to three directors over 17 years. Her devotion to observatory projects, attention to detail and public-relations skills have seen the Perth Observatory through some turbulent times and played a large part in the successful interaction the observatory enjoys with the local community. She has also played a major role in preserving many of the observatory's historical documents and artifacts. (M 27734)
Name suggested and citation provided by J. Biggs and P. Birch.

(3930) Vasilev

1982 UV_{10}. Discovered 1982 October 25 by L. V. Zhuravleva at Nauchnyj.

Named in memory of Konstantin Alekseevich Vasil'ev (1942-1976), a well known Russian painter. (M 17029)

(3931) Batten

1984 EN. Discovered 1984 March 1 by E. Bowell at Anderson Mesa.

Named in honor of Alan H. Batten, an astronomer at the Dominion Astrophysical Observatory in Victoria, British Columbia. Batten's research has centered on close binary stars and radial velocities. He has served as a vice president of the IAU, as president of two of its commissions and as editor of the *Journal of the Royal Astronomical Society of Canada.* (M 14208)
Name suggested and citation provided by C. E. Spratt.

(3932) 1984 SC_5

Discovered 1984 September 27 by M. Nolan at Palomar.

(3933) Portugal

1986 EN_4. Discovered 1986 March 12 by R. M. West at La Silla.

Named in honor of the European country whose famous navigators studied the skies with great skill and discovered many new routes to distant shores under southern stars. Its recent association with ESO now opens new, exciting celestial paths for its modern astronomers. (M 17029)

(3934) Tove

1987 DF_1. Discovered 1987 February 23 by P. Jensen and K. Augustesen and H. J. Fogh Olsen at Brorfelde.
Named in honor of Tove Augustesen, wife of the second discoverer. (M 29143)

(3935) Toatenmongakkai

1987 PB. Discovered 1987 August 14 by T. Seki at Geisei.

Named in honor of the Oriental Astronomical Association, founded in 1920 by Issei Yamamoto {see planet (2249)}, then professor at the Astronomical Institute at Kyoto Imperial University. The O.A.A. is mainly composed of amateur astronomers in Japan. The discoverer joined the O.A.A. in 1947 and has served as director of the Comet Section since 1970. (M 14633)

(3936) Elst

2321 T-3. Discovered 1977 October 16 by C. J. van Houten and I. van Houten-Groeneveld at Palomar.

Named in honor of Eric W. Elst (1936-), Belgian astronomer at the Observatory at Uccle. For many years he has been enthusiastically searching for minor planets, especially Trojans, and several of his discoveries have now been permanently numbered. (M 18138)

(3937) Bretagnon

1932 EO. Discovered 1932 March 14 by K. Reinmuth at Heidelberg.

Named in honor of Pierre Bretagnon, astronomer at the Bureau des Longitudes. He is the author of a modern, high-precision analytical theory for the motion of the planets from Mercury to Neptune, called 'VSOP'. (M 22500)
Name proposed by J. Meeus, endorsed by G. Klare and L. D. Schmadel.

(3938) Chapront

1949 PL. Discovered 1949 August 2 by K. Reinmuth at Heidelberg.

Named in honor of Jean Chapront and Michelle Chapront-Touzé, astronomers at the Bureau des Longitudes. They have constructed a high-precision analytical theory for the orbital motion of the moon, called 'ELP 2000-82'. (M 22500)

Name proposed and citation prepared by J. Meeus, endorsed by G. Klare and L. D. Schmadel.

(3939) Huruhata

1953 GO. Discovered 1953 April 7 by K. Reinmuth at Heidelberg.

Named in memory of Masaaki Huruhata (1912-1988), director of the Tokyo Astronomical Observatory from 1968 to 1973, known for his work on meteors and variable stars and the photoelectric photometry of the zodiacal light and airglow. One of his early research projects involved the light variation of (433) Eros in 1930-1931. He served as president of IAU Commission 21 (1967-1970), vice president of the Astronomical Society of Japan (1961-1963) and a member of the Science Council of Japan (1969-1972). Following his retirement he carried out photographic observations of variable stars at his home and served as an advisor to amateur astronomers in the Variable Star Observers League in Japan. (M 15261)

Name proposed by S. Nakano, who found the identifications involving this minor planet, following a suggestion by Y. Kushida, M. Inoue and O. Muramatsu. Citation prepared by S. Sakuma.

Obituaries published in Variable Star Bull., No. 9, p. 36 (1989); IAU Colloquium No. 126, p. xxv-xxvi (1991).

(3940) Larion

1973 FE_1. Discovered 1973 March 27 by L. V. Zhuravleva at Nauchnyj.

Named for Larisa Ivanovna Golubkina, dramatic actress, brilliant performer of romances. (M 22500)

(3941) Haydn

1973 UU_5. Discovered 1973 October 27 by F. Börngen at Tautenburg.
Named for the great composer Franz Joseph Haydn (1732-1809). (M 14208)

(3942) Churivannia

1977 RH_7. Discovered 1977 September 11 by N. S. Chernykh at Nauchnyj.

Named in memory of the father and the older brother of the astronomer K. I. Churyumov {see planet (2627)}, friend of the discoverer. Ivan Ivanovich Churyumov (1907-1942) perished in World War II near Kharkov, and Ivan Ivanovich Churyumov (1929-1988) was a philosopher and poet. (M 19694)

(3943) Silbermann

1981 RG_1. Discovered 1981 September 3 by F. Börngen at Tautenburg.

Named for the famous Saxon organ builder Gottfried Silbermann (1683-1753), several of whose instruments have been carefully preserved. The specific Silbermann timbre is the prototype for numerous twentieth-century organs. (M 14209)

(3944) Halliday

1981 WP_1. Discovered 1981 November 24 by E. Bowell at Anderson Mesa.

Named in honor of Ian Halliday, an astronomer at the Herzberg Institute of Astrophysics, National Research Council of Canada. Halliday has achieved international recognition in three major areas of research in meteoritics: spectroscopy, dynamics and meteorite recovery. He was the first to identify the auroral green line in the spectra of meteors; he planned and directed a large Canadian camera network called the Meteorite Observation and Recovery Project (MORP) and together with his staff determined the definitive photographic orbits of 360 fireballs; and while leading a search party Halliday found the first and largest piece of the Innisfree meteorite, one of only three meteorites recovered entirely on the basis of network photography. Halliday was able to identify six photographic trails with the corresponding six largest Innisfree fragments discovered in the fall area - a unique achievement. He also used MORP data on Geminids to study the minor planet (3200) Phaethon. (M 14634)

Name suggested by C. J. Cunningham; citation prepared by Cunningham and P. M. Millman.

(3945) Gerasimenko

1982 PL. Discovered 1982 August 14 by N. S. Chernykh at Nauchnyj.

Named in honor of Svetlana Ivanovna Gerasimenko, staff member of the Institute of Astrophysics in Dushanbe, researcher on comets and codiscoverer of periodic comet Churyumov-Gerasimenko. (M 19694)

(3946) Shor

1983 EL_2. Discovered 1983 March 5 by L. G. Karachkina at Nauchnyj.

Named in honor of Viktor Abramovich Shor {1929- }, member of the staff of the Institute for Theoretical Astronomy in Leningrad, on the occasion of his sixtieth birthday, 1989 Sept. 29. Best known for his extensive work in connection with editing and compiling the annual volume of *Ephemerides of Minor Planets* {see planet (5001)}, he has also carried out significant research on a theory of the motion of the satellites of Mars. (M 15090)

(3947) Swedenborg

1983 XD. Discovered 1983 December 1 by E. Bowell at Anderson Mesa.

Named for Emanuel Swedenborg (1688-1772), Swedish scientist, philosopher, poet and theologian. Swedenborg began publication of Sweden's first scientific journal in 1715. His work on the philosophy of nature included a cosmological theory that was a precursor to the now widely accepted Kant-Laplace nebular theory. Soon after his death, Swedenborg societies were formed to study his

thoughts, which he had published in numerous Latin volumes. His ideas have been a source of inspiration for many prominent writers, including Honoré de Balzac, Charles Baudelaire, Ralph Waldo Emerson and William Butler Yeats. (M 14972)
Name suggested and citation provided by C. J. Cunningham.

(3948) Bohr

1985 RF. Discovered 1985 September 15 by P. Jensen and K. Augustesen at Brorfelde.

Named in memory of the world-famous Danish physicist Niels Bohr (1885-1962), who received the Nobel prize for his atomic theory in 1922. (M 15427)
Bohr is also honored by a lunar crater.

(3949) Mach

1985 UL. Discovered 1985 October 20 by A. Mrkos at Kleť.

Named in memory of Ernst Mach (1838-1916), professor of physics at the universities of Graz and Prague and later professor of philosophy in Vienna. He made investigations on supersonic motion, explosions, electric sparks and philosophical positivism, and the Mach number and Mach's principle are forever associated with his name. (M 27734)
Name suggested by J. Tichá and M. Šolc.

(3950) 1986 CH

Discovered 1986 February 8 by S. Inoda and T. Urata at Karasuyama.

(3951) Zichichi

1986 CK_1. Discovered 1986 February 13 at the Osservatorio San Vittore at Bologna.

Named in honor of Antonio Zichichi, former president of the European Society of Physics and one of the protagonists of CERN. In 1963 he established and since then has directed the Centro Interdisciplinare Ettore Maiorana at Erice, Sicily. (M 14634)

(3952) Russellmark

1986 EM_2. Discovered 1986 March 14 at the Bulgarian National Observatory at Rozhen.

Named for the Russell Mark Group, of Albany, California, "Specialists in Written Communication", in grateful appreciation of their enormous assistance in editing the citations for the name proposals submitted to the Minor Planet Center. Over the past two years, the Group's president, Susan Russell, has taken a personal interest in this effort, thereby saving the Minor Planet Center about a day's work each month. (M 34619)

(3953) Perth

1986 VB_6. Discovered 1986 November 6 by E. Bowell at Anderson Mesa.

Named for the Perth Observatory in recognition of its many notable contributions to astronomy. These have included the Perth 70 catalogue, co-discovery of the rings of Uranus, and extensive work on Comet Halley during both its 1910 and 1986 apparition. Founded in 1896 on Mount Eliza, overlooking the city of Perth, Western Australia, the observatory initially undertook timekeeping, weather reporting and seismographic monitoring, as well as telescopic observations for such programs as the Carte du Ciel and the Astrographic Catalogue. City expansion and a need for darker skies led to a move to the Darling Range, east of Perth, in 1965. Current work includes astrometry and photometry of solar system objects, particularly minor planets and comets. (M 14972)
Citation prepared by P. V. Birch at the request of the discoverer.

(3954) Mendelssohn

1987 HU. Discovered 1987 April 24 by F. Börngen at Tautenburg.

Named for the celebrated composer Felix Mendelssohn-Bartholdy (1809-1847). (M 14209)

(3955) Bruckner

1988 RF_3. Discovered 1988 September 9 by F. Börngen at Tautenburg.

Named for the Austrian composer Anton Bruckner (1824-1896), chiefly known for his nine monumental symphonies. (M 14209)

(3956) Caspar

1988 VL_1. Discovered 1988 November 3 by P. Jensen at Brorfelde.

Named in honor of Caspar Karstensen, grandson of the discoverer. (M 22500)

(3957) Sugie

1933 OD. Discovered 1933 July 24 by K. Reinmuth at Heidelberg.

Named in honor of Atsushi Sugie, who works at the Dynic Astronomical Observatory and who has discovered numerous minor planets there since 1988. He is an active confirmer of new objects in collaboration with S. Nakano {see planet (3431)}, who proposed the name and made the identifications involving this object. (M 22500)

(3958) Komendantov

1953 TC. Discovered 1953 October 10 by P. F. Shajn at Simeïs.

Named in memory of Nikolaj Vasil'evich Komendantov (1895-1937), who was an active researcher of minor-planet motion. A staff member of the Astronomical Institute in Leningrad, he was arrested 1936 on spurious charges, and the date of his death in prison is uncertain. (M 34340)

Name suggested and citation prepared by Yu. V. Batrakov.

(3959) Irwin

1954 UN_2. Discovered 1954 October 28 at the Goethe Link Observatory at Brooklyn, Indiana.

Named in honor of John B. Irwin, professor of astronomy at Indiana University from 1948 to 1964. His paper published in *Science* on 1952 Feb. 29 was the first step in the series of events that led to the establishment of a national observatory in the U.S., Kitt Peak {see planet (2322)} being chosen as the site on 1958 Mar. 1. (M 25652)

Name proposed by F. K. Edmondson.

(3960) Chaliubieju

1955 BG. Discovered 1955 January 20 at the Purple Mountain Observatory at Nanking.

Named in honor of Cha Liubieju, a friend of the discoverer. She devotes her every effort to caring for and educating sick and destitute mothers and children in China. (M 32787)

(3961) Arthurcox

1962 OB. Discovered 1962 July 31 at the Goethe Link Observatory at Brooklyn, Indiana.

Named in honor of Arthur N. Cox, one of the graduate student observers in the early years of the Indiana minor planet program and the first recipient of a Ph.D. in the graduate program established after the gift of the Goethe Link Observatory {see planet (1728)} to Indiana University in 1948. His four years at Indiana University included a six-month expedition, initiated by John B. Irwin {see planet (3959)}, to the South African observatories to introduce photometry by photoelectric methods to the southern hemisphere. Cox then moved to Los

Alamos National Laboratory, becoming the leader of a scientific research group studying both nuclear test results and stellar astrophysics. His calculations of stellar opacities in the early 1960s enabled considerable progress to be made by many stellar astrophysicists, and they are used even now some 30 years later. (M 19335)
Name proposed by H. R. Johnson and F. K. Edmondson.

(3962) Valyaev

1967 CC. Discovered 1967 February 8 by T. M. Smirnova at Nauchnyj.
Named in honor of Valerij Ivanovich Valyaev (1944-), noted scientist in the field of ephemeris astronomy, head of the Ephemeris Astronomy Department of the ITA {see planet (1735)} and editor-in-chief of *Morskoj Astronomicheskij Ezhegodnik* and *Aviatsionnyj Astronomicheskij Ezhegodnik*. (M 24410)
Name proposed by the Institute of Theoretical Astronomy.

(3963) Paradzhanov

1969 TP_2. Discovered 1969 October 8 by L. I. Chernykh at Nauchnyj.
Named in memory of Sergej Iosifovich Paradzhanov (1924-1990), outstanding Soviet film producer. He directed several popular films, including *Ghosts of Forgotten Ancestors*, winner of several international film festivals. (M 27126)

(3964) Danilevskij

1974 RG_1. Discovered 1974 September 12 by L. V. Zhuravleva at Nauchnyj.
Named in memory of Grigorij Petrovich Danilevskij (1829-1890), Russian and Ukrainian writer. (M 24121)

(3965) Konopleva

1975 VA_9. Discovered 1975 November 8 by N. S. Chernykh at Nauchnyj.
Named in honor of Valentina Petrovna Konopleva, astronomer at the Main Astronomical Observatory at Golosseevo, near Kiev, known for her works on the physics and statistics of comets. (M 19694)

(3966) Cherednichenko

1976 SD_3. Discovered 1976 September 24 by N. S. Chernykh at Nauchnyj.
Named in honor of Vladimir Ivanovich Cherednichenko, docent of Kiev Polytechnical Institute, known for his research on physical and chemical processes in comets. (M 19694)

(3967) Shekhtelia

1976 YW_2. Discovered 1976 December 16 by L. I. Chernykh at Nauchnyj.
Named in memory of Fyodor Osipovich Shekhtel' (1859-1926), a famous Russian architect and a bright representative of the "Russian modern style". He designed many important buildings in Moscow and elsewhere. (M 18455)

(3968) Koptelov

1978 TU_5. Discovered 1978 October 8 by L. I. Chernykh at Nauchnyj.
Named in honor of the Soviet writer Afanasij Lazarevich Koptelov, whose works depicted life of the people of the Altaj and Siberia. (M 18645)

(3969) Rossi

1978 TQ_8. Discovered 1978 October 9 by L. V. Zhuravleva at Nauchnyj.
Named in memory of the outstanding architect Karl Ivanovich Rossi (1775-1849), designer of splendid architectural ensembles in St. Petersburg. (M 24121)

(3970) 1979 ME_9

Discovered 1979 June 28 by C. Torres at Cerro El Roble.

(3971) Voronikhin

1979 YM_8. Discovered 1979 December 23 by L. V. Zhuravleva at Nauchnyj.

Named in memory of Andrej Nikiforovich Voronikhin (1759-1814), a Russian architect, representative of classicism. (M 24121)

(3972) Richard

1981 JD_3. Discovered 1981 May 6 by C. S. Shoemaker at Palomar.

Named in honor of Richard Arthur Spellmann, brother of the discoverer. Spellmann, a chemical engineer, was a pioneer in the application of computers to the control of petroleum refinery processes. He has also dedicated much of his time to the needs of his community and served as the mayor of El Cerrito, California. (M 15575)

(3973) 1981 UC_1

Discovered 1981 October 30 by L. G. Taff at Socorro.

(3974) Verveer

1982 FS. Discovered 1982 March 28 by E. Bowell at Anderson Mesa.

Named in honor of Arie J. B. Verveer, technical manager of the Perth Observatory, in recognition of his many contributions to research. Since 1974, he has provided wide-ranging support for staff and visiting astronomers. His electronic, mechanical, and computer expertise, together with his broad astronomical knowledge and experience, have made him a great asset to the Observatory. (M 14972)

Name proposed by the discoverer, following a suggestion by M. P. Candy, who also prepared the citation.

(3975) Verdi

1982 UR_3. Discovered 1982 October 19 by F. Börngen at Tautenburg.

Named for the Italian composer Giuseppe Verdi (1813-1901), renowned the world over for his brilliant operas. (M 14634)

(3976) Lise

1983 JM. Discovered 1983 May 6 by N. G. Thomas at Anderson Mesa.

Named for the discoverer's daughter-in-law, Lise Melinda Breakey Thomas. Lise is the author of *Furry Outlaws*, a fantasy role-playing game. (M 26929)

(3977) Maxine

1983 LM. Discovered 1983 June 14 by C. S. Shoemaker and E. M. Shoemaker at Palomar.

Named in honor of Maxine Shoemaker Heath, sister of the second discoverer. An entomologist at the University of Illinois, Heath is a leading authority on the cicadas of North America and Argentina. Her research on thermoregulation in cicadas, in collaboration with James E. Heath, led to their discovery of several species of warm-blooded cicadas in the forests and thorn scrub of Argentina. (M 15575)

(3978) Klepešta

1983 VP_1. Discovered 1983 November 7 by Z. Vávrová at Kleť.

Named in memory of Josef Klepešta (1895-1976), outstanding Czech expert in astronomical photography and one of the founders of the Czech Astronomical Association. (M 22246)

(3979) Brorsen

1983 VV_1. Discovered 1983 November 8 by A. Mrkos at Kleť.

Named in memory of Theodor Brorsen (1819-1895), Danish astronomer, known for his discoveries of five comets and his studies of the gegenschein. After studying

and working in Kiel, Heidelberg and Altona he worked at the private observatory of baron John Parish in Senftenberg (Žamberk) in eastern Bohemia from 1847 to 1870. (M 27734)
Name suggested by J. Tichá.

(3980) Hviezdoslav

1983 XU. Discovered 1983 December 4 by A. Mrkos at Kleť.

Named in memory of P. O. Hviezdoslav (1849-1921), the best known Slovak poet. (M 25976)

Name proposed by the discoverer following a suggestion by astronomers of the Astronomical Institute at Tatranská Lomnica.

(3981) Stodola

1984 BL. Discovered 1984 January 26 by A. Mrkos at Kleť.

Named in memory of A. Stodola (1859-1942), professor at ETH Zürich who taught on the construction and design of various devices, such as the artificial hand he designed in 1915. (M 25976)

Name proposed by the discoverer following a suggestion by astronomers of the Astronomical Institute at Tatranská Lomnica.

(3982) Kastel

1984 JP_1. Discovered 1984 May 2 by L. G. Karachkina at Nauchnyj.

Named in honor of Galina Richardovna Kastel', well-known expert on the study of the motions of minor planets and comets, and a staff member of the Institute for Theoretical Astronomy since 1962. She contributed much to the study of the 1886 close approach of periodic comet Brooks 2 to Jupiter. For more than a quarter of a century she has headed the ITA service on comets and fast-moving minor planets, maintaining contacts with the IAU Central Bureau for Astronomical Telegrams and supplying observers in the U.S.S.R. with the necessary information on orbits and ephemerides. She also put much effort into the identification and orbit determination of minor planets observed in the course of the program at the Crimean Astrophysical Observatory, and the success of this program is in no small measure due to her help. (M 18306)

(3983) Sakiko

1984 SX. Discovered 1984 September 20 by A. Mrkos at Kleť.

Named in honor of Sakiko Nakano, sister of orbit computer Syuichi Nakano {see planet (3431)}, in remembrance of the discoverer's meeting her during the IAU General Assembly in New Delhi in 1985. (M 25976)

(3984) 1984 SB_6

Discovered 1984 September 21 by H. Debehogne at La Silla.

(3985) Raybatson

1985 CX. Discovered 1985 February 12 by C. S. Shoemaker and E. M. Shoemaker at Palomar.

Named in honor of Raymond M. Batson, planetary cartographer with the U.S. Geological Survey. Batson was responsible for detailed maps and mosaics of the lunar surface derived from the television images returned from five Surveyor spacecraft landed on the Moon in the 1960s. Later he organized and led a group to carry out systematic cartography of Mercury, Venus, Mars and the satellites of the outer planets. About 500 published maps and photomosaics have been prepared by Batson's group. (M 15575)

(3986) Rozhkovskij

1985 SF_2. Discovered 1985 September 19 by N. S. Chernykh and L. I. Chernykh at Nauchnyj.

Named in memory of Dmitrij Aleksandrovich Rozhkovskij (1915-1991), a leading astronomer at the Astrophysical Institute of the Kazakh Academy of Sciences for many decades. He was an outstanding observer and researcher on several types of celestial objects, especially reflection nebulae and new comets. (M 19694) Obituary published in Zemlya Vselennaya, No. 1, p. 41 (1992).

(3987) Wujek

1986 EL_1. Discovered 1986 March 5 by E. Bowell at Anderson Mesa.

Named in honor of Joseph H. Wujek, staff engineer in the Advanced Technology Group of Apple Computer, Inc. An electrical engineer by training, Wujek has worked on NASA's Orbiting Geophysical Observatory, the U.S.A.-U.S.S.R. Apollo-Soyuz mission and as a guest engineer at the Very Large Array of the National Radio Astronomy Observatory. His professional interests include ethics in engineering, space technology, engineering reliability, computing and the social implications of technology. In 1989 he was honored as "Volunteer of the Year" by the Astronomical Society of the Pacific. A member of the Board of Advisors of Lowell Observatory, he has given freely of his time and expertise, particularly by coordinating the donation of much-needed computer hardware and software from Apple Computer. (M 15261)

(3988) 1986 LA

Discovered 1986 June 4 by E. F. Helin at Palomar.

(3989) Odin

1986 RM. Discovered 1986 September 8 by P. Jensen and K. Augustesen at Brorfelde.

Named after the first and mightiest god in Norse mythology, who is the god of battle and victory, but also of wisdom and poetry. He rules the world from his home in Asgaard, where he gathers all the heroes who fell in battle to the great hall of Valhal {see planet (1260)}, where they enjoy feasting and fighting until Ragnarok, the end of the world. (M 22500)
The Germanic form of Odin is Wodan {see planet (2155)}.

(3990) Heimdal

1987 SO_3. Discovered 1987 September 25 by P. Jensen and K. Augustesen at Brorfelde.

Named after the god of dawn and light in Norse mythology, son of Odin {see planet (3989)}, born of nine virgins (all sisters). He is the watchman of the Aesir and guards Bifrost, the bridge between heaven and earth, against the giants. He is famous for his eyesight and can hear the grass grow. When be blows his lure Gjallarhorn it is heard throughout the world. (M 22500)

(3991) Basilevsky

1987 SW_3. Discovered 1987 September 26 by E. Bowell at Anderson Mesa.

Named in honor of Alexandr T. Basilevsky {1937- }, a planetary geologist at the Vernadsky Institute, Moscow. Basilevsky has worked for the U.S.S.R. Geological Survey on mapping the central part of the Russian platform, and for the Space Research Institute of the U.S.S.R. Academy of Sciences on potential landing sites for the Lunakhod 1 and 2 probes and the Luna 16, 20 and 24 probes, which returned lunar rock samples to the Earth. He was involved in the photogeological analysis of television images from the Mars 4 and 5 and the Venera 9, 10, 13 and 14 missions. An expert on impact cratering, Basilevsky is a coauthor of the book "Impact Craters on the Moon and Planets". He has recently worked on the interpretation of Venera 15 and 16 images of Venus. (M 16042)

(3992) Wagner

1987 SA_7. Discovered 1987 September 29 by F. Börngen at Tautenburg.

Named for the composer Richard Wagner (1813-1883), the founder of modern music-drama, known for both the music and the texts of his operas. (M 14634)

(3993) Šorm

1988 VV_5. Discovered 1988 November 4 by A. Mrkos at Kleť.

Named in memory of František Šorm (1913-1980), president of the Czechoslovak Academy of Sciences during the International Geophysical Year, founder and long-time director of the Institute for Macromolecular Chemistry in Prague. (M 25976)

(3994) Ayashi

1988 XF. Discovered 1988 December 2 by M. Koishikawa at Sendai.

Named for the western part of the city of Sendai. The observing station at which this discovery was made was set up there in 1975 in the graveyard of the Buddhist temple An-yo-ji. (M 15090)

(3995) Sakaino

1988 XM. Discovered 1988 December 5 by T. Kojima at Chiyoda.

Named in honor of Teruo Sakaino (1917-), a glass and ceramics chemist, professor emeritus of the Tokyo Institute of Technology and of the Wuhan (China) University of Engineering, known for his research on amorphous materials. In particular, he developed quantitative methods for estimating the quality of optical glass. He is also an amateur astronomer and a former teacher of the discoverer at the T.I.T. (M 28089)

(3996) Fugaku

1988 XG_1. Discovered 1988 December 5 by M. Arai and H. Mori at Yorii.

One of the ancient names of Mt. Fuji {see planet (1584)}, the highest mountain in Japan. (M 34619)

(3997) Taga

1988 XP_1. Discovered 1988 December 6 by A. Sugie at Taga.

Named for the town in which the Dynic Observatory is located. Known for its ancient shrine, Taga is located 10 km east of Lake Biwa, the largest lake in Japan, and is surrounded by the Suzuka mountain range. (M 14634)

(3998) Tezuka

1989 AB. Discovered 1989 January 1 by T. Kojima at Chiyoda.

Named in memory of Osamu Tezuka (1928-1989), celebrated Japanese master of animation. A doctor of medicine, yet already a writer of comic books at the age of 18, he made particular use of movie techniques in his works. His writings therefore differed from the usual comics or caricatures, and the style that he initiated is known in Japanese as "Manga". His works, amounting to more than 100 million copies of over 400 books, included *Tetsuwan Atom* (Astro Boy), *Jungle Taitei* (Emperor of the Jungle) and *Hi no Tori* (Fire Bird). They present both children and adults with dreams and hopes and cover a very wide range of subjects, from science fiction to human nature, philosophy, pacifism and faith. (M 19694)

(3999) Aristarchus

1989 AL. Discovered 1989 January 5 by T. Kojima at Chiyoda.

Named by the Minor Planet Names Committee for one of the earliest (3rd century B.C.) astronomers of the Alexandrine school, known for his belief that the

earth revolved around the sun. (M 19335)
Aristarchus is also honored by a lunar crater.

(4000) Hipparchus

1989 AV. Discovered 1989 January 4 by S. Ueda and H. Kaneda at Kushiro.

Named by the Minor Planet Names Committee for the greatest astronomer of ancient times, renowned for his introduction of systematic and critical procedures with regard both to observations and to theoretical concepts. (M 19335)
Hipparchus is also honored by craters on Mars and on the Moon.

(4001) Ptolemaeus

1949 PV. Discovered 1949 August 2 by K. Reinmuth at Heidelberg.

Named by the Minor Planet Names Committee for the celebrated author of the *Almagest*, whose ideas dominated astronomical thought in Europe and the Middle East from the second to the sixteenth centuries. (M 19335)
Ptolemaeus is also honored by craters on Mars and on the Moon.

(4002) Shinagawa

1950 JB. Discovered 1950 May 14 by K. Reinmuth at Heidelberg.

Named in honor of Seishi Shinagawa (1944-), who was the first to use electronic computers for orbit computation in Japan, having written a computer program for this in 1965. This program was used by Yoshiaki Banno (1952-1991) {see planet (3394)}. (M 22500)

Name proposed by S. Nakano, who found the identifications involving this object and learnt orbit computation and programming from Shinagawa and Banno.

(4003) Schumann

1964 ED. Discovered 1964 March 8 by F. Börngen at Tautenburg.

Named for the famous romanticist Robert Schumann (1810-1856), born in Zwikkau, not far from Tautenburg, and known for his song creations, piano compositions, chamber music and cello concerti. (M 14634)

(4004) List'ev

1971 SN_1. Discovered 1971 September 16 at the Crimean Astrophysical Observatory at Nauchnyj.

Named in memory of Vladislav Nikolaevich List'ev (1957-1995), a talented Russian television journalist who fell victim to hired assasins. His clever, benevolent, analytical series of telecasts 'Vzglyad' (View), 'Tema' (Subject) and 'Chas pik' (Peak hour) were popular among millions of people. They had a great influence on the public consciousness of Russians in the post-totalitarian era. (M 25229)
Name proposed by the Institute of Theoretical Astronomy.

(4005) Dyagilev

1972 TC_2. Discovered 1972 October 8 by L. V. Zhuravleva at Nauchnyj.

Named in memory of the Russian impressario Sergej Pavlovich Dyagilev (1872-1929). He is mainly known as the patron of art and literature and the founder of the travelling Russian ballet company. (M 24121)

(4006) Sandler

1972 YR. Discovered 1972 December 29 by T. M. Smirnova at Nauchnyj.

Named in memory of Grigorij Moiseevich Sandler (1912-1994), an outstanding Russian musician who made an enormous contribution to the development of choral performance. He was the main conductor of the Leningrad (now St. Petersburg) University chorus for more than 40 years. (M 24410)
Name proposed by the Institute of Theoretical Astronomy.

(4007) Euryalos

1973 SR. Discovered 1973 September 19 by C. J. van Houten at Palomar.

Named for the commander of the troops from Argos during the siege of Troy. (M 16042)

(4008) Corbin

1977 BY. Discovered 1977 January 22 at the Felix Aguilar Observatory at El Leoncito.

Named in honor of Thomas E. Corbin, head of the Meridian Circle Division at the U.S. Naval Observatory, and his wife, Brenda Groves Corbin, since 1973 the Observatory's librarian. Serving early in his career as astronomer-in-charge of the Observatory's southern station in Argentina, Tom Corbin has since been responsible for several important astrometric programs, including the *Astrographic Catalogue Reference Stars* catalogue. Brenda Corbin is one of the most renowned astronomical librarians in the world, known for her work in the Special Libraries Association, the IAU Commission 5 Working Group on Nomenclature and for her enthusiasm in tracking down copies of rare works. (M 31295)

(4009) Drobyshevskij

1977 EN_1. Discovered 1977 March 13 by N. S. Chernykh at Nauchnyj.

Named in honor of Ehduard Mikhajlovich Drobyshevskij, physicist and astrophysicist at the Ioffe Physical and Technical Institute in St. Petersburg, author of some original cosmological ideas and theories of the origin of the planets and the minor bodies of the solar system, also known for his research on the magnetic fields of the sun and other stars. (M 19694)

(4010) Nikol'skij

1977 QJ_2. Discovered 1977 August 21 by N. S. Chernykh at Nauchnyj.

Named in memory of Gennadij Mikhajlovich Nikol'skij (1929-1982), Soviet astronomer, known for his research on the sun and the solar corona and as a codiscoverer of the solar wind. (M 19695)

Obituaries published in Zemlya Vselennaya, No. 3, p. 33-34 (1983); Properties and interactions of interplanetary dust, p. XXIII-XXIV (1985).

(4011) Bakharev

1978 SC_6. Discovered 1978 September 28 by N. S. Chernykh at Nauchnyj.

Named in memory of Anatolij Mikhailovich Bakharev (1918-1979), observer and researcher on comets and meteors, one of the discoverers of comet Bakharev-Macfarlane-Krienke (1955 IV). (M 19695)
Obituary published in Byull. Inst. Astrofiz., No. 71, p. 40-41 (1982).

(4012) 1978 VK_9

Discovered 1978 November 7 by E. F. Helin and S. J. Bus at Palomar.

(4013) Ogiria

1979 OM_{15}. Discovered 1979 July 21 by N. S. Chernykh at Nauchnyj.

Named in memory of Maiya Borisovna Ogir' (1933-1991), solar physicist and staff member of the Crimean Astrophysical Observatory for more than 30 years, known for her research on the active processes on the Sun. (M 22500)

(4014) Heizman

1979 SG_{10}. Discovered 1979 September 28 by N. S. Chernykh at Nauchnyj.

Named in honor of Leonie A. Heizman, docent of the historical museum at the San Juan Capistrano mission, and Charles L. Heizman, master of technical sciences working in the computer business. They served as hostess and host to the discoverer and K. I. Churyumov {see planet (2627)} during the conference on Near Earth Asteroids in San Juan Capistrano during 1991 June 30-July 3. (M 20159)

(4015) Wilson-Harrington

1979 VA. Discovered 1979 November 15 by E. F. Helin at Palomar.

This Apollo object, which is identical with the single-apparition periodic comet 1949 III = 1949g (which appeared cometary on its discovery night of 1949 Nov. 19), is being given the name of that comet, which was discovered by A. G. Wilson and R. G. Harrington early in the first Palomar Sky Survey. (M 22246)

Named by the Minor Planet Names Committee, in the knowledge that the name violates the normal 16-character maximum.

This Apollo object was identified by E. Bowell on Palomar Sky Survey plates of 1949 Nov. 19.1 while looking for prediscovery images of minor planets. B. A. Skiff found a very slightly fanned tail on the trailed images. Subsequently, B. G. Marsden (IAUC 5585) established the identity of (4015) with the periodic comet 1949 III Wilson-Harrington.

(4016) Sambre

1979 XK. Discovered 1979 December 15 by H. Debehogne and E. R. Netto at La Silla.

Named for a tributary of the river Meuse {see planet (3016)} in Namur {see planet (3374)}, site of Caesar's victory over Boduognat {see planet (3458)}, leader of the Belgian tribes, in 59 B.C. (M 19336)

(4017) Disneya

1980 DL_5. Discovered 1980 February 21 by L. G. Karachkina at Nauchnyj.

Named in memory of Walt Disney (1901-1966), the outstanding American movie producer and master of animation. (M 17029)

(4018) Bratislava

1980 YM. Discovered 1980 December 30 by A. Mrkos at Kleť.

Named for the capital of Slovakia. (M 25976)

(4019) Klavetter

1981 EK_{14}. Discovered 1981 March 1 by S. J. Bus at Siding Spring.

Named in memory of James Jay Klavetter (1960-1997), professor of physics at California State University at Sacramento. His enthusiasm for teaching undergraduate astronomy will serve as a model for communicating the value and excitement of science. Klavetter studied the sources and behavior of cometary dust and gas tails through telescoping imaging, and he also established the chaotic rotation state of Hyperion. Klavetter's positive outlook and thirst for new experiences remained undimmed even in the final days of his battle with leukemia. (M 30475)

Name suggested and citation prepared by J. Harrington.

(4020) Dominique

1981 ET_{38}. Discovered 1981 March 1 by S. J. Bus at Siding Spring.

Named in honor of Dominique Bockelée-Morvan (1957-), researcher at the Observatoire de Paris. Bockelée-Morvan has developed excitation models for cometary parent molecules with a view to preparing and interpreting radio and infrared observations, both from space and the ground. She has been involved in the identification of new cometary species and has contributed to a better understanding of the origin of the 3.4-μmband observed in comets. (M 27458)

Citation provided by M. A. Barucci following a suggestion by E. Bowell.

(4021) Dancey

1981 QD_2. Discovered 1981 August 30 by E. Bowell at Anderson Mesa.

Named in honor of Roy Dancey and Bruce D. Dancey, father and son, who successively headed the optical shop at the Dominion Astrophysical Observatory in Victoria from 1965 until 1986. Under their hands and direction, the primary and secondary mirrors for the 3.6-m Canada-France-Hawaii Telescope were figured and polished. They also produced new primary mirrors for the 1.8-m and 1.2-m telescopes at Victoria, as well as a host of smaller telescope optics, spectrograph optics and associated test optics. Bruce Dancey's career was tragically terminated by blindness in 1986. However, the Dancey's work has given the astronomers of Canada, France and Hawaii a superb eye with which to explore the mysteries of space. (M 16042)

Name proposed by the discoverer, following a suggestion by the staff of the Dominion Astrophysical Observatory. Citation prepared by C. Aikman.

(4022) Nonna

1981 TL_4. Discovered 1981 October 8 by L. I. Chernykh at Nauchnyj.

Named in honor of Nonna (Noyabrina) Viktorovna Mordyukova, a popular Soviet cinema actress. (M 18645)

(4023) Jarník

1981 UN. Discovered 1981 October 25 by L. Brožek at Kleť.

Named in memory of Vojtěch Jarník (1897-1970), Czech mathematician and professor at the Charles University in Prague. He studied number theory and was also well known for his work on the theory of real variable functions. His excellent textbooks on the differential and integral calculus have been used by several generations of mathematicians and physicists. (M 24121)

(4024) Ronan

1981 WQ. Discovered 1981 November 24 by E. Bowell at Anderson Mesa.

Named in honor of Colin A. Ronan {1920-1995}, author and specialist in the history and philosophy of science. Particularly interested in the moon and planets, he has written more than 20 books on astronomy. He has played key roles in the administration of the British Astronomical Association and is its current president. In addition he has been director of the Association's Historical Section (1953-1965) and editor of its Journal (1965-1985). (M 17222)
Name suggested and citation material provided by S. A. Mitton.

Obituaries published in J. Br. Astron. Assoc., Vol. 105, No. 5, p. 262 (1995); Astron. Now, Vol. 9, No. 9, p. 7 (1995); Q.J.R. Astron. Soc., Vol. 37, No. 1, p. 93-94 (1996).

(4025) Ridley

1981 WU. Discovered 1981 November 24 by E. Bowell at Anderson Mesa.

Named in honor of Harold B. Ridley {1919-1995}, distinguished British astrophotographer specializing in comet and asteroid astrometry and photography. For many years, Ridley has been a very active and dedicated member of the British Astronomical Association, having served as its president (1976-1978), vice president (1978-1981, 1982-1985, 1986-1987), director of the Meteor Section (1954-1968), assistant to the director of Comet Section (1975) and currently as a Council member. (M 17222)

Name independently suggested by S. A. Mitton and B. G. Marsden, with citation material provided by the former.

Obituaries published in J. Br. Astron. Assoc., Vol. 105, No. 4, p. 189-191 (1995).

(4026) Beet

1982 BU_1. Discovered 1982 January 30 by E. Bowell at Anderson Mesa.

Named in honor of Ernest A. Beet {1904-1997}, a member of the British Astronomical Association for half of its existence. Having served in various capacities on the B.A.A.'s council between 1946 and 1987, notably as president during 1962-1964 and for twelve years as secretary, he still attends the London meetings. A schoolteacher by profession, Beet has made outstanding contributions to the advancement of astronomy education in England and has been heavily involved in the B.A.A. education committee. He has written several popular books, including *The Teaching of Astronomy in Schools* (1948). (M 17222)

Name suggested by B. G. Marsden and citation material provided by S. A. Mitton.

Obituary published in J. Br. Astron. Assoc., Vol. 108, No. 2, p. 121-122 (1998).

(4027) Mitton

1982 DN. Discovered 1982 February 21 by E. Bowell at Anderson Mesa.

Named in honor of Simon and Jacqueline Mitton, astronomers in Cambridge, England. Originally at the Mullard Radio Astronomy Observatory, Simon made comprehensive high-resolution observations of Cygnus A with the Cambridge One-Mile Telescope. Interested in the correlation of radio and optical observations, he moved to the Institute of Astronomy in 1972 and collaborated with Jacqueline on a study of the optical spectrum of Cygnus A. She also worked on equivalent widths and curves of growth. Both subsequently turned their interests to the dissemination of astronomy, he as editor of the *Cambridge Encyclopedia of Astronomy* and astronomy publisher for Cambridge University Press, she as editor of the *Journal of the British Astronomical Association*. (M 17222)

Citation provided by B. G. Marsden at the request of the discoverer.

(4028) 1982 DV_2

Discovered 1982 February 18 by L. G. Taff at Socorro.

(4029) Bridges

1982 KC_1. Discovered 1982 May 24 by C. S. Shoemaker and S. J. Bus at Palomar.

Named in honor of Patricia M. Bridges, planetary cartographer with the U.S. Geological Survey. Based on intimate familiarity with the Moon's surface, gained from long hours at the eyepiece of the 0.6-m Clark refractor at Lowell Observatory, and later on minutely detailed knowledge gleaned from spacecraft images, Bridges' shaded relief maps of the Moon, other satellites and the terrestrial planets, rendered with extraordinary skill by means of the airbrush, are generally regarded as unsurpassed. (M 15576)

(4030) Archenhold

1984 EO_1. Discovered 1984 March 2 by H. Debehogne at La Silla.

Named in memory of the German astronomer Friedrich Simon Archenhold (1861-1939). In 1896 he made it possible to build the world's longest refractor (focal length 21 meters), which is still in use at the Archenhold Observatory in Berlin - Europe's first public observatory. As its first director Archenhold had the aim of introducing astronomy to everyone. (M 34340)

Name suggested by A. Gnädig and A. Doppler and endorsed by L. D. Schmadel.

(4031) Mueller

1985 CL. Discovered 1985 February 12 by C. S. Shoemaker and E. M. Shoemaker at Palomar.

Named in honor of Jean Mueller, observer for the Second Palomar Sky Survey with the 1.2-m Oschin Telescope at Palomar Observatory. She has made numerous discoveries of supernovae and earth-approaching asteroids in the course of this survey. (M 15576)

(4032) Chaplygin

1985 UT_4. Discovered 1985 October 22 by L. V. Zhuravleva at Nauchnyj.

Named in memory of Sergej Alexeevich Chaplygin (1869-1942), a notable ex-

pert in theoretical mechanics. He was one of the founders of the aerodynamics. (M 24121)
Chaplygin is also honored by a lunar crater.

(4033) Yatsugatake

1986 FA. Discovered 1986 March 16 by M. Inoue and O. Muramatsu at Kobuchizawa.

Named for a mountain in central Japan, famous for its scenic beauty. The Kobuchizawa {see planet (3432)} observing station stands at the foot of this mountain. (M 14634)

(4034) 1986 PA

Discovered 1986 August 2 by E. F. Helin at Palomar.

(4035) 1986 WD

Discovered 1986 November 22 by K. Suzuki and T. Urata at Toyota.

(4036) 1987 DW_5

Discovered 1987 February 21 by H. Debehogne at La Silla.

(4037) Ikeya

1987 EC. Discovered 1987 March 2 by K. Suzuki and T. Urata at Toyota.

Named in honor of Kaoru Ikeya, one of the most eminent Japanese amateur astronomers. Since 1963 he has discovered five comets, including the sungrazer Comet Ikeya-Seki (1965 VIII), and two extragalactic supernovae. He has ground many large telescopic mirrors, which are in good use at observatories in Japan. (M 16443)
Name proposed by T. Urata. Citation prepared by Y. Murai and I. Hasegawa.

(4038) Kristina

1987 QH_2. Discovered 1987 August 21 by E. W. Elst at La Silla.

Named in honor of Kristina Leterme, professor of French and Russian literature, life partner to the discoverer, for her encouragement and love. (M 15090)

(4039) Souseki

1987 SH. Discovered 1987 September 17 by T. Seki at Geisei.

Named in memory of Souseki Natsume (1867-1916), a distinguished Japanese scholar in English literature who has left an indelible mark on the history of Japanese modern literature with Ougai Mori in the Meiji era, and whose brain is still kept in the medical department of Tokyo University. Many of his articles are still widely read and have been made into movies. One of the best known articles, Botchan, which means a greenhorn in English, is set in Shikoku, the island on which the Geisei {see planet (2571)} station is located. (M 16246)

(4040) Purcell

1987 SN_1. Discovered 1987 September 21 by E. Bowell at Anderson Mesa.

Named for Henry Purcell (1659-1695), English composer, a prolific and immensely gifted writer of many forms of music. Purcell's astonishing fantasias for viols, written when he was barely out of his teens, have had a great influence on the discoverer. (M 16592)

(4041) Miyamotoyohko

1988 DN_1. Discovered 1988 February 19 by T. Kojima at Chiyoda.

Named in honor of Yohko Miyamoto (1922-), wife of the amateur astronomer Yukio Miyamoto {see planet (6020)}. Mrs. Miyamoto has assisted her husband in his work, directly and indirectly, for half a century. Her sage advice to young members of the Kumamoto Astronomical Society was beneficial to the discoverer in his youth. (M 28621)

(4042) Okhotsk

1989 AT_1. Discovered 1989 January 15 by K. Endate and K. Watanabe at Kitami.

Named for the sea that extends between the northeast coast of Hokkaido and the Kamchatka Peninsula. The sea of Okhotsk, which is completely covered by ice floes from January to March, holds rich gifts of salmon, scallops and crabs for the people of the north. (M 15427)

(4043) Perolof

1175 T-3. Discovered 1977 October 17 by C. J. van Houten and I. van Houten-Groeneveld at Palomar.

Named in honor of Per Olof Lindblad, Swedish astronomer, director of the Stockholm Observatory at Saltsjöbaden. He has served as president of the council of the European Southern Observatory and is currently a vice president of the IAU. He is the son of the astronomer Bertil Lindblad (1895-1965) {see planet (1448)}. (M 18455)

(4044) Erikhøg

5142 T-3. Discovered 1977 October 16 by C. J. van Houten and I. van Houten-Groeneveld at Palomar.

Named in honor of Erik Høg (1932-), Danish astronomer at the Copenhagen Observatory. He is meridian circle astrometrist and designer of several instruments, including a micrometer for the Carlsberg Transit Circle. He has been involved with Hipparcos since the mission evaluation study in 1975 as a member of the science team, of the scientific advisory group for MATRA (leader of MESH-Consortium for construction of HIPPARCOS) and as team leader of the data reduction consortium of the Tycho part of the Hipparcos mission. (M 18455)

(4045) Lowengrub

1953 RG. Discovered 1953 September 9 at the Goethe Link Observatory at Brooklyn, Indiana.

Named in honor of Morton Lowengrub in celebration of the completion of the WIYN Observatory. As dean for Research and Graduate Development at Indiana University, and later as dean of the College of Arts and Sciences, he played a major role in the planning and construction of the WIYN 3.5-meter telescope as a charter member of the WIYN Board of Governors. He has served as a professor and administrator at Indiana University since 1967. An applied mathematician, he has authored five books on mathematics and through his research has contributed critically to problems of the propagation of cracks in solids. (M 25229)
Name proposed by R. K. Honeycutt.

(4046) Swain

1953 TV. Discovered 1953 October 7 at the Goethe Link Observatory at Brooklyn, Indiana.

Named in memory of Joseph Swain (1857-1927), successor of Daniel Kirkwood {see planet (1578)} as professor of mathematics at Indiana University in 1886. In 1891 he became professor of mathematics on the original Stanford University faculty. He returned to Indiana University as president in 1893. Kirkwood Hall, a classroom building, and the Kirkwood Observatory were built during Swain's nine years as president, and the department of mechanics and astronomy was created in 1895 with John A. Miller {see planet (1826)} as head. Swain became president

of Swarthmore College in 1902. He persuaded Miller to join him in 1906 after he raised money for a telescope twice the size of the Kirkwood Observatory 0.30-m refractor. (M 25653)
Name proposed by F. K. Edmondson.

(4047) 1964 TT_2

Discovered 1964 October 8 at the Purple Mountain Observatory at Nanking.

(4048) Samwestfall

1964 UC. Discovered 1964 October 30 at the Goethe Link Observatory at Brooklyn, Indiana.

Named in memory of Richard Samuel Westfall (1924-1996), distinguished professor of history and philosophy of science at Indiana University from 1976 to 1989. Westfall is regarded as the foremost scholar of the Scientific Revolution, primarily because of his extraordinary scholarship on Newton {see planet (8000)}, exemplified in his 1971 treatise on the development of dynamics in the seventeenth century and his 1980 biography of Newton, *Never at Rest*, both of which won the coveted Pfizer Award of the History of Science Society. (M 34619)
Name proposed by F. K. Edmondson. Citation prepared by E. Grant.

(4049) Noragal'

1973 QD_2. Discovered 1973 August 31 by T. M. Smirnova at Nauchnyj.

Named in memory of Eleonora Yakovlevna Gal'perina (1912-1992), literary critic and translator, well-known under her pseudonym Nora Gal' for her Russian translations of foreign works, namely those by the authors Theodor Dreiser, Ethel Lilian Voynich, Jack London {see planet (2625)}, Antoine de Saint-Exupéry {see planet (2578)} and others. (M 24916; M 26439)
Name proposed by E. A. Taratuta.

(4050) Mebailey

1976 SF. Discovered 1976 September 20 by C.-I. Lagerkvist and H. Rickman at Kvistaberg.

Named in honor of Mark E. Bailey, a British astronomer at the University of Manchester well known for his work on the origin of comets, the dynamics of the Oort cloud and the capture of comets into short-period orbits. (M 16042)

(4051) Hatanaka

1978 VP. Discovered 1978 November 1 by K. Tomita at Caussols.

Named in memory of Takeo Hatanaka (1914-1963), professor of astrophysics at the University of Tokyo and on the staff of the Tokyo Astronomical Observatory. Hatanaka's theoretical work covered a wide variety of astrophysical problems, involving planetary nebulae, the solar atmosphere and stellar evolution. He also pioneered research on radio astronomy in Japan and wrote enlightening books. As a member of the Japanese Science Council he was very closely involved in the early development of the national space program. (M 16043)

Name proposed by the discoverer, whose wife, Tomoko, worked for Hatanaka as a secretary before her marriage. Citation prepared by F. Moriyama.
Hatanaka is also honored by a lunar crater.

(4052) Crovisier

1981 DP_2. Discovered 1981 February 28 by S. J. Bus at Siding Spring.

Named in honor of Jacques Crovisier (1948-), an astronomer at Paris-Meudon Observatory. Crovisier has observed the OH lines of more than 40 comets using the Nançay radio telescope, as well as millimetric molecular lines of a number

of comets using the IRAM telescope. He was co-investigator for the infrared spectrometer on the VEGA mission to comet 1P/Halley and is currently involved in several cometary observation programs with ISO. Crovisier is an expert in modeling molecular processes in comets and especially in molecular spectra at radio and infrared wavelengths. (M 27458)
Citation provided by M. A. Barucci following a suggestion by E. Bowell.

(4053) Cherkasov

1981 TQ_1. Discovered 1981 October 2 by L. V. Zhuravleva at Nauchnyj.
Named in memory of Nikolaj Konstantinovich Cherkasov (1903-1966), a well-known Russian actor. (M 24121)

(4054) Turnov

1983 TL. Discovered 1983 October 5 by A. Mrkos at Kleť.
Named for a small city in a region of eastern Bohemia known as "Czech Paradise". The city is famous for the production of excellent optics and precious stones. (M 26424)

(4055) Magellan

1985 DO_2. Discovered 1985 February 24 by E. F. Helin at Palomar.
Named for the leader of the first expedition to circumnavigate the Earth and the present highly successful automated expedition to Venus. Ferdinand Magellan {1480?-1521} did not survive the epic voyage of 1519-1522, but the expedition was a triumph of navigation and yielded discoveries (including the closest galaxies to our own) that sustained generations of explorers. The Magellan spacecraft, unlike its namesake, completed its voyage in timely fashion and only then met its real goal, the mapping of the surface of our planetary twin, a task being carried out with stunning success. (M 17466)
Magellan is also honored by craters on Mars and the Moon.

(4056) Timwarner

1985 FZ_1. Discovered 1985 March 22 by E. Bowell at Anderson Mesa.
Named in memory of Timothy Warner (1961-1990), chief mission planner for the Cosmic Background Explorer's Diffuse Infrared Background Experiment (DIRBE). Tim skillfully integrated the science and engineering of DIRBE, translating the principal investigator's observing programs into common loads for the instrument that will perform an all-sky infrared survey and reveal the glow of the first stars and galaxies formed after the Big Bang. Prior to his work on COBE, he was an operations engineer for the Solar Maximum Mission's Coronagraph/Polarimeter experiment, scheduling the instrument and performing preliminary data analysis. He co-authored a paper on the instrument's observations of space debris, research that was consulted by Hubble Space Telescope engineers. Tim's artistic talents found expression in cartoons, paintings and contributions to Paramount Pictures' films, including special-effects work. He was also involved in rationalist causes, particularly in the exposure of flimflammery and fraudulent claims made in the name of science. Tim was found murdered on the night of 1990 Apr. 24, a day he had started with bright enthusiasm over the launch of the Space Telescope. (M 16885)
Name suggested and citation provided by S. T. Snell.

(4057) Demophon

1985 TQ. Discovered 1985 October 15 by E. Bowell at Anderson Mesa.
Named for a Greek warrior, the second son of Theseus and Phaedra {see planet (174)}. After fighting in the Trojan War, Demophon visited Thrace, where he

caused the queen, Phyllis {see planet (556)}, to fall in love with him. On his return to Athens, Demophon's neglect of Phyllis' kindness and love led to her suicide. (M 29669)

(4058) Cecilgreen

1986 JV. Discovered 1986 May 4 by E. Bowell at Anderson Mesa.

Named in honor of Cecil H. Green, co-founder of Texas Instruments and philanthropist extraordinary, to acknowledge a lifelong commitment to improve the quality of education and scientific research. Green and his late wife Ida funded educational and scientific projects throughout the world, such as academic buildings, health care centers, hospitals, libraries, scholarships and professorships. The Greens assisted in founding new colleges at the University of Texas at Dallas, at Oxford University in England and, in March 1990, Green College at the University of British Columbia. A graduate in electrical engineering, Green spent many years in geophysical field work before becoming a partner of Geophysical Service, Inc., in 1941. He and his partners created a division of that company that ultimately became Texas Instruments. In 1954 Texas Instruments pioneered the first successful commercialization of the silicon transistor. Green is a noted salmon fisherman. (M 17029)

Name suggested by and citation prepared at the University of British Columbia.

(4059) Balder

1987 SB_5. Discovered 1987 September 29 by P. Jensen and K. Augustesen at Brorfelde.

Named after the kindest Norse god, handsome and wise, son of Odin {see planet (3989)}, peacemaker among the Aesir. Through the treachery of Loke {see planet (4862)}, Balder was killed with a mistletoe shaft by his blind brother Høder {see planet (4669)}. (M 22500)

(4060) Deipylos

1987 YT_1. Discovered 1987 December 17 by E. W. Elst at La Silla.

Named for the Greek hero ordered by Sthenelos {see planet (3794)} to bring the horses captured from Aeneas {see planet (1172)} to the Greek vessels. (M 15090)

(4061) Martelli

1988 FF_3. Discovered 1988 March 19 by W. Ferreri at La Silla.

Named in honor of Giuseppe Martelli {1923-1994}, head of the Space and Plasma Physics Group at the University of Sussex from 1964 to 1986. Martelli pioneered the development and use of explosive accelerating techniques for the study of macroscopic hypervelocity impacts. His experimental and theoretical work constitutes a major contribution to the understanding of impact magnetization of ferromagnetic rocks, a problem particularly relevant to the origin of lunar magnetism. Over the years he has organized numerous series of experiments to study catastrophic fragmentation of free falling bodies, thereby providing experimental input to theories of asteroidal evolution and other impact phenomena of astrophysical interest. (M 14972)

Obituaries published in Planet. Space Sci., Vol. 42, No. 12, p. 1011-1012 (1994); Icarus, Vol. 112, No. 2, p. 302 (1994); Q. J. R. Astron. Soc., Vol. 36, No. 4, p. 459-460 (1995).

(4062) Schiaparelli

1989 BF. Discovered 1989 January 28 at the Osservatorio San Vittore at Bologna.

Named in memory of Giovanni Virgilio Schiaparelli (1835-1910), discoverer of the connection between comets and meteor streams. A great observer of Mars, he

discovered the famous "canali" and drew some fine maps. He made measurements of double stars and of the rotational periods of Mercury and Venus, and he discovered (69) Hesperia. From 1862 to 1900 he was director of the Brera Observatory in Milan. (M 15090)
Schiaparelli is also honored by craters on Mars and the Moon.

(4063) Euforbo

1989 CG_2. Discovered 1989 February 1 at the Osservatorio San Vittore at Bologna.

Named for the Trojan hero who wounded the Greek Patroclus in the breast before being killed by the Greek Hektor {see also the citations for planets (617) and (624)}. (M 19347)

The original citation (M 15091) was corrected by the Minor Planet Names Committee. According to the orbital elements, Euphorbos worked as a 'spy' in the foreign camp.

(4064) Marjorie

2126 P-L. Discovered 1960 September 24 by C. J. van Houten and I. van Houten-Groeneveld at Palomar.

Named in honor of Marjorie Meinel, who with her husband Aden {see planet (4065)} has formed a lifelong team, working closely together with great success. (M 19695)

(4065) Meinel

2820 P-L. Discovered 1960 September 24 by C. J. van Houten and I. van Houten-Groeneveld at Palomar.

Named in honor of U.S. astronomer Aden Meinel, originator of new techniques and facilities, including the Kitt Peak National Observatory, the Optical Sciences Center of the University of Arizona, and future space missions at the Jet Propulsion Laboratory. (M 19695)

(4066) Haapavesi

1940 RG. Discovered 1940 September 7 by H. Alikoski at Turku.

Named for the birthplace of the discoverer's father. Haapavesi has been known for centuries for the manufacture of skis. (M 18455)

(4067) Mikhel'son

1966 TP. Discovered 1966 October 11 by N. S. Chernykh at Nauchnyj.

Named in honor of Nikolaj Nikolaevich Mikhel'son (1918-), outstanding Soviet scientist known for his astronomical optics and telescopes. (M 27126)

(4068) Menestheus

1973 SW. Discovered 1973 September 19 by C. J. van Houten at Palomar.

Named for the commander of the troops from Athens during the siege of Troy. (M 16043)

(4069) Blakee

1978 VL_7. Discovered 1978 November 7 by E. F. Helin and S. J. Bus at Palomar.

Named in honor of Lawrence E. Blakee, who for 37 years has worked diligently in many capacities at the Palomar Observatory and at the Mount Wilson Observatory to support the researchers using those facilities. The many astronomers who have worked at Mount Wilson and Palomar have benefited from his conscientiousness and dedication. (M 16247)

(4070) Rozov

1980 RS_2. Discovered 1980 September 8 by L. V. Zhuravleva at Nauchnyj.

Named in honor of the Russian playwright Victor Sergeevich Rozov (1913-). (M 24121)

(4071) Rostovdon

1981 RD_2. Discovered 1981 September 7 by L. G. Karachkina at Nauchnyj.

Named for the city of Rostov, the discoverer's birthplace, a large industrial and railroad center located on the picturesque banks of the Don. (M 17029)

(4072) Yayoi

1981 UJ_4. Discovered 1981 October 30 by H. Kosai and K. Hurukawa at Kiso.

Named for the Yayoi Era, from the third century B.C. to the fourth century A.D., when the Japanese began to be influential beyond Asia. (M 21608)

(4073) 1981 UE_{10}

Discovered 1981 October 23 at the Purple Mountain Observatory at Nanking.

(4074) Sharkov

1981 UN_{11}. Discovered 1981 October 22 by N. S. Chernykh at Nauchnyj.

Named in honor of Viktor Ivanovich Sharkov (1935-), an experimental astrophysicist known for his laboratory simulation of the physical phenomena in icy cometary nuclei. A staff member of the Ioffe Physical and Technical Institute in St. Petersburg for many years, he has been director of the St.-Petersburg Planetarium since 1993. (M 26762)

(4075) Sviridov

1982 TL_1. Discovered 1982 October 14 by L. G. Karachkina at Nauchnyj.

Named in honor of Georgij Vasil'evich Sviridov {1915-1998}, a well-known Soviet composer. (M 17029)

(4076) Dörffel

1982 UF_4. Discovered 1982 October 19 by F. Börngen at Tautenburg.

Named in memory of Georg Samuel Dörffel (1643-1688), whose computations on the orbit of the great comet of 1680/1681 were the first to be made on the assumption of parabolic motion with the Sun at the focus. (M 14972)

(4077) Asuka

1982 XV_1. Discovered 1982 December 13 by H. Kosai and K. Hurukawa at Kiso.

Named of the Asuka Era, from the fifth and sixth centuries, when the capital was in the Asuka area in the central part of Japan. (M 21608)

(4078) Polakis

1983 AC. Discovered 1983 January 9 by B. A. Skiff at Anderson Mesa.

Named in honor of Thomas A. Polakis (1961-), mechanical engineer and friend of the discoverer. As an observer of the sky, he exemplifies all that is best in an amateur astronomer. (M 27126)

(4079) Britten

1983 CS. Discovered 1983 February 15 by E. Bowell at Anderson Mesa.

Named for Benjamin Britten (1913-1976), English composer. Best known for his operas and other vocal music, Lord Britten was also an accomplished pianist and conductor. His music is imbued with a rare beauty of line and limpidity of texture. (M 16592)

(4080) Galinskij

1983 PW. Discovered 1983 August 4 by L. G. Karachkina at Nauchnyj.

Named in honor of Nikolaj Dmitrievich Galinskij, Soviet radio engineer, creator of high-sensitivity image tubes for astronomical observations. The high quantum efficiency and excellent quality of these tubes were significant to the successful development of modern observational astronomy, and as long ago as 1964 stars of twentieth magnitude were being recorded with the 0.5-m telescope at the Crimean

Astrophysical Observatory with an exposure of only 4 seconds. Galinskij is a coauthor of *Television Astronomy*, published in 1984. (M 17030)
Citation provided by V. V. Prokof'eva at the request of the discoverer.

(4081) Tippett

1983 RC_2. Discovered 1983 September 14 by E. Bowell at Anderson Mesa.

Named for Sir Michael Tippett (1905-1998), English composer. Tippett's music, particularly that for voice, is marked by a compassionate humanity. (M 16592)

(4082) Swann

1984 SW_3. Discovered 1984 September 27 by C. S. Shoemaker and E. M. Shoemaker at Palomar.

Named in honor of Gordon A. Swann, geologist at Northern Arizona University. As principal investigator for the geological field investigations conducted at the Apollo 14 and 15 lunar landing sites, Swann forged a close knit and effective exploration effort that linked geologists, mission control engineers and the astronaut crew. The result was a rich return of geologic data from the first manned lunar missions planned explicitly for scientific exploration. (M 15576)

(4083) Jody

1985 CV. Discovered 1985 February 12 by C. S. Shoemaker and E. M. Shoemaker at Palomar.

Named in honor of Joan D. (Jody) Swann, planetary data librarian at the U.S. Geological Survey. Her 25 years of experience in nearly every phase of lunar and planetary exploration have enabled her to establish the world's most complete and functional archive of photographs, maps, spacecraft images and supporting data on the solid bodies of the solar system. Recently, she has utilized this archive to prepare outstanding color mosaics of Mars. (M 15576)

(4084) Hollis

1985 GM. Discovered 1985 April 14 by E. Bowell at Anderson Mesa.

Named in honor of Andrew J. Hollis, a member of the British Astronomical Association since 1961. As director of the Asteroids and Remote Planets section, he has encouraged amateur astronomers to make photoelectric photometric observations of minor planets and has published several light curves and phase curves. (M 17223)

(4085) Weir

1985 JR. Discovered 1985 May 13 by C. S. Shoemaker and E. M. Shoemaker at Palomar.

Named in honor of Doris Blackman Weir, geologist with the U.S. Geological Survey. As technical editor for the branch of astrogeology, she has made significant contributions to the quality of hundreds of maps and scientific papers published on the planets and satellites. (M 15576)

(4086) Podalirius

1985 VK_2. Discovered 1985 November 9 by L. V. Zhuravleva at Nauchnyj.

Named for the Greek physician, son of Aesculapius {see planet (1027)} and Epione and brother of Hygiea, Machaon and Panacea {see planets (10), (3063), (2878), respectively}, who stopped a pestilence during the Trojan War. (M 22246)

(4087) Pärt

1986 EM_1. Discovered 1986 March 5 by E. Bowell at Anderson Mesa.

Named for Arvo Pärt (1935-), Estonian-born composer. Until 1968, Pärt composed mainly serially. Since that year his music has evolved into what he terms a "tintinnabular" style, which is much influenced by medieval plainsong and organum. The calmness and linear clarity of the composer's recent music, which is often compared to that of Benjamin Britten {see planet (4079)}, have much affected the discoverer. (M 16593)

(4088) 1986 GG

Discovered 1986 April 3 by P. Jensen at Brorfelde.

(4089) 1986 JG

Discovered 1986 May 2 at the Palomar Observatory at Palomar.

(4090) Říšehvězd

1986 RH_1. Discovered 1986 September 2 by A. Mrkos at Kleť.

Named for Říše hvězd (*the Realm of Stars*), Czech popular journal for astronomy. Originally published by the Czech Astronomical Society in Prague, the journal has described new findings from astronomy, astrophysics and space exploration and provided information about Czech and Slovak amateur and professional astronomy for more than three-quarters of a century. (M 27734)
Name suggested by J. Tichá, M. Tichý and Z. Moravec.

(4091) Lowe

1986 TL_2. Discovered 1986 October 7 by E. Bowell at Anderson Mesa.

Named in honor of Andrew Lowe, a Canadian professional geophysicist and amateur astronomer in Calgary who has particular interests in computational astronomy involving minor planets and occultations. In recent years, he has opened up a new area of identification research by establishing several cases of linkages of orbits where the observations of the minor planets involved were made on only two nights at individual oppositions. (M 16247)

(4092) Tyr

1986 TJ_4. Discovered 1986 October 8 by P. Jensen and K. Augustesen at Brorfelde.

Named after one of the bravest gods in Norse mythology and a great warrior. It is unclear whether he is a son of Odin {see planet (3989)} or of the giant Ymer. Tyr lost his right hand when Fenrir was chained by the Aesir. (M 22500)

(4093) Bennett

1986 VD. Discovered 1986 November 4 by R. H. McNaught at Siding Spring.

Named in honor of Jack Caister Bennett {1914-1990}, prominent South African amateur astronomer. Jack Bennett's interest in comets developed from stories told by his mother about seeing Halley's comet in 1910. His dedication to comet searching resulted in an excellent catalogue of southern comet-like objects, discovery of the bright comet 1970 II and also 1974 XV. During routine comet searches he discovered the bright supernova 1968l in M83. His interests covered a wide range from involvement in Moonwatch to holding many posts within the South African Astronomical Society, including its presidency. The sight of 1970 II was one of the first and most spectacular astronomical observations made by the discoverer of this minor planet. (M 14972)

Obituaries published in Mon. Notes Astron. Soc. S. Afr., Vol. 49, No. 9-10, p. 113-115, p. 115 (1990); Komet. Tsirk., No. 420, p. 15-16 (1990); Q.J.R. Astron. Soc., Vol. 35, No. 3, p. 353 (1994).

(4094) Aoshima

1987 QC. Discovered 1987 August 26 by W. Kakei and M. Kizawa at Shizuoka.

Named in memory of Masaki Aoshima (1947-1987), founder of the Shizuoka City Amateur Astronomers' Society and an active advisor to beginning amateur astronomers. He died of a heart attack on his way home from an observing session

just one month after this minor planet was discovered. (M 15427)
Name also suggested by many amateur astronomers in Shizuoka.

(4095) Ishizuchisan

1987 SG. Discovered 1987 September 16 by T. Seki at Geisei.

Named for the highest mountain on Shikoku, the island on which the Geisei Station is located {see planets (4223) and (2571)}. The mountain has an altitude of 1982 m and is sacred to many Japanese. (M 15427)

(4096) Kushiro

1987 VC. Discovered 1987 November 15 by S. Ueda and H. Kaneda at Kushiro.

Named for a city of population of 220,000 on the Pacific coast of Hokkaido {see planet (3720)}. Its main industries are paper making, coal mining and fishing, and the citizens are especially proud of the fact that they have the largest fish catch in Japan. Behind the city area there are huge, wild fields that form Kushiro's Marsh National Park. In spite of the prevalence of fog during the summer months the first discoverer maintains his observing station there. (M 15427)

(4097) Tsurugisan

1987 WW. Discovered 1987 November 18 by T. Seki at Geisei.

Named for the second highest mountain on Shikoku {see planet (4223)} island. There is a meteorological observatory on the 1955-m summit. (M 15427)

(4098) Thraen

1987 WQ_1. Discovered 1987 November 26 by F. Börngen at Tautenburg.

Named in memory of Anton Karl Thraen (1843-1902), an astronomer and catholic priest in the German district of Eichsfeld {see planet (442)}. In his time Thraen was one of the most versatile and successful computers of orbits and ephemerides, for both comets and minor planets. Many of his articles are published in the A.N., and he also participated in the publication of the B.A.J. He was the first to show that the hyperbolic orbits of some comets are due to planetary perturbations. (M 15091)
Name proposed by the discoverer, following a suggestion by J. Dorschner.

(4099) 1988 AB_5

Discovered 1988 January 13 by H. Debehogne at La Silla.

(4100) Sumiko

1988 BF. Discovered 1988 January 16 by T. Hioki and N. Kawasato at Okutama.

Named in honor of Sumiko Hioki (1965-), wife of the first discoverer. (M 34619)

(4101) Ruikou

1988 CE. Discovered 1988 February 8 by T. Seki at Geisei.

Named in memory of Ruikou Kuroiwa (1862-1920), a great scholar, translator and commentator of the Meiji and Taisho eras. Among the best of his translations of western works are "Mask of red-hot iron", "Ghost tower" and "King of Cavern". He was born in Aki city, on the western side of which the Geisei Station is situated. The name Geisei {see planet (2571)} actually means "West side of Aki city". (M 16247)

(4102) 1988 TE_3

Discovered 1988 October 15 by V. G. Shkodrov at Rozhen.

(4103) Chahine

1989 EB. Discovered 1989 March 4 by E. F. Helin at Palomar.

Named in honor of Moustafa Chahine, chief scientist of the Jet Propulsion

Laboratory, where he coordinates the research of scientists on the world's foremost unmanned space probes. His own scientific research is in the field of atmospheric science and of remote sensing, both of which have advanced by his innovations and insight. His most recent involvement is in the project Mission to Planet Earth. Chahine possesses that rare quality that combines exceptional scientific and administrative skills. His friends and colleagues have the greatest respect and admiration for his remarkable sensitivity, vision and outstanding leadership. (M 15427)

(4104) Alu

1989 ED. Discovered 1989 March 5 by E. F. Helin at Palomar.

Named for Jeff Alu, a talented musician and composer, as well as a participant in the Palomar Planet Crossing Asteroid Survey. He has shared the long cold winter nights at Palomar and the pleasure of discovery. Jeff himself found several near-earth asteroids, as well as main-belt objects. (M 15261)

This name is endorsed by Brian P. Roman, a fellow PCAS member and good friend.

(4105) Tsia

1989 EK. Discovered 1989 March 5 by E. F. Helin at Palomar.

Named in honor of the ancient sun symbol used by Indians of the Zia Pueblo in central New Mexico (one of the Seven Golden Cities of Cibola sought by Coronado). Although the symbol's name is normally written "Zia", "Tsia" is the spelling in Keresan, the native language of the Zia Pueblo Indians. The symbol now adorns the New Mexico state flag and is often taken as an emblem of the state. It represents first and foremost the sun, the giver of life. From this symbolic sun there radiate four rays consisting of four tongues each; these represent the four cardinal directions (north, south, east and west), the four seasons (spring, summer, fall and winter) and the four stages of life (childhood, youth, adulthood and old age). Also, as ascribed in the official salute to the New Mexico state flag, the Zia is the "symbol of perfect friendship among united cultures". (M 16443)

Name proposed by the discoverer, following a suggestion of Louie V. Burke as part of a project during an undergraduate astronomy class at New Mexico State University. Citation prepared by Alan Hale, instructor of the class.

(4106) Nada

1989 EW. Discovered 1989 March 6 by T. Nomura and K. Kawanishi at Minami-Oda.

Named in honor of the Nada Junior and Senior High School in Kobe, which is known as one of the most excellent high schools in Japan. The first discoverer teaches science, especially astronomy and geology, there. Established in 1927, the school has the mottoes "make good use of your energies" and "live and let live prosperously". These are the same as the mottoes of Kodokan Judo Institute. "Nada" means "the sea that is difficult to cross". (M 15091)

(4107) Rufino

1989 GT. Discovered 1989 April 7 by E. F. Helin at Palomar.

Named in honor of Rufus J. Walker, an ardent supporter of the U.S. space program and a man of exceptional kindness and gentleness. As both friend and doctor he has often made the difference in enabling the discoverer to accomplish her goals by his sincere interest and restorative qualities. As manager of medical services at the Jet Propulsion Laboratory for the past decade, he has maintained a high level of comprehensive and compassionate medical services for employees. He has introduced, staffed and enriched a broad spectrum of counseling services that have aided countless employees and their families in coping with the inherent problems of emotional stress, substance abuse and alcoholism. (M 15428)

Name endorsed by his longtime JPL associate and personal friend, Richard House.

(4108) Rakos

3439 T-3. Discovered 1977 October 16 by C. J. van Houten and I. van Houten-Groeneveld at Palomar.

Named in honor of Karl D. Rakos (1925-), Croatian-born Austrian astronomer, since 1973 professor of astronomy at the University of Vienna and former director of the Vienna Observatory. His main field is double stars, and he served as president of IAU Commission 26 from 1985 to 1988. (M 18138)

(4109) Anokhin

1969 OW. Discovered 1969 July 17 by B. A. Burnasheva at Nauchnyj.

Named in memory of Sergej Nikolaevich Anokhin (1910-1986), test pilot and sportsman. During one of his flights he lost an eye, but he later reverted to test work. During the last years of his life he was a trainer of the Soviet cosmonauts. (M 28089)

(4110) Keats

1977 CZ. Discovered 1977 February 13 by E. Bowell at Palomar.

Named after the English poet John Keats (1795-1821), who wrote of the thrill he experienced when he opened a folio edition of Chapman's translation of Homer {see planet (5700)}. His poem *On First Looking Into Chapman's Homer*, written in 1816, includes the words: "Then felt I like some watcher of the skies/ When a new planet swims into his ken". (M 16247)

Name suggested and citation provided by J. B. Tatum.

(4111) Lamy

1981 EN_{12}. Discovered 1981 March 1 by S. J. Bus at Siding Spring.

Named in honor of Philippe Lamy, since 1973 a staff scientist at the Laboratoire d'Astronomie Spatiale, Marseilles. Lamy has been involved in studies of solar-system dust for more than 20 years. He has specialized in circumsolar, interplanetary and cometary dust, particularly in their optical and light-scattering properties. He is currently studying cometary nuclei using HST and ISO and is involved in a number of space missions (SOHO, Cassini and Rosetta). (M 27458)

Citation provided by M. A. Barucci following a suggestion by E. Bowell.

(4112) Hrabal

1981 ST. Discovered 1981 September 25 by M. Mahrová at Kleť.

Named in honor of Bohumil Hrabal (1914-), one of the most reputable contemporary Czech writers, the author of many stories and non-traditional prose (*I Attend The English King, Too Loud A Solitude, Harlequin's Millions*). Some of his works have been adapted for film screenplays, e.g., *A Close Watch On The Trains*. (M 22500)

(4113) Rascana

1982 BQ. Discovered 1982 January 18 by E. Bowell at Anderson Mesa.

Named to commemorate the centenary of the Royal Astronomical Society of Canada. Incorporated in 1890 under the name The Astronomical and Physical Society of Toronto, the RASC received the Royal Charter in March 1903. Its Journal has been published continuously since Volume 1 in 1907. The members of the Society have always included both professional and amateur astronomers. The masthead of the Journal reads "devoted to the advancement of astronomy and allied sciences". (M 16247)

Citation prepared by the Council of the RASC at the request of the discoverer.

(4114) Jasnorzewska

1982 QB_1. Discovered 1982 August 19 by Z. Vávrová at Kleť.

Named in memory of Maria Jasnorzewska-Pawlikowska (1910-1945), important Polish poet. (M 22247)

(4115) Peternorton

1982 QS_3. Discovered 1982 August 29 by N. S. Chernykh at Nauchnyj.

Named in honor of Peter Norton, internationally known programer and expert on personal computers, creator of many service programs for the PC and author of several books on the subject. He is also known as a lover of the arts, a collector of modern art and a sponsor for many young artists. (M 21608)

(4116) Elachi

1982 SU. Discovered 1982 September 20 by E. F. Helin at Palomar.

Named in honor of Charles Elachi, assistant laboratory director for Space Science and Instruments, Jet Propulsion Observatory. His expertise is in radar sensing from space, including the successful Shuttle imaging radar experiments. He has particular interest in the assessment of the surfaces of the Earth and other planets, from the standpoints of both measurement (his Ph.D. is in electrical sciences) and interpretation (he also holds an advanced degree in geology). His accomplishments include almost 200 papers, two textbooks, numerous honors and participation in almost every radar survey from space since the early 1970s. (M 15428)

(4117) Wilke

1982 SU_3. Discovered 1982 September 24 by F. Börngen at Tautenburg.

Named in memory of Alfred Wilke (1893-1972), an optician of the highest rank, who constructed a workshop and observatory at Falkensee, near Berlin. Right up to his death he made optics by the thousands for astronomical institutes and popular and school observatories, as well as for numerous amateurs. Among his works are the 0.50-m Schmidt cameras for both Sonneberg and Babelsberg, the 0.70-m Cassegrain reflector for Potsdam and the 0.70-m and 0.54-m Cassegrains for Babelsberg. (M 15261)

Name proposed by the discoverer following a suggestion by M. Gressmann, who also prepared the citation.

Obituary published in Sterne, 49. Jahrg., p. 52-53 (1973).

(4118) Sveta

1982 TH_3. Discovered 1982 October 15 by L. V. Zhuravleva at Nauchnyj.

Named for Svetlana Evgen'evna Savitskaya (1948-), 1970 world aerobatics champion, the second woman to fly in space (1982) and the first woman to fly in space twice (1984). On her second flight she became the first woman to walk in space. (M 22500)

(4119) Miles

1983 BE. Discovered 1983 January 16 by E. Bowell at Anderson Mesa.

Named in honor of Howard G. Miles, founder director of the Artificial Satellite Section of the British Astronomical Association, in which capacity he has served continuously since 1960. Miles has given sterling service to the advancement of amateur astronomy in Britain. A lecturer at Lanchester Polytechnic, Coventry, he was awarded the Member of the Order of the British Empire for his civic service in education. Miles is a former business secretary of the B.A.A. and for many years has acted as coordinator for observations of fireballs and other transient phenomena. (M 17223)

Name suggested and citation material provided by S. A. Mitton.

(4120) Denoyelle

1985 RS_4. Discovered 1985 September 14 by H. Debehogne at La Silla.

Named in honor of Jozef Denoyelle (1937-) on the occasion of his retirement from the Uccle {see planet (1276)} Observatory. His principal research concerned early-type stars and their relation to galactic structure. Early in his career he observed minor planets and comets, and his continuing interest in such work stimulated the discoverer to start observations at La Silla. Denoyelle is very active in the popularization of astronomy and has served as treasurer of the Flemish Association of Astronomy for some 20 years. (M 30095)
Name proposed by the discoverer, endorsed by E. W. Elst.

(4121) Carlin

1986 JH. Discovered 1986 May 2 at the Palomar Observatory at Palomar.

Named by Steve Singer-Brewster in honor of his daughter, Carlin Singer-Brewster, on the occasion of her seventh birthday. (M 17030)

(4122) Ferrari

1986 OA. Discovered 1986 July 28 at the Osservatorio San Vittore at Bologna.

Named in memory of Enzo Ferrari (1898-1988), world-famous Italian builder of racing cars and sports cars. After working some 20 years for Alfa Romeo, Ferrari built his factory at Maranello, near Modena, in 1940. A mechanical engineer 'honoris causa' at Bologna University, he produced racing cars that have won practically all the sporting competitions in the world. A benefactor, he gave medical apparatus to the university and to the hospital in Modena for research on muscular dystrophy. (M 16043)

(4123) 1986 QP_1

Discovered 1986 August 27 by H. Debehogne at La Silla.

(4124) Herriot

1986 SE. Discovered 1986 September 29 by Z. Vávrová at Kleť.
Named in honor of James Herriot (1916-), veterinarian and writer. (M 21608)

(4125) Lew Allen

1987 MO. Discovered 1987 June 28 by E. F. Helin at Palomar.

Named in honor of Lew Allen, who has been vice president of the California Institute of Technology and director of the Jet Propulsion Laboratory since 1982. He has provided exceptional leadership and vision to extend the U.S. exploration of the solar system. Under his guidance, 1989 saw the successful launch of the Magellan spacecraft to the planet Venus and the Galileo spacecraft to Jupiter. Allen is highly regarded by JPL people as an excellent leader as well as a caring, accessible friend. The discoverer acknowledges his genuine interest and support over the years. (M 17223)

(4126) Mashu

1988 BU. Discovered 1988 January 19 by K. Endate and K. Watanabe at Kitami.

Named for one of the lakes in Akan National Park, which is located in eastern Hokkaido, a one-hour drive from Kitami and Kushiro {see planets (3720), (3785), and (4096), respectively}. The caldera lake is 20 km in circumference, covers an area of 19.6 square km and has a maximum depth of 212 m. No rivers flow into or out of the lake, which is one of the most transparent in the world, although its surface is almost always shrouded in fog. (M 15428)

(4127) Kyogoku

1988 BA_2. Discovered 1988 January 25 by S. Ueda and H. Kaneda at Kushiro.

Named for the second discoverer's birthplace, a town of population 4,000 located approximately 70 km southwest of Sapporo {see planet (3473)} at the foot of the most beautiful mountains in Hokkaido {see planet (3720)}. The town is known for its production of potatoes, asparagus and wheat, and nearby Fukidashi park provides high-quality mineral water, which has gained in popularity among the tourists. (M 15428)

(4128) UKSTU

1988 BM_5. Discovered 1988 January 28 by R. H. McNaught at Siding Spring.

Named after the U.K. Schmidt Telescope Unit at the Royal Observatory Edinburgh and the 1.2-m U.K. Schmidt Telescope at Siding Spring. In association with ESO, UKSTU has been undertaking a multicolor survey and atlas of the southern sky since operations commenced in 1973. The high quality and faint limits of the survey material resulted in major discoveries in the morphology of galaxies. A significant proportion of quasars with redshifts greater than 4 were discovered in UKSTU material. Other programs utilizing the telescope include the UCAS survey for faint minor planets and the LUKAS survey extending to even fainter limits. The discovery images of this minor planet and those subsequently identified at earlier oppositions were all on 1.2-m U.K. Schmidt plates archived at Siding Spring and Edinburgh. In June 1988 control of the telescope passed from ROE to the Anglo-Australian Observatory. (M 15261)

(4129) Richelen

1988 DM. Discovered 1988 February 22 by R. H. McNaught at Siding Spring.

Named in honor of Richard A. Keen and Helen C. Duran, friends of the discoverer, on the occasion of their wedding. Richard has been a research meteorologist, more recently turning to writing and popularization of meteorology. His interests in amateur astronomy include visual photometry of comets and studies of the brightness of lunar eclipses. Helen is a nurse specializing in hospice care. (M 15262)

(4130) Ramanujan

1988 DQ_1. Discovered 1988 February 17 by R. Rajamohan at Kavalur.

Named in memory of Srinivasa Ramanujan (1887-1920), the Indian mathematical genius who has been classed with Euler {see planet (2002)} and Jacobi and is regarded as one of the truly great algorists in the history of mathematics. His work on the theory of partitions, done in Cambridge in collaboration with Hardy, won him worldwide recognition, and he became the first Indian mathematician to be elected a fellow of the Royal Society. The Hardy-Ramanujan theory led to the circle method, which is today one of the most powerful tools in analytic number theory. Ramanujan gave an analytic expression for pi that has been used on powerful digital computers to generate accurate values to seventeen million decimal places. Terminally ill, he returned to India in 1919 and spent his last year in Madras on further mathematical investigations, which he recorded in a notebook. This notebook was subsequently lost, eventually to be retrieved as it was about to be incinerated in Cambridge following the death of the mathematician Watson. G. E. Andrews resurrected the notebook in 1976 and by making it available for wide study caused a revival of interest in Ramanujan's work. (M 15262)
Citation prepared by D. C. V. Mallik.

(4131) Stasik

1988 DR_4. Discovered 1988 February 23 by A. J. Noymer at Siding Spring.

Named in honor of John S. Stasik, science teacher at Weston Middle School, Massachusetts, responsible for introducing the discoverer to astronomy. (M 15091)

(4132) Bartók

1988 EH. Discovered 1988 March 12 by J. Alu at Palomar.

Named for the great composer Bela Bartók (1881-1945). Bartók was influenced by the melodic lines and rhythms found in the folk songs of his native Hungary. He transformed these songs into great works of music without altering the basic musical elements of the folk songs. (M 15091)

(4133) Heureka

1942 DB. Discovered 1942 February 17 by L. Oterma at Turku.
Named for the new science center in Vantaa, near Helsinki. (M 18455)

(4134) Schütz

1961 CR. Discovered 1961 February 15 by F. Börngen at Tautenburg.

Named in memory of Heinrich Schütz (1585-1672), the greatest German composer of the seventeenth century, born in Kostritz, 22 km from Tautenburg, and a very important precursor to J. S. Bach {see planet (1814)}. The epoch-making work of Schütz is principally documented in numerous pieces of sacred music. (M 16043)

(4135) Svetlanov

1966 PG. Discovered 1966 August 14 by L. I. Chernykh and T. M. Smirnova at Nauchnyj.

Named in honor of the famous Soviet composer and conductor Evgenij Fyodorovich Svetlanov. (M 18455)

(4136) Artmane

1968 FJ. Discovered 1968 March 28 by T. M. Smirnova at Nauchnyj.

Named in honor of Vija Artmane, famous Latvian actress who appeared in more than 60 movies. *Rodnaya krov'* (The Family Blood), in which she took the leading role, was honored with the Eric Johnston prize at the fourth international movie festival, held in Argentine in 1964. (M 28089)

(4137) 1970 WC

Discovered 1970 November 24 by L. Kohoutek at Bergedorf.

(4138) Kalchas

1973 SM. Discovered 1973 September 19 by C. J. van Houten at Palomar.

Named for a prophet on the side of the Greeks during the Trojan war. (M 16043)

(4139) Ul'yanin

1975 VE_2. Discovered 1975 November 2 by T. M. Smirnova at Nauchnyj.

Named in memory of Sergej Alekseevich Ul'yanin (1871-1921), Russian pilot and talented inventor, who graduated from the Framan pilot school in France in 1910. Two of his projects involved the design of new airplanes, one of which was realized. He also designed a radio navigation system for airplanes in flight, as well as apparatus for air photography. (M 28089)

(4140) Branham

1976 VA. Discovered 1976 November 11 at the Felix Aguilar Observatory at El Leoncito.

Named in honor of Richard L. Branham Jr. (1943-), a U.S. astronomer who works in meridian astronomy and celestial mechanics. Branham worked in the 7-inch Transit Circle Division of the U.S. Naval Observatory from 1968 to 1982, toward the end of which time he was resident director of the Yale-Columbia Southern Station at El Leoncito and also in charge of the Naval Observatory's Austral Expedition. Since 1982 he has continued his astronomical research and teaching in Argentine. (M 29143)

(4141) Nintanlena

1978 PG_3. Discovered 1978 August 8 by N. S. Chernykh at Nauchnyj.

Named in honor of Nina, Tanya and Elena, wife and daughters of Kiev astronomer K. I. Churyumov {see planet (2627)}. (M 19695)

(4142) Dersu-Uzala

1981 KE. Discovered 1981 May 28 by Z. Vávrová at Kleť.

Named after an inhabitant of Siberian Taiga and the character from the novel of the same name by the Russian writer Vladimir K. Arsenev. Dersu-Uzala was a member of the Goldy tribe and he became a friend of Arsenev's during 1902-1910. (M 21609)

(4143) 1981 QN_1

Discovered 1981 August 29 by L. G. Taff at Socorro.

(4144) Vladvasil'ev

1981 SW_6. Discovered 1981 September 28 by L. V. Zhuravleva at Nauchnyj.

Named after the brilliant Russian ballet-dancer Vladimir Victorovich Vasil'ev (1940-). (M 24121)

(4145) Maximova

1981 SJ_7. Discovered 1981 September 29 by L. V. Zhuravleva at Nauchnyj.

Named in honor of the wonderful Russian ballet-dancer Ekaterina Sergeevna Maximova (1939-). (M 24121)

(4146) Rudolfinum

1982 DD_2. Discovered 1982 February 16 by L. Brožek at Kleť.

Named for one of the most beautiful and important neo-Renaissance buildings in Prague. Constructed between 1876 and 1884, Rudolfinum originally served as a concert hall and art gallery, and from 1918 to 1938 it was the seat of the Czech National Assembly. Now it is again a center of musical life as home to the Czech Philharmonic Orchestra and the international music festival 'Prague Spring'. (M 24410)

(4147) Lennon

1983 AY. Discovered 1983 January 12 by B. A. Skiff at Anderson Mesa.

Named in memory of John Lennon (1940-1980), musician and ex-Beatle. With Paul McCartney {see planet (4148)}, he wrote the early songs that helped make The Beatles the most popular group of their generation. Perhaps the most famous Beatles album is Sergeant Pepper's Lonely Hearts Club Band, which represents the Lennon-McCartney partnership at its most fertile and innovative. From his solo recording career, the song that will probably best be remembered is 'Imagine'. Passionately anti-war, John, and his wife Yoko, also made many public demonstrations of their desire for a peaceful world. (M 16247; G. V. Williams)

(4148) McCartney

1983 NT. Discovered 1983 July 11 by E. Bowell at Anderson Mesa.

Named in honor of Paul McCartney (1942-), musician and ex-Beatle. A talented composer, he was responsible for some of The Beatles' best loved songs - both 'Yesterday' and 'Michelle' have become classics of the popular repertoire, and have each been covered by several hundred artists. Paul was instrumental in the setting up of the Apple Corps, a philantropic organization that helped to launch the careers of several young musicians. Paul was the only member of The Beatles to continue regular live performances after the group split in 1970. He is still a popular performer, and many of his solo recordings have topped the charts around the world. (M 16248)

(4149) Harrison

1984 EZ. Discovered 1984 March 9 by B. A. Skiff at Anderson Mesa.

Named in honor of George Harrison (1943-), the quiet Beatle. Although overshaded by John and Paul as composers while in The Beatles, George wrote several songs for the group, including 'While My Guitar Gently Weeps' and 'Something'. Since The Beatles disbanded, George has undertaken many projects in addition to his solo recording career; notably his film company, HandMade Films. In 1971 he organized a star-studded charity concert for the relief of famine in Bangladesh, and many of his songs have reflected his concern for the environment. (M 16248)

(4150) Starr

1984 QC_1. Discovered 1984 August 31 by B. A. Skiff at Anderson Mesa.

Named in honor of Ringo Starr (Richard Starkey: 1940-), a Liverpudlian of lively personality and deadpan humor who occasionally sat in as drummer with The Beatles during their early days in Hamburg. Ringo actually joined the group in 1962, after the original drummer, Pete Best, left. Ringo's solo career has encompassed several albums and a variety of film roles. (M 16248)

(4151) Alanhale

1985 HV_1. Discovered 1985 April 24 by C. S. Shoemaker and E. M. Shoemaker at Palomar.

Named in honor of Alan Hale for his many observations of comets. In the past several years he has published careful visual observations of more than 130 comets, several at more than one apparition. His observations include both magnitude estimates and confirmations of discoveries. He has also applied his magnitude estimating skill to asteroids, particularly the fast moving objects 1989 AC and 1989 VA, and has participated in asteroid occultation teams. He has done much to promote asteroid-comet education through articles on comets and emphasis on asteroids in his introductory astronomy classes. (M 18139)
Citation provided by David Levy.

(4152) Weber

1985 JF. Discovered 1985 May 15 by E. Bowell at Anderson Mesa.

Named for Carl Maria von Weber (1786-1826), German composer. Although Weber did not invent German Romantic opera, he was strongly influential in setting its course, both as a composer and conductor. Similarly, he was a primary builder of the nineteenth-century Romantic movement in music. His operas and overtures are often performed, but his symphonies, most of his concerti and his piano works have fallen into unjust neglect. (M 16593)

(4153) Roburnham

1985 JT_1. Discovered 1985 May 14 by C. S. Shoemaker and E. M. Shoemaker at Palomar.

Named in honor of Robert Burnham, senior editor of the magazine *Astronomy*. Over many years he has been partly responsible for its evolution into the astronomical publication with the world's largest readership. In 1983, Burnham's *The Star Book* appeared with a series of clearly designed star charts to guide beginners on their way to an understanding of the night sky. (M 18139)
Citation provided by David Levy.

(4154) 1985 NE

Discovered 1985 July 10 by A. C. Gilmore and P. M. Kilmartin at Lake Tekapo.

(4155) Watanabe

1987 UB_1. Discovered 1987 October 25 by S. Ueda and H. Kaneda at Kushiro.

Named in honor of Kazuro Watanabe (1955-), known for his positional measurements of comets, minor planets and meteors. An astronomical specialist

in the Sapporo Youth Science Museum, he is responsible for passing astronomical information to the public and is known throughout Hokkaido {see planet (3720)} as a popular leader of astronomical activities. He is also involved in the preparations of maps. (M 16593)
Name suggested by S. Nakano.

(4156) 1988 BE

Discovered 1988 January 16 by T. Kojima at Chiyoda.

(4157) Izu

1988 XD_2. Discovered 1988 December 11 by Y. Oshima at Gekko.

Named for the southeastern part of Shizuoka prefecture that consists mainly of the Izu Peninsula. With its many hot springs and resorts it is well known as part of the Fuji-Hakone-Izu National Park. Gekko {see planet (4261)} Astronomical Observatory is located just at the base of the peninsula. (M 16043)

(4158) Santini

1989 BE. Discovered 1989 January 28 at the Osservatorio San Vittore at Bologna.

Named in memory of Giovanni Santini (1786-1877), director of the Padua Observatory from 1817 to 1867, a great observer of minor planets and comets and an indefatigable computer of their orbits and perturbations. His two-volume textbook *Elementi di Astronomia* was used by virtually all the Italian astronomers of the nineteenth century. (M 16043)

(4159) Freeman

1989 GK. Discovered 1989 April 5 by E. F. Helin at Palomar.

Named in honor of Ann Freeman, executive secretary of the Seismological Laboratory of the California Institute of Technology and long-time friend of discoverer, on the occasion of her birthday, 1993 Feb. 10. She has enthusiastically supported the planet-crossing asteroid survey during its first decade at Caltech and continues to celebrate its successes. (M 21955)

(4160) Sabrina-John

1989 LE. Discovered 1989 June 3 by E. F. Helin at Palomar.

Named in memory of Sabrina M. Gonsalves (1962-1980) and John H. Riggins (1962-1980), young lovers tragically lost. In their brief lives they left a legacy of love and dedication to dreams and family. Hurling through space on its eternal journey, this minor planet symbolizes the endurance of their love. (M 21955)

Citation prepared by Carl and Teresa Atallah and endorsed by the families and friends of the young couple.

(4161) Amasis

6627 P-L. Discovered 1960 September 24 by C. J. van Houten and I. van Houten-Groeneveld at Palomar.

Named after the Egyptian pharaoh Amasis (570-526 B.C.). The Greek and Roman historians praised him greatly because he gave the Greek merchants their own town in the Nile delta. Amasis introduced the so-called demotic characters for the hieroglyphs. He died just before the Persians conquered Egypt. (M 22501)

(4162) SAF

1940 WA. Discovered 1940 November 24 by A. Patry at Nice.

Named in honor of the Société Astronomique de France, founded in 1887 by Camille Flammarion {see planet (1021)} and acknowledged as being of public utility by the French government in 1897. The SAF has both professional and amateur astronomers as members, and its presidents have included Tisserand {see planet (3663)}, Janssen, Poincaré, Danjon and Lyot {see planets (2021), (1594), and (2452) for the last three astronomers}. The society publishes the magazine

L'Astronomie and the quarterly *Observations et Travaux*. The centenary of the SAF was celebrated by holding IAU Colloquium No. 98: Contributions of Amateur Astronomers to Astronomy. (M 22501)
Name suggested and citation prepared by M.-A. Combes.

(4163) Saaremaa

1941 HC. Discovered 1941 April 19 by L. Oterma at Turku.
Named for an island in the Baltic. Some 2500 years ago a large iron meteorite impacted there, causing a lake 16 m deep and some smaller craters. (M 18455)

(4164) Shilov

1969 UR. Discovered 1969 October 16 by L. I. Chernykh at Nauchnyj.
Named in honor of the outstanding Soviet painter Aleksandr Maksovich Shilov, who has created a gallery of portraits of his contemporaries. (M 18645)

(4165) Didkovskij

1976 GS_3. Discovered 1976 April 1 by N. S. Chernykh at Nauchnyj.
Named in honor of Leonid Vladimirovich Didkovskij (1948-), astrophysicist and deputy director of the Crimean Astrophysical Observatory {see planet (1725)}. He is known for his research on the brightness oscillations of the sun, for his development of scientific instrumentation for the Soviet space telescope "Astron" and for his work with the active main mirror on the 1.7-m Space Telescope "Spectrum UV", an international project. (M 34340)

(4166) Pontryagin

1978 SZ_6. Discovered 1978 September 26 by L. V. Zhuravleva at Nauchnyj.
Named after the outstanding Russian mathematician Lev Semenovich Pontryagin (1908-). (M 24121)

(4167) Riemann

1978 TQ_7. Discovered 1978 October 2 by L. V. Zhuravleva at Nauchnyj.
Named in memory of Georg Friedrich Bernhard Riemann (1826-1866), German mathematical genius. (M 24121)
Riemann is also honored by a lunar crater.

(4168) Millan

1979 EE. Discovered 1979 March 6 at the Felix Aguilar Observatory at El Leoncito.
Named in memory of Julio Rodolfo Millan (1923-1995), first rector of the San Juan National University at the time that the old Yale-Columbia Southern Station was transferred there. Trained as a mining engineer and known for his research on the promotion and development of mining in Argentina, Millan was a professor of metallurgy and later a dean at Cuyo National University {see planet (1917)}, prior to his being appointed organizing delegate of the San Juan National University on its creation in 1973. (M 32787)

(4169) Celsius

1980 FO_3. Discovered 1980 March 16 by C.-I. Lagerkvist at La Silla.
Named in memory of the Swedish astronomer Anders Celsius (1701-1744), renowned for devising the thermometric scale. A participant in the French expedition to Lapland to measure the curvature of the Earth, he was also the first astronomer to try to determine stellar magnitudes by photometric methods. The year 1990 marks the 250th anniversary of his establishment of the Uppsala Observatory, the original building of which is still preserved in the center of Uppsala. The event has been celebrated at the Nordic-Baltic Astronomy Meeting held in Uppsala during 1990 June 17-21. (M 16044)
Celsius is also honored by a lunar crater.

(4170) Semmelweis

1980 PT. Discovered 1980 August 6 by Z. Vávrová at Kleť.

Named in memory of Ignaz Fülöp Semmelweis (1818-1865), Hungarian physician. (M 22501)

(4171) Carrasco

1982 FZ_1. Discovered 1982 March 23 by C. S. Shoemaker and Q. R. Passey at Palomar.

Named in honor of Juan Carrasco, senior night assistant at the Palomar Observatory. The safe operation of the famous 5-m Hale telescope is entrusted chiefly to Carrasco's capable hands. (M 15576)

(4172) Rochefort

1982 FC_3. Discovered 1982 March 20 by H. Debehogne at La Silla.

Named for a small town in the province of Namur {see also planet (3374)} in southern Belgium. (M 21131)

(4173) Thicksten

1982 KG_1. Discovered 1982 May 27 by C. S. Shoemaker and S. J. Bus at Palomar.

Named in honor of Robert P. Thicksten, superintendent of the Palomar Observatory, California Institute of Technology since 1981. By his technical skill and his leadership of the Observatory staff, he has maintained and improved the Palomar telescopes (the 200-inch Hale Telescope, the 48-inch Oschin Schmidt, the 60-inch reflector and the 18-inch Schmidt) and their auxiliary instruments to keep them at the peak of their operational potential and, in doing so, provides invaluable support of the progress of astronomical research. (M 18139)
Citation provided by Robert J. Brucato at the request of the discoverers.

(4174) Pikulia

1982 SB_6. Discovered 1982 September 16 by L. I. Chernykh at Nauchnyj.

Named in memory of the famous Soviet writer Valentin Savvich Pikul' (1928-1990), author of many novels on Russian history. (M 18455)

(4175) Billbaum

1985 GX. Discovered 1985 April 15 by E. Bowell at Anderson Mesa.

Named in honor of William A. Baum on the occasion of his retirement from the directorship of the Planetary Research Center at Lowell Observatory, a post he has held since 1965. Baum's astronomical research is extraordinarily diverse, covering many of the major fields of optical astronomy. He has contributed importantly to studies of atmospheric and surface phenomena on Mars, Saturn's rings, the properties of grains in cometary comae, metallicity gradients in elliptical galaxies, and testing the constancy of fundamental atomic constants over cosmological timescales. Additionally, he has been active in the development of instrumentation, including early pulse-counting equipment, the Carnegie image tube, image stabilization devices and astronomical optics in general. Baum directed the International Planetary Patrol Program, was a member of the imaging team in the Viking missions to Mars and is currently a member of the Hubble Space Telescope Wide Field/Planetary Camera team. He has served on many national panels and committees and, in 1976-1977, as president of the Division for Planetary Sciences of the American Astronomical Society. (M 18139)
Citation material provided by R. L. Millis at the request of the discoverer.

(4176) Sudek

1987 DS. Discovered 1987 February 24 by A. Mrkos at Kleť.

Named in memory of Josef Sudek (1896-1976), outstanding Czech photogra-

pher, well-known for his black-and-white series of still lifes and views of Prague. (M 31295)
Name suggested by M. Tichý.

(4177) 1987 SS_1

Discovered 1987 September 21 by E. Bowell at Anderson Mesa.

(4178) 1988 EO_1

Discovered 1988 March 13 by E. F. Helin at Palomar.

(4179) Toutatis

1989 AC. Discovered 1989 January 4 by C. Pollas at Caussols.

Named after the Gaulish god, protector of the tribe. This totemic deity is well known because of the cartoon series *Les aventures d'Asterix* by Uderzo and Goscinny. This tells the stories of two almost fearless heroes living in the late village under siege in Roman-occupied Gaul in 50 B.C., and whose only fear is that the sky may fall onto their heads one day. Since this object is the Apollo object with the smallest inclination known, it is a good candidate to fall on our heads one of these days... But as the chief of the village says: "C'est pas demain la veille..." (M 16444)

Citation written by the discoverer and A. Maury and endorsed by J. D. Mulholland, who with Maury obtained the discovery plates.

(4180) Anaxagoras

6092 P-L. Discovered 1960 September 24 by C. J. van Houten and I. van Houten-Groeneveld at Palomar.

Named after the Greek philosopher Anaxagoras (c. 500-428 B.C.). He lived in Asia minor and in Athens, where he became friend of Pericles. He taught that solar and lunar eclipses originated in a scientific way and not as an act of the gods. This was an asebie (outrage against the gods) and Anaxagoras had to flee; he died in exile. (M 22501)
Anaxagoras is also honored by a lunar crater.

(4181) Kivi

1938 DK_1. Discovered 1938 February 24 by Y. Väisälä at Turku.

Named in memory of Aleksis Kivi (1834-1872), the first Finnish-language dramatist and novelist. His principal work, *Seitsemän veljestä* ('Seven brothers'), completed in 1870, was translated into many languages and also made into a movie. (M 18456)

(4182) Mount Locke

1951 JQ. Discovered 1951 May 2 at the McDonald Observatory at Fort Davis.

Named for the mountain on which the McDonald Observatory is situated. The Cook instrument was placed on Mount Locke to take all the plates for the McDonald Survey. (M 20159)

Name proposed by I. van Houten-Groeneveld, one of the people who worked on the Survey.

(4183) Cuno

1959 LM. Discovered 1959 June 5 by C. Hoffmeister at Bloemfontein.

Following the theme that several earth-approaching minor planets have four-letter masculine names, this object bears the first name of the discoverer. (M 18307)
The discoverer is also honored by minor planet (1726).

(4184) Berdyayev

1969 TJ_1. Discovered 1969 October 8 by L. I. Chernykh at Nauchnyj.

Named in honor of the famous Russian philosopher Nikolaj Aleksandrovich Berdyayev (1874-1948). (M 18456)

(4185) Phystech

1975 ED. Discovered 1975 March 4 by T. M. Smirnova at Nauchnyj.

Named in honor of the Moscow Physical and Technical Institute, the key institute in Russia for training experts and investigators in the domains of modern physics, mathematics and engineering. (M 29143)

Name proposed by the Institute of Theoretical Astronomy on the occasion of the 50th anniversary of the Institute in 1996.

(4186) Tamashima

1977 DT_1. Discovered 1977 February 18 by H. Kosai and K. Hurukawa at Kiso.

Named in honor of the city where the first discoverer was born and grew up. It is located near the Okayama {see planet (2084)} Astrophysical Observatory. For many years before the Meiji period Tamashima was one of the most important ports on the Inland Sea. (M 17979)

(4187) Shulnazaria

1978 GR_3. Discovered 1978 April 11 by N. S. Chernykh at Nauchnyj.

Named in honor of Leonid Markovich Shul'man and Galina Kirillovna Nazarchuk, husband-and-wife team of researchers on comets, staff members of the Main Ukrainian Astronomical Observatory at Golosseevo, near Kiev. (M 19695)

(4188) Kitezh

1979 HX_4. Discovered 1979 April 25 by N. S. Chernykh at Nauchnyj.

Named for the Old Russian epic town of Kitezh. As legend tells, Kitezh was a beautiful and prosperous town located near Lake Svetloyar. When troops of the Tartar khan Batyj assaulted the town, the citizens fought bravely and all fell in battle. But the town was not seized by the invaders, for it disappeared under the lake waters. Since that time, if one listens attentively, it is possible to hear the sound of the town bells under the lake waters. (M 23351)

(4189) Sayany

1979 SV_9. Discovered 1979 September 22 by N. S. Chernykh at Nauchnyj.

Named for the mountain chain in the southern part of Siberia {see planet (1094)}. (M 23351)

(4190) Kvasnica

1980 JH. Discovered 1980 May 11 by L. Brožek at Kleť.

Named in memory of Jozef Kvasnica (1930-1992), Czechoslovak theoretical physicist, professor at Charles University in Prague and head of the department of mathematical physics. Kvasnica studied nonlinear electrodynamics, quantum field theory, statistical physics and plasma theory. He was also known as the author of several textbooks, and many students remember his excellent lectures. (M 26929)

(4191) Assesse

1980 KH. Discovered 1980 May 22 by H. Debehogne at La Silla.

Named for the birthplace of the discoverer's father, a charming village in southern Belgium, site of the victory of the Austrians over the Etats Belgiques Unis in 1790. (M 19336)

(4192) Breysacher

1981 DH. Discovered 1981 February 28 by H. Debehogne and G. DeSanctis at La Silla.

Named in honor of Jacques Breysacher, astronomer in charge of the Section for Visiting Astronomers at the European Southern Observatory. His scientific work is mainly concerned with stellar spectroscopy and the classification in external galaxies, and he has carried out an extensive study of the Wolf-Rayet populations of the Magellanic Clouds. (M 19336)

(4193) 1981 SM_1

Discovered 1981 September 26 by B. A. Skiff and N. G. Thomas at Anderson Mesa.

(4194) Sweitzer

1982 RE. Discovered 1982 September 15 by E. Bowell at Anderson Mesa.

Named in honor of Paul A. Sweitzer (1936-) on the occasion of his 60th birthday. Sweitzer served as a reporter on the *Arizona Daily Sun* during 1958-1994, covering news, sports, the performing arts, religion, education, law enforcement and the courts. During that time he made a special effort to report on events and astronomical discoveries at the Lowell Observatory, and in the course of his work he developed warm relationships with many of Lowell's staff. His love of music, especially opera, has led him to become an unfailing supporter, through reviews and informative articles, of the Flagstaff Symphony Orchestra. (M 28089)
Name suggested by A.-M. Malotki, endorsed by W. L. Putnam.

(4195) Esambaev

1982 SK_8. Discovered 1982 September 19 by L. I. Chernykh at Nauchnyj.

Named in honor of Makhmud Alisultanovich Esambaev (1924-), well-known Soviet ballet dancer, an outstanding performer of the national dances of the peoples of the world. (M 26762)

(4196) Shuya

1982 SA_{13}. Discovered 1982 September 16 by L. I. Chernykh at Nauchnyj.

Named for the discoverer's birthplace, a town in Central Russia. The discoverer dedicates this name to her parents, Ekaterina Il'inichna Trushechkina (1907-1950) and Ivan Mokeevich Trushechkin (1907-). (M 18456)

(4197) 1982 TA

Discovered 1982 October 11 by E. F. Helin and E. M. Shoemaker at Palomar.

(4198) Panthera

1983 CK_1. Discovered 1983 February 11 by N. G. Thomas at Anderson Mesa.

Name derived from the Latin for 'panther', referring to a large, intimidating cat. Also, *Panthere tigris tigris*, the Bengal tiger. Many of the large cats are on the endangered species list. (M 25443)
Name suggested in a communication from the World Wildlife Fund.

(4199) Andreev

1983 RX_2. Discovered 1983 September 1 by H. Debehogne at La Silla.

Named in memory of Gennadij Andreev and his family. A professor of celestial mechanics at the University of Tomsk, Andreev is one of the leaders of the expeditions that regularly go to unravel the mysteries of the Tunguska event of 1908. (M 21609)

(4200) Shizukagozen

1983 WA. Discovered 1983 November 28 by Y. Banno and T. Urata at Karasuyama.

Named for a dancing girl, the mistress of Minamoto Yoshitsune. The orbit of this asteroid is very similar to that of (3178) Yoshitsune. (M 19695)

(4201) Orosz

1984 JA_1. Discovered 1984 May 3 by B. A. Skiff at Anderson Mesa.

Named in honor of Elizabeth M. Orosz (1970-), staff member at Melton Memorial Observatory at the University of South Carolina and friend of the discoverer. (M 27126)

(4202) 1985 CB_2

Discovered 1985 February 12 by H. Debehogne at La Silla.

(4203) Brucato

1985 FD_3. Discovered 1985 March 26 by C. S. Shoemaker and E. M. Shoemaker at Palomar.

Named for Robert J. Brucato, assistant director of the Palomar Observatory. Working with more than a dozen government agencies of the surrounding communities, he has vigorously and effectively led the effort to preserve the dark sky of Palomar. This effort has been crucial for the preservation of Palomar as a front-line astronomical observatory and has been especially important for projects such as the photographic searches for comets and minor planets. Brucato is also in charge of the new ten-year long survey of the northern sky, using the Oschin Schmidt telescope, and has directed the complete refurbishment of this telescope. He has been a strong advocate of all branches of astronomy carried out at Palomar and has helped to maintain it as one of the premier astronomical observatories in the world. (M 18456)

Citation provided by Gerry Neugebauer at the request of the discoverers.

(4204) Barsig

1985 JG_1. Discovered 1985 May 11 by C. S. Shoemaker and E. M. Shoemaker at Palomar.

Named in honor of Walter Barsig, a teacher of science and director of a major Bavarian state school board; also a director of the Annual Cultural Festival of the Ries {see planet (4327)} of Nördlingen, Germany, the location of one of the best known impact craters on Earth. Barsig has successfully promoted popular knowledge about the local impact craters, Ries and Steinheim {see planet (6563)}, and the public recognition of their scientific significance. He has been remarkable in his efforts to make the local population of the Ries area aware of the intimate relationship between the geological, cultural, and economic evolution of a natural landmark, which was created by an asteroid impact 15 million years ago. (M 18139)

Citation provided by Dieter Stöffler at the request of the discoverers.

(4205) David Hughes

1985 YP. Discovered 1985 December 18 by E. Bowell at Anderson Mesa.

Named in honor of David W. Hughes, reader in physics at Sheffield University, where he teaches courses on all aspects of astronomy. His research area concerns small solar-system bodies, particularly the relationship between comets and meteors. He has served astronomy in Britain as a vice president of both the Royal Astronomical Society and the British Astronomical Association. Hughes is a prolific reviewer of astronomy books and writes regularly on current issues in astronomy for *Nature*. (M 17223)

Name suggested and citation material provided by S. A. Mitton.

(4206) 1986 QL

Discovered 1986 August 25 by H. Debehogne at La Silla.

(4207) Chernova

1986 RO_2. Discovered 1986 September 5 by E. Bowell at Anderson Mesa.

Named in honor of Galina Pavlovna Chernova, a senior researcher at the Astrophysical Institute of the Tadjik Academy of Sciences, Dushanbe. An indefatigable observer of stars, comets and minor planets, Chernova has codiscovered a number of light-scattering effects in cometary atmospheres. She undertook an extensive program of photometric, polarimetric and astrometric observations of Comet Halley and is an active member of the Comets and Asteroids Working Groups of the U.S.S.R. Academy of Sciences. (M 16044)
Citation prepared by D. F. Lupishko at the request of the discoverer.

(4208) Kiselev

1986 RQ_2. Discovered 1986 September 6 by E. Bowell at Anderson Mesa.

Named in honor of Nikolaj Nikolaevich Kiselev, department head at the Astrophysical Institute of the Tadjik Academy of Sciences, Dushanbe. A preeminent observer of stars, comets and minor planets, Kiselev has carried out work that resulted in the discovery of a number of light-scattering effects in cometary atmospheres. He was leader of the Soviet program of polarimetric and photometric observations of Comet Halley and is one of the initiators of modern investigations of minor planets in the Soviet Union. Under his leadership the Sanglok Observatory, dark-sky site of the Astrophysical Institute, was built. (M 16044)
Citation prepared by D. F. Lupishko at the request of the discoverer.

(4209) Briggs

1986 TG_4. Discovered 1986 October 4 by E. F. Helin at Palomar.

Named in honor of Geoffrey A. Briggs, space physicist, former Director of the Solar System Exploration program at NASA headquarters, where he played a leading role in establishing the U.S.-Soviet Joint Working Group for Solar System Exploration and served as co-chairman. In the 1970s he was at JPL as an active member of the Viking Orbiter Imaging Team. Currently with the Air and Space Museum, Briggs continues to energize the presentation of Space related accomplishments to the public. The discoverer wishes to acknowledge his longterm support and enthusiasm for solar system research. (M 18456)

The naming is heartily endorsed by Jürgen Rahe and the scientific community who have known him as a serious, hardworking scientist trying to accomplish the most with limited resources.

(4210) 1987 DY_5

Discovered 1987 February 21 by H. Debehogne at La Silla.

(4211) 1987 RT

Discovered 1987 September 12 by H. Debehogne at La Silla.

(4212) 1987 SB_2

Discovered 1987 September 28 by K. Suzuki and T. Urata at Toyota.

(4213) Njord

1987 ST_4. Discovered 1987 September 25 by P. Jensen and K. Augustesen at Brorfelde.

Named after the god of winds, navigation and prosperity in Norse mythology. He belongs to the race of gods called the Vanir, who are often in conflict with the Aesir, and he was originally brought to Asgaard as a hostage. He is the father of Frej and Freja {see planet (76)}. (M 22501)

(4214) Veralynn

1987 UX_4. Discovered 1987 October 22 by L. V. Zhuravleva at Nauchnyj.
Named in honor of Vera Lynn (1917-), famous English singer. (M 24122)

(4215) Kamo

1987 VE_1. Discovered 1987 November 14 by S. Ueda and H. Kaneda at Kushiro.

Named in honor of Akira Kamo (1943-), who established the Comet Observers Network in Japan, "Hoshi no Hiroba" {see planet (4971)} in 1968. He played a leading role for many observers and discoverers of comets for more than 20 years when he was the first president of the network. His words "let's come together at Hoshi no Hiroba" are well known among amateur and professional astronomers in Japan. (M 22501)

Name proposed by K. Watanabe. Citation prepared by T. Nomura.

(4216) Neunkirchen

1988 AF_5. Discovered 1988 January 14 by H. Debehogne at La Silla.

Named to honor the birthplace of Hermann Haupt {see planet (2870)} and Gerhard Hahn {see planet (3676)}, two Austrian astronomers and colleagues of the discoverer, active in minor planet research during the past decades. Neunkirchen, situated 60 km south of Vienna, is a district capital and a major center for education and commerce. (M 16248)

(4217) Engelhardt

1988 BO_2. Discovered 1988 January 24 by C. S. Shoemaker and E. M. Shoemaker at Palomar.

Named in honor of Wolf von Engelhardt, professor of mineralogy, and director of the Institute of Mineralogy and Petrography, University of Tübingen, Germany, from 1957 to 1977. In Europe, von Engelhardt was a pioneer in research on impact craters and shock metamorphism of minerals and rocks. In 1962 he initiated and successfully conducted one of the most comprehensive research programs devoted to the study of the Ries {see planet (4327)} crater, Germany. Since the first return of lunar samples in 1969, he was and continues to be a Principal Investigator in NASA's lunar sample analysis program. (M 18456)

Citation provided by Dieter Stöffler at the request of the discoverers.

(4218) Demottoni

1988 BK_3. Discovered 1988 January 16 by H. Debehogne at La Silla.

Named in memory of Glauco de Mottoni y Palacios (1901-1988), Italian astronomer, engineer and collaborator at the Milan and Paris Observatories. A specialist in the visual observation of the planet Mars, he distinguished himself with a large cartographic work on the aspect of the Martian surface between 1907 and 1971. He designed telescopes, often in an unconventional manner. A keen popularizer of astronomy, he founded the "Urania" association in Genoa in 1951 and directed it until his death. (M 16444)

(4219) Nakamura

1988 DB. Discovered 1988 February 19 by M. Inoue and O. Muramatsu at Kobuchizawa.

Named in honor of Giichi Nakamura, the owner of Mitaka Koki, the leading maker of optical equipment in Japan. Mitaka Koki develops and produces observational and measuring equipment used at the National Astronomical Observatory and the Institute of Space and Astronomical Science. The company also puts an emphasis on producing equipment for Japanese amateur astronomers. (M 17223)

Name proposed by M. Inoue, who discovered this minor planet on films taken with a Mitaka Koki telescope. Citation prepared by O. Muramatsu and H. Kosai.

(4220) Flood

1988 DN. Discovered 1988 February 22 by R. H. McNaught at Siding Spring.

Named in memory of Thomas Flood (1919-1988), one of Scotland's bestknown and respected amateur astronomers. Despite little more than a basic education, poverty during the Depression and frequent ill-health, he was a well-read and

cultured man of great integrity. He joined the British Astronomical Association in 1954 and was for a time involved in the Lunar Section. A founder member of the Dundee Astronomical Society in 1955, he served in every position of responsibility, including secretary, treasurer and chairman, with distinction. After enforced early retiral from clerking work due to ill health he was assistant curator of the Mills Observatory in Dundee from 1974 to 1982, then wrote a history of the observatory. (M 16044)
Citation prepared by D. Gavine at the request of the discoverer.

(4221) Picasso

1988 EJ. Discovered 1988 March 13 by J. Alu at Palomar.

Named for Pablo Picasso (1881-1973), the most prolific artist of all time. His career spanned the entire course of modern art. Picasso's creativity in painting, drawing, sculpture, graphics and ceramic is filled with inventive powers, uncontrolled impulses, limitless ambition and a continuous drive for never ending discovery. (M 16044)
Name endorsed by Eleanor F. Helin.

(4222) Nancita

1988 EK_1. Discovered 1988 March 13 by E. F. Helin at Palomar.

Named in honor of Nancy Coker Helin, wife of Bruce Helin {see planet (2430)}, the discoverer's son. A talented singer, composer and teacher, Nancita has brought music and joy to the Helin family. (M 17466)

(4223) Shikoku

1988 JM. Discovered 1988 May 7 by T. Seki at Geisei.

Named for Japan's fourth largest island, the one on which the Geisei {see planet (2571)} Station is located. "Shikoku" means "the island that has four local counties". Three long bridges straddle the Seto Inland Sea and connect Shikoku to Honshu, the largest island. (M 16593)

(4224) Susa

1988 KG. Discovered 1988 May 19 by E. F. Helin at Palomar.

Named in honor of Susan and Sarah Hicks, daughters of Bill and Nancy Hicks {see also planet 2220)}. They have shared in their parent's interest and generosity regarding space-related research over the years, particularly the pursuit of the small bodies in the solar system - minor planets and comets. (M 17466)

(4225) 1989 BN

Discovered 1989 January 31 by T. Hioki and N. Kawasato at Okutama.

(4226) Damiaan

1989 RE. Discovered 1989 September 1 by E. W. Elst at St. Michel.

Named for the Flemish priest Jozef De Veuster (1840-1889) on the occasion of the 100th anniversary of his death. At the age of 19 he entered the Congregation of the Fathers of the Sacred Hearts (Picpus Fathers) and chose for himself the new name of Damiaan. In 1863 Pater Damiaan (Father Damiaan) was sent as a missionary to Hawaii, where he was ordained a priest one year later. After eight years on Kohala he asked to be transferred to the leper colony at Kalawao on the island of Molokai. There he devoted all his energy to the improvement of the conditions at the settlement until he finally contracted leprosy himself. (M 15576)

Name endorsed by E. Goffin, who found the identifications involving this planet, and endorsed by the discoverer.

(4227) Kaali

1942 DC. Discovered 1942 February 17 by L. Oterma at Turku.

Named for the place on the island of Saaremaa {see planet (4163)} where a meteorite impacted 2500 years ago. In the smaller craters one can still use magnets to collect iron fragments. (M 18457)

(4228) 1968 OC_1

Discovered 1968 July 25 by G. Plougin and Y. Belyaev at Cerro El Roble.

(4229) Plevitskaya

1971 BK. Discovered 1971 January 22 by L. I. Chernykh at Nauchnyj.

Named in memory of Nadezhda Vasil'evna Plevitskaya (1884-1940), a splendid Russian singer who sparkled on the stages of Russia, Europe and America. Many great connoisseurs of Russian art called her "Russian skylark" and "nightingale". Leonid Sobinov {see planet (4449)} sang with her, and Sergej Rakhmaninov {see planet (4345)} accompanied for her. (M 26762)

(4230) van den Bergh

1973 ST_1. Discovered 1973 September 19 by C. J. van Houten and I. van Houten-Groeneveld at Palomar.

Named in honor of Sidney van den Bergh (1929-), Dutch-born Canadian astronomer, former director of the Dominion Astrophysical Observatory in Victoria, B.C. He was president of IAU commission 50 from 1985 to 1988 and a vice president of the IAU during 1976-1982. (M 18307)

(4231) Fireman

1976 WD. Discovered 1976 November 20 at the Harvard College Observatory at Harvard.

Named in memory of Edward Leonhard Fireman (1922-1990), on the staff of the Brookhaven National Laboratory during 1950-1956 and subsequently a physicist at the Harvard-Smithsonian Center for Astrophysics. Best known for his pioneering measurements of radioactive isotopes to determine the cosmic-ray exposure ages of freshly fallen meteorites, Fireman applied the technique to lunar samples and debris from artificial satellites as well. He also developed a method for measuring the relationship of age and depth in cores of ice bored into Antarctica and Greenland and examined the implications for the earth's climatic record over the last 200,000 years. Long interested in the experimental search for neutrinos, he was in recent years particularly involved in measuring the argon 37 produced in the chlorine and potassium detectors to a depth of 1.5 km in the Homestake Mine, South Dakota. (M 18457)

Obituaries published in Bull. Am. Astron. Soc., Vol. 24, No. 4, p. 1322-1323 (1992); Meteorictics, Vol. 26, No. 2, p. 171 (1991).

(4232) Aparicio

1977 CD. Discovered 1977 February 13 at the Felix Aguilar Observatory at El Leoncito.

Named in memory of Emiliano Pedro Aparicio (1921-1988), mineralogist and geologist at the National University of Cuyo and the National University of San Juan, also rector of the latter university. (M 31023)

(4233) Pal'chikov

1977 RO_7. Discovered 1977 September 11 by N. S. Chernykh at Nauchnyj.

Named in memory of Nikolaj Borisovich Pal'chikov (1913-1937), a student in the astronomy department at Leningrad University. He wanted to specialize in celestial mechanics and was invited by M. Subbotin {see planet (1692)} to pursue post-graduate studies. Along with other Leningrad astronomers, Nikolaj was killed during one of Stalin's purges. (M 23351)

(4234) Evtushenko

1978 JT_1. Discovered 1978 May 6 by N. S. Chernykh at Nauchnyj.

Named in honor of Evgenij Aleksandrovich Evtushenko (1933-), well known Russian poet. (M 23351)

(4235) Tatishchev

1978 SL_5. Discovered 1978 September 27 by L. I. Chernykh at Nauchnyj.

Named in memory of Vasilij Nikitick Tatishchev (1686-1750), an outstanding Russian scientist and statesman, fellow companion of Peter I {see planet (2720)}. The first Russian historiographer, he is known also for his works in geography, cartography, philosophy and economics. (M 18457)

(4236) Lidov

1979 FV_1. Discovered 1979 March 23 by N. S. Chernykh at Nauchnyj.

Named in honor of Mikhail L'vovich Lidov {1926-1993}, celestial mechanician at the Keldysh Institute of Applied Mathematics of the Russian Academy of Sciences, known for his works in astrodynamics and astronautics. (M 22501)

Obituaries published in Astron. Lett., Vol. 20, No. 3, p. 337-338 (1994); Cosmic Res., Vol. 32, No. 2 (1994).

(4237) Raushenbakh

1979 SD_4. Discovered 1979 September 24 by N. S. Chernykh at Nauchnyj.

Named in honor of Boris Viktorovich Raushenbakh (1915-), member of the Russian Academy of Sciences and the International Academy of Astronautics, and a professor at Moscow Physical and Technical Institute from 1947. He has made valuable contributions in the field of mechanics and in the theory and practice of rocket-building and space exploration. (M 27126)

(4238) 1980 GF

Discovered 1980 April 13 by A. Mrkos at Kleť.

(4239) Goodman

1980 OE. Discovered 1980 July 17 by E. Bowell at Anderson Mesa.

Named in honor of Neville J. Goodman, a member of the British Astronomical Association since 1944. Professionally, he was commissioning editor for the British scientific publisher Adam Hilger for many years. Goodman's service to astronomy dates from 1952, when he became editor of the B.A.A. Journal, a post he filled with distinction for eight years. From 1962 to 1972 he devoted himself tirelessly to the position of business secretary of the Association. President from 1972 to 1974, he again served as secretary from 1976 to 1984. Goodman has contributed many ephemerides to the B.A.A. Handbook and currently serves as Handbook editor. (M 17223)

Name suggested independently by the discoverer and B. G. Marsden. Citation material provided by S. A. Mitton.

(4240) Grün

1981 EY_{20}. Discovered 1981 March 2 by S. J. Bus at Siding Spring.

Named in honor of Eberhard Grün (1942-),a physicist at the Max-Planck-Institut für Kernphysik, Heidelberg. Starting with the Helios missions, Grün has been responsible for dust collectors on a number of interplanetary spacecraft, including Giotto, Ulysses, Galileo and Cassini. He has also interpreted the data collected and determined the dynamical properties of dust particles. In addition to his research activities, Grün is a professor at the University of Heidelberg, where he passes on to students his knowledge and long experience in space missions. (M 27458)

Citation provided by H. Scholl following a suggestion by E. Bowell.

(4241) 1981 EX_{46}
Discovered 1981 March 2 by S. J. Bus at Siding Spring.

(4242) Brecher
1981 FQ. Discovered 1981 March 28 at the Harvard College Observatory at Harvard.

Named in honor of Aviva and Kenneth Brecher, each of whose research interests cover an enormous range. Both on the staff of the Massachusetts Institute of Technology in the 1970s, Kenneth has since then been a professor of physics at Boston University, working in theoretical high- energy astrophysics, X-ray and gamma-ray astronomy, observational tests and consequences of gravitational theories and cosmology, as well as in archaeoastronomy and the history of astronomy, these last with the particular view of attempting to use ancient records to establish results of astrophysical interest. Aviva's astronomical interests have been more in the area of planetary evolution, with particular attention to the effects of magnetic fields and electric conductivity in meteorites and lunar samples. Subsequently she moved to the Arthur D. Little company, where her activities included the safety of nuclear and geotechnical systems and of the environment generally, and since 1985 she has joined her husband at Boston University as director of Academic and Corporate Relations. (M 18457)

(4243) 1981 GF_1
Discovered 1981 April 4 by A. C. Gilmore and P. M. Kilmartin at Lake Tekapo.

(4244) Zakharchenko
1981 TO_3. Discovered 1981 October 7 by L. I. Chernykh at Nauchnyj.

Named in honor of Vasilij Dmitrievich Zakharchenko (1915-), remarkable Russian journalist and writer, as well as a great romantic. An expert in engineering, he is particularly known as the editor of the popular magazine *Tekhnika molodezhi* ("Engineering for Youth") for 40 years. He also edits a new magazine about adventures, travels, scientific hypotheses and science fiction called *Chudesa i priklyucheniya* ("Miracles and Adventures") and is the producer of telecasts about the achievements of amateur designers and inventors. (M 26762)

(4245) 1981 UC_{10}
Discovered 1981 October 29 at the Purple Mountain Observatory at Nanking.

(4246) Telemann
1982 SY_2. Discovered 1982 September 24 by F. Börngen at Tautenburg.

Named in memory of Georg Philipp Telemann (1681-1767), whose authority as a musician and composer of numerous and varied compositions was in his time comparable with the glory of Händel {see planet (3826)} and surpassed the authority of Johann Sebastian Bach {see planet (1814)}. (M 16045)

(4247) Grahamsmith
1983 WC. Discovered 1983 November 28 by E. Bowell at Anderson Mesa.

Named in honor of Sir Francis Graham-Smith, the thirteenth Astronomer Royal, and physics secretary of the Royal Society, London, on the occasion of his retirement. Sir Francis has been professor of Radio Astronomy at Manchester since 1964 and director of the Radio Astronomy Laboratories since 1981. In addition, he was director of the Royal Greenwich Observatory between 1975 and 1981 and served both as secretary of the Royal Astronomical Society from 1964 to 1971 and president from 1975 to 1977. (M 17030)
Name suggested and citation material provided by S. A. Mitton.

(4248) 1984 HX
Discovered 1984 April 23 by A. C. Gilmore and P. M. Kilmartin at Lake Tekapo.

(4249) Křemže

1984 SC_2. Discovered 1984 September 29 by A. Mrkos at Kleť.

Named for a small town in southern Bohemia north of Kleť Mountain. It was first mentioned as a village with the church of St. Michael Archangel in 1263, and as a town from the year 1365, but the site was first settled in ancient times. (M 29669)
Name proposed by J. Tichá, M. Tichý and Z. Moravec.

(4250) Perun

1984 UG. Discovered 1984 October 20 by Z. Vávrová at Kleť.
Named in honor of the Slavic god of thunder. (M 16444)

(4251) Kavasch

1985 JK_1. Discovered 1985 May 11 by C. S. Shoemaker and E. M. Shoemaker at Palomar.

Named in honor of Julius Kavasch (1920-1978) and his son Wulf-Dietrich Kavasch, both amateur geologists and popular science writers about the Ries {see planet (4327)} impact crater, Germany. Julius Kavasch, principal of the elementary school of Mönchsdeggingen, Ries, spent all his life popularizing the geology of the Ries crater to the benefit of students and the local public. He played an outstanding role in providing local support for the national and international professional research on the Ries impact crater. His son, a veterinarian at Hohenaltheim, Ries, has taken over and enlarged upon his father's activities. As a director of the Annual Cultural Festival of the Ries, he initiated the foundation of the "Rieskrater-Museum", which is not only devoted to the Ries crater but is one of the best examples of a museum that comprehensively presents impact cratering phenomena in the solar system. (M 18140)
Citation provided by Dieter Stöffler at the request of the discoverers.

(4252) 1985 RG_4

Discovered 1985 September 11 by H. Debehogne at La Silla.

(4253) Märker

1985 TN_3. Discovered 1985 October 11 by C. S. Shoemaker and E. M. Shoemaker at Palomar.

Named after Wolfgang Märker, owner and director of the Märker cement factory at Harburg, located on the edge of the Ries {see planet (4327)} impact crater, Germany. Märker and his forebears have maintained a long tradition of support for geological research in one of the world's best studied impact craters. Märker's responsible interest in the Ries geology and his involvement in the production of special cement from local impact breccias has kept quarries in operation that have yielded invaluable information about the Ries crater. This has been of fundamental importance for research opportunities in the field of impact cratering. (M 18140)
Citation provided by Dieter Stöffler at the request of the discoverers.

(4254) Kamél

1985 UT_3. Discovered 1985 October 24 by C.-I. Lagerkvist at Kvistaberg.

Named in honor of Lars Kamél, planetary astronomer and meticulous compiler and analyzer of cometary brightness data, on the occasion of the defense of his doctoral dissertation on May 24, 1991. Kamél's analyses have been mainly directed toward interpretation of the nongravitational effects in cometary motions and elucidation of physical evolutionary effects in comets. (M 18307)
Citation prepared by H. Rickman.

(4255) Spacewatch

1986 GW. Discovered 1986 April 4 by the Spacewatch at Kitt Peak.

Named for the Spacewatch program and its many supporters. The program uses a 2048 x 2048 CCD in the scanning mode on the Spacewatch Telescope, which is the 0.91-m Newtonian reflector of the Steward Observatory of the University of Arizona on Kitt Peak, to discover and provide astrometric observations for minor planets and comets. (M 16045)

(4256) Kagamigawa

1986 TX. Discovered 1986 October 3 by T. Seki at Geisei.

Named for the river that flows through Kochi {see planet (2396)}, home town of the discoverer. (M 19336)

(4257) Ubasti

1987 QA. Discovered 1987 August 23 by J. Mueller at Palomar.

Ubasti, also called Bastet or Bast, was an ancient Egyptian goddess worshipped in the form of a cat. Originally a goddess of the home, in the New Kingdom Ubasti was equated with the lioness war goddess. This name is being dedicated to observatory cats throughout the world. (M 19336)

Name suggested and citation prepared by G. V. Williams at the request of the discoverer.

This name is particularly dedicated to the discoverer's beloved companion for the past ten years, Pepper Cat (1974-1991).

(4258) Ryazanov

1987 RZ_2. Discovered 1987 September 1 by L. G. Karachkina at Nauchnyj.

Named in honor of Ehldar Aleksandrovich Ryazanov, a well-known Soviet movie producer, writer and poet. (M 17030)

(4259) 1988 SB_3

Discovered 1988 September 16 by S. J. Bus at Cerro Tololo.

(4260) Yanai

1989 AX. Discovered 1989 January 4 by S. Ueda and H. Kaneda at Kushiro.

Named in honor of Masayuki Yanai (1959-), who has since 1987 been an active observer and discoverer of minor planets. He lives in Kitami {see planet (3785)} and works for the Nippon Telephone and Telegram Company. (M 16593)

(4261) Gekko

1989 BJ. Discovered 1989 January 28 by Y. Oshima at Gekko.

Named for the observatory at which this minor planet was discovered. The name means "moonlight". Founded in 1957, the observatory belongs to the International Foundation for Cultural Harmony and is active in the education and popularization of astronomy. (M 18307)

(4262) 1989 CO

Discovered 1989 February 5 by M. Arai and H. Mori at Yorii.

(4263) Abashiri

1989 RL_2. Discovered 1989 September 7 by M. Yanai and K. Watanabe at Kitami.

Named for a small city of population 43,000 in the eastern part of Hokkaido {see planet (3720)}. Located at the Sea of Okhotsk, it has prospered for many years on account of its fishing industry, and it serves as a center for transportation, culture and commerce. (M 19336)

(4264) Karljosephine

1989 TB. Discovered 1989 October 2 by K. F. J. Cwach at Siding Spring.

Named by the discoverer in honor of his parents, Karl Wilhelm Cwach and Josephine Anna-Maria Cwach for their encouragement of his interest in science and amateur astronomy. (M 17979)

(4265) Kani

1989 TX. Discovered 1989 October 8 by Y. Mizuno and T. Furuta at Kani.

Named for the quiet, rural town where the first discoverer lives. It is situated about 30 km north of Nagoya, the third largest city in Japan. Kani is located on the banks of the beautiful river Kiso, which is said to resemble the Rhine in Germany. (M 16045)

(4266) Waltari

1940 YE. Discovered 1940 December 28 by Y. Väisälä at Turku.

Named in memory of Mika Waltari (1908-1979), Finnish writer, from 1957 a member of the Academy of Finland. He was best known internationally for his extensive historical novel *Sinuhe, egyptiläinen* ('Sinuhe, Egyptian'), which was translated into several languages and also filmed {see also planet (4512)}. (M 18457)

(4267) Basner

1971 QP. Discovered 1971 August 18 by T. M. Smirnova at Nauchnyj.

Named in memory of Veniamin Efimovich Basner (1925-1996), outstanding Russian composer, author of operas, ballets, operettas, excellent songs, symphonic and chamber compositions and music for many movies. (M 29143)
Name proposed by the Russian St. Petersburg Composers' Union.

(4268) Grebenikov

1972 TW_3. Discovered 1972 October 5 by T. M. Smirnova at Nauchnyj.

Named in honor of Evgenij Alexandrovich Grebenikov, head of the department in the Institute of Problems of Cybernetics in Moscow, well-known expert on analytical and celestial mechanics and on the qualitative theory of differential equations. In co-authorship with E. P. Aksenov and V. G. Demin {see planets (4777) and (5086), respectively}, he has investigated the problem of two fixed centers and found its general solution. He has proved a number of theorems substantiating the averaging method by Delaunay-Hill. In a series of studies he and his numerous pupils applied these results to the construction of precise theories of the motions of artificial earth satellites, natural satellites and minor planets. (M 19695)
Name proposed by the Institute of Theoretical Astronomy, St. Petersburg.

(4269) 1974 FN

Discovered 1974 March 22 by C. Torres at Cerro El Roble.

(4270) Juanvictoria

1975 TJ_6. Discovered 1975 October 1 at the Felix Aguilar Observatory at El Leoncito.

Named in memory of Juan Victoria (1906-1986), legal representative of the Yale-Columbia Southern Station when it was under construction in the early 1960s; he also supervised the construction of the dome and facilities for the U.S. Naval Observatory's station nearby. A mining engineer by training, Victoria was known for his teaching and research in the engineering department of the Cuyo {see planet (1917)} National University, and he also advised the government on the development and establishment of the Minerals Laboratory and the Mining Research Institute. One of Victoria's greatest accomplishments was the construction of the San Juan {see planet (2284)} Auditorium, considered to be one of best such buildings in Latin America. (M 31295)

(4271) Novosibirsk

1976 GQ_6. Discovered 1976 April 3 by N. S. Chernykh at Nauchnyj.

Named for the great Siberian industrial, cultural and scientific center, on the occasion of its 100th anniversary in 1993. (M 22501)

(4272) Entsuji

1977 EG_5. Discovered 1977 March 12 by H. Kosai and K. Hurukawa at Kiso.

Named for a temple near where the first discoverer lived in his youth. The famous priest Ryokan (1758-1831) {see planet (6031)} was trained in this temple. (M 17979)

(4273) 1978 UU_1

Discovered 1978 October 29 at the Purple Mountain Observatory at Nanking.

(4274) Karamanov

1980 RZ_3. Discovered 1980 September 6 by N. S. Chernykh at Nauchnyj.

Named in honor of Alemdar Sobitovich Karamanov, composer, who lives and works in the Crimea, known for his symphonies, ballets, music for performances and cinema films, piano concerti and other musical works, many of which are dedicated to the Crimea and its history. (M 22501)

(4275) Bogustafson

1981 EW_{14}. Discovered 1981 March 1 by S. J. Bus at Siding Spring.

Named in honor of Bo Å. S. Gustafson (1953-) of the University of Florida, Gainesville. Gustafson has specialized in studying the formation, evolution and fate of small particles in the solar system. Along with J. M. Greenberg, his name is associated with the "bird's nest" model of cometary grains as very porous aggregates. His work concerns both dynamics and physics, including optical properties in particular. Since 1994 he has led the Laboratory for Astrophysics in Gainesville, pursuing microwave simulation of light scattering. (M 27458)
Citation provided by H. Rickman following a suggestion by E. Bowell.

(4276) Clifford

1981 XA. Discovered 1981 December 2 by E. Bowell at Anderson Mesa.

Named in honor of Clifford J. Cunningham {1955- }, Canadian amateur astronomer and writer, whose interest in minor planets has encompassed positional measurements and photometry. Cunningham's book *Introduction to Asteroids*, published in 1988, has justifiably earned praise from both readers and reviewers. He has recently been working on a bibliographic database on minor planets that currently comprises more than 10,000 references dating back to the year 1789. (M 16248)

(4277) Holubov

1982 AF. Discovered 1982 January 15 by A. Mrkos at Kleť.

Named for a picturesque village in southern Bohemia northeast of Kleť Mountain. Near Holubov there are ruins of a Gothic castle, DívčíKámen, founded in 1349 by Rožmberks on the site of a Bronze Age settlement. (M 29669)

Name proposed by J. Tichá, M. Tichý and Z. Moravec, endorsed by M. J. Plavec.

(4278) Harvey

1982 SF. Discovered 1982 September 22 by E. Bowell at Anderson Mesa.

Named in honor of G. Roger Harvey, instructor in astronomy at the University of North Carolina in Charlotte. Using handcrafted Newtonian reflectors of up to 74-cm aperture, Harvey has visually observed more than 2,600 different minor planets, far surpassing previous efforts. His observations have at times proved scientifically valuable for determining rotation periods and identifying wayward objects. (M 16593)

Name suggested and citation provided by J. U. Gunter, who sparked Harvey's interest in minor planets; name endorsed by B. G. Marsden.

(4279) De Gasparis

1982 WB. Discovered 1982 November 19 at the Osservatorio San Vittore at Bologna.

Named in memory of Annibale De Gasparis (1819-1892), Italian astronomer, director of the Observatory of Naples from 1864 to 1889. A great observer of minor planets, he discovered visually those numbered 10, 11, 13, 15, 16, 20, 24, 63 and 83 and made an independent discovery of (14). He was also the author of several memoirs on orbit determination and other problems in theoretical astronomy. (M 17030)
De Gasparis is also honored by a lunar crater.

(4280) Simonenko

1985 PF_2. Discovered 1985 August 13 by N. S. Chernykh at Nauchnyj.

Named in memory of Alla Nikolaevna Simonenko (1935-1984), specialist on small bodies of the solar system. She made substantial contributions to the study of the physical process of meteoroid fragmentation, the distribution of meteor radiants, the influx of meteoric matter to the earth, and the dynamics of asteroid orbits. She published two books on the study of asteroids, and actively favored the development of physical investigations of asteroids in the former U.S.S.R. (M 22501)
Citation prepared by D. F. Lupishko at the request of the discoverer.

(4281) Pounds

1985 TE_1. Discovered 1985 October 15 by E. Bowell at Anderson Mesa.

Named in honor of Kenneth A. Pounds, professor of physics at Leicester University, where he has directed X-ray astronomy, in which he is a pioneer, for 25 years. Pounds leads the British participation on ROSAT, which has produced the first X-ray and XUV photographs of the Moon. An active supporter of amateur initiatives through many talks to societies, he is strongly in favour of publicising astronomy in the most general way. Pounds is currently president of the Royal Astronomical Society and was awarded the Commander of the British Empire for services to astronomy. (M 17030)
Name suggested and citation material provided by S. A. Mitton.

(4282) Endate

1987 UQ_1. Discovered 1987 October 28 by S. Ueda and H. Kaneda at Kushiro.

Named in honor of Kin Endate (1960-), a very active amateur astronomer who has been observing and discovering minor planets during the past few years. He lives in Bihoro {see planet (4460)} and works for the Endate Craft Company, where he makes wooden crafts, trays, cylinders, sugar pots, vessels, etc., along with the help of his family. (M 16593)

(4283) Stöffler

1988 BZ. Discovered 1988 January 23 by C. S. Shoemaker and E. M. Shoemaker at Palomar.

Named in honor of Dieter Stöffler, professor and director of the Institute for Planetology, University of Münster, Germany. He is especially well known for his careful and elegant calibration of the temperatures and peak shock pressures associated with different grades of shock metamorphism. At the Ries {see planet (4327)} impact crater in Germany, he solved these relationships chiefly from quantitative study of shock-formed phases of silica. He has worked extensively on the origin of the regolith, impact breccias, and magnetic history of the Moon, and on the petrology of meteorites. (M 18140)

(4284) Kaho

1988 FL_3. Discovered 1988 March 16 by S. Ueda and H. Kaneda at Kushiro.

Named in memory of Sigeru Kaho (1909-1981), an earnest researcher of variable

stars at the Tokyo Observatory. His discovery of comet 1936 III (Kaho-Kozik-Lis) was the first cometary discovery in Japan in modern times. The Kaho prize, established after his death, has been awarded annually to an active amateur astronomer in Japan. (M 19336)
Name suggested by H. Fukushima.

(4285) Hulkower

1988 NH. Discovered 1988 July 11 by E. F. Helin at Palomar.
Named in honor of Neal D. Hulkower, mathematician, scientist and good friend of the discoverer with a special interest in celestial mechanics. While at the Jet Propulsion Laboratory he developed a method of analyzing accessibility for spacecraft missions to near-Earth asteroids and ranked them by energy requirements for rendezvous. His enthusiasm and expertise have helped in stimulating greater interest in these objects. (M 16045)
Named by the discoverer. Name endorsed by R. Staehle and D. Bender.

(4286) Rubtsov

1988 PU_4. Discovered 1988 August 8 by L. I. Chernykh at Nauchnyj.
Named in memory of the remarkable Soviet lyric poet Nikolaj Mikhajlovich Rubtsov (1936-1971). (M 18458)

(4287) Třísov

1989 RU_2. Discovered 1989 September 7 by A. Mrkos at Kleť.
Named for a village in southern Bohemia northeast of Kleť Mountain. Near the village there are remnants of the ancient Celtic town of Abilunon, founded in the first century B.C. on a strategic peninsula in the Vltava {see planet (2123)} river. (M 29669)
Name proposed by J. Tichá, M. Tichý and Z. Moravec.

(4288) 1989 TQ_1

Discovered 1989 October 8 by T. Kojima at Chiyoda.

(4289) Biwako

1989 UA_2. Discovered 1989 October 29 by A. Sugie at Taga.
Named for the largest lake in Japan, 235 km in circumference and up to 104 m in depth. Like the Dynic Astronomical Observatory, the lake is located in Shiga prefecture. Lake Biwa, whose shape resembles the old Japanese instrument "Biwa", is one of the oldest lakes in the world and has played an important role in politics, economy, culture, transportation, and so on. It is a quasi-national park, surrounded by a spacious resort zone. (M 16249)

(4290) Heisei

1989 UK_3. Discovered 1989 October 30 by T. Seki at Geisei.
Named for the Japanese new era, or "Gengo", which began on 1989 Jan. 8. Heisei was taken from an old Chinese book *Shihchi* by Ssu-ma Chien and from *Shu Ching*, one of the Chinese Five Classics. The name embodies the hope that peace will be achieved inside and outside the country, on the earth and in the heavens. This minor planet was discoverered in the first year of Heisei. The Christian era 1990 corresponds to Japanese era 2650. (M 16444)

(4291) 1989 VH

Discovered 1989 November 2 by M. Arai and H. Mori at Yorii.

(4292) Aoba

1989 VO. Discovered 1989 November 4 by M. Koishikawa at Sendai.
Named for the castle Aoba-jo in Sendai. The castle was built in the early seventeenth century by Date Masamune, the feudal lord of the district. The name

means "fresh leaves" and frequently represents the city of Sendai. In April 1989, Sendai became a "designated city". One of the five wards of the city is named Aoba-ku, and it is there that both the Sendai Astronomical Observatory and the Ayashi Station are located {see also planets (3133) and (3994)}. (M 16045)

(4293) Masumi

1989 VT. Discovered 1989 November 1 by Y. Oshima at Gekko.

Named in honor of Masumi Furukawa (1917-), a board member of the International Foundation for Cultural Harmony and long an eminent popularizer of astronomy, mainly in Kyushu. (M 18307)

(4294) Horatius

4016 P-L. Discovered 1960 September 24 by C. J. van Houten and I. van Houten-Groeneveld at Palomar.

Named for the Roman poet Horatio (65-8 B.C.), whose main poems are collected in Sermones, Carmina and Epistolae. (M 16594)

(4295) Wisse

6032 P-L. Discovered 1960 September 24 by C. J. van Houten and I. van Houten-Groeneveld at Palomar.

Named in honor of Marijke Wisse-Schouten, who as a staff member of the Leiden Observatory performed the photometry for the three Palomar-Leiden Trojan surveys. (M 16594)

(4296) van Woerkom

1935 SA_2. Discovered 1935 September 28 by H. van Gent at Johannesburg.

Named in honor of Adrianus Jan J. van Woerkom {1915-1991}, Dutch-born astronomer, who worked in Leiden on the distribution of comet orbits and at Yale Observatory on problems of celestial mechanics before moving to the Electric Boat Company in Connecticut around 1960. (M 18140)

Obituaries published in Bull. Am. Astron. Soc., Vol. 23, No. 4, p. 1495 (1991); Phys. Today, Vol. 45, No. 8, p. 84 (1992); Celest. Mech. Dyn. Astron., Vol. 53, No. 1, p. 1 (1992).

(4297) Eichhorn

1938 HE. Discovered 1938 April 19 by W. Dieckvoss at Bergedorf.

Named in honor of Heinrich Karl Eichhorn (1927-1999), Austrian-American astronomer, educator and scholar, innovator in the astronomy of stellar positions and motions. He developed the central-overlap and other astrometric reduction methods that greatly improve their rigor, increasing the precision of stellar parallaxes and proper motions and thus also the distance scale of the universe. (M 28621)
Name proposed by A. R. Upgren.

(4298) 1941 WA

Discovered 1941 November 17 by I. Pólit at Barcelona.

(4299) WIYN

1952 QX. Discovered 1952 August 28 at the Goethe Link Observatory at Brooklyn, Indiana.

Named for the WIYN telescope, located at the Kitt Peak National Observatory. This 3.5-m telescope is jointly operated by the University of Wisconsin, Indiana University, Yale University and the National Optical Astronomy Observatories. (M 34619)
Name proposed by F. K. Edmondson.

(4300) Marg Edmondson

1955 SG_1. Discovered 1955 September 18 at the Goethe Link Observatory at Brooklyn, Indiana.

Named in memory of Margaret Russell Edmondson (1914-1999), the youngest child of Henry Norris Russell {see planet (1762)} and Lucy May (Cole) Russell. The breadth of her intellectual interests and the depth of her knowledge were the result of her genetic heritage and her close intellectual rapport with her father. (M 24620)

Name proposed by F. K. Edmondson, her husband for 64 wonderful years.

(4301) 1966 PM

Discovered 1966 August 7 at the Boyden Observatory at Bloemfontein.

(4302) Markeev

1968 HP. Discovered 1968 April 22 by T. M. Smirnova at Nauchnyj.

Named in honor of Anatolij Pavlovich Markeev, professor at the Moscow Aviation Institute, well-known expert on theoretical and celestial mechanics. He investigated the stability of Hamiltonian systems in the case of resonances of the third and the fourth orders, constructed theories of motion in the vicinity of the libration points of the three body problem and developed a theory of the motion of large constructions at the near-earth orbits. In addition, he studied various cases of motion of rigid bodies on smooth and rough surfaces. He contributed much to the methods of teaching and published a number of text-books and monographs. (M 19696)

Name proposed by the ITA.

(4303) Savitskij

1973 SZ_3. Discovered 1973 September 25 by L. V. Zhuravleva at Nauchnyj.

Named in honor of Evgenij Yakovlevich Savitskij (1910-1990), Russian aviator, father of cosmonaut Svetlana Savitskaya {see planet (4118)}. (M 22501)

(4304) Geichenko

1973 SW_4. Discovered 1973 September 27 by L. I. Chernykh at Nauchnyj.

Named in honor of Semyon Stepanovich Geichenko, writer and literary scholar, researcher of the life and the creative work of A. S. Pushkin {see planet (2208)}. Geichenko is especially known as director and curator of Mikhailovskoe, the Pushkin country-house museum in the Pskov region. He made a great effort to rebuild this memorial to Pushkin following its destruction in World War II. (M 22501)

(4305) Clapton

1976 EC. Discovered 1976 March 7 at the Harvard College Observatory at Harvard.

Named in honor of Eric Clapton (1945-), singer, composer and guitarist extraordinaire. He is the most remarkable bluesman ever to come out of England and his rock career has spanned more than 25 years. In the mid-60s, a graffito on a brick wall in London proclaimed: 'Eric is God'. He remains a musical megastar today, and is probably best known for the rock classics 'Layla' and the ballad 'Wonderful Tonight'. (M 16249)

(4306) Dunaevskij

1976 SZ_5. Discovered 1976 September 24 by N. S. Chernykh at Nauchnyj.

Named in memory of Isaak Osipovich Dunaevskij (1900-1955), prominent Soviet composer well known for his many operettas, songs and film scores. (M 23351)

(4307) Cherepashchuk

1976 UK_2. Discovered 1976 October 26 by T. M. Smirnova at Nauchnyj.

Named in honor of Anatolij Mikhajlovich Cherepashchuk (1941-), professor at Moscow State University and director of the Sternberg Astronomical Institute, respected worldwide for his work in theoretical and observational astrophysics. From an analysis of eclipses in close-binary systems he obtained correct estimates of the radii and temperatures of Wolf-Rayet stars, predicted x-ray emission in such systems, discovered optical eclipses in the unique system V1343 Aql and developed a commonly-adopted model of this object as a massive x-ray binary. He also discovered fast emission-line variability in the nuclei of Seyfert galaxies. (M 29143)
Name proposed by the Institute of Theoretical Astronomy.

(4308) Magarach

1978 PL_4. Discovered 1978 August 9 by N. S. Chernykh at Nauchnyj.

Named for the Research Institute of wine-making and viticulture at Magarach, near Yalta {see planet (1475)}. Founded in 1828 as a specialized school for gardening and wine-making, Magarach was one of the first such schools in what was then Russia. (M 22501)

(4309) Marvin

1978 QC. Discovered 1978 August 30 at the Harvard College Observatory at Harvard.

Named in honor of Ursula B. Marvin {1922- }, geologist at the Harvard-Smithsonian Center for Astrophysics, well-known for her outstanding mineralogical and petrologic research on meteorites and lunar samples. Her meteorite studies have focused in recent years on the rich wealth of samples from Antarctica, and she was also involved in collecting them. This work also led to her analysis of dust grains, some of them of volcanic origin, in Antarctic ice samples. Her discovery of cordierite in an Apollo 15 lunar sample strongly suggested that this particular rock must have originated more than 40 km below the lunar surface. Marvin has also been responsible for mapping sections of Ganymede, maintains a strong interest in the history of meteorites and currently serves as secretary general of the International Commission on the History of Geological Sciences. (M 18458)

(4310) Strömholm

1978 RJ_7. Discovered 1978 September 2 by C.-I. Lagerkvist at La Silla.

Named in honor of Stig Strömholm, scholar, author, professor and vice chancellor of Uppsala University, to commemorate his active and enduring effort to support and strengthen international collaboration in fundamental sciences and culture. Although a respector of tradition, he is ever open to significant new developments in science and society. (M 19336)
Citation prepared by B. Gustafson.

(4311) Zguridi

1978 SY_6. Discovered 1978 September 26 by L. V. Zhuravleva at Nauchnyj.

Named in memory of Aleksandr Mikhajlovich Zguridi (1904-1998), producer of popular scientific films and an actor who won public recognition, as well as official honor from the Soviet state. (M 34620)

(4312) 1978 WW_{11}

Discovered 1978 November 29 by S. J. Bus and C. T. Kowal at Palomar.

(4313) Bouchet

1979 HK_1. Discovered 1979 April 21 by H. Debehogne at La Silla.

Named in honor of Patrice Bouchet, a staff astronomer at the European Southern Observatory at La Silla, where he leads the infrared section. His work has ranged from the discoveries of a ring around Neptune and of ice in Halley's Comet, to the determination of the extinction law in the Small Magellanic Cloud and monitoring to study dust formation in supernova 1987A. (M 19337)

(4314) 1979 ML_3

Discovered 1979 June 25 by E. F. Helin and S. J. Bus at Siding Spring.

(4315) Pronik

1979 SL_{11}. Discovered 1979 September 24 by N. S. Chernykh at Nauchnyj.

Named in honor of Vladimir Ivanovich Pronik (1932-) and Iraida Ivanovna Pronik, husband and wife, astronomers at the Crimean Astrophysical Observatory for more than 40 years. The Proniks, members of IAU Commission 28 (Galaxies), are known for their research on the gaseous and stellar components of galaxies and the nature of active galactic nuclei. V. I. Pronik is a prominent expert on astronomical instruments. (M 23351)

(4316) Babinkova

1979 TZ_1. Discovered 1979 October 14 by N. S. Chernykh at Nauchnyj.

Named in honor of Artur Nikolaevich Babin (1936-) and Aleksandra Nikolaevna Koval', husband and wife, solar astrophysicists at the Crimean Astrophysical Observatory for more than 35 years, known for their research on the fine structure of active solar features. (M 23351)

(4317) Garibaldi

1980 DA_1. Discovered 1980 February 19 by Z. Vávrová at Kleť.

Named in memory of Giuseppe Garibaldi (1807-1882), fighter for human rights and hero of many battles. (M 21609)

(4318) Baťa

1980 DE_1. Discovered 1980 February 21 by Z. Vávrová at Kleť.

Named in honor of Tomáš Baťa (1876-1932), world-renowned Czech businessman. (M 16444)

(4319) 1981 ER_{14}

Discovered 1981 March 1 by S. J. Bus at Siding Spring.

(4320) 1981 EJ_{17}

Discovered 1981 March 1 by S. J. Bus at Siding Spring.

(4321) 1981 EH_{26}

Discovered 1981 March 2 by S. J. Bus at Siding Spring.

(4322) Billjackson

1981 EE_{37}. Discovered 1981 March 11 by S. J. Bus at Siding Spring.

Named in honor of William M. Jackson (1936-) of the Department of Chemistry, University of California at Davis, for his contributions to the study of the chemistry of comets. His laboratory studies of photodissociation processes have been important in understanding the processes in cometary comae and in understanding possible parent molecules. (M 27459)
Citation provided by M. F. A'Hearn following a suggestion by E. Bowell.

(4323) Hortulus

1981 QN. Discovered 1981 August 27 by P. Wild at Zimmerwald.

The name is Latin for a small, cozy garden, sheltering those flowers that gave their names to minor planets. For (4323) there are at least eight different such flower arrangements, among them (4323) = (1056) Azalea + (1080) Orchis + (1092) Lilium + (1095) Tulipa. (M 34620)

(4324) 1981 YA_1

Discovered 1981 December 24 by L. G. Taff at Socorro.

(4325) Guest

1982 HL. Discovered 1982 April 18 by E. Bowell at Anderson Mesa.

Named in honor of John E. Guest, reader in Planetary Studies at University College, London, and a colleague of the discoverer at the University of London Observatory during the latter's postgraduate days. A leading British planetary geologist and a specialist in vulcanology, Guest has worked on the geological mapping of the moon, Mars and many of the atmosphereless satellites of the major planets. He was a principal investigator on the Soviet Phobos missions and is currently a team member on the Magellan radar-mapping mission to Venus. Guest has not neglected research on terrestrial vulcanology, however, as witness his book on Etna. (M 16594)
Citation provided by D. McNally at the request of the discoverer.

(4326) McNally

1982 HS_1. Discovered 1982 April 28 by E. Bowell at Anderson Mesa.

Named in honor of Derek McNally, Director of the University of London Observatory (University College London). A teacher of the discoverer while he was an undergraduate student, McNally's main astronomical research areas are theoretical modeling of star formation and interstellar spectroscopy using both ground-based observations and data from the International Ultraviolet Explorer satellite. He also takes a keen interest in astronomical education. Currently {1988-1991}, he is general secretary of the International Astronomical Union. (M 16594)
Citation provided by J. E. Guest at the request of the discoverer.

(4327) Ries

1982 KB_1. Discovered 1982 May 24 by C. S. Shoemaker and S. J. Bus at Palomar.

Named for a large topographic basin in southern Germany formed by the impact of an extraterrestrial body in Miocene time. Within this crater, which lies between the Swabian and Franconian plateaus, is the medieval city of Nördlingen with a church made from the impact breccia. The Ries is host to an annual cultural festival in celebration of its geology, history, art, music, folklore, handicrafts, industry and agriculture. (M 17655)

(4328) Valina

1982 SQ_2. Discovered 1982 September 18 by H. Debehogne at La Silla.

Named in honor of Valentina Arkadievna Andreichenko (and her daughter Alina Eduardovna) for her collaboration with the discoverer in the Tomsk-Observatory-Brussels-Program, which was created in order to realize observations, in particular at the European Southern Observatory at La Silla, as well as orbital improvements and theoretical works in astrometry and celestial mechanics. Valentina has participated in expeditions to the site of the 1908 Tunguska {see planet (5471)} event. (M 22829)

(4329) 1982 SX_2

Discovered 1982 September 22 by L. G. Taff at Socorro.

(4330) Vivaldi

1982 UJ_3. Discovered 1982 October 19 by F. Börngen at Tautenburg.

Named for the great Italian violin virtuoso and baroque composer Antonio Vivaldi (1680-1743). (M 16045)

(4331) Hubbard

1983 HC. Discovered 1983 April 18 by N. G. Thomas at Anderson Mesa.

Named in memory of Ralph "Doc" Hubbard (1885-1980), of Medora, North

Dakota. The son of Elbert Hubbard, Ralph was a nationally-recognized authority on American Indian culture, an expert at crafting Indian ceremonial clothing and an author and educator regarding the preservation of Indian culture and history. He worked for many years with the Boy Scouts of America; he had rich experiences in frontier life from ranching in Montana and Colorado around 1900, and the discoverer was one of hundreds of young people who later were enriched through Doc's biology teaching. (M 24916)
Assistance with the citation was provided by J. Stewart and J. Paulson.

(4332) Milton

1983 RC. Discovered 1983 September 5 by C. S. Shoemaker and E. M. Shoemaker at Palomar.

Named for Daniel J. Milton, geologist with the U.S. Geological Survey. Milton, who has carried out geological studies of the moon and Mars, is best known for his investigations of meteorite craters and ancient impact features in Australia. His work on structural deformation at the largest Henbury meteorite craters and in the central uplift of the Gosses Bluff structure, Northern Territory, are tours de force in detailed geologic mapping. (M 17656)

(4333) Sinton

1983 RO_2. Discovered 1983 September 4 by E. Bowell at Anderson Mesa.

Named in honor of William M. Sinton, planetary astronomer at the Institute for Astronomy, University of Hawaii, on the occasion of his retirement. Sinton came to Hawaii in 1966 from the Lowell Observatory, where he had already established himself in the field of infrared astronomy and was codiscoverer of CO in Alpha Orionis. Instrumental in the development of Mauna Kea as a premier astronomical observatory, Sinton was responsible for the specification and design of the computer operating environment for the University's 2.24-m telescope, one of the first telescopes to be computer controlled. He has been a pioneer in bringing new technology and ideas to the subject of thermal infrared studies of the planets, particularly Venus and Mars. He has habitually selected technically difficult problems to work on, including some of the very earliest spectroscopy with interferometers, thermal emission spectroscopy of the stars and planets, time-resolved radiometry of the satellites of Jupiter, the first astronomical far-infrared work with a helium-3 bolometer, near-infrared studies of the dark hemisphere of Venus, and the thermal properties of the satellites of Jupiter during their mutual eclipses and occultations. (M 16444)

Citation prepared at the discoverer's request by D. P. Cruikshank and D. J. Tholen.

(4334) 1983 RO_3

Discovered 1983 September 2 by H. Debehogne at La Silla.

(4335) Verona

1983 VC_7. Discovered 1983 November 1 at the Giordano Bruno Observatory at Cavriana.

Named for the Italian city, founded in the fourth century B.C., and situated at the foot of the Alps. Verona knew moments of splendor during the ancient Roman Epoca and still conserves the Arena, as well as thirteenth-century masterpieces from the reign of the Scaligeri family. Shakespeare's {see planet (2985)} *Romeo and Juliet* was sited in Verona, which is also the city in which the Cavriana observers, Luciano Lai, Ivano Rocchetti and Giordano Vesentini, were born and raised. (M 16249)

(4336) Jasniewicz

1984 QE_1. Discovered 1984 August 31 by B. A. Skiff at Anderson Mesa.

Named in honor of Gerard Jasniewicz, astronomer at Strasbourg Observatory. As a long-time member of this observatory's Centre de Données Astronomiques, he is responsible for maintaining the SIMBAD {see planet (4692)} bibliographic database. (M 27126)

(4337) Arecibo

1985 GB. Discovered 1985 April 14 by E. Bowell at Anderson Mesa.

Named for the Arecibo Observatory in Puerto Rico, site of the largest filled-aperture telescope in the world. Arecibo is the principal research facility of the National Astronomy and Ionosphere Center and is used for radio astronomy, planetary radar astronomy and ionospheric investigations. Radar echoes obtained at Arecibo have provided valuable information about the physical and dynamical properties of about fifty minor planets, five comets, the terrestrial planets, the Galilean satellites and Saturn's rings. (M 16445)
Name suggested and citation provided by S. J. Ostro.

(4338) Velez

1985 PB_1. Discovered 1985 August 14 by E. Bowell at Anderson Mesa.

Named in honor of Reinaldo Velez, senior telescope operator at the Arecibo Observatory for the past fifteen years and currently head of AO's Department of Telescope and Computer Operations. Rey's expertise, dedication and extensive experience with the telescope have contributed critically to hundreds of radar/radio astronomy experiments. In particular, his direct assistance with radar observations of main-belt and earth-approaching minor planets has been a key element in the success of that research. (M 16445)
Name suggested and citation provided by S. J. Ostro.

(4339) Almamater

1985 UK. Discovered 1985 October 20 by A. Mrkos at Kleť.

Named in anticipation of the 650th anniversary of Charles University (Univerzita Karlova) in Prague. This oldest university in central Europe was founded in 1348 by Charles IV of Luxembourg. (M 30475)
Name suggested by J. Tichá, M. Šolc, M. Tichý and Z. Morovec.
Charles University in Prague is also honored by planet (2288).

(4340) Dence

1986 JZ. Discovered 1986 May 4 by C. S. Shoemaker and E. M. Shoemaker at Palomar.

Named for Michael R. Dence, executive director of the Royal Society of Canada and formerly a geologist with the Dominion Observatory. Dence was a pioneer in the geologic investigation of ancient impact craters on the Canadian shield. He made one of the first good estimates of the cratering rate on the earth in Phanerozoic time from studies of impact structures in Canada and Europe. (M 17656)

(4341) Poseidon

1987 KF. Discovered 1987 May 29 by C. S. Shoemaker and E. M. Shoemaker at Palomar.

In Greek mythology, Poseidon, a brother of Zeus {see planet (5731)}, was lord of the sea and of earthquakes (since the earth is shaken by waves beating on the shore). He ruled all that lived in the sea and gathered clouds and raised storms. Poseidon was the constant enemy of Troy during the Trojan War. (M 17656)

(4342) Freud

1987 QO_9. Discovered 1987 August 21 by E. W. Elst at La Silla.

Named in memory of Sigmund Freud (1865-1939), father of psychoanalysis. In 1885 he went to Paris and studied under the neurologist Jean Charcot, who stimulated him to investigate hysteria from a psychological point of view. This view was strengthened by Josef Breuer, who cured hysterical symptoms by putting the patient under hypnosis. Soon afterward Freud replaced hypnotism by the method of free association. Among his writings are *The Interpretations of Dreams* (1900) and *Moses and Monotheism* (1939), the latter an elucidation of a historical-cultural problem that had always fascinated him. (M 16045)

(4343) Tetsuya

1988 AC. Discovered 1988 January 10 by S. Ueda and H. Kaneda at Kushiro.

Named in honor of Tetsuya Fujii (1960-), active observer and discoverer of minor planets and director of the astronomical club in Kitami {see planet (3785)}, where he also works for the NHK broadcasting office. (M 20159)

(4344) Buxtehude

1988 CR_1. Discovered 1988 February 11 by E. W. Elst at La Silla.

Named in memory of the great organist and composer Dietrich Buxtehude (1637-1707), undoubtedly the most eminent master of the organ before Johann Sebastian Bach {see planet (1814)}. His inspired creative power is particularly evident in chorales such as "Wie schön leuchtet der Morgenstern". In 1668 he became the successor of Franz Tunder {see planet (7871)} as organist at the Marienkirche in Lübeck, a position h30e held until his death. (M 16046)

(4345) Rachmaninoff

1988 CM_2. Discovered 1988 February 11 by E. W. Elst at La Silla.

Named in memory of the composer and pianist Sergei Rachmaninoff (1873-1943), considered to be the last great figure in the tradition of Russian romanticism, and whose compositions embrace symphonies and piano concerti and preludes. Although his greatest productivity occurred in St. Petersburg and Moscow around the turn of the century, he composed the beautiful "Rhapsody on a Theme by Paganini" in the United States in 1934, during the exile in which he was principally a concert performer. (M 16046)

(4346) Whitney

1988 DS_4. Discovered 1988 February 23 by A. J. Noymer at Siding Spring.

Named in honor of Charles A. Whitney, professor at Harvard University. Whitney has done seminal work on Cepheid modeling, and he is the editor of the JAAVSO. He has brought astronomy to art history, using computer sky simulations to recreate, and hence date, the sky scenes of van Gogh {see planet (4457)}. Recently he has worked on physics and astronomy textbooks for high schools and has served as a science consultant to the Children's Television Workshop. Whitney is also the author of several astronomy books for the public, including *The Discovery of our Galaxy* and *Whitney's Star Finder*. (M 16046)

(4347) Reger

1988 PK_2. Discovered 1988 August 13 by F. Börngen at Tautenburg.

Named in memory of Max Reger (1873-1916), chief conductor and general music director of the renowned court orchestra in Meiningen during 1911-1914 and professor of composition and counterpoint in Leipzig until his death. Reger is well known for his monumental organ compositions and for his chamber and orchestra music. (M 16594)

(4348) Poulydamas

1988 RU. Discovered 1988 September 11 by C. S. Shoemaker and E. M. Shoemaker at Palomar.

Named for Hektor's {see planet (624)} most trusted advisor and strategist, Poulydamas, who was born on the same night as Hektor. The gods gave Hektor skill with arms and gave Poulydamas better judgment. He sensibly advised Hektor to lock the gates of Troy against Achilles {see planet (588)}, but Hektor

disregarded his friend's advice and went out of the city to his doom and to the eventual doom of Troy. (M 18141)
Name and citation provided by R. Preston at the request of the discoverers.

(4349) Tibúrcio

1989 LX. Discovered 1989 June 5 by W. Landgraf at La Silla.

Named in honor of Julio Cesar dos Santos Tibúrcio, a Brazilian amateur astronomer and student of information science. (M 16445)

(4350) Shibecha

1989 UG_1. Discovered 1989 October 26 by S. Ueda and H. Kaneda at Kushiro.

Named for a small town, located 47 km to the north of Kushiro {see planet (4096)}. Its cattle population, 45,500, is four times that of its human population. (M 19337)

(4351) Nobuhisa

1989 UR_1. Discovered 1989 October 28 by Y. Mizuno and T. Furuta at Kani.

Named in honor of Nobuhisa Kojima (1933-), who has been interested in astronomy since childhood. Kojima was the first Japanese amateur to make a Schmidt camera, and he discovered two comets (1970r and 1972j) photographically with a 0.30-m reflector of his own making. Kojima was also the first Japanese amateur to discover a minor planet that was given a provisional designation (1973 MA). Kojima is an inspiration for other astronomical discoverers. (M 20159)

(4352) Kyoto

1989 UW_1. Discovered 1989 October 29 by A. Sugie at Taga.

Named for the 1200th anniversary, in 1994, of the establishment by Emperor Kammu of Kyoto as the former capital of Japan. Home for many old national treasures and historical spots, Kyoto has been prosperous as a center of Japanese politics, economy and culture and nowadays attracts more than forty million visitors each year from all over the world. The Dynic Corporation, which built Dynic Astronomical Observatory, has its headquarter in Kyoto. (M 16249)

(4353) Onizaki

1989 WK_1. Discovered 1989 November 25 by Y. Mizuno and T. Furuta at Kani.

Named for the town where the second discoverer spent his boyhood. Since the name vanished when neighboring towns were united, it is particularly appropriate to revive it so that it can be the name of some location in the universe for ever. The town was situated on the Chita Peninsula in the central part of Japan and is famous for its beautiful yacht harbor. (M 16249)

(4354) Euclides

2142 P-L. Discovered 1960 September 24 by C. J. van Houten and I. van Houten-Groeneveld at Palomar.

Named for the Greek mathematician Euclid, who lived in Alexandria about 300 B.C. (M 16594)
Euclid is also honored by a lunar crater.

(4355) Memphis

3524 P-L. Discovered 1960 October 17 by C. J. van Houten and I. van Houten-Groeneveld at Palomar.
Named after the former capital of the old Egyptian kingdom. (M 18141)

(4356) Marathon

9522 P-L. Discovered 1960 October 17 by C. J. van Houten and I. van Houten-Groeneveld at Palomar.

Named after a village near Athens, near which the Athenian troops defeated the invading Persian army in the year 490 B.C. (M 18141)

(4357) Korinthos

2069 T-2. Discovered 1973 September 29 by C. J. van Houten at Palomar. Named after the important seaport town in classical Greece. (M 18141)

(4358) Lynn

A909 TF. Discovered 1909 October 5 by P. H. Cowell at Greenwich.

Named in memory of William Thynne Lynn (1835-1911), assistant at Greenwich Observatory from 1854 to 1880 and author of several popular books. He is best known, however, for the numerous short notes on astronomical matters he wrote in *The Observatory* for many years. Among those notes is an enigmatic one in 1902 in which he hinted that the Perseid comet P/Swift-Tuttle {see also planets (5035), (5036)} might be identical with Kegler's {see planet (5005)} comet of 1737. (M 22501)

Name proposed by B. G. Marsden, who made the identifications for this object.

(4359) Berlage

1935 TG. Discovered 1935 September 28 by H. van Gent at Johannesburg.

Named in honor of H. P. Berlage (1896-1968), Dutch meteorologist and astronomer. His main astronomical activities were investigations about the origin of the solar system. (M 18141)

Berlage is also honored by a lunar crater.

(4360) 1964 TG_2

Discovered 1964 October 9 at the Purple Mountain Observatory at Nanking.

(4361) Nezhdanova

1977 TG_7. Discovered 1977 October 9 by L. I. Chernykh at Nauchnyj.

Named in memory of Antonina Vasil'evna Nezhdanova (1873-1950), an outstanding Soviet singer, actress at the Bolshoj Theater and professor at the Moscow Conservatory. (M 18458)

(4362) Carlisle

1978 PR_4. Discovered 1978 August 1 by P. Jekabsons at Bickley.

Named in honor of Albert John Carlisle (1917-1993), who spent his life working in the Australian Bush. As a consequence, John Carlisle found more than 9000 meteorites on the Nullarbor Plain of Western Australia during the past 52 years. In 1982 he was awarded an Order of Australia Medal for his contribution to science. (M 22829)

(4363) Sergej

1978 TU_7. Discovered 1978 October 2 by L. V. Zhuravleva at Nauchnyj.

Named in honor of Sergej Vasil'evich Ezhov (1953-   ), since 1978 a talented surgeon in the district hospital at Bakhchisaraj {see planet (3242)}, Crimea. (M 30798)

(4364) Shkodrov

1978 VV_5. Discovered 1978 November 7 by E. F. Helin and S. J. Bus at Palomar.

Named for Vladimir Shkodrov, professor at the Bulgarian Academy of Sciences and an astronomer at the Bulgarian National Observatory. Dynamicist, science writer and translator of many astronomical classics, he has been a guiding light for many of his colleagues in Bulgaria and Europe. Shkodrov has collaborated with the discoverer on the International Near-Earth Asteroid Survey from Bulgaria and even through difficult times has persisted in carrying out asteroid and comet research. (M 18645)

(4365) Ivanova

1978 VH_8. Discovered 1978 November 7 by E. F. Helin and S. J. Bus at Palomar.

Named in honor of Violetta Ivanova, astronomer at the Bulgarian Academy of Sciences, Bulgarian National Observatory and a friend and colleague of the discoverer. Violetta and her associates have been active members of Helin's JPL International Near-Earth Asteroid Survey for the last ten years, contributing to its global sky coverage in search for NEAs. Violetta is a diligent and enthusiastic observer. Helin has spent two observing sessions in Bulgaria working with Ivanova and Shkodrov {see planet (4364)}. (M 18645)

(4366) Venikagan

1979 YV_8. Discovered 1979 December 24 by L. V. Zhuravleva at Nauchnyj.

Named in memory of the outstanding Soviet mathematician Veniamin Fyodorovich Kagan (1869-1953). (M 24122)

(4367) Meech

1981 EE_{43}. Discovered 1981 March 2 by S. J. Bus at Siding Spring.

Named in honor of Karen J. Meech (1959-) of the Institute for Astronomy of the University of Hawaii for her pioneering studies of comets very far from the sun. Her work following new and long-period comets to great distance has been a major factor in changing our ideas about water as the predominant driver of cometary activity in most comets. Meech's studies of (2060) = 95P/Chiron have likewise been critical in changing our understanding of the nature of the cometary coma. (M 27459)

Citation provided by M. F. A'Hearn following a suggestion by E. Bowell.

(4368) Pillmore

1981 JC_2. Discovered 1981 May 5 by C. S. Shoemaker at Palomar.

Named for Charles L. Pillmore, geologist with the U.S. Geological Survey. A specialist in the use of aerial photographs in geologic mapping, Pillmore is also an expert on the geology of the Raton Basin of Colorado and New Mexico, where he discovered the Cretaceous-Tertiary boundary claystone at about 20 different sites. Investigation of these sites has produced critical evidence that a large impact occurred at the end of the Cretaceous, coincident with a mass extinction of species. (M 17656)

(4369) Seifert

1982 OR. Discovered 1982 July 30 by L. Brožek at Kleť.

Named in memory of Jaroslav Seifert {1901-1986}, Czech poet and {1984} Nobel laureate. One of his volumes of poems had the name "Halleys's comet". (M 25443)

(4370) Dickens

1982 SL. Discovered 1982 September 22 by E. Bowell at Anderson Mesa.

Named for the great English novelist Charles Dickens (1812-1870). Dickens' works include *The Pickwick Papers, Little Dorrit, A Christmas Carol, Oliver Twist, Great Expectations* and *David Copperfield*, the last of which was based on his personal experiences. He created several comic characters, but he was most concerned with the social consequences of crime, corruption and moral evil. Most of his novels were first published in monthly installments, for popular consumption. Dickens later made successful reading tours of England and the United States, and several of his novels have been dramatized on the stage and on film. (M 16445)

Name suggested and citation provided by S. J. Ostro.

(4371) Fyodorov

1983 GC_2. Discovered 1983 April 10 by L. I. Chernykh at Nauchnyj.

Named in honor of Svyatoslav Nikolaevich Fyodorov, famous Moscow ophthalmologist, outstanding surgeon and brilliant manager who organized several clinics in Moscow and other cities for the treatment of eye diseases. (M 22502)

(4372) Quincy

1984 TB. Discovered 1984 October 3 at the Oak Ridge Observatory at Harvard.

Named in memory of John Quincy Adams (1767-1848), sixth president of the United States and a lifelong patron of astronomy. Instrumental in founding the Harvard College Observatory, he also first proposed - albeit unsuccessfully at the time - the establishment of an observatory under the aegis of the Smithsonian Institution. The minor planet is named on the occasion of the joint Harvard-Smithsonian Center for Astrophysics celebration of the sesquicentennial of the Harvard College Observatory and the centennial of the Smithsonian Astrophysical Observatory {see planet (3773)}, 1990 Nov. 2-4. (M 17224)

(4373) Crespo

1985 PB. Discovered 1985 August 14 by E. Bowell at Anderson Mesa.

Named in honor of Antonio Crespo, electrical engineer at the Arecibo {see planet (4337)} Observatory. For more than a decade, Crespo has been singularly responsible for maintaining the operational capability of the high-power transmitter used for all Arecibo planetary radar astronomy observations. The transmitter system is extremely complex; its outstanding performance throughout the 1980s is a credit to Crespo's expertise and hard work. (M 16446)
Name suggested and citation provided by S. J. Ostro.

(4374) Tadamori

1987 BJ. Discovered 1987 January 31 by K. Suzuki and T. Urata at Toyota.

Named for a Japanese commander of the Heike, Taira-no Tadamori (1096-1153), father of Taira-no Kiyomori {see planet (4375)}. Tadamori was a man well-educated in classics, and he laid the foundation for his son's assumption of control over the country. (M 19696)

(4375) Kiyomori

1987 DQ. Discovered 1987 February 28 by T. Niijima and T. Urata at Ojima.

Named for one of the most brilliant Japanese autocrats in the late Heian era, Taira-no Kiyomori (1118-1181). A hero of Tales of the Heike, Kiyomori succeeded his father, Taira-no Tadamori {see planet (4374)}, as chief of the Taira clan and rose to the Dajodaijin, which was the highest position in the Japanese court. In the year before his death, he crowned his two-year-old grandson as emperor and moved the capital to Kobe, where Japanese modern foreign trade has been developed. (M 19696)

(4376) Shigemori

1987 FA. Discovered 1987 March 20 by T. Niijima and T. Urata at Ojima.

Named for a Japanese military commander, Taira-no Shigemori (1138-1179), the eldest son of Taira-no Kiyomori {see planet (4375)}. Shigemori was a gentle and popular Naidaijin, Minister of Interior; after his death, the Taira clan quickly went to ruin. (M 19696)

(4377) Koremori

1987 GD. Discovered 1987 April 4 by T. Niijima and T. Urata at Ojima.

Named for Taira-no Koremori (1160-?), a Japanese military commander who was the eldest son of Taira-no Shigemori {see planet (4376)}. In 1183 he was beaten by the Genji (Minamoto) forces. It is believed that he drowned himself in the Pacific Ocean. (M 19696)

(4378) Voigt

1988 JF. Discovered 1988 May 14 by W. Landgraf at La Silla.

Named in honor of Hans-Heinrich Voigt {1921- }, known for his work on radiative transfer and stellar atmospheres, as well as for his studies of Cp, Ap and magnetic stars. An outstanding teacher, he has been a great inspiration to the discoverer. Voigt served as director of the Göttingen Observatory from 1975 to 1983 and as presidents of the Göttingen Academy of Sciences and IAU Commissions 38 and 46. (M 16446)

(4379) Snelling

1988 PT_1. Discovered 1988 August 13 by C. S. Shoemaker and E. M. Shoemaker at Palomar.

Named in honor of Reginald and Heather Snelling. Former owners of Carranya Station in Western Australia and now owners of Leigh Creek Station in South Australia, the Snellings have been staunch friends and supporters of the discoverers in their studies of Australian impact structures. It was "Reggie" Snelling who first called attention to another very small crater (Snelling Crater) on Carranya Station near Wolfe Creek Crater. These two Asutralians typify the independence, warmth and generosity the discoverers have come to think of as a national trait. (M 29669)

(4380) Geyer

1988 PB_2. Discovered 1988 August 14 by E. W. Elst at St. Michel.

Named in honor of Edward H. Geyer {1930- }, since 1965 the head of the Hoher List Observatory, on the occasion of his 60th birthday. Geyer has left his mark in many fields of astronomy, spanning from instrumental development via stellar systems to variable stars and solar-like activity. His work in solar system astronomy has included, for example, the detection of the splitting of the nucleus of Comet West (1976 VI). He is also an enthusiastic teacher, especially in the field of observational astronomy. Like many other visiting astronomers at Hoher List, the discoverer has often been supported by Geyer, long a friend of minor planets. (M 16249)

Citation prepared by M. Geffert at the request of the discoverer.

(4381) Uenohara

1989 WD_1. Discovered 1989 November 22 by N. Kawasato at Uenohara.

Named for the small town on the edge of which is the observing station where this minor planet was discovered. The town, in Yamanashi prefecture, is located in central Japan, about 100 km from Tokyo. (M 16446)

(4382) Stravinsky

1989 WQ_3. Discovered 1989 November 29 by F. Börngen at Tautenburg.

Named in memory of the famous composer Igor Stravinsky (1882-1972), known for his very versatile musical creations, including ballet music and operas. The discoverer finds his psalm symphony (1930) particularly impressive. Born in Russia, Stravinsky lived in Switzerland and France before moving to the United States in 1939. (M 16446)

(4383) Suruga

1989 XP. Discovered 1989 December 1 by Y. Oshima at Gekko.

Named for the district where the Gekko {see planet (4261)} Observatory is located. Suruga is the old name of the central part of Shizuoka prefecture. The district is well known for its mild climate, and the view from the observatory covers the whole of beautiful Suruga Bay and its seashore. (M 18307)

(4384) 1990 AA

Discovered 1990 January 3 by T. Hioki and S. Hayakawa at Okutama.

(4385) Elsässer

2534 P-L. Discovered 1960 September 24 by C. J. van Houten and I. van Houten-Groeneveld at Palomar.

Named in honor of Hans F. Elsässer (1929-), professor of astronomy at Heidelberg University and since 1968 first director of the Max-Planck-Institut für Astronomie. During 1962-1975 he was also director of the Heidelberg Observatory at Königstuhl. A scientific member of the Max-Planck-Gesellschaft, Elsässer was a founder of MPIA in Heidelberg and its associated observatory at Calar Alto. He has made important contributions to the study of interplanetary matter and the zodiacal light, the optics of the earth's atmosphere, the structure of the Galaxy and the Magellanic Clouds and star formation. He was deeply concerned with the design and establishment of large telescopes and their auxiliary instrumentation at Calar Alto. Elsässer served as president of the IAU Commission 21 during 1970-1973. His many honors include membership in the scientific academies of Austria, Halle (Leopoldina) and Heidelberg. (M 18141)
Name proposed and citation prepared by L. D. Schmadel.

(4386) Lüst

6829 P-L. Discovered 1960 September 26 by C. J. van Houten and I. van Houten-Groeneveld at Palomar.

Named in honor of Reimar Lüst (1923-), German astronomer, former director general of the Max-Planck-Gesellschaft and of the European Space Agency. (M 18141)

(4387) Tanaka

4829 T-2. Discovered 1973 September 19 by C. J. van Houten and I. van Houten-Groeneveld at Palomar.

Named in honor of Yasuo Tanaka (1931-), Japanese astrophysicist, who organized the first group of space scientists in the Netherlands. Tanaka is head of the space research group in Japan. He contributed to the launching of the Japanese satellites Hakucho (1979), Tenma (1983) and Ginga (1987). Ginga observed the first X-ray signals from supernova 1987A. (M 18142)

(4388) 1964 VE

Discovered 1964 November 3 at the Goethe Link Observatory at Brooklyn, Indiana.

(4389) Durbin

1976 GL_3. Discovered 1976 April 1 by N. S. Chernykh at Nauchnyj.

Named in honor of Deanna Durbin (1921-), talented Canadian-born American cinema actress, who played in many films in the 1930s and 40s. She was sensationally popular in America and Europe and later on in Russia. (M 23351)

(4390) Madreteresa

1976 GO_8. Discovered 1976 April 5 at the Felix Aguilar Observatory at El Leoncito.

Named in memory of Albanian-born Ines Gonxha Bojaxhiu (1910-1997), better known as Mother Teresa of Calcutta, who devoted her life to helping sick, destitute and homeless people, first in India and then in the rest of the world. Among her many awards, in 1979 she received the Nobel Peace Prize. (M 31609)

(4391) Balodis

1977 QW_2. Discovered 1977 August 21 by N. S. Chernykh at Nauchnyj.

Named in honor of Janis Balodis, chief of the cosmic geodesy department at the Astronomical Observatory of the Latvian University, known for his work on astrometric and laser observations of artificial satellites and on methods of mathematical reductions in photographic astrometry. A set of his computer programs has been used in the Crimean minor planet service for many years. (M 20837)

(4392) Agita

1978 RX_5. Discovered 1978 September 13 by N. S. Chernykh at Nauchnyj.

Named in honor of Agita Tarasova, a scientific worker at the Astronomical Observatory of the Latvian University, engaged in the mathematical reduction of observations and in the preparation of computer programs. She has rendered valuable assistance to the Crimean minor planet service by installing the Balodis {see planet (4391)} reduction programs on the CrAO computer. (M 20837)

(4393) Dawe

1978 VP_8. Discovered 1978 November 7 by E. F. Helin and S. J. Bus at Palomar.

Named in honor of John A. Dawe, who was astronomer-in-charge at the U.K. Schmidt in 1979 when the first discoverer proposed that plates be taken for a limited asteroid survey. Dawe and his staff of astronomers were hospitable and helpful during that first short survey, and they later collaborated in the more comprehensive and highly successful United Kingdom-Caltech Faint Asteroid Survey (UCAS) that was carried out in 1981. (M 17031)

(4394) 1981 EB_{19}

Discovered 1981 March 2 by S. J. Bus at Siding Spring.

(4395) 1981 EH_{41}

Discovered 1981 March 2 by S. J. Bus at Siding Spring.

(4396) Gressmann

1981 JH. Discovered 1981 May 3 by E. Bowell at Anderson Mesa.

Named in honor of Michael Gressmann, renowned astronomical optician of Falkensee, near Berlin. Scholar and successor to the famous Alfred Wilke {see planet (4117)}, Gressmann has for the past 20 years manufactured numerous telescope optics both for amateur astronomers and for observatories and institutes. These high-quality products also contain special laser and infrared optics of Gressmann's own design. Founder and leader of an East German working group on asteroids, he is also a distinguished popularizer of astronomy in his country. (M 16446)

Name proposed by the discoverer following a suggestion from L. D. Schmadel, who also prepared the citation.

(4397) Jalopez

1981 JS_1. Discovered 1981 May 9 at the Felix Aguilar Observatory at El Leoncito.

Named in honor of Jose Augusto Lopez (1922-), known for his work in meridian astrometry and as director of the Felix Aguilar Observatory during 1966-1993. He also served on the Science and Technology Committee of the government of San Juan province, as well as an advisor to the Argentine government. (M 31023)

(4398) Chiara

1984 HC_2. Discovered 1984 April 23 by W. Ferreri at La Silla.

Named in honor of Chiara Maria Faletti (1955-), wife of the discoverer. (M 23540)

(4399) Ashizuri

1984 UA. Discovered 1984 October 21 by T. Seki at Geisei.

Named for a famous promontory in the southwestern part of Shikoku {see planet (4223)} island. Many tourists visit the promontory for its beautiful ocean view, but it is often subjected to summer typhoons. (M 16595)

(4400) Bagryana

1985 QH_4. Discovered 1985 August 24 by V. G. Ivanova and V. G. Shkodrov at Rozhen.

Named in honor of Elisveta Bagryana {1893- }, the greatest Bulgarian poetess. Her lyric poetry seems to have developed outside literary schools, and she was the first in Bulgarian literature to give a bold portrait of women and their nature; her poems have been translated into more than 25 languages. Bagryana was a great admirer of astronomy; in 1984, she wrote: "Now, after outliving my 90th birthday, my only will is to see for a second time Halley's comet, whose appearance when I was 17 years old aroused my incessant thirst to reveal the mysteries of the vast Universe." Her wish was fulfilled. (M 19696)

(4401) Aditi

1985 TB. Discovered 1985 October 14 by C. S. Shoemaker and E. M. Shoemaker at Palomar.

Aditi was an Indian goddess of the sky and the air. She was the mother of 33 gods and of the sun, and also the mother of the 12 Adityas, described in Vedic literature as the gods of the heavenly light. (M 17656)

(4402) Tsunemori

1987 DP. Discovered 1987 February 25 by T. Niijima and T. Urata at Ojima.

Named for the Japanese military commander in the late Heian era, Taira-no Tsunemori (1125-1185), a son of Tadamori {see planet (4374)}. As the elder of the Heike after the dead of Kiyomori {see planet (4375)}, he made an effort to restore his family. However, he was beaten at the battle of Dannoura, and he drowned in the sea. (M 31023)

Name proposed by the second discoverer.

(4403) Kuniharu

1987 EA. Discovered 1987 March 2 by Y. Oshima at Gekko.

Named for a sister observatory that also belongs to the International Foundation for Cultural Harmony. The Kuniharu Observatory was founded in 1958 in Okazaki, Aichi prefecture, and is also active in the local popularization of astronomy. (M 18307)

(4404) Enirac

1987 GG. Discovered 1987 April 2 by A. Maury at Palomar.

Named in honor of the discoverer's spouse Carine, herself an amateur astronomer. Always wanting to share her interest in astronomy with others, she founded an astronomy club in Nice in 1978 and has presided over summer astronomy camps for teenagers during several subsequent summers. Married since 1982 and now the mother of two, she has long been and continues to be a support to her husband's activities. (M 16595)

The name is Carine spelled backwards. Otherwise there would exist the danger of confusion with planet (491) Carina.

(4405) Otava

1987 QD_1. Discovered 1987 August 21 by Z. Vávrová at Kleť.

Named for a Czech river that has its confluence with the Vltava {see planet (2123)} river near the Zvíkov castle. The Otava was considered to bear both gold and pearls in ancient and medieval times. (M 34620)

(4406) Mahler

1987 YD_1. Discovered 1987 December 22 by F. Börngen at Tautenburg.

Named in honor of the great Austrian composer, conductor and opera-director

Gustav Mahler (1860-1911). Mahler's creative endeavor was concentrated mainly to ten daring and unusual symphonies, supplemented by some lieder-groups, e.g. the four-part cycle "Lieder eines fahrenden Gesellen" and the cycle "Kindertotenlieder". (M 16446)

(4407) Taihaku

1988 TF_1. Discovered 1988 October 13 by M. Koishikawa at Sendai.

Named for the southwest part of the city of Sendai {see planet (3133)}, a scenic recreation area known as Taihaku-ku in which interesting Old Stone Age relics have been discovered. In ancient times, Venus was called "Taihaku" in Japan, and folklore says that this area's Mt. Taihaku was built up by the fall of Venus. (M 19697)

(4408) Zlatá Koruna

1988 TH_2. Discovered 1988 October 4 by A. Mrkos at Kleť.

Named for a Gothic monastery and a village in southern Bohemia east of Kleť Mountain. The Cistercian monastery of the Holy Crown of Thorns was founded there by the Czech king Přemysl Otakar II in 1263. The name of Kleť for the mountain was first mentioned in the monastery's thirteenth-century documents. There is now a literature museum at the monastery. (M 29670)
Name proposed by J. Tichá, M. Tichý and Z. Moravec.

(4409) 1989 MD

Discovered 1989 June 30 by A. C. Gilmore and P. M. Kilmartin at Lake Tekapo.

(4410) Kamuimintara

1989 YA. Discovered 1989 December 17 by S. Ueda and H. Kaneda at Kushiro.

Kamuimintara is the name given by the Ainu, Hokkaido's indigenous people, to Hokkaido's highest peak, Mt. Taisetsu. In the Ainu language the name means 'Playground of the Gods'. (M 22829)
Name proposed by J. Ueda.

(4411) Kochibunkyo

1990 AF. Discovered 1990 January 3 by T. Seki at Geisei.

Named for the association of volunteers active in education, inaugurated in the city of Kochi in April 1948, and which puts a great deal of effort into the popularization of astronomy. (M 25443)

(4412) Chephren

2535 P-L. Discovered 1960 September 26 by C. J. van Houten and I. van Houten-Groeneveld at Palomar.

Chephren was a pharaoh of the old Egyptian kingdom. He reigned about 2500 B.C. and built the second largest pyramid near Giza {see planet (5249)}. He is also known because of his monumental sphinx {see also planet (896)}. (M 18142)

(4413) Mycerinos

4020 P-L. Discovered 1960 September 24 by C. J. van Houten and I. van Houten-Groeneveld at Palomar.

Mycerinos was pharaoh of the old Egyptian kingdom of the fourth dynasty. He reigned about 2500 B.C. and built the third largest pyramid near Giza {see planet (5249)}. (M 18142)

Mycerinos is the Greek name of the Egyptian pharao Menkaure {see also planet (4568)}.

(4414) Sesostris

4153 P-L. Discovered 1960 September 24 by C. J. van Houten and I. van Houten-Groeneveld at Palomar.

Sesostris was the name of three pharaohs of the old Egyptian kingdom (twelfth dynasty). They reigned about 1900 B.C. During Sesostris I (1971-1926 B.C.) Egyptian architecture had one of its high points, with buildings in Karnak and Heliopolis. (M 18142)

(4415) Echnaton

4237 P-L. Discovered 1960 September 24 by C. J. van Houten and I. van Houten-Groeneveld at Palomar.

Echnaton was a pharaoh of the old Egyptian kingdom who reigned from 1375 to 1358 B.C. He started a religious reformation by replacing the worship of the traditional Egyptian gods by that of the single god Aten {see planet (2062)}. In his time Egyptian art was more colorful and more relaxed. (M 18142)

(4416) Ramses

4530 P-L. Discovered 1960 September 24 by C. J. van Houten and I. van Houten-Groeneveld at Palomar.

Ramses II, also called Ramesses, was a pharaoh of the old Egyptian kingdom who reigned from 1332 to 1298 B.C. He made war with the Hittites in Asia Minor and built large temple complexes at Abu Simbel. Ramses was one of the first names that Champollion {see planet (3414)} could read by translation of the Egyptian hieroglyphs. (M 18142)

(4417) Lecar

1931 GC. Discovered 1931 April 8 by K. Reinmuth at Heidelberg.

Named in honor of Myron Lecar (1930-), astrophysicist in the Planetary Sciences division of the Harvard-Smithsonian Center for Astrophysics. Lecar's research covers the dynamics of both solar system and stellar bodies, and the large-scale structure of the universe. (M 22502)

Named by the Minor Planet Names Committee following a suggestion by members of the Planetary Sciences division.

(4418) Fredfranklin

1931 TR_1. Discovered 1931 October 9 by K. Reinmuth at Heidelberg.

Named in honor of Fred Franklin (1932-), astronomer in the Planetary Sciences division of the Harvard-Smithsonian Center for Astrophysics. Franklin's researches have covered chaos in the solar system, the rings of Saturn, and mutual phenomena of planetary satellites. (M 22502)

Name proposed by members of the Planetary Sciences division, the identifications for this object having been made by B. G. Marsden.

(4419) Allancook

1932 HD. Discovered 1932 April 24 by K. Reinmuth at Heidelberg.

Named in honor of Allan F. Cook II (1922-), astrophysicist in the Planetary Sciences division of the Harvard-Smithsonian Center for Astrophysics, whose astronomical interests range from meteors and the earth's upper atmosphere, to Saturn's rings, to spectroscopy and photometry. (M 22502)

Named by the Minor Planet Names Committee following a suggestion by members of the Planetary Sciences division.

(4420) 1936 PB

Discovered 1936 August 15 by G. N. Neujmin at Simeïs.

(4421) Kayor

1942 AC. Discovered 1942 January 14 by K. Reinmuth at Heidelberg.

Named by G. V. Williams in honor of his parents, Kay and Roy, for their support over the years. This minor planet was discovered on the same day that the former Miss Merriman was born. (M 21131)

(4422) Jarre

1942 UA. Discovered 1942 October 17 by L. Boyer at Algiers.

Named in honor of the French composers Maurice (1924-) and Jean-Michel Jarre (1949-). Maurice studied composition at the Paris conservatoire, and wrote several music for concert works and boldly dramatic theatre music, before turning to film music, winning Oscars for his score to *Lawrence of Arabia* (1963) and for 'Lara's Theme' from *Doctor Zhivago* (1965). Jean-Michel abandoned his musical studies at the Conservatoire de Paris in 1967 to experiment with electronic music. His first commercial success was *Oxygene* (1977). Further successful recordings followed, and in 1981 he became the first Western rock artist to play concerts in China. Jean-Michel is best known for his spectacular live shows. (M 17031)

Name proposed by G. V. Williams, who made the principal identification involving this object.

(4423) 1949 GH

Discovered 1949 April 4 at the Goethe Link Observatory at Brooklyn, Indiana.

(4424) Arkhipova

1967 DB. Discovered 1967 February 16 by T. M. Smirnova at Nauchnyj.

Named in honor of Irina Konstantinovna Arkhipova (1925-), an outstanding representative of the Russian vocal school. For more than 30 years she was a soloist at the Bol'shoj Theatre and sang in many well-known opera houses around the world. In 1995 there will be a celebration of her career. A professor at the Moscow Conservatoire, she educated a constellation of outstanding singers. Arkhipova is an eminent public figure who established a special charitable fund. Since 1967 she has been chief judge of the M. I. Glinka {see planet (2205)} competition, and on several occasions she chaired the international jury for the P. I. Chajkovskij (Tchaikovsky) {see planet (2266)} vocal competition. (M 24916)
Name proposed by the Institute of Theoretical Astronomy.

(4425) 1967 UQ

Discovered 1967 October 30 by L. Kohoutek at Bergedorf.

(4426) Roerich

1969 TB_6. Discovered 1969 October 15 by L. I. Chernykh at Nauchnyj.

Named in honor of the family of outstanding Russian figures of culture whose lives were closely connected with India: Nikolaj Konstantinovich Roerich (1874-1947), painter, writer, philosopher, archaeologist, public figure and enlightener; his wife Elena Ivanovna Roerich (1879-1955), participant in all his expeditions, writer, philosopher; their elder son Yurij Nikolaevich Roerich (1902-1960), scientist in oriental studies and philology; and their younger son, Svyatoslav Nikolaevich Roerich (1904-1993), painter and public figure. (M 22502)

(4427) Burnashev

1971 QP_1. Discovered 1971 August 30 by T. M. Smirnova at Nauchnyj.

Named in honor of Vladislav Ivanovich Burnashev (1943-), staff scientist at the Crimean Astrophysical Observatory and an expert in stellar photometry and spectrophotometry. The name also honors his wife, Bella Alekseevna Burnasheva (1944-), former minor-planet observer on the staff of the Institute of Theoretical Astronomy. She is now a systems programer at the Crimean Astrophysical Observatory. (M 30475)

(4428) Khotinok

1977 SN. Discovered 1977 September 18 by N. S. Chernykh at Nauchnyj.

Named in honor of Roman L'vovich Khotinok, meteorite researcher in the Meteorite Committee of the Russian Academy of Sciences, who has investigated many sites of possible meteorite falls and has discovered more than 20 different meteorites that are now stored in the Moscow collection. (M 22502)

(4429) Chinmoy

1978 RJ_2. Discovered 1978 September 12 by N. S. Chernykh at Nauchnyj.

Named in honor of Sri Chinmoy, Bengali poet, artist and philosopher, preacher of peace, who travels the world, inspiring peace-loving peoples with his music, poetry and works of art. (M 23352)
Name proposed by some Russian admirers of Sri Chinmoy.

(4430) Govorukhin

1978 SX_6. Discovered 1978 September 26 by L. V. Zhuravleva at Nauchnyj.

Named in honor of the well-known Russian cinematic producer and playwright Stanislav Sergeevich Govorukhin (1936-). (M 24122)

(4431) 1978 WU_{14}

Discovered 1978 November 28 at the Purple Mountain Observatory at Nanking.

(4432) McGraw-Hill

1981 ER_{22}. Discovered 1981 March 2 by S. J. Bus at Siding Spring.

Named after the 1.3-m McGraw-Hill telescope located on the southwest ridge of Kitt Peak, Arizona, which was the site for the first physical observations for this minor planet. The telescope is operated by a consortium comprising the University of Michigan, Dartmouth College and the Massachusetts Institute of Technology. Originally erected at Stinchfield Woods near Dexter, Michigan, in July 1969, the telescope was moved to its current location in 1975 through the generous financial support of McGraw-Hill Incorporated and the Sloan Foundation. (M 19697)
Name proposed and citation provided by R. P. Binzel.

(4433) Goldstone

1981 QP. Discovered 1981 August 30 by E. Bowell at Anderson Mesa.

Named for the Goldstone Deep Space Communications Complex in Southern California's Mojave Desert. Goldstone is part of the National Aeronautics and Space Administration's Deep Space Network, which is managed, technically directed and operated by the Jet Propulsion Laboratory of the California Institute of Technology. Since 1962, the DSN has played a critical role in NASA's exploration of the solar system using unmanned spacecraft. The radio antennas at Goldstone and at similar facilities near Canberra, Australia, and Madrid, Spain, are used to receive spacecraft telemetry and to transmit commands to the spacecraft. The largest antenna at Goldstone, Deep Space Station 14, is also used for planetary radar astronomy. It was in 1968 that Goldstone observations of (1566) Icarus, carried out by R. M. Goldstein {see planet (5393)}, yielded the first detection of radar echoes from a minor planet. (M 16447)
Name suggested and citation provided by S. J. Ostro.

(4434) Nikulin

1981 RD_5. Discovered 1981 September 8 by L. V. Zhuravleva at Nauchnyj.

Named in honor of Yurij Vladimirovich Nikulin (1921-), a Russian circus artist, clown, film star. (M 24122)

(4435) Holt

1983 AG_2. Discovered 1983 January 13 by C. S. Shoemaker at Palomar.

Named for Henry E. Holt, planetary geologist with the U.S. Geological Survey and Northern Arizona University. As a member of the Surveyor and Apollo scientific teams, he investigated the detailed geology and photometric properties of the lunar surface. Since retiring from the Geological Survey, he has been a principal participant in the Palomar Asteroid and Comet Survey and is the discoverer or codiscoverer of six comets and numerous asteroids, including (4581) Asclepius, the numbered asteroid that has made the closest approach to the earth. (M 17656)

(4436) 1983 EX

Discovered 1983 March 9 by E. Barr at Anderson Mesa.

(4437) Yaroshenko

1983 GA_2. Discovered 1983 April 10 by L. I. Chernykh at Nauchnyj.

Named in memory of Nikolaj Aleksandrovich Yaroshenko (1846-1898), well-known Russian painter, a member and one of the leaders of the nineteenth-century Russian Society of Traveling Art Exhibitions. (M 18458)

(4438) Sykes

1983 WR. Discovered 1983 November 29 by E. Bowell at Anderson Mesa.

Named in honor of Mark V. Sykes, planetary scientist at the Steward Observatory of the University of Arizona, Tucson. Sykes was the first to suggest that the dust bands discovered in data from the Infrared Astronomical Satellite (IRAS) {see planet (3728)} were due to the catastrophic disruptions of small asteroids and comets. He has also discovered several additional dust bands, a second type of dust trail, and identified parent comets responsible for some of the IRAS dust trails. (M 18458)

Citation provided by E. F. Tedesco at the request of the discoverer.

(4439) Muroto

1984 VA. Discovered 1984 November 2 by T. Seki at Geisei.

Named for a small seaside city in the southeastern part of the Japanese island of Shikoku {see planet (4223)}, famous as a sightseeing spot for the beautiful headland. (M 25443)

(4440) Tchantchès

1984 YV. Discovered 1984 December 23 by F. Dossin at St. Michel.

Named in honor of the folklore character Tchantchès (Walloon dialect for François), the typical Walloon hero of the region of the discoverer. Tchantchès is obstinate but has a great heart. (M 33385)

(4441) Toshie

1985 BB. Discovered 1985 January 26 by T. Seki at Geisei.

Named in memory of Toshie Seki, the discoverer's mother, who encouraged her son to hunt for comets. She died on the day this minor planet was discovered. (M 17657)

(4442) Garcia

1985 RB_1. Discovered 1985 September 14 by the Spacewatch at Kitt Peak.

Named in memory of Jerry Garcia (1942-1995), lead guitarist for the "Grateful Dead". He was also a superb banjo player and steel pedal guitar player. His recorded work embrace many styles, among them rock-and-roll, bluegrass and country, and experimental electronic music. Garcia and the Grateful Dead have also sought to preserve endangered and underappreciated music. Garcia was best known for his consummate musicianship and for live performance and improvisations. Those embracing Garcia's music have now spanned several generations, attesting to the quality and timelessness of his music. (M 25976)

Name proposed by T. Gehrels, following a suggestion by S. Radford. Citation prepared by E. Olszewski and S. Radford.

(4443) 1985 RD_4

Discovered 1985 September 10 by H. Debehogne at La Silla.

(4444) 1985 SA

Discovered 1985 September 16 by H. U. Norgaard-Nielsen and L. Hansen and P. R. Christensen at La Silla.

(4445) 1985 TC

Discovered 1985 October 15 by K. Suzuki and T. Urata at Toyota.

(4446) Carolyn

1985 TT. Discovered 1985 October 15 by E. Bowell at Anderson Mesa.

Named in honor of Carolyn Spellmann Shoemaker, comet and asteroid discoverer. Shoemaker began searching for asteroids in 1980, using plates taken at the U.K. Schmidt Telescope at Siding Spring. She helped develop a new photographic survey program using the 0.46-m Schmidt camera at Palomar Mountain and a newly designed stereomicroscope, which greatly increased the efficiency of film scanning. In 1983 Shoemaker found her first near-earth asteroid, the Amor object (3199) Nefertiti, and later that year she found her first comet, 1983p. By February 1991 she had discovered 22 comets, at a rate of about one per 100 hours of scanning, and for discoveries recognized in the names of the comets she thus surpassed the tally of W. R. Brooks {see planet (2773)} and moved into the all-time second place behind J.-L. Pons. Shoemaker already holds the record for finding new periodic comets: 9 by early 1991. (M 18458)

Citation prepared by D. H. Levy and J. Mueller at the request of the discoverer.

(4447) Kirov

1985 VE_1. Discovered 1985 November 7 by E. Bowell at Anderson Mesa.

Named in honor of the Kirov Ballet and State Theater of St. Petersburg (formerly Leningrad). Directly descended from the Russian Imperial Ballet of St. Petersburg, the Kirov has always preserved the highest values of classical dance. In the early nineteenth century, an extraordinary synthesis of French grace, Italian virtuosity and Russian temperament produced within this school a standard of artistry admired by the world and imitated by generations of dancers. In the late nineteenth century, the spectacular choreography of Marius Petipa, particularly in his *Sleeping Beauty* and *Swan Lake*, made St. Petersburg the capital of the ballet world. The Kirov has produced some of the greatest dancers, including Tamara Karsavina, Anna Pavlova {see planet (3055)} and Vaslav Nijinsky. Ever a company of supreme elegance and impeccable taste, the Kirov performs with grace, technical ease, sensitivity and refined dramatic intensity. (M 17979)

Name suggested and citation provided by J. M. Ostro and S. J. Ostro.

(4448) Phildavis

1986 EO. Discovered 1986 March 5 by C. S. Shoemaker and E. M. Shoemaker at Palomar.

Named for Philip A. Davis, planetary geologist with the U.S. Geological Survey and chief of the Branch of Astrogeology since 1989. Davis has worked on the global geochemistry and petrology of the moon and the geology of Mars, Venus and Arabia. He is known especially for the techniques he has developed for the extraction of quantitative topographic information from spacecraft images and for analysing global patterns of geological and geophysical data. (M 17657)

(4449) Sobinov

1987 RX_3. Discovered 1987 September 3 by L. I. Chernykh at Nauchnyj.

Named in memory of Leonid Vital'evich Sobinov (1872-1934), a remarkable Russian singer, leading tenor soloist of the Bolshoj Theatre in Moscow. (M 18459)

(4450) Pan

1987 SY. Discovered 1987 September 25 by C. S. Shoemaker and E. M. Shoemaker at Palomar.

Originally a shepherd god of Arcady, Pan developed into a hunter, fisherman and warrior. He was worshipped at Athens after the Battle of Marathon {see planet (4356)}, where he aided the Athenians by spreading panic through the Persian armies. He has been represented in art as a horned half-man, half goat. (M 17657)

The name Pan has also been given to the satellite Saturn XVIII.

(4451) Grieve

1988 JJ. Discovered 1988 May 9 by C. S. Shoemaker and E. M. Shoemaker at Palomar.

Named for Richard A. F. Grieve, geologist with the Geological Survey of Canada and chief of their geophysical surveys. Grieve is a leading investigator of impact craters on the earth and moon. He maintains the global census of recognized terrestrial impact structures and has spearheaded the effort to obtain reliable ages for these structures. He also has been a leader in classifying the impacting bodies on the basis of siderophile element abundance patterns in impact melt rocks. (M 4451)

(4452) 1988 RN

Discovered 1988 September 7 by P. Jensen at Brorfelde.

(4453) 1988 VC

Discovered 1988 November 3 by P. Jensen at Brorfelde.

(4454) Kumiko

1988 VW. Discovered 1988 November 2 by S. Ueda and H. Kaneda at Kushiro.

Named in honor of Kumiko Kaneda (1951-), wife of the second discoverer. (M 19337)

(4455) Ruriko

1988 XA. Discovered 1988 December 2 by S. Ueda and H. Kaneda at Kushiro.

Named in honor of Ruriko Ueda (1957-), wife of the first discoverer. (M 19337)

(4456) Mawson

1989 OG. Discovered 1989 July 27 by R. H. McNaught at Siding Spring.

Named in memory of Sir Douglas Mawson (1882-1958), Australian geologist and Antarctic explorer. After graduating in engineering from the University of Sydney in 1902, Mawson gained valuable geological field experience on an expedition to the New Hebrides, returning to Sydney to take a geology degree in 1905. From that year onward he was associated with the University of Adelaide, being professor of geology from 1921 to 1952, making pioneering studies of uranium ores and, later, the Precambrian rocks of the Flinders Ranges. During 1907-1909 he took part in the British Antarctic Expedition led by Shackleton, and from 1911 to 1914 he led the Australasian Antarctic Expedition; on the latter was found the "Adelie Land" meteorite, the first to be discovered on the Antarctic continent. *The Home of the Blizzard*, his account of that expedition and his near-death, has become a classic. Throughout his career he promoted active fieldwork and continued his involvement in Antarctic exploration; after his retirement he was involved in planning for the IGY. (M 17979)

Named by the discoverer and endorsed by D. I. Steel, who, along with R. G. Ferguson, provided additional citation material.

(4457) van Gogh

1989 RU. Discovered 1989 September 3 by E. W. Elst at St. Michel.

Named in memory of the famous Dutch painter Vincent van Gogh (1853-1890) on the 100th anniversary of his death in Auvers sur Oise by his own hand. Deeply concerned with human suffering, van Gogh went in 1879 to the Borinage, a poor mining district in southern Belgium, to do some missionary work. It was there that he discovered his true vocation, not as a preacher but as a painter. This first great spiritual crisis in his life is testified by his early dark realistic pictures, which bear a social-critical message. In 1886 he went to Paris, where he met the "pointillists". In his later works color acquired a special function, providing a vision that influenced the upcoming expressionistic generation in the twentieth century. During his period in the Provence his landscapes expressed more and more his inner emotions. The colors become extremely vivid, although he was always in search of rest and harmony. (M 15595)

Citation prepared by Kristina Leterme, at the request of the discoverer.

(4458) Oizumi

1990 BY. Discovered 1990 January 21 by Y. Kushida and O. Muramatsu at Yatsugatake.

Named for the beautiful village in which Yatsugatake {see planet (4033)} South Base Observatory is located. The name means "great spring". The village, surrounded by forest and located at the southern end of the Yatsugatake-renpo mountain range in central Japan, has many springs, the greatest of which produces 7 million liters of pure water per day. (M 17657)

(4459) Nusamaibashi

1990 BP_2. Discovered 1990 January 30 by M. Matsuyama and K. Watanabe at Kushiro.

Named for the bridge spanning the Kushiro {see planet (4096)} river and connecting the northern and southern parts of Kushiro city. When it was first built in 1889 the bridge was named Aihoku Bashi and was the longest in Hokkaido {see planet (3720)}. The bridge has been rebuilt five times, most recently in 1976, and it now has a length of 124 m and a width of 34 m. The elegant Nusamai Bashi is known for its statues, by four well-known Japanese sculptors, of female representations of the seasons. (M 16885)

Name suggested by the first discoverer.

(4460) Bihoro

1990 DS. Discovered 1990 February 28 by K. Endate and K. Watanabe at Kitami.

Named for the city, home of the first discoverer, at the junction of the Bihoro and Abashiri {see planet (4263)} rivers, the latter running near the Kitami {see planet (3785)} station. Some 30 km from the sea of Okhotsk {see planet (4042)}, the city has a population of approximately 26,000, and its main industries are beets and potatoes. The Bihoro pass, 525 m above sea level, is in the Akan {see planet (4584)} national park 20 km from the nearest town and is renowned for its grand panorama that includes the Kussharo lake. (M 16595)

(4461) Sayama

1990 EL. Discovered 1990 March 5 by A. Sugie at Taga.

Named in honor of a city located in the Musashino hills in the center of the Kantou {see planet (3249)} plain. In the southeastern part of Saitama prefecture, Sayama is on the Iruma river and famous for producing good Japanese green tea. Since 1954 Sayama has developed with Irumagawa city and some villages into a new industrial area and as a residential area for the commuters of Tokyo. Sayama observes and promotes various traditional cultures and is noted in particular for its Star Festival ("Tanabata"), when the main streets are beautifully decorated with bamboo trees. (M 17031)

(4462) Vaughan

1952 HJ_2. Discovered 1952 April 24 at the McDonald Observatory at Fort Davis.

Named in honor of Curtis T. Vaughan, Jr., a lifelong resident, prominent businessman, and community leader of San Antonio, Texas. His early interests in science led to a physics degree at Harvard, but the need to direct a family enterprise precluded science as a career. Nevertheless, his interests and activities have remained strongly scientific, especially astronomical, as he built an outstanding collection of early astronomical instruments and constructed one of the finest private observatories in the United States, with now a fully computer-controlled 0.40-m reflector. For more than 20 years his uniquely strong encouragement and support, including financial, of astronomy at the University of Texas have played a major role in helping the astronomy program there grow into one of the preeminent such programs in the country. (M 20521)
Name proposed by the late Harlan J. Smith.

(4463) 1954 UO_2

Discovered 1954 October 28 at the Goethe Link Observatory at Brooklyn, Indiana.

(4464) Vulcano

1966 TE. Discovered 1966 October 11 by N. S. Chernykh at Nauchnyj.

Named for one of the Aeolian Islands, site of the "original" volcano, a splendid place where the workshop "Beginning the Spaceguard Survey" was held by the IAU Working Group on Near-Earth Objects during 1995 Sept. 18-22. (M 26762)

(4465) Rodita

1969 TD_5. Discovered 1969 October 14 by B. A. Burnasheva at Nauchnyj.

Named in memory of the Soviet art critic Tat'yana Mikhajlovna Rodina (1914-1989). She was the author of *Russian Theatre Art at the Beginning of the 19th Century* and *Blok and Russian Theatre of the Beginning of the 20th Century*, as well as of numerous papers in the *Great Soviet Encyclopedia* and the *Theatre Encyclopedia*. (M 28089)

(4466) Abai

1971 SX_1. Discovered 1971 September 23 at the Crimean Astrophysical Observatory at Nauchnyj.

Named in honor of Abai Kunanbaev (1845-1904), Kazakh poet, philosopher and enlightener, founder of Kazakh literature, and translator of the works of I. A. Krylov, A. S. Pushkin {see planet (2208)} and M. Yu. Lermontov {see planet (2222)}. (M 23352)

(4467) Kaidanovskij

1975 VN_2. Discovered 1975 November 2 by T. M. Smirnova at Nauchnyj.

Named in honor of Naum L'vovich Kaidanovskij (1907-), a pioneer of radio astronomy, who began research in solar and galactic radio astronomy as early as 1948. In 1955 he co-founded the radio astronomy department at Pulkovo Observatory, the first in Russia. Kaidanovskij is well known for his designs of radiotelescopes and he was principal designer of the largest radio telescope, RATAN-600, at Zelenchukskaya. (M 30475)
Name suggested by the Institute of Applied Astronomy.

(4468) Pogrebetskij

1976 SZ_3. Discovered 1976 September 24 by N. S. Chernykh at Nauchnyj.

Named in memory of Mikhail Timofeevich Pogrebetskij (1892-1956), well-known Ukrainian traveller and alpinist. He was the first to climb Khan Tengri, one of the highest mountains of Tien Shan. (M 23352)
The name was suggested by the Alpinist Club of the Crimea.

(4469) Utting

1978 PS_4. Discovered 1978 August 1 by P. Jekabsons at Bickley.
Named in honor of Muriel Janet Utting (1914-), the Perth Observatory's honorary historian since 1985. Her researches include a history of the observatory since its foundation to the present day, and these will form theses for higher degrees at Murdoch University in 1994 and 1996, the observatory's centenary year. (M 22829)

(4470) Sergeev-Censkij

1978 QP_1. Discovered 1978 August 31 by N. S. Chernykh at Nauchnyj.
Named in memory of Sergej Nikolaevich Sergeev-Censkij (real name Sergeev) (1875-1958), famous Russian writer. He lived and worked in the Crimea and was the author of a multi-volume work on Russian history. (M 23352)

(4471) Graculus

1978 VB. Discovered 1978 November 8 by P. Wild at Zimmerwald.
Pyrrhocorax graculus is the Alpine species of choughs. In great flocks they are sometimes a bit of a nuisance, but to watch their masterly elegant flight, especially in turbulent air, is a great delight. (M 34340)

(4472) Navashin

1980 TY_{14}. Discovered 1980 October 15 by N. S. Chernykh at Nauchnyj.
Named in honor of Mikhail Sergeevich Navashin (1896-1973), prominent Soviet cytologist, who was one of the first to undertake detailed research of human chromosomes. An outstanding amateur astronomer, Navashin was the father of amateur telescope making in the Soviet Union. He made several telescopes and wrote several books to aid amateur observers and telescope makers. (M 23352)

(4473) 1981 DE_2

Discovered 1981 February 28 by S. J. Bus at Siding Spring.

(4474) Proust

1981 QZ_2. Discovered 1981 August 24 by H. Debehogne at La Silla.
Named in honor of Dominique Proust, astrophysicist at the Meudon Observatory who works on observational cosmology. By means of extensive spectroscopic observations he has carried out dynamical studies of clusters of galaxies, large-scale structures and high-redshift objects. He is also a church and concert organist, whose public and broadcast performances include the compositions of astronomer-musicians such as Galileo and Herschel {see planets (697) and (2000), respectively}. The name of this minor planet also honors the French writer Marcel Proust {1871-1922}. (M 21131)

(4475) Voitkevich

1982 UQ_5. Discovered 1982 October 20 by L. G. Karachkina at Nauchnyj.
Named in honor of George Vitol'dovich Voitkevich (1920-), the 1967 Karpiński prizewinner for his outstanding geological research. He is the author of more than 190 publications on geochemistry, geophysics and cosmochemistry, including 32 books on different problems in geology. (M 22247)

(4476) Bernstein

1983 DE. Discovered 1983 February 19 by E. Bowell at Anderson Mesa.
Named in memory of Leonard Bernstein (1918-1990), American composer, con-

ductor, pianist, educator, and author. A colorful, multifaceted musician, Bernstein wrote music in many forms: symphonies, ballet music, operas, sacred music, chamber music and musicals, the last including *West Side Story*, an innovative recasting of *Romeo and Juliet* in terms of New York street gang rivalries. From 1958 to 1969, Bernstein was chief conductor of the New York Philharmonic Orchestra; during this tenure he devised a series of brilliant televised lecture demonstrations. More recently, he conducted orchestras worldwide and became particularly celebrated for his interpretation of the music of Haydn, Brahms and Mahler {these composers are not only immortalized by their music but also by minor planets (3941), (1818), and (4406)}. (M 17466)

(4477) 1983 SB

Discovered 1983 September 28 at the Bulgarian National Observatory at Rozhen.

(4478) Blanco

1984 HG_1. Discovered 1984 April 23 by W. Ferreri at La Silla.

Named in honor of Carlo Blanco, professor of astronomy at Catania University. Known for his intense activity in the observation and analysis of the mutual eclipses of the major satellites of Jupiter and Saturn, he is also involved in international campaigns devoted to observations of minor planets and the Pluto-Charon system. Furthermore, he has contributed to the study of solar-type stellar activity, in particular to analyses of stellar chromospheres and coronas. (M 17224)

(4479) 1985 CP_1

Discovered 1985 February 10 by H. Debehogne at La Silla.

(4480) Nikitibotania

1985 QM_4. Discovered 1985 August 24 by N. S. Chernykh at Nauchnyj.

Named for the Nikitian State Botanical Gardens, located in southern Crimea and founded in 1812 by the well-known Russian botanist C. C. Steven. (M 23352)

(4481) Herbelin

1985 RR. Discovered 1985 September 14 by E. Bowell at Anderson Mesa.

Named in honor of Claude Herbelin (1931-), of Neuchâtel, on the occasion of his 65th birthday (1996 Dec. 13). A friend of the discoverer and brother of the discoverer's companion Anne-Marie Malotki {see planet (3667)}, Herbelin is an electrotechnical engineer by profession. He is also an enthusiastic telecommunications specialist (frequency standards, chronometry) and mountain skier and hiker. (M 28089)

Citation provided by S. Herbelin and C. Gassmann-Herbelin.

(4482) Frèrebasile

1986 RB. Discovered 1986 September 1 by A. Maury at Palomar.

Named in honor of Nicolas Dupont, in religious life Frère Basile {1905-1999} of the Frères des Ecoles Chrétiennes, professor of mathematics and amateur astronomer. Soon after World War II he was assigned to teach cosmography at the Institution Saint Joseph in Nancy. Many students considered the subject too dry and it was subsequently removed from French curricula. To interest his puplis further by giving them "hand-on" experience with the sky, Frère Basile designed and built with a small group a 26-cm telescope and observatory, founded the Cercle Orion and in 1965 co-founded the Société Lorraine d'Astronomie. Since then he has organized many trips to several European observatories and at the age of 84 continues to teach astronomy, observe, arrange astronomy camps in the summer and publish the *Echo d'Orion*. Several professional astronomers owe him their vocations. Frère Basile is a laureate of the Société Astronomique de France, a

recipient of the gold medal of the city of Nancy, and he received the Academic Palms in 1989. (M 17224)
Citation written by J. Blondelet at the request of the discoverer.

(4483) Petöfi

1986 RC_2. Discovered 1986 September 9 by L. G. Karachkina at Nauchnyj.
Named in memory of Shandor Petöfi (1823-1849), outstanding Hungarian poet and public figure. (M 22247)
The name for this Hungaria-type planet is proposed by V. A. Shor.

(4484) Sif

1987 DD. Discovered 1987 February 25 by P. Jensen and K. Augustesen at Brorfelde.
Named after Tor's wife, who after Odin's {see planet (3989)} wife, is the highest ranked of the Asynjur, the goddesses of the Aesir. She is much pursued by the other gods, though they fear the anger of her strong husband. (M 22502)

(4485) Radonezhskij

1987 QQ_{11}. Discovered 1987 August 27 by L. I. Chernykh at Nauchnyj.
Named in honor of Reverend Sergij Radonezhskij (c. 1321-1391), prominent figure in the Russian Church, highly respected and honored person who was considered in Russia a protector of learners and learning. He actively supported Moscow Prince Dmitrij Donskoj in the policy of the consolidation of Russian principalities and in the national struggle for liberation against the Tartar-Mongolian yoke. (M 22502)

(4486) Mithra

1987 SB. Discovered 1987 September 22 by E. W. Elst and V. G. Shkodrov at Rozhen.
Named for the Indo-Iranian god of the heavenly light that led to mithraism, one of the last oriental mystery cults to reach the west, where it became the chief rival to and opponent of christianity. In Babylonia, Chaldaean astrology was incorporated, while Greek art, religion and philosophy provided the models for mithraic iconography and the mithraic mysteries. The two religions have much in common: a divine lord by whom man was assured of elevation, a sacramental meal and a ritual of baptism. Many ruins of mithraic sanctuaries are still to be found in Europe, near Frankfurt and Heidelberg, for example. This minor planet is of Apollo type, and in Asia Minor around 330 B.C. the god Mithra was identified with the god Apollo {see planet (1862)}. (M 16885)
Citation prepared by E. W. Elst.

(4487) Pocahontas

1987 UA. Discovered 1987 October 17 by C. S. Shoemaker and E. M. Shoemaker at Palomar.
An Indian princess and daughter of Powhatan, celebrated sachem and chief of the Chickahominy tribe of Virginia, Pocahontas (c. 1595-1617) developed a warm friendship for the English colonists and rendered them many services, including the rescue of their leader, Captain John Smith, from her father. Her marriage to John Rolfe pioneered the peace between the colonists and the native Americans. (M 17657)

(4488) Tokitada

1987 UK. Discovered 1987 October 21 by K. Suzuki and T. Urata at Toyota.
Named for Taira-no Tokitada (1130-1189), a Japanese military commander in the late Heian era who was the uncle of Taira-no Kiyomori {see planet (4375)}. After the battle of Dannoura, Tokitada was sent into exile to the Noto peninsula, where Percival Lowell {see planet (1886)} visited in 1889 and wrote "Noto, an unexplored corner of Japan". (M 19697)

(4489) 1988 AK

Discovered 1988 January 15 by E. Bowell at Anderson Mesa.

(4490) Bambery

1988 ND. Discovered 1988 July 14 by E. F. Helin and B. Roman at Palomar.

Named in honor of Raymond J. Bambery, chemist and image-processing scientist at the Jet Propulsion Laboratory with a great interest in all aspects of astronomy. He has spent several sessions at Palomar helping search for near-earth minor planets and has been instrumental in upgrading measuring and data reduction techniques. (M 17224)

(4491) Otaru

1988 RP. Discovered 1988 September 7 by K. Endate and K. Watanabe at Kitami.

Named for a city of population 165,000, located in Hokkaido {see planet (3720)} some 30 km west of Sapporo {see planet (3473)} and overlooking the sea of Japan. Once a prosperous port, its trade has declined in recent years, but beautiful streets and houses, an old canal, warehouses and other stone buildings still remain. This combination makes Otaru a romantic city, and nowadays its main industry is artistic glassware. (M 16886)

Name proposed by the second discoverer, following a suggestion by T. Hurukawa, president of the Otaru-Hoshiono-Kai group of amateur astronomers.

(4492) Debussy

1988 SH. Discovered 1988 September 17 by E. W. Elst at St. Michel.

Named in memory of Claude Debussy (1862-1918), famous French impressionistic composer, known particularly for his *Clair de Lune, Jardins sous la Pluie* and *Feux d'Artifice*, and more generally for his brilliant suites for the piano, such as *Estampes, Bergamasque* and *Images*. While his music is the spontaneous expression, the reflection and the image of sensation, it reaches the innermost part of one's subconscious. This effect is obtained by using free harmonies, different scales (from other cultures, e.g., in "Jardins sous la Pluie") and daring mixtures of tones. Debussy was a fervent admirer of Chopin {see planet (3784)}, even to the extent of also composing two books of twelve *Préludes* and an album of *Etudes*. (M 17031)

(4493) 1988 TG_1

Discovered 1988 October 14 by T. Kojima at Chiyoda.

(4494) Marimo

1988 TK_1. Discovered 1988 October 13 by S. Ueda and H. Kaneda at Kushiro.

Named for spherically-shaped green algae that inhabit Lake Akan {see planet (4584)} in Hokkaido, Japan. A protected species since 1894, Marimo can grow as big as a baby's head: a Marimo with a diameter of 6 cm is estimated to be between 150 and 200 years old. (M 20521)

(4495) 1988 VS

Discovered 1988 November 6 by M. Arai and H. Mori at Yorii.

(4496) Kamimachi

1988 XM_1. Discovered 1988 December 9 by T. Seki at Geisei.

Named for the street on which the discoverer lives and from where he discovered six comets, including the sungrazer 1965 VIII (Ikeya-Seki). (M 19337)

(4497) Taguchi

1989 AE_1. Discovered 1989 January 4 by K. Endate and K. Watanabe at Kitami.

Named in honor of Takeo Taguchi (1950-), famous polisher of telescope mirrors and supporter of the development of amateur astronomy in Japan. He is chief of the optics polishing room at Takahashi Seisakujo, Ltd., which makes excellent mirrors, including the one in the astrocamera Ipusiron with which this discovery was made. (M 16886)

(4498) Shinkoyama

1989 AG_1. Discovered 1989 January 5 by T. Seki at Geisei.

Named in honor of the solar physicist Shin Koyama, who has served as a professor at Kagawa University for 30 years. Born in Kyoto in 1927, Koyama has retired from public life in March 1991. (M 18001)

This planet was first (MPC 17980) accidentally named Koyama. This name, however, was already assigned to planet (3383).

(4499) Davidallen

1989 AO_3. Discovered 1989 January 4 by R. H. McNaught at Siding Spring.

Named in honor of David {Anthony} Allen {1946-1994}, staff astronomer at the Anglo-Australian Observatory (AAO). Following his Ph.D. from Cambridge University, Allen held research fellowships at the Hale Observatories and the Royal Greenwich Observatory. In 1975 he became one of the "founding members" of the scientific staff of the AAO, initially as a research fellow. He has remained as one of the pillars of that establishment ever since, having become the only permanently-appointed research astronomer. Notable for his extraordinarily wide interests across all astronomy, from the solar system to observational cosmology, his main contributions have been in the field of infrared instrumentation and its applications. He developed the radiometric method for determining asteroid diameters and recently discovered several new infrared "windows" in the atmosphere of Venus. Allen is also a leading figure in public education in astronomy, contributing to many radio and TV programs and the author of many popular articles and several books. (M 17980)

Citation prepared by R. D. Cannon at the request of the discoverer.

Obituaries published in Publ. Astron. Soc. Aust., Vol. 12, No. 1, p. 139-141 (1995); Observatory, Vol. 114, No. 1122, p. 250-252 (1994); J. Br. Astron. Assoc., Vol. 104, No. 5, p. 259 (1994); Q.J.R. Astron. Soc., Vol. 36, No. 2, p. 173-174 (1995).

(4500) Pascal

1989 CL. Discovered 1989 February 3 by S. Ueda and H. Kaneda at Kushiro.

Named by the Minor Planet Names Committee for the French mathematician and philosopher Blaise Pascal (1623-1662). (M 19337)

Pascal is also honored by a lunar crater.

(4501) Eurypylos

1989 CJ_3. Discovered 1989 February 4 by E. W. Elst at La Silla.

Named for the legendary king of Thessalon, who directed forty vessels at the siege of Troy. He was hit by an arrow from Paris {see planet (3317)} but was rescued by Patroclus {see planet (617)}. (M 17031)

(4502) Elizabethann

1989 KG. Discovered 1989 May 29 by H. E. Holt at Palomar.

Named in honor of Elizabeth Ann Holt, daughter of the discoverer. (M 18459)

(4503) Cleobulus

1989 WM. Discovered 1989 November 28 by C. S. Shoemaker at Palomar.

Named for Cleobulus (fl. c. 560 B.C.), one of the Seven Sages of Greece. A native and tyrant of Lindus in Rhodes, Cleobulus was renowned for the wisdom of his sayings, the beauty of his lyric poetry and his skill with riddles. (M 17467)

Name proposed by the discoverer, following a suggestion by K. Williams, who provided material for the citation, and G. V. Williams, who solved the riddle of the identification involving this object.

(4504) Jenkinson

1989 YO. Discovered 1989 December 21 by R. H. McNaught at Siding Spring.

Named for Nora Jenkinson, Scottish astronomy educator. In the town of Haddington, near Edinburgh, she conducted weekly classes in astronomy for youngsters for the past 22 years (missing only six weeks because of severe weather). In her retirement, she has expanded the classes, and her boundless cheerfulness and enthusiasm are communicated to all those whom she teaches, many of them having gone on to study astronomy at a university. (M 29143)

(4505) Okamura

1990 DV_1. Discovered 1990 February 20 by T. Seki at Geisei.

Named in honor of Keiichiro Okamura (1926-), a retired junior high school teacher who now gives guidance in astronomy to visitors to the Geisei Observatory. (M 27126)

(4506) Hendrie

1990 FJ. Discovered 1990 March 24 by B. G. W. Manning at Stakenbridge.

Named in honor of Michael J. Hendrie, an English amateur with a longstanding interest in astronomy and skilled in the construction of his own instruments. Hendrie has constructed several observatories, most recently one housing a 0.25 m folded reflector for astrometric comet photography. A co-discoverer of periodic comet Crommelin in 1956, he began systematic comet photography in 1952, an interest that stimulated him to take an important role in promoting professional-amateur collaboration in Britain during the International Halley Watch. He served as director of the Comet Section of the British Astronomical Association for ten years. (M 17225)

Name proposed by the discoverer and endorsed by B. G. Marsden. Citation material provided by S. A. Mitton.

(4507) 1990 FV

Discovered 1990 March 19 by H. Shiozawa and M. Kizawa at Fujieda.

(4508) Takatsuki

1990 FG_1. Discovered 1990 March 27 by K. Endate and K. Watanabe at Kitami.

torial staff of the astronomical periodical *Tenmon Guide*. He continues to make excellent astronomical tools and materials and is an advocate of the use of the Schmidt telescope as an astrocamera. The discoverers have introduced such instruments to observing stations at Tsubetsu and Sapporo {see planets (4845) and (3473), respectively}, and both were used to observe this minor planet. (M 16886)

(4509) A917 SG

Discovered 1917 September 23 by S. I. Belyavskij at Simeïs.

(4510) Shawna

1930 XK. Discovered 1930 December 13 by C. W. Tombaugh at Flagstaff.

Named in honor of Shawna Willoughby, granddaughter of the discoverer. (M 18308)

(4511) Rembrandt

1935 SP_1. Discovered 1935 September 28 by H. van Gent at Johannesburg.

Named after Rembrandt Harmensz van Rijn, the greatest Dutch painter of the 17th century, born in 1606 in Leiden. He died in Amsterdam in 1669. (M 20159)

(4512) Sinuhe

1939 BM. Discovered 1939 January 20 by Y. Väisälä at Turku.

Named for Mika Waltari's {see planet (4266)} world-renowned historical novel *Sinuhe, egyptiläinen*. (M 18459)

(4513) Louvre

1971 QW_1. Discovered 1971 August 30 by T. M. Smirnova at Nauchnyj.

Named in honor of the stately architectural ensemble in Paris, the former residence of French kings, now one of the most famous museums and largest collections of works of art in the world. (M 29143)
Name suggested by M. B. Piotrovsky.

(4514) Vilen

1972 HX. Discovered 1972 April 19 by T. M. Smirnova at Nauchnyj.

Named in honor of Vilen Valentinovich Nesterov (1935-), head of the Astrometry Department of the Sternberg Astronomical Institute in Moscow, respected for his work in astrometry, on the rotation of the earth and the analysis of observations of artificial and natural satellites. He is one of scientific leaders of the space-borne astrometric project 'Lomonosov'. His monograph, *Common Astrometry*, written with V. V. Podobed {see planet (3311)}, is used extensively by professional astronomers, students and amateurs. (M 30475)
Name suggested by the Institute of Theoretical Astronomy.

(4515) Khrennikov

1973 SD_6. Discovered 1973 September 28 by N. S. Chernykh at Nauchnyj.

Named in honor of well-known Soviet composer Tikhon Nikolaevich Khrennikov (1913-), who wrote many operas, operettas, ballets, symphony scores, and music for theater and films. He is a professor at the Moscow Conservatory and led the Union of Soviet Composers for many years. (M 34340)

(4516) Pugovkin

1973 SN_6. Discovered 1973 September 28 by N. S. Chernykh at Nauchnyj.

Named in honor of Mikhail Ivanovich Pugovkin (1923-), outstanding and popular cinema actor who has appeared in almost 200 films since 1941. He is an honorary citizen of the Crimean city of Yalta {see planet (1475)} and also has the title honorary artist of Russia. (M 30095)

(4517) 1975 SV

Discovered 1975 September 30 by S. J. Bus at Palomar.

(4518) Raikin

1976 GP_3. Discovered 1976 April 1 by N. S. Chernykh at Nauchnyj.

Named in memory of Arkadij Isaakovich Raikin (1911-1987), outstanding actor and performer of satirical sketches and monologues, master of quick transformations, founder and leader of the State Theater of Miniature. (M 23352)

(4519) Voronezh

1976 YO_4. Discovered 1976 December 18 by N. S. Chernykh at Nauchnyj.

Named for the Russian city, founded before 1177. A large industrial and cultural center, Voronezh was the site where Peter the First built the Azov Fleet in 1695-1696, and the birthplace of the well-known Russian poets A. V. Kol'tsov (1809-1842) and I. S. Nikitin (1824-1861). (M 23352)

(4520) Dovzhenko

1977 QJ_3. Discovered 1977 August 22 by N. S. Chernykh at Nauchnyj.

Named in honor of Aleksandr Petrovich Dovzhenko (1894-1956), well-known

Ukrainian cinema producer and writer, one of the founders of Soviet cinematography. (M 23352)

(4521) Akimov

1979 FU_2. Discovered 1979 March 29 by N. S. Chernykh at Nauchnyj.

Named in honor of Nikolaj Pavlovich Akimov (1901-1968), prominent Soviet theater producer and painter, an art advisor at the Leningrad Theatre of Comedy for many years. (M 23352)

(4522) Britastra

1980 BM. Discovered 1980 January 22 by E. Bowell at Anderson Mesa.

Named to commemorate the centenary of the founding of the British Astronomical Association. Intended to be a less technical counterpart of the Royal Astronomical Society, the BAA was established with the objectives: "The association of Observers, especially the possessors of small telescopes, for mutual help, and their organization in the work of Astronomical observation. The circulation of current Astronomical information. The encouragement of a popular interest in Astronomy." (M 17225)

(4523) 1981 DM_1

Discovered 1981 February 28 by S. J. Bus at Siding Spring.

(4524) Barklajdetolli

1981 RV_4. Discovered 1981 September 8 by L. V. Zhuravleva at Nauchnyj.

Named in memory of Mikhail Bogdanovich Barklaj de Tolli (Barclay de Tolly; 1761-1818), general field marshal, hero of Russian patriotic war of 1812. His family tree had roots in Scotland. During 1810-1812 he was military minister of Russia, and for two months during Napoleon's invasion he was commander-in-chief of the Russian armies. (M 34620)

(4525) 1982 JB_3

Discovered 1982 May 15 by E. F. Helin and E. M. Shoemaker and P. D. Wilder at Palomar.

(4526) Konko

1982 KN_1. Discovered 1982 May 22 by H. Kosai and K. Hurukawa at Kiso.

Named for a city in the area where the first discoverer grew up. Before the Okayama {see planet (2084)} Astrophysical Observatory was established, the Tokyo Astronomical Observatory operated a small station in Konko for the observation of variable stars. (M 17980)

(4527) Schoenberg

1982 OK. Discovered 1982 July 24 by E. Bowell at Anderson Mesa.

Named for Arnold Schönberg (1874-1951), Austrian composer. One of the twentieth century's seminal composers, Schönberg developed the dodecaphonic system of composition. He was founder of the so-called Second Viennese School, whose followers included Alban Berg and Anton Webern {see, respectively, planets (4528) and (4529)}. (M 16886)

(4528) Berg

1983 PP. Discovered 1983 August 13 by E. Bowell at Anderson Mesa.

Named for Alban Berg (1885-1935), Austrian composer. A pupil of Arnold Schönberg {see planet (4527)}, Berg did not produce a great quantity of music, but his two operas are pinnacles of twentieth-century music. (M 16886)

(4529) Webern

1984 ED. Discovered 1984 March 1 by E. Bowell at Anderson Mesa.

Named for Anton Webern (1883-1945), Austrian composer. Associated with Schönberg and Berg {see planets (4527) and (4528)}, Webern refined the use of serialism in music. His mature compositions are terse and concentrated miniatures. (M 16886)

(4530) Smoluchowski

1984 EP. Discovered 1984 March 1 by E. Bowell at Anderson Mesa.

Named in honor of Roman Smoluchowski, solid state physicist and planetary astronomer. A native of Austria, Smoluchowski became head of the physics and metals sections of the Warsaw Institute of Technology. In 1940 he was able to reach the U.S., where he did research with the General Electric Company, the Carnegie Institute of Technology and as professor of solid state physics and director of the solid-state laboratory at Princeton University. Retiring in 1978 from Princeton, he moved to the University of Texas at Austin and continued his work on solids in astrophysics. During his long career Smoluchowski has served on or headed a large number of commissions, particularly those of the National Research Council and the Space Science Board concerning physical sciences, solids and magnetism. For the past several decades he has been one of the founders and primary developers of the field of the solid state behavior of matter in astrophysical situations. He has made important contributions to the fields of planetary interiors, dynamics of rings, asteroids and comets and the structure and thermal properties of cometary nuclei. (M 17980)

Name suggested by W. D. Cochran, and citation provided by Cochran and H. J. Smith.

(4531) Asaro

1985 FC. Discovered 1985 March 20 by C. S. Shoemaker and E. M. Shoemaker at Palomar.

Named for Frank Asaro, nuclear chemist at the Lawrence Berkeley Laboratory of the University of California. With his colleagues Luis and Walter Alvarez {see planet (3581)} and Helen Michel, he discovered the global noble metal anomaly at the Cretaceous-Tertiary boundary, which is now accepted as evidence for impact of a large comet or asteroid. He also discovered noble metal anomalies in late Eocene and Miocene strata that are thought to indicate other major impacts related to the mass extinction of species. (M 17657)

(4532) Copland

1985 GM_1. Discovered 1985 April 15 by E. Bowell at Anderson Mesa.

Named in memory of American composer Aaron Copland (1900-1990). Although his earliest works show the influence of European composers, most of his later compositions make inimitable use of American folk tunes and Jazz melodies. Copland's quintessential ballet scores "Billy the Kid", "Rodeo" and "Apalachian Spring" have long been concert-hall staples, and his "Fanfare for the Common Man", which is also part of the Third Symphony, represents one of the world's most recognizable melodies. (M 18142)

Name suggested and citation provided, in part, by D. J. Tholen.

(4533) Orth

1986 EL. Discovered 1986 March 7 by C. S. Shoemaker and E. M. Shoemaker at Palomar.

Named for Charles J. Orth, nuclear chemist at the Los Alamos National Laboratory, New Mexico. Renowned for his precise analyses for elements by neutron activation methods, Orth has engaged in a world-wide search for evidence of impact of extraterrestrial bodies with the earth at the times of mass extinction of species. He discovered noble metal anomalies at the Cretaceous-Tertiary boundary in continental sedimentary deposits of North America and also near the Frasnian-Fammenian boundary in the Upper Devonian and the Cenomanian-Turonian boundary in the Upper Cretaceous. (M 17658)

(4534) Rimskij-Korsakov

1986 PV_4. Discovered 1986 August 6 by N. S. Chernykh at Nauchnyj.

Named in honor of Nikolaj Andreevich Rimskij-Korsakov (1844-1908), famous Russian composer. (M 23352)

(4535) 1986 QV_2

Discovered 1986 August 28 by H. Debehogne at La Silla.

(4536) 1987 DA_6

Discovered 1987 February 22 by H. Debehogne at La Silla.

(4537) Valgrirasp

1987 RR_3. Discovered 1987 September 2 by L. I. Chernykh at Nauchnyj.
Named in honor of the Soviet writer Valentin Grigorjevich Rasputin. (M 22502)

(4538) 1988 TP

Discovered 1988 October 10 by K. Suzuki at Toyota.

(4539) Miyagino

1988 VU_1. Discovered 1988 November 8 by M. Koishikawa at Sendai.

Named for the eastern part of the city of Sendai {see planet (3133)}, known as "Miyagino" since the seventh century. This area, where the Sendai International Port is located, sustains industry and agriculture. (M 19697)

(4540) Oriani

1988 VY_1. Discovered 1988 November 6 at the Osservatorio San Vittore at Bologna.

Named in memory of Barnaba Oriani (1752-1832), famous Italian astronomer and director of the Brera Observatory, Milan, from 1802 to 1832. Oriani studied the orbits of the planets, particularly Uranus, for which he published orbital elements and tables. A friend of Piazzi {see planet (1000)}, he was involved with the early work on Ceres and published analytical formulae for the calculation of perturbations by Jupiter on objects with highly-inclined orbits. Oriani also inspired Plana and Carlini to carry out their research on the motion of the moon. (M 17981)

(4541) Mizuno

1989 AF. Discovered 1989 January 1 by K. Suzuki and T. Furuta at Toyota.

Named in honor of Yoshikane Mizuno (1954-), who has discovered many minor planets at the Kani {see planet (4265)} Observatory. (M 26762)

(4542) Mossotti

1989 BO. Discovered 1989 January 30 at the Osservatorio San Vittore at Bologna.

Named in memory of Ottaviano Fabrizio Mossotti (1791-1863), who studied at the Brera Observatory and then went as an exile to Geneva and London. In 1827 he was named engineer astronomer and professor of calculus and physics at the University of Buenos Aires. He then went to the University of Corfu (1838-1841) and in 1848 became professor of celestial mechanics and geodesy at the University of Pisa, where he remained until his death. He is celebrated for his memoir *Nuova analisi del problema di determinazione delle orbite* and for his books *Nuova teoria degli strumenti ottici*, *Lezioni di Fisica Matematica* and *Meccanica Razionale*. (M 17981)

(4543) Phoinix

1989 CQ_1. Discovered 1989 February 2 by C. S. Shoemaker at Palomar.

Named for a wise old commander of the Greeks in the Trojan war. Phoinix, who had no sons, raised Achilles {see planet (588)} as if he were his own son. When Achilles refused to fight for the Greeks, Phoinix pleaded with Achilles, "Many a time you wet my shirt, hiccuping wine-bubbles in distress, when you were small". (M 18143)

Name and citation provided by R. Preston at the request of the discoverer.

(4544) Xanthus

1989 FB. Discovered 1989 March 31 by H. E. Holt and N. G. Thomas at Palomar.

This is another Greek name for Apollo {see planet (1862)} meaning the 'Fair'; as such, Xanthus delighted in high places, such as the peaks of high mountains and wavelapped promontaries in the seas. (M 18459)

(4545) 1989 SB_{11}

Discovered 1989 September 28 by H. Debehogne at La Silla.

(4546) Franck

1990 EW_2. Discovered 1990 March 2 by E. W. Elst at La Silla.

Named in memory of the great Belgian composer César Franck (1822-1890), well known for his piano and organ works and beautiful symphony in D minor. His ancestry included members of the famous school of Wallonian painters, and his admiration for them influenced his way of composing - as a "musician painter". After only a year's study in Paris he received "grand prix d'honneur" for piano, and a few years later the first prize for fugue, the art of which in France he restored following a lengthy period of discredit. For organ he achieved only the second prize, because the jury was not inclined to accept his genial and daring way of combining the theme of the fugue with the free theme cyclic principle. Appointed organist at St. Clothilde in 1859, he became a master in the art of improvisation. (M 17467)

Name endorsed by J. Vanvinckenroye.

(4547) Massachusetts

1990 KP. Discovered 1990 May 16 by K. Watanabe and K. Endate at Sapporo.

Named for the U.S. state in which the Minor Planet Center is located. In 1876, the Sapporo School of Agriculture (now Hokkaido University), located in what was then a very undeveloped region, invited William S. Clark, head of the Massachusetts University of Agriculture, to teach them about American agriculture and technology, which at that time was far more advanced. Before leaving to return to the United States, Clark uttered the pioneer spirit of Hokkaido's people. On 1990 Feb. 7 Hokkaido {see planet (3720)} and Massachusetts established a sister-state relationship. (M 19337)

(4548) Wielen

2538 P-L. Discovered 1960 September 24 by C. J. van Houten and I. van Houten-Groeneveld at Palomar.

Named in honor of Roland Wielen (1938-), German astronomer, director of the Astronomisches Rechen-Institut in Heidelberg, professor of theoretical astronomy at Heidelberg University since 1985, and from 1978 to 1985 professor of astronomy and astrophysics at the Berlin Technical University. His main work covers the fields of the stellar dynamics (the luminosity function of nearby stars, kinematics and dynamics of galaxies, dynamical evolution of star clusters and of clusters of galaxies) and astrometry. He has also worked on problems of galactic structure and served as president of IAU Commission 33 from 1982 to 1985. (M 18143)

Name proposed and citation prepared by L. D. Schmadel.

(4549) Burkhardt

1276 T-2. Discovered 1973 September 29 by C. J. van Houten and I. van Houten-Groeneveld at Palomar.

Named in honor of Gernot Burkhardt (1951-), astronomer at the Astronomisches Rechen-Institut in Heidelberg. Burkhardt is an expert in computer hardware and a skilled programmer whose work is close to indispensable to almost all of his colleagues. He also serves as an editor of *Astronomy and Astrophysics Abstracts* and is responsible for the rather complex data handling procedures. (M 18143)
Name proposed and citation prepared by L. D. Schmadel.

(4550) 1977 HH_1

Discovered 1977 April 24 by S. J. Bus at Palomar.

(4551) Cochran

1979 MC. Discovered 1979 June 28 by E. Bowell at Anderson Mesa.

Named in honor of William D. Cochran and Anita L. Cochran, husband and wife astronomers at the University of Texas at Austin. William's broad range of research has concerned planetary (including cometary) atmospheres, Raman scattering, stellar radial velocity variations and motions in stellar chromospheres and photospheres. Using a spectroscopic radial velocity meter, he is currently surveying several dozen stars to search for reflex motions (as small as about 2 m/s) that would indicate the presence of planetary companions. Anita is a specialist in the chemistry of cometary comae and in particular how the chemistry changes with changing heliocentric distance. An assiduous observer, she has used spatially resolved spectra to help transform the photometry of comets into a quantitative discipline. She has also developed sophisticated cometary models to understand how the observed atoms, molecules, and radicals are related to the larger parent molecules present in cometary nuclei. Anita is currently a team member of the Imaging Science Subsystem of the Comet Rendezvous-Asteroid Flyby mission. (M 18459)

(4552) Nabelek

1980 JC. Discovered 1980 May 11 by A. Mrkos at Kleť.

Named in honor of Jan Nabelek (1908-), now the most senior member of a group of teachers in Nove Mesto, Moravia, at the beginning of World War II. (M 26424)

(4553) Doncampbell

1982 RH. Discovered 1982 September 15 by E. Bowell at Anderson Mesa.

Named in honor of Donald B. Campbell, planetary scientist at Cornell University, on the occasion of his 50th birthday. Campbell is responsible for the development of the Arecibo Observatory's high-power radar system and the considerable success of the Arecibo radar astronomy program during the past quarter century. He has carried out radar observations of every class of solar system target during this period. He is the discoverer of a large number of Venus surface features whose geologic character is now being clarified by the Magellan radar. Campbell's contributions to small-body astronomy include the first radar ranging to an asteroid ((433) Eros in 1975) and observations that revealed the existence of large-particle clouds around comets. (M 20159)
Name suggested and citation written by S. J. Ostro.

(4554) Fanynka

1986 UT. Discovered 1986 October 28 by A. Mrkos at Kleť.

Named in memory of Frantiska ("Fanynka") Burian (1911-1980), associated with the discoverer during the difficult times at the beginning of World War II. (M 26424)
Name endorsed by J. Nabelek.

(4555) 1987 QL

Discovered 1987 August 24 by S. Singer-Brewster at Palomar.

(4556) Gumilyov

1987 QW_{10}. Discovered 1987 August 27 by L. G. Karachkina at Nauchnyj.

Named in memory of Nikolaj Stepanovich Gumilyov (1886-1921), famous poet and dramatist. He was one of the founders of acmeism, a poetic tendency that was popular among the Russian intelligentsia at the beginning of the current century. (M 20837)

(4557) Mika

1987 XD. Discovered 1987 December 14 by M. Yanai and K. Watanabe at Kitami.

Named in honor of Mika Watanabe (1963-), wife of the second discoverer. (M 20160)

(4558) Janesick

1988 NF. Discovered 1988 July 12 by A. Maury and J. Mueller at Palomar.

Named in honor of James R. Janesick of the Jet Propulsion Laboratory. Janesick has been instrumental in the development of CCDs for astronomy and, of equal importance, in the education of astronomers in the use of these devices and of industry in establishing requirements for good scientific imagers. His contributions have ranged from making the detectors useful in the blue and ultraviolet to the development of techniques for excellent charge transfer at very low signal levels, of amplification methods with sub-electron noise and of procedures to avoid damaging effects in high-energy radiation. (M 19337)

Citation written by J. E. Gunn at the request of the first discoverer. Name endorsed by E. F. Helin.

(4559) Strauss

1989 AP_6. Discovered 1989 January 11 by F. Börngen at Tautenburg.

Named in memory of the Austrian composer Johann Strauss (1825-1899), who represented Viennese dance music at its zenith. The leading master of Viennese operettas (*Die Fledermaus, Zigeunerbaron*, etc.), he was also the creator of immortal waltzes, such as the *An der schönen blauen Donau* (Blue Danube), *Geschichten aus dem Wiener Wald* (Tales of the Vienna Woods) and *Kaiserwalzer* (Kaiser Waltz). (M 17031)

(4560) Klyuchevskij

1976 YD_2. Discovered 1976 December 16 by L. I. Chernykh at Nauchnyj.

Named in memory of Vasilij Osipovich Klyuchevskij (1841-1911), a famous Russian historian. (M 18459)

(4561) Lemeshev

1978 RY_5. Discovered 1978 September 13 by N. S. Chernykh at Nauchnyj.

Named in honor of Sergej Yakovlevich Lemeshev (1902-1977), outstanding Russian opera-singer, artist of the Bolshoi Theatre in Moscow for many years. (M 23352)

(4562) 1979 UD_2

Discovered 1979 October 21 at the Purple Mountain Observatory at Nanking.

(4563) Kahnia

1980 OG. Discovered 1980 July 17 by E. Bowell at Anderson Mesa.

Named in honor of Franz D. Kahn, professor of astronomy and head of the department of astronomy at the University of Manchester. Kahn has made important contributions to many areas of astrophysics, including studies of H II regions, supernovae and cosmical gas dynamics, and he has shown great interest in and

support for research on comets and minor planets at Manchester. (M 19338)
Name proposed and citation prepared by G. J. Hahn.

(4564) 1981 ET_{16}

Discovered 1981 March 6 by S. J. Bus at Siding Spring.

(4565) 1981 EZ_{17}

Discovered 1981 March 2 by S. J. Bus at Siding Spring.

(4566) Chaokuangpiu

1981 WM_4. Discovered 1981 November 27 at the Purple Mountain Observatory at Nanking.

Named in honor of Kuang-piu Chao, consulting professor at Tsinghua University and Zhejiang University. For many years, he made remarkable contributions to the development of Chinese educational and cultural programs. (M 32788)

(4567) Bečvář

1982 SO_1. Discovered 1982 September 17 by M. Mahrová at Kleť.

Named in memory of the Czech astronomer Antonín Bečvář (1901-1965), founder and first director of the Skalnaté Pleso {see planet (2619)} Observatory in Slovakia. Bečvář made observations of comets, meteors and the solar photosphere, but he is best known as the author of *Atlas Coeli* and the more detailed *Atlases Eclipticalis, Borealis* and *Australis.* (M 21956)
Bečvář is also honored by a lunar crater.

(4568) Menkaure

1983 RY_3. Discovered 1983 September 2 by N. G. Thomas at Anderson Mesa.

Menkaure (c. 2530 B.C.), the son of Chephren {see planet (4412)}, reigned for 18 years and built the third and smallest of the three pyramids at Giza {see planet (5249)}. His mortuary temple was finished by his successor Shepseskaf and contained some of the finest sculptures of the pyramid age. (M 26929)
Pharao Menkaure is named in Greek Mycerinos {see planet (4413)}.

(4569) Baerbel

1985 GV_1. Discovered 1985 April 15 by C. S. Shoemaker and E. M. Shoemaker at Palomar.

Named for Baerbel K. Lucchitta, planetary geologist with the U.S. Geological Survey. She is a leading investigator of the geology of Mars and the satellites Europa and Ganymede. She is especially well known for her work on the effects of landslides and ice movement on Mars and on mapping the ice cap and the flow of ice in Antarctica. (M 17658)

(4570) Runcorn

1985 PR. Discovered 1985 August 14 by E. Bowell at Anderson Mesa.

Named in memory of Stanley Keith Runcorn (1922-1995), British geophysicist. Runcorn's research centered on the magnetic properties of rocks. He used their paleomagnetic signatures to make an important contribution to the emerging theory of plate tectonics, and he also studied the magnetism of the moon, convection in planetary interiors, and geomagnetic polarity reversals. For many years head of the department of physics at the University of Newcastle-upon-Tyne, Runcorn was celebrated for his organization of his NATO-sponsored international scientific discussion meetings. (M 28089)

Obituaries published in Nature, Vol. 379, No. 6561, p. 119 (1996); Q.J.R. Astron. Soc., Vol. 37, No. 3, p. 463-465 (1996).

(4571) Grumiaux

1985 RY_3. Discovered 1985 September 8 by H. Debehogne at La Silla.

Named in memory of Arthur Grumiaux (1921-1986), Belgian violinist who studied at the conservatories of Charleroi and Brussels. After winning the Vieuxtemps Prize in 1939 he went to Paris to study with Georges Enesco. Of particular brilliance are his recordings of the Mozart and Beethoven sonatas with Clara Haskil at the piano and of Bach's sonatas and partitas for unaccompanied violin. {The great composers mentioned are all on the minor planet sky - see planets (1034), (1815), and (1814), respectively.} (M 19338, E. Goffin)

Name proposed in consultation with E. Goffin, who found the identifications involving this minor planet.

(4572) Brage

1986 RF. Discovered 1986 September 8 by P. Jensen and K. Augustesen at Brorfelde.

Named after the god of poetry and music in Norse mythology, husband of Idun {see planet (176)} and son and principal counsellor of Odin {see planet (3989)}. (M 22502)

(4573) Piešťany

1986 TP_6. Discovered 1986 October 5 by M. Antal at Piwnice.

Named for the city located on the river Váh in the middle Považic region of western Slovakia, 70 km northwest of Bratislava {see planet (4018)}. Piešťany is the home of the discoverer and many generations of his ancestors, as well as the site of a famous spa. Evidence of settlement at Piešťany dates back to the Paleolithic era. (M 27126)

(4574) Yoshinaka

1986 YB. Discovered 1986 December 20 by T. Niijima and T. Urata at Ojima.

Named for Kiso Yoshinaka (1154-1184), the Japanese military commander in the late Heian era. He was a first cousin of Minamoto Yoritomo {see planet (3902)}. Once Yoshinaka broke the Taira forces and invaded Kyoto, then capital of Japan, and he was called "General Asahi, the rising sun". (M 19697)

(4575) Broman

1987 ME_1. Discovered 1987 June 26 by E. F. Helin at Palomar.

Named for Brian P. Roman, astronomer, chef extraordinaire and humorist, in recognition for his diligence and dedication in the pursuit of near-earth minor planets. His participation in the planet-crossing asteroid survey at Palomar has been a major contribution to its continuing success. This particular object was discovered on his birthday. (M 17225)

Name endorsed by K. Lawrence and J. Alu.

(4576) 1988 CC

Discovered 1988 February 10 by T. Kojima at Chiyoda.

(4577) Chikako

1988 WG. Discovered 1988 November 30 by Y. Kushida and M. Inoue at Yatsugatake.

Named in honor of Chikako Mihashi (1950-), who has for many years worked for the advancement of astronomical education in Japan. Her support greatly contributed to the "Astro Village" at Yatsugatake {see planet (4033)} South Base Observatory. The "Astro Village" was established for the purpose of astronomical education for children. (M 17658)

(4578) Kurashiki

1988 XL_1. Discovered 1988 December 7 by T. Seki at Geisei.

Named for a beautiful city on the shore of the Seto inland sea in western Japan. Minoru Honda (1917-1990) {see planet (3904)}, discoverer of 12 comets and 12 novae, carried out work at the Kurashiki Observatory from 1950 until his death. The city is also the home of the Ohara art museum, one of the most famous of such museums in Japan. (M 17658)

(4579) Puccini

1989 AT_6. Discovered 1989 January 11 by F. Börngen at Tautenburg.

Named in memory of the Italian composer Giacomo Puccini (1858-1924), whose magnificent operas *La Bohème, Tosca, Madame Butterfly* and *Turandot* have received the highest appreciation in the music world {see also planet (530)}. (M 17981)

(4580) Child

1989 EF. Discovered 1989 March 4 by E. F. Helin at Palomar.

Named in honor of Jack B. Child, software engineer at the Jet Propulsion Laboratory, Asteroid Project director of the World Space Foundation, and past president of the Orange County Astronomers. His generous helpfulness over the years has been greatly appreciated. He has played a very important role in introducing team members to the Palomar planet-crossing asteroid survey project. (M 17225) Name endorsed by R. L. Staehle.

(4581) Asclepius

1989 FC. Discovered 1989 March 31 by H. E. Holt and N. G. Thomas at Palomar.

Named for the Greek god of healing and medicine, son of Apollo by Coronis {see planets (1862) and (158)}. Physician to the Argonauts, many of whom he restored to life, he was killed by Zeus {see planet (5731)} with a thunderbolt for attempting to resuscitate Orion, who had been murdered by Artemis {see planet (105)}, Apollo's twin sister. For avenging Asclepius' murder by killing the Cyclops who had made the thunderbolt, Apollo was banished by Zeus. (M 18459)

(4582) Hank

1989 FW. Discovered 1989 March 31 by H. E. Holt and N. G. Thomas at Palomar. Named in honor of Henry Reid Holt, son of the first discoverer. (M 18460)

(4583) 1989 RL_4

Discovered 1989 September 1 at the Bulgarian National Observatory at Rozhen.

(4584) Akan

1990 FA. Discovered 1990 March 16 by M. Matsuyama and K. Watanabe at Kushiro.

Named for an extensive national park in eastern Hokkaido {see planet (3720)}. Designated a national park in 1934, Akan has an area of 905 square km, within which are located the three large lakes Akan, Mashu {see planet (4126)} and Kussyaro, as well as the volcanic mountains Meakandake, Oakandake and Masyudake. (M 19338)

(4585) Ainonai

1990 KQ. Discovered 1990 May 16 by K. Endate and K. Watanabe at Kitami.

Named for a small town (population less than 3,000), situated 10 km west of Kitami in eastern Hokkaido {see planets (3785) and (3720)}. (M 19338)

(4586) Gunvor

6047 P-L. Discovered 1960 September 24 by C. J. van Houten and I. van Houten-Groeneveld at Palomar.

Named in honor of Mrs. Gunvor Ulla Marie Ollongren-Lundgren (1942-), Swedish-born wife of Dutch astronomer and mathematician Alexander Ollongren. She has recently helped provide explanations for several Scandinavian names of minor planets. (M 18143)

(4587) Rees

3239 T-2. Discovered 1973 September 30 by C. J. van Houten and I. van Houten-Groeneveld at Palomar.

Named in honor of Martin J. Rees (1942-), English astronomer, professor of astronomy at the University of Cambridge and director of the Institute of Astronomy. His main research covers galactic evolution and related problems. (M 18143)

Name proposed by J. H. Oort.

(4588) 1931 EE

Discovered 1931 March 13 by M. Wolf at Heidelberg.

(4589) McDowell

1933 OB. Discovered 1933 July 24 by K. Reinmuth at Heidelberg.

Named in honor of Jonathan Christopher McDowell (1960-), astrophysicist, space enthusiast and feminist activist. His astrophysical interests include theoretical studies of background light and quasars. He shares his encyclopedic knowledge of spaceflight with the world by regularly releasing an electronic space newsletter. (M 22502)

Citation written by A. Prestwich at the request of B. G. Marsden and G. V. Williams, who appreciate Jonathan's assistance in trying to identify 1991 VG with manmade space debris. Identifications for this object were made by C. M. Bardwell.

(4590) 1968 OG_1

Discovered 1968 July 25 by G. Plougin and Y. Belyaev at Cerro El Roble.

(4591) Bryantsev

1975 VZ. Discovered 1975 November 1 by T. M. Smirnova at Nauchnyj.

Named in memory of Aleksandr Aleksandrovich Bryantsev (1883-1961), producer, People's artist of the USSR and the founder of the Theater for Young Audiences in Petrograd. He staged 48 plays, laying down the techniques of children's theater production. (M 30475)

Named at the request of the staff of the St. Petersburg Theater of Young Audiences.

(4592) Alkissia

1979 SQ_{11}. Discovered 1979 September 24 by N. S. Chernykh at Nauchnyj.

Named in honor of Aleksej Alekseevich Kisselev (1927-), astronomer at the Pulkovo Astronomical Observatory, skilled observer of double stars and leader in the field of photographic astrometry of planets and satellites. He has devised new methods of astrometric plate reduction and of orbit determination. His comprehensive treatise on photographic astrometry is a standard reference for all observers at the observatories of the former Soviet Union. After World War II, Kisselev spent seven years in prison as a victim of political repression. (M 30095)

(4593) Reipurth

1980 FV_1. Discovered 1980 March 16 by C.-I. Lagerkvist at La Silla.

Named in honor of Bo Reipurth, staff astronomer at the European Southern Observatory in charge of the Schmidt telescope. He is mainly working on stellar formation and related problems, but he also shows a very keen interest in meteorites and has a fine collection of them. (M 20521)

(4594) Dashkova

1980 KR_1. Discovered 1980 May 17 by L. I. Chernykh at Nauchnyj.

Named in honor of princess Ekaterina Romanovna Dashkova (1744-1810), one of the most educated women in Russia of her time, well-known figure in Russian culture, director of the St. Petersburg Academy of Sciences and president of the Russian Academy during 1783-1796. (M 22502)

(4595) 1981 EZ_2

Discovered 1981 March 2 by S. J. Bus at Siding Spring.

(4596) 1981 QB

Discovered 1981 August 28 by C. T. Kowal at Palomar.

(4597) 1983 UA_1

Discovered 1983 October 30 by S. J. Bus at Palomar.

(4598) Coradini

1985 PG_1. Discovered 1985 August 15 by E. Bowell at Anderson Mesa.

Named in honor of sister and brother Angioletta and Marcello Coradini, who have very actively promoted the development of planetary sciences in Europe, particularly in Italy, during the past twenty years. At the beginning of her scientific career, Angioletta carried out theoretical work on impact processes on planetary surfaces. Later, she studied the formation of planetesimals and the origin of satellite systems. Currently at the Istituto di Astrofisica Spaziale, she is involved in a number of proposed space missions. Marcello, also an IAS member but currently at the European Space Agency, has worked on Mars, participating in the Viking Lander Imaging Team, and he has carried out laboratory experiments on hypervelocity impacts. He was involved in the exploration of Halley's Comet as a co-investigator of the Giotto Multicolor Camera Team. As a member of the ESA scientific directorate, he is contributing to both the development of the European community of planetary scientists and to planetary exploration in general. (M 18460)

Citation provided by M. Fulchignoni at the request of the discoverer. Name endorsed by A. W. Harris.

(4599) 1985 RZ_2

Discovered 1985 September 5 by H. Debehogne at La Silla.

(4600) 1985 RE_4

Discovered 1985 September 10 by H. Debehogne at La Silla.

(4601) Ludkewycz

1986 LB. Discovered 1986 June 3 by M. Rudnyk at Palomar.

Named in honor of Romana Ludkewycz, mother of the discoverer. Her unending support of her son's interest in astronomy has been a constant source of inspiration since his childhood. Associated with the Palomar planet-crossing asteroid program during 1985-1987, he is currently involved in JPL's Planetary Image Facility. (M 17467)

(4602) Heudier

1986 UD_3. Discovered 1986 October 28 by the CERGA at Caussols.

Named in honor of Jean-Louis Heudier, astronomer in charge of the operations of the Schmidt telescope at Calern with which this object was discovered. Having initially gained experience in wide-field photography with the Antares satellite-tracking camera at the Observatoire de Nice, he played a crucial role in the development of the Schmidt at Calern and in convincing the French authorities to invest in modern photographic techniques. Under his leadership there were several discoveries of minor planets and supernovae. He has served as secretary and president of the IAU working group on photographic techniques. An authority on the history of astronomy, which he teaches at the University of Nice, he has always been active in the popularization of astronomy, serving in particular as the leading force behind the "Astrorama" center for public observations in the heights of Nice. (M 17981)
Citation provided by C. Pollas and A. Maury.

(4603) Bertaud

1986 WM_3. Discovered 1986 November 25 by C. Pollas at Caussols.

Named in honor of Charles Bertaud, under whose leadership at Meudon the discoverer began his astronomical work. Well known as an observer of comets, supernovae and particular stars, he was one of the initiators of the project that led to the construction of the Schmidt telescope at Caussols. (M 17467)

(4604) 1987 SK

Discovered 1987 September 18 by K. Suzuki and T. Urata at Toyota.

(4605) Nikitin

1987 SV_{17}. Discovered 1987 September 18 by L. I. Chernykh at Nauchnyj.

Named for Afanasij Nikitin (?-1472), a merchant from the town of Tver'. Nikitin, who was the first Russian to journey to India, described the trip in his diary, later published as *Walking Beyond Three Seas*. (M 23136)

(4606) Saheki

1987 UM_1. Discovered 1987 October 27 by T. Seki at Geisei.

Named in honor of Tsuneo Saheki {1916-1996}, president of the Toa Astronomical Society. An observer of Mars, Saheki has extensive records of his results covering half a century. (M 18308)
Obituary published in Yamamoto Circ., No. 1607, p. 2 (1996).

(4607) Seilandfarm

1987 WR. Discovered 1987 November 25 by K. Endate and K. Watanabe at Kitami.

Named for a farm, established by Akio Seino in 1942, located about 10 km from Kitami {see planet (3785)}. Covering an area of about 50 hectares of hilly terrain, this modern, large-scale dairy farm, with 150 cattle, is operated by four members of the Seino family. (M 20521)

(4608) 1988 BW_3

Discovered 1988 January 19 by H. Debehogne at La Silla.

(4609) Pizarro

1988 CT_3. Discovered 1988 February 13 by E. W. Elst at La Silla.

Named in honor of Guido and Oscar Pizarro, who operate the 1-m Schmidt telescope and who exposed the plates on which this minor planet was discovered. For almost 20 years the two brothers have been renowned for their patient and effective work with the telescope. They took the plates for the ESO sky surveys and have taken several thousand plates for general programs, including many specifi-

cally for the detection and follow-up of comets and minor planets. (M 17658)
Citation prepared by H.-E. Schuster at the request of the discoverer.

(4610) Kájov

1989 FO. Discovered 1989 March 26 by A. Mrkos at Kleť.

Named for a village in southern Bohemia south of Kleť Mountain that is the pilgrimage site of a Gothic church of the Assumption, founded in the fourteenth century. (M 29670)
Name proposed by J. Tichá, M. Tichý and Z. Moravec.

(4611) Vulkaneifel

1989 GR_6. Discovered 1989 April 5 by M. Geffert at La Silla.

Named in honor of the landscape and people of a region to the southwest of Bonn. The landscape of that region is full of volcanic relics such as craters and lava streams, as well as "Maare", which are small lakes. The observatory of the University of Bonn, Hoher List, where the discoverer works, is located in the Vulkaneifel near one of these lakes. (M 17467)

(4612) Greenstein

1989 JG. Discovered 1989 May 2 by E. F. Helin at Palomar.

Named in honor of Jesse Greenstein (1909-), an expert in stellar spectroscopy with many discoveries concerning low-luminosity stars, white dwarfs, stellar chemical compositions and stellar evolution. In addition, Greenstein established and guided, during its first two decades, the astronomy department at the California Institute of Technology. As a statesman of science policy he was influential in the early history of radioastronomy in the United States, as well as in the early history of AURA. (M 27734)

His friends and colleagues heartily endorse this naming. Citation prepared by J. Cohen.

(4613) Mamoru

1990 OM. Discovered 1990 July 22 by K. Watanabe at Sapporo.

Named in honor of Mamoru Mohri (1948-), a professor in the department of technology at Hokkaido University and the first payload specialist of the Japanese National Space Development Agency to travel into space. After lifting off in the space shuttle Endeavor on 1992 Sept. 12 he successfully performed 34 experiments ("Fuwatto '92") and lectured from space to Japanese school children. (M 21131)

(4614) Masamura

1990 QN. Discovered 1990 August 21 by Y. Mizuno and T. Furuta at Kani.

Named in honor of Kazutada Masamura (1920-1998), the secretary of the Oriental Astronomical Association (O.A.A.). Masamura has been observing sunspots since 1935 and has made a great effort to diffuse astronomical knowledge. In addition to establishing the Gifu Astronomical Observatory in 1971, Masamura provides financial assistance for awards given by the O.A.A. to Japanese amateur discoverers of comets, novae and minor planets. (M 19697)
Obituary published in Heavens, Vol. 79, No. 12, p. 1-3 (1998).

(4615) Zinner

A923 RH. Discovered 1923 September 13 by K. Reinmuth at Heidelberg.

Named in memory of Ernst Zinner (1886-1970), assistant astronomer (1910-1914) and director (1926-1953) of the Remeis-Sternwarte, Bamberg. In addition to his observational work on variable stars, Zinner was the first professor of astronomy to investigate the history of astronomy in the Middle Ages quantitatively. Giacobini's {see planet (1756)} comet 1900 I was rediscovered by Zinner in October 1913 and is now known as P/Giacobini-Zinner, parent of the Draconid meteors. (M 22502) Name proposed by G. Klare and L. D. Schmadel, citation prepared by R. Knigge.

(4616) Batalov

1975 BF. Discovered 1975 January 17 by L. I. Chernykh at Nauchnyj.

Named in honor of Aleksej Vladimirovich Batalov, well-known contemporary cinema actor and producer. (M 22503; M 22519)

(4617) 1976 DK

Discovered 1976 February 22 at the Felix Aguilar Observatory at El Leoncito.

(4618) Shakhovskoj

1977 RJ_3. Discovered 1977 September 12 by N. S. Chernykh at Nauchnyj.

Named in honor of Nikolaj Mikhailovich Shakhovskoj (1930-), astrophysicist at the Crimean Astrophysical Observatory, an authority in stellar polarimetry and photometry, known for his research of variable cosmic objects of diverse types, including eruptive stars, active galaxies, x-ray binaries and flare stars, as well as comets and minor planets. (M 27126)

(4619) Polyakhova

1977 RB_7. Discovered 1977 September 11 by N. S. Chernykh at Nauchnyj.

Named in honor of Elena Nikolaevna Polyakhova, assistant professor of astronomy at St. Petersburg University, well-known specialist on celestial mechanics who obtained, in particular, new insights into the problems of using solar sails in interplanetary space flight. (M 23352; M 26439)

(4620) Bickley

1978 OK. Discovered 1978 July 28 at the Perth Observatory at Bickley.

Named for the present site (since 1966) of the Perth Observatory. In 1915 the locality, in the Darling Range 23 km east-southeast of the center of the city, was officially named for Samuel Wallace Alexander Walsh Bickley (1810-1876), a pioneer in the area who was also a successful merchant and a nominated member of the Legislative Council of the colony (as it was at that time) of Western Australia. (M 31609)

(4621) Tambov

1979 QE_{10}. Discovered 1979 August 27 by N. S. Chernykh at Nauchnyj.

Named for the city of Tambov, one of the industrial and cultural centers in Russia, center of the Tambov region. (M 23352)

(4622) Solovjova

1979 WE_2. Discovered 1979 November 16 by L. I. Chernykh at Nauchnyj.

Named in memory of Sergej Mikhailovich Solovjov (1820-1879), famous Russian historian and the author of many works on the history of Russia. The name also honors his son, Vladimir Sergeevich Solovjov (1853-1900), religious philosopher, poet and publicist. (M 23136)

(4623) Obraztsova

1981 UT_{15}. Discovered 1981 October 24 by L. I. Chernykh at Nauchnyj.

Named in honor of Elena Vasil'evna Obraztsova, outstanding singer of the Bolshoi Theatre in Moscow. (M 22503)

(4624) Stefani

1982 FV_2. Discovered 1982 March 23 by C. S. Shoemaker and Q. R. Passey at Palomar.
Named for J. Stefani Salazar, granddaughter of the first discoverer. (M 17658)

(4625) Shchedrin

1982 UG_6. Discovered 1982 October 20 by L. G. Karachkina at Nauchnyj.

Named for the outstanding Russian composer Rodion Konstantinovich Shchedrin. (M 22247)

(4626) Plisetskaya

1984 YU_1. Discovered 1984 December 23 by L. G. Karachkina at Nauchnyj.
Named for the brilliant ballet-dancer Majya Mikhailovna Plisetskaya. (M 22247)

(4627) 1985 RT_2

Discovered 1985 September 5 by H. Debehogne at La Silla.

(4628) Laplace

1986 RU_4. Discovered 1986 September 7 by E. W. Elst at Rozhen.

Named in memory of the great French mathematician, astronomer and physicist, Pierre-Simon Marquis de Laplace (1749-1827). Although Newton {see planet (8000)} had concluded that divine intervention was periodically required to preserve the solar system in equilibrium, Laplace managed to prove in 1773, by applying Newtonian gravitation, the invariability of planetary mean motions to the cubes of the eccentricities and inclinations. In 1786 he showed that the effects of planetary perturbations were conservative and periodic, not cumulative and disruptive; the eccentricities and inclinations of planetary orbits to each other will remain small, constant and self-correcting. In 1796 he published his famous *Exposition du système du monde*, which treats also his "nebular hypothesis", which ascribes the origin of the solar system to the contraction of a gaseous nebula. His monumental *Traité de mécanique céleste*, published between 1798 and 1827, offered a complete mechanical interpretation of the solar system. Laplace is also well known for his investigations on probability. (M 18460)

(4629) Walford

1986 TD_7. Discovered 1986 October 7 by E. F. Helin at Palomar.

Named in honor of Roy L. Walford, professor of pathology at the UCLA School of Medicine, noted for his work in gerontology. His numerous publications include not only scientific papers, but also popular books on the subject of aging in humans. A scientist, writer, actor and explorer, he has given inspiration and encouragement to many, including the discoverer and her family. (M 18460)

(4630) Chaonis

1987 WA. Discovered 1987 November 18 by J. M. Baur at Chions.

Named in honor of the Chions, originally Chaonis, where the observatory of that name is located. The origins of this town in the province of Pordenone {see also planet (3896)} date back to the Roman empire and to the division of the Julia Concordia land, where important archaeological remains can still be found. The first settlement dates back to the time of the Longobardic king Autari, when it was founded by some of the survivors of the disastrous flood that swept over lower Friuli. Around the year 100 the town came under the jurisdiction of the Aquileia patriarchy. In 1420 it was passed to the republic of Venice and later to the 'free dominion' under the counts of Panigai. (M 17658)

(4631) Yabu

1987 WE_1. Discovered 1987 November 22 by S. Ueda and H. Kaneda at Kushiro.

Named in honor of Yasuo Yabu (1932-), long-standing meteor observer, vice-president of the Nippon Meteor Society and editor of its Astronomical Circular since 1969. (M 21131)

Name suggested by K. Watanabe. Citation prepared by I. Hasegawa.

(4632) 1987 YB

Discovered 1987 December 17 by T. Kojima at Chiyoda.

(4633) 1988 AJ_5

Discovered 1988 January 14 by H. Debehogne at La Silla.

(4634) Shibuya

1988 BA. Discovered 1988 January 16 by M. Inoue and O. Muramatsu at Kobuchizawa.

Named for a cultural district of Tokyo that contains a concert hall, two theaters, two art museums and nine other educational facilities. Among them is the Gotoh planetarium and Astronomical Museum, which has been visited by 14 million people since its establishment in 1957. (M 17659)

Name proposed by the discoverers following a suggestion by six lecturers in the museum.

(4635) Rimbaud

1988 BJ_1. Discovered 1988 January 21 by E. W. Elst at St. Michel.

Named in memory of the French poet Arthur Rimbaud (1854-1891) on the occasion of the hundredth anniversary of his death. At the age of 17 he was already known for his *Dormeur du Val* and *Le Bateau ivre*, the latter, together with *Voyelles*, probably being his most famous work. In 1872 he traveled with Paul Verlaine {see planet (6871)} to England and Belgium, their friendship ending with two gunshots fired by Verlaine at Rimbaud. In 1873 Rimbaud published *Une Saison en Enfer*, an autobiographical and psychological work. After *Les Illuminations*, written at the age of 19 and issued by Verlaine only in 1886, nothing remains of the work of this great poet. As a precursor of symbolism Rimbaud enormously influenced Verlaine and the following generation. (M 18645)

Name proposed and citation prepared by Kristina Leterme at the request of the discoverer.

(4636) Chile

1988 CJ_5. Discovered 1988 February 13 by E. W. Elst at La Silla.

Named for the beautiful South American country in which the European Southern Observatory is located. Noted for its great wines, Chile is chiefly mountainous, with the Andes dominating the landscape. The extension of Chile across some 38 degrees of latitude embraces nearly all climates. The fascinating Chilean people are racially a mixture of Europeans (the conquistadores from Spain, Basque families) and indigenous tribes (Atacamenos, Diaguitas, Picunches, Araucanians, Huilliches, Pehuenches and Cuncos). Today the proud Araucanian Indians form the only significant ethnic minority. (M 19697)

(4637) Odorico

1989 CT. Discovered 1989 February 8 by J. M. Baur at Chions.

Named for the renowned solitary traveler and missionary, the Franciscan monk Odorico (1265-1331). Born Odorico Mattiussi in Pordenone {see planet (3896)} and pronounced blessed in 1775, between 1314 and 1330 he traveled in southern Asia, stopping in Ceylon, Sumatra, Java and Borneo as a messenger of the Christian faith and of the light of civilization. He went to Tibet, starting from

Peking to Lhasa. He visited the 'canati' di Boccara and Afghanistan and crossed Mesopotamia. On his return to Italy he dictated the famous 'Relatio' (report) of his travels to his brother-monk, Guglielmo di Solagna, in St. Anthony's monastery in Padua. Studies on the Relatio, which continue even today, can be found in major European libraries and archives. Odorico was a contemporary of Dante Alighieri {see planet (2999)} and has been linked to Marco Polo. (M 17659)

(4638) Estens

1989 EG. Discovered 1989 March 2 by R. H. McNaught at Siding Spring.

Named for John (Jack) Locke Estens (1919-), Australian astronomy educator. Following a life of farming, he devoted his retirement to bring astronomy to the public with his handsomely constructed observatory and lecture room in Gilgandra, N.S.W. Opened in 1975, the observatory saw a steady stream of visitors until his second retirement in 1996. Always friendly and enthusiastic, he provided a service missing at the professional observatories, with his personal tours of the night sky. (M 29143)

(4639) Minox

1989 EK_2. Discovered 1989 March 5 by T. Seki at Geisei.

Named in honor of the Minox Club, a group of Japanese miniature-camera enthusiasts established in 1968. The discoverer is a member of the club, which holds an exhibition of still works and 9-mm movies each year in Tokyo. (M 18308)

(4640) Hara

1989 GA. Discovered 1989 April 1 by Y. Kushida and O. Muramatsu at Yatsugatake.

Named in honor of Megumi Hara, professor at Aoyama-Gakuin University, and an authority on star names, myths and history of the constellations. An observer of variable stars for many years, and long very active in the popularization of astronomy through lectures and articles, he is a leading member of the advisory committee of the Gotoh planetarium and Astronomical Museum. (M 17659) Named by the discoverers following a suggestion by S. Murayama.

(4641) 1990 QT_3

Discovered 1990 August 30 by K. Endate and K. Watanabe at Kitami.

(4642) 1990 QG_4

Discovered 1990 August 23 by H. E. Holt at Palomar.

(4643) 1990 QD_6

Discovered 1990 August 23 by H. E. Holt at Palomar.

(4644) Oumu

1990 SR_3. Discovered 1990 September 16 by A. Takahashi and K. Watanabe at Kitami.

Named for a town, noted for its fishing industry, situated on the Sea of Okhotsk some 50 km of Kitami {see planets (4042), (3785)}. (M 19338)

(4645) Tentaikojo

1990 SP_4. Discovered 1990 September 16 by T. Fujii and K. Watanabe at Kitami.

Named for a museum of stars and their images, scheduled to be opened in Sapporo in 1993. Originally built in 1892 as a brewery, the historic and beautiful building is becoming a "star factory", where visitors can experience displays and stories about constellations and the birth of stars. (M 20837)

(4646) Kwee

4009 P-L. Discovered 1960 September 24 by C. J. van Houten and I. van Houten-Groeneveld at Palomar.

Named in honor of the Leiden astronomer Kiem Keng Kwee (1927-), who works intensively on eclipsing variable stars, using their lightcurves to determine the characteristics of those systems by means of the Wilson-Devinney program. While observing at Palomar in 1963 he codiscovered a well-known short-period comet. (M 20160)

(4647) Syuji

1931 TU_1. Discovered 1931 October 9 by K. Reinmuth at Heidelberg.

Named in honor of Syuji Hayakawa (1958-), an active observer of minor planets and comets. (M 23540)

Name proposed by T. Kobayashi, who found the identifications involving this minor planet.

(4648) Tirion

1931 UE. Discovered 1931 October 18 by K. Reinmuth at Heidelberg.

Named in honor of the Dutch cartographer Wil Tirion, author of the *Sky Atlas 2000.0* and of many other modern celestial maps. (M 22503)

Name proposed and citation prepared by J. Meeus, endorsed by G. Klare and L. D. Schmadel.

(4649) Sumoto

1936 YD. Discovered 1936 December 20 by M. Laugier at Nice.

Named for the chief city of Awaji island on the Seto inland sea of Japan. The Oriental Astronomical Association's Computing Service is based in this city. Many observations of comets and minor planets by Japanese amateurs, as well as identifications and orbits, are collected there before being reported to the Minor Planet Center. (M 22503)

Name proposed by S. Nakano, who found the identifications involving this object and who lives in Sumoto.

(4650) Mori

1950 TF. Discovered 1950 October 5 by K. Reinmuth at Heidelberg.

Named in memory of Kiyoshi Mori (1906-1976), who calculated more than 400 circular, elliptical and parabolic orbits of minor planets and comets using an abacus and logarithm tables. After carrying out a statistical study of orbits of minor planets at Tokyo Imperial University in 1929 under the guidance of Hirayama {see planet (1999)}, he was a professor of physics, mathematics and statistics at many high schools and colleges. (M 18308)

Name proposed by H. Oishi, who found the identifications involving this minor planet.

(4651) Wongkwancheng

1957 UK_1. Discovered 1957 October 31 at the Purple Mountain Observatory at Nanking.

Named in memory of Kwan-cheng Wong (1907-1986), a leader in Chinese industrial and commercial circles. Wong was devoted to educational and social welfare projects all his life. The K. C. Wong Education Foundation he established has supported thousands of Chinese scholars studying abroad. (M 32788)

(4652) Iannini

1975 QO. Discovered 1975 August 30 at the Felix Aguilar Observatory at El Leoncito.

Named in honor of Gualberto Mario Iannini (1917-), Argentinean astronomer who has worked in the field of astrometry for almost 40 years. As a student, Iannini studied the orbit of C/1942 C1 (Whipple-Bernasconi-Kulin). In 1962 he joined the faculty of the School of Mathematics, Astronomy and Physics of the National University of Córdoba and became head of the Department of Astrometry of the Córdoba Observatory, where he finished the reduction of the meridian circle catalogue of the Southern Polar Cap, known as the Córdoba E catalogue. (M 33786)

(4653) Tommaso

1976 GJ_2. Discovered 1976 April 1 by N. S. Chernykh at Nauchnyj.

Named for Tommaso Campanella (1568-1639), Italian monk who was a philosopher, poet and political figure. He is known primarily as the author of *Civitās Soleil* and as a founder of Utopian Communism. Campanella was imprisoned for 27 years by the Spanish Inquisition. (M 23352)

(4654) Gor'kavyj

1977 RJ_6. Discovered 1977 September 11 by N. S. Chernykh at Nauchnyj.

Named in honor of Nikolaj Nikolaevich Gor'kavyj (1959-), staff member of the Crimean Astrophysical Observatory, celestial mechanician and cosmogonist. He has constructed a single model for the formation of the satellite systems of Jupiter, Saturn and Neptune and has offered an explanation for the origin of the retrograde satellites of the giant planets and the features of Neptune's rings. (M 27127)

(4655) Marjoriika

1978 RS. Discovered 1978 September 1 by N. S. Chernykh at Nauchnyj.

Named in honor of Marjo Riika Kuusela (1964-), specialist in Russian literature, whose perfect knowledge of the Russian language, lively wit and amiable disposition won the respect and friendship of all Russian-speaking participants at the Asteroids, Comets, Meteors 1991 conference in Flagstaff, Arizona. (M 20160)

(4656) 1978 VZ_3

Discovered 1978 November 7 by E. F. Helin and S. J. Bus at Palomar.

(4657) Lopez

1979 SU_9. Discovered 1979 September 22 by N. S. Chernykh at Nauchnyj.

Named in honor of Alvaro Lopez Garcia (1941-), professor of astronomy at Valencia University, director of the Valencia University Observatory and a specialist in astrometry and the dynamics of minor planets. In recent years he has initiated and organized several astronomical conferences in Spain that have attracted growing international participation. (M 26762)

(4658) Gavrilov

1979 SO_{11}. Discovered 1979 September 24 by N. S. Chernykh at Nauchnyj.

Named in honor of Aleksej Evgenievich Gavrilov, friend of the discoverer, famous Ukrainian architect of bridges, director of the "Kievinterproekt" Institute of the International Academy of Architecture. He designed the south bridge over the Dnieper river at Kiev and the bridge over the Daugava river at Riga. He is the great-grandson of the astronomer and physicist S. V. Shcherbakov {see planet (3886)} and the son of the geographer M. V. Shcherbakova. (M 27127; M 27147)

(4659) Roddenberry

1981 EP_{20}. Discovered 1981 March 2 by S. J. Bus at Siding Spring.

Named in memory of Eugene "Gene" W. Roddenberry (1921-1991), creator and producer of the television series "Star Trek", "Star Trek: The Next Generation", and six Star Trek motion pictures. Roddenberry explored the human condition

through the medium of science fiction, often circumventing television network censors to expound on controversial social and political topics. His unique vision of a positive future for a united humanity inspired a variety of devoted fans. Today Roddenberry, the starship Enterprise, and its crews are known worldwide, and fans number in the millions. (M 19698)
Citation provided by I. Heyer at the request of the discoverer.

(4660) Nereus

1982 DB. Discovered 1982 February 28 by E. F. Helin at Palomar.

Named for the ancient Greek god Nereus, a son of Pontus and Gaia {see planet (1184)}. A benevolent sea-god associated with ancient origins (mythology, if not science), he had the power of prophecy. (M 19338)

Name proposed by Robert M. Cutler, a Planetary Society member interested in near-earth asteroids. To English-language speakers the name seems particularly appropriate for such an object.

(4661) 1982 WM

Discovered 1982 November 17 at the Centro Astronómico de Yebes at Yebes.

(4662) 1984 HL

Discovered 1984 April 19 by A. Mrkos at Kleť.

(4663) 1984 SM_1

Discovered 1984 September 27 by A. Mrkos at Kleť.

(4664) Hanner

1985 PJ. Discovered 1985 August 14 by E. Bowell at Anderson Mesa.

Named in honor of Martha S. Hanner, planetary scientist at the Jet Propulsion Laboratory who specializes in groundbased and space studies of particulate matter in the solar system, especially cometary dust. Her publications attest to her work as an enthusiastic observer of comets at infrared wavelengths, as an organizer of the P/Halley infrared monitoring program at the Infrared Telescope Facility on Mauna Kea, as a champion of improvements in infrared photometric standards, and on the nature of zodiacal light. She has investigated the formation of infrared spectral features in dust and the scattering properties of dust. Hanner is an energetic supporter of better science education for young people, to which end she has been a teacher and organizer of space science seminars for science teachers. (M 18461)
Citation provided by Ray L. Newburn at the request of the discoverer.

(4665) Muinonen

1985 TZ_1. Discovered 1985 October 15 by E. Bowell at Anderson Mesa.

Named in honor of Karri O. Muinonen, planetary scientist at the Lowell Observatory. Formerly at the observatory and astrophysics laboratory of the University of Helsinki, Muinonen worked on the light-scattering properties of inhomogeneous media and independently originated the suggestion that coherent backscattering is responsible for both the opposition spike and the negative polarization of light from the surfaces of many atmosphereless bodies. He has since turned to the study of the spatial distribution of minor planets, particularly the population of planet crossers that might pose a hazard to the earth, and he has devised a new way of assessing the uncertainty of minor planet orbits, and hence ephemerides. (M 18461)

(4666) Dietz

1986 JA_1. Discovered 1986 May 4 by C. S. Shoemaker and E. M. Shoemaker at Palomar.

Named in honor of Robert S. Dietz, geologist and professor emeritus at Arizona State University. Dietz is a pioneer in the study and recognition of terrestrial impact structures, and he was one of the early proponents of the impact origin of the craters of the moon. He championed the concept of shatter cones as diagnostic features of terrestrial impact structures and discovered shatter cones at many of the impact sites now known. On the basis of evidence from shatter cones, Dietz revived the hypothesis of impact origin of the great Vredefort Ring of South Africa and proposed an impact origin for the Sudbury Basin, Ontario. (M 18461)

(4667) Robbiesh

1986 VC. Discovered 1986 November 4 by R. H. McNaught at Siding Spring.

Named for Hans-Christian Robert (Robbie) Wade Schmidt-Harms (1986-), son of Soo Tan and stepson of the discoverer. This minor planet was discovered three weeks after Robbie's birth. (M 29670)

(4668) 1987 DX_5

Discovered 1987 February 21 by H. Debehogne at La Silla.

(4669) Høder

1987 UF_1. Discovered 1987 October 27 by P. Jensen and K. Augustesen at Brorfelde.

Named after the blind god in Norse mythology, son of Odin {see planet (3989)} and brother of Balder {see planet (4059)}, whom he killed with a mistletoe shaft with the treacherous help of Loke {see planet (4862)}. (M 22503)

(4670) Yoshinogawa

1987 YJ. Discovered 1987 December 19 by T. Seki at Geisei.

Named for the longest river on Shikoku {see planet (4223)} island, flowing east-west for a distance of 194 km. (M 19338)

(4671) Drtikol

1988 AK_1. Discovered 1988 January 10 by A. Mrkos at Kleť.

Named in memory of František Drtikol (1883-1961), outstanding Czech photographer well-known for his black-and-white portraits in the Art Nouveau style and fine-art nude studies using geometric patterns of shadows and light. (M 34340)
Name suggested by M. Tichý.

(4672) Takuboku

1988 HB. Discovered 1988 April 17 by S. Ueda and H. Kaneda at Kushiro.

Named in honor of the poet Takuboku Ishikawa (1886-1912), who lived in the late Meiji period in Japan. He published his first collection of poems, *Akogale*, when he was 20. His most famous work is *Ichiakunosuna*, containing 551 poems, published in 1910. (M 21131)

(4673) Bortle

1988 LF. Discovered 1988 June 8 by C. S. Shoemaker at Palomar.

Named for John E. Bortle, an American astronomer specializing in comets and variable stars. One of the great visual observers, he has made more than 2,000 observations of approximately 150 cometary apparitions over the course of three decades. He has often been first to report unusual cometary activity and is well-known for his analyses of visual magnitude estimates of comets. He has also reported tens of thousands of variable star observations. Bortle has written the *Comet Digest* column in *Sky and Telescope* for more than a decade and has edited the American Association of Variable Star Observers Circular since its founding in 1970. (M 17659)
Name suggested and citation prepared by C. S. Morris.

(4674) Pauling

1989 JC. Discovered 1989 May 2 by E. F. Helin at Palomar.

Named in honor of Professor Linus Pauling on the occasion of his ninetieth birthday, 1991 Feb. 28. Pauling has had a long and distinguished career, spending 37 years as a Caltech faculty member, including 22 years as chairman of Caltech's Division of Chemistry and Chemical Engineering. He is the recipient of Nobel prizes for both Chemistry {1954} and Peace {1962}. The discoverer, and her husband Ronald, a Caltech graduate, are long time admirers of Pauling. (M 17981)
Asteroid tribute endorsed by the Caltech community.

(4675) Ohboke

1990 SD. Discovered 1990 September 19 by T. Seki at Geisei.

Named for one of the valleys of the Yoshino river, famous for its tinted autumn leaves. (M 19338)

(4676) Uedaseiji

1990 SD_4. Discovered 1990 September 16 by T. Fujii and K. Watanabe at Kitami.

Named in honor of Seiji Ueda (1952-), well known for his observations of comets and minor planets. He lives in Kushiro {see planet (4096)} and has been an enthusiastic astronomer since his junior high school days. He played an active part in ice hockey as a Japanese-league player for six years. After retirement, he renewed his interest in astronomy. Since 1987 he has used a Wright-Schmidt Camera of 16 cm aperture and has discovered many minor planets. (M 22503)

(4677) Hiroshi

1990 SQ_4. Discovered 1990 September 26 by A. Takahashi and K. Watanabe at Kitami.

Named in honor of Hiroshi Kaneda (1953-), well known for his orbit calculations of comets and minor planets. He lives in Sapporo {see planet (3473)} and is a computer programmer who has written much astronomical software. Recently, he has been finding many identifications of minor planets. He has also put much energy into the astrometric measurement of films he has obtained at the Kushiro {see planet (4096)} Observatory. (M 22503)

(4678) Ninian

1990 SS_4. Discovered 1990 September 24 by R. H. McNaught at Siding Spring.
Named in honor of Ninian T. McNaught, father of the discoverer. (M 17982)

(4679) Sybil

1990 TR_4. Discovered 1990 October 9 by R. H. McNaught at Siding Spring.
Named in honor of Sybil McNaught, mother of the discoverer. (M 17982)

(4680) Lohrmann

1937 QC. Discovered 1937 August 31 by H.-U. Sandig at Bergedorf.

Named for Wilhelm Gotthelf Lohrmann, an amateur astronomer and author of the famous lunar topographic atlas *Topographie der sichtbaren Mondoberfläche* (1824). Lohrmann was director of the Dresden "Mathematisch-Physikalischer Salon". When H.-U. Sandig founded the Institute of Geodetical Astronomy at the Dresden University of Technology in 1961, he chose the name of Lohrmann for his new institute, now known as the Lohrmann Observatory. (M 18461)
Lohrmann is honored for his monumental selenographic work by a lunar crater.

(4681) Ermak

1969 TC_2. Discovered 1969 October 8 by L. I. Chernykh at Nauchnyj.

Named for Ermak Timofeevich (?-1585), hero of Russian folklore and the Cossack chieftain who led the campaign to join Siberia to Russia. (M 23136)

(4682) Bykov

1973 SO_4. Discovered 1973 September 27 by L. I. Chernykh at Nauchnyj.

Named in memory of Leonid Fyodorovich Bykov (1928-1979), talented Ukrainian cinematic actor and producer. (M 23137)

(4683) Veratar

1976 GJ_1. Discovered 1976 April 1 by N. S. Chernykh at Nauchnyj.

Named in honor of Vera Petrovna Tarashchuk, an astrophysicist at the Astronomical Observatory of Kiev University. An active observer of major planets, minor planets and comets, she is known for her contribution to photometric and spectroscopic research on minor bodies. She also studied the association of cometary processes with solar activity, as well as the structure and rotation of minor planets. (M 30095)

(4684) Bendjoya

1978 GJ. Discovered 1978 April 10 by H. Debehogne at La Silla.

Named in honor of Philippe Bendjoya, astronomer at the Nice Observatory, who has carried out fruitful research on the identification and physical analysis of asteroid families, as well as the dynamics of planetary rings. Bendjoya was the first to apply to these fields the so-called wavelet analysis technique. (M 23137)

(4685) Karetnikov

1978 SP_6. Discovered 1978 September 27 by N. S. Chernykh at Nauchnyj.

Named in honor of Valentin Grigorievich Karetnikov (1938-), professor at Odessa University and director of the astronomical observatory there. His fields of research include light variations of artificial earth satellites, the structure and evolution of eclipsing double stars and the motion of gas in close-binary systems. (M 27127)

(4686) Maisica

1979 SX_2. Discovered 1979 September 22 by N. S. Chernykh at Nauchnyj.

Named in honor of Maria Luisa Grima Garcia, a charming Spanish woman who is the soul of her family, a hospitable hostess, and founder and curator of a small home museum. Maisica is a pet name for Maria Luisa. (M 22503)

(4687) Brunsandrej

1979 SJ_{11}. Discovered 1979 September 24 by N. S. Chernykh at Nauchnyj.

Named in honor of Andrej Vladimirovich Bruns (1931-), staff member of the Crimean Astrophysical Observatory, an authority on space astrophysics who has originated several unique instruments for ultraviolet observations of the sun, stars and galaxies from satellites and spacecraft. He designed the large Orbiting Solar Telescope controlled by cosmonauts on Salyut 4 in 1975. A relative of Ernst Heinrich Bruns {see planet (901)}, director of the Leipzig Observatory at the beginning of this century, he is related on his mother's side to Euler {see planet (2002)}. (M 26762; M 27749)

(4688) 1980 WF

Discovered 1980 November 29 by C. T. Kowal at Palomar.

(4689) Donn

1980 YB. Discovered 1980 December 30 by E. Bowell at Anderson Mesa.

Named in honor of Bertram D. Donn for his fundamental contributions to our understanding of solar system astrophysics. In addition to proposing an innovative theory of cometary composition, formation and evolution, Donn succeeded in deriving generally accepted fundamental properties of cometary nuclei that form the basis of current theories. He also carried out very original research on cometary

structures using fractals and on the condensation of pre-cometary and interstellar materials. (M 18461)
Citation provided by J. H. Rahe at the request of the discoverer.

(4690) Strasbourg

1983 AJ. Discovered 1983 January 9 by B. A. Skiff at Anderson Mesa.

Named for the French city of Strasbourg, capital of Alsace {see planet (971)}. This city near the Rhine is known for its long history, majestic cathedral, and as a center of international cooperation. It is also home of the Centre de Données Astronomiques, which maintains the SIMBAD {see planet (4692)} stellar database. (M 27127)

(4691) Toyen

1983 TU. Discovered 1983 October 7 by A. Mrkos at Kleť.

Named in memory of Toyen (1902-1980), Czech painter and graphic artist, born Marie Čermínová. She is known for her imaginative abstraction, called artificialism, resulting in a surrealistic style. (M 34340)
Name suggested by J. Tichá.

(4692) SIMBAD

1983 VM_7. Discovered 1983 November 4 by B. A. Skiff at Anderson Mesa.

Named in recognition of all the staff members of the Centre de Données Astronomiques, situated at the Observatoire de Strasbourg, for their efforts to create and maintain the SIMBAD (Set of Identifications, Measurements, and Bibliography for Astronomical Data) stellar bibliographic database and its associated on-line catalogue retrieval services. (M 27127)

(4693) Drummond

1983 WH. Discovered 1983 November 28 by E. Bowell at Anderson Mesa.

Named in honor of Jack D. Drummond of the Steward Observatory, University of Arizona. Drummond's analysis of orbital similarities led to the identification of a cometary parent for the Epsilon Geminids and to the identification of streams among near-earth asteroids. He has done extensive analysis of asteroid lightcurves to find pole directions and shapes for more than 25 objects, as well as studies of phase curves which suggest the existence of both rough and smooth surfaces among the asteroids. As one of the first to apply speckle interferometry to these bodies, he developed many theoretical contributions to the analysis of speckle data and produced the first speckle images showing features on the surface of an asteroid, namely, that of (4) Vesta. His enthusiasm for studies of asteroids, comets and meteors has made him a pleasurable colleague for collaborative efforts. (M 18462)
Citation provided by Donald R. Davis at the request of the discoverer.

(4694) Festou

1985 PM. Discovered 1985 August 14 by E. Bowell at Anderson Mesa.

Named in honor of Michel C. Festou of the Observatoire Midi-Pyrénées, Toulouse, for his extensive and wide-ranging studies of comets, both theoretical and observational. Festou was the first to calculate a vectorial model for the dissociation of water in cometary comae, and he has made major contributions to the theoretical understanding of emission by various species, including, for example, [O I] and CN. He has been a preeminent practitioner of ultraviolet spectroscopy of comets, identifying features due to OH^+, CO_2^+ and [O I] , and he has also contributed numerous ground-based observations of comets. (M 18462)
Citation provided by Michael F. A'Hearn at the request of the discoverer.

(4695) 1985 RU_3

Discovered 1985 September 7 by H. Debehogne at La Silla.

(4696) Arpigny

1985 TP. Discovered 1985 October 15 by E. Bowell at Anderson Mesa.

Named in honor of Claude Arpigny of the Université de Liège, to recognize his preeminence in high-resolution cometary spectroscopy. Arpigny's fundamental studies of the excitation of molecular spectra in comets are unparalleled for care and detailed understanding. He has made significant contributions to our understanding of the emission by NH_2, [O I] , CN and CH. He has also shown care and diligence by encouraging the work of others both in spectroscopy and other types of cometary studies, in his service to the community in preparing the forthcoming "Atlas of Cometary Spectra" and in various activities of IAU Commission 15. (M 18462)

Citation provided by Michael F. A'Hearn at the request of the discoverer.

(4697) 1986 QO

Discovered 1986 August 26 by H. Debehogne at La Silla.

(4698) Jizera

1986 RO_1. Discovered 1986 September 4 by A. Mrkos at Kleť.

Named for a Czech river rising in the Jizerské hory mountains and continuing through the towns of Turnov {see planet (4054)} and Mladá Boleslav to its confluence with the Labe river. (M 34340)

Name proposed by Z. Moravec.

(4699) Sootan

1986 VE. Discovered 1986 November 4 by R. H. McNaught at Siding Spring.

Named for Soo Hoay Tan (1956-), partner of the discoverer and high-school science teacher. This asteroid was discovered two days before Soo's thirtieth birthday. (M 29670)

(4700) Carusi

1986 VV_6. Discovered 1986 November 6 by E. Bowell at Anderson Mesa.

Named in honor of Andrea Carusi, Istituto di Astrofisica Spaziale, in recognition of his work on the dynamics of cometary orbits and studies of the identification of asteroid families. Carusi was one of the first to use numerical techniques to study the effects of close planetary encounters in changing cometary orbits and was a principal contributor to the atlas of orbital patterns at close encounters. He identified many cases of temporary satellite capture about Jupiter and was the first, together with Giovanni Valsecchi {see planet (3725)}, to recognize the importance of near-tangent orbits in the case of close planetary encounters. He was also one of the first to apply the technique of numerical cluster analysis to the problem of the identification of asteroid families, and his early result on the number of families has withstood the test of time. His enthusiasm for solar system studies has led him to deliver many popular talks on a variety of topics in planetary science for the general public. (M 18462)

Citation provided by Donald R. Davis at the request of the discoverer.

(4701) Milani

1986 VW_6. Discovered 1986 November 6 by E. Bowell at Anderson Mesa.

Named in honor of Andrea Milani of the University of Pisa for his contributions to understanding the dynamics of asteroid orbits and the use of sophisticated numerical techniques to investigate the long-term evolution of orbits in the solar system. Milani was a pioneer in applying massively parallel computers to a variety

of problems in solar system dynamics, ranging from integration of outer planet orbits over timescales of hundreds of millions of years to the evolution of hundreds of asteroids moving on orbits crossing that of the earth and other planets. He contributed to the development of refined proper elements for more than 4000 mainbelt asteroids and to the identification of secular resonances throughout the solar system. His very intense and energetic style of doing research has resulted in an impressive number of contributions so far in his career. (M 18463)
Citation provided by Donald R. Davis at the request of the discoverer.

(4702) Berounka

1987 HW. Discovered 1987 April 23 by A. Mrkos at Kleť.

Named for a Czech river originating in Plzeň {see planet (2613)} as a confluence of four rivers: Mže, Radbúza, Úhlava and Úslava. It continues through the town of Beroun to its confluence with the Vltava river near Prague {see planets (2123) and (2367), respectively}. (M 34340)
Name proposed by Z. Moravec.

(4703) Kagoshima

1988 BL. Discovered 1988 January 16 by M. Mukai and M. Takeishi at Kagoshima.

Named for the prefecture and its capital city, home of the discoverers, located at the southern tip of Kyushu Island in southwestern Japan. The rocket launching centers of the Institute of Space and Astronautical Science and the National Space Development Agency are located in Kagoshima prefecture. The city, often referred to as the "Naples of the East", is dominated by the presence of Sakurajima, one of the most active volcanoes in Japan, in the adjoining Kinko Bay. (M 18143)

(4704) Sheena

1988 BE_5. Discovered 1988 January 28 by R. H. McNaught at Siding Spring.

Named for Sheena Fleming Phillips (1952-), sister of the discoverer and a landscape artist in Haddington, Scotland. (M 29143)

(4705) Secchi

1988 CK. Discovered 1988 February 13 at the Osservatorio San Vittore at Bologna.

Named in memory of Angelo Secchi (1818-1878), Italian astronomer, director of the observatory of the Collegio Romano in Rome from 1848 to 1878. Famous for his work on stellar spectroscopy, he made the first spectroscopic survey of the heavens, and his classification scheme divided the spectra of the stars into four groups. Secchi also made an extensive study of solar phenomena and was a co-founder of the Società degli Spettroscopisti Italiani, now the Società Astronomica Italiana. (M 20160)
Secchi is also honored by craters on Mars and on the Moon.

(4706) 1988 DR

Discovered 1988 February 16 by R. Rajamohan at Kavalur.

(4707) Khryses

1988 PY. Discovered 1988 August 13 by C. S. Shoemaker and E. M. Shoemaker at Palomar.

Named for the priest of Apollo whose daughter, Khryseïs, was abducted by Agamemnon {see planets (1862), (202), and (911)}. Apollo, angered, sent a plague among the Greeks, until Agamemnon then took Achilles' girl, Briseïs {see planet (655)}, thus sparking the quarrel between Agamemnon and Achilles that led to the events described in the *Iliad*. (M 18144)
Name and citation provided by R. Preston at the request of the discoverers.

(4708) Polydoros

1988 RT. Discovered 1988 September 11 by C. S. Shoemaker and E. M. Shoemaker at Palomar.

Named for the brother of Lykaon {see planet (4792)} and the youngest son of Priam {see planet (884)}; born of the queen Laothoe. Although Polydoros was the swiftest of the sons of Priam, Priam refused to let the boy fight; but Polydoros rushed Achilles {see planet (588)} anyway, and Achilles killed him beside the River Skamander, near his brother Lykaon. (M 18144; M 18323)
Name and citation provided by R. Preston at the request of the discoverers.

(4709) Ennomos

1988 TU_2. Discovered 1988 October 12 by C. S. Shoemaker and E. M. Shoemaker at Palomar.

Named for a warrior of Mysia who was a prophet and a seer of birdflight. Ennomos was one of many Trojans butchered by Achilles {see planet (588)} in the waters of the River Skamander. (M 18144)
Name and citation provided by R. Preston at the request of the discoverers.

(4710) Wade

1989 AX_2. Discovered 1989 January 4 by R. H. McNaught at Siding Spring.

Named for Wade Richard Butler (1949-), a friend of the discoverer. Son of the famous "bare-foot" bushwalker Dot Butler, he had a long-time interest in the bush and remote places. After graduating in engineering, he overwintered in Antarctica working in cosmic-ray physics. His romantic passion for astronomy brought him to the Warrumbungle Mountains, in the shadow of Siding Spring, where he and his wife Margaret, with many friends, built a home and an observatory. In December 1995 he disappeared during a solo bushwalk in a very rugged part of southeastern Tasmania. This minor planet places Wade Butler in the sky with his children Lyra, Rigel, Leo and Eric Antares Butler, as he would have wished. (M 29143)

(4711) Kathy

1989 KD. Discovered 1989 May 31 by H. E. Holt at Palomar.

Named in honor of Kathleen Garnette Moeller, daughter of the discoverer. (M 18463)

(4712) Iwaizumi

1989 QE. Discovered 1989 August 25 by K. Endate and K. Watanabe at Kitami.

Named for a forestry town of population 15,000 in the eastern part of Iwate Prefecture. Birthplace of the first discoverer, Iwaizumi is famous for Ryusen-Do Cave, one of the Big Three Limestone Caves in Japan. (M 20838)

(4713) Steel

1989 QL. Discovered 1989 August 26 by R. H. McNaught at Siding Spring.

Named in honor of Duncan Steel, Anglo-Australian astronomer who has conducted research on the origin and evolution of asteroids, comets and meteoroids. In particular, he has shown that several Apollo asteroids are the parents of meteor showers, indicating that these Apollos are likely to be extinct or moribund cometary nuclei. He has also worked extensively on radar observations of the meteoric influx to the atmosphere, planetary impact rates, and the dynamics of small solar system bodies. (M 17982)

(4714) Toyohiro

1989 SH. Discovered 1989 September 29 by T. Fujii and K. Watanabe at Kitami.

Named in honor of the journalist Toyohiro Akiyama (1942-), the first Japanese citizen to travel into space. He lifted off on 1990 Dec. 1 in Soyuz to travel to the space station Mir, and he described the effect of weightlessness and space sickness. (M 21131)

(4715) 1989 TS_1

Discovered 1989 October 9 by Y. Oshima at Gekko.

(4716) 1989 UL_5

Discovered 1989 October 30 by S. J. Bus at Cerro Tololo.

(4717) Kaneko

1989 WX. Discovered 1989 November 20 by Y. Mizuno and T. Furuta at Kani.

Named in honor of Isao Kaneko (1918-), who worked toward social education during reconstruction in postwar Japan. He promoted cultural activities in mountain villages, established a private observatory and is now director of the Misono Space Education Service. Awarded the Chunichi Social Meritorious Prize by the Chubu-Nippon Press in 1976, he was also commended for his meritorious cultural service by the Board Education of Aichi Prefecture in 1979. (M 26762)

(4718) Araki

1990 VP_3. Discovered 1990 November 13 by T. Fujii and K. Watanabe at Kitami.

Named in honor of Chikara Araki (1946-), an astronomical photographer in Hokkaido. He has formed a group called "Konpeito", and has been taking a lot of impressive astronomical photographs. (M 22503)

(4719) Burnaby

1990 WT_2. Discovered 1990 November 21 by S. Ueda and H. Kaneda at Kushiro.

Named for the city in British Columbia, Canada, with a population of 150,000. Burnaby and Kushiro {see planet (4096)}, which are at almost the same latitude, have been sister cities since 1965 and exchange students and share cultural experiences with each other. (M 21132)

(4720) Tottori

1990 YG. Discovered 1990 December 19 by S. Ueda and H. Kaneda at Kushiro.

Named for a central city of the Sanin area, located on the Japan Sea, sister city of Kushiro {see planet (4096)} since 1963. (M 22503)

(4721) Atahualpa

4239 T-2. Discovered 1973 September 29 by C. J. van Houten and I. van Houten-Groeneveld at Palomar.

On the 500th anniversary of the discovery of the Americas by Columbus {see planet (327)}, we want to remember the last king of the Incas in Peru: Atahualpa. He was born c. 1502 and was killed - even after he paid a room full of jewels and gold as a ransom - by the conquistador Pizarro in 1533. (M 20160)

(4722) Agelaos

4271 T-3. Discovered 1977 October 16 by C. J. van Houten and I. van Houten-Groeneveld at Palomar.

Agelaos was a shepherd, who got the order to expose the baby Paris {see planet (3317)}, son of Priamus {see planet (884)}, king of Troy. He did not fulfill this order but raised and educated the boy. When Paris was still working for Agelaos as a shepherd, he had to give the 'judgement of Paris'. (M 18308)

(4723) Wolfgangmattig

1937 TB. Discovered 1937 October 11 by K. Reinmuth at Heidelberg.

Named in honor of Wolfgang Mattig (1927-), German solar physicist and cosmologist at the Freiburg Kiepenheuer-Institut, on the occasion of his retirement.

In his thesis, Mattig worked on relativistic cosmology and, in 1957, he discovered an analytical relation between the redshift and the apparent magnitude of galaxies. He took an active part in the development of the Teide Observatory, Canary Islands. Since 1980, Mattig has been the German representative in the Solar Physics Commission of COSPAR. (M 22503)

Name proposed and citation prepared by J. Schubart, endorsed by G. Klare and L. D. Schmadel.

(4724) Brocken

1961 BC. Discovered 1961 January 18 by C. Hoffmeister and J. Schubart at Tautenburg.

Named for the highest summit in the Harz Mountains that offers a wide view on a large part of northern Germany. (M 18144)

Name proposed by the second discoverer, following a suggestion of F. Börngen.

(4725) Milone

1975 YE. Discovered 1975 December 31 at the Felix Aguilar Observatory at El Leoncito.

Named in honor of Luis Ambrosio Milone (1933-), Argentinean astronomer who has been working in the field of stellar astrophysics for more than 30 years. Since 1964 he has been on the faculty of the School of Mathematics, Astronomy and Physics of the National University of Córdoba, where he teaches graduate and post-graduate courses in astronomy and astrophysics. Milone was responsible for building a photoelectric photometer for the Bosque Alegre Observatory 1.5-m telescope. He was the initiator of photoelectric photometry in Córdoba and one of the first astronomers to apply these techniques. (M 33786)

(4726) Federer

1976 SV_{10}. Discovered 1976 September 25 at the Harvard College Observatory at Harvard.

Named in honor of Charles A. Federer, Jr., founding editor of "Sky and Telescope" {see planet (3243)}. In January 1940, Federer became editor of "The Sky", a monthly popular astronomy magazine published since 1936 by the Hayden Planetarium in New York City. Then in November 1941, "The Sky" was combined with a similar publication, "The Telescope", which was founded at Perkins Observatory in 1931 but since 1934 had been published by Harvard College Observatory. Federer became editor-in-chief of "Sky and Telescope" in 1964, a post he held until his retirement a decade later. (M 19339)

(4727) Ravel

1979 UD_1. Discovered 1979 October 24 by F. Börngen at Tautenburg.

Named for the French composer Maurice Ravel (1875-1937), a significant representative of impressionism in music. His preference for Spanish music is manifested in his works *Rhapsodie espagnole* and *Bolero*. (M 18144)

(4728) Lyapidevskij

1979 VG. Discovered 1979 November 11 by N. S. Chernykh at Nauchnyj.

Named in memory of Anatolij Vasil'evich Lyapidevskij (1908-1983), Soviet airman who participated in the search and rescue of the crew members of the "Chelyuskin", a ship crushed by ice in the Arctic Ocean on 1934 Feb. 13. (M 27127)

(4729) Mikhailmil'

1980 RO_2. Discovered 1980 September 8 by L. V. Zhuravleva at Nauchnyj.

Named in memory of Mikhail Leont'evich Mil' (1909-1970), scientist and designer of helicopters with which 60 official world records were set. (M 34620)

(4730) 1980 XZ
Discovered 1980 December 7 at the Purple Mountain Observatory at Nanking.

(4731) 1981 EE_9
Discovered 1981 March 1 by S. J. Bus at Siding Spring.

(4732) Froeschlé
1981 JG. Discovered 1981 May 3 by E. Bowell at Anderson Mesa.

Named in honor of Claude and Christiane Froeschlé, Observatoire de la Côte d'Azur, Nice. Husband and wife, the Froeschlés have both contributed to our understanding of the dynamics of minor bodies (minor planets, comets, meteorites and dust particles) by numerical analysis of their orbital evolution. Claude was one of the first to investigate ergodic and chaotic regions in mean-motion resonances with Jupiter. He tested the so-called gravitational hypothesis for the formation of the Kirkwood gaps and showed that meteor streams situated in mean motion resonances may break into seperate parts. He also modeled the dynamical evolution of short-period comets using a Monte Carlo method. Christiane was one of the first to explore orbital evolution at secular resonances. She has shown that the principal secular resonances are good candidate sources for meteorites. (M 18463)
Citation provided by H. Scholl at the request of the discoverer.

(4733) ORO
1982 HB_2. Discovered 1982 April 19 at the Oak Ridge Observatory at Harvard.

Named in honor of the Oak Ridge Observatory, established in Harvard, Massachusetts, by the Harvard College Observatory in 1932 and since 1982 operated by the Smithsonian Astrophysical Observatory. This minor planet is one of the 39 discovered in the course of the Oak Ridge (also known for many years as the Agassiz Station) astrometric program, which has been conducted with the 1.5-m Wyeth reflector, the largest telescope in the eastern United States, for almost 20 years. (M 19339)

(4734) Rameau
1982 UQ_3. Discovered 1982 October 19 by F. Börngen at Tautenburg.

Named for Jean Philippe Rameau (1683-1764), the greatest French composer of the eighteenth century. He is well known for some operas, very charming piano pieces like *Le Rappel des Oiseaux* and for his fundamental writings on the theory of music. (M 18144)

(4735) Gary
1983 AN. Discovered 1983 January 9 by E. Bowell at Anderson Mesa.

Named in honor of George Gary Shoemaker, manager of Meteor Crater Enterprises, Inc., on the occasion of the Asteroids, Comets, Meteors 1991 Conference in Flagstaff, Arizona, with which his company is associated. As manager of the visitor facilities and museum at Meteor Crater, Arizona, Shoemaker has played a leading role in the preservation of this National Natural Landmark and its associated meteorites and in developing the presentation of the scientific story of the crater to the public. He has consistently encouraged and supported research at the crater and the use of the crater for educational purposes. (M 18463)
Citation provided by E. M. Shoemaker at the request of the discoverer.

(4736) Johnwood
1983 AF_2. Discovered 1983 January 13 by C. S. Shoemaker and E. M. Shoemaker at Palomar.

Named for John A. Wood {1932- }, geologist at the Harvard-Smithsonian

Center for Astrophysics, associate director of the planetary sciences division from 1981 to 1986. For more than 30 years, Wood has been a leading investigator of the petrography and petrology of chondritic meteorites as samples of primordial planetary materials. He also studied lunar rocks, beginning with those returned from the Apollo 11 mission. Wood is renowned for his theory that lunar anorthosites accumulated from a global magma ocean that developed very early in the history of the moon. (M 18464)

(4737) Kiladze

1985 QO_6. Discovered 1985 August 24 by N. S. Chernykh at Nauchnyj.

Named in honor of Rolan Il'ich Kiladze (1931-), astronomer at the Abastumani Observatory. His research has covered solar-system cosmogony and the dynamics and astrometry of Pluto, minor planets and artificial satellites. (M 27127; M 27147)

(4738) 1985 RZ_4

Discovered 1985 September 15 by D. B. Goldstein at Palomar.

(4739) Tomahrens

1985 TH_1. Discovered 1985 October 15 by E. Bowell at Anderson Mesa.

Named in honor of Thomas J. Ahrens, professor of geophysics at the California Institute of Technology. Ahrens has built and directed the premier laboratory for experimental impact and shock physics, using light-gas guns. He has used this laboratory to investigate a broad array of problems, including the equations of state of minerals at very hight pressure, shock metamorphism and cratering mechanics. He has also addressed problems of impact cratering by computer code calculations and investigated the accumulation and early bombardment of the earth and the evolution of the atmosphere. (M 18645)

Citation provided by E. M. Shoemaker at the request of the discoverer.

(4740) Veniamina

1985 UV_4. Discovered 1985 October 22 by L. V. Zhuravleva at Nauchnyj.

Named in memory of Veniamin Vasil'evich Somov (1945-1991), the discoverer's brother. (M 22503)

(4741) Leskov

1985 VP_3. Discovered 1985 November 10 by L. G. Karachkina at Nauchnyj.

Named in memory of Nikolaj Semenovich Leskov (1831-1895), well-known Russian writer. (M 22247)

(4742) Caliumi

1986 WG. Discovered 1986 November 26 at the Osservatorio San Vittore at Bologna.

Named in memory of Ferdinando Caliumi (1915-1993), well-known Italian amateur astronomer, generous constructor and restorer of astronomical telescopes. He was a great friend to the amateur astronomers at the discovery site, and he was also a member of both the Italian Astronomical Society and of the Unione Astrofili Italiani. (M 22829)

(4743) Kikuchi

1988 DA. Discovered 1988 February 16 by T. Fujii and K. Watanabe at Kitami.

Named in honor of Ryoko Kikuchi (1964-), who was backup to the first Japanese astronaut, Toyohiro Akiyama {see planet (4714)}, and helped to report details of environmental disruption. (M 21132)

(4744) 1988 RF_5

Discovered 1988 September 2 by H. Debehogne at La Silla.

(4745) Nancymarie

1989 NG_1. Discovered 1989 July 9 by H. E. Holt at Palomar.
Named in honor of Nancy Marie Martinez, daughter of the discoverer. (M 18464)

(4746) Doi

1989 TP_1. Discovered 1989 October 9 by A. Takahashi and K. Watanabe at Kitami.

Named in honor of Takao Doi (1954-), a backup payload specialist to Mamoru Mohri {see planet (4613)}. He has worked in the department of technology of the University of Tokyo and at the NASA Lewis Research Center. (M 21132)

(4747) Jujo

1989 WB. Discovered 1989 November 19 by S. Ueda and H. Kaneda at Kushiro.
Named for the place where the first discoverer works. (M 22829)

(4748) Tokiwagozen

1989 WV. Discovered 1989 November 20 by K. Suzuki and T. Urata at Toyota.

Named for Tokiwagozen (1138-?), beautiful mother of Yoshitsune {see planet (3178)}. She was captured by the Heike at the battle of Heiji in 1159 and was later in favor with Kiyomori {see planet (4375)}. (M 21132)

(4749) 1989 WE_1

Discovered 1989 November 22 by N. Kawasato at Uenohara.

(4750) Mukai

1990 XC_1. Discovered 1990 December 15 by T. Fujii and K. Watanabe at Kitami.

Named in honor of Chiaki Mukai (1952-), a backup payload specialist to Mamoru Mohri {see planet (4613)}. She has worked in the medical department of Keio University. (M 21132)

(4751) Alicemanning

1991 BG. Discovered 1991 January 17 by B. G. W. Manning at Stakenbridge.

Named in honor of Alice K. Manning, wife of the discoverer, for her support and encouragement of the discoverer's hobby over many years. The attention given to practical details such as warm clothing on cold nights, and the push needed to persuade him to finish the construction and fitting of measuring-machine encoders has contributed greatly to cometary astrometry and success in discovering minor planets. (M 18144)

(4752) Myron

1309 T-2. Discovered 1973 September 29 by C. J. van Houten and I. van Houten-Groeneveld at Palomar.

Myron of Eleutherae (fl. 470 B.C.) in Attica, one of the most celebrated Greek artists, was a pupil of Ageladus and an older contemporary of Phidias {see planet (4753)} and Polyclitus. His works, mostly in bronze, include his statue of the 'Argive runner Ladas' and the Discobolus (discus-thrower), several marble copies of which are known. (M 20160)

(4753) Phidias

4059 T-3. Discovered 1977 October 16 by C. J. van Houten and I. van Houten-Groeneveld at Palomar.

Named for the famous Greek artist, born c. 500 B.C. in Athens, a pupil of Ageladus. Eminent as an architect, a sculptor in bronze and a painter, Phidias was a friend of Pericles and his work is said to have included sculptures for the Acropolis. Phidias died in prison in 432 B.C. (M 20160)

(4754) Panthoos

5010 T-3. Discovered 1977 October 16 by C. J. van Houten and I. van Houten-Groeneveld at Palomar.

A member of the Trojan senate, Panthoos warned against a marriage of a Trojan prince and a Greek princess and felt that disaster for Troy was inevitable. (M 18308)

(4755) Nicky

1931 TE_4. Discovered 1931 October 6 by C. W. Tombaugh at Flagstaff.

Named in honor of Nichole Tombaugh, granddaughter of the discoverer. (M 18308)

(4756) Asaramas

1950 HJ. Discovered 1950 April 21 at the La Plata Observatory at La Plata.

Named in honor of the Asociación Argentina Amigos de la Astronomia, an amateur astronomical association in Argentina, founded 1929 January 4. The A.A.A.A. was the first contact with astronomy for a large number of professional astronomers at the La Plata Observatory. (M 22503)

(4757) Liselotte

1973 ST. Discovered 1973 September 19 by C. J. van Houten and I. van Houten-Groeneveld at Palomar.

Named after Elisabeth Charlotte von der Pfalz (1652-1722), known under her penname of Liselotte. A daughter of Kurfürst (Count Palatine) Karl Ludwig von der Pfalz, Liselotte married the brother of Louis XIV, Count Philipp I of Orleans. Her correspondence with her aunt, Kurfürstin Sophie von der Pfalz (wife of Herzog Ernst August von Hannover) and other famous contemporaries is preserved and gives a vivid insight into life at the time. (M 20161)

(4758) Hermitage

1978 SN_4. Discovered 1978 September 27 by L. I. Chernykh at Nauchnyj.

Named for one of the largest museums in the world, the unique depository in St. Petersburg of works of the world's art and culture. (M 29143)

Name proposed by the discoverer following a suggestion by the Institute of Theoretical Astronomy.

(4759) 1978 VG_{10}

Discovered 1978 November 7 by E. F. Helin and S. J. Bus at Palomar.

(4760) Jia-xiang

1981 GN_1. Discovered 1981 April 1 at the Harvard College Observatory at Harvard.

Named in honor of Zhang Jia-xiang, an astronomer at the Purple Mountain Observatory and the director of its Planetary Laboratory. A student of the late Y. C. Chang {see planet (2051)}, Zhang has devoted himself to observations and orbital studies of comets and minor planets for more than three decades. He and his colleagues have discovered 115 numbered minor planets and four comets. (M 19339)

(4761) Urrutia

1981 QC. Discovered 1981 August 27 by H.-E. Schuster at La Silla.

Named in honor of Antonio Urrutia A., Chilean lawyer in Santiago and for three decades legal advisor to the European Southern Observatory. The discoverer appreciates Urrutia's friendship and recalls his fruitful work. (M 34340)

(4762) Dobrynya

1982 SC_6. Discovered 1982 September 16 by L. I. Chernykh at Nauchnyj.
Named for the Russian epic hero Dobrynya Nikitich. (M 23137)

(4763) Ride

1983 BM. Discovered 1983 January 22 by E. Bowell at Anderson Mesa.

Named in honor of Sally K. Ride, the first American woman to fly in space. From 1978 to 1987 Ride was a Space Shuttle astronaut, flying on two missions that launched scientific satellites. She has served the NASA administration, the National Research Council, and the U.S. Congress in both an advisory and administrative capacity. She has conducted studies of the Soviet Space Program and matters related to International Security and Arms Control. Ride is currently professor of physics and director of the California Space Institute at the University of California, San Diego, where she is actively involved in physics education and attracting and encouraging students to enter physics-related fields of study. Her research centers on free-electron lasers. (M 19698)
Citation provided by L. A. McFadden at the request of the discoverer.

(4764) Joneberhart

1983 CC. Discovered 1983 February 11 by E. Bowell at Anderson Mesa.

Named in honor of Jonathan Eberhart, space science writer for the weekly publication *Science News*. For more than two decades Eberhart has reported advances and discoveries in planetary science through his coverage of planetary spacecraft encounters, the annual Lunar and Planetary Science Conference, and the annual meetings of the Division for Planetary Sciences (DPS) of the American Astronomical Society. He is known and respected for his attention to detail and accuracy in science reporting. Eberhart is also a talented composer and performer of folk music. As such, he is a founder of and performer in the "Titan Equatorial Band", an impromptu musical group that originated during the Voyager spacecraft encounters of Saturn. (M 18646)

Citation prepared by the Committee of the DPS at the request of the discoverer.

(4765) Wasserburg

1986 JN_1. Discovered 1986 May 5 by C. S. Shoemaker and E. M. Shoemaker at Palomar.

Named for Gerald J. Wasserburg, professor of geology and geophysics at the California Institute of Technology. One of the pioneers in the development of techniques of isotopic age determination, Wasserburg is famous for the very high precision ages obtained in his laboratory for lunar rocks. These ages have provided much of the primary chronological control for deciphering the history of evolution of the moon. He has also carried out extensive investigations of the ages of meteorites and developed key evidence bearing on the earliest history of the solar system and of nucleosynthesis events that predate the solar system. (M 18464)

(4766) Malin

1987 FF_1. Discovered 1987 March 28 by E. F. Helin at Palomar.

Named in honor of David F. Malin, chemist and astrophotographer at the Anglo-Australian Observatory. He has developed new methods of extracting information from astronomical plates, an enhancement technique that has led to the discovery of extremely faint but large-scale features associated with otherwise normal galaxies. His special photographic and developing procedures have brought him worldwide recognition and acclaim. The discoverer has long been an admirer of his innovative photographic methods, which produce extraordinary photographs of celestial subjects in glorious color and detail. Malin is a well-known lecturer on his photographic techniques and has written many scientific papers and books on the subject. (M 18646)

(4767) 1987 GC

Discovered 1987 April 4 by T. Niijima and T. Urata at Ojima.

(4768) Hartley

1988 PH_1. Discovered 1988 August 11 by A. J. Noymer at Siding Spring.

Named in honor of Malcolm Hartley, deputy astronomer in charge of the U.K. Schmidt telescope at Siding Spring, with which this minor planet was discovered. Among his own discoveries are the Amor object 1988 SM and ten comets, eight of which are of short period. (M 18464)

(4769) Castalia

1989 PB. Discovered 1989 August 9 by E. F. Helin at Palomar.

Named for Castalia, a nymph pursued by Apollo {see planet (1862)}. Fleeing his attention, she dived into the earth, whence a spring burst forth and was given her name. The mythical spring, on Mount Parnassus at Delphi, was the site of the most important oracle to ancient Greece. Castalia was sacred to the muses and was considered a divine source of poetic inspiration. The name also refers to a genus of aquatic plants of the water-lily family, distinguished by rounded, floating leaves and large, fragrant flowers of various colors. (M 18464)

Name suggested by S. Ostro, who captured the stunning radar images of 1989 PB soon after its discovery.

The name Castalia is another spelling of Kastalia - see planet (646).

(4770) Lane

1989 PC. Discovered 1989 August 9 by E. F. Helin at Palomar.

Named in honor of Arthur Lonne Lane, manager of the geology and planetary section at the Jet Propulsion Laboratory, a physical chemist and planetary scientist with special interests in ultraviolet physics and photochemistry. He has been active in the field of ultraviolet planetary astronomy of atmospheres and solid surfaces, planetary rings, spectroscopic instrumentation, remote sensing pollution problems and science interactions with space-mission design. His most recent activity is as principal investigator of the delta star ultraviolet imaging experiment. "Lonne" is a diligent, innovative and enthusiastic scientist. He is a longtime friend and colleague of the discoverer who has appreciated his perception and guidance. (M 18646)

(4771) Hayashi

1989 RM_2. Discovered 1989 September 7 by M. Yanai and K. Watanabe at Kitami.

Named in honor of Kousuke Hayashi (1934-), currently serving in the Exhibit Planning and Astronomy Section of the Sapporo Youth Science Museum, and Chairman of the Japan Planetarium Research Association since 1990. He served at the Sapporo City Astronomical Observatory for 20 years and has been a central figure in providing astronomical information to the public. One of his greatest achievements was the invention of a mobile observatory. (M 20521)

(4772) 1989 VM

Discovered 1989 November 2 by T. Hioki and N. Kawasato at Okutama.

(4773) Hayakawa

1989 WF. Discovered 1989 November 17 by K. Endate and K. Watanabe at Kitami.

Named in honor of Kazuo Hayakawa (1919-), an expert on mineralogy and petrology and professor of engineering at Hokkaigakuen University in Sapporo, Japan. His fields of interest include meteorites, craters and astroblemes. He has written a biography of William S. Clark (1826-1886) and many astrogeological essays. (M 20521)

Name proposed by K. Watanabe. Citation prepared by I. Hasegawa.

(4774) Hobetsu

1991 CV_1. Discovered 1991 February 14 by S. Ueda and H. Kaneda at Kushiro.

Named for a city, located 100 km southeast of Sapporo {see planet (3473)}, famous for the fossil of a sea dinosaur that was excavated there. Hobetsu is also well known as a production center for rice and melons. (M 22503)

(4775) Hansen

1927 TC. Discovered 1927 October 3 by M. Wolf at Heidelberg.

Named in memory of Peter Andreas Hansen (1795-1874), leading theoretical astronomer of the nineteenth century, at the occasion of the 200th anniversary of his birth. Born in Tondern, Schleswig, he became director of the Seeberg Observatory near Gotha {see planet (1346)} in 1825. His most important work was the improvement of the theories and tables of the orbits of the principal bodies in the solar system. Hansen's lunar theory – yielding a very accurate new value of the astronomical unit – was used for the *Nautical Almanac* until 1922. He invented the method of perturbations in coordinates. Simon Newcomb {see planet (855)}, who continued his work, considered Hansen the greatest master of celestial mechanics since Laplace {see planet (4628)}. (M 25229)
Name proposed by L. K. Kristensen and endorsed by L. D. Schmadel.
Hansen is also honored by a lunar crater.

(4776) Luyi

1975 VD. Discovered 1975 November 3 at the Harvard College Observatory at Harvard.

Named for a town in the eastern Henan {see planet (2085)} province of China. Luyi is the birthplace of the legendary philosopher Laotze (born circa 640 B.C.) of the western Zhou dynasty and founder of Taoism. Luyi is the original hometown of C.-Y. Shao {see planet (1881)}, who has been involved with the minor planet program at Harvard since its inception, and it is also the name of his son. (M 19339)

(4777) Aksenov

1976 SM_2. Discovered 1976 September 24 by N. S. Chernykh at Nauchnyj.

Named in memory of a well-known authority on celestial mechanics, Evgenij Petrovich Aksenov (1933-1995), director of the Sternberg Astronomical Institute in Moscow for many years. With Grebenikov and Demin {see planets (4268) and (5086), respectively}, he investigated and found a general solution for the generalized problem of two fixed centers, carrying out a qualitative investigation of all possible types of motion within the problem and considering their stability. He also developed an analytical theory for the motions of artificial satellites based on a non-Keplerian intermediate orbit, and he proved the existence of several new classes of periodic solutions within the circular restricted three-body problem. (M 29143)
Name suggested by the Institute of Theoretical Astronomy.
Obituary published in Astron. Rep., Vol. 39, No. 3, p. 382 (1995).

(4778) Fuss

1978 TV_8. Discovered 1978 October 9 by L. V. Zhuravleva at Nauchnyj.

Named in honor of the mathematicians Nikolaj Ivanovich Fuss (1755-1825) and his son Pavel Nikolaevich Fuss (1798-1855). Nikolaj Fuss contributed greatly to the development of mathematical education in Russia. Pavel Fuss published a bibliography of the more than 800 scientific papers of Euler {see planet (2002)}. (M 34620)

(4779) Whitley

1978 XQ. Discovered 1978 December 6 by E. Bowell and A. Warnsch at Palomar.

Named in memory of Keith Whitley (1958-1989), American country music singer who died tragically at the zenith of his profession. A powerful and emotional performer, Whitley began his professional career as a teenager, playing bluegrass music with Ralph Stanley and the Clinch Mountain Boys. His later success as a solo artist was a source of joy and inspiration for the many friends he made over the years. He helped rekindle an interest in traditional country music, reviving an entire industry. (M 18646)

(4780) Polina

1979 HE_5. Discovered 1979 April 25 by N. S. Chernykh at Nauchnyj.

Named in honor of Polina Evgen'evna Zakharova, stellar astronomer and director of the Kourovka Astronomical Observatory of the Ural University. (M 27127)

(4781) Sládkovič

1980 TP. Discovered 1980 October 3 by Z. Vávrová at Kleť.

Named in honor of Andrej Sládkovič (1820-1872), Slovak poet and author of the poem "Marina". (M 21609)

(4782) Gembloux

1980 TH_3. Discovered 1980 October 14 by H. Debehogne and L. Houziaux at St. Michel.

Named for an industrial city on the river Orneau in southern Belgium. (M 19339)

(4783) Wasson

1983 AH_1. Discovered 1983 January 12 by C. S. Shoemaker and E. M. Shoemaker at Palomar.

Named for John T. Wasson, cosmochemist and professor at the University of California at Los Angeles. Wasson is the leading investigator of the chemistry of iron meteorites and developed the chemical classification scheme for irons now in general use. He has written extensively on the origin and evolution of meteorites, their asteroid parent bodies and the solar nebula. With his student and colleague Frank T. Kyte, Wasson has also contributed important observations and insights on the geochemical signature left in the stratigraphic record by the impact of large extraterrestrial bodies on the earth. (M 18464)

(4784) 1984 DF_1

Discovered 1984 February 28 by H. Debehogne at La Silla.

(4785) Petrov

1984 YH_1. Discovered 1984 December 17 by L. G. Karachkina at Nauchnyj.

Named for the contemporary Russian composer Andrej Pavlovich Petrov (1930-). (M 22247)

(4786) Tatianina

1985 PE_2. Discovered 1985 August 13 by N. S. Chernykh at Nauchnyj.

Named in honor of Tatiana Aleksandrovna Somova, friend of the discoverer, nursery-school teacher in St. Petersburg, an unusual and interesting person. (M 26762)

(4787) Shul'zhenko

1986 RC_7. Discovered 1986 September 6 by L. V. Zhuravleva at Nauchnyj.

Named in memory of Klavdiya Ivanovna Shul'zhenko (1906-1984), a famous Russian variety singer. (M 24122)

(4788) Simpson

1986 TL_1. Discovered 1986 October 4 by E. Bowell at Anderson Mesa.

Named in honor of Robert W. L. Simpson, English composer, who has celebrated his 70th birthday in 1991. Simpson's musical character has been mainly expressed in his symphonies and string quartets, which are influenced by Beethoven, Bruckner and Sibelius {which are also present at the minor planet sky with numbers (1815), (3955), and (1405)}. The discoverer has long regarded Simpson as one of the greatest musical structuralists. An astronomer by avocation, Simpson is a fellow of the Royal Astronomical Society. (M 18465)

(4789) Sprattia

1987 UU_2. Discovered 1987 October 20 by D. D. Balam at Victoria.

Named in honor of Christopher E. Spratt, an active amateur member of the Royal Astronomical Society of Canada for 20 years, with special interests in comets, minor planets, meteors and variable stars. He has written numerous articles on these subjects, including some on Canadian meteorites, in the Journal of the Royal Astronomical Society of Canada, and he has been of much practical help to the astrometric program carried out at the University of Victoria. Spratt maintains a carefully-chosen representative collection of meteorites and was co-ordinator for the northwest North American section of the Amateur Observation Network of the International Halley Watch. He has contributed more than 32,000 observations of long-period and semiregular variable stars and dwarf novae to the American Association of Variable Star Observers and in 1976 was the most prolific observer of that year. The Chant Medal of the Royal Astronomical Society of Canada was bestowed upon him in 1988. Spratt holds an honors diploma in horticulture and, as a professional member of the gardening staff at the University of Victoria, does much to make the university a pleasant place in which to work. (M 18465)

Citation provided by J. B. Tatum at the request of the discoverer.

(4790) Petrpravec

1988 PP. Discovered 1988 August 9 by E. F. Helin at Palomar.

Named in honor of Petr Pravec (1967-), an astronomer at the Ondřejov Observatory well known for both astrometric and photometric work on minor planets and comets. He specializes in near-earth objects and has often been the first person to observe objects found in the course of the discoverer's NEAT program following their tentative announcement in the Minor Planet Center "NEO Confirmation Page". In his Ph.D. dissertation Pravec pointed out that the lightcurve of 1994 AW1, an NEO found by the discoverer, strongly suggests that the object is binary. (M 30095)

This minor planet is being named on the occasion of the marriage of Petr Pravec and Kateřina Macháčová, 1997 June 21 {see also planet (7492)}.

(4791) Iphidamas

1988 PB_1. Discovered 1988 August 14 by C. S. Shoemaker and E. M. Shoemaker at Palomar.

Named for the huge, stalwart son of the Trojan counselor Antenor {see planet (2207)} and brother of Agenor {see planet (1873)}. During the battle near the Greek encampment, Iphidamas thrust his spear into Agamemnon {see planet (911)}, who was saved by his war belt. Pulling the spear toward himself, Agamemnon then killed Iphidamas with a sword stroke to the neck. (M 18465)

(4792) Lykaon

1988 RK_1. Discovered 1988 September 10 by C. S. Shoemaker and E. M. Shoemaker at Palomar.

Named for the brother of Polydoros and a son of Priam {see planets (4708) and (844)}; born of the queen Laothoe. Some years before the Trojan war, Achilles {see planet (588)} captured Lykaon and sold him into slavery, but Lykaon made his way back to Troy, only to be killed by Achilles beside the River Skamander. (M 18465)

Name and citation provided by Richard Preston at the request of the discoverers.

(4793) 1988 RR_4

Discovered 1988 September 1 by H. Debehogne at La Silla.

(4794) 1988 SO_2

Discovered 1988 September 16 by S. J. Bus at Cerro Tololo.

(4795) Kihara

1989 CB_1. Discovered 1989 February 7 by A. Takahashi and K. Watanabe at Kitami.

Named in honor of Hideo Kihara (1911-1993), a founder of the Nayoro Astronomical Club, who established the Kihara Observatory, where he made observations of sunspots for 20 years. The observatory was presented to the city of Nayoro before his death and is now known as the Nayoro-Kihara Observatory. (M 22503)
Name suggested and citation prepared by H. Fukushima.

(4796) Lewis

1989 LU. Discovered 1989 June 3 by E. F. Helin at Palomar.

Named in honor of Joseph Walter Lewis, Jr. and his wife Anne Beech Lewis, good friends of the discoverer, on the occasion of their fiftieth wedding anniversary. Joe and Anne, an alumnus and alumna of Caltech and Stanford, respectively, have pursued distinguished civic as well as professional careers. Joe is president of DB Products, an electronic manufacturing company, and formerly vice president of Beckman Instruments. He is also a member of the Caltech Associates and has served as alumni president. Anne has been active in various civic and philanthropic organizations. Their three children, Jeff, Jack and Taffy, along with their many friends, warmly endorse this tribute. (M 21956)

(4797) Ako

1989 SJ. Discovered 1989 September 30 by T. Nomura and K. Kawanishi at Minami-Oda.

Named for a city in Hyogo Prefecture and its ancient castle on the Seto inland sea. Ako is famous for its salt production and the story of "Chushin-Gura", which shows the loyalty of samurais to their local lord. Three centuries ago 47 loyal samurais committed harakari after avenging their master. The second discoverer, by profession a dentist, lives in Ako and observes comets and minor planets at his private observatory there. (M 18465)

(4798) Mercator

1989 SU_1. Discovered 1989 September 26 by E. W. Elst at La Silla.

Named in memory of the Flemish cartographer Gerard De Kremer (1512-1594), known under the Latinised form of his name Gerardus Mercator. After receiving a master's degree from the University of Leuven in 1532, he studied mathematics, geography and astronomy and became a skillful engraver. He established himself in that city as an independent scientific-instrument maker and cartographer in 1537 and began to build his reputation as the most important geographer of his time. Appointed court cosmographer to the Duke of Cleve in 1564, he perfected his cylindrical map projection now known as "Mercator projection" in 1569. To him we owe the term "atlas" for a collection of geographical maps. (M 19339)

Name proposed by the discoverer following a suggestion by E. Goffin, who also prepared the citation.
Mercator is also honored by a lunar crater.

(4799) Hirasawa

1989 TC_1. Discovered 1989 October 8 by Y. Mizuno and T. Furuta at Kani.

Named in honor of Yasuo Hirasawa (1927-), a great source of inspiration for the discoverers. An eager observer of variable stars, he has also done much to impart a knowledge of astronomy in the planetarium at the Nagoya City Science Museum. (M 26762)

(4800) 1989 TG_{17}

Discovered 1989 October 9 by H. Debehogne at La Silla.

(4801) Ohře

1989 UR_4. Discovered 1989 October 22 by A. Mrkos at Kleť.

Named for a Czech river rising in Germany near the Czech border and continuing through the towns of Karlovy Vary, Žatec and Louny to its confluence with the Labe river. (M 34340)
Name proposed by Z. Moravec.

(4802) Khatchaturian

1989 UA_7. Discovered 1989 October 23 by F. Börngen at Tautenburg.

Named in memory of the Armenian composer Aram Khatchaturian (1903-1978), well known for his ballets *Gajaneh* and *Spartakus*, as well as for instrumental pieces such as the *Sabre Dance*. His compositions combine in melody, rhythm and orchestration the national folk music of the Transcaucasus with the traditions of classical symphonic music. (M 18466)

(4803) Birkle

1989 XA. Discovered 1989 December 1 by J. M. Baur at Chions.

Named in honor of Kurt Birkle {1939- }, since 1974 the local director of the Calar Alto Observatory, which is managed in close cooperation with the Max-Planck-Institut für Astronomie in Heidelberg. As an MPI collaborator Birkle performed pioneering work on site testing in mediterranean and other countries for a new large observatory. He is also an expert in astrophotography, well known for his observations of comets. (M 18466)

(4804) Pasteur

1989 XC_1. Discovered 1989 December 2 by E. W. Elst at La Silla.

Named in memory of the great French chemist and microbiologist Louis Pasteur (1822-1895), who proved that fermentation and disease are caused by microorganisms. His invention of the principle of immunization was successfully applied for the first time against rabies in 1885. In 1888 the celebrated Pasteur Institute was established in Paris, and the process of pasteurization is well known throughout the whole world. (M 19340)
Pasteur is also honored by craters on Mars and on the Moon.

(4805) Asteropaios

1990 VH_7. Discovered 1990 November 13 by C. S. Shoemaker and E. M. Shoemaker at Palomar.

Named for the leader of the Paionians, among the greatest of the Trojan allies. Spear-throwing Asteropaios, considered by Sarpedon {see planet (2223)} to be one of the bravest warriors, challenged and wounded Achilles {see planet (588)} but was slain in the encounter. (M 18466)

(4806) Miho

1990 YJ. Discovered 1990 December 22 by A. Natori and T. Urata at Yakiimo.

Named for a resort located in suburban Shimizu and about 3 km east of Yakiimo Station. Miho is known for its beautiful seashore and white sands, pine trees and its proximity to Mt. Fuji. It has interesting topography due to its sand-split, and it is most famous for the legend of angels bathing and dancing in the sky. (M 19698)

(4807) Noboru

1991 AO. Discovered 1991 January 10 by T. Kobayashi at Oizumi.

Named in memory of Noboru Yamada (1950-1989), one of the greatest Japanese alpinists. Yamada had intended to climb all the peaks in the 8000-m class on the earth and climbed nine of them in twelve expeditions. (M 23540)

(4808) 1925 BA

Discovered 1925 January 21 by K. Reinmuth at Heidelberg.

(4809) 1928 RB

Discovered 1928 September 5 by M. Wolf at Heidelberg.

(4810) Ruslanova

1972 GL. Discovered 1972 April 14 by L. I. Chernykh at Nauchnyj.

Named in memory of Lidiya Andreevna Ruslanova (1900-1973), Russian variety singer, popular for her renditions of ethnic folk songs. (M 23137)

(4811) Semashko

1973 SO_3. Discovered 1973 September 25 by L. V. Zhuravleva at Nauchnyj.

Named in memory of Nikolaj Alekseevich Semashko (1874-1949), organizer of the health-service system in the U.S.S.R. in the 1920s. He took the lead in the organization of mother-and-child health institutions, developed the network of medical research establishments and initiated the formation of the Central Medical Library and the House of Scientists in Moscow. During 1927-1936 he was editor-in-chief of the *Large Medical Encyclopedia*. (M 34620)

(4812) Hakuhou

1977 DL_3. Discovered 1977 February 18 by H. Kosai and K. Hurukawa at Kiso.

Named for the Hakuhou Era, from the seventh and eighth centuries, when the influence of Buddhism in Japan was very high. (M 21609)

(4813) Terebizh

1977 RR_7. Discovered 1977 September 11 by N. S. Chernykh at Nauchnyj.

Named in honor of Valerij Yuzefovich Terebizh (1941-), theoretical astrophysicist and observer at the Crimean station of the Sternberg Astronomical Institute. He found formal solutions to some important nonstationary and nonlinear problems of radiative transfer and time-series analysis. He developed the Occam approach to inversion problems of mathematical physics, in particular, to the problem of image restoration. (M 27127)

(4814) Casacci

1978 RW. Discovered 1978 September 1 by N. S. Chernykh at Nauchnyj.

Named in honor of Claudio Casacci (1958-), an Italian amateur astronomer who has been deeply involved in the popularization of astronomy and in the organization of meetings concerned, in particular, with minor bodies of the solar system. He was a member of the local organizing committee of the Asteroids, Comets, Meteors 1993 conference (IAU Symposium No. 160) and recently favored contacts between specialists of Italy and those of the Former Soviet Union, being particularly efficient in organizing the 1994 Belgirate Workshop on Near-Earth Asteroids. (M 25229)

Name proposed by the discoverer jointly with A. Sokolsky and V. Shor.

(4815) 1981 EA_{28}

Discovered 1981 March 2 by S. J. Bus at Siding Spring.

(4816) Connelly

1981 PK. Discovered 1981 August 3 by E. Bowell at Anderson Mesa.

Named in honor of Robert Connelly, mathematician at Cornell University. A specialist in geometry, Connelly has made significant contributions to the theoretical interpretation of optical lightcurves and radar echo power spectra. His work has provided an important foundation for asteroid radar imaging and astrometry, and it has established a valuable framework for using radar data to derive joint constraints on an asteroid's shape, rotation and delay-Doppler trajectory. (M 19698)
Name suggested and citation written by S. J. Ostro.

(4817) 1984 DC_1

Discovered 1984 February 27 by H. Debehogne at La Silla.

(4818) Elgar

1984 EM. Discovered 1984 March 1 by E. Bowell at Anderson Mesa.

Named for Sir Edward Elgar (1857-1934), English composer. Arguably the progenitor of the English musical renaissance of the 20th century, Elgar is best known for his orchestral and choral works. (M 19340)

(4819) 1985 KC

Discovered 1985 May 24 by A. C. Gilmore and P. M. Kilmartin at Lake Tekapo.

(4820) Fay

1985 RZ. Discovered 1985 September 15 by C. S. Shoemaker and E. M. Shoemaker at Palomar.

Named for Fay Gillis Wells, charter member, historian and guiding spirit of the Ninety-Nines, an international organization now numbering nearly 7000 women pilots from 35 countries. She was the first woman member of the Caterpillar Club, having jumped from a disabled aircraft to save her life in 1929, and being the first American woman pilot to fly in the Soviet Union. Still active at the age of 86, she continues to spearhead the recognition of the contribution of women to aviation. (M 24916)

(4821) Bianucci

1986 EE_5. Discovered 1986 March 5 by W. Ferreri at La Silla.

Named in honor of Piero Bianucci, scientific journalist and member of the editorial staff of "La Stampa" of Turin, one of the most important newspapers in Italy. One of the best known popular writers on astronomy in Italy, Bianucci has written more than ten popular astronomical books and thousands of scientific articles. He has inspired a growing interest in astronomy in Italy, especially among young people. (M 19340)

(4822) Karge

1986 TC_1. Discovered 1986 October 4 by E. Bowell at Anderson Mesa.

Named in honor of Orville B. Karge (1919-1990), instructor of physics at San Dieguito and Torrey Pines High Schools, near San Diego, California. Karge was known to be selflessly devoted to teaching and the development of opportunities for students to independently explore their interests beyond the classroom. His encouragement to enter science fairs, sponsorship of a long-lived rocketry and astronomy club, and an open-lab policy provided educational challenges and adventures for his physics students and many others. (M 19340)
Citation prepared by J. Bytof at the request of the discoverer.

(4823) Libenice

1986 TO_3. Discovered 1986 October 4 by A. Mrkos at Kleť.

Named for the site of remnants of an ancient Celtic sanctuary in central Bohemia that has astronomical significance and dates back to the fourth century B.C. (M 34620)
Name suggested by M. Tichý.

(4824) Stradonice

1986 WL_1. Discovered 1986 November 25 by A. Mrkos at Kleť.

Named for the site of an ancient Celtic town in central Bohemia above the Berounka {see planet (4702)} river. It was settled in the first century B.C. and is known for its treasures of gold and silver Celtic coins. (M 34620)
Name suggested by M. Tichý.

(4825) Ventura

1988 CS_2. Discovered 1988 February 11 by E. W. Elst at La Silla.

Named for the city near the coast of the Pacific, north of Los Angeles. After the conference "Near-Earth Asteroids" at San Juan Capistrano in July 1991 the discoverer and his family spent a most enjoyable time there at the home of friends. (M 19340)

(4826) Wilhelms

1988 JO. Discovered 1988 May 11 by C. S. Shoemaker and E. M. Shoemaker at Palomar.

Named for Don E. Wilhelms, a leading student of the moon. Wilhelms coordinated a 20-year-long U.S. Geological Survey program to map the geology of the moon. His masterpiece, *The Geologic History of the Moon*, was published as USGS Professional paper 1348. In his book, *To a Rocky Moon*, he recounts the human story of lunar geology from early research to the detailed exploration by the Apollo astronauts. (M 24916)

(4827) Dares

1988 QE. Discovered 1988 August 17 by C. S. Shoemaker and E. M. Shoemaker at Palomar.

Named for one of Aeneas' {see planet (1172)} wandering companions at arms after the fall of Troy and the only Trojan who had dared to box with Paris {see planet (3317)}, according to Virgil's *Aeneid*. When Aeneas' men stopped in Sicily and had games with the Sicilians, Dares boxed with a Sicilian named Entellus, using rawhide gloves. Entellus nearly beat Dares, but Aeneas called off the fight. (M 18647)
Name and citation provided by R. Preston at the request of the discoverers.

(4828) Misenus

1988 RV. Discovered 1988 September 11 by C. S. Shoemaker and E. M. Shoemaker at Palomar.

Named for Aeneas' {see planet (1172)} herald and trumpeter. As Virgil tells it, Misenus unluckily blew on a conch shell just as the Trojan wanderers were making landfall in Italy, near Cumae, and the clear note of the shell so enraged the god Triton that he killed Misenus and threw his body on the beach. In later years a Roman fleet was stationed near the legendary place, at Misenum, named after Misenus. (M 18647)
Name and citation prepared by R. Preston at the request of the discoverers.

(4829) Sergestus

1988 RM_1. Discovered 1988 September 10 by C. S. Shoemaker and E. M. Shoemaker at Palomar.

Named for one of Aeneas' {see planet (1172)} Trojan companions after the fall

of Troy, commander of the galley named "Centaur". During the Sicilian games, as described in the *Aeneid*, Sergestus took the Centaur in a race against three other galleys. He drove his ship upon the rocks, lost the race and was mocked by the crowds, but to console him Aeneas gave him a reward of a slave woman named Pholoe. (M 18647)
Name and citation provided by R. Preston at the request of the discoverers.

(4830) 1988 RG_4

Discovered 1988 September 1 by H. Debehogne at La Silla.

(4831) 1988 RX_{11}

Discovered 1988 September 14 by S. J. Bus at Cerro Tololo.

(4832) Palinurus

1988 TU_1. Discovered 1988 October 12 by C. S. Shoemaker and E. M. Shoemaker at Palomar.

Named for Aeneas' {see planet (1172)} great helmsman and navigator, who led the remnant of the Trojan fleet across unknown seas from Troy to Carthage and Sicily, and finally to Italy, as told by Virgil in the *Aeneid*. As the fleet was making its final passage to Italy, the god Neptune put Palinurus to sleep while he held the tiller in his hands; he fell overboard and was drowned, and his naked body washed up on the shores of Italy. (M 18647)
Name and citation provided by R. Preston at the request of the discoverers.

(4833) Meges

1989 AL_2. Discovered 1989 January 8 by C. S. Shoemaker and E. M. Shoemaker at Palomar.

Named for the commander of the Greek islanders who dwelt off Elis, on the west coast of Peloponnese; he sailed for Troy with forty ships, and killed Amphiklos. Later Meges, with Thoas {see planet (4834)}, was part of the Greek delegation that persuaded Achilles {see planet (588)} to return to battle. (M 18647)
Name and citation provided by R. Preston at the request of the discoverers.

(4834) Thoas

1989 AM_2. Discovered 1989 January 11 by C. S. Shoemaker and E. M. Shoemaker at Palomar.

Named for Thoas, the commander of the Aitolians, a tough man at spear-throwing, and eloquent in argument. The god Poseidon {see planet (4341)} (who favored the Greeks) must have respected Thoas for both his warlike skills and for his eloquence, because Poseidon chose to take the shape of Thoas, and, in Thoas's voice, rallied the Greeks when the Trojans were close to the Greek ships, and fought beside the Greeks in the shape of Thoas. (M 18647)
Name and citation provided by R. Preston at the request of the discoverers.

(4835) 1989 BQ

Discovered 1989 January 29 by M. Iwamoto and T. Furuta at Tokushima.

(4836) Medon

1989 CK_1. Discovered 1989 February 2 by C. S. Shoemaker and E. M. Shoemaker at Palomar.

Named for an illegitimate son of Oileus, and half-brother of Ajax Oiliades (also known as Little Ajax). Medon lived in exile for having killed a relative of his stepmother. He fought bravely at Troy, but was killed before the Greek ships by Aeneas {see planet (1172)}. (M 18648)
Name and citation provided by R. Preston at the request of the discoverers.

(4837) 1989 ME

Discovered 1989 June 30 by A. C. Gilmore and P. M. Kilmartin at Lake Tekapo.

(4838) Billmclaughlin

1989 NJ. Discovered 1989 July 2 by E. F. Helin at Palomar.

Named in honor of William I. McLaughlin (1935-) on the occasion of his retirement from the Jet Propulsion Laboratory, where he made innovative contributions to many projects, including Viking, Seasat, the Infrared Astronomical Satellite and Voyager (Uranus), and to the management of two technical sections. He also served as deputy manager of the astrophysics and fundamental physics program office. For 14 years McLuaghlin wrote a column in *Spaceflight* called "Space at JPL". He takes a special interest in philosophy, mathematics, astronomy and epistemology, as well as space exploration and has written numerous papers. (M 34340)

(4839) Daisetsuzan

1989 QG. Discovered 1989 August 25 by K. Endate and K. Watanabe at Kitami.

Named for the Daisetsuzan Mountains, which rise to the height of about 2,000 meters in central Hokkaido {see planet (3720)}. The mountains are a treasury of the beauty of nature. A quasi-national park, it contains many marshy districts and great stocks of alpine plants. (M 20838)

(4840) Otaynang

1989 UY. Discovered 1989 October 23 by Y. Oshima at Gekko.

Named in honor of Hanne Otaynang (1935-), who appealed for the environmental preservation of the Earth at the 1992 Earth Summit in Brazil. (M 24122)
Name proposed by the Gekko Astronomical Observatory.

(4841) Manjiro

1989 UO_3. Discovered 1989 October 28 by T. Seki at Geisei.

Named in memory of Manjiro Nakahama, a fisherman from Kochi prefecture who was rescued by an American whaling ship at the age of 16 when, in 1843, he was drifting in the Pacific. Beginning a new life in the U. S. under the name John Manjiro, he served as first officer on several voyages and played a very important part in the early stages of cultural exchange between Japan and the United States. (M 19340)

(4842) Atsushi

1989 WK. Discovered 1989 November 21 by S. Ueda and H. Kaneda at Kushiro.

Named in honor of Atsushi Takahashi (1965-). A resident of Kitami {see planet (3785)}, he makes observations of minor planets and comets as a member of a group named "Hokkaido Showakusei Suisei Kaigi". (M 22503)
Name proposed by K. Watanabe.

(4843) Mégantic

1990 DR_4. Discovered 1990 February 28 by H. Debehogne at La Silla.

Named for Mont Mégantic, the largest observatory in Québec. Founded in 1978, its mission is to promote research in astrophysics, to train students and to promote astronomy in the local culture. The observatory is jointly operated by the Université de Montréal and the Université Laval. (M 32344)

Name proposed by the discoverer, following a suggestion by P. Bastien and Y. Dutil.

(4844) Matsuyama

1991 BA_2. Discovered 1991 January 23 by S. Ueda and H. Kaneda at Kushiro.

Named in honor of Masanori Matsuyama (1950-), who started observing

minor planets in 1987 and has discovered many new minor planets since then. He lives in Kushiro {see planet (4096)} and makes observations of minor planets and comets as a member of "Hokkaido Showakusei Suisei Kaigi". (M 22503)
Name proposed by K. Watanabe.

(4845) Tsubetsu

1991 EC_1. Discovered 1991 March 5 by K. Endate and K. Watanabe at Kitami.
Named for a timber town, with a population of less than 9,000, in eastern Hokkaido {see planet (3720)}. The first discoverer spent his boyhood in Tsubetsu between 1966 and 1972. (M 20838)

(4846) Tuthmosis

6575 P-L. Discovered 1960 September 24 by C. J. van Houten and I. van Houten-Groeneveld at Palomar.
Tuthmosis (also written Thothmose) was the name of four Egyptian pharaohs of the 18th dynasty. The name means 'child of the god Thot'. Tuthmosis I was the first king to be buried in the Valley of the Tombs of the Kings near Thebes. Tuthmosis III was the successor of his stepmother and regent, Hatshepsut {see planet 2436)}. (M 20161)

(4847) Amenhotep

6787 P-L. Discovered 1960 September 24 by C. J. van Houten and I. van Houten-Groeneveld at Palomar.
Named after pharaohs of the 18th dynasty. Amenhotep (Greek name Amenophis) means 'Amun is merciful'. Amenhotep II was buried in the Valley of the Kings and his tomb records his military successes. Amenhotep III built one of the most famous temples at Luxor, and his mummy is now in the Cairo Museum. (M 20161)

(4848) Tutenchamun

3233 T-2. Discovered 1973 September 30 by C. J. van Houten and I. van Houten-Groeneveld at Palomar.
Named after the 18th dynasty pharaoh Tut-ench-Amun or Tutanchamun, 1355-1337 B.C., son-in-law and probably also son of Echnaton {see planet (4415)} and Nofretete (also named Nefertiti, see planets (1068) and (3199), respectively). He died unexpectedly, probably violently, at about 18 years of age. His fabulously decorated tomb was undisturbed when discovered in 1922 by Howard Carter. (M 20161)

(4849) Ardenne

1936 QV. Discovered 1936 August 17 by K. Reinmuth at Heidelberg.
Named in memory of Manfred von Ardenne (1907-1997), German inventor and researcher in various fields, especially known as a pioneer in television and radio engineering. His early interests in astronomy stimulated his enthusiasm for science and technology, and he emphasized the educational value of astronomy for young people. Ardenne established three private observatories, which were all open to the public. (M 30798)
Name proposed and citation prepared by D. B. Herrmann, endorsed by L. D. Schmadel.

(4850) Palestrina

1973 UJ_5. Discovered 1973 October 27 by F. Börngen at Tautenburg.
Named for the famous Italian Renaissance composer Giovanni Pierluigi da Palestrina (1525?-1594). His works fill 34 volumes and are marked by quiet elevation, dignity, clearness and melodious sound. (M 19341)

(4851) Vodop'yanova

1976 US_1. Discovered 1976 October 26 by T. M. Smirnova at Nauchnyj.

Named in honor of Galina Petrovna Vodop'yanova (1939-), a talented surgeon of the Russian Academy of Sciences Hospital, St. Petersburg, since 1978. (M 30475)
Name suggested by N. A. Bokhan.

(4852) Pamjones

1977 JD. Discovered 1977 May 15 by N. S. Chernykh at Nauchnyj.

Named in honor of Pamela Ann Jones, of the Lunar and Planetary Institute, Houston, in appreciation of her organization of many conferences in planetary sciences, particularly two international conferences held in June-July 1991, "Asteroids, Comets, Meteors 91" in Flagstaff and "Near-Earth Asteroids" in San Juan Capistrano. (M 18648)

(4853) 1979 ML

Discovered 1979 June 28 by C. Torres at Cerro El Roble.

(4854) 1981 ED_{27}

Discovered 1981 March 2 by S. J. Bus at Siding Spring.

(4855) Tenpyou

1982 VM_5. Discovered 1982 November 14 by H. Kosai and K. Hurukawa at Kiso.

Named for the Tenpyou Era, from the eighth century, when the Japanese capital was at Nara and many grand Buddhist temples were built. (M 21609)

(4856) Seaborg

1983 LJ. Discovered 1983 June 11 by C. S. Shoemaker and E. M. Shoemaker at Palomar.

Named in honor of Glenn T. Seaborg (1912-), American nuclear chemist, academic administrator and public servant. Seaborg participated in the discovery of most of the known transuranian elements, including plutonium, and led the chemical effort to isolate and characterize that element and produce it in large quantities from nuclear reactors. He served as professor and chancellor of the University of California at Berkeley and as a member and chairman of the U.S. Atomic Energy Commission. He shared with Edwin McMillan the 1951 Nobel prize for chemistry. (M 24916)
Name proposed and citation prepared by T. P. Kohman.

(4857) Altgamia

1984 FM. Discovered 1984 March 29 by C. S. Shoemaker at Palomar.

Named for Andrew L. T. and Angela Maria Chiarappa Green, son and wife of D. W. E. Green {see planet (2068)}, who was involved in taking the discovery films and who found the identifications for this minor planet. Angela, known as 'Lina' to family and friends, did some volunteer work with the monthly subscription book-keeping of the Minor Planet Center in 1984-85, and occasionally helps the Center and the Central Telegram Bureau with her expert linguistic knowledge of Italian and Spanish. (M 18648)

(4858) 1985 UA

Discovered 1985 October 23 by J. Gibson at Palomar.

(4859) Fraknoi

1986 TJ_2. Discovered 1986 October 7 by E. Bowell at Anderson Mesa.

Named for Andrew Fraknoi on the occasion of his resignation as executive director of the Astronomical Society of the Pacific. Under his leadership, from 1978

to 1992, the ASP {see planet (2848)} has grown into a nationally and internationally recognized science education organization, serving teachers, amateurs and the public. In addition to editing *Mercury*, Fraknoi founded and edited The *Universe in the Classroom*, a newsletter for teachers. He also organized workshops and wrote a nationally syndicated newspaper column. He hosted a radio talk show on science for two years and regularly explains astronomical developments on radio and television. An outspoken critic of astrology, he has given more than 300 public lectures on astronomical and skeptical topics. (M 20522, M 20542)
Citation provided by J. Wujek.

(4860) Gubbio

1987 EP. Discovered 1987 March 3 by E. Bowell at Anderson Mesa.

Named for the city of Gubbio, Umbria, Italy. Dating from the Umbrian civilization in the 7th century B.C., Gubbio flourished in the Middle Ages and is one of the best-preserved Italian medieval cities. It was on a hill above the city that geologist Walter Alvarez {see planet (3581)} found an iridium-rich layer originating at the transition between the Cretaceous and Tertiary periods. His father, Luis Alvarez, physicist and Nobel laureate, interpreted the layer's presence as the result of a catastrophic impact of an asteroid with the earth. Gubbio is also famous for the story of St. Francis and the wolf, and for St. Francis' "Ceri Marathon". (M 22247)
Name suggested and citation prepared by M. A. Barucci.

(4861) Nemirovskij

1987 QU_{10}. Discovered 1987 August 27 by L. G. Karachkina at Nauchnyj.

Named in honor of Lev Ruful'evich Nemirovskij (1937-), engineer, chief-constructor and artistic director of the St. Petersburg House of Satire and Humor, and one of the authors of the miscellany *Gold Ostap*. (M 23137)

(4862) Loke

1987 SJ_5. Discovered 1987 September 30 by P. Jensen and K. Augustesen at Brorfelde.

Named after the strangest character in Norse mythology. Loke is one of the giants, but he lives among the Aesir in Asgaard. He is the symbol of falseness and intrigue - at the same time the enemy and the cunning helper of the gods. He caused the death of Balder {see planet (4059)}, and he is the father of Fenrir, Hel {see planet (949)} and the Midgard serpent. (M 22504)

(4863) Yasutani

1987 VH_1. Discovered 1987 November 13 by S. Ueda and H. Kaneda at Kushiro.

Named in honor of Keiki Yasutani (1958-), an astronomical photographer renowned for his photographs of the moon and Mars. He is a member of the Sapporo Observatory. (M 22504)
Name proposed by K. Watanabe.

(4864) 1988 RA_5

Discovered 1988 September 2 by H. Debehogne at La Silla.

(4865) Sor

1988 UJ. Discovered 1988 October 18 by T. Seki at Geisei.
Named for the great Spanish guitarist Fernando Sor. (M 19341)

(4866) 1988 VB_3

Discovered 1988 November 10 by T. Kojima at Chiyoda.

(4867) Polites

1989 SZ. Discovered 1989 September 27 by C. S. Shoemaker and E. M. Shoemaker at Palomar.

Polites was a son of Priam, as was Deiphobus {see, respectively, planets (884) and (1867)}. They were both killed during the last moments of Troy: Deiphobus was butchered and mutilated by Menelaus {see planet (1647)}, and Polites was transfixed with a spear handled by Pyrrhus {see planet (5283)}. Pyrrhus, the most ruthless of the Greeks, broke open and invaded Priam's great house of fifty chambers, and he chased Polites along empty colonnades until he cornered and slaughtered him in front of his father and mother, Priam and Hecuba {see planet (108)}. (M 22248)

Name and citation provided by Richard Preston at the request of the discoverers.

(4868) 1989 UN_2

Discovered 1989 October 27 by E. F. Helin at Palomar.

(4869) Piotrovsky

1989 UE_8. Discovered 1989 October 26 by L. I. Chernykh at Nauchnyj.

Named in memory of the outstanding archaeologist Boris Borisovich Piotrovsky (1908-1990), director of the Hermitage {see planet (4758)} museum in St. Petersburg for many years. The name also honors his son Mikhail Borisovich Piotrovsky (1944-), a well-known orientalist who succeeded his father as Hermitage director. (M 29144)

Name proposed by the discoverer following a suggestion by the Institute of Theoretical Astronomy.

(4870) Shcherban'

1989 UK_8. Discovered 1989 October 25 by L. V. Zhuravleva at Nauchnyj.

Named in honor of Vladimir Onufrievich Shcherban' (1938-), who was director of an oil-extraction plant in Prikolotnoe for 30 years. He created an exceptional museum at the birthplace of Konstantin Ol'shanskij {see planet (2310)}. The plant organized a computer class and contained a beautiful garden for children and an indoor pool. (M 27329)

(4871) Riverside

1989 WH_1. Discovered 1989 November 24 by M. Koishikawa at Sendai.

Named to celebrate the 35th anniversary of the "sister city" affiliation of Sendai {see planet (3133)}, Japan, and Riverside, California, which began on 1957 Mar. 9. Riverside, which is the county seat of Riverside County and is located some 80 km east of Los Angeles, has many parks and is known as a health resort. (M 19699)

(4872) Grieg

1989 YH_7. Discovered 1989 December 25 by F. Börngen at Tautenburg.

Named for the Norwegian composer Edvard Grieg (1843-1907). In Grieg's work, the precipitous, harsh beauty of the Scandinavian landscape is reflected. Particularly popular is the orchestral suite *Peer Gynt*, a piano and violin concerto to accompany Ibsen's {see planet (5696)} play. Also well known are many short charming lyric compositions, for example the *wedding march of Troldhaugen*. (M 18648)

(4873) Fukaya

1990 EC. Discovered 1990 March 4 by A. Sugie at Taga.

Named for a city 70 km northwest of Tokyo, not far from the National Astronomical Observatory's Dodaira Station. Originally a stop on one of the main routes between Tokyo and Kyoto {see planets (498) and (4352)}, the city received

its present name in 1955, since when it has shown remarkable growth. Fukaya, home to one of the factories of the Dynic Corporation, is now one of the leading industrial centers in Saitama prefecture. (M 19341)
Name proposed by the discoverer following a suggestion by S. Sakabe.

(4874) Burke

1991 AW. Discovered 1991 January 12 by E. F. Helin at Palomar.

Named in honor of James D. Burke on the occasion of his retirement from the Jet Propulsion Laboratory. Jim served as manager of the Ranger project, the first American mission to the moon, and as manager of advanced technical studies. He invented the SNAKE guide-rope for the Soviet Mars '96 mission. In 1991 Jim served as managing director of the International Space University. He made impressive contributions to humanity's efforts to reach the planets. (M 19699)
Name endorsed by the Planetary Society and the Jet Propulsion Laboratory.

(4875) Ingalls

1991 DJ. Discovered 1991 February 19 by Y. Kushida and R. Kushida at Yatsugatake.

Named in honor of Laura Ingalls Wilder (1867-1957) and the Ingalls family. Her books in the "Little House" series illustrate the love and strong mind of the family through the American pioneer life of her girlhood and still make a deep impression on all the people of the world. (M 18648)

(4876) Strabo

1133 T-2. Discovered 1973 September 29 by C. J. van Houten and I. van Houten-Groeneveld at Palomar.

Named after the Greek geographer Strabo (Greek Strabon: 63 B.C.-c. A.D. 23), who devoted himself to historical and geographical studies and who took long journeys through Asia Minor, Egypt, Greece and Italy. His 47-volume *Historical Sketches* exists today only as fragments, but his 17-volume *Geographica* survives almost intact. (M 20161)

(4877) Humboldt

5066 T-2. Discovered 1973 September 25 by C. J. van Houten and I. van Houten-Groeneveld at Palomar.

Named after the scientist Friedrich Heinrich Alexander Freiherr von Humboldt (1769-1859), who made extended expeditions to Siberia and South America to study the flora and geology of each region. His main publication was *Cosmos, Entwurf einer physikalischen Weltbeschreibung*. (M 20161)

(4878) 1968 OF

Discovered 1968 July 18 by C. Torres and S. Cofre at Cerro El Roble.

(4879) Zykina

1974 VG. Discovered 1974 November 12 by L. I. Chernykh at Nauchnyj.

Named in honor of Lyudmila Georgievna Zykina, a Russian singer who enjoys widespread popularity. She is a distinguished performer of Russian folk songs and of songs by Russian composers. (M 26762)

(4880) Tovstonogov

1975 TR_4. Discovered 1975 October 14 by L. I. Chernykh at Nauchnyj.

Named in memory of Georgij Aleksandrovich Tovstonogov (1913-1989), outstanding Russian theatre producer, writer and art critic. As the chief producer at the Bolshoj Dramatic Theater in Leningrad (now St. Petersburg) for 40 years, he was the creator of numerous unique theatrical performances. (M 26762)

(4881) 1975 XJ

Discovered 1975 December 1 by C. Torres at Cerro El Roble.

(4882) Divari

1977 QU_2. Discovered 1977 August 21 by N. S. Chernykh at Nauchnyj.

Named in memory of Nikolaj Borisovich Divari (1921-1993), professor at the Odessa Polytechnical Institute and prominent in the study of the nature of the zodiacal light and interplanetary dust. (M 27127; M 27147)

(4883) Korolirina

1978 RJ_1. Discovered 1978 September 5 by N. S. Chernykh at Nauchnyj.

Named in honor of Irina Leonidovna Korol', a friend of the discoverer and teacher of philology in St. Petersburg who displays a keen interest in astronomy. (M 26763)

(4884) Bragaria

1979 OK_{15}. Discovered 1979 July 21 by N. S. Chernykh at Nauchnyj.

Named in honor of Luka Fedorovich Bragar' (1938-), a senior lecturer on the staff of Tiraspol Pedagogical Institute (Moldova), known for his contribution to the rotational dynamics of minor planets and as a splendid educator and popularizer of astronomical knowledge. (M 32345)

(4885) Grange

1980 LU. Discovered 1980 June 10 by C. S. Shoemaker at Palomar.

Named in honor of Alice Shoemaker Grange (1908-), aunt of E. M. Shoemaker. (M 24917)

(4886) 1981 EZ_{14}

Discovered 1981 March 1 by S. J. Bus at Siding Spring.

(4887) 1981 EV_{26}

Discovered 1981 March 2 by S. J. Bus at Siding Spring.

(4888) Doreen

1981 JX_1. Discovered 1981 May 5 by C. S. Shoemaker at Palomar.

Named in honor of Doreen Vingness Spellmann, sister-in-law of the discoverer. Spellmann led a sustained effort to beautify the city of El Cerrito, California, with flowers and shrubs. (M 24917)

(4889) Praetorius

1982 UW_3. Discovered 1982 October 19 by F. Börngen at Tautenburg.

Named for Michael Praetorius (1571-1621), outstanding German composer and musical writer of the early baroque. He published a fundamental three-volume composition theory (Syntagma musicum) and about 25 volumes with sacred and secular compositions and collections of church songs. Particularly well known is his beautiful melody for the Christmas carol "Es ist ein Ros entsprungen". (M 19341)

(4890) Shikanosima

1982 VE_4. Discovered 1982 November 14 by H. Kosai and K. Hurukawa at Kiso.

Named for the small island located off the northern coast of Kyushu. In 1784, a farmer found the golden seal that was the present from the emperor of ancient China to the king of Na, one of the local states of ancient Japan, when the King sent the first delegation to China under Han Dynasty in AD 57. This association of China and Japan is the first one appearing in the Chinese historical records. (M 22504)

(4891) Blaga

1984 GR. Discovered 1984 April 4 by V. G. Shkodrov and V. G. Ivanova at Rozhen.

Named in honor of Blaga Dimitrova, talented Bulgarian poetess, novelist and translator. Her intellectual works are profoundly philosophical, revealing the intrinsic psychology of modern mankind and the struggle for morality. The poetess herself maintains an open and courageous civic stand. Some of her works are translated into many languages. (M 19699)

(4892) Chrispollas

1985 TV_2. Discovered 1985 October 11 by the CERGA at Caussols.

Named in honor of Christian Pollas, currently head of the Schmidt telescope unit at the Observatoire de la Côte d'Azur. Christian is a dedicated observer who has discovered several asteroids and more than 60 supernovae in the last seven years. (M 23137)

Citation written by A. Maury and endorsed by J.-L. Heudier.

(4893) Seitter

1986 PT_4. Discovered 1986 August 9 by E. W. Elst and V. G. Ivanova at Rozhen.

Named in honor of Waltraud C. Seitter {1930- }, director of the Münster (Westfalen) Astronomical Institute, famous for her spectroscopic researches on novae and a good friend of the discoverer. (M 19341)

(4894) Ask

1986 RJ. Discovered 1986 September 8 by P. Jensen and K. Augustesen at Brorfelde.

Named after the first man, who, according to Norse mythology, was made from an ash tree by Odin {see planet (3989)} and his two brothers, Vile and Ve. (M 22504)

(4895) Embla

1986 TK_4. Discovered 1986 October 13 by P. Jensen and K. Augustesen at Brorfelde.

Named after the first woman, who, according to Norse mythology, was made from an ash tree by Odin {see planet (3989)}, Vile and Ve. (M 22504)

(4896) Tomoegozen

1986 YA. Discovered 1986 December 20 by T. Niijima and T. Urata at Ojima.

Named for Tomoegozen, mistress of Yoshinaka, known as a brave strategist on Yoshinaka's staff. After his death, Tomoegozen became a Buddhist nun. This planet was discovered on the same films as (4574) Yoshinaka. (M 21132)

(4897) 1987 QD_6

Discovered 1987 August 22 by E. F. Helin at Palomar.

(4898) Nishiizumi

1988 FJ. Discovered 1988 March 19 by C. S. Shoemaker and E. M. Shoemaker at Palomar.

Named for Kunihiko Nishiizumi, nuclear chemist at the University of California at Berkeley. Nishiizumi is renowned for his work on cosmogenic nuclides in meteorites, particularly meteorites from the moon, and the determination of the terrestrial ages of meteorites. He also led the determination of surface exposure ages and erosion rates of terrestrial rocks, from ^{10}Be and ^{26}Al produced by cosmic rays in quartz, and used this technique to determine ages of ancient surfaces in Antarctica and Australia and of meteorite craters. (M 24917)

(4899) Candace

1988 JU. Discovered 1988 May 9 by C. S. Shoemaker and E. M. Shoemaker at Palomar.

Named for Candace P. Kohl, American chemist and a leading investigator of ancient solar activity through analysis of solar cosmic-ray-produced nuclides in lunar samples. She has also contributed importantly in the development of techniques for dating surface exposure of materials on the earth from cosmic-ray-produced nuclides. Through her popular lectures on meteorites, the moon and the solar system, Kohl has reached a wide audience ranging from primary-school children to high-school students and the lay community. (M 25443)
Citation provided by K. Nishiizumi at the request of the discoverers.

(4900) Maymelou

1988 ME. Discovered 1988 June 16 by E. F. Helin at Palomar.

Named in honor of Mayme Lou "Stevey" Stevens Bruce, a graduate of Pomona College, California. Mother of three daughters, she and her husband, Stuart Bruce, have had a long fascination with travel to remote corners of the world. The recently published book *Beyond the Ranges*, authored by her husband, documents their research and travels. "Stevey" is an enthusiastic supporter and spokesperson for solar system research and discovery, contributing to the expansion of basic research in Helin's {see planet (3267)} PCAS and DSSS programs. (M 20161)

(4901) 1988 VJ

Discovered 1988 November 3 by M. Arai and H. Mori at Yorii.

(4902) Thessandrus

1989 AN_2. Discovered 1989 January 9 by C. S. Shoemaker and E. M. Shoemaker at Palomar.

Thessandrus was one of the Greek captains who hid inside the Trojan horse, which was devised by Epeios, who himself hid inside the horse. This minor planet is paired dynamically with (2148) Epeios. Virgil {see planet (2798)}, in the *Aeneid*, seems to imply that Thessandrus was the first warrior to slide down the lowered rope from the belly of the horse, thus beginning the sack of Troy. (M 22248)

Name and citation provided by Richard Preston at the request of the discoverers.

(4903) Ichikawa

1989 UD. Discovered 1989 October 20 by Y. Mizuno and T. Furuta at Kani.

Named in honor of the amateur astronomer Kiyotaka Ichikawa (1955-). This particular minor planet is among the many for which he established identifications and computed orbits. (M 26763)

(4904) Makio

1989 WZ. Discovered 1989 November 21 by Y. Mizuno and T. Furuta at Kani.

Named in honor of Makio Akiyama (1950-), who has discovered several minor planets at the Susono {see planet (6419)} Observatory. (M 26763)

(4905) Hiromi

1991 JM_1. Discovered 1991 May 15 by A. Takahashi and K. Watanabe at Kitami.

Named in honor of Hiromi Takahashi (1964-), wife of the first discoverer. (M 22504)

(4906) Seneferu

2533 P-L. Discovered 1960 September 24 by C. J. van Houten and I. van Houten-Groeneveld at Palomar.

Seneferu (also written Snefru) was the first king of the 4th dynasty and built two pyramids near Daschur. (M 20162, 20191)

(4907) Zoser

7618 P-L. Discovered 1960 October 17 by C. J. van Houten and I. van Houten-Groeneveld at Palomar.

Named after the Egyptian pharaoh of the third dynasty. Zoser, also written Djoser, erected the first stone pyramid - the step-pyramid at Sakkara, near Memphis. (M 20162)

(4908) Ward

1933 SD. Discovered 1933 September 17 by F. Rigaux at Uccle.

Named in honor of Steven Ward, electronics technician in the Computation Facility at the Harvard-Smithsonian Center for Astrophysics. Steve's assistance to the Minor Planet Center mainly involves system programming. (M 22504) Name proposed by B. G. Marsden, who made the identifications for this object.

(4909) Couteau

1949 SA_1. Discovered 1949 September 28 by M. Laugier at Nice.

Named in honor of Paul Couteau, double-star observer at the Observatoire de Nice since 1959, on the occasion of his retirement. Using the large refractors at Lick, Yerkes and Nice, and the telescopes of Almeria and Pic du Midi, he has discovered more than 2700 new binaries, achieving 135,000 individual measurements. Couteau, having measured star separations down to the 60-milliarcsecond resolution limit of the 2-m telescope at Pic du Midi, is one of the sharpest 'eyes' on Earth. (M 23137)

(4910) Kawasato

1953 PR. Discovered 1953 August 11 by K. Reinmuth at Heidelberg.

Named in honor of Nobuhiro Kawasato, who co-discovered 1988 VG_2 , which was identified with the lost minor planet (724) Hapag. (M 22504)

Name proposed by S. Nakano, who found the identifications involving this object.

(4911) 1953 UD

Discovered 1953 October 16 at the Goethe Link Observatory at Brooklyn, Indiana.

(4912) 1953 VX_1

Discovered 1953 November 11 at the Goethe Link Observatory at Brooklyn, Indiana.

(4913) 1965 SO

Discovered 1965 September 20 at the Purple Mountain Observatory at Nanking.

(4914) Pardina

1969 GD. Discovered 1969 April 9 at the Felix Aguilar Observatory at El Leoncito.

Named in honor of Elsa Gutierrez Rodriguez-Pardina (1921-), Argentinian astronomer who has worked in the field of celestial mechanics for more than 30 years. Form 1952 to 1956, she worked at the La Plata Observatory in the Department of Astrometry and Celestial Mechanics. In 1954 she obtained a scholarship from the French government to further her studies in celestial mechanics at the Sorbonne Henri Poincaré Institute. From 1978 to 1986, Pardina led the Celestial Mechanics Department of the Astronomical Observatory of Córdoba. For many years, she was responsible for the celestial mechanics courses taught to undergraduate students. (M 33786)

(4915) Solzhenitsyn

1969 TJ_2. Discovered 1969 October 8 by L. I. Chernykh at Nauchnyj.

Named in honor of Aleksandr Isaevich Solzhenitsyn (1918-), world-renowned Russian writer and Nobel prize-winner {1970}. (M 23137)

(4916) Brumberg

1970 PS. Discovered 1970 August 10 at the Crimean Astrophysical Observatory at Nauchnyj.

Named in honor of Victor Aleksandrovich Brumberg (1933-), staff member at the Institute of Theoretical Astronomy from 1958 to 1987 and the Institute of Applied Astronomy since 1988. His main scientific results are related to analytical and relativistic celestial mechanics (the three-body problem, general planetary theory, lunar theory, relativistic reduction of observations, reference systems and time scales, and the relativistic definitions of astronomical concepts and constants). (M 30475)

Name proposed by the Institute of Theoretical Astronomy and the Institute of Applied Astronomy.

(4917) Yurilvovia

1973 SC_6. Discovered 1973 September 28 at the Crimean Astrophysical Observatory at Nauchnyj.

Named in memory of Yurij Alekseevich L'vov (1932-1994), deputy director of the Institute of Biology and Biophysics at the Tomsk State University. A prominent geobotanist and ecologist, L'vov was for 30 years an organizer of the regular expeditions to Tunguska {see planet (5471)}, and he made important contributions to the study of the 1908 impact. (M 27127)

(4918) Rostropovich

1974 QU_1. Discovered 1974 August 24 by L. I. Chernykh at Nauchnyj.

Named in honor of Mstislav Leopol'dovich Rostropovich (1927-), outstanding Russian musician and pedagogue, world-famous cellist and conductor. (M 23137)

(4919) Vishnevskaya

1974 SR_1. Discovered 1974 September 19 by L. I. Chernykh at Nauchnyj.

Named in honor of Galina Pavlovna Vishnevskaya, outstanding Russian opera singer, actress of the Bolshoi Theatre in Moscow for many years and performer in many theaters around the world. (M 23137)

(4920) Gromov

1978 PY_2. Discovered 1978 August 8 by N. S. Chernykh at Nauchnyj.

Named in memory of Mikhail Mikhajlovich Gromov (1899-1985), well-known Russian pilot who set the world record for long-distance flight in 1934 when his aircraft flew more than 12 000 kilometers. In 1937, together with A. Yumashev and S. Danilin, he completed a nonstop flight from Moscow, over the North Pole, to the United States. He was the first Russian pilot to win the de Lawo medal of the Federation Aeronautique Internationale. Gromov also tested a great number of new aircraft. He was the founder of the State Research Center of Russia "Flying Research Institute", now named for him. (M 34340)

(4921) Volonté

1980 SJ. Discovered 1980 September 29 by Z. Vávrová at Kleť.

Named for the popular Italian actor Gian Maria Volonté, who portrayed the title role in a film about Giordano Bruno. (M 21956)

(4922) 1981 EH_4

Discovered 1981 March 2 by S. J. Bus at Siding Spring.

(4923) Clarke

1981 EO_{27}. Discovered 1981 March 2 by S. J. Bus at Siding Spring.

Named in honor of Arthur C. Clarke (1917-), renowned author and visionary. His optimistic portrayal of the future and confidence in the human spirit are themes that pervade his writing. Clarke is best known for his novel *2001: A Space Odyssey,* which was simultaneously developed into a landmark film. In his 1973 science fiction novel *Rendezvous with Rama,* Clarke describes the development of Project *Spaceguard,* a means for detecting and mitigating potentially hazardous objects in near-earth space. In recognition of Clarke's foresight, the name *Spaceguard* was adopted in a 1992 NASA proposal to the U.S. Congress for a dedicated program to survey the NEO population, as well as in the international foundation established in Rome on 1996 Mar. 26. (M 27127; M 27147)

Citation prepared with assistance from R. P. Binzel. Name independently suggested by D. I. Steel.

(4924) Hiltner

1981 EQ_{40}. Discovered 1981 March 2 by S. J. Bus at Siding Spring.

Named in memory of William Albert Hiltner (1914-1991), discoverer of the interstellar polarization of starlight, an early practitioner of precision stellar photometry, and pioneering observer of the optical counterparts of celestial x-ray sources. Director of the Yerkes Observatory for many years, while there he designed and built a rotatable telescope for polarization studies and developed photometric instrumentation. As director of the University of Michigan's observatory, he established the Michigan-Dartmouth-M.I.T. Observatory, led the construction of the 2.4-m telescope that is now named for him, and designed and constructed astronomical instrumentation. He also served as acting director of the Cerro Tololo Interamerican Observatory and as president of AURA. (M 20162)

Name suggested by R. P. Binzel and citation prepared by R. G. Teske.

Obituary published in Bull. Am. Astron. Soc., Vol. 24, No. 4, p. 1326-1327 (1992).

(4925) 1981 XH_2

Discovered 1981 December 3 at the Purple Mountain Observatory at Nanking.

(4926) Smoktunovskij

1982 ST_6. Discovered 1982 September 16 by L. I. Chernykh at Nauchnyj.

Named in honor of Innokentij Mikhailovich Smoktunovskij (1925-), outstanding Russian theater and cinema actor. (M 23137)

(4927) O'Connell

1982 UP_2. Discovered 1982 October 21 by Z. Vávrová at Kleť.

Named for the Irish freedom fighter Daniel O'Connell (1775-1847). (M 21609)

(4928) Vermeer

1982 UG_7. Discovered 1982 October 21 by L. G. Karachkina at Nauchnyj.

Named in memory of the famous Dutch painter Vermeer van Delft (1632-1675). (M 22504)

(4929) Yamatai

1982 XV. Discovered 1982 December 13 by H. Kosai and K. Hurukawa at Kiso.

Named for a region of ancient Japan. According to a Wei-dynasty Chinese historical book, there were more than 70,000 houses in Yamatai. Himiko, the queen of Yamatai, dispatched an envoy to the ancient China in 238. There is no record of Yamatai (or Himiko) in any Japanese historical document, and the site of the city has not yet been located. (M 22504)

(4930) Rephiltim

1983 AO_2. Discovered 1983 January 10 by S. L. Salyards at Palomar.

A composite name to honor the children of the discoverer: Rebecca Anne Salyards (1988-), Philip Stephen Salyards (1991-) and Timothy Theodore Salyards (1995-), all of whom enjoy studying the night sky. (M 29144)

(4931) Tomsk

1983 CN_3. Discovered 1983 February 11 by H. Debehogne and G. DeSanctis at La Silla.

Named by the first discoverer in honor of the Siberian city and its 500,000 inhabitants. (M 21609)

(4932) Texstapa

1984 EA_1. Discovered 1984 March 9 by B. A. Skiff at Anderson Mesa.

Named for the Texas Star Party on the occasion of its 18th anniversary in May 1996. The TSP is held annually on the Prude Ranch, Fort Davis, near McDonald Observatory {see planet (991)}. Founded in 1979 by Deborah Byrd {see planet (3505)}, the event has since become one of the largest meetings of amateur astronomers in North America. It provides amateur astronomers from around the world with the opportunity for visual observing, astrophotography and fellowship under clear, dark skies. (M 27127)

(4933) 1984 EN_1

Discovered 1984 March 2 by H. Debehogne at La Silla.

(4934) Rhôneranger

1985 JJ. Discovered 1985 May 15 by E. Bowell at Anderson Mesa.

Named in honor of Randall Grahm, eclectic and irrepressible owner and winemaker of Bonny Doon Vineyard, near Santa Cruz, California. Often referred to as the Rhône Ranger, Grahm is best known as the creator of Le Cigare Volant, a Rhône-style wine that in some vintages has reached celestial heights. He is also renowned for Clos de Gilroy, a grenache wine containing not a hint of garlic; Big House Red, whose untrammeled grapes leap from the glass; and a number of fruit infusions, including Prunus, which plumbs the depths of intense flavor. (M 20522)

(4935) Maslachkova

1985 PD_2. Discovered 1985 August 13 by N. S. Chernykh at Nauchnyj.

Named in memory of Iya Mikhailovna Maslachkova (1937-1996), a poet and teacher of Russian literature from St. Petersburg who worked for some years at the Crimean Astrophysical Observatory and had many friends there. Her poems are mainly dedicated to the Crimea and St. Petersburg. (M 30096)

(4936) Butakov

1985 UY_4. Discovered 1985 October 22 by L. V. Zhuravleva at Nauchnyj.

Named in memory of Grigorij Ivanovich Butakov (1820-1882), Russian admiral who drew up the first systematic directions for sailing in the Black Sea. In 1856 he became commander-in-chief of the Black Sea fleet and the military governor of Nikolaev and Sevastopol {see planet (2121)}. (M 34620)

(4937) 1986 CL_1

Discovered 1986 February 1 by H. Debehogne at La Silla.

(4938) 1986 CQ_1

Discovered 1986 February 5 by H. Debehogne at La Silla.

(4939) 1986 QL_1

Discovered 1986 August 27 by H. Debehogne at La Silla.

(4940) Polenov

1986 QY_4. Discovered 1986 August 18 by L. G. Karachkina at Nauchnyj.

Named in memory of the well-known Russian painter Vasilij Dmitrievich Polenov (1844-1927). (M 22504)

(4941) 1986 UA

Discovered 1986 October 25 by K. Suzuki and T. Urata at Toyota.

(4942) 1987 DU_6

Discovered 1987 February 24 by H. Debehogne at La Silla.

(4943) Lac d'Orient

1987 OQ. Discovered 1987 July 27 by E. W. Elst at St. Michel.

Named for the lake Lac de la Forêt d'Orient, situated about 20 km east of the city of Troyes (in the département de l'Aube, France). The region has great historical interest for its "commanderies" of the so-called Knights of the Temple. The discoverer favors this place very much for vacations, especially in the summer. (M 20838)

(4944) Kozlovskij

1987 RP_3. Discovered 1987 September 2 by L. I. Chernykh at Nauchnyj.

Named in honor of Ivan Semyonovich Kozlovskij, distinguished opera-singer who has rare lyric tenor and is popular for his many opera parties and for his rendering of Russian folk-songs. (M 22504)

(4945) Ikenozenni

1987 SJ. Discovered 1987 September 18 by K. Suzuki and T. Urata at Toyota.

Named for Ikenozenni, the second wife of (4374) Tadamori. She asked (4375) Kiyomori to spare the life of (3902) Yoritomo, who had come to destroy the Heike, family of Kiyomori. (M 21132)

(4946) Askalaphus

1988 BW_1. Discovered 1988 January 21 by C. S. Shoemaker and E. M. Shoemaker at Palomar.

Askalaphus, a son of the god Ares, became one of the Argonauts, and also participated in the Trojan War. With his brother, Ialmenos, he drew up thirty ships in line of battle. Askalaphus was killed when Deiphobus {see planet (1867)} ran him through the shoulder with his heavy spear. (M 25443)

(4947) Ninkasi

1988 TJ_1. Discovered 1988 October 12 by C. S. Shoemaker and E. M. Shoemaker at Palomar.

Named for the Sumerian goddess of wine and beer. She aided the god Lugalbanda in rescuing the tablets of fate from the demon Zu. In Semitic mythology, Ninkasi was also considered god of the Vine, corresponding to the Greek Dionysus {see planet (3671)}. An ode to Ninkasi, inscribed in cuneiform on 4000-year old clay tablets, provides a recipe that was used successfully by Fritz Maytag and Solomon H. Katz to recreate Sumerian beer. (M 20522)

(4948) 1988 VF_1

Discovered 1988 November 3 by W. Kakei and M. Kizawa and T. Urata at Oohira.

(4949) 1988 WE

Discovered 1988 November 29 by T. Kojima at Chiyoda.

(4950) House

1988 XO_1. Discovered 1988 December 7 by E. F. Helin at Palomar.

Named in honor of R. C. House, western novelist and journalist who has served, for the last 23 years, as editor of the Jet Propulsion Laboratory's internal publication 'Universe'. He has been responsible for this highly regarded chronicle of life at "the Lab" until his recent retirement. He captured for his readers the true spirit of the world's leading space exploration center. His natural warmth

endeared him to all he had contact. (M 20162)
Name endorsed by Phil Neuhauser, a good friend for many years.

(4951) Iwamoto

1990 BM. Discovered 1990 January 21 by Y. Mizuno and T. Furuta at Kani.

Named in honor of Masayuki Iwamoto (1954-), who has discovered several minor planets at the Tokushima Observatory. (M 26763)

(4952) Kibeshigemaro

1990 FC_1. Discovered 1990 March 26 by A. Sugie at Taga.

Named in memory of Shigemaro Kibe (1912-1990), foremost amateur telescope maker in Japan and well known observer of the sun, planets and variable stars. The telescope mirrors made by him are used all over Japan and include the mirror for the 0.60-m reflector at Kyoto University's Hida Observatory. (M 20838)

(4953) 1990 MU

Discovered 1990 June 23 by R. H. McNaught at Siding Spring.

(4954) Eric

1990 SQ. Discovered 1990 September 23 by B. Roman at Palomar.

The name dates back to ancient Norse times, and was also prominent among Scandinavian monarchs. Eric the Red was the first Norse explorer of Greenland and father of Leif Ericson, Eric IX was king of Sweden and is revered as Sweden's patron saint (although he was never formally canonized), and Eric of Pomerania was king of Denmark, Norway and Sweden. This Amor {see planet (1221)} object also honors Eric Dale Roman, the discoverer's first child, in his first year of life. (M 19341)
Name endorsed by E. F. Helin.

(4955) 1990 SF_2

Discovered 1990 September 17 by H. E. Holt at Palomar.

(4956) Noymer

1990 VG_1. Discovered 1990 November 12 by R. H. McNaught at Siding Spring.

Named in honor of Andrew J. Noymer on the occasion of his twentieth birthday, 1991 Nov. 4. A student at Harvard University, he has since the age of 14 been an enthusiastic assistant in a number of astronomical activities. In addition to those at the Harvard-Smithsonian Center for Astrophysics, where he is currently working at the Minor Planet Center and Central Bureau for Astronomical Telegrams, he spent several weeks at Siding Spring in 1988 (where he discovered several minor planets) and at the U.K. Infrared Telescope in 1989. (M 19341)

Name proposed by the discoverer, following a suggestion by C. M. Bardwell, D. W. E. Green, B. G. Marsden and G. V. Williams.

(4957) Brucemurray

1990 XJ. Discovered 1990 December 15 by E. F. Helin at Palomar.

Named in honor of Bruce C. Murray, professor of planetary science at the California Institute of Technology and former director of the NASA/Caltech Jet Propulsion Laboratory. As a co-founder of The Planetary Society, he has been instrumental in making science accessible and understandable to the general public. This minor planet is a deep Mars-crosser, coming within 0.2 AU of the earth at perihelion, and it is a particularly appropriate object for Murray, who has so diligently championed a mission to Mars. Future spacecraft missions could flyby or rendezvous with (4957) on the way to Mars. This tribute to 'BC' is hailed by his family, friends and colleagues. (M 22829)

(4958) 1991 NT_1

Discovered 1991 July 13 by H. E. Holt at Palomar.

(4959) Niinoama

1991 PA_1. Discovered 1991 August 15 by A. Natori and T. Urata at Yakiimo.

Named for Tokiko (?-1185), wife of (4375) Kiyomori. She is called Niinoama because she took the rank of secondary-nii after Kiyomori's death. After the battle of Dannoura, she drowned herself holding the 7-year-old Emperor Antoku {see planet (3686)} in her arms. (M 21132)

(4960) Mayo

4657 P-L. Discovered 1960 September 24 by C. J. van Houten and I. van Houten-Groeneveld at Palomar.

Named in honor of Mayo Greenberg on the occasion of his 70th birthday. Well known for his model of cometary and interstellar grains, he worked at the State University at Albany and later as head of the Astrophysical Laboratory in Leiden. (M 20162)

(4961) 1958 TH_1

Discovered 1958 October 8 at the Lowell Observatory at Flagstaff.

(4962) Vecherka

1973 TP. Discovered 1973 October 1 by T. M. Smirnova at Nauchnyj.

Named for the popular St. Petersburg evening newspaper *Vechernij Petersburg*. The newspaper regularly publishes astronomical information and articles popularizing astronomical knowledge. (M 31023)
Name proposed by the Institute of Theoretical Astronomy (ITA).

(4963) Kanroku

1977 DR_1. Discovered 1977 February 18 by H. Kosai and K. Hurukawa at Kiso.

Named for a Pekche priest who presented, by way of tribute, books on calendar-making, astronomy, geography, divination, etc., to the Japanese Government in 602. He taught the Japanese the algorithm of the luni-solar calendar used in China. (M 22504)

(4964) Kourovka

1979 OD_{15}. Discovered 1979 July 21 by N. S. Chernykh at Nauchnyj.

Named for the astronomical observatory of the Ural State University, situated near the village of Kourovka, in the Ekaterinburg region of Russia. Founded in 1965, the Kourovka Observatory is known for research in stellar astronomy and the astrometry of comets, minor planets and artificial satellites. Since 1972 it has held an annual Student Winter Astronomical School and is now a mecca for Russian students in astronomy. (M 26763)

(4965) 1981 EP_{28}

Discovered 1981 March 6 by S. J. Bus at Siding Spring.

(4966) 1981 EO_{34}

Discovered 1981 March 2 by S. J. Bus at Siding Spring.

(4967) Glia

1983 CF_1. Discovered 1983 February 11 by N. G. Thomas at Anderson Mesa.

Name derived from the Latin word for 'glue'. Glia are the cells providing support for the components of nervous tissue in the brain. According to one theory, the intricate structure of the glia may be involved in quantum processes that generate 'new' ideas, i.e. invention. (M 25443)

(4968) Suzamur

1986 PQ. Discovered 1986 August 1 by E. F. Helin at Palomar.

Named in honor of Suzanne Moss Murray, a charming and compassionate person, on the occasion of her birthday, 1993 Apr. 2. Suzanne, a special friend of the discoverer, has always shown enthusiasm and given encouragement to Helin's pursuit of minor planets. (M 22504)

(4969) Lawrence

1986 TU. Discovered 1986 October 4 by E. F. Helin at Palomar.

Named in honor of Kenneth J. Lawrence (1964-), astronomer and valuable member of the Palomar Planet-Crossing Asteroid Survey team, on the occasion of his 28th birthday, 1992 May 30. Discoverer of near-earth-asteroids and comets, his congenial personality and wry wit contribute to the smooth operation of the highly-focused NEA discovery program. (M 20522)

This recognition is heartily endorsed by his friends and coworkers, Jeff Alu, Perry Rose and Brian Roman.

(4970) Druyan

1988 VO_2. Discovered 1988 November 12 by E. F. Helin at Palomar.

Named as a tribute to Ann Druyan for her many contributions to the public understanding of the humane applications of planetary exploration and her creative directorship of the Voyager interstellar record, humanity's gift to the universe. (M 24122)

Presented on the occasion of astronomer Carl Sagan's {see planet (2709)} 60th birthday in recognition of his love and admiration for his wife and colleague.

(4971) Hoshinohiroba

1989 BY. Discovered 1989 January 30 by T. Fujii and K. Watanabe at Kitami.

Named after the Comet Observers Network in Japan, "Hoshi no Hiroba", established by Akira Kamo {see planet (4215)} in 1968. About 350 earnest observers made 8805 visual magnitude estimates of 89 comets in the first 20 years. "Hoshi no Hiroba" means an open space where everybody can come and enjoy watching stars. (M 22504)

Citation prepared by T. Nomura.

(4972) Pachelbel

1989 UE_7. Discovered 1989 October 23 by F. Börngen at Tautenburg.

Named for the composer Johann Pachelbel (1653-1706) of Nürnberg, an important representative of the art of organ music whose free-form choral preludes and fuges had a considerable influence on J. S. Bach {see planet (1814)}. (M 19699)

(4973) Showa

1990 FT. Discovered 1990 March 18 by K. Endate and K. Watanabe at Kitami.

Named for an industrial-pump company that started the production of astronomical telescopes in 1988. The first discoverer uses a Showa equatorial telescope for his observations of minor planets. (M 21132)

(4974) Elford

1990 LA. Discovered 1990 June 14 by R. H. McNaught at Siding Spring.

Named in honor of W. Graham Elford (1926-), past president of IAU Commission 22. Since entering radar meteor research soon after taking his bachelor's degree in 1949, he has spent his entire career at the University of Adelaide working in that area, apart from periods spent as a visiting scientist at Cambridge, Lund and with the Harvard Radio Meteor Project in the 1960s. Elford has made a number of major contributions to our understanding of meteor astronomy, meteor physics and atmospheric physics. (M 29144)

Name proposed by the discoverer following a suggestion by D. I. Steel, who prepared the citation.

(4975) Dohmoto

1990 SZ_1. Discovered 1990 September 16 by T. Fujii and K. Watanabe at Kitami.

Named in honor of Yoshio Dohmoto (1914-), the former head of the Asahikawa Observatory in Hokkaido. Over a period of about 30 years, he carried out many observations of comets, occultations, artificial satellites, sunspots, etc. He is now an active leader of the Asahikawa Astronomical Club. (M 22504)
Name suggested and citation prepared by H. Fukushima.

(4976) 1991 PM

Discovered 1991 August 9 by K. Watanabe at Sapporo.

(4977) Rauthgundis

2018 P-L. Discovered 1960 September 24 by C. J. van Houten and I. van Houten-Groeneveld at Palomar.

Named in honor of Rauthgundis Seitz, a friend of the discoverers, on the occasion of her 70th birthday. (M 21956)

(4978) Seitz

4069 T-2. Discovered 1973 September 29 by C. J. van Houten and I. van Houten-Groeneveld at Palomar.

Named in honor of Horstmar Seitz (1920-1943), a close friend of one of the discoverers. (M 22505)

(4979) Otawara

1949 PQ. Discovered 1949 August 2 by K. Reinmuth at Heidelberg.

Named in honor of Akira Otawara (1950-), photo artist and writer. In 1983 he published two books, *Field Star Map 2000* and *Star & Planet Catalogue 2000*, in collaboration with S. Nakano {see planet (3431)}, who proposed the name and found the identifications involving this object. (M 22505)

(4980) Magomaev

1974 SP_1. Discovered 1974 September 19 by L. I. Chernykh at Nauchnyj.

Named in honor of Muslim Magometovich Magomaev (1942-), famous Soviet singer, a soloist at the Azerbaijan Theatre of opera and ballet and later at the Bolshoj Theatre in Moscow. A remarkable performer of the songs of Russian and Azerbaijani composers and of arias in classical operas, he is also known as a songwriter and composer of movie scores. (M 26763)

(4981) Sinyavskaya

1974 VS. Discovered 1974 November 12 by L. I. Chernykh at Nauchnyj.

Named in honor of Tamara Il'inichna Sinyavskaya, distinguished Russian singer, soloist at the Bolshoj Theatre in Moscow and performer of many roles in classical and modern operas. (M 26763)

(4982) Bartini

1977 PE_1. Discovered 1977 August 14 by N. S. Chernykh at Nauchnyj.

Named in memory of Robert Lyudvigovich Bartini (1897-1974), Italian baron by birth who lived and worked in the Soviet Union. The designer of several original aircraft, Bartini was also known for his research in aerodynamics, theoretical physics and cosmology. (M 30096)

(4983) Schroeteria

1977 RD_7. Discovered 1977 September 11 by N. S. Chernykh at Nauchnyj.

Named in memory of Johann Hieronymus Schroeter (1745-1816), outstanding German amateur astronomer and telescope maker. At his private observatory in Lilienthal, Schroeter made numerous observations of the sun, the moon, and major and minor planets, comets and stars, with a special emphasis on the physical constitution and surface characteristics of celestial bodies. In particular, Schroeter correctly identified the strip-shaped details on Mars as sand storms. His drawings were important for an understanding of the physical nature of the comets of 1799, 1807 and 1811. Schroeter was one of the founders of the Astronomische Gesellschaft and host for the meeting of the famous "celestial police" in 1800. (M 26763)

Name suggested and citation provided by V. K. Abalakin.

Schroeter is also honored by a lunar crater.

(4984) 1978 VU_{10}

Discovered 1978 November 7 by E. F. Helin and S. J. Bus at Palomar.

(4985) Fitzsimmons

1979 QK_4. Discovered 1979 August 23 by C.-I. Lagerkvist at La Silla.

Named after Alan Fitzsimmons, who works on the relationships between minor planets and comets and has collaborated with the discoverer for several years. His enthusiasm and good spirit when observing on La Palma has always been very much appreciated by the discoverer. (M 21609)

(4986) Osipovia

1979 SL_7. Discovered 1979 September 23 by N. S. Chernykh at Nauchnyj.

Named in memory of Valerij Ivanovich Osipov (1945-1993), distinguished Russian historian, archivist, archaeologist and linguist in the St. Petersburg Department of Archives of the Russian Academy of Sciences. He made a valuable contribution to the study of the Russo-German scientific relations in the eighteenth century, discovered some new documents on the founding and work of the Pulkovo Astronomical Observatory, studied the peculiarities of the language of eighteenth-century Russian writers and carried out archaeological investigations in the countries of Middle Asia. (M 26763)

Name suggested and citation material provided by V. K. Abalakin.

(4987) 1980 FH_{12}

Discovered 1980 March 20 at the Perth Observatory at Bickley.

(4988) 1980 VU_1

Discovered 1980 November 6 at the Purple Mountain Observatory at Nanking.

(4989) 1981 DX_1

Discovered 1981 February 28 by S. J. Bus at Siding Spring.

(4990) 1981 ET_{26}

Discovered 1981 March 2 by S. J. Bus at Siding Spring.

(4991) 1981 EU_{29}

Discovered 1981 March 1 by S. J. Bus at Siding Spring.

(4992) Kálmán

1982 UX_{10}. Discovered 1982 October 25 by L. V. Zhuravleva at Nauchnyj.

Named in memory of the Hungarian composer Imre Kálmán (1882-1953). (M 24765)

(4993) 1983 GR

Discovered 1983 April 11 by H. Debehogne and G. DeSanctis at La Silla.

(4994) 1983 RK_3

Discovered 1983 September 1 by H. Debehogne at La Silla.

(4995) 1984 QR

Discovered 1984 August 28 by S. Swanson at Palomar.

(4996) Veisberg

1986 PX_5. Discovered 1986 August 11 by L. G. Karachkina at Nauchnyj.

Named for the Russian painter Vladimir Grigor'evich Veisberg (1924-1985). (M 22505)

(4997) Ksana

1986 TM. Discovered 1986 October 6 by L. G. Karachkina at Nauchnyj.

Named in honor of Kseniya Andreevna Nessler, chemist by profession and fighter against environmental pollution by avocation, friend of the discoverer. Ksana is a familiar variant of Kseniya, the Russian form of Xenia. (M 22505; M 22609)

(4998) 1986 VG

Discovered 1986 November 5 by K. Suzuki and T. Urata at Toyota.

(4999) MPC

1987 CJ. Discovered 1987 February 2 by E. W. Elst at La Silla.

Named by the Minor Planet Names Committee for the *Minor Planet Circulars*, the series established in 1947 for the publication of astrometric observations, orbital elements and limited ephemerides of minor planets - as well as for the announcement of new names. The abbreviation also honors the Minor Planet Center, which operates through IAU Commission 20 to issue the Circulars. Originally located at the Cincinnati Observatory, the Minor Planet Center moved to the Smithsonian Astrophysical Observatory in 1978. Data on comets also now appear in the publication, which has the alternate title *Minor Planets and Comets*. (M 19342)

(5000) IAU

1987 QN_7. Discovered 1987 August 23 by E. F. Helin at Palomar.

Named in honor of the International Astronomical Union, established in 1919 as the international association of professional astronomers. Most of its activity is concentrated in some 40 commissions, each devoted to some particular area of astronomical or astrophysical research. At a competition held in July 1991 among members of Commission 20, "Positions and Motions of Minor Planets, Comets and Satellites", for the selection of a name for the five-thousandth minor planet to be given a permanent number the name "IAU" received the largest number of votes, including that of the discoverer. (M 19342)

(5001) EMP

1987 SB_1. Discovered 1987 September 19 by E. Bowell at Anderson Mesa.

Named by the Minor Planet Names Committee for the *Ehfemeridy Malykh Planet* or *Ephemerides of Minor Planets*, the annual publication of ephemerides of the numbered minor planets and updates of the orbital elements of these bodies to a current epoch. Many new orbit improvements are also included in the EMP each year. Under the auspices of the IAU Commission 20, the EMP, successor to the old *Kleine Planeten* volumes, has been published since 1948 by the Minor Planets and Comets department of the Institute of Theoretical Astronomy in St. Petersburg (formerly Leningrad). The EMP is also now available in a computer-diskette edition. (M 19342)

(5002) Marnix

1987 SS_3. Discovered 1987 September 20 by E. W. Elst at Rozhen.

Named for Philips Marnix van Sint Aldegonde (1538-1598), mayor of Antwerp during 1583-1585 and player of a major role in defending the city against the Spanish troops. He was concerned with the religious struggle between catholics and protestants and is considered to be the composer of the beautiful Dutch national hymn 'Wilhelmus'. (M 20162)
Citation based on information supplied by R. Grignard.

(5003) 1988 ER_2

Discovered 1988 March 15 by W. Ferreri at La Silla.

(5004) Bruch

1988 RR_3. Discovered 1988 September 8 by F. Börngen at Tautenburg.

Named for the German composer Max Bruch (1838-1920), particularly well known for his grandiose first violin concerto. (M 19699)

(5005) Kegler

1988 UB. Discovered 1988 October 16 by S. Ueda and H. Kaneda at Kushiro.

Named for Ignatius Kegler (1680-1746), a German Jesuit, who worked for 29 years as Astronomer Royal in Beijing. The identity of a comet he observed in 1737 with P/Swift-Tuttle, briefly suggested by W. T. Lynn {see planet (4358)} and discussed in considerable detail by B. G. Marsden {see planet (1877)}, was proven by the comet's recovery in 1992. (M 22505)
Name suggested by I. Hasegawa.

(5006) Teller

1989 GL_5. Discovered 1989 April 5 by E. F. Helin at Palomar.

Named in honor of the distinguished Hungarian-born, U.S. physicist Edward Teller {1908- }. Known for his seminal work in physics and astrophysics, notably on the Gamow-Teller relationship, he has also made significant contributions in chemical physics, molecular physics and quantum theory. He has been a professor of physics at George Washington University, the University of Chicago and the University of California. He inspired the creation of the Lawrence Livermore National Laboratory and has served as its director. Senior research associate at the Hoover Institute, Stanford University, since 1975, he has been highly effective in promoting open international science free of secrecy. Having entered the world a few months before the Tunguska {see planet (5471)} event, Teller was honored by the naming of this minor planet by his colleagues on his eighty-fourth birthday, 1992 Jan. 15, during his participation in a NASA Workshop on procedures for the interception of threatening near-earth asteroids and comets. (M 19699)

(5007) Keay

1990 UH_2. Discovered 1990 October 20 by R. H. McNaught at Siding Spring.

Named in honor of Colin Stewart Lindsay Keay (1930-), past president of IAU Commission 22 and chairman of the IAU Working Group on the Prevention of Interplanetary Pollution, who has made several major contributions to our understanding of the meteoroidal flux to the earth. His well-controlled radar patrol from the University of Canterbury during 1960-65 in collaboration with Clifton Ellyett {see planet (5378)} remains our best knowledge of the southern-hemisphere influx. Since 1965 Keay has worked at the University of Newcastle (N.S.W.), one of his most noteworthy achievements being the development of a physical theory for the production of electrophonic sounds by bright fireballs. (M 29144)

Name proposed by the discoverer following a suggestion by D. I. Steel, who prepared the citation, and an endorsement by B. G. Marsden.

(5008) Miyazawakenji

1991 DV. Discovered 1991 February 20 by A. Sugie at Taga.

Named in memory of Kenji Miyazawa (1896-1933), a poet and fairy-tale writer with few parallels in the history of literature in Japan. His great works show a love for nature and a unique sense and expression cultivated through his life as a farmer, while he discharged his duties devotedly as an agricultural leader in Iwate Prefecture. (M 26763; M 27477)

(5009) Sethos

2562 P-L. Discovered 1960 September 24 by C. J. van Houten and I. van Houten-Groeneveld at Palomar.

Named after the Egyptian pharaoh Sethos I (reigned 1304-1290 B.C.) of the 19th dynasty. His father was Ramses I, who reigned for only two years. Sethos I built the Great Hall of Columns of the Amun {see planet (3554)} temple in Karnak and the Osiris {see planet (1923)} temple in Abydos. Sethos I was the father of Ramses II {see planet (4416)}. (M 22505)

(5010) Amenemhêt

4594 P-L. Discovered 1960 September 24 by C. J. van Houten and I. van Houten-Groeneveld at Palomar.

Named after the Egyptian pharaoh Amenemhêt III (1844-1797 B.C.), son of Sesostris III {see planet (4414)}. He regulated the oasis Fajum which became rich agricultural land. There he built an immense temple which Herodotus {see planet (3092)} later visited and called a 'labyrinth'. (M 22505)

(5011) Ptah

6743 P-L. Discovered 1960 September 24 by C. J. van Houten and I. van Houten-Groeneveld at Palomar.

In Egyptian religion Ptah was the creator of the universe and a patron of craftsmen, especially sculptors. Ptah was originally the local deity of Memphis, capital of Egypt from the 1st dynasty; the political importance of Memphis led to the expansion of Ptah's cult throughout Egypt. Ptah was always represented in purely human form, often swathed in a winding sheet. (M 20163)

(5012) Eurymedon

9507 P-L. Discovered 1960 October 17 by C. J. van Houten and I. van Houten-Groeneveld at Palomar.

Eurymedon was servant to the Greek king Nestor {see planet (659)} during the siege of Troy. (M 20163)

(5013) 1964 VT_1

Discovered 1964 November 9 at the Purple Mountain Observatory at Nanking.

(5014) Gorchakov

1974 ST. Discovered 1974 September 19 by L. I. Chernykh at Nauchnyj.

Named in memory of Aleksandr Mikhailovich Gorchakov (1798-1883), an outstanding statesman and one of the eminent diplomats of the nineteenth century. He was Minister of Foreign Affairs of Russia from 1856 to 1882. (M 32345)
Named on the occasion of the 200th anniversary of his birth.

(5015) Litke

1975 VP. Discovered 1975 November 1 by T. M. Smirnova at Nauchnyj.

Named for Fedor Petrovich Litke (1797-1882), Russian navigator and geographer, investigator of the Arctic regions, admiral and scientist, president of the Imperial Academy of Sciences in St. Petersburg during 1864-1882. His name is often found on world maps. (M 31295)

(5016) Migirenko

1976 GX_3. Discovered 1976 April 2 by N. S. Chernykh at Nauchnyj.

Named in honor of Georgij Sergeevich Migirenko (1916-), academician and professor at the Siberian Department of Russian Academy of Sciences in Novosibirsk, known for his work in applied and theoretical mechanics, hydrodynamics, mathematics and ship-building. In recent years he has worked on design problems of ecology-safe machines and transport for northern regions. Migirenko also takes an active part in research on the possibility of asteroids colliding with the earth. He is also a talented singer who graduated from the Odessa Conservatory. (M 30096)

(5017) Tenchi

1977 DS_2. Discovered 1977 February 18 by H. Kosai and K. Hurukawa at Kiso.

Named for the emperor (626-671) who made the first clepsydra in Japan in 660. On 671 June 10 (Gregorian style) he placed a new clepsydra on a new pavilion, and it sounded out the hours with bells and drums. In commemoration of this, June 10 is designated as the "Time Day" in Japan. (M 22505)

(5018) Tenmu

1977 DY_8. Discovered 1977 February 19 by H. Kosai and K. Hurukawa at Kiso.

Named for a Japanese Emperor (?-686), younger brother of the emperor Tenchi {see planet (5017)}, skilled in astronomy and in divination. He erected the first Japanese astronomical observatory in 675. (M 22505; M 22519)

(5019) 1979 MS_6

Discovered 1979 June 25 by E. F. Helin and S. J. Bus at Siding Spring.

(5020) Asimov

1981 EX_{19}. Discovered 1981 March 2 by S. J. Bus at Siding Spring.

Named in memory of Isaac Asimov (1920-1992), author and scientist. A prolific writer, Asimov produced nearly 500 works of science fiction and nonfiction, covering topics ranging from mathematics and physics to Shakespeare and history. His *Robot* stories revolutionized modern ideas about intelligent machines by introducing his "Three Laws of *Robot*ics", an ethical code of robot behavior. While a professor of biochemistry at Boston University, Asimov also devoted himself to making science accessible to the general public through writing and lecturing. (M 27128; M 27147)

Name independently proposed by B. A. Skiff.

Obituary published in Nature, Vol. 357, No. 6374, p. 113 (1992).

(5021) Krylania

1982 VK_{12}. Discovered 1982 November 13 by L. G. Karachkina at Nauchnyj.

Named in honor of Anna Alexeevna Kapitsa (1903-), the daughter of Alexei Nikolaevich Krylov. She was an outstanding Russian mathematician and the wife of famous physicist Pyotr Leonidovich Kapitsa {see planet (3437)}. (M 22505)

Name proposed by the discoverer following a suggestion by Maria Sergeevna Kapitsa.

(5022) 1984 HE_1

Discovered 1984 April 23 by W. Ferreri and V. Zappalà at La Silla.

(5023) Agapenor

1985 TG_3. Discovered 1985 October 11 by C. S. Shoemaker and E. M. Shoemaker at Palomar.

King and leader of the Arcadians {see planet (1020)} who went with the Greeks to the Trojan War, Agapenor commanded sixty ships loaned to him by Agamemnon {see planet (911)}. The Arcadians, who were able in war, thronged to go aboard as they had no ships or knowledge of seafaring themselves. (M 25443)

(5024) 1985 VP

Discovered 1985 November 14 by P. Jensen at Brorfelde.

(5025) 1986 TS_6

Discovered 1986 October 5 by M. Antal at Piwnice.

(5026) Martes

1987 QL_1. Discovered 1987 August 22 by A. Mrkos at Kleť.

Named for Martes martes and Martes foina, or pine marten and beech marten, delightful small animals of the Mustelidae family that live in the forests on Kleť Mountain. (M 34920)

Name suggested by J. Tichá and M. Tichý.

(5027) Androgeos

1988 BX_1. Discovered 1988 January 21 by C. S. Shoemaker and E. M. Shoemaker at Palomar.

Androgeos was a Greek warrior who led a detachment of Greeks into the burning city of Troy. There he encountered Aeneas {see planet (1172)} in the company of other Trojan defenders. Androgeos and his detachment were set upon and killed, after which they were stripped of their armor by the Trojans, who then disguised themselves with it. It can be said that Androgeos supplied Aeneas and his men with tickets to safety. (M 22248)

Name and citation provided by Richard Preston at the request of the discoverers.

(5028) Halaesus

1988 BY_1. Discovered 1988 January 23 by C. S. Shoemaker and E. M. Shoemaker at Palomar.

Halaesus was a son of Agamemnon {see planet (911)}. This minor planet is dynamically paired with (2759) Idomeneus, and like the Greek warrior Idomeneus, Halaesus found a home in Italy, settling on Mount Massiscus in Campania. Unlike Idomeneus, Halaesus missed joining the Trojan War and the sack of Troy. He made up for this later, for he fought against Aeneas {see planet (1172)} in the war for Italy, where he killed a large number of Trojans on Italian soil. (M 22248)

Name and citation provided by Richard Preston at the request of the discoverers.

(5029) Ireland

1988 BL_2. Discovered 1988 January 24 by C. S. Shoemaker and E. M. Shoemaker at Palomar.

Named in honor of Ireland, known poetically as Erin or Hibernia, to celebrate a year of anniversaries in 1995: the 150th anniversary of the Geological Survey of Ireland, the 150th anniversary of the Queen's Colleges (now University Colleges, Cork and Galway, and Queen's University, Belfast), and the 750th anniversary of the town of Sligo. (M 25653)

Citation prepared by P. M. Bruck at the request of the discoverers.

(5030) 1988 VK_4

Discovered 1988 November 3 by P. Jensen at Brorfelde.

(5031) Švejcar

1990 FW_1. Discovered 1990 March 16 by Z. Vávrová at Kleť.

Named in honor of Josef Švejcar (1897-), Czech physician. (M 22505)

(5032) 1990 OO

Discovered 1990 July 18 by E. F. Helin at Palomar.

(5033) Mistral

1990 PF. Discovered 1990 August 15 by E. W. Elst at St. Michel.

Named in memory of the great poet from the Provence, Frederic Mistral (1830-1914), whose entire life was dedicated to the restoration of the original dialect of the Langue d'Oc, the language of the 'troubadours'. In 1859 he published the poem *Mir entire Provence.* In 1886 he finished his *Lou tresor dou felibrige*, a Provencal-French dictionary. He was honored with the Nobel prize for literature {in 1904 together with J. Echegaray}. The Northern wind that blows through the Rhone valley and sweeps all the clouds from the sky bears the same name. (M 20163)

Citation prepared by Kristina Leterme at the request of the discoverer.

(5034) 1991 PW_{10}

Discovered 1991 August 7 by H. E. Holt at Palomar.

(5035) Swift

1991 UX. Discovered 1991 October 18 by S. Ueda and H. Kaneda at Kushiro.

Named in memory of Lewis Swift (1820-1913), famous U.S. comet hunter. The first of his 13 discoveries, P/Swift-Tuttle (1862 III) {see planet (5036) for the co-discoverer}, parent of the Perseid meteors, is also the most famous. He discovered his last comet at the age of 79. (M 22505)

Name suggested by I. Hasegawa.

(5036) Tuttle

1991 US_2. Discovered 1991 October 31 by S. Ueda and H. Kaneda at Kushiro.

Named for Horace P. Tuttle (1837-1923), astronomer at Harvard and - after serving in the U.S. Navy during the Civil War - at the Naval Observatory in Washington. Co-discoverer of both the Perseid comet P/Swift-Tuttle and the Leonid comet P/Tempel-Tuttle {see planets (5035) and (3808) for the co-discoverers}, he also found four other comets and the minor planets (66) Maja and (73) Klytia. (M 22505)

Name suggested and citation prepared by I. Hasegawa.

(5037) Habing

6552 P-L. Discovered 1960 September 24 by C. J. van Houten and I. van Houten-Groeneveld at Palomar.

Named in honor of Harm Habing (1937-), professor of astronomy at the University of Leiden and president of IAU Commission 34 (Interstellar Matter). Habing was the Dutch co-chairman of the Joint Science Team of the Infrared Astronomical Satellite (IRAS) {see planet (3728)}. Since 1983 he has supervised many publications based on IRAS observations of galaxies. Habing is a stimulating supervisor for his students and young scientists. (M 22505)

(5038) 1948 KF

Discovered 1948 May 31 by E. L. Johnson at Johannesburg.

(5039) Rosenkavalier

1967 GM_1. Discovered 1967 April 11 by F. Börngen at Tautenburg.

Named in memory of Richard Strauss (1864-1949), one of the most important twentieth-century composers of opera, notably *Der Rosenkavalier.* He also wrote numerous song compositions and symphonic poems. (M 20163)

Name endorsed by Werner Tscharnuter.

(5040) Rabinowitz

1972 RF. Discovered 1972 September 15 by T. Gehrels at Palomar.

Named in honor of the U.S. astronomer David Rabinowitz and his work in the Spacewatch program. (M 22505)

(5041) Theotes

1973 SW_1. Discovered 1973 September 19 by C. J. van Houten and I. van Houten-Groeneveld at Palomar.

Theotes was a Greek herald during the siege of Troy. (M 20163)

(5042) 1974 ME

Discovered 1974 June 20 at the Felix Aguilar Observatory at El Leoncito.

(5043) Zadornov

1974 SB_5. Discovered 1974 September 19 by L. I. Chernykh at Nauchnyj.

Named in honor of Mikhail Nikolaevich Zadornov (1948-), Russian satirist widely known for his masterful readings of his own short stories. (M 34340)

(5044) Shestaka

1977 QH_4. Discovered 1977 August 18 by N. S. Chernykh at Nauchnyj.

Named in memory of Ivan Sofronovich Shestaka (1937-1994), senior scientist and head of the Comet and Meteor Department at the Odessa University Astronomical Observatory. He was known for his research on the physics and dynamics of meteoritic and cometary matter and its interaction with other minor bodies in the solar system. (M 27128)

(5045) 1978 UL_2

Discovered 1978 October 29 at the Purple Mountain Observatory at Nanking.

(5046) 1981 DQ

Discovered 1981 February 28 by S. J. Bus at Siding Spring.

(5047) 1981 EO_{42}

Discovered 1981 March 2 by S. J. Bus at Siding Spring.

(5048) Moriarty

1981 GC. Discovered 1981 April 1 by E. Bowell at Anderson Mesa.

Named for Professor James Moriarty, whom some would describe as a fictional character in the Sherlock Holmes {see planet (5049)} stories of Sir Arthur Conan Doyle. Described by Holmes as "the Napoleon of crime", and also as "the celebrated author of *The Dynamics of an Asteroid* - a book that ascends to such rarefied heights of pure mathematics that it is said that there was no man in the scientific press capable of criticizing it", the "most dangerous and capable criminal in Europe" died in a struggle with Holmes at the Reichenbach Falls in Switzerland. (M 22505)

Name suggested by J. F. Bowers, citation prepared by J. B. Tatum.

(5049) Sherlock

1981 VC_1. Discovered 1981 November 2 by E. Bowell at Anderson Mesa.

Named for Sherlock Holmes, the famous detective in the stories by Sir Arthur Conan Doyle. Holmes's knowledge of astronomy was nil; he was ignorant of the Copernican Theory and of the composition of the solar system. "You say that we go round the sun. If we went round the moon it would not make a pennyworth of difference to me or my work." (M 22506)

Name suggested by J. F. Bowers, citation prepared by J. B. Tatum.

(5050) Doctorwatson

1983 RD_2. Discovered 1983 September 14 by E. Bowell at Anderson Mesa.

Named for Dr. John H. Watson, M.D., friend, confidant and chronicler of the detective Sherlock Holmes {see planet (5049)} in the stories by Sir Arthur Conan Doyle. (M 22506)

Name suggested by J. F. Bowers, citation prepared by J. B. Tatum.

(5051) 1984 SM

Discovered 1984 September 24 by K. Augustesen and P. Jensen and H. J. Fogh Olsen at Brorfelde.

(5052) Nancyruth

1984 UT_3. Discovered 1984 October 23 by C. S. Shoemaker and E. M. Shoemaker at Palomar.

Named in honor of Nancy R. Lebofsky of the University of Arizona's ARTIST and ACCESS! astronomy education programs, for her contributions to education. Through teacher workshops and presentations at national meetings, Lebofsky has brought the excitement of astronomy to thousands of teachers and their students. She has developed hands-on activities and produced a video that have been distributed to educators throughout the United States and other countries. (M 25443, 25457)

Name proposed by the discoverers, following a suggestion by L. Lebofsky and D. H. Levy. Citation prepared by L. Lebofsky.

(5053) Chladni

1985 FB_2. Discovered 1985 March 22 by E. Bowell at Anderson Mesa.

Named in honor of Ernst Florens Friedrich Chladni (1756-1827), German meteoriticist. Generally regarded as the 'Father of Meteoritics', Chladni was initially trained as a lawyer. He never held a permanent position, but he travelled among the great cities of Europe giving lectures. It was on these travels that Chladni's interest turned to meteorites, and in 1794 he published a pioneering book in which he presented convincing evidence that meteorites originate in space. Never before or since has the birth of a scientific discipline been so closely associated with the publication of a single book. (M 22829)

Name suggested and citation prepared by T. J. McCoy, who also recently recognized Chladni by the naming of a mineral - chladniite.

Chladni is also honored by a lunar crater.

(5054) Keil

1986 AO_2. Discovered 1986 January 12 by E. Bowell at Anderson Mesa.

Named in honor of Klaus Keil (1934-), American meteoriticist. An authority on the petrology of enstatite meteorites, Keil has also studied almost every other group of meteorites. He has headed active meteorite research groups at the University of New Mexico and the University of Hawaii. He was awarded the Leonard Medal of the Meteoritical Society, the highest award bestowed by that body, in 1988. Recently, Keil has been actively involved in spectral studies of meteorites as asteroid analogues. His work on shock-darkened meteorites and gas-rich regolith breccias has increased our understanding of the spectral properties of asteroidal surfaces. (M 22830)

Name suggested and citation prepared by T. J. McCoy.

(5055) Opekushin

1986 PB_5. Discovered 1986 August 13 by L. I. Chernykh at Nauchnyj.

Named in memory of Aleksandr Mikhailovich Opekushin (1838-1923), outstanding Russian sculptor. He has produced memorials and sculptures of many historical persons and is especially famous for the Pushkin monument in Moscow and the Lermontov monument in Pyatigorsk {see planet (2192)}. (M 30096)

(5056) 1986 RQ_5
Discovered 1986 September 9 by H. Debehogne at La Silla.

(5057) 1987 DC_6
Discovered 1987 February 22 by H. Debehogne at La Silla.

(5058) Tarrega
1987 OM. Discovered 1987 July 28 by T. Seki at Geisei.
Named in honor of Francisco Tarrega, modern Spanish guitarist. (M 22830)

(5059) Saroma
1988 AF. Discovered 1988 January 11 by K. Endate and K. Watanabe at Kitami.

Named for a lake in Abashiri {see planet (4263)} National Park in eastern Hokkaido {see planet (3720)}. With an area of 151.2 square kilometers Saroma is the third largest lake in Japan, and it is famous for its harvest of scallops and oysters. Each autumn many visitors come to see its large stock of "sango-so" plants. (M 20838)

(5060) Yoneta
1988 BO_5. Discovered 1988 January 24 by S. Ueda and H. Kaneda at Kushiro.

Named in honor of Katsuhiko Yoneta (1904-1957), doctor of engineering and a graduate of Hokkaido Imperial University. A pioneer of astronomical observation by amateur astronomers in Hokkaido, he established the Hokkaido branch of the Oriental Astronomical Association in 1924 and inspired many young students. (M 22506)
Name suggested and citation prepared by H. Fukushima.

(5061) 1988 DJ
Discovered 1988 February 22 by R. H. McNaught at Siding Spring.

(5062) 1989 CZ
Discovered 1989 February 6 by E. F. Helin at Palomar.

(5063) Monteverdi
1989 CJ_5. Discovered 1989 February 2 by F. Börngen at Tautenburg.

Named in memory of Claudio Monteverdi (1567-1643), with Schütz {see planet (4134)} the greatest musician of the seventeenth century and reputed to be one of the first composers of opera. (M 20163)

Name endorsed by Ingeborg Stein, director of the Heinrich-Schütz-Haus, Bad Köstritz, Thuringia.

(5064) Tanchozuru
1990 FS. Discovered 1990 March 16 by M. Matsuyama and K. Watanabe at Kushiro.

Named for the crane that mainly inhabits Kushiro Marsh Nationalpark in eastern Hokkaido. This bird is designated as a Special Natural Species. The white-bodied bird is about 1 meter in height and characterized by partial black plumage and a red spotted head. It is widely regarded in Japan as a lucky bird. Although the bird was once nearly exterminated, conservation means that there are now more than 400 individuals. (M 22506)

(5065) Johnstone
1990 FP_1. Discovered 1990 March 24 by E. F. Helin at Palomar.

Named in memory of Paul Johnstone (?-1976), the first director and producer of the well-known British TV program "The Sky at Night", now celebrating its 40th anniversary. (M 30475)

(5066) Garradd

1990 MA. Discovered 1990 June 22 by R. H. McNaught at Siding Spring.

Named for Gordon John Garradd (1959-), amateur astronomer and photographer who has wide-ranging interests in astronomy and natural history, amply demonstrated by his discovery of four novae in the Large Magellanic Cloud, a photographic fireball patrol and outstanding photographs of wildlife and storms. He has made extensive astrometric observations of minor planets and comets that have otherwise often gone unobserved in the southern hemisphere. He has produced some spectacular astronomical photographs and CCD images, been an observer with Spaceguard Australia and served as coordinator of a storm-spotter network.
(M 29144)

Named by the discoverer, endorsed by D. I. Steel, B. G. Marsden and G. V. Williams.

(5067) Occidental

1990 OX. Discovered 1990 July 19 by E. F. Helin at Palomar.

Named for Occidental College, a prestigious institution established in Los Angeles in 1887. Occidental is widely known for its outstanding liberal arts curricula of music, art, and political and physical science. The discoverer wishes to honor her alma mater on the occasion of Founders Day (1995 Apr. 19) and to pay special tribute to president John Brooks Slaughter, the faculty and students. May Oxy continue to provide its students with an exemplary educational experience, always inspiring excellence, equity, community and service for and from its student body with its celestial namesake elevated to heavenly rank. (M 24917)

(5068) Cragg

1990 TC. Discovered 1990 October 9 by R. H. McNaught at Siding Spring.

Named in honor of Thomas A. Cragg, a quintessential amateur astronomer. In 1945 at age 17 he joined the American Association of Variable Star Observers; by 1992 he had contributed 120,702 brightness estimates, many being of variables at their faintest, as well as sunspot counts made every clear day with a 0.15-m Newtonian reflector. After the Association of Lunar and Planetary Observers was formed in 1947, Cragg served for many years as its Saturn recorder. In 1949 he was one of the founders of the Western Amateur Astronomers. After 24 years at Mount Wilson Observatory, he resigned his position as solar observer in 1976 and joined the Anglo-Australian Observatory as chief night assistant, retiring in 1992.
(M 20163)
Citation provided by Leif J. Robinson.

(5069) Tokeidai

1991 QB. Discovered 1991 August 16 by K. Watanabe at Sapporo.

Named for a wooden clock tower, built in 1878 as a drill house for Sapporo Agricultural College. The clock is still ticking more than a century later, and the pleasant chimes still mark the hours. The building is one of Sapporo's most famous landmarks. (M 26929)

(5070) Arai

1991 XT. Discovered 1991 December 9 by S. Ueda and H. Kaneda at Kushiro.

Named in honor of Ikunosuke Arai (1836-1909), the first director of the Central Meteorological Observatory. In his early days, he contributed to the triangulation of Hokkaido. In 1887 he observed the total eclipse of the sun at Sanjo, Niigata,

and successfully photographed the event. (M 22506)
Name suggested and citation prepared by H. Fukushima.

(5071) Schoenmaker

3099 T-2. Discovered 1973 September 30 by C. J. van Houten and I. van Houten-Groeneveld at Palomar.

Named in honor of Ton A. Schoenmaker (1945-), for many years technical officer at Leiden Observatory. He wrote computer programs for the van Houtens {see planets (1673) and (1674)}, so that they were able to make the first identifications for their three Trojan Surveys. The help of Ton is very much appreciated. Ton is also a visual comet observer. (M 22506)

(5072) Hioki

1931 TS_1. Discovered 1931 October 9 by K. Reinmuth at Heidelberg.

Named in honor of Tsutomu Hioki, who co-discovered 1988 VG_2 , which was identified with the lost minor planet (724) Hapag. (M 22506)

Name proposed by S. Nakano, who found the identifications involving this object.

(5073) 1943 EN

Discovered 1943 March 3 by Y. Väisälä at Turku.

(5074) 1949 QQ_1

Discovered 1949 August 24 at the Goethe Link Observatory at Brooklyn, Indiana.

(5075) Goryachev

1969 TN_4. Discovered 1969 October 13 by B. A. Burnasheva at Nauchnyj.

Named in memory of Nikolaj Nikanorovich Goryachev (1883-1940), professor of astronomy at Tomsk University, who in 1920 founded and until 1940 headed the department of astronomy. He is the author of numerous papers on celestial mechanics, comets, meteors and astrometry. In particular, he was known for his work on Halphen's method for computing secular perturbations and his application of this to (1) Ceres. For more than 20 years Goryachev delivered lectures on astronomy and geodesy at Tomsk University and Tomsk Polytechnic Institute. (M 28089)

(5076) Lebedev-Kumach

1973 SG_4. Discovered 1973 September 26 by L. I. Chernykh at Nauchnyj.

Named in memory of Vasilij Ivanovich Lebedev-Kumach (1898-1949), prominent poet and song-writer, known for his lyrical and patriotic verses for songs for many Soviet films. (M 30096)

(5077) 1974 MG

Discovered 1974 June 17 at the Felix Aguilar Observatory at El Leoncito.

(5078) Solovjev-Sedoj

1974 SW. Discovered 1974 September 19 by L. I. Chernykh at Nauchnyj.

Named in memory of Vasilij Pavlovich Solovjev-Sedoj (1907-1979), distinguished composer, author of about 400 songs that are still popular, seven operettas and musical comedies, two ballets and music for several dozen films. (M 30096)

(5079) 1975 DB

Discovered 1975 February 16 at the Felix Aguilar Observatory at El Leoncito.

(5080) Oja

1976 EB. Discovered 1976 March 2 by C.-I. Lagerkvist at Kvistaberg.

Named in honor of Tarmo Oja, professor in astronomy at Uppsala University working on galactic structure and variable stars. During the last few years he has also enthusiastically participated in photometric observations of (4) Vesta and (10) Hygiea at both Kvistaberg and La Palma. (M 20522)

(5081) 1976 WC_1

Discovered 1976 November 18 at the Felix Aguilar Observatory at El Leoncito.

(5082) Nihonsyoki

1977 DN_4. Discovered 1977 February 18 by H. Kosai and K. Hurukawa at Kiso.

Named for the chronicles of Japan from the earliest times to 697, written in Chinese and completed in 720. It was the first historical record compiled by the Japanese government and contained records of various astronomical phenomena, such as appearances of seven comets (including the 684 return of P/Halley), 13 solar eclipses (e.g. in 628), occultations of stars and planetary phenomena. It was translated into English by W. G. Aston and published under the title of *Nihongi*. {See also the citation for planet (5454)}. (M 22506)

(5083) Irinara

1977 EV. Discovered 1977 March 13 by N. S. Chernykh at Nauchnyj.

Named in honor of Irina Evgen'evna Raksha, well-known writer in Moscow and a friend of the discoverer's family. (M 22506)

(5084) Gnedin

1977 FN_1. Discovered 1977 March 26 by N. S. Chernykh at Nauchnyj.

Named in honor of Yurij Nikolaevich Gnedin (1935-), well-known Russian astrophysicist, deputy director of the Pulkovo Astronomical Observatory and professor at St. Petersburg Technical University. Gnedin has studied the transfer of polarized radiation in cosmic media, the physics of neutron stars, black holes and comets, cosmomicrophysics and the physics of x-ray sources. (M 30475)
Name suggested by the Institute of Theoretical Astronomy.

(5085) Hippocrene

1977 NN. Discovered 1977 July 14 by N. S. Chernykh at Nauchnyj.

Named for the ancient Greek mythological spring of inspiration. The name is given in connection with the discovery of many new minor planets at the Crimean Astrophysical Observatory during the second half of 1977. (M 34620)

(5086) Demin

1978 RH_1. Discovered 1978 September 5 by N. S. Chernykh at Nauchnyj.

Named in memory of Vladimir Grigor'evich Demin (1929-1996), well-known expert on celestial mechanics and dynamics of rigid bodies, professor at Moscow University for many years and member of the editorial board of *Kosmicheskie Issledovaniya* (Cosmic Researches). With Aksenov and Grebenikov {see planets (4777) and (4268), respectively} he investigated the generalized problem of two fixed centers, and he was among the first, to develop theories of the motions of artificial satellites. He was also known for numerous monographs and textbooks on celestial mechanics, and he blazed the way to science for many students. (M 29144)

Name proposed by the discoverer following a suggestion by the Institute of Theoretical Astronomy.

(5087) Emel'yanov

1978 RM_2. Discovered 1978 September 12 by N. S. Chernykh at Nauchnyj.

Named in honor of Nikolaj Vladimirovich Emel'yanov (1946-), head of the Celestial Mechanics Department of the Sternberg Astronomical Institute in Moscow. He constructed a practical analytical theory for artificial-satellite motion based

on non-keplerian intermediate orbits. He has derived precise values for the orbital parameters of Phobos and Deimos and the dynamical parameters of Mars. He also developed ephemeris software for the majority of natural satellites. (M 30475)
Name suggested by the Institute of Theoretical Astronomy.

(5088) Tancredi

1979 QZ_1. Discovered 1979 August 22 by C.-I. Lagerkvist at La Silla.

Named in honor of Gonzalo Tancredi, Uruguayan astronomer who spent several years in Uppsala, working with the planetary group there. He has specialized in a broad range of problems related to the dynamical and physical evolution of comets and their interrelations with minor planets, using both observations and theoretical modeling. (M 22506)

(5089) Nádherná

1979 SN. Discovered 1979 September 25 by A. Mrkos at Kleť.

Named in memory of Sidonie Nádherná (1885-1950), baroness of Borutín and author of *The Chronicle of Vrchotovy Janovice.* She created a splendid park around her château of Vrchotovy Janovice in central Bohemia and became widely known as a friend of German poet Rainer Maria Rilke and Austrian writer Karl Kraus. She died a resident of Great Britain. (M 34340)
Name suggested by J. Tichá.

(5090) Wyeth

1980 CG. Discovered 1980 February 9 at the Harvard College Observatory at Harvard.

Named in memory of Stuart Wyeth, who provided the means for the construction, sixty years ago, of the Wyeth 1.5-m reflector, with which this minor planet was discovered. (M 20164)

(5091) Isakovskij

1981 SD_4. Discovered 1981 September 25 by L. I. Chernykh at Nauchnyj.

Named in memory of Mikhail Vasil'evich Isakovskij (1900-1973), a poet whose verses are remarkable for their tunefullness. Many of them were set to music and became popular songs. (M 30096)

(5092) Manara

1982 FJ. Discovered 1982 March 21 by E. Bowell at Anderson Mesa.

Named in honor of Alessandro Manara, since 1963 an astronomer at the Brera Astronomical Observatory in Milan. His fields of interest include astrometry, and the physics and dynamics of minor solar system bodies. In particular, he has studied the dynamics of comets and has carried out polarimetric analyses on several minor planets. He has recently written a study of Schiaparelli {see planet (4062)}. (M 21956)
Named suggested and citation prepared by V. Zappalà.

(5093) Svirelia

1982 TG_1. Discovered 1982 October 14 by L. G. Karachkina at Nauchnyj.

Named for Elsa Gustavovna Sviridova, the wife of Georgij Vasil'evich Sviridov {see planet (4075)}. (M 22506)

(5094) Seryozha

1982 UT_6. Discovered 1982 October 20 by L. G. Karachkina at Nauchnyj.

Named in honor of Serguei Pyotrovich Kapitsa (1928-), professor of nuclear physics, active in public matters of science, the history of science and moderator of the principal Russian television program on science and society since 1973. He has been vice-president of the European Physical Society (1977-1980) and president of the Physical Society of the U.S.S.R. (1989-1992). He received the UNESCO Kalinga prize in 1979. (M 22506)

(5095) Escalante

1983 NL. Discovered 1983 July 10 by E. Bowell at Anderson Mesa.

Named in honor of Jaime Escalante, a mathematics teacher who has inspired thousands of underprivileged high school students to use academic accomplishment to rise above the harsh realities of the inner city. In 1982, his work at Garfield High School in Los Angeles attracted national attention when fourteen of his students passed the mathematics Advanced Placement exam, were accused of cheating, were retested, and passed again. Escalante, an immigrant from Bolivia, had been acclaimed as a teacher in his native country, but in the U.S. he was required to repeat college to obtain teaching credentials. (M 22830)

Citation by S. Ostro, who assisted the making of the movie *Stand and Deliver*, which dramatized the story of Escalante's calculus students at Garfield.

(5096) Luzin

1983 RC_5. Discovered 1983 September 5 by L. V. Zhuravleva at Nauchnyj.

Named in memory of the mathematician Nikolaj Nikolaevich Luzin (1883-1950), professor at Moscow University, who made fundamental contributions to the theory of real functions. (M 34620)

(5097) Axford

1983 TW_1. Discovered 1983 October 12 by E. Bowell at Anderson Mesa.

Named in honor of William Ian Axford, director of the Max-Planck-Institut fur Aeronomie, Lindau, and current president of the Committee on Space Research, on the occasion of his sixtieth birthday. Axford's seminal contributions in the areas of magnetospheric, heliospheric, cometary and cosmic ray physics and interstellar gas dynamics have greatly enhanced our understanding and stimulated further investigation. While providing strong leadership in promoting space missions, including the highly successful Giotto mission to comet Halley in 1986, he has also been a leading advocate for international cooperation in space. (M 21956) Name suggested and citation prepared by D. A. Mendis.

(5098) 1985 CH_2

Discovered 1985 February 14 by H. Debehogne at La Silla.

(5099) 1985 DY_1

Discovered 1985 February 16 by H. Debehogne at La Silla.

(5100) Pasachoff

1985 GW. Discovered 1985 April 15 by E. Bowell at Anderson Mesa.

Named in honor of Jay M. Pasachoff, Field Memorial professor of astronomy, director of the Hopkins Observatory and chair of the astronomy department of Williams College, Williamstown, Massachusetts. Pasachoff's broad range of astronomical research has centered on the sun, and especially on studies of solar eclipses. He is also well known for an extensive series of college-level textbooks and popular-astronomy textbooks and articles. Besides being an indefatigable public lecturer, Pasachoff has served as chairperson of the astronomy section of the American Association for the Advancement of Science, as a committee member of the American Association of Physics Teachers and on the Astrophysics Council of the National Aeronautics and Space Administration. (M 21956)

(5101) Akhmerov

1985 UB_5. Discovered 1985 October 22 by L. V. Zhuravleva at Nauchnyj.

Named in honor of Vadim Zinov'evich Akhmerov (1929-), who has returned health to many people during his 45-year medical service. Since 1980 he has worked in the Alushta (Crimea) maternity hospital. (M 34620)

(5102) Benfranklin

1986 RD_1. Discovered 1986 September 2 by A. Mrkos at Kleť.

Named in memory of the American scientist, philosopher and statesman Benjamin Franklin (1706-1790), who was one of the brightest and most creative scholars of his time. He contributed in many areas of natural science: electricity, aurorae, earthquakes and atmospheric storms. Franklin proposed the two kinds of electric charge, positive and negative, and his famous experiments in 1752 with kites flown into storm clouds led to the invention of the lightning rod. (M 34621)
Name suggested by J. Tichá and M. Šolc.

(5103) Diviš

1986 RP_1. Discovered 1986 September 4 by A. Mrkos at Kleť.

Named in memory of the Czech scientist, Praemonstrat monk and priest Prokop Václav Diviš (1698-1765), who was known for his experiments with electricity. On 1754 June 15 he erected his "meteorological machine" on the grounds of his vicarage near Znojma in South Moravia. This was a kind of lightning rod consisting of some 400 iron peaks on plates on a high bar. Its aim was to exhaust the electricity from storm clouds and lead it groundward along chains, thus preventing lightning strikes. Diviš also used electricity in medical treatment. (M 34621)

Name suggested by J. Tichá and M. {Š}olc on the tricentennial of his {Diviš'} birth.

(5104) Skripnichenko

1986 RU_5. Discovered 1986 September 7 by L. I. Chernykh at Nauchnyj.

Named in honor of Vladimir Il'ich Skripnichenko (1942-), staff member of the Institute of Theoretical Astronomy since 1968 and its deputy director since 1994, well known expert on computational methods of celestial mechanics. He has studied the application of Hansen's {see planet (4775)} method of partial anomalies to the investigation of cometary motions, the development of the Universal Poissonian processor and the development of the programing system ERA. (M 30475)
Name proposed by the Institute of Theoretical Astronomy.

(5105) Westerhout

1986 TM_1. Discovered 1986 October 4 by E. Bowell at Anderson Mesa.

Named in honor of Gart Westerhout, radio astronomer, on the occasion of his retirement as scientific director of the U.S. Naval Observatory. Westerhout developed the astronomy program at the University of Maryland into one of the best in the U.S. At the Naval Observatory, his incisive leadership and pioneering work in interferometric astrometry, including development of the first radio interferometer dedicated to astrometry, led to substantial advances in precise position, motion and time determination. (M 22830)
Name suggested and citation prepared by R. H. McCracken.

(5106) 1987 DJ

Discovered 1987 February 19 by P. Jensen at Brorfelde.

(5107) 1987 DS_6

Discovered 1987 February 24 by H. Debehogne at La Silla.

(5108) Lübeck

1987 QG_2. Discovered 1987 August 21 by E. W. Elst at La Silla.

Named in memory of the famous organist and composer Vincent Lübeck (1654-1740), born in Paddingbüttel, near Dorum (Bremen area). In 1675 he became organist at St. Casmae et Damiani in Stade, retaining this post for more than 27 years. There he had one of the most beautiful north German organs made by Arp Schnitger at his disposal. In 1702 he went to Hamburg and became organist at St. Nicolai, which housed Schnitger's largest organ (four manuals, pedal and 66 voices). In 1721 the composer and organist Johann Mattheson wrote: "This unusual organ has an unusual organist. I need only say the name Vincent Lübeck and the whole panegyric is complete." (M 20164)

(5109) 1987 RM_1

Discovered 1987 September 13 by H. Debehogne at La Silla.

(5110) Belgirate

1987 SV. Discovered 1987 September 19 by E. Bowell at Anderson Mesa.

Named for the pleasant village on the western shore of Lake Maggiore, close to the town of Novara, on the occasion of the first IAU Symposium dedicated to the small bodies of the solar system (June 1993). (M 22248)
Name suggested and citation written by V. Zappalà and M. Di Martino.

(5111) Jacliff

1987 SE_4. Discovered 1987 September 29 by E. Bowell at Anderson Mesa.

Named in honor of Clifford (1929-1993) and Jackie (1935-) Holmes. With his wife Jackie's backing and support, Cliff was for decades an astronomical observer, organizer and educator. Cliff and Jackie founded the Riverside Telescope Makers' Conference in 1969; from small beginnings it has become a leading event for amateur astronomers and is known worldwide. After retirement, Cliff began a second career as an educator, teaching astronomy courses at Riverside City College. (M 22830)

Name suggested by D. H. Levy, J. Young and S. J. Edberg, citation prepared by Edberg.

(5112) Kusaji

1987 SM_{13}. Discovered 1987 September 23 by S. Ueda and H. Kaneda at Kushiro.

Named in honor of Shigeji Kusaji (1879-1956), agricultural worker in Asahikawa. He was the first head of the Asahikawa Astronomical Club and was an earnest observer of sunspots. His observations were made from 1928 to 1955, and observations are still being made by amateur astronomers at his club. (M 22506)
Name suggested by H. Kohsai. Citation prepared by H. Fukushima.

(5113) Kohno

1988 BN. Discovered 1988 January 19 by T. Seki at Geisei.

Named in honor of Masaru Kohno (1926-), Japanese guitarist. The discoverer delights in playing a handmade guitar made by Kohno. (M 23137)

(5114) Yezo

1988 CO. Discovered 1988 February 15 by S. Ueda and H. Kaneda at Kushiro.

Named for Japan's northern island of Hokkaido {see planet (3720)}, which was known as Yezo until 1869. (M 27734)

(5115) Frimout

1988 CD_4. Discovered 1988 February 13 by E. W. Elst at La Silla.

Named in honor of Dirk Frimout, the first Belgian astronaut. On 1992 Mar. 24 he went into orbit with his American colleagues on board of the space shuttle Atlantis. He is a member of BIRA, the Belgian Institute of Space Aeronomy at Uccle. The main purpose of this flight was the study of the ozone layer. (M 20164)

(5116) 1988 EU

Discovered 1988 March 13 by P. Jensen at Brorfelde.

(5117) Mokotoyama

1988 GH. Discovered 1988 April 8 by K. Endate and K. Watanabe at Kitami.

Named for a mountain in eastern Hokkaido {see planet (3720)}. Rising to 1,000 feet, the mountain gives a view from Lake Kussharo to the southwest to the Sea of Okhotsk {see planet (4042)} to the north. A hut at its eighth station has "Ginryosui", the only spring in the area. (M 20838)

(5118) 1988 RB

Discovered 1988 September 7 by P. Jensen at Brorfelde.

(5119) 1988 RA_1

Discovered 1988 September 8 by P. Jensen at Brorfelde.

(5120) Bitias

1988 TZ_1. Discovered 1988 October 13 by C. S. Shoemaker and E. M. Shoemaker at Palomar.

Bitias was a great Trojan warrior and a companion of Aeneas {see planet (1172)}. He was the first Trojan to drink a toast to the ill-fated Queen Dido {see planet (209)}, at the feast when Aeneas narrated his adventures to Dido, and she began to fall in love with him. Later, in Italy, during the Latin war, Bitias made a mistake that cost him and many other Trojans their lives, when he opened the gates of the Trojan's fortress in an effort to coax Latin warriors inside, where they could be trapped. Instead, the Latin warrior Turnus rushed inside the Trojan fortress and made slaughter, killing Bitias with a whirling pike. (M 22249)

Name and citation provided by Richard Preston at the request of the discoverers.

(5121) Numazawa

1989 AX_1. Discovered 1989 January 15 by M. Yanai and K. Watanabe at Kitami.

Named in honor of Shigemi Numazawa (1958-), a Japanese space artist with an international reputation. Relying on his rich experience as an amateur astronomer and space photographer, he has produced a large body of work, including a collection of astrophotographs taken with a Schmidt camera. (M 25653)

(5122) Mucha

1989 AZ_1. Discovered 1989 January 3 by A. Mrkos at Kleť.

Named in memory of Alfons Mucha (1860-1939), Czech painter, graphic artist and decorative artist, who also lived in Paris and the U.S. His work symbolizes the full flowering of the Art Nouveau style, as is evidenced by his posters of Sarah Bernhardt, illustrations and jewelry. (M 34621)

Name suggested by J. Tichá.

(5123) 1989 BL

Discovered 1989 January 28 by Y. Oshima at Gekko.

(5124) Muraoka

1989 CW. Discovered 1989 February 4 by T. Seki at Geisei.

Named in honor of Kenji Muraoka (1955-), a Japanese amateur astronomer who plays an active role in calculating comet orbits. (M 24765)

(5125) Okushiri

1989 CN_1. Discovered 1989 February 10 by S. Ueda and H. Kaneda at Kushiro.

Named for an island, some 30 km to the southwest of Hokkaido {see planet (3720)}, noted for its rich fishing grounds for squid and scallops. (M 28621)

Name suggested by H. Ueda.

(5126) Achaemenides

1989 CH_2. Discovered 1989 February 1 by C. S. Shoemaker and E. M. Shoemaker at Palomar.

Born in Ithaca, Achaemenides was a Greek warrior who eventually became an ally of Aeneas' {see planet (1172)} Trojans. Originally he was a companion of Odysseus {see planet (1143)} and fought on the Greek side at Troy. Later, he was left behind in the land of the Cyclops when Odysseus blinded the Cyclops and escaped. There Aeneas found him - squalid, dressed in rags held together with thorns - and rescued him. (M 22249)

Name and citation provided by Richard Preston at the request of the discoverers.

(5127) Bruhns

1989 CO_3. Discovered 1989 February 4 by E. W. Elst at La Silla.

Named in memory of the famous organist and composer Nicolaus Bruhns (1665-1697), born at Schwabstedt, near Husum (Schleswig in the north of Germany), as a descendant of an old family of musicians. He got his first lessons at the organ from his father Paul Bruhns, a pupil of Franz Tunder {see planet (7871)}. In 1681 he went to the city of Lübeck, to study the organ and composition with Dietrich Buxtehude {see planet (4344)} and the violin and viola da gamba with his uncle Peter Bruhns. He was noted for his virtuosity, sometimes playing the upper parts on the violin - with accompaniment from an appropriate pedal bass part with his feet. (M 20522)

(5128) Wakabayashi

1989 FJ. Discovered 1989 March 30 by M. Koishikawa at Sendai.

Wakabayashi is named for Wakabayashi-ku, a ward in the southeastern part of the city of Sendai {see planet (3133)}. About 400 years ago the warlord who founded Sendai city, Date Masamune {see planet (6859)}, spent his last days in this area. (M 25230)

(5129) Groom

1989 GN. Discovered 1989 April 7 by E. F. Helin at Palomar.

Named in honor of Steven L. Groom, a computer specialist at the Jet Propulsion Laboratory, who designed and assembled the Near-Earth Asteroid Tracking (NEAT) autonomous observing system, the first of its kind to search for near-earth objects. He used his expertise in systems-level software (including device drivers), parallel processing architectures, network programming and device integration to make NEAT a success. The NEAT team wishes to salute his ingenuity and dedication to the project. (M 26929)

(5130) Ilioneus

1989 SC_7. Discovered 1989 September 30 by C. S. Shoemaker and E. M. Shoemaker at Palomar.

Ilioneus, a ship commander of Aeneas {see planet (1172)}, was the official spokesman for Aeneas' band of wandering Trojans. He was the first to speak to strangers on behalf of Aeneas, and his soft words often preceded conflict. First he advised Dido {see planet (209)}, the queen of Carthage, that the Trojans and Aeneas posed no threat to her realm /when, in fact, Aeneas would break her heart and drive her to suicide). Later he spoke to King Latinus, seeking permission for the Trojans to stay in Italy, and that request ultimately led to devastating war, when the Latins tried to drive Aeneas away. (M 22249)

Name and citation provided by Richard Preston at the request of the discoverers.

(5131) 1990 BG

Discovered 1990 January 21 by E. F. Helin and B. Roman at Palomar.

(5132) Maynard

1990 ME. Discovered 1990 June 22 by H. E. Holt at Palomar.

Named in honor of Owen E. Maynard, who played a major role in the achievement of the first manned landing on the moon. An engineer in the Apollo program from the beginning, he helped design the Apollo spacecraft and oversaw the design of the command-service and lunar modules. He led the team that developed the complex lunar mission flight plans. As chief of the Systems Engineering Division, he was responsible for all aspects of the construction and testing of the Apollo hardware. He planned the early Apollo test flights, including that of the first lunar landing, Apollo 11. (M 27734)

Name suggested and citation prepared by R. Attwood.

(5133) Phillipadams

1990 PA. Discovered 1990 August 12 by R. H. McNaught at Siding Spring.

Named in honor of Phillip Adams (1939-), Australian broadcaster, writer, journalist, wit and philosopher. Amongst his multifarious good works, he helped promote the Australian search and follow-up program on near-earth asteroids from 1990 until its cessation in 1996. (M 29144)

Name proposed by the discoverer following a suggestion by D. I. Steel, who prepared the citation.

(5134) 1990 SM_2

Discovered 1990 September 17 by H. E. Holt at Palomar.

(5135) Nibutani

1990 UE. Discovered 1990 October 16 by S. Ueda and H. Kaneda at Kushiro.

Named for a valley sacred to the indigenous Ainu people of Hokkaido {see planet (3720)}. (M 28621)

Name suggested by H. Ueda.

(5136) Baggaley

1990 UG_2. Discovered 1990 October 20 by R. H. McNaught at Siding Spring.

Named in honor of W. Jack Baggaley (1938-), radar meteor researcher at the University of Canterbury. Born and educated in Sheffield (England), where he took his Ph.D. in the mid-1960s under Tom Kaiser, he then moved to Christchurch, where he currently has the rank of professor of physics. Baggaley's research has encompassed many distinct areas of radio studies of meteors, and during the 1990s he has measured more meteoroid orbits than in all previous experiments combined. (M 29144)

Name proposed by the discoverer following a suggestion by D. I. Steel, who prepared the citation.

(5137) Frevert

1990 VC. Discovered 1990 November 8 by J. M. Baur at Chions.

Named in honor of Friedrich Frevert, an amateur astronomer from Wetzlar, well-known for his work on astrometric observations and on mathematical methods of reduction in photographic astrometry. For many years he was the chief of the Vereinigung der Sternfreunde in Germany. In 1965 he established the Astronomical Group in Wetzlar and later organized a very successful Minor Planet Working Group of amateur astronomers. (M 21610)

(5138) Gyoda

1990 VD_2. Discovered 1990 November 13 by T. Hioki and S. Hayakawa at Okutama.

Named for an industrial city, home of the second discoverer, located some 60 km north of Tokyo, between the Ara and the Tone {see planet (1266)} rivers. The city contains one of the three most celebrated ancient burial grounds in Japan, and an old sword with golden Kanji characters was found there. Gyoda is sister city of Kuwana {see planet (5629)}. (M 20523)

(5139) Rumoi

1990 VH_4. Discovered 1990 November 13 by M. Mukai and M. Takeishi at Kagoshima.

Named for the city of Rumoi, situated in the northwest part of Hokkaido {see planet (3720)}, where one of the discoverers was born. (M 20523)

(5140) Kida

1990 XH. Discovered 1990 December 8 by S. Ueda and H. Kaneda at Kushiro.

Named in memory of Kinjiro Kida (1893-1962), Hokkaido-born painter, known for his landscapes, and whose work has been compared to that of Cézanne {see planet (6674)} and other impressionists. (M 28621; M 28637)

Name suggested by H. Ueda.

(5141) Tachibana

1990 YB. Discovered 1990 December 16 by T. Seki at Geisei.

Tachibana is a Kendo club formed in 1950 in Kochi prefecture. Kendo is a sport similar to fencing in the same category as judo in Japan. The discoverer is a member of this club, the name of which means 'mandarin orange'. (M 22830)

(5142) Okutama

1990 YD. Discovered 1990 December 18 by T. Hioki and S. Hayakawa at Okutama.

Named for the observatory at which this minor planet was discovered. Okutama is located to the west of Tokyo and is famous for its beautiful lake and its cherry blossom. (M 20523)

(5143) Heracles

1991 VL. Discovered 1991 November 7 by C. S. Shoemaker and E. M. Shoemaker and D. H. Levy at Palomar.

Heracles, son of Zeus and the Theban princess Alcmena {see planets (5731) and (82)}, united the finest qualities of mind and heart, according to the traditions of the heroic age. His perseverance, crowned with victories, showed the triumph of the divine part of man's nature over the earthly. Through his fabled twelve labors, he achieved immortality and took his seat among the gods. (M 20523)

(5144) Achates

1991 XX. Discovered 1991 December 2 by C. S. Shoemaker and E. M. Shoemaker at Palomar.

Achates was the best friend and companion-at-arms of Aeneas {see planet (1172)} in Virgil's {see planet (2798)} "Aeneid". Known as "fidus Achates" - faithful Achatus - he commanded his own ship in the wandering fleet of Trojans as they sought to establish a new city after the fall of Troy. Achates wandered far from known paths with Aeneas, traveling at Aeneas' side during their passage through the underworld, as the two friends threaded their way through "broad, silent tracts of night" in a journey to meet Aeneas' father, as described in the great Book VI of Virgil's poem. (M 22249)

Name and citation provided by Richard Preston at the request of the discoverers.

(5145) Pholus

1992 AD. Discovered 1992 January 9 by the Spacewatch at Kitt Peak.

This Chiron-type object is being named by the Minor Planet Names Committee for a centaur who, like Chiron {see planet (2060)}, was hospitable, charitable and shunned violence. Pholus entertained Heracles {see planet (5143)} and gave him wine, the smell of which attracted and maddened the other centaurs, who were known for their brutal behavior and who began to fight with Heracles. Heracles used his poisoned arrows and killed many of them. As Pholus buried them, he wounded himself in extracting one of the arrows. Heracles tried to save him, but the poison was too deadly. Heracles buried Pholus on a mountain that he named Pholoe. (M 20523)

(5146) Moiwa

1992 BP. Discovered 1992 January 28 by S. Ueda and H. Kaneda at Kushiro.

Named for the mountain that offers an outstanding panorama of the city of Sapporo {see planet (3473)} and is popular both to skiers and to hikers in its virgin forest. (M 28621)
Name suggested by H. Ueda.

(5147) Maruyama

1992 BQ. Discovered 1992 January 28 by S. Ueda and H. Kaneda at Kushiro.

Named for a small hill, situated near Mt. Moiwa {see planet (5146)} in the southwestern part of Sapporo {see planet (3473)} and known for a beautiful park and zoo, as well as the Hokkaido Shrine. (M 28621)
Name suggested by H. Ueda.

(5148) Giordano

5557 P-L. Discovered 1960 October 17 by C. J. van Houten and I. van Houten-Groeneveld at Palomar.

Named in honor of Giordano Bruno (1548-1600), Dominican priest, who spent 16 years travelling through Europe, but who was mostly in London, where many of his papers were published. Bruno assumed the existence of other worlds on which people could live, and he was convinced of the correctness of the heliocentric system. This brought him in conflict with the church and he was condemned and burned on the Campo dei Fiori in Rome. (M 22507)
Giordano Bruno is also honored by a lunar crater.

(5149) Leibniz

6582 P-L. Discovered 1960 September 24 by C. J. van Houten and I. van Houten-Groeneveld at Palomar.

Named after Gottfried Wilhelm Leibniz (1646-1716), German philosopher and mathematician, co-inventor of the calculus. (M 22507; M 25993)
Leibniz is also honored by a lunar crater.

(5150) Fellini

7571 P-L. Discovered 1960 October 17 by C. J. van Houten and I. van Houten-Groeneveld at Palomar.

Named in memory of Frederico Fellini (1920-1993), one of the world's greatest film directors. He was called 'the maestro of the Italian cinema' and was an inspiration to a whole generation of film makers. Fellini's first film was *Lo Sceicco Bianco*, released in 1951. His first success came with *I vitelloni*, and in 1954 he won an Oscar for *La strada*. Further Oscars followed in 1957 for *Le notti di Cabiria*, in 1964 for *Otto e mezzo* and in 1974 for *Amarcord*. In April 1993 Fellini was honored with an Oscar for his lifetime achievements in film. (M 23137)
Name proposed by M.-R. Visscher.

(5151) Weerstra

2160 T-2. Discovered 1973 September 29 by C. J. van Houten and I. van Houten-Groeneveld at Palomar.

Named in honor of Klaas Weerstra, administrative officer and longtime programmer at the Leiden Observatory. He is an enthusiastic visual comet observer and is always very helpful when the van Houtens {see planets (1673), (1674)} need help with new programs for their minor planet surveys. (M 22507; M 24421)

(5152) 1931 UD

Discovered 1931 October 18 by K. Reinmuth at Heidelberg.

(5153) 1940 GO

Discovered 1940 April 9 by Y. Väisälä at Turku.

(5154) Leonov

1969 TL_1. Discovered 1969 October 8 by L. I. Chernykh at Nauchnyj.

Named in memory of Evgenij Pavlovich Leonov (1926-1994), prominent Soviet artist, an actor of the Moscow Lenin Komsomol Theatre. Well known for his many brilliant and impressive roles on stage and screen, he was a favorite actor of children and the public generally. (M 34621)

(5155) Denisyuk

1972 HR. Discovered 1972 April 18 by T. M. Smirnova at Nauchnyj.

Named in honor of academician Yurij Nikolaevich Denisyuk (1927-), head of a laboratory at the Ioffe Physical and Technical Institute in St. Petersburg and a member of the Royal Photographic Society, known worldwide for his discoveries in holography. (M 32788)

Name suggested by the Institute of Applied Astronomy.

(5156) Golant

1972 KL. Discovered 1972 May 18 by T. M. Smirnova at Nauchnyj.

Named in honor of academician Victor Evgen'evich Golant (1928-), director of the department of plasma physics, atomic physics and astrophysics at the Ioffe Physical and Technical Institute in St. Petersburg. Golant is widely known for his work in the fields of controlled thermonuclear fusion, interaction of electromagnetic waves with plasma, transfer processes, plasma diagnostics and heating of plasma in tokamaks. (M 32788)

Name suggested by the Institute of Applied Astronomy.

(5157) Hindemith

1973 UB_5. Discovered 1973 October 27 by F. Börngen at Tautenburg.

Named in memory of the German composer Paul Hindemith (1895-1963), whose extensive creations encompassed almost all areas of music. Composer of the symphonies *Mathis der Maler* and *Harmonie der Welt*, he wrote concerning the latter that "it deals with the life and work of Johannes Kepler {see planet (1134)} and with the search for harmony, ruling the universe undubiously". He defined a system of free tonality, on the other side of major and minor, in contrast to twelve-tone music. (M 20523)

(5158) Ogarev

1976 YY. Discovered 1976 December 16 by L. I. Chernykh at Nauchnyj.

Named in memory of Nikolaj Platonovich Ogarev (1813-1877), poet and publicist, friend and associate of A. I. Herzen {see planet (3052)}, and a firm upholder of democratic ideas. He initiated and co-edited the Russian democratic newspaper *Kolokol* (The Bell), and on emigrating in 1856 became one of the leaders of *The Free Russian Printing House* in London. (M 30096)

(5159) Burbine

1977 RG. Discovered 1977 September 9 at the Harvard College Observatory at Harvard.

Named in honor of Thomas Burbine in appreciation of his assistance at the Minor Planet Center during January to May 1992. He has recently drawn attention to the similarity of (387) Aquitania and (980) Anacostia, S-type minor planets that share unusual features in their near-infrared spectra and have similar mean distances and proper inclinations - but different proper eccentricities. (M 20523)

(5160) Camoes

1979 YO. Discovered 1979 December 23 by H. Debehogne and E. R. Netto at La Silla.

Named in memory of Luiz de Camoes (Camoens, 1524-1580), the greatest of the Portuguese poets, whose epic Os Lusiadas displays an extraordinary knowledge of astronomy. (M 21610)

(5161) Wightman

1980 TX_3. Discovered 1980 October 9 by C. S. Shoemaker at Palomar.

Named in honor of Kingsley W. Wightman, loved and respected teacher of astronomy, who inspired generations of students, teachers and the lay public at the Chabot Science Center in Oakland, California. He served as director of the Center from 1976 to 1986 and played a major role in securing its future. (M 21133)

Name and citation provided by the Eastbay Astronomical Society at the request of the discoverer.

(5162) Piemonte

1982 BW. Discovered 1982 January 18 by E. Bowell at Anderson Mesa.

Named for the northwestern region of Italy. Its capital, Torino, houses a historic university famous worldwide for its scientific research. Piemonte is also well known for its natural beauty, for its manufacturing industries, and for the production of a number of wines prized by connoisseurs. (M 22249)

Name suggested and citation written by V. Zappalà and M. Di Martino.

(5163) 1983 TD_2

Discovered 1983 October 9 by J. F. Wagner at Anderson Mesa.

(5164) Mullo

1984 WE_1. Discovered 1984 November 20 by C. Pollas at Caussols.

Named for a Celtic divinity venerated in western France, especially in Mayenne, where the discoverer's family originates. The Romans associated Mullo with Mars. Several local sanctuaries with circular towers have been found, particularly in the goldmine area of Athee. (M 27128)

Citation material prepared from the work of J. Naveau.

(5165) 1985 CG

Discovered 1985 February 11 at the Copenhagen University Observatory at Brorfelde.

(5166) Olson

1985 FU_1. Discovered 1985 March 22 by E. Bowell at Anderson Mesa.

Named in memory of Irvin Edward ('Ole') Olson (1910-1993), telescope-dome manufacturer. After an unsuccessful request to a silo maker to construct telescope domes suitable for amateur astronomers, Olson quit his job to found the Ash Manufacturing Company of Plainfield, Illinois. The company's first 'Ash Dome', erected at a junior-high school in 1961, was one of more than a thousand so far made for customers in 36 countries. (M 23137)

Name suggested by J. W. Young and citation prepared by Young and R. Olson.

(5167) Joeharms

1985 GU_1. Discovered 1985 April 11 by C. S. Shoemaker and E. M. Shoemaker at Palomar.

Named in honor of John (Joe) Eric Harms, former chief geologist of Broken Hill Pty., a large Australian mineral and petroleum-producing company. During his pioneering geological exploration in Western Australia, Harms discovered and recognized the Spider, Goat Paddock and Teague Ring impact structures. In the course of a 30-year career with BHP, he led in the discovery of major iron-ore, nickel and zinc-lead deposits, which have contributed importantly to the economy of Australia. (M 27329)

(5168) Jenner

1986 EJ. Discovered 1986 March 6 by C. S. Shoemaker and E. M. Shoemaker at Palomar.

Named in memory of English country doctor Edward Jenner (1749-1823), who on 1796 May 14 vaccinated a boy against and later inoculated him with smallpox, the first successful step in a nearly 200-year journey that eventually resulted in the worldwide elimination of this disease. (M 27329)
Name suggested by B. G. Marsden.

(5169) Duffell

1986 RU_2. Discovered 1986 September 6 by E. Bowell at Anderson Mesa.

Named in honor of Stephen Duffell (1943-), a long-time friend of the discoverer, on the occasion of his 50th birthday in November 1993. (M 22830)

(5170) Sissons

1987 EH. Discovered 1987 March 3 by E. Bowell at Anderson Mesa.

Named in honor of Anthony Sissons (1943-), a long-time friend of the discoverer, on the occasion of his 50th birthday in November 1993. (M 22830)

(5171) 1987 SQ_3

Discovered 1987 September 25 by P. Jensen at Brorfelde.

(5172) 1987 UX_1

Discovered 1987 October 28 by S. Ueda and H. Kaneda at Kushiro.

(5173) 1988 EM_1

Discovered 1988 March 13 by P. Jensen at Brorfelde.

(5174) Okugi

1988 HF. Discovered 1988 April 16 by M. Yanai and K. Watanabe at Kitami.

Named in honor of Susumu Okugi (1952-), director of the software division of Goto Optical Laboratory {see planet (2621)}. Under his direction numerous automated planetarium programs have been developed. In this way, he has contributed much to the popularization of astronomy and space science. (M 25653)

(5175) Ables

1988 VS_4. Discovered 1988 November 4 by C. S. Shoemaker and E. M. Shoemaker at Palomar.

Named in honor of Harold D. Ables (1938-), an astronomer and former director at the U.S. Naval Observatory's Flagstaff Station. Ables led the Flagstaff Station in its development of CCD capabilities for astrometry and spearheaded the transition from photographic to CCD techniques in parallax determinations. (M 27459)

Name proposed by the discoverers following a suggestion by the JPL ephemeris group.

(5176) 1989 AU

Discovered 1989 January 4 by S. Ueda and H. Kaneda at Kushiro.

(5177) Hugowolf

1989 AY_6. Discovered 1989 January 10 by F. Börngen at Tautenburg.

Named in memory of the Austrian composer Hugo Wolf (1860-1903), the creator of more than 200 songs from the late-romantic era. Many of these glorious songs had texts by Goethe {see planet (3047)}, Mörike and Eichendorff. (M 50524)

(5178) 1989 CD_4

Discovered 1989 February 1 by R. Rajamohan at Kavalur.

(5179) Takeshima

1989 EO_1. Discovered 1989 March 1 by T. Seki at Geisei.

Named in honor of Toshio Takeshima (1930-), friend of the discoverer. He is a grand master of the Japanese martial art of Iai. Takeshima is also a nature photographer, producing many excellent pictures of birds. (M 23352)

(5180) Ohno

1989 GF. Discovered 1989 April 6 by T. Fujii and K. Watanabe at Kitami.

Named in honor of Keiko Ohno (1959-) for her activities in promoting the public awareness of the study of astronomy and space science. As a software developer at Goto Optical Laboratory {see planet (2621)} she has produced many computerized planetarium programs. (M 25653)

(5181) SURF

1989 GO. Discovered 1989 April 7 by E. F. Helin at Palomar.

Named to honor Caltech's Summer Undergraduate Research Fellowship (SURF) program, which was begun in 1979 by professor of chemical engineering, Frederick H. Shair. The purpose of SURF is to encourage collegial interaction between a research sponsor - Caltech faculty and JPL technical staff - and undergraduate students around a research problem of mutual interest. This tribute pays special note to those SURF students who have participated in asteroid discovery and physical observation programs at JPL. The SURF program offers a hands-on learning experience for the students and a mentor-challenging role for the research sponsors. (M 21133)

(5182) Bray

1989 NE. Discovered 1989 July 1 by E. F. Helin at Palomar.

Named in honor of Olin D. Bray on the occasion of his 85th birthday, 1992 August 28. A medical doctor formerly in the U.S. Army and a lifetime caregiver to innumerable patients, Dr. Bray has been the healer and longtime friend of four generations of the families of the discoverer and her husband. He has always found time to live life to its fullest while doctoring and rendering comfort to others wherever needed. (M 20838)

(5183) Robyn

1990 OA_1. Discovered 1990 July 22 by E. F. Helin at Palomar.

Named in honor of Laurie Robyn Ernst Yeomans, wife of current IAU Commission 20 president Donald K. Yeomans {see planet (2956)}. Her kindness and good nature have given Robyn a wide circle of friends and the love of her family. (M 24410)

(5184) Cavaillé-Coll

1990 QY_7. Discovered 1990 August 16 by E. W. Elst at La Silla.

Named in memory of Aristide Cavaillé-Coll (1811-1899), the most famous member of a family of organ builders and considered the initiator of the orchestral style of French organ building and composing. His first large organ, that at the basilica of Saint-Denis (completed in 1841), became a model for many later French organs. Napoleon III put him in charge of rebuilding a number of important cathedral organs; more than 600 instruments bore his name, a number of them in England. He made important improvements in mechanism and pipework, aiming at making the organ as expressive as symphony orchestra, and the typical romantic sound influenced a new school of composers, such as César Franck and Charles-Marie Widor {see, respectively planets (4546) and (6829)}. One of the most beautiful organs by Cavaillé-Coll is at Saint-Sulpice in Paris. (M 20524)

(5185) 1990 RV_2

Discovered 1990 September 15 by H. E. Holt at Palomar.

(5186) Donalu

1990 SB_4. Discovered 1990 September 22 by B. Roman at Palomar.

Named for Dona(lu) Wheeler Roman, wife of the discoverer. Dona is an actress, teacher, administrator and mother. Her love and support have touched everyone around her. (M 20839)

(5187) Domon

1990 TK_1. Discovered 1990 October 15 by K. Endate and K. Watanabe at Kitami.

Named in memory of Ken Domon (1911-1991), renowned in the field of art and news photography. Born in Sakata City, he developed what came to be known as "Domon realism". Among his works, collections of photographs including "Hiroshima" and "Children in Chikuno" are especially famous. (M 20839)

(5188) Paine

1990 TZ_2. Discovered 1990 October 15 by E. F. Helin at Palomar.

Named in memory of Thomas O. Paine, administrator of NASA from 1968 to 1970. He led the American space agency to its greatest achievement - the Apollo 11 mission, in which humans first walked on the moon. During his tenure at NASA, Paine laid the groundwork for the Apollo-Soyuz {see planet (2228)} test project, which marked the first joint flight of American and Soviet space travelers. In 1985 he was called again to serve the space program as chairman of the National Commission of Space, where he laid out his vision of humanity's future as a spacefaring species. Tom Paine's most enduring contribution may be the inspiration he has provided that will take humans beyond the moon to the frontier of Mars. (M 21133)
Citation prepared by Charlene Anderson of The Planetary Society.

(5189) 1990 UQ

Discovered 1990 October 20 by R. H. McNaught at Siding Spring.

(5190) 1990 UR_2

Discovered 1990 October 16 by S. Ueda and H. Kaneda at Kushiro.

(5191) 1990 VO_3

Discovered 1990 November 13 by S. Ueda and H. Kaneda at Kushiro.

(5192) Yabuki

1991 CC. Discovered 1991 February 4 by T. Fujii and K. Watanabe at Kitami.

Named in honor of Hiroshi Yabuki (1960-), one of the leaders in developing automated planetarium programs at Goto Optical Laboratory {see planet (2621)}. (M 25653)

(5193) 1992 ET

Discovered 1992 March 7 by S. Ueda and H. Kaneda at Kushiro.

(5194) Böttger

4641 P-L. Discovered 1960 September 24 by C. J. van Houten and I. van Houten-Groeneveld at Palomar.

Named after Johann Friedrich Böttger (1682-1719). He was supposed to have made gold as alchemist to King August II the Strong of Saxonia. But instead he developed, together with mathematician and physician Ehrenfried Walter von Tschirnhaus, the first European porcelain in 1707-08. In 1710, Böttger became the head of the Dresden (later Meissen) Porcelain Manufacture. Böttger tried to sell the secret for the production of porcelain to the king of Prussia, so August II put Böttger into prison, where he died a short time later. (M 22507)

(5195) Kaendler

3289 T-1. Discovered 1971 March 26 by C. J. van Houten and I. van Houten-Groeneveld at Palomar.

Named after Johann Joachim Kaendler (1706-1775), sculptor and porcelain-modeller at the Meissen Porcelain Manufacture. Kaendler was the first to develop the fine baroque porcelain figures, which are still imitated today. (M 22507)

(5196) Bustelli

3102 T-2. Discovered 1973 September 30 by C. J. van Houten and I. van Houten-Groeneveld at Palomar.

Named after Franz Anton Bustelli (1723-1763), Italian-Swiss artist. Bustelli worked first in Vienna and then became famous as modeller for the Bavarian Porcelain Manufacture in Nymphenburg. His figures from the Commedia dell'arte are especially fine. (M 22507)

(5197) Rottmann

4265 T-2. Discovered 1973 September 29 by C. J. van Houten and I. van Houten-Groeneveld at Palomar.

Named after Friedrich Rottmann (1797-1850), German Romantic landscape painter. He travelled on the order of King Ludwig I of Bavaria to Salzburg, Tirol, Rome and especially to Greece. His main work is kept in museums in Heidelberg and Munich. (M 22507)

(5198) Fongyunwah

1975 BP_1. Discovered 1975 January 16 at the Purple Mountain Observatory at Nanking.

Named in honor of Fong Yunwah, friend and fellow villager of the discoverer Yang Jiexing. Yang was introduced to astronomy as a child by Fong, who is a folk educationalist and has won the respect of all who know him. (M 23792)

Traditionally, the Purple Mountain Observatory did not announce the discoverers of minor planets. The citation therefore is exceptional.

(5199) Dortmund

1981 RP_2. Discovered 1981 September 7 by L. G. Karachkina at Nauchnyj.

Named for the capital of Nordrhein-Westfalen and the sister city of Rostov-on-Don {see planet (4071)}. Founded more than 1100 years ago, Dortmund is now one of the largest industrial, financial and cultural centers in Germany. (M 22507)

(5200) Pamal

1983 CM. Discovered 1983 February 11 by E. Bowell at Anderson Mesa.

Named for Patrick Michael Malotki (1974-), friend of the discoverer, on the occasion of his 21st birthday. For many Francophones, the phrase "pas mal" is the highest form of compliment. (M 24765)

(5201) 1983 XF

Discovered 1983 December 1 by E. Bowell at Anderson Mesa.

(5202) 1983 XX

Discovered 1983 December 5 by A. Mrkos at Kleť.

(5203) Pavarotti

1984 SF_1. Discovered 1984 September 27 by Z. Vávrová at Kleť.

Named in honor of Luciano Pavarotti (1935-), world-famous Italian opera singer. (M 22830)

(5204) Herakleitos

1988 CN_2. Discovered 1988 February 11 by E. W. Elst at La Silla.

Named for the Greek philosopher Herakleitos (540-480 B.C.) of Ephesus. By means of his metaphoric clash ("life and death are the same so opponents that become metaphorical are identified with each other") he constructed and visualized relations of a higher conceptual order. He was thereby able to give a unique and decisive contribution to the history of thought. He is well known for his sayings such as "no one can step twice into the same river" and his famous "panta rhei" (everything changes). (M 21133)
Herakleitos is also honored by a lunar crater.

(5205) 1988 CU_7

Discovered 1988 February 11 by S. Ueda and H. Kaneda at Kushiro.

(5206) Kodomonomori

1988 ED. Discovered 1988 March 7 by Y. Oshima at Gekko.

Named for the Children's Forest Program, operated by OISCA International {see planet (3843)}, the Organization for Industrial, Spiritual and Cultural Development, which is affiliated with the Gekko Observatory {see planet (4261)}. The aim of the program is to encourage children all over the world to plant more trees. (M 25653)

(5207) 1988 HE

Discovered 1988 April 15 by A. C. Gilmore and P. M. Kilmartin at Lake Tekapo.

(5208) Royer

1989 CH_1. Discovered 1989 February 6 by E. F. Helin at Palomar.

Named in honor of Monsignor Ronald Royer, whose work in astronomical photography and variable-star observation has been a life-long passion. For over twenty years he has contributed to astronomical research and has exemplified the ideals of amateur astronomy by sharing his telescopes and enthusiasm with hundreds of students, children of his parish and the general public. Royer is one of the true pioneers of tricolor astrophotography, having first applied this technology to illustrate the different colors of the gas and dust tails of Comet West (1976 VI) {see also planet (2022)}. (M 21957)

This naming is enthusiastically endorsed by Jack Child, Steve Padilla and all his many friends.

(5209) 1989 CW_1

Discovered 1989 February 13 by T. Seki at Geisei.

(5210) Saint-Saëns

1989 EL_6. Discovered 1989 March 7 by F. Börngen at Tautenburg.

Named in memory of the French composer Charles Camille Saint-Saëns (1835-1921), whose extensive musical creations ranged from church music to neo-classicism. Among his best known works are the symphonic poem *La danse macabre*, the opera *Samson and Delilah* and his third organ symphony. (M 20524)

(5211) Stevenson

1989 NX. Discovered 1989 July 8 by C. S. Shoemaker and E. M. Shoemaker at Palomar.

Named in honor of David J. Stevenson (1948-) of the California Institute of Technology. A citizen of New Zealand, Stevenson displays scientific breadth and depth to an unusual degree; his research interests span much of theoretical geophysics and planetary science. Stevenson's contributions include novel insights into our understanding of planetary interiors and the origin of planetary magnetic fields. (M 27459)

Citation provided by P. M. Goldreich following a suggestion by the discoverers and A. W. Harris.

(5212) 1989 SS

Discovered 1989 September 29 by S. Ueda and H. Kaneda at Kushiro.

(5213) Takahashi

1990 FU. Discovered 1990 March 18 by K. Endate and K. Watanabe at Kitami.

Named in memory of Kiichiro Takahashi (1922-), whose father founded the company Takahashi Seisakujo in 1932, and who started the production of astronomical telescopes in 1967. Many Japanese amateur astronomers have discovered minor planets using telescopes made by the company. (M 21133)

(5214) Oozora

1990 VN_3. Discovered 1990 November 13 by A. Takahashi and K. Watanabe at Kitami.

Oozora, or 'Big Sky', is the name of the express train that connects Hakodate, the rail gateway to Hokkaido, with Kushiro {see planets (3720) and (4096), respectively}, the largest city on the eastern side of the island. The 580-km journey takes less than eight hours. (M 26930)

(5215) Tsurui

1991 AE. Discovered 1991 January 9 by M. Matsuyama and K. Watanabe at Kushiro.

Named for a small town situated 20 km north of Kushiro in eastern Hokkaido {see planets (4096) and (3720), respectively}. (M 22507)

(5216) 1941 HA

Discovered 1941 April 16 by L. Oterma at Turku.

(5217) 1966 CL

Discovered 1966 February 13 at the Purple Mountain Observatory at Nanking.

(5218) 1969 TB_3

Discovered 1969 October 9 by B. A. Burnasheva at Nauchnyj.

(5219) Zemka

1976 GU_3. Discovered 1976 April 2 by N. S. Chernykh at Nauchnyj.

Named in honor of Aleksandr Grigorjevich Zemka (1947-), friend of the discoverer, electrotechnics engineer in Zaporozhje, both a prominent specialist and a good organizer who wins the respect of his colleagues and acquaintances. He provided valuable help to the discoverer in improving the 0.64-m telescope used for the Crimean NEA Survey. (M 32345)

(5220) Vika

1979 SA_8. Discovered 1979 September 23 by N. S. Chernykh at Nauchnyj.

Named in honor of Victoriya Semenovna Vinogradova (1928-), doctor at the Crimean Astrophysical Observatory in Nauchnyj who has attended to the staff members of the observatory for 40 years. (M 30096)

(5221) Fabribudweis

1980 FB. Discovered 1980 March 16 by L. Brožek at Kleť.

Named for the medieval Czech astronomer, physician and theologian Wenceslaus (Václav) Fabri de Budweis (c.1460-1518), mainly known as the author of almanacs. After a university career in Leipzig he returned to his native town, now České Budějovice, where he is considered one of the most distinguished medieval intellectual personages. (M 27734)

(5222) Ioffe

1980 TL_{13}. Discovered 1980 October 11 by N. S. Chernykh at Nauchnyj.

Named in memory of Abram Fedorovich Ioffe (1880-1960), one of the originators of the Soviet physics school, a pioneer in the investigation of semiconductors and an academician of the U.S.S.R. Academy of Sciences. Ioffe was the organizer and the first director of the Physical and Technical Institute, of the Institute of Semiconductors and of the Physical and Agricultural Institute in Leningrad. He also initiated the creation of physical and technical institutes in Kharkov, Dnepropetrovsk, Sverdlovsk and Tomsk. (M 26763)
Name proposed by the Institute of Theoretical Astronomy.

(5223) 1981 EX_6

Discovered 1981 March 6 by S. J. Bus at Siding Spring.

(5224) Abbe

1982 DX_3. Discovered 1982 February 21 by F. Börngen at Tautenburg.

Named in memory of Ernst Karl Abbe (1840-1905), director of the Jena Observatory from 1877 to 1900, known for his fundamental contributions to optics. These includes work on microscopes, the Abbe number and comparators - the last being very useful for the detection of minor planets. Abbe, a long-time collaborator and friend of Carl Zeiss {see planet (851)}, secured considerable financial support for the University of Jena from the Zeiss foundation. The minor planet is being named on the occasion of the first meeting of the Astronomische Gesellschaft in the {eastern part of the} reunited Germany. (M 20839)
Abbe is also honored by a lunar crater.

(5225) Loral

1983 TS_1. Discovered 1983 October 12 by E. Bowell at Anderson Mesa.

Named for Loral, Inc., an American manufacturer of charge-coupled device chips. Several of Loral's large-format chips are to be used in a search for near-earth minor planets and comets to be conducted by the discoverer. (M 23137)

(5226) Pollack

1983 WL. Discovered 1983 November 28 by E. Bowell at Anderson Mesa.

Named in memory of James B. Pollack (1938-1994), a planetary scientist who spent much of his career at the NASA Ames Research Center. He was a world leader in the study of planetary atmospheres and particles using radiative transfer techniques. His work was highly creative and interdisciplinary, and it led to numerous advances in our understanding of the solar system. He played major roles in many NASA flight missions from Mariner 9 to Cassini and guided numerous young planetary scientists into productive careers. Work on the effects on the earth's atmosphere and surface biology of the Cretaceous-Tertiary impactor led Pollack and others to the concept of "nuclear winter". (M 25653)
Citation prepared by J. Cuzzi.

Obituaries published in Bull. Am. Astron. Soc., Vol. 26, No. 4, p. 1606-1608 (1994); Icarus, Vol. 113, No. 2, p. 227-231 (1995).

(5227) 1986 PE

Discovered 1986 August 4 at the Palomar Observatory at Palomar.

(5228) Máca

1986 VT. Discovered 1986 November 3 by Z. Vávrová at Kleť.

Named in honor of Jan Máca, one-time schoolmate and old friend of the discoverer, for his contribution to the protection of nature. (M 27128)

(5229) 1987 DE_6

Discovered 1987 February 23 by H. Debehogne at La Silla.

(5230) 1988 EF

Discovered 1988 March 10 by J. Alu at Palomar.

(5231) Verne

1988 JV. Discovered 1988 May 9 by C. S. Shoemaker and E. M. Shoemaker and H. E. Holt at Palomar.

Named for Jules Verne (1828-1905), French novelist and playwright. One of the founding fathers of modern science fiction, he was also the author of numerous works of mainstream adventure fiction. His best-known tales include *From the Earth to the Moon* (1865), *Twenty Thousand Leagues under the Sea* (1870) and *Around the World in Eighty Days* (1873). (M 24765)

The discoverers are great fans of Verne's work, as is M. M. Dworetsky, who suggested the name and prepared the citation.

Jules Verne is also honored by a lunar crater.

(5232) Jordaens

1988 PR_1. Discovered 1988 August 14 by E. W. Elst at St. Michel.

Named for Jacob Jordaens (1593-1678) on the 400th anniversary of his birth. This famous Flemish painter of Antwerp was a pupil of Adam van Noort and a contemporary of P. P. Rubens. His paintings were inspired by family life and popular situations, like the joy of daily life in the villages. His well-known painting "So als de ouden songhen so pijpen de jonghen" is typically nationalistic and representative of the revival of Flemish art, putting it on an equal level with more internationalistic tendencies. (M 21610)

Citation prepared by Kristina Leterme at the request of the discoverer.

(5233) 1988 RL_{10}

Discovered 1988 September 14 by S. J. Bus at Cerro Tololo.

(5234) Sechenov

1989 VP. Discovered 1989 November 4 by L. G. Karachkina at Nauchnyj.

Named in honor of the outstanding Russian naturalist Ivan Mikhailovich Sechenov (1829-1905). (M 22507)

Sechenov is also honored by a lunar crater.

(5235) Jean-Loup

1990 SA_1. Discovered 1990 September 16 by H. E. Holt at Palomar.

Named in honor of Jean-Loup Bertaux (1942-), French planetary scientist and head of the Department of Solar System Studies at the Service d'Aéronomie of the Centre National de Recherche Spatiale. Since 1964 Bertaux has played a crucial role in space-mission exploration of the solar system, acting in collaboration with his American and Russian colleagues. He was the first to map a hydrogen cloud around a comet, and in 1975 he discovered the interstellar wind in the solar system. Bertaux has been involved in several experiments on the Vega mission to comet 1P/Halley and in other missions to study Venus, Mars, the earth and the sun. (M 27459)

Name suggested and citation prepared by M. A. Barucci.

(5236) Yoko

1990 TG_3. Discovered 1990 October 10 by Y. Mizuno and T. Furuta at Kani.

Named in honor of Yoko Furuta (1946-), wife of the second discoverer. (M 27128)

(5237) Yoshikawa

1990 UF_3. Discovered 1990 October 26 by T. Urata at Oohira.

Named in honor of Katsunori Yoshikawa (1942-), owner of the land on which the Nihondaira {see planet (2880)} Observatory stands. He and his family appreciate the discoverer's surveys for minor planets and protect the observatory, which is located in one of the green-tea producing areas in Japan. (M 21134)

(5238) Naozane

1990 VE_2. Discovered 1990 November 13 by T. Hioki and S. Hayakawa at Okutama.

Named for Naozane Jiro Kumagaya (1141-1208), respected for his courage as a soldier during the turbulent Heian-Kamakura era. (M 22507)

(5239) Reiki

1990 VC_4. Discovered 1990 November 14 by S. Izumikawa and O. Muramatsu at Yatsugatake.

Named in honor of Reiki Kushida, amateur astronomer and discoverer of supernova 1991bg, the first visual discovery of a supernova by a woman. (M 21610)

(5240) Kwasan

1990 XE. Discovered 1990 December 7 by K. Suzuki and T. Urata at Toyota.

Named for the Kyoto University's Kwasan Observatory, which is located at the top of Kwasan hill on the eastern side of Kyoto city {see planet (4352)}. Kwasan Observatory was established in 1929 and is known for its observational studies of the solar system, especially of Mars, Pluto, minor planets, comets and natural satellites. (M 21610)

Name proposed by T. Mitani, who observed this planet at Kyoto on 1953 Nov. 3.

(5241) 1990 YL

Discovered 1990 December 23 by S. Ueda and H. Kaneda at Kushiro.

(5242) Kenreimonin

1991 BO. Discovered 1991 January 18 by S. Inoda and T. Urata at Karasuyama.

Named for Tokuko Kenreimon-in (1155-1213), daughter of Kiyorami and mother of Emperor Antoku {see planet (3686)}. She was rescued from drowning after the battle of Dannoura in 1185. (M 21134, 21155)

(5243) Clasien

1246 T-2. Discovered 1973 September 29 by C. J. van Houten and I. van Houten-Groeneveld at Palomar.

Named in honor of Clasien Shane, wife of astronomy professor W. W. Shane (1928-), of Nijmegen and the Leiden Observatory. From 1982 to 1987 the second discoverer lodged with the Shanes during the measuring for the T-1, T-2 and T-3 Trojan Surveys, and she is very grateful for their charming hospitality. (M 22507)

(5244) Amphilochos

1973 SQ_1. Discovered 1973 September 29 by C. J. van Houten and I. van Houten-Groeneveld at Palomar.

Like his father Amphiaraos, Amphilochos could foretell the future. As the Greek fleet departed from Troy he was warned by a vision, and in the final moment before the ship departed he jumped to the shore. In this way he saved his life and lived together with Kalchas {see planet (4138)} in Asia Minor. (M 23137)

(5245) Maslyakov

1976 GR_2. Discovered 1976 April 1 by N. S. Chernykh at Nauchnyj.

Named in honor of Aleksandr Vasil'evich Maslyakov, Moscow television journalist, one of the founders and for many years presenter of the program *KVN* (Club of Merry and Resourceful Persons), which is very popular not only in the countries of the former Soviet Union but in others too. (M 30096)

(5246) Migliorini

1979 OB. Discovered 1979 July 26 by E. Bowell at Anderson Mesa.

Named in memory of Fabio Migliorini (1971-1997), a young researcher tragically killed in a mountain accident. Migliorini obtained important results relating to both dynamical and physical studies of minor planets. In particular, he investigated the various processes in the main belt responsible for replenishing the NEA population and was the first to point out the importance of Mars-crossing objects. He also made essential contributions to the physical study of minor planet families (among which are reconstruction of ejection velocity fields and assessment of the number of family interlopers). (M 32345)

Name suggested by A. Cellino, D. Davis, M. Di Martino, P. Farinella, V. Zappalà and endorsed by many colleagues and friends in the research community on minor planets.

Obituary published in Ir. Astron. J., Vol. 25, No. 1, p. 5 (1998).

(5247) Krylov

1982 UP_6. Discovered 1982 October 20 by L. G. Karachkina at Nauchnyj.

Named in the memory of famous Russian mathematician and naval architect Alexei Nikolaevich Krylov (1863-1945). (M 22507)

Krylov is also honored by a lunar crater.

(5248) Scardia

1983 GQ. Discovered 1983 April 6 by H. Debehogne and G. DeSanctis at La Silla.

Named in honor of Marco Scardia (1948-), astrometrist at the Merate branch of the Brera Astronomical Observatory. His principal fields of interest are visual double stars and minor solar-system bodies. As a specialist on visual binaries, in the past fifteen years he has carried out more than four thousand measurements and calculated approximately 80 orbits of binary stars. He is also active in the astrometric observation of minor planets and comets and in the calculation of their orbits. (M 23138)

Name proposed and citation prepared by H. Debehogne.

(5249) Giza

1983 HJ. Discovered 1983 April 18 by N. G. Thomas at Anderson Mesa.

Named for the Egyptian town and regional capital on the west bank of the Nile. This strip of land contains some of Egypt's greatest antiquities: the step pyramid; the sphinx {see planet (896)}; Abu Sir pyramids; the great pyramids; and Memphis, the ancient capital of Egypt. (M 26930)

(5250) Jas

1984 QF. Discovered 1984 August 21 by A. Mrkos at Kleť.

The Czech word for brightness, Jas symbolizes the South Bohemian Astronomical Society, known in Czech as Jihočeská Astronomická Společnost. Inaugurated in 1928, the society has been concerned mainly with the education and popularization of astronomy, and for this purpose it founded the České Budějovice Observatory in 1937. The establishment of an observatory on the top of the Kleť mountain was inspired by JAS members. State-aided since the 1950s, the České Budějovice Observatory oversees the directorship of the Kleť Obserservatory and

is the home of the Kleť archives. (M 27734)
Name suggested by M. Tichý.

(5251) 1985 KA

Discovered 1985 May 18 by A. C. Gilmore and P. M. Kilmartin at Lake Tekapo.

(5252) Vikrymov

1985 PZ_1. Discovered 1985 August 13 by N. S. Chernykh at Nauchnyj.

Named in honor of Viktor Aleksandrovich Krymov (1929-) deputy director of the Institute of Theoretical Astronomy for many years. His skillful and energetic work contributes much to the proper functioning of the Institute. (M 29144)

Name proposed by the discoverer following a suggestion by the Institute of Theoretical Astronomy.

(5253) 1985 XB

Discovered 1985 December 15 by S. Singer-Brewster at Palomar.

(5254) Ulysses

1986 VG_1. Discovered 1986 November 7 by E. W. Elst at St. Michel.

Named for the Latin form of Odysseus {see planet (1143)}, hero of Homer's {see planet (5700)} Iliad and Odyssey. In the Iliad Ulysses is characterised by intelligence, experience (as a builder of machines) and endurance. He caused the chief commander Agamemnon {see planet (911)} to be restored and rallied the disaffected Greeks. The killing of the Trojan Dolon by Ulysses (which means "giver of pain") and Diomedes {see planet (1437)} was a tragic episode. Ulysses invented the strategem of the wooden horse, and it was first mentioned in the Odyssey, his journey of more than nine years after the battle of Troy. The pleasant episode of Ulysses and Nausikaa {see planet (192)} at Scheria has inspired many writers and composers. In James Joyce's {see planet (5418)} famous novel "Ulysses" the man of hostility becomes a man of peace. (M 21134)

(5255) Johnsophie

1988 KF. Discovered 1988 May 19 by E. F. Helin at Palomar.

Named in honor of John and Sophie Karayusuf, parents of Alford S. Karayusuf {see planet (3800)}, a friend of the discoverer. Under the starry skies of the Syrian Desert, they inspired their children to study the stars and planets and to wonder in amazement at the ability of mankind to explore the heavens. (M 32093)

(5256) Farquhar

1988 NN. Discovered 1988 July 11 by E. F. Helin and C. Mikolajczak and R. Coker at Palomar.

Named in honor of Robert Farquhar, who has been active in the design of spacecraft and missions for low-cost solar system exploration. His work has been carried out at the Goddard Space Flight Center and NASA Headquarters (1970-1990) and more recently at Johns Hopkins University Applied Physics Laboratory. Of particular interest are his efforts to utilize international cooperation and innovative trajectory design to plan missions to comets and minor planets. He was instrumental in designing and bringing to fruition the retargeting of the third International Sun Earth Explorer spacecraft to fly through the tail of periodic comet Giacobini-Zinner {see planets (1756) and (4615)} in September 1985 thus allowing the first in situ measurements of cometary phenomena. (M 21134)

Name endorsed heartily by D. K. Yeomans, who also contributed to the preparation of the citation.

(5257) 1988 RS_{10}

Discovered 1988 September 14 by S. J. Bus at Cerro Tololo.

(5258) 1989 AU_1

Discovered 1989 January 1 by Y. Oshima at Gekko.

(5259) Epeigeus

1989 BB_1. Discovered 1989 January 30 by C. S. Shoemaker and E. M. Shoemaker at Palomar.

A noble Myrmidon fighter, Epeigeus was a son of Agakles. Before the Trojan War, he had ruled in the flourishing town of Boudeion, but there he slew a kinsman. As a suppliant to Peleus and Thetis {see planet (17)}, he was enlisted along with Achilles {see planet (588)} to make war against the Trojans. At Troy he was killed when Hector {see planet (624)} hit him upon the crest of his helmet with a great stone. (M 25443)

(5260) Philvéron

1989 RH. Discovered 1989 September 2 by E. W. Elst at St. Michel.

Named in honor of Philippe Véron, director of the Observatoire de Haute Provence. In addition to conducting research on quasars and active galaxies, Véron is a student of early cometary books and broadsides. (M 21957)

(5261) Eureka

1990 MB. Discovered 1990 June 20 by H. E. Holt and D. H. Levy at Palomar.

Named for Eureka, an expression of joy after making a discovery. According to a story that is probably apocryphal, Hiero II asked the Greek physicist Archimedes (287-212 B.C.) {see planet (3600)} to judge whether a crown contained pure gold or gold mixed with silver. After some consternation he found the answer while taking a bath: the weight of the crown in and out of water would allow the density to be determined, and hence the proportions of silver and gold. Thus Archimedes' principle was born, that a solid body surrounded by liquid is supported by a force equal to the weight of the liquid displaced. Archimedes was so thrilled by the revelation that he leaped out of the bath and dashed down the street, presumably still nude and shouting "Eureka! Eureka!" (I have found it!) (M 21134)

This planet is the first detected Martian Trojan. It moves near the Lagrangean libration point L5.

(5262) Brucegoldberg

1990 XB_1. Discovered 1990 December 14 by E. F. Helin at Palomar.

Named in honor of Bruce A. Goldberg, scientist and astronomer at Caltech's Jet Propulsion Laboratory and the USAF Phillips Laboratory, as well as friend and colleague of the discoverer. He received his Ph.D. from the University of British Columbia and has conducted research in modeling and spectroscopy of Io and comet 21P/Giacobini-Zinner {see also planets (1756), (4615)}, on adaptive optics and as guest observer at a variety of telescopes throughout the world for over twenty years. (M 25653)

Name endorsed by R. Bambery, K. J. Lawrence and his many friends at JPL.

(5263) Arrius

1991 GY_9. Discovered 1991 April 13 by D. I. Steel at Siding Spring.

Named for Harrison Callum Bertram Steel, first-born child of the discoverer. Harrison, born on 1992 December 16, shares his birth-date with several luminaries, including Christopher P. McKay (space scientist at NASA-Ames Research Center and friend of the discoverer), Arthur C. Clarke and Ludwig van Beethoven {see planets (4923) and (1815), respectively}. Their first initials provide the beginning of Harrison's second name. The 'Arrius' name is inspired by the Latin form of 'Harry', and in particular the poem 'Harrius' by the Roman poet Catullus, concerning a man of that appellation who dropped his h's. (M 23138)

(5264) Telephus

1991 KC. Discovered 1991 May 17 by C. S. Shoemaker and E. M. Shoemaker at Palomar.

Telephus, a son of Heracles {see planet (5143)} and Auge, was exposed on Mount Parthenius, nursed by a goat, and nurtured by kindly shepherds. Telephus had married a daughter of Priam {see planet (884)}, but was so grateful to the Greeks for healing a severe wound that he fought with the Greeks against his father-in-law. An oracle had declared that Troy could not be taken against the aid of a son of Heracles. (M 25444)

(5265) Schadow

2570 P-L. Discovered 1960 September 24 by C. J. van Houten and I. van Houten-Groeneveld at Palomar.

Named after Johann Gottfried Schadow (1764-1850), German sculptor in the classicistic style. His best-known sculpture is the Quadriga at the Brandenburg Gate in Berlin. (M 22507; M 25993)

(5266) Rauch

4047 T-2. Discovered 1973 September 29 by C. J. van Houten and I. van Houten-Groeneveld at Palomar.

Named after Christian Daniel Rauch (1777-1857), German sculptor, first servant of King Friedrich Wilhelm III of Prussia. He received a grant from the king to become a pupil of Schadow {see planet (5265)}. Rauch worked in the classic style in Rome from 1804 to 1811, and his main work is the sarcophagus of Queen Luise of Prussia. (M 22507)

(5267) 1966 CF

Discovered 1966 February 13 at the Purple Mountain Observatory at Nanking.

(5268) 1971 US_1

Discovered 1971 October 26 by L. Kohoutek at Bergedorf.

(5269) 1978 SL_6

Discovered 1978 September 28 by N. S. Chernykh at Nauchnyj.

(5270) 1979 KR

Discovered 1979 May 19 by R. M. West at La Silla.

(5271) 1979 MH_7

Discovered 1979 June 25 by E. F. Helin and S. J. Bus at Siding Spring.

(5272) Dickinson

1981 QH_2. Discovered 1981 August 30 by E. Bowell at Anderson Mesa.

Named in honor of Terence Dickinson (1943-), Canada's foremost popularizer of astronomy. Dickinson is the author of several books, notably *Night Watch, The Universe and Beyond, Exploring the Sky by Day* and *Exploring the Night Sky*, the last of which received the New York Academy of Sciences Children's Science Book Award in 1988. He reaches a wide Canadian audience every week with his astronomy column in *The Toronto Star* newspaper and on Canadian Broadcasting Corporation programs. From 1973 to 1975 he was editor of *Astronomy* magazine, and he has held scientific positions with the Ontario Science Centre in Toronto, the Strasenburgh Planetarium in Rochester and the McLaughlin Planetarium in Toronto. Dickinson teaches astronomy at St. Lawrence College in Kingston, Ontario. (M 23138)

Name suggested and citation prepared by C. J. Cunningham.

(5273) 1982 DQ_6

Discovered 1982 February 16 at the Beijing Observatory at Xinglong.

(5274) Degewij

1985 RS. Discovered 1985 September 14 by E. Bowell at Anderson Mesa.

Named in honor of Johan Degewij (1944-), of Utrecht Polytechnic Institute, for his contributions to small-bodies research. Degewij was among the first to use spectroscopy to search for cometary activity on near-earth asteroids and to detect color and polarization variations with rotation. His discovery of rotational albedo variegation on Vesta {see planet (4)} was the crucial observation that led to the recognition of its true rotation period as 5.34 hours, rather than 10.68 hours, the value accepted at the time. Degewij has also made significant contributions to the photometric study of comets, distant asteroids and the small satellites of Jupiter and Saturn. (M 24917)

Name suggested and citation provided by E. F. Tedesco.

(5275) Zdislava

1986 UU. Discovered 1986 October 28 by Z. Vávrová at Kleť.

Named for Saint Zdislava (c. 1220-1252), a noblewoman who devoted her life to performing charity for the poor. She is the beloved saint of Marie Mácová, wife of an old schoolmate of the discoverer. (M 27128)

(5276) Gulkis

1987 GK. Discovered 1987 April 1 by E. F. Helin at Palomar.

Named in honor of Samuel Gulkis of the Jet Propulsion Laboratory. He has over twenty-five years of research experience in radio and submillimeter astronomy, specializing in studies of Jovian magnetospheric physics, planetary atmospheres and experimental cosmology. Since 1978 he has been the project scientist in the JPL program for the search for Extra-Terrestrial Intelligence. Currently, he heads the Office of Science and Information Systems within the Space and Earth Science Programs Directorate. The discoverer and members of the Near-Earth Asteroid Tracking program wish to recognize his interest and support in the establishment of the JPL NEAT program in collaboration with the Air Force GEODSS site in Maui. (M 26930)

(5277) 1988 DO

Discovered 1988 February 22 by R. H. McNaught at Siding Spring.

(5278) Polly

1988 EJ_1. Discovered 1988 March 12 by E. F. Helin at Palomar.

Named in honor of "Polly" Blanton Brooks (1917-), a fascinating, charming Texan, whose childhood was spent on the family's Yates Ranch, where oil was discovered in 1926. She married and raised three children on the Circle C Ranch, now known as Circle C City, Texas. Polly has been a member of the The Planetary Society's New Millennium Committee, providing support and inspiration for many projects. An enthusiastic proponent of space exploration, she has fostered international communication and cooperation over the years. (M 30096)

Name proposed on the occasion of her eightieth birthday, 1997 March 11.

(5279) Arthuradel

1988 LA. Discovered 1988 June 8 by T. Rodriquez at Palomar.

Named in honor of Arthur Adel {1908-1994}, a pioneer in the field of infrared spectroscopy and upper atmospheric research. Much of his work was pursued at the Atmospheric Research Observatory, built for his use in 1952 on the campus of Northern Arizona University at Flagstaff. In the 1950s and 1960s he discovered atmospheric nitrous oxide, atmospheric heavy water and the 20-micron window,

and he prepared the first maps of the solar-telluric spectrum from 7 to 14 microns. (M 22830)
Name endorsed by E. M. and C. S. Shoemaker.

Obituary published in Bull. Am. Astron. Soc., Vol. 26, No. 4, p. 1600-1601 (1994).

(5280) 1988 PT

Discovered 1988 August 11 by C. Mikolajczak and R. Coker at Palomar.

(5281) 1988 SO_1

Discovered 1988 September 16 by S. J. Bus at Cerro Tololo.

(5282) Yamatotakeru

1988 VT. Discovered 1988 November 2 by Y. Oshima at Gekko.

Named after the tragic hero, Yamatotakeru, who is well known by many Japanese people. He appears in Japanese ancient literature such as *Kojiki* and *Nihon Shoki*. (M 24122; M 24251)
Name proposed by the Gekko Astronomical Observatory.

(5283) Pyrrhus

1989 BW. Discovered 1989 January 31 by C. S. Shoemaker and E. M. Shoemaker at Palomar.

Pyrrhus received his name from the yellow color of his hair. Also known as Neoptolemus {see planet (2260)}, he was the king of Epirus and a son of Achilles {see planet (588)}. After Achilles was killed, a Greek delegation led by Odysseus {see planet (1143)} went to Epirus and brought Pyrrhus to the Trojan War. Enraged at the death of his father, he became the most ruthless of all the Greeks. Using a battle-axe, he broke the doors leading into King Priam's {see planet (884)} house, which he sacked and burned, showing no mercy to women and children. After killing Priam's son Polites {see planet (4867)}, he also slew Priam on an altar - a sacrilege. He took away Hector's {see planet (624)} wife, Andromache {see planet (175)} as his prize. (M 22250)

Name and citation provided by Richard Preston at the request of the discoverers.

(5284) Orsilocus

1989 CK_2. Discovered 1989 February 1 by C. S. Shoemaker and E. M. Shoemaker at Palomar.

Orsilocus, one of twin sons of Diokles, was a great man among the Danaans. As he grew up, he and his brother followed along with the Argives to Troy. He helped to win honor for Agamemnon and Menelaus {see planets (911) and (1647), respectively}, before being killed by Aeneas {see planet (1172)}. (M 25444)

(5285) Krethon

1989 EO_{11}. Discovered 1989 March 9 by C. S. Shoemaker and E. M. Shoemaker at Palomar.

Krethon, one of two great Danaans, was one of the twin sons of Diokles. As a very young man he went with his brother Orsilocus {see planet (5284)} in the service of Agamemnon and Menelaus {see planets (911) and (1647), respectively}. Fighting with the Argives at Troy, he was killed by Aeneas {see planet (1172)}. (M 25444)

(5286) Haruomukai

1989 VT_1. Discovered 1989 November 4 by M. Mukai and M. Takeishi at Kagoshima.

Named in memory of Haruo Mukai (1949-1986), younger brother of one of the discoverers. (M 21957)

(5287) Heishu

1989 WE. Discovered 1989 November 20 by Y. Mizuno and T. Furuta at Kani.

Named in memory of Heishu Hosoi (1728-1801), beloved scholar during the Edo era, born in the present-day city of Tokai, home of the second discoverer. {Tokai already is a member of the minor planet sky - see planet (2478)}. (M 22507)

(5288) Nankichi

1989 XD. Discovered 1989 December 3 by Y. Mizuno and T. Furuta at Kani.

Named in memory of Nankichi Ni-imi (1913-1943), teacher and writer of fairy tales. (M 22508; M 22519)

(5289) 1990 KG_2

Discovered 1990 May 28 at the Carlos U. Cesco Observatory at El Leoncito.

(5290) 1990 OD_4

Discovered 1990 July 30 by H. E. Holt at Palomar.

(5291) Yuuko

1990 YT. Discovered 1990 December 20 by M. Matsuyama and K. Watanabe at Kushiro.

Named in honor of Yuuko Matsuyama (1950-), wife of the first discoverer. (M 22508)

(5292) 1991 AJ_1

Discovered 1991 January 12 by H. Shiozawa and M. Kizawa at Fujieda.

(5293) Bentengahama

1991 BQ_2. Discovered 1991 January 23 by M. Matsuyama and K. Watanabe at Kushiro.

Named for a beautiful sandy beach between Kushirozaki lighthouse and Chiyonoura, in the southern part of Kushiro {see planet (4096)} City. The second discoverer spent his childhood there. The beach is being eroded and is gradually disappearing. (M 22508)

(5294) Onnetoh

1991 CB. Discovered 1991 February 3 by K. Endate and K. Watanabe at Kitami.

Named for the small lake in Akan Nationalpark {see planet (4584)} in eastern Hokkaido. The lake, surrounded by volcanic mountains, is 620 meters above sea level and 4 km in circumference. (M 22508)

(5295) Masayo

1991 CE. Discovered 1991 February 5 by Y. Mizuno and T. Furuta at Kani.

Named in honor of Masayo Mizuno (1955-), wife of the first discoverer. (M 27128)

(5296) Friedrich

9546 P-L. Discovered 1960 October 17 by C. J. van Houten and I. van Houten-Groeneveld at Palomar.

Named after Caspar David Friedrich (1774-1840), German painter of the Romantic school. Most of his work has a melancholic but peaceful expression. (M 22508)

(5297) Schinkel

4170 T-2. Discovered 1973 September 29 by C. J. van Houten and I. van Houten-Groeneveld at Palomar.

Named after Karl Friedrich Schinkel (1781-1841), German architect and painter. He lived in Italy for two years. His style was mostly classicistic, but he also admired the Gothic style. His best known buildings are Schloss Charlottenhof in the park of Sanssouci and the mausoleum for Queen Luise of Prussia. (M 22508; M 24781)

(5298) 1966 PK

Discovered 1966 August 7 at the Boyden Observatory at Bloemfontein.

(5299) Bittesini

1969 LB. Discovered 1969 June 8 by C. U. Cesco at El Leoncito.

Named in honor of Luciano Bittesini (1950-), Italian amateur astronomer and astrometrist. (M 30096)

Name suggested by B. G. Marsden, who made the identifications involving this object, and by G. V. Williams, who was hosted by Luciano and family during a 1995 visit to speak at a convention of Italian astrometrists.

(5300) Sats

1974 SX_1. Discovered 1974 September 19 by L. I. Chernykh at Nauchnyj.

Named in memory of Nataliya Il'inichna Sats (1903-1993). An outstanding figure in Soviet theatre culture for children, she was the founder and art advisor of the Moscow Musical Children's Theatre, the first institution of its type in the world. She was also the author of many plays, librettos, operas and ballets for children. (M 34621)

(5301) Novobranets

1974 SD_3. Discovered 1974 September 20 by L. V. Zhuravleva at Nauchnyj.

Named in honor of Vasilij Andreevich Novobranets (1904-1984), Russian and Ukrainian writer. (M 28089)

(5302) Romanoserra

1976 YF_5. Discovered 1976 December 18 by N. S. Chernykh at Nauchnyj.

Named in honor of Romano Serra (1954-), enthusiastic president of the amateur observatory of San Giovanni in Persiceto near Bologna. A great specialist in meteorites, he has gathered the largest collection of meteorite specimens in Italy. He participated in two Tunguska {see planet (5471)} expeditions, in 1991 and 1998, and built a planetarium, the third largest in Italy. (M 34341)
Name suggested by M. di Martino and supported by the discoverer.

(5303) Parijskij

1978 TT_2. Discovered 1978 October 3 by N. S. Chernykh at Nauchnyj.

Named in honor of Yurij Nikolaevich Parijskij (1932-), radio astronomer and cosmologist. He was principal scientific investigator during the design and construction of the largest radio telescope, RATAN-600, at Zelenchukskaya, and he is now head of the radio astronomy department of the Special Astrophysical Observatory. (M 30476)

Name proposed by the Institute of Theoretical Astronomy and the Institute of Applied Astronomy.

(5304) Bazhenov

1978 TA_7. Discovered 1978 October 2 by L. V. Zhuravleva at Nauchnyj.

Named in memory of Vasilij Ivanovich Bazhenov (1738-1799), Russian architect, teacher and architectural theorist. (M 34621)

(5305) 1978 VS_5

Discovered 1978 November 7 by E. F. Helin and S. J. Bus at Palomar.

(5306) Fangfen

1980 BB. Discovered 1980 January 25 at the Harvard College Observatory at Harvard.

Named in honor of Fang Fen, a native of Zhangzhou in southern Fujian and wife of astronomer C.-Y. Shao {see planet (1881)}, one of the Agassiz {see planet (2267)} observers. (M 22508)

(5307) Paul-André

1980 YC. Discovered 1980 December 30 by E. Bowell at Anderson Mesa.

Named in memory of Paul-André Herbelin (1933-1994) of Ayent, near Sion, Switzerland, friend of the discoverer and brother of the discoverer's companion Anne-Marie Malotki {see planet (3667)}. By profession a surveyor, Herbelin was by avocation a winemaker and raclette chef extraordinaire. (M 23540)

(5308) 1981 DC_2

Discovered 1981 February 28 by S. J. Bus at Siding Spring.

(5309) 1981 ED_{25}

Discovered 1981 March 2 by S. J. Bus at Siding Spring.

(5310) 1981 EP_{26}

Discovered 1981 March 2 by S. J. Bus at Siding Spring.

(5311) 1981 GD_1

Discovered 1981 April 3 by A. C. Gilmore and P. M. Kilmartin at Lake Tekapo.

(5312) Schott

1981 VP_2. Discovered 1981 November 3 by F. Börngen at Tautenburg.

Named in memory of Friedrich Otto Schott (1851-1935), analytical chemist and glass engineer. In 1882 he founded a glass-technical laboratory and, together with E. Abbe {see planet (5224)} two years later, also the Jenaer Glaswerk Schott und Genossen. Among the glass products Schott devised are filters, thermometers and the famous "Jenaer Glas". (M 21610)

(5313) Nunes

1982 SC_2. Discovered 1982 September 18 by H. Debehogne at La Silla.

Named in memory of Pedro Nunes, Portuguese mathematician and geographer, called "the peak figure in Portuguese nautical science". He invented the instrument later known as the vernier, and he translated Sacrobosco into Portuguese. (M 21610)

(5314) Wilkickia

1982 SG_4. Discovered 1982 September 20 by N. S. Chernykh at Nauchnyj.

Named in memory of Andrej Ippolitovich Wilkitzky (Wilkicki; 1858-1913) and his son Boris Andreevich Wilkitzky (1885-1961), outstanding Russian hydrographers and Arctic explorers, who made an invaluable contribution to mastering the navigation of the Northern Maritime Way. In 1913 Boris Wilkitzky discovered a new archipelago, named by him "The Land of Emperor Nicholas II" (now Severnaya Zemlya), as well as a strait named for him. (M 30096)

This minor planet is named by the discoverer following a suggestion by V. K. Abalakin.

(5315) Bal'mont

1982 SV_5. Discovered 1982 September 16 by L. I. Chernykh at Nauchnyj.

Named in memory of Konstantin Dmitrievich Bal'mont (1867-1942), a Russian poet who lived in exile after 1920, one of the most famous of the symbolists. Known for his Russian translations of the works of many European writers, he also collected and published examples of the creative works of Egypt, Mexico, Peru, India and other countries. (M 30097)

(5316) Filatov

1982 UB_7. Discovered 1982 October 21 by L. G. Karachkina at Nauchnyj.

Named for the well-known ophthalmologist and surgeon Vladimir Petrovich Filatov (1875-1956). (M 22508)

(5317) Verolacqua

1983 CE. Discovered 1983 February 11 by C. S. Shoemaker at Palomar.

Named in honor of Veronica Lynn Passalacqua, on the occasion of her twenty-fifth birthday, by her long-time friend D. W. E. Green, who made the identifications of this object. She has done tremendous volunteer work over several years, while a student at Harvard University, toward compiling the *International Comet Quarterly* archive of photometric data on comets. (M 21957)

(5318) 1985 HG_1

Discovered 1985 April 21 by A. Mrkos at Kleť.

(5319) Petrovskaya

1985 RK_6. Discovered 1985 September 15 by N. S. Chernykh at Nauchnyj.

Named in honor of Margarita Sergeevna Petrovskaya, staff member of the Institute of Theoretical Astronomy since 1960. She determined the domains of convergence of series representing periodic solutions in planetary and satellite problems and obtained a new expansion for the perturbing function that converges even in the case of overlapping orbits. She was the first to construct convergent global expansions for the earth's potential. (M 30476)
Name proposed by the Institute of Theoretical Astronomy.

(5320) Lisbeth

1985 VD. Discovered 1985 November 14 by P. Jensen and K. Augustesen and H. J. Fogh Olsen at Brorfelde.

Named in honor of Lisbeth Fogh Olsen, daughter of the third discoverer, currently graduating in astronomy at the Copenhagen Observatory. (M 29144)

(5321) Jagras

1985 VN. Discovered 1985 November 14 by P. Jensen and K. Augustesen and H. J. Fogh Olsen at Brorfelde.

Named in honor of Jakob Grove Rasmussen, fiancé of the daughter of the third discoverer, currently graduating in astronomy at the Copenhagen Observatory. (M 29144)

(5322) 1986 QB_1

Discovered 1986 August 26 by H. Debehogne at La Silla.

(5323) 1986 TL_4

Discovered 1986 October 13 by P. Jensen at Brorfelde.

(5324) Lyapunov

1987 SL. Discovered 1987 September 22 by L. G. Karachkina at Nauchnyj.

This unusual object is named in honor of the outstanding Russian mathematician Alexandr Mikhailovich Lyapunov (1857-1918), famous for his theory of the stability of motion and the 'Lyapunov characteristic exponents'. (M 22508)

(5325) Silver

1988 JQ. Discovered 1988 May 12 by C. S. Shoemaker and E. M. Shoemaker at Palomar.

Named for Leon T. Silver, professor of geology at the California Institute of Technology. Famed for his studies of the regional geology and precise determinations of ages of precambrian and mesozoic crystalline rocks of the southwestern

United States, Silver also made important contributions to the study of the moon. He played a major role in the geologic training of astronauts for the Apollo 15, 16 and 17 missions to the moon and in the planning and conduct of these missions. (M 24917)

(5326) 1988 RT_6
Discovered 1988 September 8 by H. Debehogne at La Silla.

(5327) 1989 EX_1
Discovered 1989 March 5 by Z. Vávrová at Kleť.

(5328) 1989 UH_1
Discovered 1989 October 26 by S. Ueda and H. Kaneda at Kushiro.

(5329) 1989 YP
Discovered 1989 December 21 by R. H. McNaught at Siding Spring.

(5330) Senrikyu
1990 BQ_1. Discovered 1990 January 21 by A. Sugie at Taga.

Named in memory of Sen-no Rikyu (1522-1591), a celebrated Japanese tea master who perfected the tea ceremony and was the founder of Senke style during the Azuchi-Momoyama period. He was given an important position by both Oda-Nobunaga and Toyotomi-Hideyoshi, who were two of the greatest warriors in history. Rikyu was known as "the supreme tea master" and left his mark on the culture of the tea ceremony. (M 25653)

(5331) Erimomisaki
1990 BT_1. Discovered 1990 January 27 by K. Endate and K. Watanabe at Kitami.

Named for the promontory at the southern end of Hokkaido. Warm and cold currents, crossing each other near the promontory, bring strong winds and thick fog. (M 22508)

(5332) 1990 DA
Discovered 1990 February 16 by A. Sugie at Taga.

(5333) Kanaya
1990 UH. Discovered 1990 October 18 by M. Akiyama and T. Furuta at Mishima.

Named for the first discoverer's native town, an old station on the ancient national road "Tokai-do", some 200 km west of Tokyo. The Malinohara tablelands to the south of this town on the west bank of the Oi river are famous as one of the largest tea fields in Japan. (M 21610)

(5334) Mishima
1991 CF. Discovered 1991 February 8 by M. Akiyama and T. Furuta at Mishima.

Named for the first discoverer's second home, a city some 100 km west of Tokyo and famous for the great amount of clear water from Mt. Fuji {see planet (1584)}, 30 km to the northwest. The city's old shrine, Mishima Taisya, is famous as the point where Yoritomo Minamoto {see planet (3902)}, founder of the Kamakura shogunate, began his battle to unify the nation. (M 21611)

(5335) Damocles
1991 DA. Discovered 1991 February 18 by R. H. McNaught at Siding Spring.

Legend has it that Damocles, a courtier of the tyrant Dionysius the Elder of Syracuse, was invited to a feast, where he was seated below a sword suspended by a thread. Intended to demonstrate the impermanence of power, the 'sword of Damocles' has become a symbol of impending disaster. This minor planet, in an unusual orbit of very high eccentricity and inclination and crossing the orbits of

Mars to Uranus, could itself be seen as such a sword during periods of its orbital evolution when it becomes an Earth crosser. (M 22508)

(5336) 1991 JE_1

Discovered 1991 May 7 by N. Kawasato at Uenohara.

(5337) Aoki

1991 LD. Discovered 1991 June 6 by S. Otomo and O. Muramatsu at Kiyosato.

Named in memory of Masahiro Aoki (1920-1984), amateur astronomer and director of the variable-star section of the Oriental Astronomical Association during 1969-1984. He contributed to the popularization of astronomy and to the protection of the natural environment and was active in fighting light pollution. (M 21135)

(5338) 1991 RJ_5

Discovered 1991 September 13 by H. E. Holt at Palomar.

(5339) 1992 CD

Discovered 1992 February 4 by T. Hioki and S. Hayakawa at Okutama.

(5340) Burton

4027 P-L. Discovered 1960 September 24 by C. J. van Houten and I. van Houten-Groeneveld at Palomar.

Named in honor of William Butler Burton (1940-), American born and since 1981 professor of astronomy at the University of Leiden. He was a student of Peter van de Kamp {see planet (1965)} at Sproul Observatory. His main interest is the structure of the Galactic system at different wavelengths. He served as president of IAU Commission 33 (Structure and Dynamics of the Galactic System) from 1985 to 1987 and is chairman of the Science Team of the DENIS (Deep Near Infrared Survey of the Southern Sky) collaboration. (M 22508)

(5341) Purgathofer

6040 P-L. Discovered 1960 September 24 by C. J. van Houten and I. van Houten-Groeneveld at Palomar.

Named in memory of Alois Purgathofer (1925-1984), Austrian astronomer. Purgathofer mainly worked on photometry of galactic clusters, and he also observed special minor planets, e.g., (51) Nemausa. He put much effort into establishing an observatory with two telescopes at a remote site outside Vienna. Purgathofer was a member of the Committee on Image Tubes for Telescopes of the Carnegie Institution. He investigated planetary nebulae after building an image-tube spectrograph of his own design. (M 22508)

Name proposed and citation prepared by A. Schnell, endorsed by L. D. Schmadel.

Obituaries published in Sternenbote, 27. Jahrg., Nr. 4, p. 63-66 (1984); Mitt. Astron. Ges., Nr. 61, p. 7-8 (1984).

(5342) Le Poole

3129 T-2. Discovered 1973 September 30 by C. J. van Houten and I. van Houten-Groeneveld at Palomar.

Named in honor of Rudolf Le Poole (1942-), who spent several years in Tucson, working with G. P. Kuiper {see planet (1776)}, before moving to the Leiden Observatory. Le Poole is a member of the HIPPARCOS Science Team. He is also developing the Astroscan II, a very accurate measuring instrument for photographic plates, for the Leiden Observatory. (M 22508)

(5343) Ryzhov

1977 SG_3. Discovered 1977 September 23 by N. S. Chernykh at Nauchnyj.

Named in honor of Yurij Alekseevich Ryzhov (1930-), a member of the Russian Academy of Sciences, outstanding specialist on aero- and hydrodynamics, former long-time rector of the Moscow Aviation Institute. (M 30476; M 32561)
Name proposed by A. G. Sokolsky.

(5344) Ryabov

1978 RN. Discovered 1978 September 1 by N. S. Chernykh at Nauchnyj.

Named in honor of Yurij Aleksandrovich Ryabov (1923-), professor at the Moscow Road-Transport Institute. He developed the Poincaré-Lyapunov method of small parameters for the investigation of fine-resonance effects. He is also the author of monographs on the modern problems of celestial mechanics, as well as of popular books on astronomy. (M 30476)
Name proposed by the Institute of Theoretical Astronomy.

(5345) 1981 EY_8

Discovered 1981 March 1 by S. J. Bus at Siding Spring.

(5346) 1981 QE_3

Discovered 1981 August 24 by H. Debehogne at La Silla.

(5347) 1985 DX_2

Discovered 1985 February 24 by E. F. Helin at Palomar.

(5348) 1988 BB

Discovered 1988 January 16 by T. Kojima at Chiyoda.

(5349) Paulharris

1988 RA. Discovered 1988 September 7 by E. F. Helin at Palomar.

Named in honor of Paul Harris (1868-1947) who, in 1905, founded Rotary International as a forum for business leaders to meet and provide services to benefit their local communities and contribute to international understanding. Now, with 1.2 million members worldwide, Rotarians support a variety of worthy causes, including the Polio Plus program that has immunized millions of children around the globe. In addition, as the world's largest privately sponsored scholarship organization, Rotary provides 1350 scholarships each year to deserving students. (M 24917)

Name proposed by Theodore Combes and endorsed by Joseph Lewis and the international members of the Rotary.

(5350) Epetersen

1989 GL_1. Discovered 1989 April 3 by E. W. Elst at La Silla.

Named in honor of Erik V. Petersen (1911-), Danish amateur astronomer, who has contributed to astronomy through his study of minor planets, especially of (51) Nemausa. During 1955-1965 Petersen used the refractor at the Copenhagen University Observatory to take 712 plates, on which he measured 2,562 positions of minor planets. This is the largest single project ever carried out with that telescope. The data were used in an international collaboration, and Petersen's work received a special commendation at the IAU General Assembly in Prague in 1967. (M 21611)

Name proposed by B. Reipurth, who prepared the citation at the request of the discoverer.

(5351) Diderot

1989 SG_5. Discovered 1989 September 26 by E. W. Elst at La Silla.

Named in memory of Denis Diderot (1713-1784), French author of dramas, novels and philosophical essays. Together with d'Alembert {see planet (5956)}, he will be remembered for his Encyclopedia, a huge task that took almost 20 years of his life and resulted in more than one thousand articles on philosophy, literature, religion, politics, economics and applied sciences. (M 23540)
Citation by K. Leterme at the request of the discoverer.

(5352) Fujita

1989 YN. Discovered 1989 December 27 by Y. Kushida and O. Muramatsu at Yatsugatake.

Named in honor of Yoshio Fujita, professor emeritus of the University of Tokyo, known for his pioneering work on the spectra of late-type stars, in particular of cool, carbon stars, with his determination of the carbon isotope ratio. He served as president of the Astronomical Society of Japan during 1961-1963 and as president of IAU Commission 29 during 1970-1973. He has been a member of the Japan Academy since 1965 and of the Société Royale des Sciences de Liège since 1969. (M 21611)

Name proposed by the discoverers following a suggestion by T. Tsuji and Y. Yamashita.

(5353) 1989 YT

Discovered 1989 December 20 by Y. Oshima at Gekko.

(5354) Hisayo

1990 BJ_2. Discovered 1990 January 30 by S. Ueda and H. Kaneda at Kushiro.

Named in honor of Hisayo Kaneda (1977-), daughter of the second discoverer. (M 27734)

(5355) Akihiro

1991 CA. Discovered 1991 February 3 by S. Ueda and H. Kaneda at Kushiro.

Named in honor of Akihiro Ueda (1983-), son of the first discoverer. (M 27734)

(5356) 1991 FF_1

Discovered 1991 March 21 by K. Endate and K. Watanabe at Kitami.

(5357) 1992 EL

Discovered 1992 March 2 by T. Fujii and K. Watanabe at Kitami.

(5358) 1992 QH

Discovered 1992 August 26 by S. Ueda and H. Kaneda at Kushiro.

(5359) Markzakharov

1974 QX_1. Discovered 1974 August 24 by L. I. Chernykh at Nauchnyj.

Named in honor of Mark Anatolievich Zakharov (1933-), chief producer of the Moscow Lenin Komsomol Theatre, People's Artist of the U.S.S.R. and Russian Federation State Prize laureate. He has produced many musical plays, performances in the genre of political drama and films for television. (M 34621)

(5360) Rozhdestvenskij

1975 VD_9. Discovered 1975 November 8 by N. S. Chernykh at Nauchnyj.

Named in memory of Robert Ivanovich Rozhdestvenskij (1932-1994), outstanding poet and journalist, one of the most popular Russian poet-songwriters. For 15 years he wrote and presented the television program *The Documentary Screen*. His books were translated into many languages, and he won many awards for literature. In his last poems he appeared, not only as a lyric poet, but also as a philosopher who showed the grandeur and tragedy of his time. He arranged for the publication of Osip Mandelshtam's {see planet (3461)} works, edited the first collection of Vladimir Vysotskij's poems and organized Marina Tsvetaeva's {see planet (3511)} Memorial Museum. (M 30097)

(5361) 1976 YC_2

Discovered 1976 December 16 by L. I. Chernykh at Nauchnyj.

(5362) 1978 CH

Discovered 1978 February 2 by J. Gibson at Palomar.

(5363) Kupka

1979 UQ. Discovered 1979 October 19 by A. Mrkos at Kleť.

Named in memory of František Kupka (1871-1957), Czech painter and graphic artist, resident in France after 1906. He is considered one of the initiators of abstract art, as his *Stories of Black and White*, *Amorpha* or *Cosmic prime* show. (M 34621)

Name suggested by J. Tichá.

(5364) 1980 RC_1

Discovered 1980 September 2 by Z. Vávrová at Kleť.

(5365) Fievez

1981 EN_1. Discovered 1981 March 7 by H. Debehogne and G. DeSanctis at La Silla.

Named in memory of Charles Fievez (1844-1890), the pioneer of astrophysics in Belgium. His scientific career at the Observatoire Royal de Bruxelles was short (1877-1890) but very fruitful. In 1880 he started the first spectroscopic laboratory in Belgium. He published two dozen papers on spectroscopy, including an atlas of the solar spectrum, and in 1885 he observed the broadening effect of spectral lines due to the presence of a magnetic field (but without finding the correct interpretation) eleven years before Zeeman. (M 23138)

(5366) 1981 EY_{30}

Discovered 1981 March 2 by S. J. Bus at Siding Spring.

(5367) Sollenberger

1982 TT. Discovered 1982 October 13 by E. Bowell at Anderson Mesa.

Named in memory of Paul Sollenberger (1891-1995), the U.S. Naval Observatory's first civilian director of Time Service (from 1928 until his retirement in 1953). Sollenberger's contributions to the design of quartz-crystal clocks, chronographs and the photographic zenith tube greatly increased the precision of timekeeping during his tenure. By 1934 he had not only introduced quartz-crystal clocks, but also arranged for the automatic transmission of time by them, accurate to a millisecond. In 1949 he established the USNO's Time Service station using a PZT in Richmond, Florida. Sollenberger served as president of IAU Commission 19 during 1948-1955. (M 25653)

Citation prepared by D. D. McCarthy. Name endorsed by P. K. Seidelmann.

Obituary published in Bull. Am. Astron. Soc., Vol. 27, No. 4, p. 1482-1483 (1995).

(5368) 1984 SW_5

Discovered 1984 September 21 by H. Debehogne at La Silla.

(5369) Virgiugum

1985 SE_1. Discovered 1985 September 22 by P. Wild at Zimmerwald.

Latinization of Jungfraujoch, at 3572 m, the site of the highest astronomical observatory in the Swiss Alps. (5369) = (2040) + (3329), the numbers of the minor planets named for the astronomers Chalonge and Golay, who, with their teams did extensive observing there, contributing much to the fame of the High-Alpine Research Station, run by a foundation of the academies of eight European countries. In his student years, the discoverer was several times an assistant in Chalonge's group, thus getting the most impressive introduction to practical astronomy. (M 22508)

(5370) Taranis

1986 RA. Discovered 1986 September 2 by A. Maury at Palomar.

Named after the Gaulish god of thunder (Irish "torann" and Gallic "Taran"). His symbols are the wheel and the lightning. His cult was very common in ancient Gaul. Taranis is also the name of the first computer network in France for amateur astronomers, established by J. C. Merlin in 1988. (M 22509)
Citation by the discoverer and J. C. Merlin.

(5371) 1987 VG_1

Discovered 1987 November 15 by S. Ueda and H. Kaneda at Kushiro.

(5372) Bikki

1987 WS. Discovered 1987 November 29 by K. Endate and K. Watanabe at Kitami.

Named in honor of Bikki Sunazawa (1931-1989), self-taught Japanese sculptor. In 1954 his sculpture was honored at the Japanese Modern Art Exhibition. His major works are *Four Winds* and *Nitsunekamui*. (M 22509)

(5373) 1988 VV_3

Discovered 1988 November 14 by S. Ueda and H. Kaneda at Kushiro.

(5374) Hokutosei

1989 AM_1. Discovered 1989 January 4 by M. Yanai and K. Watanabe at Kitami.

Hokutosei, or 'Big Dipper', is the luxury all-sleeper train that connects Sapporo and Tokyo {see planets (3473) and (498)}, a 1000-km distance in 16 hours. The train passes through Seikan, the longest tunnel in the world. (M 26930)

(5375) Siedentopf

1989 AN_6. Discovered 1989 January 11 by F. Börngen at Tautenburg.

Named in memory of Heinrich Siedentopf (1906-1963), professor of astronomy and director of the observatories in Jena (1933-1945) and in Tübingen (1949-1963). His research activities covered stellar atmospheres, in particular the theory of convection and solar granulation, interplanetary matter, optics of the earth's atmosphere, and astronomical instruments. He introduced the iris-diaphragm method for photographic stellar photometry. Together with O. Heckmann and W. Fricke {see planets (1650) and (1561), respectively} he was one of the German promoters of the European Southern Observatory. (M 21611)
Citation prepared by H. Elsässer at the request of the discoverer.
Siedentopf is also honored by a lunar crater.

(5376) 1990 DD

Discovered 1990 February 16 by S. Ueda and H. Kaneda at Kushiro.

(5377) Komori

1991 FM. Discovered 1991 March 17 by S. Otomo and O. Muramatsu at Kiyosato.

Named in honor of Yukimasa Komori, owner of the Astro-Dome Company and a committee member of the Gotoh Planetarium and Astronomical Museum. Born in 1900, he is the oldest known amateur astronomer in Japan, and he had the pleasure of watching Halley's Comet in both 1910 and 1986. His main interests are in observing lunar occultations and solar eclipses. Long active in the popularization of astronomy, particularly over Japanese national radio, he planned and carried out a minute-by-minute broadcast of the total solar eclipse in 1936, linking observers at various stations. (M 21957)

Name proposed by the discoverers following suggestions by S. Kimura and E. Kobayashi.

(5378) Ellyett

1991 GD. Discovered 1991 April 9 by R. H. McNaught at Siding Spring.

Named in honor of Clifton Darfield Ellyett (1915-), pioneer of radar meteor research in New Zealand. After taking his Ph.D. at Jodrell Bank in the late 1940s, Ellyett returned to his native country and began a research program in radar meteors at what is now the University of Canterbury. In the early 1960s he carried out with Colin Keay {see planet (5007)} the important southern-hemisphere radar patrol of meteors from near Christchurch. In 1965 Ellyett left to become professor of physics in the University of Newcastle (N.S.W.), starting there a new research group in atmospheric science. (M 29144)

Name proposed by the discoverer following a suggestion by D. I. Steel, who prepared the citation.

(5379) 1991 HG

Discovered 1991 April 16 by S. Otomo and O. Muramatsu at Kiyosato.

(5380) Sprigg

1991 JT. Discovered 1991 May 7 by R. H. McNaught at Siding Spring.

Named for Reginald C. Sprigg (1919-1994), Australian exploration geologist, oceanographer, biologist, author and conservationist who in 1946 discovered the diverse pre-Cambrian Ediacaran soft-body fauna fossils representing the earliest known multicellular life on the earth. Sprigg's recognition of hydrocarbons in central Australia led to the discovery of the vast Cooper Basin gas and oil fields. His work on the formation of Australian deserts and dunes has resulted in greater understanding of present-day climate patterns. In 1968 he founded the Arkaroola Wilderness Sanctuary in the northern Flinders Ranges, and this includes an astronomical observatory. (M 31609)

Name suggested by D. I. Steel and citation prepared by the Sprigg family.

(5381) Sekhmet

1991 JY. Discovered 1991 May 14 by C. S. Shoemaker at Palomar.

The daughter of the Egyptian god Ra and wife of Ptah {see planets (2100) and (5011), respectively}, Sekhmet was a lion-headed sun goddess. Her role was that as defender of the divine order, not as creator of it. Her title was the "Mighty One", and she was a fierce goddess of war and strife and bringer of destruction to the enemies of Ra. She was considered the Eye of Ra, representing the scorching, destructive power of the sun. (M 24917)

(5382) McKay

1991 JR_2. Discovered 1991 May 8 by R. H. McNaught at Siding Spring.

Named for Christopher P. McKay (1954-), space scientist and exobiologist at NASA-Ames Research Center. McKay has made numerous seminal contributions to planetary science and the search for zones of life in the solar system, with a special interest in Mars, Titan, comets and microbial life in hostile environments on the earth. (M 31610)

Name suggested and citation prepared by D. I. Steel.

(5383) Leavitt

4293 T-2. Discovered 1973 September 29 by C. J. van Houten and I. van Houten-Groeneveld at Palomar.

Named in memory of Henriette Swan Leavitt (1868-1921) of the Harvard College Observatory, discoverer of the period-luminosity law for the classical Cepheids. (M 22509)
Leavitt is also honored by a lunar crater.

(5384) 1957 VA

Discovered 1957 November 11 by C.-H. Chang at Nanking.

(5385) 1975 TS_3

Discovered 1975 October 3 by L. I. Chernykh at Nauchnyj.

(5386) 1975 TH_6

Discovered 1975 October 1 at the Felix Aguilar Observatory at El Leoncito.

(5387) 1980 NB

Discovered 1980 July 11 at the University of Chile Observatory at Cerro El Roble.

(5388) Mottola

1981 ED_1. Discovered 1981 March 5 by H. Debehogne and G. DeSanctis at La Silla.

Named in honor of Stefano Mottola, of the Planetary Remote Sensing Section at the Deutsche Forschungsanstalt für Luft- und Raumfahrt, Wesseling. He has investigated the photometric properties of surfaces of minor planets by applying numerical computer modeling to modern CCD observations, particularly of Trojans and other objects in the outer belt. He is also an active participant in the campaigns on (951) Gaspra and (243) Ida in connection with the Galileo flybys of these objects. (M 21958)

(5389) Choikaiyau

1981 UB_{10}. Discovered 1981 October 29 at the Purple Mountain Observatory at Nanking.

Named in honor of Choi Kaiyau, honorary director of the Zhongshan Scientific Center and chairman of the Choi Educational Foundation. He made a great effort to wipe out illiteracy in China and won a high reputation for popularizing children's education. (M 25230)

(5390) 1981 YO_1

Discovered 1981 December 19 at the Purple Mountain Observatory at Nanking.

(5391) 1985 RE_2

Discovered 1985 September 13 by E. F. Helin at Palomar.

(5392) Parker

1986 AK. Discovered 1986 January 12 by C. S. Shoemaker and E. M. Shoemaker at Palomar.

This Mars-crossing Phocaea {see planet (25) for the prototype} is named in honor of the American amateur astronomer Don Parker, who specializes in high-resolution photography and CCD imaging of Mars and Jupiter. Through his precise imaging, new features have been recognized on Mars. His investigations of planetary atmospheres have improved the understanding of weather phenomena there. (M 23138)

Name proposed and citation written by P. L. Dombrowski, endorsed by D. H. Levy.

(5393) Goldstein

1986 ET. Discovered 1986 March 5 by E. Bowell at Anderson Mesa.

Named in honor of Richard M. Goldstein (1927-), radar astronomer at the Jet Propulsion Laboratory, on the occasion of the twenty-fifth anniversary of his detection of the first radar echoes from a minor planet ((1566) Icarus in June 1968). Goldstein also led observations that resulted in the first radar detections of Mars and of Saturn's rings and in the discovery of the anomalous rotation of Venus. (M 22830)

Name suggested and citation prepared by S. J. Ostro.

(5394) Jurgens

1986 EZ_1. Discovered 1986 March 6 by E. Bowell at Anderson Mesa.

Named in honor of Raymond F. Jurgens (1937-), a radar astronomer at the Jet Propulsion Laboratory, who developed the data-acquisition system used for high-resolution delay-Doppler imaging of (4179) Toutatis in 1992. Jurgens led the dual-polarization, dual-wavelength radar investigations of (433) Eros in 1975 (the first of a minor planet). He also pioneered techniques for using radar spectra to constrain small-body shapes and led the first three-station radar interferometric observations of Venus. (M 23138)

Name suggested and citation material provided by S. J. Ostro.

(5395) 1988 RK_{11}

Discovered 1988 September 14 by S. J. Bus at Cerro Tololo.

(5396) 1988 SH_1

Discovered 1988 September 20 by H. Debehogne at La Silla.

(5397) 1988 VB_5

Discovered 1988 November 14 by Y. Oshima at Gekko.

(5398) 1989 AK_1

Discovered 1989 January 13 by S. Ueda and H. Kaneda at Kushiro.

(5399) Awa

1989 BT. Discovered 1989 January 29 by M. Iwamoto and T. Furuta at Tokushima.

Named by the first discoverer for the town in which he lives in the northern part of Shikoku island {see planet (4223)}. The "Awa-odori" dancing festival, held every August, is as famous as it is unique. (M 22250)

(5400) 1989 CM

Discovered 1989 February 4 by S. Ueda and H. Kaneda at Kushiro.

(5401) Minamioda

1989 EV. Discovered 1989 March 6 by T. Nomura and K. Kawanishi and M. Sugano at Minami-Oda.

Named for the Minami-Oda district, inhabitants of which cooperated in establishing the observatory in 1971. Minami-Oda is located in the middle of Hyogo {see planet (6879)} prefecture and has the biggest hydroelectric power plant in Japan. The district has made efforts to limit light pollution. (M 22509)

(5402) Kejosmith

1989 UK_2. Discovered 1989 October 27 by E. F. Helin at Palomar.

Named in honor of Keith C. Smith and Joan Furlong of University College, London, on the occasion of their wedding. Keith is a stellar spectroscopist whose work on abundance analyses of ultraviolet and optical spectra of HgMn stars and on non-LTE effects in OB stars is well-known to his colleagues worldwide. Joan is a laboratory spectroscopist and has recently moved from UCL to the metrology

division of the U.K. National Physical Laboratory. (M 29670)
Name suggested and citation prepared by M. M. Dworetsky.

(5403) Takachiho

1990 DM. Discovered 1990 February 20 by Y. Kushida and M. Inoue at Yatsugatake.

Named for the home town of the wife of the second discoverer, Takachiho is located at the center of Miyazaki prefecture in Kyusyu, some 900 km southwest of Tokyo, and surrounded by mountains. Takachiho is famous for its legends and myths on the root of gods. The most famous is the legend of Amano-Iwato of Amaterasu-Ohmikami (the god of the sun). This story has been handed down by Yokagura (sacred music and dance) as performed by farmers. (M 22250)

(5404) Uemura

1991 EE_1. Discovered 1991 March 15 by K. Endate and K. Watanabe at Kitami.

Named in honor of Naomi Uemura (1941-1984), a famous explorer, born in Hidaka, Hyogo {see planet (6879)} Prefecture. He was the first to climb the highest mountains in five continents and the first Japanese to reach the summit of Mount Everest. From 1976 to 1978 he devoted himself to exploration of the north polar regions and Greenland by dogsled. He disappeared in an attempt to reach the summit of Mount McKinley in 1984. (M 22509)

(5405) Neverland

1991 GY. Discovered 1991 April 11 by Y. Kushida and O. Muramatsu at Yatsugatake.

Named after the world famous play *Peter Pan*, written by Sir James M. Barrie (1860-1937) and later adapted into an animated film by Walt Disney {see planet (4017)}. "Never Land" is a fantastic island floating among stars, where *Peter Pan*, Tinker Bell, and many others dwell. "*Peter Pan*" has enthralled generations of children and the young at heart throughout the world. (M 22509)

(5406) 1991 PH_{11}

Discovered 1991 August 9 by H. E. Holt at Palomar.

(5407) 1992 AX

Discovered 1992 January 4 by S. Ueda and H. Kaneda at Kushiro.

(5408) Thé

1232 T-1. Discovered 1971 March 25 by C. J. van Houten and I. van Houten-Groeneveld at Palomar.

Named in honor of Pik Sin Thé (1927-) on the occasion of his 65th birthday. Pik Sin Thé studied in the US, became director of Lembang Observatory, Indonesia, and later professor at the University of Amsterdam. His main field of interest is young variable stars, on which he has done much important research. (M 22509; M 22519)

(5409) Saale

1962 SR. Discovered 1962 September 30 by F. Börngen at Tautenburg.

Named for the river that originates in the Fichtelgebirge of Bavaria, runs not far from Tautenburg {see planet (2424)}, crosses the cities of Jena {see planet (526)} and Halle (the latter the birthplace of the discoverer) and flows into the Elbe. (M 22250)

(5410) Spivakov

1967 DA. Discovered 1967 February 16 by T. M. Smirnova at Nauchnyj.

Named in honor of the outstanding Russian musician, violinist Vladimir Theo-

dorovich Spivakov (1944-), leader of the State Chamber Orchestra "Virtuosos of Moscow" since 1979, laureate of many international competitions, artistic director of the International Musical Festival in Colmar, France, and president of the International Charitable Fund. (M 24410)
Name proposed by the Institute of Theoretical Astronomy.

(5411) Liia

1973 AT_3. Discovered 1973 January 2 by N. S. Chernykh at Nauchnyj.
Named in honor of Liia Forrer-Tsiganovskaja, wife of a friend of the discoverer, on the occasion of her 25th birthday. A student of art at Saratov and of philosophy at St. Gallen, she also takes an interest in literature and architecture. (M 30097)

(5412) Rou

1973 SR_3. Discovered 1973 September 25 by L. V. Zhuravleva at Nauchnyj.
Named in memory of Aleksandr Arturovich Rou (1906-1973), an actor and film producer specializing in Russian fairy tales. (M 34621)

(5413) Smyslov

1977 EC_2. Discovered 1977 March 13 by N. S. Chernykh at Nauchnyj.
Named in honor of Vasilij Vasil'evich Smyslov (1921-), Soviet chess grandmaster and world chess champion during 1957-1958. He is known also as a gifted amateur singer, performing operatic arias and Russian folk songs. (M 30097)

(5414) Sokolov

1977 RW_6. Discovered 1977 September 11 by N. S. Chernykh at Nauchnyj.
Named in honor of Viktor Georgievich Sokolov (1946-), specialist in analytical methods of celestial mechanics and scientific secretary of the Institute of Theoretical Astronomy. His scientific and organizational work is an essential constituent of the Institute's activity and contributes significantly to its successful operation. (M 29145)
Name proposed by the discoverer following a suggestion by the Institute of Theoretical Astronomy.

(5415) Lyanzuridi

1978 TB_2. Discovered 1978 October 3 by N. S. Chernykh at Nauchnyj.
Named in honor of Konstantin Petrovich Lyanzuridi (1934-), engineer in vacuum technology and optics who has worked at the Crimean Astrophysical Observatory {see planet (1725)} for more than 30 years. A talented experimenter, he created a new method of ion etching of optical surfaces for making precision astronomical optics for ground-based and space telescopes. (M 34341)

(5416) 1978 VE_5

Discovered 1978 November 7 by E. F. Helin and S. J. Bus at Palomar.

(5417) Solovaya

1981 QT. Discovered 1981 August 24 by L. Brožek at Kleť.
Named in honor of Nina A. Solovaya (1940-), an astronomer in the celestial mechanics department of the Sternberg State Astronomical Institute in Moscow. Her general field of work involves analytical and qualitative methods of celestial mechanics, especially the three-body problem. She has applied the results to triple-star systems, to planetary satellites and to minor planets in the outer part of the main belt. (M 29145)
Name proposed by the Kleť Observatory following a suggestion by E. M. Pittich.

(5418) Joyce

1981 QG_1. Discovered 1981 August 29 by A. Mrkos at Kleť.

Named in memory of James Joyce (1882-1941), Irish novelist noted for his experimental use of language and exploration of new literary methods in large works, such as *Ulysses*. Gell-Mann adopted the fanciful term 'quark' from a passage in his novel *Finnegans Wake*. (M 34621)
Name suggested by J. Tichá.

(5419) Benua

1981 SW_7. Discovered 1981 September 29 by L. V. Zhuravleva at Nauchnyj.
Named in memory of the Russian architect Nikolaj Leont'evich Benua (Benois; 1813-1898), as well as of his sons, the architect Leontij Nikolaevich Benua (1856-1928) and the painter Aleksandr Nikolaevich Benua (1870-1960). (M 34621)

(5420) 1982 JR_1

Discovered 1982 May 15 by E. F. Helin and E. M. Shoemaker and P. D. Wilder at Palomar.

(5421) Ulanova

1982 TD_2. Discovered 1982 October 14 by L. G. Karachkina and L. V. Zhuravleva at Nauchnyj.
Named in honor of the brilliant Russian ballet-dancer Galina Sergeevna Ulanova (1910-), whose picturesque dancing in the ballets *Giselle, Swan Lake* and others are unmatched. (M 24122)

(5422) Hodgkin

1982 YL_1. Discovered 1982 December 23 by L. G. Karachkina at Nauchnyj.
Named for the famous English chemist and biochemist Dorothy Crowfoot-Hodgkin, 1964 Nobel prize-winner. (M 22509)

(5423) 1983 DC

Discovered 1983 February 16 by Z. Vávrová at Kleť.

(5424) Covington

1983 TN_1. Discovered 1983 October 12 by E. Bowell at Anderson Mesa.
Named in honor of Arthur Covington (1913-), Canada's first radio astronomer. His discovery, during the partial solar eclipse of 1946 Nov. 23, that microwave emission was far more intense from the vicinity of sunspots than elsewhere on the sun, was the first indicator that magnetic fields were important in the generation of nonthermal cosmic radio emission. In 1947 Covington inaugurated at the National Research Council of Canada daily measurements of the solar microwave flux at 10.7 cm. (M 23541)
Name suggested and citation prepared by C. J. Cunningham.

(5425) Vojtěch

1984 SA_1. Discovered 1984 September 20 by A. Mrkos at Kleť.
Named in memory of Václav Vojtěch (1901-1932), Czech traveler and Antarctic explorer. In 1929 he became the first Czechoslovak to step on the Antarctic continent. This minor planet commemorates the seventieth anniversary of his participation with the Byrd expedition. (M 34621)
Name suggested by J. Tichá and M. Tichý.

(5426) Sharp

1985 DD. Discovered 1985 February 16 by C. S. Shoemaker and E. M. Shoemaker at Palomar.
Named for Robert P. Sharp, emeritus professor of geology at the California Institute of Technology. A leading American geomorphologist, Sharp is best known for his work on glaciers and the movement of desert sand. He played a prominent

role in the interpretation of the surface features of Mars from images returned by early spacecraft missions. As chairman for 15 years he built the Division of Geological Sciences at Caltech into one of the finest research groups of its kind. (M 24917)

(5427) 1986 JQ

Discovered 1986 May 13 at the Copenhagen University Observatory at Brorfelde.

(5428) 1987 RA_1

Discovered 1987 September 13 by H. Debehogne at La Silla.

(5429) 1988 BZ_1

Discovered 1988 January 25 by S. Ueda and H. Kaneda at Kushiro.

(5430) Luu

1988 JA_1. Discovered 1988 May 12 by C. S. Shoemaker and E. M. Shoemaker at Palomar.

Named in honor of Jane X. Luu (1963-) for her research on the small bodies of the solar system. Luu is best known for her work with David Jewitt in discovering the first and subsequent members of the Kuiper belt, as well as in following up with physical studies of those bodies. She has also contributed the most stringent upper limits on the existence of dusty comae around minor planets that might be dormant or extinct comets. (M 27459)

Citation provided by M. F. A'Hearn following a suggestion by E. Bowell and endorsed by the discoverers.

(5431) Maxinehelin

1988 MB. Discovered 1988 June 19 by E. F. Helin at Palomar.

Named in honor of Maxine Anne Helin, mother-in-law of the discoverer, on the occasion of the 90th anniversary of her birth. A resident of Colorado until 1928, she moved to California in search of a more hospitable climate. During her early years an occupational/psychiatric therapist, she has relied on that experience to provide the wisdom to help guide her family and friends along the trail of life. (M 26425)

(5432) Imakiire

1988 VN. Discovered 1988 November 3 by T. Kojima at Chiyoda.

Named in honor of Kyoko Imakiire (1965-), a Kagoshima yachtswoman who made a solo round-trip voyage between Kagoshima {see planet (4703)} and San Francisco in 1988 and a solo non-stop voyage round the world in 1991-1992. (M 26763)

Name proposed by the discoverer following a suggestion by T. Sato.

(5433) Kairen

1988 VZ_2. Discovered 1988 November 10 by T. Kojima at Chiyoda.

Named for the yacht in which Kyoko Imakiire {see planet (5432)} made her solo non-stop voyage round the world in 1991-1992. 'Kai' means 'Sea' or 'Ocean' and 'Ren' means 'Connection', Kairen being a yacht that therefore connects the seas and oceans of the world. Kairen also symbolizes Imakiire's parents, the name of her mother Umiko meaning 'a child of the ocean' and the Chinese character for the name of her father Muraji also being pronounced as 'Ren'. (M 26763)

Name proposed by the discoverer following a suggestion by T. Sato.

(5434) 1989 ES

Discovered 1989 March 6 by E. F. Helin at Palomar.

(5435) Kameoka

1990 BS_1. Discovered 1990 January 21 by A. Sugie at Taga.

Named for a city with a castle and a good deal of natural beauty, situated just west of Kyoto {see planet (4352)}. (M 24917)
Name proposed by S. Sakabe.

(5436) Eumelos

1990 DK. Discovered 1990 February 20 by C. S. Shoemaker and E. M. Shoemaker at Palomar.

Eumelos was a charioteer who had the swiftest horses in the Greek army. When the riders gathered for the funeral games for Patroclus {see planet (617)}, Eumelos was the first to rise up. He competed against Diomedes, Menelaus, Antilochus and Meriones {see planets (1437), (1647), (1583), and (3596), respectively}. He was in the lead until the goddess Athena {see planet (881)} smashed the yoke of his chariot, spinning him out. Though he came in last, Achilles {see planet (588)} awarded Eumelos the bronze corselet stripped from Asteropaios {see planet (4805)}, and he accepted it joyfully. (M 25444)

(5437) 1990 DU_3

Discovered 1990 February 26 by H. Debehogne at La Silla.

(5438) Lorre

1990 QJ. Discovered 1990 August 18 by E. F. Helin at Palomar.

Named in honor of Jean Lorre, one of the original members of the Image Processing Laboratory at the Jet Propulsion Laboratory. He has been working on JPL flight projects since 1971 and was the first to apply image processing techniques to astronomical images outside of the NASA flight projects. His expertise encompasses map projections, correlation, photometric functions, color reproduction, camera calibration, SAR, stereo, morphing and filtering. His latest astronomical project has been to optimize the NEAT processing algorithms. The NEAT team whiches to acknowledge his important contributions in software development. (M 29670)

(5439) 1990 RW

Discovered 1990 September 14 by H. E. Holt at Palomar.

(5440) Terao

1991 HD. Discovered 1991 April 16 by A. Sugie at Taga.

Named in memory of Hisashi Terao (1855-1923), who was the first Japanese professor of astronomy (1884-1919) at Tokyo Imperial University, the first director (1888-1919) of the Tokyo Astronomical Observatory and the first president of the Astronomical Society of Japan. Soon after he finished the course of physics at the newly-established Tokyo University he was sent to Paris to study celestial mechanics under Tisserand {see planet (3663)}, his main interest being in transits of Venus. He was also one of the founders of Tokyo Science University and later served as its president. (M 25653)

(5441) 1991 JZ_1

Discovered 1991 May 8 by R. H. McNaught at Siding Spring.

(5442) Drossart

1991 NH_1. Discovered 1991 July 12 by H. E. Holt at Palomar.

Named in honor of Pierre Drossart (1956-), researcher of the CNRS at Paris-Meudon Observatory. Drossart is a specialist on the atmospheres of the giant planets and has worked on the interpretation of the infrared spectra of Jupiter from Voyager and groundbased observations (Jupiter H_2O measurements, first

detection of H_3^+ on Jupiter, observations of aurorae, etc.). He has been involved as a co-investigator in several space-mission imaging spectrometers used to study Mars and the giant planets. (M 27459)
Name suggested and citation provided by M. A. Barucci.

(5443) Encrenaz

1991 NX_1. Discovered 1991 July 14 by H. E. Holt at Palomar.

Named in honor of Thérèse Encrenaz (1946-), director of the Department of Space Research at the Paris Observatory. Encrenaz is an expert on planetary atmospheres, in particular the study of the chemical composition of planetary and cometary atmospheres, mostly by means of infrared and millimeter spectroscopy. Her main achievements include determining abundance ratios in the giant planets, the detection of several planetary minor species and the detection of parent molecules in comet 1P/Halley. She has been a co-investigator on several space missions, and she is Mission Scientist on the ISO mission. (M 27459)
Citation provided by M. A. Barucci following a suggestion by E. Bowell.

(5444) Gautier

1991 PM_8. Discovered 1991 August 5 by H. E. Holt at Palomar.

Named in honor of Daniel Gautier (1936-), senior scientist in the Department of Space Research at the Paris Observatory. Gautier has played an important role in the development of European planetary space research. His work concerns the origin of the solar system through the study of planetary atmospheres. He has played a particular crucial role in promoting planetary science in France, and his enthusiasm has contributed much to the French planetary community's present-day level of excellence. (M 27459)
Citation provided by M. A. Barucci following a suggestion by E. Bowell.

(5445) Williwaw

1991 PA_{12}. Discovered 1991 August 7 by H. E. Holt at Palomar.

Named after a dramatic mountain on the skyline of Anchorage. A williwaw is a sudden seaward gust of cold air common along mountainous coasts at high latitudes. Mt. Williwaw was named by the Mountaineering Club of Alaska in 1963 and stands 5445 feet above sea level. It is the highest point in the Campbell Creek drainage. (M 34341)
Name suggested by D. Hamilton.

(5446) 1991 PB_{13}

Discovered 1991 August 5 by H. E. Holt at Palomar.

(5447) 1991 PO_{14}

Discovered 1991 August 6 by H. E. Holt at Palomar.

(5448) Siebold

1992 SP. Discovered 1992 September 26 by A. Sugie at Taga.

Named in memory of Philipp Franz Balthasar von Siebold (1796-1866), who went to Japan as a ship's doctor in the Dutch East India Company in 1823 and remained there for six years, studying in particular the natural history and ethnography of the country in an unprecedented manner. On returning to Europe he published this research. (M 27329)

(5449) 1992 US_5

Discovered 1992 October 28 by S. Ueda and H. Kaneda at Kushiro.

(5450) Sokrates

2780 P-L. Discovered 1960 September 24 by C. J. van Houten and I. van Houten-Groeneveld at Palomar.

Named after the Greek philosopher Sokrates (c. 470-399 B.C.), who taught that one should always say the truth. He was forced to drink hemlock, because the state was afraid that Sokrates influenced the youth and denied the official religion. As a good citizen he took the cup in the presence of all his pupils, without protesting against the government. As he never wrote anything down, all we know about Sokrates is from his pupil Plato {see planet (5451)}. (M 22509)

(5451) Plato

4598 P-L. Discovered 1960 September 24 by C. J. van Houten and I. van Houten-Groeneveld at Palomar.

Named after Plato (c. 428-348 B.C.), the most famous pupil of Sokrates {see planet (5450)}. He was the founder of the philosophical Academy near Athens, which existed for over 900 years. His preserved papers were written in dialogue form as the 'lessons' of Sokrates. His philosophy has strongly influenced the western world to the present day. (M 22509)
Plato is also honored by a lunar crater.

(5452) 1937 NN

Discovered 1937 July 5 by C. Jackson at Johannesburg.

(5453) Zakharchenya

1975 VS_5. Discovered 1975 November 3 by T. M. Smirnova at Nauchnyj.

Named in honor of academician Boris Petrovich Zakharchenya (1928-), outstanding Russian scientist, director of a department in the Ioffe Physical and Technical Institute in St. Petersburg. Zakharchenya is well known for his investigations in the fields of solid state physics, magneto-optics and optical orientation of electrons and nuclei in semiconductors. (M 32788)
Name suggested by the Institute of Applied Astronomy.

(5454) Kojiki

1977 EW_5. Discovered 1977 March 12 by H. Kosai and K. Hurukawa at Kiso.

Named for the first book of historical stories written in Japanese. Completed in 712, it contains myths, legends and historical accounts of the Imperial court from the earliest days of the creation up to 628. *Kojiki* is based on the oral shaman tradition and was written in the Chinese characters used to represent Japanese phonetics. Together with *Nihonsyoki* {see planet (5082)}, it is an important source book for the mythology, history and ethnology of ancient Japan. (M 22509)

(5455) Surkov

1978 RV_5. Discovered 1978 September 13 by N. S. Chernykh at Nauchnyj.

Named in honor of Vladimir Vasil'evich Surkov (1945-), well-known Russian specialist on databases and staff member of the Moscow Aviation Institute. (M 30476)

Name proposed by the discoverer following a suggestion by the Institute of Theoretical Astronomy.

(5456) Merman

1979 HH_3. Discovered 1979 April 25 by N. S. Chernykh at Nauchnyj.

Named in honor of Grigorij (Hirsh) Aronovich Merman (1921-), staff member of the Institute of Theoretical Astronomy from 1951 to 1985. Merman has obtained important results in different areas of celestial mechanics. His work on the three-body problem involved the construction of a rigorous theory of capture, establishment of criteria for the feasibility of various types of end motions and derivation of a general solution in the form of a convergent series of polynomials. He also investigated the stability of canonical equations and estimated the remainders of the infinite series used in celestial mechanics. (M 29145)

Name proposed by the discoverer following a suggestion by the Institute of Theoretical Astronomy.

(5457) Queen's

1980 TW_5. Discovered 1980 October 9 by C. S. Shoemaker at Palomar.

Named in honor of Queen's University at Kingston, Ontario. Founded by Royal Charter in 1841, its first classes were held the following spring. In more than 150 years Queen's has evolved to become one of Canada's strongest universities in many fields. The home of A. Vibert Douglas {see planet (3269)}, one of Canada's pioneering astronomers, Queen's has had a long tradition of support to professional and amateur astronomical groups. (M 24917)

Name proposed and citation prepared by D. H. Levy.

(5458) Aizman

1980 TB_{12}. Discovered 1980 October 10 by N. S. Chernykh at Nauchnyj.

Named in honor of Mikhael Iosifovich Aizman (1947-), telecommunications specialist and president of MTU-INFORM. One of the largest systems for telephone communication and data transfer in Russia, this company was developed under his supervision in Moscow. Support from MTU-INFORM enabled ITA to create a communication channel allowing the integration of the ITA computer system in the World Wide Web. (M 31023)

Name proposed by the discoverer following a suggestion by ITA.

(5459) Saraburger

1981 QP_3. Discovered 1981 August 26 by H. Debehogne at La Silla.

Named in honor of Sara Schöffer-Burger {1894- }, who celebrated her 100th birthday on 1994 March 1. She provided in her house in Amsterdam a hide-out for Jewish compatriots who had to go underground during the German occupation of The Netherlands during World War II. She and her husband were awarded the Jad. Vashem distinction. (M 24410)

Name proposed by the discoverer following a suggestion by Uccle astronomer Marijke Burger, niece of the honoree.

(5460) Tsénaat'a'í

1983 AW. Discovered 1983 January 12 by B. A. Skiff at Anderson Mesa.

This name, 'Flying Rock' or 'Rock Which Flies' in the Navajo language, signifies the motion of the minor planet through space and gives recognition to the contribution of the Navajo culture and language to northern Arizona. (M 28621)

Name suggested by Derekson Bert, eighth-grade student at Rocky Ridge School in Dinnebito, Arizona, winner of a contest to name this object in conjunction with the 1996 Flagstaff Festival of Science.

(5461) 1983 HB_1

Discovered 1983 April 18 by N. G. Thomas at Anderson Mesa.

(5462) 1984 SX_5

Discovered 1984 September 21 by H. Debehogne at La Silla.

(5463) Danwelcher

1985 TO. Discovered 1985 October 15 by E. Bowell at Anderson Mesa.

Named in honor of Dan Welcher (1948-), American composer and conductor, on the occasion of the first performance of his Symphony No. 2 for large orchestra by the Flagstaff Symphony Orchestra on 1994 Nov. 9. Subtitled *Night Watchers*, the four-movement work marks the centennial of the founding of Lowell Observatory and of the city of Flagstaff. Welcher was trained as a pianist and bassoonist and became principal bassoonist of the Louisville Orchestra. From 1990 to 1992, he was composer in residence of the Honolulu Symphony. Welcher's wide-ranging compositions encompass orchestral, chamber, instrumental and vocal works, and he specializes in conducting the music of living American composers. He is professor of composition at the University of Texas at Austin. (M 24122)

(5464) Weller

1985 VC_1. Discovered 1985 November 7 by E. Bowell at Anderson Mesa.

Named in honor of Harold Weller (1941-). Conductor of the Flagstaff Symphony Orchestra since 1981, Weller received his musical training at the Oberlin and Cincinnati Conservatories of Music, Miami University and Ohio State University. He has held positions as musical director of the Hamilton (Ohio) and Ashland (Ohio) Symphony Orchestras and has been a faculty member at Ashland College in Ohio and Old Dominion University in Norfolk, Virginia. (M 24122)

(5465) Chumakov

1986 RF_{13}. Discovered 1986 September 9 by L. G. Karachkina at Nauchnyj.

Named in honor of the well-known Russian microbiologist Mikhail Petrovich Chumakov (1909-), who founded an institute for the study of poliomyelitis and tick encephalitis. (M 22509)

(5466) Makibi

1986 WP_8. Discovered 1986 November 30 by H. Kosai and K. Hurukawa at Kiso.

Named for Kibi-no Makibi (695-775), a Japanese scholar and statesman, who travelled to T'ang-dynasty China to study in 717. Upon his return to Japan 17 years later, he presented many books on various fields (including astronomy and surveying) to the Japanese government and introduced Chinese technology into Japan. He subsequently served as an ambassador to China and also achieved the high post of minister. (M 22509)

(5467) 1988 AG

Discovered 1988 January 11 by T. Hioki and N. Kawasato at Okutama.

(5468) Hamatonbetsu

1988 BK. Discovered 1988 January 16 by M. Mukai and M. Takeishi at Kagoshima.

Located near the northern extremity of Japan, Hamatonbetsu is the town where the second discoverer lives. The lake and swamp around the town are very important for waterfowl, and this site has been registered with the Convention on Wetlands of International Importance Especially as Waterfowl Habitat (Ramusar Convention). The name also honors the inhabitants of the town, who have supported the astronomical activities of the discoverers. (M 22509; M 22519)

(5469) 1988 BK_4

Discovered 1988 January 21 by H. Debehogne at La Silla.

(5470) 1988 BK_5

Discovered 1988 January 28 by R. H. McNaught at Siding Spring.

(5471) Tunguska

1988 PK_1. Discovered 1988 August 13 by E. W. Elst at St. Michel.

Named for the site, in Siberia, of the presumed impact of a 60-meter minor planet on the 85th anniversary of the great explosion that occurred there on 1908 June 30. (M 22250)

(5472) 1988 RR

Discovered 1988 September 13 by S. Ueda and H. Kaneda at Kushiro.

(5473) Yamanashi

1988 VR. Discovered 1988 November 5 by Y. Kushida and O. Muramatsu at Yatsugatake.

Named for the prefecture in which Yatsugatake South Base Observatory and many other amateur astronomical observatories are located. This prefecture is in central Japan about 150 km west of Tokyo and is famous for Mt. Fuji {see planet (1584)} and the mountain's five lakes. This prefecture promotes environmental protection. (M 22510)

(5474) Gingasen

1988 XE_1. Discovered 1988 December 3 by T. Fujii and K. Watanabe at Kitami.

Gingasen, or 'Milky Way', is a railroad track in Hokkaido {see planet (3720)}. This 150-km public railroad connects the island's eastern cities. Each station along the line is named for a constellation. (M 26930)

(5475) 1989 QO

Discovered 1989 August 26 by R. H. McNaught at Siding Spring.

(5476) 1989 TO_{11}

Discovered 1989 October 2 by S. J. Bus at Cerro Tololo.

(5477) 1989 UH_2

Discovered 1989 October 27 by E. F. Helin at Palomar.

(5478) Wartburg

1989 UE_4. Discovered 1989 October 23 by F. Börngen at Tautenburg.

Named for a castle on a hill near the town of Eisenach, a jewel of the German territories, earlier the dwelling-place of the Thuringian landgraves and of Saint Elisabeth (1211-1227). In a special way, Wartburg was the scene of historical events in the past. In 1521-1522 Martin Luther there translated the New Testament into German, defended from his pursuers. In 1817 German students came together to express their striving for unity and liberty. In 1845 Wagner {see planet (3992)} digested in the opera *Tannhäuser* the contest of the minnesingers in the late Romanic palace of Wartburg. (M 22250)

(5479) 1989 UT_5

Discovered 1989 October 30 by S. J. Bus at Cerro Tololo.

(5480) 1989 YK_8

Discovered 1989 December 23 by S. Ueda and H. Kaneda at Kushiro.

(5481) Kiuchi

1990 CH. Discovered 1990 February 15 by K. Endate and K. Watanabe at Kitami.

Named in honor of Tsuruhiko Kiuchi (1954-), Japanese amateur astronomer, who recovered the Perseid comet P/Swift-Tuttle {see planets (5035) and (5036)} with binoculars on 1992 Sept. 26 in accordance with a prediction by B. G. Marsden. Kiuchi also co-discovered two new comets in 1990. He is employed in the manufacture of motor parts and is on the staff of a community school for star lovers. (M 22510)

Name suggested by I. Hasegawa.

(5482) 1990 DX

Discovered 1990 February 27 by K. Suzuki and T. Urata at Toyota.

(5483) 1990 UQ_{11}

Discovered 1990 October 17 by L. I. Chernykh at Nauchnyj.

(5484) Inoda

1990 VH_1. Discovered 1990 November 7 by T. Urata at Oohira.

Named in honor of Shigeru Inoda (1955-), ophthalmic vitreo-retinal surgeon and associate professor in the ophthalmology department of the Jichi Medical School. He is one of the most active amateur astronomers in the minor planet survey at Karasuyama Astronomical Observatory and is the discoverer of several minor planets. He has been an technical adviser at the Nihondaira {see planet (2880)} Observatory. (M 22510)

(5485) Kaula

1991 RQ_{21}. Discovered 1991 September 11 by H. E. Holt at Palomar.

Named in honor of William M. Kaula (1926-), professor of earth and space science at the University of California at Los Angeles. Kaula pioneered the use of artificial satellites in geodesy, producing global gravity maps of the earth, the moon, Venus and Mars. He has used the gravity, cratering record and surface topography of Venus to infer current and historical tectonics and properties of the interior. He also played an active role in establishing the Apollo Lunar Laser Ranging Experiment and has had a longstanding and wide-ranging interest in the origin and evolution of the solar system. (M 27359)

Name suggested and citation prepared by A. W. Harris, with help from S. J. Peale.

(5486) 1991 UT_2

Discovered 1991 October 31 by S. Ueda and H. Kaneda at Kushiro.

(5487) 1991 UM_4

Discovered 1991 October 18 by S. Ueda and H. Kaneda at Kushiro.

(5488) Kiyosato

1991 VK_5. Discovered 1991 November 13 by S. Otomo at Kiyosato.

Named for the town where the discoverer lives and his observatory is located. Kiyosato is famous for its natural beauty and is located on the southern slope of the Yatsugatake mountains, about 150 km west of Tokyo. Millions of visitors enjoy their vacations in Kiyosato, play sports and go picnicking and camping. (M 22250)

(5489) Oberkochen

1993 BF_2. Discovered 1993 January 17 by Y. Kushida and O. Muramatsu at Yatsugatake.

Named for the town in Germany in which the Zeiss {see also planet (851)} Optical works are located. This minor planet was discovered on the night that two Zeiss engineers visited the {Yatsugatake South Base} observatory. (M 22510)

(5490) Burbidge

2019 P-L. Discovered 1960 September 24 by C. J. van Houten and I. van Houten-Groeneveld at Palomar.

Named in honor of E. Margaret Burbidge, astronomer at the University of California at San Diego. She is well-known for her investigations of galaxies and quasars. From 1972-73 she served as head of the Royal Greenwich Observatory, the only woman director in its 300-year history. (M 22510)

(5491) Kaulbach

3128 T-1. Discovered 1971 March 26 by C. J. van Houten and I. van Houten-Groeneveld at Palomar.

Named after Wilhelm von Kaulbach (1805-1874), German painter of portraits, historical scenes and natural scenes of country life. From 1849 he was for many years director of the Art Academy of Munich. (M 22510)

(5492) Thoma

3227 T-1. Discovered 1971 March 26 by C. J. van Houten and I. van Houten-Groeneveld at Palomar.

Named after Hans Thoma (1839-1924), German painter of mostly peaceful scenes depicting people in harmony with nature. He also painted many portraits. After a visit to Paris he was influenced by Courbet. Thoma worked in Munich, Frankfurt and Karlsruhe, where he was director of the art academy and the art museum. (M 22510)

(5493) Spitzweg

1617 T-2. Discovered 1973 September 24 by C. J. van Houten and I. van Houten-Groeneveld at Palomar.

Named after Carl Spitzweg (1808-1885), German painter of the Biedermeier art period. His anecdotal and humoristic paintings mostly criticise his time through fine irony. Spitzweg travelled a lot through Europe, and his later work shows the influence of the coming impressionists. (M 22510)

(5494) 1933 UM_1

Discovered 1933 October 19 by K. Reinmuth at Heidelberg.

(5495) Rumyantsev

1972 RY_3. Discovered 1972 September 6 by L. V. Zhuravleva at Nauchnyj.

Named in memory of the Russian count Nikolaj Petrovich Rumyantsev (1754-1826), the Minister for Foreign Affairs and President of the State Council in Russia from 1807 to 1814. He amassed great collections of books, manuscripts, ethnographic and numismatic materials that became basis for Moscow's Rumyantsev museum, inaugurated in 1862. The book collection later became the foundation of the Russian State Library. (M 30476)

(5496) 1973 NA

Discovered 1973 July 4 by E. F. Helin at Palomar.

(5497) 1975 SS

Discovered 1975 September 30 by S. J. Bus at Palomar.

(5498) Gustafsson

1980 FT_3. Discovered 1980 March 16 by C.-I. Lagerkvist at La Silla.

Named in honor of Bengt Gustafsson, professor of theoretical astrophysics at and director of the Uppsala Astronomical Observatory, on the completion of his first half centennial. (M 22830)

(5499) 1981 SU_2

Discovered 1981 September 29 at the Haute Provence Observatory at St. Michel.

(5500) Twilley

1981 WR. Discovered 1981 November 24 by E. Bowell at Anderson Mesa.

Named in honor of Royston C. Twilley (1914-), primary school teacher of the discoverer in Tooting, London, during 1954-1955. Twilley was the most inspiring of teachers, leading his pupils not only through the required curriculum, but also into areas of academic discipline not usually encountered by pre-teenagers. Especially noteworthy was a thorough exploration of the past and contemporaneous River Wandle, a short river nowadays flowing mostly underground through south

London that in the eighteenth century sported a wide variety of industrial mills, a few of which were still in operation in the 1950s. During 1959 to 1977, Twilley was headmaster of two London schools, and after retiring he served as a School Governor of a Dorset primary school (1993-1994). (M 25976)

Citation material provided by B. Twilley. Name endorsed by M. C. Bradley, S. J. Duffell and A. J. Sissons.

(5501) 1982 FF_2

Discovered 1982 March 30 by L. G. Taff at Socorro.

(5502) Brashear

1984 EC. Discovered 1984 March 1 by E. Bowell at Anderson Mesa.

Named in memory of John A. Brashear (1840-1920), maker of astronomical telescopes and scientific instruments, popularizer of astronomy and university administrator. Brashear contributed much to the siting, design and fundraising for the Allegheny Observatory {see also planet (457)}, and his firm constructed its 0.76-m refractor and 0.79-m Keeler reflector. He figured the 0.4-m photographic doublet with which Max Wolf {see planets (827) and (1217), respectively} discovered many minor planets. He also had connections with the Lowell Observatory, notably in the design and construction of a fast spectrograph that was ultimately used to record the first recessional velocities of galaxies, and through the provision of a 0.13-m objective used in Lowell's {see planet (1886)} first search for a transneptunian planet. (M 25977; M 26439)

Citation provided by T. P. Kohman and H. L. Giclas following a suggestion by Kohman on behalf of the Amateur Astronomer Association of Pittsburgh.

See also the citations for planets (457) and (484). Brashear is also honored by craters on Mars and the Moon.

(5503) 1985 CE_2

Discovered 1985 February 13 by H. Debehogne at La Silla.

(5504) Lanzerotti

1985 FC_2. Discovered 1985 March 22 by E. Bowell at Anderson Mesa.

Named in honor of Louis J. Lanzerotti, American space physicist, on the occasion of his completion of six years as chair of the U.S. National Academy of Sciences Space Studies Board. Lanzerotti helped lay the experimental foundation of our understanding of energetic particles and waves in the Van Allen Belts and in the earth's magnetosphere, and his measurements of solar energetic particles elucidated their propagation in the interplanetary medium and solar corona. His experiments on proton bombardment of ices were applied to spectroscopic characteristics of interstellar grains, comet nuclei, asteroids and icy satellites in planetary magnetospheres, and he showed that the dark color of some of these objects resulted from interactions with particulate radiation. (M 23792)

Name suggested and citation prepared by M. S. Allen.

(5505) 1986 VD_1

Discovered 1986 November 6 by P. Jensen at Brorfelde.

(5506) 1987 SV_{11}

Discovered 1987 September 24 by H. Debehogne at La Silla.

(5507) Niijima

1987 UJ. Discovered 1987 October 21 by K. Suzuki and T. Urata at Toyota.

Named in honor of Tsuneo Niijima (1955-), co-discoverer of periodic comet Urata-Niijima {see minor planet (3722) for the co-discoverer} and discoverer of several minor planets. (M 22510)

Name proposed by T. Urata.

(5508) 1988 EB

Discovered 1988 March 9 by W. Kakei and M. Kizawa and T. Urata at Oohira.

(5509) Rennsteig

1988 RD_3. Discovered 1988 September 8 by F. Börngen at Tautenburg.

Named for a narrow forest road, 168 km in length, that crosses the ridge-like hill of the chain of mountains of the Thuringian Forest. It is a favorite place for wanderers. (M 22510)

(5510) 1988 RF_7

Discovered 1988 September 2 by H. Debehogne at La Silla.

(5511) Cloanthus

1988 TH_1. Discovered 1988 October 8 by C. S. Shoemaker and E. M. Shoemaker at Palomar.

Racing the sea-blue Scylla {see planet (155)}, Cloanthus was the winner of the boat race at the funeral games for Anchises {see planet (1173)}. Although the weight of the ship made it slower, Cloanthus called upon the gods of the sea to help him and came in first. (M 25444)

(5512) 1988 VD_7

Discovered 1988 November 10 by T. Hioki and N. Kawasato at Okutama.

(5513) Yukio

1988 WB. Discovered 1988 November 27 by W. Kakei and M. Kizawa and T. Urata at Oohira.

Named in honor of Yukio Hasegawa (1950-), amateur astronomer and an excellent telescope maker. Periodic comet Urata-Niijima {see planets (3722) and (5507)}, (2090) Mizuho, (3394) Banno and this minor planet were discovered using a telescope made by him. (M 22510)
Name proposed by T. Urata.

(5514) 1989 BN_1

Discovered 1989 January 29 by Z. Vávrová at Kleť.

(5515) 1989 EL_1

Discovered 1989 March 5 by E. F. Helin at Palomar.

(5516) Jawilliamson

1989 JK. Discovered 1989 May 2 by E. F. Helin at Palomar.

Named in honor of Jack Williamson, science-fiction writer, who has helped define the development of contemporary science fiction. For 65 years, Jack has written science fiction of every conceivable sort, from space opera to the most modern technologically oriented variety. He has influenced many people in their pursuit of careers in science and science writing. (M 23138)
Name endorsed by D. Brin and Williamson's many friends and colleagues.

(5517) Johnerogers

1989 LJ. Discovered 1989 June 4 by E. F. Helin at Palomar.

Named in honor of John E. Rogers, Range Safety Officer at Point Mugu, California, and an amateur astronomer long known for his computer software and for his computation of orbits. More recently he has developed an astrometric system using a CCD camera and related software that he expertly uses at his home in Camarillo. The discoverer acknowledges his unstinting assistance in using this equipment to obtain crucial observations of newly-discovered near-earth objects, and from time to time he has also joined the discoverer's team at Palomar to search for them. (M 23138; M 26947)

Name proposed by the discoverer, endorsed by K. Lawrence, B. G. Marsden and G. V. Williams.

(5518) Mariobotta

1989 YF. Discovered 1989 December 30 by J. M. Baur at Chions.

Named in honor of Mario Botta, Swiss architect. His buildings are influenced by his early meetings with Le Corbusier and Louis Kahn and are inspired by the belief that 'every building has its own individual surroundings'. Botta also lectures and is an honorary fellow of the American Institute of Architects. (M 22831)

(5519) 1990 QB_4

Discovered 1990 August 23 by H. E. Holt at Palomar.

(5520) Natori

1990 RB. Discovered 1990 September 12 by T. Urata at Oohira.

Named in honor of Akira Natori (1956-), discoverer of several minor planets. He has collaborated with the Nihondaira {see planet (2880)} Observatory for follow-up observations, and this minor planet was numbered as a result of this cooperation. (M 22510)

(5521) Morpurgo

1991 PM_1. Discovered 1991 August 15 by E. F. Helin at Palomar.

Named in honor of Pieter Morpurgo, who has been the producer for the last 18 years of "The Sky at Night", a British TV program celebrating its 40th anniversary featuring Patrick Moore {see planet (2602)} as host. Morpurgo and Moore have worked closely and successfully creating this upbeat informative program about astronomy for public education in an imaginative, entertaining forum. (M 30476)

(5522) De Rop

1991 PJ_5. Discovered 1991 August 3 by E. W. Elst at La Silla.

Named in honor of Willy De Rop (1933-), astronomer at the Royal Observatory, Uccle, on the occasion of his retirement. Besides his professional work in positional astronomy, timekeeping and rotation of the earth, De Rop has been involved in the popularization of astronomy and several other cultural societies in Belgium. For several years he was most helpful in taking care of the communication of telexes between the discoverer and observation sites, and also the Minor Planet Center. (M 22251)

Citation prepared by J. Denoyelle at the request of the discoverer.

(5523) 1991 PH_8

Discovered 1991 August 5 by H. E. Holt at Palomar.

(5524) Lecacheux

1991 RA_{30}. Discovered 1991 September 15 by H. E. Holt at Palomar.

Named in honor of Jean Lecacheux, a planetary scientist at Paris Observatory. Lecacheux is one of the most assiduous observers of solar system objects. He discovered the Saturnian satellite Helène. At Pic du Midi Observatory he developed a planetary observational station with which he fruitfully imaged Venus, Jupiter, Saturn, the jovian impacts of comet D/1993 F2, as well as other comets and minor planets. (M 27459)

Citation provided by M. A. Barucci following a suggestion by E. Bowell.

(5525) 1991 TS_4

Discovered 1991 October 15 by N. Kawasato at Uenohara.

(5526) Kenzo

1991 UP_1. Discovered 1991 October 18 by T. Urata at Oohira.

Named in honor of Kenzo Suzuki (1950-), a leading amateur observer of minor planets and comets in Japan, who has been performing astrometry since 1972 and has discovered several minor planets. (M 22510)

(5527) 1991 UQ_3

Discovered 1991 October 31 by S. Ueda and H. Kaneda at Kushiro.

(5528) 1992 AJ

Discovered 1992 January 2 by S. Ueda and H. Kaneda at Kushiro.

(5529) Perry

2557 P-L. Discovered 1960 September 24 by C. J. van Houten and I. van Houten-Groeneveld at Palomar.

Named in honor of Marcus Perry and his work in the Spacewatch program. (M 22510)

(5530) Eisinga

2835 P-L. Discovered 1960 September 24 by C. J. van Houten and I. van Houten-Groeneveld at Palomar.

Named in honor of Eise Eisinga (1744-1821), Dutch amateur astronomer, on the 250th anniversary of his birth. Eisinga lived in the northern town of Franeker, where he constructed a large (and still functioning) mechanical planetarium. (M 23541)

(5531) Carolientje

1051 T-2. Discovered 1973 September 29 by C. J. van Houten and I. van Houten-Groeneveld at Palomar.

Named for Caroline van Houten, daughter of Karel and Thea van Houten, and granddaughter of the discoverers. (M 22510)

(5532) Ichinohe

1932 CY. Discovered 1932 February 14 by K. Reinmuth at Heidelberg.

Named in memory of Naozo Ichinohe (1872-1920), instructor at the University of Tokyo and astronomer at Tokyo Observatory. He was a pioneer of astrophysics in Japan. He studied under E. B. Frost, E. E. Barnard and S. W. Burnham {see respectively planets (854), (819), and (834)} during 1905-07 at the Yerkes Observatory, where he measured the radial velocities of some spectroscopic binaries. He left about 12,000 visual observations of variable stars (including three stars discovered by himself) that were carried out in the U.S.A. and Japan. He insisted on the necessity of a large telescope in Japan. He was also a pioneer of science journalism in Japan. After he retired from the University of Tokyo, he was the editor of the monthly magazine *Gendai no Kagaku*, which he founded on the model of *Nature*. (M 25230)

Name proposed by S. Nakano, one of the identifiers involving this minor planet, following a suggestion by M. Hara and S. Sakuma.

(5533) Bagrov

1935 SC. Discovered 1935 September 21 by P. F. Shajn at Simeïs.

Named in honor of Nikolaj Vasil'evich Bagrov (1937-), geographer, professor at the Simferopol State University and author of research on geographical peculiarities of the Crimea. (M 34621)

Name proposed by the Crimean Astrophysical Observatory.

(5534) 1941 UN

Discovered 1941 October 15 by L. Oterma at Turku.

(5535) Annefrank

1942 EM. Discovered 1942 March 23 by K. Reinmuth at Heidelberg.

Named in memory of Anne Frank (1929-1945), whose life and diary form a poignant record of the ravages of war and racism, as seen through the eyes of a young person. Notebooks and papers in Anne's handwriting were found in the family's hiding place on the Prinsengracht in Amsterdam by Miep Gies, a friend who had protected and sustained the group during their isolation. They appeared in published form in 1947 and have since been widely read in more than thirty languages. Despite the incredible adversity it records, the message is inherently one of hope. Proposed on the fiftieth anniversary of the cessation of hostilities in Europe. (M 25230)

Name proposed and citation prepared by G. C. L. Aikman.

(5536) 1955 QN

Discovered 1955 August 23 at the Goethe Link Observatory at Brooklyn, Indiana.

(5537) 1964 TA_2

Discovered 1964 October 9 at the Purple Mountain Observatory at Nanking.

(5538) Luichewoo

1964 TU_2. Discovered 1964 October 9 at the Purple Mountain Observatory at Nanking.

Named in honor of Lui Chewoo, an expert in mineralogy. Lui served as a director of the Mineralogical Institute of South China for many years. He was also recently engaged as an honorary consultant at the Purple Mountain Observatory and has made a great contribution to the cause of astronomy in China. (M 25230)

(5539) 1965 UA_1

Discovered 1965 October 16 at the Purple Mountain Observatory at Nanking.

(5540) Smirnova

1971 QR_1. Discovered 1971 August 30 by T. M. Smirnova at Nauchnyj.

Named in honor of Tamara Mikhajlovna Smirnova (1935-), devoted observer of minor planets and a staff member of the Institute of Theoretical Astronomy from 1966 to 1988. During her long-lasting activity she made about 9,000 observations and measured numerous positions of minor planets at the Crimean Astrophysical Observatory. She discovered over one hundred minor planets and a comet, this object being her hundredth discovery. (M 24917)

Name proposed by the Institute of Theoretical Astronomy.

(5541) Seimei

1976 UH_{16}. Discovered 1976 October 22 by H. Kosai and K. Hurukawa at Kiso.

Named for Abe-no Seimei (921-1005), an authority of astronomy, astrology and divination. At the request of the Emperor and nobles he told their fortune. There are many mysterious legends regarding his abilities. His descendants played important roles in compiling the Japanese annotated calendar for about 800 years. (M 22511)

(5542) 1978 PT_4

Discovered 1978 August 6 at the Perth Observatory at Bickley.

(5543) Sharaf

1978 TW_2. Discovered 1978 October 3 by N. S. Chernykh at Nauchnyj.

Named in honor of Shafika Gil'mievna Sharaf (1915-), well known expert on

celestial mechanics and staff member of the Institute of Theoretical Astronomy from 1939 to 1986. She co-developed an analytical theory of Pluto using the Laplace-Newcomb method and determined new orbital elements for the planet. Later she investigated the secular variations of solar radiation incident upon given area of the earth's surface caused by perturbations of the earth's orbit. (M 30476)
Name suggested by the Institute of Theoretical Astronomy.

(5544) Kazakov

1978 TH_6. Discovered 1978 October 2 by L. V. Zhuravleva at Nauchnyj.

Named in memory of Matvej Fedorovich Kazakov (1738-1812), Russian architect, one of the founders of classicism in Russian architecture in the eighteenth century. (M 34621)

(5545) Makarov

1978 VY_{14}. Discovered 1978 November 1 by L. V. Zhuravleva at Nauchnyj.

Named in honor of Askol'd Anatol'evich Makarov (1925-), an outstanding Russian choreographer and professor of the St. Petersburg Conservatoire. From 1942 to 1970 he was the leading soloist at the Mariinskij Theatre, and since 1976 he has been artistic director at the St. Petersburg State Academic Ballet Theatre. (M 24918)
Name proposed by the Institute of Theoretical Astronomy.

(5546) Salavat

1979 YS. Discovered 1979 December 18 by H. Debehogne at La Silla.

Named in honor of the industrial city of Salavat, located in Bashkiria {see planet (2657)}, just west of the Ural mountains. (M 23138)

(5547) Acadiau

1980 LE_1. Discovered 1980 June 11 by C. S. Shoemaker at Palomar.

Named in honor of Acadia University, Wolfville, Nova Scotia. Founded in 1838, Acadia U. has become one of Canada's finest liberal arts institutions. It is located near the Minas Basin, which boasts some of the most dramatic tides on the earth, and it is also located under some of Canada's darkest night skies. Acadia's academic excellence and small student population provide a fertile environment for a good undergraduate education. (M 24918)
Name proposed and citation prepared by D. H.Levy.

(5548) 1980 TH

Discovered 1980 October 3 by Z. Vávrová at Kleť.

(5549) Bobstefanik

1981 GM_1. Discovered 1981 April 1 at the Harvard College Observatory at Harvard.

Named in honor of Robert Stefanik (1938-), director of the Oak Ridge Observatory, where this minor planet was discovered. Stefanik's stellar research includes the determination of a new set of radial-velocity standards, the search for low-mass stellar companions and the determination of binary-star orbits. (M 22511)

Name proposed by members of the Harvard-Smithsonian Center for Astrophysics Planetary Sciences division.

(5550) 1981 UB_1

Discovered 1981 October 30 by L. G. Taff at Socorro.

(5551) Glikson

1982 BJ. Discovered 1982 January 24 by C. S. Shoemaker and E. M. Shoemaker at Palomar.

Named in honor of Andrew Y. Glikson, retired senior research scientist with the Australian Geological Survey Organization. Glikson, a geologist of vision and creativity, is one of the leading investigators of the early Precambrian. He is especially noted for his work on the Archean of Western Australia and on the complex rocks of the Musgrave Block. He is also well known for his work on impact craters and on the possible role of impacts in the geology of the Archean. (M 29670)

(5552) Studnička

1982 SJ_1. Discovered 1982 September 16 by A. Mrkos at Kleť.

Named in memory of František Josef Studnička (1836-1903), professor of mathematics at Charles University in Prague, also active in astronomy and meteorology. He was known as the author of several textbooks and popular articles. He started his work as a lecturer in České Budějovice (1862-1864). (M 31295)
Name suggested by J. Tichá and M. Šolc.

(5553) Chodas

1984 CM_1. Discovered 1984 February 6 by E. Bowell at Anderson Mesa.

Paul W. Chodas (1952-) is a member of the Solar System Dynamics Group at the Jet Propulsion Laboratory. His work has contributed to the understanding of cometary nongravitational forces, the use of radar data in asteroidal and cometary orbits, orbital error analyses, earth close approaches and impact probabilities. He developed the system that allowed the use of radar landmark data on the surface of Venus for improved orbital solutions of the Magellan spacecraft. (M 22831)
Citation prepared by D. K. Yeomans, a colleague of the honoree.

(5554) Keesey

1985 TW_1. Discovered 1985 October 15 by E. Bowell at Anderson Mesa.

Michael S. W. Keesey (1937-) is a member of the Solar System Dynamics Group at the Jet Propulsion Laboratory. He began his career at JPL in 1970, and his knowledge of fundamental astronomy and celestial mechanics has been put to good use in the ephemeris development efforts for those planets, comets and minor planets that have been mission candidates or targets of ground-based observations. (M 22831)
Citation prepared by D. K. Yeomans, a colleague of the honoree.

(5555) Wimberly

1986 VF_5. Discovered 1986 November 5 by E. Bowell at Anderson Mesa.

Ravenel N. Wimberly (1946-) is a member of the Solar System Dynamics Group at the Jet Propulsion Laboratory. His intimate knowledge of various computer systems has proven invaluable in helping organize and carry out the ephemeris development efforts for the comet and asteroid targets of several flyby missions. (M 22831)
Citation prepared by D. K. Yeomans, a colleague of the honoree.

(5556) 1988 AL

Discovered 1988 January 15 by S. Ueda and H. Kaneda at Kushiro.

(5557) Chimikeppuko

1989 CM_1. Discovered 1989 February 7 by K. Endate and K. Watanabe at Kitami.

Named for the small lake in Tsubetsu {see planet (4845)}, a town in the eastern part of Hokkaido {see planet (3720)}. Some 12 km in circumference, the lake is surrounded by a deep green virgin forest. The lake is the nesting place for many wild birds and famous as bird-watching spot. (M 25977)

(5558) 1989 WL_2

Discovered 1989 November 24 by R. H. McNaught at Siding Spring.

(5559) 1990 MV

Discovered 1990 June 27 by E. F. Helin at Palomar.

(5560) Amytis

1990 MX. Discovered 1990 June 27 by E. F. Helin at Palomar.

Named in honor of Amytis Barrett (1909-), in celebration of her 85th birthday, in recognition of her many contributions to the Caltech community, including all aspects of undergraduates' activities and her significant influence on the Associates' program. (M 23352)

Name proposed by Dr. and Mrs. J. Bonner and endorsed by her many Caltech friends.

(5561) Iguchi

1991 QD. Discovered 1991 August 17 by S. Otomo at Kiyosato.

Named in honor of Masatoshi Iguchi (1928-), president of the Photovoltaic Popularization Association in Japan and publisher of *The Photovoltaic News*. (M 25654)

(5562) 1991 VS

Discovered 1991 November 4 by S. Ueda and H. Kaneda at Kushiro.

(5563) 1991 VZ_1

Discovered 1991 November 9 by S. Ueda and H. Kaneda at Kushiro.

(5564) 1991 VH_2

Discovered 1991 November 9 by S. Ueda and H. Kaneda at Kushiro.

(5565) Ukyounodaibu

1991 VN_2. Discovered 1991 November 10 by A. Natori and T. Urata at Yakiimo.

Named for the Japanese poetess, Kenreimon-in Ukyounodaibu (1157-?). She wrote a diary with 359 poems that describe her life at court and reminiscences of her young lover Taira-no Sukemori, who died at the battle of Dannoura. She worked for Kenreimonin {see planet (5242)}. (M 30476)

Name suggested and citation provided by A. Sato.

(5566) 1991 VY_3

Discovered 1991 November 11 by S. Ueda and H. Kaneda at Kushiro.

(5567) 1953 FK_1

Discovered 1953 March 21 at the Goethe Link Observatory at Brooklyn, Indiana.

(5568) 1953 TS_2

Discovered 1953 October 14 at the Goethe Link Observatory at Brooklyn, Indiana.

(5569) 1974 FO

Discovered 1974 March 22 by C. Torres at Cerro El Roble.

(5570) Kirsan

1976 GM_7. Discovered 1976 April 4 by N. S. Chernykh at Nauchnyj.

Named in honor of Kirsan Nikolaevich Ilyumzhinov (1962-), chess grandmaster and president of the International Chess Federation, elected in 1996 for a second term. (M 30097)

(5571) 1978 LG

Discovered 1978 June 1 by K. W. Kamper at La Silla.

(5572) Bliskunov

1978 SS_2. Discovered 1978 September 26 by L. V. Zhuravleva at Nauchnyj.

Named in memory of Aleksandr Ivanovich Bliskunov (1938-1996), well-known Crimean orthopedic surgeon and professor. (M 34621)

(5573) 1981 QX

Discovered 1981 August 24 by A. Mrkos at Kleť.

(5574) 1984 FS

Discovered 1984 March 20 by Z. Vávrová at Kleť.

(5575) 1985 RP_2

Discovered 1985 September 4 by H. Debehogne at La Silla.

(5576) Albanese

1986 UM_1. Discovered 1986 October 26 by the CERGA at Caussols.

Named in honor of Dominique Albanese, photographer and observer at the Schmidt telescope of the Observatoire de la Côte d'Azur. Dominique has also installed a 30-cm Schmidt telescope with which he regularly photographs the skies. (M 23138)

Citation written by A. Maury and endorsed by J.-L. Heudier.

(5577) Priestley

1986 WQ_2. Discovered 1986 November 21 by J. D. Waldron at Siding Spring.

Named for Joseph Priestley (1733-1804), an English clergyman who was the first to publish an account of his discovery in 1774 of the element oxygen. The number 5577, corresponding to the number of this minor planet, is also the wavelength in Ångstroms of the main emission line of green auroral light, due to triply-ionized oxygen. (M 22207; M 33786)

(5578) 1987 BC

Discovered 1987 January 28 by T. Niijima and T. Urata at Ojima.

(5579) Uhlherr

1988 JL. Discovered 1988 May 11 by C. S. Shoemaker and E. M. Shoemaker at Palomar.

Named in honor of engineer H. Ralph Uhlherr, a founder of the Siemens plant in Melbourne. An indefatigable collector of tektites, Uhlherr has meticulously recorded and mapped over a thousand australites from the Port Campbell Embayment of Victoria, thereby helping to resolve the conflict between the apparent young stratigraphic age of the tektites and the middle Pleistocene ages obtained by various chronometric methods. (M 29670; M 29691)

(5580) Sharidake

1988 RP_1. Discovered 1988 September 10 by K. Endate and K. Watanabe at Kitami.

Named for a mountain in eastern Hokkaido {see planet (3720)}, 1545 meters high and known as "Japan's Matterhorn". It is in the Shiretoko {see planet (3867)} range, part of the Chishima volcanic belt. (M 25977)

(5581) Mitsuko

1989 CY_1. Discovered 1989 February 10 by M. Iwamoto and T. Furuta at Tokushima.

Named in honor of Mitsuko Iwamoto (1956-), wife of the first discoverer. (M 27128)

(5582) 1989 CU_8

Discovered 1989 February 13 by H. Debehogne at La Silla.

(5583) 1989 EY_1

Discovered 1989 March 5 by A. Mrkos at Kleť.

(5584) 1989 KK

Discovered 1989 May 31 by H. E. Holt at Palomar.

(5585) Parks

1990 MJ. Discovered 1990 June 28 by E. F. Helin at Palomar.

Named in honor of Robert J. Parks, a leader in the exploration of the solar system. Early in his career at the Caltech Jet Propulsion Laboratory, he developed radio control and telemetry for the first guided missiles. As these techniques matured to become the basis for the control of and communication with interplanetary spacecraft, Parks was assigned overall responsibility for the first missions to Mercury, Venus and Mars, as well as for the first U.S. lunar lander, "Surveyor" and the "Voyager" missions to Jupiter, Saturn, Uranus and Neptune. His final position at the Jet Propulsion Laboratory was deputy director, but throughout all his assignments his task was the same: to make sure the flight missions worked. Because of his expertise and dedication, they did. (M 25977)

(5586) 1990 RE_6

Discovered 1990 September 9 by H. Debehogne at La Silla.

(5587) 1990 SB

Discovered 1990 September 16 by H. E. Holt and J. A. Brown at Palomar.

(5588) 1990 SW_3

Discovered 1990 September 23 by B. Roman at Palomar.

(5589) 1990 SD_{14}

Discovered 1990 September 23 by H. Debehogne at La Silla.

(5590) 1990 VA

Discovered 1990 November 9 by the Spacewatch at Kitt Peak.

(5591) Koyo

1990 VF_2. Discovered 1990 November 10 by T. Urata at Oohira.

Named in honor of Koyo Kawanishi (1959-), a dentist living in Ako-city, Hyogo. He has been observing comets and minor planets using his home-made 0.20-m reflector at his private observatory with his wife Kumi and daughter Saki. Familiar with electronics and mechanics, he has developed his own CCD instrumentation. (M 22511)

(5592) Oshima

1990 VB_4. Discovered 1990 November 14 by K. Suzuki and T. Urata at Toyota.

Named in honor of Yoshiaki Oshima (1952-), discoverer of the Apollo-type minor planet 1988 XB and several other minor planets. He contributed to the development of the instrumentation at the Nihondaira {see planet (2880)} Observatory. (M 22511)
Name proposed by T. Urata.

(5593) Jonsujatha

1991 JN_1. Discovered 1991 May 9 by E. F. Helin at Palomar.

Named in honor of Jonathan Brian Marsden and Sujatha Nagarajan on the occasion of their wedding on 1993 Nov. 14 in Lexington, Massachusetts. Jon and

Sujatha are talented young professionals who are friends and neighbors of the discoverer in California. Hearty good wishes are extended to them for a long, healthy and happy life together. (M 22831)

(5594) 1991 NK_1

Discovered 1991 July 12 by H. E. Holt at Palomar.

(5595) 1991 PJ

Discovered 1991 August 5 by H. E. Holt at Palomar.

(5596) Morbidelli

1991 PQ_{10}. Discovered 1991 August 7 by H. E. Holt at Palomar.

Named in honor of Alessandro Morbidelli (1966-). From Cuneo, in extreme western Italy, he joined the solar system dynamics group in Nice (50 km to the south) as an H. Poincaré {see planet (2021)} postdoctoral fellow and became a permanent staff member in 1993. Morbidelli's main interests are the theory of dynamical systems and resonances in the solar system. For his pioneering work on the overlapping of mean-motion and secular resonances, Morbidelli was awarded the bronze medal of the CNRS in 1995. (M 27460)
Name proposed and citation provided by H. Scholl.

(5597) 1991 PC_{13}

Discovered 1991 August 5 by H. E. Holt at Palomar.

(5598) Carlmurray

1991 PN_{18}. Discovered 1991 August 8 by H. E. Holt at Palomar.

Named in honor of Carl Desmond Murray (1955-), of Queen Mary and Westfield College, London, in recognition of his immense contribution to our understanding of the dynamics of the minor bodies of the solar system. In particular, he has investigated the effects of various resonances on a number of sub-systems within the solar system; for example, the Kirkwood {see planet (1578)} gaps and the interrelationships between ring structures and small satellites around the giant planets. (M 27460)
Citation prepared by I. P. Williams.

(5599) 1991 SG_1

Discovered 1991 September 29 by S. Ueda and H. Kaneda at Kushiro.

(5600) 1991 UY

Discovered 1991 October 18 by S. Ueda and H. Kaneda at Kushiro.

(5601) 1991 VR

Discovered 1991 November 4 by S. Ueda and H. Kaneda at Kushiro.

(5602) 1991 VM_1

Discovered 1991 November 4 by S. Ueda and H. Kaneda at Kushiro.

(5603) Rausudake

1992 CE. Discovered 1992 February 5 by K. Endate and K. Watanabe at Kitami.

Named for the mountain in the Shiretoko range in Hokkaido {see planets (3867) and (3720), respectively}. Known as the "Shiretoko Fuji", this 1661-meter mountain is in the Shiretoko peninsula that juts into the Sea of Okhotsk {see planets (3867), (4042)}. (M 25977)

(5604) 1992 FE

Discovered 1992 March 26 by R. H. McNaught at Siding Spring.

(5605) 1993 DB

Discovered 1993 February 17 by S. Otomo at Kiyosato.

(5606) 1993 EH

Discovered 1993 March 1 by S. Otomo at Kiyosato.

(5607) 1993 EN

Discovered 1993 March 12 by S. Ueda and H. Kaneda at Kushiro.

(5608) 1993 EO

Discovered 1993 March 12 by S. Ueda and H. Kaneda at Kushiro.

(5609) Stroncone

1993 FU. Discovered 1993 March 22 by A. Vagnozzi at Stroncone.

Named for the suburb where the Santa Lucia observatory is based. The name Stroncone is thought to come from the Greek 'astronicos', from 'astron' (star) and 'icons' (image), so it is a very appropriate name for the site of an astronomical observatory. (M 23353)

(5610) Balster

2041 T-3. Discovered 1977 October 16 by C. J. van Houten and I. van Houten-Groeneveld at Palomar.

Named in honor of Harry A. M. Balster (1946-) and his younger sister Yvonne. Harry is a very active amateur astronomer in Nijmegen, The Netherlands. His main fields of interest are sunspot research and making the general public enthusiastic for astronomy. By profession he is an electronic engineer. When I. van Houten was measuring her plates at Nijmegen, Harry always helped when there was something wrong with the measuring engine. His sister was a programmer and general assistant in the Nijmegen astronomy department, and she was just as great a help for I. van Houten as her brother. (M 23541; M 25351)

(5611) 1943 DL

Discovered 1943 February 26 by L. Oterma at Turku.

(5612) Nevskij

1975 TX_2. Discovered 1975 October 3 by L. I. Chernykh at Nauchnyj.

Named for Novgorod prince Aleksandr Yaroslavich (1220-1263), astute politician and skilled military leader of Russian troops who defended northwestern Russia from annexation by Swedish and German feudal lords. He was named Nevskij for his victory over Swedish troops at the Neva River in 1240, and he is also famous for his defeat of Teutonic knights at Chudskoe Lake in 1242. (M 30097)

(5613) Donskoj

1976 YP_1. Discovered 1976 December 16 by L. I. Chernykh at Nauchnyj.

Named for Dmitrij Donskoj (1350-1389), Alexandr Nevskij's {see planet (5612)} great-grandson, grand prince of Moscow and Vladimir principalities. He led the Russian struggle against the Tatar-Mongolian yoke and was named Donskoj for his victory over the Tatar-Mongolians in the 1380 battle at Kulikovo field between the Don and the Nepryadva {see planet (2869)} rivers. This historic victory of joint Russian troops under Dmitrij's leadership began the liberation of Russian lands from the Tatars. (M 30097)

(5614) Yakovlev

1979 VN. Discovered 1979 November 11 by N. S. Chernykh at Nauchnyj.

Named in honor of Konstantin Karol'evich Yakovlev (1954-), director of the St. Petersburg scientific-production firm "Blok". Yakovlev takes an interest in astronomy and is rendering technical support in the restoration of the Crimean 64-

cm telescope and equipping it for the observation of near-earth objects. (M 27330; M 27749)

(5615) Iskander

1983 PZ. Discovered 1983 August 4 by L. G. Karachkina at Nauchnyj.

Named in honor of Fazil' Abdulovich Iskander (1929-), well-known writer, poet, essayist, author of the humorous *Sandro from Chegem* and the winner of such literature prizes as the Pushkin {see also planet (2208)}. He is a vice-president of the Russian Pen Club. (M 23139)

Name suggested by L. R. Nemirovskij.

(5616) Vogtland

1987 ST_{10}. Discovered 1987 September 29 by F. Börngen at Tautenburg.

Named after the mountainous district of Germany that lies partly in Saxony, Thuringia, Bavaria and Bohemia. (M 22831)

(5617) 1989 EL

Discovered 1989 March 5 by E. F. Helin at Palomar.

(5618) Saitama

1990 EA. Discovered 1990 March 4 by A. Sugie at Taga.

Named for the Japanese prefecture. (M 27734)

(5619) Shair

1990 HC_1. Discovered 1990 April 26 by E. F. Helin at Palomar.

Named in honor of Fredrick H. Shair, manager of the Educational Affairs Office at the Jet Propulsion Laboratory and former dean of the College of Natural Sciences and Mathematics at California State University, Long Beach, and professor at the California Institute of Technology. While at CALTECH, he established the Summer Undergraduate Research Fellowship (SURF) {see planet (5181)} Program, which integrates students' classroom experience with a research environment. Shair has been honored with numerous awards for his teaching excellence. The discoverer has been a SURF sponsor since the early years of the program. (M 23541)

Name heartily endorsed by Shair's many friends and colleagues.

(5620) 1990 OA

Discovered 1990 July 19 by B. Roman and E. F. Helin at Palomar.

(5621) Erb

1990 SG_4. Discovered 1990 September 23 by K. J. Lawrence at Palomar.

Named in honor of Bryan and Dona Erb, friends of the discoverer. Brian is assistant director of the Canadian Space Agency, Canadian liaison officer to NASA's Johnson Space Center and assistant program manager for the space station. Dona is a software engineer working on ground control facilities. Both are currently working on the space station. The Erbs played an important role in generating the discoverer's interest in space physics. (M 25654; M 26599)

(5622) 1990 TL_4

Discovered 1990 October 14 by E. F. Helin at Palomar.

(5623) Iwamori

1990 UY. Discovered 1990 October 20 by A. Sugie at Taga.

Named in memory of Yasuke Iwamori, late principal of Kyoto {see planet (4352)} city Rakuyou {see planet (5825)} technical high school who taught physics and astronomy there. (M 27735)

Name proposed by the discoverer following a suggestion by S. Sakabe.

(5624) Shirley

1991 AY_1. Discovered 1991 January 11 by E. F. Helin at Palomar.

Named in honor of Mr. and Mrs. William J. Shirley in recognition of their generous, enthusiastic support of Caltech and Mt. Wilson Observatory. They have preserved and restored the Hale Solar Observatory in San Marino, a landmark of scientific and historical significance that they share generously with the community and educational groups for visits and study. (M 22831)
Name endorsed by their many friends.

(5625) 1991 AO_2

Discovered 1991 January 7 by R. H. McNaught at Siding Spring.

(5626) 1991 FE

Discovered 1991 March 18 by the Spacewatch at Kitt Peak.

(5627) 1991 MA

Discovered 1991 June 16 by R. H. McNaught at Siding Spring.

(5628) Preußen

1991 RP_7. Discovered 1991 September 13 by L. D. Schmadel and F. Börngen at Tautenburg.

Named for the former kingdom and German state of Prussia and for the so-called Prussian virtues of sense of duty, austerity, punctuality, order and unselfishness. Prussia, originally the name of the region at the southern end of the Baltic Sea, became a duchy in 1525. In 1701, Frederick I became king of Prussia and established his capital at Berlin. The defeat of Napoleon yielded a great territorial enlargement of Prussia. The victory over France in 1870 resulted in the creation of the German Empire, which declined in 1918. The name Prussia was abolished in World War II; the Prussian virtues, however, continue to be desirable. (M 27330)
Name proposed by the first discoverer, whose birthplace is Berlin.

(5629) Kuwana

1993 DA_1. Discovered 1993 February 20 by T. Hioki and S. Hayakawa at Okutama.

Kuwana is a city located in the delta of the Nagara and Ibe rivers, near the city of Nagoya and the Yoro and Suzuka mountains. Kuwana is the sister city of Gyoda {see planet (5138)}, where the second discoverer lives. (M 25230)

(5630) Billschaefer

1993 FZ. Discovered 1993 March 21 by J. B. Child at Palomar.

Named in honor of William Schaefer, telescope maker and amateur astronomer. For over forty years Bill Schaefer has been producing precision telescopes for amateur astronomers throughout the United States. His enthusiasm, skill and creativity have sparked the interest and involvement of hundreds of people, young and old, and have raised the quality of amateur astronomy in the Southern California community. He has patiently shared his knowledge with other telescope makers, and his award-winning telescopes will serve serious observers throughout the world for many years to come. (M 22831)

(5631) Sekihokutouge

1993 FE_1. Discovered 1993 March 20 by K. Endate and K. Watanabe at Kitami.

Named for a pass in central Hokkaido {see planet (3720)}. At an altitude of 1050 meters, the pass connects the Taisetsu mountain range, Hokkaido's highest peak, with the city of Kitami {see planet (3785)}. (M 25977)

(5632) Ingelehmann

1993 GG. Discovered 1993 April 15 by C. S. Shoemaker and E. M. Shoemaker at Palomar.

Named in honor of Inge Lehmann (1888-1993), Danish seismologist and a pioneer in the field when the science of seismology was still young and unexplored. She made fundamental contributions to the study of geophysics and was the first chief of the seismology department of the newly established Royal Danish Geodetic Institute. Her most important contribution suggested a new discontinuity in the seismic structure of the earth, now known as the Lehmann discontinuity, a region that divides the core into inner and outer parts. In later years, Lehmann became an authority on the structure of the upper mantle. (M 29670; M 29691)
Obituary published in Q.J.R. Astron. Soc., Vol. 35, No. 2, p. 231-234 (1994).

(5633) 1978 UL_7

Discovered 1978 October 27 by C. M. Olmstead at Palomar.

(5634) 1978 VT_6

Discovered 1978 November 7 by E. F. Helin and S. J. Bus at Palomar.

(5635) Cole

1981 ER_5. Discovered 1981 March 2 by S. J. Bus at Siding Spring.

Named for the fictional character Joshua Cole, amateur astronomer and star of Arthur Preston Hankin's 1923 novel *Cole of Spyglass Mountain*. A character reminiscent of Oliver Twist, Cole studied variable stars while incarcerated in a reformatory in which the boys do not use names but numbers; Cole's is 5635. He survived one crisis after another, inlcuding a near-fatal shooting in his observatory at the moment he discovered evidence of life on Mars. (M 31024)

Name proposed by the discoverer following a suggestion by D. H. Levy, who prepared the citation.

(5636) Jacobson

1985 QN. Discovered 1985 August 22 by E. Bowell at Anderson Mesa.

Robert A. Jacobson (1944-) is a noted authority on spacecraft navigation techniques, and he is currently developing ephemerides for natural satellites at the Jet Propulsion Laboratory. He was instrumental in generating the accurate satellite ephemerides used by the Voyager 2 spacecraft project during its encounters with Uranus and Neptune. These ephemerides were developed using both ground-based astrometry and spacecraft optical navigation data. (M 22831)
Citation prepared by D. K. Yeomans, a colleague of the honoree.

(5637) Gyas

1988 RF_1. Discovered 1988 September 10 by C. S. Shoemaker and E. M. Shoemaker at Palomar.

A companion of Aeneas {see planet (1172)}, Gyas took part in the boat race at the funeral games for Anchises {see planet (1173)}. He captained the Chimaera {see also the citation for planet (623)}, a boat huge in length and weight. He came in third when he was slowed by the loss of his helmsman, whom he threw overboard in anger because he sailed too wide of a dangerous rock. (M 25444)

(5638) Deikoon

1988 TA_3. Discovered 1988 October 10 by C. S. Shoemaker and E. M. Shoemaker at Palomar.

Deikoon was a son of Pergasos and a friend of Aeneas {see planet (1172)}. He was a spear fighter and a man much honored by the Trojans, because he was quick to join the battle line. He was killed by the thrust of Agamemnon's {see planet (911)} spear through his shield. (M 25444)

(5639) 1989 PE

Discovered 1989 August 9 by J. Alu and E. F. Helin at Palomar.

(5640) Yoshino

1989 UR_3. Discovered 1989 October 21 by M. Mukai and M. Takeishi at Kagoshima.
Named for the hometown of the first discoverer. Part of Kagoshima {see planet (4703)} City, it lies at the southern tip of Kyushu island. (M 33786)

(5641) McCleese

1990 DJ. Discovered 1990 February 27 by E. F. Helin at Palomar.
Named in honor of Daniel J. McCleese, whose research is primarily in atmospheric physics and infrared instrumentation for remote sensing of planetary atmospheres. As manager of the Jet Propulsion Laboratory's Earth and Space Science Division, he includes in his purview all science aspects of the Near-Earth Asteroid Tracking program. The discoverer and other NEAT team members wish to recognize the important role he played in the establishment of the program. (M 26930)

(5642) 1990 OK_1

Discovered 1990 July 27 by H. E. Holt at Palomar.

(5643) Roques

1990 QC_2. Discovered 1990 August 22 by H. E. Holt at Palomar.
Named in honor of Françoise Roques (1956-) for her work on the dynamics of planetary and circumstellar disks. Roques has numerically explored the interaction of circumstellar dust disks with embedded planets and has applied her work both to the βPictoris system and to binary stars. She has also observed stellar occultations by the giant planets with the principal goal of analyzing the dynamics of the planetary atmospheres. (M 27460)
Citation provided by M. A. Barucci following a suggestion by E. Bowell.

(5644) 1990 QG_2

Discovered 1990 August 22 by H. E. Holt at Palomar.

(5645) 1990 SP

Discovered 1990 September 20 by R. H. McNaught at Siding Spring.

(5646) 1990 TR

Discovered 1990 October 11 by S. Ueda and H. Kaneda at Kushiro.

(5647) 1990 TZ

Discovered 1990 October 14 by E. F. Helin at Palomar.

(5648) 1990 VU_1

Discovered 1990 November 11 by K. Endate and K. Watanabe at Kitami.

(5649) Donnashirley

1990 WZ_2. Discovered 1990 November 18 by E. F. Helin at Palomar.
Named in honor of Donna Shirley, manager of the highly successful Mars Exploration program at NASA's Jet Propulsion Laboratory. She managed the robotics development program that produced Sojourner, the automated rover that explored the surface of Mars in mid-1997. She was also project engineer for the Cassini mission, the spacecraft now on its way to explore Saturn, as well as the project engineer for Mariner 10, which flew by Mercury in the 1970s. In her thirty years at JPL she has had many other successful assignments in space exploration and in the development of non-space systems using space technology. She has balanced a demanding career with parenthood, writing, acting and music. (M 31024)

(5650) 1990 XK

Discovered 1990 December 10 by A. Natori and T. Urata at Yakiimo.

(5651) Traversa

1991 CA_2. Discovered 1991 February 14 by E. W. Elst at St. Michel.

In honor of Gilles Traversa, technical night-assistant at the Observatory of Haute Provence. He has been involved mainly in the Fehrenbach {see planet (3433)} Program of Radial Velocities and has made observations at Zeekoegat (South Africa), La Silla (Chile) and Haute Provence, where he has observed with the Grand Prisme Objectif (GPO), the PPO (Petit Prisme Objectif) and the Schmidt telescope. From 1986 to 1993 he has been of irreplaceable help to, and has become a very good friend of, the discoverer during the observations at Haute Provence. (M 23353)

(5652) Amphimachus

1992 HS_3. Discovered 1992 April 24 by C. S. Shoemaker and E. M. Shoemaker at Palomar.

Amphimachus was the great-hearted son of Aktorian Kteatos. He had been a suitor of Helen {see planet (101)}. He was a leader of the Epeians and was accidentally killed by Hector's {see planet (624)} spear, which had been thrown at Teucer {see planet (2797)} and had missed. (M 25444)

(5653) 1992 WD_5

Discovered 1992 November 21 by E. F. Helin and K. J. Lawrence at Palomar.

(5654) Terni

1993 KG. Discovered 1993 May 20 by A. Vagnozzi at Stroncone.

Named for the city located in a natural amphitheater, at the confluence of the Serra and Nera rivers. Founded in A.D. 672, the city (ancient name Interamna), {see also the citation for planet (704)} has long been famous for the nearby Marmore falls and is now an important industrial center. (M 23353)

(5655) Barney

1159 T-2. Discovered 1973 September 29 by C. J. van Houten and I. van Houten-Groeneveld at Palomar.

Named in memory of Ida Barney (1886-1982), astronomer at the Yale Observatory from 1924 to 1959 and supervisor of the Yale Zone Catalogue program of positions and proper motions of stars. (M 23541)

(5656) Oldfield

A920 TA. Discovered 1920 October 8 by W. Baade at Bergedorf.

Named in honor of Mike Oldfield (1953-), English composer and multi-instrumentalist, best known for *Tubular Bells*, his 1973 album that has influenced a generation of contemporary musicians. In addition to numerous studio recordings, Oldfield wrote the moving soundtrack to the 1984 film *The Killing Fields*. (M 23353)

Name proposed by G. V. Williams, who made the identifications involving this object.

(5657) 1936 QE_1

Discovered 1936 August 28 by K. Reinmuth at Heidelberg.

(5658) Clausbaader

1950 DO. Discovered 1950 February 17 by K. Reinmuth at Heidelberg.

Named in memory of Claus Baader (1824-1995), German manufacturer of planetaria, domes and telescopes, and well-known mentor of amateur astronomers in the German-speaking countries. A design engineer by profession and self-taught in astronomy, Baader constructed a new type of a small desktop planetarium that is in use in schools in many countries throughout the world. In his later years Baader, together with his son and successor Thomas, was deeply concerned

with the construction and development of observatory domes, astronomical telescopes and their auxiliary instrumentation for amateur and professional astronomers alike. (M 25977)
Name proposed and citation prepared by L. D. Schmadel.

Obituaries published in Sterne Weltraum, Jahrg. 35, Nr. 2, p. 152-153 (1996); Mitt Astron. Ges., Nr. 80, p. 5-7 (1997).

(5659) 1968 OA_1

Discovered 1968 July 18 by C. Torres and S. Cofre at Cerro El Roble.

(5660) 1974 MA

Discovered 1974 June 26 by C. T. Kowal at Palomar.

(5661) Hildebrand

1977 PO_1. Discovered 1977 August 14 by N. S. Chernykh at Nauchnyj.

This Hilda group member is named in honor of the well-known Canadian geologist Alan R. Hildebrand (1955-), who discovered that the Chicxulub Crater in the Yucatan region resulted from an asteroid impact that coincided with the terminal Cretaceous extinctions. His linking of a "smoking gun" to a major extinction event in geologic history provided evidence of fundamental importance to the resolution of a centuries-old debate on the demise of the dinosaurs and the occurrence of catastrophes in the earth's history. Hildebrand also has been highly supportive of the asteroid research program at the University of Victoria. (M 24410)

Named by the discoverer at the suggestion of J. B. Tatum. Citation provided by D. Russell.

(5662) 1981 EL_4

Discovered 1981 March 2 by S. J. Bus at Siding Spring.

(5663) 1981 EQ_{12}

Discovered 1981 March 1 by S. J. Bus at Siding Spring.

(5664) 1981 EX_{43}

Discovered 1981 March 6 by S. J. Bus at Siding Spring.

(5665) 1982 BD_{13}

Discovered 1982 January 30 by S. J. Bus at Palomar.

(5666) Rabelais

1982 TP_1. Discovered 1982 October 14 by L. G. Karachkina at Nauchnyj.

Named in memory of François Rabelais (c.1494-1553), great French writer, outstanding representative of European humanism and the author of the immortal literary works *La vie inestimable du grand Gargantua*, *Père de Pantagruel* and *Pantagruel*. (M 23139)

The name is given on the occasion of the five-hundredth anniversary of his birth, following a suggestion by the St. Petersburg House of Satire and Humor.

(5667) Nakhimovskaya

1983 QH_1. Discovered 1983 August 16 by T. M. Smirnova at Nauchnyj.

Named on the occasion of the 50th anniversary of the Nakhimov Nautical College (St. Petersburg) named after an outstanding Russian naval commander, Pavel Stepanovich Nakhimov (1802-1855). Founded to train young navigators, many of the colleges former students have become experienced captains and prominent scientists, who have made considerable contributions to the development of Russian science and the Russian navy. (M 23793)

The name is given by the Institute of Theoretical Astronomy, following a suggestion by the college.

(5668) Foucault

1984 FU. Discovered 1984 March 22 by A. Mrkos at Kleť.

Named in memory of Jean Bernard Léon Foucault (1819-1868), French physicist and astronomer, well known for his pendulum demonstration of the earth's rotation, first in Paris in 1851. (M 30476)

Name suggested by J. Tichá, M. Tichý and Z. Moravec.

(5669) 1985 CC_2

Discovered 1985 February 12 by H. Debehogne at La Silla.

(5670) Rosstaylor

1985 VF_2. Discovered 1985 November 7 by C. S. Shoemaker and E. M. Shoemaker at Palomar.

Named in honor of Stuart Ross Taylor, one of the world's leading geochemists, who has worked extensively on the abundances of the elements in the earth, moon and meteorites. His high-precision analyses have taught us much about the origin and evolution of the earth's continental crust, the sources of magmas, and the origin and evolution of the moon. As an author, he has presented comprehensive overviews of lunar and planetary science. (M 29670)

(5671) Chanal

1985 XR. Discovered 1985 December 13 by the CERGA at Caussols.

Named in honor of Roger Chanal, French amateur astronomer and friend of the discoverer. Roger has over the years built a well-equipped observatory and is a contributor to the *MPCs*. Apart from his interest in asteroid astrometry, he is also the discoverer of V1118 Orionis, temporarily named "Chanal's object". (M 23793)

Name suggested by the discoverer, Robert Chemin, and endorsed by the entire OCA Schmidt telescope team.

Officially, the institution CERGA serves as corporate discoverer.

(5672) Libby

1986 EE_2. Discovered 1986 March 6 by E. Bowell at Anderson Mesa.

Named in honor of Eleanor W. Libby, founder and president of the Donald Ware Waddell Foundation, which has provided funds to arts organizations, museums, and to cancer research. Libby's generosity has done much to enhance the visitors' program at Lowell Observatory. (M 23541)

Name suggested by W. L. Putnam, citation provided by Putnam and M. L. Evans.

(5673) McAllister

1986 RT_2. Discovered 1986 September 6 by E. Bowell at Anderson Mesa.

Named in honor of Frances McAllister, humanitarian and leading philanthropist of Flagstaff, Arizona. Founder of The Arboretum at Flagstaff, a center of horticulture and plant research, McAllister sponsored a program that allowed high-school students to observe P/Halley in 1985-86 and has helped considerably to improve Lowell Observatory's library and scientific facilities. (M 23541)

Names suggested by W. L. Putnam, citation provided by Putnam, O. G. Franz and M. L. Evans.

(5674) Wolff

1986 RW_2. Discovered 1986 September 6 by E. Bowell at Anderson Mesa.

Named in honor of John M. Wolff, a long-time trustee of the Wolff Foundation.

Endowed by Wolff's father, the foundation has provided support for children's hospitals, for children with special needs, and for arts and music organizations in Phoenix, Arizona, and St. Louis, Missouri. Generous donations to Lowell Observatory have enabled the construction of automated offset guiders and other instrumentation. (M 23541)

Name suggested by W. L. Putnam, citation provided by Putnam and M. L. Evans.

(5675) Evgenilebedev

1986 RY_5. Discovered 1986 September 7 by L. I. Chernykh at Nauchnyj.

Named in honor of Evgenij Alekseevich Lebedev (1917-1997), an actor at the Tovstonogov Bolshoi Dramatic Theater in St. Petersburg and a People's Artist of the U.S.S.R. The summit of his art was the unusual role of the horse in the performance *History of a horse*, after a story by Tolstoj {see planet (3771)}. (M 34622)

(5676) Voltaire

1986 RH_{12}. Discovered 1986 September 9 by L. G. Karachkina at Nauchnyj.

Named in memory of Marie François Arouet (Voltaire) (1694-1778), the great French writer, philosopher, historian, inspirer of French enlightenment during the eighteenth century, and the acknowledged leader of the age. (M 23139)

The name is given on the occasion of the three-hundredth anniversary of his birth, following a suggestion by the St. Petersburg House of Satire and Humor. Voltaire is also honored by a crater on the Mars satellite Deimos.

(5677) Aberdonia

1987 SQ_1. Discovered 1987 September 21 by E. Bowell at Anderson Mesa.

Named in honor of the University of Aberdeen on the occasion of the quincentenary of its founding, 1995 Feb. 10. The University is noted as having been home to the first chair of medicine in the English-speaking world. The teaching of natural philosophy was established more than 400 years ago, and the earliest record of the teaching of astronomy dates back to 1593. Occupants of chairs of natural philosophy include James Clerk Maxwell and more recently the Nobel laureate George Paget Thomson. (M 24765)

(5678) DuBridge

1989 TS. Discovered 1989 October 1 by E. F. Helin at Palomar.

Named in memory of Lee Alvin DuBridge (1901-1994), noted nuclear physicist and former president of the California Institute of Technology. DuBridge set an extraordinary example of excellence in scientific research and the administration of science. Included among his many accomplishments was the establishment and direction of the Massachusetts Institute of Technology's radiation laboratory for the development of radar. In the field of science he was a key advisor to several U.S. presidents. (M 23139)

This tribute, initiated before DuBridge's death, is endorsed by his many friends and colleagues and the discoverer's husband, Ron Helin, who was a student during DuBridge's presidency at Caltech.

(5679) Atsukadou

1989 VR. Discovered 1989 November 2 by K. Endate and K. Watanabe at Kitami.

Named for one of Japan's largest stalactite caves, located in Iwate Prefecture. (M 25977)

(5680) 1989 YZ_1

Discovered 1989 December 30 by R. H. McNaught at Siding Spring.

(5681) Bakulev

1990 RS_{17}. Discovered 1990 September 15 by L. V. Zhuravleva at Nauchnyj.

Named in memory of Aleksandr Nikolaevich Bakulev (1890-1967), one of the pioneers of neurosurgery in the U.S.S.R. He made advances in kidney surgery, bone surgery and a surgical treatment for stomach ulcers. (M 34622)

(5682) Beresford

1990 TB. Discovered 1990 October 9 by R. H. McNaught at Siding Spring.

Named in honor of Anthony Charles Beresford (1942-), prominent Australian amateur astronomer. Amongst his wide ranging astronomical interests he is an active artificial satellite observer, having been part of Operation Moonwatch from 1960 to 1975. He plays an important role in the dissemination of astronomical information and discoveries in South Australia. Always knowledgable about current events, Tony Beresford has been of considerable help to the discoverer on many occasions. (M 34341)

Name suggested and citation endorsed by D. I. Steel.

(5683) 1990 UD

Discovered 1990 October 19 by T. Urata at Oohira.

(5684) 1990 UB_2

Discovered 1990 October 21 by T. Urata at Oohira.

(5685) 1990 XA

Discovered 1990 December 8 by T. Nomura and K. Kawanishi at Minami-Oda.

(5686) Chiyonoura

1990 YQ. Discovered 1990 December 20 by M. Matsuyama and K. Watanabe at Kushiro.

Named for a beautiful sandy beach between Bentengahama {see planet (5293)} and Shiundai, in the southern part of the city of Kushiro {see planet (4096)}. A new port is under construction, and the beach is gradually disappearing as a result of erosion. (M 26930)

(5687) Yamamotoshinobu

1991 AB_1. Discovered 1991 January 13 by Y. Kushida and O. Muramatsu at Yatsugatake.

Named in honor of Shinobu Yamamoto (1911-), who served as director of the Gotoh Planetarium in Tokyo from 1987 to 1990. In the 1950s, he had made a great effort to establish that planetarium, following the destruction of its predecessor during World War II. Completed in 1957, the Gotoh Planetarium immediately attracted many people, as the first artificial satellites were launched at that time. (M 29145)

(5688) Kleewyck

1991 AD_2. Discovered 1991 January 12 by E. F. Helin at Palomar.

The name Klee Wyck was given by Vancouver Island's coastal people to the famous Victoria artist Emily Carr on one of her many painting expeditions into the coastal wilderness in the early 1900s. The name in the Nuu-Chah-Nulth people's language translates to "the smiling one". The name was also used for the friendly orca that served as the official mascot to the fifteenth Commonwealth Games held in Victoria, British Columbia, 1994 Aug. 18-28. By way of compensation for a small and temporary amount of light interference during the games, the Victoria Commonwealth Games Society has been strongly supportive of the asteroid research program at the University of Victoria, and the name chosen for this asteroid reflects the gratitude of the university's astronomers to the society. (M 24410)

Name proposed by the discoverer at the suggestion of J. B. Tatum. Citation prepared by A. Tanner.

(5689) Rhön

1991 RZ_2. Discovered 1991 September 9 by F. Börngen and L. D. Schmadel at Tautenburg.

Named for the range of young volcanic mountains that lie across Bavaria, Hesse and Thuringia. Name proposed by the first discoverer, who considers this region to be one of the most beautiful in Germany. (M 22831)

(5690) 1992 EU

Discovered 1992 March 7 by S. Ueda and H. Kaneda at Kushiro.

(5691) 1992 FD

Discovered 1992 March 26 by R. H. McNaught at Siding Spring.

(5692) Shirao

1992 FR. Discovered 1992 March 23 by K. Endate and K. Watanabe at Kitami.

Named in honor of Motomaro Shirao (1953-), a geologist and photographer whose specialties are volcanoes, geological features and the moon. (M 26930)

(5693) 1993 EA

Discovered 1993 March 3 by the Spacewatch at Kitt Peak.

(5694) Berényi

3051 P-L. Discovered 1960 September 24 by C. J. van Houten and I. van Houten-Groeneveld at Palomar.

Named in honor of Dénes Berényi (1928-), professor of the Institute of Nuclear Research of the Hungarian Academy of Sciences in Debrecen and for many years its director. His research activities include nuclear physics, atomic collision processes and electron spectrometry. In 1990 Berényi became vice-president of the Hungarian Academy of Sciences. (M 24918)

Name proposed by K.-O. Groeneveld, who works in the same fields.

See also the citation for planet (5699).

(5695) Remillieux

4577 P-L. Discovered 1960 September 24 by C. J. van Houten and I. van Houten-Groeneveld at Palomar.

Named in honor of Joseph Remillieux (1940-), professor at the Institut de Physique Nucléaire de l'Université de Lyon and vice-president of that university. His research interest is atomic collision processes in solids. (M 24918)

Name proposed by K.-O. Groeneveld.

See also the citation for planet (5699).

(5696) Ibsen

4582 P-L. Discovered 1960 September 24 by C. J. van Houten and I. van Houten-Groeneveld at Palomar.

Named in memory of the famous Norwegian writer Henrik Ibsen (1828-1906), whose most important tragedies are *Peer Gynt*, *Nora*, *Ghosts*, and *Hedda Gabler*. Ibsen lived in Germany for 20 years, before returning to Norway in 1891. His influence on the modern theater was very great. (M 25977)

See also the citation for planet (5699).

(5697) Arrhenius

6766 P-L. Discovered 1960 September 24 by C. J. van Houten and I. van Houten-Groeneveld at Palomar.

Named in memory of the Swedish physicochemist Svante August Arrhenius (1859-1927). A professor in Stockholm, he received the Nobel Prize for chemistry in 1903. He was the first to explain, in 1900, the formation of the tails of comets as a result of radiation pressure. (M 25977)

Arrhenius is also honored by craters on Mars and on the Moon. See also the citation for planet (5699).

(5698) Nolde

4121 T-1. Discovered 1971 March 26 by C. J. van Houten and I. van Houten-Groeneveld at Palomar.

Named in memory of the German expressionist painter Emil Nolde (1867-1957, real name Emil Hansen). He was a member of a group of artists called 'Die Brücke' and after 1926 lived in Seebüll together with his Danish wife. In 1937 his work was declared to be 'Entartete Kunst', and many of his paintings were therefore confiscated. In 1941 the government even forbade him to paint, although he secretly created his so-called 'unpainted paintings'. (M 25977)

See also the citation for planet (5699).

(5699) Munch

2141 T-3. Discovered 1977 October 16 by C. J. van Houten and I. van Houten-Groeneveld at Palomar.

Named in memory of Edvard Munch (1863-1944), a Norwegian expressionist painter. In his paintings the artist expresses his loneliness and anxiety. Some of his most famous paintings are 'The Cry', 'Jealousy', and 'The Girls on the Bridge'. (M 25977)

The discoverers of planets (5694) - (5699) stated in an e-mail from November 11, 1995 to Brian G. Marsden: "In honor of Brian Marsden the first letters of the names of minor planets (5694), (5695), (5696), (5697), (5698), (5699) form the name BRIAN M. Because we never, till now, had enough minor planets numbered after each other, to give your full name, we had to make it in this way." See also the addition to planet (1234).

(5700) Homerus

5166 T-3. Discovered 1977 October 16 by C. J. van Houten and I. van Houten-Groeneveld at Palomar.

Named for the legendary Greek poet Homerus, who depicted in his epic poem *Ilias* an episode of the battle around Troy. Without Homerus, we would know nothing of the heroes of Troy and Greece, including those whose names are given to minor planets. (M 23139)

(5701) 1929 VS

Discovered 1929 October 26 by C. W. Tombaugh at Flagstaff.

(5702) Morando

1931 FC. Discovered 1931 March 16 by M. Wolf at Heidelberg.

Named in honor of Bruno Morando (1931-1995), astronomer at the Bureau des Longitudes, Paris, where he has been director for 13 years. A former president of the Société Astronomique de France, he is now responsible for the reduction of the minor planet data from the Hipparcos satellite. (M 24122)

Name proposed and citation prepared by D. Savoie and J. Meeus, endorsed by L. D. Schmadel.

Obituaries published in Astronomie, Vol. 109. p. 324 (1995); Int. Astron. Union Symp., No. 172, p. 1-2 (1996); J. Astron. Fr., No. 50, p. 2-4 (1996).

(5703) Hevelius

1931 VS. Discovered 1931 November 15 by K. Reinmuth at Heidelberg.

Named in memory of the eminent astronomer Johannes Hevelius (1611-1687), brewer and city councillor of Danzig, now Gdansk. After studies at Leiden, London, Paris and Avignon he erected a private observatory and was an ardent observer. His *Selenographia* (1647), which contains detailed lunar maps obtained with long refractors, made him well-known. Further important works include *Cometographia*, *Machina coelestis* (astrometric observations of stars and planets), *Annus climactericus*, the Star Atlas and the Star Catalogue based on measurements with Tychonic-type quadrants and sextants. Hevelius' second wife, Catharina Elisabeth, was one of the first female astronomers. (M 24122)
Name proposed by J. Wünsch and L. D. Schmadel.

(5704) Schumacher

1950 DE. Discovered 1950 February 17 by K. Reinmuth at Heidelberg.

Named for the astronomer and geodesist Heinrich Christian Schumacher (1780-1850), who in 1821 founded the *Astronomische Nachrichten*. Living in Altona, he was professor in Copenhagen and leader of the Danish geodetic survey. Schumacher was the center of a lively correspondence with the leading astronomers of his day, and this ensured a collaboration across frontiers of the European geodetic surveys. His greatest service to astronomers was the institution of the *AN* {see also planet (1155)}. Its purpose was to disseminate, as quickly as possible, information about discoveries, ephemerides, short articles, reviews, and so forth. This was achieved by mailing single sheets of eight pages as soon as they could be completed. The spirit of this journal was truly international and astronomers of all nations could contribute in their own language. (M 24122)

Name proposed and citation prepared by L. K. Kristensen and endorsed by L. D. Schmadel.

Originally, this name was proposed for minor planet (6000) but received not enough votes by IAU Commission 20.

(5705) Ericsterken

1965 UA. Discovered 1965 October 21 by H. Debehogne at Uccle.

Named in memory of Eric Sterken (1948-1998), professional gardener and landscaper who took care of the gardens of the Brussels Planetarium. He is remembered by hundreds of people for his friendliness and modesty, and at the same time he was respected for his great efforts to preserve plantlife and wildlife. He was proud to have been the brother of a professional astronomer, Chris Sterken. (M 31610)

(5706) Finkelstein

1971 SS_1. Discovered 1971 September 23 at the Crimean Astrophysical Observatory at Nauchnyj.

Named in honor of Andrej Mikhajlovich Finkelstein (1942-), expert in relativistic celestial mechanics and radioastrometry, founder and director of the Russian Academy of Sciences' Institute of Applied Astronomy in St. Petersburg. (M 29145)
Name proposed by the Institute of Theoretical Astronomy.

(5707) Shevchenko

1976 GY_3. Discovered 1976 April 2 by N. S. Chernykh at Nauchnyj.

Named in honor of Vladislav Vladimirovich Shevchenko (1940-), head of the Lunar and Planetary Research Department of the Sternberg Astronomical Institute in Moscow and the Chairman of the Lunar Task Group of the IAU Working Group for Planetary System Nomenclature. He was a participant of the

'Zond' and 'Lunokhod' space missions and was co-author and scientific leader of the Global Mapping of the Moon and Mars projects. (M 30476)
Name suggested by the Institute of Theoretical Astronomy.

(5708) Melancholia

1977 TC_1. Discovered 1977 October 12 by P. Wild at Zimmerwald.

One of the four humors or human temperaments. The gloom besetting astronomers when the sky remains overcast for weeks on end. (M 30097)

(5709) 1977 TS_3

Discovered 1977 October 12 at the Purple Mountain Observatory at Nanking.

(5710) Silentium

1977 UP. Discovered 1977 October 18 by P. Wild at Zimmerwald.
Silence. (M 30097)
This is - by far - the shortest official naming citation ever published.

(5711) 1978 SO_4

Discovered 1978 September 27 by L. I. Chernykh at Nauchnyj.

(5712) 1979 SR

Discovered 1979 September 25 by A. Mrkos at Kleť.

(5713) 1982 FF_3

Discovered 1982 March 21 by H. Debehogne at La Silla.

(5714) Krasinsky

1982 PR. Discovered 1982 August 14 by N. S. Chernykh at Nauchnyj.

Named in honor of Georgij Al'bertovich Krasinskij [Krasinsky] (1939-), staff member of the Institute of Theoretical Astronomy from 1964 to 1987, now chief of the Ephemeris Astronomy Laboratory of the Institute of Applied Astronomy in St. Petersburg. He has investigated planetary motions by analytical and numerical methods, and he was one of the creators of the global relativistic theory of the inner planets and the moon in the 1970s. (M 30798)

Name suggested by Institute of Theoretical Astronomy and the Institute of Applied Astronomy.

(5715) Kramer

1982 SE_1. Discovered 1982 September 22 by E. Bowell at Anderson Mesa.

Named in honor of Kathryn Xymena Kramer, Development Director at Lowell Observatory. Kramer has been central in securing funding for the Observatory's Steel Visitor Center and for research purposes. Additionally she has been instrumental in greatly expanding the membership of the Friends of Lowell Observatory (currently to more than 1000 members) and was named Lowell Observatory employee of the year for 1994. (M 24411)
Citation prepared with the help of E. D. Pickard.

(5716) Pickard

1982 UH. Discovered 1982 October 17 by E. Bowell at Anderson Mesa.

Named in honor of Elizabeth D. Pickard, philanthropist and community volunteer in Flagstaff, Arizona, for her long-time support of Lowell Observatory, and in particular for spearheading local fundraising efforts that resulted in the construction of a major new visitors' center. (M 24411)
Citation prepared with the help of K. X. Kramer.

(5717) Damir

1982 UM_6. Discovered 1982 October 20 by L. G. Karachkina at Nauchnyj.

Named in memory of Alim Matveevich Damir (1894-1982), well-known physician, professor at the First and Second Medical Institutes in Moscow. (M 23139)
Name suggested by his son-in-law, S. P. Kapitsa.

(5718) 1983 PB

Discovered 1983 August 4 by A. C. Gilmore and P. M. Kilmartin at Lake Tekapo.

(5719) 1983 RX

Discovered 1983 September 7 by A. Mrkos at Kleť.

(5720) Halweaver

1984 FN. Discovered 1984 March 29 by C. S. Shoemaker and E. M. Shoemaker at Palomar.

Named in honor of Harold A. Weaver (1953-) for his work on the chemical composition of comets and for pioneering the use of the Hubble Space Telescope for cometary studies. His combination of infrared and ultraviolet spectroscopy has demonstrated the way in which different wavelength regimes must be combined to obtain a proper picture of the composition of comets. (M 27460)

Citation provided by M. F. A'Hearn following a suggestion by E. Bowell and endorsed by the discoverers.

(5721) 1984 SO_5

Discovered 1984 September 18 by H. Debehogne at La Silla.

(5722) 1986 JS

Discovered 1986 May 2 at the Palomar Observatory at Palomar.

(5723) Hudson

1986 RR_2. Discovered 1986 September 6 by E. Bowell at Anderson Mesa.

Named in honor of R. Scott Hudson (1959-) of the School of Electrical Engineering and Computer Science at Washington State University. Hudson has pioneered techniques for using delay-doppler radar images of an asteroid to estimate its shape, rotation and radar scattering properties, as well as the delay-doppler trajectory of the target's center of mass, a result that can improve orbit accuracy by several orders of magnitude. His reconstruction of (4769) Castalia is the first reasonably detailed model of the shape of an earth-crossing asteroid. For (4179) Toutatis, Hudson's inversion defined the non-principal-axis spin state and the ratios of the principal moments of inertia. Hudson's techniques are a cornerstone of radar investigation of small bodies and open the door to a variety of theoretical studies of these objects. (M 25654)
Name suggested and citation prepared by S. J. Ostro.

(5724) 1986 WE

Discovered 1986 November 22 by K. Suzuki and T. Urata at Toyota.

(5725) Nördlingen

1988 BK_2. Discovered 1988 January 23 by C. S. Shoemaker and E. M. Shoemaker at Palomar.

Named for the attractive, medieval, walled city located in the heart of the Ries {see planet (4327)} Basin, site of an impact crater formed by a comet or a minor planet 15 million years ago. Nördlingen is the central town of the Danube-Ries district of the Bavarian region of Swabia {see planet (6209)}. The first records of this city date back to A.D. 750-800, and its position on important trade routes allowed it to develop into a market town and cultural center. Its St. Georges Cathedral is the largest building ever constructed from suevite, the local stone created by the impact. (M 29670)

(5726) Rubin

1988 BN_2. Discovered 1988 January 24 by C. S. Shoemaker and E. M. Shoemaker at Palomar.

Named in honor of American astronomer Vera Cooper Rubin (1928-), who has studied the motions of gas and stars in galaxies and the motions of galaxies in clusters. Her observations have been fundamental to our understanding of the distribution of mass in the universe. A member of the National Academy of Sciences, she received the National Medal of Science in 1993 and in 1996 became only the second woman to receive the Gold Medal of the Royal Astronomical Society. (M 29670)

(5727) 1988 BB_4

Discovered 1988 January 19 by H. Debehogne at La Silla.

(5728) 1988 BJ_4

Discovered 1988 January 20 by H. Debehogne at La Silla.

(5729) 1988 TA_1

Discovered 1988 October 13 by S. Ueda and H. Kaneda at Kushiro.

(5730) 1988 TP_1

Discovered 1988 October 13 by Y. Oshima at Gekko.

(5731) Zeus

1988 VP_4. Discovered 1988 November 4 by C. S. Shoemaker and E. M. Shoemaker at Palomar.

Originally Zeus was the god of the sky and of atmospheric phenomena, of winds, clouds, rain and thunder. Later Zeus, father of gods (Athene {see planet (881)}, Artemis {see planet (105)}, Apollo {see planet (1862)}, Ares and Dionysus {see planet (3671)}) and men, became the supreme god of the Greeks, the protector of laws and morals and the dispenser of good and evil. (M 25230)

(5732) 1988 WC

Discovered 1988 November 29 by M. Arai and H. Mori at Yorii.

(5733) 1989 AQ

Discovered 1989 January 4 by S. Ueda and H. Kaneda at Kushiro.

(5734) Noguchi

1989 AL_1. Discovered 1989 January 15 by K. Endate and K. Watanabe at Kitami.

Named in honor of Souichi Noguchi (1965-), an astronaut working for the Japanese Aerospace Development Agency NASDA, chosen among 572 applicants for the position in 1996. He obtained a master's degree in aeronautics at the University of Tokyo and has been training as a mission specialist for NASA. (M 29145)

Name proposed by the second discoverer following a suggestion by K. Tomita.

(5735) Loripaul

1989 LM. Discovered 1989 June 4 by E. F. Helin at Palomar.

Named in honor of Lori L. Paul, Friend of All Animals, assistant director of Telescopes in Education (TIE) at the Mount Wilson Institute and Jet Propulsion Laboratory, environmentalist, veterinary technician, part time educator and artist, former zookeeper and laser operator. Her role in acquiring important sponsorship and recognition for TIE is particularly acknowledged. Like her many eclectic pursuits, her accomplishments have resulted from her curiosity, an aesthetic sense of excellence and attention to detail. This minor planet is being named to celebrate her wedding engagement to Rob Staehle {see planet (3875)}. (M 31024)

(5736) Sanford

1989 LW. Discovered 1989 June 6 by E. F. Helin at Palomar.

Named in honor of John Sanford, teacher, author, former president of the Orange County Astronomers, and a source of inspiration for hundreds of amateur astronomers. Recipient of the Western Amateur Astronomers' Bruce Blair Award, he has served amateur astronomers worldwide with the unselfish sharing of his photographic and observational skills for nearly twenty years. (M 24411)
Name proposed by the discoverer following a suggestion by J. B. Child.

(5737) 1989 SK

Discovered 1989 September 30 by T. Nomura and K. Kawanishi at Minami-Oda.

(5738) Billpickering

1989 UY_3. Discovered 1989 October 27 by E. F. Helin at Palomar.

Named in honor of William H. Pickering (1910-), former director of the Jet Propulsion Laboratory. Under his guidance instrumented probes were first sent to the inner planets, and work was begun on the "Grand Tour" of the outer planets. He was a pioneer in the development of telemetry, which was to become the essential tool of the space exploration program. In his years as an administrator he encouraged major advances in space science and technology. Many of these led to new inventions, and some spawned new industries. His inspiration and leadership established the U.S. program for the exploration of the solar system. (M 27128)

(5739) 1989 WK_2

Discovered 1989 November 24 by R. H. McNaught at Siding Spring.

(5740) Toutoumi

1989 WM_3. Discovered 1989 November 29 by Y. Oshima at Gekko.

Named for the area in the western part of Shizuoka prefecture that is the location of the Nakano-Gakuen Astro-Geological College. Several of its graduates have worked at the observatories belonging to the International Foundation for Cultural Harmony. (M 33786)

(5741) 1989 XC

Discovered 1989 December 2 by W. Kakei and M. Kizawa and T. Urata at Oohira.

(5742) 1990 TN_4

Discovered 1990 October 9 by R. H. McNaught at Siding Spring.

(5743) Kato

1990 UW. Discovered 1990 October 19 by M. Akiyama and T. Furuta at Mishima.

Named in honor of Yasuo Kato (1949-1982), a famous climber, born in Omiya, Saitama Prefecture. After reaching the summit of Mt. Everest in 1973, he lost from frostbite all of his toes and three finger tips of his right hand. In spite of that, he continued climbing. In 1982 he made the first winter ascent of Everest but was reported missing on the way down. (M 25230; M 25993; M 32127)

(5744) Yorimasa

1990 XP. Discovered 1990 December 14 by A. Natori and T. Urata at Yakiimo.

Named for a Japanese military commander and well-known poet in the late Heian era, Minomoto-no Yorimasa (1104-1180). He advised Prince Mochihitoou to fight the Heike. However, word of the plan leaked out, and Yorimasa took his own life. (M 31024)
Name proposed by the second discoverer.

(5745) 1991 AN
Discovered 1991 January 9 by T. Hioki and S. Hayakawa at Okutama.

(5746) 1991 CK
Discovered 1991 February 5 by M. Arai and H. Mori at Yorii.

(5747) 1991 CO_3
Discovered 1991 February 10 by R. H. McNaught at Siding Spring.

(5748) Davebrin
1991 DX. Discovered 1991 February 19 by E. F. Helin at Palomar.

Named in honor of David Brin, astrophysicist and popular science-fiction writer. His scientific training has given credibility and authority to his exploration of new ideas and story themes. He is the recipient of many prestigious awards, among them the Hugo, Nebula and LOCUS. (M 23139)

(5749) 1991 FV
Discovered 1991 March 17 by E. F. Helin at Palomar.

(5750) Kandatai
1991 GG_1. Discovered 1991 April 11 by A. Takahashi and K. Watanabe at Kitami.

Named in honor of Tai Kanda (1938-), a staff member of the Japanese National Astronomical Observatory who works for the Public Service Division and as the editor of *Rika Nenpyo* (Chronological Science Tables), published by the Observatory. He grew up in a family of astronomers as a son of Kiyoshi Kanda and a nephew of Shigera Kanda {see planet (2248)}. (M 29145)
Name proposed by the second discoverer following a suggestion by K. Tomita.

(5751) Zao
1992 AC. Discovered 1992 January 5 by M. Koishikawa at Sendai.

Mt. Zao is the general name of the volcanic mountains that lie in the northern part of mainland Japan. The Zao range forms the border between Yamagata and Miyagi Prefectures and was designated as a National Park in 1963. With its abundance of nature and clear skies, Mt. Zao is one of the most attractive mountains for mountaineers, skiers and starwatchers. (M 25230)

(5752) 1992 CJ
Discovered 1992 February 10 by N. Kawasato at Uenohara.

(5753) Yoshidatadahiko
1992 EM. Discovered 1992 March 4 by K. Endate and K. Watanabe at Kitami.

Named in honor of Tadahiko Yoshida (1941-), the director of the Tsukuba regional office of the company Advanced Engineering Service, which supports space development for NASDA. (M 29145)

(5754) 1992 FR_2
Discovered 1992 March 24 by S. Ueda and H. Kaneda at Kushiro.

(5755) 1992 OP_7
Discovered 1992 July 20 by H. Debehogne and A. López G. at La Silla.

(5756) Wassenbergh
6034 P-L. Discovered 1960 September 24 by C. J. van Houten and I. van Houten-Groeneveld at Palomar.

Named in honor of Or Wassenbergh (1924-), from 1977 to 1994 professor of air and space law, on the occasion of the valedictory address at Leiden University. During World War II Or joined the Dutch underground forces. Later he obtained

his doctorate, working first at KLM (Royal Dutch Air Line) and becoming more and more involved, not only with aviation policy and air law, but also with space law. He was vice-president of foreign relations and cooperation at KLM until 1990. He has published several books and articles about this subject. (M 24123)
The name was proposed by his secretary and his colleagues in Leiden.

(5757) Tichá

1967 JN. Discovered 1967 May 6 by C. U. Cesco and A. R. Klemola at El Leoncito.

Named in honor of Jana Tichá, Czech astronomer, director of the Kleť Observatory since 1992. Under her leadership, the Kleť program for astrometric observations of minor planets and comets was modernized with the introduction of a CCD camera, computers and catalogues. (M 27128; M 27147)
Name proposed following a suggestion by her husband, Miloš Tichý.

(5758) 1976 QZ_1

Discovered 1976 August 20 by M. R. Cesco at El Leoncito.

(5759) Zoshchenko

1980 BJ_4. Discovered 1980 January 22 by L. G. Karachkina at Nauchnyj.

Named in honor of the outstanding Russian writer-satirist Mikhail Mikhailovich Zoshchenko (1894-1958). (M 23541)

The name is given on the occasion of the one-hundredth anniversary of his birth, following a suggestion by L. R. Nemirovskij.

(5760) 1981 EX_{13}

Discovered 1981 March 1 by S. J. Bus at Siding Spring.

(5761) 1981 ED_{21}

Discovered 1981 March 2 by S. J. Bus at Siding Spring.

(5762) 1981 EG_{28}

Discovered 1981 March 2 by S. J. Bus at Siding Spring.

(5763) 1982 MA

Discovered 1982 June 23 by A. C. Gilmore and P. M. Kilmartin at Lake Tekapo.

(5764) 1985 CS_1

Discovered 1985 February 10 by H. Debehogne at La Silla.

(5765) Izett

1986 GU. Discovered 1986 April 4 by C. S. Shoemaker and E. M. Shoemaker at Palomar.

Named in honor of Glen A. Izett, an American geologist of many talents. Izett has carried out the most detailed studies of the Cretaceous-Tertiary boundary clay layers in the western United States and their implications for the great impact event that terminated the Cretaceous. He has also contributed the most precise determinations of ages of tektites from the extensive Australasian strewn field. His work with John Obradovich on the ages of volcanic ashes and tektites has led to the most highly refined time scale for the reversals of the geomagnetic field in the late Quaternary. (M 29671)

(5766) 1986 QR_3

Discovered 1986 August 29 by H. Debehogne at La Silla.

(5767) Moldun

1986 RV_2. Discovered 1986 September 6 by E. Bowell at Anderson Mesa.

Named for the town of Meudon, Hauts-de-Seine, location of the Astrophysics

Section of the Paris Observatory, where the discoverer had the pleasure of working for about five years. Moldun is the old Gaelic form of the names and means "sand hill". Meudon has in the past been home to Auguste Rodin {see planet (6258)}, Jean Arp, Richard Wagner {see planet (3992)}, Marcel Dupré and Georges Enescu. (M 27460)

Proposed by the discoverer following suggestions by M. A. Barucci and M. Fulchignoni.

(5768) Pittich

1986 TN_1. Discovered 1986 October 4 by E. Bowell at Anderson Mesa.

Named in honor of Eduard M. Pittich (1940-) of the Astronomical Institute of the Slovak Academy of Sciences, Bratislava. Pittich has worked on the dynamics and evolution of comets, and also on the motions of dust particles in heliocentric orbits, as influenced by sunlight. He participated in setting up the Astronomical and Geophysical Observatory of the Comenius University in Modra-Piesky and has been active in the popularization of astronomy as a successful book author. (M 27460)

Citation provided by H. Rickman.

(5769) Michard

1987 PL. Discovered 1987 August 6 by the CERGA at Caussols.

Named in honor of Raymond Michard (1925-), administrator of the Observatoire de la Côte d'Azur since 1987 and ardent supporter of the CERGA {see planet (2252)} Schmidt. He had earlier studied the sun at Sacramento Peak and led the Service Solaire at Meudon during 1971-76. He later studied the morphology of elliptical galaxies in Texas and at Nice, where he was director in 1981. (M 23541)

(5770) 1987 RY

Discovered 1987 September 12 by H. Debehogne at La Silla.

(5771) Somerville

1987 ST_1. Discovered 1987 September 21 by E. Bowell at Anderson Mesa.

Named for Mary Somerville (née Fairfax, 1780-1872), one of Europe's most distinguished women scientists of her time. Her treatise on Laplace's {see planet (4628)} *Mécanique Céleste* brought her widespread recognition, including an Honorary Fellowship of the Royal Astronomical Society. The Royal Society, unable to elect a woman as a Fellow, commissioned a sculpted bust to be displayed in their premises. J. C. Adams {see planet (1996)} is said to have told her that it was the statement, in the sixth edition of her *On the connexion of the physical sciences*, that the departure of Uranus from its predicted path might reveal the existence of an undiscovered planet that minded him to carry out his calculations that successfully predicted the position of Neptune. She campaigned for education rights and voting rights for women. Somerville College in the University of Oxford was named in her honor when founded in 1878 for the education of women undergraduates. (M 25444)

Name suggested and citation prepared by S. A. and J. Mitton.

(5772) 1988 LB

Discovered 1988 June 15 by E. F. Helin at Palomar.

(5773) 1989 NO

Discovered 1989 July 2 by E. F. Helin at Palomar.

(5774) 1989 NR

Discovered 1989 July 2 by E. F. Helin at Palomar.

(5775) Inuyama

1989 SP. Discovered 1989 September 29 by Y. Mizuno and T. Furuta at Kani.

Named for a city in the northern part of Aichi Prefecture. Rich in natural beauty and featuring the Hida-Kiso River National Park, Inuyama is also known for its traditional events, such as the Nihon-Rhine Rapid Ride and Inuyama cormorant fishing. It also boasts some historically valuable architecture, including Japan's oldest donjon (in the Inuyama Castle) and the Tea Arbor JO-AN. Both are national treasures. (M 33385)

Name proposed by the discoverers following a suggestion by their teacher, T. Yamada.

(5776) 1989 UT_2

Discovered 1989 October 29 by T. Hioki and N. Kawasato at Okutama.

(5777) 1989 XF

Discovered 1989 December 3 by Y. Mizuno and T. Furuta at Kani.

(5778) 1989 YF_5

Discovered 1989 December 28 by E. W. Elst at St. Michel.

(5779) 1990 BC_1

Discovered 1990 January 23 by S. Ueda and H. Kaneda at Kushiro.

(5780) Lafontaine

1990 EJ_2. Discovered 1990 March 2 by E. W. Elst at La Silla.

Named in memory of the French lyric poet Jean de la Fontaine (1621-1695), on the occasion of the 300th anniversary of his death. He is well known from the "fables" (first published in six books in 1668, with a total of 12 books by 1694) and his "contes". The fables were inspired by Greek and Latin writers such as Aesop and Phaedrus, and the contes, mainly by the French writers Rabelais {see planet (5666)} and Marguerite de Navarre. Although la Fontaine continued the classical tradition, he created his own style and language that exhibits versatility and fecundity. He is considered one of the greatest classical French writers. (M 23541)
Citation material provided by K. Leterme at the request of the discoverer.

(5781) Barkhatova

1990 SM_{28}. Discovered 1990 September 24 by G. R. Kastel' and L. V. Zhuravleva at Nauchnyj.

Named in memory of Claudia Alexandrovna Barkhatova (1917-1990), founder of the Kourovka {see planet (4964)} Observatory, Ural State University, Ekaterinburg. Under her guidance Kourovka became one of the best university observatories in Russia. Barkhatova made valuable contributions to stellar astronomy and to our understanding of Galactic structure. In particular, she investigated the dependence of angular diameters of open clusters on interstellar extinction, and this led to a revision of the distance scale. She also established that the orbital eccentricities of the clusters depend on their age. Her disciples work in practically all the astronomical institutions of Russia. (M 24123)
Name proposed by the first of the discoverers.

(5782) 1991 AF

Discovered 1991 January 7 by R. H. McNaught at Siding Spring.

(5783) Kumagaya

1991 CO. Discovered 1991 February 5 by T. Hioki and S. Hayakawa at Okutama.

Kumagaya, also known as the 'City of Cherry Blossoms' and 'Rugby Town', is an industrial city located northwest of Tokyo. Kumagaya is famous for the

Yukoku Temple, built for the warrior Naozane {see planet (5238)}, and for the Uchiwa Festival. (M 25230)

(5784) 1991 CY

Discovered 1991 February 9 by A. Natori and T. Urata at Yakiimo.

(5785) Fulton

1991 FU. Discovered 1991 March 17 by E. F. Helin at Palomar.

Named in honor of Joseph A. Fulton, hardware engineer specializing in electromechanical devices. Over the years he assisted in the digitization of the Planet Crossing Asteroid Survey films and Oschin Schmidt plates for the discoverer's research on minor planets. This naming is made on the occasion of Fulton's retirement after 37 years of dedicated work at the Jet Propulsion Laboratory. (M 30097)

(5786) Talos

1991 RC. Discovered 1991 September 3 by R. H. McNaught at Siding Spring.

A cousin of Icarus {see planet (1566)}, Talos was murdered by his uncle, Daedalus {see planet (1864)}, through jealousy of his inventiveness. There is another Talos in legend, a living bronze work of art created by the god Hephaistos {see planet (2212)}. Talos would run around the island of Crete and ward off strangers from its shores by throwing stones. Unwanted visitors unlucky enough to be caught would be held to his bosom, which he made red hot. It is believed that Talos is the winged figure, running with large strides and holding a stone in each hand, depicted in coins from the town of Phaestos. (M 23542)
Named by the discoverer following a suggestion by D. J. Asher.

The semimajor axis, eccentricity and inclination of this asteroid show a remarkable similarity to (1566) Icarus.

(5787) 1992 FA$_1$

Discovered 1992 March 26 by S. Ueda and H. Kaneda at Kushiro.

(5788) 1992 NJ

Discovered 1992 July 1 by R. H. McNaught at Siding Spring.

(5789) Sellin

4018 P-L. Discovered 1960 September 24 by C. J. van Houten and I. van Houten-Groeneveld at Palomar.

Named in honor of Ivan A. Sellin (1939-), professor at Oak Ridge National Laboratory and at the University of Tennessee in Knoxville. In 1976 he received the Humboldt Award. His research interests include photon and atomic collision processes, and the results of his research are given in more than 200 publications. (M 24918)
Name proposed by K.-O. Groeneveld.

(5790) Nagasaki

9540 P-L. Discovered 1960 October 17 by C. J. van Houten and I. van Houten-Groeneveld at Palomar.

Named for the Japanese city of Nagasaki in the hope that this minor planet may be a symbol for world peace. (M 25230)
Name proposed by T. Gehrels.

(5791) Comello

4053 T-2. Discovered 1973 September 29 by C. J. van Houten and I. van Houten-Groeneveld at Palomar.

Named in honor of Georg Comello (1942-), Dutch amateur astronomer. His

interests in astronomy include variable stars, comets and traveling around the world to observe solar eclipses. Since 1960 he has been employed at the Kapteyn Institute of the University of Groningen, currently as a librarian. He assisted L. Plaut {see planet (1986)} in measuring and analyzing his plates of variable stars and is still on the board of the Working Group for Variable Star Observers. For several years he has been the draftsman of the *Sterrengids* (the Dutch yearbook on astronomical events), for which he prepares in particular finding charts for minor planets. He has written articles for *Hemel en Dampkring* and *Zenit* and contributes to newspapers, radio and television. (M 24918)
Name proposed and citation prepared by T. Jurriens.

(5792) Unstrut

1964 BF. Discovered 1964 January 18 by F. Börngen at Tautenburg.

Named for the Thuringian river that flows for 189 km from Eichsfeld {see planet (442)}, past the cathedral town of Naumburg, into the Saale {see planet (5409)}. (M 23353)

(5793) 1975 TK_6

Discovered 1975 October 5 at the Felix Aguilar Observatory at El Leoncito.

(5794) Irmina

1976 SW_3. Discovered 1976 September 24 by N. S. Chernykh at Nauchnyj.

Named in memory of Irma Mikhailovna Golodyaevskaya (1931-1956), a student of the Moscow Conservatory and talented singer who tragically perished at the very beginning of her development as an artist. (M 30097)

(5795) Roshchina

1978 SH_1. Discovered 1978 September 27 by L. I. Chernykh at Nauchnyj.

Named in memory of Elena Olegovna Roshchina (1966-1994), a journalist and cinema critic from the Ivanovo region of Russia and a young poet tragically killed. She took an interest in astronomy, and some of her poems were dedicated to cosmic themes. (M 30097)

(5796) 1978 VK_5

Discovered 1978 November 7 by E. F. Helin and S. J. Bus at Palomar.

(5797) Bivoj

1980 AA. Discovered 1980 January 13 by A. Mrkos at Kleť.

This Amor-type object is named for the honorable hero of ancient Bohemian myths, known for his great courage and other good qualities. Bivoj conquered a colossal wild boar barehanded to save the crops and lives of his people. He later became acquainted with and married Kazi, the oldest daughter of prince Krok {see planet (3102)}. (M 27735)
Name suggested by J. Tichá, M. Tichý and Z. Moravec.

(5798) 1980 RL_7

Discovered 1980 September 13 by S. J. Bus at Palomar.

(5799) Brewington

1980 TG_4. Discovered 1980 October 9 by C. S. Shoemaker at Palomar.

Named in honor of Howard Brewington, discoverer of four comets beginning in 1989. To improve his chances of making a find, he moved with his wife Trudy from South Carolina to the mountaintop town of Cloudcroft, New Mexico, and built an observatory to house his homebuilt telescopes. (M 24918)

(5800) 1982 UV_1

Discovered 1982 October 16 by A. Mrkos at Kleť.

(5801) 1984 BK

Discovered 1984 January 26 by A. Mrkos at Kleť.

(5802) 1984 HL_1

Discovered 1984 April 27 by W. Ferreri and V. Zappalà at La Silla.

(5803) 1984 OA

Discovered 1984 July 21 by A. Mrkos at Kleť.

(5804) 1985 RL_1

Discovered 1985 September 9 by A. Mrkos at Kleť.

(5805) Glasgow

1985 YH. Discovered 1985 December 18 by E. Bowell at Anderson Mesa.

Named for the city of Glasgow, Scotland, and for the Astronomical Society of Glasgow, whose 100th anniversary in this year being marked by a civic dinner hosted by the Glasgow City Council. (M 24123)
Citation provided by M. Kennedy.

(5806) Archieroy

1986 AG_1. Discovered 1986 January 11 by E. Bowell at Anderson Mesa.

Named in honor of Archibald Edmiston Roy (1924-), Scottish astrophysicist, teacher and writer. Although Roy has studied stellar interiors, neural networks, the history of astronomy, as well as psychical phenomena, he is best known as a celestial mechanician. He has worked on mean-motion commensurabilities, the restricted and general three-body problems, high-order Taylor series and the long-term stability of the solar system. While teaching in the Department of Astronomy, Glasgow University (a department he headed for five years), Roy had occasion to write four textbooks. In addition, he has published six works of fiction. (M 24123)

(5807) Mshatka

1986 QA_4. Discovered 1986 August 30 by L. I. Chernykh at Nauchnyj.

Named for the country estate of Nikolaj Yakovlevich Danilevskij (1822-1885), thinker, philosopher, sociologist, publicist and natural scientist, whose book *Russia and Europe* contains ideas that are still vital today. Many well-known Russian writers stayed at Mshatka, and Danilevskij is buried there. (M 34622)

(5808) Babel'

1987 QV_{10}. Discovered 1987 August 27 by L. G. Karachkina at Nauchnyj.

Named in honor of the outstanding Russian writer and dramatist Isaac Emmanuilovich Babel' (1894-1940). (M 23542)

The name is given on the occasion of the one-hundredth anniversary of his birth, following a suggestion by L. R. Nemirovskij.

(5809) Kulibin

1987 RG_6. Discovered 1987 September 4 by L. V. Zhuravleva at Nauchnyj.

Named in memory of Ivan Petrovich Kulibin (1735-1818), brilliant Russian self-educated mechanical engineer. He designed original clockwork pieces, planned a bridge over the Neva {see planet (1603)} river and constructed various labor-saving devices. (M 34622)

(5810) 1988 EN

Discovered 1988 March 10 by Y. Oshima at Gekko.

(5811) Keck

1988 KC. Discovered 1988 May 19 by E. F. Helin at Palomar.

Named in honor of Howard B. Keck, chairman and president emeritus of the W. M. Keck Foundation. The foundation was established in 1954 by his father, William M. Keck, Sr., to support higher education, medical research and science. Under Howard Keck's leadership, the Foundation provided the grants to build the giant twin telescopes of the W. M. Keck Observatory on the summit of Mauna Kea. (M 27128)

At the suggestion of E. C. Stone, this minor planet is being named on the occasion of the dedication of the second Keck Telescope on 1996 May 8.

(5812) Jayewinkler

1988 PJ_1. Discovered 1988 August 11 by A. J. Noymer at Siding Spring.

Named in honor of Jaye Scott Winkler, a friend of the discoverer. A resident of Cambridge, Massachusetts, she is an alumna of the Harvard School of Public Health, and also an accomplished equestrienne. (M 24411)

(5813) 1988 VL

Discovered 1988 November 3 by T. Kojima at Chiyoda.

(5814) 1988 XW_1

Discovered 1988 December 11 by S. Ueda and H. Kaneda at Kushiro.

(5815) Shinsengumi

1989 AH. Discovered 1989 January 3 by T. Seki at Geisei.

Shinsengumi was the name of a 300-member group of samurai that organized the defense of the Tokugawa shogunate and kept order in Kyoto {see planet (4352)} during 1863-1869. (M 24765; M 24781)

(5816) Potsdam

1989 AO_6. Discovered 1989 January 11 by F. Börngen at Tautenburg.

Named for the capital of the German state of Brandenburg. Besides Berlin, Potsdam was the residence of Brandenburgian Electors and Prussian kings (Friedrich the Great, Friedrich Wilhelm I). Several famous castles, e.g. Sanssouci and the New Palace are located there. After World War II the Potsdam Conference took place at the Cecilienhof Palace. The Astrophysikalisches Observatorium Potsdam, devoted to the introduction and development of new research methods in astrophysics, was founded in 1874 at the Telegrafenberg. The well-known Einstein Tower for solar studies was erected by the architect Erich Mendelsohn in 1922. (M 23542)

This planet is named on the occasion of the 1994 meeting of the Astronomische Gesellschaft in Potsdam.

(5817) 1989 RZ

Discovered 1989 September 5 by E. F. Helin at Palomar.

(5818) 1989 RC_1

Discovered 1989 September 5 by A. C. Gilmore and P. M. Kilmartin at Lake Tekapo.

(5819) 1989 UZ_4

Discovered 1989 October 29 by S. J. Bus at Cerro Tololo.

(5820) Babelsberg

1989 UF_7. Discovered 1989 October 23 by F. Börngen at Tautenburg.

Named for a residential district of the city of Potsdam {see planet (5816)}, known as the summer residence of Emperor Wilhelm I and for its famous film studios.

The Berlin Observatory, founded in 1700, was transferred because of bad seeing conditions to Babelsberg before World War I. Therefore, the Babelsberg Observatory is the oldest still-existing astronomical institute in the German-speaking world. Former directors have included J. E. Bode {see planet (998)}, J. F. Encke, W. Foerster and K. H. Struve. The assistant astronomer J. G. Galle {see planet (2097)} discovered Neptune at this observatory in 1846. (M 23542)

This planet is named on the occasion of the 1994 meeting of the Astronomische Gesellschaft in Potsdam.

(5821) 1989 VV

Discovered 1989 November 4 by W. Kakei and M. Kizawa and T. Urata at Oohira.

(5822) 1989 WL

Discovered 1989 November 21 by T. Hioki and S. Hayakawa at Okutama.

(5823) Oryo

1989 YH. Discovered 1989 December 20 by T. Seki at Geisei.

Named for Oryo Narasaki (1842-1913), wife of the Japanese revolutionary Ryoma Sakamoto {see planet (2835)}. (M 24765)

(5824) Inagaki

1989 YM. Discovered 1989 December 24 by T. Seki at Geisei.

Named in honor of Minoru Inagaki (1958-), well-known guitarist from Akashi city. (M 24765)

(5825) Rakuyou

1990 BR_1. Discovered 1990 January 21 by A. Sugie at Taga.

Named for the Kyoto {see planet (4352)} city Rakuyou technical high school, originally established in 1894 as Kyoto city dyeing and weaving school. Since its foundation the school has consistently been concerned with industrial technology and has graduated many young technicians. (M 27735)

Name proposed by the discoverer following a suggestion by S. Sakabe.

(5826) 1990 DB

Discovered 1990 February 16 by S. Ueda and H. Kaneda at Kushiro.

(5827) 1990 VB_{15}

Discovered 1990 November 15 by L. I. Chernykh at Nauchnyj.

(5828) 1991 AM

Discovered 1991 January 14 by the Spacewatch at Kitt Peak.

(5829) Ishidagoro

1991 CT_1. Discovered 1991 February 11 by S. Otomo and O. Muramatsu at Kiyosato.

Named in memory of Goro Ishida (1924-1992), who was associate director of the Okayama Astrophysical Observatory {see planet (2084)} (a branch of the Tokyo Astronomical Observatory) for over a quarter of a century and contributed much to astronomical research. After retirement he was actively engaged in literary work, notably as president of the Herschel Society of Japan. He also led several study tours sponsored by that society to Europe and South Africa. He was a man of wide tastes and well known as a researcher of *Kabuki* (traditional drama). (M 25230)

Named by the discoverers following a suggestion by S. Kimura and T. Ohtani.
Obituary published in Bull. Am. Astron. Soc., Vol. 25, No. 4, p. 1497 (1993).

(5830) 1991 EG
Discovered 1991 March 9 by T. Niijima and T. Urata at Ojima.

(5831) 1991 JG
Discovered 1991 May 4 by S. Ueda and H. Kaneda at Kushiro.

(5832) 1991 LE_1
Discovered 1991 June 15 by E. F. Helin at Palomar.

(5833) 1991 PQ
Discovered 1991 August 5 by H. E. Holt at Palomar.

(5834) 1992 SZ_{14}
Discovered 1992 September 28 by S. Ueda and H. Kaneda at Kushiro.

(5835) Mainfranken
1992 SP_{24}. Discovered 1992 September 21 by F. Börngen at Tautenburg.

Named for the German district of Franken in the northern part of Bavaria, marked by the Main river that originates both in the Fichtelgebirge and in the Fränkische Schweiz. The discoverer's mother and her ancestors are from this district. (M 23353)

(5836) 1993 MF
Discovered 1993 June 22 by E. F. Helin and K. J. Lawrence at Palomar.

(5837) Hedin
2548 P-L. Discovered 1960 September 24 by C. J. van Houten and I. van Houten-Groeneveld at Palomar.

Named in memory of Sven Hedin (1865-1952), famous Swedish explorer of Central Asia. In his days there were still many unexplored spots on the map of our planet. Hedin made many expeditions and discovered, *inter alia,* the old Silk route through Asia, the Takla Makan and Gobi deserts, Lake Lop Nor and the Tarim river, as well as the sources of the Brahmaputra and Indus rivers. He was renowned, not only for his scientific papers, but also for writing articles that were interesting for laymen to read. (M 25978)
Hedin is also honored by a lunar crater.

(5838) Hamsun
2170 T-2. Discovered 1973 September 29 by C. J. van Houten and I. van Houten-Groeneveld at Palomar.

Named in memory of Knut Hamsun, pseudonym of Knut Pedersen (1859-1952), the Norwegian novelist, dramatist, poet and winner of the Nobel Prize for literature in 1920. He achieved his greatest success with the book *Sult* (*Hunger*) in 1890. Among his other best known works are *Growth of the Soil, Pan, Mysteries* and *Victoria.* (M 25978)
Name suggested by N. Arend.

(5839) GOI
1974 SJ_3. Discovered 1974 September 21 by N. S. Chernykh at Nauchnyj.

Named for Gosudarstvennyj Opticheskij Institut, the State Optical Institute, and its first director, academician Dmitrij Sergeevich Rozhdestvenskij (1876-1940), an outstanding physicist-optician. The institute has undertaken important investigations in the field of physical and applied optics. (M 32345)

The name is given on the occasion of the 80th anniversary of its founding (1918 Dec. 15).

(5840) 1978 ON

Discovered 1978 July 28 at the Perth Observatory at Bickley.

(5841) Stone

1982 ST. Discovered 1982 September 19 by E. F. Helin at Palomar.

Named in honor of Edward C. Stone for his distinguished achievements in expanding the frontiers of robotic space flight, planetary exploration, interplanetary science and the remote sensing of the earth. While highly respected for his scientific contributions, he has also achieved outstanding success as director of the Jet Propulsion Laboratory, primarily through his perspicacity and an approach to management that blends diverse individuals and ideas into a productive organization. He is also Morrison professor of physics and a vice president at the California Institute of Technology. Among his many awards and honors he received the National Medal of Science. (M 26425)

Citation prepared by M. S. Reid with the enthusiastic endorsement of the JPL staff.

(5842) 1986 CV_1

Discovered 1986 February 8 by H. Debehogne at La Silla.

(5843) 1986 UG

Discovered 1986 October 30 by K. Suzuki and T. Urata at Toyota.

(5844) 1986 UQ

Discovered 1986 October 28 by Z. Vávrová at Kleť.

(5845) 1988 QP

Discovered 1988 August 19 by R. H. McNaught at Siding Spring.

(5846) Hessen

1989 AW_6. Discovered 1989 January 11 by F. Börngen at Tautenburg.

Named for the mountainous German state. The most important towns in the region are Frankfurt/Main, Wiesbaden (the capital, see also planets (717) and (765)), Kassel and Darmstadt. (M 23793)

(5847) Wakiya

1989 YB. Discovered 1989 December 18 by K. Endate and K. Watanabe at Kitami.

Named in honor of Nanayo Wakiya (1953-), who joined the Japan Planetarium Laboratory in 1985 with the space artist Numazawa {see planet (5121)}, after working for the Mito and Niigata planetaria. Among her many activities she writes articles for astronomical magazines and directs programs about space on television. (M 29145)

(5848) Harutoriko

1990 BZ_1. Discovered 1990 January 30 by M. Matsuyama and K. Watanabe at Kushiro.

Named for a small lake in Kushiro {see planet (4096)}. The peaceful and popular park area around the lake is designated as a National Natural Monument and contains historical spots, science museums and cultural facilities. The lake is renowned for the mutated Crucian carp, Hibuna. (M 26930)

(5849) 1990 HF_1

Discovered 1990 April 27 by E. F. Helin at Palomar.

(5850) Masaharu

1990 XM. Discovered 1990 December 8 by K. Endate and K. Watanabe at Kitami.

Named in honor of Masaharu Suzuki (1949-), who is in charge of the Planetarium Programs Production Division of the Gotoh Optical Laboratory. He is also known for his production of 70-mm large-format films. (M 29145)

(5851) 1991 DM_1

Discovered 1991 February 23 by S. Inoda and T. Urata at Karasuyama.

(5852) Nanette

1991 HO. Discovered 1991 April 19 by C. S. Shoemaker and D. H. Levy at Palomar.

Named in honor of Nanette and Mark Vigil, daughter and son-in-law of Wendee and David Levy {see, respectively, planets (6485) and (3673)}, at their suggestion. (M 30476)

(5853) 1992 QG

Discovered 1992 August 26 by S. Ueda and H. Kaneda at Kushiro.

(5854) 1992 UP

Discovered 1992 October 19 by S. Ueda and H. Kaneda at Kushiro.

(5855) Yukitsuna

1992 UO_2. Discovered 1992 October 26 by A. Natori and T. Urata at Yakiimo.

Named for a Japanese military commander in the late Heian era, Minamoto-no Yukitsuna. When Yoshinaka {see planet (4574)} invaded Kyoto in 1183, Yukitsuna also fought against the Heike, and in 1185 he obstructed Yoshitsune's {see planet (3178)} departure from the city. (M 31024)

Name proposed by the second discoverer.

(5856) 1994 AL_2

Discovered 1994 January 5 by S. Ueda and H. Kaneda at Kushiro.

(5857) Neglinka

1975 TM_2. Discovered 1975 October 3 by L. I. Chernykh at Nauchnyj.

Named for a small river that flows into the Moskva river. On Borovitskij {see planet (5858)} hill, near the mouth of the Neglinka, a settlement appeared in the eleventh century. (M 30097)

This name and the three next are given on the occasion of the celebration of the 850th anniversary of the city of Moscow in 1997. See also planets (5858), (5859) and (5887), respectively.

(5858) Borovitskia

1978 SU_5. Discovered 1978 September 28 by L. I. Chernykh at Nauchnyj.

Named for Borovitskij hill (from the Russian word *bor* or pine forest), the location of the first buildings in the ancient settlement of Moscow. In the fourteenth century a stone fortress, known as the Kremlin, was erected on the hill. (M 30097)

(5859) Ostozhenka

1979 FD_2. Discovered 1979 March 23 by N. S. Chernykh at Nauchnyj.

Named for one of the oldest streets in the center of Moscow. Ostozhenka, derived from the Russian word *stog*, or haystack, shows that the street was built on a former hayfield. (M 30098)

(5860) 1981 QE_1

Discovered 1981 August 28 by Z. Vávrová at Kleť.

(5861) Glynjones

1982 RW. Discovered 1982 September 15 by E. Bowell at Anderson Mesa.

Named in memory of Kenneth Glyn Jones (1915-1995), a British historian of astronomy. Glyn Jones is the author of the *Messier's Nebulae and Star Clusters*

(1968). He was one of the founders of the Webb Society, a British-based deep-sky society formed in 1967, and served as its president until 1991. Membership of the Webb Society is now 440 in more than twenty countries worldwide. (M 25654)

Name suggested and citation prepared by R. W. Argyle, endorsed by the executive committee of the Webb Society.
Obituary published in Q.J.R. Astron. Soc., Vol. 37, No. 2, p. 265-266 (1996).

(5862) Sakanoue

1983 AB. Discovered 1983 January 13 by T. Seki at Geisei.

Named in honor of Tsutomu Sakanoue (1921-), professor emeritus of Kyushu University whose specialties included agricultural meteorology, countermeasures against meteorological disasters, medical meteorology and rainmaking. An amateur astronomer with particular interests in atmospheric seeing, the green flash and shadow bands, he contributed to the popularization of astronomy as an advisor at several science museums. He also served as vice president and president of the Oriental Astronomical Association. (M 32788)

Name proposed by the discoverer following a suggestion by S. Murayama, T. Sato and A. Fujii.

(5863) Tara

1983 RB. Discovered 1983 September 7 by C. S. Shoemaker and E. M. Shoemaker at Palomar.

In Indian Hinduism, the star-goddess Tara is a manifestation of the queen of time. She is the absolute, unquenchable hunger that propels all life. Among the Buddhists and Jains, as well as in Tibetan Lamaism, she became the symbol of spiritual hunger for release from the purely physical world. As such, Tara is the goddess of self-mastery and mysticism. She is a compassionate goddess, who sees life for the game it is. (M 25230)

(5864) Montgolfier

1983 RC_4. Discovered 1983 September 2 by N. G. Thomas at Anderson Mesa.

Named in memory of the Montgolfier brothers, Jacques Etienne (1745-1799) and Joseph Michal (1740-1810), who engineered the first manned free flight in a hot-air balloon, sailing nine km over Paris in 25 minutes on 1783 Nov. 25. Their father owned and they helped manage a paper-manufacturing business. While burning some waste paper they noticed how a paper bag was sent aloft by the hot air. (M 26930)

(5865) Qualytemocrina

1984 QQ. Discovered 1984 August 31 by A. Mrkos at Kleť.

Named for the *International Comet Quarterly*, which serves as an international archive of photometric data on comets, striving to maintain a useful bridge between amateur and professional astronomers; the naming occurs with the journal's century issue. Qualytemocrina is a play on the journal's title, each letter of the title being used and with a combination of word-order and letter-order reversal. The discoverer of this object was one of the most prolific comet discoverers and astrometrists of recent decades. (M 28089)

Name suggested by D. W. E. Green, who made the key identification for this minor planet.

(5866) Sachsen

1988 PM_2. Discovered 1988 August 13 by F. Börngen at Tautenburg.

Named for the state of Sachsen, the richest German region in the 15th and 16th centuries. The zenith of the Saxon electors (Dresden baroque) was during the reign of Friedrich August I (August the Strong, 1694-1733). The most important towns in the territory are Dresden (the capital, see planets (263) and (3053)), Leipzig, Chemnitz and Zwickau. (M 23793)

(5867) 1988 RE

Discovered 1988 September 11 by J. Phinney at Palomar.

(5868) Ohta

1988 TQ. Discovered 1988 October 13 by K. Endate and K. Watanabe at Kitami.

Named in honor of Kentaro Ohta (1942-), who was in charge of designing optical telescopes at the Gotoh Optical Laboratory for many years and who now leads the large-format film section. He has also written many articles in astronomical magazines for amateur astronomers. (M 29145)

(5869) Tanith

1988 VN_4. Discovered 1988 November 4 by C. S. Shoemaker and E. M. Shoemaker at Palomar.

Seen as a heavenly goddess by the conquering Romans who called her Caelestis, Tanith was worshipped by the Carthaginians. This winged sky-goddess of the Punic people wore a zodiac around her head and held the sun and moon in either hand. (M 24918)

(5870) Baltimore

1989 CC_1. Discovered 1989 February 11 by E. F. Helin at Palomar.

Named for the city located on the Chesapeake Bay in Maryland. Celebrating the bicentennial of its 1797 incorporation, Baltimore is home to the Maryland Academy of Sciences, founding organization of the Maryland Science Center, also celebrating its 200th year. Baltimore is also home to the Space Telescope Science Institute and Johns Hopkins University. (M 29671)

Named by the discoverer following a suggestion by J. O'Leary.

(5871) Bobbell

1989 CE_2. Discovered 1989 February 11 by E. F. Helin at Palomar.

Named in honor of Robert L. Bell, friend and associate of the discoverer's husband. Bob Bell graduated from the California Institute of Technology with a B.S. in physics and an M.S. in applied mechanics, which he promptly utilized to invent pressure-sensing devices to monitor automobile-engine functions, enabling us to drive efficiently down our highways. More recently he has joined the astronomical community with his own private dome and CCD-equipped telescope. Bob is the quintessential entrepreneur, applying and promoting new ideas to solve problems. (M 28090)

(5872) 1989 SL

Discovered 1989 September 30 by T. Nomura and K. Kawanishi at Minami-Oda.

(5873) Archilochos

1989 SB_3. Discovered 1989 September 26 by E. W. Elst at La Silla.

Named for the Greek poet Archilochos (fl. 700 B.C.) who, through his provocative attitude against the established moral values of his time, confronted his subjective comprehension of reality, his personal experience and his own notion of morality to the heroic myths of his epoch. (M 23793)

Name proposed and citation prepared by Antonia Svarna on the request of the discoverer.

(5874) 1989 XB

Discovered 1989 December 2 by N. Kawasato at Uenohara.

(5875) Kuga

1989 XO. Discovered 1989 December 5 by K. Endate and K. Watanabe at Kitami.

Named in honor of Naoto Kuga (1964-), who produces planetarium programs at the Gotoh Optical Laboratory. (M 29145)

(5876) 1990 DM_2

Discovered 1990 February 24 by H. Debehogne at La Silla.

(5877) 1990 FP

Discovered 1990 March 23 by E. F. Helin at Palomar.

(5878) Charlene

1991 CC_1. Discovered 1991 February 14 by E. F. Helin at Palomar.

Named in honor of Charlene Marie Anderson, director of publications for the *Planetary Report*, who has brought the planetary sciences to hundreds of thousands of people worldwide. As an editor she has made invaluable contributions to the public understanding of space science and planetary exploration. Her inviting, natural style of writing makes science come to life. (M 24123)

(5879) Almeria

1992 CH_1. Discovered 1992 February 8 by K. Birkle and U. Hopp at Calar Alto.

Named for the Spanish city and province where the Calar Alto Observatory of the German-Spanish Astronomical Center is located. The observatory is operated jointly by the Max-Planck-Institut für Astronomie in Heidelberg and the Spanish National Commission for Astronomy. The naming acknowledges the great Spanish hospitality experienced by German astronomers at the observatory. (M 34341)

(5880) 1992 MA

Discovered 1992 June 22 by S. Ueda and H. Kaneda at Kushiro.

(5881) 1992 SR_{12}

Discovered 1992 September 27 by M. Sugano and T. Nomura at Minami-Oda.

(5882) 1992 WW_5

Discovered 1992 November 18 by N. Kawasato at Uenohara.

(5883) 1993 VM_5

Discovered 1993 November 6 by R. H. McNaught at Siding Spring.

(5884) Dolezal

6045 P-L. Discovered 1960 September 24 by C. J. van Houten and I. van Houten-Groeneveld at Palomar.

Named in memory of Erich Dolezal (1902-1990). He was a talented writer and popularizer of astronomy and space science. He wrote many books and later became Science Advisor of Radio Austria in Vienna, where he gave numerous lectures to schoolchildren and adults. In 1949, he was a co-founder of the Austrian Society for Space Research that was eventually merged into the International Astronautical Federation. (M 24765)

Name proposed by the discoverers following a suggestion by H. Haupt.

(5885) Apeldoorn

3137 T-2. Discovered 1973 September 30 by C. J. van Houten and I. van Houten-Groeneveld at Palomar.

Named in honor of Berend Caspar Jan Apeldoorn (1944-), Dutch amateur astronomer, on the occasion of his 50th birthday. Since 1961, Ben has specialized in meteors and meteorites, observing meteors both visually and photographically. He has written many articles on astronomy for astronomical periodicals and yearbooks, as well as for general magazines and newspapers. Apeldoorn still makes important contributions to the popularization of astronomy and is a member of the Meteor Section of the Dutch Society for Meteorology and Astronomy. (M 24765)

Name proposed by the discoverers following a suggestion by F. Bettonvil, chairman of the Meteor Section.

(5886) 1975 LR

Discovered 1975 June 13 at the Felix Aguilar Observatory at El Leoncito.

(5887) Yauza

1976 SG_2. Discovered 1976 September 24 by N. S. Chernykh at Nauchnyj.

Named for one of Moscow's little rivers. Between the place where it flows into Moscow river and the mouth of the Neglinka {see planet (5858)} was the site of the first buildings of the ancient settlement of Moscow. (M 30098)

(5888) 1978 VU_7

Discovered 1978 November 7 by E. F. Helin and S. J. Bus at Palomar.

(5889) Mickiewicz

1979 FA_3. Discovered 1979 March 31 by N. S. Chernykh at Nauchnyj.

Named in memory of Adam Mickiewicz (1798-1855), great Polish poet. (M 27128)

(5890) Carlsberg

1979 KG. Discovered 1979 May 19 by R. M. West at La Silla.

Named in honor of Carlsbergfondet, established in 1876 by Jacob Christian Jacobsen (1811-1887), philantropic founder of the first Carlsberg Brewery in 1847, in order to ensure the continuation of scientific work at the Carlsberg Laboratories and to provide support for the natural sciences, mathematics, philosophy, history and linguistics. The impact of Carlsbergfondet on Danish scientific and cultural life has been immense, and thousands of research projects, including many in astronomy and astrophysics, have become possible. This minor planet was numbered in 1994 when the discoverer was living in Munich, just 150 years after Jacobsen's visit to the breweries of that city, when he obtained a famous sample of yeast that contributed much to his subsequent success. (M 27330)

(5891) Gehrig

1981 SM. Discovered 1981 September 22 by A. Mrkos at Kleť.

Named in memory of Henry Louis (Lou) Gehrig (1903-1941), professional U.S. baseball player. From 1925 to 1939 Gehrig appeared in 2130 consecutive games for the New York Yankees, a reliability record that remained unbroken until 1995. Gehrig was a consistent left-handed batsman, with a career average of 0.340, but his career was ended prematurely as a result of amyothrophic lateral sclerosis ("Lou Gehrig's Disease"). (M 27735)

Name suggested by G. V. Williams, who made the identifications involving this object, and T. B. Spahr.

(5892) 1981 YS_1

Discovered 1981 December 23 at the Purple Mountain Observatory at Nanking.

(5893) 1982 EF

Discovered 1982 March 15 by Z. Vávrová at Kleť.

(5894) 1982 RM_1

Discovered 1982 September 14 by A. Mrkos at Kleť.

(5895) 1982 UF_2

Discovered 1982 October 16 by Z. Vávrová at Kleť.

(5896) Narrenschiff

1982 VV_{10}. Discovered 1982 November 12 by L. G. Karachkina at Nauchnyj.

Named in honor of Sebastian Brant (c.1458-1521), an outstanding German writer and humanist, and the author of the immortal satiric poem *Narrenschiff*, published in 26 editions in several European languages in the sixteenth century alone. (M 23793)

The name is given on the occasion of the quincentenary of the publication in Basel in 1494 of the first edition of *Narrenschiff*. Name proposed by L. R. Nemirovskij.

(5897) Novotná

1984 SZ_1. Discovered 1984 September 29 by A. Mrkos at Kleť.

Named in memory of Jarmila Novotná-Daubková (1907-1994), Czech opera singer, who lived in the U.S. from 1948, where she was engaged at the Metropolitan Opera in New York for 16 seasons. She is also known for *The Songs of Lidice*, a selection of Czech and Moravian folk songs, which she recorded with Jan Masaryk in 1942. (M 29671)

Name suggested by J. Tichá on the occasion of ninetieth anniversary of Novotná's birth.

(5898) 1985 KE

Discovered 1985 May 23 by A. C. Gilmore and P. M. Kilmartin at Lake Tekapo.

(5899) Jedicke

1986 AH. Discovered 1986 January 9 by C. S. Shoemaker and E. M. Shoemaker at Palomar.

Named in honor of the Jedicke family, notably Peter, Robert and June. One of Canada's best known amateur astronomers, Peter Jedicke is a superb teacher of astronomy and physics, especially to young people. His brother Robert, a physicist, observes with the Spacewatch survey for near-earth objects. Although sister June Zehr has not followed in her siblings' astronomical footsteps, she often shares observing sessions with her brothers. (M 25444)

Name suggested and citation prepared by D. H. Levy.

(5900) Jensen

1986 TL. Discovered 1986 October 3 by P. Jensen and K. Augustesen and H. J. Fogh Olsen at Brorfelde.

Named in honor of Poul B. Jensen, who served in the Meridian Circle Department at the Brorfelde Observatory for 35 years, and his wife, Bodil Jensen. During the past ten years Poul also took part in the minor planet program carried out with the Schmidt telescope. (M 23793)

Name proposed by K. Augustesen and H. J. Fogh Olsen.

(5901) 1986 WB_1

Discovered 1986 November 25 by Z. Vávrová at Kleť.

(5902) Talima

1987 QY_{10}. Discovered 1987 August 27 by L. G. Karachkina at Nauchnyj.

Named for Tatiana Alimovna Damir (1923-), friend of the discoverer, daughter of Alim Matveevich Damir and wife of Sergej Petrovich Kapitsa {see, respectively, planets (5717) and (5094)}. (M 24765)

(5903) 1989 AN_1

Discovered 1989 January 6 by S. Ueda and H. Kaneda at Kushiro.

(5904) Württemberg

1989 AE_7. Discovered 1989 January 10 by F. Börngen at Tautenburg.

Named for a region in southwestern Germany that includes both parts of the Schwarzwald, the Schwäbische Alb and the Alpenvorland. Both Johannes Kepler and Albert Einstein {see, respectively, planets (1134) and (2001)} were born in this region of Baden-Württemberg, as were the poets Martin Wieland, Friedrich Schiller, Friedrich Hölderlin and Eduard Mörike {see, respectively, planets (8108), (3079), (9189) and (9204)}. The major towns in Württemberg include Stuttgart, Tübingen, Ulm and Heilbronn. (M 24123; M 25351)

(5905) Johnson

1989 CJ_1. Discovered 1989 February 11 by E. F. Helin at Palomar.

Named for Lindley N. Johnson, who is responsible for planning the potential use of space-surveillance assets to detect and track near-earth objects. With degrees in astronomy and engineering management, he has been a space enthusiast since the age of 12. The discoverer and the Near-Earth Asteroid Tracking team wish to express their appreciation for his contributions to the development of the NASA-JPL NEAT program, which became operational on the Maui GEODSS telescope in Dec. 1995. (M 27128)

(5906) 1989 SN_5

Discovered 1989 September 24 by A. C. Gilmore and P. M. Kilmartin at Lake Tekapo.

(5907) 1989 TU_5

Discovered 1989 October 2 by S. J. Bus at Cerro Tololo.

(5908) Aichi

1989 UF. Discovered 1989 October 20 by Y. Mizuno and T. Furuta at Kani.

Named for the prefecture, Japan's fourth largest by population, in which both discoverers were born and raised. It is the main industrial prefecture. In olden times it nurtured the Karakuri (intricate puppet). Nowadays the making of clocks and cars, as well as the robotic and aerospace industries, thrive there. (M 33385)

(5909) Nagoya

1989 UT. Discovered 1989 October 23 by Y. Mizuno and T. Furuta at Kani.

Named for Japan's fourth largest city, where the first discoverer was born and raised. From the time of the Tokugawa period, when the first shogun Tokugawa Ieyasu built a castle in this region, Nagoya has been an important area for land and sea travel, connecting Tokyo with Kyoto and Osaka. It is blessed with a rich cultural history and many cultural attractions, such as Nagoya Castle and the Tokugawa Art Museum, which displays artifacts of the Samurai warrior. (M 33385)

(5910) Zátopek

1989 WH_4. Discovered 1989 November 29 by A. Mrkos at Kleť.

Named in honor of the great Czechoslovak long-distance runner, Emil Zátopek (1922-). In a twelve-year career he set eighteen world records and won five Olympic medals: silver in 1948 and gold in 1952 at 5000 m, gold in 1948 and 1952 at 10 000 m and gold in the 1952 marathon. The name also honors his wife Dana (1922-), who won Olympic gold for the javelin in 1952. (M 27735)

Name suggested by G. V. Williams, who made the identifications involving this object.

(5911) 1989 WO_7

Discovered 1989 November 25 by S. Ueda and H. Kaneda at Kushiro.

(5912) 1989 YR
Discovered 1989 December 20 by T. Niijima and T. Urata at Ojima.

(5913) 1990 BU
Discovered 1990 January 21 by M. Arai and H. Mori at Yorii.

(5914) 1990 WK
Discovered 1990 November 20 by R. H. McNaught at Siding Spring.

(5915) Yoshihiro
1991 EU. Discovered 1991 March 9 by T. Seki at Geisei.

Named for Yoshihiro Yamada (1946-), astronomical scholar and keen popularizer of astronomy in Japan. (M 25231)

(5916) van der Woude
1991 JD_1. Discovered 1991 May 8 by E. F. Helin at Palomar.

Named in honor of Jurrie van der Woude, Jet Propulsion Laboratory Public Information Representative and the co-author of the "25 Years of Space Photography" Exhibit. He joined JPL in 1976 and became an active member of the photo lab and Public Information Office. He develops special products for the Director's Office and the Laboratory's Community Relations Program. (M 30477)

(5917) 1991 NG
Discovered 1991 July 7 by E. F. Helin at Palomar.

(5918) 1991 NV_3
Discovered 1991 July 6 by H. Debehogne at La Silla.

(5919) 1991 PW_{12}
Discovered 1991 August 5 by H. E. Holt at Palomar.

(5920) 1992 SX_{17}
Discovered 1992 September 30 by H. E. Holt at Palomar.

(5921) 1992 UL
Discovered 1992 October 19 by S. Ueda and H. Kaneda at Kushiro.

(5922) Shouichi
1992 UV. Discovered 1992 October 21 by S. Otomo at Kiyosato.

Named in honor of Shouichi Sato (1950-), an electric engineer who developed an inexpensive and efficient photovoltaic inverter that can be connected to commercial power supply lines. (M 25654)

(5923) Liedeke
1992 WC_8. Discovered 1992 November 26 by the Spacewatch at Kitt Peak.

Named in honor of Liedeke Gehrels-de Stoppelaar on the occasion of her completion of 21 years teaching Art History and French at University High School, a magnet school in Tucson that gives students the opportunity to pass advanced-placement examinations for credit at universities. She guided these programs through their formative years when this type of school was new in the United States, after her own preparation with a Ph.D. on the theme of justice in the writings of Albert Camus and with 13 years of teaching elsewhere. (M 29145) Named following a suggestion by C. J. van Houten and I. van Houten-Groeneveld.

(5924) Teruo
1994 CH_1. Discovered 1994 February 7 by T. Kobayashi at Oizumi.

Named in memory of Teruo Saegusa (1957-1989), one of the greatest Japanese alpinists, who climbed Mount Everest twice. (M 23542)

(5925) 1994 CP_1

Discovered 1994 February 5 by S. Ueda and H. Kaneda at Kushiro.

(5926) Schönfeld

1929 PB. Discovered 1929 August 4 by M. Wolf at Heidelberg.

Named in memory of Eduard Schönfeld (1828-1891), famous German astronomer and director of the Mannheim and Bonn observatories. Pupil and, from 1853, assistant of F. W. A. Argelander {see planet (1551)}, Schönfeld took part in the huge *Bonner Durchmusterung* project. His time as director of the Mannheim observatory mainly was devoted to the detection of variable stars. The results were published in two parts in 1866 and 1874. After Argelander's death in 1875, Schönfeld started the task to enlarge the BD with the zone between -2° and -23°. In 1886 his momumental *Südliche Durchmusterung*, containing 24 charts with 133,659 stars, was published, completing the greatest task of the Bonn observatory. Schönfeld also served as an important promoter of the foundation and development of the Astronomische Gesellschaft. (M 23793)
Name proposed and citation prepared by L. D. Schmadel.
Schönfeld is also honored by a lunar crater.

(5927) 1938 HA

Discovered 1938 April 19 by W. Dieckvoss at Bergedorf.

(5928) Pindarus

1973 SK_1. Discovered 1973 September 19 by C. J. van Houten and I. van Houten-Groeneveld at Palomar.

Named for the Greek lyric poet Pindarus (c. 520-446 B.C.), long the mainstay of Greek poetry. He lived in Greece and Sicily and united the Greek people with his poems and tragic and victory songs during festivals. He had great influence on the work of the German poet Friedrich Hölderlin {see planet (9189)}. (M 26764)

(5929) 1974 XT

Discovered 1974 December 14 at the Felix Aguilar Observatory at El Leoncito.

(5930) Zhiganov

1975 VW_2. Discovered 1975 November 2 by T. M. Smirnova at Nauchnyj.

Named in memory of Nazib Gayazovich Zhiganov (1911-1988), outstanding Soviet Tatar composer and public figure, founder of the Tatarian professional musical school. Zhiganov was the organizer and chairman of the Composers Union of Tatarstan during 1939-1976, as well as of the Kazan State Conservatory from 1945 until his death. His most significant musical compositions are the opera *Dzhalil,* the ballet *Two legends,* and his Second, Seventh and Ninth symphonies, in which he combined classical musical traditions with Tatar folk melodies. (M 25978)
Name suggested by the Institute of Theoretical Astronomy.

(5931) Zhvanetskij

1976 GK_3. Discovered 1976 April 1 by N. S. Chernykh at Nauchnyj.

Named in honor of the Russian writer Mikhail Zhvanetskij (1934-), whose satirical talent has won nation-wide recognition. He won the international festival "Gold Ostap" in 1993 and the independent prize "Triumph" in 1995. He is also an honorary citizen of Odessa, president of the World-Wide Club of Odessa Citizens and holder of the order "Friendship of peoples". (M 30098)

(5932) Prutkov

1976 GO_3. Discovered 1976 April 1 by N. S. Chernykh at Nauchnyj.

Named for Koz'ma Prutkov, a parody character, fictitious poet and author of many fables, humorous poems and aphorisms. Prutkov is the collective pen-name invented by the nineteenth-century poets A. K. Tolstoj and the brothers A. M. and V. M. Zhemchuzhnikov. Many of Prutkov's utterances have become by-words in Russian. (M 30098)

(5933) Kemurdzhian

1976 QN. Discovered 1976 August 26 by N. S. Chernykh at Nauchnyj.

Named in honor of Aleksandr Leonovich Kemurdzhian (1921-), an expert on the design of cosmic transportation and specialized robots for the investigation of the surfaces of planets and their satellites. He was responsible for designing the self-propulsion system of Lunokhod 1 and Lunokhod 2 as these vehicles explored the moon's surface in the early 1970s. Subsequently he headed the design of mobile apparatus for the investigation of Venus, Mars and Phobos. (M 31024) Name proposed by the discoverer following a suggestion by ITA.

(5934) Mats

1976 SJ. Discovered 1976 September 20 by C.-I. Lagerkvist and H. Rickman at Kvistaberg.

Named for Mats Lindgren of Uppsala Astronomical Observatory, whose research into Jupiter's role in shaping the fate of comets led to spectacular results, in particular relating to the evolution and demise of comet D/1993 F2 (Shoemaker-Levy 9). He obtained some of the best ground-based images of the impact plumes and scars in July 1994. His contributions were condensed into his doctoral thesis at Uppsala University in June 1995. (M 25444)

(5935) Ostankino

1977 EF_1. Discovered 1977 March 13 by N. S. Chernykh at Nauchnyj.

Named for a district of Moscow that included the village Ostankino, the eighteenth-century county seat of the counts Sheremetev. Among places of interest there are a county-house museum, television tower and center, and the Memorial Museum of Cosmonautics. (M 30098)

(5936) Khadzhinov

1979 FQ_2. Discovered 1979 March 29 by N. S. Chernykh at Nauchnyj.

Named in honor of Leonid Petrovich Khadzhinov (1927-), prominent Ukrainian electrical engineer involved with the construction of electrotechnical machines. (M 30098)

(5937) Lodén

1979 XQ. Discovered 1979 December 11 by C.-I. Lagerkvist at Kvistaberg.

Named for Kerstin and Lars Olof Lodén. Kerstin, an astronomer at Stockholm Observatory, is a coauthor (with the discoverer) of two books in introductory astronomy. Lars Olof is professor of astronomy at Uppsala Observatory. Both have devoted most of their research to studies of the Milky Way, in particular by making a large survey of the Southern Milky Way. (M 25444)

(5938) Keller

1980 FH_2. Discovered 1980 March 16 by C.-I. Lagerkvist at La Silla.

Named for Horst Uwe Keller, well-known scientist at the Max Planck Institute in Lindau. Uwe was the driving force of the science team for the ESA study on 'The Orbiting Planetary Telescope', in which the discoverer also participated. (M 25444)

(5939) 1981 EU_8

Discovered 1981 March 1 by S. J. Bus at Siding Spring.

(5940) Feliksobolev

1981 TJ_4. Discovered 1981 October 8 by L. I. Chernykh at Nauchnyj.

Named in memory of the Ukraine film producer Feliks Mikhailovich Sobolev (1931-1984), great master of scientific films, Honored Artist of the Ukraine and U.S.S.R. State Prize Winner. His films, made at the Kiev studio for scientific films, provided a fresh outlook at popular science, obtained the appreciation of audience and critics alike and won awards at prestigious festivals in many countries. (M 34622)

(5941) Valencia

1982 UQ_6. Discovered 1982 October 20 by L. G. Karachkina at Nauchnyj.

Named for the large Mediterranean seaport city in Spain, administrative center of the province of the same name and an important industrial, cultural and scientific center. (M 24765)

Name proposed by the Institute of Theoretical Astronomy, which collaborates with the Astronomical Observatory of the University of Valencia.

(5942) 1983 AN_2

Discovered 1983 January 10 by B. Behymer and M. Marley at Palomar.

(5943) Lovi

1984 EG. Discovered 1984 March 1 by E. Bowell at Anderson Mesa.

Named in memory of George Lovi (1939-1993), astronomical cartographer, planetarium lecturer and popularizer of sky lore. Lovi was staff astronomer at Vanderbilt Planetarium and lecturer at the American Museum-Hayden Planetarium and at the Jones Planetarium Theater, all in the New York area. Author of two books (*Men, Monsters, and the Modern Universe* and *Uranometria 2000.0*), Lovi created (in 1968) and maintained (until 1992) a monthly series of star charts for *Sky and Telescope* {see planet (3243)} magazine. In addition, he was author of the column "Rambling Through the Skies" for 260 consecutive issues of that magazine (1971-1993). A noted bibliophile and independent scholar, Lovi's diverse work ranged from planetarium installation to presentation, and from tours of the night sky to tours of major subways and railways around the world. (M 25654)
Name suggested and citation prepared by J. Rao and S. Storch.

Obituaries published in Sky Telesc., Vol. 86, No. 2, p. 59, 74 (1993); J. Br. Astron. Assoc., Vol. 103, No. 4, p. 201 (1993).

(5944) Utesov

1984 JA_2. Discovered 1984 May 2 by L. G. Karachkina at Nauchnyj.

Named in memory of Leonid Osipovich Utesov (1895-1982), famous Russian singer, musician, actor, founder and artistic leader of the first Russian theatricalized jazz band (1929). (M 24765)

The name is given by the discoverer on the occasion of onehundredth anniversary of Utesov's birth, following a suggestion by L. R. Nemirovskij.

(5945) Roachapproach

1984 SQ_3. Discovered 1984 September 28 by B. A. Skiff at Anderson Mesa.

Named in honor of Steve Roach, musician and composer of 'space music'. Among the most progressive performers in the genre, his prolific output includes seminal works such as "Structures from Silence" (1984), "Dreamtime Return" (1988) and "The Magnificent Void" (1996). His mixture of heavily-processed electronic sounds with primitive woodwinds and percussion conveys powerful impressions of spatial immensity. These have inspired the discoverer through endless nights at the telescope. (M 31296)
Name suggested by L. Kohanov.

(5946) 1984 UC_1

Discovered 1984 October 28 by A. Mrkos at Kleť.

(5947) Bonnie

1985 FD. Discovered 1985 March 21 by C. S. Shoemaker and E. M. Shoemaker at Palomar.

Named in memory of Bonnie Gail Farquhar (1936-1993), late wife of the mission director for the Near-Earth Asteroid Rendezvous (NEAR) mission, Robert Farquhar {see planet (5256)}. Rendezvous operations at (433) Eros will begin on Bonnie's birthday, 1998 Dec. 20. (M 30477)

Name proposed by the discoverers following a suggestion by R. Farquhar.

(5948) Longo

1985 JL. Discovered 1985 May 15 by E. Bowell at Anderson Mesa.

Named in honor of Giuseppe Longo (1920-), a physicist at the University of Bologna. For most of his professional life a nuclear physicist, in the late 1980s Longo became interested in the Tunguska {see planet (5471)} event. He organized an expedition to the site of the explosion in 1991, at which time microscopic particles, trapped in the resin of surviving trees, were collected. Later detailed analysis showed that the annual distribution of the particles peaked in 1908, and elements were identified that probably originated in meteoritic material. Longo recently organized in Bologna the first workshop where Russian and other modelers of the Tunguska event met and compared their views and results. (M 28621)

Name suggested and citation prepared by M. Di Martino and P. Farinella.

(5949) 1985 RL_3

Discovered 1985 September 6 by H. Debehogne at La Silla.

(5950) Leukippos

1986 PS_4. Discovered 1986 August 9 by E. W. Elst and V. G. Ivanova at Rozhen.

Named for the great Greek philosopher Leukippos, born around 450 B.C., probably at Miletus, on the west coast of Asia Minor. His main concern was to harmonize Ionic cosmology with Greek ontology. Although it was Demokritos {see planet (6129)} who brilliantly elaborated many of his ideas, the atomistic theory of matter originated with Leukippos. Only a few fragments of his work remain, and these show that he made a clear distinction between empty space and matter. (M 24918)

(5951) Alicemonet

1986 TZ_1. Discovered 1986 October 7 by E. Bowell at Anderson Mesa.

Named in honor of Alice K. B. Monet (1954-), an astronomer at the U.S. Naval Observatory's Flagstaff Station and former chair of the Division on Dynamical Astronomy of the American Astronomical Society. Monet has made valuable astrometric observations of solar-system bodies in support of NASA's Galileo and Near Earth Asteroid Rendezvous (NEAR) missions. Using the 0.20-m transit telescope in an autonomous operating mode, she has generated, in part in collaboration with the discoverer, accurate astrometric data for many asteroids and comets. (M 27460; M 27477)

Name proposed by the discoverer following a suggestion by the JPL ephemeris group.

(5952) Davemonet

1987 EV. Discovered 1987 March 4 by E. Bowell at Anderson Mesa.

Named in honor of David G. Monet (1951-), an astronomer at the U.S. Naval Observatory's Flagstaff Station. Monet received the USNO's Newcomb Award for his development of a program of high-precision differential CCD astrometry, which was implemented on the 1.55-m astrometric reflector. Monet is also the principal investigator for the Precision Measuring Microdensitometer (PMM) project. Over the years, he has been extremely generous with technical advice to the discoverer. (M 27460)

Name proposed by the discoverer following a suggestion by the JPL ephemeris group.

(5953) Shelton

1987 HS. Discovered 1987 April 25 by C. S. Shoemaker and E. M. Shoemaker at Palomar.

Named in honor of Ian Shelton, Canadian astronomer, best known for his discovery of the supernova in the Large Magellanic Cloud on 1987 Feb. 24. During his four years as resident observer at the University of Toronto Southern Observatory, he participated in programs with astronomers worldwide, and more recently he has generously contributed to the astronomical education of school children, parents and teachers. (M 31296)

Name suggested and citation prepared by D. MacCormack and R. Garrison.

(5954) Epikouros

1987 QS_1. Discovered 1987 August 19 by E. W. Elst at La Silla.

Named for the great philosopher Epikouros (341-270 B.C.), well known for his exposition of the atomistic theory of physics, inspired by the teachings of Demokritos {see planet (6129)}. He was also celebrated for his ethical teaching, to which we owe "epicurism". Living in an age when the Greeks had lost their political freedom in Macedonia, Epikouros wanted to restore mental freedom by means of his physics to ensure "quietude of the mind". (M 24765)

(5955) 1987 RT_3

Discovered 1987 September 2 by L. I. Chernykh at Nauchnyj.

(5956) d'Alembert

1988 CF_5. Discovered 1988 February 13 by E. W. Elst at La Silla.

Named in memory of Jean Le Rond d'Alembert (1717-1783), French philosopher and mathematician, famous for his mechanical principle (1742). He introduced the calculus of partial differences and solved the problem of the precession of the equinoxes. He is perhaps best remembered for his association with Diderot {see planet (5351)} in the preparation of the *Encyclopedia*. D'Alembert was much interested in music, both as a science and as an art. (M 23793)

d'Alembert is also honored by a lunar crater.

(5957) Irina

1988 JN. Discovered 1988 May 11 by C. S. Shoemaker and E. M. Shoemaker at Palomar.

Named in honor of Irina Victorovna Farquhar, wife of the NEAR mission director, Robert Farquhar {see planet (5256)}. NEAR's initial close pass by Eros will occur on 1999 January 10, their fifth wedding anniversary. Since earning a doctoral degree in economics from Leningrad State University in 1983, Irina has made several important contributions in the fields of labor, regional and health economics. (M 30477)

Name proposed by the discoverers following a suggestion by R. Farquhar, who prepared the citation.

(5958) 1989 BS_1

Discovered 1989 January 29 by A. Mrkos at Kleť.

(5959) Shaklan

1989 NB_1. Discovered 1989 July 2 by E. F. Helin at Palomar.

Named in honor of Stuart B. Shaklan, an optical engineer, who performed the imaging design and analysis required for the Near-Earth Asteroid Tracking program. He developed the models needed to understand the performance of the NEAT optical system and analyzed the imaging results to determine the optimal instrument configuration. His expertise in CCD characterization and telescope design for high-precision astrometry was an important element of the NEAT success. (M 26930)

(5960) Wakkanai

1989 US. Discovered 1989 October 21 by M. Mukai and M. Takeishi at Kagoshima.

Wakkanai, situated at the northern tip of Japan, is a city known for its marine industry, dairy farming and tourist industry. The second discoverer lived in Wakkanai for four years, and this name is proposed as a token of his gratitude to the people of the town. (M 24411)

(5961) 1989 YH_1

Discovered 1989 December 30 by R. H. McNaught at Siding Spring.

(5962) Shikokutenkyo

1990 HK. Discovered 1990 April 18 by T. Seki at Geisei.

Named for an amateur astronomical society with 50 members on Shikoku Island {see planet (4223)}. (M 25231)

(5963) 1990 QP_2

Discovered 1990 August 24 by H. E. Holt at Palomar.

(5964) 1990 QN_4

Discovered 1990 August 23 by H. E. Holt at Palomar.

(5965) 1990 SV_{15}

Discovered 1990 September 16 by H. E. Holt at Palomar.

(5966) Tomeko

1990 VS_6. Discovered 1990 November 15 by T. Seki at Geisei.

Named for Tomeko Goto (1899-), who supported her husband, Seizo Goto {see planet (2621)}, president of the Goto Optical Laboratory. (M 27330)

(5967) Edithlevy

1991 CM_5. Discovered 1991 February 9 by C. S. Shoemaker and D. H. Levy at Palomar.

Named in honor of Edith Pailet Levy (1918-), mother of the second discoverer. Born in New Orleans, she married Nathaniel Lewis Levy in 1939 and headed to Montreal to become one of McGill University's first women medical students. Levy excelled in hematology, but once her children were grown she went back to McGill to launch a career in genetics. She spent several years studying Alzheimer's patients at Montreal's Jewish Hospital of Hope. (M 24918)

(5968) Trauger

1991 FC. Discovered 1991 March 17 by E. F. Helin at Palomar.

Named in honor of John T. Trauger, widely known for his remarkable achievements as principal investigator for the Wide Field/Planetary Camera II on the Hubble Space Telescope. For defining, guiding and leading both scientific and technical aspects of the entire project, he was awarded the NASA Outstanding Leadership Medal in 1994. He is an outstanding instrumentalist, noted for his development and use of Fabry-Perot technology. He has also provided critical assistance to the Air Force Maui Optical Site (AMOS) program that has acquired follow-up astrometry for new discoveries, particularly for JPL's PCAS and NEAT programs. (M 27128)

(5969) Ryuichiro

1991 FT. Discovered 1991 March 17 by T. Seki at Geisei.

Named in honor of Ryuichiro Goto (1938-), grandson of Seizo Goto {see planet (2621)}. He directs the Goto Optical Laboratory and is very active in the development of new planetaria and the popularization of astronomy. (M 27330)

(5970) Ohdohrikouen

1991 JS_1. Discovered 1991 May 13 by K. Watanabe at Sapporo.

Named for a belt-shaped park, 105 m wide and 1.6 km long, that runs east-west through the center of Sapporo {see planet (3473)}. Statues and fountains are situated among the walkways and grassy areas. Citizens and tourists gather here throughout the year, and it is the site of the world-famous Sapporo Snow Festival in the winter. (M 26930)

(5971) 1991 NT_2

Discovered 1991 July 12 by H. E. Holt at Palomar.

(5972) 1991 PS_{12}

Discovered 1991 August 5 by H. E. Holt at Palomar.

(5973) 1991 QC

Discovered 1991 August 17 by S. Otomo at Kiyosato.

(5974) 1991 UZ_2

Discovered 1991 October 31 by S. Ueda and H. Kaneda at Kushiro.

(5975) Otakemayumi

1992 SG. Discovered 1992 September 21 by K. Endate and K. Watanabe at Kitami.

Named in honor of Mayumi Otake (1967-), well known for her work as sound producer of planetarium programs. (M 29146)

(5976) Kalatajean

1992 SR_2. Discovered 1992 September 25 at the Oak Ridge Observatory at Harvard.

Named in honor of Jean Marie Kalata, a social science analyst in the Institutional Studies Office of the Smithsonian Institution. Raised in New Jersey, she lives in Alexandria, Virginia, and among other interests sings with the Congressional Chorus. She served as a volunteer with the Christian Appalachian Project in Kentucky for 18 months. (M 27460)

Named as part of the Great Smithsonian Giveaway on the occasion of the sesquicentennial of the Smithsonian Institution.

(5977) 1992 TH_1

Discovered 1992 October 1 by H. E. Holt at Palomar.

(5978) Kaminokuni

1992 WT. Discovered 1992 November 16 by K. Endate and K. Watanabe at Kitami.

Named for one of the oldest towns in Hokkaido {see planet (3720)}, with a population of 8000. The history of the town can be traced for 800 years, which is unusually long in Hokkaido. (M 29146)

(5979) 1992 XF

Discovered 1992 December 15 by S. Ueda and H. Kaneda at Kushiro.

(5980) 1993 FP$_2$

Discovered 1993 March 26 by S. Ueda and H. Kaneda at Kushiro.

(5981) Kresilas

2140 P-L. Discovered 1960 September 24 by C. J. van Houten and I. van Houten-Groeneveld at Palomar.

Named for the Greek sculptor Kresilas, who lived in the 5th century B.C. He is mostly known for his sculpture of Pericles, of which several copies still exist. (M 26764)

(5982) Polykletus

4862 T-1. Discovered 1971 May 13 by C. J. van Houten and I. van Houten-Groeneveld at Palomar.

Named for the Greek sculptor Polykletus of Argos (c. 480-423 B.C.). Together with Phidias {see planet (4753)}, he was the most important sculptor of his time. Nothing of his original work remains, but many old copies and images on coins are handed down to us, as well as detailed written descriptions. (M 26764)

(5983) Praxiteles

2285 T-2. Discovered 1973 September 29 by C. J. van Houten and I. van Houten-Groeneveld at Palomar.

Named for the great Athenian sculptor Praxiteles (c. 370-325 B.C.). Antique sources indicate that Praxiteles worked in Athens and created at least 50 statues, although only Roman copies now exist. His figures no longer had the classical character of Phidias and Polykletus {see planets (4753) and (5982)} but had become ideal beings of smooth beauty. (M 26764)

(5984) Lysippus

4045 T-3. Discovered 1977 October 16 by C. J. van Houten and I. van Houten-Groeneveld at Palomar.

Named for Lysippus of Sikyon (c. 370-300 B.C.), a great Greek sculptor in the time of Alexander the Great. He worked in the late-classical style and had an influence on the later hellenistic style. Most of his work survives in Roman copies and descriptions. His statues have smaller heads and simulate more movement than those sculpted in the time of Polykletus {see planet (5982)}. (M 26764)

(5985) 1942 RJ

Discovered 1942 September 7 by L. Oterma at Turku.

(5986) Xenophon

1969 TA. Discovered 1969 October 2 by P. Wild at Zimmerwald.

Named for the Athenian nobleman, pupil and interpreter of Socrates {see planet (5450)}, historian, agriculturist, and military officer who lived from about 440 to 354 B.C. Because of his support of Spartan culture and politics he was exiled from Athens. He participated in a military expedition against the Persian king Artaxerxes {see planet (7212)} and, after his defeat near Babylon in 401 B.C., he prevailed to lead the army of 10 000 Greek mercenaries safely home through the mountains of Asia Minor, a great feat described in his famous book *Anabasis*. (M 34341)

(5987) 1975 LQ

Discovered 1975 June 6 at the Felix Aguilar Observatory at El Leoncito.

(5988) Gorodnitskij

1976 GN_2. Discovered 1976 April 1 by N. S. Chernykh at Nauchnyj.

Named in honor of Aleksandr Moiseevich Gorodnitskij (1933-), Russian scientist and bard. A prominent geologist and oceanologist, who works at the Russian Academy of Science's Institute of Oceanology, he has participated in many nautical research expeditions. He is also well known as an author and performer of lyric songs, filled with the romance of voyages and the exploration of novel lands. (M 34622)

(5989) Sorin

1976 QC_1. Discovered 1976 August 26 by N. S. Chernykh at Nauchnyj.

Named in memory of Sergej Ivanovich Sorin (1916-1995), astronomer and teacher who participated in the choice of a site for the astronomical observatory of the Azerbaijan Academy of Sciences and in equipping it with telescopes and other instruments. Head of the astronomical circle at the Baku Young Pioneers' Palace for some forty years, he was an outstanding educator of young amateur astronomers. Many of his pupils became prominent astronomers and are working now at various observatories in the former Soviet Union. (M 34622)

Name proposed by the discoverer, following a suggestion by some of Sorin's former pupils.

(5990) Panticapaeon

1977 EO. Discovered 1977 March 9 by N. S. Chernykh at Nauchnyj.

Named for an ancient town, the capital of a Bospor realm dating back to the sixth century B.C., located at the site of the present town of Kerch {see also planet (2216)} in the eastern Crimea. This minor planet is dedicated to the Kerch Historical-Archaeological Museum, the main repository of the Panticapaeon archaeological materials. Founded in 1826, the Kerch Museum makes a valuable contribution to the investigation of the ancient history of the region and acquaints the general public with the remote past of the region. (M 34622)

(5991) Ivavladis

1979 HE_3. Discovered 1979 April 25 by N. S. Chernykh at Nauchnyj.

Named in honor of Vladislav Aleksandrovich Ivanov (1936-), professor at the St. Petersburg Institute of Fine Mechanics and Optics. An electromechanical engineer and metrologist, he discovered a new imaging technique on the basis of magnetic resonance and invented a number of instruments for space, air, marine and subterranean investigations. (M 34622)

(5992) 1981 DZ

Discovered 1981 February 28 by S. J. Bus at Siding Spring.

(5993) 1981 EU_{22}

Discovered 1981 March 2 by S. J. Bus at Siding Spring.

(5994) Yakubovich

1981 SZ_7. Discovered 1981 September 29 by L. V. Zhuravleva at Nauchnyj.

Named in honor of Leonid Arkad'evich Yakubovich (1945-), writer, dramatist and television journalist. His high professional skill and fine sense of humor led to his becoming director of the most popular all-Russian television programs, such as *Field of miracles* and *Wheel of history*. (M 34622)

(5995) Saint-Aignan

1982 DK. Discovered 1982 February 20 by E. Bowell at Anderson Mesa.

Named in honor of Charles P. de Saint-Aignan (1977-) on the occasion of his 20th birthday, 1997 February 16. During two seasons as a summer student of the discoverer, de Saint-Aignan worked on measuring minor planet positions for orbit improvement. He has also worked at the Minor Planet Center and is currently a student at Brown University. (M 29146)

Name proposed by the discoverer, following a suggestion by B. G. Marsden, endorsed by G. V. Williams.

(5996) Julioangel

1983 NR. Discovered 1983 July 11 by E. Bowell at Anderson Mesa.

Named in honor of Julio Angel Fernández (1946-) of the Universidad de la República, Montevideo. Fernández is a noted dynamicist who has worked on the evolution of comet orbits and planetesimal scattering in the outer solar system, including the formation of the Oort {see planet (1691)} Cloud. His work has led to some of the first clear indications for the existence of the transneptunian belt. Since 1985 he has contributed to the reestablishment of Uruguayan astronomy by educating a vigorous group of young planetary scientists and dynamicists. (M 27460)

Citation procided by H. Rickman.

(5997) 1983 TH

Discovered 1983 October 1 by A. Mrkos at Kleť.

(5998) 1986 RK_1

Discovered 1986 September 2 by A. Mrkos at Kleť.

(5999) Plescia

1987 HA. Discovered 1987 April 23 by C. S. Shoemaker and E. M. Shoemaker at Palomar.

Named in honor of Jeffrey B. Plescia, a versatile scientist at the Jet Propulsion Laboratory who works with equal ease in geology, geophysics and planetary science. He has been a leading investigator of the cratering history of the satellites of Jupiter, Saturn and Uranus. More recently, Plescia has pursued gravity studies of terrestrial impact craters, including Upheaval Dome in Utah, the Manson structure of Iowa, and the Mt. Toondina, Kelly West and Teague Ring structures in Australia. (M 29671)

(6000) United Nations

1987 UN. Discovered 1987 October 27 by P. Jensen at Brorfelde.

Named by vote of IAU Commission 20 at its 1994 meeting in The Hague on the recommendation of the Minor Planet Names Committee. (M 24123)

The name was proposed by L. D. Schmadel (according to the letters UN of the provisional designation).

(6001) Thales

1988 CP_2. Discovered 1988 February 11 by E. W. Elst at La Silla.

Named for the famous Greek philosopher Thales of Miletus (c.625-547 B.C.). None of Thales' writings has come down to us, but from Aristoteles {see planet (6123)} we know that he was the first to suggest a single substratum (water) for the Universe. The correct prediction of the solar eclipse of -584 May 28 contributed considerably to his reputation as an astronomer. Thales' significance, however, lies in the fact that he attempted to explain natural phenomena by causes within nature itself, rather than by caprices of anthropomorphic gods. He must be credited with at least five important geometrical theorems. (M 24766)

Thales is also honored by a lunar crater.

(6002) 1988 RO

Discovered 1988 September 8 by P. Jensen at Brorfelde.

(6003) 1988 VO_1

Discovered 1988 November 2 by S. Ueda and H. Kaneda at Kushiro.

(6004) 1988 XY_1

Discovered 1988 December 11 by S. Ueda and H. Kaneda at Kushiro.

(6005) 1989 BD

Discovered 1989 January 29 by S. Ueda and H. Kaneda at Kushiro.

(6006) Anaximandros

1989 GB_4. Discovered 1989 April 3 by E. W. Elst at La Silla.

Named for the great Greek philosopher Anaximander of Miletus (610-546 B.C.). A pupil of Thales {see planet (6001)}, he wrote treatises on geography, astronomy and cosmology. From all this only the so-called B 1 fragment remains. Although it barely embraces nine sentences, it may be considered as the oldest philosophical citation. Anaximander derived the world from the *apeiron* (unlimited), which is the *arche* (beginning) and principal element, from which all existing things owe their birth and to which they will eventually return. Although a rationalist, he described the emergence of particular substances in metaphors, drawn from human society, in which physical injustices (hot or cold may not prevail forever) are penalized. (M 24918)

Anaximandros is also honored by a lunar crater.

(6007) 1990 BE_2

Discovered 1990 January 28 by S. Ueda and H. Kaneda at Kushiro.

(6008) 1990 BF_2

Discovered 1990 January 30 by S. Ueda and H. Kaneda at Kushiro.

(6009) 1990 FQ_1

Discovered 1990 March 24 by E. F. Helin at Palomar.

(6010) Lyzenga

1990 OE. Discovered 1990 July 19 by E. F. Helin at Palomar.

Named in honor of Gregory Lyzenga, professor of physics at Harvey Mudd College. His keen interest in minor planets and expertise in observational astronomy are combined with an aptitude for relating scientific topics to diverse audiences. His generous gifts of time and talent in support of asteroid research are greatly appreciated. (M 25231)

Citation prepared by J. B. Child.

(6011) Tozzi

1990 QU_5. Discovered 1990 August 29 by H. E. Holt at Palomar.

Named in honor of Gian Paolo Tozzi (1949-) of the Arcetri Astrophysical Observatory. Tozzi's research, primarily in cometary physics, is focused on the study of dust in comae and on the problem of atomic carbon production. Part of his activity has concerned infrared observations of asteroids. (M 27461)

Citation prepared by M. DiMartino.

(6012) 1990 SK_4

Discovered 1990 September 22 by R. H. McNaught at Siding Spring.

(6013) 1991 OZ

Discovered 1991 July 18 by H. E. Holt at Palomar.

(6014) 1991 PO_{10}

Discovered 1991 August 7 by H. E. Holt at Palomar.

(6015) 1991 PR_{10}

Discovered 1991 August 7 by H. E. Holt at Palomar.

(6016) 1991 PA_{11}

Discovered 1991 August 7 by H. E. Holt at Palomar.

(6017) 1991 PY_{11}

Discovered 1991 August 7 by H. E. Holt at Palomar.

(6018) 1991 PS_{16}

Discovered 1991 August 7 by H. E. Holt at Palomar.

(6019) 1991 RO_6

Discovered 1991 September 3 by R. H. McNaught at Siding Spring.

(6020) Miyamoto

1991 SL_1. Discovered 1991 September 30 by K. Endate and K. Watanabe at Kitami.

Named in honor of Yukio Miyamoto (1921-), who has contributed much to the popularization of astronomy and took part in establishing the Kumamoto Astronomical Society in 1968. He studied mirror-making under Jiro Hoshino {see planet (3828)} and in 1971 became the first amateur to make a Wright-Schmidt telescope. He was central to the establishment in 1982 of the society's observatory, the Kumamoto Civil Astronomical Observatory. (M 26930)

(6021) 1991 TM

Discovered 1991 October 1 by R. H. McNaught at Siding Spring.

(6022) Jyuro

1992 UB_4. Discovered 1992 October 26 by K. Endate and K. Watanabe at Kitami.

Named in honor of Jyuro Kobayashi (1949-), Japanese amateur astronomer and comet observer. He is a member of the Kumamoto Astronomical Society and participated in the establishment of the Kumamoto Civil Astronomical Observatory in 1982. (M 27128)

(6023) Tsuyashima

1992 UQ_4. Discovered 1992 October 26 by K. Endate and K. Watanabe at Kitami.

Named in honor of Takaaki Tsuyashima (1949-), popularizer of astronomy who planned the Kumamoto Civil Astronomical Observatory, which is unique in that it is open to the public every night, thanks to the volunteer activity of amateur astronomers. (M 26930)

(6024) Ochanomizu

1992 UT_4. Discovered 1992 October 27 by A. Sugie at Taga.

Named for the area in Tokyo that includes Kanda-Surugadai and Yushima. The name means 'water of tea', because during the Edo period high-quality water sprang out from the cliff there and was used for the tea ceremony by Tokugawa Shogunate. (M 29146)

(6025) Naotosato

1992 YA_3. Discovered 1992 December 30 by T. Urata at Oohira.

Named in honor of Naoto Sato (1953-), active amateur astronomer and junior high school science teacher. His outstanding educational achievement has been creating life-long interest in space and the universe among his students and followers. At his own Chichibu Astronomical Observatory, he uses a CCD camera to observe minor planets. He made a prediscovery observation of comet C/1989 Y2 (McKenzie-Russell). (M 33786)

(6026) Xenophanes

1993 BA_8. Discovered 1993 January 23 by E. W. Elst at La Silla.

Named for the Greek philosopher Xenophanes of Colophon (570-475 B.C.). He was in the first place a poet, using poetry to express his reflections about philosophical problems. Although he continued the tradition of Ionic philosophy (e.g., on nature), he spent much of his time attacking the idea of anthropomorphic gods, such as we know them from the work of Homer {see planet (5700)} and Hesiod. From this the idea of a more abstract god, in combination with ethic principles, has been evolved. Much later, Plato {see planet (5451)} took up this picture. However, the great achievement of Xenophanes remains his introduction of reflection on knowledge into philosophy. (M 24918)

Xenophanes is also honored by a lunar crater.

(6027) 1993 SS_2

Discovered 1993 September 23 by G. J. Garradd at Siding Spring.

(6028) 1994 ER_1

Discovered 1994 March 11 by S. Ueda and H. Kaneda at Kushiro.

(6029) Edithrand

1948 AG. Discovered 1948 January 14 by E. Wirtanen at Mount Hamilton.

Named for Edith (Rand) Wirtanen, who found this minor planet on an astrographic plate taken by her husband, the Lick astronomer Carl A. Wirtanen {see planet (2044)}. As a long-term resident at Mt. Hamilton, she performed numerous jobs, from library assistant to observing assistant under F. J. Neubauer and G. {see planet (2796)} and K. Kron. (M 26764)

Name endorsed by A. R. Klemola.

(6030) 1981 EG_{36}

Discovered 1981 March 7 by S. J. Bus at Siding Spring.

(6031) Ryokan

1982 BQ_4. Discovered 1982 January 26 by H. Kosai and K. Hurukawa at Kiso.

Named for Ryokan (1758-1831), a famous Buddhist monk of the Zen sect and a poet during the Edo era. He was born in Echigo (present-day Niigata Prefecture). When young, he was trained at the temple of Entsuji {see planet (4272)}, near the birthplace of the first discoverer. (M 26764)

(6032) Nobel

1983 PY. Discovered 1983 August 4 by L. G. Karachkina at Nauchnyj.

Named in memory of Alfred Bernhard Nobel (1833-1896), Swedish inventor of dynamite. He was founder of the famous Nobel International Fund, organized after his death according to his stipulation. Originally, the annual profit from this fund was divided into five parts and awarded annually for outstanding studies in physics, chemistry, physiology (including medicine), literature and peace. The centenary of the honored Nobel procedure, which includes participation by Swedish royalty, will be in 2000. (M 25231)

The name was suggested by S. P. Kapitsa, the son of P. L. Kapitsa, who won the Nobel prize in physics in 1978.

Nobel is also honored by a lunar crater.

(6033) 1984 SQ_4

Discovered 1984 September 24 by H. Debehogne at La Silla.

(6034) 1987 JA

Discovered 1987 May 5 by A. C. Gilmore and P. M. Kilmartin at Lake Tekapo.

(6035) 1987 OR
Discovered 1987 July 27 by E. W. Elst at St. Michel.

(6036) Weinberg
1988 CV_3. Discovered 1988 February 13 by E. W. Elst at La Silla.

Named in honor of Steven Weinberg, elementary-particle physicist and recipient of the 1979 Nobel Prize for Physics. Weinberg is considered one of the world's most creative scientists. He is also well known for his writings about science. In his latest book, *Dreams of a Final Theory*, he discusses quantum mechanics, beautiful theories, the weakness of philosophy and the honor of accepting a world without God. (M 24766)

(6037) 1988 EG
Discovered 1988 March 12 by J. Alu at Palomar.

(6038) 1989 EQ
Discovered 1989 March 4 by R. H. McNaught at Siding Spring.

(6039) Parmenides
1989 RS. Discovered 1989 September 3 by E. W. Elst at St. Michel.

Named for Parmenides of Elea, born around 515 B.C. and the founder of "Eleaticism". From his lengthy poem *On Nature*, a hexametric work of which only a small part has been preserved, one learns that he considered the plurality of things as the appearance of only one eternal reality. Greek philosophers before him, such as Thales and Anaximenes {see planets (6001) and (6051), respectively}, tried to explain the physical world by means of sometimes very remarkable hypotheses. Parmenides, probably influenced by the skeptical Xenophanes {see planet (6026)}, wanted to be absolutely certain about the theory he put forward. He therefore investigated the validity of theories, not by experiments, but by means of the logical soundness of laws and concepts that had been incorporated. (M 24919)

(6040) 1990 DK_3
Discovered 1990 February 24 by H. Debehogne at La Silla.

(6041) 1990 KL
Discovered 1990 May 21 by E. F. Helin at Palomar.

(6042) Cheshirecat
1990 WW_2. Discovered 1990 November 23 by A. Natori and T. Urata at Yakiimo.

Named for a cat appearing in Lewis Carroll's {see planet (6984)} famous fairy tale *Alice in Wonderland*. Its unique characteristics is its laughing voice that lingers after the cat fades away. (M 33786)
Name and citation proposed by T. Urata.

(6043) 1991 RK_2
Discovered 1991 September 9 by S. Otomo at Kiyosato.

(6044) Hammer-Purgstall
1991 RW_4. Discovered 1991 September 13 by L. D. Schmadel and F. Börngen at Tautenburg.

Named in memory of Joseph Freiherr von Hammer-Purgstall (1774-1856) on the occasion of the 150th anniversary of the Austrian Academy of Sciences. As a profound orientalist, poet and historian, he was the founder and first president (1847-1849) of the (then Imperial) Academy of Sciences in Vienna. His most famous work is the ten-volume *History of the Osman Empire*. (M 30798)

Name proposed by the first discoverer following a suggestion from the presidency of the Austrian Academy of Sciences, citation prepared by H. Haupt.

(6045) 1991 RG_9
Discovered 1991 September 11 by H. E. Holt at Palomar.

(6046) 1991 RF_{14}
Discovered 1991 September 13 by H. E. Holt at Palomar.

(6047) 1991 TB_1
Discovered 1991 October 10 by P. Rose at Palomar.

(6048) 1991 UC_1
Discovered 1991 October 18 by S. Ueda and H. Kaneda at Kushiro.

(6049) Toda
1991 VP. Discovered 1991 November 2 by A. Takahashi and K. Watanabe at Kitami.

Named in memory of Kojun Toda (1879-1951), an observer of the Tokyo Astronomical Observatory during 1902-1940 who assisted Shin Hirayama in making a photographic survey of the Milky May. He discovered the minor planet (804) Hispania at its 1903 prediscovery opposition. (M 29146)

(6050) 1992 AE
Discovered 1992 January 10 by the Spacewatch at Kitt Peak.

(6051) Anaximenes
1992 BX_1. Discovered 1992 January 30 by E. W. Elst at La Silla.

Named for the great Greek philosopher Anaximenes of Miletus. He was born in 545 B.C. and is to be considered, with Thales and Anaximander {see, respectively, planets (6001) and (6006)}, one of the three first philosophers in the western world. His principal element was "aer" (vapor, air), from which the various types of matter can be derived by condensation. His thought is typical of the transition from mythology to science. However, he was not completely liberated from mystical tendencies (orphism), since he believed that there is a kind of overarching principle between microcosmos and macrocosmos. (M 24919)
Anaximenes is also honored by a lunar crater.

(6052) Junichi
1992 CE_1. Discovered 1992 February 9 by K. Endate and K. Watanabe at Kitami.

Named in honor of Junichi Watanabe (1960-), division chief of the Public Information Office of the National Astronomical Observatory and, from 1994, president of the Japanese comet conference. Actively involved in research on minor planets, comets and meteors, he has organized an effective observational team and has contributed to the development of the planetary sciences infrastructure in Japan. He also plays an important role in the popularization of astronomy and planetary sciences. (M 26930)

(6053) 1993 BW_3
Discovered 1993 January 30 by R. H. McNaught at Siding Spring.

(6054) Ghiberti
4019 P-L. Discovered 1960 September 24 by C. J. van Houten and I. van Houten-Groeneveld at Palomar.

Named for Lorenzo Ghiberti (1378-1455), one of the most famous Italian sculptors. In the pestilence year of 1400 he vowed to create new porches for the Baptisterium, the oldest church in Florence. Donatello, Brunelleschi {see planets (6056) and (6055)} and Ghiberti together created the bronze door - the so-called Paradise door - in 41 years. In 1414 Ghiberti created the first large renaissance bronze statue of St. John the Baptist. He also worked in Siena. (M 26764)

(6055) Brunelleschi

2158 T-3. Discovered 1977 October 16 by C. J. van Houten and I. van Houten-Groeneveld at Palomar.

Named for Filippo Brunelleschi (Brunellesco, 1377-1446), a famous Italian sculptor and architect of the early renaissance. He built the 45-m dome of the cathedral of Florence. He also worked in Milan, Rimini and Mantua. (M 26764)

(6056) Donatello

2318 T-3. Discovered 1977 October 16 by C. J. van Houten and I. van Houten-Groeneveld at Palomar.

Named for Donato di Nivolo di Betto Bardi, generally known as Donatello (c. 1386-1466), a great Italian sculptor. His work in the Early Renaissance style had much influence on following generations of Italian artists. His statues and reliefs are done in marble, terracotta and wood. His statue of Gattamelata was the first equestrian statue since the antique statue of Hadrian {see planet (7446)}. Most of his work is found in Florence and Siena. (M 26764)

(6057) Robbia

5182 T-3. Discovered 1977 October 16 by C. J. van Houten and I. van Houten-Groeneveld at Palomar.

Named for Luca della Robbia (1400-1482) and his family of most important Italian sculptors. Apart from Donatello and Ghiberti {see planets (6056) and (6054)}, he was the greatest sculptor in the Early Renaissance. His first sculpture was the marble pulpit in the cathedral of Florence, and he also made reliefs for its campanile. He is mostly recognized by his clay reliefs with colored enamel. (M 26764)

(6058) 1978 VL_5

Discovered 1978 November 7 by E. F. Helin and S. J. Bus at Palomar.

(6059) 1979 TA

Discovered 1979 October 11 by Z. Vávrová at Kleť.

(6060) Doudleby

1980 DX. Discovered 1980 February 19 by A. Mrkos at Kleť.

Named for a village in southern Bohemia, south of České Budějovice. It was first mentioned as a site of a Slavonic fortified settlement above the Malše {see planet (7669)} river in 981. The Gothic church of St. Vincent was built on this site. The region of Doudleby is known for its lively traditional folk customs, including carnivals. (M 34341)

Name suggested by J. Tichá and M. Tichý.

(6061) 1981 SQ_2

Discovered 1981 September 20 by H. Debehogne at La Silla.

(6062) Vespa

1983 JQ. Discovered 1983 May 6 by N. G. Thomas at Anderson Mesa.

Named for the popular motor scooter used throughout the world. June 1996 marks the occasion of the 50th anniversary of its introduction. (M 26931)

(6063) Jason

1984 KB. Discovered 1984 May 27 by C. S. Shoemaker and E. M. Shoemaker at Palomar.

Jason, the son of Aeson of Thessaly, was educated by Chiron {see planet (2060)}, the wisest of the centaurs. He became leader of the Argonauts in the quest for the Golden Fleece, an expedition celebrated in the annals of all Greece. Many of the famous heroes of Greece accompanied Jason. The voyage of the Argo was full of struggle against the elements as well as against men. Jason succeeded in seizing the Golden Fleece, and the Argonauts made their return after a long and perilous voyage across the Danube, the Ocean, the Libyan deserts, the Red Sea and the Mediterranean. (M 25445)

(6064) Holašovice

1987 HE_1. Discovered 1987 April 23 by A. Mrkos at Kleť.

Named for a village in southern Bohemia north of Kleť Mountain. Holašovice is an exceptionally intact and well-preserved example of traditional Central European village architecture, containing a number of vernacular buildings from the eighteenth and nineteenth centuries in a style known as the South Bohemian Folk Baroque. Since 1998 Holašovice has been listed on the UNESCO World Heritage roster of cultural and natural sites. (M 34341)
Name proposed by J. Tichá.

(6065) 1987 OC

Discovered 1987 July 27 by E. F. Helin and R. S. Dunbar at Palomar.

(6066) Hendricks

1987 SZ_3. Discovered 1987 September 26 by E. Bowell at Anderson Mesa.

Named in honor of John Hendricks (1952-), founder of Discovery Communications, Inc., which owns and operates the Discovery Channel and the Learning Channel. Hendricks is a strong supporter of astronomy through his service on the Lowell Observatory Advisory Board, his sponsorship of astronomical documentaries broadcast on the Discovery Channel and his direct support of various astronomical projects. Hendricks founded the Discovery Channel to fill a void. He believes that to be human is to be curious about the world around us, and he tries to fill that need with informative and educational television programming. (M 26931)
Name suggested and citation prepared by K. X. Kramer.

(6067) 1990 QR_{11}

Discovered 1990 August 28 by Z. Vávrová at Kleť.

(6068) Brandenburg

1990 TJ_2. Discovered 1990 October 10 by F. Börngen and L. D. Schmadel at Tautenburg.

Named for the district in Germany around Berlin {see planet (422)}. The medieval capital of this region was the town of Brandenburg, and the current capital is Potsdam {see planet (5816)}. Brandenburg is a region with many lakes, waterways and pine forests. The poets Th. Fontane {see planet (8667)} and H. von Kleist, the scientist W. von Humboldt and the architect K. Fr. Schinkel {see planet (5297)} were all born in the region. Six well-known compositions by J. S. Bach {see planet (1814)} have the title *Brandenburgische Konzerte*. (M 24919)
Name proposed by the first discoverer.

(6069) Cevolani

1991 PW_{17}. Discovered 1991 August 8 by H. E. Holt at Palomar.

Named in honor of Giordano Cevolani (1945-), researcher of the Italian National Research Council (CNR) in Bologna. His main fields of interest are the physics of the earth's atmosphere, astronomy of minor bodies and space science. His scientific activity mainly concerns radar observation of meteors, the phenomena connected with their interaction with the atmosphere, and study of the wind structure (prevailing components: gravity, tidal and planetary waves)

in the mesosphere and in the lower troposphere of the polar regions (katabatic winds). (M 27461)
Name suggested and citation prepared by M. DiMartino.

(6070) Rheinland

1991 XO_1. Discovered 1991 December 10 by F. Börngen at Tautenburg.

The name "Rheinland" strictly speaking denotes the territory on both sides of the river Rhine from the town of Bingen to the border with the Netherlands. The Rheinland people (the Rheinländer) enjoy merry celebrations at carnival time and during the grape harvesting. Rheinländer is also the name of a Polka-like dance. The composer L. van Beethoven {see planet (1815)}, the physicist W. C. Röntgen and the surgeon F. Sauerbruch had their roots in the Rheinland. (M 24919)

(6071) 1992 AS_1

Discovered 1992 January 4 by T. Hioki and S. Hayakawa at Okutama.

(6072) Hooghoudt

1280 T-1. Discovered 1971 March 25 by C. J. van Houten and I. van Houten-Groeneveld at Palomar.

Named in memory of Bernard G. Hooghoudt (1924-1995), key engineer for Dutch radio telescopes, having designed the Dwingeloo telescope (1956) and the Synthesis Radio Telescope at Westerbork (1970). As a private consulting engineer he designed and led the construction of some of the world's larger telescopes, not only for radio wavelengths, but also for millimeter and optical wavelengths. (M 25978)
Name suggested and citation prepared by R. LePoole.

(6073) 1939 UB

Discovered 1939 October 18 by Y. Väisälä at Turku.

(6074) Bechtereva

1968 QE. Discovered 1968 August 24 by T. M. Smirnova at Nauchnyj.

Named in honor of Natalia Petrovna Bechtereva (1924-), outstanding neurophysiologist, director of the Institute of Experimental Medicine in St. Petersburg for many years and founder of the Institute of the Human Brain of the Russian Academy of Sciences. Bechtereva laid the foundation for basic research into the physiology of healthy and diseased human brains. She discovered the cerebral mechanisms that optimize cognitive processes - the error detector. The theory of the brain's stable pathological state as the adaptation background for many chronic diseases was developed by Bechtereva, opening up new opportunities for their treatment. (M 33786)

Name suggested by the Institute of the Human Brain and the Institute of Applied Astronomy.

(6075) Zajtsev

1976 GH_2. Discovered 1976 April 1 by N. S. Chernykh at Nauchnyj.

Named in honor of Aleksandr Leonidovich Zajtsev (1945-), staff member of the Moscow Institute of Radio Electronics, noted expert on radar observations of the planets. Under his leadership, observations of (4179) Toutatis were successfully carried out in December 1992 using the Evpatorian planetary radar, Crimea, as the signal transmitter and the radio telescope in Effelsberg, Germany, as the receiver of the radar echo from the minor planet's surface. Zajtsev is also responsible for initiating the world's first intercontinental radar astronomy experiment, Goldstone to Evpatoria observations of (6489) 1991 JX in June 1995. (M 25445)

Name proposed by the discoverer following a suggestion by the Institute of Theoretical Astronomy.

(6076) Plavec

1980 CR. Discovered 1980 February 14 by L. Brožek at Kleť.

Named in honor of Mirek J. Plavec (1925-), Czech astronomer living in the U.S. since 1969, professor of astronomy at the University of California at Los Angeles. His early works dealt with meteor showers, but he is also widely known for his studies of close binaries, especially of mass transfer in binary systems. (M 25654)

Name suggested by J. Tichá, M. Tichý and Z. Moravec, who observed this minor planet at Kleť at the 1994 opposition, just prior to its numbering.

(6077) Messner

1980 TM. Discovered 1980 October 3 by Z. Vávrová at Kleť.

Named in honor of Reinhold Messner (1944-), Italian mountaineer, adventurer and writer. (M 33385)

(6078) Burt

1980 TC_5. Discovered 1980 October 10 by C. S. Shoemaker at Palomar.

Named in honor of Burton G. Shoemaker (1912-), uncle of E. M. Shoemaker {see planet (2074)}. (M 24919)

(6079) 1981 DG_3

Discovered 1981 February 28 by S. J. Bus at Siding Spring.

(6080) 1981 EY_{26}

Discovered 1981 March 2 by S. J. Bus at Siding Spring.

(6081) 1981 EE_{35}

Discovered 1981 March 2 by S. J. Bus at Siding Spring.

(6082) Timiryazev

1982 UH_8. Discovered 1982 October 21 by L. V. Zhuravleva at Nauchnyj.

Named in memory of Russian physiologist Kliment Arkad'evich Timiryazev (1843-1920), a corresponding member of the Imperial St. Petersburg Academy of Sciences and one of the founders of the Russian school of plant physiology. (M 34622)

(6083) Janeirabloom

1984 SQ_2. Discovered 1984 September 25 by B. A. Skiff at Anderson Mesa.

Named in honor of soprano saxophonist and jazz composer Jane Ira Bloom. Bloom is known for her chromatic, lyrical playing and compositions for orchestra, unusual ensembles and dance troupes. Her performances with traditional jazz groups appear on several acclaimed recordings. As the first musician in the NASA Art Program, she wrote "Most Distant Galaxy" and "Einstein's Red/Blue Universe", the latter on commission for the American Composers' Orchestra. (M 31296)

(6084) Bascom

1985 CT. Discovered 1985 February 12 by C. S. Shoemaker and E. M. Shoemaker at Palomar.

Named in memory of Florence Bascom (1862-1945), the first woman geologist in the United States. She was an expert in crystallography, mineralogy and petrography and worked in the fields of metamorphism and crystallography in their infancy. Bascom was also an educator who trained a generation of young women as professionals at Bryn Mawr, where she founded the geology department. She was the first woman hired by the U.S. Geological Survey and the first woman elected to the Council of the Geological Society of America. Her pioneering work earned her a position among the country's hundred leading geologists in the early twentieth century. (M 31610)

(6085) 1987 SN_3

Discovered 1987 September 25 by P. Jensen at Brorfelde.

(6086) 1987 VU

Discovered 1987 November 15 by Z. Vávrová at Kleť.

(6087) Lupo

1988 FK. Discovered 1988 March 19 by C. S. Shoemaker and E. M. Shoemaker at Palomar.

Named in honor of Bob Lupo. Wearing a black hat and a genial smile, Lupo typifies the Western cowboy of today's imagination. Born in Boston, he sought his fortune in the West, becoming a registered farrier and at one time taking part in team roping in Colorado. Today he ownes and manages a fine Western-style eatery, Horsemen Lodge and Restaurant, in Flagstaff, Arizona, where he has served outstanding seafood (a heritage from Boston), steaks and barbecue to the many renowned visiting scientists and others who have entered his doors. (M 32345; M 32377)

(6088) Hoshigakubo

1988 UH. Discovered 1988 October 18 by T. Seki at Geisei.

The name refers to a section of low ground at Choja, a mountainous place in the western part of Kochi prefecture. Legend has it sthat a meteorite fell there in olden times. Now there is a pond 20 meters across. (M 28621)

(6089) Izumi

1989 AF_1. Discovered 1989 January 5 by M. Koishikawa at Sendai.

Named for Izumi-ku, a ward in the northern part of the city of Sendai {see planet (3133)}. The symbol of the city, Mt. Izumigatake, is a popular recreational area. (M 25654)

(6090) 1989 DJ

Discovered 1989 February 27 by H. Debehogne at La Silla.

(6091) Mitsuru

1990 DA_1. Discovered 1990 February 28 by K. Endate and K. Watanabe at Kitami.

Named in honor of Mitsuru Soma (1954-), of the National Astronomical Observatory of Japan, a specialist in meridian astrometry who also analyses eclipses and occultations involving various celestial objects. (M 26931)

(6092) 1990 MN

Discovered 1990 June 27 by E. F. Helin at Palomar.

(6093) Makoto

1990 QP_5. Discovered 1990 August 30 by K. Endate and K. Watanabe at Kitami.

Named in honor of Makoto Yoshikawa (1962-), senior researcher in the Communications Research Laboratory in Japan. Known for his active research in celestial mechanics, he works on the orbital analysis of minor planets, comets, meteors, artificial satellites and space debris. He is perhaps best known for his long-term orbital analyses of minor planets in mean-motion and secular resonances. (M 26931)

(6094) 1990 VQ_1

Discovered 1990 November 10 by T. Hioki and S. Hayakawa at Okutama.

(6095) 1991 UU

Discovered 1991 October 18 by S. Ueda and H. Kaneda at Kushiro.

(6096) 1991 UB_2

Discovered 1991 October 29 by S. Ueda and H. Kaneda at Kushiro.

(6097) Koishikawa

1991 UK_2. Discovered 1991 October 29 by K. Endate and K. Watanabe at Kitami.

Named in honor of Masahiro Koishikawa (1952-), a staff member of the Sendai Astronomical Observatory since 1972. He has made great efforts to observe major planets, minor planets and comets at the observatory's Ayashi station. He has also attempted to develop public awareness of astronomy as a social educational project. (M 26931)

(6098) 1991 UW_3

Discovered 1991 October 31 by M. Matsuyama and K. Watanabe at Kushiro.

(6099) Saarland

1991 UH_4. Discovered 1991 October 30 by F. Börngen at Tautenburg.

Named for the German state of Saarland, a district of beautiful landscapes situated in the middle course of the river Saar. The modern capital Saarbrücken, a center for the coal and iron industry, is named for the first stone bridge across the river and arose from a Celtic settlement, a Roman castle and the Franconian court Villa Sarabrucca. (M 25445)

(6100) Kunitomoikkansai

1991 VK_4. Discovered 1991 November 9 by A. Sugie at Taga.

Named in memory of Ikkansai Kunimoto (1778-1840). Born into a family of famous gun makers in the Edo period in Japan, he made several Gregorian reflecting telescopes entirely of his own design and observed the sun, moon and planets. For 14 months during 1835-1836 he made a continuous series of observations of sunspots. (M 33385)

Name proposed by the discoverer, endorsed by S. Murayama, K. Sato and A. Fujii.

(6101) 1993 EG

Discovered 1993 March 1 by T. Urata at Oohira.

(6102) 1993 FQ_{25}

Discovered 1993 March 21 by the Uppsala-ESO Survey at La Silla.

(6103) 1993 HV

Discovered 1993 April 16 by S. Ueda and H. Kaneda at Kushiro.

(6104) Takao

1993 HZ. Discovered 1993 April 16 by K. Endate and K. Watanabe at Kitami.

Named in honor of Takao Saito (1930-), retired professor of astrogeophysics at Tôhoku University known for his study of solar-terrestrial physics. He has investigated the various effects of the solar wind on cometary ion tails. He has also cooperated with Japanese amateur astronomers on cometary observations. (M 26931)

(6105) Verrocchio

4580 P-L. Discovered 1960 September 24 by C. J. van Houten and I. van Houten-Groeneveld at Palomar.

Named for the Italian sculptor and painter Andrea del Verrocchio (1435-1488). Originally named Andrea di Cione, he adopted the name of his teacher, Giuliano

Verrocchio, as his own. With Donatello {see planet (6056)}, he was the leading sculptor in bronze of the Florentine Renaissance. His best-known work is the statue of Bartolomeo Colleoni in Venice. (M 26764)

(6106) Stoss

6564 P-L. Discovered 1960 September 24 by C. J. van Houten and I. van Houten-Groeneveld at Palomar.

Named for the German sculptor and wood-carver Veit Stoss (c. 1445-1553), who worked in Nürnberg and Cracow. In Cracow he worked for 13 years on the 11-m by 13-m Mary altar in late-gothic/baroque style. After 1496 he lived again in Nürnberg and made many wonderful altars for the church and city. (M 26764)

(6107) Osterbrock

1948 AF. Discovered 1948 January 14 by C. A. Wirtanen at Mount Hamilton.

Named in honor of Donald E. Osterbrock (1924-), professor of astronomy and astrophysics at the University of Wisconsin-Madison (1958-1973) and at the University of California at Santa Cruz and UCO/Lick Observatory (1972-1992) and director of the Lick Observatory (1973-1981). His research spans gaseous nebulae, interstellar matter, galactic structure, and quasars and active nuclei of galaxies, and he has authored a historical series on the life and work of several astronomers who played vital roles in the progress of American astronomy. Osterbrock served as president of the American Astronomical Society during 1987-1989. He is a member of the National Academy of Sciences and the National Academy of Arts and Sciences. (M 27735)

Citation prepared by A. R. Klemola.

(6108) Glebov

1971 QN. Discovered 1971 August 18 by T. M. Smirnova at Nauchnyj.

Named in honor of academician Igor' Alekseevich Glebov (1914-), outstanding scientist in the field of electrical engineering and power engineering, since 1975 director of the Science Research Institute of Electric Machine Engineering in St. Petersburg. Glebov is renowned for his work in control systems for power turbo-generators and his application of superconductivity in electric machine engineering. (M 34341)

Name suggested by the Institute of Applied Astronomy and the Institute of Problems of Electrophysics of RAS {*Russian Academy of Sciences*}.

(6109) 1975 QC

Discovered 1975 August 29 at the Felix Aguilar Observatory at El Leoncito.

(6110) 1978 NQ_1

Discovered 1978 July 4 by L. I. Chernykh at Nauchnyj.

(6111) 1979 SP_{13}

Discovered 1979 September 20 by S. J. Bus at Palomar.

(6112) 1981 DB_1

Discovered 1981 February 28 by S. J. Bus at Siding Spring.

(6113) 1982 SX_5

Discovered 1982 September 16 by L. I. Chernykh at Nauchnyj.

(6114) 1984 HS_1

Discovered 1984 April 28 by W. Ferreri and V. Zappalà at La Silla.

(6115) Martinduncan

1984 SR_2. Discovered 1984 September 25 by B. A. Skiff at Anderson Mesa.

Named in honor of Martin J. Duncan (1950-) of Queen's University, Kingston, Ontario. Duncan has made several important contributions to the understanding of the origin and dynamical evolution of small bodies in the solar system, particularly comets and the likelihood that they originated in the Kuiper {see planet (1776)} Belt. He has been involved in the development of two important numerical algorithms that have led to orbital integrations of unprecedented duration. (M 27461)
Citation provided by H. Levison.

(6116) Still

1984 UB_3. Discovered 1984 October 26 by E. Bowell at Anderson Mesa.

Named in memory of William Grant Still (1895-1978), American composer, also known as the "Dean of American Negro Composers". Although he was very much at home in the musical world of blues and spirituals, Still's major contribution to music is as an outstanding and prolific composer of symphonies, operas, ballets, chamber music, piano music and other works in the tradition of classical musical forms. His style is basically neo-romantic, even when featuring structural complexities. Still always considered himself an American composer first and a colored composer last. (M 25654)

Name proposed and citation provided by M. J. Shott, endorsed by the composer's daughter J. A. Still and her family.

(6117) 1985 CZ_1

Discovered 1985 February 12 by H. Debehogne at La Silla.

(6118) 1986 QX_3

Discovered 1986 August 31 by H. Debehogne at La Silla.

(6119) 1986 XH

Discovered 1986 December 6 by P. Jensen at Brorfelde.

(6120) Anhalt

1987 QR. Discovered 1987 August 21 by F. Börngen at Tautenburg.

Named for the former principality of Anhalt, at the northern border of the Harz mountains. About 1770, Prince Leopold Franz of Anhalt built the Wörlitz castle in a great park that was much praised by Goethe {see planet (3047)}. Anhalt was united with the Prussian province of Sachsen {see planet (5866)} to form the state of Sachsen-Anhalt. (M 24919)

(6121) 1987 RU_3

Discovered 1987 September 2 by L. I. Chernykh at Nauchnyj.

(6122) Henrard

1987 SW_1. Discovered 1987 September 21 by E. Bowell at Anderson Mesa.

Named in honor of Jacques Henrard (1940-), professor of mathematics at the University of Namur. Henrard has applied algorithms based on Hamiltonian and Lie formalisms, particularly to the motions of the moon, the Galilean satellites of Jupiter and artificial satellites. With his celestial mechanics research group, which he formed in 1971, Henrard extended the adiabatic-invariant and action-angle theory and introduced it as a practical tool. He has also worked on the theory of slow chaotic motion and has been intensively involved in modeling mean-motion resonances in the asetroid belt. Since 1989 he has been editor-in-chief of the journal *Celestial Mechanics.* (M 27461)
Citation provided by A. Lemaître.

(6123) Aristoteles

1987 SH_2. Discovered 1987 September 19 by E. W. Elst at Rozhen.

Named for Aristoteles (384-322 B.C.), one of the most significant Greek philosophers, charging the demerits of his philosophy on the level of the knowledge at his era and ascribing the misuse of his ideas to human weakness. He was the first, and possibly the greatest, theoretician of the mechanism of thought and deduction, being at the same time well aware of the importance of what happens in the real world and in nature. (M 24766)

Citation prepared by A. Svarna and D. Sinachopoulos at the request of the discoverer. Proposal endorsed by V. Shkodrov.
Aristoteles is also honored by a lunar crater.

(6124) Mecklenburg

1987 SL_{10}. Discovered 1987 September 29 by F. Börngen at Tautenburg.

Named for a district in northern Germany, characterized by deposits of glacial moraines and hundreds of lakes rich in fishes and birds of many kinds. The towns of Wismar and Rostock are important Baltic Sea ports. The poet F. Reuter {see planet (8666)}, the archaeologist H. Schliemann {see planet (3302)} and the sculptor and poet E. Barlach {see planet (6428)} were born in Mecklenburg. Today Mecklenburg is part of the state of Mecklenburg-Vorpommern. (M 24919)
Named on the occasion of the thousandth anniversary of Mecklenburg in 1995.

(6125) 1989 CN

Discovered 1989 February 4 by S. Ueda and H. Kaneda at Kushiro.

(6126) 1989 EW_1

Discovered 1989 March 5 by Z. Vávrová at Kleť.

(6127) Hetherington

1989 HD. Discovered 1989 April 25 by E. F. Helin at Palomar.

Named in honor of Ernest Hetherington, a highly respected authority on Orchids (Orchidaceae). He was a good friend and mentor to Kay Francis {see planet (2050)}, the discoverer's mother, who grew and hybridized many exceptional "new" orchids. "Ernie" is an orchid judge, gifted writer on orchidaceae and recipient of many honors, awards and special recognition for his role as a grower, educator and popularizer of orchid culture. (M 30477)

(6128) 1989 LA

Discovered 1989 June 3 by E. F. Helin at Palomar.

(6129) Demokritos

1989 RB_2. Discovered 1989 September 4 by E. W. Elst at St. Michel.

Named for the Greek philosopher Demokritos of Abdera (460-380 B.C.), well-known for his atomistic theory of matter. He was a pupil of Leukippos {see planet (5950)}, to whom he owed many ideas. His great concern was to harmonize Ionic cosmology with Greek ontology, and he is credited with at least 57 works on ethics, physics, nature, mathematics and art. He considered an observation to be the result of an interaction between observer and observed. (M 24919)
Demokritos is also honored by a lunar crater.

(6130) 1989 SL_5

Discovered 1989 September 24 by R. H. McNaught at Siding Spring.

(6131) Towen

1990 OO_3. Discovered 1990 July 27 by H. E. Holt at Palomar.

Named in honor of Tobias C. (Toby) Owen (1936-) of the Institute for Astron-

omy of the University of Hawaii. Known for his broad work in planetary science, Owen has productively used chemical ratios and isotopic ratios from planetary atmospheres, from meteorites and from comets to understand the early solar system and the processes of formation of comets and planetary atmospheres. He has also played an important role in searching for and identifying new species in reflection spectra of ices on planetary surfaces. (M 27461)
Citation provided by M. F. A'Hearn following a suggestion by D. Morrison.

(6132) Danielson

1990 QY_3. Discovered 1990 August 22 by H. E. Holt at Palomar.

Named in honor of G. Edward Danielson (1939-) of the Division of Geological and Planetary Sciences at California Institute of Technology. After working in industry in high-energy laser research, Danielson came to the Jet Propulsion Laboratory and Caltech, where he has been involved since 1967 in the design, calibration, testing and data analysis of remote-sensing imagers. He has worked on numerous space missions, including Voyager, Pioneer Venus, Mariner Venus Mercury, the Hubble Space Telescope and Mars Observer. In 1982 Danielson was co-recoverer of comet 1P/Halley. (M 27461)
Name suggested by A. W. Harris and citation provided by W. A. Baum.

(6133) 1990 RC_3

Discovered 1990 September 14 by H. E. Holt at Palomar.

(6134) 1990 RA_5

Discovered 1990 September 15 by H. E. Holt at Palomar.

(6135) Billowen

1990 RD_9. Discovered 1990 September 14 by H. E. Holt at Palomar.

Named in honor of William Mann Owen, Jr. (1954-), an astronomer at the Jet Propulsion Laboratory. Owen developed the initial orbits for the small satellites of Uranus and Neptune discovered during the Voyager space mission. He also participated in the first determination of the mass ratio of Pluto and Charon using the Hubble Space Telescope. Using his extensive knowledge of astrometry, Owen played a significant role in reducing the ephemeris uncertainties for some flight project target bodies, including (951) Gaspra and (243) Ida. (M 27461; M 27477)

Name proposed by the discoverer following a suggestion by the JPL ephemeris group.

(6136) 1990 YH

Discovered 1990 December 22 by A. Natori and T. Urata at Yakiimo.

(6137) 1991 BY

Discovered 1991 January 25 by A. Natori and T. Urata at Yakiimo.

(6138) 1991 JH_1

Discovered 1991 May 14 by S. Otomo and O. Muramatsu at Kiyosato.

(6139) 1992 AD_1

Discovered 1992 January 10 by A. Sugie at Taga.

(6140) Kubokawa

1992 AT_1. Discovered 1992 January 6 by K. Endate and K. Watanabe at Kitami.

Named in memory of Kazuo Kubokawa (1903-1943), a staff member of Tokyo Astronomical Observatory who made photographic observations of minor planets and comets with the Brashear {see planet (5502)} astrograph at Mitaka {see planet

(1088)}. He and O. Oikawa {see planet (2667)} discovered (1139) Atami in 1929. (M 29146)

(6141) Durda

1992 YC_3. Discovered 1992 December 26 by the Spacewatch at Kitt Peak.

Named in honor of Daniel D. Durda, who has studied the generation and evolution of dust from and the size distribution of minor planets. As a member of the Galileo imaging team, he has carried out fragmentation studies of minor planets and a detailed analysis of the formation of their satellites - in particular, of (243) Ida's moon Dactyl. An avid pilot, he is also a noted astronomical artist whose paintings are inspired by his scientific work. (M 32345)

(6142) 1993 FP

Discovered 1993 March 23 by A. C. Gilmore and P. M. Kilmartin at Lake Tekapo.

(6143) Pythagoras

1993 JV. Discovered 1993 May 14 by E. W. Elst at La Silla.

Named for the great Greek philosopher and mathematician Pythagoras of Samos (580-500 B.C.). He contributed to the development of mathematics and is generally credited with the first mathematical foundation of theories about harmony in physics and the arts. Although it is difficult to distinguish his teaching from those of his disciples, Pythagorean principles strongly influenced the thought of Plato and to a somewhat lesser extent that of Aristoteles {see, respectively, planets (5451) and (6123)}. (M 24919; M 25351)

Citation written by D. Sinachopoulos at the request of the discoverer.

Pythagoras is also honored by a lunar crater.

(6144) 1994 EQ_3

Discovered 1994 March 14 by K. Endate and K. Watanabe at Kitami.

(6145) Riemenschneider

2630 P-L. Discovered 1960 September 26 by C. J. van Houten and I. van Houten-Groeneveld at Palomar.

Named for Tilman Riemenschneider (c. 1460-1531), German sculptor in stone and in wood. He lived in Würzburg and served as the city's mayor. During the 'Bauernkrieg' of 1525 he was on the side of the peasants, who lost the war. This probably resulted in his being tortured, and there is no record that he continued to sculpt after that time. In southern Germany and Austria there exist many wonderful altars from his hands, notably in Rothenburg, Creglingen, Heidelberg and Würzburg. (M 26764)

(6146) Adamkrafft

3262 T-2. Discovered 1973 September 30 by C. J. van Houten and I. van Houten-Groeneveld at Palomar.

Named for the German sculptor Adam Krafft (also written Kraft, 1460-c.1508), who lived in Nürnberg and its vicinity. His most important works are to be seen in the Sebaldus and Lorenz churches and the museum of Nürnberg. (M 26765)

(6147) Straub

1081 T-3. Discovered 1977 October 17 by C. J. van Houten and I. van Houten-Groeneveld at Palomar.

Named for the German sculptor Johann Baptist Straub (1704-1784). His most famous work is in the south-German rococo style, with transition to the classicistic style. His white and golden altars can be seen in the churches of Schäftlarn and Ettal. He worked together with his pupil Ignaz Günther {see planet (6148)}. (M 26765)

(6148) Ignazgünther

5119 T-3. Discovered 1977 October 16 by C. J. van Houten and I. van Houten-Groeneveld at Palomar.

Named for Ignaz Günther (also written Ginter and Ginther, 1725-1775). In 1743 he became a pupil of Straub {see planet (6147)} and in 1754 opened his own workshop. He is regarded as one of the most important rococo sculptors. His works can be seen in many churches and also in the Bürgersaal in München. (M 26765)

(6149) 1979 SS

Discovered 1979 September 25 by A. Mrkos at Kleť.

(6150) Neukum

1980 FR_1. Discovered 1980 March 16 by C.-I. Lagerkvist at La Silla.

Named in honor of Gerhard Neukum, director of the DLR Institute of Planetary Exploration in Berlin-Adlershof, which he has led with great dedication and enthusiasm through the challenging years following German reunification. His research on the impact record of the moon, the terrestrial planets and the minor planets has led to a method of age determination that is applicable to all planetary surfaces in the solar system. (M 26931)

Name proposed by the discoverer and G. Hahn, who prepared the citation.

(6151) Viget

1987 WF. Discovered 1987 November 19 by E. Bowell at Anderson Mesa.

Named in honor of Princeton University on the occasion of its 250th birthday 1996 Oct. 22. The name derives from Princeton's official motto: "Dei Sub Numine Viget", or "Under the Power of God She Flourishes". Princeton first became involved with astronomy when its sixth president, John Witherspoon, purchased the Rittenhouse Orrery in 1771. This miniature planetarium was one of the finest scientific instruments of the eighteenth century, and it still survives in working order. Viget joins (508) Princetonia and (534) Nassovia, both discovered by R. S. Dugan {see planet (2772)}, as the third minor planet named for Princeton. (M 27461)

Name suggested and citation prepared by J. T. Miller.

(6152) Empedocles

1989 GB_3. Discovered 1989 April 3 by E. W. Elst at La Silla.

Named for the Greek philosopher Empedocles (490-430 B.C.). Only some 400 lines remain of his poem *On Nature*, which considered that matter is composed of the four basic elements of fire, air, water and earth. Nothing comes into being or is destroyed, but things are merely transformed. Like Herakleitos {see planet (5204)} he believed that two forces, love and hate, interact to bring together and to separate the four ingredients. Fewer than 100 verses are left of *Purifications*, another poem of Empedocles. (M 24919)

(6153) Hershey

1990 OB. Discovered 1990 July 19 by E. F. Helin at Palomar.

Named in memory of Wesley Lamar Hershey (1913-1989), director of the Caltech "Y" at the California Institute of Technology for 30 years. Always ready to help students with sympathy and understanding, he developed the Y into a major factor of campus life. He provided an environment where exploration of ideas and concerns, controversial or not, was not only acceptable but encouraged. He encouraged students to face the issues of values in their own lives and for society as a whole. (M 25978)

(6154) Stevesynnott

1990 QP_1. Discovered 1990 August 22 by H. E. Holt at Palomar.

Named in honor of Stephen P. Synnott (1946-), a space scientist at the Jet Propulsion Laboratory. Synnott is a noted authority on spacecraft optical navigation techniques. He discovered the Jovian satellites Thebe and Metis and was a member of the Voyager 2 imaging team, which discovered the small Uranian and Neptunian satellites. (M 27461)

Name proposed by the discoverer following a suggestion by the JPL ephemeris group.

(6155) 1990 VY_2

Discovered 1990 November 11 by T. Nomura and K. Kawanishi at Minami-Oda.

(6156) Dall

1991 AF_1. Discovered 1991 January 12 by B. G. W. Manning at Stakenbridge.

Named in memory of Horace E. Dall (1901-1986), a talented craftsman known internationally for his skill in instrument making, optics and optical design. Inventor of the Dall 'null' test for parabolic telescope mirrors and coinventor of the Dall-Kirkham Cassegrain-type telescope, Dall was well known for his Maksutov telescopes, in particular a six-inch-aperture folding model that he used to take on his world travels. Dall refigured many telescope objective lenses to give optimum resolution, and he was also well known for making high-power microscope objectives. Dall joined the British Astronomical Association in 1925 and was awarded its Walter Goodacre Medal in 1967. (M 25231)

Obituary published in J. Br. Astron. Assoc., Vol. 97, No. 2, p. 76-80 (1987).

(6157) Prey

1991 RX_2. Discovered 1991 September 9 by L. D. Schmadel and F. Börngen at Tautenburg.

Named in memory of Adalbert Prey (1873-1949), professor of astronomy in Innsbruck, Prague and, eventually, Vienna, where he was elected to membership in the Austrian Academy of Sciences. After World War II until his death he served as one of the two secretaries of the Academy. He worked on the motion of 70 Oph (sometimes called "Prey's star"), as well as on the dynamics of minor planets and the moon. He thoroughly investigated the field of isostasy and gravitation, to which he devoted a series of papers. (M 31296)

Named by the first discoverer following a suggestion by H. Haupt, who also prepared the citation.

(6158) 1991 VB_3

Discovered 1991 November 12 by T. Niijima and T. Urata at Ojima.

(6159) 1991 YH

Discovered 1991 December 30 by S. Ueda and H. Kaneda at Kushiro.

(6160) Minakata

1993 JF. Discovered 1993 May 15 by Y. Shimizu and T. Urata at Nachi-Katsuura.

Named in memory of Kumagusu Minakata (1867-1941), a renowned Japanese naturalist who, as a child, studied mycology. Employed from 1895 as a temporary specialist by the British Museum, he returned to Japan in 1900 to continue his studies of biology and other fields. He promoted the movement to protect the forest of Kumano in the Wakayama prefecture from destruction by the government. (M 30477)

(6161) Vojno-Yasenetsky

1971 TY_2. Discovered 1971 October 14 by L. I. Chernykh at Nauchnyj.

Named in memory of Valentin Feliksovich Vojno-Yasenetsky (1877-1961), surgeon and bishop. As a physician, he practised in a number of district hospitals in central Russia, serving as head physician in the Tashkent urban hospital and as a professor at Central Asia University. In the early 1920s he took vows and was ordained as a bishop. During his life he published 55 scientific works on surgery and anatomy, winning the state prize in 1946. Although he was arrested many times and sent into exile for his views, when he died he had dignity as the archbishop of the Crimea and Simferopol. (M 32345)

(6162) Prokhorov

1973 SR_6. Discovered 1973 September 25 by L. V. Zhuravleva at Nauchnyj.

Named in honor of Yurij Vasil'evich Prokhorov (1929-), Russian mathematician noted for his work on probability theory and a member of the editorial board of the *Soviet Encyclopedia.* (M 34622)

(6163) 1977 FT

Discovered 1977 March 16 by H.-E. Schuster at La Silla.

(6164) Gerhardmüller

1977 RF_2. Discovered 1977 September 9 by N. S. Chernykh at Nauchnyj.

Named in memory of academician Gerhard Friedrich Müller (miller, according to traditional spelling in Russian; 1705-1783), first rector of St. Petersburg University and editor of the first Russian academic journal. He is considered the father of St. Petersburg's historical school, and his works were the foundation for research on the history, ethnography, archeology and geography of Russia and Siberia. (M 34341)

(6165) Frolova

1978 PD_3. Discovered 1978 August 8 by N. S. Chernykh at Nauchnyj.

Named in honor of Natalia Borisovna Frolova, assistant professor of astronomy at Ural State University in Ekaterinburg. She worked on a detailed catalogue of stars along the path of comet 1P/Halley, and this contributed to the success of the space missions Vega and Giotto. As a leader in educating local schoolchildren about astronomy, she takes an active part in organizing the annual Winter Astronomical Students' School at Kourovskaya Observatory. (M 34341)

(6166) Univsima

1978 SP_4. Discovered 1978 September 27 by L. I. Chernykh at Nauchnyj.

Named for the Simferopol State University on the eightieth anniversary of the establishment of this first institution of higher education in the Crimea. Academicians A. P. Pavlov, I. P. Pavlov, V. I. Vernadskij {see planet (2809)} and N. I. Andrusov were founding faculty members. Academicians V. I. Obruchev, A. F. Ioffe {see, respectively, planets (3128) and (5222)}, N. M. Krylov and V. I. Smirnov worked there. Among its 65,000 graduates are academicians I. V. Kurchatov {see planet (2352)} and K. I. Shchelkin. (M 34622)

(6167) Narmanskij

1979 QB_{10}. Discovered 1979 August 27 by N. S. Chernykh at Nauchnyj.

Named in honor of Vladimir Yakovlevich Narmanskij (1948-), Crimean amateur astronomer and founder of the amateur laboratory "Heliorythm", where research is conducted on the resonance structure of the solar system and on solar-terrestrial relations. He is also a poet, known for writing the text for *Mass for Chersones* by Crimean composer A. Karamanov {see planet (4274)}. (M 34341)

(6168) 1981 EB_1

Discovered 1981 March 5 by H. Debehogne and G. DeSanctis at La Silla.

(6169) 1981 EX_4

Discovered 1981 March 2 by S. J. Bus at Siding Spring.

(6170) Levasseur

1981 GP. Discovered 1981 April 5 by E. Bowell at Anderson Mesa.

Named in honor of Anny-Chantal Levasseur-Regourd (1945-), professor at Paris VI University. An incisive planetary scientist, Levasseur-Regourd works at the Service d'Aéronomie of the CNRS. She was selected as an astronaut, though her main scientific interest concerns the interplanetary medium, the physics of comets and the light-scattering properties of dust. Levasseur-Regourd has been principal investigator for several space experiments, inlcuding the Halley Optical Probe of the Giotto mission. Through her research and teaching, she actively encourages future generations of scientists. (M 27461)
Citation provided by M. A. Barucci.

(6171) 1981 UT

Discovered 1981 October 26 by L. G. Taff at Socorro.

(6172) Prokofeana

1982 TX. Discovered 1982 October 14 by L. G. Karachkina at Nauchnyj.

Named in honor of Valentina Vladimirovna Prokof'eva (1929-), prominent astrophysicist at the Crimean Astrophysical Observatory, a pioneer in the use of television photometry in astronomy and one of the authors of the monograph *Televisionnaya Astronomiya*, published in Russian in 1974 and 1984. She devised methods for receiving and calibrating television photometric data from such objects as galaxies, binaries, major and minor planets, comets and artificial satellites. Prokof'eva has obtained significant results in the interpretation of the blue clearings on Mars, nonradial pulsations of red giants, the components of symbiotic stars, and periodic light variations of the x-ray source V1055 Ori. She discovered multiple periodicities in light curves of a number of minor planets and interpreted them as a consequence of their duplicity and rotation of their components. (M 25231)
Name proposed by the discoverer following a suggestion of Yu. V. Batrakov.

(6173) Jimwestphal

1983 AD. Discovered 1983 January 9 by B. A. Skiff at Anderson Mesa.

Named in honor of James A. Westphal (1930-), professor of planetary science at the California Institute of Technology and director of the Palomar Observatory. After leading a geological research group in the petroleum industry, Westphal went to Caltech in 1961, where he developed modern infrared detectors for use in astronomy, and (with Bruce Murray, see planet (4957)) made important early observations of stars and planets. In later work (with Kristian and Sandage) he pioneered SIT spectroscopy and two-dimensional photometry of faint galaxy clusters. As principal investigator (from 1977) for the Hubble Space Telescope WFPC camera system, Westphal has played a key role in the development of CCD detctors suitable for astronomical and space research. (M 27462)
Citation provided by W. A. Baum following a suggestion by A. W. Harris.

(6174) Polybius

1983 TR_2. Discovered 1983 October 4 by N. G. Thomas at Anderson Mesa.

Named for Polybius (c. 200-118 B.C.), Greek historian and father of pragmatic history, one of 1000 eminent Archeans deported to Rome after Perseus' defeat at Pydna in 168 B.C. Making favorable friends in Rome, he traveled with them to Spain, North Africa and through the Alps back to Rome. He mediated Roman troubles in Greece and collected works of Greek historians. His own narratives cover the Punic Wars and contemporary events but with such conciseness that various problems of chronology and strategy remain. (M 26931)

(6175) 1983 XW

Discovered 1983 December 4 by A. Mrkos at Kleť.

(6176) 1985 BH

Discovered 1985 January 16 by Z. Vávrová at Kleť.

(6177) 1986 CE_2

Discovered 1986 February 12 by H. Debehogne at La Silla.

(6178) 1986 DA

Discovered 1986 February 16 by M. Kizawa at Shizuoka.

(6179) Brett

1986 EN. Discovered 1986 March 3 by C. S. Shoemaker and E. M. Shoemaker at Palomar.

Named in honor of Robin Brett (1935-), an Australian-born, Harvard-educated American citizen active in the planetary-science program from its early days. His research has included geochemical studies of meteoritic and lunar samples, modeling the cores of the terrestrial planets and geologic mapping of impact craters. He was responsible for much of the planning for the massive study of the Apollo lunar samples. A born organizer, Brett has directed his activities toward advancing international cooperation in research; the fact that he is known as a "nice guy" with a great sense of humor may have helped. (M 34623)

(6180) Bystritskaya

1986 PX_4. Discovered 1986 August 8 by L. I. Chernykh at Nauchnyj.

Named in honor of Ehlina Avraamovna Bystritskaya, distinguished Soviet actress. An actress with Moscow Malyj Theatre since 1958, she is especially popular for her brilliant roles in several films. (M 34342)

(6181) 1986 RW

Discovered 1986 September 6 by E. F. Helin at Palomar.

(6182) Katygord

1987 SC_4. Discovered 1987 September 21 by E. Bowell at Anderson Mesa.

Named in honor of Katherine Carson Gordon Kron (1917-) on the occasion of her 80th birthday, 1997 May 24. Kron devoted much of her professional life to observational astronomy at the Lick Observatory. During the second world war she served as a physicist for the U.S. Navy and for much of the 1960s was editor of the *Publications of the Astronomical Society of the Pacific.* (M 30098)

Name suggested, with love and affection, by her husband G. E. Kron {see planet (2796)}, and warmly endorsed by the discoverer.

(6183) Viscome

1987 SF_7. Discovered 1987 September 26 by C. S. Shoemaker and E. M. Shoemaker at Palomar.

Named in honor of George R. Viscome (1956-), of Lake Placid, New York. A broadcast technician responsible for master control and videotape operations at a television station in Albany, Viscome is an accomplished astrometrist. His observations of this minor planet at the 1994 opposition were sufficient for numbering it. (M 24766)

Name proposed by the discoverers, following a suggestion by G. V. Williams (who made the idenfication involving this object).

(6184) 1987 UQ_3

Discovered 1987 October 26 by P. Jensen at Brorfelde.

(6185) 1987 YD

Discovered 1987 December 20 by T. Kojima at Chiyoda.

(6186) Zenon

1988 CC_2. Discovered 1988 February 11 by E. W. Elst at La Silla.

Named for the great Greek philosopher and mathematician Zenon of Elea (494-430 B.C.). As a friend and pupil of Parmenides {see planet (6039)}, he continued his teacher's abstract and analytic thought, taking the theses of his opponents and refuting them by *reductio ad absurdum*. He tried to show that the assumption of the existence of a plurality of things in time and space carried with it more serious inconsistencies. He is especially known for the paradoxes he used for this purpose. (M 24919)

Citation by D. Sinachopoulos at the request of the discoverer.

(6187) 1988 RD_5

Discovered 1988 September 2 by H. Debehogne at La Silla.

(6188) 1988 SW_2

Discovered 1988 September 16 by S. J. Bus at Cerro Tololo.

(6189) Völk

1989 EY_2. Discovered 1989 March 2 by E. W. Elst at La Silla.

Named in honor of Elisabeth Völk (1946-), secretary at the European Southern Observatory's headquarters in Garching, where she is chiefly responsible for the administration of the ESO Schmidt plates. During the July 1994 observing campaign on comet D/1993 F2 (Shoemaker-Levy 9) she did an excellent job solving so many problems that occurred at that hectic time. During all the years (1987-1994) of the Uccle-ESO observing program on minor planets she has become a very good friend to the discoverer. (M 24766)

Proposal endorsed by L. D. Schmadel, who independently suggested naming a minor planet for Mrs. Völk.

(6190) Rennes

1989 TJ_1. Discovered 1989 October 8 by M. Koishikawa at Sendai.

Named for the city in France, Rennes is the international sister city of Sendai {see planet (3133)}, Japan, affiliated since 1997. It is the center of administration and culture in Brittany. (M 33786)

(6191) Eades

1989 WN_1. Discovered 1989 November 22 by B. G. W. Manning at Stakenbridge.

Named in honor of George Eades, currently celebrating his fiftieth year of membership in the British Astronomical Association. A structural engineer by profession, Eades constructed two 15-cm reflecting telescopes, including the mirrors and eyepiece lenses. An expert microscopist, he also constructed a high-power binocular microscope, including all the optics except for the objective lenses. Eades was of great help to the discoverer many years ago by introducing him to the BAA and scientific literature that led eventually to an interest in astrometry and the discovery of (6191) and other minor planets. (M 25445)

Name suggested by A. Manning.

(6192) 1990 KB_1

Discovered 1990 May 21 by E. F. Helin at Palomar.

(6193) Manabe

1990 QC_1. Discovered 1990 August 18 by K. Endate and K. Watanabe at Kitami.

Named in memory of Ryonosuke Manabe (1926-1983), who worked for many years as a computer assistant for H. Hirose {see planet (1612)} at the Tokyo Astronomical Observatory. In 1974 he went to work for NASDA, where he was involved with the Japanese geodesy satellite. (M 29146)

Name proposed by the second discoverer following a suggestion by K. Tomita.

(6194) Denali

1990 TN. Discovered 1990 October 12 by R. H. McNaught at Siding Spring.

Mt. Denali, rising 6194 meters above the sea level, is the tallest peak in North America. The name Denali originates from the Tanana Indian language and translates into English as "The Great One". The Tanana Indians are a subgroup of the Athabaskans; the Athabaskans were the original inhabitants of central and south central Alaska. (M 34342)

Name proposed by D. Hamilton.

(6195) Nukariya

1990 VL_2. Discovered 1990 November 13 by K. Endate and K. Watanabe at Kitami.

Named in honor of Motoi Nukariya (1943-), who was active in software development at the Tokyo Astronomical Observatory before moving to NASDA in 1971 to develop their orbit-determination system for satellites. (M 29146)

Name proposed by the second discoverer following a suggestion by K. Tomita.

(6196) 1991 UO_4

Discovered 1991 October 28 by S. Ueda and H. Kaneda at Kushiro.

(6197) 1992 AB_1

Discovered 1992 January 10 by S. Inoda and T. Urata at Karasuyama.

(6198) 1992 AF_1

Discovered 1992 January 10 by T. Hioki and S. Hayakawa at Okutama.

(6199) 1992 BK_1

Discovered 1992 January 26 by A. Sugie at Taga.

(6200) Hachinohe

1993 HL. Discovered 1993 April 16 by K. Endate and K. Watanabe at Kitami.

Named for the amateur astronomer Akio Hachinohe (1958-), who has contributed much to amateur astronomy in Hokkaido {see planet (3720)}. He is also active in sky sports such as hot-air ballooning and hang-gliding. (M 25978)

(6201) Ichiroshimizu

1993 HY. Discovered 1993 April 16 by K. Endate and K. Watanabe at Kitami.

Named in honor of Ichiro Shimizu (1923-1996), who began work for the Tokyo Astronomical Observatory in 1945 and was heavily involved in the construction of the Corona Observing Station on Mt. Norikura. He was later a key astronomer in the solar physics division of the Observatory until he retired in 1984. (M 29146)

Name proposed by the second discoverer following a suggestion by K. Tomita.

(6202) Georgemiley

3332 T-1. Discovered 1971 March 26 by C. J. van Houten and I. van Houten-Groeneveld at Palomar.

Named for George Kildare Miley (1942-) to honor his 25-year association with the Leiden Observatory. A senior scientist at the Space Telescope Science Institute during 1984-1988, he became a professor in Leiden in 1988. His research has included investigations of radio, optical, infrared and x-ray properties of active galaxies. He instigated an extensive new radio survey of the northern sky and in 1989 discovered the galaxy 4C 41.17 with the largest known redshift. (M 26765)

(6203) 1981 EC_{23}

Discovered 1981 March 3 by S. J. Bus at Siding Spring.

(6204) MacKenzie

1981 JB_3. Discovered 1981 May 6 by C. S. Shoemaker at Palomar.

Named in honor of Norman MacKenzie, a leading scholar of the poetry of Gerard Manley Hopkins. Professor of English at the University of Rhodesia from 1954 to 1964, he has subsequently been at Queen's University {see planet (5457)} in Canada. His works include co-editing the fourth edition of Hopkins' *Poems* (1967) and most recently the Oxford English Texts *Poetical Works of Gerard Manley Hopkins* (1990). MacKenzie has followed Hopkins' close observation of nature, especially astronomy, and has maintained an active interest in the sky throughout his life. (M 24919)

Name proposed and citation prepared by D. H. Levy.

(6205) Menottigalli

1983 OD. Discovered 1983 July 17 by E. Bowell at Anderson Mesa.

Named in honor of Menotti Galli (1922-), a physicist at the University of Bologna. Galli's research activity began with the study of cosmic rays and meteorites and was subsequently extended to the physics of the sun and the heliosphere. He has studied sun-earth relationships, cosmogenic isotopes and radiocarbon in tree rings. He suggested searching for microparticles embedded in the resin of trees surviving the Tunguska {see planet (5471)} explosion, and he participated in the 1991 expedition to collect the necessary wood samples, with the result that a group of elements was identified as possible constituents of the Tunguska body. (M 33786)

Name suggested and citation provided by M. di Martino.

(6206) Corradolamberti

1985 TB_1. Discovered 1985 October 15 by E. Bowell at Anderson Mesa.

Named in honor of Corrado Lamberti, an excellent writer and popularizer of astronomy and one of the founders and now director of the Italian popular astronomy review *l'Astronomia*. With his accurate work over the last 20 years, he has made a fundamental contribution to the dissemination of knowledge of astronomy in Italy. (M 33787)

Name suggested and citation provided by M. di Martino.

(6207) 1988 BV

Discovered 1988 January 24 by S. Ueda and H. Kaneda at Kushiro.

(6208) Wakata

1988 XT. Discovered 1988 December 3 by K. Endate and K. Watanabe at Kitami.

Named in honor of Kouichi Wakata, chosen by NASDA as the astronaut in 1992 among 372 applicants, after receiving a master's degree in applied dynamics at the University of Kyushu and working for Japan Air Lines. In January 1996 he was on board the space shuttle Endeavor and succeeded in recovering a Japanese satellite. (M 29146)

(6209) Schwaben

1990 TF_4. Discovered 1990 October 12 by F. Börngen and L. D. Schmadel at Tautenburg.

Schwaben was a German dukedom, in 1079 passing over to the Swabian race

of the Staufer. From 1138 to 1254, this race supplied the German kings and emperors. In the middle ages Schwaben was a rich trading place. Today, it is a Bavarian governmental district with its capital in Augsburg. It extends from the Allgäuer Alps and Lake Constance (Swabian Sea) to the Swabian-Franconian Alb. (M 25654)
Name proposed by the first discoverer.

(6210) 1991 AX_1

Discovered 1991 January 14 by M. Matsuyama and K. Watanabe at Kushiro.

(6211) 1991 DO

Discovered 1991 February 19 by S. Inoda and T. Urata at Karasuyama.

(6212) 1993 MS_1

Discovered 1993 June 23 by M. Nassir at Palomar.

(6213) Zwiers

2196 P-L. Discovered 1960 September 24 by C. J. van Houten and I. van Houten-Groeneveld at Palomar.

Named in memory of Hendrikus Johannes Zwiers (1865-1923), astronomer at the Leiden Observatory who worked in positional astronomy. He is best known for his method for calculating the orbit of visual binaries and for his computation of the orbit of comet 17P/Holmes. (M 26765)
Obituary published in Hemel en Dampkring, Vol. 22, p. 37 (1924).

(6214) Mikhailgrinev

1971 SN_2. Discovered 1971 September 26 by T. M. Smirnova at Nauchnyj.

Named in honor of Mikhail Vasil'evich Grinev (1929-), outstanding Russian surgeon, director of St. Petersburg Djanelidze Research Institute of Emergency Medicine from 1984 to 1998, member of the board of directors of the European Association of Trauma and Emergency Surgery, professor of St. Petersburg Medical University and medical academy for postgraduate training. Grinev is internationally renowned for his classis research in the field of emergency surgery, polytrauma and shock. (M 34342)

Name suggested by the Institute of Applied Astronomy and Djanelidze Research Institute of Emergency Medicine.

(6215) 1973 EK

Discovered 1973 March 7 by L. Kohoutek at Bergedorf.

(6216) San Jose

1975 SJ. Discovered 1975 September 30 by S. J. Bus at Palomar.

Named for the city in northern California, to acknowledge it as the Capital of Silicon Valley and to honor it as the world's leader in innovation and advanced technology research. The San Jose area, home of the Ohlone Indians, was settled by the Spanish in 1777. In 1849 San Jose became the first capital of the state of California. Since then it has grown into a major metropolitan area and is now the eleventh largest city in the United States. San Jose has made a major contribution to astronomy through its cooperative and friendly relationship with the Lick Observatory, which lies 20 miles to the east at the summit of Mount Hamilton. (M 31024)

Named by the discoverer following a suggestion by S. M. Faber. Citation prepared by B. F. Jones.

(6217) 1975 XH

Discovered 1975 December 1 by C. Torres and S. Barros at Cerro El Roble.

(6218) Mizushima

1977 EG_7. Discovered 1977 March 12 by H. Kosai and K. Hurukawa at Kiso.

Named for a site in the city of Kurashiki {see planet (4578)} near the birthplace of the first discoverer. Mizushima is a scenic spot in the Inland Sea and includes an old battlefield from the conflict between the Genji and Heike samurai families. This battle occurred during an annular eclipse of the sun in 1183, and Heike gained a great victory. (M 26765)

(6219) Demalia

1978 PX_2. Discovered 1978 August 8 by N. S. Chernykh at Nauchnyj.

Named in memory of Aleksandra Alekseevna Demenko (1930-1983), Ukrainian astronomer, staff member of the Astronomical Observatory of Kiev University, known for her research on comets. (M 30098)
Obituary published in Komet. Tsirk., No. 316 (1983).

(6220) Stepanmakarov

1978 SN_7. Discovered 1978 September 26 by L. V. Zhuravleva at Nauchnyj.

Named in memory of Stepan Osipovich Makarov (1848-1904), Russian vice-admiral and naval commander, oceanographer, polar researcher and ship builder. (M 34623)

(6221) Ducentesima

1980 GO. Discovered 1980 April 13 by A. Mrkos at Kleť.

The naming commemorates the 200th Kleť numbered minor planet. Its Latin name was inspired by classical female names from the pioneer days of the searching for minor planets in the nineteenth century. (M 27735)
Name suggested by J. Tichá.

(6222) 1980 PB_3

Discovered 1980 August 8 at the Royal Observatory Edinburgh at Siding Spring.

(6223) Dahl

1980 RD_1. Discovered 1980 September 3 by A. Mrkos at Kleť.

Named in honor of the Welsh-born author Roald Dahl (1916-). His *Willy Wonka and the Chocolate Factory* and *James and the Giant Peach* are classics of children's literature. (M 27735)

Name suggested by G. V. Williams, who made the identifications involving this object.

(6224) 1981 EK_8

Discovered 1981 March 1 by S. J. Bus at Siding Spring.

(6225) 1981 EK_{12}

Discovered 1981 March 1 by S. J. Bus at Siding Spring.

(6226) 1981 EY_{18}

Discovered 1981 March 2 by S. J. Bus at Siding Spring.

(6227) 1981 EQ_{42}

Discovered 1981 March 2 by S. J. Bus at Siding Spring.

(6228) 1982 BA

Discovered 1982 January 17 by T. Furuta at Tokai.

(6229) Tursachan

1983 VN_7. Discovered 1983 November 4 by B. A. Skiff at Anderson Mesa.

The name is "Standing Stones" in Gaelic, a term used to refer to the stones placed during neolithic times into groups - often circles - throughout the British Isles. Many of these arrangements exhibit astronomical alignments and may have been used to track the progression of seasons and mark the occurrence of other significant astronomical events. (M 31296)

The name was suggested by Alice Cathryne Dennis, a seventh-grade student at the Mountain School in Flagstaff, Arizona, as winner of a contest to name this minor planet in conjunction with the 1997 Flagstaff Festival of Science. Citation prepared by C. B. Luginbuhl.

(6230) 1984 SG_1

Discovered 1984 September 27 by Z. Vávrová at Kleť.

(6231) 1985 FH

Discovered 1985 March 20 by A. Mrkos at Kleť.

(6232) Zubitskia

1985 SJ_3. Discovered 1985 September 19 by N. S. Chernykh and L. I. Chernykh at Nauchnyj.

Named in honor of Danila Nikiforovich Zubitskij and Natalia Petrovna Zubitska, doctors and psycho-therapists in Kiev, who have developed original methods of treatment using herbal medicines. The authors of several books, they are well known in the Ukraine, and their methods and preparations are also being used in other countries. Natalia Petrovna, who has a broadcast program in Kiev about herbs and "people's medicine", has also published several books of poetry. (M 32345)

(6233) Kimura

1986 CG. Discovered 1986 February 8 by S. Inoda and T. Urata at Karasuyama.

Named in memory of Hisashi Kimura (1870-1943), the first director of the International Latitude Observatory of Mizusawa {see planet (7530)}. Serving in this capacity from 1899 to 1941, he was also director of the International Latitude Service from 1922 to 1934. In 1902 Kimura discovered the Z term of polar motion, the first astronomical clue for inferring physical states and processes in the earth's deep interior, in particular at the core-mantle boundary. (M 34623)

Name proposed by the second discoverer, following a suggestion by K. Hurukawa and K. Yokoyama, and in commemoration of the Mizusawa Observatory's centennial. Citation prepared by K. Yokoyama.

(6234) 1986 SF

Discovered 1986 September 30 by Z. Vávrová at Kleť.

(6235) 1987 VB

Discovered 1987 November 14 by S. Ueda and H. Kaneda at Kushiro.

(6236) 1988 WF

Discovered 1988 November 29 by W. Kakei and M. Kizawa and T. Urata at Oohira.

(6237) Chikushi

1989 CV. Discovered 1989 February 4 by T. Seki at Geisei.

This is the old name of the Fukuoka region in northern Kyushu. During 1944-1947 the Chikushi Astronomical Association was active. Several of its members, including Isao Ebihara, Takeo Mori, Mineo Nishiyama, Kiichiro Hurukawa {see planet (3425)} and Ichiro Hasegawa {see planet (3227)}, made their first studies of the orbits of comets and minor planets. (M 28621)

(6238) 1989 NM

Discovered 1989 July 2 by E. F. Helin at Palomar.

(6239) Minos

1989 QF. Discovered 1989 August 31 by C. S. Shoemaker and E. M. Shoemaker at Palomar.

Minos was the son of Zeus and Europa {see planets (5731) and (52)}. As king of Crete, he distinguished himself by the wisdom of his laws, which remained in force for nearly 1000 years, and by his sense of justice and moderation approved by all the Greeks and gods. These earned Minos, after his death, promotion to the dignity of judge of the Underworld. (M 25445)

(6240) Lucretius Carus

1989 SL_1. Discovered 1989 September 26 by E. W. Elst at La Silla.

Named for the Latin philosopher and poet Titus Lucretius Carus. He was born around 90 B.C., probably in Rome, and he is known from his *De rerum natura*, a long poem written in Latin hexameters. In this he expounds on the physical theory of the Greek philosopher Epicurus {see planet (5954)}, of whom he speaks with great admiration. The third part of the poem deals with atomic structure and the mortality of the soul, the latter with the famous words 'Death means nothing to us'. (M 25445)

Lucretius is also honored by a lunar crater.

(6241) Galante

1989 TG. Discovered 1989 October 4 at the Osservatorio San Vittore at Bologna.

Named in honor of Maria Pia Galante (1916-) wife of Ciro' Vacchi, proprietor and director of Osservatorio San Vittore. (M 25978)

(6242) 1990 OJ_2

Discovered 1990 July 29 by H. E. Holt at Palomar.

(6243) Yoder

1990 OT_3. Discovered 1990 July 27 by H. E. Holt at Palomar.

Named in honor of Charles F. Yoder (1943-), of Los Angeles, California. Yoder developed the first analysis of capture into orbit-orbit mean-motion resonances among the satellites of the giant planets, as well as the explanation of the origin and evolution of the three-body Laplacian orbital resonance among the Galilean satellites of Jupiter. He has connected the variations of the rotation of the earth with properties of the core and of the core-mantle interface, found evidence for the dissipative effects of a possible fluid core-solid mantle interaction for the moon and put constraints on the core of Venus and Venus' rotational history required by Venus' obliquity. (M 27462)

Name suggested by A. W. Harris, citation provided by S. J. Peale.

(6244) Okamoto

1990 QF. Discovered 1990 August 20 by T. Seki at Geisei.

Named in honor of Hiroshi Okamoto (1915-), an entomologist and elementary-school teacher who inspired the discoverer with a love of the stars. (M 32093)

(6245) Ikufumi

1990 SO_4. Discovered 1990 September 27 by T. Urata at Oohira.

Named in honor of Ikufumi Makino (1944-), a telecommunications systems engineer and active amateur astronomer. His extensive knowledge of computer technologies has enabled him to contribute to astronomical magazines many articles about systems for making observations. He has played an important role in introducing and spreading the data-reduction tool Astrometrica in Japan. He is a collaborating member of the Nihondaira Observatory. (M 32094)

(6246) Komurotoru
1990 VX_2. Discovered 1990 November 13 by T. Fujii and K. Watanabe at Kitami.

Named in memory of the sculptor Toru Komuro (1899-1953), on the hundredth anniversary of his birth. Komuro's work was first accepted for the Teiten (Imperial Art Exhibition, now known as Nitten) in 1924. His most important work is the equestrian statue of Datemasamune {see planet (6859)}, placed in the ruins of the castle of Sendai {see planet (3133)} as a symbol of the city. (M 34623)

(6247) Amanogawa
1990 WY_3. Discovered 1990 November 21 by K. Endate and K. Watanabe at Kitami.

Named for a river that runs through the Hokkaido town of Kaminokuni {see planet (5978)}. It is also the Japanese word for the Milky Way. (M 29146)

(6248) 1991 BM_2
Discovered 1991 January 17 by Z. Vávrová at Kleť.

(6249) Jennifer
1991 JF_1. Discovered 1991 May 7 by E. F. Helin at Palomar.

Named in honor of Jennifer Jones Simon, an actress whose creativity captivated hearts in memorable performances. She has shared with the world her passionate delight in the arts in guiding the destiny of the Norton Simon Museum of Art in Pasadena. Her Pasadena friends and admirers worldwide heartfully endorse this naming, a celebratory tribute to her talents and beautiful persona. (M 26425)

(6250) 1991 VX_1
Discovered 1991 November 2 by E. F. Helin at Palomar.

(6251) Setsuko
1992 DB. Discovered 1992 February 25 by M. Akiyama and T. Furuta at Mishima.

Named in honor of Setsuko Akiyama (1953-), wife of the first discoverer. (M 27129)

(6252) Montevideo
1992 EV_{11}. Discovered 1992 March 6 by the Uppsala-ESO Survey at La Silla.

Named for the capital of Uruguay and city of the birth of G. Tancredi {see planet (5088)}, one of the discoverers of this minor planet. The meaning of the name comes from the fact that the city is "watched over" by a bowl-shaped hill. Founded in 1726 by the Spanish crown in order to stop Portuguese expansion, Montevideo has a long tradition in astronomy, starting in 1789 when the first observatory was built to study the transit of Mercury, the observations of which were later used by Leverrier {see planet (1997)} to discover the secular advance of Mercury's perihelion. In 1955 the first planetarium in the Spanish-speaking world was built there. (M 26425)

Named on the occasion of the eighth Latino-american regional meeting of the IAU, held in Montevideo in November 1995.

(6253) 1992 FJ
Discovered 1992 March 24 by S. Ueda and H. Kaneda at Kushiro.

(6254) 1993 UM_3
Discovered 1993 October 20 by S. Ueda and H. Kaneda at Kushiro.

(6255) Kuma
1994 XT. Discovered 1994 December 5 by A. Nakamura at Kuma.

Named for the small Japanese town on Shikoku {see planet (4223)} where this planet was discovered. Kuma town, known for forestry, agriculture and as an important destination for pilgrims, also encourages local cultural activities. Kuma

Kogen Astronomical Observatory was built in 1992 for astronomical education and tourism. (M 25231)

(6256) Canova

4063 P-L. Discovered 1960 September 24 by C. J. van Houten and I. van Houten-Groeneveld at Palomar.

Named for Antonio Canova (1757-1822), an Italian sculptor in the classicistic style. An opponent of the wild form of the baroque style, he worked in a realistic style with a quiet and smooth form. He made idealized portraits of Napoleon and his family. One of his most famous sculptures is 'Amor and Psyche'. (M 26765)

(6257) Thorvaldsen

4098 T-1. Discovered 1971 March 26 by C. J. van Houten and I. van Houten-Groeneveld at Palomar.

Named for the Danish sculptor Bertel Thorvaldsen (1768-1844). With Canova {see planet (6256)} he was the greatest representative of the classicistic style. His work is found in Copenhagen, in St. Peter's in Rome, in Munich, Lucerne and Warsaw. (M 26765)

(6258) Rodin

3070 T-2. Discovered 1973 September 30 by C. J. van Houten and I. van Houten-Groeneveld at Palomar.

Named for the French sculptor Auguste Rodin (1840-1917). In his sculptures Rodin tried to express personal inner feelings with bearing and gestures. He was one of the first sculptors to abandon the classicistic style of Thorvaldsen and Canova {see planets (6257) and (6256)}. Among his most famous works are 'The Citizens of Calais', 'The Kiss' and 'The Thinker'. (M 26765)

(6259) Maillol

3236 T-2. Discovered 1973 September 30 by C. J. van Houten and I. van Houten-Groeneveld at Palomar.

Named for the French sculptor and graphic artist Aristide Maillol (1861-1944). He first studied painting with Paul Gauguin but because of a disease of his eyes became a sculptor. After Rodin {see planet (6258)} he is generally regarded as the greatest French sculptor. (M 26765)

(6260) Kelsey

1949 PN. Discovered 1949 August 2 by K. Reinmuth at Heidelberg.

Named in honor of Frances Oldham Kelsey (1914-), research pharmacologist at the U.S. Food and Drug Administration in Washington, D.C. Kelsey's prescient research led to the recognition that fetal malformation can result from drugs that appear harmless to adults. Her integrity as a scientist and as a citizen prevented the marketing of the sedative drug thalidomide in the United States, thereby preventing an occurrence of an alarming number of limbless infants being born, as happened in countries where the drug was marketed between the late 1950s and 1962. For this action, she received in 1962 the U.S. president's award for distinguished civilian service for (in the words of president J. F. Kennedy) "sparing the nation a human tragedy". (M 25654)
Name proposed and citation prepared by G. C. L. Aikman.

(6261) Chione

1976 WC. Discovered 1976 November 30 by H.-E. Schuster at La Silla.

Named for the Greek princess Chione, who had the misfortune of having both Apollo {see planet (1862)} and Hermes fall in love with her. She was the mother of Autolykos by Hermes and of Philammon by Apollo. She was arrogant enough to compare herself with Artemis {see planet (105)}, who killed her in revenge. (M 27129)

(6262) Javid

1978 RZ. Discovered 1978 September 1 by N. S. Chernykh at Nauchnyj.

Named in memory of Hussein Javid (1882-1941), Azerbaijani poet, playwright and historian, author of several historical plays, such as *Sheikh Sanan*, *Teimur, the lame* and *Khayyam*. In his works he preached the ideas of progressive romanticism. In 1937 he was branded as a dissident and exiled to Siberia. (M 30098)

(6263) 1980 PX

Discovered 1980 August 6 by Z. Vávrová at Kleť.

(6264) 1980 SQ

Discovered 1980 September 29 by Z. Vávrová at Kleť.

(6265) 1985 TW_3

Discovered 1985 October 11 by T. F. Fric and R. J. Gilbrech at Palomar.

(6266) Letzel

1986 TB_3. Discovered 1986 October 4 by A. Mrkos at Kleť.

Named in memory of Jan Letzel (1880-1925), Czech architect inspired by the Art Nouveau style and who worked in Japan from 1907 to 1923. He designed the Hiroshima Prefectural Commercial Exhibition Hall in 1915. This building, the most elegant in the city, became world famous as the Genbaku Dome or A-bomb Dome, which was left in ruins as an eternal symbol of the tragedy of Hiroshima on 1945 Aug. 6. (M 34342)

Name suggested by J. Tichá.

(6267) 1987 SO_9

Discovered 1987 September 20 by E. W. Elst at Rozhen.

(6268) Versailles

1990 SS_5. Discovered 1990 September 22 by E. W. Elst at La Silla.

Named for the city, site of the grandiose palace and start of the French revolution in 1789. In 1624 Louis XIII had a small castle built there, some 17 km from Paris, to serve as a lodge after his hunts in the surrounding woods. The present palace was built by Louis XIV, who moved his entire staff there in 1682, while carefully maintaining the old hunting lodge. The palace was substantially altered by Louis XV, and it is known for its magnificent gardens and 'Salle du jeu de paume'. (M 25445)

(6269) Kawasaki

1990 UJ. Discovered 1990 October 20 by T. Urata at Oohira.

Named in memory of Shun'ichi Kawasaki (1896-1943), the second director of the International Latitude Observatory of Mizusawa {see planet (7530)} and brother-in-law of the astronomer Issei Yamamoto {see planet (2249)}. He introduced the floating zenith telescope to supplement the visual zenith telescope for routine observations. (M 34623)

Name proposed by the discoverer following a suggestion by K. Hurukawa and K. Yokoyama. Citation prepared by K. Yokoyama.

(6270) 1991 BD

Discovered 1991 January 18 by S. Inoda and T. Urata at Karasuyama.

(6271) Farmer

1991 NF. Discovered 1991 July 9 by E. F. Helin at Palomar.

Named in honor of Crofton Bernard Farmer, distinguished visiting scientist at the Jet Propulsion Laboratory whose research has been primarily in earth and planetary atmospheres. He has been awarded the NASA Exceptional Scientific Achievement Medal on three occasions. The discoverer and members of the Near-

Earth Asteroid Tracking program wish to recognize his efforts in providing crucial assistance to the NEAT task as motivator, organizer and representative to the sponsor. (M 26931)

(6272) 1992 EB

Discovered 1992 March 2 by S. Ueda and H. Kaneda at Kushiro.

(6273) 1992 ER_{31}

Discovered 1992 March 1 by the Uppsala-ESO Survey at La Silla.

(6274) Taizaburo

1992 FV. Discovered 1992 March 23 by K. Endate and K. Watanabe at Kitami.

Named in honor of Taizaburo Koyama (1927-), a pioneer amateur astrophotographer who in 1973 invented the portable automatic equatorial telescope Sky Memo. (M 29146)

(6275) 1993 VQ

Discovered 1993 November 14 by T. Kobayashi at Oizumi.

(6276) 1994 AB

Discovered 1994 January 1 by T. Kobayashi at Oizumi.

(6277) 1949 QC_1

Discovered 1949 August 24 by H. L. Giclas and R. D. Schaldach at Flagstaff.

(6278) 1971 TF

Discovered 1971 October 10 by B. A. Burnasheva at Nauchnyj.

(6279) 1977 UO_5

Discovered 1977 October 18 by K. L. Faul at Palomar.

(6280) Sicardy

1980 RJ. Discovered 1980 September 2 by E. Bowell at Anderson Mesa.

Named in honor of Bruno Sicardy (1958-), professor of astronomy at Paris VI University. Sicardy's research concerns solar-system dynamics. Using both theoretical and numerical approaches, he has studied the confining mechanism of shepherd satellites on planetary rings and arcs. He has also taken part in the observation of the giant planets using various techniques to search for new satellites, analyze the ring systems and determine atmopsheric temperature profiles. (M 27462)

Citation provided by M. A. Barucci.

(6281) 1980 SD

Discovered 1980 September 16 by A. Mrkos at Kleť.

(6282) Edwelda

1980 TS_4. Discovered 1980 October 9 by C. S. Shoemaker at Palomar.

Named in honor of Edwin L. Aguirre and Imelda B. Joson, who have popularized astronomy in the Philippines for more than 15 years. Their 1985 book about Comet Halley, *The Second Coming*, is the first and only astronomical book ever written by Filipino authors. They designed two astronomical stamps issued by the Philippine Postal Service, and in 1990 the Philippine Department of Science and Technology commissioned them to construct a 0.44-m reflector, the largest operational telescope in the country, for a public observatory in Manila. That same year, the couple obtained the approval of then-president Corazon Aquino for an executive proclamation declaring an annual National Astronomy Week. (M 25978)

Citation prepared by S. J. O'Meara and D. H. Levy.

(6283) 1980 VX_1

Discovered 1980 November 6 at the Purple Mountain Observatory at Nanking.

(6284) 1981 EM_{19}

Discovered 1981 March 2 by S. J. Bus at Siding Spring.

(6285) 1981 EA_{26}

Discovered 1981 March 2 by S. J. Bus at Siding Spring.

(6286) 1983 EU

Discovered 1983 March 10 by E. Barr at Anderson Mesa.

(6287) Lenham

1984 AR. Discovered 1984 January 8 by E. Bowell at Anderson Mesa.

Named in memory of Alan P. Lenham (1930-1996), British research scientist at the Royal College of Science in Shrivenham and an amateur astronomer whose careful observations of the moon and planets caught the attention of Gerard Kuiper {see planet (1776)}, who invited him to spend some time at Yerkes Observatory. In 1958, using the 2.1-m telescope at the McDonald Observatory, Lenham made a superb drawing of a region on Mars now called Vallis Marineris. (M 30098)

Citation written and name endorsed by S. F. Dermott following a suggestion by the discoverer.

(6288) 1984 ER_1

Discovered 1984 March 2 by H. Debehogne at La Silla.

(6289) 1984 HP_1

Discovered 1984 April 28 by W. Ferreri and V. Zappalà at La Silla.

(6290) 1985 CA_2

Discovered 1985 February 12 by H. Debehogne at La Silla.

(6291) Renzetti

1985 TM_1. Discovered 1985 October 15 by E. Bowell at Anderson Mesa.

Named in honor of Nicholas A. Renzetti (1914-) on the occasion of his retirement from the Jet Propulsion Laboratory after 37 years of contributions to NASA's Deep Space Network. As manager of the DSN's Science Office, Renzetti revitalized the Goldstone {see planet (4433)} Solar System Radar, which has been used for investigations of 34 minor planets. He also successfully advanced the use of DSN antennas for geodynamics research, radio interferometry, microwave spectroscopy and SETI. (M 30477)

Name suggested and citation prepared by S. J. Ostro.

(6292) 1986 QQ_2

Discovered 1986 August 28 by H. Debehogne at La Silla.

(6293) Oberpfalz

1987 WV_1. Discovered 1987 November 26 by F. Börngen at Tautenburg.

Oberpfalz is a governmental district in Bavaria on the river Danube. In the Middle Ages its capital, Regensburg {see planet (927)}, was an arena of German history. Now the Oberpfalz extends from the Fir Mountains to the Franconian Alb, and from the Upper Palatinate forest to the northern part of the Bavarian forest. (M 25655)

(6294) Czerny

1988 CX_1. Discovered 1988 February 11 by E. W. Elst at La Silla.

Named in memory of the well-known German piano teacher and composer Karl

Czerny (1791-1857). His first piano lessons were from his father, his later lessons from Beethoven {see planet (1815)}. He became a piano teacher at 14 and was soon world famous, counting Liszt {see planet (3910)} and Thalberg among his pupils. Occupied with composing in the evening, he earned a living by giving piano lessons during the day. Among his enormous number of compositions the 848 *etudes* are in the arsenal of every pianist. (M 25655)

(6295) Schmoll

1988 CF_3. Discovered 1988 February 11 by E. W. Elst at La Silla.

Named for the German piano teacher Antoine Schmoll, who at the age of seven received his first piano lessons and tuition in the art of composing. He left his native country to become a piano teacher, first in Toulouse (1864), later in Brussels (1873) and Paris (1875). In 1881 he published his famous *Nouvelle méthode de piano, théorique, pratique et récréative* (with 16 editions in less than 14 years), in which the difficulty of the lessons increases gradually. (M 25445)

(6296) Cleveland

1988 NC. Discovered 1988 July 12 by E. F. Helin at Palomar.

Named in honor of the city in Ohio as a tribute to its bicentennial celebration. NASA's Lewis Research Center has been in Cleveland since 1941 performing aeronautical research and contributing to the exploration of space. In the early nineteenth century, Warner and Swasey {see planet (992)} made numerous telescopes in Cleveland that are still in use today throughout the U.S. In addition, it was at the Case Institute in Cleveland that Michelson and Morley performed the famous experiment that disproved the existence of ether. (M 25655)
Name proposed by R. L. Bondurant, Jr.

(6297) 1988 VZ_1

Discovered 1988 November 2 by S. Ueda and H. Kaneda at Kushiro.

(6298) 1988 XC

Discovered 1988 December 1 by T. Kojima at Chiyoda.

(6299) 1988 XQ_1

Discovered 1988 December 5 by M. Arai and H. Mori at Yorii.

(6300) 1988 YB

Discovered 1988 December 30 by T. Hioki and N. Kawasato at Okutama.

(6301) 1989 BR_1

Discovered 1989 January 29 by Z. Vávrová at Kleť.

(6302) 1989 CF

Discovered 1989 February 2 by T. Seki at Geisei.

(6303) 1989 EL_2

Discovered 1989 March 12 by S. Ueda and H. Kaneda at Kushiro.

(6304) Josephus Flavius

1989 GT_3. Discovered 1989 April 2 by E. W. Elst at La Silla.

Named for the Jewish historian Josephus Bar Mattheus, born around A.D. 37 in Jerusalem. He was a member of the Pharisees, an ancient Jewish sect noted for the strict observance of rites and ceremonies of the traditional law. In 66 he organised a revolt against the Roman occupation, but he was defeated and led before Vespasianus. Because of a correct prophesy he acquired his liberty and added the last part of his name. He left several Greek writings, such as The *Jewish War* and *Jewish Antiquities*. (M 25445)

(6305) Helgoland

1989 GE_8. Discovered 1989 April 6 by F. Börngen at Tautenburg.

Named for a small island in the North Sea, 45 km distant from the German coast, commanding the mouths of the rivers Elbe, Weser and Eider. It consists of a red-brown sandstone rock, about 60 m high, towering up nearly vertically from the ocean. There is a green plain upland, and the rock Long Anna is at the northern point. Helgoland has been owned successively by the dukedom of Schleswig, Denmark, Great Britain and – since 1890 – Germany. (M 25445)

(6306) 1989 UL_3

Discovered 1989 October 30 by A. Sugie at Taga.

(6307) 1989 WL_7

Discovered 1989 November 22 at the Carlos U. Cesco Observatory at El Leoncito.

(6308) 1990 BK

Discovered 1990 January 17 by Y. Kushida and O. Muramatsu at Yatsugatake.

(6309) Elsschot

1990 EM_3. Discovered 1990 March 2 by E. W. Elst at La Silla.

Named in memory of the famous Flemish writer Willem Elsschot, pseudonym of Alfons de Ridder (1882-1960). His work is very personal and deeply human. He never belonged to any literary school. In his poems and stories he shows great sincerity, sometimes evolving into bitter cynism about human mediocrity. His style is sober and shows a remarkable pureness. The cosmopolitical character of his first novel *Villa des Roses* was a complete change from the Flemish literature of that time, still trapped in a regional mentality. Because of his obvious atheism, he was for a long time despised by the catholic critics. (M 25445)
Citation by K. Leterme at the request of the discoverer.

(6310) Jankonke

1990 KK. Discovered 1990 May 21 by E. F. Helin at Palomar.

Named in honor of Janis L. Konke, space advocate and good friend of the discoverer. Born in Wisconsin and educated in Minnesota and Illinois, she has pursued a career in psychology, studying human capabilities and desires. Inspired by a friend and the starry skies of Palomar Mountain, she developed an interest in space travel and exploration, believing that humankind's greatest adventure is in the heavens. Thus began her enthusiastic support for the discoverer's Deep Solar System Survey. (M 25978)

(6311) Porubčan

1990 RQ_2. Discovered 1990 September 15 by H. E. Holt at Palomar.

Named in honor of Vladimír Porubčan (1940-   ) of the Astronomical Institute of the Slovak Academy of Sciences, Bratislava. Porubčan has worked on obtaining and analyzing observations of meteor showers. His work on the Taurid complex, together with Ján Štohl {see planet (3715)}, is widely recognized. Porubčan took an active part in the organization of the international conferences *Meteoroids and their Parent Bodies* in Smolenice in 1992, and in particular *Meteoroids* in Bratislava in 1994. (M 27462)
Name suggested and citation provided by H. Rickman.

(6312) 1990 RH_4

Discovered 1990 September 14 by H. E. Holt at Palomar.

(6313) 1990 RC_8

Discovered 1990 September 14 by H. Debehogne at La Silla.

(6314) 1990 SQ_{16}

Discovered 1990 September 17 by H. E. Holt at Palomar.

(6315) 1990 TS

Discovered 1990 October 11 by S. Ueda and H. Kaneda at Kushiro.

(6316) 1990 TL_6

Discovered 1990 October 9 by R. H. McNaught at Siding Spring.

(6317) Dreyfus

1990 UP_3. Discovered 1990 October 16 by E. W. Elst at La Silla.

Named for the infamous Dreyfus case, which dominated French politics, reinforced religious feelings, disrupted the old party system and divided the nation for more than 10 years. In 1894 a French Jewish army officer, Alfred Dreyfus {1859-1935}, was wrongfully convicted of high treason. The famous French writer Emile Zola came to his defense by writing the emotional article "J'accuse" in the journal *l'Aurore*. In 1906 Dreyfus was rehabilitated and decorated with the légion d'honneur. (M 25446)

(6318) Cronkite

1990 WA. Discovered 1990 November 18 by E. F. Helin at Palomar.

Named in honor of Walter Cronkite (1916-), whose integrity and unflappable character during his 60-year career in radio and television journalism have won him the acclaim and trust of his colleagues and the American people alike. Cronkite worked for the United Press during World War II, taking part in the Normandy beachhead assults and flying in B-17 raids. In 1950, he joined CBS News, becoming anchor of the then-15-minute-long 'CBS Evening News' in 1962. The following year this became U.S. network television's first half-hour weeknight news broadcast, marking the debut of the new format with Cronkite's interview of President Kennedy. Since retiring as news anchor in the early 1980s, Cronkite has produced and narrated dozens of award-winning programs. (M 28090)

(6319) Beregovoj

1990 WJ_3. Discovered 1990 November 19 by E. W. Elst at La Silla.

Named in memory of the Russian cosmonaut Georgij Beregovoj (1921-1995). In October 1968 he orbited the earth 64 times in a Soyuz 3 spacecraft and was safely recovered on land at Karaganda. He also performed space maneuvers near an unmanned Soyuz 2 spacecraft. Author of more than 300 scientific articles, he always put an emphasis on the importance of the role of the human factor in cosmic flights. Beregovoj was a member of the organizing committee for the 1995 international meeting "ecological consequences of the collision of the earth with small bodies of the solar system". His efforts led to the acquisition from the Russian government of a military airplane for use in the 37th expedition to Vanavara-Tunguska, and he hoped to participate in this exploration of the site of the 1908 impact. Sadly, he died unexepectedly during medical treatment shortly beforehand. (M 25655)

(6320) Bremen

1991 AL_3. Discovered 1991 January 15 by F. Börngen at Tautenburg.

Named for the important commercial seaport on the Lower Weser, seat of the shipping company Norddeutscher Lloyd and of a navigation school. In 787, Bremen became an episcopate. At the end of the Middle Ages, it was one of the most important Hanse cities. Olbers {see planet (1002)} discovered the minor planets (2) Pallas and (4) Vesta from Bremen, and Harding {see planet (2003)} discovered (3) Juno from nearby Lilienthal. In 1920 the Olbers Gesellschaft was founded in Bremen to encourage enthusiastic amateur astronomers. (M 25446)

(6321) 1991 BV
Discovered 1991 January 19 by A. Sugie at Taga.

(6322) 1991 CQ
Discovered 1991 February 10 by R. H. McNaught at Siding Spring.

(6323) Karoji
1991 CY_1. Discovered 1991 February 14 by K. Endate and K. Watanabe at Kitami.

Named in honor of Hiroshi Karoji (1946-), a professor at the Japanese National Astronomical Observatory and one of the leading astronomers in the team to construct the 8-meter Subaru telescope at Mauna Kea. He spent more than ten years at the Institut d'Astrophysique in Paris before joining the Subaru team. His main interests are the redshift distribution of galaxies and activity in infrared galaxies. (M 29146)

(6324) 1991 DN_1
Discovered 1991 February 23 by S. Inoda and T. Urata at Karasuyama.

(6325) 1991 EA_1
Discovered 1991 March 14 by M. Arai and H. Mori at Yorii.

(6326) 1991 FJ_1
Discovered 1991 March 18 by A. Sugie at Taga.

(6327) 1991 GP_1
Discovered 1991 April 9 by E. F. Helin at Palomar.

(6328) 1991 NL_1
Discovered 1991 July 12 by H. E. Holt at Palomar.

(6329) 1992 EU_1
Discovered 1992 March 12 by A. Sugie at Taga.

(6330) Koen
1992 FN. Discovered 1992 March 23 by K. Endate and K. Watanabe at Kitami.

Named in honor of Koen Yanagiya (1954-), a professional comic story teller, 'rakugoka' in Japanese. His style is different from that of other 'rakugokas', as he is very interested in astronomy, and his topics include astronomical events. He often appears at star parties to entertain the participants. (M 29146)

(6331) 1992 FZ_1
Discovered 1992 March 28 by S. Ueda and H. Kaneda at Kushiro.

(6332) Vorarlberg
1992 FP_3. Discovered 1992 March 30 by F. Börngen at Tautenburg.

Named for the westernmost Austrian province, until 1918 part of the state of Tirol {see planet (6439)}. Its capital is Bregenz (Brigantium), on Lake Constance. Vorarlberg is a mountainous region and an important center for cattle-breeding and dairy-farming. (M 25978)

(6333) Helenejacq
1992 LG. Discovered 1992 June 3 by G. J. Leonard at Palomar.

Named in honor of the discoverer's mother Helene Jacquelin, whose love and inspiration have enabled her son to discover new worlds both without and within. (M 25978)

(6334) Robleonard

1992 MM. Discovered 1992 June 27 by G. J. Leonard at Palomar.

Named in honor of the discoverer's father Robert David Leonard Sr., who has been a ceaseless source of support, inspiration and laughter. (M 25978)

(6335) 1992 NR

Discovered 1992 July 5 by E. F. Helin and J. Alu at Palomar.

(6336) 1992 UU

Discovered 1992 October 21 by S. Otomo at Kiyosato.

(6337) Shiota

1992 UC_4. Discovered 1992 October 26 by K. Endate and K. Watanabe at Kitami.

Named in honor of Kazuo Shiota (1949-), a Japanese amateur astronomer who developed image-processing technology for astronomical photographs. He is also interested in total solar eclipses and developed a special filter for use in observing these eclipses. (M 29146)

(6338) Isaosato

1992 UO_4. Discovered 1992 October 26 by K. Endate and K. Watanabe at Kitami.

Named in honor of Isao Sato (1963-), Japanese expert on occultations. He succeeded in making the first photographic observation in Japan of an occultation by a minor planet, that of γ Geminorum by (381) Myrrha on 1991 Jan. 13. He promotes observations of occultations by minor planets in Japan generally and made the first use of a color video to obtain flash spectra at the northern and southern limits of total solar eclipses. (M 27462; M 27477)

(6339) Giliberti

1993 SG. Discovered 1993 September 20 by V. S. Casulli at Colleverde di Guidonia.

Named in honor of Giuseppina Giliberti (1947-), wife of the discoverer. (M 25655)

(6340) Kathmandu

1993 TF_2. Discovered 1993 October 15 by K. Endate and K. Watanabe at Kitami.

Named for the capital city of Nepal, a favorite destination for trekkers and climbers in the Himalayas. (M 27462)

Name proposed by I. Sato.

(6341) 1993 UN_3

Discovered 1993 October 20 by S. Ueda and H. Kaneda at Kushiro.

(6342) 1993 VG

Discovered 1993 November 7 by S. Ueda and H. Kaneda at Kushiro.

(6343) 1993 VK

Discovered 1993 November 7 by S. Ueda and H. Kaneda at Kushiro.

(6344) 1993 VM

Discovered 1993 November 7 by S. Ueda and H. Kaneda at Kushiro.

(6345) Hideo

1994 AX_1. Discovered 1994 January 5 by K. Endate and K. Watanabe at Kitami.

Named in honor of Hideo Fukushima (1953-), a staff member in the public information office of the Japanese National Astronomical Observatory. Long active as an amateur astronomer, he has experimented extensively with CCD techniques and has published a book on the subject. (M 29146)

(6346) 1995 AY

Discovered 1995 January 6 by T. Kobayashi at Oizumi.

(6347) 1995 BM_4

Discovered 1995 January 28 by S. Ueda and H. Kaneda at Kushiro.

(6348) 1995 CH_1

Discovered 1995 February 3 by S. Ueda and H. Kaneda at Kushiro.

(6349) Acapulco

1995 CN_1. Discovered 1995 February 8 by M. Koishikawa at Sendai.

Named for the city in Mexico, Acapulco is the international sister city of Sendai {see planet (3133)}, Japan, affiliated since 1973. Acapulco is a famous resort and well known as one of the three most beautiful ports in the world. Hasekura Tsunenaga, the retainer of the famous Japanese feudal lord Date Masamune, made a stop there after he crossed the Pacific Ocean on his way to Rome. (M 33787)

(6350) Schlüter

3526 P-L. Discovered 1960 October 17 by C. J. van Houten and I. van Houten-Groeneveld at Palomar.

Named for Andreas Schlüter (c. 1660-1714), German sculptor and architect. Most of his work is found in Berlin, and his style was north-German baroque. His masterpieces include the sarcophagus of Queen Sophie Charlotte and King Friedrich I, as well as the equestrian statue of Friedrich Wilhelm, the Great Elector. He also designed the Berlin Castle, which was damaged during World War II and subsequently demolished. (M 26765)

(6351) Neumann

4277 T-1. Discovered 1971 March 26 by C. J. van Houten and I. van Houten-Groeneveld at Palomar.

Named for Johann Balthasar Neumann (1687-1753), one of the famous German baroque architects. He worked for the princes of Schönborn, who were the bishops of Würzburg, Bamberg {see planet (324)}, Mainz, Speyer and also of Trier. He developed the Würzburg Castle, in which he combined elements of Italian and French style with south-German late baroque. He also designed the Bruchsal {see planet (455)} Castle, destroyed during World War II but later reconstructed in its original glory. He also built churches in Vierzehnheiligen, Neresheim and Gössweinstein. (M 26765)

(6352) Schlaun

2400 T-3. Discovered 1977 October 16 by C. J. van Houten and I. van Houten-Groeneveld at Palomar.

Named for Johann Conrad Schlaun (1694-1773), baroque architect who worked mostly for the prince-bishop Klemens August of Köln. Schlaun travelled through Europe to see for himself the main buildings of his time. So he saw Neumann {see planet (6351)} in Würzburg. Schlaun constructed the Augustusburg Castle in Brühl and several other churches and castles in the style of Westphalian baroque. (M 26765)

(6353) Semper

3107 T-3. Discovered 1977 October 16 by C. J. van Houten and I. van Houten-Groeneveld at Palomar.

Named for the German architect Gottfried Semper (1803-1873). On the recommendation of Schinkel {see planet (5297)} he became a professor at Dresden Academy. He built the Dresden {see planet (3053)} museum and synagogue, as well as the most famous "Semper Operahouse", destroyed during World War II but rebuilt. Semper also worked in Zürich and Wien, where many buildings designed by him - including the Zürich Observatory - are still in use. (M 26766)

(6354) Vangelis

1934 GA. Discovered 1934 April 3 by E. Delporte at Uccle.

Named in honor of the Greek composer and multi-instrumentalist Vangelis Papathanassiou (1943-). Vangelis learned to play the piano at age four and later developed his unique musical style of multi-layered keyboard melodies, heavily influenced by both contemporary Western and Greek ethnic music. In addition to numerous studio albums and scores for theater and ballet, Vangelis has also composed numerous film soundtracks, including *1492-Conquest of Paradise* (1993), *The Bounty* (1984), *Bladerunner* (1983) and *Chariots of Fire* (1981), the last of which won him an Oscar for Best Original Score. (M 25978; M 34089)

Name proposed by G. V. Williams, who made the identification involving this object.

(6355) 1969 TX_5

Discovered 1969 October 15 by L. I. Chernykh at Nauchnyj.

(6356) Tairov

1976 QR. Discovered 1976 August 26 by N. S. Chernykh at Nauchnyj.

Named in honor of Vasilij Egorovich Tairov (1859-1938), Russian viticulturist and oenologist, author of a famous catalogue of wines (including counterfeits). He was founder and head of the first Russian research institute on viticulture, near Odessa (now the Ukrainian Tairov Research Institute of Viticulture). (M 34623)

(6357) Glushko

1976 SK_3. Discovered 1976 September 24 by N. S. Chernykh at Nauchnyj.

Named in memory of Valentin Petrovich Glushko (1908-1989), Soviet pioneer in liquid-fuel rocket-engine construction. The engines of many Soviet space rockets were designed under his leadership. (M 34623)

(6358) Chertok

1977 AL_1. Discovered 1977 January 13 by N. S. Chernykh at Nauchnyj.

Named in honor of Boris Evseevich Chertok (1912-), an authority on automation and control systems for space vehicles who has written several books on the history of Soviet rocketry and cosmonautics. (M 34623)

(6359) Dubinin

1977 AZ_1. Discovered 1977 January 13 by N. S. Chernykh at Nauchnyj.

Named in honor of Yurij Vladimirovich Dubinin (1930-), Russian diplomat, scientist, author and translator of several books from the French. (M 34342)

(6360) 1978 UA_7

Discovered 1978 October 27 by C. M. Olmstead at Palomar.

(6361) 1978 VL_{11}

Discovered 1978 November 7 by E. F. Helin and S. J. Bus at Palomar.

(6362) 1979 KO

Discovered 1979 May 19 by R. M. West at La Silla.

(6363) Doggett

1981 CB_1. Discovered 1981 February 6 by E. Bowell at Anderson Mesa.

Named in memory of LeRoy E. Doggett (1941-1996), a staff astronomer at the U.S. Naval Observatory known for his application of Chebyshev polynomials to celestial mechanics and positional astronomy. An expert in calendars, astronomical phenomena, the history of astronomy, and planetary theory, he was chief of the U.S. Nautical Almanac Office, associate editor of *Archaeoastronomy*, organizer of the nationwide "Moon Watch" program to determine earliest visibility of the lunar crescent, and secretary-treasurer of the Historical Astronomy Division of the American Astronomical Society. (M 27330)

Name proposed by the discoverer upon a suggestion by A. K. B. Monet. Citation by G. Kaplan, endorsed by D. DeVorkin and B. G. Marsden.

Obituaries published in Bull. Am. Astron. Soc., Vol. 28, No. 4, p. 1450-1451 (1996); Phys. Today, Vol. 49, No. 9, p. 110-111 (1996).

(6364) 1981 ET

Discovered 1981 March 2 by H. Debehogne and G. DeSanctis at La Silla.

(6365) 1981 ES_{29}

Discovered 1981 March 1 by S. J. Bus at Siding Spring.

(6366) 1981 UM_{22}

Discovered 1981 October 24 by S. J. Bus at Palomar.

(6367) 1982 FY_2

Discovered 1982 March 18 by H. Debehogne at La Silla.

(6368) 1983 RM_3

Discovered 1983 September 1 by H. Debehogne at La Silla.

(6369) 1983 UC

Discovered 1983 October 16 by Z. Vávrová at Kleť.

(6370) 1984 EY

Discovered 1984 March 9 by B. A. Skiff at Anderson Mesa.

(6371) Heinlein

1985 GS. Discovered 1985 April 15 by E. Bowell at Anderson Mesa.

Named in honor of Dieter Heinlein (1956-). A resident of Augsburg, Germany, Heinlein in 1989 took over direction of the western part of the European All Sky Meteor Camera Network, in collaboration with the Czech network. Since then, he has increased the network's scope to include observing stations all over Germany, Austria and the Benelux countries. Heinlein is an enthusiastic meteorite and tektite collector, and he and his wife Gabriele are friends of the discoverer. (M 27462; M 32127)

Name suggested by H. Fechtig, citation provided by K. Hopf.

(6372) Walker

1985 JW_1. Discovered 1985 May 13 by C. S. Shoemaker and E. M. Shoemaker at Palomar.

Named in honor of Robert M. Walker (1929-), a leading figure in the study of radiation damage in solids and in the investigation of interstellar grains recovered from meteorites. His work on fission tracks with others opened a new approach to the dating of rocks on the earth and the moon and led to new information on the energy spectrum, composition and flux of solar and galactic cosmic rays. He founded whole new methods of scientific study and created a premier center for research in these new fields at Washington University, where he has provided inspiring leadership to a host of students and post-doctoral fellows. (M 34623)

(6373) Stern

1986 EZ. Discovered 1986 March 5 by E. Bowell at Anderson Mesa.

Named in honor of S. Alan Stern (1957-) of the Boulder, Colorado, office of

Southwest Research Institute. Stern's research has focused on both observational and theoretical studies of the satellites of the outer planets, Pluto, comets, the Oort Cloud and the Kuiper Belt {see, respectively, planets (1691) and (1776)}. He is also active in instrument development, with a strong concentration in ultraviolet and imaging technologies. He has participated in ten planetary sounding-rocket missions, two Space Shuttle mid-deck experiments and a Shuttl -deployable satellite. He was chair of NASA's Outer Planets Science Working Group during 1991-1994. (M 27462)
Citation provided by H. Levison and J. Parker.

(6374) 1986 PY_4

Discovered 1986 August 8 by L. I. Chernykh at Nauchnyj.

(6375) Fredharris

1986 TB_5. Discovered 1986 October 1 by the CERGA at Caussols.

Named in honor of Frederick Harlan Harris, CCD guru. Harris worked at Caltech for the Palomar Observatory for many years before moving to the U.S. Naval Observatory, where he is involved in the Sloan Digital Sky Survey. Harris is a good friend of Caussols astronomer Alain Maury {see planet (3780)}, and his patient help has been essential in the development of the CCD cameras for the Schmidt telescope with which this minor planet was discovered. (M 25979)
Citation endorsed by M. Carr.

(6376) Schamp

1987 KD_1. Discovered 1987 May 29 by C. S. Shoemaker and E. M. Shoemaker at Palomar.

Named in honor of Americans Larry and Becky Schamp of Alice Springs. Larry, an electrical engineer by training, is currently the manager of a TRW high-tech aerospace group located in this remote outpost of Northern Territory. Becky is renowned for her outback hospitality. This couple took in members of the Shoemaker family on their arrival from the United States after the car collision in which Eugene {see planet (2074)} died, and then they hosted Carolyn {see planet (4446)} when she left the hospital. Larry's 6 feet, 8 inches of gentleness and know-how and Becky's warmth, understanding and enthusiasm for life helped to sustain this family in a time of great need. (M 31610)

(6377) Cagney

1987 ML_1. Discovered 1987 June 25 by A. Mrkos at Kleť.

Named in memory of the movie actor James Cagney (1899-1986). Although he played tough-guy roles in numerous gangster movies, Cagney was also a talented singer and dancer, and in 1942 he won an Academy Award for his portrayal of showman George M. Cohan in *Yankee Doodle Dandy.* (M 30098)

Name suggested by G. V. Williams, who made the identifications for this minor planet.

(6378) 1987 SE_{13}

Discovered 1987 September 27 by H. Debehogne at La Silla.

(6379) 1987 VA_1

Discovered 1987 November 15 by A. Mrkos at Kleť.

(6380) 1988 CG

Discovered 1988 February 10 by M. Arai and H. Mori at Yorii.

(6381) 1988 DO_1

Discovered 1988 February 21 by T. Fujii and K. Watanabe at Kitami.

(6382) 1988 EL

Discovered 1988 March 14 by J. Alu at Palomar.

(6383) Tokushima

1988 XU_1. Discovered 1988 December 12 by M. Iwamoto and T. Furuta at Tokushima.

Named for the first discoverer's observatory and the prefecture in which he lives. (M 33385)

(6384) Kervin

1989 AM. Discovered 1989 January 3 by E. F. Helin at Palomar.

Named in honor of Paul W. Kervin, chief scientist for the Phillips Laboratory's Air Force Maui Optical Station (AMOS). He has played a leading role in establishing and conducting the JPL/AMOS program, which is primarily focused on obtaining crucial follow-up observations of near-earth and other interesting minor planets. Before coming to AMOS, he was the chief scientist for the Relay Mirror Experiment, which received the SPIE Technology Achievement Award in 1991. (M 26931)

(6385) 1989 EC_2

Discovered 1989 March 5 by A. Mrkos at Kleť.

(6386) 1989 NK_1

Discovered 1989 July 10 by H. E. Holt at Palomar.

(6387) 1989 WC

Discovered 1989 November 19 by S. Ueda and H. Kaneda at Kushiro.

(6388) 1989 WL_1

Discovered 1989 November 25 by S. Ueda and H. Kaneda at Kushiro.

(6389) Ogawa

1990 BX. Discovered 1990 January 21 by K. Endate and K. Watanabe at Kitami.

Named in honor of Shigeo Ogawa (1932-), president of Seibundo Shinkosha Company, which publishes the monthly astronomical magazine *Tenmon Guide*, and many astronomical books. (M 29146)

(6390) Hirabayashi

1990 BG_1. Discovered 1990 January 26 by K. Endate and K. Watanabe at Kitami.

Named in honor of Shigeto Hirabayashi (1953-), amateur astronomer known for his expertise in astronomical photography. His method for increasing the sensitivity of photographic films has been used by many amateurs. (M 29147)

(6391) Africano

1990 BN_2. Discovered 1990 January 21 by E. F. Helin at Palomar.

Named in honor of John L. Africano, astronomer with the Air Force Maui Optical Station (AMOS), who has played a central role in the JPL/AMOS program on minor planets. Previously he was involved with numerous photometric projects, ranging from analyses of cool stars to the timing of occultations, as a staff observer at Kitt Peak National Observatory, and before that at the Cloudcroft Observatory. (M 26931)

(6392) 1990 HR

Discovered 1990 April 29 by Y. Mizuno and T. Furuta at Kani.

(6393) 1990 HM_1

Discovered 1990 April 29 by H. Shiozawa and M. Kizawa at Fujieda.

(6394) 1990 QM_2

Discovered 1990 August 22 by H. E. Holt at Palomar.

(6395) Hilliard

1990 UE_1. Discovered 1990 October 21 by Y. Kushida and O. Muramatsu at Yatsugatake.

Named in honor of Elizabeth (1904-) and Leslie Hilliard (1905-), who acquired the house in Bath, England, from the garden of which Herschel {see planet (2000)} discovered Uranus. The Hilliards restored the historic building and gave it to the William Herschel Society. For many years they worked tirelessly for the society following the opening of the Herschel Museum in the house on 1981 Mar. 13, the 200th anniversary of the discovery of the planet. The eighteenth-century house was thus saved from the threat of "development" and continues to please visitors from all over the world. The Hilliards also purchased the Lansdown Tower, which had been built in Bath in 1827 by William Beckford, a famous writer and art connoisseur. (M 25655)

Named by the discoverers following a suggestion by S. Kimura.

(6396) Schleswig

1991 AO_3. Discovered 1991 January 15 by F. Börngen at Tautenburg.

Schleswig, formerly an independent dukedom, is the most northern territory of Germany, situated between the North Sea and the Baltic Sea. Since 1386 it has been united with Holstein {see planet (6402)}, a duchy that borders it to the south. Prominent towns are Flensburg, the "Gate to the North", and Schleswig, site of Gottorp castle and Petri cathedral. The poet Theodor Storm was born in North Frisia, a part of Schleswig. (M 25655)

(6397) 1991 BJ

Discovered 1991 January 17 by T. Hioki and S. Hayakawa at Okutama.

(6398) Timhunter

1991 CD_1. Discovered 1991 February 10 by C. S. Shoemaker and E. M. Shoemaker and D. H. Levy at Palomar.

Named in honor of Tim Hunter, president of the International Dark Sky Association. A radiologist by profession, Hunter has followed the increasing effects of light pollution over the earth's night sky. By founding with David Crawford an association that is devoted to raising public awareness about the need for safe and sensible lighting, Hunter has performed an invaluable service to astronomy. (M 27330)

(6399) Harada

1991 GA. Discovered 1991 April 3 by T. Seki at Geisei.

Named in honor of Shoji Harada (1953-), amateur photographer and student of the orbits of comets and minor planets. (M 30798)

(6400) 1991 GQ_1

Discovered 1991 April 10 by E. F. Helin at Palomar.

(6401) Roentgen

1991 GB_2. Discovered 1991 April 15 by C. S. Shoemaker and E. M. Shoemaker and D. H. Levy at Palomar.

Named to commemorate the centennial of the discovery of x-rays by the German physicist Wilhelm Conrad Roentgen {1845-1923} in November 1895. His first x-ray of a human being was of his wife's hand. Roentgen's research revolutionized medicine by providing a diagnostic tool still unrivaled in accuracy and simplicity. X-rays quickly became an indispensable part of patient care. Early in 1896 x-ray

therapy came into being and has remained a standard treatment for cancer. This research has spread into many fields, especially astronomy, where x-rays provide a unique window on the universe. (M 27330)
Citation prepared by D. Patton, T. Hunter and D. Levy.

Roentgen was the first physicist who received the Nobel Prize in physics in 1901.

(6402) Holstein

1991 GQ_{10}. Discovered 1991 April 9 by F. Börngen at Tautenburg.

Holstein, the southern part of the German state Schleswig-Holstein {see also planet (6396)}, developed from a county to a dukedom. Well known towns are Kiel, the capital, and Lübeck, one of the leading and most properous Hansa cities, with its old city hall, Marien church, Holsten gate and ancient salt lofts. In the church of Mölln is the sepulchre of the famous buffoon and jeerer, Till Eulenspiegel {see also planet (1547)}. Holstein is the birthplace of the composer Carl Maria von Weber {see planet (4152)}, the dramatist Friedrich Hebbel, the writers Heinrich and Thomas Mann and the physicist Max Planck {see planet (1069)}. (M 25655)

(6403) Steverin

1991 NU. Discovered 1991 July 8 by E. F. Helin at Palomar.

Named for Steven Newburn and Erin Fischer on the occasion of their marriage. Steven is the son of Ray Newburn {see planet (2955)}, science coordinator for near-earth objects at the Jet Propulsion Laboratory and former co-leader of the International Halley Watch. Both young people greatly enjoyed their college astronomy courses, although neither plans a career in science. (M 27330)

(6404) Vanavara

1991 PS_6. Discovered 1991 August 6 by E. W. Elst at La Silla.

Named for the town near the Podkammenaya (stony) Tunguska river in Siberia {see planet (5471)}. On 1908 June 30 a large bolide, probably an asteroid fragment, exploded in the atmosphere, almost 70 km north of Vanavara. Almost 2000 square kilometers of the taiga forest were suddenly flattened. Tungus reindeer herders, living in the forests, were thrown to the ground, and many of them lost their herds and tepees. Several houses in Vanavara were damaged. On the occasion of the 37th expedition to Tunguska, the discoverer was very warmly received by the people of Vanavara. (M 25979)

(6405) 1992 HJ

Discovered 1992 April 30 by Y. Kushida and O. Muramatsu at Yatsugatake.

(6406) 1992 MJ

Discovered 1992 June 28 by H. E. Holt at Palomar.

(6407) 1992 PF_2

Discovered 1992 August 2 by H. E. Holt at Palomar.

(6408) Saijo

1992 UT_5. Discovered 1992 October 28 by K. Endate and K. Watanabe at Kitami.

Named in honor of Yoshihiro Saijo (1959-), amateur astronomer and editor of a monthly magazine, known for his work in astrophotography. (M 29147)

(6409) 1992 VC

Discovered 1992 November 2 by N. Kawasato at Uenohara.

(6410) 1992 WO_4

Discovered 1992 November 29 by S. Otomo at Kiyosato.

(6411) Tamaga

1993 TA. Discovered 1993 October 8 by R. H. McNaught at Siding Spring.

Named for *The Astronomer*, a British magazine in which active amateur astronomers publish and discuss their observations. Founded in 1964 by James Muirden as *The Casual Astronomer*, the magazine has developed over the years to include rapid discovery announcements and many other initiatives by the current editor, Guy Hurst {see planet (3697)}. The search programs for novae and supernovae and the observational programs for cataclysmic variables initiated through the magazine have been most successful. The subscribers include many of the most active and talented amateurs in the U.K. and overseas. The discoverer's own interest in observational astronomy was spurred by *TA*. (M 29671)

(6412) Kaifu

1993 TL_2. Discovered 1993 October 15 by K. Endate and K. Watanabe at Kitami.

Named in honor of Norio Kaifu (1943-), recently appointed the first director of the 8.2-m Subaru Telescope in Hawaii, following his direction of the construction of that telescope during the past six years. He also played an important role in the construction of the 45-m millimeter-wave radio telescope at the Nobeyama Radio Observatory, in particular, by his development of the acousto-optical radio spectrometer, a powerful instrument for very high resolution able to identify many interstellar molecular lines. Kaifu served as chairman of the Radio Astronomy Division of the National Astronomical Observatory at Mitaka during 1988-1990 and as associate director during 1992-1996. (M 30098)

(6413) Iye

1993 TJ_3. Discovered 1993 October 15 by K. Endate and K. Watanabe at Kitami.

Named in honor of Masanori Iye (1949-), a professor at the National Astronomical Observatory who works mainly on structure and evolution of galaxies and executive secretary of the Japanese National Committee of Astronomy since 1994. As project scientist for Subaru, he led the engineering core team to design and manufacture the 8.2-m telescope, and he also led teams to develop a CCD camera, active and adaptive optics, as well as a faint-object spectrograph. (M 30098)

(6414) 1993 UX

Discovered 1993 October 24 by T. Kobayashi at Oizumi.

(6415) 1993 VR_3

Discovered 1993 November 11 by S. Ueda and H. Kaneda at Kushiro.

(6416) Nyukasayama

1993 VY_3. Discovered 1993 November 14 by M. Hirasawa and S. Suzuki at Nyukasa.

Named for a mountain in central Japan, famous for many beautiful flowers and a wonderful view. Mt. Nyukasa station, where this minor planet was discovered, is located near the top of this mountain. (M 32094)

(6417) Liberati

1993 XA. Discovered 1993 December 4 by A. Vagnozzi at Stroncone.

Named in memory of Libero Liberati (1926-1962), motorcyclist who won the Absolute Italian Championship in 1955 and 1956 riding a four-cylinder Gilera. In 1957 he won the Absolute World Championship and was given the Athlete of the Year Award by the Italian Sport Press. (M 28621)

(6418) 1993 XJ

Discovered 1993 December 8 by T. Kobayashi at Oizumi.

(6419) Susono

1993 XX. Discovered 1993 December 7 by M. Akiyama and T. Furuta at Mishima.
Named by the first discoverer for the town in which he lives, some 100 km west of Tokyo. The name of this town means "foot" in English. Mt. Fuji {see planet (1584)} can be seen 25 km to the northwest and Mt. Hakone {see planet (1098)} 10 km to the east of the town. The Kise river, which reflects the beauty of the changing seasons, runs through the center. (M 27330)

(6420) 1993 XG_1

Discovered 1993 December 14 by T. Kobayashi at Oizumi.

(6421) 1993 XS_1

Discovered 1993 December 6 by S. Ueda and H. Kaneda at Kushiro.

(6422) 1994 CD_1

Discovered 1994 February 7 by T. Kobayashi at Oizumi.

(6423) 1994 CP_2

Discovered 1994 February 13 by T. Kobayashi at Oizumi.

(6424) Ando

1994 EN_3. Discovered 1994 March 14 by K. Endate and K. Watanabe at Kitami.
Named in honor of Hiroyasu Ando (1946-), chairman of the Optical and Infrared Astronomy Division of the National Astronomical Observatory since 1994 and an authority on stellar oscillations and high-resolution spectroscopy. His theoretical work with Yoji Osaki on five-minute solar oscillations played an important role in the development of the exciting field of helioseismology. A core member of the Subaru project from its beginning, Ando now leads the group constructing the high-dispersion spectrograph. (M 30099)

(6425) 1994 WZ_3

Discovered 1994 November 28 by S. Ueda and H. Kaneda at Kushiro.

(6426) Vanýsek

1995 ED. Discovered 1995 March 2 by M. Tichý at Kleť.
Named in honor of Vladimír Vanýsek (1926-1997), Czech astronomer, professor emeritus of astrophysics at Charles University in Prague. His works deal with physics and chemistry of comets, interstellar matter, and also with cosmology. From 1972 to 1986 he headed the Department of Astronomy and Astrophysics of Charles University, after which he moved to Bamberg for a three-year visiting professorship at the University of Erlangen-Nürnberg. He was president of IAU Commission 15 during 1970-1973. During the latest passage of Halley's Comet he was a member of the International Halley Watch Steering Group and a discipline specialist for the Photometry and Polarimetry Network. (M 25979)
Name proposed by the discoverer, endorsed by J. Tichá and Z. Moravec.
Obituaries published in Earth, Moon, Planets, Vol. 76, No. 1-2, p. 1-3 (1997/98); Mitt. Astron. Ges., Nr. 81, p. 9-12 (1998).

(6427) 1995 FY

Discovered 1995 March 28 by S. Ueda and H. Kaneda at Kushiro.

(6428) Barlach

3513 P-L. Discovered 1960 October 17 by C. J. van Houten and I. van Houten-Groeneveld at Palomar.
Named for Ernst Barlach (1870-1938), German sculptor, graphic artist and poet. In his work he expressed human sorrow - but also the element of hope. Sometimes he could even express humor and gladness. His sculptures have mostly the form

of a pyramid. During the Nazi era his work was regarded as 'entartete Kunst' and removed from museums and other public places. (M 26766)

(6429) Brancusi

4050 T-1. Discovered 1971 March 26 by C. J. van Houten and I. van Houten-Groeneveld at Palomar.

Named for the Roumanian sculptor Constantin Brancusi (1876-1957). He initially worked in the style of Rodin {see planet (6258)} but later became increasingly abstract. Much of his work is found in Paris, where he lived after 1904. (M 26766)

(6430) 1964 UP

Discovered 1964 October 30 at the Purple Mountain Observatory at Nanking.

(6431) 1967 UT

Discovered 1967 October 30 by L. Kohoutek at Bergedorf.

(6432) Temirkanov

1975 TR_2. Discovered 1975 October 3 by L. I. Chernykh at Nauchnyj.

Named in honor of Yurij Khatuevich Temirkanov (1938-), an outstanding conductor. From 1976 to 1988 he was the conductor of the Kirov theater {see planet (4447)} in Leningrad, where a number of splendid operas and ballets were produced under his leadership. Since 1988 Temirkanov has been the artistic director and orchestra conductor of the St. Petersburg Philharmonic Society. He is also a conductor of the Royal Philharmonic Orchestra and a main invited conductor for the Dresden Philharmonic. (M 32345; M 32377)

Name suggested by the Union of Concert Workers of Russia and supported by the discoverer.

(6433) Enya

1978 WC. Discovered 1978 November 18 by A. Mrkos at Kleť.

Named in honor of the Irish singer and songwriter Eithne NíBhraonáin (1961-), professional name Enya). (M 30099)

Name suggested by G. V. Williams, who made the identifications involving this object.

(6434) Jewitt

1981 OH. Discovered 1981 July 26 by E. Bowell at Anderson Mesa.

Named in honor of David Jewitt (1958-) of the Institute for Astronomy, University of Hawaii. The consummate astronomer, Jewitt has been devoted to astronomy from a very early age. He has made several important contributions to planetary astronomy, starting with his discovery of the Jovian satellite Adrastea from Voyager data in 1979. He was co-recoverer of comet 1P/Halley in 1982. Jewitt is perhaps best known for co-discovering the first body in the Kuiper {see planet (1776)} Belt in 1992, thus providing that accretion occurs beyond the planetary region. Jewitt's main area of research is comets, but his wide-ranging interests have also produced work on planetary rings, minor planets, Pluto and circumstellar disks. (M 27462)
Citation provided by J. Luu.

(6435) 1984 DA

Discovered 1984 February 24 by E. F. Helin and R. S. Dunbar at Palomar.

(6436) Coco

1985 JX_1. Discovered 1985 May 13 by C. S. Shoemaker and E. M. Shoemaker at Palomar.

Named in honor of Mark and Colleen Coco and their children Kymberly, Jen-

nifer, Lisa and Travis. An author of astronomical articles and an active observer, Mark has written about his many interests, including astrophotography, the Green Flash, and the relation between astronomy and religion. During his five years as manager of customer relations for Celestron International that began in 1988, Mark helped thousands of new telescope users find their way to enjoy the sky. The discoverers wish to honor the Coco family for their courage and love in dealing with Mark's long illness. (M 25979)
Name proposed and citation prepared by D. H. Levy.

(6437) 1987 QS_7

Discovered 1987 August 28 by E. W. Elst at La Silla.

(6438) 1988 BS_3

Discovered 1988 January 18 by H. Debehogne at La Silla.

(6439) Tirol

1988 CV. Discovered 1988 February 13 by F. Börngen at Tautenburg.

Named for the Austrian province, the capital of which is Innsbruck. The North Tirolian Limestone Alps, Ötztaler, Stubaier and Zillertaler Alps are popular tourist attractions. The beauty of the region is immortalized in the song "The dear God must be a Tiroler!" (M 25979)

(6440) Ransome

1988 RA_2. Discovered 1988 September 8 by A. Mrkos at Kleť.

Named in memory of Arthur Ransome (1884-1967), one of the best loved of all children's writers. His novels depict the adventures of boys and girls, *Swallows and Amazons*, who explore the lakes and mountains of England. (M 34342)
Name suggested by J. Tichá.

(6441) Milenajesenská

1988 RR_2. Discovered 1988 September 9 by A. Mrkos at Kleť.

Named in memory of Milena Jesenská (1896-1944), outstanding Czech journalist and mediator between the Czech and German cultures in Bohemia. She was detained for her political involvement and resistance in the concentration camp at Ravensbrück, where she died. She is widely known for her correspondence with Prague writer Franz Kafka {see planet (3412)} (*Briefe an Milena*). (M 34342)
Name suggested by J. Tichá.

(6442) Salzburg

1988 RU_3. Discovered 1988 September 8 by F. Börngen at Tautenburg.

Named for the Austrian province and its capital on the river Salzach. Visitors flock to the province's high mountains and the celebrated festivals in the capital, well-known as the birthplace of Mozart and Doppler {see planets (1034) and (3905)}. On 1818 December 24, the curate Joseph Mohr and his organist Franz Gruber created the well-known Christmas carol *Stille Nacht, Heilige Nacht* near the town of Hallein. (M 25979)

(6443) 1988 RH_{12}

Discovered 1988 September 14 by S. J. Bus at Cerro Tololo.

(6444) 1989 WW

Discovered 1989 November 20 by K. Suzuki and T. Urata at Toyota.

(6445) Bellmore

1990 FS_1. Discovered 1990 March 23 by E. F. Helin at Palomar.

Named in honor of Tamara Bell and Michael More on the occasion of their wedding. They are recent graduates of the University of Arizona with degrees in political science and geology. The discoverer and her husband wish the newlyweds a harmonious marriage and rewarding careers. (M 28090)

(6446) Lomberg

1990 QL. Discovered 1990 August 18 by E. F. Helin at Palomar.

Named in honor of Jon Lomberg (1948-), graphic arts virtuoso, who has excelled in his renditions of space and astronomical subjects. One of the world's leading astronomical artists, he worked with Carl Sagan {see planet (2709)} and on many television documentaries about space science, including the *Cosmos* series. He was also the design director for the Voyager Interstellar Record. His works can be found in museums and science centers all over the world. His technical research, artistic interpretation and precision in the planning of each painting is characteristic of this dedicated, accomplished artist in portraying space as it is, to the extent we know it. (M 33385)

(6447) Terrycole

1990 TO_1. Discovered 1990 October 14 by E. F. Helin at Palomar.

Named in honor of Terry Cole, chief technologist at the Jet Propulsion Laboratory and senior faculty associate in the Caltech Division of Chemistry and Chemical Engineering. In addition to his numerous technical contributions to JPL and Caltech, Cole played a central role in the development of the Caltech/JPL Summer Undergraduate Research Fellowship (SURF) {see planet (5181)} program through his vision while chairing the SURF Fellowship Executive Committee. (M 32788)

(6448) 1991 CW

Discovered 1991 February 8 by K. Suzuki and T. Urata at Toyota.

(6449) Kudara

1991 CL_1. Discovered 1991 February 7 by T. Seki at Geisei.

Named in memory of Kyoyu Kudara (1894-1964), Japanese astronomer and Buddhist priest. A staff member of the Tokyo Astronomical Observatory from 1917 to 1925, he recovered comet 10P/Tempel 2 in 1920 according to his own prediction, and he observed this comet again in 1925. Later he taught positional astronomy and the science of the calendar at Kyoto University. One of the cofounders of the Oriental Astronomical Association, he served as its president from 1959 to 1964. (M 32788)

Name proposed by the discoverer following a suggestion by I. Hasegawa and T. Sato. Citation by I. Hasegawa.

(6450) 1991 GV_1

Discovered 1991 April 9 by E. F. Helin at Palomar.

(6451) Kärnten

1991 GP_{10}. Discovered 1991 April 9 by F. Börngen at Tautenburg.

Named for the southernmost Austrian province. The capital Klagenfurt is situated on a plain beside the Wörther Lake. The name is derived from the Celtish people of the Karner. The region was part of the Roman province of Noricum. (M 25979)

(6452) Johneuller

1991 HA. Discovered 1991 April 17 by T. J. Balonek at Hamilton.

Named in honor of John E. Euller, a dedicated and highly respected teacher of high-school physics. Educated at Pennsylvania State University and the University of Buffalo, he taught physics and chemistry over a 33-year career. From 1959 to 1984 he taught physics at Eastridge High School in Irondequoit, NY, where his animated demonstrations inspired generations of students. He has had life-long interests in science and travel. (M 27129; M 27147)

(6453) 1991 NY

Discovered 1991 July 13 by H. E. Holt at Palomar.

(6454) 1991 UG$_1$

Discovered 1991 October 29 by R. H. McNaught at Siding Spring.

(6455) 1992 HE

Discovered 1992 April 25 by R. H. McNaught at Siding Spring.

(6456) Golombek

1992 OM. Discovered 1992 July 27 by E. F. Helin and K. J. Lawrence at Palomar.

Named in honor of Matthew P. Golombek, research scientist in geology and planetary geology at the Jet Propulsion Laboratory. He has devoted the past five of his 14 years there to the Pathfinder project, distinguishing himself by his highly effective performance as the Mars Pathfinder project scientist, responsible for all aspects of the mission's science. The discoverers wish to salute this popular spokesman for his understandable science and Pathfinder mission status reports. His descriptions and explanations captured the attention of a public excited by the Mars landing and the rocky excursions of the small rover, Sojourner. (M 31024)

(6457) Kremsmünster

1992 RT. Discovered 1992 September 2 by L. D. Schmadel and F. Börngen at Tautenburg.

Named as a celestial tribute to the famous Benedictine monastery, educational center and observatory in Austria. A place of scientific effort in many disciplines, the Kremsmünster Observatory remains one of the most outstanding cultural institutions in Austria – even a quarter of a millenium after its founding. This planet is also named in memory of Abbot Augustin Reslhuber (1808-1875), an ardent observer to whom it was not granted to discover a minor planet himself. (M 31610)

Named by the first discoverer and citation prepared with the support of Father Amand Kraml.

(6458) Nouda

1992 TD$_1$. Discovered 1992 October 2 by T. Seki at Geisei.

Named in memory of Tadasuke Nouda (1901-1989), Japanese astronomer. A pioneer in the field of ancient Chinese astronomy, he made important contributions to the study of early luni-solar calendars. He was a professor at the Osaka University of Education and Kyoto Sangyo University. He served as president of the Japan Calendar Association (1959-1989) and of the Oriental Astronomical Association (1965-1989). (M 32788)

Name proposed by the discoverer following a suggestion by I. Hasegawa and T. Sato. Citation by I. Hasegawa.

(6459) Hidesan

1992 UY$_5$. Discovered 1992 October 28 by K. Endate and K. Watanabe at Kitami.

Named in honor of Hideo Sato (1940-), staff member of the National Astronomical Observatory (formerly Tokyo Astronomical Observatory) who first worked on the solar corona, later moving to the Sky Patrol Section as a night observer. His life's work is the photometry of close binaries. He is also one of the leading members of the observatory's baseball team. (M 30099)

Name proposed by the second discoverer following a suggestion by K. Tomita.

(6460) Bassano

1992 UK_6. Discovered 1992 October 26 by U. Quadri and L. Strabla at Bassano Bresciano.

Named for Bassano Bresciano, an ancient village in the Italian province of Brescia. Situated on the Padana plain, in the early Middle Ages it was under Longobard and Frank dominion. Later it was ruled by Sforza and the Venice republic. In the sixteenth century the Brescian agronomists Camillo Tarello and Agostino Gallo reclaimed this marshland. The most important monuments are Luzzago's sixteenth-century palace and Brunelli's seventeenth-century villa. In the church there is a Via Crucis attributed to Tiepolo's school. (M 25655)

(6461) 1993 VB_5

Discovered 1993 November 4 by R. H. McNaught at Siding Spring.

(6462) 1994 AF_2

Discovered 1994 January 9 by T. Kobayashi at Oizumi.

(6463) Isoda

1994 AG_3. Discovered 1994 January 13 by K. Endate and K. Watanabe at Kitami.

Named in memory of Sachiko Isoda (1912-1993), who joined the Tokyo Astronomical Observatory in 1943 as a computing assistant. After the war she moved to the Department of Astronomy at University of Tokyo and was for many years secretary to Y. Hagihara {see planet (1971)}. After her formal retirement she stayed on as a part-time employee, and her fine work impressed many astronomers who graduated from the Univerity of Tokyo. (M 30099)
Name proposed by the second discoverer following a suggestion by K. Tomita.

(6464) Kaburaki

1994 CK. Discovered 1994 February 1 by Y. Kushida and O. Muramatsu at Yatsugatake.

Named in memory of Masaki Kaburaki (1902-1987), a professor of astronomy at the University of Tokyo before he retired in 1963. He taught spherical and practical astronomy there, and his main interest was in kinematics of our Galaxy. He served as the president of the Japanese Astronomical Society around 1955 and was the director of Gotoh Astronomical Museum during 1974-1987. (M 29147)

(6465) Zvezdotchet

1995 EP. Discovered 1995 March 3 by T. V. Kryachko at Zelenchukskaya.

Named for the Russian amateur astronomers' magazine *Zvezdotchet* (Stargazer) and its readers. (M 29147)

(6466) 1979 MU_8

Discovered 1979 June 25 by E. F. Helin and S. J. Bus at Siding Spring.

(6467) Prilepina

1979 TS_2. Discovered 1979 October 14 by N. S. Chernykh at Nauchnyj.

Named in honor of Svetlana Semenovna Prilepina, a graduate of the Astronomical Department at the Ural State University in Ekaterinburg and one of the organizers of the annual Winter Astronomical School at the Kourovka {see planet (4964)} Observatory. She is also a gifted amateur singer. (M 30099)

(6468) 1981 ED_{19}

Discovered 1981 March 2 by S. J. Bus at Siding Spring.

(6469) Armstrong

1982 PC. Discovered 1982 August 14 by A. Mrkos at Kleť.

Named in honor of Neil Alden Armstrong (1930-), U.S. astronaut, the first man to set foot on the moon. In 1966 he and David Scott aboard Gemini 8 conducted the first docking in space. In 1969, along with Aldrin and Collins {see planets (6470) and (6471)}, he blasted off in Apollo 11 toward the moon, and on July 20 the Eagle module landed on a plain near the southwestern edge of Mare Tranquillitatis. This minor planet commemorates the 30th anniversary of the first manned lunar-landing mission. (M 34623)

Name suggested by J. Tichá, M. Tichý and Z. Moravec, who observed this minor planet at its 1995 opposition, just prior to numbering.

(6470) Aldrin

1982 RO_1. Discovered 1982 September 14 by A. Mrkos at Kleť.

Named in honor of Edwin Eugene (Buzz) Aldrin Jr. (1930-), U.S. astronaut, the second man to set foot on the moon. In 1966 he made a record 5.5-hour spacewalk in the course of the Gemini 12 mission. In 1969 he joined Armstrong {see planet (6469)} on the surface of the moon. This minor planet commemorates the thirtieth anniversary of the first manned lunar-landing mission. (M 34623)

Name suggested by J. Tichá, M. Tichý and Z. Moravec, who observed this minor planet at its 1995 opposition, just prior to numbering.

(6471) Collins

1983 EB_1. Discovered 1983 March 4 by A. Mrkos at Kleť.

Named in honor of Michael Collins (1930-), U.S. astronaut. In 1966 he served as pilot of the Gemini 10 mission. In 1969 he was assigned to Apollo 11 as the command-module pilot. While Armstrong and Aldrin {see planets (6469) and (6470)} took their first walk on the lunar surface, Collins flew a lonely lifeguard assignment for more than 24 hours, waiting for them to launch their lunar craft and rejoin him in lunar orbit. This minor planet commemorates the thirtieth anniversary of the first manned lunar-landing mission. (M 34623)

Name suggested by J. Tichá, M. Tochý and Z. Moravec, who observed this minor planet at its 1995 opposition, just prior to its numbering.

(6472) Rosema

1985 TL. Discovered 1985 October 15 by E. Bowell at Anderson Mesa.

Named in honor of Keith D. Rosema (1967-), computer scientist at the Jet Propulsion Laboratory. In 1989, Rosema completed the first California Institute of Technology senior thesis in asteroid radar astronomy. Since then he has used his formidable computer-programming expertise and analytic talents to make crucial contributions to observations of near-earth and main-belt asteroids. His library of software has proved to be an essential tool for reduction and analysis of echoes from most of the radar-detected asteroids, including (4769) Castalia, (4179) Toutatis, (1620) Geographos and (6489) Golevka. Rosema is also one of a handful of people who have devoted extensive time to asteroid radar astronomy at both Arecibo and Goldstone {see, respectively, planets (4337) and (4433)}. (M 25655)
Name suggested and citation prepared by S. J. Ostro.

(6473) Winkler

1986 GM. Discovered 1986 April 9 by E. Bowell at Anderson Mesa.

Named in honor of Ron Winkler (1954-), digital engineer in the radio astronomy and radar group at NASA's Goldstone {see planet (4433)} deep space communications complex. Since the mid-1980s, Winkler has been singularly important in Goldstone radar astronomy and especially in observations of near-earth asteroids. His understanding of the interface between data-acquisition, antenna-pointing and transmitting systems is unique. His energy, endurance and mastery of procedures for observing close asteroids, demanding frequent switching between

transmit and receive configurations, was critical to the success of the radar imaging of (4179) Toutatis, (1620) Geographos and (6489) Golevka. (M 25655)
Name suggested and citation prepared by S. J. Ostro.

(6474) Choate

1987 SG_1. Discovered 1987 September 21 by E. Bowell at Anderson Mesa.

Named in honor of Dennis Choate (1952-), senior engineer at NASA's Goldstone {see planet (4433)} deep space communications complex. His profound understanding of the high-power planetary radar transmitter system and his awesome trouble-shooting capabilities have been a mainstay of Goldstone {see planet (4433)} radar astronomy during the past decade. In 1995, Choate's dedication and expertise were manifest in the flawless transmission of many continuous hours of radio signals toward (6489) Golevka during the Goldstone-Evpatoria and Goldstone-Kashima {see planet (6489)} intercontinental radar astronomy experiments. (M 25656)
Name suggested and citation prepared by S. J. Ostro.

(6475) Refugium

1987 SZ_6. Discovered 1987 September 29 by P. Wild at Zimmerwald.

The name is Latin for a refuge, such as might come from examining the prime factors of (6475): (5) Astraea, goddess of justice; (7) Iris, the rainbow; and (37) Fides, goddess of faith and honesty. (M 34623)

(6476) 1987 VT

Discovered 1987 November 15 by Z. Vávrová at Kleť.

(6477) 1988 AE_5

Discovered 1988 January 14 by H. Debehogne at La Silla.

(6478) 1988 JC_1

Discovered 1988 May 12 by C. S. Shoemaker at Palomar.

(6479) Leoconnolly

1988 LC. Discovered 1988 June 15 by E. F. Helin at Palomar.

Named in honor of Leo P. Connolly, astronomer and teacher at California State University, San Bernardino, and director (1985-1995) of the Summer Science Program, held for six weeks each summer at the Thacher School in Ojai. In this tranquil setting, and with a rigorous curriculum requiring high-school students to take photographic plates of minor planets and learn how to determine their orbits, Leo and his wife Jackie have nurtured the minds and spirits of nearly 400 bright and motivated young men and women. (M 25979)

Name proposed by the staff and students of SSP '95. Citation prepared at the request of the discoverer, who has lectured over the years to the summer science students at Thacher.

(6480) Scarlatti

1988 PM_1. Discovered 1988 August 12 by E. W. Elst at St. Michel.

Named in memory of the great Italian composer Domenico Scarlatti (1685-1757). He spent the first half of his life under the shadow of his father, Alessandro Scarlatti. After leaving Italy for Portugal in 1719, and particularly after his father's death in 1725, he developed a style that made him one of the greatest keyboard composers of all time. At the Portuguese court he served as music master for the young Princess Maria Barbara. In 1729 she moved with him to Spain, where he spent the rest of his life. Most of his known 550 harpsichord sonatas were dedicated to this princess, who herself was unusually gifted musically. (M 26766)

(6481) 1988 RH_2

Discovered 1988 September 9 by A. Mrkos at Kleť.

(6482) Steiermark

1989 AF_7. Discovered 1989 January 10 by F. Börngen at Tautenburg.

Named for the Austrian province, also known as Styria, and the capital of which is Graz {see planet (2806)}. The territory was colonized by the Illyrians and then by the Celts. Later it was part of the Roman provinces of Noricum and Pannonia. The writer Peter Rosegger (1843-1918) and the conductor Karl Böhm (1894-1981) were born in Styria. The white Lippizaner horses of the Spanish Court riding school in Vienna are bred in this region. (M 25979)

(6483) Nikolajvasil'ev

1990 EO_4. Discovered 1990 March 2 by E. W. Elst at La Silla.

Named in honor of Nikolaj Vasil'ev, scientific director of the Interdisciplinary Independent Tunguska Expeditions. A professor of medicine and a well-known oncologist at the Institute for Microbiology in Kharkov, he has for many years been interested in the history and scientific investigation of the Tunguska event {see planet (5471)}. In his capacity as IITE director he and other members of the group succeeded in having 4000 square kilometers of the Tunguska region set aside as a national reserve for the next 20 years. (M 26425)

(6484) Barthibbs

1990 FT_1. Discovered 1990 March 23 by E. F. Helin at Palomar.

Named in honor of Bart Hibbs, whom the discoverer has known since his earliest days, through his years of education and into his maturity as an expert and inventive aeronautical engineer. Bart was responsible for the aerodynamic design for the "Sunracer", which won the first race across central Australia for solar-powered cars. More recently, he designed a large, remote-controlled airplane, also solar powered, designed to stay at extremely high altitudes, night and day, for an extended period. (M 30099)

(6485) Wendeesther

1990 UR_1. Discovered 1990 October 25 by C. S. Shoemaker and E. M. Shoemaker and D. H. Levy at Palomar.

Named in honor of Wendee Esther Wallach on the occasion of her engagement to David Levy {see planet (3673)}. For 26 years beginning in September 1970, Wendee taught physical education in three public schools of Las Cruces, New Mexico. As an instructor-trainer in swimming with the American Red Cross for almost three decades, Wendee has taught water safety to students and to swimming instructors. Through her inspired teaching, leadership and example, she has touched the lives of thousands of young people. (M 27330)

(6486) 1991 FO

Discovered 1991 March 17 by E. F. Helin at Palomar.

(6487) Tonyspear

1991 GA_1. Discovered 1991 April 8 by E. F. Helin at Palomar.

Named in honor of Anthony Spear, manager of the successful Pathfinder Project. In his 37 years at the Jet Propulsion Laboratory he has been associated with several successful space projects, usually in a management position. These include the Magellan and VOIR projects to orbit Venus, and Seasat, the first earth satellite to make images of the surface of the oceans using radar. His association with Mars exploration goes back to the beginning of the program, when he was Cognizant Engineer for the command system of Mariner 4, which in 1964 became the first

spacecraft to reach Mars successfully. (M 31024)
Citation prepared by A. Hibbs.

(6488) Drebach

1991 GU_9. Discovered 1991 April 10 by F. Börngen at Tautenburg.

Named for the German village in the Erzgebirge mountains, 20 km south of the Saxon town of Chemnitz. It is noted for its crocuses, blooming over an area of seven hectares. The Volkssternwarte Drebach, originally established as the school observatory in 1969, is widely known, especially after the completion of a planetarium and, in 1986, of a new main building. This minor planet was numbered as a result of observations made at Drebach in April-May 1995. (M 25979)

(6489) Golevka

1991 JX. Discovered 1991 May 10 by E. F. Helin at Palomar.

Named to honor the success of the multinational radar observations of this minor planet in June 1985. The 70-m Goldstone {see planet (4433)} antenna in California provided the transmitted signal, and echoes from the object were detected by the Russian 70-m antenna near the Crimean city of Evpatoria, and also by the Japanese 34-m antenna at Kashima. This experiment, which produced the first international radar astronomy observations and also initiated planetary radar astronomy in Japan, was done during the closest approach of this minor planet for at least the next several centuries. The discoverer has encouraged radar and remote sensing observations for many years and is pleased with the results. (M 26425; M 26439)

Name proposed by the discoverer following a suggestion by A. L. Zajtsev. Citation prepared by S. J. Ostro.

The name is composed by the leading letters of the names of the observing places: GOLdstone - EVpatoria - KAshima.

(6490) 1991 NR_2

Discovered 1991 July 12 by H. E. Holt at Palomar.

(6491) 1991 OA

Discovered 1991 July 16 by H. E. Holt at Palomar.

(6492) 1991 OH_1

Discovered 1991 July 18 by H. Debehogne at La Silla.

(6493) Cathybennett

1992 CA. Discovered 1992 February 2 by E. F. Helin at Palomar.

Named in honor of Catherine A. Bennett, who with her vision, wisdom and actions has played a key role in the establishment of a successful NEO detection system. The discoverer and her JPL colleagues wish to pay tribute to Bennett with asteroid 1992 CA, discovered in the year of their first interaction, which led to new NEO discoveries by NEAT/GEODSS in early 1996. (M 27331)

(6494) 1992 NM

Discovered 1992 July 8 by S. Otomo at Kiyosato.

(6495) 1992 UB_1

Discovered 1992 October 19 by S. Ueda and H. Kaneda at Kushiro.

(6496) Kazuko

1992 UG_2. Discovered 1992 October 19 by K. Endate and K. Watanabe at Kitami.

Named in honor of Kazuko Otsuka (1934-), who joined the staff of the Tokyo Astronomical Observatory in 1957 on the occasion of the International

Geophysical Year. She did secretarial work relating to observations with the Markowitz Moon Camera, by the Baker-Nunn Satellite-Tracking Camera and the Moonwatch project. Later she worked for the Sky Patrol Section and the Dodaira Station, and she retired in 1995. (M 30099)
Name proposed by the second discoverer following a suggestion by K. Tomita.

(6497) Yamasaki

1992 UR_3. Discovered 1992 October 27 by T. Seki at Geisei.

Named in memory of Masamitsu Yamasaki (1886-1959), who worked at the International Latitude Observatory in Mizusawa {see planet (7530)}. As a codiscoverer of comet 27P/Crommelin (then named Pons-Coggia-Winnecke-Forbes) in 1928, he was the first Japanese to receive the Astronomical Society of the Pacific's Donohoe medal. (M 30798)

(6498) Ko

1992 UJ_4. Discovered 1992 October 26 by K. Endate and K. Watanabe at Kitami.

Named in honor of Ko Nagasawa (1932-), who retired from the University of Tokyo's Earthquake Research Institute in 1994 and now works for the Public Information Office at the National Astronomical Observatory of Japan. He has long been interested in the study of meteors, and he obtained many high-quality photographic spectra of the 1965 Leonids at the Dodaira Station. (M 30099)
Name proposed by the second discoverer following a suggestion by K. Tomita.

(6499) Michiko

1992 UV_6. Discovered 1992 October 27 by M. Hirasawa and S. Suzuki at Nyukasa.

Named in honor of Michiko Hirasawa (1955-), wife of the first discoverer. She is manager of Mt. Nyukasa station, where this minor planet was discovered. (M 32094)

(6500) Kodaira

1993 ET. Discovered 1993 March 15 by K. Endate and K. Watanabe at Kitami.

Named in honor of Keiichi Kodaira (1937-), who works mainly in stellar and galactic physics and served as president of IAU Commission 36 during 1982–1985. Scientific director of the Japanese National Large Telescope (SUBARU) project from its beginning, he played an essential role on its completion. He has been director of the National Astronomical Observatory of Japan since 1994. (M 27331)

(6501) Isonzo

1993 XD. Discovered 1993 December 5 at the Farra d'Isonzo Observatory at Farra d'Isonzo.

Named for the river near which the town of Farra d'Isonzo and its observatory are located. Long known as a landmark in northeastern Italy, the river has been a crossing point to eastern Europe since Roman times. (M 30099)

(6502) 1993 XR_1

Discovered 1993 December 6 by S. Ueda and H. Kaneda at Kushiro.

(6503) 1994 CP

Discovered 1994 February 4 by S. Ueda and H. Kaneda at Kushiro.

(6504) Lehmbruck

4630 P-L. Discovered 1960 September 24 by C. J. van Houten and I. van Houten-Groeneveld at Palomar.

Named for the German sculptor Wilhelm Lehmbruck (1881-1919). After his education at the Düsseldorf Academy he went to Paris to study with Rodin and Maillol {see planets (6258) and (6259)}. All his sculptures are elongated and have thin limbs. Most of his work can be seen in the museum in Duisburg, his birthplace. (M 26766)

(6505) 1976 AH

Discovered 1976 January 3 by M. R. Cesco at El Leoncito.

(6506) 1978 EN_{10}

Discovered 1978 March 15 by S. J. Bus at Palomar.

(6507) 1982 QD

Discovered 1982 August 18 by Z. Vávrová at Kleť.

(6508) 1982 QM

Discovered 1982 August 22 by A. Mrkos at Kleť.

(6509) 1983 CQ_3

Discovered 1983 February 12 by H. Debehogne and G. DeSanctis at La Silla.

(6510) Tarry

1987 DF. Discovered 1987 February 23 by C. S. Shoemaker and E. M. Shoemaker at Palomar.

Named in honor of Americans William and Nancy Tarry, whose kindness, thoughtfulness and friendship have left a lasting impression on Carolyn Shoemaker {see planet (4446)} and her family and good friends. When Eugene Shoemaker {see planet (2074)} was killed in an outback car collision far from Alice Springs, the Tarrys unhesitatingly provided their hospitality and assistance to them for several weeks. Bill is general manager for Boeing Constructors with the Boeing Defence and Space Group in Alice Springs. (M 31610)

(6511) Furmanov

1987 QR_{11}. Discovered 1987 August 27 by L. I. Chernykh at Nauchnyj.

Named in honor of Rudol'f Davidovich Furmanov (1938-), artist, producer and impresario, founder and art manager of the Andrej Mironov {see planet (3624)} Theater "Russian enterprise" in St. Petersburg, where many of the stars of the Russian theater perform. Furmanov is originator and author of the theater programs *Russian seasons*, which are widely popular even outside the country. (M 34624)

(6512) de Bergh

1987 SR_1. Discovered 1987 September 21 by E. Bowell at Anderson Mesa.

Named in honor of Catherine de Bergh (1945-), planetary scientist at the Paris Observatory who has studied the composition of the giant and terrestrial planets. She found CO on Titan and HDO in the Martian atmosphere. She has described the chemistry of Venus' lower atmosphere and has also been interested in the icy surfaces of Io, Triton and Pluto. She enthusiastically works for international cooperation in studying the solar system. (M 27462)
Citation provided by M. A. Barucci.

(6513) 1987 UW_1

Discovered 1987 October 28 by S. Ueda and H. Kaneda at Kushiro.

(6514) Torahiko

1987 WY. Discovered 1987 November 25 by T. Seki at Geisei.

Named in memory of Torahiko Terada (1878-1935), experimental physicist, geophysicist and author. His literary works are still popular today. (M 27129)

(6515) 1988 MG
Discovered 1988 June 16 by E. F. Helin at Palomar.

(6516) 1988 TC_2
Discovered 1988 October 3 by A. Mrkos at Kleť.

(6517) 1990 BW
Discovered 1990 January 21 by E. F. Helin at Palomar.

(6518) Vernon
1990 FR. Discovered 1990 March 23 by E. F. Helin at Palomar.

Named in honor of Robert and Esther Vernon, longtime friends and, for more than 35 years, neighbors of the discoverer and her parents, Fred and Kay Francis {see planet (2050)}. Their wise counsel, advice and solace over the years have been a source of comfort and renewed strength. Bob, now "retired", travels the world from Slovakia to Mongolia, sharing his wealth of experience and expertise with emerging democracies. (M 31296)

(6519) Giono
1991 CX_2. Discovered 1991 February 12 by E. W. Elst at St. Michel.

Named in memory of the Provençal writer Jean Giono (1895-1970). Born in Manosque, he left this small town only to make short trips around Provence and to Italy. His natural writing style makes his work easy to read, showing remarkable strength and a sense of humor. The wealth of poetic images, inspired by the splendid and at the same time rough nature of the Provence, evokes his personal ideas on the irreplacable things of life. In *Refus d'obeissance* (1937) rural anarchism and courageous pacifism is proclaimed, an attitude that led to difficulties with the authorities, who imprisoned him. Giono's well-known writings include the *Trilogy of Pan* and *Que ma joie demeure.* (M 25979)

Citation prepared by C. Leterme at the request of the discoverer.

(6520) Sugawa
1991 HH. Discovered 1991 April 16 by S. Otomo and O. Muramatsu at Kiyosato.

Named in honor of Chikara Sugawa (1916-), who worked for the International Latitude Observatory in Mizusawa {see planet (7530)} during 1943-1981. In 1984 he was nominated as the first president of the Herschel Society of Japan, and the following year he realized a Herschel concert in Tokyo and led a study tour to Europe. Since then he has done much to promote knowledge of the Herschels {see planet (2000)} and their contemporaries. (M 26766)

Name proposed by the discoverers following a suggestion by S. Kimura.

(6521) Pina
1991 LC_1. Discovered 1991 June 15 by E. F. Helin at Palomar.

Named in honor of Pina Toscano Blanco, regular "accompanying person" at astronomical meetings during the last 25 years. Her continuous presence has made her a supporting member of the "wives' committees" that do so much to improve such meetings. (M 28090)

Name suggested by Pina's husband Carlo Blanco {see planet (4478)}, with the blessing of the discoverer.

(6522) Aci
1991 NQ. Discovered 1991 July 9 by E. F. Helin at Palomar.

Named for a river southeast of Mount Etna, and also honoring the many towns and villages along it that contain the name: Acicastello, Acitrezza, Acireale, Acibonaccorsi, Acicatena, Aci Sant'Antonio and Aci Santo Filippo. The modern-day river evokes the myth of Aci (or Acis), the young Sicilian shepherd who was in love

with Galatea, a Nereid {see planet (74)}. The jealous cyclops Polyphemus hurled a large rock and killed Aci, whose blood was transformed into an underground river that plunged into the Ionian Sea to hug his beloved Galatea. (M 28090)
Name proposed by the discoverer, following a suggestion by Carlo Blanco.

(6523) Clube

1991 TC. Discovered 1991 October 1 by R. H. McNaught at Siding Spring.

Named for the English astronomer S. Victor M. Clube (1934-), renowned for his heterodox views on various areas of astrophysics and planetary science. Clube was a student at Oxford, where he was a cricketer of note, his bowling bamboozling various famous West Indian batsmen before the young Gary Sobers {see planet (6581)} hit him for six. After working for many years at the Royal Observatory, Edinburgh, Clube returned to Oxford, where he continues research on the small bodies in the solar system, their putative giant comet origin, and their influence upon the terrestrial climate and human civilizations. He also has interests in various other areas of astrophysics and general relativity. (M 29147)

Name proposed by the discoverer following a suggestion by D. I. Steel, who prepared the citation.

(6524) Baalke

1992 AO. Discovered 1992 January 9 by E. F. Helin at Palomar.

Named in honor of Ron Baalke, software engineer at the Jet Propulsion Laboratory who works on telemetry systems for the Deep Space Network. Since 1989 he has been performing public outreach services on the Internet on the behalf of NASA. His numerous homepages on the World Wide Web have reached millions of people around the world and have been among the most popular home pages from NASA. These pages have been instrumental in promoting the World Wide Web as a vehicle for distributing information and images. (M 29147)
Name proposed by the discoverer following a suggestion by A. Beresford.

(6525) Ocastron

1992 SQ_2. Discovered 1992 September 20 by J. B. Child and G. Fisch at Wrightwood.

Named in recognition and appreciation of the Orange County Astronomers, a non-profit, public-service organization that has served the southern California area for more than 20 years. The group has conducted public star parties, telescope making classes, observing classes and public lectures. It has been, and continues to be, a valuable resource to the community and to fledgling astronomers through the efforts of its dedicated members. (M 27331)

(6526) Matogawa

1992 TY. Discovered 1992 October 1 by K. Endate and K. Watanabe at Kitami.

Named in honor of Yasunori Matogawa (1942-), a specialist in rocket trajectory engineering, and now the chief of the External Cooperation Office at the Institute of Space and Astronautical Science in Tokyo, working to inform the public about space technology. (M 30099)
Name proposed by the second discoverer following a suggestion by K. Tomita.

(6527) 1992 UF_6

Discovered 1992 October 31 by A. Natori and T. Urata at Yakiimo.

(6528) 1993 FL_{24}

Discovered 1993 March 21 by the Uppsala-ESO Survey at La Silla.

(6529) Rhoads

1993 XR_2. Discovered 1993 December 14 at the Palomar Observatory at Palomar.

Named in honor of R. Rhoads Stephenson for his significant contributions to the advancement of technology needed and used for the exploration of outer space by robotic spacecraft. A mechanical engineer by training, he was a deputy director at the California Institute of Technology Jet Propulsion Laboratory, where he was responsible for the development and application of technology which broadly supports the United States space exploration mission. He has published in the areas of earth-based tracking and orbit determination, orbit evaluation techniques, spacecraft attitude and articulation control, flight computers, telerobotics, space power and electric propulsion. (M 32788)

(6530) Adry

1994 GW. Discovered 1994 April 12 by V. S. Casulli at Colleverde di Guidonia. Named in honor of Adriano Casulli (1974-), son of the discoverer. (M 31610)

(6531) 1994 YY

Discovered 1994 December 28 by T. Kobayashi at Oizumi.

(6532) Scarfe

1995 AC. Discovered 1995 January 4 by D. D. Balam at Victoria.

Named in honor of Colin D. Scarfe (1940-), professor of astronomy at the University of Victoria since 1965. President of Commission 30 (Radial Velocities) and of Division IX (Optical Techniques) of the IAU since 1994, Scarfe is an authority on spectroscopic and eclipsing binary stars and has made fundamental contributions to the very difficult problem of interpreting the radial-velocity curves of triple systems, as well as the determination of the orbits of systems that are simultaneously spectroscopic and eclipsing binary stars or spectroscopic and visual binary stars. This has led to the determination of stellar masses with unprecedented precision. (M 25980)

Citation prepared by J. B. Tatum at the request of the discoverer.

(6533) 1995 DM$_1$

Discovered 1995 February 24 by C. W. Hergenrother at Tucson.

(6534) 1995 DT$_1$

Discovered 1995 February 24 by T. B. Spahr at Tucson.

(6535) Archipenko

3535 P-L. Discovered 1960 October 17 by C. J. van Houten and I. van Houten-Groeneveld at Palomar.

Named for Alexander Archipenko (1887-1964), Ukrainian-American sculptor. He was the first to transform the cubism and abstract style of paintings into figurative sculptures (plastic art). He studied first in Kiev, then in Paris and Berlin, emigrating to the U.S. in 1923. Among his works are *Queen of Saba* and *King Salomon*. (M 26766)

(6536) Vysochinska

1977 NK. Discovered 1977 July 14 by N. S. Chernykh at Nauchnyj.

Named in honor of Lyudmilla Iosifovna Vysochinska, composer, pianist and music critic in Kiev and a public figure throughout Ukraine. She has written many songs based on the poetry of Ukrainian, Russian and Bulgarian poets. The founding director of the Ukraine's first Theater of Songs, she has research interests that include the connections between Ukrainian classical composers and writers. (M 32345)

(6537) Adamovich

1979 QK$_6$. Discovered 1979 August 19 by N. S. Chernykh at Nauchnyj.

Named in memory of Aleksandr Mikhajlovich Adamovich (Ales' Adamovich; 1927-1994), Byelorussian-Russian writer, literary scholar and publicist. He exhibited a harmonious combination of talent and civic responsibility. (M 34624)

(6538) 1981 SA_5

Discovered 1981 September 25 by L. I. Chernykh at Nauchnyj.

(6539) Nohavica

1982 QG. Discovered 1982 August 19 by Z. Vávrová at Kleť.

Named in honor of Jaromír Nohavica (1945-), Czech singer, poet and composer. (M 33385)

(6540) Stepling

1982 SL_1. Discovered 1982 September 16 by A. Mrkos at Kleť.

Named in memory of Joseph Stepling (1716-1778), founder and the first director of the astronomical observatory at the Jesuit college in Prague called Klementinum {see planet (3386)} (1751). He was also known for introducing Newtonian physics to Prague. (M 31296)

Name suggested by J. Tichá and M. Šolc.

(6541) 1984 DY

Discovered 1984 February 26 by H. Debehogne at La Silla.

(6542) 1985 CH_1

Discovered 1985 February 15 by A. Mrkos at Kleť.

(6543) Senna

1985 TP_3. Discovered 1985 October 11 by C. S. Shoemaker and E. M. Shoemaker at Palomar.

Named in memory of Ayrton Senna Da Silva (1960-1994), Brazilian race-car driver. One of the greatest race-car drivers of all time, he competed in the top category of motor sport, Formula One. His career highlights in this top echelon include 65 pole positions, 19 fastest laps, 41 victories and three championships. He was tragically killed while competing in his 161st race. (M 25980)

Name suggested by the Astronomical Society of South Australia and the Ayrton Senna Fan Club of Australia, the 1993 Australian Grand Prix having been his last victory.

(6544) 1986 SD

Discovered 1986 September 29 by Z. Vávrová at Kleť.

(6545) 1986 TR_6

Discovered 1986 October 5 by M. Antal at Piwnice.

(6546) Kaye

1987 DY_4. Discovered 1987 February 24 by A. Mrkos at Kleť.

Named in memory of Danny Kaye (1913-1987, born David Daniel Kominski), actor and comedian on stage and screen. Kaye's trademark screen role was the mild mannered bumbler who triumphs in the end, as in *The Court Jester* (1956). (M 30099)

Name suggested by G. V. Williams, who made the identifications involving this object.

(6547) 1987 RO_3

Discovered 1987 September 2 by L. I. Chernykh at Nauchnyj.

(6548) 1988 BO_4
Discovered 1988 January 22 by H. Debehogne at La Silla.

(6549) Skryabin
1988 PX_1. Discovered 1988 August 13 by E. W. Elst at St. Michel.

Named in memory of the Russian composer Alexandr Nikolaevich Skryabin (1872-1915). Though one of the most fascinating phenomena at the beginning of the twentieth century, his music was largely unappreciated because of his contradictory philosophical ideas. His most important symphonies, sonatas and other pieces for the piano were composed to an ever-increasing degree following his bold aim to create a *mystery* that should unite all the arts in a grand liturgic-artistical action to uplift and redeem humanity above itself into a condition of supreme ecstasy. (M 26766)

(6550) 1988 VO_5
Discovered 1988 November 4 by A. Mrkos at Kleť.

(6551) 1988 XP
Discovered 1988 December 5 by T. Kojima at Chiyoda.

(6552) 1989 GH
Discovered 1989 April 5 by E. F. Helin at Palomar.

(6553) Seehaus
1989 GP_6. Discovered 1989 April 5 by M. Geffert at La Silla.

Named in memory of the painter Paul A. Seehaus (1891-1919) and the other painters of the "Rheinische Expressionisten". Born in Bonn, Seehaus was from 1910 to 1914 a student of August Macke. In 1913 Macke, Seehaus and other members of the group had an exhibition in Bonn, and this established them as expressionist painters in Germany in their own right – in addition to the existing group of "Die Brücke" and "Der blaue Reiter". In his work Seehaus was always interested in a balance between traditional and modern painting. (M 26425)

(6554) 1989 UO_1
Discovered 1989 October 28 by Y. Mizuno and T. Furuta at Kani.

(6555) 1989 UU_1
Discovered 1989 October 29 by T. Kojima at Chiyoda.

(6556) 1989 YS_6
Discovered 1989 December 29 by A. Mrkos at Kleť.

(6557) 1990 VR_3
Discovered 1990 November 11 by T. Nomura and K. Kawanishi at Minami-Oda.

(6558) Norizuki
1991 GZ. Discovered 1991 April 14 by K. Endate and K. Watanabe at Kitami.

Named in memory of Sojiro Norizuki (1912-1995), founder of Norizuki Technical Works. Under the guidance of H. Tanaka, he constructed the first parabolic antenna for solar observations in Japan in 1949. He was later engaged in the construction of the interferometer at the Nobeyama Solar Radio Observatory and for other radio telescopes. After 1972 he extended his work to infrared and optical telescopes. (M 30099)

Name proposed by the second discoverer following a suggestion by K. Tomita.

(6559) 1991 JP
Discovered 1991 May 3 by K. Kawanishi and M. Sugano at Minami-Oda.

(6560) Pravdo

1991 NP. Discovered 1991 July 9 by E. F. Helin at Palomar.

Named in honor of Steven H. Pravdo, who works in the fields of X-ray astrophysics and extrasolar planet detection, as well as in mission and instrument development for space programs at the Jet Propulsion Laboratory. As the task manager for the Near-Earth Asteroid Tracking system, he has contributed to the design, fabrication, installation and operations of NEAT. His expertise in systems engineering, analysis and management contributed significantly to the success of the project. (M 26931)

(6561) 1991 TC_4

Discovered 1991 October 10 by K. J. Lawrence at Palomar.

(6562) 1991 VR_3

Discovered 1991 November 9 by M. Yanai and K. Watanabe at Kitami.

(6563) Steinheim

1991 XZ_5. Discovered 1991 December 11 by F. Börngen at Tautenburg.

Named for the impact crater in southwestern Germany, 10 km west of the town of Heidenheim on the river Brenz. The rim of the crater has a diameter of 4 km, and its central hillock is named Steinhirt-Klosterberg. The larger and more famous Ries crater {see planet (4327)} is located some 40 km away. Both impacts have an age of 14.5 million years, and it is very probable that they were formed simultaneously by an impactor that fragmented. (M 26425)

(6564) Asher

1992 BB. Discovered 1992 January 25 by R. H. McNaught at Siding Spring.

Named in honor of David John Asher (1966-), researcher on the dynamics of the small bodies in the solar system. Born in Scotland, Asher took degrees at the universities of Cambridge, Oxford and Edinburgh. Following this, he took the words of Douglas Adams in *The Hitchhiker's Guide to the Galaxy* to heart – "With a degree in maths, and another in astrophysics, it was either that or back to the dole queue on Monday" - and moved to Australia, where he spent two years at the Anglo-Australian Observatory. He is currently working at the National Astronomical Observatory in Japan. (M 29147)

Name proposed by the discoverer following a suggestion by D. I. Steel, who prepared the citation.

(6565) Reiji

1992 FT. Discovered 1992 March 23 by K. Endate and K. Watanabe at Kitami.

Named in honor of Reiji Shin Matumoto (1938-), whose many scientific cartoons are held in high regard among young people in Japan. As chief director of the Young Astronauts Club of Japan, he endeavors to educate future generations to participate in the development of space. (M 30099)

Name proposed by the second discoverer following a suggestion by K. Tomita.

(6566) 1992 UB_2

Discovered 1992 October 25 by T. Urata at Oohira.

(6567) Shigemasa

1992 WS. Discovered 1992 November 16 by K. Endate and K. Watanabe at Kitami.

Named in honor of Shigemasa Suzuki (1920-), an instrumentalist whose career began at the Tokyo Astronomical Observatory in Mitaka {see planet (1088)}, where he assisted in the development of a multiphase radio interferometer. In 1960 he took up a fellowship offered by the Commonwealth Scientific and Industrial Organization and joined its division of radiophysics. In 1966 he became the first person of Japanese birth naturalized as an Australian citizen. (M 30100)

Name proposed by the second discoverer, following a suggestion by K. Tomita, who was Suzuki's neighbor in Mitaka.

(6568) 1993 DT

Discovered 1993 February 21 by S. Ueda and H. Kaneda at Kushiro.

(6569) 1993 MO

Discovered 1993 June 22 by J. Mueller at Palomar.

(6570) Tomohiro

1994 JO. Discovered 1994 May 6 by K. Endate and K. Watanabe at Kitami.

Named in honor of Tomohiro Hirayama (1938-), who joined the Tokyo Astronomical Observatory in 1960 as an observer with the Baker-Nunn Satellite-Tracking Camera. He later joined the observatory's computer center and developed many programs for satellite laser ranging and other astronomical purposes. Between 1967 and 1987 all but one of the telex messages from the observatory to the Central Bureau for Astronomical Telegrams were typed by his hands. (M 30100)

Name proposed by the second discoverer K. Watanabe following a suggestion by K. Tomita.

(6571) Sigmund

3027 P-L. Discovered 1960 September 24 by C. J. van Houten and I. van Houten-Groeneveld at Palomar.

Named in honor of Peter Sigmund (1936-), professor at the Institute of Physics at Odense University and since 1970 one of the leading Danish physicists. His research as a theoretical physicist ranges from particle penetration and collision cascade phenomena to charge exchange and electron emission. He has inspired Danish and foreign colleagues immensely in these areas. (M 26932)

Name proposed by K.-O. Groeneveld. Citation prepared by J. Schou.

(6572) Carson

1938 SX. Discovered 1938 September 22 by Y. Väisälä at Turku.

Named in memory of Rachel Louise Carson (1907-1964). While a biologist with the U.S. Fish and Wildlife Service, she autored *Under the Sea Wind, The Sea Around Us* and *The Edge of the Sea.* But she is best remembered for *Silent Spring,* a landmark book that created a wide awareness of the dangers of environmental pollution and fundamentally altered humankind's perception of our global environment. (M 27129)

Name proposed and citation prepared by G. C. L. Aikman.

(6573) 1974 SK_1

Discovered 1974 September 19 by L. I. Chernykh at Nauchnyj.

(6574) Gvishiani

1976 QE_1. Discovered 1976 August 26 by N. S. Chernykh at Nauchnyj.

Named in honor of Jermen Mikhailovich Gvishiani (1928-), known for his many works in philosophy, sociology and theory of management. He serves as president of the Foundation for Prospective Research and the Moscow Institute of Economics, Politics and Law. He is a member of the Russian Academy of Sciences, the Rome Club and many foreign academies, foundations and associations. His work promotes the use of foreign achievements in science, technology and culture in present-day Russia. (M 32346)

(6575) Slavov

1978 PJ_2. Discovered 1978 August 8 by N. S. Chernykh at Nauchnyj.

Named in honor of Nikolaj Antonovich Slavov (1926-), Ukrainian river fleet engineer and sportsman. He helped lead the clean-up effort after the 1986 disaster at Chernobyl, where he had been working for several months. A national boxing champion, he is president of the Professional Boxing League of Ukraine. His interests extend to the arts, and he is president of the All-Ukrainian Foundation for the artist Leonid Bykov {see planet (4682)}. He actively contributes to the development of culture and sport in Ukraine. (M 32346)

(6576) Kievtech

1978 RK_1. Discovered 1978 September 5 by N. S. Chernykh at Nauchnyj.

Named for the Kiev Polytechnical Institute-National Technical University of Ukraine, on the hundredth anniversary of its founding in 1898 as the model for other institutions of higher education in Ukraine. Many famous people have taught there, including S. P. Korolov and B. E. Paton {see planets (1855) and (2727), respectively}. The famous aircraft and helicopter designer I. I. Sikorskij was one of the 120,000 students who have graduated from the Institute. (M 34624)

Name proposed by the discoverer following a suggestion by the Simferopol State University.

(6577) 1978 VB_6

Discovered 1978 November 7 by E. F. Helin and S. J. Bus at Palomar.

(6578) Zapesotskij

1980 TQ_{14}. Discovered 1980 October 13 by T. M. Smirnova at Nauchnyj.

Named in honor of Alexander Sergeevich Zapesotskij (1954-), president of St. Petersburg University of Humanities and Social Sciences, Russian culturologist, known for his public activity as one of the leaders of the Congress of the Russian Intelligentsia. (M 34624)

Name suggested by the Institute of Applied Astronomy.

(6579) 1981 ES_4

Discovered 1981 March 2 by S. J. Bus at Siding Spring.

(6580) 1981 EW_{21}

Discovered 1981 March 2 by S. J. Bus at Siding Spring.

(6581) Sobers

1981 SO. Discovered 1981 September 22 by A. Mrkos at Kleť.

Named in honor of Garfield St. Aubrun (Gary) Sobers (1936-), West Indian cricketer, considered by many as the best all-rounder of all time. As a batsman, Sobers scored 8032 runs in 160 Test innings, including 26 centuries and 365 (not out) against Pakistan in 1957-1958. As a bowler, he took 235 Test wickets, averaging one wicket for each 34 runs given up, and was also an excellent close-fielder. (M 27735)

Name suggested by G. V. Williams, who made the identifications involving this object.

(6582) 1981 VS

Discovered 1981 November 5 by E. Bowell at Anderson Mesa.

(6583) Destinn

1984 DE. Discovered 1984 February 21 by A. Mrkos at Kleť.

Named in memory of Ema Destinn (1878-1930), Czech opera singer, considered

one of the finest sopranos of the early twentieth century, particularly together with the excellent tenor of Enrico Caruso. (M 29671)
Name suggested by J. Tichá.

(6584) 1984 FK

Discovered 1984 March 31 by E. Bowell at Anderson Mesa.

(6585) O'Keefe

1984 SR. Discovered 1984 September 26 by C. S. Shoemaker and E. M. Shoemaker at Palomar.

Named in honor of John Aloysius O'Keefe (1916-), American astronomer and a leading figure in establishing the Theoretical Division of NASA's Goddard Space Flight Center. He is renowned for his discovery of the "pear shape" of the gravitational field of the earth from observations of the first artificial satellites. He is one of the leading students of tektites and has stimulated much interest and research on these natural glasses and played a key role in helping to found the astrogeology program of the U.S. Geological Survey. (M 29671)

(6586) Seydler

1984 UK_1. Discovered 1984 October 28 by A. Mrkos at Kleť.

Named in memory of August Seydler (1849-1891), professor of astronomy and theoretical physics at Charles University in Prague. He founded the Astronomical Institute of the Czech part of Charles University in 1886. Several of his works deal with celestial mechanics and orbit computations for minor planets and comets. (M 30477)
Name suggested by J. Tichá and M. Šolc.

(6587) Brassens

1984 WA_4. Discovered 1984 November 27 by the CERGA at Caussols.

Named in memory of George Brassens (1921-1981), a French poet born on the Mediterranean coast. This gruff artist with a soft heart sang of human nature through texts that were often nonconformist. Henceforth, accompanied by his faithful guitar, he will belong to the heritage of French culture. (M 28621)

Citation provided by members of the Groupement d'Astronomie Populaire de la Région d'Antibes, translated by C. Hayter.

(6588) 1985 RC_4

Discovered 1985 September 10 by H. Debehogne at La Silla.

(6589) 1985 SL_3

Discovered 1985 September 19 by N. S. Chernykh and L. I. Chernykh at Nauchnyj.

(6590) 1985 TA_2

Discovered 1985 October 15 by E. Bowell at Anderson Mesa.

(6591) 1986 RT_5

Discovered 1986 September 7 by L. I. Chernykh at Nauchnyj.

(6592) Goya

1986 TB_{12}. Discovered 1986 October 3 by L. G. Karachkina at Nauchnyj.

Named for Francisco José de Goya y Lucientes (1746-1828) on the 250th anniversary of the birth of this great Spanish painter, etcher and graphic artist. (M 26766)
Name proposed by the discoverer following a suggestion by L. R. Nemirovskij.

(6593) 1986 UV

Discovered 1986 October 28 by Z. Vávrová at Kleť.

(6594) Tasman

1987 MM_1. Discovered 1987 June 25 by A. Mrkos at Kleť.

Named for the Dutch navigator and explorer Abel Janszoon Tasman (c.1603-c.1659), who explored the south Pacific and discovered Tasmania, New Zealand, Tonga and the Fiji Islands. (M 30100)

Name suggested by G. V. Williams, who made the identifications involving this object.

(6595) Munizbarreto

1987 QZ_1. Discovered 1987 August 21 by E. W. Elst at La Silla.

Named in honor of former director Luiz Muniz Barreto of the National Observatory in Rio de Janeiro, who stimulated the development of astrophysics in Brazil and was responsible for the creation in Itajubá of the Laboratorio Nacional de Astrofisica. He also established research groups in astronomy and geophysics, such as the Observatory of Piedade in Minas Gerais. In 1994 he went to the Universidade Nacional Autonoma de Mexico, where he is currently involved in geophysical research. In the 1970s the discoverer was regularly invited by Muniz Barreto to carry out research at the National Observatory and to teach astrophysics at the Universidade Federal of Belo Horizonte. (M 26766)
Citation material supplied by R. Tarsia.

(6596) 1987 VC_1

Discovered 1987 November 15 by A. Mrkos at Kleť.

(6597) 1988 AF_1

Discovered 1988 January 9 by A. Mrkos at Kleť.

(6598) Modugno

1988 CL. Discovered 1988 February 13 at the Osservatorio San Vittore at Bologna.

Named in memory of Domenico Modugno (1928-1994), famous Italian singer, composer and actor. Modugno won the famous festival of Italian song at San Remo three times; in 1958 with "Nel blu dipinto di blu" (also known as "Volare"), in 1959 with "Piove" and in 1966 with "Dio, come ti amo!". (M 27129)

(6599) Tsuko

1988 PV. Discovered 1988 August 8 by K. Endate and K. Watanabe at Kitami.

Named in honor of Tsuko Nakamura (1943-), who studied celestial mechanics in the University of Tokyo's Department of Astronomy and whose research at the National Astronomical Observatory includes observations of the mutual phenomena of Jupiter's Galilean satellites and statistical investigations of the long-term orbital behavior of minor planets and comets. His interests also include the study of old Japanese archives for observations of astronomical phenomena. (M 30100)
Name proposed by the second discoverer following a suggestion by K. Tomita.

(6600) 1988 QW

Discovered 1988 August 17 by A. Mrkos at Kleť.

(6601) 1988 XK_1

Discovered 1988 December 7 by S. Ueda and H. Kaneda at Kushiro.

(6602) Gilclark

1989 EC. Discovered 1989 March 4 by E. F. Helin at Palomar.

Named in honor of Gilbert A. Clark, creator of Telescopes in Education, the first opportunity for students in classrooms to operate a research-grade telescope in real time. Since TIE saw first light in 1994, several thousand students and their teachers have observed using the 0.6-m reflector on Mount Wilson, thereby providing inspiration in astronomy, science and critical thinking. With untiring leadership, Clark reassembled and automated a telescope others thought unusable, and he created a team of volunteers, sponsors and staff to keep it operating for students everywhere. Now ready to expand to other telescopes around the world, the educational inspiration of TIE has reached and connected students in inner cities, rural communities and eight different countries. (M 31024)

(6603) 1990 KG

Discovered 1990 May 19 by E. F. Helin at Palomar.

(6604) Ilias

1990 QE_8. Discovered 1990 August 16 by E. W. Elst at La Silla.

Named for the greatest of the Greek epics, the Iliad of Homer {see planet (5700)}. This work was later divided, more or less arbitrarily, by Alexandrian scientists into 24 books of some 500-800 verses each. Although the Trojan War raged for ten years, Homer reviewed but a small episode of it. King Agamemnon {see planet (911)} was to render the captured Chryseis {see planet (202)} to her father, and he demanded Briseis {see planet (655)}, then mistress of Achilles {see planet (588)}, for himself. Thereupon Achilles withdrew himself and all his men from the battle. However, after his friend Patroclus {see planet (617)} was killed by Hektor {see planet (624)}, Achilles resumed fighting and killed Hektor. The epic ends with the funeral of Patroclus. The name also honors Ilias, born 1995 Oct. 28, the first grandson of the discoverer and son of Sigyn and Philip de Jager-Elst. (M 26425)

(6605) 1990 SM_9

Discovered 1990 September 22 by E. W. Elst at La Silla.

(6606) Makino

1990 UF. Discovered 1990 October 16 by T. Seki at Geisei.

Named in memory of Tomitaro Makino (1862-1957), Kochi-born botanist of international repute. He became interested in plants in his childhood and is credited with the discovery of 500 new species. (M 28621)

(6607) Matsushima

1991 UL_2. Discovered 1991 October 29 by K. Endate and K. Watanabe at Kitami.

Named in honor of Koichi Matsushima (1938-), head of the Instrumentation Laboratory, Control Systems Division, National Aerospace Laboratory. He has played a leading role in the design of Japan's minor-planet exploration program. He has also promoted research into minor planets, comets and meteors. (M 30477)

(6608) 1991 VC_4

Discovered 1991 November 2 by E. F. Helin at Palomar.

(6609) 1992 BN

Discovered 1992 January 28 by S. Ueda and H. Kaneda at Kushiro.

(6610) 1993 BL_3

Discovered 1993 January 28 by A. Natori and T. Urata at Yakiimo.

(6611) 1993 VW

Discovered 1993 November 9 by E. F. Helin and J. Alu at Palomar.

(6612) 1994 EM_1

Discovered 1994 March 10 by Y. Kushida and O. Muramatsu at Yatsugatake.

(6613) 1994 LK

Discovered 1994 June 2 by C. W. Hergenrother at Tucson.

(6614) Antisthenes

6530 P-L. Discovered 1960 September 24 by C. J. van Houten and I. van Houten-Groeneveld at Palomar.

Named for the Greek philosopher Antisthenes (c.455-c.360 B.C.). In ethics he said that everyone could learn "Arete" (virtue, see also planet (197)), and he founded the cynic school of philosophy. (M 26932)

(6615) Plutarchos

9512 P-L. Discovered 1960 October 17 by C. J. van Houten and I. van Houten-Groeneveld at Palomar.

Named for the great Greek philosopher and writer Plutarchos (c.45-125), who studied at the Academia of Plato {see planet (5451)} in Athens. About half of his philosophical work is preserved. His parallel biographies of 46 prominent Greeks and Romans is one of his best known works. In his *Moralia* he speaks about the correct faith and style of living in a form as Plato did. (M 26932)

(6616) Plotinos

1175 T-1. Discovered 1971 March 25 by C. J. van Houten and I. van Houten-Groeneveld at Palomar.

Named for the Greek philosopher Plotinos (205-270). In Alexandria Plotin studied under the philosopher Ammonios Sakkas and participated in the war of Gordianus III against the Persians. During 246-268 he operated a school in neo-Platonic philosophy. He combined the school of Plato {see planet (5451)} with gnostic elements and oriental mysticism, and he strongly influenced Augustine – and therefore also christianity. (M 26932)

(6617) Boethius

2218 T-1. Discovered 1971 March 25 by C. J. van Houten and I. van Houten-Groeneveld at Palomar.

Named for the western Roman politician, poet and philosopher Anicius Manilius Severinus Boethius (c. 480-524). An Advisor to the Ostrogoth king Theoderich the Great, he is called "the last Roman and the first scholastic". He translated the works of the Greek philosophers, especially Aristotle {see planet (6123)}, into Latin. He was accused of high treason, put into prison and executed without due process. In prison he wrote his most famous book, *Consolation of Philosophy*. (M 26932; M 26947)

(6618) 1936 SO

Discovered 1936 September 16 by C. W. Tombaugh at Flagstaff.

(6619) 1973 SS_4

Discovered 1973 September 27 by L. I. Chernykh at Nauchnyj.

(6620) 1973 UC

Discovered 1973 October 25 by P. Wild at Zimmerwald.

(6621) 1975 VN_5

Discovered 1975 November 2 by T. M. Smirnova at Nauchnyj.

(6622) Matvienko

1978 RG_1. Discovered 1978 September 5 by N. S. Chernykh at Nauchnyj.

Named in honor of Vladimir Pavlovich Matvienko (1938-), Ukrainian economist, author of many works on economics and banking and a member of the Ukrainian Academy of Ecological Studies. He is also a poet, and some of his poetic works have been set to music. (M 32346)

(6623) 1979 MY_2

Discovered 1979 June 25 by E. F. Helin and S. J. Bus at Siding Spring.

(6624) 1980 SG

Discovered 1980 September 16 by Z. Vávrová at Kleť.

(6625) 1981 EX_{41}

Discovered 1981 March 2 by S. J. Bus at Siding Spring.

(6626) 1981 EZ_{46}

Discovered 1981 March 2 by S. J. Bus at Siding Spring.

(6627) 1981 FT

Discovered 1981 March 27 by Z. Vávrová at Kleť.

(6628) Dondelia

1981 WA_1. Discovered 1981 November 24 by E. Bowell at Anderson Mesa.

Named in honor of Donald S. C. and Delia West of Wareham, Dorset. Friends of the discoverer, Don and Delia have had the perspicacity to make their home in one of the loveliest parts of England. (M 27463)

(6629) Kurtz

1982 UP. Discovered 1982 October 17 by E. Bowell at Anderson Mesa.

Named in honor of Paul Kurtz (1925-) of the State University of New York at Buffalo. Kurtz is professor emeritus of philosophy, chairman and founder of the Committee for the Scientific Investigation of Claims of the Paranormal (CSICOP), chairman of the Council for Secular Humanism, and founder and chairman of Prometheus Books. A noted author and philosopher, he has written more than 30 books and 650 articles on philosophy, humanism and skepticism. He has been influential in bringing philosophy back from a pure analytical study to a more activist stance. Kurtz actively promotes science and the refutation of pseudoscience. (M 27735)

Name suggested and citation provided by J. McGaha.

(6630) Skepticus

1982 VA_1. Discovered 1982 November 15 by E. Bowell at Anderson Mesa.

Named for the Committee for the Scientific Investigation of Claims of the Paranormal and its journal *The Skeptical Enquirer*. CSICOP is nonprofit educational organization, founded in 1976 to investigate paranormal, occult and fringe claims using rigorous scientific methodology. CSICOP has spawned a grassroots movement of some 70 autonomous groups worldwide. Throught public education and outreach, CSICOP has become a voice of reason to counter pseudoscience and superstition. (M 27735)

Named and citation provided by J. McGaha.

(6631) Pyatnitskij

1983 RQ_4. Discovered 1983 September 4 by L. V. Zhuravleva at Nauchnyj.

Named in memory of Mitrofan Efimovich Pyatnitskij (1864-1927), founder of the Russian National Chorus, performer and collector of Russian folk-songs, and Honored Artist of Russia (1925). (M 34624)

(6632) Scoon

1984 UX_1. Discovered 1984 October 29 by E. Bowell at Anderson Mesa.

Named in honor of George E. N. Scoon (1936-). Born in Grenada, West Indies, Scoon studied in The Netherlands and the U.K. in the fields of telecommunications engineering, general management, and business administration. He now works at the Future Scientific Project Division of the European Space Agency

in Noordwijk. With his intelligence and energy, he has promoted and continues to promote solar-system exploration in Europe and cooperation among the major space agencies. Missions such as Vesta, Cassini-Huyghens, Moro, Marsnet and the Mercury Orbiter ESA cornerstone found an enthusiastic and competent supporter in Scoon, to whom the space-science community owes a debt of gratitude. (M 27463)
Name suggested and citation prepared by M. Fulchignoni.

(6633) 1986 TR_4

Discovered 1986 October 11 by P. Jensen at Brorfelde.

(6634) 1987 KB

Discovered 1987 May 23 at the Observatorio do Valongo at Campiñas.

(6635) Zuber

1987 SH_3. Discovered 1987 September 26 by C. S. Shoemaker and E. M. Shoemaker at Palomar.

Named in honor of Maria T. Zuber, American geophysicist whose research interests have taken her into theoretical modeling of geophysical processes and the relationships between gravity, topography and tectonic features in planetary lithospheres. She has contributed to various space missions, including Mars Observer, the Clementine mission, Mars Global Surveyor and the Near Earth Asteroid Rendezvous mission. (M 32346)

(6636) 1988 RK_8

Discovered 1988 September 11 by V. G. Shkodrov at Rozhen.

(6637) Inoue

1988 XZ. Discovered 1988 December 3 by K. Endate and K. Watanabe at Kitami.

Named in honor of Keisuke Inoue (1928-), who worked for Tentai Ichiyou, nautical almanacs, at Japan's Hydrographic Department for 38 years. He was an expert in calculating the orbits of artificial satellites. After retirement, he worked for a software company and is now on the editorial staff of the astronomical almanac *Tenmon Nenkan*. (M 30477)

(6638) 1989 CA

Discovered 1989 February 2 by M. Arai and H. Mori at Yorii.

(6639) 1989 SO_8

Discovered 1989 September 25 by H. Debehogne at La Silla.

(6640) Falorni

1990 DL. Discovered 1990 February 24 at the Osservatorio San Vittore at Bologna.

Named in memory of Marco Falorni (1944-1995), Italian amateur astronomer, president of Unione Astrofili Italiani, the leading Italian association of amateur astronomers. A great visual observer of planets, principally Mars, he used in this work the refractor Amici at the Arcetri Astrophysical Observatory in Florence and the 0.83-m refractor at Meudon. Falorni also collaborated with the group of amateurs at the Osservatorio San Vittore {see planet (2235)} in Bologna, examining and measuring hundreds of photographs of planetary surfaces taken around 1970 by C. Vacchi, G. Sassi and G. Sette. (M 27331)

(6641) 1990 OK_2

Discovered 1990 July 29 by H. E. Holt at Palomar.

(6642) 1990 UE_3

Discovered 1990 October 26 by T. Urata at Oohira.

(6643) Morikubo

1990 VZ. Discovered 1990 November 7 by Y. Kushida and O. Muramatsu at Yatsugatake.

Named in honor of Shigeru Morikubo (1913-), since 1930 an amateur astronomer, observer of variable stars, sunspots, meteors and occultations, and a pioneer in observing meteoric dust. Since 1955 he has been continuing to study meteoric dust through his medical microscope. In 1967 he became the first chairman of the Japan Amateur Astronomers' Convention, and he continues to edit a history of amateur astronomy in Japan. (M 29147)
Name proposed by the discoverers following a suggestion by S. Sakuma.

(6644) Jugaku

1991 AA. Discovered 1991 January 5 by A. Takahashi and K. Watanabe at Kitami.

Named in honor of Jun Jugaku (1927-), a professor at Tokai University, formerly a professor at the Tokyo Astronomical Observatory. He is known for his work on early-type stars, particularly chemically-peculiar stars and B stars. As an associate editor of the *Publications of the Astronomical Society of Japan* from 1965 to 1989, he made every effort to upgrade the scientific level of that journal. He served as president of IAU Commission 29 during 1982-1985. (M 30798; M 30819)
Name proposed by the second discoverer, following a suggestion by W. Liller.

(6645) Arcetri

1991 AR_1. Discovered 1991 January 11 by E. F. Helin at Palomar.

Named for the Florence Observatory, moved in 1872 from the center of the city to Arcetri, near the house in which Galileo died. The original observatory, La Specola, was also associated with Galileo, and nineteenth-century directors included the comet hunters Pons, Donati and Tempel {see planet (3808)}. It was Donati who moved the observatory to its present location. (M 28090)

(6646) 1991 CA_3

Discovered 1991 February 14 by E. F. Helin at Palomar.

(6647) Josse

1991 GG_5. Discovered 1991 April 8 by E. W. Elst at La Silla.

Named in honor of Raymond Josse (1914-), a graduate of the Ecole Militaire de l'Air and the Ecole Nationale Supérieure de l'Aéronautique who has occupied important positions in the French aeronautical administration. He is also a member of the Société des Amis de Jean de La Fontaine, one of his ancestors, as well as of the Société Astronomique de France. He has authored about 100 articles on historical and on technical subjects. During the festivities in 1995 on the occasion of the 300th anniversary of the death of de La Fontaine {see planet (5780)} at Château Thierry he became a very good friend of the Elst family. (M 26766; E. W. Elst)

(6648) 1991 PM_{11}

Discovered 1991 August 9 by H. E. Holt at Palomar.

(6649) 1991 RN

Discovered 1991 September 5 by A. Natori and T. Urata at Yakiimo.

(6650) Morimoto

1991 RS_1. Discovered 1991 September 7 by K. Endate and K. Watanabe at Kitami.

Named in honor of Masaki Morimoto (1932-), Japanese radio astronomer who started his career at Mitaka {see planet (1088)} and participated in the construction of the Culgoora radioheliograph in Australia. He successfully completed and

used the Mitaka 6-m millimeter-wavelength telescope, then the 45-m telescope and 10-m interferometer at Nobeyama. His ideas led to the launch of the HALCA (VSOP) satellite with an 8-m telescope for space VLBI. He is now professor at Kagoshima University and established the astronomy program there. He has also served as president of IAU Commission 40. (M 30798; M 30819)
Name proposed by the second discoverer following a suggestion by J. Watanabe.

(6651) 1991 RV_9

Discovered 1991 September 10 by H. E. Holt at Palomar.

(6652) 1991 SJ_1

Discovered 1991 September 16 by H. E. Holt at Palomar.

(6653) Feininger

1991 XR_1. Discovered 1991 December 10 by F. Börngen at Tautenburg.

Named for the American painter and graphic artist Lyonel Feininger (1871-1956). Descended from a family of German musician, he lived in Germany (Hamburg, Berlin, Weimar and Dessau) from 1887 to 1937, almost without interruption. Initially working as a caricaturist, he began in 1907 to create architectural and landscape paintings in a very personal style similar to cubism. Feininger was a master of the Bauhaus and particularly fond of Weimar and numerous Thüringian villages. Some 120 of his works alone are of the church of Gelmeroda. Outlawed by the Nazis, he returned to the U.S. and, beginning in 1939, painted his series Skyscrapers of Manhattan. In 1947 he became the president of the Federation of American Painters and Sculptors. (M 26766)

(6654) 1992 DT_6

Discovered 1992 February 29 by the Uppsala-ESO Survey at La Silla.

(6655) 1992 EL_1

Discovered 1992 March 8 by A. Sugie at Taga.

(6656) Yokota

1992 FF. Discovered 1992 March 23 by K. Endate and K. Watanabe at Kitami.

Named in honor of Hiroshi Yokota (1927-), Japanese amateur astronomer. For almost half a century he has delivered lectures on astronomy and has guided many able amateur astronomers. He organized several national astronomical gatherings in Yamaguchi, and he also played an important role in the investigation of a previously-unknown type of meteorite. (M 30799)

Name proposed by the discoverers following a suggestion by M. Koishikawa and T. Sato.

(6657) 1992 WY

Discovered 1992 November 17 by A. Sugie at Taga.

(6658) Akiraabe

1992 WT_2. Discovered 1992 November 18 by K. Endate and K. Watanabe at Kitami.

Named in honor of Akira Abe (1934-), managing editor of *Hoshino Techo* ("Star Handbook"), a quarterly journal that existed from 1978 to 1993 and popularized astronomy with the help of high-quality articles by professional astronomers. He has played a pivotal role in introducing many amateurs and professionals to astronomy. (M 31610)
Name proposed by the discoverers following a suggestion by A. Fujii and A. Tanno.

(6659) 1992 YN

Discovered 1992 December 24 by T. Urata at Oohira.

(6660) Matsumoto

1993 BC. Discovered 1993 January 16 by T. Seki at Geisei.

Named in honor of Tatsujiro Matsumoto (1930-), popular Japanese telescope maker, known also for his observations of Mars and Jupiter. (M 30799)

(6661) 1993 BO

Discovered 1993 January 17 by Y. Mizuno and T. Furuta at Kani.

(6662) 1993 BP$_{13}$

Discovered 1993 January 22 by S. Ueda and H. Kaneda at Kushiro.

(6663) Tatebayashi

1993 CC. Discovered 1993 February 12 by T. Kobayashi at Oizumi.

Named for a city in Kanto Plain in the eastern part of Gunma {see planet (3829)} prefecture. Famous for the Tsutsujigaoka Park, where some of the trees are 1000 years old and 10 000 azalea blossoms bloom every year, the city is blessed by its beautiful natural surroundings. There are several lakes, such as Jonuma and Tataranuma, where swans and other birds come regularly. Tatebayashi is also the birthplace of nature writer Katai Tayama and Japan's first female astronaut, Chiaki Mukai {see planet (4750)}. The Tatebayashi Children's Science Exploratorium held a "star" naming competition to commemorate her second space shuttle flight, and this was a winning entry. (M 33787)

(6664) Tennyo

1993 CK. Discovered 1993 February 14 by T. Kobayashi at Oizumi.

A maiden who dances in the heavens scattering flowers, playing music and perfuming the atmosphere, this angelic figure can be seen in Buddhist sculptures and paintings traditionally found all over Japan. During astronaut Chiaki Mukai's {see planet (4750)} first space shuttle flight, she cummunicated by radio with junior high school students in Tatebayashi {see planet (6663)}. In the radio exchange, one of the students asked how it felt to be in space. She answered, "It feels like I have become a 'tennyo' and am flying." The word became famous all over Japan as another name for Mukai. This naming was also a winning entry in the Exploratorium "star" naming competition. (M 33787)

(6665) 1993 CN

Discovered 1993 February 14 by T. Urata at Oohira.

(6666) Frö

1993 FG$_{20}$. Discovered 1993 March 19 by the Uppsala-ESO Survey at La Silla.

Frö, son of Njord {see planet (4213)}, was the fertility god in the old Nordic religion. He was also the god of peace, good crops and marriage. According to the old tales he had the largest ship in the world, Skidbladner, able to sail both on land and sea. His holy animal was the pig, playing a central role in this religion. In the Nordic countries we still can see some remains of the worship to Frö. The Midsummer pole may have originated as a tribute to Frö. His sister Fröja {see planet (76)}, tremendously beautiful, who drove a carriage pulled by cats. (M 30100)

(6667) 1994 EK$_2$

Discovered 1994 March 14 by Y. Kushida and O. Muramatsu at Yatsugatake.

(6668) 1994 GY$_8$

Discovered 1994 April 11 by S. Ueda and H. Kaneda at Kushiro.

(6669) Obi

1994 JA_1. Discovered 1994 May 5 by K. Endate and K. Watanabe at Kitami.

Named in honor of Shinya Obi (1925-), professor emeritus of the University of Tokyo. In addition to his numerous research papers, he has published many popular astronomy books. He has just retired of the University of the Air, which educates via communication satellites. (M 31610)

Name proposed by the discoverers following a suggestion by A. Fujii and A. Abe.

(6670) Wallach

1994 LL_1. Discovered 1994 June 4 by C. S. Shoemaker and D. H. Levy at Palomar.

Named for Annette and Leonard Wallach, in honor of their monumental effort in building Treasure Island, a unique day camp and school on Long Island, N.Y. More than a day camp, the facilities were donated to the American Red Cross each year for a small craft and safety school. (M 30477)

Citation and name suggested by Wendee and David Levy.

(6671) 1994 NC_1

Discovered 1994 July 5 by E. F. Helin at Palomar.

(6672) Corot

1213 T-1. Discovered 1971 March 24 by C. J. van Houten and I. van Houten-Groeneveld at Palomar.

Named for the French painter and etcher Camille Corot (1796-1875). In his early years his style was classicistic, and he painted especially fine landscapes in the style of Claude Lorrain and Nicolas Poussin. Later, through his treatment of the light in his paintings, Corot became a forerunner of impressionism. (M 27331)

(6673) Degas

2246 T-1. Discovered 1971 March 25 by C. J. van Houten and I. van Houten-Groeneveld at Palomar.

Named for the French painter and sculptor Edgar Degas (1834-1917). His early paintings and sculptures were influenced by Ingres and Japanese woodcuts. Following his acquaintance with Edouard Manet, he changed to the impressionistic style. Beginning around 1900 he modeled about 70 statuettes of women, dancers and horses, and these were cast in bronze after his death. (M 26932)

(6674) Cézanne

4272 T-1. Discovered 1971 March 26 by C. J. van Houten and I. van Houten-Groeneveld at Palomar.

Named for the French impressionist Paul Cézanne (1839-1906). Cézanne was self-taught and influenced by the baroque and romantic styles. Later he came in contact with Pissarro and became an impressionist. At the end of his life he was the forerunner of expressionism, cubism and fauvism. (M 26932)

(6675) Sisley

1493 T-2. Discovered 1973 September 29 by C. J. van Houten and I. van Houten-Groeneveld at Palomar.

Named for the French impressionist painter Alfred Sisley (1839-1899). He was influenced by Monet and Renoir {see planets (6676) and (6677)}. Most of his paintings are landscapes in wonderful gleaming colors. (M 26932)

(6676) Monet

2083 T-2. Discovered 1973 September 29 by C. J. van Houten and I. van Houten-Groeneveld at Palomar.

Named for the French impressionist painter Claude Monet (1840-1926). He became acquainted with the navy painter Eugéne Boudin, who encouraged him to paint directly in natural surroundings, rather than to make a sketch in the open air and to paint it later in the studio. Monet, Manet and the other impressionist painters influenced each other strongly. (M 26932)

(6677) Renoir

3045 T-3. Discovered 1977 October 16 by C. J. van Houten and I. van Houten-Groeneveld at Palomar.

Named for the French impressionist painter Pierre Auguste Renoir (1841-1919). He had a great influence on the other impressionist painters. Renoir's great paintings show people in natural surroundings and in wonderful colors. (M 26932)

(6678) Seurat

3422 T-3. Discovered 1977 October 16 by C. J. van Houten and I. van Houten-Groeneveld at Palomar.

Named for the French impressionist painter Georges Seurat (1859-1891). Together with his friend, the painter Signac {see planet (8239)}, Seurat invented the pointillistic style. (M 26932)

(6679) 1969 UP_1

Discovered 1969 October 16 by L. I. Chernykh at Nauchnyj.

(6680) 1970 WD

Discovered 1970 November 24 by L. Kohoutek at Bergedorf.

(6681) Prokopovich

1972 RU_3. Discovered 1972 September 6 by L. V. Zhuravleva at Nauchnyj.

Named in memory of Feofan Prokopovich (1681-1736), Ukrainian and Russian writer, archbishop and associate of Peter the Great. (M 34624)

(6682) Makarij

1973 ST_3. Discovered 1973 September 25 by L. V. Zhuravleva at Nauchnyj.

Named in memory of Metropolitan Makarij (1482-1563), who had great influence on education, the church and politics in Russia. He promoted the creation of the first Russian printing-house and led a circle of educated booklovers. (M 34624)

(6683) 1976 GQ_2

Discovered 1976 April 1 by N. S. Chernykh at Nauchnyj.

(6684) 1977 QU

Discovered 1977 August 19 by N. S. Chernykh at Nauchnyj.

(6685) Boitsov

1978 QG_2. Discovered 1978 August 31 by N. S. Chernykh at Nauchnyj.

Named in memory of Vasilij Vasil'evich Boitsov (Bojtsov, 1908-1997), specialist on the technology of mechanical engineering and standardization. From 1963 to 1984 he headed the U.S.S.R. State Committee for standards and represented his country in the International Organization for Standardization, of which he served as president (1977-1979). Boitsov was an initiator and active participant in fundamental research on the creation of standard measurement systems. (M 31296)

Name suggsted by the Institute of Theoretical Astronomy and supported by the discoverer.

(6686) Hernius

1979 QC_2. Discovered 1979 August 22 by C.-I. Lagerkvist at La Silla.

Named in honor of Olof Hernius, who participated in the Uppsala-ESO Survey of Asteroids and Comets (UESAC) as an undergraduate student. (M 26932)

(6687) Lahulla

1980 FN_1. Discovered 1980 March 16 by C.-I. Lagerkvist at La Silla.

Named in honor of José Felix Lahulla, astronomer at the Observatorio Astronómico de Madrid. Lahulla has participated in photometric surveys of Hilda asteroids from Calar Alto, La Palma and ESO. (M 27129)

(6688) 1981 ER_{17}

Discovered 1981 March 2 by S. J. Bus at Siding Spring.

(6689) 1981 EQ_{24}

Discovered 1981 March 2 by S. J. Bus at Siding Spring.

(6690) 1981 SY_1

Discovered 1981 September 25 by B. A. Skiff at Anderson Mesa.

(6691) 1984 DX

Discovered 1984 February 26 by H. Debehogne at La Silla.

(6692) 1985 HL

Discovered 1985 April 18 by Z. Vávrová at Kleť.

(6693) 1986 CC_2

Discovered 1986 February 12 by H. Debehogne at La Silla.

(6694) 1986 PF

Discovered 1986 August 4 at the Palomar Observatory at Palomar.

(6695) 1986 PD_1

Discovered 1986 August 1 by E. F. Helin at Palomar.

(6696) Eubanks

1986 RC_1. Discovered 1986 September 1 at the Oak Ridge Observatory at Harvard.

Named in honor of Marshall Eubanks of the U.S. Naval Observatory. His contributions to astrometry and geodesy, through his development and imaginative use of the navy radio interferometry network and through his astute analyses of data relating to the earth's rotation and polar motion, have resulted in new insights on the causes of those complex phenomena. (M 26767)

Name proposed by I. I. Shapiro.

(6697) Celentano

1987 HM_1. Discovered 1987 April 24 by Z. Vávrová at Kleť.

Named in honor of Adriano Celentano (1938-), Italian singer and film actor. (M 33787)

(6698) Malhotra

1987 SL_1. Discovered 1987 September 21 by E. Bowell at Anderson Mesa.

Named in honor of Renu Malhotra (1961-), accomplished dynamicist and celestial mechanician at the Lunar and Planetary Institute in Houston. Born and raised in India, she has made major contributions to our understanding of how resonances affect satellite systems, the asteroid belt, and particularly Pluto. Malhotra was awarded the Harold C. Urey Prize by the Division for Planetary Sciences of the American Astronomical Society in 1997. Her talents and good spirits are much enjoyed by her colleagues. (M 31025)

Name proposed by the discoverer following a suggestion by S. A. Stern, who also wrote the citation.

(6699) Igaueno

1987 YK. Discovered 1987 December 19 by T. Seki at Geisei.

Named for a small town in Mie prefecture. A branch of the Oriental Astronomical Association was inaugurated there around the time this minor planet was discovered. (M 32094)

(6700) Kubišová

1988 AO_1. Discovered 1988 January 12 by Z. Vávrová at Kleť.

Named in honor of Marta Kubišová (1942-), Czech singer. (M 33787; M 34089)

(6701) 1988 AW_1

Discovered 1988 January 14 by A. Mrkos at Kleť.

(6702) 1988 BP_3

Discovered 1988 January 18 by H. Debehogne at La Silla.

(6703) 1988 CH

Discovered 1988 February 10 by M. Arai and H. Mori at Yorii.

(6704) 1988 CJ

Discovered 1988 February 10 by M. Arai and H. Mori at Yorii.

(6705) 1988 RK_5

Discovered 1988 September 2 by H. Debehogne at La Silla.

(6706) 1988 VD_3

Discovered 1988 November 11 by T. Kojima at Chiyoda.

(6707) Shigeru

1988 VZ_3. Discovered 1988 November 13 by M. Yanai and K. Watanabe at Kitami.

Named in honor of Shigeru Nakano (1918-), doctor of medicine specializing in obstetrics and gynecology and in medical jurisprudence. An amateur astronomer, he has published many observational guidebooks and star atlases, including the most detailed one in Japan. (M 31610)

Name proposed by the discoverers following a suggestion by A. Fujii and T. Sato.

(6708) Bobbievaile

1989 AA_5. Discovered 1989 January 4 by R. H. McNaught at Siding Spring.

Named in memory of Roberta (Bobbie) Vaile (1959-1996), lecturer in physics at the University of Western Sydney. Bobbie Vaile excelled in communicating the beauty and complexity of science to students and to the wider community. The SETI Australia Centre was founded at the UWS as a result of Bobbie's participation in the Search for Extraterrestrial Intelligence and her innovative use of SETI for teaching science. A profound faith sustained her during the achievements of her last seven years, while a brain tumor took its toll. To many colleagues and students Bobbie was a uniquely understanding and close friend, and she remains an inspiration. (M 29671)

Name proposed by P. Cass and the discoverer, citation prepared by J. Caswell.

(6709) 1989 CD

Discovered 1989 February 2 by M. Arai and H. Mori at Yorii.

(6710) Apostel

1989 GF_4. Discovered 1989 April 3 by E. W. Elst at La Silla.

Named in memory of the well-known Flemish philosopher Leo Apostel (1925-1995). He was a pupil of Chaim Perelman (Brussels), Jean Piaget (Geneva) and Rudolf Carnap (Chicago). In 1956-57 he joined the universities of Brussels and

Ghent, teaching logic and epistemology there for many years. In 1990 he established the independent interdisciplinary investigation center, Worldviews, in which the discoverer has been asked to guide the group "cosmology". (M 27129)

(6711) Holliman

1989 HG. Discovered 1989 April 30 by E. F. Helin at Palomar.

Named in memory of John Holliman (1948-1998), a national correspondent for CNN, the U.S. Cable News Network. Holliman reported extensively on the role of the Jet Propulsion Laboratory in space exploration. Specifically, he served as the network's lead reporter for the Pathfinder mission to Mars, anchoring CNN's continous coverage of the landing and mission as the spacecraft beamed live video of the planet's surface. Holliman possessed a special ability to make the complex and complicated process of space exploration understandable and accessible to his viewers. (M 32788)

(6712) 1990 DS_1

Discovered 1990 February 23 by A. Mrkos at Kleť.

(6713) 1990 KM

Discovered 1990 May 21 by E. F. Helin at Palomar.

(6714) 1990 OE_2

Discovered 1990 July 29 by H. E. Holt at Palomar.

(6715) 1990 QS_1

Discovered 1990 August 22 by H. E. Holt at Palomar.

(6716) 1990 RO_1

Discovered 1990 September 14 by H. E. Holt at Palomar.

(6717) Antal

1990 TU_{10}. Discovered 1990 October 10 by F. Börngen and L. D. Schmadel at Tautenburg.

Named in honor of the Slovak astronomer Milan Antal (1935-), on the staff of the observatories at Skalnaté Pleso {see planet (2619)} and Hurbanovo for over 38 years. Known as an excellent observer of comets and minor planets, Antal determined accurate photographic astrometric positions by the thousand. In 1973 he observed two outbursts of comet 41P/Tuttle-Giacobini-Kresák. Six of his minor planet discoveries have been numbered so far. He is a founder member and vice president of the General Štefánik Society, founded in 1989 in honor of Milan Rastislav Štefánik (1880-1919), some time assistant at the Meudon Observatory. (M 26767)

Name proposed by the first discoverer.

(6718) Beiglböck

1990 TT_{12}. Discovered 1990 October 14 by L. D. Schmadel and F. Börngen at Tautenburg.

Named in honor of Wolf D. Beiglböck (1939-), German mathematician and professor at the University of Heidelberg. His main research as a mathematical physicist ranges from the theory of Lie groups and harmonic analysis to a discussion of the equations of motion of extended bodies in the framework of Einstein's gravitation theory. For 30 years, Beiglböck has served as scientific advisor to Springer-Verlag in Heidelberg. He is the founder and co-editor of the well-known series "Lecture Notes in Physics" and "Texts and Monographs in Physics". His persistent support made it possible to realize the publication of the *Dictionary of Minor Planet Names*. (M 33386)

Name proposed and citation prepared by the first discoverer.

(6719) Gallaj

1990 UL_{11}. Discovered 1990 October 16 by L. V. Zhuravleva and G. R. Kastel' at Nauchnyj.

Named in honor of Mark Lazarevich Gallaj (1914-), a distinguished Soviet test pilot who tested 125 types of aircraft and helicopter. He pioneered in testing rocket technology and was one of the first to succeed in gaining insight into the flutter phenomenon that emerges when aircraft travel near Mach 1. Gallaj participated in training the first generation of Soviet cosmonauts and is the author of excellent books on aviation and the conquest of space. (M 26767)

Name proposed by G. R. Kastel'.

(6720) Gifu

1990 VP_2. Discovered 1990 November 11 by T. Seki at Geisei.

Named for the city of 400,000, capital of the prefecture of the same name, site of ancient battlefields and a modern public observatory. (M 32094)

(6721) 1990 VY_6

Discovered 1990 November 10 by T. Urata at Oohira.

(6722) Bunichi

1991 BG_2. Discovered 1991 January 23 by K. Endate and K. Watanabe at Kitami.

Named in honor of Bunichi Saito (1925-), professor emeritus at Niigata University and an expert on the earth's upper atmosphere, particularly the airglow. He is also an authority on the poet and novelist Kenji Miyazawa {see planet (5008)} and currently directs the museum at the Kenji Miyazawa Iihatobu Center. (M 31610)

Name proposed by the discoverers following a suggestion by A. Fujii and Y. Yoneda.

(6723) Chrisclark

1991 CL_3. Discovered 1991 February 14 by E. F. Helin at Palomar.

Named in honor of Christopher C. Clark, an electro-optical engineer at the Jet Propulsion Laboratory who designed and assembled the Near-Earth Asteroid Tracking camera system. This NEAT camera contains a large-format CCD sensor that provides digital image data, allowing for computer-automated data analysis and Near-Earth Object discoveries. He used his expertise in the development, integration and testing of focal plane arrays to make NEAT a success. (M 26932)

(6724) 1991 CX_5

Discovered 1991 February 4 by S. Ueda and H. Kaneda at Kushiro.

(6725) 1991 DS

Discovered 1991 February 21 by S. Inoda and T. Urata at Karasuyama.

(6726) 1991 PS

Discovered 1991 August 5 by H. E. Holt at Palomar.

(6727) 1991 TF_4

Discovered 1991 October 10 by K. J. Lawrence at Palomar.

(6728) 1991 UM

Discovered 1991 October 18 by S. Ueda and H. Kaneda at Kushiro.

(6729) 1991 VV_2

Discovered 1991 November 4 by S. Otomo at Kiyosato.

(6730) Ikeda

1992 BH. Discovered 1992 January 24 by T. Urata at Oohira.

Named in honor of Tetsuro Ikeda (1894-1981), the third director of the International Latitude Observatory of Mizusawa {see planet (7530)}. During his tenure (1943-1963) he established a measurement system for understanding meteorological effects on observations of the variation of latitude. (M 34624)

Name proposed by the discoverer following a suggestion by K. Hurukawa. Citation prepared by K. Yokoyama.

(6731) Hiei

1992 BK. Discovered 1992 January 24 by Y. Kushida and O. Muramatsu at Yatsugatake.

Named in honor of Eijiro Hiei (1931-), professor at Meisei University and professor emeritus of the National Astronomical Observatory of Japan. A solar physicist best known for his research on white-light flares, Hiei was the fourth director (1982-1992) of the Norikura Solar Observatory, where he conducted coronagraphic studies. (M 28090)

Name proposed by the discoverers following a suggestion by T. Sakurai.

(6732) 1992 CG_1

Discovered 1992 February 8 by S. Ueda and H. Kaneda at Kushiro.

(6733) 1992 EF

Discovered 1992 March 2 by S. Ueda and H. Kaneda at Kushiro.

(6734) 1992 FB

Discovered 1992 March 23 by S. Ueda and H. Kaneda at Kushiro.

(6735) 1992 WM_3

Discovered 1992 November 23 by T. Urata at Oohira.

(6736) 1993 EF

Discovered 1993 March 1 by T. Urata at Oohira.

(6737) Okabayashi

1993 ER. Discovered 1993 March 15 by K. Endate and K. Watanabe at Kitami.

Named in memory of Shigeki Okabayashi (1913-1944), a self-taught Japanese astronomer. The first in Japan to discover more than one kind of new celestial object, he discovered Nova V630 Sgr in 1936 and comet C/1940 S1 (Okabayashi-Honda), and also independently discovered C/1939 H1 and C/1939 V1. He was employed by Kurashiki Observatory from 1939 to 1941. Tragically, he was killed on his way home from a geological survey in Sumatra when his hospital ship was torpedoed. (M 31610)

Name proposed by the discoverers following a suggestion by T. Sato and A. Fujii.

(6738) Tanabe

1993 FD_1. Discovered 1993 March 20 by K. Endate and K. Watanabe at Kitami.

Named in honor of Hiroyoshi Tanabe (1928-), astronomer at the National Astronomical Observatory (formerly known as the Tokyo Astronomical Observatory) during 1953-1989. Tanabe's research fields is the light of the night sky. He made extensive ground-based observations of the gegenschein, the first rocket observation of the inner zodiacal light and outer corona, and he estimated the background light by making star counts in various regions of the sky. He made regular airglow observations for 33 years, Tanabe served as president of IAU Commission 21 during 1979-1982 and was an officer of the Astronomical Society of Japan for many years. (M 32094)

(6739) 1993 FU_{38}

Discovered 1993 March 19 by the Uppsala-ESO Survey at La Silla.

(6740) Goff

1993 GY. Discovered 1993 April 14 by C. S. Shoemaker and E. M. Shoemaker at Palomar.

Named in honor of Robert and Valerie Goff, of Tucson, Arizona. One of the finest masters of optics in the world, Bob Goff makes mirrors that are used in research and educational institutions worldwide. The Goffs are active in trying to increase a public interest in the sky, both through observing sessions and through the manufacture of good optics. (M 30100)

Name proposed by the discoverers following a suggestion by D. H. Levy, who prepared the citation.

(6741) Liyuan

1994 FX. Discovered 1994 March 31 by K. Endate and K. Watanabe at Kitami.

Named in honor of Li Yuan (1925-), popularizer of astronomy in the People's Republic of China. He played an important role in the opening of Beijing Planetarium in 1957 and served as a leader of planetarium activities in China. He has published more than 50 books on science, including astronomy, and has frequently contributed to domestic and foreign publications. (M 31611)

Name proposed by the discoverers following a suggestion by A. Fujii, H. Tomioka and Y. Shiono.

(6742) Biandepei

1994 GR. Discovered 1994 April 8 by K. Endate and K. Watanabe at Kitami.

Named in honor of Bian Depei (1926-), popularizer of science and astronomy in the People's Republic of China. He has published more than 60 books, many winning national prizes, and nearly 800 articles. He played an important role in establishing the Beijing Planetarium in 1954 and launching *Amateur Astronomer* magazine in 1958, both the first of their kind in China. (M 31611)

Name proposed by the discoverers following a suggestion by A. Fujii and T. Sato.

(6743) Liu

1994 GS. Discovered 1994 April 8 by K. Endate and K. Watanabe at Kitami.

Named in honor of Joseph H. C. Liu (1931-), who played a leading role in establishing the Hong Kong Space Museum and who became its first chief curator. In 1982 he received the Chiro Astronomical Award in Japan, and in 1984 he became a member of the Most Excellent Order of the British Empire, both for his promotion of popular astronomy in Hong Kong. Well known as an astrophotographer, he now lives in California. (M 31611)

Name proposed by the discoverers following a suggestion by A. Fujii and T. Sato

(6744) Komoda

1994 JL. Discovered 1994 May 6 by K. Endate and K. Watanabe at Kitami.

Named in memory of Kazuyoshi Komoda (1915-1967), amateur astronomer in Japan. Though handicapped by poliomyelitis, he enthusiastically observed Jupiter from 1944 to 1961. (M 31611)

Name proposed by the discoverers following a suggestion by T. Sato, I. Hirabayashi and A. Fujii.

(6745) Nishiyama

1994 JD_1. Discovered 1994 May 7 by K. Endate and K. Watanabe at Kitami.

Named in honor of Minewo Nishiyama (1925-), amateur astronomer and president of the Chikushi Astronomical Association {see also planet (6237)} from 1944 to 1947. Beginning in 1944 and under the direction of H. Hirose {see planet (1612)}, Nishiyama was the first amateur in Japan to make orbital calculations on minor planets. More recently he has studied the astronomical histories of Asian countries, particularly the old Burmese constellations. (M 32094)

Name proposed by the discoverers following a suggestion from H. Oishi.

(6746) Zagar

1994 NP. Discovered 1994 July 9 at the Osservatorio San Vittore at Bologna.

Named in memory of Francesco Zagar (1900-1976), professor of astronomy at the University of Milan and director of the Brera Astronomical Observatory, president of IAU Commission 31 (1967-1970) and of the Italian Astronomical Society. A lover of theoretical astronomy, Zagar studied the problem of three bodies, the orbit of Pluto, stellar statistics, stellar dynamics and cosmogony. His book *Astronomia Sferica e Teorica* is still widely studied and was recently reprinted. (M 27331)

(6747) 1995 UT_3

Discovered 1995 October 20 by T. Kobayashi at Oizumi.

(6748) Bratton

1995 UV_{30}. Discovered 1995 October 20 by the Spacewatch at Kitt Peak.

Named in honor of Durley H. Bratton (1923-), who has been a mentor and inspiration for amateur astronomers and telescope makers in the Memphis, Tennessee, area since the late 1960s. An electronics technician by profession, he became a highly skilled optical craftsman, constructing or advising in the construction of many telescopes. His designs often involved remarkable resourcefulness and ingenuity, such as using borosilicate glass cut from old color television picture-tube faceplates as f/2 Schmidt camera primary mirror blanks. (M 33787) Name suggested and citation by R. Tucker.

(6749) Ireentje

7068 P-L. Discovered 1960 October 17 by C. J. van Houten and I. van Houten-Groeneveld at Palomar.

Named in honor of the discoverers' second granddaughter, Irene Cornelia Francisca van Houten, daughter of Karel and Thea van Houten. (M 26933)

(6750) Katgert

1078 T-1. Discovered 1971 March 24 by C. J. van Houten and I. van Houten-Groeneveld at Palomar.

Named in honor of Peter Katgert (1944-), astronomer at the Leiden Observatory, and his wife, Leiden astronomer Jet Katgert-Merkelijn (1943-). Peter's main research interests are the statistics of radio sources and also the formation and dynamics of clusters of galaxies. Jet is now working on the archive of J. H. Oort {see planet (1691)}. (M 26933)

(6751) van Genderen

1114 T-1. Discovered 1971 March 25 by C. J. van Houten and I. van Houten-Groeneveld at Palomar.

Named in honor of Arnout van Genderen (1936-), astronomer at the Leiden Observatory, whose main interest is photometry of η Carinae, eclipsing variable stars, VBLUW photometry of OB stars and the photometric variability of hypergiants. (M 26933)

(6752) Ashley

4150 T-1. Discovered 1971 March 26 by C. J. van Houten and I. van Houten-Groeneveld at Palomar.

Named in honor of Ashley Thomas McDermott, professor of astronomy at the College of the Desert, Palm Desert, California, for the past 34 years. During this time he has taught astronomy to well over 10,000 students with his own special wisdom, love, enthusiasm and humor for the subject. He has also contributed many years of valuable service to the Riverside Telescope Makers' Conference, the Astronomical Society of the Pacific, the Western Amateur Astronomers and the History of Astronomy Society. He founded the Astronomical Society of the Desert and the Andromeda Astronomical Society. (M 27331)

Name proposed by T. Gehrels, following a suggestion by J. R. Crisman, who wrote the citation.

(6753) Fursenko

1974 RV_1. Discovered 1974 September 14 by N. S. Chernykh at Nauchnyj.

Named in honor of Margarita Aleksandrovna Fursenko (1931-), staff member of the Institute of Theoretical Astronomy during 1955-1997 and a well-known expert on ephemeris astronomy. She made major contribution to the calculation and publication of 40 volumes of the annual *Astronomicheskij Ezhegodnik*, as well as to the development of special ephemerides for space missions, lunar laser ranging and radioastronomical observations with RATAN-600. (M 31296)
Name suggested by ITA and supported by the discoverer.

(6754) Burdenko

1976 UD_4. Discovered 1976 October 28 by L. V. Zhuravleva at Nauchnyj.

Named in memory of Nikolaj Nilovich Burdenko (1876-1946), one of the founders of neurosurgery in the U.S.S.R. and from 1944 the first president of the Academy of Medical Sciences. (M 34624)

(6755) 1976 YE_1

Discovered 1976 December 16 by L. I. Chernykh at Nauchnyj.

(6756) 1978 VX_3

Discovered 1978 November 7 by E. F. Helin and S. J. Bus at Palomar.

(6757) 1979 SE_{15}

Discovered 1979 September 20 by S. J. Bus at Palomar.

(6758) Jesseowens

1980 GL. Discovered 1980 April 13 by A. Mrkos at Kleť.

Named in memory of the American athlete James Cleveland (Jesse) Owens (1913-1980), one of the finest track-and-field competitors in history. In the space of 45 minutes on 1935 May 26, Owens equalled the world record for the 100-yard dash and set new world records for the 220-yard dash, the 220-yard low hurdles and the long jump. At the 1936 Olympics he won four gold medals, equalling or setting new records in every event. (M 27735)

Name suggested by G. V. Williams, who made the identifications involving this object.

(6759) 1980 KD

Discovered 1980 May 21 by H. Debehogne at La Silla.

(6760) 1980 KM

Discovered 1980 May 22 by H. Debehogne at La Silla.

(6761) 1981 EV_{19}

Discovered 1981 March 2 by S. J. Bus at Siding Spring.

(6762) 1981 EC_{25}

Discovered 1981 March 2 by S. J. Bus at Siding Spring.

(6763) Kochiny

1981 RA_2. Discovered 1981 September 7 by L. G. Karachkina at Nauchnyj.

Named for Nikolaj Evgrafovich Kochin (1901-1944) and his wife Pelageya Yakovlevna Kochina (1899-), both academicians and professors of mathematics. One of the founders of modern dynamical meteorology, N. E. Kochin made great contributions to various branches of aerodynamics, hydrodynamics and theoretical meteorology. The scientific interests of P. Y. Kochina have been in subterranean

hydrodynamics; she is also the author of biographies of Kochin, Kovalevskaya {see planet (1859)}, Weierstrass and Mittag-Leffler. (M 27331; M 30309)
Name proposed by A. A. Kapitsa.

(6764) 1981 TM_3

Discovered 1981 October 7 by L. I. Chernykh at Nauchnyj.

(6765) Fibonacci

1982 BQ_2. Discovered 1982 January 20 by L. Brožek at Kleť.

Named in memory of Leonardo Fibonacci of Pisa (c. 1170-1240), the first "western" mathematician, whose *Liber abaci* (1202) contains the earliest mention of the famous sequence of numbers that bears his name. The number of this minor planet is the twentieth in that sequence, the first five members of which are 1, 1, 2, 3 and 5, and every fifth member of which is divisible by five. (M 26767)
Name proposed following a suggestion by B. G. Marsden.

(6766) Kharms

1982 UC_6. Discovered 1982 October 20 by L. G. Karachkina at Nauchnyj.

Named in honor of the outstanding Russian poet, writer and dramatist Daniil Ivanovich Yuvachev (1905-1942), whose pen-name is Kharms. He was one of the founders of the literary school known in twentieth-century European literature as "absurd". (M 27129)
Name suggested by L. R. Nemirovskij.

(6767) Shirvindt

1983 AA_3. Discovered 1983 January 6 by L. G. Karachkina at Nauchnyj.

Named in honor of the remarkable actor, People's Artist of Russia and professor at the Shchukin Theatre Institute in Moscow, Aleksandr Anatol'evich Shirvindt (1934-). He is one of the founders of the International Festival of Satire and Humor "Golden Ostap" and the first president of the "Academy of Authorities" (Academy of Humor). (M 27129)
Name suggested and citation prepared by L. R. Nemirovskij.

(6768) 1983 RY

Discovered 1983 September 7 by A. Mrkos at Kleť.

(6769) 1985 CJ_1

Discovered 1985 February 15 by A. Mrkos at Kleť.

(6770) Fugate

1985 QR. Discovered 1985 August 22 by E. Bowell at Anderson Mesa.

Named in honor of Robert Q. Fugate (1943-), of the U.S. Air Force's Starfire Optical Range (SOR), near Albuquerque. Under Fugate's leadership, the SOR has developed and operates the most advanced adaptive optics system worldwide. Fugate performed the first experiment to demonstrate the feasibility of using laser beacons for adaptive optics. He has chaired many committees and received numerous awards, including the 1997 Distinguished Civilian Award, the highest given to a civilian by the U.S. Department of Defense. (M 31025)

Named by the discoverer following a suggestion by J. D. Drummong, who prepared the citation.

(6771) Foerster

1986 EZ_4. Discovered 1986 March 9 by C.-I. Lagerkvist at Siding Spring.

Named in honor of Wilhelm Julius Foerster (1832-1921), director of the Berlin Observatory for nearly 40 years. He made measurements of star positions and in 1860 was a codiscoverer of (62) Erato. He was a strong campaigner for the organized rapid and reliable telegraphic dissemination of urgent astronomical news

and arranged for the establishment of the Astronomische Zentralstelle in Kiel. In 1889 he founded the Berlin "Urania", the first large public observatory, equipped with a 0.30-m refractor. This minor planet is being named on the occasion of the 50th anniversary of the "Wilhelm-Foerster-Sternwarte e.V.". (M 28090)
Citation prepared in part by W. Tost.

(6772) 1988 BG_4
Discovered 1988 January 20 by H. Debehogne at La Silla.

(6773) 1988 LK
Discovered 1988 June 15 by E. F. Helin at Palomar.

(6774) 1988 VH_5
Discovered 1988 November 4 by A. Mrkos at Kleť.

(6775) Giorgini
1989 GJ. Discovered 1989 April 5 by E. F. Helin at Palomar.

Named in honor of Jon D. Giorgini, who has contributed significantly to radar astronomy of minor planets through his development of an On-Site Orbit Determination software system. This allows radar astrometric measurements to be used immediately to improve a target's orbit, and hence yields ephemerides for pointing, time delay and doppler frequency. Since the system became operational at Goldstone {see planet (4433)} in 1993, there has been an order-of-magnitude improvement in the efficiency with which radar observations can progress. In particular, Giorgini's software was key to the success of the radar imaging of (1620) Geographos in Aug. 1994 and (6489) Golevka in June 1995. (M 26933)
Citation prepared by D. K. Yeomans and S. J. Ostro.

(6776) Dix
1989 GF_8. Discovered 1989 April 6 by F. Börngen at Tautenburg.

Named for the painter and graphic artist Otto Dix (1891-1969), born in the Thuringian town of Gera-Untermhaus. He was a representative of the new functionalism and is ranked with the great critical realists of the twentieth century. His pictures of the horrors of war are shocking and deeply moving. Dix was a member of the Prussian academy of arts and a master of the academy of arts in Dresden until he was dismissed by the Nazis in 1933. He was then outlawed as a "degenerate" artist, and 260 of his works were confiscated. (M 27129)

(6777) Balakirev
1989 SV_1. Discovered 1989 September 26 by E. W. Elst at La Silla.

Named in memory of the famous Russian composer and pianist Milij Alekseevich Balakirev (1837-1910). With Borodin {see planet (6780)}, Kyui, Musorgskij and Rimskij-Korsakov {see planets (1059) and (4534)}, he formed a group known as "The Innovators". Although Balakirev started to study mathematics, he changed his field to music under the influence of the diplomat A. D. Ulybyshev. He was the ruling spirit of the group and its - sometimes dictatorial - teacher of composition. Balakirev composed two symphonies and many other pieces for orchestra and piano. His fantasy for piano *Islamei* has become world-famous, and although he was himself a brilliant pianist, he admitted - with complete modesty - that there were certain passages in it that he couldn't manage. (M 26767)

(6778) Tosamakoto
1989 TX_{10}. Discovered 1989 October 4 by A. Takahashi and K. Watanabe at Kitami.

Named in honor of Makoto Tosa (1944-), a professor at Tohoku University, known for his research on galaxies. He has served as vice president of Astronomical Society of Japan and as a member of the advisory councils of both the

National Astronomical Observatory and the Sendai Astronomical Observatory. He encourages and assists the activities of amateur astronomers in the Sendai Astronomical Society and the volunteer group "Utyu-sen" (Space Ship), which popularizes astronomy for children and laymen. (M 34624)

(6779) Perrine

1990 DM_1. Discovered 1990 February 20 by Z. Vávrová at Kleť.

Named in memory of Charles Dillon Perrine (1867-1951), American astronomer working at Lick Observatory from 1893, an enthusiastic observer, who discovered the satellites Jupiter VI (Himalia) and VII (Elara) and nine comets during 1895-1902. He was appointed director of the Córdoba Observatory in Argentina in 1909 and retired in 1936. The discoverer of this minor planet rediscovered Perrine's periodic comet, lost for six returns, visually at Lomnický Štít {see planet (3168)} in 1955. After 1968 the comet was again lost. (M 34624)

Name suggested by J. Tichá, who searched unsuccessfully for this comet, now known as 18D/Perrine-Mrkos, at Kleť in 1995.

(6780) Borodin

1990 ES_3. Discovered 1990 March 2 by E. W. Elst at La Silla.

Named in memory of the famous Russian composer Alexandr Porfirevich Borodin (1833-1887). After studies in medicine and chemistry and a brief stay in the army as a doctor, he became a professor at the Medical Academy in St. Petersburg. But he was a musician at heart. He became acquainted with Balakirev {see planet (6777)}, to whom he owed his further musical education. Among his principal works are two symphonies, his well-known *In the Steppes of Central Asia* and several pieces for piano. His unfinished opera *Prince Igor* was completed by Glazunov and Rimskij-Korsakov {see planet (4534)}. (M 26767)

(6781) 1990 OD

Discovered 1990 July 19 by E. F. Helin at Palomar.

(6782) 1990 SU_{10}

Discovered 1990 September 16 by H. E. Holt at Palomar.

(6783) 1990 SO_{28}

Discovered 1990 September 24 by L. I. Chernykh at Nauchnyj.

(6784) 1990 UN_{13}

Discovered 1990 October 28 by L. I. Chernykh at Nauchnyj.

(6785) 1990 VA_7

Discovered 1990 November 12 by H. Shiozawa and M. Kizawa at Fujieda.

(6786) 1991 DT

Discovered 1991 February 21 by S. Inoda and T. Urata at Karasuyama.

(6787) 1991 PF_{15}

Discovered 1991 August 7 by H. E. Holt at Palomar.

(6788) 1991 PH_{15}

Discovered 1991 August 7 by H. E. Holt at Palomar.

(6789) 1991 RM_6

Discovered 1991 September 4 by E. F. Helin at Palomar.

(6790) 1991 SF_1

Discovered 1991 September 28 by S. Otomo at Kiyosato.

(6791) 1991 UC_2

Discovered 1991 October 29 by S. Ueda and H. Kaneda at Kushiro.

(6792) Akiyamatakashi

1991 WC. Discovered 1991 November 30 by M. Akiyama and T. Furuta at Mishima.

Named in honor of Takashi Akiyama (1923-), a leader in youth education for many years in Kanaya, Shizuoka. Long interested in science, he taught the young people of this small town to appreciate the wonders of animals, plants, and especially the cosmos. He introduced the first discoverer to astronomy and taught him a great deal. (M 31611)

(6793) Palazzolo

1991 YE. Discovered 1991 December 30 by U. Quadri and L. Strabla at Bassano Bresciano.

Named for the village of Palazzolo sull'Oglio, situated between Brescia and Bergamo and crossed by the river Oglio. The village is noted for its industries, including the first Italian factories for making cement and buttons. With a current population of 16,000, Palazzolo is famous all over the world for its production of spinning machines and zippers. Discoveries of Roman relics attest to its very ancient origin. (M 26933)

(6794) 1992 DK

Discovered 1992 February 26 by A. Sugie at Taga.

(6795) 1993 FZ_{12}

Discovered 1993 March 17 by the Uppsala-ESO Survey at La Silla.

(6796) 1993 FH_{24}

Discovered 1993 March 21 by the Uppsala-ESO Survey at La Silla.

(6797) 1993 FG_{25}

Discovered 1993 March 21 by the Uppsala-ESO Survey at La Silla.

(6798) Couperin

1993 JK_1. Discovered 1993 May 14 by E. W. Elst at La Silla.

Named for the French family that produced a succession of musicians from the early seventeenth to the mid-nineteenth century. Louis Couperin (1626-1661) was born in Chaumes and was the first of the many Couperins to become organist at the church of St. Gervais in Paris. His style was characterized by an almost aggressive use of dissonance and of baroque ornamentation. His nephew, François Couperin (1668-1733), often known as "Le Grand", is especially remembered for his keyboard pieces and for his somewhat ambiguous and obscure, theoretical work *L'art de toucher le clavecin*. François Couperin's chamber and church music is also very important, well-known examples being his two organ masses and the four volumes for the harpsichord. He was held in great esteem by J. S. Bach {see planet (1814)}. (M 26767)

(6799) 1993 KM

Discovered 1993 May 17 by E. F. Helin at Palomar.

(6800) Saragamine

1994 UC. Discovered 1994 October 29 by A. Nakamura at Kuma.

Named for a mountain (height 1271 m) located in the north of Kuma {see planet (6255)}, the town where this minor planet was discovered. Situated in a park, Mt. Saragamine is popular with campers and hikers. (M 26933)

(6801) Střekov

1995 UM_1. Discovered 1995 October 22 by Z. Moravec at Kleť.

Named for a castle situated near the town Ústínad Labem in northern Bohemia. The castle, founded in 1318, towers above the river Labe on a steep rock 100-m high and dominates the town. Thanks to its advantageous position, the castle was never conquered. Střekov is also the name of a district in the town, the discoverer's birthplace. (M 26933; M 34089)

(6802) Černovice

1995 UQ_2. Discovered 1995 October 24 by M. Tichý at Kleť.

Named for the small town where the discoverer lived in his youth. Founded in the fourteenth century, Černovice is situated amid the pleasant landscape of the Czech-Moravian Highlands in southern Bohemia. (M 27736)
Name dedicated to the parents of the discoverer.

(6803) 1995 UK_7

Discovered 1995 October 27 by S. Ueda and H. Kaneda at Kushiro.

(6804) Maruseppu

1995 WV. Discovered 1995 November 16 by A. Nakamura at Kuma.

Named for a small town (population 2400) in eastern Hokkaido. Since 1994, Maruseppu has been promoting inter-town friendship with Kuma {see planet (6255)}, where this minor planet was discovered. (M 26933)

(6805) Abstracta

4600 P-L. Discovered 1960 September 24 by C. J. van Houten and I. van Houten-Groeneveld at Palomar.

Named for the astronomical bibliography *Astronomy and Astrophysics Abstracts*. Founded in 1969, the AAA present a comprehensive documentation of all aspects of astronomy, astrophysics and related fields. The AAA are prepared under the auspices of the IAU by a special department of the Astronomisches Rechen-Institut, Heidelberg. The department, headed by Lutz D. Schmadel {see planet (2234)} for the last 20 years, has recorded, abstracted and indexed more than 500,000 documents. AAA is the direct successor of the *Astronomischer Jahresbericht*, which was founded in 1900. The AAA already amount to more than 60 volumes, occupying more than 3 meters of shelf space. (M 27331)

(6806) Kaufmann

6048 P-L. Discovered 1960 September 24 by C. J. van Houten and I. van Houten-Groeneveld at Palomar.

Named in honor of Horst W. Kaufmann (1929-), well-known German astronomical optician. After spending several years of optical apprenticeship with famous firms such as Fecker and Zeiss {see planet (851)}, Kaufmann founded his own company. He manufactured numerous telescope optics for observatories and institutes as well as for amateur astronomers. He produced several medium-sized systems for the Bonn and Heidelberg observatories and tackled very difficult light-weight infrared secondary mirrors for the 2.2-m telescope of the Max-Planck-Institut for Astronomy and for UKIRT. (M 27331)
Name suggested and citation prepared by L. D. Schmadel.

(6807) Brünnow

6568 P-L. Discovered 1960 September 24 by C. J. van Houten and I. van Houten-Groeneveld at Palomar.

Named in memory of Franz Friedrich Ernst Brünnow (1821-1891), who started his scientific career in Berlin. He continued as the director of the observatory at Bilk, near Düsseldorf, but in 1851 returned to Berlin to become the first assistant at the observatory. He worked on comets and minor planets and published a textbook on spherical astronomy. He later moved to the United States to take up the post of director of the new observatory at Ann Arbor. In 1865 he was appointed Royal Astronomer for Ireland. In that capacity he published in two parts his astronomical observations and researches made at Dunsink. (M 27736)

Name suggested by L. D. Schmadel and J. Schubart, citation prepared by J. Schubart.

(6808) Plantin

1932 CP. Discovered 1932 February 5 by K. Reinmuth at Heidelberg.

Named in memory of Christophe Plantin (c.1520-1589), printer and founder of an important printing house in Antwerp. As a youth he learned the profession of bookbinder. In 1549 he settled in Antwerp and printed his first book there in 1555. Throughout the ensuing years he established himself as a printer of great fame. His "Officina Plantiniana" published the works of contemporary scientists (among others Dodoens, Mercator {see planet (4798)}, Stevin and Vesalius {see planets (2831) and (2642), respectively}) and also many religious works. At the height of his career he had sixteen presses running, producing publications of exceptional quality. His greatest venture was the *Biblia Regia*. The "Officina" was sold in 1876 to the city of Antwerp and is now one of its most important museum pieces. (M 27736)

Name proposed and citation written by E. Goffin.

(6809) Sakuma

1938 DM_1. Discovered 1938 February 20 by K. Reinmuth at Heidelberg.

Named in honor of Seiichi Sakuma (1929-   ), a Japanese amateur astronomer and an eminent observer of variable stars since 1942. Sakuma played the principal role in founding the Variable Star Observers League in Japan, publishing its *Variable Star Bulletin* and constructing its database of more than one million Japanese observations of some 4000 variable stars. This database contains observations back to 1906 and includes more than 30,000 magnitude estimates by Sakuma. (M 28090)

Name proposed by H. Oishi, who made the principal identifications involving this minor planet.

(6810) 1969 GC

Discovered 1969 April 9 by C. U. Cesco at El Leoncito.

(6811) Kashcheev

1976 QP. Discovered 1976 August 26 by N. S. Chernykh at Nauchnyj.

Named in honor of Boris Leonidovich Kashcheev (1920-   ), professor of radioelectronics at Kharkov Technical University, researcher in the physics and dynamics of meteors. In 1956 he founded the laboratory for the exploration of meteors by radio techniques. He was the chief designer of MARS, the first automatic meteor radio system in Ukraine. More than 20 million meteors were detected with this system during 1972-1978, and many thousands of meteor orbits were determined and studied. He was also chair of the Meteor Commission of the Soviet Geophysical Committee. (M 32346)

(6812) 1978 VJ_8

Discovered 1978 November 7 by E. F. Helin and S. J. Bus at Palomar.

(6813) 1978 VV_9

Discovered 1978 November 7 by E. F. Helin and S. J. Bus at Palomar.

(6814) 1979 MC_2

Discovered 1979 June 25 by E. F. Helin and S. J. Bus at Siding Spring.

(6815) 1979 MM_5

Discovered 1979 June 25 by E. F. Helin and S. J. Bus at Siding Spring.

(6816) 1981 EB_{28}

Discovered 1981 March 2 by S. J. Bus at Siding Spring.

(6817) Pest

1982 BP_2. Discovered 1982 January 20 by A. Mrkos at Kleť.

Named for the largest of the three towns, Buda {see planet (908)}, Pest and Óbuda, that form the modern capital of Hungary. The 125th anniversary of the establishment of Budapest as Hungary's political, cultural and scientific center is being celebrated this year. The University Observatory played a major part in the life of both Pest and Buda, with its buildings dominating the panorama for three-quarters of a century after it was founded in 1780. Since 1921 the headquarters of the Konkoly {see also planet (1445)} Observatory has also been located in the city, which provided support for the construction of its large dome. (M 33386)

(6818) Sessyu

1983 EM_1. Discovered 1983 March 11 at the Kiso Observatory at Kiso.

Born at Soja-Akahama in Okayama prefecture, Sessyu (1420-1508) is called the "saint of painting". According to tradition, he drew a mouse with his own tears on a corridor when he was a novitiate at Houhukuji-Temple in Soja City. Afterwards he went to Ming (now People's Republic of China) and was initiated into the mysteries of ink painting-monochrome painting characterized by the use of sumi (black ink). In Japan he then laid the foundation for Japanese ink painting. (M 33787)

(6819) 1983 LL

Discovered 1983 June 14 by S. Smrekar at Palomar.

(6820) Buil

1985 XS. Discovered 1985 December 13 by the CERGA at Caussols.

Named in honor of Christian Buil, French amateur astronomer and long-term CCD enthusiast. Buil has been one of the first amateur astronomers to build a CCD camera – and use it. An avid observer, as well as electronician and computer programer, he has co-discovered supernovae and with his colleagues has imaged more than 5000 celestial objects incorporated in the Buil-Thouvenot Atlas, the first digital atlas available to the amateur community. He has also spent much time helping others and has for many years been the president of the T60 association that allows amateurs to use a 0.60-m telescope at Pic du Midi. (M 27463)

Name endorsed by his many friends, among them Eric and Guylaine Thouvenot and Alain Klotz.

(6821) Ranevskaya

1986 SZ_1. Discovered 1986 September 29 by L. G. Karachkina at Nauchnyj.

Named in honor of the outstanding Russian actress Faïna Grigor'evna Ranevskaya (1896-1984). With her uncommon sense of humor, Ranevskaya payed dozens of splendid comedy and dramatic parts for the theater and cinema. (M 27331; M 30309)

Name proposed by L. R. Nemirovskij on the occasion of the centennial of her birth.

(6822) 1986 UO

Discovered 1986 October 28 by Z. Vávrová at Kleť.

(6823) 1988 ED_1

Discovered 1988 March 12 by M. Arai and H. Mori at Yorii.

(6824) 1988 RE_2

Discovered 1988 September 8 by A. Mrkos at Kleť.

(6825) 1988 TJ_2

Discovered 1988 October 4 by A. Mrkos at Kleť.

(6826) Lavoisier

1989 SD_1. Discovered 1989 September 26 by E. W. Elst at La Silla.

Named for Antoine-Laurent Lavoisier (1743-1794), father of modern chemistry. In a memoir presented to the Paris Academy in 1777 he explained combustion as the result of the combination of a burning substance with oxygen (that name being due to Lavoisier). In his 1789 *Traité élémentaire de Chimie* he gave a list of simple substances that could not be further decomposed by any known process, thus providing the concept of a chemical element. He was also associated with committees on hygiene, coinage and public education. His membership in the Ferme Générale caused the authorities to be suspicious of him during the French Revolution, and he was condemned to the guillotine. The day after the execution, Lagrange {see planet (1006)} lamented: "It required only a moment to sever that head, and perhaps a century will not be sufficient to produce another like it". (M 26933)

(6827) 1990 SN_4

Discovered 1990 September 27 by T. Urata at Oohira.

(6828) Elbsteel

1990 VC_1. Discovered 1990 November 12 by D. I. Steel at Siding Spring.

Named in honor of Elliot Lewis Barnaby Steel, second child of the discoverer and Helen M. Steel. Elliot was born in Adelaide on 1995 Feb. 28, and his first home was within sight of the Siding Spring Observatory. Building on this auspicious start in life, at the age of 15 months he is already taking an interest in the night sky and other aspects of the universe around him. (M 27332)

(6829) Charmawidor

1991 BM_1. Discovered 1991 January 18 by E. W. Elst at St. Michel.

Named in memory of the famous French composer and organist Charles-Marie Widor (1845-1937). At the age of 19, by recommendation of Aristide Cavaillé-Coll {see planet (5184)}, Widor went to Brussels to study organ and composition with N. J. Lemmens and F. J. Fétis. After returning from Belgium, and after having inaugurated many organs built by Cavaillé-Coll, he was finally appointed to St. Sulpice in Paris in 1869. There he had at his disposal the most magnificient organ ever built by Cavaillé-Coll. Widor, who remained there for more than sixty years, is especially remembered for his ten *Symphonies for organ*, a genre he created. (M 34624)

(6830) 1991 JB_1

Discovered 1991 May 5 by S. Otomo and O. Muramatsu at Kiyosato.

(6831) 1991 UM_1

Discovered 1991 October 28 by S. Ueda and H. Kaneda at Kushiro.

(6832) Kawabata

1992 FP. Discovered 1992 March 23 by K. Endate and K. Watanabe at Kitami.

Named in memory of Yasunari Kawabata (1899-1972), Japanese novelist and author of "Izu no Odoriko". He was awarded the 1968 Nobel Prize in literature. (M 32346)

(6833) 1993 FC_1

Discovered 1993 March 19 by S. Shirai and S. Hayakawa at Hidaka.

(6834) 1993 JH

Discovered 1993 May 11 by Y. Shimizu and T. Urata at Nachi-Katsuura.

(6835) Molfino

1994 HT_1. Discovered 1994 April 30 at the Santa Lucia Observatory at Stroncone.

Named in memory of Alberto Molfino (1906-1977), Italian wrestler. He took part in the 1936 Berlin Olympics and won the Italian title in the 62-kg category six times, the last time being at the age of 43. (M 30799)

(6836) Paranal

1994 PW_5. Discovered 1994 August 10 by E. W. Elst at La Silla.

Named for the 2600-m high mountain in northern Chile, approximately 130 km south of the city of Antofagasta, upon which the Very Large Telescope of the European Southern Observatory is being erected. Although La Silla {see planet (2187)} ranks among the best astronomical sites in the world, subsequent studies have shown that Paranal has a substantially lower frequency of clouds and very much lower humidity. The mean value of the seeing is less than 1 arcsecond. (M 27332)

(6837) Bressi

1994 XN_4. Discovered 1994 December 8 by the Spacewatch at Kitt Peak.

Named in honor of Terrence H. Bressi, a dedicated and multi-talented engineer at the University of Arizona's Lunar and Planetary Laboratory. He has applied his expertise and ingenuity to the construction and improvement of telescopes and equipment used by the Spacewatch search for minor planets and comets. (M 33787)

(6838) Okuda

1995 UD_9. Discovered 1995 October 30 by Y. Shimizu and T. Urata at Nachi-Katsuura.

Named in honor of Toyozo Okuda (1908-1983), the fourth director of International Latitude Observatory at Mizusawa {see planet 7530)} and a former director of the Geographical Survey Institute of Japan. During his 1963-1976 tenure at Mizusawa he established a modern system of geophysical measurements of gravity and earth tides for research on the earth's rotation. (M 34624)

Name proposed by the second discoverer following a suggestion by K. Hurukawa. Citation prepared by K. Yokoyama.

(6839) 1995 WB_2

Discovered 1995 November 18 by T. Kobayashi at Oizumi.

(6840) 1995 WW_5

Discovered 1995 November 18 by S. Ueda and H. Kaneda at Kushiro.

(6841) Gottfriedkirch

2034 P-L. Discovered 1960 September 24 by C. J. van Houten and I. van Houten-Groeneveld at Palomar.

Named in memory of Gottfried Kirch (1639-1710), who was called to Berlin in 1700 to become the first astronomer at the Berlin Academy. He was recognized as the first discoverer of the great comet of 1680, and he made the sole observation of comet 55P/Tempel-Tuttle in 1699. His wife Maria Margaretha and son Christfried were also known as astronomers. (M 30799)

Name suggested and citation prepared by J. Schubart.

(6842) Krosigk

3016 P-L. Discovered 1960 September 24 by C. J. van Houten and I. van Houten-Groeneveld at Palomar.

Named in memory of Bernhard Friedrich Baron von Krosigk (1656-1714), an enthusiastic German amateur astronomer who financed astronomical expeditions to Cape Town and to Archangelsk. He owned private observatories at Poplitz and in Berlin. The latter was well equipped and used for observations by members of the Kirch {see also planet (6841)} family. (M 30799)

Name suggested by A. von Alvensleben and J. Schubart, the citation being prepared by the latter.

(6843) Heremon

1975 TC_6. Discovered 1975 October 9 by J. D. Mulholland at Fort Davis.

Named for the seventh son of Milesius, Celtic ruler of Iberia. Heremon was the first of the discoverer's Gaelic ancestors to set foot on Ireland some 25 centuries ago. (M 27129)

(6844) 1975 VR_5

Discovered 1975 November 3 by T. M. Smirnova at Nauchnyj.

(6845) Mansurova

1976 JG_2. Discovered 1976 May 2 by N. S. Chernykh at Nauchnyj.

Named in memory of Kira Sergeevna Mansurova (1931-1990), an astronomer in Irkutsk known for her observations with the zenith telescope. She was director of the Astronomical Observatory of Irkutsk University for many years and lecturer of astronomy at the University and Pedagogical Institute in Irkutsk. An active popularizer of astronomy, she published a number of works on astrometry and methods of teaching astronomy. (M 32346)

(6846) Kansazan

1976 UG_{15}. Discovered 1976 October 22 by H. Kosai and K. Hurukawa at Kiso.

Named for the writer Kansazan (1748-1827), who published many books in Chinese verse during the latter part of the Edo era. To him are attributed many popular sayings on "astronomical phenomena". (M 33787)

(6847) 1977 RL

Discovered 1977 September 5 by H.-E. Schuster at La Silla.

(6848) 1978 VG_5

Discovered 1978 November 7 by E. F. Helin and S. J. Bus at Palomar.

(6849) 1979 MX_6

Discovered 1979 June 25 by E. F. Helin and S. J. Bus at Siding Spring.

(6850) 1981 QT_3

Discovered 1981 August 28 by H. Debehogne at La Silla.

(6851) 1981 RO_1

Discovered 1981 September 1 by H. Debehogne at La Silla.

(6852) 1985 CN_2

Discovered 1985 February 14 by H. Debehogne at La Silla.

(6853) 1986 CD_2

Discovered 1986 February 12 by H. Debehogne at La Silla.

(6854) 1987 UG

Discovered 1987 October 20 by K. W. Zeigler at Anderson Mesa.

(6855) Armellini

1989 BG. Discovered 1989 January 29 at the Osservatorio San Vittore at Bologna.

Named in memory of Giuseppe Armellini (1887-1958), professor of astronomy at the University of Rome and director of the astronomical observatory there. Principally interested in celestial mechanics and classical astronomy and astrophysics, he observed and measured positions of double stars and minor planets. He is also known for his textbooks *Trattato di Astronomia Siderale*, *I Fondamenti Scientifici dell'Astronomia* and *I Fondamenti Scientifici dell'Astrofisica*. (M 27332)

(6856) 1989 EM

Discovered 1989 March 5 by E. F. Helin at Palomar.

(6857) 1990 QQ

Discovered 1990 August 19 by E. F. Helin at Palomar.

(6858) 1990 ST$_{10}$

Discovered 1990 September 16 by H. E. Holt at Palomar.

(6859) Datemasamune

1991 CZ. Discovered 1991 February 13 by M. Koishikawa at Sendai.

Named in honor of Date Masamune (1567-1636), the Japanese feudal lord who built the Sendai castle in 1601 and established the foundation of Date Clan. He made efforts to increase international relationships and trade with foreign countries, sending a mission led by his retainer Hasekuro Tsunega to Mexico, Spain and Rome. Sendai {see planet (3133)} will celebrate the 400th anniversary of its founding in 2001. (M 33787)

(6860) Sims

1991 CS$_1$. Discovered 1991 February 11 by S. Otomo and O. Muramatsu at Kiyosato.

Named in honor of Alan Sims (1920-1995), a naval officer who moved to Dublin in 1953 and became chaiman of the Dublin Astronomy Association. On his retirement in 1983 he moved to Bath, where he took an active part in the William Herschel {see planet (2000)} Society, serving first as secretary and later as vice chairman. He was also editor of the society's bulletin. Historians and librarians around the world corresponded with him, as he answered queries about the Herschels. (M 27736)

Named by the discoverers following suggestions by F. Ring and S. Kimura.

(6861) 1991 FA$_3$

Discovered 1991 March 20 by H. Debehogne at La Silla.

(6862) Virgiliomarcon

1991 GL. Discovered 1991 April 11 at the Osservatorio San Vittore at Bologna.

Named in memory of Virgilio Marcon (1903-1976), painter and teacher of art at San Donà del Piave in Venice. Also famous for building optics for amateur and professional astronomers, during the last 30 years of his life he constructed hundreds of astronomical mirrors that are still in use and famous for their quality. In 1968 he improved the technique for making aluminum mirrors and successfully worked the 1.2-meter telescope at the Merate Observatory. This technique has been carried on by his son Gianfranco, who built mirrors up to 3 meters in diameter for infrared astronomy. (M 28621; M 28637)

(6863) 1991 PX$_8$

Discovered 1991 August 5 by H. E. Holt at Palomar.

(6864) Starkenburg

1991 RC$_4$. Discovered 1991 September 12 by F. Börngen and L. D. Schmadel at Tautenburg.

Named for a castle erected around 1064 near the town of Heppenheim. The name also applies to a region in the southern part of the German state of Hes-

sen {see planet (5846)}. As a private initiative, the Starkenburg Sternwarte was founded in 1970 on the Schlossberg just below the castle. This observatory is heavily involved in educating the public and providing astronomical knowledge to people far beyond the region. This minor planet was numbered as a result of astrometric CCD observations made at Starkenburg in Dec. 1995. (M 27332)
Name proposed by the first discoverer, endorsed by the second.

(6865) Dunkerley

1991 TE_2. Discovered 1991 October 2 by Y. Kushida and O. Muramatsu at Yatsugatake.

Named in honor of Charlotte Herschel Dunkerley, a descendant of William Herschel {see planet (2000)} who recently completed the Herschel family genealogy as a result of the family reunion that took place in Bath in 1988 to celebrate the 250th anniversary of William Herschel's birth. More than 80 descendants attended the occasion and amended the family records upon which the revised genealogy is based. The record of the Herschel family began around the year 1600 and is a valuable resource for researchers and historians in the field of astronomy. (M 29147)
Name proposed by the discoverers following a suggestion by S. Kimura.

(6866) 1992 CO

Discovered 1992 February 12 by S. Otomo at Kiyosato.

(6867) Kuwano

1992 FP_1. Discovered 1992 March 28 by K. Endate and K. Watanabe at Kitami.

Named in memory of Yoshiyuki Kuwano (1931-1998), librarian and later director of Hita Municipal Tanso Memorial Library, Oita Prefecture. As an amateur astronomer, he discovered five novae (IV Cep, V3888 Sgr, V3889 Sgr, V2104 Oph, V4021 Sgr) and the peculiar variable PU Vul. An organizer and leader of local amateur astronomers, he actively popularized astronomy to the public through his frequent lectures and star parties. (M 32346)

Name proposed by the discoverers following a suggestion by T. Funada, T. Sato and A. Fujii.

Obituaries published in Yamamoto Circ., No. 2291, p. 2 (1998); Heavens, Vol. 79, No. 3, p. 16-20 (1998).

(6868) 1992 HD

Discovered 1992 April 22 by Y. Kushida and O. Muramatsu at Yatsugatake.

(6869) Funada

1992 JP. Discovered 1992 May 2 by K. Endate and K. Watanabe at Kitami.

Named in honor of Takumi Funada (1932-), retired schoolmaster and science teacher who is currently director of Sekizaki Kaisei Kan (Sea and Star Museum) and honorary president of the Oita Astronomical Society. He has greatly contributed to amateur astronomy and has helped popularize astronomy to the public. Also well known as an astrophotographer, he has published several books on astronomy. (M 32346)

Name proposed by the discoverers following a suggestion by T. Sato and A. Fujii.

(6870) Pauldavies

1992 OG. Discovered 1992 July 28 by R. H. McNaught at Siding Spring.

Named for Paul C. W. Davies (1946-), a British mathematical physicist resident in Australia, with research interests in cosmology, quantum gravity, astrophysics and the origin of life, who also has an international reputation as an author and broadcaster. His books range from the scholarly *Quantum Fields in Curved Space* (with N. D. Birrell) to the popular *About Time*. He has a long-

standing interest in the relationship between science and theology, summarized in his well-known book *The Mind of God.* His contributions to this field were recognized in 1995 with the award of the Templeton Prize for Progress in Religion. Davies has been a vociferous supporter of the need for search programs for near-earth objects. (M 33787)
Name proposed and citation prepared by D. I. Steel and the discoverer.

(6871) Verlaine

1993 BE_8. Discovered 1993 January 23 by E. W. Elst at La Silla.

Named in memory of the well-known French poet-symbolist Paul Verlaine (1844-1896). He was born in the grey and rainy Ardennes, a landscape that would never be far from his sad life. In his early work he declared himself as doomed, a "poète maudit", torn between a mystic search for pureness and demonic sensuality. Alcohol ruined his health and mind, during a libertine and bohemian existence with his friend, the poet Arthur Rimbaud {see planet (4635)}. Verlaine will be remembered for his splendid verse: "Il pleure dans mon coeur/ comme il pleut sur la ville./ Quelle est cette langueur/ qui pénétre mon coeur?". (M 27129; M 27147)
Citation written by C. Leterme at the request of the discoverer.

(6872) 1993 CN_1

Discovered 1993 February 15 by S. Ueda and H. Kaneda at Kushiro.

(6873) Tasaka

1993 HT_1. Discovered 1993 April 21 by K. Endate and K. Watanabe at Kitami.

Named in honor of Ichiro Tasaka (1929-), Japanese farmer, Mars observer and astronomical optician. He has made hundreds of precision telescope mirrors currently being used by observatories and amateur astronomers. He operates a tourist house that has an observatory. Its main telescope is a 0.73-m equatorial reflector, both the optical and mechanical parts of which were made by Tasaka himself. (M 32346)

Name proposed by the discoverers following a suggestion by T. Sato and A. Fujii.

(6874) 1994 JO_1

Discovered 1994 May 9 by G. J. Garradd at Siding Spring.

(6875) 1994 NG_1

Discovered 1994 July 4 by E. F. Helin at Palomar.

(6876) Beppeforti

1994 RK_1. Discovered 1994 September 5 by A. Boattini and M. Tombelli at Cima Ekar.

Named in honor of Giuseppe (Beppe) Forti, astronomer at the Arcetri Observatory in Florence. His encouragement and advice continuously help the discoverers to refine their observational work. Forti was trained as a solar physicist and worked for six years with the Harvard-Smithsonian Radio Meteor Project from 1965. He has done some work on nongravitational forces acting on comets and now has interests in the dynamics of minor planets. (M 27129)

(6877) Giada

1994 TB_2. Discovered 1994 October 10 by V. S. Casulli at Colleverde di Guidonia. Named in honor of Giada Casulli (1978-), daughter of the discoverer. (M 29671)

(6878) Isamu

1994 TN_2. Discovered 1994 October 2 by K. Endate and K. Watanabe at Kitami.

Named in honor of Isamu Hirabayashi (1941-), amateur astronomer who founded the Japan Lunar and Planetary Observers Network when he was a high-

school student in 1959 and who has been its president ever since. This organization soon became one of the most active groups of lunar and planetary observers and theorists in Japan. He also served as director of the Jupiter-Saturn Section of the Oriental Astronomical Association from 1971 to 1989. (M 32346)
Name proposed by the discoverers following a suggestion by T. Sato and A. Fujii.

(6879) Hyogo

1994 TC_{15}. Discovered 1994 October 14 by K. Ito at Sengamine.

Named for the prefecture where this planet was discovered. Hyogo prefecture, which includes Kobe City and the island of Awaji {see planet (3380)}, was severely damaged by the Great Hanshin Earthquake on 1995 Jan. 17, when more than 6000 people were killed and numerous houses were destroyed. (M 27130)

(6880) 1994 TG_{15}

Discovered 1994 October 13 by S. Otomo at Kiyosato.

(6881) 1994 UP

Discovered 1994 October 31 by T. Kobayashi at Oizumi.

(6882) Sormano

1995 CC_1. Discovered 1995 February 5 by P. Sicoli and V. Giuliani at Sormano.

Named in honor of the mountain village of Sormano and its nearby observatory where this object was discovered. The facility, built in 1986 by amateur astronomers of the Gruppo Astrofili Brianza, is located in the pre-Alps of Lombardy, between the southern arms of Lake Como. Since its foundation, the observatory has been involved in astrometric work on minor planets and comets, in particular the follow-up of near-earth objects. (M 27130)

(6883) 1996 AF

Discovered 1996 January 10 by T. Kobayashi at Oizumi.

(6884) Takeshisato

9521 P-L. Discovered 1960 October 17 by C. J. van Houten and I. van Houten-Groeneveld at Palomar.

Named in honor of Takeshi (Ken) Sato, a devoted educator who is currently director of the planetarium at Hiroshima's Children's Museum. He is an amateur planetary observer who served as director of the Jupiter-Saturn Section of the Oriental Astronomical Association during 1960-1971. He initiated the ordinance to minimize light pollution in the town of Bisei, Okayama Prefecture. Enacted in 1989, this ordinance was the first of its kind in Japan. (M 29147)

(6885) Nitardy

9570 P-L. Discovered 1960 October 17 by C. J. van Houten and I. van Houten-Groeneveld at Palomar.

Named in honor of John H. Nitardy, who has worked in communications engineering for the Boeing Company since 1965 and is currently a consultant to them. Together with his wife Ilene, he saved the Spacewatch {see planet (4255)} program in an intricate financial situation in early 1996. They are also known for their hospitality for and endorsement of foreign exchange students. (M 29147)

(6886) Grote

1942 CG. Discovered 1942 February 11 by L. Oterma at Turku.

Named in honor of Grote Reber (1911-), pioneer radioastronomer. A radio engineer by profession, he became intrigued by Jansky's {see planet (1932)} 1932 detection of radio waves of extraterrestrial origin. From a series of homemade radio receivers and a 9.4-m parabolic dish antenna in his backyard in Wheaton, Illinois, Reber obtained the first detailed quantitative maps of the radio sky and published

these in a series of papers during the 1940s. His stunning results gradually opened the consciousness of others to this very powerful method of observing the universe. (M 68622)
Name suggested and citation prepared by G. C. L. Aikman.

(6887) 1951 WH

Discovered 1951 November 24 by M. Laugier at Nice.

(6888) 1971 BD_3

Discovered 1971 January 27 by C. Torres and J. Petit at Cerro El Roble.

(6889) 1971 RA

Discovered 1971 September 15 by C. Torres and J. Petit at Cerro El Roble.

(6890) 1975 RP

Discovered 1975 September 3 by L. I. Chernykh at Nauchnyj.

(6891) Triconia

1976 SA. Discovered 1976 September 23 at the Harvard College Observatory at Harvard.

Named in honor of the Tri-Con Barber Shop in Lexington, Massachusetts, and for Paul, Joe, Charlie and Art who work there. (M 30100)

Name obviously propoposed by Brian G. Marsden, who, in a note to the Small Bodies Names Committee stated:"No, I am not getting free haircuts..."

(6892) 1978 VG_8

Discovered 1978 November 7 by E. F. Helin and S. J. Bus at Palomar.

(6893) 1983 RS_3

Discovered 1983 September 2 by H. Debehogne at La Silla.

(6894) Macreid

1986 RE_2. Discovered 1986 September 5 by E. F. Helin at Palomar.

Named in honor of Macgregor S. Reid, a highly regarded manager at the Jet Propulsion Laboratory. For the past ten years he has been technical executice assistant to the director, responsible for planning and identifying issues of significance to the national space program and the laboratory's future. Reid is internationally recognized for his activities concerning international standard-setting. This object is being named to honor him on his retirement after a 30-year career at JPL. (M 32094)
Citation prepared by E. C. Stone.

(6895) 1987 DG_6

Discovered 1987 February 23 by H. Debehogne at La Silla.

(6896) 1987 RE_1

Discovered 1987 September 13 by H. Debehogne at La Silla.

(6897) 1987 VQ

Discovered 1987 November 15 by A. Mrkos at Kleť.

(6898) 1988 LE

Discovered 1988 June 8 by C. S. Shoemaker at Palomar.

(6899) 1988 RP_{10}

Discovered 1988 September 14 by S. J. Bus at Cerro Tololo.

(6900) 1988 XD_1

Discovered 1988 December 2 by M. Arai and H. Mori at Yorii.

(6901) Roybishop

1989 PA. Discovered 1989 August 2 by C. S. Shoemaker and E. M. Shoemaker at Palomar.

Named in honor of Roy L. Bishop, retired professor of physics at Acadia University, Nova Scotia. Since 1982 Bishop has edited the *Observer's Handbook* of the Royal Astronomical Society of Canada, a vital reference for professional and amateur astronomers. Bishop is also known for his unique photographs, especially one of a double rainbow over Isaac Newton's {see planet (8000)} birthplace. (M 30477)

Citation provided by Carolyn and Eugene Shoemaker and Wendee and David Levy.

(6902) 1989 US_3

Discovered 1989 October 26 by Y. Mizuno and T. Furuta at Kani.

(6903) 1989 XM

Discovered 1989 December 2 by Y. Oshima at Gekko.

(6904) 1990 QW_1

Discovered 1990 August 22 by H. E. Holt at Palomar.

(6905) Miyazaki

1990 TW. Discovered 1990 October 15 by K. Endate and K. Watanabe at Kitami.

Named in honor of Isao Miyazaki (1961-), amateur astronomer well known for his high-quality visual, photographic and CCD observations of the planets. He has been director of the Jupiter-Saturn Section of the Oriental Astronomical Association (OAA) anice 1989. He is a winner of the I. Yamamoto {see planet (2249)} Award (OAA), the W. H. Steavenson Award (British Astronomical Association) and the W. H. Haas {see planet (3853)} Award (Association of Lunar and Planetary Observers). (M 32346)

Name proposed by the discoverers following a suggestion by T. Sato and A. Fujii.

(6906) 1990 WC

Discovered 1990 November 19 by R. H. McNaught at Siding Spring.

(6907) 1990 WE

Discovered 1990 November 19 by R. H. McNaught at Siding Spring.

(6908) Kunimoto

1990 WB_3. Discovered 1990 November 24 by K. Endate and K. Watanabe at Kitami.

Named in honor of Yoshihiro Kunimoto (1958-), Japanese composer and synthesizer musician. He has written and played theme and background music for more than 60 planetarium shows at the Hiroshima Children's Museum. (M 32347)

Name proposed by the discoverers following a suggestion by T. Sato and K. Kato.

(6909) Levison

1991 BY_2. Discovered 1991 January 19 by C. S. Shoemaker and E. M. Shoemaker at Palomar.

Named in honor of Harold Levison (1959-) of the Boulder, Colorado, office of the Southwest Research Institute. Since 1988 Levison has, in collaboration with Martin Duncan {see planet (6115)}, virtually revolutionized our view of the dynamics of short-period comets. The work uses sophisticated numerical models of test particles perturbed by the planets, and it has revealed important details

about the Kuiper {see planet (1776)} Belt, Centaurs, Pluto-Charon and short-period comets. Levison has also contributed revealing insights into the dynamics of perturbers in the β Pictoris system and was a leading member of the team that used Hubble Space Telescope to discover possible observational evidence for small comets in the transneptunian region. (M 27463)
Citation prepared by S. A. Stern.

(6910) 1991 FJ

Discovered 1991 March 17 by S. Otomo and O. Muramatsu at Kiyosato.

(6911) Nancygreen

1991 GN. Discovered 1991 April 10 by E. F. Helin at Palomar.

Named in honor of Nancy Green Hicks, enthusiastic horsewoman and talented artist. A graduate of Scripps College, she is with her family a principal benefactor and longterm volunteer at the Scripps Home for the elderly. Her personal contributions have made the difference for the continuation of many of the discoverer's astronomical pursuits. Her loyalty and encouragement have given support during times of diminishing governmental funds, allowing the work on observing NEOs to survive. She was most generous in her leadership role of fundraising for the planet-crossing asteroid program carried out photographically at Palomar for many years and has most recently enabled a successful transition to the CCD NEAT program. (M 27332)

(6912) Grimm

1991 GQ_2. Discovered 1991 April 8 by E. W. Elst at La Silla.

Named in memory of Friedrich Melchior Baron von Grimm (1723-1807). After studying at the University of Leipzig he went to France, as delegate of the young Prince of Saxe-Gotha. There he became associated with writers of the *Encyclopédie* such as Rousseau, Voltaire and d'Alembert {see, respectively, planets (2950), (5676) and (5956)}, but especially with Diderot {see planet (5351)}, for whom he felt great affection. The 17 volumes of his *Correspondance littéraire, philosophique et critique*, published a few years after his death, show that no eighteenth-century foreigner in France has known the country, its people and its language better than him. This correspondence, characterized by a dry and skeptical philosophy, remains an inexhaustible mine of anecdotes and judgements on men and their works. (M 27332)

(6913) Yukawa

1991 UT_3. Discovered 1991 October 31 by K. Endate and K. Watanabe at Kitami.

Named in memory of Hideki Yukawa (1907-1981), Japanese theoretical physicist who was awarded the 1949 Nobel Prize in physics. He predicted the existence of the π meson, which causes the nuclear force. (M 32347)

(6914) Becquerel

1992 GZ. Discovered 1992 April 3 by C. S. Shoemaker and D. H. Levy and H. E. Holt at Palomar.

Named for the French physicist Henri Becquerel {1852-1908} on the centennial of his discovery of radioactivity in March 1896. Becquerel's work led to the discovery of radium, nuclear transmutation and nuclear fission, and it has had a myriad applications in medicine, the sciences and industry. Radionuclides are used in diagnosis (nuclear medicine), therapy and research. Radiodating has revolutionized geology, archaeology, anthropology and history. Radioactivity explains how the earth has stayed warm; nuclear reactions explain how the sun and other stars maintain their immense energy output. (M 27332; M 27347)
Citation provided by D. Patton, T. Hunter and D. Levy.

Becquerel, together with Marie and Pierre Curie {see planet (7000)}, received the 1903 physics Nobel prize.

(6915) 1992 HH

Discovered 1992 April 30 by Y. Kushida and O. Muramatsu at Yatsugatake.

(6916) Lewispearce

1992 OJ. Discovered 1992 July 27 by R. H. McNaught at Siding Spring.

Named in memory of Lewis Percival Pearce, infant son of amateur astronomer and comet observer Andrew Pearce. Born in Nedlands, Western Australia, on 1999 Jan. 23, Lewis suffered oxygen deprivation during delivery and never gained consciousness. Before his death twelve days later, he shared some experiences with his parents, including "observing the stars" with his dad. (M 34342)

(6917) 1993 FR_2

Discovered 1993 March 29 by S. Otomo at Kiyosato.

(6918) Manaslu

1993 FV_3. Discovered 1993 March 20 by M. Hirasawa and S. Suzuki at Nyukasa.

Named for an 8156-m mountain in the Himalayas. The summit was first reached by the Japanese party led by Aritsune Maki in 1956. (M 34342)

(6919) Tomonaga

1993 HP. Discovered 1993 April 16 by K. Endate and K. Watanabe at Kitami.

Named in memory of Shinichiro Tomonaga (1906-1979), Japanese theoretical physicist who was awarded the 1965 Nobel Prize in physics. He made great contributions to the field of quantum electrodynamics, especially in the renormalization theory to remove singularities. (M 32347)

(6920) Esaki

1993 JE. Discovered 1993 May 14 by K. Endate and K. Watanabe at Kitami.

Named in honor of Leo Esaki (1925-), Japanese experimental physicist who was awarded the 1973 Nobel Proze in physics. He devised the Esaki diode using the tunnel effect. (M 32347)

(6921) 1993 JJ

Discovered 1993 May 14 by S. Ueda and H. Kaneda at Kushiro.

(6922) 1993 KY_1

Discovered 1993 May 27 by S. Otomo at Kiyosato.

(6923) Borzacchini

1993 SD. Discovered 1993 September 16 at the Santa Lucia Observatory at Stroncone.

Named in memory of Baconin Borzacchini (1898-1933), Italian automobile racing champion. (M 30799)

(6924) Fukui

1993 TP. Discovered 1993 October 8 by K. Endate and K. Watanabe at Kitami.

Named in memory of Ken'ichi Fukui (1918-1997), Japanese chemist who was awarded the 1981 Nobel Prize in chemistry. He proposed the frontier orbital theory. (M 32347; M 32561)

(6925) Susumu

1993 UW_2. Discovered 1993 October 24 by T. Seki at Geisei.

Named in honor of Augustin Susumu Yamamoto (1915-), agronomist and director of the Yamamoto Observatory and, from 1988 to 1994, chairman of the board of directors of the Oreintal Astronomical Association, which maintains its headquarters there. He published a *Catalogue of Cometary Orbits* in 1936. His father was the astronomer Issei Yamamoto {see planet (2249)}. (M 32788)

Name proposed by the discoverer following a suggestion by I. Hasegawa and T. Sato.

(6926) 1994 RO_{11}

Discovered 1994 September 1 by S. Ueda and H. Kaneda at Kushiro.

(6927) Tonegawa

1994 TE_1. Discovered 1994 October 2 by K. Endate and K. Watanabe at Kitami.

Named in honor of Susumu Tonegawa (1939-), Japanese biologist who was awarded the 1987 Nobel Prize for medicine and physiology. He revealed the mechanism of genes that produce various antibodies. (M 32347)

(6928) Lanna

1994 TM_3. Discovered 1994 October 11 by M. Tichý at Kleť.

Named in memory of Vojtěch Lanna (1805-1866), well-known Czech entrepreneur. He contributed to the development of trade and business in southern and central Bohemia. Most of his activities – shipbuilding, sailing and bridge building – were closely connected with the Vltava {see planet (2123)} river. (M 30477)

(6929) Misto

1994 UE. Discovered 1994 October 31 by V. S. Casulli at Colleverde di Guidonia.

Named in memory of Angela Misto (1902-1993), mother of the discoverer. (M 31611)

(6930) 1994 VJ_3

Discovered 1994 November 7 by S. Ueda and H. Kaneda at Kushiro.

(6931) Kenzaburo

1994 VP_6. Discovered 1994 November 4 by K. Endate and K. Watanabe at Kitami.

Named in honor of Kenzaburo Ohe (1935-), Japanese writer who was awarded the 1994 Nobel Prize in literature. His important works include "Sexual Human" and "Football on the first year of Man-en era". (M 32347)

(6932) 1994 YK

Discovered 1994 December 24 by T. Kobayashi at Oizumi.

(6933) 1994 YW

Discovered 1994 December 28 by T. Kobayashi at Oizumi.

(6934) 1994 YN_2

Discovered 1994 December 25 by S. Ueda and H. Kaneda at Kushiro.

(6935) Morisot

4524 P-L. Discovered 1960 September 24 by C. J. van Houten and I. van Houten-Groeneveld at Palomar.

Named in memory of Berthe Morisot (1841-1895), important French impressionistic painter. Her teachers were Guichart and Camille Corot {see planet (6672)} and later Edouard Manet. In turn, she influenced Manet to paint landscapes outside, rather than to make sketches first and paint the landscapes in the studio. Morisot was a specialist in oil and water colors of natural family life. (M 27463)

(6936) Cassatt

6573 P-L. Discovered 1960 September 24 by C. J. van Houten and I. van Houten-Groeneveld at Palomar.

Named in memory of Mary Cassatt (1845-1927), American impressionistic painter, who studied in Italy, Spain, Holland and especially France. She and Degas {see planet (6673)} became great friends and influenced each other. The Japanese print makers, whose work became known during her lifetime, also made an impression on her. (M 27463)

(6937) Valadon

1010 T-2. Discovered 1973 September 29 by C. J. van Houten and I. van Houten-Groeneveld at Palomar.

Named in memory of the French painter Suzanne Valadon (1865-1938), the first model for Chavannes and the impressionists Degas and Renoir {see, respectively, planets (6673) and (6677)}. Valadon learned drawing and painting from observing these painters. Until 1909 she mainly made drawings. Later she painted nudes, still life and landscapes. Her son was the well-known painter Maurice Utrillo (1883-1955). (M 27463)

(6938) Soniaterk

5140 T-2. Discovered 1973 September 25 by C. J. van Houten and I. van Houten-Groeneveld at Palomar.

Named in memory of Sonia Delaunay-Terk (1885-1979), Russian-born French painter and wife of the painter Robert Delaunay (1885-1941). In addition to her abstract painting she set the trend for modern design in her own fashion house. (M 27463)

(6939) Lestone

1952 SW_1. Discovered 1952 September 22 by L. E. Cunningham at Mount Wilson.

Named for Leighton Buzzard, a small market town in south Bedfordshire, England. The town is recorded in the Domesday Book (1086) as Lestone, and later variations of the name include Leyton and Leighton. The suffix Buzzard is believed to have been added in the twelfth century and to be a corruption of the family name of the first town prebendary, Thoebald de Busar. (M 29147)

Name suggested by G. V. Williams, who made the identifications for this object and who grew up in the town.

(6940) 1972 HL_1

Discovered 1972 April 19 by C. Torres at Cerro El Roble.

(6941) Dalgarno

1976 YA. Discovered 1976 December 16 at the Harvard College Observatory at Harvard.

Named in honor of Alexander Dalgarno (1928-   ), on the occasion of his 70th birthday. Renowned for his seminal research on collisions and radiative processes involving atoms, molecules, ions and electrons, Dalgarno has led the exploration of atmomic and molecular phenomena occurring in the interstellar medium, planetary and cometary atmospheres, supernova ejecta, the early universe and the terrestrial thermosphere. Due to his intellectual leadership, the Institute for Theoretical Atomic and Molecular Physics was established at the Harvard-Smithsonian Center for Astrophysics. He served as director of the Institute for five years and as acting director of the Harvard College Observatory for three years. (M 32347)

(6942) 1976 YB_2

Discovered 1976 December 16 by L. I. Chernykh at Nauchnyj.

(6943) 1978 VR_4

Discovered 1978 November 7 by E. F. Helin and S. J. Bus at Palomar.

(6944) 1979 MR_3

Discovered 1979 June 25 by E. F. Helin and S. J. Bus at Siding Spring.

(6945) Dahlgren

1980 FZ_3. Discovered 1980 March 16 by C.-I. Lagerkvist at La Silla.

Named in honor of Mats Dahlgren (1966-) in celebration of the completion of his Ph.D. thesis on Hilda-type minor planets. He has made a spectroscopic and photometric survey of a substantial fraction of the numbered Hildas and has studied the collisional evolution and population of the Hildas. (M 30799)

(6946) 1980 RX_1

Discovered 1980 September 15 by H. Debehogne and L. Houziaux at La Silla.

(6947) 1981 ET_8

Discovered 1981 March 1 by S. J. Bus at Siding Spring.

(6948) 1981 ET_{22}

Discovered 1981 March 2 by S. J. Bus at Siding Spring.

(6949) 1982 RZ

Discovered 1982 September 11 at the Oak Ridge Observatory at Harvard.

(6950) Simonek

1982 YQ. Discovered 1982 December 22 by F. Dossin at St. Michel.

Named in honor of Simone Ek, wife of the discoverer, in acknowledgment of the patience and understanding of a dedicated astronomer's wife. (M 33386)

Name proposed by the late discoverer in 1996. Citation prepared by J. Denoyelle.

(6951) 1985 DW_1

Discovered 1985 February 16 by H. Debehogne at La Silla.

(6952) Niccolò

1986 JT. Discovered 1986 May 4 by E. Bowell at Anderson Mesa.

Named for Niccolò Fulchignoni (1991-), second son of Marcello Fulchignoni and M. Antonietta Barucci {see, respectively, planets (3486) and (3485)}, planetary scientists at the Observatoire de Paris. Niccolò is the "Renaissance" Italian form of the name Nikolaus (northern Europe's Santa Claus) and has been given to distinguished personalities of the arts, literature and philosphy - Paganini {see planet (2859), Tommaseo and Machiavelli, for example. (M 31296)
Name suggested and citation provided by M. Fulchignoni.

(6953) 1986 PC_1

Discovered 1986 August 1 by E. F. Helin at Palomar.

(6954) 1987 RB_6

Discovered 1987 September 4 by L. V. Zhuravleva at Nauchnyj.

(6955) Ekaterina

1987 SP_{15}. Discovered 1987 September 25 by L. V. Zhuravleva at Nauchnyj.

Named in honor of the empress Ekaterina Vtoraya (Catherine II; Catherine the Great; 1729-1796), whose reign was noted for the strengthening of autocracy and the position of Russia in the world. Born the princess of Anhalt-Zerbst, she was an educated person who wrote historical and dramatic compositions, comic operas and fairy tales for children. (M 34625)

(6956) Holbach

1988 CX_3. Discovered 1988 February 13 by E. W. Elst at La Silla.

Named in memory of the famous German philosopher Paul Heinrich Dietrich von Holbach (1723-1780), with Diderot, d'Alembert, Grimm, de Jaucourt and Helvetius {see, respectively, planets (5351), (5956), (6912), (6977), and (6972)} one of the Encyclopedists. He spent almost his entire life in Paris, where his residence became the assembly of the greatest free-thinkers of his time. One of his numerous works, *Systéme de la Nature* (1770), often called "The bible of materialism", became well known and had a great impact on the philosophy that there is nothing but material and motion. According to Holbach, the greatest enemy of natural morality is religion, a product of fantasy. (M 27463)

(6957) 1988 HA

Discovered 1988 April 16 by S. Ueda and H. Kaneda at Kushiro.

(6958) 1988 TX_1

Discovered 1988 October 13 by S. Ueda and H. Kaneda at Kushiro.

(6959) 1988 VD_1

Discovered 1988 November 3 by P. Jensen at Brorfelde.

(6960) 1989 AL_5

Discovered 1989 January 4 by R. H. McNaught at Siding Spring.

(6961) Ashitaka

1989 KA. Discovered 1989 May 26 by M. Akiyama and T. Furuta at Mishima.

Named for a dormant volcano in Numazu-city, close to Mt. Fuji {see planet (1584)}. (M 34625)

Name proposed by the first discoverer following a suggestion by the Numazu Astronomical Association.

(6962) 1990 OT

Discovered 1990 July 22 by E. F. Helin at Palomar.

(6963) 1990 OQ_3

Discovered 1990 July 27 by H. E. Holt at Palomar.

(6964) Kunihiko

1990 TL_1. Discovered 1990 October 15 by K. Endate and K. Watanabe at Kitami.

Named in memory of Kunihiko Kodaira (1915-1997), Japanese mathematician who was awarded the Fields Medal in 1954. He contributed to the harmonic integral theory. (M 32347)

(6965) Niyodogawa

1990 VS_2. Discovered 1990 November 11 by T. Seki at Geisei.

Named for a river, 100 km in length, that has its source at the center of Shikoku Island {see planet (4223)}. (M 32094)

(6966) Vietoris

1991 RD_5. Discovered 1991 September 13 by L. D. Schmadel and F. Börngen at Tautenburg.

Named in honor of Leopold Vietoris (1891-), professor emeritus of mathematics of the Leopold-Franzens-University in Innsbruck, on the occasion of his forthcoming 107th birthday. Vietoris is the oldest full member of the Austrian Academy of Sciences and a highly decorated scientist. He made fundamental contributions to algebraic and set-theory topology, and he also wrote outstanding papers on the theory of real functions and on applied mathematics. (M 32094)

Named by the first discoverer following a suggestion by H. Haupt, who prepared the citation.

(6967) 1991 VJ_3

Discovered 1991 November 11 by K. Suzuki and T. Urata at Toyota.

(6968) 1991 VX_3

Discovered 1991 November 11 by S. Ueda and H. Kaneda at Kushiro.

(6969) 1991 VF_5

Discovered 1991 November 4 by S. Otomo at Kiyosato.

(6970) 1992 AL_1

Discovered 1992 January 10 by S. Otomo at Kiyosato.

(6971) Omogokei

1992 CT. Discovered 1992 February 8 by T. Seki at Geisei.

Named for a ravine, 18 hm long, at the foot of Mt. Ishizuti {see planet (4095)}, the highest mountain in western Japan. (M 32094)

(6972) Helvetius

1992 GY_3. Discovered 1992 April 4 by E. W. Elst at La Silla.

Named in memory of the famous French philosopher and Encyclopedist Claude Adrien Helvetius (1715-1771). While still very young he read Locke's *Essay concerning human understanding*, which influenced his life. In 1758 Helvetius' *De l'Esprit* was published. The book was immediately banned by parliament on the grounds that it was dangerous to the state and to religion. Another famous work, *Les progrès de la raison dans la recherche* (1775), espoused the idea that all knowledge comes from our sentences and that morality has to be based on a rational, moderate hedonism. He proposed a workday of only seven to eight hours and supported education and culture for all. (M 27463)

(6973) 1992 HK

Discovered 1992 April 27 by S. Ueda and H. Kaneda at Kushiro.

(6974) 1992 MC

Discovered 1992 June 27 by H. E. Holt at Palomar.

(6975) 1992 QM

Discovered 1992 August 25 by S. Otomo at Kiyosato.

(6976) 1993 KD_2

Discovered 1993 May 23 by S. Otomo at Kiyosato.

(6977) Jaucourt

1993 OZ_4. Discovered 1993 July 20 by E. W. Elst at La Silla.

Named in memory of the French writer Chevalier Louis de Jaucourt (1704-1779), who studied theology at Geneva and medicine at Cambridge and Leiden. He came into contact with the Encyclopedists and wrote for the *Encyclopédie* articles on physiology, chemistry, botany, pathology, history and politics. Diderot {see planet (5351)} expressed his great esteem for de Jaucourt in a letter to Sophie Volland, remarking on his prodigious labors, often reading, dictating to six or seven secretaries for thirteen or fourteen hours each day, and on his great unhappiness when the work neared completion. (M 27463)

(6978) Hironaka

1993 RD. Discovered 1993 September 12 by K. Endate and K. Watanabe at Kitami.

Named in honor of Heisuke Hironaka (1931-), Japanese mathematician who was awarded the Fields Medal in 1970. He proved the theorem of resolution of singularity. (M 32347)

(6979) Shigefumi

1993 RH. Discovered 1993 September 12 by K. Endate and K. Watanabe at Kitami.

Named in honor of Shigefumi Mori (1951-), Japanese mathematician who was awarded the Fields Medal in 1990. He resolved Hartshorne's expectation. (M 32347)

(6980) Kyusakamoto

1993 SV_1. Discovered 1993 September 16 by K. Endate and K. Watanabe at Kitami.

Named in memory of Kyu Sakamoto (1941-1985), a popular singer in Japan who was beloved as Kyu-chan. His songs "Let's walk turning our faces upward", "Look up at the stars at night" and others were big hits. The first song (English title "Sukiyaki") became one of the world's bestsellers. This planet also symbolizes the famous trio of which Kyu Sakamoto was a member, with songwriter Rokusuke Ei and pianist Hachidai Nakamura – in Japanese, Roku is 6, Kyu 9 and Hachi 8. (M 32789)

Name proposed by the discoverers following a suggestion by T. Sato, A. Fujii and Y. Katagiri.

(6981) Chirman

1993 TK_2. Discovered 1993 October 15 at the Bassano Observatory at Bassano Bresciano.

Named for the Surgical Department of the hospital in the Italian village of Manerbio, located a few kilometers from the Bassano {see planet (6460)} Observatory. The name, a contraction of Chirurgia (Surgery) and the name of the village, is presented in honor of the doctors, nurses and all the personnel of the hospital for the competence, absorption and passion they devote to the care of patients. (M 27736)

(6982) 1993 UA_3

Discovered 1993 October 16 by E. F. Helin at Palomar.

(6983) 1993 YC

Discovered 1993 December 17 by T. Kobayashi at Oizumi.

(6984) Lewiscarroll

1994 AO. Discovered 1994 January 4 by H. Shiozawa and T. Urata at Fujieda.

Named after the English fairy-tale author and mathematician Lewis Carroll (Charles Lutwidge Dodgson, 1832-1898). His great works *Alice in Wonderland* and *Through the Looking Glass* were loved not only by children, but by all generations throughout the world for over a century. (M 33788)
Name and citation proposed by T. Urata.

(6985) 1994 UF_2

Discovered 1994 October 31 by S. Ueda and H. Kaneda at Kushiro.

(6986) 1994 WE

Discovered 1994 November 24 by T. Kobayashi at Oizumi.

(6987) 1994 WZ

Discovered 1994 November 25 by T. Kobayashi at Oizumi.

(6988) 1994 WE_3

Discovered 1994 November 28 by S. Ueda and H. Kaneda at Kushiro.

(6989) 1994 XH_1

Discovered 1994 December 6 by T. Kobayashi at Oizumi.

(6990) 1994 XU_4

Discovered 1994 December 9 by T. Kobayashi at Oizumi.

(6991) 1995 AX

Discovered 1995 January 6 by T. Kobayashi at Oizumi.

(6992) 1995 BT_1

Discovered 1995 January 27 by T. Kobayashi at Oizumi.

(6993) 1995 BJ_4

Discovered 1995 January 28 by S. Ueda and H. Kaneda at Kushiro.

(6994) 1995 BV_4

Discovered 1995 January 28 by S. Ueda and H. Kaneda at Kushiro.

(6995) 1996 BZ_1

Discovered 1996 January 24 by T. Kobayashi at Oizumi.

(6996) Alvensleben

2222 T-2. Discovered 1973 September 29 by C. J. van Houten and I. van Houten-Groeneveld at Palomar.

Named in memory of Bertha von Alvensleben (1859-1912), wife of Friedrich Winfried Schubart. Her ancestor Joachim von Alvensleben (1514-1588) amassed a library and was known as a scientist in his own time. His son Gebhard Johann was interested in astronomical observations. (M 27464)

Name suggested and citation prepared by her grandson, Joachim Schubart {see planet (1911)}.

(6997) Laomedon

3104 T-3. Discovered 1977 October 16 by C. J. van Houten and I. van Houten-Groeneveld at Palomar.

Named for Laomedon, king of Troy and father of Priam and Tithonus {see planets (884) and (6998)}. (M 27736)

(6998) Tithonus

3108 T-3. Discovered 1977 October 16 by C. J. van Houten and I. van Houten-Groeneveld at Palomar.

Named for Tithonus, son of Laomedon and brother of Priam {see, respectively, planets (6997) and (884)}, kings of Troy. (M 27736)

(6999) Meitner

4379 T-3. Discovered 1977 October 16 by C. J. van Houten and I. van Houten-Groeneveld at Palomar.

Named in memory of Lise Meitner (1878-1968), Austrian nuclear physicist. Meitner was only the second woman to receive a doctorate from the University of Vienna, where she had been much inspired by Boltzmann. In 1912 she joined the Kaiser-Wilhelm-Institut in Berlin. Her collaboration with the director, Otto Hahn, resulted in the discovery of protactinium, thereby demonstrating the existence of uranium-235. In 1938 Meitner moved to Stockholm, where, with her nephew, Otto Frisch, she explained the presence of barium in the neutron-

bombardment experiments of Hahn and Strassmann as due to fission, a term they coined. (M 27464)
Named by the Small Bodies Names Committee.

(7000) Curie

1939 VD. Discovered 1939 November 6 by F. Rigaux at Uccle.

Named in memory of Marie Curie (1867-1934), the only person to receive Nobel prizes for both physics (1903) and chemistry (1911), famous for her work on radioactivity, a word she invented. The name also honors Pierre Curie (1859-1906), who shared the Nobel prize for physics with his wife and Becquerel {see planet (6914)}. Best known for the discovery of radium, the Curies also discovered polonium, an element named in honor of Marie's native land, where she was born Maria Skłodowska in Warsaw. After Pierre's death his widow was appointed to his professorship and thereby became the first woman to teach at the Sorbonne. She was also involved with the development of radiology for the medical profession. (M 27464)
Named by the Small Bodies Names Committee.

(7001) Noether

1955 EH. Discovered 1955 March 14 at the Goethe Link Observatory at Brooklyn, Indiana.

Named in memory of Emmy Noether (1882-1935), extraordinarily talented mathematician who exerted a great influence on the development of mathematics and laid the foundations of modern algebra. Educated in Erlangen, she moved to Göttingen in 1915 and was for many years the central figure in mathematical studies there, before leaving for positions at Princeton and Bryn Mawr in 1933. From axiomatic principles she developed a theory of ideals, relating them to groups and rings. She also contributed extensively to the study of noncommutative algebras. (M 27464)
Named by the Small Bodies Names Committee.

(7002) 1971 OV

Discovered 1971 July 26 by N. S. Chernykh at Nauchnyj.

(7003) Zoyamironova

1976 SZ_9. Discovered 1976 September 25 by N. S. Chernykh at Nauchnyj.

Named in honor of Zoya Sergeevna Mironova (1913-), well-known surgeon and traumatologist. During World War II, she saved the lives and health of hundreds of wounded soldiers. Since 1952 she has worked at the Central Research Institute for Orthopedics and Traumatology in Moscow. During her long career, she has performed more than 10,000 surgical operations, many of them unique. Many world-famous sports figures and ballet dancers have benefited from her skill. She has received many honors, including recognition from the International Olympic Committee. (M 32347)

Name suggested by a group of scientists and the Olympic Committee of Russia and supported by the discoverer.

(7004) 1979 OB_9

Discovered 1979 July 24 by S. J. Bus at Siding Spring.

(7005) 1981 ET_{25}

Discovered 1981 March 2 by S. J. Bus at Siding Spring.

(7006) 1981 ER_{31}
Discovered 1981 March 2 by S. J. Bus at Siding Spring.

(7007) 1981 EK_{34}
Discovered 1981 March 2 by S. J. Bus at Siding Spring.

(7008) Pavlov
1985 QH_5. Discovered 1985 August 23 by N. S. Chernykh at Nauchnyj.

Named in memory of Nikolaj Nikiforovich Pavlov (1902-1985), professor at Leningrad University and head of the Pulkovo Observatory Time Service. He made a great contribution to the study of the earth's rotation and was the first to apply photoelectric, rather than visual, registration to meridian observations. (M 32347)

(7009) Hume
1987 QU_1. Discovered 1987 August 21 by E. W. Elst at La Silla.

Named in memory of the famous British philosopher David Hume (1711-1776). He started to study law but found it distasteful. In 1744 he became a candidate for the chair of moral philosophy at Edinburgh, but he was not successful, since his opponents found evidence for heresy and even atheism in his *Treatise of Human Nature*. In 1761 the Vatican put his writings on the *Index*. Hume regarded himself chiefly as a moralist: "It is our nature to find certain human qualities intrinsically good. However, we can not explain this, since any attempt would take us into the vacuum of metaphysics." (M 27736)

(7010) Locke
1987 QH_3. Discovered 1987 August 28 by E. W. Elst at La Silla.

Named in memory of the famous British philosopher John Locke (1632-1704), initiator of the Age of Enlightenment and Reason in England and France. Although he worked extensively on political ideas, his main concern was with epistemology, noting that human knowledge rests on experience of the external world and on reflection. Mathematical reasoning is deductive and is to be understood in terms of an intellectual intuition of relations between ideas. In his famous *Essay concerning Human Understanding* (1690), he set down the foundations of an epistemology of modern science. (M 28622)
Citation prepared by S. Elst at the request of the discoverer.

(7011) Worley
1987 SK_1. Discovered 1987 September 21 by E. Bowell at Anderson Mesa.

Named in memory of Charles Edmund Worley (1935-1997), indefatigable observer and cataloguer of double-star positions. Following two years at the Lick Observatory, he was on the staff of the U.S. Naval Observatory from 1961. His work has significantly enhanced our knowledge of the motion of double-star and multiple-star systems and of stellar masses. (M 31296)
Name suggsted by H. Eichhorn.

Obituaries published in Inf. Circ. IAU Comm. 26, No. 134 (1998); Bull. Am. Astron. Soc., Vol. 30, No. 4, p. 1470-1471 (1998).

(7012) Hobbes
1988 CH_2. Discovered 1988 February 11 by E. W. Elst at La Silla.

Named in memory of the great British political philosopher Thomas Hobbes (1588-1679). In 1651 he published his masterpiece *Leviathan*, wherein he insisted that the first requirement of political and moral institutions is that they should provide citizens with security. He analyzed the conditions for peace and security and gave a recipe for constructing an ideal state: men can only live together in

peace if they agree to subject themselves to an absolute and individual sovereign. He made a severe attack on the attempt of papists and presbyterians to challenge the rights of this sovereign. (M 28622)
Citation prepared by S. Elst at the request of the discoverer.

(7013) 1988 RS_4

Discovered 1988 September 1 by H. Debehogne at La Silla.

(7014) Nietzsche

1989 GT_4. Discovered 1989 April 3 by E. W. Elst at La Silla.

Named in memory of the great German philosopher Friedrich Nietzsche (1844-1900), renowned for his criticisms on religion, philosphy and morality. One of his early writings, *The Birth of Tragedy from the Spirit of Music* (1872), remains a classic in the history of esthetics. His best-known work is *Thus spoke Zarathustra* (1883-1885), in which he noted that most men can not accept the intrinsic meaninglessness of existence: they therefore seek supplanting absolutes to invest life with meaning (philosphy and religion). But now "God is dead". The slaughter of rivals and the conquest of the earth will proceed, under the banners of universal brotherhood, democracy and socialism. (M 29148)

(7015) Schopenhauer

1990 QC_8. Discovered 1990 August 16 by E. W. Elst at La Silla.

Named for the celebrated German philosopher of pessimism, Arthur Schopenhauer (1788-1860). In 1813 he went to Weimar and came in intimate association with Goethe {see planet (3047)}. Later he considered that Hindu scriptures, together with Plato and Kant {see, respectively, planets (5451) and (7083)}, should constitute the foundations on which to erect his own philosophical system. In *On Vision and Colours* (1816), he supported Goethe against Newton {see planet (8000)}. His main work was *The World as Will and Idea* (1819), wherein he develops the fundamental idea that "the world is my representation". The world is only comprehensible with the aid of the constructs of man's intellect - space, time and causality. (M 29148)

(7016) 1991 YG

Discovered 1991 December 30 by T. Urata at Oohira.

(7017) Uradowan

1992 CE_2. Discovered 1992 February 1 by T. Seki at Geisei.

Named for the bay that is the principal entrance by sea to Kochi prefecture {see also planet (2396)}. (M 32094)

(7018) 1992 DF

Discovered 1992 February 25 by S. Ueda and H. Kaneda at Kushiro.

(7019) 1992 EM_1

Discovered 1992 March 8 by A. Sugie at Taga.

(7020) Yourcenar

1992 GR_2. Discovered 1992 April 4 by E. W. Elst at La Silla.

Named in memory of the French-Belgian-American writer Marguerite Yourcenar (Marguerite de Crayencour, 1903-1987). Well known for her novels, essays and short stories, she also did translational work. In the historical novel *Mémoires d'Hadrien* (1951), she explored the world and thoughts of the fascinating second-century Roman emperor and meditated on human destiny, morality and power. Her 1968 *L'oeuvre au noir* (The Abyss) is an erudite evocation of the medieval spirit in Flanders and Italy by means of the life of a fictitious sixteenth-century

alchemist and philosopher. In 1980 she became the first woman writer to be elected to the pretigious French Academy. (M 28622)
Citation prepared by K. Leterme at the request of the discoverer.

(7021) 1992 JN_1
Discovered 1992 May 6 by A. Sugie at Taga.

(7022) 1992 JN_4
Discovered 1992 May 2 by S. Ueda and H. Kaneda at Kushiro.

(7023) 1992 KE
Discovered 1992 May 25 by A. Sugie at Taga.

(7024) 1992 PA_4
Discovered 1992 August 2 by H. E. Holt at Palomar.

(7025) 1993 QA
Discovered 1993 August 16 by the Spacewatch at Kitt Peak.

(7026) 1993 QB_1
Discovered 1993 August 19 by E. F. Helin at Palomar.

(7027) 1993 XT
Discovered 1993 December 11 by T. Kobayashi at Oizumi.

(7028) 1993 XC_1
Discovered 1993 December 5 by M. Hirasawa and S. Suzuki at Nyukasa.

(7029) 1993 XT_2
Discovered 1993 December 14 at the Palomar Observatory at Palomar.

(7030) Colombini
1993 YU. Discovered 1993 December 18 at the Santa Lucia Observatory at Stroncone.

Named in honor of Ermes Colombini (1956-), since 1974 a member of the staff at the Osservatorio San Vittore, near Bologna. He was the first Italian amateur to obtain precise positions of comets and minor planets. He was also involved in the computation of orbits and in 1981 was awarded the Cerulli prize of the Italian Astronomical Society for his work on (2235) Vittore. (M 30799)

(7031) 1994 UU
Discovered 1994 October 31 by Y. Shimizu and T. Urata at Nachi-Katsuura.

(7032) 1994 VC_2
Discovered 1994 November 3 by Y. Shimizu and T. Urata at Nachi-Katsuura.

(7033) 1994 WN_2
Discovered 1994 November 28 by S. Ueda and H. Kaneda at Kushiro.

(7034) 1994 YT_2
Discovered 1994 December 25 by S. Ueda and H. Kaneda at Kushiro.

(7035) Gomi
1995 BD_3. Discovered 1995 January 28 by K. Endate and K. Watanabe at Kitami.

Named in honor of Kazuaki Gomi (1911-), amateur astronomer and long-time observer of variable stars. In 1936 he discovered the nova CP Lac. Although there were several earlier novae independently discovered by Japanese observers,

this was the first nova of which a Japanese observer was the first discoverer. (M 33386)
Name proposed by the discoverers following a suggestion by T. Sato and A. Fujii.

(7036) 1995 BH_3

Discovered 1995 January 29 by Y. Shimizu and T. Urata at Nachi-Katsuura.

(7037) 1995 BK_3

Discovered 1995 January 29 by Y. Shimizu and T. Urata at Nachi-Katsuura.

(7038) 1995 DJ_2

Discovered 1995 February 22 by N. Sato and T. Urata at Chichibu.

(7039) Yamagata

1996 GO_2. Discovered 1996 April 14 by T. Okuni at Nanyou.

Named for one of the six prefectures in the Tohoku (northeast) district in Japan, its capital being Yamagata city with a population of 1.25 million. The prefecture is famous for fruit production, notably cherries, apples and grapes. The Zaou and Tukigata areas of Yamagata are popular for skiing. (M 29148)

(7040) Harwood

2642 P-L. Discovered 1960 September 24 by C. J. van Houten and I. van Houten-Groeneveld at Palomar.

Named in memory of the American astronomer Margaret Harwood (1885-1979). Appointed the first director of the Maria Mitchell {see planet (1455)} Observatory in 1916, she maintained that post for 41 years. She had previously been an assistant at the Harvard College Observatory, where she worked again for 20 more years after her formal retirement. Known for her research on variable stars and the lightcurves of minor planets, she has been accorded only limited credit for her discovery of (886) Washingtonia four days before its official discovery at the U.S. Naval Observatory in November 1917. At that time, her mentors at Harward did not consider it appropriate for a woman to discover a minor planet. (M 30799)

(7041) Nantucket

4081 P-L. Discovered 1960 September 24 by C. J. van Houten and I. van Houten-Groeneveld at Palomar.

Named for the Massachusetts island and the Maria Mitchell {see planet (1455)} Observatory located there. The observatory was founded by the Maria Mitchell Association, which in 1912 created an astronomical fellowship for women to work at the observatory during the summer months. Margaret Harwood {see planet (7040)} was the first fellow and director and had wished to give the name of the island to minor planet (886), when she independently discovered it just 80 years ago. This belated new assignment of that name also comes on the sesquicentennial of the appearance of Nantucket on the astronomical map, when Maria Mitchell discovered a comet on 1847 Oct. 1 from the roof of her father's house. (M 30799)

(7042) Carver

1933 FE_1. Discovered 1933 March 24 by K. Reinmuth at Heidelberg.

Named in memory of George Washington Carver (1860-1943), credited by many as the first black American scientist. Born into slavery, he was largely self-educated, but he earned a master's degree in science in 1896 and devoted the remainder of his life to agricultural research, especially the development of food and household products from peanuts, soybeans and sweet potatoes. His scientific stature earned him election to the Royal Society in 1916, but Carver was also an outstanding teacher, artist and humanitarian. He is perhaps best remembered for his most enduring and beloved invention, peanut butter. (M 28622)
Name suggested and citation prepared by G. C. L. Aikman.

(7043) Godart

1934 RB. Discovered 1934 September 2 by E. Delporte at Uccle.

Named in honor of Odon Godart (1913-1996), Belgian mathematician who published distinguished papers on cosmic rays, theoretical and applied meteorology and cosmology. He was a professor at the Catholic University of Louvain and worked in the team of Georges Lemaître {see planet (1565)}. His main contribution was the introduction of isobaric coordinates in meteorology (1942), and this generated new developments in applied meteorology. He served as president of the Belgian Society for Astronomy, Meteorology and Physics of the Earth. (M 27736) Proposal made and citation provided by P. Pâquet.

(7044) 1971 UK

Discovered 1971 October 26 by L. Kohoutek at Bergedorf.

(7045) 1974 FJ

Discovered 1974 March 22 by C. Torres at Cerro El Roble.

(7046) 1977 QG_2

Discovered 1977 August 20 by N. S. Chernykh at Nauchnyj.

(7047) Lundström

1978 RZ_9. Discovered 1978 September 2 by C.-I. Lagerkvist at La Silla.

Named in honor of Magnus Lundström (1967-), a former student of astronomy at Uppsala and at the German Aerospace Research Establishment (DLR), Berlin, who was involved with a numerical study of the 4:1 Jupiter resonance and with several asteroid surveys, including the Uppsala-DLR Trojan Survey and the European Near-Earth Asteroids Search Observatories (EUNEASCO). (M 30799)

(7048) 1981 EH_{34}

Discovered 1981 March 2 by S. J. Bus at Siding Spring.

(7049) 1981 UV_{21}

Discovered 1981 October 24 by S. J. Bus at Palomar.

(7050) 1982 FE_3

Discovered 1982 March 20 by H. Debehogne at La Silla.

(7051) Sean

1985 JY. Discovered 1985 May 13 by C. S. Shoemaker and E. M. Shoemaker at Palomar.

Named for Sean Colin Woodard, oldest grandson of the discoverers. (M 32347)

(7052) 1988 VQ_2

Discovered 1988 November 12 by E. F. Helin at Palomar.

(7053) 1989 FA

Discovered 1989 March 28 by A. Sugie at Taga.

(7054) Brehm

1989 GL_8. Discovered 1989 April 6 by F. Börngen at Tautenburg.

Named in memory of the Thuringian naturalists Christian Ludwig Brehm (1787-1864) and Alfred Edmund Brehm (1829-1884), father and son. Ludwig Brehm was a renowned ornithologist who gathered a unique scientific collection of 15 000 bird skins and was the author of several ornithological books. Alfred Brehm,a zoologist, went on several expeditions across Europe and to Siberia and North Africa, as a result of which he published the famous *Brehms Tierleben* (1863-1870), which has been translated into many languages. Both father and son were buried in the village of Renthendorf, near Jena {see planet (526)}, where Ludwig was parson for more than 50 years and Alfred was born and died. (M 27736)

(7055) 1989 KB

Discovered 1989 May 31 by H. E. Holt at Palomar.

(7056) Kierkegaard

1989 SE_2. Discovered 1989 September 26 by E. W. Elst at La Silla.

Named in memory of the great Danish religious philosopher Søren Kierkegaard (1813-1855), known as the "father of existentialism". All his life he suffered from the influence of the overpowering personality of his father, who educated him in strict adherence to orthodox Lutheranism. His unfortunate relationship with Regine Olsen also severely influenced his life and writings. Kierkegaard believed in the necessity for each individual to make a fully conscious, responsible choice among the alternatives that life offers. His well-known *Fear and Trembling* and *Repetition* dealt with faith and the idea of sacrifice. In *The Sickness unto Death* he anticipated aspects of Freudian {see planet (4342)} psychoanalysis. He made also severe attacks on Hegel's attempt to systematize the whole of existence. (M 28622)

(7057) 1990 QL_2

Discovered 1990 August 22 by H. E. Holt at Palomar.

(7058) 1990 SN_1

Discovered 1990 September 16 by H. E. Holt at Palomar.

(7059) 1990 SK_3

Discovered 1990 September 18 by H. E. Holt at Palomar.

(7060) 1990 SF_{11}

Discovered 1990 September 16 by H. E. Holt at Palomar.

(7061) 1991 PE_1

Discovered 1991 August 15 by E. F. Helin at Palomar.

(7062) Meslier

1991 PY_5. Discovered 1991 August 6 by E. W. Elst at La Silla.

Named for the French priest and philosopher Jean Meslier (1664-1729). After his death, a 1200-page manuscript *Lettres aux Curés du Voisinage* was discovered. This showed him as a fervent atheist, materialist and revolutionist: "The real original sin of men is to be born in poverty, in misery, in dependence and tyranny of the mighty one. We must do everything to free them from this disgusting and damned sin." Although some copies of the manuscript circulated throughout the eighteenth century (eliciting comments from Voltaire and Holbach {see planets (5676) and (6956)}), the work was finally published only in 1864. (M 29148)

(7063) 1991 UK

Discovered 1991 October 18 by S. Ueda and H. Kaneda at Kushiro.

(7064) Montesquieu

1992 OC_5. Discovered 1992 July 26 by E. W. Elst at La Silla.

Named for the French philosopher and writer Charles-Louis de Secondat (1689-1755). Better known by the name of his birthplace, Montesquieu is remembered especially for his *De l'Esprit des lois* (1748), an immense work on the laws and customs of mankind. (M 30799)

(7065) 1992 PU_2

Discovered 1992 August 2 by H. E. Holt at Palomar.

(7066) Nessus

1993 HA_2. Discovered 1993 April 26 by the Spacewatch at Kitt Peak.

Named for the centaur responsible for the death of Heracles {see planet (5143)}. Driven from his home in Arcadia {see planet (1020)} by Heracles, Nessus established himself as a ferryman who later encountered Heracles on travels with his new bride Dejanira {see planet (157)}. Entrusted with ferrying Dejanira across the river Evenus, Nessus attempted to violate her and was as a result killed by Heracles. Before dying, Nessus gave Dejanira a poisonous love charm that she later unwittingly administered to Heracles. (M 29671)

(7067) Kiyose

1993 XE. Discovered 1993 December 4 by M. Hirasawa and S. Suzuki at Nyukasa.

Named for Kiyose City, a western suburb of Tokyo. The city was once home to many sanitaria because of its clean air and good environment. This minor planet was observed at Kiyose station under heavy light pollution. (M 34342)

(7068) Minowa

1994 WD_1. Discovered 1994 November 26 by Y. Kushida and O. Muramatsu at Yatsugatake.

Named in honor of Toshiyuki Minowa (1918-), a Japanese amateur astronomer. During ten years beginning in 1949, a meteor camera of the Tokyo Astronomical Observatory was installed in his garden as the end of a baseline. He was one of the founders, in 1953, of Kawasaki Astronomy Club. As a schoolmaster, he was eager to popularize astronomy among children, and he contributed a telescope that can be operated by a wheelchair user. He was the second chairman of the Japan Amateur Astronomers' Convention (1980-1983). (M 29148)
Name proposed by the discoverers following a suggestion by S. Sakuma.

(7069) 1994 YG_2

Discovered 1994 December 30 by S. Ueda and H. Kaneda at Kushiro.

(7070) 1994 YO_2

Discovered 1994 December 25 by S. Ueda and H. Kaneda at Kushiro.

(7071) 1995 BH_4

Discovered 1995 January 28 by S. Ueda and H. Kaneda at Kushiro.

(7072) Beijingdaxue

1996 CB_8. Discovered 1996 February 3 at the Beijing Observatory at Xinglong.

Named for the 100th anniversary of the Peking University (Beijing Daxue), the oldest national university in China. Founded in 1898, Peking University distinguishes itself by its rigorous style of study, prominent academic standing and fruitful education achievements. (M 30477)

(7073) 1972 RU_1

Discovered 1972 September 11 by N. S. Chernykh at Nauchnyj.

(7074) 1977 RD_3

Discovered 1977 September 10 by N. S. Chernykh at Nauchnyj.

(7075) Sadovnichij

1979 SN_4. Discovered 1979 September 24 by N. S. Chernykh at Nauchnyj.

Named in honor of Viktor Antonovich Sadovnichij (1939-), prominent mathematician, head of Moscow State University. An authority on differential equations, he has made valuable contributions to the study of mechanics and to mathematical methods for deciphering images. (M 34625)

(7076) 1980 UC

Discovered 1980 October 30 by Z. Vávrová at Kleť.

(7077) Shermanschultz

1982 VZ. Discovered 1982 November 15 by E. Bowell at Anderson Mesa.

Named in honor of Sherman W. Schultz (1922-), retired lecturer of astronomy and telescope making who taught for more than 40 years at Macalester College, St. Paul, Minnesota. An optometrist by profession, Schultz figured the mirrors and built an array of telescopes for the college observatory, and these enabled undergraduates to perform photometric observations of minor planets and variable stars that are published in the professional literature. He guided more than 250 students and college faculty to make their own telescopes. The naming of this minor planet commemorates the May 1998 dedication of the Sherman W. Schultz Observatory at Macalester College. (M 32094)

Name proposed by the discoverer following a suggestion by R. P. Binzel, who prepared the citation.

(7078) Unojönsson

1985 UH_3. Discovered 1985 October 17 by C.-I. Lagerkvist at Kvistaberg.

Named in honor of Uno Jönsson (1937-), a friend of the discoverer, to celebrate his 60th birthday. Long interested in astronomy, Jönsson has written several books for a general audience. He has also edited textbooks on astronomy, in one case with the discoverer as a contributing author. (M 31297)

(7079) 1986 RR

Discovered 1986 September 5 by E. W. Elst and V. G. Ivanova at Rozhen.

(7080) 1986 RS_1

Discovered 1986 September 5 by A. Mrkos at Kleť.

(7081) 1987 QF_7

Discovered 1987 August 30 by P. Wild at Zimmerwald.

(7082) 1987 YL_1

Discovered 1987 December 17 by E. W. Elst and G. Pizarro at La Silla.

(7083) Kant

1989 CL_3. Discovered 1989 February 4 by E. W. Elst at La Silla.

Named for the great German philosopher Immanuel Kant (1724-1804), foremost thinker of the Enlightenment. His comprehensive and systematic works in the theory of knowledge, ethics and esthetics have greatly influenced all subsequent philosophy. Although he started his study with theology, he was principally attracted to mathematics and physics. In 1781 his most famous work, *Kritik der reinen Vernunft*, was published. This deals with the roots of knowledge and the conditions of possible experience. The human mind cannot arrive, by pure thought, at truths about entities that, by their very nature, can never be objects of experience. (M 29148)

(7084) 1991 BR

Discovered 1991 January 19 by A. Sugie at Taga.

(7085) 1991 PE

Discovered 1991 August 5 by H. E. Holt at Palomar.

(7086) Bopp

1991 TA_1. Discovered 1991 October 5 by C. S. Shoemaker and E. M. Shoemaker at Palomar.

Named in honor of Tom Bopp, codiscoverer of comet C/1995 O1 (Hale-Bopp), and his father, Frank Bopp. During the last year the two Bopps have enjoyed observing the comet and in lecturing about it as a team. Tom Bopp is an accomplished observer and Frank a superb machinist who has fashioned materials for the Shoemaker-Levy observing program. (M 30100)

Name proposed by the discoverers following a suggestion by D. H. Levy, who prepared the citation.

(7087) Lewotsky

1991 TG_4. Discovered 1991 October 13 by E. F. Helin at Palomar.

Named in honor of the Lewotsky family, especially Kristin and Gretchen, friends of the discoverer. Kristin, a classmate at the California State University at Fullerton, is currently a writing editor for an optics industry trade magazine; with a degree in optical science and engineering from the University of Central Florida, she worked on mirror fabrication for NASA's Advanced X-Ray Astrophysics Facility program. Gretchen is executive director of environmental operations with a major film studio and is currently a board member of a non-profit environmental group. (M 30100)

(7088) Ishtar

1992 AA. Discovered 1992 January 1 by C. S. Shoemaker and E. M. Shoemaker at Palomar.

Named for the chief goddess of the Mesopotamian pantheon and a principal goddess of the Assyrians and Babylonians. She was a goddess of love, fertility, sex and war. Not only did she rule the moon, but Ishtar owned the morning and evening stars. As the morning star, the goddess arrayed herself in armor and hitched her chariot to seven lions before setting off in the dawn to hunt animals or humans. As the evening star, she was seen as the goddess "whose song is sweeter than honey and wine, sweeter than sprouts and herbs, superior even to pure cream". (M 29671; M 29691)

(7089) 1992 FX_1

Discovered 1992 March 23 by S. Ueda and H. Kaneda at Kushiro.

(7090) 1992 HY_4

Discovered 1992 April 23 by H. Debehogne at La Silla.

(7091) 1992 JA

Discovered 1992 May 1 by K. J. Lawrence and E. F. Helin at Palomar.

(7092) Cadmus

1992 LC. Discovered 1992 June 4 by C. S. Shoemaker and E. M. Shoemaker at Palomar.

Named for the son of the Phoenician king Agenor {see planet (1873)}, and the brother of Europa {see planet (52)}, Cilix and Phoenix. He was famous for the search for his sister Europa, whom Zeus {see planet (5731)} carried away. Cadmus overcame a dragon at Thebes with the assistance of Athena {see planet (881)} and sowed the dragon's teeth on the plain. Immediately, a company of armed men sprang up, ready for attack. Cadmus cast a stone into their midst and, in the resulting confusion, all but five of the warriors slew each other. The survivors became allies of Cadmus and helped him build Thebes. Cadmus married Harmonia {see planet (40)}, daughter of Aphrodite {see planet (1388)}. It was he who introduced the use of letters into Greece. (M 29671; M 29691)

(7093) 1992 OT

Discovered 1992 July 26 by E. F. Helin at Palomar.

(7094) Godaisan

1992 RJ. Discovered 1992 September 4 by T. Seki at Geisei.

Named for a small mountain on the eastern side of Kochi City {see planet (2396)}. (M 32094)

(7095) Lamettrie

1992 SB_{22}. Discovered 1992 September 22 by E. W. Elst at La Silla.

Named in memory of the French medical doctor and philosopher Julien Offray de La Mettrie (1709-1751). After publishing *Traité de l'Ame* in 1745, he had to flee to Holland, where he published *L'Homme-Machine* (1748). "Soul' is just an empty word. Matter by its own has the faculty of perception." In his last work *Discours préliminaire* (1751), he wrote: "How can one possibly think that reasoning is dangerous? It has never led to fanatics, sects or even theologians." (M 29672)

(7096) Napier

1992 VM. Discovered 1992 November 3 by R. H. McNaught at Siding Spring.

Named for the Scottish astronomer William (Bill) M. Napier (1940-), who has made seminal contributions in several areas of astrophysics. In collaboration with Victor Clube {see planet (6523)}, Napier was one of the first to suggest that impacts by comets and minor planets have had a significant influence upon life on the earth, and that those impacts are temporally linked to cyclic disturbances of the Oort Cloud {see planet (1691)} as the sun moves about the galactic plane, as well as to the sporadic arrival of giant comets in the inner solar system. Napier is currently investigating the effect upon the terrestrial climate of the varying influx of cometary dust in such a picture, although he also actively works on the apparent quantization of galactic redshifts. (M 29148)

Name proposed by the discoverer following a suggestion by D. I. Steel, who prepared the citation.

(7097) Yatsuka

1993 TF. Discovered 1993 October 8 by H. Abe and S. Miyasaka at Yatsuka.

Named for the first discoverer's home town, in the eastern part of Shimane prefecture, known for its production of ginseng and peonies. (M 28090)

(7098) 1993 TK_{39}

Discovered 1993 October 9 by E. W. Elst at La Silla.

(7099) Feuerbach

1996 HX_{25}. Discovered 1996 April 20 by E. W. Elst at La Silla.

Named in memory of the famous German philosopher and moralist Ludwig Andreas Feuerbach (1804-1872). In his *Über Philosophie und Christentum* (1839), he claimed that "Christianity has in fact long vanished not only from the reason but from the life of mankind and is nothing more than an obsession". But it is in his *Das Wesen des Christentums* (1841) that he put forward his ideas on religion: "God is merely the outward projection of man's inward nature". His views were later endorsed in Germany by extremists in the struggle between church and state. (M 29672)

(7100) Martin Luther

1360 T-2. Discovered 1973 September 29 by C. J. van Houten and I. van Houten-Groeneveld at Palomar.

Named in memory of Martin Luther (1483-1546), German theologian and religious reformer. After studying law, he became an Augustinian monk and priest in 1507 and a professor of biblical exegesis in Wittenberg in 1508. On 1517 Oct. 31 he posted his famous 95 theses at the door of the university church. These theses were against the misuse of indulgences by the Catholic Church, and here the Reformation started. Luther also influenced the German language through his translations of the Old and the New Testaments into German. (M 33386)

(7101) 1930 UX

Discovered 1930 October 17 by C. W. Tombaugh at Flagstaff.

(7102) 1936 NB

Discovered 1936 July 12 by C. Jackson at Johannesburg.

(7103) 1953 GH

Discovered 1953 April 7 by K. Reinmuth at Heidelberg.

(7104) Manyousyu

1977 DU. Discovered 1977 February 18 by H. Kosai and K. Hurukawa at Kiso.

Named for the earliest collection of Japanese poetry, by many poets spanning about 400 years. Celestial phenomena are mentioned in several of the poems. (M 34625)

(7105) Yousyozan

1977 DB_1. Discovered 1977 February 18 by H. Kosai and K. Hurukawa at Kiso.

Named for a 400-m mountain to the south of the Okayama {see planet (2084)} station of the National Astronomical Observatory of Japan. Site testing for the observatory was conducted for about a year on this mountain, resulting in the observatory's construction. It was opened in 1960. (M 33788)

(7106) Kondakov

1978 PM_3. Discovered 1978 August 8 by N. S. Chernykh at Nauchnyj.

Named in honor of Aleksandr Konstantinovich Kondakov (1928-), researcher on agricultural chemistry and fertilizers for fruit plants, and head of the laboratory of agrochemistry at the Michurin All-Russian Research Institute of Gardening in Michurinsk. He is well known for his new concepts involving the fertilization of garden soil and for diagnostics concerning the nourishment of garden plants. (M 30799)

(7107) Peiser

1980 PB_1. Discovered 1980 August 15 by A. Mrkos at Kleť.

Named in honor of Benny Josef Peiser (1957-), social anthropologist with a particular research interest in neocatastrophism and its implications for human, societal and cultural evolution. A senior lecturer at John Moores University in Liverpool, he launched and now moderates the Cambridge-Conference Network, a scholarly network of some 300 researchers from around the world. (M 34342)

Name suggested by J. Tichá, M. Tichý and Z. Moravec, who observed this minor planet at Kleť at the 1994 and 1996 oppositions, prior to its numbering.

(7108) Nefedov

1981 RM_3. Discovered 1981 September 2 by N. S. Chernykh at Nauchnyj.

Named in honor of Oleg Matveevich Nefedov (1931-), vice president of the Russian Academy of Sciences and chairman of the National Committee of Russian chemists. An outstanding scientist in the field of physical, synthetic and technical organic chemistry, he is best known for his fundamental research on highly reactive intermediates and small cycles. (M 32789)

Name proposed by the discoverer following a suggestion by the Institute of Applied Astronomy.

(7109) Heine

1983 RT_4. Discovered 1983 September 1 by L. G. Karachkina at Nauchnyj.

Named in memory of Heinrich Heine (1797-1856), the great German poet and writer, an outstanding master of lyric poems and political satire. The name is proposed on the occasion of the two-hundredth anniversary of his birth. (M 29148)

(7110) 1983 XH_1

Discovered 1983 December 7 at the Perth Observatory at Bickley.

(7111) 1985 QA_1

Discovered 1985 August 17 by E. F. Helin at Palomar.

(7112) Ghislaine

1986 GV. Discovered 1986 April 3 by C. S. Shoemaker and E. M. Shoemaker at Palomar.

Named in honor of the cosmochemist Ghislaine Crozaz (1939-), who began her immensely productive career three decades ago, with studies of fission tracks in lunar samples and meteorites. Since then she has contributed significantly to our understanding of the early history of the solar system and to the formation histories of various meteorite types through innovative studies of trace element microdistributions and extinct radionuclides in these objects. As a professor in the Department of Earth and Planetary Sciences and McDonnell Center for the Space Sciences at Washington University, she actively participates in the training of the next generation of planetary scientists with her characteristic nurturing and enthusiastic spirit. (M 34625)

Citation prepared by M. Wadhwa at the request of the first discoverer.

(7113) Ostapbender

1986 SD_2. Discovered 1986 September 29 by L. G. Karachkina at Nauchnyj.

Named for the character Ostap Bender, celebrated hero of the satirical novels *Twelve Chairs* (1928) and *Golden Calf* (1931) by the Soviet writers Il'ya Il'f (penname of I. Feinsilberg) and Eugeny Petrov (penname of E. Kataev). The International Festival of Satire and Humor in St. Petersburg 'Gold Ostap' (founded in 1992) bears the name of this witty and resourceful character. (M 29148)

Name proposed by L. R. Nemirovskij on the occasion of the one-hundredth anniversary of the birth of Il'ya Il'f.

(7114) 1986 WN_7

Discovered 1986 November 29 by A. Mrkos at Kleť.

(7115) 1986 WO_7

Discovered 1986 November 29 by A. Mrkos at Kleť.

(7116) Mentall

1986 XX. Discovered 1986 December 2 by E. Bowell at Anderson Mesa.

Named in honor of E. Talmadge Mentall (1927-), who retired from a drafting career to join Sky Publishing Corporation in 1994. With his artistic skill and lifelong enthusiasm for astronomy, "Tal" soon played a pivotal role in the *Millennium Star Atlas* (1997), a joint undertaking with the European Space Agency's Hipparcos project. He meticulously drafted all the outlines of nebulae and measured the orientations of 2000 galaxies not available from the literature. (M 31297)

Name proposed by R. W. Sinnott, endorsed by B. G. Marsden and the discoverer.

(7117) Claudius

1988 CA_1. Discovered 1988 February 14 by F. Börngen at Tautenburg.

Named for the popular German poet and writer Matthias Claudius (1740-1815). He was the editor of the elitist journal *Messenger of Wandsbeck* and wrote lyric poetry and prose that is distinguished by simplicity, naturalness and heartful piety. Among this work is his evening song *Der Mond ist aufgegangen.* (M 29148)

(7118) 1988 VD_5

Discovered 1988 November 4 by A. Mrkos at Kleť.

(7119) Hiera

1989 AV_2. Discovered 1989 January 11 by C. S. Shoemaker and E. M. Shoemaker at Palomar.

A famous woman warrior and general of the Mysians, Hiera fought in the Trojan War but was edited from Homer's {see planet (5700)} account because, Philostratus says, "this greatest and finest women would have outshone his heroine Helen {see planet (101)}". (M 34625)

(7120) 1989 AD_3

Discovered 1989 January 4 by R. H. McNaught at Siding Spring.

(7121) Busch

1989 AL_7. Discovered 1989 January 10 by F. Börngen at Tautenburg.

Named for the German poet and draughtsman Wilhelm Busch (1832-1908), who became world-renowned for his popular humorous and satiric picture stories. His drawing style is full of verve and movement and is of the highest originality. As creator of immortal comic types such as Max and Moritz, as well as Pious Helen, he spread much laughter. Einstein {see planet (2001)} wrote of him that "He is one of the greatest masters of the stylistic accuracy of aim". (M 29149)

(7122) Iwasaki

1989 EN_2. Discovered 1989 March 12 by K. Endate and K. Watanabe at Kitami.

Named in honor of Kazuaki Iwasaki (1935-), amateur astronomer and world-renowned space artist. His work has been frequently exhibited and he has published many books on the subject of space art. In July 1998 he opened his own space art gallery in Ito City, Shizuoka Prefecture. (M 32789)

Name proposed by the discoverers following a suggestion by T. Sato and A. Fujii.

(7123) 1989 TT_1

Discovered 1989 October 9 by T. Hioki and N. Kawasato at Okutama.

(7124) 1990 OJ_4

Discovered 1990 July 24 by H. E. Holt at Palomar.

(7125) Eitarodate

1991 CN_1. Discovered 1991 February 7 by T. Seki at Geisei.

Named in memory of Eitaro Date (1912-1953), Japanese amateur astronomer who observed the planets, particularly Mars, and who taught other planetary observers. (M 32789)

Name proposed by the discoverer following a suggestion by T. Sato and A. Fujii.

(7126) Cureau

1991 GJ_4. Discovered 1991 April 8 by E. W. Elst at La Silla.

Named in memory of Marin Cureau de la Chambre (1594-1669), a doctor of medicine and author of several scientific works. In 1635 he became one of the first members of the prestigious Académie Francaise and will be remembered for his *Traité de la Connaisance des Animaux* (1647), a work that made clear that animals, like humans, have feelings, knowledge and thoughts. (M 30478)

(7127) Stifter

1991 RD_3. Discovered 1991 September 9 by F. Börngen and L. D. Schmadel at Tautenburg.

Named for the most famous Austrian narrator Adalbert Stifter (1805-1868). After formative years spent in the Bohemian Forest, he studied near the Benedictine Abbey in Kremsmünster, later living in Vienna and Linz {see planets (397) and (1469)}. In his brillant novels and epics (*The Timber Forest, Rock Crystal, Indian Summer* and *Witiko*) landscapes were described in a superb manner. Stifter described the correlation of man and nature in a subtle manner, full of feeling. He dealt with questions of education, love and piety, and he was also engaged in painting and science. He gave full details of the total solar eclipse of 1842 July 8 as observed in Vienna. (M 29149)

(7128) Misawa

1991 SM_1. Discovered 1991 September 30 by K. Endate and K. Watanabe at Kitami.

Named in memory of Katsue Misawa (1885-1937). Although he received only a limited education, through his own efforts he became a junior-high-school geography teacher. His unique teaching practices are highly appreciated in the history of Japanese education. In astronomy, he was the real pioneer of sunspot observation in Japan, systematically observing sunspots from 1921 until 1934, when failing eyesight prevented it. His data were invaluable because in those days foreign data arrived in Japan only after a long delay. (M 34342)

Name proposed by the discoverers following a suggestion by K. Gomi, T. Sato, K. Fujimori and A. Fujii.

(7129) 1991 VE_1

Discovered 1991 November 4 by S. Ueda and H. Kaneda at Kushiro.

(7130) Klepper

1992 HR_4. Discovered 1992 April 30 by F. Börngen at Tautenburg.

Named for the German novelist and lyric poet Jochen Klepper (1903-1942), with *The Father* an important representative of the Christian historical novel. He was dismissed from his post at the Berlin Radio already in 1933 because his wife was Jewish. By courageously raising his voice against Hitler's terror he consoled people who suffered from the system. His letters and diary *Under the Shadows of your Wings* are moving. Pressed by the circumstances, he decided to die voluntarily with his wife and her daughter in order to protect them from the concentration camp. (M 29149)

(7131) 1992 YL

Discovered 1992 December 23 by A. Natori and T. Urata at Yakiimo.

(7132) Casulli

1993 SE. Discovered 1993 September 17 at the Santa Lucia Observatory at Stroncone.

Named in honor of Silvano Casulli (1944-), who was the first amateur astronomer in the world to obtain precise astrometric positions of minor planets using a CCD camera. He lives and works near Rome, and his great enthusiasm played a significant role in organizing the amateur astrometrists in Italy. (M 30800)

(7133) Kasahara

1993 TX_1. Discovered 1993 October 15 by K. Endate and K. Watanabe at Kitami.

Named in honor of Shin Kasahara (1953-), doctor of dentistry and senior lecturer at Tohoku University. An amateur astronomer, he has been enthusiastically engaged in positional observation and orbit calculation of minor planets and comets since 1973. He also played an important role in designing and constructing the biggest portable telescope in Japan, the 84-cm Chiro Memorial Telescope, which went on a national tour to allow the observation of comet 1P/Halley. (M 32789)

Name proposed by the discoverers following a suggestion by M. Koishikawa and A. Fujii.

(7134) 1993 UY
Discovered 1993 October 24 by T. Kobayashi at Oizumi.

(7135) 1993 VO
Discovered 1993 November 5 by Y. Shimizu and T. Urata at Nachi-Katsuura.

(7136) 1993 VK_2
Discovered 1993 November 14 by H. Shiozawa and T. Urata at Fujieda.

(7137) 1994 AQ_1
Discovered 1994 January 4 by S. Otomo at Kiyosato.

(7138) 1994 AK_{15}
Discovered 1994 January 15 by S. Ueda and H. Kaneda at Kushiro.

(7139) Tsubokawa
1994 CV_2. Discovered 1994 February 14 by T. Niijima and T. Urata at Ojima.

Named in memory of Ietsune Tsubokawa (1918-1994), the fifth director of the International Latitude Observatory of Mizusawa {see planet (7530)} (1976-1986) and a former director of the Earthquake Research Institute of the University of Tokyo. He strove to automate the instruments used for the geodetic survey and astronomical observations. (M 34625)

Name proposed by the second discoverer following a suggestion by K. Hurukawa and K. Yokoyama. Citation prepared by K. Yokoyama.

(7140) 1994 EE_1
Discovered 1994 March 4 by T. Kobayashi at Oizumi.

(7141) Bettarini
1994 EZ_1. Discovered 1994 March 12 by A. Boattini and M. Tombelli at Cima Ekar.

Named in memory of Otello Bettarini (1905-1982), a pioneer who left great footprints in our way of life. The astronomical observatory, built by him, allowed the discoverers to "cut their teeth" in astronomy. (M 30478)

(7142) Spinoza
1994 PC_{19}. Discovered 1994 August 12 by E. W. Elst at La Silla.

Named for the Dutch-Jewish philosopher Benedict de Spinoza (1632-1677). He incurred the disapproval of the synagogue authorities, and in 1656 he was excommunicated. Although acknowledging Descartes {see planet (3587)} to be the father of modern philosophy, he did not agree with his metaphysics. In his masterpiece, *Ethica* (1663-1675), he constructed a metaphysical system entirely deductively in terms of Euclidean geometry. Spinoza was also an expert in grinding and polishing lenses, but the inhalation of glass dust contributed to his death. (M 30800)

(7143) 1995 WU_{41}
Discovered 1995 November 17 by S. Otomo at Kiyosato.

(7144) Dossobuono

1996 KQ. Discovered 1996 May 20 by L. Lai at Madonna di Dossobuono.

Named for the village of Madonna di Dossobuono, known for its sanctuary for "700 dedicated to the virgin Mary". Located 7 km from Verona, it is the site of the observatory at which this minor planet was discovered. (M 31025)

(7145) Linzexu

1996 LO. Discovered 1996 June 7 at the Beijing Observatory at Xinglong.

Named in honor of Lin Zexu (1785-1850) for his pioneering campaign against drug abuse and drug-related crimes. Born in Minhou, Fujian Province, China, his illustrious work was crowned by the burning of 1200 tons of opium in June 1839 on the beach of Humen in southern China. As an official of Qing dynasty, he is also remembered for his achievements in water conservancy, from which his native province and other districts in China greatly benefited. (M 33788)

(7146) Konradin

3034 P-L. Discovered 1960 September 24 by C. J. van Houten and I. van Houten-Groeneveld at Palomar.

Named in honor of Konradin Ferrari d'Occhieppo (1907-), professor emeritus of astronomy of Vienna University and full member of the Austrian Academy of Sciences, on his forthcoming ninetieth birthday. Well known in Austria as a profound teacher and scientist, he is also a nobleman of extraordinary personal modesty and generosity. He has worked on variable stars, as well as on astrometric and calendrical problems and the history of astronomy. His main interest has been in the astronomical aspects of the Star of Bethlehem, a subject to which he devoted a book and many articles. (M 28622)

Name suggested by H. F. Haupt (who prepared the citation) and L. D. Schmadel.

(7147) Feijth

4015 P-L. Discovered 1960 September 24 by C. J. van Houten and I. van Houten-Groeneveld at Palomar.

Named in honor of Hendrik (Henk) Feijth (1944-1997), a devoted and true amateur astronomer: making variable star observations is his passion. Since the early 1960s he has been an observer in the variable star observing group Nederlandse Vereniging voor Weer en Sterrenkunde, Werkgroep Veranderlijke Sterren. Since 1981 he has been this group's representative to the American Association of Variable Star Observers. Feijth has made nearly 100,000 observations of variable stars and has lectured and written articles for journals and books. (M 29672)
Name proposed and citation prepared by T. Jurriens.

Obituaries published in Radiant, Jaarg. 19, Nr. 3, p. 49-50 (1997); Zenit, Jaarg. 24, Nr. 7/8 (1997); Int. Comet Q., Vol. 20, No. 2, p. 53 (1993).

(7148) Reinholdbien

1047 T-1. Discovered 1971 March 25 by C. J. van Houten and I. van Houten-Groeneveld at Palomar.

Named in honor of Reinhold Bien (1947-), astronomer at the Astronomisches Rechen-Institut in Heidelberg. Bien's fields of interest include numerical studies on celestial mechanics, stellar dynamics and involving interacting galaxies. He is known in particular for his work on the long-period evolution of Trojan orbits. (M 30800)
Name proposed and citation prepared by J. Schubart.

(7149) Bernie

3220 T-3. Discovered 1977 October 16 by C. J. van Houten and I. van Houten-Groeneveld at Palomar.

Named in honor of Hans-Heinrich "Bernie" Bernstein (1953-), senior astronomer at the Astronomisches Rechen-Institut in Heidelberg. He analyzed pulse-timing data on pulsars and was instrumental in developing an algorithm for the detection of astrometric binaries with ESA's satellite HIPPARCOS. He detected 110 new double-star systems, together with a few brown-dwarf candidates. Bernstein is an expert in numerical mathematics, with many special applications in astrometry. (M 30800)
Name proposed and citation prepared by L. D. Schmadel.

(7150) 1929 TD_1
Discovered 1929 October 11 by C. W. Tombaugh at Flagstaff.

(7151) 1971 SX_3
Discovered 1971 September 26 by C. Torres at Cerro El Roble.

(7152) Euneus
1973 SH_1. Discovered 1973 September 19 by C. J. van Houten and I. van Houten-Groeneveld at Palomar.

Named for the king of Lemnos and a son of Jason {see planet (6063)}. Euneus supplied the Greek forces with wine during the Trojan war. (M 30478)

(7153) 1975 XP_3
Discovered 1975 December 2 by T. M. Smirnova at Nauchnyj.

(7154) 1979 MJ_5
Discovered 1979 June 25 by E. F. Helin and S. J. Bus at Siding Spring.

(7155) 1979 YN
Discovered 1979 December 23 by H. Debehogne and E. R. Netto at La Silla.

(7156) 1981 EC_2
Discovered 1981 March 4 by H. Debehogne and G. DeSanctis at La Silla.

(7157) 1981 EC_8
Discovered 1981 March 1 by S. J. Bus at Siding Spring.

(7158) 1981 ES_8
Discovered 1981 March 1 by S. J. Bus at Siding Spring.

(7159) 1981 EN_{17}
Discovered 1981 March 1 by S. J. Bus at Siding Spring.

(7160) 1981 UQ_{29}
Discovered 1981 October 24 by S. J. Bus at Palomar.

(7161) 1982 UY_{10}
Discovered 1982 October 25 by L. V. Zhuravleva at Nauchnyj.

(7162) Sidwell
1982 VB_1. Discovered 1982 November 15 by E. Bowell at Anderson Mesa.

Named in honor of Daniel R. Sidwell (1932-), the facility operations manager of the Table Mountain Facility, upon his retirement after 42 years of dedicated service to the Jet Propulsion Laboratory. Sidwell was hired in 1956, and he worked with the wind-tunnel group at JPL's Pasadena facility before moving to Table Mountain in 1976. He has been a source of dedicated and sustained help beyond the call of duty for all aspects of the Table Mountain Facility. The success of many astronomy-related science assignments conducted at Table Mountain is

a direct consequence of his eagerness to serve. (M 32789)
Name suggested and citation prepared by J. W. Young.

(7163) 1984 DB

Discovered 1984 February 24 by E. F. Helin and R. S. Dunbar at Palomar.

(7164) Babadzhanov

1984 ET. Discovered 1984 March 6 by E. Bowell at Anderson Mesa.

Named in honor of Pulat Babadzhanov (1930-), director of the Institute of Astrophysics, Tajik Academy of Sciences, Dushanbe. Babadzhanov's research includes the photographic investigation of meteors, the physics and dynamics of meteors, the evolution of meteoroid streams and the association of comets and minor planets with meteor showers. He has served as rector of the Tajik State University, vice president of the Tajik Academy of Sciences and as president of IAU Commission 22. (M 32347)

(7165) Pendleton

1985 RH. Discovered 1985 September 14 by E. Bowell at Anderson Mesa.

Named for astronomer Yvonne Jean Pendleton (1957-) of the NASA Ames Research Center. She specializes in the spectroscopic study of the organic components of interstellar dust and the incorporation of that dust into bodies in the solar system and extrasolar planetary systems. She has pursued the identification of interstellar organics through comparative studies using materials produced in the laboratory and those derived from meteorites. Her research is focused on the connection of interstellar organic matter to the origin of life on the earth and other planets. Pendleton is a codiscoverer of the organic matter in interstellar dust in external galaxies by infrared spectroscopy. (M 29149)

Name proposed by the discoverer following a suggestion by D. P. Cruikshank, who prepared the citation.

(7166) Kennedy

1985 TR. Discovered 1985 October 15 by E. Bowell at Anderson Mesa.

Named in memory of Malcolm Kennedy (1944-1997), tireless and popular secretary of the Astronomical Society of Glasgow. Born and raised in New Zealand, Kennedy became a civil engineer in Scotland. He was an energetic member of the Free Church in Cumbernauld, near Glasgow. As secretary of the ASG, he ensured that meetings ran smoothly, enlivening them with wordplay and puns. (M 32348)

(7167) Laupheim

1985 TD_3. Discovered 1985 October 12 by C. S. Shoemaker and E. M. Shoemaker at Palomar.

Named in honor of Robert Clausen (1951-) and his team at the public observatory in Laupheim, a town in southern Germany. In 1975 Clausen founded an association of amateur astronomers Volkssternwarte Laupheim e.V. and became its president. By organizing traveling astronomical exhibitions and numerous international astrofests, he expanded Laupheim's reputation far beyond Germany. Since 1990 Clausen, assisted by his coworkers, has excelled in running, with great professionalism, a Zeiss planetarium and observatory that were mostly built through the members' own initiative. (M 34625)

Citation provided by G. and D. Heinlein at the request of C. Shoemaker, who visited the Laupheim facilities in 1998.

(7168) 1986 QE_2

Discovered 1986 August 28 by H. Debehogne at La Silla.

(7169) Linda

1986 TK_1. Discovered 1986 October 4 by E. Bowell at Anderson Mesa.

Named in memory of Linda McCartney (1941-1998), wife of Beatle Paul McCartney {see planet (4148)}, member of the musical group Wings, photographer, and author of vegetarian cookbooks. McCartney's strength of purpose as a friend of the environment, of animals, and of all humanity, together with her devotion to her family, provided a shining example to others. (M 32789)

Name proposed by the discoverer following a suggestion by J. Dunne, who prepared the citation.

(7170) 1987 MK

Discovered 1987 June 30 by R. H. McNaught at Siding Spring.

(7171) 1988 AT_1

Discovered 1988 January 13 by A. Mrkos at Kleť.

(7172) Multatuli

1988 DE_2. Discovered 1988 February 17 by E. W. Elst at La Silla.

Named in memory of the great Dutch writer Multatuli ("I have suffered much", pseudonym of Eduard Douwes Dekker, 1820-1887). In 1838 he went to the Durch East Indies, where he held a number of governmental posts. In 1856 he resigned as assistant commissioner of Lebak, Java, because he was not supported by the government in his struggle to protect the Javanese from exploitation by their own chiefs. Back in Europe, he soon became internationally known with his novel *Max Havelaar* (1860), which enabled him to plead for justice in Java and to satirize the Dutch middle-class mentality. (M 28622)

Name proposed by the discoverer, endorsed by C.-F. Merks and J. Meeus.

(7173) 1988 PL_1

Discovered 1988 August 15 by C. S. Shoemaker at Palomar.

(7174) 1988 SQ

Discovered 1988 September 18 by H. Debehogne at La Silla.

(7175) 1988 TN_2

Discovered 1988 October 11 by Z. Vávrová at Kleť.

(7176) Kuniji

1989 XH. Discovered 1989 December 1 by A. Takahashi and K. Watanabe at Kitami.

Named in honor of Kuniji Saito (1913-), who joined the Tokyo Astronomical Observatory in 1936 and was engaged mainly in research on the solar corona. Following his retirement in 1974, he has collected historical materials from Japan, China and other countries to analyze them from the viewpoint of modern astronomy, using computers. He named this field of research "paleoastronomy" and hopes that many other reearchers will enter into this kind of research. He also served as president of the Astronomical Society of Japan. (M 32789; M 33151)

Name proposed by the discoverers following a suggestion by A. Fujii and A. Tanno.

(7177) 1990 TF

Discovered 1990 October 9 by R. H. McNaught at Siding Spring.

(7178) 1990 VA_3

Discovered 1990 November 11 by T. Nomura and K. Kawanishi at Minami-Oda.

(7179) Gassendi

1991 GQ_6. Discovered 1991 April 8 by E. W. Elst at La Silla.

Named for the French philosopher and scientist Pierre Gassendi (1592-1655). Originally trained in theology, he came under the influence of the mathematician-theologian Mersenne. On 1631 Nov. 7 he made the first known observation of a transit of Mercury, thereby confirming the work of Kepler {see planet (1134)}. As a philosopher, he revived Epicureanism as a substitute for Aristotelianism. At Mersenne's request, he refuted Descartes' {see planet (3587)} *Meditations* in his *Disquisitio Metaphysica* (1644). In his great *Syntagma Philosophicum*, finally published as part of his *opera omnia* in 1658, he rejected the innate ideas of Descartes and favored the senses as the primary sources of knowledge. (M 31025)

(7180) 1991 NG_1

Discovered 1991 July 12 by H. E. Holt at Palomar.

(7181) 1991 PH_{12}

Discovered 1991 August 7 by H. E. Holt at Palomar.

(7182) 1991 RV_1

Discovered 1991 September 8 by E. F. Helin at Palomar.

(7183) 1991 RE_{16}

Discovered 1991 September 15 by H. E. Holt at Palomar.

(7184) 1991 RB_{25}

Discovered 1991 September 11 by H. E. Holt at Palomar.

(7185) 1991 VN_1

Discovered 1991 November 4 by S. Ueda and H. Kaneda at Kushiro.

(7186) Tomioka

1991 YF. Discovered 1991 December 26 by K. Endate and K. Watanabe at Kitami.

Named in honor of Hiroyuki Tomioka (1942-), director of weather information in Hitachi City. An amateur astronomer, he is a charter member of the Nippon Meteor Society and has observed meteors for more than 40 years, both visually and photographically. In addition to his observations from Japan, he often visits the Chiro Observatory Southern Station in Australia to observe southern meteor showers. (M 32789; M 33151)

Name proposed by the discoverers following a suggestion by A. Fujii and T. Sato.

(7187) 1992 BW

Discovered 1992 January 30 by E. F. Helin at Palomar.

(7188) Yoshii

1992 SF_1. Discovered 1992 September 23 by K. Endate and K. Watanabe at Kitami.

Named in honor of Koichi Yoshii (1914-), retired postmaster and amateur astronomer. He was a pioneer in meteor photography in Japan. In the 1930s, when the sensitivity of photographic emulsion was still very low, he succeeded in photographing 86 meteors with his homemade camera. (M 32789)

Name proposed by the discoverers following a suggestion by A. Fujii, Y. Yabu and T. Sato.

(7189) Kuniko

1992 SX_{12}. Discovered 1992 September 28 by K. Endate and K. Watanabe at Kitami.

Named in honor of Kuniko Fujita (1923-1992), née Sofue, amateur astronomer and poet. She became interested in astronomy at an early age and made an important discovery – a nova, later named CP Pup – when she was a 19-year-old schoolgirl; she was awarded a prize from the Japan Astronomical Society as the first discoverer. She became a member of the Kawasaki Astronomical Club in 1972, and she visited Australia and New Zealand in 1986 to observe comet 1P/Halley. She joined expeditions to the Okinawa annular eclipse in 1986 and to the total eclipse in Mexico in 1991. A talented poet and calligrapher, she published a book of petry, *Hoshinagisa*. (M 32789; M 32819; M 32819)

Named by the discoverers following suggestions from S. Morikubo and T. Minowa.

(7190) 1993 GB_1

Discovered 1993 April 15 by H. E. Holt at Palomar.

(7191) 1993 MA_1

Discovered 1993 June 18 by H. E. Holt at Palomar.

(7192) Cieletespace

1993 RY_1. Discovered 1993 September 12 by K. Endate and K. Watanabe at Kitami.

Named for the monthly astronomical magazine edited in Paris by Alain Cirou and published by the Association Française d'Astronomie. Created in 1947 by a small group of amateurs under the name *Le ciel normand*, *Ciel et Espace* is one of the leading magazines of astronomy and space science in Europe. Scientists, journalists and amateurs all write for the magazine, and important celestial phenomena, such as eclipses and comets, provide an opportunity for the editorial staff to invite their readers to observe the sky. (M 33386)

Name proposed by the discoverers following a suggestion by A. Fujii.

(7193) Yamaoka

1993 SE_2. Discovered 1993 September 19 by K. Endate and K. Watanabe at Kitami.

Named in honor of Hitoshi Yamaoka (1965-), astrophysicist at Kyushu University who spezializes in the study of supernovae and novae. He promotes a supernova search project at public observatories in Japan. (M 33386)

(7194) 1993 SR_3

Discovered 1993 September 18 by H. E. Holt at Palomar.

(7195) 1994 AJ

Discovered 1994 January 2 by T. Kobayashi at Oizumi.

(7196) Baroni

1994 BF. Discovered 1994 January 16 by A. Boattini and M. Tombelli at Cima Ekar.

Named in honor of the amateur astronomer Sandro Baroni (1939-). Since 1953 he has pursued many different kinds of astronomical observation, especially of variable stars, comets, lunar occultations and asteroidal occultations. He gives lectures on a regular basis at the Civic Planetarium in Milan, where he lives. Baroni has been an inspiration for generations of amateur astronomers and has shared his knowledge by publishing many informative articles, especially on the history of astronomy. (M 29149)

(7197) Pieroangela

1994 BH. Discovered 1994 January 16 by A. Boattini and M. Tombelli at Cima Ekar.

Named in honor of Piero Angela (1928-). A journalist and science writer, he has created hundreds of science programs for Italian television, including the weekly science magazine "Quark". He has produced and hosted many popular series, such as "The Marvellous Machine" (a journey inside the human body) and "The Planet of the Dinosaurs", both of which have been shown in more than 40 countries. He has written numerous books on a variety of subjects, including astronomy, biology, psychology and economics, and he has received two honorary doctorates from Italian universities in natural and biological sciences. (M 29672)

(7198) Montelupo

1994 BJ. Discovered 1994 January 16 by A. Boattini and M. Tombelli at Cima Ekar.

Named for a small but busy town very near Florence with a tradition in ceramics that goes back to the Middle Ages. During the Renaissance, Montelupo was the center of the majolica production under the Medici family. Today, the town is well known for its artistic production and for its industrial and handicraft activities, so that it is also called the "City of the ceramics". (M 29149)

Name proposed by the second discoverer, who was born in Montelupo, where most of the follow-up observations of this minor planet were made, and where she and the first discoverer both started their astrometric work on minor planets and comets.

(7199) Brianza

1994 FR. Discovered 1994 March 28 by M. Cavagna and V. Giuliani at Sormano.

Named for the area located among the cities of Milan, Como and Lecco. The name is thought to be derived from the Celtic word "brig", i.e., height. In fact, Brianza is a verdant land, with lakes, hills and prealpine mountains. It is also one of the most industrial and productive parts of Italy. The name also honors the "Gruppo Astrofili Brianza", a group of amateur astronomers to which the discoverers belong, as it celebrates 15 years of activity. (M 29149)

(7200) 1994 NO

Discovered 1994 July 8 by T. B. Spahr at Tucson.

(7201) 1994 UF$_1$

Discovered 1994 October 25 by S. Otomo at Kiyosato.

(7202) 1995 DX$_1$

Discovered 1995 February 19 by T. Niijima and T. Urata at Ojima.

(7203) 1995 DG$_2$

Discovered 1995 February 27 by S. Otomo at Kiyosato.

(7204) Ondřejov

1995 GH. Discovered 1995 April 3 by P. Pravec at Ondřejov.

Named for the village, 35 km to the southeast of Prague, where the oldest active astronomical observatory in the Czech Republic is located. The Ondřejov Observatory was founded in 1898, and it merged into the Astronomical Institute in 1953. The observatory also belongs to the Academy of Sciences of the Czech Republic. This is the first numbered minor planet to be discovered from the Ondřejov Observatory. (M 29149)

(7205) 1995 YE$_1$

Discovered 1995 December 21 by T. Kobayashi at Oizumi.

(7206) Shiki

1996 QT. Discovered 1996 August 18 by A. Nakamura at Kuma.

Named in memory of Shiki Masaoka (1867-1902), one of the most famous Haiku poets of the Meiji period, born in Matsuyama City. He proposed a revolutionary movement to adopt a new style of Haiku in terms of sketches from life, instead of using the fictitious and imaginary expressions that were common before him. His style attracted many followers and has since become common. In Japan Standard Time this minor planet was discovered on Aug. 19, which is the day of Haiku. (M 29149)

(7207) Hammurabi

2133 P-L. Discovered 1960 September 24 by C. J. van Houten and I. van Houten-Groeneveld at Palomar.

Named for the Babylonian king Hammurabi, who reigned from 1792 to 1750 B.C. and united the different small states of Ur into a single state covering much of Mesopotamia, with Babylon as its capital. Many cuneiform tablets praise his justice. Famous is the stela from Susa containing the Code of Hammurabi. (M 30800)

(7208) Ashurbanipal

2645 P-L. Discovered 1960 September 24 by C. J. van Houten and I. van Houten-Groeneveld at Palomar.

Named for the Assyrian king Ashurbanipal (c.693-626 B.C.), who became king in 669 B.C. and is historically important because of his collection of Babylonian and Assyrian documents and literature on clay tablets in cuneiform characters. The great epic poem of Gilgamesh {see planet (1812)} is found in his library. Also impressive are the large stone reliefs at his palace, where Ashurbanipal's deeds are detailed. (M 30800)

(7209) Cyrus

3523 P-L. Discovered 1960 October 17 by C. J. van Houten and I. van Houten-Groeneveld at Palomar.

Named for the Persian king Cyrus II, the Great, who reigned from 559 to 529 B.C. He conquered the Medean capital of Ecbatana and subjugated the Lydian king Croesus and the Greek colonies of Asia Minor. In 539 B.C. he conquered Babylon, then the greatest city of the ancient world. He gave the Jews permission to return to Canaan and to rebuild their temple. (M 30800)

(7210) Darius

6555 P-L. Discovered 1960 September 24 by C. J. van Houten and I. van Houten-Groeneveld at Palomar.

Named for the Persian king Darius I, the Great (550-486 B.C.), who became king after the death of Cambyses II and married two daughters of Cyrus II {see planet (7209)}. His splendid palace in Persepolio was destroyed by Alexander the Great. Darius was defeated by the Greeks at the battle of Marathon in 490 B.C., and he died as he was preparing a new campaign against them. (M 30800)

(7211) Xerxes

1240 T-1. Discovered 1971 March 25 by C. J. van Houten and I. van Houten-Groeneveld at Palomar.

Named for the Persian king Xerxes I (519-465 B.C.), son of Darius and Atossa {see, respectively, planets (7210) and (810)}. He continued the war against the Greeks but was defeated by them in naval battles at Salamis (480 B.C.) and Mycale (479 B.C.) and on land at Plateaa (479 B.C.). He was killed during a palace revolution. Aeschylus {see planet (2876)} made him the main character in his tragedy *Persai*. (M 30800)

(7212) Artaxerxes

2155 T-2. Discovered 1973 September 29 by C. J. van Houten and I. van Houten-Groeneveld at Palomar.

Named for the Persian kings Artaxerxes. Artaxerxes I, Macrocheir (484-424 B.C.) was the younger son of Xerxes I {see planet (7211)} and the father of Xerxes II. A peace treaty with Athens in 449 B.C. marked the end of Persian expansion, with Persia recognizing the independence of the Greek cities in Asia Minor. Artaxerxes II, Mnemon, reacquired those Greek cities and also received Cyprus in a peace treaty in 386 B.C. (M 30800)

(7213) 1967 KB

Discovered 1967 May 31 at the Felix Aguilar Observatory at El Leoncito.

(7214) Antielus

1973 SM_1. Discovered 1973 September 19 by C. J. van Houten and I. van Houten-Groeneveld at Palomar.

Named for one of the Greek heroes during the Trojan war. Antielus was one of the heroes hidden in the wooden horse. (M 30478)

(7215) Gerhard

1977 FS. Discovered 1977 March 16 by H.-E. Schuster at La Silla.

Named in memory of Gerhard Bachmann (1931-1996), who came to the European Southern Observatory in 1970 and was head of administration at the organization from 1972 to 1996. Throughout this time he greatly contributed to ESO's success. His unusual diplomatic abilities and thorough knowledge of the interaction of science and politics ensured the smooth running of the observatory and efficient interaction with the member countries and the European Union. The naming marks his retirement from ESO and subsequent untimely death. (M 28622)
Name proposed by the discoverer. Citation prepared by R. M. West.

(7216) 1977 QQ_2

Discovered 1977 August 21 by N. S. Chernykh at Nauchnyj.

(7217) Dacke

1979 QX_3. Discovered 1979 August 22 by C.-I. Lagerkvist at La Silla.

Named for Nils Dacke (?-1543), born in Torsås in the Swedish province Småland. Dacke was the leader of the peasants who rebelled against king Gustav Wasa in 1542. This uprising, mainly against clerical policy, tax policy and prohibition of trade across the Danish border, was the last of the large rebellions in Sweden against the king and centralized power. Initially, Dacke and his army had great success and forced the king to sign a peace treaty after he lost most of the southern provinces to the rebels. The next year the king hired foreign mercenaries, and Dacke's army was beaten. Dacke was shot, and the uprising was at an end. (M 31297)

(7218) 1979 SK

Discovered 1979 September 19 by J. Květoň at Kleť.

(7219) 1981 EZ_{47}

Discovered 1981 March 3 by S. J. Bus at Siding Spring.

(7220) Philnicholson

1981 QE. Discovered 1981 August 30 by E. Bowell at Anderson Mesa.

Named in honor of Philip D. Nicholson (1951-), professor of astronomy at Cornell University, an Australian by birth. Nicholson's research centers on the orbital dynamics of planetary ring systems and natural satellites, and infrared observational studies of planets, their satellites and rings. He has also investigated the dynamics of a planetary system around the pulsar PSR 1257+12 and has studied the zodiacal dust bands. Nicholson is codiscoverer of two irregular satellites of Uranus and is currently editor of the journal *Icarus*, which is devoted to planetary science. (M 32790)

(7221) 1981 SJ

Discovered 1981 September 22 by Z. Vávrová at Kleť.

(7222) 1981 TJ_3

Discovered 1981 October 7 by T. M. Smirnova at Nauchnyj.

(7223) Dolgorukij

1982 TF_2. Discovered 1982 October 14 by L. V. Zhuravleva and L. G. Karachkina at Nauchnyj.

Named in memory of Yurij Dolgorukij (c.1095-1157), prince of Suzdal and Kiev's grand prince, son of Vladimir Monomakh. The first reference to Moscow, made in connection with Dolgorukij's visit there, is in a monastic chronicle dated 1147. Later on, in 1156, Moscow was fortified by him. This is the reason to consider 1147 as the year of the foundation of the city and Dolgorukij as its founder, although archaeological evidence indicates that the site had been settled as early as the ninth century. (M 30100)

The name is given on the occasion of the 850th anniversary of the founding of Moscow.

(7224) Vesnina

1982 TK_3. Discovered 1982 October 15 by L. V. Zhuravleva at Nauchnyj.

Named in memory of the brothers Leonid Aleksandrovich Vesnin (1880-1933), Viktor Aleksandrovich Vesnin (1882-1950) and Aleksandr Aleksandrovich Vesnin (1883-1959), Russian architects who worked in creative cooperation. (M 34625)

(7225) Huntress

1983 BH. Discovered 1983 January 22 by E. Bowell at Anderson Mesa.

Named in honor of Wesley T. Huntress, Jr. (1942-), planetary cosmochemist and highly regarded director of NASA space science programs during the 1990s. Gaining international recognition for pioneering studies of chemical evolution in interstellar clouds, comets and planetary atmospheres, Huntress was instrumental in developing the astrochemical research group at the Jet Propulsion Laboratory. During his six years as NASA Associate Administrator for Space Science, the rate at which science missions were launched increased dramatically, along with the public awareness of space science. The naming honors Huntress on his departure from NASA after an illustrious 29-year career with the agency. (M 32348)

Name proposed by the discoverer following a suggestion by M. S. Allen, who prepared the citation.

(7226) Kryl

1984 QJ. Discovered 1984 August 21 by A. Mrkos at Kleť.

Named in memory of Karel Kryl (1944-1994), Czech singer and songwriter, from 1969 a resident of Germany working at Radio Free Europe. His songs were appreciated as a symbol of freedom by many people in the former Czechoslovakia. (M 31297)

Name suggsted by J. Tichá and M. Tichý.

(7227) 1984 SH_6

Discovered 1984 September 22 by H. Debehogne at La Silla.

(7228) MacGillivray

1985 GO. Discovered 1985 April 15 by E. Bowell at Anderson Mesa.

Named in honor of Sally M. MacGillivray of Sky Publishing Corporation. With an early bent for music and philosophy, followed by 20 years' experience in book publishing, she brought to the *Millennium Star Atlas* (1997) an elegance that is rare in a scientific work and evokes bibliophilic traditions. As publication manager of this collaborative venture with the European Space Agency's Hipparcos project, she masterfully orchestrated the many iterations by two teams of illustrators needed to complete the 1548 charts in the atlas. (M 31297)
Name suggested by R. W. Sinnott and endorsed by the discoverer.

(7229) Tonimoore

1985 RV. Discovered 1985 September 12 by the Spacewatch at Kitt Peak.

Named in honor of Toni L. Moore, who has worked at the University of Arizona's Lunar and Planetary Laboratory since 1986 as an observer, programer and data analyst on radial velocity observations of stars. She has devoted thousands of hours of observations to a search for planets orbiting other stars, a study of the stability of the solar spectrum and a unique investigation of *p*-mode oscillations in a star other than the sun. (M 33788)

(7230) Lutz

1985 RZ_1. Discovered 1985 September 12 by E. Bowell at Anderson Mesa.

Named in honor of Barry L. Lutz (1944-), professor of physics and astronomy and currently department chair at Northern Arizona University. Raised in H. L. Mencken's "Red Lion, PA", Lutz entered a career in astrophysical sciences in which he applied his training in molecular spectroscopy to astronomical objects ranging from comets to planets to stars. He was responsible for the discovery of diatomic carbon in the interstellar medium, providing important confirmation of ion-molecule chemistry in interstellar clouds. He was also codiscoverer of the pressured-induced Lutz-Dressler band of molecular nitrogen and of deuterated methane and carbon monoxide in the atmosphere of Titan. (M 33788)

(7231) Porco

1985 TQ_1. Discovered 1985 October 15 by E. Bowell at Anderson Mesa.

Named in honor of Carolyn C. Porco (1953-), planetary scientist at the University of Arizona. A pioneer in the study of planetary ring systems, Porco has made important contributions to our knowledge of spokes in Saturn's rings, eccentric ring features in the systems of Saturn and Uranus and the azimuthal structure in the rings of Neptune. Through her contributions to the Voyager project and as the team leader for the Cassini imaging system, Porco has been a leader in spacecraft exploration of the outer solar system. In 1997, she originated the idea of sending a capsule containing ashes of E. M. Shoemaker {see planet (2074)} to the moon aboard the Lunar Prospector spacecraft. (M 32348; M 32377)
Name andorsed by the Shoemaker family.

(7232) 1985 UQ

Discovered 1985 October 20 by A. Mrkos at Kleť.

(7233) 1986 EQ_5

Discovered 1986 March 7 by G. DeSanctis at La Silla.

(7234) 1986 QV_3

Discovered 1986 August 29 by H. Debehogne at La Silla.

(7235) 1986 UY

Discovered 1986 October 30 by T. Seki at Geisei.

(7236) 1987 PA

Discovered 1987 August 1 by J. Phinney at Palomar.

(7237) 1988 VH

Discovered 1988 November 3 by K. Suzuki and T. Furuta at Toyota.

(7238) 1989 OA

Discovered 1989 July 27 by Y. Mizuno and T. Furuta at Kani.

(7239) Mobberley

1989 TE. Discovered 1989 October 4 by B. G. W. Manning at Stakenbridge.

Named in honor of Martin P. Mobberley, photographer and astrometrist of comets, minor planets, variable stars, novae and supernovae. His fine photographs and CCD images have appeared in the *Journal* of the British Astronomical Association for many years. Since 1990 he has occupied the position of Papers Secretary to the association, and as such he is responsible for arranging, refereeing and recommending articles for publication. For the past six years he has prepared and given an entertaining short talk on currently visible sky phenomena at every monthly meeting of the association. Mobberley is also assistant editor of *The Astronomer* magazine. (M 30100, 30117)

Name proposed by the discoverer following a suggestion by H. McGee, who also wrote the citation.

(7240) 1989 YG

Discovered 1989 December 19 by Y. Mizuno and T. Furuta at Kani.

(7241) Kuroda

1990 VF_3. Discovered 1990 November 11 by K. Endate and K. Watanabe at Kitami.

Named in honor of Takehiko Kuroda (1946-), since 1990 the first director of the Nishi-Harima Astronomical Observatory (NHAO) and one of the leading astronomers in Japanese public observatories. He played an important role in determining activities for public education, along with astronomical research, at public observatories. His main interests include interstellar dust distribution and surveys of our galaxy in infrared light. Kuroda is currently heavily engaged in the implementation of Japan's soon-to-be largest telescope, a 2.0-m at NHAO. (M 33788)

Name proposed by the discoverers following a suggestion by Y. Yamada.

(7242) Okyudo

1990 VG_3. Discovered 1990 November 11 by K. Endate and K. Watanabe at Kitami.

Named in honor of Masami Okyudo (1961-), Japanese astronomer and currently director of the Misato Observatory, Wakayama Prefecture. Since 1995, he has been a pioneer in astronomical education over the Internet, using remotely controlled telescopes to broadcast live images of celestial objects through cyberspace. (M 33788)

Name proposed by the discoverers following a suggestion by Y. Yamada.

(7243) 1990 VV_3

Discovered 1990 November 12 by S. Ueda and H. Kaneda at Kushiro.

(7244) Villa-Lobos

1991 PQ_1. Discovered 1991 August 5 by E. W. Elst at La Silla.

Named in memory of the well-known Brazilian musician Heitor Villa-Lobos (1887-1959), one of the foremost Latin-American composers of the twentieth century. After traveling throughout his country he returned to Rio de Janeiro with a large collection of manuscripts and an intimate knowledge of Afro-Brazilian music. In 1930 he was appointed director of musical education in São Paulo, and in 1945 he founded the Brazilian Academy of Music. His 2000 compositions include operas, ballets, symphonies and pieces for solo instrument. *Bachianas brasileiras*, one of his most characteristic works, shows the influence of Bach {see planet (1814)} and some of the French composers by his use of contrapunctal techniques applied to themes of Brazilian origin. (M 28623)

(7245) 1991 RN_{10}

Discovered 1991 September 10 by H. E. Holt at Palomar.

(7246) 1991 RP_{25}

Discovered 1991 September 12 by H. E. Holt at Palomar.

(7247) 1991 TD_1

Discovered 1991 October 12 by R. H. McNaught at Siding Spring.

(7248) 1992 EV_{21}

Discovered 1992 March 1 by the Uppsala-ESO Survey at La Silla.

(7249) 1992 SN

Discovered 1992 September 26 by A. Sugie at Taga.

(7250) Kinoshita

1992 SG_1. Discovered 1992 September 23 by K. Endate and K. Watanabe at Kitami.

Named in honor of Hiroshi Kinoshita (1941-), a celestial mechanician at the National Astronomical Observatory (formerly known as the Tokyo Astronomical Observatory) who works mainly on the dynamics of the solar system bodies. The present IAU Nutation Series is based on his nutation theory of the rigid earth. He served as president of Commission 4 during 1994-1997. (M 32790)
Name proposed by the discoverers following a suggestion by K. Hurukawa.

(7251) Kuwabara

1992 SF_{13}. Discovered 1992 September 30 by K. Endate and K. Watanabe at Kitami.

Named in honor of Syoji Kuwabara (1927-). A retired school teacher in Japan, he was superintendent of the board of education in Himeji City (1981-1993) and director of the city's Science Museum (1993-1996). An amateur astronomer, he was once an enthusiastic observer of occultations; now he studies local names and legends of stars and asterisms. (M 33386)

Name proposed by the discoverers following a suggestion by T. Sato, A. Fujii and K. Kitao.

(7252) Kakegawa

1992 UZ. Discovered 1992 October 21 by T. Urata at Oohira.

Named for an ancient city in central Japan famous for the wooden castle that was reconstructed there in 1993 following old carpentry methods. Also, a well-known Japanese green tea is produced in this area. (M 33788)
Name suggested and citation prepared by S. Kasahara.

(7253) Nara

1993 CL. Discovered 1993 February 13 by F. Uto at Kashihara.

Named for the prefecture and city where the oldest capital in Japan was located, and where the first Japanese astronomical observatory was established in 675.

Nara held a central position in politics, economics and culture from the latter half of the sixth to the end of the eighth century. (M 30800)

(7254) Kuratani

1993 TN_1. Discovered 1993 October 15 by K. Endate and K. Watanabe at Kitami.

Named in honor of Hiroshi Kuratani (1934-), astronomer and educator at Toyama Observatory since 1956. He played an important role in establishing Toyama Science Museum, which opened in 1979 and from which he retired in 1998. In the 1960s he was a pioneer in improving observational techniques, such as the application of TV and development of cooled emulsion cameras. Currently he is developing a new generation space simulator. (M 33386)

Name proposed by the discoverers following a suggestion by T. Sato and A. Fujii.

(7255) 1993 VY_1

Discovered 1993 November 11 by S. Ueda and H. Kaneda at Kushiro.

(7256) Bonhoeffer

1993 VJ_5. Discovered 1993 November 11 by F. Börngen at Tautenburg.

Named in memory of the German theologian Dietrich Bonhoeffer (1906-1945). Already the author of numerous papers and well-traveled, he was shocked by the offenses of the Nazi Regime and therefore grew to a revolutionary. He was pressed by two questions: "What is the Will of God?" and "What is the Task of the Church in this time?". Bonhoeffer was expelled from Berlin in 1938 and not allowed to discourse, print or publish. In 1943 he was arrested and on 1945 Apr. 9 was executed at Flossenbürg. (M 29149)

(7257) 1994 AH_1

Discovered 1994 January 7 by T. Kobayashi at Oizumi.

(7258) 1994 EF

Discovered 1994 March 5 by A. Vagnozzi at Stroncone.

(7259) 1994 EG_1

Discovered 1994 March 6 by Y. Shimizu and T. Urata at Nachi-Katsuura.

(7260) Metelli

1994 FN. Discovered 1994 March 18 at the Santa Lucia Observatory at Stroncone.

Named in memory of Orneore Metelli (1872-1938). He is internationally acknowledged as the founder of the school for naïve painting. He started painting well past the age of 50 in the back of his shoemaker's shop. In his more than 200 works, he represented a town that is a mirror of the world, a reflection of thousands of aspects of daily life. (M 33386)

(7261) 1994 GZ

Discovered 1994 April 14 by T. Kobayashi at Oizumi.

(7262) 1995 BX_1

Discovered 1995 January 27 by T. Kobayashi at Oizumi.

(7263) 1995 DP

Discovered 1995 February 21 by T. Kobayashi at Oizumi.

(7264) 1995 FK

Discovered 1995 March 26 by Y. Shimizu and T. Urata at Nachi-Katsuura.

(7265) Edithmüller

2908 T-2. Discovered 1973 September 30 by C. J. van Houten and I. van Houten-Groeneveld at Palomar.

Named in memory of the Spanish-born Swiss astronomer Edith A. Müller (1918-1995). She was well known for her publications on theoretical astrophysics, including supernovae. She worked at many different observatories in Europe and the United States. Müller served as IAU General Secretary during 1976-1979 and was the first woman to hold this post. (M 30800)

Obituaries published in Q.J.R. Astron. Soc., Vol. 37, No. 2, p. 267-268 (1996); JOSO Annu. Rep., 1995, p. VI (1996); IAU Inf. Bull., No. 76, p. V (1996); Bull. Am. Astron. Soc., Vol. 28, No. 4, p. 1457-1458 (1996).

(7266) Trefftz

4270 T-2. Discovered 1973 September 29 by C. J. van Houten and I. van Houten-Groeneveld at Palomar.

Named in honor of the German astronomer Eleonore Trefftz (1920-). Known for her work on physical processes of astronomical interest, she worked from 1948 until her retirement at the Max-Planck-Institute for Physics and Astrophysics in Munich. (M 30800)

(7267) 1943 DF

Discovered 1943 February 23 by L. Oterma at Turku.

(7268) 1972 TF

Discovered 1972 October 3 by L. V. Zhuravleva at Nauchnyj.

(7269) 1975 VK_2

Discovered 1975 November 2 by T. M. Smirnova at Nauchnyj.

(7270) Punkin

1978 NY_7. Discovered 1978 July 7 by E. Bowell at Palomar.

Named in honor of Erica Ann Broman (1956-) on the occasion of her 41st birthday 1997 August 25. Punkin is an affectionate family name for Broman, who originated the logo used for the Lowell Observatory during its centennial in 1994 and thereafter. She has been a significant donor to the observatory, as well as a patron of the deaf, the aged and God's lesser creatures. An educator and supporter of public education, she is a most gracious lady. (M 30100)

Name proposed by the discoverer following a suggestion by W. L. Putnam, who provided the citation.

(7271) 1979 SR_2

Discovered 1979 September 22 by N. S. Chernykh at Nauchnyj.

(7272) 1980 DD_1

Discovered 1980 February 21 by Z. Vávrová at Kleť.

(7273) 1981 EK_4

Discovered 1981 March 2 by S. J. Bus at Siding Spring.

(7274) Washioyama

1982 FC. Discovered 1982 March 21 by T. Seki at Geisei.

Named for a small mountain in the southern part of Kochi City {see planet (2396)}. (M 32094)

(7275) 1983 CY_2

Discovered 1983 February 15 by N. G. Thomas at Anderson Mesa.

(7276) Maymie

1983 RE. Discovered 1983 September 4 at the Oak Ridge Observatory at Harvard.

Named in honor of Marion R. Aymie, manager of the Human Resources Department at the Smithsonian Astrophysical Observatory from 1990 to 1997. A shrewd negotiator and trusted counselor, she is noted for her creativity, humanity and wisdom, as well as for having the courage and integrity to give her best advice, even when what she has to say may be difficult for the listener to hear. (M 29672)

(7277) Klass

1983 RM_2. Discovered 1983 September 4 by E. Bowell at Anderson Mesa.

Named in honor of Philip J. Klass (1919-), for more than 30 years the senior avionics editor of *Aviation Week and Space Technology* magazine. Klass is best known to the public for his work in skeptically evaluating sensationalist claims about UFOs, making numerous TV appearances and writing five books on the subject. He has also written extensively on surveillance satellites. (M 33788) Name and citation provided by R. Sheaffer.

(7278) 1985 UW_4

Discovered 1985 October 22 by L. V. Zhuravleva at Nauchnyj.

(7279) Hagfors

1985 VD_1. Discovered 1985 November 7 by E. Bowell at Anderson Mesa.

Named in honor of Tor Hagfors (1930-), in celebration of his 68th birthday and his retirement as director of the Max-Planck-Institut für Aeronomie. In the 1960s, at the Massachusetts Institute of Technology's Lincoln Laboratory, Hagfors conducted extensive radar investigations of the moon. The radar backscattering function now called the Hagfors Law, which he derived to support his analysis of lunar echoes, remains the most widely applied model for interpreting quasispecular echoes from the moon, Mercury, Mars and Venus. Hagfors served as director of operations of the National Astronomy and Ionosphere Center's Arecibo {see planet (4337)} Observatory from 1971 to 1973 and as director of NAIC during the time, 1982-1992, when Arecibo conducted 68 successful radar experiments on minor planets. (M 34342)

(7280) Bergengruen

1988 RA_3. Discovered 1988 September 8 by F. Börngen at Tautenburg.

Named in memory of the German writer Werner Bergengruen (1892-1964) a representative of Christian humanism in his numerous reports, novels, short stories and about 400 poems. He refused the despotism through his novels *The Great Tyrant* and *On Earth as in Heaven.* He gave many people strength by his distinct conviction against the Nazi terror. (M 29149)

(7281) 1988 RX_4

Discovered 1988 September 2 by H. Debehogne at La Silla.

(7282) 1989 BC

Discovered 1989 January 29 by S. Ueda and H. Kaneda at Kushiro.

(7283) 1989 TX_{15}

Discovered 1989 October 4 by H. Debehogne at La Silla.

(7284) 1989 VW

Discovered 1989 November 4 by Y. Oshima at Gekko.

(7285) Seggewiss

1990 EX_2. Discovered 1990 March 2 by E. W. Elst at La Silla.

Named in honor of Wilhelm Seggewiss (1937-), since 1995 head of the Hoher List Observatory, on the occasion of his 60th birthday. He works mainly on Wolf-Rayet stars and related topics. In 1981 he became a professor of astronomy at the University of Bonn. Seggewiss regularly gives lectures about subjects such as the history of the calendar and astrology, in order to put them in a proper perspective. (M 30801)

Name endorsed and citation written by M. Geffert at the request of the discoverer.

(7286) 1990 QZ_4

Discovered 1990 August 24 by H. E. Holt at Palomar.

(7287) Yokokurayama

1990 VN_2. Discovered 1990 November 10 by T. Seki at Geisei.

Named for a mountain in Kochi prefecture that is a treasure house of rare plants. A museum was built on its summit in 1997. (M 32348)

(7288) 1991 FE_1

Discovered 1991 March 18 by A. Sugie at Taga.

(7289) Kamegamori

1991 JU. Discovered 1991 May 5 by T. Seki at Geisei.

Named for a grassy mountain in the center of the mountainous region of Shikoku Island {see planet (4223)}. (M 32348)

(7290) Johnrather

1991 JY_1. Discovered 1991 May 11 by E. F. Helin at Palomar.

Named in honor of John D. Rather (1938-), astrophysicist, on the completion of his 60th revolution around the sun on 1998 Jan. 25. Long associated with special projects at NASA and in industry, he served in the early 1990s as chair of the Near-Earth Object Interception Committee, which examined possible mitigation processes against the threat of minor planets and comets on collision courses with the earth. As a direct consequence of his urging the use of an aggressive treatment that allowed his wife to triumph over cancer, he has in recent years transferred his interest in mitigation to a new career in leading-edge research to fight this scourge. (M 31297)

(7291) 1991 XC_1

Discovered 1991 December 13 by S. Otomo at Kiyosato.

(7292) 1992 EM_7

Discovered 1992 March 1 by the Uppsala-ESO Survey at La Silla.

(7293) Kazuyuki

1992 FH. Discovered 1992 March 23 by K. Endate and K. Watanabe at Kitami.

Named in honor of Kazuyuki Saitoh (1957-), associate director of the Nichihara Observatory since 1985 and president of the Shimane Society of Astronomy, Shimane Prefecture. He is a widely respected astronomy scholar and a keen popularizer of astronomy, well known throughout his home province. (M 33788)
Name proposed by the discoverers following a suggestion by Y. Yamada.

(7294) 1992 LM

Discovered 1992 June 3 by G. J. Leonard at Palomar.

(7295) 1992 MB

Discovered 1992 June 22 by S. Ueda and H. Kaneda at Kushiro.

(7296) Lamarck

1992 PW_1. Discovered 1992 August 8 by E. W. Elst and C. Pollas at Caussols.

Named for Jean-Baptiste Lamarck (1744-1829), French biologist, renowned for his idea that acquired traits are inheritable. In 1778, after nine years of botanical field study, he published his three volume *Flore française*. As a result of his urging, the Muséum National d'Histoire Naturelle was founded in 1793. (M 30801)

(7297) 1992 UG

Discovered 1992 October 21 by A. Sugie at Taga.

(7298) 1992 WM_5

Discovered 1992 November 26 by K. Suzuki and T. Urata at Toyota.

(7299) 1992 WZ_5

Discovered 1992 November 21 by E. F. Helin at Palomar.

(7300) Yoshisada

1992 YV_2. Discovered 1992 December 26 by T. Urata at Oohira.

Named for Yoshisada Shimizu (1943-), a Japanese orthopaedist and an active amateur astronomer. In addition to publishing many outstanding astrophotographies, he has been searching for minor planets since 1993. (M 30478)

(7301) 1993 AB

Discovered 1993 January 2 by A. Natori and T. Urata at Yakiimo.

(7302) 1993 CQ

Discovered 1993 February 10 by S. Ueda and H. Kaneda at Kushiro.

(7303) 1993 FS_1

Discovered 1993 March 25 by S. Ueda and H. Kaneda at Kushiro.

(7304) 1994 AE_2

Discovered 1994 January 9 by T. Kobayashi at Oizumi.

(7305) Ossakajusto

1994 CX_1. Discovered 1994 February 8 by K. Endate and K. Watanabe at Kitami.

Named in memory of Justo Ossaka (1922-1998), emeritus director of the Sendai {see planet (3133)} Astronomical Observatory. Graduating from the Geophysical Institute, Tohoku University, in 1950, he worked at the Sendai Observatory from 1956 to 1994. He contributed greatly to attracting amateur astronomers and planetarium visitors. (M 33788)

Name proposed by A. Watanabe and M. Koishikawa of the Sendai Astronomical Observatory.

(7306) Panizon

1994 EH. Discovered 1994 March 6 at the Santa Lucia Observatory at Stroncone.

Named in honor of Franco Panizon (1925-). Head of the pediatric department of the University of Trieste, he is one of the key figures in Italian pediatrics, endowed with a deep knowledge of all scientific disciplines focusing on human beings, a true master of science and life. From Trieste he has influenced pediatricians all over the country and has founded a pediatric school that teaches care for children through mind and heart. (M 33386)

(7307) 1994 GT_9

Discovered 1994 April 13 by Y. Shimizu and T. Urata at Nachi-Katsuura.

(7308) Hattori

1995 BQ_4. Discovered 1995 January 31 by Y. Shimizu and T. Urata at Nachi-Katsuura.

Named in honor of Tadahiko Hattori (1908-1962), who worked at the Tokyo Astronomical Observatory and International Latitude Observatory and was appointed the first director of the central bureau of the International Polar Motion Service two months before his death. (M 34625)

Name proposed by the second discoverer following a suggestion by K. Hurukawa. Citation prepared by K. Yokoyama.

(7309) 1995 FU

Discovered 1995 March 28 by T. Kobayashi at Oizumi.

(7310) 1995 OL_1

Discovered 1995 July 19 at the Beijing Observatory at Xinglong.

(7311) 1995 TU

Discovered 1995 October 14 by D. di Cicco at Sudbury.

(7312) 1996 AT_3

Discovered 1996 January 13 by S. Ueda and H. Kaneda at Kushiro.

(7313) Pisano

6207 P-L. Discovered 1960 September 24 by C. J. van Houten and I. van Houten-Groeneveld at Palomar.

Named for the family of Italian sculptors. Nicoló Pisano (1225-c.1280) combined in his sculptures the Italian Romanesque and French Gothic styles. His work is found in the pulpits of the baptistery in Pisa and the dome in Siena, in the fountain of Perugia and tombs in Bologna. Giovanni Pisano (c.1250-c.1314), the son of Nicoló, produced work found in Siena, Pisa and Padua. Andrea Pisano (c.1290-c.1348) created the bronze south door of the baptistery in Florence. (M 30801)

(7314) Pevsner

2146 T-1. Discovered 1971 March 25 by C. J. van Houten and I. van Houten-Groeneveld at Palomar.

Named in memory of Antoine Pevsner (1886-1962), Russian-born French sculptor and painter. He and his brother, who used the name Naum Gabo (1890-1977) and later moved to the United States, used mathematical figures as a basis for their many modern sculptures. They worked with bronze, copper and various plastics. They were inspired by the Russian icons and later by Braque and Picasso {see planet (4221)}. In turn, they had a marked influence on younger generations of sculptors. (M 30801)

(7315) Kolbe

1136 T-2. Discovered 1973 September 29 by C. J. van Houten and I. van Houten-Groeneveld at Palomar.

Named in memory of the German sculptor Georg Kolbe (1877-1947). Kolbe was educated as a painter, but in 1911 he became famous for his bronze figure of the dancer. In 1947 he created the monumental statue of Beethoven {see planet (1815)} in Frankfurt. (M 30801)

(7316) Hajdu

3145 T-2. Discovered 1973 September 30 by C. J. van Houten and I. van Houten-Groeneveld at Palomar.

Named in memory of Etienne Hajdu (1907-1996), Roumanian-born sculptor who studied in Budapest and Vienna and who lived and worked in Paris from 1927.

His abstract sculptures are carved directly in marble of various colors, schist and onyx, and the free-standing works in stone, bronze, aluminum and wood combine a refined effect of light and shadow on the surfaces of volumes with a controlled tension of outlines both of inner open spaces and other contours. Reliefs stamped into thick paper, designs for coins, tapestries, stage-settings and customs, as well as decorations for Sèvres porcelain, show Hajdu's many-sided talent. (M 30801)

Name proposed by the discoverers following a suggestion by C. Portheine, who also prepared the citation.

(7317) Cabot

1940 ED. Discovered 1940 March 12 by G. Kulin at Budapest.

Named for Giovanni Caboto or John Cabot (1449/50-1498/99), who made the first recorded landfall in North America since the Norse voyages. Genoese by birth and Venetian by citizenship, Cabot moved to England in 1484, apparently motivated by his idea of reaching Asia by sailing westward across the Atlantic. In 1497 he and his crew of 18 sailed from Bristol in a tiny vessel, the 'Matthew', arriving off Newfoundland on June 24, after which he explored the coast southward to Nova Scotia. His account of the sea swarming with fish led to great interest in the rich fishery around the 'New found land'. Named to commemorate 500 years of continued contact between Europe and Canada. (M 30101)
Name suggested and citation prepared by G. C. L. Aikman.

(7318) 1969 OX

Discovered 1969 July 17 by B. A. Burnasheva at Nauchnyj.

(7319) Katterfeld

1976 SA_6. Discovered 1979 September 24 by N. S. Chernykh at Nauchnyj.

Named in honor of Gennadij Nikolaevich Katterfeld (1927-), geologist and planetologist in St. Petersburg, author of investigations on the earth, the moon, Mercury and Mars, as well as a specialist in the history of science and culture. (M 34625)

(7320) 1978 TP_6

Discovered 1978 October 2 by L. V. Zhuravleva at Nauchnyj.

(7321) 1979 MZ_2

Discovered 1979 June 25 by E. F. Helin and S. J. Bus at Siding Spring.

(7322) 1979 SW_2

Discovered 1979 September 22 by N. S. Chernykh at Nauchnyj.

(7323) 1979 SD_9

Discovered 1979 September 22 by N. S. Chernykh at Nauchnyj.

(7324) Carret

1981 BC. Discovered 1981 January 31 at the Harvard College Observatory at Harvard.

Named in honor of Philip L. Carret (1896-), on the occasion of his 101st birthday and the 80th anniversary of his graduation from Harvard University. Passionately interested in solar eclipses, Carret has travelled the globe for most of the century in search of them - from Borneo to Siberia, from Baja to Kenya and from Prince Edward Island to Indonesia. Dean of American investment management firms and legendary stock picker, he created one of the first mutual funds in the U.S., Pioneer Fund, in 1928 and helped to found the mutual fund industry. He has been generously concerned about education, and about the environment and wildlife. (M 31025)

(7325) 1981 QA_1
Discovered 1981 August 28 by Z. Vávrová at Kleť.

(7326) 1981 UK_{22}
Discovered 1981 October 24 by S. J. Bus at Palomar.

(7327) Crawford
1983 RZ_1. Discovered 1983 September 6 by E. Bowell at Anderson Mesa.

Named in honor of David L. Crawford (1931-), astronomer at the Kitt Peak National Observatory recognized for his fundamental contributions to protecting dark skies around observatories from light pollution through good outdoor lighting practices. In 1972 he was instrumental in establishing the first comprehensive lighting code, which regulated outdoor lighting in southern Arizona, and the code has become a model for such regulation throughout the world. Crawford was cofounder and executive director of the International Dark-Sky Association, a worldwide organization dedicated to the preservation of dark skies. He was also responsible for establishing the fundamental standards of the Strömgren photometric system. (M 34343)

Name suggested and citation provided by D. R. Davis.

(7328) 1984 SC_1
Discovered 1984 September 20 by A. Mrkos at Kleť.

(7329) 1985 GK
Discovered 1985 April 14 by E. Bowell at Anderson Mesa.

(7330) 1985 TD
Discovered 1985 October 15 by E. Bowell at Anderson Mesa.

(7331) 1985 TV
Discovered 1985 October 15 by E. Bowell at Anderson Mesa.

(7332) 1986 XJ_5
Discovered 1986 December 4 by A. Mrkos at Kleť.

(7333) 1987 SM_4
Discovered 1987 September 29 by E. Bowell at Anderson Mesa.

(7334) 1988 QV
Discovered 1988 August 17 by A. Mrkos at Kleť.

(7335) 1989 JA
Discovered 1989 May 1 by E. F. Helin at Palomar.

(7336) 1989 RS_1
Discovered 1989 September 6 by E. F. Helin at Palomar.

(7337) 1990 QH_1
Discovered 1990 August 22 by H. E. Holt at Palomar.

(7338) 1990 VJ_3
Discovered 1990 November 12 by H. Shiozawa and M. Kizawa at Fujieda.

(7339) 1991 RA_{16}
Discovered 1991 September 15 by H. E. Holt at Palomar.

(7340) 1991 UA_2
Discovered 1991 October 29 by S. Ueda and H. Kaneda at Kushiro.

(7341) 1991 VK

Discovered 1991 November 1 by E. F. Helin and K. J. Lawrence at Palomar.

(7342) Uchinoura

1992 FB_1. Discovered 1992 March 23 by K. Endate and K. Watanabe at Kitami.

Named for the town in southern Kyushu where the Kagoshima {see planet (4703)} Space Center of the Institute of Space and Astronautical Science (ISAS) is located. A small fishing village, the town entered the limelight in 1962, when it was selected as the launching site for ISAS (now ISAS Ministry of Education, Science, Sports and Culture). ISAS satellites are launched toward the Pacific Ocean from a plateau in the southeastern part of the town. The facilities are getting larger as the rockets also grow in size. The usual population of the town is about 10 000, but at the time of a launch staff and reporters typically cause a 20-percent increase. (M 34343)

(7343) Ockeghem

1992 GE_2. Discovered 1992 April 4 by E. W. Elst at La Silla.

Named for the Flemish composer Johannes Ockeghem (1420-1497), the most renowned polyphonist of the second half of the fifteenth century. He started career at the Antwerp cathedral and later served at the court of several French kings. Among his compositions were motets, chansons, canons and a *Deo gratias* for 36 voices. He had a great influence on his fellow musicians and still fascinates contemporary composers. (M 30801)

Citation written by K. Leterme at the request of the discoverer.

(7344) Summerfield

1992 LU. Discovered 1992 June 4 by C. S. Shoemaker and D. H. Levy at Palomar.

Named in honor of Robert and Lisa Summerfield, for their exhaustive effort to spread astronomy to the public. The couple brings their collection of telescopes, one a 0.9-m reflector, to public star parties across the United States, and they complete several cross-country marathon drives in this effort. Their organization, "Astronomy To Go", sets up education sessions in dozens of schools each year. (M 30478)

Citation prepared by David and Wendee Levy.

(7345) Happer

1992 OF. Discovered 1992 July 28 by R. H. McNaught at Siding Spring.

Named for Felix Happer, a character in Bill Forsyth's wonderful motion picture *Local Hero*. Happer desperately wanted a comet to be named for him; this minor planet will have to do. (M 29672)

Name suggested by D. I. Steel, who prepared the citation.

(7346) 1993 DQ_2

Discovered 1993 February 20 by E. W. Elst at Caussols.

(7347) 1993 EW

Discovered 1993 March 12 by S. Ueda and H. Kaneda at Kushiro.

(7348) 1993 FJ_{22}

Discovered 1993 March 21 by the Uppsala-ESO Survey at La Silla.

(7349) 1993 QK_4

Discovered 1993 August 18 by E. W. Elst at Caussols.

(7350) 1993 VA

Discovered 1993 November 7 by R. H. McNaught at Siding Spring.

(7351) 1993 XB_1
Discovered 1993 December 12 by T. Kobayashi at Oizumi.

(7352) 1994 CO
Discovered 1994 February 4 by S. Ueda and H. Kaneda at Kushiro.

(7353) Kazuya
1995 AC_1. Discovered 1995 January 6 by M. Hirasawa and S. Suzuki at Nyukasa.

Named in honor of Kazuya Yoshida (1960-). An authority in robot engineering at Tohoku University, Yoshida is now developing a robotic system to explore the minor planets. He is an excellent astrophotographer and a member of Mt. Nyukasa Station, where this minor planet was discovered. (M 34343)

(7354) 1995 BR_1
Discovered 1995 January 27 by T. Kobayashi at Oizumi.

(7355) Bottke
1995 HN_2. Discovered 1995 April 25 by the Spacewatch at Kitt Peak.

Named in honor of William F. Bottke, Jr., known for his research on the collisional and dynamical evolution of minor planets. He has also contributed to the study of the origin and evolution of NEAs, in particular by analyzing the formation of doublet craters found on the terrestrial planets. (M 32348)

(7356) Casagrande
1995 SK_5. Discovered 1995 September 27 at the Santa Lucia Observatory at Stroncone.

Named in memory of Alessandro Casagrande (1922-1964). Composer and orchestra conductor, he skillfully headed the music school in Terni and served as artistic manager of the city's symphonic bureau "S. Falchi". He was also a painter. In 1965 an international piano competition was named for him. (M 33386)

(7357) 1995 UJ_7
Discovered 1995 October 27 by S. Ueda and H. Kaneda at Kushiro.

(7358) 1995 YA_3
Discovered 1995 December 27 by T. Kobayashi at Oizumi.

(7359) Messier
1996 BH. Discovered 1996 January 16 by M. Tichý at Kleť.

Named in memory of Charles Messier (1730-1817), French astronomer and first "comet hunter". He found more than a dozen comets and independently recovered Halley's Comet at its 1759 return. Nowadays, Messier is best known for his catalogue of 103 nebulae, star clusters and galaxies - a compilation he made in order to reduce confusion to comet hunters. (M 28672)

(7360) Moberg
1996 BQ_{17}. Discovered 1996 January 30 by C.-I. Lagerkvist at La Silla.

Named in memory of Vilhelm Moberg (1898-1973), Swedish novelist and writer. Moberg mainly wrote about society's lower classes and always fought for the individual against the authorities. He also strove unsuccessfully to make Sweden a republic. Among his best-known novels are *Utvandrarna* ("The Emigrants") and *Invandrarna* ("The Immigrants"), about a familiy moving from Småland to Minnesota during the nineteenth century. (M 31297)

(7361) Endres
1996 DN_1. Discovered 1996 February 16 by the JPL NEAT Program at Haleakala.

Named for Michael Hart Endres, who as former site manager of the GEODSS project played a seminal role in the early discussions, meetings and concepts that ultimately resulted in the establishment of the Near-Earth Asteroid Tracking (NEAT) program in collaboration with the U.S. Air Force using the Ground-Based Electro Optical Deep Space Surveillance (GEODSS) 1-m telescope. "Mike" for forty years thrived in a wide-ranging career from design, installation, operation of technical equipment and management of personnel associated with technical services at various companies. (M 29672)

The NEAT team and GEODSS personnel heartily endorse the naming of the first NEAT numbered minor planet in his honor.

(7362) Rogerbyrd

1996 EY. Discovered 1996 March 15 by the JPL NEAT Program at Haleakala.

Named in honor of Roger Byrd, who has 16 years experience in project management, systems research and development, computer systems, and electronics maintenance and repair. He has extensive management experience with the GEODSS Project. He is currently the site manager at the GEODSS site in Maui and he has obtained research and development experiences working on radar projects and various space-related projects, as well as hands-on technical experience of hardware and software. (M 29672)

The NEAT and GEODSS personnel salute his dedication by honoring him with one of the first numbered NEAT minor planets.

(7363) Esquibel

1996 FA_1. Discovered 1996 March 18 by the JPL NEAT Program at Haleakala.

Named in honor of Albert Esquibel, who has for 19 years been active in project management and electronic maintenance. For the last 13 years he has been with the GEODSS project at the four U.S. Air Force sites. He has worked closely with the Jet Propulsion Laboratory as they first tested the NEAT instrument with the GEODSS telescope system in Maui. Through his diligence on site, the NEAT/GEODSS program has enjoyed unparalleled success with its autonomous NEO discovery program. (M 29672)

The NEAT and GEODSS personnel salute his dedication by honoring him with one of the first numbered NEAT minor planets.

(7364) 1996 KS

Discovered 1996 May 22 at the Višnjan Observatory at Višnjan.

(7365) Sejong

1996 QV_1. Discovered 1996 August 18 by K. Watanabe at Sapporo.

Named on the occasion of the 600th anniversary of the birth of the Korean king Sejong (1397-1450), renowned for his many achievements in astronomy. He built the Royal Observatory of Choson in Seoul in 1437, equipping it with a whole series of astronomical instruments and installing an automatic striking clepsydra. He revised the Chinese calendar for Korean use and invented the rain gauge. He established the Hangul alphabet, which is still used throughout Korea. (M 30801) Name suggested by K. Hurukawa, citation material provided by I.-S. Nha.

(7366) 1996 UY

Discovered 1996 October 20 by T. Kobayashi at Oizumi.

(7367) Giotto

3077 T-1. Discovered 1971 March 26 by C. J. van Houten and I. van Houten-Groeneveld at Palomar.

Named for Giotto di Bondone (c.1266-1337), great Italian painter and architect. Giotto liberated art in general from Byzantine art and painted mainly religious themes. His most famous works are in Assisi and Florence, and he constructed the campanile of the Florence cathedral. Giotto saw comet 1P/Halley in 1301 and is supposed to have incorporated it into his fresco, the Adoration of the Magi, in Padua. For this reason, the European spacecraft that intercepted that same comet at its return in 1986 bore his name. (M 30801)

(7368) 1966 BB

Discovered 1966 January 20 at the Goethe Link Observatory at Brooklyn, Indiana.

(7369) 1975 AN

Discovered 1975 January 13 by T. M. Smirnova at Nauchnyj.

(7370) 1978 SM_5

Discovered 1978 September 27 by L. I. Chernykh at Nauchnyj.

(7371) 1978 VA_6

Discovered 1978 November 7 by E. F. Helin and S. J. Bus at Palomar.

(7372) 1979 HH

Discovered 1979 April 19 by J. C. Muzzio at Cerro Tololo.

(7373) 1979 QX_9

Discovered 1979 August 27 by N. S. Chernykh at Nauchnyj.

(7374) 1980 DL

Discovered 1980 February 19 by Z. Vávrová at Kleť.

(7375) 1980 PZ

Discovered 1980 August 14 by Z. Vávrová at Kleť.

(7376) 1980 UU_1

Discovered 1980 October 31 by S. J. Bus at Palomar.

(7377) 1981 EW_9

Discovered 1981 March 1 by S. J. Bus at Siding Spring.

(7378) 1981 EK_{18}

Discovered 1981 March 2 by S. J. Bus at Siding Spring.

(7379) 1981 EC_{29}

Discovered 1981 March 1 by S. J. Bus at Siding Spring.

(7380) 1981 RF

Discovered 1981 September 3 by N. G. Thomas at Anderson Mesa.

(7381) Mamontov

1981 RG_5. Discovered 1981 September 8 by L. V. Zhuravleva at Nauchnyj.

Named in memory of Savva Ivanovich Mamontov (1841-1918), manufacturer and patron of the arts, a prominent figure in Russian art, theater and music. (M 34625)

(7382) 1981 RJ_5

Discovered 1981 September 8 by L. V. Zhuravleva at Nauchnyj.

(7383) Lassovszky

1981 SE. Discovered 1981 September 30 at the Oak Ridge Observatory at Harvard.

Named in memory of Károly Lassovszky (1897-1961), a staff member of the Konkoly {see planet (1445)} Observatory in Budapest from 1921 and director from 1938 to 1943. His main research was on variable stars, but he initiated the Konkoly observational program on minor planets that was largely carried out by

G. Kulin {see planet (3019)}. In 1958 Lassovszky was invited to join the staff of the Smithsonian Astrophysical Observatory (which operates the Oak Ridge Observatory), where he became chief of the Satellite Tracking Program's Photoreduction Division. (M 30801)

Named following a proposal made at the 1997 General Assembly of the Hungarian Astronomical Association.

(7384) 1981 TJ

Discovered 1981 October 6 by Z. Vávrová at Kleť.

(7385) 1981 UQ_{11}

Discovered 1981 October 22 by N. S. Chernykh at Nauchnyj.

(7386) Paulpellas

1981 WM. Discovered 1981 November 25 at the Oak Ridge Observatory at Harvard.

Named in memory of Paul Pellas (1924-1997), physicist, meteoriticist, and curator at the Muséum d'Histoire Naturelle in Paris. He timed the early cooling histories of minor planets by studying the pattern of thermal annealing of fission tracks, created by the decay of the short-lived isotope plutonium 244, in the minerals of meteorites that once resided in these larger bodies. He also measured cosmic-ray track densities in lunar rocks to determine their residence times on the surface of the moon. (M 30101)

Name proposed by J. A. Wood.

Obituary published in Meteorit. Planet. Sci., Vol. 32, No. 6, p. 983-984 (1997).

(7387) 1982 BS_1

Discovered 1982 January 30 by E. Bowell at Anderson Mesa.

(7388) 1982 FS_3

Discovered 1982 March 23 by H. Debehogne at La Silla.

(7389) 1982 UE

Discovered 1982 October 17 by E. Bowell at Anderson Mesa.

(7390) Kundera

1983 QE. Discovered 1983 August 31 at the Kleť Observatory at Kleť.

Named in honor of Milan Kundera (1929-), Czech novelist, playwright and poet. He has written various works combining the comedy of living with political criticism. His novel *The Unbearable Lightness of Being* is well known and received wide international exposure as a very successful movie. Since 1975 he has lived in France. (M 33386)

Name proposed by J. Tichá following a suggestion by M. Šidlichovský.

(7391) Strouhal

1983 VS_1. Discovered 1983 November 8 by A. Mrkos at Kleť.

Named in memory of Vincenc Strouhal (1850-1922), professor of experimental physics. He founded and built up the Institute of Physics of the Czech part of Charles University (1907). (M 31297)

Name suggsted by J. Tichá and M. Šolc.

(7392) Kowalski

1984 EX. Discovered 1984 March 6 by E. Bowell at Anderson Mesa.

Named in honor of Richard Kowalski (1963-) of Quail Hollow Observatory, near Tampa, Florida. Kowalski is engaged in follow-up and discovery work on minor planets and comets. In 1998 he founded and has energetically maintained

the *Minor Planet Mailing List*, a web-based information service and chat group for amateur and professional observers. Kowalski has also been the main organizer of the 1999 Amateur-Professional Minor Planet Workshop at the Lowell Observatory. (M 34343)
Citation written by the discoverer and P. G. Comba.

(7393) 1984 SL_3
Discovered 1984 September 28 by B. A. Skiff at Anderson Mesa.

(7394) 1985 QX_4
Discovered 1985 August 18 by N. S. Chernykh at Nauchnyj.

(7395) 1985 RP_1
Discovered 1985 September 10 by Z. Vávrová at Kleť.

(7396) Brusin
1986 EQ_2. Discovered 1986 March 4 by W. Ferreri at La Silla.

Named in honor of Silvia Rosa Brusin, head of one of the principal Italian scientific programs. She is strongly involved in popular astronomy and is in close contact with scientists around the world. Her programs on Italian television constitute a serious reference for interested people. (M 34343)

(7397) 1986 QS
Discovered 1986 August 26 by H. Debehogne at La Silla.

(7398) Walsh
1986 VM. Discovered 1986 November 3 by A. Mrkos at Kleť.

Named in honor of Martin F. Walsh (1966-) on the occasion of his marriage to Aurélie Machon near Paris on 1998 June 6. (M 32094)

Name suggested by G. V. Williams, who made the identifications for this object and has known the bridegroom for more than 20 years.

(7399) 1987 BC_2
Discovered 1987 January 29 by E. W. Elst at La Silla.

(7400) 1987 QW_1
Discovered 1987 August 21 by E. W. Elst at La Silla.

(7401) 1987 QW_7
Discovered 1987 August 21 by E. W. Elst at La Silla.

(7402) 1987 YH
Discovered 1987 December 25 by T. Kojima at Chiyoda.

(7403) 1988 AV_1
Discovered 1988 January 14 by A. Mrkos at Kleť.

(7404) 1988 AA_5
Discovered 1988 January 13 by H. Debehogne at La Silla.

(7405) 1988 FF
Discovered 1988 March 16 by S. Ueda and H. Kaneda at Kushiro.

(7406) 1988 TD
Discovered 1988 October 3 by S. Ueda and H. Kaneda at Kushiro.

(7407) 1988 TL
Discovered 1988 October 3 by S. Ueda and H. Kaneda at Kushiro.

(7408) 1989 SB
Discovered 1989 September 23 by Y. Mizuno and T. Furuta at Kani.

(7409) 1990 BS
Discovered 1990 January 21 by M. Arai and H. Mori at Yorii.

(7410) Kawazoe
1990 QG. Discovered 1990 August 20 by T. Seki at Geisei.

Named in honor of Akira Kawazoe (1934-), a member of the Geisei Observatory staff with particular interests in meteorites and geology. (M 32095)

(7411) 1990 QQ_1
Discovered 1990 August 22 by H. E. Holt at Palomar.

(7412) Linnaeus
1990 SL_9. Discovered 1990 September 22 by E. W. Elst at La Silla.

Named in memory of the great Swedish botanist Carl von Linné (1707-1778). At an early age, Linnaeus developed a great love for flowers and herbs. This led him to develop the first major systematic system of nomenclature for the flora, and this became the internationally accepted standard. He organized the large-scale collection of botanical specimens in far-away countries and had a lively correspondence with his comtemporaries. (M 30801)

(7413) Galibina
1990 SH_{28}. Discovered 1990 September 24 by L. V. Zhuravleva and G. R. Kastel' at Nauchnyj.

Named in honor of Irina Vladimirovna Galibina (1929-), a celestial mechanician known for her research on the motions of long-period comets, minor planets and meteor streams. In recent years, in coauthorship with her husband, G. N. Katterfeld, Galibina published a series of papers devoted to the principal problems in astronomical geology. (M 34343)
Name proposed by G. R. Kastel'.

(7414) Bosch
1990 TD_8. Discovered 1990 October 13 by L. D. Schmadel and F. Börngen at Tautenburg.

Named in memory of Carl Bosch (1874-1940), outstanding German chemist, inventor of high-pressure ammonia synthesis, 1931 Nobel laureate in chemistry and enthusiastic amateur astronomer. During the 1920s he built a well-equipped private observatory on his estate at Heidelberg, where he carried out spectroscopic and photometric studies. He was an important patron of science, supporting especially the Heidelberg-Königstuhl Observatory, the Astronomische Gesellschaft and the Einstein Foundation. (M 34343)
Name proposed by the first discoverer, endorsed by G. Klare and H. Mandel.

(7415) 1990 VL_8
Discovered 1990 November 14 by T. Seki at Geisei.

(7416) Linnankoski
1990 WV_4. Discovered 1990 November 16 by E. W. Elst at La Silla.

Named in memory of the Finnish writer Johannes Linnankoski, pseudonym of Vihtori Peltonen (1869-1913). He promoted Finnish independence from Russia. He became well known for his novel *Laulu tulipunaisesta kukasta* (1905), or "Song of the blood-red flower". In 1960 the discoverer set much of this novel to music. (M 31025)

(7417) 1990 YE
Discovered 1990 December 19 by M. Arai and H. Mori at Yorii.

(7418) 1991 EJ_1
Discovered 1991 March 11 by T. Fujii and K. Watanabe at Kitami.

(7419) 1991 PN_{13}
Discovered 1991 August 5 by H. E. Holt at Palomar.

(7420) Buffon
1991 RP_{11}. Discovered 1991 September 4 by E. W. Elst at La Silla.

Named for the French naturalist Georges-Louis Leclerc de Buffon (1707-1788). In 1749 he started his project *Histoire naturelle*, an immense work of 36 volumes on nature, in which he treated in succession the history of the formation of the earth, the life of animals and plants and the world of minerals. In 1779 the work was completed with his famous *Epoques de la nature*. Buffon insisted on building his theories on facts, rather than on hypotheses, and he made experience the clue for scientific investigation. His elegant style raised writing on science to the ranks of great literature. (M 31025)

(7421) 1992 HL
Discovered 1992 April 30 by Y. Kushida and O. Muramatsu at Yatsugatake.

(7422) 1992 LP
Discovered 1992 June 3 by G. J. Leonard at Palomar.

(7423) 1992 PT_2
Discovered 1992 August 2 by H. E. Holt at Palomar.

(7424) 1992 PS_6
Discovered 1992 August 6 by H. E. Holt at Palomar.

(7425) Lessing
1992 RO_5. Discovered 1992 September 2 by E. W. Elst at La Silla.

Named for Gotthold Ephraim Lessing (1729-1781), German dramatist and writer on philosophy and aesthetics. He abandoned an early interest in theology for literature, philosophy and art. In 1766, his great treatise on aesthetics, *Laokoon: oder über die Grenzen der Malerei und Poesie*, was published. As a philosopher he is remembered for *Nathan der Weise* (1779), and especially for *Die Erziehung des Menschengeschlechts* (1780), in which he expressed his belief on the perfectability of the human race. Lessing's last years were lonely and poor, and upon his death he was buried at public expense. (M 31025)

(7426) 1992 US_4
Discovered 1992 October 27 by A. Sugie at Taga.

(7427) 1992 VD
Discovered 1992 November 2 by N. Kawasato at Uenohara.

(7428) 1992 YM
Discovered 1992 December 24 by T. Urata at Oohira.

(7429) 1992 YB_1
Discovered 1992 December 24 by T. Hioki and S. Hayakawa at Okutama.

(7430) Kogure
1993 BV_2. Discovered 1993 January 23 by K. Endate and K. Watanabe at Kitami.

Named in honor of Tomokazu Kogure (1925-   ), professor emeritus of Kyoto

University. His specialties include early-type peculiar stars, emission nebulae, interstellar medium, and nuclei of active galaxies. He served as president of the Astronomical Society of Japan and is currently director of Bisei Observatory in Okayama Prefecture. He is a frequent contributor to research and education on astronomy in Asian countries, especially Indonesia and Sri Lanka. (M 34343)

Name proposed by the discoverers following a suggestion by K. Sakka and T. Sato.

(7431) 1993 FN_{41}

Discovered 1993 March 19 by the Uppsala-ESO Survey at La Silla.

(7432) 1993 HL_5

Discovered 1993 April 23 by A. C. Gilmore and P. M. Kilmartin at Lake Tekapo.

(7433) Pellegrini

1993 KD. Discovered 1993 May 21 at the Farra d'Isonzo Observatory at Farra d'Isonzo.

Named in memory of Guglielmo Pellegrini (1937-1990), an amateur astronomer who built or modified many of the telescopes used by the Italian astrometric community. His ability, skill and kind temperament will long be remembered. His epigraph reads "In the stars you loved so much we'll find your light". (M 31025)

(7434) 1994 AB_3

Discovered 1994 January 14 by T. Kobayashi at Oizumi.

(7435) Sagamihara

1994 CZ_1. Discovered 1994 February 8 by K. Endate and K. Watanabe at Kitami.

Named for a city of 571 000 located some 50 km west of Tokyo on the vast riverside of the river Sagamigawa. In 1892 the Sagamino Base Line for the nationwide triangulation was set up there. In recent years the city has grown into a large industrial center. In 1989 the campus of the Institute of Space and Astronautical Science, which develops Japanese scientific satellites, moved to Sagamihara, where the satellites Yoko and Haruka were built. (M 34343)

(7436) Kuroiwa

1994 CB_2. Discovered 1994 February 8 by K. Endate and K. Watanabe at Kitami.

Named in memory of Goro Kuroiwa (1912-1990), Japanese astronomer and observer of variable stars. A student in the department of astronomy at the University of Tokyo on the occasion of the total solar eclipse on 1936 June 19, he independently discovered the nova CP Lac, along with Kazuaki Gomi {see planet (7035)}. While serving with the Japanese army in 1942 he independently discovered the nova CP Pup. He represented Japan in the geodetic survey program using photoelectric observations of lunar occultations, carried out from 1950 to the 1960s by the U.S. Army Map Service Far East. (M 34343)

(7437) Torricelli

1994 EF_3. Discovered 1994 March 12 by V. Goretti and A. Boattini at Cima Ekar.

Named in memory of Evangelista Torricelli (1608-1647), mathematician and physicist. In 1641 he met and worked with Galileo {see planet (697)} in Florence. Torricelli continued Galileo's activities after the latter's death, making telescopes, microscopes and improving methods of lens-making. Torricelli is known for geometrical studies that presaged the development of calculus, for a theorem in fluid mechanics and for the invention of the barometer. (M 30802)

(7438) Misakatouge

1994 JE_1. Discovered 1994 May 12 by A. Nakamura at Kuma.

Named for a pass, 720 meters above sea level, in the north of the town of Kuma {see planet (6255)}, where this minor planet was discovered. Because of its altitude and steepness, Misakatouge is often covered by thick snow in winter, and it suffers from heavy fog from spring to fall. Nevertheless, much traffic uses the pass, because it is a key route connceting the cities of Matsuyama and Kochi {see, respectively, planets (4844) and (2396)}, on the island of Shikoku {see planet (4223)}. (M 30101)

(7439) 1994 XG_1

Discovered 1994 December 6 by T. Kobayashi at Oizumi.

(7440) Závist

1995 EA. Discovered 1995 March 1 by M. Tichý at Kleť.

Named for the site of the most important ancient Celtic town in Bohemia. Situated south of Prague on a hill above Vltava river {see planets (2367) and (2123), respectively}, it consists of two parts named Hradiště and Šance. It was founded in the sixth century B.C. and was destroyed five centuries later. The ramparts, stone walls, moats and an acropolis are the most impressive remnants showing the Celtic history of the discoverer's native country found up to this time. (M 34625)

(7441) Láska

1995 OZ. Discovered 1995 July 30 by J. Tichá and M. Tichý at Kleť.

Named in memory of Václav Láska (1862-1943), Czech astronomer, geophysicist, geodesist and mathematician, professor of Charles University in Prague and founder of geophysical research in Czechoslovakia. (M 30478)

(7442) Inouehideo

1995 SC_5. Discovered 1995 September 20 by K. Endate and K. Watanabe at Kitami.

Named in honor of Hideo Inoue (1917-), Japanese astronomer. An astronomy enthusiast while still a child, he studied at the Tokyo College of Physics and at the Institute of Cosmical Physics in Kyoto. While participating in Kyoto University's expedition to the solar eclipse on 1941 Sept. 21 he obtained color photographs of the solar corona, the first in Japan. He later worked at the Peking Observatory, where he calculated the national ephemeris. After the war he taught at technical high schools in Japan. For the International Geophysical Year he led the Higasimatuyama Moonwatch Team. He is also an enthusiastic ham radio operator. (M 34343)

(7443) 1996 BR_2

Discovered 1996 January 26 by T. Kobayashi at Oizumi.

(7444) 1996 TM_{10}

Discovered 1996 October 9 by S. Ueda and H. Kaneda at Kushiro.

(7445) Trajanus

4116 P-L. Discovered 1960 September 24 by C. J. van Houten and I. van Houten-Groeneveld at Palomar.

Named for the Roman emperor Marcus Ulpius Trajanus (53-117). Trajan was adopted by emperor Nerva and became his successor in 98. He expanded the Roman empire to Mesopotamia, Assyria and Armenia. In his time the Roman empire had its greatest expansion. The Trajan column in the Forum contains his ashes in a golden urn. (M 30802)

(7446) Hadrianus

2249 T-2. Discovered 1973 September 29 by C. J. van Houten and I. van Houten-Groeneveld at Palomar.

Named for Publius Aelius Hadrianus (76-138), Roman emperor, who was adopted by Trajan {see planet (7445)} and who became his successor in 117. Hadrian stopped the expansion of the Roman empire, putting his energy instead into the construction of excellent roads, aquaducts and new cities. He also fortified the borders of the immense empire, notably with the wall between England and Scotland and the limes in Germany. He reformed the economy, administration and the law. An admirer of Greek culture, Hadrian built the Castello de Sant' Angelo in Rome. (M 30802)

(7447) Marcusaurelius

1142 T-3. Discovered 1977 October 17 by C. J. van Houten and I. van Houten-Groeneveld at Palomar.

Named for the Roman emperor Marcus Aurelius Antoninus (121-180), who became emperor in 161 upon the death of his adoptive father Antoninus Pius. He tried to reign with justice but could not prevent the decay of the empire. Marcus Aurelius was often far away from Rome, suppressing secessionist border lands. He died in Vienna during a fight against the Marcomanni. His equestrian statue in Rome is the only genuine bronze monument to an emperor. (M 30802)

(7448) 1948 AA

Discovered 1948 January 14 by W. Baade at Mount Wilson.

(7449) 1949 QL

Discovered 1949 August 21 by K. Reinmuth at Heidelberg.

(7450) 1968 OZ

Discovered 1968 July 24 by L. I. Chernykh at Nauchnyj.

(7451) 1978 PU_2

Discovered 1978 August 8 by N. S. Chernykh at Nauchnyj.

(7452) 1978 QU_2

Discovered 1978 August 31 by N. S. Chernykh at Nauchnyj.

(7453) 1978 RV_1

Discovered 1978 September 5 by N. S. Chernykh at Nauchnyj.

(7454) 1981 EW_{20}

Discovered 1981 March 2 by S. J. Bus at Siding Spring.

(7455) 1981 EQ_{26}

Discovered 1981 March 2 by S. J. Bus at Siding Spring.

(7456) 1982 OD

Discovered 1982 July 17 by E. Bowell at Anderson Mesa.

(7457) 1982 SL_6

Discovered 1982 September 16 by L. I. Chernykh at Nauchnyj.

(7458) 1984 DE_1

Discovered 1984 February 28 by H. Debehogne at La Silla.

(7459) 1984 HR_1

Discovered 1984 April 28 by W. Ferreri and V. Zappalà at La Silla.

(7460) 1984 JN
Discovered 1984 May 9 by J. Gibson at Palomar.

(7461) Kachmokiam
1984 TD. Discovered 1984 October 3 at the Oak Ridge Observatory at Harvard.
Named in honor of Katherine Galindo, Christine Galindo, Molly Thompson, Kimberley Galindo and Amy Galindo, nieces of Donna Thompson {see planet (3085)}, administrator for the Minor Planet Center. (M 29673)

(7462) 1984 WM_1
Discovered 1984 November 20 by E. Bowell at Anderson Mesa.

(7463) Oukawamine
1985 SB. Discovered 1985 September 20 by T. Seki at Geisei.
Named for a high plateau at the boundary of Kochi and Ehime prefectures. (M 32348)

(7464) 1987 VB_1
Discovered 1987 November 15 by A. Mrkos at Kleť.

(7465) Munkanber
1989 UA_3. Discovered 1989 October 31 by B. G. W. Manning at Stakenbridge.
Named in honor of John Munger, Veikko Kanto and Richard Berry, the authors of *The CCD Camera Cookbook*. By providing detailed instructions and software, they have made it possible for the amateur astronomer to construct and operate an efficient, low-cost CCD camera. The final imaging and astrometry of this minor planet was carried out with such a camera made by the discoverer. (M 32348; M 32377)

(7466) 1989 VC_2
Discovered 1989 November 2 by T. Hioki and N. Kawasato at Okutama.

(7467) 1989 WQ_1
Discovered 1989 November 25 by S. Ueda and H. Kaneda at Kushiro.

(7468) 1990 UP_{11}
Discovered 1990 October 17 by L. I. Chernykh at Nauchnyj.

(7469) 1990 VU_{14}
Discovered 1990 November 15 by L. I. Chernykh at Nauchnyj.

(7470) 1991 JA
Discovered 1991 March 2 by T. Urata at Oohira.

(7471) 1991 YD
Discovered 1991 December 28 by N. Kawasato at Uenohara.

(7472) Kumakiri
1992 CU. Discovered 1992 February 13 by M. Akiyama and T. Furuta at Mishima.
Named in honor of Kazuo Kumakiri (1923-), an amateur astronomer in the Susono {see planet (6419)} area who has done much to develop a local interest in astronomy. (M 33788)

(7473) 1992 EC_4
Discovered 1992 March 1 by the Uppsala-ESO Survey at La Silla.

(7474) 1992 TC
Discovered 1992 October 1 by R. H. McNaught at Siding Spring.

(7475) 1992 UX_5
Discovered 1992 October 28 by K. Endate and K. Watanabe at Kitami.

(7476) 1993 GE
Discovered 1993 April 14 by T. B. Spahr at Tucson.

(7477) 1993 LC
Discovered 1993 June 13 by H. E. Holt at Palomar.

(7478) Hasse
1993 OA_4. Discovered 1993 July 20 by E. W. Elst at La Silla.

Named for Peter Hasse (1585-1640), the first well-known organist to be appointed at the famous Marienkirche in Lübeck. On his death, he was succeeded by Franz Tunder {see planet (7871)}, who, according to the Lübeck custom, married Hasse's eldest daughter. Another of his daughters married the famous organist Nicolaus Bruhns {see planet (5127)}. (M 31025)

(7479) 1994 EC_1
Discovered 1994 March 4 by S. Ueda and H. Kaneda at Kushiro.

(7480) Norwan
1994 PC. Discovered 1994 August 1 at the Palomar Observatory at Palomar.

Norwan, as the "dancing porcupine woman" of the California Wintuns, was a goddess of light who brought food to earth. A daughter of earth and sun, Norwan danced, light as warm air, above growing plants; she kept dancing every day until sunset. (M 29673)

(7481) San Marcello
1994 PA_1. Discovered 1994 August 11 by A. Boattini and M. Tombelli at Cima Ekar.

Named for San Marcello Pistoiese, the mountain village where the Pian dei Termini Observatory is located and an important all-year holiday Tuscan resort. Founded by the Romans in 224 B.C., its name comes from the Roman Consul Marcus Claudius Marcellus. A pedestrian suspension bridge on the river Lima is the longest in the world. (M 29673)

(7482) 1994 PC_1
Discovered 1994 August 9 by R. H. McNaught at Siding Spring.

(7483) Sekitakakazu
1994 VO_2. Discovered 1994 November 1 by K. Endate and K. Watanabe at Kitami.

Named in memory of Takakazu Seki (1642-1708), Japanese mathematician and calendar scientist in Japan. His many achievements include the formulation of determinants ten years before Leibniz {see planet (5149)}. In 1674 he developed a theory of differential and integral calculus. (M 34626)

Name proposed by the discoverers following a suggestion by T. Sato and A. Fujii.

(7484) Dogo Onsen
1994 WF_4. Discovered 1994 November 30 by A. Nakamura at Kuma.

Named for a hot spring spa in Ehime, the prefecture in which this minor planet was discovered. Already known 3,000 years ago, Dogo Onsen is the oldest and one of the most famous hot spring spas in Japan. It is visited by some 1.3 million people each year. (M 30101)

(7485) Changchun
1994 XO. Discovered 1994 December 4 by M. Koishikawa at Sendai.

Named after the city in China, Changchun is the international sister city of Sendai {see planet (3133)}, Japan, affiliated since 1980. Changchun is the capital of the Kitsurin province and the center of political and economical activities of the province. (M 33788)

(7486) 1994 XJ_1
Discovered 1994 December 6 by T. Kobayashi at Oizumi.

(7487) 1994 YM
Discovered 1994 December 28 by T. Kobayashi at Oizumi.

(7488) 1995 KB_1
Discovered 1995 May 27 by C. W. Hergenrother at Tucson.

(7489) 1995 MX
Discovered 1995 June 26 by C. W. Hergenrother at Tucson.

(7490) Babička
1995 OF_1. Discovered 1995 July 31 by P. Pravec at Ondřejov.

Named in memory of Marie Macháčová (1922-1998) and Marie Petrželová (1912-1998), the grandmothers of the discoverer's wife; and Emilie Dudková (1914-1977) and Aloisie Pravcová (1917-), the grandmothers of the discoverer. This naming is also devoted to all grandmothers, who traditionally have a major role in bringing up children in the Czech nation. This role is also reflected in the novel "Babička" ("Granny"), written by Božena Němcová, based on her own childhood experience. (M 32790)

(7491) Linzerag
1995 SD_2. Discovered 1995 September 23 at the Osservatorio San Vittore at Bologna.

Named on the occasion of 50th anniversary of the Linzer Astronomische Gemeinschaft, a very active association of amateur astronomers in Austria. The number of this minor planet, written backwards, corresponds to the year when the association was founded. (M 30101)

Name proposed by the discoverers following a suggestion from Herbert Raab, president of the association.

(7492) Kačenka
1995 UX. Discovered 1995 October 21 by P. Pravec at Ondřejov.

Named in honor of Kateřina Macháčová (1975-), who has been a constant source of support to the discoverer during many stages of his research on near-earth objects. This naming comes on the eve of her marriage to the discoverer, 1997 June 21 {see also planet (4790)}. (M 30101)

(7493) Hirzo
1995 US_2. Discovered 1995 October 24 by J. Tichá at Kleť.

Named after Hirzo (?-1275), the knight of Bohemian king Přemysl {see planet (7695)} Otakar II and the efficient surveyor of royal towns and castles. Hirzo founded the royal town of České Budějovice at the confluence of Vltava {see planet (2123)} and Malše rivers in 1265. České Budějovice is the home town of the discoverer. (M 29673)

(7494) Xiwanggongcheng
1995 UV_{48}. Discovered 1995 October 28 at the Beijing Observatory at Xinglong.

Named for the tenth anniversary of Project Hope (Xiwang Gongcheng), a non-profit social welfare program operated by the China Youth Development Foundation. The project was initiated on 1989 Oct. 30, with the aim of helping school dropouts in poverty-stricken areas return to school and finish primary education. Using contributions from people of all walks of life at home and abroad, more than 1.85 million children have been brought back to school and 6000 Hope Primary Schools have been built in the rural areas across the country. Project Hope has greatly promoted the development of elementary education in China's poor areas and has become China's most effective social welfare program in the 1990s. (M 33789)

(7495) Feynman

1995 WS_4. Discovered 1995 November 22 by M. Tichý and Z. Moravec at Kleť.

Named in memory of Richard P. Feynman (1918-1988), outstanding theoretical physicist, 1965 Nobel prizewinner for physics, known for his works on liquid helium, quantum electrodynamics, elementary-particle physics and solid-state physics. Feynman was also an outstanding lecturer, and his three-volume introductory university course, *The Feynman lectures on physics*, has been an inspiration and a source of delight for generations of physicists and other scientists all over the world. His adventures, manner of thinking and curious character are recorded in *Surely you're joking, Mr. Feynman* and *What do you care what other people think*? (M 29673)

Obituaries published in Spaceflight, Vol. 30, No. 5, p. 218 (1988); Nature, Vol. 332, No. 6165, p. 588 (1988).

(7496) Miroslavholub

1995 WN_6. Discovered 1995 November 27 by M. Tichý at Kleť.

Named in memory of Miroslav Holub (1923-1998), Czech immunologist, poet and essayist. His research resulted in the monograph *Immunology of Nude Mice*. His fine poetry combines the world of science and the world of everyday human experience, as shown, for example, in his book *Supposed to Fly*. (M 33387) Name proposed by the discoverer following a suggestion by J. Tichá.

(7497) 1995 YY_{21}

Discovered 1995 December 17 at the Beijing Observatory at Xinglong.

(7498) Blaník

1996 BF. Discovered 1996 January 16 by Z. Moravec at Kleť.

Named for a hill in central Bohemia located about 50 km southeast of Prague. According to myth, there sleeps inside the hill the army of St. Václav, which will come and help the Czech nation in its darkest hours. (M 29673)

(7499) 1996 OO_2

Discovered 1996 July 24 by A. Boattini and A. Di Paola at Campo Imperatore.

(7500) Sassi

1996 TN. Discovered 1996 October 3 at the Farra d'Isonzo Observatory at Farra d'Isonzo.

Named in honor of Giorgio Sassi (1918-), Italian amateur astronomer, cofounder of the Osservatorio San Vittore {see planet (2235)}, in operation since 1969. He conceived and realized the optical and electronic elements involved in constructing an automated telescope and is still very active today. He wrote the "bible" for Italian amateurs who want to make optical instruments. (M 31025)

(7501) Farra

1996 VD_3. Discovered 1996 November 9 at the Farra d'Isonzo Observatory at Farra d'Isonzo.

Named for the small town in northeastern Italy that is the site of the observatory where this minor planet was discovered. A Roman village built in defense of the strategic bridge over the river Isonzo, it later became a Lombard stronghold. Located in the lowlands of Isonzo, Farra is well known for the excellent quality of its wines. (M 33789)

(7502) 1996 VP_7

Discovered 1996 November 15 by Y. Shimizu and T. Urata at Nachi-Katsuura.

(7503) 1996 VJ_{38}

Discovered 1996 November 7 by S. Ueda and H. Kaneda at Kushiro.

(7504) 1997 AF_1

Discovered 1997 January 2 by T. Kobayashi at Oizumi.

(7505) 1997 AM_2

Discovered 1997 January 3 by T. Kobayashi at Oizumi.

(7506) Lub

4837 P-L. Discovered 1960 September 24 by C. J. van Houten and I. van Houten-Groeneveld at Palomar.

Named in honor of Jan Lub (1946-), Dutch astronomer in Leiden, on the occasion of his 51st birthday. After studying in Amsterdam he worked for the European Southern Observatory in Chile and Munich before moving to the Leiden Observatory. While in Leiden he was director of the Observatório Roque de los Muchachos for several years. Lub's fields of interest are photoelectric observations of RR Lyrae variables and other objects. (M 31026)

(7507) Israel

7063 P-L. Discovered 1960 October 17 by C. J. van Houten and I. van Houten-Groeneveld at Palomar.

Named in honor of Frank P. Israel (1946-), Dutch astronomer in Leiden, on the occasion of his 51st birthday. Before moving to the Leiden Observatory he worked at the California Institute of Technology and at the European Space Agency. His fields of interest are radio astronomical observations of dense interstellar clouds in galaxies, in particular the study of mass concentrations in the center of galaxies. (M 31026)

(7508) Icke

2327 T-3. Discovered 1977 October 16 by C. J. van Houten and I. van Houten-Groeneveld at Palomar.

Named in honor of Vincent Icke (1946-), Dutch astronomer in Leiden, on the occasion of his 51st birthday. After studies in Utrecht and Leiden, he worked at the Universities of Sussex and Cambridge, at the California Institute of Technology and at the University of Minnesota in Minneapolis. His fields of interest are theoretical investigations of celestial bodies and cosmology. He is also an ardent popularizer of astronomy. (M 31026)

(7509) 1977 EL

Discovered 1977 March 9 by N. S. Chernykh at Nauchnyj.

(7510) 1978 UF_6

Discovered 1978 October 27 by C. M. Olmstead at Palomar.

(7511) 1981 EX_{24}

Discovered 1981 March 2 by S. J. Bus at Siding Spring.

(7512) 1983 CA_1
Discovered 1983 February 15 by E. Bowell at Anderson Mesa.

(7513) 1985 RU_2
Discovered 1985 September 5 by H. Debehogne at La Silla.

(7514) 1986 ED
Discovered 1986 March 7 by M. Inoue and O. Muramatsu and T. Urata at Kobuchizawa.

(7515) 1986 EF_5
Discovered 1986 March 5 by G. DeSanctis at La Silla.

(7516) Kranjc
1987 MC. Discovered 1987 June 18 at the Osservatorio San Vittore at Bologna.

Named in memory of Aldo Kranjc (1919-1994), Italian astronomer at the observatories in Naples, Milan and Bologna. He succeeded F. Zagar {see planet (6746)} as director of Brera-Merate Observatory in 1971. His many activities and studies included general relativity and applications of electronic techniques to astronomy and spectroscopy. He was a pioneer in Italy on numerical electronic calculus. He wrote several programs and articles on orbit determination. (M 32348)

(7517) 1989 AD
Discovered 1989 January 3 by T. Kojima at Chiyoda.

(7518) 1989 FG
Discovered 1989 March 29 by K. Suzuki and T. Furuta at Toyota.

(7519) 1989 UN_3
Discovered 1989 October 31 by B. G. W. Manning at Stakenbridge.

(7520) 1990 BV
Discovered 1990 January 21 by T. Hioki and S. Hayakawa at Okutama.

(7521) 1990 QS_2
Discovered 1990 August 24 by H. E. Holt at Palomar.

(7522) 1991 AJ
Discovered 1991 January 9 by M. Arai and H. Mori at Yorii.

(7523) 1991 PF_{18}
Discovered 1991 August 8 by H. E. Holt at Palomar.

(7524) 1991 RW_{19}
Discovered 1991 September 14 by H. E. Holt at Palomar.

(7525) Kiyohira
1992 YE. Discovered 1992 December 18 by A. Natori and T. Urata at Yakiimo.

Named in memory of Kiyohira Fujiwara (1056-1128), founder of the Fujiwara clan in the northern part of Japan. He built a castle in Esashi {see planet (9368)}, near Mizusawa {see planet (7530)}, and accumulated the means to reign over the wider area. He then moved to Hiraizumi, about 30 km south of Esashi, which rapidly became one of the most prosperous independent realms in Japan during the eleventh and twelfth centuries. Unburnt temples and remains tell us that the cultural level of the realm was comparable with that of Kyoto {see planet (4352)}. (M 34626)

Name proposed by the second discoverer following a suggestion by the citizens of Esashi City, where there is a geophysical station operated by the Mizusawa Observatory. Citation prepared by K. Yokoyama.

(7526) 1993 AA
Discovered 1993 January 2 by T. Urata at Oohira.

(7527) 1993 BJ
Discovered 1993 January 20 by T. Urata at Oohira.

(7528) 1993 FS_{39}
Discovered 1993 March 19 by the Uppsala-ESO Survey at La Silla.

(7529) Vagnozzi
1994 BC. Discovered 1994 January 16 at the Colleverde di Guidonia Observatory at Colleverde di Guidonia.

Named in honor of Antonio Vagnozzi (1950-), who in 1993 became the first amateur astronomer in Italy to discover a minor planet with a CCD camera and, after Spacewatch {see planet (4255)}, the first person to have a CCD discovery numbered. In addition to his work on minor planets, he has more recently been engaged in a systematic search for supernovae and was a codiscoverer of SN 1996ae. (M 31611)

(7530) Mizusawa
1994 GO_1. Discovered 1994 April 15 by K. Endate and K. Watanabe at Kitami.

Named in commemoration of the hundredth anniversary of the International Latitude Observatory of Mizusawa, established by the Japanese government according to a resolution adopted by the International Association of Geodesy in 1899 as one of the six stations of the International Latitude Service, at a latitude of $+39° 08'$. In 1988 the Observatory was transferred to the Division of Earth Rotation of the National Astronomical Observatory. The observations have never been interrupted, even during the two World Wars. Mizusawa is located about 500 km north of Tokyo, and its citizens show a strong interest in the observatory's astronomical and geophysical research. (M 34626)

Name proposed by the discoverers following a suggestion by K. Hurukawa, who was an astronomer there during 1960-1969. Citation prepared by K. Yokoyama.

(7531) 1994 SC
Discovered 1994 September 24 at the Santa Lucia Observatory at Stroncone.

(7532) 1995 UR_1
Discovered 1995 October 22 by M. Tichý at Kleť.

(7533) 1995 UE_6
Discovered 1995 October 25 by Y. Shimizu and T. Urata at Nachi-Katsuura.

(7534) 1995 UA_7
Discovered 1995 October 26 by Y. Shimizu and T. Urata at Nachi-Katsuura.

(7535) 1995 WU_2
Discovered 1995 November 16 by S. Ueda and H. Kaneda at Kushiro.

(7536) Fahrenheit
1995 WB_7. Discovered 1995 November 21 by Y. Shimizu and T. Urata at Nachi-Katsuura.

Named in memory of Daniel Gabriel Fahrenheit (1686-1736), German physicist who invented both the alcohol and mercury thermometers. He is best known for developing the temperature scale that bears his name. The number of this minor planet is related to 4169, the number of the minor planet honoring the inventor of the Celsius temperature scale, in the same manner (to the nearest integer) as the relationship between the Celsius and Fahrenheit scales. (M 29673; M 29691) Named by the Small Bodies Names Committee following a suggestion by P. Wild.

(7537) Solvay

1996 HS_8. Discovered 1996 April 17 by E. W. Elst at La Silla.

Named for the Belgian chemist and philantropist Ernest Solvay (1838-1922), prodigious industrialist, scientist and engaged citizen with audacious ideas on politics and society. Today he is considered a symbol of industrial Belgium. (M 31026)

(7538) 1996 VE_6

Discovered 1996 November 15 by T. Kobayashi at Oizumi.

(7539) 1996 XS_{32}

Discovered 1996 December 6 by S. Ueda and H. Kaneda at Kushiro.

(7540) 1997 AK_{21}

Discovered 1997 January 9 by S. Ueda and H. Kaneda at Kushiro.

(7541) Nieuwenhuis

4019 T-3. Discovered 1977 October 16 by C. J. van Houten and I. van Houten-Groeneveld at Palomar.

Named in honor of Henk Nieuwenhuis (1938-), curator of the Eise Eisinga {see planet (5530)} Planetarium in Franeker, The Netherlands. Since the 1960s he has been an enthusiastic organizer of exhibitions, meetings and lectures for amateur astronomers and the general public. He also has a keen interest in space research and in observing planets at the observatory he constructed. (M 31026)

Name proposed by the discoverers following a suggestion by T. Schoenmaker, who prepared the citation.

(7542) 1953 GN

Discovered 1953 April 7 by K. Reinmuth at Heidelberg.

(7543) Prylis

1973 SY. Discovered 1973 September 19 by C. J. van Houten and I. van Houten-Groeneveld at Palomar.

Named for a son of Hermes. Inspired by Athene {see planet (881)}, Prylis suggested that entry to Troy could be gained by means of a wooden horse. Epeius {see planet (2148)}, a carpenter, volunteered to build the horse. Afterwards, of course, Odysseus {see planet (1143)} claimed all the credit for his stratagem. (M 30478)

(7544) 1976 UB_2

Discovered 1976 October 26 by T. M. Smirnova at Nauchnyj.

(7545) Smaklösa

1978 OB. Discovered 1978 July 28 by C.-I. Lagerkvist at Mount Stromlo.

Named after the music group Smaklösa, which performs mainly on the island of Gotland off the Swedish coast during the summer. Celebrating 25 years together in 1998, they made at the beginning of their career the now-famous statement that only cowards rehearse before a performance. The lyrics of their songs are very intelligently written and are even more enjoyable to those more accustomed to the lifestyle and humor of Gotland. The live performances of the group are unforgettable, and the discoverer is very thankful for those he was able to attend during the summer of 1998. (M 33387)

(7546) 1979 MB_4

Discovered 1979 June 25 by E. F. Helin and S. J. Bus at Siding Spring.

(7547) 1979 MO_4

Discovered 1979 June 25 by E. F. Helin and S. J. Bus at Siding Spring.

(7548) Engström

1980 FW_2. Discovered 1980 March 16 by C.-I. Lagerkvist at La Silla.

Named in memory of Albert Engström (1869-1940), Swedish artist and writer. Born in Lönneberga in the Swedish province of Småland, he went as a student to Uppsala in 1889, where he studied Greek and Latin before his artistic ambitions made him move to an art school in Gothenburg. Probably Sweden's best painter of caricatures, he is best known for his black and white drawings illustrating very short stories. He was also an outstanding writer, specializing in short stories. Appointed professor of drawing at the Swedish academy of art, in 1922 he became a member of the prestigious Swedish academy. (M 31611)

(7549) Woodard

1980 TO_5. Discovered 1980 October 9 by C. S. Shoemaker and E. M. Shoemaker at Palomar.

Named for Adrian Russell Woodard, youngest grandson of the discoverers. (M 32348)

(7550) 1981 EV_8

Discovered 1981 March 1 by S. J. Bus at Siding Spring.

(7551) 1981 EF_{26}

Discovered 1981 March 2 by S. J. Bus at Siding Spring.

(7552) 1981 EB_{27}

Discovered 1981 March 2 by S. J. Bus at Siding Spring.

(7553) 1981 FG

Discovered 1981 March 30 by E. Bowell at Anderson Mesa.

(7554) 1981 GQ

Discovered 1981 April 5 by E. Bowell at Anderson Mesa.

(7555) 1981 SZ_6

Discovered 1981 September 28 by L. V. Zhuravleva at Nauchnyj.

(7556) 1982 FX_2

Discovered 1982 March 18 by H. Debehogne at La Silla.

(7557) 1982 FK_3

Discovered 1982 March 21 by H. Debehogne at La Silla.

(7558) Yurlov

1982 TB_2. Discovered 1982 October 14 by L. G. Karachkina at Nauchnyj.

Named in memory of Aleksandr Aleksandrovich Yurlov (1927-1963), Russian choir master and choral conductor. His innovative interpretations of sacred music defined his contribution to Russian art. He led what later became Yurlov's Chapel Choir from 1958 to 1963, performing compositions of Prokof'ev, Shostakovich {see planet (2669)} and Sviridov {see planet (4075)}. (M 32348)
Name suggested by G. Sviridov (1915-1998) and supported by the discoverer.

(7559) 1985 VF

Discovered 1985 November 14 by P. Jensen at Brorfelde.

(7560) 1986 AJ

Discovered 1986 January 10 by C. S. Shoemaker at Palomar.

(7561) 1986 TR_2

Discovered 1986 October 7 by E. Bowell at Anderson Mesa.

(7562) Kagiroino-Oka

1986 WO_9. Discovered 1986 November 30 by H. Kosai and K. Hurukawa at Kiso.

Named for a place in one of the poems in the collection Manyousyu {see planet (7104)}. The place is said to be the present-day Ouda, Nara prefecture. (M 34626)

(7563) 1988 BC

Discovered 1988 January 16 by T. Kojima at Chiyoda.

(7564) 1988 CA

Discovered 1988 February 7 by R. Rajamohan at Kavalur.

(7565) 1988 RD_{11}

Discovered 1988 September 14 by S. J. Bus at Cerro Tololo.

(7566) 1988 SP

Discovered 1988 September 18 by H. Debehogne at La Silla.

(7567) 1988 TC_1

Discovered 1988 October 13 by S. Ueda and H. Kaneda at Kushiro.

(7568) 1988 VJ_2

Discovered 1988 November 7 by T. Hioki and N. Kawasato at Okutama.

(7569) 1989 BK

Discovered 1989 January 28 by Y. Oshima at Gekko.

(7570) 1989 CP

Discovered 1989 February 5 by M. Arai and H. Mori at Yorii.

(7571) Weisse Rose

1989 EH_6. Discovered 1989 March 7 by F. Börngen at Tautenburg.

Named in memory of the members of Weisse Rose {white rose}, a circle of students, academics and artists around Hans and Sophie Scholl in Munich. They appealed for resistance against the Nazi regime in 1942-43, motivated by ethical and Christian arguments. Numerous members of the group were later comdemned to death. (M 30478)

(7572) Znokai

1989 SF. Discovered 1989 September 23 by K. Endate and K. Watanabe at Kitami.

Named for the Z society (Z-no-kai in Japanese), a cultural and social organization of the staff of the Mizusawa {see planet (7530)} International Latitude Observatory. The name commemorates the Z-term in the latitude variation, discovered by Kimura {see planet (6233)}, the first director. (M 34626)

Name proposed by the discoverers following a suggestion of K. Hurukawa, who is a member of Z-no-kai. Citation prepared by K. Yokoyama.

(7573) 1989 VX

Discovered 1989 November 4 by B. G. W. Manning at Stakenbridge.

(7574) 1989 WO_1

Discovered 1989 November 20 by W. Kakei and M. Kizawa and T. Urata at Oohira.

(7575) Kimuraseiji

1989 YK. Discovered 1989 December 22 by Y. Kushida and O. Muramatsu at Yatsugatake.

Named in honor of Seiji Kimura (1932-), an amateur astronomer who first suggested sending a solar eclipse expedition to the U.S.S.R. in 1968. Since then he has organized several overseas expeditions to observe total solar eclipses. Secretary

of the committee of the Japan Amateur Astronomers' Convention for over 14 years, Kimura established the Herschel Society of Japan in 1984 and has been editing bimonthly newsletters promoting knowledge of the Herschels and keeping contact with the William Herschel Society in the U.K. (M 33789)
Name proposed by the discoverers following a suggestion by S. Morikubo.

(7576) 1990 BN

Discovered 1990 January 21 by M. Arai and H. Mori at Yorii.

(7577) 1990 QV$_4$

Discovered 1990 August 24 by H. E. Holt at Palomar.

(7578) Georgböhm

1990 SP$_7$. Discovered 1990 September 22 by E. W. Elst at La Silla.

Named for the North German organist Georg Böhm (1661-1733), who studied under Buxtehude {see planet (4344)} in Lübeck. In 1698 Böhm was appointed organist at the St. Johannis church in Lüneburg. There he wrote his imposing Prelude and Fugue in C major. This opens with a virtuoso pedal solo, highly characteristic of North German organ composers. (M 31026)

(7579) 1990 TN$_1$

Discovered 1990 October 14 by E. F. Helin at Palomar.

(7580) Schwabhausen

1990 TM$_7$. Discovered 1990 October 13 by F. Börngen and L. D. Schmadel at Tautenburg.

Named for the Jena University Observatory's outstation, located some 10 km to the west of the city. The station was established in 1962 on the watershed of the rivers Saale {see planet (5409)} and Ilm. It is located in a small preserved wood between the villages of Gross- and Kleinschwabhausen, and the discoverer's principal residence is nearby. This minor planet was numbered as the result of astrometric CCD observations made at Schwabhausen in April 1997. (M 30101)
Name proposed by the first discoverer.

(7581) Yudovich

1990 VY$_{13}$. Discovered 1990 November 14 by L. G. Karachkina at Nauchnyj.

Named in honor of Viktor Iosifovich Yudovich (1934-), a professor at Rostov University and notable specialist in the mathematical hydrodynamics of convective fluid and mathematical geophysics. Yudovich developed convective models for earthquakes and volcanic eruptions. (M 31026)

(7582) 1990 WL

Discovered 1990 November 20 by R. H. McNaught at Siding Spring.

(7583) Rosegger

1991 BA$_3$. Discovered 1991 January 17 by F. Börngen at Tautenburg.

Named for the Styrian author Peter Rosegger (1843-1918), in his lifetime extremely popular in Austria. Particularly known are his novels *Die Schriften des Waldschulmeisters* and *Als ich noch der Waldbauernbub war* describing the people and manners in the villages of his homeland. (M 30101)

(7584) Ossietzky

1991 GK$_{10}$. Discovered 1991 April 9 by F. Börngen at Tautenburg.

Named in memory of Carl von Ossietzky (1889-1938), German publicist and editor of the weekly paper Weltbühne from 1927 to 1933. A staunch republican, he warned against militarism and National Socialism. In 1931 he received an 18-month prison sentence for disclosing the rearmament of the Reich in violation of the treaty of Versailles. Imprisoned by the Nazis in 1933, Ossietzky received the 1935 Nobel Peace Prize but was not allowed to accept it. He was released in 1936 following worldwide protests but died shortly afterwards due to the ill-treatment he had received while in prison. (M 30478)

(7585) 1991 PK_8

Discovered 1991 August 5 by H. E. Holt at Palomar.

(7586) Bismarck

1991 RH_7. Discovered 1991 September 13 by L. D. Schmadel and F. Börngen at Tautenburg.

Named in memory of Prince Otto Eduard Leopold von Bismarck (1815-1898), Prussian prime minister and minister of foreign affairs and, from 1871, the so-called Great Chancellor of the German Reich. His name is closely connected with the unification of Germany and with his politics of a balance of power among the European states. Bismarck introduced a comprehensive collection of civil laws and promoted an extensive social welfare system which - in its essential parts - remains important even a century after his death. (M 32348)
Name proposed by the first discoverer.

(7587) Weckmann

1992 CF_3. Discovered 1992 February 2 by E. W. Elst at La Silla.

Named for the North German organist Matthias Weckmann (1619-1674). Appointed organist at St. Jacobi's church in Hamburg in 1655, he had the mighty Arp Schnitger organ at his disposal until his death. A pupil of Heinrich Schütz {see planet (4134)}, he worked under the direction of Praetorius {see planet (4889)} and Scheidemann. In 1668 he founded the important *Collegium musicum*. Many of his organ and cembalo compositions (toccatas, canzones, suites and sonatas) survive. His Fantasia in D minor is a multisectional piece that is thoroughly contrapunctal in nature, and it reflects the great seriousness of Weckmann's style. (M 31026)

(7588) 1992 FJ_1

Discovered 1992 March 24 by R. H. McNaught at Siding Spring.

(7589) 1992 SR_1

Discovered 1992 September 26 by A. Sugie at Taga.

(7590) 1992 UP_4

Discovered 1992 October 26 by K. Endate and K. Watanabe at Kitami.

(7591) 1992 WG_3

Discovered 1992 November 18 by S. Ueda and H. Kaneda at Kushiro.

(7592) 1992 WR_3

Discovered 1992 November 23 by S. Otomo at Kiyosato.

(7593) 1992 WP_4

Discovered 1992 November 21 by E. F. Helin at Palomar.

(7594) Shotaro

1993 BH_2. Discovered 1993 January 19 by T. Seki at Geisei.

Named in memory of Shotaro Miyamoto (1912-1992), professor of astrophysics at Kyoto University. His interests included relativistic cosmology, planetary nebulae, solar physics, lunar geology and Martian meteorology. He also contributed greatly to the popularization of astronomy through his many books and radio and television broadcasts. Director of the Kwasan and Hida Observatories, he also served as president of the International Lunar Society, vice president of the

International Association of Planetology, president of the Astronomical Society of Japan and vice president of the Oriental Astronomical Association. (M 32790)
Name proposed by the discoverer following a suggestion by K. Sakka and T. Sato. Obituary published in Yamamoto Circ., No. 2185, p. 2 (1992).

(7595) 1993 FN_{26}

Discovered 1993 March 21 by the Uppsala-ESO Survey at La Silla.

(7596) Yumi

1993 GH. Discovered 1993 April 10 by K. Endate and K. Watanabe at Kitami.

Named in honor of Shigeru Yumi (1916-), the second director of the central bureau of the International Polar Motion Survey. During his 1962-1980 tenure he directed and saw the completion of the comprehensive work of rereducing the complete set of latitude observations. (M 34626)

Name proposed by the discoverers following a suggestion by K. Hurukawa. Citation prepared by K. Yokoyama.

(7597) 1993 GM

Discovered 1993 April 14 by S. Otomo at Kiyosato.

(7598) 1994 CS

Discovered 1994 February 4 by S. Ueda and H. Kaneda at Kushiro.

(7599) Munari

1994 PB. Discovered 1994 August 3 by A. Boattini and M. Tombelli at San Marcello Pistoiese.

Named in honor of Ulisse Munari (1960-), staff astronomer at the Astronomical Observatories of Padua and Asiago {see planets (363) and (7679), respectively} since 1990. His researches are concerned with symbiotic stars, cataclysmic variables, novae and open clusters as well as minor planets. As a high-school student he founded the Minor Planet Section of the Italian National Amateur Organization, and he still maintains his links with amateurs, allowing them access to some of the Italian professional telescopes. The discoverers found their first asteroid with one of these instruments. (M 30478)

(7600) Vacchi

1994 RB_1. Discovered 1994 September 9 by V. S. Casulli at Colleverde di Guidonia.

Named in honor of Ciro Vacchi (1916-), amateur astronomer and co-founder of San Vittore {see planet (2235)} Observatory, near Bologna. In 1974 he began the San Vittore search for asteroids, and he has many new discoveries to his credit. Today he is still a leader in the Italian community of amateur astrometrists. (M 31026)

(7601) 1994 US_1

Discovered 1994 October 25 by S. Ueda and H. Kaneda at Kushiro.

(7602) 1994 YW_1

Discovered 1994 December 31 by T. Kobayashi at Oizumi.

(7603) Salopia

1995 OA_2. Discovered 1995 July 25 by S. P. Laurie at Church Stretton.

Named for the English county of Shropshire (or Salop), where Church Stretton is located. Situated in the once-tumultuous borderland of the Welsh Marches, the earliest major settlement was the Roman garnison town of Uriconium built near the Wrekin, the county's prominent hill. The current county town of Shrewsbury is situated on a meandering bend of the river Severn and was the birthplace of the naturalist Charles Darwin {see planet (1991)}. The other main town, Telford, includes the Coalbrookdale and Ironbridge areas that were the birthplace of the Industrial Revolution in the eighteenth century. The county was celebrated in A. E. Housman's *A Shropshire Lad* and is the original 'land of lost content'. (M 34626)

(7604) 1995 QY_2

Discovered 1995 August 31 by R. H. McNaught at Siding Spring.

(7605) 1995 SR_1

Discovered 1995 September 21 by T. B. Spahr at Tucson.

(7606) 1995 SV_2

Discovered 1995 September 20 by S. Ueda and H. Kaneda at Kushiro.

(7607) Billmerline

1995 SB_{13}. Discovered 1995 September 18 by the Spacewatch at Kitt Peak.

Named in honor of William J. Merline (1954-), planetary scientist at the Southwest Research Institute in Boulder, Colorado, and a dedicated instrumentalist and observer. While a graduate student at the University of Arizona's Lunar and Planetary Laboratory he codiscovered periodic variations of the radial velocity of the star Arcturus. He made thousands of hours of observations for a search for planets orbiting other stars and a study of the stability of the solar spectrum. He is now a member of the imaging team of the Galileo mission, studying cratering records from spacecraft imagery. He is also conducting a search for satellites of minor planets using coronagraphic imaging with groundbased telescopes. (M 33789)

(7608) Telegramia

1995 UO_1. Discovered 1995 October 22 by J. Tichá at Kleť.

Named on the 75th anniversary of the publication of the first *IAU Circular*, which was issued on 1922 Oct. 22 by the IAU Central Bureau for Astronomical Telegrams, then at the Copenhagen Observatory. The Bureau moved to the Smithsonian Astrophysical Observatory in 1965 and has now issued 6756 *IAU Circulars*, relating generally to transient astronomical objects and phenomena in urgent need of further observation. (M 30802)

Name proposed by the discoverer and endorsed by her colleagues M. Tichý and Z. Moravec.

(7609) 1995 WX_3

Discovered 1995 November 18 by Y. Shimizu and T. Urata at Nachi-Katsuura.

(7610) 1995 XB

Discovered 1995 December 3 by D. di Cicco at Sudbury.

(7611) 1996 BW_1

Discovered 1996 January 23 by T. Kobayashi at Oizumi.

(7612) 1996 CN_2

Discovered 1996 February 12 by S. Ueda and H. Kaneda at Kushiro.

(7613) 1996 DK

Discovered 1996 February 16 by the JPL NEAT Program at Haleakala.

(7614) 1996 EA

Discovered 1996 March 2 by T. Kobayashi at Oizumi.

(7615) 1996 TA_{11}

Discovered 1996 October 9 by S. Ueda and H. Kaneda at Kushiro.

(7616) 1996 VF_2

Discovered 1996 November 6 by T. Kobayashi at Oizumi.

(7617) 1996 VF_{30}

Discovered 1996 November 7 by S. Ueda and H. Kaneda at Kushiro.

(7618) 1997 AU_4

Discovered 1997 January 6 by T. Kobayashi at Oizumi.

(7619) 1997 AP_{21}

Discovered 1997 January 13 by Y. Shimizu and T. Urata at Nachi-Katsuura.

(7620) Willaert

4077 P-L. Discovered 1960 September 24 by C. J. van Houten and I. van Houten-Groeneveld at Palomar.

Named for the Flemish composer Adrian Willaert (1485-1562). From 1527 Willaert was a musician, conductor and organist at St: Mark's cathedral in Venice. He combined Dutch polyphony with Italian concertante music and started the Venetian school. His main works are motets and madrigals. (M 30802)

(7621) Sweelinck

4127 P-L. Discovered 1960 September 24 by C. J. van Houten and I. van Houten-Groeneveld at Palomar.

Named for the Dutch composer and organist Jan Pieterszoon Sweelinck (1562-1621). At the age of 18, Sweelinck succeeded his father as organist of the Old Church of Amsterdam, a position he held to his death. He composed many choral songs, motets and madrigals. His fantasias and variations influenced North German organ music. Praetorius {see planet (4889)} was one of his pupils. (M 30802)

(7622) Pergolesi

6624 P-L. Discovered 1960 September 24 by C. J. van Houten and I. van Houten-Groeneveld at Palomar.

Named for the Italian composer Giovanni Battista Pergolesi (1710-1736). He was a conductor in Naples and Rome and died in Pozzuoli in extreme poverty. He did not live to see the glory of his intermezzo *La serva pedrona*, which especially in France became the beginning of *opera buffa*. Pergolesi's *Stabat mater* for soprano, alto, string instruments and organ shows his ability to combine large choral and instrumental components. (M 30802)

(7623) Stamitz

9508 P-L. Discovered 1960 October 17 by C. J. van Houten and I. van Houten-Groeneveld at Palomar.

Named for the German-Bohemian violinist and composer Johann Wenzel Anton Stamitz (1717-1757). He played in the orchestra during the coronation of Karl VII in Frankfurt in 1742. The Elector Palatine Carl Theodor invited Stamitz to Mannheim to conduct his orchestra. The composer of 74 symphonies, Stamitz started the Mannheim school of musicians and influenced Haydn, Mozart and Beethoven {see planets (3941), (1034) and (1815), respectively}. (M 30802)

(7624) Gluck

1251 T-1. Discovered 1971 March 25 by C. J. van Houten and I. van Houten-Groeneveld at Palomar.

Named of the German composer Ritter Christoph Willibald von Gluck (1714-1787). After studying in Prague and Milan he traveled to London to meet Handel {see planet (3826)}. He also worked with the traveling opera companies of Mingotti and Locatelli. In 1752 Gluck became conductor in Sachsen-Hildburghausen and in

Vienna. He composed 17 symphonies and 107 operas, including the three Italian reform operas *Orfeo ed Euridice*, *Alceste* and *Paride ed Elena*. (M 30802)

(7625) Louisspohr

2150 T-2. Discovered 1973 September 29 by C. J. van Houten and I. van Houten-Groeneveld at Palomar.

Named for the German composer and violinist Louis Spohr (1784-1859). At the age of 15 he became a violinist in the orchestra of the duke of Brunswick. He made many journeys throughout Europe to give concerts and founded a school for violinists. His eleven operas included *Faust* and *Jessonda*, and he also composed nine symphonies, 15 violin concerti and much chamber music. (M 30802)

(7626) 1976 QL_2

Discovered 1976 August 20 at the Felix Aguilar Observatory at El Leoncito.

(7627) Wakenokiyomaro

1977 DS_4. Discovered 1977 February 18 by H. Kosai and K. Hurukawa at Kiso.

Named for the person, born in Okayama prefecture in the Nara Era, who moved the Japanese capital from Nara {see planet (7253)} to Kyoto {see planet (4352)}. (M 34626)

(7628) 1977 QY

Discovered 1977 August 19 by N. S. Chernykh at Nauchnyj.

(7629) Foros

1977 QK_1. Discovered 1977 August 19 by N. S. Chernykh at Nauchnyj.
Named for the health resort on the south coast of the Crimea. (M 34626)

(7630) 1979 MR_2

Discovered 1979 June 25 by E. F. Helin and S. J. Bus at Siding Spring.

(7631) 1981 WH

Discovered 1981 November 20 by E. Bowell at Anderson Mesa.

(7632) Stanislav

1982 UT_5. Discovered 1982 October 20 by L. G. Karachkina at Nauchnyj.

Named in memory of the Ukrainian writer, poet, literary scholar and public figure Stanislav Volodymyrovych Tel'nyuk (1935-1990). His daughters Lesya and Galnya are well known Ukrainian singers who have set his verses to music. (M 31026)

(7633) Volodymyr

1982 UD_7. Discovered 1982 October 21 by L. G. Karachkina at Nauchnyj.

Named in honor of the vice president of the Ukrainian Astronomical Association and director of the Kiev University Observatory, Volodymyr Volodymyrovich Tel'nyuk-Adamchuk (1936-). He is an expert in astrometry and is the author of observational and summary catalogues of precise stellar positions. (M 31026)

(7634) Shizutani-Kou

1982 VO_3. Discovered 1982 November 14 by H. Kosai and K. Hurukawa at Kiso.

Named for a school in Bizen-city, Okayama prefecture. Founded in 1668, it is the oldest Japanese school building in existence. (M 34626)

(7635) 1983 VH_1

Discovered 1983 November 6 by A. Mrkos at Kleť.

(7636) Comba

1984 CM. Discovered 1984 February 5 by E. Bowell at Anderson Mesa.

Named in honor of Paul G. Comba (1926-), who gave up a career in celestial mechanics and astronomy for one in mathematics and computer science. After teaching at the University of Hawaii during the 1950s, he worked for three decades at IBM in New York and Massachusetts, where he carried out software research and development in many areas, including database management, geometric modeling, elasticity and cryptography. Following his formal retirement to Prescott, Arizona, he undertook a program of astrometry of minor planets in 1995. He has already been credited with the discovery of 20 numbered objects. (M 34344)

Name proposed by the discoverer following a suggestion by B. G. Marsden, who prepared the citation.

(7637) 1984 DN

Discovered 1984 February 23 by H. Debehogne at La Silla.

(7638) 1984 UX

Discovered 1984 October 26 by E. Bowell at Anderson Mesa.

(7639) Offutt

1985 DC_1. Discovered 1985 February 21 at the Oak Ridge Observatory at Harvard.

Named in honor of Warren Offutt (1928-), on the occasion of his 70th birthday, 1998 Feb. 13. After a career as an engineering executive, he turned in his retirement to the astronomical applications of CCDs, considering in particular the contributions that can be made by amateur astronomers. At his observatory in New Mexico he has made key observations of several of the objects discovered in the Kuiper Belt in recent years, as well as of other comets and minor planets as faint as 22nd magnitude. His follow-up of S/1997 U 2, one of the two recently discovered satellites of Uranus, played a crucial role in the establishment of its orbit. (M 31297)

(7640) 1985 PX

Discovered 1985 August 14 by E. Bowell at Anderson Mesa.

(7641) 1986 TT_6

Discovered 1986 October 5 by M. Antal at Piwnice.

(7642) 1988 TZ

Discovered 1988 October 13 by S. Ueda and H. Kaneda at Kushiro.

(7643) 1988 VQ_1

Discovered 1988 November 6 by M. Arai and H. Mori at Yorii.

(7644) 1988 VR_5

Discovered 1988 November 4 by A. Mrkos at Kleť.

(7645) Pons

1989 AC_2. Discovered 1989 January 4 by A. Mrkos at Kleť.

Named in memory of Jean-Louis Pons (1761-1831), whose name appears on 26 different comets and who claimed to discover as many as 37. Initially hired as the doorkeeper of the Marseilles Observatory, he soon showed his proficiency for both constructing and using telescopes, and at the end of his career he was directing observatories in Tuscany. He discovered comet 2P/Encke in both 1805 and 1818, and although it is named for the person who recognized this and other apparitions of it as being one and the same body, Encke himself always referred to it as "Pons' comet". (M 31611)

Name suggested by J. Tichá, citation prepared by B. G. Marsden.

(7646) 1989 KE

Discovered 1989 May 29 by H. E. Holt at Palomar.

(7647) Etrépigny

1989 SR_2. Discovered 1989 September 26 by E. W. Elst at La Silla.

Named for a small village in the French Ardennes, close to the Belgian border. (M 30478)

(7648) 1989 TB_1

Discovered 1989 October 8 by Y. Mizuno and T. Furuta at Kani.

(7649) Bougainville

1990 SV_5. Discovered 1990 September 22 by E. W. Elst at La Silla.

Named in memory of the French navigator Louis-Antoine de Bougainville (1729-1811), best remembered for his circumnavigation on the Bondeuse (1766-1769) and his discovery of several archipelagos in Polynesia. His adventures are recounted in his *Voyage autour du monde* (1771), wherein he expounds theories about the goodness and the moral virtues of man living in harmony with nature. (M 30802)

(7650) Kaname

1990 UG. Discovered 1990 October 16 by T. Seki at Geisei.

Named in memory of Kaname Nakamura (1904-1932), Japanese amateur astronomer and volunteer at the Kwasan Observatory. As a member of W. H. Pickering's {see planet (784)} League of Mars Observers, he was the first in Japan to make systematic observations of that planet. He was also a pioneer in astronomical optics and made many precision mirrors and lenses. (M 32790)
Name proposed by the discoverer following a suggestion by T. Sato and A. Fujii.

(7651) Villeneuve

1990 VD_6. Discovered 1990 November 15 by E. W. Elst at La Silla.

Named in honor of Don Villeneuve, anthropologist and friend of the discoverer. (M 30478)

(7652) 1991 RL_5

Discovered 1991 September 13 by H. E. Holt at Palomar.

(7653) 1991 UV

Discovered 1991 October 18 by S. Ueda and H. Kaneda at Kushiro.

(7654) 1991 VV_3

Discovered 1991 November 11 by S. Ueda and H. Kaneda at Kushiro.

(7655) Adamries

1991 YM_1. Discovered 1991 December 28 by F. Börngen at Tautenburg.

Named for the famous German arithmetician Adam Ries (1492-1559), author of the first German arithmetic books, which were used over more than two hundred years and which explain the procedure of calculations in a clear manner. Ries made an effort to help the man in the street to appreciate arithmetic, which was considered at that time to be difficult. In 1539, Ries was appointed to the *Churfürstlich Sächsischen Hofarithmeticus.* (M 30478)
This is the discoverer's 100th numbered discovery.

(7656) Joemontani

1992 HX. Discovered 1992 April 24 by the Spacewatch at Kitt Peak.

Named in honor of Joseph L. Montani (1952-), instrumentalist, optics expert and observer with the Spacewatch project of the University of Arizona's Lunar and Planetary Laboratory. While making observations for Spacewatch he discovered comets P/1997 G1, C/1997 G2 and C/1998 M6 and more than 16 near-earth objects. Prior to joining the Spacewatch group, he made millimeter-wave observations of CO in the Magellanic Clouds, Orion and Monoceros, did high-resolution infrared spectroscopy of solar-system objects from the Kuiper Airborne Observatory and made a laboratory study of volatiles in asteroidal surface minerals. (M 33789)

(7657) Jefflarsen

1992 HK_1. Discovered 1992 April 25 by the Spacewatch at Kitt Peak.

Named in honor of astronomer Jeffrey A. Larsen (1967-), whose software expertise has allowed the Spacewatch project to upgrade its computational capability in the automatic detection of small solar system bodies. While making observations for Spacewatch he discovered comets P/1997 V1 and C/1998 M3 and more than seven near-earth objects. Prior to joining the Spacewatch group, he participated in the Automated Plate Scanner project at the University of Minnesota, where he studied a large asymmetry in the distribution of faint halo/thick disk stars in the Milky Way. (M 33789)

(7658) 1993 BM_{12}

Discovered 1993 January 22 by S. Ueda and H. Kaneda at Kushiro.

(7659) 1993 CP_1

Discovered 1993 February 15 by S. Ueda and H. Kaneda at Kushiro.

(7660) 1993 VM_1

Discovered 1993 November 5 by R. H. McNaught at Siding Spring.

(7661) Reincken

1994 PK_{38}. Discovered 1994 August 10 by E. W. Elst at La Silla.

Named for the North German organist Johann Adam Reincken (1623-1722), a central figure of Hamburg's musical life who lived to a legendary age of 99 years. A pupil of Scheidemann, he was appointed organist at the Katherinenkirche in Hamburg in 1663. In 1678 he was co-founder of Hamburg's famous opera. Reincken was one of the principal representatives of the art of playing the organ in North Germany, although his style was sometimes too vituoistic. J. S. Bach {see planet (1814)} went to Hamburg several times to hear the master play. In addition to his *Hortus musicus*, there survives his famous Prelude and Fugue in G minor, a piece that every organist likes to play, because of its extremely brilliant and fluent style. (M 31026)

(7662) 1994 RM_1

Discovered 1994 September 3 by Y. Shimizu and T. Urata at Nachi-Katsuura.

(7663) 1994 RX_1

Discovered 1994 September 2 by E. F. Helin at Palomar.

(7664) 1994 TE_3

Discovered 1994 October 2 by K. Endate and K. Watanabe at Kitami.

(7665) 1994 TK_3

Discovered 1994 October 11 by V. S. Casulli and S. Valentini at Colleverde di Guidonia.

(7666) Keyaki

1994 VC_1. Discovered 1994 November 4 by K. Cross at Sendai.
Named for the Sendai "city tree". (M 31026)

The name, zelkova in English, was selected from many ideas submitted by Sendai citizens, among them K. and G. Miyashita.

(7667) 1995 BL_3

Discovered 1995 January 29 by Y. Shimizu and T. Urata at Nachi-Katsuura.

(7668) 1995 BR_3

Discovered 1995 January 31 by T. Kobayashi at Oizumi.

(7669) Malše

1995 PB. Discovered 1995 August 4 by M. Tichý and Z. Moravec at Kleť.

Named for a Czech river. The České Budějovice Observatory is situated at the confluence of the Malše and Vltava {see planet (2123)} rivers. (M 31297)

(7670) Kabeláč

1995 QJ. Discovered 1995 August 20 by L. Šarounová at Ondřejov.

Named in memory of Miloslav Kabeláč (1908-1979), outstanding Czech composer and teacher of contemporary music. He displayed high moral qualities during difficult periods in the history of the discoverer's country. (M 31027; M 31045)

Name proposed by the discoverer's father, Jaroslav Šaroun, who was Kabeláč's student of composition.

(7671) Albis

1995 UK_1. Discovered 1995 October 22 by Z. Moravec at Kleť.

The Latin name for one of Europe's principal rivers, known in German as the Elbe and in Czech as the Labe. Rising in the Krkonoše mountains of the Czech Republic, the river continues into Germany through Dresden {see planet (263) or (3053)} to its mouth beyond Hamburg {see planet (449) or (723)}. Its most significant tributary is the Vltava (or Moldau) {see planet (2123)}, on which Prague {see planet (2367)} is located. (M 31297)

(7672) Hawking

1995 UO_2. Discovered 1995 October 24 at the Kleť Observatory at Kleť.

Named in honor of Stephen Hawking (1942-), theoretical physicist, professor of mathematics at Cambridge University. He is renowned for his research on cosmology and quantum gravitation theory, as well as on the physics of black holes. Besides his scientific work, he is also the author of the well-known popularizations *A Brief History of Time* and *Black Holes and Baby Universes and Other Essays*. (M 31297)

(7673) 1995 UY_3

Discovered 1995 October 20 by T. Kobayashi at Oizumi.

(7674) Kasuga

1995 VO_1. Discovered 1995 November 15 by K. Endate and K. Watanabe at Kitami.

Named in honor of Ryo Kasuga (1950-), Japanese Buddhist priest, opera singer, professional magician, television and radio personality. He is also an amateur astronomer, who operates a planetarium at his temple and is very active in the movement against light pollution. (M 34626)

Name proposed by the discoverers following a suggestion by T. Sato and A. Fujii.

(7675) Gorizia

1995 WT_5. Discovered 1995 November 23 at the Farra d'Isonzo Observatory at Farra d'Isonzo.

Named for the Italian town on the 1000th anniversary of the first mention of its actual name in an official document. Gorizia is the main city in the lowlands of the river Isonzo, on which Farra d'Isonzo {see planet (7501)} and its astronomical observatory rise. Under Hapsburg dominion for over four centuries, Gorizio is now a green city where buildings and streets continuously remind one of this Austro-Hungarian influence. (M 33789)

(7676) 1995 WN_8

Discovered 1995 November 18 by Y. Shimizu and T. Urata at Nachi-Katsuura.

(7677) 1995 YP_3

Discovered 1995 December 27 by T. Kobayashi at Oizumi.

(7678) Onoda

1996 CW_2. Discovered 1996 February 15 by A. Nakamura at Kuma.

Named for the city in the southwest of Yamaguchi Prefecture where the discoverer was born. Long known for its coal mines, Onoda is now famous for its chemical industry and cement production. (M 30803)

(7679) Asiago

1996 CA_9. Discovered 1996 February 15 by U. Munari and M. Tombelli at Cima Ekar.

Named for the mountain town and region hosting the largest astronomical observatory on Italian territory. When dedicated in 1942, the 1.2-m telescope was the largest in Europe; a 1.8-m Cassegrain reflector was added in 1973. The observatory has been particularly active in the discovery and photometric and spectroscopic monitoring of all types of variables, including supernovae. A program to discover new minor planets has been conducted with Asiago Schmidts since 1993. (M 30478)

(7680) Cari

1996 HB. Discovered 1996 April 16 at the Santa Lucia Observatory at Stroncone.

Named in memory of Amleto Cari (1904-1982). A versatile athlete of rare talent, from 1919 to 1933 he was captain of the Terni {see planet (5654)} football team that, under his guidance, reached the national league. (M 33789)

(7681) 1996 YK_2

Discovered 1996 December 24 at the Beijing Observatory at Xinglong.

(7682) 1997 CY_{19}

Discovered 1997 February 12 by T. Kobayashi at Oizumi.

(7683) 1997 DE

Discovered 1997 February 19 at the Beijing Observatory at Xinglong.

(7684) Marioferrero

1997 EY. Discovered 1997 March 3 by P. G. Comba at Prescott.

Named in memory of Mario Ferrero (1904-1965) who encouraged and nurtured the discoverer's interest in astronomy when the latter was in his early teens. Ferrero was the co-discoverer of (1169) Alwine and (1218) Aster. He was an astronomer at Pino Torinese {see planet (2694)} in the 1930s. Because of his interest in teaching, Ferrero later joined the physics faculty at the Polytechnic Institute in Turin. He was also a music lover and accomplished organist. (M 30479)

(7685) 1997 EP_{17}

Discovered 1997 March 1 by S. Ueda and H. Kaneda at Kushiro.

(7686) Wolfernst

2024 P-L. Discovered 1960 September 24 by C. J. van Houten and I. van Houten-Groeneveld at Palomar.

Named in honor of Wolfgang Ernst (1947-), amateur astronomer who initiated

the astrometric program at the Starkenburg {see planet (6864)} Observatory in Heppenheim. Since 1995, the major part of his work has consisted of making follow-up observations of objects discovered in the course of the 1990-1993 KSO-ARI minor planet surveys at Tautenburg. (M 33387)
Name proposed by L. D. Schmadel.

(7687) Matthias

2099 P-L. Discovered 1960 September 24 by C. J. van Houten and I. van Houten-Groeneveld at Palomar.

Named in honor of Matthias Busch (1968-), amateur astronomer at the Starkenburg {see planet (6864)} Observatory in Heppenheim who has made a substantial contribution to the astrometry of minor planets. He is an ardent follow-up observer of near-earth objects and a keen developer of astronomical software. (M 33387)
Name proposed by L. D. Schmadel.

(7688) Lothar

2536 P-L. Discovered 1960 September 24 by C. J. van Houten and I. van Houten-Groeneveld at Palomar.

Named in honor of Lothar Kurtze (1972-), German amateur astronomer at the Starkenburg Observatory, Heppenheim, and a cofounder of the minor planet astrometry group in 1995. Kurtze played a major role in both the planning and the construction of the student observatory at the Werner Heisenberg senior high school in Weinheim. His lectures on astronomical topics inspired many young students. (M 34344)
Name proposed by L. D. Schmadel.

(7689) Reinerstoss

4036 P-L. Discovered 1960 September 24 by C. J. van Houten and I. van Houten-Groeneveld at Palomar.

Named in honor of Reiner Stoss (1975-), German amateur astronomer at the Starkenburg Observatory, Heppenheim. Stoss is an ardent observer who detected most of the minor planets found there. His main interests are the establishment of an NEO search program and to make follow-up observations of known NEOs. (M 34344)
Name proposed by L. D. Schmadel.

(7690) Sackler

2291 T-1. Discovered 1971 March 25 by C. J. van Houten and I. van Houten-Groeneveld at Palomar.

Named in honor of Raymond and Beverly Sackler, wise, perceptive and generous supporters of science and the arts. Their special interest in stimulating astronomical research through direct support, and the fostering of collaborations and exchange of knowledge among astronomers worldwide, has been a great boon to this exciting field, now in a truly "golden age". (M 31297)

(7691) Brady

3186 T-3. Discovered 1977 October 16 by C. J. van Houten and I. van Houten-Groeneveld at Palomar.

Named in honor of Charles E. "Chuck" Brady, Jr. (1951-), captain in the U.S. Navy, NASA astronaut, surgeon, experienced amateur radio operator and close friend to the name proposer. As a mission specialist, Chuck flew on the Life and Microgravity Spacelab mission aboard "Columbia" and has logged over 405 hours in space. In 1997, Chuck organized a radio expedition to Palmyra Island and the

Kingman Reef in the Pacific. His ham call sign, N4BQW, could have been heard around the globe. (M 32790)

Name proposed and citation prepared by L. D. Schmadel, DK8UH, who contacted him on Palmyra.

(7692) 1981 EZ_{25}

Discovered 1981 March 2 by S. J. Bus at Siding Spring.

(7693) 1982 WE

Discovered 1982 November 20 by T. Seki at Geisei.

(7694) 1983 SF

Discovered 1983 September 29 by A. Mrkos at Kleť.

(7695) Přemysl

1984 WA_1. Discovered 1984 November 27 by A. Mrkos at Kleť.

Named for a mythical hero Přemysl, called Ploughman. He married princess Libuše {see planet (264)}, and they became the mythical founders of the royal dynasty of the Přemyslids, ruling over Bohemia until 1306. Bohemian kings Přemysl Otakar I (?-1230) and Přemysl Otakar II (?-1278) are considered the most reputable members of this dynasty. (M 32095)

Name suggested by J. Tichá, M. Tichý and Z. Moravec.

(7696) 1988 JD

Discovered 1988 May 10 by W. Landgraf at La Silla.

(7697) 1989 AE

Discovered 1989 January 3 by T. Kojima at Chiyoda.

(7698) Schweitzer

1989 AS_6. Discovered 1989 January 11 by F. Börngen at Tautenburg.

Named in memory of Albert Schweitzer (1875-1965), German theologian, doctor and philosopher, organist and musicologist. Early in his life he felt deep "reverence for life" and studied medicine as a professor of theology. In 1913 he founded a hospital in the jungle near Lambaréné, when he was working in Africa as a missionary-doctor. His philosophy culminated in a universal affirmative ethics of an active charity. His accomplishments involving organ music included an interpretation of Bach {see planet (1814)}. Schweitzer was awarded the 1952 Nobel Peace Prize for his efforts on behalf of the Brotherhood of Nations. (M 30803)

(7699) 1989 CB_4

Discovered 1989 February 2 by A. Mrkos at Kleť.

(7700) Rote Kapelle

1990 TE_8. Discovered 1990 October 13 by F. Börngen and L. D. Schmadel at Tautenburg.

Named for the members of Rote Kapelle, a left-wing and communist resistance group who fought against the fascist regime, operating mainly in Berlin and Hamburg. Faced with the genocide of the European Jews, the group worked to accelerate both the breakdown of the fascist regime and the end of the war. Members of the group were, among others, Arvid Harnack, Harro Schulze-Boysen and Erika Gräfin Brockdorf. In 1942, the group was suppressed by the Geheime Staatspolizei, with the arrest of about 600 people, almost 60 of whom were executed. (M 31611)

Name proposed by the first discoverer.

(7701) 1990 TX_8

Discovered 1990 October 14 by A. Mrkos at Kleť.

(7702) 1991 PO_{13}

Discovered 1991 August 5 by H. E. Holt at Palomar.

(7703) 1991 RW

Discovered 1991 September 7 by E. F. Helin at Palomar.

(7704) 1992 EB_7

Discovered 1992 March 1 by the Uppsala-ESO Survey at La Silla.

(7705) 1993 FU_7

Discovered 1993 March 17 by the Uppsala-ESO Survey at La Silla.

(7706) 1993 FZ_{36}

Discovered 1993 March 19 by the Uppsala-ESO Survey at La Silla.

(7707) 1993 HM_1

Discovered 1993 April 17 by C. W. Hergenrother at Tucson.

(7708) 1994 GF_9

Discovered 1994 April 11 by S. Ueda and H. Kaneda at Kushiro.

(7709) 1994 RN_1

Discovered 1994 September 8 by Y. Shimizu and T. Urata at Nachi-Katsuura.

(7710) 1994 WT_2

Discovered 1994 November 30 by T. Kobayashi at Oizumi.

(7711) Říp

1994 XF. Discovered 1994 December 2 by Z. Moravec at Kleť.

Named for a hill in central Bohemia (459 m above sea level) located about 30 km north of Prague. When a mythological ancestor of Čech, the Father of the Czech nation, came to Bohemia, he is said to have climbed this hill, looked around and made the decision to stay there because of the wealth and fertility of the country. (M 31298)

(7712) 1995 TB_1

Discovered 1995 October 12 by Y. Shimizu and T. Urata at Nachi-Katsuura.

(7713) 1995 YE

Discovered 1995 December 17 by T. Kobayashi at Oizumi.

(7714) Briccialdi

1996 CC_1. Discovered 1996 February 9 at the Santa Lucia Observatory at Stroncone.

Named in memory of Gulio Briccialdi (1818-1881). Flutist and composer, he became a famous concert musician at the age of 15. He taught in Milan and Vienna and gave concerts in Europe and America. He invented a particular kind of flute and created the musical scales for it. (M 33387)

(7715) Leonidarosino

1996 CR_7. Discovered 1996 February 14 by U. Munari and M. Tombelli at Cima Ekar.

Named in memory of Leonida Rosino (1915-1997), director of the Padua and Asiago Observatories from 1953 to 1985. His principal research was on globular clusters, variable stars of all kinds, and novae and supernovae. At Padua he was instrumental in establishing the first Italian department of astronomy in 1968. Rosino's undisputed great teaching ability, keen scientific guidance and everlasting enthusiasm benefited several generations of Italian astronomers. (M 30803)

Obituaries published in G. Astrom., Vol. 23, N. 3, p. 2 (1997); Mem. Soc. Astron. Ital., Vol. 69, N. 1, p. 11-14 (1998).

(7716) Ube

1996 DA_3. Discovered 1996 February 22 by A. Nakamura at Kuma.

Named for the city in Yamaguchi prefecture where the discoverer attended high school. Famous for its chemical industry, Ube is also known as a 'sculpture city', a large sculpture exhibition being held there every two years, after which some of the pieces are purchased by the city and displayed along the downtown sidewalks in a unique 'open air museum'. (M 30803)

(7717) 1997 AL_5

Discovered 1997 January 7 by T. Kobayashi at Oizumi.

(7718) Desnoux

1997 EP_{30}. Discovered 1997 March 10 by C. Buil at Ramonville Saint Agne.

Named in honor of Valerie Desnoux, talented amateur astronomer. Together with J. P. Rozelot and J. Rosch, she assisted in the development of a scanning heliometer at the Pic du Midi Observatory. Her work on the evolution of the spectra of Be stars with the T60 amateur telescope has also been very much appreciated. (M 31027)

(7719) 1997 GT_{36}

Discovered 1997 April 7 by the Lincoln NEA Research Team at Socorro.

(7720) Lepaute

4559 P-L. Discovered 1960 September 26 by C. J. van Houten and I. van Houten-Groeneveld at Palomar.

Named after Nicole-Reine Lepaute (1723-1788), extraordinarily talented French mathematician. The astronomers Lalande and Clairaut sought her help for the computations needed to predict the 1759 return of comet 1P/Halley. Lepaute's computations for the annular eclipse of 1764 were used throughout Europe. Unfortunately, years of calculation left her nearly blind. (M 31298)

(7721) Andrillat

6612 P-L. Discovered 1960 September 24 by C. J. van Houten and I. van Houten-Groeneveld at Palomar.

Named in honor of Yvette Marie Josette Andrillat (1925-), French spectroscopist who works mainly on infrared spectra of early-type stars. She has also contributed to the spectroscopy of galaxies, novae, planetary nebulae and comets. Near the end of her renowned career she became acting director of the Observatoire de Haute Provence. The planet is named also in honor of her husband, Henri Andrillat, who is a retired professor of astronomy and a prominent cosmologist. The couple now live in Toulouse. (M 31298; M 31457)
Name suggested and citation prepared by H. Haupt.

(7722) Firneis

2240 T-2. Discovered 1973 September 29 by C. J. van Houten and I. van Houten-Groeneveld at Palomar.

Named in honor of Maria Gertrude Firneis (1947-), a professor of astronomy at the University of Vienna. She has worked on delicate astrometric and statistical problems and is an authority on the history of astronomy. She is also engaged in dating ancient events and correlating them with celestial phenomena. (M 31298; M 31457)
Name suggested and citation prepared by H. Haupt.

(7723) 1952 QW
Discovered 1952 August 28 at the Goethe Link Observatory at Brooklyn, Indiana.

(7724) 1970 OB
Discovered 1970 July 24 at the Felix Aguilar Observatory at El Leoncito.

(7725) Sel'vinskij
1972 RX_1. Discovered 1972 September 11 by N. S. Chernykh at Nauchnyj.
Named in memory of Il'ya L'vovich Sel'vinskij (1899-1968), poet and publicist, born in the Crimea. As a journalist, he witnessed and participated in many events in the history of the U.S.S.R. and he depicted these events in his poetical works. (M 34626)

(7726) 1974 QM_2
Discovered 1974 August 27 by L. I. Chernykh at Nauchnyj.

(7727) 1975 EA_3
Discovered 1975 March 8 by N. S. Chernykh at Nauchnyj.

(7728) 1977 AW_2
Discovered 1977 January 12 by E. Bowell at Palomar.

(7729) 1977 QY_3
Discovered 1977 August 24 by N. S. Chernykh at Nauchnyj.

(7730) 1978 NN_1
Discovered 1978 July 4 by L. I. Chernykh at Nauchnyj.

(7731) 1978 UV
Discovered 1978 October 28 by H. L. Giclas at Anderson Mesa.

(7732) 1978 VE_9
Discovered 1978 November 7 by E. F. Helin and S. J. Bus at Palomar.

(7733) 1979 MH_4
Discovered 1979 June 25 by E. F. Helin and S. J. Bus at Siding Spring.

(7734) 1979 MZ_6
Discovered 1979 June 25 by E. F. Helin and S. J. Bus at Siding Spring.

(7735) 1980 UL_1
Discovered 1980 October 31 by S. J. Bus at Palomar.

(7736) 1981 RC_5
Discovered 1981 September 8 by L. V. Zhuravleva at Nauchnyj.

(7737) 1981 VU
Discovered 1981 November 5 by E. Bowell at Anderson Mesa.

(7738) Heyman
1981 WS_1. Discovered 1981 November 24 at the Oak Ridge Observatory at Harvard.
Named in honor of I. Michael Heyman (1930-), secretary and linchpin of the Smithsonian Institution's diverse and incomparably rich programs. Educator, legal scholar, civil rights champion, he has unfailingly supported research at the frontiers of astronomy. (M 32348)

(7739) Čech

1982 CE. Discovered 1982 February 14 by L. Brožek at Kleť.

Named in memory of Eduard Čech (1893-1960), Czech mathematician, professor at Charles University in Prague and Masaryk University in Brno. One of his papers, *On bicompact spaces*, introduced what today is called the Stone-Čech compactification of regular topological spaces. His research in topology is fundamental, but his work was also important in the development of functional analysis. (M 34344)

Name proposed by J. Tichá and Z. Moravec following a number of suggestions from Charles University.

(7740) 1983 RR_2

Discovered 1983 September 6 by E. Bowell at Anderson Mesa.

(7741) Fedoseev

1983 RR_4. Discovered 1983 September 1 by L. G. Karachkina at Nauchnyj.

Named in honor of Vladimir Ivanovich Fedoseev (1932-), outstanding Russian conductor. He has been artistic director of Tchaikovsky's symphony orchestra in Moscow since 1974. He is also a principal conductor of the Vienna Symphony Orchestra and guest conductor in Tokyo and in many other cities. Fedoseev promotes the Russian classical music of Glinka, Mussorgsky, Rakhmaninov, Sviridov {see, respectively, planets (2205), (1059), (4345) and (4075)} and others. (M 32348) Name suggested by G. Sviridov and supported by the discoverer.

(7742) 1985 US

Discovered 1985 October 20 by A. Mrkos at Kleť.

(7743) 1986 JA

Discovered 1986 May 2 at the Copenhagen University Observatory at Brorfelde.

(7744) 1986 QA_1

Discovered 1986 August 26 by H. Debehogne at La Silla.

(7745) 1987 DB_6

Discovered 1987 February 22 by H. Debehogne at La Silla.

(7746) 1987 RC_1

Discovered 1987 September 13 by H. Debehogne at La Silla.

(7747) 1987 SO

Discovered 1987 September 19 by E. Bowell at Anderson Mesa.

(7748) 1987 TA

Discovered 1987 October 12 by T. Niijima and T. Urata at Ojima.

(7749) Jackschmitt

1988 JP. Discovered 1988 May 12 by C. S. Shoemaker and E. M. Shoemaker at Palomar.

Named in honor of Harrison Schmitt (1935-), the last Apollo astronaut to walk on the moon and, as a planetary geologist, the only Apollo scientist. Schmitt later served as a member of the United States Senate from New Mexico. (M 32349)

(7750) McEwen

1988 QD_1. Discovered 1988 August 18 by C. S. Shoemaker and E. M. Shoemaker at Palomar.

Named in honor of Alfred McEwen for his significant contributions to planetary geology in the fields of photogeology, photometry, spectroscopy, photoclinometry,

instrument calibration and mission planning. His research has involved diverse topics concerning the earth, the moon, Mars, minor planets, Galilean satellites and outer planets. He has been (or is) a member of science teams on Cassini, Galileo, Clementine and Mars Global Surveyor. (M 33789)
Citation prepared by M. S. Robinson at the request of C. S. Shoemaker.

(7751) 1988 UA
Discovered 1988 October 16 by S. Ueda and H. Kaneda at Kushiro.

(7752) 1988 US
Discovered 1988 October 31 by T. Niijima and K. Kanai at Ojima.

(7753) 1988 XB
Discovered 1988 December 5 by Y. Oshima at Gekko.

(7754) 1989 TT_{11}
Discovered 1989 October 2 by S. J. Bus at Cerro Tololo.

(7755) 1989 YO_5
Discovered 1989 December 28 by E. W. Elst at St. Michel.

(7756) 1990 FR_1
Discovered 1990 March 27 by C. S. Shoemaker at Palomar.

(7757) 1990 KO
Discovered 1990 May 22 by E. F. Helin at Palomar.

(7758) 1990 KT
Discovered 1990 May 21 by E. F. Helin at Palomar.

(7759) 1990 QD_2
Discovered 1990 August 22 by H. E. Holt at Palomar.

(7760) 1990 RW_3
Discovered 1990 September 14 by H. E. Holt at Palomar.

(7761) 1990 SL
Discovered 1990 September 20 by R. H. McNaught at Siding Spring.

(7762) 1990 SY_2
Discovered 1990 September 18 by H. E. Holt at Palomar.

(7763) 1990 UT_5
Discovered 1990 October 16 by E. W. Elst at La Silla.

(7764) 1991 AB
Discovered 1991 January 7 by S. Inoda and T. Urata at Karasuyama.

(7765) 1991 AD
Discovered 1991 January 8 by Y. Kushida and O. Muramatsu at Yatsugatake.

(7766) Jododaira
1991 BH_2. Discovered 1991 January 23 by K. Endate and K. Watanabe at Kitami.

Named for the place where the first really big star parties were held in Japan, the "Chiro's Star Festivals", from 1975 to 1984. Jododaira (1600 m above the sea) is near the top of Mt. Azuma, a famous volcano in Fukushima Prefecture in northeastern Japan. After the violent eruption of the volcano in 1893, Percival Lowell {see planet (1886)} climbed the mountain to investigate the result of the eruption. One hundred years later, an astronomical observatory was established there and is open to the public. (M 32790)

Name proposed by the discoverers following a suggestion by M. Koishikawa and T. Sato.

(7767) Tomatic

1991 RB_5. Discovered 1991 September 13 by L. D. Schmadel and F. Börngen at Tautenburg.

Named in honor of A. U. Tomatic (1997-), collaborator at the Minor Planet Center. An ardent computer of orbits and distributor of observational data of minor planets and comets, Tomatic published his first contribution to this field on *MPEC* 1997-Y01 (1997 Dec. 16). Tomatic is a godchild of the *MPC* astronomers B. G. Marsden and G. V. Williams. (M 34344)

Name proposed by the first discoverer in gratitude for Tomatic's indefatigable service.

(7768) 1991 SX_1

Discovered 1991 September 16 by H. E. Holt at Palomar.

(7769) 1991 VF_4

Discovered 1991 November 4 by S. Otomo at Kiyosato.

(7770) 1992 EQ_8

Discovered 1992 March 2 by the Uppsala-ESO Survey at La Silla.

(7771) 1992 EZ_9

Discovered 1992 March 2 by the Uppsala-ESO Survey at La Silla.

(7772) 1992 EQ_{15}

Discovered 1992 March 1 by the Uppsala-ESO Survey at La Silla.

(7773) 1992 FS

Discovered 1992 March 23 by K. Endate and K. Watanabe at Kitami.

(7774) 1992 UU_2

Discovered 1992 October 19 by K. Endate and K. Watanabe at Kitami.

(7775) Taiko

1992 XD. Discovered 1992 December 4 by Y. Kushida and O. Muramatsu at Yatsugatake.

Named in honor of Taiko Takeuchi, a Japanese amateur astronomer who started observing the stars more than half a century ago. She then became a schoolteacher, was active in popularizing astronomy and computed the orbits of some minor planets. (M 30803)

Name proposed by the discoverers, following a suggestion by S. Kimura.

(7776) Takeishi

1993 BF. Discovered 1993 January 20 by T. Urata at Oohira.

Named for Masanori Takeishi (1950-), a Japanese amateur astronomer who discovered a number of minor planets and was chief editor of the *Japan Astronomical Circular* from 1975 to 1993. (M 31027)

(7777) 1993 CO_1

Discovered 1993 February 15 by S. Ueda and H. Kaneda at Kushiro.

(7778) Markrobinson

1993 HK_1. Discovered 1993 April 17 by C. S. Shoemaker and E. M. Shoemaker at Palomar.

Named in honor of Mark Robinson, a planetary geologist who has worked on

M-type bodies (Mercury, moon, Mars, and minor planets). He began his career in Alaska, where he prospected for mineral ores and earned his B.S. in geology at the University of Alaska. In Hawaii, he completed his doctorate in 1993 on lunar and Martian volcanism and then began his outstanding work on the Clementine mission to the moon as a member of the U.S. Geological Survey Astrogeology Branch. His efforts, which he continues from his new base at Northwestern University, are now directed to the NEAR spacecraft mission and to future exploration of Mercury. (M 33789)
Cittaion prepared with the assistance of A. McEwen.

(7779) Susanring

1993 KL. Discovered 1993 May 19 by J. B. Child at Palomar.

Named in honor of Susan Ivanka Ring, inspired and dedicated member of the Canberra Astronomical Society. Susan's love of astronomy has expressed itself through committed support of school observing programs and leadership in the Society, as well as through her devoted managing editorship of the magazine *Southern Sky*. (M 31027)

Citation enthusiastically endorsed by J. Mould and the Canberra Astronomical Society.

(7780) 1993 NJ

Discovered 1993 July 15 by J. B. Child at Palomar.

(7781) Townsend

1993 QT. Discovered 1993 August 19 by E. F. Helin at Palomar.

Named in honor of Charles Townsend, a retired laser physicist and community-college instructor of astronomy and mathematics, and a former president of the Ventura County Astronomical Society and the Western Amateur Astronomers. He has co-authored two books, *Patrolling Night Skies* (1984) and *Observational Astronomy* (1989), the latter being used to teach an advanced astronomy course at Moorpark College. One of his students, John E. Rogers {see planet (5517)}, is extremely grateful for Twonsend's encouragement and leadership. This led to the existence of the Camarillo Observatory, which has contributed to the astrometry of near-earth objects and comets generally since 1992. (M 31027)

(7782) Mony

1994 CY. Discovered 1994 February 7 at the Santa Lucia Observatory at Stroncone.

Named in memory of Monica De Magistris (1977-1998), a student of physics at Perugia University and a great lover of astronomy who sometimes visited the Santa Lucia observatory. She organized an astronomy exposition as part of Italy's "Scientific and Technological Culture Week" in 1995, giving lectures on the subject to visitors. In her struggle with illness, she was an example of strength, intelligence and passion throughout her short life. (M 31611)

(7783) 1994 JD

Discovered 1994 May 4 by T. B. Spahr at Tucson.

(7784) 1994 PL

Discovered 1994 August 5 by T. B. Spahr at Tucson.

(7785) 1994 QW

Discovered 1994 August 29 by Y. Shimizu and T. Urata at Nachi-Katsuura.

(7786) 1994 TB_{15}

Discovered 1994 October 14 by Y. Shimizu and T. Urata at Nachi-Katsuura.

(7787) Annalaura

1994 WW. Discovered 1994 November 23 by L. Tesi and A. Boattini at San Marcello Pistoiese.

Named in honor of Annalaura Calvani Tesi, wife of the first discoverer. (M 30803)

(7788) Tsukuba

1994 XS. Discovered 1994 December 5 by A. Nakamura at Kuma.

Named for the city located in the southwest of Ibaraki prefecture. Originally a region of pastoral farms at the foot of Mt. Tsukuba, it dramatically changed with the establishment in 1973 of the University of Tsukuba, from which the discoverer graduated. The city is now also home to 32 national laboratories and is known as the 'academic capital' of Japan. (M 30803)

(7789) 1994 XE_6

Discovered 1994 December 2 by E. Bowell at Palomar.

(7790) Miselli

1995 DK_2. Discovered 1995 February 28 at the Santa Lucia Observatory at Stroncone.

Named in memory of Furio Miselli (1867-1949). Poet, singer, songwriter and singing master in Terni, he promoted "Cantamaggio", a folk style that is currently a popular tourist draw. He also founded a humor magazine, *Sborbottu*, which parodied city life at the end of the nineteenth century. (M 33387)

(7791) 1995 EB

Discovered 1995 March 1 by M. Tichý at Kleť.

(7792) 1995 WZ_3

Discovered 1995 November 18 by Y. Shimizu and T. Urata at Nachi-Katsuura.

(7793) 1995 YC_3

Discovered 1995 December 27 by the JPL NEAT Program at Haleakala.

(7794) Sanvito

1996 AD_4. Discovered 1996 January 15 by U. Munari and M. Tombelli at Cima Ekar.

Named in honor of Roberto di San Vito, amateur astronomer. Strongly committed to astronomy and astrometry, he is supporting a new observatory in Montelupo {see planet (7198)} that will bear his name, the "San Vito Observatory". (M 32790)

(7795) 1996 AN_{15}

Discovered 1996 January 14 at the Beijing Observatory at Xinglong.

(7796) Járacimrman

1996 BG. Discovered 1996 January 16 by Z. Moravec at Kleť.

Named for Jára Cimrman, a fictitious Czech genius. An analogue to Leonardo da Vinci {see planet (3000)}, he was a playwright, composer, poet, painter, versatile scientist, inventor, polar explorer, sportsman, first man on the moon, etc. Although his name is not mentioned in any encyclopedia, his work is explored at the Jára Cimrman Theatre in Prague. This theater is headed by the famous cimrmanologists Z. Svěrák and L. Smoljak, who endorsed the name proposal. (M 31298)

(7797) 1996 BK_2

Discovered 1996 January 26 by T. Kobayashi at Oizumi.

(7798) 1996 CL
Discovered 1996 February 1 at the Beijing Observatory at Xinglong.

(7799) Martinšolc
1996 DW_1. Discovered 1996 February 24 at the Kleť Observatory at Kleť.
Named in honor of Martin Šolc (1949-), Czech astronomer and head of the Astronomical Institute of Charles University in Prague since 1987. His work deals with cosmic dust, dust in comets and isotopes. He participated in data evaluation from the Vega and Giotto missions to comet 1P/Halley in 1986-1987. At present he collaborates on the ISO project with the Institut für Kernphysik in Heidelberg. Much of his time is devoted to teaching. He is also working on a history of Czech astronomy, together with his wife Alena Šolcová (1950-). (M 34344)
Name proposed by discoverers J. Tichá, M. Tichý and Z. Moravec.

(7800) Zhongkeyuan
1996 EW_2. Discovered 1996 March 11 at the Beijing Observatory at Xinglong.
Named for the Chinese Academy of Sciences (zhong guo ke xue yuan) on the occasion of its 50th anniversary. Founded on 1949 Nov. 1, it is China's most prestigious academic institution and comprehensive research and development center in science and technology. Over the past half century, the Chinese Academy of Sciences has made tremendous contributions to science and technology in China, to the country's economic construction and development and to human civilization and progress. (M 33789)

(7801) Goretti
1996 GG_2. Discovered 1996 April 12 by L. Tesi and A. Boattini at San Marcello Pistoiese.
Named in honor of Vittorio Goretti (1939-), a high-school teacher of physics and mathematics in Bologna for almost three decades. As an amateur astronomer, he has focused primarily on the study of minor planets, and in 1976 he was one of the founders of the survey at San Vittore {see planet (2235)}. In 1995 he started a new program from his home in Pianoro that is mainly devoted to follow-up observations. (M 30803)

(7802) 1996 XG_1
Discovered 1996 December 2 by T. Kobayashi at Oizumi.

(7803) 1997 EW_2
Discovered 1997 March 4 by T. Kobayashi at Oizumi.

(7804) Boesgaard
3083 P-L. Discovered 1960 September 24 by C. J. van Houten and I. van Houten-Groeneveld at Palomar.
Named in honor of Ann Merchant Boesgaard (1939-), astronomer at the University of Hawaii, well-known for her investigations of the lithium and beryllium content of stellar atmospheres. (M31298)

(7805) Moons
7610 P-L. Discovered 1960 October 17 by C. J. van Houten and I. van Houten-Groeneveld at Palomar.
Named in memory of Michèle Moons (1951-1998), well known for her work on celestial mechanics done at the department of mathematics of Facultés Universitaires N. D. de la Paix, Namur, Belgium. She was assistant editor of the journal *Celestial Mechanics and Dynamical Astronomy* for nearly ten years. In the early 1980s she developed an analytical theory of the libration of the moon that is widely used by several centers interested in analyzing the moon's motion. Her

later work included studies on the effects of resonant motion in the minor planet belt. Her early death is a great loss to her family and to science. (M 32095)

Citation prepared by J. Henrard and J. Schubart, the name being suggested by the latter.

Obituary published in Celest. Mech. Dyn. Astron., Vol. 68, No. 3, p. 3-9 (1997/98).

(7806) 1971 UM

Discovered 1971 October 26 by L. Kohoutek at Bergedorf.

(7807) 1975 SJ_1

Discovered 1975 September 30 by S. J. Bus at Palomar.

(7808) 1976 GL_8

Discovered 1976 April 5 by M. R. Cesco at El Leoncito.

(7809) 1979 ML_1

Discovered 1979 June 25 by E. F. Helin and S. J. Bus at Siding Spring.

(7810) 1981 DE

Discovered 1981 February 26 by H. Debehogne and G. DeSanctis at La Silla.

(7811) 1982 DT_6

Discovered 1982 February 23 at the Beijing Observatory at Xinglong.

(7812) 1984 UT

Discovered 1984 October 26 by E. Bowell at Anderson Mesa.

(7813) 1985 UF_3

Discovered 1985 October 16 by C.-I. Lagerkvist at Kvistaberg.

(7814) 1986 CF_2

Discovered 1986 February 13 by H. Debehogne at La Silla.

(7815) Dolon

1987 QN. Discovered 1987 August 21 by E. W. Elst at La Silla.

Named for the sad Trojan hero who figures in *Doloneia*, the tenth book of the *Iliad*. Dolon penetrated the Greek's camp at night to learn of their intentions but was caught and killed without pity by Ulysses and Diomedes {see, respectively, planets (5254) and (1437)}. (M 30803)

(7816) Hanoi

1987 YA. Discovered 1987 December 18 by M. Koishikawa at Sendai.

Named for the capital of Vietnam. The discoverer visited the city in 1997 with Yoshihide Kozai {see planet (3040)} to help Vietnamese astronomers install a 0.40-m Schmidt-Cassegrain telescope with a CCD camera and a 0.10-m refractor. The equipment, installed at the National Pedagogical University, was obtained using funds from the Sumitomo Foundations, Japan, with the hope that astronomical research in Vietnam will thereby be promoted. (M 33790)

(7817) 1988 RH_{10}

Discovered 1988 September 14 by S. J. Bus at Cerro Tololo.

(7818) Muirhead

1990 QO. Discovered 1990 August 19 by E. F. Helin at Palomar.

Named in honor of Brian K. Muirhead, flight system manager and deputy project manager of the Jet Propulsion Laboratory's Pathfinder mission. He was responsible for the design, development, test and launch of the Pathfinder spacecraft. He has been commended for his leading role in the successful Pathfinder project and became a highly recognized scientist in the media blitz reporting the status of this remarkable Mars landing. (M 31027)

(7819) 1990 RR_3

Discovered 1990 September 14 by H. E. Holt at Palomar.

(7820) 1990 TU_8

Discovered 1990 October 14 by A. Mrkos at Kleť.

(7821) 1991 AC

Discovered 1991 January 8 by Y. Kushida and O. Muramatsu at Yatsugatake.

(7822) 1991 CS

Discovered 1991 February 13 by R. H. McNaught at Siding Spring.

(7823) 1991 PF_{10}

Discovered 1991 August 7 by H. E. Holt at Palomar.

(7824) Lynch

1991 RM_2. Discovered 1991 September 7 by E. F. Helin at Palomar.

Named in honor of William Lynch III of the Jet Propulsion Laboratory and an outstanding model of efficiency, friendliness and dedication to his work and responsibilities. He is the epitome of the NASA/JPL motto, "Faster, better and cheaper", in the sense that anyone who works in a highly-charged, positive fashion, streamlined for action, leads to a more successful and economical operation. Always with a big smile and a cheerful greeting, he is one of JPL's biggest assets. (M 31298)

(7825) 1991 TL_1

Discovered 1991 October 10 by J. Alu at Palomar.

(7826) Kinugasa

1991 VO. Discovered 1991 November 2 by A. Takahashi and K. Watanabe at Kitami.

Named in honor of Sachio Kinugasa (1947-), Japanese professional baseball player who was called the "Iron Man". From 1970 to 1987, he played 2215 games without interruption - the world record until it was broken by Cal Ripken, Jr. of the Baltimore Orioles. Throughout his 22-year professional career for the Hiroshima Toyo Carp team he had 2543 hits, including 504 home runs. Kinugasa is now a television and newspaper commentator, as well as a guest professor at the Hiroshima prefectural university. (M 34626)

Name proposed by the discoverers following a suggestion by T. Sato and A. Fujii.

(7827) 1992 QE_2

Discovered 1992 August 22 by H. E. Holt at Palomar.

(7828) 1992 SD_{13}

Discovered 1992 September 28 by M. Yanai and K. Watanabe at Kitami.

(7829) Jaroff

1992 WY_4. Discovered 1992 November 21 by E. F. Helin at Palomar.

Named in honor of Leon Jaroff (1927-), science journalist with a long, distinguished association with *Time* magazine. Jaroff is internationally known for his well-researched, insightful articles and essays on scientific subjects combining factual reporting and intelligent commentary. He has won many awards and honors for his fine journalism. Through his writing, he has drawn attention to the issue of NEOs and the potentially catastrophic consequences for our civilization should a large comet or asteroid strike the earth. (M 32349; M 32561)

(7830) Akihikotago

1993 DC_1. Discovered 1993 February 24 by Y. Kushida and O. Muramatsu at Yatsugatake.

Named in honor of Akihiko Tago (1932-), credited with the discovery of comets C/1968 H1, C/1969 T1, C/1987 B1 (and an independent discovery of C/1970 U1) and the 1994 nova V2313 Oph. At the Satsuki Municipal Observatory in Yanahara, Okayama, Tago works as an instructor, popularizing astronomy, particularly among the young. (M 31298)

Name proposed by the discoverers, following a suggestion by H. Kosai, who prepared the citation.

(7831) François-Xavier

1993 FQ. Discovered 1993 March 21 by E. F. Helin at Palomar.

Named in memory of François-Xavier Bagnoud (1961-1986), whose passion for flying made him, at the age of 23, the youngest professional IFR pilot in Europe of both fixed-wing airplanes and helicopters. That passion, combined with compassion for others in need, led him to join his father at Air Glaciers and to lead some 300 successful rescue missions in the Alps. On 1986 Jan. 14, François died in a tragic helicopter accident in Mali. The François-Xavier Bagnoud Foundation was born out of the desire of his family and close friends to continue the loving concern for others shown by François during his life. An observatory established in his name in Switzerland provides astronomers with many clear nights for observation. (M 34626)

Name proposed by M. Wilson, a classmate of François at the American School in Paris.

(7832) 1993 FA_{27}

Discovered 1993 March 21 by the Uppsala-ESO Survey at La Silla.

(7833) 1993 FV_{32}

Discovered 1993 March 19 by the Uppsala-ESO Survey at La Silla.

(7834) 1993 JL

Discovered 1993 May 14 by S. Ueda and H. Kaneda at Kushiro.

(7835) 1993 MC

Discovered 1993 June 16 by T. B. Spahr at Tucson.

(7836) 1993 TG

Discovered 1993 October 9 by N. Kawasato at Uenohara.

(7837) Mutsumi

1993 TX. Discovered 1993 October 11 by H. Abe and S. Miyasaka at Yatsuka.

Named in honor of Mutsumi Abe (1957-), wife of the first discoverer. (M 30803)

(7838) 1993 WA

Discovered 1993 November 16 at the Farra d'Isonzo Observatory at Farra d'Isonzo.

(7839) 1994 ND

Discovered 1994 July 3 by R. H. McNaught at Siding Spring.

(7840) Hendrika

1994 TL_3. Discovered 1994 October 5 by G. C. L. Aikman at Victoria.

Named in honor of Hendrika Cornelia Marshall Aikman (née Grootendorst), beloved wife of the discoverer. (M 31027)

(7841) 1994 UE_1

Discovered 1994 October 31 by Y. Shimizu and T. Urata at Nachi-Katsuura.

(7842) 1994 XQ

Discovered 1994 December 1 by K. Endate and K. Watanabe at Kitami.

(7843) 1994 YE_1

Discovered 1994 December 22 by S. Ueda and H. Kaneda at Kushiro.

(7844) 1995 YL_1

Discovered 1995 December 21 by T. Kobayashi at Oizumi.

(7845) 1996 AC

Discovered 1996 January 1 by T. Kobayashi at Oizumi.

(7846) Setvák

1996 BJ. Discovered 1996 January 16 by M. Tichý at Kleť.

Named in honor of Martin Setvák (1958-), meteorologist and head of the Satellite Department of the Czech Hydrometeorological Institute in Prague, on the occasion of his 40th birthday, 1998 Feb. 14. His main professional interest is in convective storms, but his long-time hobby is astrophotography. This minor planet also honors his wife Stáňa Setváková (1967-), a staff member of the Prague Planetarium. (M 31298)

Name proposed by the discoverer and J. Tichá.

(7847) Mattiaorsi

1996 CS_8. Discovered 1996 February 14 by U. Munari and M. Tombelli at Cima Ekar.

Named for Mattia Orsi (1995-), nephew of Arcetri astronomer Giuseppe Forti. (M 31611)

(7848) Bernasconi

1996 DF_1. Discovered 1996 February 22 by M. Cavagna and A. Testa at Sormano.

Named in memory of the brothers Giovanni (1901-1965) and Angelo (1911-1990) Bernasconi, renowned Italian amateur astronomers. Giovanni was a codiscoverer of comets C/1941 K1 (although his discovery was not registered because of a delay in communications in wartime Europe), C/1942 C1 and C/1948 L1. Angelo developed a method for determining the real shape and orientation of cometary tails. Both brothers were active observers of comets, meteors and variable stars. In 1965 they established in Saronno a club of amateur astronomers, now named 'Gruppo Astrofili Giovanni e Angelo Bernasconi'. (M 30803)

Citation provided by L. Pansecchi at the request of the discoverers.

Obituary (A.B.) published in Universo, Vol. 12, No. 36, p. 51 (1992).

(7849) Janjosefrič

1996 HR. Discovered 1996 April 18 by P. Pravec and L. Šarounová at Ondřejov.

Named in memory of Jan Frič (1863-1897) and Josef Frič (1861-1945), the two brothers who founded the Ondřejov Observatory. On 1898 Jan 21, one year after Jan's sudden death, Josef started to bring to fruition their shared idea of establishing an astronomical institution. He purchased the land, and he adopted Jan's name as his own second name. The observatory grew extensively during the twentieth century and became internationally renowned. It was at Ondřejov that Josef Jan Frič, together with František Nušl {see planet (3424)}, discovered an anomalous refraction effect and developed astrogeodetic instruments, such as the well-known Nušl-Frič circumzenithal, for the determination of geographical coordinates. (M 31027)

Name proposed by M. Kopecký, astronomer and archivist of the Ondřejov Observatory.

(7850) Buenos Aires

1996 LH. Discovered 1996 June 10 by L. Macri at Mount Hopkins.

Named for the capital city of Argentina. First founded in 1536 by Don Pedro de Mendoza, who named it Santa Maria de los Buenos Aires, the city was soon destroyed by the natives. It was founded anew in 1580 by Don Juan de Garay. (M 30803)

(7851) Azumino

1996 YW_2. Discovered 1996 December 29 by N. Sato at Chichibu.

Named for an area in Nagano prefecture, Azumino is the rice field area east of the Japanese Northern Alps. In various places, natural spring water surfaces to form clear mountain streams. Azumino is famous throughout Japan for its beautiful scenery year-round. (M 33387)

(7852) Itsukushima

7604 P-L. Discovered 1960 October 17 by C. J. van Houten and I. van Houten-Groeneveld at Palomar.

Named for Itsukushima Island near Hiroshima City {see planet (2247)}. Known as "one of the Scenic Trio of Japan", the island is also called "Miyajima", which means "Shrine Island". The origin of the shrine is not known, but it goes back at least to the sixth century; in the twelfth century its building complex was much enlarged. The shrine and its giant torii gate stand in the sea at high tide. In 1996 the shrine was assigned "World Heritage" recognition by UNESCO. (M 32349)
Name proposed by Takeshi Sato.

(7853) Confucius

2086 T-2. Discovered 1973 September 29 by C. J. van Houten and I. van Houten-Groeneveld at Palomar.

Named after Confucius (551-471 B.C.), the famous Chinese philosopher. Nothing written by Confucius has survived, but his pupils compiled a collection of his sayings and anecdotes in the *Master Cung*. Confucius was convinced that the self-discipline and virtue shown by society's leaders would influence everyone. During the Han dynasty, Confucius was venerated as a god, and in the Sung dynasty his rules of living filtered into politics and education. (M 32790; M 32819; M 33151)

(7854) Laotse

1076 T-3. Discovered 1977 October 17 by C. J. van Houten and I. van Houten-Groeneveld at Palomar.

Named after Laotse (sixth century B.C.), the famous Chinese philosopher. He worked at the court of the emperor as a registrar, and it is said that he became a hermit and went to the West. Before he left, he was asked to record his doctrine. The result is the famous book of Tao, *Tao-Te-King*, which describes the origin of the universe and of Te, the strength possessed by all human beings. (M 32790)

(7855) Tagore

4092 T-3. Discovered 1977 October 16 by C. J. van Houten and I. van Houten-Groeneveld at Palomar.

Named in memory of Rabindranath Tagore (1861-1941), Indian poet and writer. Opposed to the Indian caste system, he created a school (*Viśva-Bharati*) in Santiniketan, and it became a state university in 1951. Tagore tried to unit the positive elements of European and Eastern philosophy. He received the 1913 Nobel Prize in literature for his lyric work *Gitanjali.* (M 32790)

(7856) 1975 VB_1

Discovered 1975 November 1 by T. M. Smirnova at Nauchnyj.

(7857) Lagerros

1978 QC_3. Discovered 1978 August 22 by C.-I. Lagerkvist at Mount Stromlo.

Named in honor of Johan S. V. Lagerros (1968-) to celebrate the completion of his Ph.D. thesis titled "Thermal Physics of Asteroids". In a series of papers, he extended and improved the Standard Thermal Model of minor planets. As a result, his model was used for the official calibration of ISO data. The discoverer has also benefited from his experience with computers on numerous occasions. (M 31611)

(7858) 1978 SB_3

Discovered 1978 September 26 by L. V. Zhuravleva at Nauchnyj.

(7859) 1979 US

Discovered 1979 October 19 by A. Mrkos at Kleť.

(7860) 1980 PF

Discovered 1980 August 6 by E. Bowell at Anderson Mesa.

(7861) 1981 EK_{25}

Discovered 1981 March 2 by S. J. Bus at Siding Spring.

(7862) 1981 EE_{28}

Discovered 1981 March 2 by S. J. Bus at Siding Spring.

(7863) 1981 VK

Discovered 1981 November 2 by B. A. Skiff at Anderson Mesa.

(7864) 1982 EE

Discovered 1982 March 14 by A. Mrkos at Kleť.

(7865) 1982 FG_3

Discovered 1982 March 21 by H. Debehogne at La Silla.

(7866) 1982 TK

Discovered 1982 October 13 by E. Bowell at Anderson Mesa.

(7867) 1984 SB_1

Discovered 1984 September 20 by A. Mrkos at Kleť.

(7868) 1984 UX_2

Discovered 1984 October 26 by E. Bowell at Anderson Mesa.

(7869) 1987 RV_3

Discovered 1987 September 2 by L. I. Chernykh at Nauchnyj.

(7870) 1987 UP_2

Discovered 1987 October 25 by P. Jensen at Brorfelde.

(7871) Tunder

1990 SW_4. Discovered 1990 September 22 by E. W. Elst at La Silla.

Named for the organist Franz Tunder (1614-1667), founder of the North German school of organ composition. In 1641 he became organist in Lübeck at the

Marienkirche, with its beautiful organ that was destroyed in 1945. It was Tunder also who started the famous "Abendmusiken", which his successor, Dietrich Buxtehude {see planet (4344)}, continued and raised to further glory. (M 31027)

(7872) 1990 UC

Discovered 1990 October 18 by T. Urata at Oohira.

(7873) Böll

1991 AE_3. Discovered 1991 January 15 by F. Börngen at Tautenburg.

Named in honor of Heinrich Böll (1917-1985), outstanding German writer. In his work he described the reality of World War II and the postwar era, mostly through the fate of ordinary people. A democrat with great historical farsightedness, he defended the liberty of the mind whenwever it was endangered. In 1971-1974, he was president of the International PEN Club. His receipt of the Nobel Prize in literature in 1972 was the visible sign of his appreciation worldwide. (M 31612)

(7874) 1991 BE

Discovered 1991 January 18 by S. Inoda and T. Urata at Karasuyama.

(7875) 1991 ES_1

Discovered 1991 March 7 by S. Ueda and H. Kaneda at Kushiro.

(7876) 1991 VW_3

Discovered 1991 November 11 by S. Ueda and H. Kaneda at Kushiro.

(7877) 1992 AH_1

Discovered 1992 January 10 by N. Kawasato at Uenohara.

(7878) 1992 DZ

Discovered 1992 February 27 by N. Kawasato at Uenohara.

(7879) 1992 EX_{17}

Discovered 1992 March 3 by the Uppsala-ESO Survey at La Silla.

(7880) 1992 OM_7

Discovered 1992 July 19 by H. Debehogne and A. López G. at La Silla.

(7881) Schieferdecker

1992 RC_7. Discovered 1992 September 2 by E. W. Elst at La Silla.

Named for the German organist Johann Christian Schieferdecker (1679-1732), who became Dietrich Buxtehude's {see planet (4344)} successor at the Marienkirche in Lübeck, by marrying the rather undesirable eldest daughter of his predecessor, apparently one of the side conditions for getting the job. Schieferdecker was a fine harpsichordist and had come to Lübeck in order to learn from Buxtehude the art of sacred composition. (M 31027)

(7882) 1993 FL_6

Discovered 1993 March 17 by the Uppsala-ESO Survey at La Silla.

(7883) 1993 GD_1

Discovered 1993 April 15 by H. E. Holt at Palomar.

(7884) 1993 HH_7

Discovered 1993 April 24 by H. Debehogne at La Silla.

(7885) 1993 KQ_2

Discovered 1993 May 17 by T. B. Spahr at Tucson.

(7886) Redman

1993 PE. Discovered 1993 August 12 by D. D. Balam at Victoria.

Named in memory of Roderick Oliver Redman (1905-1975), professor of astronomy and longtime director of the University of Cambridge Observatories, and in honor of Russell Ormond Redman (1951-), a radio astronomer on the staff of the Dominion Astrophysical Observatory. The senior Redman established a lifelong association with the DAO when he obtained the observational material for his doctoral thesis there; he became renowned for his superlative observational techniques and designs for astronomical instrumentation. The junior Redman, whose association with the DAO began as a summer student assistant in 1970, has pioneered the study of minor planets at submillimeter wavelengths and is an authority in using the thermal emission spectra to study their surfaces. (M 32095)
Citation prepared by F. A. Feldman at the request of the discoverer.

(7887) 1993 SU_2

Discovered 1993 September 18 by C. W. Hergenrother at Tucson.

(7888) 1993 UC

Discovered 1993 October 20 by R. H. McNaught at Siding Spring.

(7889) 1994 LX

Discovered 1994 June 15 by the Spacewatch at Kitt Peak.

(7890) 1994 TC_3

Discovered 1994 October 2 by K. Endate and K. Watanabe at Kitami.

(7891) Fuchie

1994 VJ_7. Discovered 1994 November 11 by M. Hirasawa and S. Suzuki at Nyukasa.

Named in honor of the Tokyo Metropolitan Fuchie Senior High School, where the first discoverer teaches astronomy and geology. Fuchie means the bank of a large river. This place has been called Fuchie for nearly a thousand years. (M 32095)

(7892) 1994 WQ_{12}

Discovered 1994 November 27 by M. Hirasawa and S. Suzuki at Nyukasa.

(7893) 1994 XY

Discovered 1994 December 2 by Y. Shimizu and T. Urata at Nachi-Katsuura.

(7894) 1994 XC_1

Discovered 1994 December 6 by T. Kobayashi at Oizumi.

(7895) Kaseda

1995 DK_1. Discovered 1995 February 22 by F. Uto at Kashihara.

Named for a city in the southwest of Kagoshima prefecture, where the discoverer was born. This town is fully blessed with natural beauty and is relatively free of light pollution. It once won the highest commendation for the most beautiful starry sky in Japan. (M 33387)

(7896) Švejk

1995 EC. Discovered 1995 March 1 by Z. Moravec at Kleť.

Named for a literary character created by Jaroslav Hašek {see planet (2734)}, Czech writer and humorist (1883-1923). His "good soldier" Josef Švejk is devoted to the disintegrating Austrian-Hungarian monarchy regime, but by his simplicity (or perhaps his ingenuity) he helps to reveal its flaws. (M 34344)

(7897) Bohuška

1995 EL_1. Discovered 1995 March 12 by L. Šarounová at Ondřejov.

Named in honor of the discoverer's mother, Bohumila Šarounová, in hearty thanks for her tolerance and support of the life and work of the discoverer. This minor planet was discovered on the discoverer's mother's birthday. (M 31027; M 31045)

(7898) 1995 XR_1

Discovered 1995 December 15 by T. Kobayashi at Oizumi.

(7899) 1996 BV_3

Discovered 1996 January 30 by T. Kobayashi at Oizumi.

(7900) Portule

1996 CV_8. Discovered 1996 February 14 by U. Munari and M. Tombelli at Cima Ekar.

Named for the highest mountain peak in the Asiago {see planet (7679)} tableland, near the Asiago Astrophysical Observatory. The wild mountain range is a trekking favorite for local astronomers. (M 32349)

(7901) 1996 DP

Discovered 1996 February 19 by T. Kobayashi at Oizumi.

(7902) Hanff

1996 HT_{17}. Discovered 1996 April 18 by E. W. Elst at La Silla.

Named in memory of the German organist Johann Nicolaus Hanff (1665-1711). In 1696 he became organist and conductor to the Bishop of Lübeck. Hanff's style, with the melody moving slowly but with rich ornamentation above a slow-moving and not very clearly individualized accompaniment, was favored by Buxtehude {see planet (4344)}. (M 31027)

(7903) Albinoni

1996 HV_{24}. Discovered 1996 April 20 by E. W. Elst at La Silla.

Named for the Italian composer Tomaso Albinoni (1671-1750). Although he considered himself an amateur composer, his 50 or so operas and instrumental works achieved wide popularity. Particularly notable are his concerti for solo violin and for one or two oboes, as well as his famous "Adagio" for strings. (M 31027)

(7904) Morrow

1997 JL_4. Discovered 1997 May 1 by the Lincoln NEA Research Team at Socorro.

Named in honor of Walter W. Morrow, Jr. (1928-), for his exceptional leadership as director of the Massachusetts Institute of Technology's Lincoln Laboratory for over 20 years, on the occasion of his retirement. (M 31028)

(7905) Juzoitami

1997 OX. Discovered 1997 July 24 by A. Nakamura at Kuma.

Named in memory of Juzo Itami (1933-1997), actor, translator, essayist, editor and well-known director since the great success of his first work *Osoushiki* (*Funeral*). His last work *Shizukana Seikatsu* (*Quiet Life*) is based on the novel written by Kenzaburo Oe, winner of the Nobel Prize in literature in 1994 and brother-in-law of Itami. (M 32095)

(7906) Melanchton

3081 P-L. Discovered 1960 September 24 by C. J. van Houten and I. van Houten-Groeneveld at Palomar.

Named after Philipp Melanchton (1497-1560), German reformer who worked closely with Martin Luther {see planet (7100)}. At the early age of twelve, he took lectures at the University of Heidelberg. His personal synthesis of humanism and the Christian faith left a strong mark on German Protestantism, expressed in the *Confessio Augustana* in 1530. (M 32790)

(7907) Erasmus

4047 P-L. Discovered 1960 September 24 by C. J. van Houten and I. van Houten-Groeneveld at Palomar.

Erasmus of Rotterdam (1469-1536) was the son of a priest and became a priest himself. In 1517 he was released by the pope from his holy vows, becoming an advisor to the emperor Charles V. He published papers critical of the practices of the Church; eventually he became the father of European humanism. He translated the classics, books of the church fathers and the New Testament into Latin in a critical way. His collection of proverbs helped shed new light on ancient literature and influenced religion, art and sciences. (M 32791)

(7908) Zwingli

4192 T-1. Discovered 1971 March 26 by C. J. van Houten and I. van Houten-Groeneveld at Palomar.

Named in memory of Ulrich Zwingli (1484-1531), Swiss reformer of the church. He was originally a Catholic priest, but under the influence of the books of Erasmus and Luther {see, respectively, planets (7907) and (7100)} he felt the need for reform. In 1523 he broke with the bishop, married and became a fervent puritan. For him the holy communion was only a symbolic celebration, and in this respect he did not agree with Luther. Zwingli died in a fight against the catholic cantons as a clergyman of the reformed soldiers. (M 33387)

(7909) 1975 SK

Discovered 1975 September 30 by S. J. Bus at Palomar.

(7910) 1976 GD_2

Discovered 1976 April 1 by N. S. Chernykh at Nauchnyj.

(7911) 1977 RZ_8

Discovered 1977 September 8 by E. Bowell at Palomar.

(7912) 1978 PO_3

Discovered 1978 August 8 by N. S. Chernykh at Nauchnyj.

(7913) 1978 TU_8

Discovered 1978 October 9 by L. V. Zhuravleva at Nauchnyj.

(7914) 1978 UW_7

Discovered 1978 October 27 by C. M. Olmstead at Palomar.

(7915) 1979 MA_6

Discovered 1979 June 25 by E. F. Helin and S. J. Bus at Siding Spring.

(7916) 1981 EN

Discovered 1981 March 1 by H. Debehogne and G. DeSanctis at La Silla.

(7917) 1981 EG_5

Discovered 1981 March 2 by S. J. Bus at Siding Spring.

(7918) 1981 EJ_{22}

Discovered 1981 March 2 by S. J. Bus at Siding Spring.

(7919) Prime

1981 EZ_{27}. Discovered 1981 March 2 by S. J. Bus at Siding Spring.

Recognizing the importance that mathematics has played in science, this naming acknowledges that 7919 is the one-thousandth prime number. (M 34344)
Name proposed by the discoverer following a suggestion by D. Hamilton.

(7920) 1981 XM_2

Discovered 1981 December 3 at the Purple Mountain Observatory at Nanking.

(7921) 1982 RF

Discovered 1982 September 15 by E. Bowell at Anderson Mesa.

(7922) 1983 CO_3

Discovered 1983 February 12 by H. Debehogne and G. DeSanctis at La Silla.

(7923) 1983 WJ

Discovered 1983 November 28 by E. Bowell at Anderson Mesa.

(7924) Simbirsk

1986 PW_4. Discovered 1986 August 6 by N. S. Chernykh and L. I. Chernykh at Nauchnyj.

Named for the city of Simbirsk (Ul'yanovsk since 1924), a large industrial and cultural center on the Volga river. Founded in 1648, it is connected with many great events and people in Russian history. (M 32349)

(7925) 1986 RX_2

Discovered 1986 September 6 by E. Bowell at Anderson Mesa.

(7926) 1986 RD_5

Discovered 1986 September 3 by H. Debehogne at La Silla.

(7927) 1986 WV_1

Discovered 1986 November 29 by A. Mrkos at Kleť.

(7928) 1986 WM_5

Discovered 1986 November 27 by the CERGA at Caussols.

(7929) 1987 SK_{12}

Discovered 1987 September 16 by H. Debehogne at La Silla.

(7930) 1987 VD

Discovered 1987 November 15 by S. Ueda and H. Kaneda at Kushiro.

(7931) 1988 EB_1

Discovered 1988 March 13 by P. Jensen at Brorfelde.

(7932) 1989 GP

Discovered 1989 April 7 by E. F. Helin at Palomar.

(7933) Magritte

1989 GP_4. Discovered 1989 April 3 by E. W. Elst at La Silla.

Named for the Belgian painter René Magritte Lessines (1898-1967), well kown for his magical surrealistic style. In the 1930s he composed his *Magritte dictionary*, placing ordinary objects, such as apples, stones and pipes, in a surrealistic context, thereby aiming to surprise and alienate. The philosophy of his work may best be summarized as "creating the unknown with known things". (M 31298)
Citation written by K. Leterme at the request of the discoverer.

(7934) Sinatra

1989 SG_1. Discovered 1989 September 26 by E. W. Elst at La Silla.

Named in memory of Frank Sinatra (1915-1998), American singer. Beginning in 1935 when he won a radio talent show, Sinatra showed impeccable taste as an American song stylist for over 50 years. He won an "Oscar" in 1954 for his role in the movie "From Here to Eternity", and in 1985 he was awarded the Presidential Medal of Freedom. Throughout his career, he "did it his way". (M 32791)

(7935) 1990 EZ_5

Discovered 1990 March 1 by H. Debehogne at La Silla.

(7936) 1990 OW_2

Discovered 1990 July 30 by H. E. Holt at Palomar.

(7937) 1990 QA_2

Discovered 1990 August 22 by H. E. Holt at Palomar.

(7938) 1990 SL_2

Discovered 1990 September 17 by H. E. Holt at Palomar.

(7939) Asphaug

1991 AP_1. Discovered 1991 January 14 by E. F. Helin at Palomar.

Named in honor of Erik Asphaug (1961-), currently in the Earth Sciences Department at the University of California at Santa Cruz. Asphaug has pioneered the application of fracture mechanics and hydrodynamics to numerous problems in planetary science. Using a radar-derived model of (4769) Castalia, he has explored how effects of a hypervelocity impact depend on the projectile's energy, the equation of state of the target material and the target's internal configuration of fractures and voids. His calculations have important implications for our understanding of the collisional evolution of minor planets and for the mitigation of hazards due to comets and minor planets. (M 32791)

(7940) Erichmeyer

1991 EO_1. Discovered 1991 March 13 at the Oak Ridge Observatory at Harvard.

Named in honor of Erich Meyer (1951-), Austrian amateur astronomer, on the occasion of his 20th anniversary as an astrometricist. Using a measuring engine he constructed himself, Meyer measured about 250 precise positions of minor planets and comets from photographic plates. Amoung the 2,600 positions he derived after switching to CCD equipment in 1993 are some for the 1997 opposition of this object, thereby rendering it appropriate for numbering. An electrical engineer by profession, Meyer is also a well-known astrophotographer and popularizer of astronomy. (M 32095)

(7941) 1991 NE_1

Discovered 1991 July 12 by H. E. Holt at Palomar.

(7942) 1991 OK_1

Discovered 1991 July 18 by H. Debehogne at La Silla.

(7943) 1991 PQ_{12}

Discovered 1991 August 5 by H. E. Holt at Palomar.

(7944) 1991 PR_{12}

Discovered 1991 August 5 by H. E. Holt at Palomar.

(7945) Kreisau

1991 RK_7. Discovered 1991 September 13 by F. Börngen and L. D. Schmadel at Tautenburg.

Named in memory of an anti-Nazi resistance group formed in 1942 in Silesia at the Kreisau farm, the family estate of H. J. Graf von Moltke. The members of the group, called Kreisauer Kreis, were mainly conservative people from a variety of social backgrounds and political attitudes. They considered the possibility of eliminating the fascist regime and revising Germany's ethical and political character. Most of the members were executed following the assassination attempt on the fascist leader in July 1944. (M 31612)
Name proposed by the first discoverer.

(7946) 1991 RV_{13}

Discovered 1991 September 13 by H. E. Holt at Palomar.

(7947) Toland

1992 BE_2. Discovered 1992 January 30 by E. W. Elst at La Silla.

Named in memory of the controversial freethinker John Toland (1670-1722). In 1696 he published his celebrated *Christianity not Mysterious*, a classic exposition on deism. Later, in his *Letters to Serena* (1704), he severely attacked the deism of both Spinoza {see planet (7142)} and Newton {see planet (8000)} with the remark "That after admitting the Activity of Matter, there seems to be no need of presiding Intelligence", thus anticipating Holbach's {see planet (6956)} *Système de la Nature*. (M 31028)

(7948) 1992 HY

Discovered 1992 April 24 by the Spacewatch at Kitt Peak.

(7949) 1992 SU

Discovered 1992 September 23 by E. F. Helin at Palomar.

(7950) Berezov

1992 SS_{26}. Discovered 1992 September 28 by L. V. Zhuravleva at Nauchnyj.

Named for a Siberian town, founded in 1593, situated on the Sos'va river in the Khanty-Mansi national district. In recent times a rich gas field was discovered that favored the growth of industry and the development of the town. Russian prince Aleksandr Danilovich Menshikov and his family were banished there in 1728. (M 34626)

(7951) 1992 WC_2

Discovered 1992 November 18 by S. Ueda and H. Kaneda at Kushiro.

(7952) 1992 XB

Discovered 1992 December 3 by A. Natori and T. Urata at Yakiimo.

(7953) 1993 KP

Discovered 1993 May 20 by S. Otomo at Kiyosato.

(7954) 1993 SQ_2

Discovered 1993 September 19 by K. Endate and K. Watanabe at Kitami.

(7955) 1993 WE

Discovered 1993 November 18 by T. Urata at Oohira.

(7956) 1993 YH

Discovered 1993 December 17 by T. Kobayashi at Oizumi.

(7957) Antonella

1994 BT. Discovered 1994 January 17 by A. Boattini and M. Tombelli at Cima Ekar.

Named in honor of Antonella Bartolini (1956-), amateur astronomer and friend of the discoverers. (M 31612)

(7958) Leakey

1994 LE_3. Discovered 1994 June 5 by C. S. Shoemaker and E. M. Shoemaker at Palomar.

Named for Mary Leakey (1913-1996), her husband Louis Leakey (1903-1972) and her son Richard Leakey (1944-), all major figures in the paleoanthropology of Africa. Mary and Louis worked together for 30 years and proved that human evolution was centered in Africa, rather than in Asia. Many of their most important discoveries were made at Olduvai Gorge in Tanzania. With his team in Kenya, Richard Leakey discovered that the Koobi Fora site contained one of the richest and most varied assemblages of early human remains anywhere in the world. (M 31612)

(7959) 1994 PK

Discovered 1994 August 2 by C. W. Hergenrother at Tucson.

(7960) Condorcet

1994 PW_{16}. Discovered 1994 August 10 by E. W. Elst at La Silla.

Named for the French philosopher of the Enlightenment, Marie-Jean-Antoine-Nicolas de Caritat, marquis de Condorcet (1743-1794), an advocate of educational reform and a believer in the indefinite perfectability of mankind. At an early age, Condorcet showed remarkable mathematical abilities, notably in probability theory, and as a friend of d'Alembert {see planet (5956)} he took an active part in the preparation of the *Encyclopédie*. But he is mainly remembered for his *Esquisse d'un tableau historique des progrès de l'esprit humain* (1795), with its fundamental idea of the continuous progress of humankind to an ultimate perfection. Suspected as a *Girondin*, he has to flee, but eventually he was captured and imprisoned at Bourg-La-Reine, where he was poisoned. (M 31028)

(7961) 1994 TD_2

Discovered 1994 October 10 by V. S. Casulli and S. Valentini at Colleverde di Guidonia.

(7962) 1994 WG_3

Discovered 1994 November 28 by S. Ueda and H. Kaneda at Kushiro.

(7963) Falcinelli

1995 CA. Discovered 1995 February 1 at the Santa Lucia Observatory at Stroncone.

Named in memory of Amleto Falcinelli (1921-1996), Italian bantam-weight boxing champion, who fought from 1936 to 1956 in Italian, European and international rings. Legendary are his five matches against the Spanish champion Romero for the European title, which he won several times. (M 33387)

(7964) 1995 DD_2

Discovered 1995 February 23 by Y. Shimizu and T. Urata at Nachi-Katsuura.

(7965) 1996 BD_1

Discovered 1996 January 17 by K. Endate and K. Watanabe at Kitami.

(7966) 1996 DA

Discovered 1996 February 18 by T. Kobayashi at Oizumi.

(7967) Beny

1996 DV_2. Discovered 1996 February 28 by Z. Moravec at Kleť.

Named in memory of Michal 'Beny' Böhm (1968-1998), schoolmate, roommate at Charles University and great friend of the discoverer. (M 34344)

(7968) Elst-Pizarro

1996 N2. Discovered 1996 July 14 by E. W. Elst and G. Pizarro at La Silla.

Originally classified as a comet because of the persistent appearance of a tail, this newly-numbered Themis-family minor planet is being given the same name it has had as a comet. (M 30803)

Normally, minor planets are not named after their discoverers. In this unusual case this decision was made by the Small Bodies Names Committee of Division III of the IAU.

(7969) 1997 RP_3

Discovered 1997 September 5 by Y. Shimizu and T. Urata at Nachi-Katsuura.

(7970) Lichtenberg

6065 P-L. Discovered 1960 September 24 by C. J. van Houten and I. van Houten-Groeneveld at Palomar.

Named in memory of Georg Christoph Lichtenberg (1742-1799). In his time he was one of Europe's best-known experimental physicists. He worked in Göttingen, Germany, and in Brighton, England. Even today, he is known for his aphorisms, such as: "it is almost impossible to carry the torch of the truth through a crowd without singeing someone's beard", "the healthy scholar is the man for whom thinking is not an illness" and "I always feel pain when a gifted person dies – the world has a greater need for these than the heavens". (M 33387)
Name proposed by K.-O. Groeneveld.

(7971) Meckbach

9002 P-L. Discovered 1960 October 17 by C. J. van Houten and I. van Houten-Groeneveld at Palomar.

Named in memory of Wolfgang Meckbach (1919-1998), microwave spectroscopist who emigrated from Germany to Argentina in 1951, where he became the co-founder of the Centro Atómico Bariloche and founder of the Bariloche school of atomic collision physics. In spite of difficult times, he excelled as a teacher and researcher in atomic collision physics using ion penetration through matter. By sad coincidence, the name was proposed just as Meckbach suffered a fatal heart attack while skiing. (M 33387)
Name proposed and citation prepared by K.-O. Groenveld.

(7972) Mariotti

1174 T-1. Discovered 1971 March 25 by C. J. van Houten and I. van Houten-Groeneveld at Palomar.

Named in memory of Jean-Marie Mariotti (1955-1998), French astronomer, most recently in Garching at the European Southern Observatory. He led the pioneering project to establish optical interferometry with the new Very Large Telescope as a breakthrough new astronomical instrument for the next century. His interest was in the area of high angular resolution, and he was involved in interferometric projects both on the ground (e.g., FLUOR and VLTI) and in space (DARWIN). He hoped with these techniques to find low-mass companions, and ultimately planets, outside our solar system. He had an extraordinarily effective mix of technical expertise and scientific eagerness. (M 33387)
Name proposed and citation prepared by R. Le Poole.

Obituaries published in Messenger, No. 93, p. 47 (1998); J. Astron. Fr., No. 57-58, p. 3-5 (1998).

(7973) 1344 T-2
Discovered 1973 September 29 by C. J. van Houten and I. van Houten-Groeneveld at Palomar.

(7974) 2218 T-2
Discovered 1973 September 29 by C. J. van Houten and I. van Houten-Groeneveld at Palomar.

(7975) 1974 FD
Discovered 1974 March 22 by C. Torres at Cerro El Roble.

(7976) 1977 QT_2
Discovered 1977 August 21 by N. S. Chernykh at Nauchnyj.

(7977) 1977 QQ_5
Discovered 1977 August 21 by R. H. McNaught at Siding Spring.

(7978) 1978 SR_4
Discovered 1978 September 27 by L. I. Chernykh at Nauchnyj.

(7979) 1978 SV_7
Discovered 1978 September 26 by L. V. Zhuravleva at Nauchnyj.

(7980) 1978 TD_2
Discovered 1978 October 3 by N. S. Chernykh at Nauchnyj.

(7981) 1978 VL_{10}
Discovered 1978 November 7 by E. F. Helin and S. J. Bus at Palomar.

(7982) 1979 MX_5
Discovered 1979 June 25 by E. F. Helin and S. J. Bus at Siding Spring.

(7983) Festin
1980 FY. Discovered 1980 March 16 by C.-I. Lagerkvist at La Silla.

Named in honor of Leif Festin (1967-) to celebrate the completion of his Ph.D. thesis on the faint end of the luminosity function. He assisted with photometric observations of minor planets while he was working at the Nordic Optical Telescope {see planet (2857)} on La Palma. He is co-author of several publications on lightcurves of minor planets. (M 31612)

(7984) 1980 SM
Discovered 1980 September 29 by Z. Vávrová at Kleť.

(7985) 1981 EK_{10}
Discovered 1981 March 1 by S. J. Bus at Siding Spring.

(7986) 1981 EG_{15}
Discovered 1981 March 1 by S. J. Bus at Siding Spring.

(7987) 1981 EV_{22}
Discovered 1981 March 2 by S. J. Bus at Siding Spring.

(7988) 1981 EX_{30}
Discovered 1981 March 2 by S. J. Bus at Siding Spring.

(7989) 1981 EW_{41}
Discovered 1981 March 2 by S. J. Bus at Siding Spring.

(7990) 1981 SN_1
Discovered 1981 September 26 by N. G. Thomas at Anderson Mesa.

(7991) Kaguyahime
1981 UT_7. Discovered 1981 October 30 by H. Kosai and K. Hurukawa at Kiso.
Named for the main female character in the old Japanese romance *Taketori-monogatari.* (M 34627)

(7992) 1981 WC
Discovered 1981 November 28 by T. Furuta at Tokai.

(7993) 1982 UD_2
Discovered 1982 October 16 by A. Mrkos at Kleť.

(7994) 1983 CQ_2
Discovered 1983 February 15 by E. Bowell at Anderson Mesa.

(7995) Khvorostovsky
1983 PX. Discovered 1983 August 4 by L. G. Karachkina at Nauchnyj.
Named in honor of the outstanding Russian baritone Dmitrij Aleksandrovich Khvorostovsky (1962-), a people's artist of Russia. His extensive opera and chamber repertoire includes both Russian and world music. The last composition of G. Sviridov {see planet (4075)}, the poem 'Petersburg', is dedicated to him. (M 32349)
Name proposed by G. Sviridov and supported by the discoverer.

(7996) Vedernikov
1983 RX_3. Discovered 1983 September 1 by L. G. Karachkina at Nauchnyj.
Named in honor of brilliant Russian bass Aleksandr Filippovich Vedernikov (1927-), a people's artist of Russia. He was the first to perform such works of G. Sviridov {see planet (4075)} as "Passionate oratorio", "Voice from the choir" and the solo part in "Pushkin's garland". (M 32349)
Name suggested by G. Sviridov and supported by the discoverer.

(7997) 1985 CN_1
Discovered 1985 February 13 by H. Debehogne at La Silla.

(7998) 1985 JK
Discovered 1985 May 15 by E. Bowell at Anderson Mesa.

(7999) 1986 RA_3
Discovered 1986 September 11 by E. Bowell at Anderson Mesa.

(8000) Isaac Newton
1986 RL_5. Discovered 1986 September 5 by H. Debehogne at La Silla.
Named for Isaac Newton (1643-1727), hailed by some as the greatest universal genius of all time. Newton formulated the laws of motion, with the specific application to gravitation, and is known for his 1687 publication *Principia Mathematica.* He also made seminal contributions to optics and the construction of the reflecting telescope. Newton was the last to observe, from his rooms at Trinity College, Cambridge, the great comet of 1860 - the first comet to have its motion explained on gravitational principles. He also served as master of the Royal Mint in London. (M 31298)
Named by the Small Bodies Names Committee.

(8001) 1986 TR_3
Discovered 1986 October 4 by A. Mrkos at Kleť.

(8002) 1986 XF_5
Discovered 1986 December 4 by A. Mrkos at Kleť.

(8003) 1987 RJ
Discovered 1987 September 1 by E. W. Elst at La Silla.

(8004) 1987 RX
Discovered 1987 September 12 by H. Debehogne at La Silla.

(8005) Albinadubois
1988 MJ. Discovered 1988 June 16 by E. F. Helin at Palomar.

Named in honor of Albina du Boisrouvray (1939-), daughter of Count Guy du Boisrouvray and Luz Mila Patino. She was a journalist, publisher and film producer in Paris in the 1960s and 1970s. When her only son, François-Xavier Bagnoud {see planet (7831)}, died in a helicopter accident in 1986, she sold most of her possessions and began to support and participate in projects to rescue children in need. This extended to 17 countries, and the rights of such children are championed through a major center she funded at Harvard University. She continues "hands-on" philantropy through an endless schedule of visits to the projects and participation in advocacy forums. (M 34627)
Name proposed by the discoverer, following a suggestion by M. Wilson.

(8006) Tacchini
1988 QU. Discovered 1988 August 22 at the Osservatorio San Vittore at Bologna.

Named in memory of Pietro Tacchini (1838-1905), Italian astronomer. He was a pioneer, with Secchi {see planet (4705)}, of astrophysics in Italy and cofounded the Società degli Spettroscopisti Italiani (1871), now the Italian Astronomical Society. Director of the Modena Observatory from 1859 to 1863, he moved to the Palermo Observatory in 1879. In 1902, he was called to Rome to become director of the Central Office of Meteorology and of the Astronomical Observatory of Collegio Romano. He also studied meteorology and seismology and founded the Società Sismologica Italiana (1895). (M 32349)

(8007) 1988 RU_6
Discovered 1988 September 8 by H. Debehogne at La Silla.

(8008) 1988 TQ_4
Discovered 1988 October 10 by Y. Oshima at Gekko.

(8009) 1989 BA_1
Discovered 1989 January 25 by C. Pollas at Caussols.

(8010) Böhnhardt
1989 GB_1. Discovered 1989 April 3 by E. W. Elst at La Silla.

Named in honor of Hermann Böhnhardt (1955-), astronomer at the European Southern Observatory, known for his observational and theoretical studies of comets. He took a particular interest in (7968) Elst-Pizarro, a Themis-family {see planet (24)} minor planet that also appeared in the guise of comet 133P/1996 N2. In addition to the main, sunward tail temporariliy exhibited by this strange object, Böhnhardt detected a faint anti-sunward tail, which he had predicted by means of Finson-Probstein modeling. (M 31298)

With its numbering, the comet received the final designation 133P/Elst-Pizarro. See the citation of planet (7968) for more details.

(8011) 1989 WG_7
Discovered 1989 November 29 by K. Endate and K. Watanabe at Kitami.

(8012) 1990 HO_3
Discovered 1990 April 29 by A. Zytkow and M. J. Irwin at Siding Spring.

(8013) 1990 KA
Discovered 1990 May 18 by E. F. Helin at Palomar.

(8014) 1990 MF
Discovered 1990 June 26 by E. F. Helin at Palomar.

(8015) 1990 QT_2
Discovered 1990 August 24 by H. E. Holt at Palomar.

(8016) 1990 QW_{10}
Discovered 1990 August 27 by H. E. Holt at Palomar.

(8017) 1990 RM_5
Discovered 1990 September 15 by H. E. Holt at Palomar.

(8018) 1990 SW
Discovered 1990 September 16 by H. E. Holt at Palomar.

(8019) 1990 TH_{12}
Discovered 1990 October 14 by L. D. Schmadel and F. Börngen at Tautenburg.

(8020) Erzgebirge
1990 TV_{13}. Discovered 1990 October 14 by F. Börngen and L. D. Schmadel at Tautenburg.

Named for the low mountain range along the border of Saxony and Bohemia, about 150 km in length and rich in forests. This region is relatively densely populated, owing to its past richness of silver and tin ore ("Erz", in German). Pitchblende ("Joachimsthaler Pechblende"), first found in the Erzgebirge, was used by the Curies {see planet (7000)} in 1898 to isolate the radioactive elements polonium and radium. The Sternwarte Drehbach, which is situated in Middle Erzgebirge, contributed an extensive series of observations to the minor planet in 1997, thereby enabling it to be numbered. (M 31299)
Name proposed by the first discoverer.

(8021) 1990 UO_2
Discovered 1990 October 22 by C. S. Shoemaker at Palomar.

(8022) 1990 VD_7
Discovered 1990 November 10 by A. Mrkos at Kleť.

(8023) 1991 DD
Discovered 1991 February 17 by T. Urata at Oohira.

(8024) 1991 FN
Discovered 1991 March 17 by E. F. Helin at Palomar.

(8025) 1991 FB_4
Discovered 1991 March 22 by H. Debehogne at La Silla.

(8026) 1991 JA_1
Discovered 1991 May 8 by E. F. Helin at Palomar.

(8027) 1991 PB_{12}
Discovered 1991 August 7 by H. E. Holt at Palomar.

(8028) 1991 QE
Discovered 1991 August 30 by R. H. McNaught at Siding Spring.

(8029) 1991 RR_{30}
Discovered 1991 September 15 by H. E. Holt at Palomar.

(8030) 1991 SK
Discovered 1991 September 29 by R. H. McNaught at Siding Spring.

(8031) 1992 ER
Discovered 1992 March 7 by S. Ueda and H. Kaneda at Kushiro.

(8032) 1992 ES_1
Discovered 1992 March 8 by S. Ueda and H. Kaneda at Kushiro.

(8033) 1992 FY_1
Discovered 1992 March 26 by S. Ueda and H. Kaneda at Kushiro.

(8034) 1992 LR
Discovered 1992 June 3 by C. S. Shoemaker at Palomar.

(8035) 1992 TB
Discovered 1992 October 2 by the Spacewatch at Kitt Peak.

(8036) 1992 UG_4
Discovered 1992 October 26 by K. Endate and K. Watanabe at Kitami.

(8037) 1993 HO_1
Discovered 1993 April 20 by R. H. McNaught at Siding Spring.

(8038) 1993 JG
Discovered 1993 May 11 by Y. Shimizu and T. Urata at Nachi-Katsuura.

(8039) 1993 RB_{16}
Discovered 1993 September 15 by H. Debehogne and E. W. Elst at La Silla.

(8040) 1993 SY_3
Discovered 1993 September 16 by K. Endate and K. Watanabe at Kitami.

(8041) Masumoto
1993 VR_2. Discovered 1993 November 15 by F. Uto at Kashihara.
Named in honor of Takeji Masumoto, who built the discoverer's observa tory. Initially, he had to make a great effort to secure the land for the observatory. (M 33388)

(8042) 1994 AX_2
Discovered 1994 January 12 by S. Ueda and H. Kaneda at Kushiro.

(8043) 1994 XE_1
Discovered 1994 December 6 by T. Kobayashi at Oizumi.

(8044) 1994 YT
Discovered 1994 December 28 by T. Kobayashi at Oizumi.

(8045) 1995 AW
Discovered 1995 January 6 by T. Kobayashi at Oizumi.

(8046) 1995 BU
Discovered 1995 January 25 by T. Kobayashi at Oizumi.

(8047) 1995 BT_3
Discovered 1995 January 31 by T. Kobayashi at Oizumi.

(8048) Andrle

1995 DB_1. Discovered 1995 February 22 by J. Tichá and Z. Moravec at Kleť.

Named in memory of Pavel Andrle (1936-1991), Czech astronomer and celestial mechanician. Methods described in his textbook were used at Kleť for the preliminary orbit determination of this minor planet. (M 31299)
Name proposed by the discoverers and endorsed by J. Tichá.

Obituaries published in Vesmír, Vol. 71, No. 2, p. 115 (1992); Bull. Astron. Inst. Czech., Vol. 42, No. 6, p. 408 (1991); Ríse hvezd, Vol. 73, No. 1, p. 10 (1992).

(8049) 1996 FL_2

Discovered 1996 March 17 by the JPL NEAT Program at Haleakala.

(8050) 1996 ST

Discovered 1996 September 18 at the Beijing Observatory at Xinglong.

(8051) Pistoria

1997 PP_4. Discovered 1997 August 13 by L. Tesi and G. Cattani at San Marcello Pistoiese.

Named for the Italian city of Pistoria, situated about 30 km from Florence at the foot of the western slope of the Central Apennines between Tuscany and Aemilia. Pistoria is the ancient name of the settlement dating to its founding in the second century B.C. by the Romans. (M 32349)

(8052) Novalis

2093 P-L. Discovered 1960 September 24 by C. J. van Houten and I. van Houten-Groeneveld at Palomar.

Friedrich Leopold Freiherr von Hardenberg, known under his pseudonym Novalis (1772-1801), is one of the greatest Romantic writers in German literature. The death of his fiancée and his pietistic thinking brought him to his romantic way of writing. In one of his novels he uses the image of a blue flower for the magical task poets have. Novalis had strong connections with the other Romantic writers of his time, such as Tieck {see planet (8056)} and Schlegel. (M 34344)

(8053) Kleist

4082 P-L. Discovered 1960 September 25 by C. J. van Houten and I. van Houten-Groeneveld at Palomar.

Heinrich von Kleist (1777-1811) was a prolific writer who was not known during his lifetime. He traveled throughout Europe, living in Dresden {see planet (3053)} and Königsberg, where he studied the philosophy of Kant {see planet (7083)}. In 1811 he committed suicide together with his girlfriend. His plays, including *Das Kätchen von Heilbronn*, *Prinz von Homburg* and the popular *Der zerbrochene Krug*, are highly regarded nowadays, as are his novels and short stories. (M 34344)

(8054) Brentano

4581 P-L. Discovered 1960 September 24 by C. J. van Houten and I. van Houten-Groeneveld at Palomar.

Named for the German Romantic poet Clemens Brentano (1778-1842), esteemed in his time by important German poets and writers. Together with his friend Achim von Arnim {see planet (8055)} he collected about 600 German folksongs in *Des Knaben Wunderhorn*. He also published famous fairy tales such as *Gockel, Hinkel und Gackeleia*. (M 34344)

(8055) Arnim

5004 P-L. Discovered 1960 October 17 by C. J. van Houten and I. van Houten-Groeneveld at Palomar.

Achim (Ludwig Joachim) von Arnim (1781-1831) is one of the well-known German Romantics. Although he studied law and natural sciences, he dedicated his life to literature, especially German literature. With his friend Brentano {see planet (8054)} he published about 600 German folksongs in *Des Knaben Wunderhorn*. German poets from the "Heidelberger Romantik", such as the brothers Grimm, Tieck {see planet (8056)} and Runge, worked together with Arnim and Brentano. (M 34345)

(8056) Tieck

6038 P-L. Discovered 1960 September 24 by C. J. van Houten and I. van Houten-Groeneveld at Palomar.

Ludwig Tieck (1773-1853) was one of the great Romantic poets and writers. He became friends with Brentano {see planet (8054)}, Fichte, Novalis {see planet (8052)}, the brothers Grimm and many other poets of his time. He wrote the ironic fairy tales *Ritter Blaubart, Der gestiefelte Kater* and other sensitive and imaginative stories. (M 34345)

(8057) Hofmannsthal

4034 T-1. Discovered 1971 March 26 by C. J. van Houten and I. van Houten-Groeneveld at Palomar.

Hugo von Hofmannsthal (1874-1929) began writing poems at the age of 16. His lyrical and dramatic work reflects Austrian impressionism and symbolism. Together with Richard Strauss {see planet (5039)} and Max Reinhardt, he founded the Salzburger Festspiele. His best-known play is *Jedermann.* (M 34345)

(8058) Zuckmayer

3241 T-3. Discovered 1977 October 16 by C. J. van Houten and I. van Houten-Groeneveld at Palomar.

Named for Jewish playwright Carl Zuckmayer (1896-1977). He wrote many comedies, his most famous play being *Der Hauptmann von Köpenick.* As a young soldier he took part in World War I. His plays were later banned in Germany, so he emigrated to Austria and in 1938 moved via Switzerland to the U.S. During 1940-1946 he earned his living on a farm. There he wrote the much-discussed drama *Des Teufels General.* Although he returned to Germany in 1947, he lived in Switzerland from 1958 onward. (M 34345)

(8059) 1957 JP

Discovered 1957 May 6 at the Goethe Link Observatory at Brooklyn, Indiana.

(8060) Anius

1973 SD_1. Discovered 1973 September 19 by C. J. van Houten and I. van Houten-Groeneveld at Palomar.

Named after Anius, son and priest to Apollo {see planet (1862)} on the isle of Delos. His three daughters could change anything into wine, grain or soil. So it was easy for Anius to supply the Greek army with these things. As a result, Agamemnon {see planet (911)} kidnapped the three daughters. But after they prayed to Dionysos {see planet (3671)}, the god transformed the girls into birds. (M 31612)

(8061) 1975 UF

Discovered 1975 October 27 by P. Wild at Zimmerwald.

(8062) 1977 EZ

Discovered 1977 March 13 by N. S. Chernykh at Nauchnyj.

(8063) 1977 XP_2

Discovered 1977 December 7 by S. J. Bus at Palomar.

(8064) 1978 RR

Discovered 1978 September 1 by N. S. Chernykh at Nauchnyj.

(8065) 1979 FD_3
Discovered 1979 March 31 by N. S. Chernykh at Nauchnyj.

(8066) 1980 PB_2
Discovered 1980 August 6 by R. M. West at La Silla.

(8067) 1980 RU
Discovered 1980 September 7 by E. Bowell at Anderson Mesa.

(8068) 1981 EQ_{28}
Discovered 1981 March 6 by S. J. Bus at Siding Spring.

(8069) 1981 EF_{30}
Discovered 1981 March 2 by S. J. Bus at Siding Spring.

(8070) 1981 EM_{30}
Discovered 1981 March 2 by S. J. Bus at Siding Spring.

(8071) 1981 GO
Discovered 1981 April 5 by E. Bowell at Anderson Mesa.

(8072) 1981 GO_1
Discovered 1981 April 1 at the Harvard College Observatory at Harvard.

(8073) 1982 BS
Discovered 1982 January 24 by E. Bowell at Anderson Mesa.

(8074) 1984 WC_2
Discovered 1984 November 20 by E. Bowell at Palomar.

(8075) 1985 PE
Discovered 1985 August 14 by E. Bowell at Anderson Mesa.

(8076) 1985 RV_4
Discovered 1985 September 15 by H. Debehogne at La Silla.

(8077) 1986 AW_2
Discovered 1986 January 12 by E. Bowell at Anderson Mesa.

(8078) 1986 RS_2
Discovered 1986 September 6 by E. Bowell at Anderson Mesa.

(8079) 1986 XF_1
Discovered 1986 December 4 by E. Bowell at Anderson Mesa.

(8080) 1987 WU_2
Discovered 1987 November 17 by the CERGA at Caussols.

(8081) Leopardi
1988 DD. Discovered 1988 February 17 at the Osservatorio San Vittore at Bologna.

Named in memory of Giacomo Leopardi (1798-1837), Italian poet and prose writer famous for his collection of poems *I Canti* and the collection of prose *Operette Morali*. When he was only 15 years old he wrote an erudite history of astronomy, recently reprinted in Italy. (M 32349)

(8082) 1988 NR
Discovered 1988 July 12 by E. F. Helin at Palomar.

(8083) Mayeda

1988 VB. Discovered 1988 November 1 by T. Seki at Geisei.

Named in memory of Shizuo (Haruhisa) Mayeda (1914-1952), amateur astronomer and pioneer Mars observer who studied changes in the colors of albedo features on that planet. (M 32791)

Name proposed by the discoverer following a suggestion by T. Sato and A. Fujii

(8084) Dallas

1989 CL_1. Discovered 1989 February 6 by M. Koishikawa at Sendai.

Named for the city in Texas, Dallas is the international sister city of Sendai {see planet (3133)}, Japan, affiliated since 1997. Dallas, popularly called "BIG D", is among the principal cities of the American southwest. (M 33790)

(8085) 1989 CD_8

Discovered 1989 February 7 by H. Debehogne at La Silla.

(8086) 1989 RB_6

Discovered 1989 September 1 by E. Bowell at Palomar.

(8087) 1989 WA_2

Discovered 1989 November 29 by K. Endate and K. Watanabe at Kitami.

(8088) Australia

1990 SL_{27}. Discovered 1990 September 23 by G. R. Kastel' and L. V. Zhuravleva at Nauchnyj.

Named for the continent in the Southern Hemisphere. With this naming the first discoverer pays tribute to the feelings she experienced during her recent visit to this marvelous country. (M 32791)

(8089) 1990 TW_7

Discovered 1990 October 13 by L. D. Schmadel and F. Börngen at Tautenburg.

(8090) 1991 RO_{23}

Discovered 1991 September 15 by H. E. Holt at Palomar.

(8091) 1992 BG

Discovered 1992 January 24 by T. Urata at Oohira.

(8092) 1992 DC_{10}

Discovered 1992 February 29 by the Uppsala-ESO Survey at La Silla.

(8093) 1992 UZ_2

Discovered 1992 October 25 by N. Kawasato at Uenohara.

(8094) 1992 UG_3

Discovered 1992 October 24 by A. Sugie at Taga.

(8095) 1992 WS_2

Discovered 1992 November 18 by S. Ueda and H. Kaneda at Kushiro.

(8096) Emilezola

1993 OW_3. Discovered 1993 July 20 by E. W. Elst at La Silla.

Named in memory of the French writer Emile Zola (1840-1902), founder of the naturalist movement in literature. His writings show a credulous faith in science and an uncritical acceptance of scientific determinism. Although he believed that human nature was completely determined by heredity, he thought that it could be perfected. The controversial views expressed in his famous "J'accuse", published in the newspaper *L'Aurore* on 1898 Jan. 13, forced his flight to England. Nevertheless, he received a public funeral in France, and his remains are preserved in the Panthéon, although he never became a member of the Académie. (M 31299)

(8097) 1993 RE

Discovered 1993 September 12 by K. Endate and K. Watanabe at Kitami.

(8098) 1993 SH_2

Discovered 1993 September 19 by K. Endate and K. Watanabe at Kitami.

(8099) 1993 TE

Discovered 1993 October 8 by H. Abe and S. Miyasaka at Yatsuka.

(8100) Nobeyama

1993 XF. Discovered 1993 December 4 by M. Hirasawa and S. Suzuki at Nyukasa.

Named for a village in central Japan, famous for its beautiful scenery and night sky. Nobeyama Radio Observatory is located there. (M 32095)

(8101) 1993 XK_1

Discovered 1993 December 15 by T. Kobayashi at Oizumi.

(8102) 1994 AQ_2

Discovered 1994 January 14 by T. Kobayashi at Oizumi.

(8103) Fermi

1994 BE. Discovered 1994 January 19 at the Farra d'Isonzo Observatory at Farra d'Isonzo.

Named in memory of the Italian physicist Enrico Fermi (1901-1954), winner of the 1938 Nobel Prize in physics for his demonstration of the existence of new radioactive elements produced by neutron irradiation and for the related discovery of nuclear reactions primed by slow neutrons. A many-sided personality, Fermi has been a star in the heavens of this century's physics, as shown in his studies of nuclear and particle physics. He is the second Italian Nobel laureate in physics, joining Guglielmo Marconi {see planet (1332)} (1909). (M 33790)

(8104) 1994 BW_4

Discovered 1994 January 19 by T. Kobayashi at Oizumi.

(8105) 1994 WH_2

Discovered 1994 November 28 by S. Ueda and H. Kaneda at Kushiro.

(8106) Carpino

1994 YB. Discovered 1994 December 23 by M. Cavagna and P. Sicoli at Sormano.

Named in honor of Mario Carpino (1957-), Italian astronomer at Brera Astronomical Observatory in Milan. From his initial studies of satellite geodesy, he acquired a taste for extreme accuracy in orbit determination, applying these skills to the study of the dynamics of solar-system bodies in projects such as LONGSTOP and SPACEGUARD. When the Spaceguard Foundation was established in Rome in 1996 he became its secretary. (M 31299)
Citation prepared by A. Milani at the request of the discoverers.

(8107) 1995 BR_4

Discovered 1995 January 31 by Y. Shimizu and T. Urata at Nachi-Katsuura.

(8108) Wieland

1995 BC_{16}. Discovered 1995 January 30 by F. Börngen at Tautenburg.

Named for the author Christoph Martin Wieland (1733-1813), a representative of the Rococo period, as well as of the German Enlightenment. Founder of modern

German storytelling prose, he translated the works of Shakespeare {see planet (2985)} and other writers and paved the way to the German classical period. The complete edition of his works contains 42 volumes. From 1773 to 1810 he was editor of the journal *Der Teutsche Merkur*. In 1772, he was appointed educator of the young Prince Carl August at the court in Weimar {see planet (3539)}, where he was a highly respected member of Goethe's {see planet (3047)} sphere. (M 31612)

(8109) 1995 DU_1

Discovered 1995 February 25 by C. W. Hergenrother at Tucson.

(8110) 1995 DE_2

Discovered 1995 February 27 by T. Kobayashi at Oizumi.

(8111) Hoepli

1995 GE. Discovered 1995 April 2 by A. Testa and V. Giuliani at Sormano.

Named in memory of Ulrico Hoepli (1847-1935), the founder of Hoepli Publishing Company and Hoepli International Bookstore in Milan. One of the greatest publishers of his time, Ulrico Hoepli published several important astronomy works within the 8000 titles the company printed; he was the publisher of G. V. Schiaparelli {see planet (4062)}. Today the company is still managed by the Hoepli family. A city patron and benefactor, in 1930 he donated a planetarium to the city of Milan, among the first in the world. (M 32791)
Citation prepared by Hoepli family.

(8112) Cesi

1995 JJ. Discovered 1995 May 3 at the Santa Lucia Observatory at Stroncone.

Named in memory of Federico Cesi (1585-1630), a scientist and humanist of the Italian Renaissance who devoted his life and property to the Academy of the Lincei, which he founded at the age of eighteen and of which Galileo {see planet (697)} was a member. He published numerous works about nature, including Galileo's "History and evidence about sunspots" and tables of classification for the animal, vegetable and mineral kingdoms. (M 33388)

(8113) Matsue

1996 HD_1. Discovered 1996 April 21 by R. H. McNaught and H. Abe at Yatsuka.

Named for a city in the ancient Izumo district in western Japan. A beautiful city rich in culture, history and archeological remains, Matsue has been designated as an International Cultural Center. (M 31299)

(8114) Lafcadio

1996 HZ_1. Discovered 1996 April 24 by H. Abe at Yatsuka.

Named in memory of Lafcadio Hearn (1850-1904). Born of an Irish father and Greek mother on the Ionian isle of Lefkás, he worked 20 years as a newspaper reporter in the United States before moving to Japan in 1890, changing his name to Yakumo Koizumi and becoming a naturalized Japanese citizen. He was commissioned by the Japanese Ministry of Education to teach in a Shimane prefectural junior high school in Matsue {see planet (8113)}, and from insights and impressions gleaned during his stay there he wrote his first and perhaps most famous major work on Japan, "Glimpses of Unfamiliar Japan". (M 31299)

(8115) Sakabe

1996 HB_2. Discovered 1996 April 24 by R. H. McNaught and Y. Ikari at Moriyama.

Named in honor of Sanjirou Sakabe (1923-), an amateur astronomer who studied under the late Issei Yamamoto {see planet (2249)}. Sakabe is the founder of the Dynic Astronomical Observatory, which contributes to the spread of astronomy in the surrounding area. (M 32349)

(8116) Jeanperrin

1996 HA_{15}. Discovered 1996 April 17 by E. W. Elst at La Silla.

Named in memory of the French physicist Jean Perrin (1870-1942), who studied Brownian motion and thereby confirmed the atomic nature of matter. At the same time he formed his ideas about the atom and its electrons, envisaging an atom as a small solar system, thereby anticipating the model by Rutherford. In 1926 he was honored with the Nobel prize in physics. The Centre National de la Recherche Scientifique and the Palais de la Découverte in Paris were created through his efforts. French astronomers especially honor him for his creation of the Observatoire de Haute Provence in 1936. (M 31299)

(8117) 1996 SD_1

Discovered 1996 September 18 at the Beijing Observatory at Xinglong.

(8118) 1996 WG_3

Discovered 1996 November 26 at the Beijing Observatory at Xinglong.

(8119) 1997 TP_{25}

Discovered 1997 October 12 at the Beijing Observatory at Xinglong.

(8120) Kobe

1997 VT. Discovered 1997 November 2 by H. Abe at Yatsuka.

Named for the primary port on the Seto Island Sea since the eighth century and one of Japan's most cosmopolitan cities, where the discoverer lived for five years during his student days. Kobe has made a remarkable recovery from the great earthquake in 1995. (M 33790)
Name proposed by the discoverer and citation proposed by I. Hasegawa.

(8121) Altdorfer

2572 P-L. Discovered 1960 September 24 by C. J. van Houten and I. van Houten-Groeneveld at Palomar.

Albrecht Altdorfer (c.1480-1538) is one of the painters of the "Donauschool". He created the first landscape paintings without figures. He was also an architect in the city of Regensburg {see planet (927)} and a member of the municipality. In addition to his paintings, many of his copper engravings and etchings have been preserved. (M 34345)

(8122) Holbein

4038 P-L. Discovered 1960 September 24 by C. J. van Houten and I. van Houten-Groeneveld at Palomar.

Named after Hans Holbein Sr. (c.1465-1524) and Hans Holbein Jr. (c.1497-1543), German painters. Holbein Sr. created mostly religious paintings in the transition style from Gothic to Renaissance. Holbein Jr. is more famous, working in Basel, Switzerland and England. He became court painter to king Henry VIII and made well-known portraits of the English aristocracy and other famous people of the time. (M 34345)

(8123) Canaletto

3138 T-1. Discovered 1971 March 26 by C. J. van Houten and I. van Houten-Groeneveld at Palomar.

Canaletto, whose birth name was Giovanni Antonio Canal (1697-1768), developed the so-called Veduten style of painting, characterized by detailed and precise views of cities. Best known for his paintings of Venice, he also painted London and Dresden. (M 34345)

(8124) Guardi

4370 T-1. Discovered 1971 March 26 by C. J. van Houten and I. van Houten-Groeneveld at Palomar.

Painter Francesco Guardi (1712-1793) was influenced by Pietro Longhi and Canaletto {see planet (8123)}. Guardi's style is more impressionistic, however, and he plays with the colors of light in the Venetian atmosphere more than Canelotto. His paintings are typical of the Rococo style. (M 34345)

(8125) Tyndareus

5493 T-2. Discovered 1973 September 30 by C. J. van Houten and I. van Houten-Groeneveld at Palomar.

Named for a king of Sparta, father of Clytemnestra and Helena {see planets (179) and (101)}. Often it is said, however, that Tyndareus was the foster-father of Helena, because Helena is generally considered to have been the daughter of Zeus and Leda {see, respectively, planets (5731) and (38)}. (M 31612)

(8126) 1966 BL

Discovered 1966 January 20 at the Purple Mountain Observatory at Nanking.

(8127) 1967 HA

Discovered 1967 April 27 by C. U. Cesco at El Leoncito.

(8128) Nicomachus

1967 JP. Discovered 1967 May 6 by C. U. Cesco and A. R. Klemola at El Leoncito.

Named for Nicomachus of Gerasa, first-century arithmetician and numerologist who has been credited with the discovery that 496 and 8128 are the third and fourth perfect numbers, i.e., numbers that are the sums of their factors (including unity but excluding the numbers themselves). The perfection of 6 and 28 was known to even earlier Greek and Hindu mathematicians. In 1644, Mersenne {see planet (8191)} showed that the sequence of perfect numbers is given by the product of $2^n - 1$ and 2^{n-1}, where $2^n - 1$ is prime - as then is n, which takes the values 2, 3, 5 and 7 for the first four cases. (M 31612)

Name proposed by the second discoverer, following a suggestion by B. G. Marsden, who wrote the citation.

(8129) 1975 SK_1

Discovered 1975 September 30 by S. J. Bus at Palomar.

(8130) Seeberg

1976 DJ_1. Discovered 1976 February 27 by F. Börngen at Tautenburg.

Named for the Seeberg Observatory, situated upon a well-marked hill close to the town of Gotha {see planet (1346)}. In the time of the duke Ernst II von Sachsen-Gotha-Altenburg (1745-1804) and the astronomer F. X. von Zach {see planet (999)}, it was an important center for astronomy. The first meeting of European astronomers took place at the Seeberg Observatory in August 1789. This naming honors the 200th anniversary of that conference, as well as the 1998 International Spring Meeting of the Astronomische Gesellschaft in Gotha, held on the occasion of this anniversary. (M 31612)
Named endorsed by P. Brosche.

(8131) Scanlon

1976 SC. Discovered 1976 September 27 by E. F. Helin at Palomar.

Named in honor of Leo J. Scanlon (1903-), prominent amateur whose interest in astronomy was sparked by seeing comet 1P/Halley in 1910. A plumber by profession, he built his own astronomical equipment and instructed others in telescope making. With Chester B. Roe he co-founded the Amateur Astronomers

Association of Pittsburgh on 1929. The following year he built the first aluminum observatory dome. He observed variable stars, and he helped organize the Moonwatch team at the Allegheny Observatory in 1957. In 1985 he was happy to be able to observe comet 1P/Halley again. (M 32095; M 33817)

Name proposed by the discoverer, following a suggestion by T. P. Kohman, who prepared the citation.

(8132) 1976 YA_6

Discovered 1976 December 18 by L. I. Chernykh at Nauchnyj.

(8133) 1977 DX_3

Discovered 1977 February 18 by H. Kosai and K. Hurukawa at Kiso.

(8134) 1978 SQ_7

Discovered 1978 September 26 by L. V. Zhuravleva at Nauchnyj.

(8135) 1978 VP_{10}

Discovered 1978 November 7 by E. F. Helin and S. J. Bus at Palomar.

(8136) 1979 MH_2

Discovered 1979 June 25 by E. F. Helin and S. J. Bus at Siding Spring.

(8137) Kvíz

1979 SJ. Discovered 1979 September 19 at the Kleť Observatory at Kleť.

Named in memory of Zdeněk Kvíz (1932-1993), Czech astronomer. His early works dealt with meteor showers, although his main interest was in light curves of close binaries and eclipsing binaries. Beginning in 1968 he lived in Australia, working also in Switzerland. The Czech Astronomical Society prize for young astronomers bears his name. (M 32349)

Name suggested by J. Tichá

(8138) 1980 FF_{12}

Discovered 1980 March 20 at the Perth Observatory at Bickley.

(8139) 1980 UM_1

Discovered 1980 October 31 by S. J. Bus at Palomar.

(8140) 1981 EO_{15}

Discovered 1981 March 1 by S. J. Bus at Siding Spring.

(8141) 1982 SO_4

Discovered 1982 September 20 by N. S. Chernykh at Nauchnyj.

(8142) Zolotov

1982 UR_6. Discovered 1982 October 20 by L. G. Karachkina at Nauchnyj.

Named in honor of Andrej Andreevich Zolotov (1937-), Russian screenwriter and art and music critic. He is the author of more than 30 documentary films about Russian musicians, composers and conductors. (M 32349)

Name suggested by G. Spiridov and supported by the discoverer.

(8143) 1982 VN

Discovered 1982 November 11 by A. Mrkos at Kleť.

(8144) 1982 VY_2

Discovered 1982 November 14 by H. Kosai and K. Hurukawa at Kiso.

(8145) 1983 RY_4
Discovered 1983 September 5 by L. V. Zhuravleva at Nauchnyj.

(8146) 1983 WG
Discovered 1983 November 28 by E. Bowell at Anderson Mesa.

(8147) 1984 SU_3
Discovered 1984 September 28 by B. A. Skiff at Anderson Mesa.

(8148) 1985 CR_2
Discovered 1985 February 15 by H. Debehogne at La Silla.

(8149) 1985 JN_1
Discovered 1985 May 11 by C. S. Shoemaker at Palomar.

(8150) 1985 QL_4
Discovered 1985 August 24 by N. S. Chernykh at Nauchnyj.

(8151) 1986 PK_6
Discovered 1986 August 12 by L. V. Zhuravleva at Nauchnyj.

(8152) 1986 VY
Discovered 1986 November 3 by A. Mrkos at Kleť.

(8153) 1986 WO_1
Discovered 1986 November 25 by A. Mrkos at Kleť.

(8154) Stahl
1988 CQ_7. Discovered 1988 February 15 by E. W. Elst at La Silla.

Named in memory of the German physician and chemist Georg Ernst Stahl (1660-1734), who developed the phlogiston theory of combustion, which dominated chemical thought for almost a century. Contrary to the view of his friend Friedrich Hoffmann at the University of Halle, who considered living organisms as machines to be explained by the laws of mechanics, Stahl insisted that neither mechanical nor chemical laws alone were sufficient to account for the phenomenon of life. Most likely influenced by his pietism, he insisted that life required a force for which he reserved the Latin word *anima*, which in turn gave rise to the theory of animism (vitalism). (M 34345)

(8155) Battaglini
1988 QA. Discovered 1988 August 17 at the Osservatorio San Vittore at Bologna.

Named in memory of Giuseppe Battaglini (1826-1894), Italian mathematician. University professor of superior geometry at Naples and Rome, he was founder of the *Giornale di Matematiche ad uso degli studenti delle Università Italiane*. He introduced Italy to the non-Euclidean geometry of Lobachevskij {see planet (1858)}, and his translations of some of Todhunter's texts were used for many years by Italian university students. (M 33790)

(8156) 1988 TR
Discovered 1988 October 13 by K. Endate and K. Watanabe at Kitami.

(8157) 1988 XG_2
Discovered 1988 December 15 by Y. Oshima at Gekko.

(8158) Herder
1989 UH_7. Discovered 1989 October 23 by F. Börngen at Tautenburg.

Named for Johann Gottfried Herder (1744-1803), German philosopher, theologian, writer, translator and collector of folk songs. Through his study of history, language, culture, and literature, he became a great authority on the history of the European mind. Through Goethe, he was appointed head clergyman at the court of Weimar {see planet (3539) in 1776. With Goethe, Schiller and Wieland {see, respectively, planets (3047), (3079), (8108)}, Herder is one of the four "Greats" of the Weimar classical period. (M 31612)

(8159) 1990 BE_1

Discovered 1990 January 24 by K. Endate and K. Watanabe at Kitami.

(8160) 1990 MG

Discovered 1990 June 21 by H. E. Holt at Palomar.

(8161) Newman

1990 QP_3. Discovered 1990 August 19 at the Oak Ridge Observatory at Harvard.

Named in honor of Constance B. Newman (1935-), Smithsonian Institution undersecretary whose unwavering devotion to the principles of exemplary management and diversity has enabled the Smithsonian to flourish and the Astrophysical Observatory to thrive. (M 32349)

(8162) 1990 SK_{11}

Discovered 1990 September 16 by H. E. Holt at Palomar.

(8163) Ishizaki

1990 UF_2. Discovered 1990 October 27 by T. Seki at Geisei.

Named in honor of Masako Ishizaki (1902-), long-time Japanese amateur astronomer and pioneer in the study of meteoric dust. In particular, she studied dust collected for her at the Showa Base in Antarctica, where terrestrial contamination is low. (M 32791)

Name proposed by the discoverer following a suggestion by S. Morikubo, I. Hasegawa and T. Sato.

(8164) Andreasdoppler

1990 UO_3. Discovered 1990 October 16 by E. W. Elst at La Silla.

Named in honor of Andreas Doppler (1963-), German amateur astronomer and identifier of minor planets. (M 34345)

This name was independently proposed by F. Börngen and L. D. Schmadel.

(8165) Gnädig

1990 WQ_3. Discovered 1990 November 21 by E. W. Elst at La Silla.

Named in honor of Arno Gnädig (1956-), German amateur astronomer and identifier of minor planets. (M 34345)

This name was independently proposed by F. Börngen and L. D. Schmadel.

(8166) 1991 AH_1

Discovered 1991 January 12 by B. G. W. Manning at Stakenbridge.

(8167) 1991 CM_3

Discovered 1991 February 14 by K. Endate and K. Watanabe at Kitami.

(8168) 1991 FK_1

Discovered 1991 March 18 by E. F. Helin at Palomar.

(8169) Mirabeau

1991 PO_2. Discovered 1991 August 2 by E. W. Elst at La Silla.

Named in memory of French politician and orator Honoré-Gabriel Mirabeau (1749-1791), one of the greatest figures of the early French Revolution. In 1789 he was elected deputy at the States General to represent Marseille and Aix-en-

Provence. From May to October 1789 he played a decisive part in the battle between the Third Estate (common people) and the privileged orders (clergy and nobility). This made him very popular in Paris, but intrigues with the court and his Machiavellian game in general led him to political defeat, since he was too much a monarchist for the revolution and too revolutionary for the monarchy. However, he was given a magnificent funeral at the new church of Sainte-Geneviève, which, in honor of him, was converted into the Panthèon, for the burial of great men. (M 34345)

(8170) 1991 PZ_{11}

Discovered 1991 August 7 by H. E. Holt at Palomar.

(8171) Stauffenberg

1991 RV_3. Discovered 1991 September 5 by F. Börngen and L. D. Schmadel at Tautenburg.

Named in honor of the German count Claus Graf Schenk von Stauffenberg (1907-1944), the driving force behind a conspiracy of German military and politicians to overthrow the fascist regime. An unsuccessful attempt on the fascist leader's life was carried out personally by Stauffenberg on 1994 July 20. This had far-ranging consequences that can not always be fully appreciated, even though the immediate outcome was the arrest of approximately 1000 and the execution of more than 200, including Stauffenberg, the very same day. (M 31612)
Name proposed by the first discoverer.

(8172) 1991 RP_{15}

Discovered 1991 September 15 by H. E. Holt at Palomar.

(8173) 1991 RX_{23}

Discovered 1991 September 11 by H. E. Holt at Palomar.

(8174) 1991 SL_2

Discovered 1991 September 17 by H. E. Holt at Palomar.

(8175) Boerhaave

1991 VV_5. Discovered 1991 November 2 by E. W. Elst at La Silla.

Named in memory of the Dutch physician and professor of medicine Herman {Hermannus} Boerhaave (1668-1738). He was a brilliant teacher of botany, medicine and chemistry at the University of Leiden, where students came from all parts of Europe to hear him. Baron d'Holbach and de Lamettrie {see planets (6956) and (7095), respectively} studied with him. He is credited with founding the modern system of teaching medical students at the patient's bedside. He is the author of *Institutiones Medicae* and *Aphorismi de Cognoscendis et Curandis Morbis*, two textbooks widely used during and after his lifetime. (M 34346)

(8176) 1991 WA

Discovered 1991 November 29 by R. H. McNaught at Siding Spring.

(8177) 1992 BO

Discovered 1992 January 28 by S. Ueda and H. Kaneda at Kushiro.

(8178) 1992 DQ_{10}

Discovered 1992 February 29 by the Uppsala-ESO Survey at La Silla.

(8179) 1992 EA_7

Discovered 1992 March 1 by the Uppsala-ESO Survey at La Silla.

(8180) 1992 PY_2
Discovered 1992 August 6 by H. E. Holt at Palomar.

(8181) Rossini
1992 ST_{26}. Discovered 1992 September 28 by L. V. Zhuravleva at Nauchnyj.

Named in memory of the famous Italian composer Gioacchino Antonio Rossini (1792-1868). (M 34627)

(8182) 1992 TX
Discovered 1992 October 1 by M. Yanai and K. Watanabe at Kitami.

(8183) 1992 UE_3
Discovered 1992 October 22 by S. Ueda and H. Kaneda at Kushiro.

(8184) 1992 WL
Discovered 1992 November 16 by S. Ueda and H. Kaneda at Kushiro.

(8185) 1992 WR_2
Discovered 1992 November 18 by S. Ueda and H. Kaneda at Kushiro.

(8186) 1992 WP_3
Discovered 1992 November 17 by A. Sugie at Taga.

(8187) 1992 XL
Discovered 1992 December 15 by S. Otomo at Kiyosato.

(8188) 1992 YE_3
Discovered 1992 December 18 by Y. Mizuno and T. Furuta at Kani.

(8189) 1992 YG_3
Discovered 1992 December 30 by T. Hioki and S. Hayakawa at Okutama.

(8190) Bouguer
1993 ON_9. Discovered 1993 July 20 by E. W. Elst at La Silla.

Named in memory of the French scientist Pierre Bouguer (1698-1758), the founder of astronomical photometry. In 1720 he made some of the earliest measurements in astronomical photometry by comparing the apparent brightness of celestial objects to the light of a candle. In 1735 he followed de la Condamine to measure an arc of the meridian near the equator in Peru. During his stay he investigated the absorption of light in the atmosphere, which led to the formulation of "Bouguer's absorption law". (M 34346)

(8191) Mersenne
1993 OX_9. Discovered 1993 July 20 by E. W. Elst at La Silla.

Named for Marin Mersenne (1588-1648), French mathematician and philosopher, best known for his discovery of prime numbers of the form $2^n - 1$ and that n is then necessarily also prime. The number 8191 = $2^{13} - 1$ is the fifth Mersenne prime. Mersenne correctly verified that the numbers are prime also for n = 17, 19 and 31 - and that they are *not* prime for n = 11, 23 and 29. The largest prime Mersenne correctly established was for n = 127. (M 31612)

Name proposed by the discoverer, following a suggestion by B. G. Marsden, who wrote the citation.

(8192) Tonucci
1993 RB. Discovered 1993 September 10 at the Santa Lucia Observatory at Stroncone.

Named in memory of Giuseppe Tonucci (1938-1988), Italian road and track cycling champion. During his long career he won numerous competitions throughout

Europe, distinguishing himself in the Rome Olympic Games in 1960 and in the World Championship in the Netherlands. (M 33388)

(8193) Ciaurro

1993 SF. Discovered 1993 September 17 at the Santa Lucia Observatory at Stroncone.

Named in memory of Ilario Ciaurro (1889-1992), an art teacher and ceramist, but most famous as a painter. His favorite subject was Terni, his adopted town, and he loved using etchings, poems and stories to explore its innermost aspects. (M 33388)

(8194) 1993 SB_1

Discovered 1993 September 16 by K. Endate and K. Watanabe at Kitami.

(8195) 1993 UC_1

Discovered 1993 October 19 by E. F. Helin at Palomar.

(8196) 1993 UB_3

Discovered 1993 October 16 by E. F. Helin at Palomar.

(8197) 1993 VX

Discovered 1993 November 15 by T. Kobayashi at Oizumi.

(8198) 1993 VE_2

Discovered 1993 November 11 by S. Ueda and H. Kaneda at Kushiro.

(8199) 1993 XR

Discovered 1993 December 9 by T. Kobayashi at Oizumi.

(8200) Souten

1994 AY_1. Discovered 1994 January 7 by M. Hirasawa and S. Suzuki at Nyukasa.

Named for the Waseda University Astronomy Association, nicknamed Souten. Established in 1959 by Hidetaka Tojo, Souten has produced an astronaut and many astrophotographers and observers. The discoverers of this minor planet were members of Souten. (M 34346)

(8201) 1994 AH_2

Discovered 1994 January 5 by G. J. Garradd at Siding Spring.

(8202) 1994 CX_2

Discovered 1994 February 11 by K. Endate and K. Watanabe at Kitami.

(8203) Jogolehmann

1994 CP_{10}. Discovered 1994 February 7 by E. W. Elst at La Silla.

Named in memory of the German geologist Johann Gottlob Lehmann (1719-1767), who made important contributions to stratigraphy, the study of order and sequence in bedded sedimentary rocks. Educated originally at the University of Wittenberg as a medical doctor, he soon discovered that his real field of interest was mines and mining. He successfully explained the origin of mineral deposits in mines and recognized that rocks do not fall in haphazard positions but form a historical sequence, thereby supplying a strong argument against the biblical diluvian theory. (M 34346)

(8204) 1994 GC_1

Discovered 1994 April 8 by K. Endate and K. Watanabe at Kitami.

(8205) Van Dijck

1994 PE_{10}. Discovered 1994 August 10 by E. W. Elst at La Silla.

Named in memory of the Flemish painter Anthony Van Dijck (Van Dyck or Vandyke; 1599-1641), after Rubens the most prolific painter of the seventeenth century. His work embraces portraits of the European aristocracy, depictions of religious subjects and paintings of enchanting mythological stories. In contrast to Rubens, the lights and shades of Van Dijck are more abrupt, the figures less harmoniously proportioned and their expressions exaggerated. In 1632 Van Dijck was appointed court painter of king Charles I of England and was knighted the same year. His style influenced the great English portrait painters of the eighteenth century. (M 34627)

(8206) 1994 WK_1

Discovered 1994 November 27 by T. Kobayashi at Oizumi.

(8207) 1994 YS_1

Discovered 1994 December 31 by T. Kobayashi at Oizumi.

(8208) Volta

1995 DL_2. Discovered 1995 February 28 by P. Sicoli and P. Ghezzi at Sormano.

Named in memory of Alessandro Volta (1745-1827), physicist at the University of Livia. His discovery of the significance of separating silver-zinc disk pairs with disks of moist cardbord was the key to his invention of the battery, just 200 years ago. Volta also made important contributions to meteorology and the study of gases, notably with his discovery of methane. (M 34627)

(8209) 1995 DM_2

Discovered 1995 February 28 by P. Sicoli and P. Ghezzi at Sormano.

(8210) 1995 EH

Discovered 1995 March 5 by T. Kobayashi at Oizumi.

(8211) 1995 EB_1

Discovered 1995 March 5 by S. Ueda and H. Kaneda at Kushiro.

(8212) 1995 EF_1

Discovered 1995 March 6 by S. Otomo at Kiyosato.

(8213) 1995 FE

Discovered 1995 March 26 by Y. Shimizu and T. Urata at Nachi-Katsuura.

(8214) 1995 FH

Discovered 1995 March 29 by S. Mottola at La Silla.

(8215) 1995 FZ

Discovered 1995 March 31 by Y. Shimizu and T. Urata at Nachi-Katsuura.

(8216) 1995 FX_{14}

Discovered 1995 March 27 by the Spacewatch at Kitt Peak.

(8217) Dominikhašek

1995 HC. Discovered 1995 April 21 by P. Pravec and L. Šarounová at Ondřejov.

Named in honor of Dominik Hašek (1965-), goalkeeper, and the Czech Olympic hockey team, winners of the gold medal at the 1998 Winter Olympic Games in Nagano. Although the success was basically due to teamwork, Hašek's contribution was the most significant. This great win in "the hockey tournament of the century" was widely appreciated by the Czech people and recognized as one of the greatest achievements in the history of Czech sport. (M 31613)

(8218) 1996 JH
Discovered 1996 May 8 by R. H. McNaught at Siding Spring.

(8219) 1996 JL
Discovered 1996 May 10 by R. H. McNaught and T. Kojima at Chiyoda.

(8220) Nanyou
1996 JD_1. Discovered 1996 May 13 by T. Okuni at Nanyou.

Named for the city where the discoverer lives and his observatory is located. Nanyou City lies to the north of Tokyo, two and a half hours away on the Yamagata {see planet (7039)} line (the new express train). It is famous for Akayu-onsen (hot springs), the Kumano shrine and Kikuningyou (a doll made of chrysanthemums). (M 32791)

(8221) 1996 NA_4
Discovered 1996 July 14 by E. W. Elst at La Silla.

(8222) Gellner
1996 OX. Discovered 1996 July 22 by M. Tichý and Z. Moravec at Kleť.

Named in memory of Czech poet František Gellner (1881-1914), a Bohemian anarchist. His poetry depicted, sometimes in shocking ways, night life in a big city. He died in the front lines during World War I. (M 34346)

(8223) Bradshaw
1996 PD. Discovered 1996 August 6 by P. G. Comba at Prescott.

Named for the Bradshaw mountains, which lie to the south of Prescott, Arizona. This area of sparsely populated pine-covered mountains provides a dark southern sky for the Prescott Observatory. The mountains themselves were named for William D. Bradshaw, a Western adventurer and miner. (M 31613)

(8224) Fultonwright
1996 PE. Discovered 1996 August 6 by P. G. Comba at Prescott.

Named in honor of Fulton Wright, Jr. (1937-), a friend of the discoverer, professor at Yavapai College, active amateur astronomer and telescope maker, and accomplished double bassist. (M 31613)

(8225) Emerson
1996 QC. Discovered 1996 August 16 by C. F. Durman and B. M. Ewen-Smith at Portimão.

Named in memory of David Emerson (1943-1996), lecturer in astrophysics at the Royal Observatory, Edinburgh. His research interests were in the interaction of matter and radiation and star formation. He was director of studies of the first discoverer. Emerson was also an ordained lay-preacher in the Scottish Episcopal Church. He died at a relatively young age on 1996 Sept. 26, just a few weeks after the discovery was made. (M 31613)
Obituary published in Astron. Geophys., Vol. 38, Issue 4, p. 38 (1997).

(8226) 1996 TF_7
Discovered 1996 October 5 by Y. Shimizu and T. Urata at Nachi-Katsuura.

(8227) 1996 VD_4
Discovered 1996 November 8 at the Beijing Observatory at Xinglong.

(8228) 1996 YB_2
Discovered 1996 December 22 at the Beijing Observatory at Xinglong.

(8229) Kozelský

1996 YU_2. Discovered 1996 December 28 by M. Wolf (Marek) and L. Šarounová at Ondřejov.

Named in honor of František Kozelský (1913-), a Czech telescope maker well-known for his work in collaboration with V. Gajdušek {see planet (3603)}. Kozelský made a set of unique fine-mechanics tools that allowed Gajdušek to produce a number of telescope mirrors, resulting finally in the production of six excellent 0.6-m mirrors, two of them now used at the Ondřejov and the Kleť observatories in telescopes devoted to observations of minor planets. Kozelský also made a series of eight 0.2-m refracting telescopes, twenty 0.3-m reflecting telescopes, and several coelostats for small observatories in Czech Republic and Slovakia. (M 32791)

Name proposed by the discoverers, following a suggestion by G. Polášek.

(8230) Perona

1997 TW_{16}. Discovered 1997 October 8 at the Santa Lucia Observatory at Stroncone.

Named in memory of Renato Perona (1927-1984), a track cycling champion who for more than a decade was the undisputed ruler in national and international competitions. He won the gold medal in the London Olympic Games in 1948. (M 33388)

(8231) 1997 TX_{17}

Discovered 1997 October 6 by K. Endate and K. Watanabe at Kitami.

(8232) 1997 UW_3

Discovered 1997 October 26 by T. Kobayashi at Oizumi.

(8233) 1997 VZ_2

Discovered 1997 November 5 by T. Kobayashi at Oizumi.

(8234) Nobeoka

1997 VK_8. Discovered 1997 November 3 by T. Seki at Geisei.

Named for a town in Miyazaki {see planet (6905)} prefecture, at the mouth of the Gokase River. Although it is the site of one of the largest chemical factories in Japan, it remains rich in natural beauty. The sound of the bell at the top of the castle hill is described in a poem by Bokusui Wakayama {see planet (8367)}. (M 33388)

Name proposed by the discoverer following a suggestion by T. Sato and R. Ukishima.

(8235) Fragonard

2096 P-L. Discovered 1960 September 24 by C. J. van Houten and I. van Houten-Groeneveld at Palomar.

Jean-Honoré Fragonard (1732-1806) was a student of Boucher. Both teacher and pupil painted the colorful world of the Rococo aristocracy. In his landscape paintings Fragonard shows the influence of the Venetian artists. (M 34346)

(8236) Gainsborough

4040 P-L. Discovered 1960 September 24 by C. J. van Houten and I. van Houten-Groeneveld at Palomar.

Named for Thomas Gainsborough (1727-1788), British portrait and landscape painter. He was strongly influenced by the Rococo style and Dutch landscape painting, and his work is characterized by luminous colors. His portraits often feature fine landscapes, and he is notable for painting not only the aristocracy of England but also its farmers. (M 34346)

(8237) Constable

7581 P-L. Discovered 1960 October 17 by C. J. van Houten and I. van Houten-Groeneveld at Palomar.

Named for the English painter John Constable (1776-1837), known for his landscapes painted directly from nature. A forerunner of the Impressionists of the late nineteenth century, he is notable for his use of color. For some time he made portraits, but later he went back to landscape painting. (M 34346)

(8238) Courbet

4232 T-1. Discovered 1971 March 26 by C. J. van Houten and I. van Houten-Groeneveld at Palomar.

Gustave Courbet (1819-1877) was an autodidact who studied nature and the great paintings in the Louvre. Initially his paintings were not accepted in the salons of Paris; he then held his own exhibition, which was a great success. His harmonious colors influenced the French and German Impressionists. (M 34346)

(8239) Signac

1153 T-2. Discovered 1973 September 29 by C. J. van Houten and I. van Houten-Groeneveld at Palomar.

Paul Signac (1863-1935) worked together with Seurat {see planet (6678)} to develop the technique of pointillism. For his pointillistic paintings Signac mostly chose water, ships and ports as subjects. After Seurat and Signac not many painters worked in this time-consuming method. (M 34346)

(8240) Matisse

4172 T-2. Discovered 1973 September 29 by C. J. van Houten and I. van Houten-Groeneveld at Palomar.

Henri Matisse (1869-1954) studied at the School of Fine Arts in Paris. After traveling through Europe and the Pacific islands, he initially developed an Impressionistic style of painting. Later he became the leader of a new school, Fauvism, characterized by bold use of color and distorted forms. (M 34346)

(8241) Agrius

1973 SE_1. Discovered 1973 September 19 by C. J. van Houten and I. van Houten-Groeneveld at Palomar.

Named after Agrius, a Greek from Aetolia and father of Thersites {see planet (1868)}. Thersites was the ugliest Greek in Troy. (M 32791)

(8242) 1975 SA_1

Discovered 1975 September 30 by S. J. Bus at Palomar.

(8243) 1975 SF_1

Discovered 1975 September 30 by S. J. Bus at Palomar.

(8244) 1975 TO_2

Discovered 1975 October 3 by L. I. Chernykh at Nauchnyj.

(8245) 1977 RC_9

Discovered 1977 September 8 by S. J. Bus at Palomar.

(8246) 1979 QT_8

Discovered 1979 August 20 by N. S. Chernykh at Nauchnyj.

(8247) 1979 SP_{14}

Discovered 1979 September 20 by S. J. Bus at Palomar.

(8248) 1979 TV_2
Discovered 1979 October 14 by N. S. Chernykh at Nauchnyj.

(8249) 1980 GG
Discovered 1980 April 13 by A. Mrkos at Kleť.

(8250) 1980 RP
Discovered 1980 September 2 by E. Bowell at Anderson Mesa.

(8251) 1980 VA
Discovered 1980 November 8 by T. Furuta at Tokai.

(8252) 1981 EY_{14}
Discovered 1981 March 1 by S. J. Bus at Siding Spring.

(8253) 1981 EU_{15}
Discovered 1981 March 1 by S. J. Bus at Siding Spring.

(8254) 1981 EF_{18}
Discovered 1981 March 2 by S. J. Bus at Siding Spring.

(8255) 1981 EZ_{18}
Discovered 1981 March 2 by S. J. Bus at Siding Spring.

(8256) 1981 UZ_9
Discovered 1981 October 25 at the Purple Mountain Observatory at Nanking.

(8257) 1982 HO_1
Discovered 1982 April 28 by E. Bowell at Anderson Mesa.

(8258) 1982 RW_1
Discovered 1982 September 15 by A. Mrkos at Kleť.

(8259) 1983 UG
Discovered 1983 October 16 by Z. Vávrová at Kleť.

(8260) 1984 SH
Discovered 1984 September 23 at the Bulgarian National Observatory at Rozhen.

(8261) 1985 RD
Discovered 1985 September 11 at the Brorfelde Observatory at Brorfelde.

(8262) 1985 RG
Discovered 1985 September 14 by E. Bowell at Anderson Mesa.

(8263) 1986 QT
Discovered 1986 August 26 by H. Debehogne at La Silla.

(8264) 1986 QA_3
Discovered 1986 August 29 by H. Debehogne at La Silla.

(8265) 1986 RB_5
Discovered 1986 September 1 by H. Debehogne at La Silla.

(8266) Bertelli
1986 TC. Discovered 1986 October 1 at the Osservatorio San Vittore at Bologna.

Named in memory of Francesco Bertelli (1794-1844), Italian astronomer at the observatory of Bologna and successor to P. Caturegli (1786-1833) as professor of astronomy at the University of Bologna. Only the first volume of his book, *Elementi di Meccanica Celeste*, was published at Bologna (1841) before his untimely death. He collaborated in the calculation of the *Effemeridi*, the annual publication initiated in 1715 by E. Manfredi (1674-1739) and that ceased with Bertelli's death in 1844. (M 33790)

(8267) 1986 TX_3

Discovered 1986 October 4 by A. Mrkos at Kleť.

(8268) Goerdeler

1987 SQ_{10}. Discovered 1987 September 29 by F. Börngen at Tautenburg.

Named in memory of Carl Friedrich Goerdeler (1884-1945), mayor of the towns of Königsberg and Leipzig, respectively (1920-1937). Early on he opposed the fascist dictatorship. After the outbreak of World War II, he became the leading force behind a Resistance group of conservative-minded Germans, planning a coup d'état against the fascist leader through contacts with a military opposition. Goerdeler's main goal was to restore and institutionalize a state of law. In September 1944, he was condemned to death by the Volksgerichtshof. (M 31613)

(8269) Calandrelli

1988 QB. Discovered 1988 August 17 at the Osservatorio San Vittore at Bologna.

Named in memory of Ignazio Calandrelli (1792-1866), professor of optics and astronomy in Bologna and Rome. He served as director of the Bologna Observatory during 1845-1848 and of Campidoglio at Rome from 1848 to his death. He was an observer of the major planets, minor planets and comets, as well as an orbit computer. (M 34627)

(8270) Winslow

1989 JF. Discovered 1989 May 2 by E. F. Helin at Palomar.

Named in honor of John Seymour Winslow, a friend of the discoverer's husband, Ron, since grade school. John graduated from Caltech with a degree in physics and later returned for a master's degree in electrical engineering. He developed the equipment used to master the MCA video disc, and later became a partner in Optical Disc Corporation, a firm that provides mastering equipment for compact discs. This minor planet was presented on the occasion of the 45th reunion of John and Ron's Caltech Class of '53. (M 32350)

(8271) 1989 NY

Discovered 1989 July 2 by E. F. Helin at Palomar.

(8272) 1989 SG

Discovered 1989 September 24 by Y. Mizuno and T. Furuta at Kani.

(8273) 1989 WB_2

Discovered 1989 November 29 by M. Akiyama and T. Furuta at Mishima.

(8274) 1990 TJ_1

Discovered 1990 October 15 by K. Endate and K. Watanabe at Kitami.

(8275) 1990 VR_8

Discovered 1990 November 11 by E. W. Elst at La Silla.

(8276) Shigei

1991 FL. Discovered 1991 March 17 by S. Otomo and O. Muramatsu at Kiyosato.

Named in honor of Mika Shigei (1968-   ), lecturer and curator of the Gotoh Planetarium and Astronomical Museum in Tokyo, where she was in charge of an exhibition on meteorites. (M 34346)

(8277) 1991 GV_8
Discovered 1991 April 8 by E. W. Elst at La Silla.

(8278) 1991 JJ
Discovered 1991 May 4 by Y. Mizuno and T. Furuta at Kani.

(8279) 1991 PN_7
Discovered 1991 August 6 by E. W. Elst at La Silla.

(8280) 1991 PG_{16}
Discovered 1991 August 7 by H. E. Holt at Palomar.

(8281) 1991 PC_{18}
Discovered 1991 August 8 by H. E. Holt at Palomar.

(8282) Delp
1991 RR_{40}. Discovered 1991 September 10 by F. Börngen at Tautenburg.
Named in memory of the German theologian Alfred Delp (1907-1945). Beginning in 1942, he was a member of the illegal "Kreisauer Kreis" {see planet (7945)}, a group of political conservatives who worked to eliminate the Nazi regime. With this group, Delp collaborated on a first draft of a Christian social order for his country. Following the 1944 attempt on the fascist leader's life, he was condemned to death and executed in February 1945. (M 32350)

(8283) 1991 SV
Discovered 1991 September 30 by R. H. McNaught at Siding Spring.

(8284) Cranach
1991 TT_{13}. Discovered 1991 October 8 by F. Börngen at Tautenburg.
Named in honor of the German painter, draftsman and copper engraver Lucas Cranach the Elder (1472-1553). As a friend of Luther and Melanchton {see planets (7100) and (7906)}, he is reputed to be the primary master of the German Reformation period. His first work is ascertainable in Vienna; beginning in 1505, he was an esteemed court painter for the elector of Saxony Friedrich der Weise in Wittenberg. One of his best-known works is the altarpiece in the main church of Weimar {see planet (3539)}, which was completed by his son Lucas Cranch the Younger. (M 32095)

(8285) 1991 UK_3
Discovered 1991 October 31 by S. Ueda and H. Kaneda at Kushiro.

(8286) 1992 EK_1
Discovered 1992 March 8 by K. Endate and K. Watanabe at Kitami.

(8287) 1992 EJ_4
Discovered 1992 March 1 by the Uppsala-ESO Survey at La Silla.

(8288) 1992 ED_{17}
Discovered 1992 March 1 by the Uppsala-ESO Survey at La Silla.

(8289) An-Eefje
1992 JQ_3. Discovered 1992 May 3 by H. Debehogne at La Silla.
Named in memory of the two young Belgian women An and Eefje, whose young and hopeful lives came dramatically to an end in 1996. May their memory stand for all women throughout the world who suffer from abuse. (M 32095)

(8290) 1992 NP
Discovered 1992 July 2 by E. F. Helin and L. Lee at Palomar.

(8291) 1992 RV_1
Discovered 1992 September 2 by E. W. Elst at Palomar.

(8292) 1992 SU_{14}
Discovered 1992 September 30 by H. E. Holt at Palomar.

(8293) 1992 UQ
Discovered 1992 October 19 by S. Ueda and H. Kaneda at Kushiro.

(8294) 1992 UM_3
Discovered 1992 October 26 by K. Endate and K. Watanabe at Kitami.

(8295) 1992 UN_4
Discovered 1992 October 26 by K. Endate and K. Watanabe at Kitami.

(8296) 1993 AD
Discovered 1993 January 13 by K. Endate and K. Watanabe at Kitami.

(8297) 1993 QJ_4
Discovered 1993 August 18 by E. W. Elst at Caussols.

(8298) Loubna
1993 SQ_{10}. Discovered 1993 September 22 by H. Debehogne and E. W. Elst at La Silla.

Named in memory of the Moroccan-Belgian girl Loubna, whose young life came abruptly to an end. May she stand as a symbol for a hopeful life in a multicultural society. (M 32095)

(8299) 1993 TP_{24}
Discovered 1993 October 9 by E. W. Elst at La Silla.

(8300) 1994 AO_2
Discovered 1994 January 9 by T. Kobayashi and H. Fujii at Oizumi.

(8301) 1995 BG_2
Discovered 1995 January 30 by T. Kobayashi at Oizumi.

(8302) 1995 CY
Discovered 1995 February 3 by T. Kobayashi at Oizumi.

(8303) 1995 CO_1
Discovered 1995 February 9 by T. Kobayashi at Oizumi.

(8304) 1995 DJ_1
Discovered 1995 February 22 by T. Kobayashi at Oizumi.

(8305) 1995 DQ_1
Discovered 1995 February 22 by T. Kobayashi at Oizumi.

(8306) Shoko
1995 DY_1. Discovered 1995 February 24 by A. Nakamura at Kuma.

Named in honor of Shoko Sawada (1962-), Japanese singer and songwriter. Since her debut in 1979, she has released 52 singles and 22 albums, ranging from touching ballads to rhythmical pop songs. Her recordings have fascinated many fans in Japan, including the discoverer of this minor planet. (M 32095)

(8307) Peltan

1995 EN. Discovered 1995 March 5 by J. Tichá at Kleť.

Named for the discoverer's family, including her mother Marie (nee Kosová) Peltanová (1930-), her father Bohuslav Peltan (1927-1983), her brother Jiří Peltan (1953-), her sister-in-law Adéla (1951-) and her nephews Petr (1983-) and Libor (1989-). (M 34346)

(8308) Julie-Mélissa

1996 HD_{13}. Discovered 1996 April 17 by E. W. Elst at La Silla.

Named in memory of the two Belgian children Julie and Mélissa, who were murdered tragically in 1996, symbolizing all innocent children suffering from abuse. (M 32096)

(8309) 1996 NL_1

Discovered 1996 July 14 by the JPL NEAT Program at Haleakala.

(8310) Seelos

1996 PL_2. Discovered 1996 August 9 by the JPL NEAT Program at Haleakala.

Named in honor of Frank P. Seelos IV who, as a student at Wolford College, participated in the 1998 CalTech Summer Undergraduate Research Fellow program. He assisted and carried out research with the Near-Earth Asteroid Tracking program at the Jet Propulsion Laboratory and developed the software package HAVANA, which rapidly accesses images of specified objects from the extensive observational archive available. Frank is an outstanding student with a double major in physics and mathematics. (M 33790)

(8311) Zhangdaning

1996 TV_1. Discovered 1996 October 3 at the Beijing Observatory at Xinglong.

Named for Daning Zhang (1944-), physician and vice-president of Tianjin Traditional Chinese Medicine (TCM) Hospital, professor of Tianjin Medical University, guest professor of Taiwan Chinese Academic College, chairman of the International TCM Kidney Diseases Conference. As the founder of the kidney diseases practice of TCM, he has won many international awards. He has written more than 60 publications on kidney disease. (M 32350)

(8312) 1996 TJ_{12}

Discovered 1996 October 15 by Y. Shimizu and T. Urata at Nachi-Katsuura.

(8313) Christiansen

1996 YU_1. Discovered 1996 December 19 at the Beijing Observatory at Xinglong.

Named in honor of Wilbur N. Christiansen (1913-), foreign member of the Chinese Academy of Sciences and unfailing friend of Chinese astronomers. A pioneer in radio astronomy, he invented and developed a series of radio telescopes that in their time provided the highest angular resolution. These were the "grating telescope", the "grating cross" and the "rotational synthesis telescope". His textbook on radio telescopes, with Högbom, was translated into Russian and Chinese. Immediately after the discovery of the 21-cm hydrogen line in space, he confirmed this and went on to make the first map that showed we live in a spiral galaxy. He served as a vice president of the IAU and as president of URSI, and he now is an honorary president of URSI. (M 33388)

(8314) 1997 US_8

Discovered 1997 October 25 by K. Endate and K. Watanabe at Kitami.

(8315) 1997 WA_{22}

Discovered 1997 November 25 at the Beijing Observatory at Xinglong.

(8316) Wolkenstein

3002 P-L. Discovered 1960 September 24 by C. J. van Houten and I. van Houten-Groeneveld at Palomar.

Named for Oswald von Wolkenstein (?1377-1445), one of the last minstrels. We know much about his life and times from his own songs, which have been preserved. At the age of ten he became shield-bearer for a knight and traveled throughout the world as a groom, cook and singer, eventually returning to his home castle of Wolkenstein in the Grödner valley in Tirol. He was also ambassador to the emperor Sigismund and traveled to England and the Iberian peninsula in his name. Especially interesting are his financial documents, which have all been saved and show the difficult position of knights around 1400. (M 31791)

(8317) Eurysaces

4523 P-L. Discovered 1960 September 24 by C. J. van Houten and I. van Houten-Groeneveld at Palomar.

Named for the son of Ajax by Tecmessa {see, respectively planets (1404) and (604)}, known for his enormous shield. (M 32791)

(8318) Averroës

1306 T-2. Discovered 1973 September 29 by C. J. van Houten and I. van Houten-Groeneveld at Palomar.

Named for Averroës (1126-1198), whose Arabian name was Ibn Roschd Abdul Walid and who studied law and medicine in Córdoba. He brought together the philosophy of Aristotle {see planet (6123)}, Islam and Christianity. Both Islam and the Christian church, especially Thomas Aquinas, condemned him. Most of his publications have been translated into Latin. (M 32792)

(8319) Antiphanes

3365 T-2. Discovered 1973 September 25 by C. J. van Houten and I. van Houten-Groeneveld at Palomar.

Named after the Greek comic poet Antiphanes (408-330 B.C.). Today 119 complete titles and about 300 fragments are known. He wrote parodies of Sophocles, Euripides {see planets (2921) and (2930), respectively} and many different characters and professions. (M 34346)

(8320) 1955 RV

Discovered 1955 September 13 at the Goethe Link Observatory at Brooklyn, Indiana.

(8321) 1977 EX

Discovered 1977 March 13 by N. S. Chernykh at Nauchnyj.

(8322) 1978 RL_1

Discovered 1978 September 5 by N. S. Chernykh at Nauchnyj.

(8323) 1979 UH

Discovered 1979 October 17 by E. Bowell at Anderson Mesa.

(8324) 1981 DF_2

Discovered 1981 February 28 by S. J. Bus at Siding Spring.

(8325) 1981 EM_{26}

Discovered 1981 March 2 by S. J. Bus at Siding Spring.

(8326) 1981 JS_2

Discovered 1981 May 6 by C. S. Shoemaker at Palomar.

(8327) 1981 JE_3
Discovered 1981 May 6 by C. S. Shoemaker at Palomar.

(8328) 1981 QQ_2
Discovered 1981 August 23 by H. Debehogne at La Silla.

(8329) 1982 FP_3
Discovered 1982 March 22 by H. Debehogne at La Silla.

(8330) 1982 FX_3
Discovered 1982 March 28 by H. Debehogne at La Silla.

(8331) 1982 KK_1
Discovered 1982 May 27 by C. S. Shoemaker at Palomar.

(8332) Ivantsvetaev
1982 TL_2. Discovered 1982 October 14 by L. V. Zhuravleva and L. G. Karachkina at Nauchnyj.

Named in honor of Ivan Vladimirovich Tsvetaev (1847-1913), art critic, professor of philology at the universities in Warsaw, Kiev, Moscow and St. Petersburg. He was the founder and first director of the Moscow Museum of Fine Arts and the director of the Rumyantsev {see planet (5495)} Museum. (M 34627)

(8333) 1982 VF
Discovered 1982 November 7 by A. Mrkos at Kleť.

(8334) 1984 CF
Discovered 1984 February 10 by J. Gibson at Palomar.

(8335) 1984 DD_1
Discovered 1984 February 28 by H. Debehogne at La Silla.

(8336) Šafařík
1984 SK_1. Discovered 1984 September 27 by A. Mrkos at Kleť.

Named in memory of Vojtěch Šafařík (1829-1902), known for his work in inorganic chemistry, Czech chemical nomenclature and textbooks, as well as for his observations of variable stars. He obtained about 20 000 observations of variable stars. This minor planet also honors the memory of his wife and co-worker Paulína Šafaříková (1836-1920), who was interested in the history and popularization of astronomy. (M 34347)
Name suggested by J. Tichá and M. Tichý.

(8337) 1984 SF_6
Discovered 1984 September 22 by H. Debehogne at La Silla.

(8338) 1985 FE_3
Discovered 1985 March 27 at the Brorfelde Observatory at Brorfelde.

(8339) 1985 RM_6
Discovered 1985 September 15 by N. S. Chernykh at Nauchnyj.

(8340) 1985 TS_1
Discovered 1985 October 15 by E. Bowell at Anderson Mesa.

(8341) 1986 QQ
Discovered 1986 August 26 by H. Debehogne at La Silla.

(8342) 1986 QN_3
Discovered 1986 August 29 by H. Debehogne at La Silla.

(8343) 1986 TG_3
Discovered 1986 October 4 by A. Mrkos at Kleť.

(8344) 1987 BB
Discovered 1987 January 25 by T. Niijima and T. Urata at Ojima.

(8345) 1987 BO_1
Discovered 1987 January 22 by E. W. Elst at La Silla.

(8346) 1987 DW_6
Discovered 1987 February 26 by H. Debehogne at La Silla.

(8347) 1987 HK
Discovered 1987 April 21 by C. S. Shoemaker at Palomar.

(8348) 1988 BX
Discovered 1988 January 26 by R. Rajamohan at Kavalur.

(8349) 1988 DH_1
Discovered 1988 February 19 by Y. Oshima at Gekko.

(8350) 1989 AG
Discovered 1989 January 2 by T. Hioki and N. Kawasato at Okutama.

(8351) 1989 EH_1
Discovered 1989 March 10 by K. Suzuki and T. Furuta at Toyota.

(8352) 1989 GE
Discovered 1989 April 6 by S. Ueda and H. Kaneda at Kushiro.

(8353) 1989 GC_4
Discovered 1989 April 3 by E. W. Elst at La Silla.

(8354) 1989 RF
Discovered 1989 September 1 by E. W. Elst at St. Michel.

(8355) 1989 RQ_1
Discovered 1989 September 5 by E. F. Helin at Palomar.

(8356) Wadhwa
1989 RO_2. Discovered 1989 September 3 by C. S. Shoemaker and E. M. Shoemaker at Palomar.

Named in honor of Meenakshi Wadhwa, known as Mini to her friends. Mini's studies of the microdistribution of trace elements in all known martian meteorites have led to a better understanding of the origin of these rocks and have contributed to the igneous history of planet Mars. Her use of extinct and long-lived radioisotopes to decipher the chronology of meteorites is bringing new insights into the formation of a variety of objects from different asteroidal bodies. As Curator of Meteorites, Minerals and Gems at the Field Museum of Natural History, she is actively involved in public education about meteorites and their parent asteroids. (M 33790)

Citation prepared by M. S. Robinson at the request of C. S. Shoemaker.

(8357) O'Connor
1989 SC_1. Discovered 1989 September 25 at the Oak Ridge Observatory at Harvard.

Named in honor of J. Dennis O'Connor (1942-), internationally acknowledged biological scientist, provost and staunch proponent of research excellence at the Smithsonian Institution, and resolute champion of the programs of the Astrophysical Observatory. (M 32350)

(8358) 1989 VN_5
Discovered 1989 November 4 by C. S. Shoemaker at Palomar.

(8359) 1989 WD
Discovered 1989 November 19 by S. Ueda and H. Kaneda at Kushiro.

(8360) 1990 FD_1
Discovered 1990 March 26 by A. Sugie at Taga.

(8361) 1990 JN_1
Discovered 1990 May 1 by A. Zytkow and M. J. Irwin at Siding Spring.

(8362) 1990 QM_1
Discovered 1990 August 22 by H. E. Holt at Palomar.

(8363) 1990 RV
Discovered 1990 September 13 by C. M. Olmstead at Palomar.

(8364) 1990 RE_5
Discovered 1990 September 15 by H. E. Holt at Palomar.

(8365) 1990 RR_5
Discovered 1990 September 15 by H. E. Holt at Palomar.

(8366) 1990 UL_1
Discovered 1990 October 20 by A. Sugie at Taga.

(8367) Bokusui
1990 UL_2. Discovered 1990 October 23 by T. Seki at Geisei.

Named in memory of Bokusui (Shigeru) Wakayama (1885-1928), one of the most beloved of Japanese poets. Born in a village in Miyazaki {see planet (6905)} prefecture, he graduated from the nearby Nobeoka {see planet (8234)} Middle School and later from Waseda University. Fond of both travel and sake, he wrote many poems about the joys and sorrows of life and nature. After his death, his complete works were published in 13 volumes. (M 33388)

Name proposed by the discoverer following a suggestion by R. Ukishima and T. Sato.

(8368) 1991 DM
Discovered 1991 February 20 by R. H. McNaught at Siding Spring.

(8369) 1991 GR
Discovered 1991 April 8 by E. F. Helin at Palomar.

(8370) 1991 RK_{11}
Discovered 1991 September 4 by E. W. Elst at La Silla.

(8371) 1991 TJ_{14}
Discovered 1991 October 2 by C. P. de Saint-Aignan at Palomar.

(8372) 1991 VC_2
Discovered 1991 November 9 by S. Ueda and H. Kaneda at Kushiro.

(8373) 1992 AB
Discovered 1992 January 1 by C. S. Shoemaker at Palomar.

(8374) 1992 AK_1
Discovered 1992 January 10 by S. Otomo at Kiyosato.

(8375) Kenzokohno
1992 AP_1. Discovered 1992 January 12 by T. Seki at Geisei.

Named in honor of Kenzo Kohno (1934-), staff member of Akashi Planetarium since 1960 and its director from 1982 to 1995. He served as president of the Japan Planetarium Society from 1983 to 1984. In 1981, he was awarded the Minister's Prize by the Japanese Ministry of Education. (M 33388)
Name proposed by the discoverer following a suggestion by T. Sato and A. Fujii.

(8376) 1992 OZ_9
Discovered 1992 July 30 by H. Debehogne and A. López G. at La Silla.

(8377) 1992 SD_1
Discovered 1992 September 23 by K. Endate and K. Watanabe at Kitami.

(8378) 1992 SN_1
Discovered 1992 September 23 by E. F. Helin at Palomar.

(8379) Straczynski
1992 SW_{10}. Discovered 1992 September 27 by the Spacewatch at Kitt Peak.

Named in honor of J. Michael Straczynski (1954-), creator, executive producer, and writer of *Babylon 5*, an award-winning science-fiction novel for television that chronicles the story of the last of a series of space stations, the last best hope for peace in the galaxy. Straczynski has been involved in genre television for many years, editing and writing for a number of popular series. He has also published many short stories, an anthology and two fantasy/horror novels. Straczynski's primary criterion for a good science-fiction television series is that it must be good science fiction and good television. With *Babylon 5* he succeeded superbly. (M 33790)
Name proposed by J. Scotti, W. Bottke and D. Durda. Citation by D. Durda.

(8380) Tooting
1992 SW_{17}. Discovered 1992 September 29 by H. E. Holt at Palomar.
Named for a suburb of London. Tooting's postal code is SW17. (M 32350)
Name suggested by E. Bowell, who lived in Tooting for 13 years as a youngster.

(8381) Hauptmann
1992 SO_{24}. Discovered 1992 September 21 by F. Börngen at Tautenburg.

Named for the dramatist and writer Gerhart Hauptmann (1862-1946), a leading personality in the field of German naturalism. His work frequently depicts social problems and the ups and downs of life, often viewed from his homeland of Silesia. Beginning in 1904, he lived on Hiddensee, a small island in the Baltic Sea, where he was also buried. Hauptmann was honored with the 1912 Nobel Prize in literature. (M 32096)

(8382) Mann
1992 SQ_{26}. Discovered 1992 September 23 by F. Börngen at Tautenburg.

Named in memory of Heinrich (1871-1950) and Thomas Mann (1875-1955), born in the Hanseatic town of Lübeck, both prominent German writers and social critics. In 1930, Heinrich became director of the section "Art of Creative Writing" of the Prussian Academy of Sciences. Thomas received the 1929 Nobel Prize for literature. After emigrating in 1933, they later settled in the United States. (M 32096)

(8383) 1992 UA_3

Discovered 1992 October 25 by T. Hioki and S. Hayakawa at Okutama.

(8384) 1992 YB

Discovered 1992 December 16 by T. Urata at Oohira.

(8385) 1993 AN

Discovered 1993 January 13 by S. Ueda and H. Kaneda at Kushiro.

(8386) 1993 BB_6

Discovered 1993 January 27 by E. W. Elst at Caussols.

(8387) Fujimori

1993 DO. Discovered 1993 February 19 by T. Seki at Geisei.

Named in honor of Kenichi Fujimori (1934-), an amateur astronomer who observes sunspots, faculae and prominences. A formal observer designated by the Sunspot Index Data Center, he served as director of the solar section of the Oriental Astronomical Association from 1971 to 1978. (M 33388)
Name proposed by the discoverer following a suggestion by T. Sato and A. Fujii.

(8388) 1993 FO_6

Discovered 1993 March 17 by the Uppsala-ESO Survey at La Silla.

(8389) 1993 FT_{37}

Discovered 1993 March 19 by the Uppsala-ESO Survey at La Silla.

(8390) 1993 FE_{48}

Discovered 1993 March 19 by the Uppsala-ESO Survey at La Silla.

(8391) 1993 HH_3

Discovered 1993 April 20 by the Spacewatch at Kitt Peak.

(8392) 1993 OP

Discovered 1993 July 18 by E. F. Helin at Palomar.

(8393) 1993 TJ_1

Discovered 1993 October 15 by K. Endate and K. Watanabe at Kitami.

(8394) 1993 TM_{12}

Discovered 1993 October 13 by H. E. Holt at Palomar.

(8395) Rembaut

1993 TQ_{23}. Discovered 1993 October 9 by E. W. Elst at La Silla.

Named in memory of Peter Rembaut (1966-1997), industrial engineer at the electronic laboratory at the Royal Observatory at Uccle, who died tragically in 1997. Of great help to the discoverer for improving the software for the Zeiss comparator-measuring device, he was esteemed by everyone at the observatory for his kindness and ability. (M 34347)

(8396) 1993 UR_2

Discovered 1993 October 19 by E. F. Helin at Palomar.

(8397) 1993 XO

Discovered 1993 December 8 by S. Otomo at Kiyosato.

(8398) Rubbia

1993 XY. Discovered 1993 December 12 at the Farra d'Isonzo Observatory at Farra d'Isonzo.

Named in honor of the Italian physicist Carlo Rubbia (1934-), winner of the 1984 Nobel Prize in physics for his decisive contribution to the large project that led to the discovery of the field particles W and Z, communicators of weak interaction. His discovery has supplied a fundamental check of the unified theory of electro-weak interactions. He is the fourth Italian Nobel laureate in physics. (M 33790)

(8399) 1994 AD

Discovered 1994 January 2 by T. Kobayashi at Oizumi.

(8400) 1994 AQ

Discovered 1994 January 4 by T. Kobayashi at Oizumi.

(8401) 1994 DA

Discovered 1994 February 16 at the Farra d'Isonzo Observatory at Farra d'Isonzo.

(8402) 1994 GH_9

Discovered 1994 April 11 by E. F. Helin at Palomar.

(8403) 1994 JG

Discovered 1994 May 6 by T. Kobayashi at Oizumi.

(8404) 1995 AN

Discovered 1995 January 1 by T. B. Spahr at Tucson.

(8405) 1995 GO

Discovered 1995 April 5 by the Spacewatch at Kitt Peak.

(8406) 1995 HJ

Discovered 1995 April 20 by K. Endate and K. Watanabe at Kitami.

(8407) Houlahan

1995 ON. Discovered 1995 July 25 by P. G. Comba at Prescott.

Named in honor of Padraig Houlahan (1958-). Although Houlahan studied astronomy, he has worked mainly as a computer expert at various institutions, lately at Lowell Observatory. When the discoverer was first attempting to do astrometry by measuring photographic prints, Houlahan suggested putting the prints through a digital scanner and measuring them with a computer. The method, although slow, gave satisfactory results. Thereafter Houlahan and the discoverer, working jointly, measured more than 200 prints, including 20 with images of this minor planet. (M 32096)

(8408) 1995 SX_{12}

Discovered 1995 September 18 by the Spacewatch at Kitt Peak.

(8409) 1995 WB_{43}

Discovered 1995 November 28 by R. Weber at Socorro.

(8410) 1996 QZ_1

Discovered 1996 August 24 by S. Ueda and H. Kaneda at Kushiro.

(8411) 1996 TO

Discovered 1996 October 3 at the Farra d'Isonzo Observatory at Farra d'Isonzo.

(8412) 1996 TM_6
Discovered 1996 October 7 at the Beijing Observatory at Xinglong.

(8413) 1996 TV_{10}
Discovered 1996 October 9 by S. Ueda and H. Kaneda at Kushiro.

(8414) 1996 TW_{10}
Discovered 1996 October 9 by S. Ueda and H. Kaneda at Kushiro.

(8415) 1996 UT
Discovered 1996 October 16 by Y. Shimizu and T. Urata at Nachi-Katsuura.

(8416) 1996 VB_8
Discovered 1996 November 3 by S. Ueda and H. Kaneda at Kushiro.

(8417) 1996 VG_8
Discovered 1996 November 7 by K. Endate and K. Watanabe at Kitami.

(8418) Mogamigawa
1996 VS_{30}. Discovered 1996 November 10 by T. Okuni at Nanyou.

Named for the longest river in the discoverer's home prefecture Yamagata {see planet (7039)}. One of the three wildest rivers in Japan, Mogamigawa has its source in Mt. Azuma, south of Yamagata. It flows north and pours into the Sea of Japan in Sakata City. Called "Mother River", it is very popular in this area. (M 32792)

(8419) 1996 VK_{38}
Discovered 1996 November 7 by S. Ueda and H. Kaneda at Kushiro.

(8420) Angrogna
1996 WQ. Discovered 1996 November 17 by P. G. Comba at Prescott.

Named for a village nestled in the Cottian Alps, in the Piedmont region of Italy, where the discoverer's paternal ancestors were born and lived for at least six generations. (M 32096)

(8421) Montanari
1996 XA_9. Discovered 1996 December 2 at the Osservatorio San Vittore at Bologna.

Named in memory of Geminiano Montanari (1633-1687), professor of mathematics at university of Bologna during 1664-1678. He invented and improved several astronomical instruments, discovered the variability of Algol and drew fine lunar maps. (M 34627)

(8422) 1996 XJ_{26}
Discovered 1996 December 5 at the Farra d'Isonzo Observatory at Farra d'Isonzo.

(8423) Macao
1997 AO_{22}. Discovered 1997 January 11 at the Beijing Observatory at Xinglong.

Named in honor of Macao, the beautiful peninsula near Zhuhai {see planet (2903)}. An international tourist attraction as well as a free trading point, Macao is famous for its combination of Chinese and Western cultures. (M 32792)

(8424) 1997 CP
Discovered 1997 February 1 by T. Kobayashi at Oizumi.

(8425) 1997 CJ_{29}
Discovered 1997 February 14 at the Beijing Observatory at Xinglong.

(8426) 1997 ST

Discovered 1997 September 16 at the Beijing Observatory at Xinglong.

(8427) 1997 TH_{17}

Discovered 1997 October 6 by Y. Shimizu and T. Urata at Nachi-Katsuura.

(8428) 1997 VJ_8

Discovered 1997 November 3 by T. Seki at Geisei.

(8429) 1997 YK_4

Discovered 1997 December 23 at the Beijing Observatory at Xinglong.

(8430) Florey

1997 YB_5. Discovered 1997 December 25 by F. B. Zoltowski at Woomera.

Named in memory of Howard Walter Florey (1898-1968), celebrated pathologist, born and educated in South Australia. He and the biochemist Ernst Chain isolated penicillin and formulated procedures for its extraction and production, thereby enabling its large-scale production during the latter part of World War II. For his work, he shared the 1945 Nobel Prize in physiology and medicine with Chain and Fleming. (M 34627)

(8431) 1997 YQ_{13}

Discovered 1997 December 31 by T. Kobayashi at Oizumi.

(8432) Tamakasuga

1997 YD_{18}. Discovered 1997 December 27 by A. Nakamura at Kuma.

Named in honor of Ryoji Matsumoto (1972-), Japanese sumo wrestler whose professional name is Tamakasuga. Born in Ehime prefecture, where this minor planet was discovered, he became a professional wrestler in 1994. In 1996 he was promoted to "Makunouchi", the top division in the official listing of rank. Famous for his powerful "tsuki-oshi", or pushing-thrusting style, Tamakasuga has won the Outstanding Performance Award, the Technique Prize and the Fighting Spirit Prize (twice) in his 26 tournaments. (M 32096)

(8433) Brachyrhynchus

2561 P-L. Discovered 1960 September 24 by C. J. van Houten and I. van Houten-Groeneveld at Palomar.

Named for Anser brachyrhynchus, or pink-footed goose. {It belongs to the family of the Anatidae.} (M 33790; M 34089)

This species as well as many of those described in the citations that follow, are species whose presence in Europe is endangered. This bird is on the Dutch Blue List of endangered birds. It does not breed in the Netherlands, but stays there during the winter. Therefore it is of international importance for it to find feeding places there. The Netherlands is a very important feeding ground for many migratory birds. In particular the birds find food in the Wadden Sea and along the Dutch coast.

(8434) Columbianus

6571 P-L. Discovered 1960 September 24 by C. J. van Houten and I. van Houten-Groeneveld at Palomar.

Named for Cygnus columbianus, also called Cygnus bewickii, or Bewick's swan. {It belongs to the family of the Anatidae. Bewick's Swan comes from Northern Russia and stays along the North Sea in winter. In 1994 only 3,000 pairs of Cygnus columbianus were counted.} (M 33790; M 34089)

(8435) Anser

6643 P-L. Discovered 1960 September 26 by C. J. van Houten and I. van Houten-Groeneveld at Palomar.

Named for Anser anser, or grey-lag goose. {It belongs to the family of the Anatidae. The bird is on the Dutch Blue List of endangered birds, coming from the East and staying in the Netherlands in great numbers during the winter. It is of international importance to find feeding places for them.} (M 33790; M 34089)

(8436) Leucopsis

2259 T-1. Discovered 1971 March 25 by C. J. van Houten and I. van Houten-Groeneveld at Palomar.

Named for Branta leucopsis, or barnacle goose. {It belongs to the family of the Anatidae. The bird is on the Dutch Blue List of endangered birds. It comes from the Arctic and stays during the winter in the Netherlands.} (M 33791; M 34089)

(8437) Bernicla

3057 T-1. Discovered 1971 March 26 by C. J. van Houten and I. van Houten-Groeneveld at Palomar.

Named for Branta bernicla, or Brant goose. {It belongs to the family of Anatidae. In 1994 only 1,000 pairs were counted. It is also on the 'List of Vulnerable Birds in Europe' and on the Dutch Blue List of endangered birds.} (M 33791; M 34089)

(8438) Marila

4825 T-1. Discovered 1971 May 13 by C. J. van Houten and I. van Houten-Groeneveld at Palomar.

Named for Aythya marila, or greater scaup. {It belongs to the family of Anatidae. The bird does not breed in the Netherlands, but stays there during the winter.} (M 33791; M 34089)

(8439) Albellus

2034 T-2. Discovered 1973 September 29 by C. J. van Houten and I. van Houten-Groeneveld at Palomar.

Named for Mergus albellus, or smew. {It belongs to the family of the Anatidae. The bird does not breed in the Netherlands, but stays there during the winter.} (M 33791; M 34089)

(8440) Wigeon

1017 T-3. Discovered 1977 October 17 by C. J. van Houten and I. van Houten-Groeneveld at Palomar.

Named after the bird Anas penelope; wigeon is its English name. It belongs to the family of the Anatidae. It is on the Dutch Blue List of endangered birds. Although this bird does not breed in the Netherlands, it stays there during the winter. (M 34347; M 34089)

(8441) Lapponica

4008 T-3. Discovered 1977 October 16 by C. J. van Houten and I. van Houten-Groeneveld at Palomar.

Named for Limosa lapponica, or bar-tailed godwit. {It belongs to the family of the Scolopacidae. In 1994 only 2,200 pairs were counted. The bird does not breed in the Netherlands, but stays there during the winter.} (M 33791; M 34089)

(8442) Ostralegus

4237 T-3. Discovered 1977 October 16 by C. J. van Houten and I. van Houten-Groeneveld at Palomar.

Named for Haematopus ostralegus, or oyster catcher. {It belongs to the family of the Haematopodidae and is on the Dutch Blue List of endangered birds.} (M 33791; M 34089)

(8443) Svecica

4343 T-3. Discovered 1977 October 16 by C. J. van Houten and I. van Houten-Groeneveld at Palomar.

Named for Luscinia svecica, or bluethroat. {It belongs to the family of the Turdidae.} (M 33791; M 34089)

(8444) 1969 TR_1

Discovered 1969 October 8 by L. I. Chernykh at Nauchnyj.

(8445) 1973 QG_2

Discovered 1973 August 31 by T. M. Smirnova at Nauchnyj.

(8446) 1973 SB_6

Discovered 1973 September 28 by N. S. Chernykh at Nauchnyj.

(8447) 1974 OE

Discovered 1974 July 16 at the Felix Aguilar Observatory at El Leoncito.

(8448) Belyakina

1976 UT_1. Discovered 1976 October 26 by T. M. Smirnova at Nauchnyj.

Named in honor of Tamara Sergeevna Belyakina (1934-), astrophysicist and stellar photometrist who worked at the Crimean Astrophysical Observatory from 1955 to 1990. She is known for her multicolor photometric observations of symbiotic stars. She first discovered nonradial pulsations of red giants in such systems and proposed the interpretation of this phenomenon. (M 32350)

(8449) 1977 EO_1

Discovered 1977 March 13 by N. S. Chernykh at Nauchnyj.

(8450) 1977 QL_1

Discovered 1977 August 19 by N. S. Chernykh at Nauchnyj.

(8451) 1977 RY_6

Discovered 1977 September 11 by N. S. Chernykh at Nauchnyj.

(8452) Clay

1978 WB. Discovered 1978 November 27 at the Harvard College Observatory at Harvard.

Named in honor of Landon and Lavinia Clay and their four sons, steadfast friends of science and the arts, Harvard University and the Smithsonian Institution. They are in particular enthusiastic and penetratingly knowledgeable supporters of astronomy, conservation biology and mathematics. (M 32350; M 32561)

(8453) 1981 EQ

Discovered 1981 March 1 by H. Debehogne and G. DeSanctis at La Silla.

(8454) 1981 EG_1

Discovered 1981 March 5 by H. Debehogne and G. DeSanctis at La Silla.

(8455) 1981 ER_6

Discovered 1981 March 6 by S. J. Bus at Siding Spring.

(8456) 1981 EJ_7

Discovered 1981 March 1 by S. J. Bus at Siding Spring.

(8457) 1981 EO_8
Discovered 1981 March 1 by S. J. Bus at Siding Spring.

(8458) 1981 EY_9
Discovered 1981 March 1 by S. J. Bus at Siding Spring.

(8459) 1981 EQ_{18}
Discovered 1981 March 2 by S. J. Bus at Siding Spring.

(8460) 1981 EP_{19}
Discovered 1981 March 2 by S. J. Bus at Siding Spring.

(8461) 1981 EC_{21}
Discovered 1981 March 2 by S. J. Bus at Siding Spring.

(8462) 1981 ED_{22}
Discovered 1981 March 2 by S. J. Bus at Siding Spring.

(8463) 1981 EM_{27}
Discovered 1981 March 2 by S. J. Bus at Siding Spring.

(8464) 1981 EF_{28}
Discovered 1981 March 2 by S. J. Bus at Siding Spring.

(8465) 1981 EQ_{31}
Discovered 1981 March 2 by S. J. Bus at Siding Spring.

(8466) 1981 EV_{34}
Discovered 1981 March 2 by S. J. Bus at Siding Spring.

(8467) 1981 ES_{35}
Discovered 1981 March 2 by S. J. Bus at Siding Spring.

(8468) 1981 EA_{40}
Discovered 1981 March 2 by S. J. Bus at Siding Spring.

(8469) 1981 TZ
Discovered 1981 October 5 by N. G. Thomas at Anderson Mesa.

(8470) 1982 SA_4
Discovered 1982 September 17 by N. S. Chernykh at Nauchnyj.

(8471) 1983 RX_4
Discovered 1983 September 5 by L. V. Zhuravleva at Nauchnyj.

(8472) Tarroni
1983 TC. Discovered 1983 October 12 at the Osservatorio San Vittore at Bologna.
Named in memory of the Italian amateur astronomer Gino Tarroni (1958-1986), a member of the Sezione Astrofili dell'Università Popolare Sestrese. A fine observer of the sun, he was in charge of the solar section of the Unione Astrofili Italiani, and he served as secretary of the Unione for the four years preceding his tragic death in a road accident. Tarroni also had interests in speleology and mountain climbing. (M 34627)

(8473) 1984 SS_5
Discovered 1984 September 21 by H. Debehogne at La Silla.

(8474) 1985 GA_1
Discovered 1985 April 15 by E. Bowell at Anderson Mesa.

(8475) 1985 PC_2
Discovered 1985 August 13 by N. S. Chernykh at Nauchnyj.

(8476) 1986 QT_2
Discovered 1986 August 28 by H. Debehogne at La Silla.

(8477) 1986 RF_7
Discovered 1986 September 6 by L. V. Zhuravleva at Nauchnyj.

(8478) 1987 DO_6
Discovered 1987 February 23 by H. Debehogne at La Silla.

(8479) 1987 HD_2
Discovered 1987 April 29 by A. Mrkos at Kleť.

(8480) 1987 RD_1
Discovered 1987 September 13 by H. Debehogne at La Silla.

(8481) 1988 LH
Discovered 1988 June 14 by A. C. Gilmore and P. M. Kilmartin at Lake Tekapo.

(8482) 1988 RA_{11}
Discovered 1988 September 14 by S. J. Bus at Cerro Tololo.

(8483) 1988 SY_1
Discovered 1988 September 16 by S. J. Bus at Cerro Tololo.

(8484) 1988 VM_2
Discovered 1988 November 10 by M. Arai and H. Mori at Yorii.

(8485) 1989 FL
Discovered 1989 March 29 by T. Seki at Geisei.

(8486) 1989 QV
Discovered 1989 August 26 by R. H. McNaught at Siding Spring.

(8487) 1989 SQ
Discovered 1989 September 29 by Y. Mizuno and T. Furuta at Kani.

(8488) 1989 SR_1
Discovered 1989 September 26 by E. W. Elst at La Silla.

(8489) Boulder
1989 TA_3. Discovered 1989 October 7 by E. W. Elst at La Silla.

Named for the city of Boulder, Colorado, about 48 km northwest of Denver. Settled in 1859 by miners and named for the large stones in the area, the University of Colorado was founded there in 1876, making Boulder a center for scientific and environmental research. The water supply for the city is unique, coming from the Arapahoe glacier high in the Rocky Mountains. The discoverer and his wife spent a year there at the JILA institution during 1967-1968. (M 34347)

(8490) 1989 TU_{10}
Discovered 1989 October 4 by Y. Mizuno and T. Furuta at Kani.

(8491) 1989 YL_5
Discovered 1989 December 28 by E. W. Elst at St. Michel.

(8492) 1990 BZ
Discovered 1990 January 21 by T. Seki at Geisei.

(8493) 1990 BY_1

Discovered 1990 January 30 by M. Matsuyama and K. Watanabe at Kushiro.

(8494) 1990 OT_4

Discovered 1990 July 25 by H. E. Holt at Palomar.

(8495) 1990 QV_1

Discovered 1990 August 22 by H. E. Holt at Palomar.

(8496) Jandlsmith

1990 QO_3. Discovered 1990 August 16 at the Oak Ridge Observatory at Harvard.

Named in honor of Jim and Laurie Smith in recognition of their generosity and wisdom in the support of forefront tools to explore the heavens to the benefit of all of humanity. Most especially, their support was crucial for Harvard University's participation in the Magellan Project. (M 33388)

(8497) 1990 RE_7

Discovered 1990 September 13 by H. Debehogne at La Silla.

(8498) 1990 RM_{17}

Discovered 1990 September 15 by L. V. Zhuravleva at Nauchnyj.

(8499) 1990 SC_{13}

Discovered 1990 September 22 by H. Debehogne at La Silla.

(8500) Hori

1990 TU. Discovered 1990 October 10 by K. Endate and K. Watanabe at Kitami.

Named in honor of Gen-Ichiro Hori (1930-), professor emeritus at Tokyo University who gave lectures on celestial mechanics at Tokyo University, Yale University and the University of Texas at Austin. He developed the famous perturbation theory based on Lie transformations that has been adopted by many scientists to solve a set of differential equations by perturbation methods. In fact, his publication on this in 1966 is the most frequently cited paper during the 50-year history of the *Publication of the Astronomical Society of Japan*. (M 32792)

Name proposed by the discoverers, following a suggestion by S. Murayama, K. Hurukawa and A. Fujii, and endorsed by Y. Kozai, K. Aksnes and B. G. Marsden.

(8501) 1990 TK_8

Discovered 1990 October 13 by L. D. Schmadel and F. Börngen at Tautenburg.

(8502) Bauhaus

1990 TR_{12}. Discovered 1990 October 14 by F. Börngen and L. D. Schmadel at Tautenburg.

Named for the school of design founded in 1919 by the German architect W. Gropius (1883-1969) in Weimar {see planet (3539)}. The school's philosophy emphasized the unity of fine art and trade as the basis for artistic work. By incorporating engineering, the way was open for industrial design. The Bauhaus moved to Dessau in 1925 and was closed by the Nazis for its "decadence" in 1932. Many Bauhaus masters emigrated to the United States. In 1937, the New Bauhaus was founded by L. Moholy-Nagy (1895-1946) in Chicago. (M 32350)
Name proposed by the first discoverer.

(8503) 1990 WX_3

Discovered 1990 November 21 by K. Endate and K. Watanabe at Kitami.

(8504) 1990 YC
Discovered 1990 December 17 by S. Ueda and H. Kaneda at Kushiro.

(8505) 1990 YK
Discovered 1990 December 19 by S. Ueda and H. Kaneda at Kushiro.

(8506) 1991 CN
Discovered 1991 February 5 by M. Arai and H. Mori at Yorii.

(8507) 1991 CB_1
Discovered 1991 February 15 by the Spacewatch at Kitt Peak.

(8508) 1991 CU_1
Discovered 1991 February 14 by S. Ueda and H. Kaneda at Kushiro.

(8509) 1991 FV_2
Discovered 1991 March 20 by H. Debehogne at La Silla.

(8510) 1991 PT_8
Discovered 1991 August 5 by H. E. Holt at Palomar.

(8511) 1991 PY_{10}
Discovered 1991 August 7 by H. E. Holt at Palomar.

(8512) 1991 PC_{11}
Discovered 1991 August 7 by H. E. Holt at Palomar.

(8513) 1991 PK_{11}
Discovered 1991 August 9 by H. E. Holt at Palomar.

(8514) 1991 PK_{15}
Discovered 1991 August 7 by H. E. Holt at Palomar.

(8515) 1991 RJ
Discovered 1991 September 4 by R. H. McNaught at Siding Spring.

(8516) 1991 TW_1
Discovered 1991 October 13 by T. Hioki and S. Hayakawa at Okutama.

(8517) 1992 BB_5
Discovered 1992 January 28 by S. Ueda and H. Kaneda at Kushiro.

(8518) 1992 DM_6
Discovered 1992 February 29 by the Uppsala-ESO Survey at La Silla.

(8519) 1992 DB_{10}
Discovered 1992 February 29 by the Uppsala-ESO Survey at La Silla.

(8520) 1992 EC_{12}
Discovered 1992 March 6 by the Uppsala-ESO Survey at La Silla.

(8521) Boulainvilliers
1992 GF_4. Discovered 1992 April 4 by E. W. Elst at La Silla.

Named in memory of Henri de Boulainvilliers (1658-1722), French historian and political writer who influenced intellectual developments in the French Enlightenment. Claiming that historical studies can supply the tools for analyzing the present state of society, he worked out a theory of comparative historical study which approached the later writings of Montesquieu {see planet (7064)}. In 1683 he published his *L'Ideé d'un système genéral de la Nature*, which anticipated Holbach's {see planet (6956)} *Système de la Nature* (1770). His *Histoire de la religion et de la philosophie ancienne* was published around 1700. (M 34347)

(8522) 1992 ML

Discovered 1992 June 25 by G. J. Leonard at Palomar.

(8523) Bouillabaisse

1992 PX. Discovered 1992 August 8 by E. W. Elst at Caussols.

Named for the famous French fish soup, the glory of Provençal cooking. It contains fish, shellfish, olive oil, onions, tomatoes, garlic, parsley, saffron, fennel, thyme, bay leaf and orange peel (according to the Marseille recipe). All ingredients must be boiled together quickly. (M 34347)

(8524) Paoloruffini

1992 RJ_3. Discovered 1992 September 2 by E. W. Elst at La Silla.

Named in memory of the famous Italian mathematician and physician Paolo Ruffini (1765-1822). In 1799 he published a book on the theory of equations, with the claim that the solution by radicals of a general equation of degree greater of four is impossible. Initially the mathematical community showed no interest in his work. However, in 1821 his work was acknowledged by Cauchy, who was influenced by his investigations and had generalized some of Ruffini's results. Due to political problems he had to leave his chair in mathematics at Modena and begin a career in medicine, tending to patients from the poorest to the richest. (M 34347)

(8525) Nielsabel

1992 RZ_5. Discovered 1992 September 2 by E. W. Elst at La Silla.

Named in memory of the famous Norwegian mathematician Niels Henrik Abel (1802-1829). In 1824 he proved the impossibility of solving a general equation of the fifth degree by radicals. Through his friendship with the editor Crelle, who encouraged him in his work, he wrote his masterpiece *Recherches sur les fonctions elliptiques* (1827), from which he could prove that Jacobi's work on elliptic integrals were consequences of his own work. It is interesting to note that neither Gauss {see planet (1001)} nor Cauchy showed interest in Abel's work. (M 34347)

(8526) 1992 SM_{12}

Discovered 1992 September 23 by K. Endate and K. Watanabe at Kitami.

(8527) 1992 SV_{12}

Discovered 1992 September 28 by K. Endate and K. Watanabe at Kitami.

(8528) 1992 SC_{24}

Discovered 1992 September 29 by H. E. Holt at Palomar.

(8529) 1992 UH_2

Discovered 1992 October 19 by K. Endate and K. Watanabe at Kitami.

(8530) 1992 UK_5

Discovered 1992 October 25 by M. Hirasawa and S. Suzuki at Nyukasa.

(8531) 1992 WX_2

Discovered 1992 November 16 by K. Endate and K. Watanabe at Kitami.

(8532) 1992 YW_3

Discovered 1992 December 29 by Y. Kushida and O. Muramatsu at Yatsugatake.

(8533) 1993 BM

Discovered 1993 January 20 by T. Urata at Oohira.

(8534) Knutsson

1993 FJ_{10}. Discovered 1993 March 17 by C.-I. Lagerkvist at La Silla.

Named in memory of Gösta Knutsson (1908-1973), Swedish author and radio producer who introduced quiz programs to Sweden. His children's stories about the cat Pelle Svanslös {see planet (8535)} and his adventures in Uppsala have been very popular. The author of twelve books, the first in 1939, Knutsson lived not far from the locations where many of the adventures take place. (M 33791)

(8535) Pellesvanslös

1993 FH_{22}. Discovered 1993 March 21 by C.-I. Lagerkvist at La Silla.

Named for the fictional character Pelle Svanslös, a brave cat that appears in Gösta Knutsson's {see planet (8534)} children's stories. Some of the adventures of this cat, whose tail was bitten off by a rat when he was only a few days old, take place in the section of Uppsala where the astronomical observatory is located. (M 33791)

(8536) Måns

1993 FK_{23}. Discovered 1993 March 21 by C.-I. Lagerkvist at La Silla.

Named for the fictional cat Måns, the eternal "bad guy" in the Knutsson {see planet (8534)} stories, always devising new ways of ridiculing Pelle Svanslös {see planet (8535)} over his nonexistent tail. (M 33791)

(8537) Billochbull

1993 FG_{24}. Discovered 1993 March 21 by C.-I. Lagerkvist at La Silla.

Named for the fictional cats Bill and Bull, dim and fawning cronies of the bad cat Måns {see planet (8536)} of the Knutsson {see planet (8534)} stories. (M 33791)

(8538) Gammelmaja

1993 FR_{26}. Discovered 1993 March 21 by C.-I. Lagerkvist at La Silla.

Named for Gammelmaja, an old and wise cat who seldom fails to notice when Pelle {see planet (8535)} is being treated unfairly and often takes his side in arguments. She lives in the belfry of the Uppsala cathedral. (M 33791)

(8539) Laban

1993 FT_{32}. Discovered 1993 March 19 by C.-I. Lagerkvist at La Silla.

Named for the Knutsson's {see planet (8534)} fictional cat Laban. He lived in the Observatory park in Uppsala and gave his name to one of the first modern computers at the Astronomical Observatory. (M 33791)

(8540) 1993 FK_{80}

Discovered 1993 March 17 by the Uppsala-ESO Survey at La Silla.

(8541) 1993 TZ_{32}

Discovered 1993 October 9 by E. W. Elst at La Silla.

(8542) 1993 VB_2

Discovered 1993 November 11 by S. Ueda and H. Kaneda at Kushiro.

(8543) 1993 XO_1

Discovered 1993 December 15 by T. Kobayashi at Oizumi.

(8544) 1993 YE

Discovered 1993 December 17 by T. Kobayashi at Oizumi.

(8545) 1994 AM_1
Discovered 1994 January 2 by B. G. W. Manning at Stakenbridge.

(8546) 1994 AH_3
Discovered 1994 January 13 by K. Endate and K. Watanabe at Kitami.

(8547) 1994 CQ
Discovered 1994 February 4 by S. Ueda and H. Kaneda at Kushiro.

(8548) 1994 ER_3
Discovered 1994 March 14 by K. Endate and K. Watanabe at Kitami.

(8549) Alcide
1994 FS. Discovered 1994 March 30 at the Farra d'Isonzo Observatory at Farra d'Isonzo.

Named in memory of Alcide Bittesini (1913-1981), father of Luciano Bittesini, one of the Farra d'Isonzo amateur astronomers who discovered this minor planet. A natural sciences high-school teacher in Italy, Alcide Bittesini kindled his then-nine-year-old son's interest in astronomy by showing him a comet, using a hand-made telescope constructed from a tin can, a pair of glasses and an eyepiece from his microscope. (M 33791)

(8550) 1994 PV_{24}
Discovered 1994 August 12 by E. W. Elst at La Silla.

(8551) 1994 VC_7
Discovered 1994 November 11 by M. Hirasawa and S. Suzuki at Nyukasa.

(8552) Hyoichi
1995 HE. Discovered 1995 April 20 by A. Nakamura at Kuma.

Named in honor of Hyoichi Kohno (1958-), Japanese adventurer, born in Ehime prefecture, where this minor planet was discovered. Since 1980, he has boated down the Yukon River; climbed Mt. McKinley (6194 m) and Cerro Aconcagua (6959 m), the highest mountains in North and South America; walked across Patagonia; walked from Los Angeles to New York; walked from Algeria to Togo across the Sahara desert; and so on. In 1997, he became the first Japanese to walk to the North Pole alone. (M 32096)

(8553) 1995 HG
Discovered 1995 April 20 by K. Endate and K. Watanabe at Kitami.

(8554) 1995 KH
Discovered 1995 May 25 by M. Tichý at Kleť.

(8555) Mirimao
1995 LD. Discovered 1995 June 3 at the Santa Lucia Observatory at Stroncone.

Named in memory of Guido Mirimao (1909-1990), internationally known painter and draftsman. A graphic artist who contributed regularly to newspapers and magazines, from 1931 to 1940 he received a great number of prizes in national exhibitions. He also created art works and murals on sacred subjects in Italy and abroad. (M 33388)

(8556) Jana
1995 NB. Discovered 1995 July 7 by Z. Moravec at Kleť.

Named in honor of Jana Moravcová, wife of the discoverer, on the occasion of her thirtieth birthday, 1999 July 7. (M 34627)

(8557) Šaroun

1995 OK. Discovered 1995 July 23 by L. Šarounová at Ondřejov.

Named in honor of the discoverer's father, Jaroslav Šaroun (1943-). A teacher at the Prague Academy of Musical Arts and a member of the Czech Philharmonic Orchestra, he is a pianist who is always in great demand as an accompanist for Czech and foreign singers. As a lover of astronomy, he influenced and supported his daughter in her desire to become an astronomer. This minor planet was discovered on the day after his birthday. (M 32350)

(8558) Hack

1995 PC. Discovered 1995 August 1 by L. Tesi and A. Boattini at San Marcello Pistoiese.

Named in honor of Margherita Hack (1922-), director of the Trieste Astronomical Observatory (1964-1987), director of the astronomy department of Trieste University (1985-1991 and 1996-1997) and a former president of IAU Commission 29. Although her studies have ranged from optics and solar physics to radioastronomy (galactic 21-cm emission), her main fields of research remain stellar spectroscopy, stellar atmospheres and observable effects of stellar evolution. Her present interests are the ultraviolet and optical spectroscopy of close interacting binaries, atmospheric eclipsing binaries and symbiotic stars. (M 32350)

(8559) 1995 QM_2

Discovered 1995 August 25 by Y. Shimizu and T. Urata at Nachi-Katsuura.

(8560) 1995 SD_5

Discovered 1995 September 20 by K. Endate and K. Watanabe at Kitami.

(8561) 1995 SO_{29}

Discovered 1995 September 26 by T. V. Kryachko at Zelenchukskaya.

(8562) 1995 SK_{53}

Discovered 1995 September 28 at the Beijing Observatory at Xinglong.

(8563) 1995 US

Discovered 1995 October 19 by T. B. Spahr at Tucson.

(8564) 1995 UL_3

Discovered 1995 October 17 by Y. Shimizu and T. Urata at Nachi-Katsuura.

(8565) 1995 WB_6

Discovered 1995 November 24 by T. Niijima and T. Urata at Ojima.

(8566) 1996 EN

Discovered 1996 March 15 by the JPL NEAT Program at Haleakala.

(8567) 1996 HW_1

Discovered 1996 April 23 by the Spacewatch at Kitt Peak.

(8568) 1996 RU_2

Discovered 1996 September 10 by the JPL NEAT Program at Haleakala.

(8569) 1996 TG

Discovered 1996 October 1 by V. S. Casulli at Colleverde di Guidonia.

(8570) 1996 TN_{10}

Discovered 1996 October 9 by S. Ueda and H. Kaneda at Kushiro.

(8571) 1996 UX

Discovered 1996 October 20 by T. Kobayashi at Oizumi.

(8572) Nijo

1996 UG_1. Discovered 1996 October 19 by J. Tichá and M. Tichý at Kleť.

Named for Nijo Castle (Nijo-jo) in Kyoto {see planet (4352)}, built in 1603 as the official residence of the first Tokugawa shogun Ieyasu. The site is famous for its wooden architecture and the screen paintings of trees and birds of Ninomaru Palace, as well as for the beautiful Ninomaru garden, designed by the tea master and landscape architect Kobori Enshu. The discoverers visited and admired Ninomaru and several other famous gardens in Kyoto during the IAU General Assembly in 1997. (M 34627)

(8573) Ivanka

1996 VQ. Discovered 1996 November 4 by Z. Moravec at Kleť.

Named for Ivanka Moravcová (1943-), the discoverer's mother. (M 34628)

(8574) 1996 VC_2

Discovered 1996 November 6 by T. Kobayashi at Oizumi.

(8575) 1996 VL_8

Discovered 1996 November 7 by K. Endate and K. Watanabe at Kitami.

(8576) 1996 VN_8

Discovered 1996 November 7 by S. Ueda and H. Kaneda at Kushiro.

(8577) 1996 VX_8

Discovered 1996 November 7 by K. Endate and K. Watanabe at Kitami.

(8578) 1996 WZ

Discovered 1996 November 19 by T. Kobayashi at Oizumi.

(8579) 1996 XV_{19}

Discovered 1996 December 11 by T. Kobayashi at Oizumi.

(8580) Pinsky

1996 XZ_{25}. Discovered 1996 December 14 by P. G. Comba at Prescott.

Named in honor of Robert Pinsky (1940-), poet laureate of the United States since 1997. Besides several books of poetry, Pinsky has produced a much-acclaimed new English translation of Dante's {see planet (2999)} *Inferno*. Written in slant rhyme, Pinsky's version captures the rhythm and grandeur of the great Italian poet's masterpiece. (M 32096)

(8581) Johnen

1996 YO_2. Discovered 1996 December 28 by N. Sato at Chichibu.

Named for a mountain in Nagano prefecture, part of the Japanese Northern Alps. Popular with climbers, the 2857-m peak is especially famous because an Englishman, Walter Weston (1861-1940), climbed it in 1894 and spread the word of its beauty all over the world. (M 33388)

(8582) 1997 AY

Discovered 1997 January 2 by T. Kobayashi at Oizumi.

(8583) Froberger

1997 AK_6. Discovered 1997 January 8 by P. G. Comba at Prescott.

Named for Johann Jacob Froberger (1616-1667), organist and the foremost German keyboard composer of his day. A pupil of Frescobaldi, he combined features of many national styles. His toccate, full of imaginative chromatic harmonies, were copied out and imitated by J. S. Bach {see planet (1814)}. (M 32096)

(8584) 1997 AN_{22}

Discovered 1997 January 11 at the Beijing Observatory at Xinglong.

(8585) Purpurea

2025 P-L. Discovered 1960 September 24 by C. J. van Houten and I. van Houten-Groeneveld at Palomar.

Named for Ardea purpurea, or purple heron. {It belongs to the family of the Ardeidae. The bird is on the Dutch Red List of birds endangered in the Netherlands and also on the 'List of Vulnerable Birds in Europe'.} (M 33791; M 34089)
See also the remarks to planet (8600).

(8586) Epops

2563 P-L. Discovered 1960 September 24 by C. J. van Houten and I. van Houten-Groeneveld at Palomar.

Named for Upupa epops, or hoopoe. {It belongs to the family of the Upupidae.} (M 33791; M 34089)
See also the remarks to planet (8600).

(8587) Ruficollis

3078 P-L. Discovered 1960 September 25 by C. J. van Houten and I. van Houten-Groeneveld at Palomar.

Named for Tachybaptus ruficollis, or little grebe. {It belongs to the family of the Podicipidae and is on the Dutch Red List of endangered birds.} (M 33791; M 34089)
See also the remarks to planet (8600).

(8588) Avosetta

4025 P-L. Discovered 1960 September 24 by C. J. van Houten and I. van Houten-Groeneveld at Palomar.

Named for Recurvirostra avosetta, or avocet. {It belongs to the family of the Recurvirostridae and is on the Dutch Red List of birds endangered in the Netherlands.} (M 33791; M 34089)
See also the remarks to planet (8600).

(8589) Stellaris

4068 P-L. Discovered 1960 September 24 by C. J. van Houten and I. van Houten-Groeneveld at Palomar.

Named for Botaurus stellaris, or bittern. {It belongs to the family of the Ardeidae. It is on the Dutch Red List of birds endangered in the Netherlands as well as on the 'List of Vulnerable Birds in Europe'.} (M 33791; M 34089)
See also the remarks to planet (8600).

(8590) Pygargus

6533 P-L. Discovered 1960 September 24 by C. J. van Houten and I. van Houten-Groeneveld at Palomar.

Named for Circus pygargus, or Montagu's harrier. {It belongs to the family of the Accipitridae and is on the Dutch Red List of birds endangered in the Netherlands.} (M 33792; M 34089)
See also the remarks to planet (8600).

(8591) Excubitor

6543 P-L. Discovered 1960 September 24 by C. J. van Houten and I. van Houten-Groeneveld at Palomar.

Named for Lanius excubitor, or great grey or northern shrike. {It belongs to the family of the Laniidae and is on the Dutch Red List of birds endangered in the Netherlands.} (M 33792; M 34089)
See also the remarks to planet (8600).

(8592) Rubetra

1188 T-1. Discovered 1971 March 25 by C. J. van Houten and I. van Houten-Groeneveld at Palomar.

Named for Saxicola rubetra, or whinchat. {It belongs to the family of the Turdidae and is on the Dutch Red List of endangered birds.} (M 33792; M 34089)
See also the remarks to planet (8600).

(8593) Angustirostris

2186 T-1. Discovered 1971 March 25 by C. J. van Houten and I. van Houten-Groeneveld at Palomar.

Named for Marmaronetta angustirostris, also called anas angustirostris, marbled teal or marbled duck. {It belongs to the Anatidae family and is one of the birds on the 'List of Endangered Birds in Europe'. In 1994 only 200 pairs were counted.} (M 33792; M 34089)
See also the remarks to planet (8600).

(8594) Albifrons

2245 T-1. Discovered 1971 March 25 by C. J. van Houten and I. van Houten-Groeneveld at Palomar.

Named for Sterna albifrons, or little tern. {It belongs to the family of the Sternidae.} (M 33792; M 34089)
See also the remarks to planet (8600).

(8595) Dougallii

3233 T-1. Discovered 1971 March 26 by C. J. van Houten and I. van Houten-Groeneveld at Palomar.

Named for Sterna dougallii, or roseate tern. {It belongs to the family of the Sternidae. In 1994 only 1,600 pairs were counted.} (M 33792; M 34089)
See also the remarks to planet (8600).

(8596) Alchata

1298 T-2. Discovered 1973 September 29 by C. J. van Houten and I. van Houten-Groeneveld at Palomar.

Named for Pterocles alchata, or pin-tailed sandgrouse. {It belongs to the Pteroclidae family and is one of the birds on the 'List of Endangered Birds in Europe'.} (M 33792; M 34089)
See also the remark to planet (8600).

(8597) Sandvicensis

2045 T-2. Discovered 1973 September 29 by C. J. van Houten and I. van Houten-Groeneveld at Palomar.

Named for Sterna sandvicensis, or sandwich tern. {It belongs to the Sternidae family and is on the Dutch Red List of endangered birds.} (M 33792; M 34089)
See also the remarks to planet (8600).

(8598) Tetrix

2202 T-2. Discovered 1973 September 29 by C. J. van Houten and I. van Houten-Groeneveld at Palomar.

Named for Tetrao tetrix, also named lyrurus tetrix, or black grouse. {It belongs to the family of the Tetraonidae. The bird is on the Dutch as well as on the

European list of endangered birds.} (M 33792; M 34089)
See also the remarks to planet (8600).

(8599) Riparia

2277 T-2. Discovered 1973 September 29 by C. J. van Houten and I. van Houten-Groeneveld at Palomar.

Named for Riparia riparia, or sand martin. {It belongs to the family of the Hirundinidae and is on the Dutch Red List of endangered birds.} (M 33792; M 34089)
See also the remarks to planet (8600).

(8600) Arundinaceus

3060 T-2. Discovered 1973 September 30 by C. J. van Houten and I. van Houten-Groeneveld at Palomar.

Named for Acrocephalus arundinaceus, or great reed warbler. {It belongs to the family of the Sylviidae and is on the Dutch Red List of birds endangered in the Netherlands.} (M 33792; M 34089)

The first letters of the names of minor planets (8585) to (8600) result in the well-known Latin motto "per aspera ad astra" {Through difficulties to the Stars}. This arrangement by L. D. Schmadel remained unnoticed after the publication of the new names in MPC itself.

(8601) Ciconia

3155 T-2. Discovered 1973 September 30 by C. J. van Houten and I. van Houten-Groeneveld at Palomar.

Named for Ciconia ciconia, or white stork. {It belongs to the family of the Ciconiidae and is on both the Dutch and the European list of endangered birds.} (M 33792; M 34089)

(8602) Oedicnemus

2480 T-3. Discovered 1977 October 16 by C. J. van Houten and I. van Houten-Groeneveld at Palomar.

Named for Burhinus oedicnemus, or stone curlew. {It belongs to the family of the Burhinidae and is endangered in the Netherlands and in Euope.} (M 33792; M 34089)

(8603) Senator

3134 T-3. Discovered 1977 October 16 by C. J. van Houten and I. van Houten-Groeneveld at Palomar.

Named for Lanius senator, or woodchat shrike. {It belongs to the family of the Laniidae. The bird is on the Dutch Red List of birds endangered in the Netherlands and also on the 'List of Vulnerable Birds in Europe'.} (M 33792; M 34089)

(8604) 1929 PK

Discovered 1929 August 12 by C. J. Krieger at Mount Hamilton.

(8605) 1968 OH

Discovered 1968 July 18 by C. Torres and S. Cofre at Cerro El Roble.

(8606) 1971 UG

Discovered 1971 October 26 by L. Kohoutek at Bergedorf.

(8607) 1971 UT

Discovered 1971 October 26 by L. Kohoutek at Bergedorf.

(8608) 1976 YO_2
Discovered 1976 December 16 by L. I. Chernykh at Nauchnyj.

(8609) 1977 QH_3
Discovered 1977 August 22 by N. S. Chernykh at Nauchnyj.

(8610) 1977 UD
Discovered 1977 October 22 at the Harvard College Observatory at Harvard.

(8611) 1977 UM_4
Discovered 1977 October 18 by S. J. Bus at Palomar.

(8612) 1978 SS_7
Discovered 1978 September 26 by L. V. Zhuravleva at Nauchnyj.

(8613) 1978 VE_{10}
Discovered 1978 November 7 by E. F. Helin and S. J. Bus at Palomar.

(8614) 1978 VP_{11}
Discovered 1978 November 7 by E. F. Helin and S. J. Bus at Palomar.

(8615) 1979 MB_2
Discovered 1979 June 25 by E. F. Helin and S. J. Bus at Siding Spring.

(8616) Fogelquist
1980 FY_4. Discovered 1980 March 16 by C.-I. Lagerkvist at La Silla.

Named in honor of Rune Fogelquist (1924-) for his inspiring activities in astronomy popularization within the Mariestad Astronomy Club, lovated near Lake Vänern in southern Sweden, and the building and running of the nearby Bifrost Observatory, the main instrument at which is a 0.60-m reflector. The observatory has about 1000 visitors annually. The naming commemorates the twentieth anniversary of the Mariestad Astronomy Club, celebrated in August 1998. (M 32350)
Name proposed and citation prepared by H. Rickman.

(8617) 1980 PW
Discovered 1980 August 6 by Z. Vávrová at Kleť.

(8618) 1981 DX
Discovered 1981 February 28 by S. J. Bus at Siding Spring.

(8619) 1981 EH_1
Discovered 1981 March 6 by H. Debehogne and G. DeSanctis at La Silla.

(8620) 1981 EK_5
Discovered 1981 March 2 by S. J. Bus at Siding Spring.

(8621) 1981 EK_7
Discovered 1981 March 1 by S. J. Bus at Siding Spring.

(8622) 1981 EM_8
Discovered 1981 March 1 by S. J. Bus at Siding Spring.

(8623) 1981 EQ_9
Discovered 1981 March 1 by S. J. Bus at Siding Spring.

(8624) 1981 ES_9
Discovered 1981 March 1 by S. J. Bus at Siding Spring.

(8625) 1981 EX_{15}
Discovered 1981 March 1 by S. J. Bus at Siding Spring.

(8626) 1981 EC_{18}
Discovered 1981 March 2 by S. J. Bus at Siding Spring.

(8627) 1981 EU_{20}
Discovered 1981 March 2 by S. J. Bus at Siding Spring.

(8628) 1981 EX_{21}
Discovered 1981 March 2 by S. J. Bus at Siding Spring.

(8629) 1981 EU_{26}
Discovered 1981 March 2 by S. J. Bus at Siding Spring.

(8630) 1981 EY_{35}
Discovered 1981 March 2 by S. J. Bus at Siding Spring.

(8631) 1981 EK_{41}
Discovered 1981 March 2 by S. J. Bus at Siding Spring.

(8632) 1981 FR
Discovered 1981 March 28 at the Harvard College Observatory at Harvard.

(8633) 1981 FC_1
Discovered 1981 March 16 by S. J. Bus at Siding Spring.

(8634) 1981 GG
Discovered 1981 April 5 by E. Bowell at Anderson Mesa.

(8635) 1985 PG_2
Discovered 1985 August 13 by N. S. Chernykh at Nauchnyj.

(8636) 1985 UH_2
Discovered 1985 October 17 by the CERGA at Caussols.

(8637) 1986 CS_1
Discovered 1986 February 6 by H. Debehogne at La Silla.

(8638) 1986 QY
Discovered 1986 August 26 by H. Debehogne at La Silla.

(8639) 1986 VB_1
Discovered 1986 November 3 by A. Mrkos at Kleť.

(8640) 1986 VX_5
Discovered 1986 November 6 by E. Bowell at Anderson Mesa.

(8641) 1987 BM_1
Discovered 1987 January 27 by P. Jensen at Brorfelde.

(8642) 1988 RZ_{11}
Discovered 1988 September 14 by S. J. Bus at Cerro Tololo.

(8643) Quercus
1988 SC. Discovered 1988 September 16 by E. W. Elst at St. Michel.

Named for Quercus, a genus belonging to the family Fagaceae. The tree Quercus robur (English oak) reaches a height of 30-40 m and an age of more than a thousand years. (M 34347)

(8644) Betulapendula

1988 SD. Discovered 1988 September 16 by E. W. Elst at St. Michel.

Named for Betula pendula (silver birch), a species in the genus Betula belonging to the family Betulaceae. It is a beautiful tree with an almost white bark. It grows fast and reaches a height of about 25 m and an age of 60-80 years. (M 34347)

(8645) 1988 TN

Discovered 1988 October 5 by S. Ueda and H. Kaneda at Kushiro.

(8646) 1988 TB_1

Discovered 1988 October 13 by S. Ueda and H. Kaneda at Kushiro.

(8647) Populus

1989 RG. Discovered 1989 September 2 by E. W. Elst at St. Michel.

Named for Populus, a genus belonging to the family Salicaceae. Populus nigra (black poplar) is a fast-growing tree with a height of about 30 m, whereas Populus tremula (trembling aspen) is easily recognizable by its shimmering appearance. (M 34347)

(8648) Salix

1989 RJ. Discovered 1989 September 2 by E. W. Elst at St. Michel.

Named for Salix, a genus belonging to the family Salicaceae. Salix alba (white willow) is a small tree with long, thin leaves. This fast-growing tree is ised for windbreaks and screens. Salix caprea (goat or pussy willow) is a fast-growing small tree with striking catkins in early spring. (M 34347)

(8649) Juglans

1989 SS_2. Discovered 1989 September 26 by E. W. Elst at La Silla.

Named for Juglans, a genus belonging to the family Juglandaceae. Juglans regia (English walnut) produces a particularly delicious fruit. (M 34347)

(8650) 1989 TJ_2

Discovered 1989 October 5 by A. Mrkos at Kleť.

(8651) 1989 YU_5

Discovered 1989 December 29 by E. W. Elst at St. Michel.

(8652) Acacia

1990 EA_5. Discovered 1990 March 2 by E. W. Elst at La Silla.

Named for Acacia, belonging to the family Mimosaceae. Acacia mearnsii (mimosa) is a shrub with feathery leaves composed of many small leaflets. The strongly scented tiny flowers are grouped in loose, rounded clusters. (M 34347)

(8653) 1990 KE

Discovered 1990 May 20 by R. H. McNaught at Siding Spring.

(8654) 1990 KC_1

Discovered 1990 May 20 by R. H. McNaught at Siding Spring.

(8655) 1990 QJ_1

Discovered 1990 August 22 by H. E. Holt at Palomar.

(8656) Cupressus

1990 QY_8. Discovered 1990 August 16 by E. W. Elst at La Silla.

Named for Cupressus, belonging to the family Cupressaceae. Cupressus macrocarpia (Monterrey cypress) produces a durable wood. Cupressus leylandii (Leyland cypress), hybridized from the macrocarpia, has scale-like green leaves that are arranged at various angles to the shoot. These trees are typical of the landscape in Tuscany. (M 34348)

(8657) Cedrus

1990 QE_9. Discovered 1990 August 16 by E. W. Elst at La Silla.

Named for Cedrus, belonging to the family Pinaceae. Cedrus libani (cedar of Lebanon) has a fruit that is an ovoid upright cone (M 34348)

(8658) 1990 RG_3

Discovered 1990 September 14 by H. E. Holt at Palomar.

(8659) 1990 SE_{11}

Discovered 1990 September 17 by H. E. Holt at Palomar.

(8660) 1990 TM_1

Discovered 1990 October 15 by K. Endate and K. Watanabe at Kitami.

(8661) 1990 TA_{13}

Discovered 1990 October 14 by L. D. Schmadel and F. Börngen at Tautenburg.

(8662) 1990 UT_{10}

Discovered 1990 October 22 by S. Ueda and H. Kaneda at Kushiro.

(8663) 1991 DJ_1

Discovered 1991 February 18 by E. F. Helin at Palomar.

(8664) 1991 GR_1

Discovered 1991 April 10 by E. F. Helin at Palomar.

(8665) 1991 GA_9

Discovered 1991 April 8 by E. W. Elst at La Silla.

(8666) Reuter

1991 GG_{10}. Discovered 1991 April 9 by F. Börngen at Tautenburg.

Named for the German poet and social critic Fritz Reuter (1810-1874), co-founder of the new Low German literature and one of its greatest humorists. He often expressed worldly wisdom by descriptions of celestial bodies, especially the sun. His work, most of which takes place in Mecklenburg {see planet (6124)}, criticizes unjust social conditions. As a student he was imprisoned for seven years because of his views in favor of democracy and the unity of Germany. Although written in a regional dialect, his work has been widely translated. (M 32792)

(8667) Fontane

1991 GH_{10}. Discovered 1991 April 9 by F. Börngen at Tautenburg.

Named for Theodor Fontane (1819-1898) on the occasion of the 100th anniversary of his death. He brought the German novel to worldwide importance. Descended from a Huguenot family, he worked as an apothecary, travel writer, editor, drama critic and finally as a full-time writer. Besides ballads and the multivolume *Wanderungen durch die Mark Brandenburg*, his novels are especially significant (e.g. *Effi Briest, Der Stechlin*). He chose his material mostly from Prussian history and the Berlin of his times. He combined love of tradition with open-mindedness for the new and skepticism with religious faith. (M 32792)

(8668) 1991 HM

Discovered 1991 April 16 by S. Otomo and O. Muramatsu at Kiyosato.

(8669) 1991 NS_1

Discovered 1991 July 13 by H. E. Holt at Palomar.

(8670) 1991 OM_1

Discovered 1991 July 18 by H. Debehogne at La Silla.

(8671) 1991 PW
Discovered 1991 August 5 by H. E. Holt at Palomar.

(8672) 1991 PW_{16}
Discovered 1991 August 6 by E. W. Elst at La Silla.

(8673) 1991 RN_5
Discovered 1991 September 13 by H. E. Holt at Palomar.

(8674) 1991 VA_1
Discovered 1991 November 4 by S. Ueda and H. Kaneda at Kushiro.

(8675) 1991 YZ
Discovered 1991 December 30 by S. Ueda and H. Kaneda at Kushiro.

(8676) 1992 CT_2
Discovered 1992 February 2 by E. W. Elst at La Silla.

(8677) 1992 ES_5
Discovered 1992 March 2 by the Uppsala-ESO Survey at La Silla.

(8678) 1992 ER_6
Discovered 1992 March 1 by the Uppsala-ESO Survey at La Silla.

(8679) 1992 EG_8
Discovered 1992 March 2 by the Uppsala-ESO Survey at La Silla.

(8680) 1992 EJ_9
Discovered 1992 March 2 by the Uppsala-ESO Survey at La Silla.

(8681) 1992 EN_9
Discovered 1992 March 2 by the Uppsala-ESO Survey at La Silla.

(8682) 1992 ER_9
Discovered 1992 March 2 by the Uppsala-ESO Survey at La Silla.

(8683) 1992 EE_{13}
Discovered 1992 March 2 by the Uppsala-ESO Survey at La Silla.

(8684) Reichwein
1992 FO_3. Discovered 1992 March 30 by F. Börngen at Tautenburg.

Named for the German reform pedagogue Adolf Reichwein (1898-1944), who ran the Adult Education Courses in Jena during 1925-1929 and became professor of history and civics in Halle in 1930. An early opponent of facism, he was dismissed from Halle in 1933 and worked as teacher in an outlying village school. Reichwein took part in founding the illegal "Kreisauer Kreis" {see planet (7945)}, was among its closest advisors and designated to be the minister of education in a new liberated German State. He was arrested early in July 1944, condemned to death by the Volksgerichtshof three months later and executed the same day. (M 34628)

(8685) 1992 GG_3
Discovered 1992 April 4 by E. W. Elst at La Silla.

(8686) 1992 OX_1
Discovered 1992 July 26 by E. W. Elst at La Silla.

(8687) 1992 PV
Discovered 1992 August 8 by E. W. Elst at Caussols.

(8688) 1992 PV_1

Discovered 1992 August 8 by E. W. Elst at Caussols.

(8689) 1992 PU_3

Discovered 1992 August 5 by H. E. Holt at Palomar.

(8690) 1992 SW_3

Discovered 1992 September 24 by the Spacewatch at Kitt Peak.

(8691) Etsuko

1992 UZ_1. Discovered 1992 October 21 by Y. Kushida and O. Muramatsu at Yatsugatake.

Named in honor of Etsuko Kobayashi (1926-), the first female lecturer at the Gotoh Planetarium und Astronomical Museum in Tokyo, where she worked for 30 years until 1986. Famous for her lecturing style, which left many listeners spellbound, she has educated many young lecturers who are active in the field today. She also worked for the publisher of a science journal and is now a well-known author of astronomy books herself. (M 34348)
Name proposed by the discoverers following a suggestion by S. Kimura.

(8692) 1992 WH

Discovered 1992 November 16 by S. Ueda and H. Kaneda at Kushiro.

(8693) 1992 WH_1

Discovered 1992 November 16 by K. Endate and K. Watanabe at Kitami.

(8694) 1993 CO

Discovered 1993 February 10 by S. Ueda and H. Kaneda at Kushiro.

(8695) 1993 FW_8

Discovered 1993 March 17 by the Uppsala-ESO Survey at La Silla.

(8696) 1993 FM_{16}

Discovered 1993 March 17 by the Uppsala-ESO Survey at La Silla.

(8697) 1993 FT_{23}

Discovered 1993 March 21 by the Uppsala-ESO Survey at La Silla.

(8698) 1993 FT_{41}

Discovered 1993 March 19 by the Uppsala-ESO Survey at La Silla.

(8699) 1993 FO_{48}

Discovered 1993 March 19 by the Uppsala-ESO Survey at La Silla.

(8700) 1993 JL_1

Discovered 1993 May 14 by E. W. Elst at La Silla.

(8701) 1993 LG_2

Discovered 1993 June 15 by H. E. Holt at Palomar.

(8702) 1993 VX_3

Discovered 1993 November 14 by M. Hirasawa and S. Suzuki at Nyukasa.

(8703) 1993 XP_1

Discovered 1993 December 15 by T. Kobayashi at Oizumi.

(8704) 1993 YJ

Discovered 1993 December 17 by T. Kobayashi at Oizumi.

(8705) 1994 AL_3
Discovered 1994 January 8 by H. Shiozawa and T. Urata at Fujieda.

(8706) 1994 CM
Discovered 1994 February 3 by T. Kobayashi at Oizumi.

(8707) 1994 CE_2
Discovered 1994 February 12 by T. Kobayashi at Oizumi.

(8708) 1994 DD
Discovered 1994 February 17 by S. Otomo at Kiyosato.

(8709) Kadlu
1994 JF_1. Discovered 1994 May 14 by C. S. Shoemaker and E. M. Shoemaker at Palomar.

The Eskimo thunder-goddess was originally a little girl who played so noisily that her parents told her and her sisters to go outside to play. They did, and they invented a game in which Kadlu jumped on hollow ice, causing a thunderous sound. Transported to the sky, the goddess lived in a whale-bone house far in the west. Some legends say that Kadlu made thunder by rubbing dry sealskins together, or by singing. (M 32350)

(8710) Hawley
1994 JK_9. Discovered 1994 May 15 by C. P. de Saint-Aignan at Palomar.

Named in honor of the discoverer's friend and mentor, Walter N. Hawley, a physics and astronomy teacher at Saint Paul's School in Concord, New Hampshire. He is also director of the Saint Paul's Astronomy Center, an observatory of unequaled quality at the high-school level. Hawley has observed the night sky with his students since 1972, acquainting them with such wonders as the Aurora Borealis and Messier {see planet (7359)} Objects, as well as lesser wonders, such as Murphy's Law – and frostbite. (M 32792)

(8711) 1994 LL
Discovered 1994 June 5 by C. W. Hergenrother at Tucson.

(8712) 1994 TH_2
Discovered 1994 October 2 by K. Endate and K. Watanabe at Kitami.

(8713) 1995 BT_2
Discovered 1995 January 26 by K. Endate and K. Watanabe at Kitami.

(8714) 1995 OT
Discovered 1995 July 24 by Y. Shimizu and T. Urata at Nachi-Katsuura.

(8715) 1995 OX_1
Discovered 1995 July 26 by Y. Shimizu and T. Urata at Nachi-Katsuura.

(8716) Ginestra
1995 SB_2. Discovered 1995 September 23 by V. S. Casulli at Colleverde di Guidonia.

Named for a poem by Giacomo Leopardi {see planet (8081)} (1798-1837), the great poet and philosopher from the Italian Romantic period. Full of astronomical references, his poetry expresses the great sense of bewilderment of post-Copernican man, faced with an infinite variety of worlds of which he is no longer the center, but only infinitesimal and marginal. Nevertheless, the "Ginestra" becomes the symbol of man/flower in the middle of the cosmos/desert, a cosmos sustained by rigid mechanistic laws, indifferent to every desire and human

sentiment, existing only to perpetuate the cycle of production and universal destruction. (M 32351)
Name suggested and citation prepared by M. Vicoli.

(8717) 1995 SN_{29}
Discovered 1995 September 26 by T. V. Kryachko at Zelenchukskaya.

(8718) 1995 UC_8
Discovered 1995 October 27 by Y. Shimizu and T. Urata at Nachi-Katsuura.

(8719) 1995 VR
Discovered 1995 November 11 at the Kleť Observatory at Kleť.

(8720) Takamizawa
1995 WE_1. Discovered 1995 November 16 by A. Nakamura at Kuma.

Named in honor of Kesao Takamizawa (1952-), one of the most renowned amateur astronomers in Japan. During his 25-year visual search for comets, Takamizawa made two new discoveries and one independent discovery. In 1994, he moved from visual to photographic surveying. Since then, he has discovered two more comets, one supernova, two galactic novae, and many variable stars. (M 32792)

(8721) AMOS
1996 AO_3. Discovered 1996 January 14 by the JPL NEAT Program at Haleakala.

Named for the Air Force Maui Optical Station (AMOS), the research and development arm of the Maui Space Surveillance System at Haleakala on Maui, Hawaii. The AMOS project for observations of minor planets is the result of a collaboration between the Air Force Research Laboratory and Eleanor Helin's {see planet (3267)} observing programs at the Jet Propulsion Laboratory. This collaboration began in 1993 and continues today. AMOS performs follow-up observations for several programs, primarily the highly successful JPL NEAT program. (M 32351)

(8722) 1996 QU_1
Discovered 1996 August 19 by R. G. Davis at Granville.

(8723) 1996 SL_7
Discovered 1996 September 23 by T. Okuni at Nanyou.

(8724) 1996 SK_8
Discovered 1996 September 17 by S. Otomo at Kiyosato.

(8725) Keiko
1996 TG_5. Discovered 1996 October 5 by H. Abe at Yatsuka.

Named in honor of Keiko Morinaga (1969-), a member of the Matsue Astronomy Club and observing partner of the discoverer. (M 32351; M 32377)

(8726) 1996 VP_5
Discovered 1996 November 14 by T. Kobayashi at Oizumi.

(8727) 1996 VZ_7
Discovered 1996 November 3 by S. Ueda and H. Kaneda at Kushiro.

(8728) 1996 VF_9
Discovered 1996 November 7 by K. Endate and K. Watanabe at Kitami.

(8729) 1996 VZ_{12}
Discovered 1996 November 5 by the Spacewatch at Kitt Peak.

(8730) 1996 VT_{30}
Discovered 1996 November 10 by T. Okuni at Nanyou.

(8731) 1996 WY
Discovered 1996 November 19 by T. Kobayashi at Oizumi.

(8732) 1996 XR_{25}
Discovered 1996 December 8 by T. Seki at Geisei.

(8733) 1996 YB_1
Discovered 1996 December 20 by T. Kobayashi at Oizumi.

(8734) Warner
1997 AA. Discovered 1997 January 1 by P. G. Comba at Prescott.

Named in honor of Brian D. Warner (1952-). For more than eight years he has published the *Minor Planet Observer*, a monthly newsletter with minor-planet finder charts and other news of interest to minor planet observers. In recent years he has taken up CCD imaging and contributed hundreds of astrometric observations. (M 34348)

(8735) 1997 AA_1
Discovered 1997 January 2 by T. Kobayashi at Oizumi.

(8736) 1997 AD_7
Discovered 1997 January 9 by T. Kobayashi at Oizumi.

(8737) 1997 AL_{13}
Discovered 1997 January 11 by T. Kobayashi at Oizumi.

(8738) Saji
1997 AQ_{16}. Discovered 1997 January 5 at the Saji Observatory at Saji.

Saji observatory, with its 1.03-m telescope, is situated on a hill overlooking Saji village. Saji's 3200 residents constructed their observatory in 1994, and the village assembly adopted a rule of keeping the skies dark. (M 33792)

(8739) 1997 BE_3
Discovered 1997 January 30 by T. Kobayashi at Oizumi.

(8740) Václav
1998 AS_8. Discovered 1998 January 12 by M. Tichý and Z. Moravec at Kleť.

Named for Czech prince Václav (c.908-c.929), Czech kings Václav I (1205-1253), Václav II (1271-1305) and his son Václav III (1289-1306), all from royal dynasty of Přemyslids {see planet (7695)}, and Czech king and Roman emperor Václav IV of Luxembourg (1361-1419), the son of Charles IV. Prince Václav was killed by his brother, Boleslav I. Later he was worshipped as a saint and from the eleventh century he became patron of the Czech lands and the symbol of Czech statehood. King Václav III was also killed, and he was the last male descendant of the Přemyslids dynasty. Václav is still the most common Slavonic name in the Czech Republic. (M 34628)
Name endorsed by J. Tichá.

(8741) 1998 BR_8
Discovered 1998 January 25 by T. Kobayashi at Oizumi.

(8742) 1998 CB_2
Discovered 1998 February 14 by V. S. Casulli at Colleverde di Guidonia.

(8743) 1998 EH_{12}
Discovered 1998 March 1 by E. W. Elst at La Silla.

(8744) Cilla
1998 FE_{59}. Discovered 1998 March 20 by the Lincoln NEA Research Team at Socorro.

Named in memory of Priscilla Annette (1994-1998), niece of Lincoln Laboratory staff member Colleen Cilley. (M 33792)

(8745) Delaney
1998 FO_{65}. Discovered 1998 March 20 by the Lincoln NEA Research Team at Socorro.

Named in honor of William P. Delaney, director's fellow at M.I.T. Lincoln Laboratory. This is a donation of the fruit of an optical search program to an expert in radar. (M 33792)

(8746) 1998 FL_{68}
Discovered 1998 March 20 by the Lincoln NEA Research Team at Socorro.

(8747) 1998 FS_{73}
Discovered 1998 March 24 by T. Okuni at Nanyou.

(8748) 1998 FV_{113}
Discovered 1998 March 31 by the Lincoln NEA Research Team at Socorro.

(8749) 1998 GJ_{10}
Discovered 1998 April 3 by J. Broughton at Reedy Creek.

(8750) Nettarufina
2197 P-L. Discovered 1960 September 24 by C. J. van Houten and I. van Houten-Groeneveld at Palomar.

Named for Netta rufina, or red-crested pochard. {It belongs to the family of the Anatidae and is on the Dutch Red List of birds endangered in the Netherlands.} (M 33792; M 34089)

(8751) Nigricollis
2594 P-L. Discovered 1960 September 24 by C. J. van Houten and I. van Houten-Groeneveld at Palomar.

Named for Podiceps nigricollis, or black-necked grebe. {It belongs to the family of the Podicipidae and is on the Dutch Red List.} (M 33792; M 34089)

(8752) Flammeus
2604 P-L. Discovered 1960 September 24 by C. J. van Houten and I. van Houten-Groeneveld at Palomar.

Named for Asio flammeus, or short-eared owl. {It belongs to the family of the Strigidae and is on the Dutch and European lists of endangered birds.} (M 33792; M 34089)

(8753) Nycticorax
2636 P-L. Discovered 1960 September 24 by C. J. van Houten and I. van Houten-Groeneveld at Palomar.

Named for Nycticorax nycticorax, or night heron. {It belongs to the family of the Ardeidae and is endangered in the Netherlands.} (M 33792; M 34089)

(8754) Leucorodia
4521 P-L. Discovered 1960 September 24 by C. J. van Houten and I. van Houten-Groeneveld at Palomar.

Named for Platalea leucorodia, or spoonbill. {It belongs to the family of the Plataleidae. The bird is mentioned on the Dutch and European lists of endangered species.} (M 33792; M 34089)

(8755) Querquedula

4586 P-L. Discovered 1960 September 24 by C. J. van Houten and I. van Houten-Groeneveld at Palomar.

Named for Anas querquedula, or garganey. {It belongs to the family of the Anatidae. It is on the Dutch Red List of birds endangered in the Netherlands and also on the 'List of Vulnerable Birds in Europe'.} (M 33792; m 34089)

(8756) Mollissima

6588 P-L. Discovered 1960 September 24 by C. J. van Houten and I. van Houten-Groeneveld at Palomar.

Named for Somateria mollissima, or common eider. {It belongs to the Anatidae family and is on the Dutch Red List of endangered birds.} (M 33793; M 34089)

(8757) Cyaneus

6600 P-L. Discovered 1960 September 24 by C. J. van Houten and I. van Houten-Groeneveld at Palomar.

Named for Circus cyaneus, or hen or northern harrier. {It belongs to the family of the Accipitridae and is endangered in the Netherlands as well as in Europe.} (M 33793; M 34089)

(8758) Perdix

6683 P-L. Discovered 1960 September 24 by C. J. van Houten and I. van Houten-Groeneveld at Palomar.

Named for Perdix perdix, or partridge. {It belongs to the family of the Phasianidae. It is on the Dutch Red List of birds endangered in the Netherlands and also on the 'List of Vulnerable Birds in Europe'.} (M 33793; M 34089)

(8759) Porzana

7603 P-L. Discovered 1960 October 17 by C. J. van Houten and I. van Houten-Groeneveld at Palomar.

Named for Porzana porzana, or spotted crake. {This endangered bird belongs to the family of the Rallidae.} (M 33793; M 34089)

(8760) Crex

1081 T-1. Discovered 1971 March 25 by C. J. van Houten and I. van Houten-Groeneveld at Palomar.

Named for Crex crex, or corncrake. {It belongs to the family of the Rallidae and is endangered in the Netherlands as well as in Europe. In 1994 only 3,800 pairs were counted.} (M 33793; M 34089)

(8761) Crane

1163 T-1. Discovered 1971 March 25 by C. J. van Houten and I. van Houten-Groeneveld at Palomar.

Named for the bird Grus grus; crane is its English name. It belongs to the family of the Gruidae and is an endangered species in Europe. (M 34348)

(8762) Hiaticula

3196 T-1. Discovered 1971 March 26 by C. J. van Houten and I. van Houten-Groeneveld at Palomar.

Named for Charadrius hiaticula, or ringed plover. {It belongs to the family of the Charadriidae and is on the Dutch Red List.} (M 33793; M 34089)

(8763) Pugnax

3271 T-1. Discovered 1971 March 26 by C. J. van Houten and I. van Houten-Groeneveld at Palomar.

Named for Philomachus pugnax, or ruff (male) and reeve (female). {It belongs to the family of the Scolopacidae and is on the list of endangered birds in the Netherlands.} (M 33793; M 34089)

(8764) Gallinago

1109 T-2. Discovered 1973 September 29 by C. J. van Houten and I. van Houten-Groeneveld at Palomar.

Named for Gallinago gallinago, or snipe. {It belongs to the family of the Scolopacidae and is on the Dutch Red List.} (M 33793; M 34089)

(8765) Limosa

1274 T-2. Discovered 1973 September 29 by C. J. van Houten and I. van Houten-Groeneveld at Palomar.

Named for Limosa limosa, or black-tailed godwit. {It belongs to the family of the Scolopacidae. It is on the Dutch Red List of birds endangered in the Netherlands and also on the 'List of Vulnerable Birds in Europe.} (M 33793; M 34089)

(8766) Niger

1304 T-2. Discovered 1973 September 29 by C. J. van Houten and I. van Houten-Groeneveld at Palomar.

Named for Chlidonias niger, or black tern. {It belongs to the family of the Sternidae and is on the Dutch Red List of endangered birds.} (M 33793; M 34089)

(8767) Commontern

1335 T-2. Discovered 1973 September 29 by C. J. van Houten and I. van Houten-Groeneveld at Palomar.

Named for the bird Sterna hirundo; common tern is its English name. It belongs to the family of the Sternidae. (M 34348)

(8768) Barnowl

2080 T-2. Discovered 1973 September 29 by C. J. van Houten and I. van Houten-Groeneveld at Palomar.

Named for the bird Tyto alba; barn owl is its English name. It belongs to the family of Tytonidae. (M 34348)

(8769) Arctictern

2181 T-2. Discovered 1973 September 29 by C. J. van Houten and I. van Houten-Groeneveld at Palomar.

Named for the bird Sterna paradisaea; arctic tern is its English name. It belongs to the family of the Sternidae. (M 34348)

(8770) Totanus

3076 T-2. Discovered 1973 September 30 by C. J. van Houten and I. van Houten-Groeneveld at Palomar.

Named for Tringa totanus, or redshank. {It belongs to the family of the Scolopacidae and is endangered in the Netherlands.} (M 33793; M 34089)

(8771) Biarmicus

3187 T-2. Discovered 1973 September 30 by C. J. van Houten and I. van Houten-Groeneveld at Palomar.

Named for Panurus biarmicus, or bearded tit. {It belongs to the family of the Timaliidae and is on the Dutch Red List.} (M 33793; M 34089)

(8772) Minutus

4254 T-2. Discovered 1973 September 29 by C. J. van Houten and I. van Houten-Groeneveld at Palomar.

Named for Ixobrychus minutus, or little bittern. {It belongs to the family of the Ardeidae. It is on the Dutch Red List of birds endangered in the Netherlands and also on the 'List of Vulnerable Birds in Europe.} (M 33793; M 34089)

(8773) Torquilla

5006 T-2. Discovered 1973 September 25 by C. J. van Houten and I. van Houten-Groeneveld at Palomar.

Named for Jynx torquilla, or wryneck. {It belongs to the family of the Picidae and is on the Dutch list of endangered birds.} (M 33793; M 34089)

(8774) Viridis

5162 T-2. Discovered 1973 September 25 by C. J. van Houten and I. van Houten-Groeneveld at Palomar.

Named for Picus viridis, or green woodpecker. {It belongs to the family of the Picidae and is endangered in the Netherlands.} (M 33793; M 34089)

(8775) Cristata

5490 T-2. Discovered 1973 September 30 by C. J. van Houten and I. van Houten-Groeneveld at Palomar.

Named for Galerida cristata, or crested lark. {It belongs to the Alaudidae family and is on the Dutch Red List.} (M 33793; M 34089)

(8776) Campestris

2287 T-3. Discovered 1977 October 16 by C. J. van Houten and I. van Houten-Groeneveld at Palomar.

Named for Anthus campestris, or tawny pipit. {It belongs to the family of the Motacillidae. It is on the Dutch Red List of birds endangered in the Netherlands and also on the 'List of Vulnerable Birds in Europe.} (M 33793; M 34089)

(8777) Torquata

5016 T-3. Discovered 1977 October 16 by C. J. van Houten and I. van Houten-Groeneveld at Palomar.

Named for Saxicola torquata, or stonechat. {It belongs to the family of the Turdidae and is on the Dutch list of endangered birds.} (M 33793; M 34089)

(8778) 1931 TD_3

Discovered 1931 October 10 by C. W. Tombaugh at Flagstaff.

(8779) 1971 UH_1

Discovered 1971 October 26 by L. Kohoutek at Bergedorf.

(8780) 1975 LT

Discovered 1975 June 13 by M. R. Cesco at El Leoncito.

(8781) 1976 GA_2

Discovered 1976 April 1 by N. S. Chernykh at Nauchnyj.

(8782) 1976 UG_2

Discovered 1976 October 26 by T. M. Smirnova at Nauchnyj.

(8783) 1977 EK_1

Discovered 1977 March 13 by N. S. Chernykh at Nauchnyj.

(8784) 1977 RQ_{19}

Discovered 1977 September 9 by C. M. Olmstead at Palomar.

(8785) 1978 RR_1

Discovered 1978 September 5 by N. S. Chernykh at Nauchnyj.

(8786) Belskaya

1978 RA_8. Discovered 1978 September 2 by C.-I. Lagerkvist at La Silla.

Named in honor of Irina N. Belskaya, senior astronomer at the Kharkov Astronomical Observatory and a friend of the discoverer. Her research focuses on the spin and surface properties of main-belt minor planets, in particular, objects of the taxonomic type M. She has made important contributions in this field by combining photometric and polarimetric observations with optical laboratory measurements of the analogous asteroidal materials. (M 33389)
Citation prepared by A. Erikson.

(8787) 1978 TL_4

Discovered 1978 October 4 by T. M. Smirnova at Nauchnyj.

(8788) Labeyrie

1978 VP_2. Discovered 1978 November 1 by K. Tomita at Caussols.

Named in honor of Catherine and Antoine Labeyrie. Catherine was the hypersensitization expert at the 0.9-m Schmidt telescope with which this object was discovered; in recent years, she has devoted her energy to the preservation of the ecology of the Provence region. Antoine is an optician as well as an astronomer; he participated in the invention of holographic diffraction and is the inventor of speckle interferometry. The speckle technique, which has given birth to optical multi-telescope interferometry, has been used to obtain images of several minor planets. Antoine currently holds the chair of observational astrophysics at the Collège de France and is currently director of the Observatoire de Haute Provence. (M 34628)
Citation prepared by A. Morely.

(8789) 1978 VZ_7

Discovered 1978 November 7 by E. F. Helin and S. J. Bus at Palomar.

(8790) 1978 VN_9

Discovered 1978 November 7 by E. F. Helin and S. J. Bus at Palomar.

(8791) 1978 VG_{11}

Discovered 1978 November 7 by E. F. Helin and S. J. Bus at Palomar.

(8792) 1978 VH_{11}

Discovered 1978 November 7 by E. F. Helin and S. J. Bus at Palomar.

(8793) Thomasmüller

1979 QX. Discovered 1979 August 22 by C.-I. Lagerkvist at La Silla.

Named in honor of Thomas G. Müller for his contribution to the study of minor planets in the thermal infrared. His observational work includes mid- to far-infrared photometry, spectroscopy and polarimetry with the Infrared Space Observatory. He has developed and applied various thermophysical models and techniques, in order to derive physical properties of minor planets. Because of the great accuracy he achieved, future telescope projects in the infrared will also benefit from these efforts, since minor planets are well suited as calibration targets. (M 33389)
Name proposed and citation prepared by J. S. V. Lagerros.

(8794) 1981 EA_7
Discovered 1981 March 6 by S. J. Bus at Siding Spring.

(8795) 1981 EO_9
Discovered 1981 March 1 by S. J. Bus at Siding Spring.

(8796) 1981 EA_{12}
Discovered 1981 March 7 by S. J. Bus at Siding Spring.

(8797) 1981 EU_{18}
Discovered 1981 March 2 by S. J. Bus at Siding Spring.

(8798) 1981 EF_{24}
Discovered 1981 March 7 by S. J. Bus at Siding Spring.

(8799) 1981 ER_{25}
Discovered 1981 March 2 by S. J. Bus at Siding Spring.

(8800) 1981 EB_{26}
Discovered 1981 March 2 by S. J. Bus at Siding Spring.

(8801) 1981 EQ_{29}
Discovered 1981 March 1 by S. J. Bus at Siding Spring.

(8802) 1981 EW_{31}
Discovered 1981 March 2 by S. J. Bus at Siding Spring.

(8803) 1981 EL_{34}
Discovered 1981 March 2 by S. J. Bus at Siding Spring.

(8804) 1981 JB_2
Discovered 1981 May 5 by C. S. Shoemaker at Palomar.

(8805) 1981 UM_{11}
Discovered 1981 October 22 by N. S. Chernykh at Nauchnyj.

(8806) 1981 UU_{11}
Discovered 1981 October 22 by N. S. Chernykh at Nauchnyj.

(8807) 1981 UD_{23}
Discovered 1981 October 24 by S. J. Bus at Palomar.

(8808) 1981 UH_{28}
Discovered 1981 October 24 by S. J. Bus at Palomar.

(8809) 1981 WE_1
Discovered 1981 November 24 by E. Bowell at Anderson Mesa.

(8810) 1982 JM_1
Discovered 1982 May 15 by E. F. Helin and E. M. Shoemaker at Palomar.

(8811) 1982 UX_5
Discovered 1982 October 20 by L. G. Karachkina at Nauchnyj.

(8812) Kravtsov
1982 UY_6. Discovered 1982 October 20 by L. G. Karachkina at Nauchnyj.
Named for Ukrainian pilot and writer Yurij Fedorovich Kravtsov (1924-1994). The author and the heroes in his books are examples of great human courage. (M 34348)

(8813) 1983 WF_1
Discovered 1983 November 29 by E. Bowell at Anderson Mesa.

(8814) 1983 XG
Discovered 1983 December 1 by E. Bowell at Anderson Mesa.

(8815) 1984 DR
Discovered 1984 February 23 by H. Debehogne at La Silla.

(8816) Gamow
1984 YN_1. Discovered 1984 December 17 by L. G. Karachkina at Nauchnyj.

Named in memory of the scientist George (Georgij Antonovich) Gamow (1904-1968). His main scientific achievements include the creation of alpha and beta decay theory and the theory of the exploding Universe. Gamow was also the first to decipher the genetic code. He worked at institutions around the world, in Odessa, Leningrad, Göttingen, Copenhagen, Cambridge, as well as in the U.S. Through his popular lectures, articles and books he promoted public interest in science. In 1956 he received the Calling Prize awarded by UNESCO for the popularization of science. (M 34348)
Name suggested by S. P. Kapitza and supported by the discoverer.

(8817) 1985 JU_1
Discovered 1985 May 13 by C. S. Shoemaker at Palomar.

(8818) 1985 RW_2
Discovered 1985 September 5 by H. Debehogne at La Silla.

(8819) 1985 RR_4
Discovered 1985 September 14 by H. Debehogne at La Silla.

(8820) 1985 VG
Discovered 1985 November 14 by P. Jensen at Brorfelde.

(8821) 1987 DP_6
Discovered 1987 February 23 by H. Debehogne at La Silla.

(8822) Shuryanka
1987 RQ_2. Discovered 1987 September 1 by L. G. Karachkina at Nauchnyj.

Named in honor of Aleksandra Semenova Morozova (1917-), mother of the discoverer. Shuryanka is a pet name used by her parents. Before her retirement she worked as a barley selector. (M 34628)

(8823) 1987 WS_3
Discovered 1987 November 24 by S. McDonald at Anderson Mesa.

(8824) 1988 BH
Discovered 1988 January 18 by M. Matsuyama and K. Watanabe at Kushiro.

(8825) 1988 MF
Discovered 1988 June 16 by E. F. Helin at Palomar.

(8826) 1988 PZ_1
Discovered 1988 August 13 by E. W. Elst at St. Michel.

(8827) Kollwitz
1988 PO_2. Discovered 1988 August 13 by F. Börngen at Tautenburg.

Named for the distinguished German graphic artist and sculptor Käthe Kollwitz (1867-1945). She studied art in Berlin and Munich and was married to a physician.

She became a professor and a member of the Prussian Academy of Arts in 1919 but was ousted in 1933. In her pictures, woodcuts, lithographs and sculptures, Kollwitz represented the destitution and misery of the poor, the unhappy fate of poor mothers and children, and the death of young men in war. She was deeply influenced by the tragedies of both world wars and the horrors of fascism, always hoping for a peaceful and humane society. (M 32792)

(8828) 1988 RC_7

Discovered 1988 September 10 by H. Debehogne at La Silla.

(8829) 1988 RV_{10}

Discovered 1988 September 14 by S. J. Bus at Cerro Tololo.

(8830) 1988 VZ

Discovered 1988 November 7 by Y. Kushida and M. Inoue at Yatsugatake.

(8831) Brändström

1989 CO_5. Discovered 1989 February 2 by F. Börngen at Tautenburg.

Named in memory of the uncommon Swede Elsa Brändström (1888-1948). Born in St. Petersburg, she was a daughter of the diplomat general Edvard Brändström. During World War I, she stood up beyond all measure for indigent German prisoners in Russia. Under the sign of the Red Cross, she organized numerous relief missions to Siberian camps; from 1914 to 1920, about 700 000 prisoners benefited of her help, which gave her the honorary title "Angle of Siberia". After 1918, she founded several homes for war orphans and repatriates of war in Germany. After the fascists' seizure of power, she left Berlin together with her husband and emigrated in the U.S. There she stood up anew for needy refugees from Europe, persecuted for racial or political motives. (M 32792)

(8832) Altenrath

1989 EC_3. Discovered 1989 March 2 by E. W. Elst at La Silla.

Named in memory of Henricus Hubertus Altenrath (1832-1892), initiator and first director of the "Nijverheidsschool", a well-known Antwerp school for the teaching of technical professions. Under his direction, the school began teaching in Flemish, which was not common at that time. His name is still honored by the Association "Henric Altenrath". The discoverer has taught for many years at this school. (M 33389)

(8833) Acer

1989 RW. Discovered 1989 September 3 by E. W. Elst at St. Michel.

Named for Aceraceae, the maple family, with two genera and more than 100 species. One well-known species is Acer saccharum (sugar maple), the sap of which is used for maple syrup and maple sugar. (M 34348)

(8834) Anacardium

1989 SX_2. Discovered 1989 September 26 by E. W. Elst at La Silla.

Named for Anacardiaceae, the cashew or mango family, with 80 genera and over 800 species of evergreen and deciduous trees, shrubs and climbing plants. Anacardium occidentale (western cashew) yields a delicious fruit. (M 34348)

(8835) Annona

1989 SA_3. Discovered 1989 September 26 by E. W. Elst at La Silla.

Named for Annonaceae, the custard apple family, with more than 2000 species. The trees are mainly tropical and include the species Annona squamose (sweetsop), which has a sweet, pulpy fruit. (M 34348)

(8836) Aquifolium

1989 SU_3. Discovered 1989 September 26 by E. W. Elst at La Silla.

Named for Aquifoliaceae, the holly family, with 700 evergreens and deciduous species. Ilex aquifolium (English holly) belongs to this family, as does Ilex paraguariensis (Yerba maté), which makes a tasteful tea. (M 34348)

(8837) 1989 TF_4

Discovered 1989 October 7 by E. W. Elst at La Silla.

(8838) 1989 UW_2

Discovered 1989 October 29 by T. Hioki and N. Kawasato at Okutama.

(8839) 1989 UB_8

Discovered 1989 October 24 by L. I. Chernykh at Nauchnyj.

(8840) 1989 WT

Discovered 1989 November 20 by S. Ueda and H. Kaneda at Kushiro.

(8841) 1990 EA_7

Discovered 1990 March 2 by H. Debehogne at La Silla.

(8842) 1990 KF

Discovered 1990 May 20 by R. H. McNaught at Siding Spring.

(8843) 1990 OH

Discovered 1990 July 22 by E. F. Helin at Palomar.

(8844) 1990 QR_2

Discovered 1990 August 24 by H. E. Holt at Palomar.

(8845) 1990 RD

Discovered 1990 September 14 by H. E. Holt at Palomar.

(8846) 1990 RK_7

Discovered 1990 September 13 by H. Debehogne at La Silla.

(8847) Huch

1990 TO_3. Discovered 1990 October 12 by F. Börngen and L. D. Schmadel at Tautenburg.

Named in honor of the German novelist and poet Ricarda Huch (1864-1947). Born into a prosperous commercial family, she earned a doctorate in history in Zurich in 1891, afterwards working as a librarian and teacher. In 1897 she decided to become a writer. Her novels and stories depict historical figures and events. She was a critic of the fascist regime, and her last unfinished work was to portray the German resistance movement. In 1933, she retired from the Prussian Academy of Arts. (M 32792)

Name proposed by the first discoverer.

(8848) 1990 VK_1

Discovered 1990 November 12 by S. Ueda and H. Kaneda at Kushiro.

(8849) 1990 VZ_4

Discovered 1990 November 15 by E. W. Elst at La Silla.

(8850) Bignonia

1990 VQ_6. Discovered 1990 November 15 by E. W. Elst at La Silla.

Named for Bignoniaceae, the catalpa family, with about 100 genera and 700 species with tubular flowers. Among them are Bignonia capreolata (trumpet flower) and Crescentia cujete (calabash). (M 34348)

(8851) 1990 XB
Discovered 1990 December 8 by Y. Mizuno and T. Furuta at Kani.

(8852) Buxus
1991 GG_6. Discovered 1991 April 8 by E. W. Elst at La Silla.

Named for Buxaceae, the box family, with four or five genera and some 60 species of evergreen trees and shrubs. Buxus sempervirens (English boxwood) has very small leaves and is used for hedges and borders. (M 34349)

(8853) Gerdlehmann
1991 GC_{10}. Discovered 1991 April 9 by F. Börngen at Tautenburg.

Named in honor of Gerhard Lehmann (1960-), an amateur astronomer who with Jens Kandler {see planet (8861)} measured some 70 precise positions of minor planets and comets from photographic plates obtained at Drebach {see planet (6488)}. Among the 1500 positions he has derived after switching to CCD equipment in 1994 are some for the 1998 opposition of this object. The Drebach positions have made more than 30 Tautenburg objects appropriate for numbering. A teacher of physics and astronomy, Lehmann is also a popularizer of astronomy and since 1997 the head of the minor planets section of the Vereinigung der Sternfreunde. (M 33389)

(8854) 1991 HC
Discovered 1991 April 16 by N. Kawasato at Uenohara.

(8855) Miwa
1991 JL. Discovered 1991 May 3 by S. Otomo and O. Muramatsu at Kiyosato.

Named in honor of Miwa Saito, a junior high school science teacher. She has made an effort to popularize science and astronomy on her World Wide Web site. (M 33793)

(8856) Celastrus
1991 LH_1. Discovered 1991 June 6 by E. W. Elst at La Silla.

Named for Celastraceae, the staff-tree family, with 100 genera and over 1000 species including many climbing plants. These include Celastrus candens (bittersweet) and Euonymus europaeus (European spindle tree), the latter having pink fruit and orange seeds. (M 34349)

(8857) Cercidiphyllum
1991 PA_7. Discovered 1991 August 6 by E. W. Elst at La Silla.

Named for Cercidiphyllaceae, a family with only one member, Cercidiphyllum japonicum (katsura), regarded as a plant of primitive origin. It was originally classified as belonging to the magnolias, but it seems more related to the planes. (M 34349)

(8858) Cornus
1991 PT_7. Discovered 1991 August 6 by E. W. Elst at La Silla.

Named for Cornaceae, the dogwood family, with about 12 genera and 100 species of evergreens, deciduous trees and shrubs. Species include Cornus sanguinea (red dogwood) and Cornus florida (flowering dogwood). (M 34349)

(8859) 1991 PQ_{11}
Discovered 1991 August 9 by H. E. Holt at Palomar.

(8860) 1991 TE_5
Discovered 1991 October 5 by L. D. Schmadel and F. Börngen at Tautenburg.

(8861) Jenskandler

1991 TF_7. Discovered 1991 October 3 by F. Börngen and L. D. Schmadel at Tautenburg.

Named in honor of Jens Kandler (1973-), who has been an observer at Drebach {see planet (6488)} since the age of 14. Beginning in 1993, he has been observing minor planets and comets, at first photographically and then using a CCD. Among the Drebach observations are many from both the 1996 and 1998 oppositions of this object. A skilled industrial mechanic and tile setter, Kandler is an enthusiastic proponent of astronomy. (M 33389)
Name proposed by the first discoverer.

(8862) 1991 UZ

Discovered 1991 October 18 by S. Otomo at Kiyosato.

(8863) 1991 UV_2

Discovered 1991 October 31 by S. Ueda and H. Kaneda at Kushiro.

(8864) 1991 VU

Discovered 1991 November 4 by Y. Mizuno and T. Furuta at Kani.

(8865) 1992 AF

Discovered 1992 January 1 by A. Natori and T. Urata at Yakiimo.

(8866) Tanegashima

1992 BR. Discovered 1992 January 26 by M. Mukai and M. Takeishi at Kagoshima.

Named for the small, long and narrow island to the south of Kyushu Island in southeastern Japan. It contains the launch site for the National Space Development Agency. (M 33793)

(8867) 1992 BF_4

Discovered 1992 January 29 by the Spacewatch at Kitt Peak.

(8868) 1992 EE_7

Discovered 1992 March 1 by the Uppsala-ESO Survey at La Silla.

(8869) 1992 EE_{11}

Discovered 1992 March 6 by the Uppsala-ESO Survey at La Silla.

(8870) 1992 EQ_{11}

Discovered 1992 March 6 by the Uppsala-ESO Survey at La Silla.

(8871) 1992 EA_{22}

Discovered 1992 March 1 by the Uppsala-ESO Survey at La Silla.

(8872) Ebenum

1992 GA_4. Discovered 1992 April 4 by E. W. Elst at La Silla.

Named for Ebenaceae, the ebony family, with only two genera and 500 species. Male and female flowers are usually borne on separate plants. The Diospyros genus dominates, the fruit of the Diospyros virginiana (persimmon) being good to eat. Diospyros ebenum (Macassar ebony) is a valuable wood. (M 34349)

(8873) 1992 UM_2

Discovered 1992 October 21 by Y. Mizuno and T. Furuta at Kani.

(8874) 1992 UY_3

Discovered 1992 October 26 by K. Endate and K. Watanabe at Kitami.

(8875) 1992 UP_{10}

Discovered 1992 October 22 by E. Bowell at Palomar.

(8876) 1992 WU_3

Discovered 1992 November 23 by Y. Kushida and O. Muramatsu at Yatsugatake.

(8877) 1993 BK_2

Discovered 1993 January 19 by T. Seki at Geisei.

(8878) 1993 FN_{16}

Discovered 1993 March 17 by the Uppsala-ESO Survey at La Silla.

(8879) 1993 FN_{20}

Discovered 1993 March 19 by the Uppsala-ESO Survey at La Silla.

(8880) 1993 FT_{33}

Discovered 1993 March 19 by the Uppsala-ESO Survey at La Silla.

(8881) 1993 FW_{36}

Discovered 1993 March 19 by the Uppsala-ESO Survey at La Silla.

(8882) 1994 AP_2

Discovered 1994 January 10 by K. Endate and K. Watanabe at Kitami.

(8883) 1994 BS_4

Discovered 1994 January 16 by T. Kobayashi at Oizumi.

(8884) 1994 CM_2

Discovered 1994 February 12 by A. C. Gilmore and P. M. Kilmartin at Lake Tekapo.

(8885) Sette

1994 EL_3. Discovered 1994 March 13 by M. Tombelli and V. Goretti at Cima Ekar.

Named in honor of Giancarlo Sette (1927-), Italian amateur astronomer. Since 1950 he has focused his attention mainly on astronomical photography with several groups of amateur astronomers in Bologna. From 1975 to 1981, at San Vittore Observatory, he was involved in the International Planetary Patrol Program of the Lowell Observatory. He still devotes his time to wide-field Schmidt astronomical photography. (M 32792)

(8886) Elaeagnus

1994 EG_6. Discovered 1994 March 9 by E. W. Elst at Caussols.

Named for Elaeagnaceae, the oleaster family with three genera and about 50 species. In many cases the flowers develop into edible fruits. Elaeagnus augustifolia (Russian olive) has important commercial value for its fruits. (M 34349)

(8887) Scheeres

1994 LK_1. Discovered 1994 June 9 by E. F. Helin at Palomar.

Named in honor of Daniel J. Scheeres (1963-), currently in the Department of Aerospace Engineering Mechanics at Iowa State University. Scheeres has pioneered the investigation of the dynamics of orbits close to small, irregularly shaped minor planets. His research has included studies of the short-term evolution and the long-term stability of orbits around radar-derived models of (4179) Toutatis and (4769) Castalia. His work has far-reaching implications for the operation of spacecraft orbiting minor planets, for the cosmogony of satellites of minor planets and for understanding the distribution of non-escaping impact ejecta on small bodies. (M 32793)

(8888) 1994 NT_1

Discovered 1994 July 8 by E. W. Elst at Caussols.

(8889) 1994 OC
Discovered 1994 July 31 by Y. Shimizu and T. Urata at Nachi-Katsuura.

(8890) 1994 PS_{37}
Discovered 1994 August 10 by E. W. Elst at La Silla.

(8891) 1994 RC_1
Discovered 1994 September 1 by K. Endate and K. Watanabe at Kitami.

(8892) 1994 RC_{11}
Discovered 1994 September 11 by M. Sugano and T. Nomura at Minami-Oda.

(8893) 1995 KZ
Discovered 1995 May 23 by T. B. Spahr at Tucson.

(8894) 1995 PV
Discovered 1995 August 2 by Y. Shimizu and T. Urata at Nachi-Katsuura.

(8895) Nha
1995 QN. Discovered 1995 August 21 by K. Watanabe at Sapporo.

Named in honor of Nha Il-Seong (1932-), professor emeritus at Yonsei University, Korea, on the occasion of the dedication of his Museum of Astronomy in Yecheon. Professor Nha's major works include photometry of close binary stars, the discovery of the apsidal motion in the CW Cep system and contributions to the history of Asian astronomy. He has served as president of the Korean Astronomical Society and director of Yonsei University Observatory. (M 34349)

Name proposed by the discoverer following a suggestion by K. Hurukawa. Citation by I. Hasegawa.

(8896) 1995 QG_2
Discovered 1995 August 24 by Y. Shimizu and T. Urata at Nachi-Katsuura.

(8897) Defelice
1995 SX. Discovered 1995 September 22 at the Santa Lucia Observatory at Stroncone.

Named in memory of Aurelio De Felice (1915-1996), Italian sculptor known for going "against the flow" in his creations, which appear in museums around the world. (M 34628)

(8898) 1995 SL_5
Discovered 1995 September 29 by G. Emerson at Golden.

(8899) 1995 SX_{29}
Discovered 1995 September 22 by R. H. McNaught at Siding Spring.

(8900) 1995 UD_2
Discovered 1995 October 24 by D. di Cicco at Sudbury.

(8901) 1995 UJ_4
Discovered 1995 October 20 by T. Kobayashi at Oizumi.

(8902) 1995 UK_4
Discovered 1995 October 20 by T. Kobayashi at Oizumi.

(8903) 1995 UB_7
Discovered 1995 October 26 by Y. Shimizu and T. Urata at Nachi-Katsuura.

(8904) 1995 VY
Discovered 1995 November 15 by T. Kobayashi at Oizumi.

(8905) 1995 WJ
Discovered 1995 November 16 by T. Kobayashi at Oizumi.

(8906) 1995 WF_2
Discovered 1995 November 18 by T. Kobayashi at Oizumi.

(8907) 1995 WM_5
Discovered 1995 November 24 by T. Kobayashi at Oizumi.

(8908) 1995 WY_6
Discovered 1995 November 18 by S. Ueda and H. Kaneda at Kushiro.

(8909) 1995 WL_7
Discovered 1995 November 27 by T. Kobayashi at Oizumi.

(8910) 1995 WV_{42}
Discovered 1995 November 25 by S. Ueda and H. Kaneda at Kushiro.

(8911) 1995 YA
Discovered 1995 December 17 by T. Kobayashi at Oizumi.

(8912) 1995 YN_1
Discovered 1995 December 21 by T. Kobayashi at Oizumi.

(8913) 1995 YB_2
Discovered 1995 December 22 by the JPL NEAT Program at Haleakala.

(8914) 1995 YP_2
Discovered 1995 December 25 by B. G. W. Manning at Stakenbridge.

(8915) 1995 YK_3
Discovered 1995 December 27 by T. Kobayashi at Oizumi.

(8916) 1996 CC
Discovered 1996 February 1 at the Beijing Observatory at Xinglong.

(8917) 1996 EU_2
Discovered 1996 March 9 at the Beijing Observatory at Xinglong.

(8918) 1996 OR_1
Discovered 1996 July 20 at the Beijing Observatory at Xinglong.

(8919) 1996 TU_{13}
Discovered 1996 October 9 at the Beijing Observatory at Xinglong.

(8920) 1996 VZ_{29}
Discovered 1996 November 7 by S. Ueda and H. Kaneda at Kushiro.

(8921) 1996 VH_{30}
Discovered 1996 November 7 by S. Ueda and H. Kaneda at Kushiro.

(8922) 1996 VQ_{30}
Discovered 1996 November 10 by T. Okuni at Nanyou.

(8923) 1996 WQ_1
Discovered 1996 November 30 by T. Kobayashi at Oizumi.

(8924) 1996 XA_{32}
Discovered 1996 December 14 by N. Sato at Chichibu.

(8925) Boattini
1996 XG_{32}. Discovered 1996 December 4 by M. Tombelli and U. Munari at Cima Ekar.

Named in honor of Andrea Boattini (1969-). After developing a growing interest in minor planets, he graduated in 1996 from the University of Bologna with a thesis on near-earth objects. He is involved in various projects related to NEO follow-up and search programs, with special interest in the NEO class known as Atens {see planet (2062)}. He currently works at the Istituto di Astrofisica Spaziale of the National Research Council in Rome. The discoverers have started their astrometric activity on minor planets together with him. (M 33793)

(8926) 1996 YK
Discovered 1996 December 20 by T. Kobayashi at Oizumi.

(8927) 1996 YT
Discovered 1996 December 20 by T. Kobayashi at Oizumi.

(8928) 1996 YH_2
Discovered 1996 December 23 at the Beijing Observatory at Xinglong.

(8929) 1996 YQ_2
Discovered 1996 December 29 by T. Kobayashi at Oizumi.

(8930) 1997 AX_3
Discovered 1997 January 6 by T. Kobayashi at Oizumi.

(8931) 1997 AC_4
Discovered 1997 January 6 by T. Kobayashi at Oizumi.

(8932) 1997 AR_4
Discovered 1997 January 6 by T. Kobayashi at Oizumi.

(8933) 1997 AU_6
Discovered 1997 January 6 by N. Sato at Chichibu.

(8934) 1997 AQ_{12}
Discovered 1997 January 10 by T. Kobayashi at Oizumi.

(8935) 1997 AV_{13}
Discovered 1997 January 11 by P. Sicoli and M. Cavagna at Sormano.

(8936) 1997 AS_{17}
Discovered 1997 January 14 at the Farra d'Isonzo Observatory at Farra d'Isonzo.

(8937) 1997 AK_{19}
Discovered 1997 January 13 by T. Okuni at Nanyou.

(8938) 1997 AF_{21}
Discovered 1997 January 9 by S. Ueda and H. Kaneda at Kushiro.

(8939) 1997 BU_1
Discovered 1997 January 29 by T. Kobayashi at Oizumi.

(8940) 1997 BA_2
Discovered 1997 January 29 by T. Kobayashi at Oizumi.

(8941) 1997 BL_2
Discovered 1997 January 30 by T. Kobayashi at Oizumi.

(8942) 1997 BR_2
Discovered 1997 January 30 by T. Kobayashi at Oizumi.

(8943) 1997 BH_3
Discovered 1997 January 30 at the Santa Lucia Observatory at Stroncone.

(8944) Ortigara
1997 BF_9. Discovered 1997 January 30 by U. Munari and M. Tombelli at Cima Ekar.

Mount Ortigara, located near the Asiago Astrophysical Observatory, is one of the highest peaks in the Asiago Tableland and was the location of one of the most famous and bloodiest alpine battles of World War I. (M 32793)

(8945) Cavaradossi
1997 CM. Discovered 1997 February 1 by P. G. Comba at Prescott.

Named for a character in Puccini's {see planet (4579)} opera *Tosca*. Cavaradossi, a revolutionary and fighter against despotism, is imprisoned and condemned to die. While awaiting execution, he sings the nostalgic aria "E lucevan le stelle", recalling the sweet times he had with his mistress *Tosca* and bemoaning his impending death. Meanwhile, *Tosca* tries to gain his release by bargaining with the lecherous chief of police Scarpia. They double-cross each other, and everyone dies a violent death. (M 32793)

(8946) 1997 CO
Discovered 1997 February 1 by T. Kobayashi at Oizumi.

(8947) 1997 CH_{26}
Discovered 1997 February 14 by T. Kobayashi at Oizumi.

(8948) 1997 CW_{27}
Discovered 1997 February 6 at the Beijing Observatory at Xinglong.

(8949) 1997 CM_{28}
Discovered 1997 February 13 at the Beijing Observatory at Xinglong.

(8950) 1997 EG_{46}
Discovered 1997 March 15 at the Beijing Observatory at Xinglong.

(8951) 1997 FO
Discovered 1997 March 19 at the Beijing Observatory at Xinglong.

(8952) 1998 EG_2
Discovered 1998 March 2 by the OCA-DLR Survey at Caussols.

(8953) 1998 FC_{61}
Discovered 1998 March 20 by the Lincoln NEA Research Team at Socorro.

(8954) 1998 FK_{62}
Discovered 1998 March 20 by the Lincoln NEA Research Team at Socorro.

(8955) 1998 FR_{79}
Discovered 1998 March 24 by the Lincoln NEA Research Team at Socorro.

(8956) 1998 FN_{119}
Discovered 1998 March 31 by the Lincoln NEA Research Team at Socorro.

(8957) 1998 FM_{125}
Discovered 1998 March 22 by T. Seki at Geisei.

(8958) 1998 FJ_{126}

Discovered 1998 March 23 by J. Broughton at Reedy Creek.

(8959) Oenanthe

2550 P-L. Discovered 1960 September 24 by C. J. van Houten and I. van Houten-Groeneveld at Palomar.

Named for Oenanthe oenanthe, or wheatear. {It belongs to the family of the Turdidae. It is on the Dutch Red List of birds endangered in the Netherlands.} (M 33793; M 34089)

(8960) Luscinioides

2575 P-L. Discovered 1960 September 24 by C. J. van Houten and I. van Houten-Groeneveld at Palomar.

Named for Locustella luscinioides, or Savi's warbler. {It belongs to the family of the Sylviidae and is endangered in the Netherlands.} (M 33794; M 34089)

(8961) Schoenobaenus

2702 P-L. Discovered 1960 September 24 by C. J. van Houten and I. van Houten-Groeneveld at Palomar.

Named for Acrocephalus schoenobaenus, or sedge warbler. {It belongs to the family of Sylviidae and is on the Dutch Red List.} (M 33794; M 34089)

(8962) Noctua

2771 P-L. Discovered 1960 September 24 by C. J. van Houten and I. van Houten-Groeneveld at Palomar.

Named for Athene noctua, or little owl. {It belongs to the family of the Strigidae. It is on the Dutch Red List of endangered birds.} (M 33794; M 34089)

(8963) Collurio

4651 P-L. Discovered 1960 September 24 by C. J. van Houten and I. van Houten-Groeneveld at Palomar.

Named for Lanius collurio, or red-backed shrike. {This bird from the Dutch Red List of birds endangered in the Netherlands belongs to the family of the Laniidae.} (M 33794; M 34089)

(8964) Corax

7643 P-L. Discovered 1960 October 17 by C. J. van Houten and I. van Houten-Groeneveld at Palomar.

Named for Corvus corax, or raven. {It belongs to the family of the Corvidae and is an endangered bird in the Netherlands.} (M 33794; M 34089)

(8965) Citrinella

9511 P-L. Discovered 1960 October 17 by C. J. van Houten and I. van Houten-Groeneveld at Palomar.

Named for Emberiza citrinella, or yellowhammer. {It belongs to the family of the Emberizidae and is on the Dutch Red List.} (M 33794; M 34089)

(8966) Hortulana

3287 T-1. Discovered 1971 March 26 by C. J. van Houten and I. van Houten-Groeneveld at Palomar.

Named for Emberiza hortulana, or ortolan bunting. {It belongs to the family of the Emberizidae. It is on the Dutch Red List of birds endangered in the Netherlands and also on the 'List of Vulnerable Birds in Europe'.} (M 33794; M 34089)

(8967) Calandra

4878 T-1. Discovered 1971 May 13 by C. J. van Houten and I. van Houten-Groeneveld at Palomar.

Named for Miliaria calandra, or corn bunting. {It belongs to the family of the Emebrizidae and is on the Dutch Red List.} (M 33794; M 34089)

(8968) Europaeus

1212 T-2. Discovered 1973 September 29 by C. J. van Houten and I. van Houten-Groeneveld at Palomar.

Named for Caprimulgus europaeus, or nightjar. {It belongs to the Caprimulgidae family and is an endangered bird in the Netherlands.} (M 33794; M 34089)

(8969) Alexandrinus

1218 T-2. Discovered 1973 September 29 by C. J. van Houten and I. van Houten-Groeneveld at Palomar.

Named for Charadrius alexandrinus, or Kentish plover. {It belongs to the family of the Charadriidae and is on the Dutch Red List.} (M 33794; M 34089)

(8970) Islandica

1355 T-2. Discovered 1973 September 29 by C. J. van Houten and I. van Houten-Groeneveld at Palomar.

Named for Bucephala islandica, or Barrow's goldeneye. {It belongs to the family of the Anatidae and is one of the birds on the 'List of E ndangered Birds in Europe'. In 1994 only 200 pairs were counted.} (M 33794; M 34089)

(8971) Leucocephala

2256 T-2. Discovered 1973 September 29 by C. J. van Houten and I. van Houten-Groeneveld at Palomar.

Named for Oxyura leucocephala, or white-headed duck. {It belongs to the Anatidae family and is endangered in Europe. Im 1994 only 210 pairs were counted.} (M 33794; M 34089)

(8972) Sylvatica

2319 T-2. Discovered 1973 September 29 by C. J. van Houten and I. van Houten-Groeneveld at Palomar.

Named for Turnix sylvatica, or Andalusian hemipode. {It belongs to the Turnicidae family and it is one of the birds on the 'List of Endangered Birds in Europe'. In 1994, only five pairs were counted.} (M 33794; M 34089)

(8973) Pratincola

3297 T-2. Discovered 1973 September 30 by C. J. van Houten and I. van Houten-Groeneveld at Palomar.

Named for Glareola pratincola, or collared pratincole. {It belongs to the Glareolidae family and is endangered in Europe.} (M 33794; M 34089)

(8974) Gregaria

3357 T-2. Discovered 1973 September 25 by C. J. van Houten and I. van Houten-Groeneveld at Palomar.

Named for Chettusia gregaria, or sociable plover. {It belongs to the Chardadriidae family and is endangered in Europe. In 1994 only 1,000 pairs were counted.} (M 33794; M 34089)

(8975) Atthis

4076 T-2. Discovered 1973 September 29 by C. J. van Houten and I. van Houten-Groeneveld at Palomar.

Named for Alcedo atthis, or kingfisher. {It belongs to the family of the Alcedinidae and it is on the Dutch Red List of birds endangered in the Netherlands.} (M 33794; M 34089)

(8976) Leucura

4221 T-2. Discovered 1973 September 29 by C. J. van Houten and I. van Houten-Groeneveld at Palomar.

Named for Oenanthe leucura, or black wheatear. {It belongs to the Turdidae family and is endangered in Europe. In 1994 only 4,100 pairs were counted.} (M 33794; M 34089)

(8977) Paludicola

4272 T-2. Discovered 1973 September 29 by C. J. van Houten and I. van Houten-Groeneveld at Palomar.

Named for Acrocephalus paludicola, or aquatic warbler. {It belongs to the Sylviidae family. It is one of the birds on the 'List of Endangered Birds in Europe'. In 1994 only 3,700 pairs were counted.} (M 33794; M 34089)

(8978) Barbatus

3109 T-3. Discovered 1977 October 16 by C. J. van Houten and I. van Houten-Groeneveld at Palomar.

Named for the bird Gypaetus barbatus, or lammergeier. {It belongs to the Accipitridae family and it is an endangered bird in Europe. In 1994 only 200 pairs were counted.} (M 34349)

(8979) Clanga

3476 T-3. Discovered 1977 October 16 by C. J. van Houten and I. van Houten-Groeneveld at Palomar.

Named for the bird Aquila clanga, or greater spotted eagle. {It belongs to the family of the Accipitridae and it is endangered in Europe. In 1994 only 860 pairs were counted.} (M 34349)

(8980) Heliaca

4190 T-3. Discovered 1977 October 16 by C. J. van Houten and I. van Houten-Groeneveld at Palomar.

Named for the bird Aquila heliaca, or imperial eagle. {It belongs to the Accipitridae family. It is one of the birds on the 'List of Endangered Birds in Europe'. In 1994 only 320 pairs were counted.} (M 34349)

(8981) 1964 YJ

Discovered 1964 December 31 at the Purple Mountain Observatory at Nanking.

(8982) 1973 SQ_3

Discovered 1973 September 25 by L. V. Zhuravleva at Nauchnyj.

(8983) 1977 ED_1

Discovered 1977 March 13 by N. S. Chernykh at Nauchnyj.

(8984) 1977 QD_3

Discovered 1977 August 22 by N. S. Chernykh at Nauchnyj.

(8985) 1978 PV_3

Discovered 1978 August 9 by N. S. Chernykh and L. I. Chernykh at Nauchnyj.

(8986) Kineyayasuyo

1978 VN_2. Discovered 1978 November 1 by K. Tomita at Caussols.

Named in honor of Yasuyo Kineya, the stage name of Nobuko Shimamura (1920-), elder sister of the discoverer. Ever since her youth, she has devoted herself to Japanese traditional music, especially to Nagauta and Shamisen, and she is still active as a performer. The discoverer was inspired by her to become an astronomer. (M 34628)

(8987) 1978 VD_4
Discovered 1978 November 7 by E. F. Helin and S. J. Bus at Palomar.

(8988) 1979 MA_4
Discovered 1979 June 25 by E. F. Helin and S. J. Bus at Siding Spring.

(8989) 1979 XJ
Discovered 1979 December 15 by H. Debehogne and E. R. Netto at La Silla.

(8990) 1980 DN
Discovered 1980 February 19 by Z. Vávrová at Kleť.

(8991) 1980 PV_1
Discovered 1980 August 6 by R. M. West at La Silla.

(8992) 1980 TE_7
Discovered 1980 October 14 at the Purple Mountain Observatory at Nanking.

(8993) 1980 UL
Discovered 1980 October 30 by R. M. West at La Silla.

(8994) 1980 VG
Discovered 1980 November 6 by B. A. Skiff at Anderson Mesa.

(8995) 1981 EB_9
Discovered 1981 March 1 by S. J. Bus at Siding Spring.

(8996) 1981 EC_{10}
Discovered 1981 March 1 by S. J. Bus at Siding Spring.

(8997) 1981 ES_{14}
Discovered 1981 March 1 by S. J. Bus at Siding Spring.

(8998) 1981 EG_{23}
Discovered 1981 March 3 by S. J. Bus at Siding Spring.

(8999) 1981 EJ_{28}
Discovered 1981 March 2 by S. J. Bus at Siding Spring.

(9000) Hal
1981 JO. Discovered 1981 May 3 by E. Bowell at Anderson Mesa.

Named in honor of the computer Hal 9000 that is the central character of both Clarke's {see planet (4923)} novel and Kubrick's movie *2001: A Space Odyssey*. Hal serves to this day, more than three decades later, as an icon for artificial intelligence and a beacon that has motivated an incalculable number of careers in computing, computer science, electrical engineering and space exploration. (M 34628)

(9001) 1981 QE_2
Discovered 1981 August 30 by E. Bowell at Anderson Mesa.

(9002) 1981 QV_2
Discovered 1981 August 23 by H. Debehogne at La Silla.

(9003) 1981 UW_{21}
Discovered 1981 October 24 by S. J. Bus at Palomar.

(9004) 1982 UZ_2
Discovered 1982 October 22 by G. Aldering at Kitt Peak.

(9005) Sidorova

1982 UU_5. Discovered 1982 October 20 by L. G. Karachkina at Nauchnyj.

Named in honor of Sophia Ivanovna Sidorova (1943-), tireless worker on behalf of public education in Ukraine and leader of the Crimean Republic Committee for the education of workers. She is an educator of great ability, a brilliant organizer and a thoughtful wife and mother. (M 34349)

(9006) Voytkevych

1982 UA_7. Discovered 1982 October 21 by L. G. Karachkina at Nauchnyj.

Named in honor of Vanda Georgievna Voytkevych (1949-), a friend of the discoverer and specialist in analytical chemistry, particularly in its application to ecology and the metallurgy of welding. During 1971-1994 she was a researcher at the E. O. Paton Welding Institute. Now she lives in New Zealand with her family. (M 34628)

(9007) James Bond

1983 TE_1. Discovered 1983 October 5 by A. Mrkos at Kleť.

Named for British Secret Service agent James Bond (007), the creation of novelist Ian Fleming, a former British naval intelligence officer. Between 1952 and 1964, Fleming churned out twelve novels and two collections of short stories. Eighteen movies were based on the Bond legend, including *Dr. No*, *Goldeneye* and *Tomorrow Never Dies*, all with the compelling ingredients of action, seduction, humor and high technology. Two of the best-known Bond players are Sean Connery and Pierce Brosnan. (M 34349)
Name suggested by J. Tichá, M. Tichý and Z. Moravec.

(9008) Bohšternberk

1984 BS. Discovered 1984 January 27 by A. Mrkos at Kleť.

Named in memory of Bohumil Šternberk (1897-1983), outstanding Czech astronomer and astrophysicist. He worked in a wide variety of fields, including photometry of variable stars, and he established the modern time service in Czechoslovakia. On the evening of 1930 Mar. 18, at the Stará Ďala Observatory (O'Gyalla {see planet (1259)}, now Hurbanovo), he made the first deliberate astrometric observations of Pluto in Europe. Šternberk served as long-time director of the Astronomical Institute of the Czechoslovak Academy of Sciences (1954-1968) and as a vice-president of the IAU (1958-1964). (M 34628)

Name suggested by J. Tichá, M. Tichý and Z. Moravec, who made observations of this minor planet at Kleť during its oppositions in 1997 and 1998.

Obituaries published in Říse hvezd, Vol. 64, p. 106-107 (1983); Vesmír, Vol. 62, p. 156 (1983); Q.J.R. Astron. Soc., Vol. 25, No. 3, p. 379-380 (1984); Bull. Astron. Inst. Czech., Vol. 34, No. 4, p. 193-194 (1983).

(9009) 1984 HJ_1

Discovered 1984 April 23 by W. Ferreri and V. Zappalà at La Silla.

(9010) 1984 HM_1

Discovered 1984 April 27 by W. Ferreri and V. Zappalà at La Silla.

(9011) 1984 SU

Discovered 1984 September 20 by A. Mrkos at Kleť.

(9012) 1984 UW

Discovered 1984 October 26 by E. Bowell at Anderson Mesa.

(9013) 1985 PA_1

Discovered 1985 August 14 by E. Bowell at Anderson Mesa.

(9014) 1985 UG_5
Discovered 1985 October 22 by L. V. Zhuravleva at Nauchnyj.

(9015) 1985 VK
Discovered 1985 November 14 by P. Jensen at Brorfelde.

(9016) Henrymoore
1986 AE. Discovered 1986 January 10 by C. S. Shoemaker and E. M. Shoemaker at Palomar.

Named in memory of Henry J. Moore (1928-1998). A geologist with the U.S. Geological Survey since 1960, Moore took part in the fledling astrogeology program helping to train the Apollo astronauts in impact geology prior to the moon landings. He participated in the analysis of the lunar samples they brought back and worked with the photographs of the moon's surface taken from orbit. His experience in the selection of lunar landing sites transferred to the Mars Viking project. Moore's work on the Mars Pathfinder mission and with its Mars rover, Sojourner, were among his last contributions. (M 34349)

(9017) 1986 TW_9
Discovered 1986 October 2 by L. V. Zhuravleva at Nauchnyj.

(9018) 1987 JG
Discovered 1987 May 5 by A. C. Gilmore and P. M. Kilmartin at Lake Tekapo.

(9019) Eucommia
1987 QF_3. Discovered 1987 August 28 by E. W. Elst at La Silla.

Named for Eucommiaceae, a famliy with only one member, the elmlike Eucommia ulmoides, a vigorous and decorative plant. It is the only tree from temperate regions that produces latex - in small quantities. (M 34349)

(9020) Eucryphia
1987 SG_2. Discovered 1987 September 19 by E. W. Elst at Rozhen.

Named for Eucryphiaceae, a family with five species in a single genus. Eucryphia cordifolia (Chilean elm) may reach a height of 12m. Hybrids between the different species are cultivated for the garden. (M 34349)

(9021) Fagus
1988 CT_5. Discovered 1988 February 14 by E. W. Elst at La Silla.

Named for Fagaceae, the beech family, with eight genera and over 1000 species, including oaks and chestnuts. Well-known species are Fagus grandifolia (American beech) and Fagus sylvatica (European beech). (M 34349)

(9022) 1988 PC_1
Discovered 1988 August 14 by C. S. Shoemaker at Palomar.

(9023) 1988 RG_1
Discovered 1988 September 10 by C. S. Shoemaker at Palomar.

(9024) 1988 RF_9
Discovered 1988 September 5 by H. Debehogne at La Silla.

(9025) 1988 SM_2
Discovered 1988 September 16 by S. J. Bus at Cerro Tololo.

(9026) 1988 ST_2
Discovered 1988 September 16 by S. J. Bus at Cerro Tololo.

(9027) 1988 VP_5
Discovered 1988 November 4 by A. Mrkos at Kleť.

(9028) 1989 BE_1
Discovered 1989 January 26 by A. Mrkos at Kleť.

(9029) 1989 GM
Discovered 1989 April 6 by E. F. Helin at Palomar.

(9030) 1989 UX_5
Discovered 1989 October 30 by S. J. Bus at Cerro Tololo.

(9031) 1989 WG_4
Discovered 1989 November 29 by A. Mrkos at Kleť.

(9032) 1989 WK_4
Discovered 1989 November 23 by T. Seki at Geisei.

(9033) 1990 AD
Discovered 1990 January 4 by M. Akiyama and T. Furuta at Mishima.

(9034) 1990 QZ_{17}
Discovered 1990 August 26 by L. V. Zhuravleva at Nauchnyj.

(9035) 1990 SH_1
Discovered 1990 September 16 by H. E. Holt at Palomar.

(9036) 1990 SJ_{16}
Discovered 1990 September 17 by H. E. Holt at Palomar.

(9037) 1990 UJ_2
Discovered 1990 October 20 by A. Sugie at Taga.

(9038) Helensteel
1990 VE_1. Discovered 1990 November 12 by D. I. Steel at Siding Spring.
Named in honor of Helen Margaret Steel (née Pope; 1959-), wife of the discoverer and a native of Adelaide, South Australia. (M 34628)

(9039) 1990 WB_4
Discovered 1990 November 16 by Y. Mizuno and T. Furuta at Kani.

(9040) Flacourtia
1991 BH_1. Discovered 1991 January 18 by E. W. Elst at St. Michel.
Named for Flacourtiaceae, the Indian plum family, with almost 90 genera and 900 species. Flacourtia indica (governor's plum) is one of the species, and the southeast Asian genus Hydnocarpus produces an oil that is used for treating some skin diseases. (M 34350)

(9041) 1991 CX
Discovered 1991 February 9 by S. Otomo and O. Muramatsu at Kiyosato.

(9042) 1991 EN_2
Discovered 1991 March 11 by H. Debehogne at La Silla.

(9043) 1991 EJ_4
Discovered 1991 March 12 by H. Debehogne at La Silla.

(9044) Kaoru
1991 KA. Discovered 1991 May 18 by S. Otomo and O. Muramatsu at Kiyosato.
Named in honor of Kaoru Kimura (1964-), lecturer and curator of the Gotoh Planetarium and Astronomical Museum in Tokyo, where she was in charge of an

exhibition about the history of constellations. She traveled abroad several times to observe solar eclipses. (M 34350)

(9045) 1991 PG_{15}
Discovered 1991 August 7 by H. E. Holt at Palomar.

(9046) 1991 PG_{17}
Discovered 1991 August 9 by H. E. Holt at Palomar.

(9047) 1991 QF
Discovered 1991 August 30 by R. H. McNaught at Siding Spring.

(9048) 1991 RD_{24}
Discovered 1991 September 12 by H. E. Holt at Palomar.

(9049) 1991 RQ_{27}
Discovered 1991 September 12 by H. E. Holt at Palomar.

(9050) 1991 RF_{29}
Discovered 1991 September 13 by H. E. Holt at Palomar.

(9051) 1991 UG_3
Discovered 1991 October 31 by S. Ueda and H. Kaneda at Kushiro.

(9052) Uhland
1991 UJ_4. Discovered 1991 October 30 by F. Börngen at Tautenburg.

Named for the German poet Johann Ludwig Uhland (1787-1862), best known for lyric poetry and ballads in which historical and legendary themes dominate, and which were set to music by well-known composers. Uhland is the most famous representative of Swabian Romanticism. In his lifetime, his "Gedichte" (1815) were printed in 42 editions. He also engaged himself in political issues, and in 1848-49 was a member of the Frankfurt parliament. (M 32793)

(9053) Hamamelis
1991 VW_5. Discovered 1991 November 2 by E. W. Elst at La Silla.

Named for Hamamelidaceae, the witch hazel family, with about 25 genera and 100 species, typical in South America and Africa. Hamamelis virginiana (witch hazel) is not a true hazel, although the leaves are similar. (M 34350)

(9054) Hippocastanum
1991 YO. Discovered 1991 December 30 by E. W. Elst at St. Michel.

Named for Hippocastanaceae, the horse-chestnut family, with only two genera and 15 species. Aesculus hippocastanum (common horse-chestnut) is known for its creamy yellow flowers, in large, conical, upright panicles. (M 34350)

(9055) 1992 DP_8
Discovered 1992 February 29 by the Uppsala-ESO Survey at La Silla.

(9056) 1992 EQ_{14}
Discovered 1992 March 1 by the Uppsala-ESO Survey at La Silla.

(9057) 1992 HA_5
Discovered 1992 April 24 by H. Debehogne at La Silla.

(9058) 1992 JB
Discovered 1992 May 1 by J. Alu and K. J. Lawrence at Palomar.

(9059) 1992 PJ
Discovered 1992 August 8 by E. W. Elst at Caussols.

(9060) 1992 RM
Discovered 1992 September 4 by S. Otomo at Kiyosato.

(9061) 1992 WC$_3$
Discovered 1992 November 18 by A. Sugie at Taga.

(9062) 1992 WO$_5$
Discovered 1992 November 27 by T. Seki at Geisei.

(9063) 1992 YS
Discovered 1992 December 17 by T. Seki at Geisei.

(9064) 1993 BH$_8$
Discovered 1993 January 21 by the Spacewatch at Kitt Peak.

(9065) 1993 FN$_1$
Discovered 1993 March 25 by S. Ueda and H. Kaneda at Kushiro.

(9066) 1993 FR$_{34}$
Discovered 1993 March 19 by the Uppsala-ESO Survey at La Silla.

(9067) 1993 HR
Discovered 1993 April 16 by K. Endate and K. Watanabe at Kitami.

(9068) 1993 OD
Discovered 1993 July 16 by E. F. Helin at Palomar.

(9069) 1993 OV
Discovered 1993 July 16 by E. F. Helin at Palomar.

(9070) 1993 OZ$_2$
Discovered 1993 July 23 by C. S. Shoemaker at Palomar.

(9071) 1993 OB$_{13}$
Discovered 1993 July 19 by E. W. Elst at La Silla.

(9072) 1993 RX$_3$
Discovered 1993 September 12 at the Palomar Observatory at Palomar.

(9073) 1994 ER
Discovered 1994 March 4 by T. Kobayashi at Oizumi.

(9074) 1994 FZ
Discovered 1994 March 31 by K. Endate and K. Watanabe at Kitami.

(9075) 1994 GD$_9$
Discovered 1994 April 14 by E. F. Helin at Palomar.

(9076) Shinsaku
1994 JT. Discovered 1994 May 8 by A. Nakamura at Kuma.

Named in memory of Shinsaku Takasugi (1839-1867), Japanese revolutionary, who though a lower-ranking samurai, attracted many followers with his innovative ideas and extraordinary drive to modernize Japan. As a primary leader of anti-shogunate forces, he played the most important role in the success of the Meiji Restoration. (M 33389)

(9077) Ildo
1994 NC. Discovered 1994 July 3 at the Farra d'Isonzo Observatory at Farra d'Isonzo.

Named in memory of Ildo Lombardi (1934-1954), Italian gymnast and brother of Giuseppe Lombardi, a member of the team of amateur astronomers who discovered this minor planet. Ildo died by a fall from the parallel bars while training with the Italian team for the Melbourne Olympic games in 1956. (M 33794)

(9078) 1994 PB_2
Discovered 1994 August 9 at the Palomar Observatory at Palomar.

(9079) 1994 PC_{34}
Discovered 1994 August 10 by E. W. Elst at La Silla.

(9080) 1994 TP
Discovered 1994 October 2 by K. Endate and K. Watanabe at Kitami.

(9081) Hideakianno
1994 VY. Discovered 1994 November 3 by A. Nakamura at Kuma.

Named in honor of Hideaki Anno (1960-), Japanese animator and director, born in Ube City {see planet (7716)}, Yamaguchi Prefecture. Since 1981, he has contributed to the production of several highly successful works of animation. (M 33794)

(9082) 1994 VR_6
Discovered 1994 November 4 by C. S. Shoemaker at Palomar.

(9083) 1994 WC_4
Discovered 1994 November 28 by C. S. Shoemaker at Palomar.

(9084) 1995 CS_1
Discovered 1995 February 3 by D. J. Asher at Siding Spring.

(9085) 1995 QH_2
Discovered 1995 August 24 by Y. Shimizu and T. Urata at Nachi-Katsuura.

(9086) 1995 SA_3
Discovered 1995 September 20 by S. Ueda and H. Kaneda at Kushiro.

(9087) Neff
1995 SN_3. Discovered 1995 September 29 at the Kleť Observatory at Kleť.

Named for Vladimír Neff and Ondřej Neff. Vladimír Neff (1909-1983) was an outstanding Czech writer, well-known for his five-part chronicle, spanning the mid-nineteenth to the mid-twentieth century, of the Prague entrepreneur family, the Borns and the Nedobyls. His last work was a great ironic novel *The Vesture of Mr. de Balzac*. Ondřej Neff (1945-) is a Czech writer of science fiction, a journalist and son of Vladimír Neff. In 1996 he founded the first Czech independent Internet daily newspaper called *Neviditelný pes*, or "The invisible Dog". He continues this work today. (M 34350)

Name propsoed by discoverers J. Tichá, M. Tichý and Z. Moravec.

(9088) 1995 SX_3
Discovered 1995 September 20 by K. Endate and K. Watanabe at Kitami.

(9089) 1995 UC_7
Discovered 1995 October 26 by Y. Shimizu and T. Urata at Nachi-Katsuura.

(9090) Chirotenmondai
1995 UW_8. Discovered 1995 October 28 by K. Endate and K. Watanabe at Kitami.

Named for Chiro Astronomical Observatory (Chiro tenmondai) in Western Australia. Founded in 1995 by the joint efforts of Australian and Japanese amateur astronomers, it is also operated by the two groups. It is considered a sister observatory to the Japanese amateurs' Shirakawa Observatory, the two sites being at

approximately the same latitude in opposite hemispheres. (M 33389)
Name proposed by the discoverers following a suggestion by L. Taylor and T. Sato.

(9091) 1995 VK
Discovered 1995 November 2 by T. Kobayashi at Oizumi.

(9092) 1995 VU_{18}
Discovered 1995 November 4 at the Beijing Observatory at Xinglong.

(9093) 1995 WA
Discovered 1995 November 16 by T. Kobayashi at Oizumi.

(9094) 1995 WH
Discovered 1995 November 16 by T. Kobayashi at Oizumi.

(9095) 1995 WT_2
Discovered 1995 November 16 by S. Ueda and H. Kaneda at Kushiro.

(9096) 1995 XE_1
Discovered 1995 December 15 by T. Kobayashi at Oizumi.

(9097) Davidschlag
1996 AU_1. Discovered 1996 January 14 at the Davidschlag Observatory at Linz.
Named for a small rural village, some 10 km to the north of Linz {see planet (1469)}, at the entrance to a region known as "Sterngartl", or "small garden of stars". This object is the first minor planet discovered at the amateur astronomical observatory that is located in this village. (M 34628)

(9098) 1996 BQ_3
Discovered 1996 January 27 by T. Kobayashi at Oizumi.

(9099) 1996 VN_3
Discovered 1996 November 6 by T. Kobayashi at Oizumi.

(9100) 1996 XU_1
Discovered 1996 December 2 by T. Kobayashi at Oizumi.

(9101) 1996 XG_2
Discovered 1996 December 3 at the Farra d'Isonzo Observatory at Farra d'Isonzo.

(9102) 1996 XS_{18}
Discovered 1996 December 12 at the Kleť Observatory at Kleť.

(9103) 1996 XW_{30}
Discovered 1996 December 14 by T. Kobayashi at Oizumi.

(9104) 1996 YB
Discovered 1996 December 20 by T. Kobayashi at Oizumi.

(9105) 1997 AU
Discovered 1997 January 2 by T. Kobayashi at Oizumi.

(9106) 1997 AY_1
Discovered 1997 January 3 by T. Kobayashi at Oizumi.

(9107) 1997 AE_4
Discovered 1997 January 6 by T. Kobayashi at Oizumi.

(9108) 1997 AZ_6
Discovered 1997 January 9 by T. Kobayashi at Oizumi.

(9109) 1997 AH_7
Discovered 1997 January 9 by T. Kobayashi at Oizumi.

(9110) 1997 AM_{19}
Discovered 1997 January 13 by T. Okuni at Nanyou.

(9111) 1997 BD_2
Discovered 1997 January 28 by P. Sicoli and F. Manca at Sormano.

(9112) 1997 BU_3
Discovered 1997 January 31 by T. Kobayashi at Oizumi.

(9113) 1997 CN_5
Discovered 1997 February 3 by Y. Shimizu and T. Urata at Nachi-Katsuura.

(9114) 1997 CU_{19}
Discovered 1997 February 12 by T. Kobayashi at Oizumi.

(9115) Battisti
1997 DG. Discovered 1997 February 27 by P. Sicoli and F. Manca at Sormano.

Named in memory of Lucio Battisti (1943-1998), innovative Italian composer and singer. He wrote many songs for himself and other singers, among them "Emozioni" and "I giardini di marzo". Beginning in 1973, he lived in a small village in the neighborhood of the observatory where this minor planet was discovered. (M 33389)

(9116) 1997 ES_{40}
Discovered 1997 March 7 by M. W. Buie at Anderson Mesa.

(9117) Aude
1997 FR_1. Discovered 1997 March 27 by D. Morata and S. Morata at Martigues.

Named in honor of the French "Association des Utilisateurs de Detecteurs Electroniques" (Electronic Detectors User's Association). Founded by Christian Buil and Alain Maury {see, respectively, planets (6820) and (3780)} with the aim of assisting amateur astronomers in the use of electronic detectors, the organization is now heavily involved with the discovery and follow-up of solar-system objects and photometry of supernovae. (M 33794)

(9118) 1997 GD_{20}
Discovered 1997 April 5 by the Lincoln NEA Research Team at Socorro.

(9119) Georgpeuerbach
1998 DT. Discovered 1998 February 18 at the Davidschlag Observatory at Linz.

Named in memory of Georg Aunpekh von Peuerbach (1423-1461), professor at the University of Vienna, astronomer at the court of emperor Friedrich III, mathematician, poet, early humanist and teacher of Regiomontanus {see planet (9307)}. He discovered the magnetic declination, introduced sines into trigonometry and invented the foldable sundial, the first trustworthy pocket timepiece. (M 34350)

Named on the occasion of the unveiling of a memorial tablet on St. Stephan's cathedral in Vienna, where Peuerbach is buried.

(9120) 1998 DR_8
Discovered 1998 February 22 at the Beijing Observatory at Xinglong.

(9121) 1998 DJ_{11}
Discovered 1998 February 24 by V. S. Casulli at Colleverde di Guidonia.

(9122) 1998 FZ_8
Discovered 1998 March 22 by the Spacewatch at Kitt Peak.

(9123) 1998 FQ_{11}
Discovered 1998 March 24 by T. Kagawa at Gekko.

(9124) 1998 FR_{60}
Discovered 1998 March 20 by the Lincoln NEA Research Team at Socorro.

(9125) 1998 FT_{62}
Discovered 1998 March 20 by the Lincoln NEA Research Team at Socorro.

(9126) 1998 FR_{64}
Discovered 1998 March 20 by the Lincoln NEA Research Team at Socorro.

(9127) Brucekoehn
1998 HX_{51}. Discovered 1998 April 30 by the LONEOS at Anderson Mesa.

Named in honor of Bruce W. Koehn (1948-), who has developed much of the software used in the operation of LONEOS, the Lowell Observatory Near-Earth Object Search program. Among his many contributions, he devised moving-object detection and field-sequencing algorithms and has been central to the increasingly automated operation of LONEOS. Koehn has also written much of the code used in the minor-planet services at the Lowell Observatory's website. (M 34628)
Name suggested and citation written by E. Bowell.

(9128) 1998 HQ_{52}
Discovered 1998 April 30 by T. Okuni at Nanyou.

(9129) 1998 HU_{144}
Discovered 1998 April 21 by the Lincoln NEA Research Team at Socorro.

(9130) Galois
1998 HQ_{148}. Discovered 1998 April 25 by E. W. Elst at La Silla.

Named in memory of the famous French mathematician Evariste Galois (1811-1832), well known for his contributions to group theory. At the age of only 16, he launched himself into a deeper understanding of the essential conditions that an equation must satisfy to be solvable by radicals (Galois theory). He died in Paris in a duel, under circumstances that have never been fully explained, although his republican activities and sympathies were well known. (M 33794)

(9131) 1998 JV
Discovered 1998 May 1 by the JPL NEAT Program at Haleakala.

(9132) Walteranderson
2821 P-L. Discovered 1960 September 24 by C. J. van Houten and I. van Houten-Groeneveld at Palomar.

Named in honor of Walter Anderson, chief executive officer of Entreé International and long-time supporter of space exploration, most notably via the International Space University. (M 34350)
Name proposed and citation prepared by SBNC.

(9133) d'Arrest
3107 P-L. Discovered 1960 September 25 by C. J. van Houten and I. van Houten-Groeneveld at Palomar.

Named in memory of Heinrich Louis d'Arrest (1822-1875), German astronomer. As an assistant astronomer at the Berlin Observatory, he, together with J. G. Galle {see planet (2097)}, discovered the planet Neptune in 1846. From 1852 d'Arrest served as an astronomer at the Leipzig Observatory and was appointed

professor of astronomy. In 1857 he went to Denmark as director of the Copenhagen Observatory, where he discovered (79) Freia. He detected several comets, among them a famous short-period comet, and he was a prolific observer of minor planets and nebulae. (M 34350; M 34385)
Name proposed and citation prepared by L. D. Schmadel.

(9134) Encke

4822 P-L. Discovered 1960 September 24 by C. J. van Houten and I. van Houten-Groeneveld at Palomar.

Named in memory of Johann Franz Encke (1791-1865), eminent German astronomer. Beginning in 1816, Encke was assistant professor and director of the Seeberg Observatory, near Gotha {see, respectively planets (8130) and (1346)}. He computed the orbit of a short-period comet discovered by Pons {see planet (7645)} and demonstrated that this object had been observed repeatedly. Later it was named comet 2P/Encke. In 1825 Encke was appointed director of the Berlin Observatory and member of the Berlin Academy of Sciences. Encke is most famous for editing the *Berliner Astronomisches Jahrbuch* from 1830 to 1866. This contained several very important papers on orbit determination and perturbation computations. (M 34350; M 34385)
Name proposed and citation prepared by L. D. Schmadel.

(9135) Lacaille

7609 P-L. Discovered 1960 October 17 by C. J. van Houten and I. van Houten-Groeneveld at Palomar.

Named in memory of Nicholas-Louis de Lacaille (1713-1762), French astronomer known as the "father of southern astronomy" for his naming of southern constellations and his countless astrometric observations of planets and stars. In 1750, the abbé Lacaille went to an expedition to the Cape for trigonometric determinations of the lunar and solar parallax, his observations being partly supported by simultaneous measurements by Lalande {see planet (9136)} in Berlin. Lacaille's observations of the southern heavens resulted in a famous catalogue of 9766 stars, a work that was eventually published in 1847 by Baily {see planet (3115)}. Lacaille is also known as a precise observer of comet 1P/Halley in 1759. (M 34350; M 34385)
Name proposed and citation prepared by L. D. Schmadel.

(9136) Lalande

4886 T-1. Discovered 1971 May 13 by C. J. van Houten and I. van Houten-Groeneveld at Palomar.

Named in memory of Joseph-Jérôme Lefrançais de Lalande (1732-1807), famous French astronomer who computed many useful tables for practical astronomical work and authored many papers and textbooks. After some years as an assistant to Delisle, Le Monnier and Lacaille {see planet (9135)}, Lalande went to Berlin in order to prepare measurements for determining the lunar parallax. He was a member of the Berlin and Paris academies of science and was editor of many volumes of the almanac *Connaissance des temps* during 1760-1776 and 1794-1807. His monumental *Bibliographie astronomique* (1802) has lasted for two centuries. (M 34350; M 34385)
Name proposed and citation prepared by L. D. Schmadel.

(9137) Remo

2114 T-2. Discovered 1973 September 29 by C. J. van Houten and I. van Houten-Groeneveld at Palomar.

Named in honor of John L. Remo (1941-), U.S. physicist who has contributed to laser resonator optics, quantum optics instrumentation, energy systems and materials interactions. He has also helped promote an international interest in policy toward the detction of earth-threatening minor planets and comets and their mitigation, notably with his organization of a conference on the subject at the United Nations {see planet (6000)} in New York in 1995 and publication of the proceedings. (M 34629)

(9138) Murdoch

2280 T-2. Discovered 1973 September 29 by C. J. van Houten and I. van Houten-Groeneveld at Palomar.

Named in memory of Jean Iris Murdoch (1919-1999), prolific novelist and philosopher, whose 26 moralist-existentialist fictional works in 41 years earned her a place among the leading twentieth-century writers in the English language. Already suffering from Alzheimer's disease when her last novel was published in 1995, Murdoch was lovingly cared for by her husband, the critic John Bayley, whose recent *Elegy for Iris* is a penetrating study of conflicting emotions. (M 34350)
Name proposed and citation prepared by SBNC.

(9139) Barrylasker

4180 T-2. Discovered 1973 September 29 by C. J. van Houten and I. van Houten-Groeneveld at Palomar.

Named in memory of U.S. astronomer Barry M. Lasker (1939-1999), whose long-time association with the Space Telescope Science Institute includes the writing of the proposal to establish the Institute. Under his leadership the *Guide Star Catalogue* and the *Digitized Sky Survey* were created for Space Telescope operations. By leading the construction of these momentous enterprises, he provided tools of very broad astronomical use, including the astrometry of minor planets. (M 34351)
Name proposed and citation prepared by SBNC.

(9140) Deni

4195 T-3. Discovered 1977 October 16 by C. J. van Houten and I. van Houten-Groeneveld at Palomar.

Named after the Department of Education for Northern Ireland for its support of the Armagh Observatory, where Ernst Öpik {see planet (2099)} carried out seminal work on the collisional evolution and dynamics of small bodies. The DENI has promoted life-long learning and research across many disciplines, encompassing higher education, schools, museums, recreation and culture. It has sought for many years to advance knowledge and understanding of the natural world, and to provide the means by which as many people as possible may appreciate and continue to develop Northern Ireland's rich cultural heritage, its diversity and contributions to the scientific, intellectual and social life of the community. (M 34351)
Name proposed and citation prepared by SBNC.

(9141) Kapur

5174 T-3. Discovered 1977 October 16 by C. J. van Houten and I. van Houten-Groeneveld at Palomar.

Named in honor of Shekhar Kapur (1945-), Indian actor and Bombay film director whose recent acclaim has put "Bollywood" on the international map. The 800 feature films produced annually in Bombay represent nearly 20 percent of the world's total – and are twice as many as come from Hollywood. Kapur's 1998 movie *Elizabeth* received seven Oscar and 12 British Academy nominations. The showing of his 1994 movie *The Bandit Queen*, a true story about a low-caste woman who fought back after being gang-raped and was eventually elected to the Indian parliament, was initially banned in India. (M 34629)

(9142) Rhesus

5191 T-3. Discovered 1977 October 16 by C. J. van Houten and I. van Houten-Groeneveld at Palomar.

Named for the king of the Thracians and ally of the Trojans. He and his twelve companions were stealthily assassinated in their sleep by Odysseus and Diomedes {see, respectively, planets (1143) and (1437)}. (M 33389)

(9143) 1955 SF

Discovered 1955 September 16 at the Goethe Link Observatory at Brooklyn, Indiana.

(9144) 1955 UN_1

Discovered 1955 October 25 at the Goethe Link Observatory at Brooklyn, Indiana.

(9145) 1976 GG_3

Discovered 1976 April 1 by N. S. Chernykh at Nauchnyj.

(9146) 1976 YG_1

Discovered 1976 December 16 by L. I. Chernykh at Nauchnyj.

(9147) 1977 DD_1

Discovered 1977 February 18 by H. Kosai and K. Hurukawa at Kiso.

(9148) 1977 EL_1

Discovered 1977 March 13 by N. S. Chernykh at Nauchnyj.

(9149) 1977 TD_1

Discovered 1977 October 12 by P. Wild at Zimmerwald.

(9150) 1978 SE_1

Discovered 1978 September 27 by L. I. Chernykh at Nauchnyj.

(9151) 1979 MQ_8

Discovered 1979 June 25 by E. F. Helin and S. J. Bus at Siding Spring.

(9152) 1980 VZ_2

Discovered 1980 November 1 by S. J. Bus at Palomar.

(9153) 1981 UD_2

Discovered 1981 October 30 by H. Kosai and K. Hurukawa at Kiso.

(9154) 1982 SP_6

Discovered 1982 September 16 by L. I. Chernykh at Nauchnyj.

(9155) 1982 SM_7

Discovered 1982 September 18 by N. S. Chernykh at Nauchnyj.

(9156) Malanin

1982 TQ_2. Discovered 1982 October 15 by L. G. Karachkina at Nauchnyj.

Named in honor of Ivan Ivanovich Malanin (1897-1969), talented Russian accordionist, well known for his solo concerts throughout Siberia. He has also been an accompanist at performances of the Russian singers Lemeshev, Nezhdanova {see, respectively, planets (4561) and (4361)}, Obukhova and Pantofel'-Nechetskaya. (M 34629)

(9157) 1983 RB_4

Discovered 1983 September 2 by N. G. Thomas at Anderson Mesa.

(9158) 1984 MR

Discovered 1984 June 25 by T. M. Smirnova at Nauchnyj.

(9159) 1984 UD_3
Discovered 1984 October 26 by E. Bowell at Anderson Mesa.

(9160) 1986 UH_3
Discovered 1986 October 28 by Z. Vávrová at Kleť.

(9161) 1987 BZ_1
Discovered 1987 January 26 by E. W. Elst at La Silla.

(9162) 1987 OA
Discovered 1987 July 29 by J. Mueller at Palomar.

(9163) 1987 RB_1
Discovered 1987 September 13 by H. Debehogne at La Silla.

(9164) 1987 SQ
Discovered 1987 September 19 by E. Bowell at Anderson Mesa.

(9165) 1987 SJ_3
Discovered 1987 September 27 by C. S. Shoemaker and E. M. Shoemaker at Palomar.

(9166) 1987 SC_6
Discovered 1987 September 21 by Z. Vávrová at Kleť.

(9167) 1987 SS_{17}
Discovered 1987 September 18 by L. I. Chernykh at Nauchnyj.

(9168) 1987 ST_{17}
Discovered 1987 September 18 by L. I. Chernykh at Nauchnyj.

(9169) 1988 TL_1
Discovered 1988 October 13 by S. Ueda and H. Kaneda at Kushiro.

(9170) 1988 TG_5
Discovered 1988 October 3 by S. Ueda and H. Kaneda at Kushiro.

(9171) 1989 GD_5
Discovered 1989 April 4 by E. W. Elst at La Silla.

(9172) 1989 OB
Discovered 1989 July 29 by C. S. Shoemaker at Palomar.

(9173) 1989 TZ_{15}
Discovered 1989 October 4 by H. Debehogne at La Silla.

(9174) 1989 WC_3
Discovered 1989 November 27 by Y. Oshima at Gekko.

(9175) 1990 OO_2
Discovered 1990 July 29 by H. E. Holt at Palomar.

(9176) 1990 VC_{15}
Discovered 1990 November 15 by L. I. Chernykh at Nauchnyj.

(9177) 1990 YA
Discovered 1990 December 18 by E. F. Helin at Palomar.

(9178) 1991 DU
Discovered 1991 February 23 by S. Inoda and T. Urata at Karasuyama.

(9179) 1991 EM_1
Discovered 1991 March 13 at the Oak Ridge Observatory at Harvard.

(9180) 1991 GQ
Discovered 1991 April 8 by E. F. Helin at Palomar.

(9181) 1991 NP_2
Discovered 1991 July 14 by H. E. Holt at Palomar.

(9182) 1991 NB_4
Discovered 1991 July 8 by H. Debehogne at La Silla.

(9183) 1991 OW
Discovered 1991 July 18 by H. E. Holt at Palomar.

(9184) 1991 PJ_3
Discovered 1991 August 2 by E. W. Elst at La Silla.

(9185) 1991 PX_{17}
Discovered 1991 August 7 by H. E. Holt at Palomar.

(9186) 1991 RZ_1
Discovered 1991 September 7 by E. F. Helin at Palomar.

(9187) 1991 RD_4
Discovered 1991 September 12 by L. D. Schmadel and F. Börngen at Tautenburg.

(9188) 1991 RM_{15}
Discovered 1991 September 15 by H. E. Holt at Palomar.

(9189) Hölderlin
1991 RH_{41}. Discovered 1991 September 10 by F. Börngen at Tautenburg.
Named for the German poet Friedrich Hölderlin (1770-1843), who lived mainly in the Swabian town Tübingen {see planet (1481)}. In 1795, he studied in Jena {see planet (526)} and there wrote some of his patriotic hymns. He is one of the most original and well-known representatives of German idealism. His philosophy and vision shaped much of the poetry of the German language and were not fully appreciated until the twentieth century. (M 33795)

(9190) Masako
1991 VR_1. Discovered 1991 November 4 by Y. Kushida and O. Muramatsu at Yatsugatake.
Named in honor of Masako Muramatsu, wife of the second discoverer. (M 33795)

(9191) 1991 XU
Discovered 1991 December 13 by S. Otomo at Kiyosato.

(9192) 1992 AR_1
Discovered 1992 January 14 by S. Ueda and H. Kaneda at Kushiro.

(9193) Geoffreycopland
1992 ED_1. Discovered 1992 March 10 by D. I. Steel at Siding Spring.
Named for Geoffrey Malcolm Copland (1942-), rector and vice-chancellor of the University of Westminster. Copland, an Oxford-educated physicist, has dedicated his career to the benefit of the students under his care, initially within the University of London, but more recently at the above institution. (M 34629) Name proposed by the discoverer, who owes much to Copland's tutelage.

(9194) 1992 OV_2
Discovered 1992 July 26 by E. W. Elst at La Silla.

(9195) 1992 OF_9
Discovered 1992 July 26 by H. Debehogne and A. López G. at La Silla.

(9196) 1992 WP_5
Discovered 1992 November 27 by T. Seki at Geisei.

(9197) Endo
1992 WH_8. Discovered 1992 November 24 by M. Hirasawa and S. Suzuki at Nyukasa.

Named in honor of Shu Endo (1953-), one of Japan's leading astrophotographers. Endo's astrophotographs are famous for their beauty and detail. Many astrophotographers have been inspired by his ideas and advice. The films used to discover this minor planet were hypersensitized by him. (M 34351)

(9198) 1993 BJ_3
Discovered 1993 January 25 by T. Seki at Geisei.

(9199) 1993 FO_1
Discovered 1993 March 25 by S. Ueda and H. Kaneda at Kushiro.

(9200) 1993 FK_{21}
Discovered 1993 March 21 by the Uppsala-ESO Survey at La Silla.

(9201) 1993 FU_{39}
Discovered 1993 March 19 by the Uppsala-ESO Survey at La Silla.

(9202) 1993 PB
Discovered 1993 August 13 by the Spacewatch at Kitt Peak.

(9203) Myrtus
1993 TM_{16}. Discovered 1993 October 9 by E. W. Elst at La Silla.

Named for Myrtaceae, the myrtle family of aromatic trees and shrubs, with more than 100 genera and nearly 4000 species. This family is typical of the Southern Hemisphere, where the Eucalyptus genus is prevalent. Myrtus communis (common myrtle) is a species native to the Mediterranean and is cultivated as far north as England. (M 34351)

(9204) Mörike
1994 PZ_1. Discovered 1994 August 4 by F. Börngen at Tautenburg.

Named for the Swabian poet Eduard Mörike (1804-1875), one of the great German lyric poets. The wonders of nature were revealed to him everywhere, even in the smallest thing. More than 50 of his poems were set to music by Hugo Wolf {see planet (5177)}, notably "At midnight". His most beautiful work in prose is the short novel "Mozarts Reise nach Prag", written in a fluid, breathtaking style similar to Mozart's {see planet (1034)} music. (M 33795)

(9205) Eddywally
1994 PO_9. Discovered 1994 August 10 by E. W. Elst at La Silla.

Named in honor of Eduard Van de Walle (1930-), Flemish singer known as the "Voice of Europe". In a career that spans more than half a century, he recently scored a record sale of one million singles of his song "Chérie". He is also the first well-known Western musician to sing in Chinese. In Ertvelde (near Ghent) he operates the club "Paris-Vegas", which each weekend is visited by many fans. (M 33795)

(9206) 1994 RQ
Discovered 1994 September 1 by K. Endate and K. Watanabe at Kitami.

(9207) 1994 SF_{12}
Discovered 1994 September 29 by the Spacewatch at Kitt Peak.

(9208) 1994 TX_2
Discovered 1994 October 2 by K. Endate and K. Watanabe at Kitami.

(9209) 1994 UK_1
Discovered 1994 October 25 by S. Ueda and H. Kaneda at Kushiro.

(9210) 1995 BW_2
Discovered 1995 January 27 by S. Ueda and H. Kaneda at Kushiro.

(9211) 1995 SB_{27}
Discovered 1995 September 19 by the Spacewatch at Kitt Peak.

(9212) 1995 UR_3
Discovered 1995 October 20 by T. Kobayashi at Oizumi.

(9213) 1995 UX_5
Discovered 1995 October 21 by S. Ueda and H. Kaneda at Kushiro.

(9214) 1995 UC_6
Discovered 1995 October 21 by S. Ueda and H. Kaneda at Kushiro.

(9215) 1995 UB_{45}
Discovered 1995 October 28 by K. Endate and K. Watanabe at Kitami.

(9216) 1995 VS
Discovered 1995 November 1 by S. Otomo at Kiyosato.

(9217) 1995 WN
Discovered 1995 November 16 by T. Kobayashi at Oizumi.

(9218) 1995 WV_2
Discovered 1995 November 20 by T. Kobayashi at Oizumi.

(9219) 1995 WO_8
Discovered 1995 November 18 by Y. Shimizu and T. Urata at Nachi-Katsuura.

(9220) 1995 XL_1
Discovered 1995 December 15 by T. Kobayashi at Oizumi.

(9221) 1995 XP_2
Discovered 1995 December 2 at the Beijing Observatory at Xinglong.

(9222) 1995 YM
Discovered 1995 December 19 by T. Kobayashi at Oizumi.

(9223) 1995 YY_7
Discovered 1995 December 18 by the Spacewatch at Kitt Peak.

(9224) 1996 AE
Discovered 1996 January 10 at the Kleť Observatory at Kleť.

(9225) 1996 AU
Discovered 1996 January 10 by T. Kobayashi at Oizumi.

(9226) 1996 AB_1
Discovered 1996 January 12 by T. Kobayashi at Oizumi.

(9227) 1996 BO_2
Discovered 1996 January 26 by T. Kobayashi at Oizumi.

(9228) 1996 CG_1
Discovered 1996 February 11 by T. Kobayashi at Oizumi.

(9229) 1996 DJ_1
Discovered 1996 February 20 by K. Endate and K. Watanabe at Kitami.

(9230) 1996 YY_2
Discovered 1996 December 29 by N. Sato at Chichibu.

(9231) 1997 BB_2
Discovered 1997 January 29 by T. Kobayashi at Oizumi.

(9232) Miretti
1997 BG_8. Discovered 1997 January 31 by V. Goretti at Pianoro.

Named in memory of Manlio Miretti (1928-1996), who, blessed with a powerful and warm baritone voice, devoted his whole life to the opera, even though he achieved a very high position in a bank. After studying with Gino Bechi's maestro, Raul Frazzi, and Valiano Natali, he performed the entire baritone repertoire. He loved Giuseppe Verdi's {see planet (3975)} music and characters best, especially "Rigoletto". He also successfully sang Toschi's classical songs, the Neapolitan ones and the "musica proibita". He sang in the theaters of Losanna and Bologna and at La Scala of Milan, and he also had a very busy concert career. (M 33795)

(9233) 1997 CC_1
Discovered 1997 February 1 by T. Kobayashi at Oizumi.

(9234) 1997 CH_4
Discovered 1997 February 3 by T. Kobayashi at Oizumi.

(9235) Shimanamikaido
1997 CT_{21}. Discovered 1997 February 9 by A. Nakamura at Kuma.

Named for a toll road, the Nishiseto Expressway, whose nickname is Shimanami Kaido. The third road connecting the islands of Honshu and Shikoku {see planet (4223)}, Shimanami Kaido is 60 km long and includes ten long-span bridges. Scheduled for use starting 1999 May 1, it is expected to encourage economic activity by increasing the interchange of goods and personnel in the Nishiseto area. (M 34629)

(9236) Obermair
1997 EV_{32}. Discovered 1997 March 12 by E. Meyer at Linz.

Named in honor of the Austrian amateur astronomer Erwin Obermair (1946-), who, together with the discoverer, is co-owner of the private observatory in Davidschlag, near Linz {see, respectively, planets (9097) and (1469)}. A technician by profession, Obermair is also a well-known astrophotographer and popularizer of astronomy. (M 34629)

(9237) 1997 GY_7
Discovered 1997 April 2 by the Lincoln NEA Research Team at Socorro.

(9238) Yavapai
1997 HO_2. Discovered 1997 April 28 by P. G. Comba at Prescott.

Named for the county in Arizona of which Prescott is the county seat. The county is named for the Yavapai tribe of American Indians that has lived in the region since the times preceding the European conquest. (M 33795)

(9239) 1997 JP_{15}

Discovered 1997 May 3 by E. W. Elst at La Silla.

(9240) 1997 KR_3

Discovered 1997 May 31 by the Spacewatch at Kitt Peak.

(9241) 1997 PE_6

Discovered 1997 August 10 by J. Broughton at Reedy Creek.

(9242) Olea

1998 CS_3. Discovered 1998 February 6 by E. W. Elst at La Silla.

Named for Oleaceae, the olive family, with about 25 genera and nearly 1000 species. Genera include ash, lilac, jasmine and forsythia. Olea europea (common olive) is profitable as a fruit and source of oil. (M 34351)

(9243) 1998 FF_{68}

Discovered 1998 March 20 by the Lincoln NEA Research Team at Socorro.

(9244) 1998 HV_7

Discovered 1998 April 21 at the Višnjan Observatory at Višnjan.

(9245) 1998 HF_{101}

Discovered 1998 April 21 by the Lincoln NEA Research Team at Socorro.

(9246) 1998 HB_{149}

Discovered 1998 April 25 by E. W. Elst at La Silla.

(9247) 1998 MO_{19}

Discovered 1998 June 23 by the Lincoln NEA Research Team at Socorro.

(9248) Sauer

4593 P-L. Discovered 1960 September 24 by C. J. van Houten and I. van Houten-Groeneveld at Palomar.

Named in honor of Carl G. Sauer, Jr., a principal flight mechanics engineer at NASA's Jet Propulsion Laboratory. Since 1952, when he began work at JPL, Carl has been a leader in the development of the analysis tools used for designing the ballistic and low-thrust spacecraft missions that explore the solar system. His work has included mission designs for advanced spacecraft propulsion systems, including solar electric ion propulsion, nuclear propulsion and solar sailing. His database of potential spacecraft trajectories to hundreds of comets and minor planets is an invaluable resource for mission design work at JPL. (M 34351)
Name proposed and citation prepared by SBNC.

(9249) Yen

4606 P-L. Discovered 1960 September 24 by C. J. van Houten and I. van Houten-Groeneveld at Palomar.

Named in honor of Chen-wan L. Yen, a senior analyst within the Mission and Systems Architecture Section of NASA's Jet Propulsion Laboratory. Chen-wan has done seminal work in the development and application of optimization techniques to interplanetary trajectories. Using multiple-impulse and gravity-assist techniques, her optimized interplanetary trajectories have allowed significant payloads to be launched within current launch vehicle capabilities. Her work is evident in the interplanetary trajectories designed for the Galileo mission to Jupiter, the Magellan mission to Venus, the Cassini mission to Saturn, and the Stardust mis-

sion to return a dust sample from comet 81P/Wild 2. (M 34351)
Name proposed and citation prepared by SBNC.

(9250) Chamberlin

4643 P-L. Discovered 1960 September 24 by C. J. van Houten and I. van Houten-Groeneveld at Palomar.

Named in honor of Alan B. Chamberlin, a senior engineer within the Navigation and Flight Mechanics Section of NASA's Jet Propulsion Laboratory. Since joining the staff of JPL in 1996, Alan has shown his expertise in asrtonomical observing and celestial mechanics. His research work has centered on the search for gaseous emissions from suspected defunct comets and the uncertainties associated with the close approaches to the earth of the so-called potentially hazardous objects. For the New Millennium Deep Space One spacecraft that was launched on 1998 Oct. 25, Alan is responsible for the ephemeris development for both the mission targets and the two dozen well-observed minor planets that will be used by this spacecraft to navigate autonomously to its targets. (M 34351)
Name proposed and citation prepared by SBNC.

(9251) Harch

4896 P-L. Discovered 1960 September 26 by C. J. van Houten and I. van Houten-Groeneveld at Palomar.

Named in honor of the Cornell University scientist Ann P. Harch, an expert in the design and execution of spacecraft camera sequencing and pointing commands. Ann played a major role in obtaining the first spacecraft images of (951) Gaspra and (243) Ida during the Galileo spacecraft flybys in Oct. 1991 and Aug. 1993. During the latter encounter, she was the first to note evidence for Ida's moon, Dactyl. In addition, Ann was instrumental in the successful imaging of (253) Mathilde and (433) Eros during the NEAR spacecraft flybys in June 1997 and Dec. 1998. (M 34351)
Name proposed and citation prepared by SBNC.

(9252) Goddard

9058 P-L. Discovered 1960 October 17 by C. J. van Houten and I. van Houten-Groeneveld at Palomar.

Named in memory of the American rocket pioneer Robert H. Goddard (1882-1945). Beginning in 1906, Goddard combined theory and practice in a long career of building and testing rockets. While he dedicated himself to the development of rockets, including the first test of a liquid fuel rocket in 1926, Goddard never lost sight of its ultimate purpose - the exploration of space. (M 34351)
Name proposed and citation prepared by SBNC.

(9253) Oberth

1171 T-1. Discovered 1971 March 25 by C. J. van Houten and I. van Houten-Groeneveld at Palomar.

Named in memory of the German pioneer of space flight Hermann J. Oberth (1894-1989). In his 1923 work entitled *Die Rakete zu den Planetenräumen*, Oberth gave a thorough discussion of many phases of rocket travel, including the launching of payloads into earth orbit and the abnormal effects of pressure on the human body. (M 34351)
Name proposed and citation prepared by SBNC.
Obituary published in Spaceflight, Vol. 32, No. 4, p. 144 (1990).

(9254) Shunkai

2151 T-1. Discovered 1971 March 25 by C. J. van Houten and I. van Houten-Groeneveld at Palomar.

Named in memory of Shibukawa Shunkai (1639-1715), who was an expert on the calendar. Born into a family of "go" teachers in Kyoto, he studied the calendar system and astronomy and found that the lunisolar calendar, which had been used without any modification for more than 800 years after it was imported from China, did not match observations. He revised the calendar using observations of the sun and moon, and his calendar system was adopted by the Tokugawa government in 1684. He was then nominated as "The Astronomer", which was a new title for him. The title was inherited by his family. The Tokugawa government also established the observatory with Shunkai as the director. (M 34352)
Name proposed and citation prepared by SBNC.

(9255) Inoutadataka

3174 T-1. Discovered 1971 March 26 by C. J. van Houten and I. van Houten-Groeneveld at Palomar.

Named in memory of Inou Tadataka (1745-1818), a very able geographer and surveyor. He began to study astronomy and surveying methods after he retired from his official job at the age of 50. He began surveying in Hokkaido {see planet (3720)} at his own expense. Since the map he made was found to be excellent, the Tokugawa government asked him to continue the survey in Honshu. He completed his survey of the Japanese coastal area by measuring astronomical latitudes at 1200 places, and he derived the length of the meridional arc of one degree to be 110.8 km. His map, smuggled to Europe on a Durch ship, won wide respect there. (M 34352)
Name proposed and citation prepared by SBNC.

(9256) Tsukamoto

1324 T-2. Discovered 1973 September 29 by C. J. van Houten and I. van Houten-Groeneveld at Palomar.

Named in memory of Tsukamoto Akitake (1833-1885), a geographer who worked for both the Tokugawa and Meiji governments. He recommended that the Japanese government adopt the Gregorian calendar to replace the lunisolar calendar. The government accepted his advice, and Japan adopted the Gregorian calendar on 1873 Jan. 1. (M 34352)
Name proposed and citation prepared by SBNC.

(9257) Kunisuke

1552 T-2. Discovered 1973 September 24 by C. J. van Houten and I. van Houten-Groeneveld at Palomar.

Named in memory of Kunisuke Kinoshita (1901-1931), an astronomer at the Tokyo Astronomical Observatory from 1924 to his death. At the observatory he worked with Oikawa {see planet (2667)} and discovered 1927 EB and 1927 FC. However, after he was sent to Malaysia in 1929 to observe the total eclipse, he became ill and stopped most of his work in 1930. He also wrote excellent articles on modern astronomy in the *Astronomical Herald*, the monthly journal of the Astronomical Society of Japan, for which he served as editor after 1929. (M 34352)
Name proposed and citation prepared by SBNC.

(9258) 2137 T-2

Discovered 1973 September 29 by C. J. van Houten and I. van Houten-Groeneveld at Palomar.

(9259) 2189 T-2

Discovered 1973 September 29 by C. J. van Houten and I. van Houten-Groeneveld at Palomar.

(9260) 1953 TA_1
Discovered 1953 October 8 at the Goethe Link Observatory at Brooklyn, Indiana.

(9261) 1953 TD_1
Discovered 1953 October 8 at the Goethe Link Observatory at Brooklyn, Indiana.

(9262) 1973 RF
Discovered 1973 September 6 by T. M. Smirnova at Nauchnyj.

(9263) 1976 SX_5
Discovered 1976 September 24 by N. S. Chernykh at Nauchnyj.

(9264) 1978 OQ
Discovered 1978 July 28 at the Perth Observatory at Bickley.

(9265) 1978 RC_9
Discovered 1978 September 2 by C.-I. Lagerkvist at La Silla.

(9266) 1978 RD_{10}
Discovered 1978 September 2 by C.-I. Lagerkvist at La Silla.

(9267) 1978 RL_{10}
Discovered 1978 September 2 by C.-I. Lagerkvist at La Silla.

(9268) 1978 VZ_2
Discovered 1978 November 7 by E. F. Helin and S. J. Bus at Palomar.

(9269) 1978 VW_6
Discovered 1978 November 7 by E. F. Helin and S. J. Bus at Palomar.

(9270) 1978 VO_8
Discovered 1978 November 7 by E. F. Helin and S. J. Bus at Palomar.

(9271) 1978 VT_8
Discovered 1978 November 7 by E. F. Helin and S. J. Bus at Palomar.

(9272) 1979 KQ
Discovered 1979 May 19 by R. M. West at La Silla.

(9273) 1979 QW_3
Discovered 1979 August 22 by C.-I. Lagerkvist at La Silla.

(9274) 1980 FF_3
Discovered 1980 March 16 by C.-I. Lagerkvist at La Silla.

(9275) 1980 FS_3
Discovered 1980 March 16 by C.-I. Lagerkvist at La Silla.

(9276) 1980 RB_8
Discovered 1980 September 13 by S. J. Bus at Palomar.

(9277) 1980 TT_3
Discovered 1980 October 9 by C. S. Shoemaker at Palomar.

(9278) 1981 EM_1
Discovered 1981 March 7 by H. Debehogne and G. DeSanctis at La Silla.

(9279) 1981 EY_{12}
Discovered 1981 March 1 by S. J. Bus at Siding Spring.

(9280) 1981 EQ_{14}
Discovered 1981 March 1 by S. J. Bus at Siding Spring.

(9281) 1981 EJ_{15}
Discovered 1981 March 1 by S. J. Bus at Siding Spring.

(9282) 1981 EP_{16}
Discovered 1981 March 6 by S. J. Bus at Siding Spring.

(9283) 1981 EY_{17}
Discovered 1981 March 2 by S. J. Bus at Siding Spring.

(9284) 1981 ED_{24}
Discovered 1981 March 7 by S. J. Bus at Siding Spring.

(9285) 1981 EL_{24}
Discovered 1981 March 2 by S. J. Bus at Siding Spring.

(9286) 1981 ED_{35}
Discovered 1981 March 2 by S. J. Bus at Siding Spring.

(9287) 1981 ER_{43}
Discovered 1981 March 6 by S. J. Bus at Siding Spring.

(9288) 1981 EV_{46}
Discovered 1981 March 2 by S. J. Bus at Siding Spring.

(9289) 1981 QR_3
Discovered 1981 August 26 by H. Debehogne at La Silla.

(9290) 1981 TT
Discovered 1981 October 6 by Z. Vávrová at Kleť.

(9291) 1982 QO
Discovered 1982 August 17 at the Oak Ridge Observatory at Harvard.

(9292) 1982 UE_2
Discovered 1982 October 16 by A. Mrkos at Kleť.

(9293) 1982 XQ_1
Discovered 1982 December 13 by H. Kosai and K. Hurukawa at Kiso.

(9294) 1983 EV
Discovered 1983 March 10 by E. Barr at Anderson Mesa.

(9295) 1983 RT_1
Discovered 1983 September 2 by E. Bowell at Anderson Mesa.

(9296) 1983 RB_2
Discovered 1983 September 5 by Z. Vávrová at Kleť.

(9297) 1984 MP
Discovered 1984 June 25 by T. M. Smirnova at Nauchnyj.

(9298) 1985 JM
Discovered 1985 May 15 by E. Bowell at Anderson Mesa.

(9299) 1985 JG_2
Discovered 1985 May 13 by C. S. Shoemaker at Palomar.

(9300) 1985 PS
Discovered 1985 August 14 by E. Bowell at Anderson Mesa.

(9301) 1985 RB_4
Discovered 1985 September 10 by H. Debehogne at La Silla.

(9302) 1985 TB_3
Discovered 1985 October 12 by P. Wild at Zimmerwald.

(9303) 1986 QH_3
Discovered 1986 August 29 by H. Debehogne at La Silla.

(9304) 1986 RA_5
Discovered 1986 September 1 by H. Debehogne at La Silla.

(9305) 1986 TR_1
Discovered 1986 October 7 by E. Bowell at Anderson Mesa.

(9306) Pittosporum
1987 CG. Discovered 1987 February 2 by E. W. Elst at La Silla.

Named for Pittosporaceae, the Australian laurel family, with nine genera and more than 200 species, mainly in tropical Africa and Pacific islands. Among the species is Pittosporum eugenioides (lemonwood). (M 34352)

(9307) Regiomontanus
1987 QS. Discovered 1987 August 21 by F. Börngen at Tautenburg.

Named for the German mathematician and astronomer Johannes Regiomontanus (1436-1476), originally called J. Müller, one of the most famous scholars of his time and trailblazer of the new world view. He improved mathematical methods and created modern trigonometry. He was sure that calculations of the orbits of celestial bodies could be improved essentially by new, more exact systematic observations. So in 1472 he founded the first German observatory in Nürnberg. His plan was interrupted by his early death. Calendars and ephemeris he calculated and published helped the sailors Columbus, da Gama and Vespucci. (M 33795)

(9308) 1987 SD_4
Discovered 1987 September 21 by E. Bowell at Anderson Mesa.

(9309) Platanus
1987 SS_9. Discovered 1987 September 20 by E. W. Elst at Rozhen.

Named ofr Platanaceae, the plane family, with one genus and seven species. They grow wild mainly in the United States and Mexico, but they are also typical of southern France. Platanus acerifolia (London plane) reaches a height of almost 30m and is widely planted in cities because of its resistance to air pollution. The brown, bristly fruits, hanging two to four together on a single stalk, persist over the winter. (M 34352)

(9310) 1987 SV_{12}
Discovered 1987 September 18 by H. Debehogne at La Silla.

(9311) 1987 UV_1
Discovered 1987 October 25 by S. Ueda and H. Kaneda at Kushiro.

(9312) 1987 VE_2
Discovered 1987 November 15 by S. Ueda and H. Kaneda at Kushiro.

(9313) Protea
1988 CH_3. Discovered 1988 February 13 by E. W. Elst at La Silla.

Named for Proteaceae, the firebush family, with some 75 genera and over 1000 species. They are native to the Southern Hemisphere, where they grow wild. Some species, such as Protea mellifera (honeyflower), extend also to the warm regions of the Northern Hemisphere. The family is best known for its ornamental plants, such as Embothrium coccineum (Chilean firebush), which are cultivated for their edible nuts. (M 34352)

(9314) 1988 DJ_1

Discovered 1988 February 19 by Y. Oshima at Gekko.

(9315) Weigel

1988 PP_2. Discovered 1988 August 13 by F. Börngen at Tautenburg.

Named for the German mathematician and astronomer Erhard Weigel (1625-1699) on the occasion of the 300th anniversary of his death. A lecturer at the University of Jena {see planet (526)} beginning in 1652, his larger-than-life personality and inventive mind contributed essentially to the prominence of this university in his time. He created, for example, giant globes with diameters of several meters, in which one could set foot and admire the artificial firmament in the daytime. In 1670, his seven-storeyed dwelling-house was completed, built according to his plans and supporting an observation platform. This unusual house was one of the "Seven Miracles of Jena". (M 33795)

(9316) Rhamnus

1988 PX_2. Discovered 1988 August 12 by E. W. Elst at St. Michel.

Named for Rhamnaceae, the buckthorn family of deciduous and evergreen trees, shrubs and climbing plants, with 60 genera and around 900 species. They grow wild in all parts of the world. Some of the species yield dyes. The berrylike fruits appear in dense clusters. Rhamnus purshina (bearwood) produces a dry bark (cascara sagrada) that is used as a laxative. (M 34352)

(9317) 1988 RO_4

Discovered 1988 September 1 by H. Debehogne at La Silla.

(9318) 1988 RG_9

Discovered 1988 September 6 by H. Debehogne at La Silla.

(9319) 1988 RV_{11}

Discovered 1988 September 14 by S. J. Bus at Cerro Tololo.

(9320) 1988 VN_3

Discovered 1988 November 11 by Y. Oshima at Gekko.

(9321) 1989 AK

Discovered 1989 January 5 by T. Kojima at Chiyoda.

(9322) Lindenau

1989 AC_7. Discovered 1989 January 10 by F. Börngen at Tautenburg.

Named for the German scientist, statesman and promoter of the arts Bernhard August von Lindenau (1779-1854). From 1808 to 1819, he was the director of the Seeberg {see planet (8130)} Observatory. He communicated closely with contemporaries such as Goethe, A. von Humboldt, von Zach, Gauss and Bessel {see, respectively, planets (3047), (4877), (999), (1001) and (1552)}. At the beginning of the nineteenth century, he fostered the founding of several astronomical periodicals. Lindenau was a member of numerous renowned European academies. From 1830 to 1843, he served as a high public official in the Kingdom of Saxony. He donated his art collection, including a unique collection of early Italian panel paintings, to his hometown of Altenburg {see planet (9336)} to be displayed in the Lindenau Museum. (M 33795)

(9323) 1989 CV_1
Discovered 1989 February 11 by T. Seki at Geisei.

(9324) 1989 CH_4
Discovered 1989 February 7 by S. Ueda and H. Kaneda at Kushiro.

(9325) 1989 GG_4
Discovered 1989 April 3 by E. W. Elst at La Silla.

(9326) Ruta
1989 SP_2. Discovered 1989 September 26 by E. W. Elst at La Silla.

Named for Rutaceae, the rue family, with over 150 genera and 1500 species, including citrus fruits. Ruta graviolensis (common rue) is a perennial herb with a particularly strong scent. The family thrives in tropical and warm, temperate regions. (M 34352)

(9327) 1989 SW_2
Discovered 1989 September 26 by E. W. Elst at La Silla.

(9328) 1990 DL_3
Discovered 1990 February 24 by H. Debehogne at La Silla.

(9329) 1990 EO
Discovered 1990 March 2 by E. W. Elst at La Silla.

(9330) 1990 EF_7
Discovered 1990 March 3 by H. Debehogne at La Silla.

(9331) 1990 QM_9
Discovered 1990 August 16 by E. W. Elst at La Silla.

(9332) 1990 SB_1
Discovered 1990 September 16 by H. E. Holt at Palomar.

(9333) 1990 TK_3
Discovered 1990 October 15 by K. Endate and K. Watanabe at Kitami.

(9334) 1990 UU_3
Discovered 1990 October 16 by E. W. Elst at La Silla.

(9335) 1991 AA_1
Discovered 1991 January 10 by Y. Kushida and O. Muramatsu at Yatsugatake.

(9336) Altenburg
1991 AY_2. Discovered 1991 January 15 by F. Börngen at Tautenburg.

Named for the town of Altenburg, noted for its outstanding buildings (e.g. the theater and Lindenau {see planet (9322)} Museum) and its eventful 1000-year history. It was first the palace of German kings and emperors (Ottonian, Hohenstaufen), later an imperial city and the capital of the duchy Sachsen-Altenburg. The Saxonian princes resided in this place for 600 years, until 1918. Altenburg is also renowned as the "town of playing cards", which have been produced here for 450 years. In 1813, "Skat", the chief card game of the Germans, was created here by impassioned players associated with the chancellor Hans Carl Leopold von der Gabelentz. (M 33795)

(9337) 1991 FO_1
Discovered 1991 March 17 by H. Debehogne at La Silla.

(9338) 1991 FL_4
Discovered 1991 March 25 by H. Debehogne at La Silla.

(9339) 1991 GT_5
Discovered 1991 April 8 by E. W. Elst at La Silla.

(9340) 1991 LW_1
Discovered 1991 June 6 by E. W. Elst at La Silla.

(9341) 1991 PH_2
Discovered 1991 August 2 by E. W. Elst at La Silla.

(9342) 1991 PJ_7
Discovered 1991 August 6 by E. W. Elst at La Silla.

(9343) 1991 PO_{11}
Discovered 1991 August 9 by H. E. Holt at Palomar.

(9344) Klopstock
1991 RB_4. Discovered 1991 September 12 by F. Börngen and L. D. Schmadel at Tautenburg.

Named for Friedrich Gottfried Klopstock (1724-1803), one of the most famous German poets between the late Baroque and the classical period. He studied theology in Jena {see planet (526)} and Leipzig. In his poems he proclaimed a new mentality – the Age of Sensibility. Klopstock's "Messias", a cycle of 20 religious poems, is the first great epic of the Low High German literature. His inspired odes have love, friendship, fatherland and the experience of nature as their subjects. (M 33795; M 33817)
Name proposed and citation prepared by the first discoverer.

(9345) 1991 RA_{10}
Discovered 1991 September 12 by H. E. Holt at Palomar.

(9346) 1991 RN_{11}
Discovered 1991 September 4 by E. W. Elst at La Silla.

(9347) 1991 RY_{21}
Discovered 1991 September 15 by H. E. Holt at Palomar.

(9348) 1991 RH_{25}
Discovered 1991 September 11 by H. E. Holt at Palomar.

(9349) Lucas
1991 SX. Discovered 1991 September 30 by R. H. McNaught at Siding Spring.

Named in memory of François Edouard Anatole Lucas (1842-1891), French number theorist, best known for his test for Mersenne {see planet (8191)} primes and the demonstration that 2^{127}-1 is prime. His interests included magic squares and Fibonacci {see planet (6765)} numbers, and while studying the latter he devised the similar sequence 2, 1, 3, 4, 7,..., now known as Lucas numbers. The number of this minor planet is the twentieth member of this sequence. (M 33389)
Name proposed by the discoverer following a suggestion by B. G. Marsden.

(9350) Waseda
1991 TH_2. Discovered 1991 October 13 by M. Hirasawa and S. Suzuki at Nyukasa.

Named for one of the most prestigious private universities in Japan. It was established in 1882 by Shigenobu Okuma, who later became the eighth and the seventeenth prime minister of Japan. One of the primary objectives of the university was to promote the independence of study and spirit of progress known as Waseda Spirits. Both discoverers of this minor planet graduated from Waseda. (M 34352)

(9351) 1991 TH_6
Discovered 1991 October 2 by L. D. Schmadel and F. Börngen at Tautenburg.

(9352) 1991 UB_4
Discovered 1991 October 31 by S. Ueda and H. Kaneda at Kushiro.

(9353) 1991 VM_4
Discovered 1991 November 9 by A. Sugie at Taga.

(9354) 1991 VF_7
Discovered 1991 November 11 by S. Ueda and H. Kaneda at Kushiro.

(9355) 1991 XO_2
Discovered 1991 December 5 by S. Ueda and H. Kaneda at Kushiro.

(9356) 1991 YV
Discovered 1991 December 30 by E. W. Elst at St. Michel.

(9357) 1992 AT_3
Discovered 1992 January 11 by O. A. Naranjo at Mérida.

(9358) 1992 DN_7
Discovered 1992 February 29 by the Uppsala-ESO Survey at La Silla.

(9359) 1992 ED_{11}
Discovered 1992 March 6 by the Uppsala-ESO Survey at La Silla.

(9360) 1992 EV_{13}
Discovered 1992 March 2 by the Uppsala-ESO Survey at La Silla.

(9361) 1992 EM_{18}
Discovered 1992 March 3 by the Uppsala-ESO Survey at La Silla.

(9362) 1992 FE_1
Discovered 1992 March 23 by K. Endate and K. Watanabe at Kitami.

(9363) 1992 GR
Discovered 1992 April 3 by S. Ueda and H. Kaneda at Kushiro.

(9364) Clusius
1992 HZ_3. Discovered 1992 April 23 by E. W. Elst at La Silla.

Named for Carolus Clusius (Charles de l'Escluse; 1526-1609), a Flemish botanist well-known for his catalogue, published in Antwerp in 1576, of fine drawings of more than 200 plants he had studied in southern France, Portugal and Spain. His system of classification proposed the idea of natural families. The Clusiaceae, a family of plants and trees, was named after him. It is said that he introduced the tulip into Holland. (M 34352)

(9365) Chinesewilson
1992 RU_3. Discovered 1992 September 2 by E. W. Elst at La Silla.

Named for Ernest "Chinese" Wilson (1876-1930) a prolific hunter of plants, who introduced more than 1000 new plants from China into Europe. Because of him Europeans have the Azalea {see planet (1056)}, Hortensia and Magnolia wilsonii {see also planet (1060)}. He lost his leg while hunting the lily Lilium regale. (M 34352)

(9366) 1992 WR_1

Discovered 1992 November 17 by A. Sugie at Taga.

(9367) 1993 BO_3

Discovered 1993 January 30 by A. Natori and T. Urata at Yakiimo.

(9368) Esashi

1993 BS_3. Discovered 1993 January 26 by M. Mukai and M. Takeishi at Kagoshima.

Named for a small town in northern Hokkaido, on the sea of Okhotsk {see planets (3720) and (4042), respectively}, not far from the home of the second discoverer. Known as "The Town of the Solar Eclipse", it was visited by astronomers from all over the world during the total solar eclipses of 1896 Aug. 9 and 1936 June 19. After the former eclipse the American astronomer David P. Todd gave many books to the town, and these became the nucleus of the first public library to be established in Hokkaido. (M 34353)

(9369) 1993 DB_1

Discovered 1993 February 20 by T. Hioki and S. Hayakawa at Okutama.

(9370) 1993 FC_{22}

Discovered 1993 March 21 by the Uppsala-ESO Survey at La Silla.

(9371) 1993 FV_{31}

Discovered 1993 March 19 by the Uppsala-ESO Survey at La Silla.

(9372) 1993 FK_{37}

Discovered 1993 March 19 by the Uppsala-ESO Survey at La Silla.

(9373) 1993 FY_{43}

Discovered 1993 March 19 by the Uppsala-ESO Survey at La Silla.

(9374) 1993 FJ_{46}

Discovered 1993 March 19 by the Uppsala-ESO Survey at La Silla.

(9375) 1993 HK

Discovered 1993 April 16 by K. Endate and K. Watanabe at Kitami.

(9376) Thionville

1993 OU_7. Discovered 1993 July 20 by E. W. Elst at La Silla.

Named for the French city in northeastern France, near the Luxembourg border. It lies in the center of an iron-mining district. (M 34353)

(9377) Metz

1993 PJ_7. Discovered 1993 August 15 by E. W. Elst at Caussols.

Named for the French city in the Lorraine region of France, at the confluence of the Moselle and Seille rivers. It derives its name from the Gallic tribe Mediomatrici. In 843 it became the capital of Lorraine {see planet (1114)}. The Metz cathedral has remarkable stained glass windows from the thirteenth and fourteenth centuries. (M 34353)

(9378) Nancy-Lorraine

1993 QF_3. Discovered 1993 August 18 by E. W. Elst at Caussols.

Named for the French city Nancy, in the Lorraine {see planet (1114)} region of France, near the left bank of the river Meuse {see planet (3016)}. The city has a very handsome town hall. (M 34353)

(9379) Dijon

1993 QH_3. Discovered 1993 August 18 by E. W. Elst at Caussols.

Named for the capital of Côte d'Or, in east-central France. In the ninth century the city was known as Castrum Divionense. Dijon is a major communication center and offers many tourist attractions. It lies among beautiful, fertile vineyards. The city is famous for its strong mustard, vinegar and gingerbread. (M 34353)

(9380) Mâcon

1993 QZ_5. Discovered 1993 August 17 by E. W. Elst at Caussols.

Named for the French city Mâcon in the Bourgogne, on the right side of the river Saône. Although a very old city (Matiscô), it has been ruined by several religious wars, and little of its original beauty is left. (M 34353)

(9381) Lyon

1993 RT_{19}. Discovered 1993 September 15 by H. Debehogne and E. W. Elst at La Silla.

Named for the French city Lyon, capital of the Rhône department in east-central France, at the confluence of the Rhône and Saône rivers. Originally a Roman military colony (Lugdunum), it subsequently became the capital for the Gauls. As early as 1473 Lyon was the most active printing center in Europe. (M 34353)

(9382) 1993 TK_{11}

Discovered 1993 October 11 by K. Endate and K. Watanabe at Kitami.

(9383) Montélimar

1993 TP_{15}. Discovered 1993 October 9 by E. W. Elst at La Silla.

Named for the city in southeastern France, near the confluence of the Roubion and Rhône rivers. The city was called Acunum by the Romans and Monteil d'Adhémar in medieval times. It was a papal posession from 1340 to 1470. The Chateau des Papes serves now as a prison. The city is famous for its candy. (M 34353)

(9384) Aransio

1993 TP_{26}. Discovered 1993 October 9 by E. W. Elst at La Silla.

Named for the city of Orange in southeastern France in the department Vaucluse. In early times the city was called Aransio and was a well-known Roman colony. In 406 and in 410 the city was pillaged, and during the seventeenth century it suffered from religious wars. In spite of this, some Roman architecture remains, such as the triumphal arch and the old theater. (M 34353)

(9385) Avignon

1993 TJ_{30}. Discovered 1993 October 9 by E. W. Elst at La Silla.

Named for the city in southeastern France, at the east bank of the Rhône river. In 1309 it became the capital of the Roman Catholic papacy and harbored heretics and criminals. The city was well-known for its Saint-Bénézet bridge, which everyone knows from the song "Sur le pont d'Avignon". (M 34353)

(9386) Hitomi

1993 XD_1. Discovered 1993 December 5 by M. Hirasawa and S. Suzuki at Nyukasa.

Named in honor of Hitomi Doi (1955-), wife of Takao Doi, the first Japanese space walker. Hitomi is the author of many articles and essays, mostly about astronomy. She is also an enthusiastic amateur astronomer. (M 34353)

(9387) 1994 CA

Discovered 1994 February 2 by H. Shiozawa and T. Urata at Fujieda.

(9388) 1994 EH_2
Discovered 1994 March 10 by T. Kobayashi at Oizumi.

(9389) Condillac
1994 ET_6. Discovered 1994 March 9 by E. W. Elst at Caussols.

Named in memory of the philosopher and priest Etienne Bonnot de Condillac (1715-1780), the leading advocate in France of the ideas of John Locke {see planet (7010)}. In 1740 he became acquainted with the Encyclopaedists, establishing his position there with his first book *Essai sur l'origine des connaissances humaines* in 1746. In his *Traité des sensations* (1754), which inspired Holbach and Diderot {see planets (6956) and (5351), respectively}, he expressed his views on sensations: "Human knowledge is entirely and exclusively based on observations made by the sense perception". (M 34629)

(9390) 1994 NJ_1
Discovered 1994 July 12 by Y. Shimizu and T. Urata at Nachi-Katsuura.

(9391) 1994 PH_1
Discovered 1994 August 14 by R. H. McNaught at Siding Spring.

(9392) 1994 PK_7
Discovered 1994 August 10 by E. W. Elst at La Silla.

(9393) 1994 PT_{14}
Discovered 1994 August 10 by E. W. Elst at La Silla.

(9394) 1994 PV_{16}
Discovered 1994 August 10 by E. W. Elst at La Silla.

(9395) 1994 PC_{39}
Discovered 1994 August 10 by E. W. Elst at La Silla.

(9396) 1994 QT
Discovered 1994 August 17 by T. Kobayashi at Oizumi.

(9397) 1994 RJ
Discovered 1994 September 6 at the Santa Lucia Observatory at Stroncone.

(9398) 1994 SH_3
Discovered 1994 September 28 by the Spacewatch at Kitt Peak.

(9399) 1994 ST_{12}
Discovered 1994 September 29 by the Spacewatch at Kitt Peak.

(9400) 1994 TW_1
Discovered 1994 October 9 by E. F. Helin and K. J. Lawrence at Palomar.

(9401) 1994 TS_3
Discovered 1994 October 13 by Y. Shimizu and T. Urata at Nachi-Katsuura.

(9402) 1994 UN_1
Discovered 1994 October 25 by S. Ueda and H. Kaneda at Kushiro.

(9403) 1994 UJ_{11}
Discovered 1994 October 31 by the Spacewatch at Kitt Peak.

(9404) 1994 UQ_{11}
Discovered 1994 October 26 by S. Ueda and H. Kaneda at Kushiro.

(9405) 1994 WQ_1
Discovered 1994 November 27 by T. Kobayashi at Oizumi.

(9406) 1994 WG_2
Discovered 1994 November 28 by S. Ueda and H. Kaneda at Kushiro.

(9407) 1994 WS_3
Discovered 1994 November 28 by S. Otomo at Kiyosato.

(9408) 1995 BC
Discovered 1995 January 20 by T. Kobayashi at Oizumi.

(9409) 1995 BG_1
Discovered 1995 January 25 by T. Seki at Geisei.

(9410) 1995 BJ_1
Discovered 1995 January 26 by T. Urata at Oohira.

(9411) 1995 CF
Discovered 1995 February 1 by T. Kobayashi at Oizumi.

(9412) 1995 GZ_8
Discovered 1995 April 4 by S. Ueda and H. Kaneda at Kushiro.

(9413) Eichendorff
1995 SQ_{54}. Discovered 1995 September 21 by F. Börngen at Tautenburg.

Named for the Silesian poet Joseph Freiherr von Eichendorff (1788-1857). He studied philosophy and law in Halle and Heidelberg. From 1816 to 1844, he was in the Prussian civil service. The most popular writer among the German romanticists, he sings of nature, walking, the woods and the night. The highlights of his work, some of which have been set to music, include "In einem kühlen Grunde", "O Täler weit, o Höhen", "Wer hat dich, du schöner Wald"). His most beautiful short story is "Aus dem Leben eines Taugenichts". (M 33795)

(9414) 1995 UV_4
Discovered 1995 October 25 by T. Kobayashi at Oizumi.

(9415) 1995 VE
Discovered 1995 November 1 by T. Kobayashi at Oizumi.

(9416) 1995 WS
Discovered 1995 November 17 by T. Kobayashi at Oizumi.

(9417) 1995 WU
Discovered 1995 November 17 by T. Kobayashi at Oizumi.

(9418) 1995 WX_5
Discovered 1995 November 18 by N. Sato and T. Urata at Chichibu.

(9419) 1995 XS
Discovered 1995 December 12 by T. Kobayashi at Oizumi.

(9420) 1995 XP_4
Discovered 1995 December 14 by the Spacewatch at Kitt Peak.

(9421) Violilla
1995 YM_2. Discovered 1995 December 24 by S. P. Laurie at Church Stretton.

Named in memory of the discoverer's mother, Violet Lilian Laurie (1921-1995), who died the same year that (9421) was discovered. (M 34629)

(9422) 1996 AO_2
Discovered 1996 January 13 by T. Kobayashi at Oizumi.

(9423) 1996 AT_7
Discovered 1996 January 12 by the Spacewatch at Kitt Peak.

(9424) 1996 BN
Discovered 1996 January 16 by T. Kobayashi at Oizumi.

(9425) 1996 CM_7
Discovered 1996 February 14 by U. Munari and M. Tombelli at Cima Ekar.

(9426) Aliante
1996 CO_7. Discovered 1996 February 14 by U. Munari and M. Tombelli at Cima Ekar.

Italian name for a glider, the silent plane. The Latin root of the name emphasizes the key role of the wings in the absence of an engine. (M 33796)

(9427) 1996 CV_7
Discovered 1996 February 14 by U. Munari and M. Tombelli at Cima Ekar.

(9428) Angelalouise
1996 DW_2. Discovered 1996 February 26 by S. P. Laurie at Church Stretton.

Named in honor of the discoverer's wife, Angela Louise Laurie (née Freeman; 1962-). This naming is in recognition of the inspiration, encouragement and support she has given to her husband's astronomical endeavors. (M 34629)

(9429) 1996 EW_1
Discovered 1996 March 14 at the Višnjan Observatory at Višnjan.

(9430) 1996 HU_{10}
Discovered 1996 April 17 by E. W. Elst at La Silla.

(9431) 1996 PS_1
Discovered 1996 August 12 at the Farra d'Isonzo Observatory at Farra d'Isonzo.

(9432) 1997 CQ
Discovered 1997 February 1 by T. Kobayashi at Oizumi.

(9433) 1997 CF_3
Discovered 1997 February 3 by the JPL NEAT Program at Haleakala.

(9434) 1997 CJ_{20}
Discovered 1997 February 12 by T. Kobayashi at Oizumi.

(9435) 1997 CK_{20}
Discovered 1997 February 12 by T. Kobayashi at Oizumi.

(9436) 1997 EB
Discovered 1997 March 1 by T. Kobayashi at Oizumi.

(9437) 1997 EA_3
Discovered 1997 March 4 by T. Kobayashi at Oizumi.

(9438) 1997 EE_{16}
Discovered 1997 March 5 by the Spacewatch at Kitt Peak.

(9439) 1997 EB_{42}
Discovered 1997 March 10 by the Lincoln NEA Research Team at Socorro.

(9440) 1997 FZ_1
Discovered 1997 March 29 at the Beijing Observatory at Xinglong.

(9441) 1997 GN_8
Discovered 1997 April 2 by the Lincoln NEA Research Team at Socorro.

(9442) 1997 GQ_{27}
Discovered 1997 April 2 at the Beijing Observatory at Xinglong.

(9443) 1997 HR_9
Discovered 1997 April 30 by the Lincoln NEA Research Team at Socorro.

(9444) 1997 JA
Discovered 1997 May 1 at the Kleť Observatory at Kleť.

(9445) Charpentier
1997 JA_8. Discovered 1997 May 8 by P. G. Comba at Prescott.

Named for the French musician Marc-Antoine Charpentier (1643-1704), composer of operas and church music, best known for his voluminous and inventive choral music that combines Italian influences and French refinement. (M 33796)

(9446) Cicero
1997 JT_{11}. Discovered 1997 May 3 by E. W. Elst at La Silla.

Named in memory of Marcus Tullius Cicero (106-43 B.C.), Roman statesman, lawyer, scholar and writer. Perhaps the greatest orator of Roman times, he made his reputation in politics and in the law courts. He was trained by Molon of Rhodes. In his *Brutus* he gives a description of the equipment of an orator: a thorough knowledge of literature, a grounding of philosophy, legal expertise, a storehouse of history – as well as the capacity to tie up an opponent and reduce the jury to laughter. The more than 900 of his letters that have survived constitute a primary historical source of the ancient world. In his philosophical writings his aim was to provide Rome with a kind of philosophic encyclopedia. (M 34629)

(9447) 1997 JJ_{18}
Discovered 1997 May 3 by E. W. Elst at La Silla.

(9448) 1997 LJ_3
Discovered 1997 June 5 by the Spacewatch at Kitt Peak.

(9449) 1997 VU_2
Discovered 1997 November 4 by L. Šarounová at Ondřejov.

(9450) 1998 BT_1
Discovered 1998 January 19 by T. Kobayashi at Oizumi.

(9451) 1998 BE_2
Discovered 1998 January 20 by the Lincoln NEA Research Team at Socorro.

(9452) Rogerpeeters
1998 DY_{33}. Discovered 1998 February 27 by E. W. Elst at La Silla.

Named in honor of Roger Peeters (1948-), head of the mechanical workshop at the Royal Observatory at Uccle {see planet (1276)} and responsible for maintaining all the telescopes. He was always available for keeping the Zeiss measuring instrument in good repair, and this was used in particular for measuring this minor planet. (M 34353)

(9453) Mallorca
1998 FO_1. Discovered 1998 March 19 by A. Lopez and R. Pacheco at Mallorca.

Named for the largest of the Balearic Islands, site of the discovery of this minor planet. (M 33796)

(9454) 1998 FX_{54}
Discovered 1998 March 20 by the Lincoln NEA Research Team at Socorro.

(9455) 1998 FJ_{56}
Discovered 1998 March 20 by the Lincoln NEA Research Team at Socorro.

(9456) 1998 FQ_{67}
Discovered 1998 March 20 by the Lincoln NEA Research Team at Socorro.

(9457) 1998 FB_{75}
Discovered 1998 March 24 by the Lincoln NEA Research Team at Socorro.

(9458) 1998 FF_{97}
Discovered 1998 March 31 by the Lincoln NEA Research Team at Socorro.

(9459) 1998 FW_{113}
Discovered 1998 March 31 by the Lincoln NEA Research Team at Socorro.

(9460) 1998 HS_{30}
Discovered 1998 April 29 by the JPL NEAT Program at Haleakala.

(9461) 1998 HV_{33}
Discovered 1998 April 20 by the Lincoln NEA Research Team at Socorro.

(9462) 1998 HC_{37}
Discovered 1998 April 20 by the Lincoln NEA Research Team at Socorro.

(9463) 1998 HW_{38}
Discovered 1998 April 20 by the Lincoln NEA Research Team at Socorro.

(9464) 1998 HL_{117}
Discovered 1998 April 23 by the Lincoln NEA Research Team at Socorro.

(9465) 1998 HJ_{121}
Discovered 1998 April 23 by the Lincoln NEA Research Team at Socorro.

(9466) 1998 KR_{46}
Discovered 1998 May 22 by the Lincoln NEA Research Team at Socorro.

(9467) 1998 KQ_{47}
Discovered 1998 May 22 by the Lincoln NEA Research Team at Socorro.

(9468) Brewer
1998 LT_2. Discovered 1998 June 1 by E. W. Elst at La Silla.

Named in honor of James Brewer, ESO astronomer working on stellar populations and carbon stars. Using intermediate band photometry, he has identified C- and M-type stars in M31 and demonstrated that the ratio of these stars tracks the metallicity gradient in the disk. He is currently working on the globular clusters of the Fornax dwarf spheroidal galaxy. During 1998 he was of great help to the discoverer in obtaining observations with the ESO Schmidt. (M 34353)

(9469) 1998 MY_{34}
Discovered 1998 June 24 by the Lincoln NEA Research Team at Socorro.

(9470) Jussieu

1998 OS_{10}. Discovered 1998 July 26 by E. W. Elst at La Silla.

Named in memory of Bernard (1699-1777), Joseph (1704-1779), Antoine-Laurent (1748-1836) and Adrien-Laurent-Henri (1797-1853) de Jussieu, a family of French botanists. Bernard found a method of plant classification based on the anatomical characters of the plant embryo, while Antoine-Laurent laid down the principles for a natural system of plant classification. Adrien-Laurent-Henri wrote a treatise on botany, and Joseph is well known from the voyage to Peru with de la Condamine to measure a meridional arc; Joseph remained in South America for 35 years, returning to Paris in 1771, when he introduced the common "garden heliotrope" into Europe. (M 34629)

(9471) Ostend

1998 OU_{13}. Discovered 1998 July 26 by E. W. Elst at La Silla.

Named for a town (Flemish Oostende, French Ostende) on the Belgian coast. A fishing village since the ninth century, it was fortified in 1583 as a Dutch stronghold in Belgium. However, in 1604, it fell to the Spanish troops after a three-year siege. With emperor Charles VI of Austria, who founded the "Ostend Company", Ostend entered into a period of prosperity. After Belgian independence in 1830 it became a fashionable seaside resort. Connected to England by sea and air services, Ostend is often called the "Gateway to Europe". (M 34629)

(9472) Bruges

1998 OD_{14}. Discovered 1998 July 26 by E. W. Elst at La Silla.

Named for Bruges (or Brugge), a beautiful small town in West Flanders that was already mentioned in the seventh century as "Municipium Brug ense", the name being derived from a Roman bridge over the Reie river. In the thirteenth century, together with Ghent {see planet (9473)} and Ypres, the town held a monopoly on English wool. During the fifteenth century the city began to decline when silt clogged the way to the port of Zeebrugge. However, the city remained a strong center for the Flemish school of painting, until the religious struggles of the sixteenth century completed its ruin. (M 34630)

(9473) Ghent

1998 OO_{14}. Discovered 1998 July 26 by E. W. Elst at La Silla.

Named for the Flemish city Ghent (or Gand), capital of the province of East Flanders. One of the oldest cities (seventh century) in Belgium, united in wealthy guilds, it was virtually independent until 1584. Its prosperity was and is still based on the cloth industry. The city will be always commemorated in the Pacification of Ghent (1576) and the Treaty of Ghent (1814). It is now the second largest Belgian port. It is also a horticultural center and every five years holds a great flower show known as "Les Floralies". (M 34630)

(9474) 1998 QK_{15}

Discovered 1998 August 17 by the Lincoln NEA Research Team at Socorro.

(9475) 1998 QC_{19}

Discovered 1998 August 17 by the Lincoln NEA Research Team at Socorro.

(9476) 1998 QQ_{36}

Discovered 1998 August 17 by the Lincoln NEA Research Team at Socorro.

(9477) 1998 QK_{41}

Discovered 1998 August 17 by the Lincoln NEA Research Team at Socorro.

(9478) Caldeyro

2148 P-L. Discovered 1960 September 24 by C. J. van Houten and I. van Houten-Groeneveld at Palomar.

Named in memory of Roberto Caldeyro Barcia (1921-1996), Uruguayan physiologist who led a team of researchers, together with Hermógenes Alvarez, in a pioneering study of the physiological aspects of perinatology. Their investigations on uterine contractility were seminal for the development of procedures for the control of high-risk pregnancies and deliveries. Their unit system ("Montevideo Units"), introduced to measure the intensity of the uterine contractions, is used worldwide. During his term as president of the Program for the Development of Basic Science (1986-1996), Caldeyro Barcia played a fundamental role in restoring Uruguaian scientific research, which had been severely damaged during the previous period of military dictatorship. (M 34353)
Name proposed and citation prepared by SBNC.

(9479) Madresplazamayo

2175 P-L. Discovered 1960 September 26 by C. J. van Houten and I. van Houten-Groeneveld at Palomar.

Named in honor of the group of women who became a symbol of human rights activism and courage by denouncing the crimes and atrocities commited by the military junta that ruled Argentina during 1976-1983. Dressed in black, they demonstrated for years every Thursday afternoon in the famous Plaza de Mayo in Buenos Aires, demanding to know the fates of their disappeared sons and daughters. During the course of their movement, many Mothers became themselves victims of the military regime. The Mothers of Plaza de Mayo stand out as a shining example to the world that courage and dignity can still blossom in the hearts of many people, even under the most oppressive regimes. (M 34630)

(9480) Inti

2553 P-L. Discovered 1960 September 24 by C. J. van Houten and I. van Houten-Groeneveld at Palomar.

Named for the sun god in Inca religion, also called Apu-Punchau, believed to be the ancestor of the Incas. Inti was at the head of the state cult. He was usually represented in human form, his face portrayed as a gold disk from which rays and flames extended. Inti's sister was the moon, Mama-Kilya, who was portrayed as a silver disk with human features. (M 34353)
Name proposed and citation prepared by SBNC.

(9481) Menchú

2559 P-L. Discovered 1960 September 24 by C. J. van Houten and I. van Houten-Groeneveld at Palomar.

Named in honor of Rigoberta Menchú Tum (1959-), a Mayan Indian from Guatemala who has campaigned for human rights, especially for indigenous people. Menchú, whose parents and brother were killed by government troops, became a symbol of peace and reconciliation in a country devastated by civil war. She was awarded the Nobel Peace prize in 1992. According to Nobel Committee Chairman Sejersted, Menchú belongs to "...the shining individual examples of people who manage to preserve their humanity in brutal and violent surroundings. Such people give us hope that there are ways out of a vicious circle." (M 34354)
Name proposed and citation prepared by SBNC.

(9482) Rubéndarío

4065 P-L. Discovered 1960 September 24 by C. J. van Houten and I. van Houten-Groeneveld at Palomar.

Named in memory of Rubén Darío (1867-1916), pseudonym of Félix Rubén Gar-

cia Sarmiento, Nicaraguan poet, journalist and diplomat who became a leader of the Spanish-American literary movement known as Modernism. Darío's poetry is notable for its remarkable musicality, grace and sonority, and he is indeed widely considered one of the greatest poets who ever wrote in Spanish. (M 34354)
Name proposed and citation prepared by SBNC.

(9483) Chagas

4121 P-L. Discovered 1960 September 24 by C. J. van Houten and I. van Houten-Groeneveld at Palomar.

Named in memory of Carlos Chagas (1879-1934), Brazilian epidemiologist who first described the agent, vectors and clinical signs of a new disease, the American Trypanosomiasis. Also known as Chagas' disease, this is an infection produced by the flagellate protozoan Trypanosoma Cruzi that is transmitted to humans by blood-sucking insects. It is endemic in most rural areas of Central and South America. (M 34354)
Name proposed and citation prepared by SBNC.

(9484) Wanambi

4590 P-L. Discovered 1960 September 24 by C. J. van Houten and I. van Houten-Groeneveld at Palomar.

Named for the Australian Aborigines' best known spirit, the Rainbow Snake, which is the rainbow and the revealer of truth. Wanambi is one of the rainbow's names. (M 34354)
Name proposed and citation prepared by SBNC.

(9485) Uluru

6108 P-L. Discovered 1960 September 24 by C. J. van Houten and I. van Houten-Groeneveld at Palomar.

Named for the Australian Aboriginal sacred place, the center or navel of the island continent. It is also known as Ayers Rock. (M 34354)
Name proposed and citation prepared by SBNC.

(9486) Utemorrah

6130 P-L. Discovered 1960 September 24 by C. J. van Houten and I. van Houten-Groeneveld at Palomar.

Named for the Australian Aboriginal poet Daisy Utemorrah (1922-), teller of Dreamtime stories in the Kimberley: "Words are my gun and my spear". (M 34354)
Name proposed and citation prepared by SBNC.

(9487) Kupe

7633 P-L. Discovered 1960 October 17 by C. J. van Houten and I. van Houten-Groeneveld at Palomar.

Named for the legendary sailor and explorer whose discovery of the islands now known as New Zealand led to the coming there of the Maori people around the ninth century. (M 34354)
Name proposed and citation prepared by SBNC.

(9488) Huia

9523 P-L. Discovered 1960 September 24 by C. J. van Houten and I. van Houten-Groeneveld at Palomar.

Named for the bird, extinct since 1907, with splendid green and black plumage and a long white-tipped tail (Heteralocha acutirostris). The tail feathers of the huia were highly prized by the Maori people as symbols of rank. Carved boxes, waka huia, were made to keep them in. (M 34354)
Name proposed and citation prepared by SBNC.

(9489) Tanemahuta

1146 T-1. Discovered 1971 March 25 by C. J. van Houten and I. van Houten-Groeneveld at Palomar.

Named for the god of the forests of the Maori people in New Zealand and for the great kauri tree in the Waipoua forest that also bears this name. (M 34354) Name proposed and citation prepared by SBNC.

(9490) 1181 T-1

Discovered 1971 March 25 by C. J. van Houten and I. van Houten-Groeneveld at Palomar.

(9491) 1205 T-1

Discovered 1971 March 25 by C. J. van Houten and I. van Houten-Groeneveld at Palomar.

(9492) 2066 T-1

Discovered 1971 March 25 by C. J. van Houten and I. van Houten-Groeneveld at Palomar.

(9493) 3100 T-1

Discovered 1971 March 26 by C. J. van Houten and I. van Houten-Groeneveld at Palomar.

(9494) 3212 T-1

Discovered 1971 March 26 by C. J. van Houten and I. van Houten-Groeneveld at Palomar.

(9495) 4177 T-1

Discovered 1971 March 26 by C. J. van Houten and I. van Houten-Groeneveld at Palomar.

(9496) 4260 T-1

Discovered 1971 March 26 by C. J. van Houten and I. van Houten-Groeneveld at Palomar.

(9497) 1001 T-2

Discovered 1973 September 29 by C. J. van Houten and I. van Houten-Groeneveld at Palomar.

(9498) 1197 T-2

Discovered 1973 September 29 by C. J. van Houten and I. van Houten-Groeneveld at Palomar.

(9499) 1269 T-2

Discovered 1973 September 29 by C. J. van Houten and I. van Houten-Groeneveld at Palomar.

(9500) 1281 T-2

Discovered 1973 September 29 by C. J. van Houten and I. van Houten-Groeneveld at Palomar.

(9501) 2071 T-2

Discovered 1973 September 29 by C. J. van Houten and I. van Houten-Groeneveld at Palomar.

(9502) 2075 T-2

Discovered 1973 September 29 by C. J. van Houten and I. van Houten-Groeneveld at Palomar.

(9503) 2180 T-2

Discovered 1973 September 29 by C. J. van Houten and I. van Houten-Groeneveld at Palomar.

(9504) 2224 T-2
Discovered 1973 September 29 by C. J. van Houten and I. van Houten-Groeneveld at Palomar.

(9505) 4131 T-2
Discovered 1973 September 29 by C. J. van Houten and I. van Houten-Groeneveld at Palomar.

(9506) 5200 T-2
Discovered 1973 September 25 by C. J. van Houten and I. van Houten-Groeneveld at Palomar.

(9507) 5447 T-2
Discovered 1973 September 30 by C. J. van Houten and I. van Houten-Groeneveld at Palomar.

(9508) 3395 T-3
Discovered 1977 October 16 by C. J. van Houten and I. van Houten-Groeneveld at Palomar.

(9509) 3453 T-3
Discovered 1977 October 16 by C. J. van Houten and I. van Houten-Groeneveld at Palomar.

(9510) 5022 T-3
Discovered 1977 October 16 by C. J. van Houten and I. van Houten-Groeneveld at Palomar.

(9511) 5051 T-3
Discovered 1977 October 16 by C. J. van Houten and I. van Houten-Groeneveld at Palomar.

(9512) 1966 CM
Discovered 1966 February 13 at the Purple Mountain Observatory at Nanking.

(9513) 1971 UN
Discovered 1971 October 26 by L. Kohoutek at Bergedorf.

(9514) 1973 SG$_5$
Discovered 1973 September 27 by L. V. Zhuravleva at Nauchnyj.

(9515) 1975 RA$_2$
Discovered 1975 September 5 by M. R. Cesco at El Leoncito.

(9516) 1976 YL$_3$
Discovered 1976 December 16 by L. I. Chernykh at Nauchnyj.

(9517) 1977 VL$_1$
Discovered 1977 November 3 at the Purple Mountain Observatory at Nanking.

(9518) 1978 GA
Discovered 1978 April 7 at the Harvard College Observatory at Harvard.

(9519) 1978 VK$_3$
Discovered 1978 November 6 by E. F. Helin and S. J. Bus at Palomar.

(9520) 1978 VV$_6$
Discovered 1978 November 7 by E. F. Helin and S. J. Bus at Palomar.

(9521) 1980 FS$_1$
Discovered 1980 March 16 by C.-I. Lagerkvist at La Silla.

(9522) 1981 DS
Discovered 1981 February 28 by S. J. Bus at Siding Spring.

(9523) 1981 EE_1
Discovered 1981 March 5 by H. Debehogne and G. DeSanctis at La Silla.

(9524) 1981 EJ_5
Discovered 1981 March 2 by S. J. Bus at Siding Spring.

(9525) 1981 EF_{11}
Discovered 1981 March 1 by S. J. Bus at Palomar.

(9526) 1981 EC_{13}
Discovered 1981 March 1 by S. J. Bus at Siding Spring.

(9527) 1981 EH_{23}
Discovered 1981 March 3 by S. J. Bus at Siding Spring.

(9528) 1981 EH_{24}
Discovered 1981 March 7 by S. J. Bus at Siding Spring.

(9529) 1981 EF_{25}
Discovered 1981 March 2 by S. J. Bus at Siding Spring.

(9530) 1981 EO_{26}
Discovered 1981 March 2 by S. J. Bus at Siding Spring.

(9531) 1981 QK
Discovered 1981 August 30 by E. Bowell at Anderson Mesa.

(9532) Abramenko
1981 RQ_2. Discovered 1981 September 7 by L. G. Karachkina at Nauchnyj.
Named in honor of Aleksandr Nikolaevich Abramenko (1921-), chief engineer at the Crimean Astrophysical Observatory. He created a special television system that has been used for observations of astronomical objects since 1963. This has produced unique data on novae, supernovae, cataclysmic variables, minor planets and comets. Abramenko is one of the authors of a monograph on television astronomy (1974, 1984). (M 34630)

(9533) 1981 SA_7
Discovered 1981 September 28 by L. V. Zhuravleva at Nauchnyj.

(9534) 1981 TP
Discovered 1981 October 4 by N. G. Thomas at Anderson Mesa.

(9535) 1981 UO_{11}
Discovered 1981 October 22 by N. S. Chernykh at Nauchnyj.

(9536) 1981 UR_{27}
Discovered 1981 October 24 by S. J. Bus at Palomar.

(9537) 1982 BM
Discovered 1982 January 18 by E. Bowell at Anderson Mesa.

(9538) 1982 UM_2
Discovered 1982 October 20 by A. Mrkos at Kleť.

(9539) Prishvin
1982 UE_7. Discovered 1982 October 21 by L. G. Karachkina at Nauchnyj.
Named in memory of the writer Mikhail Mikhajlovich Prishvin (1873-1954),

famous for his deep philosophical descriptions of Russian nature. His affecting stories about animals are thrilling, especially for children. (M 34630)

(9540) Mikhalkov

1982 UJ_7. Discovered 1982 October 21 by L. G. Karachkina at Nauchnyj.

Named in honor of the Russian writer and poet Sergej Vladimirovich Mikhalkov (1913-), whose wonderful verses and fables are the best companion for any child. (M 34630)

(9541) 1983 CH

Discovered 1983 February 11 by E. Bowell at Anderson Mesa.

(9542) 1983 TU_1

Discovered 1983 October 12 by E. Bowell at Anderson Mesa.

(9543) 1983 XN_1

Discovered 1983 December 4 by M. Antal at Piszkéstető.

(9544) 1984 EL

Discovered 1984 March 1 by E. Bowell at Anderson Mesa.

(9545) 1984 MQ

Discovered 1984 June 25 by T. M. Smirnova at Nauchnyj.

(9546) 1984 SD_6

Discovered 1984 September 22 by H. Debehogne at La Silla.

(9547) 1985 AE

Discovered 1985 January 15 by K. Suzuki and T. Urata at Toyota.

(9548) 1985 CN

Discovered 1985 February 13 by the Spacewatch at Kitt Peak.

(9549) 1985 SM_2

Discovered 1985 September 19 by N. S. Chernykh and L. I. Chernykh at Nauchnyj.

(9550) 1985 TY_1

Discovered 1985 October 15 by E. Bowell at Anderson Mesa.

(9551) Kazi

1985 UJ. Discovered 1985 October 20 by A. Mrkos at Kleť.

Named for the oldest daughter of prince Krok {see planet (3102)} and wife of honorable hero Bivoj {see planet (5797)}, known from Bohemian myths. Kazi was experienced in magic, herbs, formulas and signs. (M 34354)
Name suggested by J. Tichá, M. Tichý and Z. Moravec.

(9552) 1985 UY

Discovered 1985 October 24 by A. Mrkos at Kleť.

(9553) 1985 UG_2

Discovered 1985 October 17 by the CERGA at Caussols.

(9554) 1985 XA

Discovered 1985 December 13 by R. Chemin at Caussols.

(9555) 1986 GC

Discovered 1986 April 2 by P. Jensen at Brorfelde.

(9556) 1986 GF

Discovered 1986 April 8 at the Palomar Observatory at Palomar.

(9557) 1986 QL_2
Discovered 1986 August 28 by H. Debehogne at La Silla.

(9558) 1986 QB_3
Discovered 1986 August 29 by H. Debehogne at La Silla.

(9559) 1987 DH_6
Discovered 1987 February 23 by H. Debehogne at La Silla.

(9560) 1987 EQ
Discovered 1987 March 3 by E. Bowell at Anderson Mesa.

(9561) 1987 QT_1
Discovered 1987 August 19 by E. W. Elst at La Silla.

(9562) 1987 RG
Discovered 1987 September 1 by E. W. Elst at La Silla.

(9563) 1987 SJ_1
Discovered 1987 September 21 by E. Bowell at Anderson Mesa.

(9564) 1987 SG_3
Discovered 1987 September 26 by C. S. Shoemaker and E. M. Shoemaker at Palomar.

(9565) 1987 SU_{17}
Discovered 1987 September 18 by L. I. Chernykh at Nauchnyj.

(9566) 1987 SX_{17}
Discovered 1987 September 18 by L. I. Chernykh at Nauchnyj.

(9567) 1987 US_4
Discovered 1987 October 22 by L. V. Zhuravleva at Nauchnyj.

(9568) 1988 AX_4
Discovered 1988 January 13 by H. Debehogne at La Silla.

(9569) 1988 CL_2
Discovered 1988 February 11 by E. W. Elst at La Silla.

(9570) 1988 RQ_5
Discovered 1988 September 2 by H. Debehogne at La Silla.

(9571) 1988 RR_5
Discovered 1988 September 2 by H. Debehogne at La Silla.

(9572) 1988 RS_6
Discovered 1988 September 8 by H. Debehogne at La Silla.

(9573) 1988 UC
Discovered 1988 October 16 by K. Endate and K. Watanabe at Kitami.

(9574) Taku
1988 XB_5. Discovered 1988 December 5 by T. Nakamura at Kiso.

Named in memory of Hiroshi Nakamura (1891-1974), Japanese medical biochemist and researcher of old maps. He was often called Taku, since the Chinese character representing his first name can also be pronounced as Taku. He is well-known in the community of historical cartography as twice the winner of the Imago Mundi prize, awarded by the authoritative international journal on old maps. Taku was a long-time member of the editorial board of this journal. He wrote several important books and papers on far-east Asian maps of the sixteenth and seventeenth centuries. He showed that, contrary to Western understanding, European sea charts and maps of far-east Asia then being made were largely influenced by the knowledge of contemporary Japanese and Chinese sailors. (M 34354)

(9575) 1989 BW_1
Discovered 1989 January 29 by A. Mrkos at Kleť.

(9576) 1989 CX_2
Discovered 1989 February 4 by E. W. Elst at La Silla.

(9577) Gropius
1989 CE_5. Discovered 1989 February 2 by F. Börngen at Tautenburg.

Named for the German architect Walter Gropius (1883-1969), born in Berlin {see planet (422)} and one of the most fascinating representatives of twentieth-century architecture. He turned away from all historical influence and favored expedient, glass and concrete constructions. In 1919, he founded the Bauhaus in Weimar {see planets (8502) and (3139)} and was its director until 1928. In 1933 he emigrated to England. From 1937 to 1952 he taught at Harvard University, where he founded an architecture school and the Architects Collaborative and built the Harvard Graduate Center. His public works can be found in Berlin, Dessau, Chicago, Athens, and New York (notably the Pan Am Building). (M 33976)

(9578) 1989 GA_3
Discovered 1989 April 3 by E. W. Elst at La Silla.

(9579) 1989 GO_4
Discovered 1989 April 3 by E. W. Elst at La Silla.

(9580) 1989 TB_{11}
Discovered 1989 October 4 by T. Nomura and K. Kawanishi at Minami-Oda.

(9581) 1990 DM_3
Discovered 1990 February 24 by H. Debehogne at La Silla.

(9582) 1990 EL_7
Discovered 1990 March 3 by H. Debehogne at La Silla.

(9583) 1990 HL_1
Discovered 1990 April 28 by R. H. McNaught at Siding Spring.

(9584) 1990 OL_4
Discovered 1990 July 25 by H. E. Holt at Palomar.

(9585) 1990 QY_2
Discovered 1990 August 28 by H. E. Holt at Palomar.

(9586) 1990 SG_{11}
Discovered 1990 September 16 by H. E. Holt at Palomar.

(9587) 1990 UG_4
Discovered 1990 October 16 by E. W. Elst at La Silla.

(9588) 1990 WE_2
Discovered 1990 November 18 by E. W. Elst at La Silla.

(9589) 1990 WU_5
Discovered 1990 November 21 by E. W. Elst at La Silla.

(9590) 1991 DK_1
Discovered 1991 February 21 by the Spacewatch at Kitt Peak.

(9591) 1991 FH_2
Discovered 1991 March 20 by H. Debehogne at La Silla.

(9592) 1991 GK_4
Discovered 1991 April 8 by E. W. Elst at La Silla.

(9593) 1991 PZ_{17}
Discovered 1991 August 7 by H. E. Holt at Palomar.

(9594) 1991 RG
Discovered 1991 September 4 by R. H. McNaught at Siding Spring.

(9595) 1991 RE_{11}
Discovered 1991 September 13 by H. E. Holt at Palomar.

(9596) 1991 RC_{22}
Discovered 1991 September 15 by H. E. Holt at Palomar.

(9597) 1991 UF
Discovered 1991 October 18 by S. Ueda and H. Kaneda at Kushiro.

(9598) 1991 UQ
Discovered 1991 October 18 by S. Ueda and H. Kaneda at Kushiro.

(9599) 1991 UP_2
Discovered 1991 October 29 by K. Endate and K. Watanabe at Kitami.

(9600) 1991 UB_3
Discovered 1991 October 31 by S. Ueda and H. Kaneda at Kushiro.

(9601) 1991 UE_3
Discovered 1991 October 18 by S. Ueda and H. Kaneda at Kushiro.

(9602) 1991 UU_3
Discovered 1991 October 31 by T. Fujii and K. Watanabe at Kitami.

(9603) 1991 VG_2
Discovered 1991 November 9 by S. Ueda and H. Kaneda at Kushiro.

(9604) 1991 YW
Discovered 1991 December 30 by E. W. Elst at St. Michel.

(9605) 1992 AP_3
Discovered 1992 January 11 by O. A. Naranjo at Mérida.

(9606) 1992 BZ
Discovered 1992 January 28 by S. Ueda and H. Kaneda at Kushiro.

(9607) 1992 DS_6
Discovered 1992 February 29 by the Uppsala-ESO Survey at La Silla.

(9608) 1992 PD_2
Discovered 1992 August 2 by H. E. Holt at Palomar.

(9609) 1992 QL_2
Discovered 1992 August 26 by L. I. Chernykh at Nauchnyj.

(9610) Vischer

1992 RQ. Discovered 1992 September 2 by F. Börngen and L. D. Schmadel at Tautenburg.

Named for Peter Vischer the Elder (c.1460-1529), the greatest German brass-caster master in the Middle Ages. Ledgers, tombs, artistic parcloses, baptismal fonts and statues made of bronze from his workshop in Nürnberg can be found throughout Germany. His father and sons also worked in this foundry. His masterpiece is the world-renowned Sebaldusgrab in St. Sebald church at Nürnberg, made during 1508-1519. (M 34354)
Name proposed by the first discoverer.

(9611) 1992 RF_7

Discovered 1992 September 2 by E. W. Elst at La Silla.

(9612) 1992 RT_7

Discovered 1992 September 4 by L. V. Zhuravleva at Nauchnyj.

(9613) 1993 BN_3

Discovered 1993 January 26 by T. J. Balonek at Kitt Peak.

(9614) 1993 BQ_4

Discovered 1993 January 27 by E. W. Elst at Caussols.

(9615) 1993 BX_{13}

Discovered 1993 January 23 by E. W. Elst at La Silla.

(9616) 1993 FR_3

Discovered 1993 March 21 by E. F. Helin at Palomar.

(9617) 1993 FA_5

Discovered 1993 March 17 by the Uppsala-ESO Survey at La Silla.

(9618) 1993 FQ_8

Discovered 1993 March 17 by the Uppsala-ESO Survey at La Silla.

(9619) 1993 FS_9

Discovered 1993 March 17 by the Uppsala-ESO Survey at La Silla.

(9620) 1993 FU_{13}

Discovered 1993 March 17 by the Uppsala-ESO Survey at La Silla.

(9621) 1993 FT_{26}

Discovered 1993 March 21 by the Uppsala-ESO Survey at La Silla.

(9622) 1993 FV_{26}

Discovered 1993 March 21 by the Uppsala-ESO Survey at La Silla.

(9623) 1993 FU_{28}

Discovered 1993 March 21 by the Uppsala-ESO Survey at La Silla.

(9624) 1993 FH_{38}

Discovered 1993 March 19 by the Uppsala-ESO Survey at La Silla.

(9625) 1993 HF

Discovered 1993 April 16 by S. Ueda and H. Kaneda at Kushiro.

(9626) 1993 JF_1

Discovered 1993 May 14 by E. W. Elst at La Silla.

(9627) 1993 LU_1
Discovered 1993 June 15 by H. E. Holt at Palomar.

(9628) 1993 OB_2
Discovered 1993 July 16 by E. F. Helin at Palomar.

(9629) 1993 PU_7
Discovered 1993 August 15 by E. W. Elst at Caussols.

(9630) 1993 PW_7
Discovered 1993 August 15 by E. W. Elst at Caussols.

(9631) 1993 SL_6
Discovered 1993 September 17 by E. W. Elst at La Silla.

(9632) 1993 TK_3
Discovered 1993 October 15 by K. Endate and K. Watanabe at Kitami.

(9633) Cotur
1993 UP_8. Discovered 1993 October 20 by E. W. Elst at La Silla.

Named in honor of Peter Cotur (1944-), scientific journalist at the newspaper *Het Laatste Nieuws* in Brussels {see planet (2689)} and a good friend of the discoverer. Educated in biology at the University of Ghent {see planet (9473)}. he is especially interested in medicine, environmental problems, space technology and astronomy in general. Fascinated by history, he is currently occupied in discovering the earliest facts about technical and scientific inventions, as well as about peculiarities in the lives of the inventors. (M 34630)

(9634) 1993 XB
Discovered 1993 December 4 at the Farra d'Isonzo Observatory at Farra d'Isonzo.

(9635) 1993 XS
Discovered 1993 December 9 by T. Urata at Oohira.

(9636) 1993 YO
Discovered 1993 December 17 at the Farra d'Isonzo Observatory at Farra d'Isonzo.

(9637) 1994 PJ_2
Discovered 1994 August 9 at the Palomar Observatory at Palomar.

(9638) Fuchs
1994 PO_7. Discovered 1994 August 10 by E. W. Elst at La Silla.

Named in memory of the German botanist and physician Leonhard Fuchs (1501-1566), whose botanical work *Historia Stirpium commentarii insignes* (1542) remains a landmark in the organized presentation of botanical observations. The book contains beautiful woodcuts and accurate descriptions (and a glossary) of more than 550 plants, most of them of medical use. His name is also commemorated in the Fuchsia genus of flowering plants. (M 34630)

(9639) Scherer
1994 PS_{11}. Discovered 1994 August 10 by E. W. Elst at La Silla.

Named in honor of Marc Scherer (1944-), a well-known scientist at The Belgian Institute for Space Aeronomy in Uccle. His main interest is the development of kinetic models for polar and solar winds, using the kinetic theory of gases. This naming also honors his sisters Elisabeth and Trixie Scherer, in sweet remembrance of a time at the Atheneum of Kapellen in Antwerp. (M 34630)

(9640) Lippens

1994 PP_{26}. Discovered 1994 August 12 by E. W. Elst at La Silla.

Named in honor of Carlos Lippens (1945-), well known for his work on trace substances in the earth's atmosphere by means of infrared spectroscopy. Working at the Belgian Institute for Space Aeronomy, he was responsible for the GRILLE spectrometer on board Spacelab 1 (1983) and Atlas 1 (1992). Lippens was also closely involved with the MIRAS instrumentation on board MIR. As a computer specialist he has frequently helped solve data-communication problems in connection with the discoverer's work on minor planets. (M 34630)

(9641) Demazière

1994 PB_{30}. Discovered 1994 August 12 by E. W. Elst at La Silla.

Named in honor of Martine De Mazière (1960-), a scientist at the Belgian Institute for Space Aeronomy involved with the optical remote sensing of the lower atmosphere. Her main interest is in the atmospheric composition, including aerosols, and how it changes. She has also made a quantitative evaluation of the post-Pinatubo NO_2 reduction and recovery, based on ten years of ultraviolet and optical spectroscopic measurements at the Jungfraujoch station. She shares with the discoverer a great love for music. (M 34630)

(9642) 1994 RU

Discovered 1994 September 1 by K. Endate and K. Watanabe at Kitami.

(9643) 1994 RX

Discovered 1994 September 2 by Y. Shimizu and T. Urata at Nachi-Katsuura.

(9644) 1994 WQ_3

Discovered 1994 November 26 by Y. Shimizu and T. Urata at Nachi-Katsuura.

(9645) Grünewald

1995 AO_4. Discovered 1995 January 5 by F. Börngen at Tautenburg.

Named for the German painter and watercolor artist Matthias Grünewald (ca 1460-1528). His original name was Mathis Gothard Neithardt (or Niethart). His masterpiece is the Isenheimer altar in Colmar, one of the great works of Western art, featuring spiritual and realistic scenes juxtaposed and notable for the variety and beauty of the colors. Paul Hindemith {see planet (5157)} made Grünewald's life the subject of his opera and symphony *Mathis der Maler*. (M 34354)

(9646) 1995 BV

Discovered 1995 January 25 by T. Kobayashi at Oizumi.

(9647) 1995 UM_8

Discovered 1995 October 27 by T. Kobayashi at Oizumi.

(9648) 1995 UB_9

Discovered 1995 October 30 by F. Uto at Kashihara.

(9649) 1995 XG

Discovered 1995 December 2 by T. Kobayashi at Oizumi.

(9650) 1995 YG

Discovered 1995 December 17 by T. Kobayashi at Oizumi.

(9651) 1996 AJ

Discovered 1996 January 7 by the JPL NEAT Program at Haleakala.

(9652) 1996 AF_2

Discovered 1996 January 12 by S. Ueda and H. Kaneda at Kushiro.

(9653) 1996 AL_2
Discovered 1996 January 13 by T. Urata at Oohira.

(9654) 1996 AQ_2
Discovered 1996 January 13 by T. Kobayashi at Oizumi.

(9655) 1996 CH_1
Discovered 1996 February 11 by T. Kobayashi at Oizumi.

(9656) 1996 DK_1
Discovered 1996 February 23 by T. Kobayashi at Oizumi.

(9657) 1996 DG_2
Discovered 1996 February 24 at the Višnjan Observatory at Višnjan.

(9658) Imabari
1996 DD_3. Discovered 1996 February 28 by A. Nakamura at Kuma.

Named for a city in eastern Ehime prefecture, famous for the shipbuilding and textile industries, especially the production of towels. Imabari is the terminal city on the Shikoku {see planet (4223)} Island side of the Nishiseto Expressway (Shimanami Kaido) {see planet (9235)} and expects to welcome many visitors after this road opens on 1999 May 1. (M 34630)

(9659) 1996 EJ
Discovered 1996 March 10 by S. Ueda and H. Kaneda at Kushiro.

(9660) 1996 FW_4
Discovered 1996 March 22 by the JPL NEAT Program at Haleakala.

(9661) 1996 FU_{13}
Discovered 1996 March 18 by the Spacewatch at Kitt Peak.

(9662) Frankhubbard
1996 GS. Discovered 1996 April 12 by P. G. Comba at Prescott.

Named for the American harpsichord maker Frank Hubbard (1920-1976). After graduating from Harvard he traveled in Europe and gained a profound knowledge of early keyboard instruments, setting this down in his book *Three Centuries of Harpsichord Making*. In 1949, jointly with William Dowd, he set up a harpsichord workshop in Boston, which, through a process of fission, became the progenitor of several such workshops and led to the establishment of a tradition of harpsichord making in North America. (M 34355)

(9663) 1996 GC_{18}
Discovered 1996 April 15 by E. W. Elst at La Silla.

(9664) Brueghel
1996 HT_{14}. Discovered 1996 April 17 by E. W. Elst at La Silla.

Named in memory of Pieter Brueghel (Bruegel, Breughel; c.1525-1569), the greatest Flemish painter of the sixteenth century. He was apprenticed to Pieter Coecke, a leading Antwerp artist, whose daughter he married. Although Brueghel was little influenced by Coecke's italianate art, the apprenticeship provided an important contact with a humanistic milieu, sensitizing Brueghel for human life. Perhaps for this reason his landscapes show, not only trees, mountains and snow, but also humans suffering from one side and enjoying life from the other. Through his sons Jan and Pieter, Brueghel became the ancestor of a dynasty of painters that survived into the eighteenth century. (M 34631)

(9665) 1996 LA
Discovered 1996 June 5 at the Kleť Observatory at Kleť.

(9666) 1997 GM_{22}
Discovered 1997 April 6 by the Lincoln NEA Research Team at Socorro.

(9667) 1997 HC_{16}
Discovered 1997 April 29 by the Spacewatch at Kitt Peak.

(9668) 1997 LN
Discovered 1997 June 3 at the Beijing Observatory at Xinglong.

(9669) Symmetria
1997 NC_3. Discovered 1997 July 8 by P. G. Comba at Prescott.

Name suggested by the palindromic shape of the numeral 9669, and the fact that each pair of its digit is invariant under a rotation by 180 degrees. (M 33796)

(9670) Magni
1997 NJ_{10}. Discovered 1997 July 10 by A. Boattini at Campo Imperatore.

Named in honor of Gianfranco Magni (1943-), who joined the Istituto di Astrofisica Spaziale in Rome in 1975. He has worked on the physics of stellar interiors, in particular on the equation of state of gas at high pressure and temperature. Currently, his main field of interest is the origin of the solar system and of planetary systems in general, with special attention to the structure and evolution of circumstellar disks and the formation of giant planets. Involved also in the study of the origin and structure of comets, he is a team member of the cometary mission Rosetta. A lover of early music, he plays the recorder and belongs to two musical groups ,"L'Amoroso Cantar" (Medieval music) and "Il Martellato" (Renaissance music). (M 34355)

(9671) 1997 TU_9
Discovered 1997 October 5 by L. Šarounová at Ondřejov.

(9672) Rosenbergerezek
1997 TA_{10}. Discovered 1997 October 5 by P. Pravec at Ondřejov.

Named in honor of Kamila Rosenbergerová (1976-) and Tomáš Rezek (1974-). Tomáš, a friend of the discoverer, has worked as a research assistant at the Ondřejov Observatory for two years. His exceptional skill with various operating systems have often amazed his colleagues. This minor planet is being named on the occasion of the marriage of Kamila and Tomáš on 1999 June 19. (M 34631)

(9673) 1997 UC_{25}
Discovered 1997 October 25 by S. Otomo at Kiyosato.

(9674) Slovenija
1998 QU_{15}. Discovered 1998 August 23 at the Crni Vrh Observatory at Crni Vrh.

Named for the European nation in which the Crni Vrh Observatory is situated. This is the first minor planet credited to Slovenian astronomers. (M 33796)

(9675) 1998 QK_{36}
Discovered 1998 August 17 by the Lincoln NEA Research Team at Socorro.

(9676) 2023 P-L
Discovered 1960 September 24 by C. J. van Houten and I. van Houten-Groeneveld at Palomar.

(9677) 2532 P-L
Discovered 1960 September 24 by C. J. van Houten and I. van Houten-Groeneveld at Palomar.

(9678) 2584 P-L
Discovered 1960 September 24 by C. J. van Houten and I. van Houten-Groeneveld at Palomar.

(9679) 2600 P-L
Discovered 1960 September 24 by C. J. van Houten and I. van Houten-Groeneveld at Palomar.

(9680) 3557 P-L
Discovered 1960 October 22 by C. J. van Houten and I. van Houten-Groeneveld at Palomar.

(9681) 4069 P-L
Discovered 1960 September 24 by C. J. van Houten and I. van Houten-Groeneveld at Palomar.

(9682) 4073 P-L
Discovered 1960 September 24 by C. J. van Houten and I. van Houten-Groeneveld at Palomar.

(9683) 4099 P-L
Discovered 1960 September 24 by C. J. van Houten and I. van Houten-Groeneveld at Palomar.

(9684) 4113 P-L
Discovered 1960 September 24 by C. J. van Houten and I. van Houten-Groeneveld at Palomar.

(9685) 4247 P-L
Discovered 1960 September 24 by C. J. van Houten and I. van Houten-Groeneveld at Palomar.

(9686) 4604 P-L
Discovered 1960 September 24 by C. J. van Houten and I. van Houten-Groeneveld at Palomar.

(9687) 4614 P-L
Discovered 1960 September 24 by C. J. van Houten and I. van Houten-Groeneveld at Palomar.

(9688) 4665 P-L
Discovered 1960 September 24 by C. J. van Houten and I. van Houten-Groeneveld at Palomar.

(9689) 4831 P-L
Discovered 1960 September 24 by C. J. van Houten and I. van Houten-Groeneveld at Palomar.

(9690) 6039 P-L
Discovered 1960 September 24 by C. J. van Houten and I. van Houten-Groeneveld at Palomar.

(9691) 6053 P-L
Discovered 1960 September 24 by C. J. van Houten and I. van Houten-Groeneveld at Palomar.

(9692) 6354 P-L
Discovered 1960 September 24 by C. J. van Houten and I. van Houten-Groeneveld at Palomar.

(9693) 6547 P-L
Discovered 1960 September 24 by C. J. van Houten and I. van Houten-Groeneveld at Palomar.

(9694) Lycomedes
6581 P-L. Discovered 1960 September 26 by C. J. van Houten and I. van Houten-Groeneveld at Palomar.

Lycomedes was king of Scyros during the Trojan war. At his court Thetis hid her son Achilles {see, respectively, planets (17) and (588)}, dressed in girl's clothes to save him from the Trojan war. He was found by Odysseus {see planet (1143)}. (M 34355)

(9695) 6583 P-L
Discovered 1960 September 24 by C. J. van Houten and I. van Houten-Groeneveld at Palomar.

(9696) 6628 P-L
Discovered 1960 September 24 by C. J. van Houten and I. van Houten-Groeneveld at Palomar.

(9697) 1295 T-1
Discovered 1971 March 25 by C. J. van Houten and I. van Houten-Groeneveld at Palomar.

(9698) 2205 T-1
Discovered 1971 March 25 by C. J. van Houten and I. van Houten-Groeneveld at Palomar.

(9699) 3036 T-1
Discovered 1971 March 26 by C. J. van Houten and I. van Houten-Groeneveld at Palomar.

(9700) 3058 T-1
Discovered 1971 March 26 by C. J. van Houten and I. van Houten-Groeneveld at Palomar.

(9701) 1157 T-2
Discovered 1973 September 29 by C. J. van Houten and I. van Houten-Groeneveld at Palomar.

(9702) 2108 T-2
Discovered 1973 September 29 by C. J. van Houten and I. van Houten-Groeneveld at Palomar.

(9703) 3146 T-2
Discovered 1973 September 30 by C. J. van Houten and I. van Houten-Groeneveld at Palomar.

(9704) 5469 T-2
Discovered 1973 September 30 by C. J. van Houten and I. van Houten-Groeneveld at Palomar.

(9705) 3137 T-3
Discovered 1977 October 16 by C. J. van Houten and I. van Houten-Groeneveld at Palomar.

(9706) 3176 T-3
Discovered 1977 October 16 by C. J. van Houten and I. van Houten-Groeneveld at Palomar.

(9707) 3226 T-3

Discovered 1977 October 16 by C. J. van Houten and I. van Houten-Groeneveld at Palomar.

(9708) 4140 T-3

Discovered 1977 October 16 by C. J. van Houten and I. van Houten-Groeneveld at Palomar.

(9709) 5192 T-3

Discovered 1977 October 16 by C. J. van Houten and I. van Houten-Groeneveld at Palomar.

(9710) 1964 VN1

Discovered 1964 November 9 at the Purple Mountain Observatory at Nanking.

(9711) 1972 PA

Discovered 1972 August 7 by P. Wild at Zimmerwald.

(9712) Nauplius

1973 SO_1. Discovered 1973 September 19 by C. J. van Houten and I. van Houten-Groeneveld at Palomar.

Nauplius, a son of Poseidon {see planet (4341)} and Amymone, was king of Euboea and father of Palamedes {see, respectively, planets (1119) and (2456)}. After a false accusation by Odysseus {see planet (1143)} caused the death of Palamedes, Nauplius sailed to Troy to seek vengeance. But this was denied by Agamemnon {see planet (911)}, so Nauplius returned to Greece, together with his surviving son Oceax {see planet (9713)}. However, Nauplius earned his revenge. He misled the Greeks returning from Troy by placing lights on the cliffs of his land, so that many ships ran aground. (M 34355)

(9713) Oceax

1973 SP_1. Discovered 1973 September 19 by C. J. van Houten and I. van Houten-Groeneveld at Palomar.

Oceax was the son of Nauplius of Euboea and brother of Palamedes {see planets (9712), (1119) and (2456), respectively}. He traveled with his father from Troy to Greece. (M 34355)

(9714) 1975 LF_1

Discovered 1975 June 1 by R. H. McNaught at Siding Spring.

(9715) 1975 SB_1

Discovered 1975 September 30 by S. J. Bus at Palomar.

(9716) 1975 UE

Discovered 1975 October 27 by P. Wild at Zimmerwald.

(9717) 1976 SR_5

Discovered 1976 September 24 by N. S. Chernykh at Nauchnyj.

(9718) 1976 YR_1

Discovered 1976 December 16 by L. I. Chernykh at Nauchnyj.

(9719) 1977 DF_2

Discovered 1977 February 18 by H. Kosai and K. Hurukawa at Kiso.

(9720) 1980 FH_1

Discovered 1980 March 16 by C.-I. Lagerkvist at La Silla.

(9721) 1980 GB
Discovered 1980 April 14 by E. Bowell at Anderson Mesa.

(9722) 1981 EZ
Discovered 1981 March 4 by H. Debehogne and G. DeSanctis at La Silla.

(9723) 1981 EP_{13}
Discovered 1981 March 1 by S. J. Bus at Siding Spring.

(9724) 1981 EW_{17}
Discovered 1981 March 2 by S. J. Bus at Siding Spring.

(9725) 1981 EE_{19}
Discovered 1981 March 2 by S. J. Bus at Siding Spring.

(9726) 1981 EY_{19}
Discovered 1981 March 2 by S. J. Bus at Siding Spring.

(9727) 1981 EW_{24}
Discovered 1981 March 2 by S. J. Bus at Siding Spring.

(9728) 1981 EX_{38}
Discovered 1981 March 2 by S. J. Bus at Siding Spring.

(9729) 1981 RQ
Discovered 1981 September 7 by A. Mrkos at Kleť.

(9730) 1982 FA
Discovered 1982 March 23 by M. L. Sitko and W. A. Stern at Mount Lemmon.

(9731) 1982 JD_1
Discovered 1982 May 15 at the Palomar Observatory at Palomar.

(9732) 1984 SJ_7
Discovered 1984 September 24 at the Bulgarian National Observatory at Rozhen.

(9733) 1985 SC_3
Discovered 1985 September 19 by N. S. Chernykh and L. I. Chernykh at Nauchnyj.

(9734) 1986 CB_2
Discovered 1986 February 12 by H. Debehogne at La Silla.

(9735) 1986 JD
Discovered 1986 May 2 at the Palomar Observatory at Palomar.

(9736) 1986 QP_2
Discovered 1986 August 28 by H. Debehogne at La Silla.

(9737) Dudarova
1986 SC_2. Discovered 1986 September 29 by L. G. Karachkina at Nauchnyj.

Named in honor of Veronica Borisovna Dudarova (1916-), the only woman conductor in Russia. She is a People's Artist of the U.S.S.R. (1977) and the state prize laureate of Russia (1980). She was artistic director and principal conductor of the Moscow State Symphony Orchestra from 1960 to 1989 and has been artistic director and principal conductor of the Veronica Dudarova Orchestra since 1991. Her innovative interpretations of such large musical forms as symphonies and oratorios with choruses has defined her contribution to Russian art. (M 34355)

Name suggested by G. V. Sviridov and S. P. Kapitza and supported by the discoverer.

(9738) 1987 DF_6
Discovered 1987 February 23 by H. Debehogne at La Silla.

(9739) 1987 SH_7
Discovered 1987 September 26 by C. S. Shoemaker at Palomar.

(9740) 1987 ST_{11}
Discovered 1987 September 23 by H. Debehogne at La Silla.

(9741) 1987 UU_4
Discovered 1987 October 22 by L. V. Zhuravleva at Nauchnyj.

(9742) Worpswede
1987 WT_1. Discovered 1987 November 26 by F. Börngen at Tautenburg.

Named for the German colony of artists near the city of Bremen {see planet (6320)}. Beginning in 1889, many painters settled in this lonely region, trying to live in harmony with nature. For their subjects they chose the heath and marshland, the farmhouses and the simple people. The poet R. M. Rilke {see planet (9833)} also lived here for some time. The fame of Worpswede as a paradise for painters, unique in nineteenth-century Europe, continues to fascinate art lovers all over the world. (M 34355)

(9743) 1988 GD
Discovered 1988 April 8 by E. F. Helin at Palomar.

(9744) 1988 JW
Discovered 1988 May 9 by C. S. Shoemaker at Palomar.

(9745) 1988 VY
Discovered 1988 November 2 by T. Seki at Geisei.

(9746) Kazukoichikawa
1988 VS_1. Discovered 1988 November 7 by Y. Kushida and M. Inoue at Yatsugatake.

Named in honor of Kazuko Ichikawa, who has loved dolls since her childhood. While working for a toy company as a designer of "Jenny", she was charmed by the lovely figures and started making teddy bears in 1987. She now attracts attention as one of the leading teddy-bear artists in Japan, and she teaches students the joy of making teddy bears themselves. Her works are introduced at Teddy Bear Museum as well as through her books, the best known of which is *My Friend Teddy Bear*. (M 34631)
Citation prepared by the second discoverer.

(9747) 1989 AT
Discovered 1989 January 4 by S. Ueda and H. Kaneda at Kushiro.

(9748) van Ostaijen
1989 CS_2. Discovered 1989 February 4 by E. W. Elst at La Silla.

Named in memory of Paul van Ostaijen (1896-1928), Flemish poet who influenced Belgian and Dutch poetry with his avant-garde writings. His second volume of poetry, *Het Sienjaal* (1918), inspired the Humanitarian Expressionist movement in Flanders. However, compromised as a political activist, he had to leave the country. He went to Berlin, where he was very disappointed by the artistic and political climate. Soon he developed his own poetic system, aiming at "pure poetry", without any personal and humanitarian confessions. His *First book of Schmoll* (1928) contains his best and most original poems. (M 34631)

(9749) 1989 GC_1
Discovered 1989 April 3 by E. W. Elst at La Silla.

(9750) 1989 NE_1
Discovered 1989 July 8 by A. C. Gilmore and P. M. Kilmartin at Lake Tekapo.

(9751) 1990 QM
Discovered 1990 August 20 by T. Seki at Geisei.

(9752) 1990 QZ_1
Discovered 1990 August 22 by H. E. Holt at Palomar.

(9753) 1990 QL_3
Discovered 1990 August 28 by H. E. Holt at Palomar.

(9754) 1990 QJ_4
Discovered 1990 August 23 by H. E. Holt at Palomar.

(9755) 1990 RR_2
Discovered 1990 September 15 by H. E. Holt at Palomar.

(9756) 1991 CC_3
Discovered 1991 February 12 by T. Seki at Geisei.

(9757) 1991 GA_6
Discovered 1991 April 8 by E. W. Elst at La Silla.

(9758) Dainty
1991 GZ_9. Discovered 1991 April 13 by D. I. Steel at Siding Spring.

Named in honor of J. Christopher Dainty (1947-), Pilkington Professor of Optics at Imperial College, London. Dainty has made numerous seminal contributions in the fields of stellar speckle interferometry, the characterisation of atmospheric turbulence, scattering and propagation of light, and adaptive optics. (M 34631)

Name proposed by the discoverer in recognition of the inspiration and guidance afforded him by Dainty whilst an undergraduate at the University of London.

(9759) 1991 NE_7
Discovered 1991 July 12 by H. Debehogne at La Silla.

(9760) 1991 PJ_{13}
Discovered 1991 August 5 by H. E. Holt at Palomar.

(9761) 1991 RR_4
Discovered 1991 September 13 by L. D. Schmadel and F. Börngen at Tautenburg.

(9762) Hermannhesse
1991 RA_5. Discovered 1991 September 13 by F. Börngen and L. D. Schmadel at Tautenburg.

Named in memory of Hermann Hesse (1877-1962), born in Swabia and since 1923 a citizen of Switzerland. As a lyric poet, writer, essayist, critic and illustrator, he was strongly impressed by Indian philosophy. Translated into many languages, his books were well received in the U.S.A. Hesse received numerous honors, among them the Nobel Prize for Literature in 1946 and the Peace Prize of the German Buchhandel in 1955. (M 34631)

Name proposed by the first discoverer.

(9763) 1991 RU_{17}
Discovered 1991 September 13 by H. E. Holt at Palomar.

(9764) Morgenstern

1991 UE_5. Discovered 1991 October 30 by F. Börngen at Tautenburg.

Named in memory of the German poet, writer and translator Christian Morgenstern (1871-1914). His most successful poems are the witty, grotesque and thoughtful *Galgenlieder*. His pensive contemplative poetry was influenced by Nietzsche {see planet (7014)}, later by Buddhism and the anthroposophy of his friend R. Steiner. (M 34631)

(9765) 1991 XZ

Discovered 1991 December 14 by H. Shiozawa and M. Kizawa at Fujieda.

(9766) 1992 DZ_2

Discovered 1992 February 24 by the Spacewatch at Kitt Peak.

(9767) Midsomer Norton

1992 EB_1. Discovered 1992 March 10 by D. I. Steel at Siding Spring.

Named for the small but historic English town where the discoverer was born and bred. Formerly a coal-mining center, Midsomer Norton lies on the fringe of the Mendip Hills, eight miles from the city of Bath,where William Herschel {see planet (2000)} discovered Uranus in 1781, and on the Fosseway, the great Roman road stretching from central to southwestern England. The town was also a hiding place of king Charles II in the Civil War during the 1650s. (M 34631)

(9768) 1992 GB_1

Discovered 1992 April 5 by C. S. Shoemaker at Palomar.

(9769) 1993 DG_2

Discovered 1993 February 24 by A. Natori and T. Urata at Yakiimo.

(9770) 1993 EE

Discovered 1993 March 1 by T. Urata at Oohira.

(9771) 1993 FU_{17}

Discovered 1993 March 17 by the Uppsala-ESO Survey at La Silla.

(9772) 1993 MB

Discovered 1993 June 16 by T. B. Spahr at Tucson.

(9773) 1993 MG_1

Discovered 1993 June 23 by E. F. Helin at Palomar.

(9774) 1993 NO

Discovered 1993 July 12 by E. W. Elst at La Silla.

(9775) 1993 OH_{12}

Discovered 1993 July 19 by E. W. Elst at La Silla.

(9776) 1993 VL_3

Discovered 1993 November 11 by S. Ueda and H. Kaneda at Kushiro.

(9777) 1994 OB

Discovered 1994 July 31 by Y. Shimizu and T. Urata at Nachi-Katsuura.

(9778) 1994 PA_{19}

Discovered 1994 August 12 by E. W. Elst at La Silla.

(9779) 1994 RA_{11}

Discovered 1994 September 1 by S. Ueda and H. Kaneda at Kushiro.

(9780) 1994 SB
Discovered 1994 September 25 by Y. Shimizu and T. Urata at Nachi-Katsuura.

(9781) 1994 UB_1
Discovered 1994 October 31 by Y. Shimizu and T. Urata at Nachi-Katsuura.

(9782) 1994 WM
Discovered 1994 November 25 by T. Kobayashi at Oizumi.

(9783) 1994 YD_1
Discovered 1994 December 28 by T. Kobayashi at Oizumi.

(9784) 1994 YJ_1
Discovered 1994 December 31 by T. Kobayashi at Oizumi.

(9785) 1994 YX_1
Discovered 1994 December 31 by T. Kobayashi at Oizumi.

(9786) 1995 BB
Discovered 1995 January 19 by T. Kobayashi at Oizumi.

(9787) 1995 BA_3
Discovered 1995 January 27 by S. Ueda and H. Kaneda at Kushiro.

(9788) 1995 EQ_1
Discovered 1995 March 11 by T. Kobayashi at Oizumi.

(9789) 1995 GO_7
Discovered 1995 April 4 by S. Ueda and H. Kaneda at Kushiro.

(9790) 1995 OK_8
Discovered 1995 July 25 by the Spacewatch at Kitt Peak.

(9791) 1995 YD_1
Discovered 1995 December 21 by T. Kobayashi at Oizumi.

(9792) 1996 BX_1
Discovered 1996 January 23 by T. Kobayashi at Oizumi.

(9793) 1996 BW_4
Discovered 1996 January 16 by the Spacewatch at Kitt Peak.

(9794) 1996 FO_5
Discovered 1996 March 25 at the Beijing Observatory at Xinglong.

(9795) 1996 GJ_{19}
Discovered 1996 April 15 by E. W. Elst at La Silla.

(9796) 1996 HW
Discovered 1996 April 19 by F. Manca and P. Chiavenna at Sormano.

(9797) 1996 HR_{21}
Discovered 1996 April 18 by E. W. Elst at La Silla.

(9798) 1996 JK
Discovered 1996 May 8 by S. Ueda and H. Kaneda at Kushiro.

(9799) 1996 RJ
Discovered 1996 September 8 by T. B. Spahr at Tucson.

(9800) 1997 ES_2
Discovered 1997 March 4 by T. Kobayashi at Oizumi.

(9801) 1997 FX_3
Discovered 1997 March 31 by the Lincoln NEA Research Team at Socorro.

(9802) 1997 GQ_6
Discovered 1997 April 2 by the Lincoln NEA Research Team at Socorro.

(9803) 1997 GL_8
Discovered 1997 April 2 by the Lincoln NEA Research Team at Socorro.

(9804) 1997 NU
Discovered 1997 July 1 by E. O. Ofek at Mitzpeh Ramon.

(9805) 1997 NZ
Discovered 1997 July 1 at the Beijing Observatory at Xinglong.

(9806) 1997 NR_6
Discovered 1997 July 10 at the Beijing Observatory at Xinglong.

(9807) 1997 SJ_4
Discovered 1997 September 27 by T. Kobayashi at Oizumi.

(9808) 1998 QS_{70}
Discovered 1998 August 24 by the Lincoln NEA Research Team at Socorro.

(9809) 1998 RZ_5
Discovered 1998 September 13 by the LONEOS at Anderson Mesa.

(9810) 1998 RJ_{65}
Discovered 1998 September 14 by the Lincoln NEA Research Team at Socorro.

(9811) 1998 ST
Discovered 1998 September 16 by the OCA-DLR Survey at Caussols.

(9812) 1998 SJ_{144}
Discovered 1998 September 18 by E. W. Elst at La Silla.

(9813) 1998 TP_5
Discovered 1998 October 13 by K. Korlević at Višnjan.

(9814) 1998 UU_{18}
Discovered 1998 October 23 by K. Korlevic at Višnjan.

(9815) 2079 P-L
Discovered 1960 September 24 by C. J. van Houten and I. van Houten-Groeneveld at Palomar.

(9816) 2643 P-L
Discovered 1960 September 24 by C. J. van Houten and I. van Houten-Groeneveld at Palomar.

(9817) Thersander
6540 P-L. Discovered 1960 September 24 by C. J. van Houten and I. van Houten-Groeneveld at Palomar.

Thersander, a son of the Theban Polyneices, was wounded by Telephus and later was one of the heroes in the Wooden Horse. (M 34355)

(9818) Eurymachos

6591 P-L. Discovered 1960 September 24 by C. J. van Houten and I. van Houten-Groeneveld at Palomar.

Eurymachos was one of the heroes hidden in the Wooden Horse. (M 34355)

(9819) 2172 T-1

Discovered 1971 March 25 by C. J. van Houten and I. van Houten-Groeneveld at Palomar.

(9820) 3064 T-1

Discovered 1971 March 26 by C. J. van Houten and I. van Houten-Groeneveld at Palomar.

(9821) 4033 T-1

Discovered 1971 March 26 by C. J. van Houten and I. van Houten-Groeneveld at Palomar.

(9822) 4114 T-1

Discovered 1971 March 26 by C. J. van Houten and I. van Houten-Groeneveld at Palomar.

(9823) 4271 T-1

Discovered 1971 March 26 by C. J. van Houten and I. van Houten-Groeneveld at Palomar.

(9824) 3033 T-2

Discovered 1973 September 30 by C. J. van Houten and I. van Houten-Groeneveld at Palomar.

(9825) 1214 T-3

Discovered 1977 October 17 by C. J. van Houten and I. van Houten-Groeneveld at Palomar.

(9826) 2114 T-3

Discovered 1977 October 16 by C. J. van Houten and I. van Houten-Groeneveld at Palomar.

(9827) 1958 TL_1

Discovered 1958 October 8 at the Lowell Observatory at Flagstaff.

(9828) Antimachos

1973 SS. Discovered 1973 September 19 by C. J. van Houten and I. van Houten-Groeneveld at Palomar.

Antimachos was one of the heroes hidden in the Wooden Horse. (M 34355)

(9829) Murillo

1973 SJ_1. Discovered 1973 September 19 by C. J. van Houten and I. van Houten-Groeneveld at Palomar.

Bartolomé Estéban Murillo (1618-1682) created many religious paintings for churches. The painter from Seville is famous for his genre pictures of children that were poor but happy, and these are imitated even today. (M 34355)

(9830) 1978 VE_{11}

Discovered 1978 November 7 by E. F. Helin and S. J. Bus at Palomar.

(9831) 1979 QZ

Discovered 1979 August 22 by C.-I. Lagerkvist at La Silla.

(9832) 1981 EH_3

Discovered 1981 March 2 by S. J. Bus at Siding Spring.

(9833) Rilke

1982 DW_3. Discovered 1982 February 21 by F. Börngen at Tautenburg.

Named for the Austrian poet Rainer Maria Rilke (1875-1926), husband of the sculptor Clara Westhoff. His monographs *Auguste Rodin* (1903) and *Das Marienleben* (1913), set to music by Hindemith {see planet (5157)}, were written in Paris. *Duineser Elegien* was his greatest late work. In his lyric creations he always strove for sonorous language. Some of his work is strongly influenced by religious longings. (M 34631)

(9834) Kirsanov

1982 TS_1. Discovered 1982 October 14 by L. G. Karachkina at Nauchnyj.

Named in memory of the brilliant Russian poet Semen Isaakovich Kirsanov (1906-1972). The innovative form of his verses and poems is a reflection of his deep lyrical-philosophical world outlook, while his popular verses are enriched by his peerless humor. He was given an enthusiastic welcome at the Crimean Astrophysical Observatory. (M 34631)
Named by the discoverer, following a suggestion by R. E. Gershberg.

(9835) 1984 UD

Discovered 1984 October 17 by Z. Vávrová at Kleť.

(9836) 1985 TU

Discovered 1985 October 15 by E. Bowell at Anderson Mesa.

(9837) 1986 AA_2

Discovered 1986 January 12 by I. Horowitz at Anderson Mesa.

(9838) 1987 RN_6

Discovered 1987 September 4 by L. V. Zhuravleva at Nauchnyj.

(9839) 1988 CT_2

Discovered 1988 February 11 by E. W. Elst at La Silla.

(9840) 1988 RQ_2

Discovered 1988 September 8 by P. Jensen at Brorfelde.

(9841) 1988 UT

Discovered 1988 October 18 by Z. Vávrová at Kleť.

(9842) 1989 AS_1

Discovered 1989 January 15 by K. Endate and K. Watanabe at Kitami.

(9843) 1989 AL_3

Discovered 1989 January 4 by R. H. McNaught at Siding Spring.

(9844) Otani

1989 WF_1. Discovered 1989 November 23 by Y. Kushida and O. Muramatsu at Yatsugatake.

Named in honor of Toyokazu Otani (1928-), first lecturer at the Gotoh Planetarium and Astronomical Museum in Tokyo, where he worked for 32 years until 1988. His deep knowledge of acoustics and classical music was instrumental in establishing the well-known BMG sound system, appreciated by planetarium audiences. Otani is also known as an eminent observer of the planets. (M 34355)

(9845) 1990 FM_1

Discovered 1990 March 27 by K. Endate and K. Watanabe at Kitami.

(9846) 1990 OS_1
Discovered 1990 July 29 by H. E. Holt at Palomar.

(9847) 1990 QJ_5
Discovered 1990 August 25 by H. E. Holt at Palomar.

(9848) 1990 QX_{17}
Discovered 1990 August 26 by L. V. Zhuravleva at Nauchnyj.

(9849) 1990 RF_2
Discovered 1990 September 14 by H. E. Holt at Palomar.

(9850) 1990 TM_5
Discovered 1990 October 9 by R. H. McNaught at Siding Spring.

(9851) 1990 UG_3
Discovered 1990 October 24 by K. Endate and K. Watanabe at Kitami.

(9852) 1990 YX
Discovered 1990 December 24 by T. Seki at Geisei.

(9853) 1991 AN_2
Discovered 1991 January 7 by R. H. McNaught at Siding Spring.

(9854) 1991 AC_3
Discovered 1991 January 15 by F. Börngen at Tautenburg.

(9855) 1991 CU
Discovered 1991 February 7 by R. H. McNaught at Siding Spring.

(9856) 1991 EE
Discovered 1991 March 13 by the Spacewatch at Kitt Peak.

(9857) 1991 EN
Discovered 1991 March 10 by R. H. McNaught at Siding Spring.

(9858) 1991 OL_1
Discovered 1991 July 18 by H. Debehogne at La Silla.

(9859) 1991 PE_5
Discovered 1991 August 3 by E. W. Elst at La Silla.

(9860) 1991 PW_9
Discovered 1991 August 6 by E. W. Elst at La Silla.

(9861) 1991 RB_3
Discovered 1991 September 9 by L. D. Schmadel and F. Börngen at Tautenburg.

(9862) 1991 RA_6
Discovered 1991 September 13 by H. E. Holt at Palomar.

(9863) Reichardt
1991 RJ_7. Discovered 1991 September 13 by F. Börngen and L. D. Schmadel at Tautenburg.

Named in memory of the German composer Johann Friedrich Reichardt (1752-1814), who for 20 years was conductor of the orchestra of the Royal Prussian court in Potsdam {see planet (5816)}. Dismissed because of his positive attitude towards the French revolution, he lived in Halle-Giebichenstein after 1794. For a time he was a collaborator of Goethe {see planet (3047)}. Numerous young poets and artists received encouragement from him. His extensive compositions include some 1000 songs in 30 collections using words from nearly all the German poets. He is considered the precursor of Schubert {see planet (3917)}. (M 34631) Name proposed by the first discoverer.

(9864) 1991 RT_{17}
Discovered 1991 September 13 by H. E. Holt at Palomar.

(9865) 1991 TP_1
Discovered 1991 October 3 by K. Suzuki and T. Urata at Toyota.

(9866) 1991 TV_4
Discovered 1991 October 15 by S. Otomo at Kiyosato.

(9867) 1991 VM
Discovered 1991 November 3 by A. Natori and T. Urata at Yakiimo.

(9868) 1991 VP_1
Discovered 1991 November 4 by S. Ueda and H. Kaneda at Kushiro.

(9869) 1992 CD_1
Discovered 1992 February 9 by K. Endate and K. Watanabe at Kitami.

(9870) 1992 DA
Discovered 1992 February 24 by T. Seki at Geisei.

(9871) 1992 DG_1
Discovered 1992 February 28 by T. Fujii and K. Watanabe at Kitami.

(9872) Solf
1992 DJ_4. Discovered 1992 February 27 by F. Börngen at Tautenburg.

Named in honor of Josef K. M. Solf (1934-), professor of astronomy at Heidelberg and Jena, and since 1994 director of the Thüringer Landessternwarte Tautenburg. During 1969-1994 he served as staff member of the Max-Planck-Institut für Astronomie at Heidelberg and its Calar Alto Observatory. He is primarily interested in instrumentation for large optical telescopes. A pioneer in using high-resolution long-slit spectroscopic techniques, Solf made important contributions to the study of collimated mass outflows from both young and evolved stars, such as jets from pre-main sequence stars, Herbig-Haro objects, protoplanetary nebulae, nova shells and jets from symbiotic stars and central stars of planetary nebulae. (M 34555)

(9873) 1992 GH
Discovered 1992 April 9 by R. H. McNaught at Siding Spring.

(9874) 1993 FG_{23}
Discovered 1993 March 21 by the Uppsala-ESO Survey at La Silla.

(9875) 1993 FH_{25}
Discovered 1993 March 21 by the Uppsala-ESO Survey at La Silla.

(9876) 1993 FY_{37}
Discovered 1993 March 19 by the Uppsala-ESO Survey at La Silla.

(9877) 1993 ST_3
Discovered 1993 September 18 by H. E. Holt at Palomar.

(9878) 1994 FQ
Discovered 1994 March 17 at the Farra d'Isonzo Observatory at Farra d'Isonzo.

(9879) 1994 PZ_{29}
Discovered 1994 August 12 by E. W. Elst at La Silla.

(9880) 1994 PQ_{31}
Discovered 1994 August 12 by E. W. Elst at La Silla.

(9881) 1994 SE
Discovered 1994 September 25 by R. H. McNaught at Siding Spring.

(9882) 1994 SS_9
Discovered 1994 September 28 by the Spacewatch at Kitt Peak.

(9883) 1994 TU_1
Discovered 1994 October 8 by J. E. Rogers at Camarillo.

(9884) 1994 TN_3
Discovered 1994 October 12 by M. Tichý and Z. Moravec at Kleť.

(9885) 1994 TM_{14}
Discovered 1994 October 12 by the Spacewatch at Kitt Peak.

(9886) 1994 VM_7
Discovered 1994 November 8 by S. Otomo at Kiyosato.

(9887) 1995 AH
Discovered 1995 January 2 by T. Kobayashi at Oizumi.

(9888) 1995 CD
Discovered 1995 February 1 by T. Kobayashi at Oizumi.

(9889) 1995 FG_1
Discovered 1995 March 28 by S. Ueda and H. Kaneda at Kushiro.

(9890) 1995 SY_2
Discovered 1995 September 20 by S. Ueda and H. Kaneda at Kushiro.

(9891) 1995 XN_1
Discovered 1995 December 15 by T. Kobayashi at Oizumi.

(9892) 1995 YN_3
Discovered 1995 December 27 by T. Kobayashi at Oizumi.

(9893) 1996 AA_1
Discovered 1996 January 12 by T. Kobayashi at Oizumi.

(9894) 1996 BS_1
Discovered 1996 January 23 by T. Kobayashi at Oizumi.

(9895) 1996 BR_3
Discovered 1996 January 27 by T. Kobayashi at Oizumi.

(9896) 1996 BL_{17}
Discovered 1996 January 22 by the Lincoln Laboratory ETS at Socorro.

(9897) 1996 CX_7
Discovered 1996 February 14 by U. Munari and M. Tombelli at Cima Ekar.

(9898) 1996 DF
Discovered 1996 February 18 by T. Kobayashi at Oizumi.

(9899) 1996 EH

Discovered 1996 March 12 by R. H. McNaught at Siding Spring.

(9900) 1997 LL_6

Discovered 1997 June 13 by M. Blasco at Mallorca.

(9901) 1997 NV

Discovered 1997 July 1 at the Kleť Observatory at Kleť.

(9902) Kirkpatrick

1997 NY. Discovered 1997 July 3 by P. G. Comba at Prescott.

Named for Ralph Kirkpatrick (1911-1984), eminent American harpsichordist, clavichordist, musicologist and teacher. He traveled extensively in Europe, where he studied seventeenth-century and eighteenth-century performance practices in chamber music. He wrote a biography of Domenico Scarlatti {see planet (6480)} and prepared a facsimile edition of his more than 500 sonatas. He gave many concerts, specializing in the performance of works by Scarlatti and J. S. Bach {see planet (1814)}. (M 34356)

(9903) Leonhardt

1997 NA_1. Discovered 1997 July 4 by P. G. Comba at Prescott.

Named for Gustav Leonhardt (1928-), eminent Dutch harpsichordist, conductor and pedagogue. As a harpsichordist, he has made many concert tours in Europe and North America and recorded a large number of baroque works. (M 34356)

(9904) Mauratombelli

1997 OC_1. Discovered 1997 July 29 by A. Boattini and L. Tesi at San Marcello Pistoiese.

Named in honor of Maura Tombelli (1952-). Initially trained as an observer of variable stars, in 1994 she started a five-year survey of minor planets at Asiago {see planet (7679)} Astrophysical Observatory with Ulisse Munari {see planet (7599)} and Giuseppe Forti in Arcetri {see planets (6876) and (6645)}. She also shared a lot of observing with the discoverers, especially the follow-up of NEOs, and she contributed to the discovery of 1994 QC, the first NEA found from Italy. She is currently involved in a project to build a new observatory near the town of Montelupo, where she lives. She is still the only female astrometrist in Italy. (M 34632)

(9905) Tiziano

4611 P-L. Discovered 1960 September 24 by C. J. van Houten and I. van Houten-Groeneveld at Palomar.

Tiziano Vecellio (Titian; c.1477 or 1490-1576) was a typical Renaissance painter. He loved colors, the contrast of light and dark, and gave the people in his paintings a personal expression. Titian painted portraits of Charles V, Phillip II, Franz I and pope Paul III. (M 34356)

(9906) Tintoretto

6523 P-L. Discovered 1960 September 26 by C. J. van Houten and I. van Houten-Groeneveld at Palomar.

Named after Venetian painter Jacopo Robusti (1518-1594), called Tintoretto. He was influenced by Titian {see planet (9905)}. His main work can be seen in the Scuola di S. Rocco and the palace of the Doge in Venice. (M 34356)

(9907) Oileus

6541 P-L. Discovered 1960 September 24 by C. J. van Houten and I. van Houten-Groeneveld at Palomar.

Oileus was the father of Ajax the Little. Next to Achilles {see planet (588)}, Ajax was the fastest-running Greek hero. (M 34356)

(9908) Aue

2140 T-1. Discovered 1971 March 25 by C. J. van Houten and I. van Houten-Groeneveld at Palomar.

Hartmann von Aue (c.1160-1210) was educated in a cloister and took part in the crusade of Barbarossa {see also planet (1860)}. He wrote epic works *Erec*, *Iwein* and *Der arme Heinrich* after the model of Chrèstien de Troyes. He also created minstrel songs. (M 34356)

(9909) Eschenbach

4355 T-1. Discovered 1971 March 26 by C. J. van Houten and I. van Houten-Groeneveld at Palomar.

Wolfram von Eschenbach (1168-1220) was, with von Aue {see planet (9908)} and Gottfried von Strassburg, one of the greatest poets of medieval times. Beginning as a song writer, he later wrote three great epics: *Parzifal*, with the fragmentarian Gralsepos of Chrèstien de Troyes as a basis, and the two unfinished works *Titurel* and *Willehalm*. (M 34356)

(9910) Vogelweide

3181 T-2. Discovered 1973 September 30 by C. J. van Houten and I. van Houten-Groeneveld at Palomar.

Walther von der Vogelweide (c.1170-1230) was probably born in Niederösterreich and died near Würzburg, Germany. A famous minstrel, his portrait appears in the *Manesse-Handschrift* from 1320. Quite a number of his songs with several verses have been preserved. (M 34356)

(9911) Quantz

4129 T-2. Discovered 1973 September 29 by C. J. van Houten and I. van Houten-Groeneveld at Palomar.

Johan Joachim Quantz (1697-1773), the son of a smith, became city piper of Merseburg, Germany, and later came to the court in Dresden {see planet (3053)} as an oboe player. Thirty years later he took up the flute, studying throughout Europe and introducing new techniques for the flute. In 1741 he became the teacher and musical conductor for the Prussian king Friedrich II the Great. Quantz composed about 500 pieces for flute and published a book about playing the flute and on Baroque music in general. (M 34356)

(9912) Donizetti

2078 T-3. Discovered 1977 October 16 by C. J. van Houten and I. van Houten-Groeneveld at Palomar.

The Italian composer Gaetano Donizetti (1797-1848) wrote symphonies, church and chamber music. After 1818 he composed 74 operas. He is still much esteemed today. (M 34356)

(9913) Humperdinck

4071 T-3. Discovered 1977 October 16 by C. J. van Houten and I. van Houten-Groeneveld at Palomar.

Engelbert Humperdinck (1854-1921) was first assistant to Richard Wagner {see planet (3992)}. Later he traveled throughout Europe and Northern Africa and became a teacher in different conservatories and art academies. His greatest success was the opera *Hänsel und Gretel*, which remains a favorite today, with its folksong melodies. (M 34356)

(9914) 1976 UJ_4

Discovered 1976 October 28 by L. V. Zhuravleva at Nauchnyj.

(9915) 1977 RD_2
Discovered 1977 September 8 by N. S. Chernykh at Nauchnyj.

(9916) 1978 TR_2
Discovered 1978 October 3 by N. S. Chernykh at Nauchnyj.

(9917) 1979 MK
Discovered 1979 June 26 by C. Torres at Cerro El Roble.

(9918) 1979 MK_3
Discovered 1979 June 25 by E. F. Helin and S. J. Bus at Siding Spring.

(9919) 1979 QF_1
Discovered 1979 August 22 by C.-I. Lagerkvist at La Silla.

(9920) 1981 EZ_{10}
Discovered 1981 March 1 by S. J. Bus at Siding Spring.

(9921) 1981 EO_{18}
Discovered 1981 March 2 by S. J. Bus at Siding Spring.

(9922) 1981 EO_{21}
Discovered 1981 March 2 by S. J. Bus at Siding Spring.

(9923) 1981 EB_{24}
Discovered 1981 March 7 by S. J. Bus at Siding Spring.

(9924) 1981 EM_{24}
Discovered 1981 March 2 by S. J. Bus at Siding Spring.

(9925) 1981 EU_{24}
Discovered 1981 March 2 by S. J. Bus at Siding Spring.

(9926) 1981 EU_{41}
Discovered 1981 March 2 by S. J. Bus at Siding Spring.

(9927) Tyutchev
1981 TW_1. Discovered 1981 October 3 by L. G. Karachkina at Nauchnyj.
Named in memory of the Russian poet Fedor Ivanovich Tyutchev (1803-1873). One of the greatest representatives of Russian lyricism of the nineteenth century, he was a master of Russian poetry. Many of his verses were set to music and translated into other languages. (M 34632)

(9928) 1981 WE_9
Discovered 1981 November 16 at the Perth Observatory at Bickley.

(9929) 1982 DP_1
Discovered 1982 February 24 at the Oak Ridge Observatory at Harvard.

(9930) 1984 CP
Discovered 1984 February 5 by E. Bowell at Anderson Mesa.

(9931) 1985 HH
Discovered 1985 April 18 by A. Mrkos at Kleť.

(9932) 1985 QP_5
Discovered 1985 August 23 by N. S. Chernykh at Nauchnyj.

(9933) 1985 SM_3
Discovered 1985 September 19 by N. S. Chernykh and L. I. Chernykh at Nauchnyj.

(9934) 1985 UC
Discovered 1985 October 20 by E. Bowell at Anderson Mesa.

(9935) 1986 CP_1
Discovered 1986 February 4 by H. Debehogne at La Silla.

(9936) 1986 PN_4
Discovered 1986 August 8 by E. W. Elst and V. G. Ivanova at Rozhen.

(9937) 1988 DJ_2
Discovered 1988 February 17 by E. W. Elst at La Silla.

(9938) 1988 KA
Discovered 1988 May 18 by W. Landgraf at La Silla.

(9939) 1988 VK
Discovered 1988 November 3 by T. Kojima at Chiyoda.

(9940) 1988 VM_3
Discovered 1988 November 11 by Y. Oshima at Gekko.

(9941) 1989 CB_3
Discovered 1989 February 4 by E. W. Elst at La Silla.

(9942) 1989 TM_1
Discovered 1989 October 8 by T. Hioki and N. Kawasato at Okutama.

(9943) 1989 UG_3
Discovered 1989 October 29 by M. Iwamoto and T. Furuta at Tokushima.

(9944) 1990 DA_3
Discovered 1990 February 24 by H. Debehogne at La Silla.

(9945) 1990 KX
Discovered 1990 May 21 by E. F. Helin at Palomar.

(9946) 1990 ON_2
Discovered 1990 July 29 by H. E. Holt at Palomar.

(9947) 1990 QB
Discovered 1990 August 17 by E. F. Helin at Palomar.

(9948) 1990 QB_2
Discovered 1990 August 22 by H. E. Holt at Palomar.

(9949) 1990 SK_6
Discovered 1990 September 22 by E. W. Elst at La Silla.

(9950) 1990 VB
Discovered 1990 November 8 by C. Pollas at Caussols.

(9951) 1990 VK_5
Discovered 1990 November 15 by E. W. Elst at La Silla.

(9952) 1991 AK
Discovered 1991 January 9 by M. Arai and H. Mori at Yorii.

(9953) 1991 EB
Discovered 1991 March 7 by S. Ueda and H. Kaneda at Kushiro.

(9954) 1991 GX_7
Discovered 1991 April 8 by E. W. Elst at La Silla.

(9955) 1991 PU_{11}
Discovered 1991 August 7 by H. E. Holt at Palomar.

(9956) 1991 TX_4
Discovered 1991 October 5 by L. D. Schmadel and F. Börngen at Tautenburg.

(9957) 1991 TO_{13}
Discovered 1991 October 6 by F. Börngen at Tautenburg.

(9958) 1991 VL_1
Discovered 1991 November 4 by S. Ueda and H. Kaneda at Kushiro.

(9959) 1991 VF_2
Discovered 1991 November 9 by S. Ueda and H. Kaneda at Kushiro.

(9960) 1991 VE_4
Discovered 1991 November 4 by S. Otomo at Kiyosato.

(9961) 1991 XK
Discovered 1991 December 4 by S. Ueda and H. Kaneda at Kushiro.

(9962) 1991 YL_1
Discovered 1991 December 28 by F. Börngen at Tautenburg.

(9963) 1992 AN
Discovered 1992 January 9 by E. F. Helin at Palomar.

(9964) 1992 CF_1
Discovered 1992 February 13 by T. Seki at Geisei.

(9965) 1992 EF_2
Discovered 1992 March 5 by the Spacewatch at Kitt Peak.

(9966) 1992 ES_{13}
Discovered 1992 March 2 by the Uppsala-ESO Survey at La Silla.

(9967) 1992 FV_1
Discovered 1992 March 31 by K. Endate and K. Watanabe at Kitami.

(9968) 1992 JS_2
Discovered 1992 May 4 by H. Debehogne at La Silla.

(9969) 1992 KD
Discovered 1992 May 27 by E. F. Helin and K. J. Lawrence at Palomar.

(9970) 1992 ST_1
Discovered 1992 September 26 by A. Sugie at Taga.

(9971) 1993 HS
Discovered 1993 April 16 by K. Endate and K. Watanabe at Kitami.

(9972) 1993 KQ
Discovered 1993 May 26 by S. Otomo at Kiyosato.

(9973) 1993 NB_2
Discovered 1993 July 12 by E. W. Elst at La Silla.

(9974) 1993 OG_{13}
Discovered 1993 July 19 by E. W. Elst at La Silla.

(9975) 1993 RZ_1
Discovered 1993 September 12 by K. Endate and K. Watanabe at Kitami.

(9976) 1993 TQ
Discovered 1993 October 9 by S. Shirai and S. Hayakawa at Hidaka.

(9977) 1994 AH
Discovered 1994 January 2 by T. Kobayashi at Oizumi.

(9978) 1994 AJ_1
Discovered 1994 January 7 by T. Kobayashi at Oizumi.

(9979) 1994 VT
Discovered 1994 November 3 by T. Kobayashi at Oizumi.

(9980) 1995 BQ_3
Discovered 1995 January 31 by T. Kobayashi at Oizumi.

(9981) 1995 BS_3
Discovered 1995 January 31 by T. Kobayashi at Oizumi.

(9982) 1995 CH
Discovered 1995 February 1 by T. Kobayashi at Oizumi.

(9983) 1995 DA
Discovered 1995 February 19 by D. di Cicco at Sudbury.

(9984) 1996 HT
Discovered 1996 April 18 by R. H. McNaught and J. B. Child at Macquarie.

(9985) Akiko
1996 JF. Discovered 1996 May 12 by R. H. McNaught and H. Abe at Yatsuka.

Named in honor of Akiko Yamamoto (1963-), a member of the Yatsuka Observatory and observing partner of the second discoverer. (M 34632)

(9986) 1996 NX
Discovered 1996 July 12 by H. Shiozawa and T. Urata at Nachi-Katsuura.

(9987) Peano
1997 OO_1. Discovered 1997 July 29 by P. G. Comba at Prescott.

Named for Giuseppe Peano (1858-1932), Italian mathematician, professor at the University of Turin and a founder of symbolic logic. Peano's *Formulario Mathematico* was intended to develop mathematics in its entirety from a set of basic postulates by use of a logic notation, and it had a profound influence on later studies in the foundations of mathematics. Peano was also the creator of the artificial language "Latino sine flexione". Later renamed "Interlingua", it had a simple grammar and a lexicon based on the Latin roots of words common to the major European languages. (M 34632)

(9988) Erictemplebell
1997 RX_6. Discovered 1997 September 9 by P. G. Comba at Prescott.

Named for Eric Temple Bell (1883-1960), Scottish-American mathematician and professor at the California Institute of Technology, where the discoverer attended his algebra class and was exposed to his acerbic wit. Bell wrote more than 250 scholarly papers, mainly in number theory, and several influential books of popularization, including *Mathematics, Queen and Servant of Science.* he also wrote several books of science fiction under the pseudonym John Taine. (M 34632)

(9989) 1997 SG_{16}

Discovered 1997 September 27 by N. Kawasato at Uenohara.

(9990) 1997 SO_{17}

Discovered 1997 September 30 by T. Okuni at Nanyou.

(9991) Anežka

1997 TY_7. Discovered 1997 October 5 by Z. Moravec at Kleť.

Named in honor of Anežka Moravcová (1924-), the discoverer's grandmother, on the occasion of her 75th birthday. (M 34632)

(9992) 1997 TG_{19}

Discovered 1997 October 8 by T. Kagawa and T. Urata at Gekko.

(9993) 1997 VX_5

Discovered 1997 November 6 by J. Kobayashi at Kumamoto.

(9994) 4028 P-L

Discovered 1960 September 24 by C. J. van Houten and I. van Houten-Groeneveld at Palomar.

(9995) 4805 P-L

Discovered 1960 September 24 by C. J. van Houten and I. van Houten-Groeneveld at Palomar.

(9996) 9070 P-L

Discovered 1960 October 17 by C. J. van Houten and I. van Houten-Groeneveld at Palomar.

(9997) 1217 T-1

Discovered 1971 March 25 by C. J. van Houten and I. van Houten-Groeneveld at Palomar.

(9998) 1293 T-1

Discovered 1971 March 25 by C. J. van Houten and I. van Houten-Groeneveld at Palomar.

(9999) Wiles

4196 T-2. Discovered 1973 September 29 by C. J. van Houten and I. van Houten-Groeneveld at Palomar.

Named in honor of Andrew J. Wiles (1953-), British mathematician and world-renowned for his 1993 proof of Fermat's Last Theorem, one of the most difficult mathematical problems, which had remained unsolved for more than 350 years. A gap in the proof was closed by Wiles in 1994. His achievement came too late for him to receive the Field Medal, because Wiles was then over the age limit of 40. The International Mathematical Union therefore gave him its most prestigious award, the Silver Plaque. Wiles' fields of research are number theory and arithmetic geometry. He is a professor at Princeton University. (M 34356) Name proposed and citation prepared by L. D. Schmadel.

(10000) Myriostos

1951 SY. Discovered 1951 September 30 by A. G. Wilson at Palomar.

The Greek word for ten-thousandth, Myriostos honors all the astronomers, past and present, from all around the world, professional and amateur, observer and orbit computer, who participated, over an interval of 198 years, in the achievement of accumulating 10,000 minor planets with orbit determinations of the highest quality. (M 34632)

This name was proposed by the majority of members of the Small Bodies Names Committee. Citation prepared by B. G. Marsden.

The naming of this planet was very controversial. The Minor Planet Center in a Editorial Notice (M.P.C. 33615) discussed the naming procedure during the nearly two centuries of minor planet discoveries: "... It has been traditional to have a special celebration with each thousandth numbering... Obviously, it would be appropriate to have some very special celebration to acknowledge (10000)... Most readers... will be aware of recent discussions in the press concerning a proposal that the number (10000) should be given to Pluto. The principal reasoning for this is the recognition during the past few years that Pluto was the first discovered and largest known member of the "Transneptunian Belt" of small objects beyond Neptune that possess some similarity, at least dynamically, to bodies in the Cisjovian Belt. Although as many as 95 members of the Transneptunian Belt are now listed, most of the orbital solutions are very weak... Although it is not unlikely that further Transneptunian Objects as large as Pluto will be discovered in the future, Pluto obviously holds a very special place in our appreciation of this new population, and by assigning to it the number (10000), we should guarantee that Pluto will be at the head of the Transneptunian list. It is also very important to affirm that there is absolutely no implied "demotion" or "reclassification" of Pluto from its position in the list of the "planets"..." All users of the M.P.C.s were invited to vote on this proposal or to make an alternative suggestion. An examination of the votes was published in M.P.C. 33817. Although the messages received were from 22 different countries, some 57 percent of the vote originated in the U.S. Respondents from outside the U.S. voted overwhelmingly - 82 percent - in favor of (10000) Pluto. Inside the U.S., the vote in favor of (10000) Pluto was 51 percent. A slight majority in the Small Bodies Names Committee was in favor, but this "...had a rather sudden change of heart following agitation by a group of planetary astronomers, mainly located in the U.S. Acting on this sudden decision, the IAU Secretariat announced on Feb. 3 that the Small Bodies Names Committee had "decided against assigning any Minor Planet number to Pluto"..."

(10001) 1969 TM_1

Discovered 1969 October 8 by L. I. Chernykh at Nauchnyj.

(10002) 1969 TQ_1

Discovered 1969 October 8 by L. I. Chernykh at Nauchnyj.

(10003) 1971 UD_1

Discovered 1971 October 26 by L. Kohoutek at Bergedorf.

(10004) 1975 VV_2

Discovered 1975 November 2 by T. M. Smirnova at Nauchnyj.

(10005) 1976 SS_2

Discovered 1976 September 24 by N. S. Chernykh at Nauchnyj.

(10006) 1976 UR_{15}

Discovered 1976 October 22 by H. Kosai and K. Hurukawa at Kiso.

(10007) 1976 YF_3

Discovered 1976 December 16 by L. I. Chernykh at Nauchnyj.

(10008) 1977 DT_2

Discovered 1977 February 18 by H. Kosai and K. Hurukawa at Kiso.

(10009) 1977 EA_6
Discovered 1977 March 12 by H. Kosai and K. Hurukawa at Kiso.

(10010) 1978 PW_3
Discovered 1978 August 9 by N. S. Chernykh and L. I. Chernykh at Nauchnyj.

(10011) 1978 QY_1
Discovered 1978 August 31 by N. S. Chernykh at Nauchnyj.

(10012) 1978 RE_3
Discovered 1978 September 3 by N. S. Chernykh and L. G. Karachkina at Nauchnyj.

(10013) 1978 RR_8
Discovered 1978 September 2 by C.-I. Lagerkvist at La Silla.

(10014) 1978 SE_3
Discovered 1978 September 26 by L. V. Zhuravleva at Nauchnyj.

(10015) 1978 SA_5
Discovered 1978 September 27 by L. I. Chernykh at Nauchnyj.

(10016) 1978 SW_7
Discovered 1978 September 26 by L. V. Zhuravleva at Nauchnyj.

(10017) 1978 UP_2
Discovered 1978 October 30 at the Purple Mountain Observatory at Nanking.

(10018) 1979 MG_4
Discovered 1979 June 25 by E. F. Helin and S. J. Bus at Siding Spring.

(10019) 1979 MK_7
Discovered 1979 June 25 by E. F. Helin and S. J. Bus at Siding Spring.

(10020) 1979 OQ_5
Discovered 1979 July 24 by S. J. Bus at Palomar.

(10021) 1979 QC_1
Discovered 1979 August 22 by C.-I. Lagerkvist at La Silla.

(10022) 1979 SU_2
Discovered 1979 September 22 by N. S. Chernykh at Nauchnyj.

(10023) 1979 WX_3
Discovered 1979 November 17 by L. I. Chernykh at Nauchnyj.

(10024) 1980 EB
Discovered 1980 March 10 at the Harvard College Observatory at Harvard.

(10025) 1980 FO_1
Discovered 1980 March 16 by C.-I. Lagerkvist at La Silla.

(10026) 1980 RE_1
Discovered 1980 September 3 by A. Mrkos at Kleť.

(10027) 1981 FL
Discovered 1981 March 30 by E. Bowell at Anderson Mesa.

(10028) 1981 JM_2
Discovered 1981 May 5 by C. S. Shoemaker at Palomar.

(10029) 1981 QF
Discovered 1981 August 30 by E. Bowell at Anderson Mesa.

(10030) 1981 QG
Discovered 1981 August 30 by E. Bowell at Anderson Mesa.

(10031) 1981 RB_2
Discovered 1981 September 7 by L. G. Karachkina at Nauchnyj.

(10032) 1981 RF_7
Discovered 1981 September 3 by S. J. Bus at Palomar.

(10033) 1981 UJ_{23}
Discovered 1981 October 24 by S. J. Bus at Palomar.

(10034) 1981 YG
Discovered 1981 December 30 by E. Bowell at Anderson Mesa.

(10035) 1982 DC_2
Discovered 1982 February 16 by L. Brožek at Kleť.

(10036) 1982 OF
Discovered 1982 July 24 by E. Bowell at Anderson Mesa.

(10037) 1984 BQ
Discovered 1984 January 26 by A. Mrkos at Kleť.

(10038) 1984 HO_1
Discovered 1984 April 28 by W. Ferreri and V. Zappalà at La Silla.

(10039) 1984 LK
Discovered 1984 June 2 by B. A. Skiff at Anderson Mesa.

(10040) 1984 QM
Discovered 1984 August 24 by Z. Vávrová at Kleť.

(10041) 1985 HS_1
Discovered 1985 April 24 by C. S. Shoemaker at Palomar.

(10042) 1985 PL
Discovered 1985 August 14 by E. Bowell at Anderson Mesa.

(10043) 1985 PN
Discovered 1985 August 14 by E. Bowell at Anderson Mesa.

(10044) 1985 RU
Discovered 1985 September 15 by C. S. Shoemaker at Palomar.

(10045) 1985 RJ_3
Discovered 1985 September 6 by H. Debehogne at La Silla.

(10046) 1986 JC
Discovered 1986 May 2 at the Palomar Observatory at Palomar.

(10047) 1986 QK_2
Discovered 1986 August 28 by H. Debehogne at La Silla.

(10048) 1986 TQ
Discovered 1986 October 3 by P. Jensen at Brorfelde.

(10049) 1986 TZ_{11}
Discovered 1986 October 3 by L. G. Karachkina at Nauchnyj.

(10050) 1987 MA_1
Discovered 1987 June 28 by E. F. Helin at Palomar.

(10051) 1987 QG_6
Discovered 1987 August 23 by E. F. Helin at Palomar.

(10052) 1987 SM_{12}
Discovered 1987 September 16 by H. Debehogne at La Silla.

(10053) 1987 SR_{12}
Discovered 1987 September 16 by H. Debehogne at La Silla.

(10054) 1987 SQ_{17}
Discovered 1987 September 17 by L. I. Chernykh at Nauchnyj.

(10055) Silcher
1987 YC_1. Discovered 1987 December 22 by F. Börngen at Tautenburg.

Named in memory of the German composer of popular songs Philipp Friedrich Silcher (1789-1860), musical director at the University of Tübingen {see planet (1481)} from 1817. He was a famous music educator, promoter of music for the laity and founder of several choirs. Silcher published numerous collections of folksongs and composed, transcribed and arranged music for male and mixed choirs. Among his best-known compositions, still very popular today, are *Ännchen von Tharau*, *Ich weiss nicht, was soll es bedeuten* and *Alle Jahre wieder*. (M 34632)

(10056) 1988 BX_3
Discovered 1988 January 19 by H. Debehogne at La Silla.

(10057) 1988 CO_1
Discovered 1988 February 11 by E. W. Elst at La Silla.

(10058) 1988 DD_5
Discovered 1988 February 25 by R. H. McNaught at Siding Spring.

(10059) 1988 FS_2
Discovered 1988 March 21 at the Bulgarian National Observatory at Rozhen.

(10060) 1988 GL
Discovered 1988 April 12 by C. S. Shoemaker at Palomar.

(10061) 1988 PG_1
Discovered 1988 August 11 by A. J. Noymer at Siding Spring.

(10062) 1988 RV_4
Discovered 1988 September 1 by H. Debehogne at La Silla.

(10063) 1988 SZ_2
Discovered 1988 September 16 by S. J. Bus at Cerro Tololo.

(10064) 1988 UO
Discovered 1988 October 31 by T. Kojima at Chiyoda.

(10065) 1988 XK
Discovered 1988 December 3 by Y. Oshima at Gekko.

(10066) 1988 XV2
Discovered 1988 December 1 by P. Jensen at Brorfelde.

(10067) 1989 AL_6
Discovered 1989 January 11 by F. Börngen at Tautenburg.

(10068) 1989 CT_2
Discovered 1989 February 4 by E. W. Elst at La Silla.

(10069) 1989 CW_2
Discovered 1989 February 4 by E. W. Elst at La Silla.

(10070) 1989 CB_8
Discovered 1989 February 7 by H. Debehogne at La Silla.

(10071) 1989 EZ_2
Discovered 1989 March 2 by E. W. Elst at La Silla.

(10072) 1989 GF_1
Discovered 1989 April 3 by E. W. Elst at La Silla.

(10073) 1989 GJ_2
Discovered 1989 April 3 by E. W. Elst at La Silla.

(10074) 1989 GH_4
Discovered 1989 April 3 by E. W. Elst at La Silla.

(10075) 1989 GR_4
Discovered 1989 April 3 by E. W. Elst at La Silla.

(10076) 1989 PK
Discovered 1989 August 9 by E. F. Helin at Palomar.

(10077) 1989 UL_1
Discovered 1989 October 26 by S. Ueda and H. Kaneda at Kushiro.

(10078) 1989 UJ_3
Discovered 1989 October 30 by T. Seki at Geisei.

(10079) 1989 XD_2
Discovered 1989 December 2 by E. W. Elst at La Silla.

(10080) 1990 OF_1
Discovered 1990 July 18 by E. F. Helin at Palomar.

(10081) 1990 OW_1
Discovered 1990 July 29 by H. E. Holt at Palomar.

(10082) 1990 OF_2
Discovered 1990 July 29 by H. E. Holt at Palomar.

(10083) 1990 QE_2
Discovered 1990 August 22 by H. E. Holt at Palomar.

(10084) 1990 QC_5
Discovered 1990 August 25 by H. E. Holt at Palomar.

(10085) 1990 QF_5
Discovered 1990 August 25 by H. E. Holt at Palomar.

(10086) 1990 SZ
Discovered 1990 September 16 by H. E. Holt at Palomar.

(10087) 1990 SG$_3$
Discovered 1990 September 18 by H. E. Holt at Palomar.

(10088) 1990 SG$_8$
Discovered 1990 September 22 by E. W. Elst at La Silla.

(10089) 1990 SS$_9$
Discovered 1990 September 22 by E. W. Elst at La Silla.

(10090) 1990 TK$_{15}$
Discovered 1990 October 13 by L. G. Karachkina and G. R. Kastel' at Nauchnyj.

(10091) 1990 VD$_3$
Discovered 1990 November 11 by T. Seki at Geisei.

(10092) 1990 VD$_4$
Discovered 1990 November 15 by K. Endate and K. Watanabe at Kitami.

(10093) 1990 WX$_1$
Discovered 1990 November 18 by E. W. Elst at La Silla.

(10094) 1991 DK
Discovered 1991 February 20 by T. Seki at Geisei.

(10095) Carlloewe
1991 RP$_2$. Discovered 1991 September 9 by F. Börngen and L. D. Schmadel at Tautenburg.

Named in memory of the German composer Johann Carl Friedrich Loewe (1796-1869), born in Löbejün, near Halle. At the age of 23 he visited Goethe in Weimar {see planet (3539)}. From 1820 to 1866 he lived in Stettin and was organist, precentor, music teacher and director for music in the town. There he directed the Pomeranian music festivals. His compositional work includes about 400 ballads, for which he favored dramatic-epic and romantic words, particularly by Goethe, Uhland and Herder {see, respectively, planets (3047), (9052) and (8158)}. This form of song owes its real popularity to him. He performed his ballads himself at concerts. (M 34632; M 34673)
Name proposed by the first discoverer.

(10096) 1991 RK$_5$
Discovered 1991 September 13 by H. E. Holt at Palomar.

(10097) 1991 RV$_{16}$
Discovered 1991 September 15 by H. E. Holt at Palomar.

(10098) 1991 SC$_1$
Discovered 1991 September 30 by R. H. McNaught at Siding Spring.

(10099) 1991 VB$_9$
Discovered 1991 November 4 by the Spacewatch at Kitt Peak.

(10100) 1991 XH$_1$
Discovered 1991 December 10 by F. Börngen at Tautenburg.

(10101) 1992 BM$_2$
Discovered 1992 January 30 by E. W. Elst at La Silla.

(10102) 1992 DA_6
Discovered 1992 February 29 by the Uppsala-ESO Survey at La Silla.

(10103) 1992 DB_9
Discovered 1992 February 29 by the Uppsala-ESO Survey at La Silla.

(10104) 1992 EY_9
Discovered 1992 March 2 by the Uppsala-ESO Survey at La Silla.

(10105) 1992 EM_{12}
Discovered 1992 March 6 by the Uppsala-ESO Survey at La Silla.

(10106) 1992 EV_{15}
Discovered 1992 March 1 by the Uppsala-ESO Survey at La Silla.

(10107) 1992 FW_1
Discovered 1992 March 27 by D. I. Steel at Siding Spring.

(10108) 1992 HM
Discovered 1992 April 26 by C. S. Shoemaker at Palomar.

(10109) 1992 KQ
Discovered 1992 May 29 by E. F. Helin at Palomar.

(10110) 1992 LJ
Discovered 1992 June 3 by G. J. Leonard at Palomar.

(10111) 1992 OO_1
Discovered 1992 July 25 by E. W. Elst at Caussols.

(10112) 1992 OP_1
Discovered 1992 July 31 by H. E. Holt at Palomar.

(10113) 1992 PX_2
Discovered 1992 August 6 by H. E. Holt at Palomar.

(10114) 1992 RZ
Discovered 1992 September 4 by L. D. Schmadel and F. Börngen at Tautenburg.

(10115) 1992 SK
Discovered 1992 September 24 by E. F. Helin and J. Alu at Palomar.

(10116) Robertfranz
1992 SJ_2. Discovered 1992 September 21 by F. Börngen and L. D. Schmadel at Tautenburg.

Named in memory of Robert Franz (1815-1892), German composer of songs. An organist in his native city of Halle, he was also conductor of the Singakademie and the university director of music. Franz set more than 350 songs to music, using in particular the words by Heine {see planet (7109)} (*Im Rhein, im heiligen Strome, Wie des Mondes Abbild*). His arrangements and editions of compositions of Bach and Handel {see planets (1814) and (3826), respectively} were very important. He was appointed Knight of the Order of Maximilian for arts and sciences in 1878 by the king of Bavaria. (M 34632)
Name proposed by the first discoverer.

(10117) 1992 TW
Discovered 1992 October 1 by M. Yanai and K. Watanabe at Kitami.

(10118) 1992 UK_1
Discovered 1992 October 19 by S. Ueda and H. Kaneda at Kushiro.

(10119) 1992 YC_1
Discovered 1992 December 18 by E. W. Elst at Caussols.

(10120) 1992 YH_2
Discovered 1992 December 18 by E. W. Elst at Caussols.

(10121) 1993 BS_4
Discovered 1993 January 27 by E. W. Elst at Caussols.

(10122) 1993 BC_5
Discovered 1993 January 27 by E. W. Elst at Caussols.

(10123) 1993 FJ_{16}
Discovered 1993 March 17 by the Uppsala-ESO Survey at La Silla.

(10124) 1993 FE_{23}
Discovered 1993 March 21 by the Uppsala-ESO Survey at La Silla.

(10125) 1993 FB_{24}
Discovered 1993 March 21 by the Uppsala-ESO Survey at La Silla.

(10126) 1993 FW_{24}
Discovered 1993 March 21 by the Uppsala-ESO Survey at La Silla.

(10127) 1993 FF_{26}
Discovered 1993 March 21 by the Uppsala-ESO Survey at La Silla.

(10128) 1993 FT_{31}
Discovered 1993 March 19 by the Uppsala-ESO Survey at La Silla.

(10129) 1993 FO_{40}
Discovered 1993 March 19 by the Uppsala-ESO Survey at La Silla.

(10130) 1993 FJ_{50}
Discovered 1993 March 19 by the Uppsala-ESO Survey at La Silla.

(10131) 1993 FP_{73}
Discovered 1993 March 21 by the Uppsala-ESO Survey at La Silla.

(10132) 1993 FL_{84}
Discovered 1993 March 20 by the Uppsala-ESO Survey at La Silla.

(10133) 1993 GC_1
Discovered 1993 April 15 by H. E. Holt at Palomar.

(10134) 1993 HL_6
Discovered 1993 April 17 by H. Debehogne at La Silla.

(10135) 1993 LZ_1
Discovered 1993 June 13 by H. E. Holt at Palomar.

(10136) 1993 OM_3
Discovered 1993 July 20 by E. W. Elst at La Silla.

(10137) 1993 PV_6
Discovered 1993 August 15 by E. W. Elst at Caussols.

(10138) 1993 SS_1
Discovered 1993 September 16 by K. Endate and K. Watanabe at Kitami.

(10139) 1993 ST_4
Discovered 1993 September 19 by E. W. Elst at Caussols.

(10140) 1993 SX_4
Discovered 1993 September 19 by E. W. Elst at Caussols.

(10141) 1993 VE
Discovered 1993 November 5 by S. Otomo at Kiyosato.

(10142) 1993 VG_1
Discovered 1993 November 15 by A. Sugie at Taga.

(10143) 1994 AP_1
Discovered 1994 January 8 by A. Sugie at Taga.

(10144) 1994 AB_2
Discovered 1994 January 9 by Y. Kushida and O. Muramatsu at Yatsugatake.

(10145) 1994 CK_1
Discovered 1994 February 10 by the Spacewatch at Kitt Peak.

(10146) 1994 CV_1
Discovered 1994 February 8 by K. Endate and K. Watanabe at Kitami.

(10147) 1994 CK_2
Discovered 1994 February 13 by T. Kobayashi at Oizumi.

(10148) 1994 GR_9
Discovered 1994 April 14 by S. Otomo at Kiyosato.

(10149) 1994 PA
Discovered 1994 August 3 by M. Tombelli and A. Boattini at San Marcello Pistoiese.

(10150) 1994 PN
Discovered 1994 August 7 by G. J. Garradd at Siding Spring.

(10151) 1994 PF_{22}
Discovered 1994 August 12 by E. W. Elst at La Silla.

(10152) 1994 RJ_{11}
Discovered 1994 September 11 by S. Otomo at Kiyosato.

(10153) 1994 UB
Discovered 1994 October 26 by D. di Cicco at Sudbury.

(10154) 1994 UH
Discovered 1994 October 31 by T. Kobayashi at Oizumi.

(10155) 1994 VZ_2
Discovered 1994 November 4 by K. Endate and K. Watanabe at Kitami.

(10156) 1994 VQ_7
Discovered 1994 November 7 by S. Ueda and H. Kaneda at Kushiro.

(10157) 1994 WE_1
Discovered 1994 November 27 by T. Kobayashi at Oizumi.

(10158) 1994 XK
Discovered 1994 December 3 by T. Kobayashi at Oizumi.

(10159) 1994 XS_4
Discovered 1994 December 9 by T. Kobayashi at Oizumi.

(10160) 1994 YQ_1
Discovered 1994 December 31 by T. Kobayashi at Oizumi.

(10161) 1994 YZ_1
Discovered 1994 December 31 by T. Kobayashi at Oizumi.

(10162) 1995 AL
Discovered 1995 January 2 by T. Niijima and T. Urata at Ojima.

(10163) Onomichi
1995 BH_1. Discovered 1995 January 26 by A. Nakamura at Kuma.
Named for a city in eastern Hiroshima {see planet (2247)} prefecture. Blessed with abundant natural beauty, Onomichi has been artistically captured by many Japanese novellists and artists. It has also been featured on film, thanks to the *Onomichi Trilogy* directed by Nobuhiko Obayashi. Onomichi is the terminal city on the Honshu side of the new Nishiseto Expressway {see planet (9235)}. (M 34632)

(10164) 1995 BS_1
Discovered 1995 January 27 by T. Kobayashi at Oizumi.

(10165) 1995 BL_2
Discovered 1995 January 31 by the Spacewatch at Kitt Peak.

(10166) 1995 BN_3
Discovered 1995 January 30 by T. Kobayashi at Oizumi.

(10167) 1995 BQ_{15}
Discovered 1995 January 31 by T. Seki at Geisei.

(10168) 1995 CN
Discovered 1995 February 4 by J. B. Child and J. E. Rogers at Stony Ridge.

(10169) 1995 DK
Discovered 1995 February 21 by T. Kobayashi at Oizumi.

(10170) Petrjakeš
1995 DA_1. Discovered 1995 February 22 by M. Tichý and Z. Moravec at Kleť.
Named in honor of Petr Jakeš (1940-), Czech geologist, geochemist and lecturer at Charles University. He was a member of the team that made a preliminary analysis of the lunar rocks returned by Apollo 14 and 15. He also studies meteorites and has searched for them in desert regions. (M 34632)
Name endorsed by J. Tichá.

(10171) 1995 EE_8
Discovered 1995 March 7 by M. Hirasawa and S. Suzuki at Nyukasa.

(10172) 1995 FW_{19}
Discovered 1995 March 31 by the Spacewatch at Kitt Peak.

(10173) 1995 HA
Discovered 1995 April 21 by P. Pravec and L. Šarounová at Ondřejov.

(10174) 1995 JD
Discovered 1995 May 2 by Z. Moravec at Kleť.

(10175) 1996 CR_1
Discovered 1996 February 14 at the Višnjan Observatory at Višnjan.

(10176) 1996 CW_7
Discovered 1996 February 14 by U. Munari and M. Tombelli at Cima Ekar.

(10177) 1996 CK_9
Discovered 1996 February 10 by the Spacewatch at Kitt Peak.

(10178) 1996 DD
Discovered 1996 February 18 by T. Kobayashi at Oizumi.

(10179) 1996 DE
Discovered 1996 February 18 by T. Kobayashi at Oizumi.

(10180) 1996 EE_2
Discovered 1996 March 15 by G. J. Garradd at Loomberah.

(10181) 1996 FP_3
Discovered 1996 March 26 by P. G. Comba at Prescott.

(10182) 1996 FL_5
Discovered 1996 March 20 by K. Endate and K. Watanabe at Kitami.

(10183) 1996 GV_{20}
Discovered 1996 April 15 by E. W. Elst at La Silla.

(10184) 1996 HC_{19}
Discovered 1996 April 18 by E. W. Elst at La Silla.

(10185) 1996 HD_{21}
Discovered 1996 April 18 by E. W. Elst at La Silla.

(10186) 1996 HD_{24}
Discovered 1996 April 20 by E. W. Elst at La Silla.

(10187) 1996 JV
Discovered 1996 May 12 by T. B. Spahr at Tucson.

(10188) 1996 JY
Discovered 1996 May 14 by R. H. McNaught and Y. Ikari at Moriyama.

(10189) 1996 JK_{16}
Discovered 1996 May 15 by the Spacewatch at Kitt Peak.

(10190) 1996 NC
Discovered 1996 July 14 by the JPL NEAT Program at Haleakala.

(10191) 1996 NU_1
Discovered 1996 July 14 by the JPL NEAT Program at Haleakala.

(10192) 1996 OQ_1
Discovered 1996 July 20 at the Beijing Observatory at Xinglong.

(10193) 1996 PR_1
Discovered 1996 August 8 by the JPL NEAT Program at Haleakala.

(10194) 1996 QN_1
Discovered 1996 August 18 by G. R. Viscome at Lake Placid.

(10195) Nebraska
1996 RS_5. Discovered 1996 September 13 by R. Linderholm at Lime Creek.
Named for the U.S. state of Nebraska, this is the first minor planet to be discovered from an observatory there. (M 34632)

(10196) 1996 TJ_{15}
Discovered 1996 October 9 by S. Ueda and H. Kaneda at Kushiro.

(10197) 1996 UO
Discovered 1996 October 18 by V. Goretti at Pianoro.

(10198) 1996 XN_{26}
Discovered 1996 December 6 by U. Munari and M. Tombelli at Cima Ekar.

(10199) 1997 CU_{26}
Discovered 1997 February 15 by the Spacewatch at Kitt Peak.

(10200) 1997 NZ_2
Discovered 1997 July 7 by V. Goretti at Pianoro.

(10201) 1997 NL_6
Discovered 1997 July 12 at the Farra d'Isonzo Observatory at Farra d'Isonzo.

(10202) 1997 PE
Discovered 1997 August 1 by the JPL NEAT Program at Haleakala.

(10203) Flinders
1997 PQ. Discovered 1997 August 1 by F. B. Zoltowski at Woomera.
Named in memory of Matthew Flinders (1774-1814), British navigator and explorer who charted the Australian coastline. A self-educated navigator, he explored the Bass Strait between Australia and Tasmania and circumnavigated Australia, charting large sections of its coast. His use of the term Australia led to its acceptance as the name of the continent. (M 34632)

(10204) Turing
1997 PK_1. Discovered 1997 August 1 by P. G. Comba at Prescott.
Named in memory of Alan Mathison Turing (1912-1954), English mathematician and logician, a pioneer in the study of computability. In a fundamental paper published in 1936, he introduced the concept of an abstract computing machine and showed how such a machine can be programed to simulate the behavior of any other computing device. This concept, now referred to as a "universal Turing machine", was introduced years before the advent of programable computers, and it was used to demonstrate the existence of noncomputable numbers and undecidable mathematical propositions. (M 34633)

(10205) Pokorný
1997 PX_1. Discovered 1997 August 7 by M. Tichý and Z. Moravec at Kleť.
Named in honor of Zdeněk Pokorný (1947-), Czech astronomer who works at the Nicolas Copernicus Observatory and Planetarium, as well as at Masaryk University in Brno {see planets (1841) and (2889), respectively}. He started his career by studying the giant planets, but his main work consists of education and the popularization of astronomy. He created many programs for the Brno Planetarium and is the author of the first Czech astronomical multimedia CD-ROM "ASTRO 2001". In 1991 Pokorný founded and served as the first president

of the Association of Observatories and Planetaria in Czechoslovakia. (M 34633)
Name endorsed and citation prepared by J. Tichá.

(10206) 1997 PC_2

Discovered 1997 August 7 by L. L. Amburgey at Fitchburg.

(10207) 1997 QA

Discovered 1997 August 16 by P. Kolény and L. Kornoš at Modra.

(10208) 1997 QN_1

Discovered 1997 August 30 at the Santa Lucia Observatory at Stroncone.

(10209) 1997 QY_1

Discovered 1997 August 24 by Y. Shimizu and T. Urata at Nachi-Katsuura.

(10210) 1997 QV_3

Discovered 1997 August 30 by the OCA-DLR Survey at Caussols.

(10211) 1997 RG_3

Discovered 1997 September 6 at the Monte Viseggi Observatory at Monte Viseggi.

(10212) 1997 RA_7

Discovered 1997 September 3 by S. P. Laurie at Church Stretton.

(10213) Koukolík

1997 RK_7. Discovered 1997 September 10 by M. Tichý and Z. Moravec at Kleť.

Named in honor of František Koukolík (1941-), Czech neuropathologist who works on the relationship between the brain and human behavior. He is also known as an excellent popularizer of science, being the author of many articles and several books of essays, such as *An Ant and the Universe*. (M 34633)
Name endorsed by J. Tichá.

(10214) 1997 RT_9

Discovered 1997 September 10 by T. Pauwels at Uccle.

(10215) 1997 SQ

Discovered 1997 September 20 by L. Šarounová at Ondřejov.

(10216) 1997 SN_3

Discovered 1997 September 22 by S. P. Laurie at Church Stretton.

(10217) 1997 SN_4

Discovered 1997 September 27 by the JPL NEAT Program at Haleakala.

(10218) 1997 SJ_{23}

Discovered 1997 September 29 by the Spacewatch at Kitt Peak.

(10219) 1997 UJ_5

Discovered 1997 October 25 by L. Tesi and A. Boattini at San Marcello Pistoiese.

(10220) Pigott

1997 UG_7. Discovered 1997 October 20 by R. A. Tucker at Tucson.

Named in memory of the English amateur astronomer Edward Pigott (1753-1825), discoverer of several comets and the variable stars η Aql, R Sct and R CrB. A skilled user of the transit instrument, he determined the proper motions of several stars. Pigott was a friend and observing collaborator of John Goodricke {see planet (3116)}. (M 34633)

(10221) Kubrick

1997 UM_9. Discovered 1997 October 28 by P. Pravec at Ondřejov.

Named in memory of Stanley Kubrick (1928-1999), filmmaking genius, whose 1968 rendition of Arthus Clarke's {see planet (4923)} novel, *2001: A Space Odyssey*, remains the definitive science-fiction movie. At least half of his 12 other feature films, from *The Killing* (1956) and *Paths of Glory* (1957), through Peter Sellers' performance in the brilliant *Dr. Strangelove* (1964), and on to *A Clockwork Orange* (1971), *The Shining* (1980) and *Full Metal Jacket* (1987), explored twentieth-century culture in a manner that was unprecedented in its artistry and its longevity. (M 34633)

(10222) 1997 UV_{10}

Discovered 1997 October 29 by C. Buil at Ramonville Saint Agne.

(10223) 1997 UD_{11}

Discovered 1997 October 31 by T. Urata at Oohira.

(10224) 1997 UK_{22}

Discovered 1997 October 26 by N. Sato at Chichibu.

(10225) 1997 VQ_1

Discovered 1997 November 1 by S. Ueda and H. Kaneda at Kushiro.

(10226) 1997 VK_5

Discovered 1997 November 8 by T. Kobayashi at Oizumi.

(10227) 1997 VO_6

Discovered 1997 November 4 by T. Kagawa and T. Urata at Gekko.

(10228) 1997 VY_8

Discovered 1997 November 1 by S. Ueda and H. Kaneda at Kushiro.

(10229) 1997 WR_3

Discovered 1997 November 19 by Y. Shimizu and T. Urata at Nachi-Katsuura.

(10230) 1997 WU_{35}

Discovered 1997 November 29 by the Lincoln NEA Research Team at Socorro.

(10231) 1997 WQ_{37}

Discovered 1997 November 29 by the Lincoln NEA Research Team at Socorro.

(10232) 1997 WR_{49}

Discovered 1997 November 26 by the Lincoln NEA Research Team at Socorro.

(10233) Le Creusot

1997 XQ_2. Discovered 1997 December 5 by J.-C. Merlin at Le Creusot.

Named for the home city and observatory site of the discoverer. It is in the Burgundy region of France, some 75 km southwest of Dijon {see planet (9379)}. (M 34633)

(10234) 1997 YB_8

Discovered 1997 December 27 at the Kleť Observatory at Kleť.

(10235) 1998 QR_{37}

Discovered 1998 August 17 by the Lincoln NEA Research Team at Socorro.

(10236) 1998 QA_{93}

Discovered 1998 August 28 by the Lincoln NEA Research Team at Socorro.

(10237) 1998 SJ_{119}
Discovered 1998 September 26 by the Lincoln NEA Research Team at Socorro.

(10238) 1998 SO_{140}
Discovered 1998 September 26 by the Lincoln NEA Research Team at Socorro.

(10239) 1998 TY_{30}
Discovered 1998 October 10 by the LONEOS at Anderson Mesa.

(10240) 1998 VW_{34}
Discovered 1998 November 12 by S. Ueda and H. Kaneda at Kushiro.

(10241) 1999 AU_{6}
Discovered 1999 January 9 by K. Korlevic at Višnjan.

(10242) 2808 P-L
Discovered 1960 September 24 by C. J. van Houten and I. van Houten-Groeneveld at Palomar.

(10243) 3553 P-L
Discovered 1960 October 22 by C. J. van Houten and I. van Houten-Groeneveld at Palomar.

(10244) 4668 P-L
Discovered 1960 September 26 by C. J. van Houten and I. van Houten-Groeneveld at Palomar.

(10245) 6071 P-L
Discovered 1960 September 24 by C. J. van Houten and I. van Houten-Groeneveld at Palomar.

(10246) 6381 P-L
Discovered 1960 September 24 by C. J. van Houten and I. van Houten-Groeneveld at Palomar.

(10247) 6629 P-L
Discovered 1960 September 24 by C. J. van Houten and I. van Houten-Groeneveld at Palomar.

(10248) 7639 P-L
Discovered 1960 October 17 by C. J. van Houten and I. van Houten-Groeneveld at Palomar.

(10249) 9515 P-L
Discovered 1960 October 17 by C. J. van Houten and I. van Houten-Groeneveld at Palomar.

(10250) 1252 T-1
Discovered 1971 March 25 by C. J. van Houten and I. van Houten-Groeneveld at Palomar.

(10251) 3089 T-1
Discovered 1971 March 26 by C. J. van Houten and I. van Houten-Groeneveld at Palomar.

(10252) 4164 T-1
Discovered 1971 March 26 by C. J. van Houten and I. van Houten-Groeneveld at Palomar.

(10253) 2116 T-2
Discovered 1973 September 29 by C. J. van Houten and I. van Houten-Groeneveld at Palomar.

(10254) 2314 T-2
Discovered 1973 September 29 by C. J. van Houten and I. van Houten-Groeneveld at Palomar.

(10255) 3398 T-3
Discovered 1977 October 16 by C. J. van Houten and I. van Houten-Groeneveld at Palomar.

(10256) 4157 T-3
Discovered 1977 October 16 by C. J. van Houten and I. van Houten-Groeneveld at Palomar.

(10257) 4333 T-3
Discovered 1977 October 16 by C. J. van Houten and I. van Houten-Groeneveld at Palomar.

(10258) 1940 AB
Discovered 1940 January 6 by G. Kulin at Budapest.

(10259) 1972 HL
Discovered 1972 April 18 by T. M. Smirnova at Nauchnyj.

(10260) 1972 TC
Discovered 1972 October 4 by L. Kohoutek at Bergedorf.

(10261) 1974 QF$_1$
Discovered 1974 August 22 by L. V. Zhuravleva at Nauchnyj.

(10262) 1975 TQ$_3$
Discovered 1975 October 3 by L. I. Chernykh at Nauchnyj.

(10263) 1976 SE$_5$
Discovered 1976 September 24 by N. S. Chernykh at Nauchnyj.

(10264) 1978 PH$_3$
Discovered 1978 August 8 by N. S. Chernykh at Nauchnyj.

(10265) 1978 RY$_6$
Discovered 1978 September 2 by C.-I. Lagerkvist at La Silla.

(10266) 1978 SA$_7$
Discovered 1978 September 26 by L. V. Zhuravleva at Nauchnyj.

(10267) 1978 VD$_7$
Discovered 1978 November 7 by E. F. Helin and S. J. Bus at Palomar.

(10268) 1979 HW$_6$
Discovered 1979 April 26 at the Perth Observatory at Bickley.

(10269) 1979 SU$_{11}$
Discovered 1979 September 24 by N. S. Chernykh at Nauchnyj.

(10270) 1980 FX$_3$
Discovered 1980 March 16 by C.-I. Lagerkvist at La Silla.

(10271) 1980 TV_2
Discovered 1980 October 14 by H. Debehogne and L. Houziaux at St. Michel.

(10272) 1981 EF_{13}
Discovered 1981 March 1 by S. J. Bus at Siding Spring.

(10273) 1981 ED_{14}
Discovered 1981 March 1 by S. J. Bus at Siding Spring.

(10274) 1981 ET_{15}
Discovered 1981 March 1 by S. J. Bus at Siding Spring.

(10275) 1981 EC_{16}
Discovered 1981 March 1 by S. J. Bus at Siding Spring.

(10276) 1981 EK_{23}
Discovered 1981 March 3 by S. J. Bus at Siding Spring.

(10277) 1981 EC_{27}
Discovered 1981 March 2 by S. J. Bus at Siding Spring.

(10278) 1981 EW_{30}
Discovered 1981 March 2 by S. J. Bus at Siding Spring.

(10279) 1981 ET_{42}
Discovered 1981 March 2 by S. J. Bus at Siding Spring.

(10280) 1981 EA_{43}
Discovered 1981 March 2 by S. J. Bus at Siding Spring.

(10281) 1981 EE_{45}
Discovered 1981 March 11 by S. J. Bus at Siding Spring.

(10282) 1981 ET_{46}
Discovered 1981 March 2 by S. J. Bus at Siding Spring.

(10283) 1981 JE_2
Discovered 1981 May 5 by C. S. Shoemaker at Palomar.

(10284) 1981 QY_2
Discovered 1981 August 24 by H. Debehogne at La Silla.

(10285) 1982 QX_1
Discovered 1982 August 17 by C.-I. Lagerkvist at La Silla.

(10286) 1982 SM_6
Discovered 1982 September 16 by L. I. Chernykh at Nauchnyj.

(10287) 1982 UK_7
Discovered 1982 October 21 by L. G. Karachkina at Nauchnyj.

(10288) 1983 WN
Discovered 1983 November 28 by E. Bowell at Anderson Mesa.

(10289) 1984 QS
Discovered 1984 August 24 at the Oak Ridge Observatory at Harvard.

(10290) 1985 SR
Discovered 1985 September 17 at the Oak Ridge Observatory at Harvard.

(10291) 1985 UT
Discovered 1985 October 20 by A. Mrkos at Kleť.

(10292) 1986 PM
Discovered 1986 August 2 at the Palomar Observatory at Palomar.

(10293) 1986 TU_6
Discovered 1986 October 5 by M. Antal at Piwnice.

(10294) 1988 AA_2
Discovered 1988 January 14 by A. Mrkos at Kleť.

(10295) 1988 GB
Discovered 1988 April 12 by C. S. Shoemaker and E. M. Shoemaker at Palomar.

(10296) 1988 RQ_{12}
Discovered 1988 September 14 by S. J. Bus at Cerro Tololo.

(10297) 1988 RJ_{13}
Discovered 1988 September 14 by S. J. Bus at Cerro Tololo.

(10298) 1988 SU_2
Discovered 1988 September 16 by S. J. Bus at Cerro Tololo.

(10299) 1988 VS_3
Discovered 1988 November 13 by Y. Oshima at Gekko.

(10300) 1989 EG_1
Discovered 1989 March 6 by T. Seki at Geisei.

(10301) 1989 FH
Discovered 1989 March 30 by K. Endate and K. Watanabe at Kitami.

(10302) 1989 ML
Discovered 1989 June 29 by E. F. Helin and J. Alu at Palomar.

(10303) 1989 RD_2
Discovered 1989 September 2 by E. W. Elst at St. Michel.

(10304) 1989 SY
Discovered 1989 September 30 by K. Endate and K. Watanabe at Kitami.

(10305) 1989 YP_5
Discovered 1989 December 29 by E. W. Elst at St. Michel.

(10306) 1990 QY
Discovered 1990 August 21 by E. W. Elst at St. Michel.

(10307) 1990 QX_1
Discovered 1990 August 22 by H. E. Holt at Palomar.

(10308) 1990 QC_3
Discovered 1990 August 28 by H. E. Holt at Palomar.

(10309) 1990 QC_6
Discovered 1990 August 23 by H. E. Holt at Palomar.

(10310) 1990 QZ_8
Discovered 1990 August 16 by E. W. Elst at La Silla.

(10311) 1990 QL_9
Discovered 1990 August 16 by E. W. Elst at La Silla.

(10312) 1990 QT_9
Discovered 1990 August 23 by H. E. Holt at Palomar.

(10313) 1990 QW_{17}
Discovered 1990 August 26 by L. V. Zhuravleva at Nauchnyj.

(10314) 1990 RF
Discovered 1990 September 14 by H. E. Holt at Palomar.

(10315) 1990 SC_4
Discovered 1990 September 23 by E. F. Helin at Palomar.

(10316) 1990 SF_9
Discovered 1990 September 22 by E. W. Elst at La Silla.

(10317) 1990 SA_{15}
Discovered 1990 September 17 by H. E. Holt at Palomar.

(10318) 1990 TX
Discovered 1990 October 15 by T. Nomura and K. Kawanishi at Minami-Oda.

(10319) 1990 TB_1
Discovered 1990 October 11 by A. Takahashi and K. Watanabe at Kitami.

(10320) 1990 TR_1
Discovered 1990 October 14 by E. F. Helin at Palomar.

(10321) 1990 UN_2
Discovered 1990 October 26 by T. Seki at Geisei.

(10322) 1990 VT_1
Discovered 1990 November 11 by K. Endate and K. Watanabe at Kitami.

(10323) 1990 VW_6
Discovered 1990 November 14 by E. W. Elst at La Silla.

(10324) 1990 VB_{14}
Discovered 1990 November 14 by L. G. Karachkina at Nauchnyj.

(10325) 1990 WB_2
Discovered 1990 November 18 by E. W. Elst at La Silla.

(10326) 1990 WS_2
Discovered 1990 November 21 by K. Endate and K. Watanabe at Kitami.

(10327) 1990 WQ_6
Discovered 1990 November 21 by E. W. Elst at La Silla.

(10328) 1991 GC_1
Discovered 1991 April 10 by E. F. Helin at Palomar.

(10329) 1991 GJ_1
Discovered 1991 April 11 by S. Ueda and H. Kaneda at Kushiro.

(10330) 1991 GH_3
Discovered 1991 April 8 by E. W. Elst at La Silla.

(10331) 1991 GM_{10}
Discovered 1991 April 9 by F. Börngen at Tautenburg.

(10332) 1991 JT_1
Discovered 1991 May 13 by C. S. Shoemaker at Palomar.

(10333) 1991 NZ_6
Discovered 1991 July 12 by H. Debehogne at La Silla.

(10334) 1991 PG_5
Discovered 1991 August 3 by E. W. Elst at La Silla.

(10335) 1991 PG_9
Discovered 1991 August 15 by E. F. Helin at Palomar.

(10336) 1991 PJ_{12}
Discovered 1991 August 7 by H. E. Holt at Palomar.

(10337) 1991 RO_1
Discovered 1991 September 10 by A. Sugie at Taga.

(10338) 1991 RB_{11}
Discovered 1991 September 10 by H. E. Holt at Palomar.

(10339) 1991 RK_{17}
Discovered 1991 September 11 by H. E. Holt at Palomar.

(10340) 1991 RT_{40}
Discovered 1991 September 10 by F. Börngen at Tautenburg.

(10341) 1991 SC_2
Discovered 1991 September 16 by H. E. Holt at Palomar.

(10342) 1991 TQ
Discovered 1991 October 1 by R. H. McNaught at Siding Spring.

(10343) 1991 VW_8
Discovered 1991 November 4 by the Spacewatch at Kitt Peak.

(10344) 1992 CA_2
Discovered 1992 February 12 by O. A. Naranjo and J. Stock at Mérida.

(10345) 1992 DC_{11}
Discovered 1992 February 29 by the Uppsala-ESO Survey at La Silla.

(10346) 1992 GA_1
Discovered 1992 April 2 by C. S. Shoemaker at Palomar.

(10347) 1992 HG_4
Discovered 1992 April 23 by E. W. Elst at La Silla.

(10348) 1992 HL_4
Discovered 1992 April 29 by F. Börngen at Tautenburg.

(10349) 1992 LN
Discovered 1992 June 3 by G. J. Leonard at Palomar.

(10350) 1992 OG_2
Discovered 1992 July 26 by E. W. Elst at La Silla.

(10351) 1992 SE_1
Discovered 1992 September 23 by K. Endate and K. Watanabe at Kitami.

(10352) 1992 UO_3
Discovered 1992 October 26 by K. Endate and K. Watanabe at Kitami.

(10353) 1992 YS_2
Discovered 1992 December 20 by S. Otomo at Kiyosato.

(10354) 1993 BU_5
Discovered 1993 January 27 by E. W. Elst at Caussols.

(10355) 1993 EQ
Discovered 1993 March 15 by K. Endate and K. Watanabe at Kitami.

(10356) 1993 RQ_4
Discovered 1993 September 15 by E. W. Elst at La Silla.

(10357) 1993 SL_3
Discovered 1993 September 19 by H. E. Holt at Palomar.

(10358) 1993 TH_{32}
Discovered 1993 October 9 by E. W. Elst at La Silla.

(10359) 1993 TU_{36}
Discovered 1993 October 13 by H. E. Holt at Palomar.

(10360) 1993 VN
Discovered 1993 November 7 by S. Ueda and H. Kaneda at Kushiro.

(10361) 1994 PR_{20}
Discovered 1994 August 12 by E. W. Elst at La Silla.

(10362) 1994 UC_2
Discovered 1994 October 31 by S. Ueda and H. Kaneda at Kushiro.

(10363) 1994 UP_{11}
Discovered 1994 October 31 at the Palomar Observatory at Palomar.

(10364) 1994 VR_1
Discovered 1994 November 3 by T. Kobayashi at Oizumi.

(10365) 1994 WL_1
Discovered 1994 November 27 by T. Kobayashi at Oizumi.

(10366) 1994 WD_4
Discovered 1994 November 24 by K. Endate and K. Watanabe at Kitami.

(10367) 1994 YL_1
Discovered 1994 December 31 by T. Kobayashi at Oizumi.

(10368) 1995 CM_1
Discovered 1995 February 7 by T. Kobayashi at Oizumi.

(10369) 1995 CE_2
Discovered 1995 February 8 by D. J. Asher at Siding Spring.

(10370) 1995 DW_2
Discovered 1995 February 27 by D. Jewitt and J. X. Luu at Mauna Kea.

(10371) 1995 DU_3
Discovered 1995 February 27 by L. Tesi and A. Boattini at San Marcello Pistoiese.

(10372) 1995 FO_{10}
Discovered 1995 March 26 by the Spacewatch at Kitt Peak.

(10373) 1996 ER
Discovered 1996 March 14 by D. di Cicco at Sudbury.

(10374) 1996 GN_{19}
Discovered 1996 April 15 by E. W. Elst at La Silla.

(10375) 1996 HM_1
Discovered 1996 April 21 by A. Nakamura at Kuma.

(10376) 1996 KW
Discovered 1996 May 16 at the Osservatorio San Vittore at Bologna.

(10377) 1996 NN_4
Discovered 1996 July 14 by E. W. Elst at La Silla.

(10378) 1996 NE_5
Discovered 1996 July 14 by E. W. Elst at La Silla.

(10379) 1996 OH
Discovered 1996 July 18 by G. R. Viscome at Lake Placid.

(10380) 1996 PY_7
Discovered 1996 August 8 by E. W. Elst at La Silla.

(10381) 1996 RB
Discovered 1996 September 3 by B. G. W. Manning at Stakenbridge.

(10382) 1996 RJ_3
Discovered 1996 September 15 by P. G. Comba at Prescott.

(10383) 1996 SR_7
Discovered 1996 September 16 by S. P. Laurie at Church Stretton.

(10384) 1996 TQ_{10}
Discovered 1996 October 9 by S. Ueda and H. Kaneda at Kushiro.

(10385) 1996 TL_{12}
Discovered 1996 October 15 by Y. Shimizu and T. Urata at Nachi-Katsuura.

(10386) 1996 TS_{15}
Discovered 1996 October 12 by V. S. Casulli at Colleverde di Guidonia.

(10387) 1996 UQ
Discovered 1996 October 18 by P. Sicoli and F. Manca at Sormano.

(10388) 1996 YH_3
Discovered 1996 December 25 at the Beijing Observatory at Xinglong.

(10389) 1997 LD
Discovered 1997 June 1 by the JPL NEAT Program at Haleakala.

(10390) 1997 QD_1
Discovered 1997 August 27 by P. Pravec and M. Wolf (Marek) at Ondřejov.

(10391) 1997 RR_3
Discovered 1997 September 5 by Y. Shimizu and T. Urata at Nachi-Katsuura.

(10392) 1997 RP_7
Discovered 1997 September 11 by R. Linderholm at Lime Creek.

(10393) 1997 RF_8
Discovered 1997 September 4 by T. Kagawa and T. Urata at Gekko.

(10394) 1997 SG_1
Discovered 1997 September 22 by P. Sala at Giesing.

(10395) 1997 SZ_1
Discovered 1997 September 23 by M. Wolf (Marek) and P. Pravec at Ondřejov.

(10396) 1997 SW_{33}
Discovered 1997 September 17 at the Beijing Observatory at Xinglong.

(10397) 1997 SX_{33}
Discovered 1997 September 17 at the Beijing Observatory at Xinglong.

(10398) 1997 UP_8
Discovered 1997 October 23 by S. Ueda and H. Kaneda at Kushiro.

(10399) 1997 UZ_8
Discovered 1997 October 29 by T. Kobayashi at Oizumi.

(10400) 1997 VX
Discovered 1997 November 1 by T. Kobayashi at Oizumi.

(10401) 1997 VD_3
Discovered 1997 November 6 by T. Kobayashi at Oizumi.

(10402) 1997 VS_5
Discovered 1997 November 8 by T. Kobayashi at Oizumi.

(10403) 1997 WU_3
Discovered 1997 November 22 at the Kleť Observatory at Kleť.

(10404) 1997 WP_{14}
Discovered 1997 November 22 by the Spacewatch at Kitt Peak.

(10405) 1997 WT_{23}
Discovered 1997 November 19 by T. Okuni at Nanyou.

(10406) 1997 WZ_{29}
Discovered 1997 November 24 by S. Ueda and H. Kaneda at Kushiro.

(10407) 1997 WS_{32}
Discovered 1997 November 29 by the Lincoln NEA Research Team at Socorro.

(10408) 1997 WL_{44}
Discovered 1997 November 29 by the Lincoln NEA Research Team at Socorro.

(10409) 1997 WP_{44}
Discovered 1997 November 29 by the Lincoln NEA Research Team at Socorro.

(10410) 1997 XR_9
Discovered 1997 December 4 at the Beijing Observatory at Xinglong.

(10411) 1997 XO_{11}
Discovered 1997 December 15 at the Beijing Observatory at Xinglong.

(10412) 1997 YO_4
Discovered 1997 December 21 by Y. Shimizu and T. Urata at Nachi-Katsuura.

(10413) 1997 YG_{20}
Discovered 1997 December 29 at the Osservatorio San Vittore at Bologna.

(10414) 1998 QJ_{37}
Discovered 1998 August 17 by the Lincoln NEA Research Team at Socorro.

(10415) 1998 UT_{15}
Discovered 1998 October 23 by K. Korlevic at Višnjan.

(10416) 1998 VA_{32}
Discovered 1998 November 14 by the Lincoln NEA Research Team at Socorro.

(10417) 1998 WA_{23}
Discovered 1998 November 18 by the Lincoln NEA Research Team at Socorro.

(10418) 1998 WZ_{23}
Discovered 1998 November 25 by the Lincoln NEA Research Team at Socorro.

(10419) 1998 XB_4
Discovered 1998 December 11 by T. Kobayashi at Oizumi.

(10420) 1998 YB_{12}
Discovered 1998 December 27 by T. Kobayashi at Oizumi.

(10421) 1999 AY_6
Discovered 1999 January 9 by K. Korlevic at Višnjan.

(10422) 1999 AN_{22}
Discovered 1999 January 14 at the Beijing Observatory at Xinglong.

(10423) 1999 BB
Discovered 1999 January 16 by K. Korlevic at Višnjan.

(10424) 1999 BD_5
Discovered 1999 January 20 by the OCA-DLR Survey at Caussols.

(10425) 1999 BE_6
Discovered 1999 January 20 by the OCA-DLR Survey at Caussols.

(10426) 1999 BB_{27}
Discovered 1999 January 16 by the Spacewatch at Kitt Peak.

(10427) 2017 P-L
Discovered 1960 September 24 by C. J. van Houten and I. van Houten-Groeneveld at Palomar.

(10428) 2073 P-L
Discovered 1960 September 24 by C. J. van Houten and I. van Houten-Groeneveld at Palomar.

(10429) 2546 P-L
Discovered 1960 September 24 by C. J. van Houten and I. van Houten-Groeneveld at Palomar.

(10430) 4030 P-L

Discovered 1960 September 24 by C. J. van Houten and I. van Houten-Groeneveld at Palomar.

(10431) 4042 P-L

Discovered 1960 September 24 by C. J. van Houten and I. van Houten-Groeneveld at Palomar.

(10432) 4623 P-L

Discovered 1960 September 24 by C. J. van Houten and I. van Houten-Groeneveld at Palomar.

(10433) 4716 P-L

Discovered 1960 September 24 by C. J. van Houten and I. van Houten-Groeneveld at Palomar.

(10434) 4722 P-L

Discovered 1960 September 24 by C. J. van Houten and I. van Houten-Groeneveld at Palomar.

(10435) 6064 P-L

Discovered 1960 September 24 by C. J. van Houten and I. van Houten-Groeneveld at Palomar.

(10436) 6073 P-L

Discovered 1960 September 24 by C. J. van Houten and I. van Houten-Groeneveld at Palomar.

(10437) 6085 P-L

Discovered 1960 September 24 by C. J. van Houten and I. van Houten-Groeneveld at Palomar.

(10438) 6615 P-L

Discovered 1960 September 24 by C. J. van Houten and I. van Houten-Groeneveld at Palomar.

(10439) 6676 P-L

Discovered 1960 September 24 by C. J. van Houten and I. van Houten-Groeneveld at Palomar.

(10440) 7636 P-L

Discovered 1960 October 17 by C. J. van Houten and I. van Houten-Groeneveld at Palomar.

(10441) 9076 P-L

Discovered 1960 October 17 by C. J. van Houten and I. van Houten-Groeneveld at Palomar.

(10442) 4062 T-1

Discovered 1971 March 26 by C. J. van Houten and I. van Houten-Groeneveld at Palomar.

(10443) 1045 T-2

Discovered 1973 September 29 by C. J. van Houten and I. van Houten-Groeneveld at Palomar.

(10444) 3290 T-2

Discovered 1973 September 30 by C. J. van Houten and I. van Houten-Groeneveld at Palomar.

(10445) 4090 T-2
Discovered 1973 September 29 by C. J. van Houten and I. van Houten-Groeneveld at Palomar.

(10446) 3006 T-3
Discovered 1977 October 16 by C. J. van Houten and I. van Houten-Groeneveld at Palomar.

(10447) 3357 T-3
Discovered 1977 October 16 by C. J. van Houten and I. van Houten-Groeneveld at Palomar.

(10448) 4314 T-3
Discovered 1977 October 16 by C. J. van Houten and I. van Houten-Groeneveld at Palomar.

(10449) 1936 UD
Discovered 1936 October 16 by M. Laugier at Nice.

(10450) 1967 JQ
Discovered 1967 May 6 by C. U. Cesco and A. R. Klemola at El Leoncito.

(10451) 1975 SE
Discovered 1975 September 28 by H. L. Giclas at Anderson Mesa.

(10452) 1976 SQ_7
Discovered 1976 September 25 by N. S. Chernykh at Nauchnyj.

(10453) 1977 DY_3
Discovered 1977 February 18 by H. Kosai and K. Hurukawa at Kiso.

(10454) 1978 NY
Discovered 1978 July 9 by H.-E. Schuster at La Silla.

(10455) 1978 NU_3
Discovered 1978 July 9 by C.-I. Lagerkvist at Mount Stromlo.

(10456) 1978 PS_2
Discovered 1978 August 8 by N. S. Chernykh at Nauchnyj.

(10457) 1978 QE_2
Discovered 1978 August 31 by N. S. Chernykh at Nauchnyj.

(10458) 1978 RM_7
Discovered 1978 September 2 by C.-I. Lagerkvist at La Silla.

(10459) 1978 SJ_5
Discovered 1978 September 27 by L. I. Chernykh at Nauchnyj.

(10460) 1978 VK_8
Discovered 1978 November 7 by E. F. Helin and S. J. Bus at Palomar.

(10461) 1978 XU
Discovered 1978 December 6 by E. Bowell and A. Warnsch at Palomar.

(10462) 1979 KM
Discovered 1979 May 19 by R. M. West at La Silla.

(10463) 1979 MB_9
Discovered 1979 June 25 by E. F. Helin and S. J. Bus at Siding Spring.

(10464) 1979 SC
Discovered 1979 September 17 at the Harvard College Observatory at Harvard.

(10465) 1980 WE_5
Discovered 1980 November 29 by S. J. Bus at Palomar.

(10466) 1981 ET_7
Discovered 1981 March 1 by S. J. Bus at Siding Spring.

(10467) 1981 EZ_7
Discovered 1981 March 1 by S. J. Bus at Siding Spring.

(10468) 1981 EH_9
Discovered 1981 March 1 by S. J. Bus at Siding Spring.

(10469) 1981 EE_{14}
Discovered 1981 March 1 by S. J. Bus at Siding Spring.

(10470) 1981 EW_{18}
Discovered 1981 March 2 by S. J. Bus at Siding Spring.

(10471) 1981 EH_{20}
Discovered 1981 March 2 by S. J. Bus at Siding Spring.

(10472) 1981 EO_{20}
Discovered 1981 March 2 by S. J. Bus at Siding Spring.

(10473) 1981 EL_{21}
Discovered 1981 March 2 by S. J. Bus at Siding Spring.

(10474) 1981 EJ_{23}
Discovered 1981 March 3 by S. J. Bus at Siding Spring.

(10475) 1981 EX_{28}
Discovered 1981 March 1 by S. J. Bus at Siding Spring.

(10476) 1981 EY_{38}
Discovered 1981 March 2 by S. J. Bus at Siding Spring.

(10477) 1981 ET_{41}
Discovered 1981 March 2 by S. J. Bus at Siding Spring.

(10478) 1981 WO
Discovered 1981 November 24 by E. Bowell at Anderson Mesa.

(10479) 1982 HJ
Discovered 1982 April 18 by M. Watt at Anderson Mesa.

(10480) 1982 JB_2
Discovered 1982 May 15 at the Palomar Observatory at Palomar.

(10481) 1982 QK_3
Discovered 1982 August 23 by N. S. Chernykh at Nauchnyj.

(10482) 1983 RG_2
Discovered 1983 September 14 by E. Bowell at Anderson Mesa.

(10483) 1983 RP_2
Discovered 1983 September 4 by E. Bowell at Anderson Mesa.

(10484) 1983 WM
Discovered 1983 November 28 by E. Bowell at Anderson Mesa.

(10485) 1984 SY_5
Discovered 1984 September 21 by H. Debehogne at La Silla.

(10486) 1985 CS_2
Discovered 1985 February 15 by H. Debehogne at La Silla.

(10487) 1985 GP_1
Discovered 1985 April 14 by C. S. Shoemaker at Palomar.

(10488) 1985 RS_1
Discovered 1985 September 12 by P. Wild at Zimmerwald.

(10489) 1985 TJ_1
Discovered 1985 October 15 by E. Bowell at Anderson Mesa.

(10490) 1985 VL
Discovered 1985 November 14 by P. Jensen at Brorfelde.

(10491) 1986 QS_1
Discovered 1986 August 27 by H. Debehogne at La Silla.

(10492) 1986 QZ_1
Discovered 1986 August 28 by H. Debehogne at La Silla.

(10493) 1986 QH_2
Discovered 1986 August 28 by H. Debehogne at La Silla.

(10494) 1986 QO_3
Discovered 1986 August 29 by H. Debehogne at La Silla.

(10495) 1986 RD
Discovered 1986 September 8 by P. Jensen at Brorfelde.

(10496) 1986 RK
Discovered 1986 September 11 by P. Jensen at Brorfelde.

(10497) 1986 RQ
Discovered 1986 September 11 by P. Jensen at Brorfelde.

(10498) 1986 RG_3
Discovered 1986 September 11 by E. Bowell at Anderson Mesa.

(10499) 1986 RN_5
Discovered 1986 September 7 by H. Debehogne at La Silla.

(10500) 1987 GA
Discovered 1987 April 3 by M. Koishikawa at Sendai.

(10501) 1987 OT
Discovered 1987 July 19 by E. F. Helin at Palomar.

(10502) 1987 QF_6
Discovered 1987 August 22 by E. F. Helin at Palomar.

(10503) 1987 SG_{13}
Discovered 1987 September 27 by H. Debehogne at La Silla.

(10504) 1987 UF_5
Discovered 1987 October 22 by L. V. Zhuravleva at Nauchnyj.

(10505) 1988 BN_4
Discovered 1988 January 22 by H. Debehogne at La Silla.

(10506) 1988 CW_4
Discovered 1988 February 13 by E. W. Elst at La Silla.

(10507) 1988 ER_1
Discovered 1988 March 13 by P. Jensen at Brorfelde.

(10508) 1988 RM_4
Discovered 1988 September 1 by H. Debehogne at La Silla.

(10509) 1989 GD_4
Discovered 1989 April 3 by E. W. Elst at La Silla.

(10510) 1989 GQ_4
Discovered 1989 April 3 by E. W. Elst at La Silla.

(10511) 1989 OD
Discovered 1989 July 21 by R. H. McNaught at Siding Spring.

(10512) 1989 TP_{11}
Discovered 1989 October 2 by S. J. Bus at Cerro Tololo.

(10513) 1989 TJ_{14}
Discovered 1989 October 2 by H. Debehogne at La Silla.

(10514) 1989 TD_{16}
Discovered 1989 October 4 by H. Debehogne at La Silla.

(10515) 1989 UB_3
Discovered 1989 October 31 by B. G. W. Manning at Stakenbridge.

(10516) 1989 VQ
Discovered 1989 November 1 by M. Mukai and M. Takeishi at Kagoshima.

(10517) 1990 BH_1
Discovered 1990 January 28 by S. Ueda and H. Kaneda at Kushiro.

(10518) 1990 MC
Discovered 1990 June 18 by H. E. Holt at Palomar.

(10519) 1990 RO_2
Discovered 1990 September 15 by H. E. Holt at Palomar.

(10520) 1990 RS_2
Discovered 1990 September 15 by H. E. Holt at Palomar.

(10521) 1990 RW_7
Discovered 1990 September 14 by H. Debehogne at La Silla.

(10522) 1990 SN_3
Discovered 1990 September 18 by H. E. Holt at Palomar.

(10523) 1990 SM_6
Discovered 1990 September 22 by E. W. Elst at La Silla.

(10524) 1990 SZ_7
Discovered 1990 September 22 by E. W. Elst at La Silla.

(10525) 1990 TO
Discovered 1990 October 12 by R. H. McNaught at Siding Spring.

(10526) 1990 UK_1
Discovered 1990 October 19 by T. Hioki and S. Hayakawa at Okutama.

(10527) 1990 UN_1
Discovered 1990 October 20 by A. Sugie at Taga.

(10528) 1990 VX_3
Discovered 1990 November 12 by S. Ueda and H. Kaneda at Kushiro.

(10529) 1990 WQ_4
Discovered 1990 November 16 by E. W. Elst at La Silla.

(10530) 1991 EA
Discovered 1991 March 7 by S. Ueda and H. Kaneda at Kushiro.

(10531) 1991 GB_1
Discovered 1991 April 8 by E. F. Helin at Palomar.

(10532) 1991 NA_2
Discovered 1991 July 14 by H. E. Holt at Palomar.

(10533) 1991 PT_{12}
Discovered 1991 August 5 by H. E. Holt at Palomar.

(10534) 1991 PV_{16}
Discovered 1991 August 7 by H. E. Holt at Palomar.

(10535) 1991 RB_1
Discovered 1991 September 10 by A. Sugie at Taga.

(10536) 1991 RZ_8
Discovered 1991 September 11 by H. E. Holt at Palomar.

(10537) 1991 RY_{16}
Discovered 1991 September 15 by H. E. Holt at Palomar.

(10538) 1991 VP_2
Discovered 1991 November 11 by B. G. W. Manning at Stakenbridge.

(10539) 1991 VH_4
Discovered 1991 November 9 by S. Ueda and H. Kaneda at Kushiro.

(10540) 1991 VP_4
Discovered 1991 November 13 by S. Otomo at Kiyosato.

(10541) 1991 YX
Discovered 1991 December 31 by E. W. Elst at St. Michel.

(10542) 1992 CN_3
Discovered 1992 February 2 by E. W. Elst at La Silla.

(10543) 1992 DL_4
Discovered 1992 February 27 by F. Börngen at Tautenburg.

(10544) 1992 DA_9
Discovered 1992 February 29 by the Uppsala-ESO Survey at La Silla.

(10545) 1992 EQ_9
Discovered 1992 March 2 by the Uppsala-ESO Survey at La Silla.

(10546) 1992 FS_1
Discovered 1992 March 28 by K. Endate and K. Watanabe at Kitami.

(10547) 1992 JF
Discovered 1992 May 2 by T. Seki at Geisei.

(10548) 1992 PJ_2
Discovered 1992 August 2 by H. E. Holt at Palomar.

(10549) 1992 RM_2
Discovered 1992 September 2 by E. W. Elst at La Silla.

(10550) 1992 RK_7
Discovered 1992 September 2 by E. W. Elst at La Silla.

(10551) 1992 YL_2
Discovered 1992 December 18 by E. W. Elst at Caussols.

(10552) 1993 BH_{13}
Discovered 1993 January 22 by E. W. Elst at La Silla.

(10553) 1993 FZ_4
Discovered 1993 March 17 by the Uppsala-ESO Survey at La Silla.

(10554) 1993 FO_{34}
Discovered 1993 March 19 by the Uppsala-ESO Survey at La Silla.

(10555) 1993 HH
Discovered 1993 April 16 by K. Endate and K. Watanabe at Kitami.

(10556) 1993 QS
Discovered 1993 August 19 by E. F. Helin at Palomar.

(10557) 1993 RL_5
Discovered 1993 September 15 by E. W. Elst at La Silla.

(10558) 1993 RB_7
Discovered 1993 September 15 by E. W. Elst at La Silla.

(10559) 1993 SJ_1
Discovered 1993 September 16 by K. Endate and K. Watanabe at Kitami.

(10560) 1993 TN
Discovered 1993 October 8 by K. Endate and K. Watanabe at Kitami.

(10561) 1993 TE_2
Discovered 1993 October 15 by K. Endate and K. Watanabe at Kitami.

(10562) 1993 UB_1
Discovered 1993 October 19 by E. F. Helin at Palomar.

(10563) 1993 WD
Discovered 1993 November 19 by C. S. Shoemaker at Palomar.

(10564) 1993 XQ_2
Discovered 1993 December 14 at the Palomar Observatory at Palomar.

(10565) 1994 AT_1
Discovered 1994 January 9 by H. Shiozawa and T. Urata at Fujieda.

(10566) 1994 AZ_2
Discovered 1994 January 14 by Y. Kushida and O. Muramatsu at Yatsugatake.

(10567) 1994 CV
Discovered 1994 February 7 at the Farra d'Isonzo Observatory at Farra d'Isonzo.

(10568) 1994 CF_1
Discovered 1994 February 2 by S. Otomo at Kiyosato.

(10569) 1994 GQ
Discovered 1994 April 8 by K. Endate and K. Watanabe at Kitami.

(10570) 1994 GT
Discovered 1994 April 8 by K. Endate and K. Watanabe at Kitami.

(10571) 1994 LA_1
Discovered 1994 June 5 by C. W. Hergenrother at Tucson.

(10572) 1994 VO_7
Discovered 1994 November 8 by S. Otomo at Kiyosato.

(10573) 1994 WU_1
Discovered 1994 November 29 at the Santa Lucia Observatory at Stroncone.

(10574) 1994 YH_1
Discovered 1994 December 31 by T. Kobayashi at Oizumi.

(10575) 1994 YV_1
Discovered 1994 December 31 by T. Kobayashi at Oizumi.

(10576) 1995 GF
Discovered 1995 April 3 by T. Kobayashi at Oizumi.

(10577) 1995 JC
Discovered 1995 May 2 by M. Tichý at Kleť.

(10578) 1995 LH
Discovered 1995 June 5 by G. J. Garradd at Siding Spring.

(10579) 1995 OE
Discovered 1995 July 20 at the Osservatorio San Vittore at Bologna.

(10580) 1995 OV
Discovered 1995 July 24 by Y. Shimizu and T. Urata at Nachi-Katsuura.

(10581) 1995 OD_1
Discovered 1995 July 30 by P. Pravec at Ondřejov.

(10582) 1995 TG
Discovered 1995 October 3 by Y. Ikari at Moriyama.

(10583) 1995 WC_4
Discovered 1995 November 21 by T. Okuni at Nanyou.

(10584) 1996 GJ_2
Discovered 1996 April 14 by L. Tesi and A. Boattini at San Marcello Pistoiese.

(10585) 1996 GD_{21}
Discovered 1996 April 13 by the Spacewatch at Kitt Peak.

(10586) 1996 KY_4
Discovered 1996 May 22 by E. W. Elst at La Silla.

(10587) 1996 NF_3
Discovered 1996 July 14 by E. W. Elst at La Silla.

(10588) 1996 OE
Discovered 1996 July 18 by P. G. Comba at Prescott.

(10589) 1996 OM_2
Discovered 1996 July 23 by A. Boattini and A. Di Paola at Campo Imperatore.

(10590) 1996 OP_2
Discovered 1996 July 24 by A. Boattini and A. Di Paola at Campo Imperatore.

(10591) 1996 PD_3
Discovered 1996 August 13 by M. Tombelli and G. Forti at Montelupo.

(10592) 1996 PN_5
Discovered 1996 August 10 by the JPL NEAT Program at Haleakala.

(10593) 1996 QQ_1
Discovered 1996 August 25 by R. G. Sandness at King City.

(10594) 1996 RE_4
Discovered 1996 September 10 at the Beijing Observatory at Xinglong.

(10595) 1996 SS_6
Discovered 1996 September 21 at the Beijing Observatory at Xinglong.

(10596) 1996 TS
Discovered 1996 October 4 by D. di Cicco at Sudbury.

(10597) 1996 TR_{10}
Discovered 1996 October 9 by S. Ueda and H. Kaneda at Kushiro.

(10598) 1996 TT_{11}
Discovered 1996 October 13 by P. G. Comba at Prescott.

(10599) 1996 TK_{15}
Discovered 1996 October 9 by S. Ueda and H. Kaneda at Kushiro.

(10600) 1996 TK_{48}
Discovered 1996 October 9 by S. Ueda and H. Kaneda at Kushiro.

(10601) 1996 UC
Discovered 1996 October 16 by A. Nakamura at Kuma.

(10602) 1996 UG_3
Discovered 1996 October 16 by S. Otomo at Kiyosato.

(10603) 1996 UF_4
Discovered 1996 October 29 at the Beijing Observatory at Xinglong.

(10604) 1996 VJ
Discovered 1996 November 3 by T. Urata at Oohira.

(10605) 1996 VC$_1$
Discovered 1996 November 3 by V. Giuliani and F. Manca at Sormano.

(10606) 1996 VD$_1$
Discovered 1996 November 3 by V. Giuliani and F. Manca at Sormano.

(10607) 1996 VQ$_6$
Discovered 1996 November 13 by P. G. Comba at Prescott.

(10608) 1996 VB$_9$
Discovered 1996 November 7 by K. Endate and K. Watanabe at Kitami.

(10609) 1996 WC$_3$
Discovered 1996 November 28 by A. Nakamura at Kuma.

(10610) 1996 XR$_1$
Discovered 1996 December 2 by T. Kobayashi at Oizumi.

(10611) 1997 BB$_1$
Discovered 1997 January 23 at the Beijing Observatory at Xinglong.

(10612) 1997 JR$_{17}$
Discovered 1997 May 3 by E. W. Elst at La Silla.

(10613) 1997 RO$_3$
Discovered 1997 September 4 by Y. Shimizu and T. Urata at Nachi-Katsuura.

(10614) 1997 UH$_1$
Discovered 1997 October 21 by Y. Shimizu and T. Urata at Nachi-Katsuura.

(10615) 1997 UK$_3$
Discovered 1997 October 26 by T. Kobayashi at Oizumi.

(10616) 1997 UW$_8$
Discovered 1997 October 25 by K. Endate and K. Watanabe at Kitami.

(10617) 1997 UK$_{24}$
Discovered 1997 October 25 by M. Hirasawa and S. Suzuki at Nyukasa.

(10618) 1997 VU$_3$
Discovered 1997 November 6 by T. Kobayashi at Oizumi.

(10619) 1997 WO$_{13}$
Discovered 1997 November 27 by T. Kagawa and T. Urata at Gekko.

(10620) 1997 WQ$_{34}$
Discovered 1997 November 29 by the Lincoln NEA Research Team at Socorro.

(10621) 1997 XN
Discovered 1997 December 3 by T. Kobayashi at Oizumi.

(10622) 1997 XA$_{12}$
Discovered 1997 December 5 by the Lincoln NEA Research Team at Socorro.

(10623) 1997 YP$_7$
Discovered 1997 December 27 by T. Kobayashi at Oizumi.

(10624) 1997 YR_{13}
Discovered 1997 December 31 by T. Kobayashi at Oizumi.

(10625) 1998 AC_8
Discovered 1998 January 2 by the Lincoln NEA Research Team at Socorro.

(10626) 1998 AP_8
Discovered 1998 January 10 by L. Šarounová at Ondřejov.

(10627) 1998 BW_2
Discovered 1998 January 19 by Y. Shimizu and T. Urata at Nachi-Katsuura.

(10628) 1998 BD_5
Discovered 1998 January 18 by the OCA-DLR Survey at Caussols.

(10629) 1998 BK_{11}
Discovered 1998 January 23 by the Lincoln NEA Research Team at Socorro.

(10630) 1998 BV_{12}
Discovered 1998 January 23 by the Lincoln NEA Research Team at Socorro.

(10631) 1998 BM_{15}
Discovered 1998 January 24 by the JPL NEAT Program at Haleakala.

(10632) 1998 CV_1
Discovered 1998 February 1 at the Beijing Observatory at Xinglong.

(10633) 1998 DP_1
Discovered 1998 February 20 by P. Pravec at Ondřejov.

(10634) 1998 GM_1
Discovered 1998 April 8 by L. Šarounová at Ondřejov.

(10635) 1998 QH_8
Discovered 1998 August 17 by the Lincoln NEA Research Team at Socorro.

(10636) 1998 QK_{56}
Discovered 1998 August 28 by the Lincoln NEA Research Team at Socorro.

(10637) 1998 QP_{104}
Discovered 1998 August 26 by E. W. Elst at La Silla.

(10638) 1998 SV_{54}
Discovered 1998 September 16 by the LONEOS at Anderson Mesa.

(10639) 1998 VV_{41}
Discovered 1998 November 14 by the Spacewatch at Kitt Peak.

(10640) 1998 WU_{19}
Discovered 1998 November 25 by the Lincoln NEA Research Team at Socorro.

(10641) 1998 XS_{52}
Discovered 1998 December 14 by the Lincoln NEA Research Team at Socorro.

(10642) 1999 BF_8
Discovered 1999 January 19 by A. Boattini and L. Tesi at San Marcello Pistoiese.

(10643) 1999 CE_{78}
Discovered 1999 February 12 by the Lincoln NEA Research Team at Socorro.

(10644) 1999 DM_2
Discovered 1999 February 19 by T. Kobayashi at Oizumi.

(10645) 1999 ES_4
Discovered 1999 March 14 by K. Korlević at Višnjan.

(10646) 2077 P-L
Discovered 1960 September 26 by C. J. van Houten and I. van Houten-Groeneveld at Palomar.

(10647) 3074 P-L
Discovered 1960 September 25 by C. J. van Houten and I. van Houten-Groeneveld at Palomar.

(10648) 4089 P-L
Discovered 1960 September 24 by C. J. van Houten and I. van Houten-Groeneveld at Palomar.

(10649) 4098 P-L
Discovered 1960 September 24 by C. J. van Houten and I. van Houten-Groeneveld at Palomar.

(10650) 4110 P-L
Discovered 1960 September 24 by C. J. van Houten and I. van Houten-Groeneveld at Palomar.

(10651) 4522 P-L
Discovered 1960 September 24 by C. J. van Houten and I. van Houten-Groeneveld at Palomar.

(10652) 4599 P-L
Discovered 1960 September 24 by C. J. van Houten and I. van Houten-Groeneveld at Palomar.

(10653) 6030 P-L
Discovered 1960 September 24 by C. J. van Houten and I. van Houten-Groeneveld at Palomar.

(10654) 6673 P-L
Discovered 1960 September 24 by C. J. van Houten and I. van Houten-Groeneveld at Palomar.

(10655) 9535 P-L
Discovered 1960 October 17 by C. J. van Houten and I. van Houten-Groeneveld at Palomar.

(10656) 2213 T-1
Discovered 1971 March 25 by C. J. van Houten and I. van Houten-Groeneveld at Palomar.

(10657) 2251 T-1
Discovered 1971 March 25 by C. J. van Houten and I. van Houten-Groeneveld at Palomar.

(10658) 2281 T-1
Discovered 1971 March 25 by C. J. van Houten and I. van Houten-Groeneveld at Palomar.

(10659) 3266 T-1
Discovered 1971 March 26 by C. J. van Houten and I. van Houten-Groeneveld at Palomar.

(10660) 4348 T-1

Discovered 1971 March 26 by C. J. van Houten and I. van Houten-Groeneveld at Palomar.

(10661) 1211 T-2

Discovered 1973 September 29 by C. J. van Houten and I. van Houten-Groeneveld at Palomar.

(10662) 3201 T-2

Discovered 1973 September 30 by C. J. van Houten and I. van Houten-Groeneveld at Palomar.

(10663) 4283 T-2

Discovered 1973 September 29 by C. J. van Houten and I. van Houten-Groeneveld at Palomar.

(10664) 5187 T-2

Discovered 1973 September 25 by C. J. van Houten and I. van Houten-Groeneveld at Palomar.

(10665) 3019 T-3

Discovered 1977 October 16 by C. J. van Houten and I. van Houten-Groeneveld at Palomar.

(10666) 4171 T-3

Discovered 1977 October 16 by C. J. van Houten and I. van Houten-Groeneveld at Palomar.

Hermes

1937 UB. Discovered 1937 October 28 by K. Reinmuth at Heidelberg.

Named after the messenger of the gods (named Mercury by the Romans), son of Zeus and Maia {see planets (5731) and (66), respectively}. His attributes are the most complex and varied of those of any of the major gods. He was a deity of wealth, god of trade and travelers, of commerce, manual skill, oratory and eloquence, of thieves, and of the wind - with whose speed he was able to move; he was also patron of athletes. Within a few hours of birth, he had stolen Apollo's {see planet (1862)} cattle. Hermes invented the lyre and gave it to Apollo who, in turn gave him the caduceus, a golden staff with wings at the top, intertwined with serpents - symbol of today's medical profession. His son Autolycus became the champion thief of the world. The name "Hermes" means hastener, and representations of him are symbolic of the messenger or of speed and majesty in flight. (Z 124; LDS)

Named by the Astronomisches Rechen-Institut (AN 264, 343 (1938)): "Da das von K. Reinmuth entdeckte Objekt wegen des ungewöhnlichen Charakters der Bahn dauernd in die Literatur eingehen wird, soll es mit Zustimmung des Entdeckers den Namen "Hermes" erhalten."

Hermes is the only planet which received a name but which got no number. The planet was very near the Earth, moving at an hourly rate of more than 20 minutes of arc in the discovery night. Hermes passed from opposition to conjunction within only two days. He reached closest approach on Oct. 30 with only 0.005 a.u. Unfortunately, the planet was lost immediately after discovery.

Discoverers in Alphabetical Order

Abe, H.	8 (5)	1993-1997
Aikman, G. C. L.	1	1994
Akiyama, M.	10 (10)	1989-1993
Albitskij, V. A.	10	1923-1925
Aldering, G.	1	1982
Alikoski, H.	13	1938-1953
Alu, J.	12 (6)	1988-1993
Amburgey, L. L.	1	1997
Andrews, A. D.	1	1965
Antal, M.	9	1971-1986
Arai, M.	27 (27)	1988-1991
Arend, S.	51	1929-1961
Asher, D. J.	2	1995
Augustesen, K.	27 (27)	1982-1987
Baade, W.	10	1920-1949
Bailey, S. I.	1	1902
Balam, D. D.	3	1987-1995
Balonek, T. J.	2	1991-1993
Bancilhon, O.	1	1934
Banno, Y.	1 (1)	1983
Barr, E.	5	1983
Barros, S.	2 (1)	1975-1977
Barucci, M. A.	2 (2)	1984-1985
Baur, J. M.	6	1987-1990
Beatty, D.	1 (1)	1980
Behymer, B.	1 (1)	1983
Belyaev, Y.	2 (2)	1968
Belyavskij, S. I.	36 (3)	1912-1927
Bigourdan, G.	1	1894
Birkle, K.	1 (1)	1992
Blasco, M.	1	1997
Boattini, A.	22 (21)	1994-1999
Bohrmann, A.	9	1936-1938
Börngen, F.	165 (40)	1961-1995
Borrelly, A.	19	1866-1894
Bourgeois, P.	1	1929
Bowell, E.	499 (2)	1977-1994
Boyer, L.	40	1930-1952
Brady, J. L.	1	1952
Broughton, J.	3	1997-1998
Brown, J. A.	1 (1)	1990
Brožek, L.	21	1979-1982
Bruwer, J. A.	4	1953-1970
Buchar, E.	1	1925
Buie, M. W.	1	1997
Buil, C.	2	1997
Burnasheva, B. A.	13	1969-1971
Burnham, R. A.	1 (1)	1964
Bus, E. S.	1	1977
Bus, S. J.	426 (99)	1975-1989

Cameron, R. C.	1	1950
Candy, M. P.	5 (3)	1980-1984
Candy, V. M.	2 (2)	1984
Carnera, L.	16 (3)	1901-1902
Carrasco, R.	1	1935
Casulli, V. S.	12 (2)	1993-1998
Cattani, G.	1 (1)	1997
Cavagna, M.	4 (4)	1994-1997
Cerulli, V.	1	1910
Cesco, C. U.	19 (13)	1965-1971
Cesco, M. R.	5	1975-1976
Chacornac, J.	6	1853-1860
Chang, C.-H.	1	1957
Chang, Y. C.	2	1928-1957
Charlois, A.	99	1887-1904
Chemin, R.	1	1985
Chernykh, L. I.	226 (13)	1966-1992
Chernykh, N. S.	464 (13)	1966-1986
Chiavenna, P.	1 (1)	1996
Child, J. B.	6 (3)	1992-1996
Christensen, P. R.	1 (1)	1985
Churms, J.	2	1953
Coddington, E. F.	3	1898-1899
Cofre, S.	3 (3)	1968
Coggia, J.	5	1868-1899
Coker, R.	2 (2)	1988
Comas Solá, J.	11	1915-1930
Comba, P. G.	23	1995-1997
Cottenot, P.	1	1878
Courty, F.	2	1894
Cowell, P. H.	1	1909
Cross, K.	1	1994
Cunningham, L. E.	2	1951-1952
Cwach, K. F. J.	1	1989
d'Arrest, H.	1	1862
Davis, R. G.	1	1996
de Ball, L.	1	1882
de Gasparis, A.	9	1849-1865
de Saint-Aignan, C. P.	2	1991-1994
Debehogne, H.	333 (46)	1965-1993
Delporte, E.	66	1925-1942
DeSanctis, G.	32 (30)	1981-1986
Deutsch, A. N.	1	1929
di Cicco, D.	7	1994-1996
Di Paola, A.	3 (3)	1996
Dieckvoss, W.	2	1938
Djurkovic, P.	2	1936-1940
Dobrzycki, J.	1 (1)	1949
Dossin, F.	4	1981-1984
Dugan, R. S.	16	1902-1904
Dunbar, R. S.	6 (5)	1981-1987
Durman, C. F.	1 (1)	1996

Elst, E. W.	393 (14)	1986-1998
Emerson, G.	1	1995
Endate, K.	236 (236)	1987-1997
Ernst, E.	1	1910
Ewen-Smith, B. M.	1 (1)	1996
Faul, K. L.	1	1977
Ferguson, J.	3	1854-1860
Ferreri, W.	16 (8)	1984-1988
Ferrero, M.	2 (2)	1930
Fisch, G.	1 (1)	1992
Foerster, W.	1 (1)	1860
Fogh Olsen, H. J.	10 (10)	1984-1987
Forti, G.	1 (1)	1996
Fric, T. F.	1 (1)	1985
Frost, R. H.	1	1902
Fujii, H.	1 (1)	1994
Fujii, T.	17 (17)	1988-1992
Furuta, T.	59 (53)	1980-1993
Garradd, G. J.	6	1993-1996
Geffert, M.	2	1989
Gehrels, T.	6	1971-1973
Georgieva, A.	1 (1)	1983
Gessner, H.	1	1953
Ghezzi, P.	2 (2)	1995
Gibson, J.	13 (3)	1971-1985
Giclas, H. L.	15 (2)	1934-1978
Gilbrech, R. J.	1 (1)	1985
Gilmore, A. C.	29 (29)	1981-1994
Giuliani, V.	5 (5)	1994-1996
Goldschmidt, H.	14	1852-1861
Goldstein, D. B.	1	1985
Gonnessiat, F.	2	1918-1920
Gonzalez, L. E.	2	1981-1982
Goretti, V.	5 (2)	1994-1997
Götz, P.	20 (2)	1903-1905
Graham, A.	1	1848
Groeneveld, I. *	2 (1)	1955-1956
Grossman, A.	1	1984
Gyllenberg, K.	1	1916
Hansen, L.	1 (1)	1985
Harding, K.	1	1804
Harlan, E. A.	1 (1)	1984
Hartmann, J.	3	1921-1932
Hayakawa, S.	22 (22)	1989-1993
Helffrich, J.	13	1909-1911
Helin, E. F.	388 (122)	1973-1994
Hencke, K. L.	2	1845-1847
Henry, P. M.	7	1872-1878
Henry, P. P.	7	1872-1882

* See also van Houten-Groeneveld

Hergenrother, C. W.	10	1993-1995
Herkenhoff, K.	2 (2)	1983
Hertzsprung, E.	2	1924-1929
Hind, J. R.	10	1847-1854
Hioki, T.	33 (33)	1988-1993
Hirasawa, M.	18 (18)	1991-1997
Hoffmeister, C.	5 (1)	1938-1963
Holt, H. E.	261 (7)	1988-1993
Hopp, U.	1 (1)	1992
Horowitz, I.	1	1986
Houziaux, L.	3 (3)	1980
Hubble, E.	1	1935
Huchra, J. P.	1 (1)	1975
Hunaerts, J.	2	1936
Hurukawa, K.	48 (48)	1976-1986
Ikari, Y.	3 (2)	1995-1996
Inkeri, K.	1	1937
Inoda, S.	15 (15)	1986-1992
Inoue, M.	9 (9)	1986-1990
Irwin, M. J.	2 (2)	1990
Ito, K.	1	1994
Itzigsohn, M.	15	1948-1954
Ivanov, N.	3 (3)	1927
Ivanova, V. G.	9 (9)	1983-1987
Iwamoto, M.	5 (5)	1988-1989
Izumikawa, S.	1 (1)	1990
Jackisch, G.	1 (1)	1959
Jackson, C.	72 (1)	1929-1939
Jekabsons, P.	4 (1)	1978-1981
Jekhovsky, B.	12	1921-1927
Jensen, K. S.	1 (1)	1982
Jensen, P.	64 (26)	1984-1998
Jewitt, D.	1 (1)	1995
Johnson, E. L.	18	1946-1951
Kagawa, T.	5 (4)	1997-1998
Kaiser, F.	21	1911-1914
Kakei, W.	8 (8)	1987-1989
Kamper, K. W.	2	1963-1978
Kanai, K.	1 (1)	1988
Kaneda, H.	330 (330)	1987-1998
Karachkina, L. G.	99 (8)	1978-1990
Kastel', G. R.	5 (5)	1990
Kawanishi, K.	12 (12)	1989-1991
Kawasato, N.	29 (13)	1988-1997
Keeler, J. E.	1	1899
Kilmartin, P. M.	29 (29)	1981-1994
Kirk, R.	1 (1)	1983
Kizawa, M.	15 (14)	1986-1991
Klemola, A. R.	16 (10)	1965-1984
Knežević, Z.	1	1980
Knorre, V.	4	1876-1887

Kobayashi, J.	1	1997
Kobayashi, T.	256 (1)	1991-1999
Kohoutek, L.	62 (1)	1967-1981
Koishikawa, M.	15	1987-1995
Kojima, T.	31 (1)	1987-1996
Kolény, P.	1 (1)	1997
König, A.	1 (1)	1959
Kopff, A.	67 (1)	1904-1909
Korlevic, K.	7	1998-1999
Kornoš, L.	1 (1)	1997
Kosai, H.	49 (49)	1976-1986
Kowal, C. T.	16 (1)	1970-1981
Krieger, C. J.	1	1929
Kriete, A.	1 (1)	1967
Kryachko, T. V.	3	1995
Kubokawa, K.	1 (1)	1929
Kulin, G.	21	1936-1941
Kurochkin, N. E.	2 (1)	1972-1983
Kushida, R.	1 (1)	1991
Kushida, Y.	38 (38)	1988-1994
Květon, J.	2	1979
Kwiek, A.	1 (1)	1949
Lagerkvist, C.-I.	65 (2)	1975-1996
Lagrula, J.	1	1914
Lai, L.	1	1996
Landgraf, W.	5	1987-1989
Lauer, T.	1 (1)	1977
Laugier, M.	20	1932-1955
Laurent, A.	1	1858
Laurie, S. P.	6	1995-1997
Lawrence, K. J.	11 (8)	1990-1994
Le Morvan, C.	1	1913
Lee, L.	1 (1)	1992
Leonard, G. J.	7	1992
Lesser, O.	1 (1)	1860
Levy, D. H.	10 (10)	1990-1994
Liller, W.	2	1978-1979
Linderholm, R.	2	1996-1997
Lohnert, K.	4	1906-1907
Lokalov, A.	1 (1)	1970
López G., A.	4 (4)	1992
Lopez, A.	1 (1)	1998
Lorenz, W.	4	1908-1909
Lovas, M.	2	1977-1982
Lowell, P.	1	1907
Luther, R.	24	1852-1890
Luu, J. X.	1 (1)	1995
Macri, L.	1	1996
Mahrová, M.	4	1981-1985
Manca, F.	6 (6)	1996-1997
Manning, B. G. W.	14	1989-1996
Marley, M.	1 (1)	1983

Marth, A.	1	1854
Massinger, A.	7	1912-1914
Matsuyama, M.	12 (12)	1988-1991
Maury, A.	5 (1)	1986-1988
McDonald, S.	1	1987
McLeish, D.	1	1964
McNaught, R. H.	130 (6)	1975-1996
McNay, A.	1	1980
Melotte, P. J.	1	1909
Merlin, J.-C.	1	1997
Metcalf, J. H.	41	1905-1914
Metlova, N. V.	1 (1)	1983
Meyer, E.	1	1997
Mikolajczak, C.	2 (2)	1988
Millosevich, E.	2	1891
Minkowski, R.	1 (1)	1951
Mitani, T.	1	1953
Miyasaka, S.	3 (3)	1993
Mizuno, Y.	34 (34)	1989-1993
Morata, D.	1 (1)	1997
Morata, S.	1 (1)	1997
Moravec, Z.	20 (9)	1994-1998
Mori, H.	27 (27)	1988-1991
Mottola, S.	1	1995
Mrkos, A.	212	1978-1990
Mueller, J.	4 (1)	1987-1993
Mukai, M.	9 (9)	1988-1993
Mulholland, J. D.	1	1975
Munari, U.	13 (13)	1996-1997
Muramatsu, O.	53 (53)	1986-1994
Muzzio, J. C.	2	1979
Nakamura, A.	22	1994-1997
Nakamura, T.	1	1988
Naranjo, O. A.	3 (1)	1992
Nassir, M.	1	1993
Natori, A.	21 (21)	1990-1993
Netto, E. R.	6 (6)	1979
Neujmin, G. N.	74	1913-1939
Nicholson, S. B.	2	1916-1957
Niijima, T.	21 (21)	1986-1995
Nolan, M.	1	1984
Nomura, T.	13 (13)	1989-1994
Norgaard-Nielsen, H. U.	1 (1)	1985
Noymer, A. J.	5	1988
Ofek, E. O.	1	1997
Oikawa, O.	8 (1)	1927-1929
Ojakangas, G.	2 (2)	1983
Okuni, T.	13	1995-1998
Olbers, H. W.	2	1802-1807
Olmstead, C. M.	6	1977-1990
Oshima, Y.	30	1987-1989
Oterma, L.	53	1938-1953

Otomo, S.	67 (14)	1991-1997
Pacheco, R.	1 (1)	1998
Palisa, J.	121	1874-1923
Parchomenko, P.	2	1929-1930
Paroubek, A.	1 (1)	1955
Passey, Q. R.	2 (2)	1982
Patry, A.	9	1936-1940
Pauwels, T.	1	1997
Pereyra, Z. M.	1	1966
Perrotin, J.	6	1874-1885
Peters, C. H. F.	48	1861-1889
Peters, G. H.	3	1904-1921
Petit, J.	2 (2)	1971
Phinney, J.	2	1987-1988
Piazzi, G.	1	1801
Pizarro, G.	3 (3)	1979-1996
Pizarro, O.	1 (1)	1979
Platt, J.	2	1984
Plougin, G.	2 (2)	1968
Podstanicka, R.	1 (1)	1955
Pogson, N. R.	8	1856-1885
Pólit, I.	1	1941
Pollas, C.	6 (1)	1984-1992
Potter, H.	1 (1)	1970
Pratt, C.	1 (1)	1978
Pravec, P.	12 (5)	1995-1998
Protitch, M. B.	7	1936-1952
Quadri, U.	2 (2)	1991-1992
Rajamohan, R.	5	1988-1989
Reinmuth, K.	389 (1)	1914-1957
Reiss, G.	5	1931-1935
Rheden, J.	3	1913-1916
Rickman, H.	2 (2)	1976
Rigaux, F.	7	1933-1941
Rodriquez, T.	1	1988
Roemer, E.	2	1964-1975
Rogers, J. E.	2 (1)	1994-1995
Roman, B.	6 (3)	1988-1990
Rose, P.	1	1991
Rudnyk, M.	1	1986
Rudy, D.	1 (1)	1983
Sala, P.	1	1997
Salyards, S. L.	1	1983
Samuel, A. G.	1 (1)	1968
Sandig, H.-U.	1	1937
Sandness, R. G.	1	1996
Šarounová, L.	12 (4)	1995-1998
Sasaki, G.	1 (1)	1979
Sato, N.	8 (2)	1995-1997
Sause, G.	1 (1)	1988

Schaldach, R. D.	2 (2)	1949
Schaumasse, A.	2	1921-1928
Schiaparelli, G.	1	1861
Schildknecht, T.	3	1985
Schmadel, L. D.	40 (40)	1990-1992
Schmitt, A.	4	1932-1953
Schorr, R.	2	1917-1932
Schubart, J.	2 (1)	1960-1961
Schulhof, L.	1	1875
Schuster, H.-E.	17	1976-1982
Schwassmann, A.	22 (13)	1898-1932
Searle, G.	1	1858
Sebok, W.	2	1977
Sei, S.	1	1983
Seki, T.	112	1981-1998
Shajn, G.	3	1925
Shajn, P. F.	19	1928-1953
Shimizu, Y.	67 (67)	1993-1998
Shiozawa, H.	12 (12)	1990-1996
Shirai, S.	2 (2)	1993
Shkodrov, V. G.	7 (5)	1983-1988
Shoemaker, C. S.	252 (170)	1980-1994
Shoemaker, E. M.	168 (168)	1977-1994
Sicoli, P.	8 (8)	1994-1997
Singer-Brewster, S.	2	1985-1987
Sitko, M. L.	1 (1)	1982
Skiff, B. A.	44 (4)	1980-1984
Skvortsov, E.	3	1929-1930
Smirnova, T. M.	124 (1)	1966-1984
Smrekar, S.	1	1983
Soulie, G.	2	1967-1968
Spahr, T. B.	14	1993-1996
Staus, A.	1	1892
Steel, D. I.	7	1990-1992
Stephan, E.	1	1866
Stern, W. A.	1 (1)	1982
Stewart, B.	1 (1)	1978
Stewart, D.	1	1901
Stock, J.	1 (1)	1992
Strabla, L.	2 (2)	1991-1992
Strommer, G.	1	1940
Struve, O.	2	1922
Sugano, M.	4 (4)	1989-1994
Sugie, A.	50	1988-1994
Suvanto, R.	1	1936
Suzuki, K.	33 (32)	1984-1992
Suzuki, S.	18 (18)	1991-1997
Swanson, S.	1	1984
Sy, F.	2	1916
Taff, L. G.	11 (1)	1980-1982
Takahashi, A.	13 (13)	1989-1991
Takeishi, M.	9 (9)	1988-1993

Tempel, E. W.	5	1861-1868
Tesi, L.	9 (9)	1994-1999
Testa, A.	2 (2)	1995-1996
Thiele, H.	4	1914-1916
Tholen, D. J.	1	1981
Thomas, N. G.	50 (8)	1964-1989
Tichá, J.	6 (3)	1995-1996
Tichý, M.	21 (10)	1994-1998
Tietjen, F.	1	1866
Tombaugh, C. W.	15	1929-1936
Tombelli, M.	24 (24)	1994-1997
Tomita, K.	6	1978-1982
Torres, C.	31 (6)	1968-1982
Tucker, R. A.	1	1997
Tuttle, H. P.	2	1861-1862
Ueda, S.	330 (330)	1987-1998
Urata, T.	214 (173)	1978-1998
Uto, F.	4	1993-1995
Vagnozzi, A.	4	1993-1994
Väisälä, M.	2	1942
Väisälä, Y.	127	1935-1944
Valentini, S.	2 (2)	1994
Van Biesbroeck, G.	16	1922-1939
van Gent, H.	39	1929-1935
van Houten, C. J.	602 (597)	1960-1977
van Houten-Groeneveld, I. *	598 (597)	1960-1977
Vávrová, Z.	103	1978-1991
Villiger, W.	1	1897
Viscome, G. R.	2	1996
Vogt, H.	1	1912
Volta, L.	5	1928-1934
Wachmann, A. A.	3	1938-1939
Wagner, J. F.	5	1983-1984
Waldron, J. D.	2	1986
Wallenquist, A. A. E.	1 (1)	1950
Warnsch, A.	2 (2)	1978
Watanabe, K.	297 (291)	1987-1997
Watson, J. C.	22	1863-1877
Watt, M.	4	1982
Weber, R.	1	1995
Wenzel, W.	1 (1)	1959
West, R. M.	26	1976-1986
Whipple, F. L.	1	1933
Wild, P.	90	1961-1987
Wilder, P. D.	2 (2)	1982
Willis, J. E.	1	1941
Wilson, A. G.	4 (2)	1950-1953
Wirtanen, C. A.	7	1947-1950
Wirtanen, E.	1	1948
Witt, G.	2	1896-1898

* See also Groeneveld

Wolf, M. (Marek)	3 (3)	1996-1997
Wolf, M. (Max)	248 (19)	1891-1932
Wood, H. E.	12 (1)	1911-1932
Wroblewski, H.	2	1968-1975
Yanai, M.	13 (13)	1987-1992
Zappalà, V.	8 (8)	1984
Zeigler, K. W.	2	1987-1988
Zhuravleva, L. V.	158 (10)	1972-1992
Zoltowski, F. B.	2	1997
Zytkow, A.	2 (2)	1990

Discoverers Ranking List

1.	van Houten, C. J.	602 (597)	1960-1977
2.	van Houten-Groeneveld, I.	598 (597)	1960-1977
3.	Bowell, E.	499 (2)	1977-1994
4.	Chernykh, N. S.	464 (13)	1966-1986
5.	Bus, S. J.	426 (99)	1975-1989
6.	Elst, E. W.	393 (14)	1986-1998
7.	Reinmuth, K.	389 (1)	1914-1957
8.	Helin, E. F.	388 (122)	1973-1994
9.	Debehogne, H.	333 (46)	1965-1993
10.	Kaneda, H.	330 (330)	1987-1998
	Ueda, S.	330 (330)	1987-1998
12.	Watanabe, K.	297 (291)	1987-1997
13.	Holt, H. E.	261 (7)	1988-1993
14.	Kobayashi, T.	256 (1)	1991-1999
15.	Shoemaker, C. S.	252 (170)	1980-1994
16.	Wolf, M. (Max)	248 (19)	1891-1932
17.	Endate, K.	236 (236)	1987-1997
18.	Chernykh, L. I.	226 (13)	1966-1992
19.	Urata, T.	214 (173)	1978-1998
20.	Mrkos, A.	212	1978-1990
21.	Shoemaker, E. M.	168 (168)	1977-1994
22.	Börngen, F.	165 (40)	1961-1995
23.	Zhuravleva, L. V.	158 (10)	1972-1992
24.	McNaught, R. H.	130 (6)	1975-1996
25.	Väisälä, Y.	127	1935-1944
26.	Smirnova, T. M.	124 (1)	1966-1984
27.	Palisa, J.	121	1874-1923
28.	Seki, T.	112	1981-1998
29.	Vávrová, Z.	103	1978-1991
30.	Charlois, A.	99	1887-1904
	Karachkina, L. G.	99 (8)	1978-1990
32.	Wild, P.	90	1961-1987
33.	Neujmin, G. N.	74	1913-1939
34.	Jackson, C.	72 (1)	1929-1939
35.	Kopff, A.	67 (1)	1904-1909
	Otomo, S.	67 (14)	1991-1997
	Shimizu, Y.	67 (67)	1993-1998
38.	Delporte, E.	66	1925-1942
39.	Lagerkvist, C.-I.	65 (2)	1975-1996
40.	Jensen, P.	64 (26)	1984-1998
41.	Kohoutek, L.	62 (1)	1967-1981
42.	Furuta, T.	59 (53)	1980-1993
43.	Muramatsu, O.	53 (53)	1986-1994
	Oterma, L.	53	1938-1953
45.	Arend, S.	51	1929-1961
46.	Sugie, A.	50	1988-1994
	Thomas, N. G.	50 (8)	1964-1989
48.	Kosai, H.	49 (49)	1976-1986
49.	Hurukawa, K.	48 (48)	1976-1986
	Peters, C. H. F.	48	1861-1889

51.	Skiff, B. A.	44 (4)	1980-1984
52.	Metcalf, J. H.	41	1905-1914
53.	Boyer, L.	40	1930-1952
	Schmadel, L. D.	40 (40)	1990-1992
55.	van Gent, H.	39	1929-1935
56.	Kushida, Y.	38 (38)	1988-1994
57.	Belyavskij, S. I.	36 (3)	1912-1927
58.	Mizuno, Y.	34 (34)	1989-1993
59.	Hioki, T.	33 (33)	1988-1993
	Suzuki, K.	33 (32)	1984-1992
61.	DeSanctis, G.	32 (30)	1981-1986
62.	Kojima, T.	31 (1)	1987-1996
	Torres, C.	31 (6)	1968-1982
64.	Oshima, Y.	30	1987-1989
65.	Gilmore, A. C.	29 (29)	1981-1994
	Kawasato, N.	29 (13)	1988-1997
	Kilmartin, P. M.	29 (29)	1981-1994
68.	Arai, M.	27 (27)	1988-1991
	Augustesen, K.	27 (27)	1982-1987
	Mori, H.	27 (27)	1988-1991
71.	West, R. M.	26	1976-1986
72.	Luther, R.	24	1852-1890
	Tombelli, M.	24 (24)	1994-1997
74.	Comba, P. G.	23	1995-1997
75.	Boattini, A.	22 (21)	1994-1999
	Hayakawa, S.	22 (22)	1989-1993
	Nakamura, A.	22	1994-1997
	Schwassmann, A.	22 (13)	1898-1932
	Watson, J. C.	22	1863-1877
80.	Brožek, L.	21	1979-1982
	Kaiser, F.	21	1911-1914
	Kulin, G.	21	1936-1941
	Natori, A.	21 (21)	1990-1993
	Niijima, T.	21 (21)	1986-1995
	Tichý, M.	21 (10)	1994-1998
86.	Götz, P.	20 (2)	1903-1905
	Laugier, M.	20	1932-1955
	Moravec, Z.	20 (9)	1994-1998

Corporate Discoveries

1.	Purple Mountain Observatory	140
2.	Goethe Link Observatory	116
3.	Uppsala-ESO Survey	107
4.	Lincoln NEA Research Team	72
	Spacewatch	72
6.	Beijing (Peking) Observatory	59
7.	Felix Aguilar Observatory (now C. U. Cesco Observatory)	48
8.	Harvard College Observatory (now Oak Ridge Observatory)	45
9.	Osservatorio San Vittore	32
10.	Oak Ridge Observatory	26
11.	JPL NEAT Program	25
12.	Santa Lucia Observatory	23
13.	Farra d'Isonzo Observatory	21
14.	Palomar Observatory	19
15.	Perth Observatory	17
16.	CERGA (Caussols)	13
	Kleť Observatory	13
18.	Crimean Astrophysical Observatory	11
19.	Bulgarian National Observatory	6
	Carlos U. Cesco Observatory	6
	OCA-DLR Survey	6
22.	La Plata Observatory	5
	Višnjan Observatory	5
24.	LONEOS	4
25.	Copenhagen University Observatory	3
	Lowell Observatory	3
27.	Boyden Observatory	2
	Brorfelde Observatory	2
	Davidschlag Observatory	2
	Haute Provence Observatory	2
	Infrared Astronomical Satellite	2
	McDonald Observatory	2
	University of Chile Observatory	2
34.	Bassano Observatory	1
	Centro Astronomico de Yebes	1
	Colleverde di Guidonia Observatory	1
	Crni Vrh Observatory	1
	Giordano Bruno Observatory	1
	Kiso Observatory	1
	Lincoln Laboratory ETS	1
	Monte Viseggi Observatory	1
	Observatorio do Valongo	1
	Royal Observatory Edinburgh	1
	Saji Observatory	1

Discovery Places in Alphabetical Order

Place	Number
Albany, NY, U.S.A.	1
Algiers, Algeria	64
Anderson Mesa, AZ, U.S.A.	604
Ann Arbor, MI, U.S.A.	21
Arequipa, Peru	3
Barcelona, Spain	12
Bassano Bresciano, Italy	3
Belgrade, Yugoslavia	8
Bergedorf, Germany	90
Berlin, Germany	8
Bickley, Australia	27
Bloemfontein, South Africa	5
Bologna, Italy	32
Bordeaux, France	4
Bothkamp, Germany	1
Bremen, Germany	2
Brooklyn, IN, U.S.A.	117
Brorfelde, Denmark	70
Budapest, Hungary	22
Calar Alto, Spain	2
Camarillo, CA, U.S.A.	1
Cambridge, MA, U.S.A.	3
Campiñas, Brazil	1
Campo Imperatore, Italy	4
Caussols, France	61
Cavriana, Italy	1
Cerro El Roble, Chile	41
Cerro Tololo, Chile	35
Chichibu, Japan	8
Chions, Italy	6
Chirorin, Japan	1
Chiyoda, Japan	31
Church Stretton, United Kingdom	6
Cima Ekar, Italy	22
Clinton, NY, U.S.A.	48
Colleverde di Guidonia, Italy	13
Copenhagen, Denmark	1
Córdoba, Argentina	2
Costitx, Spain	2
Crni vrh, Slovenia	1
Driesen, Germany	2
Düsseldorf, Germany	24
El Leoncito, Argentina	83
Farra d'Isonzo, Italy	21
Fitchburg, MA, U.S.A.	1
Flagstaff, AZ, U.S.A.	29
Fort Davis, TX, U.S.A.	3
Fujieda, Japan	11
Geisei, Japan	112
Gekko, Japan	35
Giesing, Germany	1
Golden, CO, U.S.A.	1
Granville, MA, U.S.A.	1
Greenwich, United Kingdom	2
Haleakala, HI, U.S.A.	25
Hamilton, NY, U.S.A.	1
Hartbeespoort, South Africa	1
Harvard, MA, U.S.A.	71
Heidelberg, Germany	815
Hidaka, Japan	2
Johannesburg, South Africa	147
Kagoshima, Japan	9
Kani, Japan	34
Karasuyama, Japan	16
Kashihara, Japan	4
Kavalur, India	5
King City, ON, Canada	1
Kiso, Japan	51
Kitami, Japan	278
Kitt Peak, AZ, U.S.A.	76
Kiyosato, Japan	67
Kleť, Czech Rep.	391
Kobuchizawa, Japan	5
Kuma, Japan	22
Kumomoto, Japan	1
Kushiro, Japan	342
Kvistaberg, Sweden	9
Kwasan, Japan	1
La Plata, Argentina	23
La Silla, Chile	862
Lake Placid, NY, U.S.A.	2
Lake Tekapo, New Zealand	29
Le Creusot, France	1
Lilienthal, Germany	1
Lime Creek, NE, U.S.A.	2
Linz, Austria	3
London, United Kingdom	11
Loomberah, NSW, Australia	1
Macquarie, Australia	1
Madonna di Dossobuono, Italy	1
Madras, India	5
Madrid, Spain	1
Markree, Ireland	1
Marseilles, France	32
Martigues, France	1
Mauna Kea, HI, U.S.A.	1
Merida, Venezuela	3
Milan, Italy	1
Minami-Oda, Japan	14
Mishima, Japan	10
Mitzpe Ramon, Israel	1
Modra, Slovakia	1
Monte Viseggi, Italy	1
Montelupo, Italy	1
Moriyama, Japan	3
Mount Hamilton, CA, U.S.A.	19
Mount Hopkins, AZ, U.S.A.	1
Mount Lemmon, AZ, U.S.A.	1

Mount Stromlo, Australia	6
Mount Wilson, CA, U.S.A.	7
Munich, Germany	1
Nachi-Katsuura, Japan	68
Nanking, China	142
Nanyou, Japan	13
Naples, Italy	9
Nauchnyj, Ukraine	1077
Nice, France	131
Nîmes, France	1
Nyukasa, Japan	18
Oizumi, Japan	256
Ojima, Japan	21
Okutama, Japan	33
Ondřejov, Czech Rep.	21
Oohira, Japan	47
Oxford, United Kingdom	3
Palermo, Italy	1
Palomar, CA, U.S.A.	1635
Paris, France	35
Peking, China	1
Pianoro, Italy	3
Pino Torinese, Italy	5
Piszkéstetö, Hungary	6
Piwnice, Poland	5
Pola, Austria (now Croatia)	28
Portimão, Portugal	1
Poznań, Poland	1
Prescott, AZ, U.S.A.	23
Ramonville Saint Agne, France	2
Reedy Creek, QLD, Australia	3
Rome, Italy	2
Rozhen, Bulgaria	24
Saji, Japan	1
San Marcello Pistoiese, Italy	11
Sapporo, Japan	7
Sendai, Japan	16
Sengamine, Japan	1
Shizuoka, Japan	2
Siding Spring, NSW, Australia	444
Simeïs, Russia	148
Skalnaté Pleso, Slovak Rep.	2
Socorro, NM, U.S.A.	85
Sonneberg, Germany	4
Sormano, Italy	14
St. Michel, France	51
Stakenbridge, United Kingdom	14
Stony Ridge, CA, U.S.A.	1
Stroncone, Italy	27
Sudbury, MA, U.S.A.	7
Taga, Japan	50
Taunton, MA, U.S.A.	31
Tautenburg, Germany	167
Teramo, Italy	1
Tokai, Japan	6
Tokushima, Japan	5
Tokyo, Japan	8
Toulouse, France	5
Toyota, Japan	33
Tucson, AZ, U.S.A.	25
Turku, Finland	197
Uccle, Belgium	134
Uenohara, Japan	16
Victoria, BC, Canada	4
Vienna, Austria	97
Višnjan, Croatia	12
Washington, DC, U.S.A.	7
Williams Bay, WI, U.S.A.	19
Winchester, MA, U.S.A.	10
Woomera, SA, Australia	2
Wrightwood, CA, U.S.A.	1
Xinglong, China	59
Yakiimo, Japan	22
Yatsugatake, Japan	39
Yatsuka, Japan	8
Yebes, Spain	1
Yorii, Japan	27
Zelenchukskaya, Ukraine	3
Zimmerwald, Switzerland	93

Observatory Index List

Albany, NY	Dudley Obs.
Algiers	Bouzaréah Obs.
Anderson Mesa, AZ	Lowell Obs. Branch
Arequipa	Harvard Obs. Branch
Barcelona	Fabra Obs.
Bergedorf	Hamburg Obs.
Bickley	Perth Obs.
Bloemfontein	Boyden Obs.
Bologna	Oss. San Vittore
Bordeaux	Univ. Obs. Floirac

Brooklyn, IN	Goethe Link Obs.
Brorfelde	Copenhagen Univ. Obs.
Budapest	Konkoly Obs.
Calar Alto	Max-Planck-Inst. Astron. Branch
Cambridge, MA	Harvard Coll. Obs.
Campinãs	Obs. do Valongo, Rio de Janeiro
Caussols	CERGA
Cavriana	Giordano Bruno Obs.
Cerro Tololo	Inter-American Obs.
Chions	Oss. Chaonis
Cima Ekar	Asiago Astrophys. Obs.
Clinton, NY	Hamilton Coll. Obs.
Costitx	Obs. Astron. Mallorca
Düsseldorf	Bilk Obs.
El Leoncito	Carlos U. Cesco Obs.
	Félix Aguilar Obs. Branch
Flagstaff, AZ	Lowell Obs.
	U.S. Naval Obs.
Fort Davis, TX	McDonald Obs.
Golden, CO	Chamberlin Obs. Field Stn.
Greenwich	Royal Obs.
Hamilton, NY	Foggy Bottom Obs.
Hartbeespoort	Union Obs. Annexe
Harvard, MA	Harvard Obs. Agassiz Stn.
	Oak Ridge Obs.
Heidelberg	Königstuhl Obs.
Johannesburg	Union Obs.
	Yale-Columbia Stn.
Kavalur	Vainu Bappu Obs.
Kiso	Tokyo Astron. Obs. Branch
Kitt Peak, AZ	Kitt Peak Natl. Obs.
	McGraw-Hill Obs.
	Steward Obs.
La Silla	European Southern Obs.
Lake Placid	Rand Obs. (Viscome)
Lake Tekapo	Mount John Univ. Obs.
Linz	Davidschlag Obs.
Mitzpe Ramon	Wise Obs., Tel-Aviv Univ.
Mount Hamilton, CA	Lick Obs.
Mount Hopkins, AZ	F. L. Whipple Obs.
Mount Lemmon, AZ	Infrared Obs., Univ. Minnesota
Nanking	Purple Mountain Obs.
Nauchnyj	Crimean Astrophys. Obs.
	Sternberg Stn.
Oohira	Nihondaira Obs. Branch
Pola	Hydrographic Office Obs.
Portimão	Algarve Astron. Obs.
Rozhen	Bulgarian Natl. Obs.
Sendai	Sendai Obs. Ayashi Stn.
Siding Spring, NSW	Anglo-Australian Obs.
	Uppsala Southern Stn.
Smolyan	see Rozhen
Socorro, NM	Lincoln Lab. ETS

St. Michel	Haute Provence Obs.
Stroncone	Santa Lucia Obs.
Taga	Dynic Astron. Obs.
Tautenburg	Karl Schwarzschild Obs.
Tucson, AZ	Catalina Stn., Univ. Arizona
	Goodricke-Pigott Obs. (Tucker)
Uccle	Obs. Royal Belgique
Victoria	Climenhaga Obs.
	Dominion Astrophys. Obs.
Vienna	Univ. Obs.
Washington, DC	Natl. Obs.
	U.S. Naval Obs.
Williams Bay, WI	Yerkes Obs.
Wrightwood, CA	Ford Obs.
Xinglong	Peking Obs. Branch

Discovery Places Ranking List

1.	Palomar	1635
2.	Nauchnyj	1077
3.	La Silla	862
4.	Heidelberg	815
5.	Anderson Mesa	604
6.	Siding Spring	444
7.	Kleť	391
8.	Kushiro	342
9.	Kitami	278
10.	Oizumi	256
11.	Turku	197
12.	Tautenburg	167
13.	Simeïs	148
14.	Johannesburg	147
15.	Nanking	142
16.	Uccle	134
17.	Nice	131
18.	Brooklyn, Indiana	117
19.	Geisei	112
20.	Vienna	97
21.	Zimmerwald	93
22.	Bergedorf	90
23.	Socorro	85
24.	El Leoncito	83
25.	Kitt Peak	76
26.	Harvard	71
27.	Brorfelde	70
28.	Nachi-Katsuura	68
29.	Kiyosato	67
30.	Algiers	64
31.	Caussols	61
32.	Xinglong	59
33.	Kiso	51
	St. Michel	51
35.	Taga	50
36.	Clinton	48
37.	Oohira	47
38.	Cerro El Roble	41
39.	Yatsugatake	39
40.	Cerro Tololo	35
	Gekko	35
	Paris	35
43.	Kani	34
44.	Okutama	33
	Toyota	33
46.	Bologna	32
	Marseilles	32
48.	Chiyoda	31
	Taunton	31

50.	Flagstaff	29
	Lake Tekapo	29
52.	Pola	28
53.	Bickley	27
	Stroncone	27
	Yorii	27
56.	Haleakala	25
	Tucson	25
58.	Düsseldorf	24
	Rozhen	24
60.	La Plata	23
	Prescott	23
62.	Budapest	22
	Cima Ekar	22
	Kuma	22
	Yakiimo	22
66.	Ann Arbor	21
	Farra d'Isonzo	21
	Ojima	21
	Ondřejov	21
70.	Mount Hamilton	19
	Williams Bay	19
72.	Nyukasa	18
73.	Karasuyama	16
	Sendai	16
	Uenohara	16
76.	Minami-Oda	14
	Sormano	14
	Stakenbridge	14
79.	Colleverde di Guidonia	13
	Nanyou	13
81.	Barcelona	12
	Višnjan	12
83.	Fujieda	11
	London	11
	San Marcello Pistoiese	11
86.	Mishima	10
	Winchester	10

Categories of Minor Planet Names

Men	3475
Astronomers	1420
Women	870
Scientists (no astronomers)	790
Cities, harbors, buildings	601
Mythological terms	580
Writers	370
Amateur astronomers	333
Relatives of discoverers	326
Countries, provinces, islands	271
Historical or political figures	229
Composers, musicians, dancers	227
Rivers, seas, mountains	179
Plants, trees, animals	178
Figures from literature, sagas, etc.	160
Painters, sculptors	150
Institutes, observatories, universities	149
Acronyms, pseudonyms, compound terms	76
Curiosities	28

Names Classification by Countries

Former USSR	890
USA	860
Japan	599
Germany	450
France	224
Italy	215
Czech & Slovak Rep.	181
UK	151
Finland	138
Belgium	107
China	107
Netherlands	78
Sweden	62
Austria	61
Argentina	56
Greece	56
Australia	55
Switzerland	43
Denmark	38
South Africa	38
Canada	37
Spain	35
Hungary	30
Egypt	21
Poland	17
Brazil	15
New Zealand	15
Chile	13
Former Yugoslavia	11
Bulgaria	10
Algeria	9
India	9
Norway	9
Indonesia	7
Ireland	7
Portugal	4
Uruguay	4
Hong Kong	3
Kenya	3
Puerto Rico	3
Romania	3
Korea	2
Mexico	2
Peru	2
Bolivia	1
El Salvador	1
Ethiopia	1
Guatemala	1
Jamaica	1
Libya	1

Luxembourg	1
Philippines	1
Turkey	1
Uganda	1

Nobel Laureates

Literature

(5033) F. Mistral, 1904
(8381) G. Hauptmann, 1912
(7855) R. Tagore, 1913
(1269) R. Rolland, 1915
(5838) K. Hamsun, 1920
(8382) T. Mann, 1929
(3890) I. A. Bunin, 1933
(1446) F. E. Sillanpää, 1939
(9762) H. Hesse, 1946
(3656) E. Hemingway, 1954
(3508) B. L. Pasternak, 1958
(2448) M. A. Scholochow, 1965
(6832) Y. Kawabata, 1968
(4915) A. I. Solzhenitsyn, 1970
(7873) H. Böll, 1972
(4369) J. Seifert, 1984
(6931) K. Ohe, 1994

Physics

(6401) W. C. Roentgen, 1901
(6914) H. A. Becquerel, 1903
(7000) M. S. Curie, 1903
(1332) G. Marconi, 1909
(1069) M. Planck, 1918
(2001) A. Einstein, 1921
(3948) N. Bohr, 1922
(8116) J. B. Perrin, 1926
(8103) E. Fermi, 1938
(6913) H. Yukawa, 1949
(3443) T.-D. Lee, 1957
(7495) R. P. Feynman, 1965
(6919) S. Tomonaga, 1965
(3421) C. N. Yang, 1957
(3599) N. G. Basov, 1964
(1778) H. O. G. Alfvén, 1968
(3581) L. W. Alvarez, 1968
(6920) L. Esaki, 1973
(3437) P. L. Kapitsa, 1978
(6036) S. Weinberg, 1979
(1958) S. Chandrasekhar, 1983
(8398) C. Rubbia, 1984

Physiology, medicine

(1007) I. P. Pawlow, 1904
(2087) T. Kocher, 1909
(8430) H. W. Florey, 1945
(2550) B. A. Houssay, 1947
(3861) K. Z. Lorenz, 1973
(6927) S. Tonegawa, 1987

Chemistry

(5697) S. A. Arrhenius, 1903
(7000) M. S. Curie, 1911
(7414) C. Bosch, 1931
(1449) A. I. Virtanen, 1945
(4856) G. T. Seaborg, 1951
(4674) L. C. Pauling, 1954
(3069) J. Heyrovský, 1959
(5422) D. Crowfoot-Hodgkin, 1964
(2548) L. Leloir, 1970
(3316) G. Herzberg, 1971
(6924) K. Fukui, 1981

Peace

(1962) H. Dunant, 1901
(853) F. Nansen, 1922
(7584) C. von Ossietzky, 1935
(4390) A. Schweitzer, 1952
(3228) G. Pire, 1958
(4674) L. C. Pauling, 1962
(2305) M. L. King, 1964
(1979) A. D. Sacharow, 1975
(4390) Mother Teresa, 1979
(9481) R. Menchú, 1992

IAU Officers

Presidents

(1280)	B. Baillaud (France)	1919-1922
(2751)	W. W. Campbell (U.S.A.)	1922-1925
(1686)	W. de Sitter (Netherlands)	1925-1928
(1241)	F. W. Dyson (U.K.)	1928-1932
(1770)	F. Schlesinger (U.S.A.)	1932-1935
(1509)	E. Esclangon (France)	1935-1938
(2761)	A. A. Eddington (U.K.)	1938-1943
(3282)	H. Spencer Jones (U.K.)	1943-1948
(1448)	B. Lindblad (Sweden)	1948-1952
(2227)	O. Struve (U.S.A.)	1952-1955
(1594)	A. Danjon (France)	1955-1958
(1691)	J. H. Oort (Netherlands)	1958-1961
(1905)	V. A. Ambartsumian (U.S.S.R.)	1961-1964
(1637)	P. Swings (Belgium)	1964-1967
(1650)	O. Heckmann (Germany)	1967-1970
(1846)	B. Strömgren (Denmark)	1970-1973
(3572)	L. Goldberg (U.S.A.)	1973-1976
(2145)	A. Blaauw (Netherlands)	1976-1979
(2596)	M. K. V. Bappu (India)	1979-1982
(2605)	J. Sahade (Argentina)	1985-1988
(3040)	Y. Kozai (Japan)	1988-1991
(2563)	A. A. Boyarchuk (U.S.S.R)	1991-1994
(3377)	L. Woltjer (Netherlands)	1994-1997
(3712)	R. P. Kraft (U.S.A.)	1997-2000

General Secretaries

(1560)	F. J. M. Stratton (U.K.)	1925-1938
(1691)	J. H. Oort (Netherlands)	1938-1948
(1846)	B. Strömgren (Denmark)	1948-1952
(1738)	P. T. Oosterhoff (Netherlands)	1952-1958
(1629)	J.-C. Pecker (France)	1964-1967
(2900)	L. Perek (Czechoslovakia)	1967-1970
(3798)	C. de Jager (Netherlands)	1970-1973
(7265)	E. A. Müller (Switzerland)	1976-1979
(2022)	R. M. West (Denmark)	1982-1985
(4326)	D. McNally (U.K.)	1988-1991
(2373)	I. Appenzeller (Germany)	1994-1997

Small Bodies Names Committee

(3192)	M. F. A'Hearn (U.S.A.	1997-2000
(2067)	K. Aksnes (Norway	1991-2000
(5996)	J. Fernandez (Uruguay	1997-2000
(3907)	P. M. Kilmartin (New Zealand	1994-2000
(3040)	Y. Kozai (Japan)	1991-2000
(1877)	B. G. Marsden (U.S.A.	1991-2000
(3692)	H. Rickman (Sweden	1994-2000
(2234)	L. D. Schmadel (Germany	1991-2000
(3946)	V. A. Shor (U.S.S.R	1991-2000
(2022)	R. M. West (Denmark	1994-2000
(2956)	D. K. Yeomans (U.S.A.	1991-2000

Special Type Numbered Minor Planets

Amor planets

(4401) Aditi
(719) Albert
(887) Alinda
(5879) Almeria
(1221) Amor
(1943) Anteros
(2061) Anza
(2059) Baboquivari
(2368) Beltrovata
(1580) Betulia
(5797) Bivoj
(1916) Boreas
(4957) Brucemurray
(4503) Cleobulus
(1917) Cuyo
(3671) Dionysus
(3552) Don Quixote
(4954) Eric
(433) Eros
(3122) Florence
(1036) Ganymed
(6489) Golevka
(6456) Golombek
(7088) Ishtar
(1627) Ivar
(3102) Krok
(5324) Lyapunov
(4055) Magellan
(3352) McAuliffe
(3553) Mera
(3199) Nefertiti
(4947) Ninkasi
(7480) Norwan
(3908) Nyx
(2202) Pele
(4487) Pocahontas
(1915) Quetzalcoatl
(3288) Seleucus
(2608) Seneca
(5869) Tanith
(5863) Tara
(5370) Taranis
(1980) Tezcatlipoca
(3271) Ul
(3551) Verenia
(4015) Wilson-Harrington
(5751) Zao
(7977) 1977 QQ_5
(4688) 1980 WF
(4596) 1981 QB
(3691) 1982 FT
(3757) 1982 XB
(6178) 1986 DA
(3988) 1986 LA
(7236) 1987 PA
(10302) 1989 ML
(9172) 1989 OB
(7336) 1989 RS_1
(5332) 1990 DA
(8013) 1990 KA
(5620) 1990 OA
(5587) 1990 SB
(5646) 1990 TR
(9950) 1990 VB
(5626) 1991 FE
(6491) 1991 OA
(6050) 1992 AE
(8034) 1992 LR
(7474) 1992 TC
(5653) 1992 WD_5
(6053) 1993 BW_3
(8037) 1993 HO_1
(5836) 1993 MF
(6569) 1993 MO
(7025) 1993 QA
(7839) 1994 ND
(10150) 1994 PN
(9400) 1994 TW_1
(7358) 1995 YA_3
(8567) 1996 HW_1

Apollo planets

(2101) Adonis
(1863) Antinous
(1862) Apollo
(2135) Aristaeus
(4581) Asclepius
(2063) Bacchus
(7092) Cadmus
(3752) Camillo
(4769) Castalia
(1865) Cerberus
(4183) Cuno
(1864) Daedalus
(3103) Eger
(3838) Epona
(1620) Geographos
(2212) Hephaistos
(5143) Heracles
Hermes
(1566) Icarus
(6063) Jason
(1981) Midas
(6239) Minos
(4486) Mithra
(4660) Nereus
(2201) Oljato
(3361) Orpheus
(2329) Orthos
(4450) Pan
(3200) Phaethon
(4341) Poseidon
(5011) Ptah
(1866) Sisyphus
(5786) Talos
(2102) Tantalus
(1685) Toro
(4179) Toutatis
(4257) Ubasti
(4544) Xanthus
(5731) Zeus
(5496) 1973 NA
(5660) 1974 MA
(3360) 1981 VA
(4197) 1982 TA
(4034) 1986 PA
(9162) 1987 OA
(6037) 1988 EG
(7753) 1988 XB
(7335) 1989 JA
(5131) 1990 BG
(8014) 1990 MF
(4953) 1990 MU

(5645) 1990 SP
(5189) 1990 UQ
(5828) 1991 AM
(8507) 1991 CB_1
(7822) 1991 CS
(9856) 1991 EE
(6047) 1991 TB_1
(7341) 1991 VK
(8176) 1991 WA
(6455) 1992 HE
(9058) 1992 JB
(10115) 1992 SK
(8035) 1992 TB
(5693) 1993 EA
(9202) 1993 PB
(7888) 1993 UC
(7350) 1993 VA
(6611) 1993 VW
(10563) 1993 WD
(8201) 1994 AH_2
(10145) 1994 CK_1
(7889) 1994 LX
(7482) 1994 PC_1
(10165) 1995 BL_2
(8566) 1996 EN
(10636) 1998 QK_{56}

Aten planets

(3554) Amun
(2062) Aten
(3753) Cruithne
(2340) Hathor
(3362) Khufu
(2100) Ra-Shalom
(5381) Sekhmet
(5590) 1990 VA
(5604) 1992 FE

Hilda planets

(2067) Aksnes
(8721) AMOS
(958) Asplinda
(3290) Azabu
(361) Bononia
(2246) Bowell
(1746) Brouwer
(3254) Bus
(4446) Carolyn
(334) Chicago
(6237) Chikushi
(1754) Cunningham
(3415) Danby
(3561) Devine
(2312) Duboshin
(3655) Eupraksia
(1212) Francette
(4317) Garibaldi
(3202) Graff
(2483) Guinevere
(3990) Heimdal
(153) Hilda
(5661) Hildebrand
(190) Ismene
(2760) Kacha
(1578) Kirkwood
(3134) Kostinsky
(1162) Larissa
(6984) Lewiscarroll
(1268) Libya
(4757) Liselotte
(1202) Marina
(1877) Marsden
(1748) Mauderli
(6124) Mecklenburg
(3571) Milanštefánik
(9829) Murillo
(1256) Normannia
(3843) OISCA
(1529) Oterma
(1512) Oulu
(5928) Pindarus
(1345) Potomac
(3577) Putilin
(3923) Radzievskij
(5603) Rausudake
(1180) Rita
(1269) Rollandia
(2624) Samitchell
(2959) Scholl
(1911) Schubart
(8130) Seeberg
(1902) Shaposhnikov
(3694) Sharon
(4196) Shuya
(748) Simeïsa
(3557) Sokolsky
(4255) Spacewatch
(1038) Tuckia
(4230) van den Bergh
(499) Venusia
(1439) Vogtia
(1941) Wild
(3514) 1971 UJ
(5711) 1978 SO_4
(7458) 1984 DE_1
(5368) 1984 SW_5
(7394) 1985 QX_4
(7174) 1988 SQ
(10063) 1988 SZ_2
(4495) 1988 VS
(8086) 1989 RB_6
(7284) 1989 VW
(5439) 1990 RW
(8376) 1992 OZ_9
(7027) 1993 XT
(8550) 1994 PV_{24}
(8551) 1994 VC_7
(8913) 1995 YB_2
(8915) 1995 YK_3
(9661) 1996 FU_{13}
(9121) 1998 DJ_{11}
(8743) 1998 EH_{12}

Trojan planets

The Trojan planets are named for Greek and Trojan heroes of the Trojan War. The Greeks, which are marked by an asterisk, move around the Lagrange libration point L4 of Jupiter, the Trojans around L5. There are, unfortunately, some errors concerning the assignment of Homer's heroes to the correct class. Astronomers interpret those warriors as being "spies" in the enemies' camp.

(2594) Acamas
(5126) Achaemenides *
(5144) Achates
(588) Achilles *
(911) Agamemnon *
(5023) Agapenor *
(4722) Agelaos
(1873) Agenor
(8241) Agrius *
(1404) Ajax *
(2241) Alcathous
(5244) Amphilochos *
(5652) Amphimachus *
(1173) Anchises
(5027) Androgeos *
(1172) Äneas
(8060) Anius *
(2207) Antenor
(7214) Antielus *
(1583) Antilochus *
(9828) Antimachos *
(4946) Askalaphus *
(4805) Asteropaios
(1871) Astyanax
(2920) Automedon *
(5120) Bitias
(2363) Cebriones
(5511) Cloanthus
(4827) Dares
(5638) Deikoon
(1867) Deiphobus
(4060) Deipylos *
(4057) Demophon *
(1437) Diomedes *
(7815) Dolon
(4709) Ennomos
(5259) Epeigeus *
(2148) Epeios *
(4063) Euforbo *
(5436) Eumelos *
(7152) Euneus *
(4007) Euryalos *
(3548) Eurybates *
(9818) Eurymachos *
(4501) Eurypylos *
(8317) Eurysaces *
(1870) Glaukos
(5637) Gyas
(5028) Halaesus *
(624) Hektor *
(1872) Helenos
(7119) Hiera *
(2759) Idomeneus *
(5130) Ilioneus
(4791) Iphidamas
(4138) Kalchas *
(4707) Khryses
(5285) Krethon *
(3240) Laocoon
(6997) Laomedon
(3793) Leonteus *
(9694) Lycomedes *
(4792) Lykaon
(3063) Makhaon *
(4836) Medon *
(4833) Meges *
(2895) Memnon
(1647) Menelaus *
(4068) Menestheus *
(3451) Mentor
(3596) Meriones *
(4828) Misenus
(9712) Nauplius *
(2260) Neoptolemus *
(659) Nestor *
(9713) Oceax *
(1143) Odysseus *
(9907) Oileus *
(5284) Orsilocus *
(2456) Palamedes *
(4832) Palinurus
(2674) Pandarus
(4754) Panthoos
(3317) Paris
(617) Patroclus
(2893) Peiroos
(2357) Phereclos
(1869) Philoctetes *
(4086) Podalirius *
(4867) Polites
(4708) Polydoros
(3709) Polypoites *
(4348) Poulydamas
(884) Priamus
(3540) Protesilaos *
(7543) Prylis *
(5283) Pyrrhus *
(9142) Rhesus
(2223) Sarpedon
(4829) Sergestus
(3391) Sinon *
(2146) Stentor *
(3794) Sthenelos *
(3564) Talthybius *
(1749) Telamon *
(5264) Telephus *
(2797) Teucer *
(5041) Theotes *
(9817) Thersander *
(1868) Thersites *
(4902) Thessandrus *
(4834) Thoas *
(3801) Thrasymedes *
(6998) Tithonus
(1208) Troilus
(8125) Tyndareus *
(5254) Ulysses *
(3708) 1974 FV_1
(6545) 1986 TR_6 *
(5025) 1986 TS_6 *
(7641) 1986 TT_6 *
(4035) 1986 WD *
(4489) 1988 AK *
(5119) 1988 RA_1
(9023) 1988 RG_1
(6443) 1988 RH_{12}
(5233) 1988 RL_{10}
(6002) 1988 RO
(5257) 1988 RS_{10}
(5258) 1989 AU_1 *
(5123) 1989 BL *
(4835) 1989 BQ *

(5012) Eurymedon *
(6090) 1989 DJ *
(5476) 1989 TO_{11}
(4715) 1989 TS_1
(5907) 1989 TU_5
(9030) 1989 UX_5
(5648) 1990 VU_1
(4543) Phoinix *
(9590) 1991 DK_1 *
(9857) 1991 EN *
(7352) 1994 CO
(9790) 1995 OK_8 *
(9430) 1996 HU_{10}
(9431) 1996 PS_1 *
(5209) 1989 CW_1 *
(9799) 1996 RJ *
(9807) 1997 SJ_4 *
(10664) 5187 T-2 *
(10247) 6629 P-L *

Martian Trojan at L5

(5261) Eureka

Minor Planet Names with Unknown Meaning

(164) Eva	1876 P. P. Henry (Paris)
(177) Irma	1877 P. P. Henry (Paris)
(186) Celuta	1878 P. M. Henry (Paris)
(210) Isabella	1879 J. Palisa (Pola)
(223) Rosa	1882 J. Palisa (Vienna)
(245) Vera	1885 N. R. Pogson (Madras)
(252) Clementina	1885 J. Perrotin (Nice)
(262) Valda	1886 J. Palisa (Vienna)
(272) Antonia	1888 A. Charlois (Nice)
(276) Adelheid	1888 J. Palisa (Vienna)
(278) Paulina	1888 J. Palisa (Vienna)
(283) Emma	1889 A. Charlois (Nice)
(284) Amalia	1889 A. Charlois (Nice)
(285) Regina	1889 A. Charlois (Nice)
(291) Alice	1890 J. Palisa (Vienna)
(292) Ludovica	1890 J. Palisa (Vienna)
(294) Felicia	1890 A. Charlois (Nice)
(295) Theresia	1890 J. Palisa (Vienna)
(297) Caecilia	1890 A. Charlois (Nice)
(298) Baptistina	1890 A. Charlois (Nice)
(300) Geraldina	1890 A. Charlois (Nice)
(302) Clarissa	1890 A. Charlois (Nice)
(310) Margarita	1891 A. Charlois (Nice)
(311) Claudia	1891 A. Charlois (Nice)
(312) Pierretta	1891 A. Charlois (Nice)
(314) Rosalia	1891 A. Charlois (Nice)
(316) Goberta	1891 A. Charlois (Nice)
(318) Magdalena	1891 A. Charlois (Nice)
(319) Leona	1891 A. Charlois (Nice)
(331) Etheridgea	1892 A. Charlois (Nice)
(332) Siri	1892 M. Wolf (Heidelberg)
(337) Devosa	1892 A. Charlois (Nice)
(338) Budrosa	1892 A. Charlois (Nice)
(345) Tercidina	1892 A. Charlois (Nice)
(347) Pariana	1892 A. Charlois (Nice)
(351) Yrsa	1892 M. Wolf (Heidelberg)
(354) Eleonora	1893 A. Charlois (Nice)
(357) Ninina	1893 A. Charlois (Nice)
(360) Carlova	1893 A. Charlois (Nice)
(368) Haidea	1893 A. Charlois (Nice)
(375) Ursula	1893 A. Charlois (Nice)
(383) Janina	1894 A. Charlois (Nice)
(391) Ingeborg	1894 M. Wolf (Heidelberg)
(413) Edburga	1896 M. Wolf (Heidelberg)
(419) Aurelia	1896 M. Wolf (Heidelberg)

(435) Ella	1898 M. Wolf and A. Schwassmann (Heidelberg)
(436) Patricia	1898 M. Wolf and A. Schwassmann (Heidelberg)
(441) Bathilde	1898 A. Charlois (Nice)
(448) Natalie	1899 M. Wolf and A. Schwassmann (Heidelberg)
(450) Brigitta	1899 M. Wolf and A. Schwassmann (Heidelberg)
(453) Tea	1900 A. Charlois (Nice)
(481) Emita	1902 L. Carnera (Heidelberg)
(491) Carina	1902 M. Wolf (Heidelberg)
(509) Iolanda	1903 M. Wolf (Heidelberg)
(520) Franziska	1903 M. Wolf and P. Götz (Heidelberg)
(522) Helga	1904 M. Wolf (Heidelberg)
(532) Herculina	1904 M. Wolf (Heidelberg)
(574) Reginhild	1905 M. Wolf (Heidelberg)
(575) Renate	1905 M. Wolf (Heidelberg)
(591) Irmgard	1906 A. Kopff (Heidelberg)
(599) Luisa	1906 J. H. Metcalf (Taunton)
(602) Marianna	1906 J. H. Metcalf (Taunton)
(610) Valeska	1906 M. Wolf (Heidelberg)
(611) Valeria	1906 J. H. Metcalf (Taunton)
(612) Veronika	1906 A. Kopff (Heidelberg)
(618) Elfriede	1906 K. Lohnert (Heidelberg)
(625) Xenia	1907 A. Kopff (Heidelberg)
(628) Christine	1907 A. Kopff (Heidelberg)
(629) Bernardina	1907 A. Kopff (Heidelberg)
(633) Zelima	1907 A. Kopff (Heidelberg)
(636) Erika	1907 J. H. Metcalf (Taunton)
(641) Agnes	1907 M. Wolf (Heidelberg)
(647) Adelgunde	1907 A. Kopff (Heidelberg)
(649) Josefa	1907 A. Kopff (Heidelberg)
(660) Crescentia	1908 J. H. Metcalf (Taunton)
(663) Gerlinde	1908 A. Kopff (Heidelberg)
(665) Sabine	1908 W. Lorenz (Heidelberg)
(667) Denise	1908 A. Kopff (Heidelberg)
(684) Hildburg	1909 A. Kopff (Heidelberg)
(685) Hermia	1909 W. Lorenz (Heidelberg)
(687) Tinette	1909 J. Palisa (Vienna)
(688) Melanie	1909 J. Palisa (Vienna)
(695) Bella	1909 J. H. Metcalf (Taunton)
(795) Fini	1914 J. Palisa (Vienna)
(796) Sarita	1914 K. Reinmuth (Heidelberg)
(802) Epyaxa	1915 M. Wolf (Heidelberg)
(817) Annika	1916 M. Wolf (Heidelberg)
(820) Adriana	1916 M. Wolf (Heidelberg)
(821) Fanny	1916 M. Wolf (Heidelberg)
(822) Lalage	1916 M. Wolf (Heidelberg)

(825) Tanina	1916 G. N. Neujmin (Simeis)
(826) Henrika	1916 M. Wolf (Heidelberg)
(833) Monica	1916 M. Wolf (Heidelberg)
(835) Olivia	1916 M. Wolf (Heidelberg)
(838) Seraphina	1916 M. Wolf (Heidelberg)
(842) Kerstin	1916 M. Wolf (Heidelberg)
(845) Naëma	1916 M. Wolf (Heidelberg)
(860) Ursina	1917 M. Wolf (Heidelberg)
(868) Lova	1917 M. Wolf (Heidelberg)
(873) Mechthild	1917 M. Wolf (Heidelberg)
(882) Swetlana	1917 G. N. Neujmin (Simeis)
(891) Gunhild	1918 M. Wolf (Heidelberg)
(969) Leocadia	1921 S. I. Belyavskij (Simeis)
(1015) Christa	1924 K. Reinmuth (Heidelberg)
(1043) Beate	1925 K. Reinmuth (Heidelberg)
(1049) Gotho	1925 K. Reinmuth (Heidelberg)
(1050) Meta	1925 K. Reinmuth (Heidelberg)
(1053) Vigdis	1925 M. Wolf (Heidelberg)
(1109) Tata	1929 K. Reinmuth (Heidelberg)
(1156) Kira	1928 K. Reinmuth (Heidelberg)
(1169) Alwine	1930 M. Wolf and M. Ferrero (Heidelberg)
(1175) Margo	1930 K. Reinmuth (Heidelberg)
(1180) Rita	1931 K. Reinmuth (Heidelberg)
(1182) Ilona	1927 K. Reinmuth (Heidelberg)
(1183) Jutta	1930 K. Reinmuth (Heidelberg)
(1187) Afra	1929 K. Reinmuth (Heidelberg)
(1214) Richilde	1932 M. Wolf (Heidelberg)
(1219) Britta	1932 M. Wolf (Heidelberg)
(1288) Santa	1933 E. Delporte (Uccle)
(1293) Sonja	1933 E. Delporte (Uccle)
(1438) Wendeline	1937 K. Reinmuth (Heidelberg)
(1514) Ricouxa	1906 M. Wolf (Heidelberg)

References

Bauschinger, J.: Astron. Nachr., Band 149, Nr. 3570, p. 318 (1899)
Bauschinger, J.: Astron. Nachr., Band 154, Nr. 3683, p. 227 (1901)
Bowell, E., Chernykh, N. S., Marsden, B. G.: in 'Asteroids II', R. P. Binzel, T. Gehrels, M. S. Matthews (eds.), Univ. Arizona Press, Tucson, AZ (U.S.A.), p. 21 (1989)
Bower, E. C.: Astron. Nachr., Band 223, Nr. 5337, p. 149 (1924)
Bruhns, C.: Circ. Berliner Astron. Jahrb., No. 96 (1878)
Coenen, D.: Herder Lexikon - Griechische und römische Mythologie, Herder, Freiburg i. Br., Germany, 235 p. (1981)
Combes, M.-A.: Astronomie, Vol. 107, p. 327 (1993)
Combes, M.-A.: Statistiques sur les déconvertes des 6000 premiérs astéroïdes numérotés, private print, 23 p. (1994)
de Freitas Mourão, R. R.: Dicionário Enciclopédico de Astronomia e Astronáutica, Editora Nova Fronteira, Rio de Janeiro, XXXV + 914 p. (1987)
Denoyelle, J.: Ciel Terre, Vol. 111, p. 14 (1995); Vol. 112, p. 18 (1996)
Deutsch, A. N.: Byull. Inst. Teor. Astron. (Leningrad), Tom 13, No. 9 (152), p. 551 (1974)
Encke, J. F.: Berliner Astronomisches Jahrbuch für 1854, Berlin (1851)
Ferguson, J.: Astron. Nachr., Band 35, Nr. 820, p. 51 (1852)
Förster, W.: Astron. Nachr., Band 55, Nr. 1304, p. 116 (1861)
Goldschmidt, H.: Astron. Nachr., Band 55, Nr. 1311, p. 235 (1861)
Grimal, P.: The Dictionary of Classical Mythology, Blackwell, New York, X + 603 p. (1987)
Herget, P.: The Names of the Minor Planets, Cincinnati Observatory, p. 1-38 (1955); p. 39-138 (1968)
Herget, P.: Minor Planet Circ. 813 (1952)
Holden, E. S.: Publ. Astron. Soc. Pac., Vol. 8, No. 47, p. 28 (1896)
Keller, H. L.: Reclams Lexikon der Heiligen und der biblischen Gestalten, Reclam, Stuttgart, Germany, 655 p. (1987)
Kopff, A.: Astron. Nachr., Band 223, Nr. 5337, p. 149 (1924)
Kreutz, H.: Astron. Nachr., Band 149, Nr. 3570, p. 317 (1899)
Krueger, A.: Astron. Nachr., Band 130, Nr. 3106, p. 159 (1892)
Krueger, A.: Astron. Nachr., Band 134, Nr. 3194, p. 31 (1893)
Laugier, E.: Astron. Nachr., Band 50, Nr. 1178, p. 27 (1859)
Luther, R.: Astron. Nachr., Band 42, Nr. 991, p. 107 (1855)
Luther, R.: Astron. Nachr., Band 55, Nr. 1308, p. 179 (1861)
Luther, R.: Circ. Berliner Astron. Jahrb., No. 94 (1878)
Marsden, B. G.: in 'Asteroids', T. Gehrels (ed.), Univ. Arizona Press, Tucson, AZ (U.S.A.), p. 77 (1979)
Marsden, B. G.: Annu. Rep. Minor Planet Cent. 1990, 3 p. (1991)

Marsden, B. G.: in 'Dynamics, Ephemerides and Astrometry of the Solar System', S. Ferraz-Mello, B. Morando, J.-E. Arlot (eds.), Kluwer, Dordrecht, The Netherlands, p. 153 (1996)

Menzel, D. H., Minnaert, M., Levin, B., Dollfus, A., Bell, B.: Space Sci. Rev., Vol. 12, p. 136 (1971)

Meeus, J.: The Names of the Minor Planets, private print, Kessel-Lo, Belgium, 47 p. (1963)

Meeus, J.: Heelal, Vol. 33, p. 157 (1988)

Neyachenko, I. I.: Istoriko-astronomicheskie issledovaniya, Vyp. XIII, p. 43-116, Nauka, Moskva (1977)

Paluzíe-Borrell, A.: The Names of the Minor Planets and their Meanings, published privately by J. Meeus, Kessel-Lo, Belgium, 120 p., (1963)

Peters, C. A. F.: Astron. Nachr., Band 55, Nr. 1311, p. 236 (1861)

Pilcher, F.: in 'Asteroids', T. Gehrels (ed.), Univ. Arizona Press, Tucson, AZ (U.S.A.), p. 1130 (1979)

Pilcher, F., Meeus, J.: Tables of Minor Planets, private print, 104 p. (1973)

Schmadel, L. D.: Messenger, No. 69, p. 32 (1992)

Schnell, A., Haupt, H.: Kleine Planeten, deren Namen einen Österreichbezug aufweisen, Inst. Astron. Vienna, 63 p. (1996)

Steinheil, C. A.: Astron. Nachr., Band 55, Nr. 1311, p. 225 (1861)

van Houten, C. J., van Houten-Groeneveld, I., Herget, P., Gehrels. T.: Astron. Astrophys., Suppl. Ser., Vol. 2, p. 339 (1970)

van Houten, C. J., Herget, P., Marsden, B. G.: Icarus, Vol. 59, p. 1 (1984)

van Houten, C. J., van Houten-Groeneveld, I., Wisse-Schouten, M., Bardwell, C., Green, D. W. E., Gehrels, T.: Icarus, Vol. 91, p. 326 (1991)

van Houten-Groeneveld, I., van Houten, C. J., Wisse-Schouten, M., Bardwell, C., Gehrels, T.: Astron. Astrophys., Vol. 224, p. 299 (1989)

von Littrow, J. J.: Die Wunder des Himmels oder Gemeinfassliche Darstellung des Weltsystemes, K. von Littrow (ed.), 6. ed., Hempel, Berlin (1878)

von Littrow, J. J.: Die Wunder des Himmels oder Gemeinfassliche Darstellung des Weltsystemes, E. Weiß (ed.), 7. ed., Hempel, Berlin (1886)

Wolf, R.: Astron. Nachr., Band 50, Nr. 1185, p. 144 (1859)

Wolf, R.: Handbuch der Astronomie, ihrer Geschichte und Literatur. Zrich (1892)

Zimmermann, J. E.: Dictionary of Classical Mythology, Harper & Row, New York, XX + 300 p. (1964)

Index of Principal Designations

Index of Minor Planet Names